Selected Keys of the Graphing Calculator*

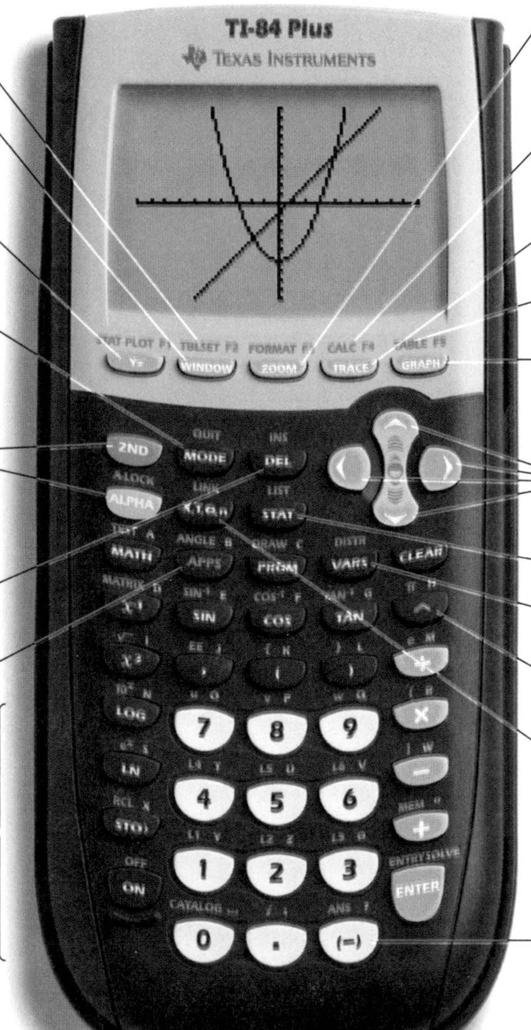

Controls the values that are used when creating a table.

Determines the portion of the curve(s) shown and the scale of the graph.

Used to enter the equation(s) that is to be graphed.

Controls whether graphs are drawn sequentially or simultaneously and if the window is split.

Activates the secondary functions printed above many keys in blue or green.

Used to delete previously entered characters.

Accesses preprogrammed applications and tutorials.

These keys are similar to those found on a scientific calculator.

Magnifies or reduces a portion of the curve being viewed and can "square" the graph to reduce distortion.

Used to determine certain important values associated with a graph.

Used to display the coordinates of points on a curve.

Used to display x- and y-values in a table.

Used to graph equations that were entered using the Y= key.

Used to move the cursor and adjust contrast.

Used to fit curves to data.

Used to access a previously named function or equation.

Used to raise a base to a power.

Used to write the variable, x.

Used as a negative sign.

*Key functions and locations are the same for the TI-83 Plus.

4TH EDITION

Elementary and Intermediate Algebra: Graphs and Models

Marvin L. Bittinger

Indiana University Purdue University Indianapolis

David J. Ellenbogen

Community College of Vermont

Barbara L. Johnson

Indiana University Purdue University Indianapolis

Addison-Wesley

Boston Columbus Indianapolis New York San Francisco Upper Saddle River
Amsterdam Cape Town Dubai London Madrid Milan Munich Paris Montréal Toronto
Delhi Mexico City São Paulo Sydney Hong Kong Seoul Singapore Taipei Tokyo

Editorial Director	Christine Hoag
Editor in Chief	Maureen O'Connor
Executive Editor	Cathy Cantin
Executive Content Editor	Kari Heen
Associate Content Editor	Christine Whitlock
Assistant Editor	Jonathan Wooding
Production Manager	Ron Hampton
Composition	PreMediaGlobal
Production and Editorial Services	Martha K. Morong/Quadrata, Inc.
Art Editor and Photo Researcher	Geri Davis/The Davis Group, Inc.
Associate Media Producer	Nathaniel Koven
Content Development Manager	Eric Gregg (MathXL)
Senior Content Developer	Mary Durnwald (TestGen)
Executive Marketing Manager	Michelle Renda
Marketing Coordinator	Alicia Frankel
Manufacturing Manager	Evelyn Beaton
Senior Manufacturing Buyer	Carol Melville
Senior Media Buyer	Ginny Michaud
Text Designer	Geri Davis/The Davis Group, Inc.
Senior Designer/Cover Design	Beth Paquin
Cover Photograph	© Y-tea/Shutterstock

Photo Credits

Photo credits appear on page G-8.

Library of Congress Cataloging-in-Publication Data

Bittinger, Marvin L.
Elementary & intermediate algebra : graphs & models / Marvin L. Bittinger, David J. Ellenbogen, Barbara L. Johnson. — 4th ed.
p. cm.
Includes index.
1. Algebra—Textbooks. 2. Algebra—Graphic methods—Textbooks. I. Ellenbogen, David. II. Johnson, Barbara L. III. Title. IV. Title: Elementary and intermediate algebra.
QA152.3.B55 2012
512.9—dc22 2010023188

4 5 6 7 8 9 10—V057—15

Addison-Wesley
is a imprint of

www.pearsonhighered.com

ISBN-13: 978-0-321-72634-6
ISBN-10: 0-321-72634-0

Contents

ONLINE CHAPTER*

13 Conic Sections 937

ONLINE CHAPTER*

14 Sequences, Series, and the Binomial Theorem 981

R Elementary Algebra Review R-1

*Available upon request.

Preface

The Bittinger Graphs and Models series helps students "see the math" and learn algebra by making connections between mathematical concepts and their real-world applications. The authors use a variety of tools and techniques—including side-by-side algebraic and graphical solutions and graphing calculators, when appropriate—to engage and motivate all types of learners. The authors have included an abundance of applications, many of which use real data, giving students a lens through which to learn the math.

Appropriate for two consecutive courses or a course combining the study of elementary and intermediate algebra, *Elementary and Intermediate Algebra: Graphs and Models,* Fourth Edition, covers both elementary and intermediate topics without the repetition necessary in two separate texts. This text is more interactive than most other elementary and intermediate texts. Our goal is to enhance the learning process by encouraging students to visualize the mathematics and by providing as much support as possible to help students in their study of algebra. *Intermediate Algebra: Graphs and Models,* Fourth Edition, is also part of this series.

New and Updated Features in This Edition

Responding to both user and reviewer feedback, we have refined the pedagogy, content, design, and art and have expanded the supplements package for this edition. The following is a list of major changes:

New!
- **Chapter openers** pose real-world questions that relate to the chapter contents. Real-data graphs help students consider the answers to the questions. (See pp. 81, 423, and 647.)

New!
- **"Try" exercises** conclude nearly every example by pointing students to one or more parallel exercises in the corresponding exercise set so they can immediately reinforce the skills and concepts presented in the examples. Answers to these exercises appear at the end of each exercise set, as well as in the answer section. (See pp. 139, 244, and 521.)

New!
- **Your Turn** exercises encourage immediate practice of graphing calculator features discussed in the text. Many of these exercises include keystrokes, and answers are included where appropriate. (See pp. 88, 344, and 701.)

New!
- A **Mid-Chapter Review** gives students the opportunity to reinforce their understanding of the mathematical skills and concepts before moving on to new material. Section and objective references are included for convenient studying. (See pp. 214, 537, and 811.)
  - **Guided Solutions** are worked-out problems with blanks for students to fill in the correct expression to complete the solution. (See pp. 214, 538, and 812.)
  - **Mixed Review** provides free-response exercises, similar to those in the preceding sections in the chapter, reinforcing mastery of skills and concepts. (See pp. 214, 538, and 812.)

New!
- **Skill Review** exercises in the section exercise sets offer just-in-time review of previously presented skills that students will need to learn before moving on to the next section. (See pp. 131, 297, and 337.)

- In total, **27% of the exercises are new or updated.** All exercises have been carefully reviewed and revised to improve their grading and to update applications.
- The **Study Summary** at the end of each chapter is expanded to provide more comprehensive in-text practice and review. Important Concepts are paired with a worked-out example for reference and review and a similar practice exercise for students to solve. (See pp. 153, 261, and 324.)
- **New design!** While incorporating a new layout, a fresh palette of colors, and new features, we have increased the page dimension for an open look and a typeface that is easy to read. As always, it is our goal to make the text look mature without being intimidating. In addition, we continue to pay close attention to the pedagogical use of color to make sure that it is used to present concepts in the clearest possible manner.

Content Changes

- Section 10.7 now contains the distance formula and the midpoint formula.
- Sections 11.3 and 11.4 from the previous edition have switched order.
- The appendixes have been removed.

New and Updated Student and Instructor Resources

- The *Instructor's Resource Manual* includes new **Mini-Lectures** for every section of the text.
- **Enhancements to the Bittinger MyMathLab course** include the following:
  - **Increased exercise coverage** provides more practice options for students.
  - **New math games** help students practice math skills in a fun, interactive environment.
  - **Premade homework assignments** are available for each section of the text. In addition to the section-level premade assignments, the Bittinger MyMathLab courses include premade **mid-chapter review assignments** for each chapter.
  - **Interactive Translating for Success** matching activities help students learn to associate word problems (through translation) with their appropriate mathematical equations. These activities are now assignable.
  - **Interactive Visualizing for Success** activities ask students to match equations and inequalities with their graphs, allowing them to recognize the important characteristics of the equation and visualize the corresponding attributes of its graph. These activities are now assignable.
  - **The English/Spanish Audio Glossary** allows students to see key mathematical terms and hear their definitions in either English or Spanish.

Hallmark Features

Problem Solving

One distinguishing feature of our approach is our treatment of and emphasis on problem solving. We use problem solving and applications to motivate the material wherever possible, and we include real-life applications and problem-solving techniques throughout the text. Problem solving not only encourages students to think about how mathematics can be used, it helps to prepare them for more advanced material in future courses.

In Chapter 2, we introduce the five-step process for solving problems: (1) *Familiarize*, (2) *Translate*, (3) *Carry out*, (4) *Check*, and (5) *State* the answer. These steps are then used consistently throughout the text whenever we encounter a problem-solving situation. Repeated use of this problem-solving strategy gives students a sense that they have a starting point for any type of problem they encounter, and frees them to focus on the mathematics necessary to successfully translate the problem situation. We often use estimation and carefully checked guesses to help with the *Familiarize* and *Check* steps. (See pp. 117, 120, and 305–306.)

Algebraic/Graphical Side-by-Sides

Algebraic/graphical side-by-sides give students a direct comparison between these two problem-solving approaches. They show the connection between algebraic and graphical or visual solutions and demonstrate that there is more than one way to obtain a result. This feature also illustrates the comparative efficiency and accuracy of the two methods. (See pp. 303, 551, and 611.) Instructors using this text have found that it works superbly both in courses where the graphing calculator is required and in courses where it is optional.

Applications

Interesting applications of mathematics help motivate both students and instructors. Solving applied problems gives students the opportunity to see their conceptual understanding put to use in a real way. In the fourth edition of *Elementary and Intermediate Algebra: Graphs and Models*, the number of applications and source lines has been increased, and effort has been made to present the most current and relevant applications. As in the past, art is integrated into the applications and exercises to aid the student in visualizing the mathematics. (See pp. 128, 212, and 483.)

Interactive Discoveries

Interactive Discoveries invite students to develop analytical and reasoning skills while taking an active role in the learning process. These discoveries can be used as lecture launchers to introduce new topics at the beginning of a class and quickly guide students through a concept, or as out-of-class concept discoveries. (See pp. 216, 356, and 426.)

Pedagogical Features

Concept Reinforcement Exercises. This feature is designed to help students build their confidence and comprehension through true/false, matching, and fill-in-the-blank exercises at the beginning of most exercise sets. Whenever possible, special attention is devoted to increasing student understanding of the new vocabulary and notation developed in that section. (See pp. 39, 207, and 281.)

Translating for Success. These problem sets give extra practice with the important *Translate* step of the process for solving word problems. After translating each of ten problems into its appropriate equation or inequality, students are asked to choose from fifteen possible translations, encouraging them to comprehend the problem before matching. (See pp. 10 and 147.)

Visualizing for Success. These matching exercises provide students with an opportunity to match an equation with its graph by focusing on the characteristics of the equation and the corresponding attributes of the graph. This feature occurs once in each chapter at the end of a related section. (See pp. 280, 358, and 485.)

Student Notes. These comments, strategically located in the margin within each section, are specific to the mathematics appearing on that page. Remarks are often more casual in format than the typical exposition and range from suggestions on how to avoid common mistakes to how to best read new mathematical notation. (See pp. 189, 276, and 523.)

Connecting the Concepts. To help students understand the big picture, Connecting the Concepts subsections relate the concept at hand to previously learned and upcoming concepts. Because students occasionally lose sight of the forest because of the trees, this feature helps students keep their bearings as they encounter new material. (See pp. 168, 315, and 630.)

Study Tips. These remarks, located in the margin near the beginning of each section, provide suggestions for successful study habits that can be applied to both this and other college courses. Ranging from ideas for better time management to suggestions for test preparation, these comments can be useful even to experienced college students. (See pp. 9, 110, and 448.)

Skill Review Exercises. Retention of skills is critical to a student's success in this and future courses. Thus, beginning in Section 1.2, every exercise set includes Skill Review exercises that review skills and concepts from preceding sections of the text. Often, these exercises provide practice with specific skills needed for the next section of the text. (See pp. 142, 297, and 337.)

Synthesis Exercises. Following the Skill Review section, every exercise set ends with a group of Synthesis exercises that offers opportunities for students to synthesize skills and concepts from earlier sections with the present material, and often provides students with deeper insights into the current topic. Synthesis exercises are generally more challenging than those in the main body of the exercise set and occasionally include Aha! exercises (exercises that can be solved more quickly by reasoning than by computation). (See pp. 116, 213, and 387.)

Thinking and Writing Exercises. Writing exercises have been found to aid in student comprehension, critical thinking, and conceptualization. Thus every set of exercises includes at least four Thinking and Writing exercises. Two of these appear just before the Skill Review exercises. The others are more challenging and appear as Synthesis exercises. All are marked with TW and require answers that are one or more complete sentences. Because some instructors may collect answers to writing exercises, and because more than one answer may be correct, answers to the Thinking and Writing exercises are listed at the back of the text only when they are within review exercises. (See pp. 107, 201, and 349.)

Collaborative Corners. Studies have shown that students who work together generally outperform those who do not. Throughout the text, we provide optional Collaborative Corner features that require students to work in groups to explore and solve problems. There is at least one Collaborative Corner per chapter, each one appearing after the appropriate exercise set. (See pp. 99, 173, and 314.)

Study Summary. Each three-column study summary contains a list of key terms and concepts from the chapter, with definitions, as well as formulas from the chapter. Examples of important concepts are shown in the second column, with a practice exercise relating to that concept appearing in the third column. Section references are provided so students can reference the corresponding exposition in the chapter. The Summary provides a terrific point from which to begin reviewing for a chapter test. (See pp. 153, 261, and 324.)

Ancillaries

The following ancillaries are available to help both instructors and students use this text more effectively.

STUDENT SUPPLEMENTS

Student's Solutions Manual
(ISBN 978-0-321-72660-5)

- By Math Made Visible
- Contains completely worked-out solutions for all the odd-numbered exercises in the text, with the exception of the Thinking and Writing exercises, as well as completely worked-out solutions to all the exercises in the Chapter Reviews, Chapter Tests, and Cumulative Reviews.

Worksheets for Classroom or Lab Practice
(ISBN 978-0-321-72666-7)

These classroom- and lab-friendly workbooks offer the following resources for every section of the text: a list of learning objectives, vocabulary practice problems, and extra practice exercises with ample workspace.

Graphing Calculator Manual
(ISBN 978-0-321-73726-7)

- By Math Made Visible
- Uses actual examples and exercises from the text to help teach students to use the graphing calculator.
- Order of topics mirrors the order of the text, providing a just-in-time mode of instruction.

Video Resources on DVD
(ISBN 978-0-321-72662-9)

- Complete set of digitized videos on DVDs for student use at home or on campus.
- Presents a series of lectures correlated directly to the content of each section of the text.
- Features an engaging team of instructors, including authors Barbara Johnson and David Ellenbogen, who present material in a format that stresses student interaction, often using examples and exercises from the text.
- Ideal for distance learning or supplemental instruction.
- Includes an expandable window that shows text captioning. Captions can be turned on or off.

INSTRUCTOR SUPPLEMENTS

Annotated Instructor's Edition
(ISBN 978-0-321-72663-6)

- Includes answers to all exercises printed in blue on the same page as those exercises.

Instructor's Solutions Manual
(ISBN 978-0-321-72664-3)

- By Math Made Visible
- Contains full, worked-out solutions to all the exercises in the exercise sets, including the Thinking and Writing exercises, and worked-out solutions to all the exercises in the Chapter Reviews, Chapter Tests, and Cumulative Reviews.

Instructor's Resource Manual with Printable Test Forms

- By Math Made Visible
- Download at www.pearsonhighered.com
- Features resources and teaching tips designed to help both new and adjunct faculty with course preparation and classroom management, including general/first-time advice, sample syllabi, and teaching tips arranged by textbook section.
- New! Includes mini-lectures, one for every section of the text, with objectives, key examples, and teaching tips.
- Provides 5 revised test forms for every chapter and revised test forms for the final exam.
- For the chapter tests, test forms are organized by topic order following the chapter tests in the text, and 2 test forms are multiple choice.
- Resources include extra practice sheets, conversion guide, video index, and transparency masters.
- Available electronically so course/adjunct coordinators can customize material specific to their schools.

TestGen® (www.pearsoned.com/testgen)
(ISBN 978-0-321-72661-2)

- Enables instructors to build, edit, print, and administer tests.
- Features a computerized bank of questions developed to cover all text objectives.
- Algorithmically based content allows instructors to create multiple but equivalent versions of the same question or test with a click of a button.
- Instructors can also modify test-bank questions or add new questions.
- The software and testbank are available for download from Pearson Education's online catalog.

MathXL® Online Course (access code required). MathXL is a powerful online homework, tutorial, and assessment system that accompanies Pearson Education's textbooks in mathematics or statistics. With MathXL, instructors can create, edit, and assign online homework and tests using algorithmically generated exercises correlated at the objective level to the textbook. They can also create and assign their own online exercises and import TestGen tests for added flexibility. All student work is tracked and records maintained in MathXL's online gradebook. Students can take chapter tests in MathXL and receive personalized study plans and/or personalized homework assignments based on their test results. The study plan diagnoses weaknesses and links students directly to tutorial exercises for the objectives they need to study and retest. Students can also access supplemental animations and video clips directly from selected exercises. MathXL is available to qualified adopters. For more information, visit our Web site at www.mathxl.com or contact your Pearson representative.

MyMathLab® Online Course (access code required). MyMathLab is a text-specific, easily customizable online course that integrates interactive multimedia instruction with textbook content. MyMathLab gives you the tools you need to deliver all or a portion of your course online, whether your students are in a lab setting or working from home.

- **Interactive homework exercises,** correlated to your textbook at the objective level, are algorithmically generated for unlimited practice and mastery. Most exercises are free-response and provide guided solutions, sample problems, and tutorial learning aids for extra help.
- **Personalized homework** assignments can be designed to meet the needs of your class. MyMathLab tailors the assignment for each student on the basis of their test or quiz scores. Each student receives a homework assignment that contains only the problems he or she still needs to master.
- **Personalized Study Plan,** generated when students complete a test or quiz or homework, indicates which topics have been mastered and links to tutorial exercises for topics students have not mastered. You can customize the Study Plan so that the topics available match your course content.
- **Multimedia learning aids,** such as video lectures and podcasts, animations, interactive games, and a complete multimedia textbook, help students independently improve their understanding and performance. You can assign these multimedia learning aids as homework to help your students grasp the concepts.
- **Homework and Test Manager** lets you assign homework, quizzes, and tests that are automatically graded. Select just the right mix of questions from the MyMathLab exercise bank, instructor-created custom exercises, and/or TestGen® test items.
- **Gradebook,** designed specifically for mathematics and statistics, automatically tracks students' results, lets you stay on top of student performance, and gives you control over how to calculate final grades. You can also add offline (paper-and-pencil) grades to the gradebook.
- **MathXL Exercise Builder** allows you to create static and algorithmic exercises for your online assignments. You can use the library of sample exercises as an easy starting point, or you can edit any course-related exercise.
- **Pearson Tutor Center** (www.pearsontutorservices.com) access is automatically included with MyMathLab. The Tutor Center is staffed by qualified math instructors who provide textbook-specific tutoring for students via toll-free phone, fax, email, and interactive Web sessions.

Students do their assignments in the Flash®-based MathXL Player, which is compatible with almost any browser (Firefox®, Safari™, or Internet Explorer®) on almost any platform (Macintosh® or Windows®). MyMathLab is powered by CourseCompass™, Pearson Education's online teaching and learning environment, and by MathXL®, our online homework, tutorial, and assessment system. MyMathLab is available to qualified adopters. For more information, visit www.mymathlab.com or contact your Pearson representative.

InterAct Math Tutorial Web site, www.interactmath.com. Get practice and tutorial help online! This interactive tutorial Web site provides algorithmically generated practice exercises that correlate directly to the exercises in the textbook. Students can retry an exercise as many times as they like with new values each time for unlimited practice and mastery. Every exercise is accompanied by an interactive guided solution that provides helpful feedback for incorrect answers, and students can also view a worked-out sample problem that steps them through an exercise similar to the one they're working on.

Pearson Math Adjunct Support Center. The Pearson Math Adjunct Support Center (http://www.pearsontutorservices.com/math-adjunct.html) is staffed by qualified instructors with over 100 years of combined experience at both the community college and university levels. Assistance is provided for faculty in the following areas:

- Suggested syllabus consultation
- Tips on using materials packed with your book
- Book-specific content assistance
- Teaching suggestions including advice on classroom strategies

Acknowledgments

No book can be produced without a team of professionals who take pride in their work and are willing to put in long hours. Thanks to Math Made Visible for their work on the print (or printable) supplements. Holly Martinez, Mindy Pergl, Jeremy Pletcher, Art Bouvier, David Johnson, Beverly Fusfield, Christine Verity, Carrie Green, and Gary Williams provided enormous help, often in the face of great time pressure, as accuracy checkers. Michael Avidon, Ebony Harvey, and Robert Pierce from the Pearson Tutor Center also made valuable contributions. We are also indebted to Michelle Lanosga for her help with applications research.

Martha Morong, of Quadrata, Inc., provided editorial and production services of the highest quality imaginable—she is simply a joy to work with. Geri Davis, of the Davis Group, Inc., performed superb work as designer, art editor, and photo researcher, and always with a disposition that can brighten an otherwise gray day. Network Graphics generated the graphs, the charts, and many of the illustrations. Not only are the people at Network reliable, but they clearly take pride in their work. The many representational illustrations appear thanks to Bill Melvin, a gifted artist with true mathematical sensibilities.

Our team at Pearson deserves special thanks. Assistant Editor Jonathan Wooding managed many of the day-to-day details—always in a pleasant and reliable manner. Executive Content Editor Kari Heen expertly provided information and a steadying influence along with gentle prodding at just the right moments. Executive Editor Cathy Cantin provided many fine suggestions along with unflagging support. Production Manager Ron Hampton exhibited careful supervision and an eye for detail throughout production. Executive Marketing Manager Michelle Renda and Marketing Coordinator Alicia Frankel skillfully kept us in touch with the needs of faculty. Associate Media Producer Nathaniel Koven provided us with the technological guidance so necessary for our many supplements and our fine video series. To all of these people, we owe a real debt of gratitude.

Reviewers

Reviewers of *Elementary and Intermediate Algebra: Graphs and Models,* Fourth Edition

Susan Bradley, *Angelina College*
Ben Franklin, *Indiana University–Kokomo*
Julie Mays, *Angelina College*
Melissa Reid, *Rowan-Cabarrus Community College*
Pansy Waycaster, *Virginia Highlands Community College*
Kevin Yokoyama, *College of the Redwoods*

Reviewers of *Intermediate Algebra: Graphs and Models*, Fourth Edition

Clark Brown, *Mohave Community College–Kingman*
Steve Drucker, *Santa Rosa Junior College*
Diana Hunt, *Arapahoe Community Colleg*

M.L.B.
D.J.E.
B.L.J.

1

Introduction to Algebraic Expressions

How Do You Like Your Music?

Do you download single tracks or do you buy complete albums? According to the Nielsen Company, one billion single tracks were sold online in 2008. The relatively low cost of purchasing a single track is one reason for this high volume of sales. In Example 11 of Section 1.1, an algebraic model is developed for the cost of music downloads.

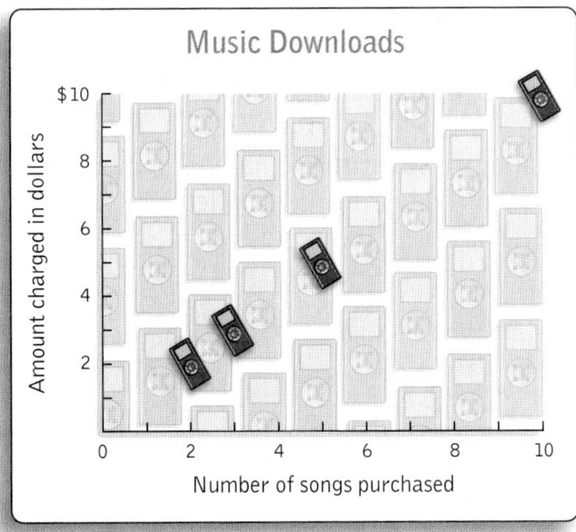

Problem solving using algebra is the focus of this text.

In Chapter 1, we begin working with the algebra that we will later apply to solving problems. The chapter includes a review of arithmetic, a discussion of real numbers and their properties, and an examination of how real numbers are added, subtracted, multiplied, divided, and raised to powers. The graphing calculator is also introduced as a problem-solving tool.

1.1 Introduction to Algebra

- Algebraic Expressions
- Translating to Algebraic Expressions
- Translating to Equations
- Models

This section introduces some basic concepts and expressions used in algebra. Solving real-world problems is an important part of algebra, so we will focus on expressions that can arise in applications.

ALGEBRAIC EXPRESSIONS

Probably the greatest difference between arithmetic and algebra is the use of *variables*. Suppose that n represents the number of tickets sold in one day for a U2 concert and that each ticket costs \$60. Then a total of 60 times n, or $60 \cdot n$, dollars will be collected for tickets.

The letter n is a **variable** because it can represent any one of a set of numbers.

The number 60 is a **constant** because it does not change.

The expression $60 \cdot n$ is a **variable expression** because it contains a variable.

An **algebraic expression** consists of variables and/or numerals, often with operation signs and grouping symbols. In the algebraic expression $60 \cdot n$, the **operation** is multiplication.

To **evaluate** an algebraic expression, we **substitute** a number for each variable in the expression and calculate the result. This result is called the **value** of the expression. The table below lists several values of the expression $60 \cdot n$.

Cost per Ticket (in dollars) 60	Number of Tickets Sold n	Total Collected (in dollars) $60n$
60	150	9,000
60	200	12,000
60	250	15,000

Other examples of algebraic expressions are:

$t + 37$; This contains the variable t, the constant 37, and the operation of addition.

$a \cdot c - b$; This contains the variables a, b, and c and the operations multiplication and subtraction.

$(s + t) \div 2$. This contains the variables s and t, the constant 2, grouping symbols, and the operations addition and division.

Multiplication can be written in several ways. For example, "60 times n" can be written as $60 \cdot n$, $60 \times n$, $60(n)$, $60 * n$, or simply (and usually) $60n$. Division can also be represented by a fraction bar: $\frac{9}{7}$, or $9/7$, means $9 \div 7$.

Student Notes

As we will see later, it is sometimes necessary to use parentheses when substituting a number for a variable. It is safe (and often wise) to use parentheses whenever you substitute. In Example 1, we could write

$$x + y = (37) + (28) = 65$$

and

$$5ab = 5(2)(3) = 30.$$

EXAMPLE 1 Evaluate each expression for the given values.

a) $x + y$ for $x = 37$ and $y = 28$ **b)** $5ab$ for $a = 2$ and $b = 3$

SOLUTION

a) We substitute 37 for x and 28 for y and carry out the addition:

$$x + y = 37 + 28 = 65.$$

The value of the expression is 65.

b) We substitute 2 for a and 3 for b and multiply:

$$5ab = 5 \cdot 2 \cdot 3 = 10 \cdot 3 = 30.$$ **$5ab$ means 5 times a times b.**

Try Exercise 13.

EXAMPLE 2 The area A of a rectangle of length l and width w is given by the formula $A = lw$. Find the area when l is 17 in. and w is 10 in.

SOLUTION We evaluate, substituting 17 in. for l and 10 in. for w and carrying out the multiplication:

$$A = lw = (17 \text{ in.})(10 \text{ in.})$$
$$= (17)(10)(\text{in.})(\text{in.})$$
$$= 170 \text{ in}^2, \text{ or } 170 \text{ square inches.}$$

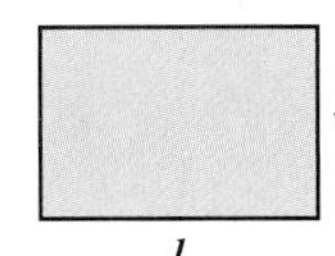

Try Exercise 25.

Note that we always use square units for area and (in.)(in.) = in^2. Exponents like the 2 in the expression in^2 are discussed further in Section 1.8.

EXAMPLE 3 The area A of a triangle with a base of length b and a height of length h is given by the formula $A = \frac{1}{2}bh$. Find the area when b is 8 m and h is 6.4 m.

SOLUTION We substitute 8 m for b and 6.4 m for h and then multiply:

$$A = \tfrac{1}{2}bh = \tfrac{1}{2}(8 \text{ m})(6.4 \text{ m})$$
$$= \tfrac{1}{2}(8)(6.4)(\text{m})(\text{m})$$
$$= 4(6.4) \text{ m}^2$$
$$= 25.6 \text{ m}^2, \text{ or } 25.6 \text{ square meters.}$$

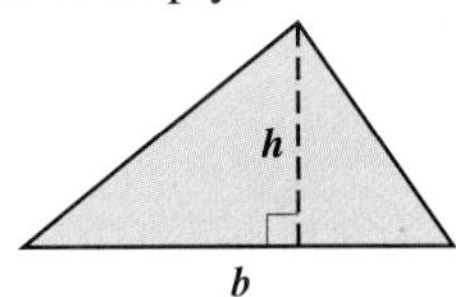

Try Exercise 27.

Student Notes

Using a graphing calculator is not a substitute for understanding mathematical procedures. Be sure that you understand every new procedure before you rely on a graphing calculator. Even when you are working a problem by hand, a calculator can provide a useful check of your answers.

Introduction to the Graphing Calculator

Graphing calculators and graphing software can be valuable aids in understanding and applying algebra. In this text, we will reference features that are common to most graphing calculators. Specific keystrokes and instructions for certain calculators are included in the Graphing Calculator Manual that accompanies this book. For other procedures, consult your instructor or a user's manual.

There are two important things to keep in mind as you proceed through this text:

1. You will not learn to use a graphing calculator by simply reading about it; you must in fact *use* the calculator. Press the keys on your calculator as you

(*continued*)

read the text, do the calculator exercises in the exercise set, and experiment as you learn new procedures.

2. Your user's manual contains more information about your calculator than appears in this text. If you need additional explanation and examples, be sure to consult the manual.

Keypad A diagram of the keypad of a graphing calculator appears at the front of this text. The organization and labeling of the keys are not the same for all calculators. Note that there are options written above keys as well as on the keys. To access the options shown above the keys, press 2ND or ALPHA, depending on the color of the desired option, and then the key below the desired option.

Screen After you have turned the calculator on, you should see a blinking rectangle, or **cursor,** at the top left corner of the screen. If you do not see anything, try adjusting the **contrast.** On many calculators, this is done by pressing 2ND and the up or down arrow keys. The up key darkens the screen; the down key lightens it. To perform computations, you should be in the **home screen.** Pressing QUIT (often the 2nd option associated with the MODE key) will return you to the home screen.

Performing Operations To perform addition, subtraction, multiplication, and division using a graphing calculator, key in the expression as it would be written. The entire expression should appear on the screen, and you can check your keystrokes. A multiplication symbol appears on the screen as *, and division is usually shown by the symbol /. Grouping symbols such as brackets or braces are entered as parentheses. Once the expression appears correctly, press ENTER. At that time, the calculator will evaluate the expression and display the result.

Catalog A graphing calculator's catalog lists all the functions of the calculator in alphabetical order. On many calculators, CATALOG is the 2nd option associated with the 0 key. To copy an item from the catalog to the screen, press CATALOG, and then scroll through the list using the up and down arrow keys until the desired item is indicated. To move through the list more quickly, press the key associated with the first letter of the item. The indicator will move to the first item beginning with that letter. When the desired item is indicated, press ENTER.

Error Messages When a calculator cannot complete an instruction, a message similar to the one shown below appears on the screen.

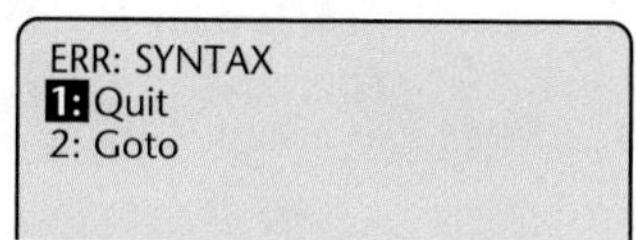

Press 1 to return to the home screen, and press 2 to go to the instruction that caused the error. Not all errors are operator errors; ERR:OVERFLOW indicates a result too large for the calculator to handle.

Your Turn

1. Turn your calculator on.
2. Adjust the contrast.
3. Identify the home screen and the cursor.
4. Calculate $4 \cdot 5$ and $10 \div 2$. The results should be 20 and 5.
5. Find how many commands in the catalog begin with the letter Q.
6. Create an error message by trying to calculate $10 \div 0$.
7. Return to the home screen.
8. Turn your calculator off.

We can use a graphing calculator to evaluate algebraic expressions.

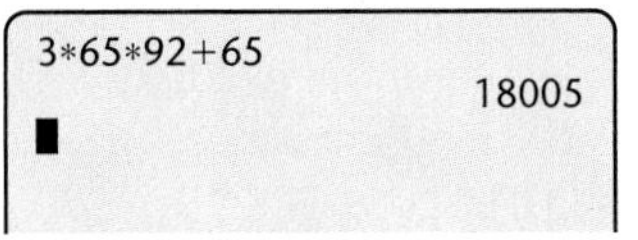

EXAMPLE 4 Use a graphing calculator to evaluate $3xy + x$ for $x = 65$ and $y = 92$.

SOLUTION We enter the expression in the graphing calculator as it is written, replacing x with 65 and y with 92. Note that since $3xy$ means $3 \cdot x \cdot y$, we must supply a multiplication symbol $(*)$ between 3, 65, and 92. We then press ENTER after the expression is complete. The value of the expression appears on the right of the screen as shown here.

We see that $3xy + x = 18{,}005$ for $x = 65$ and $y = 92$.

Try Exercise 23.

TRANSLATING TO ALGEBRAIC EXPRESSIONS

Before attempting to translate problems to equations, we must be able to translate certain phrases to algebraic expressions. Any variable can be used to represent an

Important Words	Sample Phrase or Sentence	Variable Definition	Translation
Addition (+)			
added to	700 lb was added to the car's weight.	Let w represent the car's weight, in pounds.	$w + 700$
sum of	The sum of a number and 12	Let n represent the number.	$n + 12$
plus	53 plus some number	Let x represent "some number."	$53 + x$
more than	8000 more than Detroit's population	Let p represent Detroit's population.	$p + 8000$
increased by	Alex's original guess, increased by 4	Let n represent Alex's original guess.	$n + 4$
Subtraction (−)			
subtracted from	2 oz was subtracted from the weight.	Let w represent the weight, in ounces.	$w - 2$
difference of	The difference of two scores	Let m represent the larger score and n represent the smaller score.	$m - n$
minus	A team of size s, minus 2 injured players	Let s represent the number of players.	$s - 2$
less than	9 less than the number of volunteers	Let v represent the number of volunteers.	$v - 9$
decreased by	The car's speed, decreased by 8 mph	Let s represent the car's speed, in miles per hour.	$s - 8$
Multiplication (•)			
multiplied by	The number of guests, multiplied by 3	Let g represent the number of guests.	$g \cdot 3$
product of	The product of two numbers	Let m and n represent the numbers.	$m \cdot n$
times	5 times the dog's weight	Let w represent the dog's weight, in pounds.	$5w$
twice	Twice the wholesale cost	Let c represent the wholesale cost.	$2c$
of	$\frac{1}{2}$ of Rita's salary	Let s represent Rita's salary.	$\frac{1}{2}s$
Division (÷)			
divided by	A 2-lb coffee cake, divided by 3	*No variables are required for translation.*	$2 \div 3$
quotient of	The quotient of 14 and 7	*No variables are required for translation.*	$14 \div 7$
divided into	4 divided into the delivery fee	Let f represent the delivery fee.	$f \div 4$
ratio of	The ratio of \$500 to the cost of a new car	Let n represent the cost of a new car, in dollars.	$500/n$
per	There were 18 students per teacher.	Let t represent the number of teachers.	$18/s$

unknown quantity; however, it is helpful to choose a descriptive letter. For example, w suggests weight or width and p suggests population or price. It is important to write down what each variable represents, as well as the unit in which it is measured.

EXAMPLE 5 Translate each phrase to an algebraic expression.

a) Four less than Gwen's height, in inches

b) Eighteen more than a number

c) A day's pay, in dollars, divided by eight

SOLUTION To help think through a translation, we sometimes begin with a specific number in place of a variable.

a) If the height were 60 in., then 4 less than 60 would mean $60 - 4$. If we use h to represent "Gwen's height, in inches," the translation of "Four less than Gwen's height, in inches" is $h - 4$.

b) If we knew the number to be 10, the translation would be $10 + 18$, or $18 + 10$. If we use t to represent "a number," the translation of "Eighteen more than a number" is $t + 18$, or $18 + t$.

c) We let d represent "a day's pay, in dollars." If the pay were \$78, the translation would be $78 \div 8$, or $\frac{78}{8}$. Thus our translation of "a day's pay, in dollars, divided by eight" is $d \div 8$, or $\frac{d}{8}$.

Try Exercise 33.

CAUTION! The order in which we subtract and divide affects the answer! Answering $4 - h$ or $8 \div d$ in Examples 5(a) and 5(c) is incorrect.

Student Notes

Try looking for "than" or "from" in a phrase and writing what follows it first. Then add or subtract the necessary quantity. (See Examples 6c and 6d.)

EXAMPLE 6 Translate each phrase to an algebraic expression.

a) Half of some number

b) Seven more than twice Malcolm's weight

c) Six less than the product of two numbers

d) Nine times the difference of a number and 10

e) Eighty-two percent of last year's enrollment

SOLUTION

Phrase	Variable(s)	Algebraic Expression
a) Half of some number	Let n represent the number.	$\frac{1}{2}n$, or $\frac{n}{2}$, or $n \div 2$
b) Seven more than twice Malcolm's weight	Let w represent Malcolm's weight, in pounds.	$2w + 7$, or $7 + 2w$
c) Six less than the product of two numbers	Let m and n represent the numbers.	$mn - 6$
d) Nine times the difference of a number and 10	Let a represent the number.	$9(a - 10)$
e) Eighty-two percent of last year's enrollment	Let r represent last year's enrollment.	82% of r, or $0.82r$

Try Exercise 47.

TRANSLATING TO EQUATIONS

The symbol $=$ ("equals") indicates that the expressions on either side of the equals sign represent the same number. An **equation** is a number sentence with the verb $=$. Equations may be true, false, or neither true nor false.

EXAMPLE 7 Determine whether each equation is true, false, or neither.

a) $8 \cdot 4 = 32$ **b)** $7 - 2 = 4$ **c)** $x + 6 = 13$

SOLUTION

a) $8 \cdot 4 = 32$ The equation is *true*.

b) $7 - 2 = 4$ The equation is *false*.

c) $x + 6 = 13$ The equation is *neither* true nor false, because we do not know what will replace the variable x.

Solution A replacement or substitution that makes an equation true is called a *solution*. Some equations have more than one solution, and some have no solution. When all solutions have been found, we have *solved* the equation.

To determine whether a number is a solution, we evaluate all expressions in the equation. If the values on both sides of the equation are the same, the number is a solution.

EXAMPLE 8 Determine whether 7 is a solution of $x + 6 = 13$.

SOLUTION

$x + 6 = 13$	**Writing the equation**
$7 + 6 \mid 13$	**Substituting 7 for x**
$13 \stackrel{?}{=} 13$	**$13 = 13$ is TRUE.**

Since the left-hand and the right-hand sides are the same, 7 is a solution.

Try Exercise 55.

Although we do not study solving equations until Chapter 2, we can translate certain problem situations to equations now. The words "is the same as," "equal," "is," "are," "was," and "were" translate to "=."

Words indicating equality, =: "is the same as," "equal," "is," "are," "was," "were."

When translating a problem to an equation, we translate phrases to algebraic expressions and the entire statement to an equation containing those expressions.

EXAMPLE 9 Translate the following problem to an equation.

What number plus 478 is 1019?

SOLUTION We let y represent the unknown number. The translation then comes almost directly from the English sentence.

What number	plus	478	is	1019?
↓	↓	↓	↓	↓
y	$+$	478	$=$	1019

Note that "plus" translates to "+" and "is" translates to "=."

Try Exercise 63.

Sometimes it helps to reword a problem before translating.

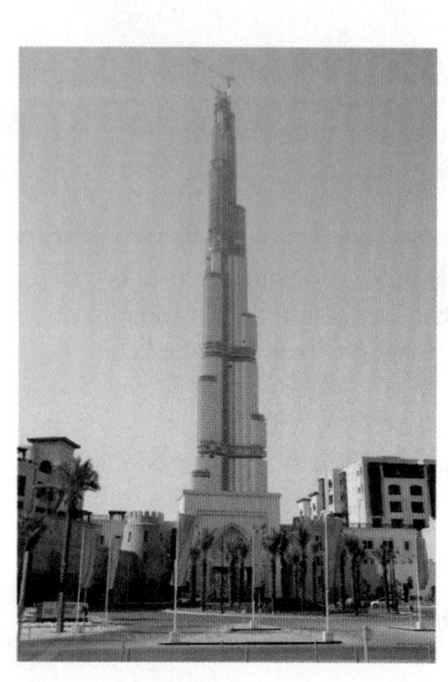

EXAMPLE 10 Translate the following problem to an equation.

The Burj Dubai in the United Arab Emirates is the world's tallest building. At 818 m, it is 310 m taller than Taipei 101 in Taiwan. How tall is Taipei 101?
Source: www.infoplease.com

SOLUTION We let h represent the height, in meters, of Taipei 101. A rewording and translation follow:

Rewording:	The height of the Burj Dubai	is	310 m more than the height of Taipei 101.
	↓	↓	↓
Translating:	818	$=$	$h + 310$

Try Exercise 69.

MODELS

When we translate a problem into mathematical language, we say that we **model** the problem. A **mathematical model** is a mathematical representation of a real-world situation.

Information about a problem is often given as a set of numbers, called **data.** Sometimes data follow a pattern that can be modeled using an equation.

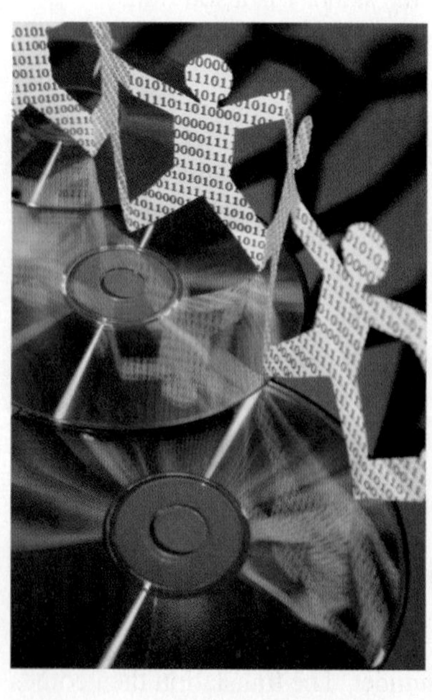

EXAMPLE 11 Music. The table below lists the amount charged for several purchases from an online music store. We let a represent the amount charged, in dollars, and n the number of songs. Find an equation giving a in terms of n.

Number of Songs Purchased, n	Amount Charged, a
2	$1.98
3	2.97
5	4.95
10	9.90

SOLUTION To write an equation for a **in terms of** n means that a will be on one side of the equals sign and an expression involving n will be on the other side.

We look for a pattern in the data. Since the amount charged increases as the number of songs increases, we can try dividing the amount by the number of songs:

$$1.98/2 = 0.99;$$
$$2.97/3 = 0.99;$$
$$4.95/5 = 0.99;$$
$$9.90/10 = 0.99.$$

The quotient is the same, 0.99, for each pair of numbers. Thus each song costs \$0.99. We reword and translate as follows:

Rewording: The amount charged is 0.99 times the number of songs.

Translating: $a = 0.99 \cdot n$

Try Exercise 79.

STUDY TIP

Get the Facts

In each section of this textbook, you will find a Study Tip. These tips are intended to help improve your math study skills. As the first Study Tip, we suggest that you complete this chart.

Instructor

Name ______

Office hours and location ______

Phone number ______

E-mail address ______

Math lab on campus

Location ______

Phone ______

Hours ______

Tutoring

Campus location ______

Phone ______

Hours ______

Classmates

1. Name ______

Phone number ______

E-mail address ______

2. Name ______

Phone number ______

E-mail address ______

Important supplements

See the preface for a complete list of available supplements.

Supplements recommended by the instructor

______

______

______

______

Translating for Success

1. Twice the difference of a number and 11

2. The product of a number and 11 is 2.

3. Twice the difference of two numbers is 11.

4. The quotient of twice a number and 11

5. The quotient of 11 and the product of two numbers

6. Eleven times the sum of a number and twice the number

7. Twice the sum of two numbers is 11.

8. Two more than twice a number is 11.

9. Twice the sum of 11 times a number and 2

10. Twenty percent of some number is 11.

Translate to an expression or an equation and match that translation with one of the choices A–O below. Do not solve.

A. $x = 0.2(11)$

B. $\frac{2x}{11}$

C. $2x + 2 = 11$

D. $2(11x + 2)$

E. $11x = 2$

F. $0.2x = 11$

G. $11(2x - y)$

H. $2(x - 11)$

I. $11 + 2x = 2$

J. $2x + y = 11$

K. $2(x - y) = 11$

L. $11(x + 2x)$

M. $2(x + y) = 11$

N. $2 + \frac{x}{11}$

O. $\frac{11}{xy}$

Answers on page A-1

An additional, animated version of this activity appears in MyMathLab. To use MyMathLab, you need a course ID and a student access code. Contact your instructor for more information.

1.1 Exercise Set

FOR EXTRA HELP MyMathLab Math XL PRACTICE WATCH DOWNLOAD READ REVIEW  

Concept Reinforcement *Classify each of the following as either an expression or an equation.*

1. $4x + 7$
2. $3x = 21$
3. $2x - 5 = 9$
4. $5(x - 2)$
5. $38 = 2t$
6. $45 = a - 1$
7. $4a - 5b$
8. $3t + 4 = 19$
9. $2x - 3y = 8$
10. $12 - 4xy$
11. $r(t + 7) + 5$
12. $9a + b$

To the student and the instructor: The Try Exercises *for examples are indicated by a shaded block on the exercise number. Answers to these exercises appear at the end of the exercise set as well as at the back of the book.*

Evaluate.

13. $3a$, for $a = 9$
14. $8x$, for $x = 7$
15. $t + 6$, for $t = 2$
16. $13 - r$, for $r = 9$
17. $\dfrac{x + y}{4}$, for $x = 2$ and $y = 14$
(*Hint*: Add $x + y$ before dividing by 4.)
18. $\dfrac{c + d}{7}$, for $c = 15$ and $d = 20$
19. $\dfrac{m - n}{2}$, for $m = 20$ and $n = 6$
20. $\dfrac{x - y}{6}$, for $x = 23$ and $y = 5$
21. $\dfrac{9m}{q}$, for $m = 6$ and $q = 18$
22. $\dfrac{5z}{y}$, for $z = 9$ and $y = 15$

To the student and the instructor: The calculator symbol, ▦, indicates those exercises designed to be solved with a calculator.

▦ *Evaluate using a calculator.*

23. $27a - 18b$, for $a = 136$ and $b = 13$
24. $19xy - 9x + 13y$, for $x = 87$ and $y = 29$

Substitute to find the value of each expression.

25. *Basketball.* The area of a rectangle with base b and height h is bh. A regulation basketball backboard is 6 ft wide and $3\frac{1}{2}$ ft high. Find the area of the backboard.
Source: NBA

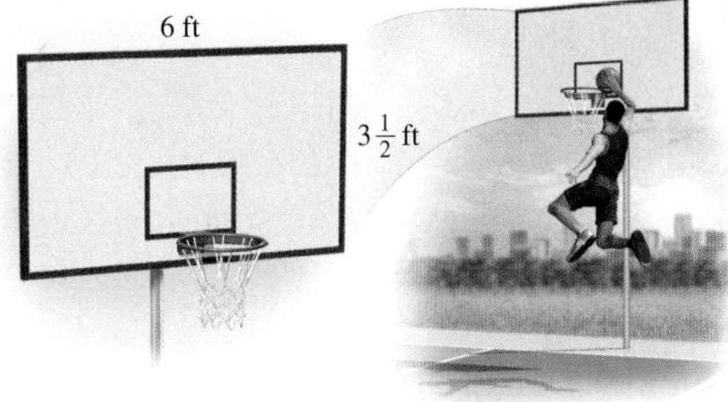

26. *Travel Time.* The length of a flight from Seattle, Washington, to St. Paul, Minnesota, is approximately 1400 mi. The time, in hours, for the flight is

$$\frac{1400}{v},$$

where v is the velocity, in miles per hour. How long will a flight take at a velocity of 400 mph?

27. *Zoology.* A great white shark has triangular teeth. Each tooth measures about 5 cm across the base and has a height of 6 cm. Find the surface area of the front side of one such tooth. (See Example 3.)

28. *Work Time.* Alan takes twice as long to do a job as Connor does. Suppose t represents the time it takes Connor to do the job. Then $2t$ represents the time it takes Alan. How long does it take Alan if Connor takes **(a)** 30 sec? **(b)** 35 min? **(c)** $2\frac{1}{2}$ hr?

29. *Area of a Parallelogram.* The area of a parallelogram with base b and height h is bh. Edward Tufte's sculpture *Spring Arcs* is in the shape of a parallelogram with base 67 ft and height 12 ft. What is the area of the parallelogram?
Source: edwardtufte.com

Spring Arcs (2004), Edward Tufte. Solid stainless steel, footprint 12′ × 67′/www.tufte.com

30. *Olympic Softball.* A softball player's batting average is h/a, where h is the number of hits and a is the number of "at bats." In the 2008 Summer Olympics, Jessica Mendoza had 8 hits in 22 at bats. What was her batting average? Round to the nearest thousandth.
Source: sports_reference.com

Translate to an algebraic expression.

31. 5 more than Ron's age
32. 8 times Luke's speed
33. The product of 4 and a
34. 7 more than Lou's weight
35. 9 less than c
36. 4 less than d
37. 6 increased by q
38. 11 increased by z
39. The difference of m and n
40. t subtracted from p
41. x less than y
42. 2 less than Lorrie's age
43. x divided by w
44. The quotient of two numbers
45. The sum of the box's length and height
46. The sum of d and f
47. Panya's speed minus twice the wind speed
48. The product of 9 and twice m
49. Twelve less than a quarter of some number
50. Twenty less than six times a number
51. Eight times the difference of two numbers
52. One third of the sum of two numbers
53. 64% of the women attending
54. 38% of a number

Determine whether the given number is a solution of the given equation.

55. 15; $x + 17 = 32$
56. 75; $y + 28 = 93$
57. 93; $a - 28 = 75$
58. 12; $8t = 96$
59. 63; $\frac{t}{7} = 9$
60. 52; $\frac{x}{8} = 6$
61. 3; $\frac{108}{x} = 36$
62. 7; $\frac{94}{y} = 12$

Translate each problem to an equation. Do not solve.

63. Seven times what number is 1596?
64. What number added to 73 is 201?
65. When 42 is multiplied by a number, the result is 2352. Find the number.
66. When 345 is added to a number, the result is 987. Find the number.
67. *Chess.* A chess board has 64 squares. If pieces occupy 19 squares, how many squares are unoccupied?
68. *Hours Worked.* A carpenter charges \$35 an hour. How many hours did she work if she billed a total of \$3150?
69. *Recycling.* Currently, Americans recover 33.4% of all municipal solid waste. This is the same as recovering 85 million tons. What is the total amount of waste generated?
Source: Environmental Protection Agency
70. *Travel to Work.* For U.S. cities with populations greater than 5000, the longest average commute is 59.8 min in Indian Wells, Arizona. This is 51.2 min longer than the shortest average commute, which is in Fort Bliss, Texas. How long is the average commute in Fort Bliss?
Source: www.city-data.com

In each of Exercises 71–78, match the phrase or sentence with the appropriate expression or equation from the column on the right.

71. ____ Twice the sum of two numbers

72. ____ Five less than a number is nine.

73. ____ Two more than a number is five.

74. ____ Half the product of two numbers

75. ____ Three times the sum of a number and five

76. ____ Twice the sum of two numbers is 48.

77. ____ One less than the product of two numbers is 49.

78. ____ Six more than the quotient of two numbers

a) $\frac{x}{y} + 6$

b) $2(x + y) = 48$

c) $\frac{1}{2} \cdot a \cdot b$

d) $t + 2 = 5$

e) $ab - 1 = 49$

f) $2(m + n)$

g) $3(t + 5)$

h) $x - 5 = 9$

79. *Nutrition.* The number of grams f of dietary fiber recommended daily for children depends on the age a of the child, as shown in the table below. Find an equation for f in terms of a.

Age of Child, a (in years)	Grams of Dietary Fiber Recommended Daily, f
3	8
4	9
5	10
6	11
7	12
8	13

Source: The American Health Foundation

80. *Tuition.* The table below lists the tuition costs for students taking various numbers of hours of classes. Find an equation for the cost c of tuition for a student taking h hours of classes.

Number of Class Hours, h	Tuition, c
12	$1200
15	1500
18	1800
21	2100

81. *Postage Rates.* The U.S. Postal Service charges extra for packages that must be processed by hand. The table below lists machinable and nonmachinable costs for certain packages. Find an equation for the nonmachinable cost n in terms of the machinable cost m.

Weight (in pounds)	Machinable Cost, m	Nonmachinable Cost, n
1	$2.74	$4.95
2	3.08	5.29
3	3.42	5.63

Source: pe.usps.gov

82. *Foreign Currency.* On Emily's trip to France, she used her debit card to withdraw money. The table below lists the amounts r that she received and the amounts s that were subtracted from her account. Find an equation for r in terms of s.

Amount Received, r (in U.S. dollars)	Amount Subtracted, s (in U.S. dollars)
$150	$153
75	78
120	123

83. *Number of Drivers.* The table below lists the number of vehicle miles v traveled annually per household by the number of drivers d in the household. Find an equation for v in terms of d.

Number of Drivers, d	Number of Vehicle Miles Traveled, v
1	10,000
2	20,000
3	30,000
4	40,000

Source: Energy Information Administration

84. *Meteorology.* The table below lists the number of centimeters of water w to which various amounts of snow s will melt under certain conditions. Find an equation for w in terms of s.

Depth of Snow, s (in centimeters)	Depth of Water, w (in centimeters)
120	12
135	13.5
160	16
90	9

To the student and the instructor: Thinking and writing exercises, denoted by TW, are meant to be answered using one or more English sentences. Because answers to many writing exercises will vary, solutions are not listed in the answers at the back of the book.

TW **85.** What is the difference between a variable, a variable expression, and an equation?

TW **86.** What does it mean to evaluate an algebraic expression?

SYNTHESIS

To the student and the instructor: Synthesis exercises are designed to challenge students to extend the concepts or skills studied in each section. Many synthesis exercises require the assimilation of skills and concepts from several sections.

TW **87.** If the lengths of the sides of a square are doubled, is the area doubled? Why or why not?

TW **88.** Write a problem that translates to $1998 + t = 2006$.

89. Signs of Distinction charges $120 per square foot for handpainted signs. The town of Belmar commissioned a triangular sign with a base of 3.0 ft and a height of 2.5 ft. How much will the sign cost?

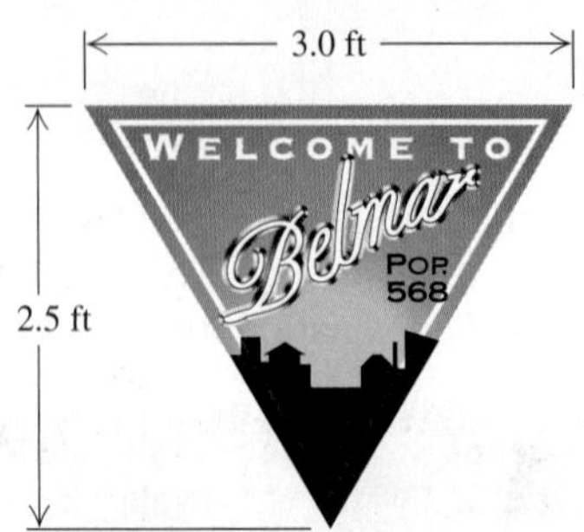

90. Find the area that is shaded in the figure below.

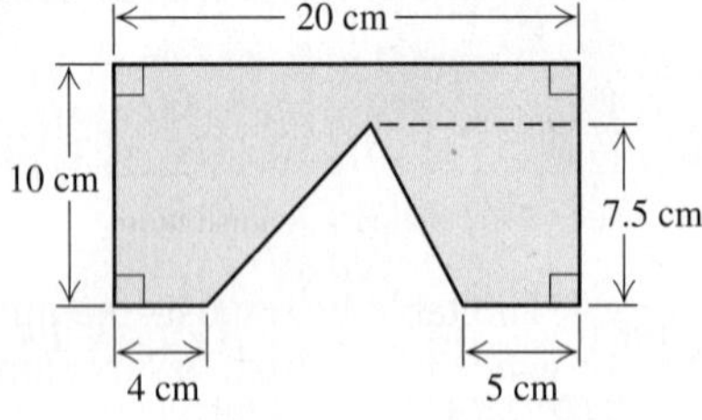

91. Evaluate $\frac{x + y}{2}$ when y is twice x and $x = 6$.

92. Evaluate $\frac{a - b}{3}$ when a is three times b and $a = 18$.

Answer each question with an algebraic expression.

93. If $w + 3$ is a whole number, what is the next whole number after it?

94. If $d + 2$ is an odd number, what is the preceding odd number?

Translate to an algebraic expression.

95. The perimeter of a rectangle with length l and width w (perimeter means distance around)

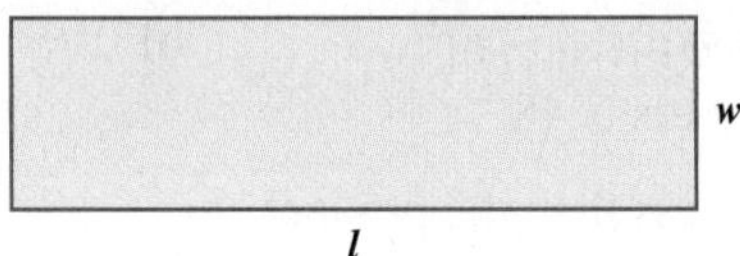

96. The perimeter of a square with side s (perimeter means distance around)

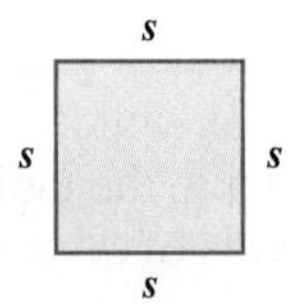

97. Ella's race time, assuming she took 5 sec longer than Kyle and Kyle took 3 sec longer than Amy. Assume that Amy's time was t seconds.

98. Ray's age 7 years from now if he is 2 years older than Monique and Monique is a years old

TW **99.** If the height of a triangle is doubled, is its area also doubled? Why or why not?

Try Exercise Answers: Section 1.1

13. 27 **23.** 3438 **25.** 21 ft^2 **27.** 15 cm^2 **33.** $4a$, or $a \cdot 4$ **47.** Let p represent Panya's speed and w the wind speed; $p - 2w$ **55.** Yes **63.** Let x represent the unknown number; $7x = 1596$ **69.** Let w represent the amount of solid waste generated, in millions of tons; 33.4% of $w = 85$, or $0.334w = 85$ **79.** $f = a + 5$

Collaborative Corner

Teamwork

Focus: Group problem solving; working collaboratively
Time: 15 minutes
Group Size: 2

Are two (or more) heads better than one? Try this activity and decide whether group work can help us solve problems.

ACTIVITY

1. The left-hand column below contains the names of 12 colleges. A scrambled list of the names of their sports teams is on the right. As a group, match the names of the colleges to the teams.

1. University of Texas	**a)** Antelopes
2. Western State College of Colorado	**b)** Fighting Banana Slugs
3. University of North Carolina	**c)** Sea Warriors
4. University of Massachusetts	**d)** Gators
5. Hawaii Pacific University	**e)** Mountaineers
6. University of Nebraska	**f)** Sailfish
7. University of California, Santa Cruz	**g)** Longhorns
8. University of Louisiana at Lafayette	**h)** Tar Heels
9. Grand Canyon University	**i)** Seawolves
10. Palm Beach Atlantic College	**j)** Ragin' Cajuns
11. University of Alaska, Anchorage	**k)** Cornhuskers
12. University of Florida	**l)** Minutemen

2. After working for 5 min, confer with another group and reach mutual agreement.
3. Does the class agree on all 12 pairs?
4. Do you agree that group collaboration increases our ability to solve problems?

1.2 The Commutative, Associative, and Distributive Laws

- Equivalent Expressions
- The Commutative Laws
- The Associative Laws
- The Distributive Law
- The Distributive Law and Factoring

In order to solve equations, we must be able to work with algebraic expressions. The commutative, associative, and distributive laws discussed in this section allow us to write *equivalent expressions* that will simplify our work. Indeed, much of this text is devoted to finding equivalent expressions.

EQUIVALENT EXPRESSIONS

The expressions $4 + 4 + 4$, $3 \cdot 4$, and $4 \cdot 3$ all represent the same number, 12. Expressions that represent the same number are said to be **equivalent.** The equivalent expressions $t + 18$ and $18 + t$ were used on p. 6 when we translated "eighteen more

STUDY TIP 

Learn by Example

The examples in each section are designed to prepare you for success with the exercise set. Study the step-by-step solutions of the examples, noting that color is used to indicate substitutions and to call attention to the new steps in multistep examples. The time you spend studying the examples will save you valuable time when you do your assignment.

than a number." These expressions are equivalent because they represent the same number for any value of t. We can illustrate this by making some choices for t.

When $t = 3$, $t + 18 = 3 + 18 = 21$
and $18 + t = 18 + 3 = 21$.
When $t = 40$, $t + 18 = 40 + 18 = 58$
and $18 + t = 18 + 40 = 58$.

THE COMMUTATIVE LAWS

Recall that changing the order in addition or multiplication does not change the result. Equations like $3 + 78 = 78 + 3$ and $5 \cdot 14 = 14 \cdot 5$ illustrate this idea and show that addition and multiplication are **commutative.**

The Commutative Laws *For Addition.* For any numbers a and b,

$$a + b = b + a.$$

(Changing the order of addition does not affect the answer.)

For Multiplication. For any numbers a and b,

$$ab = ba.$$

(Changing the order of multiplication does not affect the answer.)

EXAMPLE 1 Use the commutative laws to write an expression equivalent to each of the following: **(a)** $y + 5$; **(b)** $9x$; **(c)** $7 + ab$.

SOLUTION

a) $y + 5$ is equivalent to $5 + y$ by the commutative law of addition.

b) $9x$ is equivalent to $x \cdot 9$ by the commutative law of multiplication.

c) $7 + ab$ is equivalent to $ab + 7$ by the commutative law of *addition.*

$7 + ab$ is also equivalent to $7 + ba$ by the commutative law of *multiplication.*

$7 + ab$ is also equivalent to $ba + 7$ by the two commutative laws, used together.

Try Exercise 11.

THE ASSOCIATIVE LAWS

Parentheses are used to indicate groupings. We normally simplify within the parentheses first. For example,

$3 + (8 + 4) = 3 + 12 = 15$
and $(3 + 8) + 4 = 11 + 4 = 15$.

Similarly,

$4 \cdot (2 \cdot 3) = 4 \cdot 6 = 24$
and $(4 \cdot 2) \cdot 3 = 8 \cdot 3 = 24$.

Note that, so long as only addition or only multiplication appears in an expression, changing the grouping does not change the result. Equations such as $3 + (7 + 5) = (3 + 7) + 5$ and $4(5 \cdot 3) = (4 \cdot 5)3$ illustrate that addition and multiplication are **associative.**

Student Notes

Examine and compare the statements of the commutative laws and the associative laws. Note that the order of the variables changes in the commutative laws. In the associative laws, the order does not change, but the grouping does.

The Associative Laws *For Addition.* For any numbers a, b, and c,

$$a + (b + c) = (a + b) + c.$$

(Numbers can be grouped in any manner for addition.)

For Multiplication. For any numbers a, b, and c,

$$a \cdot (b \cdot c) = (a \cdot b) \cdot c.$$

(Numbers can be grouped in any manner for multiplication.)

EXAMPLE 2 Use an associative law to write an expression equivalent to each of the following: **(a)** $y + (z + 3)$; **(b)** $(8x)y$.

SOLUTION

a) $y + (z + 3)$ is equivalent to $(y + z) + 3$ by the associative law of addition.

b) $(8x)y$ is equivalent to $8(xy)$ by the associative law of multiplication.

Try Exercise 33.

When only addition or only multiplication is involved, parentheses do not change the result. For that reason, we sometimes omit them altogether. Thus,

$$x + (y + 7) = x + y + 7, \quad \text{and} \quad l(wh) = lwh.$$

A sum such as $(5 + 1) + (3 + 5) + 9$ can be simplified by pairing numbers that add to 10. The associative and commutative laws allow us to do this:

$$\begin{aligned}(5 + 1) + (3 + 5) + 9 &= 5 + 5 + 9 + 1 + 3\\ &= 10 + 10 + 3 = 23.\end{aligned}$$

EXAMPLE 3 Use the commutative and/or associative laws of addition to write two expressions equivalent to $(7 + x) + 3$. Then simplify.

SOLUTION

$$(7 + x) + 3 = (x + 7) + 3$$ **Using the commutative law; $(x + 7) + 3$ is one equivalent expression.**

$$= x + (7 + 3)$$ **Using the associative law; $x + (7 + 3)$ is another equivalent expression.**

$$= x + 10$$ **Simplifying**

Try Exercise 39.

EXAMPLE 4 Use the commutative and/or associative laws of multiplication to write two expressions equivalent to $2(x \cdot 3)$. Then simplify.

SOLUTION

$$2(x \cdot 3) = 2(3x)$$ **Using the commutative law; $2(3x)$ is one equivalent expression.**

$$= (2 \cdot 3)x$$ **Using the associative law; $(2 \cdot 3)x$ is another equivalent expression.**

$$= 6x$$ **Simplifying**

Try Exercise 41.

Student Notes

To remember the names *commutative*, *associative*, and *distributive*, first understand the concept. Next, use everyday life to link the word to the concept. For example, think of commuting to and from college as changing the order of appearance.

THE DISTRIBUTIVE LAW

The *distributive law* is probably the single most important law for manipulating algebraic expressions. Unlike the commutative and associative laws, the distributive law uses multiplication together with addition.

The distributive law relates expressions like $5(x + 2)$ and $5x + 10$, which involve both multiplication and addition. When two numbers are multiplied, the result is a **product.** The parts of the product are called **factors.** When two numbers are added, the result is a **sum.** The parts of the sum are called **terms.**

$5\ (x + 2)$ is a product. The factors are 5 and $(x + 2)$.

This factor is a sum. The terms are x and 2.

$5x + 10$ is a sum. The terms are $5x$ and 10.

This term is a product. The factors are 5 and x.

In general, a term is a number, a variable, or a product or a quotient of numbers and/or variables. Terms are separated by plus signs.

EXAMPLE 5 List the terms in the expression $3s + st + \frac{2s}{t}$.

SOLUTION Terms are separated by plus signs, so the terms in $3s + st + \frac{2s}{t}$ are $3s$, st, and $\frac{2s}{t}$.

Try Exercise 63.

EXAMPLE 6 List the factors in $x(3 + y)$.

SOLUTION Factors are parts of products, so the factors in $x(3 + y)$ are x and $(3 + y)$.

Try Exercise 81.

You have already used the distributive law although you may not have realized it at the time. To illustrate, try to multiply $3 \cdot 21$ mentally. Many people find the product, 63, by thinking of 21 as $20 + 1$ and then multiplying 20 by 3 and 1 by 3. The sum of the two products, $60 + 3$, is 63. Note that if the 3 does not multiply *both* 20 and 1, the result will not be correct.

EXAMPLE 7 Compute in two ways: $4(7 + 2)$.

SOLUTION

a) As in the discussion of $3(20 + 1)$ above, to compute $4(7 + 2)$, we can multiply both 7 and 2 by 4 and add the results:

$$\begin{aligned} 4(7 + 2) &= 4 \cdot 7 + 4 \cdot 2 && \textbf{Multiplying both 7 and 2 by 4} \\ &= 28 + 8 = 36. && \textbf{Adding} \end{aligned}$$

b) By first adding inside the parentheses, we get the same result in a different way:

$$\begin{aligned} 4(7 + 2) &= 4(9) && \textbf{Adding; } 7 + 2 = 9 \\ &= 36. && \textbf{Multiplying} \end{aligned}$$

The Distributive Law For any numbers a, b, and c,

$$a(b + c) = ab + ac.$$

(The product of a number and a sum can be written as the sum of two products.)

EXAMPLE 8 Multiply: $3(x + 2)$.

SOLUTION Since $x + 2$ cannot be simplified unless a value for x is given, we use the distributive law:

$$3(x + 2) = 3 \cdot x + 3 \cdot 2$$ **Using the distributive law**

$$= 3x + 6.$$ **$3 \cdot x$ is the same as $3x$**

Try Exercise 47.

The distributive law can also be used when more than two terms are inside the parentheses.

EXAMPLE 9 Multiply: $6(s + 2 + 5w)$.

SOLUTION

$$6(s + 2 + 5w) = 6 \cdot s + 6 \cdot 2 + 6 \cdot 5w$$ **Using the distributive law**

$$= 6s + 12 + (6 \cdot 5)w$$ **Using the associative law for multiplication**

$$= 6s + 12 + 30w$$

Try Exercise 57.

Because of the commutative law of multiplication, the distributive law can be used on the "right": $(b + c)a = ba + ca$.

EXAMPLE 10 Multiply: $(c + 4)5$.

SOLUTION

$$(c + 4)5 = c \cdot 5 + 4 \cdot 5$$ **Using the distributive law on the right**

$$= 5c + 20$$

Try Exercise 51.

CAUTION! To use the distributive law for removing parentheses, be sure to multiply *each* term inside the parentheses by the multiplier outside.

THE DISTRIBUTIVE LAW AND FACTORING

If we use the distributive law in reverse, we have the basis of a process called **factoring:** $ab + ac = a(b + c)$. Factoring involves multiplication: To **factor** an expression means to write an equivalent expression that is a *product*. Recall that the parts of the product are called factors. Note that "factor" can be used as either a verb or a noun. A **common factor** is a factor that appears in every term in an expression.

EXAMPLE 11 Use the distributive law to factor each of the following.

a) $3x + 3y$

b) $7x + 21y + 7$

SOLUTION

a) By the distributive law,

$$3x + 3y = 3(x + y).$$ **The *common factor* for 3x and 3y is 3.**

b) $7x + 21y + 7 = 7 \cdot x + 7 \cdot 3y + 7 \cdot 1$ **The common factor is 7.**

$= 7(x + 3y + 1)$ **Using the distributive law**

Be sure to include both the 1 and the common factor, 7.

Try Exercise 69.

To check our factoring, we multiply to see if the original expression is obtained. For example, to check the **factorization** in Example 11(b), note that

$$7(x + 3y + 1) = 7x + 7 \cdot 3y + 7 \cdot 1$$
$$= 7x + 21y + 7.$$

Since $7x + 21y + 7$ is what we began with in Example 11(b), we have a check.

1.2 Exercise Set

FOR EXTRA HELP MyMathLab

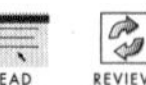

PRACTICE WATCH DOWNLOAD READ REVIEW

Concept Reinforcement *Complete each sentence using one of these terms:* commutative, associative, *or* distributive.

1. $8 + t$ is equivalent to $t + 8$ by the ____________ law for addition.

2. $3(xy)$ is equivalent to $(3x)y$ by the ____________ law for multiplication.

3. $(5b)c$ is equivalent to $5(bc)$ by the ____________ law for multiplication.

4. mn is equivalent to nm by the ____________ law for multiplication.

5. $x(y + z)$ is equivalent to $xy + xz$ by the ____________ law.

6. $(9 + a) + b$ is equivalent to $9 + (a + b)$ by the ____________ law for addition.

7. $a + (6 + d)$ is equivalent to $(a + 6) + d$ by the ____________ law for addition.

8. $t + 4$ is equivalent to $4 + t$ by the ____________ law for addition.

9. $x \cdot 7$ is equivalent to $7 \cdot x$ by the ____________ law for multiplication.

10. $2(a + b)$ is equivalent to $2 \cdot a + 2 \cdot b$ by the ____________ law.

Use the commutative law of addition to write an equivalent expression.

11. $7 + x$

12. $a + 2$

13. $x + 3y$

14. $9x + 3y$

15. $ab + c$

16. $uv + xy$

17. $5(a + 1)$

18. $9(x + 5)$

Use the commutative law of multiplication to write an equivalent expression.

19. $2 \cdot a$

20. xy

21. st

22. $4x$

23. $5 + ab$

24. $x + 3y$

25. $5(a + 1)$

26. $9(x + 5)$

Use the associative law of addition to write an equivalent expression.

27. $(a + 5) + b$

28. $(5 + m) + r$

29. $r + (t + 7)$

30. $x + (2 + y)$

31. $(ab + c) + d$

32. $(m + np) + r$

Use the associative law of multiplication to write an equivalent expression.

33. $(7m)n$

34. $(13x)y$

35. $2(ab)$

36. $9(rp)$

37. $3[2(a + b)]$

38. $5[x(2 + y)]$

Use the commutative and/or associative laws to write two equivalent expressions. Then simplify. Answers may vary.

39. $2 + (t + 6)$

40. $(11 + v) + 4$

41. $(3a) \cdot 7$

42. $5(x \cdot 8)$

Use the commutative and/or associative laws to show why the expression on the left is equivalent to the expression on the right. Write a series of steps with labels, as in Example 4.

43. $(5 + x) + 2$ is equivalent to $x + 7$

44. $(2a)4$ is equivalent to $8a$

45. $(m \cdot 3)7$ is equivalent to $21m$

46. $4 + (9 + x)$ is equivalent to $x + 13$

Multiply.

47. $4(a + 3)$

48. $3(x + 5)$

49. $6(1 + x)$

50. $6(v + 4)$

51. $(n + 5)2$

52. $(1 + t)3$

53. $8(3x + 5y)$

54. $7(4x + 5y)$

55. $9(2x + 6)$

56. $9(6m + 7)$

57. $5(r + 2 + 3t)$

58. $4(5x + 8 + 3p)$

59. $(a + b)2$

60. $(x + 2)7$

61. $(x + y + 2)5$

62. $(2 + a + b)6$

List the terms in each expression.

63. $x + xyz + 19$

64. $9 + 17a + abc$

65. $2a + \frac{a}{b} + 5b$

66. $3xy + 20 + \frac{4a}{b}$

Use the distributive law to factor each of the following. Check by multiplying.

67. $2a + 2b$

68. $5y + 5z$

69. $7 + 7y$

70. $13 + 13x$

71. $18x + 3$

72. $20a + 5$

73. $5x + 10 + 15y$

74. $3 + 27b + 6c$

75. $12x + 9$

76. $25y + 30$

77. $3a + 9b$

78. $5a + 15b$

79. $44x + 88y + 66z$

80. $24a + 48b + 60$

List the factors in each expression.

81. st

82. $5x$

83. $3(x + y)$

84. $(a + b)6$

85. $7 \cdot a \cdot b$

86. $m \cdot n \cdot 2$

87. $(a - b)(x - y)$

88. $(3 - a)(b + c)$

TW **89.** Explain how you can determine the terms and the factors in an expression.

TW **90.** Explain how the distributive, commutative, and associative laws can be used to show that $2(3x + 4y)$ is equivalent to $6x + 8y$.

SKILL REVIEW

To the student and the instructor: *Exercises included for Skill Review include skills previously studied in the text. Often these exercises provide preparation for the next section of the text. The numbers in brackets immediately following the directions or exercise indicate the section in which the skill was introduced. For example, the notation* [1.1] *refers to Chapter* 1, *Section* 1. *The answers to all Skill Review exercises appear at the back of the book. If a Skill Review exercise gives you difficulty, review the material in the indicated section of the text.*

Translate to an algebraic expression. [1.1]

91. Half of Kylie's salary

92. Twice the sum of m and 7

SYNTHESIS

TW **93.** Is subtraction commutative? Why or why not?

TW **94.** Is division associative? Why or why not?

Tell whether the expressions in each pairing are equivalent. Then explain why or why not.

95. $8 + 4(a + b)$ and $4(2 + a + b)$

96. $5(a \cdot b)$ and $5 \cdot a \cdot 5 \cdot b$

97. $7 \div (3m)$ and $m \cdot 3 \div 7$

98. $(rt + st)5$ and $5t(r + s)$

99. $30y + x \cdot 15$ and $5[2(x + 3y)]$

100. $[c(2 + 3b)]5$ and $10c + 15bc$

TW **101.** Evaluate the expressions $3(2 + x)$ and $6 + x$ for $x = 0$. Do your results indicate that $3(2 + x)$ and $6 + x$ are equivalent? Why or why not?

TW **102.** Factor $15x + 40$. Then evaluate both $15x + 40$ and the factorization for $x = 4$. Do your results *guarantee* that the factorization is correct? Why or why not? (*Hint*: See Exercise 101.)

Try Exercise Answers: Section 1.2

11. $x + 7$ **33.** $7(mn)$ **39.** $2 + (6 + t)$; $(2 + 6) + t$; $8 + t$ **41.** $(a \cdot 3) \cdot 7$; $a(3 \cdot 7)$; $a \cdot 21$, or $21a$ **47.** $4a + 12$ **51.** $2n + 10$ **57.** $5r + 10 + 15t$ **63.** x, xyz, 19 **69.** $7(1 + y)$ **81.** s, t

1.3 Fraction Notation

- Factors and Prime Factorizations
- Fraction Notation
- Multiplication and Simplification
- Division
- Addition and Subtraction

This section covers multiplication, division, addition, and subtraction with fractions. Although much of this may be review, note that fraction expressions that contain variables are also included.

FACTORS AND PRIME FACTORIZATIONS

We first review how *natural numbers* are factored. **Natural numbers** can be thought of as the counting numbers:

$$1, 2, 3, 4, 5, \ldots.^*$$

(The dots indicate that the pattern continues without ending.)

Since factors are parts of products, to factor a number, we express it as a product of two or more numbers. Several factorizations of 12 are $1 \cdot 12$, $2 \cdot 6$, $3 \cdot 4$, and $2 \cdot 2 \cdot 3$.

To list all factors of a number, we carefully list the factorizations of the number.

EXAMPLE 1 List all factors of 18.

SOLUTION Beginning at 1, we check all natural numbers to see if they are factors of 18. If they are, we write the factorization. We stop when we have already included the next natural number in a factorization.

1 is a factor of every number.	$1 \cdot 18$
2 is a factor of 18.	$2 \cdot 9$
3 is a factor of 18.	$3 \cdot 6$
4 is *not* a factor of 18.	
5 is *not* a factor of 18.	

6 is a factor of 18, but we have already listed it in the product $3 \cdot 6$.

We need check no additional numbers, because any natural number greater than 6 must be paired with a factor less than 6.

*A similar collection of numbers, the **whole numbers,** includes 0: 0, 1, 2, 3, ...

We now write the factors of 18 beginning with 1, going down the list of factorizations writing the first factor, then up the list of factorizations writing the second factor:

$$1,\quad 2,\quad 3,\quad 6,\quad 9,\quad 18.$$

Try Exercise 5.

Some numbers have only two factors, the number itself and 1. Such numbers are called **prime.**

Prime Number A *prime number* is a natural number that has exactly two different factors: the number itself and 1. The first several primes are 2, 3, 5, 7, 11, 13, 17, 19, and 23.

If a natural number other than 1 is not prime, we call it **composite.**

EXAMPLE 2 Label each number as prime, composite, or neither: 29, 4, 1.

SOLUTION

29 is prime. It has exactly two different factors, 29 and 1.

4 is not prime. It has three different factors, 1, 2, and 4. It is composite.

1 is not prime. It does not have two *different* factors. It is neither prime nor composite.

Try Exercise 9.

Every composite number can be factored into a product of prime numbers. Such a factorization is called the **prime factorization** of that composite number.

Student Notes

When writing a factorization, you are writing an equivalent expression for the original number. Some students do this with a tree diagram:

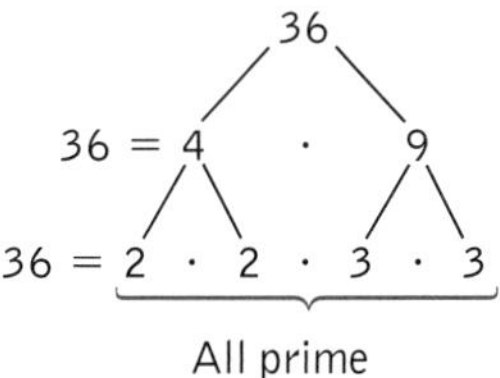

EXAMPLE 3 Find the prime factorization of 36.

SOLUTION We first factor 36 in any way that we can. One way is like this:

$$36 = 4 \cdot 9.$$

The factors 4 and 9 are not prime, so we factor them:

$$36 = 4 \cdot 9$$
$$= 2 \cdot 2 \cdot 3 \cdot 3. \qquad \textbf{2 and 3 are both prime.}$$

The prime factorization of 36 is $2 \cdot 2 \cdot 3 \cdot 3$.

Try Exercise 21.

FRACTION NOTATION

An example of **fraction notation** for a number is

$$\frac{2}{3}. \quad \begin{matrix} \leftarrow \textbf{Numerator} \\ \leftarrow \textbf{Denominator} \end{matrix}$$

The top number is called the **numerator,** and the bottom number is called the **denominator.** When the numerator and the denominator are the same nonzero number, we have fraction notation for the number 1.

Student Notes

Fraction notation for 1 appears in many contexts in algebra. Keep in mind that when the (nonzero) numerator of a fraction is the same as the denominator, the fraction is equivalent to 1.

Fraction Notation for 1 For any number a, except 0,

$$\frac{a}{a} = 1.$$

(Any nonzero number divided by itself is 1.)

Note that in the definition for fraction notation for the number 1, we have excluded 0. In fact, 0 cannot be the denominator of *any* fraction. Later in this chapter, we will discuss why denominators cannot be 0.

MULTIPLICATION AND SIMPLIFICATION

Recall from arithmetic that fractions are multiplied as follows.

Multiplication of Fractions For any two fractions $\frac{a}{b}$ and $\frac{c}{d}$,

$$\frac{a}{b} \cdot \frac{c}{d} = \frac{ac}{bd}.$$

(The numerator of the product is the product of the two numerators. The denominator of the product is the product of the two denominators.)

EXAMPLE 4 Multiply: **(a)** $\frac{2}{3} \cdot \frac{5}{7}$; **(b)** $\frac{4}{x} \cdot \frac{8}{y}$.

SOLUTION We multiply numerators as well as denominators.

a) $\frac{2}{3} \cdot \frac{5}{7} = \frac{2 \cdot 5}{3 \cdot 7} = \frac{10}{21}$

b) $\frac{4}{x} \cdot \frac{8}{y} = \frac{4 \cdot 8}{x \cdot y} = \frac{32}{xy}$

Try Exercise 53.

When one of the fractions being multiplied is 1, multiplying yields an equivalent expression because of the *identity property of* 1.

The Identity Property of 1 For any number a,

$$a \cdot 1 = 1 \cdot a = a.$$

(Multiplying a number by 1 gives that same number.) The number 1 is called the *multiplicative identity*.

When simplifying fraction notation, we use the identity property of 1. First, note that since $\frac{6}{6} = 1$, the expression $\frac{4}{5} \cdot \frac{6}{6}$ is equivalent to $\frac{4}{5} \cdot 1$. Thus we can say that

$$\frac{4}{5} = \frac{4}{5} \cdot 1 = \frac{4}{5} \cdot \frac{6}{6} = \frac{24}{30}.$$

This tells us that $\frac{24}{30}$ is equivalent to $\frac{4}{5}$.

The steps above are reversed by "removing a factor equal to 1"—in this case, $\frac{6}{6}$. By removing a factor equal to 1, we can *simplify* an expression like $\frac{24}{30}$ to an equivalent expression like $\frac{4}{5}$.

To simplify, we factor the numerator and the denominator, looking for the largest factor common to both. This is sometimes made easier by writing prime factorizations. After identifying common factors, we can express the fraction as a product of two fractions, one of which is in the form $\dfrac{a}{a}$.

Student Notes

The following rules can help you quickly determine whether 2, 3, or 5 is a factor of a number.

2 is a factor of a number if the number is even (the ones digit is 0, 2, 4, 6, or 8).

3 is a factor of a number if the sum of its digits is divisible by 3.

5 is a factor of a number if its ones digit is 0 or 5.

EXAMPLE 5 Simplify: **(a)** $\dfrac{15}{40}$; **(b)** $\dfrac{36}{24}$.

SOLUTION

a) Note that 5 is a factor of both 15 and 40:

$$\frac{15}{40} = \frac{3 \cdot 5}{8 \cdot 5}$$ **Factoring the numerator and the denominator, using the common factor, 5**

$$= \frac{3}{8} \cdot \frac{5}{5}$$ **Rewriting as a product of two fractions; $\frac{5}{5} = 1$**

$$= \frac{3}{8} \cdot 1 = \frac{3}{8}.$$ **Using the identity property of 1 (removing a factor equal to 1)**

b)

$$\frac{36}{24} = \frac{2 \cdot 2 \cdot 3 \cdot 3}{2 \cdot 2 \cdot 2 \cdot 3}$$ **Writing the prime factorizations and identifying common factors; 12/12 could also be used.**

$$= \frac{3}{2} \cdot \frac{2 \cdot 2 \cdot 3}{2 \cdot 2 \cdot 3}$$ **Rewriting as a product of two fractions; $\frac{2 \cdot 2 \cdot 3}{2 \cdot 2 \cdot 3} = 1$**

$$= \frac{3}{2} \cdot 1 = \frac{3}{2}$$ **Using the identity property of 1**

Try Exercise 35.

It is always wise to check your result to see if any common factors of the numerator and the denominator remain. (This will never happen if prime factorizations are used correctly.) If common factors remain, repeat the process by removing another factor equal to 1 to simplify your result.

Student Notes

Canceling is the same as removing a factor equal to 1; each pair of slashes indicates a factor of 1. Thus numerators and denominators must be factored, or written as products, before canceling is used.

You may have used a shortcut called "canceling" to remove a factor equal to 1 when working with fraction notation. With *great* concern, we mention it as a possible way to speed up your work. Canceling can be used only when removing common factors in numerators and denominators. Canceling *cannot* be used in sums or differences. Our concern is that "canceling" be used with understanding. Example 5(b) might have been done faster as follows:

$$\frac{36}{24} = \frac{\not{2} \cdot \not{2} \cdot 3 \cdot \not{3}}{\not{2} \cdot \not{2} \cdot 2 \cdot \not{3}} = \frac{3}{2}, \quad \text{or} \quad \frac{36}{24} = \frac{3 \cdot \not{12}}{2 \cdot \not{12}} = \frac{3}{2}, \quad \text{or} \quad \frac{\overset{\overset{3}{\not{18}}}{\not{36}}}{\underset{\underset{2}{\not{12}}}{\not{24}}} = \frac{3}{2}.$$

CAUTION! Unfortunately, canceling is often performed incorrectly:

$$\cancel{\frac{2+3}{2} = 3,} \quad \cancel{\frac{4-1}{4-2} = \frac{1}{2},} \quad \cancel{\frac{15}{54} = \frac{1}{4}.}$$

The above cancellations are incorrect because the expressions canceled are *not* factors. Correct simplifications are as follows:

$$\frac{2+3}{2} = \frac{5}{2}, \quad \frac{4-1}{4-2} = \frac{3}{2}, \quad \frac{15}{54} = \frac{5 \cdot \cancel{3}}{18 \cdot \cancel{3}} = \frac{5}{18}.$$

***Remember*: If you can't factor, you can't cancel! If in doubt, don't cancel!**

Sometimes it is helpful to use 1 as a factor in the numerator or the denominator when simplifying.

EXAMPLE 6 Simplify: $\frac{9}{72}$.

SOLUTION

$$\frac{9}{72} = \frac{1 \cdot 9}{8 \cdot 9}$$ Factoring and using the identity property of 1 to write 9 as $1 \cdot 9$

$$= \frac{1 \cdot \cancel{9}}{8 \cdot \cancel{9}} = \frac{1}{8}$$ Simplifying by removing a factor equal to 1; $\frac{9}{9} = 1$

Try Exercise 39.

DIVISION

Two numbers whose product is 1 are **reciprocals,** or **multiplicative inverses,** of each other. All numbers, except zero, have reciprocals. For example,

the reciprocal of $\frac{2}{3}$ is $\frac{3}{2}$ because $\frac{2}{3} \cdot \frac{3}{2} = \frac{6}{6} = 1$;
the reciprocal of 9 is $\frac{1}{9}$ because $9 \cdot \frac{1}{9} = \frac{9}{9} = 1$; and
the reciprocal of $\frac{1}{4}$ is 4 because $\frac{1}{4} \cdot 4 = 1$.

Reciprocals are used to rewrite division using multiplication.

Division of Fractions To divide two fractions, multiply by the reciprocal of the divisor:

$$\frac{a}{b} \div \frac{c}{d} = \frac{a}{b} \cdot \frac{d}{c}.$$

EXAMPLE 7 Divide: $\frac{1}{2} \div \frac{3}{5}$.

SOLUTION Note that the *divisor* is $\frac{3}{5}$:

$$\frac{1}{2} \div \frac{3}{5} = \frac{1}{2} \cdot \frac{5}{3} \qquad \textbf{$\frac{5}{3}$ is the reciprocal of $\frac{3}{5}$}$$
$$= \frac{5}{6}.$$

Try Exercise 73.

ADDITION AND SUBTRACTION

When denominators are the same, fractions are added or subtracted by adding or subtracting numerators and keeping the same denominator.

Addition and Subtraction of Fractions For any two fractions $\frac{a}{d}$ and $\frac{b}{d}$,

$$\frac{a}{d} + \frac{b}{d} = \frac{a+b}{d} \quad \text{and} \quad \frac{a}{d} - \frac{b}{d} = \frac{a-b}{d}.$$

EXAMPLE 8 Add and simplify: $\frac{4}{8} + \frac{5}{8}$.

SOLUTION We add the numerators and keep the common denominator:

$$\frac{4}{8} + \frac{5}{8} = \frac{4+5}{8} = \frac{9}{8}.$$

You can think of this as 4 eighths + 5 eighths = 9 eighths, or $\frac{9}{8}$.

Try Exercise 63.

In arithmetic, we often write $1\frac{1}{8}$ rather than the "improper" fraction $\frac{9}{8}$. In algebra, $\frac{9}{8}$ is generally more useful and is quite "proper" for our purposes.

When denominators are different, we use the identity property of 1 and multiply to obtain a common denominator. Then we add, as in Example 8.

Student Notes

See Section 7.3 for an explanation of determining a common denominator. Although it may not be the *best* choice, the product of the denominators can always be used as a common denominator.

EXAMPLE 9 Add or subtract as indicated: **(a)** $\frac{7}{8} + \frac{5}{12}$; **(b)** $\frac{9}{8} - \frac{4}{5}$.

SOLUTION

a) The number 24 is divisible by both 8 and 12. We multiply both $\frac{7}{8}$ and $\frac{5}{12}$ by suitable forms of 1 to obtain two fractions with denominators of 24:

$$\frac{7}{8} + \frac{5}{12} = \frac{7}{8} \cdot \frac{3}{3} + \frac{5}{12} \cdot \frac{2}{2}$$

Multiplying by 1.
Since $8 \cdot 3 = 24$, we multiply $\frac{7}{8}$ by $\frac{3}{3}$.
Since $12 \cdot 2 = 24$, we multiply $\frac{5}{12}$ by $\frac{2}{2}$.

$$= \frac{21}{24} + \frac{10}{24} = \frac{31}{24}. \qquad \textbf{Adding fractions}$$

b) $\frac{9}{8} - \frac{4}{5} = \frac{9}{8} \cdot \frac{5}{5} - \frac{4}{5} \cdot \frac{8}{8}$ **Using 40 as a common denominator**

$= \frac{45}{40} - \frac{32}{40} = \frac{13}{40}$ **Subtracting fractions**

Try Exercise 69.

After adding, subtracting, multiplying, or dividing, we may still need to simplify the answer.

EXAMPLE 10 Perform the indicated operation and, if possible, simplify.

a) $\frac{7}{10} - \frac{1}{5}$ b) $8 \cdot \frac{5}{12}$ c) $\frac{\frac{5}{6}}{\frac{25}{9}}$

SOLUTION

a) $\frac{7}{10} - \frac{1}{5} = \frac{7}{10} - \frac{1}{5} \cdot \frac{2}{2}$ **Using 10 as the common denominator**

$= \frac{7}{10} - \frac{2}{10}$

$= \frac{5}{10} = \frac{1 \cdot \cancel{5}}{2 \cdot \cancel{5}} = \frac{1}{2}$ **Removing a factor equal to 1: $\frac{5}{5} = 1$**

b) $8 \cdot \frac{5}{12} = \frac{8 \cdot 5}{12}$ **Multiplying numerators and denominators. Think of 8 as $\frac{8}{1}$.**

$= \frac{2 \cdot 2 \cdot 2 \cdot 5}{2 \cdot 2 \cdot 3}$ **Factoring; $\frac{4 \cdot 2 \cdot 5}{4 \cdot 3}$ can also be used.**

$= \frac{\cancel{2} \cdot \cancel{2} \cdot 2 \cdot 5}{\cancel{2} \cdot \cancel{2} \cdot 3}$ **Removing a factor equal to 1: $\frac{2 \cdot 2}{2 \cdot 2} = 1$**

$= \frac{10}{3}$ **Simplifying**

c) $\frac{\frac{5}{6}}{\frac{25}{9}} = \frac{5}{6} \div \frac{25}{9}$ **Rewriting horizontally. Remember that a fraction bar indicates division.**

$= \frac{5}{6} \cdot \frac{9}{25}$ **Multiplying by the reciprocal of $\frac{25}{9}$**

$= \frac{5 \cdot 3 \cdot 3}{2 \cdot 3 \cdot 5 \cdot 5}$ **Writing as one fraction and factoring**

$= \frac{\cancel{5} \cdot \cancel{3} \cdot 3}{2 \cdot \cancel{3} \cdot \cancel{5} \cdot 5}$ **Removing a factor equal to 1: $\frac{5 \cdot 3}{3 \cdot 5} = 1$**

$= \frac{3}{10}$ **Simplifying**

Try Exercise 59.

STUDY TIP

Do the Exercises

- When you complete the odd-numbered exercises, you can check your answers at the back of the book. If you miss any, closely examine your work, and if necessary, consult the *Student's Solutions Manual* or your instructor for guidance.
- Do some even-numbered exercises, even if none is assigned. Because there are no answers given for them, you will gain practice doing exercises that are similar to quiz or test problems. Check your answers later with a friend or your instructor.

Fraction Notation and Menus

Some graphing calculators can perform operations using fraction notation. Others can convert answers to fraction notation. Often this conversion is performed using a command found in a **menu.**

Menus A menu is a list of options that appears when a key is pressed. To select an item from the menu, highlight its number using the up or down arrow keys and press ENTER or simply press the number of the item. For example, pressing MATH results in a screen like the following. Four menu titles, or **submenus,** are listed across the top of the screen. Use the left and right arrow keys to highlight the desired submenu.

```
MATH NUM CPX PRB
1:▸Frac
2:▸Dec
3:³
4:³√(
5:ˣ√
6:fMin(
7↓fMax(
```

In the screen shown above, the MATH submenu is highlighted. The options in that menu appear on the screen. We will refer to this submenu as MATH MATH, meaning that we first press MATH and then highlight the MATH submenu. Note that the submenu contains more options than can fit on a screen, as indicated by the arrow in entry 7. The remaining options will appear as the down arrow is pressed.

Fraction Notation To convert a number to fraction notation, enter the number on the home screen and choose the FRAC option from the MATH MATH submenu. After the notation ▸FRAC is copied on the home screen, press ENTER.

Reciprocals To find the reciprocal of a number, enter the number, press x^{-1}, and then press ENTER. The reciprocal will be given in decimal notation and can be converted to fraction notation using the FRAC option.

Your Turn

1. Find the LCM(entry in the MATH NUM submenu.
2. Copy the LCM(entry to the home screen, and press 6 , 8) ENTER to find the least, or smallest, common multiple of 6 and 8.
   The result should be 24.
3. Convert 0.5 to fraction notation. Press 0 . 5 and then find the ▸FRAC entry in the MATH MATH submenu. Copy this to the home screen and press ENTER. The result should be 1/2.
4. Find the reciprocal of 2/3. Press (2 ÷ 3) x^{-1}. Then copy ▸FRAC to the home screen and press ENTER. The result should be 3/2.

Student Notes

On some calculators, use of the FRAC option will result in a fraction written as $\frac{2}{15}$ rather than 2/15.

On some calculators, fractions can be entered using the N/D option in the MATH NUM submenu. After choosing this option, enter the numerator, move to the denominator by pressing ▽, and then enter the denominator. Press ▷ when you are finished entering the fraction.

$$\frac{2}{15}+\frac{7}{12} \qquad \frac{43}{60}$$

EXAMPLE 11 Use a graphing calculator to find fraction notation for $\frac{2}{15} + \frac{7}{12}$.

SOLUTION A fraction bar indicates division, so we enter $\frac{2}{15}$ as 2/15 and $\frac{7}{12}$ as 7/12. Parentheses are not necessary for this calculation, but they are for others, so we include them here. After pressing ENTER, we see the answer in decimal notation. We then convert to fraction notation by pressing MATH and selecting the FRAC option. The ANS notation on the screen indicates the result, or answer, of the most recent operation. In this case, the notation ANS ▶FRAC indicates that the calculator will convert 0.7166666667 to fraction notation.

```
(2/15)+(7/12)
                 .7166666667
Ans▶Frac
                       43/60
```

This procedure can be done in one step, using the keystrokes

2 ÷ 1 5 + 7 ÷ 1 2 MATH 1 ENTER.

```
2/15+7/12▶Frac
                       43/60
```

Try Exercise 65.

1.3 Exercise Set

FOR EXTRA HELP MyMathLab    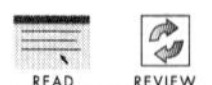

PRACTICE WATCH DOWNLOAD READ REVIEW

Concept Reinforcement *In each of Exercises 1–4, match the description with the appropriate number from the list on the right.*

1. ___ A factor of 35   a) 2
2. ___ A number that has 3 as a factor   b) 7
3. ___ An odd composite number   c) 60
4. ___ The only even prime number   d) 65

To the student and the instructor: Beginning in this section, selected exercises are marked with the symbol Aha!*. Students who pause to inspect an Aha! exercise should find the answer more readily than those who proceed mechanically. This is done to discourage rote memorization. Some "Aha!" exercises in this exercise set are unmarked, to encourage students to always pause before working a problem.*

Write all two-factor factorizations of each number. Then list all the factors of the number.

5. 50   **6.** 70   **7.** 42   **8.** 60

Label each of the following numbers as prime, composite, or neither.

9. 21   **10.** 15   **11.** 31   **12.** 35

13. 25   **14.** 37   **15.** 2   **16.** 1

17. 0   **18.** 4   **19.** 40   **20.** 75

Find the prime factorization of each number. If the number is prime, state this.

21. 26   **22.** 15   **23.** 30

24. 55   **25.** 27   **26.** 98

27. 40   **28.** 54   **29.** 43

30. 120   **31.** 210   **32.** 79

33. 115   **34.** 143

Simplify.

35. $\frac{14}{21}$ **36.** $\frac{20}{26}$ **37.** $\frac{16}{56}$

38. $\frac{72}{27}$ **39.** $\frac{6}{48}$ **40.** $\frac{18}{84}$

41. $\frac{52}{13}$ **42.** $\frac{132}{11}$ **43.** $\frac{19}{76}$

44. $\frac{17}{51}$ **45.** $\frac{150}{25}$ **46.** $\frac{170}{34}$

47. $\frac{42}{50}$ **48.** $\frac{75}{80}$ **49.** $\frac{120}{82}$

50. $\frac{75}{45}$ **51.** $\frac{210}{98}$ **52.** $\frac{140}{350}$

Perform the indicated operation and, if possible, simplify. If there are no variables, check using a calculator.

53. $\frac{1}{2} \cdot \frac{3}{7}$

54. $\frac{9}{4} \cdot \frac{3}{8}$

55. $\frac{12}{5} \cdot \frac{10}{9}$

Aha! **56.** $\frac{11}{12} \cdot \frac{12}{11}$

57. $\frac{1}{8} + \frac{3}{8}$

58. $\frac{1}{2} + \frac{1}{8}$

59. $\frac{4}{9} + \frac{13}{18}$

60. $\frac{4}{5} + \frac{8}{15}$

61. $\frac{3}{a} \cdot \frac{b}{7}$

62. $\frac{x}{5} \cdot \frac{y}{z}$

63. $\frac{4}{a} + \frac{3}{a}$

64. $\frac{7}{a} - \frac{5}{a}$

65. $\frac{3}{10} + \frac{8}{15}$

66. $\frac{7}{8} + \frac{5}{12}$

67. $\frac{9}{7} - \frac{2}{7}$

68. $\frac{12}{5} - \frac{2}{5}$

69. $\frac{13}{18} - \frac{4}{9}$

70. $\frac{13}{15} - \frac{8}{45}$

Aha! **71.** $\frac{20}{30} - \frac{2}{3}$

72. $\frac{5}{7} - \frac{5}{21}$

73. $\frac{7}{6} \div \frac{3}{5}$

74. $\frac{7}{5} \div \frac{3}{4}$

75. $\frac{8}{9} \div \frac{4}{15}$

76. $\frac{1}{8} \div \frac{1}{4}$

77. $12 \div \frac{3}{7}$

78. $\frac{10}{9} \div 10$

Aha! **79.** $\frac{7}{13} \div \frac{7}{13}$

80. $\frac{17}{8} \div \frac{5}{6}$

81. $\frac{\frac{2}{7}}{\frac{5}{3}}$

82. $\frac{\frac{3}{8}}{\frac{1}{5}}$

83. $\frac{9}{\frac{1}{2}}$

84. $\frac{\frac{3}{7}}{6}$

TW **85.** How many even numbers are there? Explain your reasoning.

TW **86.** Under what circumstances would the sum of two fractions be easier to compute than the product of the same two fractions?

SKILL REVIEW

Use a commutative law to write an equivalent expression. There can be more than one correct answer. [1.2]

87. $5(x + 3)$

88. $7 + (a + b)$

SYNTHESIS

TW **89.** Bryce insists that $\frac{2 + x}{8}$ is equivalent to $\frac{1 + x}{4}$. What mistake do you think is being made and how could you demonstrate to Bryce that the two expressions are not equivalent?

TW **90.** Why are 0 and 1 considered neither prime nor composite?

91. In the table below, the top number can be factored in such a way that the sum of the factors is the bottom number. For example, in the first column, 56 is factored as $7 \cdot 8$, since $7 + 8 = 15$, the bottom number. Find the missing numbers in each column.

Product	56	63	36	72	140	96	168
Factor	7						
Factor	8						
Sum	15	16	20	38	24	20	29

92. *Packaging.* Tritan Candies uses two sizes of boxes, 6 in. long and 8 in. long. These are packed end to end in bigger cartons to be shipped. What is the shortest-length carton that will accommodate boxes of either size without any room left over? (Each carton must contain boxes of only one size; no mixing is allowed.)

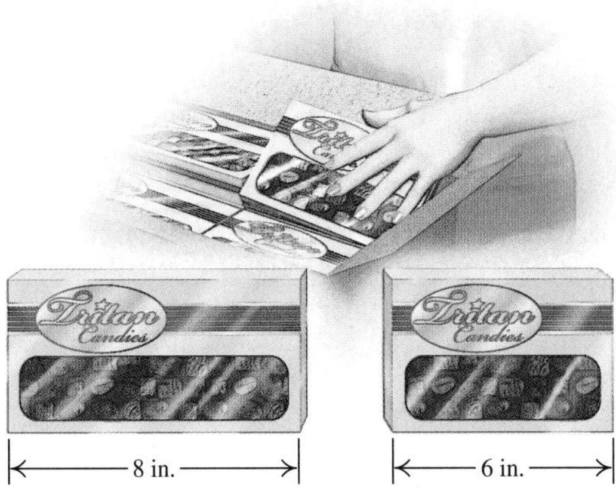

Simplify.

93. $\frac{16 \cdot 9 \cdot 4}{15 \cdot 8 \cdot 12}$

94. $\frac{9 \cdot 8xy}{2xy \cdot 36}$

95. $\frac{27pqrs}{9prst}$

96. $\frac{247}{323}$

97. $\frac{15 \cdot 4xy \cdot 9}{6 \cdot 25x \cdot 15y}$

98. $\frac{10x \cdot 12 \cdot 25y}{2z \cdot 30x \cdot 20y}$

99. $\frac{\frac{27ab}{15mn}}{\frac{18bc}{25np}}$

100. $\frac{\frac{45xyz}{24ab}}{\frac{30xz}{32ac}}$

101. $\frac{5\frac{3}{4}rs}{4\frac{1}{2}st}$

102. $\frac{3\frac{5}{7}mn}{2\frac{4}{5}np}$

Find the area of each figure.

103.

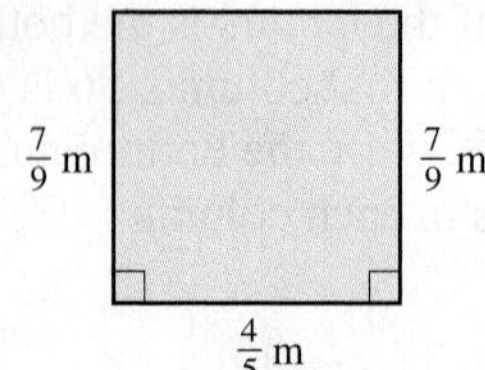

104.

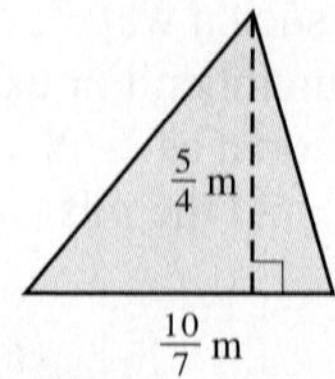

105. Find the perimeter of a square with sides of length $3\frac{5}{9}$m.

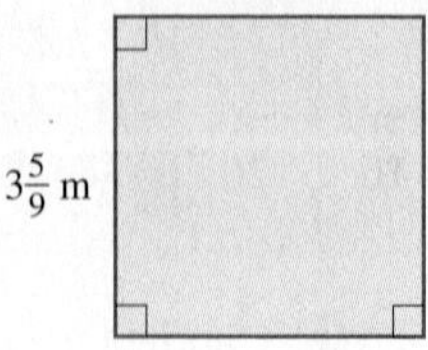

106. Find the perimeter of the rectangle in Exercise 103.

107. Find the total length of the edges of a cube with sides of length $2\frac{3}{10}$ cm.

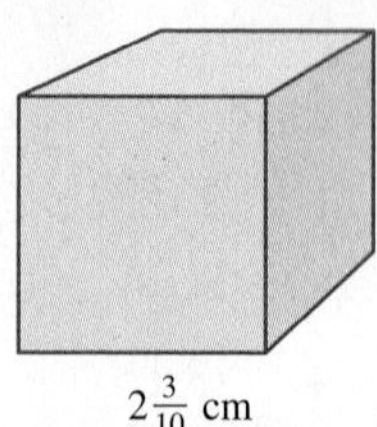

Try Exercise Answers: Section 1.3

5. $1 \cdot 50$; $2 \cdot 25$; $5 \cdot 10$; 1, 2, 5, 10, 25, 50 **9.** Composite **21.** $2 \cdot 13$ **35.** $\frac{2}{3}$ **39.** $\frac{1}{8}$ **53.** $\frac{3}{14}$ **59.** $\frac{7}{6}$ **63.** $\frac{7}{a}$ **65.** $\frac{5}{6}$ **69.** $\frac{5}{18}$ **73.** $\frac{35}{18}$

1.4 Positive and Negative Real Numbers

- The Integers
- The Rational Numbers
- Real Numbers and Order
- Absolute Value

A **set** is a collection of objects. The set containing 1, 3, and 7 is usually written $\{1, 3, 7\}$. In this section, we examine some important sets of numbers.

THE INTEGERS

Two sets of numbers were mentioned in Section 1.3. We represent these sets using dots on the number line.

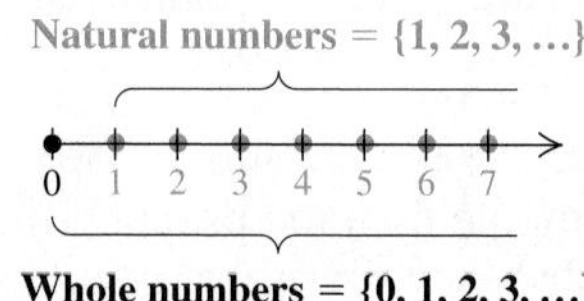

To create the set of *integers*, we include all whole numbers, along with their *opposites*. To find the opposite of a number, we locate the number that is the same distance from 0 but on the other side of the number line. For example,

the opposite of 1 is negative 1, written -1;

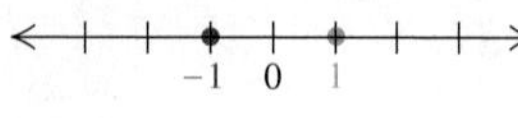

and

the opposite of 3 is negative 3, written -3.

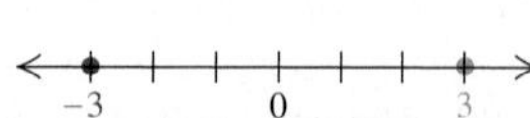

The **integers** consist of all whole numbers and their opposites.

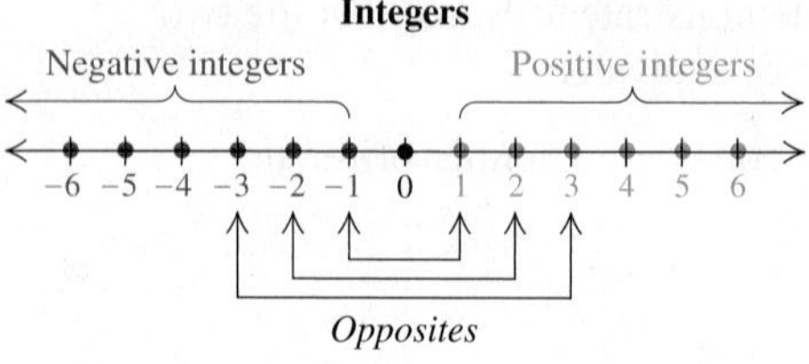

Opposites are discussed in more detail in Section 1.6. Note that, except for 0, opposites occur in pairs. Thus, 5 is the opposite of -5, just as -5 is the opposite of 5. Note that 0 acts as its own opposite.

Set of Integers

The set of integers $= \{\ldots, -4, -3, -2, -1, 0, 1, 2, 3, 4, \ldots\}$.

Integers are associated with many real-world problems and situations.

EXAMPLE 1 State which integer(s) corresponds to each situation.

a) During 2008, the U.S. economy lost 2.6 million jobs.
Source: U.S. Department of Labor

b) Badwater Basin in Death Valley, California, is 282 ft below sea level.

c) To lose one pound of fat, it is necessary for most people to create a 3500-calorie deficit.
Source: World Health Organization

SOLUTION

a) The integer $-2{,}600{,}000$ corresponds to a loss of 2.6 million jobs.

b) The integer -282 corresponds to 282 ft below sea level. The elevation is -282 ft.

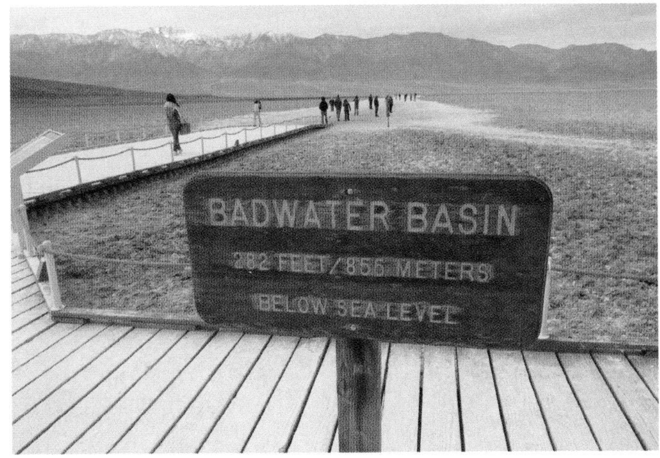

c) The integer -3500 corresponds to a deficit of 3500 calories.

Try Exercise 9.

STUDY TIP 

Try to Get Ahead

Try to keep one section ahead of your instructor. If you study ahead of your lectures, you can concentrate on what is being explained in them, rather than trying to write everything down. You can then write notes on only special points or questions related to what is happening in class. A few minutes of preparation before class can save you much more study time after class.

THE RATIONAL NUMBERS

Although built out of integers, a number like $\frac{5}{9}$ is not itself an integer. Another set of numbers, the **rational numbers,** contains fractions and decimals that repeat or terminate, as well as the integers. Some examples of rational numbers are

$$\frac{5}{9},\quad -\frac{4}{7},\quad 95,\quad -16,\quad 0,\quad \frac{-35}{8},\quad 2.4,\quad -0.31.$$

In Section 1.7, we show that $-\frac{4}{7}$ can be written as $\frac{-4}{7}$ or $\frac{4}{-7}$. Indeed, every number listed above can be written as an integer over an integer. For example, 95 can be written as $\frac{95}{1}$ and 2.4 can be written as $\frac{24}{10}$. In this manner, any *ratio*nal number can be expressed as the *ratio* of two integers. Rather than attempt to list all rational numbers, we use this idea of ratio to describe the set as follows.

Set of Rational Numbers

The set of rational numbers $= \left\{ \frac{a}{b} \,\middle|\, a \text{ and } b \text{ are integers and } b \neq 0 \right\}$.

This is read "the set of all numbers *a* over *b*, where *a* and *b* are integers and *b* does not equal zero."

To *graph* a number is to mark its location on the number line.

EXAMPLE 2 Graph each of the following rational numbers: **(a)** $\frac{5}{2}$; **(b)** -3.2; **(c)** $\frac{11}{8}$.

SOLUTION

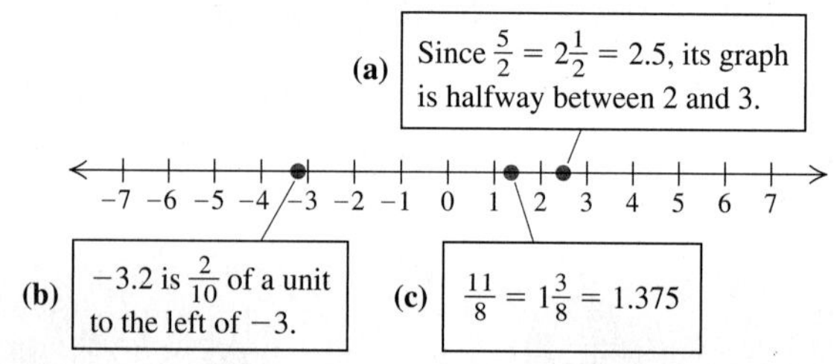

Try Exercise 19.

Every rational number can be written as a fraction or a decimal.

EXAMPLE 3 Convert to decimal notation: $-\frac{5}{8}$.

SOLUTION We first find decimal notation for $\frac{5}{8}$. Since $\frac{5}{8}$ means $5 \div 8$, we divide.

$$\begin{array}{r} 0.625 \\ 8\overline{)5.000} \\ \underline{4800} \\ 200 \\ \underline{160} \\ 40 \\ \underline{40} \\ 0 \end{array} \leftarrow \textbf{The remainder is 0.}$$

Thus, $\frac{5}{8} = 0.625$, so $-\frac{5}{8} = -0.625$.

Try Exercise 25.

Because the division in Example 3 ends with the remainder 0, we consider -0.625 a **terminating decimal.** If we are "bringing down" zeros and a remainder reappears, we have a **repeating decimal,** as shown in the next example.

EXAMPLE 4 Convert to decimal notation: $\frac{7}{11}$.

SOLUTION We divide:

$$\begin{array}{r} 0.6\,3\,6\,3\ldots \\ 11\overline{)7.0\,0\,0\,0} \\ \underline{6\,6} \\ 4\,0 \\ \underline{3\,3} \\ 7\,0 \\ \underline{6\,6} \\ 4\,0 \end{array}$$

4 reappears as a remainder, so the pattern of 6's and 3's in the quotient will continue.

We abbreviate repeating decimals by writing a bar over the repeating part—in this case, $0.\overline{63}$. Thus, $\frac{7}{11} = 0.\overline{63}$.

Try Exercise 29.

Although we do not prove it here, every rational number can be expressed as either a terminating decimal or a repeating decimal, and every terminating decimal or repeating decimal can be expressed as a ratio of two integers.

REAL NUMBERS AND ORDER

Some numbers, when written in decimal form, neither terminate nor repeat. Such numbers are called **irrational numbers.**

What sort of numbers are irrational? One example is π (the Greek letter *pi*, read "pie"), which is used to find the area and the circumference of a circle: $A = \pi r^2$ and $C = 2\pi r$.

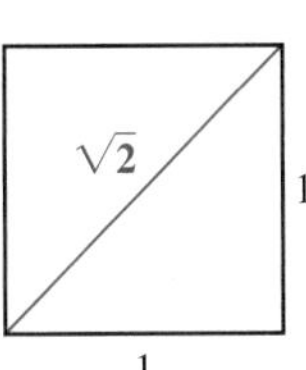

Another irrational number, $\sqrt{2}$ (read "the square root of 2"), is the length of the diagonal of a square with sides of length 1 (see the figure at left). It is also the number that, when multiplied by itself, gives 2. No rational number can be multiplied by itself to get 2, although some approximations come close:

1.4 is an *approximation* of $\sqrt{2}$ because $(1.4)(1.4) = 1.96$;

1.41 is a better approximation because $(1.41)(1.41) = 1.9881$;

1.4142 is an even better approximation because $(1.4142)(1.4142) = 1.99996164$.

To approximate $\sqrt{2}$ on most graphing calculators, we press $\sqrt{\ }$ and then enter 2 enclosed by parentheses. Some calculators will supply the left parenthesis automatically when $\sqrt{\ }$ is pressed. Other calculators use the notation $\sqrt{2}$ rather than $\sqrt{\ }(2)$. Square roots are discussed in detail in Chapter 10.

EXAMPLE 5 Graph the real number $\sqrt{3}$ on the number line.

SOLUTION We use a calculator as shown at left and approximate: $\sqrt{3} \approx 1.732$ ("$\approx$" means "approximately equals"). Then we locate this number on the number line.

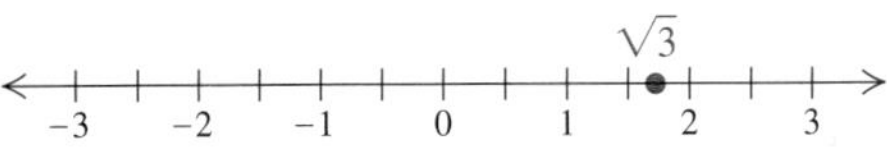

Try Exercise 37.

The rational numbers and the irrational numbers together correspond to all the points on the number line and make up what is called the **real-number system.**

Set of Real Numbers

The set of real numbers = The set of all numbers corresponding to points on the number line.

The following figure shows the relationships among various kinds of numbers.

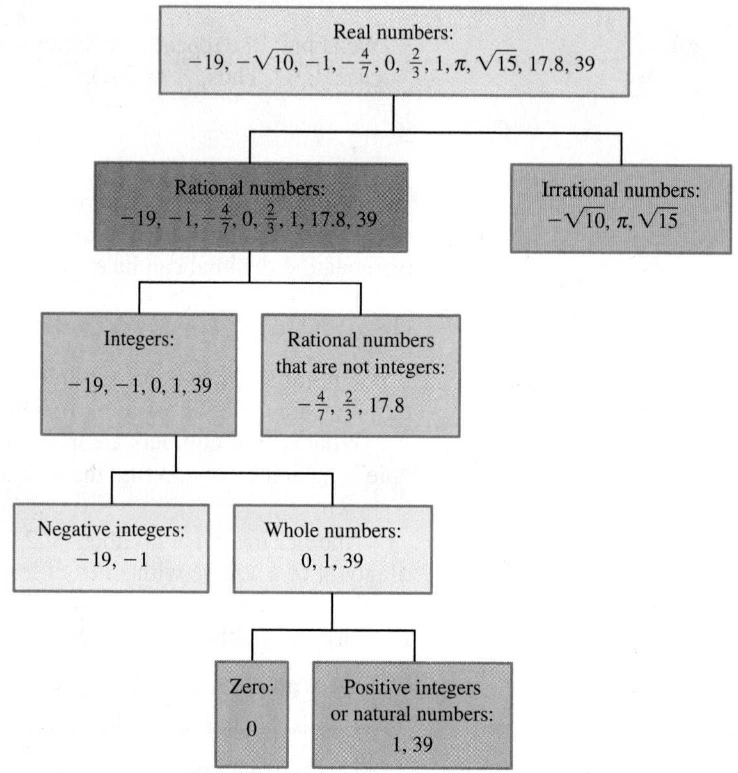

EXAMPLE 6 Which numbers in the following list are **(a)** whole numbers? **(b)** integers? **(c)** rational numbers? **(d)** irrational numbers? **(e)** real numbers?

$$-38, \quad -\frac{8}{5}, \quad 0, \quad 4.5, \quad \sqrt{30}, \quad 52, \quad \frac{10}{2}$$

SOLUTION

a) $0, 52$, and $\frac{10}{2}$ are whole numbers.

b) $-38, 0, 52$, and $\frac{10}{2}$ are integers.

c) $-38, -\frac{8}{5}, 0, 4.5, 52$, and $\frac{10}{2}$ are rational numbers.

d) $\sqrt{30}$ is an irrational number.

e) $-38, -\frac{8}{5}, 0, 4.5, \sqrt{30}, 52$, and $\frac{10}{2}$ are real numbers.

Try Exercise 75.

Real numbers are named in order on the number line, with larger numbers further to the right. For any two numbers, the one to the left is less than the one to the right. We use the symbol **<** to mean "**is less than.**" The sentence $-8 < 6$ means "-8 is less than 6." The symbol **>** means "**is greater than.**" The sentence $-3 > -7$ means "-3 is greater than -7."

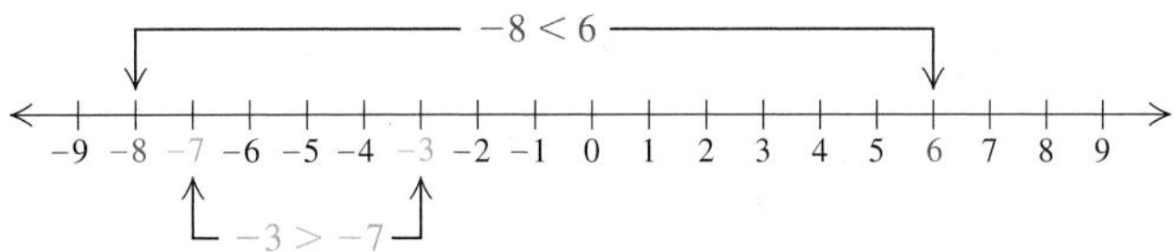

EXAMPLE 7 Use either $<$ or $>$ for $\square$ to write a true sentence.

a) $2 \square 9$ **b)** $-3.45 \square 1.32$ **c)** $6 \square -12$

d) $-18 \square -5$ **e)** $\frac{7}{11} \square \frac{5}{8}$

SOLUTION

a) Since 2 is to the left of 9 on the number line, we know that 2 is less than 9, so $2 < 9$.

b) Since -3.45 is to the left of 1.32, we have $-3.45 < 1.32$.

c) Since 6 is to the right of -12, we have $6 > -12$.

d) Since -18 is to the left of -5, we have $-18 < -5$.

e) We convert to decimal notation: $\frac{7}{11} = 0.\overline{63}$ and $\frac{5}{8} = 0.625$. Thus, $\frac{7}{11} > \frac{5}{8}$. We also could have used a common denominator: $\frac{7}{11} = \frac{56}{88} > \frac{55}{88} = \frac{5}{8}$.

Try Exercise 43.

Sentences like "$a < -5$" and "$-3 > -8$" are **inequalities.** It is useful to remember that every inequality can be written in two ways. For example,

$$-3 > -8 \quad \text{has the same meaning as} \quad -8 < -3.$$

It may be helpful to think of an inequality sign as an "arrow" with the smaller side pointing to the smaller number.

Note that $a > 0$ means that a represents a positive real number and $a < 0$ means that a represents a negative real number.

Statements like $a \leq b$ and $a \geq b$ are also inequalities. We read $a \leq b$ as "**a is less than or equal to b**" and $a \geq b$ as "**a is greater than or equal to b.**"

EXAMPLE 8 Classify each inequality as true or false.

a) $-3 \leq 5$ **b)** $-3 \leq -3$ **c)** $-5 \geq -4$

SOLUTION

a) $-3 \leq 5$ is *true* because $-3 < 5$ is true.

b) $-3 \leq -3$ is *true* because $-3 = -3$ is true.

c) $-5 \geq -4$ is *false* since neither $-5 > -4$ nor $-5 = -4$ is true.

Try Exercise 57.

Student Notes

It is important to remember that just because an equation or an inequality is written or printed, it is not necessarily *true*. For instance, $6 = 7$ is an equation and $2 > 5$ is an inequality. Of course, both statements are *false*.

ABSOLUTE VALUE

There is a convenient terminology and notation for the distance a number is from 0 on the number line. It is called the **absolute value** of the number.

Absolute Value We write $|a|$, read "the absolute value of a," to represent the number of units that a is from zero.

EXAMPLE 9 Find each absolute value: **(a)** $|-3|$; **(b)** $|7.2|$; **(c)** $|0|$.

SOLUTION

a) $|-3| = 3$ since -3 is 3 units from 0.

b) $|7.2| = 7.2$ since 7.2 is 7.2 units from 0.

c) $|0| = 0$ since 0 is 0 units from itself.

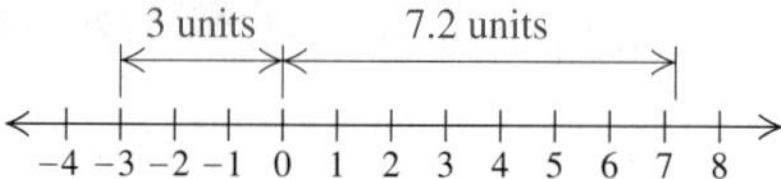

Try Exercise 63.

Since distance is never negative, numbers that are opposites have the same absolute value. If a number is nonnegative, its absolute value is the number itself. If a number is negative, its absolute value is its opposite.

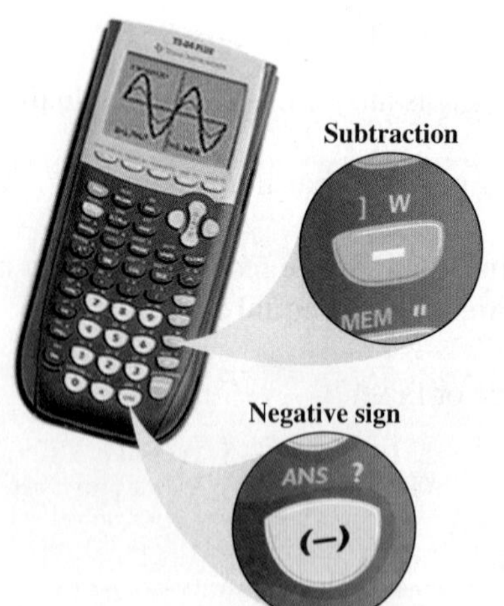

Negative Numbers and Absolute Value

Graphing calculators have different keys for writing negatives and subtracting. The key labeled (−) is used to create a negative sign whereas the one labeled − is used for subtraction. Using the wrong key may result in an ERR:SYNTAX message.

Many graphing calculators use the notation "abs" with parentheses to indicate absolute value. Thus, $\text{abs}(2) = |2| = 2$. The abs notation is often found in the MATH NUM submenu. Some calculators automatically supply the left parenthesis. Some calculators use the notation $|-3|$ rather than $\text{abs}(-3)$. For these calculators, parentheses are not used.

To check Example 9(a) using a calculator, press MATH ▷ 1 (−) 3) ENTER. We see that $|-3| = 3$.

Your Turn

1. Find the ABS(entry in the MATH NUM submenu.
2. Copy ABS(to the home screen and find $|4|$. 4
3. Use the subtraction key to calculate $10 - 4$. 6
4. Create an error message by using − instead of (−) to enter $|-6|$.
5. Return to the home screen. Find $\left|-\frac{1}{3}\right|$ in fraction notation. 1/3

1.4 Exercise Set

FOR EXTRA HELP MyMathLab

Concept Reinforcement *In each of Exercises 1–8, fill in the blank using one of the following terms:* natural number, whole number, integer, rational number, terminating, repeating, irrational number, absolute value.

1. Division can be used to show that $\frac{4}{7}$ can be written as a(n) ______________ decimal.
2. Division can be used to show that $\frac{3}{20}$ can be written as a(n) ______________ decimal.
3. If a number is a(n) ______________, it is either a whole number or the opposite of a whole number.
4. 0 is the only ______________ that is not a natural number.
5. Any number of the form a/b, where a and b are integers, with $b \neq 0$, is an example of a(n) ______________.
6. A number like $\sqrt{5}$, which cannot be written precisely in fraction notation or decimal notation, is an example of a(n) ______________.
7. If a number is a(n) ______________, then it can be thought of as a counting number.
8. When two numbers are opposites, they have the same ______________.

State which real number(s) correspond to each situation.

9. *Record Temperature.* The highest temperature recorded in Alaska is 100 degrees Fahrenheit (°F) at Fort Yukon. The lowest temperature recorded in Alaska is 80°F below zero at Prospect Creek Camp.
Source: www.netstate.com

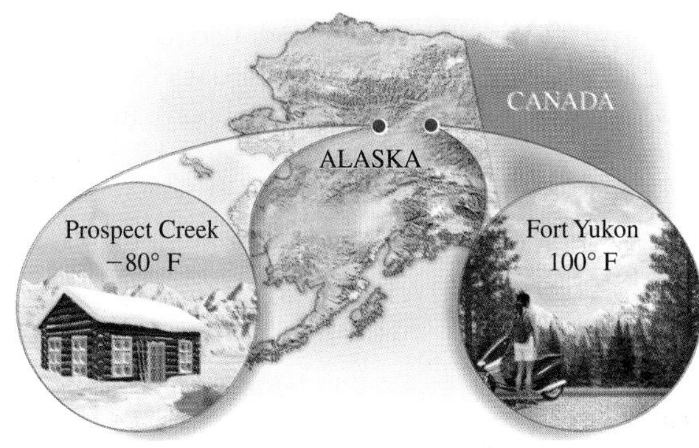

10. *Calories.* During a yoga class, Sharrita burned 250 calories. She then drank an isotonic drink containing 65 calories.

11. *Stock Market.* The Dow Jones Industrial Average is an indicator of the stock market. On September 29, 2008, the Dow Jones fell a record 777.68 points. On October 13, 2008, the Dow Jones gained a record 936.42 points.
Source: www.finfacts.com

12. *Record Elevation.* The Dead Sea is 1340 ft below sea level, and Mt. Everest is 29,035 ft above sea level.
Source: encarta.msn.com

13. *Student Loans and Scholarships.* The maximum amount that a student may borrow each year with a Stafford Loan is $12,500. The maximum annual award for the Nursing Economics Foundation Scholarship is $5000.
Sources: www.studentaid.ed.gov; www.collegezone.com

14. *Birth Rate and Death Rate.* Recently, the world birth rate was 20.18 per thousand. The death rate was 8.23 per thousand.
Source: Central Intelligence Agency, 2009

15. *Football.* The halfback gained 8 yd on the first play. The quarterback was tackled for a 5-yd loss on the second play.

16. *Golf.* In the 2009 Open Championship, golfer Ernie Els finished 1 over par. In the 2010 Farmers Insurance Open, he finished 11 under par.
Source: www.pgatour.com

17. Ignition occurs 10 sec before liftoff. A spent fuel tank is detached 235 sec after liftoff.

18. Melanie deposited $750 in a savings account. Two weeks later, she withdrew $125.

Graph each rational number on the number line.

19. -2

20. 5

21. -4.3

22. 3.87

23. $\frac{10}{3}$

24. $-\frac{17}{5}$

Write decimal notation for each number.

25. $\frac{7}{8}$

26. $-\frac{1}{8}$

27. $-\frac{3}{4}$

28. $\frac{11}{6}$

29. $-\frac{7}{6}$

30. $-\frac{5}{12}$

31. $\frac{2}{3}$

32. $\frac{1}{4}$

33. $-\frac{1}{2}$

34. $-\frac{2}{9}$

Aha! **35.** $\frac{13}{100}$

36. $-\frac{7}{20}$

Graph each irrational number on the number line.

37. $\sqrt{5}$

38. $\sqrt{92}$

39. $-\sqrt{22}$

40. $-\sqrt{54}$

Write a true sentence using either < or >.

41. $7 \square 0$

42. $8 \square -8$

43. $-6 \square 6$

44. $0 \square -7$

45. $-8 \square -5$

46. $-4 \square -3$

47. $-5 \square -11$

48. $-3 \square -4$

49. $-12.5 \square -9.4$

50. $-10.3 \square -14.5$

51. $-\frac{5}{12} \square -\frac{11}{25}$

52. $-\frac{14}{17} \square -\frac{27}{35}$

For each of the following, write a second inequality with the same meaning.

53. $-7 > x$

54. $a > 9$

55. $-10 \le y$

56. $12 \ge t$

Classify each inequality as either true or false.

57. $-3 \ge -11$

58. $5 \le -5$

59. $0 \ge 8$

60. $-5 \le 7$

61. $-8 \le -8$

62. $8 \ge 8$

Find each absolute value.

63. $|-58|$

64. $|-47|$

65. $|5.6|$

66. $|-\frac{2}{5}|$

67. $|\sqrt{2}|$

68. $|-456|$

69. $|-\frac{9}{7}|$

70. $|-\sqrt{3}|$

71. $|0|$

72. $|4.3|$

73. $|x|$, for $x = -8$

74. $|a|$, for $a = -5$

For Exercises 75–80, consider the following list:

$$18, \quad -4.7, \quad 0, \quad -\frac{5}{9}, \quad \pi, \quad \sqrt{17}, \quad 2.1\overline{6}, \quad -37.$$

75. List all rational numbers.

76. List all natural numbers.

77. List all integers.

78. List all irrational numbers.

79. List all real numbers.

80. List all nonnegative integers.

TW **81.** Is every integer a rational number? Why or why not?

TW **82.** Is every integer a natural number? Why or why not?

SKILL REVIEW

83. Evaluate $3xy$ for $x = 2$ and $y = 7$. [1.1]

84. Use a commutative law to write an expression equivalent to $ab + 5$. [1.2]

SYNTHESIS

TW **85.** Is the absolute value of a number always positive? Why or why not?

TW **86.** How many rational numbers are there between 0 and 1? Justify your answer.

TW **87.** Does "nonnegative" mean the same thing as "positive"? Why or why not?

List in order from least to greatest.

88. $13, -12, 5, -17$

89. $-23, 4, 0, -17$

90. $-\frac{2}{3}, \frac{1}{2}, -\frac{3}{4}, -\frac{5}{6}, \frac{3}{8}, \frac{1}{6}$

91. $\frac{4}{5}, \frac{4}{3}, \frac{4}{8}, \frac{4}{6}, \frac{4}{9}, \frac{4}{2}, -\frac{4}{3}$

Write a true sentence using either <, >, or =.

92. $|-5| \square |-2|$

93. $|4| \square |-7|$

94. $|-8| \square |8|$

95. $|23| \square |-23|$

Solve. Consider only integer replacements.

Aha! **96.** $|x| = 7$

97. $|x| < 3$

98. $2 < |x| < 5$

Given that $0.3\overline{3} = \frac{1}{3}$ and $0.6\overline{6} = \frac{2}{3}$, express each of the following as a ratio of two integers.

99. $0.1\overline{1}$

100. $0.9\overline{9}$

101. $5.5\overline{5}$

102. $7.7\overline{7}$

Translate to an inequality.

103. A number a is negative.

104. A number x is nonpositive.

105. The distance from x to 0 is no more than 10.

106. The distance from t to 0 is at least 20.

TW **107.** When Helga's calculator gives a decimal value for $\sqrt{2}$ and that value is promptly squared, the result is 2. Yet when that same decimal approximation is entered by hand and then squared, the result is not exactly 2. Why do you suppose this is?

TW **108.** Is the following statement true? Why or why not?

$$\sqrt{a^2} = |a| \quad \text{for any real number } a.$$

Try Exercise Answers: Section 1.4

9. $100, -80$ **19.**  **25.** 0.875 **29.** $-1.1\overline{6}$ **37.**  **43.** $<$ **57.** True **63.** 58 **75.** $18, -4.7, 0, -\frac{5}{9}, 2.1\overline{6}, -37$

Mid-Chapter Review

An introduction to algebra involves learning some basic laws and terms.

Commutative Laws: $a + b = b + a$; $ab = ba$

Associative Laws: $a + (b + c) = (a + b) + c$; $a(bc) = (ab)c$

Distributive Law: $a(b + c) = ab + ac$

GUIDED SOLUTIONS

1. Evaluate $\frac{x - y}{3}$ for $x = 22$ and $y = 10$. [1.1]*

Solution

$$\frac{x - y}{3} = \frac{\square - \square}{3}$$ **Substituting**

$$= \frac{\square}{3}$$ **Subtracting**

$$= \square$$ **Dividing**

2. Factor: $40x + 8$. [1.2]

Solution

$$40x + 8 = \square \cdot 5x + \square \cdot 1$$ **Factoring each term using a common factor**

$$= \square\,(5x + 1)$$ **Factoring out the common factor**

MIXED REVIEW

Evaluate. [1.1]

1. $x + y$, for $x = 3$ and $y = 12$

2. $\frac{2a}{5}$, for $a = 10$

Translate to an algebraic expression. [1.1]

3. 10 less than d

4. The product of 8 and the number of hours worked

5. Translate to an equation. Do not solve. [1.1]
Janine's class has 27 students. This is 5 fewer than the number originally enrolled. How many students originally enrolled in the class?

6. Determine whether 8 is a solution of $13t = 94$. [1.1]

7. Use the commutative law of addition to write an expression equivalent to $7 + 10x$. [1.2]

8. Use the associative law of multiplication to write an expression equivalent to $3(ab)$. [1.2]

Multiply. [1.2]

9. $4(2x + 8)$

10. $3(2m + 5n + 10)$

Factor. [1.2]

11. $18x + 9$

12. $8a + 24y + 20$

13. Find the prime factorization of 84. [1.3]

Simplify. [1.3]

14. $\frac{48}{40}$

15. $\frac{135}{315}$

Perform the indicated operation and, if possible, simplify. [1.3]

16. $\frac{11}{12} - \frac{3}{8}$

17. $\frac{8}{15} \div \frac{6}{11}$

18. Graph -2.5 on the number line. [1.4]

19. Write decimal notation for $-\frac{3}{20}$. [1.4]

Write a true sentence using either $<$ *or* $>$. [1.4]

20. $-16 \; \square \; -24$

21. $-\frac{3}{22} \; \square \; -\frac{2}{15}$

22. Write a second inequality with the same meaning as $x \geq 9$. [1.4]

23. Classify as true or false: $-6 \leq -5$. [1.4]

Find the absolute value. [1.4]

24. $|-5.6|$

25. $|0|$

---

*The *section reference* [1.1] refers to Chapter 1, Section 1. The concept reviewed in Guided Solution 1 was developed in this section.

1.5 Addition of Real Numbers

- Adding with the Number Line
- Adding without the Number Line
- Problem Solving
- Combining Like Terms

We now consider addition of real numbers. To gain understanding, we will use the number line first and then develop rules that allow us to work more quickly without the number line.

ADDING WITH THE NUMBER LINE

To add $a + b$ on the number line, we start at a and move according to b.

a) If b is positive, we move to the right (the positive direction).

b) If b is negative, we move to the left (the negative direction).

c) If b is 0, we stay at a.

EXAMPLE 1 Add: $-4 + 9$.

SOLUTION To add on the number line, we locate the first number, -4, and then move 9 units to the right. Note that it requires 4 units to reach 0. The difference between 9 and 4 is where we finish.

$-4 + 9 = 5$

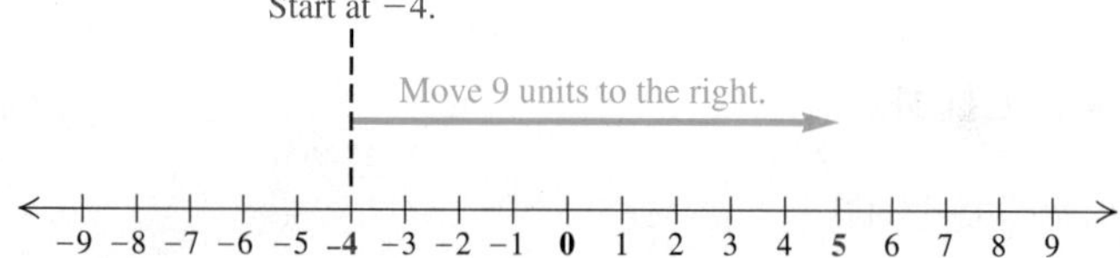

Try Exercise 9.

Student Notes

Parentheses are essential when a negative sign follows an operation. Just as we would never write $8 \div \times 2$, it is improper to write $3 + -5$.

EXAMPLE 2 Add: $3 + (-5)$.

SOLUTION We locate the first number, 3, and then move 5 units to the left. Note that it requires 3 units to reach 0. The difference between 5 and 3 is 2, so we finish 2 units to the left of 0.

$3 + (-5) = -2$

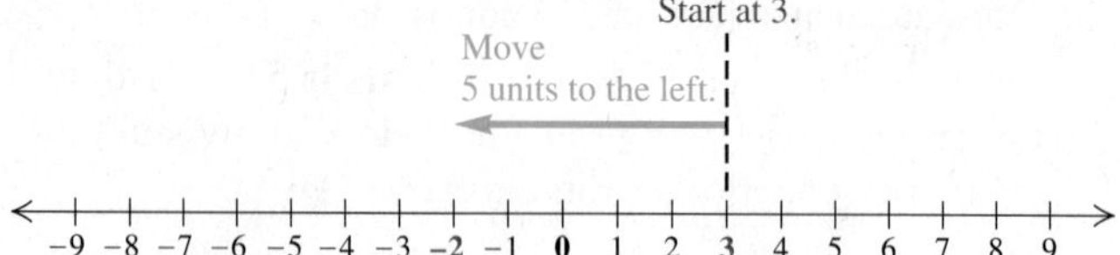

Try Exercise 7.

EXAMPLE 3 Add: $-4 + (-3)$.

SOLUTION After locating -4, we move 3 units to the left. We finish a total of 7 units to the left of 0.

$-4 + (-3) = -7$

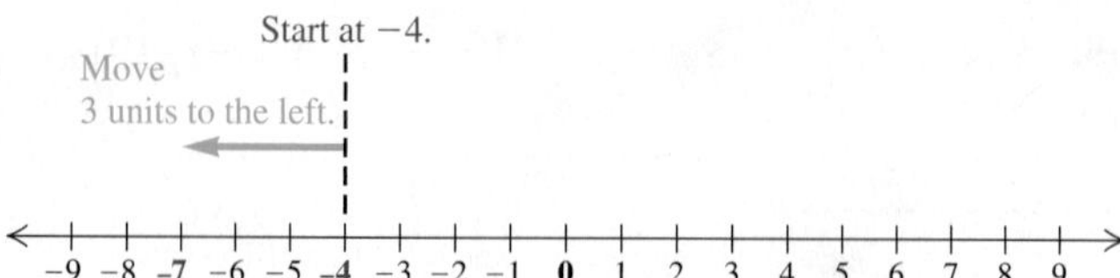

Try Exercise 13.

EXAMPLE 4 Add: $-5.2 + 0$.

SOLUTION We locate -5.2 and move 0 units. Thus we finish where we started, at -5.2.

$-5.2 + 0 = -5.2$

Start at -5.2.

Stay at -5.2.

$-9\ -8\ -7\ -6\ -5\ -4\ -3\ -2\ -1\ 0\ 1\ 2\ 3\ 4\ 5\ 6\ 7\ 8\ 9$

-5.2

Try Exercise 11.

From Examples 1–4, we observe the following rules.

Rules for Addition of Real Numbers

1. *Positive numbers*: Add as usual. The answer is positive.
2. *Negative numbers*: Add absolute values and make the answer negative (see Example 3).
3. *A positive number and a negative number*: Subtract the smaller absolute value from the greater absolute value. Then:
   a) If the positive number has the greater absolute value, the answer is positive (see Example 1).
   b) If the negative number has the greater absolute value, the answer is negative (see Example 2).
   c) If the numbers have the same absolute value, the answer is 0.
4. *One number is zero*: The sum is the other number (see Example 4).

Rule 4 is known as the **identity property of 0.**

The Identity Property of 0

For any real number a,

$$a + 0 = 0 + a = a.$$

(Adding 0 to a number gives that same number.) The number 0 is called the *additive identity.*

ADDING WITHOUT THE NUMBER LINE

The rules listed above can be used without drawing the number line.

EXAMPLE 5 Add without using the number line.

a) $-12 + (-7)$

b) $-1.4 + 8.5$

c) $-36 + 21$

d) $1.5 + (-1.5)$

e) $-\frac{7}{8} + 0$

f) $\frac{2}{3} + \left(-\frac{5}{8}\right)$

STUDY TIP 

Two (or More) Heads Are Better Than One

Consider forming a study group with some of your fellow students. Exchange telephone numbers, schedules, and any e-mail addresses so that you can coordinate study time for homework and tests.

SOLUTION

a) $-12 + (-7) = -19$ — **Two negatives.** ***Think*****: Add the absolute values, 12 and 7, to get 19. Make the answer *negative*, −19.**

b) $-1.4 + 8.5 = 7.1$ — **A negative and a positive.** ***Think*****: The difference of absolute values is 8.5 − 1.4, or 7.1. The positive number has the greater absolute value, so the answer is *positive*, 7.1.**

c) $-36 + 21 = -15$ — **A negative and a positive.** ***Think*****: The difference of absolute values is 36 − 21, or 15. The negative number has the greater absolute value, so the answer is *negative*, −15.**

d) $1.5 + (-1.5) = 0$ — **A negative and a positive.** ***Think*****: Since the numbers are opposites, they have the same absolute value and the answer is 0.**

e) $-\frac{7}{8} + 0 = -\frac{7}{8}$ — **One number is zero. The sum is the other number, $-\frac{7}{8}$.**

f) $\frac{2}{3} + \left(-\frac{5}{8}\right) = \frac{16}{24} + \left(-\frac{15}{24}\right) = \frac{1}{24}$ — **This is similar to part (b) above. We find a common denominator and then add.**

Try Exercises 15 and 17.

If we are adding several numbers, some positive and some negative, the commutative and associative laws allow us to add all the positive numbers, then add all the negative numbers, and then add the results. Of course, we can also add from left to right, if we prefer.

EXAMPLE 6 Add: $15 + (-2) + 7 + 14 + (-5) + (-12)$.

SOLUTION

$15 + (-2) + 7 + 14 + (-5) + (-12)$

$= 15 + 7 + 14 + (-2) + (-5) + (-12)$ — **Using the commutative law of addition**

$= (15 + 7 + 14) + [(-2) + (-5) + (-12)]$ — **Using the associative law of addition**

$= 36 + (-19)$ — **Adding the positives; adding the negatives**

$= 17$ — **Adding a positive and a negative**

Try Exercise 55.

A calculator can be helpful when we are adding real numbers. However, it is not a substitute for knowledge of the rules for addition. When the rules are new to you, use a calculator only when checking your work.

PROBLEM SOLVING

Problems that ask us to find a total translate to addition.

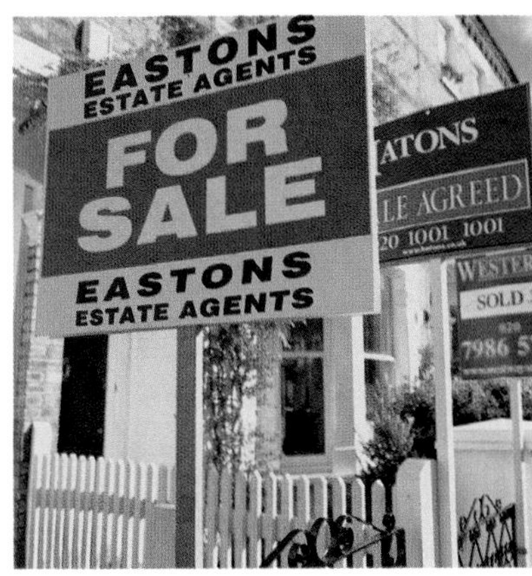

EXAMPLE 7 Median House Prices. From the second quarter of 2008 through the second quarter of 2009, median house prices in Cedar Rapids, Iowa, dropped \$5.8 thousand, rose \$1.5 thousand, dropped \$9.6 thousand, and rose \$14.4 thousand. By how much did median prices change over the year?
Source: National Association of Realtors

SOLUTION The problem translates to a sum:

Rewording:	The 1st change	plus	the 2nd change	plus	the 3rd change	plus	the 4th change	is	the total change.
	↓	↓	↓	↓	↓	↓	↓	↓	↓
Translating:	-5.8	$+$	1.5	$+$	(-9.6)	$+$	14.4	$=$	Total change

Adding from left to right, we have

$$\begin{aligned}-5.8 + 1.5 + (-9.6) + 14.4 &= -4.3 + (-9.6) + 14.4\\ &= -13.9 + 14.4 = 0.5.\end{aligned}$$

The median house price rose \$0.5 thousand, or \$500, between the second quarter of 2008 and the second quarter of 2009.

Try Exercise 59.

COMBINING LIKE TERMS

When two terms have variable factors that are exactly the same, like $5n$ and $-7n$, the terms are called **like,** or **similar, terms.*** The distributive law enables us to **combine,** or **collect, like terms.** The above rules for addition will again apply.

EXAMPLE 8 Combine like terms.

a) $-7x + 9x$ **b)** $2a + (-3b) + (-5a) + 9b$

c) $6 + y + (-3.5y) + 2$

SOLUTION

a) $-7x + 9x = (-7 + 9)x$ **Using the distributive law**
$= 2x$ **Adding -7 and 9**

b) $2a + (-3b) + (-5a) + 9b$
$= 2a + (-5a) + (-3b) + 9b$ **Using the commutative law of addition**
$= (2 + (-5))a + (-3 + 9)b$ **Using the distributive law**
$= -3a + 6b$ **Adding**

c) $6 + y + (-3.5y) + 2 = y + (-3.5y) + 6 + 2$ **Using the commutative law of addition**
$= (1 + (-3.5))y + 6 + 2$ **Using the distributive law**
$= -2.5y + 8$ **Adding**

Try Exercise 69.

With practice we can leave out some steps, combining like terms mentally. Note that numbers like 6 and 2 in the expression $6 + y + (-3.5y) + 2$ are constants and are also considered to be like terms.

*Like terms are discussed in greater detail in Section 1.8.

1.5 Exercise Set

FOR EXTRA HELP MyMathLab   

Concept Reinforcement *In each of Exercises 1–6, match the term with a like term from the column on the right.*

1. ___ $8n$ a) $-3z$
2. ___ $7m$ b) $5x$
3. ___ 43 c) $2t$
4. ___ $28z$ d) $-4m$
5. ___ $-2x$ e) 9
6. ___ $-9t$ f) $-3n$

Add using the number line.

7. $5 + (-8)$
8. $2 + (-5)$
9. $-5 + 9$
10. $-3 + 8$
11. $-4 + 0$
12. $-6 + 0$
13. $-3 + (-5)$
14. $-4 + (-6)$

Add. Do not use the number line except as a check.

15. $-6 + (-5)$
16. $-8 + (-12)$
17. $10 + (-15)$
18. $12 + (-22)$
19. $12 + (-12)$
20. $17 + (-17)$
21. $-24 + (-17)$
22. $-17 + (-25)$
23. $-13 + 13$
24. $-18 + 18$
25. $18 + (-11)$
26. $8 + (-5)$
27. $-36 + 0$
28. $0 + (-74)$
29. $-3 + 14$
30. $13 + (-6)$
31. $-14 + (-19)$
32. $11 + (-9)$
33. $19 + (-19)$
34. $-20 + (-6)$
35. $23 + (-5)$
36. $-15 + (-7)$
37. $-31 + (-14)$
38. $40 + (-8)$
39. $40 + (-40)$
40. $-25 + 25$
41. $85 + (-65)$
42. $63 + (-18)$
43. $-3.6 + 1.9$
44. $-6.5 + 4.7$
45. $-5.4 + (-3.7)$
46. $-3.8 + (-9.4)$
47. $\frac{-3}{5} + \frac{4}{5}$
48. $\frac{-2}{7} + \frac{3}{7}$
49. $\frac{-4}{7} + \frac{-2}{7}$
50. $\frac{-5}{9} + \frac{-2}{9}$
51. $-\frac{2}{5} + \frac{1}{3}$
52. $-\frac{4}{13} + \frac{1}{2}$
53. $\frac{-4}{9} + \frac{2}{3}$
54. $\frac{-1}{6} + \frac{1}{3}$
55. $35 + (-14) + (-19) + (-5)$
56. $28 + (-44) + 17 + 31 + (-94)$

Aha! 57. $-4.9 + 8.5 + 4.9 + (-8.5)$

58. $24 + 3.1 + (-44) + (-8.2) + 63$

Solve. Write your answer as a complete sentence.

59. *Gas Prices.* During one month, the price of a gallon of 87-octane gasoline dropped 5¢, dropped 3¢, and then rose 7¢. By how much did the price change during that period?

60. *Oil Prices.* During one winter, the price of a gallon of home heating oil dropped 6¢, rose 12¢, and then dropped 4¢. By how much did the price change during that period?

61. *Lake Level.* Between January 2004 and January 2008, the south end of the Great Salt Lake dropped $\frac{1}{2}$ ft, rose $\frac{6}{5}$ ft, rose $\frac{3}{4}$ ft, and dropped $\frac{3}{2}$ ft. By how much did the level change?
Source: U.S. Geological Survey

62. *Profits and Losses.* The table below lists the profits and losses of Fitness Sales over a 3-year period. Find the profit or loss after this period of time.

Year	*Profit or loss*
2008	−\$26,500
2009	−\$10,200
2010	+\$32,400

63. *Yardage Gained.* In an intramural football game, the quarterback attempted passes with the following results.

First try	13-yd gain
Second try	12-yd loss
Third try	21-yd gain

Find the total gain (or loss).

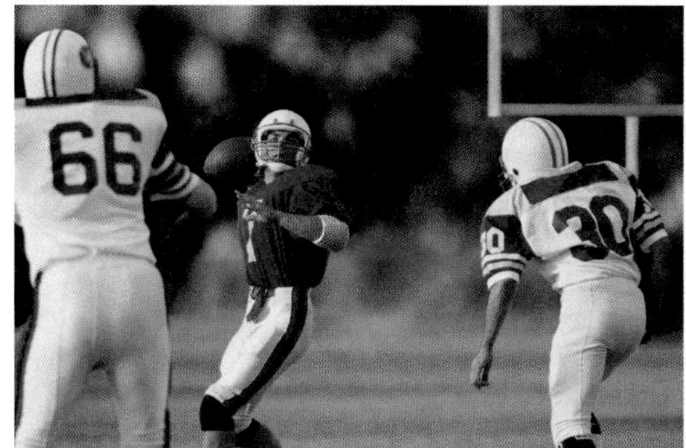

64. *Account Balance.* Lynn has \$350 in her checking account. She writes a check for \$530, makes a deposit of \$75, and then writes a check for \$90. What is the balance in her account?

65. *Telephone Bills.* Yusuf's cell-phone bill for July was \$82. He sent a check for \$50 and then ran up \$63 in charges for August. What was his new balance?

66. *Credit-Card Bills.* Ian's credit-card bill indicates that he owes \$470. He sends a check to the credit-card company for \$45, charges another \$160 in merchandise, and then pays off another \$500 of his bill. What is Ian's new balance?

67. *Peak Elevation.* The tallest mountain in the world, as measured from base to peak, is Mauna Kea in Hawaii. From a base 19,684 ft below sea level, it rises 33,480 ft. What is the elevation of its peak?
Source: *The Guinness Book of Records*

68. *Class Size.* During the first two weeks of the semester, 5 students withdrew from Meghan's algebra class, 8 students were added to the class, and 4 students were dropped as "no-shows." By how many students did the original class size change?

Combine like terms.

69. $5a + (-8a)$

70. $-3x + 8x$

71. $-3x + 12x$

72. $2m + (-7m)$

73. $-5a + (-2a)$

74. $10n + (-17n)$

75. $-3 + 8x + 4 + (-10x)$

76. $8a + 5 + (-a) + (-3)$

77. $6m + 9n + (-9n) + (-10m)$

78. $-11s + (-8t) + (-3s) + 8t$

79. $-4x + 6.3 + (-x) + (-10.2)$

80. $-7 + 10.5y + 13 + (-11.5y)$

Find the perimeter of each figure.

81.

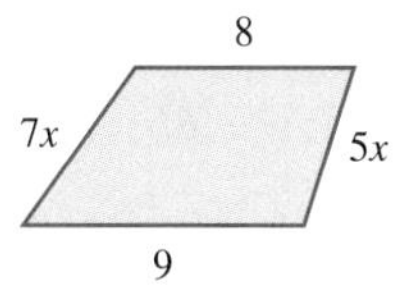

82.

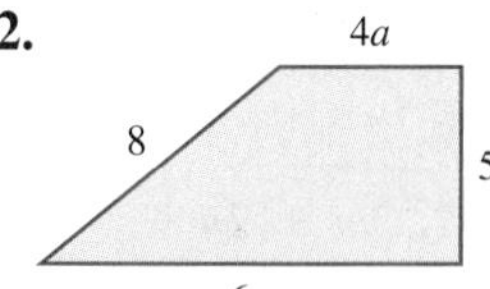

83.

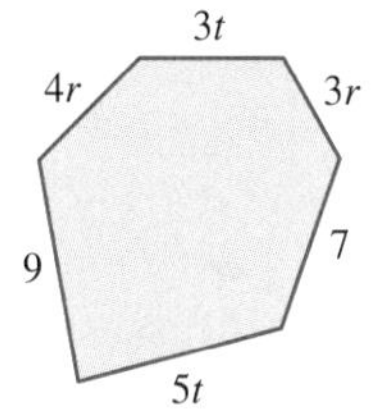

84.

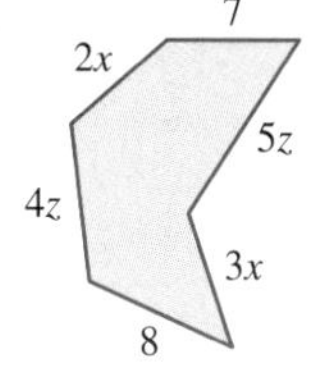

85.

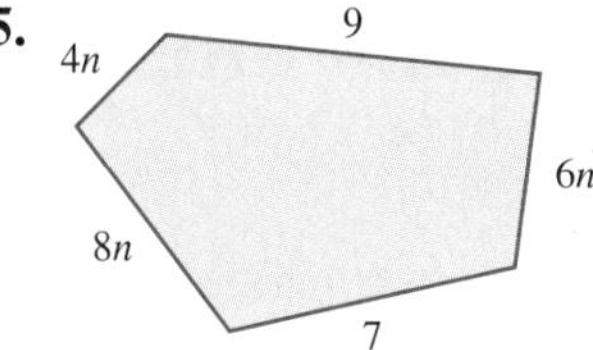

86.

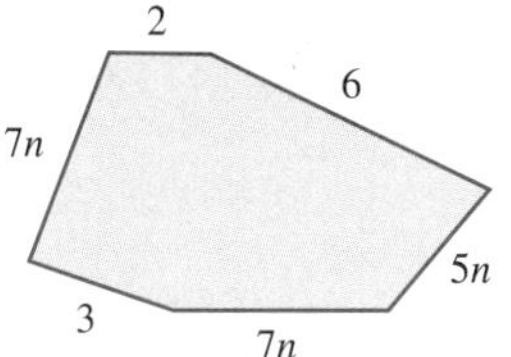

TW **87.** Explain in your own words why the sum of two negative numbers is negative.

TW **88.** Without performing the actual addition, explain why the sum of all integers from -10 to 10 is 0.

SKILL REVIEW

89. Multiply: $7(3z + y + 2)$. [1.2]

90. Divide and simplify: $\frac{7}{2} \div \frac{3}{8}$. [1.3]

SYNTHESIS

TW **91.** Under what circumstances will the sum of one positive number and several negative numbers be positive?

TW **92.** Is it possible to add real numbers without knowing how to calculate $a - b$ with a and b both nonnegative and $a \geq b$? Why or why not?

93. *Banking.* Travis had \$257.33 in his checking account. After depositing \$152 in the account and writing a check, his account was overdrawn by \$42.37. What was the amount of the check?

94. *Sports-Card Values.* The value of a sports card dropped \$12 and then rose \$17.50 before settling at \$61. What was the original value of the card?

Find the missing term or terms.

95. $4x + \_\_\_\_ + (-9x) + (-2y) = -5x - 7y$

96. $-3a + 9b + \_\_\_\_ + 5a = 2a - 6b$

97. $3m + 2n + \_\_\_\_ + (-2m) = 2n + (-6m)$

98. $\_\_\_\_ + 9x + (-4y) + x = 10x - 7y$

Aha! **99.** $7t + 23 + \_\_\_\_ + \_\_\_\_ = 0$

100. *Geometry.* The perimeter of a rectangle is $7x + 10$. If the length of the rectangle is 5, express the width in terms of x.

101. *Golfing.* After five rounds of golf, a golf pro was 3 under par twice, 2 over par once, 2 under par once, and 1 over par once. On average, how far above or below par was the golfer?

Try Exercise Answers: Section 1.5

7. −3 **9.** 4 **11.** −4 **13.** −8 **15.** −11 **17.** −5 **55.** −3 **59.** The price dropped 1¢. **69.** $-3a$

1.6 Subtraction of Real Numbers

- Opposites and Additive Inverses
- Subtraction
- Problem Solving

In arithmetic, when a number b is subtracted from another number a, the difference, $a - b$, is the number that when added to b gives a. For example, $10 - 7 = 3$ because $3 + 7 = 10$. We will use this approach to develop an efficient way of finding the value of $a - b$ for any real numbers a and b.

OPPOSITES AND ADDITIVE INVERSES

Numbers such as 6 and −6 are *opposites*, or *additive inverses*, of each other. Whenever opposites are added, the result is 0; and whenever two numbers add to 0, those numbers are opposites.

EXAMPLE 1 Find the opposite of each number: **(a)** 34; **(b)** −8.3; **(c)** 0.

SOLUTION

a) The opposite of 34 is −34: $34 + (-34) = 0.$

b) The opposite of −8.3 is 8.3: $-8.3 + 8.3 = 0.$

c) The opposite of 0 is 0: $0 + 0 = 0.$

Try Exercise 17.

To write the opposite, we use the symbol −, as follows.

Opposite The *opposite*, or *additive inverse*, of a number a is written $-a$ (read "the opposite of a" or "the additive inverse of a").

Note that if we choose a number (say, 8) and find its opposite (-8) and then find the opposite of the result, we will have the original number (8) again.

The Opposite of an Opposite

For any real number a,

$$-(-a) = a.$$

(The opposite of the opposite of a is a.)

EXAMPLE 2 Find $-x$ and $-(-x)$ when $x = 16$.

SOLUTION

If $x = 16$, then $-x = -16$. **The opposite of 16 is −16.**

If $x = 16$, then $-(-x) = -(-16) = 16$. **The opposite of the opposite of 16 is 16.**

Try Exercise 29.

EXAMPLE 3 Find $-x$ and $-(-x)$ when $x = -3$.

SOLUTION

If $x = -3$, then $-x = -(-3) = 3$. **The opposite of −3 is 3.**

If $x = -3$, then $-(-x) = -(-(-3)) = -(\ 3\ ) = -3$.

Try Exercise 23.

Student Notes

As you read mathematics, it is important to verbalize correctly the words and symbols to yourself. Consistently reading the expression $-x$ as "the opposite of x" is a good step in this direction.

Note in Example 3 that an extra set of parentheses is used to show that we are substituting the negative number -3 for x. The notation $--x$ is not used.

A symbol such as -8 is usually read "negative 8." It could be read "the additive inverse of 8," because the additive inverse of 8 is negative 8. It could also be read "the opposite of 8," because the opposite of 8 is -8.

A symbol like $-x$, which has a variable, should be read "the opposite of x" or "the additive inverse of x" and *not* "negative x," since to do so suggests that $-x$ represents a negative number.

The symbol "$-$" is read differently depending on where it appears. For example, $-5 - (-x)$ should be read "negative five minus the opposite of x."

EXAMPLE 4 Write each of the following in words.

a) $2 - 8$ **b)** $t - (-4)$ **c)** $-7 - (-x)$

SOLUTION

a) $2 - 8$ is read "two minus eight."

b) $t - (-4)$ is read "t minus negative four."

c) $-7 - (-x)$ is read "negative seven minus the opposite of x."

Try Exercise 11.

As we saw in Example 3, $-x$ can represent a positive number. This notation can be used to restate a result from Section 1.5 as *the law of opposites*.

The Law of Opposites

For any two numbers a and $-a$,

$$a + (-a) = 0.$$

(When opposites are added, their sum is 0.)

A negative number is said to have a "negative *sign*." A positive number is said to have a "positive *sign*." If we change a number to its opposite, or additive inverse, we say that we have "changed or reversed its sign."

EXAMPLE 5 Change the sign (find the opposite) of each number: **(a)** -3; **(b)** -10; **(c)** 14.

SOLUTION

a) When we change the sign of -3, we obtain 3.

b) When we change the sign of -10, we obtain 10.

c) When we change the sign of 14, we obtain -14.

Try Exercise 33.

SUBTRACTION

Opposites are helpful when subtraction involves negative numbers. To see why, look for a pattern in the following.

Subtracting		Adding the Opposite
$9 - 5 = 4$	since $4 + 5 = 9$	$9 + (-5) = 4$
$5 - 8 = -3$	since $-3 + 8 = 5$	$5 + (-8) = -3$
$-6 - 4 = -10$	since $-10 + 4 = -6$	$-6 + (-4) = -10$
$-7 - (-10) = 3$	since $3 + (-10) = -7$	$-7 + 10 = 3$
$-7 - (-2) = -5$	since $-5 + (-2) = -7$	$-7 + 2 = -5$

The matching results suggest that we can subtract by adding the opposite of the number being subtracted. This can always be done and often provides the easiest way to subtract real numbers.

Subtraction of Real Numbers For any real numbers a and b,

$$a - b = a + (-b).$$

(To subtract, add the opposite, or additive inverse, of the number being subtracted.)

EXAMPLE 6 Subtract each of the following and then check with addition.

a) $2 - 6$ **b)** $4 - (-9)$ **c)** $-4.2 - (-3.6)$
d) $-1.8 - (-7.5)$ **e)** $\frac{1}{5} - \left(-\frac{3}{5}\right)$

SOLUTION

a) $2 - 6 = 2 + (-6) = -4$ **The opposite of 6 is -6. We change the subtraction to addition and add the opposite.** ***Check:*** **$-4 + 6 = 2$.**

b) $4 - (-9) = 4 + 9 = 13$ **The opposite of -9 is 9. We change the subtraction to addition and add the opposite.** ***Check:*** **$13 + (-9) = 4$.**

c) $-4.2 - (-3.6) = -4.2 + 3.6$ **Adding the opposite of -3.6.** ***Check:*** **$-0.6 + (-3.6) = -4.2$.**

$= -0.6$

d) $-1.8 - (-7.5) = -1.8 + 7.5$ **Adding the opposite.** ***Check:*** **$5.7 + (-7.5) = -1.8$.**

$= 5.7$

e) $\frac{1}{5} - \left(-\frac{3}{5}\right) = \frac{1}{5} + \frac{3}{5}$ **Adding the opposite**

$= \frac{1 + 3}{5}$ **A common denominator exists so we add in the numerator.**

$= \frac{4}{5}$

Check: $\frac{4}{5} + \left(-\frac{3}{5}\right) = \frac{4}{5} + \frac{-3}{5} = \frac{4 + (-3)}{5} = \frac{1}{5}$.

Try Exercises 41 and 43.

EXAMPLE 7 Simplify: $8 - (-4) - 2 - (-5) + 3$.

SOLUTION

$$8 - (-4) - 2 - (-5) + 3 = 8 + 4 + (-2) + 5 + 3$$ **To subtract, we add the opposite.**

$$= 18$$

Try Exercise 99.

Recall from Section 1.2 that the terms of an algebraic expression are separated by plus signs. This means that the terms of $5x - 7y - 9$ are $5x$, $-7y$, and -9, since $5x - 7y - 9 = 5x + (-7y) + (-9)$.

EXAMPLE 8 Identify the terms of $4 - 2ab + 7a - 9$.

SOLUTION We have

$$4 - 2ab + 7a - 9 = 4 + (-2ab) + 7a + (-9),$$ **Rewriting as addition**

so the terms are 4, $-2ab$, $7a$, and -9.

Try Exercise 107.

STUDY TIP

Indicate the Highlights

Most students find it helpful to draw a star or use a color, felt-tipped highlighter to indicate important concepts or trouble spots that require further study. Most campus bookstores carry a variety of highlighters that permit you to brightly color written material while keeping it easy to read.

EXAMPLE 9 Combine like terms.

a) $1 + 3x - 7x$

b) $-5a - 7b - 4a + 10b$

c) $4 - 3m - 9 + 2m$

SOLUTION

a) $1 + 3x - 7x = 1 + 3x + (-7x)$ **Adding the opposite**

$= 1 + (3 + (-7))x$
$= 1 + (-4)x$ **Using the distributive law. Try to do this mentally.**

$= 1 - 4x$ **Rewriting as subtraction to be more concise: $1 + (-4x) = 1 - 4x$**

b) $$\begin{aligned} -5a - 7b - 4a + 10b &= -5a + (-7b) + (-4a) + 10b && \textbf{Adding the opposite} \\ &= -5a + (-4a) + (-7b) + 10b && \textbf{Using the commutative law of addition} \\ &= -9a + 3b && \textbf{Combining like terms mentally} \end{aligned}$$

c) $$\begin{aligned} 4 - 3m - 9 + 2m &= 4 + (-3m) + (-9) + 2m && \textbf{Rewriting as addition} \\ &= 4 + (-9) + (-3m) + 2m && \textbf{Using the commutative law of addition} \\ &= -5 + (-1m) && \textbf{We can write } -1m \textbf{ as } -m. \\ &= -5 - m \end{aligned}$$

Try Exercise 113.

PROBLEM SOLVING

We use subtraction to solve problems involving differences. These include problems that ask "How much more?" or "How much higher?"

EXAMPLE 10 Elevation. The Jordan River begins in Lebanon at an elevation of 550 m above sea level and empties into the Dead Sea at an elevation of 400 m below sea level. During its 360-km length, by how many meters does it fall?
Source: Brittanica Online

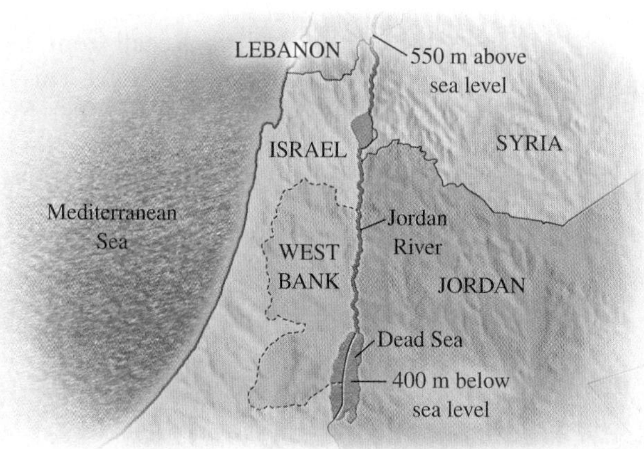

SOLUTION To find the difference between two elevations, we subtract the lower elevation from the higher elevation:

$$\begin{aligned} &\underbrace{\text{Higher elevation}} \quad - \quad \underbrace{\text{Lower elevation}} \\ &\quad 550\text{ m} \qquad\quad - \qquad (-400\text{ m}) \\ &= 550 + 400 \\ &= 950. \end{aligned}$$

The Jordan River falls 950 m.

Try Exercise 123.

1.6 Exercise Set

FOR EXTRA HELP MyMathLab MathXL PRACTICE  WATCH  DOWNLOAD  READ REVIEW

Concept Reinforcement *In each of Exercises 1–8, match the expression with the appropriate wording from the column on the right.*

1. ___ $-x$
2. ___ $12 - x$
3. ___ $12 - (-x)$
4. ___ $x - 12$
5. ___ $x - (-12)$
6. ___ $-x - 12$
7. ___ $-x - x$
8. ___ $-x - (-12)$

a) x minus negative twelve
b) The opposite of x minus x
c) The opposite of x minus twelve
d) The opposite of x
e) The opposite of x minus negative twelve
f) Twelve minus the opposite of x
g) Twelve minus x
h) x minus twelve

Write each of the following in words.

9. $4 - 10$
10. $5 - 13$
11. $2 - (-9)$
12. $4 - (-1)$
13. $-x - y$
14. $-a - b$
15. $-3 - (-n)$
16. $-7 - (-m)$

Find the opposite, or additive inverse.

17. 39
18. -17
19. $-\frac{11}{2}$
20. $\frac{7}{2}$
21. -3.14
22. 48.2

Find $-x$ when x is each of the following.

23. -45
24. 13
25. $-\frac{14}{3}$
26. $\frac{1}{328}$
27. 0.101
28. 0

Find $-(-x)$ when x is each of the following.

29. 72
30. 29
31. $-\frac{2}{5}$
32. -9.1

Change the sign. (Find the opposite.)

33. -1
34. -7
35. 7
36. 10

Subtract.

37. $6 - 8$
38. $4 - 13$
39. $0 - 5$
40. $0 - 8$
41. $-4 - 3$
42. $-5 - 6$
43. $-9 - (-3)$
44. $-9 - (-5)$
45. Aha! $-8 - (-8)$
46. $-10 - (-10)$
47. $30 - 40$
48. $20 - 27$
49. $-7 - (-9)$
50. $-8 - (-3)$
51. $-9 - (-9)$
52. $-40 - (-40)$
53. $5 - 5$
54. $7 - 7$
55. $4 - (-4)$
56. $6 - (-6)$
57. $-7 - 4$
58. $-6 - 8$
59. $6 - (-10)$
60. $3 - (-12)$
61. $-6 - (-5)$
62. $-4 - (-7)$
63. $5 - (-12)$
64. $5 - (-6)$
65. $0 - (-10)$
66. $0 - (-1)$
67. $-5 - (-2)$
68. $-3 - (-1)$
69. $7 - 14$
70. $-9 - 16$
71. $-8 - 0$
72. $-9 - 0$
73. $0 - 11$
74. $0 - 31$
75. $2 - 25$
76. $18 - 63$
77. $-4.2 - 3.1$
78. $-10.1 - 2.6$
79. $-1.8 - (-2.4)$
80. $-5.8 - (-7.3)$
81. $3.2 - 8.7$
82. $1.5 - 9.4$
83. $0.072 - 1$
84. $0.825 - 1$
85. $\frac{2}{11} - \frac{9}{11}$
86. $\frac{3}{7} - \frac{5}{7}$
87. $\frac{-1}{5} - \frac{3}{5}$
88. $\frac{-2}{9} - \frac{5}{9}$
89. $-\frac{4}{17} - \left(-\frac{9}{17}\right)$
90. $-\frac{2}{13} - \left(-\frac{5}{13}\right)$

To find the difference between a and b, we subtract b from a.

91. Subtract 37 from -21.

92. Subtract 19 from -7.

93. Subtract -25 from 9.

94. Subtract -31 from -5.

In each of Exercises 95–98, translate the phrase to mathematical language and simplify.

95. The difference between 3.8 and -5.2

96. The difference between -2.1 and -5.9

97. The difference between 114 and -79

98. The difference between 23 and -17

Simplify.

99. $25 - (-12) - 7 - (-2) + 9$

100. $22 - (-18) + 7 + (-42) - 27$

101. $-31 + (-28) - (-14) - 17$

102. $-43 - (-19) - (-21) + 25$

103. $-34 - 28 + (-33) - 44$

104. $39 + (-88) - 29 - (-83)$

Aha! **105.** $-93 + (-84) - (-93) - (-84)$

106. $84 + (-99) + 44 - (-99) - 43$

Identify the terms in each expression.

107. $-7x - 4y$

108. $7a - 9b$

109. $9 - 5t - 3st$

110. $-4 - 3x + 2xy$

Combine like terms.

111. $4x - 7x$

112. $3a - 14a$

113. $7a - 12a + 4$

114. $-9x - 13x + 7$

115. $-8n - 9 + 7n$

116. $-7 + 9n - 8n$

117. $2 - 6t - 9 + t$

118. $-5 + b - 7 - 5b$

119. $5y + (-3x) - 9x + 1 - 2y + 8$

120. $14 - (-5x) + 2z - (-32) + 4z - 2x$

121. $13x - (-2x) + 45 - (-21) - 7x$

122. $8t - (-2t) - 14 - (-5t) + 53 - 9t$

Solve.

123. *Record Elevations.* The current world records for the highest parachute jump and the lowest manned-vessel ocean dive were both set in 1960. On August 16 of that year, Captain Joseph Kittinger jumped from a height of 102,880 ft above sea level. Earlier, on January 23, Jacques Piccard and Navy Lieutenant Donald Walsh descended in a bathyscaphe 35,797 ft below sea level. What was the difference in elevation between the highest parachute jump and the lowest ocean dive?
Sources: www.firstflight.org; www.seasky.org

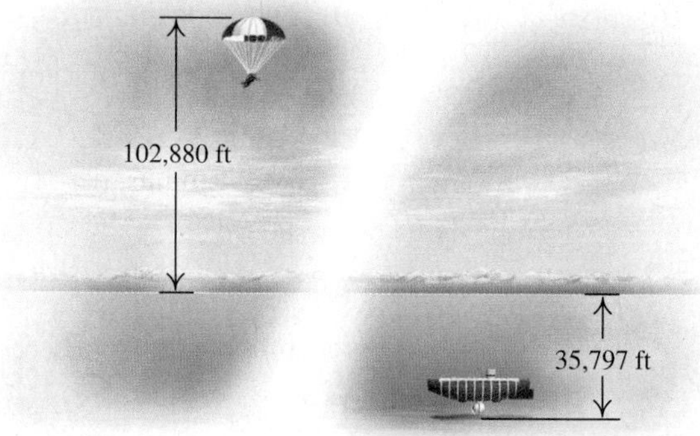

124. *Elevation Extremes.* The lowest elevation in Asia, the Dead Sea, is 1340 ft below sea level. The highest elevation in Asia, Mount Everest, is 29,035 ft. Find the difference in elevation.
Source: encarta.msn.com

125. *Temperature Change.* On January 12, 1980, the temperature in Great Falls, Montana, rose from $-32°F$ to 15°F in just 7 min. By how much did the temperature change?
Source: www.greaterfalls.com

126. *Temperature Extremes.* The highest temperature ever recorded in the United States is 134°F in Greenland Ranch, California, on July 10, 1913. The lowest temperature ever recorded is $-79.8°F$ in Prospect Creek, Alaska, on January 23, 1971. How much higher was the temperature in Greenland Ranch than that in Prospect Creek?
Source: infoplease, Pearson Education, Inc.

127. *Basketball.* A team's scoring differential is the difference between points scored and points allowed. The Cleveland Cavaliers improved their scoring differential from -0.4 in 2007–2008 to +8.5 in 2008–2009. By how many points did they improve?
Source: sports.espn.go.com

128. *Underwater Elevation.* The deepest point in the Pacific Ocean is the Challenger Deep in the Marianas Trench, with a depth of 10,911.5 m. The deepest point in the Atlantic Ocean is the Milwaukee Deep in the Puerto Rico Trench, with a depth of 8530 m. What is the difference in elevation of the two trenches?
Source: infoplease, Pearson Education, Inc.

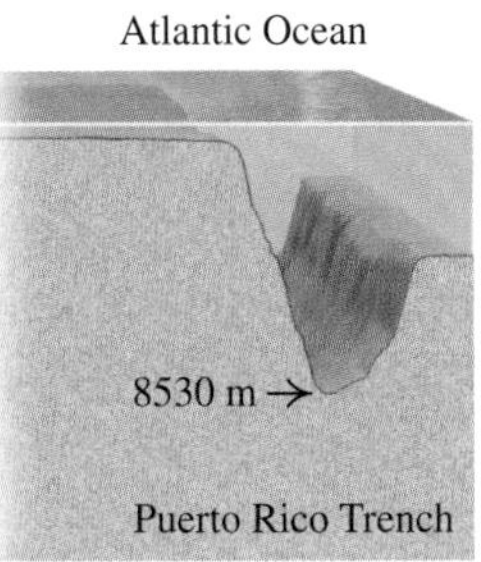

TW **129.** Brianna insists that if you can *add* real numbers, then you can also *subtract* real numbers. Do you agree? Why or why not?

TW **130.** Are the expressions $-a + b$ and $a + (-b)$ opposites of each other? Why or why not?

SKILL REVIEW

131. Find the area of a rectangle when the length is 36 ft and the width is 12 ft. [1.1]

132. Find the prime factorization of 864. [1.3]

SYNTHESIS

TW **133.** Explain the different uses of the symbol "−". Give examples of each use of the symbol.

TW **134.** If a and b are both negative, under what circumstances will $a - b$ be negative?

135. *Power Outages.* During the Northeast's electrical blackout of August 14, 2003, residents of Bloomfield, New Jersey, lost power at 4:00 P.M. One resident returned from vacation at 3:00 P.M. the following day to find the clocks in her apartment reading 8:00 A.M. At what time, and on what day, was power restored?

Tell whether each statement is true or false for all real numbers m and n. Use various replacements for m and n to support your answer.

136. If $m > n$, then $m - n > 0$.

137. If $m > n$, then $m + n > 0$.

138. If m and n are opposites, then $m - n = 0$.

139. If $m = -n$, then $m + n = 0$.

TW **140.** A gambler loses a wager and then loses "double or nothing" (meaning the gambler owes twice as much) twice more. After the three losses, the gambler's assets are −$20. Explain how much the gambler originally bet and how the $20 debt occurred.

141. List the keystrokes needed to compute $-9 - (-7)$.

TW **142.** If n is positive and m is negative, what is the sign of $n + (-m)$? Why?

Try Exercise Answers: Section 1.6

11. Two minus negative nine **17.** −39 **23.** 45 **29.** 72 **33.** 1 **41.** −7 **43.** −6 **99.** 41 **107.** $-7x, -4y$ **113.** $-5a + 4$ **123.** 138,677 ft

1.7 Multiplication and Division of Real Numbers

- Multiplication
- Division

We now develop rules for multiplication and division of real numbers. Because multiplication and division are closely related, the rules are quite similar.

MULTIPLICATION

We already know how to multiply two nonnegative numbers. To see how to multiply a positive number and a negative number, consider the following pattern in which multiplication is regarded as repeated addition:

This number decreases by 1 each time. →

$$\begin{aligned} 4(-5) &= (-5)+(-5)+(-5)+(-5) = -20 \\ 3(-5) &= (-5)+(-5)+(-5) = -15 \\ 2(-5) &= (-5)+(-5) = -10 \\ 1(-5) &= (-5) = -5 \\ 0(-5) &= 0 = 0 \end{aligned}$$

← This number increases by 5 each time.

This pattern illustrates that the product of a negative number and a positive number is negative.

The Product of a Negative Number and a Positive Number To multiply a positive number and a negative number, multiply their absolute values. The answer is negative.

STUDY TIP

How Did They Get That!?

The *Student's Solutions Manual* is an excellent resource if you need additional help with an exercise in the exercise sets. It contains step-by-step solutions to the odd-numbered exercises in each exercise set.

EXAMPLE 1 Multiply: **(a)** $8(-5)$; **(b)** $-\frac{1}{3} \cdot \frac{5}{7}$.

SOLUTION The product of a negative number and a positive number is negative.

a) $8(-5) = -40$ *Think*: **$8 \cdot 5 = 40$; make the answer negative.**

b) $-\frac{1}{3} \cdot \frac{5}{7} = -\frac{5}{21}$ *Think*: **$\frac{1}{3} \cdot \frac{5}{7} = \frac{5}{21}$; make the answer negative.**

Try Exercise 11.

The pattern developed above includes not just products of positive numbers and negative numbers, but a product involving zero as well.

The Multiplicative Property of Zero For any real number a,

$$0 \cdot a = a \cdot 0 = 0.$$

(The product of 0 and any real number is 0.)

EXAMPLE 2 Multiply: $173(-452)0$.

SOLUTION We have

$$\begin{aligned} 173(-452)0 &= 173[(-452)0] && \text{Using the associative law of multiplication} \\ &= 173[0] && \text{Using the multiplicative property of zero} \\ &= 0. && \text{Using the multiplicative property of zero again} \end{aligned}$$

Note that whenever 0 appears as a factor, the product will be 0.

Try Exercise 29.

We can extend the above pattern still further to examine the product of two negative numbers.

This number decreases by 1 each time. →

$$\begin{aligned} 2(-5) &= (-5) + (-5) = -10 \\ 1(-5) &= (-5) = -5 \\ 0(-5) &= 0 = 0 \\ -1(-5) &= -(-5) = 5 \\ -2(-5) &= -(-5) - (-5) = 10 \end{aligned}$$

← This number increases by 5 each time.

According to the pattern, the product of two negative numbers is positive.

The Product of Two Negative Numbers To multiply two negative numbers, multiply their absolute values. The answer is positive.

EXAMPLE 3 Multiply: **(a)** $(-6)(-8)$; **(b)** $(-1.2)(-3)$.

SOLUTION The product of two negative numbers is positive.

a) The absolute value of -6 is 6 and the absolute value of -8 is 8. Thus,

$$\begin{aligned} (-6)(-8) &= 6 \cdot 8 && \text{Multiplying absolute values. The answer is positive.} \\ &= 48. \end{aligned}$$

b)

$$\begin{aligned} (-1.2)(-3) &= (1.2)(3) && \text{Multiplying absolute values. The answer is positive.} \\ &= 3.6 && \text{Try to go directly to this step.} \end{aligned}$$

Try Exercise 17.

When three or more numbers are multiplied, we can order and group the numbers as we please because of the commutative and associative laws.

EXAMPLE 4 Multiply: **(a)** $-3(-2)(-5)$; **(b)** $-4(-6)(-1)(-2)$.

SOLUTION

a)

$$\begin{aligned} -3(-2)(-5) &= 6(-5) && \text{Multiplying the first two numbers. The product of two negatives is positive.} \\ &= -30 && \text{The product of a positive and a negative is negative.} \end{aligned}$$

b) $-4(-6)(-1)(-2) = 24 \cdot 2$ **Multiplying the first two numbers and the last two numbers**

$= 48$

Try Exercise 43.

We can see the following pattern in the results of Example 4.

The product of an even number of negative numbers is positive.

The product of an odd number of negative numbers is negative.

DIVISION

Recall that $a \div b$, or $\frac{a}{b}$, is the number, if one exists, that when multiplied by b gives a. For example, to show that $10 \div 2$ is 5, we need only note that $5 \cdot 2 = 10$. Thus division can always be checked with multiplication.

The rules for signs for division are the same as those for multiplication: The quotient of a positive number and a negative number is negative; the quotient of two negative numbers is positive.

Rules for Multiplication and Division To multiply or divide two nonzero real numbers:

1. Using the absolute values, multiply or divide, as indicated.
2. If the signs are the same, the answer is positive.
3. If the signs are different, the answer is negative.

EXAMPLE 5 Divide, if possible, and check your answer.

a) $14 \div (-7)$ b) $\frac{-32}{-4}$ c) $\frac{-10}{9}$ d) $\frac{-17}{0}$

SOLUTION

a) $14 \div (-7) = -2$ **We look for a number that when multiplied by -7 gives 14. That number is -2. *Check*: $(-2)(-7) = 14$.**

b) $\frac{-32}{-4} = 8$ **We look for a number that when multiplied by -4 gives -32. That number is 8. *Check*: $8(-4) = -32$.**

c) $\frac{-10}{9} = -\frac{10}{9}$ **We look for a number that when multiplied by 9 gives -10. That number is $-\frac{10}{9}$. *Check*: $-\frac{10}{9} \cdot 9 = -10$.**

d) $\frac{-17}{0}$ is **undefined.** **We look for a number that when multiplied by 0 gives -17. There is no such number because if 0 is a factor, then the product is 0, not -17.**

Student Notes

Try to regard "undefined" as a mathematical way of saying "we do not give any meaning to this expression."

Try Exercises 51 and 53.

Had Example 5(a) been written as $-14 \div 7$ or $-\frac{14}{7}$, rather than $14 \div (-7)$, the result would still have been -2. Thus from Examples 5(a)–5(c), we have the following:

$$\frac{-a}{b} = \frac{a}{-b} = -\frac{a}{b} \quad \text{and} \quad \frac{-a}{-b} = \frac{a}{b}.$$

EXAMPLE 6 Rewrite each of the following in two equivalent forms: **(a)** $\frac{5}{-2}$; **(b)** $-\frac{3}{10}$.

SOLUTION We use one of the properties just listed.

a) $\dfrac{5}{-2} = \dfrac{-5}{2}$ and $\dfrac{5}{-2} = -\dfrac{5}{2}$

b) $-\dfrac{3}{10} = \dfrac{-3}{10}$ and $-\dfrac{3}{10} = \dfrac{3}{-10}$

Since $\dfrac{-a}{b} = \dfrac{a}{-b} = -\dfrac{a}{b}$

Try Exercise 71.

When a fraction contains a negative sign, it may be helpful to rewrite (or simply visualize) the fraction in an equivalent form.

EXAMPLE 7 Perform the indicated operation: **(a)** $\left(-\frac{4}{5}\right)\left(\frac{-7}{3}\right)$; **(b)** $-\frac{2}{7} + \frac{9}{-7}$.

SOLUTION

a) $\left(-\dfrac{4}{5}\right)\left(\dfrac{-7}{3}\right) = \left(-\dfrac{4}{5}\right)\left(-\dfrac{7}{3}\right)$ **Rewriting $\dfrac{-7}{3}$ as $-\dfrac{7}{3}$**

$= \dfrac{28}{15}$ **Try to go directly to this step.**

b) Given a choice, we generally choose a positive denominator:

$-\dfrac{2}{7} + \dfrac{9}{-7} = \dfrac{-2}{7} + \dfrac{-9}{7}$ **Rewriting both fractions with a common denominator of 7**

$= \dfrac{-11}{7}$, or $-\dfrac{11}{7}$.

Try Exercise 91.

To divide with fraction notation, it is usually easiest to find a reciprocal and then multiply.

EXAMPLE 8 Find the reciprocal of each number, if it exists.

a) -27 **b)** $\frac{-3}{4}$ **c)** $-\frac{1}{5}$ **d)** 0

SOLUTION Recall from Section 1.3 that two numbers are reciprocals of each other if their product is 1.

a) The reciprocal of -27 is $\frac{1}{-27}$. More often, this number is written as $-\frac{1}{27}$.
Check: $(-27)\left(-\frac{1}{27}\right) = \frac{27}{27} = 1$.

b) The reciprocal of $\frac{-3}{4}$ is $\frac{4}{-3}$, or, equivalently, $-\frac{4}{3}$.
Check: $\frac{-3}{4} \cdot \frac{4}{-3} = \frac{-12}{-12} = 1$.

c) The reciprocal of $-\frac{1}{5}$ is $-\frac{5}{1}$, or -5.
Check: $-\frac{1}{5}(-5) = \frac{5}{5} = 1$.

d) The reciprocal of 0 does not exist. To see this, recall that there is no number r for which $0 \cdot r = 1$.

Try Exercise 79.

EXAMPLE 9 Divide: **(a)** $-\frac{2}{3} \div \left(-\frac{5}{4}\right)$; **(b)** $-\frac{3}{4} \div \frac{3}{10}$.

SOLUTION We divide by multiplying by the reciprocal of the divisor.

a) $-\frac{2}{3} \div \left(-\frac{5}{4}\right) = -\frac{2}{3} \cdot \left(-\frac{4}{5}\right) = \frac{8}{15}$ **Multiplying by the reciprocal**

Be careful not to change the sign when taking a reciprocal!

b) $-\frac{3}{4} \div \frac{3}{10} = -\frac{3}{4} \cdot \left(\frac{10}{3}\right) = -\frac{30}{12} = -\frac{5}{2} \cdot \frac{6}{6} = -\frac{5}{2}$ **Removing a factor equal to 1: $\frac{6}{6} = 1$**

Try Exercise 99.

To divide with decimal notation, it is usually easiest to carry out the division.

EXAMPLE 10 Divide: $27.9 \div (-3)$.

SOLUTION

$$27.9 \div (-3) = \frac{27.9}{-3} = -9.3$$

Dividing: $3\overline{)27.9}$ = 9.3. The answer is negative.

Try Exercise 57.

In Example 5(d), we explained why we cannot divide -17 by 0. To see why *no* nonzero number b can be divided by 0, remember that $b \div 0$ would have to be the number that when multiplied by 0 gives b. But since the product of 0 and any number is 0, not b, we say that $b \div 0$ is **undefined** for $b \neq 0$. In the special case of $0 \div 0$, we look for a number r such that $0 \div 0 = r$ and $r \cdot 0 = 0$. But, $r \cdot 0 = 0$ for *any* number r. For this reason, we say that $b \div 0$ is undefined for any choice of b.*

Finally, note that $0 \div 7 = 0$ since $0 \cdot 7 = 0$. This can be written $0/7 = 0$. It is important not to confuse division *by* 0 with division *into* 0.

EXAMPLE 11 Divide, if possible: **(a)** $\frac{0}{-2}$; **(b)** $\frac{5}{0}$.

SOLUTION

a) $\frac{0}{-2} = 0$ *Check*: $0(-2) = 0$.

b) $\frac{5}{0}$ is undefined.

Try Exercise 67.

Division Involving Zero For any real number a, $\frac{a}{0}$ is undefined, and for $a \neq 0$, $\frac{0}{a} = 0$.

*Sometimes $0 \div 0$ is said to be *indeterminate*.

CAUTION! It is important not to confuse *opposite* with *reciprocal*. Keep in mind that the opposite, or *additive inverse*, of a number is what we add to the number to get 0. The reciprocal, or *multiplicative inverse*, is what we multiply the number by to get 1.

Compare the following.

Number	Opposite (Change the sign.)	Reciprocal (Invert but do not change the sign.)
$-\frac{3}{8}$	$\frac{3}{8}$ ← $-\frac{3}{8}+\frac{3}{8}=0$	$-\frac{8}{3}$ ← $\left(-\frac{3}{8}\right)\left(-\frac{8}{3}\right)=1$
19	-19	$\frac{1}{19}$
$\frac{18}{7}$	$-\frac{18}{7}$	$\frac{7}{18}$
-7.9	7.9	$-\frac{1}{7.9}$, or $-\frac{10}{79}$
0	0	Undefined

1.7 Exercise Set

FOR EXTRA HELP MyMathLab  

Concept Reinforcement *In each of Exercises 1–10, replace the blank with either* 0 *or* 1 *to match the description given.*

1. The product of two reciprocals ____
2. The sum of a pair of opposites ____
3. The sum of a pair of additive inverses ____
4. The product of two multiplicative inverses ____
5. This number has no reciprocal. ____
6. This number is its own reciprocal. ____
7. This number is the multiplicative identity. ____
8. This number is the additive identity. ____
9. A nonzero number divided by itself ____
10. Division by this number is undefined. ____

Multiply.

11. $-3 \cdot 8$
12. $-3 \cdot 7$
13. $-8 \cdot 7$
14. $-9 \cdot 2$
15. $8 \cdot (-3)$
16. $9 \cdot (-5)$
17. $-6 \cdot (-7)$
18. $-2 \cdot (-5)$
19. $19 \cdot (-10)$
20. $12 \cdot (-10)$
21. $-12 \cdot 12$
22. $-13 \cdot (-15)$
23. $-25 \cdot (-48)$
24. $15 \cdot (-43)$
25. $4.5 \cdot (-28)$
26. $-49 \cdot (-2.1)$
27. $-5 \cdot (-2.3)$
28. $-6 \cdot 4.8$
29. $(-25) \cdot 0$
30. $0 \cdot (-4.7)$
31. $\frac{2}{3} \cdot \left(-\frac{3}{5}\right)$
32. $\frac{5}{7} \cdot \left(-\frac{2}{3}\right)$
33. $-\frac{3}{8} \cdot \left(-\frac{2}{9}\right)$
34. $-\frac{5}{8} \cdot \left(-\frac{2}{5}\right)$
35. $(-5.3)(2.1)$
36. $(-4.3)(9.5)$
37. $-\frac{5}{9} \cdot \frac{3}{4}$
38. $-\frac{8}{3} \cdot \frac{9}{4}$
39. $3 \cdot (-7) \cdot (-2) \cdot 6$
40. $9 \cdot (-2) \cdot (-6) \cdot 7$
41. $-\frac{1}{3} \cdot \frac{1}{4} \cdot \left(-\frac{3}{7}\right)$
42. $-\frac{1}{2} \cdot \frac{3}{5} \cdot \left(-\frac{2}{7}\right)$
43. $-2 \cdot (-5) \cdot (-3) \cdot (-5)$

44. $-3 \cdot (-5) \cdot (-2) \cdot (-1)$

Aha! **45.** $(-31) \cdot (-27) \cdot 0$

46. $7 \cdot (-6) \cdot 5 \cdot (-4) \cdot 3 \cdot (-2) \cdot 1 \cdot 0$

47. $(-8)(-9)(-10)$

48. $(-7)(-8)(-9)(-10)$

49. $(-6)(-7)(-8)(-9)(-10)$

50. $(-5)(-6)(-7)(-8)(-9)(-10)$

Divide, if possible, and check. If a quotient is undefined, state this.

51. $14 \div (-2)$

52. $\frac{24}{-3}$

53. $-26 \div (-13)$

54. $-32 \div (-4)$

55. $\frac{-50}{5}$

56. $\frac{-50}{25}$

57. $-10.2 \div (-2)$

58. $-2 \div 0.8$

59. $-100 \div (-11)$

60. $\frac{-64}{-7}$

61. $\frac{400}{-50}$

62. $-300 \div (-13)$

63. $\frac{28}{0}$

64. $\frac{0}{-5}$

65. $-4.8 \div 1.2$

66. $-3.9 \div 1.3$

67. $\frac{0}{-9}$

68. $0 \div (-47)$

Aha! **69.** $\frac{9.7(-2.8)0}{4.3}$

70. $\frac{(-4.9)(7.2)}{0}$

Write each expression in two equivalent forms, as in Example 6.

71. $\frac{-8}{3}$

72. $\frac{-12}{7}$

73. $\frac{29}{-35}$

74. $\frac{9}{-14}$

75. $-\frac{7}{3}$

76. $-\frac{4}{15}$

77. $\frac{-x}{2}$

78. $\frac{9}{-a}$

Find the reciprocal of each number, if it exists.

79. $-\frac{4}{5}$

80. $-\frac{13}{11}$

81. $\frac{51}{-10}$

82. $\frac{43}{-24}$

83. -10

84. 34

85. 4.3

86. -1.7

87. $\frac{-1}{4}$

88. $\frac{-1}{11}$

89. 0

90. -1

Perform the indicated operation and, if possible, simplify. If a quotient is undefined, state this.

91. $\left(\frac{-7}{4}\right)\left(-\frac{3}{5}\right)$

92. $\left(-\frac{5}{6}\right)\left(\frac{-1}{3}\right)$

93. $\frac{-3}{8} + \frac{-5}{8}$

94. $\frac{-4}{5} + \frac{7}{5}$

Aha! **95.** $\left(\frac{-9}{5}\right)\left(\frac{5}{-9}\right)$

96. $\left(-\frac{2}{7}\right)\left(\frac{5}{-8}\right)$

97. $\left(-\frac{3}{11}\right) - \left(-\frac{6}{11}\right)$

98. $\left(-\frac{4}{7}\right) - \left(-\frac{2}{7}\right)$

99. $\frac{7}{8} \div \left(-\frac{1}{2}\right)$

100. $\frac{3}{4} \div \left(-\frac{2}{3}\right)$

101. $\frac{9}{5} \cdot \frac{-20}{3}$

102. $\frac{-5}{12} \cdot \frac{7}{15}$

103. $\left(-\frac{18}{7}\right) + \left(-\frac{3}{7}\right)$

104. $\left(-\frac{12}{5}\right) + \left(-\frac{3}{5}\right)$

Aha! **105.** $-\frac{5}{9} \div \left(-\frac{5}{9}\right)$

106. $-\frac{5}{12} \div \frac{15}{7}$

107. $\frac{5}{9} - \frac{7}{9}$

108. $\frac{2}{7} - \frac{6}{7}$

109. $\frac{-3}{10} + \frac{2}{5}$

110. $\frac{-5}{9} + \frac{2}{3}$

111. $\frac{7}{10} \div \left(\frac{-3}{5}\right)$

112. $\left(\frac{-3}{5}\right) \div \frac{6}{15}$

113. $\frac{14}{-9} \div \frac{0}{3}$

114. $\frac{0}{-10} \div \frac{-3}{8}$

115. $\frac{-4}{15} + \frac{2}{-3}$

116. $\frac{3}{-10} + \frac{-1}{5}$

TW **117.** Most calculators have a key, often appearing as x^{-1}, for finding reciprocals. To use this key, we enter a number and then press x^{-1} ENTER to find its reciprocal. What should happen if we enter a number and then find the reciprocal twice? Why?

TW **118.** Multiplication can be regarded as repeated addition. Using this idea and the number line, explain why $3 \cdot (-5) = -15$.

SKILL REVIEW

119. Simplify: $\frac{264}{468}$. [1.3]

120. Determine whether 12 is a solution of $35 - a = 13$. [1.1]

SYNTHESIS

TW **121.** If two nonzero numbers are opposites of each other, are their reciprocals opposites of each other? Why or why not?

TW **122.** If two numbers are reciprocals of each other, are their opposites reciprocals of each other? Why or why not?

Translate to an algebraic expression or equation.

123. The reciprocal of a sum

124. The sum of two reciprocals

125. The opposite of a sum

126. The sum of two opposites

127. A real number is its own opposite.

128. A real number is its own reciprocal.

129. Show that the reciprocal of a sum is *not* the sum of the two reciprocals.

130. Which real numbers are their own reciprocals?

Tell whether each expression represents a positive number or a negative number when m and n are negative.

131. $\frac{m}{-n}$

132. $\frac{-n}{-m}$

133. $-m \cdot \left(\frac{-n}{m}\right)$

134. $-\left(\frac{n}{-m}\right)$

135. $(m + n) \cdot \frac{m}{n}$

136. $(-n - m)\frac{n}{m}$

137. What must be true of m and n if $-mn$ is to be **(a)** positive? **(b)** zero? **(c)** negative?

138. Jenna is a meteorologist. On December 10, she notes that the temperature is $-3°F$ at 6:00 A.M. She predicts that the temperature will rise at a rate of 2° per hour for 3 hr, and then rise at a rate of 3° per hour for 6 hr. She also predicts that the temperature will then fall at a rate of 2° per hour for 3 hr, and then fall at a rate of 5° per hour for 2 hr. What is Jenna's temperature forecast for 8:00 P.M?

139. The following is a proof that a positive number times a negative number is negative. Provide a reason for each step. Assume that $a > 0$ and $b > 0$.

$$\begin{aligned} a(-b) + ab &= a[-b + b] \\ &= a(0) \\ &= 0 \end{aligned}$$

Therefore, $a(-b)$ is the opposite of ab.

TW **140.** Is it true that for any numbers a and b, if a is larger than b, then the reciprocal of a is smaller than the reciprocal of b? Why or why not?

Try Exercise Answers: Section 1.7

11. -24 **17.** 42 **29.** 0 **43.** 150 **51.** -7 **53.** 2 **57.** 5.1 **67.** 0 **71.** $-\frac{8}{3}; \frac{8}{-3}$ **79.** $-\frac{5}{4}$ **91.** $\frac{21}{20}$ **99.** $-\frac{7}{4}$

1.8 Exponential Notation and Order of Operations

- Exponential Notation
- Order of Operations
- Simplifying and the Distributive Law
- The Opposite of a Sum

Algebraic expressions often contain *exponential notation.* In this section, we learn how to use exponential notation as well as rules for the *order of operations* in performing certain algebraic manipulations.

EXPONENTIAL NOTATION

A product like $3 \cdot 3 \cdot 3 \cdot 3$, in which the factors are the same, is called a **power.** Powers occur often enough that a simpler notation called **exponential notation** is used. For

$$\underbrace{3 \cdot 3 \cdot 3 \cdot 3}_{4 \text{ factors}}, \quad \text{we write} \quad 3^4.$$

Because $3^4 = 81$, we sometimes say that 81 "is a power of 3."

This is read "three to the fourth power," or simply "three to the fourth." The number 4 is called an **exponent** and the number 3 a **base.**

Expressions like s^2 and s^3 are usually read "s squared" and "s cubed," respectively. This is derived from the fact that a square with sides of length s has an area A given by $A = s^2$ and a cube with sides of length s has a volume V given by $V = s^3$.

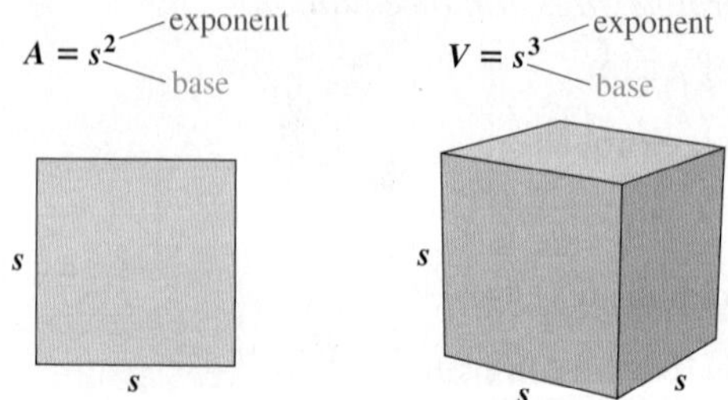

STUDY TIP

A Journey of 1000 Miles Starts with a Single Step

It is extremely important to include steps when working problems. Doing so allows you and others to follow your thought process. It also helps you to avoid careless errors and to identify specific areas in which you may have made mistakes.

EXAMPLE 1 Write exponential notation for $10 \cdot 10 \cdot 10 \cdot 10 \cdot 10$.

SOLUTION

Exponential notation is 10^5. 5 is the exponent. 10 is the base.

Try Exercise 3.

EXAMPLE 2 Simplify: **(a)** 5^2; **(b)** $(-5)^3$; **(c)** $(2n)^3$.

SOLUTION

a) $5^2 = 5 \cdot 5 = 25$ **The exponent 2 indicates two factors of 5.**

b) $(-5)^3 = (-5)(-5)(-5)$ **The exponent 3 indicates three factors of -5.**

$= 25(-5)$ **Using the associative law of multiplication**

$= -125$

c) $(2n)^3 = (2n)(2n)(2n)$ **The exponent 3 indicates three factors of $2n$.**

$= 2 \cdot 2 \cdot 2 \cdot n \cdot n \cdot n$ **Using the associative and commutative laws of multiplication**

$= 8n^3$

Try Exercise 13.

To determine what the exponent 1 will mean, look for a pattern in the following:

$$7 \cdot 7 \cdot 7 \cdot 7 = 7^4$$
$$7 \cdot 7 \cdot 7 = 7^3$$
$$7 \cdot 7 = 7^2$$
$$7 = 7^?$$

We divide by 7 each time.

The exponents decrease by 1 each time. To continue the pattern, we say that

$$7 = 7^1.$$

Exponential Notation For any natural number n,

$$b^n \text{ means } \overbrace{b \cdot b \cdot b \cdot b \cdots b}^{n \text{ factors}}.$$

ORDER OF OPERATIONS

How should $4 + 2 \times 5$ be computed? If we multiply 2 by 5 and then add 4, the result is 14. If we add 2 and 4 first and then multiply by 5, the result is 30. Since these results differ, the order in which we perform operations matters. If grouping symbols such as parentheses ( ), brackets [ ], braces { }, absolute-value symbols | |, or fraction bars are used, they tell us what to do first. For example,

$$(4 + 2) \times 5 \quad \text{indicates} \quad 6 \times 5, \quad \text{resulting in } 30,$$

and

$$4 + (2 \times 5) \quad \text{indicates} \quad 4 + 10, \quad \text{resulting in } 14.$$

In addition to grouping symbols, conventions exist for determining the order in which operations should be performed. Most scientific and graphing calculators follow these rules when evaluating expressions.

In this text, we direct exploration of mathematical concepts using a graphing calculator in Interactive Discovery features such as the one that follows. Such explorations are part of the development of the material presented, and should be performed as you read the text.

Interactive Discovery

Use a graphing calculator to compute $4 + 2 \times 5$.

1. What operation does the calculator perform first?
2. Insert parentheses in the expression $4 + 2 \times 5$ to indicate the order in which the calculator performs the addition and the multiplication.

The correct way to compute $4 + 2 \times 5$ is to first multiply 2 by 5 and then add 4. The result is 14.

Rules for Order of Operations

1. Calculate within the innermost grouping symbols, ( ), [ ], { }, | |, and above or below fraction bars.
2. Simplify all exponential expressions.
3. Perform all multiplication and division, working from left to right.
4. Perform all addition and subtraction, working from left to right.

EXAMPLE 3 Simplify: $15 - 2 \cdot 5 + 3$.

SOLUTION When no groupings or exponents appear, we *always* multiply or divide before adding or subtracting:

$$\begin{aligned} 15 - 2 \cdot 5 + 3 &= 15 - 10 + 3 && \textbf{Multiplying} \\ &= 5 + 3 && \textbf{Subtracting and adding} \\ &= 8. && \textbf{from left to right} \end{aligned}$$

Try Exercise 31.

Always calculate within parentheses first. When there are exponents and no parentheses, simplify powers before multiplying or dividing.

EXAMPLE 4 Simplify: **(a)** $(3 \cdot 4)^2$; **(b)** $3 \cdot 4^2$.

SOLUTION

a) $(3 \cdot 4)^2 = (12)^2$ **Working within parentheses first**

$= 144$

b) $3 \cdot 4^2 = 3 \cdot 16$ **Simplifying the power**

$= 48$ **Multiplying**

Note that $(3 \cdot 4)^2 \neq 3 \cdot 4^2$.

Try Exercise 45.

CAUTION! Example 4 illustrates that, in general, $(ab)^2 \neq ab^2$.

Finding the opposite of a number is the same as multiplying the number by -1. Thus we evaluate expressions like $(-7)^2$ and -7^2 differently.

$(-7)^2 = 49$		$-7^2 = -49$	
1. *Parentheses*:	The opposite of 7 is -7.	1. *Exponents*:	The square of 7 is 49.
2. *Exponents*:	The square of -7 is 49.	2. *Multiplication*:	The opposite of 49 is -49.

EXAMPLE 5 Evaluate for $x = 5$: **(a)** $(-x)^2$; **(b)** $-x^2$.

SOLUTION

a) $(-x)^2 = (-5)^2 = (-5)(-5) = 25$ **We square the opposite of 5.**

b) $-x^2 = -5^2 = -25$ **We square 5 and then find the opposite.**

Try Exercise 67.

CAUTION! Example 5 illustrates that, in general, $(-x)^2 \neq -x^2$.

EXAMPLE 6 Evaluate $-15 \div 3(6 - a)^3$ for $a = 4$.

SOLUTION

$-15 \div 3(6 - a)^3 = -15 \div 3(6 - 4)^3$ **Substituting 4 for *a***

$= -15 \div 3(2)^3$ **Working within parentheses first**

$= -15 \div 3 \cdot 8$ **Simplifying the exponential expression**

$= -5 \cdot 8$ } **Dividing and multiplying from left to right**

$= -40$

Try Exercise 71.

When combinations of grouping symbols are used, we begin with the innermost grouping symbols and work to the outside.

Student Notes

The symbols ( ), [ ], and { } are all used in the same way. Used inside or next to each other, they make it easier to locate the left and right sides of a grouping. Try doing this in your own work to minimize mistakes.

EXAMPLE 7 Simplify: $8 \div 4 + 3[9 + 2(3 - 5)^3]$.

SOLUTION

$8 \div 4 + 3[9 + 2(3 - 5)^3]$

$= 8 \div 4 + 3[9 + 2(-2)^3]$ — **Doing the calculations in the innermost parentheses first**

$= 8 \div 4 + 3[9 + 2(-8)]$ — $(-2)^3 = (-2)(-2)(-2) = -8$

$= 8 \div 4 + 3[9 + (-16)]$

$= 8 \div 4 + 3[-7]$ — **Completing the calculations within the brackets**

$= 2 + (-21)$ — **Multiplying and dividing from left to right**

$= -19$

Try Exercise 47.

Exponents and Grouping Symbols

Exponents To enter an exponential expression on most graphing calculators, enter the base and then press ^ and enter the exponent. x^2 is often used to enter an exponent of 2. Include parentheses around a negative base.

Grouping Symbols Graphing calculators follow the rules for order of operations, so expressions can be entered as they are written. Grouping symbols such as brackets or braces are entered as parentheses. For fractions, you may need to enter parentheses around the numerator and around the denominator.

Your Turn

1. Calculate 8^2 using both 8 ^ 2 ENTER and 8 x^2 ENTER. Both results should be 64.
2. Calculate and compare (−) 2 ^ 4 ENTER and ((−) 2) ^ 4 ENTER. Which keystroke sequence gives the value of $(-2)^4$? Only the second sequence yields 16.
3. Calculate $\frac{3 + 5}{2}$. Use parentheses around the numerator. 4
4. Enter the expression in Example 7 and compare your result with the solution.

EXAMPLE 8 Calculate: $\frac{12(9 - 7) + 4 \cdot 5}{2^4 + 3^2}$.

SOLUTION An equivalent expression with brackets is

$$[12(9 - 7) + 4 \cdot 5] \div [2^4 + 3^2].$$

In effect, we need to simplify the numerator, simplify the denominator, and then divide the results:

$$\frac{12(9 - 7) + 4 \cdot 5}{2^4 + 3^2} = \frac{12(2) + 4 \cdot 5}{16 + 9} = \frac{24 + 20}{25} = \frac{44}{25}.$$

```
(12(9−7)+4*5)/(2
^4+3²)►Frac
                 44/25
```

To use a calculator, we enter the expression, replacing the brackets with parentheses, choose the FRAC option of the MATH MATH submenu, and press ENTER, as shown in the figure at left.

Try Exercise 53.

Example 8 demonstrates that graphing calculators do not make it any less important to understand the mathematics being studied. A solid understanding of the rules for order of operations is necessary in order to locate parentheses properly in an expression.

SIMPLIFYING AND THE DISTRIBUTIVE LAW

Sometimes we cannot simplify within grouping symbols. When a sum or a difference is within parentheses, the distributive law provides a method for removing the grouping symbols.

EXAMPLE 9 Simplify: $5x - 9 + 2(4x + 5)$.

SOLUTION

$$
\begin{aligned}
5x - 9 + 2(4x + 5) &= 5x - 9 + 8x + 10 && \textbf{Using the distributive law} \\
&= 13x + 1 && \textbf{Combining like terms}
\end{aligned}
$$

Try Exercise 97.

Now that exponents have been introduced, we can make our definition of *like*, or *similar, terms* more precise. **Like,** or **similar, terms** are either constant terms or terms containing the same variable(s) raised to the same exponent(s). Thus, 5 and -7, $19xy$ and $2yx$, and $4a^3b$ and a^3b are all pairs of like terms.

EXAMPLE 10 Simplify: $7x^2 + 3(x^2 + 2x) - 5x$.

SOLUTION

$$
\begin{aligned}
7x^2 + 3(x^2 + 2x) - 5x &= 7x^2 + 3x^2 + 6x - 5x && \textbf{Using the distributive law} \\
&= 10x^2 + x && \textbf{Combining like terms}
\end{aligned}
$$

Try Exercise 103.

THE OPPOSITE OF A SUM

When a number is multiplied by -1, the result is the opposite of that number. For example, $-1(7) = -7$ and $-1(-5) = 5$.

The Property of -1 For any real number a,

$$-1 \cdot a = -a.$$

(Negative one times a is the opposite of a.)

An expression such as $-(x + y)$ indicates the *opposite*, or *additive inverse*, of the sum of x and y. When a sum within grouping symbols is preceded by a "$-$" symbol, we can multiply the sum by -1 and use the distributive law. In this manner, we can find an equivalent expression for the opposite of a sum.

EXAMPLE 11 Write an expression equivalent to $-(3x + 2y + 4)$ without using parentheses.

SOLUTION

$$-(3x + 2y + 4) = -1(3x + 2y + 4) \quad \text{Using the property of } -1$$
$$= -1(3x) + (-1)(2y) + (-1)4 \quad \text{Using the distributive law}$$
$$= -3x - 2y - 4 \quad \text{Using the associative law and the property of } -1$$

Try Exercise 85.

The Opposite of a Sum For any real numbers a and b,

$$-(a + b) = (-a) + (-b).$$

(The opposite of a sum is the sum of the opposites.)

To remove parentheses from an expression like $-(x - 7y + 5)$, we can first rewrite the subtraction as addition:

$$-(x - 7y + 5) = -(x + (-7y) + 5) \quad \text{Rewriting as addition}$$
$$= -x + 7y - 5. \quad \text{Taking the opposite of a sum}$$

This procedure is normally streamlined to one step in which we find the opposite by "removing parentheses and changing the sign of every term":

$$-(x - 7y + 5) = -x + 7y - 5.$$

EXAMPLE 12 Simplify: $3x - (4x + 2)$.

SOLUTION

$$3x - (4x + 2) = 3x + [-(4x + 2)] \quad \text{Adding the opposite of } 4x + 2$$
$$= 3x + [-4x - 2] \quad \text{Taking the opposite of } 4x + 2$$
$$= 3x + (-4x) + (-2)$$
$$= 3x - 4x - 2 \quad \text{Try to go directly to this step.}$$
$$= -x - 2 \quad \text{Combining like terms}$$

Try Exercise 93.

In practice, the first three steps of Example 12 are usually skipped.

EXAMPLE 13 Simplify: $5t^2 - 2t - (4t^2 - 9t)$.

SOLUTION

$$5t^2 - 2t - (4t^2 - 9t) = 5t^2 - 2t - 4t^2 + 9t \quad \text{Removing parentheses and changing the sign of each term inside}$$
$$= t^2 + 7t \quad \text{Combining like terms}$$

Try Exercise 101.

Expressions such as $7 - 3(x + 2)$ can be simplified as follows:

$7 - 3(x + 2) = 7 + [-3(x + 2)]$ **Adding the opposite of $3(x + 2)$**

$= 7 + [-3x - 6]$ **Multiplying $x + 2$ by -3**

$= 7 - 3x - 6$ **Try to go directly to this step.**

$= 1 - 3x.$ **Combining like terms**

EXAMPLE 14 Simplify.

a) $3n - 2(4n - 5)$

b) $7x^3 + 2 - [5(x^3 - 1) + 8]$

SOLUTION

a) $3n - 2(4n - 5) = 3n - 8n + 10$ **Multiplying each term inside the parentheses by -2**

$= -5n + 10$ **Combining like terms**

b) $7x^3 + 2 - [5(x^3 - 1) + 8] = 7x^3 + 2 - [5x^3 - 5 + 8]$ **Removing parentheses**

$= 7x^3 + 2 - [5x^3 + 3]$

$= 7x^3 + 2 - 5x^3 - 3$ **Removing brackets**

$= 2x^3 - 1$ **Combining like terms**

Try Exercise 99.

It is important that we be able to distinguish between the two tasks of **simplifying an expression** and **solving an equation.** In Chapter 1, we did not solve equations, but we did simplify expressions. This enabled us to write *equivalent expressions* that were simpler than the given expression. In Chapter 2, we will continue to simplify expressions, but we will also begin to solve equations.

1.8 Exercise Set

FOR EXTRA HELP MyMathLab Math XL PRACTICE

WATCH

DOWNLOAD

READ

REVIEW

Concept Reinforcement *In each part of Exercises 1 and 2, name the operation that should be performed first. Do not perform the calculations.*

1. a) $4 + 8 \div 2 \cdot 2$

b) $7 - 9 + 15$

c) $5 - 2(3 + 4)$

d) $6 + 7 \cdot 3$

e) $18 - 2[4 + (3 - 2)]$

f) $\dfrac{5 - 6 \cdot 7}{2}$

2. a) $9 - 3 \cdot 4 \div 2$

b) $8 + 7(6 - 5)$

c) $5 \cdot [2 - 3(4 + 1)]$

d) $8 - 7 + 2$

e) $4 + 6 \div 2 \cdot 3$

f) $\dfrac{37}{8 - 2 \cdot 2}$

Write exponential notation.

3. $x \cdot x \cdot x \cdot x \cdot x \cdot x \cdot x$

4. $y \cdot y \cdot y \cdot y \cdot y \cdot y$

5. $(-5)(-5)(-5)$

6. $(-7)(-7)(-7)(-7)$

7. $3t \cdot 3t \cdot 3t \cdot 3t \cdot 3t$

8. $5m \cdot 5m \cdot 5m \cdot 5m \cdot 5m$

9. $2 \cdot n \cdot n \cdot n \cdot n$

10. $8 \cdot a \cdot a \cdot a$

Simplify.

11. 3^2
12. 5^3
13. $(-4)^2$
14. $(-9)^2$
15. -4^2
16. -9^2
17. 4^3
18. 9^1
19. $(-5)^4$
20. 5^4
21. 7^1
22. $(-1)^7$
23. $(-2)^5$
24. -2^5
25. $(3t)^4$
26. $(5t)^2$
27. $(-7x)^3$
28. $(-5x)^4$
29. $5 + 3 \cdot 7$
30. $3 - 4 \cdot 2$
31. $8 \cdot 7 + 6 \cdot 5$
32. $10 \cdot 5 + 1 \cdot 1$
33. $9 \div 3 + 16 \div 8$
34. $32 - 8 \div 4 - 2$

Aha! 35. $14 \cdot 19 \div (19 \cdot 14)$

36. $18 - 6 \div 3 \cdot 2 + 7$
37. $3(-10)^2 - 8 \div 2^2$
38. $9 - 3^2 \div 9(-1)$
39. $8 - (2 \cdot 3 - 9)$
40. $(8 - 2 \cdot 3) - 9$
41. $(8 - 2)(3 - 9)$
42. $32 \div (-2)^2 \cdot 4$
43. $5 \cdot 3^2 - 4^2 \cdot 2$
44. $112 \div 28 - 112 \div 28$
45. $5 + 3(2 - 9)^2$
46. $9 - (3 - 5)^3 - 4$
47. $[2 \cdot (5 - 8)]^2 - 12$
48. $2^3 + 2^4 - 5[8 - 4(9 - 10)^2]$
49. $\dfrac{7 + 2}{5^2 - 4^2}$
50. $\dfrac{5^2 - 3^2}{2 \cdot 6 - 4}$
51. $8(-7) + |6(-5)|$
52. $|10(-5)| + 1(-1)$
53. $\dfrac{(-2)^3 + 4^2}{3 - 5^2 + 3 \cdot 6}$
54. $\dfrac{7^2 - (-1)^5}{3 - 2 \cdot 3^2 + 5}$
55. $\dfrac{-3^3 - 2 \cdot 3^2}{8 \div 2^2 - (6 - |2 - 15|)}$
56. $\dfrac{(-5)^2 - 3 \cdot 5}{3^2 + 4 \cdot |6 - 7| \cdot (-1)^5}$

In Exercises 57–60, match the algebraic expression with the equivalent rewritten expression below. Check your answer by calculating the expression by hand and by using a calculator.

a) $(5(3 - 7) + 4 \wedge 3)/(-2 - 3)^2$
b) $(5(3 - 7) + 4 \wedge 3)/(-2 - 3^2)$
c) $(5(3 - 7) + 4) \wedge 3/-2 - 3^2$
d) $5(3 - 7) + 4 \wedge 3/(-2 - 3)^2$

57. $\dfrac{5(3 - 7) + 4^3}{(-2 - 3)^2}$
58. $(5(3 - 7) + 4)^3 \div (-2) - 3^2$
59. $5(3 - 7) + 4^3 \div (-2 - 3)^2$
60. $\dfrac{5(3 - 7) + 4^3}{(-2) - 3^2}$

Simplify using a calculator. Round your answer to the nearest thousandth.

61. $\dfrac{2.5^2 - 10 \cdot 12 \div (-1.5)}{(3 + 5)^2 - 60}$
62. $\dfrac{46 - (3 - 8)^3}{2[35 - (18 - 26)^2]}$
63. $\dfrac{13.4 - 5|1.2 + 4.6|}{(9.3 - 5.4)^2}$
64. $|13.5 + 8(-4.7)|^3$

Evaluate.

65. $9 - 4x$, for $x = 5$
66. $1 + x^3$, for $x = -2$
67. $24 \div t^3$, for $t = -2$
68. $20 \div a \cdot 4$, for $a = 5$
69. $45 \div 3 \cdot a$, for $a = -1$
70. $50 \div 2 \cdot t$, for $t = -5$
71. $5x \div 15x^2$, for $x = 3$
72. $6a \div 12a^3$, for $a = 2$
73. $45 \div 3^2 x(x - 1)$, for $x = 3$
74. $-30 \div t(t + 4)^2$, for $t = -6$
75. $-x^2 - 5x$, for $x = -3$
76. $(-x)^2 - 5x$, for $x = -3$
77. $\dfrac{3a - 4a^2}{a^2 - 20}$, for $a = 5$
78. $\dfrac{a^3 - 4a}{a(a - 3)}$, for $a = -2$

Evaluate using a calculator.

79. $13 - (y - 4)^3 + 10$, for $y = 6$

80. $(t + 4)^2 - 12 \div (19 - 17) + 68$, for $t = 5$

81. $3(m + 2n) \div m$, for $m = 1.6$ and $n = 5.9$

82. $1.5 + (2x - y)^2$, for $x = 9.25$ and $y = 1.7$

83. $\frac{1}{2}(x + 5/z)^2$, for $x = 141$ and $z = 0.2$

84. $a - \frac{3}{4}(2a - b)$, for $a = 213$ and $b = 165$

Write an equivalent expression without using grouping symbols.

85. $-(9x + 1)$

86. $-(3x + 5)$

87. $-[5 - 6x]$

88. $-(6x - 7)$

89. $-(4a - 3b + 7c)$

90. $-[5x - 2y - 3z]$

91. $-(3x^2 + 5x - 1)$

92. $-(8x^3 - 6x + 5)$

Simplify.

93. $8x - (6x + 7)$

94. $2a - (5a - 9)$

95. $2x - 7x - (4x - 6)$

96. $2a + 5a - (6a + 8)$

97. $9t - 5r + 2(3r + 6t)$

98. $4m - 9n + 3(2m - n)$

99. $15x - y - 5(3x - 2y + 5z)$

100. $4a - b - 4(5a - 7b + 8c)$

101. $3x^2 + 7 - (2x^2 + 5)$

102. $5x^4 + 3x - (5x^4 + 3x)$

103. $5t^3 + t + 3(t - 2t^3)$

104. $8n^2 - 3n + 2(n - 4n^2)$

105. $12a^2 - 3ab + 5b^2 - 5(-5a^2 + 4ab - 6b^2)$

106. $-8a^2 + 5ab - 12b^2 - 6(2a^2 - 4ab - 10b^2)$

107. $-7t^3 - t^2 - 3(5t^3 - 3t)$

108. $9t^4 + 7t - 5(9t^3 - 2t)$

109. $5(2x - 7) - [4(2x - 3) + 2]$

110. $3(6x - 5) - [3(1 - 8x) + 5]$

TW **111.** Some students use the mnemonic device PEMDAS to help remember the rules for the order of operations. Explain how this can be done and how the order of the letters in PEMDAS could lead a student to a wrong conclusion about the order of some operations.

TW **112.** Jake keys $18/2 \cdot 3$ into his calculator and expects the result to be 3. What mistake is he probably making?

SKILL REVIEW

Translate to an algebraic expression. [1.1]

113. Nine more than twice a number

114. Half of the sum of two numbers

SYNTHESIS

TW **115.** Write the sentence $(-x)^2 \neq -x^2$ in words. Explain why $(-x)^2$ and $-x^2$ are not equivalent.

TW **116.** Write the sentence $-|x| \neq -x$ in words. Explain why $-|x|$ and $-x$ are not equivalent.

Simplify.

117. $5t - \{7t - [4r - 3(t - 7)] + 6r\} - 4r$

118. $z - \{2z - [3z - (4z - 5z) - 6z] - 7z\} - 8z$

119. $\{x - [f - (f - x)] + [x - f]\} - 3x$

TW **120.** Is it true that for all real numbers a and b,

$$ab = (-a)(-b)?$$

Why or why not?

TW **121.** Is it true that for all real numbers a, b, and c,

$$a|b - c| = ab - ac?$$

Why or why not?

If $n > 0$, $m > 0$, and $n \neq m$, classify each of the following as either true or false.

122. $-n + m = -(n + m)$

123. $m - n = -(n - m)$

124. $n(-n - m) = -n^2 + nm$

125. $-m(n - m) = -(mn + m^2)$

126. $-n(-n - m) = n(n + m)$

Evaluate.

Aha! **127.** $[x + 3(2 - 5x) \div 7 + x](x - 3)$, for $x = 3$

Aha! **128.** $[x + 2 \div 3x] \div [x + 2 \div 3x]$, for $x = -7$

129. $\dfrac{x^2 + 2^x}{x^2 - 2^x}$, for $x = 3$

130. $\dfrac{x^2 + 2^x}{x^2 - 2^x}$, for $x = 2$

131. In Mexico, between 500 B.C. and 600 A.D., the Mayans represented numbers using powers of 20 and certain symbols. For example, the symbols

represent $4 \cdot 20^3 + 17 \cdot 20^2 + 10 \cdot 20^1 + 0 \cdot 20^0$. Evaluate this number.

Source: National Council of Teachers of Mathematics, 1906 Association Drive, Reston, VA 22091

132. Examine the Mayan symbols and the numbers in Exercise 131. What numbers do

, , and 

each represent?

133. Calculate the volume of the tower shown below.

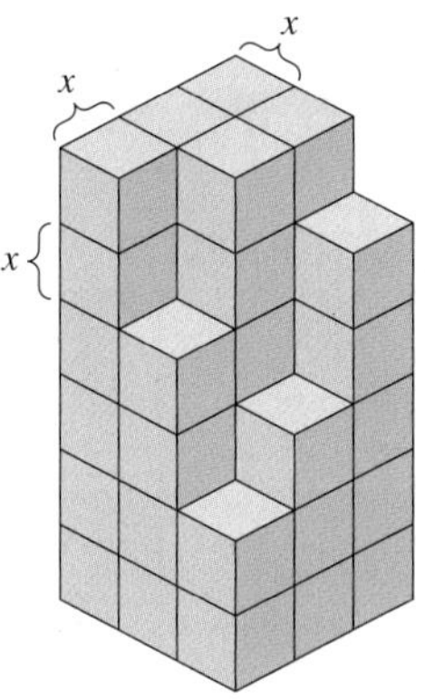

Try Exercise Answers: Section 1.8

3. x^7 **13.** 16 **31.** 86 **45.** 152 **47.** 24 **53.** -2 **67.** -3 **71.** 9 **85.** $-9x - 1$ **93.** $2x - 7$ **97.** $21t + r$ **99.** $9y - 25z$ **101.** $x^2 + 2$ **103.** $-t^3 + 4t$

Collaborative Corner

Select the Symbols

Focus: Order of operations
Time: 15 minutes
Group Size: 2

Do you understand the rules for order of operations well enough to work backwards? For example, by inserting operation signs and grouping symbols, the display

1 2 3 4 5

can be used to obtain the result 21:

$$(1 + 2) \div 3 + 4 \cdot 5.$$

ACTIVITY

1. Each group should prepare an exercise similar to the example shown above. (Exponents are not allowed.) To do so, first select five single-digit numbers for display. Then insert operation signs and grouping symbols and calculate the result.
2. Pair with another group. Each group should give the other its result along with its five-number display (without the symbols), and challenge the other group to insert symbols that will make the display equal the result given.
3. Share with the entire class the various mathematical statements developed by each group.

Study Summary

KEY TERMS AND CONCEPTS	EXAMPLES	PRACTICE EXERCISES
SECTION 1.1: INTRODUCTION TO ALGEBRA		
To **evaluate** an algebraic expression, substitute a number for each variable and carry out the operations. The result is a **value** of that expression.	Evaluate $\frac{x+y}{8}$ for $x = 15$ and $y = 9$. $\frac{x+y}{8} = \frac{15+9}{8} = \frac{24}{8} = 3$	**1.** Evaluate $3 + 5c - d$ for $c = 3$ and $d = 10$.
To find the area of a rectangle, a triangle, or a parallelogram, evaluate the appropriate formula for the given values.	Find the area of a triangle with base 3.1 m and height 6 m. $A = \frac{1}{2}bh = \frac{1}{2}(3.1\text{ m})(6\text{ m}) = \frac{1}{2}(3.1)(6)(\text{m}\cdot\text{m}) = 9.3\text{ m}^2$	**2.** Find the area of a rectangle with length 8 ft and width $\frac{1}{2}$ ft.
Many problems can be solved by **translating** phrases to algebraic expressions and then forming an equation. The table on p. 5 shows translations of many words that occur in problems.	Translate to an equation. Do not solve. When 34 is subtracted from a number, the result is 13. What is the number? Let n represent the number. *Rewording:* 34 subtracted from a number is 13 *Translating:* $n - 34 = 13$	**3.** Translate to an equation. Do not solve. 78 is 92 less than some number. What is the number?
An **equation** is a number sentence with the verb =. A substitution for the variable in an equation that makes the equation true is a **solution** of the equation.	Determine whether 9 is a solution of $47 - n = 38$. $47 - n = 38$ $47 - 9 \mid 38$ $38 \stackrel{?}{=} 38$ TRUE Since $38 = 38$ is true, 9 is a solution.	**4.** Determine whether 13 is a solution of $29 + t = 43$.
SECTION 1.2: THE COMMUTATIVE, ASSOCIATIVE, AND DISTRIBUTIVE LAWS		
The Commutative Laws $a + b = b + a$; $ab = ba$	$3 + (-5) = -5 + 3$; $8(10) = 10(8)$	**5.** Use the commutative law of addition to write an expression equivalent to $6 + 10n$.
The Associative Laws $a + (b + c) = (a + b) + c$; $a \cdot (b \cdot c) = (a \cdot b) \cdot c$	$-5 + (5 + 6) = (-5 + 5) + 6$; $2 \cdot (5 \cdot 9) = (2 \cdot 5) \cdot 9$	**6.** Use the associative law of multiplication to write an expression equivalent to $3(ab)$.
The Distributive Law $a(b + c) = ab + ac$	Multiply: $3(2x + 5y)$. $3(2x + 5y) = 3 \cdot 2x + 3 \cdot 5y = 6x + 15y$ Factor: $16x + 24y + 8$. $16x + 24y + 8 = 8(2x + 3y + 1)$	**7.** Multiply: $10(5m + 9n + 1)$. **8.** Factor: $26x + 13$.

SECTION 1.3: FRACTION NOTATION

Natural numbers: $\{1, 2, 3, \ldots\}$ **Whole numbers:** $\{0, 1, 2, 3, \ldots\}$	15, 39, and 1567 are examples of natural numbers. 0, 5, 16, and 2890 are examples of whole numbers.	**9.** Which of the following are whole numbers? $7, \frac{1}{3}, 2.8, 0$
A **prime** number has only two different factors, the number itself and 1. Natural numbers that have factors other than 1 and the number itself are **composite** numbers.	2, 3, 5, 7, 11, and 13 are the first six prime numbers. 4, 6, 8, 24, and 100 are examples of composite numbers.	**10.** Is 15 prime or composite?
The **prime factorization** of a composite number expresses that number as a product of prime numbers.	The prime factorization of 136 is $2 \cdot 2 \cdot 2 \cdot 17$.	**11.** Find the prime factorization of 84.
For any nonzero number a, $\frac{a}{a} = 1.$	$\frac{15}{15} = 1$ and $\frac{2x}{2x} = 1.$	**12.** Simplify: $\frac{t}{t}$.
The Identity Property of 1 $a \cdot 1 = 1 \cdot a = a$ The number 1 is called the **multiplicative identity.**	$\frac{2}{3} = \frac{2}{3} \cdot \frac{5}{5}$ since $\frac{5}{5} = 1.$	**13.** Simplify: $\frac{9}{10} \cdot \frac{13}{13}$.
$\frac{a}{d} + \frac{b}{d} = \frac{a+b}{d}$ $\frac{a}{d} - \frac{b}{d} = \frac{a-b}{d}$ $\frac{a}{b} \cdot \frac{c}{d} = \frac{a \cdot c}{b \cdot d}$ $\frac{a}{b} \div \frac{c}{d} = \frac{a}{b} \cdot \frac{d}{c}$	$\frac{1}{6} + \frac{3}{8} = \frac{4}{24} + \frac{9}{24} = \frac{13}{24}$ $\frac{5}{12} - \frac{1}{6} = \frac{5}{12} - \frac{2}{12} = \frac{3}{12} = \frac{1 \cdot 3}{4 \cdot 3} = \frac{1}{4} \cdot \frac{3}{3} = \frac{1}{4} \cdot 1 = \frac{1}{4}$ $\frac{2}{5} \cdot \frac{7}{8} = \frac{\cancel{2} \cdot 7}{5 \cdot \cancel{2} \cdot 4} = \frac{7}{20}$ **Removing a factor equal to 1:** $\frac{2}{2} = 1$ $\frac{10}{9} \div \frac{4}{15} = \frac{10}{9} \cdot \frac{15}{4} = \frac{\cancel{2} \cdot 5 \cdot \cancel{3} \cdot 5}{\cancel{3} \cdot 3 \cdot \cancel{2} \cdot 2} = \frac{25}{6}$ **Removing a factor equal to 1:** $\frac{2 \cdot 3}{2 \cdot 3} = 1$	**14.** Add and, if possible, simplify: $\frac{2}{3} + \frac{5}{6}$. **15.** Subtract and, if possible, simplify: $\frac{3}{4} - \frac{3}{10}$. **16.** Multiply and, if possible, simplify: $\frac{15}{14} \cdot \frac{35}{9}$. **17.** Divide and, if possible, simplify: $15 \div \frac{3}{5}$.

SECTION 1.4: POSITIVE AND NEGATIVE REAL NUMBERS

Integers: $\{\ldots, -3, -2, -1, 0, 1, 2, 3, \ldots\}$ **Rational numbers:** $\left\{\frac{a}{b} \middle\vert\ a \text{ and } b \text{ are integers and } b \neq 0\right\}$ The rational numbers and the **irrational numbers** make up the set of **real numbers.**	$-25, -2, 0, 1$, and 2000 are examples of integers. $\frac{1}{6}, \frac{-3}{7}, 0, 17, 0.758$, and $9.\overline{608}$ are examples of rational numbers. $\sqrt{7}$ and π are examples of irrational numbers.	**18.** Which of the following are integers? $\frac{9}{10}, \; 0, \; -15, \; \sqrt{2}, \; \frac{30}{3}$
Every rational number can be written using fraction notation or decimal notation. When written in decimal notation, a rational number either **repeats** or **terminates.**	$-\frac{1}{16} = -0.0625$ This is a terminating decimal. $\frac{5}{6} = 0.8333\ldots = 0.8\overline{3}$ This is a repeating decimal.	**19.** Find decimal notation: $-\frac{10}{9}$.
Every real number corresponds to a point on the number line. For any two numbers, the one to the left is less than the one to the right. The symbol **<** means "**is less than**" and the symbol **>** means "**is greater than.**"	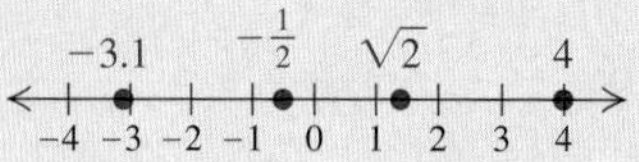 $4 > -3.1 \qquad -\frac{1}{2} < \sqrt{2}$	**20.** Classify as true or false: $-15 < -16$.
The **absolute value** of a number is the number of units that number is from zero on the number line.	$\lvert 3 \rvert = 3$ since 3 is 3 units from 0. $\lvert -3 \rvert = 3$ since -3 is 3 units from 0.	**21.** Find the absolute value: $\lvert -1.5 \rvert$.

SECTION 1.5: ADDITION OF REAL NUMBERS

To **add** two real numbers, use the rules on p. 43.	$-8 + (-3) = -11$; $-8 + 3 = -5$; $8 + (-3) = 5$; $-8 + 8 = 0$	**22.** Add: $-15 + (-10) + 20$.
The Identity Property of 0 $a + 0 = 0 + a = a$ The number 0 is called the **additive identity.**	$-35 + 0 = -35$; $0 + \frac{2}{9} = \frac{2}{9}$	**23.** Add: $-2.9 + 0$.

SECTION 1.6: SUBTRACTION OF REAL NUMBERS

The **opposite,** or **additive inverse,** of a number a is written $-a$. The opposite of the opposite of a is a. $-(-a) = a$	Find $-x$ and $-(-x)$ when $x = -11$. $-x = -(-11) = 11$; $-(-x) = -(-(-11)) = -11 \qquad -(-x) = x$	**24.** Find $-(-x)$ when $x = -12$.

To **subtract** two real numbers, add the opposite of the number being subtracted.	$-10 - 12 = -10 + (-12) = -22$; $-10 - (-12) = -10 + 12 = 2$	**25.** Subtract: $6 - (-9)$.
The **terms** of an expression are separated by plus signs. **Like terms** either are constants or have the same variable factors raised to the same power. Like terms can be **combined** using the distributive law.	In the expression $-2x + 3y + 5x - 7y$: The terms are $-2x$, $3y$, $5x$, and $-7y$. The like terms are $-2x$ and $5x$, and $3y$ and $-7y$. Combining like terms gives $-2x + 3y + 5x - 7y = -2x + 5x + 3y - 7y$ $= (-2 + 5)x + (3 - 7)y$ $= 3x - 4y$.	**26.** Combine like terms: $3c + d - 10c - 2 + 8d$.
SECTION 1.7: MULTIPLICATION AND DIVISION OF REAL NUMBERS		
To **multiply** or **divide** two real numbers, use the rules on p. 58. Division by 0 is **undefined**.	$(-5)(-2) = 10$; $30 \div (-6) = -5$; $0 \div (-3) = 0$; $-3 \div 0$ is undefined.	**27.** Multiply: $-3(-7)$. **28.** Divide: $10 \div (-2.5)$.
SECTION 1.8: EXPONENTIAL NOTATION AND ORDER OF OPERATIONS		
Exponential notation Exponent, Base, n factors: $b^n = \underbrace{b \cdot b \cdot b \cdots b}_{n \text{ factors}}$	$6^2 = 6 \cdot 6 = 36$; $(-6)^2 = (-6) \cdot (-6) = 36$; $-6^2 = -(6 \cdot 6) = -36$; $(6x)^2 = (6x) \cdot (6x) = 36x^2$	**29.** Evaluate: -10^2.
To perform multiple operations, use the rules for **order of operations** on p. 65.	$-3 + (3 - 5)^3 \div 4(-1) = -3 + (-2)^3 \div 4(-1)$ $= -3 + (-8) \div 4(-1)$ $= -3 + (-2)(-1)$ $= -3 + 2$ $= -1$	**30.** Simplify: $120 \div (-10) \cdot 2$ $- 3(4 - 5)$.
The Property of −1 For any real number a, $-1 \cdot a = -a$. **The Opposite of a Sum** For any real numbers a and b, $-(a + b) = -a - b$.	$-1 \cdot 5x = -5x$ and $-5x = -1(5x)$ $-(2x - 3y) = -(2x) - (-3y) = -2x + 3y$	**31.** Write an equivalent expression without using grouping symbols: $-(-a + 2b - 3c)$.
Expressions containing parentheses can be simplified by removing parentheses using the distributive law.	Simplify: $3x^2 - 5(x^2 - 4xy + 2y^2) - 7y^2$. $3x^2 - 5(x^2 - 4xy + 2y^2) - 7y^2$ $= 3x^2 - 5x^2 + 20xy - 10y^2 - 7y^2$ $= -2x^2 + 20xy - 17y^2$	**32.** Simplify: $2m + n - 3(5 - m - 2n)$ $- 12$.

Review Exercises 

Concept Reinforcement *Classify each of the following statements as either true or false.*

1. $4x - 5y$ and $12 - 7a$ are both algebraic expressions containing two terms. [1.2]*
2. $3t + 1 = 7$ and $8 - 2 = 9$ are both equations. [1.1]
3. The fact that $2 + x$ is equivalent to $x + 2$ is an illustration of the associative law for addition. [1.2]
4. The statement $4(a + 3) = 4 \cdot a + 4 \cdot 3$ illustrates the distributive law. [1.2]
5. The number 2 is neither prime nor composite. [1.3]
6. Every irrational number can be written as a repeating decimal or a terminating decimal. [1.4]
7. Every natural number is a whole number and every whole number is an integer. [1.4]
8. The expressions $9r^2s$ and $5rs^2$ are like terms. [1.8]
9. The opposite of x, written $-x$, never represents a positive number. [1.6]
10. The number 0 has no reciprocal. [1.3]

Evaluate.

11. $5t$, for $t = 3$ [1.1]
12. $9 - y^2$, for $y = -5$ [1.8]
13. $-10 + a^2 \div (b + 1)$, for $a = 5$ and $b = -6$ [1.8]

Translate to an algebraic expression. [1.1]

14. 7 less than z
15. 10 more than the product of x and z
16. 15 times the difference of Brent's speed and the wind speed
17. Determine whether 35 is a solution of $x/5 = 8$. [1.1]
18. Translate to an equation. Do not solve. [1.1]

    Backpacking burns twice as many calories per hour as housecleaning. If Katie burns 237 calories per hour cleaning, how many calories per hour would she burn backpacking?
    Source: www.myoptumhealth.com

19. The table below lists the number of calories Kim burns when bowling for various lengths of time. Find an equation for the number of calories burned c when Kim bowls for t hours. [1.1]

Number of Hours Spent Bowling, t	Number of Calories Burned, c
$\frac{1}{2}$	100
2	400
$2\frac{1}{2}$	500

20. Use the commutative law of multiplication to write an expression equivalent to $3t + 5$. [1.2]
21. Use the associative law of addition to write an expression equivalent to $(2x + y) + z$. [1.2]
22. Use the commutative and associative laws to write three expressions equivalent to $4(xy)$. [1.2]

Multiply. [1.2]

23. $6(3x + 5y)$
24. $8(5x + 3y + 2)$

Factor. [1.2]

25. $21x + 15y$
26. $35x + 77y + 7$
27. Find the prime factorization of 52. [1.3]

Simplify. [1.3]

28. $\dfrac{20}{48}$
29. $\dfrac{18}{8}$

Perform the indicated operation and, if possible, simplify. [1.3]

30. $\dfrac{5}{12} + \dfrac{4}{9}$
31. $\dfrac{9}{16} \div 3$
32. $\dfrac{2}{3} - \dfrac{1}{15}$
33. $\dfrac{9}{10} \cdot \dfrac{16}{5}$
34. Tell which integers correspond to this situation. [1.4]

    The world record for the highest dive, 172 ft, is held by Dana Kunze. The record for the deepest free dive, 820 ft, is held by Alexey Molchanov.
    Sources: peak.com; AIDA International

35. Graph on the number line: $\frac{-1}{3}$. [1.4]
36. Write an inequality with the same meaning as $-3 < x$. [1.4]

*The notation [1.2] refers to Chapter 1, Section 2.

37. Classify as true or false: $0 \le -1$. [1.4]

38. Find decimal notation: $-\frac{7}{8}$. [1.4]

39. Find the absolute value: $|-1|$. [1.4]

40. Find $-(-x)$ when x is -9. [1.6]

Simplify.

41. $-3 + (-7)$ [1.5]

42. $-\frac{2}{3} + \frac{1}{12}$ [1.5]

43. $10 + (-9) + (-8) + 7$ [1.5]

44. $-3.8 + 5.1 + (-12) + (-4.3) + 10$ [1.5]

45. $-2 - (-7)$ [1.6]

46. $\frac{1}{2} - \frac{9}{10}$ [1.6]

47. $-3.8 - 4.1$ [1.6]

48. $-9 \cdot (-6)$ [1.7]

49. $-2.7(3.4)$ [1.7]

50. $\frac{2}{3} \cdot \left(-\frac{3}{7}\right)$ [1.7]

51. $2 \cdot (-7) \cdot (-2) \cdot (-5)$ [1.7]

52. $35 \div (-5)$ [1.7]

53. $-5.1 \div 1.7$ [1.7]

54. $-\frac{3}{5} \div \left(-\frac{4}{5}\right)$ [1.7]

55. $|-3 \cdot 4 - 12 \cdot 2| - 8(-7)$ [1.8]

56. $16 \div (-2)^3 - 5[3 - 1 + 2(4 - 7)]$ [1.8]

57. $120 - 6^2 \div 4 \cdot 8$ [1.8]

58. $(120 - 6^2) \div 4 \cdot 8$ [1.8]

59. $(120 - 6^2) \div (4 \cdot 8)$ [1.8]

60. $\dfrac{4(18 - 8) + 7 \cdot 9}{9^2 - 8^2}$ [1.8]

Combine like terms.

61. $11a + 2b + (-4a) + (-5b)$ [1.5]

62. $7x - 3y - 9x + 8y$ [1.6]

63. Find the opposite of -7. [1.6]

64. Find the reciprocal of -7. [1.7]

65. Write exponential notation for $2x \cdot 2x \cdot 2x \cdot 2x$. [1.8]

66. Simplify: $(-5x)^3$. [1.8]

Remove parentheses and simplify. [1.8]

67. $2a - (5a - 9)$

68. $11x^4 + 2x + 8(x - x^4)$

69. $2n^2 - 5(-3n^2 + m^2 - 4mn) + 6m^2$

70. $8(x + 4) - 6 - [3(x - 2) + 4]$

SYNTHESIS

TW 71. Explain the difference between a constant and a variable. [1.1]

TW 72. Explain the difference between a term and a factor. [1.2]

TW 73. Describe at least three ways in which the distributive law was used in this chapter. [1.2]

TW 74. Devise a rule for determining the sign of a negative number raised to an exponent. [1.8]

75. Evaluate $a^{50} - 20a^{25}b^4 + 100b^8$ for $a = 1$ and $b = 2$. [1.8]

76. If $0.090909\ldots = \frac{1}{11}$ and $0.181818\ldots = \frac{2}{11}$, what rational number is named by each of the following?

a) $0.272727\ldots$ [1.4] **b)** $0.909090\ldots$ [1.4]

Simplify. [1.8]

77. $-\left|\frac{7}{8} - \left(-\frac{1}{2}\right) - \frac{3}{4}\right|$

78. $(|2.7 - 3| + 3^2 - |-3|) \div (-3)$

In each of Exercises 79–89, match the phrase in the left column with the appropriate choice from the right column.

79. ____ A number is nonnegative. [1.4]

80. ____ The reciprocal of a sum [1.7]

81. ____ A number squared [1.8]

82. ____ A sum of squares [1.8]

83. ____ The opposite of an opposite is the original number. [1.6]

84. ____ The order in which numbers are added does not change the result. [1.2]

85. ____ A number is positive. [1.4]

86. ____ The absolute value of a product [1.4]

87. ____ A sum of a number and its reciprocal [1.7]

88. ____ The square of a sum [1.8]

89. ____ The absolute value of one number is less than the absolute value of another number. [1.4]

a) a^2

b) $a + b = b + a$

c) $a > 0$

d) $a + \dfrac{1}{a}$

e) $|ab|$

f) $(a + b)^2$

g) $|a| < |b|$

h) $a^2 + b^2$

i) $a \ge 0$

j) $\dfrac{1}{a + b}$

k) $-(-a) = a$

Chapter Test 1

1. Evaluate $\frac{2x}{y}$ for $x = 10$ and $y = 5$.
2. Write an algebraic expression: Nine less than the product of two numbers.
3. Find the area of a triangle when the height h is 30 ft and the base b is 16 ft.
4. Use the commutative law of addition to write an expression equivalent to $3p + q$.
5. Use the associative law of multiplication to write an expression equivalent to $x \cdot (4 \cdot y)$.
6. Determine whether 7 is a solution of $65 - x = 69$.
7. Translate to an equation. Do not solve.

   About 1500 golden lion tamarins, an endangered species of monkey, live in the wild. This is 1050 more than live in zoos worldwide. How many golden lion tamarins live in zoos?

   Source: nationalzoo.si.edu

Multiply.

8. $7(5 + x)$
9. $-5(y - 2)$

Factor.

10. $11 + 44x$
11. $7x + 7 + 14y$
12. Find the prime factorization of 300.
13. Simplify: $\frac{10}{35}$.

Write a true sentence using either $<$ or $>$.

14. $-4 \square 0$
15. $-3 \square -8$

Find the absolute value.

16. $\left|\frac{9}{4}\right|$
17. $|-3.8|$
18. Find the opposite of $-\frac{2}{3}$.
19. Find the reciprocal of $-\frac{4}{7}$.
20. Find $-x$ when x is -10.
21. Write an inequality with the same meaning as $x \leq -5$.

Perform the indicated operations and, if possible, simplify.

22. $3.1 - (-4.7)$
23. $-8 + 4 + (-7) + 3$
24. $3.2 - 5.7$
25. $-\frac{1}{8} - \frac{3}{4}$
26. $4 \cdot (-12)$
27. $-\frac{1}{2} \cdot \left(-\frac{4}{9}\right)$
28. $-66 \div 11$
29. $-\frac{3}{5} \div \left(-\frac{4}{5}\right)$
30. $4.864 \div (-0.5)$
31. $10 - 2(-16) \div 4^2 + |2 - 10|$
32. $256 \div (-16) \cdot 4$
33. $2^3 - 10[4 - (-2 + 18)3]$
34. Combine like terms: $-18y + 30a - 9a + 4y$.
35. Simplify: $(-2x)^4$.

Remove parentheses and simplify.

36. $4x - (3x - 7)$
37. $4(2a - 3b) + a - 7$
38. $3[5(y - 3) + 9] - 2(8y - 1)$

SYNTHESIS

39. Evaluate $\frac{5y - x}{2}$ when $x = 20$ and y is 4 less than half of x.
40. Insert one pair of parentheses to make the following a true statement:

$$9 - 3 - 4 + 5 = 15.$$

Simplify.

41. $|-27 - 3(4)| - |-36| + |-12|$
42. $a - \{3a - [4a - (2a - 4a)]\}$
43. Classify the following as either true or false:

$$a|b - c| = |ab| - |ac|.$$

2

Equations, Inequalities, and Problem Solving

How Much Space Does a Seahorse Need?

Seahorses are beautiful but delicate ocean creatures. Even those that have been raised in an aquarium require special care. In Example 7 of Section 2.5, we use the data shown here to estimate an appropriate aquarium size for 15 seahorses.

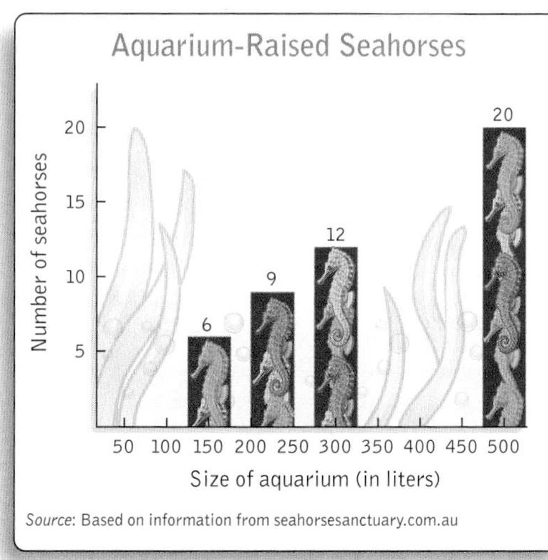

Source: Based on information from seahorsesanctuary.com.au

Solving equations and inequalities is an important theme in mathematics. In this chapter, we will study some of the principles used to solve equations and inequalities. We will then use equations and inequalities to solve applied problems.

2.1 Solving Equations

- Equations and Solutions
- The Addition Principle
- The Multiplication Principle
- Selecting the Correct Approach

Solving equations is essential for problem solving in algebra. In this section, we study two of the most important principles used for this task.

EQUATIONS AND SOLUTIONS

An equation is a number sentence stating that the expressions on either side of the equals sign represent the same number. Some equations, like $3 + 2 = 5$ or $2x + 6 = 2(x + 3)$, are *always* true and some, like $3 + 2 = 6$ or $x + 2 = x + 3$, are *never* true. In this text, we will concentrate on equations like $x + 6 = 13$ or $7x = 141$ that are *sometimes* true, depending on the replacement value for the variable.

> Any replacement for the variable that makes an equation true is called a *solution* of the equation. To *solve* an equation means to find all of its solutions.

To determine whether a number is a solution, we substitute that number for the variable throughout the equation. If the values on both sides of the equals sign are the same, then the number that was substituted is a solution.

EXAMPLE 1 Determine whether 7 is a solution of $x + 6 = 13$.

SOLUTION We have

$x + 6 = 13$		**Writing the equation**
$7 + 6 \mid 13$		**Substituting 7 for x**
$13 \stackrel{?}{=} 13$	TRUE	**$13 = 13$ is a true statement.**

Since the left-hand and the right-hand sides are the same, 7 is a solution.

Try Exercise 11.

> **CAUTION!** Note that in Example 1, the solution is 7, not 13.

EXAMPLE 2 Determine whether 3 is a solution of $7x - 2 = 4x + 5$.

SOLUTION We have

$7x - 2 = 4x + 5$			Writing the equation
$7(3) - 2$	$4(3) + 5$		Substituting 3 for x
$21 - 2$	$12 + 5$		Carrying out calculations on both sides
$19 \stackrel{?}{=} 17$		FALSE	The statement $19 = 17$ is false.

Since the left-hand and the right-hand sides differ, 3 is not a solution.

Try Exercise 15.

THE ADDITION PRINCIPLE

Consider the equation

$$x = 7.$$

We can easily see that the solution of this equation is 7. Replacing x with 7, we get

$$7 = 7, \quad \text{which is true.}$$

Now consider the equation

$$x + 6 = 13.$$

In Example 1, we found that the solution of $x + 6 = 13$ is also 7. Although the solution of $x = 7$ may seem more obvious, the equations $x + 6 = 13$ and $x = 7$ have identical solutions and are said to be **equivalent.**

Student Notes

Be sure to remember the difference between an expression and an equation. For example, $5a - 10$ and $5(a - 2)$ are *equivalent expressions* because they represent the same value for all replacements for a. The *equations* $5a = 10$ and $a = 2$ are *equivalent* because they have the same solution, 2.

Equivalent Equations Equations with the same solutions are called *equivalent equations.*

There are principles that enable us to begin with one equation and end up with an equivalent equation, like $x = 7$, for which the solution is obvious. One such principle concerns addition. The equation $a = b$ says that a and b stand for the same number. Suppose this is true, and some number c is added to a. We get the same result if we add c to b, because a and b are the same number.

The Addition Principle For any real numbers a, b, and c,

$$a = b \quad \text{is equivalent to} \quad a + c = b + c.$$

To visualize the addition principle, consider a balance similar to one a jeweler might use. When the two sides of the balance hold equal weight, the balance is level. If weight is then added or removed, equally, on both sides, the balance will remain level.

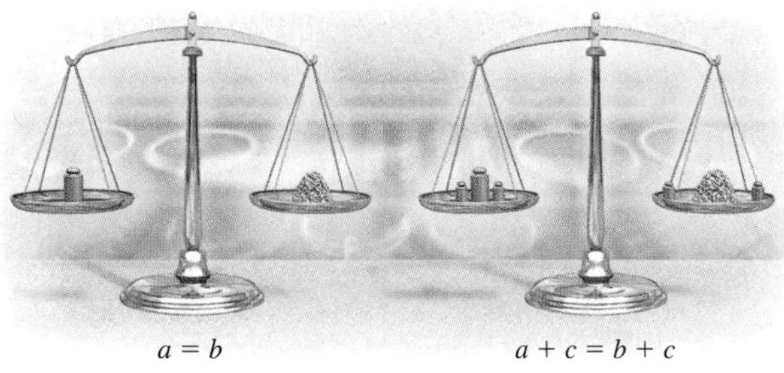

When using the addition principle, we often say that we "add the same number to both sides of an equation." We can also "subtract the same number from both sides," since subtraction can be regarded as the addition of an opposite.

EXAMPLE 3 Solve: $x + 5 = -7$.

SOLUTION We can add any number we like to both sides. Since -5 is the opposite, or additive inverse, of 5, we add -5 to each side:

$$x + 5 = -7$$

$$x + 5 - 5 = -7 - 5 \quad \text{Using the addition principle: adding } -5 \text{ to both sides or subtracting 5 from both sides}$$

$$x + 0 = -12 \quad \text{Simplifying; } x + 5 - 5 = x + 5 + (-5) = x + 0$$

$$x = -12. \quad \text{Using the identity property of 0}$$

The equation $x = -12$ is equivalent to the equation $x + 5 = -7$ by the addition principle, so the solution of $x = -12$ is the solution of $x + 5 = -7$.

It is obvious that the solution of $x = -12$ is the number -12. To check the answer in the original equation, we substitute.

Check:

$$\begin{array}{c|c} x + 5 & = -7 \\ \hline -12 + 5 & -7 \\ -7 & \stackrel{?}{=} -7 \end{array} \quad \text{TRUE} \quad -7 = -7 \text{ is true.}$$

The solution of the original equation is -12.

Try Exercise 19.

In Example 3, note that because we added the *opposite*, or *additive inverse*, of 5, the left-hand side of the equation simplified to x plus the *additive identity*, 0, or simply x. To solve $x + a = b$ for x, we add $-a$ to (or subtract a from) both sides.

Student Notes

We can also think of "undoing" operations to isolate a variable. In Example 4, we began with $y - 8.4$ on the right side. To undo the subtraction, we *add* 8.4.

EXAMPLE 4 Solve: $-6.5 = y - 8.4$.

SOLUTION The variable is on the right-hand side this time. We can isolate y by adding 8.4 to each side:

$$-6.5 = y - 8.4 \quad y - 8.4 \text{ can be regarded as } y + (-8.4).$$

$$-6.5 + 8.4 = y - 8.4 + 8.4 \quad \text{Using the addition principle: Adding 8.4 to both sides "eliminates" } -8.4 \text{ on the right-hand side.}$$

$$1.9 = y. \quad y - 8.4 + 8.4 = y + (-8.4) + 8.4 = y + 0 = y$$

Check:

$$\begin{array}{c|c} -6.5 & = y - 8.4 \\ \hline -6.5 & 1.9 - 8.4 \\ -6.5 & \stackrel{?}{=} -6.5 \end{array} \quad \text{TRUE} \quad -6.5 = -6.5 \text{ is true.}$$

The solution is 1.9.

Try Exercise 27.

Note that the equations $a = b$ and $b = a$ have the same meaning. Thus, $-6.5 = y - 8.4$ could have been rewritten as $y - 8.4 = -6.5$.

THE MULTIPLICATION PRINCIPLE

A second principle for solving equations concerns multiplying. Suppose a and b are equal. If a and b are multiplied by some number c, then ac and bc will also be equal.

The Multiplication Principle For any real numbers a, b, and c, with $c \neq 0$,

$a = b$ is equivalent to $a \cdot c = b \cdot c$.

EXAMPLE 5 Solve: $\frac{5}{4}x = 10$.

SOLUTION We can multiply both sides by any nonzero number. Since $\frac{4}{5}$ is the reciprocal of $\frac{5}{4}$, we multiply each side by $\frac{4}{5}$:

$$\frac{5}{4}x = 10$$

$$\frac{4}{5} \cdot \frac{5}{4}x = \frac{4}{5} \cdot 10 \qquad \text{Using the multiplication principle: Multiplying both sides by } \tfrac{4}{5} \text{ "eliminates" the } \tfrac{5}{4} \text{ on the left.}$$

$$1 \cdot x = 8 \qquad \text{Simplifying}$$

$$x = 8. \qquad \text{Using the identity property of 1}$$

Check:

$\frac{5}{4}x = 10$		
$\frac{5}{4} \cdot 8$	10	
$\frac{40}{4}$		Think of 8 as $\frac{8}{1}$.
$10 \stackrel{?}{=} 10$	TRUE	$10 = 10$ is true.

The solution is 8.

Try Exercise 55.

In Example 5, to get x alone, we multiplied by the *reciprocal*, or *multiplicative inverse* of $\frac{5}{4}$. We then simplified the left-hand side to x times the *multiplicative identity*, 1, or simply x.

Because division is the same as multiplying by a reciprocal, the multiplication principle also tells us that we can "divide both sides by the same nonzero number." That is,

$$\text{if } a = b, \text{ then } \frac{1}{c} \cdot a = \frac{1}{c} \cdot b \quad \text{and} \quad \frac{a}{c} = \frac{b}{c} \qquad (\text{provided } c \neq 0).$$

In a product like $3x$, the multiplier 3 is called the **coefficient.** *When the coefficient of the variable is an integer or a decimal, it is usually easiest to solve an equation by dividing on both sides. When the coefficient is in fraction notation, it is usually easier to multiply by the reciprocal.*

Student Notes

In Example 6(a), we can think of undoing the multiplication $-4 \cdot x$ by *dividing* both sides by -4.

EXAMPLE 6 Solve: **(a)** $-4x = 9$; **(b)** $-x = 7$; **(c)** $\frac{2y}{9} = \frac{8}{3}$.

SOLUTION

a) $-4x = 9$

$\frac{-4x}{-4} = \frac{9}{-4}$ **Using the multiplication principle: Dividing both sides by -4 is the same as multiplying by $-\frac{1}{4}$.**

$1 \cdot x = -\frac{9}{4}$ **Simplifying**

$x = -\frac{9}{4}$ **Using the identity property of 1**

Check:

$$\begin{array}{c|c} -4x & 9 \\ \hline -4\left(-\frac{9}{4}\right) & 9 \\ 9 & \overset{?}{=} 9 \end{array}$$

TRUE **$9 = 9$ is true.**

The solution is $-\frac{9}{4}$.

b) To solve an equation like $-x = 7$, remember that when an expression is multiplied or divided by -1, its sign is changed. Here we multiply both sides by -1 to change the sign of $-x$:

$-x = 7$

$(-1)(-x) = (-1)7$ **Multiplying both sides by -1. (Dividing by -1 would also work.) Note that the reciprocal of -1 is -1.**

$x = -7.$ **Note that $(-1)(-x)$ is the same as $(-1)(-1)x$.**

Check:

$$\begin{array}{c|c} -x & 7 \\ \hline -(-7) & 7 \\ 7 & \overset{?}{=} 7 \end{array}$$

TRUE **$7 = 7$ is true.**

The solution is -7.

c) To solve an equation like $\frac{2y}{9} = \frac{8}{3}$, we rewrite the left-hand side as $\frac{2}{9} \cdot y$ and then use the multiplication principle, multiplying by the reciprocal of $\frac{2}{9}$:

$\frac{2y}{9} = \frac{8}{3}$

$\frac{2}{9} \cdot y = \frac{8}{3}$ **Rewriting $\frac{2y}{9}$ as $\frac{2}{9} \cdot y$**

$\frac{9}{2} \cdot \frac{2}{9} \cdot y = \frac{9}{2} \cdot \frac{8}{3}$ **Multiplying both sides by $\frac{9}{2}$**

$1y = \frac{3 \cdot \cancel{3} \cdot \cancel{2} \cdot 4}{\cancel{2} \cdot \cancel{3}}$ **Removing a factor equal to 1: $\frac{3 \cdot 2}{2 \cdot 3} = 1$**

$y = 12.$

Check:

$$\frac{2y}{9} = \frac{8}{3}$$

$$\frac{2 \cdot 12}{9} \;\Big|\; \frac{8}{3}$$

$$\frac{24}{9}$$

$$\frac{8}{3} \stackrel{?}{=} \frac{8}{3} \quad \text{TRUE} \qquad \tfrac{8}{3} = \tfrac{8}{3} \text{ is true.}$$

The solution is 12.

Try Exercise 49.

SELECTING THE CORRECT APPROACH

It is important that you be able to determine which principle should be used in order to solve a particular equation.

STUDY TIP

Seeking Help?

A variety of resources are available to help make studying easier and more enjoyable.

- **Textbook supplements.** See the preface for a description of the supplements that exist for this textbook.
- **Your college or university.** Your own college or university probably has resources to enhance your math learning: a learning lab or tutoring center, study skills workshops or group tutoring sessions tailored for the course you are taking, or a bulletin board or network where you can locate the names of experienced private tutors.
- **Your instructor.** Find out your instructor's office hours and make it a point to visit when you need additional help. Many instructors also welcome student e-mail.

EXAMPLE 7 Solve: **(a)** $-8 + x = -3$; **(b)** $1.8 = 3t$.

SOLUTION

a) To undo the addition of -8, we subtract -8 from (or add 8 to) both sides. Note that the opposite of *negative* 8 is *positive* 8.

$$-8 + x = -3$$

$$-8 + x + 8 = -3 + 8 \qquad \text{Using the addition principle}$$

$$x = 5$$

Check:

$$-8 + x = -3$$

$$-8 + 5 \;\Big|\; -3$$

$$-3 \stackrel{?}{=} -3 \quad \text{TRUE} \qquad -3 = -3 \text{ is true.}$$

The solution is 5.

b) To undo the multiplication by 3, we either divide both sides by 3 or multiply both sides by $\frac{1}{3}$. Note that the reciprocal of *positive* 3 is *positive* $\frac{1}{3}$.

$$1.8 = 3t$$

$$\frac{1.8}{3} = \frac{3t}{3} \qquad \text{Using the multiplication principle}$$

$$0.6 = t \qquad \text{Simplifying}$$

Check:

$$1.8 = 3t$$

$$1.8 \;\Big|\; 3(0.6)$$

$$1.8 \stackrel{?}{=} 1.8 \quad \text{TRUE} \qquad 1.8 = 1.8 \text{ is true.}$$

The solution is 0.6.

Try Exercises 65 and 67.

Editing and Evaluating Expressions

To correct an error or change a value in an expression, use the *arrow*, *insert*, and *delete* keys.

As the expression is being typed, use the arrow keys to move the cursor to the character you want to change. Pressing INS (the 2nd option associated with the DEL key) changes the calculator between the INSERT and OVERWRITE modes. The OVERWRITE mode is often indicated by a rectangular cursor and the INSERT mode by an underscore cursor. The table below lists how to make changes.

Insert a character in front of the cursor.	In the INSERT mode, press the character you wish to insert.
Replace the character under the cursor.	In the OVERWRITE mode, press the character you want as the replacement.
Delete the character under the cursor.	Press DEL.

After you have evaluated an expression by pressing ENTER, the expression can be recalled to the screen by pressing ENTRY (the 2nd option associated with the ENTER key.) Then it can be edited as described above. Pressing ENTER will then evaluate the edited expression.

Expressions such as $2xy - x$ can be entered into a graphing calculator as written. The calculator will evaluate an expression using the values that it has stored for the variables. Variable names can be entered by pressing ALPHA and then the key associated with that letter. The variable x can also be entered by pressing the X,T,Θ,n key. Thus, for example, to store the value 1 as y, press 1 STO› ALPHA Y ENTER. The value for y can then be seen by pressing ALPHA Y ENTER.

Your Turn

1. Press 8 ÷ 2 to enter the expression $8 \div 2$. Do not press ENTER.
2. With the calculator in OVERWRITE mode, change the expression to $8 + 2$. Do not press ENTER.
3. With the calculator in INSERT mode, change the expression to $85 + 2$. Do not press ENTER.
4. Change the expression to $5 + 2$. Evaluate by pressing ENTER.
5. Store the value 10 as y.
6. Store the value -3 as x.
7. Evaluate $2xy - x$ for $x = -3$ and $y = 10$ by pressing 2 × X,T,Θ,n × ALPHA Y − X,T,Θ,n ENTER. The result should be -57.
8. Store the value 8 as y.
9. Evaluate $2xy - x$ for $x = -3$ and $y = 8$ by recalling the entry $2xy - x$ and pressing ENTER. The result should be -45.

EXAMPLE 8 Use a calculator to determine whether each of the following is a solution of $2x - 5 = -7$: 3; -1.

SOLUTION We evaluate the expression on the left-hand side of the equals sign, $2x - 5$, for each given value of x. We begin by substituting 3 for x. Since $2(3) - 5 \neq -7$, 3 is not a solution of the equation.

To check −1, we recall the last entry, $2(3) - 5$. Note that −1 has one more character than 3. We can insert a negative sign and then overwrite the 3 with a 1. We see that −1 is a solution of the equation.

Try Exercise 17.

As a check of Example 8, we store 3 to the variable x and evaluate the expression. We see that for $x = 3$, the value of $2x - 5$ is 1.

Now we store −1 to x. We then press ENTRY repeatedly until $2x - 5$ appears again on the screen. Then we press ENTER. We see that for $x = -1$, the value of $2x - 5$ is −7. Thus, −1 is a solution of $2x - 5 = -7$, but 3 is not.

```
−1 → X
                 −1
2X−5
                 −7
```

2.1 Exercise Set

FOR EXTRA HELP MyMathLab

Concept Reinforcement *For each of Exercises 1–6, match the statement with the most appropriate choice from the column on the right.*

1. ___ The equations $x + 3 = 7$ and $6x = 24$
2. ___ The expressions $3(x - 2)$ and $3x - 6$
3. ___ A replacement that makes an equation true
4. ___ The role of 9 in $9ab$
5. ___ The principle used to solve $\frac{2}{3} \cdot x = -4$
6. ___ The principle used to solve $\frac{2}{3} + x = -4$

a) Coefficient
b) Equivalent expressions
c) Equivalent equations
d) The multiplication principle
e) The addition principle
f) Solution

For each of Exercises 7–10, match the equation with the step, from the column on the right, that would be used to solve the equation.

7. $6x = 30$ a) Add 6 to both sides.
8. $x + 6 = 30$ b) Subtract 6 from both sides.
9. $\frac{1}{6}x = 30$ c) Multiply both sides by 6.
10. $x - 6 = 30$ d) Divide both sides by 6.

To the student and the instructor: The Try Exercises *for examples are indicated by a shaded block on the exercise number. Answers to these exercises appear at the end of the exercise set as well as at the back of the book.*

Determine whether the given number is a solution of the given equation.

11. $6 - x = -2;\ 4$
12. $6 - x = -2;\ 8$
13. $\frac{2}{3}t = 12;\ 18$
14. $\frac{2}{3}t = 12;\ 8$
15. $x + 7 = 3 - x;\ -2$
16. $-4 + x = 5x;\ -1$
17. $4 - \frac{1}{5}n = 8;\ -20$
18. $-3 = 5 - \frac{n}{2};\ 4$

Solve using the addition principle. Don't forget to check!

19. $x + 6 = 23$
20. $x + 5 = 8$
21. $y + 7 = -4$
22. $t + 6 = 43$
23. $-6 = y + 25$
24. $-5 = x + 8$
25. $x - 8 = 5$
26. $x - 9 = 6$
27. $12 = -7 + y$
28. $15 = -8 + z$
29. $-5 + t = -9$
30. $-6 + y = -21$
31. $r + \frac{1}{3} = \frac{8}{3}$
32. $t + \frac{3}{8} = \frac{5}{8}$
33. $x - \frac{3}{5} = -\frac{7}{10}$
34. $x - \frac{2}{3} = -\frac{5}{6}$
35. $-\frac{1}{5} + z = -\frac{1}{4}$
36. $-\frac{1}{8} + y = -\frac{3}{4}$
37. $m + 3.9 = 5.4$
38. $y + 5.3 = 8.7$
39. $-9.7 = -4.7 + y$
40. $-7.8 = 2.8 + x$

Solve using the multiplication principle. Don't forget to check!

41. $5x = 70$
42. $3x = 39$
43. $84 = 7n$
44. $56 = 7t$
45. $-x = 23$
46. $100 = -x$
47. Aha! $-t = -8$
48. $-68 = -r$
49. $2x = -5$
50. $-3x = 5$
51. $-1.3a = -10.4$
52. $-3.4t = -20.4$
53. $\frac{y}{-8} = 11$
54. $\frac{a}{4} = 13$
55. $\frac{4}{5}x = 16$
56. $\frac{3}{4}x = 27$
57. $\frac{-x}{6} = 9$
58. $\frac{-t}{5} = 9$
59. $\frac{1}{9} = \frac{z}{5}$
60. $\frac{2}{7} = \frac{x}{3}$
61. Aha! $-\frac{3}{5}r = -\frac{3}{5}$
62. $-\frac{2}{5}y = -\frac{4}{15}$
63. $\frac{-3r}{2} = -\frac{27}{4}$
64. $\frac{5x}{7} = -\frac{10}{14}$

Solve. The symbol ▦ indicates an exercise designed to give practice using a calculator.

65. $4.5 + t = -3.1$
66. $\frac{3}{4}x = 18$
67. $-8.2x = 20.5$
68. $t - 7.4 = -12.9$
69. $x - 4 = -19$
70. $y - 6 = -14$
71. $-12x = 72$
72. $-15x = 105$
73. $48 = -\frac{3}{8}y$
74. $14 = t + 27$
75. $a - \frac{1}{6} = -\frac{2}{3}$
76. $-\frac{x}{7} = \frac{2}{9}$
77. $-24 = \frac{8x}{5}$
78. $\frac{1}{5} + y = -\frac{3}{10}$
79. $-\frac{4}{3}t = -16$
80. $\frac{17}{35} = -x$
81. ▦ $-483.297 = -794.053 + t$
82. ▦ $-0.2344x = 2028.732$
83. TW When solving an equation, how do you determine what number to add, subtract, multiply, or divide by on both sides of that equation?
84. TW What is the difference between equivalent expressions and equivalent equations?

SKILL REVIEW

To prepare for Section 2.2, review the rules for order of operations (Section 1.8).

Simplify. [1.8]

85. $3 \cdot 4 - 18$
86. $14 - 2(7 - 1)$
87. $16 \div (2 - 3 \cdot 2) + 5$
88. $12 - 5 \cdot 2^3 + 4 \cdot 3$

SYNTHESIS

TW **89.** To solve $-3.5 = 14t$, Gregory adds 3.5 to both sides. Will this form an equivalent equation? Will it help solve the equation? Explain.

TW **90.** Explain why it is not necessary to state a subtraction principle: For any real numbers a, b, and c, $a = b$ is equivalent to $a - c = b - c$.

Solve for x. Assume $a, c, m \neq 0$.

91. $mx = 9.4m$

92. $x - 4 + a = a$

93. $cx + 5c = 7c$

94. $c \cdot \frac{21}{a} = \frac{7cx}{2a}$

95. $7 + |x| = 20$

96. $ax - 3a = 5a$

97. If $t - 3590 = 1820$, find $t + 3590$.

98. If $n + 268 = 124$, find $n - 268$.

99. Alayna makes a calculation and gets an answer of 22.5. On the last step, she multiplies by 0.3 when she should have divided by 0.3. What is the correct answer?

TW **100.** Are the equations $x = 5$ and $x^2 = 25$ equivalent? Why or why not?

Try Exercise Answers: Section 2.1

11. No **15.** Yes **17.** Yes **19.** 17 **27.** 19 **49.** $-\frac{5}{2}$ **55.** 20 **65.** -7.6 **67.** -2.5

2.2 Using the Principles Together

- Applying Both Principles
- Combining Like Terms
- Clearing Fractions and Decimals
- Contradictions and Identities

An important strategy for solving new problems is to find a way to make a new problem look like a problem that we already know how to solve. In this section, you will find that the last steps of the examples are nearly identical to the steps used for solving the examples of Section 2.1. What is new in this section appears in the early steps of each example.

APPLYING BOTH PRINCIPLES

The addition and multiplication principles, along with the laws discussed in Chapter 1, are our tools for solving equations.

EXAMPLE 1 Solve: $5 + 3x = 17$.

SOLUTION Were we to evaluate $5 + 3x$, the rules for the order of operations direct us to *first* multiply by 3 and *then* add 5. Because of this, we can isolate $3x$ and then x by reversing these operations: We first subtract 5 from both sides and then divide both sides by 3. Our goal is an equivalent equation of the form $x = a$.

$$5 + 3x = 17$$

$$5 + 3x - 5 = 17 - 5$$ **Using the addition principle: subtracting 5 from both sides (adding -5)**

$$5 + (-5) + 3x = 12$$ **Using a commutative law. Try to perform this step mentally.**

Isolate the x-term.

$$3x = 12$$ **Simplifying**

$$\frac{3x}{3} = \frac{12}{3}$$ **Using the multiplication principle: dividing both sides by 3 (multiplying by $\frac{1}{3}$)**

Isolate x.

$$x = 4$$ **Simplifying**

Check:

$$\begin{array}{c|c} 5 + 3x & 17 \\ \hline 5 + 3 \cdot 4 & 17 \\ 5 + 12 & \\ 17 \stackrel{?}{=} 17 & \end{array}$$ TRUE

We use the rules for order of operations: Find the product, 3 · 4, and then add.

The solution is 4.

Try Exercise 7.

EXAMPLE 2 Solve: $\frac{4}{3}x - 7 = 1$.

SOLUTION In $\frac{4}{3}x - 7$, we multiply first and then subtract. To reverse these steps, we first add 7 and then either divide by $\frac{4}{3}$ or multiply by $\frac{3}{4}$.

$$\frac{4}{3}x - 7 = 1$$

$$\frac{4}{3}x - 7 + 7 = 1 + 7 \qquad \textbf{Adding 7 to both sides}$$

$$\frac{4}{3}x = 8$$

$$\frac{3}{4} \cdot \frac{4}{3}x = \frac{3}{4} \cdot 8 \qquad \textbf{Multiplying both sides by } \tfrac{3}{4}$$

$$\left.\begin{aligned} 1 \cdot x &= \frac{3 \cdot \not{4} \cdot 2}{\not{4}} \\ x &= 6 \end{aligned}\right\} \quad \textbf{Simplifying}$$

```
4/3*6-7
                1
```

This time we check using a calculator. We replace x with 6 and evaluate the expression on the left-hand side of the equation, as shown at left. Since $\frac{4}{3} \cdot 6 - 7 = 1$, the answer checks. The solution is 6.

Try Exercise 27.

EXAMPLE 3 Solve: $45 - t = 13$.

SOLUTION We have

$$45 - t = 13$$

$$45 - t - 45 = 13 - 45 \qquad \textbf{Subtracting 45 from both sides}$$

$$\left.\begin{aligned} 45 + (-t) + (-45) &= 13 - 45 \\ 45 + (-45) + (-t) &= 13 - 45 \end{aligned}\right\} \quad \textbf{Try to do these steps mentally.}$$

$$-t = -32 \qquad \textbf{Try to go directly to this step.}$$

$$(-1)(-t) = (-1)(-32) \qquad \textbf{Multiplying both sides by } -1. \textbf{ (Dividing by } -1 \textbf{ would also work.)}$$

$$t = 32.$$

Check:

$$\begin{array}{c|c} 45 - t & 13 \\ \hline 45 - 32 & 13 \\ 13 \stackrel{?}{=} 13 & \end{array}$$ TRUE

The solution is 32.

Try Exercise 17.

As our skills improve, many of the steps can be streamlined.

STUDY TIP 

Use the Answer Section Carefully

When using the answers listed at the back of this book, try not to "work backward" from the answer. If you frequently require two or more attempts to answer an exercise correctly, you probably need to work more carefully and/or reread the section preceding the exercise set. Remember that on quizzes and tests you have only one attempt per problem and no answer section to check.

EXAMPLE 4 Solve: $16.3 - 7.2y = -8.18$.

SOLUTION We have

$$
\begin{aligned}
16.3 - 7.2y &= -8.18 \\
16.3 - 7.2y - 16.3 &= -8.18 - 16.3 && \text{Subtracting 16.3 from both sides} \\
-7.2y &= -24.48 && \text{Simplifying} \\
\frac{-7.2y}{-7.2} &= \frac{-24.48}{-7.2} && \text{Dividing both sides by } -7.2 \\
y &= 3.4. && \text{Simplifying}
\end{aligned}
$$

Check:

$$
\begin{array}{r|l}
16.3 - 7.2y & = -8.18 \\
\hline
16.3 - 7.2(3.4) & -8.18 \\
16.3 - 24.48 & \\
-8.18 & \stackrel{?}{=} -8.18 \quad \text{TRUE}
\end{array}
$$

The solution is 3.4.

Try Exercise 21.

COMBINING LIKE TERMS

If like terms appear on the same side of an equation, we combine them and then solve. Should like terms appear on both sides of an equation, we can use the addition principle to rewrite all like terms on one side.

EXAMPLE 5 Solve.

a) $3x + 4x = -14$ **b)** $-x + 5 = -8x + 6$

c) $6x + 5 - 7x = 10 - 4x + 7$ **d)** $2 - 5(x + 5) = 3(x - 2) - 1$

SOLUTION

a) $3x + 4x = -14$

$$
\begin{aligned}
7x &= -14 && \text{Combining like terms} \\
\frac{7x}{7} &= \frac{-14}{7} && \text{Dividing both sides by 7} \\
x &= -2
\end{aligned}
$$

The check is left to the student. The solution is -2.

b) To solve $-x + 5 = -8x + 6$, we must first write only variable terms on one side and only constant terms on the other. This can be done by subtracting 5 from both sides, to get all constant terms on the right, and adding $8x$ to both sides to get all variable terms on the left.

Isolate variable terms on one side and constant terms on the other side.

$$
\begin{aligned}
-x + 5 &= -8x + 6 \\
-x + 5 - 5 &= -8x + 6 - 5 && \text{Subtracting 5 from both sides} \\
-x &= -8x + 1 && \text{Simplifying} \\
-x + 8x &= -8x + 8x + 1 && \text{Adding } 8x \text{ to both sides} \\
7x &= 1 && \text{Combining like terms and simplifying} \\
\frac{7x}{7} &= \frac{1}{7} && \text{Dividing both sides by 7} \\
x &= \frac{1}{7} && \text{Simplifying}
\end{aligned}
$$

The check is left to the student. The solution is $\frac{1}{7}$.

c) $6x + 5 - 7x = 10 - 4x + 7$

$-x + 5 = 17 - 4x$ — **Combining like terms on both sides**

$-x + 5 + 4x = 17 - 4x + 4x$ — **Adding 4x to both sides**

$5 + 3x = 17$ — **Simplifying. This is identical to Example 1.**

$3x = 12$ — **Subtracting 5 from both sides and simplifying**

$\frac{3x}{3} = \frac{12}{3}$ — **Dividing both sides by 3**

$x = 4$

```
6(4)+5−7(4)
                    1
10−4(4)+7
                    1
```

To check with a calculator, we evaluate the expressions on both sides of the equation for $x = 4$. We see from the figure at left that both expressions have the same value, so 4 checks. The solution is 4.

d) $2 - 5(x + 5) = 3(x - 2) - 1$

$2 - 5x - 25 = 3x - 6 - 1$ — **Using the distributive law. This is now similar to part (c) above.**

$-5x - 23 = 3x - 7$ — **Combining like terms on both sides**

$-5x - 23 + 7 = 3x - 7 + 7$ — **Adding 7 to both sides**

$-5x - 16 = 3x$ — **Simplifying**

$-5x - 16 + 5x = 3x + 5x$ — **Adding 5x to both sides**

$-16 = 8x$ — **Simplifying**

$\frac{-16}{8} = \frac{8x}{8}$ — **Dividing both sides by 8**

$-2 = x$ — **This is equivalent to $x = -2$.**

Check:

$2 - 5(x + 5)$	$= 3(x - 2) - 1$
$2 - 5(-2 + 5)$	$3(-2 - 2) - 1$
$2 - 5(3)$	$3(-4) - 1$
$2 - 15$	$-12 - 1$
$-13 \stackrel{?}{=}$	-13 TRUE

The solution is -2.

■ Try Exercise 57.

CLEARING FRACTIONS AND DECIMALS

Equations are generally easier to solve when they do not contain fractions or decimals. The multiplication principle can be used to "clear" fractions or decimals, as shown here.

Clearing Fractions	Clearing Decimals
$\frac{1}{2}x + 5 = \frac{3}{4}$	$2.3x + 7 = 5.4$
$4\left(\frac{1}{2}x + 5\right) = 4 \cdot \frac{3}{4}$	$10(2.3x + 7) = 10 \cdot 5.4$
$2x + 20 = 3$	$23x + 70 = 54$

In each case, the resulting equation is equivalent to the original equation, but easier to solve.

An Equation-Solving Procedure

1. Use the multiplication principle to clear any fractions or decimals. (This is optional, but can ease computations. See Examples 6 and 7.)
2. If necessary, use the distributive law to remove parentheses. Then combine like terms on each side. (See Example 5.)
3. Use the addition principle, as needed, to get all variable terms on one side and all constant terms on the other. Then combine like terms. (See Examples 1–7.)
4. Multiply or divide to solve for the variable, using the multiplication principle. (See Examples 1–7.)
5. Check all possible solutions in the original equation. (See Examples 1–5.)

The easiest way to clear an equation of fractions is to multiply *both sides* of the equation by the smallest, or *least*, common denominator.

EXAMPLE 6 Solve: **(a)** $\frac{2}{3}x - \frac{1}{6} = 2x$; **(b)** $\frac{2}{5}(3x + 2) = 8$.

SOLUTION

a) We multiply both sides by 6, the least common denominator of $\frac{2}{3}$ and $\frac{1}{6}$.

$$6\left(\frac{2}{3}x - \frac{1}{6}\right) = 6 \cdot 2x$$ **Multiplying both sides by 6**

$$6 \cdot \frac{2}{3}x - 6 \cdot \frac{1}{6} = 6 \cdot 2x$$

CAUTION! Be sure the distributive law is used to multiply *all* the terms by 6.

$$4x - 1 = 12x$$ **Simplifying. Note that the fractions are cleared: $6 \cdot \frac{2}{3} = 4$, $6 \cdot \frac{1}{6} = 1$, and $6 \cdot 2 = 12$.**

$$4x - 1 - 4x = 12x - 4x$$ **Subtracting 4x from both sides**

$$-1 = 8x$$

$$\frac{-1}{8} = \frac{8x}{8}$$ **Dividing both sides by 8**

$$-\frac{1}{8} = x$$ $\frac{-1}{8} = -\frac{1}{8}$

The student can confirm that $-\frac{1}{8}$ checks and is the solution.

b) To solve $\frac{2}{5}(3x + 2) = 8$, we can multiply both sides by $\frac{5}{2}$ $\left(\text{or divide by } \frac{2}{5}\right)$ to "undo" the multiplication by $\frac{2}{5}$ on the left-hand side.

$$\frac{5}{2} \cdot \frac{2}{5}(3x + 2) = \frac{5}{2} \cdot 8$$ **Multiplying both sides by $\frac{5}{2}$**

$$3x + 2 = 20$$ **Simplifying; $\frac{5}{2} \cdot \frac{2}{5} = 1$ and $\frac{5}{2} \cdot \frac{8}{1} = 20$**

$$3x + 2 - 2 = 20 - 2$$ **Subtracting 2 from both sides**

$$3x = 18$$

$$\frac{3x}{3} = \frac{18}{3}$$ **Dividing both sides by 3**

$$x = 6$$

The student can confirm that 6 checks and is the solution.

Try Exercise 65.

To clear an equation of decimals, we count the greatest number of decimal places in any one number. If the greatest number of decimal places is 1, we multiply both sides by 10; if it is 2, we multiply by 100; and so on.

Student Notes

Compare the steps of Examples 4 and 7. Note that the two different approaches yield the same solution. Whenever you can use two approaches to solve a problem, try to do so, both as a check and as a valuable learning experience.

EXAMPLE 7 Solve: $16.3 - 7.2y = -8.18$.

SOLUTION The greatest number of decimal places in any one number is *two*. Multiplying by 100 will clear all decimals.

$$100(16.3 - 7.2y) = 100(-8.18) \quad \text{Multiplying both sides by 100}$$
$$100(16.3) - 100(7.2y) = 100(-8.18) \quad \text{Using the distributive law}$$
$$1630 - 720y = -818 \quad \text{Simplifying}$$
$$1630 - 720y - 1630 = -818 - 1630 \quad \text{Subtracting 1630 from both sides}$$
$$-720y = -2448 \quad \text{Combining like terms}$$
$$\frac{-720y}{-720} = \frac{-2448}{-720} \quad \text{Dividing both sides by } -720$$
$$y = 3.4$$

In Example 4, we found the same solution without clearing decimals. Finding the same answer in two ways is a good check. The solution is 3.4.

Try Exercise 71.

CONTRADICTIONS AND IDENTITIES

All of the equations that we have examined so far had a solution. Equations that are true for some values (solutions), but not for others, are called **conditional equations.** Equations that have no solution, such as $x + 1 = x + 2$, are called **contradictions.** If, when solving an equation, we obtain an equation that is false for any value of x, the equation has no solution.

EXAMPLE 8 Solve: $3x - 5 = 3(x - 2) + 4$.

SOLUTION

$$3x - 5 = 3(x - 2) + 4$$
$$3x - 5 = 3x - 6 + 4 \quad \text{Using the distributive law}$$
$$3x - 5 = 3x - 2 \quad \text{Combining like terms}$$
$$-3x + 3x - 5 = -3x + 3x - 2 \quad \text{Using the addition principle}$$
$$-5 = -2$$

Since the original equation is equivalent to $-5 = -2$, which is false for any choice of x, the original equation has no solution. There is no choice of x that will make $3x - 5 = 3(x - 2) + 4$ true. The equation is a contradiction. It is *never* true.

Try Exercise 45.

Some equations, like $x + 1 = x + 1$, are true for all replacements. Such an equation is called an **identity.**

EXAMPLE 9 Solve: $2x + 7 = 7(x + 1) - 5x$.

SOLUTION

$2x + 7 = 7(x + 1) - 5x$

$2x + 7 = 7x + 7 - 5x$ **Using the distributive law**

$2x + 7 = 2x + 7$ **Combining like terms**

The equation $2x + 7 = 2x + 7$ is true regardless of the replacement for x, so all real numbers are solutions. Note that $2x + 7 = 2x + 7$ is equivalent to $2x = 2x$, $7 = 7$, or $0 = 0$. All real numbers are solutions and the equation is an identity.

Try Exercise 61.

In Sections 2.1 and 2.2, we have solved *linear equations.* A **linear equation** in one variable—say, x—is an equation equivalent to one of the form $ax = b$ with a and b constants and $a \neq 0$.

We will sometimes refer to the set of solutions, or **solution set,** of a particular equation. Thus the solution set for Example 7 is $\{3.4\}$. The solution set for Example 9 is simply $\mathbb{R}$, the set of all real numbers, and the solution set for Example 8 is the **empty set,** denoted $\varnothing$ or $\{\ \}$. As its name suggests, the empty set is the set containing no elements.

2.2 Exercise Set

FOR EXTRA HELP MyMathLab MathXL PRACTICE WATCH DOWNLOAD READ REVIEW

Concept Reinforcement *In each of Exercises 1–6, match the equation with an equivalent equation from the column on the right that could be the next step in finding a solution.*

1. ___ $3x - 1 = 7$ a) $6x - 6 = 2$
2. ___ $4x + 5x = 12$ b) $4x + 2x = 3$
3. ___ $6(x - 1) = 2$ c) $3x = 7 + 1$
4. ___ $7x = 9$ d) $8x + 2x = 6 + 5$
5. ___ $4x = 3 - 2x$ e) $9x = 12$
6. ___ $8x - 5 = 6 - 2x$ f) $x = \frac{9}{7}$

Solve and check. Label any contradictions or identities.

7. $2x + 9 = 25$
8. $3x + 6 = 18$
9. $6z + 4 = -20$
10. $6z + 3 = -45$
11. $7t - 8 = 27$
12. $6x - 3 = 15$
13. $3x - 9 = 33$
14. $5x - 9 = 41$
15. $-91 = 9t + 8$
16. $-39 = 1 + 8x$
17. $12 - t = 16$
18. $9 - t = 21$
19. $-6z - 18 = -132$
20. $-7x - 24 = -129$
21. $5.3 + 1.2n = 1.94$
22. $6.4 - 2.5n = 2.2$
23. $4x + 5x = 10$
24. $13 = 5x + 7x$
25. $32 - 7x = 11$
26. $27 - 6x = 99$
27. $\frac{3}{5}t - 1 = 8$
28. $\frac{2}{3}t - 1 = 5$
29. $4 + \frac{7}{2}x = -10$
30. $6 + \frac{5}{4}x = -4$
31. $-\frac{3a}{4} - 5 = 2$
32. $-\frac{7a}{8} - 2 = 1$
33. $2x = x + x$
34. $-3z + 8z = 45$
35. $4x - 6 = 6x$
36. $4x - x = 2x + x$
37. $2 - 5y = 26 - y$
38. $6x - 5 = 7 + 2x$
39. $7(2a - 1) = 21$
40. $5(3 - 3t) = 30$
41. Aha! $8 = 8(x + 1)$
42. $9 = 3(5x - 2)$
43. $2(3 + 4m) - 6 = 48$
44. $3(5 + 3m) - 8 = 7$

45. $3(y + 4) = 3(y - 1)$

46. $5(y - 7) = 3(y - 2) + 2y$

47. $2r + 8 = 6r + 10$

48. $3p - 2 = 7p + 4$

49. $5 - 2x = 3x - 7x + 25$

50. $10 - 3x = 2x - 8x + 40$

51. $7 + 3x - 6 = 3x + 5 - x$

52. $5 + 4x - 7 = 4x - 2 - x$

53. $4y - 4 + y + 24 = 6y + 20 - 4y$

54. $5y - 10 + y = 7y + 18 - 5y$

55. $4 + 7a = 7(a - 1)$

56. $3(t + 2) + t = 2(3 + 2t)$

57. $13 - 3(2x - 1) = 4$

58. $5(d + 4) = 7(d - 2)$

59. $7(5x - 2) = 6(6x - 1)$

60. $5(t + 1) + 8 = 3(t - 2) + 6$

61. $2(7 - x) - 20 = 7x - 3(2 + 3x)$

62. $5(x - 7) = 3(x - 2) + 2x$

63. $19 - (2x + 3) = 2(x + 3) + x$

64. $13 - (2c + 2) = 2(c + 2) + 3c$

Clear fractions or decimals, solve, and check.

65. $\frac{2}{3} + \frac{1}{4}t = 2$

66. $-\frac{5}{6} + x = -\frac{1}{2} - \frac{2}{3}$

67. $\frac{2}{3} + 4t = 6t - \frac{2}{15}$

68. $\frac{1}{2} + 4m = 3m - \frac{5}{2}$

69. $\frac{1}{3}x + \frac{2}{5} = \frac{4}{15} + \frac{3}{5}x - \frac{2}{3}$

70. $1 - \frac{2}{3}y = \frac{9}{5} - \frac{1}{5}y + \frac{3}{5}$

71. $2.1x + 45.2 = 3.2 - 8.4x$

72. $0.91 - 0.2z = 1.23 - 0.6z$

73. $0.76 + 0.21t = 0.96t - 0.49$

74. $1.7t + 8 - 1.62t = 0.4t - 0.32 + 8$

75. $\frac{2}{5}x - \frac{3}{2}x = \frac{3}{4}x + 2$

76. $\frac{5}{16}y + \frac{3}{8}y = 2 + \frac{1}{4}y$

77. $\frac{1}{3}(2x - 1) = 7$

78. $\frac{4}{3}(5x + 1) = 8$

79. $\frac{3}{4}(3t - 6) = 9$

80. $\frac{3}{2}(2x + 5) = -\frac{15}{2}$

81. $\frac{1}{6}\left(\frac{3}{4}x - 2\right) = -\frac{1}{5}$

82. $\frac{2}{3}\left(\frac{7}{8} - 4x\right) - \frac{5}{8} = \frac{3}{8}$

83. $0.7(3x + 6) = 1.1 - (x + 2)$

84. $0.9(2x + 8) = 20 - (x + 5)$

85. $a + (a - 3) = (a + 2) - (a + 1)$

86. $0.8 - 4(b - 1) = 0.2 + 3(4 - b)$

TW 87. When an equation contains decimals, is it essential to clear the equation of decimals? Why or why not?

TW 88. Why must the rules for the order of operations be understood before solving the equations in this section?

SKILL REVIEW

To prepare for Section 2.3, review evaluating algebraic expressions (Section 1.8).

Evaluate. [1.8]

89. $3 - 5a$, for $a = 2$

90. $12 \div 4 \cdot t$, for $t = 5$

91. $7x - 2x$, for $x = -3$

92. $t(8 - 3t)$, for $t = -2$

SYNTHESIS

TW 93. What procedure would you follow to solve an equation like $0.23x + \frac{17}{3} = -0.8 + \frac{3}{4}x$? Could your procedure be streamlined? If so, how?

TW 94. Joseph is determined to solve the equation $3x + 4 = -11$ by first using the multiplication principle to "eliminate" the 3. How should he proceed and why?

Solve. Label any contradictions or identities.

95. $8.43x - 2.5(3.2 - 0.7x) = -3.455x + 9.04$

96. $0.008 + 9.62x - 42.8 = 0.944x + 0.0083 - x$

97. $-2[3(x - 2) + 4] = 4(5 - x) - 2x$

98. $0 = y - (-14) - (-3y)$

99. $2x(x + 5) - 3(x^2 + 2x - 1) = 9 - 5x - x^2$

100. $x(x - 4) = 3x(x + 1) - 2(x^2 + x - 5)$

101. $9 - 3x = 2(5 - 2x) - (1 - 5x)$

Aha! 102. $[7 - 2(8 \div (-2))]x = 0$

103. $\frac{x}{14} - \frac{5x + 2}{49} = \frac{3x - 4}{7}$

104. $\frac{5x + 3}{4} + \frac{25}{12} = \frac{5 + 2x}{3}$

105. $2\{9 - 3[-2x - 4]\} = 12x + 42$

106. $-9t + 2 = 2 - 9t - 5(8 \div 4(1 + 3^4))$

107. $3|x| - 2 = 10$

Try Exercise Answers: Section 2.2

7. 8 **17.** -4 **21.** -2.8 **27.** 15 **45.** No solution; contradiction **57.** 2 **61.** All real numbers; identity **65.** $\frac{16}{3}$ **71.** -4

Collaborative Corner

Step-by-Step Solutions

Focus: Solving linear equations
Time: 20 minutes
Group size: 3

In general, there is more than one correct sequence of steps for solving an equation. This makes it important that you write your steps clearly and logically so that others can follow your approach.

ACTIVITY

1. Each group member should select a different one of the following equations and, on a fresh sheet of paper, perform the first step of the solution.

$$4 - 3(x - 3) = 7x + 6(2 - x)$$
$$5 - 7[x - 2(x - 6)] = 3x + 4(2x - 7) + 9$$
$$4x - 7[2 + 3(x - 5) + x] = 4 - 9(-3x - 19)$$

2. Pass the papers around so that the second and third steps of each solution are performed by the other two group members. Before writing, make sure that the previous step is correct. If a mistake is discovered, return the problem to the person who made the mistake for repairs. Continue passing the problems around until all equations have been solved.
3. Each group should reach a consensus on what the three solutions are and then compare their answers to those of other groups.

2.3 Formulas

- Evaluating Formulas
- Solving for a Variable

An equation that shows a relationship between quantities will use letters and is known as a **formula.** Most of the letters in this book are variables, but some are constants. For example, c in $E = mc^2$ represents the speed of light.

EVALUATING FORMULAS

EXAMPLE 1 Event Promotion. Event promoters use the formula

$$p = \frac{1.2x}{s}$$

to determine a ticket price p for an event with x dollars of expenses and s anticipated ticket sales. Grand Events expects expenses for an upcoming concert to be \$80,000 and anticipates selling 4000 tickets. What should the ticket price be?
Source: *The Indianapolis Star*, 2/27/03

SOLUTION We substitute 80,000 for x and 4000 for s in the formula and calculate p:

$$p = \frac{1.2x}{s} = \frac{1.2(80{,}000)}{4000} = 24.$$

The ticket price should be \$24.

Try Exercise 1.

SOLVING FOR A VARIABLE

In the Northeast, the formula $B = 30a$ is used to determine the minimum furnace output B, in British thermal units (Btu's), for a well-insulated home with a square feet of flooring. Suppose that a contractor has an extra furnace and wants to determine the size of the largest (well-insulated) house in which it can be used. The contractor can substitute the amount of the furnace's output in Btu's—say, 63,000—for B, and then solve for a:

$$63{,}000 = 30a \quad \textbf{Replacing } B \textbf{ with 63,000}$$
$$2100 = a. \quad \textbf{Dividing both sides by 30}$$

The home should have no more than 2100 ft^2 of flooring.

Were these calculations to be performed for a variety of furnaces, the contractor would find it easier to first solve $B = 30a$ for a, and *then* substitute values for B. This can be done in much the same way that we solved equations in Sections 2.1 and 2.2.

EXAMPLE 2 Solve for a: $B = 30a$.

SOLUTION We have

$$B = 30a \quad \text{We want this letter alone.}$$
$$\frac{B}{30} = a. \quad \textbf{Dividing both sides by 30}$$

The equation $a = \frac{B}{30}$ gives a quick, easy way to determine the floor area of the largest (well-insulated) house that a furnace supplying B Btu's could heat.

Try Exercise 9.

To see how solving a formula is just like solving an equation, compare the following. In (A), we solve as usual; in (B), we show steps but do not simplify; and in (C), we *cannot* simplify since a, b, and c are unknown.

A. $5x + 2 = 12$
$$5x = 12 - 2$$
$$5x = 10$$
$$x = \frac{10}{5} = 2$$

B. $5x + 2 = 12$
$$5x = 12 - 2$$
$$x = \frac{12 - 2}{5}$$

C. $ax + b = c$
$$ax = c - b$$
$$x = \frac{c - b}{a}$$

EXAMPLE 3 Circumference of a Circle. The formula $C = 2\pi r$ gives the *circumference* C of a circle with radius r. Solve for r.

SOLUTION The **circumference** is the distance around a circle.

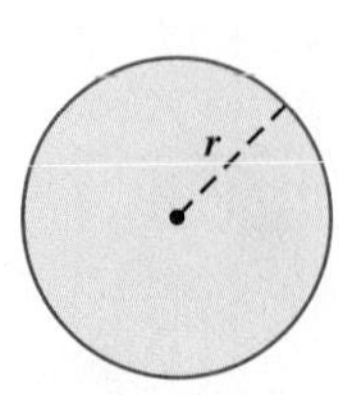

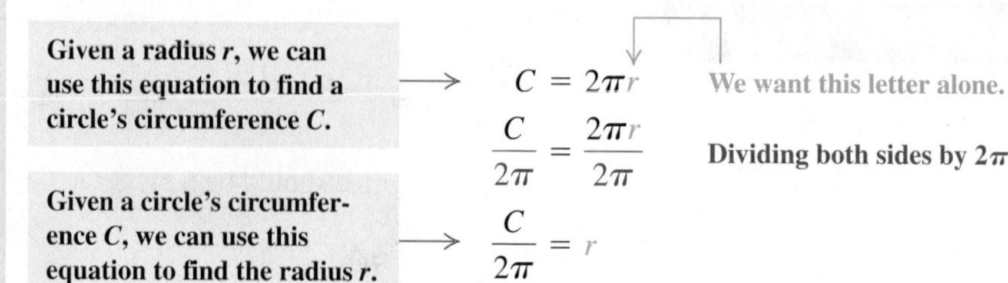

$$C = 2\pi r$$
$$\frac{C}{2\pi} = \frac{2\pi r}{2\pi}$$
$$\frac{C}{2\pi} = r$$

Try Exercise 11.

EXAMPLE 4 Motion. The rate r at which an object moves is found by dividing distance d traveled by time t, or

$$r = \frac{d}{t}.$$

Solve for t.

STUDY TIP

You Are What You Eat

Studies have shown that good nutrition is important for good learning. Try to keep healthful food and snacks readily available so you won't be tempted by food with little nutritional value.

SOLUTION We use the multiplication principle to clear fractions and then solve for t:

$$r = \frac{d}{t}$$ We want this variable alone.

$$r \cdot t = \frac{d}{t} \cdot t$$ **Multiplying both sides by t**

$$rt = \frac{dt}{t}$$ $\frac{d}{t} \cdot t = \frac{d}{t} \cdot \frac{t}{1} = \frac{dt}{t}$

$$rt = d$$ **Removing a factor equal to 1: $\frac{t}{t} = 1$. The equation is cleared of fractions.**

$$\frac{rt}{r} = \frac{d}{r}$$ **Dividing both sides by r**

$$t = \frac{d}{r}.$$

This formula can be used to find the time spent traveling when the distance and the rate are known.

Try Exercise 27.

EXAMPLE 5 Solve for y: $3x - 4y = 10$.

SOLUTION There is one term that contains y, so we begin by isolating that term on one side of the equation.

$3x - 4y = 10$	We want this variable alone.
$-4y = 10 - 3x$	**Subtracting $3x$ from both sides**
$-\frac{1}{4}(-4y) = -\frac{1}{4}(10 - 3x)$	**Multiplying both sides by $-\frac{1}{4}$**
$y = -\frac{10}{4} + \frac{3}{4}x$	**Multiplying using the distributive law**
$y = -\frac{5}{2} + \frac{3}{4}x$	**Simplifying the fraction**

Try Exercise 33.

The steps above are similar to those used in Section 2.2 to solve equations. We use the addition and multiplication principles just as before. An important difference that we will see in the next example is that we will sometimes need to factor.

To Solve a Formula for a Given Variable

1. If the variable for which you are solving appears in a fraction, use the multiplication principle to clear fractions.
2. Isolate the term(s) with the variable for which you are solving on one side of the equation.
3. If two or more terms contain the variable for which you are solving, factor the variable out.
4. Multiply or divide to solve for the variable in question.

We can also solve for a letter that represents a constant.

EXAMPLE 6 Surface Area of a Right Circular Cylinder. The formula

$$A = 2\pi rh + 2\pi r^2$$

gives the surface area A of a right circular cylinder of height h and radius r. Solve for π.

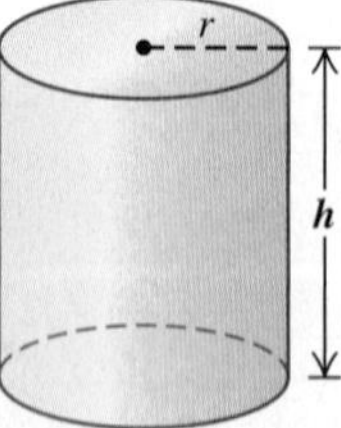

SOLUTION We have

$$A = 2\pi rh + 2\pi r^2 \quad \text{We want this letter alone.}$$

$$A = \pi(2rh + 2r^2) \quad \text{Factoring}$$

$$\frac{A}{2rh + 2r^2} = \pi. \quad \text{Dividing both sides by } 2rh + 2r^2\text{, or multiplying both sides by } 1/(2rh + 2r^2)$$

We can also write this as

$$\pi = \frac{A}{2rh + 2r^2}.$$

Try Exercise 43.

CAUTION! Had we performed the following steps in Example 6, we would *not* have solved for π:

$$A = 2\pi rh + 2\pi r^2 \quad \text{We want } \pi \text{ alone.}$$

$$A - 2\pi r^2 = 2\pi rh \quad \text{Subtracting } 2\pi r^2 \text{ from both sides}$$

Two occurrences of π

$$\frac{A - 2\pi r^2}{2rh} = \pi. \quad \text{Dividing both sides by } 2rh$$

The mathematics of each step is correct, but because π occurs on both sides of the formula, *we have not solved the formula for π*. Remember that the letter for which we are solving should be alone on one side of the equation, with no occurrence of that letter on the other side!

Tables and Entering Equations

A formula can be evaluated for several values of a variable using the TABLE feature of a graphing calculator. In order to use the table, the formula must be entered using the Y= editor.

Y= Editor To enter equations in a graphing calculator, first press MODE and make sure that the FUNC mode is selected. When the calculator is set in FUNCTION mode, pressing Y= accesses the Y= editor, as shown on the following page.

Equations containing two variables can be entered into the calculator using this editor. First solve for a particular variable. Then replace this variable with y and the other variable with x. To enter the equation, position the cursor after an = sign and enter the rest of the formula. Since all equations are written in terms of y, they are distinguished by the subscripts 1, 2, and so on. The notation Y1 is read "y sub one."

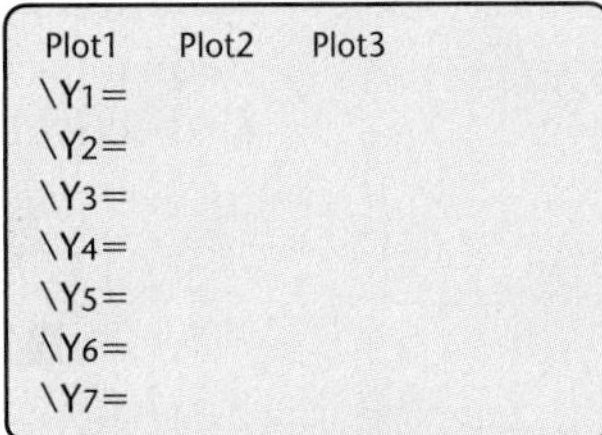

Table After we enter an equation on the Y= editor screen, we can view a table of solutions. A table is set up by pressing (TBLSET) (the 2nd option associated with (WINDOW)). Since the value of y *depends on* the choice of the value for x, we say that y is the **dependent** variable and x is the **independent** variable.

If we want to choose the values for the independent variable, we set Indpnt to Ask, as shown on the left below.

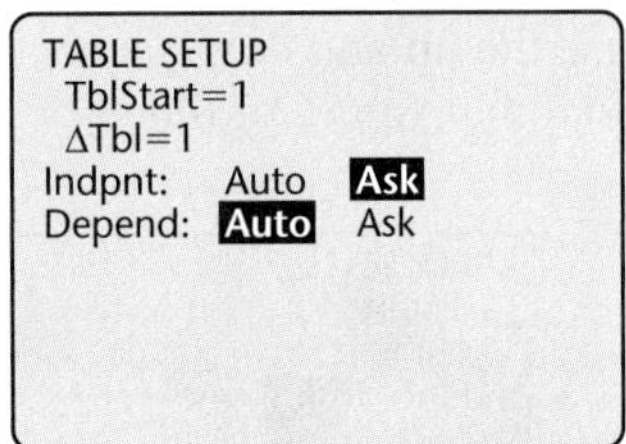

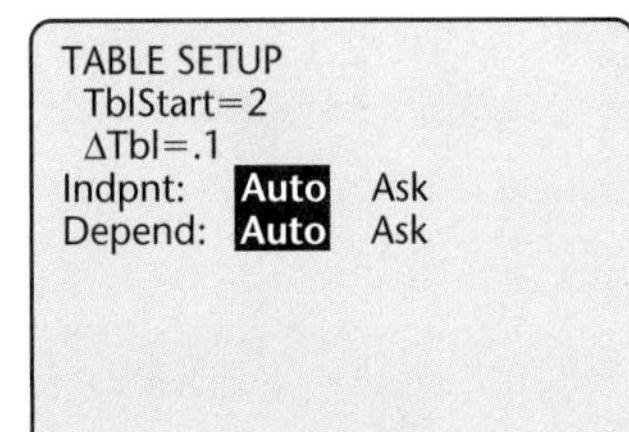

If Indpnt is set to Auto, the calculator will provide values for x, beginning with the value specified as TblStart. The symbol Δ (the Greek letter delta) often indicates a step or a change. Here, the value of ΔTbl is added to the preceding value of x. The figure on the right above shows a beginning value of 2 and an increment, or ΔTbl, of 0.1.

To view the values in a table, press (TABLE) (the 2nd option associated with (GRAPH)). The up and down arrow keys allow us to scroll up and down the table.

Your Turn

1. Make sure your calculator is set in FUNCTION mode.
2. Press (Y=) and clear any equations present by highlighting each equation and pressing (CLEAR).
3. Enter the equation $y = x + 1$.
4. Set up a table with Indpnt set to Ask.
5. Press (TABLE) and enter several values for x. Each corresponding y-value should be 1 more than the x-value.

EXAMPLE 7 Sound. The formula $d = 344t$ can be used to determine how far d, in meters, sound travels through room temperature air in t seconds. At outdoor concerts, fans far from the stage experience a time lag between the time a word is pronounced and the time the sound reaches their ears. Sound and video engineers often adjust speakers or monitors to compensate for this time lag.

a) Use a table to determine how far fans are from a stage when the time lag is 0.25 sec and 1 sec.

b) Create a table of values showing distance from the stage for time lags of 0 sec, 0.1 sec, 0.2 sec, 0.3 sec, and so on.

SOLUTION

a) We replace d with y and t with x and enter the formula as $y = 344x$, as shown on the left below. To enter specific values for x, we set up a table with Indpnt set to Ask. We then view the table, entering 0.25 and 1 for x. As shown on the right below, the distances from the stage are 86 m and 344 m, respectively.

```
Plot1   Plot2   Plot3
\Y1 = 344X
\Y2=
\Y3=
\Y4=
\Y5=
\Y6=
\Y7=
```

X	Y1	
.25	86	
1	344	

X =

b) To create this table of values, we first press TBLSET and set Indpnt to Auto. Then TblStart is 0, to indicate the start of the x-values, and ΔTbl = 0.1, to indicate the increment used for x. To view the table, we press TABLE. We can use the up and down arrow keys to scroll through the table to find distances for other values of time.

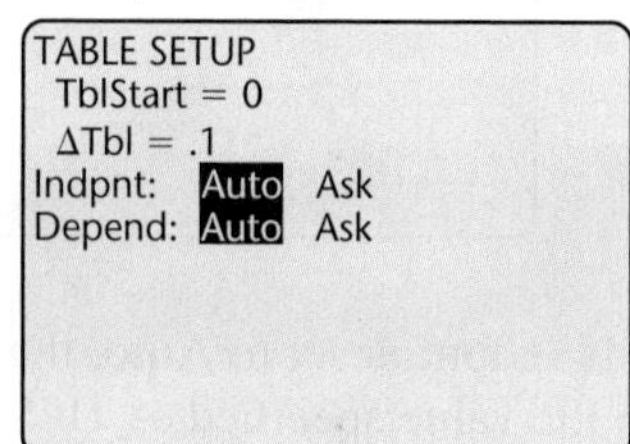

X	Y1	
0	0	
.1	34.4	
.2	68.8	
.3	103.2	
.4	137.6	
.5	172	
.6	206.4	

Y1 = 206.4

Try Exercise 7.

2.3 Exercise Set

FOR EXTRA HELP

1. *College Enrollment.* At many colleges, the number of "full-time-equivalent" students f is given by

$$f = \frac{n}{15},$$

where n is the total number of credits for which students have enrolled in a given semester. Determine the number of full-time-equivalent students on a campus in which students registered for a total of 21,345 credits.

2. *Distance from a Storm.* The formula $M = \frac{1}{5}t$ can be used to determine how far M, in miles, you are from lightning when its thunder takes t seconds to reach your ears. If it takes 10 sec for the sound of thunder to reach you after you have seen the lightning, how far away is the storm?

3. *Electrical Power.* The power rating P, in watts, of an electrical appliance is determined by

$$P = I \cdot V,$$

where I is the current, in amperes, and V is the voltage, measured in volts. If the appliances in a kitchen require 30 amps of current and the voltage in the house is 115 volts, what is the wattage of the kitchen?

4. *Wavelength of a Musical Note.* The wavelength w, in meters per cycle, of a musical note is given by

$$w = \frac{r}{f},$$

where r is the speed of the sound, in meters per second, and f is the frequency, in cycles per second. The speed of sound in air is 344 m/sec. What is the wavelength of a note whose frequency in air is 24 cycles per second?

5. *Federal Funds Rate.* The Federal Reserve Board sets a target f for the federal funds rate, that is, the interest rate that banks charge each other for overnight borrowing of Federal funds. This target rate can be estimated by

$$f = 8.5 + 1.4(I - U),$$

where I is the core inflation rate over the preceding 12 months and U is the seasonally adjusted unemployment rate. If core inflation is 0.025 and unemployment is 0.044, what should the federal funds rate be?
Source: Greg Mankiw, Harvard University, www.gregmankiw.blogspot.com/2006/06/what-would-alan-do.html

6. *Calorie Density.* The calorie density D, in calories per ounce, of a food that contains c calories and weighs w ounces is given by

$$D = \frac{c}{w}.$$

Eight ounces of fat-free milk contains 84 calories. Find the calorie density of fat-free milk.
Source: *Nutrition Action Healthletter*, March 2000, p. 9. Center for Science in the Public Interest, Suite 300; 1875 Connecticut Ave NW, Washington, D.C. 20008.

7. *Absorption of Ibuprofen.* When 400 mg of the painkiller ibuprofen is swallowed, the number of milligrams n in the bloodstream t hours later (for $0 \le t \le 6$) is estimated by

$$n = 0.5t^4 + 3.45t^3 - 96.65t^2 + 347.7t.$$

How many milligrams of ibuprofen remain in the blood 1 hr after 400 mg has been swallowed?

8. *Size of a League Schedule.* When all n teams in a league play every other team twice, a total of N games are played, where

$$N = n^2 - n.$$

If a soccer league has 7 teams and all teams play each other twice, how many games are played?

Solve each formula for the indicated letter.

9. $A = bh$, for b
(Area of parallelogram with base b and height h)

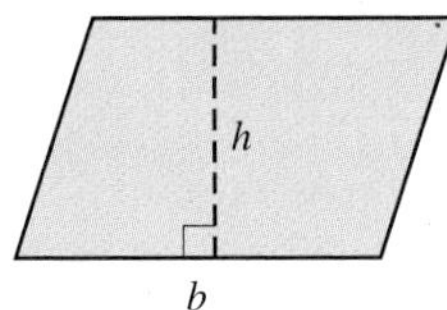

10. $A = bh$, for h

11. $I = Prt$, for P
(Simple-interest formula, where I is interest, P is principal, r is interest rate, and t is time)

12. $I = Prt$, for t

13. $H = 65 - m$, for m
(To determine the number of heating degree days H for a day with m degrees Fahrenheit as the average temperature)

14. $d = h - 64$, for h
(To determine how many inches d above average an h-inch-tall woman is)

15. $P = 2l + 2w$, for l
(Perimeter of a rectangle of length l and width w)

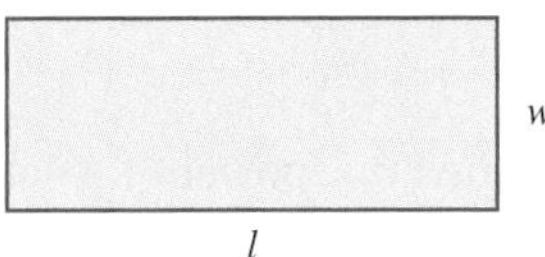

16. $P = 2l + 2w$, for w

17. $A = \pi r^2$, for π
(Area of a circle with radius r)

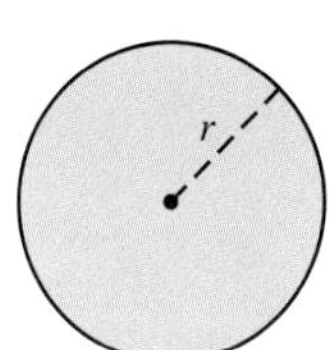

18. $A = \pi r^2$, for r^2

19. $A = \frac{1}{2}bh$, for h
(Area of a triangle with base b and height h)

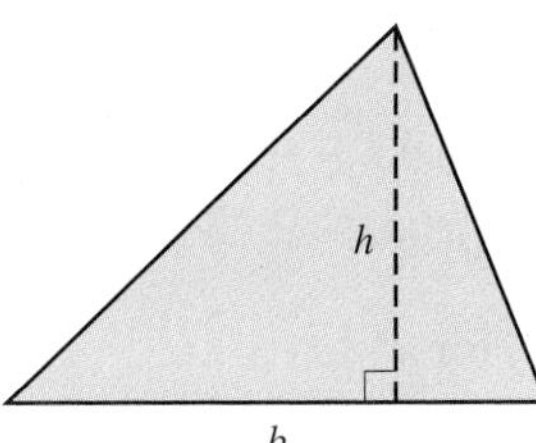

20. $A = \frac{1}{2}bh$, for b

21. $E = mc^2$, for m
(A relativity formula from physics)

22. $E = mc^2$, for c^2

23. $Q = \dfrac{c + d}{2}$, for d

24. $A = \dfrac{a + b + c}{3}$, for b

25. $p - q + r = 2$, for q
(Euler's formula from graph theory)

26. $p = \dfrac{r - q}{2}$, for q

27. $w = \dfrac{r}{f}$, for r
(To compute the wavelength w of a musical note with frequency f and speed of sound r)

28. $M = \dfrac{A}{s}$, for A
(To compute the Mach number M for speed A and speed of sound s)

29. $H = \dfrac{TV}{550}$, for T
(To determine the horsepower of an airplane propeller)

30. $P = \dfrac{ab}{c}$, for b

31. $F = \frac{9}{5}C + 32$, for C
(To convert the Celsius temperature C to the Fahrenheit temperature F)

32. $M = \frac{3}{7}n + 29$, for n

33. $2x - y = 1$, for y

34. $3x - y = 7$, for y

35. $2x + 5y = 10$, for y

36. $3x + 2y = 12$, for y

37. $4x - 3y = 6$, for y

38. $5x - 4y = 8$, for y

39. $9x + 8y = 4$, for y

40. $x + 10y = 2$, for y

41. $3x - 5y = 8$, for y

42. $7x - 6y = 7$, for y

43. $A = at + bt$, for t

44. $S = rx + sx$, for x

45. *Area of a Trapezoid.* The formula

$$A = \tfrac{1}{2}ah + \tfrac{1}{2}bh$$

can be used to find the area A of a trapezoid with bases a and b and height h. Solve for h. (*Hint*: First clear fractions.)

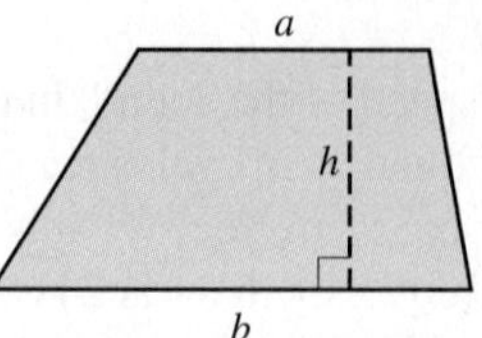

46. *Compounding Interest.* The formula

$$A = P + Prt$$

is used to find the amount A in an account when simple interest is added to an investment of P dollars. (See Exercise 11.) Solve for P.

47. $z = 13 + 2(x + y)$, for x

48. $A = 115 + \frac{1}{2}(p + s)$, for s

49. $t = 27 - \frac{1}{4}(w - l)$, for l

50. $m = 19 - 5(x - n)$, for n

51. *Chess Rating.* The formula

$$R = r + \frac{400(W - L)}{N}$$

is used to establish a chess player's rating R after that player has played N games, won W of them, and lost L of them. Here r is the average rating of the opponents. Solve for L.
Source: The U.S. Chess Federation

52. *Angle Measure.* The angle measure S of a sector of a circle is given by

$$S = \frac{360A}{\pi r^2},$$

where r is the radius, A is the area of the sector, and S is in degrees. Solve for r^2.

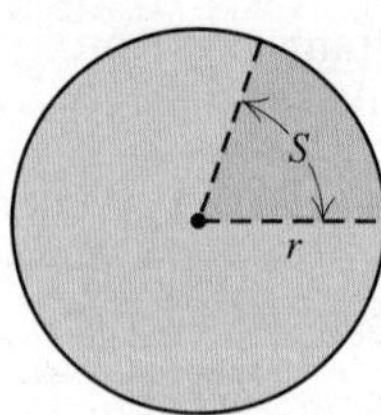

TW **53.** Audra has a formula that allows her to convert Celsius temperatures to Fahrenheit temperatures. She needs a formula for converting Fahrenheit temperatures to Celsius temperatures. What advice can you give her?

TW **54.** Under what circumstances would it be useful to solve $I = Prt$ for P? (See Exercise 11.)

SKILL REVIEW

Review simplifying expressions (Sections 1.6, 1.7, and 1.8).

Perform the indicated operations and simplify.

55. $-2 + 5 - (-4) - 17$ [1.6]

56. $-98 \div \frac{1}{2}$ [1.7]

Aha! **57.** $4.2(-11.75)(0)$ [1.7]

58. $(-2)^5$ [1.8]

59. $20 \div (-4) \cdot 2 - 3$ [1.8]

60. $5|8 - (2 - 7)|$ [1.8]

SYNTHESIS

TW **61.** The equations

$$P = 2l + 2w \quad \text{and} \quad w = \frac{P}{2} - l$$

are equivalent formulas involving the perimeter P, length l, and width w of a rectangle. Devise a problem for which the second of the two formulas would be more useful.

TW **62.** While solving $2A = ah + bh$ for h, Dee writes $\frac{2A - ah}{b} = h$. What is her mistake?

63. The Harris–Benedict formula gives the number of calories K needed each day by a moderately active man who weighs w kilograms, is h centimeters tall, and is a years old as

$$K = 21.235w + 7.75h - 10.54a + 102.3.$$

If Janos is moderately active, weighs 80 kg, is 190 cm tall, and needs to consume 2852 calories per day, how old is he?

64. *Altitude and Temperature.* Air temperature drops about 1° Celsius (C) for each 100-m rise above ground level, up to 12 km. If the ground level temperature is t°C, find a formula for the temperature T at an elevation of h meters.
Source: *A Sourcebook of School Mathematics*, Mathematical Association of America, 1980

65. *Surface Area of a Cube.* The surface area A of a cube with side s is given by

$$A = 6s^2.$$

If a cube's surface area is 54 in^2, find the volume of the cube.

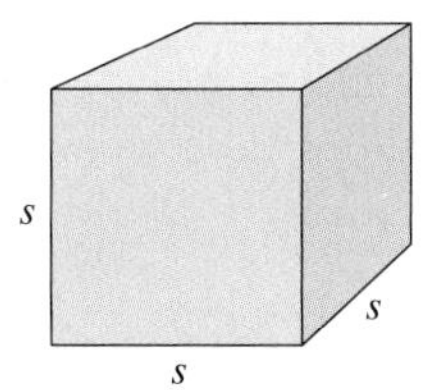

66. *Dosage Size.* Clark's rule for determining the size of a particular child's medicine dosage c is

$$c = \frac{w}{a} \cdot d,$$

where w is the child's weight, in pounds, and d is the usual adult dosage for an adult weighing a pounds. Solve for a.
Source: Olsen, June Looby, et al., *Medical Dosage Calculations.* Redwood City, CA: Addison-Wesley, 1995

Solve each formula for the given letter.

67. $\frac{y}{z} \div \frac{z}{t} = 1$, for y

68. $ac = bc + d$, for c

69. $qt = r(s + t)$, for t

70. $3a = c - a(b + d)$, for a

71. *Furnace Output.* The formula

$$B = 50a$$

is used in New England to estimate the minimum furnace output B, in Btu's, for an old, poorly insulated house with a square feet of flooring. Find an equation for determining the number of Btu's saved by insulating an old house. (*Hint*: See Example 2.)

72. Revise the formula in Exercise 63 so that a man's weight in pounds (2.2046 lb = 1 kg) and his height in inches (0.3937 in. = 1 cm) are used.

Try Exercise Answers: Section 2.3

1. 1423 students **7.** 255 mg **9.** $b = \frac{A}{h}$ **11.** $P = \frac{I}{rt}$ **27.** $r = wf$ **33.** $y = 2x - 1$ **43.** $t = \frac{A}{a + b}$

Mid-Chapter Review

We solve equations using the addition and multiplication principles.

For any real numbers a, b, and c:

a) $a = b$ is equivalent to $a + c = b + c$;

b) $a = b$ is equivalent to $ac = bc$, provided $c \neq 0$.

GUIDED SOLUTIONS

Solve. [2.2]*

1. $2x + 3 = 10$

Solution

$2x + 3 - 3 = 10 - \square$ **Using the addition principle**

$2x = \square$ **Simplifying**

$\frac{1}{2} \cdot 2x = \square \cdot 7$ **Using the multiplication principle**

$x = \square$ **Simplifying**

2. $\frac{1}{2}(x - 3) = \frac{1}{3}(x - 4)$

Solution

$6 \cdot \frac{1}{2}(x - 3) = \square \cdot \frac{1}{3}(x - 4)$ **Multiplying to clear fractions**

$\square(x - 3) = \square(x - 4)$ **The fractions are cleared.**

$3x - \square = 2x - \square$ **Multiplying**

$3x - 9 + 9 = 2x - 8 + \square$ **Using the addition principle**

$3x = 2x + \square$ **Simplifying**

$3x - \square = 2x + 1 - 2x$ **Using the addition principle**

$x = \square$ **Simplifying**

MIXED REVIEW

Solve.

1. $x - 2 = -1$ [2.1]

2. $2 - x = -1$ [2.1]

3. $3t = 5$ [2.1]

4. $-\frac{3}{2}x = 12$ [2.1]

5. $\frac{y}{8} = 6$ [2.1]

6. $0.06x = 0.03$ [2.1]

7. $3x - 7x = 20$ [2.2]

8. $9x - 7 = 17$ [2.2]

9. $4(t - 3) - t = 6$ [2.2]

10. $3(y + 5) = 8y$ [2.2]

11. $8n - (3n - 5) = 5 - n$ [2.2]

12. $\frac{9}{10}y - \frac{7}{10} = \frac{21}{5}$ [2.2]

13. $2(t - 5) - 3(2t - 7) = 12 - 5(3t + 1)$ [2.2]

14. $\frac{2}{3}(x - 2) - 1 = -\frac{1}{2}(x - 3)$ [2.2]

Solve for the indicated variable. [2.3]

15. $E = wA$, for A

16. $V = lwh$, for w

17. $Ax + By = C$, for y

18. $at + ap = m$, for a

19. $m = \frac{F}{a}$, for a

20. $v = \frac{d_2 - d_1}{t}$, for d_1

*The notation [2.2] refers to Chapter 2, Section 2.

2.4 Applications with Percent

- Converting Between Percent Notation and Decimal Notation
- Solving Percent Problems

Percent problems arise so frequently in everyday life that often we are not even aware of them. In this section, we will solve some real-world percent problems. Before doing so, however, we need to review a few basics.

CONVERTING BETWEEN PERCENT NOTATION AND DECIMAL NOTATION

Oceans cover 70% of the earth's surface. This means that of every 100 square miles on the surface of the earth, 70 square miles is ocean. Thus, 70% is a ratio of 70 to 100.

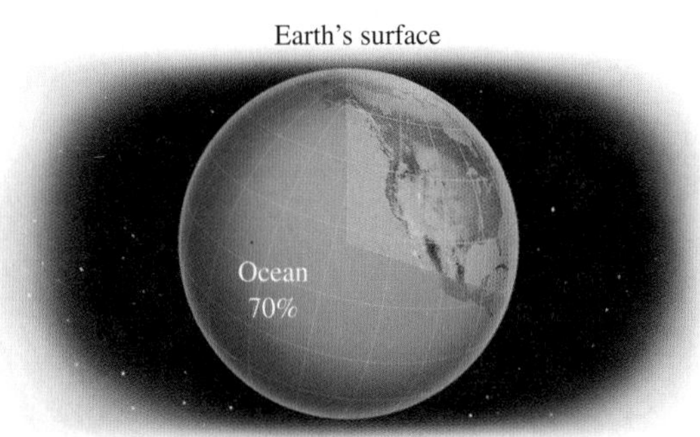

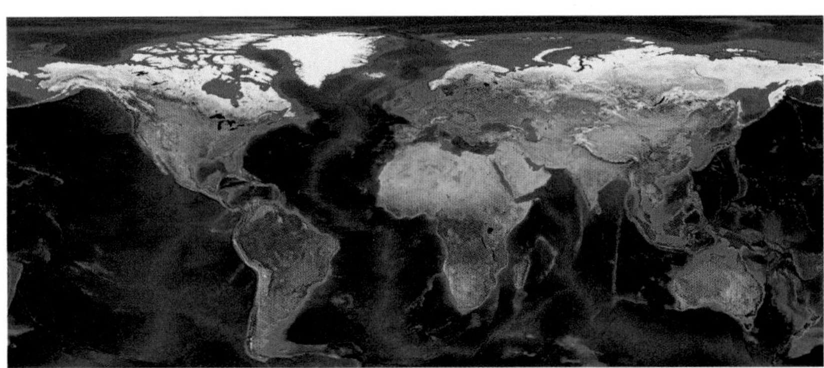

The percent symbol % means "per hundred." We can regard the percent symbol as part of a name for a number. For example,

$$70\% \quad \text{is defined to mean} \quad \frac{70}{100}, \quad \text{or} \quad 70 \times \frac{1}{100}, \quad \text{or} \quad 70 \times 0.01.$$

Percent Notation

$$n\% \quad \text{means} \quad \frac{n}{100}, \quad \text{or} \quad n \times \frac{1}{100}, \quad \text{or} \quad n \times 0.01.$$

EXAMPLE 1 Convert to decimal notation: **(a)** 78%; **(b)** 1.3%.

SOLUTION

a) $78\% = 78 \times 0.01$ **Replacing % with ×0.01**

$= 0.78$

b) $1.3\% = 1.3 \times 0.01$ **Replacing % with ×0.01**

$= 0.013$

Try Exercise 19.

As shown above, multiplication by 0.01 simply moves the decimal point two places to the left.

To convert from percent notation to decimal notation, move the decimal point two places to the left and drop the percent symbol.

STUDY TIP

Pace Yourself

Most instructors agree that it is better for a student to study for one hour four days in a week, than to study once a week for four hours. Of course, the total weekly study time will vary from student to student. It is common to expect an average of two hours of homework for each hour of class time.

EXAMPLE 2 Convert the percent notation in the following sentence to decimal notation: Plastic makes up 90% of all trash floating in the ocean.
Source: http://environment.nationalgeographic.com

SOLUTION

$$90\% = 90.0\% \qquad 0.90.0 \qquad 90\% = 0.90, \quad \text{or simply } 0.9$$

Move the decimal point two places to the left.

Try Exercise 11.

The procedure used in Examples 1 and 2 can be reversed:

$$\begin{aligned} 0.38 &= 38 \times 0.01 \\ &= 38\%. \quad \text{Replacing } \times 0.01 \text{ with } \% \end{aligned}$$

To convert from decimal notation to percent notation, move the decimal point two places to the right and write a percent symbol.

EXAMPLE 3 Convert to percent notation: **(a)** 1.27; **(b)** $\frac{1}{4}$; **(c)** 0.3.

SOLUTION

a) We first move the decimal point two places to the right: 1.27.
and then write a % symbol: 127% **This is the same as multiplying 1.27 by 100 and writing %.**

b) Note that $\frac{1}{4} = 0.25$. We move the decimal point two places to the right: 0.25.
and then write a % symbol: 25% **Multiplying by 100 and writing %**

c) We first move the decimal point two places to the right (recall that $0.3 = 0.30$): 0.30.
and then write a % symbol: 30% **Multiplying by 100 and writing %**

Try Exercise 33.

SOLVING PERCENT PROBLEMS

In solving percent problems, we first *translate* the problem to an equation. Then we *solve* the equation using the techniques discussed in Sections 2.1–2.3. The key words in the translation are as follows.

Key Words in Percent Translations

"**Of**" translates to " · " or " × ". "**Is**" or "**Was**" translates to " = ".
"**What**" translates to a variable. "**%**" translates to "$\times\frac{1}{100}$" or "×0.01".

Student Notes

A way of checking answers is by estimating as follows:

$$\begin{aligned} 11\% \times 49 &\approx 10\% \times 50 \\ &= 0.10 \times 50 = 5. \end{aligned}$$

Since 5 is close to 5.39, our answer is reasonable.

EXAMPLE 4 What is 11% of 49?

SOLUTION

Translate: What is 11% of 49?

$$a = 0.11 \cdot 49 \qquad \text{"of" means multiply; } 11\% = 0.11$$

$$a = 5.39$$

Thus, 5.39 is 11% of 49. The answer is 5.39.

Try Exercise 51.

EXAMPLE 5 3 is 16 percent of what?

SOLUTION

Translate: 3 is 16 percent of what?

$$3 = 0.16 \cdot y$$

$$\frac{3}{0.16} = y \qquad \textbf{Dividing both sides by 0.16}$$

$$18.75 = y$$

Thus, 3 is 16 percent of 18.75. The answer is 18.75.

Try Exercise 47.

EXAMPLE 6 What percent of \$50 is \$34?

SOLUTION

Translate: What percent of \$50 is \$34?

$$n \cdot 50 = 34$$

$$n = \frac{34}{50} \qquad \textbf{Dividing both sides by 50}$$

$$n = 0.68 = 68\% \qquad \textbf{Converting to percent notation}$$

Thus, \$34 is 68% of \$50. The answer is 68%.

Try Exercise 43.

Examples 4–6 represent the three basic types of percent problems. Note that in all the problems, the following quantities are present:

- a percent, expressed in decimal notation in the translation;
- a base amount, indicated by "of" in the problem; and
- a percentage of the base, found by multiplying the base times the percent.

EXAMPLE 7 Alzheimer's Disease. In 2009, there were 307 million people in the United States. About 1.7% of them had Alzheimer's disease. How many people had Alzheimer's?
Source: Alzheimer's Association, *2009 Facts and Figures*

Student Notes

Always look for connections between examples. Here you should look for similarities between Examples 4 and 7 as well as between Examples 5 and 8 and between Examples 6 and 9.

SOLUTION To solve the problem, we first reword and then translate. We let a = the number of people in the United States with Alzheimer's, in millions.

Rewording: What is 1.7% of 307?

Translating: $a = 0.017 \times 307$

The letter is by itself. To solve the equation, we need only multiply:

$$a = 0.017 \times 307 = 5.219.$$

Thus, 5.219 million is 1.7% of 307 million, so in 2009, about 5.219 million people in the United States had Alzheimer's disease.

Try Exercise 65.

EXAMPLE 8 College Enrollment. About 2.2 million students who graduated from high school in 2008 were attending college in the fall of 2008. This was 68.6% of all 2008 high school graduates. How many students graduated from high school in 2008?
Source: U.S. Bureau of Labor Statistics

SOLUTION Before translating the problem to mathematics, we reword and let S represent the total number of students, in millions, who graduated from high school in 2008.

Rewording: 2.2 is 68.6% of S.

$$\downarrow \quad \downarrow \quad \downarrow \quad \downarrow \quad \downarrow$$

Translating: $2.2 = 0.686 \cdot S$

$$\frac{2.2}{0.686} = S \qquad \textbf{Dividing both sides by 0.686}$$

$$3.2 \approx S \qquad \textbf{The symbol } \approx \textbf{ means } \textit{is approximately equal to.}$$

About 3.2 million students graduated from high school in 2008.

Try Exercise 67.

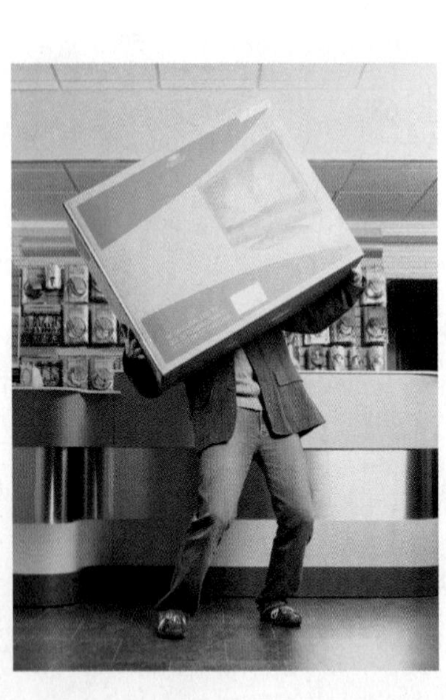

EXAMPLE 9 Television Prices. Recently, Giant Electronics reduced the price of a 47-in. LG1 LCD HDTV from \$1500 to \$1200.

a) What percent of the original price does the sale price represent?

b) What is the percent of discount?

SOLUTION

a) We reword and translate, using n for the unknown percent.

Rewording: What percent of 1500 is 1200?

$$\downarrow \quad \downarrow \quad \downarrow \quad \downarrow \quad \downarrow$$

Translating: $n \cdot 1500 = 1200$

$$n = \frac{1200}{1500} \qquad \textbf{Dividing both sides by 1500}$$

$$n = 0.8 = 80\% \qquad \textbf{Converting to percent notation}$$

The sale price is 80% of the original price.

b) Since \$1500 represents 100% of the original price, the sale price represents a discount of $(100 - 80)\%$, or 20%.

Try Exercise 69.

2.4 Exercise Set

FOR EXTRA HELP MyMathLab    

Concept Reinforcement *In each of Exercises 1–10, match the question with the most appropriate translation from the column on the right. Some choices are used more than once.*

1. ___ What percent of 57 is 23?	**a)** $a = (0.57)23$
2. ___ What percent of 23 is 57?	**b)** $57 = 0.23y$
3. ___ 23 is 57% of what number?	**c)** $n \cdot 23 = 57$
4. ___ 57 is 23% of what number?	**d)** $n \cdot 57 = 23$
5. ___ 57 is what percent of 23?	**e)** $23 = 0.57y$
6. ___ 23 is what percent of 57?	**f)** $a = (0.23)57$
7. ___ What is 23% of 57?	
8. ___ What is 57% of 23?	
9. ___ 23% of what number is 57?	
10. ___ 57% of what number is 23?	

Convert the percent notation in each sentence to decimal notation.

11. *Musical Instruments.* Of those who play Guitar Hero and Rock Band but do not currently play an actual musical instrument, 67% indicated that they are likely to begin playing an actual instrument.
Source: Guitar Center Survey, gonintendo.com

12. *Volunteering.* Of all Americans, 55% do volunteer work.
Source: *The Nonprofit Almanac in Brief*

13. *Dehydration.* A 2% drop in water content of the human body can affect one's ability to study mathematics.
Source: Gopinathan, P. M., G. Pichan, and V. M. Sharma, "Role of Dehydration in Heat Stress-Induced Variations in Mental Performance," *Archives of Environmental Health*, 1988, Jan–Feb, **43**(1): 15-7

14. *Left-Handed Golfers.* Of those who golf, 7% are left-handed.
Source: National Association of Left-Handed Golfers

15. *Plant Species.* Trees make up about 3.5% of all plant species found in the United States.
Source: South Dakota Project Learning Tree

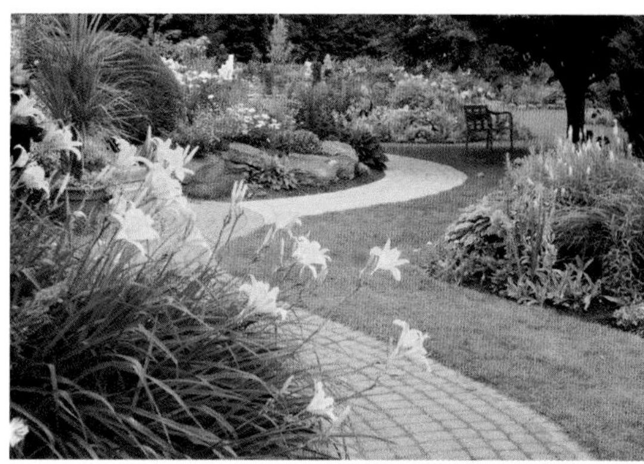

16. *Gold.* Gold that is marked 10K is 41.6% gold.

17. *Women-Owned Businesses.* Women own 40% of all privately held firms.
Source: Center for Women's Business Research

18. *Women-Owned Businesses.* Women own 20% of all firms with revenues of $1 million or more.
Source: Center for Women's Business Research

Convert to decimal notation.

19. 62.58%	**20.** 39.81%
21. 0.7%	**22.** 0.3%
23. 125%	**24.** 150%

Convert the decimal notation in each sentence to percent notation.

25. *NASCAR Fans.* Of those who are fans of NASCAR racing, 0.13 are between the ages of 18 and 24.
Source: Scarborough Research cited in *Street & Smith's Sports Business Daily*, 2/16/09

26. *The Arts.* In 2008, 0.35 of U.S. adults attended an art museum or an arts performance.
Source: National Endowment for the Arts

27. *Foreign Student Enrollment.* Of all foreign students studying in the United States, 0.014 are from Vietnam.
Source: voanews.com

28. *Salmon Population.* In 2008, sea lions ate 0.029 of the salmon passing Bonneville Dam on the Columbia River.
Source: news.yahoo.com

29. *Women in the Workforce.* Women comprise 0.326 of all lawyers.
Source: U.S. Census Bureau

30. *Women in the Workforce.* Women comprise 0.247 of all architects.
Source: U.S. Census Bureau

31. *Water in Watermelon.* Watermelon is 0.9 water.

32. *Jupiter's Atmosphere.* The atmosphere of Jupiter is 0.1 helium.

Convert to percent notation.

33. 0.0049

34. 0.0008

35. 1.08

36. 1.05

37. 2.3

38. 2.9

39. $\frac{4}{5}$

40. $\frac{3}{4}$

41. $\frac{8}{25}$

42. $\frac{3}{8}$

Solve.

43. What percent of 68 is 17?

44. What percent of 150 is 39?

45. What percent of 125 is 30?

46. What percent of 300 is 57?

47. 14 is 30% of what number?

48. 54 is 24% of what number?

49. 0.3 is 12% of what number?

50. 7 is 175% of what number?

51. What number is 35% of 240?

52. What number is 1% of one million?

53. What percent of 60 is 75?

Aha! **54.** What percent of 70 is 70?

55. What is 2% of 40?

56. What is 40% of 2?

Aha! **57.** 25 is what percent of 50?

58. 8 is 2% of what number?

59. What percent of 69 is 23?

60. What percent of 40 is 9?

In the 2007–2008 National Pet Owners Survey, dog owners reported annual expenses of $1425 for their pet. The circle graph below shows how those expenses were distributed. In each of Exercises 61–64, determine the annual cost of the item.

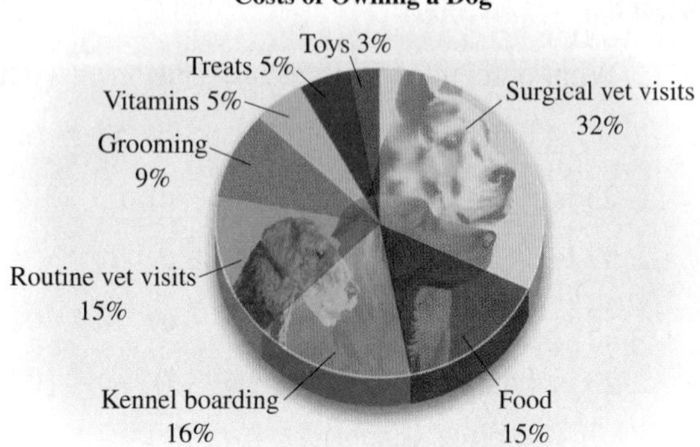

Source: The American Pet Products Manufacturers Association

61. Surgical vet visits

62. Food

63. Treats

64. Toys

65. *College Graduation.* To obtain his bachelor's degree in nursing, Clayton must complete 125 credit hours of instruction. If he has completed 60% of his requirement, how many credits has Clayton completed?

66. *College Graduation.* To obtain her bachelor's degree in journalism, Lydia must complete 125 credit hours of instruction. If 20% of Lydia's credit hours remain to be completed, how many credits does she still need to take?

67. *Batting Average.* At one point in a recent season, Ichiro Suzuki of the Seattle Mariners had 213 hits. His batting average was 0.355, or 35.5%. That is, of the total number of at-bats, 35.5% were hits. How many at-bats did he have?
Source: Major League Baseball

68. *Pass Completions.* In a recent season, Peyton Manning of the Indianapolis Colts completed 371 passes. This was 66.8% of his attempts. How many attempts did he make?
Source: National Football League

69. *Tipping.* Shane left a $4 tip for a meal that cost $25.

a) What percent of the cost of the meal was the tip?

b) What was the total cost of the meal including the tip?

70. *Tipping.* Selena left a $12.76 tip for a meal that cost $58.

a) What percent of the cost of the meal was the tip?
b) What was the total cost of the meal including the tip?

71. *Crude Oil Imports.* In August 2009, crude oil imports to the United States averaged 8.9 million barrels per day. Of this total, 3.1 million came from Canada and Mexico. What percent of crude oil imports came from Canada and Mexico? What percent came from the rest of the world?
Source: Energy Information Administration

72. *Voting.* Approximately 131.3 million Americans voted in the 2008 presidential election. In that election, Barack Obama received 69.5 million votes. What percent of the votes cast did Obama receive? What percent of the votes were cast for other candidates?
Source: Federal Election Commission

73. *Student Loans.* To finance her community college education, Irena takes out a Stafford loan for $3500. After a year, Irena decides to pay off the interest, which is 6% of $3500. How much will she pay?

74. *Student Loans.* Glenn takes out a subsidized federal Stafford loan for $2400. After a year, Glenn decides to pay off the interest, which is 5% of $2400. How much will he pay?

75. *Infant Health.* In a study of 300 pregnant women with "good-to-excellent" diets, 95% had babies in good or excellent health. How many women in this group had babies in good or excellent health?

76. *Infant Health.* In a study of 300 pregnant women with "poor" diets, 8% had babies in good or excellent health. How many women in this group had babies in good or excellent health?

77. *Cost of Self-Employment.* Because of additional taxes and fewer benefits, it has been estimated that a self-employed person must earn 20% more than a non–self-employed person performing the same task(s). If Sara earns $16 per hour working for Village Copy, how much would she need to earn on her own for a comparable income?

78. Refer to Exercise 77. Adam earns $18 per hour working for Round Edge stairbuilders. How much would Adam need to earn on his own for a comparable income?

79. *Social Networking.* The number of minutes that users spent on Facebook grew from 1.7 million in April 2008 to 13.9 million in April 2009. Calculate the percentage by which the number increased.
Source: Nielsen NetView

80. *Social Networking.* The number of minutes that users spent on Myspace.com fell from 7.3 million in April 2008 to 5.0 million in April 2009. Calculate the percentage by which the number decreased.
Source: Nielsen NetView

81. A bill at Officeland totaled $37.80. How much did the merchandise cost if the sales tax is 5%?

82. Isabella's checkbook shows that she wrote a check for $987 for building materials, including the tax. What was the price of the materials if the sales tax is 5%?

83. *Deducting Sales Tax.* A tax-exempt school group received a bill of $157.41 for educational software. The bill incorrectly included sales tax of 6%. How much should the school group pay?

84. *Deducting Sales Tax.* A tax-exempt charity received a bill of $145.90 for a sump pump. The bill incorrectly included sales tax of 5%. How much does the charity owe?

85. *Body Fat.* One author of this text exercises regularly at a local YMCA that recently offered a body-fat percentage test to its members. The author's body-fat percentage was found to be 16.5% and he weighs 191 lb. What part, in pounds, of his body weight is fat?

86. *Areas of Alaska and Arizona.* The area of Arizona is 19% of the area of Alaska. The area of Alaska is 586,400 mi^2. What is the area of Arizona?

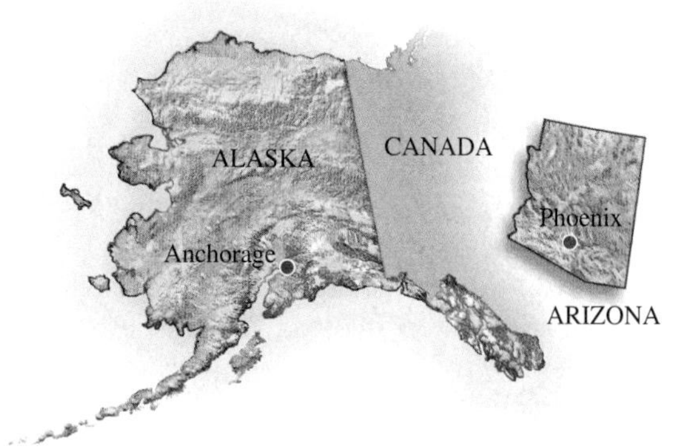

87. *Direct Mail.* Only 2.15% of mailed ads lead to a sale or a response from customers. In 2006, businesses sent out 114 billion pieces of direct mail (catalogs, coupons, and so on). How many pieces of mail led to a response from customers?
Sources: Direct Marketing Association; U.S. Postal Service

88. *Kissing and Colds.* In a medical study, it was determined that if 800 people kiss someone who has a cold, only 56 will actually catch the cold. What percent is this?

89. *Calorie Content.* A 1-oz serving of Baked! Lay's® Original Potato Crisps contains 120 calories. This is 20% less than the number of calories in a serving of Lay's Classic potato chips. How many calories are in a serving of the classic potato chips?

90. *Fat Content.* Each serving of Baked! Lay's® Original Potato Crisps contains 15 calories of fat. This is $83\frac{1}{3}\%$ less than the fat content in Lay's Classic potato chips. How many calories of fat are in a serving of the classic potato chips?

TW **91.** Campus Bookbuyers pays $30 for a book and sells it for $60. Is this a 100% markup or a 50% markup? Explain.

TW **92.** If Alexander leaves a $12 tip for a $90 dinner, is he being generous, stingy, or neither? Explain.

SKILL REVIEW

To prepare for Section 2.5, review translating to algebraic expressions and equations (Section 1.1).

Translate to an algebraic expression or equation. [1.1]

93. Twice the length plus twice the width

94. 5% of $180

95. 5 fewer than the number of points Tino scored

96. 15 plus the product of 1.5 and x

97. The product of 10 and half of a

98. 10 more than three times a number

99. The width is 2 in. less than the length.

100. A number is four times as large as a second number.

SYNTHESIS

TW **101.** How is the use of statistics in each of the following misleading?

a) A business explaining new restrictions on sick leave cited a recent survey indicating that 40% of all sick days were taken on Monday or Friday.

b) An advertisement urging summer installation of a security system quoted FBI statistics stating that over 26% of home burglaries occur between Memorial Day and Labor Day.

TW **102.** Erin is returning a tent that she bought during a 25%-off storewide sale that has ended. She is offered store credit for 125% of what she paid (not to be used on sale items). Is this fair to Erin? Why or why not?

103. The community of Bardville has 1332 left-handed females. If 48% of the community is female and 15% of all females are left-handed, how many people are in the community?

104. It has been determined that at the age of 15, a boy has reached 96.1% of his final adult height. Jaraan is 6 ft 4 in. at the age of 15. What will his final adult height be?

105. It has been determined that at the age of 10, a girl has reached 84.4% of her final adult height. Dana is 4 ft 8 in. at the age of 10. What will her final adult height be?

106. *Photography.* A 6-in. by 8-in. photo is framed using a mat meant for a 5 in. by 7 in. photo. What percentage of the photo will be hidden by the mat?

107. *Dropout Rate.* Between 2005 and 2007, the high school dropout rate in the United States decreased from 94 to 87 per thousand. Calculate the percent by which the dropout rate decreased and use that percentage to estimate dropout rates for the United States in 2008 and in 2009.
Source: U.S. Department of Education, National Center for Education Statistics

TW Aha! **108.** Would it be better to receive a 5% raise and then, a year later, an 8% raise or the other way around? Why?

TW **109.** Jose is in the 30% tax bracket. This means that 30¢ of each dollar earned goes to taxes. Which would cost him the least: contributing $50 that is tax-deductible or contributing $40 that is not tax-deductible? Explain.

Try Exercise Answers: Section 2.4

11. 0.67 **19.** 0.6258 **33.** 0.49% **43.** 25% **47.** $46\frac{2}{3}$, or $\frac{140}{3}$ **51.** 84 **65.** 75 credits **67.** 600 at-bats **69.** **(a)** 16%; **(b)** $29

2.5 Problem Solving

- Five Steps for Problem Solving
- Organizing Information Using Tables

Probably the most important use of algebra is as a tool for problem solving. In this section, we develop a problem-solving approach that is used throughout the remainder of the text.

FIVE STEPS FOR PROBLEM SOLVING

In Section 2.4, we solved several real-world problems. To solve them, we first *familiarized* ourselves with percent notation. We then *translated* each problem into an equation, *solved* the equation, *checked* the solution, and *stated* the answer.

Five Steps for Problem Solving in Algebra

1. *Familiarize* yourself with the problem.
2. *Translate* to mathematical language. (This often means writing an equation.)
3. *Carry out* some mathematical manipulation. (This often means *solving* an equation.)
4. *Check* your possible answer in the original problem.
5. *State* the answer clearly, using a complete English sentence.

Of the five steps, the most important is probably the first one: becoming familiar with the problem. Here are some hints for familiarization.

To Become Familiar with a Problem

1. Read the problem carefully. Try to visualize the problem.
2. Reread the problem, perhaps aloud. Make sure you understand all important words.
3. List the information given and the question(s) to be answered. Choose a variable (or variables) to represent the unknown and specify what the variable represents. For example, let $L =$ length in centimeters, $d =$ distance in miles, and so on.
4. Look for similarities between the problem and other problems you have already solved. Ask yourself what type of problem this is.
5. Find more information. Look up a formula in a book, at a library, or online. Consult a reference librarian or an expert in the field.
6. Make a table that uses all the information you have available. Look for patterns that may help in the translation.
7. Make a drawing and label it with known and unknown information, using specific units if given.
8. Think of a possible answer and check the guess. Note the manner in which the guess is checked.

EXAMPLE 1 Hiking. In 1957 at the age of 69, Emma "Grandma" Gatewood became the first woman to hike solo all 2100 mi of the Appalachian trail—from Springer Mountain, Georgia, to Mount Katahdin, Maine. Gatewood repeated the feat in 1960 and again in 1963, becoming the first person to hike the trail three times. When Gatewood stood atop Big Walker Mountain, Virginia, she was three times

Emma "Grandma" Gatewood (1887–1973)

as far from the northern end of the trail as from the southern end. At that point, how far was she from each end of the trail?

SOLUTION

1. **Familiarize.** It may be helpful to make a drawing.

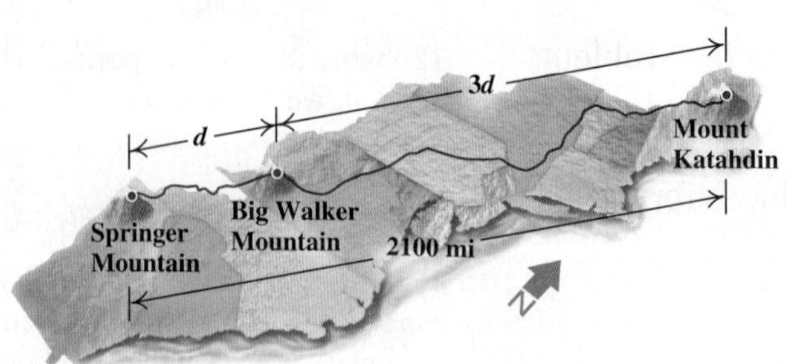

To gain some familiarity, let's suppose that Gatewood stood 600 mi from Springer Mountain. Three times 600 mi is 1800 mi. Since 600 mi + 1800 mi = 2400 mi and 2400 mi > 2100 mi, we see that our guess is too large. Rather than guess again, we let

d = the distance, in miles, to the southern end

and

$3d$ = the distance, in miles, to the northern end.

(We could also let d = the distance to the northern end and $\frac{1}{3}d$ = the distance to the southern end.)

2. **Translate.** From the drawing, we see that the lengths of the two parts of the trail must add up to 2100 mi. This leads to our translation.

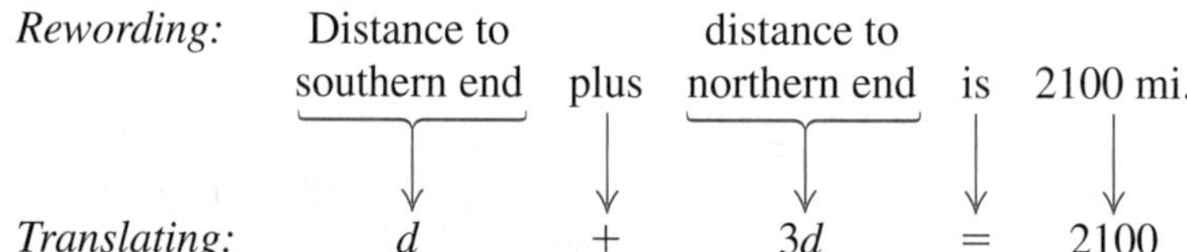

Rewording: Distance to southern end plus distance to northern end is 2100 mi.

Translating: $d \quad + \quad 3d \quad = \quad 2100$

3. **Carry out.** We solve the equation:

$$d + 3d = 2100$$

$$4d = 2100 \qquad \text{Combining like terms}$$

$$d = 525. \qquad \text{Dividing both sides by 4}$$

4. **Check.** As predicted in the *Familiarize* step, d is less than 600 mi. If d = 525 mi, then $3d$ = 1575 mi. Since 525 mi + 1575 mi = 2100 mi, we have a check.

5. **State.** Atop Big Walker Mountain, Gatewood stood 525 mi from Springer Mountain and 1575 mi from Mount Katahdin.

Try Exercise 9.

Before we solve the next problem, we need to learn some additional terminology regarding integers.

The following are examples of **consecutive integers:** 16, 17, 18, 19, 20; and −31, −30, −29, −28. Consecutive integers can be represented in the form x, $x + 1$, $x + 2$, and so on.

The following are examples of **consecutive even integers:** 16, 18, 20, 22, 24; and −52, −50, −48, −46. Consecutive even integers can be represented in the form $x, x + 2, x + 4$, and so on.

The following are examples of **consecutive odd integers:** 21, 23, 25, 27, 29; and $-71, -69, -67, -65$. Consecutive odd integers can be represented in the form $x, x + 2, x + 4$, and so on.

STUDY TIP

Set Reasonable Expectations

Do not be surprised if your success rate drops some as you work through the exercises in this section. *This is normal.* Your success rate will increase as you gain experience with these types of problems and use some of the study tips already listed.

EXAMPLE 2 Interstate Mile Markers. U.S. interstate highways post numbered markers at every mile to indicate location in case of an emergency. The sum of two consecutive mile markers on I-70 in Kansas is 559. Find the numbers on the markers.

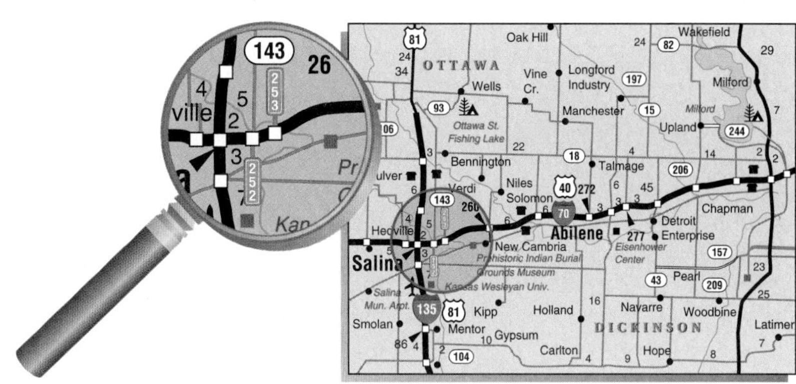

SOLUTION

1. **Familiarize.** The numbers on the mile markers are consecutive positive integers. Thus if we let $x =$ the smaller number, then $x + 1 =$ the larger number.

   The TABLE feature of a graphing calculator enables us to try many possible mile markers. We begin with a value for x. If we let $y_1 = x + 1$ represent the second mile marker, then $y_2 = x + (x + 1)$ is the sum of the markers. We enter both equations and set up a table. We will start the table at $x = 250$ and increase by 10's; that is, we set $\Delta\text{Tbl} = 10$.

   Since we are looking for a sum of 559, we see from the table at left that the value of x will be between 270 and 280. The problem could actually be solved by examining a table of values for x between 270 and 280, but let's work on developing our algebra skills.

$y_1 = x + 1, y_2 = x + (x + 1)$

X	Y1	Y2
250	251	501
260	261	521
270	271	541
280	281	561
290	291	581
300	301	601
310	311	621

X = 250

2. **Translate.** We reword the problem and translate as follows.

   Rewording: First integer plus second integer is 559.

   Translating: $x \quad + \quad (x + 1) \quad = \quad 559$

3. **Carry out.** We solve the equation:

$$x + (x + 1) = 559$$
$$2x + 1 = 559 \qquad \text{Using an associative law and combining like terms}$$
$$2x = 558 \qquad \text{Subtracting 1 from both sides}$$
$$x = 279. \qquad \text{Dividing both sides by 2}$$

   If x is 279, then $x + 1$ is 280.

4. **Check.** The possibilities for the mile markers are 279 and 280. These are consecutive positive integers and $279 + 280 = 559$, so the answers check.

5. **State.** The mile markers are 279 and 280.

Try Exercise 13.

EXAMPLE 3 Color Printers. Egads Computer Corporation rents a Xerox Phaser 8500 Color Laser Printer for \$200 per month. A new art gallery needs to lease a printer for a 2-month advertising campaign. The ink and paper for the brochures will cost an additional 21.5¢ per copy. If the gallery allots a budget of \$3000, how many brochures can they print?
Source: egadscomputer.com

SOLUTION

1. **Familiarize.** Suppose that the art gallery prints 20,000 brochures. Then the cost is the monthly charges plus ink and paper cost, or

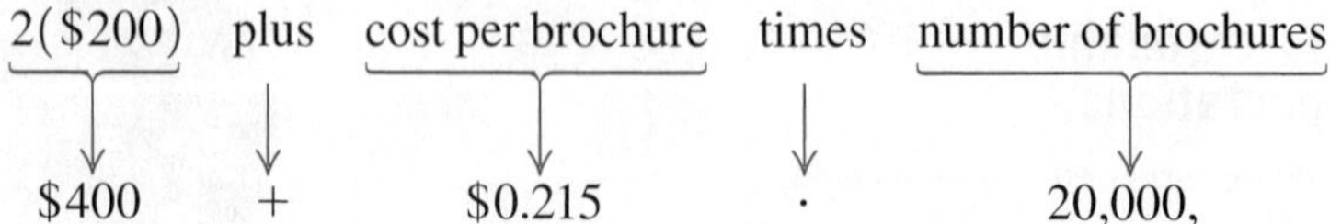

which is \$4700. Our guess of 20,000 is too large, but we have familiarized ourselves with the way in which a calculation is made. Note that we convert 21.5¢ to \$0.215 so that all information is in the same unit, dollars. We let c = the number of brochures that can be printed for \$3000.

2. **Translate.** We reword the problem and translate as follows.

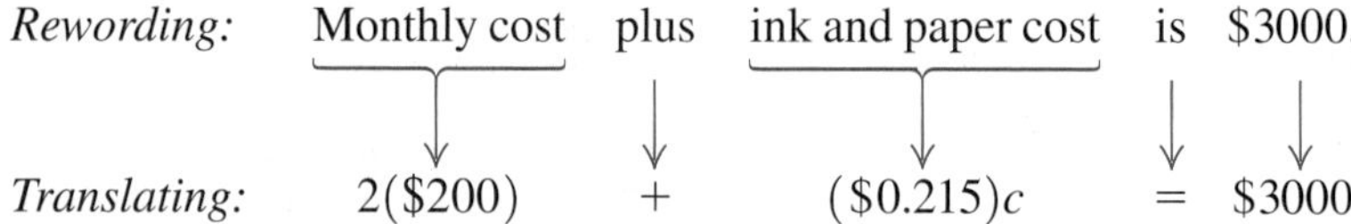

Student Notes

For many students, the most challenging step is step (2), "Translate." The table on p. 5 (Section 1.1) can be helpful in this regard.

3. **Carry out.** We solve the equation:

$$2(200) + 0.215c = 3000$$
$$400 + 0.215c = 3000$$
$$0.215c = 2600 \quad \textbf{Subtracting 400 from both sides}$$
$$c = \frac{2600}{0.215} \quad \textbf{Dividing both sides by 0.215}$$
$$c \approx 12{,}093. \quad \textbf{Rounding to the nearest one}$$

4. **Check.** We check in the original problem. The cost for 12,093 brochures is 12,093(\$0.215) = \$2599.995. The rental for 2 months is 2(\$200) = \$400. The total cost is then \$2599.995 + \$400 ≈ \$3000, which is the amount that was allotted. Our answer is less than 20,000, as we expected from the *Familiarize* step.

5. **State.** The art gallery can produce 12,093 brochures with the rental allotment of \$3000.

Try Exercise 37.

EXAMPLE 4 Perimeter of NBA Court. The perimeter of an NBA basketball court is 288 ft. The length is 44 ft longer than the width. Find the dimensions of the court.
Source: National Basketball Association

SOLUTION

1. **Familiarize.** Recall that the perimeter of a rectangle is twice the length plus twice the width. Suppose the court were 30 ft wide. The length would then be 30 + 44, or 74 ft, and the perimeter would be $2 \cdot 30$ ft + $2 \cdot 74$ ft, or 208 ft. This shows that in order for the perimeter to be 288 ft, the width must exceed 30 ft. Instead of guessing again, we let w = the width of the court, in feet.

Since the court is "44 ft longer than it is wide," we let $w + 44 =$ the length of the court, in feet.

2. **Translate.** To translate, we use $w + 44$ as the length and 288 as the perimeter. To double the length, $w + 44$, parentheses are essential.

Rewording:	Twice the length	plus	twice the width	is	288 ft.
	↓	↓	↓	↓	↓
Translating:	$2(w + 44)$	$+$	$2w$	$=$	288

3. **Carry out.** We solve the equation:

$$2(w + 44) + 2w = 288$$
$$2w + 88 + 2w = 288 \quad \text{Using the distributive law}$$
$$4w + 88 = 288 \quad \text{Combining like terms}$$
$$4w = 200$$
$$w = 50.$$

The dimensions appear to be $w = 50$ ft, and $l = w + 44 = 94$ ft.

4. **Check.** If the width is 50 ft and the length is 94 ft, then the court is 44 ft longer than it is wide. The perimeter is $2(50\text{ ft}) + 2(94\text{ ft}) = 100\text{ ft} + 188\text{ ft}$, or 288 ft, as specified. We have a check.
5. **State.** An NBA court is 50 ft wide and 94 ft long.

Try Exercise 25.

Student Notes

Get in the habit of writing what each variable represents before writing an equation. In Example 4, you might write

width $= w$,

length $= w + 44$

before translating the problem to an equation. This step becomes more important as problems become more complex.

CAUTION! Always be sure to answer the question in the original problem completely. For instance, in Example 1 we needed to find *two* numbers: the distances from *each* end of the trail to the hiker. Similarly, in Example 4 we needed to find two dimensions, not just the width. Be sure to label each answer with the proper unit.

EXAMPLE 5 Selling a Home. The McCanns are planning to sell their home. If they want to be left with \$117,500 after paying 6% of the selling price to a realtor as a commission, for how much must they sell the house?

SOLUTION

1. **Familiarize.** Suppose the McCanns sell the house for \$120,000. A 6% commission can be determined by finding 6% of \$120,000:

$$6\% \text{ of } \$120{,}000 = 0.06(\$120{,}000) = \$7200.$$

Subtracting this commission from \$120,000 would leave the McCanns with

$$\$120{,}000 - \$7200 = \$112{,}800.$$

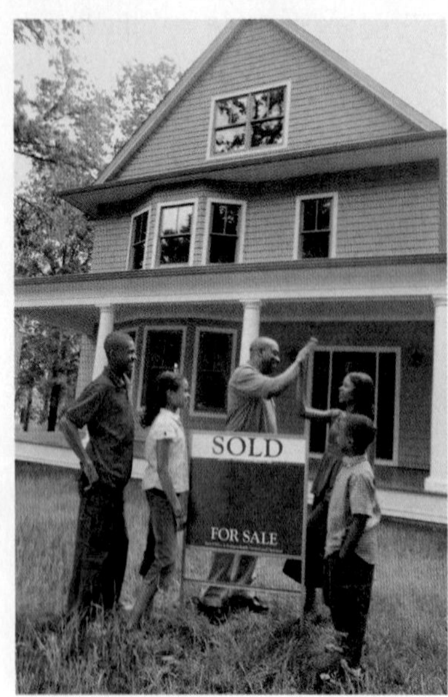

This shows that in order for the McCanns to clear \$117,500, the house must sell for more than \$120,000. To determine what the sale price must be, we let $x =$ the selling price, in dollars. With a 6% commission, the realtor would receive $0.06x$.

2. **Translate.** We reword the problem and translate as follows.

Rewording:	Selling price	less	commission	is	amount remaining.
	↓	↓	↓	↓	↓
Translating:	x	$-$	$0.06x$	$=$	117,500

3. **Carry out.** We solve the equation:

$$x - 0.06x = 117{,}500$$
$$1x - 0.06x = 117{,}500$$
$$0.94x = 117{,}500 \quad \text{Combining like terms. Had we noted that after the commission has been paid, 94\% remains, we could have begun with this equation.}$$
$$x = \frac{117{,}500}{0.94} \quad \text{Dividing both sides by 0.94}$$
$$x = 125{,}000.$$

4. **Check.** To check, we first find 6% of \$125,000:

$$6\% \text{ of } \$125{,}000 = 0.06(\$125{,}000) = \$7500. \quad \text{This is the commission.}$$

Next, we subtract the commission to find the remaining amount:

$$\$125{,}000 - \$7500 = \$117{,}500.$$

Since, after the commission, the McCanns are left with \$117,500, our answer checks. Note that the \$125,000 sale price is greater than \$120,000, as predicted in the *Familiarize* step.

5. **State.** To be left with \$117,500, the McCanns must sell their house for \$125,000.

Try Exercise 49.

EXAMPLE 6 Cross Section of a Roof. In a triangular gable end of a roof, the angle of the peak is twice as large as the angle on the back side of the house. The measure of the angle on the front side is 20° greater than the angle on the back side. How large are the angles?

SOLUTION

1. **Familiarize.** We make a drawing. In this case, the measure of the back angle is x, the measure of the front angle is $x + 20$, and the measure of the peak angle is $2x$.

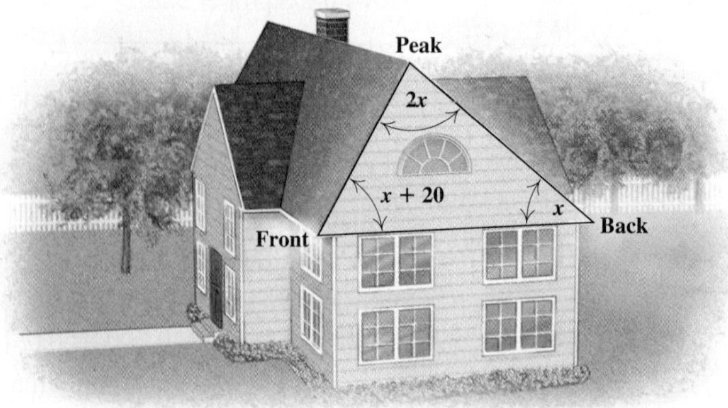

Student Notes

You may be expected to recall material that you have learned in an earlier course. If you have forgotten a formula, refresh your memory by consulting a textbook or a Web site.

2. **Translate.** To translate, we need to recall that the sum of the measures of the angles in a triangle is 180°.

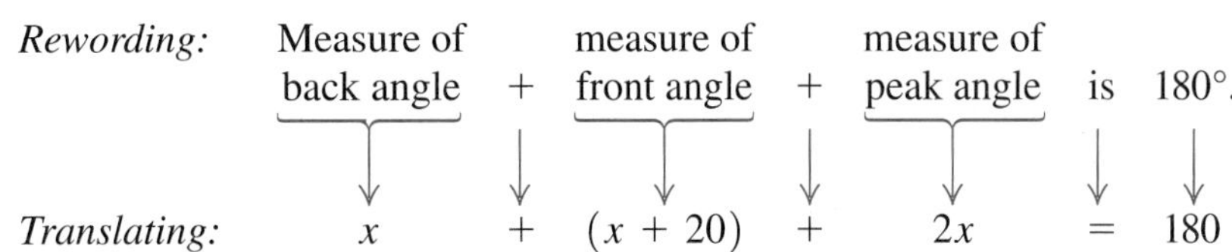

3. **Carry out.** We solve:

$$\begin{aligned} x + (x + 20) + 2x &= 180 \\ 4x + 20 &= 180 \quad \text{Combining like terms} \\ 4x &= 160 \quad \text{Subtracting 20 from both sides} \\ x &= 40. \quad \text{Dividing both sides by 4} \end{aligned}$$

The measures for the angles appear to be:

Back angle: $x = 40°$,
Front angle: $x + 20 = 40 + 20 = 60°$,
Peak angle: $2x = 2(40) = 80°$.

4. **Check.** Consider 40°, 60°, and 80°. The measure of the front angle is 20° greater than the measure of the back angle, the measure of the peak angle is twice the measure of the back angle, and the sum is 180°. These numbers check.
5. **State.** The measures of the angles are 40°, 60°, and 80°.

Try Exercise 31.

EXAMPLE 7 Zookeeping. A zookeeper is designing an aquarium for a group of 15 seahorses. She finds the following recommendations for aquarium sizes for various numbers of seahorses. What size aquarium should she design?

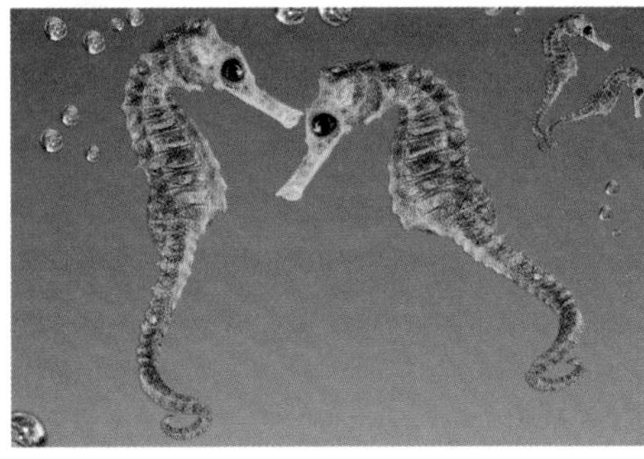

Size of Aquarium (in liters)	Number of Seahorses
150	6
225	9
300	12
500	20

Source: Based on information from seahorsesanctuary.com.au

SOLUTION

1. **Familiarize.** We look for a relationship between the size of the aquarium and the number of seahorses. We try dividing the size of the aquarium by the number of seahorses:

$$\begin{aligned} 150 \div 6 &= 25, \\ 225 \div 9 &= 25, \\ 300 \div 12 &= 25, \\ 500 \div 20 &= 25. \end{aligned}$$

Thus if we let a represent the size of the aquarium and n the number of seahorses, we see that $a/n = 25$.

2. **Translate.** We substitute 15 for the number of seahorses:

$$\frac{a}{n} = 25$$

$$\frac{a}{15} = 25. \quad \textbf{Substituting 15 for } n$$

3. **Carry out.** We solve the equation:

$$\frac{a}{15} = 25$$

$$15 \cdot \frac{a}{15} = 15 \cdot 25 \quad \textbf{Multiplying both sides by 15}$$

$$a = 375. \quad \textbf{Simplifying}$$

4. **Check.** Since $375 \div 15 = 25$, a 375-L aquarium fits the pattern. We can also see from the table that this aquarium size is between the sizes recommended for 12 seahorses and for 20 seahorses, as we would expect. The answer checks.

5. **State.** The zookeeper should design an aquarium that holds 375 L.

Try Exercise 53.

ORGANIZING INFORMATION USING TABLES

It is often helpful to organize the information given in a problem using a table. We fill in as many entries in the table as possible using the given information and then use a variable or variable expression to represent each of the remaining quantities.

When working with motion problems, we often use tables, as well as the following motion formula.

Motion Formula

$$d = r \cdot t$$

(Distance = Rate times Time)

EXAMPLE 8 Motion. Sharon drove for 3 hr on a highway and then for 1 hr on a side road. Her speed on the highway was 20 mph faster than her speed on the side road. If she traveled a total of 220 mi, how fast did she travel on the side road?

SOLUTION

1. **Familiarize.** After reading the problem carefully, we see that we are asked to find the rate of travel on the side road. We then define the variable as

x = Sharon's speed on the side road, in miles per hour.

Since we are dealing with motion, we will use the motion formula $d = r \cdot t$. We use the variables in this formula as headings for three columns in a table. The rows of the table correspond to the different motion situations—in this case, the highway driving and the side-road driving. We know the times traveled for each type of driving, and we have defined a variable representing the speed on the side road. Thus we can fill in those entries in the table.

	d	$= r$	$\cdot\ t$
Highway			3 hr
Side Road		x mph	1 hr

We want to fill in the remaining entries using the variable x. We know that Sharon's speed on the highway was 20 mph faster than her speed on the side road, or

Highway speed $= (x + 20)$ mph.

Then, since $d = r \cdot t$, we multiply to find each distance:

Highway distance $= (x + 20)(3)$ mi;
Side-road distance $= x(1)$ mi.

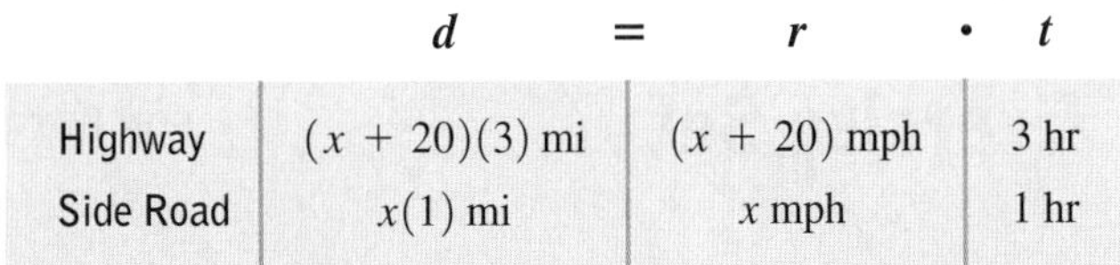

	d	$= r$	$\cdot\ t$
Highway	$(x + 20)(3)$ mi	$(x + 20)$ mph	3 hr
Side Road	$x(1)$ mi	x mph	1 hr

2. **Translate.** We are also told in the statement of the problem that Sharon traveled a total of 220 mi. This gives us the translation to an equation.

Rewording:	Highway distance	plus	side-road distance	is	total distance.
	↓	↓	↓	↓	↓
Translating:	$(x + 20)(3)$	$+$	$x(1)$	$=$	220

3. **Solve.** We now have an equation to solve:

$$(x + 20)(3) + x(1) = 220$$
$$3x + 60 + x = 220 \quad \text{Using the distributive law}$$
$$4x + 60 = 220 \quad \text{Combining like terms}$$
$$4x = 160 \quad \text{Subtracting 60 from both sides}$$
$$x = 40. \quad \text{Dividing both sides by 4}$$

4. **Check.** Since x represents Sharon's speed on the side road, we have the following.

	Speed (Rate)	Time	Distance
Side Road	$x = 40$	1	$40(1) = 40$
Highway	$x + 20 = 60$	3	$60(3) = 180$

The total distance traveled is 40 mi + 180 mi, or 220 mi. The answer checks.

5. **State.** Sharon's speed on the side road was 40 mph.

Try Exercise 57.

We close this section with some tips to aid you in problem solving.

Problem-Solving Tips

1. The more problems you solve, the more your skills will improve.
2. Look for patterns when solving problems. Each time you study an example in a text, you may observe a pattern for problems that you will encounter later in the exercise sets or in other practical situations.
3. Clearly define variables before translating to an equation.
4. Consider the dimensions of the variables and constants in the equation. The variables that represent length should all be in the same unit, those that represent money should all be in dollars or all in cents, and so on.
5. Make sure that units appear in the answer whenever appropriate and that you have completely answered the question in the original problem.

Exercise Set

FOR EXTRA HELP MyMathLab  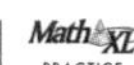  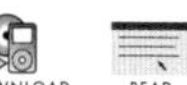

Solve. Even though you might find the answer quickly in some other way, practice using the five-step problem-solving process.

1. Two fewer than ten times a number is 78. What is the number?
2. Three less than twice a number is 19. What is the number?
3. Five times the sum of 3 and some number is 70. What is the number?
4. Twice the sum of 4 and some number is 34. What is the number?
5. *Comparing Prices.* Miles paid \$180 for a 16-GB iPod Nano. This was only 20% more than the cost of an 8-GB Nano. What was the price of the 8-GB Nano?
6. *Comparing Prices.* Eva paid \$120 for a TI-84 Plus graphing calculator. This was 20% less than the price of a TI-89 Titanium graphing calculator. How much did the TI-89 cost?
7. *Sales Tax.* Amy paid \$90.95, including 7% tax, for a pair of Nike running shoes. How much did the pair of shoes itself cost?
8. *Sales Tax.* Patrick paid \$275.60, including 6% tax, for a Canon printer. How much did the printer itself cost?
9. *Running.* In 2008, Yiannis Kouros of Australia set the record for the greatest distance run in 48 hr by running 433 km. After 16 hr, he had approximately twice as far to go as he had already run. How far had he run?
   Source: yianniskouros.com
10. *Sled-Dog Racing.* The Iditarod sled-dog race extends for 1049 mi from Anchorage to Nome. If a musher is twice as far from Anchorage as from Nome, how many miles has the musher traveled?

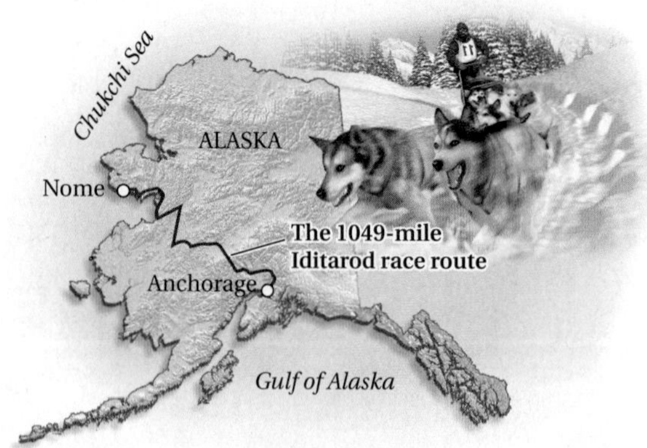

11. *Indy Car Racing.* In May 2010, Scott Dixon won the Road Runner Turbo 300 with a time of 01:50:43.1410 for the 300-mi race. At one point, Dixon was 20 mi closer to the finish than to the start. How far had Dixon traveled at that point?
12. *NASCAR Racing.* In May 2010, Kyle Busch won the Autism Speaks 400 with a 7.551-sec margin of victory

for the 400-mi race. At one point, Busch was 80 mi closer to the finish than to the start. How far had Busch traveled at that point?

13. *Apartment Numbers.* The apartments in Lara's apartment house are consecutively numbered on each floor. The sum of her number and her next-door neighbor's number is 2409. What are the two numbers?

14. *Apartment Numbers.* The apartments in Jonathan's apartment house are numbered consecutively on each floor. The sum of his number and his next-door neighbor's number is 1419. What are the two numbers?

15. *Street Addresses.* The houses on the south side of Elm Street are consecutive even numbers. Chrissy and Bryan are next-door neighbors and the sum of their house numbers is 794. Find their house numbers.

16. *Street Addresses.* The houses on the west side of Lincoln Avenue are consecutive odd numbers. Art and Colleen are next-door neighbors and the sum of their house numbers is 572. Find their house numbers.

17. The sum of three consecutive page numbers is 60. Find the numbers.

18. The sum of three consecutive page numbers is 99. Find the numbers.

19. *Oldest Bride.* The world's oldest bride was 19 yr older than her groom. Together, their ages totaled 185 yr. How old were the bride and the groom?
Source: *Guinness World Records* 2010

20. *Best Actress.* Jessica Tandy was the oldest woman to win the Oscar award for Best Actress, for *Driving Miss Daisy* in 1990. Marlee Matlin was the youngest to win a Best Actress Oscar, for *Children of a Lesser God* in 1987. Tandy was 59 years older than Matlin, and together their ages totaled 101 years. How old was each when she won the award?
Source: *Guinness World Records* 2010

21. *e-mail.* In 2008, approximately 210 billion e-mail messages were sent each day. The number of spam messages was about five times the number of nonspam messages. How many of each type of message were sent each day in 2008?
Source: Radicati Group and ICF International

22. *Laptop Computers.* Approximately 160 million laptop computers, including netbooks, were sold in 2009. The number of laptops sold that are not netbooks was four times the number of netbooks sold. How many of each type of computer were sold in 2009?
Source: IDC research estimate quoted in "How Laptops Took Over the World," in guardian.co.uk

23. *Page Numbers.* The sum of the page numbers on the facing pages of a book is 281. What are the page numbers?

24. *Perimeter of a Triangle.* The perimeter of a triangle is 195 mm. If the lengths of the sides are consecutive odd integers, find the length of each side.

25. *Hancock Building Dimensions.* The top of the John Hancock Building in Chicago is a rectangle whose length is 60 ft more than the width. The perimeter is 520 ft. Find the width and the length of the rectangle. Find the area of the rectangle.

26. *Dimensions of a State.* The perimeter of the state of Wyoming is 1280 mi. The width is 90 mi less than the length. Find the width and the length.

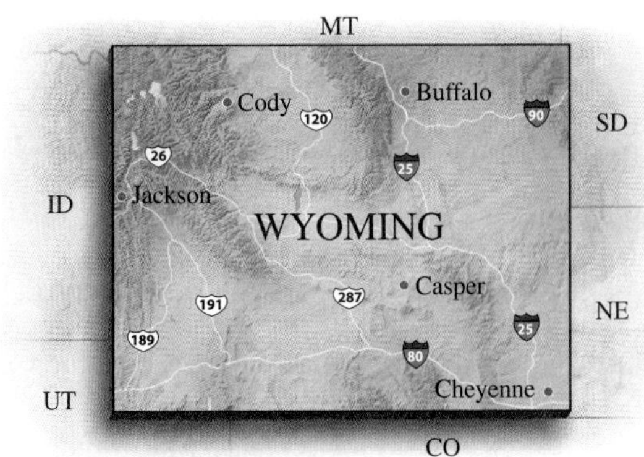

27. *Perimeter of a High School Basketball Court.* The perimeter of a standard high school basketball court is 268 ft. The length is 34 ft longer than the width. Find the dimensions of the court.
Source: Indiana High School Athletic Association

28. A rectangular community garden is to be enclosed with 92 m of fencing. In order to allow for compost storage, the garden must be 4 m longer than it is wide. Determine the dimensions of the garden.

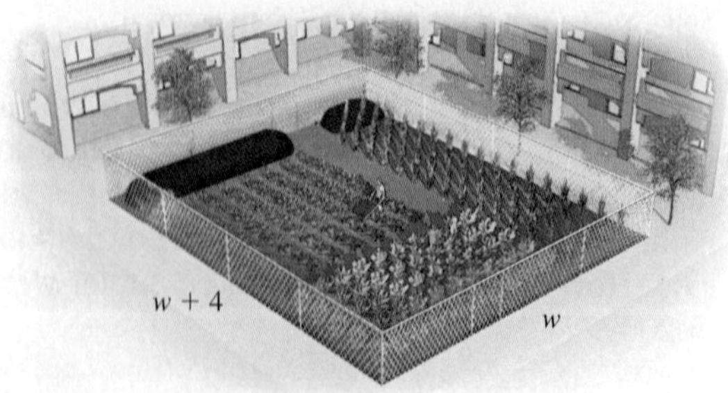

29. *Two-by-Four.* The perimeter of a cross section of a "two-by-four" piece of lumber is 10 in. The length is 2 in. longer than the width. Find the actual dimensions of the cross section of a two-by-four.

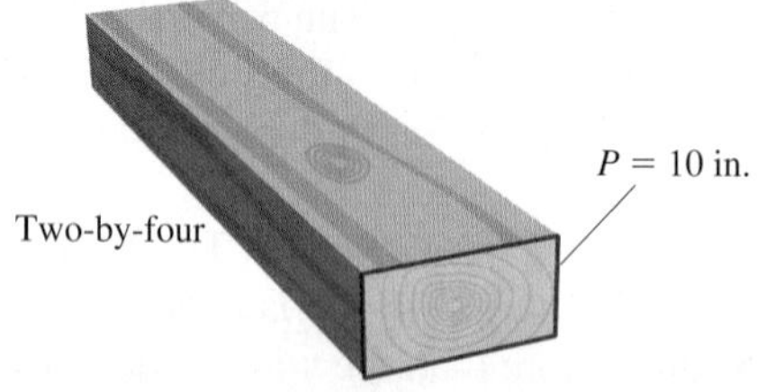

30. *Standard Billboard Sign.* A standard rectangular highway billboard sign has a perimeter of 124 ft. The length is 6 ft more than three times the width. Find the dimensions of the sign.

31. *Angles of a Triangle.* The second angle of an architect's triangle is three times as large as the first. The third angle is 30° more than the first. Find the measure of each angle.

32. *Angles of a Triangle.* The second angle of a triangular garden is four times as large as the first. The third angle is 45° less than the sum of the other two angles. Find the measure of each angle.

33. *Angles of a Triangle.* The second angle of a triangular building lot is three times as large as the first. The third angle is 10° more than the sum of the other two angles. Find the measure of the third angle.

34. *Angles of a Triangle.* The second angle of a triangular kite is four times as large as the first. The third angle is 5° more than the sum of the other two angles. Find the measure of the second angle.

35. *Rocket Sections.* A rocket is divided into three sections: the payload and navigation section in the top, the fuel section in the middle, and the rocket engine section in the bottom. The top section is one-sixth the length of the bottom section. The middle section is one-half the length of the bottom section. The total length is 240 ft. Find the length of each section.

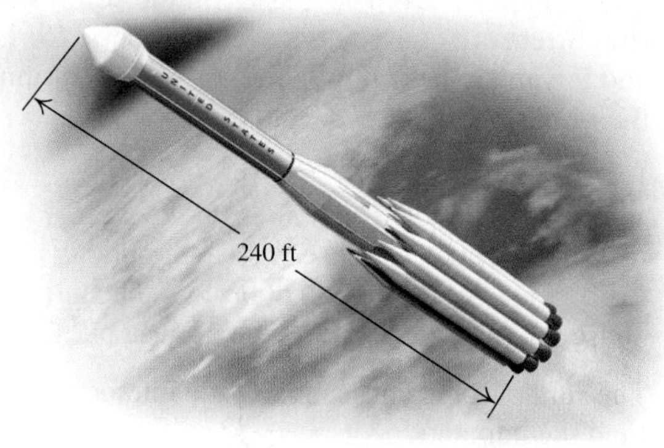

36. *Gourmet Sandwiches.* Jenny, Demi, and Shaina buy an 18-in. long gourmet sandwich and take it back to their apartment. Since they have different appetites, Jenny cuts the sandwich so that Demi gets half of what Jenny gets and Shaina gets three-fourths of what Jenny gets. Find the length of each person's sandwich.

37. *Taxi Rates.* In Chicago, a taxi ride costs \$3.25 plus \$1.80 for each mile traveled. Debbie has budgeted \$19 for a taxi ride (excluding tip). How far can she travel on her \$19 budget?
Source: cityofchicago.org

38. *Taxi Fares.* In New York City, taxis charge \$2.50 plus \$2.00 per mile for off-peak fares. How far can Oscar travel for \$17.50 (assuming an off-peak fare)?
Source: New York City Taxi and Limousine Commission

39. *Truck Rentals.* Truck-Rite Rentals rents trucks at a daily rate of \$49.95 plus 39¢ per mile. Concert Productions has budgeted \$100 for renting a truck to haul equipment to an upcoming concert. How far can they travel in one day and stay within their budget?

40. *Truck Rentals.* Fine Line Trucks rents an 18-ft truck for \$42 plus 35¢ per mile. Judy needs a truck for one day to deliver a shipment of plants. How far can she drive and stay within a budget of \$70?

41. *Complementary Angles.* The sum of the measures of two *complementary* angles is 90°. If one angle measures 15° more than twice the measure of its complement, find the measure of each angle.

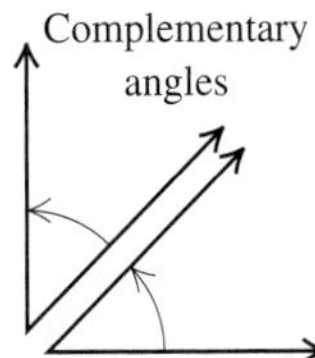

42. *Supplementary Angles.* The sum of the measures of two *supplementary* angles is 180°. If one angle measures 45° less than twice the measure of its supplement, find the measure of each angle.

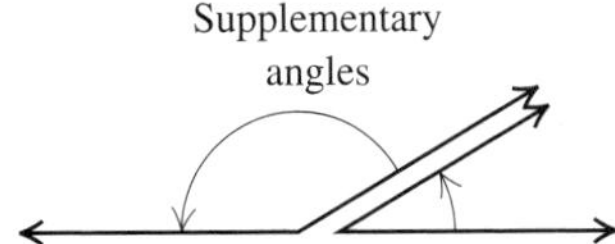

43. *Copier Paper.* The perimeter of standard-size copier paper is 99 cm. The width is 6.3 cm less than the length. Find the length and the width.

44. *Stock Prices.* Sarah's stock investment grew 28% to \$448. How much did she invest?

45. *Savings Interest.* Amber invested money in a savings account at a rate of 6% simple interest. After one year, she has \$6996 in the account. How much did Amber originally invest?

46. *Credit Cards.* The balance in Will's Mastercard® account grew 2%, to \$870, in one month. What was his balance at the beginning of the month?

47. *Scrabble®.* During the 25th Annual Phoenix Scrabble Tournament, held in Ahwatukee in February 2009, Laurie Cohen and Nigel Peltier scored a total of 1127 points, setting a world record. If Cohen scored 323 points more than Peltier, what was the winning score?
Source: "World Record Set in Ahwatukee Tournament," by Coty Dolores Miranda, 2/19/09, azcentral.com

48. *Color Printers.* The art gallery in Example 3 raises its budget to \$5000 for the 2-month period. How many brochures can they produce for \$5000?

49. *Selling at an Auction.* Thomas is selling his collection of Transformers at an auction. He wants to be left with \$1150 after paying a seller's premium of 8% on the final bid (hammer price) for the collection. What must the hammer price be in order for him to clear \$1150?

50. *Budget Overruns.* The massive roadworks project in Boston known as The Big Dig cost approximately \$14.6 billion. This cost was 484% more than the original estimate. What was the original estimate of the cost of The Big Dig?
Sources: Taxpayers for Common Sense; www.msnbc.cmsn.com

51. *Cricket Chirps and Temperature.* The equation $T = \frac{1}{4}N + 40$ can be used to determine the temperature T, in degrees Fahrenheit, given the number of times N that a cricket chirps per minute. Determine the number of chirps per minute for a temperature of 80°F.

52. *Race Time.* The equation $R = -0.028t + 20.8$ can be used to predict the world record in the 200-m dash, where R is the record, in seconds, and t is the number of years since 1920. In what year will the record be 18.0 sec?

53. *Aquariums.* The table below lists the maximum recommended stocking density for fish in aquariums of various sizes. The density is calculated by adding the lengths of all fish in the aquarium. What size aquarium would be needed for 30 in. of fish?

Size of Aquarium (in gallons)	Recommended Stocking Density (in inches of fish)
100	20
120	24
200	40
250	50

Source: Aquarium Fish Magazine, June 2002

54. *Internet Sales.* The table on the following page lists the cost, including shipping, for various orders from walgreens.com. If the cost, including shipping, for an order was $25.68, what was the cost of the items before shipping?

Cost of Order Before Shipping	Cost Including Shipping
$15.50	$20.99
36.48	41.97
4.46	9.95
45.65	51.14

55. *Agriculture.* The table below lists the weight of a prize-winning pumpkin for various days in August. On what day did the pumpkin weigh 920 lb?

Date	Weight of Pumpkin (in pounds)
August 1	380
August 2	410
August 3	440
August 4	470
August 11	680
August 25	1100

Source: *USA Weekend*, Oct. 19–21, 2001

56. *House Cleaning.* The table below lists how much Susan and Leah charge for cleaning houses of various sizes. For what size house is the cleaning cost $145?

House Size (in square feet)	Cleaning Cost
1000	$ 60
1100	65
1200	70
2000	110
3000	160

57. Samantha ran on a fitness track for 20 min and then walked for 10 min. Her running rate was 250 ft per minute faster than her walking rate. If she ran and walked a total of 15,500 ft, how fast did she run?

58. Stephanie walked 5 min to meet her jogging partner and then ran for 40 min. Her jogging rate was twice as fast as her walking rate. If she walked and jogged a total of 21,250 ft, how fast did she jog?

59. Anthony drove for 2 hr in a snowstorm and then for 5 more hr in clear weather. He drove half as fast through the snow as he did in the clear weather. If he drove 240 more mi in clear weather than he did in the snow, how fast did he drive through the snow?

60. Robert drove for 6 hr on a level highway and then for 2 more hr through the mountains. His speed through the mountains was 20 mph slower than his speed on the first part of the trip. If he drove 300 mi longer on the level road than he did through the mountains, how fast did he drive through the mountains?

61. Justin rode his bicycle at a rate of 15 mph and then slowed to 10 mph. He rode 30 min longer at 15 mph than he did at 10 mph. If he traveled a total of 25 mi, how long did he ride at the faster rate? [*Hint*: Convert minutes to hours or hours to minutes.]

62. Courtney rode her scooter to the library at a rate of 25 mph and then continued to school at a rate of 15 mph. It took her 4 min longer to get to the library than it did for her to get to the school from the library. If she traveled a total of 15 mi, how long did it take her to get to the library? [*Hint*: Convert minutes to hours or hours to minutes.]

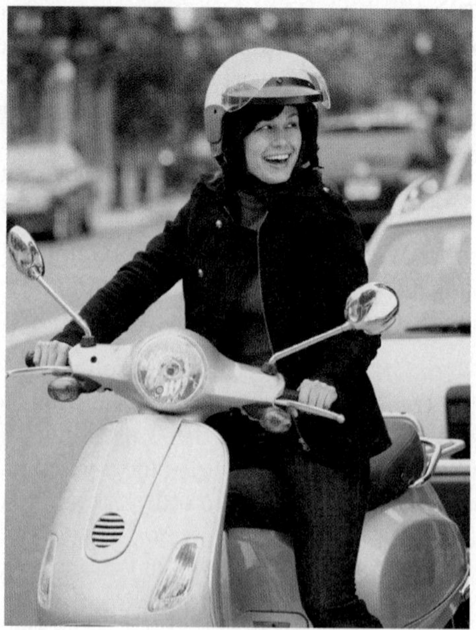

TW **63.** Ethan claims he can solve most of the problems in this section by guessing. Is there anything wrong with this approach? Why or why not?

TW **64.** When solving Exercise 19, Erica used a to represent the bride's age and Ben used a to represent the groom's age. Is one of these approaches preferable to the other? Why or why not?

SKILL REVIEW

To prepare for Section 2.6, review inequalities (Section 1.4).

Write a true sentence using either < or >. [1.4]

65. $-9 \square 5$

66. $1 \square 3$

67. $-4 \square 7$

68. $-9 \square -12$

Write a second inequality with the same meaning. [1.4]

69. $x \geq -4$

70. $x < 5$

71. $5 > y$

72. $-10 \leq t$

SYNTHESIS

TW **73.** Write a problem for a classmate to solve. Devise it so that the problem can be translated to the equation $x + (x + 2) + (x + 4) = 375$.

TW **74.** Write a problem for a classmate to solve. Devise it so that the solution is "Audrey can drive the rental truck for 50 mi without exceeding her budget."

75. *Discounted Dinners.* Kate's "Dining Card" entitles her to \$10 off the price of a meal after a 15% tip has been added to the cost of the meal. If, after the discount, the bill is \$32.55, how much did the meal originally cost?

76. *Test Scores.* Pam scored 78 on a test that had 4 fill-in questions worth 7 points each and 24 multiple-choice questions worth 3 points each. She had one fill-in question wrong. How many multiple-choice questions did Pam get right?

77. *Gettysburg Address.* Abraham Lincoln's 1863 Gettysburg Address refers to the year 1776 as "four *score* and seven years ago." Determine what a score is.

78. One number is 25% of another. The larger number is 12 more than the smaller. What are the numbers?

79. A storekeeper goes to the bank to get \$10 worth of change. She requests twice as many quarters as half dollars, twice as many dimes as quarters, three times as many nickels as dimes, and no pennies or dollars. How many of each coin did the storekeeper get?

80. *Perimeter of a Rectangle.* The width of a rectangle is three-fourths of the length. The perimeter of the rectangle becomes 50 cm when the length and the width are each increased by 2 cm. Find the length and the width.

81. *Discounts.* In exchange for opening a new credit account, Macy's Department Stores® subtracts 10% from all purchases made on the day the account is established. Julio is opening an account and has a coupon for which he receives 10% off the first day's reduced price of a camera. If Julio's final price is \$77.75, what was the price of the camera before the two discounts?

82. *Sharing Fruit.* Apples are collected in a basket for six people. One-third, one-fourth, one-eighth, and one-fifth of the apples are given to four people, respectively. The fifth person gets ten apples, and one apple remains for the sixth person. Find the original number of apples in the basket.

83. *Winning Percentage.* In a basketball league, the Falcons won 15 of their first 20 games. In order to win 60% of the total number of games, how many more games will they have to play, assuming they win only half of the remaining games?

84. *eBay Purchases.* An eBay seller charges \$9.99 for the first DVD purchased and \$6.99 for all others. For shipping and handling, he charges the full shipping fee of \$3 for the first DVD, one-half of the shipping charge for the second item, and one-third of the shipping charge per item for all remaining items. The total cost of a shipment (excluding tax) was \$45.45. How many DVDs were in the shipment?

85. *Test Scores.* Elsa has an average score of 82 on three tests. Her average score on the first two tests is 85. What was the score on the third test?

86. *Taxi Fares.* In New York City, a taxi ride costs \$2.50 plus 40¢ per $\frac{1}{5}$ mile and 20¢ per minute stopped in traffic. Due to traffic, Glenda's taxi took 20 min to complete what is, without traffic, a 10-min drive. If she is charged \$16.50 for the ride, how far did Glenda travel?

TW **87.** A school purchases a piano and must choose between paying \$2000 at the time of purchase or \$2150 at the end of one year. Which option should the school select and why?

TW Aha! **88.** Annette claims the following problem has no solution: "The sum of the page numbers on facing pages is 191. Find the page numbers." Is she correct? Why or why not?

89. The perimeter of a rectangle is 101.74 cm. If the length is 4.25 cm longer than the width, find the dimensions of the rectangle.

90. The second side of a triangle is 3.25 cm longer than the first side. The third side is 4.35 cm longer than the second side. If the perimeter of the triangle is 26.87 cm, find the length of each side.

Try Exercise Answers: Section 2.5

9. Approximately $144\frac{1}{3}$ km **13.** 1204 and 1205 **25.** Width: 100 ft; length: 160 ft; area: 16,000 ft^2 **31.** $30°, 90°, 60°$ **37.** $8\frac{3}{4}$ mi **49.** \$1250 **53.** 150 gal **57.** 600 ft per minute

2.6 Solving Inequalities

- Solutions of Inequalities
- Graphs of Inequalities
- Set-Builder Notation and Interval Notation
- Solving Inequalities Using the Addition Principle
- Solving Inequalities Using the Multiplication Principle
- Using the Principles Together

Many real-world situations translate to *inequalities*. For example, a student might need to register for *at least* 12 credits; an elevator might be designed to hold *at most* 2000 lb; a tax credit might be allowable for families with incomes of *less than* \$40,000; and so on. Before solving applications of this type, we must adapt our equation-solving principles to the solving of inequalities.

SOLUTIONS OF INEQUALITIES

Recall from Section 1.4 that an inequality is a number sentence containing $>$ (is greater than), $<$ (is less than), $\geq$ (is greater than or equal to), or $\leq$ (is less than or equal to). Inequalities like

$$-7 > x, \quad t < 5, \quad 5x - 2 \geq 9, \quad \text{and} \quad -3y + 8 \leq -7$$

are true for some replacements of the variable and false for others.

Any value for the variable that makes an inequality true is called a **solution.** The set of all solutions is called the **solution set.** When all solutions of an inequality have been found, we say that we have **solved** the inequality.

EXAMPLE 1 Determine whether the given number is a solution of $x < 2$: **(a)** -3; **(b)** 2.

SOLUTION

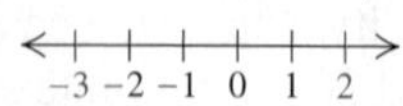

a) Since $-3 < 2$ is true, -3 is a solution.

b) Since $2 < 2$ is false, 2 is not a solution.

Try Exercise 9.

EXAMPLE 2 Determine whether the given number is a solution of $y \geq 6$: **(a)** 6; **(b)** -4.

SOLUTION

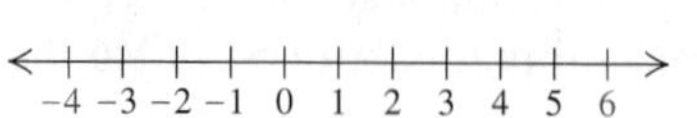

a) Since $6 \geq 6$ is true, 6 is a solution.

b) Since $-4 \geq 6$ is false, -4 is not a solution.

Try Exercise 11.

GRAPHS OF INEQUALITIES

Because the solutions of inequalities like $x < 2$ are too numerous to list, it is helpful to make a drawing that represents all the solutions. The **graph** of an inequality is such a drawing. Graphs of inequalities in one variable can be drawn on the number line by shading all points that are solutions. Parentheses are used to indicate endpoints that are *not* solutions and brackets indicate endpoints that *are* solutions.

EXAMPLE 3 Graph each inequality.

a) $x < 2$

b) $y \geq -3$

c) $-2 < x \leq 3$

SOLUTION

a) The solutions of $x < 2$ are those numbers less than 2. They are shown on the number line by shading all points to the left of 2. The right parenthesis at 2 and the shading to its left indicate that 2 is *not* part of the graph, but numbers like 1.2 and 1.99 are.

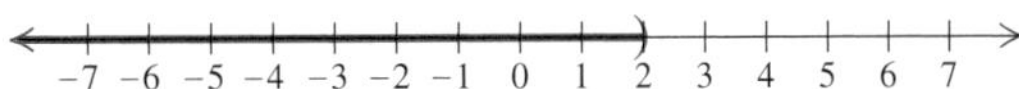

b) The solutions of $y \geq -3$ are shown on the number line by shading the point for -3 and all points to the right of -3. The left bracket at -3 indicates that -3 *is* part of the graph.

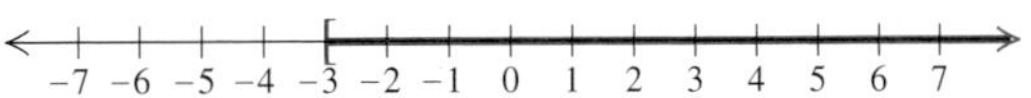

Student Notes

Note that $-2 < x < 3$ means $-2 < x$ *and* $x < 3$. Because of this, statements like $2 < x < 1$ make no sense—no number is both greater than 2 and less than 1.

c) The inequality $-2 < x \leq 3$ is read "-2 is less than x *and* x is less than or equal to 3," or "x is greater than -2 *and* less than or equal to 3." To be a solution of $-2 < x \leq 3$, a number must be a solution of both $-2 < x$ *and* $x \leq 3$. The number 1 is a solution, as are -0.5, 1.9, and 3. The parenthesis indicates that -2 is *not* a solution, whereas the bracket indicates that 3 *is* a solution. The other solutions are shaded.

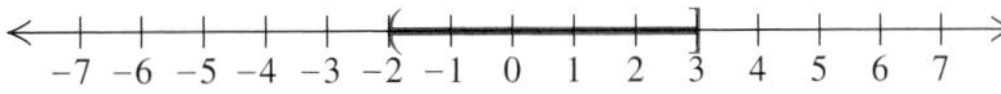

Try Exercise 15.

SET-BUILDER NOTATION AND INTERVAL NOTATION

The solutions of an inequality are numbers. In Example 3, $x < 2$, $y \geq -3$, and $-2 < x \leq 3$ are *inequalities*, not *solutions*. We will use two types of notation to write the **solution set** of an inequality: set-builder notation and interval notation.

One way to write the solution set of an inequality is **set-builder notation.** The solution set of Example 3(a), written in set-builder notation, is

$$\{x|x < 2\}.$$

This notation is read

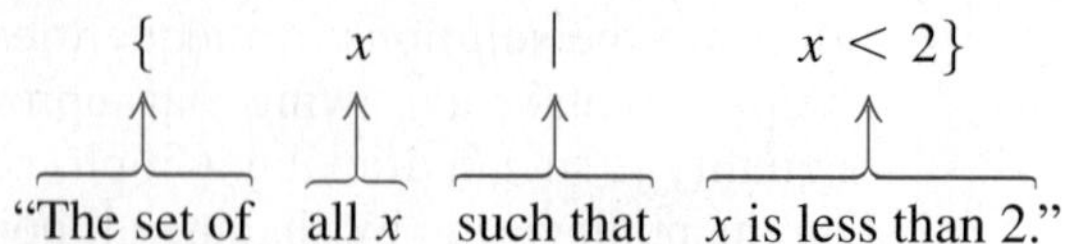

A number is in $\{x|x < 2\}$ if that number is less than 2. Thus, for example, 0 and -1 are in $\{x|x < 2\}$, but 2 and 3 are not.

Another way to write solutions of an inequality in one variable is to use **interval notation.** Interval notation uses parentheses, ( ), and brackets, [ ].

If a and b are real numbers with $a < b$, we define the **open interval** (a, b) as the set of all numbers x for which $a < x < b$. This means that x can be any number between a and b, but it cannot be either a or b.

The **closed interval** $[a, b]$ is defined as the set of all numbers x for which $a \leq x \leq b$. **Half-open intervals** $(a, b]$ and $[a, b)$ contain one endpoint and not the other.

We use the symbols ∞ and $-\infty$ to represent positive infinity and negative infinity, respectively. Thus the notation (a, ∞) represents the set of all real numbers greater than a, and $(-\infty, a)$ represents the set of all real numbers less than a.

Interval notation for a set of numbers corresponds to its graph.

Student Notes

The notation for the *interval* (a, b) is the same as that for the *ordered pair* (a, b). The context in which the notation appears should make the meaning clear.

Interval Notation	Set-Builder Notation	Graph*
(a, b) open interval	$\{x\|a < x < b\}$	(a, b)
$[a, b]$ closed interval	$\{x\|a \leq x \leq b\}$	$[a, b]$
$(a, b]$ half-open interval	$\{x\|a < x \leq b\}$	$(a, b]$
$[a, b)$ half-open interval	$\{x\|a \leq x < b\}$	$[a, b)$
(a, ∞)	$\{x\|x > a\}$	
$[a, \infty)$	$\{x\|x \geq a\}$	
$(-\infty, a)$	$\{x\|x < a\}$	
$(-\infty, a]$	$\{x\|x \leq a\}$	

*The alternative representations —o—o— (a, b) and —•—•— (a, b) are sometimes used instead of, respectively, —(—)— (a, b) and —[—]— (a, b).

CAUTION! For interval notation, the order of the endpoints should mimic the number line, with the smaller number on the left and the larger on the right.

EXAMPLE 4 Graph $t \geq -2$ on the number line and write the solution set using both interval notation and set-builder notation.

SOLUTION Using interval notation, we write the solution set as $[-2, \infty)$. Using set-builder notation, we write the solution set as $\{t | t \geq -2\}$.

To graph the solution, we shade all numbers to the right of -2 and use a bracket to indicate that -2 is also a solution.

Try Exercise 25.

SOLVING INEQUALITIES USING THE ADDITION PRINCIPLE

Consider a balance similar to one that appears in Section 2.1. When one side of the balance holds more weight than the other, the balance tips in that direction. If equal amounts of weight are then added to or subtracted from both sides of the balance, the balance remains tipped in the same direction.

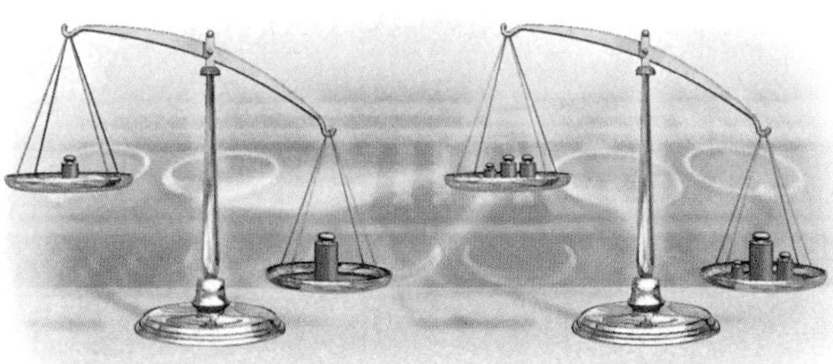

The balance illustrates the idea that when a number, such as 2, is added to (or subtracted from) both sides of a true inequality, such as $3 < 7$, we get another true inequality:

$$3 + 2 < 7 + 2, \quad \text{or} \quad 5 < 9.$$

Similarly, if we add -4 to both sides of $x + 4 < 10$, we get an *equivalent* inequality:

$$x + 4 + (-4) < 10 + (-4), \quad \text{or} \quad x < 6.$$

We say that $x + 4 < 10$ and $x < 6$ are **equivalent,** which means that both inequalities have the same solution set.

STUDY TIP

Sleep Well

Being well rested, alert, and focused is very important when studying math. Often, problems that may seem confusing to a sleepy person are easily understood after a good night's sleep. Using your time efficiently is always important, so you should be aware that an alert, wide-awake student can often accomplish more in 10 minutes than a sleepy student can accomplish in 30 minutes.

The Addition Principle for Inequalities For any real numbers a, b, and c:

$a < b$ is equivalent to $a + c < b + c$;
$a \leq b$ is equivalent to $a + c \leq b + c$;
$a > b$ is equivalent to $a + c > b + c$;
$a \geq b$ is equivalent to $a + c \geq b + c$.

As with equations, our goal is to isolate the variable on one side.

EXAMPLE 5 Solve $x + 2 > 8$ and then graph the solution.

SOLUTION We use the addition principle, subtracting 2 from both sides:

$$x + 2 - 2 > 8 - 2 \qquad \textbf{Subtracting 2 from, or adding } -2 \textbf{ to, both sides}$$
$$x > 6.$$

From the inequality $x > 6$, we can determine the solutions easily. Any number greater than 6 makes $x > 6$ true and is a solution of that inequality as well as the inequality $x + 2 > 8$. Using set-builder notation, we write the solution set as $\{x|x > 6\}$. Using interval notation, we write the solution set as $(6, \infty)$. The graph is as follows:

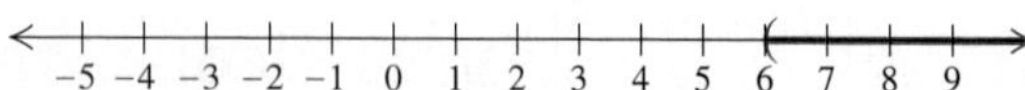

Because most inequalities have an infinite number of solutions, we cannot possibly check them all. A partial check can be made using one of the possible solutions. For this example, we can substitute any number greater than 6—say, 6.1—into the original inequality:

$x + 2 > 8$	
$6.1 + 2$	8
$8.1 \overset{?}{>} 8$	TRUE

8.1 > 8 is a true statement.

Since $8.1 > 8$ is true, 6.1 is a solution. Any number greater than 6 is a solution.

Try Exercise 41.

EXAMPLE 6 Solve $3x - 1 \le 2x - 5$ and then graph the solution.

SOLUTION We have

$$3x - 1 \le 2x - 5$$
$$3x - 1 + 1 \le 2x - 5 + 1 \qquad \textbf{Adding 1 to both sides}$$
$$3x \le 2x - 4 \qquad \textbf{Simplifying}$$
$$3x - 2x \le 2x - 4 - 2x \qquad \textbf{Subtracting } 2x \textbf{ from both sides}$$
$$x \le -4. \qquad \textbf{Simplifying}$$

The graph is as follows:

−7 −6 −5 −4 −3 −2 −1 0 1 2 3 4 5 6 7

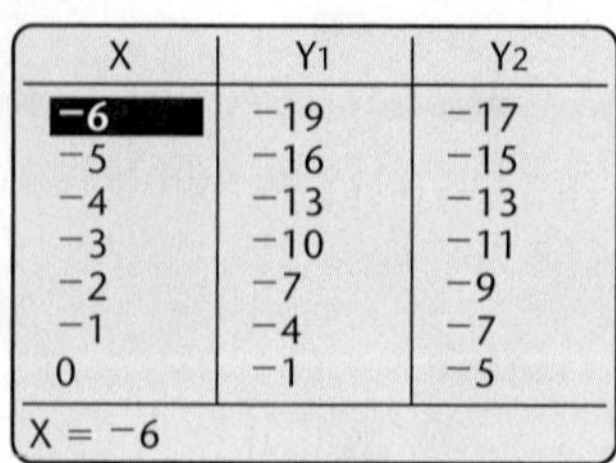

X	Y1	Y2
−6	−19	−17
−5	−16	−15
−4	−13	−13
−3	−10	−11
−2	−7	−9
−1	−4	−7
0	−1	−5

X = −6

Any number less than or equal to -4 is a solution, so the solution set is $\{x|x \le -4\}$, or, in interval notation, $(-\infty, -4]$.

As a partial check using the TABLE feature of a graphing calculator, we can let $y_1 = 3x - 1$ and $y_2 = 2x - 5$. By scrolling up or down, you can note that for $x \le -4$, we have $y_1 \le y_2$.

Try Exercise 47.

SOLVING INEQUALITIES USING THE MULTIPLICATION PRINCIPLE

There is a multiplication principle for inequalities similar to that for equations, but it must be modified when multiplying both sides by a negative number. Consider the true inequality

$$3 < 7.$$

If we multiply both sides by a *positive* number—say, 2—we get another true inequality:

$$3 \cdot 2 < 7 \cdot 2, \quad \text{or} \quad 6 < 14. \qquad \text{TRUE}$$

If we multiply both sides by a *negative* number—say, -2—we get a *false* inequality:

$$3 \cdot (-2) < 7 \cdot (-2), \quad \text{or} \quad -6 < -14. \qquad \text{FALSE}$$

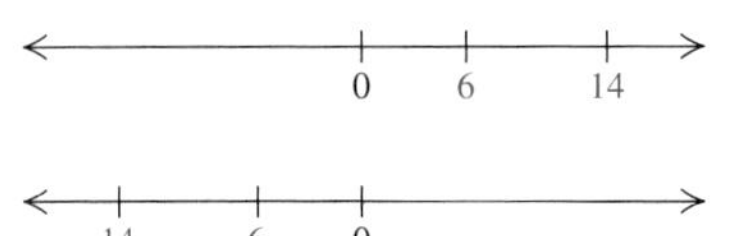

The fact that $6 < 14$ is true, but $-6 < -14$ is false, stems from the fact that the negative numbers, in a sense, mirror the positive numbers. Whereas 14 is to the *right* of 6, the number -14 is to the *left* of -6. Thus if we reverse the inequality symbol in $-6 < -14$, we get a true inequality:

$$-6 > -14. \qquad \text{TRUE}$$

The Multiplication Principle for Inequalities For any real numbers a and b:

when c is any *positive* number,

$a < b$ is equivalent to $ac < bc$, and

$a > b$ is equivalent to $ac > bc$;

when c is any *negative* number,

$a < b$ is equivalent to $ac > bc$, and

$a > b$ is equivalent to $ac < bc$.

Similar statements hold for $\leq$ and $\geq$.

CAUTION! When multiplying or dividing both sides of an inequality by a negative number, don't forget to reverse the inequality symbol!

EXAMPLE 7 Solve and graph each inequality: **(a)** $\frac{1}{4}x < 7$; **(b)** $-2y < 18$.

SOLUTION

a) $\frac{1}{4}x < 7$

$4 \cdot \frac{1}{4}x < 4 \cdot 7$ **Multiplying both sides by 4, the reciprocal of $\frac{1}{4}$**

The symbol stays the same, since 4 is positive.

$x < 28$ **Simplifying**

The solution set is $\{x | x < 28\}$, or $(-\infty, 28)$. The graph is as follows:

b) $-2y < 18$

$$\frac{-2y}{-2} > \frac{18}{-2}$$ **Multiplying both sides by $-\frac{1}{2}$, or dividing both sides by -2**

***At this step,* we reverse the inequality, because $-\frac{1}{2}$ is negative.**

$$y > -9$$ **Simplifying**

As a partial check, we substitute a number greater than -9, say -8, into the original inequality:

$$\begin{array}{c|c} -2y & < 18 \\ \hline -2(-8) & 18 \\ 16 & \overset{?}{<} 18 \end{array}$$ TRUE **$16 < 18$ is a true statement.**

The solution set is $\{y|y > -9\}$, or $(-9, \infty)$. The graph is as follows:

Try Exercise 59.

USING THE PRINCIPLES TOGETHER

We use the addition and multiplication principles together to solve inequalities much as we did when solving equations.

EXAMPLE 8 Solve: **(a)** $6 - 5y > 7$; **(b)** $2x - 9 \leq 7x + 1$.

SOLUTION

a)

$$6 - 5y > 7$$

$$-6 + 6 - 5y > -6 + 7$$ **Adding -6 to both sides**

$$-5y > 1$$ **Simplifying**

$$-\frac{1}{5} \cdot (-5y) < -\frac{1}{5} \cdot 1$$ **Multiplying both sides by $-\frac{1}{5}$, or dividing both sides by -5**

Remember to reverse the inequality symbol!

$$y < -\frac{1}{5}$$ **Simplifying**

As a partial check, we substitute a number smaller than $-\frac{1}{5}$, say -1, into the original inequality:

$$\begin{array}{c|c} 6 - 5y & > 7 \\ \hline 6 - 5(-1) & 7 \\ 6 - (-5) & \\ 11 & \overset{?}{>} 7 \end{array}$$ TRUE **$11 > 7$ is a true statement.**

The solution set is $\left\{y|y < -\frac{1}{5}\right\}$, or $\left(-\infty, -\frac{1}{5}\right)$.

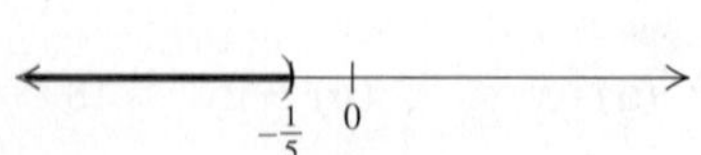

b)

$$
\begin{aligned}
2x - 9 &\le 7x + 1 \\
2x - 9 - 1 &\le 7x + 1 - 1 && \textbf{Subtracting 1 from both sides} \\
2x - 10 &\le 7x && \textbf{Simplifying} \\
2x - 10 - 2x &\le 7x - 2x && \textbf{Subtracting } 2x \textbf{ from both sides} \\
-10 &\le 5x && \textbf{Simplifying} \\
\frac{-10}{5} &\le \frac{5x}{5} && \textbf{Dividing both sides by 5} \\
-2 &\le x && \textbf{Simplifying}
\end{aligned}
$$

The solution set is $\{x|-2 \le x\}$, or $\{x|x \ge -2\}$, or $[-2, \infty)$.

Try Exercise 69.

All of the equation-solving techniques used in Sections 2.1 and 2.2 can be used with inequalities provided we remember to reverse the inequality symbol when multiplying or dividing both sides by a negative number.

EXAMPLE 9 Solve.

a) $16.3 - 7.2p \le -8.18$ **b)** $3(x - 9) - 1 < 2 - 5(x + 6)$

SOLUTION

a) The greatest number of decimal places in any one number is *two*. Multiplying both sides by 100 will clear decimals. Then we proceed as before.

$$
\begin{aligned}
16.3 - 7.2p &\le -8.18 \\
100(16.3 - 7.2p) &\le 100(-8.18) && \textbf{Multiplying both sides by 100} \\
100(16.3) - 100(7.2p) &\le 100(-8.18) && \textbf{Using the distributive law} \\
1630 - 720p &\le -818 && \textbf{Simplifying} \\
-720p &\le -818 - 1630 && \textbf{Subtracting 1630 from both sides} \\
-720p &\le -2448 && \textbf{Simplifying} \\
p &\ge \frac{-2448}{-720} && \textbf{Dividing both sides by } -720 \\
& && \textbf{Remember to reverse the symbol!} \\
p &\ge 3.4
\end{aligned}
$$

The solution set is $\{p|p \ge 3.4\}$, or $[3.4, \infty)$.

b)

$$
\begin{aligned}
3(x - 9) - 1 &< 2 - 5(x + 6) \\
3x - 27 - 1 &< 2 - 5x - 30 && \textbf{Using the distributive law to remove parentheses} \\
3x - 28 &< -5x - 28 && \textbf{Simplifying} \\
3x - 28 + 28 &< -5x - 28 + 28 && \textbf{Adding 28 to both sides} \\
3x &< -5x \\
3x + 5x &< -5x + 5x && \textbf{Adding } 5x \textbf{ to both sides} \\
8x &< 0 \\
x &< 0 && \textbf{Dividing both sides by 8}
\end{aligned}
$$

The solution set is $\{x|x < 0\}$, or $(-\infty, 0)$.

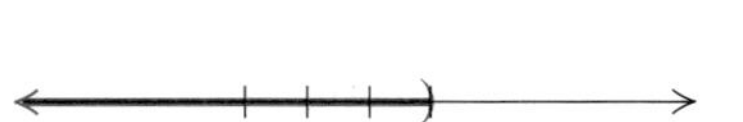

Try Exercise 89.

Connecting the Concepts

The procedure for solving inequalities is very similar to that used to solve equations. There are, however, two important differences.

- The multiplication principle for inequalities differs from the multiplication principle for equations: When we multiply or divide on both sides of an inequality by a *negative* number, we must *reverse* the direction of the inequality.
- The solution set of an equation like those we solved in this chapter typically consists of one number. The solution set of an inequality typically consists of a set of numbers and is written using either set-builder notation or interval notation.

Compare the following solutions.

Solve: $2 - 3x = x + 10$.

SOLUTION

$2 - 3x = x + 10$	
$-3x = x + 8$	**Subtracting 2 from both sides**
$-4x = 8$	**Subtracting x from both sides**
$x = -2$	**Dividing both sides by -4**

The solution is -2.

Solve: $2 - 3x > x + 10$.

SOLUTION

$2 - 3x > x + 10$	
$-3x > x + 8$	**Subtracting 2 from both sides**
$-4x > 8$	**Subtracting x from both sides**
$x < -2$	**Dividing both sides by -4 and reversing the direction of the inequality symbol**

The solution is $\{x|x < -2\}$, or $(-\infty, -2)$.

2.6 Exercise Set

FOR EXTRA HELP MyMathLab

Concept Reinforcement *Insert the symbol $<, >, \leq,$ or $\geq$ to make each pair of inequalities equivalent.*

1. $-5x \leq 30;\ x \,\square\, -6$
2. $-7t \geq 56;\ t \,\square\, -8$
3. $-2t > -14;\ t \,\square\, 7$
4. $-3x < -15;\ x \,\square\, 5$

Classify each pair of inequalities as "equivalent" or "not equivalent."

5. $x < -2;\ -2 > x$
6. $t > -1;\ -1 < t$
7. $-4x - 1 \leq 15;$ $-4x \leq 16$
8. $-2t + 3 \geq 11;$ $-2t \geq 14$

Determine whether each number is a solution of the given inequality.

9. $x > -2$
   a) 5 b) 0 c) -3
10. $y < 5$
   a) 0 b) 5 c) -13
11. $y \leq 19$
   a) 18.99 b) 19.01 c) 19
12. $x \geq 11$
   a) 11 b) $11\frac{1}{2}$ c) $10\frac{2}{3}$
13. $a \geq -6$
   a) -6 b) -6.1 c) -5.9

14. $c \leq -10$
a) 0 b) -10 c) -10.1

Graph on the number line.

15. $x \leq 7$
16. $y < 2$
17. $t > -2$
18. $y > 4$
19. $1 \leq m$
20. $0 \leq t$
21. $-3 < x \leq 5$
22. $-5 \leq x < 2$
23. $0 < x < 3$
24. $-5 \leq x \leq 0$

Graph each inequality, and write the solution set using both set-builder notation and interval notation.

25. $y < 6$
26. $x > 4$
27. $x \geq -4$
28. $t \leq 6$
29. $t > -3$
30. $y < -3$
31. $x \leq -7$
32. $x \geq -6$

Describe each graph using both set-builder notation and interval notation.

33. −7 −6 −5 −4 −3 −2 −1 0 1 2 3 4 5 6 7
34. −7 −6 −5 −4 −3 −2 −1 0 1 2 3 4 5 6 7
35. −7 −6 −5 −4 −3 −2 −1 0 1 2 3 4 5 6 7
36. −7 −6 −5 −4 −3 −2 −1 0 1 2 3 4 5 6 7
37. −7 −6 −5 −4 −3 −2 −1 0 1 2 3 4 5 6 7
38. −7 −6 −5 −4 −3 −2 −1 0 1 2 3 4 5 6 7
39. −7 −6 −5 −4 −3 −2 −1 0 1 2 3 4 5 6 7
40. −7 −6 −5 −4 −3 −2 −1 0 1 2 3 4 5 6 7

Solve using the addition principle. Graph and write both set-builder notation and interval notation for each answer.

41. $y + 2 > 9$
42. $y + 6 > 9$
43. $x - 8 \leq -10$
44. $x - 9 \leq -12$
45. $5 \leq t + 8$
46. $4 \leq t + 9$
47. $2x + 4 \leq x + 9$
48. $2x + 4 \leq x + 1$

Solve using the addition principle.

49. $y + \frac{1}{3} \leq \frac{5}{6}$
50. $x + \frac{1}{4} \leq \frac{1}{2}$
51. $t - \frac{1}{8} > \frac{1}{2}$
52. $y - \frac{1}{3} > \frac{1}{4}$
53. $-9x + 17 > 17 - 8x$
54. $-8n + 12 > 12 - 7n$

Aha! 55. $-23 < -t$
56. $19 < -x$

Solve using the multiplication principle. Graph and write both set-builder notation and interval notation for each answer.

57. $5x < 35$
58. $8x \geq 32$
59. $-24 > 8t$
60. $-16x < -64$
61. $1.8 \geq -1.2n$
62. $9 \leq -2.5a$

Solve using the multiplication principle.

63. $-2y \leq \frac{1}{5}$
64. $-2x \geq \frac{1}{5}$
65. $-\frac{8}{5} > -2x$
66. $-\frac{5}{8} < -10y$

Solve using the addition and multiplication principles.

67. $7 + 3x < 34$
68. $5 + 4y < 37$
69. $4t - 5 \leq 23$
70. $13x - 7 < -46$
71. $16 < 4 - a$
72. $22 < 6 - n$
73. $5 - 7y \geq 5$
74. $8 - 2y \geq 14$
75. $-3 < 8x + 7 - 7x$
76. $-5 < 9x + 8 - 8x$
77. $6 - 4y > 4 - 3y$
78. $7 - 8y > 5 - 7y$
79. $7 - 9y \leq 4 - 8y$
80. $6 - 13y \leq 4 - 12y$
81. $2.1x + 43.2 > 1.2 - 8.4x$
82. $0.96y - 0.79 \leq 0.21y + 0.46$
83. $0.7n - 15 + n \geq 2n - 8 - 0.4n$
84. $1.7t + 8 - 1.62t < 0.4t - 0.32 + 8$
85. $\frac{x}{3} - 4 \leq 1$
86. $\frac{2}{3} - \frac{x}{5} < \frac{4}{15}$
87. $3 < 5 - \frac{t}{7}$
88. $2 > 9 - \frac{x}{5}$
89. $4(2y - 3) \leq -44$
90. $3(2y - 3) \geq 21$
91. $3(t - 2) \geq 9(t + 2)$

92. $8(2t + 1) > 4(7t + 7)$

93. $3(r - 6) + 2 < 4(r + 2) - 21$

94. $5(t + 3) + 9 > 3(t - 2) + 6$

95. $\frac{2}{3}(2x - 1) \ge 10$

96. $\frac{4}{5}(3x + 4) \le 20$

97. $\frac{3}{4}\left(3x - \frac{1}{2}\right) - \frac{2}{3} < \frac{1}{3}$

98. $\frac{2}{3}\left(\frac{7}{8} - 4x\right) - \frac{5}{8} < \frac{3}{8}$

TW **99.** Are the inequalities $x > -3$ and $x \ge -2$ equivalent? Why or why not?

TW **100.** Are the inequalities $t > -7$ and $7 < -t$ equivalent? Why or why not?

SKILL REVIEW

Review solving equations and inequalities (Sections 2.2 and 2.6).

Solve.

101. $4 - x = 8 - 5x$ [2.2]

102. $4 - x > 8 - 5x$ [2.6]

103. $2(5 - x) = \frac{1}{2}(x + 1)$ [2.2]

104. $2(5 - x) \le \frac{1}{2}(x + 1)$ [2.6]

SYNTHESIS

TW **105.** Explain how it is possible for the graph of an inequality to consist of just one number. (*Hint*: See Example 3c.)

TW **106.** The statements of the addition and multiplication principles for inequalities begin with *conditions* set for the variables. Explain the conditions given for each principle.

Solve.

Aha! **107.** $x < x + 1$

108. $6[4 - 2(6 + 3t)] > 5[3(7 - t) - 4(8 + 2t)] - 20$

109. $27 - 4[2(4x - 3) + 7] \ge 2[4 - 2(3 - x)] - 3$

Solve for x.

110. $-(x + 5) \ge 4a - 5$

111. $\frac{1}{2}(2x + 2b) > \frac{1}{3}(21 + 3b)$

112. $y < ax + b$ (Assume $a > 0$.)

113. $y < ax + b$ (Assume $a < 0$.)

114. Graph the solutions of $|x| < 3$ on the number line.

Aha! **115.** Determine the solution set of $|x| > -3$.

116. Determine the solution set of $|x| < 0$.

Try Exercise Answers: Section 2.6

9. (a) Yes; **(b)** yes; **(c)** no **11. (a)** Yes; **(b)** no; **(c)** yes
15. 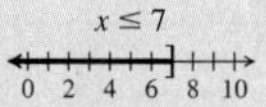

25. 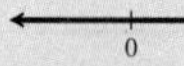

; $\{y \mid y < 6\}$, $(-\infty, 6)$
41. $\{y \mid y > 7\}$, or $(7, \infty)$; 

47. $\{x \mid x \le 5\}$, or $(-\infty, 5]$; 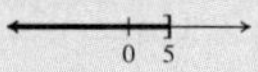

59. $\{t \mid t < -3\}$, or $(-\infty, -3)$; 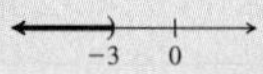

69. $\{t \mid t \le 7\}$, or $(-\infty, 7]$
89. $\{y \mid y \le -4\}$, or $(-\infty, -4]$

2.7 Solving Applications with Inequalities

- Translating to Inequalities
- Solving Problems

The five steps for problem solving can be used for problems involving inequalities.

TRANSLATING TO INEQUALITIES

Before solving problems that involve inequalities, we list some important phrases to look for. Sample translations are listed as well.

Important Words	Sample Sentence	Definition of Variables	Translation
is at least	Kelby walks at least 1.5 mi a day.	Let k represent the length of Kelby's walk, in miles.	$k \geq 1.5$
is at most	At most 5 students dropped the course.	Let n represent the number of students who dropped the course.	$n \leq 5$
cannot exceed	The cost cannot exceed $12,000.	Let c represent the cost, in dollars.	$c \leq 12{,}000$
must exceed	The speed must exceed 40 mph.	Let s represent the speed, in miles per hour.	$s > 40$
is less than	Hamid's weight is less than 130 lb.	Let w represent Hamid's weight, in pounds.	$w < 130$
is more than	Boston is more than 200 mi away.	Let d represent the distance to Boston, in miles.	$d > 200$
is between	The film is between 90 min and 100 min long.	Let t represent the length of the film, in minutes.	$90 < t < 100$
minimum	Ned drank a minimum of 5 glasses of water a day.	Let w represent the number of glasses of water.	$w \geq 5$
maximum	The maximum penalty is $100.	Let p represent the penalty, in dollars.	$p \leq 100$
no more than	Alan consumes no more than 1500 calories.	Let c represent the number of calories Alan consumes.	$c \leq 1500$
no less than	Patty scored no less than 80.	Let s represent Patty's score.	$s \geq 80$

The following phrases deserve special attention.

Translating "At Least" and "At Most"

The quantity x is at least some amount q: $x \geq q$.

(If x is *at least* q, it cannot be less than q.)

The quantity x is at most some amount q: $x \leq q$.

(If x is *at most* q, it cannot be more than q.)

SOLVING PROBLEMS

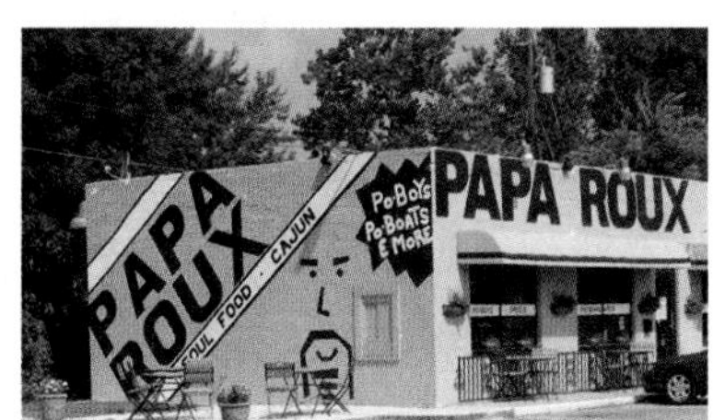

EXAMPLE 1 Catering Costs. To cater a party, Papa Roux charges a $50 setup fee plus $15 per person. The cost of Hotel Pharmacy's end-of-season softball party cannot exceed $450. How many people can attend the party?

SOLUTION

1. **Familiarize.** Suppose that 20 people were to attend the party. The cost would then be $\$50 + \$15 \cdot 20$, or $350. This shows that more than 20 people could attend without exceeding $450. Instead of making another guess, we let $n =$ the number of people in attendance.

2. **Translate.** The cost of the party will be $50 for the setup fee plus $15 times the number of people attending. We can reword as follows:

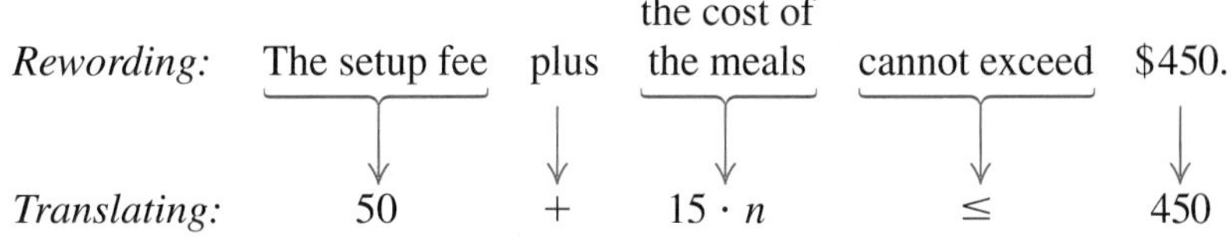

Rewording:	The setup fee	plus	the cost of the meals	cannot exceed	$450.
Translating:	50	$+$	$15 \cdot n$	$\leq$	450

3. **Carry out.** We solve for n:

$$50 + 15n \leq 450$$

$$15n \leq 400 \quad \textbf{Subtracting 50 from both sides}$$

$$n \leq \frac{400}{15} \quad \textbf{Dividing both sides by 15}$$

$$n \leq 26\frac{2}{3}. \quad \textbf{Simplifying}$$

4. **Check.** The solution set is all numbers less than or equal to $26\frac{2}{3}$. Since n represents the number of people in attendance, we round to a whole number. Since the nearest whole number, 27, is not part of the solution set, we round *down* to 26. If 26 people attend, the cost will be $\$50 + \$15 \cdot 26$, or \$440, and if 27 attend, the cost will exceed \$450.
5. **State.** At most 26 people can attend the party.

Try Exercise 47.

CAUTION! Solutions of problems should always be checked using the original wording of the problem. In some cases, answers might need to be whole numbers or integers or rounded off in a particular direction.

Some applications with inequalities involve *averages*, or *means*. You are already familiar with the concept of averages from grades in courses that you have taken.

Average, or Mean

To find the **average**, or **mean**, of a set of numbers, add the numbers and then divide by the number of addends.

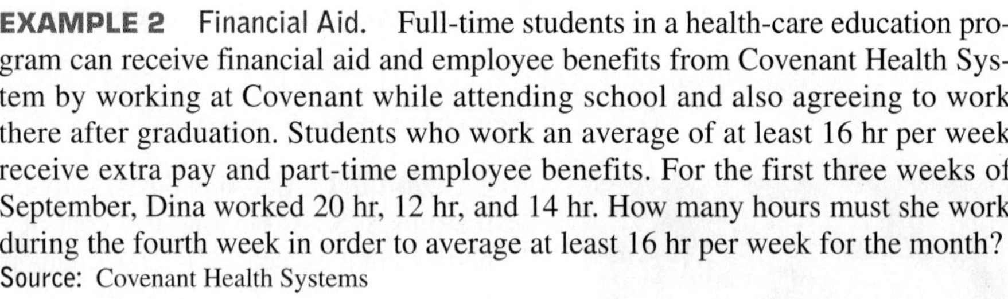

EXAMPLE 2 Financial Aid. Full-time students in a health-care education program can receive financial aid and employee benefits from Covenant Health System by working at Covenant while attending school and also agreeing to work there after graduation. Students who work an average of at least 16 hr per week receive extra pay and part-time employee benefits. For the first three weeks of September, Dina worked 20 hr, 12 hr, and 14 hr. How many hours must she work during the fourth week in order to average at least 16 hr per week for the month?
Source: Covenant Health Systems

SOLUTION

1. **Familiarize.** Suppose Dina works 10 hr during the fourth week. Her average for the month would be

$$\frac{20 \text{ hr} + 12 \text{ hr} + 14 \text{ hr} + 10 \text{ hr}}{4} = 14 \text{ hr}. \quad \textbf{There are 4 addends, so we divide by 4.}$$

This shows that Dina must work more than 10 hr during the fourth week, if she is to average at least 16 hr of work per week. We let x represent the number of hours Dina works during the fourth week.

2. **Translate.** We reword the problem and translate as follows:

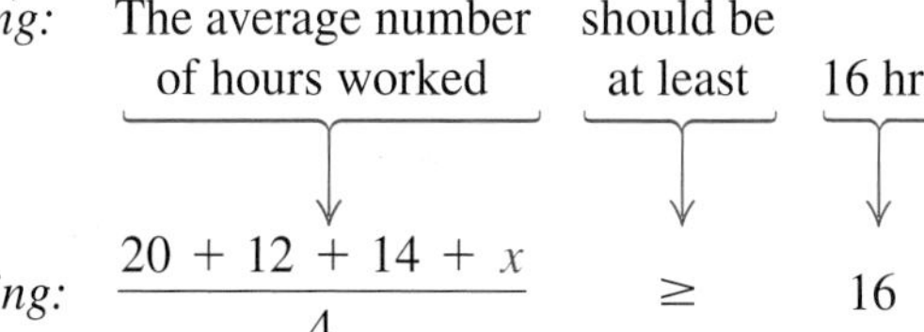

Rewording: The average number of hours worked / should be at least / 16 hr.

Translating: $\dfrac{20 + 12 + 14 + x}{4} \geq 16$

3. **Carry out.** Because of the fraction, it is convenient to use the multiplication principle first:

$$\frac{20 + 12 + 14 + x}{4} \geq 16$$

$$4\left(\frac{20 + 12 + 14 + x}{4}\right) \geq 4 \cdot 16 \qquad \textbf{Multiplying both sides by 4}$$

$$20 + 12 + 14 + x \geq 64$$

$$46 + x \geq 64 \qquad \textbf{Simplifying}$$

$$x \geq 18. \qquad \textbf{Subtracting 46 from both sides}$$

X	Y1	
16	15.5	
17	15.75	
18	16	
19	16.25	
20	16.5	
21	16.75	
22	17	
X = 18		

4. **Check.** As a partial check, we let $y = (20 + 12 + 14 + x)/4$ and create a table of values, as shown at left. We see that when $x = 18$, the average is 16 hr. For $x > 18$, the average is more than 16 hr.

5. **State.** Dina will average at least 16 hr of work per week for September if she works at least 18 hr during the fourth week.

Try Exercise 25.

STUDY TIP

Avoiding Temptation

Make the most of your study time by choosing the right location. For example, sit in a comfortable chair in a well-lit location, but stay away from a coffee shop where friends may stop by. Once you begin studying, let the answering machine answer the phone, shut off any cell phones, and do not check e-mail.

EXAMPLE 3 Job Offers. After graduation, Jessica had two job offers in sales:

Uptown Fashions: A salary of \$600 per month, plus a commission of 4% of sales;

Ergo Designs: A salary of \$800 per month, plus a commission of 6% of sales in excess of \$10,000.

If sales always exceed \$10,000, for what amount of sales would Uptown Fashions provide higher pay?

SOLUTION

1. **Familiarize.** Suppose that Jessica sold a certain amount—say, \$12,000—in one month. Which plan would be better? Working for Uptown, she would earn \$600 plus 4% of \$12,000, or

$$\$600 + 0.04(\$12{,}000) = \$1080.$$

Since with Ergo Designs commissions are paid only on sales in excess of \$10,000, Jessica would earn \$800 plus 6% of (\$12,000 − \$10,000), or

$$\$800 + 0.06(\$2000) = \$920.$$

This shows that for monthly sales of \$12,000, Uptown's rate of pay is better. Similar calculations will show that for sales of \$30,000 per month, Ergo's rate of pay is better. To determine *all* values for which Uptown pays more money, we must solve an inequality that is based on the calculations above.

We let $S =$ the amount of monthly sales, in dollars, and will assume that $S > 10{,}000$ so that both plans will pay a commission. Listing the given information in a table will be helpful.

Uptown Fashions Monthly Income	Ergo Designs Monthly Income
\$600 salary 4% of sales $= 0.04S$ *Total*: $600 + 0.04S$	\$800 salary 6% of sales over \$10,000 $= 0.06(S - 10{,}000)$ *Total*: $800 + 0.06(S - 10{,}000)$

2. **Translate.** We want to find all values of S for which

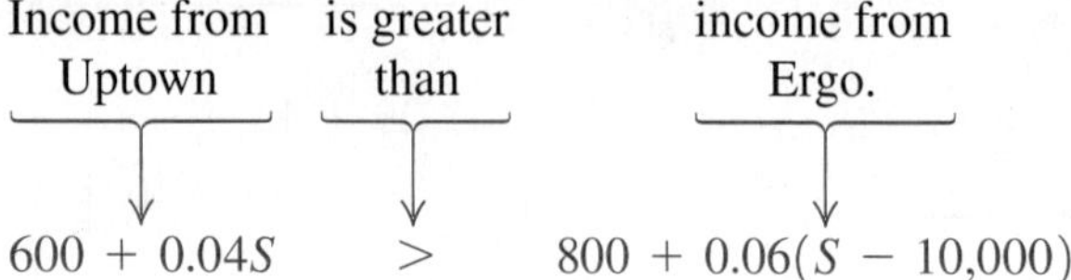

$$600 + 0.04S \quad > \quad 800 + 0.06(S - 10{,}000)$$

3. **Carry out.** We solve the inequality:

$$600 + 0.04S > 800 + 0.06(S - 10{,}000)$$

$$600 + 0.04S > 800 + 0.06S - 600 \qquad \text{Using the distributive law}$$

$$600 + 0.04S > 200 + 0.06S \qquad \text{Combining like terms}$$

$$400 > 0.02S \qquad \text{Subtracting 200 and } 0.04S \text{ from both sides}$$

$$20{,}000 > S, \text{ or } S < 20{,}000. \qquad \text{Dividing both sides by 0.02}$$

4. **Check.** The steps above indicate that income from Uptown Fashions is higher than income from Ergo Designs for sales less than \$20,000. In the *Familiarize* step, we saw that for sales of \$12,000, Uptown pays more. Since $12{,}000 < 20{,}000$, this is a partial check.

5. **State.** When monthly sales are less than \$20,000, Uptown Fashions provides the higher pay (assuming both plans pay a commission).

Try Exercise 55.

Translating for Success

1. ***Consecutive Integers.*** The sum of two consecutive even integers is 102. Find the integers.

2. ***Salary Increase.*** After Susanna earned a 5% raise, her new salary was \$25,750. What was her former salary?

3. ***Dimensions of a Rectangle.*** The length of a rectangle is 6 in. more than the width. The perimeter of the rectangle is 102 in. Find the length and the width.

4. ***Population.*** The population of Kelling Point is decreasing at a rate of 5% per year. The current population is 25,750. What was the population the previous year?

5. ***Reading Assignment.*** Quinn has 6 days to complete a 150-page reading assignment. How many pages must he read the first day so that he has no more than 102 pages left to read on the 5 remaining days?

Translate each word problem to an equation or an inequality and select a correct translation from A–O.

A. $0.05(25{,}750) = x$

B. $x + 2x = 102$

C. $2x + 2(x + 6) = 102$

D. $150 - x \leq 102$

E. $x - 0.05x = 25{,}750$

F. $x + (x + 2) = 102$

G. $x + (x + 6) > 102$

H. $x + 5x = 150$

I. $x + 0.05x = 25{,}750$

J. $x + (2x + 6) = 102$

K. $x + (x + 1) = 102$

L. $102 + x > 150$

M. $0.05x = 25{,}750$

N. $102 + 5x > 150$

O. $x + (x + 6) = 102$

Answers on page A-5

An additional, animated version of this activity appears in MyMathLab. To use MyMathLab, you need a course ID and a student access code. Contact your instructor for more information.

6. ***Numerical Relationship.*** One number is 6 more than twice another. The sum of the numbers is 102. Find the numbers.

7. ***DVD Collections.*** Together Mindy and Ken have 102 DVDs. If Ken has 6 more DVDs than Mindy, how many does each have?

8. ***Sales Commissions.*** Kirk earns a commission of 5% on his sales. One year he earned commissions totaling \$25,750. What were his total sales for the year?

9. ***Fencing.*** Jess has 102 ft of fencing that he plans to use to enclose two dog runs. The perimeter of one run is to be twice the perimeter of the other. Into what lengths should the fencing be cut?

10. ***Quiz Scores.*** Lupe has a total of 102 points on the first 6 quizzes in her sociology class. How many total points must she earn on the 5 remaining quizzes in order to have more than 150 points for the semester?

2.7 Exercise Set

FOR EXTRA HELP MyMathLab Math XL PRACTICE  WATCH  DOWNLOAD  READ 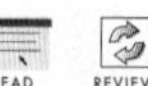 REVIEW

Concept Reinforcement *In each of Exercises 1–8, match the sentence with one of the following:*

$a < b$; $a \le b$; $b < a$; $b \le a$.

1. a is at least b.
2. a exceeds b.
3. a is at most b.
4. a is exceeded by b.
5. b is no more than a.
6. b is no less than a.
7. b is less than a.
8. b is more than a.

Translate to an inequality.

9. A number is less than 10.
10. A number is greater than or equal to 4.
11. The temperature is at most $-3°$C.
12. The credit-card debt of the average college freshman is at least \$2000.
13. The age of the Mayan altar exceeds 1200 yr.
14. The time of the test was between 45 min and 55 min.
15. Normandale Community College is no more than 15 mi away.
16. Angenita earns no less than \$12 per hour.
17. To rent a car, a driver must have a minimum of 5 years of driving experience.
18. The maximum safe-exposure limit of formaldehyde is 2 parts per million.
19. The costs of production of the software cannot exceed \$12,500.
20. The cost of gasoline was at most \$4 per gallon.

Use an inequality and the five-step process to solve each problem.

21. *Furnace Repairs.* RJ's Plumbing and Heating charges \$55 plus \$40 per hour for emergency service. Gary remembers being billed over \$100 for an emergency call. How long was RJ's there?

22. *College Tuition.* Karen's financial aid stipulates that her tuition not exceed \$1000. If her local community college charges a \$35 registration fee plus \$375 per course, what is the greatest number of courses for which Karen can register?

23. *Graduate School.* An unconditional acceptance into the Master of Business Administration (MBA) program at Arkansas State University will be given to students whose GMAT score plus 200 times the undergraduate grade point average is at least 950. Chloe's GMAT score was 500. What must her grade point average be in order to be unconditionally accepted into the program?
Source: graduateschool.astate.edu

24. *Car Payments.* As a rule of thumb, debt payments (other than mortgages) should be less than 8% of a consumer's monthly gross income. Oliver makes \$54,000 per year and has a \$100 student-loan payment every month. What size car payment can he afford?
Source: money.cnn.com

25. *Quiz Average.* Rod's quiz grades are 73, 75, 89, and 91. What scores on a fifth quiz will make his average quiz grade at least 85?

26. *Nutrition.* Following the guidelines of the Food and Drug Administration, Dale tries to eat at least 5 servings of fruits or vegetables each day. For the first six days of one week, she had 4, 6, 7, 4, 6, and 4 servings. How many servings of fruits or vegetables should Dale eat on Saturday, in order to average at least 5 servings per day for the week?

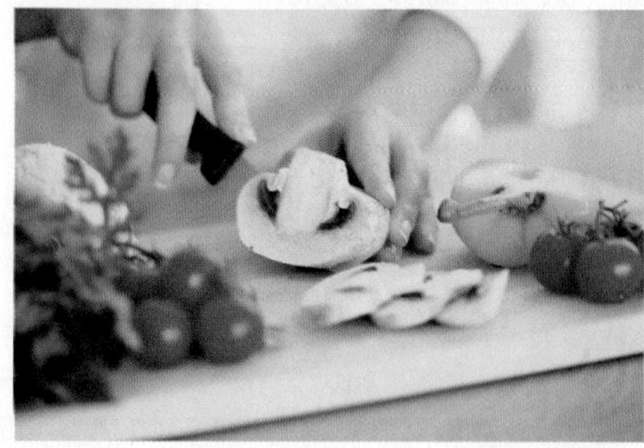

27. *College Course Load.* To remain on financial aid, Millie must complete an average of at least 7 credits per quarter each year. In the first three quarters of 2010, Millie completed 5, 7, and 8 credits. How many credits of course work must Millie complete in the fourth quarter if she is to remain on financial aid?

28. *Music Lessons.* Band members at Colchester Middle School are expected to average at least 20 min of practice time per day. One week Monroe practiced 15 min, 28 min, 30 min, 0 min, 15 min, and 25 min. How long must he practice on the seventh day if he is to meet expectations?

29. *Baseball.* In order to qualify for a batting title, a major-league baseball player must average at least 3.1 plate appearances per game. For the first nine games of the season, a player had 5, 1, 4, 2, 3, 4, 4, 3, and 2 plate appearances. How many plate appearances must the player have in the tenth game in order to average at least 3.1 per game?
Source: Major League Baseball

30. *Education.* The Mecklenberg County Public Schools stipulate that a standard school day will average at least $5\frac{1}{2}$ hr, excluding meal breaks. For the first four days of one school week, bad weather resulted in school days of 4 hr, $6\frac{1}{2}$ hr, $3\frac{1}{2}$ hr, and $6\frac{1}{2}$ hr. How long must the Friday school day be in order to average at least $5\frac{1}{2}$ hr for the week?
Source: www.meck.k12.va.us

31. *Perimeter of a Triangle.* One side of a triangle is 2 cm shorter than the base. The other side is 3 cm longer than the base. What lengths of the base will allow the perimeter to be greater than 19 cm?

32. *Perimeter of a Pool.* The perimeter of a rectangular swimming pool is not to exceed 70 ft. The length is to be twice the width. What widths will meet these conditions?

33. *Well Drilling.* All Seasons Well Drilling offers two plans. Under the "pay-as-you-go" plan, they charge $500 plus $8 per foot for a well of any depth. Under their "guaranteed-water" plan, they charge a flat fee of $4000 for a well that is guaranteed to provide adequate water for a household. For what depths would it save a customer money to use the pay-as-you-go plan?

34. *Cost of Road Service.* Rick's Automotive charges $50 plus $15 for each (15-min) unit of time when making a road call. Twin City Repair charges $70 plus $10 for each unit of time. Under what circumstances would it be more economical for a motorist to call Rick's?

35. *Insurance-Covered Repairs.* Most insurance companies will replace a vehicle if an estimated repair exceeds 80% of the "blue-book" value of the vehicle. Michelle's insurance company paid $8500 for repairs to her Subaru after an accident. What can be concluded about the blue-book value of the car?

36. *Insurance-Covered Repairs.* Following an accident, Jeff's Ford pickup was replaced by his insurance company because the damage was so extensive. Before the damage, the blue-book value of the truck was $21,000. How much would it have cost to repair the truck? (See Exercise 35.)

37. *Sizes of Envelopes.* Rhetoric Advertising is a direct-mail company. It determines that for a particular campaign, it can use any envelope with a fixed width of $3\frac{1}{2}$ in. and an area of at least $17\frac{1}{2}$ in^2. Determine (in terms of an inequality) those lengths that will satisfy the company constraints.

38. *Sizes of Packages.* The U.S. Postal Service defines a "package" as a parcel for which the sum of the length and the girth is less than 84 in. (Length is the longest side of a package and girth is the distance around the other two sides of the package.) A box has a fixed girth of 29 in. Determine (in terms of an inequality) those lengths for which the box is considered a "package."

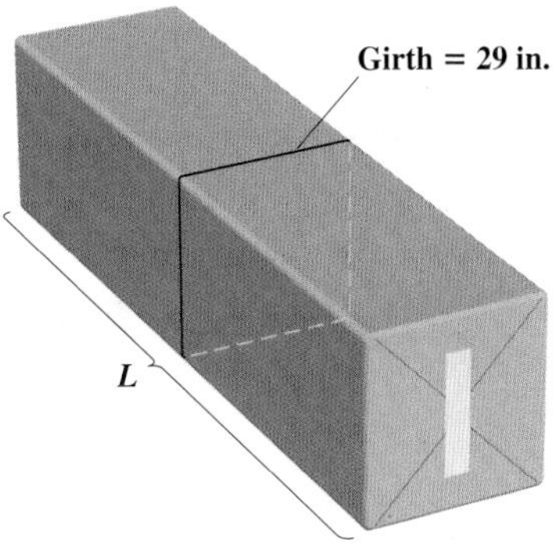

39. *Body Temperature.* A person is considered to be feverish when his or her temperature is higher than 98.6°F. The formula $F = \frac{9}{5}C + 32$ can be used to convert Celsius temperatures C to Fahrenheit temperatures F. For which Celsius temperatures is a person considered feverish?

40. *Gold Temperatures.* Gold stays solid at Fahrenheit temperatures below 1945.4°. Determine (in terms of an inequality) those Celsius temperatures for which gold stays solid. Use the formula given in Exercise 39.

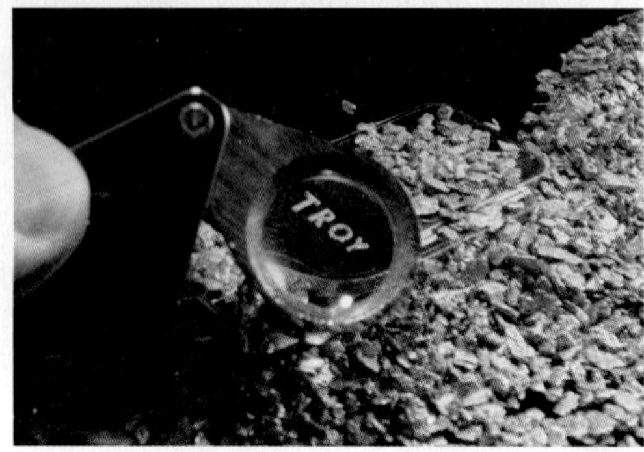

41. *Area of a Triangular Flag.* As part of an outdoor education course, Tricia needs to make a bright-colored triangular flag with an area of at least 3 ft^2. What heights can the triangle be if the base is $1\frac{1}{2}$ ft?

42. *Area of a Triangular Sign.* Zoning laws in Harrington prohibit displaying signs with areas exceeding 12 ft^2. If Flo's Marina is ordering a triangular sign with an 8-ft base, how tall can the sign be?

43. *Fat Content in Foods.* Reduced Fat Oreo® cookies contain 4.5 g of fat per serving. In order for a food to be labeled "reduced fat," it must have at least 25% less fat than the regular item. What can you conclude about the number of grams of fat in a serving of the regular Oreo cookies?
Source: Nutrition facts label

44. *Fat Content in Foods.* Reduced Fat Sargento® colby cheese contains 6 g of fat per serving. What can you conclude about the number of grams of fat in regular Sargento colby cheese? (See Exercise 43.)
Source: Nutrition facts label

45. *Weight Gain.* In the last weeks before the yearly Topsfield Weigh In, heavyweight pumpkins gain about 26 lb per day. Charlotte's heaviest pumpkin weighs 532 lb on September 5. For what dates will its weight exceed 818 lb?

46. *Pond Depth.* On July 1, Garrett's Pond was 25 ft deep. Since that date, the water level has dropped $\frac{2}{3}$ ft per week. For what dates will the water level not exceed 21 ft?

47. *Cell-Phone Budget.* Braden has budgeted \$60 per month for his cell phone. For his service, he pays a monthly fee of \$39.95, plus taxes of \$6.65, plus 10¢ for each text message sent or received. How many text messages can he send or receive and not exceed his budget?

48. *Banquet Costs.* The women's volleyball team can spend at most \$450 for its awards banquet at a local restaurant. If the restaurant charges a \$40 setup fee plus \$16 per person, at most how many can attend?

49. *World Records in the Mile Run.* The formula $R = -0.0065t + 4.3259$ can be used to predict the world record, in minutes, for the 1-mi run t years after 1900. Determine (in terms of an inequality) those years for which the world record will be less than 3.6 min.
Source: Based on information from Information Please Database 2007, Pearson Education, Inc.

50. *Women's Records in the Women's 1500-m Run.* The formula $R = -0.0026t + 4.0807$ can be used to predict the world record, in minutes, for the 1500-m run t years after 1900. Determine (in terms of an inequality) those years for which the world record will be less than 3.8 min.
Source: Based on information from *Track and Field*

51. *Toll Charges.* The equation $y = 0.06x + 0.50$ can be used to determine the approximate cost y, in dollars, of driving x miles on the Pennsylvania Turnpike. For what mileages x will the cost be at most \$14?

52. *Price of a Movie Ticket.* The average price of a movie ticket can be estimated by the equation $P = 0.169Y - 333.04$, where Y is the year and P is the average price, in dollars. For what years will the average price of a movie ticket be at least \$7? (Include the year in which the \$7 ticket first occurs.)
Source: Based on data from National Association of Theatre Owners

53. *Checking-Account Rates.* The Hudson Bank offers two checking-account plans. Their Anywhere plan charges 20¢ per check whereas their Acu-checking plan costs \$2 per month plus 12¢ per check. For what numbers of checks per month will the Acu-checking plan cost less?

54. *Moving Costs.* Musclebound Movers charges \$85 plus \$40 per hour to move households across town. Champion Moving charges \$60 per hour for cross-town moves. For what lengths of time is Champion more expensive?

55. *Wages.* Toni can be paid in one of two ways:

Plan A: A salary of \$400 per month, plus a commission of 8% of gross sales;

Plan B: A salary of \$610 per month, plus a commission of 5% of gross sales.

For what amount of gross sales should Toni select plan A?

56. *Wages.* Aiden can be paid for his masonry work in one of two ways:

Plan A: \$300 plus \$9.00 per hour;

Plan B: Straight \$12.50 per hour.

Suppose that the job takes n hours. For what values of n is plan B better for Aiden?

57. *Car Rental.* Abriana rented a compact car for a business trip. At the time of the rental, she was given the option of prepaying for an entire tank of gasoline at \$3.099 per gallon, or waiting until her return and paying \$6.34 per gallon for enough gasoline to fill the tank. If the tank holds 14 gal, how many gallons can she use and still save money by choosing the second option?

58. Refer to Exercise 57. If Abriana's rental car gets 30 mpg, how many miles must she drive in order to make the first option more economical?

TW 59. If f represents Fran's age and w represents Walt's age, write a sentence that would translate to $w + 3 < f$.

TW 60. Explain how the meanings of "Five more than a number" and "Five is more than a number" differ.

SKILL REVIEW

Review operations with real numbers (Sections 1.5–1.8).

Simplify.

61. $-2 + (-5) - 7$ [1.6]

62. $\frac{1}{2} \div \left(-\frac{3}{4}\right)$ [1.7]

63. $3 \cdot (-10) \cdot (-1) \cdot (-2)$ [1.7]

64. $-6.3 + (-4.8)$ [1.5]

65. $(3 - 7) - (4 - 8)$ [1.8]

66. $3 - 2 + 5 \cdot 10 \div 5^2 \cdot 2$ [1.8]

67. $\frac{-2 - (-6)}{8 - 10}$ [1.8]

68. $\frac{1 - (-7)}{-3 - 5}$ [1.8]

SYNTHESIS

TW 69. Write a problem for a classmate to solve. Devise the problem so the answer is "The Rothmans can drive 90 mi without exceeding their truck rental budget."

TW 70. Write a problem for a classmate to solve. Devise the problem so the answer is "At most 18 passengers can go on the boat." Design the problem so that at least one number in the solution must be rounded down.

71. *Wedding Costs.* The Arnold Inn offers two plans for wedding parties. Under plan A, the inn charges \$30 for each person in attendance. Under plan B, the inn charges \$1300 plus \$20 for each person in excess of the first 25 who attend. For what size parties will plan B cost less? (Assume that more than 25 guests will attend.)

72. *Insurance Benefits.* Bayside Insurance offers two plans. Under plan A, Giselle would pay the first \$50 of her medical bills and 20% of all bills after that. Under plan B, Giselle would pay the first \$250 of bills, but only 10% of the rest. For what amount of medical bills will plan B save Giselle money? (Assume that her bills will exceed \$250.)

73. *Parking Fees.* Mack's Parking Garage charges \$4.00 for the first hour and \$2.50 for each additional hour. For how long has a car been parked when the charge exceeds \$16.50?

74. *Ski Wax.* Green ski wax works best between 5° and 15° Fahrenheit. Determine those Celsius temperatures for which green ski wax works best. (See Exercise 39.)

Aha! **75.** The area of a square can be no more than 64 cm^2. What lengths of a side will allow this?

Aha! **76.** The sum of two consecutive odd integers is less than 100. What is the largest pair of such integers?

77. *Nutritional Standards.* In order for a food to be labeled "lowfat," it must have fewer than 3 g of fat per serving. Reduced Fat Tortilla Pops® contain 60% less fat than regular nacho cheese tortilla chips, but still cannot be labeled lowfat. What can you conclude about the fat content of a serving of nacho cheese tortilla chips?

78. *Parking Fees.* When asked how much the parking charge is for a certain car (see Exercise 73), Mack replies "between 14 and 24 dollars." For how long has the car been parked?

79. *Frequent Buyer Bonus.* Alice's Books allows customers to select one free book for every 10 books purchased. The price of that book cannot exceed the average cost of the 10 books. Neoma has bought 9 books that average $12 per book. How much should her tenth book cost if she wants to select a $15 book for free?

TW **80.** *Grading.* After 9 quizzes, Blythe's average is 84. Is it possible for Blythe to improve her average by two points with the next quiz? Why or why not?

TW **81.** *Discount Card.* Barnes & Noble offers a member card for $25 per year. This card entitles a customer to a 40% discount off list price on hardcover bestsellers, a 20% discount on adult hardcovers, and a 10% discount on other purchases. Describe two sets of circumstances for which an individual would save money by becoming a member.

Source: Barnes & Noble

Try Exercise Answers: Section 2.7

25. Scores greater than or equal to 97 **47.** No more than 134 text messages **55.** Gross sales greater than $7000

Study Summary

KEY TERMS AND CONCEPTS	EXAMPLES	PRACTICE EXERCISES
SECTION 2.1: SOLVING EQUATIONS		
The Addition Principle for Equations $a = b$ is equivalent to $a + c = b + c$.	Solve: $x + 5 = -2$. $x + 5 = -2$ $x + 5 + (-5) = -2 + (-5)$ **Adding −5 to both sides** $x = -7$	**1.** Solve: $x - 8 = -3$.
The Multiplication Principle for Equations $a = b$ is equivalent to $ac = bc$, for $c \neq 0$.	Solve: $-\frac{1}{3}x = 7$. $-\frac{1}{3}x = 7$ $(-3)(-\frac{1}{3}x) = (-3)(7)$ **Multiplying both sides by −3** $x = -21$	**2.** Solve: $\frac{1}{4}x = 1.2$.
SECTION 2.2: USING THE PRINCIPLES TOGETHER		
When solving equations, we work in the reverse order of the order of operations.	Solve: $-3x - 7 = -8$. $-3x - 7 + 7 = -8 + 7$ **Adding 7 to both sides** $-3x = -1$ $\frac{-3x}{-3} = \frac{-1}{-3}$ **Dividing both sides by −3** $x = \frac{1}{3}$	**3.** Solve: $4 - 3x = 7$.
We can **clear fractions** by multiplying both sides of an equation by the least common multiple of the denominators in the equation. We can **clear decimals** by multiplying both sides by a power of 10. If there is at most one decimal place in any one number, we multiply by 10. If there are at most two decimal places, we multiply by 100, and so on.	Solve: $\frac{1}{2}x - \frac{1}{3} = \frac{1}{6}x + \frac{2}{3}$. $6\left(\frac{1}{2}x - \frac{1}{3}\right) = 6\left(\frac{1}{6}x + \frac{2}{3}\right)$ **Multiplying by 6, the least common denominator** $6 \cdot \frac{1}{2}x - 6 \cdot \frac{1}{3} = 6 \cdot \frac{1}{6}x + 6 \cdot \frac{2}{3}$ **Using the distributive law** $3x - 2 = x + 4$ **Simplifying** $2x = 6$ **Subtracting x from and adding 2 to both sides** $x = 3$	**4.** Solve: $\frac{1}{6}t - \frac{3}{4} = t - \frac{2}{3}$.
SECTION 2.3: FORMULAS		
A **formula** uses letters to show a relationship among two or more quantities. Formulas can be solved for a given letter using the addition and multiplication principles.	Solve: $x = \frac{2}{5}y + 7$ for y. $x = \frac{2}{5}y + 7$ **We are solving for y.** $x - 7 = \frac{2}{5}y$ **Isolating the term containing y** $\frac{5}{2}(x - 7) = \frac{5}{2} \cdot \frac{2}{5}y$ **Multiplying both sides by $\frac{5}{2}$** $\frac{5}{2}x - \frac{5}{2} \cdot 7 = 1 \cdot y$ **Using the distributive law** $\frac{5}{2}x - \frac{35}{2} = y$ **We have solved for y.**	**5.** Solve $ac - bc = d$ for c.

SECTION 2.4: APPLICATIONS WITH PERCENT

Key Words in Percent Translations "Of" translates to "$\cdot$" or "$\times$" "What" translates to a variable "Is" or "Was" translates to "=" "%" translates to "$\times \frac{1}{100}$" or "$\times$ 0.01"	What percent of 60 is 7.2? $n \cdot 60 = 7.2$ $n = \frac{7.2}{60}$ $n = 0.12$ Thus, 7.2 is 12% of 60.	**6.** 12 is 15% of what number?

SECTION 2.5: PROBLEM SOLVING

Five Steps for Problem Solving in Algebra **1.** *Familiarize* yourself with the problem. **2.** *Translate* to mathematical language. (This often means writing an equation.) **3.** *Carry out* some mathematical manipulation. (This often means *solving* an equation.) **4.** *Check* your possible answer in the original problem. **5.** *State* the answer clearly.	The perimeter of a rectangle is 70 cm. The width is 5 cm longer than half the length. Find the length and the width. **1. Familiarize.** The formula for the perimeter of a rectangle is $P = 2l + 2w$. We can describe the width in terms of the length: $w = \frac{1}{2}l + 5$. **2. Translate.** *Rewording:* Twice the length plus twice the width is the perimeter. *Translating:* $2l + 2\left(\frac{1}{2}l + 5\right) = 70$ **3. Carry out.** Solve the equation: $2l + 2\left(\frac{1}{2}l + 5\right) = 70$ $2l + l + 10 = 70$ **Using the distributive law** $3l + 10 = 70$ **Combining like terms** $3l = 60$ **Subtracting 10 from both sides** $l = 20.$ **Dividing both sides by 3** If $l = 20$, then $w = \frac{1}{2}l + 5 = \frac{1}{2} \cdot 20 + 5 = 10 + 5 = 15$. **4. Check.** The width should be 5 cm longer than half the length. Since half the length is 10 cm, and 15 cm is 5 cm longer, this statement checks. The perimeter should be 70 cm. Since $2l + 2w = 2(20) + 2(15) = 40 + 30 = 70$, this statement also checks. **5. State.** The length is 20 cm and the width is 15 cm.	**7.** Deborah rode a total of 120 mi in two bicycle tours. One tour was 25 mi longer than the other. How long was each tour?

SECTION 2.6: SOLVING INEQUALITIES

An **inequality** is any sentence containing $<, >, \leq, \geq$, or $\neq$. Solution sets of inequalities can be **graphed** and written in **set-builder notation** or **interval notation.**

Interval Notation	Set-builder Notation	Graph
(a, b)	$\{x \mid a < x < b\}$	
$[a, b]$	$\{x \mid a \leq x \leq b\}$	
$[a, b)$	$\{x \mid a \leq x < b\}$	
$(a, b]$	$\{x \mid a < x \leq b\}$	
(a, ∞)	$\{x \mid a < x\}$	
$(-\infty, a)$	$\{x \mid x < a\}$	

8. Write using interval notation:

$\{x \mid x \leq 0\}$.

The Addition Principle for Inequalities

For any real numbers a, b, and c,

$a < b$ is equivalent to $a + c < b + c$;

$a > b$ is equivalent to $a + c > b + c$.

Similar statements hold for $\leq$ and $\geq$.

Solve: $x + 3 \leq 5$.

$$x + 3 \leq 5$$
$$x + 3 - 3 \leq 5 - 3 \quad \textbf{Subtracting 3 from both sides}$$
$$x \leq 2$$

9. Solve: $x - 11 > -4$.

The Multiplication Principle for Inequalities

For any real numbers a and b, and for any *positive* number c,

$a < b$ is equivalent to $ac < bc$;

$a > b$ is equivalent to $ac > bc$.

For any real numbers a and b, and for any *negative* number c,

$a < b$ is equivalent to $ac > bc$;

$a > b$ is equivalent to $ac < bc$.

Similar statements hold for $\leq$ and $\geq$.

Solve: $3x > 9$.

$$3x > 9$$
$$\frac{1}{3} \cdot 3x > \frac{1}{3} \cdot 9$$
The inequality symbol does not change because $\frac{1}{3}$ is positive.
$$x > 3$$

Solve: $-3x > 9$.

$$-3x > 9$$
$$-\frac{1}{3} \cdot -3x < -\frac{1}{3} \cdot 9$$
The inequality symbol is reversed because $-\frac{1}{3}$ is negative.
$$x < -3$$

10. Solve: $-8x \leq 2$.

SECTION 2.7: SOLVING APPLICATIONS WITH INEQUALITIES

Many real-world problems can be solved by translating the problem to an inequality and applying the five-step problem-solving strategy.

Translate to an inequality.

The test score must exceed 85.	$s > 85$
At most 15 volunteers greeted visitors.	$v \leq 15$
Ona makes no more than \$100 per week.	$w \leq 100$
Herbs need at least 4 hr of sun per day.	$h \geq 4$

11. Translate to an inequality:

Luke runs no less than 3 mi per day.

Review Exercises 2

Concept Reinforcement *Classify each of the following statements as either true or false.*

1. $5x - 4 = 2x$ and $3x = 4$ are equivalent equations. [2.1]
2. $5 - 2t < 9$ and $t > 6$ are equivalent inequalities. [2.6]
3. Some equations have no solution. [2.1]
4. Consecutive odd integers are 2 units apart. [2.5]
5. For any number a, $a \le a$. [2.6]
6. The addition principle is always used before the multiplication principle. [2.2]
7. A 10% discount results in a sale price that is 90% of the original price. [2.4]
8. Often it is impossible to list all solutions of an inequality number by number. [2.6]

Solve. Label any contradictions or identities.

9. $x + 9 = -16$ [2.1]
10. $-8x = -56$ [2.1]
11. $-\dfrac{x}{5} = 13$ [2.1]
12. $-8 = n - 11$ [2.1]
13. $\frac{2}{5}t = -8$ [2.1]
14. $x - 0.1 = 1.01$ [2.1]
15. $-\frac{2}{3} + x = -\frac{1}{6}$ [2.1]
16. $5z + 3 = 41$ [2.2]
17. $5 - x = 13$ [2.2]
18. $5t + 9 = 3t - 1$ [2.2]
19. $7x - 6 = 25x$ [2.2]
20. $\frac{1}{4}a - \frac{5}{8} = \frac{3}{8}$ [2.2]
21. $14y = 23y - 17 - 9y$ [2.2]
22. $0.22y - 0.6 = 0.12y + 3 - 0.8y$ [2.2]
23. $\frac{1}{4}x - \frac{1}{8}x = 3 - \frac{1}{16}x$ [2.2]
24. $3(5 - n) = 36$ [2.2]
25. $4(5x - 7) = -56$ [2.2]
26. $8(x - 2) = 5(x + 4)$ [2.2]
27. $3(x - 4) + 2 = x + 2(x - 5)$ [2.2]

Solve each formula for the given letter. [2.3]

28. $C = \pi d$, for d
29. $V = \dfrac{1}{3}Bh$, for B
30. $5x - 2y = 10$, for y
31. $tx = ax + b$, for x
32. Find decimal notation: 0.9%. [2.4]
33. Find percent notation: $\frac{11}{25}$. [2.4]
34. What percent of 60 is 42? [2.4]
35. 42 is 30% of what number? [2.4]

Determine whether each number is a solution of $x \le -5$. [2.6]

36. -3
37. -7
38. 0

Graph on the number line. [2.6]

39. $5x - 6 < 2x + 3$
40. $-2 < x \le 5$
41. $t > 0$

Solve. Write the answers in both set-builder notation and interval notation. [2.6]

42. $t + \frac{2}{3} \ge \frac{1}{6}$
43. $9x \ge 63$
44. $2 + 6y > 20$
45. $7 - 3y \ge 27 + 2y$
46. $3x + 5 < 2x - 6$
47. $-4y < 28$
48. $3 - 4x < 27$
49. $4 - 8x < 13 + 3x$
50. $13 \le -\frac{2}{3}t + 5$
51. $7 \le 1 - \frac{3}{4}x$

Solve.

52. About 35% of all charitable contributions are made to religious organizations. In 2008, \$106.9 billion was given to religious organizations. How much was given to charities in general? [2.4]
Source: The Giving USA Foundation/Giving Institute
53. A 32-ft beam is cut into two pieces. One piece is 2 ft longer than the other. How long are the pieces? [2.5]
54. The sum of two consecutive odd integers is 116. Find the integers. [2.5]
55. The perimeter of a rectangle is 56 cm. The width is 6 cm less than the length. Find the width and the length. [2.5]
56. After a 25% reduction, a picnic table is on sale for \$120. What was the regular price? [2.4]

57. Children ages 3–6 need about 12 hr of sleep per day. This is 25% less than infants need. How many hours of sleep do infants need? [2.4]
Source: www.helpguide.org

58. The measure of the second angle of a triangle is 50° more than that of the first. The measure of the third angle is 10° less than twice the first. Find the measures of the angles. [2.5]

59. *Magazine Subscriptions.* The table below lists the cost of gift subscriptions for *Taste of Home*, a cooking magazine, during a recent subscription campaign. Tonya spent $73 for gift subscriptions. How many subscriptions did she purchase? [2.5]
Source: *Taste of Home*

Number of Gift Subscriptions	Total Cost of Subscriptions
1	$ 13
2	23
4	43
10	103

60. Caroline has budgeted an average of $95 per month for entertainment. For the first five months of the year, she has spent $98, $89, $110, $85, and $83. How much can Caroline spend in the sixth month without exceeding her average budget? [2.7]

61. The length of a rectangular frame is 43 cm. For what widths would the perimeter be greater than 120 cm? [2.7]

SYNTHESIS

TW **62.** How does the multiplication principle for equations differ from the multiplication principle for inequalities? [2.1], [2.6]

TW **63.** Explain how checking the solutions of an equation differs from checking the solutions of an inequality. [2.1], [2.6]

64. A study of sixth- and seventh-graders in Boston revealed that, on average, the students spent 3 hr 20 min per day watching TV or playing video and computer games. This represents 108% more than the average time spent reading or doing homework. How much time each day was spent, on average, reading or doing homework? [2.4]
Source: Harvard School of Public Health

65. In June 2007, a team of Brazilian scientists exploring the Amazon measured its length as 65 mi longer than the Nile. If the combined length of both rivers is 8385 mi, how long is each river? [2.5]
Source: news.nationalgeographic.com

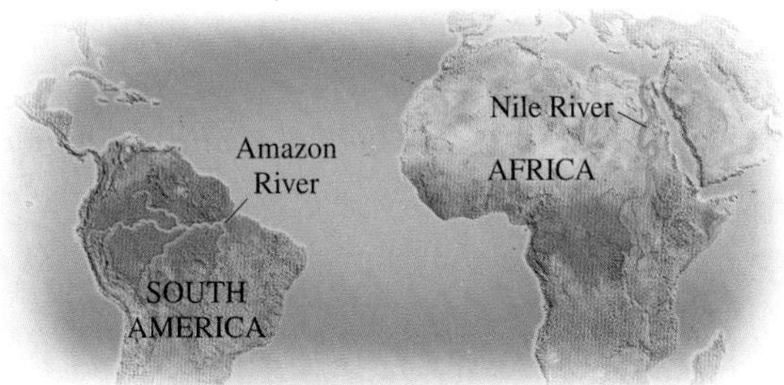

66. Consumer experts advise us never to pay the sticker price for a car. A rule of thumb is to pay the sticker price minus 20% of the sticker price, plus $200. A car is purchased for $15,080 using the rule. What was the sticker price? [2.4], [2.5]

Solve.

67. $2|n| + 4 = 50$ [1.4], [2.2]

68. $|3n| = 60$ [1.4], [2.1]

69. $y = 2a - ab + 3$, for a [2.3]

Chapter Test 2

Solve. Label any contradictions or identities.

1. $t + 7 = 16$
2. $t - 3 = 12$
3. $6x = -18$
4. $-\frac{4}{7}x = -28$
5. $3t + 7 = 2t - 5$
6. $\frac{1}{2}x - \frac{3}{5} = \frac{2}{5}$
7. $8 - y = 16$
8. $4.2x + 3.5 = 1.2 - 2.5x$
9. $4(x + 2) = 36$
10. $9 - 3x = 6(x + 4)$
11. $\frac{5}{6}(3x + 1) = 20$
12. $3(2x - 8) = 6(x - 4)$

Solve. Write the answers in both set-builder notation and interval notation.

13. $x + 6 > 1$
14. $14x + 9 > 13x - 4$
15. $-2y \leq 26$
16. $4y \leq -32$
17. $4n + 3 < -17$
18. $\frac{1}{2}t - \frac{1}{4} \leq \frac{3}{4}t$
19. $5 - 9x \geq 19 + 5x$

Solve each formula for the given letter.

20. $A = 2\pi rh$, for r
21. $w = \dfrac{P + l}{2}$, for l
22. Find decimal notation: 230%.
23. Find percent notation: 0.054.
24. What number is 32% of 50?
25. What percent of 75 is 33?

Graph on the number line.

26. $y < 4$
27. $-2 \leq x \leq 2$

Solve.

28. The perimeter of a rectangular calculator is 36 cm. The length is 4 cm greater than the width. Find the width and the length.
29. Kari is taking a 240-mi bicycle trip through Vermont. She has three times as many miles to go as she has already ridden. How many miles has she biked so far?
30. The perimeter of a triangle is 249 mm. If the sides are consecutive odd integers, find the length of each side.
31. By lowering the temperature of their electric hot-water heater from 140°F to 120°F, the Tuttles dropped their average electric bill by 7% to $60.45. What was their electric bill before they lowered the temperature of their hot water?
32. *Van Rentals.* Budget rents a moving truck at a daily rate of $14.99 plus $0.59 per mile. A business has budgeted $250 for a one-day van rental. What mileages will allow the business to stay within budget? (Round to the nearest tenth of a mile.)

SYNTHESIS

Solve.

33. $c = \dfrac{2cd}{a - d}$, for d
34. $3|w| - 8 = 37$
35. Translate to an inequality: A plant marked "partial sun" needs at least 4 hr but no more than 6 hr of sun each day.
    Source: www.yardsmarts.com
36. A concert promoter had a certain number of tickets to give away. Five people got the tickets. The first got one-third of the tickets, the second got one-fourth of the tickets, and the third got one-fifth of the tickets. The fourth person got eight tickets, and there were five tickets left for the fifth person. Find the total number of tickets given away.

3

Introduction to Graphing and Functions

Is Live Music Becoming a Luxury?

As the accompanying graph shows, concert ticket prices have more than doubled from 1997 to 2009. In Example 10 of Section 3.7, we use these data to predict the average price of a concert ticket in 2012.

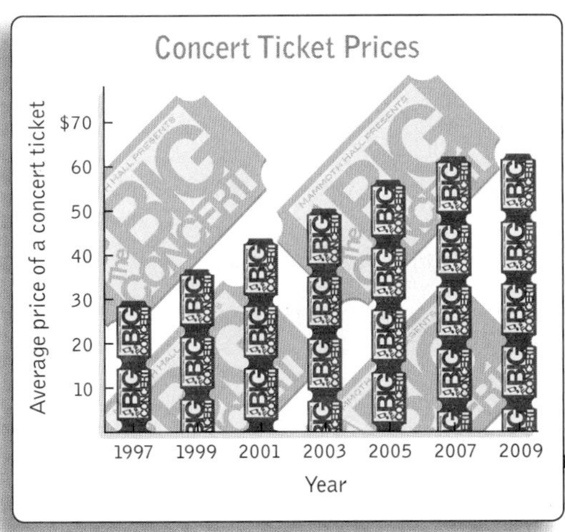

We now begin our study of graphing. From graphs of data, we move to graphs of equations and the related ideas of rate and slope. We will use the graphing calculator as a tool in graphing equations and learn how graphs can be used as a problem-solving tool. Finally, we develop the idea of a function and look at some relationships between graphs and functions.

3.1 Reading Graphs, Plotting Points, and Scaling Graphs

- Problem Solving with Bar, Circle, and Line Graphs
- Points and Ordered Pairs
- Axes and Windows

STUDY TIP

To Err Is Human

It is no coincidence that the students who experience the greatest success in this course work in pencil. We all make mistakes and by using pencil and eraser we are more willing to admit to ourselves that something needs to be rewritten. Please work with a pencil and eraser if you aren't doing so already.

Today's print and electronic media make almost constant use of graphs. In this section, we consider problem solving with bar graphs, line graphs, and circle graphs. Then we examine graphs that use a coordinate system.

PROBLEM SOLVING WITH BAR, CIRCLE, AND LINE GRAPHS

A *bar graph* is a convenient way of showing comparisons. In every bar graph, certain categories, such as levels of education in the following example, are paired with certain numbers.

EXAMPLE 1 Earnings. The bar graph below shows median income for full-time, year-round workers with various levels of education.

Median Earnings for Full-Time, Year-Round Workers Ages 25 and Older by Educational Attainment

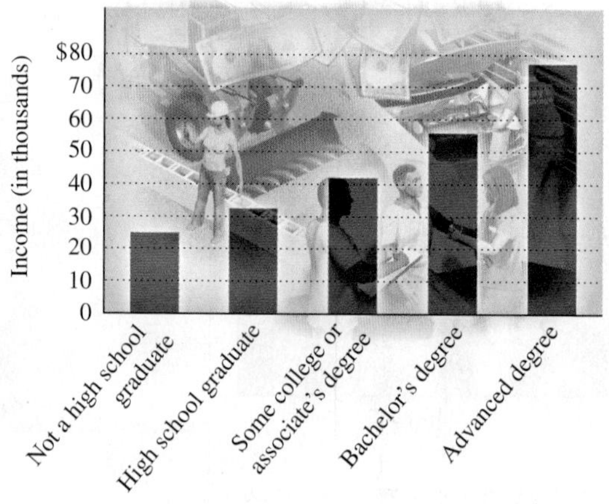

Source: Educational Attainment in the United States; http://www.census.gov/prod/2009pubs/p20-560.pdf

a) Quincy plans to earn an associate's degree. What is the median income for workers with associate's degrees?

b) Anne would like to make at least $50,000 per year. What level of education should she pursue?

SOLUTION

a) Since level of education is shown on the horizontal scale, we go to the top of the bar above the label reading "associate's degree." Then we move horizontally from the top of the bar to the vertical scale, which shows earnings. The median income for workers with associate's degrees is about $40,000.

b) By locating $50,000 on the vertical scale and then moving horizontally, we see that the bars reaching a height of $50,000 or higher correspond to a bachelor's degree and an advanced degree. Therefore, Anne should pursue a bachelor's degree or an advanced degree in order to make at least $50,000 per year.

Try Exercise 5.

Circle graphs, or *pie charts*, are often used to show what percent of the whole each particular item in a group represents.

EXAMPLE 2 Student Aid. The circle graph below shows the sources for student aid in 2008 and the percentage of aid students received from each source. In that year, the total amount of aid distributed was $143.4 billion. About 5.4 million students received a federal Pell grant. What was the average amount of aid per recipient?

Source: Trends in Student Aid 2008, www.collegeboard.com

Student Aid

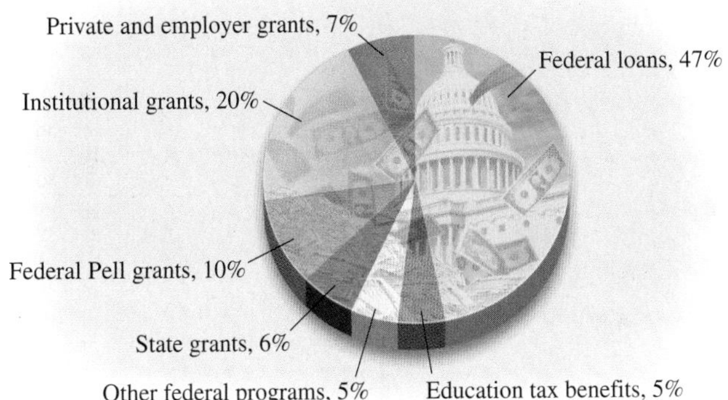

SOLUTION

1. **Familiarize.** The problem involves percents, so if we were unsure of how to solve percent problems, we might review Section 2.4.

   The solution of this problem will involve two steps. We are told the total amount of student aid distributed. In order to find the average amount of a Pell grant, we must first calculate the total of all Pell grants and then divide by the number of students.

   We let g = the average amount of a Pell grant in 2008.

2. **Translate.** From the circle graph, we see that federal Pell grants were 10% of the total amount of aid. The total amount distributed was $143.4 billion, or $143,400,000,000, so we have

Find the value of all Pell grants.

$$\text{the value of all Pell grants} = 0.10(143{,}400{,}000{,}000) = 14{,}340{,}000{,}000.$$

Calculate the average amount of a Pell grant.

Then we reword the problem and translate as follows:

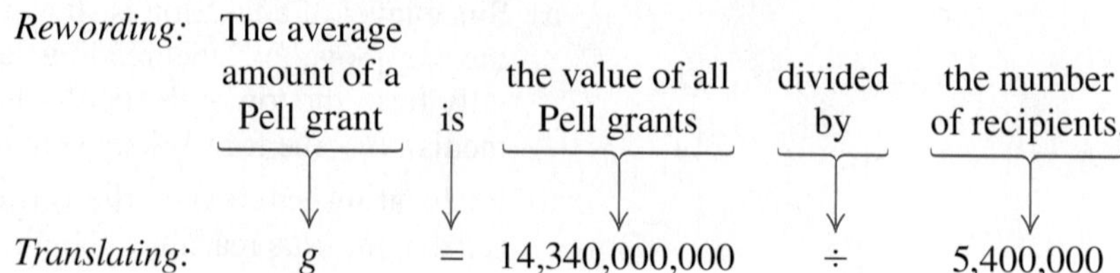

3. **Carry out.** We solve the equation:

$$g = 14{,}340{,}000{,}000 \div 5{,}400{,}000$$
$$\approx 2656. \quad \textbf{Rounding to the nearest dollar}$$

4. **Check.** If each student received \$2656, the total amount of aid distributed through Pell grants would be $\$2656 \cdot 5{,}400{,}000$, or \$14,342,400,000. Since this is approximately 10% of the total student aid for 2008, our answer checks.
5. **State.** In 2008, the average Pell grant was \$2656.

Try Exercise 9.

EXAMPLE 3 Exercise and Pulse Rate. The line graph below shows the relationship between a person's resting pulse rate and months of regular exercise.* Note that the symbol ⦚ is used to indicate that counting on the vertical axis begins at 50.

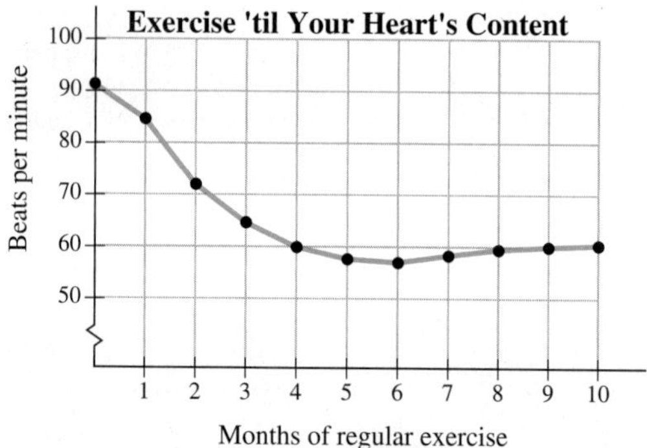

a) How many months of regular exercise are required to lower the pulse rate as much as possible?
b) How many months of regular exercise are needed to achieve a pulse rate of 65 beats per minute?

SOLUTION

a) The lowest point on the graph occurs above the number 6. Thus, after 6 months of regular exercise, the pulse rate is lowered as much as possible.
b) To determine how many months of exercise are needed to lower a person's resting pulse rate to 65, we locate 65 midway between 60 and 70 on the vertical axis. From that location, we move right until the line is reached. At that

*Data from *Body Clock* by Dr. Martin Hughes (New York: Facts on File, Inc.), p. 60.

point, we move down to the horizontal scale and read the number of months required, as shown.

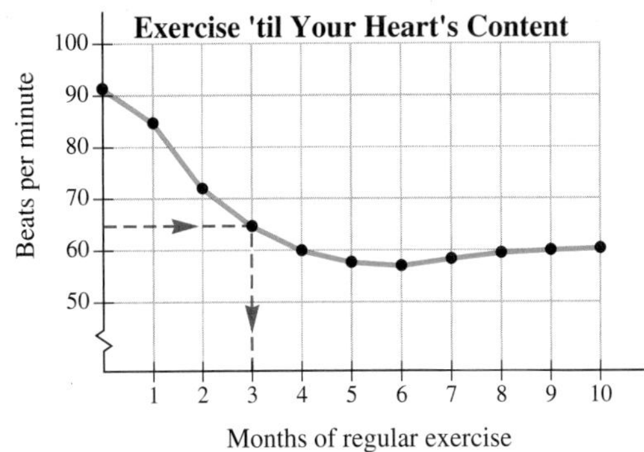

The pulse rate is 65 beats per minute after 3 months of regular exercise.

Try Exercise 17.

POINTS AND ORDERED PAIRS

The line graph in Example 3 contains a collection of points. Each point pairs up a number of months of exercise with a pulse rate. To create such a graph, we **graph**, or **plot**, pairs of numbers on a plane. This is done using two perpendicular number lines called **axes** (pronounced "ak-sēz"; singular, **axis**). The point at which the axes cross is called the **origin**. Arrows on the axes indicate the positive directions.

Consider the pair $(3, 4)$. The numbers in such a pair are called **coordinates**. The **first coordinate** in this case is 3 and the **second coordinate** is 4.* To plot $(3, 4)$, we start at the origin, move horizontally to the 3, move up vertically 4 units, and then make a "dot." Thus, $(3, 4)$ is located above 3 on the first axis and to the right of 4 on the second axis.

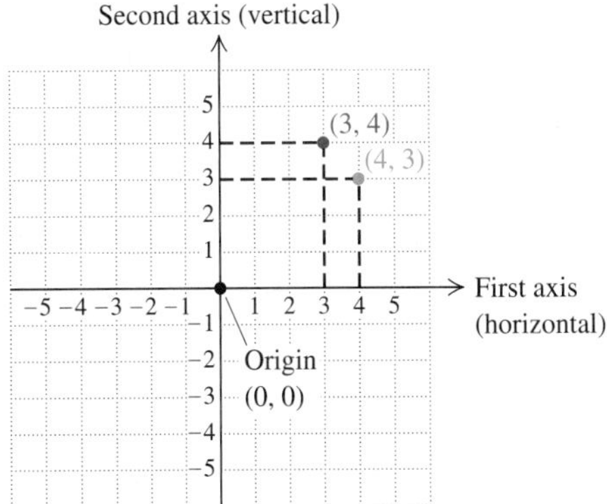

The point $(4, 3)$ is also plotted in the figure above. Note that (3, 4) and (4, 3) are different points. For this reason, coordinate pairs are called **ordered pairs**—the order in which the numbers appear is important.

*The first coordinate is called the *abscissa* and the second coordinate is called the *ordinate*. The plane is called the *Cartesian coordinate plane* after the French mathematician René Descartes (1595–1650).

EXAMPLE 4 Plot the point $(-3, 4)$.

SOLUTION The first number, -3, is negative. Starting at the origin, we move 3 units in the negative horizontal direction (3 units to the left). The second number, 4, is positive, so we move 4 units in the positive vertical direction (up). The point $(-3, 4)$ is above -3 on the first axis and to the left of 4 on the second axis.

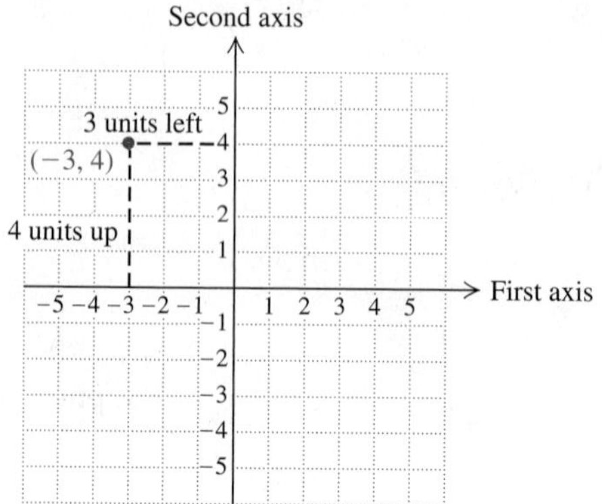

Try Exercise 21.

CAUTION! Do not confuse the *ordered pair* (*a*, *b*) with the *interval* (*a*, *b*). The context in which the notation appears usually makes the meaning clear.

To find the coordinates of a point, we see how far to the right or the left of the origin the point is and how far above or below the origin it is. Note that the coordinates of the origin itself are (0, 0).

EXAMPLE 5 Find the coordinates of points *A*, *B*, *C*, *D*, *E*, *F*, and *G*.

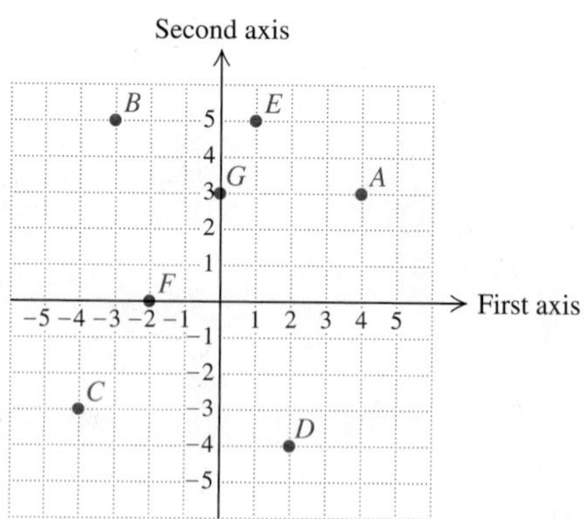

SOLUTION Point *A* is 4 units to the right of the origin and 3 units above the origin. Its coordinates are $(4, 3)$. The coordinates of the other points are as follows:

B: $(-3, 5)$; C: $(-4, -3)$; D: $(2, -4)$;
E: $(1, 5)$; F: $(-2, 0)$; G: $(0, 3)$.

Try Exercise 27.

The variables x and y are commonly used when graphing on a plane. Coordinates of ordered pairs are often labeled

(x-coordinate, y-coordinate).

The first, or horizontal, axis is then labeled the x-axis, and the second, or vertical, axis is labeled the y-axis.

The horizontal and vertical axes divide the plane into four regions, or **quadrants**, as indicated by Roman numerals in the figure below. Note that the point $(-4, 5)$ is in the second quadrant and the point $(5, -5)$ is in the fourth quadrant. The points $(3, 0)$ and $(0, 1)$ are on the axes and are not considered to be in any quadrant.

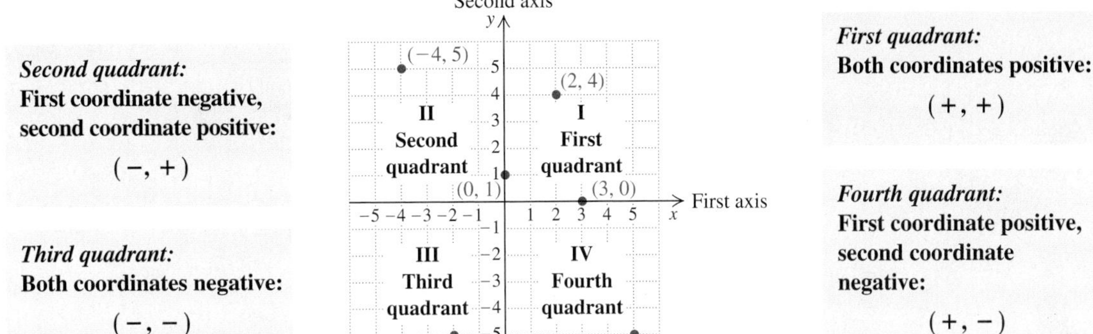

AXES AND WINDOWS

We draw only part of the plane when we create a graph. Although it is standard to show the origin and parts of all quadrants, as in the graphs in Examples 4 and 5, for some applications it may be more practical to show a different portion of the plane. Note, too, that on the graphs in Examples 4 and 5, the marks on the axes are one unit apart. In this case, we say that the **scale** of each axis is 1. Often it is necessary to use a different scale on one or both of the axes.

EXAMPLE 6 Use a grid 10 squares wide and 10 squares high to plot $(-34, 450)$, $(48, 95)$ and $(10, -200)$.

SOLUTION Since x-values vary from a low of -34 to a high of 48, the 10 horizontal squares must span $48 - (-34)$, or 82 units. Because 82 is not a multiple of 10, we round *up* to the next multiple of 10, which is 90. Dividing 90 by 10, we find that if each square is 9 units wide (has a scale of 9), we could represent all the x-values. However, since it is more convenient to count by 10's, we will instead use a scale of 10. Starting at 0, we count backward to -40 and forward to 60.

This is how we will arrange the x-axis.

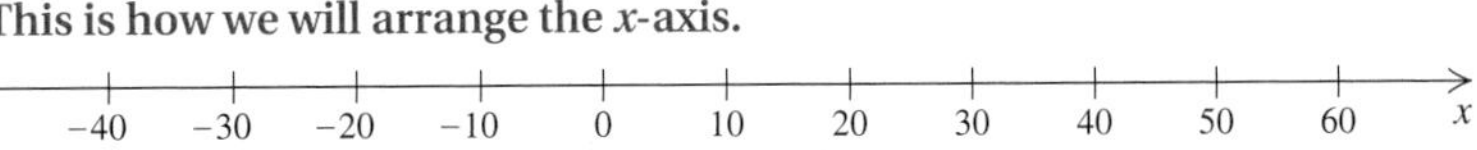

This is how we will arrange the y-axis.

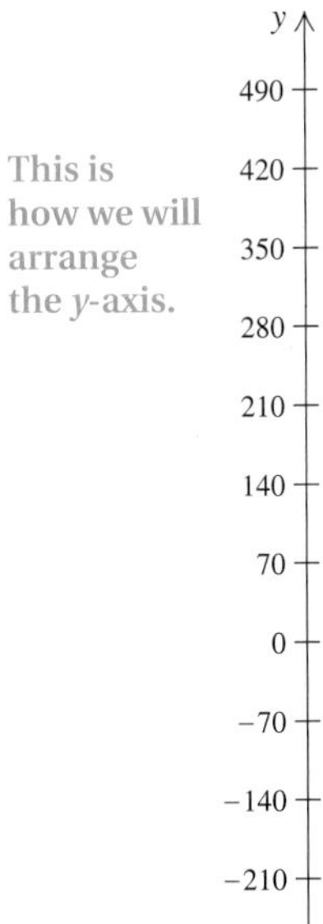

There is more than one correct way to cover the values from -34 to 48 using 10 increments. For instance, we could have counted from -60 to 90, using a scale of 15. In general, we try to use the smallest span and scale that will cover the given coordinates. Scales that are multiples of 2, 5, or 10 are especially convenient. Numbering must always begin at the origin.

Since we must be able to show y-values from -200 to 450, the 10 vertical squares must span $450 - (-200)$, or 650 units. For convenience, we round 650 *up* to 700 and then divide by 10: $700 \div 10 = 70$. Using 70 as the scale, we count *down* from 0 until we pass -200 and *up* from 0 until we pass 450, as shown at left.

Next, we use the x-axis and the y-axis that we just developed to form a grid. If the axes are to be placed correctly, the two 0's must coincide where the axes cross. Finally, once the graph has been numbered, we plot the points as shown below.

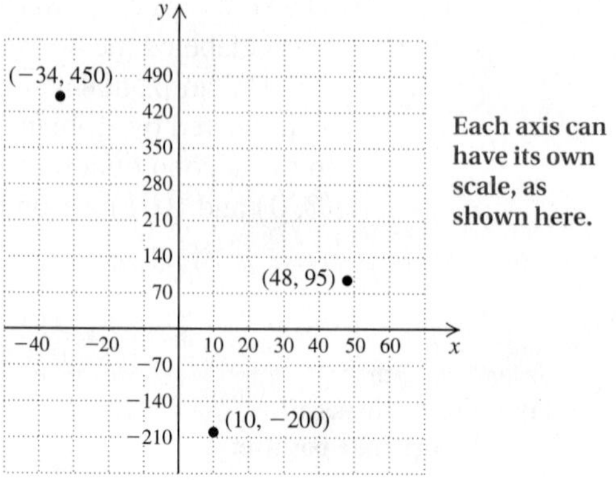

Each axis can have its own scale, as shown here.

Try Exercise 31.

Windows

On a graphing calculator, the rectangular portion of the screen in which a graph appears is called the **viewing window**. Windows are described by four numbers of the form [L, R, B, T], representing the **L**eft and **R**ight endpoints of the x-axis and the **B**ottom and **T**op endpoints of the y-axis. The **standard viewing window** is the window described by $[-10, 10, -10, 10]$. On most graphing calculators, a standard viewing window can be set up quickly using the ZStandard option in the ZOOM menu.

Press WINDOW to set the window dimensions. The scales for the axes are set using Xscl and Yscl. Press GRAPH to see the viewing window. Xres indicates the pixel resolution, which we generally set as Xres = 1.

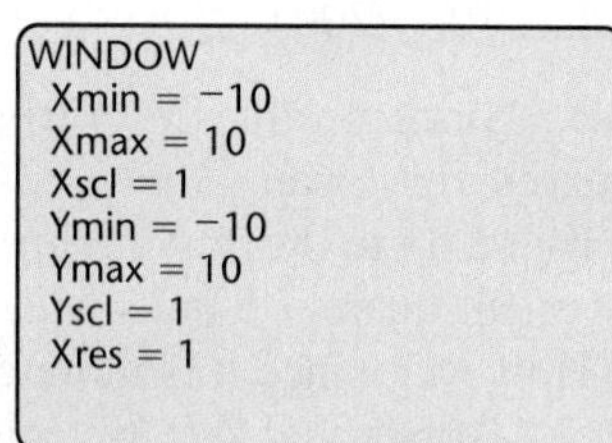

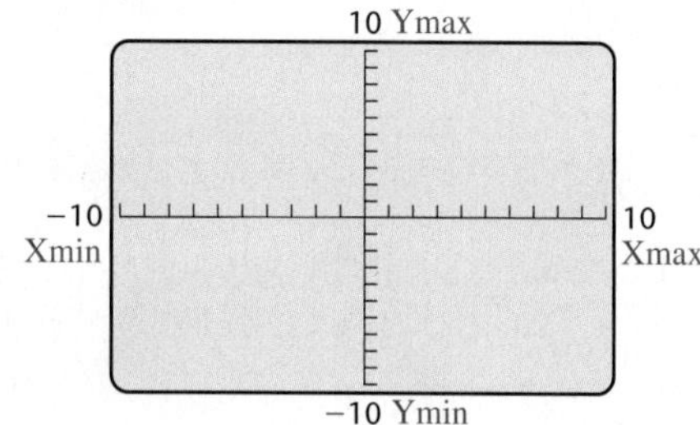

In the graph on the right above, each mark on the axes represents 1 unit. In this text, the window dimensions are written outside the graphs. A scale other than 1 is indicated below the graph.

Inappropriate window dimensions may result in an error. For example, the message ERR:WINDOW RANGE occurs when Xmin is not less than Xmax.

Your Turn

1. Press Y= and clear any equations present.
2. Press ZOOM 6 to see a standard viewing window.
3. Press WINDOW and change the dimensions to $[-20, 20, -5, 5]$. Press GRAPH to see the window.

EXAMPLE 7 Set up a $[-100, 100, -5, 5]$ viewing window on a graphing calculator, choosing appropriate scales for the axes.

SOLUTION We are to show the x-axis from -100 to 100 and the y-axis from -5 to 5. The window dimensions are set as shown in the screen on the left below. The choice of a scale for an axis may vary. If the scale chosen is too small, the marks on the axes will not be distinct. The resulting window is shown on the right below.

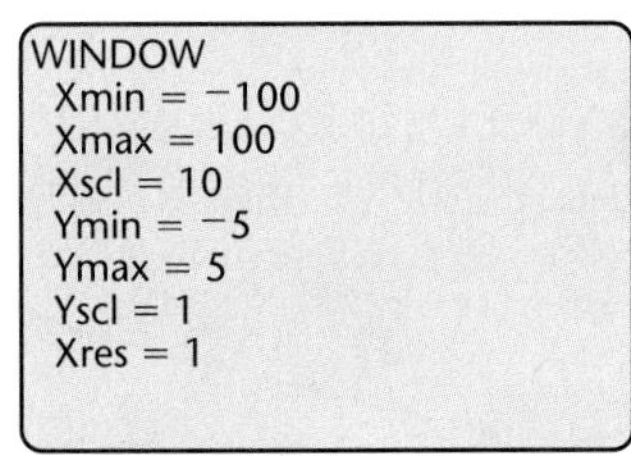

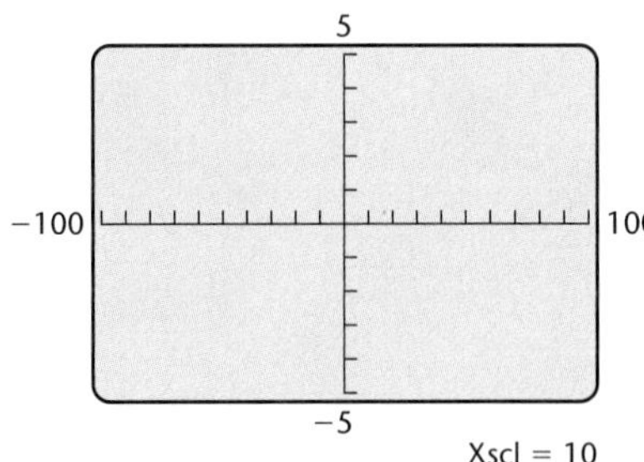

Note that the viewing window is not square. Also note that each mark on the x-axis represents 10 units, and each mark on the y-axis represents 1 unit.

Try Exercise 57.

Sometimes we do not want to start counting on an axis at 0. On a paper graph, we indicate skipped numbers by a jagged break in the axis, as shown on the left below. This grid shows the horizontal axis from -10 to 6 with a scale of 2 and the vertical axis from 2000 to 2500 with a scale of 100.

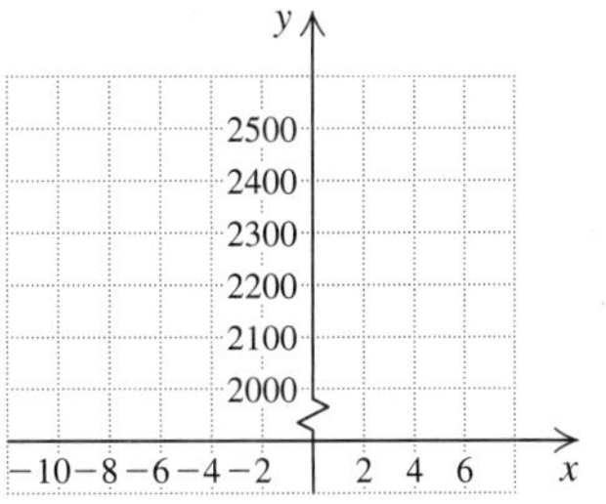

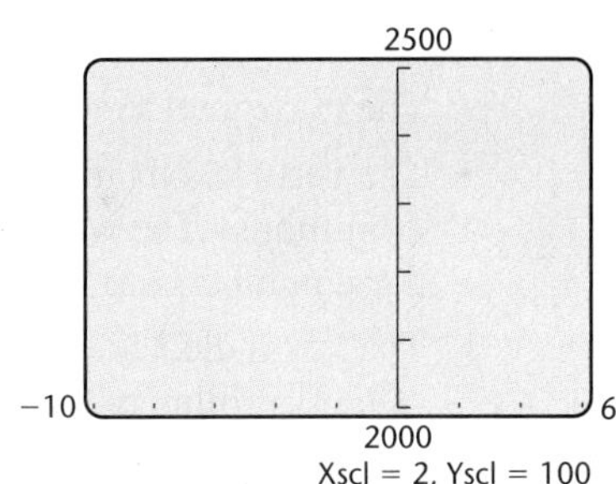

The same portion of the plane is shown in the figure on the right above in a graphing calculator's window. Note that the horizontal units are shown by dots at the bottom of the screen, but the horizontal axis does not appear.

Connecting the Concepts

Reading Graphs

A graph can present a great quantity of information in a compact form. When reading a graph, pay attention to labels, units, relationships, trends, and maximums and minimums, as well as actual values indicated.

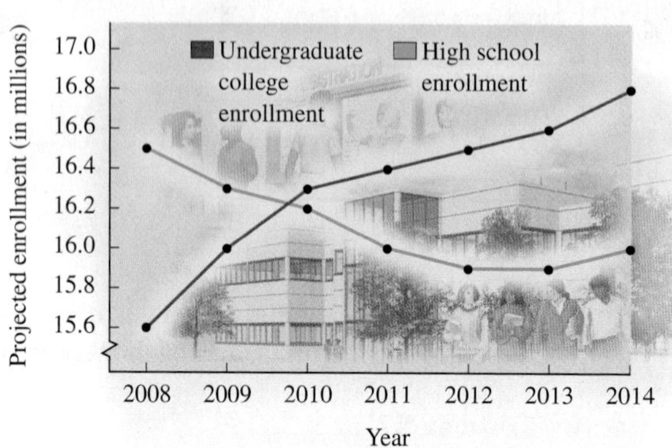

Source: U.S. National Center for Education Statistics

1. *Labels.* Examine the title, the labels on the axes, and any key.
   - The title of the graph indicates that the graph shows school enrollment in the United States.
   - The horizontal axis indicates the year.
   - The vertical axis indicates projected enrollment, in millions. The word "projected" indicates that these numbers are future estimates.
   - The key indicates which line represents high school enrollment, and which line represents undergraduate college enrollment.
2. *Units.* Understand the scales used on the axes before attempting to read any values.
   - Each unit on the horizontal axis represents one year.
   - Each unit on the vertical axis represents 0.2 million, or 200,000.
3. *Relationships.* Any one graph indicates relationships between the quantities on the horizontal axis and those on the vertical axis. When two or more graphs are shown together, we can also see relationships between the quantities these graphs represent.
   - Each graph shows the number of students enrolled that year.
   - By looking at the intersection of the graphs, we can estimate the year in which the number of high school students will be the same as the number of college undergraduates.
4. *Trends.* As we move from left to right on a graph, the quantity represented on the vertical axis may increase, decrease, or remain constant.
   - The number of college undergraduates is projected to increase every year.
   - The number of high school students is projected to decrease from 2008 to 2012, remain constant from 2012 to 2013, and increase from 2013 to 2014.
5. *Maximums and minimums.* A "high" point or a "low" point on a graph indicates a maximum value or a minimum value for the quantity on the vertical axis.
   - The maximum number of high school students shown is in 2008.
   - The minimum number of high school students shown is in 2012 and 2013.
6. *Values.* If we know one coordinate of a point on a graph, we can find the other coordinate.
   - In 2008, there were 16.5 million high school students.
   - There are projected to be 16.5 million undergraduate college students in 2012.

3.1 Exercise Set

FOR EXTRA HELP MyMathLab   

 *Concept Reinforcement* *In each of Exercises 1–4, match the set of coordinates with the graph from (a)–(d) below that would be the best for plotting the points.*

1. ___ $(-9, 3), (-2, -1), (1, 5)$
2. ___ $(-2, -1), (1, 5), (7, 3)$
3. ___ $(-2, -9), (2, 1), (4, -6)$
4. ___ $(-2, -1), (-9, 3), (-4, -6)$

a) 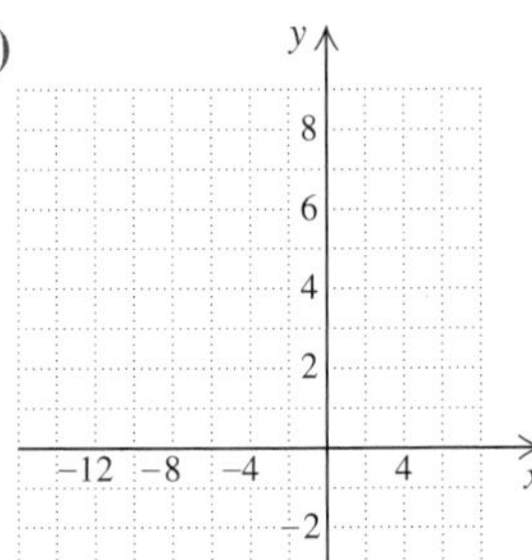

b) 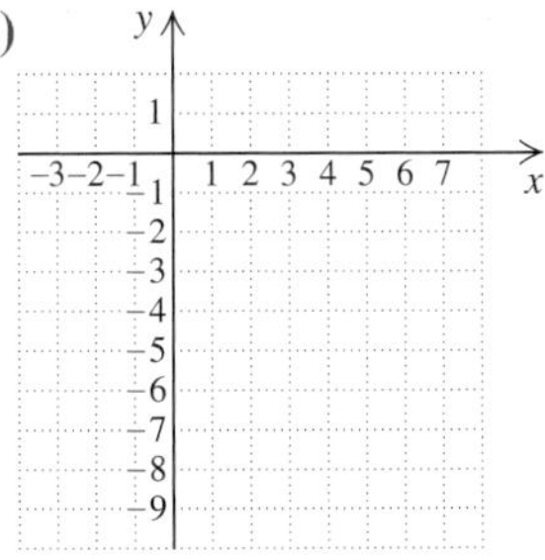

c) 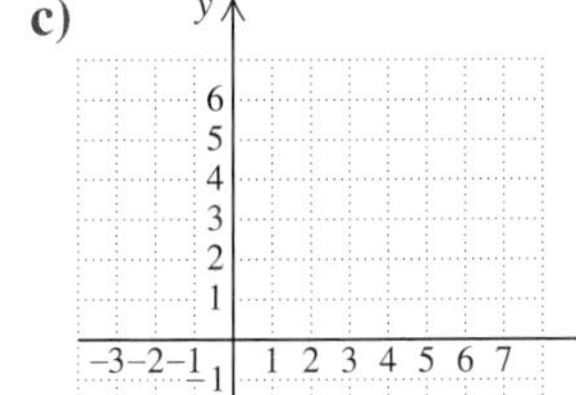

d) 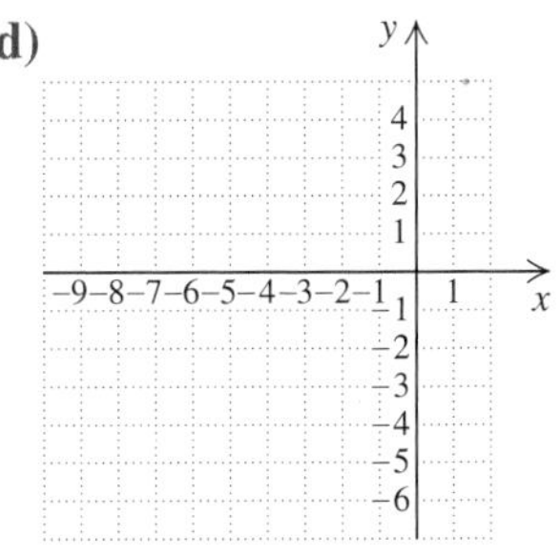

The Try Exercises *for examples are indicated by a shaded block on the exercise number. Answers to these exercises appear at the end of the exercise set as well as at the back of the book.*

Driving under the Influence. *A blood-alcohol level of 0.08% or higher makes driving illegal in the United States. This bar graph shows how many drinks a person of a certain weight would need to consume in 1 hr to achieve a blood-alcohol level of 0.08%. Note that a 12-oz beer, a 5-oz glass of wine, or a cocktail containing $1\frac{1}{2}$ oz of distilled liquor all count as one drink.*

Source: Adapted from www.medicinenet.com

Friends Don't Let Friends Drive Drunk

5. Approximately how many drinks would a 100-lb person have consumed in 1 hr to reach a blood-alcohol level of 0.08%?

6. Approximately how many drinks would a 160-lb person have consumed in 1 hr to reach a blood-alcohol level of 0.08%?

7. What can you conclude about the weight of someone who has consumed 3 drinks in 1 hr without reaching a blood-alcohol level of 0.08%?

8. What can you conclude about the weight of someone who has consumed 4 drinks in 1 hr without reaching a blood-alcohol level of 0.08%?

Student Aid. *Use the information in Example 2 to answer Exercises 9–12.*

9. In 2008, there were 13,803,201 full-time equivalent students in U.S. colleges and universities. What was the average federal loan per full-time equivalent student?

10. In 2008, there were 13,803,201 full-time equivalent students in U.S. colleges and universities. What was the average education tax benefit received per full-time equivalent student?

11. Approximately 70% of state grants are need-based. How much did students receive in 2008 in need-based state grants?

12. Approximately 19% of Pell grant dollars is given to students in for-profit institutions. How much did students in for-profit institutions receive in Pell grants in 2008?

Sorting Solid Waste. Use the pie chart below to answer Exercises 13–16.

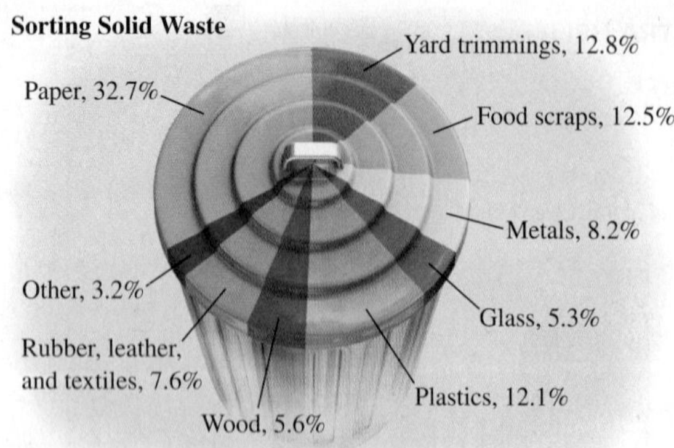

Source: Environmental Protection Agency

13. In 2007, Americans generated 254 million tons of waste. How much of the waste was glass?

14. In 2007, the average American generated 4.62 lb of waste per day. How much of that was paper?

15. Americans are recycling about 23.7% of all glass that is in the waste stream. How much glass did Americans recycle in 2007? (See Exercise 13.)

16. Americans are recycling about 54.5% of all paper. What amount of paper did the average American recycle per day in 2007? (Use the information in Exercise 14.)

Wireless-Only Households. The graph below shows the number of U.S. households with only wireless phones.

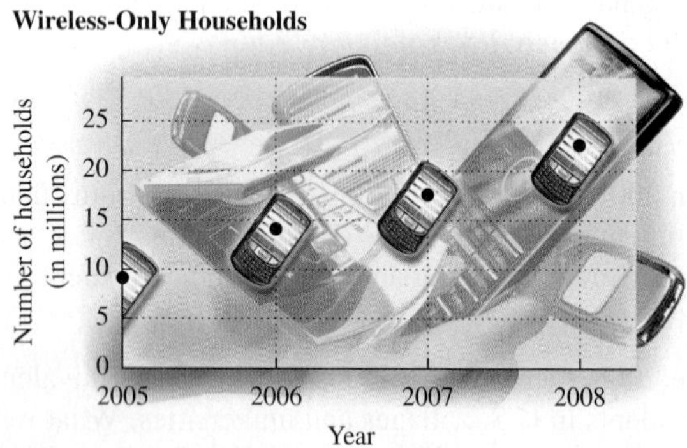

17. Approximately how many households were wireless only in 2005?

18. Approximately how many households were wireless only in 2008?

19. For what years were fewer than 15 million households wireless only?

20. By how much did the number of wireless-only households increase from 2005 to 2006?

Plot each group of points.

21. $(1,2), (-2,3), (4,-1), (-5,-3), (4,0), (0,-2)$

22. $(-2,-4), (4,-3), (5,4), (-1,0), (-4,4), (0,5)$

23. $(4,4), (-2,4), (5,-3), (-5,-5), (0,4), (0,-4), (3,0), (-4,0)$

24. $(2,5), (-1,3), (3,-2), (-2,-4), (0,4), (0,-5), (5,0), (-5,0)$

25. *Text Messaging.* Listed below are estimates of the number of text messages sent in the United States. Make a line graph of the data.

Year	Monthly Text Messages (in billions)
2003	1.2
2005	7.25
2006	12.5
2007	28.8
2008	75
2009	135

Sources: CTIA—The Wireless Association; The Nielsen Company

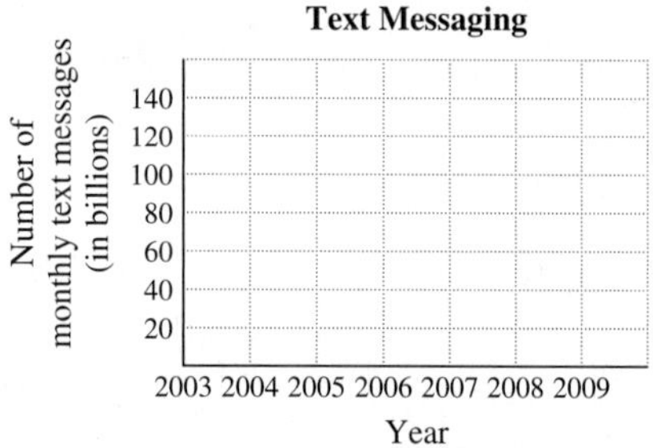

26. *Ozone Layer.* Make a line graph of the data in the table below, listing years on the horizontal scale.

Year	Ozone Level in the Southern Polar Region (in Dobson Units)
2003	108.7
2004	131.7
2005	112.8
2006	97.0
2007	115.8
2008	112.0

Source: National Aeronautics and Space Administration

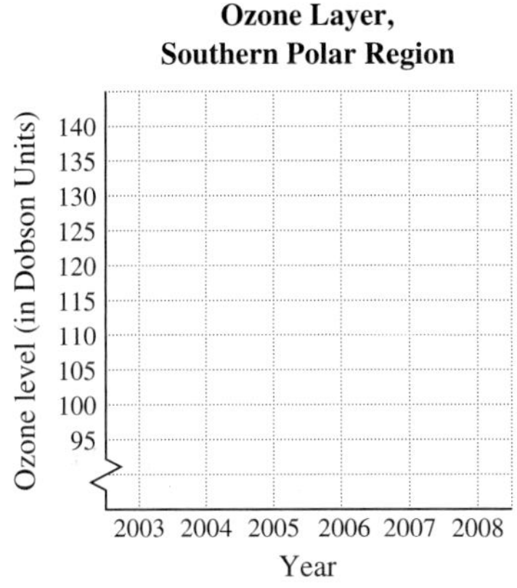

In Exercises 27–30, find the coordinates of points A, B, C, D, and E.

27.

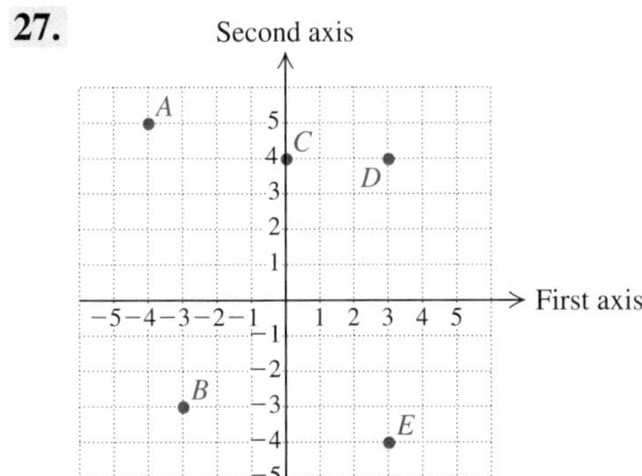

28.

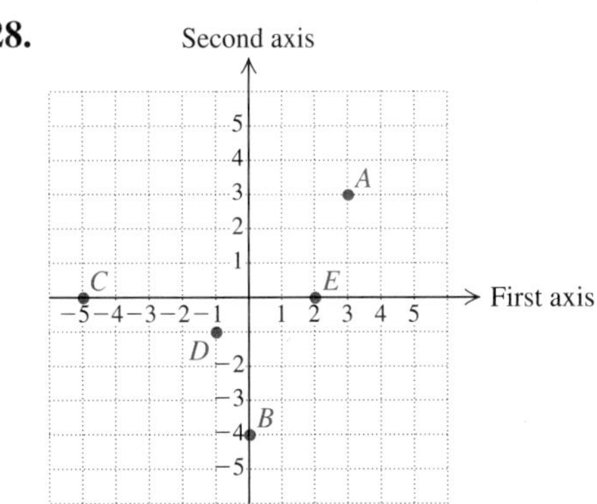

29.

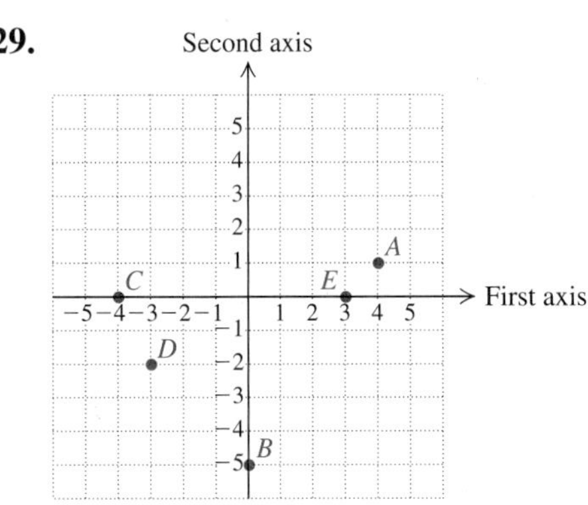

30.

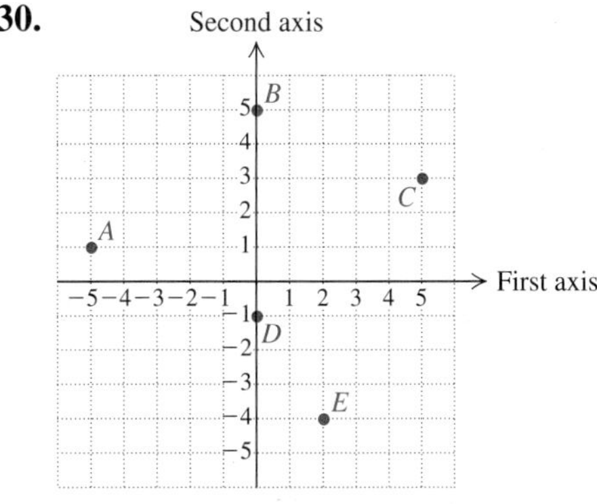

In Exercises 31–40, use a grid 10 squares wide and 10 squares high to plot the given coordinates. Choose your scale carefully. Scales may vary.

31. $(-75, 5), (-18, -2), (9, -4)$

32. $(-13, 3), (48, -1), (62, -4)$

33. $(-1, 83), (-5, -14), (5, 37)$

34. $(2, -79), (4, -25), (-4, 12)$

35. $(-10, -4), (-16, 7), (3, 15)$

36. $(5, -16), (-7, -4), (12, 3)$

37. $(-100, -5), (350, 20), (800, 37)$

38. $(750, -8), (-150, 17), (400, 32)$

39. $(-83, 491), (-124, -95), (54, -238)$

40. $(738, -89), (-49, -6), (-165, 53)$

In which quadrant or on which axis is each point located?

41. $(7, -2)$ **42.** $(-1, -4)$ **43.** $(-4, -3)$

44. $(1, -5)$ **45.** $(0, -3)$ **46.** $(6, 0)$

47. $(-4.9, 8.3)$ **48.** $(7.5, 2.9)$ **49.** $\left(-\frac{5}{2}, 0\right)$

50. $(0, 2.8)$ **51.** $(160, 2)$ **52.** $\left(-\frac{1}{2}, 2000\right)$

53. In which quadrants are the first coordinates positive?

54. In which quadrants are the second coordinates negative?

55. In which quadrants do both coordinates have the same sign?

56. In which quadrants do the first and second coordinates have opposite signs?

For Exercises 57–60, set up the indicated viewing window on a graphing calculator, choosing appropriate scales for the axes. Choice of scale may vary.

57. $[-3, 15, -50, 500]$ **58.** $[-8000, 0, -100, 600]$

59. $\left[-\frac{1}{2}, 1, 0, 0.1\right]$ **60.** $\left[-2.6, 2.6, -\frac{1}{4}, \frac{1}{4}\right]$

TW **61.** The graph below was included in a mailing sent by Agway® to their oil customers in 2000. What information is missing from the graph and why is the graph misleading?

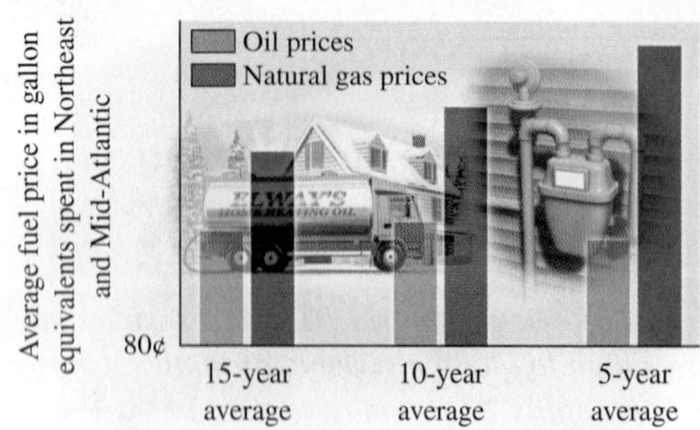

Source: Energy Research Center Inc. *3/1/99–2/29/00

TW **62.** What do all of the points on the vertical axis of a graph have in common?

SKILL REVIEW

To prepare for Section 3.2, review solving for a variable (Section 2.3).

Solve for y. [2.3]

63. $5y = 2x$

64. $2y = -3x$

65. $x - y = 8$

66. $2x + 5y = 10$

67. $2x + 3y = 5$

68. $5x - 8y = 1$

SYNTHESIS

TW **69.** Describe what the result would be if the first and second coordinates of every point in the following graph of an arrow were interchanged.

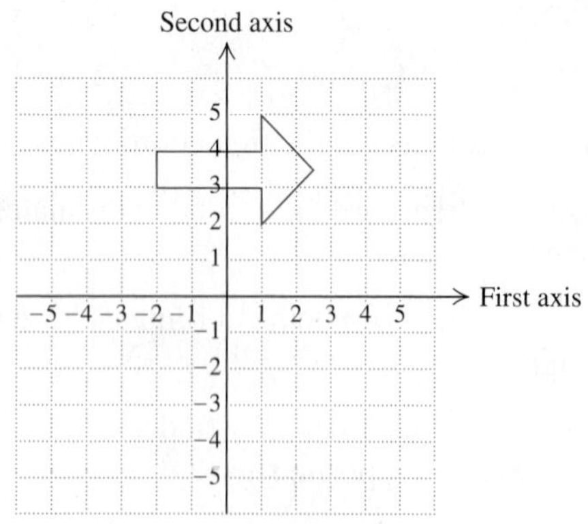

TW **70.** The graph accompanying Example 3 flattens out. Why do you think this occurs?

71. In which quadrant(s) could a point be located if its coordinates are opposites of each other?

72. In which quadrant(s) could a point be located if its coordinates are reciprocals of each other?

73. The points $(-1, 1)$, $(4, 1)$, and $(4, -5)$ are three vertices of a rectangle. Find the coordinates of the fourth vertex.

74. The pairs $(-2, -3)$, $(-1, 2)$, and $(4, -3)$ can serve as three (of four) vertices for three different parallelograms. Find the fourth vertex of each parallelogram.

75. Graph eight points such that the sum of the coordinates in each pair is 7. Answers may vary.

76. Find the perimeter of a rectangle if three of its vertices are $(5, -2)$, $(-3, -2)$, and $(-3, 3)$.

77. Find the area of a triangle whose vertices have coordinates $(0, 9)$, $(0, -4)$, and $(5, -4)$.

Coordinates on the Globe. *Coordinates can also be used to describe the location on a sphere: 0° latitude is the equator and 0° longitude is a line from the North Pole to the South Pole through France and Algeria. In the figure shown here, hurricane Clara is at a point about 260 mi northwest of Bermuda near latitude 36.0° North, longitude 69.0° West.*

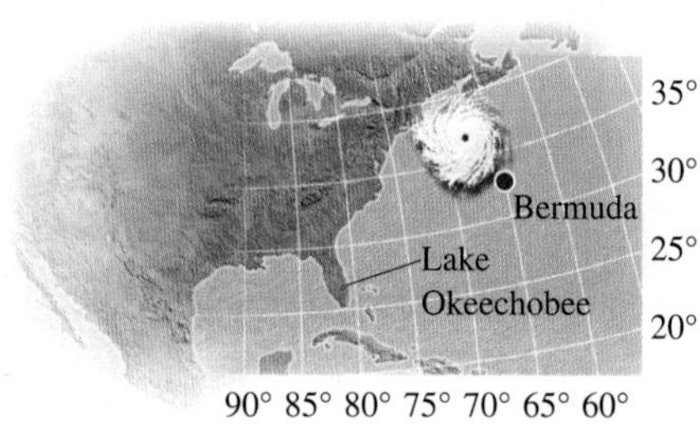

78. Approximate the latitude and the longitude of Bermuda.

79. Approximate the latitude and the longitude of Lake Okeechobee.

TW **80.** In the *Star Trek* science-fiction series, a three-dimensional coordinate system is used to locate objects in space. If the center of a planet is used as the origin, how many "quadrants" will exist? Why? If possible, sketch a three-dimensional coordinate system and label each "quadrant."

Try Exercise Answers: Section 3.1

5. 2 drinks **9.** About \$4883 **17.** About 9 million households **21.** **27.** $A(-4, 5)$; $B(-3, -3)$; $C(0, 4)$; $D(3, 4)$; $E(3, -4)$ **31.** **57.**

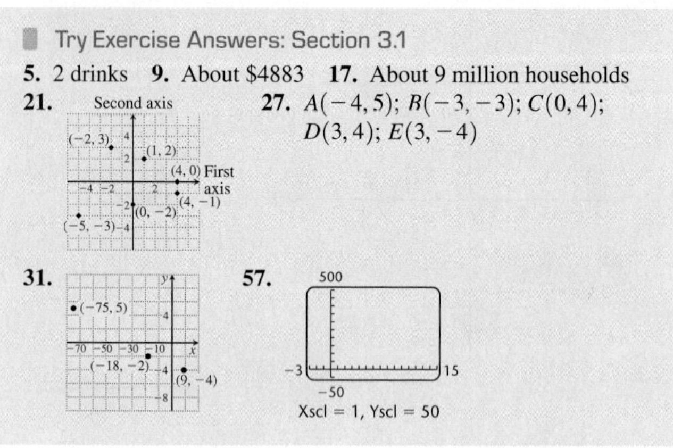

Collaborative Corner

You Sank My Battleship!

Focus: Graphing points; logical questioning
Time: 15–25 minutes
Group Size: 3–5
Materials: Graph paper

In the game Battleship®, a player places a miniature ship on a grid that only that player can see. An opponent guesses at coordinates that might "hit" the "hidden" ship. The following activity is similar to this game.

ACTIVITY

1. Using only integers from -10 to 10 (inclusive), one group member should secretly record the coordinates of a point on a slip of paper. (This point is the hidden "battleship.")
2. The other group members can then ask up to 10 "yes/no" questions in an effort to determine the coordinates of the secret point. Be sure to phrase each question mathematically (for example, "Is the x-coordinate negative?")
3. The group member who selected the point should answer each question. On the basis of the answer given, another group member should cross out the points no longer under consideration. All group members should check that this is done correctly.
4. If the hidden point has not been determined after 10 questions have been answered, the secret coordinates should be revealed to all group members.
5. Repeat parts (1)–(4) until each group member has had the opportunity to select the hidden point and answer questions.

3.2 Graphing Equations

- Solutions of Equations
- Graphing Linear Equations
- Applications
- Graphing Nonlinear Equations

We have seen how bar, line, and circle graphs can represent information. Now we begin to learn how graphs can be used to represent solutions of equations.

SOLUTIONS OF EQUATIONS

When an equation contains two variables, solutions are ordered pairs in which each number in the pair replaces a variable in the equation. Unless stated otherwise, the first number in each pair replaces the variable that occurs first alphabetically.

EXAMPLE 1 Determine whether each of the following pairs is a solution of $4b - 3a = 22$: **(a)** (2, 7); **(b)** (1, 6).

SOLUTION

a) We substitute 2 for a and 7 for b (alphabetical order of variables):

$$\begin{array}{r|l} 4b - 3a & = 22 \\ \hline 4(7) - 3(2) & 22 \\ 28 - 6 & \\ 22 & \stackrel{?}{=} 22 \quad \text{TRUE} \end{array}$$

Since $22 = 22$ is *true*, the pair (2, 7) *is* a solution.

STUDY TIP

Talk It Out

If you are finding a particular topic difficult, talk about it with a classmate. Verbalizing your questions about the material might help clarify it for you. If your classmate is also finding the material difficult, you can ask your instructor to explain the concept again.

b) In this case, we replace a with 1 and b with 6:

$$\begin{array}{c|c} 4b - 3a = & 22 \\ \hline 4(6) - 3(1) & 22 \\ 24 - 3 & \\ 21 \stackrel{?}{=} & 22 \end{array} \quad \text{FALSE} \quad 21 \neq 22$$

Since $21 = 22$ is *false*, the pair $(1, 6)$ is *not* a solution.

Try Exercise 7.

EXAMPLE 2 Show that the pairs $(3, 7)$, $(0, 1)$, and $(-3, -5)$ are solutions of $y = 2x + 1$. Then graph the three points to determine another pair that is a solution.

SOLUTION To show that a pair is a solution, we substitute, replacing x with the first coordinate and y with the second coordinate of each pair.

$$\begin{array}{c|c} y = & 2x + 1 \\ \hline 7 & 2 \cdot 3 + 1 \\ & 6 + 1 \\ 7 \stackrel{?}{=} & 7 \end{array} \quad \text{TRUE} \qquad \begin{array}{c|c} y = & 2x + 1 \\ \hline 1 & 2 \cdot 0 + 1 \\ & 0 + 1 \\ 1 \stackrel{?}{=} & 1 \end{array} \quad \text{TRUE} \qquad \begin{array}{c|c} y = & 2x + 1 \\ \hline -5 & 2(-3) + 1 \\ & -6 + 1 \\ -5 \stackrel{?}{=} & -5 \end{array} \quad \text{TRUE}$$

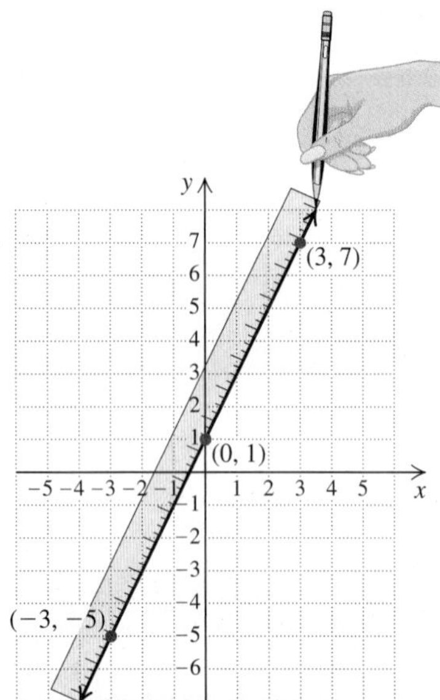

In each of the three cases, the substitution results in a true equation. Thus the pairs $(3, 7)$, $(0, 1)$, and $(-3, -5)$ are all solutions. We graph them as shown at left.

Note that the three points appear to "line up." Will other points that line up with these points also represent solutions of $y = 2x + 1$? To find out, we use a ruler and draw a line passing through $(-3, -5)$, $(0, 1)$, and $(3, 7)$.

The line appears to pass through $(2, 5)$. Let's check if this pair is a solution of $y = 2x + 1$:

$$\begin{array}{c|c} y = & 2x + 1 \\ \hline 5 & 2 \cdot 2 + 1 \\ & 4 + 1 \\ 5 \stackrel{?}{=} & 5 \end{array} \quad \text{TRUE}$$

We see that $(2, 5)$ *is* a solution. You should perform a similar check for at least one other point that appears to be on the line.

Try Exercise 13.

Example 2 leads us to suspect that *any* point on the line passing through $(3, 7)$, $(0, 1)$, and $(-3, -5)$ represents a solution of $y = 2x + 1$. In fact, every solution of $y = 2x + 1$ is represented by a point on this line and every point on this line represents a solution. The line is called the **graph** of the equation.

Connecting the Concepts

Solutions

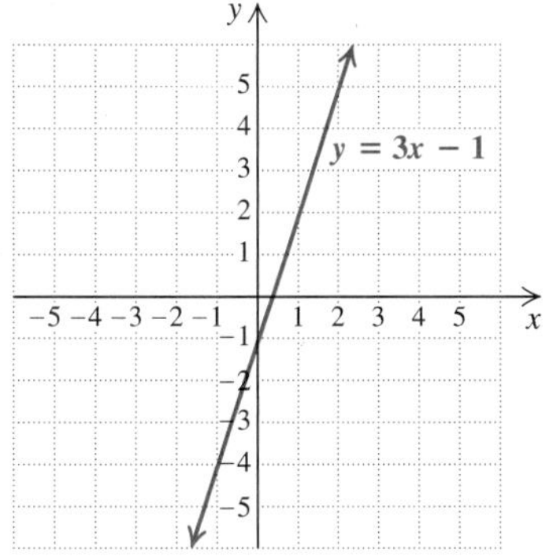

A solution of an equation like $5 = 3x - 1$ is a number. For this particular equation, the solution is 2. Were we to graph the solution, it would be a point on the number line.

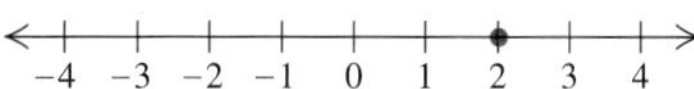

A solution of an equation like $y = 3x - 1$ is an ordered pair. One solution of this equation is $(2, 5)$. The graph of the solutions of this equation is a line in a coordinate plane, as shown in the figure at left.

GRAPHING LINEAR EQUATIONS

Equations like $y = 2x + 1$ or $4b - 3a = 22$ are said to be **linear** because the graph of each equation is a line. In general, any equation that can be written in the form $y = mx + b$ or $Ax + By = C$ (where m, b, A, B, and C are constants and A and B are not both 0) is linear. An equation of the form $Ax + By = C$ is said to be written in **standard form.**

To *graph* an equation is to make a drawing that represents its solutions. Linear equations can be graphed as follows.

To Graph a Linear Equation

1. Select a value for one coordinate and calculate the corresponding value of the other coordinate. Form an ordered pair. This pair is one solution of the equation.
2. Repeat step (1) to find a second ordered pair. A third ordered pair can be used as a check.
3. Plot the ordered pairs and draw a straight line passing through the points. The line represents all solutions of the equation.

EXAMPLE 3 Graph: $y = -3x + 1$.

SOLUTION Since $y = -3x + 1$ is in the form $y = mx + b$, the equation is linear and the graph is a straight line. We select a convenient value for x, compute y, and form an ordered pair. Then we repeat the process for other choices of x.

If $x = 2$, then $y = -3 \cdot 2 + 1 = -5$, and $(2, -5)$ is a solution.
If $x = 0$, then $y = -3 \cdot 0 + 1 = 1$, and $(0, 1)$ is a solution.
If $x = -1$, then $y = -3(-1) + 1 = 4$, and $(-1, 4)$ is a solution.

Results are often listed in a table, as shown below. The points corresponding to each pair are then plotted.

$y = -3x + 1$

Calculate ordered pairs.

x	y	(x, y)
2	−5	(2, −5)
0	1	(0, 1)
−1	4	(−1, 4)

(1) Choose x.
(2) Compute y.
(3) Form the pair (x, y).
(4) Plot the points.

Plot the points.

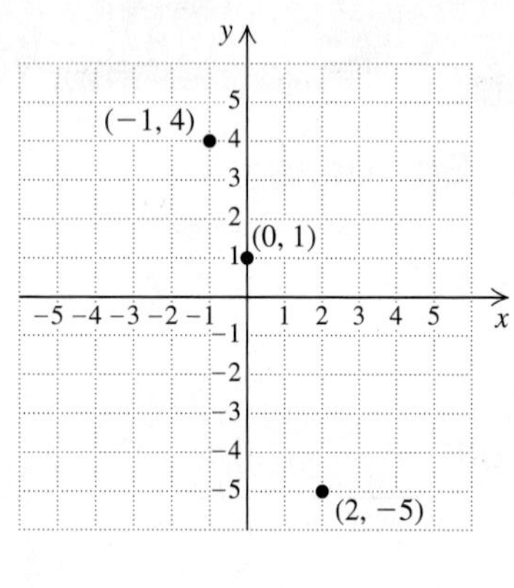

Note that all three points line up. If they didn't, we would know that we had made a mistake, because the equation is linear. When only two points are plotted, an error is more difficult to detect.

Draw the graph.

Finally, we use a ruler to draw a line. We add arrowheads to the ends of the line to indicate that the line extends indefinitely beyond the edge of the grid drawn. Every point on the line represents a solution of $y = -3x + 1$.

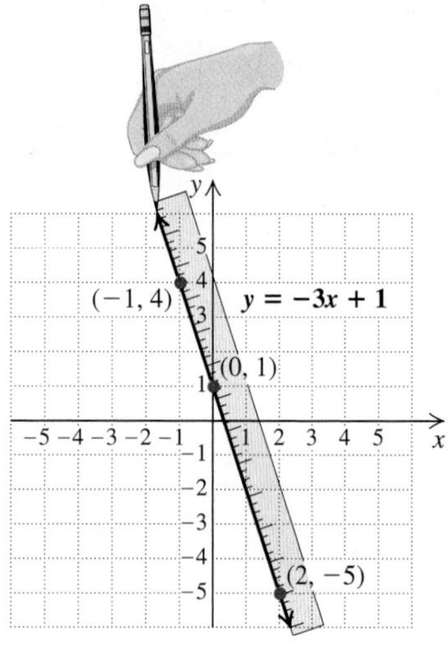

Try Exercise 21.

A graphing calculator also graphs equations by calculating solutions, plotting points, and connecting them. After entering an equation and setting a viewing window, we can view the graph of the equation.

Graphing Equations

Equations are entered using the equation-editor screen, often accessed by pressing Y=.

- The variables used are x and y. If an equation contains different variables, such as a and b, replace them with x and y before entering the equation.
- The first part of each equation, "Y=," is supplied by the calculator, including a subscript that identifies the equation. Thus we need to solve for y before entering an equation.
- Use X,T,Θ,n to enter x.
- The symbol before the Y indicates the graph style.
- A highlighted = indicates that the equation is selected to be graphed. An equation can be selected or deselected by positioning the cursor on the = and pressing ENTER.
- An equation can be cleared by positioning the cursor on the equation and pressing CLEAR.

(continued)

Selected equations are graphed by pressing GRAPH. A standard viewing window is not always the best window to use. Choosing an appropriate viewing window for a graph can be challenging. If the window is not set appropriately, the graph may not appear on the screen at all.

There is generally no one "correct" window, and a good choice often involves trial and error. For now, start with a standard window, and change the dimensions if needed. The ZOOM menu can help in setting window dimensions; for example, choosing ZStandard from the menu will graph the selected equations using the standard viewing window.

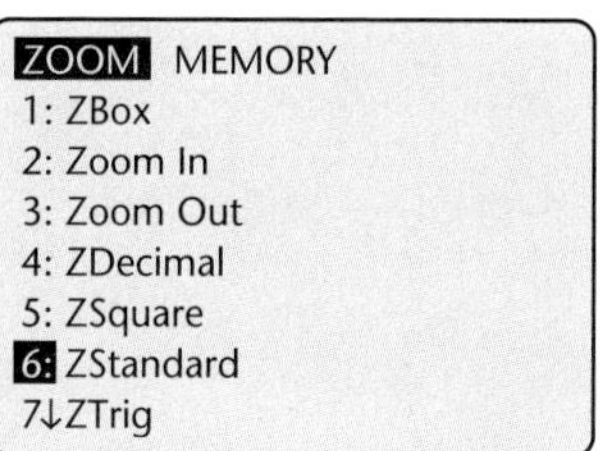

Your Turn

1. Press Y= and clear any equations present.
2. Enter $y = -3x + 1$.
3. Graph the equation on a standard viewing window by pressing ZOOM 6. The graph should look like the graph in Example 3.

The equation in the next example is graphed both by hand and by using a graphing calculator. Let's compare the processes and the results.

EXAMPLE 4 Graph the equation: $y = -\frac{1}{2}x$.

SOLUTION

BY HAND

We find some ordered pairs that are solutions. To find an ordered pair, we can choose *any* number for x and then determine y. By choosing even numbers for x, we can avoid fractions when calculating y. For example, if we choose 4 for x, we get $y = \left(-\frac{1}{2}\right)(4)$, or -2. If x is -6, we get $y = \left(-\frac{1}{2}\right)(-6)$, or 3. We find several ordered pairs, plot them, and draw the line.

$y = -\frac{1}{2}x$

x	y	(x, y)
4	-2	$(4, -2)$
-6	3	$(-6, 3)$
0	0	$(0, 0)$
2	-1	$(2, -1)$

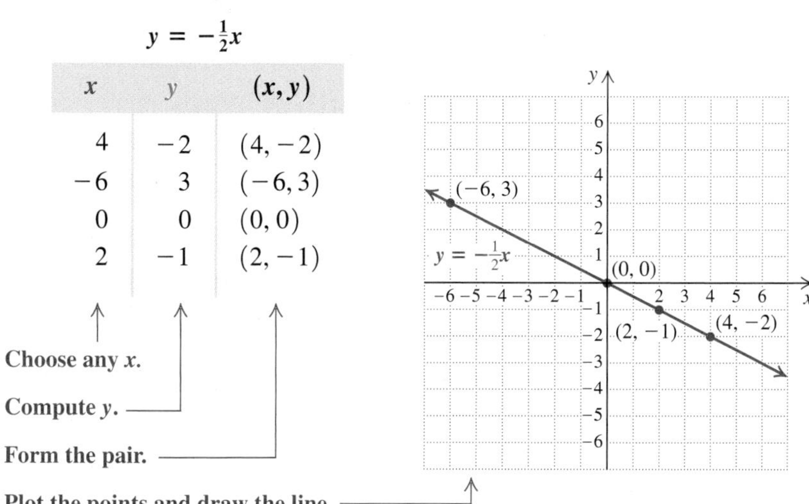

WITH A GRAPHING CALCULATOR

We press Y= and clear any equations present, and then enter the equation as $y = -(1/2)x$. Remember to use the (−) key for the negative sign. Also note that some calculators require parentheses around a fraction coefficient. The standard viewing window is a good choice for this graph.

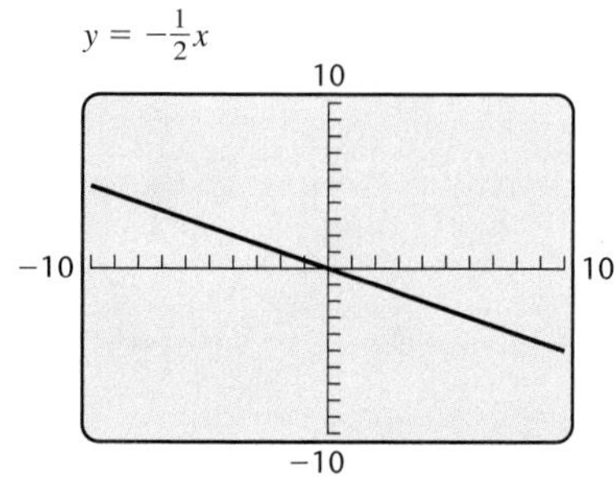

Try Exercise 55.

EXAMPLE 5 Graph: $4x + 2y = 12$.

SOLUTION To form ordered pairs, we can replace either variable with a number and then calculate the other coordinate:

If $y = 0$, we have
$$4x + 2 \cdot 0 = 12$$
$$4x = 12$$
$$x = 3. \quad \textbf{(3, 0) is a solution.}$$

If $x = 0$, we have
$$4 \cdot 0 + 2y = 12$$
$$2y = 12$$
$$y = 6. \quad \textbf{(0, 6) is a solution.}$$

If $y = 2$, we have
$$4x + 2 \cdot 2 = 12$$
$$4x + 4 = 12$$
$$4x = 8$$
$$x = 2. \quad \textbf{(2, 2) is a solution.}$$

$4x + 2y = 12$

x	y	(x, y)
3	0	(3, 0)
0	6	(0, 6)
2	2	(2, 2)

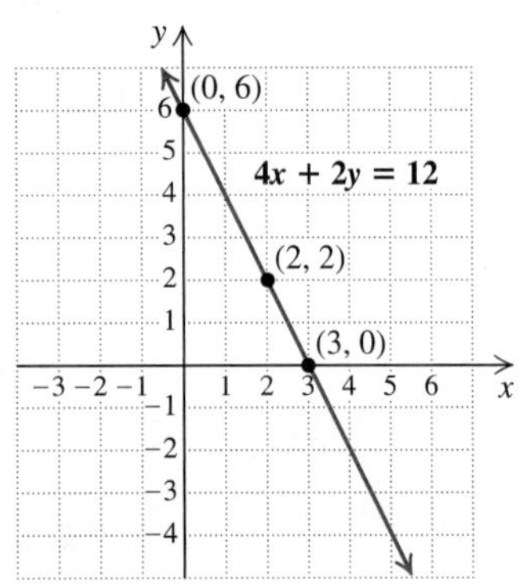

Try Exercise 43.

Note that in Examples 3 and 4 the variable y is isolated on one side of the equation. This often simplifies calculations, so it is important to be able to solve for y before graphing. Also, equations must generally be solved for y before they can be entered on a graphing calculator.

EXAMPLE 6 Graph $3y = 2x$ by first solving for y.

SOLUTION To isolate y, we divide both sides by 3, or multiply both sides by $\frac{1}{3}$:

$$3y = 2x$$
$$\frac{1}{3} \cdot 3y = \frac{1}{3} \cdot 2x \quad \textbf{Using the multiplication principle to multiply both sides by } \tfrac{1}{3}$$
$$\left.\begin{aligned} 1y &= \tfrac{2}{3} \cdot x \\ y &= \tfrac{2}{3}x. \end{aligned}\right\} \quad \textbf{Simplifying}$$

Because all the equations above are equivalent, we can use $y = \frac{2}{3}x$ to draw the graph of $3y = 2x$.

To graph $y = \frac{2}{3}x$, we can select x-values that are multiples of 3. This will allow us to avoid fractions when the corresponding y-values are computed.

$$\left.\begin{aligned} &\text{If } x = 3, && \text{then } y = \tfrac{2}{3} \cdot 3 = 2. \\ &\text{If } x = -3, && \text{then } y = \tfrac{2}{3}(-3) = -2. \\ &\text{If } x = 6, && \text{then } y = \tfrac{2}{3} \cdot 6 = 4. \end{aligned}\right\}$$

Note that when multiples of 3 are substituted for x, the y-coordinates are not fractions.

The table below lists these solutions. Next, we plot the points and see that they form a line. Finally, we draw and label the line.

$3y = 2x$, or $y = \frac{2}{3}x$

x	y	(x, y)
3	2	(3, 2)
−3	−2	(−3, −2)
6	4	(6, 4)

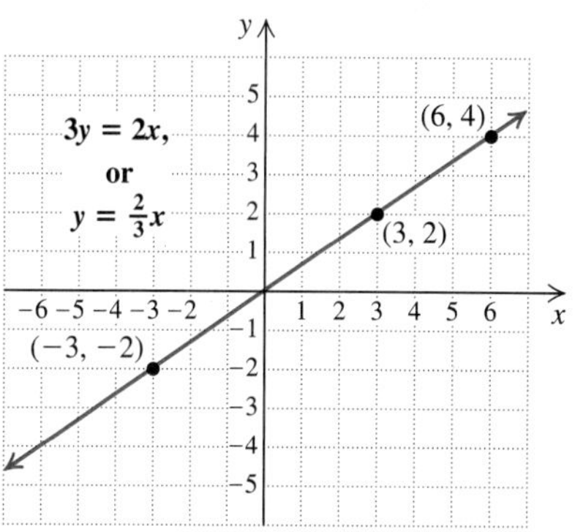

Try Exercise 41.

Student Notes

We can enter the equation in Example 7 as $y = (-1/5)x - 2$. Alternatively, we could solve for y by dividing on both sides by 5 instead of multiplying by $\frac{1}{5}$:

$$x + 5y = -10$$
$$5y = -x - 10$$
$$y = (-x - 10)/5.$$

This approach is often used for entering equations. The approach used in Example 7 is necessary for further study of linear graphs.

EXAMPLE 7 Graph $x + 5y = -10$ by first solving for y.

SOLUTION We have

$$x + 5y = -10$$
$$5y = -x - 10 \quad \text{Adding } -x \text{ to both sides}$$
$$y = \tfrac{1}{5}(-x - 10) \quad \text{Multiplying both sides by } \tfrac{1}{5}$$
$$y = -\tfrac{1}{5}x - 2. \quad \text{Using the distributive law}$$

CAUTION! You must multiply *both* $-x$ and -10 by $\frac{1}{5}$.

Thus, $x + 5y = -10$ is equivalent to $y = -\frac{1}{5}x - 2$. If we choose x-values that are multiples of 5, we can avoid fractions when calculating the corresponding y-values.

If $x = 5$, then $y = -\frac{1}{5} \cdot 5 - 2 = -1 - 2 = -3$.
If $x = 0$, then $y = -\frac{1}{5} \cdot 0 - 2 = 0 - 2 = -2$.
If $x = -5$, then $y = -\frac{1}{5}(-5) - 2 = 1 - 2 = -1$.

$x + 5y = -10$, or $y = -\frac{1}{5}x - 2$

x	y	(x, y)
5	−3	(5, −3)
0	−2	(0, −2)
−5	−1	(−5, −1)

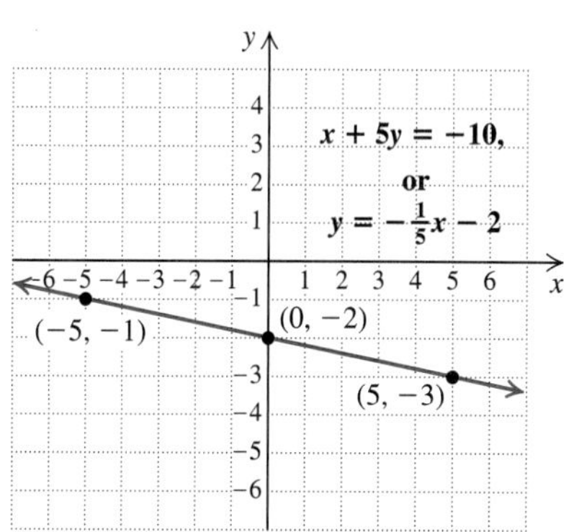

Try Exercise 37.

APPLICATIONS

Linear equations appear in many real-life situations.

EXAMPLE 8 Fuel Efficiency. A typical tractor-trailer will move 18 tons of air per mile at 55 mph. Air resistance increases with speed, causing fuel efficiency to decrease at higher speeds. At highway speeds, a certain truck's fuel efficiency t, in miles per gallon (mpg), can be given by

$$t = -0.1s + 13.1,$$

where s is the speed of the truck, in miles per hour (mph). Graph the equation and then use the graph to estimate the fuel efficiency at 66 mph.
Source: Based on data from Kenworth Truck Co.

SOLUTION We graph $t = -0.1s + 13.1$ by first selecting values for s and then calculating the associated values t. Since the equation is true for highway speeds, we use $s \geq 50$.

If $s = 50$, then $t = -0.1(50) + 13.1 = 8.1$.
If $s = 60$, then $t = -0.1(60) + 13.1 = 7.1$.
If $s = 70$, then $t = -0.1(70) + 13.1 = 6.1$.

s	t
50	8.1
60	7.1
70	6.1

Because we are *selecting* values for s and *calculating* values for t, we represent s on the horizontal axis and t on the vertical axis. Counting by 5's horizontally, beginning at 50, and by 0.5 vertically, beginning at 4, will allow us to plot all three pairs, as shown below.

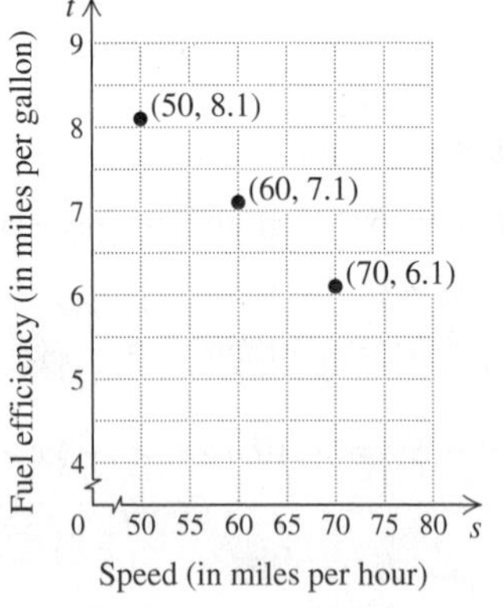

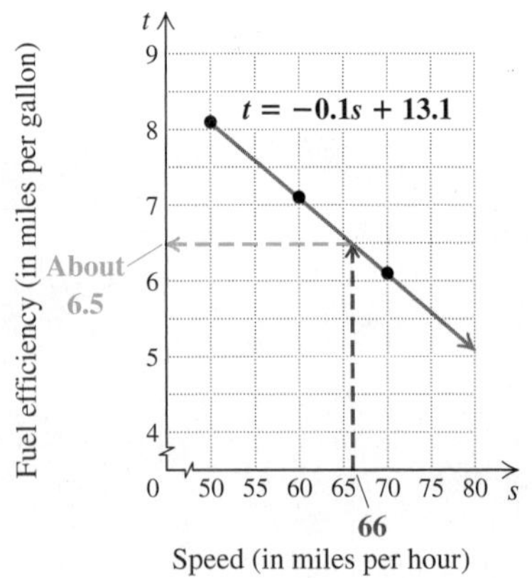

Since the three points line up, our calculations are probably correct. We draw a line, beginning at (50, 8.1). To estimate the fuel efficiency at 66 mph, we locate the point on the line that is above 66 and then find the value on the t-axis that corresponds to that point, as shown in the figure on the right above. The fuel efficiency at 66 mph is about 6.5 mpg.

Try Exercise 73.

CAUTION! When the coordinates of a point are read from a graph, as in Example 8, values should not be considered exact.

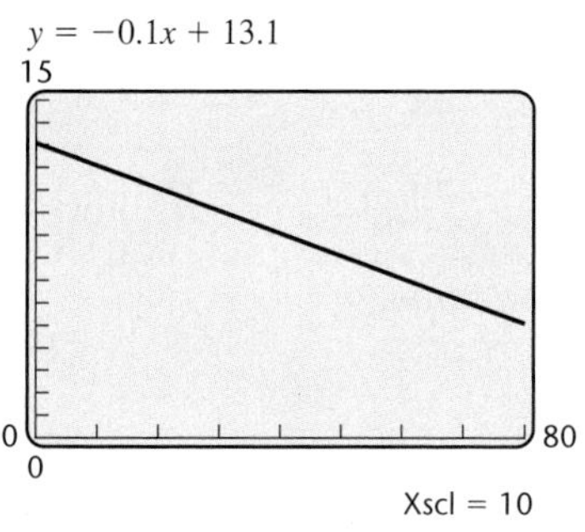

To graph an equation like the one in Example 8 using a graphing calculator, we replace the variables in the equation with x and y. To graph $t = -0.1s + 13.1$, we enter the equation $y = -0.1x + 13.1$. The graph of the equation is shown at left. We show only the first quadrant because neither x nor y can be negative. Note that the equation does not apply to speeds of less than 50 mph, but the graph shown does extend above (50, 8.1).

GRAPHING NONLINEAR EQUATIONS

We refer to any equation whose graph is a straight line as a **linear equation.** Many equations, however, are not linear. When ordered pairs that are solutions of such an equation are plotted, the pattern formed is not a straight line. Let's look at the graph of a **nonlinear equation.**

Student Notes

If you *know* that an equation is linear, you can draw the graph using only two points. If you are not sure, or if you know that the equation is nonlinear, you must calculate and plot more than two points—as many as is necessary in order for you to determine the shape of the graph.

EXAMPLE 9 Graph: $y = x^2 - 5$.

SOLUTION We select numbers for x and find the corresponding values for y. For example, if we choose -2 for x, we get $y = (-2)^2 - 5 = 4 - 5 = -1$. The table lists several ordered pairs.

$y = x^2 - 5$

x	y
0	−5
−1	−4
1	−4
−2	−1
2	−1
−3	4
3	4

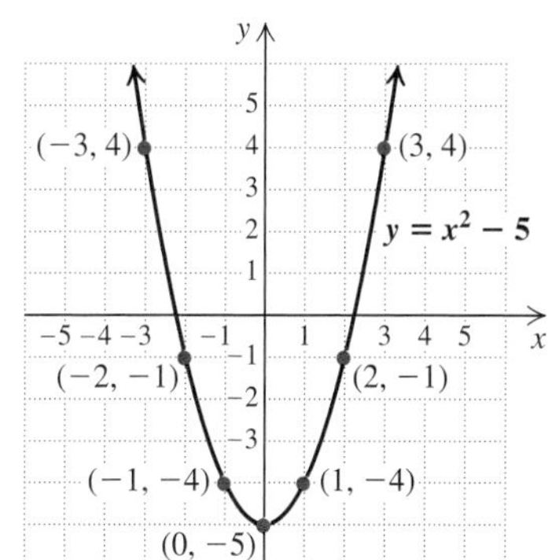

Next, we plot the points. The more points plotted, the clearer the shape of the graph becomes. Since the value of $x^2 - 5$ grows rapidly as x moves away from the origin, the graph rises steeply on either side of the y-axis.

Try Exercise 47.

Curves similar to the one in Example 9 are studied in detail in Chapter 11. Graphing calculators are especially helpful when drawing such nonlinear graphs.

3.2 Exercise Set

FOR EXTRA HELP MyMathLab

READ

Concept Reinforcement *Classify each of the following statements as either true or false.*

1. A linear equation in two variables has at most one solution.
2. Every solution of $y = 3x - 7$ is an ordered pair.
3. The graph of $y = 3x - 7$ represents all solutions of the equation.
4. If a point is on the graph of $y = 3x - 7$, the corresponding ordered pair is a solution of the equation.

5. To find a solution of $y = 3x - 7$, we can choose any value for x and calculate the corresponding value for y.

6. The graph of every equation is a straight line.

Determine whether each equation has the given ordered pair as a solution.

7. $y = 4x - 7; (2, 1)$

8. $y = 2x + 3; (0, 3)$

9. $3y + 2x = 12; (4, 2)$

10. $5x - 3y = 15; (0, 5)$

11. $4m - 5n = 7; (3, -1)$

12. $3b - 2a = -8; (1, -2)$

In Exercises 13–20, an equation and two ordered pairs are given. Show that each pair is a solution of the equation. Then graph the two pairs to determine another solution. See Example 2. Answers may vary.

13. $y = x + 3;\ (-1, 2), (4, 7)$

14. $y = x - 2;\ (3, 1), (-2, -4)$

15. $y = \frac{1}{2}x + 3;\ (4, 5), (-2, 2)$

16. $y = \frac{1}{2}x - 1;\ (6, 2), (0, -1)$

17. $y + 3x = 7;\ (2, 1), (4, -5)$

18. $2y + x = 5;\ (-1, 3), (7, -1)$

19. $4x - 2y = 10;\ (0, -5), (4, 3)$

20. $6x - 3y = 3;\ (1, 1), (-1, -3)$

Graph each equation by hand.

21. $y = x + 1$

22. $y = x - 1$

23. $y = -x$

24. $y = x$

25. $y = 2x$

26. $y = -3x$

27. $y = 2x + 2$

28. $y = 3x - 2$

29. $y = -\frac{1}{2}x$

30. $y = \frac{1}{4}x$

31. $y = \frac{1}{3}x - 4$

32. $y = \frac{1}{2}x + 1$

33. $x + y = 4$

34. $x + y = -5$

35. $x - y = -2$

36. $y - x = 3$

37. $x + 2y = -6$

38. $x + 2y = 8$

39. $y = -\frac{2}{3}x + 4$

40. $y = \frac{3}{2}x + 1$

41. $4x = 3y$

42. $2x = 5y$

43. $8x - 4y = 12$

44. $6x - 3y = 9$

45. $6y + 2x = 8$

46. $8y + 2x = -4$

47. $y = x^2 + 1$

48. $y = x^2 - 3$

49. $y = 2 - x^2$

50. $y = 3 + x^2$

Graph the solutions of each equation.

51. **(a)** $x + 1 = 4$; **(b)** $x + 1 = y$

52. **(a)** $2x = 7$; **(b)** $2x = y$

53. **(a)** $y = 2x - 3$; **(b)** $-7 = 2x - 3$

54. **(a)** $y = x - 3$; **(b)** $-2 = x - 3$

Graph each equation using a graphing calculator. Remember to solve for y first if necessary.

55. $y = -\frac{3}{2}x + 1$

56. $y = -\frac{2}{3}x - 2$

57. $4y - 3x = 1$

58. $4y - 3x = 2$

59. $y = -2$

60. $y = 5$

61. $y = -x^2$

62. $y = x^2$

63. $x^2 - y = 3$

64. $y - 2 = x^2$

65. $y = x^3$

66. $y = x^3 - 2$

Graph each equation using both viewing windows indicated. Determine which window best shows both the shape of the graph and where the graph crosses the x- and y-axes.

67. $y = x - 15$
a) $[-10, 10, -10, 10]$, Xscl = 1, Yscl = 1
b) $[-20, 20, -20, 20]$, Xscl = 5, Yscl = 5

68. $y = -3x + 30$
a) $[-10, 10, -10, 10]$, Xscl = 1, Yscl = 1
b) $[-20, 20, -20, 40]$, Xscl = 5, Yscl = 5

69. $y = 5x^2 - 8$
a) $[-10, 10, -10, 10]$, Xscl = 1, Yscl = 1
b) $[-3, 3, -3, 3]$, Xscl = 1, Yscl = 1

70. $y = \frac{1}{10}x^2 + \frac{1}{3}$
a) $[-10, 10, -10, 10]$, Xscl = 1, Yscl = 1
b) $[-0.5, 0.5, -0.5, 0.5]$, Xscl = 0.1, Yscl = 0.1

71. $y = 4x^3 - 12$
a) $[-10, 10, -10, 10]$, Xscl = 1, Yscl = 1
b) $[-5, 5, -20, 10]$, Xscl = 1, Yscl = 5

72. $y = x^2 + 10$
a) $[-10, 10, -10, 10]$, Xscl = 1, Yscl = 1
b) $[-3, 3, 0, 20]$, Xscl = 1, Yscl = 5

Solve by graphing. Label all axes, and show where each solution is located on the graph.

73. *Student Aid.* The average award a of federal student financial assistance per student is approximated by $a = 0.08t + 2.5$, where a is in thousands of dollars and t is the number of years since 1994. Graph the equation and use the graph to estimate the average amount of federal student aid per student in 2012.
Source: Based on data from U.S. Department of Education, Office of Postsecondary Education

74. *Value of a Color Copier.* The value of Duplio-graphic's color copier is given by $v = -0.68t + 3.4$, where v is the value, in thousands of dollars, t years from the date of purchase. Graph the equation and use the graph to estimate the value of the copier after $2\frac{1}{2}$ years.

75. *FedEx Mailing Costs.* Recently, the cost c, in dollars, of shipping a FedEx Priority Overnight package weighing 1 lb or more a distance of 1001 to 1400 mi was given by $c = 3.1w + 29.07$, where w is the package's weight, in pounds. Graph the equation and use the graph to estimate the cost of shipping a $6\frac{1}{2}$-lb package.
Source: Based on data from FedEx.com

76. *Increasing Life Expectancy.* A smoker is 15 times more likely to die from lung cancer than a nonsmoker. An ex-smoker who stopped smoking t years ago is w times more likely to die from lung cancer than a nonsmoker, where $w = 15 - t$. Graph the equation and use the graph to estimate how much more likely it is for Sandy to die from lung cancer than Polly, if Polly never smoked and Sandy quit $2\frac{1}{2}$ years ago.
Source: Data from *Body Clock* by Dr. Martin Hughes, p. 60. New York: Facts on File, Inc.

77. *Scrapbook Pricing.* The price p, in dollars, of an 8-in. by 8-in. assembled scrapbook is given by $p = 3.5n + 9$, where n is the number of pages in the scrapbook. Graph the equation and use the graph to estimate the price of a scrapbook containing 25 pages.
Source: www.scrapbooksplease.com

78. *Value of Computer Software.* The value v of a shopkeeper's inventory software program, in hundreds of dollars, is given by $v = -\frac{3}{4}t + 6$, where t is the number of years since the shopkeeper first bought the program. Graph the equation and use the graph to estimate what the program is worth 4 years after it was first purchased.

79. *Cost of College.* The cost T, in hundreds of dollars, of tuition and fees at many community colleges can be approximated by $T = \frac{5}{4}c + 2$, where c is the number of credits for which a student registers. Graph the equation and use the graph to estimate the cost of tuition and fees when a student registers for 4 three-credit courses.

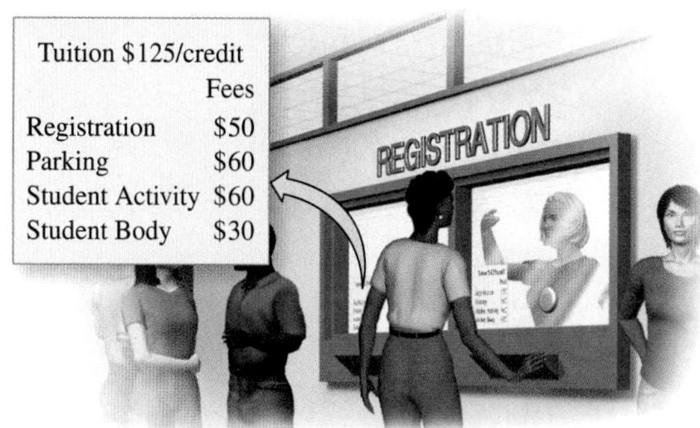

80. *Cost of College.* The cost C, in thousands of dollars, of tuition for a year at a private four-year college can be approximated by $C = \frac{6}{5}t + 14$, where t is the number of years since 1995. Graph the equation and use the graph to estimate the cost of tuition at a private four-year college in 2012.
Source: Based on information from U.S. National Center for Education Statistics

81. *Record Temperature Drop.* On January 22, 1943, the temperature T, in degrees Fahrenheit, in Spearfish, South Dakota, could be approximated by $T = -2m + 54$, where m is the number of minutes since 9:00 A.M. that morning. Graph the equation and use the graph to estimate the temperature at 9:15 A.M.
Source: Based on information from the National Oceanic and Atmospheric Administration

82. *Recycling.* The number of tons n of paper recovered in the United States, in millions, can be approximated by $n = \frac{3}{2}d + 34$, where d is the number of years since 2000. Graph the equation and use the graph to estimate the amount of paper recovered in 2010.
Source: Based on information from Franklin Associates, a division of ERG

TW **83.** The equations $3x + 4y = 8$ and $y = -\frac{3}{4}x + 2$ are equivalent. Which equation would be easier to graph and why?

TW **84.** Suppose that a linear equation is graphed by plotting three points and that the three points line up with each other. Does this *guarantee* that the equation is being correctly graphed? Why or why not?

SKILL REVIEW

To prepare for Section 3.3, review solving equations and formulas (Sections 2.2 and 2.3).

Solve and check. [2.2]

85. $5x + 3 \cdot 0 = 12$

86. $7 \cdot 0 - 4y = 10$

87. $5x + 3(2 - x) = 12$

88. $3(y - 5) - 8y = 6$

Solve. [2.3]

89. $pt + p = w$, for p

90. $Ax + By = C$, for y

91. $A = \dfrac{T + Q}{2}$, for Q

92. $\dfrac{y - k}{m} = x - h$, for y

SYNTHESIS

TW **93.** Janice consistently makes the mistake of plotting the x-coordinate of an ordered pair using the y-axis, and the y-coordinate using the x-axis. How will Janice's incorrect graph compare with the appropriate graph?

TW **94.** Explain how the graph in Example 8 can be used to determine the speed at which the fuel efficiency is 5 mpg.

95. *Bicycling.* Long Beach Island in New Jersey is a long, narrow, flat island. For exercise, Lauren routinely bikes to the northern tip of the island and back. Because of the steady wind, she uses one gear going north and another for her return. Lauren's bike has 14 gears and the sum of the two gears used on her ride is always 18. Write and graph an equation that represents the different pairings of gears that Lauren uses. Note that there are no fraction gears on a bicycle.

In Exercises 96–99, try to find an equation for the graph shown.

96. 

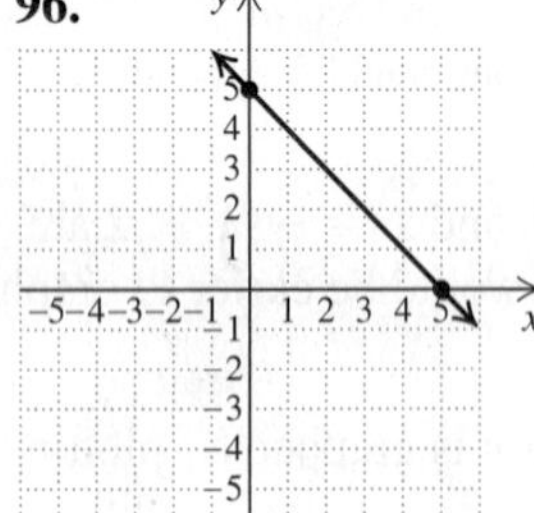

97. 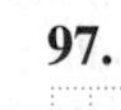

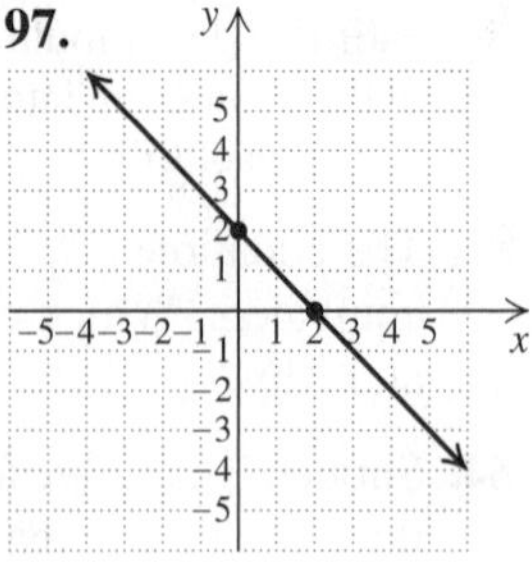

98. 

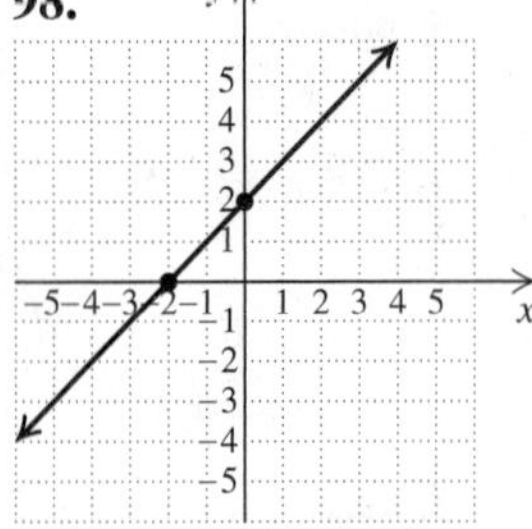

99.

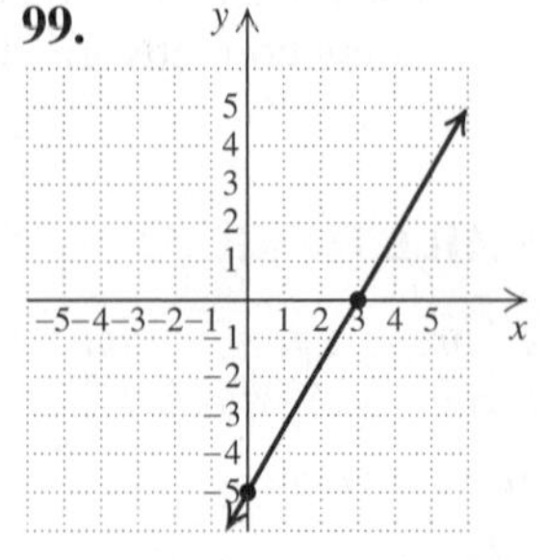

100. Translate to an equation:

d dimes and *n* nickels total $1.75.

Then graph the equation and use the graph to determine three different combinations of dimes and nickels that total $1.75. (See also Exercise 108.)

101. Translate to an equation:

d $25 dinners and *l* $5 lunches total $225.

Then graph the equation and use the graph to determine three different combinations of lunches and dinners that total $225. (See also Exercise 108.)

Use the suggested x-values $-3, -2, -1, 0, 1, 2,$ *and* 3 *to graph each equation. Check using a graphing calculator.*

102. $y = |x|$

Aha! **103.** $y = -|x|$

Aha! **104.** $y = |x| - 2$

105. $y = -|x| + 2$

106. $y = |x| + 3$

107. Example 8 discusses fuel efficiency. If fuel costs $3.50 per gallon, how much money will a truck driver save on a 500-mi trip by driving at 55 mph instead of 70 mph? How many gallons of fuel will be saved?

TW **108.** Study the graph of Exercise 100 or 101. Does *every* point on the graph represent a solution of the associated problem? Why or why not?

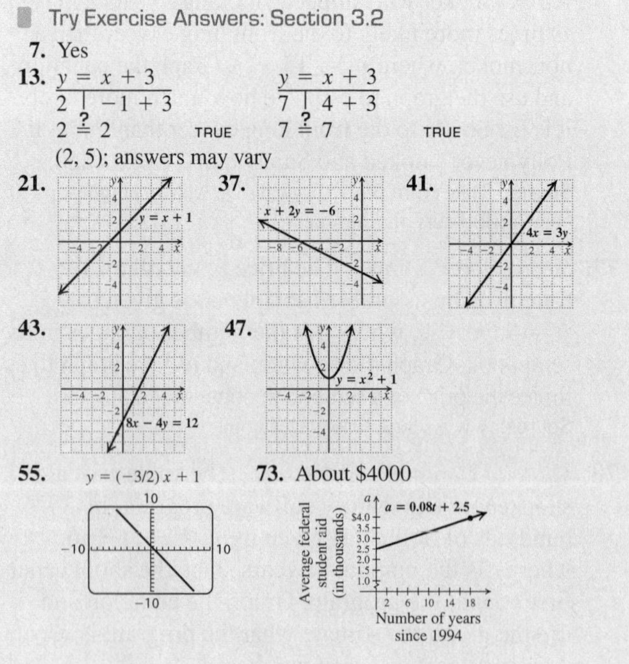

3.3 Linear Equations and Intercepts

- Recognizing Linear Equations
- Intercepts
- Using Intercepts to Graph
- Graphing Horizontal or Vertical Lines

As we saw in Section 3.2, a *linear equation* is an equation whose graph is a straight line. We can determine whether an equation is linear without graphing.

RECOGNIZING LINEAR EQUATIONS

A linear equation may appear in different forms, but all linear equations can be written in the *standard form* $Ax + By = C$.

> **The Standard Form of a Linear Equation** Any equation of the form $Ax + By = C$, where A, B, and C are real numbers and A and B are not both 0, is linear.
> Any equation of the form $Ax + By = C$ is said to be a linear equation in *standard form.*

$y = 3x - 5$

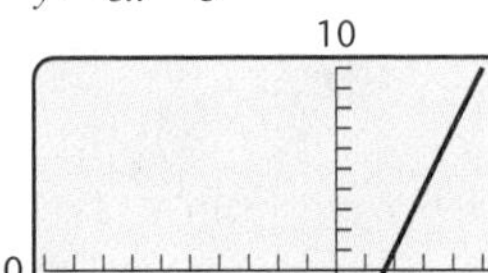

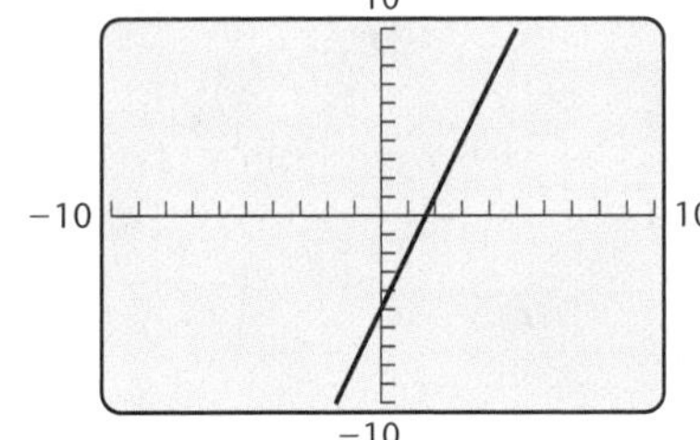

$y = x^2 - 5$

$y = 7/3$

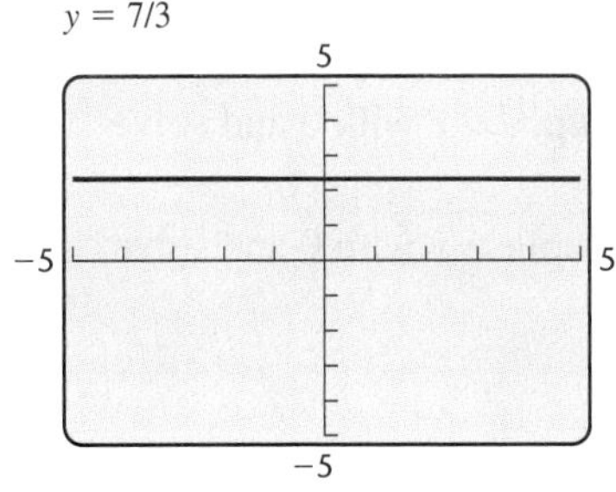

EXAMPLE 1 Determine whether each of the following equations is linear: **(a)** $y = 3x - 5$; **(b)** $y = x^2 - 5$; **(c)** $3y = 7$.

SOLUTION We attempt to write each equation in the form $Ax + By = C$.

a) We have

$$y = 3x - 5$$
$$-3x + y = -5. \qquad \textbf{Adding } -3x \textbf{ to both sides}$$
$$Ax + By = C \qquad A = -3, B = 1, C = -5$$

The equation $y = 3x - 5$ is linear, as confirmed by the first graph at left.

b) We attempt to put the equation in standard form:

$$y = x^2 - 5$$
$$-x^2 + y = -5. \qquad \textbf{Adding } -x^2 \textbf{ to both sides}$$

This last equation is not linear because it has an x^2-term.

We can see this as well from the graph of the equation shown at left.

c) Although there is no x-term in $3y = 7$, the definition of the standard form allows for either A or B to be 0. Thus we write the equation as

$$0 \cdot x + 3y = 7.$$
$$Ax + By = C \qquad A = 0, B = 3, C = 7$$

Thus, $3y = 7$ is linear, as we can see in the third graph at left.

Try Exercise 7.

Once we have determined that an equation is linear, we can use any two points to graph the line because only one line can be drawn through two given points. Unless a line is horizontal or vertical, it will cross both axes at points known as *intercepts*. Often, finding the intercepts of a graph is a convenient way to graph a linear equation.

INTERCEPTS

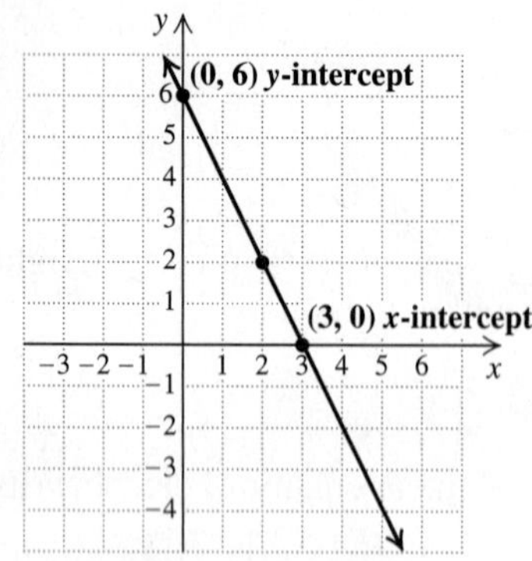

In Example 5 of Section 3.2, we graphed $4x + 2y = 12$ by plotting the points $(3, 0)$, $(0, 6)$, and $(2, 2)$ and then drawing the line.

- The point at which a graph crosses the y-axis is called the **y-intercept.** In the figure shown at left, the y-intercept is $(0, 6)$. The x-coordinate of a y-intercept is always 0.
- The point at which a graph crosses the x-axis is called the **x-intercept.** In the figure shown at left, the x-intercept is $(3, 0)$. The y-coordinate of an x-intercept is always 0.

It is possible for the graph of a nonlinear curve to have more than one y-intercept or more than one x-intercept.

EXAMPLE 2 For the graph shown below, **(a)** give the coordinates of any x-intercepts and **(b)** give the coordinates of any y-intercepts.

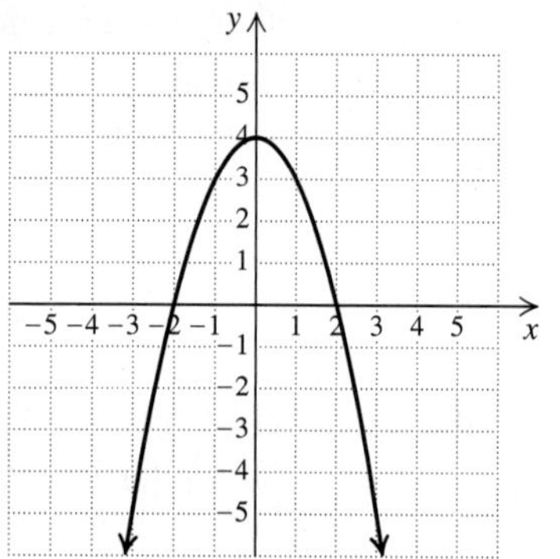

SOLUTION

a) The x-intercepts are the points at which the graph crosses the x-axis. For the graph shown, the x-intercepts are $(-2, 0)$ and $(2, 0)$.

b) The y-intercept is the point at which the graph crosses the y-axis. For the graph shown, the y-intercept is $(0, 4)$.

Try Exercise 17.

STUDY TIP 

Create Your Own Glossary

Understanding the meaning of mathematical terms is essential for success in any math course. Try writing your own glossary of important words toward the back of your notebook. Often, just the act of writing out a word's definition can help you remember what the word means.

USING INTERCEPTS TO GRAPH

It is important to know how to locate a graph's intercepts from the equation being graphed.

> **To Find Intercepts**
>
> To find the y-intercept(s) of an equation's graph, replace x with 0 and solve for y.
>
> To find the x-intercept(s) of an equation's graph, replace y with 0 and solve for x.

EXAMPLE 3 Find the y-intercept and the x-intercept of the graph of $2x + 4y = 20$.

SOLUTION To find the y-intercept, we let $x = 0$ and solve for y:

$$2 \cdot 0 + 4y = 20 \quad \text{Replacing } x \text{ with } 0$$
$$4y = 20$$
$$y = 5.$$

Thus the y-intercept is $(0, 5)$.

To find the x-intercept, we let $y = 0$ and solve for x:

$$2x + 4 \cdot 0 = 20 \quad \text{Replacing } y \text{ with } 0$$
$$2x = 20$$
$$x = 10.$$

Thus the x-intercept is $(10, 0)$.

Try Exercise 25.

Intercepts can be used to graph a linear equation.

EXAMPLE 4 Graph $2x + 4y = 20$ using intercepts.

SOLUTION In Example 3, we showed that the y-intercept is $(0, 5)$ and the x-intercept is $(10, 0)$. We plot both intercepts. Before drawing a line, we plot a third point as a check. We substitute any convenient value for x and solve for y.

If we let $x = 5$, then

$$2 \cdot 5 + 4y = 20 \quad \text{Substituting 5 for } x$$
$$10 + 4y = 20$$
$$4y = 10 \quad \text{Subtracting 10 from both sides}$$
$$y = \tfrac{10}{4}, \text{ or } 2\tfrac{1}{2}. \quad \text{Solving for } y$$

Student Notes

Since each intercept is on an axis, one of the numbers in the ordered pair is 0. If both numbers in an ordered pair are nonzero, the pair is *not* an intercept.

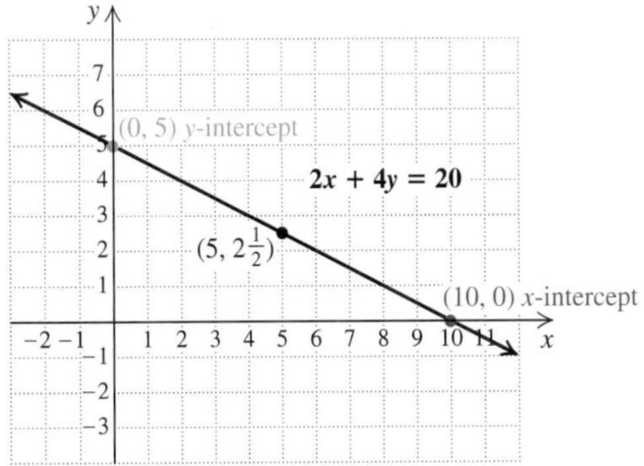

The point $\left(5, 2\tfrac{1}{2}\right)$ appears to line up with the intercepts, so our work is probably correct. To finish, we draw and label the line.

Try Exercise 35.

Note that when we solved for the y-intercept, we simplified $2x + 4y = 20$ to $4y = 20$. Thus, to find the y-intercept, we can momentarily ignore the x-term and solve the remaining equation.

In a similar manner, when we solved for the x-intercept, we simplified $2x + 4y = 20$ to $2x = 20$. Thus, to find the x-intercept, we can momentarily ignore the y-term and then solve this remaining equation.

EXAMPLE 5 Graph $3x - 2y = 60$ using intercepts.

SOLUTION To find the y-intercept, we let $x = 0$. This amounts to temporarily ignoring the x-term and then solving:

$$-2y = 60 \qquad \textbf{For } x = 0\textbf{, we have } 3 \cdot 0 - 2y\textbf{, or simply } -2y.$$
$$y = -30.$$

The y-intercept is $(0, -30)$.

To find the x-intercept, we let $y = 0$. This amounts to temporarily disregarding the y-term and then solving:

$$3x = 60 \qquad \textbf{For } y = 0\textbf{, we have } 3x - 2 \cdot 0\textbf{, or simply } 3x.$$
$$x = 20.$$

The x-intercept is $(20, 0)$.

To find a third point, we can replace x with 4 and solve for y:

$$3 \cdot 4 - 2y = 60 \qquad \textbf{Numbers other than 4 can be used for } x.$$
$$12 - 2y = 60$$
$$-2y = 48$$
$$y = -24. \qquad \textbf{This means that } (4, -24) \textbf{ is on the graph.}$$

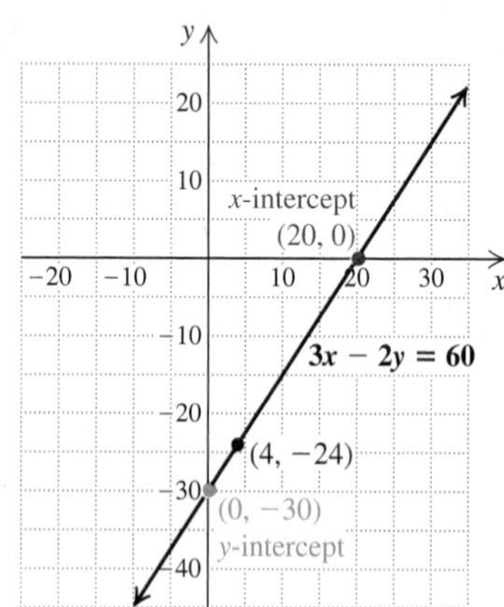

In order for us to graph all three points, the y-axis of our graph must go down to at least -30 and the x-axis must go up to at least 20. Using a scale of 5 units per square allows us to display both intercepts and $(4, -24)$, as well as the origin.

The point $(4, -24)$ appears to line up with the intercepts, so we draw and label the line, as shown at left.

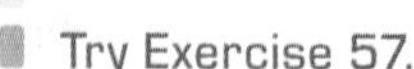

Try Exercise 57.

EXAMPLE 6 Determine a viewing window that shows the intercepts of the graph of the equation $y = 2x + 15$. Then graph the equation.

SOLUTION To find the y-intercept, we let $x = 0$:

$$y = 2 \cdot 0 + 15$$
$$y = 15.$$

The y-intercept is $(0, 15)$.

To find the x-intercept, we let $y = 0$ and solve for x:

$$0 = 2x + 15$$
$$-15 = 2x \qquad \textbf{Subtracting 15 from both sides}$$
$$-\tfrac{15}{2} = x.$$

The x-intercept is $\left(-\frac{15}{2}, 0\right)$.

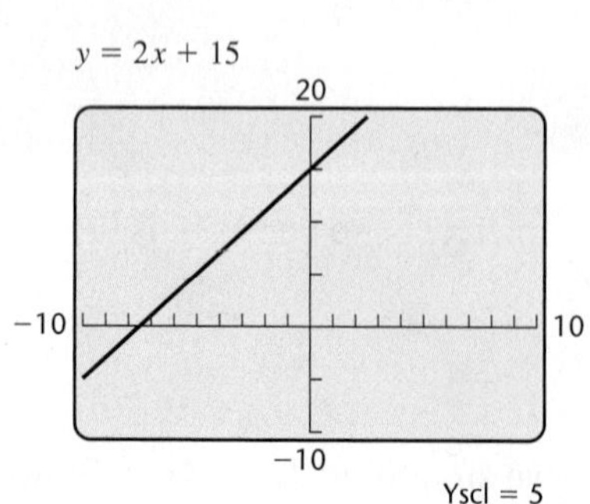

A standard viewing window will not show the y-intercept, $(0, 15)$. Thus we adjust the Ymax value and then we choose a viewing window of $[-10, 10, -10, 20]$, with Yscl $= 5$. Other choices of window dimensions are also possible.

Try Exercise 83.

GRAPHING HORIZONTAL OR VERTICAL LINES

The equations graphed in Examples 4 and 5 are both in the form $Ax + By = C$. We have already stated that any equation in the form $Ax + By = C$ is linear, provided A and B are not both zero. What if A or B (but not both) is zero? We will find that when A is zero, there is no x-term and the graph is a horizontal line. We will also find that when B is zero, there is no y-term and the graph is a vertical line.

Student Notes

Sometimes students draw horizontal lines when they should be drawing vertical lines and vice versa. To avoid this mistake, first locate the correct number on the axis whose label is given. Then draw a line perpendicular to that axis. Thus, to graph $x = 2$, we locate 2 on the x-axis and then draw a line perpendicular to that axis at that point. Note that the graph of $x = 2$ on a plane is a line, whereas the graph of $x = 2$ on the number line is a point.

EXAMPLE 7 Graph: $y = 3$.

SOLUTION We can think of the equation $y = 3$ as $0 \cdot x + y = 3$. No matter what number we choose for x, we find that y must be 3 if the equation is to be solved. Consider the following table.

$y = 3$

Choose any number for x. ⟶

x	y	(x, y)
-2	3	$(-2, 3)$
0	3	$(0, 3)$
4	3	$(4, 3)$

y must be 3.

All pairs will have 3 as the y-coordinate.

When we plot the ordered pairs $(-2, 3)$, $(0, 3)$, and $(4, 3)$ and connect the points, we obtain a horizontal line. Any ordered pair of the form $(x, 3)$ is a solution, so the line is parallel to the x-axis with y-intercept $(0, 3)$. Note that the graph of $y = 3$ has no x-intercept.

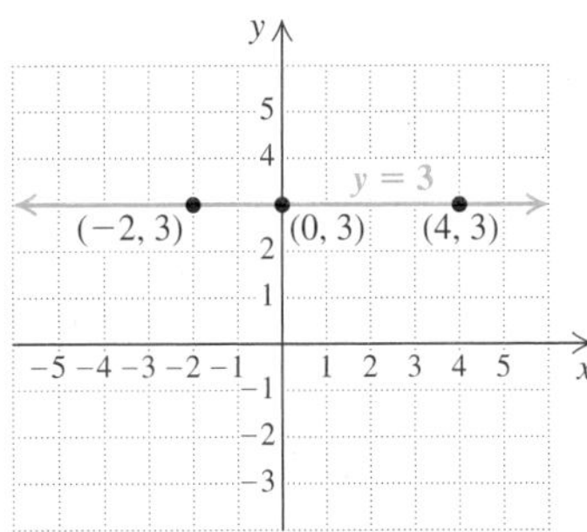

Try Exercise 63.

EXAMPLE 8 Graph: $x = -4$.

SOLUTION We can think of the equation $x = -4$ as $x + 0 \cdot y = -4$. We make up a table with all -4's in the x-column.

$x = -4$

x must be -4. ⟶

x	y	(x, y)
-4	-5	$(-4, -5)$
-4	1	$(-4, 1)$
-4	3	$(-4, 3)$

Choose any number for y.

All pairs will have -4 as the x-coordinate.

When we plot the ordered pairs $(-4, -5)$, $(-4, 1)$, and $(-4, 3)$ and connect them, we obtain a vertical line. Any ordered pair of the form $(-4, y)$ is a solution. The line is parallel to the y-axis with x-intercept $(-4, 0)$. Note that the graph of $x = -4$ has no y-intercept.

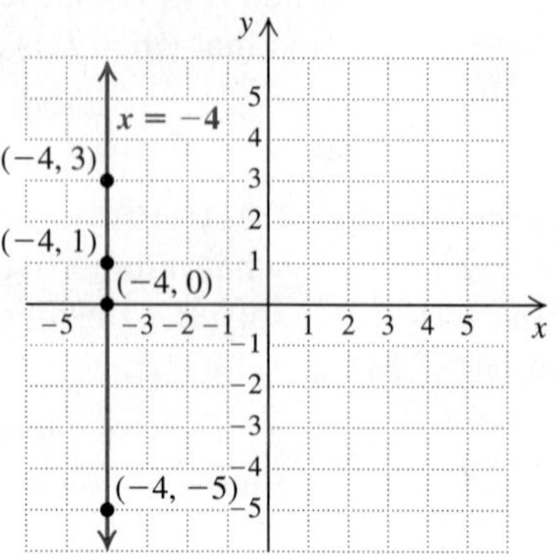

Try Exercise 65.

Linear Equations in One Variable

The graph of $y = b$ is a horizontal line, with y-intercept $(0, b)$.

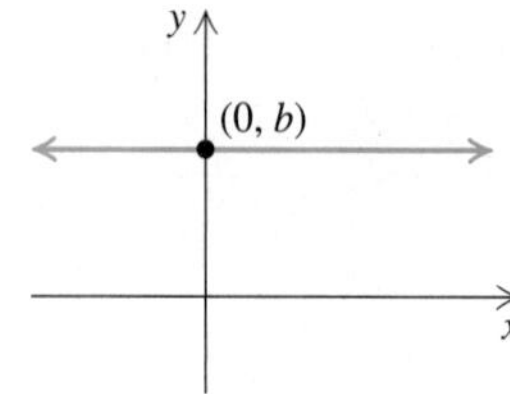

The graph of $x = a$ is a vertical line, with x-intercept $(a, 0)$.

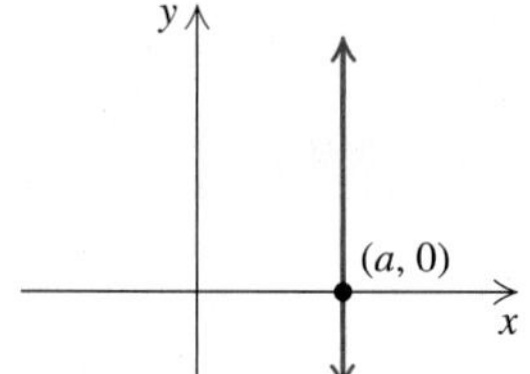

EXAMPLE 9 Write an equation for each graph.

a)

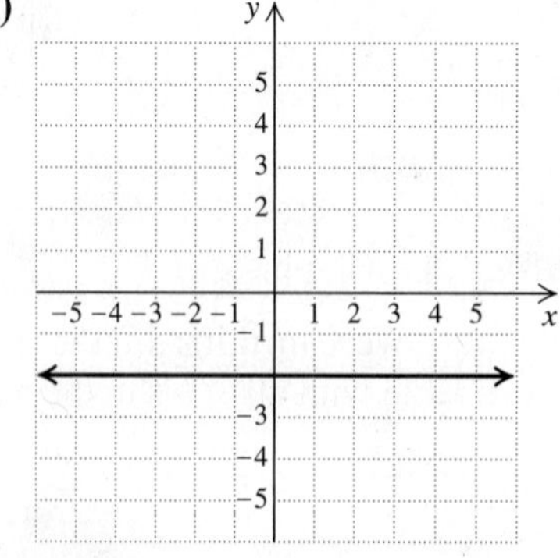

b)

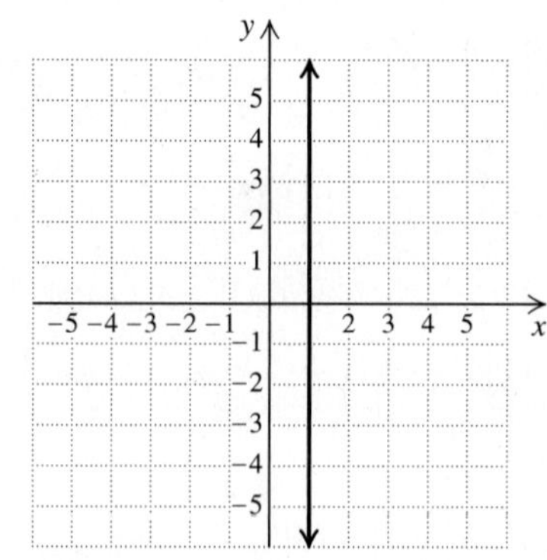

SOLUTION

a) Every point on the horizontal line passing through $(0, -2)$ has -2 as the y-coordinate. Thus the equation of the line is $y = -2$.

b) Every point on the vertical line passing through $(1, 0)$ has 1 as the x-coordinate. Thus the equation of the line is $x = 1$.

Try Exercise 77.

3.3 Exercise Set

FOR EXTRA HELP MyMathLab   

Concept Reinforcement *In each of Exercises 1–6, match the phrase with the most appropriate choice from the column on the right.*

1. ___ A vertical line — **a)** $2x + 5y = 100$
2. ___ A horizontal line — **b)** $(3, -2)$
3. ___ A y-intercept — **c)** $(1, 0)$
4. ___ An x-intercept — **d)** $(0, 2)$
5. ___ A third point as a check — **e)** $y = 3$
6. ___ Use a scale of 10 units per square. — **f)** $x = -4$

Determine whether each equation is linear.

7. $5x - 3y = 15$

8. $2y + 10x = 5$

9. $7y = x - 5$

10. $3y = 4x$

11. $xy = 7$

12. $\dfrac{3y}{4x} = 2x$

13. $16 + 4y = 0$

14. $3x - 12 = 0$

15. $2y - \dfrac{3}{x} = 5$

16. $x + 6xy = 10$

For Exercises 17–24, list **(a)** *the coordinates of the y-intercept and* **(b)** *the coordinates of all x-intercepts.*

17.

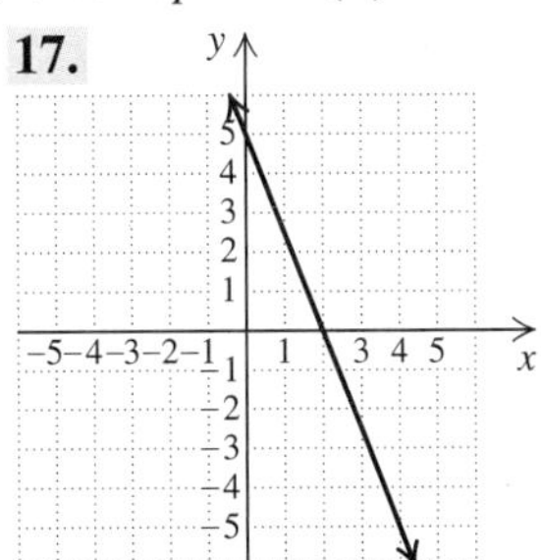

18.

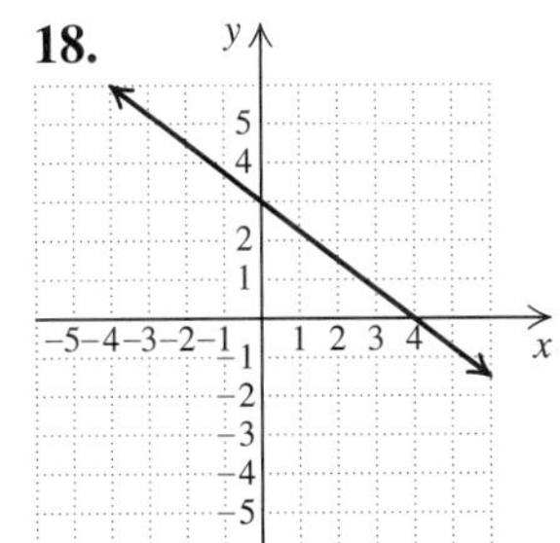

19.

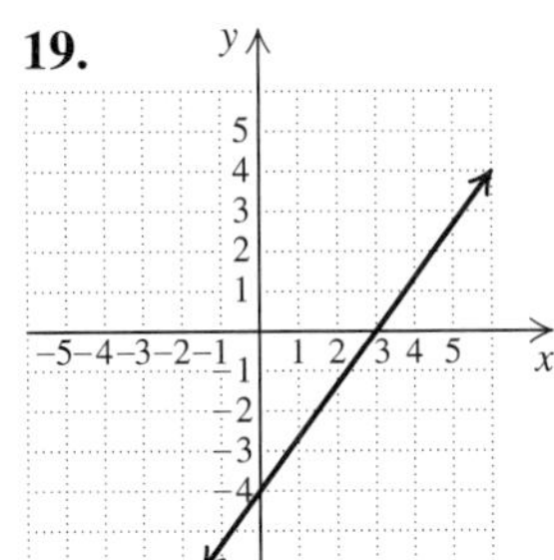

20.

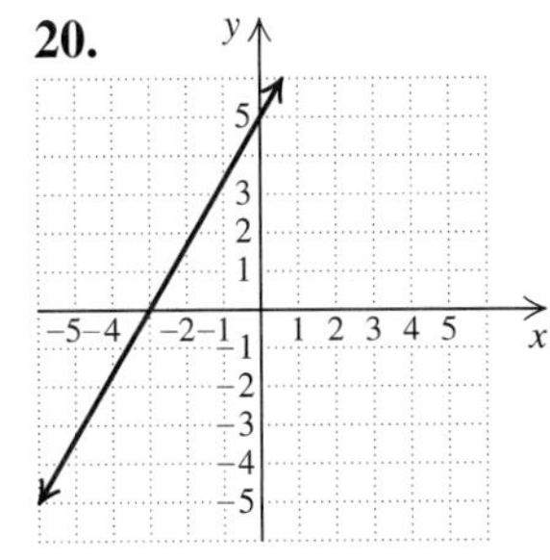

21.

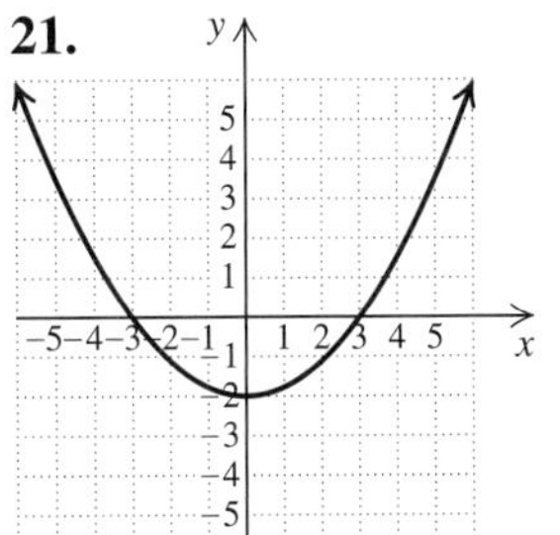

22.

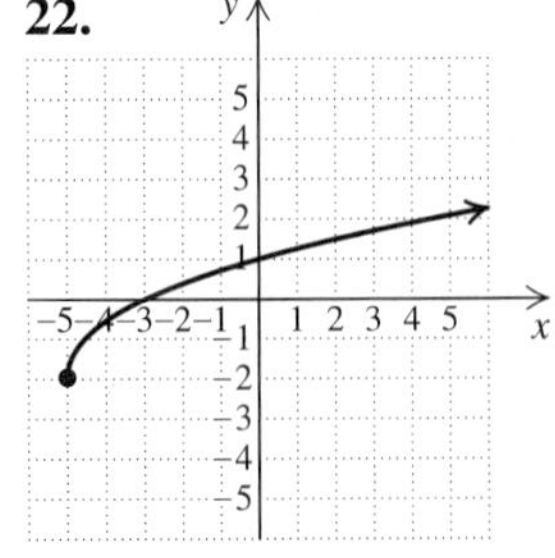

23.

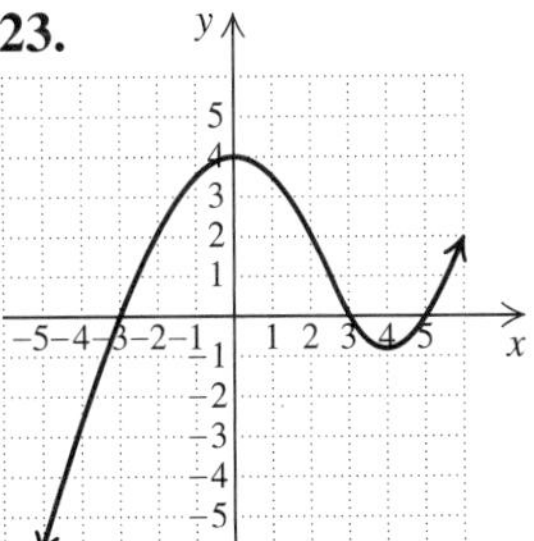

24.

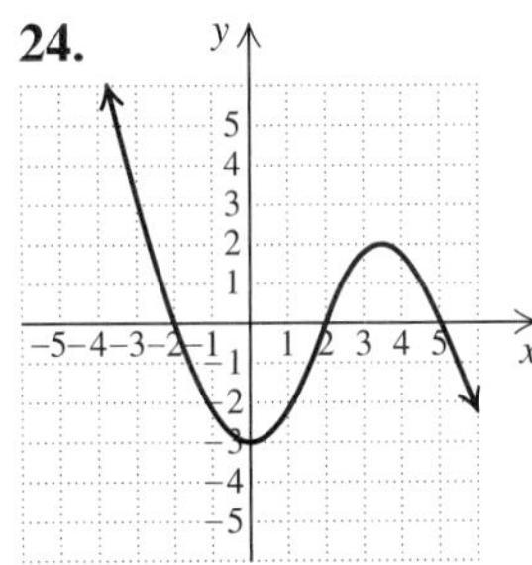

For Exercises 25–34, list **(a)** *the coordinates of any y-intercept and* **(b)** *the coordinates of any x-intercept. Do not graph.*

25. $5x + 3y = 15$

26. $5x + 2y = 20$

27. $7x - 2y = 28$

28. $4x - 3y = 24$

29. $-4x + 3y = 150$

30. $-2x + 3y = 80$

Aha! **31.** $y = 9$

32. $x = 8$

33. $x = -7$

34. $y = -1$

Find the intercepts. Then graph.

35. $3x + 2y = 12$

36. $x + 2y = 6$

37. $x + 3y = 6$

38. $6x + 9y = 36$

39. $-x + 2y = 8$

40. $-x + 3y = 9$

41. $3x + y = 9$

42. $2x - y = 8$

43. $y = 2x - 6$

44. $y = -3x + 6$

45. $3x - 9 = 3y$

46. $5x - 10 = 5y$

47. $2x - 3y = 6$

48. $2x - 5y = 10$

49. $4x + 5y = 20$

50. $6x + 2y = 12$

51. $3x + 2y = 8$

52. $x - 1 = y$

53. $2x + 4y = 6$

54. $5x - 6y = 18$

55. $5x + 3y = 180$

56. $10x + 7y = 210$

57. $y = -30 + 3x$

58. $y = -40 + 5x$

59. $-4x = 20y + 80$

60. $60 = 20x - 3y$

61. $y - 3x = 0$

62. $x + 2y = 0$

Graph.

63. $y = 5$

64. $y = -4$

65. $x = -1$

66. $x = 6$

67. $y = -15$

68. $x = 20$

69. $y = 0$

70. $y = \frac{3}{2}$

71. $x = -\frac{5}{2}$

72. $x = 0$

73. $-4x = -100$

74. $12y = -360$

75. $35 + 7y = 0$

76. $-3x - 24 = 0$

Write an equation for each graph.

77.

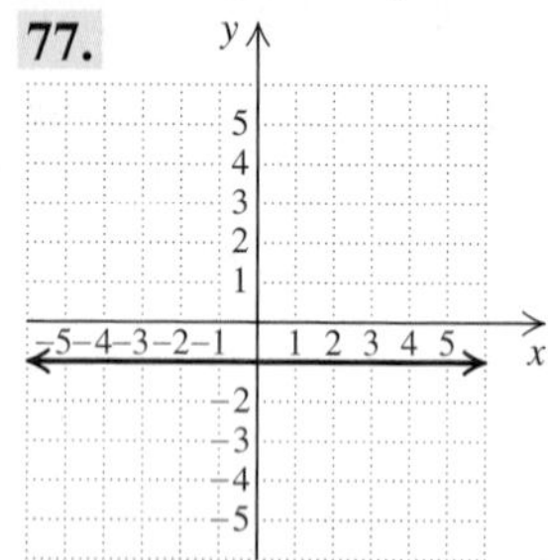

78.

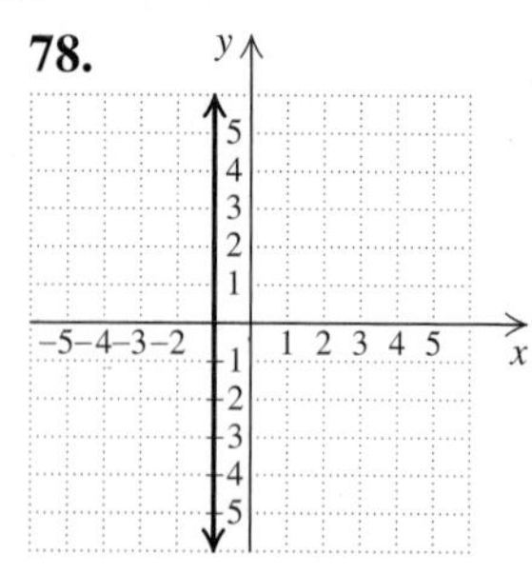

79.

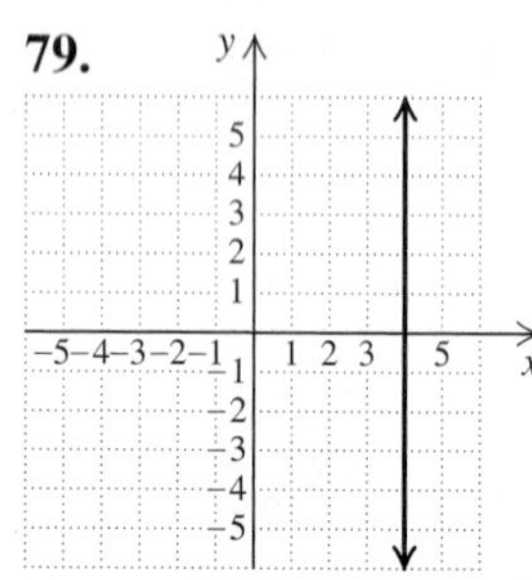

80.

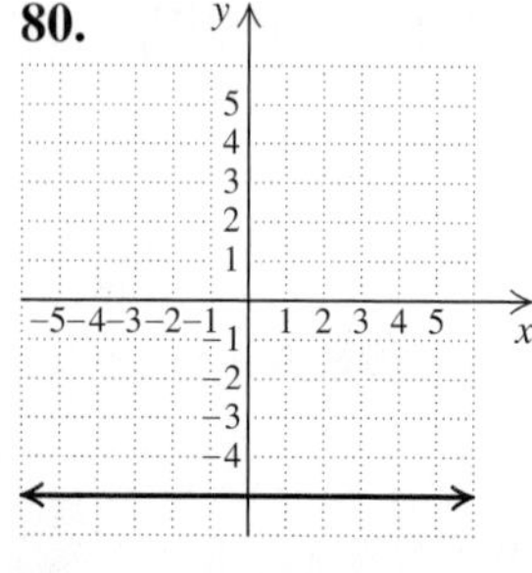

81.

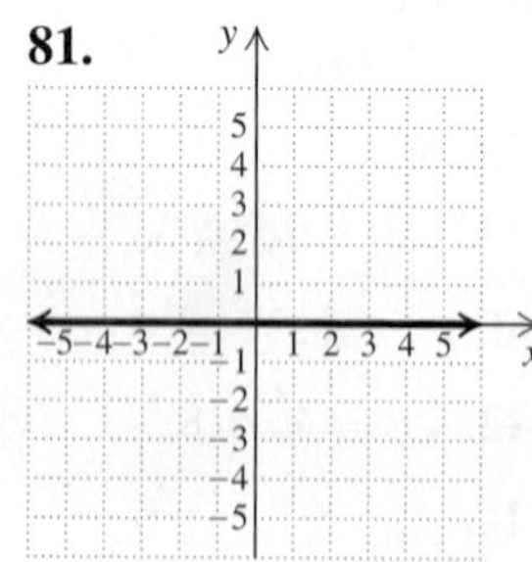

82.

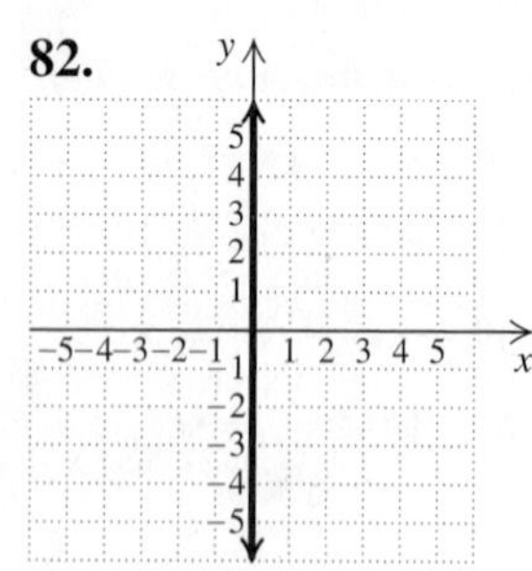

For each equation, find the x-intercept and the y-intercept. Then determine which of the given viewing windows will show both intercepts.

83. $y = 20 - 4x$

a) $[-10, 10, -10, 10]$
b) $[-5, 10, -5, 10]$
c) $[-10, 10, -10, 30]$
d) $[-10, 10, -30, 10]$

84. $y = 3x + 7$

a) $[-10, 10, -10, 10]$
b) $[-1, 15, -1, 15]$
c) $[-15, 5, -15, 5]$
d) $[-10, 10, -30, 0]$

85. $y = -35x + 7000$

a) $[-10, 10, -10, 10]$
b) $[-35, 0, 0, 7000]$
c) $[-1000, 1000, -1000, 1000]$
d) $[0, 500, 0, 10{,}000]$

86. $y = 0.2 - 0.01x$

a) $[-10, 10, -10, 10]$
b) $[-5, 30, -1, 1]$
c) $[-1, 1, -5, 30]$
d) $[0, 0.01, 0, 0.2]$

Using a graphing calculator, graph each equation so that both intercepts can be easily viewed. Adjust the window settings so that tick marks can be clearly seen on both axes.

87. $y = -0.72x - 15$

88. $y - 2.13x = 27$

89. $5x + 6y = 84$

90. $2x - 7y = 150$

91. $19x - 17y = 200$

92. $6x + 5y = 159$

TW **93.** Explain in your own words why the graph of $y = 8$ is a horizontal line.

TW **94.** Explain in your own words why the graph of $x = -4$ is a vertical line.

SKILL REVIEW

To prepare for Section 3.4, review translating to algebraic expressions (Section 1.1).

Translate to an algebraic expression. [1.1]

95. 7 less than d

96. 5 more than w

97. The sum of 7 and four times a number

98. The product of 3 and a number

99. Twice the sum of two numbers

100. Half of the sum of two numbers

SYNTHESIS

TW **101.** Describe what the graph of $x + y = C$ will look like for any choice of C.

TW **102.** If the graph of a linear equation has one point that is both the x-intercept and the y-intercept, what is that point? Why?

103. Write an equation for the x-axis.

104. Write an equation of the line parallel to the x-axis and passing through $(3, 5)$.

105. Write an equation of the line parallel to the y-axis and passing through $(-2, 7)$.

106. Find the coordinates of the point of intersection of the graphs of $y = x$ and $y = 6$.

107. Find the coordinates of the point of intersection of the graphs of the equations $x = -3$ and $y = x$.

108. Write an equation of the line shown in Exercise 17.

109. Write an equation of the line shown in Exercise 20.

110. Find the value of C such that the graph of $3x + C = 5y$ has an x-intercept of $(-4, 0)$.

111. Find the value of C such that the graph of $4x = C - 3y$ has a y-intercept of $(0, -8)$.

TW **112.** For A and B nonzero, the graphs of $Ax + D = C$ and $By + D = C$ will be parallel to an axis. Explain why.

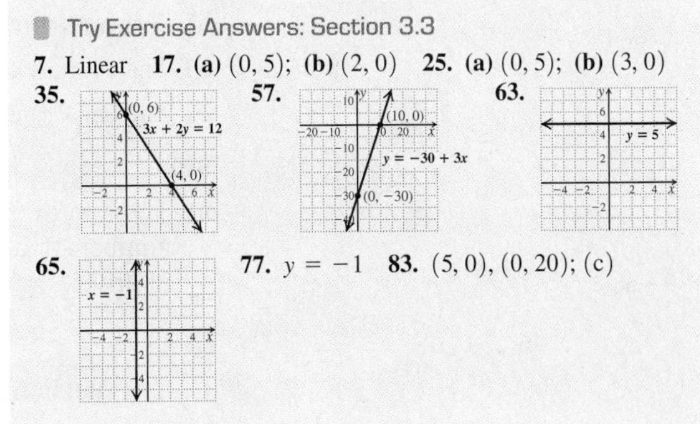

Try Exercise Answers: Section 3.3

7. Linear **17. (a)** $(0, 5)$; **(b)** $(2, 0)$ **25. (a)** $(0, 5)$; **(b)** $(3, 0)$

35. **57.** **63.**

65. **77.** $y = -1$ **83.** $(5, 0), (0, 20)$; (c)

3.4 Rates

- Rates of Change
- Visualizing Rates

RATES OF CHANGE

Because graphs make use of two axes, they allow us to visualize how two quantities change with respect to each other. A number accompanied by units is used to represent this type of change and is referred to as a *rate*.

Rate A *rate* is a ratio that indicates how two quantities change with respect to each other.

STUDY TIP 

Find a Study Buddy

It is not always a simple matter to find a study partner. Tension can develop if one of the study partners feels that he or she is working too much or too little. Try to find a partner with whom you feel compatible and don't take it personally if your first partner is not a good match. With some effort you will be able to locate a suitable partner. Often tutor centers or learning labs are good places to look for one.

Rates occur often in everyday life:

A Web site that grows by 50,000 visitors over a period of 2 months has a *growth rate* of $\frac{50,000}{2}$, or 25,000, visitors per month.

A vehicle traveling 260 mi in 4 hr is moving at a *rate* of $\frac{260}{4}$, or 65, mph (miles per hour).

A class of 25 students pays a total of \$93.75 to visit a museum. The *rate* is $\frac{\$93.75}{25}$, or \$3.75, per student.

CAUTION! To calculate a rate, it is important to keep track of the units being used.

EXAMPLE 1 On January 3, Shannon rented a Ford Focus with a full tank of gas and 9312 mi on the odometer. On January 7, she returned the car with 9630 mi on the odometer.* If the rental agency charged Shannon \$108 for the rental and the necessary 12 gal of gas to fill up the gas tank, find the following rates.

a) The car's rate of gas consumption, in miles per gallon

b) The average cost of the rental, in dollars per day

c) The car's rate of travel, in miles per day

SOLUTION

a) The rate of gas consumption, in miles per gallon, is found by dividing the number of miles traveled by the number of gallons used for that amount of driving:

$$\text{Rate, in miles per gallon} = \frac{9630 \text{ mi} - 9312 \text{ mi}}{12 \text{ gal}}$$ **The word "per" indicates division.**

$$= \frac{318 \text{ mi}}{12 \text{ gal}}$$

$$= 26.5 \text{ mi/gal}$$ **Dividing**

$$= 26.5 \text{ miles per gallon.}$$

b) The average cost of the rental, in dollars per day, is found by dividing the cost of the rental by the number of days:

$$\text{Rate, in dollars per day} = \frac{108 \text{ dollars}}{4 \text{ days}}$$ **From January 3 to January 7 is 7 − 3 = 4 days.**

$$= 27 \text{ dollars/day}$$

$$= \$27 \text{ per day.}$$

c) The car's rate of travel, in miles per day, is found by dividing the number of miles traveled by the number of days:

$$\text{Rate, in miles per day} = \frac{318 \text{ mi}}{4 \text{ days}}$$ **9630 mi − 9312 mi = 318 mi. From January 3 to January 7 is 7 − 3 = 4 days.**

$$= 79.5 \text{ mi/day}$$

$$= 79.5 \text{ mi per day.}$$

Try Exercise 7.

CAUTION! Units are a vital part of real-world problems. They must be considered in the translation of a problem and included in the answer to a problem.

Many problems involve a rate of travel, or *speed.* The **speed** of an object is found by dividing the distance traveled by the time required to travel that distance.

EXAMPLE 2 Transportation. An Atlantic City Express bus makes regular trips between Paramus and Atlantic City, New Jersey. At 6:00 P.M., the bus is at mileage marker 40 on the Garden State Parkway, and at 8:00 P.M. it is at marker 170. Find the average speed of the bus.

*For all rental problems, assume that the pickup time was later in the day than the return time so that no late fees were applied.

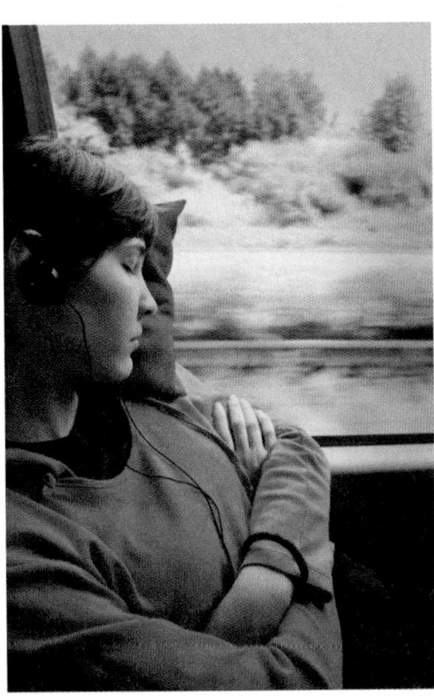

SOLUTION Speed is the distance traveled divided by the time spent traveling:

$$\text{Bus speed} = \frac{\text{Distance traveled}}{\text{Time spent traveling}}$$

$$= \frac{\text{Change in mileage}}{\text{Change in time}}$$

$$= \frac{130 \text{ mi}}{2 \text{ hr}}$$ **170 mi − 40 mi = 130 mi; 8:00 P.M. − 6:00 P.M. = 2 hr**

$$= 65\frac{\text{mi}}{\text{hr}}$$

$= 65$ miles per hour. **This *average* speed does not indicate by how much the bus speed may vary along the route.**

Try Exercise 13.

VISUALIZING RATES

Graphs allow us to visualize a rate of change. As a rule, the quantity listed in the numerator appears on the vertical axis and the quantity listed in the denominator appears on the horizontal axis.

EXAMPLE 3 Sports. In 2001, there were approximately 30 thousand USA Triathlon members. Between 2001 and 2009, this number increased at a rate of approximately 11 thousand athletes per year. Draw a graph to represent this information.

Source: Based on information from USA Triathlon

SOLUTION To label the axes, note that the rate is given as 11,000 athletes per year, or

$$11 \text{ thousand} \frac{\text{athletes}}{\text{year}}.$$ **← Numerator: vertical axis; ← Denominator: horizontal axis**

We list *Number of Triathlon members, in thousands,* on the vertical axis and *Year* on the horizontal axis. (See the figure on the left below.)

Next, we select a scale for each axis that allows us to plot the given information. If we count by increments of 20 thousand on the vertical axis, we can show 30 thousand members for 2001 and increasing amounts for later years. On the horizontal axis, we count by increments of one year. (See the figure on the right below.)

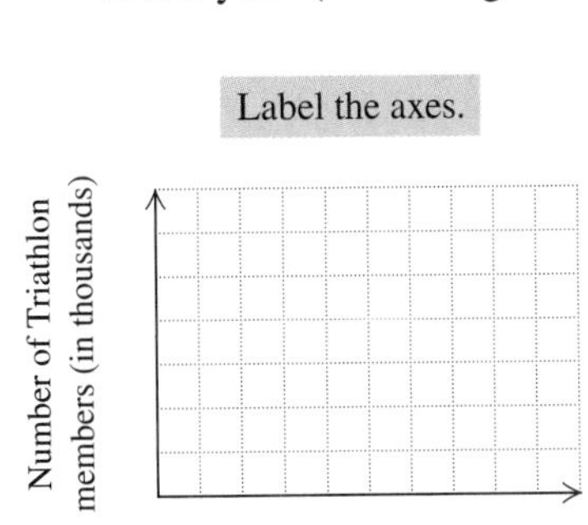

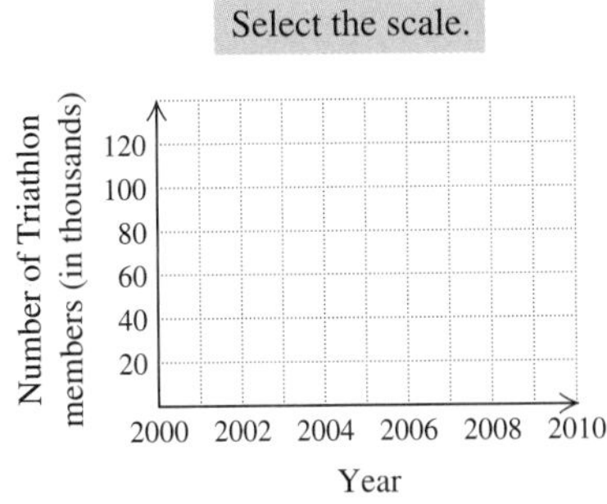

We now plot the point corresponding to (2001, 30 thousand). Then, to display the rate of growth, we move from that point to a second point that represents 11 thousand more athletes 1 year later.

(2001, 30 thousand)	**Beginning point**
(2001 + 1, 30 thousand + 11 thousand)	**11 thousand more athletes, 1 year later**
(2002, 41 thousand)	**A second point on the graph**

Similarly, we can find the coordinates for 2009. Since 2009 is 8 years after 2001, we add 8 to the year and 8(11 thousand) = 88 thousand to the amount.

(2001, 30 thousand)	**Beginning point**
(2001 + 8, 30 thousand + 88 thousand)	**8(11 thousand) more athletes 8 years later**
(2009, 118 thousand)	**A third point on the graph**

Draw the graph.

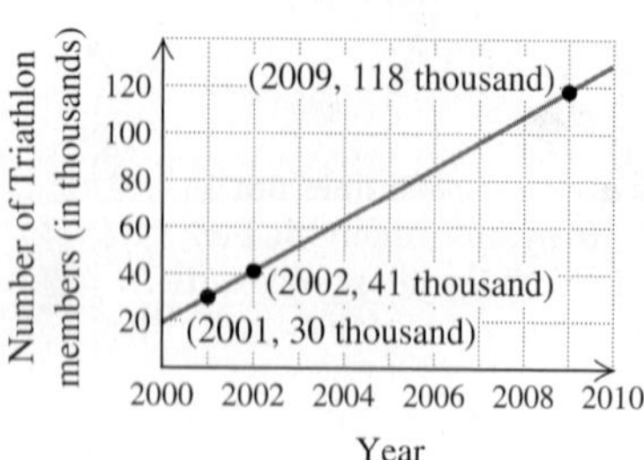

After plotting the three points, we draw a line through them, as shown in the figure at left. This gives us the graph.

Try Exercise 19.

EXAMPLE 4 Banking. Nadia prepared the graph shown below from data collected on a recent day at a branch bank.

a) What rate can be determined from the graph?

b) What is that rate?

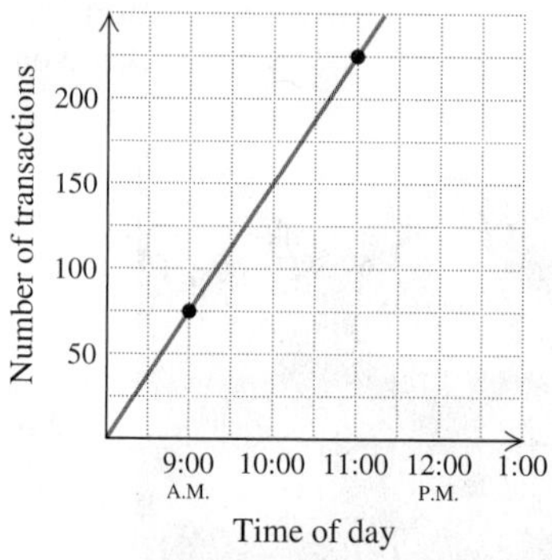

SOLUTION

a) Because the vertical axis shows the number of transactions and the horizontal axis lists the time in hour-long increments, we can determine the rate *Number of transactions per hour.*

b) The points (9:00, 75) and (11:00, 225) are both on the graph. This tells us that in the 2 hours between 9:00 and 11:00, there were 225 − 75 = 150 transactions. Thus the rate is

$$\frac{225 \text{ transactions} - 75 \text{ transactions}}{11{:}00 - 9{:}00} = \frac{150 \text{ transactions}}{2 \text{ hours}}$$
$$= 75 \text{ transactions per hour.}$$

Note that this is an *average* rate.

Try Exercise 29.

3.4 Exercise Set

FOR EXTRA HELP MyMathLab Math XL PRACTICE  WATCH  DOWNLOAD  READ REVIEW

Concept Reinforcement *For Exercises 1–6, fill in the missing units for each rate.*

1. If Jane biked 100 miles in 5 hours, her average rate was 20 ______________.
2. If it took Sabrina 18 hours to read 6 chapters, her average rate was 3 ______________.
3. If Denny's ticket cost $300 for a 150-mile flight, his average rate was 2 ______________.
4. If Geoff planted 36 petunias along a 12-foot sidewalk, his average rate was 3 ______________.
5. If Christi ran 8 errands in 40 minutes, her average rate was 5 ______________.
6. If Karl made 8 cakes using 20 cups of flour, his average rate was $2\frac{1}{2}$ ______________.

Solve. For Exercises 7–14, round answers to the nearest cent. For Exercises 7 and 8, assume that the pickup time was later in the day than the return time so that no late fees were applied.

7. *Van Rentals.* Late on June 5, Tina rented a Ford Focus with a full tank of gas and 13,741 mi on the odometer. On June 8, she returned the van with 14,014 mi on the odometer. The rental agency charged Tina $118 for the rental and the necessary 13 gal of gas to fill up the tank.
   a) Find the van's rate of gas consumption, in miles per gallon.
   b) Find the average cost of the rental, in dollars per day.
   c) Find the rate of travel, in miles per day.
   d) Find the rental rate, in cents per mile.

8. *Car Rentals.* On February 10, Rocco rented a Chevy Trailblazer with a full tank of gas and 13,091 mi on the odometer. On February 12, he returned the vehicle with 13,322 mi on the odometer. The rental agency charged $92 for the rental and the necessary 14 gal of gas to fill the tank.
   a) Find the SUV's rate of gas consumption, in miles per gallon.
   b) Find the average cost of the rental, in dollars per day.
   c) Find the rate of travel, in miles per day.
   d) Find the rental rate, in cents per mile.

9. *Bicycle Rentals.* At 2:00, Perry rented a mountain bike from The Slick Rock Cyclery. He returned the bike at 5:00, after cycling 18 mi. Perry paid $12 for the rental.
   a) Find Perry's average speed, in miles per hour.
   b) Find the rental rate, in dollars per hour.
   c) Find the rental rate, in dollars per mile.

10. *Bicycle Rentals.* At 9:00, Jodi rented a mountain bike from The Bike Rack. She returned the bicycle at 11:00, after cycling 14 mi. Jodi paid $15 for the rental.
    a) Find Jodi's average speed, in miles per hour.
    b) Find the rental rate, in dollars per hour.
    c) Find the rental rate, in dollars per mile.

11. *Proofreading.* Dylan began proofreading at 9:00 A.M., starting at the top of page 93. He worked until 2:00 P.M. that day and finished page 195. He billed the publisher $110 for the day's work.
    a) Find the rate of pay, in dollars per hour.
    b) Find the average proofreading rate, in number of pages per hour.
    c) Find the rate of pay, in dollars per page.

12. *Temporary Help.* A typist for Kelly Services reports to 3E's Properties for work at 10:00 A.M. and leaves at 6:00 P.M. after having typed from the end of page 8 to the end of page 50 of a proposal. 3E's pays $120 for the typist's services.
    a) Find the rate of pay, in dollars per hour.
    b) Find the average typing rate, in number of pages per hour.
    c) Find the rate of pay, in dollars per page.

13. *TV Prices.* The average price of a 32-in. LCD TV was $700 in January 2008 and $460 in July 2009. Find the rate at which the price was decreasing.
    Source: NPD Group's retail tracking service

14. *Four-Year-College Tuition.* The average tuition at a public four-year college was $5939 in 2005 and approximately $6836 in 2007. Find the rate at which tuition was increasing.
    Source: U.S. National Center for Education Statistics

15. *Elevators.* At 2:38, Peter entered an elevator on the 34th floor of the Regency Hotel. At 2:40, he stepped off at the 5th floor.

a) Find the elevator's average rate of travel, in number of floors per minute.
b) Find the elevator's average rate of travel, in seconds per floor.

16. *Snow Removal.* By 1:00 P.M., Shelby had already shoveled 2 driveways, and by 6:00 P.M., the number was up to 7.
a) Find Shelby's shoveling rate, in number of driveways per hour.
b) Find Shelby's shoveling rate, in hours per driveway.

17. *Mountaineering.* The fastest ascent of the Nepalese side of Mt. Everest was accomplished by Pemba Dorje of Nepal in 2004. Pemba Dorje climbed from base camp, elevation 17,552 ft, to the summit, elevation 29,035 ft, in 8 hr 10 min.
Source: "Bid for fastest Everest ascent" at www.thehimalayantimes.com, 06/02/2009
a) Find Pemba Dorje's average rate of ascent, in feet per minute.
b) Find Pemba Dorje's average rate of ascent, in minutes per foot.

18. *Mountaineering.* The fastest ascent of the northern face of Mt. Everest was accomplished by Christian Stangl of Austria in 2006. Stangl climbed from Advanced Base Camp, elevation 21,300 ft, to the summit, elevation 29,035 ft, in 16 hr 42 min.
Source: "Fastest Everest climber eats 3, 6000 m peaks in 16 hours", www.mounteverest.net/news, 11/09/2008
a) Find Stangl's rate of ascent, in feet per minute.
b) Find Stangl's rate of ascent, in minutes per foot.

In Exercises 19–28, draw a linear graph to represent the given information. Be sure to label and number the axes appropriately. (See Example 3.)

19. *Recycling.* In 2003, the amount of paper recovered for recycling in the United States was about 340 lb per person, and the figure was rising at a rate of 5 lb per person per year.
Sources: paperrecycles.org and epa.gov

20. *Health-Care Costs.* In 2009, the average cost for health insurance for a single employee was about \$4800, and the figure was rising at a rate of about \$200 per year.
Source: Kaiser/HRET Survey of Employer-Sponsored Health Benefits, 1999–2009

21. *Law Enforcement.* In 2006, there were approximately 10 million property crimes reported in the United States, and the figure was dropping at a rate of about 0.1 million per year.
Source: Based on data from the U.S. Department of Justice

22. *Retail.* In 2009, approximately 21% of shoppers said that merchandise selection was the most important factor in choosing to shop at a particular store. This percentage was dropping at a rate of 1.1% per year.
Source: 2009 BIG research; National Retail Federation

23. *Train Travel.* At 3:00 P.M., the Boston–Washington Metroliner had traveled 230 mi and was cruising at a rate of 90 miles per hour.

24. *Plane Travel.* At 4:00 P.M., the Seattle–Los Angeles shuttle had traveled 400 mi and was cruising at a rate of 300 miles per hour.

25. *Wages.* By 2:00 P.M., Diane had earned \$50. She continued earning money at a rate of \$15 per hour.

26. *Wages.* By 3:00 P.M., Arnie had earned \$70. He continued earning money at a rate of \$12 per hour.

27. *Telephone Bills.* Roberta's phone bill was already \$7.50 when she made a call for which she was charged at a rate of \$0.10 per minute.

28. *Telephone Bills.* At 3:00 P.M., Larry's phone bill was \$6.50 and increasing at a rate of 7¢ per minute.

In Exercises 29–38, use the graph provided to calculate a rate of change in which the units of the horizontal axis are used in the denominator.

29. *Call Center.* The graph below shows data from a technical assistance call center. At what rate are calls being handled?

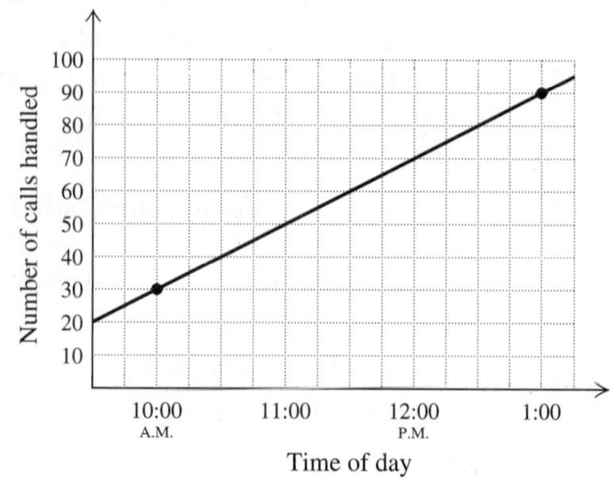

30. *Hairdresser.* Eve's Custom Cuts has a graph displaying data from a recent day of work. At what rate does Eve work?

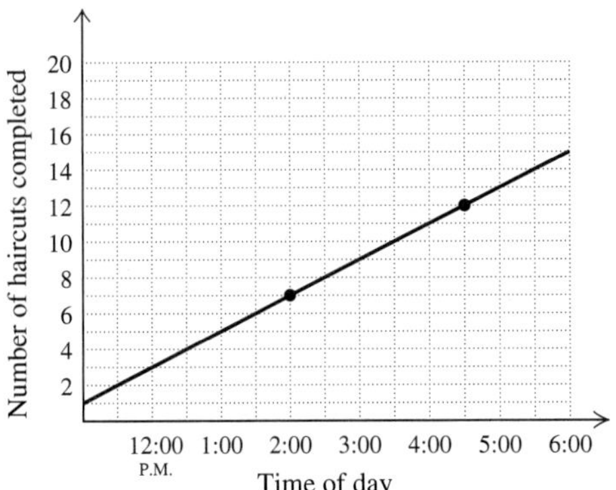

31. *Train Travel.* The graph below shows data from a recent train ride from Chicago to St. Louis. At what rate did the train travel?

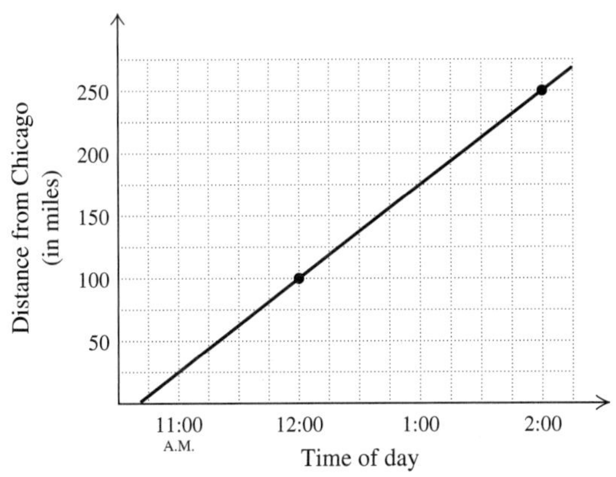

32. *Train Travel.* The graph below shows data from a recent train ride from Denver to Kansas City. At what rate did the train travel?

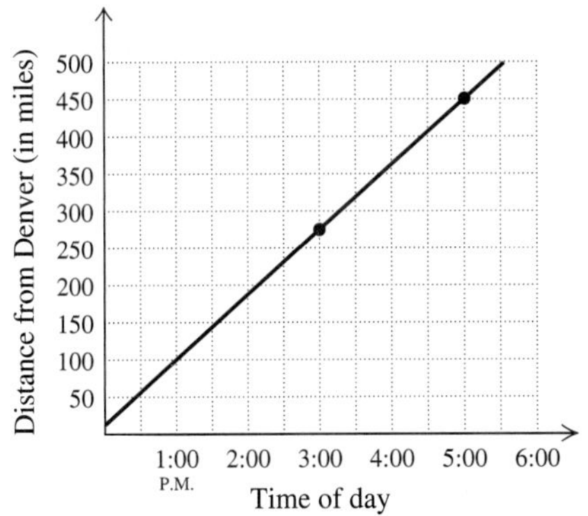

33. *Cost of a Telephone Call.* The graph below shows data from a recent phone call between the United States and the Netherlands. At what rate was the customer being billed?

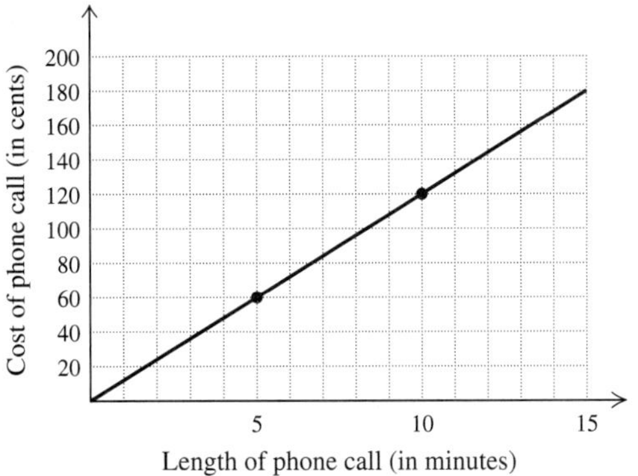

34. *Cost of a Telephone Call.* The graph below shows data from a recent phone call between the United States and South Korea. At what rate was the customer being billed?

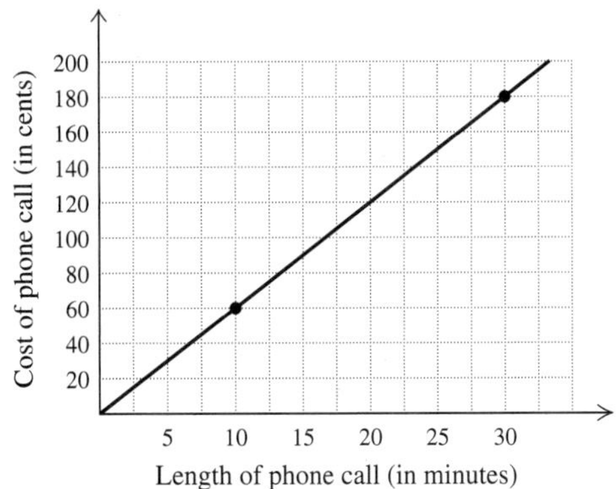

35. *Population.* The graph below shows data regarding the population of Youngstown, Ohio. At what rate was the population changing?

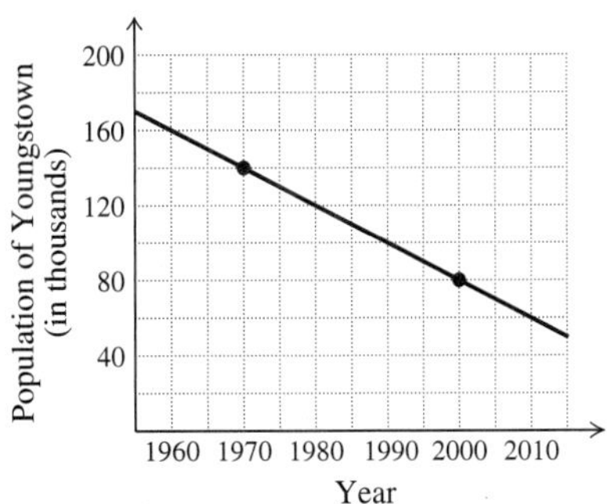

36. *Depreciation of an Office Machine.* Data regarding the value of a particular color copier is represented in the graph below. At what rate is the value changing?

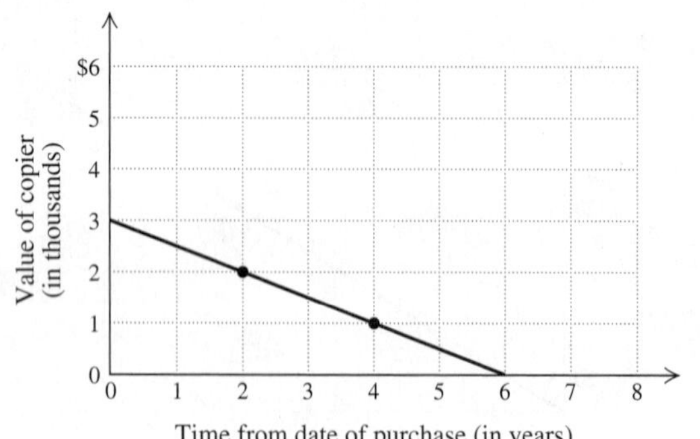

37. *Gas Mileage.* The graph below shows data for a Honda Insight (hybrid) driven on city streets. At what rate was the vehicle consuming gas?

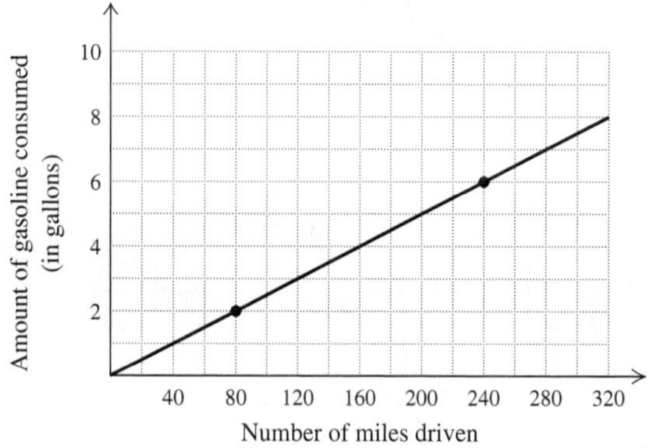

38. *Gas Mileage.* The graph below shows data for a Ford Mustang driven on highways. At what rate was the vehicle consuming gas?

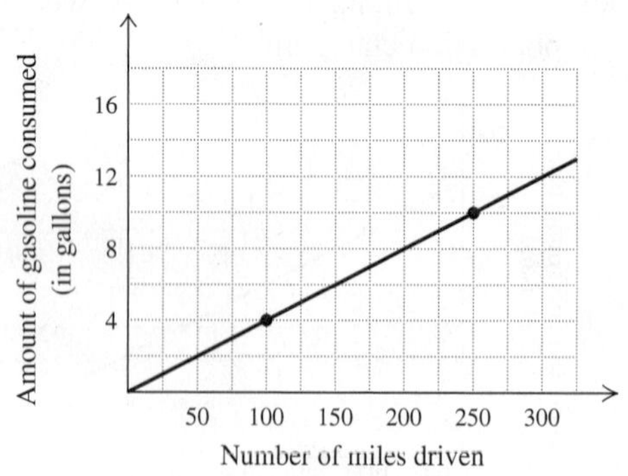

In each of Exercises 39–44, match the description with the most appropriate graph from the choices (a)–(f) below. Scales are intentionally omitted. Assume that of the three sports listed, swimming is the slowest and biking is the fastest.

39. ____ Robin trains for triathlons by running, biking, and then swimming every Saturday.

40. ____ Gene trains for triathlons by biking, running, and then swimming every Sunday.

41. ____ Shirley trains for triathlons by swimming, biking, and then running every Sunday.

42. ____ Evan trains for triathlons by swimming, running, and then biking every Saturday.

43. ____ Angie trains for triathlons by biking, swimming, and then running every Sunday.

44. ____ Mick trains for triathlons by running, swimming, and then biking every Saturday.

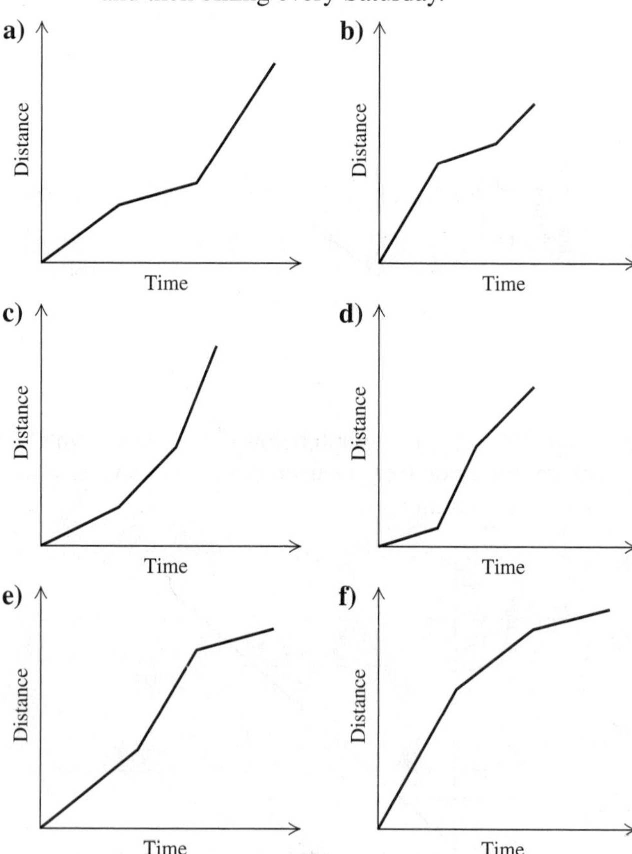

TW **45.** What does a negative rate of travel indicate? Explain.

TW **46.** Explain how to convert from kilometers per hour to meters per second.

SKILL REVIEW

To prepare for Section 3.5, review subtraction and order of operations (Sections 1.6 and 1.8).

Simplify.

47. $-2-(-7)$ [1.6]

48. $-9-(-3)$ [1.6]

49. $\frac{5-(-4)}{-2-7}$ [1.8]

50. $\frac{8-(-4)}{2-11}$ [1.8]

51. $\frac{-4-8}{7-(-2)}$ [1.8]

52. $\frac{-5-3}{6-(-4)}$ [1.8]

53. $\frac{-6-(-6)}{-2-7}$ [1.8]

54. $\frac{-3-5}{-1-(-1)}$ [1.8]

SYNTHESIS

TW **55.** How would the graphs of Jon's and Jenny's total earnings compare in each of the following situations?

a) Jon earns twice as much per hour as Jenny.
b) Jon and Jenny earn the same hourly rate, but Jenny received a bonus for a cost-saving suggestion.
c) Jon is paid by the hour, and Jenny is paid a weekly salary.

TW **56.** Write an exercise similar to Exercises 7–18 for a classmate to solve. Design the problem so that the solution is "The motorcycle's rate of gas consumption was 65 miles per gallon."

57. *Aviation.* A Boeing 737 climbs from sea level to a cruising altitude of 31,500 ft at a rate of 6300 ft/min. After cruising for 3 min, the jet is forced to land, descending at a rate of 3500 ft/min. Represent the flight with a graph in which altitude is measured on the vertical axis and time on the horizontal axis.

58. *Wages with Commissions.* Each salesperson at Mike's Bikes is paid \$140 per week plus 13% of all sales up to \$2000, and then 20% on any sales in excess of \$2000. Draw a graph in which sales are measured on the horizontal axis and wages on the vertical axis. Then use the graph to estimate the wages paid when a salesperson sells \$2700 in merchandise in one week.

59. *Taxi Fares.* The driver of a New York City Yellow Cab recently charged \$2.50 plus 40¢ for each fifth of a mile traveled. Draw a graph that could be used to determine the cost of a fare.

60. *Gas Mileage.* Suppose that a Honda motorcycle travels twice as far as a Honda Insight on the same amount of gas (see Exercise 37). Draw a graph that reflects this information.

61. *Navigation.* In 3 sec, Penny walks 24 ft, to the bow (front) of a tugboat. The boat is cruising at a rate of 5 ft/sec. What is Penny's rate of travel with respect to land?

62. *Aviation.* Tim's F-14 jet is moving forward at a deck speed of 95 mph aboard an aircraft carrier that is traveling 39 mph in the same direction. How fast is the jet traveling, in minutes per mile, with respect to the sea?

63. *Running.* Zoe ran from the 4-km mark to the 7-km mark of a 10-km race in 15.5 min. At this rate, how long would it take Zoe to run a 5-mi race? (*Hint*: 1 km ≈ 0.62 mi.)

64. *Running.* Jerod ran from the 2-mi marker to the finish line of a 5-mi race in 25 min. At this rate, how long would it take Jerod to run a 10-km race? (*Hint*: 1 mi ≈ 1.61 km.)

65. At 3:00 P.M., Camden and Natalie had already made 46 candles. By 5:00 P.M., the total reached 100 candles. Assuming a constant production rate, at what time did they make their 82nd candle?

66. Marcy picks apples twice as fast as Whitney does. By 4:30, Whitney had already picked 4 bushels of apples. Fifty minutes later, her total reached $5\frac{1}{2}$ bushels. Find Marcy's picking rate. Give your answer in number of bushels per hour.

Try Exercise Answers: Section 3.4

7. (a) 21 mpg; **(b)** \$39.33/day; **(c)** 91 mi/day; **(d)** 43¢/mi
13. \$13.33/month **19.**

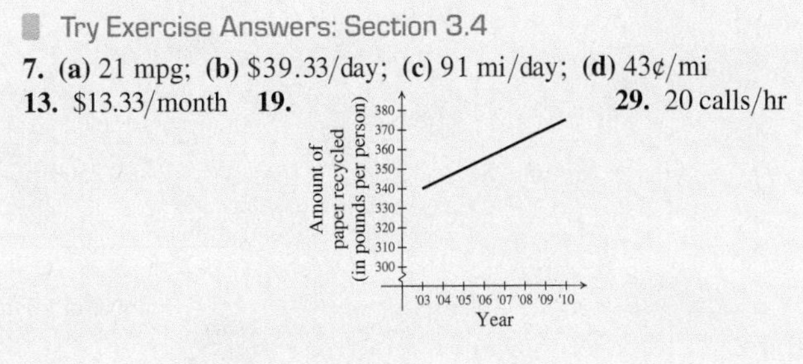

29. 20 calls/hr

3.5 Slope

- Rate and Slope
- Horizontal and Vertical Lines
- Applications

In Section 3.4, we introduced *rate* as a method of measuring how two quantities change with respect to each other. In this section, we will discuss how rate can be related to the slope of a line.

RATE AND SLOPE

A candy company owns two automatic candy-wrapping machines. The LX-269 will double-twist wrap 2500 hard candies every 3 min. The LX-266 will single-twist wrap 300 hard candies every 2 min. The tables below list the number of candies wrapped after various amounts of time for each machine.

Source: Based on information from the Labh Group of Companies

LX-269 (Double twist)	
Minutes Elapsed	Candies Wrapped
0	0
3	2,500
6	5,000
9	7,500
12	10,000

LX-266 (Single twist)	
Minutes Elapsed	Candies Wrapped
0	0
2	300
4	600
6	900
8	1200

We now graph the pairs of numbers listed in the tables, using the horizontal axis for the number of minutes elapsed and the vertical axis for the number of candies wrapped.

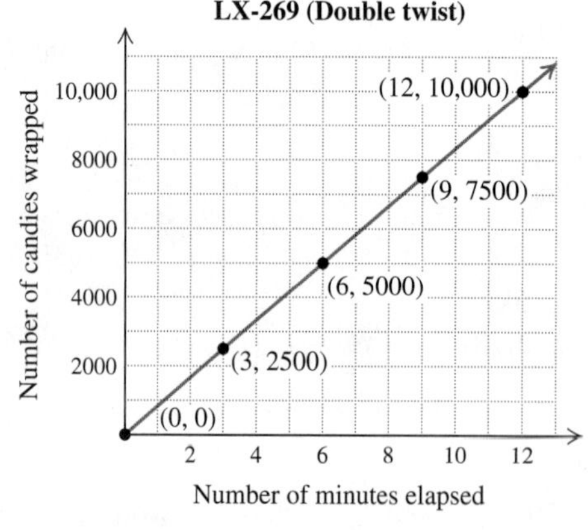

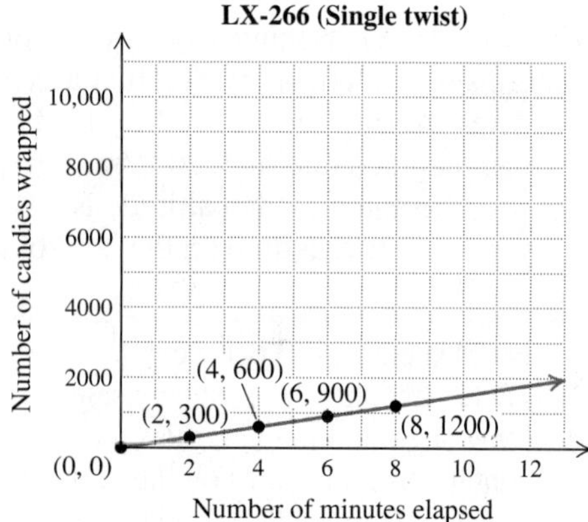

Let's compare the rates of the machines. The double-twist machine wraps 2500 candies every 3 min, so its *rate* is $2500 \div 3 = \frac{2500}{3} = 833\frac{1}{3}$ candies per minute. Since the single-twist machine wraps 300 candies every 2 min, its rate is $300 \div 2 = \frac{300}{2} = 150$ candies per minute. Note that the rate of the double-twist machine is greater so its graph is steeper.

The rates can also be found using the coordinates of two points that are on the line. Because the lines are straight, the same rates can be found using *any* pair of points on the line. For example, we can use the points (3, 2500) and (9, 7500) to find the wrapping rate for the double-twist machine:

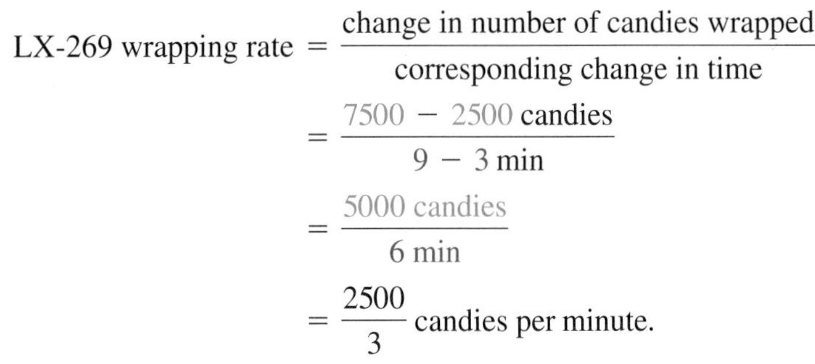

$$\text{LX-269 wrapping rate} = \frac{\text{change in number of candies wrapped}}{\text{corresponding change in time}}$$
$$= \frac{7500 - 2500 \text{ candies}}{9 - 3 \text{ min}}$$
$$= \frac{5000 \text{ candies}}{6 \text{ min}}$$
$$= \frac{2500}{3} \text{ candies per minute.}$$

Now, using the points (0, 0) and (12, 10,000), we can find the same rate:

$$\text{LX-269 wrapping rate} = \frac{10{,}000 - 0 \text{ candies}}{12 - 0 \text{ min}}$$
$$= \frac{10{,}000 \text{ candies}}{12 \text{ min}}$$
$$= \frac{2500}{3} \text{ candies per minute.}$$

> The rate is always the vertical change divided by the corresponding horizontal change.

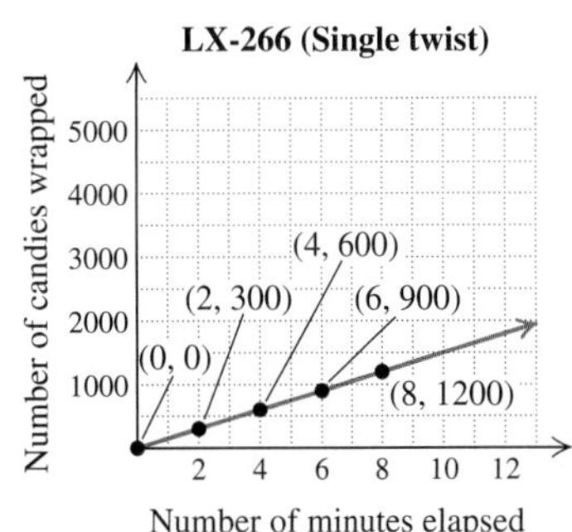

EXAMPLE 1 Use the graph of candy wrapping by the single-twist machine to find the rate at which candies are wrapped.

SOLUTION We can use any two points on the line, such as (2, 300) and (8, 1200):

$$\text{LX-266 wrapping rate} = \frac{\text{change in number of candies wrapped}}{\text{corresponding change in time}}$$
$$= \frac{1200 - 300 \text{ candies}}{8 - 2 \text{ min}}$$
$$= \frac{900 \text{ candies}}{6 \text{ min}}$$
$$= 150 \text{ candies per minute.}$$

As a check, we can use another pair of points to calculate the rate.

Try Exercise 11.

STUDY TIP

Add Your Voice to the Author's

If you own your text, consider using it as a notebook. Since many instructors' work closely parallels the book, it is often useful to make notes on the appropriate page as he or she is lecturing.

When the axes of a graph are simply labeled x and y, the ratio of vertical change to horizontal change is the rate at which y is changing with respect to x. This ratio is a measure of a line's slant, or **slope.**

Consider a line passing through $(2, 3)$ and $(6, 5)$, as shown below. We find the ratio of vertical change, or *rise*, to horizontal change, or *run*, as follows:

$$\text{Ratio of vertical change to horizontal change} = \frac{\text{change in } y}{\text{change in } x} = \frac{\text{rise}}{\text{run}}$$

$$= \frac{5 - 3}{6 - 2}$$

$$= \frac{2}{4}, \text{ or } \frac{1}{2}.$$

Note that these calculations can be performed without viewing a graph.

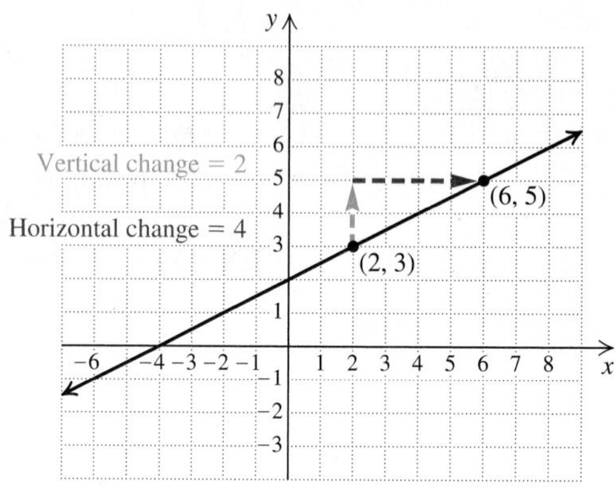

Thus the y-coordinates of points on this line increase at a rate of 2 units for every 4-unit increase in x, 1 unit for every 2-unit increase in x, or $\frac{1}{2}$ unit for every 1-unit increase in x. The slope of the line is $\frac{1}{2}$.

In the definition of *slope* below, the *subscripts* 1 and 2 are used to distinguish point 1 and point 2 from each other. That is, the slightly lowered 1's and 2's are not exponents but are used to indicate x-values (or y-values) that may differ from each other.

Student Notes

The notation x_1 is read "x sub one."

Slope The *slope* of the line containing points (x_1, y_1) and (x_2, y_2) is given by

$$m = \frac{\text{change in } y}{\text{change in } x} = \frac{\text{rise}}{\text{run}} = \frac{y_2 - y_1}{x_2 - x_1}.$$

EXAMPLE 2 Find the slope of the line containing the points $(-4, 3)$ and $(2, -6)$.

SOLUTION From $(-4, 3)$ to $(2, -6)$, the change in y, or rise, is $-6 - 3$, or -9. The change in x, or run, is $2 - (-4)$, or 6. Thus,

$$\text{Slope} = \frac{\text{change in } y}{\text{change in } x} = \frac{\text{rise}}{\text{run}}$$

$$= \frac{-6 - 3}{2 - (-4)} = \frac{-9}{6} = -\frac{9}{6}, \text{ or } -\frac{3}{2}.$$

The graph of the line is shown at left for reference.

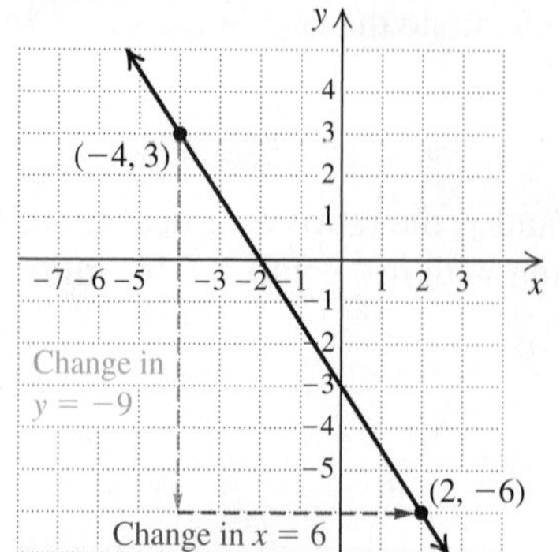

Try Exercise 37.

Student Notes

You may wonder which point should be regarded as (x_1, y_1) and which should be (x_2, y_2). To see that the math works out the same either way, perform both calculations on your own.

CAUTION! When we use the formula

$$m = \frac{y_2 - y_1}{x_2 - x_1},$$

it makes no difference which point is considered (x_1, y_1). What matters is that we subtract the y-coordinates in the same order that we subtract the x-coordinates.

To illustrate, we reverse *both* of the subtractions in Example 2. The slope is still $-\frac{3}{2}$:

$$\text{Slope} = \frac{\text{change in } y}{\text{change in } x} = \frac{3 - (-6)}{-4 - 2} = \frac{9}{-6} = -\frac{3}{2}.$$

As shown in the graphs below, a line with positive slope slants up from left to right, and a line with negative slope slants down from left to right. The larger the absolute value of the slope, the steeper the line.

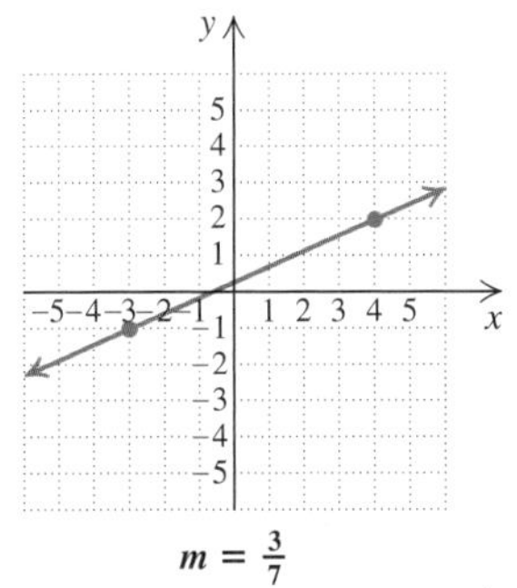

$m = \frac{3}{7}$

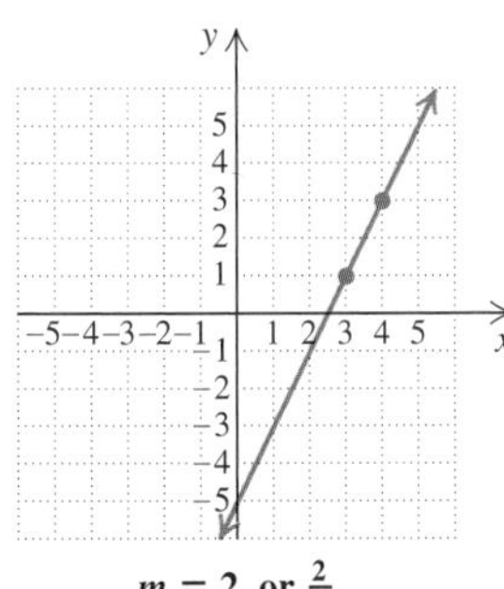

$m = 2$, or $\frac{2}{1}$

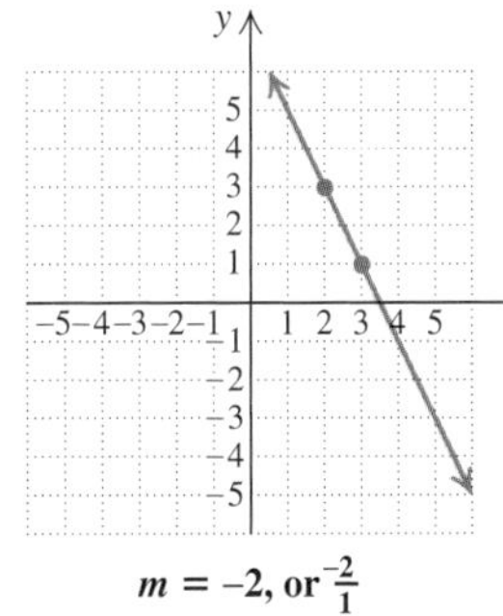

$m = -2$, or $\frac{-2}{1}$

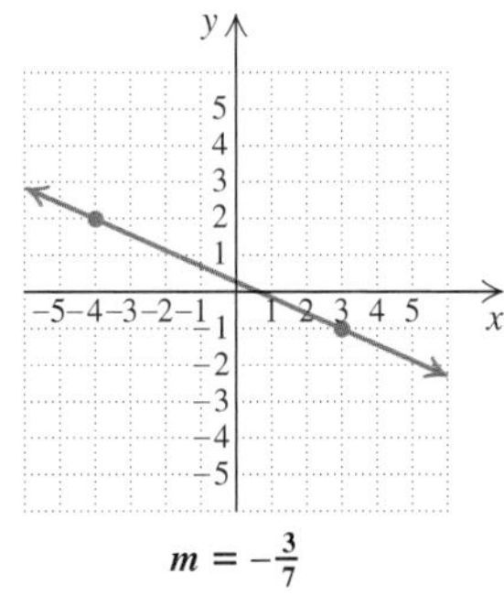

$m = -\frac{3}{7}$

HORIZONTAL AND VERTICAL LINES

What about the slope of a horizontal line or a vertical line?

EXAMPLE 3 Find the slope of the line $y = 4$.

SOLUTION Consider the points $(2, 4)$ and $(-3, 4)$, which are on the line. The change in y, or the rise, is $4 - 4$, or 0. The change in x, or the run, is $-3 - 2$, or -5. Thus,

$$\begin{aligned} m &= \frac{4 - 4}{-3 - 2} \\ &= \frac{0}{-5} \\ &= 0. \end{aligned}$$

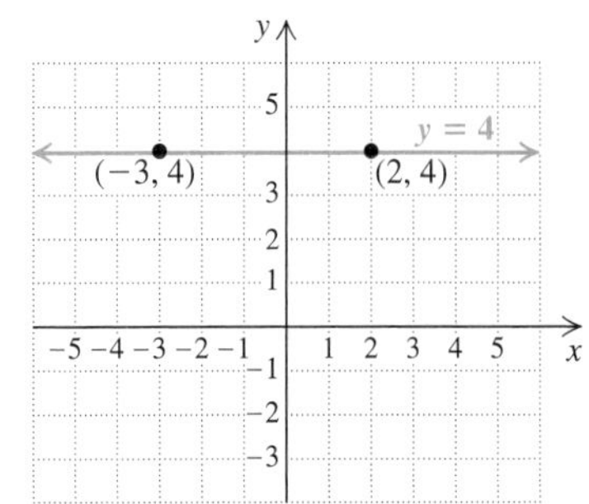

Any two points on a horizontal line have the same y-coordinate. Thus the change in y is 0, so the slope is 0.

Try Exercise 53.

A horizontal line has slope 0.

EXAMPLE 4 Find the slope of the line $x = -3$.

SOLUTION Consider the points $(-3, 4)$ and $(-3, -2)$, which are on the line. The change in y, or the rise, is $-2 - 4$ or -6. The change in x, or the run, is $-3 - (-3)$, or 0. Thus,

$$m = \frac{-2 - 4}{-3 - (-3)}$$

$$= \frac{-6}{0} \quad \text{(undefined)}.$$

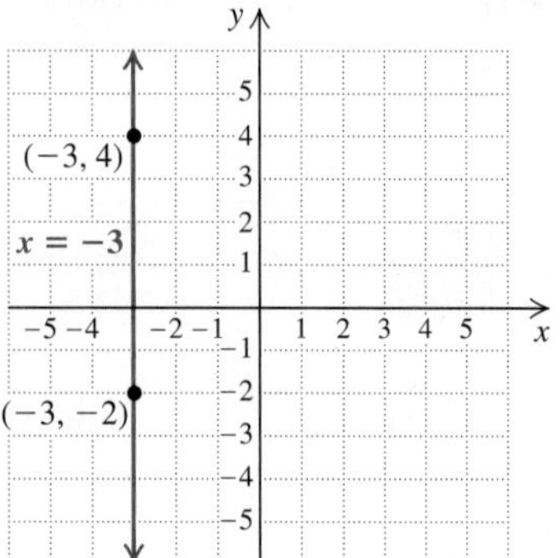

Since division by 0 is not defined, the slope of this line is not defined. The answer to a problem of this type is "The slope of this line is undefined."

Try Exercise 55.

The slope of a vertical line is undefined.

APPLICATIONS

We have seen that slope has many real-world applications, ranging from car speed to production rate. Slope can also measure steepness. For example, numbers like 2%, 3%, and 6% are often used to represent the **grade** of a road, a measure of a road's steepness. That is, since $3\% = \frac{3}{100}$, a 3% grade means that for every horizontal distance of 100 ft, the road rises or drops 3 ft. The concept of grade also occurs in skiing or snowboarding, where a 7% grade is considered very tame, but a 70% grade is considered steep.

EXAMPLE 5 Skiing. Among the steepest skiable terrain in North America, the Headwall on Mount Washington, in New Hampshire, drops 720 ft over a horizontal distance of 900 ft. Find the grade of the Headwall.

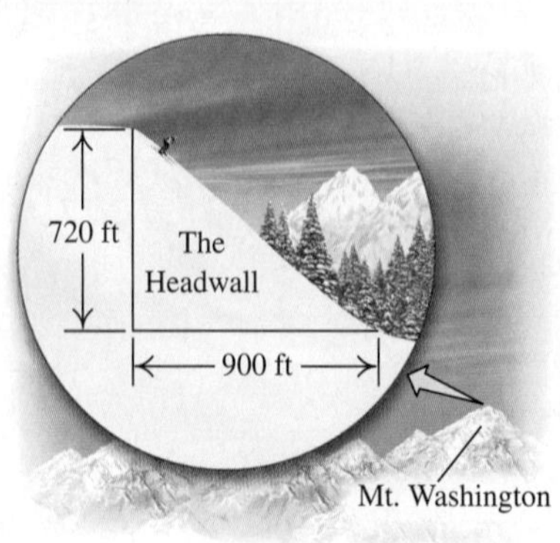

SOLUTION The grade of the Headwall is its slope, expressed as a percent:

$$m = \frac{720}{900} = \frac{8}{10} = 80\%. \qquad \textbf{Grade is slope expressed as a percent.}$$

Try Exercise 61.

Carpenters use slope when designing stairs, ramps, or roof pitches. Another application occurs in the engineering of a dam—the force or strength of a river depends on how much the river drops over a specified distance.

EXAMPLE 6 Running Speed. Kathy runs 10 km during each workout. For the first 7 km, her pace is twice as fast as it is for the last 3 km. Which of the following graphs best describes Kathy's workout?

A.

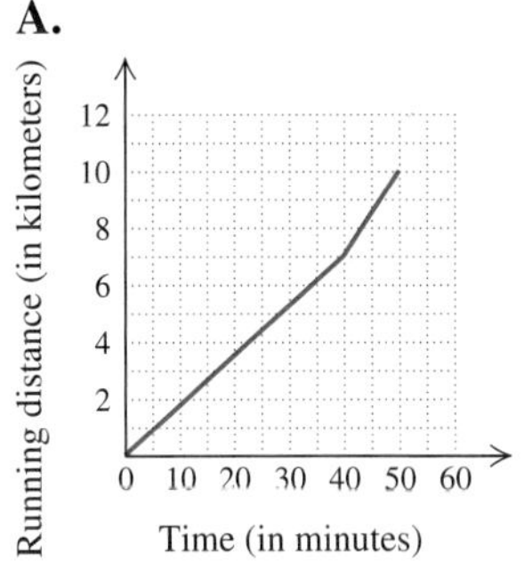

B.

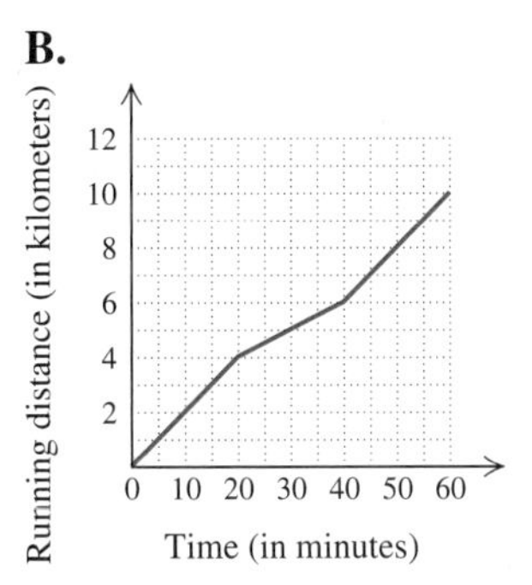

C.

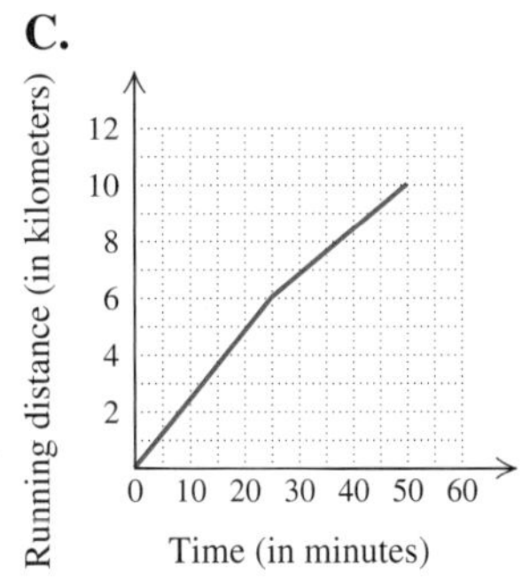

D.

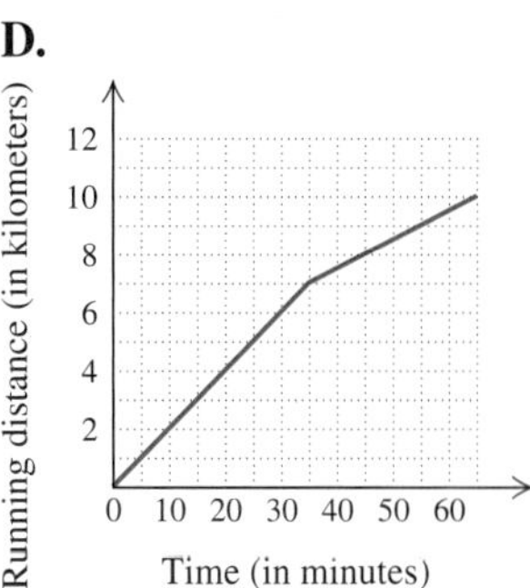

SOLUTION The slopes in graph A increase as we move to the right. This would indicate that Kathy ran faster for the *last* part of her workout. Thus graph A is not the correct one.

The slopes in graph B indicate that Kathy slowed down in the middle of her run and then resumed her original speed. Thus graph B does not correctly model the situation either.

According to graph C, Kathy slowed down not at the 7-km mark, but at the 6-km mark. Thus graph C is also incorrect.

Graph D indicates that Kathy ran the first 7 km in 35 min, a rate of 0.2 km/min. It also indicates that she ran the final 3 km in 30 min, a rate of 0.1 km/min. This means that Kathy's rate was twice as fast for the first 7 km, so graph D provides a correct description of her workout.

Try Exercise 69.

Exercise Set

FOR EXTRA HELP MyMathLab

Concept Reinforcement *State whether each of the following rates is positive, negative, or zero.*

1. The rate at which a teenager's height changes
2. The rate at which an elderly person's height changes
3. The rate at which a pond's water level changes during a drought
4. The rate at which a pond's water level changes during the rainy season
5. The rate at which the number of people in attendance at a basketball game changes in the moments before the opening tipoff
6. The rate at which the number of people in attendance at a basketball game changes in the moments after the final buzzer sounds
7. The rate at which a person's IQ changes during his or her sleep
8. The rate at which a gift shop's sales change as the holidays approach
9. The rate at which a bookstore's inventory changes during a liquidation sale
10. The rate at which the number of U.S. senators changes

11. Find the rate of change of the U.S. population.
Source: Based on information from the U.S. Census Bureau

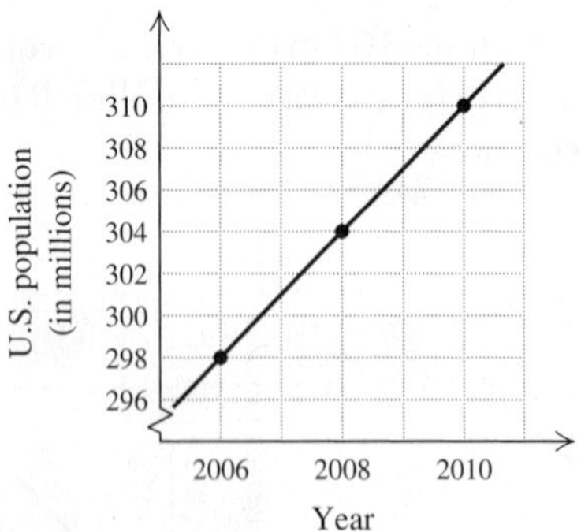

12. Find the rate at which a runner burns calories.

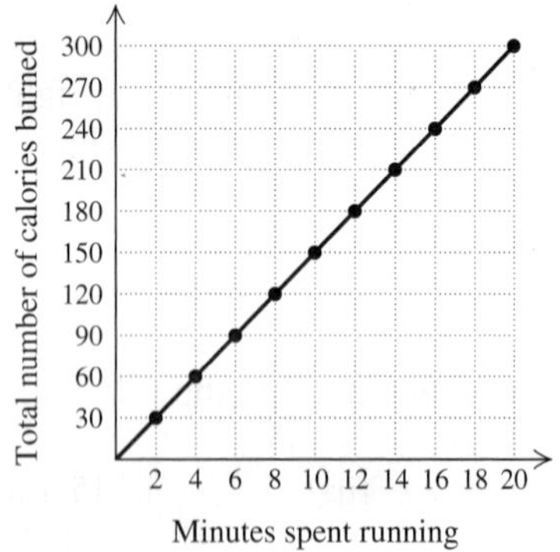

13. Find the rate of change of the men's world record for the mile run.

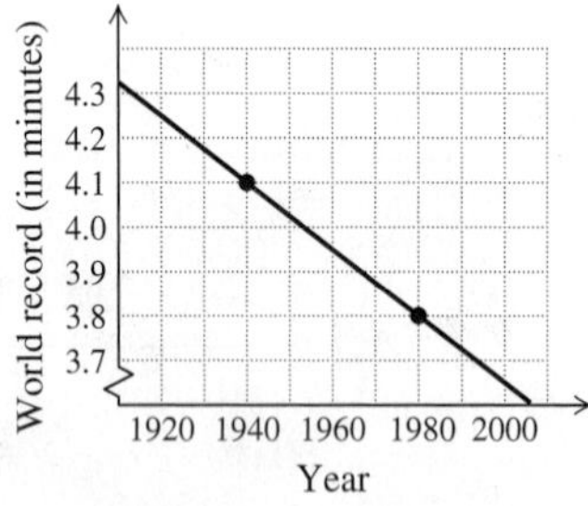

14. *Retail Sales.* Find the rate of change of the total retail sales of department stores.
Source: Based on information from the U.S. Census Bureau

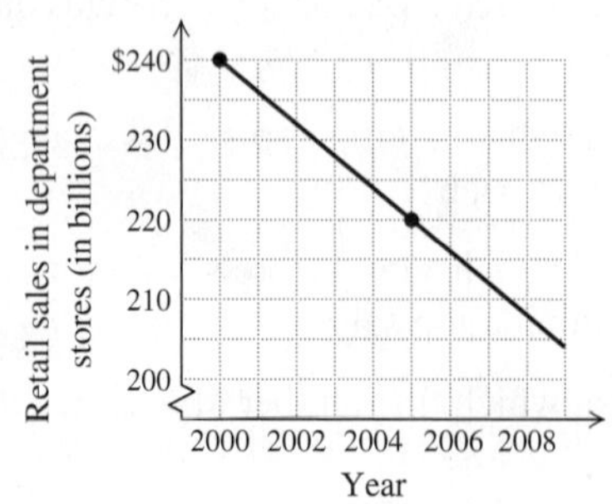

15. Find the rate of change of SAT math scores with respect to family income.
Source: College Board

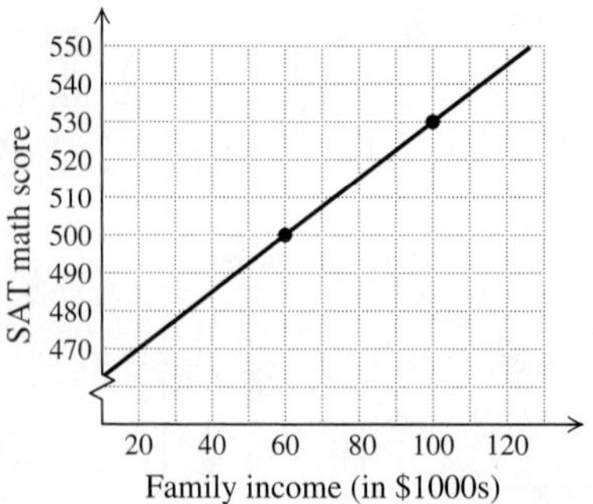

16. *Long-Term Care.* Find the rate of change of Medicaid spending on long-term care.
Source: Based on data from Thomson Medstat, prepared by AARP Public Policy Institute, and CMS, prepared for the Kaiser Commission on Medicaid and the Uninsured

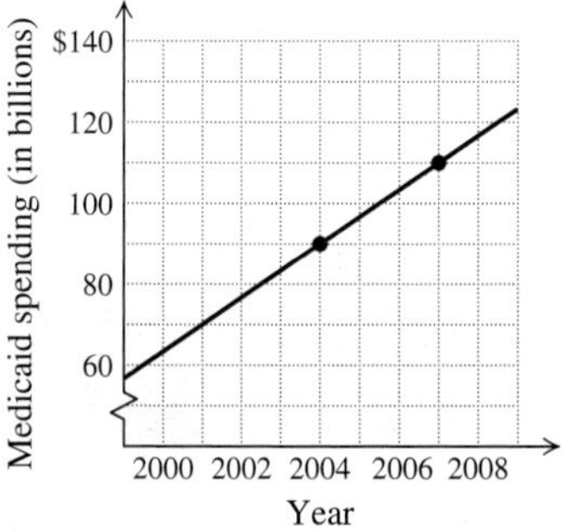

17. *Meteorology.* Find the rate of change of the temperature in Spearfish, Montana, on January 22, 1943, as shown below.
Source: National Oceanic Atmospheric Administration

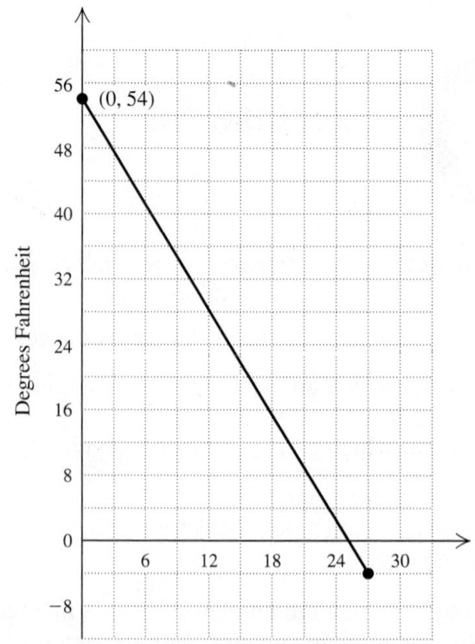

18. Find the rate of change of the birth rate among teenagers reported in the United States.
Source: Based on statistics from the U.S. National Center for Health Statistics

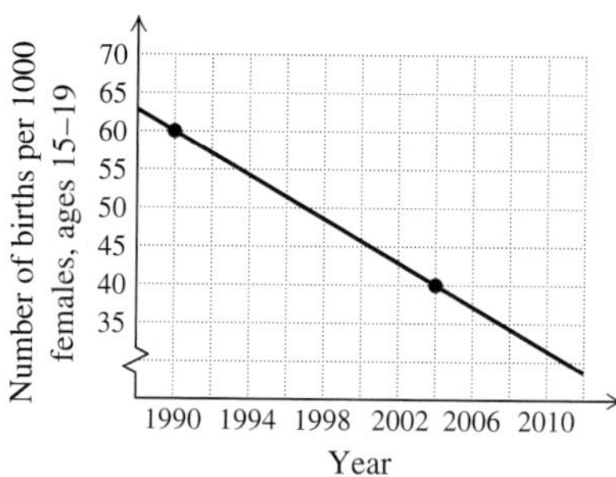

Find the slope, if it is defined, of each line. If the slope is undefined, state this.

19.

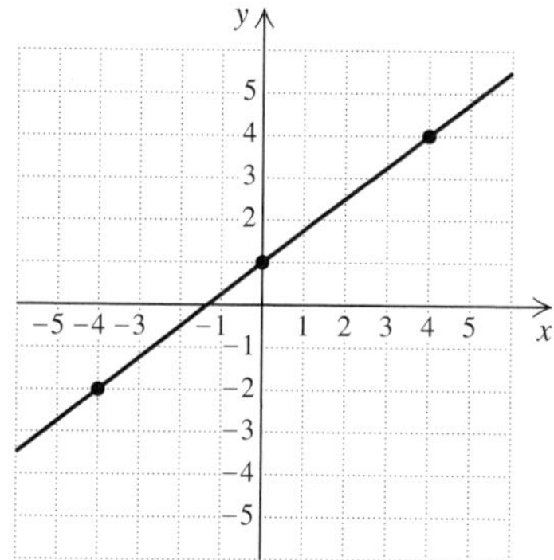

20.

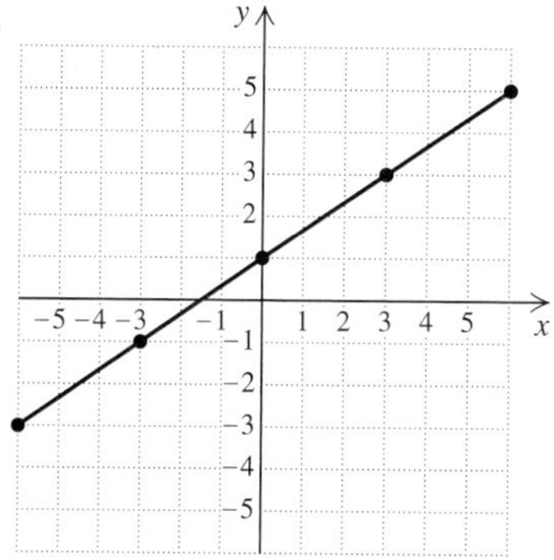

21.

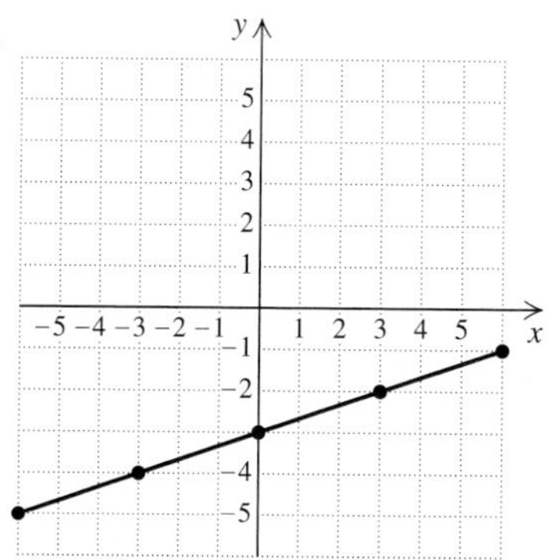

22.

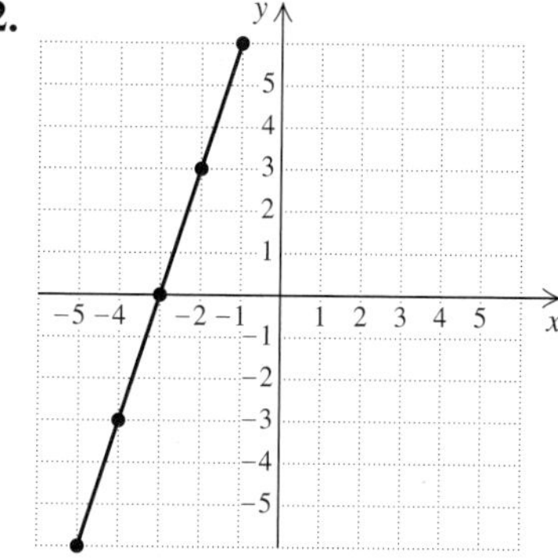

23.

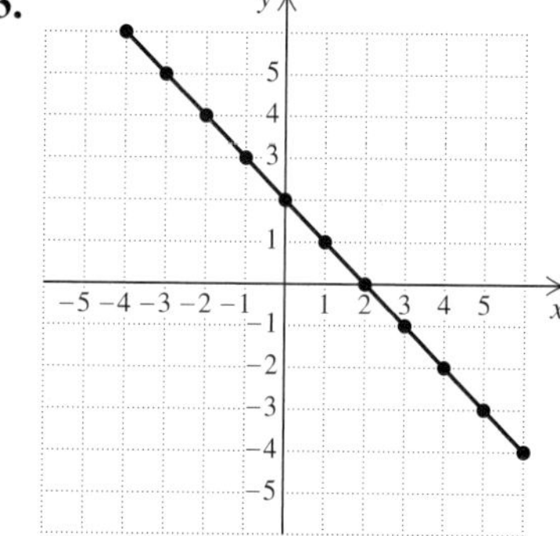

24.

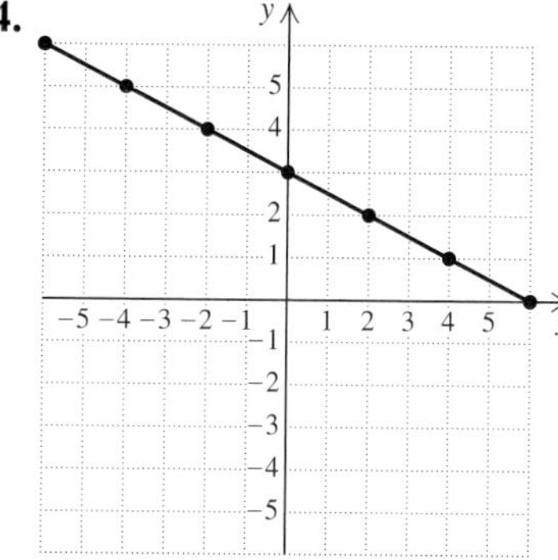

25.

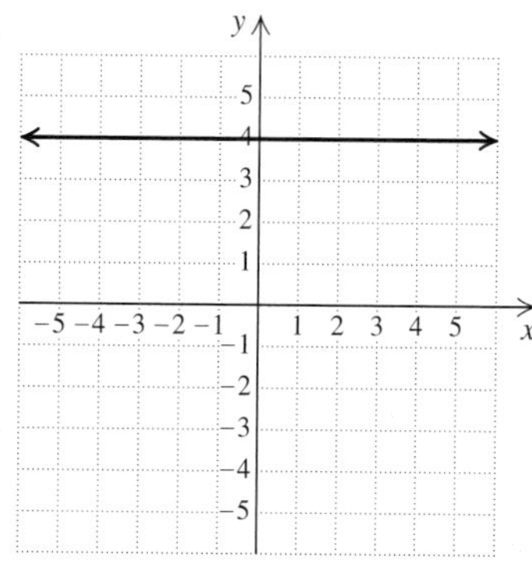

26.

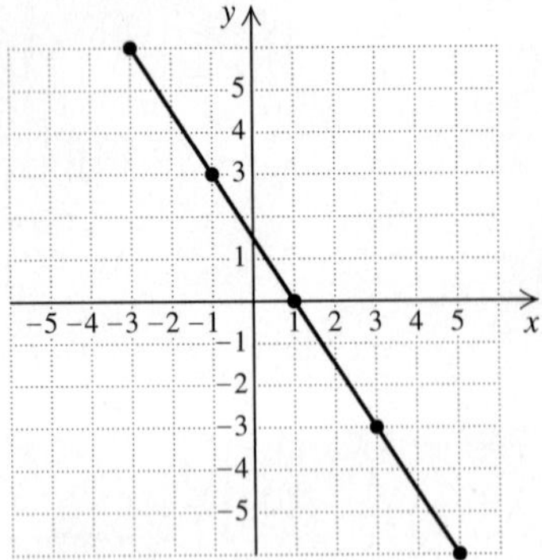

27.

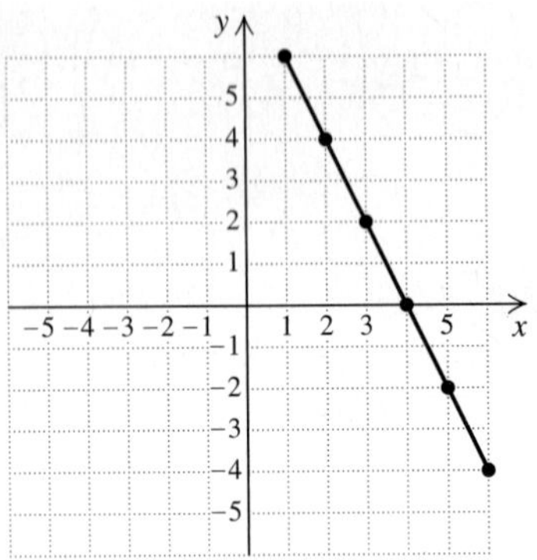

28.

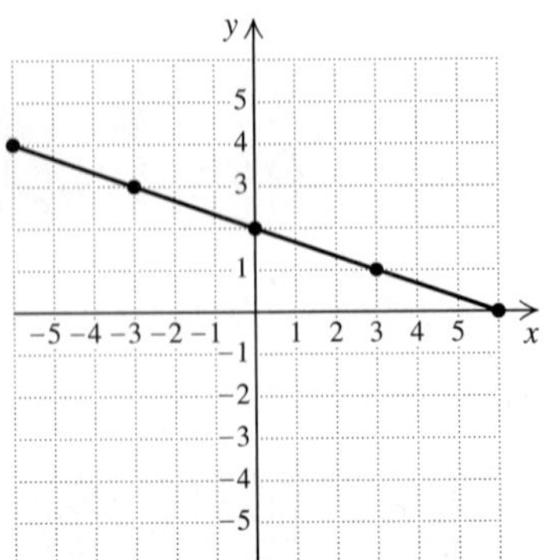

29.

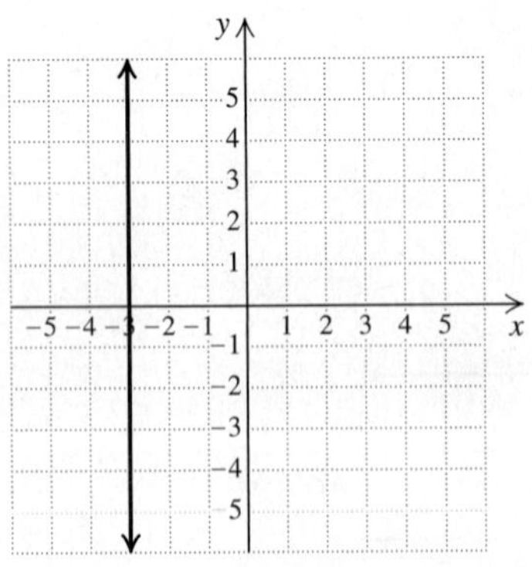

30.

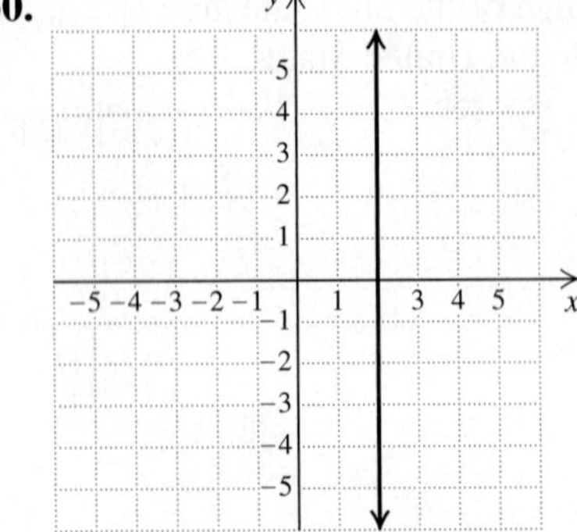

31.

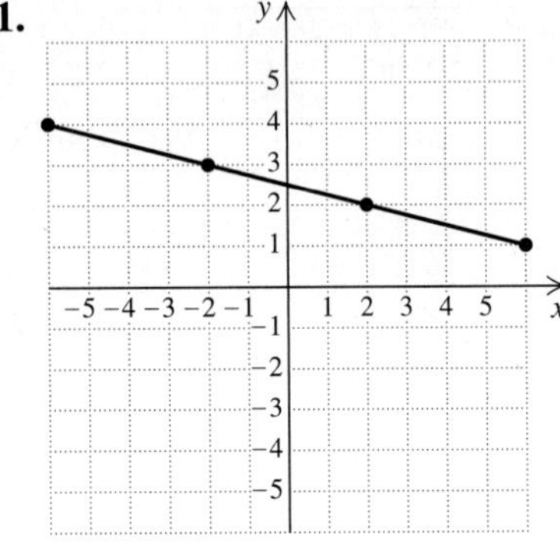

32.

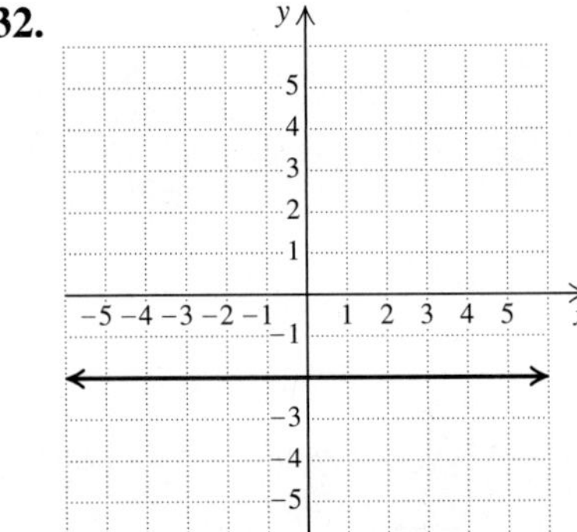

33.

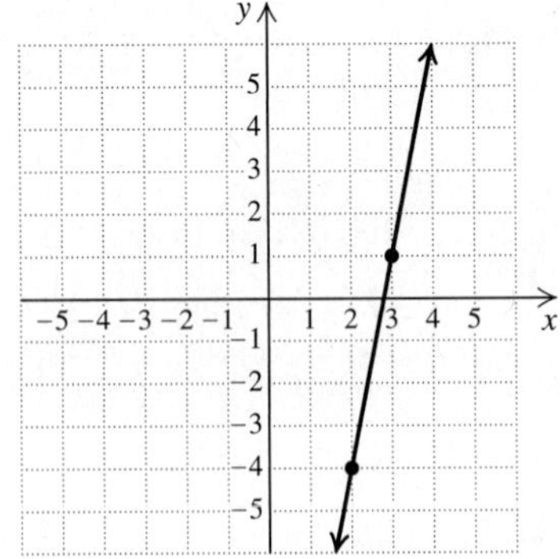

34.

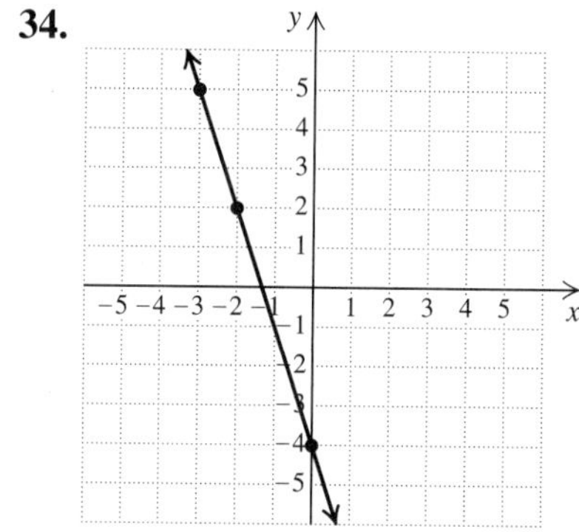

35.

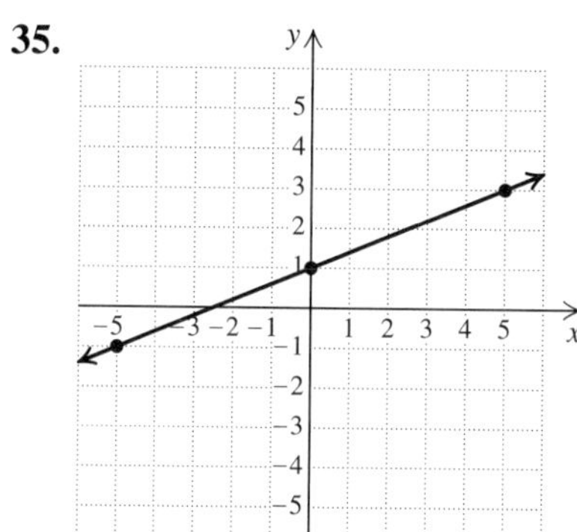

36.

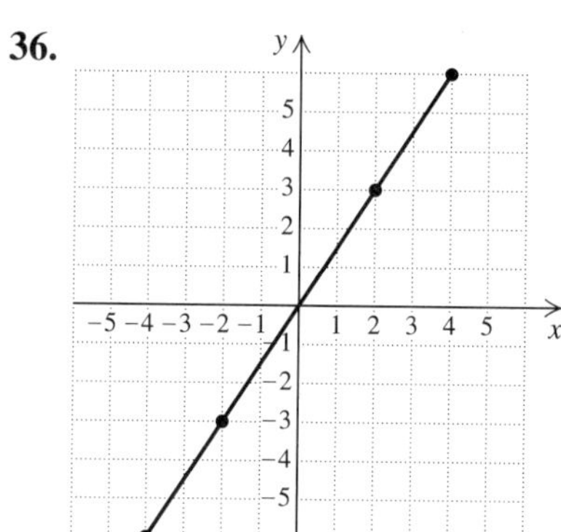

Find the slope of the line containing each given pair of points. If the slope is undefined, state this.

37. $(1, 2)$ and $(5, 8)$
38. $(1, 8)$ and $(6, 9)$
39. $(-2, 4)$ and $(3, 0)$
40. $(-4, 2)$ and $(2, -3)$
41. $(-4, 0)$ and $(2, 3)$
42. $(3, 0)$ and $(6, 9)$
43. $(0, 7)$ and $(-3, 10)$
44. $(0, 9)$ and $(-5, 0)$
45. $(-2, 3)$ and $(-6, 5)$
46. $(-1, 4)$ and $(5, -8)$

Aha! 47. $\left(-2, \frac{1}{2}\right)$ and $\left(-5, \frac{1}{2}\right)$
48. $(-5, -1)$ and $(2, 3)$
49. $(5, -4)$ and $(2, -7)$
50. $(-10, 3)$ and $(-10, 4)$
51. $(6, -4)$ and $(6, 5)$
52. $(5, -2)$ and $(-4, -2)$

Find the slope of each line whose equation is given. If the slope is undefined, state this.

53. $y = 3$
54. $y = 17$
55. $x = -1$
56. $x = 18$
57. $x = 9$
58. $x = -7$
59. $y = -9$
60. $y = -4$

61. *Surveying.* Tucked between two ski areas, Vermont Route 108 rises 106 m over a horizontal distance of 1325 m. What is the grade of the road?

62. *Navigation.* Capital Rapids drops 54 ft vertically over a horizontal distance of 1080 ft. What is the slope of the rapids?

63. *Road Design.* To meet Minnesota Department of Transportation standards, a walkway cannot rise more than 1 ft over a horizontal distance of 20 ft. Express this slope as a grade.

64. *Engineering.* At one point, Yellowstone's Beartooth Highway rises 315 ft over a horizontal distance of 4500 ft. Find the grade of the road.

65. *Carpentry.* Find the slope (or pitch) of the roof.

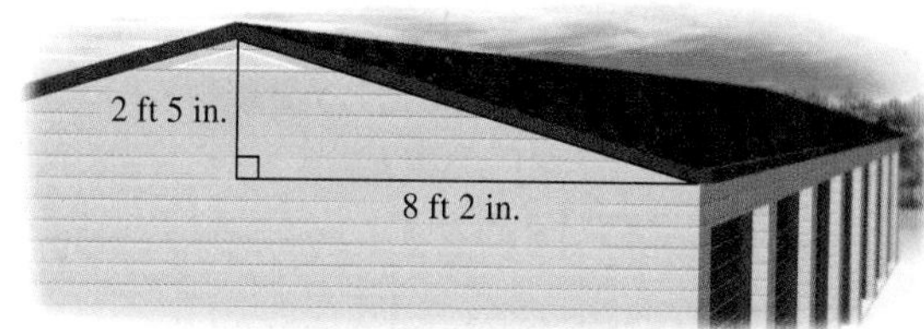

66. *Exercise.* Find the slope (or grade) of the treadmill.

67. *Bicycling.* To qualify as a rated climb on the Tour de France, a grade must average at least 4%. The ascent of Dooley Mountain, Oregon, part of the Elkhorn

Classic, begins at 3500 ft and climbs to 5400 ft over a horizontal distance of 37,000 ft. What is the grade of the road? Would it qualify as a rated climb if it were part of the Tour de France?
Source: barkercityherald.com

68. *Construction.* Public buildings regularly include steps with 7-in. risers and 11-in. treads. Find the grade of such a stairway.

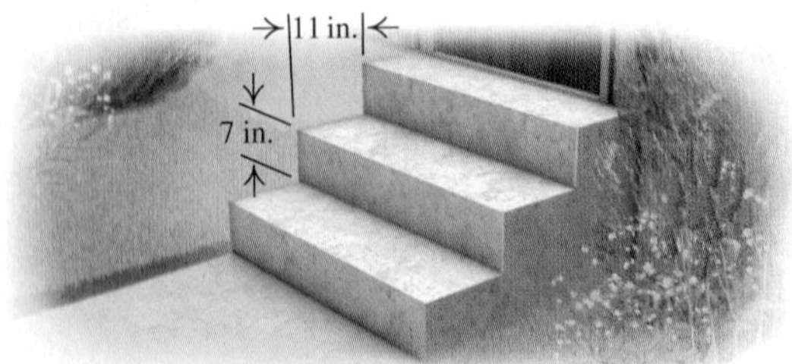

69. *Nursing.* Match each sentence with the most appropriate of the four graphs shown.

a) The rate at which fluids were given intravenously was doubled after 3 hr.
b) The rate at which fluids were given intravenously was gradually reduced to 0.
c) The rate at which fluids were given intravenously remained constant for 5 hr.
d) The rate at which fluids were given intravenously was gradually increased.

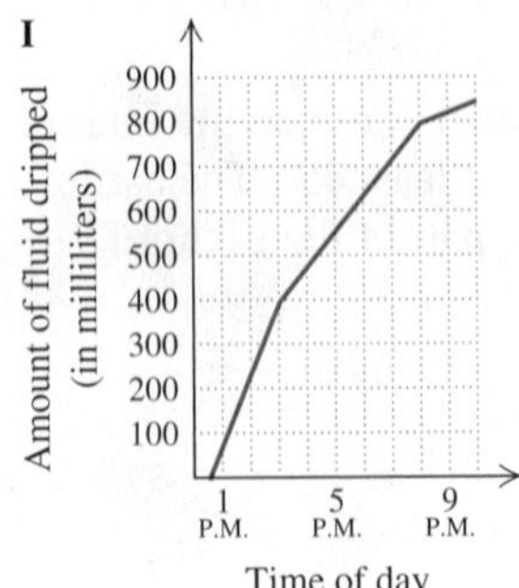

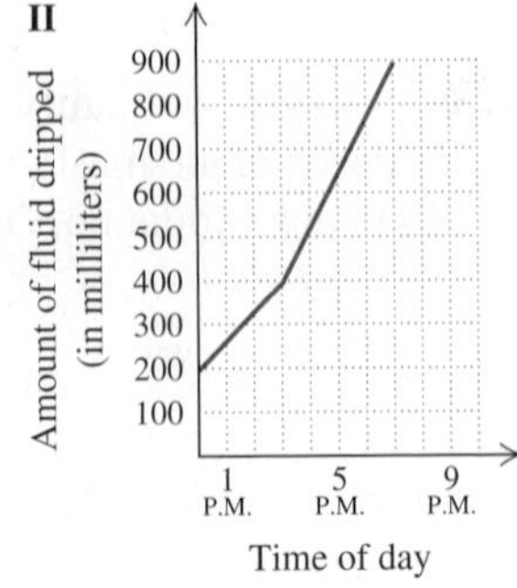

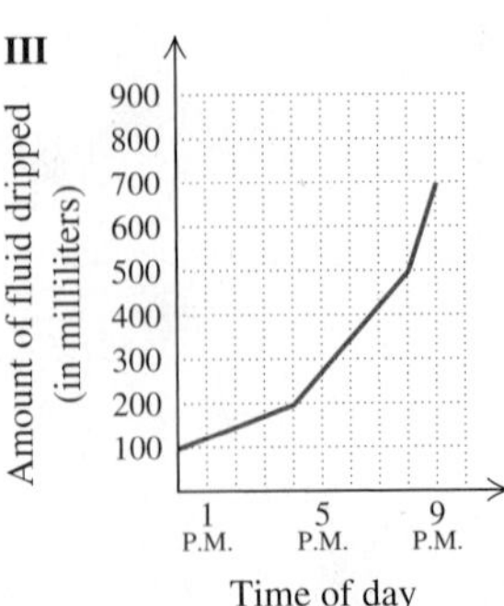

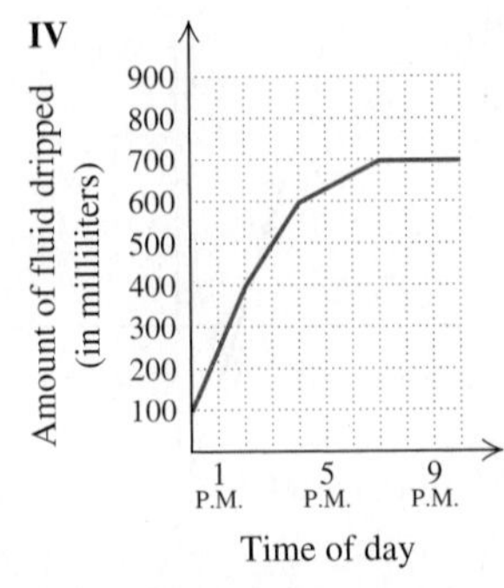

70. *Market Research.* Match each sentence with the most appropriate of the four graphs shown.

a) After January 1, daily sales continued to rise, but at a slower rate.
b) After January 1, sales decreased faster than they ever grew.
c) The rate of growth in daily sales doubled after January 1.
d) After January 1, daily sales decreased at half the rate that they grew in December.

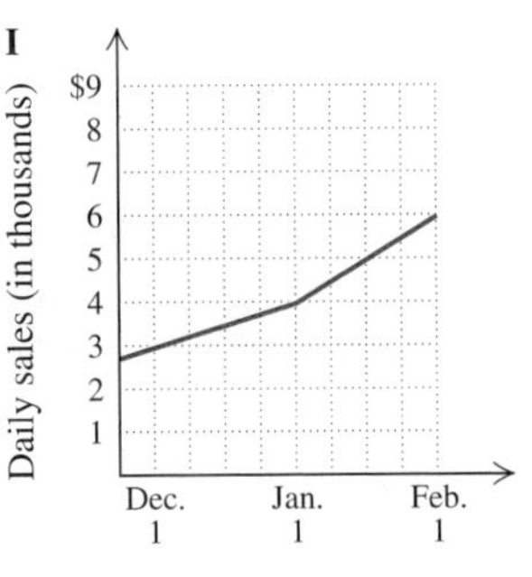

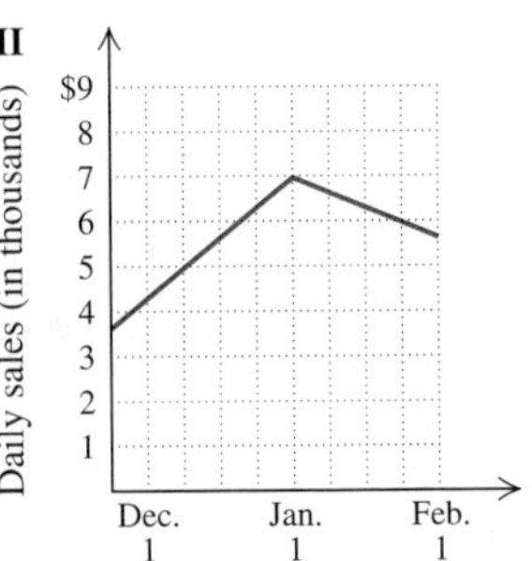

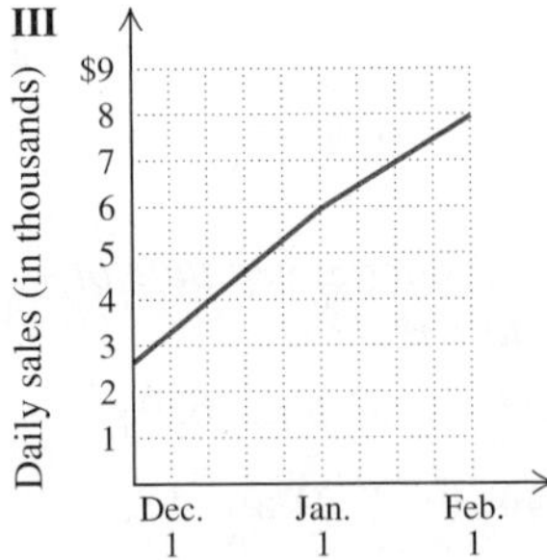

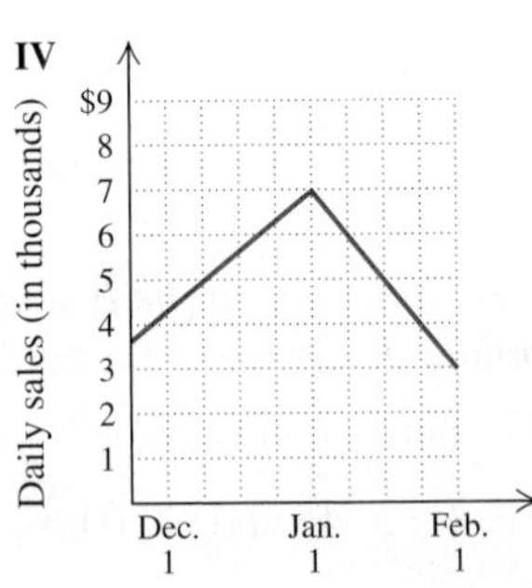

TW **71.** Explain why the order in which coordinates are subtracted to find slope does not matter so long as y-coordinates and x-coordinates are subtracted in the same order.

TW **72.** If one line has a slope of -3 and another has a slope of 2, which line is steeper? Why?

SKILL REVIEW

To prepare for Section 3.6, review solving a formula for a variable and graphing linear equations (Sections 2.3 and 3.2).

Solve. [2.3]

73. $ax + by = c$, for y

74. $rx - mn = p$, for r

75. $ax - by = c$, for y

76. $rs + nt = q$, for t

Graph. [3.2]

77. $8x + 6y = 24$

78. $3y = 4$

SYNTHESIS

TW 79. The points $(-4, -3)$, $(1, 4)$, $(4, 2)$, and $(-1, -5)$ are vertices of a quadrilateral. Use slopes to explain why the quadrilateral is a parallelogram.

TW 80. Can the points $(-4, 0)$, $(-1, 5)$, $(6, 2)$, and $(2, -3)$ be vertices of a parallelogram? Why or why not?

81. A line passes through $(4, -7)$ and never enters the first quadrant. What numbers could the line have for its slope?

82. A line passes through $(2, 5)$ and never enters the second quadrant. What numbers could the line have for its slope?

83. *Architecture.* Architects often use the equation $x + y = 18$ to determine the height y, in inches, of the riser of a step when the tread is x inches wide. Express the slope of stairs designed with this equation without using the variable y.

In Exercises 84 and 85, the slope of each line is $-\frac{2}{3}$, but the numbering on one axis is missing. How many units should each tick mark on that unnumbered axis represent?

84.

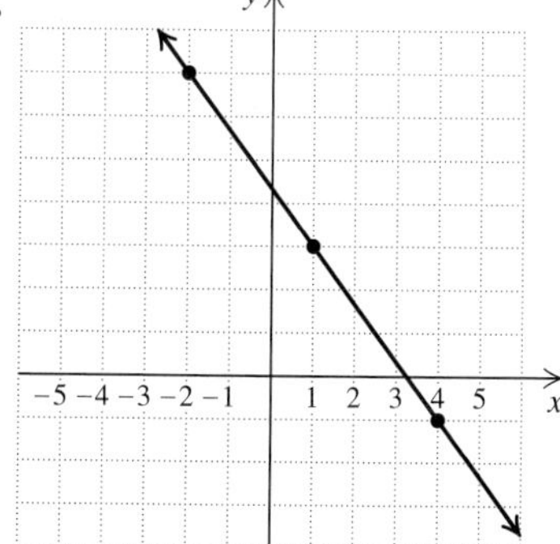

85.

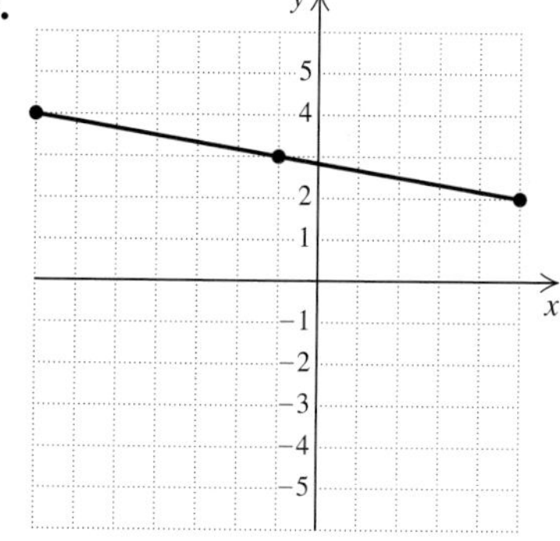

86. Match each sentence with the most appropriate of the four graphs below.

a) Annie drove 2 mi to a lake, swam 1 mi, and then drove 3 mi to a store.

b) During a preseason workout, Rico biked 2 mi, ran for 1 mi, and then walked 3 mi.

c) James bicycled 2 mi to a park, hiked 1 mi over the notch, and then took a 3-mi bus ride back to the park.

d) After hiking 2 mi, Marcy ran for 1 mi before catching a bus for the 3-mi ride into town.

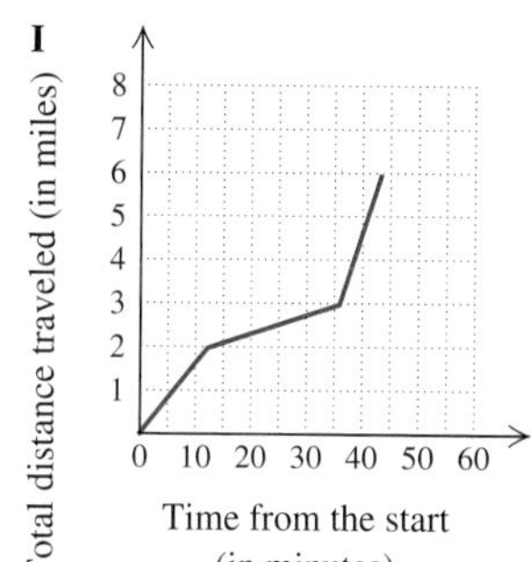

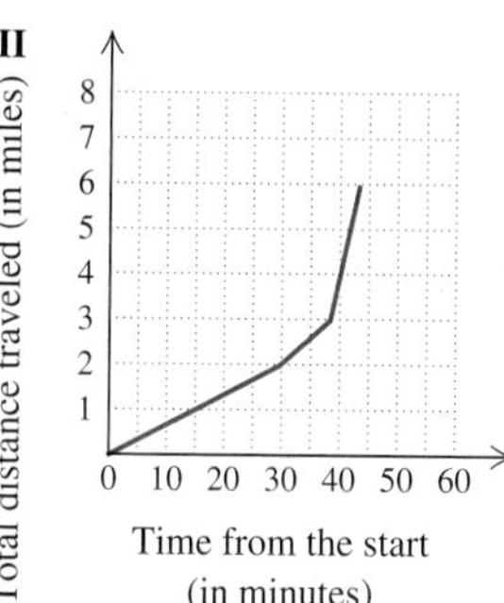

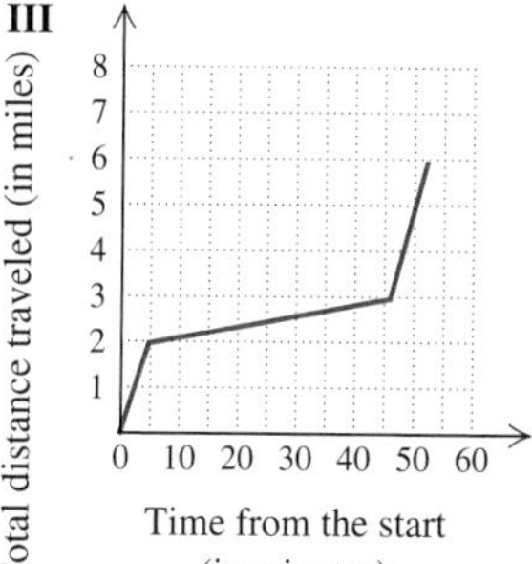

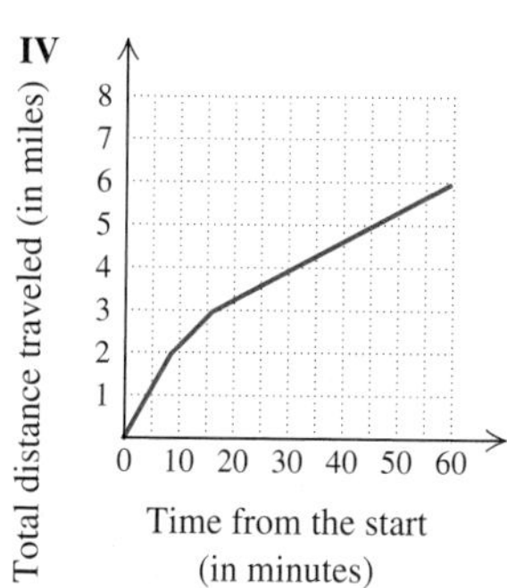

87. The plans below are for a skateboard "Fun Box." For the ramps labeled A, find the slope or grade.
Source: www.heckler.com

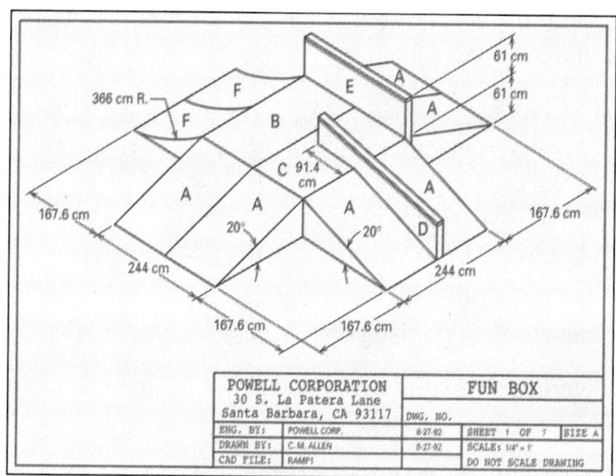

Try Exercise Answers: Section 3.5

11. 3 million people/yr **37.** $\frac{3}{2}$ **53.** 0 **55.** Undefined **61.** 8% **69.** (a) II; (b) IV; (c) I; (d) III

Mid-Chapter Review

By hand or by using a graphing calculator, we can plot points and graph equations on a *Cartesian coordinate plane.*

- A point is represented by an *ordered pair.*
- The graph of an equation represents all of its solutions.
- Equations can be *linear* or *nonlinear.*
- The *slope* of a line represents a rate of change: Slope $= m = \dfrac{\text{change in } y}{\text{change in } x} = \dfrac{y_2 - y_1}{x_2 - x_1}$.

GUIDED SOLUTIONS

1. Find the y-intercept and the x-intercept of the graph of $y - 3x = 6$. [3.3]

Solution

y-intercept: $y - 3 \cdot \square = 6$

$y = \square$

The y-intercept is $(\square, \square)$.

x-intercept: $\square - 3x = 6$

$-3x = 6$

$x = \square$

The x-intercept is $(\square, \square)$.

2. Find the slope of the line containing the points $(1, 5)$ and $(3, -1)$. [3.5]

Solution

$$m = \frac{y_2 - y_1}{x_2 - x_1} = \frac{-1 - \square}{3 - \square}$$

$$= \frac{\square}{2}$$

$$= \square$$

MIXED REVIEW

1. Plot the point $(0, -3)$. [3.1]

2. In which quadrant is the point $(4, -15)$ located? [3.1]

3. Determine whether the ordered pair $(-2, -3)$ is a solution of the equation $y = 5 - x$. [3.2]

Graph by hand.

4. $y = x - 3$ [3.2]

5. $y = -3x$ [3.2]

6. $3x - y = 2$ [3.2]

7. $4x - 5y = 20$ [3.3]

8. $y = -2$ [3.3]

9. $x = 1$ [3.3]

10. Graph using a graphing calculator: $y = 2x^2 + x$. [3.2]

11. Determine whether the equation $4x - 5y = 20$ is linear. [3.3]

12. Determine whether the equation $y = 2x^2 + x$ is linear. [3.3]

13. By the end of June, Construction Builders had winterized 10 homes. By the end of August, they had winterized a total of 38 homes. Find the rate at which the company was winterizing homes. [3.4]

14. From a base elevation of 9600 ft, Longs Peak, Colorado, rises to a summit elevation of 14,255 ft over a horizontal distance of 15,840 ft. Find the average grade of Longs Peak. [3.5]

Find the slope of the line containing the given pair of points. If the slope is undefined, state this. [3.5]

15. $(-5, -2)$ and $(1, 8)$

16. $(1, 2)$ and $(4, -7)$

17. $(0, 0)$ and $(0, -2)$

18. What is the slope of the line $y = 4$? [3.5]

19. What is the slope of the line $x = -7$? [3.5]

20. Find the x-intercept and the y-intercept of the line given by $2y - 3x = 12$. [3.3]

3.6 Slope–Intercept Form

- Using the y-intercept and the Slope to Graph a Line
- Equations in Slope–Intercept Form
- Graphing and Slope–Intercept Form
- Parallel and Perpendicular Lines

If we know the slope and the y-intercept of a line, we can graph the line. In this section, we will discover that a line's slope and its y-intercept can be read directly from the line's equation if the equation is written in a certain form.

USING THE y-INTERCEPT AND THE SLOPE TO GRAPH A LINE

Let's modify the candy wrapping situation that first appeared in Section 3.5. Suppose that as a new workshift begins, 12,000 candies have already been wrapped by the double-twist machine. Then the number wrapped during the new workshift is given by the table and the graph shown here.

LX-269 (Double twist)	
Minutes Elapsed	**Candies Wrapped**
0	12,000
3	14,500
6	17,000
9	19,500

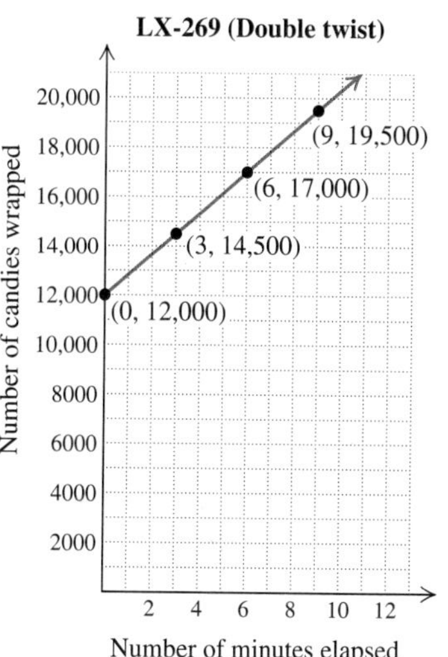

The wrapping rate is still $\frac{2500}{3}$, as we can confirm by calculating the slope of the line:

$$\text{Slope} = \frac{\text{change in } y}{\text{change in } x} = \frac{y_2 - y_1}{x_2 - x_1}$$

$$= \frac{19{,}500 - 14{,}500}{9 - 3} \qquad \textbf{Using the points (3, 14,500) and (9, 19,500)}$$

$$= \frac{5000}{6} = \frac{2500}{3}.$$

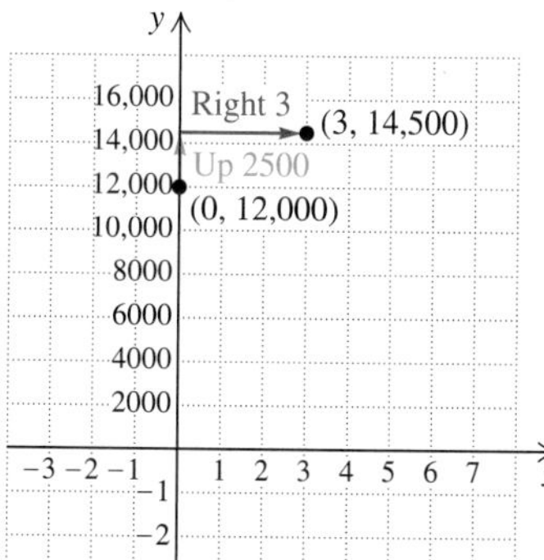

Knowing that the slope is $\frac{2500}{3}$, we could have drawn the graph by plotting $(0, 12{,}000)$ and from there moving *up* 2500 units and *to the right* 3 units. This would have located the point $(3, 14{,}500)$. Using these two points, we could then draw the line. This is the method used in the following example.

STUDY TIP

When One Just Isn't Enough

When an exercise gives you difficulty, practice solving some other exercises that are very similar to the one that gave you trouble. Usually, if the troubling exercise is odd-numbered, the next (even-numbered) exercise is quite similar.

EXAMPLE 1 Draw a line that has slope $\frac{1}{4}$ and y-intercept $(0, 2)$.

SOLUTION We plot $(0, 2)$ and from there move *up* 1 unit and *to the right* 4 units. This locates the point $(4, 3)$. We plot $(4, 3)$and draw a line passing through $(0, 2)$ and $(4, 3)$, as shown on the right below.

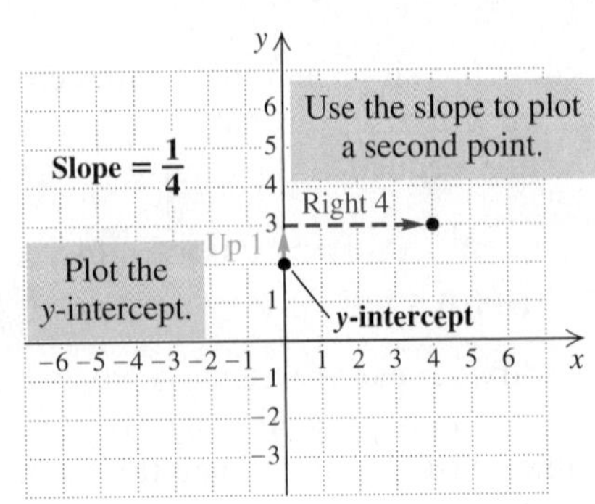

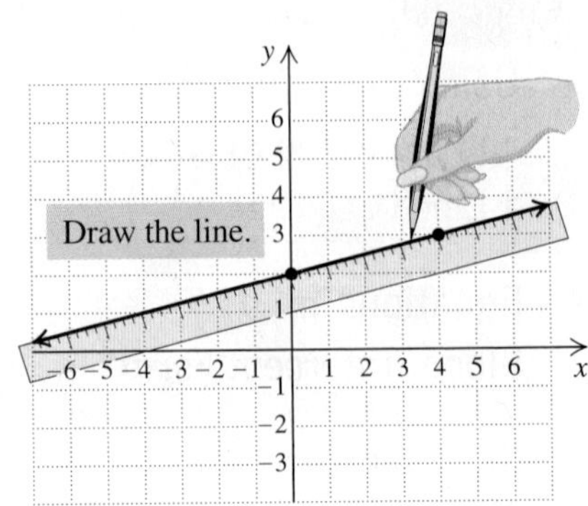

Try Exercise 7.

EQUATIONS IN SLOPE–INTERCEPT FORM

Recall from Section 3.3 that an equation of the form $Ax + By = C$ is linear. A linear equation can also be written in a form from which the slope and the y-intercept are read directly.

Interactive Discovery

Enter and graph $y_1 = 2x + 3$. Note the slope of the line and the y-intercept.

1. How does the graph change when we change the coefficient of the x-term in the equation? Enter and graph equations like $y_2 = 4x + 3$, $y_2 = \frac{1}{2}x + 3$, $y_2 = -2x + 3$, and $y_2 = -\frac{1}{2}x + 3$. If you have the Transfrm app on your calculator, enter $y_1 = Ax + 3$ and vary the coefficient A. Does the slope change? Does the y-intercept change?
2. How does the graph change when we change the constant term in the equation? Enter and graph equations like $y_2 = 2x + 1$, $y_2 = 2x + 5$, $y_2 = 2x - 1$, and $y_2 = 2x - 5$. If you have the Transfrm app on your calculator, enter $y_1 = 2x + A$ and vary the constant term A. Does the slope change? Does the y-intercept change?

Student Notes

An equation for a given line can be written in many different forms. Note that in the slope–intercept form, the equation is solved for y.

The pattern you may have observed is true in general.

The Slope–Intercept Equation The equation $y = mx + b$ is called the *slope–intercept equation.* The equation represents a line of slope m with y-intercept $(0, b)$.

The equation of any nonvertical line can be written in this form. The letter m is traditionally used for slope. This usage has its roots in the French verb *monter*, to climb.

EXAMPLE 2 Find the slope and the y-intercept of each line whose equation is given.

a) $y = \frac{4}{5}x - 8$ **b)** $2x + y = 5$ **c)** $4x - 4y = 7$

SOLUTION

a) We rewrite $y = \frac{4}{5}x - 8$ as $y = \frac{4}{5}x + (-8)$. Now we simply read the slope and the y-intercept from the equation:

$$y = \tfrac{4}{5}x + (-8).$$
$$y = mx + \quad b \qquad m = \tfrac{4}{5}, b = -8$$

The slope is $\frac{4}{5}$. The y-intercept is $(0, -8)$.

b) We first solve for y to find an equivalent equation in the form $y = mx + b$:

$$2x + y = 5$$
$$y = -2x + 5. \qquad \textbf{Adding } -2x \textbf{ to both sides}$$

The slope is -2. The y-intercept is $(0, 5)$.

c) We rewrite the equation in the form $y = mx + b$:

$$4x - 4y = 7$$
$$-4y = -4x + 7 \qquad \textbf{Adding } -4x \textbf{ to both sides}$$
$$y = -\tfrac{1}{4}(-4x + 7) \qquad \textbf{Multiplying both sides by } -\tfrac{1}{4}$$
$$y = x - \tfrac{7}{4} \qquad \textbf{Using the distributive law}$$
$$y = 1 \cdot x - \tfrac{7}{4}. \qquad \textbf{The (unwritten) coefficient of } x \textbf{ is 1.}$$

The slope is 1. The y-intercept is $\left(0, -\frac{7}{4}\right)$.

Try Exercise 19.

EXAMPLE 3 A line has slope $-\frac{12}{5}$ and y-intercept $(0, 11)$. Find an equation of the line.

SOLUTION We use the slope–intercept equation, substituting $-\frac{12}{5}$ for m and 11 for b:

$$y = mx + b = -\tfrac{12}{5}x + 11.$$

The desired equation is $y = -\frac{12}{5}x + 11$.

Try Exercise 37.

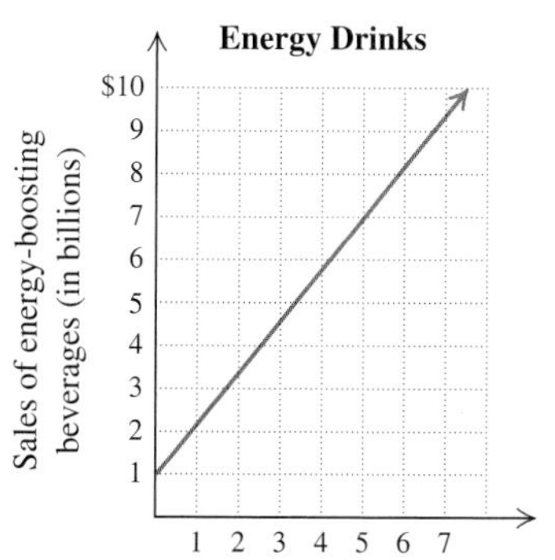

Source: Based on information from Packaged Facts and Restaurants & Institutions, Inc.

EXAMPLE 4 Determine an equation for the graph shown at left.

SOLUTION To write an equation for a line, we can use slope–intercept form, provided the slope and the y-intercept are known. From the graph, we see that $(0, 1)$ is the y-intercept. Looking closely, we see that the line passes through $(5, 7)$. We can either count squares on the graph or use the formula to calculate the slope:

$$m = \frac{\text{change in } y}{\text{change in } x} = \frac{7 - 1}{5 - 0} = \frac{6}{5}.$$

The desired equation is

$$y = \frac{6}{5}x + 1, \qquad \textbf{Using } \frac{6}{5} \textbf{ for } m \textbf{ and 1 for } b$$

where y is the sales of energy–boosting beverages, in billions of dollars, x years after 2002.

Try Exercise 45.

GRAPHING AND SLOPE-INTERCEPT FORM

In Example 1, we drew a graph, knowing only the slope and the y-intercept. In Example 2, we determined the slope and the y-intercept of a line by examining its equation. We now combine the two procedures to develop a quick way to graph a linear equation.

EXAMPLE 5 Graph: **(a)** $y = \frac{3}{4}x + 5$; **(b)** $2x + 3y = 3$.

SOLUTION To graph each equation, we plot the y-intercept and find additional points using the slope.

a) We can read the slope and the y-intercept from the equation $y = \frac{3}{4}x + 5$:

Determine the slope and the y-intercept.

Slope: $\frac{3}{4}$; y-intercept: $(0, 5)$.

Plot the y-intercept.

We plot the y-intercept $(0, 5)$. This gives us one point on the line.

Use the slope to find a second point.

Starting at $(0, 5)$, we use the slope $\frac{3}{4}$ to find another point.

We move *up* 3 units since the numerator (change in y) is *positive.*

We move *to the right* 4 units since the denominator (change in x) is *positive.*

This gives us a second point on the line, $(4, 8)$.

We can find a third point on the line by rewriting the slope $\frac{3}{4}$ as $\frac{-3}{-4}$, since these fractions are equivalent. Now, starting again at $(0, 5)$, we use the slope $\frac{-3}{-4}$ to find another point.

Use the slope to find a third point.

We move *down* 3 units since the numerator (change in y) is *negative.*

We move *to the left* 4 units since the denominator (change in x) is *negative.*

This gives us a third point on the line, $(-4, 2)$.

Draw the line.

Finally, we draw the line.

Student Notes

Recall the following:

$$\frac{3}{4} = \frac{-3}{-4};$$

$$-\frac{3}{4} = \frac{-3}{4} = \frac{3}{-4};$$

$$2 = \frac{2}{1} \text{ and } -2 = \frac{-2}{1}.$$

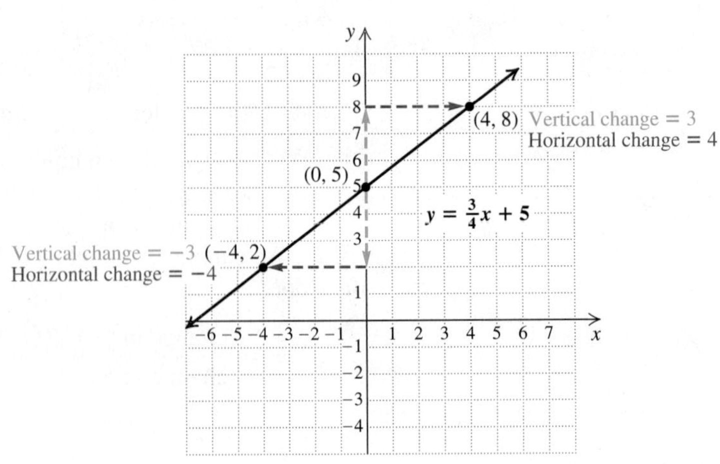

Student Notes

The signs of the numerator and the denominator of the slope indicate whether to move up, down, left, or right. Compare the following slopes.

$\frac{1}{2}$ ← 1 unit up, ← 2 units right

$\frac{-1}{-2}$ ← 1 unit down, ← 2 units left

$\frac{-1}{2}$ ← 1 unit down, ← 2 units right

$\frac{1}{-2}$ ← 1 unit up, ← 2 units left

b) To graph $2x + 3y = 3$, we first rewrite it to find the slope and the y-intercept:

$$
\begin{aligned}
2x + 3y &= 3 \\
3y &= -2x + 3 && \textbf{Adding } -2x \textbf{ to both sides} \\
y &= \tfrac{1}{3}(-2x + 3) && \textbf{Multiplying both sides by } \tfrac{1}{3} \\
y &= -\tfrac{2}{3}x + 1. && \textbf{Using the distributive law}
\end{aligned}
$$

We plot the y-intercept, $(0, 1)$.

The slope is $-\frac{2}{3}$. For graphing, we think of this slope as $\frac{-2}{3}$ or $\frac{2}{-3}$. Starting at $(0, 1)$, we use the slope $\frac{-2}{3}$ to find a second point.

We move *down* 2 units since the numerator is *negative.*

We move *to the right* 3 units since the denominator is *positive.*

We plot the new point, $(3, -1)$.

Now, starting at $(3, -1)$ and again using the slope $\frac{-2}{3}$, we move to a third point, $(6, -3)$. Alternatively, we can start at $(0, 1)$ and use the slope $\frac{2}{-3}$.

We move *up* 2 units since the numerator is *positive.*

We move *to the left* 3 units since the denominator is *negative.*

This leads to another point on the graph, $(-3, 3)$.

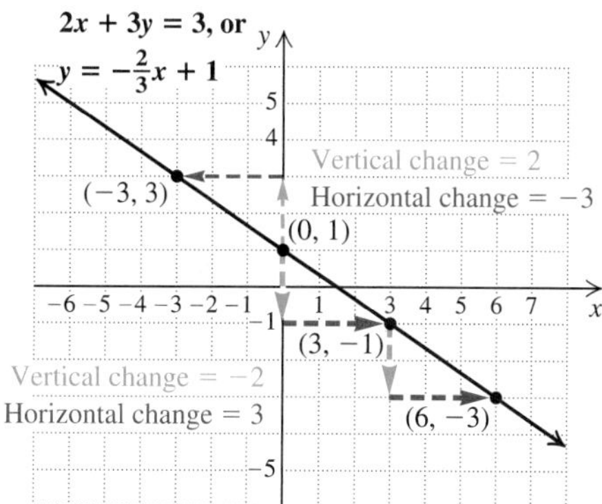

It is important to be able to use both $\frac{2}{-3}$ and $\frac{-2}{3}$ to draw the graph.

Try Exercise 49.

Slope–intercept form allows us to quickly determine the slope of a line by simply inspecting its equation. This can be especially helpful when attempting to decide whether two lines are parallel or perpendicular.

PARALLEL AND PERPENDICULAR LINES

Two lines are parallel if they lie in the same plane and do not intersect no matter how far they are extended. If two lines are vertical, they are parallel. How can we tell if nonvertical lines are parallel? The answer is simple: We look at their slopes.

> **Slope and Parallel Lines**
>
> Two lines with different y-intercepts are parallel if they have the same slope. Also, two vertical lines are parallel.

EXAMPLE 6 Determine whether the graphs of

$$y = -3x + 4 \quad \text{and} \quad 6x + 2y = -10$$

are parallel.

$y_1 = -3x + 4,\ y_2 = -3x - 5$

SOLUTION When two lines have the same slope but different y-intercepts, they are parallel. One of the two equations is given in slope–intercept form:

$y = -3x + 4.$ **The slope is -3 and the y-intercept is $(0, 4)$.**

To find the slope of the other line, we first solve for y:

$6x + 2y = -10$

$2y = -6x - 10$ **Adding $-6x$ to both sides**

$y = -3x - 5.$ **The slope is -3 and the y-intercept is $(0, -5)$.**

Since both lines have slope -3 but different y-intercepts, the graphs are parallel. The graphs of both equations are shown at left, and do appear to be parallel.

Try Exercise 65.

EXAMPLE 7 Determine whether the graphs of

$$7x + 10y = 50 \quad \text{and} \quad 4x + 5y = -25$$

are parallel.

$y_1 = -\frac{7}{10}x + 5,\ y_2 = -\frac{4}{5}x - 5$

SOLUTION We find the slope of each line:

$7x + 10y = 50$

$10y = -7x + 50$ **Adding $-7x$ to both sides**

$y = -\frac{7}{10}x + 5;$ **Multiplying both sides by $\frac{1}{10}$**

The slope is $-\frac{7}{10}$.

$4x + 5y = -25$

$5y = -4x - 25$ **Adding $-4x$ to both sides**

$y = -\frac{4}{5}x - 5.$ **Multiplying both sides by $\frac{1}{5}$**

The slope is $-\frac{4}{5}$.

Since the slopes are not the same, the lines are not parallel. In decimal form, the slopes are -0.7 and -0.8. Since these slopes are close in value, in some viewing windows the graphs may appear to be parallel, as shown at left, when in reality they are not.

Try Exercise 67.

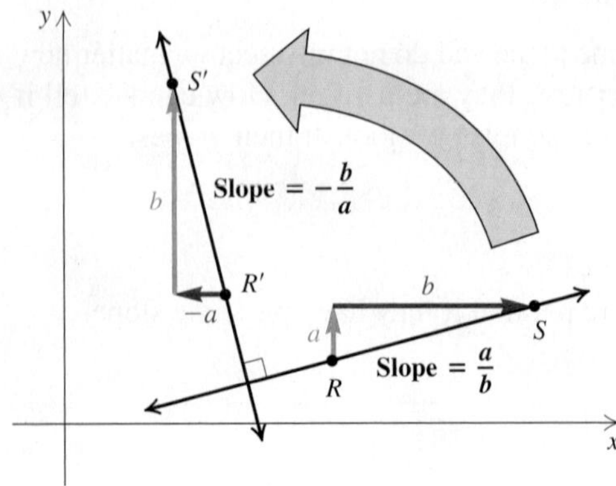

Two lines are perpendicular if they intersect at a right angle. If one line is vertical and another is horizontal, they are perpendicular. There are other instances in which two lines are perpendicular.

Consider a line $\overleftrightarrow{RS}$, as shown at left, with slope a/b. Then think of rotating the figure 90° to get a line $\overleftrightarrow{R'S'}$, perpendicular to $\overleftrightarrow{RS}$. For the new line, the rise and the run are interchanged, but the run is now negative. Thus the slope of the new line is $-b/a$. Let's multiply the slopes:

$$\frac{a}{b}\left(-\frac{b}{a}\right) = -1.$$

This can help us determine which lines are perpendicular.

> **Slope and Perpendicular Lines**
>
> Two lines are perpendicular if the product of their slopes is -1 or if one line is vertical and the other is horizontal.

Thus, if one line has slope m $(m \neq 0)$, the slope of a line perpendicular to it is $-1/m$. That is, we take the reciprocal of m $(m \neq 0)$ and change the sign.

EXAMPLE 8 Determine whether the graphs of $2x + y = 8$ and $y = \frac{1}{2}x + 7$ are perpendicular.

SOLUTION The second equation is given in slope–intercept form:

$y = \frac{1}{2}x + 7$. **The slope is $\frac{1}{2}$.**

To find the slope of the other line, we solve for y:

$2x + y = 8$

$y = -2x + 8$. **Adding $-2x$ to both sides**

The slope is -2.

The lines are perpendicular if the product of their slopes is -1. Since

$\frac{1}{2}(-2) = -1,$

the graphs are perpendicular. The graphs of both equations are shown at left, and do appear to be perpendicular.

EXAMPLE 9 For each equation, find the slope of a line parallel to its graph and the slope of a line perpendicular to its graph.

a) $y = -\frac{3}{4}x + 7$ **b)** $y = \frac{1}{9}x - 2$ **c)** $y = -5x + \frac{1}{3}$

SOLUTION For each equation, we use the following process:

1. Find the slope of the given line.
2. Find the reciprocal of the slope.
3. Find the opposite of the reciprocal.

The number found in step (1) is the slope of a line *parallel* to the given line. The number found in step (3) is the slope of a line *perpendicular* to the given line.

	1. Find the slope.	**2. Find the reciprocal.**	**3. Find the opposite of the reciprocal.**	**Slope of a parallel line**	**Slope of a perpendicular line**
a) $y = -\frac{3}{4}x + 7$	$-\frac{3}{4}$	$-\frac{4}{3}$	$\frac{4}{3}$	$-\frac{3}{4}$	$\frac{4}{3}$
b) $y = \frac{1}{9}x - 2$	$\frac{1}{9}$	$\frac{9}{1}$, or 9	-9	$\frac{1}{9}$	-9
c) $y = -5x + \frac{1}{3}$	-5	$-\frac{1}{5}$	$\frac{1}{5}$	-5	$\frac{1}{5}$

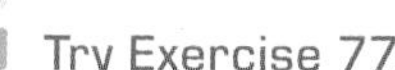

EXAMPLE 10 Write a slope–intercept equation for the line whose graph is described.

a) Parallel to the graph of $2x - 3y = 7$, with y-intercept $(0, -1)$

b) Perpendicular to the graph of $2x - 3y = 7$, with y-intercept $(0, -1)$

SOLUTION We begin by determining the slope of the line represented by $2x - 3y = 7$:

$$2x - 3y = 7$$

$$-3y = -2x + 7 \qquad \textbf{Adding } -2x \textbf{ to both sides}$$

$$y = \tfrac{2}{3}x - \tfrac{7}{3}. \qquad \textbf{Dividing both sides by } -3$$

The slope is $\frac{2}{3}$.

a) A line parallel to the graph of $2x - 3y = 7$ has a slope of $\frac{2}{3}$. Since the y-intercept is $(0, -1)$, the slope–intercept equation is

$$y = \tfrac{2}{3}x - 1. \qquad \textbf{Substituting in } y = mx + b$$

b) A line perpendicular to the graph of $2x - 3y = 7$ has a slope that is the opposite of the reciprocal of $\frac{2}{3}$, or $-\frac{3}{2}$. Since the y-intercept is $(0, -1)$, the slope–intercept equation is

$$y = -\tfrac{3}{2}x - 1. \qquad \textbf{Substituting in } y = mx + b$$

Try Exercises 85 and 87.

Student Notes

Although it is helpful to visualize parallel lines and perpendicular lines graphically, a graph cannot be used to *determine* whether two lines are parallel or perpendicular. Slopes must be used to determine this.

Squaring a Viewing Window

If the units on the x-axis are a different length than those on the y-axis, two lines that are perpendicular may not appear to be so when graphed. Finding a viewing window with units the same length on both axes is called *squaring* the viewing window.

Windows can be squared by choosing the ZSquare option in the ZOOM menu. They can be squared manually by choosing the portions of the axes shown in the correct proportion. For example, if the ratio of the length of a viewing window to its height is 3:2, a window like $[-9, 9, -6, 6]$ will be squared.

Consider the graphs of $y = 3x - 7$ and $y = -\frac{1}{3}x + \frac{1}{3}$. The lines are perpendicular, since $3 \cdot \left(-\frac{1}{3}\right) = -1$. The graphs are shown in a standard viewing window on the left below, but the lines do not appear to be perpendicular. If we press ZOOM 5, the lines are graphed in a squared viewing window as shown in the graph on the right below, and they do appear perpendicular.

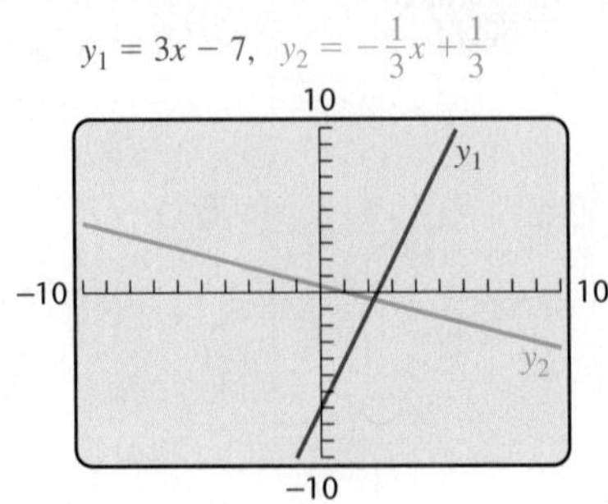

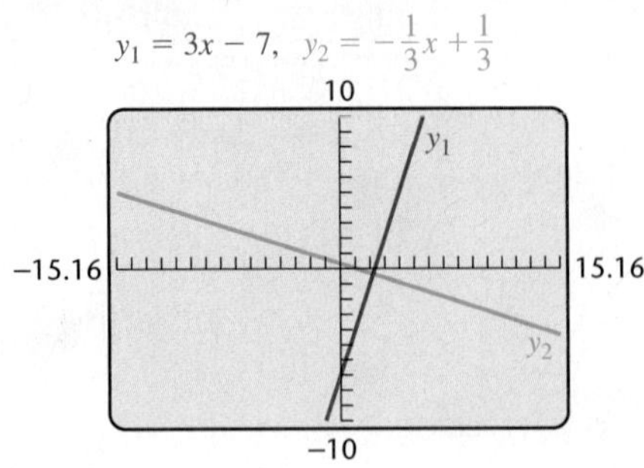

(continued)

Your Turn

1. Graph $y = x$ using a standard viewing window $[-10, 10, -10, 10]$, with Xscl = 1 and Yscl = 1. Compare the distance between marks on the axes.
2. Press ZOOM 5 to square the window. Again, compare the distance between marks on the axes. They should be the same.

3.6 Exercise Set

FOR EXTRA HELP MyMathLab

Concept Reinforcement *In each of Exercises 1–6, match the phrase with the most appropriate choice from the column on the right.*

1. ___ The slope of the graph of $y = 3x - 2$
2. ___ The slope of the graph of $y = 2x - 3$
3. ___ The slope of the graph of $y = \frac{2}{3}x + 3$
4. ___ The y-intercept of the graph of $y = 2x - 3$
5. ___ The y-intercept of the graph of $y = 3x - 2$
6. ___ The y-intercept of the graph of $y = \frac{2}{3}x + \frac{3}{4}$

a) $\left(0, \frac{3}{4}\right)$
b) 2
c) $(0, -3)$
d) $\frac{2}{3}$
e) $(0, -2)$
f) 3

Draw a line that has the given slope and y-intercept.

7. Slope $\frac{2}{5}$; y-intercept $(0, 1)$
8. Slope $\frac{3}{5}$; y-intercept $(0, -1)$
9. Slope $\frac{5}{3}$; y-intercept $(0, -2)$
10. Slope $\frac{1}{2}$; y-intercept $(0, 0)$
11. Slope $-\frac{1}{3}$; y-intercept $(0, 5)$
12. Slope $-\frac{4}{5}$; y-intercept $(0, 6)$
13. Slope 2; y-intercept $(0, 0)$
14. Slope -2; y-intercept $(0, -3)$
15. Slope -3; y-intercept $(0, 2)$
16. Slope 3; y-intercept $(0, 4)$
17. Slope 0; y-intercept $(0, -5)$
18. Slope 0; y-intercept $(0, 1)$

Find the slope and the y-intercept of each line whose equation is given.

19. $y = -\frac{2}{7}x + 5$
20. $y = -\frac{3}{8}x + 4$
21. $y = \frac{1}{3}x + 7$
22. $y = \frac{4}{5}x + 1$
23. $y = \frac{9}{5}x - 4$
24. $y = -\frac{9}{10}x - 5$
25. $-3x + y = 7$
26. $-4x + y = 7$
27. $4x + 2y = 8$
28. $3x + 4y = 12$
29. Aha! $y = 4$
30. $y - 3 = 5$
31. $2x - 5y = -8$
32. $12x - 6y = 9$
33. $9x - 8y = 0$
34. $7x = 5y$
35. Use the slope and the y-intercept of each line to match each equation with the correct graph.
    a) $y = 3x - 5$
    b) $y = 0.7x + 1$
    c) $y = -0.25x - 3$
    d) $y = -4x + 2$

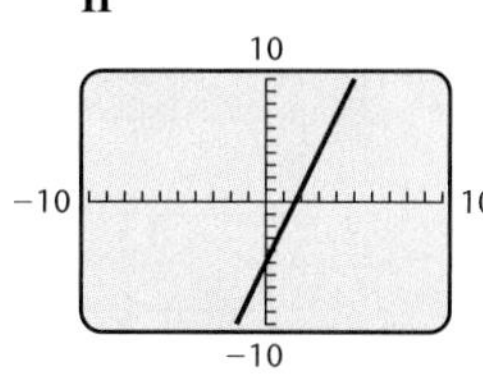

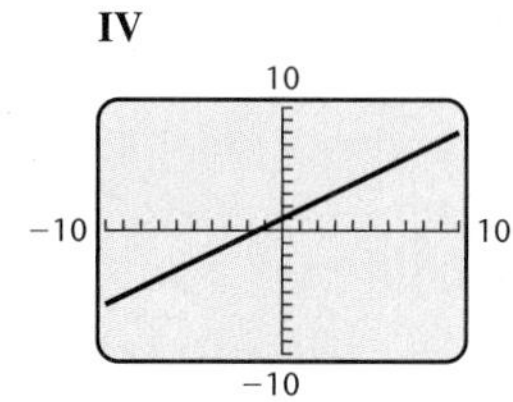

36. Use the slope and the y-intercept of each line to match each equation with the correct graph.

a) $y = \frac{1}{2}x - 5$
b) $y = 2x + 3$
c) $y = -3x + 1$
d) $y = -\frac{3}{4}x - 2$

I

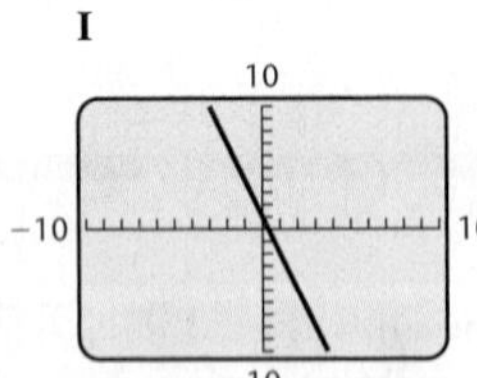

II

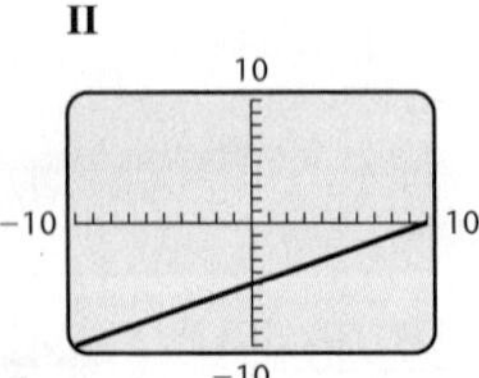

III

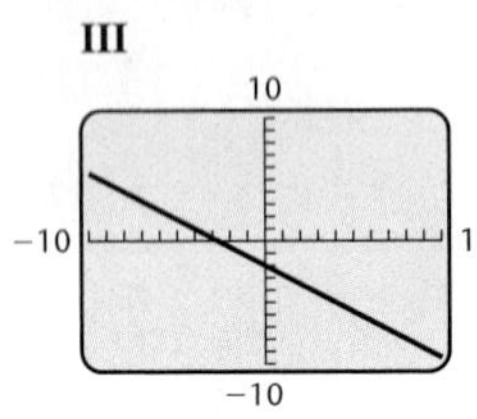

IV

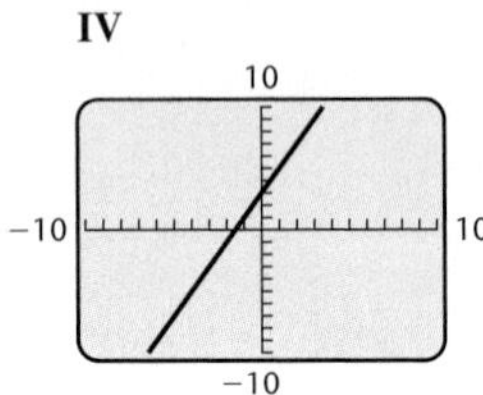

Find the slope–intercept equation for the line with the indicated slope and y-intercept.

37. Slope 3; y-intercept $(0, 7)$

38. Slope -4; y-intercept $\left(0, -\frac{3}{5}\right)$

39. Slope $\frac{7}{8}$; y-intercept $(0, -1)$

40. Slope $\frac{5}{7}$; y-intercept $(0, 4)$

41. Slope $-\frac{5}{3}$; y-intercept $(0, -8)$

42. Slope $\frac{3}{4}$; y-intercept $(0, -35)$

Aha! **43.** Slope 0; y-intercept $\left(0, \frac{1}{3}\right)$

44. Slope 7; y-intercept $(0, 0)$

Determine an equation for each graph shown.

45.

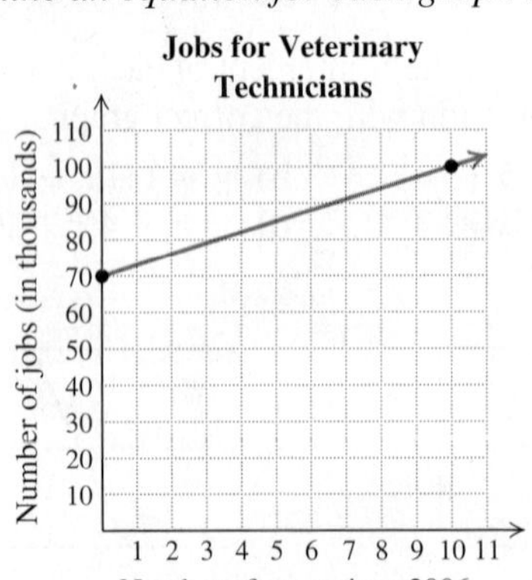

Estimates based on information from the U.S. Census Bureau, *Statistical Abstract of the United States*, 2009

46.

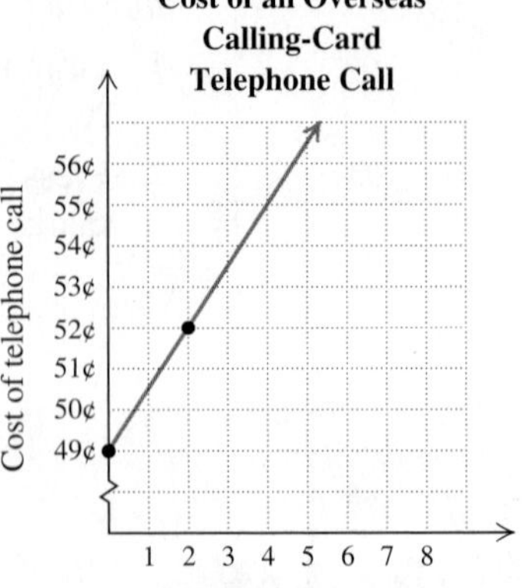

Source: www.pennytalk.com

47.

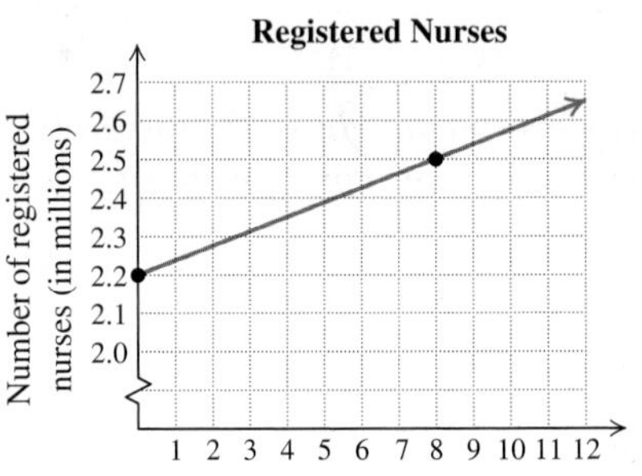

Based on information from the Bureau of Labor

48.

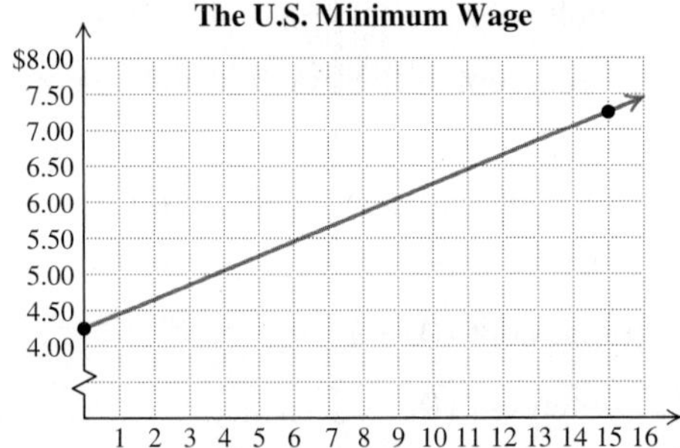

Based on information from www.infoplease.com

Graph by hand.

49. $y = \frac{3}{5}x + 2$

50. $y = -\frac{3}{5}x - 1$

51. $y = -\frac{3}{5}x + 1$

52. $y = \frac{3}{5}x - 2$

53. $y = \frac{5}{3}x + 3$

54. $y = \frac{5}{3}x - 2$

55. $y = -\frac{3}{2}x - 2$

56. $y = -\frac{4}{3}x + 3$

57. $2x + y = 1$

58. $3x + y = 2$

59. $3x + y = 0$

60. $2x + y = 0$

61. $2x + 3y = 9$

62. $4x + 5y = 15$

63. $x - 4y = 12$

64. $x + 5y = 20$

Determine whether each pair of equations represents parallel lines.

65. $y = \frac{2}{3}x + 7,$
$y = \frac{2}{3}x - 5$

66. $y = -\frac{5}{4}x + 1,$
$y = \frac{5}{4}x + 3$

67. $y = 2x - 5,$
$4x + 2y = 9$

68. $y = -3x + 1,$
$6x + 2y = 8$

69. $3x + 4y = 8,$
$7 - 12y = 9x$

70. $3x = 5y - 2,$
$10y = 4 - 6x$

Determine whether each pair of equations represents perpendicular lines.

71. $y = 4x - 5,$
$4y = 8 - x$

72. $2x - 5y = -3,$
$2x + 5y = 4$

73. $x - 2y = 5,$
$2x + 4y = 8$

74. $y = -x + 7,$
$y - x = 3$

75. $2x + 3y = 1,$
$3x - 2y = 1$

76. $y = 5 - 3x,$
$3x - y = 8$

For each equation, **(a)** *determine the slope of a line parallel to its graph, and* **(b)** *determine the slope of a line perpendicular to its graph.*

77. $y = \frac{7}{8}x - 3$

78. $y = -\frac{9}{10}x + 4$

79. $y = -\frac{1}{4}x - \frac{5}{8}$

80. $y = \frac{1}{6}x - \frac{3}{11}$

81. $20x - y = 12$

82. $y + 15x = 30$

83. $x + y = 4$

84. $x - y = 19$

Write a slope–intercept equation of the line whose graph is described.

85. Parallel to the graph of $y = 5x - 7$; y-intercept $(0, 11)$

86. Parallel to the graph of $2x - y = 1$; y-intercept $(0, -3)$

87. Perpendicular to the graph of $2x + y = 0$; y-intercept $(0, 0)$

88. Perpendicular to the graph of $y = \frac{1}{3}x + 7$; y-intercept $(0, 5)$

Aha! **89.** Parallel to the graph of $y = x$; y-intercept $(0, 3)$

Aha! **90.** Perpendicular to the graph of $y = x$; y-intercept $(0, 0)$

91. Perpendicular to the graph of $x + y = 3$; y-intercept $(0, -4)$

92. Parallel to the graph of $3x + 2y = 5$; y-intercept $(0, -1)$

TW **93.** A student makes a mistake when using a graphing calculator to draw $4x + 5y = 12$ and the following screen appears. Use algebra to show that a mistake has been made. What do you think the mistake was?

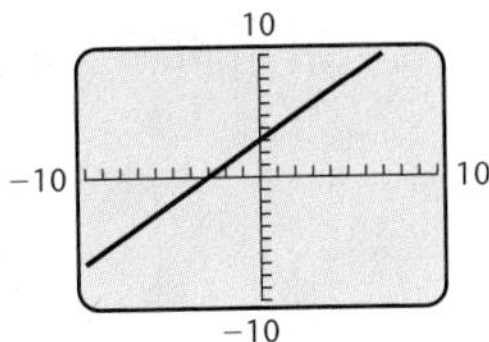

TW **94.** A student makes a mistake when using a graphing calculator to draw $5x - 2y = 3$ and the following screen appears. Use algebra to show that a mistake has been made. What do you think the mistake was?

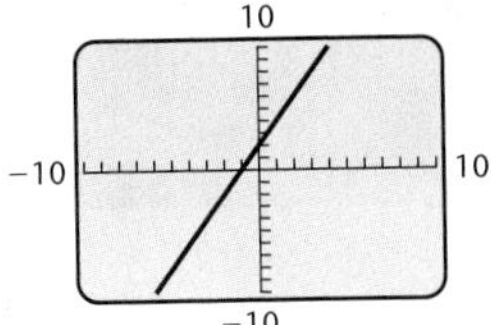

SKILL REVIEW

To prepare for Section 3.7, review solving a formula for a variable and subtracting real numbers (Sections 1.6 and 2.3).

Solve. [2.3]

95. $y - k = m(x - h)$, for y

96. $y - 9 = -2(x + 4)$, for y

Simplify. [1.6]

97. $-10 - (-3)$

98. $8 - (-5)$

99. $-4 - 5$

100. $-6 - 5$

SYNTHESIS

TW **101.** Explain how it is possible for an incorrect graph to be drawn, even after plotting three points that line up.

TW **102.** Which would you prefer, and why: graphing an equation of the form $y = mx + b$ or graphing an equation of the form $Ax + By = C$?

103. Show that the slope of the line given by $y = mx + b$ is m. (*Hint*: Substitute both 0 and 1 for x to find two pairs of coordinates. Then use the formula, Slope = change in y/change in x.)

104. Find k such that the line containing $(-3, k)$ and $(4, 8)$ is parallel to the line containing $(5, 3)$ and $(1, -6)$.

Solve.

105. *Refrigerator Size.* Kitchen designers recommend that a refrigerator be selected on the basis of the number of people in the household. For 1–2 people, a 16-ft^3 model is suggested. For each additional person, an additional 1.5 ft^3 is recommended. If x is the number of residents in excess of 2, find the slope–intercept equation for the recommended size of a refrigerator.

106. *Cost of a Speeding Ticket.* The penalty schedule shown below is used to determine the cost of a speeding ticket in certain states. Use this schedule to graph the cost of a speeding ticket given the number of miles per hour over the limit that a driver is going.

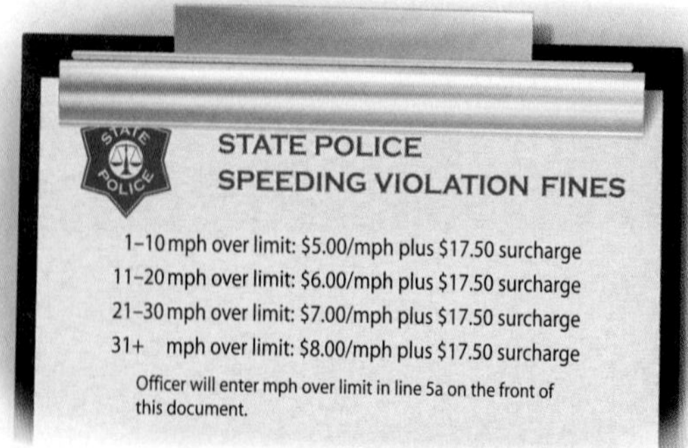

STATE POLICE
SPEEDING VIOLATION FINES

1–10 mph over limit: $5.00/mph plus $17.50 surcharge
11–20 mph over limit: $6.00/mph plus $17.50 surcharge
21–30 mph over limit: $7.00/mph plus $17.50 surcharge
31+ mph over limit: $8.00/mph plus $17.50 surcharge

Officer will enter mph over limit in line 5a on the front of this document.

In Exercises 107 and 108, assume that r, p, and s are constants and that x and y are variables. Determine the slope and the y-intercept.

107. $rx + py = s$

108. $rx + py = s - ry$

TW **109.** *Aerobic Exercise.* The formula $T = -\frac{3}{4}a + 165$ can be used to determine the *target heart rate*, in number of beats per minute, for a person, a years old, participating in aerobic exercise. Graph the equation and interpret the significance of its slope.

110. Graph the equations

$$y_1 = 1.4x + 2, \qquad y_2 = 0.6x + 2,$$
$$y_3 = 1.4x + 5, \quad \text{and} \quad y_4 = 0.6x + 5$$

using a graphing calculator. If possible, use the SIMULTANEOUS mode so that you cannot tell which equation is being graphed first. Then decide which line corresponds to each equation.

Try Exercise Answers: Section 3.6

7.

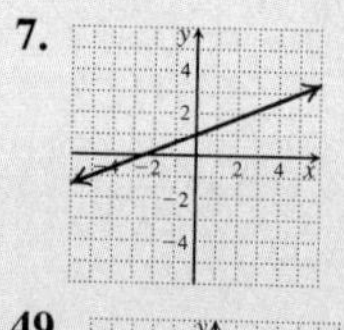

19. $-\frac{2}{7}$; $(0, 5)$ **37.** $y = 3x + 7$ **45.** $y = 3x + 70$, where y is the number of jobs, in thousands, and x is the number of years since 2006

49.

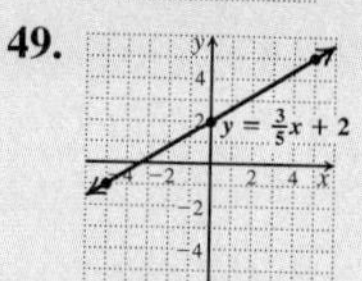

65. Yes **67.** No **71.** Yes **77.** (a) $\frac{7}{8}$; (b) $-\frac{8}{7}$ **85.** $y = 5x + 11$ **87.** $y = \frac{1}{2}x$

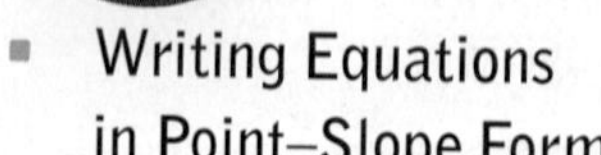

3.7 Point–Slope Form; Introduction to Curve Fitting

- Writing Equations in Point–Slope Form
- Graphing and Point–Slope Form
- Estimations and Predictions Using Two Points
- Curve Fitting

Specifying a line's slope and one point through which the line passes enables us to draw the line. In this section, we study how this same information can be used to produce an *equation* of the line.

WRITING EQUATIONS IN POINT–SLOPE FORM

Consider a line with slope 2 passing through the point $(4, 1)$, as shown in the figure on the following page. In order for a point (x, y) to be on the line, the coordinates x and y must be solutions of the slope equation

$$\frac{y - 1}{x - 4} = 2. \qquad \text{This is true only when } (x, y) \text{ is on the line.}$$

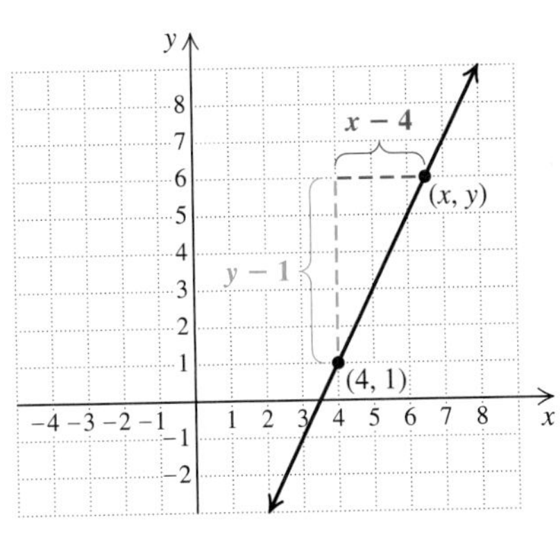

Take a moment to examine this equation. Pairs like (5, 3) and (3, −1) are solutions, since

$$\frac{3-1}{5-4}=2 \quad \text{and} \quad \frac{-1-1}{3-4}=2.$$

Note, however, that (4, 1) is not itself a solution of the equation:

$$\frac{1-1}{4-4}\neq 2.$$

To avoid this difficulty, we can use the multiplication principle:

$$(x-4)\cdot\frac{y-1}{x-4}=2(x-4) \qquad \textbf{Multiplying both sides by } x-4$$

$$y-1=2(x-4). \qquad \textbf{Removing a factor equal to 1: } \frac{x-4}{x-4}=1$$

This is considered **point–slope form** for the line shown above. A point–slope equation can be written whenever a line's slope and a point on the line are known.

Student Notes

You can remember the point–slope equation by calculating the slope m of a line containing a *general* point (x, y) and a *specific* point (x_1, y_1):

$$m=\frac{y-y_1}{x-x_1}.$$

Multiplying by $x - x_1$, we have

$$m(x-x_1)=y-y_1.$$

The Point–Slope Equation The equation $y - y_1 = m(x - x_1)$ is called the *point–slope equation* for the line with slope m that contains the point (x_1, y_1).

EXAMPLE 1 Write a point–slope equation for the line with slope $-\frac{4}{3}$ that contains the point $(1, -6)$.

SOLUTION We substitute $-\frac{4}{3}$ for m, 1 for x_1, and -6 for y_1:

$$y-y_1=m(x-x_1) \qquad \textbf{Using the point–slope equation}$$

$$y-(-6)=-\tfrac{4}{3}(x-1). \qquad \textbf{Substituting}$$

Try Exercise 19.

EXAMPLE 2 Write the slope–intercept equation for the line with slope 2 that contains the point (3, 1).

SOLUTION There are two parts to this solution. First, we write an equation in point–slope form:

$$y-y_1=m(x-x_1)$$

$$y-1=2(x-3). \qquad \textbf{Substituting}$$

Next, we find an equivalent equation of the form $y = mx + b$:

$$y-1=2(x-3)$$

$$y-1=2x-6 \qquad \textbf{Using the distributive law}$$

$$y=2x-5. \qquad \textbf{Adding 1 to both sides to get slope–intercept form}$$

Try Exercise 33.

Student Notes

There are several forms in which a line's equation can be written. For instance, as shown in Example 2, $y - 1 = 2(x - 3)$, $y - 1 = 2x - 6$, and $y = 2x - 5$ all are equations for the same line.

EXAMPLE 3 Consider the line given by the equation $8y = 7x - 24$.

a) Write the slope–intercept equation for a parallel line passing through $(-1, 2)$.

b) Write the slope–intercept equation for a perpendicular line passing through $(-1, 2)$.

SOLUTION Both parts (a) and (b) require us to find the slope of the line given by $8y = 7x - 24$. To do so, we solve for y to find slope–intercept form:

$$8y = 7x - 24$$

$$y = \tfrac{7}{8}x - 3. \quad \textbf{Multiplying both sides by } \tfrac{1}{8}$$

The slope is $\frac{7}{8}$.

a) The slope of any parallel line will be $\frac{7}{8}$. The point–slope equation yields

$$y - 2 = \tfrac{7}{8}[x - (-1)] \quad \textbf{Substituting } \tfrac{7}{8} \textbf{ for the slope and } (-1, 2) \textbf{ for the point}$$

$$y - 2 = \tfrac{7}{8}[x + 1]$$

$$y = \tfrac{7}{8}x + \tfrac{7}{8} + 2 \quad \textbf{Using the distributive law and adding 2 to both sides}$$

$$y = \tfrac{7}{8}x + \tfrac{23}{8}.$$

b) The slope of a perpendicular line is given by the opposite of the reciprocal of $\frac{7}{8}$, or $-\frac{8}{7}$. The point–slope equation yields

$$y - 2 = -\tfrac{8}{7}[x - (-1)] \quad \textbf{Substituting } -\tfrac{8}{7} \textbf{ for the slope and } (-1, 2) \textbf{ for the point}$$

$$y - 2 = -\tfrac{8}{7}[x + 1]$$

$$y = -\tfrac{8}{7}x - \tfrac{8}{7} + 2 \quad \textbf{Using the distributive law and adding 2 to both sides}$$

$$y = -\tfrac{8}{7}x + \tfrac{6}{7}.$$

Try Exercises 43 and 51.

EXAMPLE 4 Write the slope–intercept equation for the line that contains the points $(-1, -5)$ and $(3, -2)$.

SOLUTION We first determine the slope of the line and then use the point–slope equation:

Find the slope.

$$m = \frac{-5 - (-2)}{-1 - 3} = \frac{-3}{-4} = \frac{3}{4}.$$

Since the line passes through $(3, -2)$, we have

Find the point–slope form.

$$y - (-2) = \tfrac{3}{4}(x - 3) \quad \textbf{Substituting into the point–slope equation}$$

$$y + 2 = \tfrac{3}{4}x - \tfrac{9}{4}. \quad \textbf{Using the distributive law}$$

Finally, we solve for y to write the equation in slope–intercept form:

Find the slope–intercept form.

$$y = \tfrac{3}{4}x - \tfrac{9}{4} - 2 \quad \textbf{Subtracting 2 from both sides}$$

$$y = \tfrac{3}{4}x - \tfrac{17}{4}. \quad -\tfrac{9}{4} - \tfrac{8}{4} = -\tfrac{17}{4}$$

You can check that substituting $(-1, -5)$ instead of $(3, -2)$ in the point–slope equation will yield the same slope–intercept equation.

Try Exercise 59.

GRAPHING AND POINT–SLOPE FORM

When we know a line's slope and a point that is on the line, we can draw the graph, much as we did in Section 3.6. For example, the information given in the statement of Example 2 is sufficient for drawing a graph.

EXAMPLE 5 Graph the line with slope 2 that passes through $(3, 1)$.

SOLUTION We plot $(3, 1)$, move *up* 2 units and *to the right* 1 unit $\left(\text{since } 2 = \frac{2}{1}\right)$, and draw the line.

STUDY TIP 

Understand Your Mistakes

When you receive a graded quiz, test, or assignment back from your instructor, it is important to review and understand what your mistakes were. While the material is still fresh, take advantage of the opportunity to learn from your mistakes.

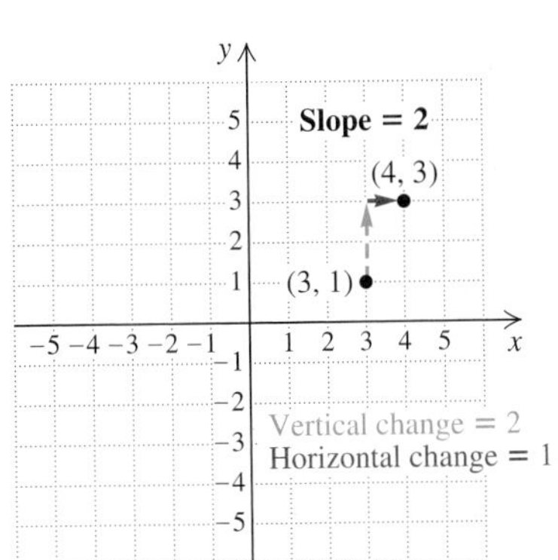

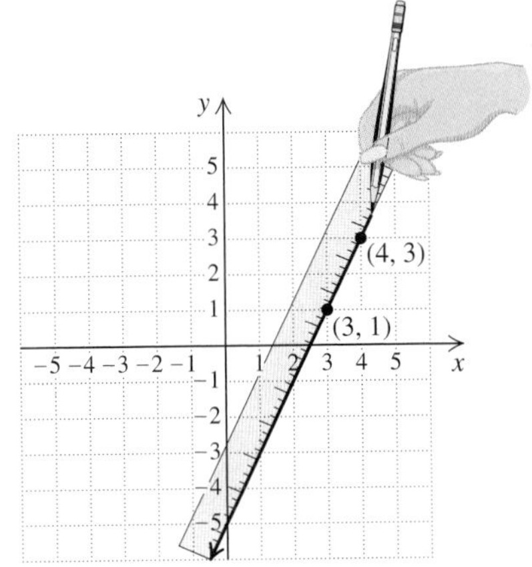

Try Exercise 67.

EXAMPLE 6 Graph: $y - 2 = 3(x - 4)$.

SOLUTION Since $y - 2 = 3(x - 4)$ is in point–slope form, we know that the line has slope 3, or $\frac{3}{1}$, and passes through the point $(4, 2)$. We plot $(4, 2)$ and then find a second point by moving *up* 3 units and *to the right* 1 unit. The line can then be drawn, as shown below.

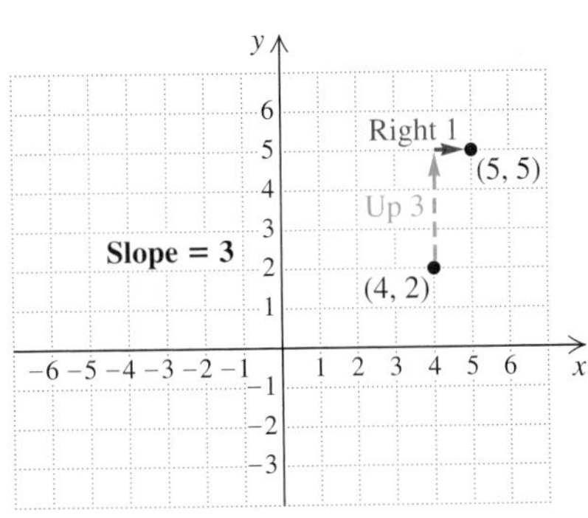

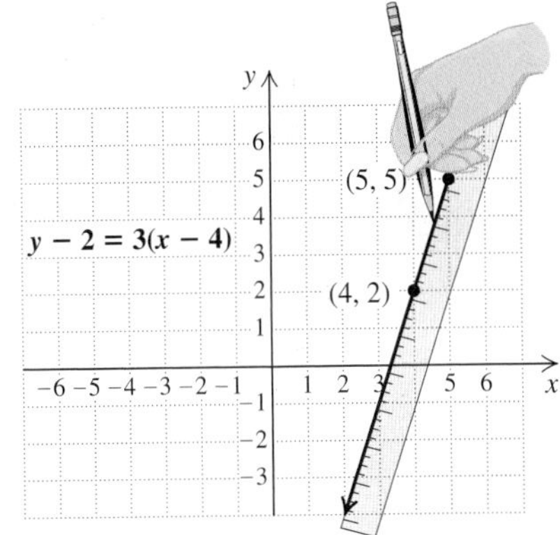

Try Exercise 71.

EXAMPLE 7 Graph: $y + 4 = -\frac{5}{2}(x + 3)$.

SOLUTION Once we have written the equation in point–slope form, $y - y_1 = m(x - x_1)$, we can proceed much as we did in Example 6. To find an equivalent equation in point–slope form, we subtract opposites instead of adding:

$$y + 4 = -\tfrac{5}{2}(x + 3)$$

$$y - (-4) = -\tfrac{5}{2}(x - (-3)).$$

Subtracting a negative instead of adding a positive. This is now in point–slope form.

From this last equation, $y - (-4) = -\frac{5}{2}(x - (-3))$, we see that the line passes through $(-3, -4)$ and has slope $-\frac{5}{2}$, or $\frac{5}{-2}$.

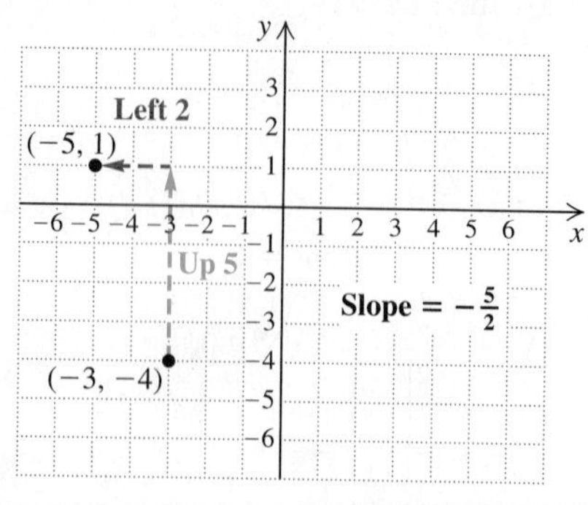

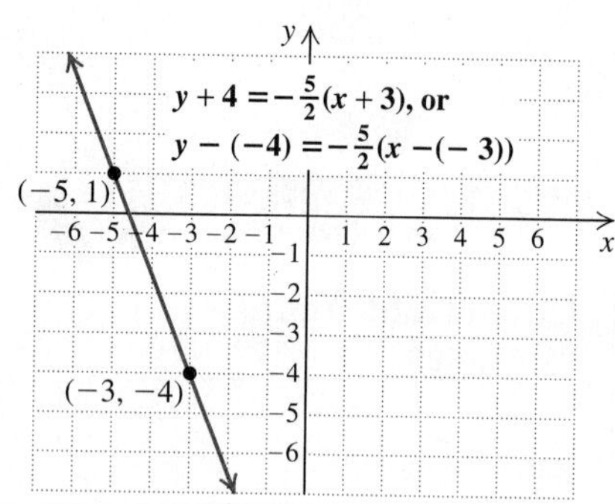

Try Exercise 79.

ESTIMATIONS AND PREDICTIONS USING TWO POINTS

We can estimate real-life quantities already known. To do so, we calculate the coordinates of an unknown point by using two points with known coordinates. When the unknown point is located *between* the two points, this process is called **interpolation.*** Sometimes a graph passing through the known points is *extended* to predict future values. Making predictions in this manner is called **extrapolation.***

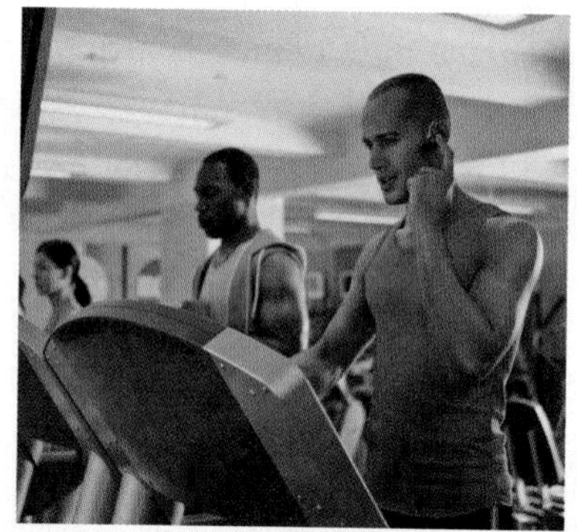

EXAMPLE 8 Aerobic Exercise. A person's target heart rate is the number of beats per minute that brings the most aerobic benefit to his or her heart. The target heart rate for a 20-year-old is 150 beats per minute and for a 60-year-old, 120 beats per minute.

a) Graph the given data and calculate the target heart rate for a 36-year-old.

b) Calculate the target heart rate for a 75-year-old.

SOLUTION

a) We first draw a horizontal axis for "Age" and a vertical axis for "Target heart rate." Next, we number the axes, using a scale that will permit us to view both the given data and the desired data. The given information allows us to plot $(20, 150)$ and $(60, 120)$ and draw a line passing through both points.

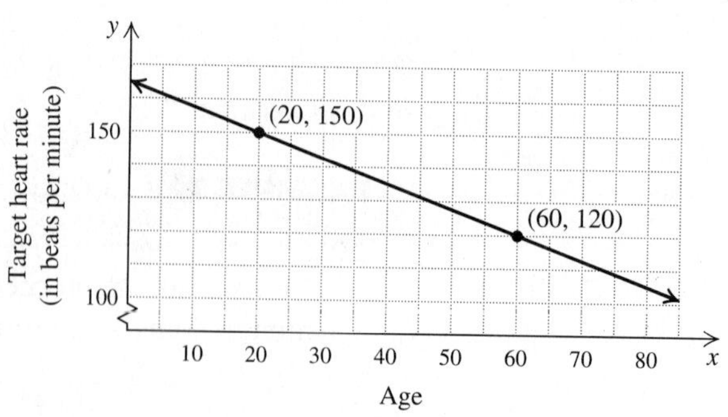

*Both interpolation and extrapolation can be performed using more than two known points and using curves other than lines.

To find an equation for the line, we first find the slope:

$$m = \frac{\text{change in } y}{\text{change in } x} = \frac{150 - 120 \text{ beats per minute}}{20 - 60 \text{ years}}$$

$$= \frac{30 \text{ beats per minute}}{-40 \text{ years}} = -\frac{3}{4} \text{ beat per minute per year.}$$

The target heart rate drops at a rate of $\frac{3}{4}$ beat per minute for each year that we age. We can use either of the given points to write a point–slope equation for the line:

$y - 150 = -\frac{3}{4}(x - 20)$ **Using the point (20, 150)**

$y - 150 = -\frac{3}{4}x + 15$ **Using the distributive law**

$y = -\frac{3}{4}x + 165.$ **Adding 150 to both sides. This is slope–intercept form.**

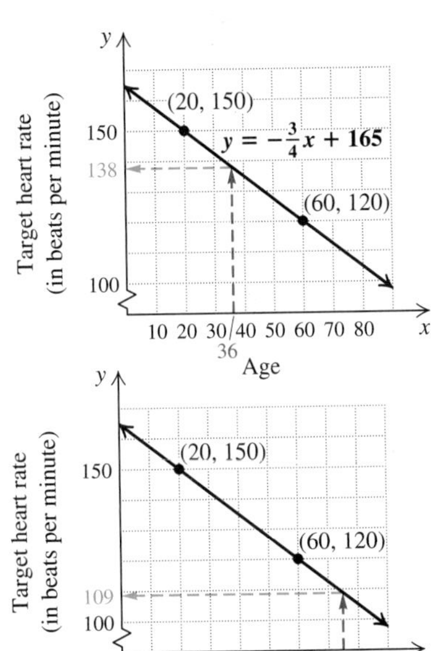

To calculate the target heart rate for a 36-year-old, we substitute 36 for x in the slope–intercept equation:

$$y = -\frac{3}{4} \cdot 36 + 165 = -27 + 165 = 138.$$

As the graph at left confirms, the target heart rate for a 36-year-old is about 138 beats per minute. Because 36 is *between* the given values of 20 and 60, we are *interpolating* here.

b) To calculate the target heart rate for a 75-year-old, we again substitute for x in the slope–intercept equation:

$$y = -\frac{3}{4} \cdot 75 + 165 = -56.25 + 165 = 108.75 \approx 109.$$

As the graph at left confirms, the target heart rate for a 75-year-old is about 109 beats per minute. Because 75 is *beyond* the given values, we are *extrapolating* here.

Try Exercise 81.

CURVE FITTING

The process of understanding and interpreting *data*, or lists of information, is called *data analysis*. One helpful tool in data analysis is **curve fitting,** or finding an algebraic equation that describes the data. We fit a linear equation to two data points in Example 8.

One of the first steps in curve fitting is to decide what kind of equation will best describe the data. In Example 8, we assumed a linear relationship. Generally, we need more than two points in order to determine the pattern of the data. We gather all the data known and plot the points. If the data appear linear, we can fit a linear equation to the data.

EXAMPLE 9 Following are three graphs of sets of data. Determine whether each appears to be linear.

a) Golf Ball Launch

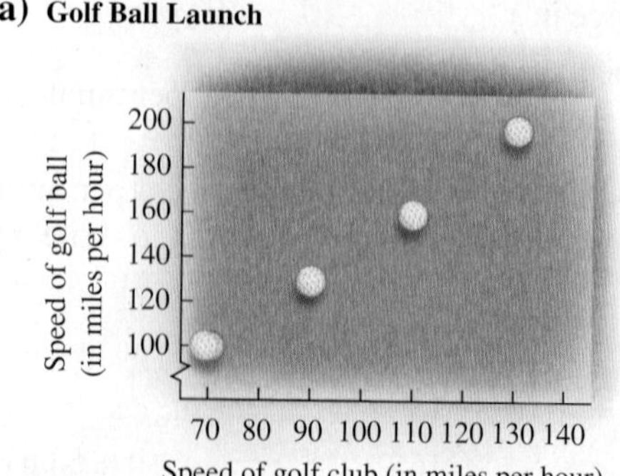

Source: Wishon

b) Average Mobile-Phone Prices

$105
100
95
90
85
Average price
2003 2004 2005 2006 2007 2008
Year

Source: U.S. Wireless Mobile Evaluation Studies

c) Concert Ticket Prices

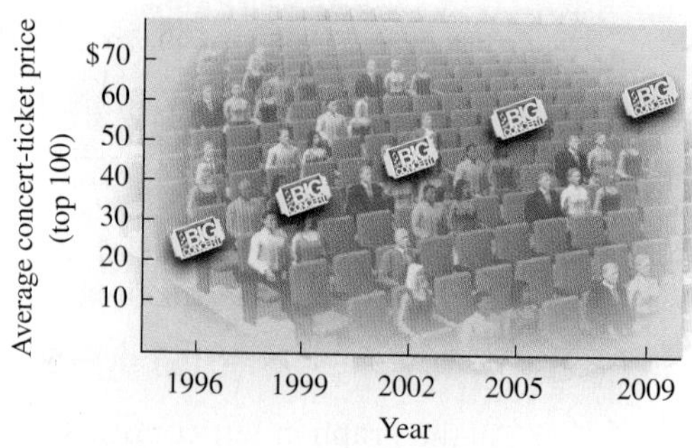

Source: Pollstar

SOLUTION In order for data to be linear, the points must lie, at least approximately, on a straight line. The rate of change is constant for a linear equation, so the change in the quantity on the vertical axis should be about the same for each unit on the horizontal axis.

a) The points lie on a straight line, so the data are linear. The rate of change is constant.

b) Note that the average price of a mobile phone increased, then decreased, and then began to increase. The points do not lie on a straight line. The data are not linear.

c) The points lie approximately on a straight line. The data appear to be linear.

Try Exercise 91.

If data are linear, we can fit a linear equation to the data by choosing two points. However, if the data do not lie exactly on a straight line, different choices of points will yield different equations. The line that *best* describes the data may not actually go through any of the given points. Equation-fitting methods generally consider all the points, not just two, when fitting an equation to data. The most commonly used method is *linear regression*.

The development of the method of linear regression belongs to a later mathematics course, but most graphing calculators offer regression as a way of fitting a line or a curve to a set of data. To use a REGRESSION feature, we must enter the data into the calculator.

Entering and Plotting Data

Data are entered in a graphing calculator in *lists* using the STAT menu.

To enter data, press STAT and choose the EDIT option. Lists appear as three columns on the screen. A number is entered by moving the cursor to the correct position, typing the number, and pressing ENTER.

To clear a list, move the cursor to the title of the list (L1, L2, and so on) and press CLEAR ENTER.

To enter a set of ordered pairs, enter the first coordinates as one list and the second coordinates as another list. The coordinates of each point should be at the same position on both lists. A DIM MISMATCH error will occur if there is not the same number of items in each list.

To define the Plot, press STAT PLOT, the 2nd option associated with the Y= key, and choose Plot 1. (See the screen on the left below.) Then turn Plot 1 on by positioning the cursor over On and pressing ENTER. Define the remaining items on the screen, using the down arrow key to move to the next item. (See the screen on the right below.)

There are six available types of graphs. To plot points, choose the first type of graph shown, a scatter diagram or scatterplot. For the second option, a line graph, the points are connected. The third type is a bar graph. The remaining types are not discussed in this text.

Now make sure that Xlist is set to the list in which the first coordinates were entered, probably L1, and Ylist to the list in which the second coordinates were entered, probably L2. List names can be selected by pressing LIST, the 2nd option associated with the STAT key.

Any of the three marks can be used to plot the points.

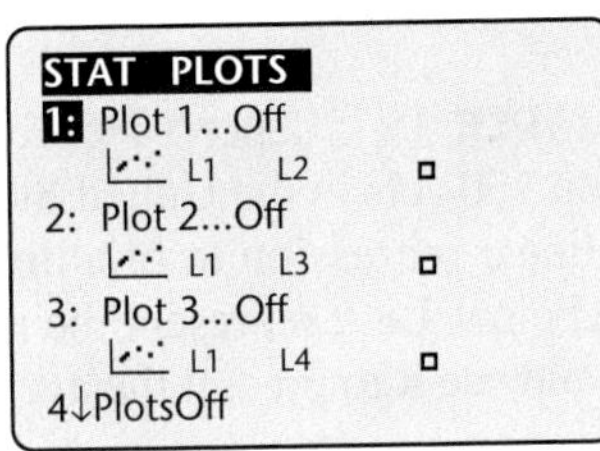

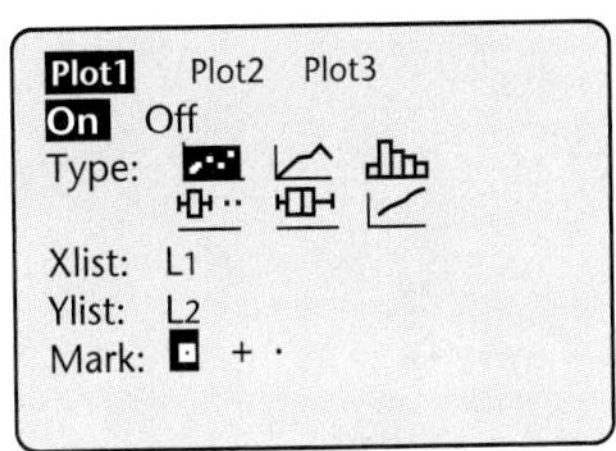

To plot the points, choose window dimensions that will allow all the points to be seen and press GRAPH. This can be done automatically by the ZoomStat option of the ZOOM menu.

When you are done, turn off the plot. A quick way to do this is to move the cursor to the highlighted PLOT name at the top of the equation-editor screen and press ENTER.

Your Turn

1. Press STAT 1. Clear the lists if there are any entries in them.
2. Enter the ordered pairs (1, 4) and (2, 6). Enter 1 and 2 in L1 and 4 and 6 in L2.
3. Clear any equations present in the equation-editor screen.
4. Press STAT PLOT 1. Make sure On is highlighted, as well as the first type of graph. Designate Xlist as L1 and Ylist as L2, and highlight any one of the marks.
5. Press ZOOM 9 to set window dimensions and plot the ordered pairs.
6. To turn off the plot, press Y=, move the cursor up to Plot1, and press ENTER.

Linear Regression

Fitting a curve to a set of data is done using the STAT menu.

Enter the data as described above.

Choose the equation by pressing STAT and selecting the CALC menu. For linear regression, choose the LinReg option. After copying LinReg(ax + b) to the home screen, enter the list names, found as 2nd options associated with the number keys 1 through 6. Enter the list containing the independent values first, and separate the list names by commas. If you wish, enter a Y-variable name, chosen in the VARS Y-VARS FUNCTION submenu, after the linear regression command. This will copy the equation found to the equation-editor screen.

Lin Reg (ax+b) L1, L2, Y1

The command below indicates that L1 contains the values for the independent variable, L2 contains the values for the dependent variable, and the equation is to be copied to Y1. If no list names are entered, L1 and L2 will be used.

The *coefficient of correlation, r,* gives an indication of how well the regression line fits the data. When r^2 is close to 1, the line is a good fit. Turn DiagnosticOn using the CATALOG to show the values of r^2 and r along with the values for the regression equation.

Your Turn

1. Enter the ordered pairs (1, 4) and (2, 6) if you have not already done so, and turn Plot1 on.
2. Press STAT ▷ 4 VARS ▷ 1 ENTER ENTER. You should see values for a and b. Both values are 2.
3. Press Y= to see the regression equation Y1 = 2X + 2. Press ZOOM 9 to graph the equation and the data.

EXAMPLE 10 Concert Ticket Prices. The average price P of a concert ticket for the top 100 acts in the United States has increased almost 150% from 1996 to 2009. Use linear regression to fit a linear equation to the average ticket price data in the table below. Let t represent the number of years since 1996. Graph the line with the data and use it to predict the average price of a concert ticket in 2012.

Year	Number of Years Since 1996, t	Average Price of a Concert Ticket, P
1996	0	$25.81
1997	1	29.81
1998	2	32.20
1999	3	36.84
2000	4	40.74
2001	5	43.68
2002	6	46.56
2003	7	50.35
2004	8	52.39
2005	9	56.88
2006	10	61.58
2007	11	62.07
2008	12	67.33
2009	13	62.57

Source: Pollstar

SOLUTION We are looking for an equation of the form $P = mt + b$. We enter the data, with the number of years since 1996 in L1 and the average ticket price in L2.

L1	L2	L3 1
0	25.81	----------
1	29.81	
2	32.2	
3	36.84	
4	40.74	
5	43.68	
6	46.56	

L1(1) = 0

Student Notes

The procedures used to calculate a regression equation involve squaring. Thus large numbers in the data lists can cause overflow errors. This is one reason we use "years since 1996" instead of the actual year.

Next, we make sure that Plot1 is turned on and clear any equations listed in the Y= screen. Since years vary from 0 to 13 and price varies from \$25.81 to \$67.33, we set a viewing window of [0, 20, 0, 100], with Xscl = 1 and Yscl = 10.

To calculate the equation, we choose the LinReg option in the STAT CALC menu and use the default list names L1 and L2. We select Y1 from the VARS Y-VARS FUNCTION menu and press ENTER.

The screen on the left below indicates that the equation is

$$y = 3.171934066x + 27.15457143.$$

The screen in the middle shows the equation copied as Y1. Pressing GRAPH gives the screen on the right below.

Lin Reg
y = ax+b
a = 3.171934066
b = 27.15457143

Plot1 Plot2 Plot3
\Y1 = 3.1719340659341X+27.154571428571
\Y2=
\Y3=
\Y4=
\Y5=

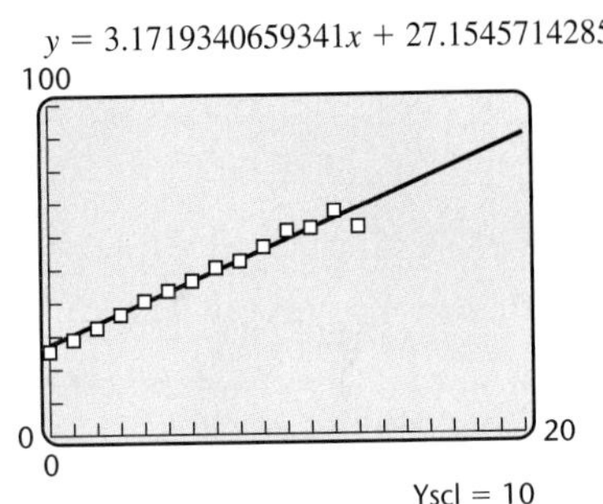

Since 2012 is 16 years after 1996, we substitute 16 for x in order to predict the average price of a concert ticket in 2012. By using either the VALUE option of the CALC menu or a table, we get a value of approximately \$77.91.

To state our answer, we will round the regression coefficients and use the variables defined in the problem.

The regression equation is $P = 3.172t + 27.155$. The average ticket price in 2012 will be about \$77.91.

Try Exercise 99.

Connecting the Concepts

Equations of Lines

Any line can be described by a number of equivalent equations. We write the equation in the form that is most useful for us. For example, all four of the equations below describe the given line.

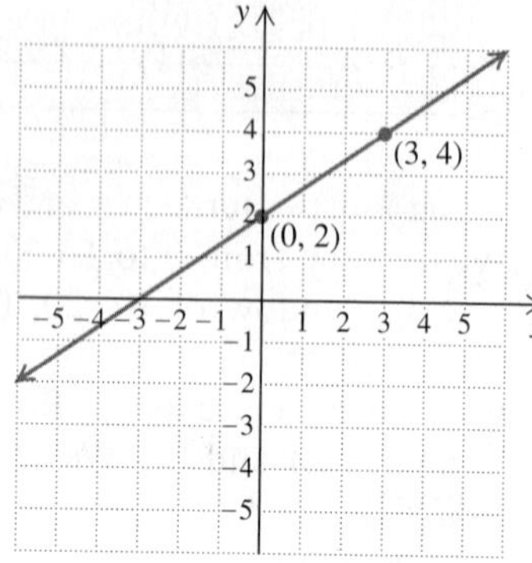

$2x - 3y = -6;$
$y = \frac{2}{3}x + 2;$
$y - 4 = \frac{2}{3}(x - 3);$
$2x + 6 = 3y$

Form of a Linear Equation	Example	Uses
Standard form: $Ax + By = C$	$2x - 3y = -6$	Finding x- and y-intercepts; Graphing using intercepts
Slope–intercept form: $y = mx + b$	$y = \frac{2}{3}x + 2$	Finding slope and y-intercept; Graphing using slope and y-intercept; Writing an equation given slope and y-intercept
Point–slope form: $y - y_1 = m(x - x_1)$	$y - 4 = \frac{2}{3}(x - 3)$	Finding slope and a point on the line; Graphing using slope and a point on the line; Writing an equation given slope and a point on the line or two points on a line

Visualizing for Success

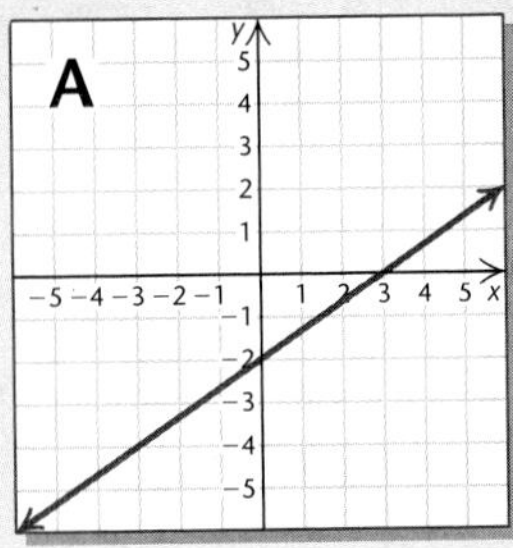

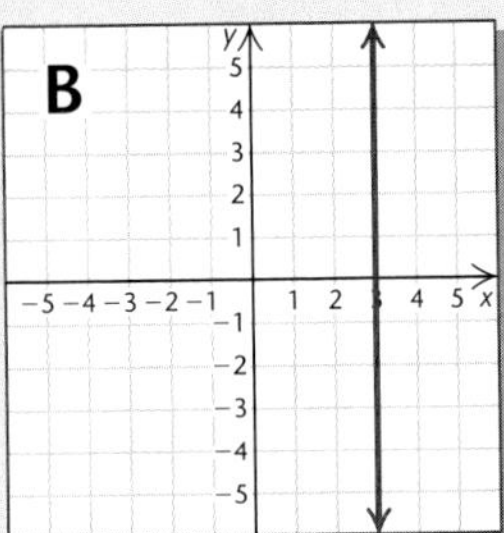

C

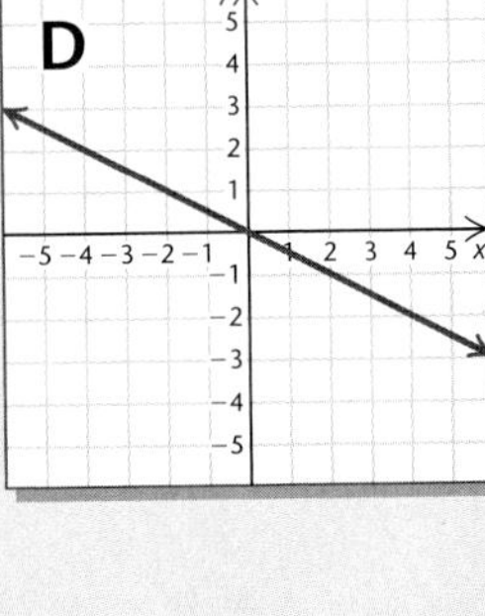

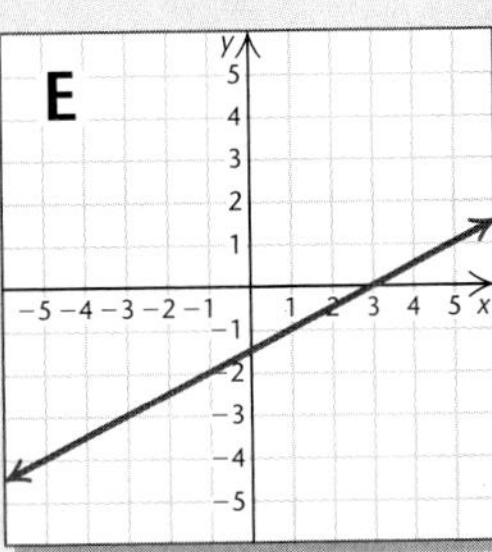

Match each equation or function with its graph.

1. $y = x + 4$

2. $y = 2x$

3. $y = 3$

4. $x = 3$

5. $y = -\frac{1}{2}x$

6. $2x - 3y = 6$

7. $y = -3x - 2$

8. $3x + 2y = 6$

9. $y - 3 = 2(x - 1)$

10. $y + 2 = \frac{1}{2}(x + 1)$

Answers on page A-11

An additional, animated version of this activity appears in MyMathLab. To use MyMathLab, you need a course ID and a student access code. Contact your instructor for more information.

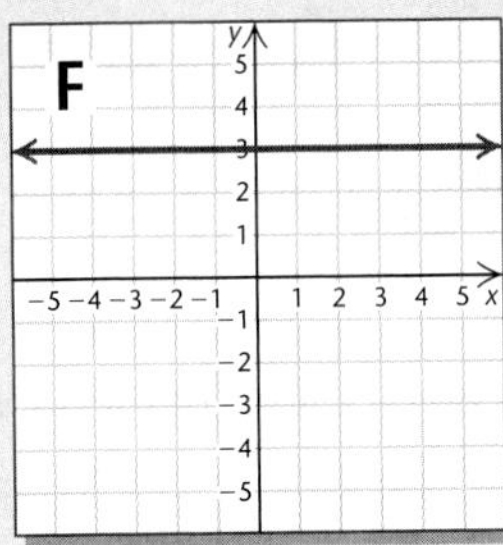

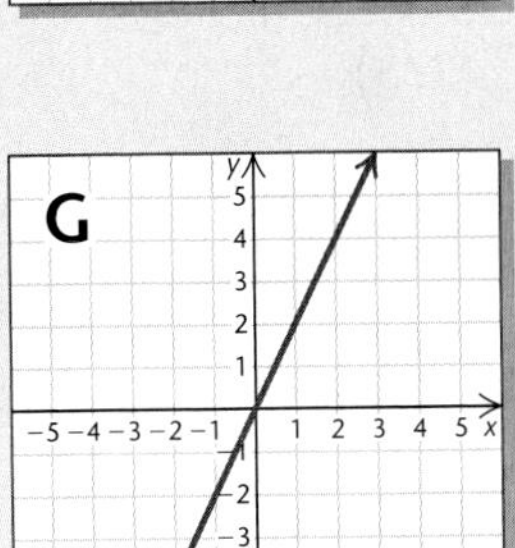

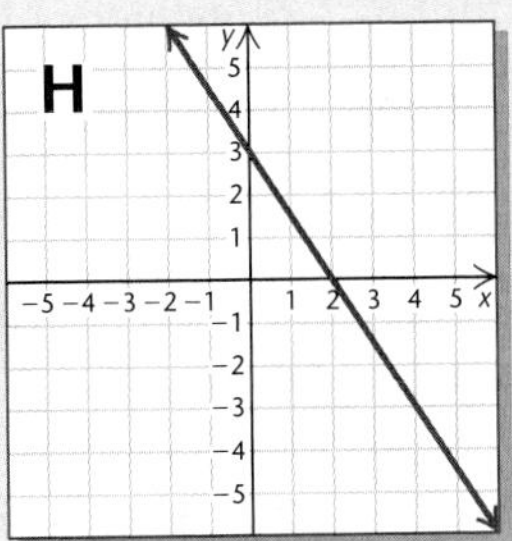

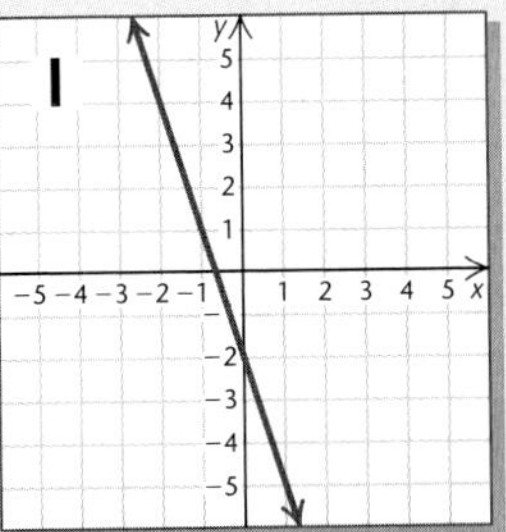

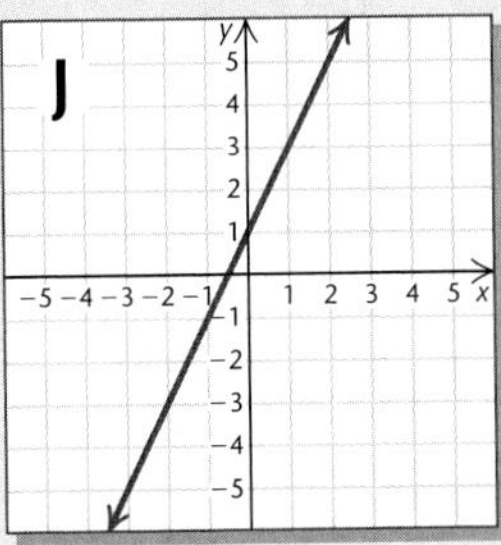

3.7 Exercise Set

FOR EXTRA HELP MyMathLab Math XL PRACTICE  WATCH  DOWNLOAD READ 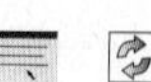 REVIEW

Concept Reinforcement In each of Exercises 1–8, match the given information about a line with the appropriate equation from the column on the right.

1. ___ Slope 5; includes $(2, 3)$
2. ___ Slope -5; includes $(2, 3)$
3. ___ Slope -5; includes $(-2, -3)$
4. ___ Slope 5; includes $(-2, -3)$
5. ___ Slope 5; includes $(3, 2)$
6. ___ Slope -5; includes $(3, 2)$
7. ___ Slope -5; includes $(-3, -2)$
8. ___ Slope 5; includes $(-3, -2)$

a) $y + 3 = 5(x + 2)$
b) $y - 2 = 5(x - 3)$
c) $y + 2 = 5(x + 3)$
d) $y - 3 = -5(x - 2)$
e) $y + 3 = -5(x + 2)$
f) $y + 2 = -5(x + 3)$
g) $y - 3 = 5(x - 2)$
h) $y - 2 = -5(x - 3)$

In each of Exercises 9–12, match the graph with the appropriate equation from the column on the right.

9.

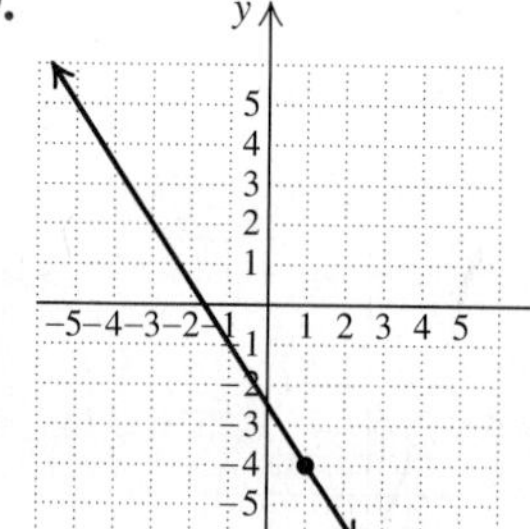

10.

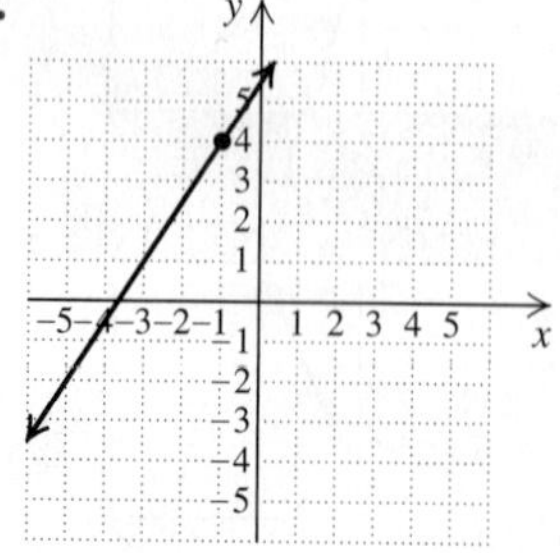

11.

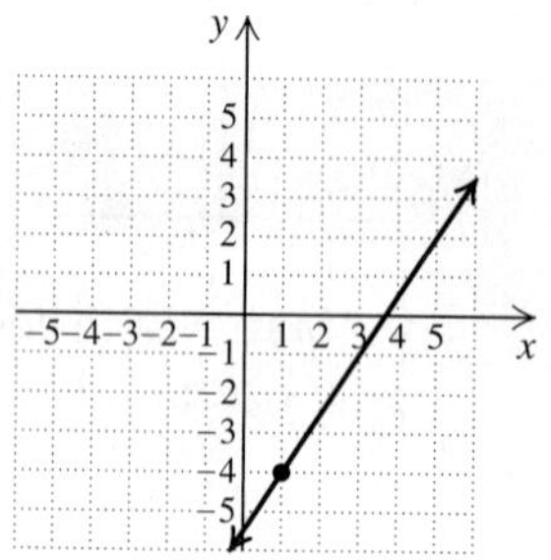

12.

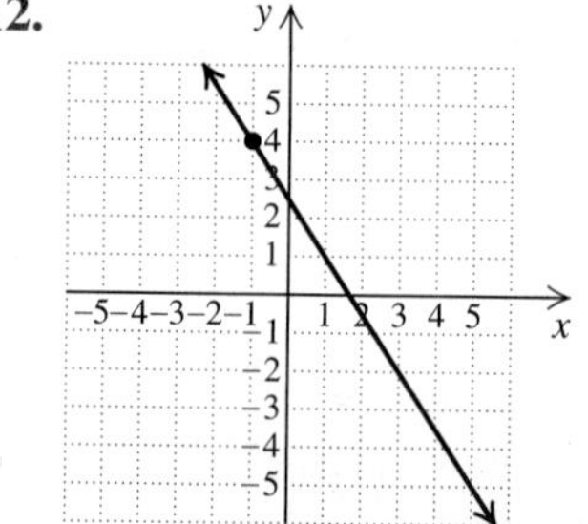

a) $y - 4 = -\frac{3}{2}(x + 1)$
b) $y - 4 = \frac{3}{2}(x + 1)$
c) $y + 4 = -\frac{3}{2}(x - 1)$
d) $y + 4 = \frac{3}{2}(x - 1)$

Write a point–slope equation for the line with the given slope and containing the given point.

13. $m = 5;\ (6, 2)$
14. $m = 4;\ (3, 5)$
15. $m = -4;\ (3, 1)$
16. $m = -5;\ (6, 2)$
17. $m = \frac{3}{2};\ (5, -4)$
18. $m = -\frac{4}{3};\ (7, -1)$
19. $m = -\frac{5}{4};\ (-2, 6)$
20. $m = \frac{7}{2};\ (-3, 4)$
21. $m = -2;\ (-4, -1)$
22. $m = -3;\ (-2, -5)$
23. $m = 1;\ (-2, 8)$
24. $m = -1;\ (-3, 6)$

For each point–slope equation given, state the slope and a point on the graph.

25. $y - 9 = \frac{2}{7}(x - 8)$
26. $y - 3 = 9(x - 2)$
27. $y + 2 = -5(x - 7)$
28. $y - 4 = -\frac{2}{9}(x + 5)$
29. $y - 4 = -\frac{5}{3}(x + 2)$
30. $y + 7 = -4(x - 9)$

Aha! 31. $y = \frac{4}{7}x$
32. $y = 3x$

Write the slope–intercept equation for the line with the given slope and containing the given point.

33. $m = 2;\ (5, 7)$
34. $m = 3;\ (6, 2)$
35. $m = \frac{7}{4};\ (4, -2)$
36. $m = \frac{8}{3};\ (3, -4)$
37. $m = -3;\ (-1, 6)$
38. $m = -2;\ (-1, 3)$
39. $m = -4;\ (-2, -1)$
40. $m = -5;\ (-1, -4)$

Aha! 41. $m = -\frac{5}{6};\ (0, 4)$
42. $m = -\frac{3}{4};\ (0, 5)$

Write an equation of the line that contains the specified point and is parallel to the indicated line.

43. $(4, 7),\ x + 2y = 6$
44. $(1, 3),\ 3x - y = 7$

Aha! 45. $(0, -7),\ y = 2x + 1$
46. $(-7, 0),\ 5x + 2y = 6$
47. $(2, -6),\ 5x - 3y = 8$
48. $(0, 2),\ y = x - 11$
49. $(5, -4),\ x = 2$
50. $(-3, 6),\ y = 7$

Write an equation of the line that contains the specified point and is perpendicular to the indicated line.

51. $(3, -2),\ 3x - 6y = 5$
52. $(-3, -5),\ 4x - 2y = 4$
53. $(-4, 2),\ x + y = 6$
54. $(-4, 5),\ 7x - 2y = 1$

Aha! 55. $(0, 6),\ 2x - 5 = y$
56. $(4, 0),\ x - 3y = 0$
57. $(-3, 7),\ y = 5$
58. $(4, -2),\ x = 1$

Write the slope–intercept equation for the line containing the given pair of points.

59. $(1, 5)$ and $(4, 2)$
60. $(3, 7)$ and $(4, 8)$
61. $(-3, 1)$ and $(3, 5)$
62. $(-2, 3)$ and $(2, 5)$
63. $(5, 0)$ and $(0, -2)$
64. $(-2, 0)$ and $(0, 3)$
65. $(-2, -4)$ and $(2, -1)$
66. $(-3, 5)$ and $(-1, -3)$

67. Graph the line with slope $\frac{4}{3}$ that passes through the point $(1, 2)$.

68. Graph the line with slope $\frac{2}{5}$ that passes through the point $(3, 4)$.

69. Graph the line with slope $-\frac{3}{4}$ that passes through the point $(2, 5)$.

70. Graph the line with slope $-\frac{3}{2}$ that passes through the point $(1, 4)$.

Graph.

71. $y - 2 = \frac{1}{2}(x - 1)$
72. $y - 5 = \frac{1}{3}(x - 2)$
73. $y - 1 = -\frac{1}{2}(x - 3)$
74. $y - 1 = -\frac{1}{4}(x - 3)$
75. $y + 4 = 3(x + 1)$
76. $y + 3 = 2(x + 1)$
77. $y + 3 = -(x + 2)$
78. $y + 3 = -1(x - 4)$
79. $y + 1 = -\frac{3}{5}(x + 2)$
80. $y + 2 = -\frac{2}{3}(x + 1)$

In Exercises 81–90, assume the relationship is linear, as in Example 8.

81. *Community Involvement.* The number of college students attending public meetings grew from 392.5 million students in 2006 to 468.2 million students in 2008.
Source: Volunteering in America, Corporation for National and Community Service

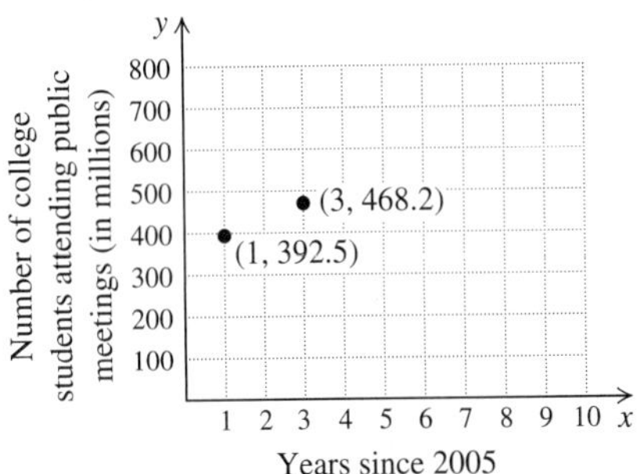

a) Find a linear equation that fits the data.
b) Calculate the number of college students attending public meetings in 2007.
c) Predict the number of students attending public meetings in 2012.

82. *SNAP Participants.* Participation in the U.S. Supplementary Nutrition Assistance Program grew from approximately 23.9 million people in 2004 to approximately 28.4 million in 2008.
Source: U.S. Department of Agriculture

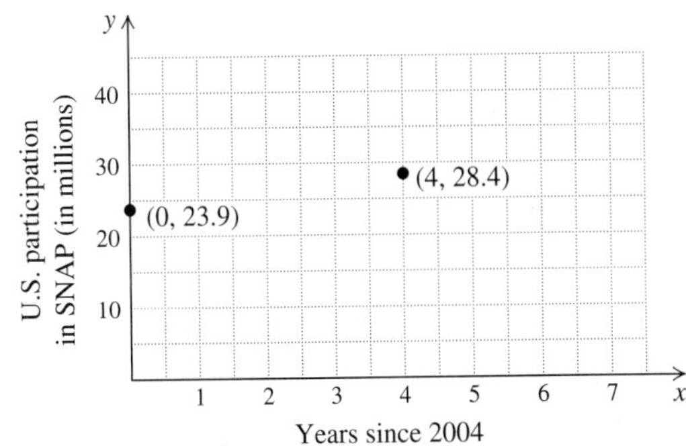

a) Find a linear equation that fits the data.
b) Calculate the number of participants in 2007.
c) Predict the number of participants in 2011.

83. *National Park Land.* The number of acres in the National Park system has grown from 78.2 million

acres in 2000 to 78.8 million acres in 2007. Let A represent the number of acres, in millions, in the system and t the number of years after 2000.
Source: U.S. National Park Service

a) Find a linear equation that fits the data.
b) Estimate the number of acres in the National Park system in 2003.
c) Predict the number of acres in the National Park system in 2010.

84. *Aging Population.* The number of U.S. residents over the age of 65 was approximately 35.6 million in 2002 and 37.9 million in 2007. Let R represent the number of U.S. residents over the age of 65, in millions, and t the number of years after 2000.
Source: U.S. Census Bureau

a) Find a linear equation that fits the data.
b) Calculate the number of U.S. residents over the age of 65 in 2006.
c) Predict the number of U.S. residents over the age of 65 in 2010.

85. *Environmental Awareness.* The percentage of Americans who are familiar with the term "carbon footprint" grew from 38 percent in 2007 to 57 percent in July 2009. Let C represent the percentage of Americans who are familiar with the term "carbon footprint" and t the number of years since 2006.
Source: National Marketing Institute

a) Find a linear equation that fits the data.
b) Predict the percentage of Americans who will be familiar with the term "carbon footprint" in 2012.
c) When will all Americans be familiar with the term "carbon footprint"?

86. *Medical Care.* In 2002, Medicaid long-term care expenses totaled \$92 billion. This figure had risen to \$109 billion by 2006. Let M represent Medicaid long-term care expenses, in billions of dollars, and t the number of years since 2000.
Source: Kaiser Commission on Medicaid and the Uninsured, Analysis of 2008 National Health Interview Survey data

a) Find a linear equation that fits the data.
b) Predict the amount of Medicaid long-term care expenses in 2010.
c) In what year will Medicaid long-term care expenses reach \$150 billion?

87. *Life Expectancy of Females in the United States.* In 1993, the life expectancy at birth of females was 78.8 years. In 2006, it was 80.2 years. Let E represent life expectancy and t the number of years since 1990.
Source: National Vital Statistics Reports

a) Find a linear equation that fits the data.
b) Predict the life expectancy of females in 2010.

88. *Life Expectancy of Males in the United States.* In 1993, the life expectancy at birth of males was 72.2 years. In 2006, it was 75.1 years. Let E represent life expectancy and t the number of years since 1990.
Source: National Vital Statistics Reports

a) Find a linear equation that fits the data.
b) Predict the life expectancy of males in 2010.

89. *PAC Contributions.* In 1996, Political Action Committees (PACs) contributed \$217.8 million to federal political candidates. In 2006, the figure rose to \$372.1 million. Let N represent the amount of political contributions, in millions of dollars, and t the number of years since 1996.
Source: Federal Election Commission

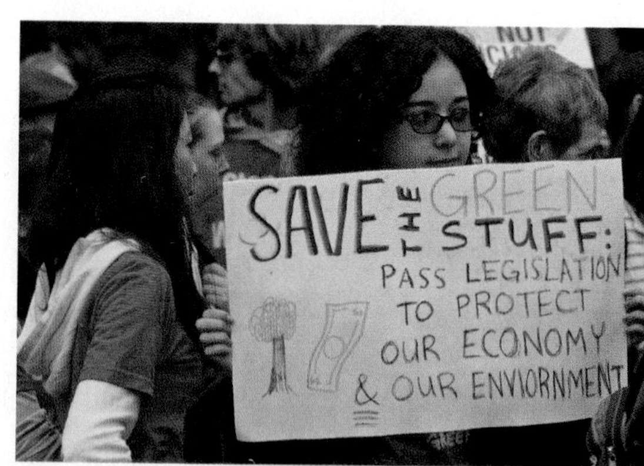

a) Find a linear equation that fits the data.
b) Predict the amount of PAC contributions in 2010.

90. *Recycling.* In 2003, Americans recovered 72.3 million tons of solid waste. In 2007, the figure grew to 85.0 million tons. Let N represent the number of tons recovered, in millions, and t the number of years since 2000.
Source: U.S. Environmental Protection Agency

a) Find a linear equation that fits the data.
b) Predict the amount recycled in 2010.

Determine whether the data in each graph appear to be linear.

91. Registered Nurses

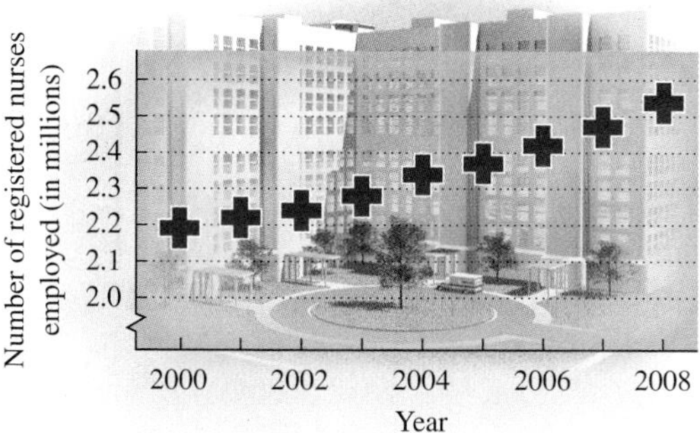

Source: Bureau of Labor Statistics, U.S. Department of Labor

92. Holiday Shopping

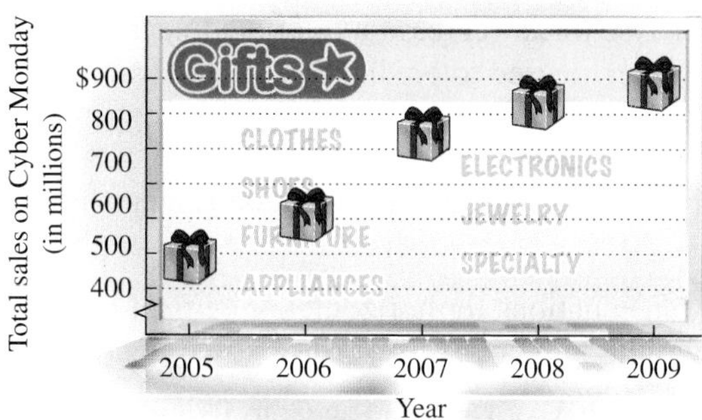

Source: comScore

93. Holiday Shopping

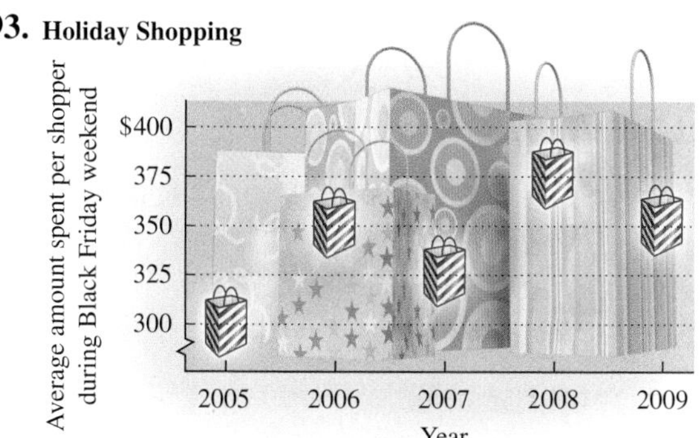

Source: Based on data from National Retail Federation

94. U.S. Farming

Source: U.S. Department of Agriculture

95. Near-Term Births

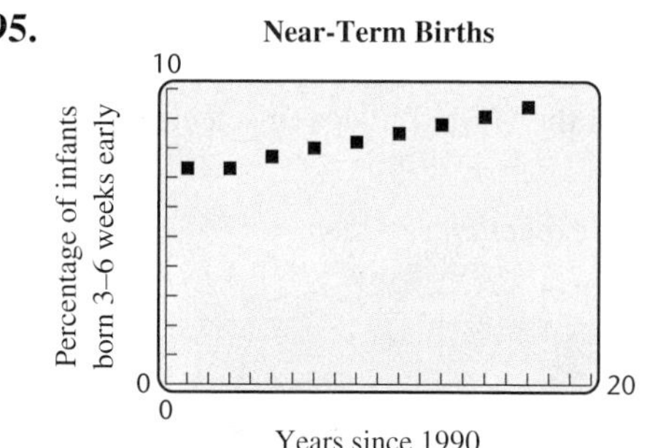

Source: U.S. Centers for Disease Control

96. Caesarean Births

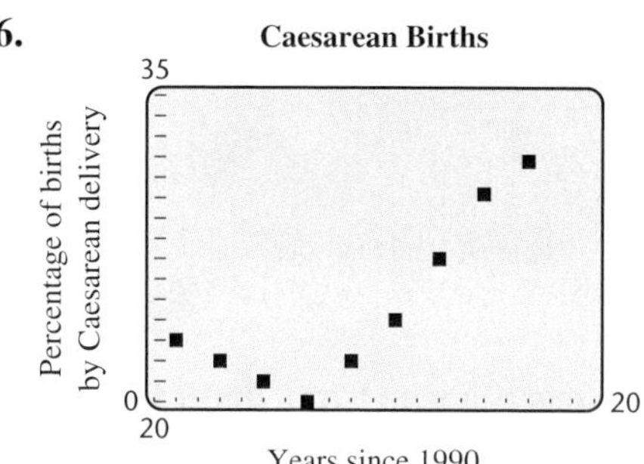

Source: U.S. Centers for Disease Control

97. *Life Expectancy of Females in the United States.* The table below lists the life expectancy at birth of women who were born in the United States in selected years.

Life expectancy of women

Year	Life Expectancy (in years)
1960	73.1
1970	74.4
1980	77.5
1990	78.8
2000	79.7
2006	80.2

Source: National Vital Statistics Reports

a) Use linear regression to find a linear function that can be used to predict the life expectancy W of a woman. Let x = the number of years since 1900.

b) Predict the life expectancy of a woman in 2010 and compare your answer with the answer to Exercise 87.

98. *Life Expectancy of Males in the United States.* The table below lists the life expectancy of males born in the United States in selected years.

Life expectancy of men

Year	Life Expectancy (in years)
1960	66.6
1970	67.1
1980	70.0
1990	71.8
2000	74.4
2006	75.1

Source: National Vital Statistics Reports

a) Use linear regression to find a linear function that can be used to predict the life expectancy M of a man. Let $x =$ the number of years since 1900.

b) Predict the life expectancy of a man in 2010 and compare your answer with the answer to Exercise 88.

99. *Nursing.* The table below lists the number of registered nurses employed in the United States for various years.

Year	Number of Registered Nurses Employed (in millions)
2000	2.19
2001	2.22
2002	2.24
2003	2.28
2004	2.34
2005	2.37
2006	2.42
2007	2.47
2008	2.54

Source: Bureau of Labor Statistics, U.S. Department of Labor

a) Use linear regression to find a linear equation that can be used to predict the number N of registered nurses, in millions, employed t years after 2000.

b) Estimate the number of registered nurses employed in 2012.

100. *Holiday Shopping.* The amount of money spent by online shoppers on "Cyber Monday," the Monday following Thanksgiving Day, is shown in the table below.

Year	Amount Spent on Cyber Monday (in millions)
2005	$486
2006	610
2007	730
2008	834
2009	887

Source: comScore

a) Use linear regression to find a linear equation that can be used to predict the amount spent A, in millions of dollars, on Cyber Monday x years after 2005.

b) Estimate the amount spent on Cyber Monday in 2011.

TW **101.** Can equations for horizontal or vertical lines be written in point–slope form? Why or why not?

TW **102.** On the basis of your answers to Exercises 87 and 88, would you predict that at some point in the future the life expectancy of males will exceed that of females? Why or why not?

SKILL REVIEW

To prepare for Section 3.8, review evaluating algebraic expressions (Section 1.8).

Evaluate. [1.8]

103. $3 - 4x$, for $x = 5$

104. $2 - x$, for $x = -3$

105. $n^2 - 5n$, for $n = -1$

106. $3n^2 + n$, for $n = 0$

107. $\dfrac{x - 6}{2x + 8}$, for $x = 4$

108. $\dfrac{x - 5}{x + 7}$, for $x = 5$

SYNTHESIS

TW **109.** Why is slope–intercept form more useful than point–slope form when using a graphing calculator? How can point–slope form be modified so that it is more easily used with graphing calculators?

TW **110.** Any nonvertical line has many equations in point–slope form, but only one in slope–intercept form. Why is this?

Graph.

Aha! **111.** $y - 3 = 0(x - 52)$ **112.** $y + 4 = 0(x + 93)$

Write the slope–intercept equation for each line shown.

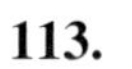

113.

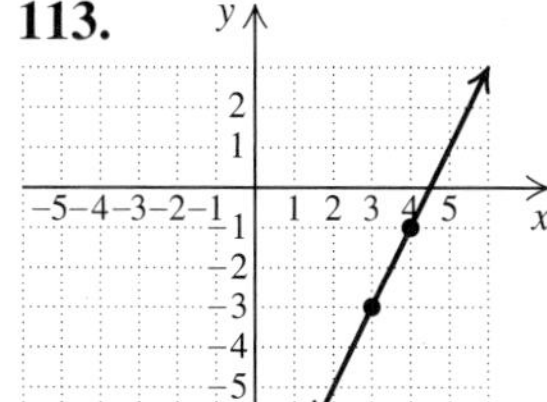

114.

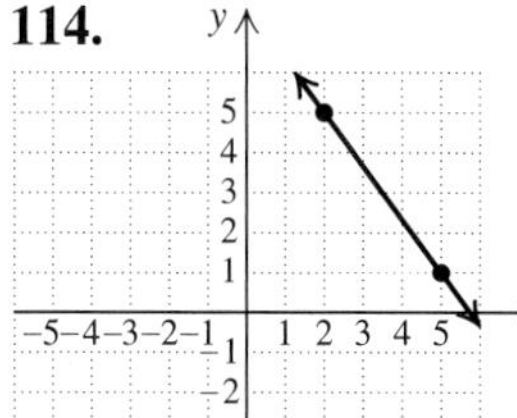

115. Write the slope–intercept equation of the line that has the same y-intercept as the line $x - 3y = 6$ and contains the point $(5, -1)$.

116. Write the slope–intercept equation of the line that contains the point $(-1, 5)$ and is parallel to the line passing through $(2, 7)$ and $(-1, -3)$.

117. Write the slope–intercept equation of the line that has x-intercept $(-2, 0)$ and is parallel to $4x - 8y = 12$.

118. *Depreciation of a Computer.* After 6 months of use, the value of Dani's computer had dropped to \$900. After 8 months, the value had gone down to \$750. How much did the computer cost originally?

119. *Temperature Conversion.* Water freezes at 32° Fahrenheit and at 0° Celsius. Water boils at 212°F and at 100°C. What Celsius temperature corresponds to a room temperature of 70°F?

120. *Cell-Phone Charges.* The total cost of Mel's cell phone was \$230 after 5 months of service and \$390 after 9 months. What costs had Mel already incurred when his service just began?

121. *Operating Expenses.* The total cost for operating Ming's Wings was \$7500 after 4 months and \$9250 after 7 months. Predict the total cost after 10 months.

Try Exercise Answers: Section 3.7

19. $y - 6 = -\frac{5}{4}(x - (-2))$ **33.** $y = 2x - 3$
43. $y = -\frac{1}{2}x + 9$ **51.** $y = -2x + 4$ **59.** $y = -x + 6$
67. **71.**

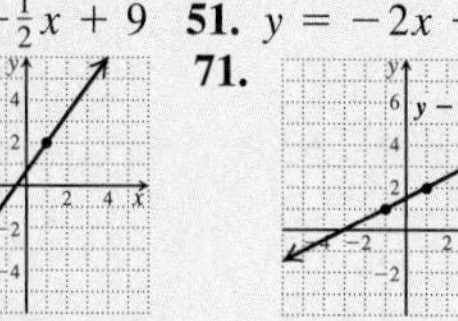

79. $y + 1 = -\frac{3}{5}(x + 2)$

81. (a) $y = 37.85x + 354.65$; **(b)** 430.35 million students; **(c)** 619.6 million students
91. Linear
99. (a) $N = 0.0433t + 2.1678$; **(b)** 2.69 million registered nurses

3.8 Functions

- Functions and Graphs
- Function Notation and Equations
- Functions Defined Piecewise
- Linear Functions and Applications

We now develop the idea of a *function*—one of the most important concepts in mathematics. A function is a special kind of correspondence between two sets. For example:

To each person in a class	there corresponds	a date of birth.
To each bar code in a store	there corresponds	a price.
To each real number	there corresponds	the cube of that number.

In each example, the first set is called the **domain.** The second set is called the **range.** For any member of the domain, there is *exactly one* member of the range to which it corresponds. This kind of correspondence is called a **function.**

STUDY TIP 

And Now for Something Completely Different

When a new topic, such as functions, is introduced, there are often new terms and notation to become familiar with. Listing definitions, verbalizing the vocabulary, and practicing the use of notation in exercises will all help you become comfortable with new concepts.

Note that although two members of a class may have the same date of birth, and two bar codes may correspond to the same price, each correspondence is still a function since every member of the domain is paired with exactly one member of the range.

EXAMPLE 1 Determine whether each correspondence is a function.

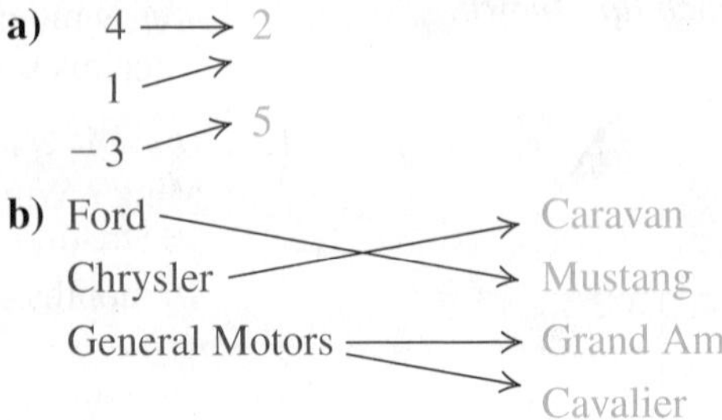

SOLUTION

a) The correspondence *is* a function because each member of the domain corresponds to *exactly one* member of the range.

b) The correspondence *is not* a function because a member of the domain (General Motors) corresponds to more than one member of the range.

Try Exercise 9.

Function

A *function* is a correspondence between a first set, called the *domain*, and a second set, called the *range*, such that each member of the domain corresponds to *exactly one* member of the range.

EXAMPLE 2 Determine whether each correspondence is a function.

a) The correspondence matching a person with his or her weight

b) The correspondence matching the numbers $-2, 0, 1$, and 2 with each number's square

c) The correspondence matching a best-selling author with the titles of books written by that author

SOLUTION

a) For this correspondence, the domain is a set of people and the range is a set of positive numbers (the weights). We ask ourselves, "Does a person have *only one* weight?" Since the answer is Yes, this correspondence *is* a function.

b) The domain is $\{-2, 0, 1, 2\}$ and the range is $\{0, 1, 4\}$. We ask ourselves, "Does each number have *only one* square?" Since the answer is Yes, the correspondence *is* a function.

c) The domain is a set of authors and the range is a set of book titles. We ask ourselves, "Has each author written *only one* book?" Since many authors have multiple titles published, the answer is No, the correspondence *is not* a function.

Try Exercise 17.

Although the correspondence in Example 2(c) is not a function, it is a *relation.*

> **Relation**
>
> A *relation* is a correspondence between a first set, called the *domain*, and a second set, called the *range*, such that each member of the domain corresponds to *at least one* member of the range.

FUNCTIONS AND GRAPHS

The functions in Examples 1(a) and 2(b) can be expressed as sets of ordered pairs. Example 1(a) can be written $\{(-3, 5), (1, 2), (4, 2)\}$ and Example 2(b) can be written $\{(-2, 4), (0, 0), (1, 1), (2, 4)\}$. We can graph these functions as follows.

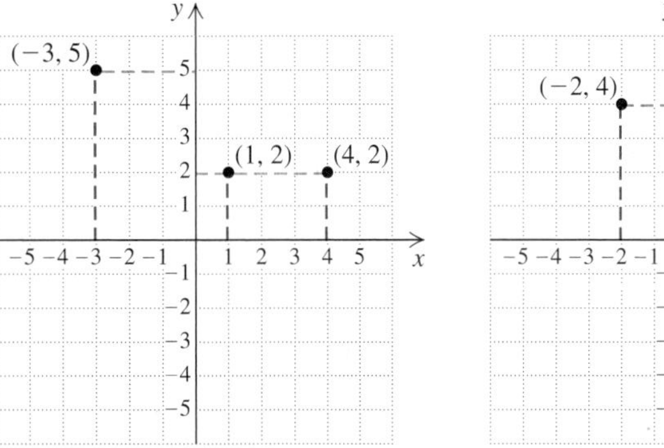

The function $\{(-3, 5), (1, 2), (4, 2)\}$
Domain is $\{-3, 1, 4\}$
Range is $\{5, 2\}$

The function $\{(-2, 4), (0, 0), (1, 1), (2, 4)\}$
Domain is $\{-2, 0, 1, 2\}$
Range is $\{4, 0, 1\}$

When a function is given as a set of ordered pairs, the domain is the set of all first coordinates and the range is the set of all second coordinates.

Functions are generally represented by lower-case or upper-case letters.

EXAMPLE 3 Find the domain and the range of the function f shown here.

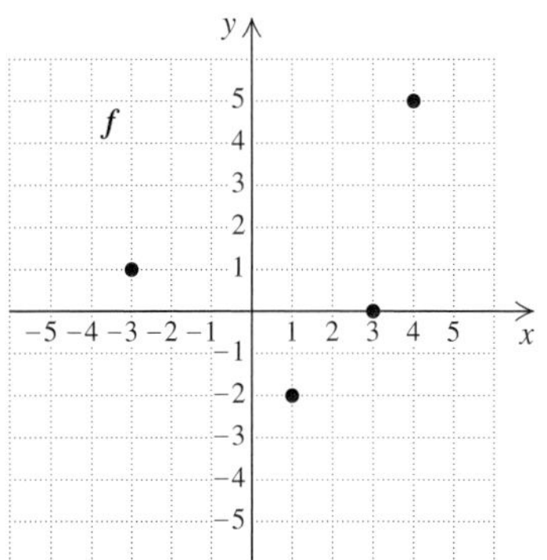

SOLUTION Here f can be written $\{(-3, 1), (1, -2), (3, 0), (4, 5)\}$. The domain is the set of all first coordinates, $\{-3, 1, 3, 4\}$, and the range is the set of all second coordinates, $\{1, -2, 0, 5\}$.

Try Exercise 29.

We can also find the domain and the range directly from the graph, without first listing all pairs.

EXAMPLE 4 For the function f shown here, determine each of the following.

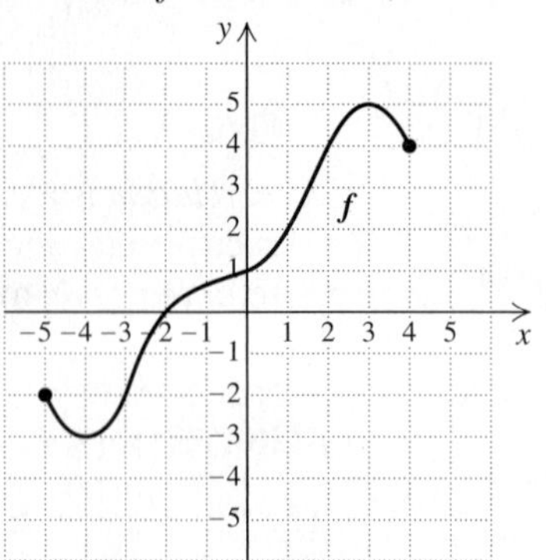

a) What member of the range is paired with 2

b) The domain of f

c) What member of the domain is paired with -3

d) The range of f

SOLUTION

a) To determine what member of the range is paired with 2, we first note that we are considering 2 in the domain. Thus we locate 2 on the horizontal axis. Next, we find the point directly above 2 on the graph of f. From that point, we can look to the vertical axis to find the corresponding y-coordinate, 4. Thus, 4 is the member of the range that is paired with 2.

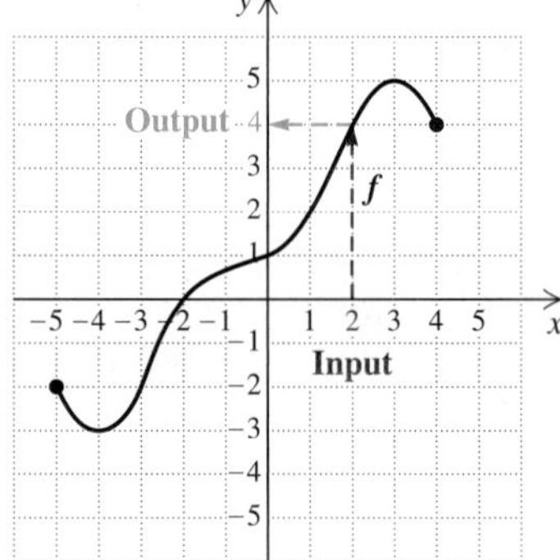

b) The domain of the function is the set of all x-values that are used in the points of the curve. Because there are no breaks in the graph of f, these extend continuously from -5 to 4 and can be viewed as the curve's shadow, or *projection*, on the x-axis. Thus the domain is $\{x|-5 \leq x \leq 4\}$, or, using interval notation, $[-5, 4]$.

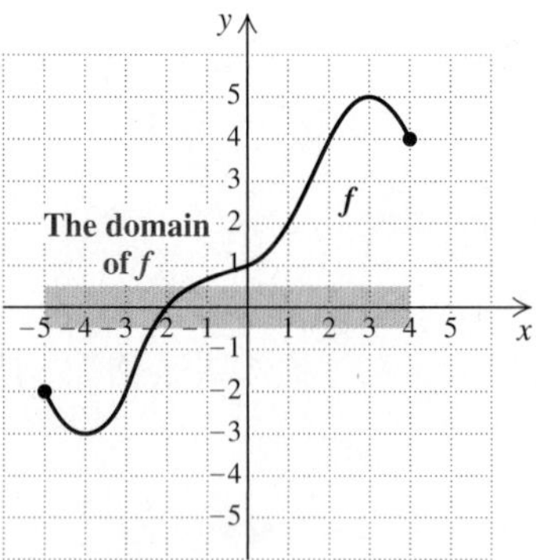

c) To determine what member of the domain is paired with -3, we note that we are considering -3 in the range. Thus we locate -3 on the vertical axis. From there we look left and right to the graph of f to find any points for which -3 is the second coordinate. One such point exists, $(-4, -3)$. We note that -4 is the only element of the domain paired with -3.

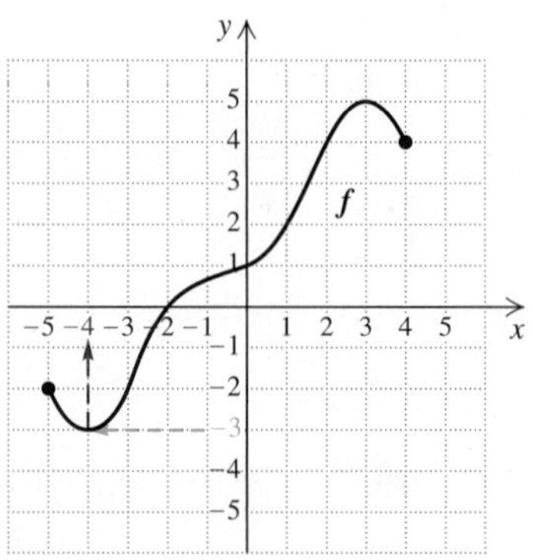

d) The range of the function is the set of all y-values that are in the graph. These extend continuously from -3 to 5, and can be viewed as the curve's projection on the y-axis. Thus the range is $\{y \mid -3 \leq y \leq 5\}$, or, using interval notation, $[-3, 5]$.

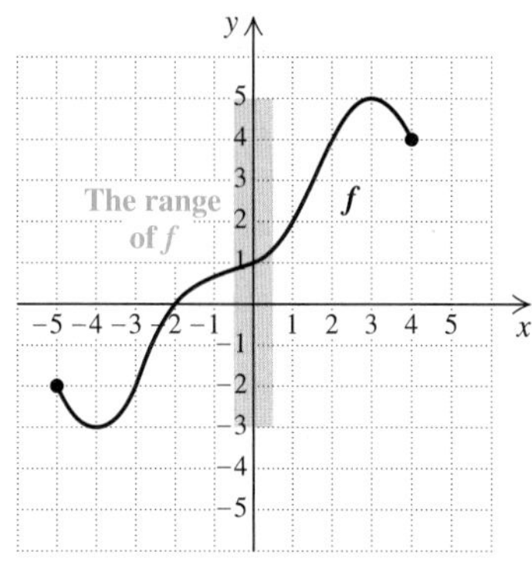

Try Exercise 27.

A closed dot, as in Example 4, emphasizes that a particular point *is* on a graph. An open dot, ∘, indicates that a particular point *is not* on a graph.

The graphs of some functions have no endpoints. For example, consider the graph of the linear function f shown below.

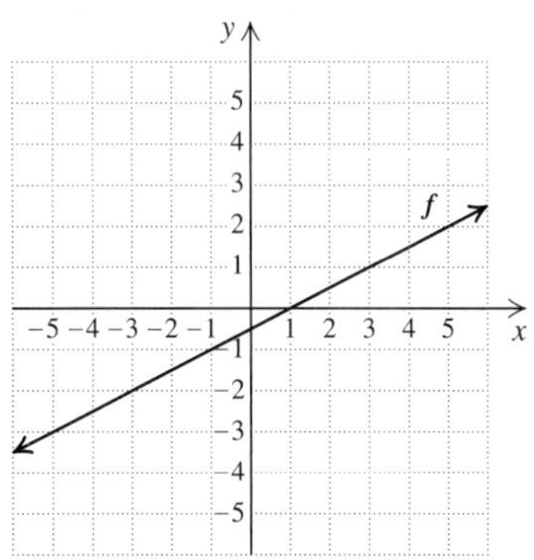

The graph extends indefinitely both left and right and up and down. Any real number can be an input, and any real number can be an output.

Both the domain and the range of this function are the set of all real numbers. This can be written $(-\infty, \infty)$, or $\mathbb{R}$.

EXAMPLE 5 Find the domain and the range of the function f shown here.

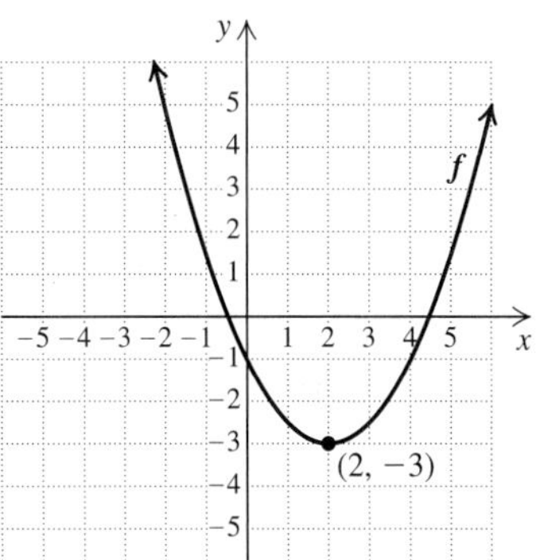

SOLUTION The domain is the set of all x-values that are used in the points on the curve. The arrows indicate that the graph extends without end. Thus the shadow, or projection, of the graph on the x-axis is the entire x-axis. (See the graph on the left below.) The domain is $\{x|x \text{ is a real number}\}$, or $(-\infty, \infty)$, or $\mathbb{R}$.

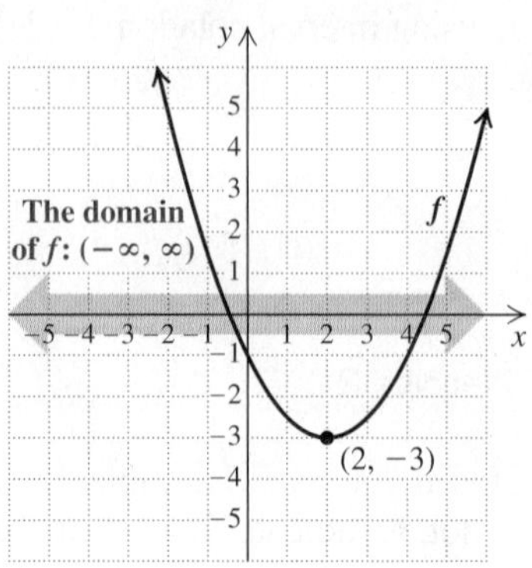

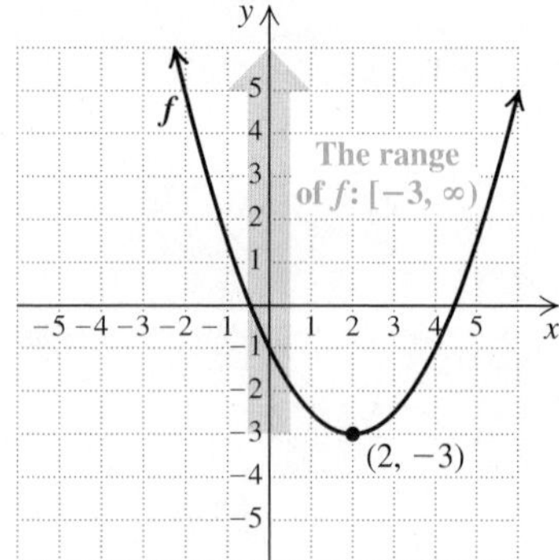

The range of f is the set of all y-values that are used in the points on the curve. The function has no y-values less than -3, and every y-value greater than or equal to -3 corresponds to at least one member of the domain. Thus the projection of the graph on the y-axis is the portion of the y-axis greater than or equal to -3. (See the graph on the right above.) The range is $\{y|y \geq -3\}$, or $[-3, \infty)$.

Try Exercise 35.

EXAMPLE 6 Find the domain and the range of the function f shown here.

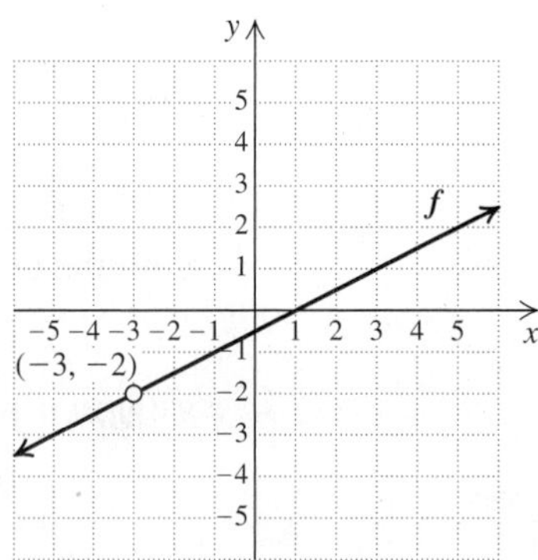

SOLUTION The domain of f is the set of all x-values that are in the graph. The open dot in the graph at $(-3, -2)$ indicates that there is no y-value that corresponds to $x = -3$; that is, the function is not defined for $x = -3$. Thus, -3 is not in the domain of the function, and

$$\text{Domain of } f = \{x|x \text{ is a real number } and\ x \neq -3\}.$$

There is no function value at $(-3, -2)$, so -2 is not in the range of the function. Thus we have

$$\text{Range of } f = \{y|y \text{ is a real number } and\ y \neq -2\}.$$

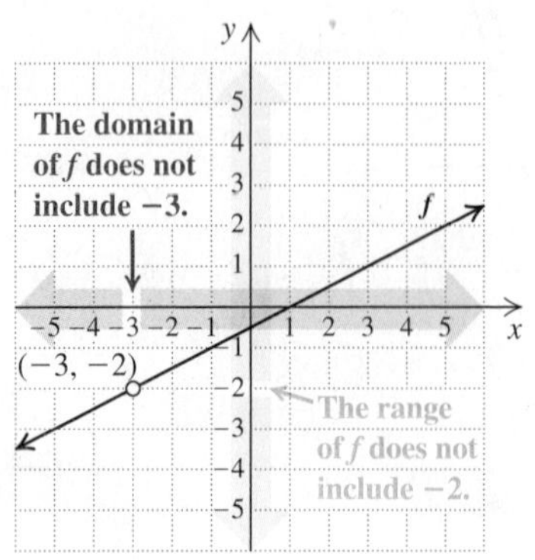

Try Exercise 41.

Note that if a graph contains two or more points with the same first coordinate, that graph cannot represent a function (otherwise one member of the domain would correspond to more than one member of the range). This observation is the basis of the *vertical-line test.*

The Vertical-Line Test If it is possible for a vertical line to cross a graph more than once, then the graph is not the graph of a function.

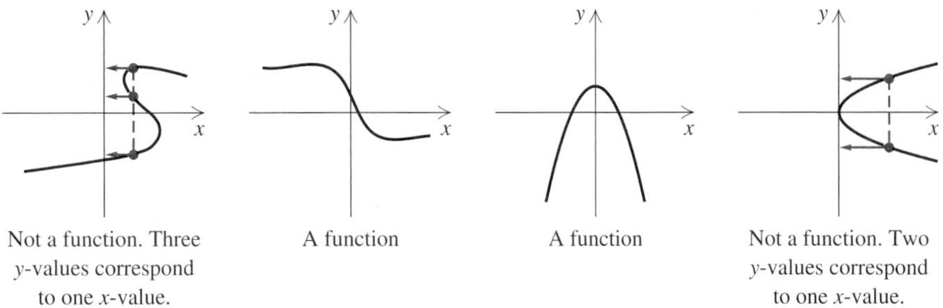

Not a function. Three y-values correspond to one x-value.

A function

A function

Not a function. Two y-values correspond to one x-value.

FUNCTION NOTATION AND EQUATIONS

We often think of an element of the domain of a function as an **input** and its corresponding element of the range as an **output.** In Example 3, the function f is written

$$\{(-3, 1), (1, -2), (3, 0), (4, 5)\}.$$

Thus, for an input of -3, the corresponding output is 1, and for an input of 3, the corresponding output is 0.

We use *function notation* to indicate what output corresponds to a given input. For the function f defined above, we write

$$f(-3) = 1, \quad f(1) = -2, \quad f(3) = 0, \quad \text{and} \quad f(4) = 5.$$

The notation $f(x)$ is read "f of x," "f at x," or "the value of f at x." If x is an input, then $f(x)$ is the corresponding output.

CAUTION! ***$f(x)$ does not mean f times x.***

Most functions are described by equations. For example, $f(x) = 2x + 3$ describes the function that takes an input x, multiplies it by 2, and then adds 3.

Input

$$f(x) = 2x + 3$$

Double Add 3

To calculate the output $f(4)$, we take the input 4, double it, and add 3 to get 11. That is, we substitute 4 into the formula for $f(x)$:

$$f(4) = 2 \cdot 4 + 3$$
$$= 11. \longleftarrow \text{Output}$$

To understand function notation, it can help to imagine a "function machine." Think of putting an input into the machine. For the function $f(x) = 2x + 3$, the machine will double the input and then add 3. The result will be the output.

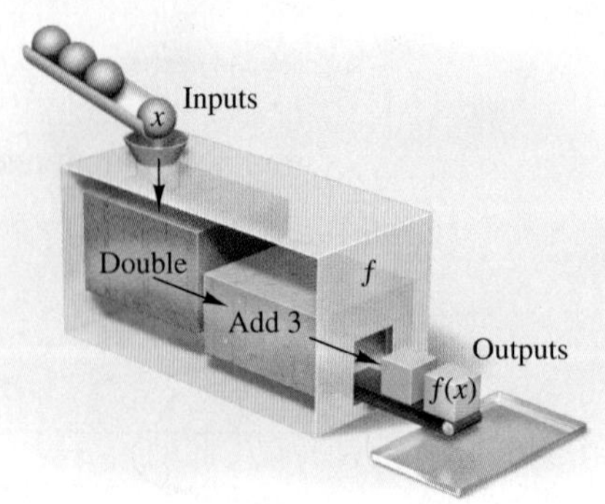

Sometimes, in place of $f(x) = 2x + 3$, we write $y = 2x + 3$, where it is understood that the value of y, the *dependent variable*, depends on our choice of x, the *independent variable.* To understand why $f(x)$ notation is so useful, consider two equivalent statements:

a) Find the member of the range that is paired with 2.

b) Find $f(2)$.

Function notation is not only more concise; it also emphasizes that x is the independent variable.

EXAMPLE 7 Find each indicated function value.

a) $f(5)$, for $f(x) = 3x + 2$

b) $g(-2)$, for $g(r) = 5r^2 + 3r$

c) $h(4)$, for $h(x) = 7$

d) $F(a) + 1$, for $F(x) = 3x + 2$

e) $F(a + 1)$, for $F(x) = 3x + 2$

SOLUTION Finding function values is much like evaluating an algebraic expression.

a) $f(x) = 3x + 2$

$f(5) = 3 \cdot 5 + 2 = 17$ **5 is the input; 17 is the output.**

b) $g(r) = 5r^2 + 3r$

$g(-2) = 5(-2)^2 + 3(-2)$ Substitute.

$= 5 \cdot 4 - 6 = 14$ Evaluate.

c) For the function given by $h(x) = 7$, all inputs share the same output, 7. Therefore, $h(4) = 7$. The function h is an example of a *constant function.*

d) $F(x) = 3(x) + 2$

$F(a) + 1 = [3(a) + 2] + 1$ **The input is a.**

$= [3a + 2] + 1 = 3a + 3$

e) $F(x) = 3(x) + 2$

$F(a + 1) = 3(a + 1) + 2$ **The input is $a + 1$.**

$= 3a + 3 + 2 = 3a + 5$

Try Exercise 53.

Student Notes

In Example 7(e), it is important to note that the parentheses on the left are for function notation, whereas those on the right indicate multiplication.

Note that whether we write $f(x) = 3x + 2$, or $f(t) = 3t + 2$, or $f(\square) = 3\square + 2$, we still have $f(5) = 17$. The variable in parentheses (the independent variable) is the same as the variable used in the algebraic expression.

Function Notation

Function values can be found directly using a graphing calculator. For example, if $f(x) = x^2 - 6x + 7$ is entered as $Y_1 = x^2 - 6x + 7$, then $f(2)$ can be calculated by evaluating $Y_1(2)$.

After entering the function, move to the home screen by pressing QUIT. The notation "Y1" can be found by pressing VARS, choosing the Y-VARS submenu, selecting the FUNCTION option, and then selecting Y1. After Y1 appears on the home screen, press (2) ENTER to evaluate Y1(2).

Function values can also be found using a table with Indpnt set to Ask. They can also be read from the graph of a function using the VALUE option of the CALC menu.

Your Turn

1. Let $f(x) = x^2 + x$. Enter $Y_1 = x^2 + x$ using the Y= key.
2. To find $f(-5)$, press QUIT and then VARS ▷ 1 1 to copy Y1 to the home screen. Then press ((−) 5) ENTER. $f(-5) = 20$

EXAMPLE 8 For $f(a) = 2a^2 - 3a + 1$, find $f(3)$ and $f(-5.1)$.

SOLUTION We first enter the function into the graphing calculator. The equation $Y_1 = 2x^2 - 3x + 1$ represents the same function, with the understanding that the Y1-values are the outputs and the x-values are the inputs; $Y_1(x)$ is equivalent to $f(a)$.

To find $f(3)$, we enter $Y_1(3)$ and find that $f(3) = 10$.

```
Plot1  Plot2  Plot3
\Y1=2X²-3X+1
\Y2=
\Y3=
```

```
Y1(3)
                 10
Y1(-5.1)
              68.32
```

X	Y1
3	10
−5.1	68.32

To find $f(-5.1)$, we can press ENTRY and edit the previous entry. Or, using a table, we can find both $f(3)$ and $f(-5.1)$, as shown on the right above. We see that $f(-5.1) = 68.32$.

Try Exercise 59.

When we find a function value, we are determining an output that corresponds to a given input. To do this, we evaluate the expression for the given value of the input.

Sometimes we want to determine an input that corresponds to a given output. To do this, we solve an equation.

EXAMPLE 9 Let $f(x) = 3x - 7$.

a) What output corresponds to an input of 5?
b) What input corresponds to an output of 5?

SOLUTION

a) We ask ourselves, "$f(5) = \square$?" Thus we find $f(5)$:

$$f(x) = 3x - 7$$
$$f(5) = 3(5) - 7 \quad \text{The input is 5. We substitute 5 for } x.$$
$$= 15 - 7 = 8. \quad \text{Carrying out the calculations}$$

The output 8 corresponds to the input 5; that is, $f(5) = 8$.

b) We ask ourselves, "$f(\square) = 5$? Thus we must find the value of x for which $f(x) = 5$:

$$f(x) = 3x - 7$$
$$5 = 3x - 7 \quad \text{The output is 5. We substitute 5 for } f(x).$$
$$12 = 3x$$
$$4 = x. \quad \text{Solving for } x$$

The input 4 corresponds to the output 5; that is, $f(4) = 5$.

Try Exercise 79.

When a function is described by an equation, we assume that the domain is the set of all real numbers for which function values can be calculated. If an x-value is not in the domain of a function, no point on the graph will have that x-value as its first coordinate.

EXAMPLE 10 For each equation, determine the domain of f.

a) $f(x) = |x|$ **b)** $f(x) = \dfrac{7}{2x - 6}$

SOLUTION

a) We ask ourselves, "Is there any number x for which we cannot compute $|x|$?" Since we can find the absolute value of *any* number, the answer is no. Thus the domain of f is $\mathbb{R}$, the set of all real numbers.

b) Is there any number x for which $\dfrac{7}{2x - 6}$ cannot be computed? Since $\dfrac{7}{2x - 6}$ cannot be computed when the denominator $2x - 6$ is 0, the answer is yes. To determine what x-value causes the denominator to be 0, we set up and solve an equation:

$$2x - 6 = 0 \quad \text{Setting the denominator equal to 0}$$
$$2x = 6 \quad \text{Adding 6 to both sides}$$
$$x = 3. \quad \text{Dividing both sides by 2}$$

$y_1 = 7/(2x - 6)$

X	Y1	
3	ERROR	

Y1 = ERROR

Thus, 3 is *not* in the domain of f, whereas all other real numbers are. The table at left indicates that $f(3)$ is not defined. The domain of f is $\{x \mid x \text{ is a real number } and\ x \neq 3\}$.

Try Exercise 81.

FUNCTIONS DEFINED PIECEWISE

Some functions are defined by different equations for various parts of their domains. Such functions are said to be **piecewise** defined. For example, the function given by $f(x) = |x|$ is described by

$$f(x) = \begin{cases} x, & \text{if } x \geq 0, \\ -x, & \text{if } x < 0. \end{cases}$$

To evaluate a piecewise-defined function for an input a, we first determine what part of the domain a belongs to. Then we use the appropriate formula for that part of the domain.

EXAMPLE 11 Find each function value for the function f given by

$$f(x) = \begin{cases} 2x, & \text{if } x < 0, \\ x + 1, & \text{if } x \geq 0. \end{cases}$$

a) $f(4)$ **b)** $f(-10)$

SOLUTION

a) The function f is defined using two different equations. To find $f(4)$, we must first determine whether to use the equation $f(x) = 2x$ or the equation $f(x) = x + 1$. To do this, we focus first on the two parts of the domain.

$$f(x) = \begin{cases} 2x, & \text{if } x < 0, \\ x + 1, & \text{if } x \geq 0. \end{cases}$$ **4 is in the second part of the domain.**

Since $4 \geq 0$, we use the equation $f(x) = x + 1$. Thus, $f(4) = 4 + 1 = 5$.

b) To find $f(-10)$, we first note that $-10 < 0$, so we must use the equation $f(x) = 2x$. Thus, $f(-10) = 2(-10) = -20$.

Try Exercise 95.

EXAMPLE 12 Find each function value for the function g given by

$$g(x) = \begin{cases} x + 2, & \text{if } x \leq -2, \\ x^2, & \text{if } -2 < x \leq 5, \\ 3x, & \text{if } x > 5. \end{cases}$$

a) $g(-2)$ **b)** $g(3)$ **c)** $g(7)$

$g(x) = x + 2$ $g(x) = x^2$ $g(x) = 3x$

−4 −3 −2 −1 0 1 2 3 4 5 6 7

$x \leq -2$ $-2 < x \leq 5$ $x > 5$

SOLUTION It may help to visualize the domain on the number line.

a) To find $g(-2)$, we note that -2 is in the part of the domain that is shaded blue. Since $-2 \leq -2$, we use the first equation, $g(x) = x + 2$:

$$g(-2) = -2 + 2 = 0.$$

b) We note that 3 is in the part of the domain that is shaded red. Since $-2 < 3 \leq 5$, we use the second equation, $g(x) = x^2$:

$$g(3) = 3^2 = 9.$$

c) We note that 7 is in the part of the domain that is shaded gray. Since $7 > 5$, we use the last equation, $g(x) = 3x$:

$$g(7) = 3 \cdot 7 = 21.$$

Try Exercise 97.

LINEAR FUNCTIONS AND APPLICATIONS

Any nonvertical line passes the vertical-line test and is the graph of a function. Such a function is called a **linear function** and can be written in the form $f(x) = mx + b$.

Linear functions arise continually in today's world. As with rate problems, it is critical to use proper units in all answers.

EXAMPLE 13 Salvage Value. Tyline Electric uses the function

$$S(t) = -700t + 3500$$

to determine the *salvage value* $S(t)$, in dollars, of a color photocopier t years after its purchase.

a) What do the numbers -700 and 3500 signify?

b) How long will it take the copier to *depreciate* completely?

c) What is the domain of S?

SOLUTION Drawing, or at least visualizing, a graph can be useful here.

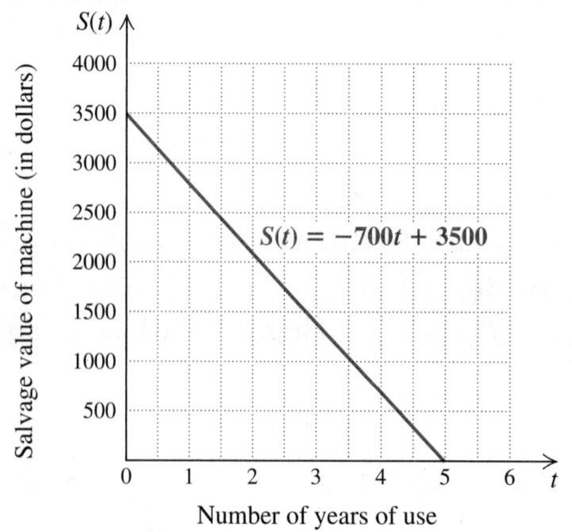

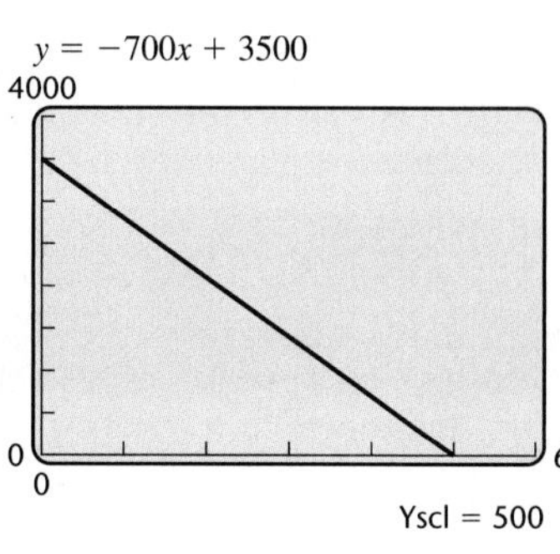

a) The point (0, 3500) is the y-intercept. Thus the number 3500 signifies the original cost of the copier, in dollars. The number -700 is the slope of the line. Since the output is measured in dollars and the input in years, the number -700 signifies that the value of the copier is declining at a rate of \$700 per year.

b) The copier will have depreciated completely when its value drops to 0:

$$S(t) = 0 \qquad \textbf{We set } S(t) = 0 \textbf{ and solve.}$$

$$-700t + 3500 = 0 \qquad \textbf{Substituting } -700t + 3500 \textbf{ for } S(t)$$

$$-700t = -3500 \qquad \textbf{Subtracting 3500 from both sides}$$

$$t = 5. \qquad \textbf{Dividing both sides by } -700$$

The copier will have depreciated completely in 5 years.

c) Neither the number of years of service nor the salvage value can be negative. In part (b) we found that after 5 years, the salvage value will have dropped to 0. Thus the domain of S is $\{t|0 \le t \le 5\}$, or $[0, 5]$. Either graph above serves as a visual check of this result.

Try Exercise 109.

3.8 Exercise Set

FOR EXTRA HELP MyMathLab   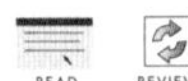

Concept Reinforcement *Complete each of the following sentences.*

1. A function is a special kind of ____________ between two sets.
2. In any function, each member of the domain is paired with ____________ one member of the range.
3. For any function, the set of all inputs, or first values, is called the ____________.
4. For any function, the set of all outputs, or second values, is called the ____________.
5. When a function is graphed, members of the domain are located on the ____________ axis.
6. When a function is graphed, members of the range are located on the ____________ axis.
7. The notation $f(3)$ is read ____________.
8. The ____________ line test can be used to determine whether or not a graph represents a function.

Determine whether each correspondence is a function.

9.

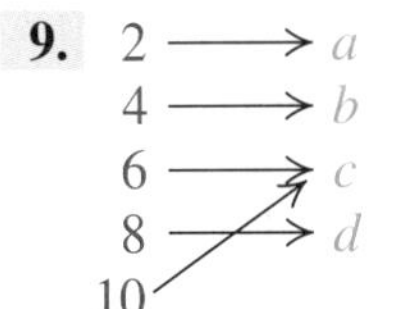

10. 3 → 9; 6 → 8; 6 → 7; 9 → 8; 9 → 7; 12 → 6

11. *Girl's age (in months)* → *Average daily weight gain (in grams)*

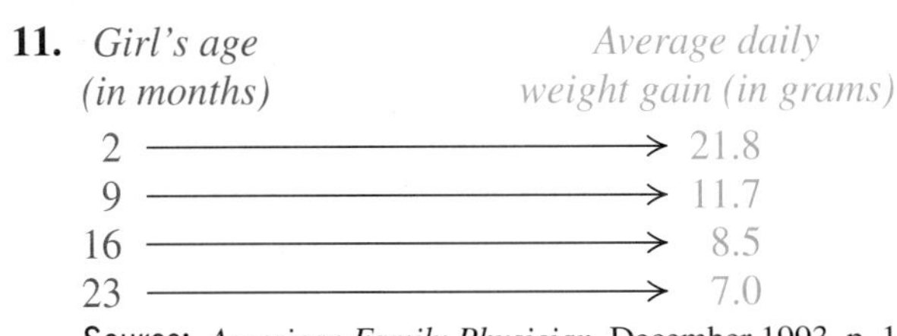

Source: *American Family Physician*, December 1993, p. 1435

12. *Boy's age (in months)* → *Average daily weight gain (in grams)*

Boy's age (in months)	Average daily weight gain (in grams)
2	24.8
9	11.7
16	8.2
23	7.0

Source: *American Family Physician*, December 1993, p. 1435

13. Year of Induction into Rock and Roll Hall of Fame → Musician

2008 → John Mellencamp
2008 → Leonard Cohen
2008 → Madonna
2009 → Jeff Beck
2009 → Metallica

Source: The Rock and Roll Hall of Fame and Museum, Inc.

14.

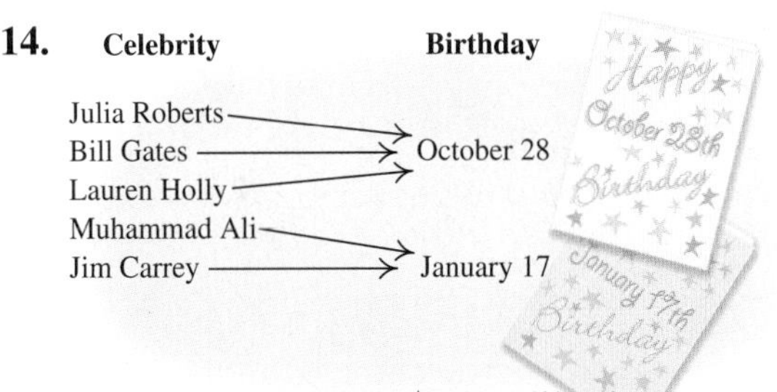

Sources: www.leannesbirthdays.com/; and www.kidsparties.com

15.

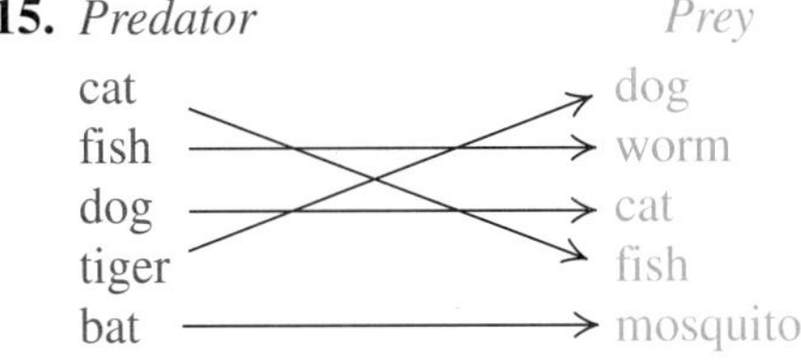

16.

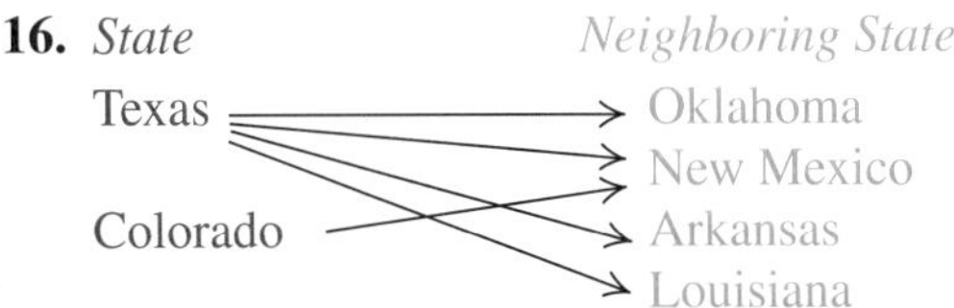

Determine whether each of the following is a function. Identify any relations that are not functions.

17. The correspondence matching a USB flash drive with its storage capacity
18. The correspondence matching a member of a rock band with the instrument the person can play
19. The correspondence matching a player on a team with that player's uniform number
20. The correspondence matching a triangle with its area

For each graph of a function, determine **(a)** $f(1)$; **(b)** *the domain;* **(c)** *any x-values for which* $f(x) = 2$; *and* **(d)** *the range.*

21.

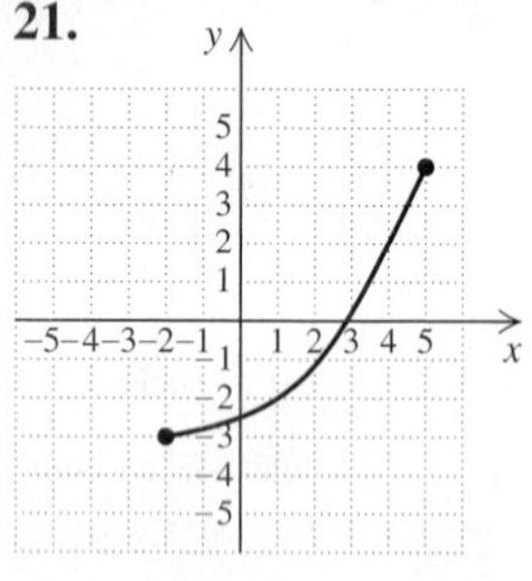

22.

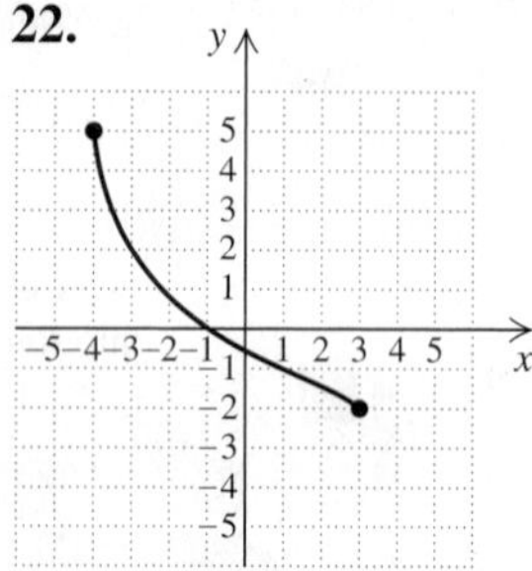

23.

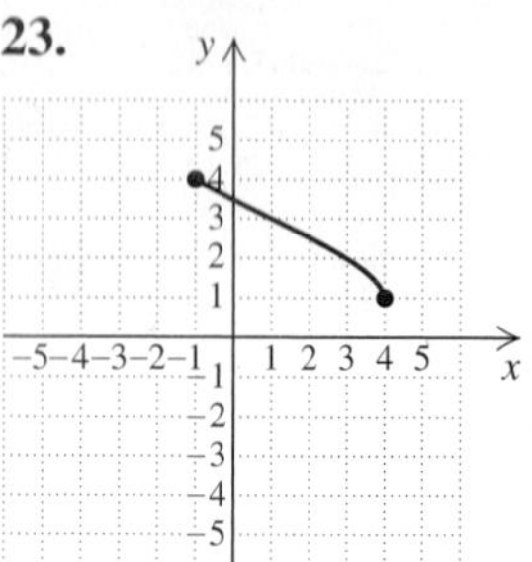

24.

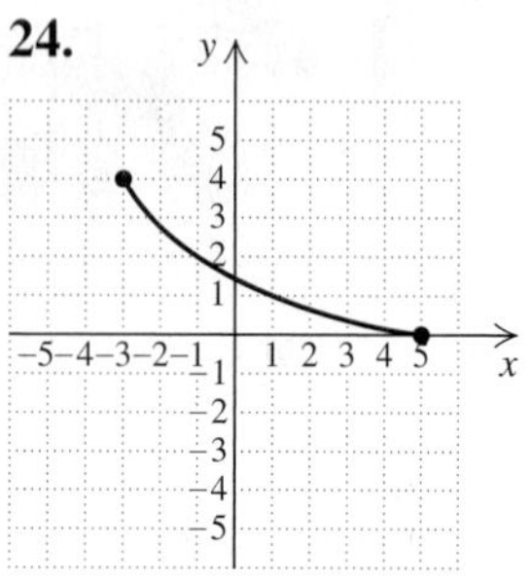

25.

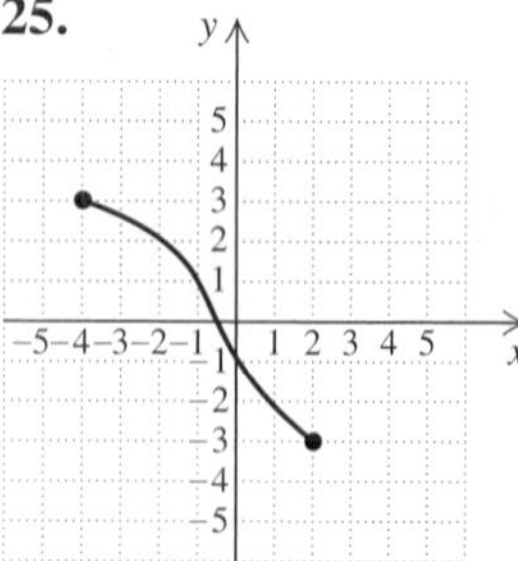

26.

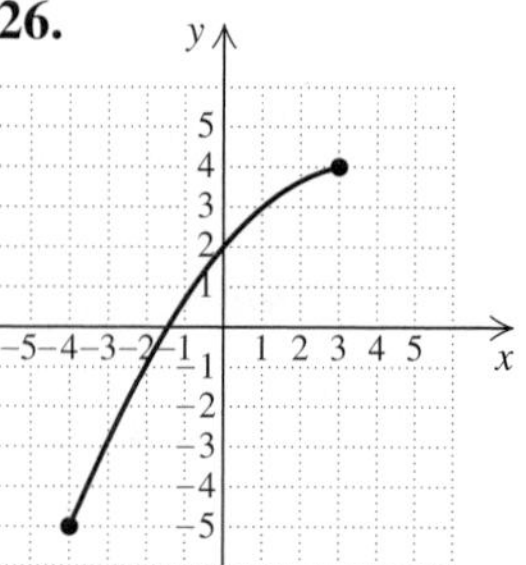

27.

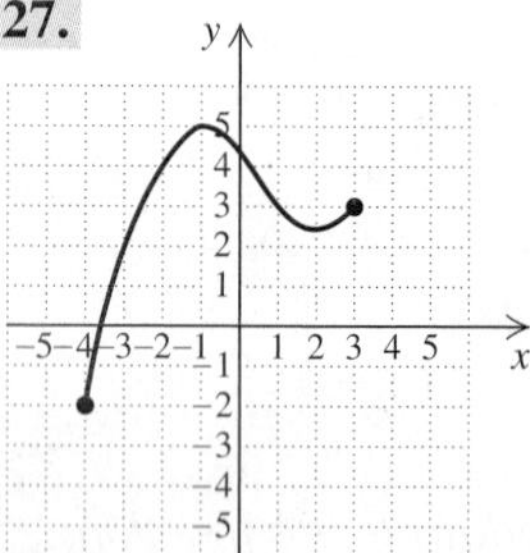

28.

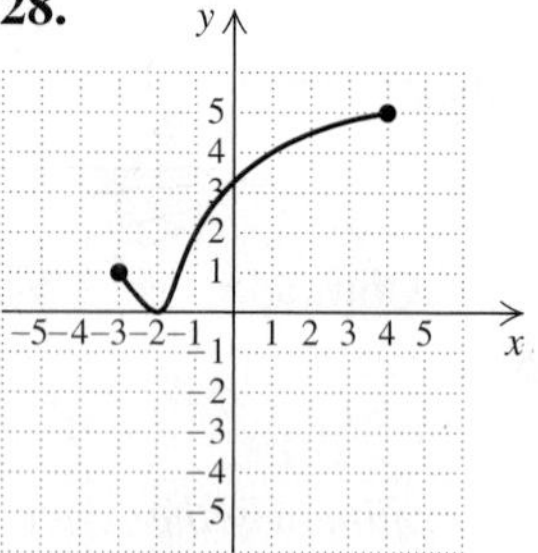

29.

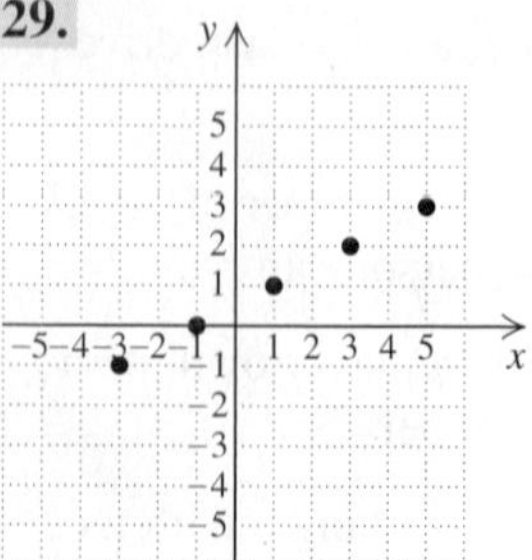

30.

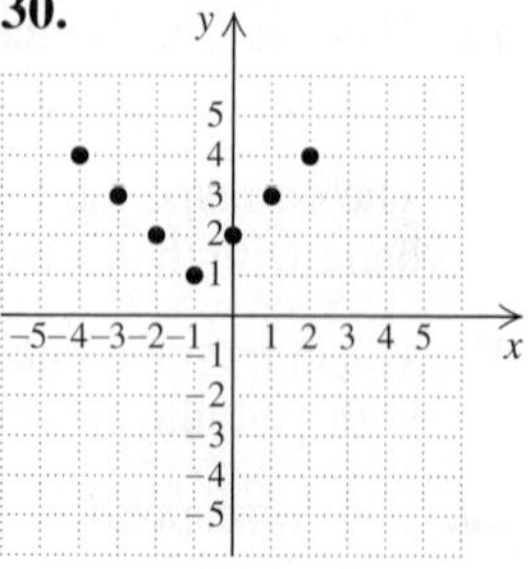

31.

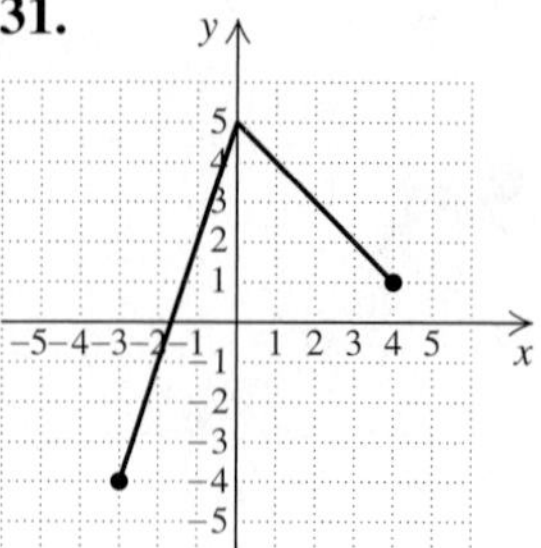

32.

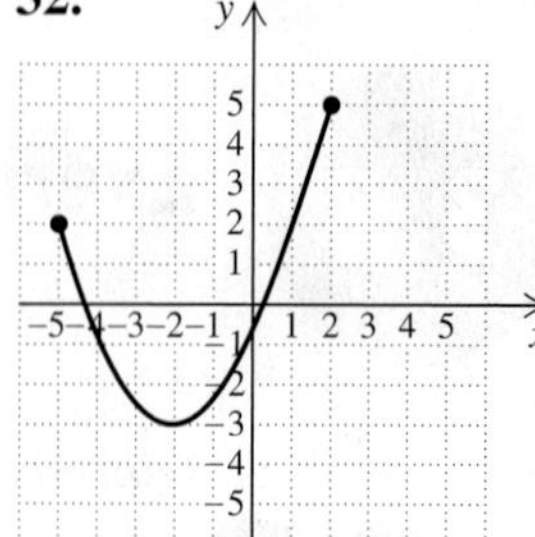

33.

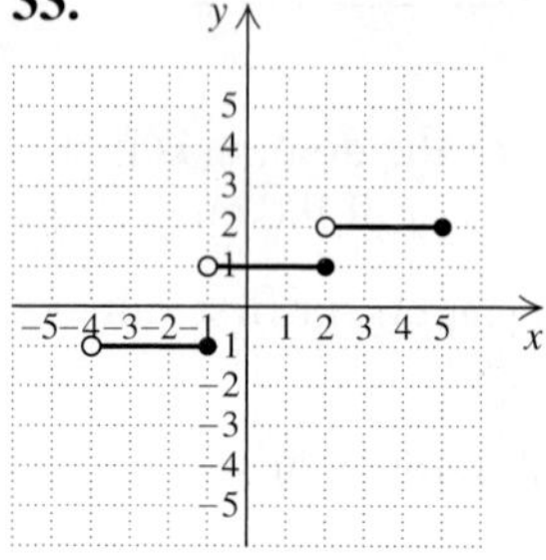

34.

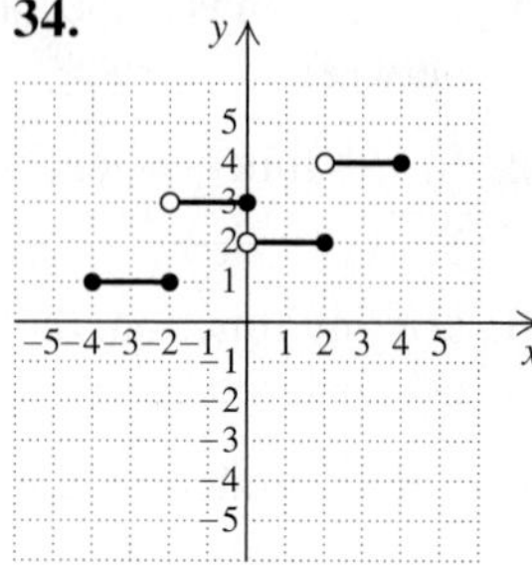

Determine the domain and the range of each function.

35.

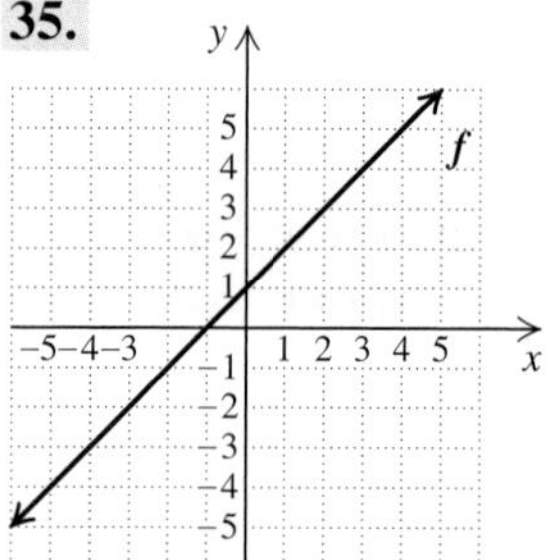

36.

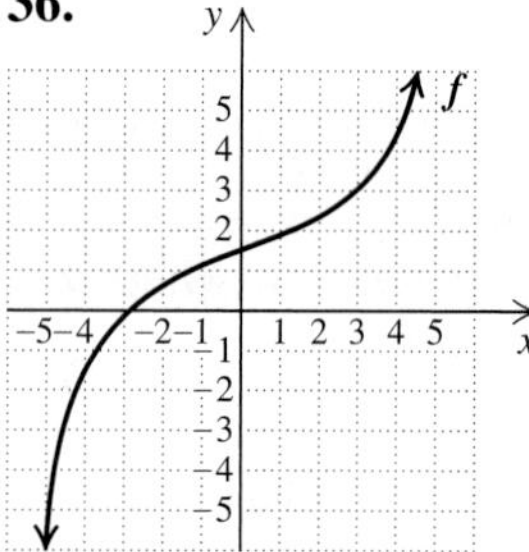

37.

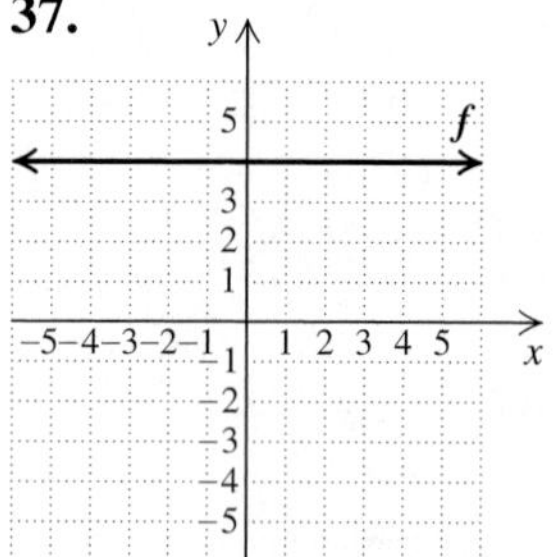

38.

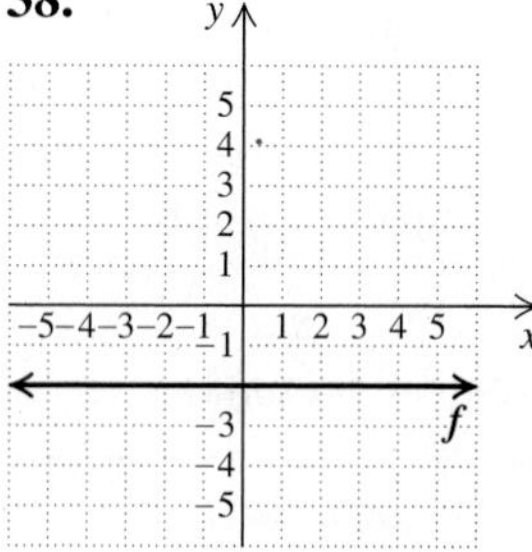

39.

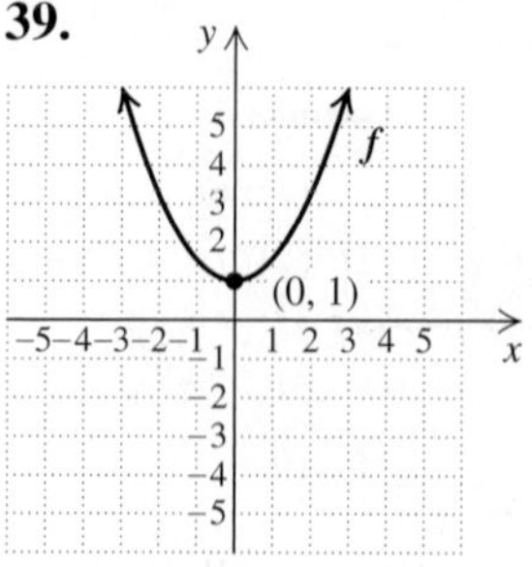

40.

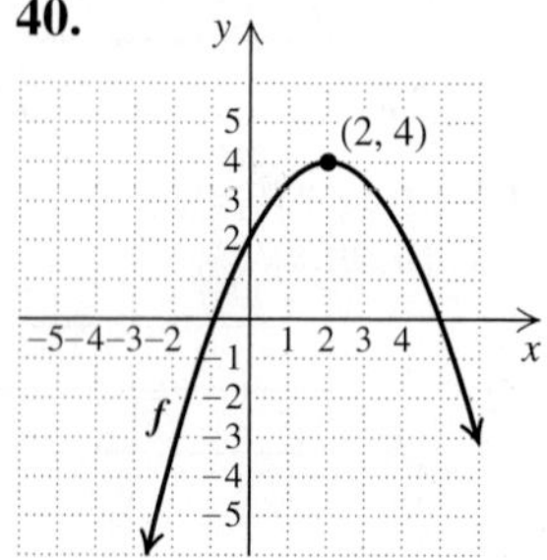

41.

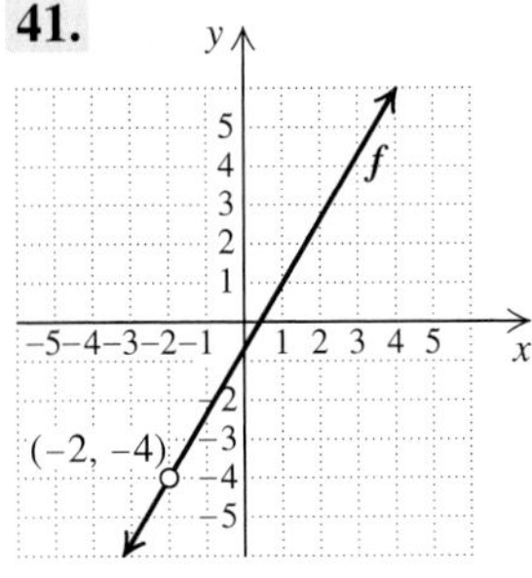

42.

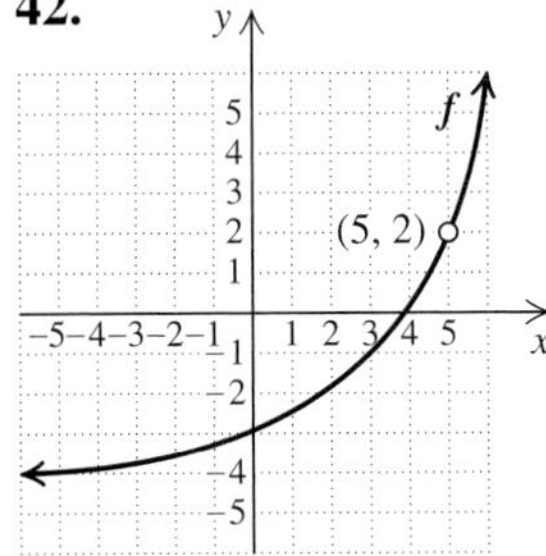

43.

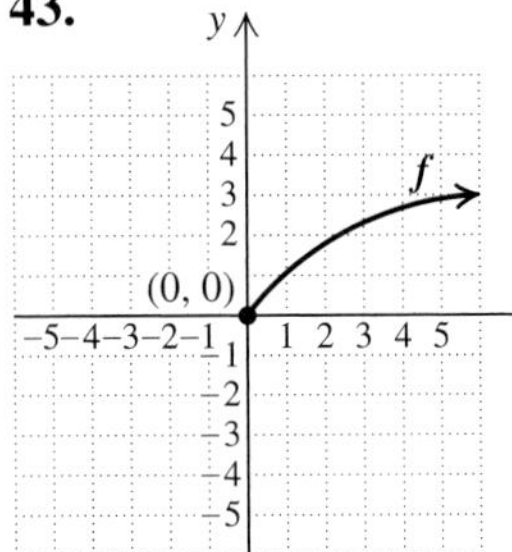

44.

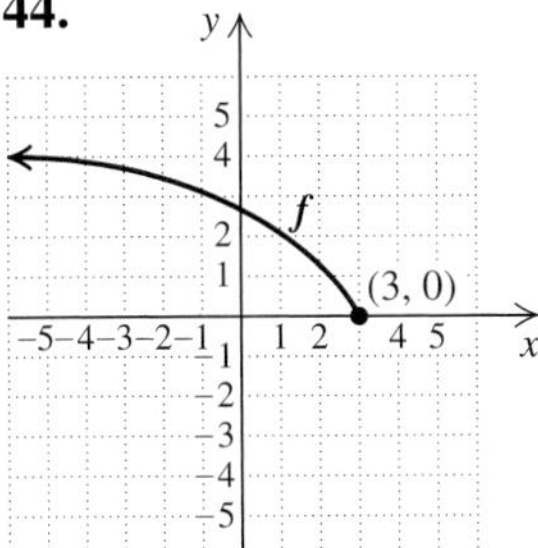

Determine whether each of the following is the graph of a function.

45.

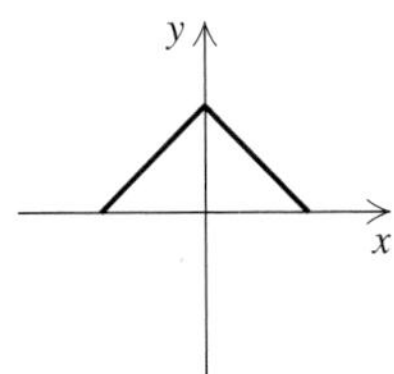

46.

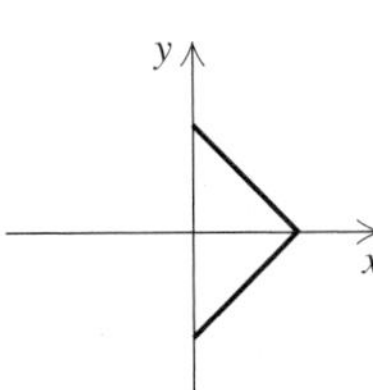

47.

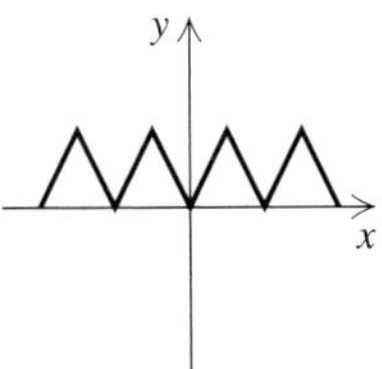

48.

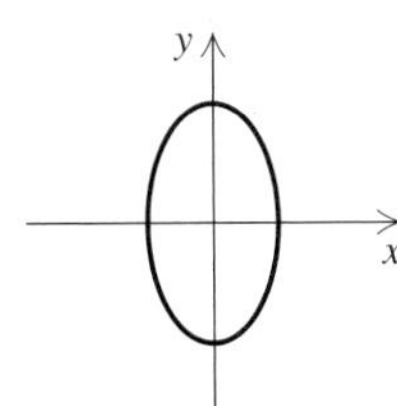

49.

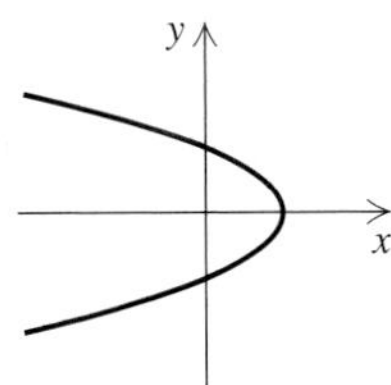

50.

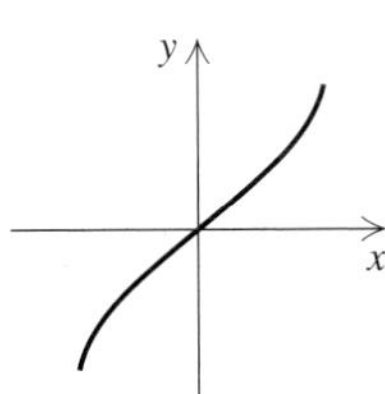

51.

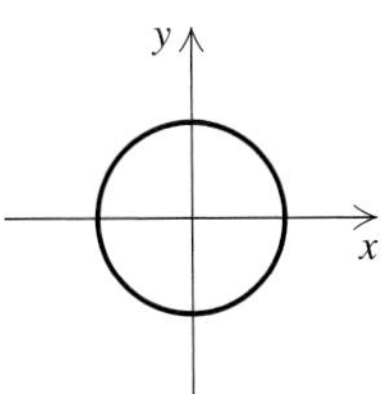

52.

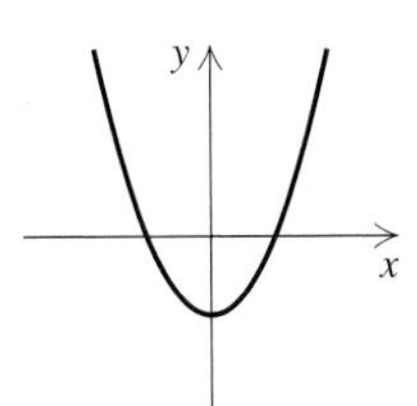

Find the function values.

53. $g(x) = 2x + 3$

a) $g(0)$ **b)** $g(-4)$ **c)** $g(-7)$
d) $g(8)$ **e)** $g(a + 2)$ **f)** $g(a) + 2$

54. $h(x) = 3x - 2$

a) $h(4)$ **b)** $h(8)$ **c)** $h(-3)$
d) $h(-4)$ **e)** $h(a - 1)$ **f)** $h(a) - 1$

55. $f(n) = 5n^2 + 4n$

a) $f(0)$ **b)** $f(-1)$ **c)** $f(3)$
Aha! **d)** $f(t)$ **e)** $f(2a)$ **f)** $2 \cdot f(a)$

56. $g(n) = 3n^2 - 2n$

a) $g(0)$ **b)** $g(-1)$ **c)** $g(3)$
d) $g(t)$ **e)** $g(2a)$ **f)** $2 \cdot g(a)$

57. $f(x) = \dfrac{x - 3}{2x - 5}$

a) $f(0)$ **b)** $f(4)$ **c)** $f(-1)$
d) $f(3)$ **e)** $f(x + 2)$

58. $s(x) = \dfrac{3x - 4}{2x + 5}$

a) $s(10)$ **b)** $s(2)$ **c)** $s\left(\frac{1}{2}\right)$
d) $s(-1)$ **e)** $s(x + 3)$

Use a graphing calculator to find the function values.

59. $f(a) = a^2 + a - 1$

a) $f(-6)$ **b)** $f(1.7)$

60. $g(t) = 3t^2 - 8$

a) $g(29)$ **b)** $g(-0.1)$

61. $h(n) = 8 - n - \dfrac{1}{n}$

a) $h(0.2)$ **b)** $h\left(-\frac{1}{4}\right)$

62. $p(a) = \dfrac{2}{a} - a^2$

a) $p\left(\frac{1}{8}\right)$ **b)** $p(-0.5)$

The function A described by $A(s) = s^2\dfrac{\sqrt{3}}{4}$ gives the area of an equilateral triangle with side s.

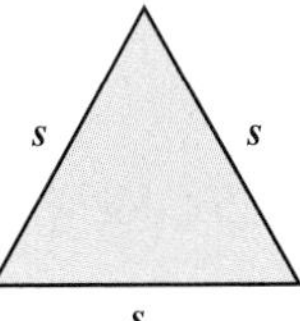

63. Find the area when a side measures 4 cm.

64. Find the area when a side measures 6 in.

The function V described by $V(r) = 4\pi r^2$ gives the surface area of a sphere with radius r.

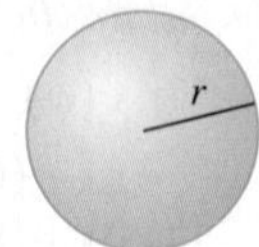

65. Find the surface area when the radius is 3 in.

66. Find the surface area when the radius is 5 cm.

67. *Pressure at Sea Depth.* The function $P(d) = 1 + (d/33)$ gives the pressure, in *atmospheres* (atm), at a depth of d feet in the sea. Note that $P(0) = 1$ atm, $P(33) = 2$ atm, and so on. Find the pressure at 20 ft, at 30 ft, and at 100 ft.

68. *Melting Snow.* The function $W(d) = 0.112d$ approximates the amount, in centimeters, of water that results from d centimeters of snow melting. Find the amount of water that results from snow melting from depths of 16 cm, 25 cm, and 100 cm.

Fill in the missing values in each table.

	$f(x) = 2x - 5$	
	x	$f(x)$
69.	8	
70.		13
71.		-5
72.	-4	

	$f(x) = \frac{1}{3}x + 4$	
	x	$f(x)$
73.		$\frac{1}{2}$
74.		$-\frac{1}{3}$
75.	$\frac{1}{2}$	
76.	$-\frac{1}{3}$	

77. If $f(x) = 4 - x$, for what input is the output 7?

78. If $f(x) = 5x + 1$, for what input is the output $\frac{1}{2}$?

79. If $f(x) = 0.1x - 0.5$, for what input is the output -3?

80. If $f(x) = 2.3 - 1.5x$, for what input is the output 10?

Find the domain of f.

81. $f(x) = \dfrac{5}{x - 3}$

82. $f(x) = \dfrac{7}{6 - x}$

83. $f(x) = \dfrac{x}{2x - 1}$

84. $f(x) = \dfrac{2x}{4x + 3}$

85. $f(x) = 2x + 1$

86. $f(x) = x^2 + 3$

87. $f(x) = |5 - x|$

88. $f(x) = |3x - 4|$

89. $f(x) = \dfrac{5}{x - 9}$

90. $f(x) = \dfrac{3}{x + 1}$

91. $f(x) = x^2 - 9$

92. $f(x) = x^2 - 2x + 1$

93. $f(x) = \dfrac{2x - 7}{5}$

94. $f(x) = \dfrac{x + 5}{8}$

Find the indicated function values for each function.

95. $f(x) = \begin{cases} x, & \text{if } x < 0, \\ 2x + 1, & \text{if } x \geq 0 \end{cases}$

a) $f(-5)$ b) $f(0)$ c) $f(10)$

96. $g(x) = \begin{cases} x - 5, & \text{if } x \leq 5, \\ 3x, & \text{if } x > 5 \end{cases}$

a) $g(0)$ b) $g(5)$ c) $g(6)$

97. $G(x) = \begin{cases} x - 5, & \text{if } x < -1, \\ x, & \text{if } -1 \leq x \leq 2, \\ x + 2, & \text{if } x > 2 \end{cases}$

a) $G(0)$ b) $G(2)$ c) $G(5)$

98. $F(x) = \begin{cases} 2x, & \text{if } x \leq 0, \\ x, & \text{if } 0 < x \leq 3, \\ -5x, & \text{if } x > 3 \end{cases}$

a) $F(-1)$ b) $F(3)$ c) $F(10)$

99. $f(x) = \begin{cases} x^2 - 10, & \text{if } x < -10, \\ x^2, & \text{if } -10 \leq x \leq 10, \\ x^2 + 10, & \text{if } x > 10 \end{cases}$

a) $f(-10)$ b) $f(10)$ c) $f(11)$

100. $f(x) = \begin{cases} 2x^2 - 3, & \text{if } x \leq 2, \\ x^2, & \text{if } 2 < x < 4, \\ 5x - 7, & \text{if } x \geq 4 \end{cases}$

a) $f(0)$ b) $f(3)$ c) $f(6)$

In Exercises 101–108, each model is of the form $f(x) = mx + b$. In each case, determine what m and b signify.

101. *Catering.* When catering a party for x people, Jennette's Catering uses the formula $C(x) = 25x + 75$, where $C(x)$ is the cost of the party, in dollars.

102. *Hair Growth.* After Ty gets a "buzz cut," the length $L(t)$ of his hair, in inches, is given by $L(t) = \frac{1}{2}t + 1$, where t is the number of months after he gets the haircut.

103. *Renewable Energy.* U.S. consumption of renewable energy, in quadrillions of Btu's, is approximated by $D(t) = \frac{2}{3}t + \frac{10}{3}$, where t is the number of years after 1960.
Source: Based on data from the U.S. Energy Information Administration

104. *Life Expectancy of American Women.* The life expectancy of American women t years after 1970 is given by $A(t) = \frac{1}{7}t + 75.5$.

105. *Landscaping.* After being cut, the length $G(t)$ of the lawn, in inches, at Great Harrington Community College is given by $G(t) = \frac{1}{8}t + 2$, where t is the number of days since the lawn was cut.

106. *Cost of Renting a Truck.* The cost, in dollars, of a one-day truck rental is given by $C(d) = 0.3d + 20$, where d is the number of miles driven.

107. *Cost of a Movie Ticket.* The average price $P(t)$, in dollars, of a movie ticket is given by $P(t) = 0.21t + 5.43$, where t is the number of years since 2000.

108. *Cost of a Taxi Ride.* The cost, in dollars, of a taxi ride during off-peak hours in New York City is given by $C(d) = 2d + 2.5$, where d is the number of miles traveled.

109. *Salvage Value.* Green Glass Recycling uses the function given by $F(t) = -5000t + 90{,}000$ to determine the salvage value $F(t)$, in dollars, of a waste removal truck t years after it has been put into use.

a) What do the numbers -5000 and 90,000 signify?
b) How long will it take the truck to depreciate completely?
c) What is the domain of F?

110. *Salvage Value.* Consolidated Shirt Works uses the function given by $V(t) = -2000t + 15{,}000$ to determine the salvage value $V(t)$, in dollars, of a color separator t years after it has been put into use.

a) What do the numbers -2000 and 15,000 signify?
b) How long will it take the machine to depreciate completely?
c) What is the domain of V?

111. *Records in the 400-m Run.* The record $R(t)$ for the 400-m run t years after 1930 is given by $R(t) = 46.8 - 0.075t$.

a) What do the numbers 46.8 and -0.075 signify?
b) When will the record be 38.7 sec?
c) What is the domain of R?

112. *Pressure at Sea Depth.* The pressure $P(d)$, in atmospheres, at a depth d feet beneath the surface of the ocean is given by $P(d) = 0.03d + 1$.

a) What do the numbers 0.03 and 1 signify?
b) Where is the pressure 4 atmospheres?
c) What is the domain of P?

TW **113.** Explain why the domain of the function given by $f(x) = \dfrac{x + 3}{2}$ is $\mathbb{R}$, but the domain of the function given by $g(x) = \dfrac{2}{x + 3}$ is not $\mathbb{R}$.

TW **114.** Abby asserts that for a function described by a set of ordered pairs, the range of the function will always have the same number of elements as there are ordered pairs. Is she correct? Why or why not?

SKILL REVIEW

To prepare for Chapter 4, review exponential notation and order of operations (Section 1.8).

Simplify. [1.8]

115. $(-5)^3$

116. $(-2)^6$

117. -2^6

118. $3 \cdot 2^4 - 5 \cdot 2^3$

119. $2 - (3 - 2^2) + 10 \div 2 \cdot 5$

120. $(5 - 7)^2(3 - 2 \cdot 2)$

SYNTHESIS

TW **121.** For the function given by $n(z) = ab + wz$, what is the independent variable? How can you tell?

TW **122.** Explain in your own words why every function is a relation, but not every relation is a function.

For Exercises 123 and 124, let $f(x) = 3x^2 - 1$ *and* $g(x) = 2x + 5$.

123. Find $f(g(-4))$ and $g(f(-4))$.

124. Find $f(g(-1))$ and $g(f(-1))$.

125. If f represents the function in Exercise 15, find $f(f(f(f(tiger))))$.

Pregnancy. For Exercises 126–129, use the following graph of a woman's "stress test." This graph shows the size of a pregnant woman's contractions as a function of time.

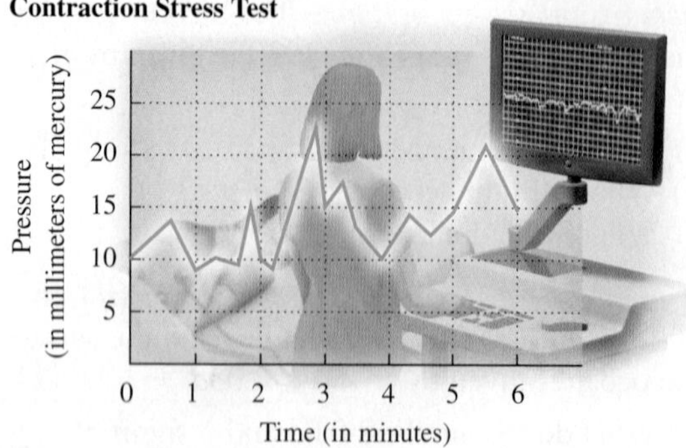

126. How large is the largest contraction that occurred during the test?

127. At what time during the test did the largest contraction occur?

TW 128. On the basis of the information provided, how large a contraction would you expect 60 sec after the end of the test? Why?

129. What is the frequency of the largest contraction?

130. The *greatest integer function* $f(x) = [\![x]\!]$ is defined as follows: $[\![x]\!]$ is the greatest integer that is less than or equal to x. For example, if $x = 3.74$, then $[\![x]\!] = 3$; and if $x = -0.98$, then $[\![x]\!] = -1$. Graph the greatest integer function for $-5 \leq x \leq 5$. (The notation $f(x) = \text{int}(x)$, used in many graphing calculators, is often found in the MATH NUM submenu.)

131. Suppose that a function g is such that $g(-1) = -7$ and $g(3) = 8$. Find a formula for g if $g(x)$ is of the form $g(x) = mx + b$, where m and b are constants.

Given that $f(x) = mx + b$, classify each of the following as true or false.

132. $f(c + d) = f(c) + f(d)$

133. $f(cd) = f(c)f(d)$

134. $f(kx) = kf(x)$

135. $f(c - d) = f(c) - f(d)$

Try Exercise Answers: Section 3.8

9. Yes **17.** Function **27.** **(a)** 3; **(b)** $\{x|-4 \leq x \leq 3\}$, or $[-4, 3]$; **(c)** -3; **(d)** $\{y|-2 \leq y \leq 5\}$, or $[-2, 5]$ **29.** **(a)** 1; **(b)** $\{-3, -1, 1, 3, 5\}$; **(c)** 3; **(d)** $\{-1, 0, 1, 2, 3\}$ **35.** Domain: $\mathbb{R}$; range: $\mathbb{R}$ **41.** Domain: $\{x|x$ is a real number *and* $x \neq -2\}$; range: $\{y|y$ is a real number *and* $y \neq -4\}$ **53.** **(a)** 3; **(b)** -5; **(c)** -11; **(d)** 19; **(e)** $2a + 7$; **(f)** $2a + 5$ **59.** **(a)** 29; **(b)** 3.59 **79.** -25 **81.** $\{x|x$ is a real number *and* $x \neq 3\}$ **95.** **(a)** -5; **(b)** 1; **(c)** 21 **97.** **(a)** 0; **(b)** 2; **(c)** 7 **109.** **(a)** -5000 signifies that the depreciation is \$5000 per year; 90,000 signifies that the original value of the truck was \$90,000; **(b)** 18 yr; **(c)** $\{t|0 \leq t \leq 18\}$

Study Summary

KEY TERMS AND CONCEPTS | **EXAMPLES** | **PRACTICE EXERCISES**

SECTION 3.1: READING GRAPHS, PLOTTING POINTS, AND SCALING GRAPHS

Ordered pairs, like $(-3, 2)$ or $(4, 3)$, can be **plotted** or **graphed** using a **coordinate system** that uses two **axes,** which are most often labeled x and y. The axes intersect at the **origin,** $(0, 0)$, and divide a plane into four **quadrants.**

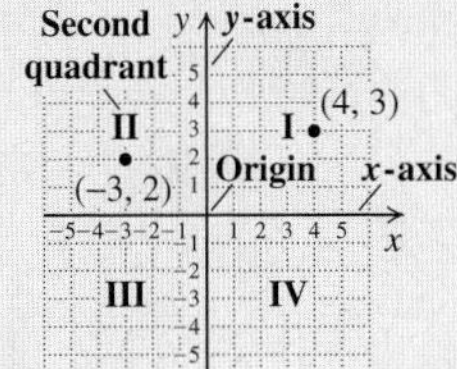

1. Plot the points $(0, -5)$ and $(-2, 1)$.
2. In which quadrant is the point $(-10, -20)$ located?

SECTION 3.2: GRAPHING EQUATIONS

To **graph** an equation means to make a drawing that represents all of its solutions.

A **linear equation,** such as $y = 2x - 7$ or $2x + 3y = 12$, has a graph that is a straight line.

$3x = y + 1$

x	y	(x, y)
1	2	$(1, 2)$
0	-1	$(0, -1)$
-1	-4	$(-1, -4)$

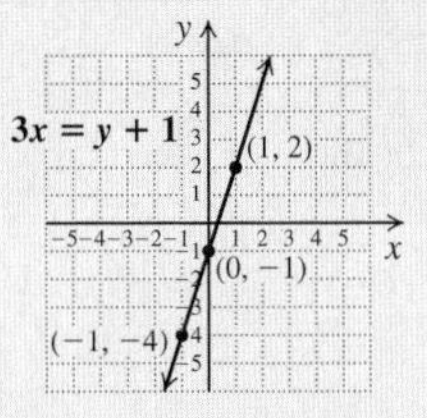

3. Graph: $y = 2x + 1$.

SECTION 3.3: LINEAR EQUATIONS AND INTERCEPTS

To find a y-intercept $(0, b)$, let $x = 0$ and solve for y.

To find an x-intercept $(a, 0)$, let $y = 0$ and solve for x.

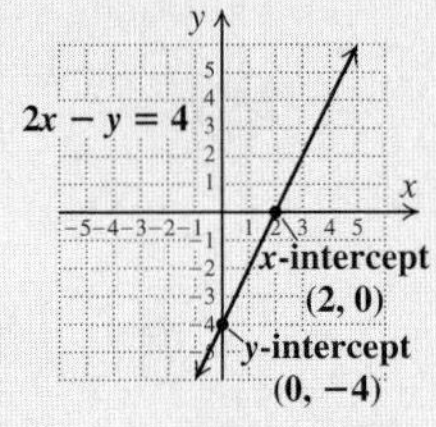

4. Find the x-intercept and the y-intercept of the line given by $10x - y = 10$.

Horizontal Lines

The graph of $y = b$ is a horizontal line, with y-intercept $(0, b)$.

Vertical Lines

The graph of $x = a$ is a vertical line, with x-intercept $(a, 0)$.

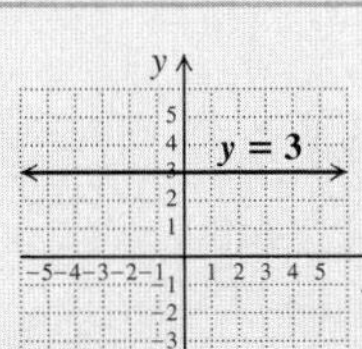

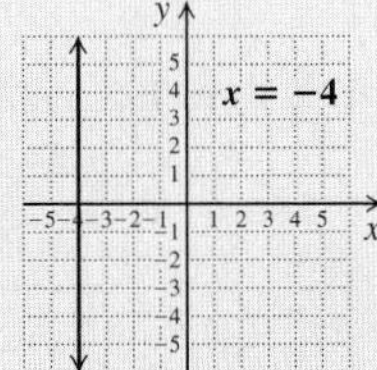

5. Graph: $y = -2$.
6. Graph: $x = 3$.

SECTION 3.4: RATES

A **rate** is a ratio that indicates how two quantities change with respect to each other.

Lara had \$1500 in her savings account at the beginning of February, and \$2400 at the beginning of May. Find the rate at which Lara is saving.

$$\text{Savings rate} = \frac{\text{Amount saved}}{\text{Number of months}} = \frac{\$2400 - \$1500}{3 \text{ months}}$$

$$= \frac{\$900}{3 \text{ months}} = \$300 \text{ per month}$$

7. At 8:30 A.M., the Buck Creek Fire Department had served 47 people at their annual pancake breakfast. By 9:15 A.M., the total served had reached 67. Find the serving rate, in number of meals per minute.

SECTION 3.5: SLOPE

Slope

$$\text{Slope} = m = \frac{\text{change in } y}{\text{change in } x} = \frac{\text{rise}}{\text{run}} = \frac{y_2 - y_1}{x_2 - x_1}$$

The slope of the line containing the points $(-1, -4)$ and $(2, -6)$ is

$$m = \frac{-6 - (-4)}{2 - (-1)} = \frac{-2}{3} = -\frac{2}{3}.$$

8. Find the slope of the line containing the points $(1, 4)$ and $(-9, 3)$.

The slope of a horizontal line is 0.

The slope of a vertical line is undefined.

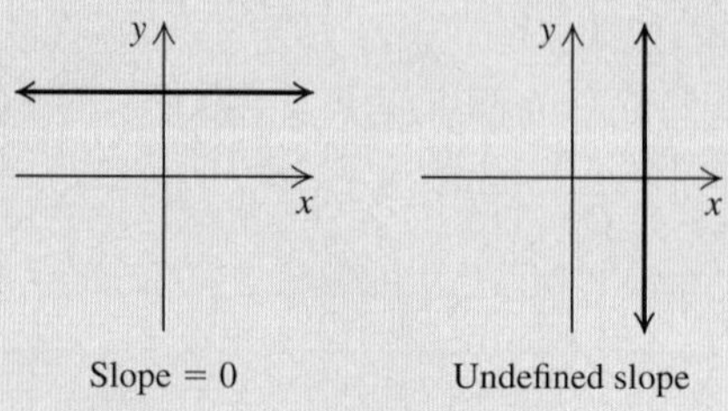

Slope = 0 Undefined slope

9. Find the slope of the line given by $y = 10$.

SECTION 3.6: SLOPE–INTERCEPT FORM

Slope–Intercept Form

$y = mx + b$

The slope of the line is m. The y-intercept of the line is $(0, b)$.

For the line given by $y = \frac{2}{3}x - 8$:

The slope is $\frac{2}{3}$ and the y-intercept is $(0, -8)$.

10. Find the slope and the y-intercept of the line given by $y = -4x + \frac{2}{5}$.

To graph a line written in slope–intercept form, plot the y-intercept and count off the slope.

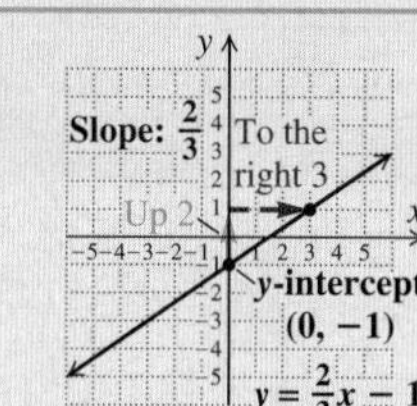

11. Graph: $y = \frac{1}{2}x + 2$.

Parallel Lines

Two lines are parallel if they have the same slope or if both are vertical.

Determine whether the graphs of $y = \frac{2}{3}x - 5$ and $3y - 2x = 7$ are parallel.

$y = \frac{2}{3}x - 5$

The slope is $\frac{2}{3}$.

$3y - 2x = 7$

$3y = 2x + 7$

$y = \frac{2}{3}x + \frac{7}{3}$

The slope is $\frac{2}{3}$.

Since the slopes are the same, the graphs are parallel.

12. Determine whether the graphs of $y = 4x - 12$ and $4y = x - 9$ are parallel.

Perpendicular Lines

Two lines are perpendicular if the product of their slopes is -1 or if one line is vertical and the other line is horizontal.

Determine whether the graphs of $y = \frac{2}{3}x - 5$ and $2y + 3x = 1$ are perpendicular.

$y = \frac{2}{3}x - 5$

The slope is $\frac{2}{3}$.

$2y + 3x = 1$

$2y = -3x + 1$

$y = -\frac{3}{2}x + \frac{1}{2}$

The slope is $-\frac{3}{2}$.

Since $\frac{2}{3}\left(-\frac{3}{2}\right) = -1$, the graphs are perpendicular.

13. Determine whether the graphs of $y = x - 7$ and $x + y = 3$ are perpendicular.

SECTION 3.7: POINT–SLOPE FORM; INTRODUCTION TO CURVE FITTING

Point–Slope Form $y - y_1 = m(x - x_1)$ The slope of the line is m. The line passes through (x_1, y_1).	Write a point–slope equation for the line with slope -2 that contains the point $(3, -5)$. $y - y_1 = m(x - x_1)$ $y - (-5) = -2(x - 3)$	**14.** Write a point–slope equation for the line with slope $\frac{1}{4}$ and containing the point $(-1, 6)$.

SECTION 3.8: FUNCTIONS

A **function** is a correspondence between a first set, called the **domain,** and a second set, called the **range,** such that each member of the domain corresponds to *exactly one* member of the range.	The correspondence f: $\{(-1, \frac{1}{2}), (0, 1), (1, 2), (2, 4), (3, 8)\}$ is a function. The domain of $f = \{-1, 0, 1, 2, 3\}$. The range of $f = \{\frac{1}{2}, 1, 2, 4, 8\}$. $f(-1) = \frac{1}{2}$ The input -1 corresponds to the output $\frac{1}{2}$.	**15.** Find $f(-1)$ for $f(x) = 2 - 3x$.
The Vertical-Line Test If it is possible for a vertical line to cross a graph more than once, then the graph is not the graph of a function.	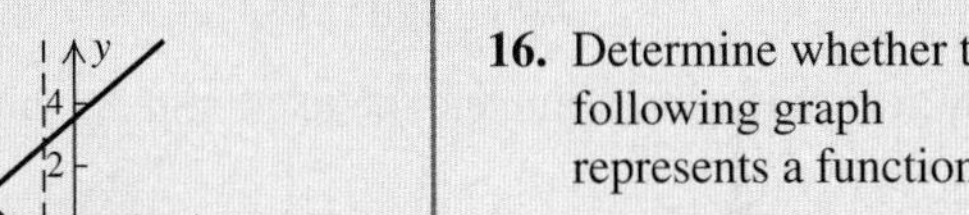 This *is* the graph of a function. This *is not* the graph of a function.	**16.** Determine whether the following graph represents a function. 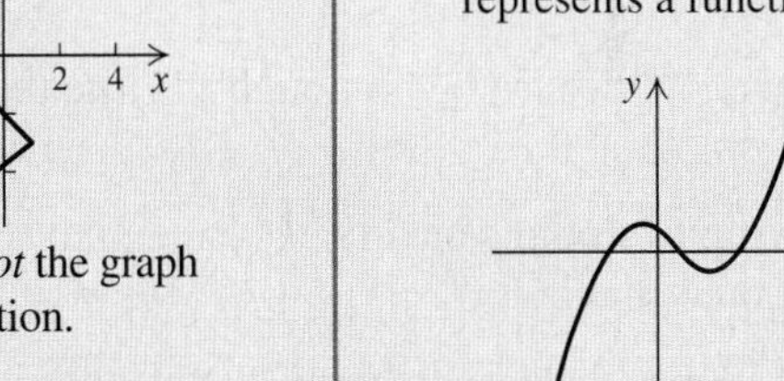
The domain of a function is the set of all x-coordinates of the points on the graph. The range of a function is the set of all y-coordinates of the points on the graph.	Consider the function given by $f(x) = \|x\| - 3$. The domain of the function is $\mathbb{R}$. The range of the function is $\{y\|y \geq -3\}$, or $[-3, \infty)$. 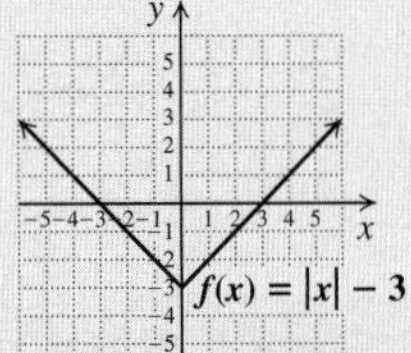 	**17.** Determine the domain and the range of the function represented in the following graph. 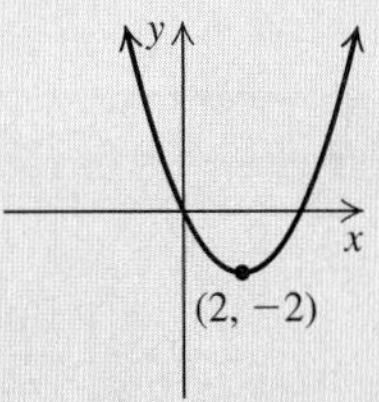
Unless otherwise stated, the domain of a function is the set of all numbers for which function values can be calculated.	Consider the function given by $f(x) = \dfrac{x + 2}{x - 7}$. Function values cannot be calculated when the denominator is 0. Since $x - 7 = 0$ when $x = 7$, the domain of f is $\{x\|x \text{ is a real number } and\ x \neq 7\}$.	**18.** Determine the domain of the function given by $f(x) = \frac{1}{4}x - 5$.

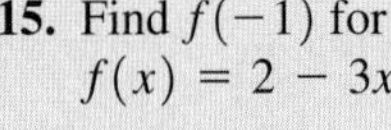

Review Exercises 3

Concept Reinforcement *Classify each of the following statements as either true or false.*

1. Every ordered pair lies in one of the four quadrants. [3.1]
2. The equation of a vertical line cannot be written in slope–intercept form. [3.6]
3. Equations for lines written in slope–intercept form appear in the form $Ax + By = C$. [3.6]
4. Every horizontal line has an x-intercept. [3.3]
5. A line's slope is a measure of rate. [3.5]
6. A positive rate of inflation means prices are rising. [3.4]
7. Any two points on a line can be used to determine the line's slope. [3.5]
8. Knowing a line's slope is enough to write the equation of the line. [3.6]
9. Knowing two points on a line is enough to write the equation of the line. [3.7]
10. Parallel lines that are not vertical have the same slope. [3.6]

The circle graph below shows the percentage of online searches done in August 2009 that were performed by a particular search engine. [3.1]
Source: Nielsen Mega View Search

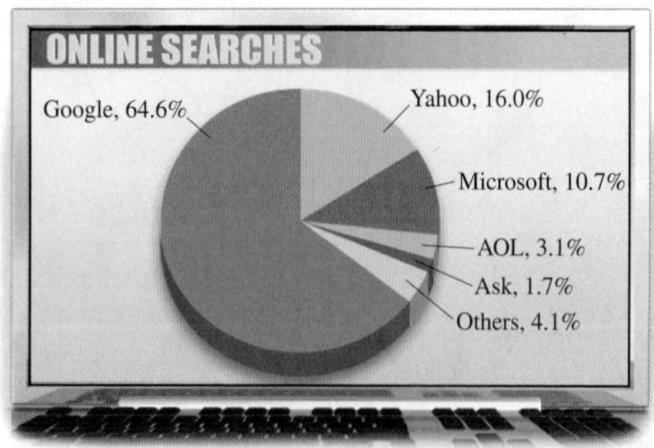

11. There were 10.8 billion searches done by home and business Internet users in August 2009. How many searches were done using Yahoo?

12. About 55% of the online searches done by Advanced Graphics are image searches. In August 2009, Advanced employees did 4200 online searches. If their search engine use is typical, how many image searches did they do using Google?

Plot each point. [3.1]

13. $(5, -1)$ 14. $(2, 3)$ 15. $(-4, 0)$

In which quadrant or on what axis is each point located? [3.1]

16. $(2, -6)$ 17. $(-0.5, -12)$

18. $(-8, 0)$

Find the coordinates of each point in the figure. [3.1]

19. A 20. B 21. C

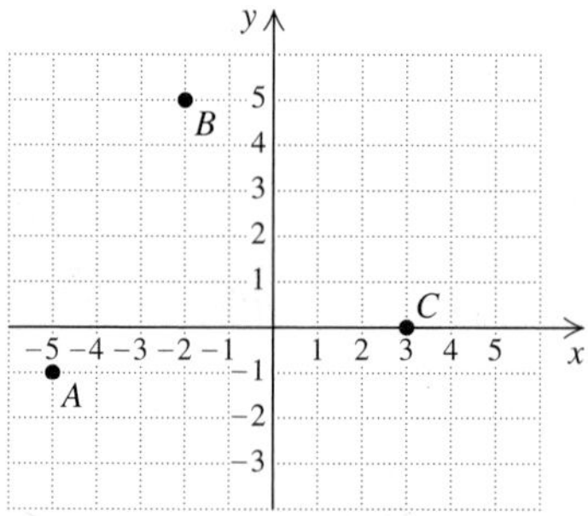

22. Use a grid 10 squares wide and 10 squares high to plot $(-65, -2)$, $(-10, 6)$, and $(25, 7)$. Choose the scale carefully. Scales may vary. [3.1]

23. Determine whether the equation $y = 2x - 5$ has *each* ordered pair as a solution: **(a)** $(3, 1)$; **(b)** $(-3, 1)$. [3.2]

24. Show that the ordered pairs $(0, -3)$ and $(2, 1)$ are solutions of the equation $2x - y = 3$. Then use the graph of the two points to determine another solution. Answers may vary. [3.2]

Graph by hand.

25. $y = x - 5$ [3.2] 26. $y = -\frac{1}{4}x$ [3.2]

27. $y = -x + 4$ [3.2] 28. $4x + y = 3$ [3.2]

29. $4x + 5 = 3$ [3.3] 30. $5x - 2y = 10$ [3.3]

Graph using a graphing calculator. [3.2]

31. $y = x^2 + 1$ 32. $2y - x = 8$

33. *Organic Gardening.* The number of U.S. households g, in millions, that use only all-natural fertilizer and pest control is given by $g = 1.75t + 5$, where t is the number of years since 2004. Graph the equation and use the graph to estimate the number of households using natural garden products in 2012. [3.2]
Source: Based on data from The National Gardening Association

Determine whether each equation is linear. [3.3]

34. $y = x^2 - 2$

35. $y = x - 2$

36. At 4:00 P.M., Jesse's Honda Civic was at mile marker 17 of Interstate 290 in Chicago. At 4:45 P.M., Jesse was at mile marker 23. [3.4]

a) Find Jesse's driving rate, in number of miles per minute.

b) Find Jesse's driving rate, in number of minutes per mile.

37. *Gas Mileage.* The graph below shows data for a Ford Explorer driven on city streets. At what rate was the vehicle consuming gas? [3.4]

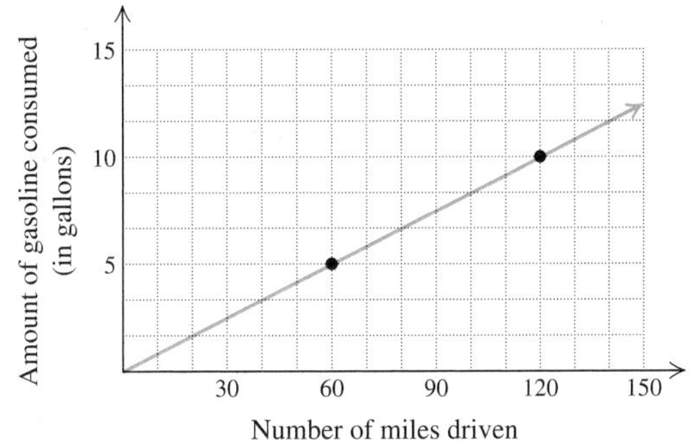

For each line in Exercises 38–40, list **(a)** *the coordinates of the y-intercept,* **(b)** *the coordinates of any x-intercepts, and* **(c)** *the slope.* [3.3], [3.5]

38.

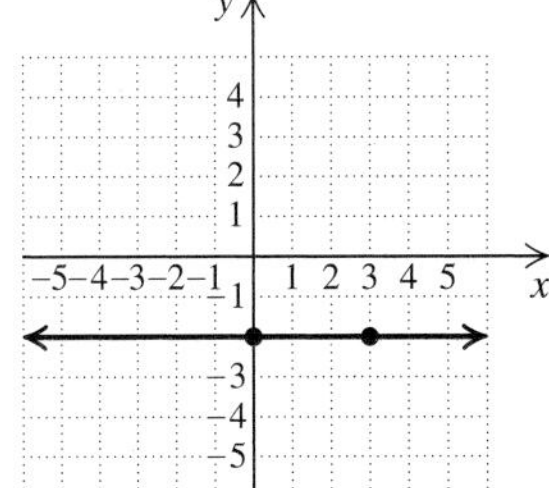

39.

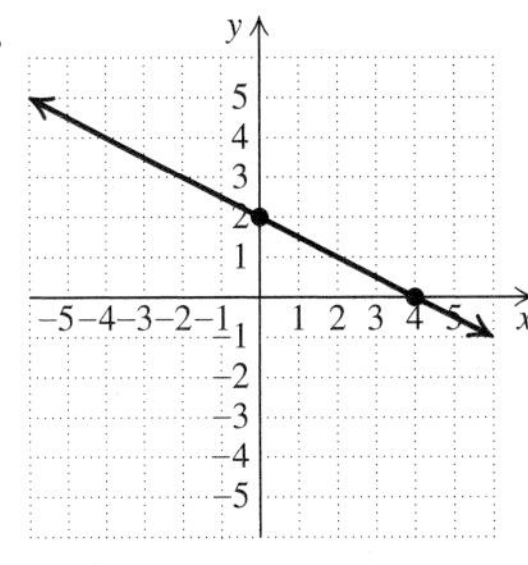

40.

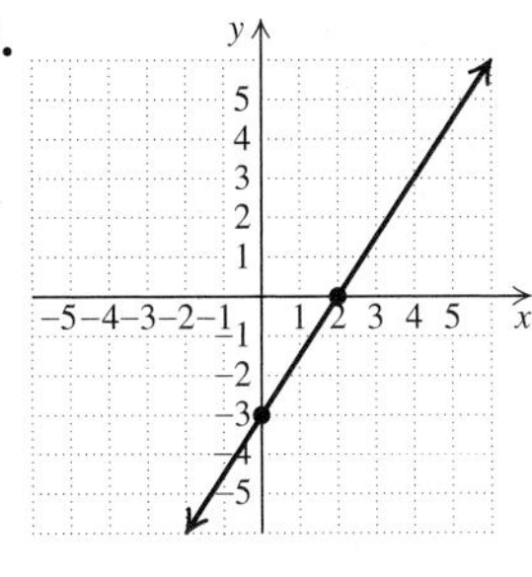

Find the slope of the line containing the given pair of points. If the slope is undefined, state this. [3.5]

41. $(-2, 5)$ and $(3, -1)$

42. $(5, 1)$ and $(-1, 1)$

43. $(-3, 0)$ and $(-3, 5)$

44. $(-8.3, 4.6)$ and $(-9.9, 1.4)$

45. *Architecture.* To meet federal standards, a wheelchair ramp cannot rise more than 1 ft over a horizontal distance of 12 ft. Express this slope as a grade. [3.5]

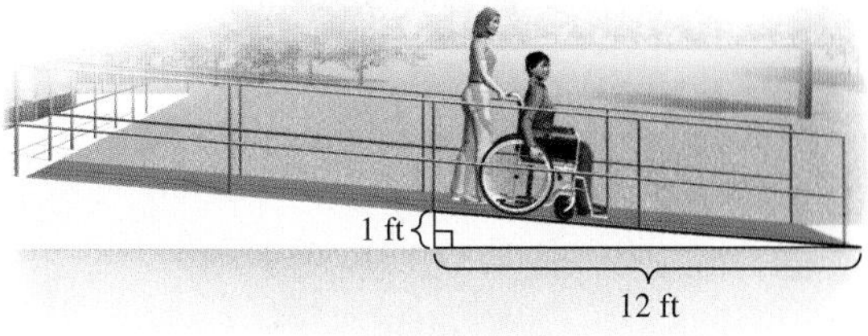

Find the slope of each line. If the slope is undefined, state this.

46. $y = -\frac{7}{10}x - 3$ [3.6]

47. $y = 5$ [3.5]

48. $x = -\frac{1}{3}$ [3.5]

49. $3x - 2y = 6$ [3.6]

50. Find the x-intercept and the y-intercept of the line given by $5x - y = 30$. [3.3]

51. Find the slope and the y-intercept of the line given by $2x + 4y = 20$. [3.6]

Determine whether each pair of lines is parallel, perpendicular, or neither. [3.6]

52. $y + 5 = -x$,
$x - y = 2$

53. $3x - 5 = 7y$,
$7y - 3x = 7$

54. Write the slope–intercept equation of the line with slope $-\frac{3}{4}$ and y-intercept $(0, 6)$. [3.6]

55. Write a point–slope equation for the line with slope $-\frac{1}{2}$ that contains the point $(3, 6)$. [3.7]

56. Write the slope–intercept equation for the line that contains the points $(1, -2)$ and $(-3, -7)$. [3.7]

57. Write the slope–intercept equation for the line that is perpendicular to the line $3x - 5y = 9$ and that contains the point $(2, -5)$. [3.7]

58. *Performing Arts.* Total attendance for U.S. arts performances increased from 11.4 million in 2003 to 12.0 million in 2006.
Source: The League of American Theaters and Producers, Inc., New York, NY

a) Assuming that the growth was linear, find a linear equation that fits the data. Let a represent the attendance, in millions, and t the number of years since 2000. [3.7]

b) Calculate the attendance in 2004. [3.7]

c) Predict the attendance in 2012. [3.7]

Graph.

59. $y = \frac{2}{3}x - 5$ [3.6]

60. $2x + y = 4$ [3.3]

61. $y = 6$ [3.3]

62. $x = -2$ [3.3]

63. $y + 2 = -\frac{1}{2}(x - 3)$ [3.7]

Fitness and Recreation. The table below lists the estimated revenue of U.S. fitness and recreation centers for various years. [3.7]

Year	Revenue (in billions)
2000	$12.5
2002	15.0
2003	16.1
2004	16.8
2005	17.5
2006	18.4

Source: U.S. Census Bureau, "2006 Service Annual Survey, Arts, Entertainment, and Recreation Services"

64. Graph the data and determine whether the relationship appears to be linear.

65. Use linear regression to find a linear equation that fits the data. Let F = the revenue, in billions of dollars, t years after 2000.

66. Use the equation from Exercise 65 to estimate the revenue in 2012.

67. For the graph of f shown here, determine **(a)** $f(2)$; **(b)** the domain of f; **(c)** any x-values for which $f(x) = 2$; and **(d)** the range of f. [3.8]

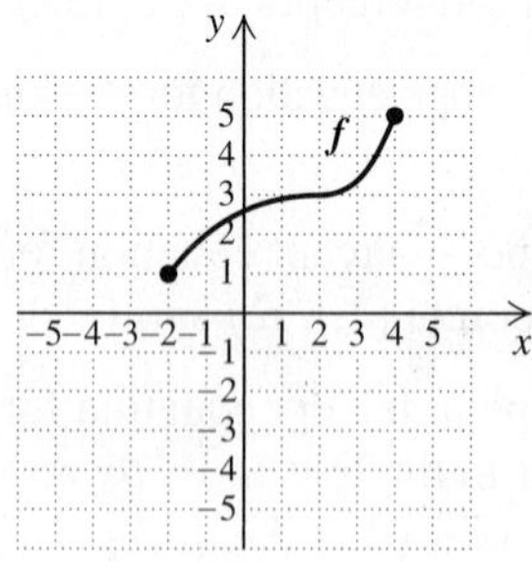

68. Find $g(-3)$ for $g(x) = x^2 - 10x$. [3.8]

69. Find $f(2a)$ for $f(x) = x^2 + 2x - 3$. [3.8]

70. The function $A(t) = 0.11t + 7.9$ can be used to estimate the median age of cars in the United States t years after 1995. (In this context, a median age of 3 years means that half the cars are more than 3 years old and half are less.)

Source: Based on information from R.L. Polk & Co.

a) Predict the median age of cars in 2015; that is, find $A(20)$. [3.8]

b) Determine what 0.11 and 7.9 signify. [3.8]

*For each of the graphs in Exercises 71–74, **(a)** determine whether the graph represents a function and **(b)** if so, determine the domain and the range of the function.* [3.8]

71.

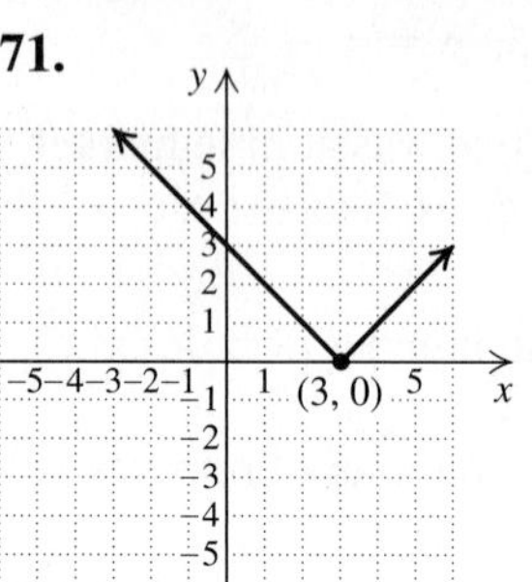

72.

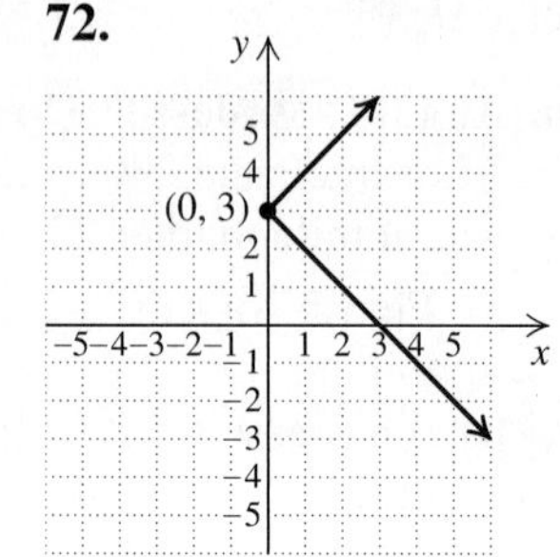

73.

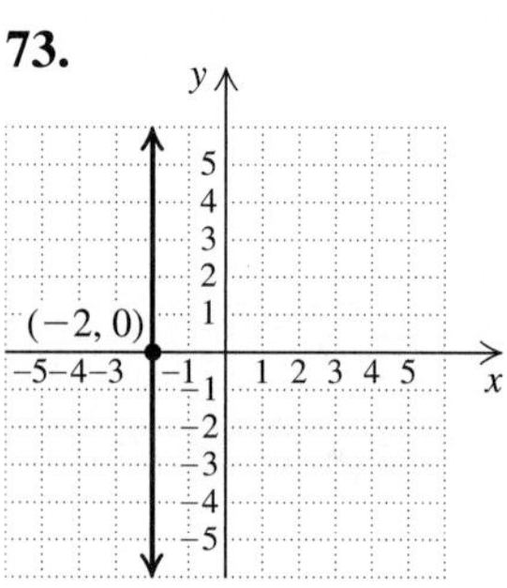

74.

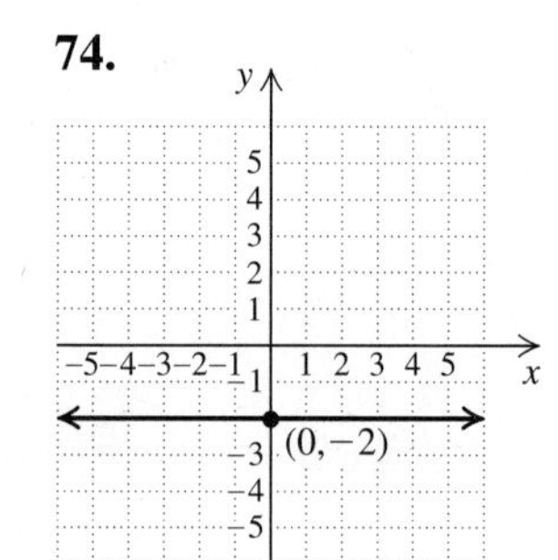

Find the domain of each function. [3.8]

75. $f(x) = 3x^2 - 7$

76. $g(x) = \dfrac{x^2}{x - 1}$

77. For the function given by

$$f(x) = \begin{cases} 2 - x, & \text{for } x \le -2, \\ x^2, & \text{for } -2 < x \le 5, \\ x + 10, & \text{for } x > 5, \end{cases}$$

find **(a)** $f(-3)$; **(b)** $f(-2)$; **(c)** $f(4)$; and **(d)** $f(25)$. [3.8]

SYNTHESIS

TW **78.** Can two perpendicular lines share the same y-intercept? Why or why not? [3.3], [3.6]

TW **79.** If two functions have the same domain and range, are the functions identical? Why or why not? [3.8]

80. Find the value of m in $y = mx + 3$ such that $(-2, 5)$ is on the graph. [3.2]

81. Find the value of b in $y = -5x + b$ such that $(3, 4)$ is on the graph. [3.2]

82. Find the area and the perimeter of a rectangle for which $(-2, 2)$, $(7, 2)$, and $(7, -3)$ are three of the vertices. [3.1]

83. Find three solutions of $y = 4 - |x|$. [3.2]

84. Determine the domain and the range of the function graphed below. [3.8]

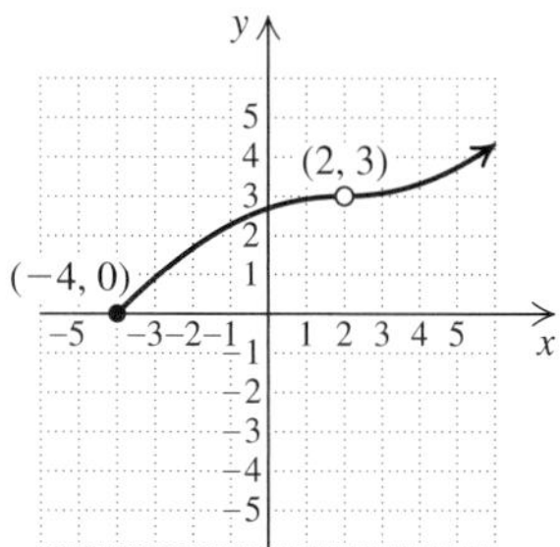

Chapter Test 3

Volunteering. The pie chart below shows the types of organizations in which people ages 16–24 volunteer. Use this pie chart to answer Exercises 1 and 2.

Volunteers, Ages 16–24

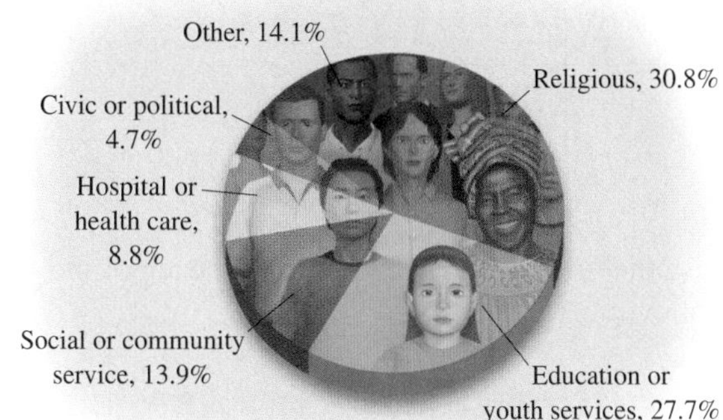

Source: U.S. Bureau of Labor Statistics

1. At Rolling Hills College, 25% of the 1200 students volunteer. How many students will volunteer in education or youth services?

2. At Valley University, $\frac{1}{3}$ of the 3900 students volunteer. How many students will volunteer in hospital or health-care services?

In which quadrant or on which axis is each point located?

3. $(0, 6)$

4. $(-1.6, 2.3)$

Find the coordinates of each point in the figure.

5. A

6. B

7. C

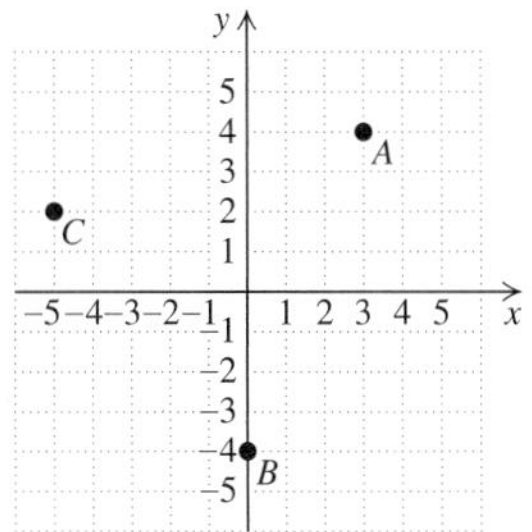

Graph by hand.

8. $y = 2x - 1$

9. $2x - 4y = -8$

10. $y + 4 = -\frac{1}{2}(x - 3)$

11. $y = \frac{3}{4}x$

12. $2x - y = 3$

13. $x = -1$

14. Graph using a graphing calculator: $1.2x - y = 5$.

15. Find the x-intercept and the y-intercept of the line given by $x - 2y = 16$.

16. Find the slope of the line containing the points $(4, -1)$ and $(6, 8)$.

17. *Running.* Blake reached the 3-km mark of a race at 2:15 P.M. and the 6-km mark at 2:24 P.M. What is his running rate, in kilometers/minute?

18. At one point Filbert Street, the steepest street in San Francisco, drops 63 ft over a horizontal distance of 200 ft. Find the road grade.

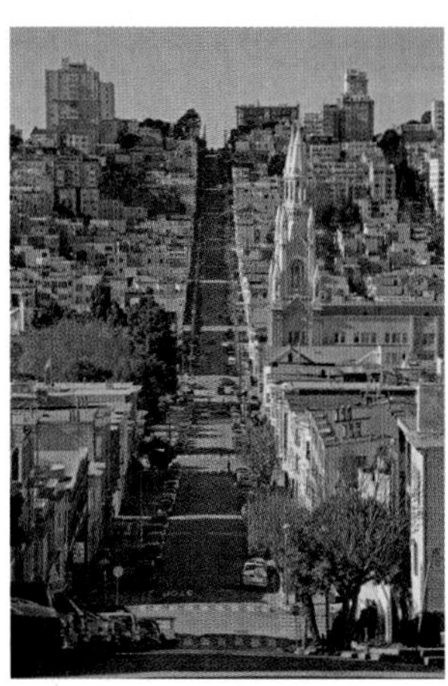

19. Find the slope and the y-intercept of the line given by $y - 8x = 10$.

Determine without graphing whether each pair of lines is parallel, perpendicular, or neither.

20. $4y + 2 = 3x$,
$-3x + 4y = -12$

21. $y = -2x + 5$,
$2y - x = 6$

22. Write a point–slope equation for the line of slope -3 that contains the point $(6, 8)$.

23. Write the slope–intercept equation of the line that is perpendicular to the line $x - y = 3$ and that contains the point $(3, 7)$.

24. The average urban commuter was sitting in traffic for 16 hr in 1982 and for 41 hr in 2007.
Source: Urban Mobility Report

a) Assuming a linear relationship, find an equation that fits the data. Let $c =$ the number of hours that the average urban commuter is sitting in traffic and t the number of years after 1982.
b) Calculate the number of hours that the average urban commuter was sitting in traffic in 2000.
c) Predict the number of hours that the average urban commuter will be sitting in traffic in 2012.

The table below lists the number of twin births in the United States for various years. Use the table for Exercises 25 and 26.

Year	Number of Twin Births (in thousands)
1980	68
1985	77
1990	94
1995	97
2000	119
2005	133
2006	137

Source: Centers for Disease Control and Prevention, *National Vital Statistics Reports*, Vol. 57, No. 7; Jan.7, 2009

25. Use linear regression to find a linear equation that fits the data. Let $B =$ the number of twin births, in thousands, and t the number of years after 1980.

26. Use the equation from Exercise 25 to estimate the number of twin births in 2012.

For each of the graphs in Exercises 27–29, **(a)** *determine whether the graph represents a function and* **(b)** *if so, determine the domain and the range of the function.*

27.

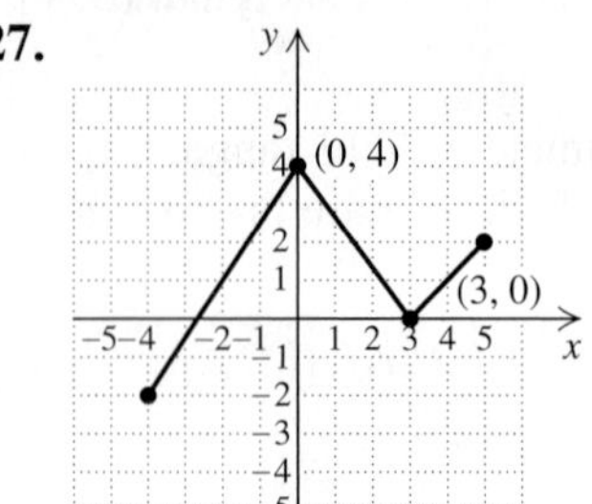

28.

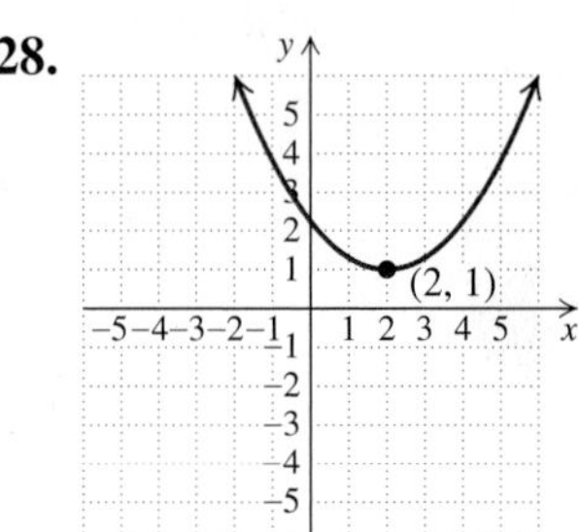

29.

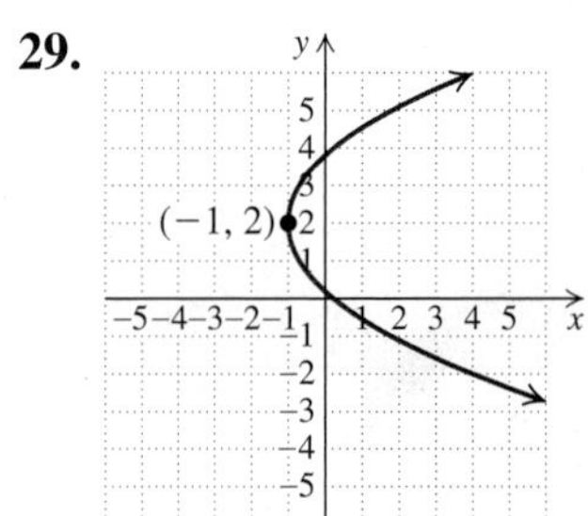

30. Use the graph in Exercise 27 to find each of the following.
a) $f(3)$
b) Any inputs x such that $f(x) = -2$

31. If $g(x) = \dfrac{4}{2x + 1}$, find each of the following.
a) $g(1)$
b) The domain of g

32. For the function given by

$$f(x) = \begin{cases} x^2, & \text{if } x < 0, \\ 3x - 5, & \text{if } 0 \le x \le 2, \\ x + 7, & \text{if } x > 2, \end{cases}$$

find **(a)** $f(0)$; **(b)** $f(3)$.

SYNTHESIS

33. Write an equation of the line that is parallel to the graph of $2x - 5y = 6$ and has the same y-intercept as the graph of $3x + y = 9$.

34. A diagonal of a square connects the points $(-3, -1)$ and $(2, 4)$. Find the area and the perimeter of the square.

Cumulative Review: Chapters 1–3

1. Evaluate $\frac{x}{5y}$ for $x = 70$ and $y = 2$. [1.1]

2. Multiply: $6(2a - b + 3)$. [1.2]

3. Factor: $8x - 4y + 4$. [1.2]

4. Find the prime factorization of 54. [1.3]

5. Find decimal notation: $-\frac{3}{20}$. [1.4]

6. Find the absolute value: $|-37|$. [1.4]

7. Find the opposite of $-\frac{1}{10}$. [1.6]

8. Find the reciprocal of $-\frac{1}{10}$. [1.7]

9. Find decimal notation: 36.7%. [2.4]

Simplify.

10. $\frac{3}{5} - \frac{5}{12}$ [1.3]

11. $(-2)(-1.4)(2.6)$ [1.7]

12. $\frac{3}{8} \div \left(-\frac{9}{10}\right)$ [1.7]

13. $1 + 6 \cdot 10 \div (-1) \cdot 2^2$ [1.8]

14. $1 - [32 \div (4 + 2^2)]$ [1.8]

15. $-5 + 16 \div 2 \cdot 4$ [1.8]

16. $y - (3y + 7)$ [1.8]

17. $3(x - 1) - 2[x - (2x + 7)]$ [1.8]

Solve.

18. $2.7 = 5.3 + x$ [2.1]

19. $\frac{5}{3}x = -45$ [2.1]

20. $3x - 7 = 41$ [2.2]

21. $\frac{3}{4} = \frac{-n}{8}$ [2.1]

22. $14 - 5x = 2x$ [2.2]

23. $3(5 - x) = 2(3x + 4)$ [2.2]

24. $\frac{1}{4}x - \frac{2}{3} = \frac{3}{4} + \frac{1}{3}x$ [2.2]

25. $y + 5 - 3y = 5y - 9$ [2.2]

26. $x - 28 < 20 - 2x$ [2.6]

27. $2(x + 2) \geq 5(2x + 3)$ [2.6]

28. Solve $A = 2\pi rh + \pi r^2$ for h. [2.3]

29. In which quadrant is the point $(3, -1)$ located? [3.1]

30. Graph on a number line: $-1 < x \leq 2$. [2.6]

31. Use a grid 10 squares wide and 10 squares high to plot $(-150, -40)$, $(40, -7)$, and $(0, 6)$. Choose the scale carefully. [3.1]

Graph.

32. $x = 3$ [3.3]

33. $2x - 5y = 10$ [3.3]

34. $y = -2x + 1$ [3.2]

35. $y = \frac{2}{3}x$ [3.2]

36. $y = -\frac{3}{4}x + 2$ [3.6]

37. $2y - 5 = 3$ [3.3]

Find the coordinates of the x- and y-intercepts. Do not graph.

38. $2x - 7y = 21$ [3.3]

39. $y = 4x + 5$ [3.3]

40. Find the slope and the y-intercept of the line given by $3x - y = 2$. [3.6]

41. Find the slope of the line containing the points $(-4, 1)$ and $(2, -1)$. [3.5]

42. Write an equation of the line with slope $\frac{2}{7}$ and y-intercept $(0, -4)$. [3.6]

43. Write a point–slope equation of the line with slope $-\frac{3}{8}$ that contains the point $(-6, 4)$. [3.7]

44. Write the slope–intercept form of the equation in Exercise 43. [3.6]

45. Determine an equation for the graph below. [3.6], [3.7]

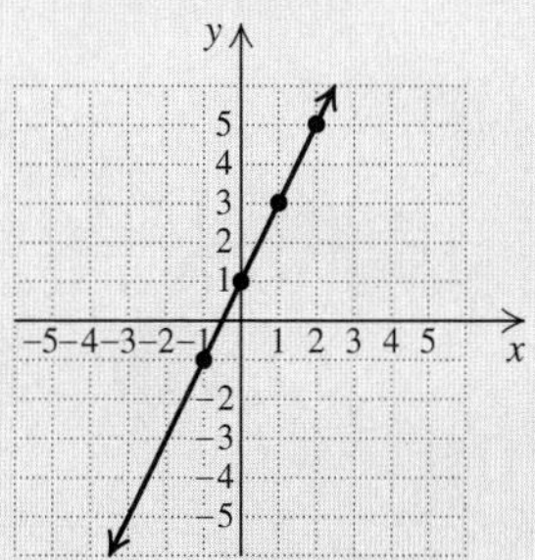

46. For the graph of f shown, determine the domain, the range, $f(-3)$, and any value of x for which $f(x) = 5$. [3.8]

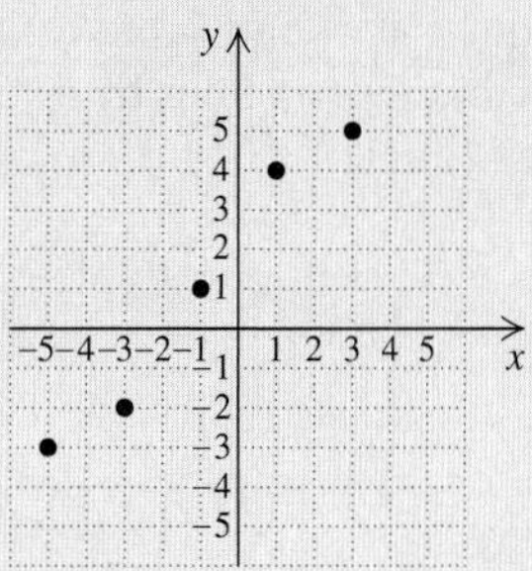

47. For the function given by $f(x) = \dfrac{7}{2x - 1}$, determine each of the following.

a) $f(0)$ [3.8]
b) The domain of f [3.8]

48. A 150-lb person will burn 240 calories per hour when riding a bicycle at 6 mph. The same person will burn 410 calories per hour when cycling at 12 mph. [3.7]
Source: American Heart Association

a) Graph the data and determine an equation for the related line. Let r = the rate at which the person is cycling and c = the number of calories burned per hour.
b) Use the equation of part (a) to estimate the number of calories burned per hour by a 150-lb person cycling at 10 mph.

49. The table below lists the number of calories burned per hour while riding a bicycle at 12 mph for persons of various weights.

Weight (in pounds)	Calories Burned Per Hour at 12 mph
100	270
150	410
200	534

Source: American Heart Association

a) Use linear regression to find an equation that fits the data. Let c = the number of calories burned and w = the weight, in pounds. [3.7]
b) Use the equation of part (a) to estimate the number of calories burned per hour by a 135-lb person cycling at 12 mph. [3.7]

50. In 2007, the mean earnings of individuals with a high school diploma was $31,286. This was about 54.7% of the mean earnings of those with a bachelor's degree. What were the mean earnings of individuals with a bachelor's degree in 2007? [2.4]
Source: U.S. Census Bureau

51. Recently there were 136 million Americans with either O-positive or O-negative blood. Those with O-positive blood outnumbered those with O-negative blood by 115 million. How many Americans had O-negative blood? [2.5]
Source: American Red Cross

52. Tara paid $126 for a cordless drill, including a 5% sales tax. How much did the drill itself cost? [2.4]

53. A 143-m wire is cut into three pieces. The second is 3 m longer than the first. The third is four-fifths as long as the first. How long is each piece? [2.5]

54. In order to qualify for availability pay, a criminal investigator must average at least 2 hr of unscheduled duty per workday. For the first four days of one week, Clint worked 1, 0, 3, and 2 extra hours. How many extra hours must he work on Friday in order to qualify for availability pay? [2.7]
Source: U.S. Department of Justice

SYNTHESIS

55. Anya's salary at the end of a year is $26,780. This reflects a 4% salary increase in February and then a 3% cost-of-living adjustment in June. What was her salary at the beginning of the year? [2.4]

Solve. If no solution exists, state this.

56. $4|x| - 13 = 3$ [1.4], [2.2]

57. $\dfrac{2 + 5x}{4} = \dfrac{11}{28} + \dfrac{8x + 3}{7}$ [2.2]

58. $5(7 + x) = (x + 6)5$ [2.2]

59. Solve $p = \dfrac{2}{m + Q}$ for Q. [2.3]

60. The points $(-3, 0)$, $(0, 7)$, $(3, 0)$, and $(0, -7)$ are vertices of a parallelogram. Find four equations of lines that intersect to form the parallelogram. [3.6]

4

Systems of Equations in Two Variables

Where Are the Wide, Open Spaces?

Some of America's National Parks are crowded with visitors, while in others it is easy to find solitude. As the accompanying graph shows, the number of visits to Gates of the Arctic National Park is increasing and the number to North Cascades National Park is decreasing. In Example 6 of Section 4.1, we will estimate the year in which the number of visits to both parks will be the same.

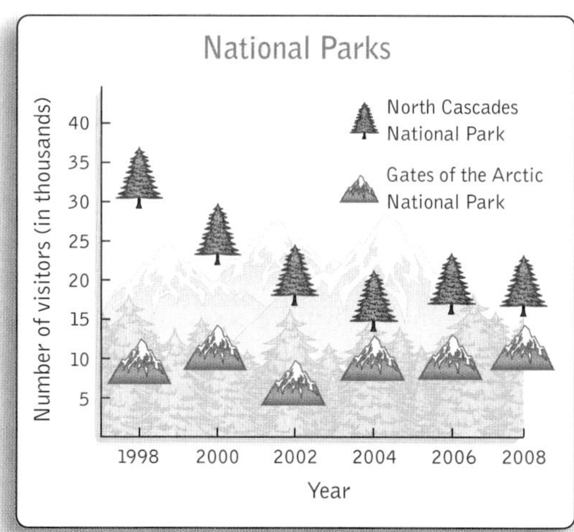

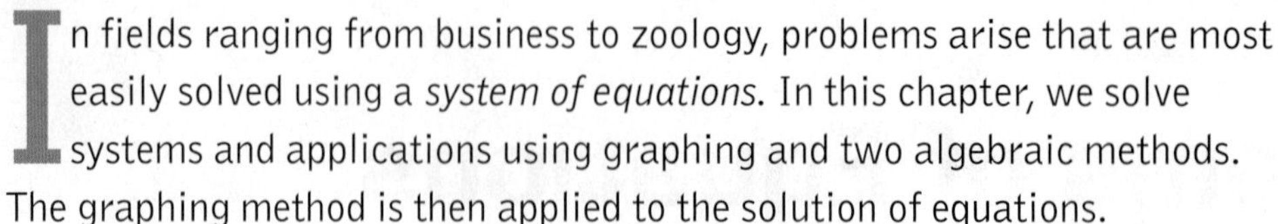

In fields ranging from business to zoology, problems arise that are most easily solved using a *system of equations*. In this chapter, we solve systems and applications using graphing and two algebraic methods. The graphing method is then applied to the solution of equations.

4.1 Systems of Equations and Graphing

- Solutions of Systems
- Solving Systems of Equations by Graphing
- Models

SOLUTIONS OF SYSTEMS

A **system of equations** is a set of two or more equations, in two or more variables, for which we seek a common solution. It is often easier to translate real-world situations to a system of two equations that use two variables, or unknowns, than it is to represent the situation with one equation using one variable. For example, the following problem from Section 2.5 can be represented by two equations using two unknowns:

> The perimeter of an NBA basketball court is 288 ft. The length is 44 ft longer than the width. Find the dimensions of the court.
> Source: National Basketball Association

If we let $w =$ the width of the court, in feet, and $l =$ the length of the court, in feet, the problem translates to the following system of equations:

$$2l + 2w = 288, \quad \text{The perimeter is 288.}$$
$$l = w + 44. \quad \text{The length is 44 more than the width.}$$

A solution of this system will be a length and a width that make *both* equations true, often written as an ordered pair of the form (l, w). Note that variables are listed in alphabetical order within an ordered pair unless stated otherwise.

EXAMPLE 1 Consider the system from above:

$$2l + 2w = 288,$$
$$l = w + 44.$$

Determine whether each pair is a solution of the system: **(a)** (94, 50); **(b)** (90, 46).

SOLUTION

a) Unless stated otherwise, we use alphabetical order of the variables. Thus we replace l with 94 and w with 50.

$$\begin{array}{c|c} 2l + 2w & = 288 \\ \hline 2 \cdot 94 + 2 \cdot 50 & 288 \\ 188 + 100 & \\ 288 & \stackrel{?}{=} 288 \quad \text{TRUE} \end{array} \qquad \begin{array}{c|c} l & = w + 44 \\ \hline 94 & 50 + 44 \\ 94 & \stackrel{?}{=} 94 \quad \text{TRUE} \end{array}$$

Since (94, 50) checks in *both* equations, it is a solution of the system.

STUDY TIP

Speak Up

Feel free to ask questions in class at appropriate times. Most instructors welcome questions and encourage students to ask them. Other students in your class probably have the same questions you do.

CAUTION! Be sure to check the ordered pair in *both* equations.

b) We substitute 90 for l and 46 for w:

$$\begin{array}{r|l} 2l + 2w = & 288 \\ \hline 2 \cdot 90 + 2 \cdot 46 & 288 \\ 180 + 92 & \\ 272 \stackrel{?}{=} & 288 \quad \text{FALSE} \end{array} \qquad \begin{array}{r|l} l = & w + 44 \\ \hline 90 & 46 + 44 \\ 90 \stackrel{?}{=} & 90 \quad \text{TRUE} \end{array}$$

Since (90, 46) is not a solution of *both* equations, it is not a solution of the system.

Try Exercise 9.

A solution of a system of two equations in two variables is an ordered pair of numbers that makes both equations true. The numbers in the ordered pair correspond to the variables in alphabetical order.

In Example 1, we demonstrated that (94, 50) is a solution of the system, but we did not show how the pair (94, 50) was found. One way to find such a solution uses graphs.

SOLVING SYSTEMS OF EQUATIONS BY GRAPHING

Recall that the graph of an equation is a drawing that represents its solution set. Each point on the graph corresponds to an ordered pair that is a solution of the equation. By graphing two equations using one set of axes, we can identify a solution of both equations by looking for a point of intersection.

Student Notes

Because it is so critical to accurately identify any point(s) of intersection, use graph paper, a ruler or other straightedge, and a pencil to graph each line as neatly as possible.

EXAMPLE 2 Solve this system of equations by graphing:

$$x + y = 7,$$
$$y = 3x - 1.$$

SOLUTION We graph the equations using any method studied earlier. The equation $x + y = 7$ can be graphed easily using the intercepts, $(0, 7)$ and $(7, 0)$. The equation $y = 3x - 1$ is in slope–intercept form, so it can be graphed by plotting its y-intercept, $(0, -1)$, and "counting off" a slope of 3.

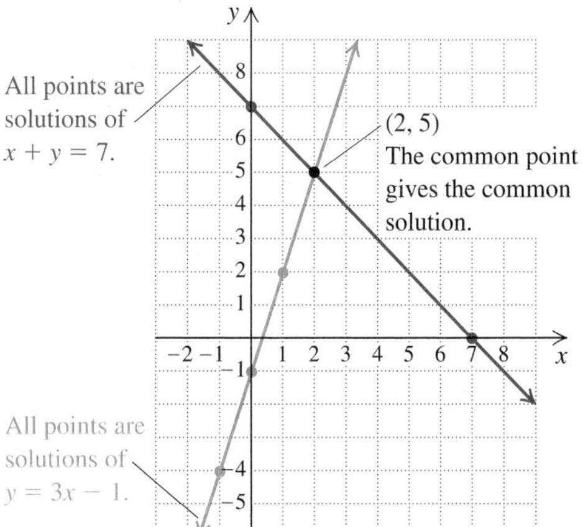

The "apparent" solution of the system, (2, 5), should be checked in both equations. Note that the solution is an ordered pair in the form (x, y).

Check:

$x + y = 7$		$y = 3x - 1$	
$2 + 5$	7	5	$3 \cdot 2 - 1$
$7 \stackrel{?}{=} 7$	TRUE	$5 \stackrel{?}{=} 5$	TRUE

Since it checks in both equations, (2, 5) is a solution of the system.

Try Exercise 17.

Point of Intersection

Often a graphing calculator provides a more accurate solution than graphs drawn by hand. We can trace along the graph of one of the functions to find the coordinates of the point of intersection. Most graphing calculators can find the point of intersection directly using an INTERSECT feature.

To find the point of intersection of two graphs, press CALC and choose the INTERSECT option. The questions FIRST CURVE? and SECOND CURVE? are used to identify the graphs with which we are concerned. Position the cursor on each graph, in turn, and press ENTER, using the up and down arrow keys, if necessary, to move to another graph.

The calculator then asks a third question, GUESS?. Since graphs may intersect at more than one point, we must visually identify the point of intersection in which we are interested and indicate the general location of that point. Enter a guess either by moving the cursor near the point of intersection and pressing ENTER or by typing a guess and pressing ENTER. The calculator then returns the coordinates of the point of intersection. At this point, the calculator variables X and Y contain the coordinates of the point of intersection. These values can be used to check an answer, and often they can be converted to fraction notation from the home screen using the FRAC option of the MATH MATH menu.

Your Turn

1. Enter the equations $y_1 = 2x - 3$ and $y_2 = 4 - x$. Press ZOOM 6 to graph them in a standard viewing window.
2. Press CALC 5 to choose the INTERSECT option.
3. Choose the curves. If these two equations are the only ones entered, simply press ENTER ENTER.
4. Make a guess by moving the cursor close to the point of intersection and pressing ENTER.
5. The coordinates of the point of intersection are written in decimal form. Return to the home screen and convert both X and Y to fraction notation. $X = \frac{7}{3}, Y = \frac{5}{3}$

Most pairs of lines have exactly one point in common. We will soon see, however, that this is not always the case.

EXAMPLE 3 Solve graphically:

$$y - x = 1,$$
$$y + x = 3.$$

SOLUTION

BY HAND

We graph each equation using any method studied in Chapter 3. All ordered pairs from line L_1 in the graph below are solutions of the first equation. All ordered pairs from line L_2 are solutions of the second equation. The point of intersection has coordinates that make *both* equations true. Apparently, $(1, 2)$ is the solution. Graphing is not always accurate, so solving by graphing may yield approximate answers. Our check below shows that $(1, 2)$ is indeed the solution.

Graph both equations.

Look for any points in common.

Check.

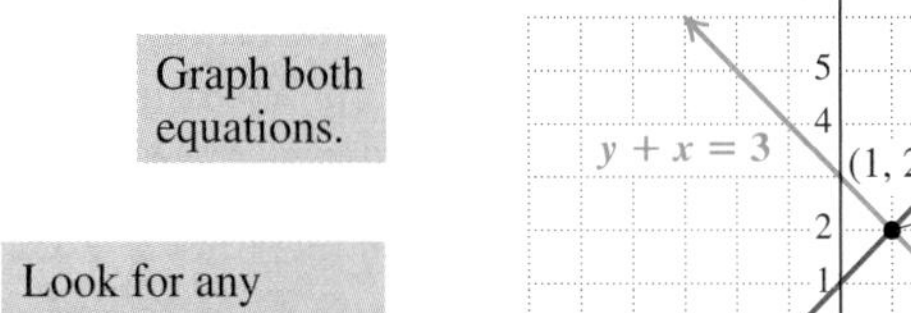

Check:

$$\begin{array}{c|c} y - x & = 1 \\ \hline 2 - 1 & 1 \\ 1 & \stackrel{?}{=} 1 \end{array} \quad \text{TRUE}$$

$$\begin{array}{c|c} y + x & = 3 \\ \hline 2 + 1 & 3 \\ 3 & \stackrel{?}{=} 3 \end{array} \quad \text{TRUE}$$

WITH A GRAPHING CALCULATOR

We first solve each equation for y:

$$y - x = 1 \qquad\qquad y + x = 3$$
$$y = x + 1; \qquad\qquad y = 3 - x.$$

Then we enter and graph $y_1 = x + 1$ and $y_2 = 3 - x$.

After selecting the INTERSECT option, we choose the two *curves* and enter a guess. The coordinates of the point of intersection then appear at the bottom of the screen.

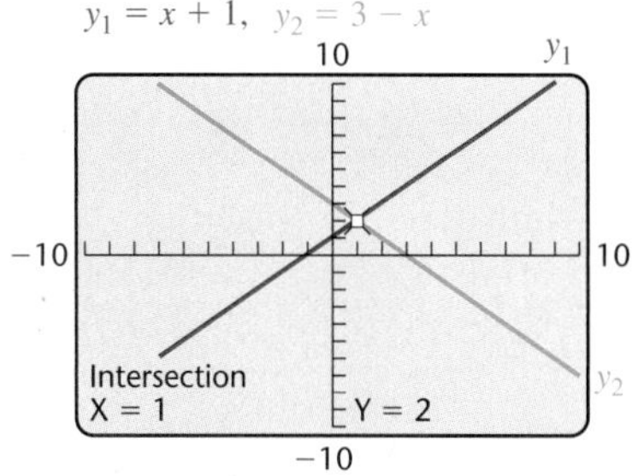

Since, in the calculator, X now has the value 1 and Y has the value 2, we can check from the home screen. The solution is $(1, 2)$.

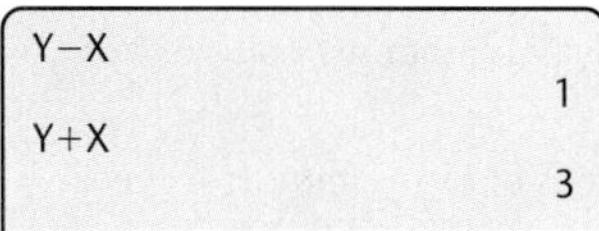

Try Exercise 19.

EXAMPLE 4 Solve each system graphically.

a) $y = -3x + 5,$
$y = -3x - 2$

b) $3y - 2x = 6,$
$-12y + 8x = -24$

SOLUTION

a) We graph the equations. The lines have the same slope, -3, and different y-intercepts, so they are parallel. There is no point at which they cross, so the system has no solution.

$$y = -3x + 5,$$
$$y = -3x - 2$$

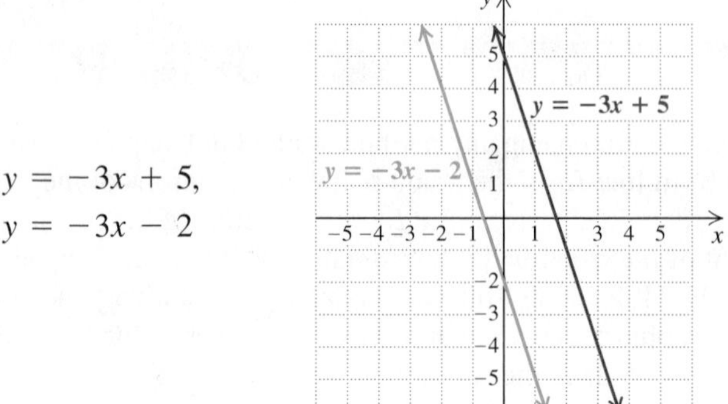

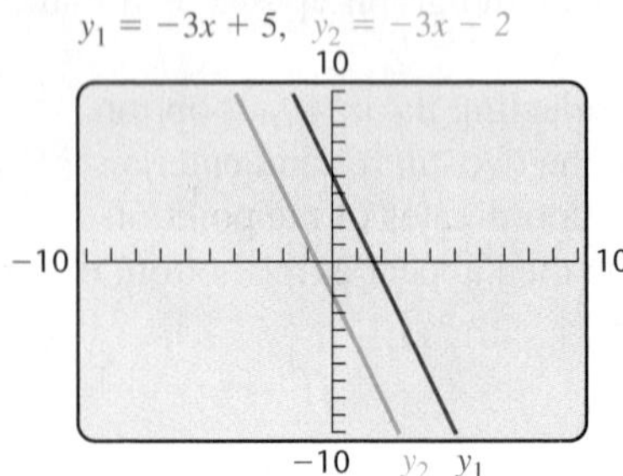

What happens when we try to solve this system using a graphing calculator? As shown at left, the graphs appear to be parallel. (This must be verified algebraically by confirming that they have the same slope.) When we attempt to find the intersection using the INTERSECT feature, we get an error message.

b) We graph the equations and find that the same line is drawn twice. Thus any solution of one equation is a solution of the other. Each equation has an infinite number of solutions, so the system itself has an infinite number of solutions. We check one solution, $(0, 2)$, which is the y-intercept of each equation.

$$3y - 2x = 6,$$
$$-12y + 8x = -24$$

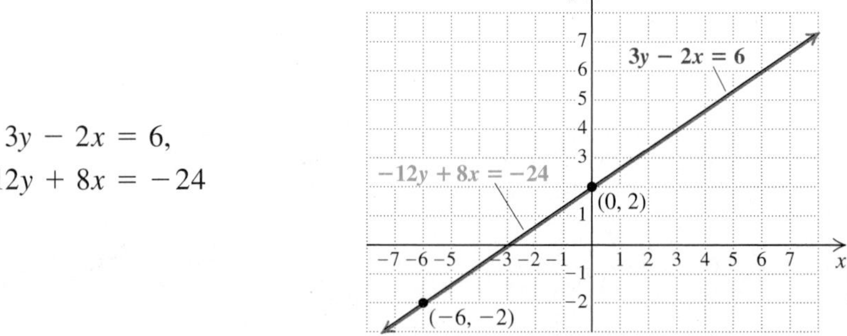

Student Notes

Although the system in Example 4(b) is true for an infinite number of ordered pairs, those pairs must be of a certain form. Only pairs that are solutions of $3y - 2x = 6$ or $-12y + 8x = -24$ are solutions of the system. It is incorrect to think that *all* ordered pairs are solutions.

Check:

$3y - 2x = 6$	
$3(2) - 2(0)$	6
$6 - 0$	
$6 \stackrel{?}{=} 6$	TRUE

$-12y + 8x = -24$	
$-12(2) + 8(0)$	-24
$-24 + 0$	
$-24 \stackrel{?}{=} -24$	TRUE

You can check that $(-6, -2)$ is another solution of both equations. In fact, any pair that is a solution of one equation is a solution of the other equation as well. Thus the solution set is

$$\{(x, y) | 3y - 2x = 6\}$$

or, in words, "the set of all pairs (x, y) for which $3y - 2x = 6$." Since the two equations are equivalent, we could have written instead

$$\{(x, y) | -12y + 8x = -24\}.$$

If we attempt to find the intersection of the graphs of the equations in this system using INTERSECT, the graphing calculator will return as the intersection whatever point we choose as the guess. This is a point of intersection, as is any other point on the graph of the lines.

Try Exercise 33.

When we graph a system of two linear equations in two variables, one of the following three outcomes will occur.

1. The lines have one point in common, and that point is the only solution of the system (see Example 3). Any system that has at least one solution is said to be **consistent.**
2. The lines are parallel, with no point in common, and the system has no solution (see Example 4a). This type of system is called **inconsistent.**
3. The lines coincide, sharing the same graph. Because every solution of one equation is a solution of the other, the system has an infinite number of solutions (see Example 4b). Since it has a solution, this type of system is also consistent.

When one equation in a system can be obtained by multiplying both sides of another equation by a constant, the two equations are said to be **dependent.** Thus the equations in Example 4(b) are dependent, but those in Examples 3 and 4(a) are **independent.** For systems of three or more equations, the definitions of dependent and independent must be slightly modified.

Connecting the Concepts

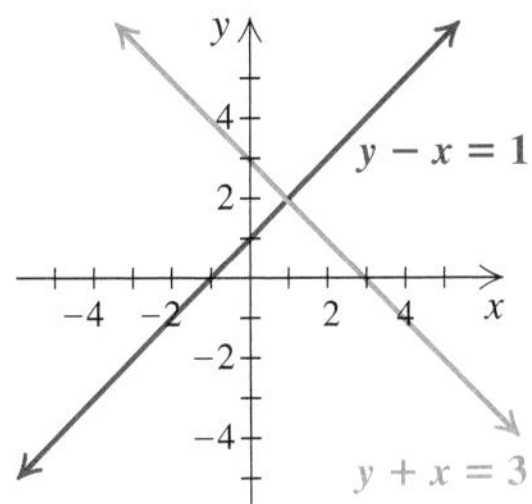

Graphs intersect at one point.

The system

$$y - x = 1,$$
$$y + x = 3$$

is *consistent* and has one solution.

Since neither equation is a multiple of the other, the equations are *independent.*

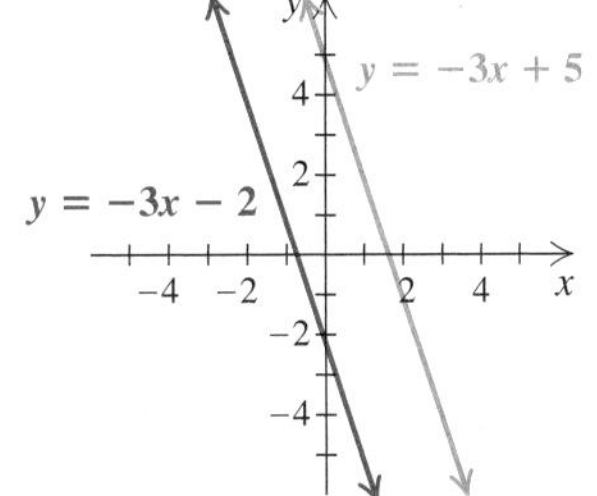

Graphs are parallel.

The system

$$y = -3x - 2,$$
$$y = -3x + 5$$

is *inconsistent* because there is no solution.

Since neither equation is a multiple of the other, the equations are *independent.*

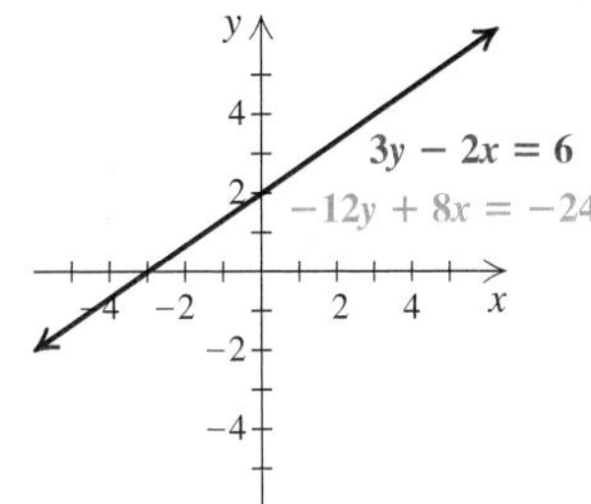

Equations have the same graph.

The system

$$3y - 2x = 6,$$
$$-12y + 8x = -24$$

is *consistent* and has an infinite number of solutions.

Since one equation is a multiple of the other, the equations are *dependent.*

Graphing calculators are especially useful when equations contain fractions or decimals or when the coordinates of the intersection are not integers.

EXAMPLE 5 Solve graphically:

$$3.45x + 4.21y = 8.39,$$
$$7.12x - 5.43y = 6.18.$$

SOLUTION First, we solve for y in each equation:

$3.45x + 4.21y = 8.39$

$4.21y = 8.39 - 3.45x$ — **Subtracting $3.45x$ from both sides**

$y = (8.39 - 3.45x)/4.21;$ — **Dividing both sides by 4.21. This can also be written $y = 8.39/4.21 - 3.45x/4.21$.**

$7.12x - 5.43y = 6.18$

$-5.43y = 6.18 - 7.12x$ — **Subtracting $7.12x$ from both sides**

$y = (6.18 - 7.12x)/(-5.43).$ — **Dividing both sides by -5.43**

It is not necessary to simplify further. We have the system

$$y = (8.39 - 3.45x)/4.21,$$
$$y = (6.18 - 7.12x)/(-5.43).$$

Next, we enter both equations and graph using the same viewing window. By using the INTERSECT feature in the CALC menu, we see that, to the nearest hundredth, the solution is $(1.47, 0.79)$. Note that the coordinates found by the calculator are approximations.

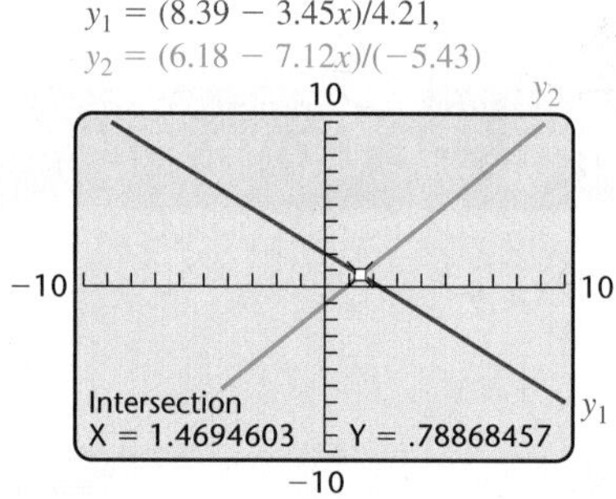

Try Exercise 37.

MODELS

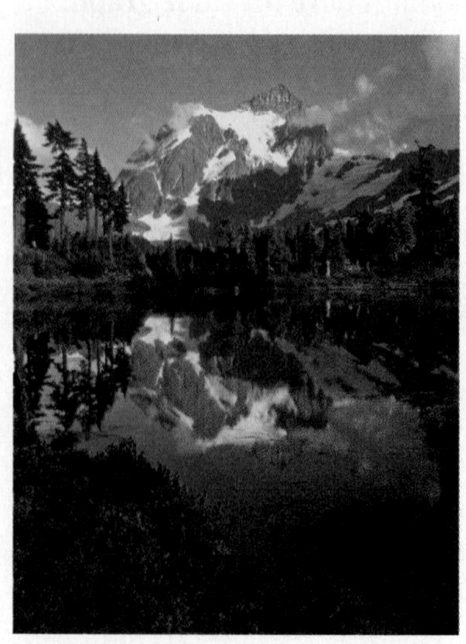

EXAMPLE 6 National Parks. The Gates of the Arctic National Park is growing in popularity while the North Cascades National Park is attracting fewer visitors. Use the data in the table below to predict when the number of visitors to the two parks will be the same.

Year	Number of Visitors to Gates of the Arctic National Park (in thousands)	Number of Visitors to North Cascades National Park (in thousands)
1996	6.4	27.9
1998	8.3	32.8
2000	11.3	25.7
2002	6.6	20.7
2004	10.3	16.9
2006	10.0	19.2
2008	11.4	18.7

Source: National Parks Service

SOLUTION We enter the number of years after 1996 as L1, the number of visitors to Gates of the Arctic as L2, and the number of visitors to North Cascades as L3. A good choice of a viewing window is $[0, 15, 0, 40]$, with Xscl = 5 and Yscl = 5.

We graph the data, using a box as the mark for Gates of the Arctic and a plus sign as the mark for North Cascades. Although neither set of data is exactly linear, a line is still a good model of each set for purposes of estimation and prediction.

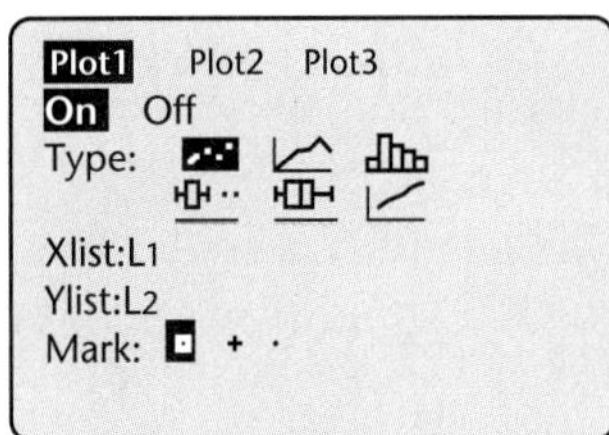

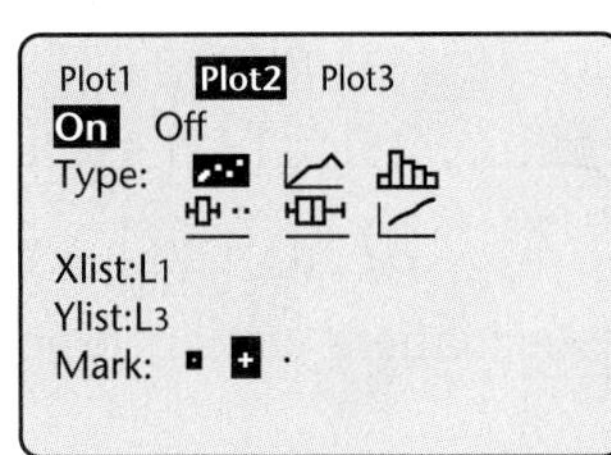

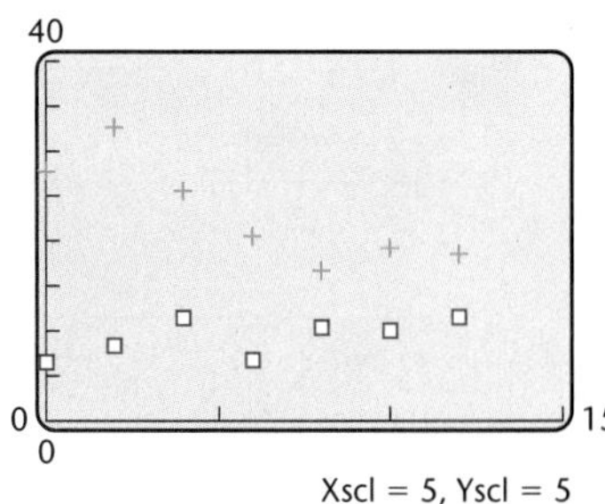

Xscl = 5, Yscl = 5

LinReg(ax+b) L1, L2, Y1

LinReg(ax+b) L1, L3, Y2

The figures at left show the commands used to find the linear regression lines for the data. The line for Gates of the Arctic, using lists L1 and L2, is stored as Y1. The line for North Cascades, using lists L1 and L3, is stored as Y2.

The screen on the left below gives us the two equations. Rounding the coefficients to four decimal places gives

$$y_1 = 0.3107x + 7.3214 \quad \text{and} \quad y_2 = -1.1357x + 29.9429.$$

Plot1 Plot2 Plot3
\Y1=.31071428571429X+7.3214285714286
\Y2=−1.1357142857143X+29.942857142857
\Y3=

$y_1 = .3107x + 7.3214,$
$y_2 = -1.1357x + 29.9429$

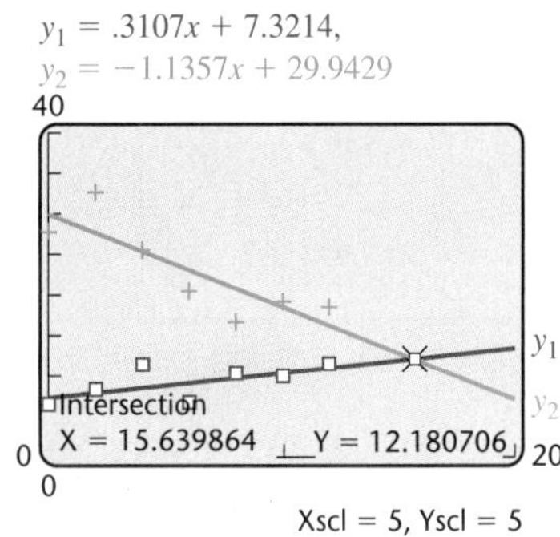

Xscl = 5, Yscl = 5

We graph the equations (using the rounded coefficients) along with the data and determine the coordinates of the point of intersection. The x-coordinate of the point of intersection must be within the window dimensions, so we extend the window, as shown on the right above. The point of intersection is approximately $(15.6, 12.2)$. Since x represents the number of years after 1996, this tells us that approximately 15.6 years after 1996, both parks will have about 12.2 thousand visitors. Rounding, we state that there will be the same number of visitors to the parks in about 2012.

Try Exercise 47.

Graphing is helpful when solving systems because it allows us to "see" the solution. It can also be used with systems of nonlinear equations, and in many applications, it provides a satisfactory answer. However, often we cannot get a precise solution of a system using graphing. In Sections 4.2 and 4.3, we will develop two algebraic methods of solving systems that will produce exact answers.

Visualizing for Success

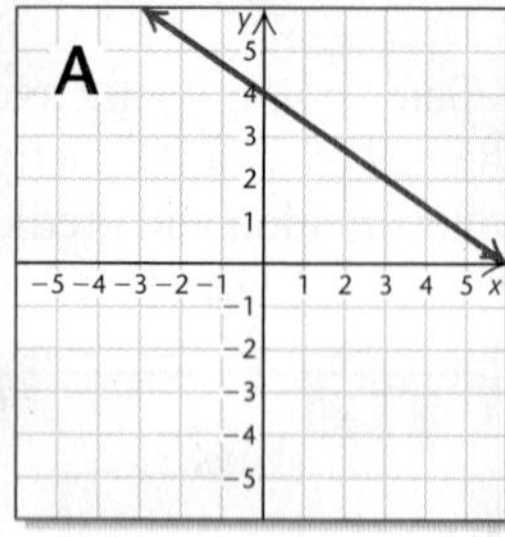

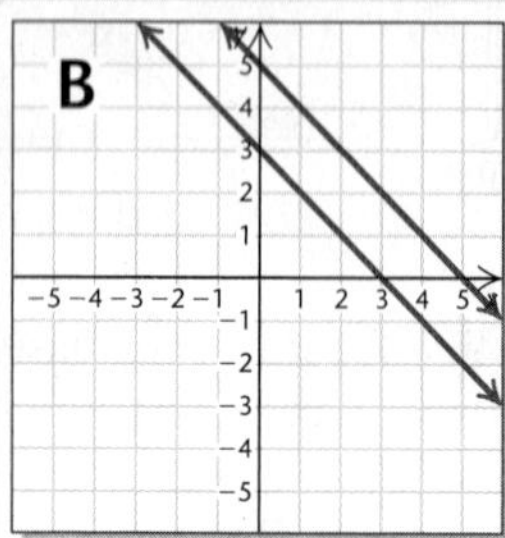

C

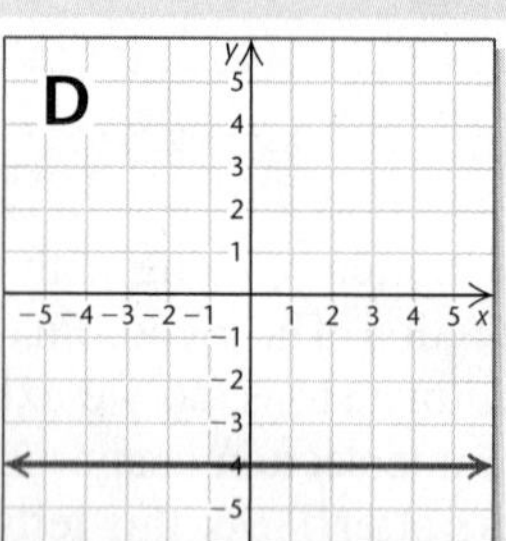

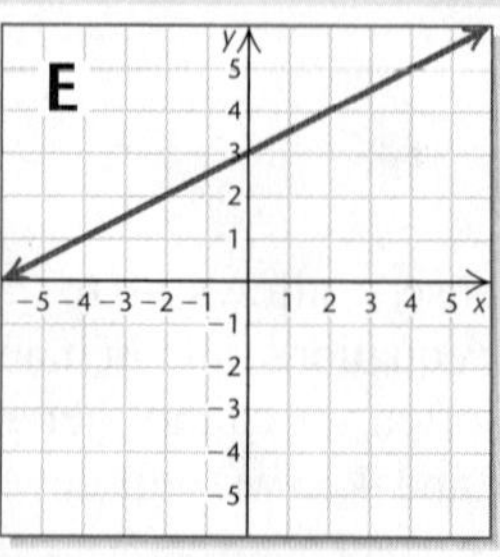

Match each equation or system of equations with its graph.

1. $x + y = 2,$
   $x - y = 2$

2. $y = \frac{1}{3}x - 5$

3. $4x - 2y = -8$

4. $2x + y = 1,$
   $x + 2y = 1$

5. $8y + 32 = 0$

6. $y = -x + 4$

7. $\frac{2}{3}x + y = 4$

8. $x = 4,$
   $y = 3$

9. $y = \frac{1}{2}x + 3,$
   $2y - x = 6$

10. $y = -x + 5,$
    $y = 3 - x$

Answers on page A-15

An additional, animated version of this activity appears in MyMathLab. To use MyMathLab, you need a course ID and a student access code. Contact your instructor for more information.

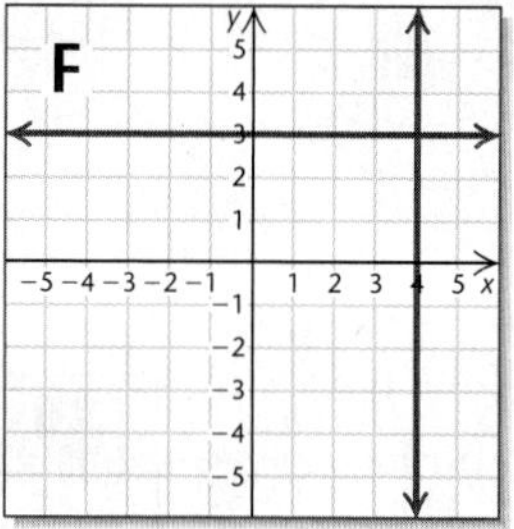

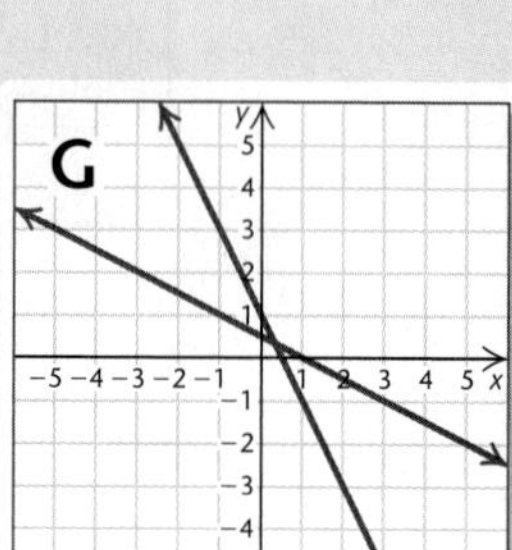

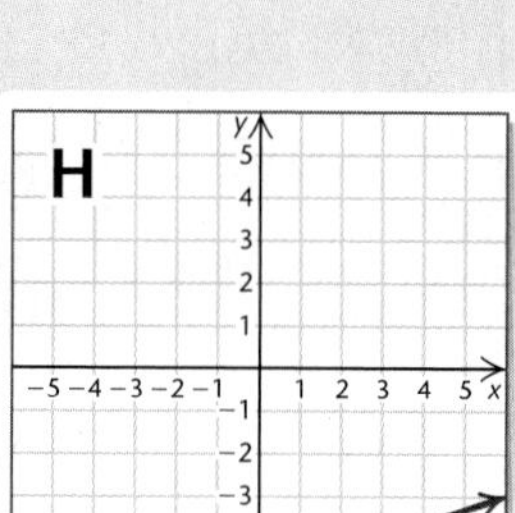

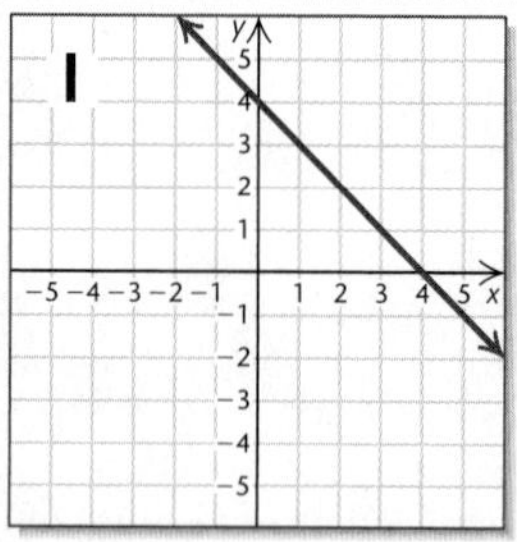

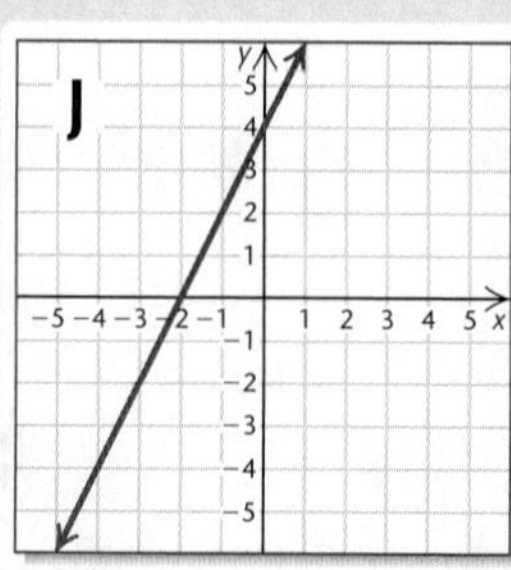

4.1 Exercise Set

FOR EXTRA HELP MyMathLab     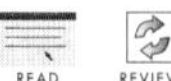

Concept Reinforcement *Classify each of the following statements as either true or false.*

1. Solutions of systems of equations in two variables are ordered pairs.
2. Every system of equations has at least one solution.
3. It is possible for a system of equations to have an infinite number of solutions.
4. The graphs of the equations in a system of two equations may coincide.
5. The graphs of the equations in a system of two equations could be parallel lines.
6. Any system of equations that has at most one solution is said to be consistent.
7. Any system of equations that has more than one solution is said to be inconsistent.
8. The equations $x + y = 5$ and $2(x + y) = 2(5)$ are dependent.

Determine whether the ordered pair is a solution of the given system of equations. Remember to use alphabetical order of variables.

9. $(1, 2)$; $4x - y = 2,$ $10x - 3y = 4$
10. $(4, 0)$; $2x + 7y = 8,$ $x - 9y = 4$
11. $(-5, 1)$; $x + 5y = 0,$ $y = 2x + 9$
12. $(-1, -2)$; $x + 3y = -7,$ $3x - 2y = 12$
13. $(0, -5)$; $x - y = 5,$ $y = 3x - 5$
14. $(5, 2)$; $a + b = 7,$ $2a - 8 = b$

Aha! 15. $(3, 1)$; $3x + 4y = 13,$ $6x + 8y = 26$

16. $(4, -2)$; $-3x - 2y = -8,$ $8 = 3x + 2y$

Solve each system graphically. Be sure to check your solution. If a system has an infinite number of solutions, use set-builder notation to write the solution set. If a system has no solution, state this. Where appropriate, round to the nearest hundredth.

17. $x - y = 3,$ $x + y = 5$
18. $x + y = 4,$ $x - y = 2$
19. $3x + y = 5,$ $x - 2y = 4$
20. $2x - y = 4,$ $5x - y = 13$
21. $4y = x + 8,$ $3x - 2y = 6$
22. $4x - y = 9,$ $x - 3y = 16$
23. $x = y - 1,$ $2x = 3y$
24. $a = 1 + b,$ $b = 5 - 2a$
25. $x = -3,$ $y = 2$
26. $x = 4,$ $y = -5$
27. $t + 2s = -1,$ $s = t + 10$
28. $b + 2a = 2,$ $a = -3 - b$
29. $2b + a = 11,$ $a - b = 5$
30. $y = -\frac{1}{3}x - 1,$ $4x - 3y = 18$
31. $y = -\frac{1}{4}x + 1,$ $2y = x - 4$
32. $6x - 2y = 2,$ $9x - 3y = 1$
33. $y - x = 5,$ $2x - 2y = 10$
34. $y = -x - 1,$ $4x - 3y = 24$
35. $y = 3 - x,$ $2x + 2y = 6$
36. $2x - 3y = 6,$ $3y - 2x = -6$
37. $y = -5.43x + 10.89,$ $y = 6.29x - 7.04$
38. $y = 123.52x + 89.32,$ $y = -89.22x + 33.76$
39. $2.6x - 1.1y = 4,$ $1.32y = 3.12x - 5.04$
40. $2.18x + 7.81y = 13.78,$ $5.79x - 3.45y = 8.94$
41. $0.2x - y = 17.5,$ $2y - 10.6x = 30$

42. $1.9x = 4.8y + 1.7,$
$12.92x + 23.8 = 32.64y$

43. For the systems in the odd-numbered exercises 17–41, which are consistent?

44. For the systems in the even-numbered exercises 18–42, which are consistent?

45. For the systems in the odd-numbered exercises 17–41, which contain dependent equations?

46. For the systems in the even-numbered exercises 18–42, which contain dependent equations?

47. *College Faculty.* The number of part-time faculty in institutions of higher learning is growing rapidly. The table below lists the number of full-time faculty and the number of part-time faculty for various years. Use linear regression to fit a line to each set of data, and use those equations to predict the year in which the number of part-time faculty and the number of full-time faculty will be the same.

Year	Number of Full-time Faculty (in thousands)	Number of Part-time Faculty (in thousands)
1980	450	236
1985	459	256
1991	536	291
1995	551	381
1999	591	437
2005	676	615

Source: U.S. National Center for Education Statistics

48. *Milk Cows.* The number of milk cows in Vermont has decreased since 2004, while the number in Colorado has increased, as shown in the table below. Use linear regression to fit a line to each set of data, and use those equations to predict the year in which the number of milk cows in the two states will be the same.

Year	Number of Milk Cows in Vermont (in thousands)	Number of Milk Cows in Colorado (in thousands)
2004	145	102
2005	143	104
2006	141	110
2007	140	118
2008	140	128

Source: U.S. Department of Agriculture

49. *Financial Advisers.* Financial advisers are leaving major national firms and setting up their own businesses, as shown in the table below. Use linear regression to fit a line to each set of data, and use those equations to predict the year in which there will be the same number of independent financial advisers as there are those in major national firms.

Year	Number of Independent Financial Advisers (in thousands)	Number of Financial Advisers with National Firms (in thousands)
2004	21	60
2005	23	62
2006	25	59
2007	28	57
2008	32	55

Source: Based on information from Cerulli Associates

50. *Recycling.* In the United States, the amount of waste being recovered is slowly catching up to the amount of waste being discarded, as shown in the table below. Use linear regression to fit a line to each set of data, and use those equations to predict the year in which the amount of waste recycled will equal the amount discarded.

Year	Amount of Waste Recovered (in pounds per person per day)	Amount of Waste Discarded (in pounds per person per day)
1980	0.35	3.24
1990	0.73	3.12
2000	1.35	2.64
2008	1.50	2.43

Source: U.S. Environmental Protection Agency

TW **51.** Suppose that the equations in a system of two linear equations are dependent. Does it follow that the system is consistent? Why or why not?

TW **52.** Why is slope–intercept form especially useful when solving systems of equations by graphing?

SKILL REVIEW

To prepare for Section 4.2, review solving equations and formulas (Sections 2.2 and 2.3).

Solve. [2.2]

53. $2(4x - 3) - 7x = 9$

54. $6y - 3(5 - 2y) = 4$

55. $4x - 5x = 8x - 9 + 11x$

56. $8x - 2(5 - x) = 7x + 3$

Solve. [2.3]

57. $3x + 4y = 7$, for y

58. $2x - 5y = 9$, for y

SYNTHESIS

Advertising Media. *For Exercises 59 and 60, consider the graph below showing the U.S. market share for various advertising media.*

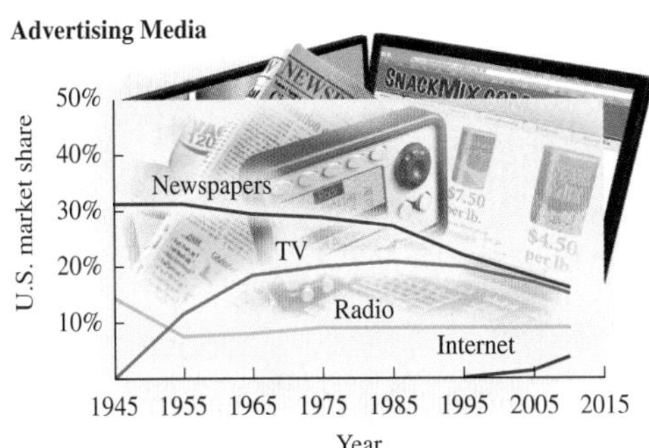

Source: *The Wall Street Journal*, 12/30/07

TW **59.** At what point in time could it have been said that no medium was third in market share? Explain.

TW **60.** Will the Internet advertising market share ever exceed that of radio? TV? newspapers? If so, when? Explain your answers.

61. For each of the following conditions, write a system of equations.

a) $(5, 1)$ is a solution.
b) There is no solution.
c) There is an infinite number of solutions.

62. A system of linear equations has $(1, -1)$ and $(-2, 3)$ as solutions. Determine:

a) a third point that is a solution, and
b) how many solutions there are.

63. The solution of the following system is $(4, -5)$. Find A and B.

$$Ax - 6y = 13,$$
$$x - By = -8.$$

Solve graphically.

64. $y = |x|,$
$x + 4y = 15$

65. $x - y = 0,$
$y = x^2$

In Exercises 66–69, match each system with the appropriate graph from the selections given.

a)
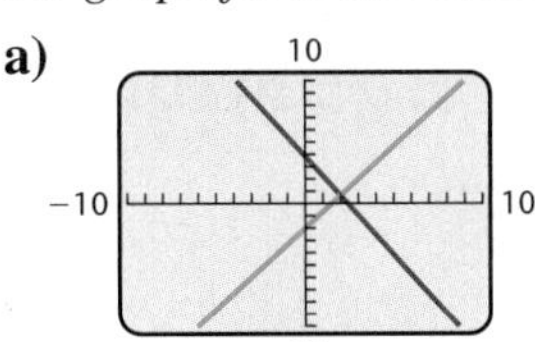

b)
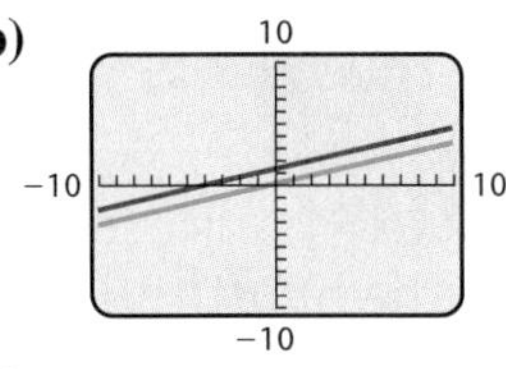

c)
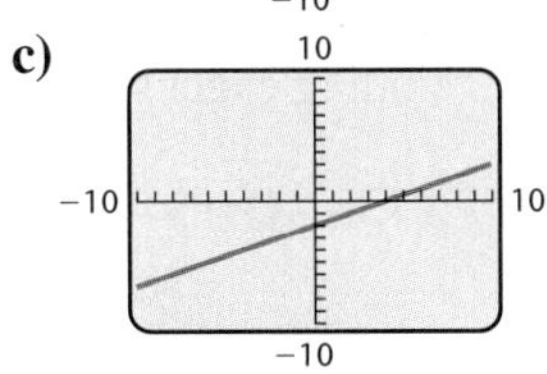

d)
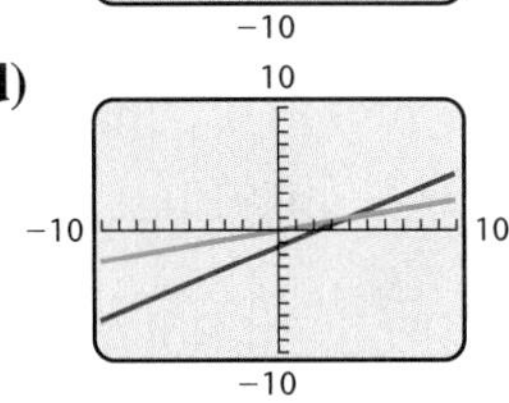

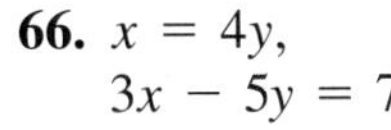

66. $x = 4y,$
$3x - 5y = 7$

67. $2x - 8 = 4y,$
$x - 2y = 4$

68. $8x + 5y = 20,$
$4x - 3y = 6$

69. $x = 3y - 4,$
$2x + 1 = 6y$

70. *Copying Costs.* Aaron occasionally goes to an office store with small copying jobs. He can purchase a "copy card" for \$20 that will entitle him to 500 copies, or he can simply pay 6¢ per page.

a) Create cost equations for each method of paying for a number (up to 500) of copies.
b) Graph both cost equations on the same set of axes.
c) Use the graph to determine how many copies Aaron must make if the card is to be more economical.

71. *Computers.* In 2004, about 46 million notebook PCs were shipped worldwide, and that number was growing at a rate of $22\frac{2}{3}$ million per year. There were 140 million desktop PCs shipped worldwide in 2004, and that number was growing at a rate of 4 million per year.
Source: iSuppli

a) Write two equations that can be used to predict n, the number of notebook PCs and desktop PCs, in millions, shipped t years after 2004.
b) Use a graphing calculator to determine the year in which the numbers of notebook PCs and desktop PCs were the same.

Try Exercise Answers: Section 4.1

9. Yes **17.** $(4, 1)$ **19.** $(2, -1)$ **33.** No solution
37. Approximately $(1.53, 2.58)$
47. Full-time faculty: $y = 9.0524x + 430.6778$; part-time faculty: $y = 14.7175x + 185.3643$; y is in thousands and x is the number of years after 1980; in about 2023

4.2 Systems of Equations and Substitution

- The Substitution Method
- Solving for the Variable First
- Problem Solving

Near the end of Section 4.1, we mentioned that graphing can be an imprecise method for solving systems. In this section and the next, we develop *algebraic* methods of finding exact solutions.

THE SUBSTITUTION METHOD

One algebraic method for solving systems, the **substitution method,** relies on having a variable isolated.

EXAMPLE 1 Solve the system

$$x + y = 7, \quad (1)$$
$$y = 3x - 1. \quad (2)$$

We have numbered the equations (1) and (2) for easy reference.

SOLUTION Equation (2) says that y and $3x - 1$ name the same number. Thus we can substitute $3x - 1$ for y in equation (1):

$x + y = 7$ **Equation (1)**

$x + (3x - 1) = 7.$ **Substituting $3x - 1$ for y**

This last equation has only one variable, for which we now solve:

$x + (3x - 1) = 7$

$4x - 1 = 7$ **Removing parentheses and combining like terms**

$4x = 8$ **Adding 1 to both sides**

$x = 2.$ **Dividing both sides by 4**

We have found the x-value of the solution. To find the y-value, we return to the original pair of equations. Substituting into either equation will give us the y-value. We choose equation (1):

$x + y = 7$ **Equation (1)**

$2 + y = 7$ **Substituting 2 for x**

$y = 5.$ **Subtracting 2 from both sides**

We now have the ordered pair (2, 5). A check assures us that it is the solution.

$x + y = 7$		$y = 3x - 1$	
$2 + 5 \mid 7$		$5 \mid 3 \cdot 2 - 1$	
$7 \stackrel{?}{=} 7$	TRUE	$5 \stackrel{?}{=} 5$	TRUE

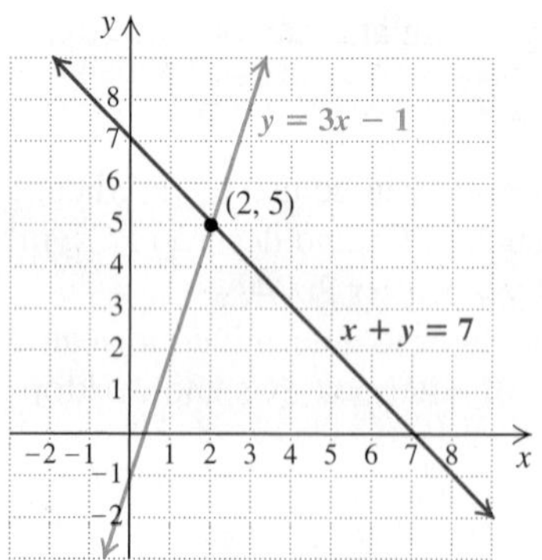

Since (2, 5) checks, it is the solution. For this particular system, we can also check by examining the graph from Example 2 in Section 4.1, as shown at left.

Try Exercise 5.

STUDY TIP

Learn from Your Mistakes

Immediately after each quiz or test, write out a step-by-step solution to any questions you missed. Visit your professor during office hours or consult with a tutor for help with problems that still give you trouble. Errors tend to repeat themselves if not corrected as soon as possible.

CAUTION! A solution of a system of equations in two variables is an ordered *pair* of numbers. Once you have solved for one variable, don't forget the other. A common mistake is to solve for only one variable.

EXAMPLE 2 Solve:

$x = 3 - 2y,$ (1) **x and $3 - 2y$ name the same number.**
$y - 3x = 5.$ (2)

SOLUTION We substitute $3 - 2y$ for x in the second equation:

$y - 3x = 5$ **Equation (2)**
$y - 3(3 - 2y) = 5.$ **Substituting $3 - 2y$ for x, making sure to use parentheses**

Now we solve for y:

$y - 9 + 6y = 5$ **Using the distributive law**
$7y - 9 = 5$
$7y = 14$ **Solving for y**
$y = 2.$

Next, we substitute 2 for y in equation (1) of the original system:

$x = 3 - 2y$ **Equation (1)**
$x = 3 - 2 \cdot 2$ **Substituting 2 for y**
$x = -1.$ **Simplifying**

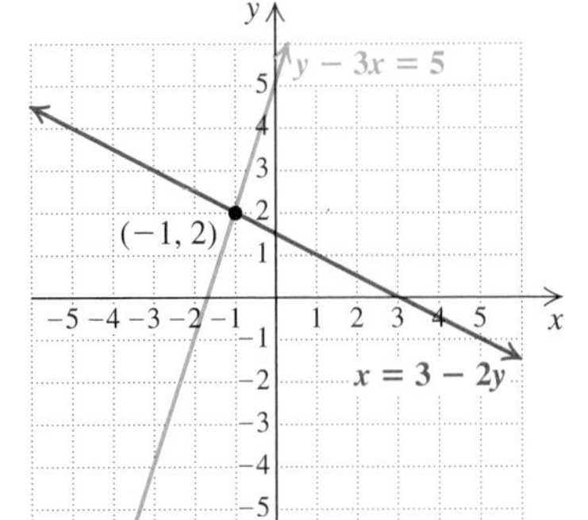

We check the ordered pair $(-1, 2)$.

Check:

$x = 3 - 2y$	
-1	$3 - 2 \cdot 2$
	$3 - 4$
$-1 \overset{?}{=} -1$	TRUE

$y - 3x = 5$	
$2 - 3(-1)$	5
$2 + 3$	
$5 \overset{?}{=} 5$	TRUE

The pair $(-1, 2)$ is the solution. A graph is shown at left as another check.

Try Exercise 9.

SOLVING FOR THE VARIABLE FIRST

Sometimes neither equation has a variable alone on one side. In that case, we solve one equation for one of the variables and then proceed as before.

EXAMPLE 3 Solve:

$x - 2y = 6,$ (1)
$3x + 2y = 4.$ (2)

SOLUTION We can solve either equation for either variable. Since the coefficient of x is 1 in equation (1), it is easier to solve that equation for x:

Solve for x in terms of y.

$x - 2y = 6$ **Equation (1)**
$x = 6 + 2y.$ (3) **Adding $2y$ to both sides**

We substitute $6 + 2y$ for x in equation (2) of the original pair and solve for y:

Substitute.

$3x + 2y = 4$ **Equation (2)**
$3(6 + 2y) + 2y = 4$ **Substituting $6 + 2y$ for x**

Remember to use parentheses when you substitute.

$$18 + 6y + 2y = 4 \quad \text{Using the distributive law}$$
$$18 + 8y = 4 \quad \text{Combining like terms}$$
$$8y = -14 \quad \text{Subtracting 18 from both sides}$$

Find the value of y.

$$y = \frac{-14}{8} = -\frac{7}{4}. \quad \text{Dividing both sides by 8}$$

To find x, we can substitute $-\frac{7}{4}$ for y in equation (1), (2), or (3). Because it is generally easier to use an equation that has already been solved for a specific variable, we decide to use equation (3):

Substitute.

Find the value of x.

$$x = 6 + 2y$$
$$x = 6 + 2\left(-\tfrac{7}{4}\right)$$
$$x = 6 - \tfrac{7}{2} = \tfrac{12}{2} - \tfrac{7}{2} = \tfrac{5}{2}.$$

We check the ordered pair $\left(\frac{5}{2}, -\frac{7}{4}\right)$.

Check:

$x - 2y = 6$	
$\frac{5}{2} - 2\left(-\frac{7}{4}\right)$	6
$\frac{5}{2} + \frac{7}{2}$	
$\frac{12}{2}$	
$6 \stackrel{?}{=} 6$	TRUE

$3x + 2y = 4$	
$3 \cdot \frac{5}{2} + 2\left(-\frac{7}{4}\right)$	4
$\frac{15}{2} - \frac{7}{2}$	
$\frac{8}{2}$	
$4 \stackrel{?}{=} 4$	TRUE

$y_1 = (6 - x)/(-2)$, $y_2 = (4 - 3x)/2$

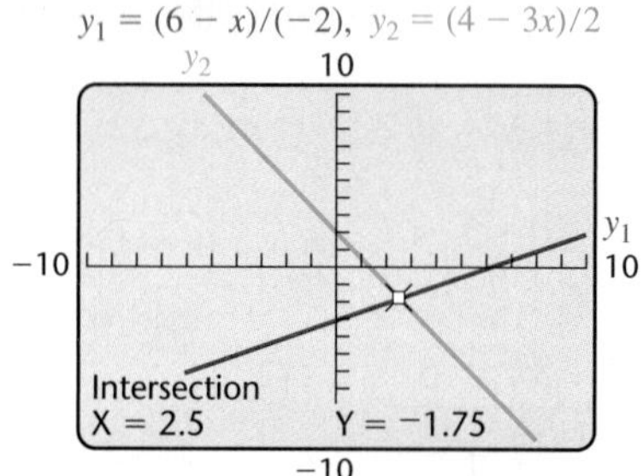

To check using a graphing calculator, we must first solve each equation for y. When we do so, equation (1) becomes $y = (6 - x)/(-2)$ and equation (2) becomes $y = (4 - 3x)/2$. The graph at left confirms the solution. Since $\left(\frac{5}{2}, -\frac{7}{4}\right)$ checks, it is the solution.

Try Exercise 19.

Some systems have no solution and some have an infinite number of solutions.

EXAMPLE 4 Solve each system.

a) $y = \frac{5}{2}x + 4,$ (1)
$y = \frac{5}{2}x - 3$ (2)

b) $2y = 6x + 4,$ (1)
$y = 3x + 2$ (2)

SOLUTION

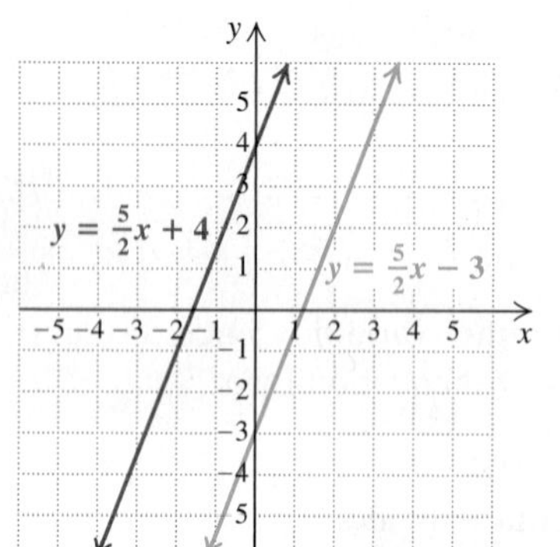

a) Since the lines have the same slope, $\frac{5}{2}$, and different y-intercepts, they are parallel. Let's see what happens if we try to solve this system by substituting $\frac{5}{2}x - 3$ for y in the first equation:

$$y = \tfrac{5}{2}x + 4 \quad \text{Equation (1)}$$
$$\tfrac{5}{2}x - 3 = \tfrac{5}{2}x + 4 \quad \text{Substituting } \tfrac{5}{2}x - 3 \text{ for } y$$
$$-3 = 4. \quad \text{Subtracting } \tfrac{5}{2}x \text{ from both sides}$$

When we subtract $\frac{5}{2}x$ from both sides, we obtain a *false* equation, or contradiction. In such a case, when solving algebraically leads to a false equation, we state that the system has no solution and thus is inconsistent.

b) If we use substitution to solve the system, we can substitute $3x + 2$ for y in equation (1):

$$2y = 6x + 4 \quad \text{Equation (1)}$$
$$2(3x + 2) = 6x + 4 \quad \text{Substituting } 3x + 2 \text{ for } y$$
$$6x + 4 = 6x + 4.$$

This last equation is true for *any* choice of x, indicating that for any choice of x, a solution can be found. When the algebraic solution of a system of two equations leads to an identity—that is, an equation that is true for all real numbers—any pair that is a solution of equation (1) is also a solution of equation (2). The equations are dependent and the solution set is infinite:

$$\{(x, y) \mid y = 3x + 2\}, \quad \text{or equivalently,} \quad \{(x, y) \mid 2y = 6x + 4\}.$$

Try Exercises 15 and 21.

To Solve a System Using Substitution

1. Solve for a variable in either one of the equations if neither equation already has a variable isolated.
2. Using the result of step (1), substitute in the *other* equation for the variable isolated in step (1).
3. Solve the equation from step (2).
4. Substitute the solution from step (3) into one of the other equations to solve for the other variable.
5. Check that the ordered pair resulting from steps (3) and (4) checks in both of the original equations.

PROBLEM SOLVING

Now let's use the substitution method in problem solving.

EXAMPLE 5 Supplementary Angles. Two angles are supplementary. One angle measures 30° more than twice the other. Find the measures of the two angles.

SOLUTION

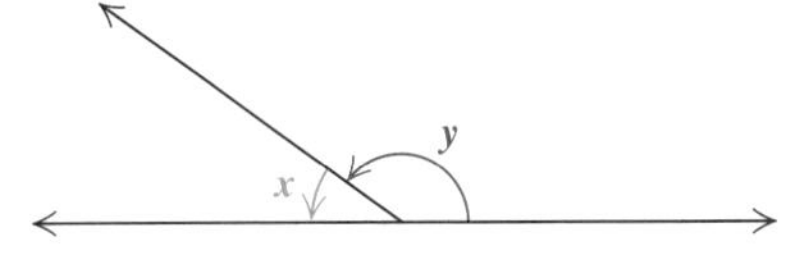

Supplementary angles

1. **Familiarize.** Recall that two angles are supplementary if the sum of their measures is 180°. We could try to guess a solution, but instead we make a drawing and translate. We let x and y represent the measures of the two angles.
2. **Translate.** Since we are told that the angles are supplementary, one equation is

   $$x + y = 180. \qquad (1)$$

   The second sentence can be rephrased and translated as follows:

   Rewording: One angle is 30° more than two times the other.

   Translating: $y = 2x + 30 \qquad (2)$

   We now have a system of two equations in two unknowns:

   $$x + y = 180, \qquad (1)$$
   $$y = 2x + 30. \qquad (2)$$

3. **Carry out.** We substitute $2x + 30$ for y in equation (1):

   $$\begin{aligned} x + y &= 180 && \textbf{Equation (1)} \\ x + (2x + 30) &= 180 && \textbf{Substituting} \\ 3x + 30 &= 180 \\ 3x &= 150 && \textbf{Subtracting 30 from both sides} \\ x &= 50. && \textbf{Dividing both sides by 3} \end{aligned}$$

Substituting 50 for x in equation (1) then gives us

$$\begin{aligned} x + y &= 180 && \textbf{Equation (1)} \\ 50 + y &= 180 && \textbf{Substituting 50 for } x \\ y &= 130. \end{aligned}$$

4. **Check.** If one angle is 50° and the other is 130°, then the sum of the measures is 180°. Thus the angles are supplementary. If 30° is added to twice the measure of the smaller angle, we have $2 \cdot 50° + 30°$, or 130°, which is the measure of the other angle. The numbers check.
5. **State.** One angle measures 50° and the other 130°.

Try Exercise 43.

4.2 Exercise Set

Concept Reinforcement *Classify each of the following statements as either true or false.*

1. When using the substitution method, we must solve for the variables in alphabetical order.
2. The substitution method often requires us to first solve for a variable, much as we did when solving for a letter in a formula.
3. When solving a system of equations algebraically leads to a false equation, the system has no solution.
4. When solving a system of two equations algebraically leads to an equation that is always true, the system has an infinite number of solutions.

Solve each system using the substitution method. If a system has an infinite number of solutions, use set-builder notation to write the solution set. If a system has no solution, state this.

5. $x + y = 5,\quad y = x + 3$

6. $x + y = 9,\quad x = y + 1$

7. $x = y + 1,\quad x + 2y = 4$

8. $y = x - 3,\quad 3x + y = 5$

9. $y = 2x - 5,\quad 3y - x = 5$

10. $y = 2x + 1,\quad x + y = 4$

11. $a = -4b,\quad a + 5b = 5$

12. $r = -3s,\quad r + 4s = 10$

13. $2x + 3y = 8,\quad x = y - 6$

14. $x = y - 8,\quad 3x + 2y = 1$

15. $x = 2y + 1,\quad 3x - 6y = 2$

16. $y = 3x - 1,\quad 6x - 2y = 2$

17. $s + t = -4,\quad s - t = 2$

18. $x - y = 2,\quad x + y = -2$

19. $x - y = 5,\quad x + 2y = 7$

20. $y - 2x = -6,\quad 2y - x = 5$

21. $x - 2y = 7,\quad 3x - 21 = 6y$

22. $x - 4y = 3,\quad 2x - 6 = 8y$

23. $y = 2x + 5,\quad -2y = -4x - 10$

24. $y = -2x + 3,\quad 3y = -6x + 9$

25. $2x + 3y = -2,\quad 2x - y = 9$

26. $x + 2y = 10,\quad 3x + 4y = 8$

27. $a - b = 6,\quad 3a - 4b = 18$

28. $3a + 2b = 2,\quad -2a + b = 8$

29. $s = \frac{1}{2}r,\quad 3r - 4s = 10$

30. $x = \frac{1}{2}y,\quad 2x + y = 12$

31. $8x + 2y = 6,\quad y = 3 - 4x$

32. $x - 3y = 7,\quad -4x + 12y = 28$

33. $x - 2y = 5,\quad 2y - 3x = 1$

34. $x - 3y = -1,\quad 5y - 2x = 4$

Aha! **35.** $2x - y = 0,\quad 2x - y = -2$

36. $5x = y - 3,\quad 5x = y + 5$

Solve using a system of equations.

37. The sum of two numbers is 83. One number is 5 more than the other. Find the numbers.

38. The sum of two numbers is 76. One number is 2 more than the other. Find the numbers.

39. Find two numbers for which the sum is 93 and the difference is 9.

40. Find two numbers for which the sum is 76 and the difference is 12.

41. The difference between two numbers is 16. Three times the larger number is seven times the smaller. What are the numbers?

42. The difference between two numbers is 18. Twice the smaller number plus three times the larger is 74. What are the numbers?

43. *Supplementary Angles.* Two angles are supplementary. One angle is 15° more than twice the other. Find the measure of each angle.

44. *Supplementary Angles.* Two angles are supplementary. One angle is 8° less than three times the other. Find the measure of each angle.

45. *Complementary Angles.* Two angles are complementary. Their difference is 18°. Find the measure of each angle. (*Complementary angles* are angles for which the sum is 90°.)

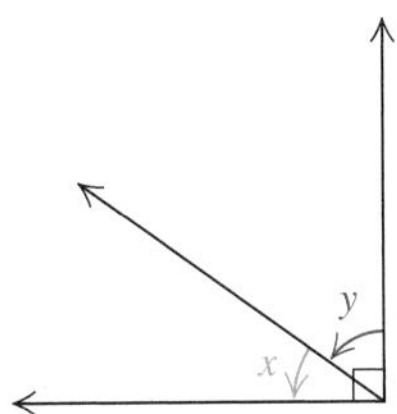

Complementary angles

46. *Complementary Angles.* Two angles are complementary. One angle is 42° more than one-half the other. Find the measure of each angle.

47. *Poster Dimensions.* Morrison ordered a poster printed from his prize-winning photograph. The poster had a perimeter of 100 in., and the length was 6 in. more than the width. Find the length and the width.

48. *Two-by-Four.* The perimeter of a cross section of a "two-by-four" piece of lumber is 10 in. The length is 2 in. longer than the width. Find the actual dimensions of the cross section of a two-by-four.

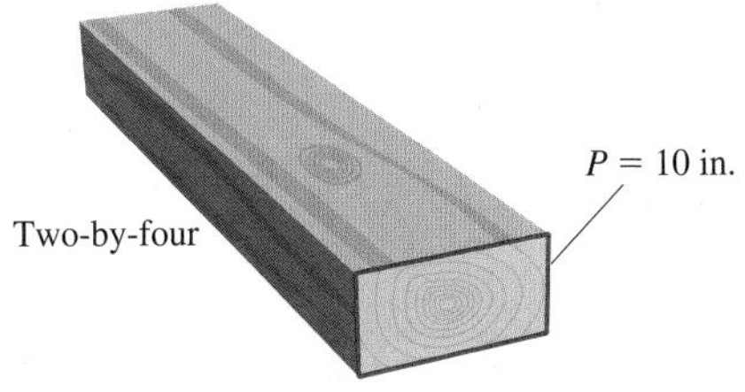

49. *Dimensions of Colorado.* The state of Colorado is roughly in the shape of a rectangle whose perimeter is 1300 mi. The width is 110 mi less than the length. Find the length and the width.

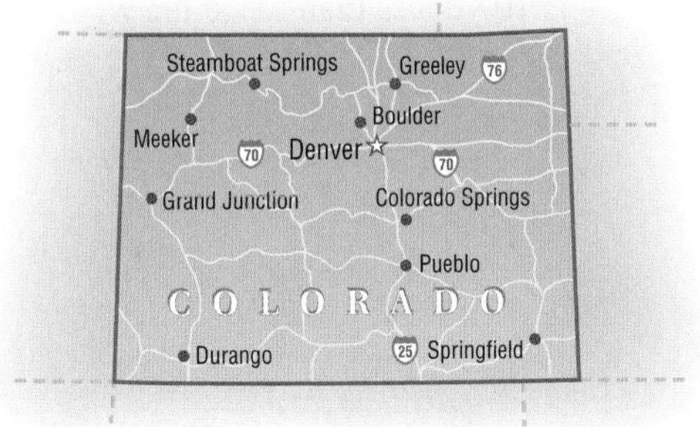

50. *Dimensions of Wyoming.* The state of Wyoming is in the shape of a rectangle with a perimeter of 1280 mi. The width is 90 mi less than the length. Find the length and the width.

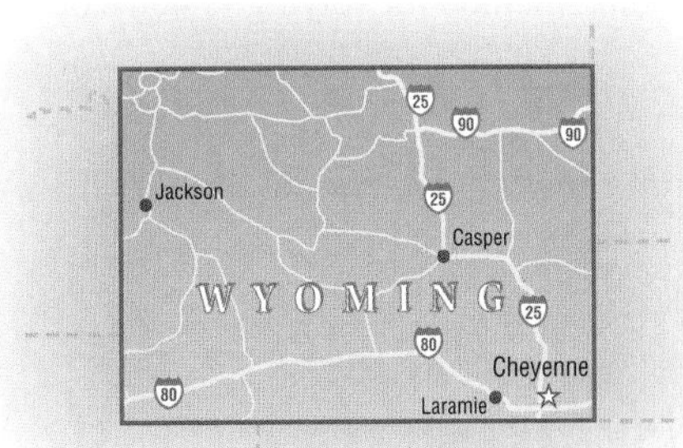

51. *Soccer.* The perimeter of a soccer field is 280 yd. The width is 5 yd more than half the length. Find the length and the width.

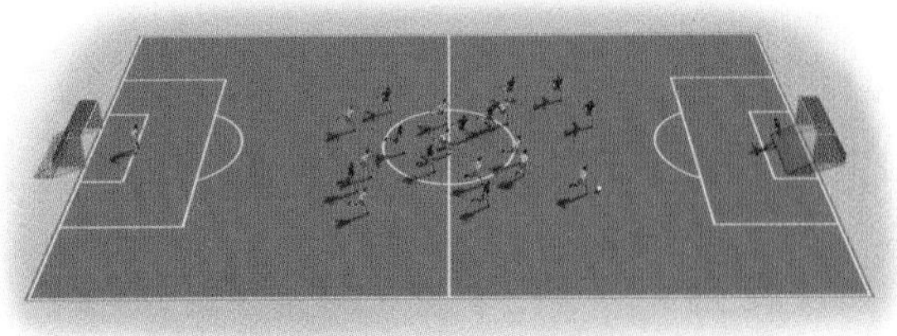

52. *Racquetball.* A regulation racquetball court has a perimeter of 120 ft, with a length that is twice the width. Find the length and the width of a court.

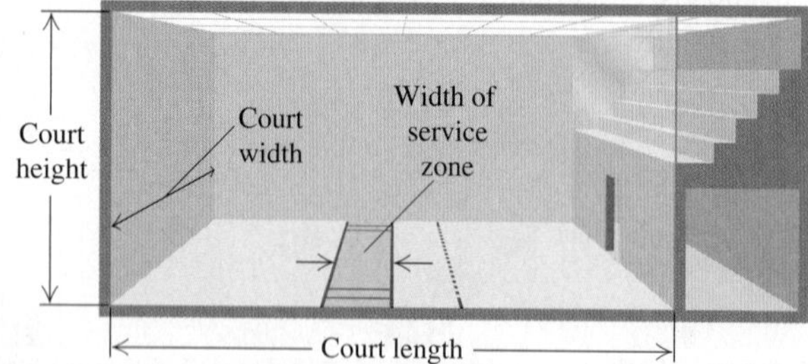

53. *Racquetball.* The height of the front wall of a standard racquetball court is four times the width of the service zone (see the figure). Together, these measurements total 25 ft. Find the height and the width.

54. *Lacrosse.* The perimeter of a lacrosse field is 340 yd. The length is 10 yd less than twice the width. Find the length and the width.

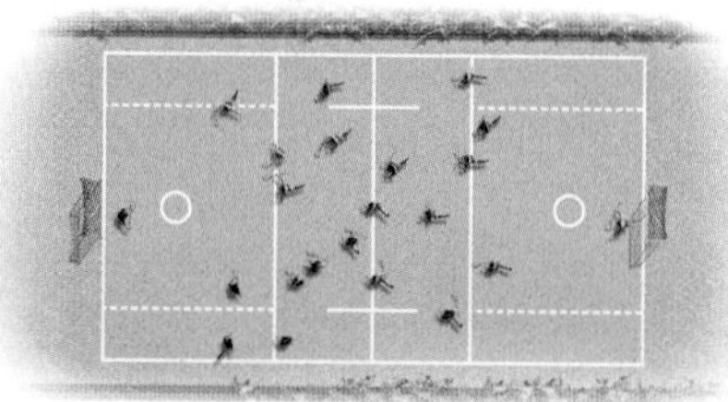

TW 55. Briley solves every system of two equations (in x and y) by first solving for y in the first equation and then substituting into the second equation. Is he using the best approach? Why or why not?

TW 56. Describe two advantages of the substitution method over the graphing method for solving systems of equations.

SKILL REVIEW

To prepare for Section 4.3, review simplifying algebraic expressions (Section 1.8).

Simplify. [1.8]

57. $2(5x + 3y) - 3(5x + 3y)$

58. $5(2x + 3y) - 3(7x + 5y)$

59. $4(5x + 6y) - 5(4x + 7y)$

60. $2(7x + 5 - 3y) - 7(2x + 5)$

61. $2(5x - 3y) - 5(2x + y)$

62. $4(2x + 3y) + 3(5x - 4y)$

SYNTHESIS

TW 63. Hilary can tell by inspection that the system

$$x = 2y - 1,$$
$$x = 2y + 3$$

has no solution. How can she tell?

TW 64. Under what circumstances can a system of equations be solved more easily by graphing than by substitution?

Solve by the substitution method.

65. $\frac{1}{6}(a + b) = 1,$
$\frac{1}{4}(a - b) = 2$

66. $\frac{x}{5} - \frac{y}{2} = 3,$
$\frac{x}{4} + \frac{3y}{4} = 1$

67. $y + 5.97 = 2.35x,$
$2.14y - x = 4.88$

68. $a + 4.2b = 25.1,$
$9a - 1.8b = 39.78$

69. *Age at Marriage.* Trudy is 20 years younger than Dennis. She feels that she needs to be 7 more than half of Dennis's age before they can marry. What is the youngest age at which Trudy can marry Dennis and honor this requirement?

Exercises 70 and 71 contain systems of three equations in three variables. A solution is an ordered triple of the form (x, y, z). Use the substitution method to solve.

70. $x + y + z = 4,$
$x - 2y - z = 1,$
$y = -1$

71. $x + y + z = 180,$
$x = z - 70,$
$2y - z = 0$

72. *Softball.* The perimeter of a softball diamond is two-thirds of the perimeter of a baseball diamond. Together, the two perimeters measure 200 yd. Find the distance between the bases in each sport.

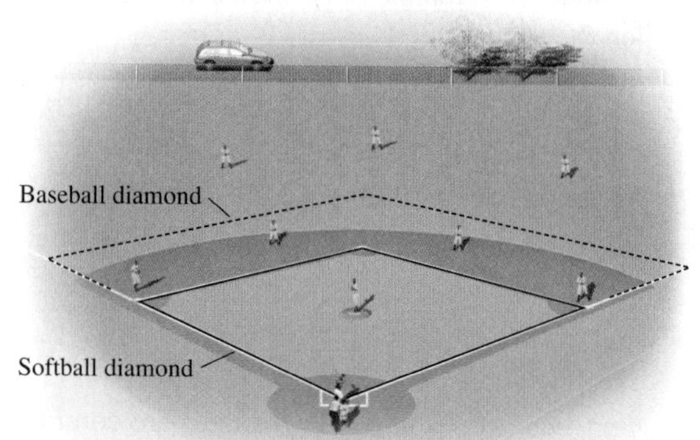

TW 73. Solve Example 3 by first solving for $2y$ in equation (1) and then substituting for $2y$ in equation (2). Is this method easier than the procedure used in Example 3? Why or why not?

74. Write a system of two linear equations that can be solved more quickly—but still precisely—by a graphing calculator than by substitution. Time yourself using both methods to solve the system.

Try Exercise Answers: Section 4.2

5. (1, 4) **9.** (4, 3) **15.** No solution **19.** $\left(\frac{17}{3}, \frac{2}{3}\right)$ **21.** $\{(x, y) | x - 2y = 7\}$ **43.** 55°, 125°

4.3 Systems of Equations and Elimination

- Solving by the Elimination Method
- Problem Solving

STUDY TIP 

Use What You Know

An excellent strategy for solving any new type of problem is to rewrite the problem in an equivalent form that we already know how to solve. When this is feasible, as in some of the problems in this section, it can make a new topic much easier to learn.

Although substitution can be used to solve any system of two equations, another method, *elimination,* is simpler to use for many problems.

SOLVING BY THE ELIMINATION METHOD

The **elimination method** for solving systems of equations makes use of the addition principle.

EXAMPLE 1 Solve the system

$$2x + 3y = 13, \qquad (1)$$
$$4x - 3y = 17. \qquad (2)$$

SOLUTION According to equation (2), $4x - 3y$ and 17 are the same number. Thus we can add $4x - 3y$ to the left side of equation (1) and 17 to the right side:

$$\begin{aligned} 2x + 3y &= 13 && (1)\\ 4x - 3y &= 17 && (2)\\ \hline 6x + 0y &= 30. && \text{Adding. Note that } y \text{ has been "eliminated."} \end{aligned}$$

The resulting equation has just one variable:

$$6x = 30.$$

Dividing both sides of this equation by 6, we find that $x = 5$.

Next, we substitute 5 for x in either of the original equations:

$$\begin{aligned} 2x + 3y &= 13 && \text{Equation (1)}\\ 2(5) + 3y &= 13 && \text{Substituting 5 for } x\\ 10 + 3y &= 13 &&\\ 3y &= 3 &&\\ y &= 1. && \text{Solving for } y \end{aligned}$$

We check the ordered pair (5, 1). The graph shown at left also serves as a check.

2x + 3y = 13

(5, 1)

4x − 3y = 17

Check:

$$\begin{array}{r|l} 2x + 3y = 13 & \\ \hline 2(5) + 3(1) & 13 \\ 10 + 3 & \\ 13 \stackrel{?}{=} 13 & \text{TRUE} \end{array} \qquad \begin{array}{r|l} 4x - 3y = 17 & \\ \hline 4(5) - 3(1) & 17 \\ 20 - 3 & \\ 17 \stackrel{?}{=} 17 & \text{TRUE} \end{array}$$

Since (5, 1) checks in both equations, it is the solution.

Try Exercise 5.

Adding in Example 1 eliminated the variable y because two terms, $-3y$ in equation (2) and $3y$ in equation (1), are opposites. Most systems have no pair of terms that are opposites. We can create such a pair of terms by multiplying one or both of the equations by appropriate numbers.

EXAMPLE 2 Solve:

$$2x + 3y = 8, \qquad (1)$$
$$x + 3y = 7. \qquad (2)$$

SOLUTION Adding these equations as they now appear will not eliminate a variable. However, if the $3y$ were $-3y$ in one equation, we could eliminate y. We multiply both sides of equation (2) by -1 to find an equivalent equation that contains $-3y$, and then add:

$$\begin{array}{rcll} 2x + 3y &=& 8 & \textbf{Equation (1)} \\ -x - 3y &=& -7 & \textbf{Multiplying both sides of equation (2) by } -1 \\ \hline x &=& 1. & \textbf{Adding} \end{array}$$

Next, we substitute 1 for x in either of the original equations:

$$\begin{array}{rcll} x + 3y &=& 7 & \textbf{Equation (2)} \\ 1 + 3y &=& 7 & \textbf{Substituting 1 for } x \\ 3y &=& 6 & \\ y &=& 2. & \textbf{Solving for } y \end{array}$$

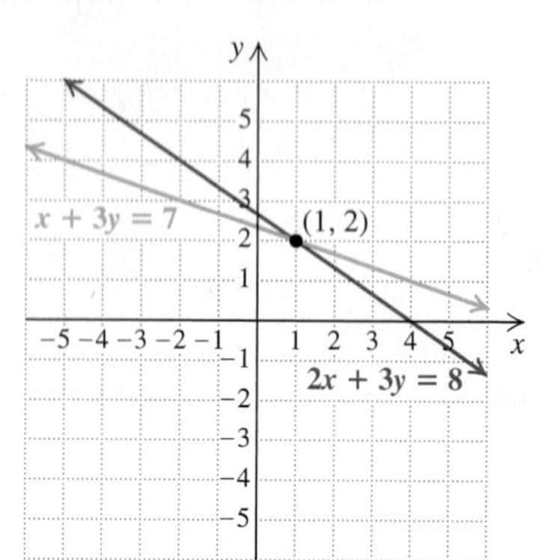

We can check the ordered pair (1, 2). The graph shown at left is also a check.

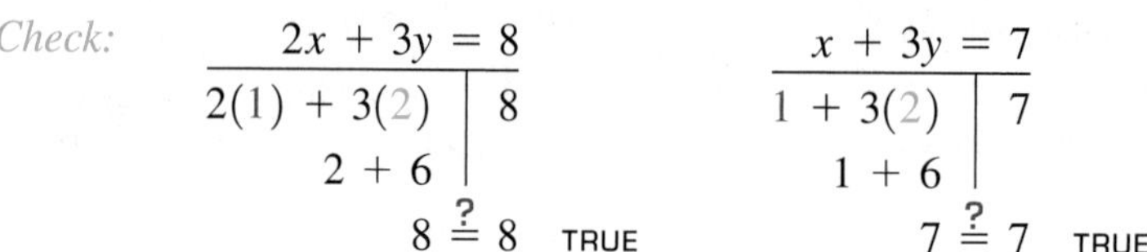

Check:

$$\begin{array}{c|c} 2x + 3y & 8 \\ \hline 2(1) + 3(2) & 8 \\ 2 + 6 & \\ 8 \stackrel{?}{=} 8 & \text{TRUE} \end{array} \qquad \begin{array}{c|c} x + 3y & 7 \\ \hline 1 + 3(2) & 7 \\ 1 + 6 & \\ 7 \stackrel{?}{=} 7 & \text{TRUE} \end{array}$$

Since (1, 2) checks in both equations, it is the solution.

Try Exercise 17.

When deciding which variable to eliminate, we inspect the coefficients in both equations. If one coefficient is a multiple of the coefficient of the same variable in the other equation, that is the easiest variable to eliminate.

EXAMPLE 3 Solve:

$$\begin{array}{rcll} 3x + 6y &=& -6, & (1) \\ 5x - 2y &=& 14. & (2) \end{array}$$

SOLUTION No terms are opposites, but if both sides of equation (2) are multiplied by 3 (or if both sides of equation (1) are multiplied by $\frac{1}{3}$), the coefficients of y will be opposites. Note that 6 is the LCM of 2 and 6:

$$\begin{array}{rcll} 3x + 6y &=& -6 & \textbf{Equation (1)} \\ 15x - 6y &=& 42 & \textbf{Multiplying both sides of equation (2) by 3} \\ \hline 18x &=& 36 & \textbf{Adding} \\ x &=& 2. & \textbf{Solving for } x \end{array}$$

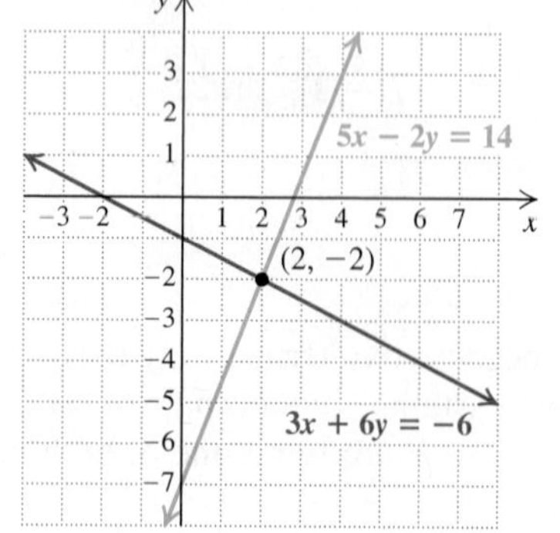

We then substitute 2 for x in either equation (1) or equation (2):

$$\begin{array}{rcll} 3(2) + 6y &=& -6 & \textbf{Substituting 2 for } x \textbf{ in equation (1)} \\ 6 + 6y &=& -6 & \\ 6y &=& -12 & \textbf{Solving for } y \\ y &=& -2. & \end{array}$$

We leave it to the student to confirm that $(2, -2)$ checks and is the solution. The graph at left also serves as a check.

Try Exercise 21.

Sometimes both equations must be multiplied to find the least common multiple of two coefficients.

EXAMPLE 4 Solve:

$$3y + 1 + 2x = 0, \qquad (1)$$
$$5x = 7 - 4y. \qquad (2)$$

SOLUTION It is often helpful to write both equations in the form $Ax + By = C$ before attempting to eliminate a variable:

$$2x + 3y = -1, \qquad (3)$$ **Subtracting 1 from both sides and rearranging the terms of the first equation**

$$5x + 4y = 7. \qquad (4)$$ **Adding 4*y* to both sides of equation (2)**

Since neither coefficient of x is a multiple of the other and neither coefficient of y is a multiple of the other, we use the multiplication principle with *both* equations. Note that we can eliminate the x-term by multiplying both sides of equation (3) by 5 and both sides of equation (4) by -2:

Multiply to get terms that are opposites.

$$10x + 15y = -5$$ **Multiplying both sides of equation (3) by 5**

$$-10x - 8y = -14$$ **Multiplying both sides of equation (4) by −2**

Solve for one variable.

$$7y = -19$$ **Adding**

$$y = \frac{-19}{7} = -\frac{19}{7}.$$ **Dividing by 7**

We substitute $-\frac{19}{7}$ for y in equation (3):

Substitute.

$$2x + 3y = -1$$ **Equation (3)**

$$2x + 3\left(-\frac{19}{7}\right) = -1$$ **Substituting $-\frac{19}{7}$ for *y***

$$2x - \frac{57}{7} = -1$$

$$2x = -1 + \frac{57}{7}$$ **Adding $\frac{57}{7}$ to both sides**

Solve for the other variable.

$$2x = -\frac{7}{7} + \frac{57}{7} = \frac{50}{7}$$

$$x = \frac{50}{7} \cdot \frac{1}{2} = \frac{25}{7}.$$ **Solving for *x***

We check the ordered pair $\left(\frac{25}{7}, -\frac{19}{7}\right)$.

Check in both equations.

Check:

$3y + 1 + 2x = 0$	
$3\left(-\frac{19}{7}\right) + 1 + 2\left(\frac{25}{7}\right)$	0
$-\frac{57}{7} + \frac{7}{7} + \frac{50}{7}$	
$0 \stackrel{?}{=} 0$	TRUE

$5x = 7 - 4y$	
$5\left(\frac{25}{7}\right)$	$7 - 4\left(-\frac{19}{7}\right)$
$\frac{125}{7}$	$\frac{49}{7} + \frac{76}{7}$
$\frac{125}{7} \stackrel{?}{=} \frac{125}{7}$	TRUE

State the solution as an ordered pair.

The solution is $\left(\frac{25}{7}, -\frac{19}{7}\right)$.

Try Exercise 33.

Next, we consider a system with no solution and see what happens when the elimination method is used.

EXAMPLE 5 Solve:

$$y - 3x = 2, \qquad (1)$$
$$y - 3x = 1. \qquad (2)$$

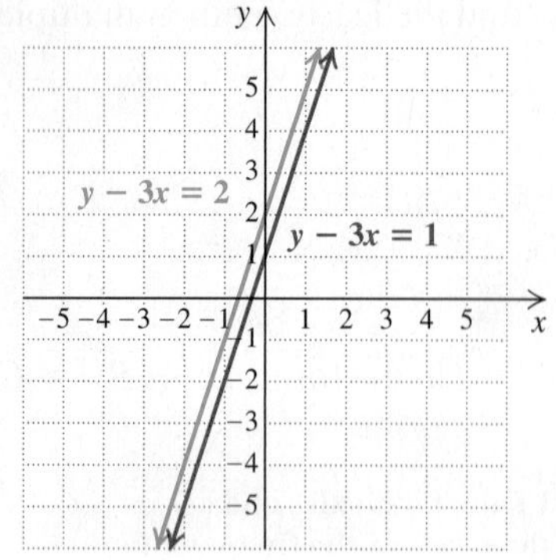

SOLUTION To eliminate y, we multiply both sides of equation (2) by -1. Then we add:

$$\begin{aligned} y - 3x &= 2 \\ -y + 3x &= -1 \qquad \textbf{Multiplying both sides of equation (2) by } -1 \\ \hline 0 &= 1. \qquad \textbf{Adding} \end{aligned}$$

Note that in eliminating y, we eliminated x as well. The resulting equation, $0 = 1$, is false for any pair (x, y), so there is *no solution*.

Try Exercise 23.

Sometimes there is an infinite number of solutions.

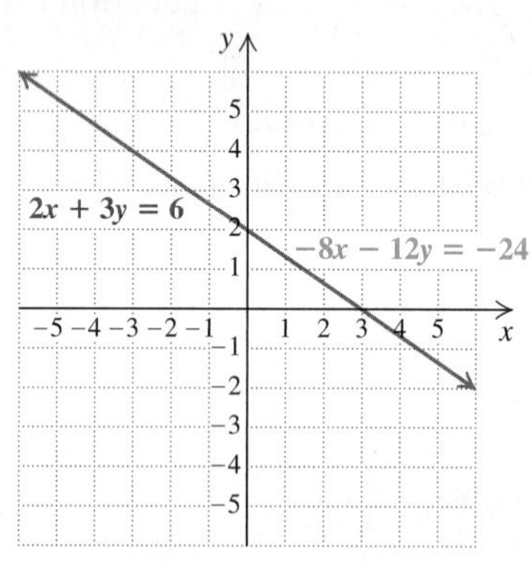

EXAMPLE 6 Solve:

$$\begin{aligned} 2x + 3y &= 6, \qquad (1) \\ -8x - 12y &= -24. \qquad (2) \end{aligned}$$

SOLUTION To eliminate x, we multiply both sides of equation (1) by 4 and then add the two equations:

$$\begin{aligned} 8x + 12y &= 24 \qquad \textbf{Multiplying both sides of equation (1) by 4} \\ -8x - 12y &= -24 \\ \hline 0 &= 0. \qquad \textbf{Adding} \end{aligned}$$

Again, we have eliminated *both* variables. The resulting equation, $0 = 0$, is always true, indicating that the equations are dependent. Such a system has an infinite solution set:

$\{(x, y) | 2x + 3y = 6\}$, or equivalently, $\{(x, y) | -8x - 12y = -24\}$.

Try Exercise 15.

When decimals or fractions appear, we can first multiply to clear them. Then we proceed as before.

EXAMPLE 7 Solve:

$$\begin{aligned} \tfrac{1}{2}x + \tfrac{3}{4}y &= 2, \qquad (1) \\ x + 3y &= 7. \qquad (2) \end{aligned}$$

SOLUTION The number 4 is the LCD for equation (1). Thus we multiply both sides of equation (1) by 4 to clear fractions:

$$\begin{aligned} 4\left(\tfrac{1}{2}x + \tfrac{3}{4}y\right) &= 4 \cdot 2 \qquad \textbf{Multiplying both sides of equation (1) by 4} \\ 4 \cdot \tfrac{1}{2}x + 4 \cdot \tfrac{3}{4}y &= 8 \qquad \textbf{Using the distributive law} \\ 2x + 3y &= 8. \end{aligned}$$

The resulting system is

$$\begin{aligned} 2x + 3y &= 8, \qquad \textbf{This equation is equivalent to equation (1).} \\ x + 3y &= 7. \end{aligned}$$

As we saw in Example 2, the solution of this system is (1, 2).

Try Exercise 35.

PROBLEM SOLVING

We now use the elimination method to solve a problem.

EXAMPLE 8 Birthday Parties. Anne plans to give a birthday party for her 5-year-old daughter at either Karate World or Fitness King. Karate World charges a \$60 setup fee plus \$10 per child for the party. Fitness King charges a \$40 setup fee plus \$12 per child. For how many children will the party prices be the same?

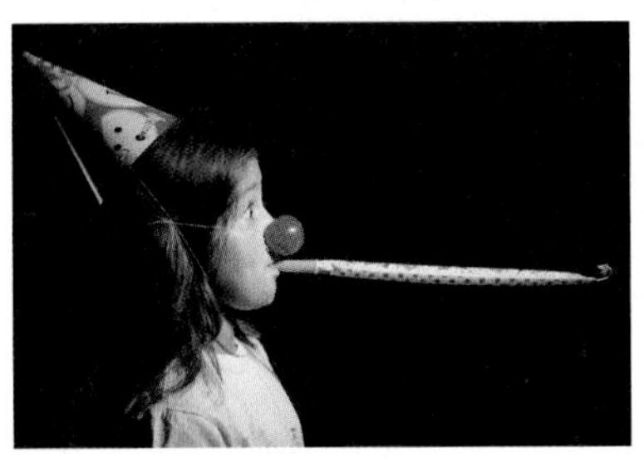

SOLUTION

1. **Familiarize.** Let's make and check a guess of 8 children. The two party prices would be:

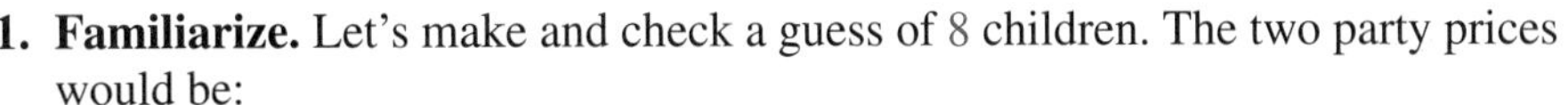

Karate World: $\$60 + \$10(8) = \$60 + \$80 = \$140;$

Fitness King: $\$40 + \$12(8) = \$40 + \$96 = \$136.$

We see that our guess is close, but that the prices are not exactly equal. From the check, we can see how equations can be written to model the situation. We let n = the number of children and p = the price of the party.

2. **Translate.** We reword the problem and translate.

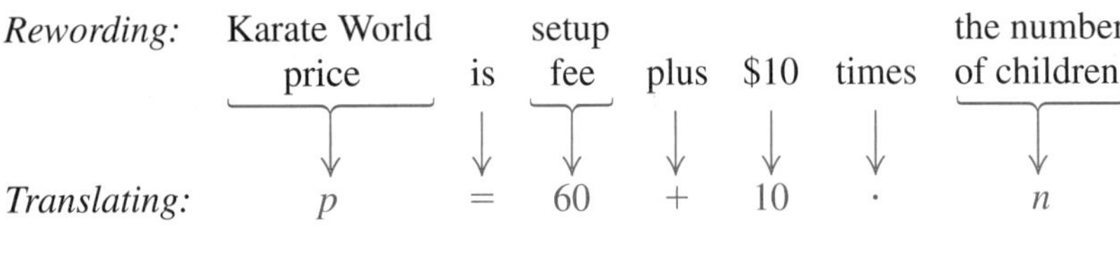

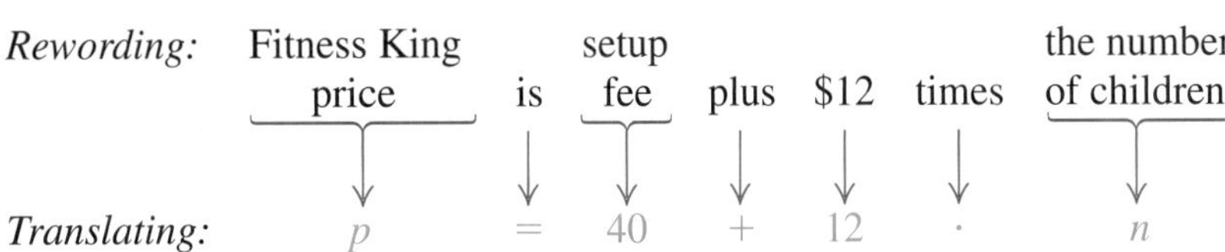

We now have the system of equations

$$p = 60 + 10n,$$
$$p = 40 + 12n.$$

3. **Carry out.** We use elimination to solve the system.

$$\begin{aligned} p &= 60 + 10n \\ -p &= -40 - 12n \quad \textbf{Multiplying by } -1 \\ \hline 0 &= 20 - 2n \quad \textbf{Adding to eliminate } p \end{aligned}$$

We can now solve for n:

$$\begin{aligned} 0 &= 20 - 2n \\ 2n &= 20 \\ n &= 10. \end{aligned}$$

4. **Check.** We compare the prices for 10 children:

Karate World: $\$60 + \$10(10) = \$60 + \$100 = \$160;$

Fitness King: $\$40 + \$12(10) = \$40 + \$120 = \$160.$

The prices are the same.

5. **State.** The prices are the same for a party of 10 children.

Try Exercise 39.

4.3 Exercise Set

FOR EXTRA HELP MyMathLab     

Concept Reinforcement *Classify each of the following statements as either true or false.*

1. The elimination method is never easier to use than the substitution method.
2. The elimination method works especially well when the coefficients of one variable are opposites of each other.
3. When the elimination method yields an equation that is always true, there is an infinite number of solutions.
4. When the elimination method yields an equation that is never true, the system has no solution.

Solve using the elimination method. If a system has an infinite number of solutions, use set-builder notation to write the solution set. If a system has no solution, state this.

5. $x - y = 6,$ $x + y = 12$
6. $x + y = 3,$ $x - y = 7$
7. $x + y = 6,$ $-x + 3y = -2$
8. $x + y = 6,$ $-x + 2y = 15$
9. $4x - y = 1,$ $3x + y = 13$
10. $3x - y = 9,$ $2x + y = 6$
11. $5a + 4b = 7,$ $-5a + b = 8$
12. $7c + 4d = 16,$ $c - 4d = -4$
13. $8x - 5y = -9,$ $3x + 5y = -2$
14. $3a - 3b = -15,$ $-3a - 3b = -3$
15. $3a - 6b = 8,$ $-3a + 6b = -8$
16. $8x + 3y = 4,$ $-8x - 3y = -4$
17. $-x - y = 8,$ $2x - y = -1$
18. $x + y = -7,$ $3x + y = -9$
19. $x + 3y = 19,$ $x - y = -1$
20. $3x - y = 8,$ $x + 2y = 5$
21. $8x - 3y = -6,$ $5x + 6y = 75$
22. $x - y = 3,$ $2x - 3y = -1$
23. $2w - 3z = -1,$ $-4w + 6z = 5$
24. $7p + 5q = 2,$ $8p - 9q = 17$
25. $4a + 6b = -1,$ $a - 3b = 2$
26. $x + 9y = 1,$ $2x - 6y = 10$
27. $3y = x,$ $5x + 14 = y$
28. $5a = 2b,$ $2a + 11 = 3b$
29. $4x - 10y = 13,$ $-2x + 5y = 8$
30. $2p + 5w = 9,$ $3p - 2w = 4$
31. $8n + 6 - 3m = 0,$ $32 = m - n$
32. $6x - 8 + y = 0,$ $11 = 3y - 8x$
33. $3x + 5y = 4,$ $-2x + 3y = 10$
34. $2x + y = 13,$ $4x + 2y = 23$
35. $0.06x + 0.05y = 0.07,$ $0.4x - 0.3y = 1.1$
36. $x - \frac{3}{2}y = 13,$ $\frac{3}{2}x - y = 17$
37. $x + \frac{9}{2}y = \frac{15}{4},$ $\frac{9}{10}x - y = \frac{9}{20}$
38. $1.8x - 2y = 0.9,$ $0.04x + 0.18y = 0.15$

Solve.

39. *Car Rentals.* The University of Oklahoma School of Drama can rent a cargo van for \$27 per day plus 22¢ per mile or a pickup truck for \$29 per day plus 17¢ per mile. For what mileage is the cost the same?
Source: fleetservices.ou.edu

40. *RV Rentals.* Stoltzfus RV rents a Class A recreational vehicle (RV) for \$1545 for one week plus 20¢ per mile. Martin RV rents a similar RV for \$1183 for one week plus 30¢ per mile. For what mileage is the cost the same?
Sources: Stoltzfus RV; Martin RV

41. *Complementary Angles.* Two angles are complementary. The difference of the angle measures is 38°. Find the measure of each angle.

42. *Complementary Angles.* Two angles are complementary. The difference of the angle measures is 86°. Find the measure of each angle.

43. *Phone Rates.* Elke makes frequent calls from the United States to Austria. She is choosing between a PowerNet Global plan that costs \$1.99 per month plus 43¢ per minute and an AT&T plan that costs \$3.99 per month plus 28¢ per minute. For what number of minutes will the two plans cost the same?
Sources: AT&T; PNG

44. *Phone Rates.* Recently, AT&T offered the One Rate 5¢ Nationwide Advantage Plan for $5.00 per month plus 5¢ per minute. AT&T also offered the Value Card Plus Plan for $1.95 per month plus 15¢ per minute. For what number of minutes per month will the two plans cost the same?
Source: AT&T

45. *Supplementary Angles.* Two angles are supplementary. The difference of the angle measures is 68°. Find the measure of each angle.

46. *Supplementary Angles.* Two angles are supplementary. The difference of the angle measures is 90°. Find the measure of each angle.

47. *Planting Grapes.* South Wind Vineyards uses 820 acres to plant Chardonnay and Riesling grapes. The vintner knows the profits will be greatest by planting 140 more acres of Chardonnay than Riesling. How many acres of each type of grape should be planted?

48. *Baking.* Maple Branch Bakers sells 175 loaves of bread each day—some white and the rest whole-wheat. Because of a regular order from a local sandwich shop, Maple Branch consistently bakes 9 more loaves of white bread than whole-wheat. How many loaves of each type of bread do they bake?

49. *Framing.* Anu has 18 ft of molding from which he needs to make a rectangular frame. Because of the dimensions of the mirror being framed, the frame must be twice as long as it is wide. What should the dimensions of the frame be?

50. *Gardening.* Patrice has 108 ft of fencing for a rectangular garden. If the garden's length is to be $1\frac{1}{2}$ times its width, what should the garden's dimensions be?

TW 51. Describe a method that could be used for writing a system that contains dependent equations.

TW 52. Describe a method that could be used for writing an inconsistent system of equations.

SKILL REVIEW

To prepare for Section 4.4, review percent notation (Section 2.4).

Convert to decimal notation. [2.4]

53. 12.2%

54. 0.5%

Solve. [2.4]

55. What percent of 65 is 26?

56. What number is 17% of 18?

Translate to an algebraic expression. [1.1]

57. 12% of the number of liters

58. 10.5% of the number of pounds

SYNTHESIS

TW 59. If a system has an infinite number of solutions, does it follow that *any* ordered pair is a solution? Why or why not?

TW 60. Explain how the multiplication and addition principles are used in this section. Then count the number of times that these principles are used in Example 4.

Solve using substitution, elimination, or graphing.

61. $x + y = 7,$
$3(y - x) = 9$

62. $y = 3x + 4,$
$3 + y = 2(y - x)$

63. $2(5a - 5b) = 10,$
$-5(2a + 6b) = 10$

64. $0.05x + y = 4,$
$\frac{x}{2} + \frac{y}{3} = 1\frac{1}{3}$

Aha! 65. $y = -\frac{2}{7}x + 3,$
$y = \frac{4}{5}x + 3$

66. $y = \frac{2}{5}x - 7,$
$y = \frac{2}{5}x + 4$

Solve for x and y.

67. $y = ax + b,$
$y = x + c$

68. $ax + by + c = 0,$
$ax + cy + b = 0$

69. *Math in History.* Several ancient Chinese books included problems that can be solved by translating to systems of equations. *Arithmetical Rules in Nine Sections* is a book of 246 problems compiled by a Chinese mathematician, Chang Tsang, who died in 152 B.C. One of the problems is: Suppose there are a number of rabbits and pheasants confined in a cage. In all, there are 35 heads and 94 feet. How many rabbits and how many pheasants are there? Solve the problem.

70. *Age.* Miguel's age is 20% of his mother's age. Twenty years from now, Miguel's age will be 52% of his mother's age. How old are Miguel and his mother now?

71. *Age.* If 5 is added to a man's age and the total is divided by 5, the result will be his daughter's age. Five years ago, the man's age was eight times his daughter's age. Find their present ages.

72. *Dimensions of a Triangle.* When the base of a triangle is increased by 1 ft and the height is increased by 2 ft, the height changes from being two-thirds of the base to being four-fifths of the base. Find the original dimensions of the triangle

Try Exercise Answers: Section 4.3

5. (9, 3) **15.** $\{(a, b) \mid 3a - 6b = 8\}$ **17.** $(-3, -5)$
21. (3, 10) **23.** No solution **33.** $(-2, 2)$
35. $(2, -1)$ **39.** 40 mi

Mid-Chapter Review

We now have three different methods for solving systems of equations. Each method has certain strengths and weaknesses, as outlined below.

Method	Strengths	Weaknesses
Graphical	Solutions are displayed visually. Can be used with any system that can be graphed.	For some systems, only approximate solutions can be found graphically. The graph drawn may not be large enough to show the solution.
Substitution	Yields exact solutions. Easy to use when a variable has a coefficient of 1.	Introduces extensive computations with fractions when solving more complicated systems. Solutions are not displayed graphically.
Elimination	Yields exact solutions. Easy to use when fractions or decimals appear in the system. The preferred method for systems of 3 or more equations in 3 or more variables. (See Chapter 9.)	Solutions are not displayed graphically.

GUIDED SOLUTIONS

Solve. [4.2], [4.3]

1. $2x - 3y = 5,$
$y = x - 1$

Solution

$2x - 3(\square) = 5$ **Substituting $x - 1$ for y**

$2x - \square + \square = 5$ **Using the distributive law**

$\square + 3 = 5$ **Combining like terms**

$-x = \square$ **Subtracting 3 from both sides**

$x = \square$ **Dividing both sides by -1**

$y = x - 1$

$y = \square - 1$ **Substituting**

$y = \square$

The solution is $(\square, \square)$.

2. $2x - 5y = 1,$
$x + 5y = 8$

Solution

$2x - 5y = 1$
$x + 5y = 8$

$\square = \square$

$x = \square$

$x + 5y = 8$

$\square + 5y = 8$ **Substituting**

$5y = \square$

$y = \square$

The solution is $(\square, \square)$.

MIXED REVIEW

Solve using the best method.

1. $x = y,$
$x + y = 2$

2. $x + y = 10,$
$x - y = 8$

3. $y = \frac{1}{2}x + 1,$
$y = 2x - 5$

4. $y = 2x - 3,$
$x + y = 12$

5. $x = 5,$
$y = 10$

6. $3x + 5y = 8,$
$3x - 5y = 4$

7. $2x - y = 1,$
$2y - 4x = 3$

8. $x = 2 - y,$
$3x + 3y = 6$

9. $x + 2y = 3,$
$3x = 4 - y$

10. $9x + 8y = 0,$
$11x - 7y = 0$

11. $10x + 20y = 40,$
$x - y = 7$

12. $y = \frac{5}{3}x + 7,$
$y = \frac{5}{3}x - 8$

13. $2x - 5y = 1,$
$3x + 2y = 11$

14. $\frac{x}{2} + \frac{y}{3} = \frac{2}{3},$
$\frac{x}{5} + \frac{5y}{2} = \frac{1}{4}$

15. $1.1x - 0.3y = 0.8,$
$2.3x + 0.3y = 2.6$

16. $y = -3,$
$x = 11$

17. $x - 2y = 5,$
$3x - 15 = 6y$

18. $12x - 19y = 13,$
$8x + 19y = 7$

19. $0.2x + 0.7y = 1.2,$
$0.3x - 0.1y = 2.7$

20. $\frac{1}{4}x = \frac{1}{3}y,$
$\frac{1}{2}x - \frac{1}{15}y = 2$

4.4 More Applications Using Systems

- Total-Value Problems
- Mixture Problems
- Motion Problems

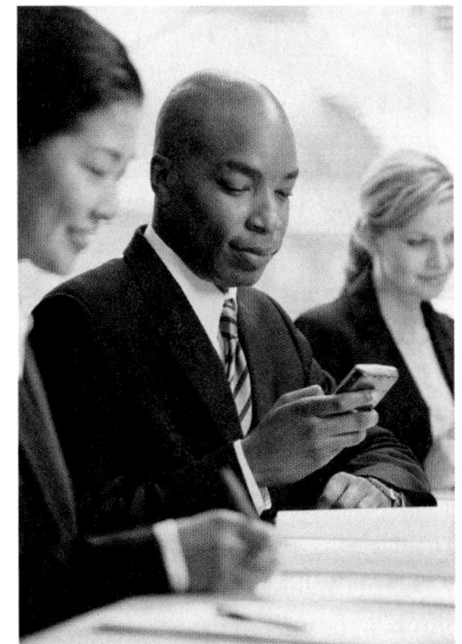

The five steps for problem solving and the methods for solving systems of equations can be used in a variety of applications.

EXAMPLE 1 Text Messaging. In 2008, the average wireless subscriber sent or received a total of 561 calls and text messages each month. The number of text messages was 153 more than the number of calls. How many calls and how many text messages were sent or received each month?
Source: Nielsen Telecom Practice Group

SOLUTION

1. **Familiarize.** We let c = the number of calls and t = the number of text messages sent or received monthly. Since we have two unknowns, we look for two relationships to translate.
2. **Translate.** There are two statements to translate. First, we know the total number of calls and text messages:

Rewording: The number of calls | plus | the number of text messages | was | 561.

Translating: $c \quad + \quad t \quad = \quad 561$

We also know how many more text messages there were than calls:

Rewording: The number of text messages | was | 153 more than the number of calls.

Translating: $t \quad = \quad c + 153$

We have now translated the problem to a system of equations:

$c + t = 561,$
$t = c + 153.$

3. **Carry out.** We solve the system of equations both algebraically and graphically.

ALGEBRAIC APPROACH

Since one equation already has a variable isolated, let's use the substitution method:

$$c + t = 561$$

$c + (c + 153) = 561$ **Substituting $c + 153$ for t**

$2c + 153 = 561$ **Combining like terms**

$2c = 408$ **Subtracting 153 from both sides**

$c = 204.$ **Dividing both sides by 2**

Next, using either of the original equations, we substitute and solve for t:

$$t = c + 153$$
$$t = 204 + 153$$
$$t = 357.$$

We have $c = 204, t = 357$.

GRAPHICAL APPROACH

First, we let $x = c$ and $y = t$ and substitute. Then we graph $y_1 = 561 - x$ and $y_2 = x + 153$. Since the total number of calls and text messages is 561, we choose a viewing window of $[0, 600, 0, 600]$, with Xscl = 50 and Yscl = 50.

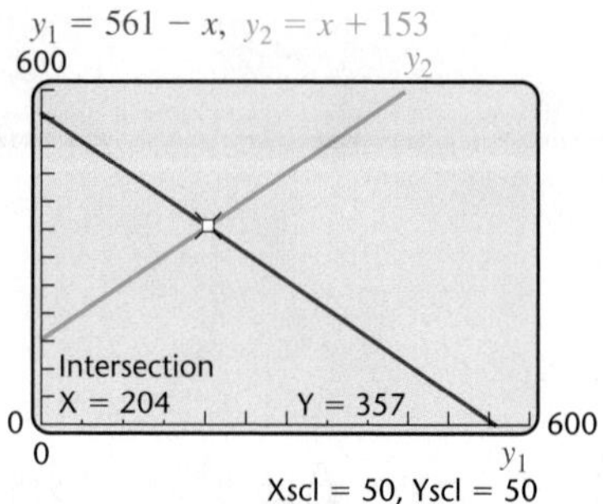

We have a solution of (204, 357).

4. **Check.** Checking using the wording in the original problem, we see that the total number of calls and text messages is $204 + 357$, or 561. Also, the number of text messages, 357, is 153 more than the number of calls: $357 - 204 = 153$. The numbers check.
5. **State.** There were 204 calls and 357 text messages sent or received monthly.

Try Exercise 1.

TOTAL-VALUE PROBLEMS

EXAMPLE 2 Purchasing. Recently the Riley Recreation Center purchased 120 stamps for \$40.80. If the stamps were a combination of 28¢ postcard stamps and 44¢ first-class stamps, how many of each type were bought?

SOLUTION

1. **Familiarize.** To familiarize ourselves with this problem, let's guess that the center bought 60 stamps at 28¢ each and 60 stamps at 44¢ each. The total cost would then be

$$60 \cdot \$0.28 + 60 \cdot \$0.44 = \$16.80 + \$26.40, \text{ or } \$43.20.$$

Since $\$43.20 \neq \$40.80$, our guess is incorrect. We let p = the number of postcard stamps and f = the number of first-class stamps purchased.

2. **Translate.** Since a total of 120 stamps was purchased, we have

$$p + f = 120.$$

To find a second equation, we reword some information, focusing on the amount of money paid:

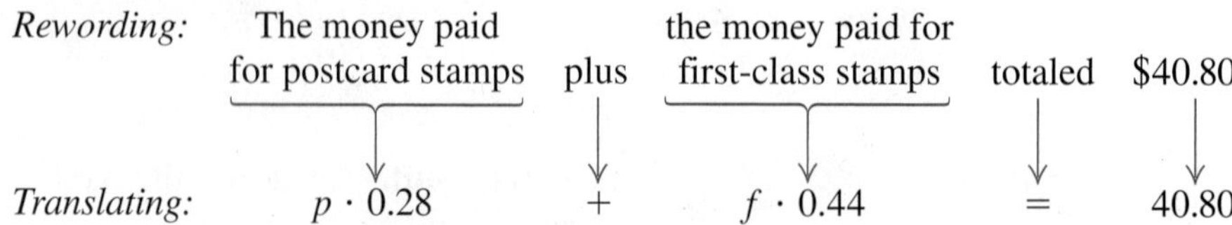

Note that we changed all the money units to dollars.

Presenting the information in a table can be helpful.

	Postcard Stamps	First-Class Stamps	Total	
Cost per Stamp	\$0.28	\$0.44		
Number of Stamps	p	f	120	$\longrightarrow p + f = 120$
Amount Paid	$0.28p$	$0.44f$	40.80	$\longrightarrow 0.28p + 0.44f = 40.80$

We have translated to a system of equations:

$$p + f = 120, \quad (1)$$
$$0.28p + 0.44f = 40.80. \quad (2)$$

3. Carry out. We now solve the system of equations

$$p + f = 120, \quad (1)$$
$$28p + 44f = 4080. \quad (2) \qquad \textbf{Working in cents rather than dollars}$$

ALGEBRAIC APPROACH

Because both equations are in the form $Ax + By = C$, let's use the elimination method to solve the system. We can eliminate p by multiplying both sides of equation (1) by -28 and adding them to the corresponding sides of equation (2):

$$-28p - 28f = -3360 \qquad \textbf{Multiplying both sides of equation (1) by } -28$$
$$28p + 44f = 4080$$
$$16f = 720 \qquad \textbf{Adding}$$
$$f = 45. \qquad \textbf{Solving for } f$$

To find p, we substitute 45 for f in equation (1) and then solve for p:

$$p + f = 120 \qquad \textbf{Equation (1)}$$
$$p + 45 = 120 \qquad \textbf{Substituting 45 for } f$$
$$p = 75. \qquad \textbf{Solving for } p$$

We obtain (45, 75), or $f = 45$, $p = 75$.

GRAPHICAL APPROACH

We replace f with x and p with y and solve for y:

$$y + x = 120 \qquad \textbf{Solving for } y \textbf{ in equation (1)}$$
$$y = 120 - x$$

and

$$28y + 44x = 4080 \qquad \textbf{Solving for } y \textbf{ in equation (2)}$$
$$28y = 4080 - 44x$$
$$y = (4080 - 44x)/28.$$

Since the number of each kind of stamp is between 0 and 120, an appropriate viewing window is [0, 120, 0, 120].

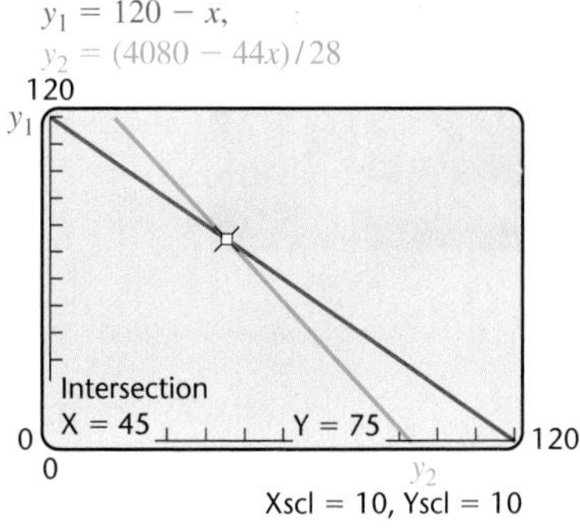

The point of intersection is (45, 75). Since f was replaced with x and p with y, we have $f = 45$, $p = 75$.

Student Notes

It is very important that you clearly label precisely what each variable represents. Not only will this help you write equations, but it will also allow you to write solutions correctly.

4. **Check.** We check in the original problem. Recall that f is the number of first-class stamps and p the number of postcard stamps.

Number of stamps: $f + p = 45 + 75 = 120$

Cost of first-class stamps: $\$0.44f = 0.44 \times 45 = \$19.80$

Cost of postcard stamps: $\$0.28p = 0.28 \times 75 = \$21.00$

Total $= \$40.80$

The numbers check.

5. **State.** The center bought 45 first-class stamps and 75 postcard stamps.

Try Exercise 9.

Example 2 involves two types of items (first-class stamps and postcard stamps), the quantity of each type bought, and the total value of the items. We refer to this type of problem as a *total-value problem.*

MIXTURE PROBLEMS

EXAMPLE 3 Blending Teas. Tea Pots n Treasures sells loose Oolong tea for \$2.15 per ounce. Donna mixed Oolong tea with shaved almonds that sell for \$0.95 per ounce to create the Market Street Oolong blend that sells for \$1.85 per ounce. One week, she made 300 oz of Market Street Oolong. How much tea and how much shaved almonds did Donna use?

SOLUTION

1. **Familiarize.** This problem is similar to Example 2. Since we know the price per ounce of the blend, we can find the total value of the blend by multiplying 300 ounces times \$1.85 per ounce. We let l = the number of ounces of Oolong tea and a = the number of ounces of shaved almonds.

2. **Translate.** Since a 300-oz batch is being made, we must have

$$l + a = 300.$$

To find a second equation, note that the total value of the 300-oz blend must match the combined value of the separate ingredients:

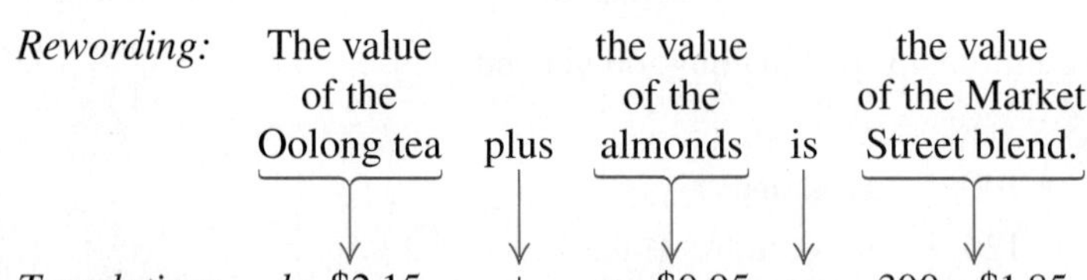

These equations can also be obtained from a table.

	Oolong Tea	Almonds	Market Street Blend	
Number of Ounces	l	a	300	→ $l + a = 300$
Price per Ounce	\$2.15	\$0.95	\$1.85	
Value of Tea	$\$2.15l$	$\$0.95a$	$300 \cdot \$1.85$, or \$555	→ $2.15l + 0.95a = 555$

Clearing decimals in the second equation, we have $215l + 95a = 55{,}500$. We have translated to a system of equations:

$$l + \quad a = 300, \qquad (1)$$
$$215l + 95a = 55{,}500. \qquad (2)$$

3. **Carry out.** We solve the system both algebraically and graphically.

ALGEBRAIC APPROACH

We can solve using substitution. When equation (1) is solved for l, we have $l = 300 - a$. Substituting $300 - a$ for l in equation (2), we find a:

$215(300 - a) + 95a = 55{,}500$	**Substituting**
$64{,}500 - 215a + 95a = 55{,}500$	**Using the distributive law**
$-120a = -9000$	**Combining like terms; subtracting 64,500 from both sides**
$a = 75.$	**Dividing both sides by −120**

We have $a = 75$ and, from equation (1) above, $l + a = 300$. Thus, $l = 225$.

GRAPHICAL APPROACH

We replace a with x and l with y and solve for y, which gives us the system of equations

$$y = 300 - x,$$
$$y = (55{,}500 - 95x)/215.$$

We choose the viewing window $[0, 300, 0, 300]$, graph the equations, and find the point of intersection.

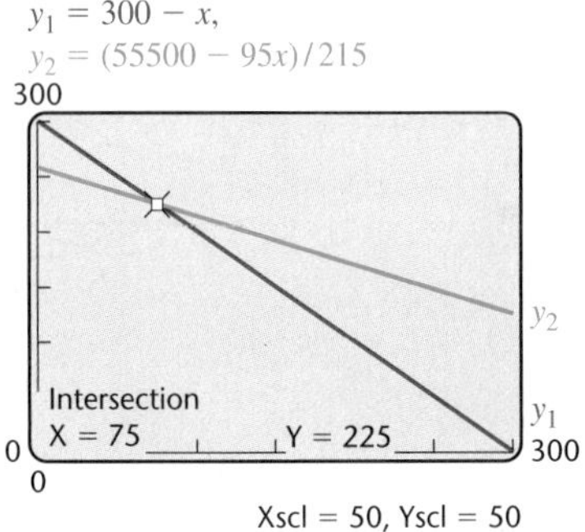

The point of intersection is (75, 225). Since a was replaced with x and l with y, we have $a = 75, l = 225$.

4. **Check.** Combining 225 oz of Oolong tea and 75 oz of almonds will give a 300-oz blend. The value of 225 oz of Oolong is 225($2.15), or $483.75. The value of 75 oz of almonds is 75($0.95), or $71.25. Thus the combined value of the blend is $483.75 + $71.25, or $555. A 300-oz blend priced at $1.85 per ounce would also be worth $555, so our answer checks.
5. **State.** The Market Street blend was made by combining 225 oz of Oolong tea and 75 oz of almonds.

Try Exercise 21.

EXAMPLE 4 Student Loans. Rani's student loans totaled $9600. Part was a PLUS loan made at 3.28% interest and the rest was a Perkins loan made at 5% interest. After one year, Rani's loans accumulated $402.60 in interest. What was the original amount of each loan?

SOLUTION

1. **Familiarize.** We begin with a guess. If \$7000 was borrowed at 3.28% and \$2600 was borrowed at 5%, the two loans would total \$9600. The interest would then be 0.0328(\$7000), or \$229.60, and 0.05(\$2600), or \$130, for a total of only \$359.60 in interest. Our guess was wrong, but checking the guess familiarized us with the problem. More than \$2600 was borrowed at the higher rate.
2. **Translate.** We let l = the amount of the PLUS loan and k = the amount of the Perkins loan. Next, we organize a table in which the entries in each column come from the formula for simple interest:

   Principal · Rate · Time = Interest.

	PLUS Loan	Perkins Loan	Total	
Principal	l	k	\$9600	$\rightarrow l + k = 9600$
Rate of Interest	3.28%	5%		
Time	1 year	1 year		
Interest	$0.0328l$	$0.05k$	\$402.60	$\rightarrow 0.0328l + 0.05k = 402.60$

The total amount borrowed is found in the first row of the table:

$$l + k = 9600.$$

A second equation, representing the accumulated interest, can be found in the last row:

$$0.0328l + 0.05k = 402.60, \quad \text{or} \quad 328l + 500k = 4{,}026{,}000. \quad \textbf{Clearing decimals}$$

3. **Carry out.** The system can be solved by elimination:

$$\begin{aligned} l + k &= 9600 \xrightarrow{\text{Multiplying both sides by } -500} & -500l - 500k &= -4{,}800{,}000 \\ 328l + 500k &= 4{,}026{,}000 & 328l + 500k &= 4{,}026{,}000 \\ & & -172l &= -774{,}000 \\ l + k &= 9600 \longleftarrow & l &= 4500 \\ 4500 + k &= 9600 & & \\ k &= 5100. & & \end{aligned}$$

We find that $l = 4500$ and $k = 5100$.

4. **Check.** The total amount borrowed is \$4500 + \$5100, or \$9600. The interest on \$4500 at 3.28% for 1 year is 0.0328(\$4500), or \$147.60. The interest on \$5100 at 5% for 1 year is 0.05(\$5100), or \$255. The total amount of interest is \$147.60 + \$255, or \$402.60, so the numbers check.
5. **State.** The PLUS loan was for \$4500 and the Perkins loan was for \$5100.

Try Exercise 27.

Problem-Solving Tip

When solving a problem, see if it is patterned or modeled after a problem that you have already solved.

Before proceeding to Example 5, briefly scan Examples 2–4 for similarities. Note that in each case, one of the equations in the system is a simple sum while the other equation represents a sum of products. Example 5 continues this pattern.

EXAMPLE 5 Mixing Fertilizers. Nature's Green Gardening, Inc., carries two brands of fertilizer containing nitrogen and water. "Gentle Grow" is 3% nitrogen and "Sun Saver" is 8% nitrogen. Nature's Green needs to combine the two types of solutions into a 90-L mixture that is 6% nitrogen. How much of each brand should be used?

SOLUTION

1. **Familiarize.** We make a drawing and note that we must consider not only the size of the mixture, but also its strength. Let's make a guess to gain familiarity with the problem.

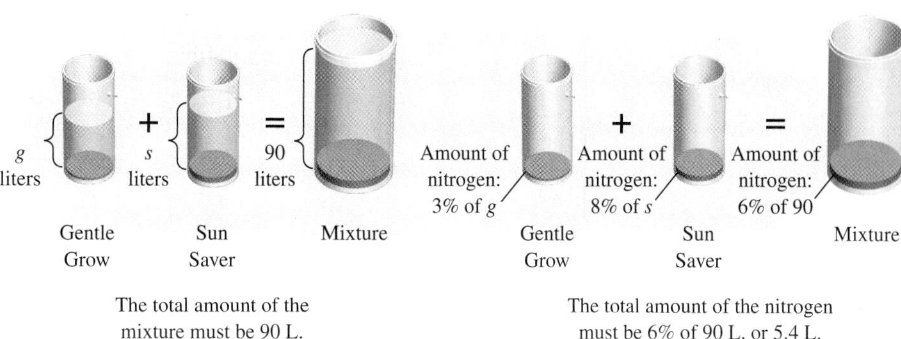

Suppose that 40 L of Gentle Grow and 50 L of Sun Saver are mixed. The resulting mixture will be the right size, 90 L, but will it be the right strength? To find out, note that 40 L of Gentle Grow would contribute $0.03(40) = 1.2$ L of nitrogen to the mixture while 50 L of Sun Saver would contribute $0.08(50) = 4$ L of nitrogen to the mixture. The total amount of nitrogen in the mixture would then be $1.2 + 4$, or 5.2 L. But we want 6% of 90, or 5.4 L, to be nitrogen. Our guess of 40 L and 50 L is close but incorrect. Checking our guess has familiarized us with the problem.

2. **Translate.** Let g = the number of liters of Gentle Grow and s = the number of liters of Sun Saver. The information can be organized in a table.

	Gentle Grow	Sun Saver	Mixture	
Number of Liters	g	s	90	→ $g + s = 90$
Percent of Nitrogen	3%	8%	6%	
Amount of Nitrogen	$0.03g$	$0.08s$	0.06×90, or 5.4 liters	→ $0.03g + 0.08s = 5.4$

If we add g and s in the first row, we get one equation. It represents the total amount of mixture: $g + s = 90$.

If we add the amounts of nitrogen listed in the third row, we get a second equation. This equation represents the amount of nitrogen in the mixture: $0.03g + 0.08s = 5.4$.

After clearing decimals, we have translated the problem to the system

$$g + s = 90, \quad (1)$$
$$3g + 8s = 540. \quad (2)$$

3. **Carry out.** We use the elimination method to solve the system:

$$\begin{aligned} -3g - 3s &= -270 & &\textbf{Multiplying both sides of equation (1) by } -3\\ 3g + 8s &= 540 & &\\ \hline 5s &= 270 & &\textbf{Adding}\\ s &= 54; & &\textbf{Solving for } s\\ g + 54 &= 90 & &\textbf{Substituting into equation (1)}\\ g &= 36. & &\textbf{Solving for } g \end{aligned}$$

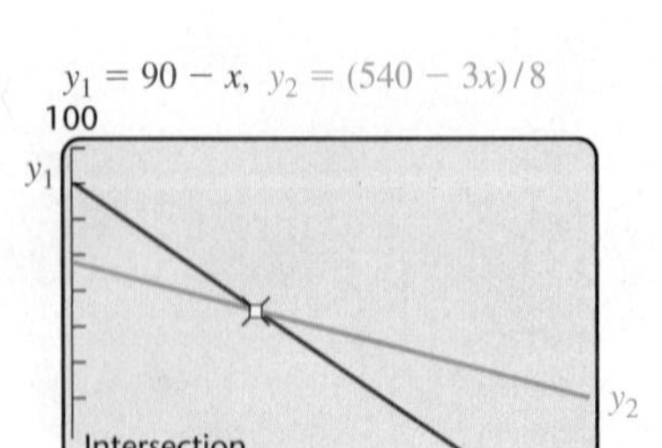

4. **Check.** Remember, g is the number of liters of Gentle Grow and s is the number of liters of Sun Saver.

Total amount of mixture: $g + s = 36 + 54 = 90$

Total amount of nitrogen: 3% of 36 + 8% of 54 = 1.08 + 4.32 = 5.4

Percentage of nitrogen in mixture: $\dfrac{\text{Total amount of nitrogen}}{\text{Total amount of mixture}} = \dfrac{5.4}{90} = 6\%$

The numbers check in the original problem. We can also check graphically, as shown at left, where x replaces g and y replaces s.

5. **State.** Nature's Green Gardening should mix 36 L of Gentle Grow with 54 L of Sun Saver.

Try Exercise 25.

MOTION PROBLEMS

When a problem deals with distance, speed (rate), and time, recall the following.

Distance, Rate, and Time Equations If r represents rate, t represents time, and d represents distance, then

$$d = rt, \quad r = \frac{d}{t}, \quad \text{and} \quad t = \frac{d}{r}.$$

Be sure to remember at least one of these equations. The others can be obtained by multiplying or dividing on both sides as needed.

EXAMPLE 6 Train Travel. A Vermont Railways freight train, loaded with logs, leaves Boston, heading to Washington, D.C., at a speed of 60 km/h. Two hours later, an Amtrak® Metroliner leaves Boston, bound for Washington, D.C., on a parallel track at 90 km/h. At what point will the Metroliner catch up to the freight train?

SOLUTION

1. **Familiarize.** Let's make a guess and check to see if it is correct. Suppose the trains meet after traveling 180 km. We can calculate the time for each train.

	Distance	Rate	Time
Freight train	180 km	60 km/h	$\frac{180}{60} = 3$ hr
Metroliner	180 km	90 km/h	$\frac{180}{90} = 2$ hr

We see that the distance cannot be 180 km, since the difference in travel times for the trains is *not* 2 hr. Although our guess is wrong, we can use a similar chart to organize the information in this problem.

The distance at which the trains meet is unknown, but we do know that the trains will have traveled the same distance when they meet. We let $d =$ this distance.

The time that the trains are running is also unknown, but we do know that the freight train has a 2-hr head start. Thus if we let $t =$ the number of hours that the freight train is running before they meet, then $t - 2$ is the number of hours that the Metroliner runs before catching up to the freight train.

2. **Translate.** We can organize the information in a chart. The formula *Distance* = *Rate* · *Time* guides our choice of rows and columns.

	Distance	Rate	Time	
Freight Train	d	60	t	→ $d = 60t$
Metroliner	d	90	$t - 2$	→ $d = 90(t - 2)$

Using *Distance* = *Rate* · *Time* twice, we get two equations:

$$d = 60t, \quad (1)$$
$$d = 90(t - 2). \quad (2)$$

3. **Carry out.** We solve the system both algebraically and graphically.

ALGEBRAIC APPROACH

We solve the system using the substitution method:

$60t = 90(t - 2)$ **Substituting $60t$ for d in equation (2)**

$60t = 90t - 180$ **Using the distributive law**

$-30t = -180$

$t = 6.$

The time for the freight train is 6 hr, which means that the time for the Metroliner is $6 - 2$, or 4 hr. Remember that it is distance, not time, that the problem asked for. Thus for $t = 6$, we have $d = 60 \cdot 6 = 360$ km.

GRAPHICAL APPROACH

We replace d with y and t with x. Since $y =$ distance and $x =$ time, we use a viewing window of $[0, 10, 0, 500]$. We graph $y_1 = 60x$ and $y_2 = 90(x - 2)$ and find the point of intersection, $(6, 360)$.

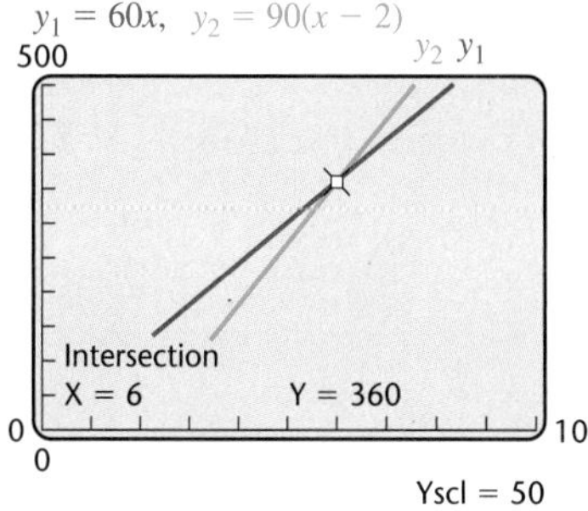

The problem asks for the distance that the trains travel. Recalling that $y =$ distance, we see that the distance that both trains travel is 360 km.

Student Notes

Always be careful to answer the question asked in the problem. In Example 6, the problem asks for distance, not time. Answering "6 hr" would be incorrect.

4. **Check.** At 60 km/h, the freight train will travel $60 \cdot 6$, or 360 km, in 6 hr. At 90 km/h, the Metroliner will travel $90 \cdot (6 - 2) = 360$ km in 4 hr. The numbers check.
5. **State.** The Metroliner will catch up to the freight train at a point 360 km from Boston.

Try Exercise 35.

EXAMPLE 7 Jet Travel. A Boeing 747-400 jet flies 4 hr west with a 60-mph tailwind. Returning *against* the wind takes 5 hr. Find the speed of the jet with no wind.

SOLUTION

1. **Familiarize.** We imagine the situation and make a drawing. Note that the wind *speeds up* the jet on the outbound flight but *slows down* the jet on the return flight.

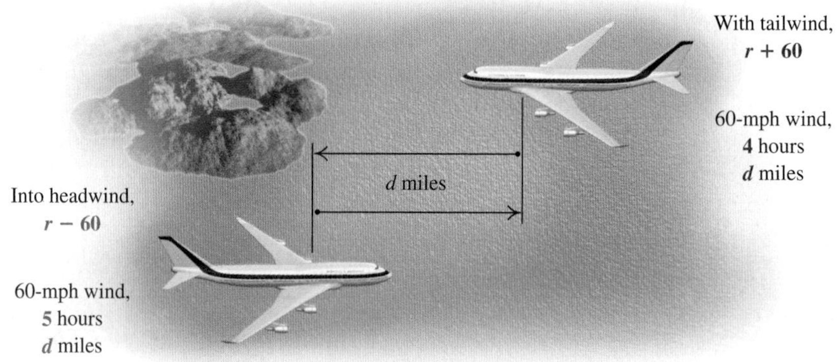

Let's make a guess of the jet's speed if there were no wind. Note that the distances traveled each way must be the same.

Speed with no wind: 400 mph
Speed with the wind: $400 + 60 = 460$ mph
Speed against the wind: $400 - 60 = 340$ mph
Distance with the wind: $460 \cdot 4 = 1840$ mi
Distance against the wind: $340 \cdot 5 = 1700$ mi

Since the distances are not the same, our guess of 400 mph is incorrect.

We let r = the speed, in miles per hour, of the jet in still air. Then $r + 60$ = the jet's speed with the wind and $r - 60$ = the jet's speed against the wind. We also let d = the distance traveled, in miles.

2. **Translate.** The information can be organized in a chart. The distances traveled are the same, so we use *Distance* = *Rate* (or *Speed*) · *Time*. Each row of the chart gives an equation.

	Distance	Rate	Time	
With Wind	d	$r + 60$	4	→ $d = (r + 60)4$
Against Wind	d	$r - 60$	5	→ $d = (r - 60)5$

The two equations constitute a system:

$$d = (r + 60)4, \quad (1)$$
$$d = (r - 60)5. \quad (2)$$

3. **Carry out.** We solve the system using substitution:

$$(r - 60)5 = (r + 60)4 \quad \text{Substituting } (r - 60)5 \text{ for } d \text{ in equation (1)}$$
$$5r - 300 = 4r + 240 \quad \text{Using the distributive law}$$
$$r = 540. \quad \text{Solving for } r$$

4. **Check.** When $r = 540$, the speed with the wind is $540 + 60 = 600$ mph, and the speed against the wind is $540 - 60 = 480$ mph. The distance with the wind, $600 \cdot 4 = 2400$ mi, matches the distance into the wind, $480 \cdot 5 = 2400$ mi, so we have a check.
5. **State.** The speed of the jet with no wind is 540 mph.

Try Exercise 37.

Tips for Solving Motion Problems

1. Draw a diagram using an arrow or arrows to represent distance and the direction of each object in motion.
2. Organize the information in a chart.
3. Look for times, distances, or rates that are the same. These often can lead to an equation.
4. Translating to a system of equations allows for the use of two variables.
5. Always make sure that you have answered the question asked.

4.4 Exercise Set

FOR EXTRA HELP MyMathLab

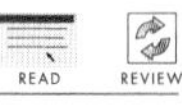

1. *Endangered Species.* In 2009, there were 1010 endangered plant and animal species in the United States. There were 192 more endangered plant species than animal species. How many plant species and how many animal species were considered endangered in 2009?
Source: U.S. Fish and Wildlife Service

2. *e-mail Usage.* In 2007, the average e-mail user sent 578 personal and business e-mails each week. The number of personal e-mails was 30 fewer than the number of business e-mails. How many of each type were sent each week?
Source: *JupiterResearch*

3. *Social Networking.* In April 2009, Facebook and MySpace together had an estimated 160 million unique users. Facebook had 8 million fewer users than twice the number of users MySpace had that same month. How many unique users did each online social network have in April 2009?
Source: www.insidefacebook.com

4. *Snowmen.* The tallest snowman ever recorded—really a snow *woman* named Olympia—was built by residents of Bethel, Maine, and surrounding towns. Her body and head together made up her total record height of 122 ft. The body was 2 ft longer than 14 times the

height of the head. What were the separate heights of Olympia's head and body?
Source: Based on information from *Guinness World Records 2010*

5. *Geometry.* Two angles are supplementary. One angle is 3° less than twice the other. Find the measures of the angles.

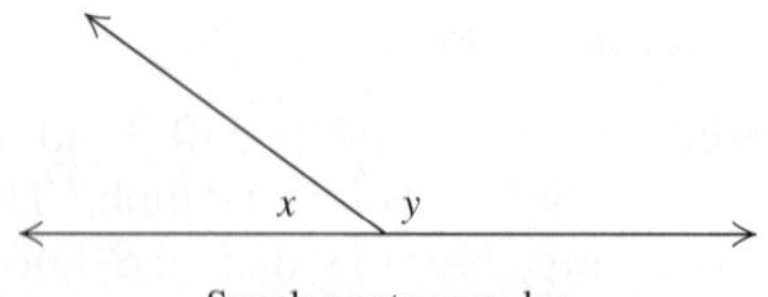

Supplementary angles

6. *Geometry.* Two angles are complementary. The sum of the measures of the first angle and half the second angle is 64°. Find the measures of the angles.

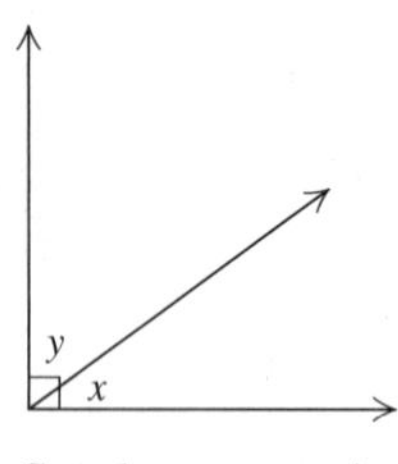

Complementary angles

7. *College Credits.* Each course at Mt. Regis College is worth either 3 or 4 credits. The members of the men's swim team are taking a total of 48 courses that are worth a total of 155 credits. How many 3-credit courses and how many 4-credit courses are being taken?

8. *College Credits.* Each course at Pease County Community College is worth either 3 or 4 credits. The members of the women's golf team are taking a total of 27 courses that are worth a total of 89 credits. How many 3-credit courses and how many 4-credit courses are being taken?

9. *Returnable Bottles.* As part of a fundraiser, the Cobble Hill Daycare collected 430 returnable bottles and cans, some worth 5 cents each and the rest worth 10 cents each. If the total value of the cans and bottles was $26.20, how many 5-cent bottles or cans and how many 10-cent bottles or cans were collected?

10. *Retail Sales.* Cool Treats sold 60 ice cream cones. Single-dip cones sold for $2.50 each and double-dip cones for $4.15 each. In all, $179.70 was taken in for the cones. How many of each size cone were sold?

11. *Yellowstone Park Admissions.* Entering Yellowstone National Park costs $25 for a car and $20 for a motorcycle. One June day, 5950 cars or motorcycles enter and pay a total of $137,625. How many motorcycles entered on that day?
Source: National Park Service, U.S. Department of the Interior

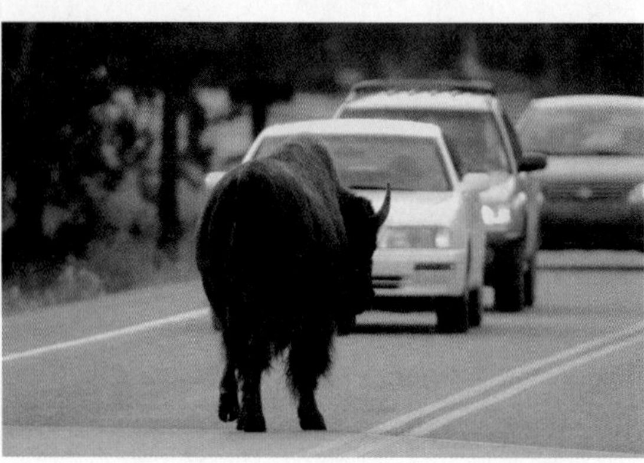

12. *Zoo Admissions.* During the summer months, the Bronx Zoo charges $15 for adults and $11 for children. One July day, a total of $11,920 was collected from 960 adults and children. How many adult admissions were there?
Source: Bronx Zoo

13. *Printing.* King Street Printing recently charged 1.9¢ per sheet of paper, but 2.4¢ per sheet for paper made of recycled fibers. Darren's bill for 150 sheets of paper was $3.41. How many sheets of each type were used?

14. *Photocopying.* Quick Copy recently charged 6¢ per page for copying pages that can be machine-fed and 18¢ per page for copying pages that must be hand-placed on the copier. If Lea's bill for 90 copies was $9.24, how many copies of each type were made?

15. *Jewelry Design.* In order to make a necklace, Alicia purchased 80 beads for a total of $39 (excluding tax). Some of the beads were sterling silver beads that cost 40¢ each and the rest were gemstone beads that cost 65¢ each. How many of each type of bead did Alicia buy?

16. *Vehicles.* 2 Your Door is purchasing new hybrid SUVs for its delivery fleet. On order are 35 Mercury Mariner and Ford Escape hybrids. Each Mercury Mariner costs $29,000, and each Ford Escape costs $32,000. If the total price is $1,042,000, how many of each did they order?

17. *Sales.* Office Depot® recently sold a black Epson® T069120-S ink cartridge for $16.99 and a black HP C4902AN cartridge for $25.99. At the start of a recent fall semester, a total of 50 of these cartridges was sold for a total of $984.50. How many of each type were purchased?

18. *Office Supplies.* Hancock County Social Services is preparing materials for a seminar. They purchase a combination of 80 large and small binders. The large binders cost $8.49 each and the small ones cost

\$5.99 each. If the total cost of the binders was \$544.20, how many of each size were purchased?

Aha! **19.** *Blending Coffees.* The Roasted Bean charges \$13.00 per pound for Fair Trade Organic Mexican coffee and \$11.00 per pound for Fair Trade Organic Peruvian coffee. How much of each type should be used to make a 28-lb blend that sells for \$12.00 per pound?

20. *Mixed Nuts.* Oh Nuts! sells pistachio kernels for \$6.50 per pound and almonds for \$8.00 per pound. How much of each type should be used to make a 50-lb mixture that sells for \$7.40 per pound?

21. *Blending Spices.* Spice of Life sells ground sumac for \$1.35 per ounce and ground thyme for \$1.85 per ounce. Aman wants to make a 20-oz Zahtar seasoning blend using the two spices that sells for \$1.65 per ounce. How much of each spice should Aman use?

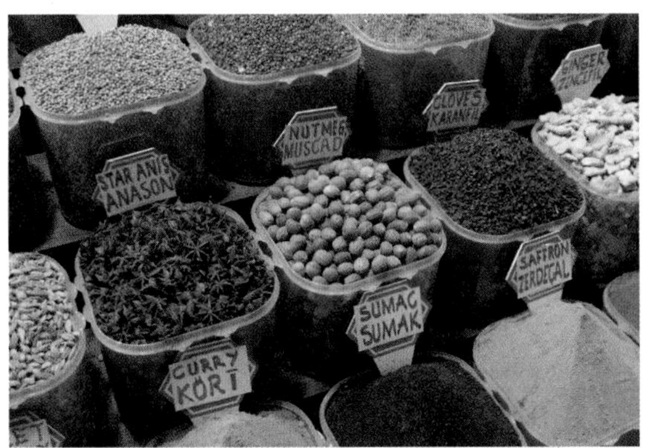

22. *Blending Coffees.* The Gathering Grounds charges \$10.00 per pound for Columbian Supreme coffee and \$12.00 per pound for Columbian Supreme decaffeinated coffee. The Lanosgas buy some of each to make their own partially-decaffeinated blend. If they pay \$53 for a 5-lb blend, how much of each did they purchase?

23. *Acid Mixtures.* Jerome's experiment requires him to mix a 50%-acid solution with an 80%-acid solution to create 200 mL of a 68%-acid solution. How much 50%-acid solution and how much 80%-acid solution should he use? Complete the following table as part of the *Translate* step.

Type of Solution	50%-Acid	80%-Acid	68%-Acid Mix
Amount of Solution	x	y	
Percent Acid	50%		68%
Amount of Acid in Solution		$0.8y$	

24. *Ink Remover.* Etch Clean Graphics uses one cleanser that is 25% acid and a second that is 50% acid. How many liters of each should be mixed to get 30 L of a solution that is 40% acid?

25. *Catering.* Cati's Catering is planning an office reception. The office administrator has requested a candy mixture that is 25% chocolate. Cati has available mixtures that are either 50% chocolate or 10% chocolate. How much of each type should be mixed to get a 20-lb mixture that is 25% chocolate?

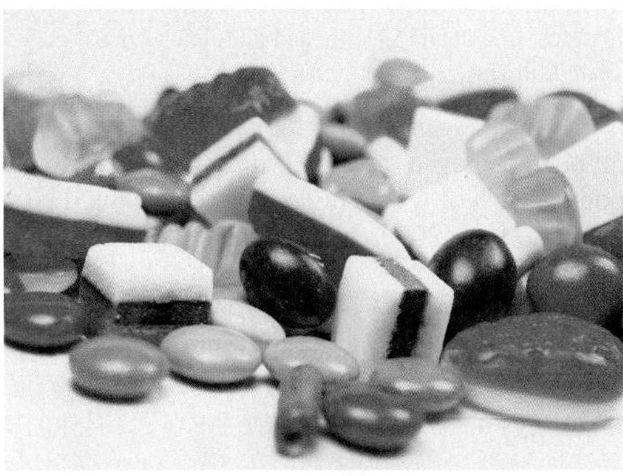

26. *Livestock Feed.* Soybean meal is 16% protein and corn meal is 9% protein. How many pounds of each should be mixed to get a 350-lb mixture that is 12% protein?

27. *Student Loans.* Asel's two student loans totaled \$12,000. One of her loans was at 6.5% simple interest and the other at 7.2%. After one year, Asel owed \$811.50 in interest. What was the amount of each loan?

28. *Investments.* A self-employed contractor nearing retirement made two investments totaling \$15,000. In one year, these investments yielded \$1023 in simple interest. Part of the money was invested at 6% and the rest at 7.5%. How much was invested at each rate?

29. *Automotive Maintenance.* "Steady State" antifreeze is 18% alcohol and "Even Flow" is 10% alcohol. How many liters of each should be mixed to get 20 L of a mixture that is 15% alcohol?

30. *Chemistry.* E-Chem Testing has a solution that is 80% base and another that is 30% base. A technician needs 150 L of a solution that is 62% base. The 150 L will be prepared by mixing the two solutions on hand. How much of each should be used?

31. *Octane Ratings.* The octane rating of a gasoline is a measure of the amount of isooctane in the gas. Manufacturers recommend using 93-octane gasoline on retuned motors. How much 87-octane gas and 95-octane gas should Yousef mix in order to make 10 gal of 93-octane gas for his retuned Ford F-150?
Source: Champlain Electric and Petroleum Equipment

32. *Octane Ratings.* Subaru recommends 91-octane gasoline for the 2008 Legacy 3.0 R. How much 87-octane gas and 93-octane gas should Kelsey mix in order to make 12 gal of 91-octane gas for her Legacy?
Sources: Champlain Electric and Petroleum Equipment: Dean Team Ballwin

33. *Food Science.* The bar graph below shows the milk fat percentages in three dairy products. How many pounds each of whole milk and cream should be mixed to form 200 lb of milk for cream cheese?

34. *Food Science.* How much lowfat milk (1% fat) and how much whole milk (4% fat) should be mixed to make 5 gal of reduced fat milk (2% fat)?

35. *Train Travel.* A train leaves Danville Union and travels north at a speed of 75 km/h. Two hours later, an express train leaves on a parallel track and travels north at 125 km/h. How far from the station will they meet?

36. *Car Travel.* Two cars leave Salt Lake City, traveling in opposite directions. One car travels at a speed of 80 km/h and the other at 96 km/h. In how many hours will they be 528 km apart?

37. *Canoeing.* Kahla paddled for 4 hr with a 6-km/h current to reach a campsite. The return trip against the same current took 10 hr. Find the speed of Kahla's canoe in still water.

38. *Boating.* Cody's motorboat took 3 hr to make a trip downstream with a 6-mph current. The return trip against the same current took 5 hr. Find the speed of the boat in still water.

39. *Point of No Return.* A plane flying the 3458-mi trip from New York City to London has a 50-mph tailwind. The flight's *point of no return* is the point at which the flight time required to return to New York is the same as the time required to continue to London. If the speed of the plane in still air is 360 mph, how far is New York from the point of no return?

40. *Point of No Return.* A plane is flying the 2553-mi trip from Los Angeles to Honolulu into a 60-mph headwind. If the speed of the plane in still air is 310 mph, how far from Los Angeles is the plane's point of no return? (See Exercise 39.)

41. *Basketball Scoring.* Wilt Chamberlain once scored a record 100 points on a combination of 64 foul shots (each worth one point) and two-pointers. How many shots of each type did he make?

42. *Basketball Scoring.* LeBron James recently scored 34 points on a combination of 21 foul shots and two-pointers. How many shots of each type did he make?

43. *Phone Rates.* Kim makes frequent calls from the United States to Korea. Her calling plan costs \$3.99 per month plus 9¢ per minute for calls made to a landline and 15¢ per minute for calls made to a wireless number. One month her bill was \$58.89. If she talked for a total of 400 min, how many minutes were to a landline and how many minutes to a wireless number?
Source: wireless.att.com

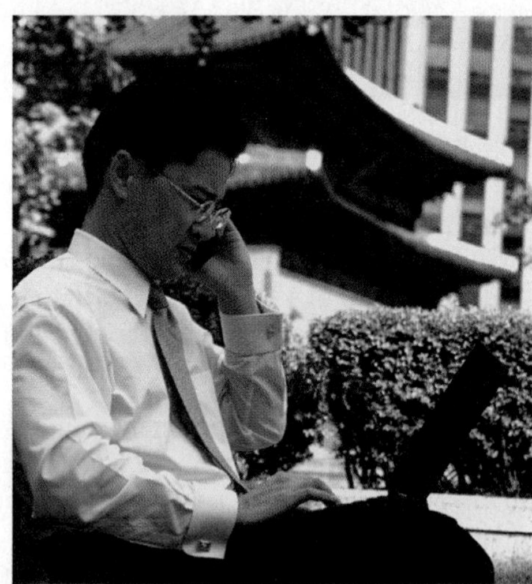

44. *Radio Airplay.* Roscoe must play 12 commercials during his 1-hr radio show. Each commercial is either 30 sec or 60 sec long. If the total commercial time during that hour is 10 min, how many commercials of each type does Roscoe play?

45. *Making Change.* Monica makes a \$9.25 purchase at the bookstore with a \$20 bill. The store has no bills and gives her the change in quarters and fifty-cent pieces. There are 30 coins in all. How many of each kind are there?

46. *Money Exchange.* Sabina goes to a bank and gets change for a $50 bill consisting of all $5 bills and $1 bills. There are 22 bills in all. How many of each kind are there?

TW **47.** In what ways are Examples 3 and 4 similar? In what sense are their systems of equations similar?

TW **48.** Write at least three study tips of your own for someone beginning this exercise set.

SKILL REVIEW

To prepare for Section 4.5, review graphing equations and functions (Sections 3.3, 3.6, and 3.8).

Graph.

49. $y = 2x - 3$ [3.6]

50. $y = -\frac{1}{3}x + 4$ [3.6]

51. $y = 2$ [3.3]

52. $y = -4$ [3.3]

53. $f(x) = -\frac{2}{3}x + 1$ [3.8]

54. $g(x) = 5x - 2$ [3.8]

SYNTHESIS

TW **55.** Suppose that in Example 3 you are asked only for the amount of almonds needed for the Market Street Blend. Would the method of solving the problem change? Why or why not?

TW **56.** Write a problem similar to Example 2 for a classmate to solve. Design the problem so that the solution is "The florist sold 14 hanging plants and 9 flats of petunias."

57. *Metal Alloys.* In order for a metal to be labeled "sterling silver," the silver alloy must contain at least 92.5% pure silver. Mitchell has 32 oz of coin silver, which is 90% pure silver. How much pure silver must he add to the coin silver in order to have a sterling-silver alloy?
Source: Hardy, R. Allen, *The Jewelry Repair Manual*. Courier Dover Publications, 1996, p. 271.

58. *Recycled Paper.* Unable to purchase 60 reams of paper that contains 20% post-consumer fiber, the Naylor School bought paper that was either 0% post-consumer fiber or 30% post-consumer fiber. How many reams of each should be purchased in order to use the same amount of post-consumer fiber as if the 20% post-consumer fiber paper were available?

59. *Automotive Maintenance.* The radiator in Michelle's car contains 6.3 L of antifreeze and water. This mixture is 30% antifreeze. How much of this mixture should she drain and replace with pure antifreeze so that there will be a mixture of 50% antifreeze?

60. *Exercise.* Cindi jogs and walks to school each day. She averages 4 km/h walking and 8 km/h jogging. From home to school is 6 km and Cindi makes the trip in 1 hr. How far does she jog in a trip?

61. *Book Sales.* *American Economic History* can be purchased as a three-volume set for $88 or each volume can be purchased separately for $39. An economics class spent $1641 for 51 volumes. How many three-volume sets were ordered?
Source: National History Day, www.nhd.org.

62. The tens digit of a two-digit positive integer is 2 more than three times the units digit. If the digits are interchanged, the new number is 13 less than half the given number. Find the given integer. (*Hint*: Let x = the tens-place digit and y = the units-place digit; then $10x + y$ is the number.)

63. *Wood Stains.* Williams' Custom Flooring has 0.5 gal of stain that is 20% brown and 80% neutral. A customer orders 1.5 gal of a stain that is 60% brown and 40% neutral. How much pure brown stain and how much neutral stain should be added to the original 0.5 gal in order to make up the order? (This problem was suggested by Professor Chris Burditt of Yountville, California.)

64. *Train Travel.* A train leaves Union Station for Central Station, 216 km away, at 9 A.M. One hour later, a train leaves Central Station for Union Station. They meet at noon. If the second train had started at 9 A.M. and the first train at 10:30 A.M., they would still have met at noon. Find the speed of each train.

65. *Fuel Economy.* Grady's station wagon gets 18 miles per gallon (mpg) in city driving and 24 mpg in highway driving. The car is driven 465 mi on 23 gal of gasoline. How many miles were driven in the city and how many were driven on the highway?

66. *Biochemistry.* Industrial biochemists routinely use a machine to mix a buffer of 10% acetone by adding 100% acetone to water. One day, instead of adding 5 L of acetone to create a vat of buffer, a machine added 10 L. How much additional water was needed to bring the concentration down to 10%?

67. Recently, Staples® charged $5.99 for a 2-count pack of Bic® Round Stic Grip mechanical pencils and $7.49 for a 12-count pack of Bic® Matic Grip mechanical pencils. Wiese Accounting purchased 138 of these two types of mechanical pencils for a total of $157.26. How many packs of each did they buy?

Try Exercise Answers: Section 4.4

1. Plant species: 601; animal species: 409 **9.** 5-cent bottles or cans: 336; 10-cent bottles or cans: 94 **21.** Sumac: 8 oz; thyme: 12 oz **25.** 50% chocolate: 7.5 lb; 10% chocolate: 12.5 lb **27.** $7500 at 6.5%; $4500 at 7.2% **35.** 375 km **37.** 14 km/h

Collaborative Corner

How Many Two's? How Many Three's?

Focus: Systems of linear equations
Time: 20 minutes
Group Size: 3

The box score at right, from the 2010 NBA All-Star game, contains information on how many field goals (worth either 2 or 3 points) and free throws (worth 1 point) each player attempted and made. For example, the line "James 10-22 4-4 25" means that the East's LeBron James made 10 field goals out of 22 attempts and 4 free throws out of 4 attempts, for a total of 25 points.

ACTIVITY

1. Work as a group to develop a system of two equations in two unknowns that can be used to determine how many 2-pointers and how many 3-pointers were made by the West.
2. Each group member should solve the system from part (1) in a different way: one person algebraically, one person by making a table and methodically checking all combinations of 2- and 3-pointers, and one person by guesswork. Compare answers when this has been completed.
3. Determine, as a group, how many 2- and 3-pointers the East made.

East (141)
James 10–22 4–4 25, Garnett 2–4 0–0 4, Howard 7–10 2–3 17, Johnson 4–8 0–0 10, Wade 12–16 4–6 28, Pierce 3–6 0–0 8, Bosh 9–16 5–7 23, Rondo 2–3 0–0 4, Wallace 1–3 0–0 2, Lee 2–3 0–0 4, Rose 4–8 0–0 8, Horford 4–5 0–1 8
Totals 60–104 15–21 141

West (139)
Duncan 1–4 1–2 3, Nowitzki 8–15 6–6 22, Stoudemire 5–10 2–2 12, Anthony 13–22 0–1 27, Nash 2–4 0–0 4, Gasol 5–9 3–3 13, Billups 6–11 0–0 17, Williams 6–11 0–0 14, Durant 7–14 0–0 15, Randolph 4–10 0–0 8, Kaman 2–4 0–0 4, Kidd 0–1 0–0 0
Totals 59–115 12–14 139

East 37 39 42 23 — 141
West 34 35 40 30 — 139

4.5 Solving Equations by Graphing

- Solving Equations Graphically: The Intersect Method
- Solving Equations Graphically: The Zero Method
- Applications

Recall that to *solve* an equation means to find all the replacements for the variable that make the equation true. We have seen how to solve algebraically; we now use a graphical method to solve.

SOLVING EQUATIONS GRAPHICALLY: THE INTERSECT METHOD

To see how solutions of equations are related to graphs, consider the graphs of the functions given by $f(x) = 2x + 5$ and $g(x) = -3$. The graphs intersect at $(-4, -3)$. At that point,

$$f(-4) = -3 \quad \text{and}$$
$$g(-4) = -3.$$

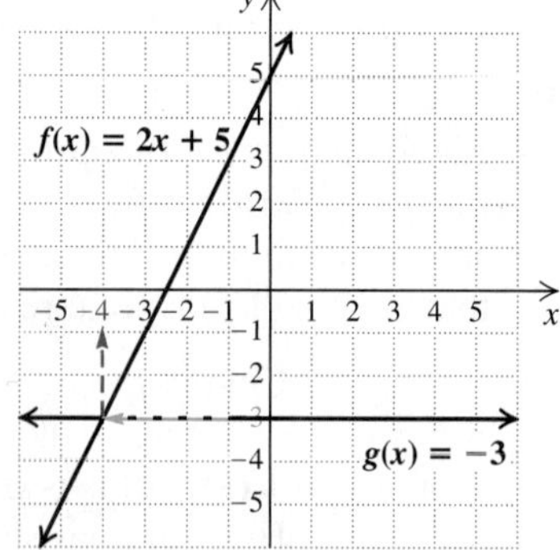

Thus,

$$f(x) = g(x) \quad \text{when } x = -4;$$
$$2x + 5 = -3 \quad \text{when } x = -4.$$

The solution of $2x + 5 = -3$ is -4, which is the x-coordinate of the point of intersection of the graphs of f and g.

EXAMPLE 1 Solve graphically: $\frac{1}{2}x + 3 = 2$.

SOLUTION First, we define $f(x)$ and $g(x)$. If $f(x) = \frac{1}{2}x + 3$ and $g(x) = 2$, then the equation becomes

$$\frac{1}{2}x + 3 = 2$$
$$f(x) = g(x). \quad \textbf{Substituting}$$

We graph both functions. The intersection appears to be $(-2, 2)$, so the solution is apparently -2.

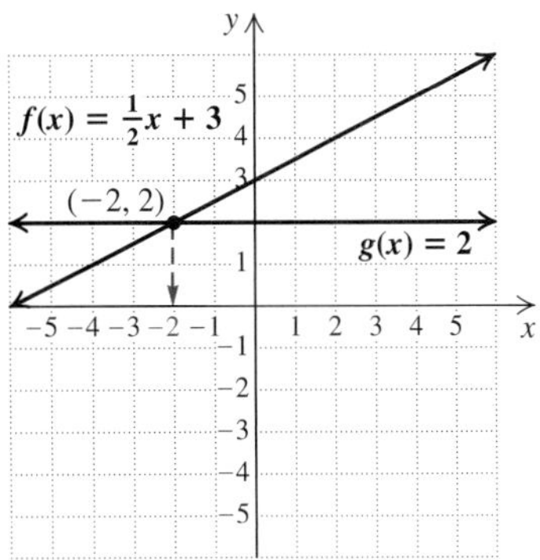

Check:

$$\begin{array}{c|c} \frac{1}{2}x + 3 & 2 \\ \hline \frac{1}{2}(-2) + 3 & 2 \\ -1 + 3 & \\ 2 & \stackrel{?}{=} 2 \quad \text{TRUE} \end{array}$$

The solution is -2.

Try Exercise 7.

Connecting the Concepts

Although graphical solutions of systems and equations look similar, different solutions are read from the graph. The solution of a system of equations is an ordered pair. The solution of a linear equation in one variable is a single number.

Solving a system of equations

$y = x - 3,$
$y = 3x - 5$

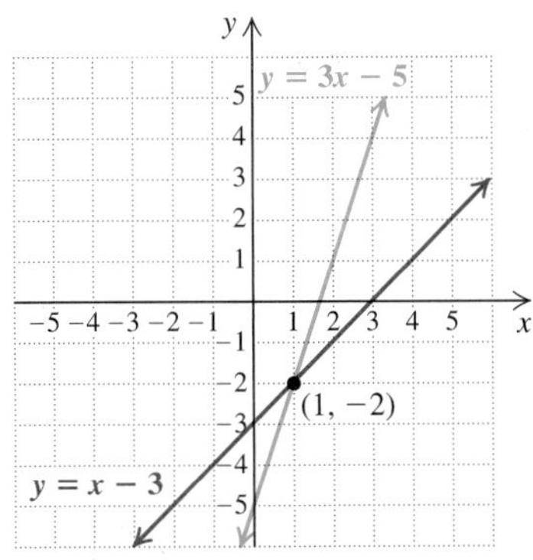

The solution of the system is $(1, -2)$.

Solving an equation

$x - 3 = 3x - 5$

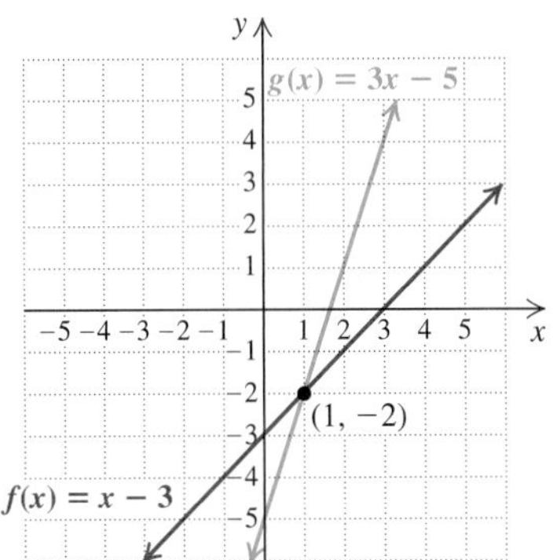

The solution of the equation is 1.

EXAMPLE 2 Solve: $-\frac{3}{4}x + 6 = 2x - 1$.

ALGEBRAIC APPROACH

We have

$$-\tfrac{3}{4}x + 6 = 2x - 1$$

$$-\tfrac{3}{4}x + 6 - 6 = 2x - 1 - 6 \quad \text{Subtracting 6 from both sides}$$

$$-\tfrac{3}{4}x = 2x - 7 \quad \text{Simplifying}$$

$$-\tfrac{3}{4}x - 2x = 2x - 7 - 2x \quad \text{Subtracting } 2x \text{ from both sides}$$

$$-\tfrac{3}{4}x - \tfrac{8}{4}x = -7 \quad \text{Simplifying}$$

$$-\tfrac{11}{4}x = -7 \quad \text{Combining like terms}$$

$$-\tfrac{4}{11}\left(-\tfrac{11}{4}x\right) = -\tfrac{4}{11}(-7) \quad \text{Multiplying both sides by } -\tfrac{4}{11}$$

$$x = \tfrac{28}{11}. \quad \text{Simplifying}$$

Check:

$-\frac{3}{4}x + 6$	$= 2x - 1$
$-\frac{3}{4}\left(\frac{28}{11}\right) + 6$	$2\left(\frac{28}{11}\right) - 1$
$-\frac{21}{11} + \frac{66}{11}$	$\frac{56}{11} - \frac{11}{11}$
$\frac{45}{11} \stackrel{?}{=}$	$\frac{45}{11}$ TRUE

The solution is $\frac{28}{11}$.

GRAPHICAL APPROACH

We graph $f(x) = -\frac{3}{4}x + 6$ and $g(x) = 2x - 1$. It appears that the lines intersect at $(2.5, 4)$.

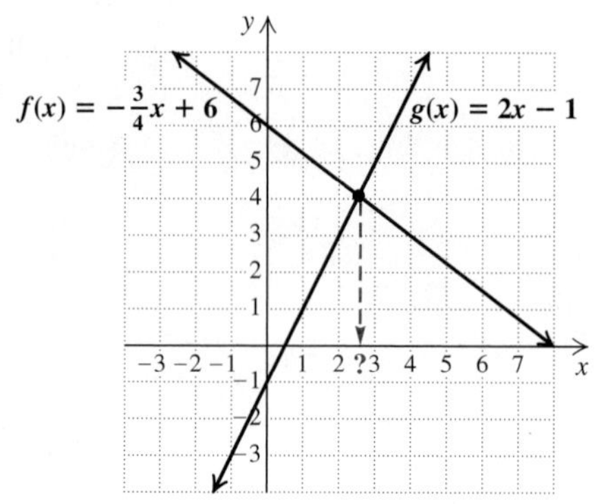

Check:

$-\frac{3}{4}x + 6$	$= 2x - 1$
$-\frac{3}{4}(2.5) + 6$	$2(2.5) - 1$
$-1.875 + 6$	$5 - 1$
$4.125 \stackrel{?}{=}$	4 FALSE

Our check shows that 2.5 is *not* the solution. To find the exact solution graphically, we need a method that will determine coordinates more precisely.

Try Exercise 37.

STUDY TIP

If You Must Miss a Class

Occasionally you may be forced to miss an upcoming class. It is usually best to alert your instructor to this situation as soon as possible. He or she may permit you to make up a missed quiz or test *if you provide enough advance notice.* Make an effort to find out what assignment will be given in your absence and try your best to learn the material on your own beforehand so that you can ask questions in advance of your absence. Sometimes it is possible to attend another section of the course or view a video for the lesson that you must miss.

CAUTION! When using a hand-drawn graph to solve an equation, it is important to use graph paper and to work as neatly as possible. Use a straightedge when drawing lines and be sure to erase any mistakes completely.

We can use the INTERSECT option of the CALC menu on a graphing calculator to find the point of intersection.

EXAMPLE 3 Solve using a graphing calculator: $-\frac{3}{4}x + 6 = 2x - 1$.

SOLUTION In Example 2, we saw that the graphs of $f(x) = -\frac{3}{4}x + 6$ and $g(x) = 2x - 1$ intersect near the point $(2.5, 4)$. Using a calculator, we now graph $y_1 = -\frac{3}{4}x + 6$ and $y_2 = 2x - 1$ and use the INTERSECT feature.

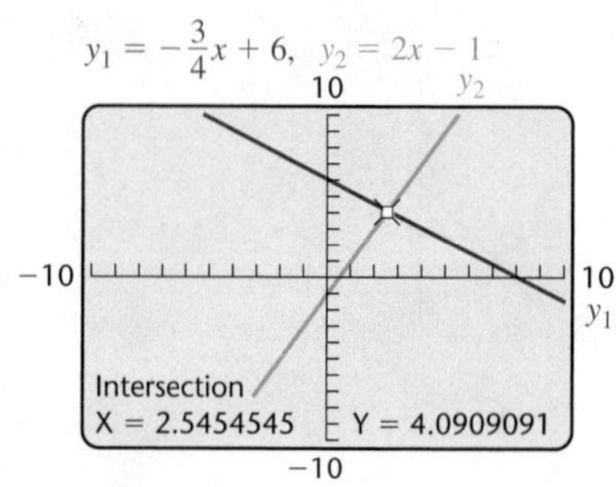

It appears from the screen above that the solution is 2.5454545. To check, we evaluate both sides of the equation $-\frac{3}{4}x + 6 = 2x - 1$ for this value of x. We evaluate Y1(X) and Y2(X) from the home screen, as shown on the left below. Converting X to fraction notation will give an exact solution. From the screen on the right below, we see that $\frac{28}{11}$ is the solution of the equation.

```
Y1(X)
                4.090909091
Y2(X)
                4.090909091
```

Try Exercise 41.

SOLVING EQUATIONS GRAPHICALLY: THE ZERO METHOD

When we are solving an equation graphically, it can be challenging to determine a portion of the x, y-coordinate plane that contains the point of intersection. The Zero method makes that determination easier because we are interested only in the point at which a graph crosses the x-axis.

To solve an equation using the Zero method, we use the addition principle to get zero on one side of the equation. Then we look for the intersection of the line and the x-axis, or the *x-intercept* of the graph. If using function notation, we look for the *zero* of the function, or the input x that makes the output zero.

Zero of a Function A zero of a function is an input whose corresponding output is 0. If a is a zero of the function f, then $f(a) = 0$.

The y-coordinate of a point is 0 when the point is on the x-axis. Thus a zero of a function is the x-coordinate of any point at which its graph crosses or touches the x-axis.

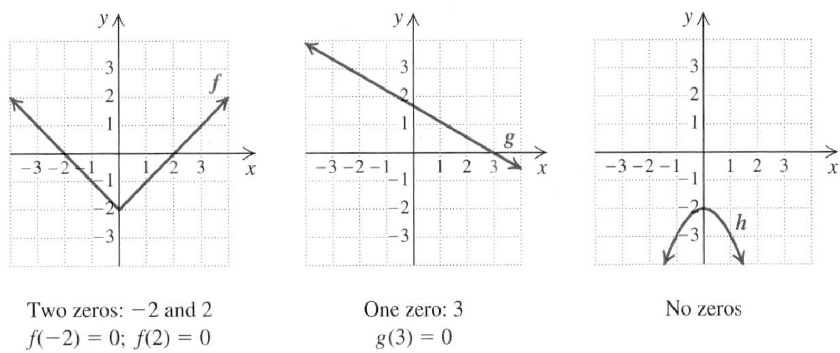

Two zeros: -2 and 2
$f(-2) = 0$; $f(2) = 0$

One zero: 3
$g(3) = 0$

No zeros

Zeros of a Function

We can determine any zeros of a function using the ZERO option in the CALC menu of a graphing calculator. After graphing the function and choosing the ZERO option, we will be prompted for a Left Bound, a Right Bound, and a Guess. Since there may be more than one zero of a function, the left and right bounds indicate which zero we are currently finding. We examine the graph to find any places where it appears that the graph touches or crosses the x-axis, and then find those x-values one at a time. By using the arrow keys or entering a value on

(continued)

the keypad, we choose an x-value less than the zero for the left bound, an x-value more than the zero for the right bound, and a value close to the zero for the Guess.

Your Turn

1. Find the zero of $f(x) = 4.5 - 0.81x$. Enter $y_1 = 4.5 - 0.81x$ and graph using a standard viewing window. The zero should appear to be about 5.
2. Press CALC 2. Press 3 ENTER 8 ENTER to choose a Left Bound of 3 and a Right Bound of 8. Then press 5 ENTER to enter a Guess of 5. The zero is 5.5555556.
3. From the home screen, press X,T,Θ,n MATH 1 ENTER to convert the solution to a fraction. $\frac{50}{9}$

EXAMPLE 4 Find the zeros of the function given by $f(x) = \frac{2}{3}x + 5$.

ALGEBRAIC APPROACH

We want to find any x-values for which $f(x) = 0$, so we substitute 0 for $f(x)$ and solve:

$$f(x) = \tfrac{2}{3}x + 5$$

$$0 = \tfrac{2}{3}x + 5 \qquad \textbf{Substituting 0 for } f(x)$$

$$-\tfrac{2}{3}x = 5 \qquad \textbf{Subtracting } -\tfrac{2}{3}x \textbf{ from both sides}$$

$$\left(-\tfrac{3}{2}\right)\left(-\tfrac{2}{3}x\right) = \left(-\tfrac{3}{2}\right)(5) \qquad \textbf{Multiplying by the reciprocal of } -\tfrac{2}{3}$$

$$x = -\tfrac{15}{2}. \qquad \textbf{Simplifying}$$

The zero of the function is $-\frac{15}{2}$, or -7.5.

GRAPHICAL APPROACH

We graph the function. There appears to be one zero of the function, approximately -7.

We press CALC 2 to choose the ZERO option of the CALC menu. We then choose -10 for a Left Bound, -5 for a Right Bound, and -7 for a Guess. The zero of the function is -7.5.

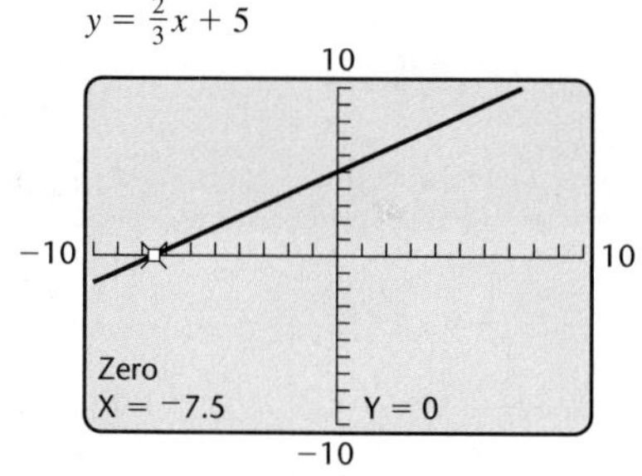

Try Exercise 25.

Note that the answer to Example 4 is a value for x. A zero of a function is an x-value, not a function value or an ordered pair.

EXAMPLE 5 Solve graphically, using the Zero method: $2x - 5 = 4x - 11$.

SOLUTION We first get 0 on one side of the equation:

$$2x - 5 = 4x - 11$$

$$-2x - 5 = -11 \qquad \textbf{Subtracting } 4x \textbf{ from both sides}$$

$$-2x + 6 = 0. \qquad \textbf{Adding 11 to both sides}$$

We then graph $f(x) = -2x + 6$, and find the x-intercept of the graph, as shown at left.

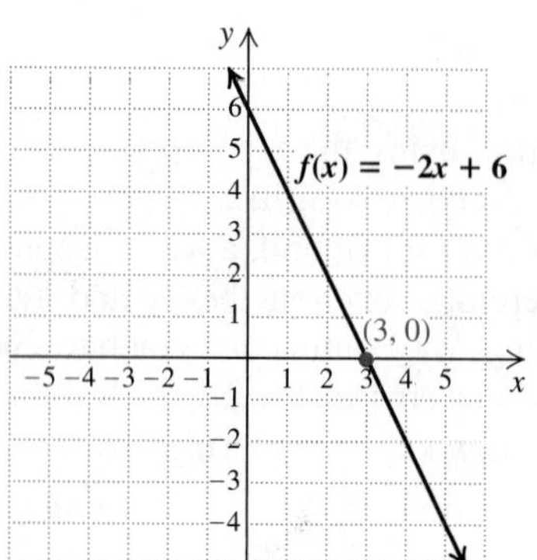

The x-intercept of the graph appears to be $(3, 0)$. We check 3 in the original equation.

Check:

$$\begin{array}{c|c} 2x - 5 & 4x - 11 \\ \hline 2 \cdot 3 - 5 & 4 \cdot 3 - 11 \\ 6 - 5 & 12 - 11 \\ 1 & 1 \end{array} \quad 1 \stackrel{?}{=} 1 \quad \text{TRUE}$$

The solution is 3.

Try Exercise 31.

EXAMPLE 6 Solve $3 - 8x = 5 - 7x$ using both the Intersect method and the Zero method.

GRAPHICAL APPROACH: INTERSECT METHOD

We graph $y_1 = 3 - 8x$ and $y_2 = 5 - 7x$ and determine the coordinates of any point of intersection.

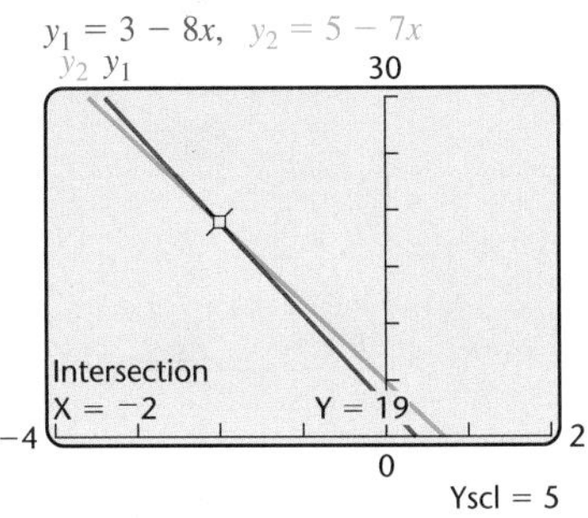

The point of intersection is $(-2, 19)$. The solution is -2.

GRAPHICAL APPROACH: ZERO METHOD

We first get zero on one side of the equation:

$$3 - 8x = 5 - 7x$$
$$-2 - 8x = -7x \quad \textbf{Subtracting 5 from both sides}$$
$$-2 - x = 0. \quad \textbf{Adding } 7x \textbf{ to both sides}$$

Then we graph $y = -2 - x$ and use the ZERO option to determine the x-intercept of the graph.

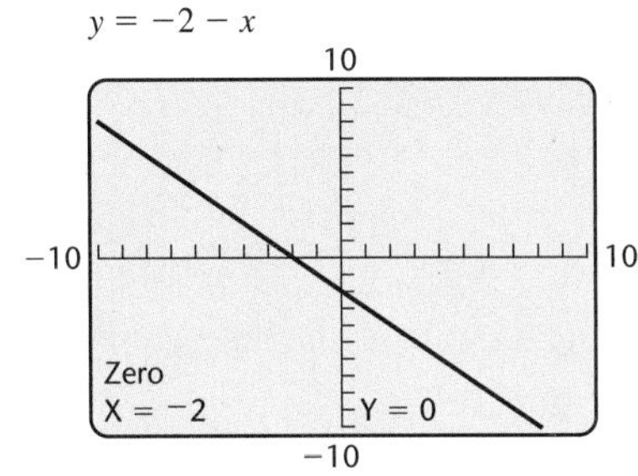

The solution is the first coordinate of the x-intercept, or -2.

Try Exercise 39.

APPLICATIONS

EXAMPLE 7 Photo-Book Prices. A medium wire-bound photo book from iPhoto costs \$10 for the book and 20 pages. Each additional page costs \$0.50. Formulate and graph a mathematical model for the price. Then use the model to estimate the size of a book for which the price is \$18.00.
Source: www.apple.com

SOLUTION

1. **Familiarize.** For a 20-page book, the price is \$10. How much would a 26-page book cost? The price is \$10 plus the cost for the additional 6 pages:

$$\$10 + 6(\$0.50) = \$10 + \$3 = \$13.$$

This can be generalized in a model if we let $P(x)$ represent the price of the book, in dollars, for a book with x pages more than the 20 included in the price.

2. **Translate.** We reword and translate as follows:

Rewording:	The total price	is	the price of the 20-page book	plus	\$0.50 per additional page.
	↓	↓	↓	↓	↓
Translating:	$P(x)$	$=$	10	$+$	$0.50x$

3. **Carry out.** To estimate the size of a book for which the price is \$18.00, we are estimating the solution of

$$0.50x + 10 = 18. \quad \textbf{Replacing } P(x) \textbf{ with 18}$$

We do this by graphing $P(x) = 0.50x + 10$ and $y = 18$ and looking for the point of intersection. On a graphing calculator, we let $y_1 = 0.5x + 10$ and $y_2 = 18$ and adjust the window dimensions to include the point of intersection. We find a point of intersection at $(16, 18)$.

Thus we estimate that an \$18.00 book will have 16 additional pages, or a total of 36 pages.

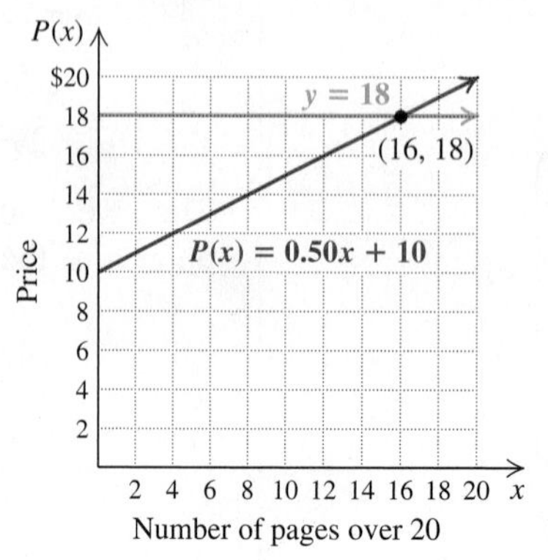

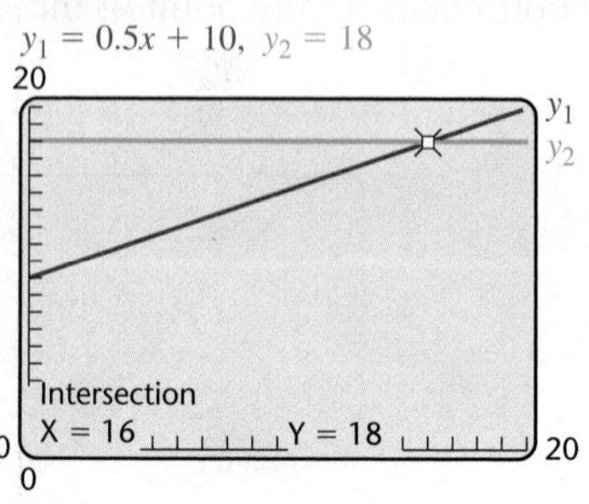

4. **Check.** For a book with 36 pages, the price is \$10 plus \$0.50 for each page over 20. Since $36 - 20 = 16$, the price will be

$$\$10 + 16(\$0.50) = \$10 + \$8 = \$18.$$

Our estimate turns out to be the exact answer.

5. **State.** A photo book that costs \$18 has 36 pages.

Try Exercise 43.

4.5 Exercise Set

FOR EXTRA HELP

Concept Reinforcement *Exercises 1–6 refer to the graph below. Match each description with the appropriate answer from the column on the right.*

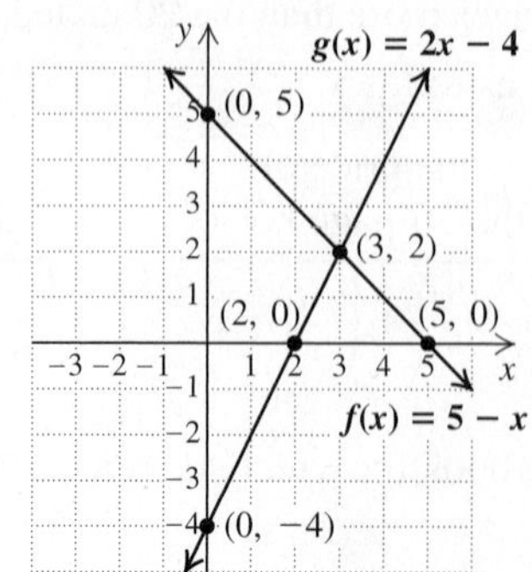

1. The solution of $5 - x = 2x - 4$
2. The zero of g
3. The solution of $5 - x = 0$
4. The x-intercept of the graph of f
5. The y-intercept of the graph of g
6. The solution of

$$y = 5 - x,$$
$$y = 2x - 4$$

a) $(0, -4)$
b) $(5, 0)$
c) $(3, 2)$
d) 2
e) 3
f) 5

Estimate the solution of each equation from the associated graph.

7. $2x - 1 = -5$

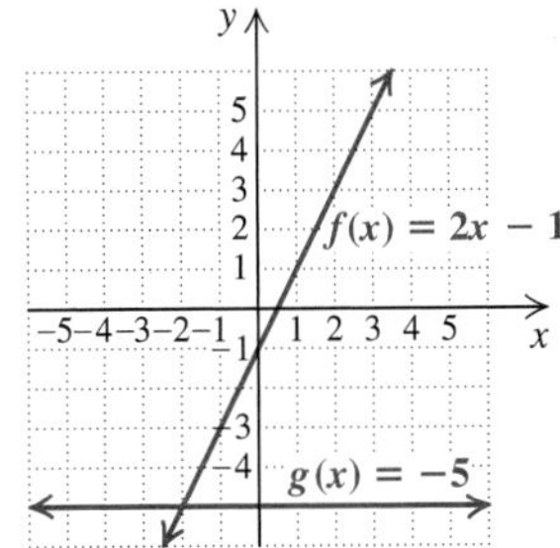

8. $\frac{1}{2}x + 1 = 3$

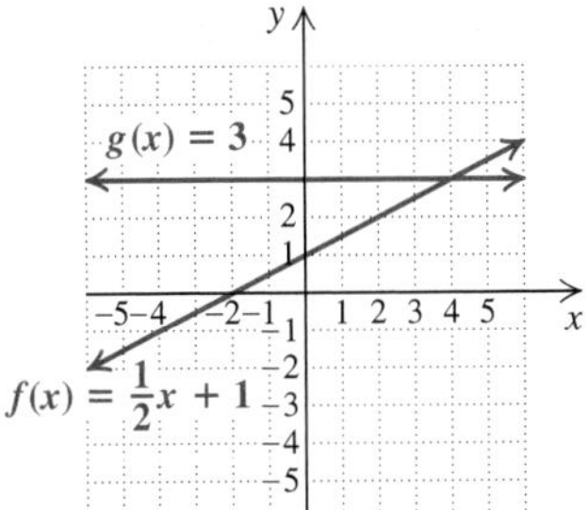

9. $2x + 3 = x - 1$

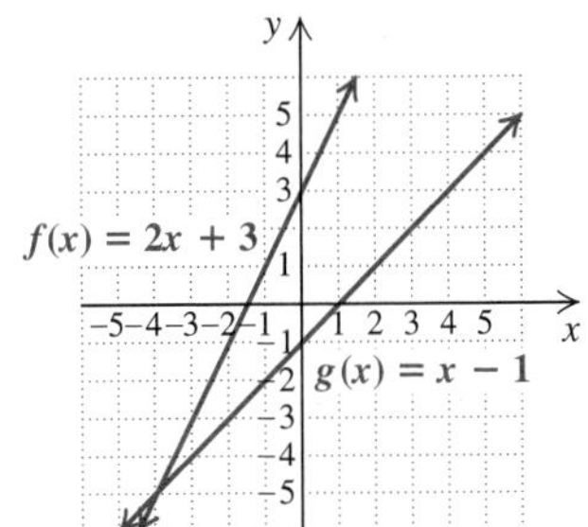

10. $2 - x = 3x - 2$

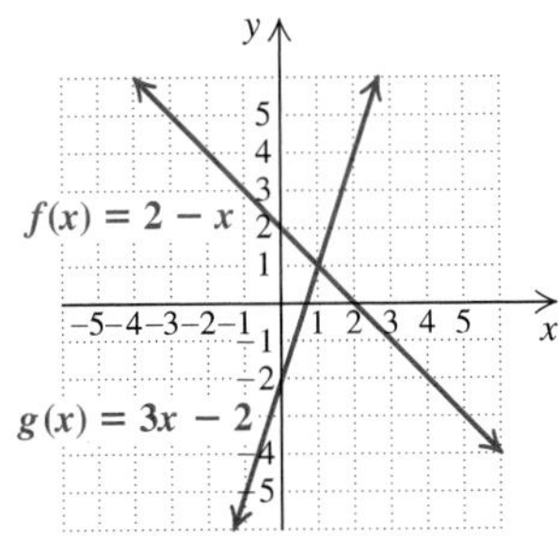

11. $\frac{1}{2}x + 3 = x - 1$

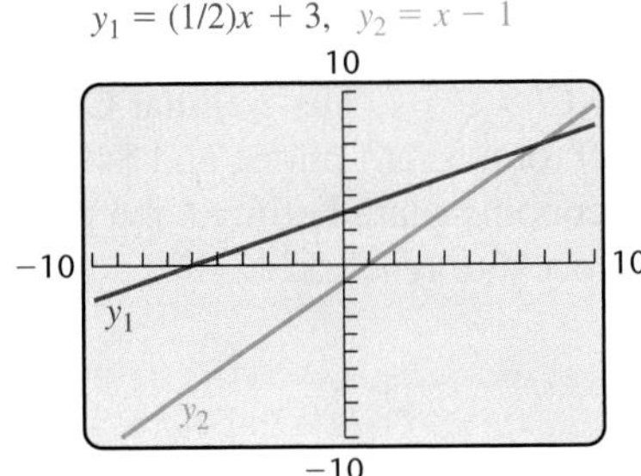

12. $2x - \frac{1}{2} = x + \frac{5}{2}$

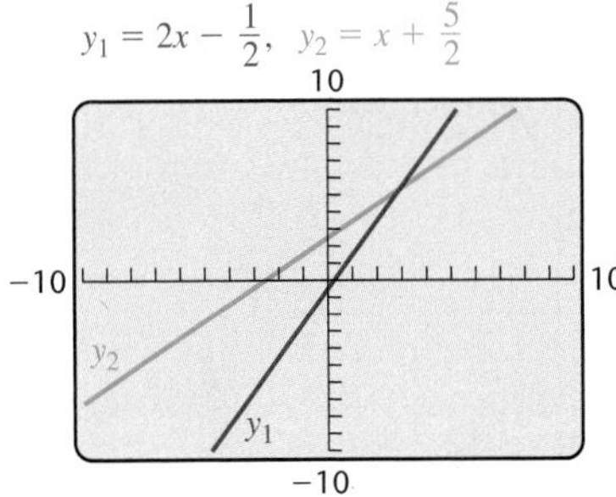

13. $f(x) = g(x)$

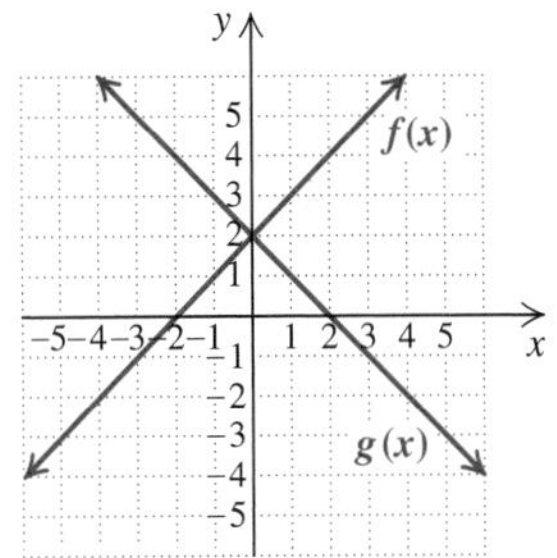

14. $f(x) = g(x)$

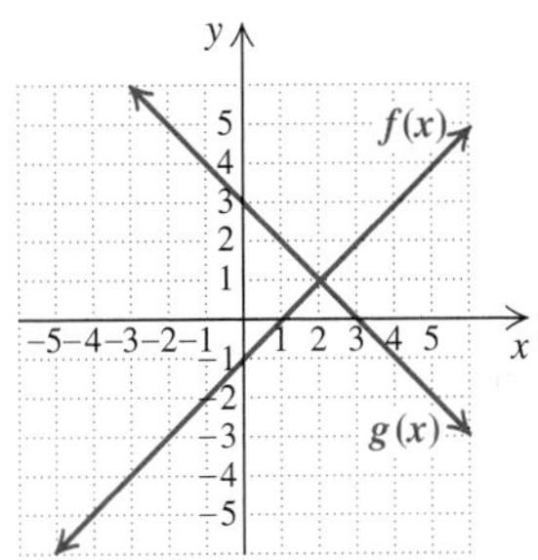

15. Estimate the value of x for which $y_1 = y_2$.

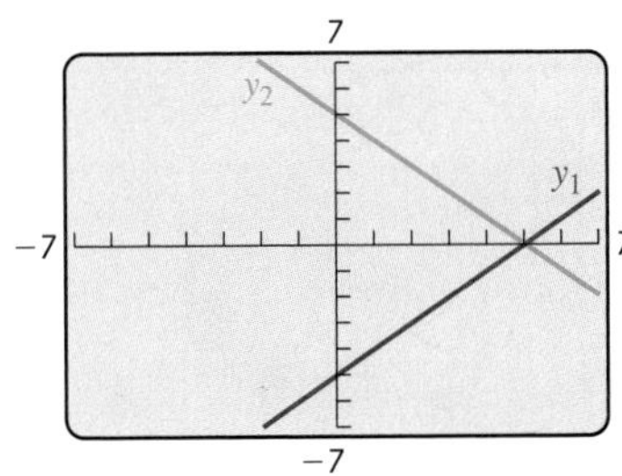

16. Estimate the value of x for which $y_1 = y_2$.

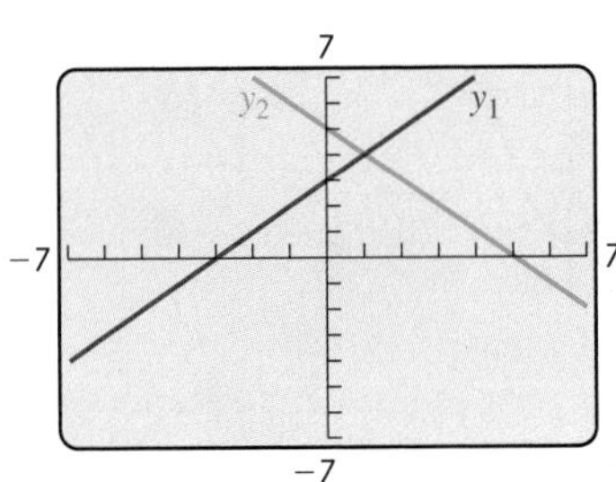

In Exercises 17–30, determine the zeros, if any, of each function.

17.

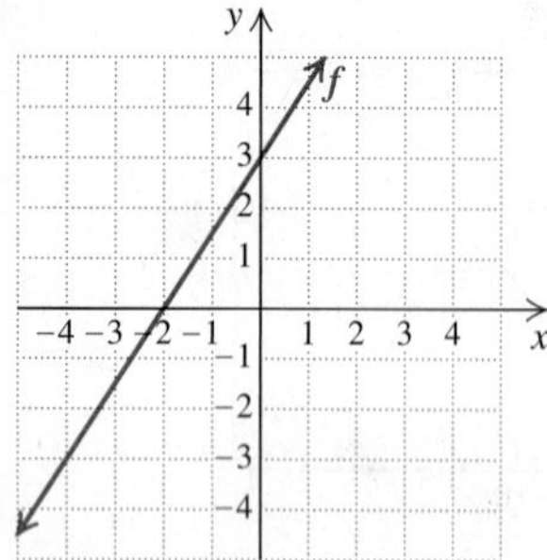

18.

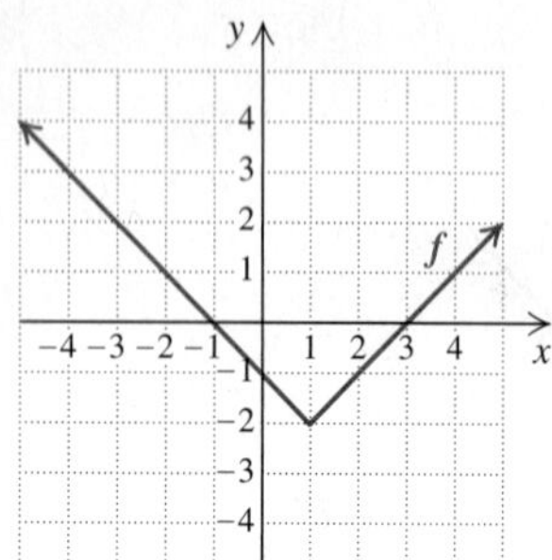

19.

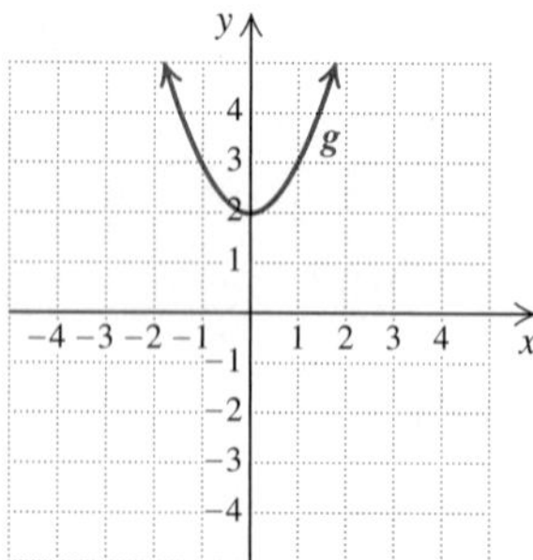

20.

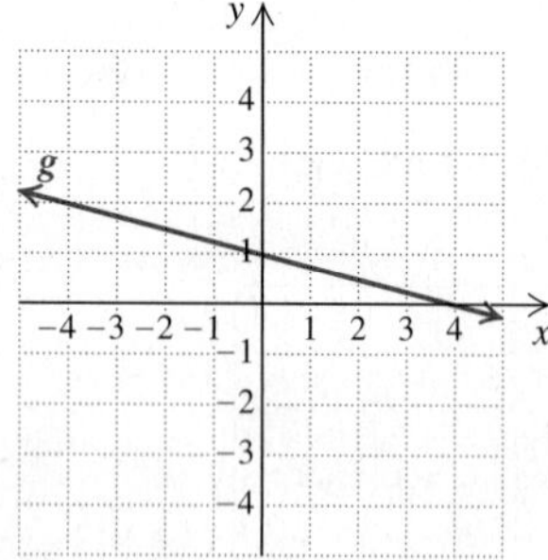

21.

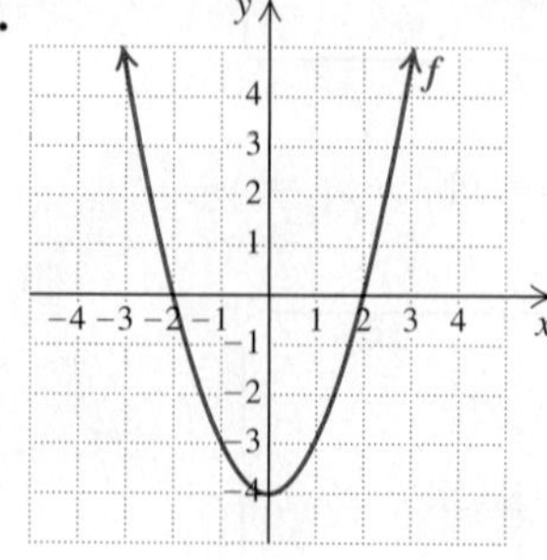

22.

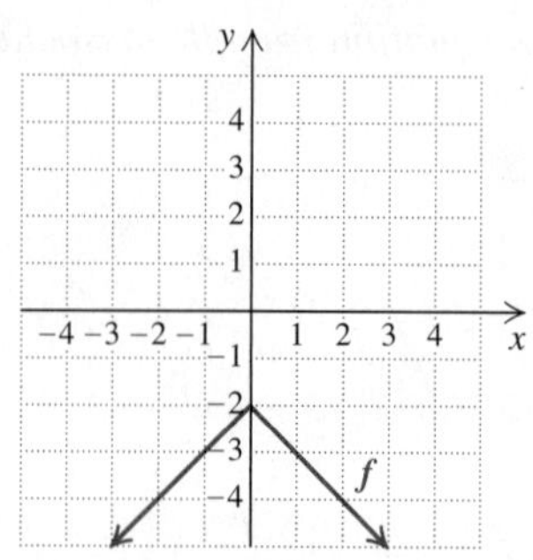

23. $f(x) = x - 5$

24. $f(x) = x + 3$

25. $f(x) = \frac{1}{2}x + 10$

26. $f(x) = \frac{2}{3}x - 6$

27. $f(x) = 2.7 - x$

28. $f(x) = 0.5 - x$

29. $f(x) = 3x + 7$

30. $f(x) = 5x - 8$

Solve graphically.

31. $x - 3 = 4$

32. $x + 4 = 6$

33. $2x + 1 = 7$

34. $3x - 5 = 1$

35. $\frac{1}{3}x - 2 = 1$

36. $\frac{1}{2}x + 3 = -1$

37. $x + 3 = 5 - x$

38. $x - 7 = 3x - 3$

39. $5 - \frac{1}{2}x = x - 4$

40. $3 - x = \frac{1}{2}x - 3$

41. $2x - 1 = -x + 3$

42. $-3x + 4 = 3x - 4$

Use a graph to estimate the solution in each of the following. Be sure to use graph paper and a straightedge if graphing by hand.

43. *Health Care.* Under a particular Anthem health-insurance plan, an individual pays the first \$5000 of hospitalization charges each year plus 30% of all charges in excess of \$5000. In 2010, Gerry's only hospitalization was for rotator cuff surgery. By approximately how much did Gerry's hospital bill exceed \$5000, if the surgery ended up costing him \$6350?
Source: www.ehealthinsurance.com

44. *Cable TV.* Gina's new TV service costs \$200 for the hardware plus \$35 per month for the service. After how many months has she spent \$480 for cable TV?

45. *Cell-Phone Charges.* Skytone Calling charges \$100 for a Smart phone and \$35 per month under its economy plan. Estimate the time required for the total cost to reach \$275.

46. *Cell-Phone Charges.* The Cellular Connection charges \$80 for a Smart phone and \$40 per month under its economy plan. Estimate the time required for the total cost to reach \$240.

47. *Parking Fees.* Karla's Parking charges $3.00 to park plus 50¢ for each 15-min unit of time. Estimate how long someone can park for $7.50.*

48. *Cost of a Road Call.* Dave's Foreign Auto Village charges $50 for a road call plus $15 for each 15-min unit of time. Estimate the time required for a road call that cost $140.*

49. *Cost of a FedEx Delivery.* In 2010, for Standard delivery to the closest zone of packages weighing from 100 to 499 lb, FedEx charged $130 plus $1.30 for each pound over 100. Estimate the weight of a package that cost $325 to ship.
Source: www.fedex.com

50. *Copying Costs.* For each copy of a research paper, a university copy center charged $2.25 for a spiral binding and 8¢ for each page that was a double-sided copy. Estimate the number of pages in a spiral-bound research paper that cost $8.25 per copy. Assume that every page in the report was a double-sided copy.

TW **51.** Explain the difference between finding $f(0)$ and finding the zeros of f.

TW **52.** Darnell used a graphical method to solve $2x - 3 = x + 4$. He stated that the solution was $(7, 11)$. Can this answer be correct? What mistake do you think Darnell is making?

SKILL REVIEW

To prepare for Chapter 5, review exponential notation and order of operations (Section 1.8).

Simplify. [1.8]

53. $(-5)^3$

54. $(-2)^6$

55. -2^6

56. $3 \cdot 2^4 - 5 \cdot 2^3$

57. $2 - (3 - 2^2) + 10 \div 2 \cdot 5$

58. $(5 - 7)^2(3 - 2 \cdot 2)$

*More precise, nonlinear models of Exercises 47 and 48 appear in Exercises 69 and 70, respectively.

SYNTHESIS

TW **59.** Explain why, when we are solving an equation graphically, the x-coordinate of the point of intersection gives the solution of the equation.

TW **60.** Explain the difference between "solving by graphing" and "graphing the solution set."

Estimate the solution(s) from the associated graph.

61. $f(x) = g(x)$

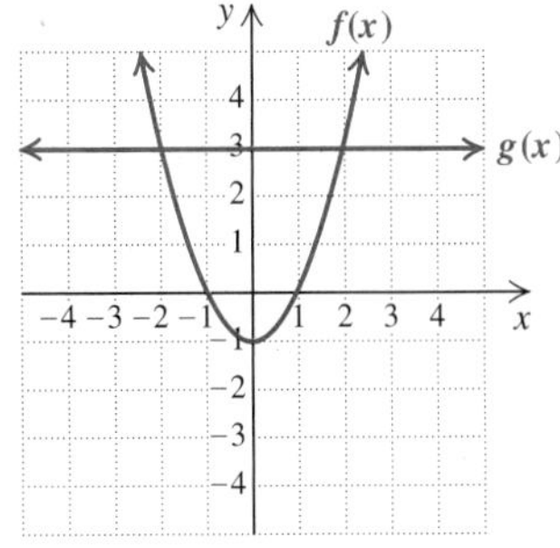

62. $f(x) = g(x)$

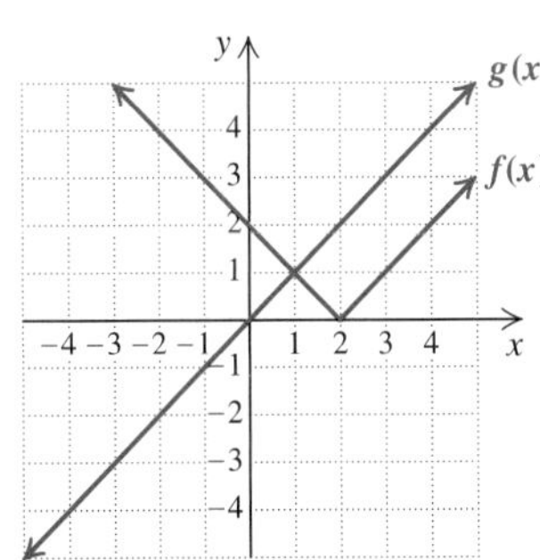

Solve graphically. Be sure to check.

63. $2x = |x + 1|$

64. $x - 1 = |4 - 2x|$

65. $\frac{1}{2}x = 3 - |x|$

66. $2 - |x| = 1 - 3x$

67. $x^2 = x + 2$

68. $x^2 = x$

69. (Refer to Exercise 47.) It costs as much to park at Karla's for 16 min as it does for 29 min. Thus the linear graph drawn in the solution of Exercise 47 is not a precise representation of the situation. Draw a graph with a series of "steps" that more accurately reflects the situation.

70. (Refer to Exercise 48.) A 32-min road call with Dave's costs the same as a 44-min road call. Thus the linear graph drawn in the solution of Exercise 48 is not a precise representation of the situation. Draw a graph with a series of "steps" that more accurately reflects the situation.

Try Exercise Answers: Section 4.5

7. −2 **25.** −20 **31.** 7 **37.** 1 **39.** 6 **41.** $1\frac{1}{3}$ **43.** $9500, or $4500 over $5000

Study Summary

KEY TERMS AND CONCEPTS	EXAMPLES	PRACTICE EXERCISES
SECTION 4.1: SYSTEMS OF EQUATIONS AND GRAPHING		
A solution of a **system of two equations** is an ordered pair that makes both equations true. The intersection of the graphs of the equations gives the solution of the system. A **consistent** system has at least one solution; an **inconsistent** system has no solution. When one equation in a system of two equations is a nonzero multiple of the other, the equations are **dependent** and there is an infinite number of solutions. Otherwise, the equations are **independent.**	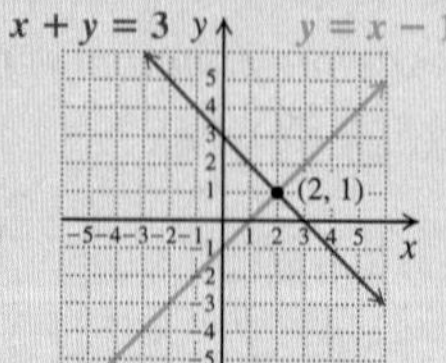 $x + y = 3,$ $y = x - 1$ The graphs intersect at (2, 1). The solution is (2, 1). The system is consistent. The equations are independent. 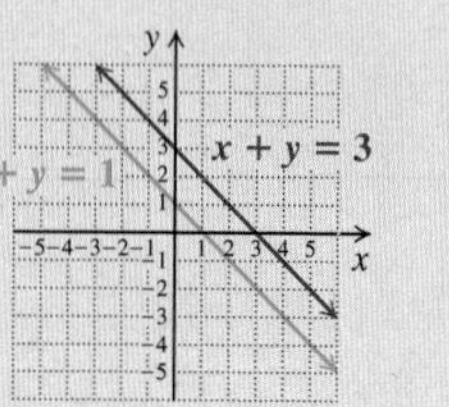 $x + y = 3,$ $x + y = 1$ The graphs do not intersect. There is no solution. The system is inconsistent. The equations are independent. 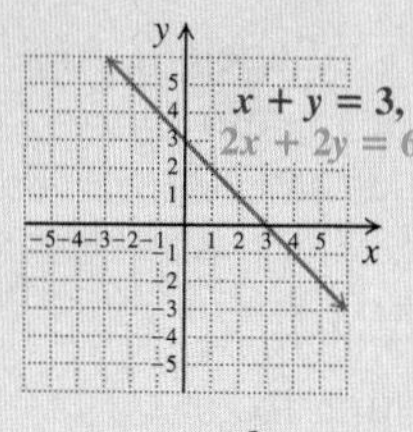 $x + y = 3,$ $2x + 2y = 6$ The graphs are the same. The solution set is $\{(x, y) \mid x + y = 3\}$. The system is consistent. The equations are dependent.	**1.** Solve by graphing: $x - y = 3,$ $y = 2x - 5.$
SECTION 4.2: SYSTEMS OF EQUATIONS AND SUBSTITUTION		
To use the **substitution method,** we solve one equation for a variable and substitute that expression for that variable in the other equation.	Solve: $2x + 3y = 8,$ $x = y + 1.$ Substitute and solve for y: $2(y + 1) + 3y = 8$ $2y + 2 + 3y = 8$ $y = \frac{6}{5}.$ Substitute and solve for x: $x = y + 1$ $x = \frac{6}{5} + 1$ $x = \frac{11}{5}.$ The solution is $\left(\frac{11}{5}, \frac{6}{5}\right)$.	**2.** Solve by substitution: $x = 3y - 2,$ $y - x = 1.$
SECTION 4.3: SYSTEMS OF EQUATIONS AND ELIMINATION		
To use the **elimination method,** we add to eliminate a variable.	Solve: $4x - 2y = 6,$ $3x + y = 7.$	**3.** Solve by elimination: $2x - y = 5,$ $x + 3y = 1.$

Eliminate y and solve for x:

$$\begin{aligned} 4x - 2y &= 6 \\ 6x + 2y &= 14 \\ \hline 10x \quad &= 20 \\ x &= 2. \end{aligned}$$

Substitute and solve for y:

$$\begin{aligned} 3x + y &= 7 \\ 3 \cdot 2 + y &= 7 \\ y &= 1. \end{aligned}$$

The solution is $(2, 1)$.

SECTION 4.4: MORE APPLICATIONS USING SYSTEMS

Total-value, mixture, and **motion problems** often translate directly to systems of equations.

Motion problems use one of the following relationships:

$$d = rt, \quad r = \frac{d}{t}, \quad t = \frac{d}{r}.$$

Simple-interest problems use the formula

Principal · Rate · Time = Interest.

Total Value

Recently the Riley Recreation Center purchased 120 stamps for \$40.80. If the stamps were a combination of 28¢ postcard stamps and 44¢ first-class stamps, how many of each type were bought? (See Example 2 on pp. 300–302 for a solution.)

Mixture

Nature's Green Gardening, Inc., carries two brands of fertilizer containing nitrogen and water. "Gentle Grow" is 3% nitrogen and "Sun Saver" is 8% nitrogen. Nature's Green needs to combine the two types of solutions into a 90-L mixture that is 6% nitrogen. How much of each brand should be used? (See Example 5 on pp. 305–306 for a solution.)

Motion

A Boeing 747-400 jet flies 4 hr west with a 60-mph tailwind. Returning against the wind takes 5 hr. Find the speed of the jet with no wind. (See Example 7 on pp. 308–309 for a solution.)

4. Barlow's Office Supply charges \$17.49 for a box of Roller Grip™ pens and \$16.49 for a box of eGEL™ pens. If Letsonville Community College purchased 120 such boxes for \$2010.80, how many boxes of each type did they purchase?

5. An industrial cleaning solution that is 40% nitric acid is added to a solution that is 15% nitric acid in order to create 2 L of a solution that is 25% nitric acid. How much 40%-acid and how much 15%-acid should be used?

6. Ruth paddled for $1\frac{1}{2}$ hr with a 2-mph current to view a rock formation. The return trip against the same current took $2\frac{1}{2}$ hr. Find the speed of Ruth's canoe in still water.

SECTION 4.5: SOLVING EQUATIONS BY GRAPHING

A **zero** of a function is an input x such that $f(x) = 0$. Graphically, it is the x-coordinate of an x-intercept of the graph of f.

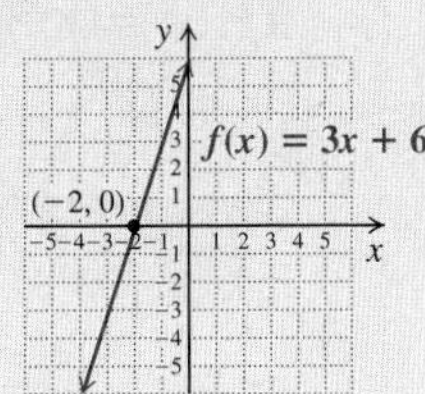

The x-intercept is $(-2, 0)$.

The zero of f is -2 because $f(-2) = 0$.

7. Find the zero of the function given by $f(x) = 8x - 1$.

Equations can be solved graphically using either the **Intersect** method or the **Zero** method.	Solve graphically: $x + 1 = 2x - 3$. **Intersect Method** Graph $y = x + 1$ and $y = 2x - 3$. 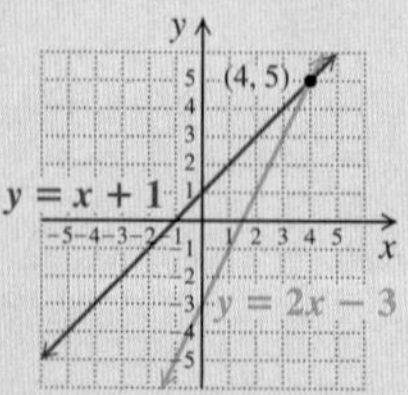 The solution is 4. **Zero Method** $x + 1 = 2x - 3$ $0 = x - 4$ Graph $y = x - 4$. 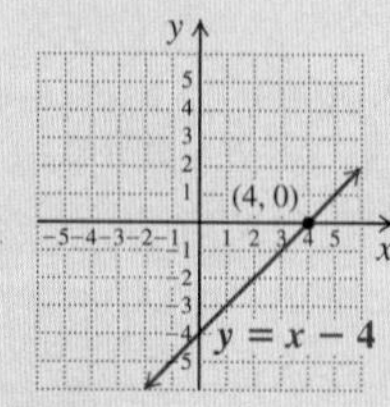 The solution is 4.	**8.** Solve graphically: $x - 3 = 5x + 1$.

Review Exercises 4

Concept Reinforcement. *Complete each of the following sentences.*

1. The system

$$5x + 3y = 7,$$
$$y = 2x + 1$$

is most easily solved using the ____________ method. [4.2]

2. The system

$$-2x + 3y = 8,$$
$$2x + 2y = 7$$

is most easily solved using the ____________ method. [4.3]

3. Of the methods used to solve systems of equations, the ____________ method may yield only approximate solutions. [4.1]

4. When one equation in a system is a multiple of another equation in that system, the equations are said to be ____________. [4.1]

5. A system for which there is no solution is said to be ____________. [4.1]

6. When we are using an algebraic method to solve a system of equations, obtaining a ____________ tells us that the system is inconsistent. [4.2]

7. When we are graphing to solve a system of two equations, if there is no solution, the lines will be ____________. [4.1]

8. If $f(10) = 0$, 10 is a ____________ of f. [4.5]

9. The numbers in an ordered pair that is a solution of a system correspond to the variables in ____________ order. [4.1]

10. When we are solving an equation graphically, the solution is the ____________ of the point of intersection of the graphs. [4.5]

For Exercises 11–19, if a system has an infinite number of solutions, use set-builder notation to write the solution set. If a system has no solution, state this.

Solve graphically. [4.1]

11. $y = x - 3,$ $y = \frac{1}{4}x$

12. $16x - 7y = 25,$ $8x + 3y = 19$

Solve using the substitution method. [4.2]

13. $x - y = 8,$ $y = 3x + 2$

14. $y = x + 2,$ $y - x = 8$

15. $x - 3y = -2,$ $7y - 4x = 6$

Solve using the elimination method. [4.3]

16. $2x - 5y = 11,$ $y - 2x = 5$

17. $4x - 7y = 18,$ $9x + 14y = 40$

18. $3x - 5y = 9,$ $5x - 3y = -1$

19. $1.5x - 3 = -2y,$ $3x + 4y = 6$

Solve.

20. *Architecture.* The rectangular ground floor of the John Hancock building has a perimeter of 860 ft. The length is 100 ft more than the width. Find the length and the width. [4.2]

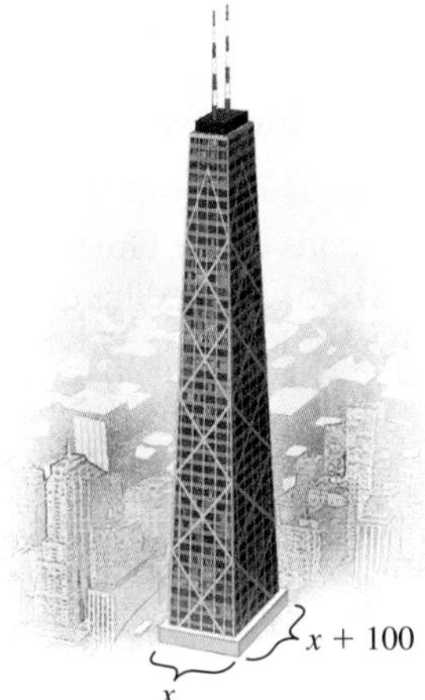

21. A nontoxic floor wax can be made from lemon juice and food-grade linseed oil. The amount of oil should be twice the amount of lemon juice. How much of each ingredient is needed to make 32 oz of floor wax? (The mix should be spread with a rag and buffed when dry.) [4.4]

22. *Music Lessons.* Jillian charges \$25 for a private guitar lesson and \$18 for a group guitar lesson. One day in August, Jillian earned \$265 from 12 students. How many students of each type did Jillian teach? [4.4]

23. *Geometry.* Two angles are supplementary. One angle is 7° less than 10 times the other. Find the measures of the angles. [4.2]

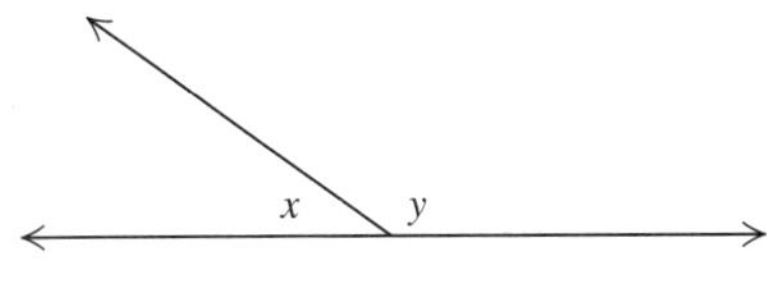

Supplementary angles

24. A freight train leaves Houston at midnight traveling north at a speed of 44 mph. One hour later, a passenger train, going 55 mph, travels north from Houston on a parallel track. How many hours will the passenger train travel before it overtakes the freight train? [4.4]

25. D'Andre wants 14 L of fruit punch that is 10% juice. At the store, he finds punch that is 15% juice and punch that is 8% juice. How much of each should he purchase? [4.4]

26. *Printing.* Using some pages that hold 1300 words per page and others that hold 1850 words per page, a typesetter is able to completely fill 12 pages with an 18,350-word document. How many pages of each kind were used? [4.4]

27. Use the graph below to solve $x - 1 = 2x + 1$. [4.5]

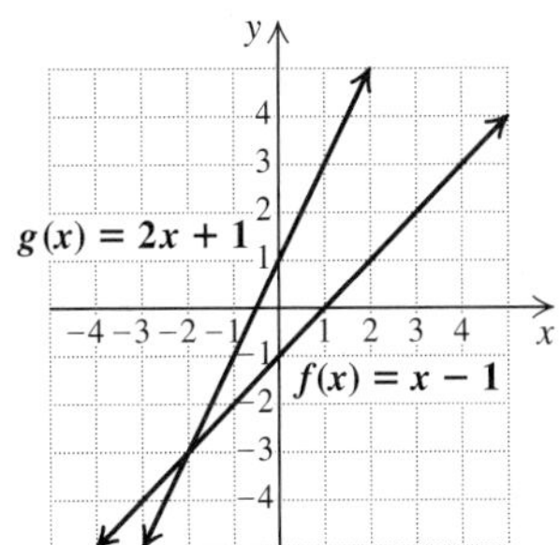

28. Use the graph in Exercise 27 to determine any zeros of f. [4.5]

29. Determine any zeros of $f(x) = 4 - 7x$. [4.5]

30. Solve graphically: $3x - 2 = x + 4$. [4.5]

SYNTHESIS

TW **31.** Explain why any solution of a system of equations is a point of intersection of the graphs of each equation in the system. [4.1]

TW **32.** Explain the difference between solving a system of equations graphically and solving an equation graphically. [4.1], [4.5]

33. Solve graphically:

$$y = x + 2,$$
$$y = x^2 + 2. \quad [4.1]$$

34. The solution of the following system is $(6, 2)$. Find C and D.

$$2x - Dy = 6,$$
$$Cx + 4y = 14 \quad [4.1]$$

35. For a two-digit number, the sum of the ones digit and the tens digit is 6. When the digits are reversed, the new number is 18 more than the original number. Find the original number. [4.4]

36. A sales representative agrees to a 1-year compensation package of \$42,000 plus a computer. After 7 months, the sales representative leaves the company and receives a prorated compensation package consisting of the computer and \$23,750. What was the value of the computer? [4.4]

Chapter Test 4

For Exercises 1–9, if a system has an infinite number of solutions, use set-builder notation to write the solution set. If a system has no solution, state this.

Solve by graphing.

1. $2x + y = 8,$
$y - x = 2$

2. $2y - x = 7,$
$2x - 4y = 4$

Solve using the substitution method.

3. $x + 3y = -8,$
$4x - 3y = 23$

4. $2x + 4y = -6,$
$y = 3x - 9$

5. $x = 5y - 10,$
$15y = 3x + 30$

Solve using the elimination method.

6. $3x - y = 7,$
$x + y = 1$

7. $4y + 2x = 18,$
$3x + 6y = 26$

8. $4x - 6y = 3,$
$6x - 4y = -3$

9. $4x + 5y = 5,$
$6x + 7y = 7$

Solve.

10. *Perimeter of a Garden.* Hassam is designing a rectangular garden with a perimeter of 66 ft. The length of the garden is 1 ft longer than three times the width. Find the dimensions of the garden.

11. *Geometry.* Two angles are complementary. The sum of the measures of the first angle and half the second angle is 64°. Find the measures of the angles.

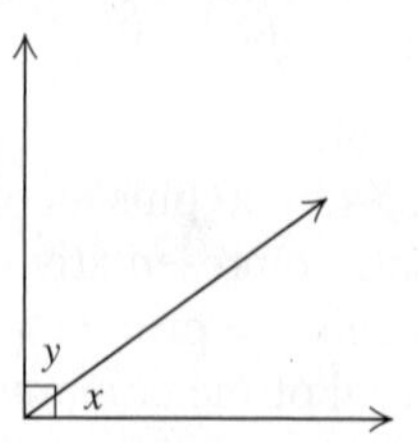

Complementary angles

12. In 2007, Nintendo Co. sold three times as many Wii game machines in Japan as Sony Corp. sold PlayStation 3 consoles. Together, they sold 4.84 million game machines in Japan. How many of each were sold?
Source: Bloomberg.com

13. *Books.* The Friends of the Greenfield Public Library charge $1.75 for used hardbacks and 75¢ for used paperbacks. Keith purchased a total of 23 books for $28.25. How many books of each type did he buy?

14. Pepperidge Farm® Goldfish is a snack food for which 40% of its calories come from fat. Rold Gold® Pretzels receive 9% of their calories from fat. How many grams of each would be needed to make 620 g of a snack mix for which 15% of the calories are from fat?

15. *Boating.* Kylie's motorboat took 3 hr to make a trip downstream on a river flowing at 5 mph. The return trip against the same current took 5 hr. Find the speed of the boat in still water.

16. *Coin Value.* A collection of quarters and nickels is worth $1.25. There are 13 coins in all. How many of each are there?

17. Solve graphically: $2x - 5 = 3x - 3$.

18. Determine any zeros of the function given by $f(x) = \frac{1}{2}x - 5$.

SYNTHESIS

Solve.

19. $3(x - y) = 4 + x,$
$x = 5y + 2$

20. $\frac{3}{2}x - y = 24,$
$2x + \frac{3}{2}y = 15$

21. You are in line at a ticket window. There are 2 more people ahead of you in line than there are behind you. In the entire line, there are three times as many people as there are behind you. How many are in the line?

22. The graph of the function $f(x) = mx + b$ contains the points $(-1, 3)$ and $(-2, -4)$. Find m and b.

Polynomials

5

Who Drives the Most Miles?

Teens? College students? Senior citizens? Since insurance premiums and even tax rates can be based on the answer, it is important to have a good measure of miles driven. One measure often used is the number of vehicle miles traveled, or VMT. In Example 8 in Section 5.3, an algebraic model of VMT is used to estimate the average number of miles driven each year on the basis of the driver's age.

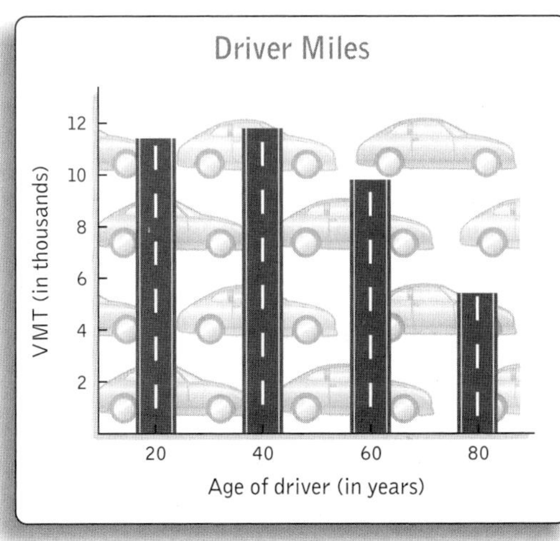

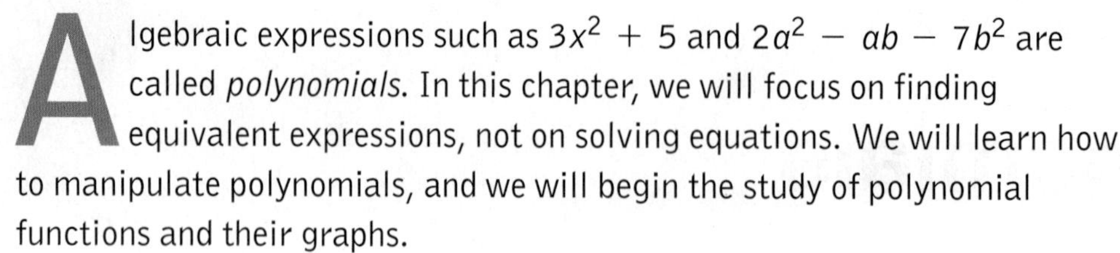

Algebraic expressions such as $3x^2 + 5$ and $2a^2 - ab - 7b^2$ are called *polynomials*. In this chapter, we will focus on finding equivalent expressions, not on solving equations. We will learn how to manipulate polynomials, and we will begin the study of polynomial functions and their graphs.

5.1 Exponents and Their Properties

- Multiplying Powers with Like Bases
- Dividing Powers with Like Bases
- Zero as an Exponent
- Raising a Power to a Power
- Raising a Product or a Quotient to a Power

In Section 5.3, we begin our study of polynomials. Before doing so, however, we must develop some rules for working with exponents.

MULTIPLYING POWERS WITH LIKE BASES

Recall from Section 1.8 that an expression like a^3 means $a \cdot a \cdot a$. We can use this fact to find the product of two expressions that have the same base:

$a^3 \cdot a^2 = (a \cdot a \cdot a)(a \cdot a)$ **There are three factors in a^3 and two factors in a^2.**

$a^3 \cdot a^2 = a \cdot a \cdot a \cdot a \cdot a$ **Using an associative law**

$a^3 \cdot a^2 = a^5.$

Note that the exponent in a^5 is the sum of the exponents in $a^3 \cdot a^2$. That is, $3 + 2 = 5$. Similarly,

$$b^4 \cdot b^3 = (b \cdot b \cdot b \cdot b)(b \cdot b \cdot b)$$
$$b^4 \cdot b^3 = b^7, \quad \text{where } 4 + 3 = 7.$$

Adding the exponents gives the correct result.

The Product Rule For any number a and any positive integers m and n,

$$a^m \cdot a^n = a^{m+n}.$$

(To multiply powers with the same base, keep the base and add the exponents.)

EXAMPLE 1 Multiply and simplify each of the following. (Here "simplify" means express the product as one base to a power whenever possible.)

a) $2^3 \cdot 2^8$ **b)** $5 \cdot 5^8 \cdot 5^3$

c) $(r + s)^7(r + s)^6$ **d)** $(a^3b^2)(a^3b^5)$

SOLUTION

a) $2^3 \cdot 2^8 = 2^{3+8}$ **Adding exponents: $a^m \cdot a^n = a^{m+n}$**

$= 2^{11}$

CAUTION! The base is unchanged: $2^3 \cdot 2^8 \neq 4^{11}$.

STUDY TIP 

Helping Yourself by Helping Others

When you feel confident in your command of a topic, don't hesitate to help classmates experiencing trouble. Your understanding and retention of a concept will deepen when you explain it to someone else and your classmate will appreciate your help.

b) $5 \cdot 5^8 \cdot 5^3 = 5^1 \cdot 5^8 \cdot 5^3$ — Recall that $x^1 = x$ for any number x.

$= 5^{1+8+3}$ — Adding exponents

$= 5^{12}$

CAUTION! $5^{12} \neq 5 \cdot 12$.

c) $(r + s)^7(r + s)^6 = (r + s)^{7+6}$ — The base here is $r + s$.

$= (r + s)^{13}$

CAUTION! $(r + s)^{13} \neq r^{13} + s^{13}$.

d) $(a^3b^2)(a^3b^5) = a^3b^2a^3b^5$ — Using an associative law

$= a^3a^3b^2b^5$ — Using a commutative law

$= a^6b^7$ — Adding exponents

Try Exercise 15.

DIVIDING POWERS WITH LIKE BASES

Recall that any expression that is divided or multiplied by 1 is unchanged. This, together with the fact that anything (besides 0) divided by itself is 1, can lead to a rule for division:

$$\frac{a^5}{a^2} = \frac{a \cdot a \cdot a \cdot a \cdot a}{a \cdot a} = \frac{a \cdot a \cdot a}{1} \cdot \frac{a \cdot a}{a \cdot a}$$

$$\frac{a^5}{a^2} = \frac{a \cdot a \cdot a}{1} \cdot 1$$

$$\frac{a^5}{a^2} = a \cdot a \cdot a = a^3.$$

Note that the exponent in a^3 is the difference of the exponents in $\frac{a^5}{a^2}$. Similarly,

$$\frac{x^4}{x^3} = \frac{x \cdot x \cdot x \cdot x}{x \cdot x \cdot x} = \frac{x}{1} \cdot \frac{x \cdot x \cdot x}{x \cdot x \cdot x} = \frac{x}{1} \cdot 1 = x^1, \text{ or } x.$$

Subtracting the exponents gives the correct result.

The Quotient Rule For any nonzero number a and any positive integers m and n for which $m > n$,

$$\frac{a^m}{a^n} = a^{m-n}.$$

(To divide powers with the same base, subtract the exponent of the denominator from the exponent of the numerator.)

EXAMPLE 2 Divide and simplify. (Here "simplify" means express the quotient as one base to an exponent whenever possible.)

a) $\frac{x^8}{x^2}$ **b)** $\frac{7^9}{7^4}$ **c)** $\frac{(5a)^{12}}{(5a)^4}$ **d)** $\frac{4p^5q^7}{6p^2q}$

SOLUTION

a) $\frac{x^8}{x^2} = x^{8-2}$ **Subtracting exponents:** $\frac{a^m}{a^n} = a^{m-n}$

$= x^6$

b) $\frac{7^9}{7^4} = 7^{9-4}$

$= 7^5$

CAUTION! The base is unchanged: $\frac{7^9}{7^4} \neq 1^5$.

c) $\frac{(5a)^{12}}{(5a)^4} = (5a)^{12-4} = (5a)^8$ **The base here is** $5a$.

d) $\frac{4p^5q^7}{6p^2q} = \frac{4}{6} \cdot \frac{p^5}{p^2} \cdot \frac{q^7}{q^1}$ **Note that the 4 and the 6 are factors, not exponents!**

$= \frac{2}{3} \cdot p^{5-2} \cdot q^{7-1} = \frac{2}{3}p^3q^6$ **Using the quotient rule twice; simplifying**

Try Exercise 33.

ZERO AS AN EXPONENT

The quotient rule can be used to help determine what 0 should mean when it appears as an exponent. Consider a^4/a^4, where a is nonzero. Since the numerator and the denominator are the same,

$$\frac{a^4}{a^4} = 1.$$

On the other hand, using the quotient rule would give us

$$\frac{a^4}{a^4} = a^{4-4} = a^0.$$ **Subtracting exponents**

Since $a^0 = a^4/a^4 = 1$, this suggests that $a^0 = 1$ for any nonzero value of a.

The Exponent Zero For any real number a, with $a \neq 0$,

$$a^0 = 1.$$

(Any nonzero number raised to the 0 exponent is 1.)

Note that in the box above, 0^0 is not defined. For this text, we will assume that expressions like a^m do not represent 0^0.

Recall that in the rules for order of operations, simplifying exponential expressions is done before multiplying.

EXAMPLE 3 Simplify: **(a)** 1948^0; **(b)** $(-9)^0$; **(c)** $(3x)^0$; **(d)** $3x^0$; **(e)** $(-1)9^0$; **(f)** -9^0.

SOLUTION

a) $1948^0 = 1$ **Any nonzero number raised to the 0 exponent is 1.**

b) $(-9)^0 = 1$ **Any nonzero number raised to the 0 exponent is 1. The base here is** -9.

c) $(3x)^0 = 1$, for any $x \neq 0$. **The parentheses indicate that the base is $3x$.**

d) $3x^0 = 3 \cdot x^0$ **The base is x.**

$= 3 \cdot 1$ **$x^0 = 1$, for any $x \neq 0$**

$= 3$

e) $(-1)9^0 = (-1)1 = -1.$ **The base here is 9.**

f) -9^0 is read "the opposite of 9^0" and is equivalent to $(-1)9^0$:

$$-9^0 = (-1)9^0 = (-1)1 = -1.$$

Note from parts (b), (e), and (f) that $-9^0 = (-1)9^0$ and $-9^0 \neq (-9)^0$.

Try Exercise 49.

CAUTION! $-9^0 \neq (-9)^0$, and, in general, $-a^n \neq (-a)^n$.

RAISING A POWER TO A POWER

Consider an expression like $(7^2)^4$:

$(7^2)^4 = (7^2)(7^2)(7^2)(7^2)$ **There are four factors of 7^2.**

$= (7 \cdot 7)(7 \cdot 7)(7 \cdot 7)(7 \cdot 7)$ **We could also use the product rule.**

$= 7 \cdot 7 \cdot 7 \cdot 7 \cdot 7 \cdot 7 \cdot 7 \cdot 7$ **Using an associative law**

$= 7^8.$

Note that the exponent in 7^8 is the product of the exponents in $(7^2)^4$. Similarly,

$(y^5)^3 = y^5 \cdot y^5 \cdot y^5$ **There are three factors of y^5.**

$= (y \cdot y \cdot y \cdot y \cdot y)(y \cdot y \cdot y \cdot y \cdot y)(y \cdot y \cdot y \cdot y \cdot y)$

$= y^{15}.$

Once again, we get the same result if we multiply exponents:

$$(y^5)^3 = y^{5 \cdot 3} = y^{15}.$$

The Power Rule For any number a and any whole numbers m and n,

$$(a^m)^n = a^{mn}.$$

(To raise a power to a power, multiply the exponents and leave the base unchanged.)

Student Notes

There are several rules for manipulating exponents in this section. One way to remember them all is to replace variables with small numbers (other than 1) and see what the results suggest. For example, multiplying $2^2 \cdot 2^3$ and examining the result is a fine way of reminding yourself that $a^m \cdot a^n = a^{m+n}$.

Remember that for this text we assume that 0^0 is not considered.

EXAMPLE 4 Simplify: $(m^2)^5$.

SOLUTION

$(m^2)^5 = m^{2 \cdot 5}$ **Multiplying exponents: $(a^m)^n = a^{mn}$**

$= m^{10}$

Try Exercise 57.

RAISING A PRODUCT OR A QUOTIENT TO A POWER

When an expression inside parentheses is raised to a power, the inside expression is the base. Let's compare $2a^3$ and $(2a)^3$:

$2a^3 = 2 \cdot a \cdot a \cdot a$ **The base is *a*.**

$$\begin{aligned}(2a)^3 &= (2a)(2a)(2a) \quad \textbf{The base is } 2a.\\ &= (2 \cdot 2 \cdot 2)(a \cdot a \cdot a)\\ &= 2^3a^3\\ &= 8a^3.\end{aligned}$$

We see that $2a^3$ and $(2a)^3$ are *not* equivalent. Note too that $(2a)^3$ can be simplified by cubing each factor in $2a$. This leads to the following rule for raising a product to a power.

Raising a Product to a Power For any numbers a and b and any whole number n,

$$(ab)^n = a^nb^n.$$

(To raise a product to a power, raise each factor to that power.)

EXAMPLE 5 Simplify: **(a)** $(4a)^3$; **(b)** $(-5x^4)^2$; **(c)** $(a^7b)^2(a^3b^4)$.

SOLUTION

a) $(4a)^3 = 4^3a^3 = 64a^3$ **Raising each factor to the third power and simplifying**

b) $(-5x^4)^2 = (-5)^2(x^4)^2$ **Raising each factor to the second power. Parentheses are important here.**

$= 25x^8$ **Simplifying $(-5)^2$ and using the power rule**

c) $(a^7b)^2(a^3b^4) = (a^7)^2b^2a^3b^4$ **Raising a product to a power**

$= a^{14}b^2a^3b^4$ **Multiplying exponents**

$= a^{17}b^6$ **Adding exponents**

Try Exercise 65.

CAUTION! The rule $(ab)^n = a^nb^n$ applies only to *products* raised to a power, not to sums or differences. For example, $(3 + 4)^2 \neq 3^2 + 4^2$ since $7^2 \neq 9 + 16$. Similarly, $(5x)^2 = 5^2 \cdot x^2$, but $(5 + x)^2 \neq 5^2 + x^2$.

There is a similar rule for raising a quotient to a power.

Raising a Quotient to a Power For any numbers a and b, $b \neq 0$, and any whole number n,

$$\left(\frac{a}{b}\right)^n = \frac{a^n}{b^n}.$$

(To raise a quotient to a power, raise the numerator to the power and divide by the denominator to the power.)

EXAMPLE 6 Simplify: **(a)** $\left(\frac{x}{5}\right)^2$; **(b)** $\left(\frac{5}{a^4}\right)^3$; **(c)** $\left(\frac{3a^4}{b^3}\right)^2$.

SOLUTION

a) $\left(\frac{x}{5}\right)^2 = \frac{x^2}{5^2} = \frac{x^2}{25}$ **Squaring the numerator and the denominator**

b) $\left(\frac{5}{a^4}\right)^3 = \frac{5^3}{(a^4)^3}$ **Raising a quotient to a power**

$= \frac{125}{a^{4\cdot3}} = \frac{125}{a^{12}}$ **Using the power rule and simplifying**

c) $\left(\frac{3a^4}{b^3}\right)^2 = \frac{(3a^4)^2}{(b^3)^2}$ **Raising a quotient to a power**

$= \frac{3^2(a^4)^2}{b^{3\cdot2}} = \frac{9a^8}{b^6}$ **Raising a product to a power and using the power rule**

Try Exercise 75.

In the following summary of definitions and rules, we assume that no denominators are 0 and 0^0 is not considered.

Definitions and Properties of Exponents

For any whole numbers m and n,

1 as an exponent:	$a^1 = a$
0 as an exponent:	$a^0 = 1$
The Product Rule:	$a^m \cdot a^n = a^{m+n}$
The Quotient Rule:	$\frac{a^m}{a^n} = a^{m-n}$
The Power Rule:	$(a^m)^n = a^{mn}$
Raising a product to a power:	$(ab)^n = a^n b^n$
Raising a quotient to a power:	$\left(\frac{a}{b}\right)^n = \frac{a^n}{b^n}$

5.1 Exercise Set

FOR EXTRA HELP MyMathLab MathXL PRACTICE  WATCH  DOWNLOAD  READ REVIEW

Concept Reinforcement *In each of Exercises 1–8, complete the sentence using the most appropriate phrase from the column on the right.*

1. To raise a product to a power, ___
2. To raise a quotient to a power, ___
3. To raise a power to a power, ___
4. To divide powers with the same base, ___
5. Any nonzero number raised to the 0 exponent ___
6. To multiply powers with the same base, ___
7. To square a fraction, ___
8. To square a product, ___

a) keep the base and add the exponents.
b) multiply the exponents and leave the base unchanged.
c) square the numerator and square the denominator.
d) square each factor.
e) raise each factor to that power.
f) raise the numerator to the power and divide by the denominator to the power.
g) is one.
h) subtract the exponent of the denominator from the exponent of the numerator.

Identify the base and the exponent in each expression.

9. $(5x)^7$
10. $(y - 3)^2$
11. $8n^0$
12. $-x^4$
13. $\frac{4y^3}{7}$
14. $-6n^0$

Simplify. Assume that no denominator is zero and that 0^0 is not considered.

15. $d^3 \cdot d^{10}$
16. $8^4 \cdot 8^3$
17. $a^6 \cdot a$
18. $y^7 \cdot y^9$
19. $8^4 \cdot 8^7$
20. $t^0 \cdot t^{16}$
21. $(3y)^4(3y)^8$
22. $(2t)^8(2t)^{17}$
23. $(7p)^0(7p)^1$
24. $(9n)(9n)^5$
25. $(x + 1)^5(x + 1)^7$
26. $(m - 3)^4(m - 3)^5$
27. $(a^2b^7)(a^3b^2)$
28. $(a^8b^3)(a^4b)$
29. $r^3 \cdot r^7 \cdot r^0$
30. $s^4 \cdot s^5 \cdot s^2$
31. $(mn^5)(m^3n^4)$
32. $(a^3b)(ab)^4$
33. $\frac{7^5}{7^2}$
34. $\frac{4^7}{4^3}$
35. $\frac{t^5}{t}$
36. $\frac{x^7}{x}$
37. $\frac{(5a)^7}{(5a)^6}$
38. $\frac{(3m)^9}{(3m)^8}$

Aha! 39. $\frac{(x + y)^8}{(x + y)^8}$
40. $\frac{(a - b)^8}{(a - b)^2}$
41. $\frac{(r + s)^{12}}{(r + s)^4}$
42. $\frac{(6x)^7}{(6x)^7}$
43. $\frac{8a^9b^7}{2a^2b}$
44. $\frac{12r^{10}s^7}{4r^2s}$
45. $\frac{12d^9}{15d^2}$
46. $\frac{10n^7}{15n^3}$
47. $\frac{m^9n^8}{m^0n^4}$
48. $\frac{a^{10}b^{12}}{a^2b^0}$

Simplify.

49. x^0 when $x = 13$
50. y^0 when $y = 38$
51. $5x^0$ when $x = -4$
52. $7m^0$ when $m = 1.7$
53. $7^0 + 4^0$
54. $(8 + 5)^0$
55. $(-3)^1 - (-3)^0$
56. $(-4)^0 - (-4)^1$

Simplify. Assume that no denominator is zero and that 0^0 is not considered.

57. $(x^4)^7$
58. $(a^3)^8$
59. $(5^8)^2$
60. $(2^5)^3$
61. $(t^{20})^4$
62. $(t^3)^9$
63. $(7x)^2$
64. $(5a)^2$

65. $(-2a)^3$

66. $(-3x)^3$

67. $(-5n^7)^2$

68. $(-4m^4)^2$

69. $(a^2b)^7$

70. $(xy^4)^9$

71. $(x^3y)^2(x^2y^5)$

72. $(a^4b^6)(a^2b)^5$

73. $(2x^5)^3(3x^4)$

74. $(5x^3)^2(2x^7)$

75. $\left(\frac{a}{4}\right)^3$

76. $\left(\frac{3}{x}\right)^4$

77. $\left(\frac{7}{5a}\right)^2$

78. $\left(\frac{5x}{2}\right)^3$

79. $\left(\frac{a^4}{b^3}\right)^5$

80. $\left(\frac{x^5}{y^2}\right)^7$

81. $\left(\frac{x^2y}{z^3}\right)^4$

82. $\left(\frac{x^3}{y^2z}\right)^5$

83. $\left(\frac{a^3}{-2b^5}\right)^4$

84. $\left(\frac{x^5}{-3y^3}\right)^4$

85. $\left(\frac{5x^7y}{-2z^4}\right)^3$

86. $\left(\frac{-4p^8}{3m^2n^3}\right)^3$

Aha! **87.** $\left(\frac{4x^3y^5}{3z^7}\right)^0$

88. $\left(\frac{5a^7}{2b^5c}\right)^0$

TW **89.** Explain in your own words why $-5^2 \neq (-5)^2$.

TW **90.** Under what circumstances should exponents be added?

SKILL REVIEW

To prepare for Section 5.2, review operations with integers (Sections 1.5–1.7).

Perform the indicated operations.

91. $-10 - 14$ [1.6]

92. $-3(5)$ [1.7]

93. $-16 + 5$ [1.5]

94. $12 - (-4)$ [1.6]

95. $-8(-10)$ [1.7]

96. $-3 + (-11)$ [1.5]

SYNTHESIS

TW **97.** Under what conditions does a^n represent a negative number? Why?

TW **98.** Using the quotient rule, explain why 9^0 is 1.

TW **99.** Suppose that the width of a square is three times the width of a second square (see the figure below). How do the areas of the squares compare? Why?

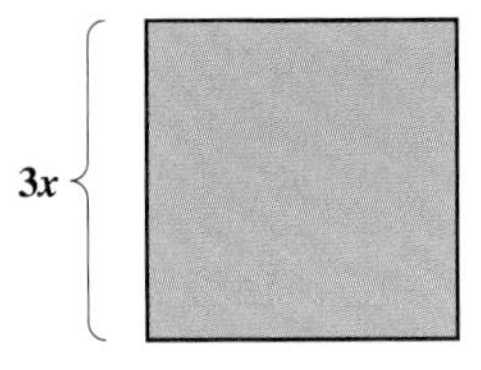

TW **100.** Suppose that the width of a cube is twice the width of a second cube. How do the volumes of the cubes compare? Why?

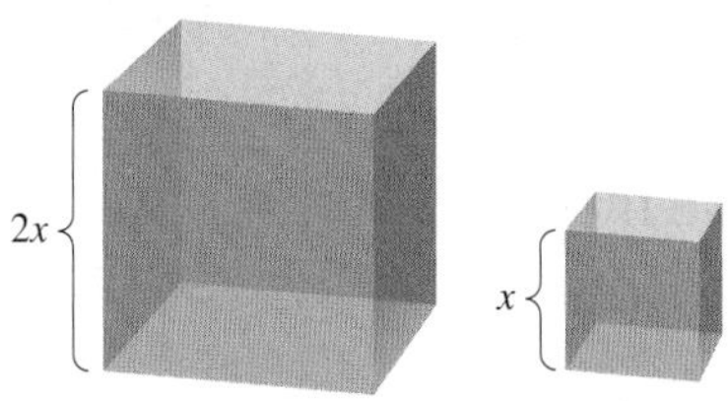

Find a value of the variable that shows that the two expressions are not *equivalent. Answers may vary.*

101. $(a + 5)^2;\ a^2 + 5^2$

102. $3x^2;\ (3x)^2$

103. $\frac{a+7}{7};\ a$

104. $\frac{t^6}{t^2};\ t^3$

Simplify.

105. $a^{10k} \div a^{2k}$

106. $y^{4x} \cdot y^{2x}$

107. $\frac{\left(\frac{1}{2}\right)^3\left(\frac{2}{3}\right)^4}{\left(\frac{5}{6}\right)^3}$

108. $\frac{x^{5t}(x^t)^2}{(x^{3t})^2}$

109. Solve for x: $\frac{t^{26}}{t^x} = t^x$.

Replace □ *with* $>$, $<$, *or* $=$ *to write a true sentence.*

110. 3^5 □ 3^4

111. 4^2 □ 4^3

112. 4^3 □ 5^3

113. 4^3 □ 3^4

114. 9^7 □ 3^{13}

115. 25^8 □ 125^5

Use the fact that $10^3 \approx 2^{10}$ to estimate each of the following powers of 2. Then compute the power of 2 with a calculator and find the difference between the exact value and the approximation.

116. 2^{14}

117. 2^{22}

118. 2^{26}

119. 2^{31}

In computer science, 1 KB of memory refers to 1 kilobyte, or 1×10^3 bytes, of memory. This is really an approximation of 1×2^{10} bytes (since computer memory uses powers of 2).

120. The TI-84 Plus graphing calculator has 480 KB of "FLASH ROM." How many bytes is this?

121. The TI-84 Plus Silver Edition graphing calculator has 1.5 MB (megabytes) of FLASH ROM, where 1 MB is 1000 KB. How many bytes of FLASH ROM does this calculator have?

Try Exercise Answers: Section 5.1

15. d^{13} **33.** 7^3 **49.** 1 **57.** x^{28} **65.** $-8a^3$ **75.** $\frac{a^3}{64}$

5.2 Negative Exponents and Scientific Notation

- Negative Integers as Exponents
- Scientific Notation
- Multiplying, Dividing, and Significant Digits
- Problem Solving Using Scientific Notation

We now attach a meaning to negative exponents. Once we understand both positive exponents and negative exponents, we can study a method of writing numbers known as *scientific notation.*

NEGATIVE INTEGERS AS EXPONENTS

Let's define negative exponents so that the rules that apply to whole-number exponents will hold for all integer exponents. To do so, consider a^{-5} and the rule for adding exponents:

$$a^{-5} = a^{-5} \cdot 1 \qquad \textbf{Using the identity property of 1}$$

$$a^{-5} = \frac{a^{-5}}{1} \cdot \frac{a^5}{a^5} \qquad \textbf{Writing 1 as } \frac{a^5}{a^5} \textbf{ and } a^{-5} \textbf{ as } \frac{a^{-5}}{1}$$

$$a^{-5} = \frac{a^{-5+5}}{a^5} \qquad \textbf{Adding exponents}$$

$$a^{-5} = \frac{1}{a^5}. \qquad -5 + 5 = 0 \textbf{ and } a^0 = 1$$

This leads to our definition of negative exponents.

Negative Exponents For any real number a that is nonzero and any integer n,

$$a^{-n} = \frac{1}{a^n}.$$

(The numbers a^{-n} and a^n are reciprocals of each other.)

EXAMPLE 1 Express using positive exponents and, if possible, simplify.

a) m^{-3} b) 4^{-2} c) $(-3)^{-2}$ d) ab^{-1}

SOLUTION

a) $m^{-3} = \frac{1}{m^3}$ **m^{-3} is the reciprocal of m^3.**

b) $4^{-2} = \frac{1}{4^2} = \frac{1}{16}$ **4^{-2} is the reciprocal of 4^2. Note that $4^{-2} \neq 4(-2)$.**

c) $(-3)^{-2} = \frac{1}{(-3)^2} = \frac{1}{(-3)(-3)} = \frac{1}{9}$ **$(-3)^{-2}$ is the reciprocal of $(-3)^2$. Note that $(-3)^{-2} \neq -\frac{1}{3^2}$.**

d) $ab^{-1} = a\left(\frac{1}{b^1}\right) = a\left(\frac{1}{b}\right) = \frac{a}{b}$ **b^{-1} is the reciprocal of b^1. Note that the base is b, not ab.**

Try Exercise 11.

STUDY TIP

Connect the Dots

Whenever possible, look for connections between concepts covered in different sections or chapters. For example, both Sections 5.1 and 5.2 discuss exponents, and both Chapters 5 and 6 cover polynomials.

CAUTION! A negative exponent does not, in itself, indicate that an expression is negative. As shown in Example 1,

$$4^{-2} \neq 4(-2) \quad \text{and} \quad (-3)^{-2} \neq -\frac{1}{3^2}.$$

The following is another way to illustrate why negative exponents are defined as they are.

On this side, we divide by 5 at each step.

$$125 = 5^3$$
$$25 = 5^2$$
$$5 = 5^1$$
$$1 = 5^0$$
$$\frac{1}{5} = 5^?$$
$$\frac{1}{25} = 5^?$$

On this side, the exponents decrease by 1.

To continue the pattern, it follows that

$$\frac{1}{5} = \frac{1}{5^1} = 5^{-1}, \quad \frac{1}{25} = \frac{1}{5^2} = 5^{-2}, \quad \text{and, in general,} \quad \frac{1}{a^n} = a^{-n}.$$

EXAMPLE 2 Express $\frac{1}{x^7}$ using negative exponents.

SOLUTION We know that $\frac{1}{a^n} = a^{-n}$. Thus, $\frac{1}{x^7} = x^{-7}$.

Try Exercise 31.

The rules for exponents still hold when exponents are negative.

EXAMPLE 3 Simplify. Do not use negative exponents in the answer.

a) $t^5 \cdot t^{-2}$ **b)** $(5x^{-2}y^3)^{-4}$ **c)** $\frac{x^{-4}}{x^{-5}}$

d) $\frac{1}{t^{-5}}$ **e)** $\frac{s^{-3}}{t^{-5}}$ **f)** $\frac{-10x^{-3}y}{5x^2y^5}$

SOLUTION

a) $t^5 \cdot t^{-2} = t^{5+(-2)} = t^3$ **Adding exponents**

b) $(5x^{-2}y^3)^{-4} = 5^{-4}(x^{-2})^{-4}(y^3)^{-4}$ **Raising each factor to the exponent of −4**

$= \frac{1}{5^4}x^8y^{-12} = \frac{x^8}{625y^{12}}$ **Multiplying exponents; writing with positive exponents**

c) $\frac{x^{-4}}{x^{-5}} = x^{-4-(-5)} = x^1 = x$ **We subtract exponents even if the exponent in the denominator is negative.**

d) Since $\dfrac{1}{a^n} = a^{-n}$, we have $\dfrac{1}{t^{-5}} = t^{-(-5)} = t^5$.

e) $\dfrac{s^{-3}}{t^{-5}} = s^{-3} \cdot \dfrac{1}{t^{-5}} = \dfrac{1}{s^3} \cdot t^5 = \dfrac{t^5}{s^3}$ **Using the result from part (d) above**

f) $\dfrac{-10x^{-3}y}{5x^2y^5} = \dfrac{-10}{5} \cdot \dfrac{x^{-3}}{x^2} \cdot \dfrac{y^1}{y^5}$ **Note that the -10 and 5 are factors.**

$= -2 \cdot x^{-3-2} \cdot y^{1-5}$ **Using the quotient rule twice**

$= -2x^{-5}y^{-4} = \dfrac{-2}{x^5y^4}$ **Simplifying**

Try Exercises 37 and 45.

The result from Example 3(e) can be generalized.

Factors and Negative Exponents For any nonzero real numbers a and b and any integers m and n,

$$\frac{a^{-n}}{b^{-m}} = \frac{b^m}{a^n}.$$

(A factor can be moved to the other side of the fraction bar if the sign of the exponent is changed.)

EXAMPLE 4 Simplify: $\dfrac{-15x^{-7}}{5y^2z^{-4}}$.

SOLUTION We can move the factors x^{-7} and z^{-4} to the other side of the fraction bar if we change the sign of each exponent:

$$\frac{-15x^{-7}}{5y^2z^{-4}} = \frac{-15}{5} \cdot \frac{x^{-7}}{y^2z^{-4}}$$ **We can simply divide the constant factors.**

$$= -3\frac{z^4}{y^2x^7}$$

$$= \frac{-3z^4}{x^7y^2}.$$

Try Exercise 27.

Another way to change the sign of the exponent is to take the reciprocal of the base. To understand why this is true, note that

$$\left(\frac{s}{t}\right)^{-5} = \frac{s^{-5}}{t^{-5}} = \frac{t^5}{s^5} = \left(\frac{t}{s}\right)^5.$$

This often provides the easiest way to simplify an expression containing a negative exponent.

Reciprocals and Negative Exponents For any nonzero real numbers a and b and any integer n,

$$\left(\frac{a}{b}\right)^{-n} = \left(\frac{b}{a}\right)^{n}.$$

(Any base to an exponent is equal to the reciprocal of the base raised to the opposite exponent.)

EXAMPLE 5 Simplify: $\left(\frac{x^4}{2y}\right)^{-3}$.

SOLUTION

$\left(\frac{x^4}{2y}\right)^{-3} = \left(\frac{2y}{x^4}\right)^{3}$ — **Taking the reciprocal of the base and changing the sign of the exponent**

$= \frac{(2y)^3}{(x^4)^3}$ — **Raising a quotient to an exponent by raising both the numerator and the denominator to the exponent**

$= \frac{2^3y^3}{x^{12}}$ — **Raising a product to an exponent; using the power rule in the denominator**

$= \frac{8y^3}{x^{12}}$ — **Cubing 2**

Try Exercise 67.

SCIENTIFIC NOTATION

When we are working with the very large or very small numbers that frequently occur in science, **scientific notation** provides a useful way of writing numbers. The following are examples of scientific notation.

The mass of the earth:

6.0×10^{24} kilograms (kg) $=$ 6,000,000,000,000,000,000,000,000 kg

The mass of a hydrogen atom:

1.7×10^{-24} g $=$ 0.0000000000000000000000017 g

Student Notes

Definitions are usually written as concisely as possible, so that every phrase included is important. The definition for scientific notation states that $1 \leq N < 10$. Thus, 2.68×10^5 is written in scientific notation, but 26.8×10^5 and 0.268×10^5 are *not* written in scientific notation.

Scientific Notation *Scientific notation* for a number is an expression of the type

$$N \times 10^m,$$

where N is at least 1 but less than 10 ($1 \leq N < 10$), N is expressed in decimal notation, and m is an integer.

Converting from scientific notation to decimal notation involves multiplying by a power of 10. Consider the following.

Scientific Notation $N \times 10^m$	Multiplication	Decimal Notation
4.52×10^2	4.52×100	452.
4.52×10^1	4.52×10	45.2
4.52×10^0	4.52×1	4.52
4.52×10^{-1}	4.52×0.1	0.452
4.52×10^{-2}	4.52×0.01	0.0452

We generally perform this multiplication mentally. Thus to convert $N \times 10^m$ to decimal notation, we move the decimal point.

- When m is positive, we move the decimal point right m places.
- When m is negative, we move the decimal point left $|m|$ places.

EXAMPLE 6 Convert to decimal notation.

a) 7.893×10^5 **b)** 4.7×10^{-8}

SOLUTION

a) Since the exponent is positive, the decimal point moves to the right:

7.89300. $7.893 \times 10^5 = 789{,}300$ **The decimal point moves 5 places to the right.**

5 places

b) Since the exponent is negative, the decimal point moves to the left:

0.00000004.7 $4.7 \times 10^{-8} = 0.000000047$ **The decimal point moves 8 places to the left.**

8 places

Try Exercise 83.

To convert from decimal notation to scientific notation, this procedure is reversed.

EXAMPLE 7 Write in scientific notation: **(a)** 83,000; **(b)** 0.0327.

SOLUTION

a) We need to find m such that $83{,}000 = 8.3 \times 10^m$. To change 8.3 to 83,000 requires moving the decimal point 4 places to the right. This can be accomplished by multiplying by 10^4. Thus,

$83{,}000 = 8.3 \times 10^4$. **This is scientific notation.**

b) We need to find m such that $0.0327 = 3.27 \times 10^m$. To change 3.27 to 0.0327 requires moving the decimal point 2 places to the left. This can be accomplished by multiplying by 10^{-2}. Thus,

$0.0327 = 3.27 \times 10^{-2}$. **This is scientific notation.**

Try Exercise 93.

Remember that positive exponents are used to represent large numbers and negative exponents are used to represent small numbers between 0 and 1.

MULTIPLYING, DIVIDING, AND SIGNIFICANT DIGITS

In the world of science, it is important to know just how accurate a measurement is. For example, the measurement 5.12×10^3 km is more precise than the measurement 5.1×10^3 km. We say that 5.12×10^3 has three **significant digits** whereas 5.1×10^3 has only two significant digits. If 5.1×10^3, or 5100, includes no rounding in the tens column, we would indicate that by writing 5.10×10^3.

> When two or more measurements written in scientific notation are multiplied or divided, the result should be rounded so that it has the same number of significant digits as the measurement with the fewest significant digits. Rounding should be performed at the *end* of the calculation.

Thus,

$$(\underbrace{3.1}_{\text{2 digits}} \times 10^{-3}\text{ mm})(\underbrace{2.45}_{\text{3 digits}} \times 10^{-4}\text{ mm}) = 7.595 \times 10^{-7}\text{ mm}^2$$

should be rounded to

$$\overbrace{7.6}^{\text{2 digits}} \times 10^{-7}\text{ mm}^2.$$

> When two or more measurements written in scientific notation are added or subtracted, the result should be rounded so that it has as many decimal places as the measurement with the fewest decimal places.

For example,

$$1.\underbrace{6354}_{\text{4 decimal places}} \times 10^4\text{ km} + 2.\underbrace{078}_{\text{3 decimal places}} \times 10^4\text{ km} = 3.7134 \times 10^4\text{ km}$$

should be rounded to

$$3.\overbrace{713}^{\text{3 decimal places}} \times 10^4\text{ km}.$$

EXAMPLE 8 Multiply and write scientific notation for the answer:

$$(7.2 \times 10^5)(4.3 \times 10^9).$$

SOLUTION We have

$$(7.2 \times 10^5)(4.3 \times 10^9) = (7.2 \times 4.3)(10^5 \times 10^9)$$ **Using the commutative and associative laws**

$$= 30.96 \times 10^{14}.$$ **Adding exponents**

To find scientific notation for this result, we convert 30.96 to scientific notation and simplify:

$$\begin{aligned} 30.96 \times 10^{14} &= (3.096 \times 10^{1}) \times 10^{14} \\ &= 3.096 \times 10^{15} \\ &\approx 3.1 \times 10^{15}. \qquad \textbf{Rounding to 2 significant digits} \end{aligned}$$

Try Exercise 103.

EXAMPLE 9 Divide and write scientific notation for the answer:

$$\frac{3.48 \times 10^{-7}}{4.64 \times 10^{6}}.$$

SOLUTION

$$\begin{aligned} \frac{3.48 \times 10^{-7}}{4.64 \times 10^{6}} &= \frac{3.48}{4.64} \times \frac{10^{-7}}{10^{6}} && \textbf{Separating factors. Our answer must have 3 significant digits.} \\ &= 0.75 \times 10^{-13} && \textbf{Subtracting exponents; simplifying} \\ &= (7.5 \times 10^{-1}) \times 10^{-13} && \textbf{Converting 0.75 to scientific notation} \\ &= 7.50 \times 10^{-14} && \textbf{Adding exponents. We write 7.50 to indicate 3 significant digits.} \end{aligned}$$

Try Exercise 109.

Exponents and Scientific Notation

To simplify an exponential expression like 3^5, we can use a calculator's exponentiation key, usually labeled ^. If the exponent is 2, we can also use the x^2 key. If it is -1, we can use the x^{-1} key. If the exponent is a single number, not an expression, we do not need to enclose it in parentheses.

Graphing calculators will accept entries using scientific notation, and will normally write very large or very small numbers using scientific notation. The EE key, the 2nd option associated with the , key, is used to enter scientific notation. On the calculator screen, a notation like E22 represents $\times\ 10^{22}$.

Calculators will display all numbers using scientific notation if they are in the SCI mode, set using MODE.

Your Turn

1. Press 1 . 2 × 1 0 ^ (−) 9 ENTER. From the screen, you should see that 1.2×10^{-9} is represented as 1.2E−9.
2. Use the EE key to enter 9×10^5. If the number is displayed as 900000, change the mode to scientific notation. (Change the mode back to NORMAL when you are no longer using scientific notation.)

EXAMPLE 10 Use a graphing calculator to calculate 3^5, $(-4.7)^2$, and $(-8)^{-1}$.

SOLUTION Keystrokes are shown for each calculation.

$(-8)^{-1}$: 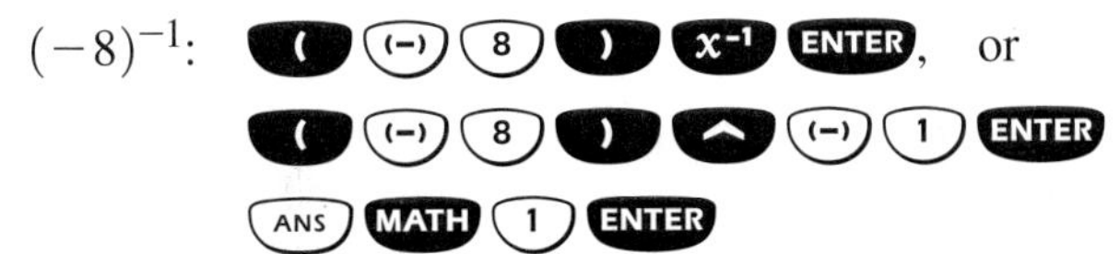

3^5	243
(−4.7)^2	22.09
(−4.7)2	22.09

$(-8)^{-1}$	−.125
(−8)^−1	−.125
Ans ▶ Frac	−1/8

We see that $3^5 = 243$, $(-4.7)^2 = 22.09$, and $(-8)^{-1} = -0.125$, or $-\frac{1}{8}$.

Try Exercise 77.

EXAMPLE 11 Use a graphing calculator to calculate

$$(7.5 \times 10^8)(1.2 \times 10^{-14}).$$

SOLUTION We press 7 . 5 EE 8 × 1 . 2 EE (−) 1 4 ENTER.

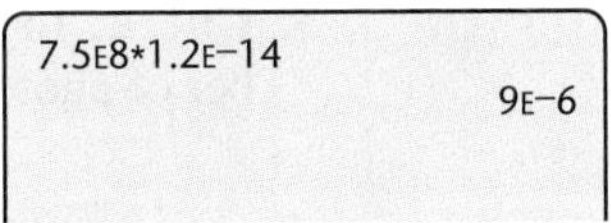

The result shown is then read as 9×10^{-6}.

Try Exercise 105.

PROBLEM SOLVING USING SCIENTIFIC NOTATION

The table below lists common names and prefixes of powers of 10, in both decimal notation and scientific notation. These values are used in many applications.

One thousand	kilo-	1000	1×10^3
One million	mega-	1,000,000	1×10^6
One billion	giga-	1,000,000,000	1×10^9
One trillion	tera-	1,000,000,000,000	1×10^{12}
One thousandth	milli-	0.001	1×10^{-3}
One millionth	micro-	0.000001	1×10^{-6}
One billionth	nano-	0.000000001	1×10^{-9}
One trillionth	pico-	0.000000000001	1×10^{-12}

EXAMPLE 12 Information Storage. By October 2008, Facebook users had uploaded 10 billion photographs to the social networking site. If one DVD can store about 1500 photographs, how many DVDs would it take to store all the photographs on Facebook?
Source: Facebook

SOLUTION

1. **Familiarize.** In order to find the number of DVDs needed to store the photographs, we need to divide the total number of photos by the number of photos each DVD can store. We first write each number using scientific notation:

$$10 \text{ billion} = 10 \times 10^9 = 1 \times 10^{10}, \quad \text{and}$$
$$1500 = 1.5 \times 10^3.$$

We also let d = the number of DVDs needed to store the photographs.

2. **Translate.** To find d, we divide:

$$d = \frac{1 \times 10^{10}}{1.5 \times 10^3}.$$

3. **Carry out.** We calculate and write scientific notation for the result:

$$d = \frac{1.0 \times 10^{10}}{1.5 \times 10^3}$$
$$d = \frac{1.0}{1.5} \times \frac{10^{10}}{10^3}$$
$$d \approx 0.67 \times 10^7 \qquad \textbf{Rounding to 2 significant digits}$$
$$d \approx 6.7 \times 10^{-1} \times 10^7 \qquad \textbf{Writing scientific notation}$$
$$d \approx 6.7 \times 10^6.$$

4. **Check.** To check, we multiply the number of DVDs, 6.7×10^6, by the number of photos that each DVD can hold:

$$(6.7 \times 10^6 \text{ DVDs})\left(1.5 \times 10^3 \frac{\text{photos}}{\text{DVD}}\right) \qquad \textbf{We also check the units.}$$
$$= (6.7 \times 1.5)(10^6 \times 10^3) \text{ DVDs} \cdot \frac{\text{photos}}{\text{DVD}}$$
$$= 10.05 \times 10^9 \text{ photos.} \qquad \textbf{This is approximately 10 billion. The unit, photos, also checks.}$$

5. **State.** It would take about 6.7×10^6, or 6.7 million, DVDs to store the photos that had been uploaded to Facebook.

Try Exercise 117.

5.2 Exercise Set

FOR EXTRA HELP MyMathLab

Concept Reinforcement *State whether scientific notation for each of the following numbers would include a positive power of 10 or a negative power of 10.*

1. The length of an Olympic marathon, in centimeters
2. The thickness of a cat's whisker, in meters
3. The mass of a hydrogen atom, in grams
4. The mass of a pickup truck, in grams
5. The time between leap years, in seconds
6. The time between a bird's heartbeats, in hours

Match each expression with an equivalent expression from the column on the right.

7. ___ $\left(\frac{x^3}{y^2}\right)^{-2}$ a) $\frac{y^6}{x^9}$

8. ___ $\left(\frac{y^2}{x^3}\right)^{-2}$ b) $\frac{x^9}{y^6}$

9. ___ $\left(\frac{y^{-2}}{x^{-3}}\right)^{-3}$ c) $\frac{y^4}{x^6}$

10. ___ $\left(\frac{x^{-3}}{y^{-2}}\right)^{-3}$ d) $\frac{x^6}{y^4}$

Express using positive exponents and, if possible, simplify.

11. 7^{-2} **12.** 10^{-4} **13.** $(-2)^{-6}$

14. $(-3)^{-4}$ **15.** a^{-3} **16.** n^{-6}

17. $\frac{1}{5^{-3}}$ **18.** $\frac{1}{2^{-6}}$ **19.** 7^{-1}

20. 3^{-1} **21.** $8x^{-3}$ **22.** xy^{-9}

23. $3a^8b^{-6}$ **24.** $5a^{-7}b^4$ **25.** $\frac{z^{-4}}{3x^5}$

26. $\frac{y^{-5}}{x^{-3}}$ **27.** $\frac{5x^{-2}y^7}{z^{-4}}$ **28.** $\frac{y^4z^{-3}}{7x^{-2}}$

29. $\left(\frac{a}{2}\right)^{-3}$ **30.** $\left(\frac{x}{3}\right)^{-4}$

Express using negative exponents.

31. $\frac{1}{8^4}$ **32.** $\frac{1}{t^5}$ **33.** $\frac{1}{x}$

34. $\frac{1}{2}$ **35.** x^5 **36.** 4^7

Simplify. If negative exponents appear in the answer, write a second answer using only positive exponents.

37. $8^{-2} \cdot 8^{-4}$ **38.** $9^{-1} \cdot 9^{-6}$

39. $b^2 \cdot b^{-5}$ **40.** $a^4 \cdot a^{-3}$

41. $a^{-3} \cdot a^4 \cdot a$ **42.** $x^{-6} \cdot x^5 \cdot x$

43. $(5a^{-2}b^{-3})(2a^{-4}b)$ **44.** $(3a^{-5}b^{-7})(2ab^{-2})$

45. $\frac{y^4}{y^{-5}}$ **46.** $\frac{a^3}{a^{-2}}$

47. $\frac{2^{-8}}{2^{-5}}$ **48.** $\frac{9^{-4}}{9^{-6}}$

49. $\frac{24a^2}{-8a^3}$ **50.** $\frac{-12x^9}{2x^{11}}$

51. $\frac{-6a^3b^{-5}}{-3a^7b^{-8}}$ **52.** $\frac{-12x^{-2}y^4}{-3xy^{-7}}$

53. $\frac{6x^{-2}y^4z^8}{24x^{-5}y^6z^{-3}}$ **54.** $\frac{8a^6b^{-4}c^8}{32a^{-4}b^5c^9}$

55. $(n^{-5})^3$ **56.** $(m^{10})^{-5}$

57. $(t^{-8})^{-5}$ **58.** $(x^{-4})^{-3}$

59. $(mn)^{-7}$ **60.** $(ab)^{-9}$

61. $(5r^{-4}t^3)^2$ **62.** $(4x^5y^{-6})^3$

63. $(3m^5n^{-3})^{-2}$ **64.** $(2y^{-4}z^2)^{-3}$

65. $(a^{-5}b^7c^{-2})(a^{-3}b^{-2}c^6)$ **66.** $(x^3y^{-4}z^{-5})(x^{-4}y^{-2}z^9)$

67. $\left(\frac{a^4}{3}\right)^{-2}$ **68.** $\left(\frac{7}{x^{-3}}\right)^{-4}$

69. $\left(\frac{m^{-1}}{n^{-4}}\right)^3$ **70.** $\left(\frac{x^2y}{z^{-5}}\right)^3$

71. $\left(\frac{-4x^4y^{-2}}{5x^{-1}y^4}\right)^{-4}$ **72.** $\left(\frac{2x^3y^{-2}}{3y^{-3}}\right)^3$

Aha! **73.** $\left(\frac{4a^3b^{-9}}{6a^{-2}b^5}\right)^0$ **74.** $\left(\frac{5x^0y^{-7}}{2x^{-2}y^4}\right)^1$

75. $\frac{(2a^3)^3 4a^{-3}}{(a^2)^5}$ **76.** $\frac{(3x^2)^3 2x^{-4}}{(x^4)^2}$

Evaluate using a calculator.

77. -8^4 **78.** $(-8)^4$ **79.** $(-2)^{-4}$

80. -2^{-4} **81.** $3^4 5^{-3}$ **82.** $\left(\frac{2}{3}\right)^{-5}$

Convert to decimal notation.

83. 4.92×10^5 **84.** 8×10^4

85. 8.02×10^{-3} **86.** 5.49×10^{-4}

87. 3.497×10^{-6} **88.** 7.034×10^{-2}

89. 9.03×10^{10} **90.** 8.001×10^7

Convert to scientific notation.

91. 47,000,000,000 **92.** 2,600,000,000,000

93. 0.00583 **94.** 0.0814

95. 407,000,000,000 **96.** 3,090,000,000,000

97. 0.000000603 **98.** 0.00000000802

Write scientific notation for the number represented on each calculator screen.

99. 5.02E18

100. 1.067E−6

101. −3.05E−10

102. −5.968E27

Simplify and write scientific notation for the answer. Use the correct number of significant digits.

103. $(2.3 \times 10^6)(4.2 \times 10^{-11})$

104. $(6.5 \times 10^3)(5.2 \times 10^{-8})$

105. $(2.34 \times 10^{-8})(5.7 \times 10^{-4})$

106. $(4.26 \times 10^{-6})(8.2 \times 10^{-6})$

Aha! 107. $(2.0 \times 10^6)(3.02 \times 10^{-6})$

108. $(7.04 \times 10^{-9})(9.01 \times 10^{-7})$

109. $\dfrac{5.1 \times 10^6}{3.4 \times 10^3}$

110. $\dfrac{8.5 \times 10^8}{3.4 \times 10^5}$

111. $\dfrac{7.5 \times 10^{-9}}{2.5 \times 10^{-4}}$

112. $\dfrac{1.26 \times 10^9}{4.2 \times 10^{-3}}$

113. $\dfrac{1.23 \times 10^8}{6.87 \times 10^{-13}}$

114. $\dfrac{4.95 \times 10^{-3}}{1.64 \times 10^{10}}$

115. $5.9 \times 10^{23} + 6.3 \times 10^{23}$

116. $7.8 \times 10^{-34} + 5.4 \times 10^{-34}$

Solve. Write the answers using scientific notation.

117. *Information Technology.* In 2003, the University of California at Berkeley estimated that in 2002 approximately 5 exabytes of information were generated by the worldwide population of 6.3 billion people. Given that 1 exabyte is 10^{12} megabytes, find the average number of megabytes of information generated per person in 2002.

118. *Printing and Engraving.* A ton of five-dollar bills is worth $4,540,000. How many pounds does a five-dollar bill weigh?

119. *Hospital Care.* In 2007, 121 million patients visited emergency rooms in the United States. If the average visit lasted 5.8 hr, how many minutes in all did people spend in emergency rooms in 2007?
Source: Kaiser Family Foundation; ScienceDaily.com, March 5, 2009

120. *Computer Technology.* In 2007, Intel Corp. began making silicon modulators that can encode data onto a beam of light at a rate of 40 gigabits per second. If 25 of these communication lasers are packed on a single chip, how many bits per second could that chip encode?
Source: *The Wall Street Journal*, 7/25/2007

121. *Astronomy.* The diameter of the Milky Way galaxy is approximately 5.88×10^{17} mi. The distance light travels in one year, or one light year, is 5.88×10^{12} mi. How many light years is it from one end of the galaxy to the other?

122. *Astronomy.* The average distance of Earth from the sun is about 9.3×10^7 mi. About how far does Earth travel in a yearly orbit about the sun? (Assume a circular orbit.)

123. *Telecommunications.* A fiber-optic cable is to be used for 125 km of transmission line. The cable has a diameter of 0.6 cm. What is the volume of cable needed for the line? (*Hint*: The volume of a cylinder is given by $V = \pi r^2 h$.)

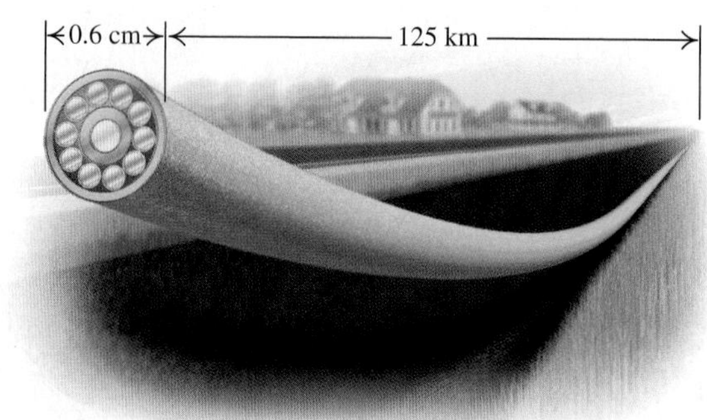

124. *High-Tech Fibers.* A carbon nanotube is a thin cylinder of carbon atoms that, pound for pound, is stronger than steel. With a diameter of about 4.0×10^{-10} in., a fiber can be made 100 yd long. Find the volume of such a fiber.
Source: www.pa.msu.edu

125. *Coral Reefs.* There are 10 million bacteria per square centimeter of coral in a coral reef. The coral reefs near the Hawaiian Islands cover 14,000 km^2. How many bacteria are there in Hawaii's coral reef?
Sources: livescience.com; U.S. Geological Survey

126. *Biology.* A human hair is about 4×10^{-5} m in diameter. A strand of DNA is 2 nanometers in diameter. How many strands of DNA laid side by side would it take to equal the width of a human hair?

127. *Home Maintenance.* The thickness of a sheet of plastic is measured in *mils,* where 1 mil $= \frac{1}{1000}$ in. To help conserve heat, the foundation of a 24-ft by 32-ft rectangular home is covered with a 4-ft high sheet of 8-mil plastic. Find the volume of plastic used.

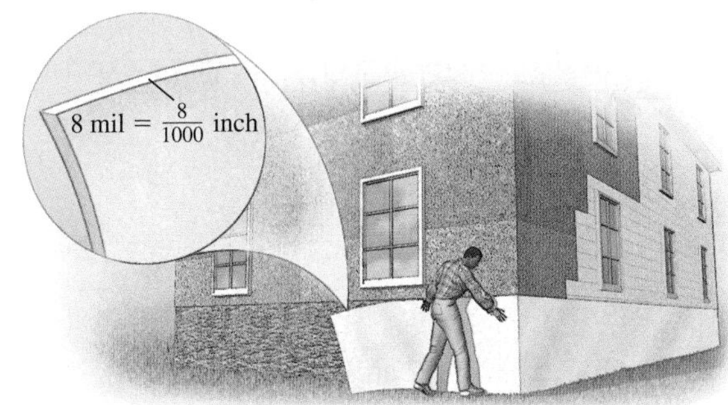

128. *Office Supplies.* A ream of copier paper weighs 2.25 kg. How much does a sheet of copier paper weigh?

TW **129.** Without performing actual computations, explain why 3^{-29} is smaller than 2^{-29}.

TW **130.** Explain why each of the following is not scientific notation: 12.6×10^8; $4.8 \times 10^{1.7}$; 0.207×10^{-5}.

SKILL REVIEW

To prepare for Section 5.3, review combining like terms and evaluating expressions (Sections 1.6 and 1.8).

Combine like terms. [1.6]

131. $9x + 2y - x - 2y$

132. $5a - 7b - 8a + b$

133. $-3x + (-2) - 5 - (-x)$

134. $2 - t - 3t - r - 7$

Evaluate. [1.8]

135. $4 + x^3$, for $x = 10$

136. $-x^2 - 5x + 3$, for $x = -2$

SYNTHESIS

TW **137.** Explain why $(-17)^{-8}$ is positive.

TW **138.** Explain what requirements must be met in order for x^{-n} to represent a negative number.

TW **139.** Some numbers exceed the limits of the calculator. Enter 1.3×10^{-1000} and 1.3×10^{1000} and explain the results.

Simplify.

140. $[(5^{-3})^2]^{-1}$

141. $5^0 - 5^{-1}$

142. $3^{-1} + 4^{-1}$

143. $(7^{-12})^2 \cdot 7^{25}$

144. $\dfrac{125^{-4}(25^2)^4}{125}$

145. $\dfrac{4.2 \times 10^8[(2.5 \times 10^{-5}) \div (5.0 \times 10^{-9})]}{3.0 \times 10^{-12}}$

146. Write $8^{-3} \cdot 32 \div 16^2$ as a power of 2.

147. Write $81^3 \cdot 27 \div 9^2$ as a power of 3.

148. Compare 8×10^{-90} and 9×10^{-91}. Which is the larger value? How much larger? Write scientific notation for the difference.

149. Write the reciprocal of 8.00×10^{-23} in scientific notation.

150. Write $\frac{4}{32}$ in decimal notation, simplified fraction notation, and scientific notation.

151. A grain of sand is placed on the first square of a chessboard, two grains on the second square, four grains on the third, eight on the fourth, and so on. Without a calculator, use scientific notation to approximate the number of grains of sand required for the 64th square. (*Hint*: Use the fact that $2^{10} \approx 10^3$.)

Try Exercise Answers: Section 5.2

11. $\frac{1}{7^2}$, or $\frac{1}{49}$ **27.** $\frac{5y^7z^4}{x^2}$ **31.** 8^{-4} **37.** 8^{-6}, or $\frac{1}{8^6}$ **45.** y^9
67. $9a^{-8}$, or $\frac{9}{a^8}$ **77.** -4096 **83.** 492,000 **93.** 5.83×10^{-3}
103. 9.7×10^{-5} **105.** 1.3×10^{-11} **109.** 1.5×10^3
117. 8×10^2 megabytes of information per person

5.3 Polynomials and Polynomial Functions

- Terms
- Types of Polynomials
- Degree and Coefficients
- Combining Like Terms
- Polynomial Functions
- Graphs of Polynomial Functions

We now examine an important algebraic expression known as a *polynomial.* Certain polynomials have appeared earlier in this text so you already have some experience working with them.

TERMS

At this point, we have seen a variety of algebraic expressions like

$$3a^2b^4, \qquad 2l + 2w, \qquad \text{and} \qquad 5x^2 + x - 2.$$

Within these expressions, $3a^2b^4$, $2l$, $2w$, $5x^2$, x, and -2 are examples of *terms.* A **term** can be a number (like -2), a variable (like x), a product of numbers and/or variables (like $3a^2b^4$, $2l$, or $5x^2$), or a quotient of numbers and/or variables (like $7/t$).

TYPES OF POLYNOMIALS

If a term is a product of constants and/or variables, it is called a **monomial.** Note that a term, but not a monomial, can include division by a variable. A **polynomial** is a monomial or a sum of monomials.

Examples of monomials: $3, \quad n, \quad 2w, \quad 5x^2y^3z, \quad \frac{1}{3}t^{10}$

Examples of polynomials: $3a + 2, \quad \frac{1}{2}x^2, \quad -3t^2 + t - 5, \quad x, \quad 0$

The following algebraic expressions are *not* polynomials:

$$\textbf{(1)}\ \frac{x+3}{x-4}, \qquad \textbf{(2)}\ 5x^3 - 2x^2 + \frac{1}{x}, \qquad \textbf{(3)}\ \frac{1}{x^3 - 2}.$$

Expressions (1) and (3) are not polynomials because they represent quotients, not sums. Expression (2) is not a polynomial because $1/x$ is not a monomial.

When a polynomial is written as a sum of monomials, each monomial is called a *term of the polynomial.*

STUDY TIP

A Text Is Not Light Reading

Do not expect a math text to read like a magazine or novel. On the one hand, most assigned readings in a math text consist of only a few pages. On the other hand, every sentence and word is important and should make sense. If they don't, ask for help as soon as possible.

EXAMPLE 1 Identify the terms of the polynomial $3t^4 - 5t^6 - 4t + 2$.

SOLUTION The terms are $3t^4$, $-5t^6$, $-4t$, and 2. We can see this by rewriting all subtractions as additions of opposites:

$$3t^4 - 5t^6 - 4t + 2 = 3t^4 + (-5t^6) + (-4t) + 2.$$

These are the terms of the polynomial.

Try Exercise 15.

A polynomial with two terms is called a **binomial,** whereas those with three terms are called **trinomials.** Polynomials with four or more terms have no special name.

Monomials	Binomials	Trinomials	No Special Name
$4x^2$	$2x + 4$	$3t^3 + 4t + 7$	$4x^3 - 5x^2 + xy - 8$
9	$3a^5 + 6bc$	$6x^7 - 8z^2 + 4$	$z^5 + 2z^4 - z^3 + 7z + 3$
$-7a^{19}b^5$	$-9x^7 - 6$	$4x^2 - 6x - \frac{1}{2}$	$4x^6 - 3x^5 + x^4 - x^3 + 2x - 1$

DEGREE AND COEFFICIENTS

The **degree of a term** of a polynomial is the number of variable factors in that term. Thus the degree of $7t^2$ is 2 because $7t^2$ has two variable factors: $7t^2 = 7 \cdot t \cdot t$. We will revisit the meaning of degree in Section 5.7 when polynomials in several variables are examined.

EXAMPLE 2 Determine the degree of each term: **(a)** $8x^4$; **(b)** $3x$; **(c)** 7.

SOLUTION

a) The degree of $8x^4$ is 4. **x^4 represents 4 variable factors: $x \cdot x \cdot x \cdot x$.**

b) The degree of $3x$ is 1. **There is 1 variable factor.**

c) The degree of 7 is 0. **There is no variable factor.**

The degree of a constant polynomial, such as 7, is 0, since there are no variable factors. The polynomial 0 is an exception, since $0 = 0x = 0x^2 = 0x^3$, and so on. We say that the polynomial 0 has *no* degree.

The part of a term that is a constant factor is the **coefficient** of that term. Thus the coefficient of $3x$ is 3, and the coefficient for the term 7 is simply 7.

EXAMPLE 3 Identify the coefficient of each term in the polynomial

$$4x^3 - 7x^2y + x - 8.$$

SOLUTION

The coefficient of $4x^3$ is 4.

The coefficient of $-7x^2y$ is -7.

The coefficient of the third term is 1, since $x = 1x$.

The coefficient of -8 is simply -8.

Try Exercise 19.

The **leading term** of a polynomial is the term of highest degree. Its coefficient is called the **leading coefficient** and its degree is referred to as the **degree of the polynomial.** To see how this terminology is used, consider the polynomial

$$3x^2 - 8x^3 + 5x^4 + 7x - 6.$$

The *terms* are	$3x^2$,	$-8x^3$,	$5x^4$,	$7x$,	and	-6.
The *coefficients* are	3,	-8,	5,	7,	and	-6.
The *degree of each term* is	2,	3,	4,	1,	and	0.

The *leading term* is $5x^4$ and the *leading coefficient* is 5.

The *degree of the polynomial* is 4.

COMBINING LIKE TERMS

Recall from Section 1.8 that *like*, or *similar*, *terms* are either constant terms or terms containing the same variable(s) raised to the same power(s). To simplify certain polynomials, we can often *combine*, or *collect*, like terms.

EXAMPLE 4 Identify the like terms in $4x^3 + 5x - 7x^2 + 2x^3 + x^2$.

SOLUTION

Like terms: $4x^3$ and $2x^3$ **Same variable and exponent**

Like terms: $-7x^2$ and x^2 **Same variable and exponent**

Student Notes

Remember that when we combine like terms, we are forming equivalent expressions. In these examples, there are no equations to solve.

EXAMPLE 5 Write an equivalent expression by combining like terms.

a) $2x^3 + 6x^3$

b) $8x^2 + 7 + 2x^4 - 9x^2 - 11 - 2x^4$

c) $7a^3 - 5a^2 + 9a^3 + a^2$

d) $\frac{2}{3}x^4 - x^3 - \frac{1}{6}x^4 + \frac{2}{5}x^3 - \frac{3}{10}x^3$

SOLUTION

a) $2x^3 + 6x^3 = (2 + 6)x^3$ **Using the distributive law**

$= 8x^3$

b) $8x^2 + 7 + 2x^4 - 9x^2 - 11 - 2x^4$

$= 8x^2 - 9x^2 + 2x^4 - 2x^4 + 7 - 11$

$= (8 - 9)x^2 + (2 - 2)x^4 + (7 - 11)$

$= -1x^2 + 0x^4 + (-4)$

$= -x^2 - 4$

These steps are often done mentally.

c) $7a^3 - 5a^2 + 9a^3 + a^2 = 7a^3 - 5a^2 + 9a^3 + 1a^2$ **$a^2 = 1 \cdot a^2 = 1a^2$**

$= 16a^3 - 4a^2$

d) $\frac{2}{3}x^4 - x^3 - \frac{1}{6}x^4 + \frac{2}{5}x^3 - \frac{3}{10}x^3 = \left(\frac{2}{3} - \frac{1}{6}\right)x^4 + \left(-1 + \frac{2}{5} - \frac{3}{10}\right)x^3$

$= \left(\frac{4}{6} - \frac{1}{6}\right)x^4 + \left(-\frac{10}{10} + \frac{4}{10} - \frac{3}{10}\right)x^3$

$= \frac{3}{6}x^4 - \frac{9}{10}x^3$

$= \frac{1}{2}x^4 - \frac{9}{10}x^3$

Try Exercise 43.

Note in Example 5 that the solutions are written so that the term of highest degree appears first, followed by the term of next highest degree, and so on. This is known as **descending order** and is the form in which answers will normally appear.

POLYNOMIAL FUNCTIONS

In a *polynomial function*, such as $P(x) = 5x^4 - 6x^2 + x - 7$, outputs are determined by evaluating a polynomial. Polynomial functions are classified by the degree of the polynomial used to define the function, as shown below.

Type of Function	Degree	Example
Linear	1	$f(x) = 2x + 5$
Quadratic	2	$g(x) = x^2$
Cubic	3	$p(x) = 5x^3 - \frac{1}{3}x + 2$
Quartic	4	$h(x) = 9x^4 - 6x^3$

We studied linear functions in Chapter 3, and we will study quadratic functions in Chapter 11.

To evaluate a polynomial, we substitute a number for the variable. The result will be a number.

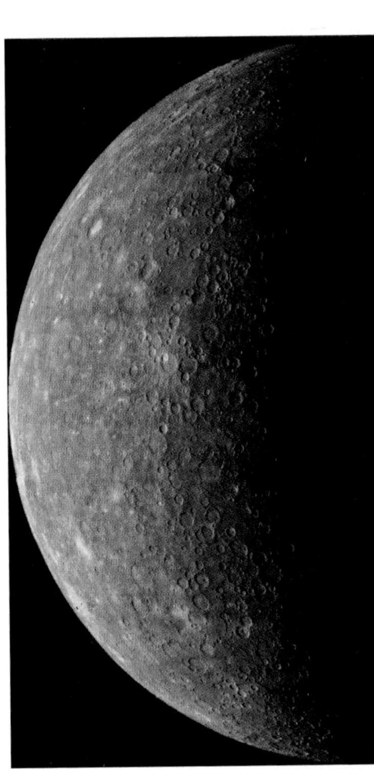

EXAMPLE 6 Astronomy. The area of a circle of radius r is given by the polynomial πr^2. The largest circular basin that has been observed on Mercury has a radius of 650 km. What is the area of the basin?
Source: NASA

SOLUTION We evaluate the polynomial πr^2 for $r = 650$:

$$\begin{aligned} \pi r^2 &= \pi(650 \text{ km})^2 && \textbf{Substituting 650 km for } r \\ &\approx 3.14(650 \text{ km})^2 && \textbf{Using 3.14 for an approximation of } \pi \\ &= 3.14 \cdot 422{,}500 \text{ km}^2 && \textbf{Evaluating the exponential expression} \\ &= 1{,}326{,}650 \text{ km}^2. && \textbf{Multiplying} \end{aligned}$$

The area of the basin is approximately 1,326,650 km^2.

Try Exercise 77.

EXAMPLE 7 Find $P(-5)$ for the polynomial function given by $P(x) = -x^2 + 4x - 1$.

SOLUTION We evaluate the function using several methods discussed in Chapter 3.

Using algebraic substitution. We substitute -5 for x and carry out the operations using the rules for order of operations:

$$\begin{aligned} P(-5) &= -(-5)^2 + 4(-5) - 1 \\ &= -25 - 20 - 1 \\ &= -46. \end{aligned}$$

CAUTION! Note that $-(-5)^2 = -25$. We square the input first and then take its opposite.

Using function notation on a graphing calculator. We let $y_1 = -x^2 + 4x - 1$. We enter the function into the graphing calculator and evaluate Y1(-5) using the Y-VARS menu.

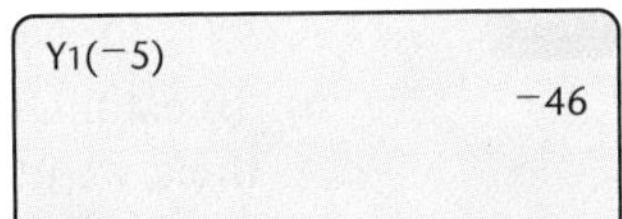

Using a table. If $y_1 = -x^2 + 4x - 1$, we can find the value of y_1 for $x = -5$ by setting Indpnt to Ask in the TABLE SETUP. If Depend is set to Auto, the y-value will appear when the x-value is entered. If Depend is set to Ask, we position the cursor in the y-column and press ENTER to see the corresponding y-value.

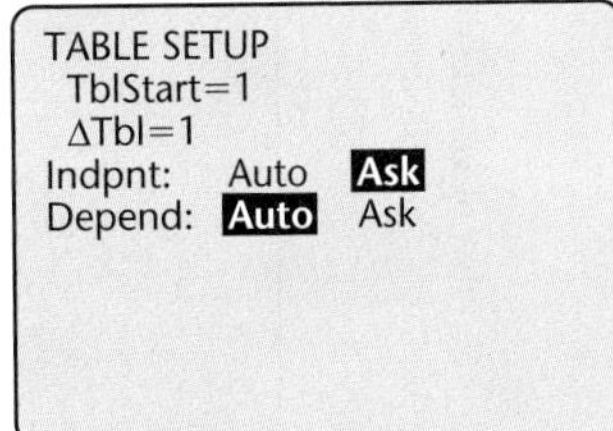

X	Y1	
−5	−46	

X =

Try Exercise 65.

EXAMPLE 8 Vehicle Miles Traveled. The table below lists the average annual number of vehicle miles traveled (VMT), in thousands, for drivers of various ages a. The data can be modeled by the polynomial function

$$v(a) = -0.003a^2 + 0.2a + 8.6.$$

a) Use the given function to estimate the number of vehicle miles traveled annually by a 25-year-old driver.

b) Use the graph to estimate $v(50)$ and tell what that number represents.

Age a of Driver	VMT (in thousands)
20	11.4
40	11.8
60	9.8
80	5.4

Source: Based on information from the Energy Information Administration

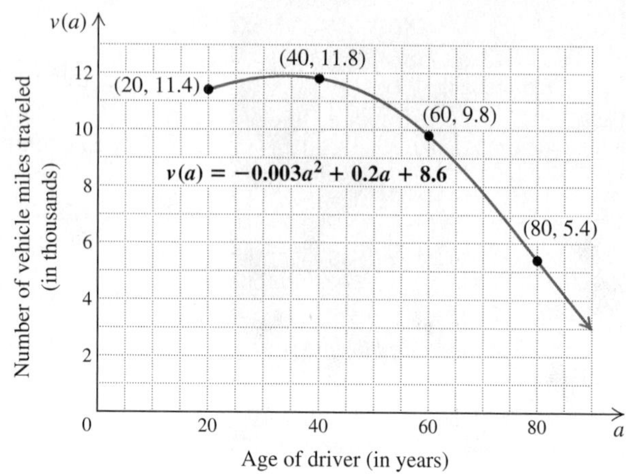

SOLUTION

a) We evalute the function for $a = 25$:

$$\begin{aligned} v(25) &= -0.003(25)^2 + 0.2(25) + 8.6 \\ &= -0.003(625) + 0.2(25) + 8.6 \\ &= -1.875 + 5 + 8.6 \\ &= 11.725. \end{aligned}$$

According to this model, the average annual number of vehicle miles traveled by a 25-year-old driver is 11.725 thousand, or 11,725.

b) To estimate $v(50)$ using a graph, we locate 50 on the horizontal axis. From there we move vertically to the graph of the function and then horizontally to the $v(a)$-axis, as shown below. This locates a value of about 11.1.

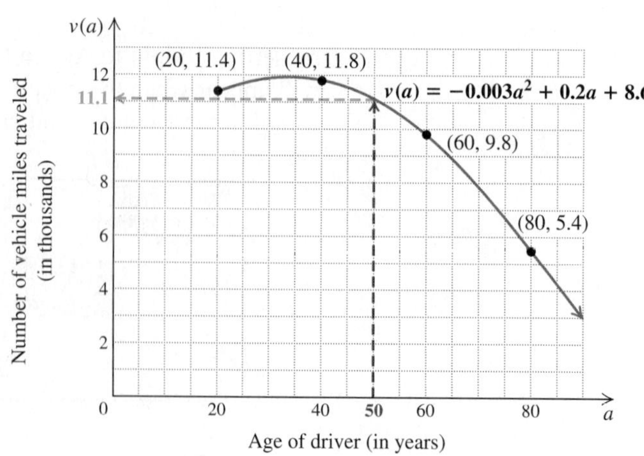

The value $v(50)$, or 11.1, indicates that the average annual number of vehicle miles traveled by a 50-year-old driver is about 11.1 thousand, or 11,100.

Try Exercise 87.

Note in Example 8 that 11.1 is a good estimate of the value of v at $a = 50$ found by evaluating the function:

$$v(50) = -0.003(50)^2 + 0.2(50) + 8.6 = 11.1.$$

GRAPHS OF POLYNOMIAL FUNCTIONS

Connecting the Concepts

Families of Graphs of Functions

Often, the shape of the graph of a function can be predicted by examining the equation describing the function. Simple functions of various types, such as those shown below, form a **library of functions.**

Linear functions can be expressed in the form $f(x) = mx + b$. They have graphs that are straight lines. Their slope, or rate of change, is constant. We can think of linear functions as a "family" of functions, with the function $f(x) = x$ being the simplest such function.

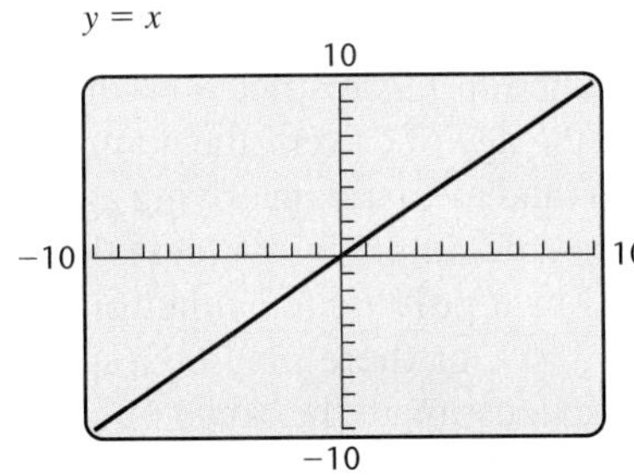

An *absolute-value function* of the form $f(x) = |ax + b|$, where $a \neq 0$, has a graph similar to that of $f(x) = |x|$. This is an example of a nonlinear function.

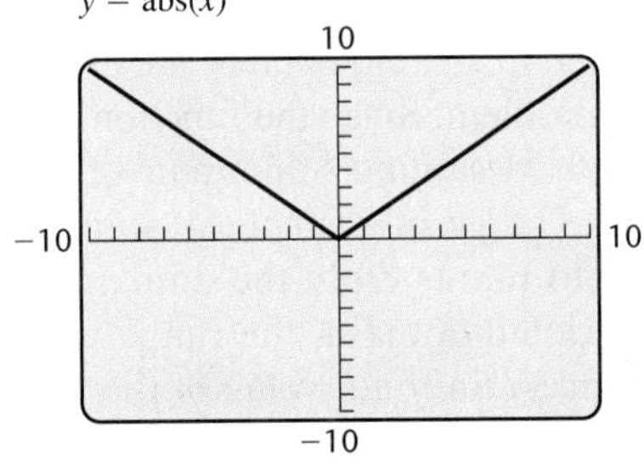

Another nonlinear function that we will consider in Chapter 10 is a square-root function. A *square-root function* of the form $f(x) = \sqrt{ax + b}$, where $a \neq 0$, has a graph similar to the graph of $f(x) = \sqrt{x}$.

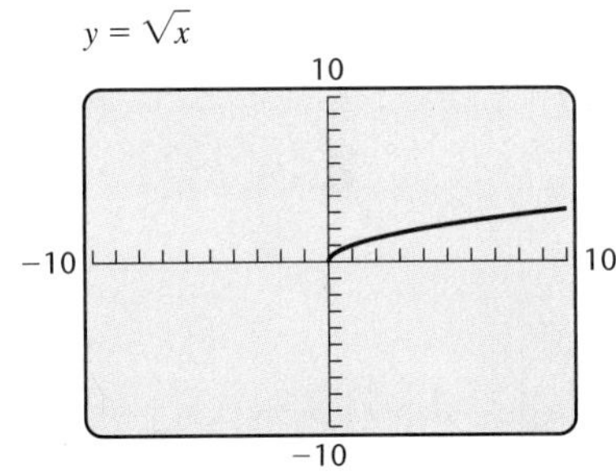

The shape of the graph of a polynomial function is largely determined by its degree. One "family" of polynomial functions, those of degree 1, have graphs that are straight lines, as discussed above. Polynomial functions of degree 2, or quadratic functions, have cup-shaped graphs called *parabolas*.

The graphs of polynomial functions have certain common characteristics.

Interactive Discovery

The table below lists some polynomial functions and some nonpolynomial functions. Graph each function and compare the graphs.

Polynomial Functions	Nonpolynomial Functions
$f(x) = x^2 + 3x + 5$	$f(x) = \lvert x - 4 \rvert$
$f(x) = 4$	$f(x) = 1 + \sqrt{2x - 5}$
$f(x) = -0.5x^3 + 5x - 2.3$	$f(x) = \dfrac{x - 7}{2x}$

1. What are some characteristics of graphs of polynomial functions?

You may have noticed the following:

- The graph of a polynomial function is "smooth," that is, there are no sharp corners.
- The graph of a polynomial function is continuous, that is, there are no holes or breaks.

Recall from Section 3.8 that the domain of a function, when not specified, is the set of all real numbers for which the function is defined. If no restrictions are given, a polynomial function is defined for all real numbers. In other words,

The domain of a polynomial function is $(-\infty, \infty)$.

Note in Example 8 that the domain of the function v is restricted by the context of the problem. Since the function is defined for the age of drivers, the domain of v is $[16, 110]$, assuming a minimum driving age of 16 and a maximum driving age of 110.

The range of a function is the set of all possible outputs (y-values) that correspond to inputs from the domain (x-values). For a polynomial function with an unrestricted domain, the range may be $(-\infty, \infty)$, or there may be a maximum value or a minimum value of the function. We can estimate the range of a function from its graph.

EXAMPLE 9 Estimate the range of each of the following functions from its graph.

a)

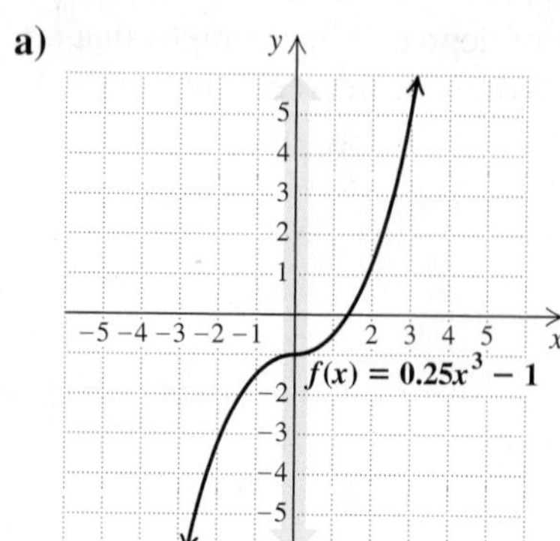

b)

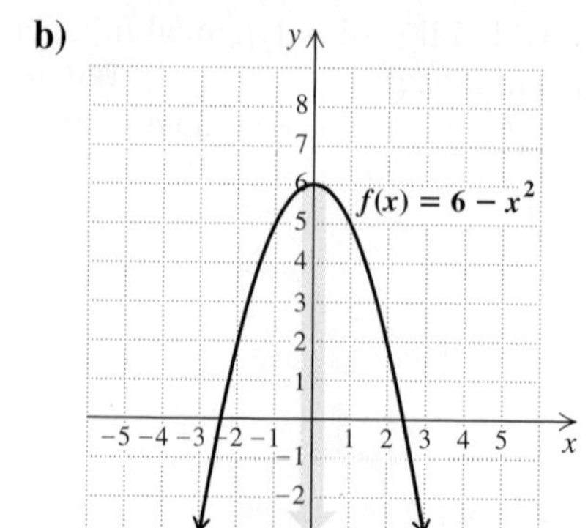

c)

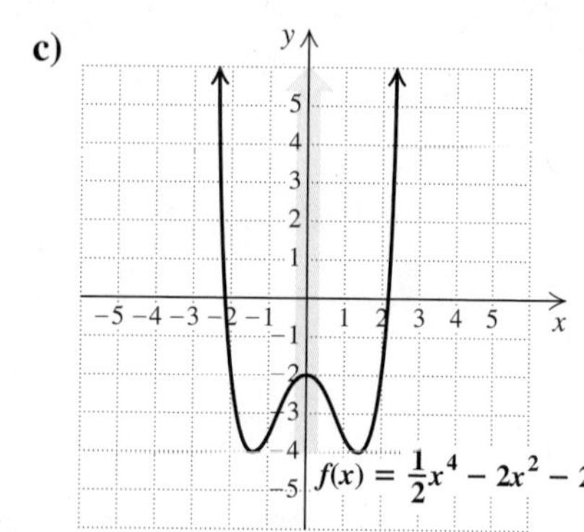

SOLUTION The domain of each function represented in the graphs is $(-\infty, \infty)$.

a) Since there is no maximum or minimum y-value indicated on the graph of the first function, we estimate its range to be $(-\infty, \infty)$. The range is indicated by the shading on the y-axis.

b) The second function has a maximum value of about 6, and no minimum is indicated, so its range is about $(-\infty, 6]$.

c) The third function has a minimum value of about -4 and no maximum value, so its range is $[-4, \infty)$.

Try Exercise 91.

We need to know the "behavior" of the graph of a polynomial function in order to estimate its range. In this chapter, we will supply the graph or an appropriate viewing window for the graph of a polynomial function. As you learn more about graphs of polynomial functions, you will be able to more readily choose appropriate scales or viewing windows.

EXAMPLE 10 Graph each of the following functions in the standard viewing window and estimate the range of the function.

a) $f(x) = x^3 - 4x^2 + 5$ **b)** $g(x) = x^4 - 4x^2 + 5$

SOLUTION

a) The graph of $y = x^3 - 4x^2 + 5$ is shown on the left below. It appears that the graph extends downward and upward indefinitely, so the range of the function is $(-\infty, \infty)$.

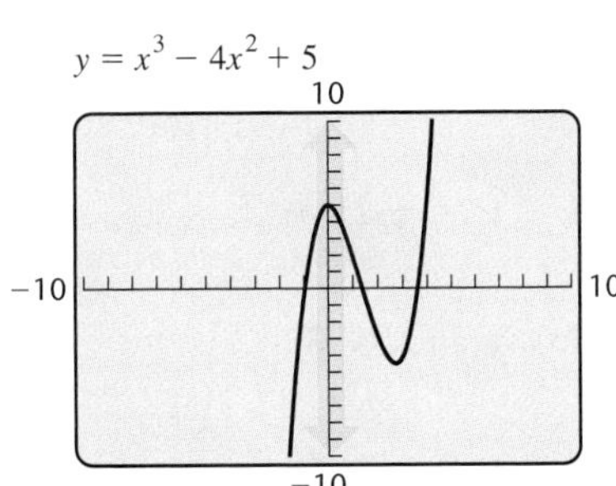

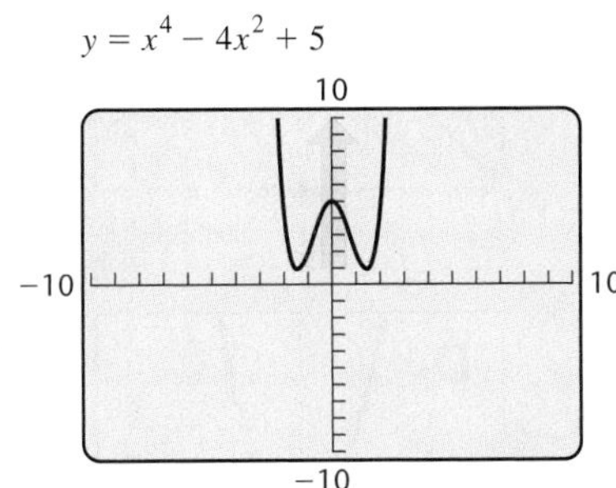

b) The graph of $y = x^4 - 4x^2 + 5$ is shown on the right above. The graph extends upward indefinitely but does not go below a y-value of 1. (This can be confirmed by tracing along the graph.) Since there is a minimum function value of 1 and no maximum function value, the range of the function is $[1, \infty)$.

Try Exercise 99.

Visualizing for Success

Match each equation with its graph.

1. $y = x$
2. $y = |x|$
3. $y = \sqrt{x}$
4. $y = x^2$
5. $y = 2x$
6. $y = -2x$
7. $y = 2x + 3$
8. $y = -2x - 3$
9. $y = x + 3$
10. $y = x - 3$

Answers on page A-18

An additional, animated version of this activity appears in MyMathLab. To use MyMathLab, you need a course ID and a student access code. Contact your instructor for more information.

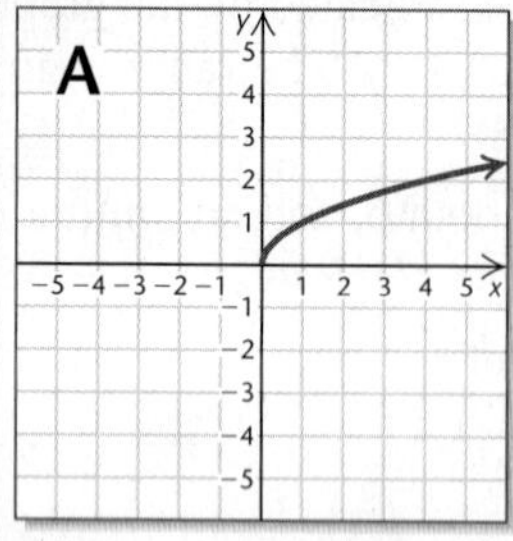

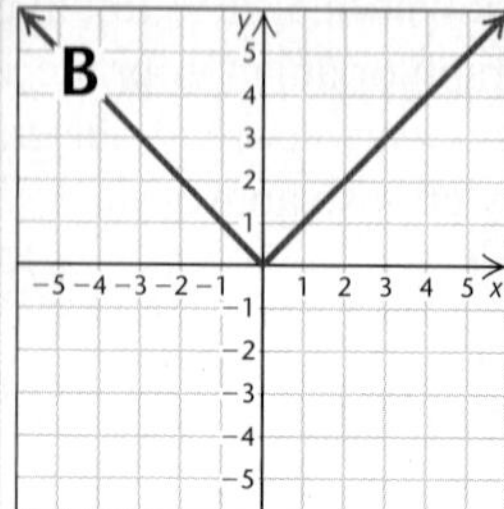

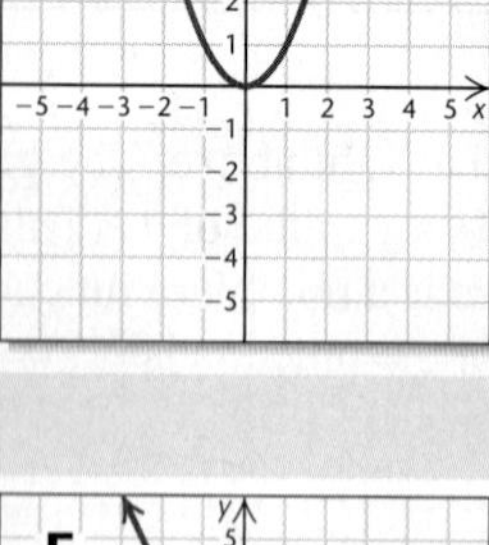

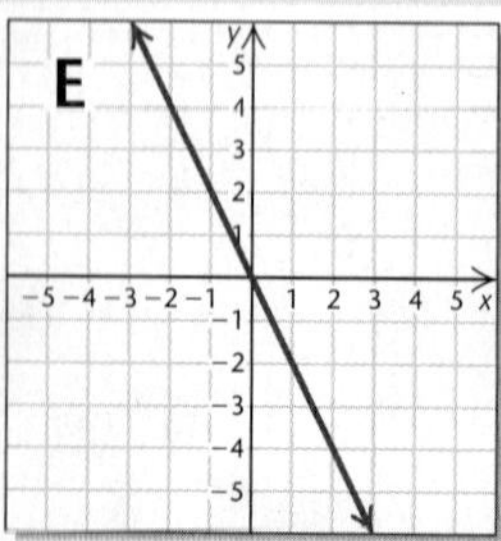

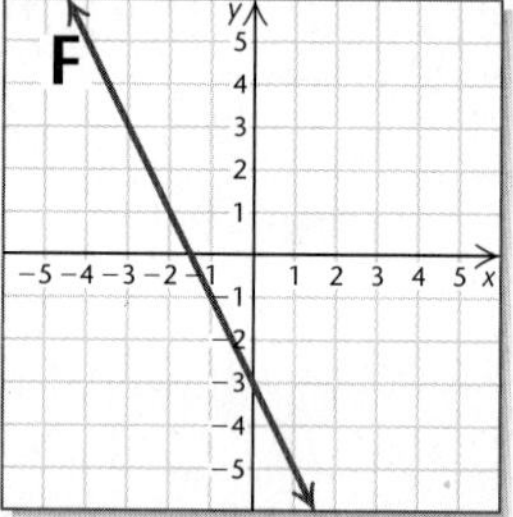

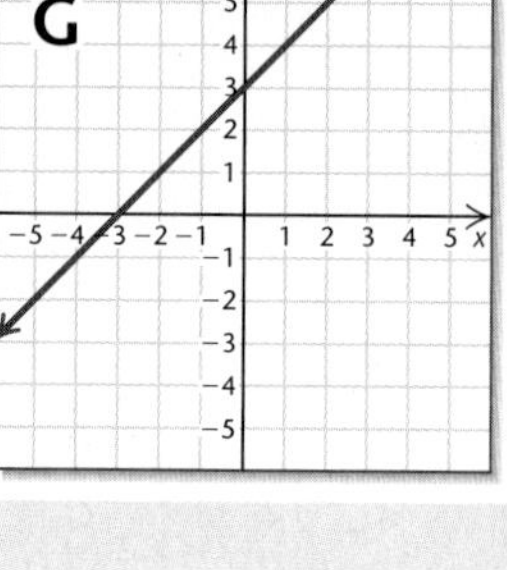

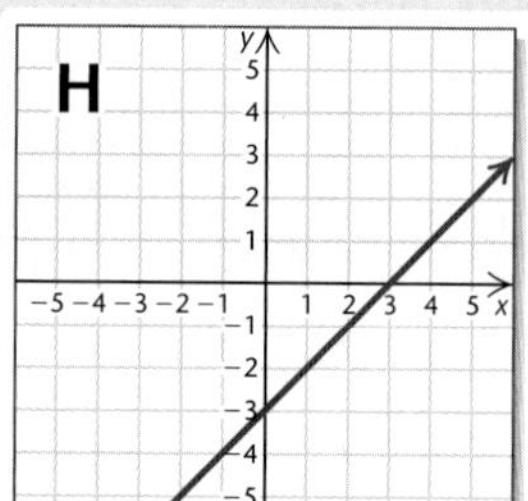

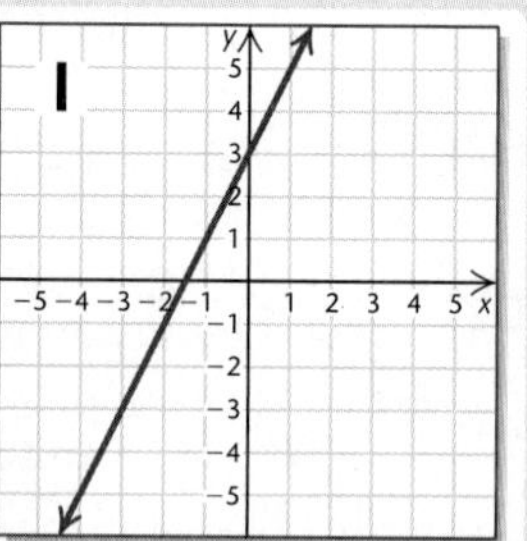

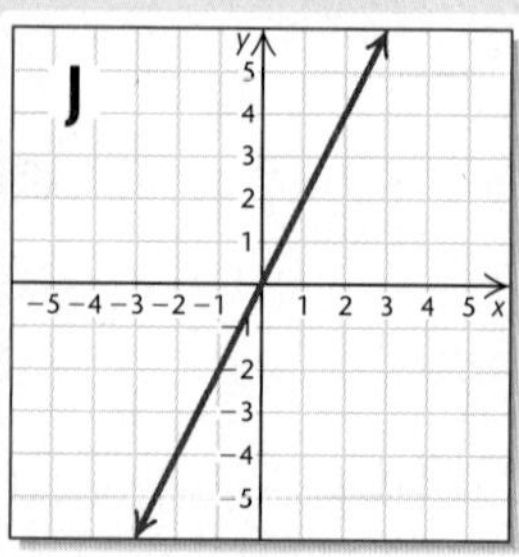

5.3 Exercise Set

FOR EXTRA HELP MyMathLab MathXL PRACTICE  WATCH  DOWNLOAD  READ REVIEW

Concept Reinforcement In each of Exercises 1–8, match the description with the most appropriate algebraic expression from the column on the right.

1. ____ A polynomial with four terms
2. ____ A polynomial with 7 as its leading coefficient
3. ____ A trinomial written in descending order
4. ____ A polynomial with degree 5
5. ____ A binomial with degree 7
6. ____ A monomial of degree 0
7. ____ An expression with two terms that is not a binomial
8. ____ An expression with three terms that is not a trinomial

a) $8x^3 + \frac{2}{x^2}$

b) $5x^4 + 3x^3 - 4x + 7$

c) $\frac{3}{x} - 6x^2 + 9$

d) $8t - 4t^5$

e) 5

f) $6x^2 + 7x^4 - 2x^3$

g) $4t - 2t^7$

h) $3t^2 + 4t + 7$

Determine whether each expression is a polynomial.

9. $3x - 7$
10. $-2x^5 + 9 - 7x^2$
11. $\dfrac{x^2 + x + 1}{x^3 - 7}$
12. -10
13. $\frac{1}{4}x^{10} - 8.6$
14. $\frac{3}{x^4} - \frac{1}{x} + 13$

Identify the terms of each polynomial.

15. $7x^4 + x^3 - 5x + 8$
16. $5a^3 + 4a^2 - a - 7$
17. $-t^6 + 7t^3 - 3t^2 + 6$
18. $n^5 - 4n^3 + 2n - 8$

Determine the coefficient and the degree of each term in each polynomial.

19. $4x^5 + 7x$
20. $9a^3 - 4a^2$
21. $9t^2 - 3t + 4$
22. $7x^4 + 5x - 3$
23. $x^4 - x^3 + 4x - 3$
24. $3a^5 - a^3 + a - 9$

*For each of the following polynomials, **(a)** list the degree of each term; **(b)** determine the leading term and the leading coefficient; and **(c)** determine the degree of the polynomial.*

25. $2a^3 + 7a^5 + a^2$
26. $5x - 9x^2 + 3x^6$
27. $9x^4 + x^2 + x^7 + 4$
28. $8 + 6x^2 - 3x - x^5$
29. $9a - a^4 + 3 + 2a^3$
30. $-x + 2x^5 - 5x^2 + x^6$
31. Complete the table below for the polynomial $7x^2 + 8x^5 - 4x^3 + 6 - \frac{1}{2}x^4$.

Term	Coefficient	Degree of the Term	Degree of the Polynomial
		5	
$-\frac{1}{2}x^4$			
	-4		
		2	
	6		

32. Complete the table below for the polynomial $-3x^4 + 6x^3 - 2x^2 + 8x + 7$.

Term	Coefficient	Degree of the Term	Degree of the Polynomial
	-3		
$6x^3$			
		2	
		1	
	7		

Classify each polynomial as a monomial, a binomial, a trinomial, or a polynomial with no special name.

33. $x^2 - 23x + 17$

34. $-9x^2$

35. $x^3 - 7x^2 + 2x - 4$

36. $t^3 + 4t$

37. $y + 5$

38. $4x^2 + 12x + 9$

39. 17

40. $2x^4 - 7x^3 + x^2 + x - 6$

Combine like terms. Write all answers in descending order.

41. $7x^2 + 3x + 4x^2$

42. $5a + 7a^2 + 3a$

43. $3a^4 - 2a + 2a + a^4$

44. $9b^5 + 3b^2 - 2b^5 - 3b^2$

45. $9t^3 - 11t + 5t + t^2$

46. $3x^4 - 7x + x^4 - 2x^2$

47. $4b^3 + 5b + 7b^3 + b^2 - 6b$

48. $6x^2 + 2x^4 - 2x^2 - x^4 - 4x^2$

49. $10x^2 + 2x^3 - 3x^3 - 4x^2 - 6x^2 - x^4$

50. $12t^6 - t^3 + 8t^6 + 4t^3 - t^7 - 3t^3$

51. $\frac{1}{5}x^4 + 7 - 2x^2 + 3 - \frac{2}{15}x^4 + 2x^2$

52. $\frac{1}{6}x^3 + 3x^2 - \frac{1}{3}x^3 + 7 + x^2 - 10$

53. $5.9x^2 - 2.1x + 6 + 3.4x - 2.5x^2 - 0.5$

54. $7.4x^3 - 4.9x + 2.9 - 3.5x - 4.3 + 1.9x^3$

Evaluate each polynomial for $x = 3$ and for $x = -3$.

55. $-7x + 4$

56. $-5x + 7$

57. $2x^2 - 3x + 7$

58. $4x^2 - 6x + 9$

59. $-2x^3 - 3x^2 + 4x + 2$

60. $-3x^3 + 7x^2 - 4x - 8$

61. $\frac{1}{3}x^4 - 2x^3$

62. $2x^4 - \frac{1}{9}x^3$

63. $-x - x^2 - x^3$

64. $-x^2 - 3x^3 - x^4$

Find the specified function values.

65. Find $P(4)$ and $P(0)$: $P(x) = 3x^2 - 2x + 7$.

66. Find $Q(3)$ and $Q(-1)$: $Q(x) = -4x^3 + 7x^2 - 6$.

67. Find $P(-2)$ and $P\left(\frac{1}{3}\right)$: $P(y) = 8y^3 - 12y - 5$.

68. Find $Q(-3)$ and $Q(0)$:

$$Q(y) = -8y^3 + 7y^2 - 4y - 9.$$

Evaluate each polynomial function for $x = -1$.

69. $f(x) = -5x^3 + 3x^2 - 4x - 3$

70. $g(x) = -4x^3 + 2x^2 + 5x - 7$

Back-to-College Expenses. The amount of money, in billions of dollars, spent on shoes for college can be estimated by the polynomial

$$0.4t + 1.13,$$

where t is the number of years since 2004. That is, $t = 0$ for 2004, $t = 1$ for 2005, and so on.

Source: Based on data from the National Retail Federation

71. Estimate the amount spent on shoes for college in 2006.

72. Estimate the amount spent on shoes for college in 2010.

73. *Skydiving.* During the first 13 sec of a jump, the number of feet that a skydiver falls in t seconds is approximated by the polynomial

$$11.12t^2.$$

In 2009, 108 U.S. skydivers fell headfirst in formation from a height of 18,000 ft. How far had they fallen 10 sec after having jumped from the plane?

Source: www.telegraph.co.uk

74. *Skydiving.* For jumps that exceed 13 sec, the polynomial

$$173t - 369$$

can be used to approximate the distance, in feet, that a skydiver has fallen in t seconds. The skydivers in Exercise 73 had 40 sec to complete their formation. How far did they fall in 40 sec?

Circumference. The circumference of a circle of radius r is given by the polynomial $2\pi r$, where π is an irrational number. For an approximation of π, use 3.14.

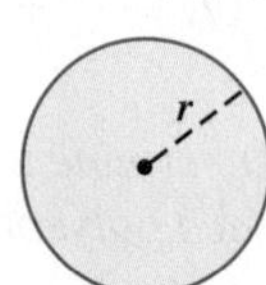

75. Find the circumference of a circle with radius 10 cm.

76. Find the circumference of a circle with radius 5 ft.

Area of a Circle. The area of a circle of radius r is given by the polynomial πr^2. Use 3.14 for π.

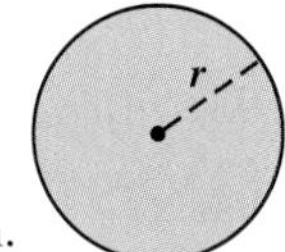

77. Find the area of a circle with radius 7 m.

78. Find the area of a circle with radius 6 ft.

79. *Kayaking.* The distance $s(t)$, in feet, traveled by a body falling freely from rest in t seconds is approximated by

$$s(t) = 16t^2.$$

On March 4, 2009, Brazilian kayaker Pedro Olivia set a world record waterfall descent on the Rio Sacre in Brazil. He was airborne for 2.9 sec. How far did he drop?

Source: www.telegraph.co.uk

80. *SCAD diving.* The SCAD thrill ride is a 2.5-sec free fall into a net. How far does the diver fall? (See Exercise 79.)

Source: "What is SCAD?", www.scadfreefall.co.uk

World Wide Web. The total number of Web sites $w(t)$, in millions, t years after 2003 can be approximated by the polynomial function given by

$$w(t) = 4.03t^2 + 6.78t + 42.86.$$

Use the following graph for Exercises 81–84.

Source: Based on information from Web Server Surveys, www.news.netcraft.com

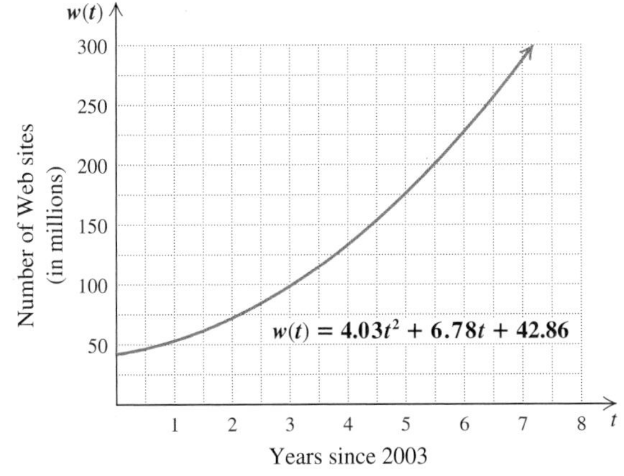

81. Estimate the number of Web sites in 2004.

82. Estimate the number of Web sites in 2009.

83. Approximate $w(5)$.

84. Approximate $w(2)$.

85. *Stacking Spheres.* In 2004, the journal *Annals of Mathematics* accepted a proof of the so-called Kepler Conjecture: that the most efficient way to pack spheres is in the shape of a square pyramid. The number N of balls in the stack is given by the polynomial function

$$N(x) = \frac{1}{3}x^3 + \frac{1}{2}x^2 + \frac{1}{6}x,$$

where x is the number of layers. Use both the function and the figure to find $N(3)$. Then calculate the number of oranges in a pyramid with 5 layers.

Source: *The New York Times* 4/6/04

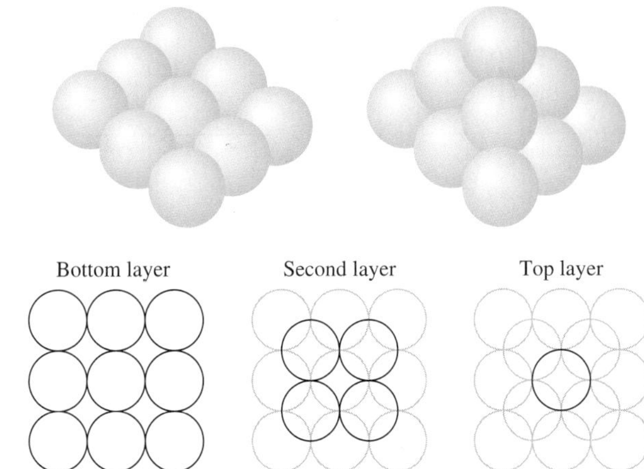

Bottom layer Second layer Top layer

86. *Stacking Cannonballs.* The function in Exercise 85 was discovered by Thomas Harriot, assistant to Sir Walter Raleigh, when preparing for an expedition at sea. How many cannonballs did they pack if there were 10 layers to their pyramid?

Source: *The New York Times* 4/7/04

Veterinary Science. Gentamicin is an antibiotic frequently used by veterinarians. The concentration, in micrograms per milliliter (mcg/mL), of Gentamicin in a horse's bloodstream t hours after injection can be approximated by the polynomial function

$$C(t) = -0.005t^4 + 0.003t^3 + 0.35t^2 + 0.5t.$$

Use the following graph for Exercises 87–90.

Source: Michele Tulis, DVM, telephone interview

C(t)
$C(t) = -0.005t^4 + 0.003t^3 + 0.35t^2 + 0.5t$
Concentration (in micrograms per milliliter)
Time (in hours)

87. Estimate the concentration, in mcg/mL, of Gentamicin in the bloodstream 2 hr after injection.

88. Estimate the concentration, in mcg/mL, of Gentamicin in the bloodstream 4 hr after injection.

89. Approximate the range of C.

90. Approximate the domain of C.

Estimate the range of each function from its graph.

91.

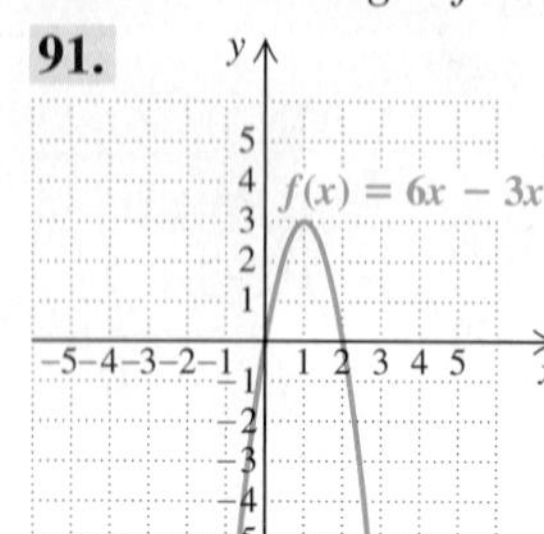

92.

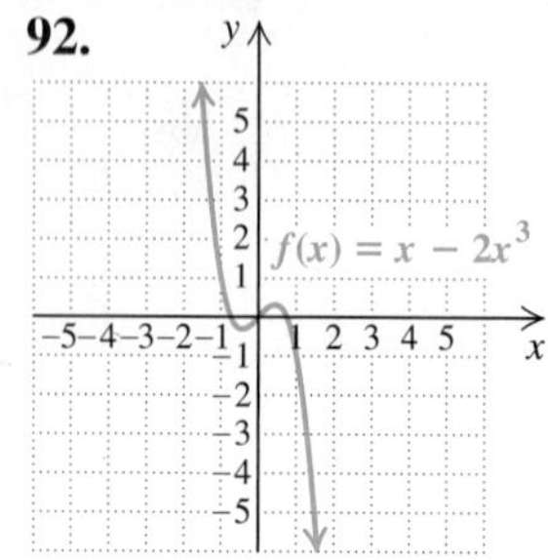

93.

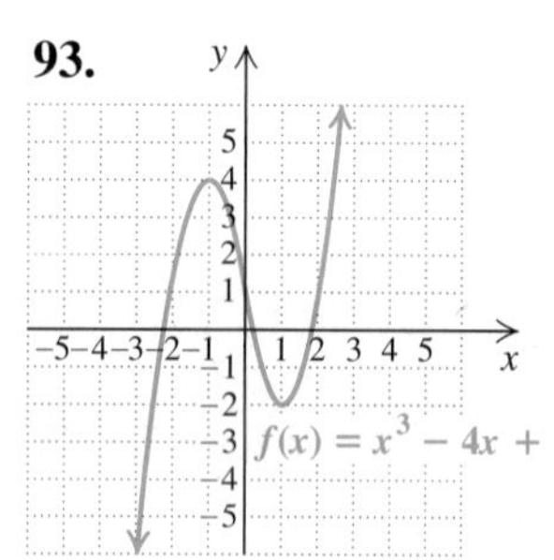

94.

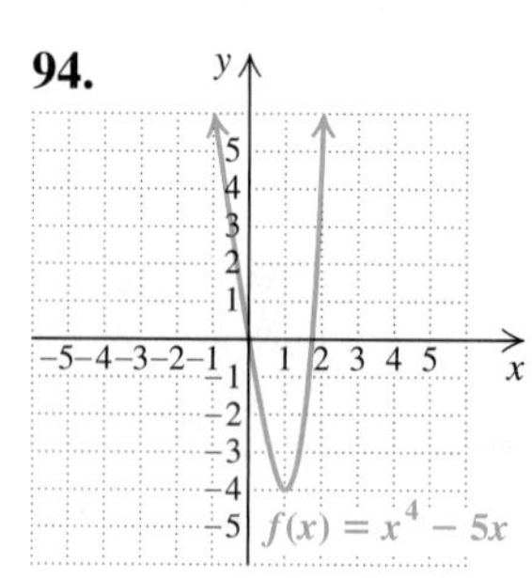

95.

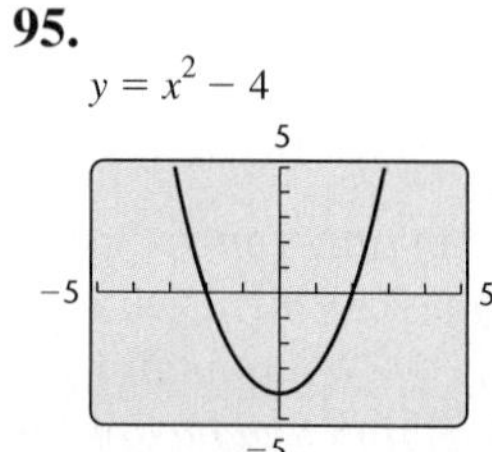

96.

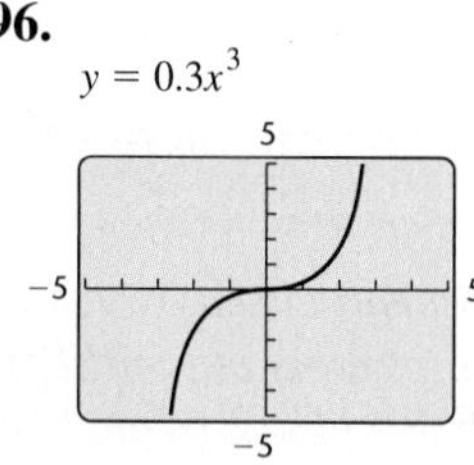

97.

$y = x^4 - 5x^3 + x - 2$

50

−10 10

−80

Yscl = 10

98.

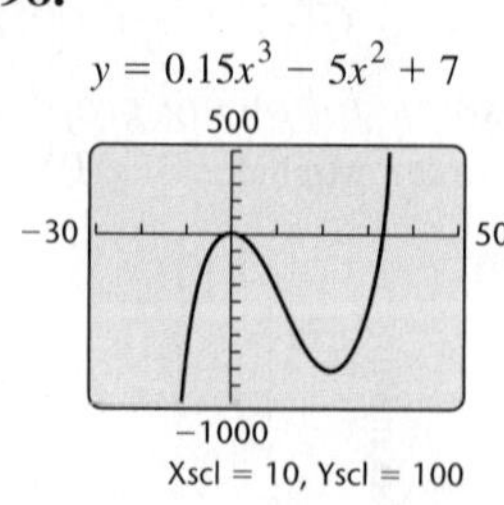

Use a graphing calculator to graph each polynomial function in the indicated viewing window, and estimate its range.

99. $f(x) = x^2 + 2x + 1, [-10, 10, -10, 10]$

100. $p(x) = x^2 + x - 6, [-10, 10, -10, 10]$

101. $q(x) = -2x^2 + 5, [-10, 10, -10, 10]$

102. $g(x) = 1 - x^2, [-10, 10, -10, 10]$

103. $p(x) = -2x^3 + x + 5, [-10, 10, -10, 10]$

104. $f(x) = -x^4 + 2x^3 - 10, [-5, 5, -30, 10]$, Yscl = 5

105. $g(x) = x^4 + 2x^3 - 5, [-5, 5, -10, 10]$

106. $q(x) = x^5 - 2x^4, [-5, 5, -5, 5]$

TW **107.** Estimate the domain and the range of the VMT function in Example 8, and explain your reasoning.

TW **108.** Is it possible to evaluate polynomials without understanding the rules for order of operations? Why or why not?

SKILL REVIEW

To prepare for Section 5.4, review simplifying expressions containing parentheses (Section 1.8).

Simplify. [1.8]

109. $3x + 7 - (x + 3)$

110. $2t - 5 - (3t - 6)$

111. $4a + 11 - (-2a - 9)$

112. $\frac{1}{2}x - \frac{1}{4} - \left(\frac{3}{2}x + \frac{3}{4}\right)$

113. $4t^4 + 3t^2 + 8t - (3t^4 + 9t^2 + 8t)$

114. $0.1n^2 + 5 - (-0.3n^2 + n - 6)$

SYNTHESIS

TW **115.** Is it easier to evaluate a polynomial before or after like terms have been combined? Why?

TW **116.** A student who is trying to graph

$$p(x) = 0.05x^4 - x^2 + 5$$

gets the following screen. How can the student tell at a glance that a mistake has been made?

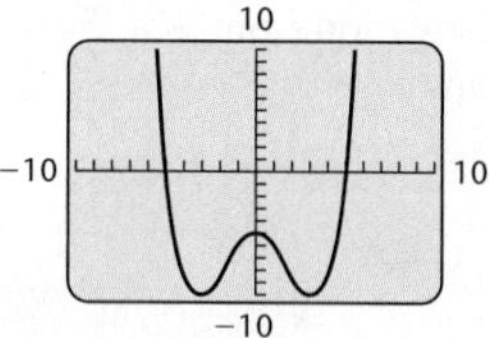

117. Construct a polynomial in x (meaning that x is the variable) of degree 5 with four terms and coefficients that are consecutive even integers.

118. Construct a trinomial in y of degree 4 with coefficients that are rational numbers.

119. What is the degree of $(5m^5)^2$?

120. Construct three like terms of degree 4.

Simplify.

121. $\frac{9}{2}x^8 + \frac{1}{9}x^2 + \frac{1}{2}x^9 + \frac{9}{2}x + \frac{9}{2}x^9 + \frac{8}{9}x^2 + \frac{1}{2}x - \frac{1}{2}x^8$

122. $(3x^2)^3 + 4x^2 \cdot 4x^4 - x^4(2x)^2 + ((2x)^2)^3 - 100x^2(x^2)^2$

123. A polynomial in x has degree 3. The coefficient of x^2 is 3 less than the coefficient of x^3. The coefficient of x is three times the coefficient of x^2. The remaining constant is 2 more than the coefficient of x^3. The sum of the coefficients is -4. Find the polynomial.

124. *Daily Accidents.* The average number of accidents per day involving drivers of age r can be approximated by the polynomial

$$0.4r^2 - 40r + 1039.$$

For what age is the number of daily accidents smallest?

Semester Averages. Professor Kopecki calculates a student's average for her course using

$$A = 0.3q + 0.4t + 0.2f + 0.1h,$$

with q, t, f, and h representing a student's quiz average, test average, final exam score, and homework average, respectively. In Exercises 125 and 126, find the given student's course average rounded to the nearest tenth.

125. Mary Lou: quizzes: 60, 85, 72, 91; final exam: 84; tests: 89, 93, 90; homework: 88

126. Nigel: quizzes: 95, 99, 72, 79; final exam: 91; tests: 68, 76, 92; homework: 86

In Exercises 127 and 128, complete the table for the given choices of t. Then plot the points and connect them with a smooth curve representing the graph of the polynomial.

127.

t	$-t^2 + 10t - 18$
3	
4	
5	
6	
7	

128.

t	$-t^2 + 6t - 4$
1	
2	
3	
4	
5	

129. *Path of the Olympic Arrow.* The Olympic flame at the 1992 Summer Olympics was lit by a flaming arrow. As the arrow moved d meters horizontally from the archer, its height h, in meters, was approximated by the polynomial

$$-0.0064d^2 + 0.8d + 2.$$

Complete the table for the choices of d given. Then plot the points and draw a graph representing the path of the arrow.

d	$-0.0064d^2 + 0.8d + 2$
0	
30	
60	
90	
120	

Try Exercise Answers: Section 5.3

15. $7x^4, x^3, -5x, 8$ **19.** Coefficients: 4, 7; degrees: 5, 1 **43.** $4a^4$ **65.** 47; 7 **77.** 153.86 m^2 **87.** About 2.3 mcg/mL **91.** $(-\infty, 3]$ **99.** $[0, \infty)$

5.4 Addition and Subtraction of Polynomials

- Addition of Polynomials
- Opposites of Polynomials
- Subtraction of Polynomials
- Problem Solving

CAUTION! Note that equations like those in Examples 1 and 2 are written to show how one expression can be rewritten in an equivalent form. This is very different from solving an equation.

ADDITION OF POLYNOMIALS

To add two polynomials, we write a plus sign between them and combine like terms.

EXAMPLE 1 Write an equivalent expression by adding.

a) $(-5x^3 + 6x - 1) + (4x^3 + 3x^2 + 2)$

b) $\left(\frac{2}{3}x^4 + 3x^2 - 7x + \frac{1}{2}\right) + \left(-\frac{1}{3}x^4 + 5x^3 - 3x^2 + 3x - \frac{1}{2}\right)$

SOLUTION

a) $(-5x^3 + 6x - 1) + (4x^3 + 3x^2 + 2)$

$= -5x^3 + 6x - 1 + 4x^3 + 3x^2 + 2$ **Writing without parentheses**

$= -5x^3 + 4x^3 + 3x^2 + 6x - 1 + 2$ **Using the commutative and associative laws to write like terms together**

$= (-5 + 4)x^3 + 3x^2 + 6x + (-1 + 2)$ **Combining like terms; using the distributive law**

$= -x^3 + 3x^2 + 6x + 1$ **Note that $-1x^3 = -x^3$.**

b) $\left(\frac{2}{3}x^4 + 3x^2 - 7x + \frac{1}{2}\right) + \left(-\frac{1}{3}x^4 + 5x^3 - 3x^2 + 3x - \frac{1}{2}\right)$

$= \left(\frac{2}{3} - \frac{1}{3}\right)x^4 + 5x^3 + (3 - 3)x^2 + (-7 + 3)x + \left(\frac{1}{2} - \frac{1}{2}\right)$ **Combining like terms**

$= \frac{1}{3}x^4 + 5x^3 - 4x$

Try Exercise 9.

After some practice, polynomial addition is often performed mentally.

EXAMPLE 2 Add: $(2 - 3x + x^2) + (-5 + 7x - 3x^2 + x^3)$.

SOLUTION We have

$(2 - 3x + x^2) + (-5 + 7x - 3x^2 + x^3)$ **Note that $x^2 = 1x^2$.**

$= (2 - 5) + (-3 + 7)x + (1 - 3)x^2 + x^3$ **You might do this step mentally.**

$= -3 + 4x - 2x^2 + x^3$. **Then you would write only this.**

Try Exercise 13.

The polynomials in the last example are written with the terms arranged according to degree, from least to greatest. Such an arrangement is called *ascending order.* As a rule, answers are written in ascending order when the polynomials in the original problem are given in ascending order. When the polynomials in the original problem are given in descending order, the answer is usually written in descending order.

Checking

A graphing calculator can be used to check whether two algebraic expressions given by y_1 and y_2 are equivalent. Some of the methods listed below have been described before; here they are listed together as a summary.

1. *Comparing graphs.* Graph both y_1 and y_2 on the same set of axes using the SEQUENTIAL mode. If the expressions are equivalent, the graphs will be identical. Using the PATH graphstyle, indicated by a small circle, for y_2 allows us to see if the graph of y_2 is being traced over the graph of y_1. To change the GRAPHSTYLE, locate the cursor on the icon before Y1= or Y2= and press ENTER. Because two different graphs may appear identical in certain windows, this check is only partial.
2. *Subtracting expressions.* Let $y_3 = y_1 - y_2$. If the expressions are equivalent, the graph of y_3 will be $y = 0$, or the x-axis. Using the PATH graphstyle or TRACE allows us to see if the graph of y_3 is the x-axis.
3. *Comparing values.* A table allows us to compare values of y_1 and y_2. If the expressions are equivalent, the values will be the same for any given x-value. Again, this is only a partial check.
4. *Comparing both graphs and values.* Many graphing calculators will split the viewing screen and show both a graph and a table. This choice is usually made using the MODE key. In the bottom line of the screen on the left below, the FULL mode indicates a full-screen table or graph. In the HORIZ mode, the screen is split in half horizontally, and in the G-T mode, the screen is split vertically. In the screen in the middle below, a graph and a table of values are shown for $y = x^2 - 1$ using the HORIZ mode. The screen on the right shows the same graph and table in the G-T mode.

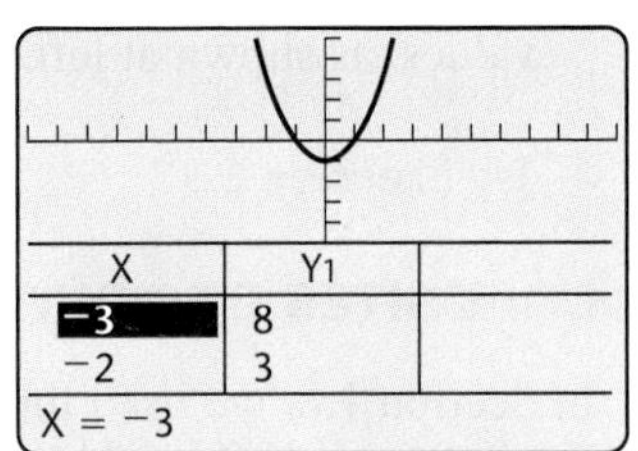

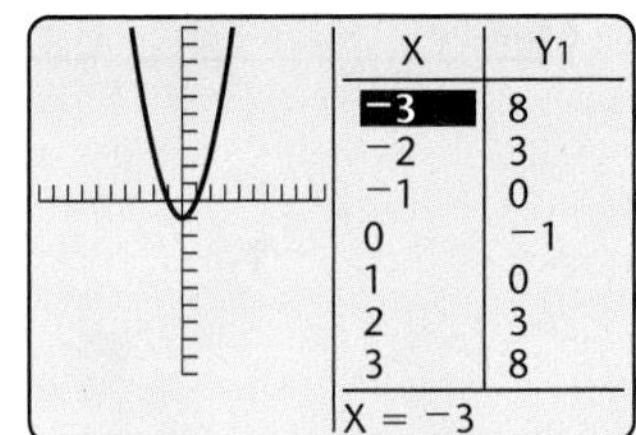

Your Turn

1. By comparing graphs, check the addition $(x^2 + x) + (x^2 - x) = 2x^2 + 2x$. Let $y_1 = (x^2 + x) + (x^2 - x)$ and $y_2 = 2x^2 + 2x$. The graphs should *not* be identical; the correct sum is $2x^2$.
2. By subtracting expressions, check the addition $(3x^2 + 5) + (1 - x^2) = 2x^2 + 6$. Let $y_1 = (3x^2 + 5) + (1 - x^2)$, $y_2 = 2x^2 + 6$, and $y_3 = y_1 - y_2$. The graph of y_3 should be the x-axis; the sum is correct.
3. By comparing values, check the addition $(3x^2 - 7) + (x^2 - x) = 4x^2 - 8x$. Let $y_1 = (3x^2 - 7) + (x^2 - x)$ and $y_2 = 4x^2 - 8x$. Except for $x = 1$, $y_1 \neq y_2$; the correct sum is $4x^2 - x - 7$.
4. Split the screen using the G-T mode and again check the addition from Exercise 3. Both the graphs and the table columns should be different.

Graphs and tables of values provide good checks of the results of algebraic manipulation. When checking by graphing, remember that two different graphs might not appear different in some viewing windows. It is also possible for two different functions to have the same value for several entries in a table. However, when enough values of two polynomial functions are the same, a table does provide a foolproof check.

If the values of two nth-degree polynomial functions are the same for $n + 1$ or more values, then the polynomials are equivalent. For example, if the values of two cubic functions ($n = 3$) are the same for 4 or more values, the cubic polynomials are equivalent.

To add using columns, we write the polynomials one under the other, listing like terms under one another and leaving spaces for missing terms.

EXAMPLE 3 Add: $9x^5 - 2x^3 + 6x^2 + 3$ and $5x^4 - 7x^2 + 6$ and $3x^6 - 5x^5 + x^2 + 5$. Check using a table of values.

SOLUTION We arrange the polynomials with like terms in columns.

$$\begin{array}{rrrrrrl} & 9x^5 & & -2x^3 & +6x^2 & +3 & \\ & & 5x^4 & & -7x^2 & +6 & \text{We leave spaces for missing terms.} \\ 3x^6 & -5x^5 & & & +1x^2 & +5 & \text{Writing } x^2 \text{ as } 1x^2 \\ \hline 3x^6 & +4x^5 & +5x^4 & -2x^3 & & +14 & \text{Adding} \end{array}$$

To check, we let

$$y_1 = (9x^5 - 2x^3 + 6x^2 + 3) + (5x^4 - 7x^2 + 6) + (3x^6 - 5x^5 + x^2 + 5)$$

and $y_2 = 3x^6 + 4x^5 + 5x^4 - 2x^3 + 14$.

X	Y1	Y2
−2	174	174
−1	20	20
0	14	14
1	24	24
2	398	398
3	3524	3524
4	17550	17550

X = −2

Because of the polynomial with degree 6, we need to show that $y_1 = y_2$ for seven different x-values. This can be accomplished quickly by creating a table of values, as shown at left. The answer is $3x^6 + 4x^5 + 5x^4 - 2x^3 + 14$.

Try Exercise 23.

OPPOSITES OF POLYNOMIALS

In Section 1.8, we used the property of -1 to show that the opposite of a sum is the sum of the opposites. This idea can be extended.

The Opposite of a Polynomial To find an equivalent polynomial for the *opposite*, or *additive inverse*, of a polynomial, change the sign of every term. This is the same as multiplying the polynomial by -1.

EXAMPLE 4 Write two equivalent expressions for the opposite of

$$4x^5 - 7x^3 - 8x + \tfrac{5}{6}.$$

SOLUTION

i) $-\left(4x^5 - 7x^3 - 8x + \frac{5}{6}\right)$ — **This is one way to write the opposite of $4x^5 - 7x^3 - 8x + \frac{5}{6}$.**

ii) $-4x^5 + 7x^3 + 8x - \frac{5}{6}$ — **Changing the sign of every term**

Thus, $-\left(4x^5 - 7x^3 - 8x + \frac{5}{6}\right)$ and $-4x^5 + 7x^3 + 8x - \frac{5}{6}$ are equivalent. Both expressions represent the opposite of $4x^5 - 7x^3 - 8x + \frac{5}{6}$.

Try Exercise 27.

STUDY TIP 

Aim for Mastery

In each exercise set, you may encounter one or more exercises that give you trouble. Do not ignore these problems, but instead work to master them. These are the problems (not the "easy" problems) that you will need to revisit as you progress through the course. Consider marking these exercises so that you can easily locate them when studying or reviewing for a quiz or a test.

EXAMPLE 5 Simplify: $-\left(-7x^4 - \frac{5}{9}x^3 + 8x^2 - x + 67\right)$.

SOLUTION

$$-\left(-7x^4 - \tfrac{5}{9}x^3 + 8x^2 - x + 67\right) = 7x^4 + \tfrac{5}{9}x^3 - 8x^2 + x - 67$$

Try Exercise 33.

SUBTRACTION OF POLYNOMIALS

We can now subtract one polynomial from another by adding the opposite of the polynomial being subtracted.

EXAMPLE 6 Write an equivalent expression by subtracting.

a) $(9x^5 + x^3 - 2x^2 + 4) - (-2x^5 + x^4 - 4x^3 - 3x^2)$

b) $(7x^5 + x^3 - 9x) - (3x^5 - 4x^3 + 5)$

SOLUTION

a) $(9x^5 + x^3 - 2x^2 + 4) - (-2x^5 + x^4 - 4x^3 - 3x^2)$

$= 9x^5 + x^3 - 2x^2 + 4 + 2x^5 - x^4 + 4x^3 + 3x^2$ **Adding the opposite**

$= 11x^5 - x^4 + 5x^3 + x^2 + 4$ **Combining like terms**

b) $(7x^5 + x^3 - 9x) - (3x^5 - 4x^3 + 5)$

$= 7x^5 + x^3 - 9x + (-3x^5) + 4x^3 - 5$ **Adding the opposite**

$= 7x^5 + x^3 - 9x - 3x^5 + 4x^3 - 5$ **Try to go directly to this step.**

$= 4x^5 + 5x^3 - 9x - 5$ **Combining like terms**

Try Exercise 39.

To subtract using columns, we first replace the coefficients in the polynomial being subtracted with their opposites. We then add as before.

EXAMPLE 7 Write in columns and subtract:

$$(5x^2 - 3x + 6) - (9x^2 - 5x - 3).$$

SOLUTION

i)
$$\begin{array}{r} 5x^2 - 3x + 6 \\ -(9x^2 - 5x - 3) \\ \hline \end{array}$$
Writing similar terms in columns

ii)
$$\begin{array}{r} 5x^2 - 3x + 6 \\ -9x^2 + 5x + 3 \\ \hline \end{array}$$
Changing signs and removing parentheses

iii)
$$\begin{array}{r} 5x^2 - 3x + 6 \\ -9x^2 + 5x + 3 \\ \hline -4x^2 + 2x + 9 \end{array}$$
Adding

Try Exercise 53.

If you can do so without error, you can arrange the polynomials in columns, mentally find the opposite of each term being subtracted, and write the answer. Lining up like terms is important and may require leaving some blanks.

EXAMPLE 8 Write in columns and subtract:

$$(x^3 + x^2 - 12) - (-2x^3 + x^2 - 3x + 8).$$

SOLUTION We have

$$\begin{array}{r} x^3 + x^2 \phantom{{}- 3x} - 12 \\ -(-2x^3 + x^2 - 3x + 8) \\ \hline 3x^3 \phantom{{}+ x^2} + 3x - 20 \end{array}$$ **Leaving a blank space for the missing term**

Try Exercise 55.

PROBLEM SOLVING

EXAMPLE 9 Find a polynomial for the sum of the areas of rectangles A, B, C, and D.

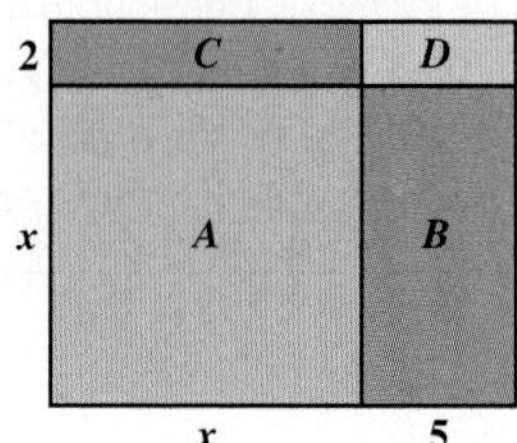

SOLUTION

1. **Familiarize.** Recall that the area of a rectangle is the product of its length and width.
2. **Translate.** We translate the problem to mathematical language. The sum of the areas is a sum of products. We find each product and then add:

Area of A	plus	area of B	plus	area of C	plus	area of D.
↓	↓	↓	↓	↓	↓	↓
$x \cdot x$	$+$	$5x$	$+$	$2x$	$+$	$2 \cdot 5$

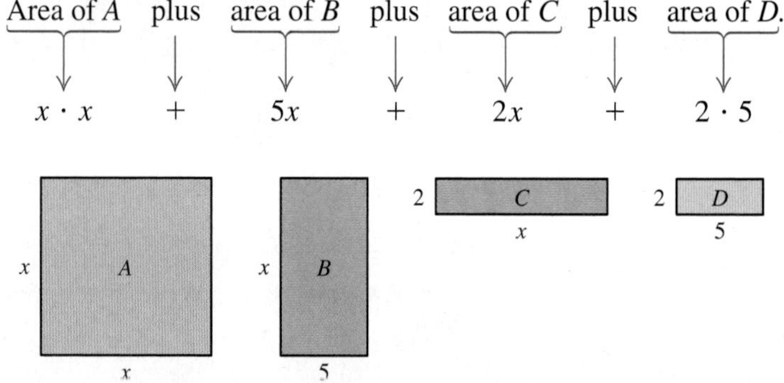

3. **Carry out.** We simplify $x \cdot x$ and $2 \cdot 5$ and combine like terms:

$$x^2 + 5x + 2x + 10 = x^2 + 7x + 10.$$

4. **Check.** A partial check is to replace x with a number, say, 3. Then we evaluate $x^2 + 7x + 10$ and compare that result with an alternative calculation:

$$3^2 + 7 \cdot 3 + 10 = 9 + 21 + 10 = 40.$$

When we substitute 3 for x and calculate the total area by regarding the figure as one large rectangle, we should also get 40:

$$\text{Total area} = (x + 5)(x + 2) = (3 + 5)(3 + 2) = 8 \cdot 5 = 40.$$

Our check is only partial, since it is possible for an incorrect answer to equal 40 when evaluated for $x = 3$. This would be unlikely, especially if a second choice of x, say $x = 5$, also checks. We leave that check to the student.

5. **State.** A polynomial for the sum of the areas is $x^2 + 7x + 10$.

Try Exercise 57.

EXAMPLE 10 A 16-ft wide round fountain is built in a square that measures x ft by x ft. Find a polynomial for the remaining area of the square.

SOLUTION

1. **Familiarize.** We make a drawing of the square and the circular fountain, and let x represent the length of a side of the square.

The area of a square is given by $A = s^2$, and the area of a circle is given by $A = \pi r^2$. Note that a circle with a diameter of 16 ft has a radius of 8 ft.

2. **Translate.** We reword the problem and translate as follows.

Rewording:	Area of square	minus	area of fountain	is	area left over.
	↓	↓	↓	↓	↓
Translating:	$x\text{ ft} \cdot x\text{ ft}$	$-$	$\pi \cdot 8\text{ ft} \cdot 8\text{ ft}$	$=$	Area left over

3. **Carry out.** We carry out the multiplication:

$$x^2\text{ ft}^2 - 64\pi\text{ ft}^2 = \text{Area left over.}$$

4. **Check.** As a partial check, note that the units in the answer are square feet (ft^2), a measure of area, as expected.

5. **State.** The remaining area of the square is $(x^2 - 64\pi)\text{ ft}^2$.

Try Exercise 69.

5.4 Exercise Set

FOR EXTRA HELP MyMathLab    

Concept Reinforcement *For Exercises 1–4, replace* □ *with the correct expression or operation sign.*

1. $(3x^2 + 2) + (6x^2 + 7) = (3 + 6)\square + (2 + 7)$
2. $(5t - 6) + (4t + 3) = (5 + 4)t + (\square + 3)$
3. $(9x^3 - x^2) - (3x^3 + x^2) = 9x^3 - x^2 - 3x^3 \square x^2$
4. $(-2n^3 + 5) - (n^2 - 2) = -2n^3 + 5 - n^2 \square 2$

Add.

5. $(3x + 2) + (x + 7)$
6. $(x + 1) + (12x + 10)$
7. $(2y - 3) + (-9y + 1)$
8. $(-8t - 5) + (-t - 3)$
9. $(-6x + 2) + (x^2 + x - 3)$

10. $(x^2 - 5x + 4) + (8x - 9)$

11. $(7t^2 - 3t + 6) + (2t^2 + 8t - 9)$

12. $(8a^2 + 4a - 5) + (6a^2 - 3a - 1)$

13. $(2m^3 - 7m^2 + m - 6) + (4m^3 + 7m^2 - 4m - 2)$

14. $(5n^3 - n^2 + 4n - 3) + (2n^3 - 4n^2 + n + 3)$

15. $(3 + 6a + 7a^2 + a^3) + (4 + 7a - 8a^2 + 6a^3)$

16. $(7 + 4t - 5t^2 + 6t^3) + (2 + t + 6t^2 - 4t^3)$

17. $(9x^8 - 7x^4 + 2x^2 + 5) + (8x^7 + 4x^4 - 2x)$

18. $(4x^5 - 6x^3 - 9x + 1) + (6x^3 + 9x^2 + 9x)$

19. $\left(\frac{1}{4}x^4 + \frac{2}{3}x^3 + \frac{5}{8}x^2 + 7\right) + \left(-\frac{3}{4}x^4 + \frac{3}{8}x^2 - 7\right)$

20. $\left(\frac{1}{3}x^9 + \frac{1}{5}x^5 - \frac{1}{2}x^2 + 7\right) + \left(-\frac{1}{5}x^9 + \frac{1}{4}x^4 - \frac{3}{5}x^5\right)$

21. $(5.3t^2 - 6.4t - 9.1) + (4.2t^3 - 1.8t^2 + 7.3)$

22. $(4.9a^3 + 3.2a^2 - 5.1a) + (2.1a^2 - 3.7a + 4.6)$

23.
$$\begin{array}{r} -3x^4 + 6x^2 + 2x - 1 \\ -3x^2 + 2x + 1 \\ \hline \end{array}$$

24.
$$\begin{array}{r} -4x^3 + 8x^2 + 3x - 2 \\ -4x^2 + 3x + 2 \\ \hline \end{array}$$

25.
$$\begin{array}{lllll} 0.15x^4 & + 0.10x^3 & - 0.9x^2 & & \\ & - 0.01x^3 & + 0.01x^2 & + x & \\ 1.25x^4 & & + 0.11x^2 & & + 0.01 \\ & 0.27x^3 & & & + 0.99 \\ -0.35x^4 & & + 15x^2 & & - 0.03 \\ \hline \end{array}$$

26.
$$\begin{array}{lllll} 0.05x^4 & + 0.12x^3 & - 0.5x^2 & & \\ & - 0.02x^3 & + 0.02x^2 & + 2x & \\ 1.5x^4 & & + 0.01x^2 & & + 0.15 \\ & 0.25x^3 & & & + 0.85 \\ -0.25x^4 & & + 10x^2 & & - 0.04 \\ \hline \end{array}$$

Write two equivalent expressions for the opposite of each polynomial, as in Example 4.

27. $-t^3 + 4t^2 - 9$

28. $-4x^3 - 5x^2 + 2x$

29. $12x^4 - 3x^3 + 3$

30. $5a^3 + 2a - 17$

Simplify.

31. $-(8x - 9)$

32. $-(-6x + 5)$

33. $-(3a^4 - 5a^2 + 9)$

34. $-(-6a^3 + 2a^2 - 7)$

35. $-\left(-4x^4 + 6x^2 + \frac{3}{4}x - 8\right)$

36. $-(-5x^4 + 4x^3 - x^2 + 0.9)$

Subtract.

37. $(7x + 4) - (2x + 1)$

38. $(5x + 6) - (2x + 4)$

39. $(-5t + 6) - (t^2 + 3t - 1)$

40. $(a^2 - 5a + 2) - (3a^2 + 2a - 4)$

41. $(8y^2 + y - 11) - (3 - 6y^3 - 8y^2)$

42. $(-4x^2 + 2x) - (-5x^2 + 2x^3 + 3)$

43. $(1.2x^3 + 4.5x^2 - 3.8x) - (-3.4x^3 - 4.7x^2 + 23)$

44. $(0.5x^4 - 0.6x^2 + 0.7) - (2.3x^4 + 1.8x - 3.9)$

Aha! **45.** $(7x^3 - 2x^2 + 6) - (7x^3 - 2x^2 + 6)$

46. $(8x^5 + 3x^4 + x - 1) - (8x^5 + 3x^4 - 1)$

47. $(3 + 5a + 3a^2 - a^3) - (2 + 3a - 4a^2 + 2a^3)$

48. $(7 + t - 5t^2 + 2t^3) - (1 + 2t - 4t^2 + 5t^3)$

49. $\left(\frac{5}{8}x^3 - \frac{1}{4}x - \frac{1}{3}\right) - \left(-\frac{1}{8}x^3 + \frac{1}{4}x - \frac{1}{3}\right)$

50. $\left(\frac{1}{5}x^3 + 2x^2 - \frac{3}{10}\right) - \left(-\frac{2}{5}x^3 + 2x^2 + \frac{7}{1000}\right)$

51. $(0.07t^3 - 0.03t^2 - 0.25t) - (0.02t^3 + 0.04t^2 - 0.3t)$

52. $(0.9a^3 + 0.2a - 5) - (0.7a^4 - 0.15a - 0.1)$

53.
$$\begin{array}{r} x^2 + 5x + 6 \\ -(x^2 + 2x + 1) \\ \hline \end{array}$$

54.
$$\begin{array}{r} x^3 + 3x^2 + 1 \\ -(x^3 + x^2 - 5) \\ \hline \end{array}$$

55.
$$\begin{array}{r} 5x^4 + 6x^3 \\ -(-6x^4 - 6x^3 + x^2) \\ \hline \end{array}$$

56.
$$\begin{array}{r} 5x^4 - 2x^3 + 6x^2 \\ -(7x^4 \qquad + 7x^2) \\ \hline \end{array}$$

57. Solve.

a) Find a polynomial for the sum of the areas of the rectangles shown in the figure.

b) Find the sum of the areas when $x = 5$ and $x = 7$.

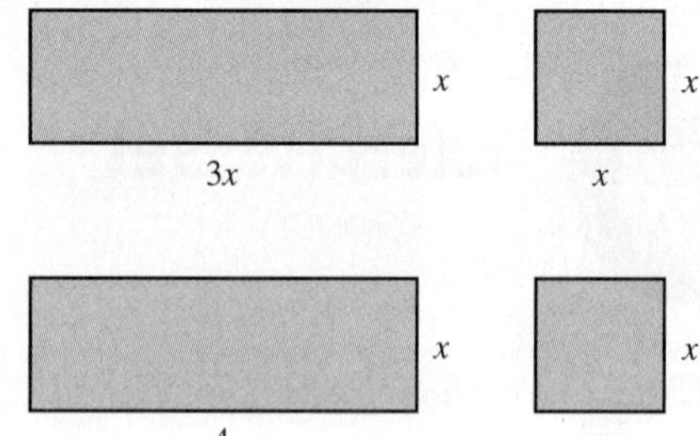

58. Solve.

a) Find a polynomial for the sum of the areas of the circles shown in the figure.

b) Find the sum of the areas when $r = 5$ and $r = 11.3$.

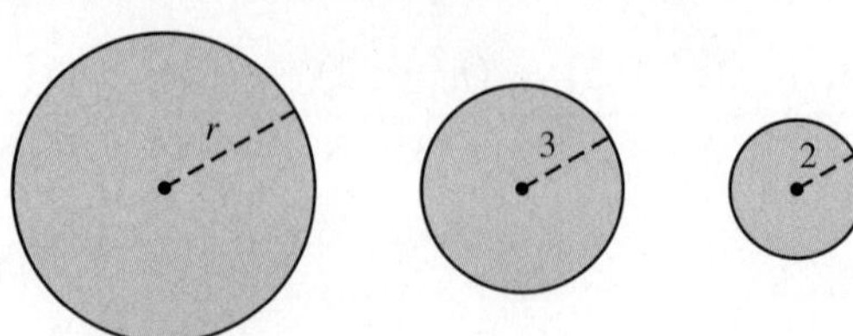

Find a polynomial for the perimeter of each figure in Exercises 59 and 60.

59.

60.

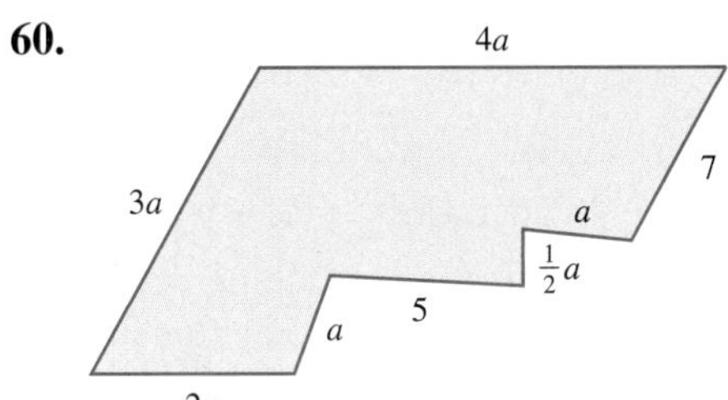

Find two algebraic expressions for the area of each figure. First, regard the figure as one large rectangle, and then regard the figure as a sum of four smaller rectangles.

61.

62.

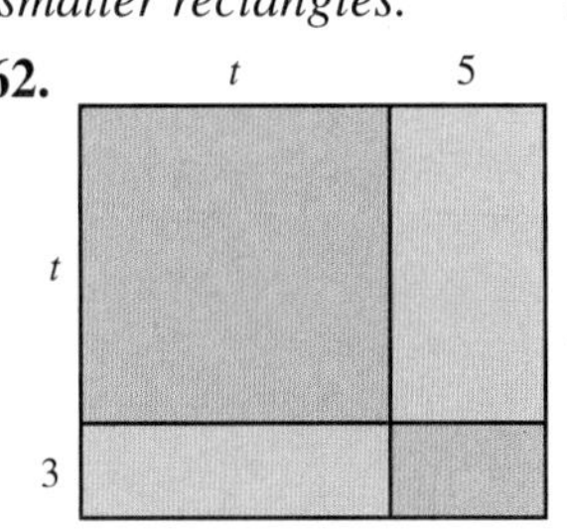

63.

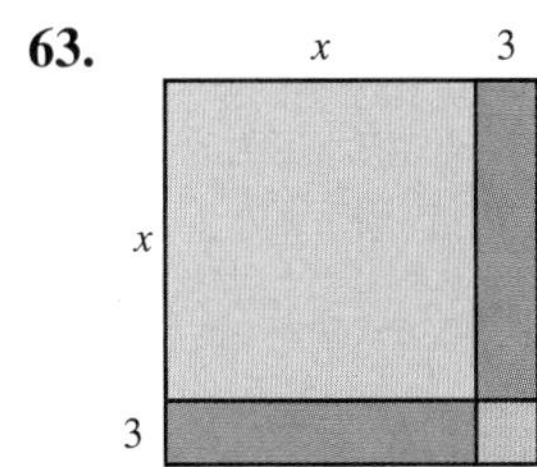

64.

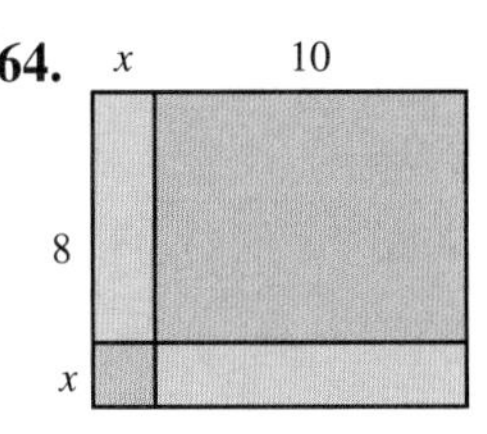

Find a polynomial for the shaded area of each figure.

65.

66.

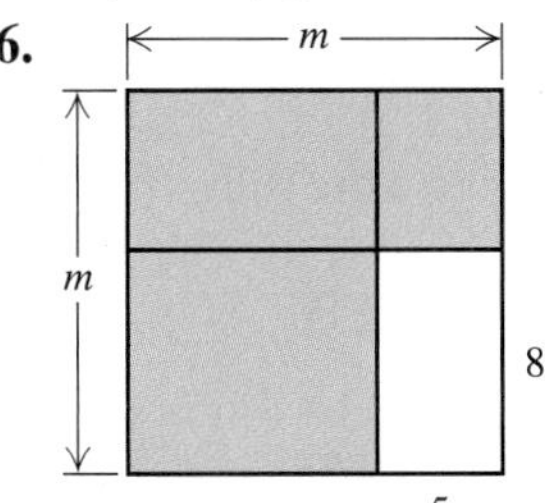

67.

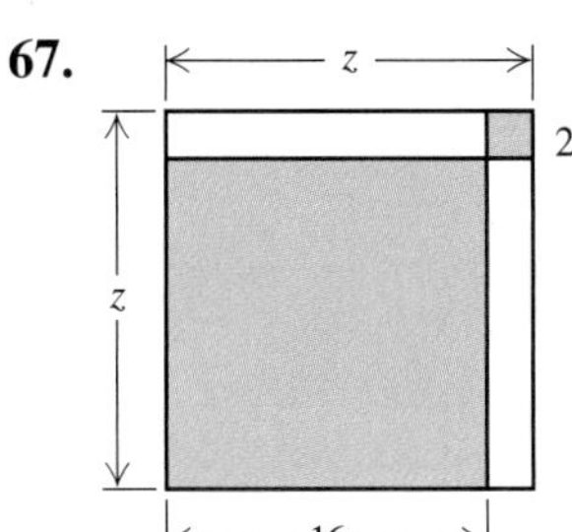

68.

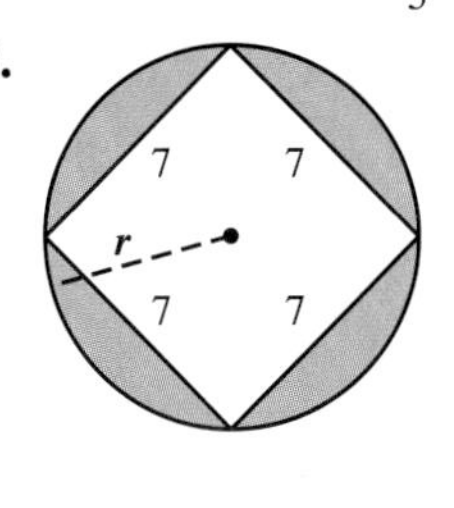

69. A 2-ft by 6-ft bath enclosure is installed in a new bathroom measuring x ft by x ft. Find a polynomial for the remaining floor area.

70. A 5-ft by 7-ft Jacuzzi™ is installed on an outdoor deck measuring y ft by y ft. Find a polynomial for the remaining area of the deck.

71. A 12-ft wide round patio is laid in a garden measuring z ft by z ft. Find a polynomial for the remaining area of the garden.

72. A 10-ft wide round water trampoline is floating in a pool measuring x ft by x ft. Find a polynomial for the remaining surface area of the pool.

73. A 12-m by 12-m mat includes a circle of diameter d meters for wrestling. Find a polynomial for the area of the mat outside the wrestling circle.

74. A 2-m by 3-m rug is spread inside a tepee that has a diameter of x meters. Find a polynomial for the area of the tepee's floor that is not covered.

Use a graphing calculator to determine whether each addition or subtraction is correct.

75. $(3x^2 - x^3 + 5x) + (4x^3 - x^2 + 7) = 7x^3 - 2x^2 + 5x + 7$

76. $(-5x + 3) + (4x - 7) + (10 - x) = -2x + 6$

77. $\left(x^2 - x - 3x^3 + \frac{1}{2}\right) - \left(x^3 - x^2 - x - \frac{1}{2}\right) = -4x^3 + 2x^2 + 1$

78. $(8x^2 - 6x - 5) - (4x^3 - 6x + 2) = 4x^2 - 7$

79. $(4x^2 + 3) + (x^2 - 5x) - (8 - 3x) = 5x^2 - 8x - 5$

80. $(4.1x^2 - 1.3x - 2.7) - (3.7x - 4.8 - 6.5x^2) = 0.4x^2 + 3.5x + 3.8$

TW **81.** Explain why parentheses are used in the statement of the solution of Example 10: $(x^2 - 64\pi)$ ft^2.

TW **82.** Is the sum of two trinomials always a trinomial? Why or why not?

SKILL REVIEW

To prepare for Section 5.5, review multiplying using the distributive law and multiplying with exponential notation (Sections 1.8 and 5.1).

Simplify.

83. $2(x^2 - x + 3)$ [1.8]

84. $-5(3x^2 - 2x - 7)$ [1.8]

85. $t^2 \cdot t^{11}$ [5.1]

86. $y^6 \cdot y$ [5.1]

87. $2n \cdot n^6$ [5.1]

88. $-9n^4 \cdot n^8$ [5.1]

SYNTHESIS

TW **89.** What can be concluded about two polynomials whose sum is zero?

TW **90.** Which, if any, of the commutative, associative, and distributive laws are needed for adding polynomials? Why?

Simplify.

91. $(6t^2 - 7t) + (3t^2 - 4t + 5) - (9t - 6)$

92. $(3x^2 - 4x + 6) - (-2x^2 + 4) + (-5x - 3)$

93. $4(x^2 - x + 3) - 2(2x^2 + x - 1)$

94. $3(2y^2 - y - 1) - (6y^2 - 3y - 3)$

95. $(345.099x^3 - 6.178x) - (94.508x^3 - 8.99x)$

Find a polynomial for the surface area of each right rectangular solid.

96.

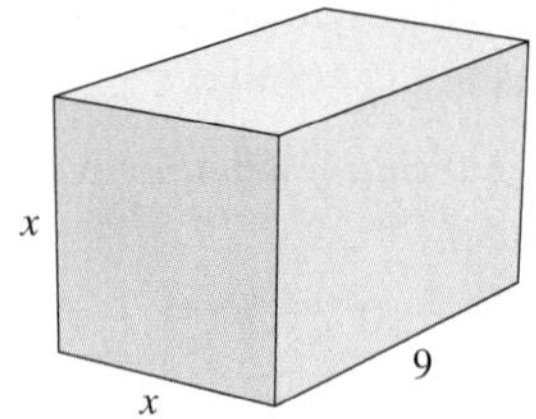

97.

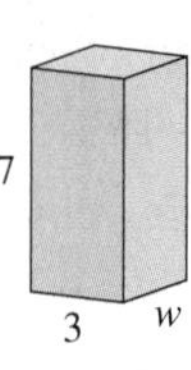

98.

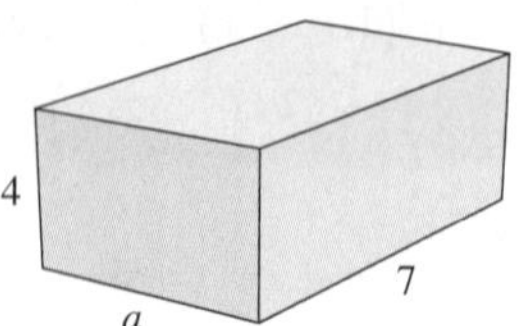

99.

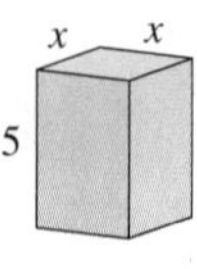

100. Find a polynomial for the total length of all edges in the figure appearing in Exercise 99.

101. *Total Profit.* Hadley Electronics is marketing a new kind of camera. Total revenue is the total amount of money taken in. The firm determines that when it sells x cameras, its total revenue is given by

$$R(x) = 175x - 0.4x^2.$$

Total cost is the total cost of producing x cameras. Hadley Electronics determines that the total cost of producing x cameras is given by

$$C(x) = 5000 + 0.6x^2.$$

The total profit $P(x)$ is

(Total Revenue) − (Total Cost) $= R(x) - C(x)$.

a) Find a polynomial function for total profit.
b) What is the total profit on the production and sale of 75 cameras?
c) What is the total profit on the production and sale of 120 cameras?

TW **102.** Does replacing each occurrence of the variable x in $4x^7 - 6x^3 + 2x$ with its opposite result in the opposite of the polynomial? Why or why not?

Try Exercise Answers: Section 5.4

9. $x^2 - 5x - 1$ **13.** $6m^3 - 3m - 8$ **23.** $-3x^4 + 3x^2 + 4x$ **27.** $-(-t^3 + 4t^2 - 9); t^3 - 4t^2 + 9$ **33.** $-3a^4 + 5a^2 - 9$ **39.** $-t^2 - 8t + 7$ **53.** $3x + 5$ **55.** $11x^4 + 12x^3 - x^2$ **57.** (a) $5x^2 + 4x$; (b) 145; 273 **69.** $(x^2 - 12)$ ft^2

5.5 Multiplication of Polynomials

- Multiplying Monomials
- Multiplying a Monomial and a Polynomial
- Multiplying Any Two Polynomials

We now multiply polynomials using techniques based on the distributive, associative, and commutative laws and the rules for exponents.

MULTIPLYING MONOMIALS

Consider $(3x)(4x)$. We multiply as follows:

$$(3x)(4x) = 3 \cdot x \cdot 4 \cdot x \quad \text{Using an associative law}$$
$$(3x)(4x) = 3 \cdot 4 \cdot x \cdot x \quad \text{Using a commutative law}$$
$$(3x)(4x) = (3 \cdot 4) \cdot x \cdot x \quad \text{Using an associative law}$$
$$(3x)(4x) = 12x^2.$$

To Multiply Monomials To find an equivalent expression for the product of two monomials, multiply the coefficients and then multiply the variables using the product rule for exponents.

Student Notes

Remember that when we compute $(3 \cdot 5)(2 \cdot 4)$, each factor is used only once, even if we change the order:

$$\begin{aligned}(3 \cdot 5)(2 \cdot 4) &= (3 \cdot 2)(5 \cdot 4)\\ &= 6 \cdot 20 = 120.\end{aligned}$$

In the same way,

$$\begin{aligned}(3 \cdot x)(2 \cdot x) &= (3 \cdot 2)(x \cdot x)\\ &= 6x^2.\end{aligned}$$

Some students mistakenly "reuse" a factor.

EXAMPLE 1 Multiply to form an equivalent expression.

a) $(5x)(6x)$
b) $(3a)(-a)$
c) $(-7x^5)(4x^3)$

SOLUTION

a) $(5x)(6x) = (5 \cdot 6)(x \cdot x)$ **Multiplying the coefficients; multiplying the variables**
$= 30x^2$ **Simplifying**

b) $(3a)(-a) = (3a)(-1a)$ **Writing $-a$ as $-1a$ can ease calculations.**
$= (3)(-1)(a \cdot a)$ **Using an associative law and a commutative law**
$= -3a^2$

c) $(-7x^5)(4x^3) = (-7 \cdot 4)(x^5 \cdot x^3)$
$= -28x^{5+3}$
$= -28x^8$ } **Using the product rule for exponents**

Try Exercise 7.

After some practice, you can try writing only the answer.

MULTIPLYING A MONOMIAL AND A POLYNOMIAL

To find an equivalent expression for the product of a monomial, such as $5x$, and a polynomial, such as $2x^2 - 3x + 4$, we use the distributive law.

EXAMPLE 2 Multiply: **(a)** $x(x + 3)$; **(b)** $5x(2x^2 - 3x + 4)$.

SOLUTION

a) $x(x + 3) = x \cdot x + x \cdot 3$ **Using the distributive law**
$= x^2 + 3x$

We can check this product using a table by letting $y_1 = x(x + 3)$ and $y_2 = x^2 + 3x$. The table of values for y_1 and y_2 shown at left indicates that the answer found is correct.

X	Y1	Y2
0	0	0
1	4	4
2	10	10
3	18	18
4	28	28
5	40	40
6	54	54
X = 0		

b) $5x(2x^2 - 3x + 4) = (5x)(2x^2) - (5x)(3x) + (5x)(4)$ **Using the distributive law**
$= 10x^3 - 15x^2 + 20x$ **Performing the three multiplications**

Try Exercise 23.

The product in Example 2(a) can be visualized as the area of a rectangle with width x and length $x + 3$.

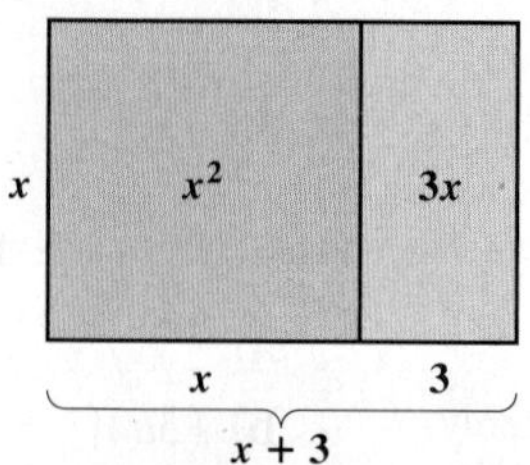

Note that the total area can be expressed as $x(x + 3)$ or, by adding the two smaller areas, $x^2 + 3x$.

The Product of a Monomial and a Polynomial To multiply a monomial and a polynomial, multiply each term of the polynomial by the monomial.

Try to do this mentally, when possible. Remember that we multiply coefficients and, when the bases match, add exponents.

EXAMPLE 3 Multiply: $2x^2(x^3 - 7x^2 + 10x - 4)$.

SOLUTION

Think: $2x^2 \cdot x^3 - 2x^2 \cdot 7x^2 + 2x^2 \cdot 10x - 2x^2 \cdot 4$

$$2x^2(x^3 - 7x^2 + 10x - 4) = 2x^5 - 14x^4 + 20x^3 - 8x^2$$

Try Exercise 31.

MULTIPLYING ANY TWO POLYNOMIALS

Before considering the product of *any* two polynomials, let's look at the product of two binomials. To multiply, we again begin by using the distributive law. This time, however, it is a *binomial* rather than a monomial that is being distributed.

EXAMPLE 4 Multiply each pair of binomials.

a) $x + 5$ and $x + 4$

b) $4x - 3$ and $x - 2$

SOLUTION

a) $(x + 5)(x + 4) = (x + 5)x + (x + 5)4$ **Using the distributive law**

$= x(x + 5) + 4(x + 5)$ **Using the commutative law for multiplication**

$= x \cdot x + x \cdot 5 + 4 \cdot x + 4 \cdot 5$ **Using the distributive law (twice)**

$= x^2 + 5x + 4x + 20$ **Multiplying the monomials**

$= x^2 + 9x + 20$ **Combining like terms**

To visualize this product, consider a rectangle of length $x + 5$ and width $x + 4$. The total area can be expressed as $(x + 5)(x + 4)$ or, by adding the four smaller areas, $x^2 + 5x + 4x + 20$.

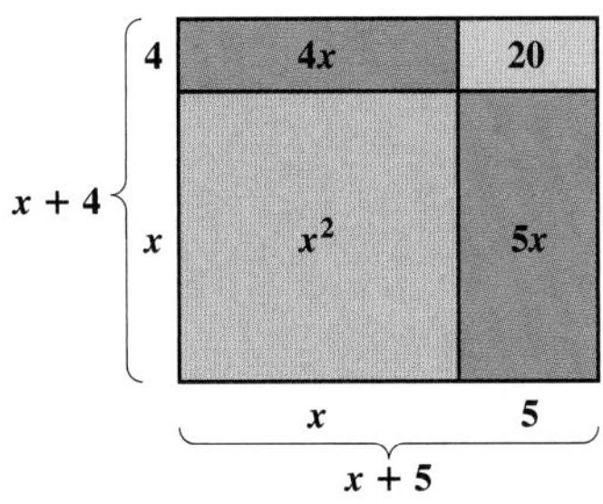

b) $(4x - 3)(x - 2) = (4x - 3)x - (4x - 3)2$ **Using the distributive law**

$= x(4x - 3) - 2(4x - 3)$ **Using the commutative law for multiplication. This step is often omitted.**

$= x \cdot 4x - x \cdot 3 - 2 \cdot 4x - 2(-3)$ **Using the distributive law (twice)**

$= 4x^2 - 3x - 8x + 6$ **Multiplying the monomials**

$= 4x^2 - 11x + 6$ **Combining like terms**

Try Exercise 35.

Let's consider the product of a binomial and a trinomial. Again we make repeated use of the distributive law.

STUDY TIP

Take a Peek Ahead

Try to at least glance at the next section of material that will be covered in class. This will make it easier to concentrate on your instructor's lecture instead of trying to write everything down.

EXAMPLE 5 Multiply: $(x^2 + 2x - 3)(x + 4)$.

SOLUTION

$(x^2 + 2x - 3)(x + 4)$

$= (x^2 + 2x - 3)x + (x^2 + 2x - 3)4$ **Using the distributive law**

$= x(x^2 + 2x - 3) + 4(x^2 + 2x - 3)$ **Using the commutative law**

$= x \cdot x^2 + x \cdot 2x - x \cdot 3 + 4 \cdot x^2 + 4 \cdot 2x - 4 \cdot 3$ **Using the distributive law (twice)**

$= x^3 + 2x^2 - 3x + 4x^2 + 8x - 12$ **Multiplying the monomials**

$= x^3 + 6x^2 + 5x - 12$ **Combining like terms**

Try Exercise 57.

Perhaps you have discovered the following in the preceding examples.

The Product of Two Polynomials To multiply two polynomials P and Q, select one of the polynomials, say P. Then multiply each term of P by every term of Q and combine like terms.

To use columns for long multiplication, multiply each term in the top row by every term in the bottom row. We write like terms in columns, and then add the results. Such multiplication is like multiplying with whole numbers:

3 2 1	300 + 20 + 1	
× 1 2	× 10 + 2	
6 4 2	600 + 40 + 2	**Multiplying the top row by 2**
3 2 1	3000 + 200 + 10	**Multiplying the top row by 10**
3 8 5 2	3000 + 800 + 50 + 2	**Adding**

EXAMPLE 6 Multiply: $(5x^4 - 2x^2 + 3x)(x^2 + 2x)$.

SOLUTION

$$\begin{array}{r} 5x^4 - 2x^2 + 3x \\ x^2 + 2x \\ \hline 10x^5 \quad - 4x^3 + 6x^2 \\ 5x^6 \quad - 2x^4 + 3x^3 \quad \\ \hline 5x^6 + 10x^5 - 2x^4 - x^3 + 6x^2 \end{array}$$

Note that each polynomial is written in descending order.

Multiplying the top row by $2x$

Multiplying the top row by x^2

Combining like terms

Line up like terms in columns.

Try Exercise 55.

Sometimes we multiply horizontally, while still aligning like terms as we write the product.

EXAMPLE 7 Multiply: $(2x^3 + 3x^2 - 4x + 6)(3x + 5)$.

SOLUTION

$$\begin{aligned}(2x^3 + 3x^2 - 4x + 6)(3x + 5) &= \overbrace{6x^4 + 9x^3 - 12x^2 + 18x}^{\text{Multiplying by } 3x} \\ &\quad \underbrace{+ 10x^3 + 15x^2 - 20x + 30}_{\text{Multiplying by } 5} \\ &= 6x^4 + 19x^3 + 3x^2 - 2x + 30\end{aligned}$$

X	Y1	Y2
−3	36	36
−2	−10	−10
−1	22	22
0	30	30
1	56	56
2	286	286
3	1050	1050
X = −3		

To check the multiplication by evaluating or using a table of values, we must compare values for at least 5 x-values, because the degree is 4. The table at left shows values of

$$y_1 = (2x^3 + 3x^2 - 4x + 6)(3x + 5) \quad \text{and}$$
$$y_2 = 6x^4 + 19x^3 + 3x^2 - 2x + 30$$

for 7 x-values. Thus the multiplication checks.

Try Exercise 59.

5.5 Exercise Set

FOR EXTRA HELP MyMathLab   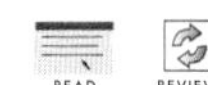

Concept Reinforcement *In each of Exercises 1–6, match the product with the correct result from the column on the right. Choices may be used more than once.*

1. ___ $3x^2 \cdot 2x^4$ — a) $6x^8$
2. ___ $3x^8 + 5x^8$ — b) $8x^6$
3. ___ $4x^3 \cdot 2x^5$ — c) $6x^6$
4. ___ $3x^5 \cdot 2x^3$ — d) $8x^8$
5. ___ $4x^6 + 2x^6$
6. ___ $4x^4 \cdot 2x^2$

Multiply.

7. $(4x^3)9$
8. $(5x^4)6$
9. $(-x^2)(-x)$
10. $(-x^3)(x^4)$
11. $(-x^6)(x^2)$
12. $(-x^3)(-x^5)$
13. $(7t^5)(4t^3)$
14. $(10a^2)(3a^2)$
15. $(-0.1x^6)(0.2x^4)$
16. $(0.3x^3)(-0.4x^6)$
17. $\left(-\frac{1}{5}x^3\right)\left(-\frac{1}{3}x\right)$
18. $\left(-\frac{1}{4}x^4\right)\left(\frac{1}{5}x^8\right)$
19. $(-1)(-19t^2)$
20. $(-5n^3)(-1)$
21. $(-4y^5)(6y^2)(-3y^3)$
22. $7x^2(-2x^3)(2x^6)$
23. $4x(x + 1)$
24. $3x(x - 2)$
25. $(a - 7)4a$
26. $(a + 9)3a$
27. $x^2(x^3 + 1)$
28. $-2x^3(x^2 - 1)$
29. $-3n(2n^2 - 8n + 1)$
30. $4n(3n^3 - 4n^2 - 5n + 10)$
31. $-5t^2(3t^3 + 6t)$
32. $-7t^2(2t^2 + t)$
33. $\frac{2}{3}a^4\left(6a^5 - 12a^3 - \frac{5}{8}\right)$
34. $\frac{3}{4}t^5\left(8t^6 - 12t^4 + \frac{12}{7}\right)$
35. $(x + 6)(x + 1)$
36. $(x + 5)(x + 2)$
37. $(x + 5)(x - 2)$
38. $(x + 6)(x - 2)$
39. $(a - 0.6)(a - 0.7)$
40. $(a - 0.4)(a - 0.8)$
41. $(x + 3)(x - 3)$
42. $(x + 6)(x - 6)$
43. $(5 - x)(5 - 2x)$
44. $(3 + x)(6 + 2x)$
45. $\left(t + \frac{3}{2}\right)\left(t + \frac{4}{3}\right)$
46. $\left(a - \frac{2}{5}\right)\left(a + \frac{5}{2}\right)$
47. $\left(\frac{1}{4}a + 2\right)\left(\frac{3}{4}a - 1\right)$
48. $\left(\frac{2}{5}t - 1\right)\left(\frac{3}{5}t + 1\right)$

Draw and label rectangles similar to those illustrating Examples 2 and 4 to illustrate each product.

49. $x(x + 5)$
50. $x(x + 2)$
51. $(x + 1)(x + 2)$
52. $(x + 3)(x + 1)$
53. $(x + 5)(x + 3)$
54. $(x + 4)(x + 6)$

Multiply and check.

55. $(x^2 - x + 5)(x + 1)$
56. $(x^2 + x - 7)(x + 2)$
57. $(2a + 5)(a^2 - 3a + 2)$
58. $(3t + 4)(t^2 - 5t + 1)$
59. $(y^2 - 7)(2y^3 + y + 1)$
60. $(a^2 + 4)(5a^3 - 3a - 1)$
61. Aha! $(3x + 2)(5x + 4x + 7)$
62. $(4x - 5x - 3)(1 + 2x^2)$
63. $(x^2 - 3x + 2)(x^2 + x + 1)$
64. $(x^2 + 5x - 1)(x^2 - x + 3)$
65. $(2t^2 - 5t - 4)(3t^2 - t + 1)$
66. $(5t^2 - t + 1)(2t^2 + t - 3)$
67. $(x + 1)(x^3 + 7x^2 + 5x + 4)$
68. $(x + 2)(x^3 + 5x^2 + 9x + 3)$
69. TW Is it possible to understand polynomial multiplication without understanding the distributive law? Why or why not?
70. TW The polynomials

$$(a + b + c + d) \quad \text{and} \quad (r + s + m + p)$$

are multiplied. Without performing the multiplication, determine how many terms the product will contain. Provide a justification for your answer.

SKILL REVIEW

Review simplifying expressions using the rules for order of operations (Section 1.8).

Simplify. [1.8]

71. $(9 - 3)(9 + 3) + 3^2 - 9^2$
72. $(7 + 2)(7 - 2) - 7^2 + 2^2$
73. $5 + \dfrac{7 + 4 + 2 \cdot 5}{7}$
74. $11 - \dfrac{2 + 6 \cdot 3 + 4}{6}$

75. $(4 + 3 \cdot 5 + 5) \div 3 \cdot 4$

76. $(2 + 2 \cdot 7 + 4) \div 2 \cdot 5$

SYNTHESIS

TW **77.** Under what conditions will the product of two binomials be a trinomial?

TW **78.** How can the figure below be used to show that $(x + 3)^2 \neq x^2 + 9$?

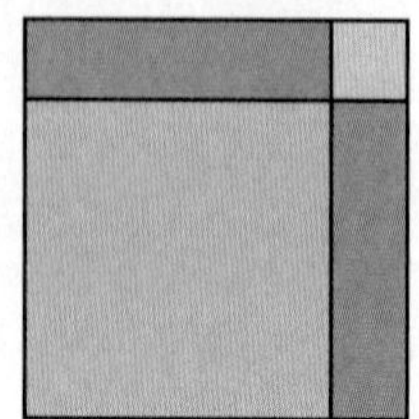

Find a polynomial for the shaded area of each figure.

79.

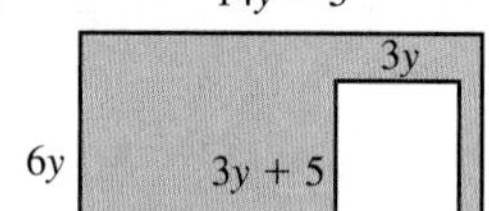

80.

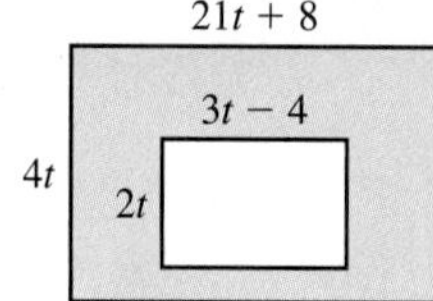

For each figure, determine what the missing number must be in order for the figure to have the given area.

81. Area is $x^2 + 7x + 10$

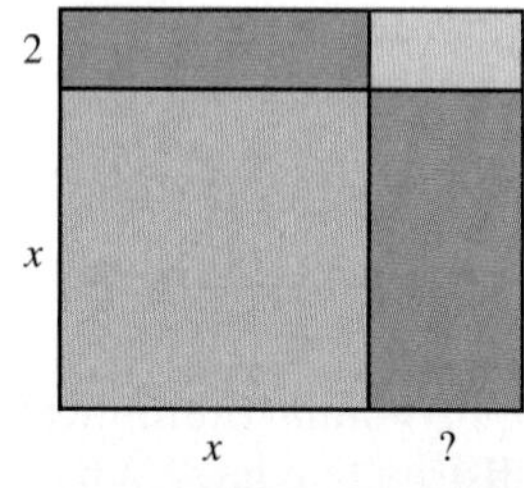

82. Area is $x^2 + 8x + 15$

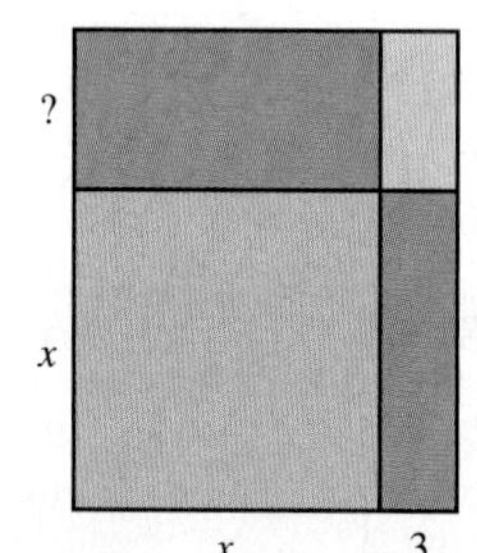

83. A box with a square bottom is to be made from a 12-in.-square piece of cardboard. Squares with side x are cut out of the corners and the sides are folded up. Find the polynomials for the volume and the outside surface area of the box.

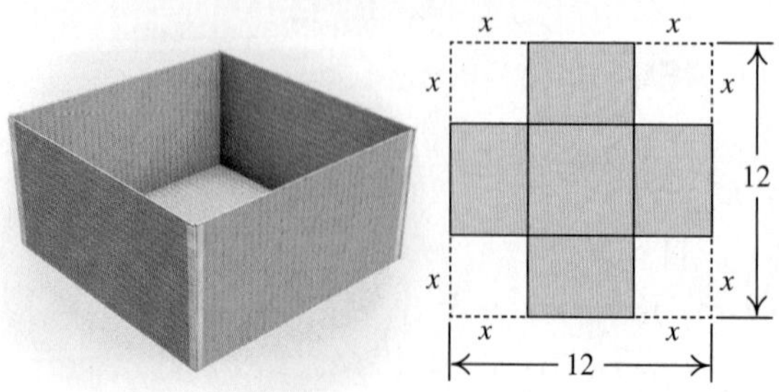

84. A rectangular garden is twice as long as it is wide and is surrounded by a sidewalk that is 4 ft wide (see the figure below). The area of the sidewalk is 256 ft^2. Find the dimensions of the garden.

85. An open wooden box is a cube with side x centimeters. The box, including its bottom, is made of wood that is 1 cm thick. Find a polynomial for the interior volume of the cube.

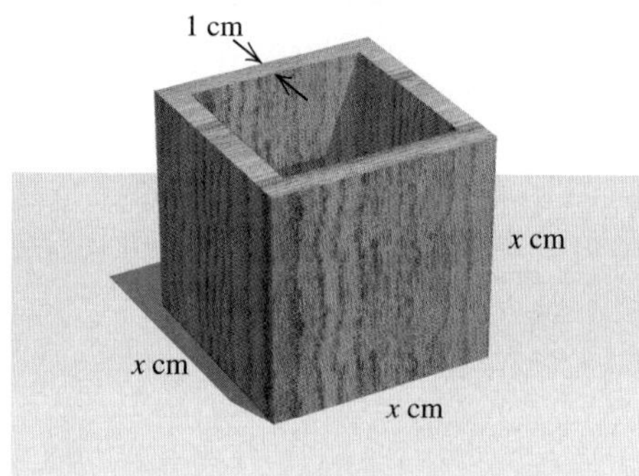

86. A side of a cube is $(x + 2)$ cm long. Find a polynomial for the volume of the cube.

87. Find a polynomial for the volume of the solid shown below.

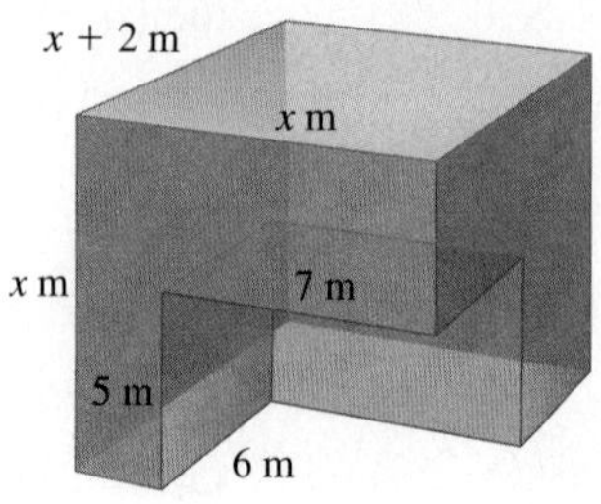

Compute and simplify.

88. $(x + 3)(x + 6) + (x + 3)(x + 6)$

Aha! **89.** $(x - 2)(x - 7) - (x - 7)(x - 2)$

90. $(x + 5)^2 - (x - 3)^2$

91. $(x + 2)(x + 4)(x - 5)$

92. $(x - 3)^3$

Aha! **93.** Extend the pattern and simplify

$(x - a)(x - b)(x - c)(x - d) \cdots (x - z).$

Try Exercise Answers: Section 5.5

7. $36x^3$ **23.** $4x^2 + 4x$ **31.** $-15t^5 - 30t^3$ **35.** $x^2 + 7x + 6$ **55.** $x^3 + 4x + 5$ **57.** $2a^3 - a^2 - 11a + 10$ **59.** $2y^5 - 13y^3 + y^2 - 7y - 7$

5.6 Special Products

- Products of Two Binomials
- Multiplying Sums and Differences of Two Terms
- Squaring Binomials
- Multiplications of Various Types

Student Notes

The special products discussed in this section are developed by recognizing patterns.

Looking for patterns often aids in understanding, remembering, and applying the material you are studying.

Patterns that you may have observed in the products of two binomials allow us to compute such products quickly.

PRODUCTS OF TWO BINOMIALS

In Section 5.5, we found the product $(x + 5)(x + 4)$ by using the distributive law a total of three times (see p. 374). Note that each term in $x + 5$ is multiplied by each term in $x + 4$. To shorten our work, we can go right to this step:

$$(x + 5)(x + 4) = x \cdot x + x \cdot 4 + 5 \cdot x + 5 \cdot 4$$
$$(x + 5)(x + 4) = x^2 + 4x + 5x + 20$$
$$(x + 5)(x + 4) = x^2 + 9x + 20.$$

Note that $x \cdot x$ is found by multiplying the *First* terms of each binomial, $x \cdot 4$ is found by multiplying the *Outer* terms of the two binomials, $5 \cdot x$ is the product of the *Inner* terms of the two binomials, and $5 \cdot 4$ is the product of the *Last* terms of each binomial:

First terms, Outer terms, Inner terms, Last terms

$$(x + 5)(x + 4) = x \cdot x + 4 \cdot x + 5 \cdot x + 5 \cdot 4.$$

To remember this shortcut for multiplying, we use the initials **FOIL.**

The FOIL Method To multiply two binomials, $A + B$ and $C + D$, multiply the First terms AC, the Outer terms AD, the Inner terms BC, and then the Last terms BD. Then combine like terms, if possible.

$$(A + B)(C + D) = AC + AD + BC + BD$$

1. Multiply First terms: AC.
2. Multiply Outer terms: AD.
3. Multiply Inner terms: BC.
4. Multiply Last terms: BD.

↓
FOIL

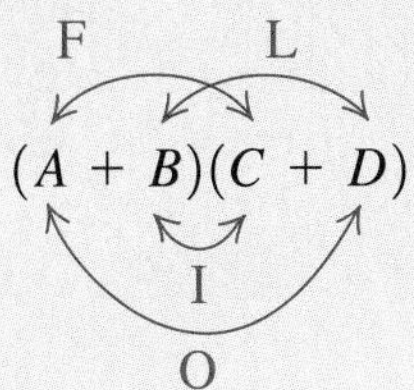

Because addition is commutative, the individual multiplications can be performed in any order. Both FLOI and FIOL yield the same result as FOIL, but FOIL is most easily remembered and most widely used.

EXAMPLE 1 Form an equivalent expression by multiplying: $(x + 8)(x^2 + 5)$.

SOLUTION

F L F O I L

$$(x + 8)(x^2 + 5) = x^3 + 5x + 8x^2 + 40 \quad \text{There are no like terms.}$$
$$= x^3 + 8x^2 + 5x + 40 \quad \text{Writing in descending order}$$

I O

Try Exercise 5.

After multiplying, remember to combine any like terms.

EXAMPLE 2 Multiply to form an equivalent expression.

a) $(x + 7)(x + 4)$ **b)** $(y + 3)(y - 2)$
c) $(4t^3 + 5t)(3t^2 - 2)$ **d)** $(3 - 4x)(7 - 5x^3)$

$y_1 = (x + 7)(x + 4)$,
$y_2 = x^2 + 11x + 28$,
$y_3 = y_1 - y_2$

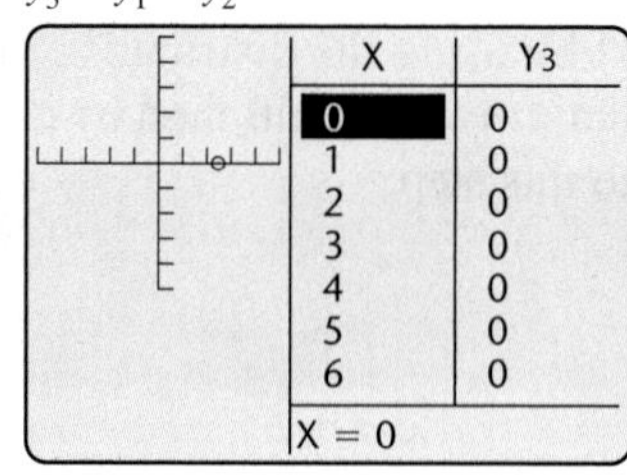

SOLUTION

a) $(x + 7)(x + 4) = x^2 + 4x + 7x + 28$ **Using FOIL**
$= x^2 + 11x + 28$ **Combining like terms**

We can check multiplication by using a table of values or by graphing. We let $y_1 = (x + 7)(x + 4)$, $y_2 = x^2 + 11x + 28$, and $y_3 = y_1 - y_2$. The graph and table at left show that the difference of the original expression and the resulting product is 0.

b) $(y + 3)(y - 2) = y^2 - 2y + 3y - 6$
$= y^2 + y - 6$

c) $(4t^3 + 5t)(3t^2 - 2) = 12t^5 - 8t^3 + 15t^3 - 10t$
Remember to add exponents when multiplying terms with the same base.
$= 12t^5 + 7t^3 - 10t$

d) $(3 - 4x)(7 - 5x^3) = 21 - 15x^3 - 28x + 20x^4$
$= 21 - 28x - 15x^3 + 20x^4$
In general, if the original binomials are in *ascending* order, we also write the answer that way.

Try Exercise 9.

MULTIPLYING SUMS AND DIFFERENCES OF TWO TERMS

Consider the product of the sum and the difference of the same two terms, such as

$$(x + 5)(x - 5).$$

Since this is the product of two binomials, we can use FOIL. In doing so, we find that the "outer" and "inner" products are opposites:

$$(x + 5)(x - 5) = x^2 - 5x + 5x - 25 \quad \text{The "outer" and "inner" terms "drop out." Their sum is zero.}$$
$$= x^2 - 25.$$

Interactive Discovery

Determine whether each of the following is an identity.

1. $(x + 3)(x - 3) = x^2 - 9$
2. $(x + 3)(x - 3) = x^2 + 9$
3. $(x + 3)(x - 3) = x^2 - 6x + 9$
4. $(x + 3)(x - 3) = x^2 + 6x + 9$

The pattern you may have observed is true in general:

$$\begin{aligned} (A + B)(A - B) &= \overset{\text{F}}{A^2} - \overset{\text{O}}{AB} + \overset{\text{I}}{AB} - \overset{\text{L}}{B^2} \\ &= A^2 - B^2. \qquad -AB + AB = 0 \end{aligned}$$

The Product of a Sum and a Difference The product of the sum and the difference of the same two terms is the square of the first term minus the square of the second term:

$$(A + B)(A - B) = \underbrace{A^2 - B^2}.$$

This is called a *difference of squares.*

STUDY TIP

Two Books Are Better Than One

Many students find it helpful to use a second book as a reference when studying. Perhaps you or a friend own a text from a previous math course that can serve as a resource. Often professors have older texts that they will happily give away. Library book sales and thrift shops can also be excellent sources for extra books. Saving your text when you finish a math course can provide you with an excellent aid for your next course.

EXAMPLE 3 Multiply.

a) $(x + 4)(x - 4)$

b) $(5 + 2w)(5 - 2w)$

c) $(3a^4 - 5)(3a^4 + 5)$

SOLUTION

$$(A + B)\,(A - B) = A^2 - B^2$$

a) $(x + 4)(x - 4) = x^2 - 4^2$ Saying the words can help: "The square of the first term, x^2, minus the square of the second, 4^2"

$= x^2 - 16$ Simplifying

b) $(5 + 2w)(5 - 2w) = 5^2 - (2w)^2$

$= 25 - 4w^2$ Squaring both 5 and $2w$

c) $(3a^4 - 5)(3a^4 + 5) = (3a^4)^2 - 5^2$

$= 9a^8 - 25$ Remember to multiply exponents when raising a power to a power.

Try Exercise 41.

SQUARING BINOMIALS

Consider the square of a binomial, such as $(x + 3)^2$. This is the same as $(x + 3)(x + 3)$. Since this is the product of two binomials, we can use FOIL. But again, this product occurs so often that a faster method has been developed. Look for a pattern in the following.

Interactive Discovery

Determine whether each of the following is an identity.

1. $(x + 3)^2 = x^2 + 9$
2. $(x + 3)^2 = x^2 + 6x + 9$
3. $(x - 3)^2 = x^2 - 6x - 9$
4. $(x - 3)^2 = x^2 - 9$
5. $(x - 3)^2 = x^2 - 6x + 9$

A fast method for squaring binomials can be developed using FOIL:

$$\begin{aligned}(A + B)^2 &= (A + B)(A + B)\\ &= A^2 + AB + AB + B^2 \quad \text{Note that } AB \text{ occurs twice.}\\ &= A^2 + 2AB + B^2;\end{aligned}$$

$$\begin{aligned}(A - B)^2 &= (A - B)(A - B)\\ &= A^2 - AB - AB + B^2 \quad \text{Note that } -AB \text{ occurs twice.}\\ &= A^2 - 2AB + B^2.\end{aligned}$$

The Square of a Binomial The square of a binomial is the square of the first term, plus twice the product of the two terms, plus the square of the last term:

$$\left.\begin{aligned}(A + B)^2 &= A^2 + 2AB + B^2;\\ (A - B)^2 &= A^2 - 2AB + B^2.\end{aligned}\right\} \quad \text{These are called } \textit{perfect-square trinomials.}^*$$

EXAMPLE 4 Write an equivalent expression for the square of a binomial.

a) $(x + 7)^2$
b) $(t - 5)^2$
c) $(3a + 0.4)^2$
d) $(5x - 3x^4)^2$

SOLUTION

$$\begin{aligned}(A + B)^2 &= A^2 + 2 \cdot A \cdot B + B^2\\ &\;\;\downarrow\\ \text{a) } (x + 7)^2 &= x^2 + 2 \cdot x \cdot 7 + 7^2\\ &= x^2 + 14x + 49\end{aligned}$$

Saying the words can help: "The square of the first term, x^2, plus twice the product of the terms, $2 \cdot 7x$, plus the square of the second term, 7^2"

*Another name for these is *trinomial squares*.

b) $(t - 5)^2 = t^2 - 2 \cdot t \cdot 5 + 5^2$
$= t^2 - 10t + 25$

c) $(3a + 0.4)^2 = (3a)^2 + 2 \cdot 3a \cdot 0.4 + 0.4^2$
$= 9a^2 + 2.4a + 0.16$

d) $(5x - 3x^4)^2 = (5x)^2 - 2 \cdot 5x \cdot 3x^4 + (3x^4)^2$
$= 25x^2 - 30x^5 + 9x^8$ **Using the rules for exponents**

Try Exercise 51.

CAUTION! Although the square of a product is the product of the squares, the square of a sum is *not* the sum of the squares. That is, $(AB)^2 = A^2B^2$, but

The term 2AB is missing.

$$(A + B)^2 \neq A^2 + B^2.$$

To confirm this inequality, note that

$$(7 + 5)^2 = 12^2 = 144,$$

whereas

$$7^2 + 5^2 = 49 + 25 = 74, \quad \text{and} \quad 74 \neq 144.$$

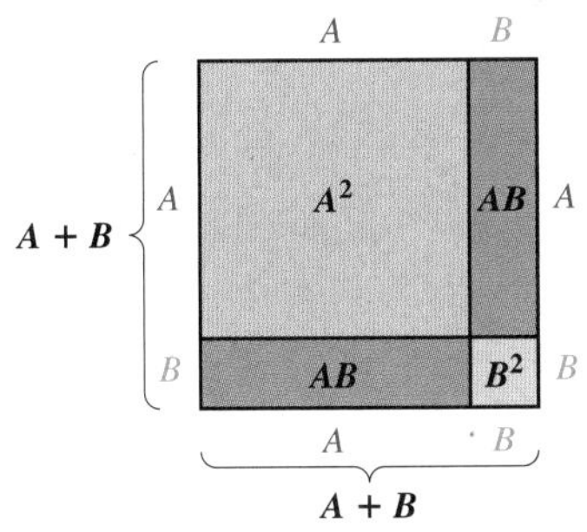

Note that the areas A^2 and B^2 do not fill the area $(A + B)^2$.

Geometrically, $(A + B)^2$ can be viewed as the area of a square with sides of length $A + B$:

$$(A + B)(A + B) = (A + B)^2.$$

This is equal to the sum of the areas of the four smaller regions:

$$A^2 + AB + AB + B^2 = A^2 + 2AB + B^2.$$

Thus,

$$(A + B)^2 = A^2 + 2AB + B^2.$$

MULTIPLICATIONS OF VARIOUS TYPES

Recognizing patterns often helps when new problems are encountered. To simplify a new multiplication problem, always examine what type of product it is so that the best method for finding that product can be used. To do this, ask yourself questions similar to the following.

Multiplying Two Polynomials

1. Is the multiplication the product of a monomial and a polynomial? If so, multiply each term of the polynomial by the monomial.
2. Is the multiplication the product of two binomials? If so:
   a) Is it the product of the sum and the difference of the *same* two terms? If so, use the pattern
   $$(A + B)(A - B) = A^2 - B^2.$$
   b) Is the product the square of a binomial? If so, use the pattern
   $$(A + B)(A + B) = (A + B)^2 = A^2 + 2AB + B^2,$$
   or
   $$(A - B)(A - B) = (A - B)^2 = A^2 - 2AB + B^2.$$
   c) If neither (a) nor (b) applies, use FOIL.
3. Is the multiplication the product of two polynomials other than those above? If so, multiply each term of one by every term of the other. Use columns if you wish.

EXAMPLE 5 Multiply.

a) $(x + 3)(x - 3)$ b) $(t + 7)(t - 5)$
c) $(x + 7)(x + 7)$ d) $2x^3(9x^2 + x - 7)$
e) $(p + 3)(p^2 + 2p - 1)$ f) $\left(3x - \frac{1}{4}\right)^2$

SOLUTION

a) $(x + 3)(x - 3) = x^2 - 9$ This is the product of the sum and the difference of the same two terms.

b) $(t + 7)(t - 5) = t^2 - 5t + 7t - 35$ Using FOIL
$= t^2 + 2t - 35$

c) $(x + 7)(x + 7) = x^2 + 14x + 49$ This is the square of a binomial, $(x + 7)^2$.

d) $2x^3(9x^2 + x - 7) = 18x^5 + 2x^4 - 14x^3$ Multiplying each term of the trinomial by the monomial

e) We multiply each term of $p^2 + 2p - 1$ by every term of $p + 3$:

$(p + 3)(p^2 + 2p - 1) = p^3 + 2p^2 - p$ Multiplying by p
$+ 3p^2 + 6p - 3$ Multiplying by 3
$= p^3 + 5p^2 + 5p - 3$

f) $\left(3x - \frac{1}{4}\right)^2 = 9x^2 - 2(3x)\left(\frac{1}{4}\right) + \frac{1}{16}$ Squaring a binomial
$= 9x^2 - \frac{3}{2}x + \frac{1}{16}$

Try Exercise 69.

5.6 Exercise Set

FOR EXTRA HELP MyMathLab MathXL PRACTICE  WATCH  DOWNLOAD 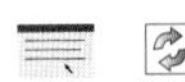 READ REVIEW

Concept Reinforcement *Identify each statement as either true or false.*

1. FOIL is simply a memory device for finding the product of two binomials.
2. Once FOIL is used, it is always possible to combine like terms.
3. The square of a binomial cannot be found using FOIL.
4. The square of $A + B$ is not the sum of the squares of A and B.

Multiply.

5. $(x + 3)(x^2 + 5)$
6. $(x^2 - 3)(x - 1)$
7. $(t^4 - 2)(t + 7)$
8. $(n^3 + 8)(n - 4)$
9. $(y + 2)(y - 3)$
10. $(a + 2)(a + 2)$
11. $(3x + 2)(3x + 5)$
12. $(4x + 1)(2x + 7)$
13. $(5x - 6)(x + 2)$
14. $(4x - 5)(4x + 5)$
15. $(1 + 3t)(2 - 3t)$
16. $(7 - a)(2 + 3a)$
17. $(x^2 + 3)(x^2 - 7)$
18. $(x^2 + 2)(x^2 - 8)$
19. $\left(p - \frac{1}{4}\right)\left(p + \frac{1}{4}\right)$
20. $\left(q + \frac{3}{4}\right)\left(q + \frac{3}{4}\right)$
21. $(x - 0.1)(x - 0.1)$
22. $(x + 0.3)(x - 0.3)$
23. $(-3n + 2)(n + 7)$
24. $(-m + 5)(2m - 9)$
25. $(a + 9)(a + 9)$
26. $(2y + 7)(2y + 7)$
27. $(1 - 3t)(1 + 5t^2)$
28. $(1 + 2t)(1 - 3t^2)$
29. $(x^2 + 3)(x^3 - 1)$
30. $(x^4 - 3)(2x + 1)$
31. $(3x^2 - 2)(x^4 - 2)$
32. $(x^{10} + 3)(x^{10} - 3)$
33. $(2t^3 + 5)(2t^3 + 3)$
34. $(5t^2 + 1)(2t^2 + 3)$
35. $(8x^3 + 5)(x^2 + 2)$
36. $(4 - 2x)(5 - 2x^2)$
37. $(10x^2 + 3)(10x^2 - 3)$
38. $(7x - 2)(2x - 7)$

Multiply. Try to recognize the type of product before multiplying.

39. $(x + 7)(x - 7)$
40. $(x + 1)(x - 1)$
41. $(2x + 1)(2x - 1)$
42. $(4n + 7)(4n - 7)$
43. $(5m^2 - 9)(5m^2 + 9)$
44. $(3x^4 + 2)(3x^4 - 2)$
45. $(6a^3 + 1)(6a^3 - 1)$
46. $(t^2 - 0.2)(t^2 + 0.2)$
47. $(x^4 + 0.1)(x^4 - 0.1)$
48. $(t^3 + 4)(t^3 - 4)$
49. $\left(t - \frac{3}{4}\right)\left(t + \frac{3}{4}\right)$
50. $\left(m - \frac{2}{3}\right)\left(m + \frac{2}{3}\right)$
51. $(x + 2)^2$
52. $(2x - 1)^2$
53. $(3x^5 - 1)^2$
54. $(4x^3 + 1)^2$
55. $\left(a - \frac{2}{5}\right)^2$
56. $\left(t - \frac{1}{5}\right)^2$
57. $(x^2 + 1)(x^2 - x + 2)$
58. $(a + 3)(a^2 + 6a + 1)$
59. $(2 - 3x^4)^2$
60. $(5 - 2t^3)^2$
61. $(5 + 6t^2)^2$
62. $(3p^2 - p)^2$
63. $(7x - 0.3)^2$
64. $(4a - 0.6)^2$
65. $5a^3(2a^2 - 1)$
66. $9x^3(2x^2 - 5)$
67. $(a - 3)(a^2 + 2a - 4)$
68. $(x^2 - 5)(x^2 + x - 1)$
69. $(7 - 3x^4)(7 - 3x^4)$
70. $(x - 4x^3)^2$
71. $-4x(x^2 + 6x - 3)$
72. $8x(-x^5 + 6x^2 + 9)$
73. $(-t^3 + 1)^2$
74. $(-x^2 + 1)^2$
75. $3t^2(5t^3 - t^2 + t)$

76. $-5x^3(x^2 + 8x - 9)$

77. $(6x^4 - 3x)^2$

78. $(8a^3 + 5)(8a^3 - 5)$

79. $(9a + 0.4)(2a^3 + 0.5)$

80. $(2a - 0.7)(8a^3 - 0.5)$

81. $\left(\frac{1}{5} - 6x^4\right)\left(\frac{1}{5} + 6x^4\right)$

82. $(3 + \frac{1}{2}t^5)(3 + \frac{1}{2}t^5)$

83. $(a + 1)(a^2 - a + 1)$

84. $(x - 5)(x^2 + 5x + 25)$

Find the total area of all shaded rectangles.

85.

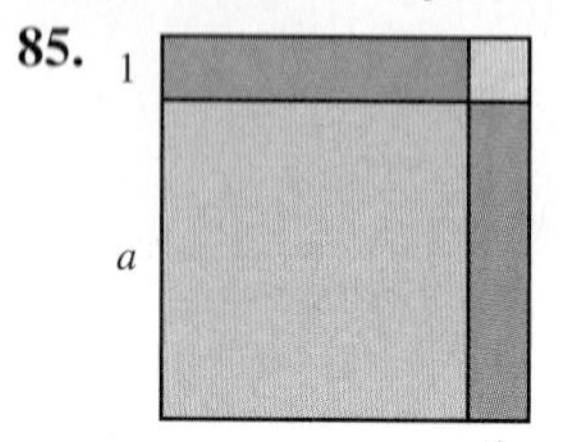

86.

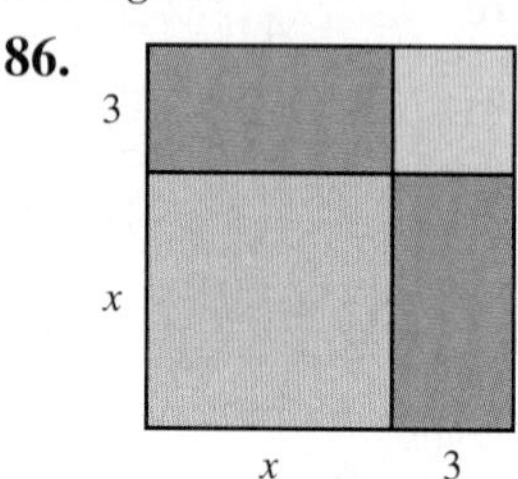

87.

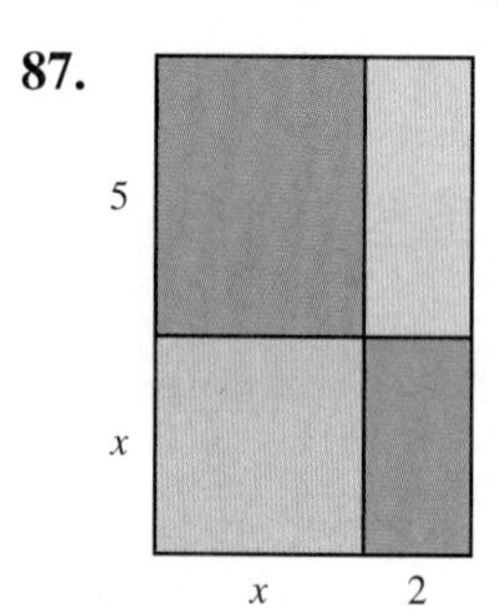

88.

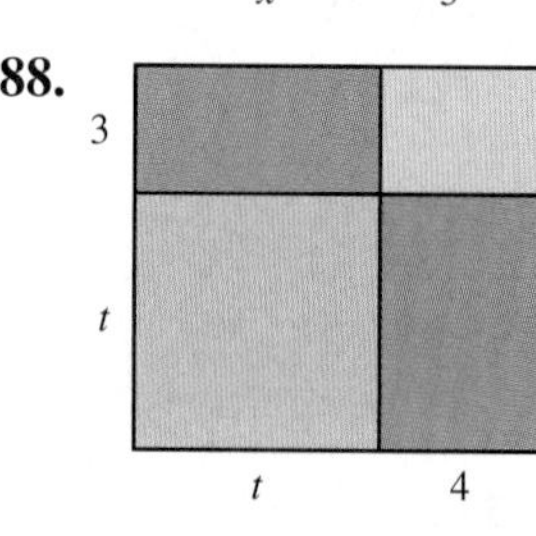

89.

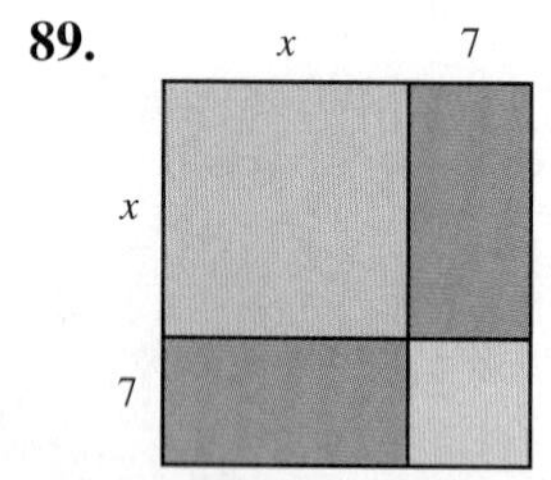

90.

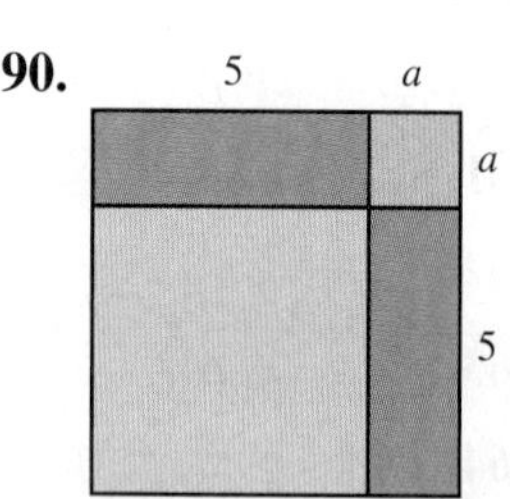

91.

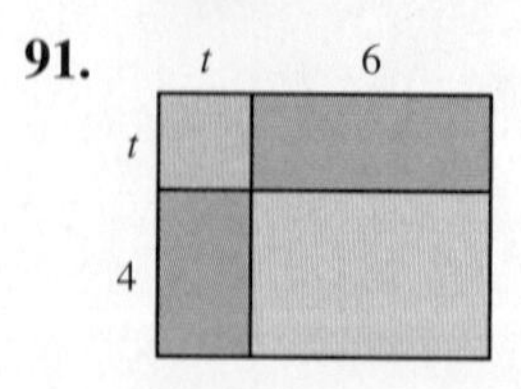

92.

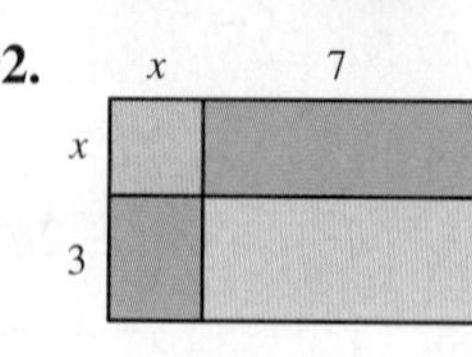

93.

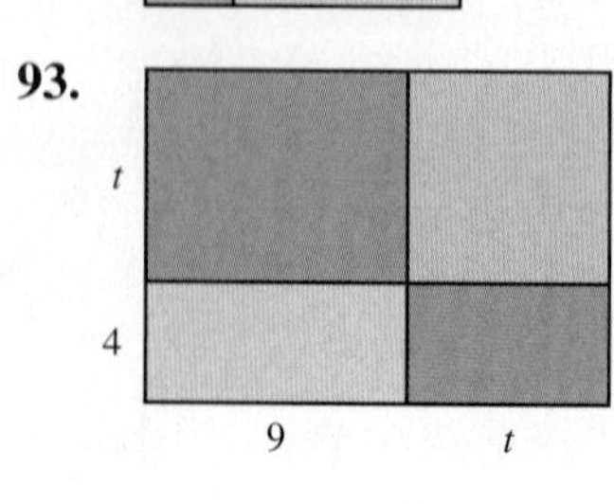

94.

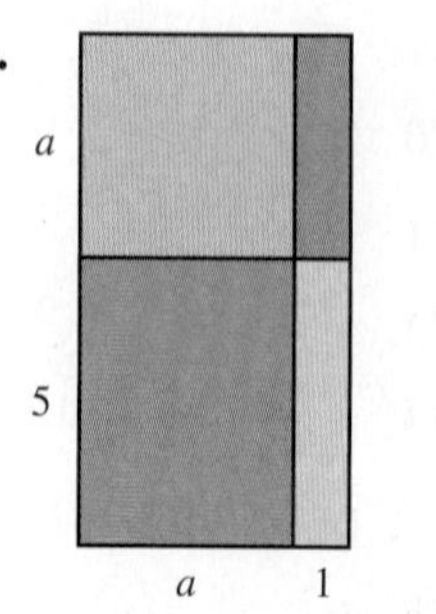

95. **96.**

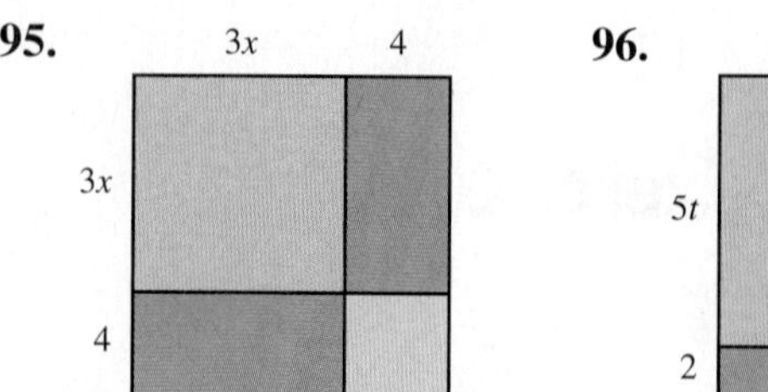

Draw and label rectangles similar to those in Exercises 85–96 to illustrate each of the following.

97. $(x + 5)^2$

98. $(x + 8)^2$

99. $(t + 9)^2$

100. $(a + 12)^2$

101. $(3 + x)^2$

102. $(7 + t)^2$

TW **103.** Gaston feels that since he can find the product of any two binomials using FOIL, he needn't study the other special products. What advice would you give him?

TW **104.** Under what conditions is the product of two binomials a binomial?

SKILL REVIEW

Review problem solving and solving a formula for a variable (Sections 2.3 and 2.5).

Solve. [2.5]

105. *Energy Use.* Under typical use, a refrigerator, a freezer, and a washing machine together use 297 kilowatt-hours per month (kWh/mo). A refrigerator uses 21 times as much energy as a washing machine, and a freezer uses 11 times as much energy as a washing machine. How much energy is used by each appliance?

106. *Searches.* Americans conducted 10.5 billion online searches in July 2009. This was a 5% increase over the number of searches in June 2009. How many searches were conducted in June 2009?
Source: The Nielsen Company

Solve. [2.3]

107. $5xy = 8$, for y

108. $3ab = c$, for a

109. $ax - by = c$, for x

110. $ax - by = c$, for y

SYNTHESIS

TW **111.** By writing $19 \cdot 21$ as $(20 - 1)(20 + 1)$, Patti can find the product mentally. How is this possible?

TW **112.** The product $(A + B)^2$ can be regarded as the sum of the areas of four regions (as shown following Example 4). How might one visually represent $(A + B)^3$? Why?

Calculate as the difference of squares.

113. 18×22 [*Hint*: $(20 - 2)(20 + 2)$.]

114. 93×107

Multiply.

Aha! **115.** $(4x^2 + 9)(2x + 3)(2x - 3)$

116. $(9a^2 + 1)(3a - 1)(3a + 1)$

Aha! **117.** $(3t - 2)^2(3t + 2)^2$

118. $(5a + 1)^2(5a - 1)^2$

119. $(t^3 - 1)^4(t^3 + 1)^4$

120. $(32.41x + 5.37)^2$

Solve.

121. $(x + 2)(x - 5) = (x + 1)(x - 3)$

122. $(2x + 5)(x - 4) = (x + 5)(2x - 4)$

123. Find $(y - 2)^2$ by subtracting the white areas from y^2.

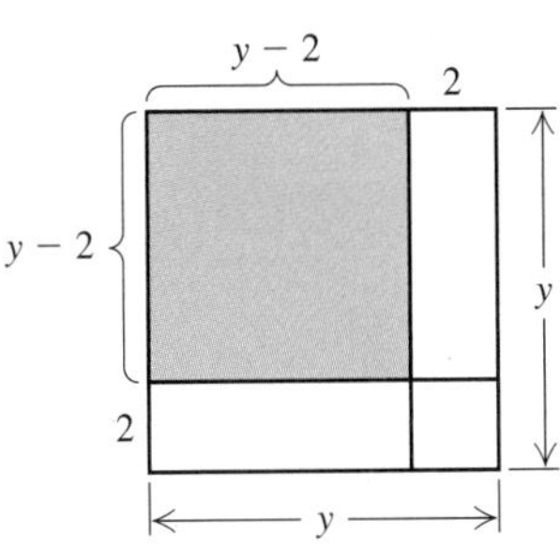

124. Find $(10 - 2x)^2$ by subtracting the white areas from 10^2.

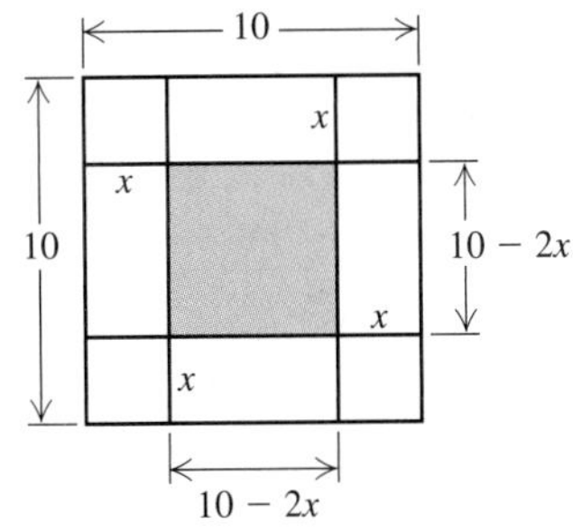

The height of a box is 1 more than its length l, and the length is 1 more than its width w. Find a polynomial for the volume V in terms of the following.

125. The length l

126. The width w

Find a polynomial for the total shaded area in each figure.

127.

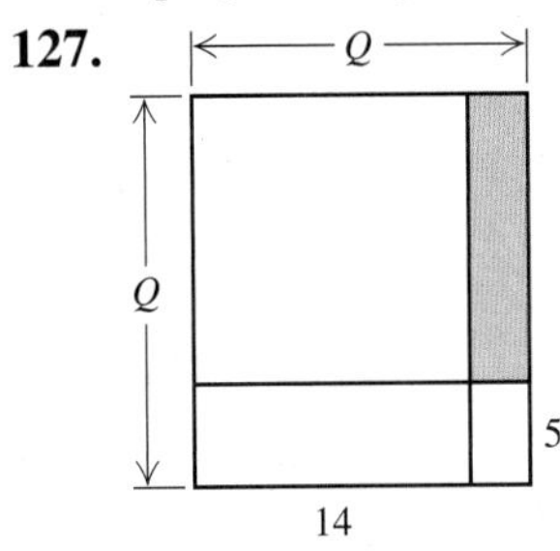

128.

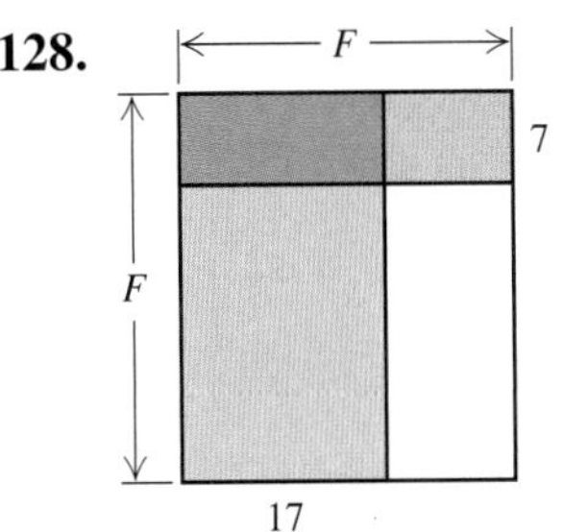

129.

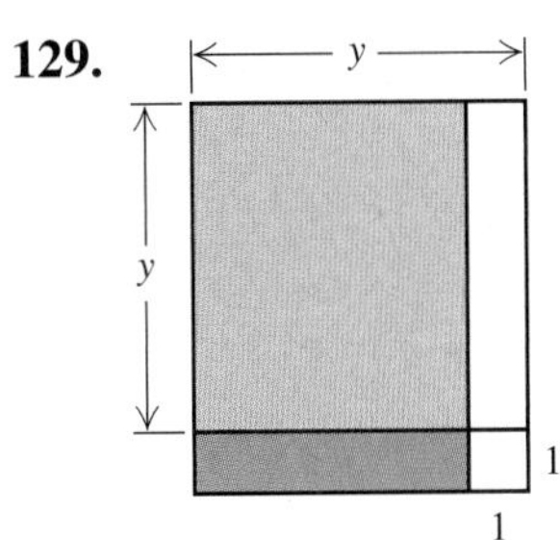

130. Find three consecutive integers for which the sum of the squares is 65 more than three times the square of the smallest integer.

Try Exercise Answers: Section 5.6

5. $x^3 + 3x^2 + 5x + 15$ **9.** $y^2 - y - 6$ **41.** $4x^2 - 1$ **51.** $x^2 + 4x + 4$ **69.** $49 - 42x^4 + 9x^8$

Mid-Chapter Review

When writing equivalent expressions, look first at the operation that you are asked to perform.

EXPONENTIAL EXPRESSIONS

Operation	Procedure
Multiply expressions with the same base.	Add exponents.
Divide expressions with the same base.	Subtract exponents.
Raise a power to a power.	Multiply exponents.

POLYNOMIAL EXPRESSIONS

Operation	Procedure
Addition	Combine like terms.
Subtraction	Add the opposite of the polynomial being subtracted.
Multiplication	Multiply each term of one polynomial by every term of the other.

When multiplying polynomials, also remember the special-product rules.

$(A + B)(A - B) = A^2 - B^2$ **The product of a sum and a difference**

$\left.\begin{array}{r}(A + B)^2 = A^2 + 2AB + B^2 \\ (A - B)^2 = A^2 - 2AB + B^2\end{array}\right\}$ **The square of a binomial**

$(A + B)(C + D) = AC + AD + BC + BD$ **FOIL**

GUIDED SOLUTIONS

1. $(2x^2y^{-5})^{-10} = 2^{\square}(x^2)^{\square}(y^{-5})^{\square}$ **Raising each factor to the exponent**

$= 2^{\square}x^{\square}y^{\square}$ **Multiplying exponents**

$= \dfrac{y^{\square}}{2^{\square}x^{\square}}$ [5.1], [5.2] **Writing without negative exponents**

2. $(x^2 + 7x)(x^2 - 7x) = (x^2)^2 - \square^2$ **The product is a difference of squares.**

$= \square - \square$ [5.6] **Simplifying**

MIXED REVIEW

Simplify. Write your answer without negative exponents.

1. $(x^2y^5)^8$ [5.1]

2. $(4x)^0$ [5.1]

3. d^{-10} [5.2]

4. $\dfrac{3a^{11}}{12a}$ [5.1]

5. $(m^{-3}y)^{-1}$ [5.2]

6. $\dfrac{-48ab^{-7}}{18a^{11}b^{-6}}$ [5.2]

Perform the indicated operation and simplify.

7. $(3x^2 - 2x + 6) + (5x - 3)$ [5.4]

8. $(9x + 6) - (2x - 1)$ [5.4]

9. $6x^3(8x^2 - 7)$ [5.5]

10. $(3x + 2)(2x - 1)$ [5.6]

11. $(4x^2 - x - 7) - (10x^2 - 3x + 5)$ [5.4]

12. $(3x + 8)(3x + 7)$ [5.6]

13. $(t^9 + 3t^6 - 8t^2) + (5t^7 - 3t^6 + 8t^2)$ [5.4]

14. $(2m - 1)^2$ [5.6]

15. $(x - 1)(x^2 + x + 1)$ [5.5]

16. $(c + 3)(c - 3)$ [5.6]

17. $(4y^3 + 7)^2$ [5.6]

18. $(3a^4 - 9a^3 - 7) - (4a^3 + 13a^2 - 3)$ [5.4]

19. $(4t^2 - 5)(4t^2 + 5)$ [5.6]

20. $(a^4 + 3)(a^4 - 8)$ [5.6]

5.7 Polynomials in Several Variables

- Evaluating Polynomials
- Like Terms and Degree
- Addition and Subtraction
- Multiplication
- Function Notation

Thus far, the polynomials that we have studied have had only one variable. Polynomials such as

$$5x - x^2y + 7, \qquad 9ab^2c - 2a^3b^2 + 8a^2b^3, \quad \text{and} \quad 4m^2 - 9n^2$$

contain two or more variables. In this section, we will add, subtract, multiply, and evaluate such **polynomials in several variables.**

EVALUATING POLYNOMIALS

To evaluate a polynomial in two or more variables, we substitute numbers for the variables. Then we compute, using the rules for order of operations.

EXAMPLE 1 Evaluate the polynomial $4 + 3x + xy^2 + 8x^3y^3$ for $x = -2$ and $y = 5$.

SOLUTION We substitute -2 for x and 5 for y:

$$\begin{aligned} 4 + 3x + xy^2 + 8x^3y^3 &= 4 + 3(-2) + (-2) \cdot 5^2 + 8(-2)^3 \cdot 5^3 \\ &= 4 - 6 - 50 - 8000 = -8052. \end{aligned}$$

Try Exercise 9.

We can evaluate polynomials in several variables using a graphing calculator. To evaluate the polynomial in Example 1 for $x = -2$ and $y = 5$, we store -2 to X and 5 to Y using the STO› key and then enter the polynomial. The result is shown in the screen below.

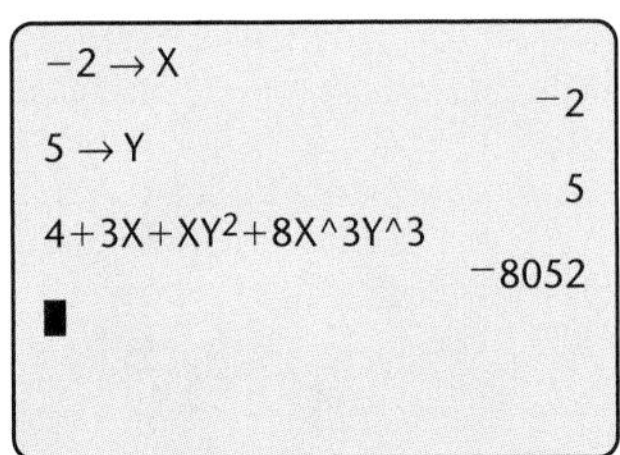

EXAMPLE 2 Surface Area of a Right Circular Cylinder. The surface area of a right circular cylinder is given by the polynomial

$$2\pi rh + 2\pi r^2,$$

where h is the height and r is the radius of the base. A 12-oz can has a height of 4.7 in. and a radius of 1.2 in. Approximate its surface area to the nearest tenth of a square inch.

SOLUTION We evaluate the polynomial for $h = 4.7$ in. and $r = 1.2$ in. If 3.14 is used to approximate π, we have

$$\begin{aligned} 2\pi rh + 2\pi r^2 &\approx 2(3.14)(1.2 \text{ in.})(4.7 \text{ in.}) + 2(3.14)(1.2 \text{ in.})^2 \\ &\approx 2(3.14)(1.2 \text{ in.})(4.7 \text{ in.}) + 2(3.14)(1.44 \text{ in}^2) \\ &\approx 35.4192 \text{ in}^2 + 9.0432 \text{ in}^2 \approx 44.4624 \text{ in}^2. \end{aligned}$$

If the π key of a calculator is used, we have

$$\begin{aligned} 2\pi rh + 2\pi r^2 &\approx 2(3.141592654)(1.2 \text{ in.})(4.7 \text{ in.}) + 2(3.141592654)(1.2 \text{ in.})^2 \\ &\approx 44.48495197 \text{ in}^2. \end{aligned}$$

Note that the unit in the answer (square inches) is a unit of area. The surface area is about 44.5 in^2 (square inches).

Try Exercise 13.

LIKE TERMS AND DEGREE

Recall that the degree of a term is the number of variable factors in the term. For example, the degree of $5x^2$ is 2 because there are two variable factors in $5 \cdot x \cdot x$. Similarly, the degree of $5a^2b^4$ is 6 because there are 6 variable factors in $5 \cdot a \cdot a \cdot b \cdot b \cdot b \cdot b$. Note that 6 can be found by adding the exponents 2 and 4.

As we learned in Section 5.3, the degree of a polynomial is the degree of the term of highest degree.

STUDY TIP

Use Your Voice

Don't hesitate to ask questions in class at appropriate times. Most instructors welcome questions and encourage students to ask them. Other students in your class probably have the same questions you do.

EXAMPLE 3 Identify the coefficient and the degree of each term and the degree of the polynomial

$$9x^2y^3 - 14xy^2z^3 + xy + 4y + 5x^2 + 7.$$

SOLUTION

Term	Coefficient	Degree	Degree of the Polynomial
$9x^2y^3$	9	5	
$-14xy^2z^3$	-14	6	6
xy	1	2	
$4y$	4	1	
$5x^2$	5	2	
7	7	0	

Try Exercise 21.

Note in Example 3 that although both xy and $5x^2$ have degree 2, they are *not* like terms. *Like*, or *similar*, *terms* either have exactly the same variables with exactly the same exponents or are constants. For example,

$8a^4b^7$ and $5b^7a^4$ are like terms

and

-17 and 3 are like terms,

but

$-2x^2y$ and $9xy^2$ are *not* like terms.

As always, combining like terms is based on the distributive law.

EXAMPLE 4 Combine like terms.

a) $9x^2y + 3xy^2 - 5x^2y - xy^2$

b) $7ab - 5ab^2 + 3ab^2 + 6a^3 + 9ab - 11a^3 + b - 1$

SOLUTION

a) $9x^2y + 3xy^2 - 5x^2y - xy^2 = (9 - 5)x^2y + (3 - 1)xy^2$

$= 4x^2y + 2xy^2$ **Try to go directly to this step.**

b) $7ab - 5ab^2 + 3ab^2 + 6a^3 + 9ab - 11a^3 + b - 1$

$= -5a^3 - 2ab^2 + 16ab + b - 1$ **We choose to write descending powers of *a*. Other, equivalent, forms can also be used.**

Try Exercise 27.

ADDITION AND SUBTRACTION

The procedure used for adding polynomials in one variable is used to add polynomials in several variables.

Student Notes

Always read the problem carefully. The difference between

$$(-5x^3 + 7) + (8x^3 + y)$$

and

$$(-5x^3 + 7)(8x^3 + y)$$

is enormous. To avoid wasting time working on an incorrectly copied exercise, be sure to double-check that you have written the correct problem in your notebook.

EXAMPLE 5 Add.

a) $(-5x^3 + 3y - 5y^2) + (8x^3 + 4x^2 + 7y^2)$

b) $(5ab^2 - 4a^2b + 5a^3 + 2) + (3ab^2 - 2a^2b + 3a^3b - 5)$

SOLUTION

a) $(-5x^3 + 3y - 5y^2) + (8x^3 + 4x^2 + 7y^2)$

$= (-5 + 8)x^3 + 4x^2 + 3y + (-5 + 7)y^2$ **Try to do this step mentally.**

$= 3x^3 + 4x^2 + 3y + 2y^2$

b) $(5ab^2 - 4a^2b + 5a^3 + 2) + (3ab^2 - 2a^2b + 3a^3b - 5)$

$= 8ab^2 - 6a^2b + 5a^3 + 3a^3b - 3$

Try Exercise 33.

When subtracting a polynomial, remember to find the opposite of each term in that polynomial and then add.

EXAMPLE 6 Subtract:

$$(4x^2y + x^3y^2 + 3x^2y^3 + 6y) - (4x^2y - 6x^3y^2 + x^2y^2 - 5y).$$

SOLUTION

$(4x^2y + x^3y^2 + 3x^2y^3 + 6y) - (4x^2y - 6x^3y^2 + x^2y^2 - 5y)$

$= 4x^2y + x^3y^2 + 3x^2y^3 + 6y - 4x^2y + 6x^3y^2 - x^2y^2 + 5y$

$= 7x^3y^2 + 3x^2y^3 - x^2y^2 + 11y$ **Combining like terms**

Try Exercise 35.

MULTIPLICATION

To multiply polynomials in several variables, multiply each term of one polynomial by every term of the other, just as we did in Sections 5.5 and 5.6.

EXAMPLE 7 Multiply: $(3x^2y - 2xy + 3y)(xy + 2y)$.

SOLUTION

$$\begin{array}{rl}
3x^2y - 2xy + 3y & \\
xy + 2y & \\
\hline
6x^2y^2 - 4xy^2 + 6y^2 & \text{Multiplying by } 2y \\
3x^3y^2 - 2x^2y^2 + 3xy^2 \qquad & \text{Multiplying by } xy \\
\hline
3x^3y^2 + 4x^2y^2 - xy^2 + 6y^2 & \text{Adding}
\end{array}$$

Try Exercise 45.

The special products discussed in Section 5.6 can speed up our work.

EXAMPLE 8 Multiply.

a) $(p + 5q)(2p - 3q)$
b) $(3x + 2y)^2$
c) $(a^3 - 7a^2b)^2$
d) $(3x^2y + 2y)(3x^2y - 2y)$
e) $(-2x^3y^2 + 5t)(2x^3y^2 + 5t)$
f) $(2x + 3 - 2y)(2x + 3 + 2y)$

SOLUTION

a) $(p + 5q)(2p - 3q) = \overset{F}{2p^2} - \overset{O}{3pq} + \overset{I}{10pq} - \overset{L}{15q^2}$

$= 2p^2 + 7pq - 15q^2$ **Combining like terms**

$(A + B)^2 = A^2 + 2 \cdot A \cdot B + B^2$

b) $(3x + 2y)^2 = (3x)^2 + 2(3x)(2y) + (2y)^2$ **Using the pattern for squaring a binomial**

$= 9x^2 + 12xy + 4y^2$

$(A - B)^2 = A^2 - 2 \cdot A \cdot B + B^2$

c) $(a^3 - 7a^2b)^2 = (a^3)^2 - 2(a^3)(7a^2b) + (7a^2b)^2$ **Squaring a binomial**

$= a^6 - 14a^5b + 49a^4b^2$ **Using the rules for exponents**

$(A + B)(A - B) = A^2 - B^2$

d) $(3x^2y + 2y)(3x^2y - 2y) = (3x^2y)^2 - (2y)^2$ **Using the pattern for multiplying the sum and the difference of two terms**

$= 9x^4y^2 - 4y^2$ **Using the rules for exponents**

e) $(-2x^3y^2 + 5t)(2x^3y^2 + 5t) = (5t - 2x^3y^2)(5t + 2x^3y^2)$ **Using the commutative law for addition twice**

$= (5t)^2 - (2x^3y^2)^2$ **Multiplying the sum and the difference of the same two terms**

$= 25t^2 - 4x^6y^4$

$$(\quad A \quad - \quad B)(\quad A \quad + \quad B) = \quad A^2 \quad - \quad B^2$$

f) $(2x + 3 - 2y)(2x + 3 + 2y) = (2x + 3)^2 - (2y)^2$ **Multiplying a sum and a difference**

$= 4x^2 + 12x + 9 - 4y^2$ **Squaring a binomial**

Try Exercise 49.

In Example 8, we recognized patterns that might not be obvious, particularly in parts (e) and (f). In part (e), we *can* use FOIL, and in part (f), we *can* use long multiplication, but doing so would be slower. By carefully inspecting a problem before "jumping in," we can often save ourselves considerable work. At least one instructor refers to this as "working smart" instead of "working hard."*

FUNCTION NOTATION

Our work with multiplying can be used when evaluating functions.

EXAMPLE 9 Given $f(x) = x^2 - 4x + 5$, find and simplify each of the following.

a) $f(a + 3)$

b) $f(a + h) - f(a)$

SOLUTION

a) To find $f(a + 3)$, we replace x with $a + 3$. Then we simplify:

$$\begin{aligned} f(a + 3) &= (a + 3)^2 - 4(a + 3) + 5 \\ &= a^2 + 6a + 9 - 4a - 12 + 5 \\ &= a^2 + 2a + 2. \end{aligned}$$

b) To find $f(a + h)$ and $f(a)$, we replace x with $a + h$ and a, respectively.

$$\begin{aligned} f(a + h) &= (a + h)^2 - 4(a + h) + 5 \\ &= a^2 + 2ah + h^2 - 4a - 4h + 5 \\ f(a) &= a^2 - 4a + 5 \end{aligned}$$

Then

$$\begin{aligned} f(a + h) - f(a) &= a^2 + 2ah + h^2 - 4a - 4h + 5 - [a^2 - 4a + 5] \\ &= a^2 + 2ah + h^2 - 4a - 4h + 5 - a^2 + 4a - 5 \\ &= 2ah + h^2 - 4h. \end{aligned}$$

Try Exercise 83.

*Thanks to Pauline Kirkpatrick of Wharton County Junior College for this language.

5.7 Exercise Set

FOR EXTRA HELP MyMathLab MathXL PRACTICE  WATCH  DOWNLOAD  READ REVIEW

Concept Reinforcement *Choose from the following list of words and phrases to complete each sentence. Choices may be used more than once.*

binomial, coefficient, degree, leading coefficient, like terms, three variables

1. The ____________ of $5ab^8$ is 5.
2. The ____________ of $5ab^8$ is 9.
3. The ____________ of $4x^2y^3 - x^9y$ is 10.
4. The ____________ of $4x^2y^3 - x^9y$ is -1.
5. The expression $9x^2y - 3xy^2$ is a ____________.
6. The expression $6 - st - rs$ is a polynomial in ____________.
7. The expressions $7x^3y^2$ and $-3y^2x^3$ are ____________.
8. The expression $(2a^5 - c)^2$ is the square of a ____________.

Evaluate each polynomial for $x = 5$ and $y = -2$.

9. $x^2 - 3y^2 + 2xy$
10. $x^2 + 5y^2 - 4xy$

Evaluate each polynomial for $x = 2$, $y = -3$, and $z = -4$.

11. $xyz^2 - z$
12. $xy - xz + yz$

Lung Capacity. The polynomial

$$0.041h - 0.018A - 2.69$$

can be used to estimate the lung capacity, in liters, of a female with height h, in centimeters, and age A, in years.

13. Find the lung capacity of a 50-year-old woman who is 160 cm tall.
14. Find the lung capacity of a 20-year-old woman who is 165 cm tall.
15. *Male Caloric Needs.* The number of calories needed each day by a moderately active man who weighs w kilograms, is h centimeters tall, and is a years old can be estimated by the polynomial

$$19.18w + 7h - 9.52a + 92.4.$$

Marv is moderately active, weighs 87 kg, is 185 cm tall, and is 59 years old. What are his daily caloric needs?

Source: Parker, M., *She Does Math.* Mathematical Association of America

16. *Female Caloric Needs.* The number of calories needed each day by a moderately active woman who weighs w pounds, is h inches tall, and is a years old can be estimated by the polynomial

$$917 + 6w + 6h - 6a.$$

Christine is moderately active, weighs 125 lb, is 64 in. tall, and is 27 years old. What are her daily caloric needs?

Source: Parker, M., *She Does Math.* Mathematical Association of America

Surface Area of a Silo. A silo is a structure that is shaped like a right circular cylinder with a half sphere on top. The surface area of a silo of height h and radius r (including the area of the base) is given by the polynomial $2\pi rh + \pi r^2$. (Note that h is the height of the entire silo.)

17. A coffee grinder is shaped like a silo, with a height of 7 in. and a radius of $1\frac{1}{2}$ in. Find the surface area of the coffee grinder. Use 3.14 for π.

18. A $1\frac{1}{2}$-oz bottle of roll-on deodorant has a height of 4 in. and a radius of $\frac{3}{4}$ in. Find the surface area of the bottle if the bottle is shaped like a silo. Use 3.14 for π.

Altitude of a Launched Object. The altitude of an object, in meters, is given by the polynomial

$$h + vt - 4.9t^2,$$

where h is the height, in meters, at which the launch occurs, v is the initial upward speed (or velocity), in meters per second, and t is the number of seconds for which the object is airborne.

19. A bocce ball is thrown upward with an initial speed of 18 m/sec by a person atop the Leaning Tower of

Pisa, which is 50 m above the ground. How high will the ball be 2 sec after it is thrown?

20. A golf ball is launched upward with an initial speed of 30 m/sec by a golfer atop the Washington Monument, which is 160 m above the ground. How high above the ground will the ball be after 3 sec?

Identify the coefficient and the degree of each term of each polynomial. Then find the degree of each polynomial.

21. $x^3y - 2xy + 3x^2 - 5$

22. $xy^2 - y^2 + 9x^2y + 7$

23. $11 - abc + a^2b + 0.5ab^2$

24. $3m - \frac{1}{2}mn - 8m^2n^3 - 4mn^6$

Combine like terms.

25. $7a + b - 4a - 3b$

26. $8r + s - 5r - 4s$

27. $3x^2y - 2xy^2 + x^2 + 5x$

28. $m^3 + 2m^2n - 3m^2 + 3mn^2$

29. $2u^2v - 3uv^2 + 6u^2v - 2uv^2 + 7u^2$

30. $3x^2 + 6xy + 3y^2 - 5x^2 - 10xy$

31. $5a^2c - 2ab^2 + a^2b - 3ab^2 + a^2c - 2ab^2$

32. $3s^2t + r^2t - 4st^2 - s^2t + 3st^2 - 7r^2t$

Add or subtract, as indicated.

33. $(4x^2 - xy + y^2) + (-x^2 - 3xy + 2y^2)$

34. $(2r^3 + 3rs - 5s^2) - (5r^3 + rs + 4s^2)$

35. $(3a^4 - 5ab + 6ab^2) - (9a^4 + 3ab - ab^2)$

36. $(2r^2t - 5rt + rt^2) - (7r^2t + rt - 5rt^2)$

Aha! **37.** $(5r^2 - 4rt + t^2) + (-6r^2 - 5rt - t^2) + (-5r^2 + 4rt - t^2)$

38. $(2x^2 - 3xy + y^2) + (-4x^2 - 6xy - y^2) + (4x^2 + 6xy + y^2)$

39. $(x^3 - y^3) - (-2x^3 + x^2y - xy^2 + 2y^3)$

40. $(a^3 + b^3) - (-5a^3 + 2a^2b - ab^2 + 3b^3)$

41. $(2y^4x^2 - 5y^3x) + (5y^4x^2 - y^3x) + (3y^4x^2 - 2y^3x)$

42. $(5a^2b + 7ab) + (9a^2b - 5ab) + (a^2b - 6ab)$

43. Subtract $7x + 3y$ from the sum of $4x + 5y$ and $-5x + 6y$.

44. Subtract $5a + 2b$ from the sum of $2a + b$ and $3a - 4b$.

Multiply.

45. $(3z - u)(2z + 3u)$

46. $(5x + y)(2x - 3y)$

47. $(xy + 7)(xy - 4)$

48. $(ab + 3)(ab - 5)$

49. $(2a - b)(2a + b)$

50. $(a - 3b)(a + 3b)$

51. $(5rt - 2)(3rt + 1)$

52. $(3xy - 1)(4xy + 2)$

53. $(m^3n + 8)(m^3n - 6)$

54. $(3 - c^2d^2)(4 + c^2d^2)$

55. $(6x - 2y)(5x - 3y)$

56. $(7a - 6b)(5a + 4b)$

57. $(aw + 0.1)(-aw + 0.1)$

58. $(rt + 0.2)(-rt + 0.2)$

59. $(x + h)^2$

60. $(a - y)^2$

61. $(4a - 5b)^2$

62. $(3x + 2y)^2$

63. $(ab + cd^2)(ab - cd^2)$

64. $(m^3 - 5n)(m^3 + 5n)$

65. $(2xy + x^2y + 3)(xy + y^2)$

66. $(5cd - c^2 - d^2)(2c - c^2d)$

Aha! **67.** $(a + b - c)(a + b + c)$

68. $(x + y + z)(x + y - z)$

69. $[a + b + c][a - (b + c)]$

70. $(a + b + c)(a - b - c)$

Find the total area of each shaded region.

71.

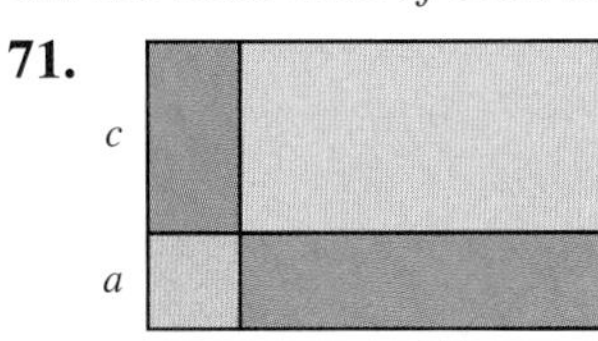

72.

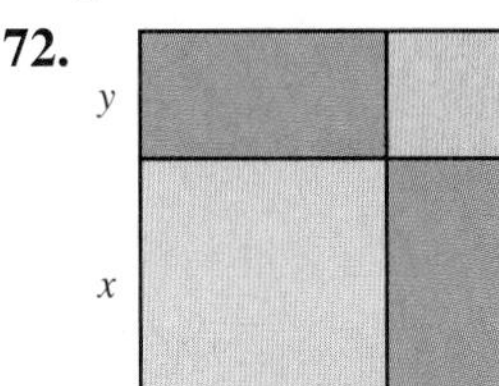

73.

74.

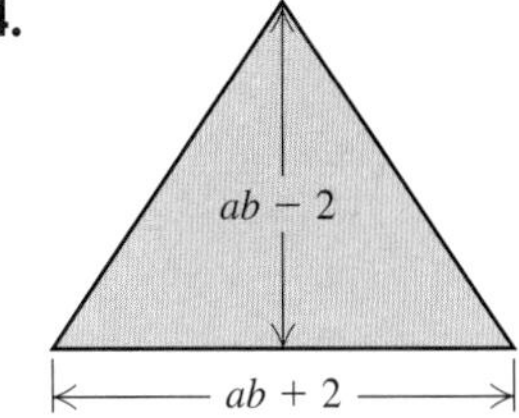

75.

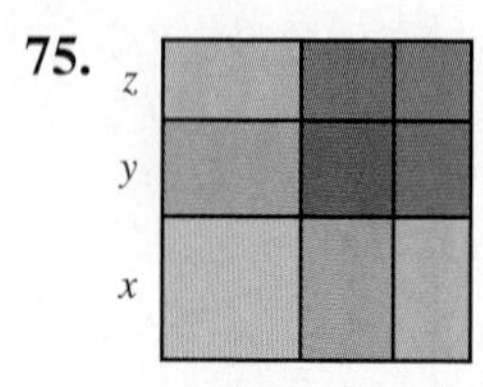

76.

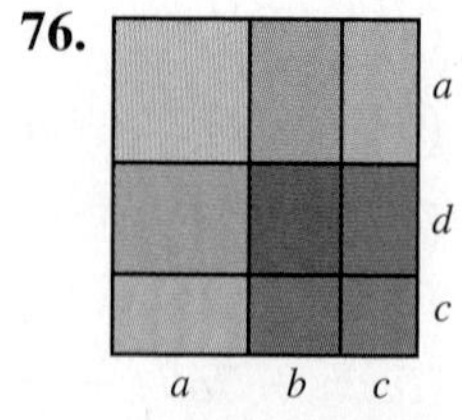

77.

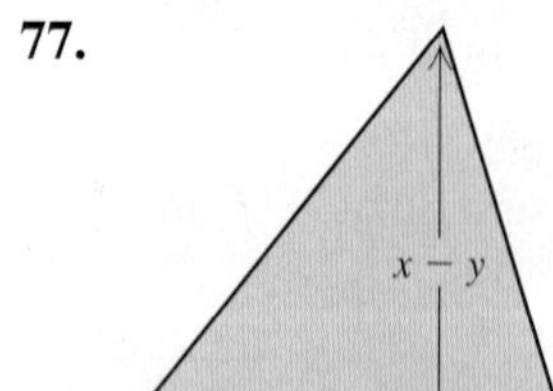

78.

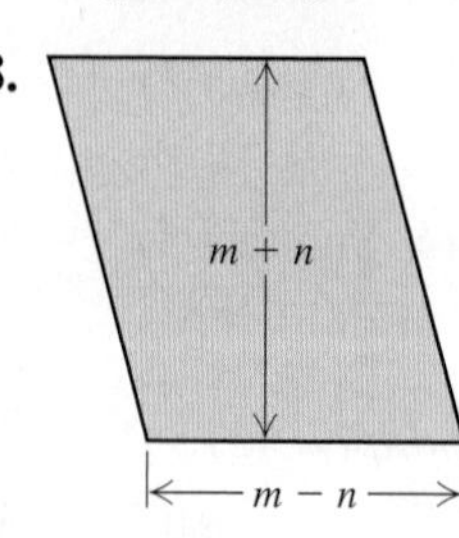

Draw and label rectangles similar to those in Exercises 71, 72, 75, and 76 to illustrate each product.

79. $(r + s)(u + v)$

80. $(m + r)(n + v)$

81. $(a + b + c)(a + d + f)$

82. $(r + s + t)^2$

83. Given $f(x) = x^2 + 5$, find and simplify.

a) $f(t - 1)$
b) $f(a + h) - f(a)$
c) $f(a) - f(a - h)$

84. Given $f(x) = x^2 + 7$, find and simplify.

a) $f(p + 1)$
b) $f(a + h) - f(a)$
c) $f(a) - f(a - h)$

TW **85.** Is it possible for a polynomial in 4 variables to have a degree less than 4? Why or why not?

TW **86.** A fourth-degree polynomial is multiplied by a third-degree polynomial. What is the degree of the product? Explain your reasoning.

SKILL REVIEW

To prepare for Section 5.8, review subtraction of polynomials using columns (Section 5.4).

Subtract. [5.4]

87. $$\begin{array}{r} x^2 - 3x - 7 \\ -(\quad 5x - 3) \\ \hline \end{array}$$

88. $$\begin{array}{r} 2x^3 \qquad\quad - x + 3 \\ -(\qquad x^2 \qquad - 1) \\ \hline \end{array}$$

89. $$\begin{array}{r} 3x^2 + x + 5 \\ -(3x^2 + 3x) \\ \hline \end{array}$$

90. $$\begin{array}{r} 4x^3 - 3x^2 + x \\ -(4x^3 - 8x^2) \\ \hline \end{array}$$

91. $$\begin{array}{r} 5x^3 - 2x^2 + 1 \\ -(5x^3 - 15x^2) \\ \hline \end{array}$$

92. $$\begin{array}{r} 2x^2 + 5x - 3 \\ -(2x^2 + 6x) \\ \hline \end{array}$$

SYNTHESIS

TW **93.** The concept of "leading term" was intentionally not discussed in this section. Why?

TW **94.** Explain how it is possible for the sum of two trinomials in several variables to be a binomial in one variable.

Find a polynomial for the shaded area. (Leave results in terms of π where appropriate.)

95.

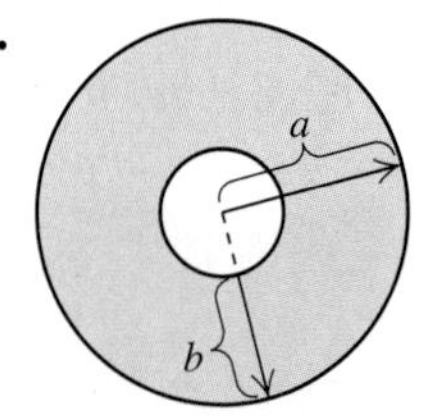

96.

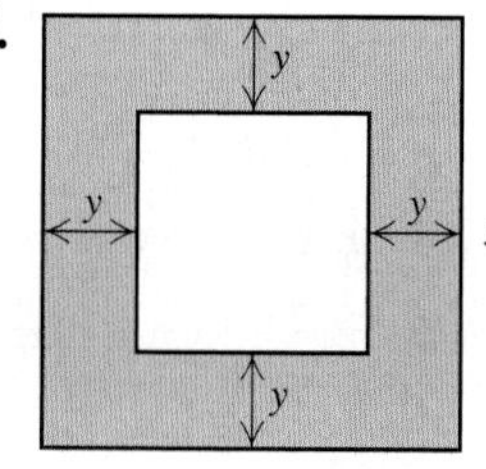

97.

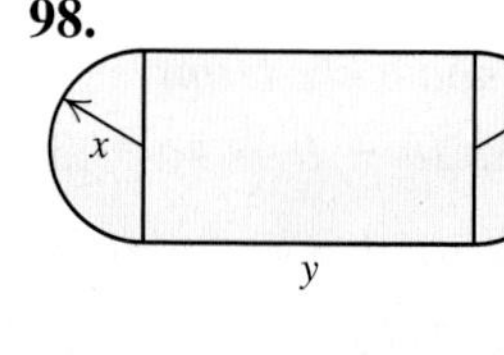

98.

99. Find a polynomial for the total volume of the figure shown.

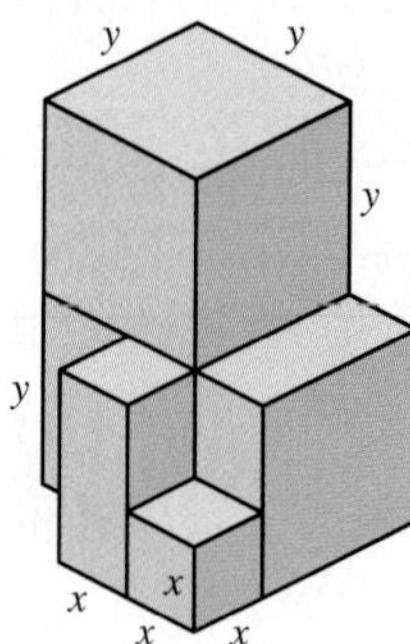

100. Find the shaded area in this figure using each of the approaches given below. Then check that both answers match.

a) Find the shaded area by subtracting the area of the unshaded square from the total area of the figure.

b) Find the shaded area by adding the areas of the three shaded rectangles.

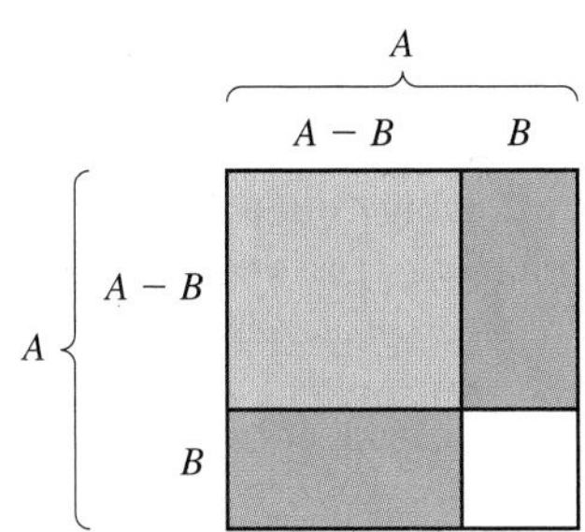

Find a polynomial for the surface area of each solid object shown. (Leave results in terms of π.)

101.

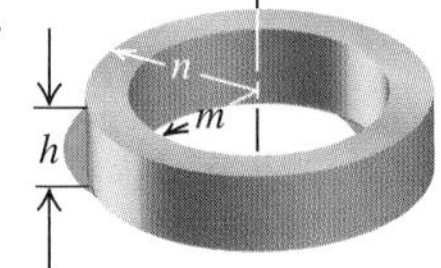

102.

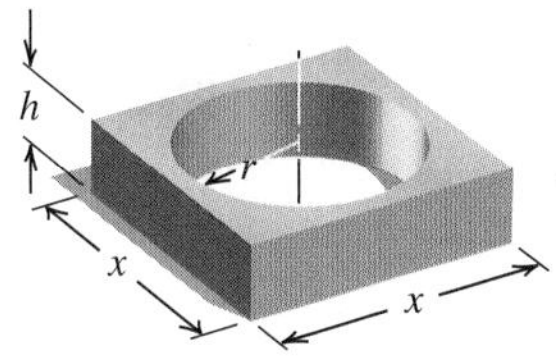

TW **103.** The observatory at Danville University is shaped like a silo that is 40 ft high and 30 ft wide (see Exercise 17). The Heavenly Bodies Astronomy Club is to paint the exterior of the observatory using paint that covers 250 ft^2 per gallon. How many gallons should they purchase? Explain your reasoning.

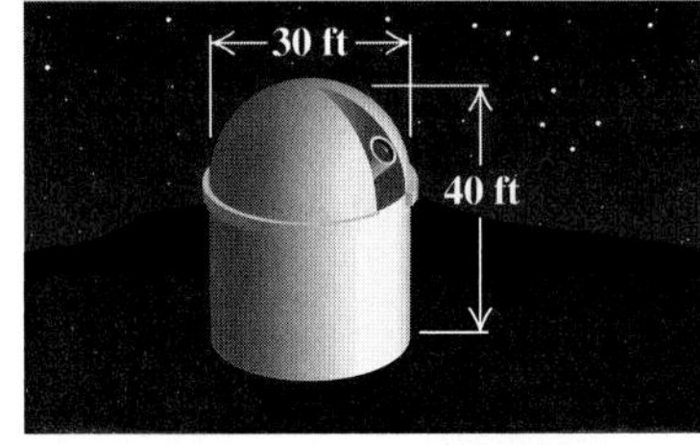

104. Multiply: $(x + a)(x - b)(x - a)(x + b)$.

Computer spreadsheet applications allow values for cells in a spreadsheet to be calculated from values in other cells. For example, if the cell C1 *contains the formula*

$$= \text{A1} + 2{*}\text{B1},$$

the value in C1 *will be the sum of the value in* A1 *and twice the value in* B1. *This formula is a polynomial in the two variables* A1 *and* B1.

105. The cell D4 contains the formula

$$= 2{*}\text{A4} + 3{*}\text{B4}.$$

What is the value in D4 if the value in A4 is 5 and the value in B4 is 10?

106. The cell D6 contains the formula

$$= \text{A1} - 0.2{*}\text{B1} + 0.3{*}\text{C1}.$$

What is the value in D6 if the value in A1 is 10, the value in B1 is -3, and the value in C1 is 30?

107. *Interest Compounded Annually.* An amount of money P that is invested at the yearly interest rate r grows to the amount

$$P(1 + r)^t$$

after t years. Find a polynomial that can be used to determine the amount to which P will grow after 2 years.

108. *Yearly Depreciation.* An investment P that drops in value at the yearly rate r drops in value to

$$P(1 - r)^t$$

after t years. Find a polynomial that can be used to determine the value to which P has dropped after 2 years.

109. Suppose that \$10,400 is invested at 8.5% compounded annually. How much is in the account at the end of 5 years? (See Exercise 107.)

110. A \$90,000 investment in computer hardware is depreciating at a yearly rate of 12.5%. How much is the investment worth after 4 years? (See Exercise 108.)

Try Exercise Answers: Section 5.7

9. -7 **13.** 2.97 L **21.** Coefficients: $1, -2, 3, -5$; degrees: 4, 2, 2, 0; 4 **27.** $3x^2y - 2xy^2 + x^2 + 5x$ **33.** $3x^2 - 4xy + 3y^2$ **35.** $-6a^4 - 8ab + 7ab^2$ **45.** $6z^2 + 7uz - 3u^2$ **49.** $4a^2 - b^2$ **83.** **(a)** $t^2 - 2t + 6$; **(b)** $2ah + h^2$; **(c)** $2ah - h^2$

Collaborative Corner

Finding the Magic Number

Focus: Evaluating polynomials in several variables
Time: 15–25 minutes
Group size: 3
Materials: A coin for each person

Can you know how soon it is possible for your team to clinch first place? A team's "magic number" is the combined number of wins by that team and losses by the second-place team that guarantee the leading team a first-place finish. For example, if the Cubs' magic number is 3 over the Reds, any combination of Cubs wins and Reds losses that totals 3 will guarantee a first-place finish for the Cubs. A team's magic number is computed using the polynomial

$$G - P - L + 1,$$

where G is the length of the season, in games, P is the number of games that the leading team has played, and L is the total number of games that the second-place team has lost minus the total number of games that the leading team has lost.

ACTIVITY

1. The standings shown are from a fictitious league with a 162-game season. Each group should calculate the Jaguars' magic number over the Catamounts as well as the Jaguars' magic number over the Wildcats.

	W	L
Jaguars	92	64
Catamounts	90	66
Wildcats	89	66

2. Each group member should play the role of one of the teams, using coin tosses to "play" the remaining games. If a group member correctly predicts the side (heads or tails) that comes up, the coin toss represents a win for that team. Should the other side appear, the toss represents a loss. Assume that these games are against other (unlisted) teams in the league. Each group member should perform three coin tosses and then update the standings.
3. Recalculate the two magic numbers, using the updated standings from part (2).
4. Slowly—one coin toss at a time—play out the remainder of the season. Record all wins and losses, update the standings, and recalculate the magic numbers each time all three group members have completed a round of coin tosses.
5. Examine the work in part (4) and explain why a magic number of 0 indicates that a team has been eliminated from contention.

5.8 Division of Polynomials

- Dividing by a Monomial
- Dividing by a Binomial
- Synthetic Division

In this section, we study division of polynomials. We will find that polynomial division is similar to division in arithmetic.

DIVIDING BY A MONOMIAL

We first consider division by a monomial. When dividing a monomial by a monomial, we use the quotient rule of Section 5.1 to subtract exponents when bases are the same. For example,

$$\frac{15x^{10}}{3x^4} = 5x^{10-4}$$
$$= 5x^6$$

CAUTION! The coefficients are divided but the exponents are subtracted.

STUDY TIP

Anticipate the Future

It is never too soon to begin reviewing for an exam. Take a few minutes each week to review important problems, formulas, and properties. Working a few problems from material covered earlier in the course will help you keep your skills fresh.

and

$$\frac{42a^2b^5}{-3ab^2} = \frac{42}{-3}a^{2-1}b^{5-2} \qquad \text{Recall that } a^m/a^n = a^{m-n}.$$
$$= -14ab^3.$$

To divide a polynomial by a monomial, we note that since

$$\frac{A}{C} + \frac{B}{C} = \frac{A+B}{C},$$

it follows that

$$\frac{A+B}{C} = \frac{A}{C} + \frac{B}{C}. \qquad \text{Switching the left and right sides of the equation}$$

This is actually how we perform divisions like $86 \div 2$:

$$\frac{86}{2} = \frac{80+6}{2} = \frac{80}{2} + \frac{6}{2} = 40 + 3.$$

Similarly, to divide a polynomial by a monomial, we divide each term by the monomial:

$$\frac{80x^5 + 6x^4}{2x^3} = \frac{80x^5}{2x^3} + \frac{6x^4}{2x^3}$$
$$= \frac{80}{2}x^{5-3} + \frac{6}{2}x^{4-3} \qquad \text{Dividing coefficients and subtracting exponents}$$
$$= 40x^2 + 3x.$$

EXAMPLE 1 Divide $x^4 + 15x^3 - 6x^2$ by $3x$.

SOLUTION We divide each term of $x^4 + 15x^3 - 6x^2$ by $3x$:

$$\frac{x^4 + 15x^3 - 6x^2}{3x} = \frac{x^4}{3x} + \frac{15x^3}{3x} - \frac{6x^2}{3x}$$
$$= \frac{1}{3}x^{4-1} + \frac{15}{3}x^{3-1} - \frac{6}{3}x^{2-1} \qquad \text{Dividing coefficients and subtracting exponents}$$
$$= \frac{1}{3}x^3 + 5x^2 - 2x. \qquad \text{This is the quotient.}$$

To check, we multiply our answer by $3x$, using the distributive law:

$$3x\left(\frac{1}{3}x^3 + 5x^2 - 2x\right) = 3x \cdot \frac{1}{3}x^3 + 3x \cdot 5x^2 - 3x \cdot 2x$$
$$= x^4 + 15x^3 - 6x^2.$$

This is the polynomial that was being divided, so our answer, $\frac{1}{3}x^3 + 5x^2 - 2x$, checks.

Try Exercise 9.

EXAMPLE 2 Divide and check: $(10a^5b^4 - 2a^3b^2 + 6a^2b) \div (-2a^2b)$.

SOLUTION We have

$$\frac{10a^5b^4 - 2a^3b^2 + 6a^2b}{-2a^2b} = \frac{10a^5b^4}{-2a^2b} - \frac{2a^3b^2}{-2a^2b} + \frac{6a^2b}{-2a^2b}$$

We divide coefficients and subtract exponents.

$$= -\frac{10}{2}a^{5-2}b^{4-1} - \left(-\frac{2}{2}\right)a^{3-2}b^{2-1} + \left(-\frac{6}{2}\right)$$

$$= -5a^3b^3 + ab - 3.$$

Check:

$$-2a^2b(-5a^3b^3 + ab - 3) = -2a^2b(-5a^3b^3) + (-2a^2b)(ab) + (-2a^2b)(-3)$$
$$= 10a^5b^4 - 2a^3b^2 + 6a^2b$$

Our answer, $-5a^3b^3 + ab - 3$, checks.

Try Exercise 17.

DIVIDING BY A BINOMIAL

The divisors in Examples 1 and 2 have just one term. For divisors with more than one term, we use long division, much as we do in arithmetic. Polynomials are written in descending order and any missing terms in the dividend are written in, using 0 for the coefficients.

EXAMPLE 3 Divide $x^2 + 5x + 6$ by $x + 3$.

SOLUTION We begin by dividing x^2 by x:

$$\begin{array}{r} x \phantom{{}+5x+6} \\ x+3\overline{)x^2+5x+6} \\ -(x^2+3x) \phantom{{}+6} \\ \hline 2x \phantom{{}+6} \end{array}$$

x ← **Divide the first term, x^2, by the first term in the divisor: $x^2/x = x$. Ignore the term 3 for the moment.**

$-(x^2 + 3x)$ ← **Multiply $x + 3$ by x, using the distributive law.**

$2x$ ← **Subtract both x^2 and $3x$: $x^2 + 5x - (x^2 + 3x) = 2x$.**

Student Notes

Since long division of polynomials requires many steps, we recommend that you double-check each step of your work as you move forward.

Now we "bring down" the next term of the dividend—in this case, 6. The current remainder, $2x + 6$, now becomes the focus of our division. We divide $2x$ by x:

$$\begin{array}{r} x+2 \phantom{{}+6} \\ x+3\overline{)x^2+5x+6} \\ -(x^2+3x) \phantom{{}+6} \\ \hline 2x+6 \\ -(2x+6) \\ \hline 0 \end{array}$$

$2x + 6$: **Divide $2x$ by x: $2x/x = 2$.**

$-(2x + 6)$ ← **Multiply 2 by the divisor, $x + 3$, using the distributive law.**

0 ← **Subtract: $(2x + 6) - (2x + 6) = 0$.**

The quotient is $x + 2$. The notation R 0 indicates a remainder of 0, although a remainder of 0 is generally not listed in an answer.

Check: To check, we multiply the quotient by the divisor and add any remainder to see if we get the dividend:

$$\underbrace{(x+3)}_{\text{Divisor}}\ \underbrace{(x+2)}_{\text{Quotient}} + \underbrace{0}_{\text{Remainder}} = \underbrace{x^2 + 5x + 6}_{\text{Dividend}}.$$

Our answer, $x + 2$, checks.

Try Exercise 21.

EXAMPLE 4 Divide: $(2x^2 + 5x - 1) \div (2x - 1)$.

> **CAUTION!** Write parentheses around the polynomial being subtracted to remind you to subtract all its terms.

SOLUTION We have

$$\begin{array}{r} x \\ 2x-1\overline{)2x^2 + 5x - 1} \\ -(2x^2 - x) \\ \hline 6x \end{array}$$

Divide the first term by the first term: $2x^2/(2x) = x$.

Multiply $2x - 1$ by x.

Subtract by changing signs and adding: $2x^2 + 5x - (2x^2 - x) = 6x$.

Now, we bring down the next term of the dividend, -1.

$$\begin{array}{r} x + 3 \\ 2x-1\overline{)2x^2 + 5x - 1} \\ -(2x^2 - x) \\ \hline 6x - 1 \\ -(6x - 3) \\ \hline 2 \end{array}$$

Divide $6x$ by $2x$: $6x/(2x) = 3$.

Multiply 3 by the divisor, $2x - 1$.

Subtract. Note that $-1 - (-3) = -1 + 3 = 2$.

The answer is $x + 3$ with R 2.

Another way to write $x + 3$ R 2 is as

$$\underbrace{x + 3}_{\text{Quotient}} + \frac{2 \leftarrow \text{Remainder}}{\underbrace{2x - 1}_{\text{Divisor}}}.$$

(This is the way answers will be given at the back of the book.)

Check: To check, we multiply the divisor by the quotient and add the remainder:

$$\begin{aligned}(2x - 1)(x + 3) + 2 &= 2x^2 + 5x - 3 + 2 \\ &= 2x^2 + 5x - 1.\end{aligned}$$ **Our answer checks.**

Try Exercise 23.

Our division procedure ends when the degree of the remainder is less than that of the divisor. Check that this was indeed the case in Example 4.

EXAMPLE 5 Divide each of the following.

a) $(x^3 + 1) \div (x + 1)$ **b)** $(9a^2 + a^3 - 5) \div (a^2 - 1)$

SOLUTION

a)
$$
\begin{array}{r}
x^2 - x + 1 \\
x + 1 \overline{) x^3 + 0x^2 + 0x + 1} \\
\underline{-(x^3 + x^2) } \\
-x^2 + 0x \\
\underline{-(-x^2 - x) } \\
x + 1 \\
\underline{-(x + 1)} \\
0
\end{array}
$$

$\longleftarrow$ **Writing in the missing terms**

$\longleftarrow$ **Subtracting $x^3 + x^2$ from $x^3 + 0x^2$ and bringing down the $0x$**

$\longleftarrow$ **Subtracting $-x^2 - x$ from $-x^2 + 0x$ and bringing down the 1**

The answer is $x^2 - x + 1$.

Check: $(x + 1)(x^2 - x + 1) = x^3 - x^2 + x + x^2 - x + 1$
$= x^3 + 1.$

b) We rewrite the problem in descending order:

$$(a^3 + 9a^2 - 5) \div (a^2 - 1).$$

Thus,

$$
\begin{array}{r}
a + 9 \\
a^2 - 1 \overline{) a^3 + 9a^2 + 0a - 5} \\
\underline{-(a^3 - a) } \\
9a^2 + a - 5 \\
\underline{-(9a^2 - 9)} \\
a + 4
\end{array}
$$

$\longleftarrow$ **Writing in the missing term**

$\longleftarrow$ **Subtracting $a^3 - a$ from $a^3 + 9a^2 + 0a$ and bringing down the -5**

The degree of the remainder is less than the degree of the divisor, so we are finished.

The answer is $a + 9 + \dfrac{a + 4}{a^2 - 1}$.

Check: $(a^2 - 1)(a + 9) + a + 4 = a^3 + 9a^2 - a - 9 + a + 4$
$= a^3 + 9a^2 - 5.$

Try Exercise 29.

SYNTHETIC DIVISION

To divide a polynomial by a binomial of the type $x - a$, we can streamline the usual procedure to develop a process called *synthetic division.*

Compare the following. When a polynomial is written in descending order, the coefficients provide the essential information:

$$
\begin{array}{r}
4x^2 + 5x + 11 \\
x - 2 \overline{) 4x^3 - 3x^2 + x + 7} \\
\underline{4x^3 - 8x^2 } \\
5x^2 + x + 7 \\
\underline{5x^2 - 10x } \\
11x + 7 \\
\underline{11x - 22} \\
29
\end{array}
\qquad\qquad
\begin{array}{r}
4 + 5 + 11 \\
1 - 2 \overline{) 4 - 3 + 1 + 7} \\
\underline{4 - 8 } \\
5 + 1 + 7 \\
\underline{5 - 10 } \\
11 + 7 \\
\underline{11 - 22} \\
29
\end{array}
$$

Because the coefficient of x is 1 in the divisor, each time we multiply the divisor by a term in the answer, the leading coefficient of that product duplicates a coefficient in the answer. The process of synthetic division eliminates the duplication. To simplify the process further, we reverse the sign of the constant in the divisor and add rather than subtract.

EXAMPLE 6 Use synthetic division to divide:

$$(x^3 + 6x^2 - x - 30) \div (x - 2).$$

SOLUTION

$$\begin{array}{r|rrrr} 2 & 1 & 6 & -1 & -30 \\ & & & & \\ \hline & 1 & & & \end{array}$$

Write the 2 of $x - 2$ and the coefficients of the dividend.

Bring down the first coefficient.

$$\begin{array}{r|rrrr} 2 & 1 & 6 & -1 & -30 \\ & & 2 & & \\ \hline & 1 & 8 & & \end{array}$$

Multiply 1 by 2 to get 2.

Add 6 and 2.

$$\begin{array}{r|rrrr} 2 & 1 & 6 & -1 & -30 \\ & & 2 & 16 & \\ \hline & 1 & 8 & 15 & \end{array}$$

Multiply 8 by 2.

Add -1 and 16.

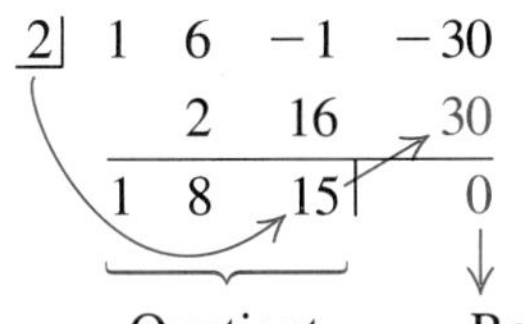

$$\begin{array}{r|rrrr} 2 & 1 & 6 & -1 & -30 \\ & & 2 & 16 & 30 \\ \hline & 1 & 8 & 15 & 0 \end{array}$$

Multiply 15 by 2 and add.

Quotient Remainder

The last number, 0, is the remainder. The other numbers are the coefficients of the quotient. The degree of the quotient is 1 less than the degree of the dividend.

1 8 15 | 0 ← This is the remainder.

This is the zero-degree coefficient.

This is the first-degree coefficient.

This is the second-degree coefficient.

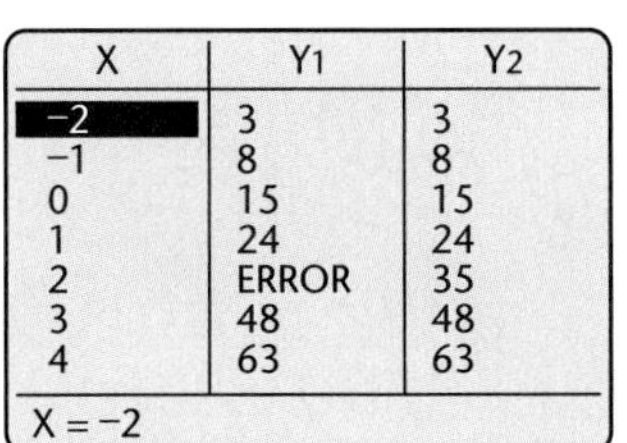

X	Y1	Y2
−2	3	3
−1	8	8
0	15	15
1	24	24
2	ERROR	35
3	48	48
4	63	63

X = −2

Since the remainder is 0, we have

$$(x^3 + 6x^2 - x - 30) \div (x - 2) = x^2 + 8x + 15.$$

We can check this with a table, letting

$$y_1 = (x^3 + 6x^2 - x - 30) \div (x - 2) \quad \text{and} \quad y_2 = x^2 + 8x + 15.$$

The values of both expressions are the same except for $x = 2$, when y_1 is not defined.

The answer is $x^2 + 8x + 15$ with R 0, or just $x^2 + 8x + 15$.

Try Exercise 43.

Remember that in order for this method to work, the divisor must be of the form $x - a$, that is, a variable minus a constant. The coefficient of the variable must be 1.

EXAMPLE 7 Use synthetic division to divide.

a) $(2x^3 + 7x^2 - 5) \div (x + 3)$

b) $(10x^2 - 13x + 3x^3 - 20) \div (4 + x)$

SOLUTION

a) $(2x^3 + 7x^2 - 5) \div (x + 3)$

The dividend has no x-term, so we need to write 0 for its coefficient of x. Note that $x + 3 = x - (-3)$, so we write -3 inside the $\rfloor$.

$$\begin{array}{r|rrrr} -3 & 2 & 7 & 0 & -5 \\ & & -6 & -3 & 9 \\ \hline & 2 & 1 & -3 & 4 \end{array}$$

The answer is $2x^2 + x - 3$, with R 4, or $2x^2 + x - 3 + \dfrac{4}{x + 3}$.

b) We first rewrite $(10x^2 - 13x + 3x^3 - 20) \div (4 + x)$ in descending order:

$(3x^3 + 10x^2 - 13x - 20) \div (x + 4)$.

Next, we use synthetic division. Note that $x + 4 = x - (-4)$.

$$\begin{array}{r|rrrr} -4 & 3 & 10 & -13 & -20 \\ & & -12 & 8 & 20 \\ \hline & 3 & -2 & -5 & 0 \end{array}$$

The answer is $3x^2 - 2x - 5$.

Try Exercise 51.

5.8 Exercise Set

FOR EXTRA HELP MyMathLab

Concept Reinforcement *Use the words from the following list to label the numbered expressions from the division shown.*

dividend, divisor, quotient, remainder

$$\begin{array}{l} \quad\quad\quad\ \ \text{①}\ x + 2 \\ \text{②}\ x - 3\overline{)x^2 - x + 9}\ \text{③} \\ \quad\quad\quad -(x^2 - 3x) \\ \quad\quad\quad\quad\quad 2x + 9 \\ \quad\quad\quad\quad -(2x - 6) \\ \quad\quad\quad\quad\quad\quad 15\ \text{④} \end{array}$$

1. ______________
2. ______________
3. ______________
4. ______________

Divide and check.

5. $\dfrac{32x^5 - 24x}{8}$

6. $\dfrac{12a^4 - 3a^2}{6}$

7. $\dfrac{u - 2u^2 + u^7}{u}$

8. $\dfrac{50x^5 - 7x^4 + x^2}{x}$

9. $(15t^3 - 24t^2 + 6t) \div (3t)$

10. $(20t^3 - 15t^2 + 30t) \div (5t)$

11. $(24t^5 - 40t^4 + 6t^3) \div (4t^3)$

12. $(18t^6 - 27t^5 - 3t^3) \div (9t^3)$

13. $(15x^7 - 21x^4 - 3x^2) \div (-3x^2)$

14. $(16x^6 + 32x^5 - 8x^2) \div (-8x^2)$

15. $\dfrac{8x^2 - 10x + 1}{2x}$

16. $\dfrac{9x^2 + 3x - 2}{3x}$

17. $\dfrac{9r^2s^2 + 3r^2s - 6rs^2}{-3rs}$

18. $\dfrac{4x^4y - 8x^6y^2 + 12x^8y^6}{4x^4y}$

19. $(10x^5y^2 + 15x^2y^2 - 5x^2y) \div (5x^2y)$

20. $(12a^3b^2 + 4a^4b^5 + 16ab^2) \div (4ab^2)$

21. $(x^2 + 10x + 21) \div (x + 7)$

22. $(y^2 - 8y + 16) \div (y - 4)$

23. $(a^2 - 8a - 16) \div (a + 4)$

24. $(y^2 - 10y - 25) \div (y - 5)$

25. $(2x^2 + 11x - 5) \div (x + 6)$

26. $(3x^2 - 2x - 13) \div (x - 2)$

Aha! 27. $(y^2 - 25) \div (y + 5)$

28. $(a^2 - 81) \div (a - 9)$

29. $\dfrac{a^3 + 8}{a + 2}$

30. $\dfrac{t^3 + 27}{t + 3}$

31. $\dfrac{t^2 - 13}{t - 4}$

32. $\dfrac{a^2 - 21}{a - 5}$

33. $\dfrac{2t^3 - 9t^2 + 11t - 3}{2t - 3}$

34. $\dfrac{8t^3 - 22t^2 - 5t + 12}{4t + 3}$

35. $(5x^2 - 14x) \div (5x + 1)$

36. $(3x^2 - 7x) \div (3x - 1)$

37. $(t^3 + t - t^2 - 1) \div (t + 1)$

38. $(x^3 + x - x^2 - 1) \div (x - 1)$

39. $(t^4 + 4t^2 + 3t - 6) \div (5 + t^2)$

40. $(t^4 - 2t^2 + 4t - 5) \div (t^2 - 3)$

41. $(4x^4 - 3 - x - 4x^2) \div (2x^2 - 3)$

42. $(x + 6x^4 - 4 - 3x^2) \div (1 + 2x^2)$

Use synthetic division to divide.

43. $(x^3 - 2x^2 + 2x - 7) \div (x + 1)$

44. $(x^3 - 2x^2 + 2x - 7) \div (x - 1)$

45. $(a^2 + 8a + 11) \div (a + 3)$

46. $(a^2 + 8a + 11) \div (a + 5)$

47. $(x^3 - 13x - 7x^2 + 3) \div (x - 2)$

48. $(x^3 - 13x - 7x^2 + 3) \div (x + 2)$

49. $(3x^3 + 7x^2 - 4x + 3) \div (x + 3)$

50. $(3x^3 + 7x^2 - 4x + 3) \div (x - 3)$

51. $(x^5 - 32) \div (x - 2)$

52. $(y^5 - 1) \div (y - 1)$

53. $(3x^3 + 1 - x + 7x^2) \div \left(x + \frac{1}{3}\right)$

54. $(8x^3 - 1 + 7x - 6x^2) \div \left(x - \frac{1}{2}\right)$

TW 55. How is the distributive law used when dividing a polynomial by a binomial?

TW 56. On an assignment, Katia *incorrectly* writes

$$\frac{12x^3 - 6x}{3x} = 4x^2 - 6x.$$

What mistake do you think she is making and how might you convince her that a mistake has been made?

SKILL REVIEW

Review graphing linear equations (Chapter 3).

Graph.

57. $3x - 4y = 12$ [3.3]

58. $y = -\frac{2}{3}x + 4$ [3.6]

59. $3y - 2 = 7$ [3.3]

60. $8x = 4y$ [3.2]

61. Find the slope of the line containing the points (3, 2) and $(-7, 5)$. [3.5]

62. Find the slope and the y-intercept of the line given by $2y = 8x + 7$. [3.6]

63. Find the slope–intercept form of the line with slope -5 and y-intercept $(0, -10)$. [3.6]

64. Find the slope–intercept form of the line containing the points (6, 3) and $(-2, -7)$. [3.7]

SYNTHESIS

TW 65. Explain how to construct a trinomial that has a remainder of 2 when divided by $x - 5$.

TW 66. Explain how the quotient of two binomials can have more than two terms.

Divide.

67. $(10x^{9k} - 32x^{6k} + 28x^{3k}) \div (2x^{3k})$

68. $(45a^{8k} + 30a^{6k} - 60a^{4k}) \div (3a^{2k})$

69. $(6t^{3h} + 13t^{2h} - 4t^h - 15) \div (2t^h + 3)$

70. $(x^4 + a^2) \div (x + a)$

71. $(5a^3 + 8a^2 - 23a - 1) \div (5a^2 - 7a - 2)$

72. $(15y^3 - 30y + 7 - 19y^2) \div (3y^2 - 2 - 5y)$

73. Divide the sum of $4x^5 - 14x^3 - x^2 + 3$ and $2x^5 + 3x^4 + x^3 - 3x^2 + 5x$ by $3x^3 - 2x - 1$.

74. Divide $5x^7 - 3x^4 + 2x^2 - 10x + 2$ by the sum of $(x - 3)^2$ and $5x - 8$.

If the remainder is 0 when one polynomial is divided by another, the divisor is a factor *of the dividend. Find the value(s) of c for which $x - 1$ is a factor of each polynomial.*

75. $x^2 - 4x + c$

76. $2x^2 - 3cx - 8$

77. $c^2x^2 + 2cx + 1$

78. Let

$$f(x) = \frac{3x + 7}{x + 2}.$$

a) Use division to find an expression equivalent to $f(x)$. Then graph f.

b) On the same set of axes, sketch both $g(x) = 1/(x + 2)$ and $h(x) = 1/x$.

c) How do the graphs of f, g, and h compare?

Try Exercise Answers: Section 5.8

9. $5t^2 - 8t + 2$ **17.** $-3rs - r + 2s$ **21.** $x + 3$ **23.** $a - 12 + \frac{32}{a + 4}$ **29.** $a^2 - 2a + 4$ **43.** $x^2 - 3x + 5 + \frac{-12}{x + 1}$ **51.** $x^4 + 2x^3 + 4x^2 + 8x + 16$

5.9 The Algebra of Functions

- The Sum, Difference, Product, or Quotient of Two Functions
- Domains and Graphs

STUDY TIP

Test Preparation

The best way to prepare for taking tests is by working consistently throughout the course. That said, here are some extra suggestions.

- Make up your own practice test.
- Ask your instructor or former students for old exams to practice on.
- Review your notes and all homework that gave you difficulty.
- Make use of the Study Summary, Review Exercises, and Test at the end of each chapter.

We now examine four ways in which functions can be combined.

THE SUM, DIFFERENCE, PRODUCT, OR QUOTIENT OF TWO FUNCTIONS

Suppose that a is in the domain of two functions, f and g. The input a is paired with $f(a)$ by f and with $g(a)$ by g. The outputs can then be added to get $f(a) + g(a)$.

EXAMPLE 1 Let $f(x) = x + 4$ and $g(x) = x^2 + 1$. Find $f(2) + g(2)$.

SOLUTION We visualize two function machines. Because 2 is in the domain of each function, we can compute $f(2)$ and $g(2)$.

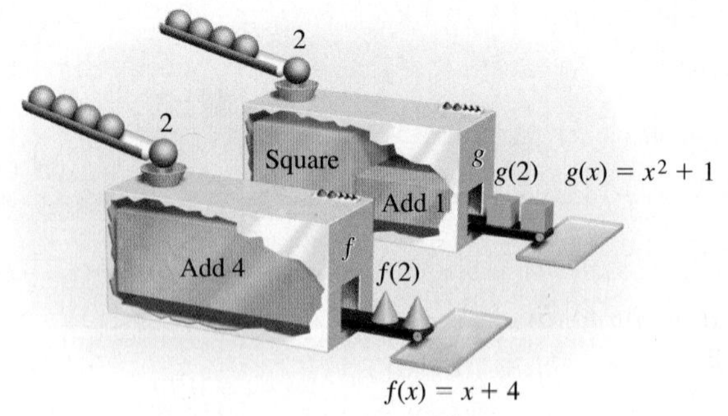

Since

$$f(2) = 2 + 4 = 6 \quad \text{and} \quad g(2) = 2^2 + 1 = 5,$$

we have

$$f(2) + g(2) = 6 + 5 = 11.$$

Try Exercise 7.

In Example 1, we could also add the expressions defining f and g:

$$f(x) + g(x) = (x + 4) + (x^2 + 1) = x^2 + x + 5.$$

This can then be regarded as a "new" function, written $(f + g)(x)$.

The Algebra of Functions If f and g are functions and x is in the domain of both functions, then:

1. $(f + g)(x) = f(x) + g(x)$;
2. $(f - g)(x) = f(x) - g(x)$;
3. $(f \cdot g)(x) = f(x) \cdot g(x)$;
4. $(f/g)(x) = f(x)/g(x)$, provided $g(x) \neq 0$.

EXAMPLE 2 For $f(x) = x^2 - x$ and $g(x) = x + 2$, find the following.

a) $(f + g)(3)$
b) $(f - g)(x)$ and $(f - g)(-1)$
c) $(f/g)(x)$ and $(f/g)(-4)$
d) $(f \cdot g)(3)$

CAUTION! Although $(f + g)(3) = f(3) + g(3)$, it is *not true* that $f(2 + 3)$ is the same as $f(2) + f(3)$. In the expression $f(2 + 3)$, the *inputs* are added; for $(f + g)(3)$, the functions themselves are added.

SOLUTION

a) Since $f(3) = 3^2 - 3 = 6$ and $g(3) = 3 + 2 = 5$, we have

$$\begin{aligned}(f + g)(3) &= f(3) + g(3)\\ &= 6 + 5 \qquad \textbf{Substituting}\\ &= 11.\end{aligned}$$

Alternatively, we could first find $(f + g)(x)$:

$$\begin{aligned}(f + g)(x) &= f(x) + g(x)\\ &= x^2 - x + x + 2\\ &= x^2 + 2. \qquad \textbf{Combining like terms}\end{aligned}$$

Then

$$(f + g)(3) = 3^2 + 2 = 11. \qquad \textbf{Our results match.}$$

b) We have

$$\begin{aligned}(f - g)(x) &= f(x) - g(x)\\ &= x^2 - x - (x + 2) \qquad \textbf{Substituting}\\ &= x^2 - 2x - 2. \qquad \textbf{Removing parentheses and combining like terms}\end{aligned}$$

Then

$$\begin{aligned}(f - g)(-1) &= (-1)^2 - 2(-1) - 2 \qquad \textbf{Using } (f - g)(x) \textbf{ is faster than using } f(x) - g(x).\\ &= 1. \qquad \textbf{Simplifying}\end{aligned}$$

c) We have

$$\begin{aligned}(f/g)(x) &= f(x)/g(x)\\ &= \frac{x^2 - x}{x + 2}. \qquad \textbf{We assume that } x \neq -2.\end{aligned}$$

Then

$$(f/g)(-4) = \frac{(-4)^2 - (-4)}{-4 + 2} \qquad \textbf{Substituting}$$

$$= \frac{20}{-2} = -10.$$

d) Using our work in part (a), we have

$$(f \cdot g)(3) = f(3) \cdot g(3)$$
$$= 6 \cdot 5$$
$$= 30.$$

Alternatively, we could first find $(f \cdot g)(x)$:

$$(f \cdot g)(x) = f(x) \cdot g(x)$$
$$= (x^2 - x)(x + 2)$$
$$= x^3 + x^2 - 2x. \qquad \textbf{Multiplying and combining like terms}$$

Then

$$(f \cdot g)(3) = 3^3 + 3^2 - 2 \cdot 3$$
$$= 27 + 9 - 6$$
$$= 30.$$

Try Exercise 17.

DOMAINS AND GRAPHS

Applications involving sums or differences of functions often appear in print. For example, the following graphs are similar to those published by the California Department of Education to promote breakfast programs in which students eat a balanced meal of fruit or juice, toast or cereal, and 2% or whole milk. The combination of carbohydrate, protein, and fat gives a sustained release of energy, delaying the onset of hunger for several hours.

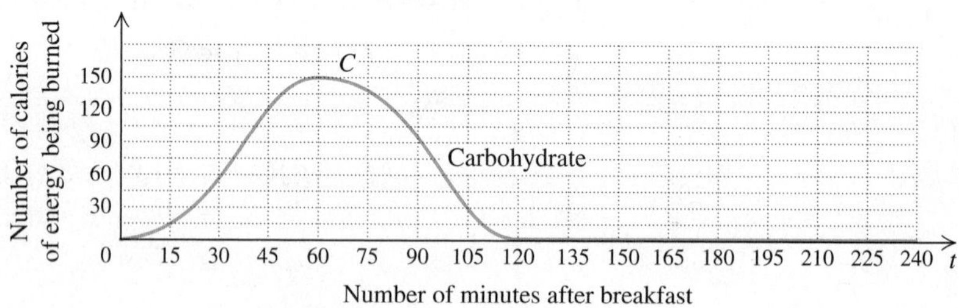

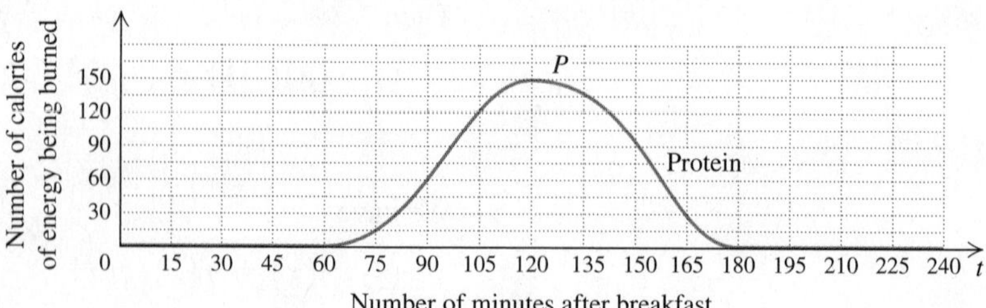

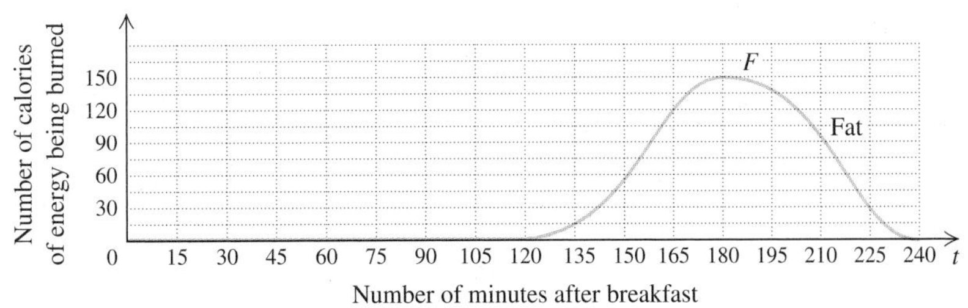

When the calorie expenditures are added, it becomes clear that a balanced meal results in a steady, sustained supply of energy.

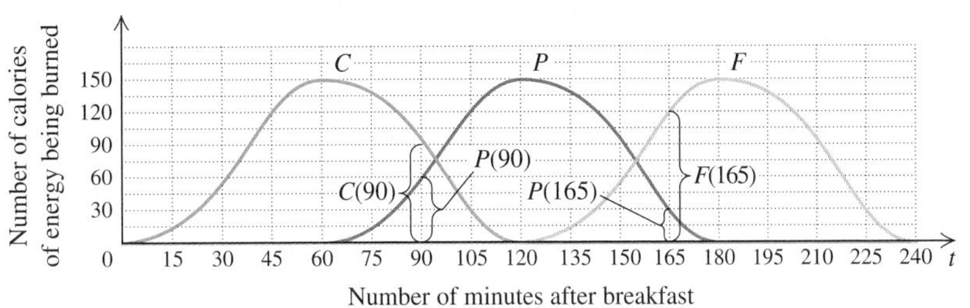

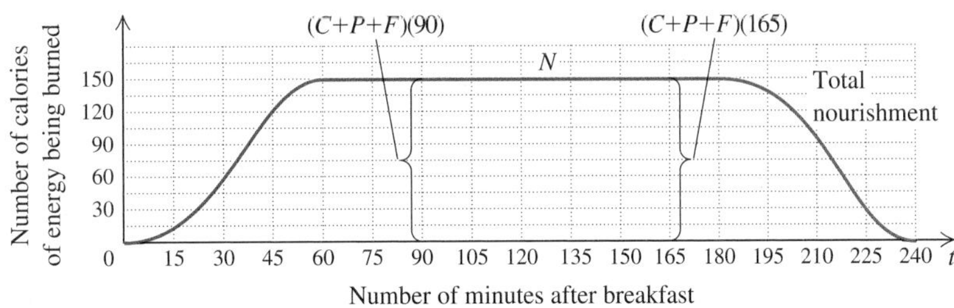

For any point $(t, N(t))$, we have

$$N(t) = (C + P + F)(t) = C(t) + P(t) + F(t).$$

To find $(f + g)(a)$, $(f - g)(a)$, $(f \cdot g)(a)$, or $(f/g)(a)$, we must know that $f(a)$ and $g(a)$ exist. This means that a must be in the domain of both f and g.

Interactive Discovery

Let $f(x) = \dfrac{5}{x}$ and $g(x) = \dfrac{2x - 6}{x + 1}$. Enter $y_1 = f(x)$, $y_2 = g(x)$, and $y_3 = y_1 + y_2$. Create a table of values for the functions with TblStart $= -3$, ΔTbl $= 1$, and Indpnt set to Auto.

1. What number is not in the domain of f?
2. What number is not in the domain of g?
3. What numbers are not in the domain of $f + g$?

Now enter $y_4 = y_1 - y_2$, $y_5 = y_1 \cdot y_2$, and $y_6 = y_1/y_2$. Create a table of values for the functions.

4. Does the domain of $f - g$ appear to be the same as the domain of $f + g$?
5. Does the domain of $f \cdot g$ appear to be the same as the domain of $f + g$?
6. Does the domain of f/g appear to be the same as the domain of $f + g$?

The table below summarizes the results of the Interactive Discovery above.

Function	Domain	Explanation
$f(x) = \frac{5}{x}$	$\{x \mid x$ is a real number *and* $x \neq 0\}$	The denominator, x, is 0 when $x = 0$.
$g(x) = \frac{2x - 6}{x + 1}$	$\{x \mid x$ is a real number *and* $x \neq -1\}$	The denominator, $x + 1$, is 0 when $x = -1$.
$f + g$ $f - g$ $f \cdot g$	$\{x \mid x$ is a real number *and* $x \neq 0$ *and* $x \neq -1\}$	0 is not in the domain of f. -1 is not in the domain of g.
f/g	$\{x \mid x$ is a real number *and* $x \neq 0$ *and* $x \neq -1$ *and* $x \neq 3\}$	0 is not in the domain of f. -1 is not in the domain of g. The denominator, g, is 0 when $x = 3$.

We can generalize the results of the Interactive Discovery. To find the domains of $f + g$, $f - g$, $f \cdot g$, and f/g, we use the domain of f, the domain of g, and any values of x for which $g(x) = 0$.

To find the domain of the sum, the difference, the product, or the quotient of two functions f and g:

1. Find the domain of f and the domain of g.
2. The functions $f + g$, $f - g$, and $f \cdot g$ have the same domain. It is the intersection of the domains of f and g, or, in other words, the set of all values common to the domains of f and g.
3. Find any values of x for which $g(x) = 0$.
4. The domain of f/g is the set found in step (2) (the set of all values common to the domains of f and g) *excluding* any values of x found in step (3).

Determining the Domain

The domain of $f + g$, $f - g$, or $f \cdot g$ is the set of all values common to the domains of f and g.

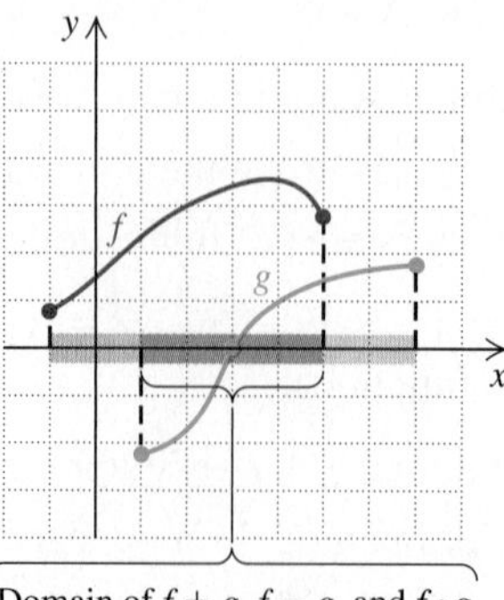

The domain of f/g is the set of all values common to the domains of f and g, excluding any values for which $g(x)$ is 0.

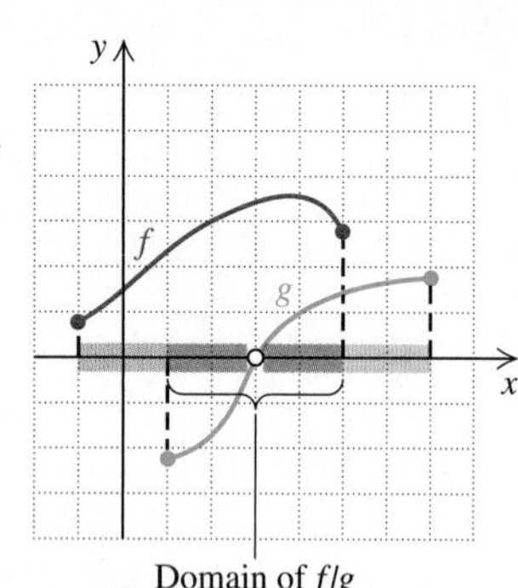

EXAMPLE 3 Given $f(x) = 1/x$ and $g(x) = 2x - 7$, find the domains of $f + g, f - g, f \cdot g$, and f/g.

SOLUTION We first find the domain of f and the domain of g:

Find the domains of f and g.

The domain of f is $\{x \mid x \text{ is a real number } \textit{and } x \neq 0\}$.

The domain of g is $\mathbb{R}$.

The domains of $f + g, f - g$, and $f \cdot g$ are the set of all elements common to the domains of f and g. This consists of all real numbers except 0.

Find the domain of $f + g, f - g$, and $f \cdot g$.

$$\text{The domain of } f + g = \text{the domain of } f - g = \text{the domain of } f \cdot g = \{x \mid x \text{ is a real number } \textit{and } x \neq 0\}.$$

Find any values of x for which $g(x) = 0$.

Because we cannot divide by 0, the domain of f/g must also exclude any values of x for which $g(x)$ is 0. We determine those values by solving $g(x) = 0$:

$$\begin{aligned} g(x) &= 0 \\ 2x - 7 &= 0 \qquad \text{Replacing } g(x) \text{ with } 2x - 7 \\ 2x &= 7 \\ x &= \tfrac{7}{2}. \end{aligned}$$

Find the domain of f/g.

The domain of f/g is the domain of the sum, the difference, and the product of f and g, found above, excluding $\frac{7}{2}$.

$$\text{The domain of } f/g = \{x \mid x \text{ is a real number } \textit{and } x \neq 0 \textit{ and } x \neq \tfrac{7}{2}\}.$$

Try Exercises 47 and 55.

EXAMPLE 4 Let $f(x) = 125x^3 - 8$ and $g(x) = 5x - 2$. If $F(x) = (f/g)(x)$, find a simplified expression for $F(x)$.

SOLUTION Since $(f/g)(x) = f(x)/g(x)$,

$$F(x) = \frac{125x^3 - 8}{5x - 2}.$$

Note that if $x = \frac{2}{5}$, $g(x) = 0$. Thus the domain of F cannot include $\frac{2}{5}$.

We can simplify the expression for $F(x)$:

$$\begin{array}{r|l} \begin{array}{r} 25x^2 + 10x + 4 \\ 5x - 2\overline{)125x^3 + 0x^2 + 0x - 8} \\ 125x^3 - 50x^2 \\ \hline 50x^2 + 0x \\ 50x^2 - 20x \\ \hline 20x - 8 \\ 20x - 8 \\ \hline 0. \end{array} & \begin{array}{l} \\ \text{Writing in the missing terms} \\ \\ \text{Subtracting } 125x^3 - 50x^2 \text{ from} \\ 125x^3 + 0x^2 \text{ and bringing down the } 0x \\ \text{Subtracting } 50x^2 - 20x \text{ from} \\ 50x^2 + 0x \text{ and bringing down the } -8 \\ \\ \end{array} \end{array}$$

Including the restriction on the domain noted above,

$$F(x) = 25x^2 + 10x + 4, \quad \text{provided } x \neq \tfrac{2}{5}.$$

Try Exercise 63.

Student Notes

The concern over a denominator being 0 arises throughout this course. Try to develop the habit of checking for any possible input values that would create a denominator of 0 whenever you work with functions.

Division by 0 is not the only condition that can force restrictions on the domain of a function. In Chapter 10, we will examine functions similar to that given by $f(x) = \sqrt{x}$, for which the concern is taking the square root of a negative number.

5.9 Exercise Set

FOR EXTRA HELP MyMathLab Math XL PRACTICE WATCH DOWNLOAD READ REVIEW

Concept Reinforcement *Make each of the following sentences true by selecting the correct word for each blank.*

1. The function $f - g$ is the ______ (sum/difference) of f and g.

2. One way to compute $(f - g)(2)$ is to ______ (erase/subtract) $g(2)$ from $f(2)$.

3. One way to compute $(f - g)(2)$ is to simplify $f(x) - g(x)$ and then ______ (evaluate/substitute) the result for $x = 2$.

4. The domain of $f + g$ is the set of all values ______ (common to/excluded from) the domains of f and g.

5. The domain of f/g is the set of all values common to the domains of f and g, ______ (including/excluding) any values for which $g(x)$ is 0.

6. The height of $(f + g)(a)$ on a graph is the ______ (product/sum) of the heights of $f(a)$ and $g(a)$.

Let $f(x) = -3x + 1$ and $g(x) = x^2 + 2$. Find each of the following.

7. $f(2) + g(2)$
8. $f(-1) + g(-1)$
9. $f(5) - g(5)$
10. $f(4) - g(4)$
11. $f(-1) \cdot g(-1)$
12. $f(-2) \cdot g(-2)$
13. $f(-4)/g(-4)$
14. $f(3)/g(3)$
15. $g(1) - f(1)$
16. $g(2)/f(2)$
17. $(f + g)(x)$
18. $(g - f)(x)$
19. $(f \cdot g)(x)$
20. $(g/f)(x)$

Let $F(x) = x^2 - 2$ and $G(x) = 5 - x$. Find each of the following.

21. $(F + G)(x)$
22. $(F + G)(a)$
23. $(F + G)(-4)$
24. $(F + G)(-5)$
25. $(F - G)(3)$
26. $(F - G)(2)$
27. $(F \cdot G)(-3)$
28. $(F \cdot G)(-4)$
29. $(F/G)(x)$
30. $(F \cdot G)(x)$
31. $(F - G)(x)$
32. $(G - F)(x)$
33. $(F/G)(-2)$
34. $(F/G)(-1)$

In 2004, a study comparing high doses of the cholesterol-lowering drugs Lipitor and Pravachol indicated that patients taking Lipitor were significantly less likely to have heart attacks or require angioplasty or surgery.

In the graph below, $L(t)$ is the percentage of patients on Lipitor (80 mg) and $P(t)$ is the percentage of patients on Pravachol (40 mg) who suffered heart problems or death t years after beginning to take the medication.

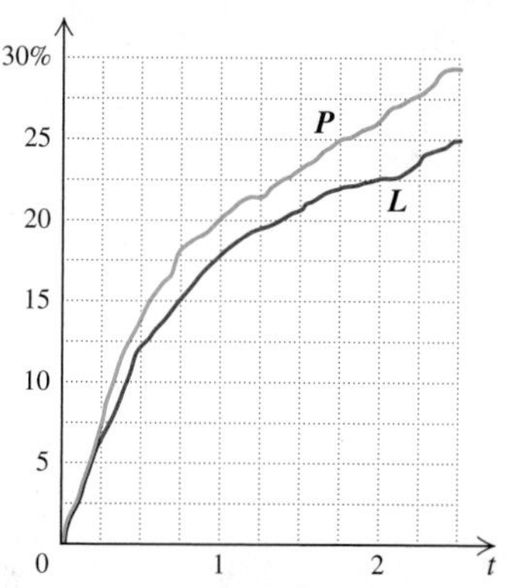

Source: New England Journal of Medicine

35. Use estimates of $P(2)$ and $L(2)$ to estimate $(P - L)(2)$.

36. Use estimates of $P(1)$ and $L(1)$ to estimate $(P - L)(1)$.

The graph below shows the number of births in the United States, in millions, from 1970–2004. Here $C(t)$ represents the number of Caesarean section births, $B(t)$ the number of non-Caesarean section births, and $N(t)$ the total number of births in year t.

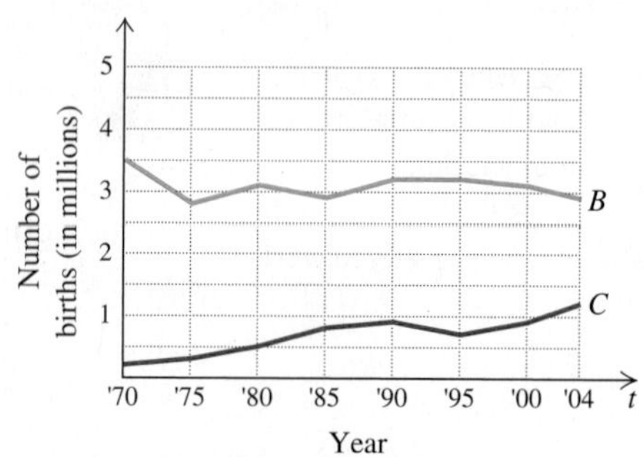

Source: National Center for Health Statistics

37. Use estimates of $C(2004)$ and $B(2004)$ to estimate $N(2004)$.

38. Use estimates of $C(1985)$ and $B(1985)$ to estimate $N(1985)$.

Often function addition is represented by stacking the individual functions directly on top of each other. The graph below indicates how U.S. municipal solid waste has been managed. The braces indicate the values of the individual functions.

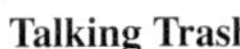

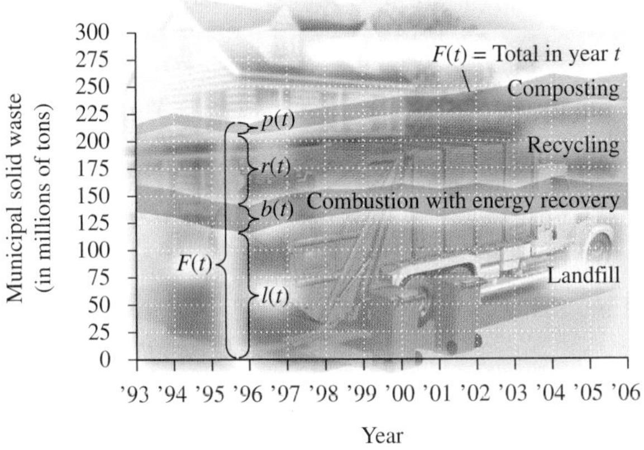

Source: Environmental Protection Agency

39. Estimate $(p + r)('05)$. What does it represent?

40. Estimate $(p + r + b)('05)$. What does it represent?

41. Estimate $F('96)$. What does it represent?

42. Estimate $F('06)$. What does it represent?

43. Estimate $(F - p)('04)$. What does it represent?

44. Estimate $(F - l)('03)$. What does it represent?

For each pair of functions f and g, determine the domain of the sum, the difference, and the product of the two functions.

45. $f(x) = x^2$, $g(x) = 7x - 4$

46. $f(x) = 5x - 1$, $g(x) = 2x^2$

47. $f(x) = \dfrac{1}{x - 3}$, $g(x) = 4x^3$

48. $f(x) = 3x^2$, $g(x) = \dfrac{1}{x - 9}$

49. $f(x) = \dfrac{2}{x}$, $g(x) = x^2 - 4$

50. $f(x) = x^3 + 1$, $g(x) = \dfrac{5}{x}$

51. $f(x) = x + \dfrac{2}{x - 1}$, $g(x) = 3x^3$

52. $f(x) = 9 - x^2$, $g(x) = \dfrac{3}{x - 6} + 2x$

53. $f(x) = \dfrac{x}{2x - 9}$, $g(x) = \dfrac{5}{1 - x}$

54. $f(x) = \dfrac{5}{3 - x}$, $g(x) = \dfrac{x}{4x - 1}$

For each pair of functions f and g, determine the domain of f/g.

55. $f(x) = x^4$, $g(x) = x - 3$

56. $f(x) = 2x^3$, $g(x) = 5 - x$

57. $f(x) = 3x - 2$, $g(x) = 2x - 8$

58. $f(x) = 5 + x$, $g(x) = 6 - 2x$

59. $f(x) = \dfrac{3}{x - 4}$, $g(x) = 5 - x$

60. $f(x) = \dfrac{1}{2 - x}$, $g(x) = 7 - x$

61. $f(x) = \dfrac{2x}{x + 1}$, $g(x) = 2x + 5$

62. $f(x) = \dfrac{7x}{x - 2}$, $g(x) = 3x + 7$

For Exercises 63–66, $f(x)$ and $g(x)$ are as given. Find a simplified expression for $F(x)$ if $F(x) - (f/g)(x)$. (See Example 4.)

63. $f(x) = 8x^3 + 27$, $g(x) = 2x + 3$

64. $f(x) = 64x^3 - 8$, $g(x) = 4x - 2$

65. $f(x) = 6x^2 - 11x - 10$, $g(x) = 3x + 2$

66. $f(x) = 8x^2 - 22x - 21$, $g(x) = 2x - 7$

For Exercises 67–74, consider the functions F and G as shown.

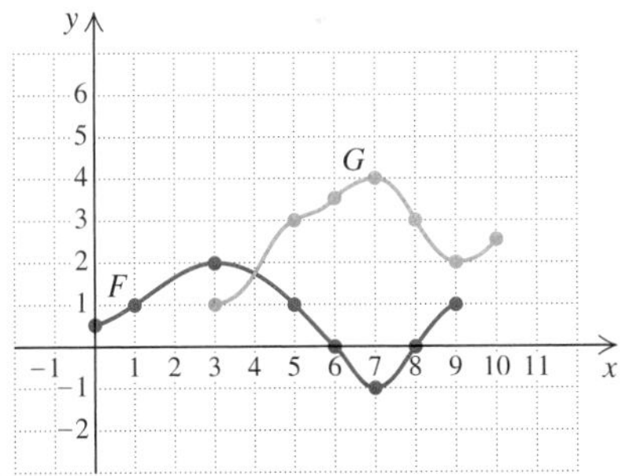

67. Determine $(F + G)(5)$ and $(F + G)(7)$.

68. Determine $(F \cdot G)(6)$ and $(F \cdot G)(9)$.

69. Determine $(G - F)(7)$ and $(G - F)(3)$.

70. Determine $(F/G)(3)$ and $(F/G)(7)$.

71. Find the domains of F, G, $F + G$, and F/G.

72. Find the domains of $F - G$, $F \cdot G$, and G/F.

73. Graph $F + G$.

74. Graph $G - F$.

In the graph below, $S(t)$ represents the number of gallons of carbonated soft drinks consumed by the average American in year t, $M(t)$ the number of gallons of milk, $J(t)$ the number of gallons of fruit juice, and $W(t)$ the number of gallons of bottled water.

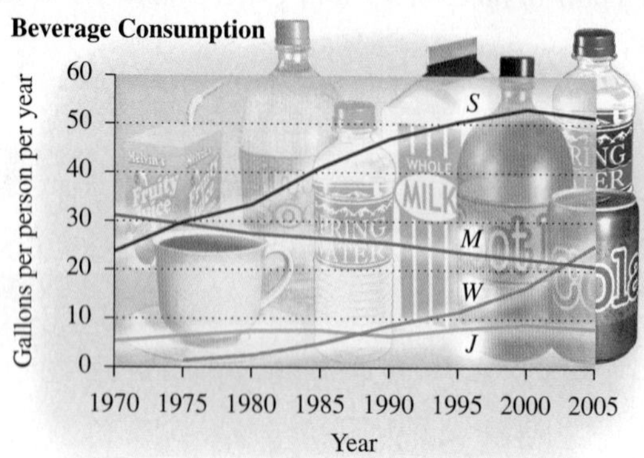

Source: Economic Research Service, U. S. Department of Agriculture

TW **75.** Between what years did the average American drink more soft drinks than juice, bottled water, and milk combined? Explain how you determined this.

TW **76.** Examine the graphs before Exercises 37 and 38. Did the total number of births increase or decrease from 1970 to 2004? Did the percent of births by Caesarean section increase or decrease from 1970 to 2004? Explain how you determined your answers.

SKILL REVIEW

To prepare for Chapter 6, review factoring expressions and solving equations (Sections 1.2 and 2.2).

Factor. [1.2]

77. $15x + 20y + 5$

78. $12m + 4n + 8$

Solve. [2.2]

79. $x + 5 = 0$

80. $4x + 9 = 0$

81. $3x - 1 = 0$

82. $4x = 0$

SYNTHESIS

TW **83.** Examine the graphs showing number of calories expended following Example 2 and explain how they might be modified to represent the absorption of 200 mg of Advil® taken four times a day.

TW **84.** If $f(x) = c$, where c is some positive constant, describe how the graphs of $y = g(x)$ and $y = (f + g)(x)$ will differ.

85. Find the domain of f/g, if

$$f(x) = \frac{3x}{2x + 5} \quad \text{and} \quad g(x) = \frac{x^4 - 1}{3x + 9}.$$

86. Find the domain of F/G, if

$$F(x) = \frac{1}{x - 4} \quad \text{and} \quad G(x) = \frac{x^2 - 4}{x - 3}.$$

87. Sketch the graph of two functions f and g such that the domain of f/g is

$$\{x \mid -2 \le x \le 3 \textit{ and } x \ne 1\}.$$

88. Find the domains of $f + g$, $f - g$, $f \cdot g$, and f/g, if

$$f = \{(-2, 1), (-1, 2), (0, 3), (1, 4), (2, 5)\}$$

and

$$g = \{(-4, 4), (-3, 3), (-2, 4), (-1, 0), (0, 5), (1, 6)\}.$$

89. Find the domain of m/n, if

$$m(x) = 3x \text{ for } -1 < x < 5$$

and

$$n(x) = 2x - 3.$$

90. For f and g as defined in Exercise 88, find $(f + g)(-2)$, $(f \cdot g)(0)$, and $(f/g)(1)$.

91. Write equations for two functions f and g such that the domain of $f + g$ is

$$\{x \mid x \text{ is a real number } \textit{and } x \ne -2 \textit{ and } x \ne 5\}.$$

92. Using the window $[-5, 5, -1, 9]$, graph $y_1 = 5$, $y_2 = x + 2$, and $y_3 = \sqrt{x}$. Then predict what shape the graphs of $y_1 + y_2$, $y_1 + y_3$, and $y_2 + y_3$ will take. Use a graph to check each prediction.

93. Let $y_1 = 2.5x + 1.5$, $y_2 = x - 3$, and $y_3 = y_1/y_2$. For many calculators, depending on whether the CONNECTED or DOT mode is used, the graph of y_3 appears as follows.

CONNECTED MODE

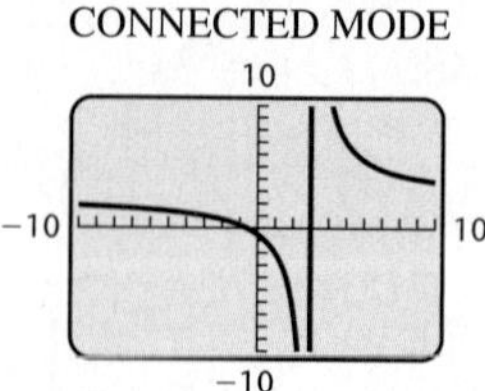

DOT MODE

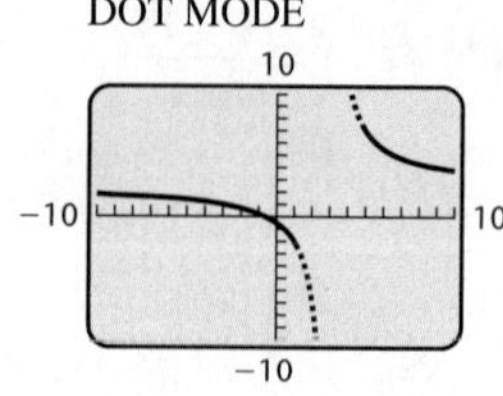

Use algebra to determine which graph more accurately represents y_3.

94. Use the graphs of f and g, shown below, to match each of $(f+g)(x)$, $(f-g)(x)$, $(f \cdot g)(x)$, and $(f/g)(x)$ with its graph.

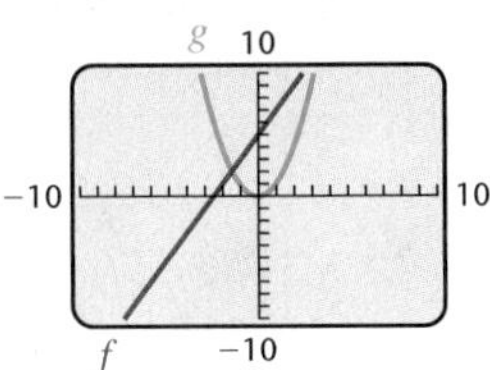

a) $(f+g)(x)$
b) $(f-g)(x)$
c) $(f \cdot g)(x)$
d) $(f/g)(x)$

I

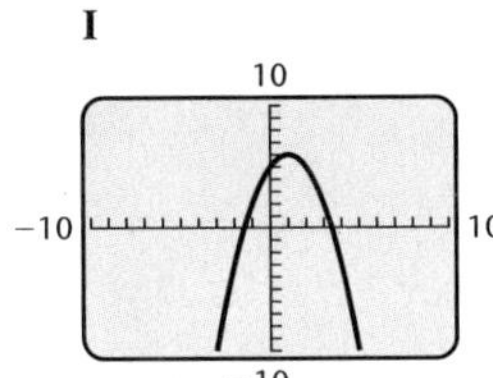

II

10

−10 10

−10

III

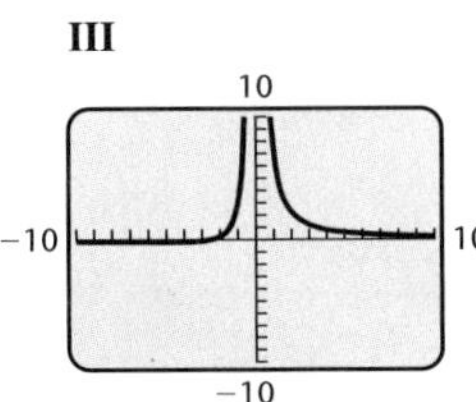

IV

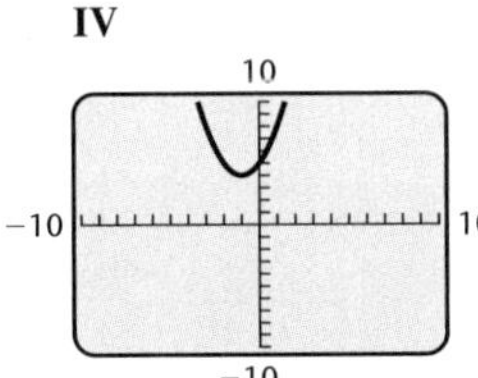

Try Exercise Answers: Section 5.9

7. 1 **17.** $x^2 - 3x + 3$
47. $\{x \mid x \text{ is a real number } and\ x \neq 3\}$
55. $\{x \mid x \text{ is a real number } and\ x \neq 3\}$
63. $4x^2 - 6x + 9, x \neq -\frac{3}{2}$

Collaborative Corner

Time On Your Hands

Focus: The algebra of functions
Time: 10–15 minutes
Group Size: 2–3

How much money do you need to retire? To answer this question, you might consider the graph and the data at right. They chart the average retirement age $R(x)$ and life expectancy $E(x)$ of U.S. citizens in year x.

ACTIVITY

1. Working as a team, perform the appropriate calculations and then graph $E - R$.
2. What does $(E - R)(x)$ represent? In what fields of study or business might the function $E - R$ prove useful?
3. Should E and R really be calculated separately for men and women? Why or why not?
4. What advice would you give to someone considering early retirement?

Year:	1955	1965	1975	1985	1995	2005
Average Retirement Age:	67.3	64.9	63.2	62.8	62.7	61.5
Average Life Expectancy:	73.9	75.5	77.2	78.5	79.1	80.1

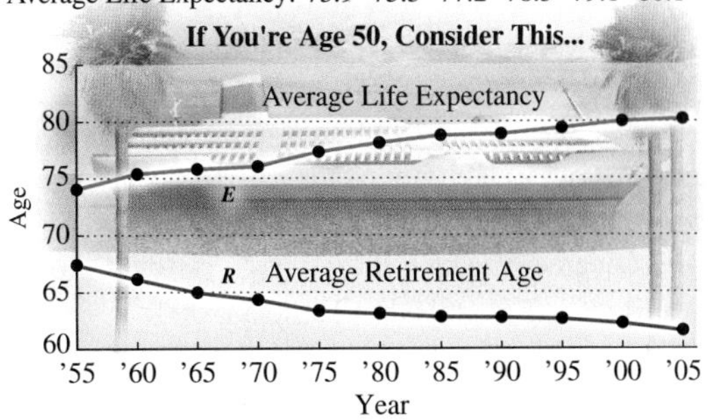

Source: U.S. Bureau of Labor Statistics

Study Summary

KEY TERMS AND CONCEPTS | **EXAMPLES** | **PRACTICE EXERCISES**

SECTION 5.1: EXPONENTS AND THEIR PROPERTIES

Key Terms and Concepts

- 1 as an exponent: $a^1 = a$
- 0 as an exponent: $a^0 = 1$
- The Product Rule: $a^m \cdot a^n = a^{m+n}$
- The Quotient Rule: $\dfrac{a^m}{a^n} = a^{m-n}$
- The Power Rule: $(a^m)^n = a^{mn}$
- Raising a product to a power: $(ab)^n = a^n b^n$
- Raising a quotient to a power: $\left(\dfrac{a}{b}\right)^n = \dfrac{a^n}{b^n}$

Examples

- $3^1 = 3$
- $3^0 = 1$
- $3^5 \cdot 3^9 = 3^{5+9} = 3^{14}$
- $\dfrac{3^7}{3} = 3^{7-1} = 3^6$
- $(3^4)^2 = 3^{4 \cdot 2} = 3^8$
- $(3x^5)^4 = 3^4(x^5)^4 = 81x^{20}$
- $\left(\dfrac{3}{x}\right)^6 = \dfrac{3^6}{x^6}$

Practice Exercises

Simplify.

1. 6^1
2. $(-5)^0$
3. x^5x^{11}
4. $\dfrac{8^9}{8^2}$
5. $(y^5)^3$
6. $(x^3y)^{10}$
7. $\left(\dfrac{x^2}{7}\right)^5$

SECTION 5.2: NEGATIVE EXPONENTS AND SCIENTIFIC NOTATION

Key Terms and Concepts

$a^{-n} = \dfrac{1}{a^n}$;

$\dfrac{a^{-n}}{b^{-m}} = \dfrac{b^m}{a^n}$

Examples

$3^{-2} = \dfrac{1}{3^2} = \dfrac{1}{9}$;

$\dfrac{3^{-7}}{x^{-5}} = \dfrac{x^5}{3^7}$

Practice Exercises

Write without negative exponents.

8. 10^{-1}
9. $\dfrac{x^{-1}}{1y^{-3}}$

Key Terms and Concepts

Scientific notation: $N \times 10^m, 1 \le N < 10.$

Examples

$4100 = 4.1 \times 10^3$;

$5 \times 10^{-3} = 0.005$

Practice Exercises

10. Convert to scientific notation: 0.000904.
11. Convert to decimal notation: 6.9×10^5.

SECTION 5.3: POLYNOMIALS AND POLYNOMIAL FUNCTIONS

Key Terms and Concepts

A **polynomial** is a monomial or a sum of monomials.

When a polynomial is written as a sum of monomials, each monomial is a **term** of the polynomial.

The **degree of a term** of a polynomial is the number of variable factors in that term.

The **coefficient** of a term is the part of the term that is a constant factor.

The **leading term** of a polynomial is the term of highest degree.

The **leading coefficient** is the coefficient of the leading term.

The **degree of the polynomial** is the degree of the leading term.

Examples

Polynomial: $10x - x^3 + 4x^5 + 7$

Term	$10x$	$-x^3$	$4x^5$	7
Degree of term	1	3	5	0
Coefficient of term	10	−1	4	7
Leading term	$4x^5$			
Leading coefficient	4			
Degree of polynomial	5			

Practice Exercises

For Exercises 12–17, consider the polynomial $x^2 - 10 + 5x - 8x^6$.

12. List the terms of the polynomial.
13. What is the degree of the term $5x$?
14. What is the coefficient of the term x^2?
15. What is the leading term of the polynomial?
16. What is the leading coefficient of the polynomial?
17. What is the degree of the polynomial?

A **monomial** has one term. A **binomial** has two terms. A **trinomial** has three terms.	Monomial: $4x^3$ Binomial: $x^2 - 5$ Trinomial: $3t^3 + 2t - 10$	**18.** Classify the polynomial $8x - 3 - x^4$ as a monomial, a binomial, a trinomial, or a polynomial with no special name.
Like terms, or **similar terms,** are either constant terms or terms containing the same variable(s) raised to the same power(s). These can be **combined** within a polynomial.	Combine like terms: $3y^4 + 6y^2 - 7 - y^4 - 6y^2 + 8.$ $3y^4 + 6y^2 - 7 - y^4 - 6y^2 + 8$ $= \underbrace{3y^4 - y^4} + \underbrace{6y^2 - 6y^2} \underbrace{- 7 + 8}$ $= \quad 2y^4 \quad + \quad 0 \quad + 1$ $= 2y^4 + 1$	**19.** Combine like terms: $3x^2 + 5x - 10x + x.$
To **evaluate** a polynomial, replace the variable with a number. The **value** is calculated using the rules for order of operations.	Evaluate $t^3 - 2t^2 - 5t + 1$ for $t = -2$. $t^3 - 2t^2 - 5t + 1$ $= (-2)^3 - 2(-2)^2 - 5(-2) + 1$ $= -8 - 2(4) - (-10) + 1$ $= -8 - 8 + 10 + 1$ $= -5$	**20.** Evaluate $2 - 3x - x^2$ for $x = -1$.
SECTION 5.4: ADDITION AND SUBTRACTION OF POLYNOMIALS		
Add polynomials by combining like terms.	$(2x^2 - 3x + 7) + (5x^3 + 3x - 9)$ $= 2x^2 + (-3x) + 7 + 5x^3 + 3x + (-9)$ $= 5x^3 + 2x^2 - 2$	**21.** Add: $(9x^2 - 3x) + (4x - x^2)$.
Subtract polynomials by adding the opposite of the polynomial being subtracted.	$(2x^2 - 3x + 7) - (5x^3 + 3x - 9)$ $= 2x^2 - 3x + 7 + (-5x^3 - 3x + 9)$ $= 2x^2 - 3x + 7 - 5x^3 - 3x + 9$ $= -5x^3 + 2x^2 - 6x + 16$	**22.** Subtract: $(9x^2 - 3x) - (4x - x^2)$.
SECTION 5.5: MULTIPLICATION OF POLYNOMIALS		
Multiply polynomials by multiplying each term of one polynomial by each term of the other.	$(x + 2)(x^2 - x - 1)$ $= x \cdot x^2 - x \cdot x - x \cdot 1 + 2 \cdot x^2 - 2 \cdot x - 2 \cdot 1$ $= x^3 - x^2 - x + 2x^2 - 2x - 2$ $= x^3 + x^2 - 3x - 2$	**23.** Multiply: $(x - 1)(x^2 - x - 2)$.
SECTION 5.6: SPECIAL PRODUCTS		
FOIL (First, Outer, Inner, Last): $(A + B)(C + D) = AC + AD + BC + BD$ 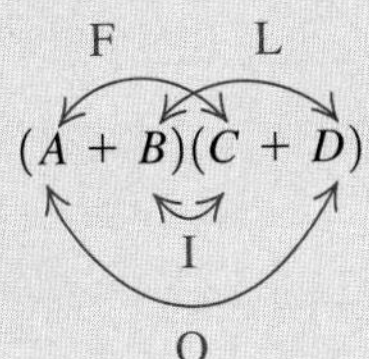	$(x + 3)(x - 2) = x^2 - 2x + 3x - 6$ $= x^2 + x - 6$	**24.** Multiply: $(x + 4)(2x + 3)$.

The product of a sum and a difference: $(A + B)(A - B) = A^2 - B^2$ $A^2 - B^2$ is called a **difference of squares.**	$(t^3 + 5)(t^3 - 5) = (t^3)^2 - 5^2$ $= t^6 - 25$	**25.** Multiply: $(5 + 3x)(5 - 3x)$.
The square of a binomial: $(A + B)^2 = A^2 + 2AB + B^2$; $(A - B)^2 = A^2 - 2AB + B^2$ $A^2 + 2AB + B^2$ and $A^2 - 2AB + B^2$ are called **perfect-square trinomials.**	$(5x + 3)^2 = (5x)^2 + 2(5x)(3) + 3^2$ $= 25x^2 + 30x + 9$; $(5x - 3)^2 = (5x)^2 - 2(5x)(3) + 3^2$ $= 25x^2 - 30x + 9$	*Multiply.* **26.** $(x + 9)^2$ **27.** $(8x - 1)^2$

SECTION 5.7: POLYNOMIALS IN SEVERAL VARIABLES

To **evaluate** a polynomial, replace each variable with a number and simplify.	Evaluate $4 - 3xy + x^2y$ for $x = 5$ and $y = -1$. $4 - 3xy + x^2y = 4 - 3(5)(-1) + (5)^2(-1)$ $= 4 - (-15) + (-25)$ $= -6$	**28.** Evaluate $xy - y^2 - 4x$ for $x = -2$ and $y = 3$.
The **degree** of a term is the number of variables in the term or the sum of the exponents of the variables.	The degree of $-19x^3yz^2$ is 6.	**29.** What is the degree of $4mn^5$?
Add, subtract, and multiply polynomials in several variables in the same way as polynomials in one variable.	$(3xy^2 - 4x^2y + 5xy) + (xy - 6x^2y)$ $= 3xy^2 - 10x^2y + 6xy$; $(3xy^2 - 4x^2y + 5xy) - (xy - 6x^2y)$ $= 3xy^2 + 2x^2y + 4xy$; $(2a^2b + 3a)(5a^2b - a)$ $= 10a^4b^2 + 13a^3b - 3a^2$	**30.** Add: $(3cd^2 + 2c) + (4cd - 9c)$. **31.** Subtract: $(8pw - p^2w) - (p^2w + 8pw)$. **32.** Multiply: $(7xy - x^2)^2$.

SECTION 5.8: DIVISION OF POLYNOMIALS

To divide a polynomial by a monomial, divide each term by the monomial. Divide coefficients and subtract exponents.	$\frac{3t^5 - 6t^4 + 4t^2 + 9t}{3t} = \frac{3t^5}{3t} - \frac{6t^4}{3t} + \frac{4t^2}{3t} + \frac{9t}{3t}$ $= t^4 - 2t^3 + \frac{4}{3}t + 3$	**33.** Divide: $\frac{4y^5 - 8y^3 + 16y^2}{4y^2}$.
To divide a polynomial by a binomial, use long division or synthetic division.	Divide: $(x^2 + 5x - 2) \div (x - 3)$. $\begin{array}{r} x + 8 \\ x - 3\overline{)x^2 + 5x - 2} \\ -(x^2 - 3x) \\ \hline 8x - 2 \\ -(8x - 24) \\ \hline 22 \end{array}$ $(x^2 + 5x - 2) \div (x - 3) = x + 8 + \frac{22}{x - 3}$	**34.** Divide: $(x^2 - x + 4) \div (x + 1)$.

SECTION 5.9: THE ALGEBRA OF FUNCTIONS

$(f+g)(x) = f(x) + g(x)$	For $f(x) = x^2 + 3x$ and $g(x) = x - 5$: $(f+g)(x) = f(x) + g(x) = x^2 + 3x + x - 5 = x^2 + 4x - 5;$	*For Exercises 35–38, consider the functions $f(x) = x - 2$ and $g(x) = x - 7$.* **35.** Find $(f+g)(x)$.
$(f-g)(x) = f(x) - g(x)$	$(f-g)(x) = f(x) - g(x) = x^2 + 3x - (x - 5) = x^2 + 2x + 5;$	**36.** Find $(f-g)(x)$. **37.** Find $(f \cdot g)(x)$.
$(f \cdot g)(x) = f(x) \cdot g(x)$	$(f \cdot g)(x) = f(x) \cdot g(x) = (x^2 + 3x)(x - 5) = x^3 - 2x^2 - 15x;$	**38.** Find $(f/g)(x)$.
$(f/g)(x) = f(x)/g(x)$, provided $g(x) \neq 0$	$(f/g)(x) = f(x)/g(x)$, provided $g(x) \neq 0$ $= \dfrac{x^2 + 3x}{x - 5}$, provided $x \neq 5$.	

Review Exercises 

Concept Reinforcement *Classify each of the following statements as either true or false.*

1. When two polynomials that are written in descending order are added, the result is generally written in ascending order. [5.4]
2. The product of the sum and the difference of the same two terms is a difference of squares. [5.6]
3. When a binomial is squared, the result is a perfect-square trinomial. [5.6]
4. FOIL can be used whenever two polynomials are being multiplied. [5.6]
5. The degree of a polynomial can exceed the value of the polynomial's leading coefficient. [5.3]
6. Scientific notation is used only for extremely large numbers. [5.2]
7. In order for $(f/g)(a)$ to exist, we must have $g(a) \neq 0$. [5.9]
8. A positive number raised to a negative exponent can never represent a negative number. [5.2]

Simplify. [5.1]

9. $y^7 \cdot y^3 \cdot y$

10. $(3x)^5 \cdot (3x)^9$

11. $t^8 \cdot t^0$

12. $\dfrac{4^5}{4^2}$

13. $\dfrac{(a+b)^4}{(a+b)^4}$

14. $\dfrac{-20m^4n^5}{10mn^3}$

15. $(-2xy^2)^3$

16. $(2x^3)(-3x)^2$

17. $(a^2b)(ab)^5$

18. $\left(\dfrac{3x^2}{2y^3}\right)^2$

19. Express using a positive exponent: m^{-7}. [5.2]

Simplify. Do not use negative exponents in the answer. [5.2]

20. $7^2 \cdot 7^{-4}$

21. $\dfrac{6a^{-5}b}{6a^8b^8}$

22. $(x^3)^{-4}$

23. $(2x^{-3}y)^{-2}$

24. $\left(\dfrac{2x}{y}\right)^{-3}$

25. Convert to decimal notation: 8.3×10^6. [5.2]

26. Convert to scientific notation: 0.0000328. [5.2]

Multiply or divide and write scientific notation for the result. Use the correct number of significant digits. [5.2]

27. $(3.8 \times 10^4)(5.5 \times 10^{-1})$

28. $\dfrac{1.28 \times 10^{-8}}{2.5 \times 10^{-4}}$

29. *Blood Donors.* Every 4–6 weeks, Nathan donates 0.5 L of blood. In one microliter of blood, there are about 4.5×10^6 red blood cells. Approximate the number of red blood cells in Nathan's typical donation. [*Hint*: 1 liter $= 10^6$ microliters.] [5.2]

Identify the terms of each polynomial. [5.3]

30. $3x^2 + 6x + \frac{1}{2}$

31. $-4y^5 + 7y^2 - 3y - 2$

List the coefficients of the terms in each polynomial. [5.3]

32. $7x^2 - x + 7$

33. $4x^3 + x^2 - 5x + \frac{5}{3}$

For each polynomial, **(a)** *list the degree of each term;* **(b)** *determine the leading term and the leading coefficient; and* **(c)** *determine the degree of the polynomial.* [5.3]

34. $4t^2 + 6 + 15t^3$

35. $-2x^5 + x^4 - 3x^2 + x$

Classify each polynomial as a monomial, a binomial, a trinomial, or a polynomial with no special name. [5.3]

36. $4x^3 - 1$

37. $4 - 9t^3 - 7t^4 + 10t^2$

38. $7y^2$

Combine like terms and write in descending order. [5.3]

39. $5x - x^2 + 4x$

40. $\frac{3}{4}x^3 + 4x^2 - x^3 + 7$

41. $-2x^4 + 16 + 2x^4 + 9 - 3x^5$

42. $-x + \frac{1}{2} + 14x^4 - 7x^2 - 1 - 4x^4$

43. Find $P(-1)$ for $P(x) = x^2 - 3x + 6$. [5.3]

44. Evaluate $3 - 5x$ for $x = -5$. [5.3]

45. Graph $p(x) = 2 - x^2$ in a standard viewing window and estimate the range of p. [5.3]

Medicine. Ibuprofen is a medication used to relieve pain. The polynomial function

$$M(t) = 0.5t^4 + 3.45t^3 - 96.65t^2 + 347.7t,$$
$$0 \le t \le 6,$$

can be used to estimate the number of milligrams of ibuprofen in the bloodstream t hours after 400 mg of the medication has been swallowed. Use the graph at the top of the next column for Exercises 46–49. [5.3]

Source: Based on data from Dr. P. Carey, Burlington, VT

Time (in hours)	Milligrams of Ibuprofen in Bloodstream
0.5	150
1	255
2.5	340
4.5	125
5.5	20

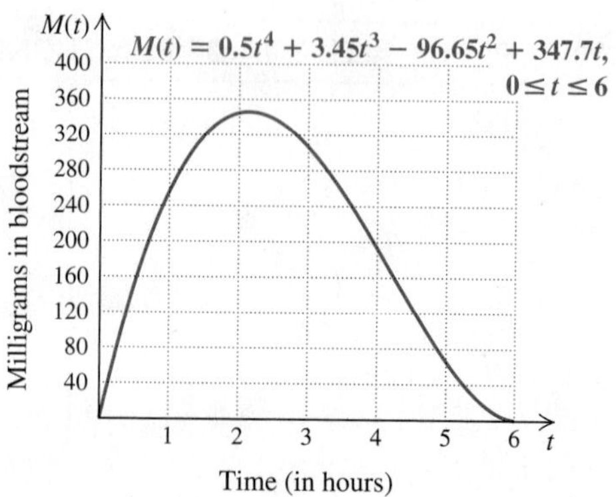

46. Use the graph above to estimate the number of milligrams of ibuprofen in the bloodstream 2 hr after 400 mg has been swallowed.

47. Use the graph above to estimate the number of milligrams of ibuprofen in the bloodstream 4 hr after 400 mg has been swallowed.

48. Approximate the range of M.

49. Give the domain of M.

Add or subtract. [5.4]

50. $(3x^4 - x^3 + x - 4) + (x^5 + 7x^3 - 3x - 5)$

51. $(5x^2 - 4x + 1) - (3x^2 + 7)$

52. $(3x^5 - 4x^4 + 2x^2 + 3) - (2x^5 - 4x^4 + 3x^3 + 4x^2 - 5)$

53.
$$\begin{array}{rrrrr} -\frac{3}{4}x^4 & +\frac{1}{2}x^3 & & & +\frac{7}{8} \\ & -\frac{1}{4}x^3 & -x^2 & -\frac{7}{4}x & \\ +\frac{3}{2}x^4 & & +\frac{2}{3}x^2 & & -\frac{1}{2} \\ \hline \end{array}$$

54.
$$\begin{array}{rrrrrr} 2x^5 & & -x^3 & & +x & +3 \\ -(3x^5 & -x^4 & +4x^3 & +2x^2 & -x & +3) \\ \hline \end{array}$$

55. The length of a rectangle is 3 m greater than its width.

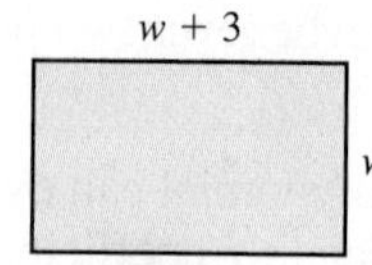

a) Find a polynomial for the perimeter. [5.4]
b) Find a polynomial for the area. [5.5]

Multiply.

56. $3x(-4x^2)$ [5.5]

57. $(7x + 1)^2$ [5.6]

58. $(a - 7)(a + 4)$ [5.6]

59. $(m + 5)(m - 5)$ [5.6]

60. $(4x^2 - 5x + 1)(3x - 2)$ [5.5]

61. $(x - 9)^2$ [5.6]

62. $3t^2(5t^3 - 2t^2 + 4t)$ [5.5]

63. $(3a + 8)(3a - 8)$ [5.6]

64. $(x - 0.3)(x - 0.5)$ [5.6]

65. $(x^4 - 2x + 3)(x^3 + x - 1)$ [5.5]

66. $(3x^4 - 5)^2$ [5.6]

67. $(2t^2 + 3)(t^2 - 7)$ [5.6]

68. $\left(a - \frac{1}{2}\right)\left(a + \frac{2}{3}\right)$ [5.6]

69. $(-7 + 2n)(7 + 2n)$ [5.6]

70. Evaluate $2 - 5xy + y^2 - 4xy^3 + x^6$ for $x = -1$ and $y = 2$. [5.7]

Identify the coefficient and the degree of each term of each polynomial. Then find the degree of each polynomial. [5.7]

71. $x^5y - 7xy + 9x^2 - 8$

72. $x^2y^5z^9 - y^{40} + x^{13}z^{10}$

Combine like terms. [5.7]

73. $y + w - 2y + 8w - 5$

74. $6m^3 + 3m^2n + 4mn^2 + m^2n - 5mn^2$

Add or subtract. [5.7]

75. $(5x^2 - 7xy + y^2) + (-6x^2 - 3xy - y^2)$

76. $(6x^3y^2 - 4x^2y - 6x) - (-5x^3y^2 + 4x^2y + 6x^2 - 6)$

Multiply. [5.7]

77. $(p - q)(p^2 + pq + q^2)$

78. $(5ab - cd^2)^2$

79. Find a polynomial for the shaded area. [5.7]

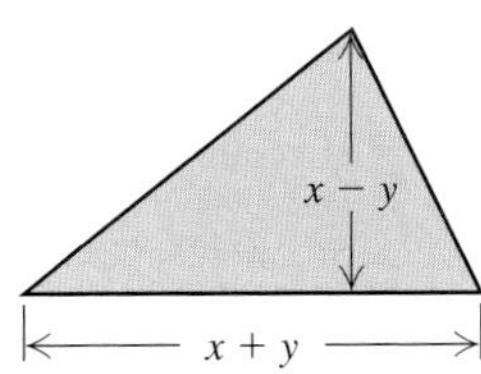

Divide. [5.8]

80. $(10x^3 - x^2 + 6x) \div (2x)$

81. $(6x^3 - 5x^2 - 13x + 13) \div (2x + 3)$

82. $\dfrac{t^4 + t^3 + 2t^2 - t - 3}{t + 1}$

Let $g(x) = 3x - 6$ and $h(x) = x^2 + 1$. Find the following. [5.9]

83. $(g \cdot h)(4)$

84. $(g - h)(-2)$

85. $(g/h)(-1)$

86. $(g + h)(x)$

87. $(g \cdot h)(x)$

88. $(h/g)(x)$

89. The domains of $g + h$ and $g \cdot h$

90. The domain of h/g

SYNTHESIS

TW **91.** Explain why $5x^3$ and $(5x)^3$ are not equivalent expressions. [5.1]

TW **92.** A binomial is squared and the result, written in descending order, is $x^2 - 6x + 9$. Is it possible to determine what binomial was squared? Why or why not? [5.6]

93. Determine, without performing the multiplications, the degree of each product. [5.5]

a) $(x^5 - 6x^2 + 3)(x + x^7 + 11)$
b) $(x^3 - 1)^5$

94. Combine like terms: [5.1], [5.3]

$$-3x^5 \cdot 3x^3 - x^6(2x)^2 + (3x^4)^2 + (2x^2)^4 - 40x^2(x^3)^2.$$

95. A polynomial has degree 4. The x^2-term is missing. The coefficient of x^4 is two times the coefficient of x^3. The coefficient of x is 3 less than the coefficient of x^4. The remaining coefficient is 7 less than the coefficient of x. The sum of the coefficients is 15. Find the polynomial. [5.3]

Aha! **96.** Multiply: $[(x - 5) - 4x^3][(x - 5) + 4x^3]$. [5.6]

97. Solve: $(x - 7)(x + 10) = (x - 4)(x - 6)$. [2.2], [5.6]

Chapter Test 

Simplify.

1. $t^2 \cdot t^5 \cdot t$
2. $(x^4)^9$
3. $\dfrac{3^5}{3^2}$
4. $\dfrac{(2x)^5}{(2x)^5}$
5. $(5x^4y)(-2x^5y)^3$
6. $\dfrac{-24a^7b^4}{8a^2b}$
7. Express using a positive exponent: 5^{-3}.

Simplify. Do not use negative exponents in the answer.

8. $t^{-4} \cdot t^{-2}$
9. $\dfrac{12x^3y^2}{15x^8y^{-3}}$
10. $(2a^3b^{-1})^{-4}$
11. $\left(\dfrac{ab}{c}\right)^{-3}$
12. Convert to scientific notation: 3,060,000,000.
13. Convert to decimal notation: 5×10^{-4}.

Multiply or divide and write scientific notation for the result. Use the correct number of significant digits.

14. $\dfrac{5.6 \times 10^6}{3.2 \times 10^{-11}}$
15. $(2.4 \times 10^5)(5.4 \times 10^{16})$
16. Every day about 12.4 billion spam e-mails are sent. If each spam e-mail wastes 4 sec, how many hours are wasted each day due to spam?
Source: spam-filter-review.toptenreviews.com
17. Classify $6t^2 - 9t$ as a monomial, a binomial, a trinomial, or a polynomial with no special name.
18. Identify the coefficient of each term of the polynomial
$$\tfrac{1}{3}x^5 - x + 7.$$
19. Determine the degree of each term, the leading term and the leading coefficient, and the degree of the polynomial:
$$2t^3 - t + 7t^5 + 4.$$
20. Find $p(-2)$ for $p(x) = x^2 + 5x - 1$.

Combine like terms and write in descending order.

21. $4a^2 - 6 + a^2$
22. $y^2 - 3y - y + \tfrac{3}{4}y^2$
23. $3 - x^2 + 2x^3 + 5x^2 - 6x - 2x + x^5$
24. Graph $f(x) = x^3 - x + 1$ in the standard viewing window and estimate the range of f.

Add or subtract.

25. $(3x^5 + 5x^3 - 5x^2 - 3) + (x^5 + x^4 - 3x^2 + 2x - 4)$
26. $(2x^4 + x^3 - 8x^2 - 6x - 3) - (6x^4 - 8x^2 + 2x)$
27. $(t^3 - 0.3t^2 - 20) - (t^4 - 1.5t^3 + 0.3t^2 - 11)$

Multiply.

28. $-3x^2(4x^2 - 3x - 5)$
29. $\left(x - \tfrac{1}{3}\right)^2$
30. $(5t - 7)(5t + 7)$
31. $(3b + 5)(b - 3)$
32. $(x^6 - 4)(x^8 + 4)$
33. $(8 - y)(6 + 5y)$
34. $(2x + 1)(3x^2 - 5x - 3)$
35. $(8a + 3)^2$
36. Combine like terms:
$$x^3y - y^3 + xy^3 + 8 - 6x^3y - x^2y^2 + 11.$$
37. Subtract:
$$(8a^2b^2 - ab + b^3) - (-6ab^2 - 7ab - ab^3 + 5b^3).$$
38. Multiply: $(3x^5 - 4y)(3x^5 + 4y)$.

Divide.

39. $(12x^4 + 9x^3 - 15x^2) \div (3x^2)$
40. $(6x^3 - 8x^2 - 14x + 13) \div (x + 2)$

Find each of the following, given that $g(x) = 1/x$ and $h(x) = 2x + 1$.

41. $(g \cdot h)(3)$
42. $(g + h)(x)$
43. The domain of g/h

SYNTHESIS

44. The height of a box is 1 less than its length, and the length is 2 more than its width. Express the volume in terms of the length.
45. Simplify: $2^{-1} - 4^{-1}$.

6

Polynomial Factorizations and Equations

How Does Experience Affect a Football Player's Salary?

As the graph shows, the NFL minimum salary increases with experience, but not linearly. In Example 6 of Section 6.7, we estimate how many seasons an athlete has played if his minimum salary is $900,000.

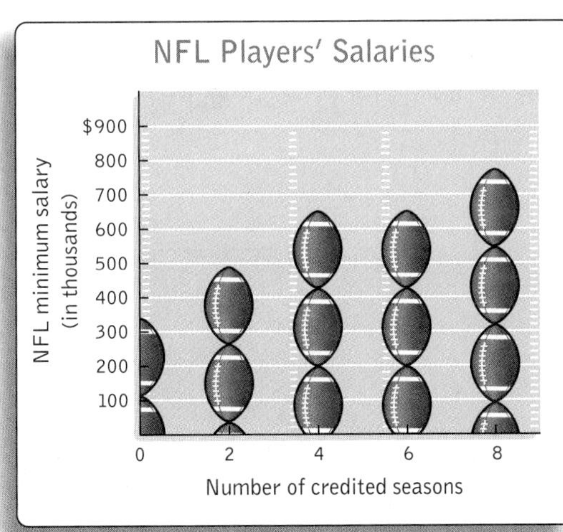

Factoring is writing an expression as a product. Thus factoring polynomials requires a solid command of the multiplication methods studied in Chapter 5. In this chapter, we factor polynomials and use factoring to solve equations.

6.1 Introduction to Polynomial Factorizations and Equations

- Graphical Solutions
- The Principle of Zero Products
- Terms with Common Factors
- Factoring by Grouping
- Factoring and Equations

STUDY TIP

Math Web Sites

If for some reason you feel uneasy about asking an instructor or a tutor for help, you may benefit from visiting math Web sites. Even if a site is not geared for adult learners, the math content can be very helpful nonetheless.

Whenever two polynomials are set equal to each other, the result is a **polynomial equation**. In this section, we discuss solving such equations both graphically and algebraically.

GRAPHICAL SOLUTIONS

EXAMPLE 1 Solve: $x^2 = 6x$.

SOLUTION We can find real-number solutions of a polynomial equation by finding the points of intersection of two graphs.

Alternatively, we can rewrite the equation so that one side is 0 and then find the x-intercepts of one graph, or the zeros of a function. Both methods were discussed in Section 4.5.

INTERSECT METHOD

To solve $x^2 = 6x$ using the first method, we graph $y_1 = x^2$ and $y_2 = 6x$ and find the coordinates of any points of intersection.

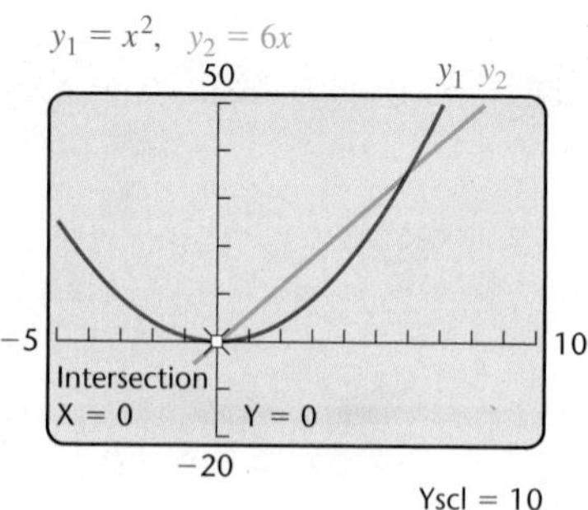

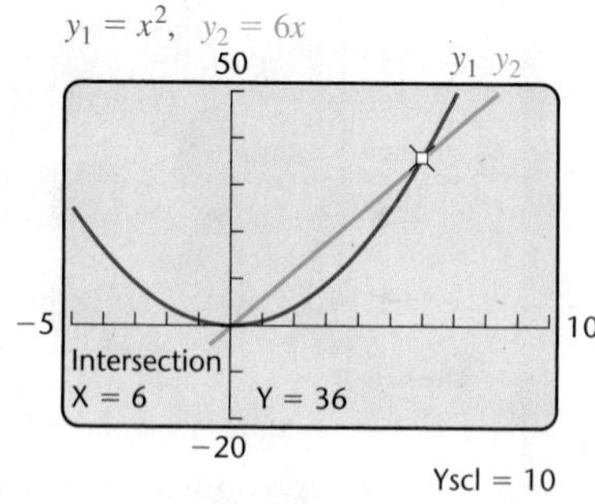

There are two points of intersection of the graphs, so there are two solutions of the equation. The x-coordinates of the points of intersection are 0 and 6, so the solutions of the equation are 0 and 6.

ZERO METHOD

Using the second method, we rewrite the equation so that one side is 0:

$$x^2 = 6x$$
$$x^2 - 6x = 0. \qquad \textbf{Adding } -6x \textbf{ to both sides}$$

To solve $x^2 - 6x = 0$, we graph the function $f(x) = x^2 - 6x$ and look for values of x for which $f(x) = 0$, or the zeros of f. These correspond to the x-intercepts of the graph.

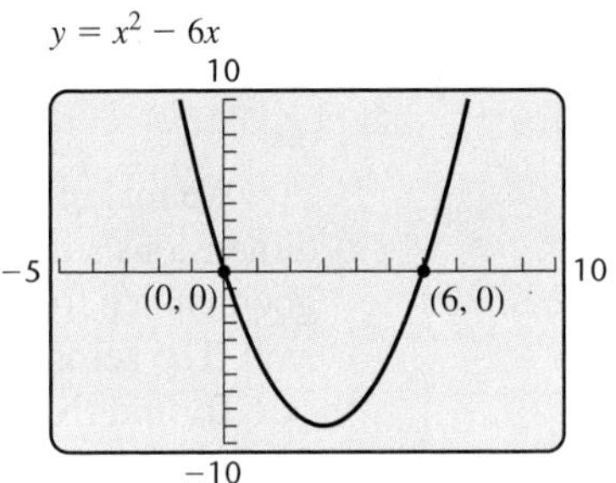

Student Notes

Recall that a zero of a function is a number, not an ordered pair. In Example 1, the zeros are the first coordinates of the x-intercepts.

The solutions of the equation are the x-coordinates of the x-intercepts of the graph, 0 and 6. Both 0 and 6 check in the original equation.

Try Exercise 19.

In Example 1, the values 0 and 6 are called **zeros** of the function $f(x) = x^2 - 6x$. They are also referred to as **roots** of the equation $f(x) = 0$.

Zeros and Roots The x-values for which a function $f(x)$ is 0 are called the *zeros* of the function.

The x-values for which an equation such as $f(x) = 0$ is true are called the *roots* of the equation.

In this chapter, we consider only the real-number zeros of functions. We can solve, or find the roots of, the equation $f(x) = 0$ by finding the zeros of the function f.

EXAMPLE 2 Find the zeros of the function given by

$$f(x) = x^3 - 3x^2 - 4x + 12.$$

SOLUTION First, we graph the equation $y = x^3 - 3x^2 - 4x + 12$, choosing a viewing window that shows the x-intercepts of the graph. It may require trial and error to choose an appropriate viewing window. In the standard viewing window, shown on the left at the top of the next page, it appears that there are three zeros. However, we cannot see the shape of the graph for x-values between -2 and 1, so Ymax should be increased. A viewing window of $[-10, 10, -100, 100]$, with Yscl = 10, as shown in the middle below, gives a better idea of the overall shape of the graph, but we cannot see all three x-intercepts clearly.

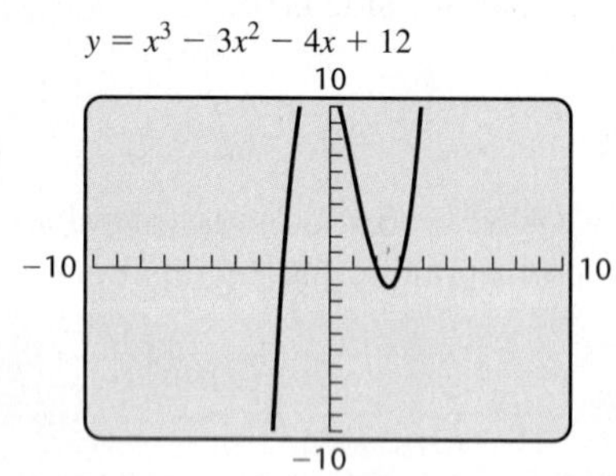

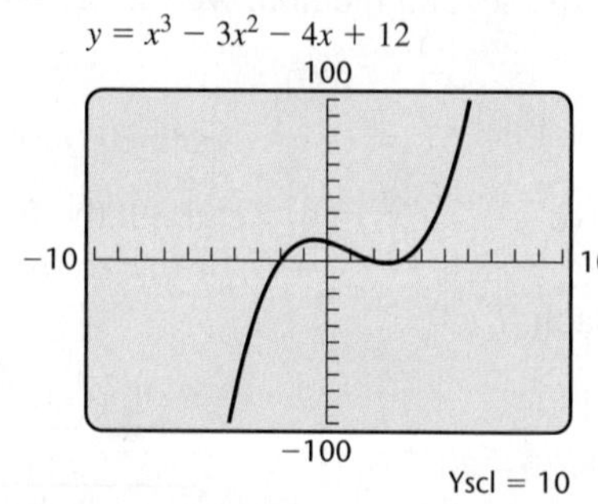

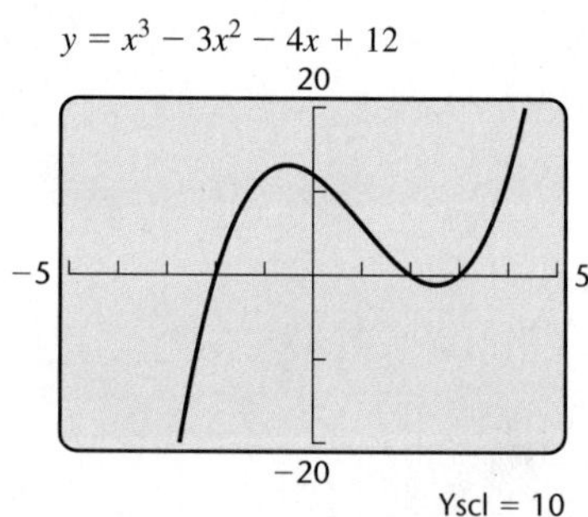

We magnify the portion of the graph close to the origin by making the window dimensions smaller. The window shown on the right above, $[-5, 5, -20, 20]$, is a good choice for viewing the zeros of this function. There are other good choices as well. The zeros of the function seem to be about -2, 2, and 3.

To find the zero that appears to be about -2, we first choose the ZERO option from the CALC menu. We then choose a Left Bound to the left of -2 on the x-axis. Next, we choose a Right Bound to the right of -2 on the x-axis. For a Guess, we choose an x-value close to -2. We see that -2 is indeed a zero of the function f.

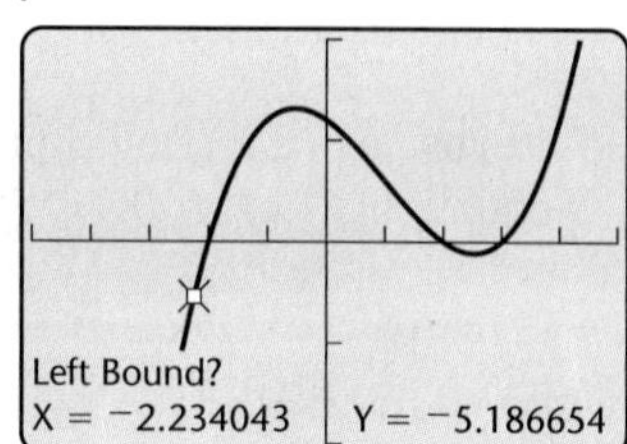

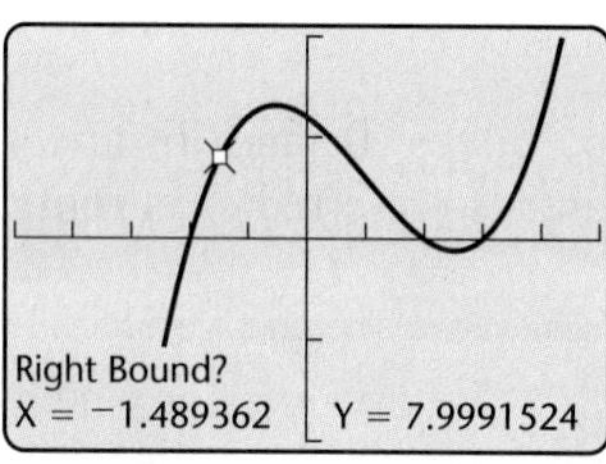

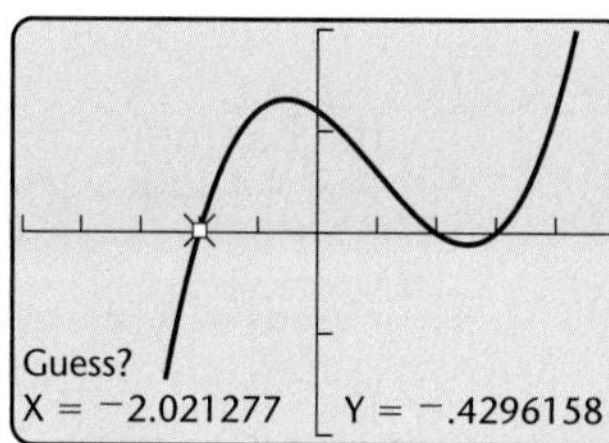

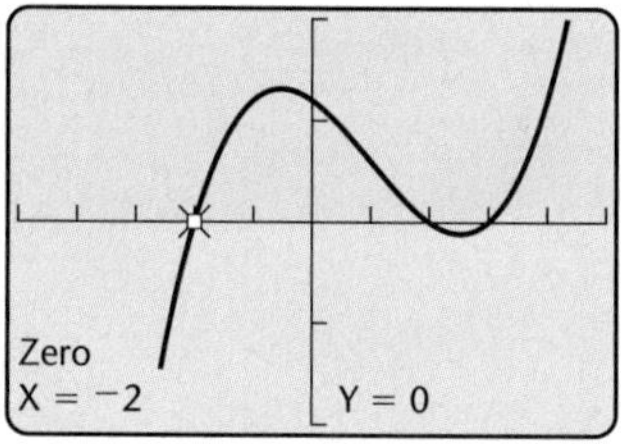

Using the same procedure for each of the other two zeros, we find that the zeros of the function $f(x) = x^3 - 3x^2 - 4x + 12$ are -2, 2, and 3.

Try Exercise 35.

From Examples 1 and 2, we see that $f(x) = x^2 - 6x$ has 2 zeros and that $f(x) = x^3 - 3x^2 - 4x + 12$ has 3 zeros. We can tell something about the number of zeros of a polynomial function from its degree.

Interactive Discovery

Each of the following is a second-degree polynomial (quadratic) function. Use a graph to determine the *number* of real-number zeros of each function.

1. $f(x) = x^2 - 2x + 1$
2. $g(x) = x^2 + 9$
3. $h(x) = 2x^2 + 7x - 4$

Each of the following is a third-degree polynomial (cubic) function. Use a graph to determine the *number* of real-number zeros of each function.

4. $f(x) = 2x^3 - 5x^2 - 3x$
5. $g(x) = x^3 - 7x^2 + 16x - 12$
6. $h(x) = x^3 - 4x^2 + 5x - 20$

7. Compare the degree of each function with the number of real-number zeros of that function. What conclusion can you draw?

A second-degree polynomial function will have 0, 1, or 2 real-number zeros. A third-degree polynomial function will have 1, 2, or 3 real-number zeros. This result can be generalized, although we will not prove it here.

An nth-degree polynomial function will have at most n zeros.

"Seeing" all the zeros of a polynomial function when solving an equation graphically depends on a good choice of viewing window. Many polynomial equations can be solved algebraically. One principle used in solving polynomial equations is the *principle of zero products*.

THE PRINCIPLE OF ZERO PRODUCTS

When we multiply two or more numbers, the product is 0 if any one of those numbers (factors) is 0. Conversely, if a product is 0, then at least one of the factors must be 0. This property of 0 gives us a new principle for solving equations.

The Principle of Zero Products For any real numbers a and b: If $ab = 0$, then $a = 0$ or $b = 0$. If $a = 0$ or $b = 0$, then $ab = 0$.

When a polynomial is written as a product, we say that it is *factored*. The product is called a *factorization* of the polynomial. For example, $3x(x + 4)$ is a factorization of $3x^2 + 12x$. The polynomials $3x$ and $x + 4$ are *factors* of $3x^2 + 12x$.

Suppose that we want to solve $3x^2 + 12x = 0$. Using the factorization of $3x^2 + 12x$, we see that the equation becomes

$$3x(x + 4) = 0.$$

The solutions of this equation are all x-values that make the equation true. We can see that $3x = 0$ when $x = 0$ and that $x + 4 = 0$ when $x = -4$. Their product is 0 when $x = 0$ *or* when $x = -4$.

For $x = 0$: $\quad 3x(x + 4) = (3 \cdot 0)(0 + 4) = 0 \cdot 4 = 0.$

For $x = -4$: $\quad 3x(x + 4) = (3(-4))(-4 + 4) = -12 \cdot 0 = 0.$

Thus the solutions of $3x^2 + 12x = 0$ are 0 and -4. We can use the principle of zero products and factorizations to solve a polynomial equation.

EXAMPLE 3 Solve: $(x - 3)(x + 2) = 0$.

SOLUTION The principle of zero products says that in order for $(x - 3)(x + 2)$ to be 0, at least one factor must be 0. Thus,

$x - 3 = 0 \quad or \quad x + 2 = 0.$ **Using the principle of zero products**

Each of these linear equations is then solved separately:

$x = 3 \quad or \quad x = -2.$

We check both algebraically and graphically, as follows.

ALGEBRAIC CHECK

For 3:

$$\begin{array}{c|c} (x-3)(x+2) = 0 & \\ \hline (3-3)(3+2) & 0 \\ 0(5) & \\ 0 \stackrel{?}{=} 0 & \text{TRUE} \end{array}$$

For $x = 3$, $x - 3 = 0$.

For -2:

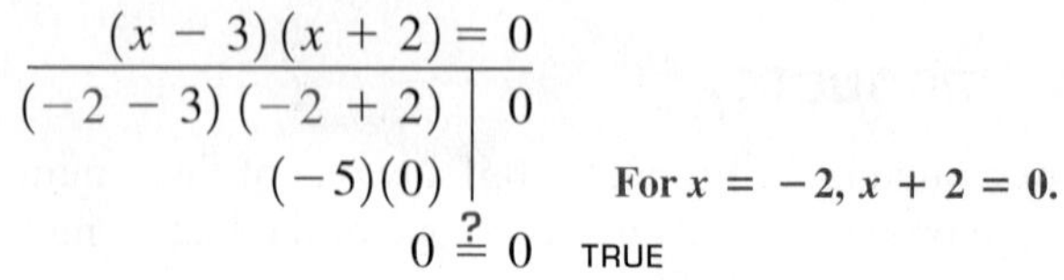

$$\begin{array}{c|c} (x-3)(x+2) = 0 & \\ \hline (-2-3)(-2+2) & 0 \\ (-5)(0) & \\ 0 \stackrel{?}{=} 0 & \text{TRUE} \end{array}$$

For $x = -2$, $x + 2 = 0$.

GRAPHICAL CHECK

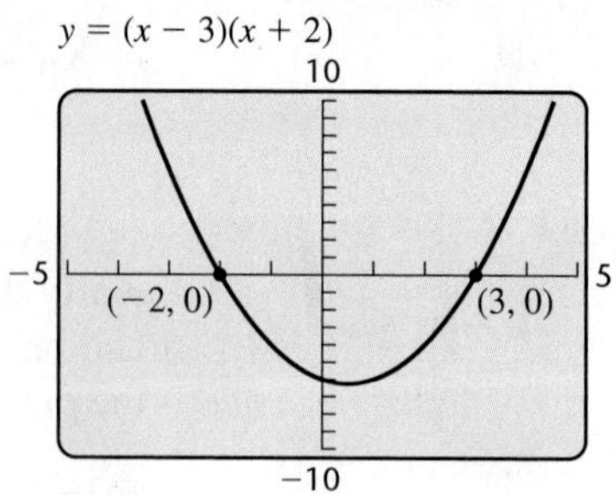

The solutions are -2 and 3.

Try Exercise 103.

EXAMPLE 4 Given that $f(x) = x(3x + 2)$, find all the values of a for which $f(a) = 0$.

SOLUTION We are looking for all numbers a for which $f(a) = 0$. Since $f(a) = a(3a + 2)$, we must have

$$a(3a + 2) = 0 \quad \textbf{Setting } f(a) \textbf{ equal to 0}$$

$$a = 0 \quad or \quad 3a + 2 = 0 \quad \textbf{Using the principle of zero products}$$

$$a = 0 \quad or \quad a = -\tfrac{2}{3}.$$

We check by evaluating $f(0)$ and $f\left(-\frac{2}{3}\right)$. One way to check with a graphing calculator is to enter $y_1 = x(3x + 2)$ and calculate Y1(0) and Y1$(-2/3)$. We can also graph $f(x) = x(3x + 2)$ and note that its zeros are $-\frac{2}{3}$ and 0. Thus, to have $f(a) = 0$, we must have $a = 0$ or $a = -\frac{2}{3}$.

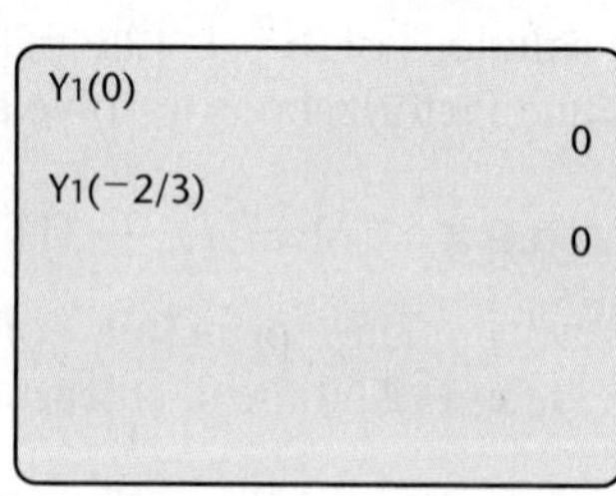

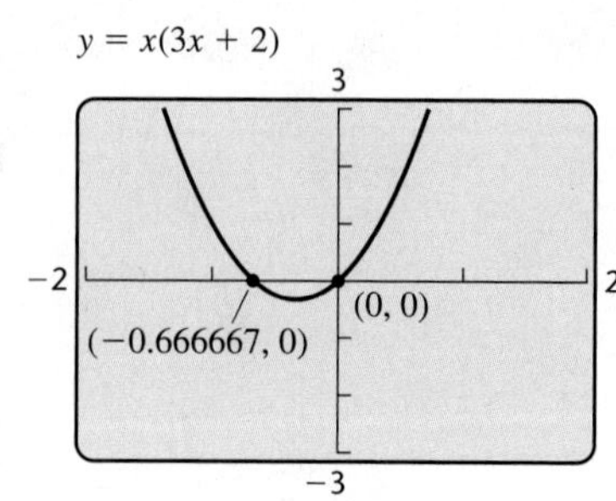

Try Exercise 113.

If the polynomial in an equation of the form $p(x) = 0$ is not in factored form, we must *factor* it before we can use the principle of zero products to solve the equation.

Factoring To *factor* a polynomial is to find an equivalent expression that is a product of polynomials. An equivalent expression of this type is called a *factorization* of the polynomial.

TERMS WITH COMMON FACTORS

When factoring a polynomial, we look for factors common to every term and then use the distributive law: $a(b + c) = ab + ac$.

EXAMPLE 5 Factor out a common factor: $6y^2 - 18$.

SOLUTION We have

$$6y^2 - 18 = 6 \cdot y^2 - 6 \cdot 3 \quad \text{Noting that 6 is a common factor}$$
$$= 6(y^2 - 3). \quad \text{Using the distributive law}$$

Check: $6(y^2 - 3) = 6y^2 - 18$.

Try Exercise 45.

Suppose in Example 5 that the common factor 2 were used:

$$6y^2 - 18 = 2 \cdot 3y^2 - 2 \cdot 9 \quad \text{2 is a common factor.}$$
$$= 2(3y^2 - 9). \quad \text{Using the distributive law}$$

Note that $3y^2 - 9$ itself has a common factor, 3. It is standard practice to factor out the *largest*, or *greatest*, *common factor*. Thus, by now factoring out 3, we can complete the factorization:

$$6y^2 - 18 = 2(3y^2 - 9)$$
$$= 2 \cdot 3(y^2 - 3) = 6(y^2 - 3). \quad \text{Remember to multiply the two common factors: } 2 \cdot 3 = 6.$$

The greatest common factor of a polynomial is the largest common factor of the coefficients times the greatest common factor of the variable(s) in all the terms. Thus, to find the greatest common factor of $30x^4 + 20x^5$, we multiply the greatest common factor of 30 and 20, which is 10, by the greatest common factor of x^4 and x^5, which is x^4:

$$30x^4 + 20x^5 = 10 \cdot 3 \cdot x^4 + 10 \cdot 2 \cdot x^4 \cdot x$$
$$= 10x^4(3 + 2x). \quad \text{The greatest common factor is } 10x^4.$$

EXAMPLE 6 Write an expression equivalent to $8p^6q^2 - 4p^5q^3 + 10p^4q^4$ by factoring out the greatest common factor.

Student Notes

To write the variable factors in a greatest common factor, write each variable that appears in *every* term and the *smallest* exponent with which it appears.

SOLUTION First, we look for the greatest positive common factor of the coefficients:

$$8, -4, 10 \longrightarrow \text{Greatest common factor} = 2.$$

Second, we look for the greatest common factor of the powers of p:

$$p^6, p^5, p^4 \longrightarrow \text{Greatest common factor} = p^4.$$

Third, we look for the greatest common factor of the powers of q:

$$q^2, q^3, q^4 \longrightarrow \text{Greatest common factor} = q^2.$$

Thus, $2p^4q^2$ is the greatest common factor of the given polynomial. Then

$$8p^6q^2 - 4p^5q^3 + 10p^4q^4 = 2p^4q^2 \cdot 4p^2 - 2p^4q^2 \cdot 2pq + 2p^4q^2 \cdot 5q^2$$
$$= 2p^4q^2(4p^2 - 2pq + 5q^2).$$

We can check a factorization by multiplying:

$$2p^4q^2(4p^2 - 2pq + 5q^2) = 2p^4q^2 \cdot 4p^2 - 2p^4q^2 \cdot 2pq + 2p^4q^2 \cdot 5q^2$$
$$= 8p^6q^2 - 4p^5q^3 + 10p^4q^4.$$

The factorization is $2p^4q^2(4p^2 - 2pq + 5q^2)$.

Try Exercise 51.

The polynomials in Examples 5 and 6 have been **factored completely**. They cannot be factored further. The factors in the resulting factorizations are said to be **prime polynomials**.

EXAMPLE 7 Factor: $15x^5 - 12x^4 + 27x^3 - 3x^2$.

SOLUTION We have

$$15x^5 - 12x^4 + 27x^3 - 3x^2$$
$$= 3x^2 \cdot 5x^3 - 3x^2 \cdot 4x^2 + 3x^2 \cdot 9x - 3x^2 \cdot 1$$ **Try to do this mentally.**
$$= 3x^2(5x^3 - 4x^2 + 9x - 1).$$ **Factoring out $3x^2$**

CAUTION! Don't forget the term -1. The check below shows why it is essential.

Since $5x^3 - 4x^2 + 9x - 1$ has no common factor, we are finished, except for a check:

$$3x^2(5x^3 - 4x^2 + 9x - 1) = 15x^5 - 12x^4 + 27x^3 - 3x^2.$$ **Our factorization checks.**

The factorization is $3x^2(5x^3 - 4x^2 + 9x - 1)$.

Try Exercise 49.

When the leading coefficient is a negative number, we generally factor out a common factor with a negative coefficient.

EXAMPLE 8 Write an equivalent expression by factoring out a common factor with a negative coefficient.

a) $-4x - 24$ **b)** $-2x^3 + 6x^2 - 2x$

SOLUTION

a) $-4x - 24 = -4(x + 6)$

b) $-2x^3 + 6x^2 - 2x = -2x(x^2 - 3x + 1)$ **The 1 is essential.**

Try Exercise 63.

EXAMPLE 9 Height of a Thrown Object. Suppose that a baseball is thrown upward with an initial velocity of 64 ft/sec and an initial height of 0 ft. Its height in feet, $h(t)$, after t seconds is given by

$$h(t) = -16t^2 + 64t.$$

Find an equivalent expression for $h(t)$ by factoring out a common factor with a negative coefficient.

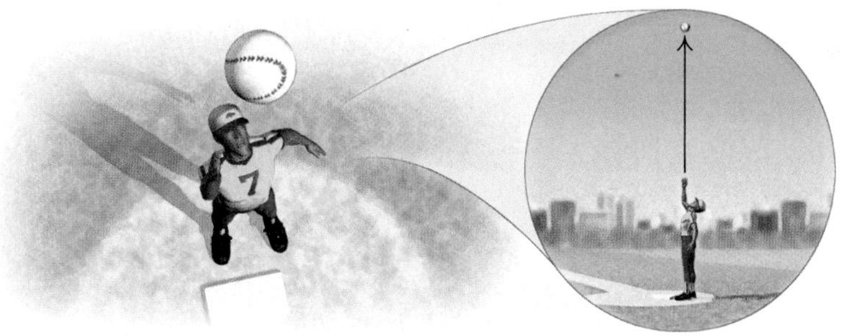

SOLUTION We factor out $-16t$ as follows:

$$h(t) = -16t^2 + 64t = -16t(t - 4).$$

Check: $-16t \cdot t = -16t^2$ and $-16t(-4) = 64t$.

Try Exercise 91.

$y_1 = -16x^2 + 64x$,
$y_2 = (-16x)(x - 4)$

X	Y1	Y2
0	0	0
1	48	48
2	64	64
3	48	48
4	0	0
5	−80	−80
6	−192	−192
X = 0		

Note in Example 9 that we can obtain function values using either expression for $h(t)$, since factoring forms equivalent expressions. For example,

$$h(1) = -16 \cdot 1^2 + 64 \cdot 1 = 48$$

and $h(1) = -16 \cdot 1(1 - 4) = 48.$ **Using the factorization**

We can evaluate the expressions $-16t^2 + 64t$ and $-16t(t - 4)$ using any value for t. The results should always match. Thus a quick partial check of any factorization is to evaluate the factorization and the original polynomial for one or two convenient replacements. The check for Example 9 becomes foolproof if three replacements are used. Recall that, in general, an nth-degree factorization is correct if it checks for $n + 1$ different replacements. The table shown at left confirms that the factorization is correct.

Tips for Factoring

1. Factor out the largest common factor, if one exists.
2. The common factor multiplies a polynomial with the same number of terms as the original polynomial.
3. Factoring can always be checked by multiplying. Multiplication should yield the original polynomial.

FACTORING BY GROUPING

The largest common factor is sometimes a binomial.

EXAMPLE 10 Factor: $(a - b)(x + 5) + (a - b)(x - y^2)$.

SOLUTION Here the largest common factor is the binomial $a - b$:

$$(a - b)(x + 5) + (a - b)(x - y^2) = (a - b)[(x + 5) + (x - y^2)]$$
$$= (a - b)[2x + 5 - y^2].$$

Try Exercise 73.

Often, in order to identify a common binomial factor in a polynomial with four terms, we must regroup into two groups of two terms each.

$y_1 = x^3 + 3x^2 + 4x + 12,$
$y_2 = (x + 3)(x^2 + 4)$

X	Y1	Y2
0	12	12
1	20	20
2	40	40
3	78	78
4	140	140
5	232	232
6	360	360

X = 0

EXAMPLE 11 Write an equivalent expression by factoring.

a) $x^3 + 3x^2 + 4x + 12$ **b)** $4t^3 - 15 + 20t^2 - 3t$

SOLUTION

a)
$$\begin{aligned} x^3 + 3x^2 + 4x + 12 &= (x^3 + 3x^2) + (4x + 12) && \text{Each grouping has a common factor.} \\ &= x^2(x + 3) + 4(x + 3) && \text{Factoring out a common factor from each binomial} \\ &= (x + 3)(x^2 + 4) && \text{Factoring out } x + 3 \end{aligned}$$

We can check by letting

$$y_1 = x^3 + 3x^2 + 4x + 12 \quad \text{and} \quad y_2 = (x + 3)(x^2 + 4).$$

The table at left indicates that the factorization is correct.

b) When we try grouping $4t^3 - 15 + 20t^2 - 3t$ as

$$(4t^3 - 15) + (20t^2 - 3t),$$

we are unable to factor $4t^3 - 15$. When this happens, we can rearrange the polynomial and try a different grouping:

$$\begin{aligned} 4t^3 - 15 + 20t^2 - 3t &= 4t^3 + 20t^2 - 3t - 15 && \text{Using the commutative law to rearrange the terms} \\ &= 4t^2(t + 5) - 3(t + 5) && \text{By factoring out } -3\text{, we see that } t + 5 \text{ is a common factor.} \\ &= (t + 5)(4t^2 - 3). \end{aligned}$$

Try Exercise 79.

We can "reverse subtraction" when necessary by factoring out -1.

Factoring out −1

$$b - a = -1(a - b) = -(a - b)$$

EXAMPLE 12 Factor: $ax - bx + by - ay$.

SOLUTION We have

$$\begin{aligned} ax - bx + by - ay &= (ax - bx) + (by - ay) && \text{Grouping} \\ &= x(a - b) + y(b - a) && \text{Factoring each binomial} \\ &= x(a - b) + y(-1)(a - b) && \text{Factoring out } -1 \text{ to reverse } b - a \\ &= x(a - b) - y(a - b) && \text{Simplifying} \\ &= (a - b)(x - y). && \text{Factoring out } a - b \end{aligned}$$

Check: To check, note that $a - b$ and $x - y$ are both prime and that

$$(a - b)(x - y) = ax - ay - bx + by = ax - bx + by - ay.$$

The factorization is $(a - b)(x - y)$.

Try Exercise 85.

Some polynomials with four terms, like $x^3 + x^2 + 3x - 3$, are *prime*. Not only is there no common monomial factor, but no matter how we group terms, there is no common binomial factor:

$$x^3 + x^2 + 3x - 3 = x^2(x + 1) + 3(x - 1);$$ **No common factor**

$$x^3 + 3x + x^2 - 3 = x(x^2 + 3) + (x^2 - 3);$$ **No common factor**

$$x^3 - 3 + x^2 + 3x = (x^3 - 3) + x(x + 3).$$ **No common factor**

FACTORING AND EQUATIONS

Factoring can help us solve polynomial equations.

EXAMPLE 13 Solve: $6x^2 = 30x$.

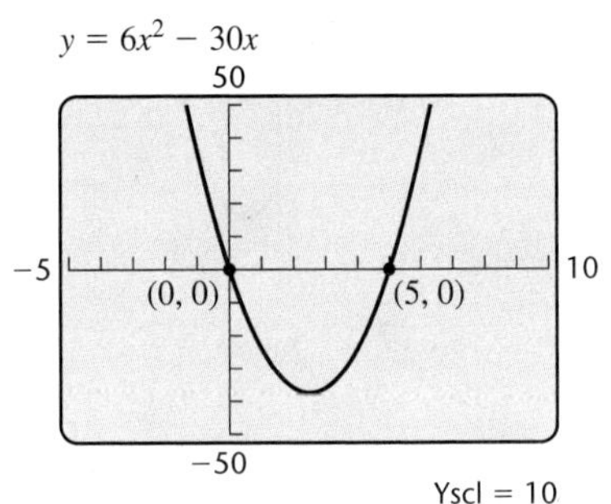

SOLUTION We can use the principle of zero products if there is a 0 on one side of the equation and the other side is in factored form:

$$6x^2 = 30x$$

$$6x^2 - 30x = 0$$ **Subtracting $30x$. One side is now 0.**

$$6x(x - 5) = 0$$ **Factoring. Do not "divide both sides by x."**

$$6x = 0 \quad or \quad x - 5 = 0$$ **Using the principle of zero products**

$$x = 0 \quad or \quad x = 5.$$

We check by substitution or graphically, as shown in the figure at left. The solutions are 0 and 5.

Try Exercise 107.

CAUTION! Remember that we can *factor expressions* and *solve equations.* In Example 13, we factored the expression $6x^2 - 30x$ in order to solve the equation $6x^2 - 30x = 0$.

The principle of zero products can be used to show that an nth-degree polynomial function can have at most n zeros. In Example 13, we wrote a quadratic polynomial as a product of two linear factors. Each linear factor corresponded to one zero of the polynomial function. In general, a polynomial function of degree n can have at most n linear factors, so a polynomial function of degree n can have at most n zeros. Thus, when solving an nth-degree polynomial equation, we need not look for more than n zeros.

To Use the Principle of Zero Products

1. Write an equivalent equation with 0 on one side, using the addition principle.
2. Factor the nonzero side of the equation.
3. Set each factor that is not a constant equal to 0.
4. Solve the resulting equations.

6.1 Exercise Set

FOR EXTRA HELP MyMathLab Math XL PRACTICE WATCH DOWNLOAD READ REVIEW

Concept Reinforcement *Classify each of the following statements as either true or false.*

1. The largest common factor of $10x^4 + 15x^2$ is $5x$.
2. The largest common factor of a polynomial always has the same degree as the polynomial itself.
3. The polynomial $8x + 9y$ is prime.
4. When the leading coefficient of a polynomial is negative, we generally factor out a common factor with a negative coefficient.
5. A polynomial is not prime if it contains a common factor other than 1 or -1.
6. All polynomials with four terms can be factored by grouping.
7. The expressions $b - a$, $-(a - b)$, and $-1(a - b)$ are all equivalent.
8. The complete factorization of $12x^3 - 20x^2$ is $4x(3x^2 - 5x)$.

Tell whether each of the following is an expression or an equation.

9. $x^2 + 6x + 9$
10. $x^3 = x^2 - x + 3$
11. $3x^2 = 3x$
12. $x^4 + 3x^3 + x^2$
13. $2x^3 + x^2 = 0$
14. $5x^4 + 5x$

In Exercises 15 and 16, use the graph to solve $f(x) = 0$.

15.

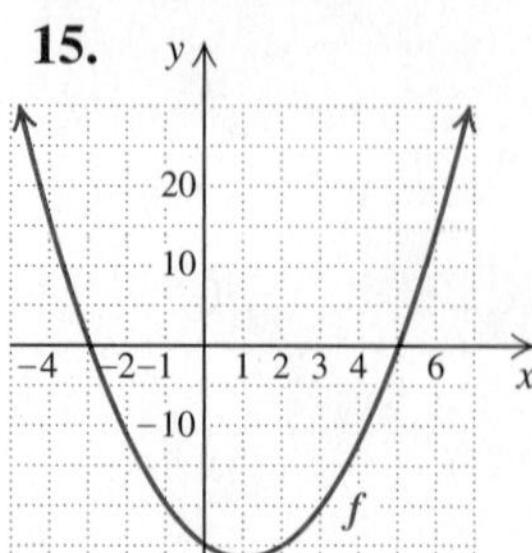

16.

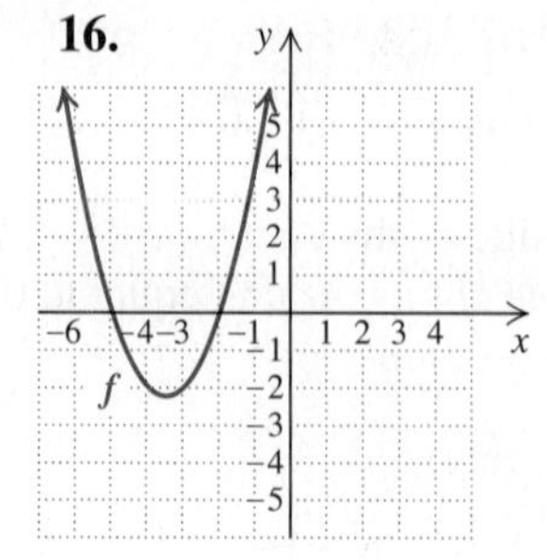

In Exercises 17 and 18, use the graph to find the zeros of the function f.

17.

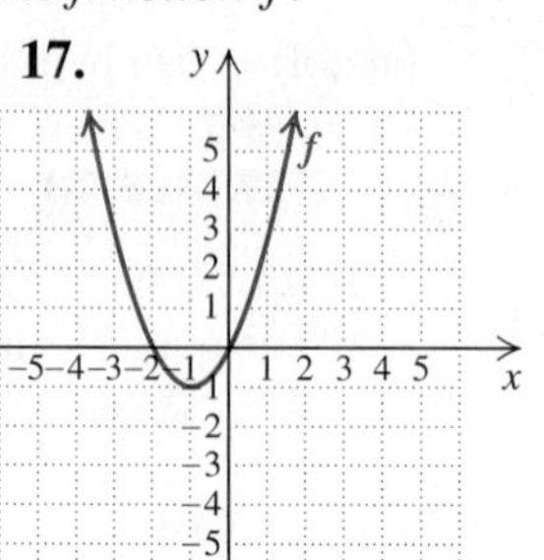

18.

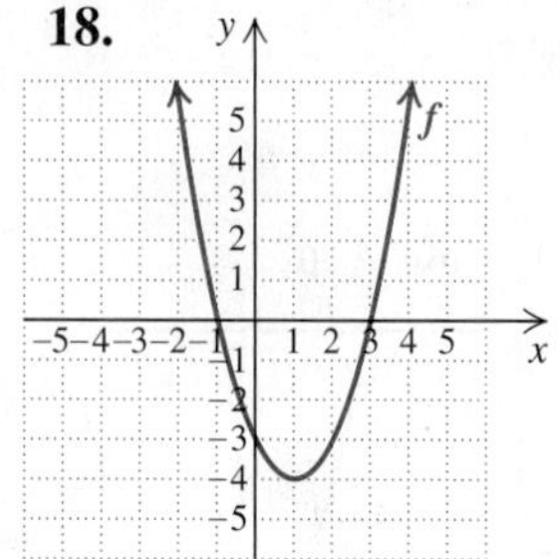

19. Use the graph below to solve $x^2 + 2x = 3$.

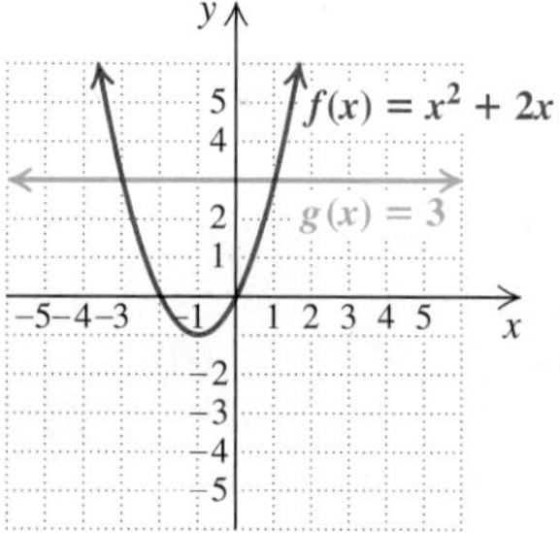

20. Use the graph below to solve $x^2 = 4$.

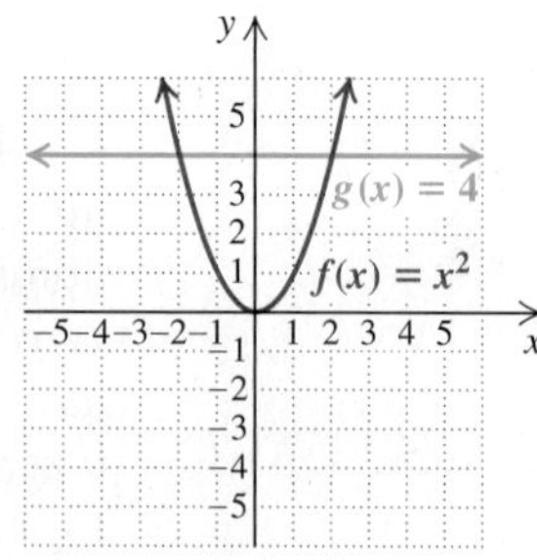

21. Use the graph below to solve $x^2 + 2x - 8 = 0$.

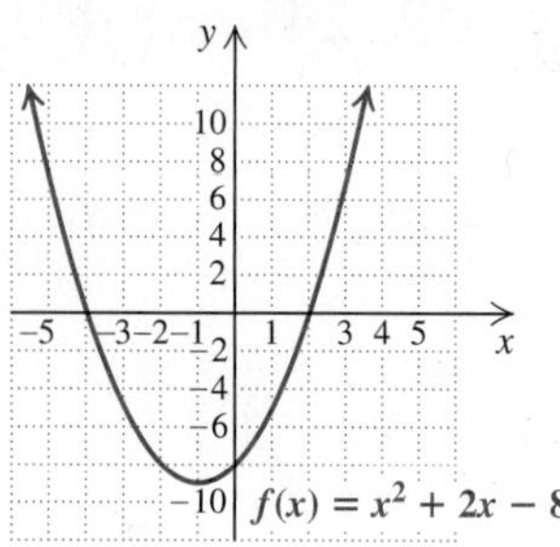

22. Use the graph below to find the zeros of the function given by $f(x) = x^2 - 2x + 1$.

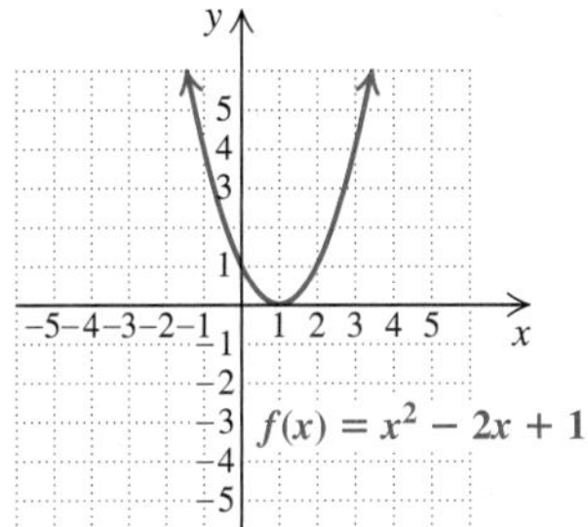

Solve using a graphing calculator.

23. $x^2 = 5x$

24. $2x^2 = 20x$

25. $4x = x^2 + 3$

26. $x^2 = 1$

27. $x^2 + 150 = 25x$

28. $2x^2 + 25 = 51x$

29. $x^3 - 3x^2 + 2x = 0$

30. $x^3 + 2x^2 = x + 2$

31. $x^3 - 3x^2 - 198x + 1080 = 0$

32. $2x^3 + 25x^2 - 282x + 360 = 0$

33. $21x^2 + 2x - 3 = 0$

34. $66x^2 - 49x - 5 = 0$

Find the zeros of each function.

35. $f(x) = x^2 - 4x - 45$

36. $g(x) = x^2 + x - 20$

37. $p(x) = 2x^2 - 13x - 7$

38. $f(x) = 6x^2 + 17x + 6$

39. $f(x) = x^3 - 2x^2 - 3x$

40. $r(x) = 3x^3 - 12x$

Aha! *Match each graph to the corresponding function in Exercises 41–44.*

I

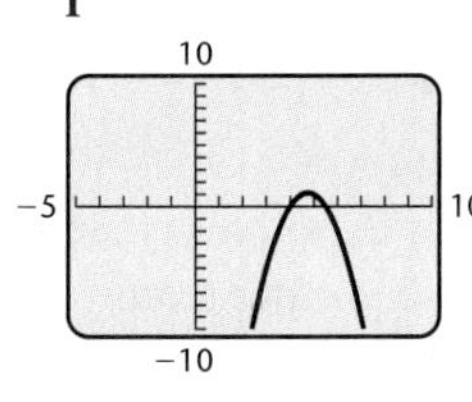

II

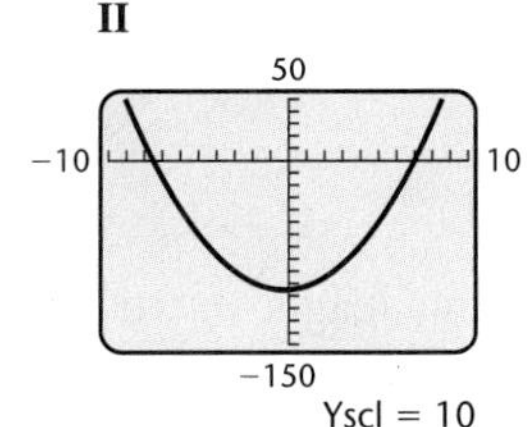

III

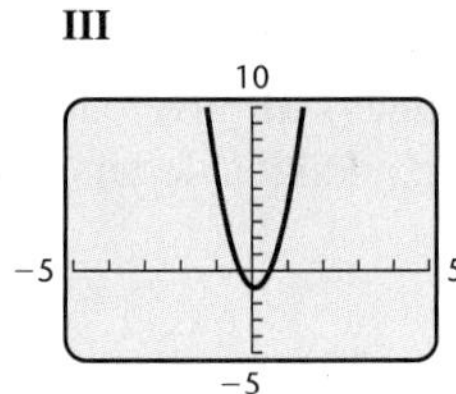

IV

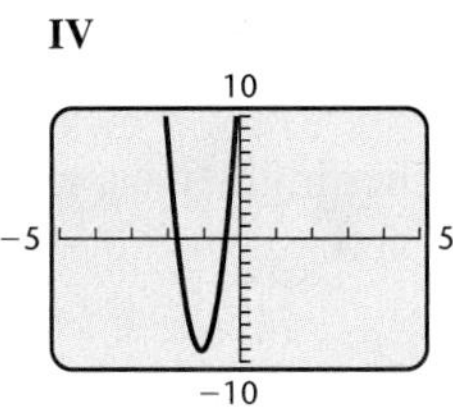

41. $f(x) = (2x - 1)(3x + 1)$

42. $f(x) = (2x + 15)(x - 7)$

43. $f(x) = (4 - x)(2x - 11)$

44. $f(x) = (5x + 2)(4x + 7)$

Write an equivalent expression by factoring out the greatest common factor.

45. $2t^2 + 8t$

46. $3y^2 - 6y$

47. $9y^3 - y^2$

48. $x^3 + 8x^2$

49. $15x^2 - 5x^4 + 5x$

50. $8y^2 + 4y^4 - 2y$

51. $4x^2y - 12xy^2$

52. $5x^2y^3 + 15x^3y^2$

53. $3y^2 - 3y - 9$

54. $15x^2 - 5x + 5$

55. $6ab - 4ad + 12ac$

56. $8xy + 10xz - 14xw$

57. $72x^3 - 36x^2 + 24x$

58. $12a^4 - 21a^3 - 9a^2$

59. $x^5y^5 + x^4y^3 + x^3y^3 - xy^2$

60. $x^9y^6 - x^7y^5 + x^4y^4 + x^3y$

61. $9x^3y^6z^2 - 12x^4y^4z^4 + 15x^2y^5z^3$

62. $14a^4b^3c^5 + 21a^3b^5c^4 - 35a^4b^4c^3$

Write an equivalent expression by factoring out a factor with a negative coefficient.

63. $-5x + 35$

64. $-6y - 72$

65. $-2x^2 + 4x - 12$

66. $-2x^2 + 12x + 40$

67. $3y - 24x$

68. $7x - 56y$

69. $-x^2 + 5x - 9$

70. $-p^3 - 4p^2 + 11$

71. $-a^4 + 2a^3 - 13a$

72. $-m^{10} - m^9 + m^8 - 2m^7$

Write an equivalent expression by factoring.

73. $a(b - 5) + c(b - 5)$

74. $r(t - 3) - s(t - 3)$

75. $(x + 7)(x - 1) + (x + 7)(x - 2)$

76. $(a + 5)(a - 2) + (a + 5)(a + 1)$

77. $a^2(x - y) + 5(y - x)$

78. $5x^2(x - 6) + 2(6 - x)$

Factor by grouping, if possible, and check.

79. $ac + ad + bc + bd$

80. $xy + xz + wy + wz$

81. $b^3 - b^2 + 2b - 2$

82. $y^3 - y^2 + 3y - 3$

83. $x^3 - x^2 - 2x + 5$

84. $p^3 + p^2 - 3p + 10$

85. $a^3 - 3a^2 + 6 - 2a$

86. $t^3 + 6t^2 - 2t - 12$

87. $x^6 - x^5 - x^3 + x^4$

88. $y^4 - y^3 - y + y^2$

89. $2y^4 + 6y^2 + 5y^2 + 15$

90. $2xy - x^2y - 6 + 3x$

91. *Height of a Baseball.* A baseball is popped up with an upward velocity of 72 ft/sec. Its height in feet, $h(t)$, after t seconds is given by

$$h(t) = -16t^2 + 72t.$$

a) Find an equivalent expression for $h(t)$ by factoring out a common factor with a negative coefficient.

b) Perform a partial check of part (a) by evaluating both expressions for $h(t)$ at $t = 1$.

92. *Height of a Rocket.* A water rocket is launched upward with an initial velocity of 96 ft/sec. Its height in feet, $h(t)$, after t seconds is given by

$$h(t) = -16t^2 + 96t.$$

a) Find an equivalent expression for $h(t)$ by factoring out a common factor with a negative coefficient.

b) Check your factoring by evaluating both expressions for $h(t)$ at $t = 1$.

93. *Airline Routes.* When an airline links n cities so that from any one city it is possible to fly directly to each of the other cities, the total number of direct routes is given by

$$R(n) = n^2 - n.$$

Find an equivalent expression for $R(n)$ by factoring out a common factor.

94. *Surface Area of a Silo.* A silo is a structure that is shaped like a right circular cylinder with a half sphere on top. The surface area of a silo of height h and radius r (including the area of the base) is given by the polynomial $2\pi rh + \pi r^2$. (Note that h is the height of the entire silo.) Find an equivalent expression by factoring out a common factor.

95. *Total Profit.* When x hundred gaming systems are sold, Rolics Electronics collects a profit of $P(x)$, where

$$P(x) = x^2 - 3x,$$

and $P(x)$ is in thousands of dollars. Find an equivalent expression by factoring out a common factor.

96. *Total Profit.* After t weeks of production, Claw Foot, Inc., is making a profit of $P(t) = t^2 - 5t$ from sales of their surfboards. Find an equivalent expression by factoring out a common factor.

97. *Total Revenue.* Urban Sounds is marketing a new MP3 player. The firm determines that when it sells x units, the total revenue R is given by the polynomial function

$$R(x) = 280x - 0.4x^2 \text{ dollars.}$$

Find an equivalent expression for $R(x)$ by factoring out $0.4x$.

98. *Total Cost.* Urban Sounds determines that the total cost C of producing x MP3 players is given by the polynomial function

$$C(x) = 0.18x + 0.6x^2.$$

Find an equivalent expression for $C(x)$ by factoring out $0.6x$.

99. *Counting Spheres in a Pile.* The number N of spheres in a triangular pile like the one shown here is a polynomial function given by

$$N(x) = \tfrac{1}{6}x^3 + \tfrac{1}{2}x^2 + \tfrac{1}{3}x,$$

where x is the number of layers and $N(x)$ is the number of spheres. Find an equivalent expression for $N(x)$ by factoring out $\frac{1}{6}$.

100. *Number of Games in a League.* If there are n teams in a league and each team plays every other team once, we can find the total number of games played by using the polynomial function $f(n) = \frac{1}{2}n^2 - \frac{1}{2}n$. Find an equivalent expression by factoring out $\frac{1}{2}$.

101. *High-fives.* When a team of n players all give each other high-fives, a total of $H(n)$ hand slaps occurs, where

$$H(n) = \tfrac{1}{2}n^2 - \tfrac{1}{2}n.$$

Find an equivalent expression by factoring out $\frac{1}{2}n$.

102. *Number of Diagonals.* The number of diagonals of a polygon having n sides is given by the polynomial function

$$P(n) = \tfrac{1}{2}n^2 - \tfrac{3}{2}n.$$

Find an equivalent expression for $P(n)$ by factoring out $\tfrac{1}{2}n$.

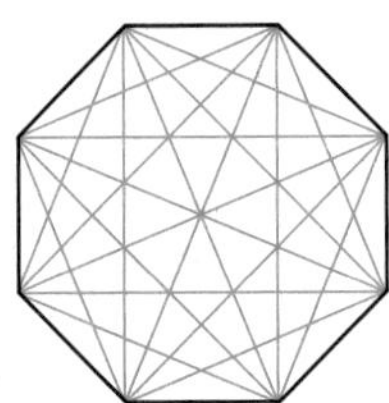

Solve using the principle of zero products.

103. $(x + 3)(x - 4) = 0$

104. $(x + 10)(x + 11) = 0$

105. $x(x + 1) = 0$

106. $5x(x - 2) = 0$

107. $x^2 - 3x = 0$

108. $2x^2 + 8x = 0$

109. $-5x^2 = 15x$

110. $2x - 4x^2 = 0$

111. $12x^4 + 4x^3 = 0$

112. $21x^3 = 7x^2$

113. Given that $f(x) = (x - 3)(x + 7)$, find all values of a for which $f(a) = 0$.

114. Given that $f(x) = (3x + 1)(x + 8)$, find all values of a for which $f(a) = 0$.

115. Given that $f(x) = 2x(5x + 9)$, find all values of a for which $f(a) = 0$.

116. Given that $f(x) = 8x(x - 1)$, find all values of a for which $f(a) = 0$.

117. Given that $f(x) = x^3 - 3x^2$, find all values of a for which $f(a) = 0$.

118. Given that $f(x) = 6x + 9x^2$, find all values of a for which $f(a) = 0$.

TW **119.** Write a two-sentence paragraph in which the word "factor" is used at least once as a noun and once as a verb.

TW **120.** Jasmine claims that the zeros of the function given by $f(x) = x^4 - 3x^2 + 7x + 20$ are -1, 1, 2, 4, and 5. How can you tell, without performing any calculations, that she cannot be correct?

SKILL REVIEW

To prepare for Section 6.2, review multiplying binomials using FOIL (Section 5.6).

Multiply. [5.6]

121. $(x + 2)(x + 7)$

122. $(x - 2)(x - 7)$

123. $(x + 2)(x - 7)$

124. $(x - 2)(x + 7)$

125. $(a - 1)(a - 3)$

126. $(t + 3)(t + 5)$

127. $(t - 5)(t + 10)$

128. $(a + 4)(a - 6)$

SYNTHESIS

TW **129.** Ashlee factors $8x^2y - 10xy^2$ as

$$2xy \cdot 4x - 2xy \cdot 5y.$$

Is this the factorization of the polynomial? Why or why not?

TW **130.** What is wrong with solving $x^2 = 3x$ by dividing both sides of the equation by x?

131. Use the results of Exercise 21 to factor $x^2 + 2x - 8$.

132. Use the results of Exercise 22 to factor $x^2 - 2x + 1$.

Complete each of the following factorizations.

133. $x^5y^4 + \_\_\_ = x^4y^4(\_\_\_ + y^2)$

134. $a^3b^7 - \_\_\_ = \_\_\_(ab^4 - c^2)$

Write an equivalent expression by factoring out the smallest power of x in each of the following.

135. $x^{-6} + x^{-9} + x^{-3}$

136. $x^{-8} + x^{-4} + x^{-6}$

137. $x^{1/3} - 5x^{1/2} + 3x^{3/4}$

138. $x^{3/4} + x^{1/2} - x^{1/4}$

Factor.

Aha! **139.** $5x^5 - 5x^4 + x^3 - x^2 + 3x - 3$

140. $ax^2 + 2ax + 3a + x^2 + 2x + 3$

Write an equivalent expression by factoring. Assume that all exponents are natural numbers.

141. $2x^{3a} + 8x^a + 4x^{2a}$

142. $3a^{n+1} + 6a^n - 15a^{n+2}$

Find a polynomial in factored form for the shaded area in each figure. (Use π in your answers where appropriate.)

143.

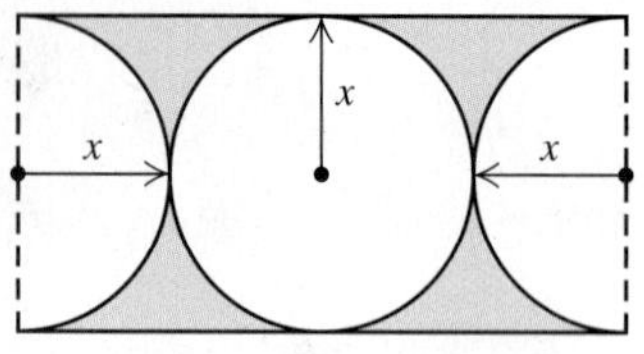

144.

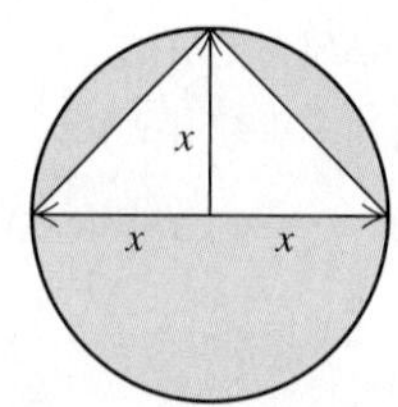

Try Exercise Answers: Section 6.1

19. $-3, 1$ **35.** $-5, 9$ **45.** $2t(t + 4)$ **49.** $5x(3x - x^3 + 1)$ **51.** $4xy(x - 3y)$ **63.** $-5(x - 7)$ **73.** $(b - 5)(a + c)$ **79.** $(c + d)(a + b)$ **85.** $(a - 3)(a^2 - 2)$ **91. (a)** $h(t) = -8t(2t - 9)$; **(b)** $h(1) = 56$ ft **103.** $-3, 4$ **107.** $0, 3$ **113.** $-7, 3$

6.2 Trinomials of the Type $x^2 + bx + c$

- Factoring Trinomials of the Type $x^2 + bx + c$
- Equations Containing Trinomials
- Zeros and Factoring

STUDY TIP

Read the Instructions First

Take the time to carefully read the instructions before beginning an exercise or a series of exercises. Not only will this give you direction in your work, it may also help you in your problem solving. For example, you may be asked to supply more than one answer, or you may be told that answers may vary.

We now learn how to factor trinomials like

$$x^2 + 5x + 4 \quad \text{or} \quad x^2 + 3x - 10$$

and how to solve equations containing such trinomials.

FACTORING TRINOMIALS OF THE TYPE $x^2 + bx + c$

When trying to factor trinomials of the type $x^2 + bx + c$, we think of FOIL in reverse.

Constant Term Positive

Recall the FOIL method of multiplying two binomials:

$$\begin{aligned} & \quad\ \ \text{F} \quad \text{O} \quad \text{I} \quad \text{L} \\ (x + 3)(x + 5) &= x^2 + \underbrace{5x + 3x}_{} + 15 \\ &= x^2 + \quad 8x \quad + 15. \end{aligned}$$

Because the leading coefficient in each binomial is 1, the leading coefficient in the product is also 1. To factor $x^2 + 8x + 15$, we think of FOIL in reverse. The x^2 resulted from x times x, which suggests that the First term in each binomial is x. The challenge is to find two numbers p and q such that

$$\begin{aligned} x^2 + 8x + 15 &= (x + p)(x + q) \\ &= x^2 + qx + px + pq. \end{aligned}$$

Note that the Outer and Inner products, qx and px, can be written as $(p + q)x$. The Last product, pq, will be a constant. Thus we need two numbers, p and q, whose product is 15 and whose sum is 8. These numbers are 3 and 5. The factorization is

$$(x + 3)(x + 5), \quad \text{or} \quad (x + 5)(x + 3). \qquad \textbf{Using a commutative law}$$

A GEOMETRIC APPROACH TO EXAMPLE 1

In Section 5.5, we saw that the product of two binomials can be regarded as the sum of the areas of four rectangles (see p. 375). Thus we can regard the factoring of $x^2 + 5x + 6$ as a search for p and q so that the sum of areas A, B, C, and D is $x^2 + 5x + 6$.

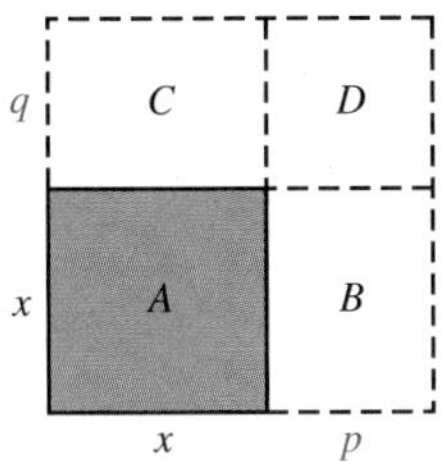

Note that area D is simply the product of p and q. In order for area D to be 6, p and q must be either 1 and 6 or 2 and 3. We illustrate both below.

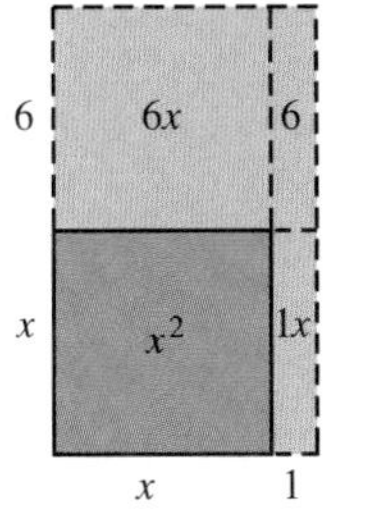

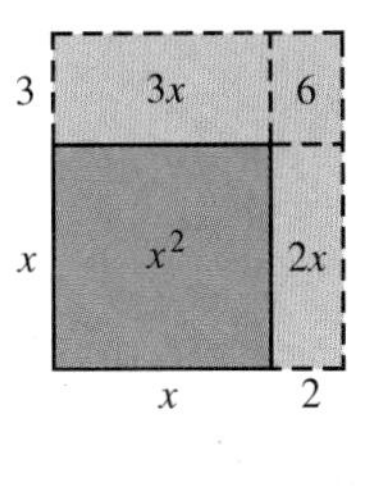

When p and q are 1 and 6, the total area is $x^2 + 7x + 6$, but when p and q are 2 and 3, as shown on the right, the total area is $x^2 + 5x + 6$, as desired. Thus the factorization of $x^2 + 5x + 6$ is $(x + 2)(x + 3)$.

EXAMPLE 1 Factor to form an equivalent expression:

$$x^2 + 5x + 6.$$

SOLUTION Think of FOIL in reverse. The first term of each factor is x:

$$(x + \quad)(x + \quad).$$

To complete the factorization, we need a constant term for each binomial. The constants must have a product of 6 and a sum of 5. We list some pairs of numbers that multiply to 6 and then check the sum of each pair of factors.

Pairs of Factors of 6	Sums of Factors
1, 6	7
2, 3	5 ← The numbers we seek are 2 and 3.
−1, −6	−7
−2, −3	−5

One pair has a sum of 5.
Every pair has a product of 6.

Since

$$2 \cdot 3 = 6 \quad \text{and} \quad 2 + 3 = 5,$$

the factorization of $x^2 + 5x + 6$ is $(x + 2)(x + 3)$.

Check: $(x + 2)(x + 3) = x^2 + 3x + 2x + 6$
$= x^2 + 5x + 6.$

Thus, $(x + 2)(x + 3)$ is a product that is equivalent to $x^2 + 5x + 6$.

Note that since 5 and 6 are both positive, when factoring $x^2 + 5x + 6$ we need not consider negative factors of 6. Note too that changing the signs of the factors changes only the sign of the sum (see the table above).

Try Exercise 9.

Compare the following:

$$(x + 2)(x + 3) = x^2 + 5x + 6;$$
$$(x - 2)(x - 3) = x^2 - 5x + 6.$$

When the constant term of a trinomial is positive, it can be the product of two positive numbers or two negative numbers. To factor such a trinomial, we use the sign of the trinomial's middle term.

To Factor $x^2 + bx + c$ When c Is Positive

When the constant term c of a trinomial is positive, look for two numbers with the same sign. Select pairs of numbers with the sign of b, the coefficient of the middle term.

$$x^2 - 7x + 10 = (x - 2)(x - 5);$$

$$x^2 + 7x + 10 = (x + 2)(x + 5)$$

EXAMPLE 2 Factor: $t^2 - 9t + 20$.

SOLUTION Since the constant term is positive and the coefficient of the middle term is negative, we look for a factorization of 20 in which both factors are negative. Their sum must be -9.

Pairs of Factors of 20	Sums of Factors
$-1, -20$	-21
$-2, -10$	-12
$-4, -5$	-9 ←

We need a sum of -9. The numbers we need are -4 and -5.

$y_1 = x^2 - 9x + 20,$
$y_2 = (x - 4)(x - 5)$

X	Y1	Y2
0	20	20
1	12	12
2	6	6
3	2	2
4	0	0
5	0	0
6	2	2

X = 0

The factorization is $(t - 4)(t - 5)$. We check by comparing values using a table, as shown at left.

Try Exercise 13.

Constant Term Negative

When the constant term of a trinomial is negative, one factor will be negative and one will be positive.

EXAMPLE 3 Factor: $x^3 - x^2 - 30x$.

SOLUTION *Always* look first for a common factor! This time there is one, x. We factor it out:

$$x^3 - x^2 - 30x = x(x^2 - x - 30).$$

Now we consider $x^2 - x - 30$. When the constant term, -30, is factored, one factor will be negative and one will be positive. Since the sum of these two numbers must be negative (specifically, -1), the negative number must have the greater absolute value.

Pairs of Factors of -30	Sums of Factors
$1, -30$	-29
$2, -15$	-13
$3, -10$	-7
$5, -6$	-1 ←

We need not consider pairs of factors such as $-1, 30$ for which the sum is positive.

The numbers we need are 5 and -6.

The factorization of $x^2 - x - 30$ is $(x + 5)(x - 6)$. *Don't forget to include the factor that was factored out earlier*! In this case, the factorization of the original trinomial is $x(x + 5)(x - 6)$.

CAUTION! When factoring involves more than one step, be careful to write out the *entire* factorization.

$y_1 = x^3 - x^2 - 30x,\ y_2 = x(x + 5)(x - 6)$

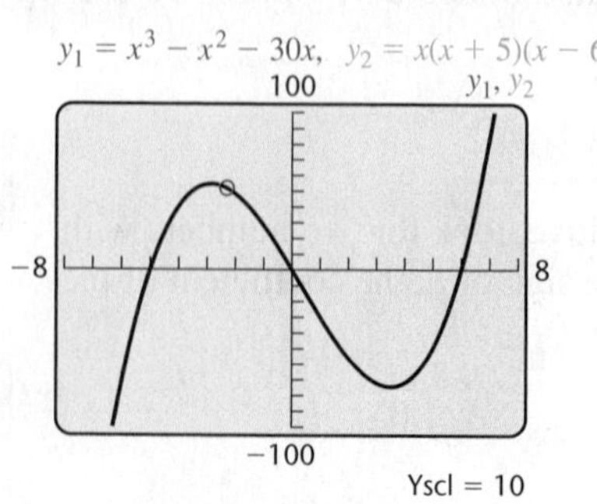

We check by graphing $y_1 = x^3 - x^2 - 30x$ and $y_2 = x(x + 5)(x - 6)$, as shown at left.

Try Exercise 21.

To Factor $x^2 + bx + c$ When c Is Negative

When the constant term c of a trinomial is negative, look for a positive number and a negative number that multiply to c. Select pairs of numbers for which the number with the larger absolute value has the same sign as b, the coefficient of the middle term.

$$x^2 - 4x - 21 = (x + 3)(x - 7);$$

$$x^2 + 4x - 21 = (x - 3)(x + 7)$$

EXAMPLE 4 Factor: $t^2 - 24 + 5t$.

SOLUTION It helps to first write the trinomial in descending order: $t^2 + 5t - 24$. The factorization of the constant term, -24, must have one factor positive and one factor negative. The sum must be 5, so the positive factor must have the larger absolute value. Thus we consider only pairs of factors in which the positive factor has the larger absolute value.

Pairs of Factors of −24	Sums of Factors	
−1, 24	23	
−2, 12	10	
−3, 8	5	← **The numbers we need are −3 and 8.**
−4, 6	2	

Check: $(t - 3)(t + 8) = t^2 + 5t - 24$.

The factorization is $(t - 3)(t + 8)$.

Try Exercise 25.

Some polynomials are not factorable using integers.

EXAMPLE 5 Factor: $x^2 - x + 7$.

SOLUTION Since 7 has very few factors, we can easily check all possibilities.

Pairs of Factors of 7	Sums of Factors	
7, 1	8	**No pair gives a sum of −1.**
−7, −1	−8	

The polynomial is not factorable using integer coefficients; it is **prime.**

Try Exercise 35.

To Factor $x^2 + bx + c$

1. *Always* factor out any common factor first. If the coefficient of x^2 is -1, factor out a -1.
2. If necessary, rewrite the trinomial in descending order.
3. Find a pair of factors that have c as their product and b as their sum.
   - If c is positive, both factors will have the same sign as b.
   - If c is negative, one factor will be positive and the other will be negative. Select the factors such that the factor with the larger absolute value is the factor with the same sign as b.
   - If the sum of the two factors is the opposite of b, changing the signs of both factors will give the desired factors whose sum is b.
4. Check by multiplying.

EXAMPLE 6 Factor: $x^2 - 2xy - 48y^2$.

SOLUTION We look for numbers p and q such that

$$x^2 - 2xy - 48y^2 = (x + py)(x + qy).$$

The x's and y's can be written in the binomials in advance.

Our thinking is much the same as if we were factoring $x^2 - 2x - 48$. We look for factors of -48 whose sum is -2. Those factors are 6 and -8. Thus,

$$x^2 - 2xy - 48y^2 = (x + 6y)(x - 8y).$$

The check is left to the student.

Try Exercise 37.

CAUTION! In Example 7, we are solving an equation. Do not try to solve an expression!

EQUATIONS CONTAINING TRINOMIALS

EXAMPLE 7 Solve: $x^2 + 9x + 8 = 0$.

ALGEBRAIC APPROACH

We use the principle of zero products:

$$x^2 + 9x + 8 = 0$$

$$(x + 1)(x + 8) = 0 \quad \textbf{Factoring}$$

$$x + 1 = 0 \quad or \quad x + 8 = 0 \quad \textbf{Using the principle of zero products}$$

$$x = -1 \quad or \quad x = -8. \quad \textbf{Solving each equation for } x$$

Check:

For -1:

$$\begin{array}{c|c} x^2 + 9x + 8 = 0 & \\ \hline (-1)^2 + 9(-1) + 8 & 0 \\ 1 - 9 + 8 & \\ 0 \stackrel{?}{=} 0 & \text{TRUE} \end{array}$$

For -8:

$$\begin{array}{c|c} x^2 + 9x + 8 = 0 & \\ \hline (-8)^2 + 9(-8) + 8 & 0 \\ 64 - 72 + 8 & \\ 0 \stackrel{?}{=} 0 & \text{TRUE} \end{array}$$

The solutions are -1 and -8.

GRAPHICAL APPROACH

The real-number solutions of $x^2 + 9x + 8 = 0$ are the first coordinates of the x-intercepts of the graph of $f(x) = x^2 + 9x + 8$.

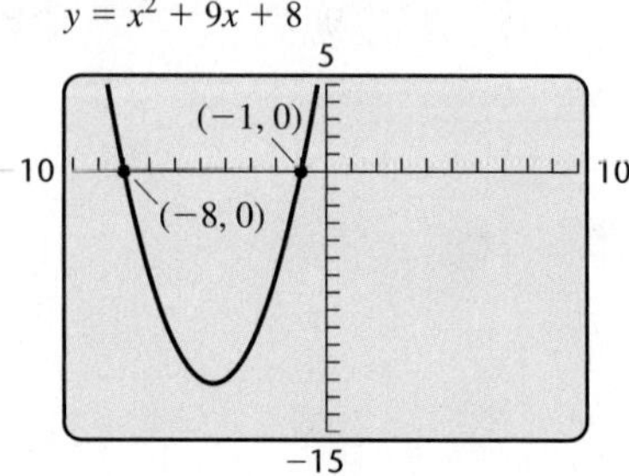

We find that -1 and -8 are solutions.

Try Exercise 43.

EXAMPLE 8 Solve: $(t - 10)(t + 1) = -24$.

ALGEBRAIC APPROACH

Note that the left side of the equation is factored. It may be tempting to set both factors equal to -24 and solve, but this is *not* correct. The principle of zero products requires 0 on one side of the equation. We begin by multiplying the left side.

$$(t - 10)(t + 1) = -24$$

$$t^2 - 9t - 10 = -24 \quad \text{Multiplying}$$

$$t^2 - 9t + 14 = 0 \quad \text{Adding 24 to both sides to get 0 on one side}$$

$$(t - 2)(t - 7) = 0 \quad \text{Factoring}$$

$$t - 2 = 0 \quad or \quad t - 7 = 0 \quad \text{Using the principle of zero products}$$

$$t = 2 \quad or \quad t = 7$$

The solutions are 2 and 7.

GRAPHICAL APPROACH

There is no need to multiply. We let $y_1 = (x - 10)(x + 1)$ and $y_2 = -24$ and look for any points of intersection of the graphs. The x-coordinates of these points are the solutions of the equation.

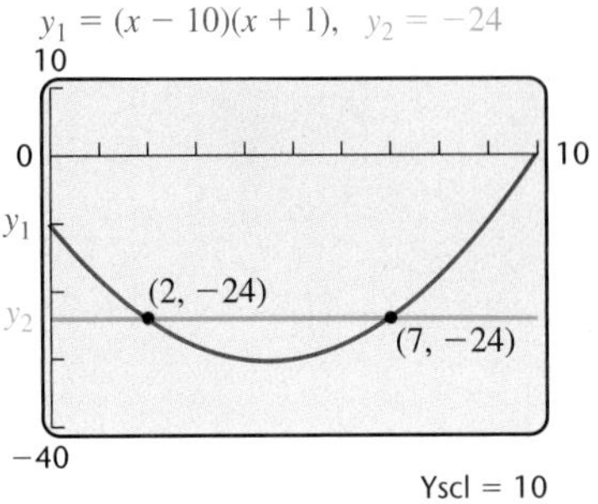

We find that 2 and 7 are solutions.

Try Exercise 61.

EXAMPLE 9 Solve: $x^3 = x^2 + 30x$.

ALGEBRAIC APPROACH

In order to solve $x^3 = x^2 + 30x$ using the principle of zero products, we must rewrite the equation with 0 on one side. We have

$$x^3 = x^2 + 30x$$

$$x^3 - x^2 - 30x = 0 \quad \text{Getting 0 on one side}$$

$$x(x - 6)(x + 5) = 0 \quad \text{Using the factorization from Example 3}$$

$$x = 0 \quad or \quad x - 6 = 0 \quad or \quad x + 5 = 0 \quad \text{Using the principle of zero products}$$

$$x = 0 \quad or \quad x = 6 \quad or \quad x = -5.$$

We check by substituting $-5, 0$, and 6 into the original equation. The solutions are $-5, 0$, and 6.

GRAPHICAL APPROACH

We let $y_1 = x^3$ and $y_2 = x^2 + 30x$ and look for points of intersection of the graphs. Since the polynomial equation is of degree 3, we know there will be no more than 3 solutions. The following viewing window shows that there are indeed three points of intersection. The x-coordinates of the points of intersection are -5, 0, and 6.

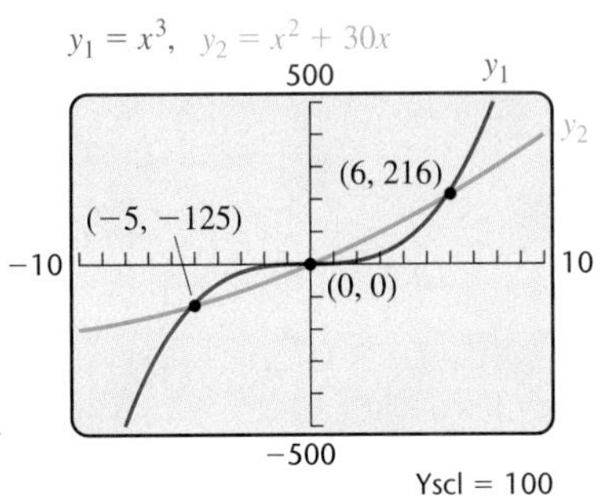

The solutions are -5, 0, and 6.

Try Exercise 59.

ZEROS AND FACTORING

We can use the principle of zero products "in reverse" to factor a polynomial and to write a function with given zeros.

EXAMPLE 10 Factor: $x^2 - 13x - 608$.

SOLUTION The factorization of the trinomial $x^2 - 13x - 608$ will be in the form

$$x^2 - 13x - 608 = (x + p)(x + q).$$

We could use lists of factors to find p and q, but there are many factors of 608. We can factor the trinomial by first finding the roots of the equation

$$x^2 - 13x - 608 = 0.$$

We graph the function $f(x) = x^2 - 13x - 608$. The zeros are -19 and 32.

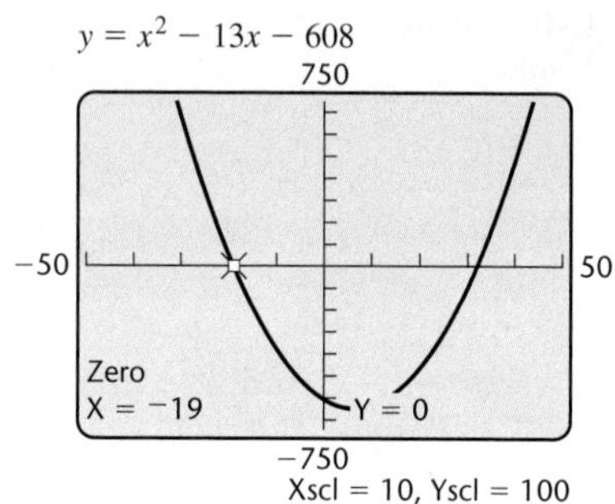

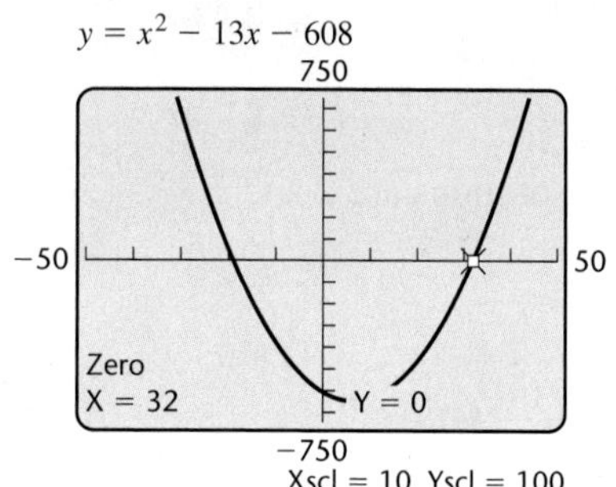

Thus, when $x^2 - 13x - 608 = 0$, we have

$$x = -19 \quad or \quad x = 32$$

$$x + 19 = 0 \quad or \quad x - 32 = 0$$

$$(x + 19)(x - 32) = 0.$$ **Using the principle of zero products "in reverse"**

Then $x^2 - 13x - 608 = (x + 19)(x - 32)$.

Multiplication indicates that the factorization is correct.

Try Exercise 67.

Not every trinomial is factorable. The roots of $x^2 - 2x - 2 = 0$ are irrational; thus $x^2 - 2x - 2$ cannot be factored using integers. The equation $x^2 - x + 1 = 0$ has no real roots, and $x^2 - x + 1$ is prime. We will study equations like these in Chapter 11.

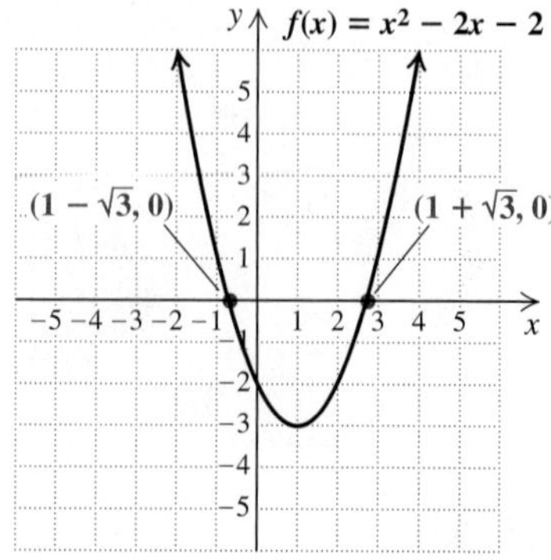

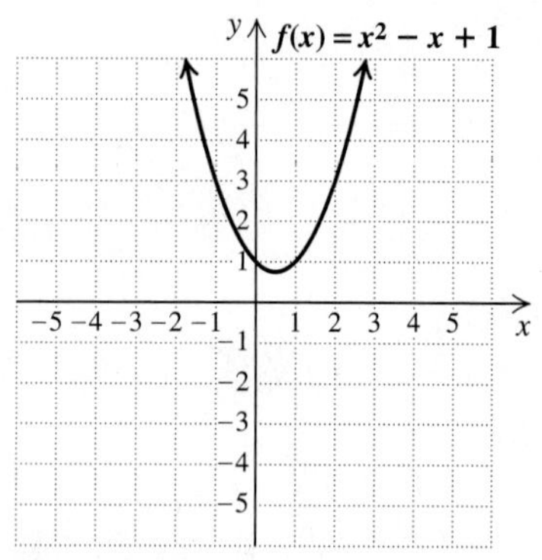

We can also use the principle of zero products to write a function whose zeros are given.

EXAMPLE 11 Write a polynomial function $f(x)$ whose zeros are -1, 0, and 3.

SOLUTION Each zero of the polynomial function f yields a linear factor of the polynomial.

For $x = -1$, $x + 1 = 0$. **Writing a linear factor for each zero**

For $x = 0$, $x = 0$.

For $x = 3$, $x - 3 = 0$.

Thus a polynomial function with zeros -1, 0, and 3 is

$$f(x) = (x + 1) \cdot x \cdot (x - 3);$$

multiplying gives us

$$f(x) = x^3 - 2x^2 - 3x.$$

Try Exercise 71.

6.2 Exercise Set

FOR EXTRA HELP MyMathLab MathXL PRACTICE WATCH DOWNLOAD READ REVIEW

Concept Reinforcement Classify each of the following statements as either true or false.

1. When factoring any polynomial, it is always best to look first for a common factor.
2. Whenever the sum of a negative number and a positive number is negative, the negative number has the greater absolute value.
3. Whenever the product of a pair of factors is negative, the factors have the same sign.
4. If $p + q = -17$, then $-p + (-q) = 17$.
5. To factor $x^2 + 16x + 60$, consider only pairs of positive factors of 60.
6. To factor $x^2 - 4x - 60$, consider only pairs of negative factors of 60.
7. If 1 is a zero of a polynomial function, then $x - 1$ is a factor of the polynomial.
8. If $x - 2$ is a factor of $p(x)$, then $p(2) = 0$.

Factor completely. Remember to look first for a common factor. If a polynomial is prime, state this.

9. $x^2 + 8x + 12$
10. $x^2 + 6x + 5$
11. $t^2 + 8t + 15$
12. $y^2 + 12y + 27$
13. $a^2 - 7a + 12$
14. $z^2 - 8z + 7$
15. $x^2 - 2x - 15$
16. $x^2 - x - 42$
17. $x^2 + 2x - 15$
18. $x^2 + x - 42$
19. $2n^2 - 20n + 50$
20. $2a^2 - 16a + 32$
21. $a^3 - a^2 - 72a$
22. $x^3 + 3x^2 - 54x$
23. $14x + x^2 + 45$
24. $12y + y^2 + 32$
25. $3x + x^2 - 10$
26. $x + x^2 - 6$
27. $3x^2 - 15x + 18$
28. $5y^2 - 40y + 35$
29. $56 + x - x^2$
30. $32 + 4y - y^2$
31. $32y + 4y^2 - y^3$
32. $56x + x^2 - x^3$
33. $x^4 + 11x^3 - 80x^2$
34. $y^4 + 5y^3 - 84y^2$
35. $x^2 + 12x + 13$
36. $x^2 - 3x + 7$
37. $p^2 - 5pq - 24q^2$
38. $x^2 + 12xy + 27y^2$
39. $y^2 + 8yz + 16z^2$
40. $x^2 - 14xy + 49y^2$
41. $p^4 - 80p^3 + 79p^2$
42. $x^4 - 50x^3 + 49x^2$
43. Use the results of Exercise 9 to solve $x^2 + 8x + 12 = 0$.
44. Use the results of Exercise 10 to solve $x^2 + 6x + 5 = 0$.
45. Use the results of Exercise 19 to solve $2n^2 + 50 = 20n$.
46. Use the results of Exercise 30 to solve $32 + 4y = y^2$.

In Exercises 47–50, use the graph to solve the given equation. Check by substituting into the equation.

Aha! **47.** $x^2 + 4x - 5 = 0$

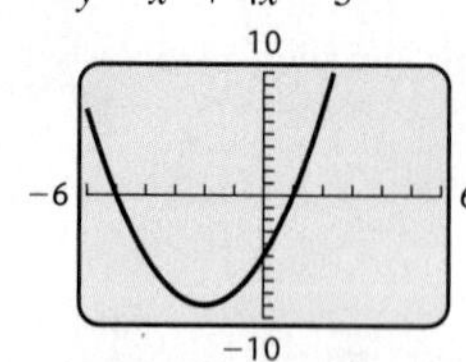

48. $x^2 - x - 6 = 0$

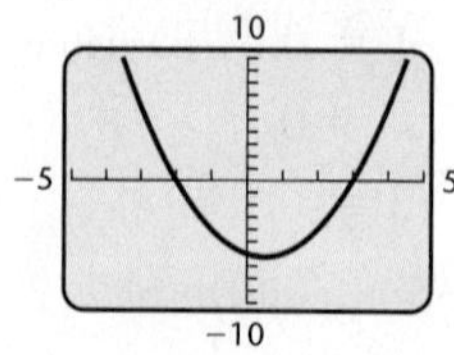

49. $x^2 + x - 6 = 0$

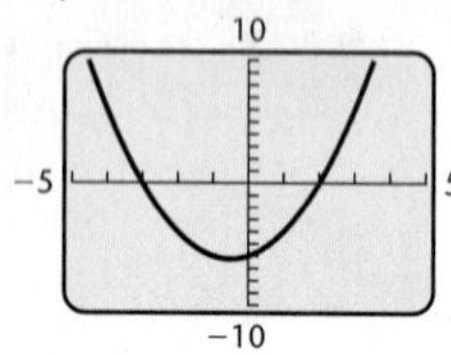

50. $x^2 + 8x + 15 = 0$

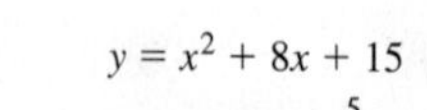

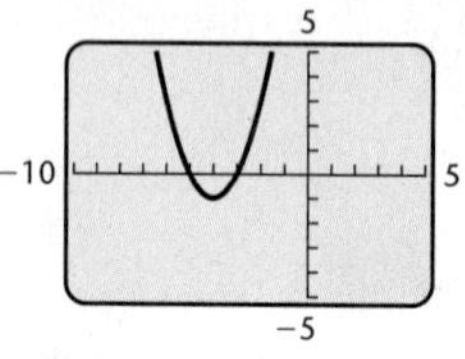

Find the zeros of each function.

51. $f(x) = x^2 - 4x - 45$

52. $f(x) = x^2 + x - 20$

53. $r(x) = x^3 + 4x^2 + 3x$

54. $g(x) = 3x^2 - 21x + 30$

Solve.

55. $x^2 + 4x = 45$

56. $t^2 - 3t = 28$

57. $x^2 - 9x = 0$

58. $a^2 + 18a = 0$

59. $a^3 + 40a = 13a^2$

60. $x^3 - 2x^2 = 63x$

61. $(x - 3)(x + 2) = 14$

62. $(z + 4)(z - 2) = -5$

63. $35 - x^2 = 2x$

64. $40 - x^2 + 3x = 0$

In Exercises 65 and 66, use the graph to factor the given polynomial.

65. $x^2 + 10x - 264$

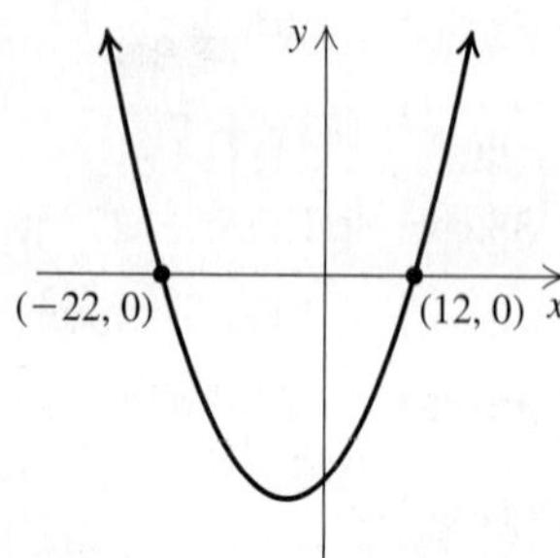

$f(x) = x^2 + 10x - 264$

66. $x^2 + 16x - 336$

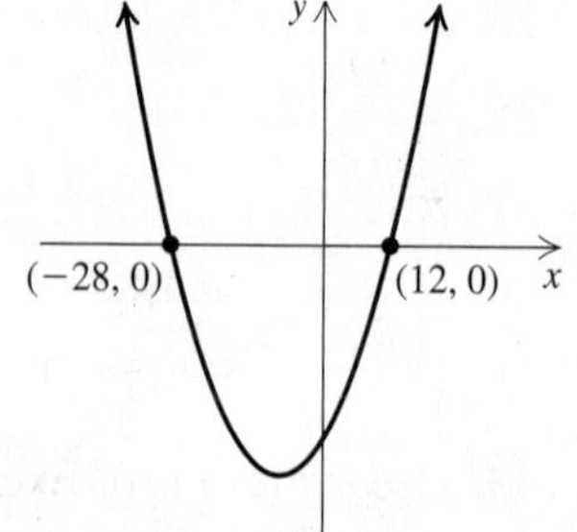

$f(x) = x^2 + 16x - 336$

In Exercises 67–70, use a graph to help factor each polynomial.

67. $x^2 + 40x + 384$

68. $x^2 - 13x - 300$

69. $x^2 + 26x - 2432$

70. $x^2 - 46x + 504$

Write a polynomial function that has the given zeros. Answers may vary.

71. $-1, 2$

72. $2, 5$

73. $-7, -10$

74. $8, -3$

75. $0, 1, 2$

76. $-3, 0, 5$

TW **77.** Allison says that she will never miss a point of intersection when solving graphically because she always uses a $[-100, 100, -100, 100]$ window. Is she correct? Why or why not?

TW **78.** Shari factors $x^3 - 8x^2 + 15x$ as $(x^2 - 5x)(x - 3)$. Is she wrong? Why or why not? What advice would you offer?

SKILL REVIEW

To prepare for Section 6.3, review multiplying binomials using FOIL (Section 5.6).

Multiply. [5.6]

79. $(2x + 3)(3x + 4)$

80. $(2x + 3)(3x - 4)$

81. $(2x - 3)(3x + 4)$

82. $(2x - 3)(3x - 4)$

83. $(5x - 1)(x - 7)$

84. $(x + 6)(3x - 5)$

SYNTHESIS

TW **85.** Explain how the following graph of

$$y = x^2 + 3x - 2 - (x - 2)(x + 1)$$

can be used to show that

$$x^2 + 3x - 2 \neq (x - 2)(x + 1).$$

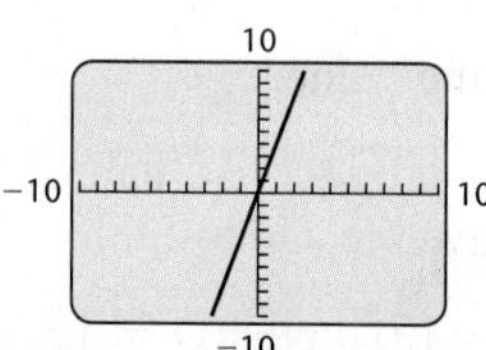

TW **86.** When searching for a factorization, why do we list pairs of numbers with the correct *product* instead of pairs of numbers with the correct *sum*?

87. Use the following graph of $f(x) = x^2 - 2x - 3$ to solve $x^2 - 2x - 3 = 0$ and to solve $x^2 - 2x - 3 < 5$.

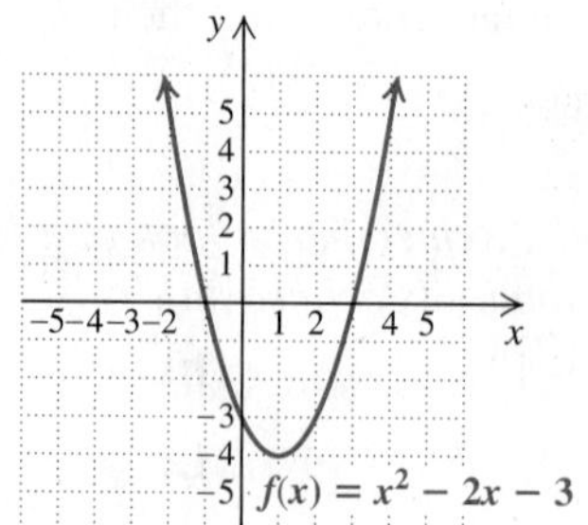

88. Use the following graph of $g(x) = -x^2 - 2x + 3$ to solve $-x^2 - 2x + 3 = 0$ and to solve $-x^2 - 2x + 3 \geq -5$.

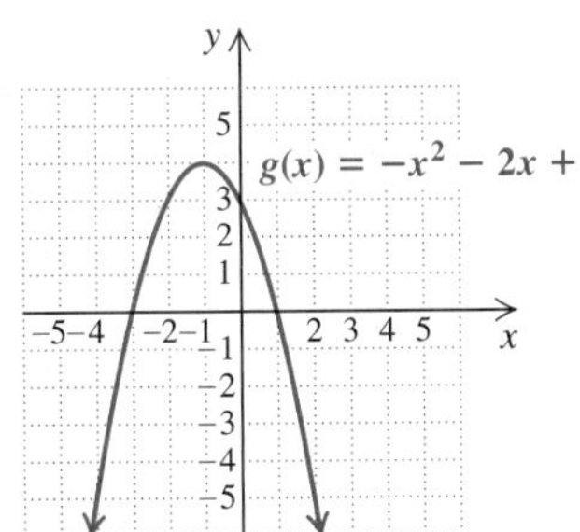

89. Find a polynomial function f for which $f(2) = 0$, $f(-1) = 0$, $f(3) = 0$, and $f(0) = 30$.

90. Find a polynomial function g for which $g(-3) = 0$, $g(1) = 0$, $g(5) = 0$, and $g(0) = 45$.

In Exercises 91–94, use a graphing calculator to find any solutions that exist accurate to two decimal places.

91. $-x^2 + 13.80x = 47.61$

92. $-x^2 + 3.63x + 34.34 = x^2$

93. $x^3 - 3.48x^2 + x = 3.48$

94. $x^2 + 4.68 = 1.2x$

Factor. Assume that variables in exponents represent positive integers.

95. $x^2 + \frac{1}{2}x - \frac{3}{16}$

96. $y^2 + 0.4y - 0.05$

97. $x^{2a} + 5x^a - 24$

98. $a^2p^{2a} + a^2p^a - 2a^2$

Aha! **99.** $(a + 1)x^2 + (a + 1)3x + (a + 1)2$

100. $ax^2 - 5x^2 + 8ax - 40x - (a - 5)9$ (*Hint*: See Exercise 99.)

Aha! **101.** $(x + 3)^2 - 2(x + 3) - 35$

102. Find all integers m for which $x^2 + mx + 75$ can be factored.

103. Find all integers q for which $x^2 + qx - 32$ can be factored.

104. One factor of $x^2 - 345x - 7300$ is $x + 20$. Find the other factor.

Find a polynomial in factored form for the shaded area in each figure.

105.

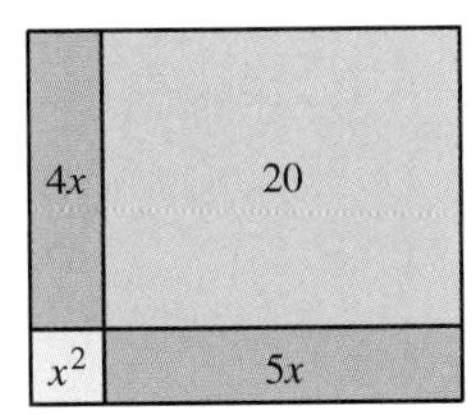

106.

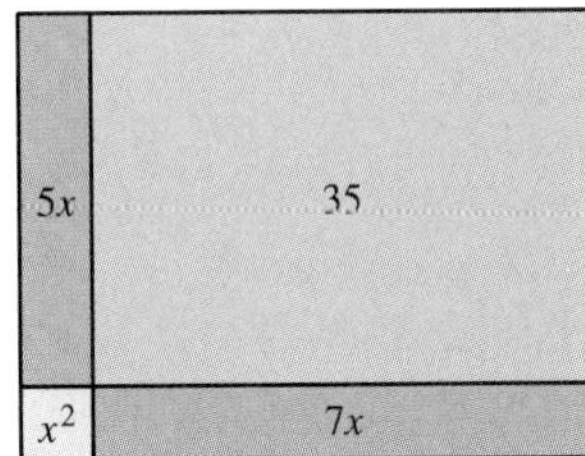

Try Exercise Answers: Section 6.2

9. $(x + 2)(x + 6)$ **13.** $(a - 3)(a - 4)$ **21.** $a(a + 8)(a - 9)$ **25.** $(x + 5)(x - 2)$ **35.** Prime **37.** $(p - 8q)(p + 3q)$ **43.** $-6, -2$ **59.** $0, 5, 8$ **61.** $-4, 5$ **67.** $(x + 24)(x + 16)$ **71.** $f(x) = (x + 1)(x - 2)$, or $f(x) = x^2 - x - 2$

6.3 Trinomials of the Type $ax^2 + bx + c$

- Factoring Trinomials of the Type $ax^2 + bx + c$
- Equations and Functions

FACTORING TRINOMIALS OF THE TYPE $ax^2 + bx + c$

Now we look at trinomials in which the leading coefficient is not 1. We consider two factoring methods. Use the method that you prefer or the one recommended by your instructor.

Method 1: Factoring with FOIL

We first consider the **FOIL method** for factoring trinomials of the type

$$ax^2 + bx + c, \quad \text{where } a \neq 1.$$

Consider the following multiplication.

$$\begin{aligned} & \quad\;\; \text{F} \quad\; \text{O} \quad\;\; \text{I} \quad\; \text{L} \\ (3x + 2)(4x + 5) &= 12x^2 + 15x + 8x + 10 \\ &= 12x^2 + \quad 23x \quad + 10 \end{aligned}$$

To factor $12x^2 + 23x + 10$, we could reverse the multiplication and look for two binomials whose product is this trinomial. The product of the First terms must be $12x^2$. The product of the Outer terms plus the product of the Inner terms must be $23x$. The product of the Last terms must be 10. How can such a factorization be found? Our first approach relies on FOIL.

STUDY TIP

Leave a Trail

Most instructors regard your reasoning as more important than your final answer. Try to organize your supporting work so that your instructor (and you as well) can follow your steps. Instead of erasing work you are not pleased with, consider simply crossing it out so that you (and others) can study and learn from these attempts.

To Factor $ax^2 + bx + c$ Using FOIL

1. Make certain that all common factors have been removed. If any remain, factor out the largest common factor.
2. Find two **F**irst terms whose product is ax^2:

$$(\square x + \quad)(\square x + \quad) = ax^2 + bx + c.$$
FOIL

3. Find two **L**ast terms whose product is c:

$$(\quad x + \square)(\quad x + \square) = ax^2 + bx + c.$$
FOIL

4. Check by multiplying to see if the sum of the **O**uter and **I**nner products is bx. If necessary, repeat steps (2) and (3) until the correct combination is found.

$$(\square x + \square)(\square x + \square) = ax^2 + bx + c.$$
I
O
FOIL

If no correct combination exists, state that the polynomial is prime.

EXAMPLE 1 Factor: $3x^2 - 10x - 8$.

SOLUTION

1. First, check for a common factor. In this case, there is none (other than 1 or -1).
2. Find two **F**irst terms whose product is $3x^2$. The only possibilities for the **F**irst terms are $3x$ and x. Thus, if a factorization exists, it must be of the form

$$(3x + \square)(x + \square).$$

3. Find two **L**ast terms whose product is -8. There are four pairs of factors of -8 and each can be listed in two ways:

$-1, \; 8$		$8, -1$
$1, -8$	and	$-8, \; 1$
$-2, \; 4$		$4, -2$
$2, -4$		$-4, \; 2.$

Important! Since the **F**irst terms are not identical, changing the order of the factors of -8 results in a different product.

4. Find a pair of factors for which the sum of the **O**uter and **I**nner products is the middle term, $-10x$.

Pair of Factors	Corresponding Trial	Product	
$-1,\ 8$	$(3x - 1)(x + 8)$	$3x^2 + 24x - x - 8$ $= 3x^2 + 23x - 8$	Wrong middle term
$1, -8$	$(3x + 1)(x - 8)$	$3x^2 - 24x + x - 8$ $= 3x^2 - 23x - 8$	Wrong middle term
$-2,\ 4$	$(3x - 2)(x + 4)$	$3x^2 + 12x - 2x - 8$ $= 3x^2 + 10x - 8$	Wrong middle term
$2, -4$	$(3x + 2)(x - 4)$	$3x^2 - 12x + 2x - 8$ $= 3x^2 - 10x - 8$	Correct middle term!
$8, -1$	$(3x + 8)(x - 1)$	$3x^2 - 3x + 8x - 8$ $= 3x^2 + 5x - 8$	Wrong middle term
$-8,\ 1$	$(3x - 8)(x + 1)$	$3x^2 + 3x - 8x - 8$ $= 3x^2 - 5x - 8$	Wrong middle term
$4, -2$	$(3x + 4)(x - 2)$	$3x^2 - 6x + 4x - 8$ $= 3x^2 - 2x - 8$	Wrong middle term
$-4,\ 2$	$(3x - 4)(x + 2)$	$3x^2 + 6x - 4x - 8$ $= 3x^2 + 2x - 8$	Wrong middle term

The correct factorization is $(3x + 2)(x - 4)$.

Try Exercise 9.

Two observations can be made from Example 1. First, we listed all possible trials even though we generally stop after finding the correct factorization. We did this to show that **each trial differs only in the middle term of the product.** Second, note that **only the sign of the middle term changes when the signs in the binomials are reversed.**

Student Notes

Keep your work organized so that you can see what you have already considered. For example, when factoring $6x^2 - 19x + 10$, we can list all possibilities and cross out those in which a common factor appears:

~~$(3x - 1)(2x - 10)$~~,
$(3x - 10)(2x - 1)$,
$(3x - 2)(2x - 5)$,
~~$(3x - 5)(2x - 2)$~~,
$(6x - 1)(x - 10)$,
~~$(6x - 10)(x - 1)$~~,
~~$(6x - 2)(x - 5)$~~,
$(6x - 5)(x - 2)$.

By being organized and not erasing, we can see that there are only four possible factorizations.

EXAMPLE 2 Factor: $6x^6 - 19x^5 + 10x^4$.

SOLUTION

1. First, factor out the common factor x^4:

$$x^4(6x^2 - 19x + 10).$$

2. Since $6x^2 = 6x \cdot x$ and $6x^2 = 3x \cdot 2x$, we have two possibilities for the factorization of $6x^2 - 19x + 10$:

$$(3x + \square)(2x + \square) \quad \text{or} \quad (6x + \square)(x + \square).$$

3. The constant term, 10, can be factored as $1 \cdot 10$, $2 \cdot 5$, $(-1)(-10)$, and $(-2)(-5)$. Since the middle term is negative, we need consider only the factorizations with negative factors:

$$-1, -10 \quad\text{and}\quad -10, -1$$
$$-2, -5 \qquad\qquad -5, -2.$$

4. There are 4 possibilities for *each* factorization in step (2). We need factors for which the sum of the Outer and Inner products is the middle term, $-19x$.

We first try these factors with $(3x + \quad)(2x + \quad)$. If none gives the correct factorization, then we will consider $(6x + \quad)(x + \quad)$.

Trial	Product	
$(3x - 1)(2x - 10)$	$6x^2 - 30x - 2x + 10$	
	$= 6x^2 - 32x + 10$	Wrong middle term
$(3x - 10)(2x - 1)$	$6x^2 - 3x - 20x + 10$	
	$= 6x^2 - 23x + 10$	Wrong middle term
$(3x - 2)(2x - 5)$	$6x^2 - 15x - 4x + 10$	
	$= 6x^2 - 19x + 10$	Correct middle term!

$y_1 = 6x^6 - 19x^5 + 10x^4$,
$y_2 = x^4(3x - 2)(2x - 5)$

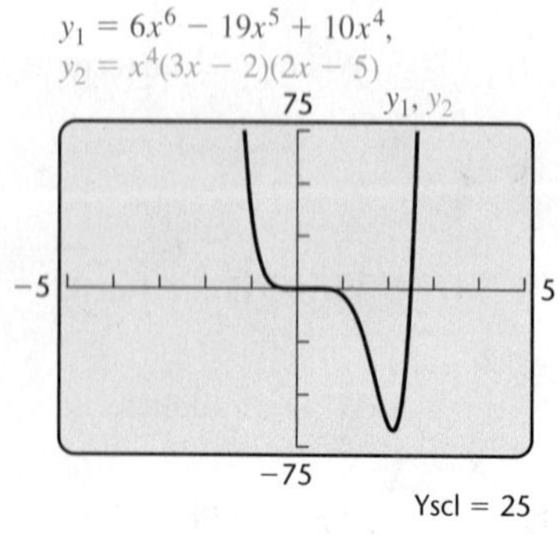

Since we have a correct factorization, we need not consider any additional trials. The factorization of $6x^2 - 19x + 10$ is $(3x - 2)(2x - 5)$. *But do not forget the common factor*! We must include it to get the complete factorization of the original trinomial:

$$6x^6 - 19x^5 + 10x^4 = x^4(3x - 2)(2x - 5).$$

The graphs of the original polynomial and the factorization coincide, as shown in the figure at left.

Try Exercise 21.

In Example 2, look again at the possibility $(3x - 1)(2x - 10)$. Without multiplying, we can reject such a possibility. To see why, note that

$$(3x - 1)(2x - 10) = (3x - 1)2(x - 5).$$

The expression $2x - 10$ has a common factor, 2. But we removed the *largest* common factor in step (1). If $2x - 10$ were one of the factors, then 2 would be *another* common factor in addition to the original, x^4. Thus, $(2x - 10)$ cannot be part of the factorization of $6x^2 - 19x + 10$. Similar reasoning can be used to reject $(3x - 5)(2x - 2)$ as a possible factorization.

Once the largest common factor is factored out, none of the remaining factors can have a common factor.

Tips for Factoring $ax^2 + bx + c$ with FOIL

1. If the largest common factor has been factored out of the original trinomial, then no binomial factor can contain a common factor (other than 1 or -1.)
2. If necessary, factor out a -1 so that a is positive. Then if c is also positive, the signs in the factors must match the sign of b.
3. Reversing the signs in the binomials reverses the sign of the middle term of their product.
4. Organize your work so that you can keep track of those possibilities that you have checked.

Method 2: The *ac*-Method

Another method for factoring trinomials of the type $ax^2 + bx + c$, $a \neq 1$, is known as the ***ac*-method,** or the **grouping method.*** This method relies on rewriting $ax^2 + bx + c$ in the form $ax^2 + px + qx + c$ and then factoring by grouping.

*The rationale behind this method is outlined in Exercise 113.

The sum of p and q is b, and the product of p and q is ac. For example, to factor $6x^2 + 23x + 20$, we form the product $6 \cdot 20 = 120$. Then we determine factors of 120 that add to 23. This process is very similar to the procedure we used to factor trinomials of the form $x^2 + bx + c$. For this trinomial, the numbers we want are 8 and 15. We factor by grouping as follows:

$$\begin{aligned} 6x^2 + 23x + 20 &= 6x^2 + 8x + 15x + 20 && \textbf{Writing } 23x \textbf{ as } 8x + 15x \\ &= 2x(3x + 4) + 5(3x + 4) && \\ &= (3x + 4)(2x + 5). && \textbf{Factoring by grouping} \end{aligned}$$

To Factor $ax^2 + bx + c$ Using the ac-Method

1. Factor out the largest common factor, if one exists.
2. Multiply the leading coefficient a and the constant c.
3. Find a pair of factors of ac whose sum is b.
4. Rewrite the middle term as a sum or a difference using the factors found in step (3).
5. Factor by grouping.
6. Include any common factor from step (1) and check by multiplying.

EXAMPLE 3 Factor: $3x^2 + 10x - 8$.

SOLUTION

1. First, note that there is no common factor (other than 1 or -1.)
2. Multiply the leading coefficient and the constant, 3 and -8.

   $3(-8) = -24.$

3. Next, look for a factorization of -24 in which the sum of the factors is the coefficient of the middle term, 10.

Pairs of Factors of -24	Sums of Factors
$1, -24$	-23
$-1, 24$	23
$2, -12$	-10
$-2, 12$	10 ← $-2 + 12 = 10$
$3, -8$	-5
$-3, 8$	5
$4, -6$	-2
$-4, 6$	2

We normally stop listing pairs of factors once we have found the one we are after.

4. Split $10x$ using the results of step (3):

   $10x = 12x - 2x.$

5. Factor by grouping:

$$3x^2 + 10x - 8 = 3x^2 + 12x - 2x - 8$$

Substituting $12x - 2x$ for $10x$. We could also use $-2x + 12x$.

$$\begin{aligned} &= 3x(x + 4) - 2(x + 4) \\ &= (x + 4)(3x - 2). \end{aligned}$$

Factoring by grouping

$y_1 = 3x^2 + 10x - 8,$
$y_2 = (x + 4)(3x - 2)$

X	Y1	Y2
0	−8	−8
1	5	5
2	24	24
3	49	49
4	80	80
5	117	117
6	160	160
X = 0		

6. We check the solution by multiplying. A second check using a table is shown at left.

$$(x + 4)(3x - 2) = 3x^2 - 2x + 12x - 8$$
$$= 3x^2 + 10x - 8$$

The factorization of $3x^2 + 10x - 8$ is $(x + 4)(3x - 2)$.

Try Exercise 17.

EXAMPLE 4 Factor: $6x^4 - 116x^3 - 80x^2$.

SOLUTION

1. Factor out the greatest common factor, $2x^2$:

$$6x^4 - 116x^3 - 80x^2 = 2x^2(3x^2 - 58x - 40).$$

2. To factor $3x^2 - 58x - 40$, multiply the leading coefficient, 3, and the constant, -40: $3(-40) = -120$.

3. Next, look for factors of -120 that add to -58. We see from the table below that the factors we need are 2 and -60.

Pairs of Factors of −120	Sum of Factors
1, −120	−119
2, −60	−58
3, −40	−37
4, −30	−26
5, −24	−19
6, −20	−14
8, −15	−7
10, −12	−2

4. Split the middle term, $-58x$, using the results of step (3): $-58x = 2x - 60x$.

5. Factor by grouping:

$$3x^2 - 58x - 40 = 3x^2 + 2x - 60x - 40 \quad \text{Substituting } 2x - 60x \text{ for } -58x$$
$$= x(3x + 2) - 20(3x + 2)$$
$$= (3x + 2)(x - 20). \quad \text{Factoring by grouping}$$

The factorization of $3x^2 - 58x - 40$ is $(3x + 2)(x - 20)$. *But don't forget the common factor*! We must include it to factor the original trinomial:

$$6x^4 - 116x^3 - 80x^2 = 2x^2(3x + 2)(x - 20).$$

6. *Check:* $2x^2(3x + 2)(x - 20) = 2x^2(3x^2 - 58x - 40)$
$= 6x^4 - 116x^3 - 80x^2.$

The complete factorization is $2x^2(3x + 2)(x - 20)$.

Try Exercise 19.

EQUATIONS AND FUNCTIONS

We now use our new factoring skill to solve polynomial equations. We factor a polynomial *expression* and use the principle of zero products to solve an *equation*.

EXAMPLE 5 Solve: $6x^6 - 19x^5 + 10x^4 = 0$.

SOLUTION We note at the outset that the polynomial is of degree 6, so there will be at most 6 solutions of the equation.

ALGEBRAIC APPROACH

We solve as follows:

$$6x^6 - 19x^5 + 10x^4 = 0$$

$$x^4(3x - 2)(2x - 5) = 0 \quad \text{Factoring as in Example 2}$$

$$x^4 = 0 \quad or \quad 3x - 2 = 0 \quad or \quad 2x - 5 = 0 \quad \text{Using the principle of zero products}$$

$$x = 0 \quad or \quad x = \tfrac{2}{3} \quad or \quad x = \tfrac{5}{2}. \quad \text{Solving for } x;\ x^4 = x \cdot x \cdot x \cdot x = 0 \text{ when } x = 0$$

The solutions are 0, $\frac{2}{3}$, and $\frac{5}{2}$.

GRAPHICAL APPROACH

We find the x-intercepts of the function

$$f(x) = 6x^6 - 19x^5 + 10x^4$$

using the ZERO option of the CALC menu.

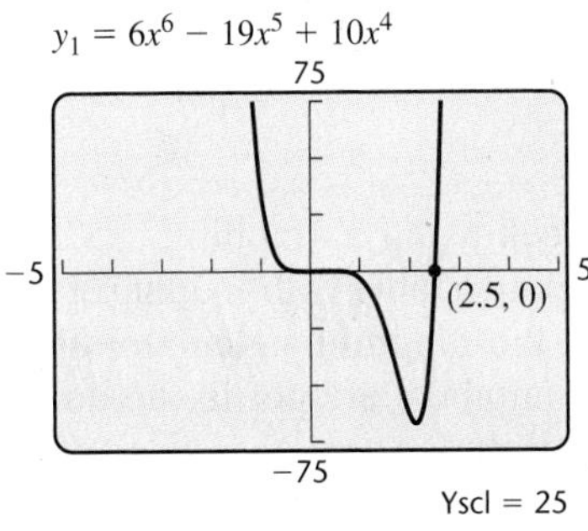

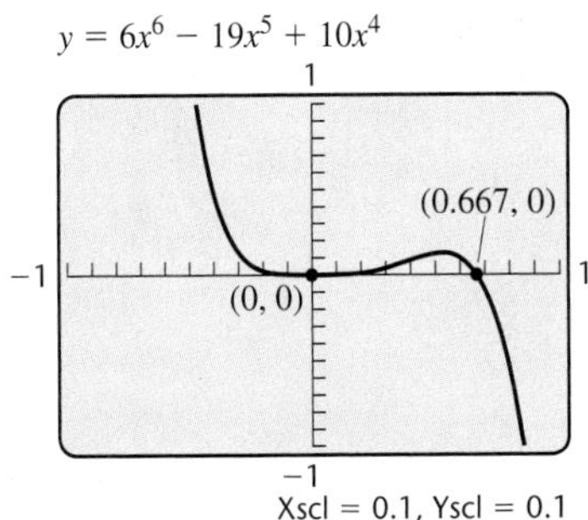

Using the viewing window $[-5, 5, -75, 75]$ shown on the left above, we see that 2.5 is a zero of the function. From this window, we cannot tell how many times the graph of the function intersects the x-axis between -1 and 1. We magnify that portion of the x-axis, as shown on the right above. The x-coordinates of the x-intercepts are 0, 0.667, and 2.5. Since $\frac{5}{2} = 2.5$ and $\frac{2}{3} \approx 0.667$, the solutions are 0, $\frac{2}{3}$, and $\frac{5}{2}$.

Try Exercise 59.

Example 5 illustrates two disadvantages of solving equations graphically: (1) It can be easy to "miss" solutions in some viewing windows and (2) solutions may be given as approximations.

Our work with factoring can help us when we are working with functions.

EXAMPLE 6 Given that $f(x) = 3x^2 - 4x$, find all values of a for which $f(a) = 4$.

SOLUTION We want all numbers a for which $f(a) = 4$. Since $f(a) = 3a^2 - 4a$, we must have

$$\begin{aligned} 3a^2 - 4a &= 4 && \text{Setting } f(a) \text{ equal to 4} \\ 3a^2 - 4a - 4 &= 0 && \text{Getting 0 on one side} \\ (3a + 2)(a - 2) &= 0 && \text{Factoring} \\ 3a + 2 = 0 \quad &or \quad a - 2 = 0 \\ a = -\tfrac{2}{3} \quad &or \quad a = 2. \end{aligned}$$

Y1(−2/3)	
	4
Y1(2)	
	4

To check using a graphing calculator, we enter $y_1 = 3x^2 - 4x$ and calculate Y1(−2/3) and Y1(2). To have $f(a) = 4$, we must have $a = -\frac{2}{3}$ or $a = 2$.

Try Exercise 69.

EXAMPLE 7 Find the domain of F if $F(x) = \dfrac{x - 2}{x^2 + 2x - 15}$.

SOLUTION The domain of F is the set of all values for which

$$\frac{x - 2}{x^2 + 2x - 15}$$

is a real number. Since division by 0 is undefined, we *exclude* any x-value for which the denominator, $x^2 + 2x - 15$, is 0. We solve:

$$\begin{aligned} x^2 + 2x - 15 &= 0 && \text{Setting the denominator equal to 0} \\ (x - 3)(x + 5) &= 0 && \text{Factoring} \\ x - 3 = 0 \quad &or \quad x + 5 = 0 \\ x = 3 \quad &or \quad x = -5. && \text{These are the values to } \textit{exclude}. \end{aligned}$$

To check using a graphing calculator, we enter $y = (x - 2)/(x^2 + 2x - 15)$ and set up a table with Indpnt set to Ask. When the x-values of 3 and −5 are supplied, the graphing calculator should return an error message, indicating that those numbers are not in the domain of the function. There will be a function value given for any other value of x.

The domain of F is $\{x \mid x \text{ is a real number } and\ x \neq -5\ and\ x \neq 3\}$.

X	Y1
3	ERROR
−5	ERROR
0	.13333
.5	.10909
8	.09231
127	.00764
−3	.41667
X = 3	

Try Exercise 71.

6.3 Exercise Set

FOR EXTRA HELP MyMathLab Math XL PRACTICE  WATCH  DOWNLOAD  READ REVIEW

Concept Reinforcement *In each of Exercises 1–8, match the polynomial with one of its factors from the column on the right.*

1. $2x^2 - 7x - 15$ a) $5x + 2$
2. $3x^2 + 4x - 7$ b) $5x + 3$
3. $6x^2 + 7x + 2$ c) $3x + 7$
4. $10x^2 - x - 2$ d) $2x + 7$
5. $3x^2 + 4x - 15$ e) $3x + 2$
6. $2x^2 + 9x + 7$ f) $2x + 3$
7. $10x^2 + 9x - 7$ g) $3x - 5$
8. $15x^2 + 14x + 3$ h) $5x + 7$

Factor completely. If a polynomial is prime, state this.

9. $2x^2 + 7x - 4$
10. $3x^2 + x - 4$
11. $3x^2 - 17x - 6$
12. $5x^2 - 19x - 4$
13. $15a^2 - 14a + 3$
14. $3a^2 - 10a + 8$
15. $6t^2 + 17t + 7$
16. $9a^2 + 18a + 8$
17. $6x^2 - 10x - 4$
18. $15t^2 + 20t - 75$
19. $8x^2 - 16 - 28x$
20. $18x^2 - 24 - 6x$
21. $14x^4 - 19x^3 - 3x^2$
22. $70x^4 - 68x^3 + 16x^2$
23. $10 - 23x + 12x^2$
24. $x - 15 + 2x^2$
25. $9x^2 + 15x + 4$
26. $6y^2 - y - 2$
27. Aha! $4x^2 + 15x + 9$
28. $2y^2 + 7y + 6$
29. $4 + 6t^2 - 13t$
30. $2t^2 - 19 - 6t$
31. $-8t^2 - 8t + 30$
32. $-36a^2 + 21a - 3$
33. $8 - 6z - 9z^2$
34. $3 + 35a - 12a^2$
35. $18xy^3 + 3xy^2 - 10xy$
36. $3x^3y^2 - 5x^2y^2 - 2xy^2$
37. $24x^2 - 2 - 47x$
38. $15z^2 - 10 - 47z$
39. $63x^3 + 111x^2 + 36x$
40. $50t^3 + 115t^2 + 60t$
41. $48x^4 + 4x^3 - 30x^2$
42. $40y^4 + 4y^2 - 12$
43. $12a^2 - 17ab + 6b^2$
44. $20a^2 - 23ax + 6x^2$
45. $2x^2 + xy - 6y^2$
46. $8m^2 - 6mn - 9n^2$
47. $8s^2 + 22st + 14t^2$
48. $10s^2 + 4st - 6t^2$
49. $9x^2 - 30xy + 25y^2$
50. $4p^2 + 12pq + 9q^2$
51. $9x^2y^2 + 5xy - 4$
52. $7a^2b^2 + 13ab + 6$
53. Use the results of Exercise 33 to solve $9z^2 + 6z = 8$.
54. Use the results of Exercise 34 to solve $3 + 35a = 12a^2$.
55. Use the results of Exercise 39 to solve $63x^3 + 111x^2 + 36x = 0$.
56. Use the results of Exercise 40 to solve $50t^3 + 115t^2 + 60t = 0$.

Solve.

57. $3x^2 - 8x + 4 = 0$
58. $9x^2 - 15x + 4 = 0$
59. $4t^3 + 11t^2 + 6t = 0$
60. $8n^3 + 10n^2 + 3n = 0$
61. $6x^2 = 13x + 5$
62. $40x^2 + 43x = 6$
63. $x(5 + 12x) = 28$
64. $a(1 + 21a) = 10$
65. Find the zeros of the function given by
$$f(x) = 2x^2 - 13x - 7.$$
66. Find the zeros of the function given by
$$g(x) = 6x^2 + 13x + 6.$$
67. Let $f(x) = x^2 + 12x + 40$. Find all values of a for which $f(a) = 8$.
68. Let $f(x) = x^2 + 14x + 50$. Find all values of a for which $f(a) = 5$.
69. Let $g(x) = 2x^2 + 5x$. Find all values of a for which $g(a) = 12$.
70. Let $g(x) = 2x^2 - 15x$. Find all values of a for which $g(a) = -7$.

Find the domain of the function f given by each of the following.

71. $f(x) = \dfrac{3}{x^2 - 4x - 5}$
72. $f(x) = \dfrac{2}{x^2 - 7x + 6}$

73. $f(x) = \dfrac{x - 5}{9x - 18x^2}$

74. $f(x) = \dfrac{1 + x}{3x - 15x^2}$

75. $f(x) = \dfrac{3x}{2x^2 - 9x + 4}$

76. $f(x) = \dfrac{-x}{6x^2 + 13x + 6}$

77. $f(x) = \dfrac{7}{5x^3 - 35x^2 + 50x}$

78. $f(x) = \dfrac{3}{2x^3 - 2x^2 - 12x}$

TW **79.** Asked to factor $4x^2 + 28x + 48$, Aziz *incorrectly* answers

$$\begin{aligned} 4x^2 + 28x + 48 &= (2x + 6)(2x + 8) \\ &= 2(x + 3)(x + 4). \end{aligned}$$

If this were a 10-point quiz question, how many points would you take off? Why?

TW **80.** Austin says that the domain of the function

$$F(x) = \frac{x + 3}{3x^2 - x - 2}$$

is $\{-\frac{2}{3}, 1\}$. Is he correct? Why or why not?

SKILL REVIEW

To prepare for Section 6.4, review the special products in Section 5.6.

Multiply. [5.6]

81. $(x - 2)^2$

82. $(x + 2)^2$

83. $(x + 2)(x - 2)$

84. $(5t - 3)^2$

85. $(4a + 1)^2$

86. $(2n + 7)(2n - 7)$

87. $(3c - 10)^2$

88. $(1 - 5a)^2$

89. $(8n + 3)(8n - 3)$

90. $(9 - y)(9 + y)$

SYNTHESIS

TW **91.** Explain how you would prove to a fellow student that a given trinomial is prime.

TW **92.** Tori has factored a polynomial as $(a - b)(x - y)$, while Tracy has factored the same polynomial as $(b - a)(y - x)$. Can both be correct? Why or why not?

Use a graph to help factor each polynomial.

93. $4x^2 + 120x + 675$

94. $4x^2 + 164x + 1197$

95. $3x^3 + 150x^2 - 3672x$

96. $5x^4 + 20x^3 - 1600x^2$

Solve.

97. $(8x + 11)(12x^2 - 5x - 2) = 0$

98. $(x + 1)^3 = (x - 1)^3 + 26$

99. $(x - 2)^3 = x^3 - 2$

Factor. Assume that variables in exponents represent positive integers. If a polynomial is prime, state this.

100. $9x^2y^2 - 12xy - 2$

101. $18a^2b^2 - 3ab - 10$

102. $16x^2y^3 + 20xy^2 + 4y$

103. $16a^2b^3 + 25ab^2 + 9$

104. $9t^8 + 12t^4 + 4$

105. $25t^{10} - 10t^5 + 1$

106. $-15x^{2m} + 26x^m - 8$

107. $20x^{2n} + 16x^n + 3$

108. $3(a + 1)^{n+1}(a + 3)^2 - 5(a + 1)^n(a + 3)^3$

109. $7(t - 3)^{2n} + 5(t - 3)^n - 2$

110. $6(x - 7)^2 + 13(x - 7) - 5$

111. $2a^4b^6 - 3a^2b^3 - 20$

112. $5x^8y^6 + 35x^4y^3 + 60$

113. To better understand factoring $ax^2 + bx + c$ by the ac-method, suppose that

$$ax^2 + bx + c = (mx + r)(nx + s).$$

Show that if $P = ms$ and $Q = rn$, then $P + Q = b$ and $PQ = ac$.

Try Exercise Answers: Section 6.3

9. $(2x - 1)(x + 4)$ **17.** $2(3x + 1)(x - 2)$ **19.** $4(x - 4)(2x + 1)$ **21.** $x^2(2x - 3)(7x + 1)$ **59.** $-2, -\frac{3}{4}, 0$ **69.** $-4, \frac{3}{2}$ **71.** $\{x \mid x \text{ is a real number } and\ x \neq 5 \text{ } and\ x \neq -1\}$

Mid-Chapter Review

In Sections 6.1–6.3, we have considered factoring out a common factor, factoring by grouping, and factoring with FOIL. The following is a good strategy to follow when you encounter a mixed set of factoring problems.

1. Factor out any common factor.
2. Try factoring by grouping for polynomials with four terms.
3. Try factoring with FOIL for trinomials. If the leading coefficient of the trinomial is not 1, you may instead try factoring by grouping.

GUIDED SOLUTIONS

Factor completely.

1. $12x^3y - 8xy^2 + 24x^2y = \square(3x^2 - 2y + 6x)$ [6.1] **Factoring out the largest common factor. No further factorization is possible.**

2. $3a^3 - 3a^2 - 90a = 3a(\square - \square - \square)$ **Factoring out the largest common factor**

   $= 3a(a - \square)(a + \square)$ [6.2] **Factoring the trinomial**

MIXED REVIEW

Factor completely. If a polynomial is prime, state this.

1. $6x^5 - 18x^2$ [6.1]
2. $x^2 + 10x + 16$ [6.2]
3. $2x^2 + 13x - 7$ [6.3]
4. $x^3 + 3x^2 + 2x + 6$ [6.1]
5. $5x^2 + 40x - 100$ [6.2]
6. $x^2 - 2x - 5$ [6.2]
7. $7x^2y - 21xy - 28y$ [6.2]
8. $15a^4 - 27a^2b^2 + 21a^2b$ [6.1]
9. $b^2 - 14b + 49$ [6.2]
10. $12x^2 - x - 1$ [6.3]
11. $x^2y^2 - xy - 2$ [6.2]
12. $2x^2 + 30x - 200$ [6.2]
13. $t^2 + t - 10$ [6.2]
14. $15d^2 - 30d + 75$ [6.1]
15. $15p^2 + 16px + 4x^2$ [6.3]
16. $-2t^3 - 10t^2 - 12t$ [6.2]
17. $x^2 + 4x - 77$ [6.2]
18. $10c^2 + 20c + 10$ [6.2]
19. $5 + 3x - 2x^2$ [6.3]
20. $2m^3n - 10m^2n - 6mn + 30n$ [6.1]

6.4 Perfect-Square Trinomials and Differences of Squares

- Perfect-Square Trinomials
- Differences of Squares
- More Factoring by Grouping
- Solving Equations

Student Notes

If you're not already quick to recognize squares, this is a good time to memorize these numbers:

$1 = 1^2$, $49 = 7^2$,
$4 = 2^2$, $64 = 8^2$,
$9 = 3^2$, $81 = 9^2$,
$16 = 4^2$, $100 = 10^2$,
$25 = 5^2$, $121 = 11^2$,
$36 = 6^2$, $144 = 12^2$.

Reversing the rules for special products provides us with shortcuts for factoring certain polynomials.

PERFECT-SQUARE TRINOMIALS

Consider the trinomial

$$x^2 + 6x + 9.$$

To factor it, we can look for factors of 9 that add to 6. These factors are 3 and 3:

$$x^2 + 6x + 9 = (x + 3)(x + 3) = (x + 3)^2.$$

Note that the result is the square of a binomial. Because of this, we call $x^2 + 6x + 9$ a **perfect-square trinomial.** Although reversing FOIL can be used to factor a perfect-square trinomial, once recognized, a perfect-square trinomial can be quickly factored.

To Recognize a Perfect-Square Trinomial

- Two terms must be squares, such as A^2 and B^2.
- Neither A^2 nor B^2 is being subtracted.
- The remaining term must be $2AB$ or its opposite, $-2AB$.

Note that in order for a term to be a square, its coefficient must be a perfect square and the exponent(s) of the variable(s) must be even.

EXAMPLE 1 Determine whether each polynomial is a perfect-square trinomial.

a) $x^2 + 10x + 25$ **b)** $4x + 16 + 3x^2$ **c)** $100y^2 + 81 - 180y$

SOLUTION

a)
- Two of the terms in $x^2 + 10x + 25$ are squares: x^2 and 25. Here, $A = x$ and $B = 5$.
- Neither x^2 nor 25 is being subtracted.
- The remaining term, $10x$, is $2 \cdot x \cdot 5$.

Thus, $x^2 + 10x + 25$ *is* a perfect square.

b) In $4x + 16 + 3x^2$, only one term, 16, is a square ($3x^2$ is not a square because 3 is not a perfect square; $4x$ is not a square because x is not a square). Thus, $4x + 16 + 3x^2$ *is not* a perfect square.

c) It can help to write the polynomial in descending order: $100y^2 - 180y + 81$.
- Two of the terms, $100y^2$ and 81, are squares.
- Neither $100y^2$ nor 81 is being subtracted.
- Twice the product of the square roots, $2 \cdot 10y \cdot 9$, is $180y$. The remaining term, $-180y$, is the opposite of this product.

Thus, $100y^2 + 81 - 180y$ *is* a perfect-square trinomial.

Try Exercise 9.

STUDY TIP

Fill In Your Blanks

Don't hesitate to write out any missing steps that you'd like to see included. For instance, in Example 1(c), we state that $100y^2$ is a square. To solidify your understanding, you may want to write $10y \cdot 10y = 100y^2$ in the margin of your textbook.

To factor a perfect-square trinomial, we reuse the patterns that we learned in Section 5.6.

Factoring a Perfect-Square Trinomial

$$A^2 + 2AB + B^2 = (A + B)^2;$$
$$A^2 - 2AB + B^2 = (A - B)^2$$

CAUTION! In Example 2, we are factoring expressions. These are *not* equations and thus we do not try to solve them.

EXAMPLE 2 Factor.

a) $x^2 - 10x + 25$ **b)** $16y^2 + 49 + 56y$ **c)** $-20xy + 4y^2 + 25x^2$

SOLUTION

a) $x^2 - 10x + 25 = (x - 5)^2$ **$A = x$ and $B = 5$. We write the square roots with a minus sign between them.**

Note the sign!

As always, any factorization can be checked by multiplying:

$$(x - 5)^2 = (x - 5)(x - 5) = x^2 - 5x - 5x + 25 = x^2 - 10x + 25.$$

b) $16y^2 + 49 + 56y = 16y^2 + 56y + 49$ **Using a commutative law**

$= (4y + 7)^2$ **$16y^2 = (4y)^2$ and $49 = 7^2$. We write the square roots with a plus sign between them.**

The check is left to the student.

c) $-20xy + 4y^2 + 25x^2 = 4y^2 - 20xy + 25x^2$ **$4y^2$ and $25x^2$ are squares.**

$= (2y - 5x)^2$

This square can also be expressed as

$$25x^2 - 20xy + 4y^2 = (5x - 2y)^2.$$

The student should confirm that both factorizations check.

Try Exercise 17.

When factoring, always look first for a factor common to all the terms.

EXAMPLE 3 Factor: $-4y^2 - 144y^8 + 48y^5$.

SOLUTION We look first for a common factor. In this case, we factor out $-4y^2$ so that the leading coefficient of the polynomial inside the parentheses is positive:

Factor out the common factor.

Factor the perfect-square trinomial.

$-4y^2 - 144y^8 + 48y^5 = -4y^2(1 + 36y^6 - 12y^3)$ **Factoring out the common factor**

$= -4y^2(36y^6 - 12y^3 + 1)$ **Changing order**

$= -4y^2(6y^3 - 1)^2.$ **$36y^6 = (6y^3)^2$ and $1 = 1^2$**

Check: $-4y^2(6y^3 - 1)^2 = -4y^2(6y^3 - 1)(6y^3 - 1)$

$= -4y^2(36y^6 - 12y^3 + 1)$

$= -144y^8 + 48y^5 - 4y^2$

$= -4y^2 - 144y^8 + 48y^5.$

The factorization is $-4y^2(6y^3 - 1)^2$.

Try Exercise 25.

DIFFERENCES OF SQUARES

Any expression, like $16x^2 - 9$, that can be written in the form $A^2 - B^2$ is called a **difference of squares.**

To Recognize a Difference of Squares

- There must be two expressions, both squares, such as A^2 and B^2.
- The terms in the binomial must have different signs.

EXAMPLE 4 Determine whether each of the following is a difference of squares.

a) $9x^2 - 64$ **b)** $25 - t^3$ **c)** $-4x^{10} + 36$

SOLUTION

a) • The first expression in $9x^2 - 64$ is a square: $9x^2 = (3x)^2$. The second expression is a square: $64 = 8^2$.

• The terms have different signs.

Thus, $9x^2 - 64$ is a difference of squares, $(3x)^2 - 8^2$.

b) The expression t^3 in $25 - t^3$ is not a square.

Thus, $25 - t^3$ is not a difference of squares.

c) • The expressions $4x^{10}$ and 36 in $-4x^{10} + 36$ are squares: $4x^{10} = (2x^5)^2$ and $36 = 6^2$.

• The terms have different signs.

Thus, $-4x^{10} + 36$ is a difference of squares, $6^2 - (2x^5)^2$. It is often useful to rewrite $-B^2 + A^2$ in the equivalent form $A^2 - B^2$.

Try Exercise 41.

To factor a difference of two squares, we can reverse a pattern from Section 5.6.

Factoring a Difference of Two Squares

$$A^2 - B^2 = (A + B)(A - B)$$

We often refer to a factorization such as $(A + B)(A - B)$ as the product of the sum and the difference of A and B.

EXAMPLE 5 Factor: **(a)** $x^2 - 9$; **(b)** $25y^6 - 49x^2$.

SOLUTION

a) $x^2 - 9 = x^2 - 3^2 = (x + 3)(x - 3)$

$$\begin{array}{ccccc} A^2 & - & B^2 & = & (A + B)(A - B) \\ \downarrow & & \downarrow & & \downarrow \quad \downarrow \quad \downarrow \quad \downarrow \end{array}$$

b) $25y^6 - 49x^2 = (5y^3)^2 - (7x)^2 = (5y^3 + 7x)(5y^3 - 7x)$

Try Exercise 47.

As always, the first step in factoring is to look for common factors. Factoring is complete when no factor can be factored further.

EXAMPLE 6 Factor: **(a)** $5 - 5x^2y^6$; **(b)** $16x^4y - 81y$.

SOLUTION

a) $5 - 5x^2y^6 = 5(1 - x^2y^6)$ **Factoring out the common factor**

$= 5[1^2 - (xy^3)^2]$ **Rewriting x^2y^6 as a quantity squared**

$= 5(1 + xy^3)(1 - xy^3)$ **Factoring the difference of squares**

Check: $5(1 + xy^3)(1 - xy^3) = 5(1 - xy^3 + xy^3 - x^2y^6)$

$= 5(1 - x^2y^6) = 5 - 5x^2y^6.$

The factorization $5(1 + xy^3)(1 - xy^3)$ checks.

Factor out a common factor.

Factor a difference of squares.

Factor another difference of squares.

b) $16x^4y - 81y = y(16x^4 - 81)$ **Factoring out the common factor**

$= y[(4x^2)^2 - 9^2]$

$= y(4x^2 + 9)(4x^2 - 9)$ **Factoring the difference of squares**

$= y(4x^2 + 9)(2x + 3)(2x - 3)$ **Factoring $4x^2 - 9$, which is itself a difference of squares**

The check is left to the student.

Try Exercise 55.

Note in Example 6(b) that the factor $4x^2 + 9$ is a *sum* of squares that cannot be factored further.

CAUTION! There is no general formula for factoring a sum of squares.

MORE FACTORING BY GROUPING

Sometimes, when factoring a polynomial with four terms, we may be able to factor further.

EXAMPLE 7 Factor: $x^3 + 3x^2 - 4x - 12$.

SOLUTION

$x^3 + 3x^2 - 4x - 12 = x^2(x + 3) - 4(x + 3)$ **Factoring by grouping**

$= (x + 3)(x^2 - 4)$ **Factoring out $x + 3$**

$= (x + 3)(x + 2)(x - 2)$ **Factoring $x^2 - 4$**

Try Exercise 71.

A difference of squares can have four or more terms. For example, one of the squares may be a trinomial. In this case, a type of grouping can be used.

EXAMPLE 8 Factor: **(a)** $x^2 + 6x + 9 - y^2$; **(b)** $a^2 - b^2 + 8b - 16$.

SOLUTION

a) $x^2 + 6x + 9 - y^2 = (x^2 + 6x + 9) - y^2$ — **Grouping as a perfect-square trinomial minus y^2 to show a difference of squares**

$= (x + 3)^2 - y^2$

$= (x + 3 + y)(x + 3 - y)$

b) Grouping $a^2 - b^2 + 8b - 16$ into two groups of two terms does not yield a common binomial factor, so we look for a perfect-square trinomial. In this case, the perfect-square trinomial is being subtracted from a^2:

$a^2 - b^2 + 8b - 16 = a^2 - (b^2 - 8b + 16)$ — **Factoring out -1 and rewriting as subtraction**

$= a^2 - (b - 4)^2$ — **Factoring the perfect-square trinomial**

$= (a + (b - 4))(a - (b - 4))$ — **Factoring a difference of squares**

$= (a + b - 4)(a - b + 4).$ — **Removing parentheses**

Try Exercise 69.

SOLVING EQUATIONS

We can now solve polynomial equations involving differences of squares and perfect-square trinomials.

EXAMPLE 9 Solve: $x^3 + 3x^2 = 4x + 12$.

SOLUTION We first note that the equation is a third-degree polynomial equation. Thus it will have 3 or fewer solutions.

ALGEBRAIC APPROACH

We factor and use the principle of zero products:

$x^3 + 3x^2 = 4x + 12$

$x^3 + 3x^2 - 4x - 12 = 0$ — **Getting 0 on one side**

$(x + 3)(x + 2)(x - 2) = 0$ — **Factoring; using the results of Example 7**

$x + 3 = 0$ *or* $x + 2 = 0$ *or* $x - 2 = 0$ — **Using the principle of zero products**

$x = -3$ *or* $x = -2$ *or* $x = 2$.

The solutions are -3, -2, and 2.

GRAPHICAL APPROACH

We let $f(x) = x^3 + 3x^2$ and $g(x) = 4x + 12$ and look for any points of intersection of the graphs of f and g.

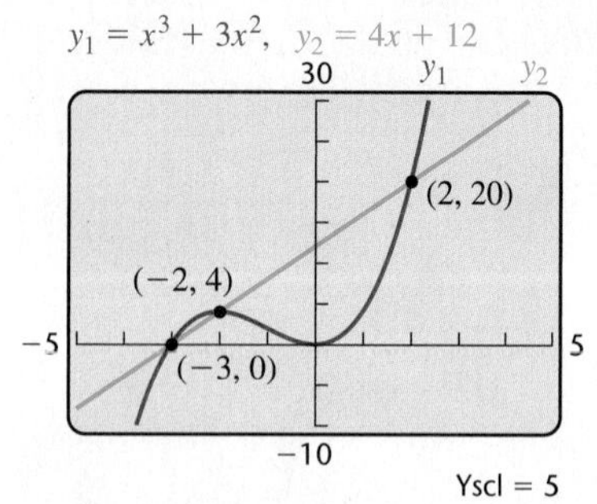

The x-coordinates of the points of intersection are -3, -2, and 2. These are the solutions of the equation.

Try Exercise 83.

EXAMPLE 10 Find the zeros of the function given by

$$f(x) = x^3 + x^2 - x - 1.$$

SOLUTION This function is a third-degree polynomial function, so there will be at most 3 zeros. To find the zeros of the function, we find the roots of the equation $x^3 + x^2 - x - 1 = 0$.

ALGEBRAIC APPROACH

We first factor the polynomial:

$$x^3 + x^2 - x - 1 = 0$$
$$x^2(x + 1) - 1(x + 1) = 0 \qquad \textbf{Factoring by grouping}$$
$$(x + 1)(x^2 - 1) = 0$$
$$(x + 1)(x + 1)(x - 1) = 0. \qquad \textbf{Factoring a difference of squares}$$

Then we use the principle of zero products:

$$x + 1 = 0 \quad or \quad x + 1 = 0 \quad or \quad x - 1 = 0$$
$$x = -1 \quad or \quad x = -1 \quad or \quad x = 1.$$

The solutions are -1 and 1. The zeros are -1 and 1.

GRAPHICAL APPROACH

We graph $f(x) = x^3 + x^2 - x - 1$ and find the zeros of the function.

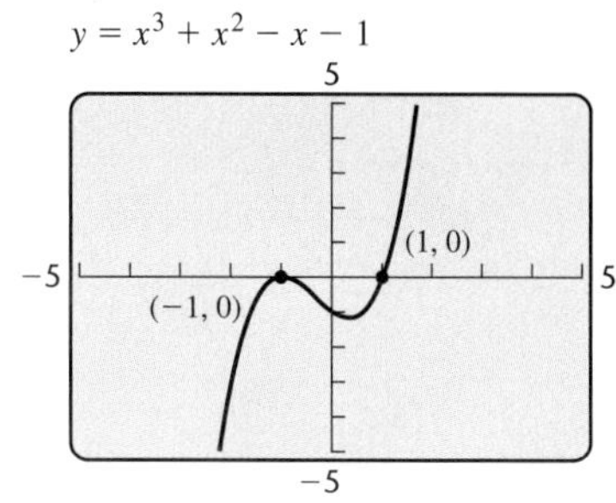

The zeros are -1 and 1.

Try Exercise 103.

In Example 10, the factor $(x + 1)$ appeared twice in the factorization. When a factor appears two or more times, we say that we have a **repeated root.** If the factor occurs twice, we say that we have a **double root,** or a **root of multiplicity two.** In Example 10, -1 is a double root.

In this chapter, we have emphasized the relationship between factoring and equation solving. There are many polynomial equations, however, that cannot be solved by factoring. For these, we can find approximations of any real-number solutions using a graphing calculator.

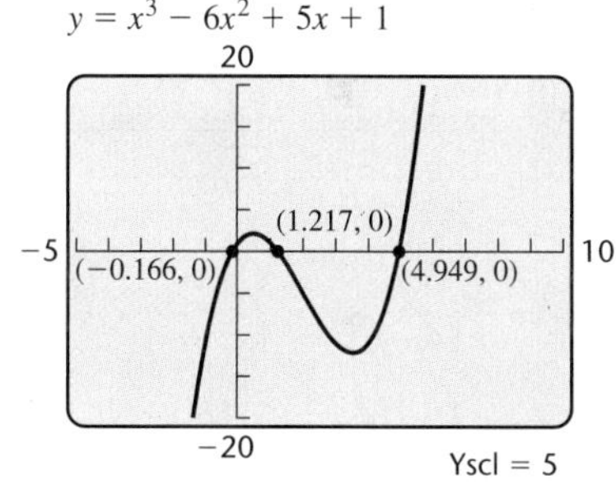

EXAMPLE 11 Solve: $x^3 - 6x^2 + 5x + 1 = 0$.

SOLUTION The polynomial $x^3 - 6x^2 + 5x + 1$ is prime; it cannot be factored using rational coefficients. We solve the equation by graphing $f(x) = x^3 - 6x^2 + 5x + 1$ and finding the zeros of the function. Since the polynomial is of degree 3, we need not look for more than 3 zeros.

There are three x-intercepts of the graph. We use the ZERO feature to find each root, or zero. Since the solutions are irrational, the calculator will give approximations. Rounding each to the nearest thousandth, we have the solutions -0.166, 1.217, and 4.949.

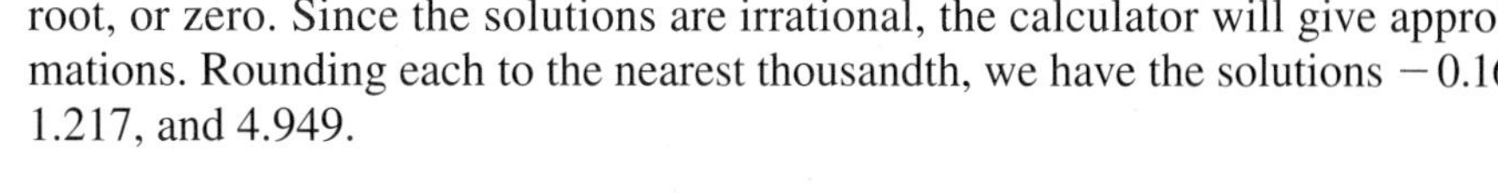

Try Exercise 95.

Connecting the Concepts

Algebraic and Graphical Methods

We have considered both algebraic and graphical methods of solving polynomial equations. It is important to understand and be able to use both methods. Some of the advantages and disadvantages of each method are given in the following table.

	Advantages	Disadvantages
Algebraic Method	• Can find exact answers • Works well when the polynomial is in factored form or can be readily factored • Can be used to find solutions that are not real numbers (see Chapter 11)	• Cannot be used if the polynomial is not factorable • Can be difficult to use if factorization is not readily apparent
Graphical Method	• Does not require the polynomial to be factored • Can visualize solutions	• Easy to miss solutions if an appropriate viewing window is not chosen • Gives approximations of solutions • For most graphing calculators, only real-number solutions can be found.

6.4 Exercise Set

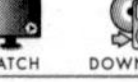

Concept Reinforcement *Identify each of the following as a perfect-square trinomial, a difference of two squares, a prime polynomial, or none of these.*

1. $9t^2 - 49$

2. $25x^2 - 20x + 4$

3. $36x^2 - 12x + 1$

4. $36a^2 - 25$

5. $4x^2 + 8x + 10$

6. $x^2 + 1$

7. $100y^2 + z^2$

8. $t^2 - 6t + 8$

Determine whether each of the following is a perfect-square trinomial.

9. $x^2 + 18x + 81$

10. $x^2 - 16x + 64$

11. $x^2 - 10x - 25$

12. $x^2 - 14x - 49$

13. $x^2 - 3x + 9$

14. $x^2 + 4x + 4$

15. $9x^2 + 25 - 30x$

16. $36x^2 + 16 - 24x$

Factor completely.

17. $t^2 + 6t + 9$

18. $a^2 + 16a + 64$

19. $a^2 - 14a + 49$

20. $x^2 - 8x + 16$

21. $4a^2 - 16a + 16$

22. $2a^2 + 8a + 8$

23. $1 - 2t + t^2$

24. $1 + t^2 + 2t$

25. $24a^2 + a^3 + 144a$

26. $-18y^2 + y^3 + 81y$

27. $20x^2 + 100x + 125$

28. $32x^2 + 48x + 18$

29. $1 + 8d^3 + 16d^6$

30. $64 + 25y^8 - 80y^4$

31. $-y^3 + 8y^2 - 16y$ **32.** $10a^2 - a^3 - 25a$

33. $0.25x^2 + 0.30x + 0.09$

34. $0.04x^2 - 0.28x + 0.49$

35. $x^2 - 2xy + y^2$

36. $m^{10} + 2m^5n^5 + n^{10}$

37. $25a^6 + 30a^3b^3 + 9b^6$

38. $49p^2 - 84pt + 36t^2$

39. $5a^2 - 10ab + 5b^2$

40. $4t^2 - 8tr + 4r^2$

Determine whether each of the following is a difference of squares.

41. $x^2 - 100$ **42.** $x^2 + 49$

43. $n^4 + 1$ **44.** $n^4 - 81$

45. $-1 + 64t^2$ **46.** $-12 + 25t^2$

Factor completely. Remember to look first for a common factor. If a polynomial is prime, state this.

47. $y^2 - 100$ **48.** $x^2 - 16$

49. $m^2 - 64$ **50.** $q^2 + 1$

51. $-49 + t^2$ **52.** $-64 + m^2$

53. $8x^2 - 8y^2$ **54.** $6x^2 - 6y^2$

55. $-80a^6 + 45$ **56.** $-81x^7 + 16x$

57. $49a^4 + 100$ **58.** $9x^4 - 25x^2$

59. $t^4 - 1$ **60.** $x^4 - 16$

61. $9a^4 - 25a^2b^4$ **62.** $16x^6 - 121x^2y^4$

63. $16x^4 - y^4$ **64.** $2a^4 - 32y^8$

65. $\frac{1}{49} - x^2$ **66.** $\frac{1}{16} - y^2$

67. $(a + b)^2 - 9$ **68.** $(p + q)^2 - 25$

69. $x^2 - 6x + 9 - y^2$ **70.** $a^2 - 8a + 16 - b^2$

71. $t^3 + 8t^2 - t - 8$ **72.** $x^3 - 7x^2 - 4x + 28$

73. $r^3 - 3r^2 - 9r + 27$ **74.** $t^3 + 2t^2 - 4t - 8$

75. $m^2 - 2mn + n^2 - 25$ **76.** $x^2 + 2xy + y^2 - 9$

77. $36 - (x + y)^2$ **78.** $49 - (a + b)^2$

Aha! **79.** $16 - a^2 - 2ab - b^2$ **80.** $9 - x^2 - 2xy - y^2$

81. $a^3 - ab^2 - 2a^2 + 2b^2$

82. $p^2q - 25q + 3p^2 - 75$

Solve. Round any irrational solutions to the nearest thousandth.

83. $a^2 + 1 = 2a$ **84.** $r^2 + 16 = 8r$

85. $2x^2 - 24x + 72 = 0$ **86.** $-t^2 - 16t - 64 = 0$

87. $x^2 - 9 = 0$ **88.** $r^2 - 64 = 0$

89. $a^2 = \frac{1}{25}$ **90.** $x^2 = \frac{1}{100}$

91. $8x^3 + 1 = 4x^2 + 2x$

92. $27x^3 + 18x^2 = 12x + 8$

93. $x^3 + 3 = 3x^2 + x$

94. $x^3 + x^2 = 16x + 16$

95. $x^2 - 3x - 7 = 0$ **96.** $x^2 - 5x + 1 = 0$

97. $2x^2 + 8x + 1 = 0$ **98.** $3x^2 + x - 1 = 0$

99. $x^3 + 3x^2 + x - 1 = 0$

100. $x^3 + x^2 + x - 1 = 0$

101. Let $f(x) = x^2 - 12x$. Find a such that $f(a) = -36$.

102. Let $g(x) = x^2$. Find a such that $g(a) = 144$.

Find the zeros of each function.

103. $f(x) = x^2 - 16$

104. $f(x) = x^2 - 8x + 16$

105. $f(x) = 2x^2 + 4x + 2$

106. $f(x) = 3x^2 - 27$

107. $f(x) = x^3 - 2x^2 - x + 2$

108. $f(x) = x^3 + x^2 - 4x - 4$

TW **109.** Explain in your own words how to determine whether a polynomial is a perfect-square trinomial.

TW **110.** Explain in your own words how to determine whether a polynomial is a difference of squares.

SKILL REVIEW

To prepare for Section 6.5, review the product and power rules for exponents and multiplication of polynomials (Sections 5.1 and 5.5).

Simplify. [5.1]

111. $(2x^2y^4)^3$ **112.** $(-5x^2y)^3$

Multiply. [5.5]

113. $(x + 1)(x + 1)(x + 1)$ **114.** $(x - 1)^3$

115. $(m + n)^3$ **116.** $(m - n)^3$

SYNTHESIS

TW **117.** Without finding the entire factorization, determine the number of factors of $x^{256} - 1$. Explain how you arrived at your answer.

TW **118.** Akio concludes that since $x^2 - 9 = (x - 3)(x + 3)$, it must follow that $x^2 + 9 = (x + 3)(x - 3)$. What mistake(s) is he making?

Factor completely. Assume that variables in exponents represent positive integers.

119. $x^8 - 2^8$

120. $x^2 - \left(\frac{1}{x}\right)^2$

121. $3x^2 - \frac{1}{3}$

122. $18x^3 - \frac{8}{25}x$

123. $0.09x^8 + 0.48x^4 + 0.64$

124. $a^2 + 2ab + b^2 - c^2 + 6c - 9$

125. $r^2 - 8r - 25 - s^2 - 10s + 16$

126. $x^{2a} - y^2$

127. $x^{4a} - 49y^{2a}$

Aha! **128.** $(y + 3)^2 + 2(y + 3) + 1$

129. $3(x + 1)^2 + 12(x + 1) + 12$

130. $5c^{100} - 80d^{100}$

131. $9x^{2n} - 6x^n + 1$

132. $m^2 + 4mn + 4n^2 + 5m + 10n$

133. $s^2 - 4st + 4t^2 + 4s - 8t + 4$

134. If $P(x) = x^2$, use factoring to simplify
$$P(a + h) - P(a).$$

135. If $P(x) = x^4$, use factoring to simplify
$$P(a + h) - P(a).$$

136. *Volume of Carpeting.* The volume of a carpet that is rolled up can be estimated by the polynomial $\pi R^2h - \pi r^2h$.

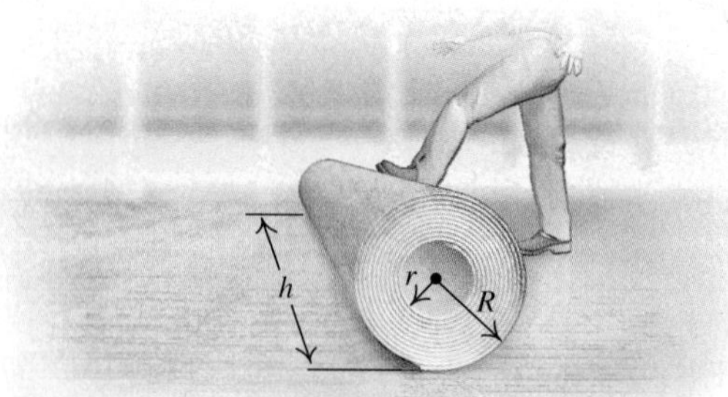

a) Factor the polynomial.

b) Use both the original and the factored forms to find the volume of a roll for which $R = 50$ cm, $r = 10$ cm, and $h = 4$ m. Use 3.14 for π.

Try Exercise Answers: Section 6.4

9. Yes **17.** $(t + 3)^2$ **25.** $a(a + 12)^2$ **41.** Yes **47.** $(y + 10)(y - 10)$ **55.** $-5(4a^3 + 3)(4a^3 - 3)$ **69.** $(x - 3 + y)(x - 3 - y)$ **71.** $(t + 8)(t + 1)(t - 1)$ **83.** 1 **95.** $-1.541, 4.541$ **103.** $-4, 4$

6.5 Sums or Differences of Cubes

- Factoring Sums or Differences of Cubes
- Solving Equations

STUDY TIP

A Good Night's Sleep

Study when you are alert. A good night's sleep or a 10-minute "power nap" can often make a problem suddenly seem much easier to solve.

FACTORING SUMS OR DIFFERENCES OF CUBES

We have seen that a difference of two squares can be factored but (unless a common factor exists) a *sum* of two squares is usually prime. The situation is different with cubes: The difference *or sum* of two cubes can always be factored. To see this, consider the following products:

$$\begin{aligned}(A + B)(A^2 - AB + B^2) &= A(A^2 - AB + B^2) + B(A^2 - AB + B^2)\\ &= A^3 - A^2B + AB^2 + A^2B - AB^2 + B^3\\ &= A^3 + B^3 \quad \text{Combining like terms}\end{aligned}$$

and

$$\begin{aligned}(A - B)(A^2 + AB + B^2) &= A(A^2 + AB + B^2) - B(A^2 + AB + B^2)\\ &= A^3 + A^2B + AB^2 - A^2B - AB^2 - B^3\\ &= A^3 - B^3. \quad \text{Combining like terms}\end{aligned}$$

These products allow us to factor a sum or a difference of two cubes. Note how the location of the + and − signs changes.

Factoring a Sum or a Difference of Two Cubes

$$A^3 + B^3 = (A + B)(A^2 - AB + B^2);$$
$$A^3 - B^3 = (A - B)(A^2 + AB + B^2)$$

Remembering this list of cubes may prove helpful when factoring.

N	0.2	0.1	0	1	2	3	4	5	6
N^3	0.008	0.001	0	1	8	27	64	125	216

We say that 2 is the *cube root* of 8, that 3 is the cube root of 27, and so on.

EXAMPLE 1 Write an equivalent expression by factoring: $x^3 + 27$.

SOLUTION We first observe that

$$x^3 + 27 = x^3 + 3^3.$$

Next, in one set of parentheses, we write the first cube root, x, plus the second cube root, 3:

$$(x + 3)(\qquad).$$

To get the other factor, we think of $x + 3$ and do the following:

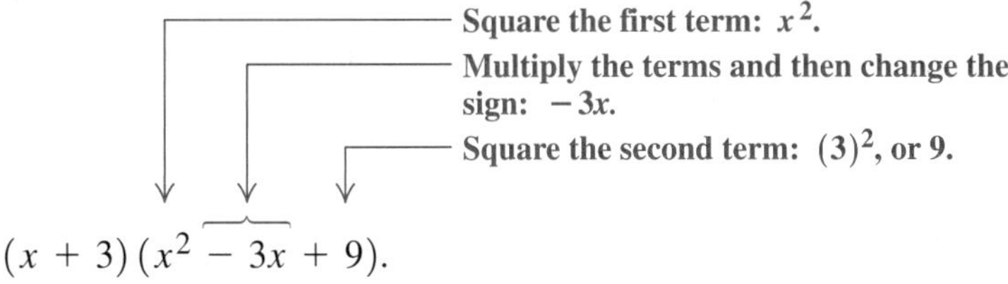

$$(x + 3)(x^2 - 3x + 9).$$

Check: $(x + 3)(x^2 - 3x + 9) = x^3 - 3x^2 + 9x + 3x^2 - 9x + 27$
$= x^3 + 27.$ **Combining like terms**

Thus, $x^3 + 27 = (x + 3)(x^2 - 3x + 9)$.

Try Exercise 11.

In Example 1, note that $x^2 - 3x + 9$ cannot be factored further.

EXAMPLE 2 Factor.

a) $125x^3 - y^3$
b) $m^6 + 64$
c) $128y^7 - 250x^6y$
d) $r^6 - s^6$

SOLUTION

a) We have

$$125x^3 - y^3 = (5x)^3 - y^3.$$ **This is a difference of cubes.**

In one set of parentheses, we write the cube root of the first term, $5x$, minus the second cube root, y:

$$(5x - y)(\quad).$$

To get the other factor, we think of $5x - y$ and do the following:

Square the first term: $(5x)^2$, or $25x^2$.
Multiply the terms and then change the sign: $5xy$.
Square the second term: $(-y)^2$, or y^2.

$$(5x - y)(25x^2 + 5xy + y^2).$$

Check: $(5x - y)(25x^2 + 5xy + y^2) = 125x^3 + 25x^2y + 5xy^2 - 25x^2y - 5xy^2 - y^3$
$= 125x^3 - y^3.$ **Combining like terms**

Thus, $125x^3 - y^3 = (5x - y)(25x^2 + 5xy + y^2)$.

Student Notes

If you think of $A^3 - B^3$ as $A^3 + (-B)^3$, you then need remember only the pattern for factoring a sum of two cubes. Be sure to simplify your result if you do this.

b) We have

$$m^6 + 64 = (m^2)^3 + 4^3. \quad \textbf{Rewriting as a sum of quantities cubed}$$

Next, we use the pattern for a sum of cubes:

$$A^3 + B^3 = (A + B)(A^2 - A \cdot B + B^2)$$
$$\downarrow \quad \downarrow$$
$$(m^2)^3 + 4^3 = (m^2 + 4)((m^2)^2 - m^2 \cdot 4 + 4^2)$$
$$= (m^2 + 4)(m^4 - 4m^2 + 16).$$

The check is left to the student.

c) We have

$$128y^7 - 250x^6y = 2y(64y^6 - 125x^6) \quad \textbf{Remember: \textit{Always} look for a common factor.}$$
$$= 2y[(4y^2)^3 - (5x^2)^3]. \quad \textbf{Rewriting as a difference of quantities cubed}$$

To factor $(4y^2)^3 - (5x^2)^3$, we use the pattern for a difference of cubes:

$$A^3 - B^3 = (A - B)(A^2 + A \cdot B + B^2)$$
$$\downarrow \quad \downarrow \quad \downarrow \quad \downarrow \quad \downarrow \quad \downarrow \quad \downarrow \quad \downarrow$$
$$(4y^2)^3 - (5x^2)^3 = (4y^2 - 5x^2)((4y^2)^2 + 4y^2 \cdot 5x^2 + (5x^2)^2)$$
$$= (4y^2 - 5x^2)(16y^4 + 20x^2y^2 + 25x^4).$$

CAUTION!
Use parentheses when evaluating A^2 and B^2:

$(4y^2)^2 = 16y^4$ and
$(5x^2)^2 = 25x^4$.

The check is left to the student. We have

$$128y^7 - 250x^6y = 2y(4y^2 - 5x^2)(16y^4 + 20x^2y^2 + 25x^4).$$

d) We have

$$r^6 - s^6 = (r^3)^2 - (s^3)^2$$
$$= (r^3 + s^3)(r^3 - s^3) \quad \textbf{Factoring a difference of two \textit{squares}}$$
$$= (r + s)(r^2 - rs + s^2)(r - s)(r^2 + rs + s^2).$$
Factoring the sum and the difference of two cubes

To check, read the steps in reverse order and inspect the multiplication.

Try Exercise 33.

In Example 2(d), suppose we first factored $r^6 - s^6$ as a difference of two cubes:

$$(r^2)^3 - (s^2)^3 = (r^2 - s^2)(r^4 + r^2s^2 + s^4)$$
$$= (r + s)(r - s)(r^4 + r^2s^2 + s^4).$$

In this case, we might have missed some factors; $r^4 + r^2s^2 + s^4$ can be factored as $(r^2 - rs + s^2)(r^2 + rs + s^2)$, but we probably would never have suspected that such a factorization exists. **Given a choice, it is generally better to factor as a difference of squares before factoring as a sum or a difference of cubes.**

Useful Factoring Facts

Sum of cubes: $A^3 + B^3 = (A + B)(A^2 - AB + B^2)$

Difference of cubes: $A^3 - B^3 = (A - B)(A^2 + AB + B^2)$

Difference of squares: $A^2 - B^2 = (A + B)(A - B)$

There is no formula for factoring a sum of squares.

SOLVING EQUATIONS

We can now solve equations involving sums or differences of cubes.

EXAMPLE 3 Solve: $x^3 = 27$.

SOLUTION We rewrite the equation to get 0 on one side and factor:

$$x^3 = 27$$
$$x^3 - 27 = 0 \quad \text{Subtracting 27 from both sides}$$
$$(x - 3)(x^2 + 3x + 9) = 0. \quad \text{Factoring a difference of cubes}$$

The second-degree factor, $x^2 + 3x + 9$, does not factor using real coefficients. When we apply the principle of zero products, we have

$$x - 3 = 0 \quad or \quad x^2 + 3x + 9 = 0. \quad \text{Using the principle of zero products}$$
$$x = 3 \quad \text{We cannot factor } x^2 + 3x + 9.$$

We have one solution, 3. In Chapter 11, we will learn how to solve equations involving prime quadratic polynomials like $x^2 + 3x + 9 = 0$. For now, we can find the real-number solutions of the equation $x^3 = 27$ graphically, by looking for any points of intersection of the graphs of $y_1 = x^3$ and $y_2 = 27$. There is only one point of intersection.

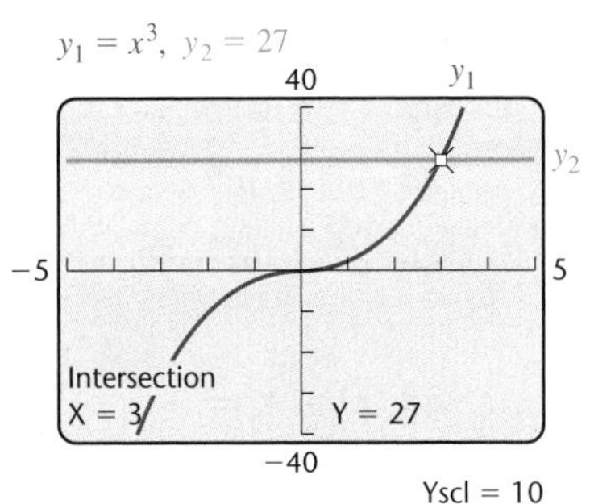

The real-number solution of $x^3 = 27$ is 3.

Try Exercise 47.

6.5 Exercise Set

FOR EXTRA HELP MyMathLab Math XL PRACTICE  WATCH  DOWNLOAD READ  REVIEW

Concept Reinforcement *Classify each binomial as either a sum of cubes, a difference of cubes, a difference of squares, or none of these.*

1. $x^3 - 1$
2. $8 + t^3$
3. $9x^4 - 25$
4. $9x^2 + 25$
5. $1000t^3 + 1$
6. $x^3y^3 - 27z^3$
7. $25x^2 + 8x$
8. $100y^8 - 25x^4$
9. $s^{12} - t^{15}$
10. $14x^3 - 2x$

Factor completely.

11. $x^3 + 64$
12. $t^3 + 27$
13. $z^3 - 1$
14. $x^3 - 8$
15. $t^3 - 1000$
16. $m^3 - 125$
17. $27x^3 + 1$
18. $8a^3 + 1$
19. $64 - 125x^3$
20. $27 - 8t^3$
21. $8y^3 + 64$
22. $8a^3 + 1000$
23. $x^3 - y^3$
24. $y^3 - z^3$
25. $a^3 + \frac{1}{8}$
26. $x^3 + \frac{1}{27}$
27. $8t^3 - 8$
28. $2y^3 - 128$
29. $y^3 - \frac{1}{1000}$
30. $x^3 - \frac{1}{125}$
31. $ab^3 + 125a$
32. $rs^3 + 64r$
33. $5x^3 - 40z^3$
34. $2y^3 - 54z^3$
35. $x^3 + 0.001$
36. $y^3 + 0.125$
37. $64x^6 - 8t^6$
38. $125c^6 - 8d^6$
39. $2y^4 - 128y$
40. $3z^5 - 3z^2$
41. $z^6 - 1$
42. $t^6 + 1$
43. $t^6 + 64y^6$
44. $p^6 - w^6$
45. $x^{12} - y^3z^{12}$
46. $a^9 + b^{12}c^{15}$

Solve.

47. $x^3 + 1 = 0$
48. $t^3 - 8 = 0$
49. $8x^3 = 27$
50. $64x^3 + 27 = 0$
51. $2t^3 - 2000 = 0$
52. $375 = 24x^3$

TW 53. How could you use factoring to convince someone that $x^3 + y^3 \neq (x + y)^3$?

TW 54. Is the following statement true or false and why? If A^3 and B^3 have a common factor, then A and B have a common factor.

SKILL REVIEW

Review graphing linear equations (Chapter 3).

55. Find the slope of the line containing the points $(-2, -5)$ and $(3, -6)$. [3.5]

56. Find the slope of the line given by $y - 3 = \frac{1}{4}x$. [3.6]

Graph.

57. $2x - 5y = 10$ [3.3]

58. $-5x = 10$ [3.3]

59. $y = \frac{2}{3}x - 1$ [3.6]

60. $y - 2 = -2(x + 4)$ [3.7]

SYNTHESIS

TW 61. Explain how the geometric model below can be used to verify the formula for factoring $a^3 - b^3$.

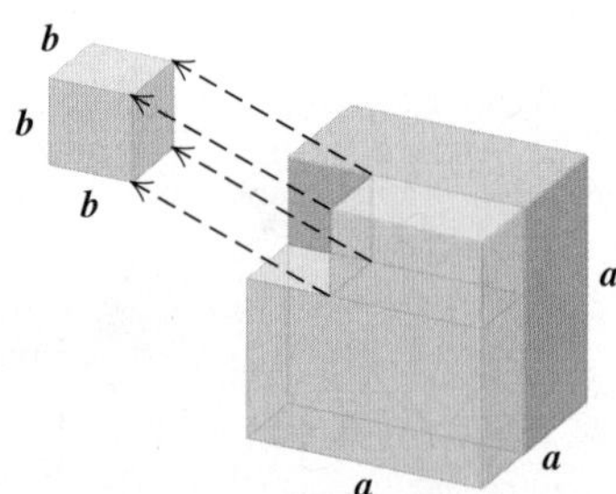

TW 62. Explain how someone could construct a binomial that is both a difference of two cubes and a difference of two squares.

Factor.

63. $x^{6a} - y^{3b}$
64. $2x^{3a} + 16y^{3b}$
Aha! 65. $(x + 5)^3 + (x - 5)^3$
66. $\frac{1}{16}x^{3a} + \frac{1}{2}y^{6a}z^{9b}$
67. $5x^3y^6 - \frac{5}{8}$
68. $x^3 - (x + y)^3$
69. $x^{6a} - (x^{2a} + 1)^3$
70. $(x^{2a} - 1)^3 - x^{6a}$
71. $t^4 - 8t^3 - t + 8$

72. If $P(x) = x^3$, use factoring to simplify $P(a + h) - P(a)$.

73. If $Q(x) = x^6$, use factoring to simplify

$$Q(a + h) - Q(a).$$

74. Using one viewing window, graph the following.

a) $f(x) = x^3$
b) $g(x) = x^3 - 8$
c) $h(x) = (x - 2)^3$

Try Exercise Answers: Section 6.5

11. $(x + 4)(x^2 - 4x + 16)$
33. $5(x - 2z)(x^2 + 2xz + 4z^2)$ **47.** -1

6.6 Factoring: A General Strategy

- Choosing the Right Method

Thus far, each section in this chapter has examined one or two different methods for factoring polynomials. In practice, when factoring a polynomial, we must decide on our own which method to use.

We can use the following guidelines to factor polynomials.

To Factor a Polynomial

A. Always look for a common factor first. If there is one, factor out the largest common factor. Be sure to include it in your final answer.

B. Then look at the number of terms.

Two terms: Try factoring as a difference of squares first:

$$A^2 - B^2 = (A - B)(A + B).$$

Next, try factoring as a sum or a difference of cubes:

$$A^3 + B^3 = (A + B)(A^2 - AB + B^2)$$

and

$$A^3 - B^3 = (A - B)(A^2 + AB + B^2).$$

Three terms: If the trinomial is a perfect-square trinomial, factor accordingly:

$$A^2 + 2AB + B^2 = (A + B)^2$$

or

$$A^2 - 2AB + B^2 = (A - B)^2.$$

If it is not a perfect-square trinomial, try using FOIL or grouping.

Four terms: Try factoring by grouping.

C. Always *factor completely*. When a factor can itself be factored, be sure to factor it. Remember that some polynomials, like $x^2 + 9$, are prime.

D. Check by multiplying.

CHOOSING THE RIGHT METHOD

STUDY TIP

Keep Your Focus

When studying with someone else, it can be very tempting to talk about topics other than what you need to study. If you see that this may happen, explain to your partner(s) that you enjoy the conversation, but would enjoy it more later—after the work has been completed.

EXAMPLE 1 Factor: $5t^4 - 80$.

SOLUTION

A. We look for a common factor:

$$5t^4 - 80 = 5(t^4 - 16).$$ **5 is the largest common factor.**

B. The factor $t^4 - 16$ is a difference of squares: $(t^2)^2 - 4^2$. We factor it, being careful to rewrite the 5 from step (A):

$$5t^4 - 80 = 5(t^2 + 4)(t^2 - 4).$$ $t^4 - 16 = (t^2 + 4)(t^2 - 4)$

C. Since $t^2 - 4$ is a difference of squares, we continue factoring:

$$5t^4 - 80 = 5(t^2 + 4)(t^2 - 4) = 5(t^2 + 4)(t - 2)(t + 2).$$

↑ **This is a sum of squares with no common factor. It cannot be factored!**

D. *Check:* $5(t^2 + 4)(t - 2)(t + 2) = 5(t^2 + 4)(t^2 - 4)$
$= 5(t^4 - 16) = 5t^4 - 80.$

The factorization is $5(t^2 + 4)(t - 2)(t + 2)$.

Try Exercise 5.

EXAMPLE 2 Factor: $2x^3 + 10x^2 + x + 5$.

SOLUTION

A. We look for a common factor. There is none (other than 1 or -1).

B. Because there are four terms, we try factoring by grouping:

$2x^3 + 10x^2 + x + 5$

$= (2x^3 + 10x^2) + (x + 5)$ **Separating into two binomials**

$= 2x^2(x + 5) + 1(x + 5)$ **Factoring out the largest common factor from each binomial. The 1 serves as an aid.**

$= (x + 5)(2x^2 + 1).$ **Factoring out the common factor, $x + 5$**

C. Nothing can be factored further, so we have factored completely.

D. *Check:* $(x + 5)(2x^2 + 1) = 2x^3 + x + 10x^2 + 5$
$= 2x^3 + 10x^2 + x + 5.$

The factorization is $(x + 5)(2x^2 + 1)$.

Try Exercise 41.

EXAMPLE 3 Factor: $-n^5 + 2n^4 + 35n^3$.

SOLUTION

A. We note that there is a common factor, $-n^3$:

$$-n^5 + 2n^4 + 35n^3 = -n^3(n^2 - 2n - 35).$$

B. The factor $n^2 - 2n - 35$ is not a perfect-square trinomial. We factor it using FOIL in reverse:

$$-n^5 + 2n^4 + 35n^3 = -n^3(n^2 - 2n - 35)$$
$$= -n^3(n - 7)(n + 5).$$

C. Nothing can be factored further, so we have factored completely.

D. *Check:* $-n^3(n - 7)(n + 5) = -n^3(n^2 - 2n - 35)$
$= -n^5 + 2n^4 + 35n^3.$

The factorization is $-n^3(n - 7)(n + 5)$.

Try Exercise 21.

EXAMPLE 4 Factor: $x^2 - 20x + 100$.

SOLUTION

A. We look first for a common factor. There is none.

B. This polynomial is a perfect-square trinomial. We factor it accordingly:

$$x^2 - 20x + 100 = x^2 - 2 \cdot x \cdot 10 + 10^2 \qquad \text{Try to do this step mentally.}$$
$$= (x - 10)^2.$$

C. Nothing can be factored further, so we have factored completely.

D. *Check:* $(x - 10)(x - 10) = x^2 - 20x + 100$.

The factorization is $(x - 10)(x - 10)$, or $(x - 10)^2$.

Try Exercise 7.

Student Notes

Quickly checking the leading and constant terms of a trinomial to see if they are squares can save you time. If they aren't both squares, the trinomial can't possibly be a perfect-square trinomial.

EXAMPLE 5 Factor: $6x^2y^4 - 21x^3y^5 + 3x^2y^6$.

SOLUTION

A. We first factor out the largest common factor, $3x^2y^4$:

$$6x^2y^4 - 21x^3y^5 + 3x^2y^6 = 3x^2y^4(2 - 7xy + y^2).$$

B. There are three terms in $2 - 7xy + y^2$. Since only y^2 is a square, we do not have a perfect-square trinomial. Note that x appears only in $-7xy$. The product of a form like $(1 - y)(2 - y)$ has no x in the middle term. Thus, $2 - 7xy + y^2$ cannot be factored.

C. Nothing can be factored further, so we have factored completely.

D. *Check:* $3x^2y^4(2 - 7xy + y^2) = 6x^2y^4 - 21x^3y^5 + 3x^2y^6$.

The factorization is $3x^2y^4(2 - 7xy + y^2)$.

Try Exercise 55.

EXAMPLE 6 Factor: $x^6 - 64$.

SOLUTION

A. We look first for a common factor. There is none (other than 1 or -1).

B. There are two terms, a difference of squares: $(x^3)^2 - (8)^2$. We factor it:

$$x^6 - 64 = (x^3 + 8)(x^3 - 8). \qquad \text{Note that } x^6 = (x^3)^2.$$

C. One factor is a sum of two cubes, and the other factor is a difference of two cubes. We factor both:

$$x^6 - 64 = (x + 2)(x^2 - 2x + 4)(x - 2)(x^2 + 2x + 4).$$

The factorization is complete because no factor can be factored further.

D. *Check:*

$$(x + 2)(x^2 - 2x + 4)(x - 2)(x^2 + 2x + 4) = (x^3 + 8)(x^3 - 8)$$
$$= x^6 - 64.$$

The factorization is $(x + 2)(x^2 - 2x + 4)(x - 2)(x^2 + 2x + 4)$.

Try Exercise 53.

EXAMPLE 7 Factor: $-25m^2 - 20mn - 4n^2$.

Student Notes

Most factoring procedures assume that the leading coefficient of a polynomial is positive. You can factor out a -1, if necessary, to make this true.

SOLUTION

A. We look first for a common factor. Since all the terms are negative, we factor out a -1:

$$-25m^2 - 20mn - 4n^2 = -1(25m^2 + 20mn + 4n^2).$$

B. There are three terms in the parentheses. Note that the first term and the last term are squares: $25m^2 = (5m)^2$ and $4n^2 = (2n)^2$. We see that twice the product of $5m$ and $2n$ is the middle term,

$$2 \cdot 5m \cdot 2n = 20mn,$$

so the trinomial is a perfect square. To factor, we write a binomial squared:

$$-25m^2 - 20mn - 4n^2 = -1(25m^2 + 20mn + 4n^2)$$
$$= -1(5m + 2n)^2.$$

C. Nothing can be factored further, so we have factored completely.

D. *Check:* $-1(5m + 2n)^2 = -1(25m^2 + 20mn + 4n^2)$
$= -25m^2 - 20mn - 4n^2.$

The factorization is $-1(5m + 2n)^2$, or $-(5m + 2n)^2$.

Try Exercise 59.

EXAMPLE 8 Factor: $x^2y^2 + 7xy + 12$.

SOLUTION

A. We look first for a common factor. There is none.

B. Since only one term is a square, we do not have a perfect-square trinomial. We use trial and error, thinking of the product xy as a single variable:

$$(xy + \quad)(xy + \quad).$$

We factor the last term, 12. All the signs are positive, so we consider only positive factors. Possibilities are 1, 12 and 2, 6 and 3, 4. The pair 3, 4 gives a sum of 7 for the coefficient of the middle term. Thus,

$$x^2y^2 + 7xy + 12 = (xy + 3)(xy + 4).$$

C. Nothing can be factored further, so we have factored completely.

D. *Check:* $(xy + 3)(xy + 4) = x^2y^2 + 7xy + 12.$

The factorization is $(xy + 3)(xy + 4)$.

Try Exercise 73.

Compare the variables appearing in Example 7 with those appearing in Example 8. Note that when one variable appears in the leading term and another variable appears in the last term, as in Example 7, each binomial contains two variable terms. When two variables appear in the leading term, as in Example 8, each binomial contains just one variable term.

EXAMPLE 9 Factor: $2a^3 + 12a^2 + 18a - 8ab^2$.

SOLUTION

A. We note that there is a common factor, $2a$:

$$2a^3 + 12a^2 + 18a - 8ab^2 = 2a(a^2 + 6a + 9 - 4b^2).$$

B. There are four terms. Grouping into two groups of two terms does not yield a common binomial factor, but we can group into a difference of squares:

$$\begin{aligned} 2a^3 + 12a^2 + 18a - 8ab^2 &= 2a[(a^2 + 6a + 9) - 4b^2] \\ &= 2a[(a + 3)^2 - (2b)^2] \\ &= 2a[(a + 3 + 2b)(a + 3 - 2b)]. \end{aligned}$$

C. Nothing can be factored further, so we have factored completely.

D. *Check:* $\begin{aligned} 2a(a + 3 + 2b)(a + 3 - 2b) &= 2a[(a + 3)^2 - (2b)^2] \\ &= 2a[a^2 + 6a + 9 - 4b^2] \\ &= 2a^3 + 12a^2 + 18a - 8ab^2. \end{aligned}$

The factorization is $2a(a + 3 + 2b)(a + 3 - 2b)$.

Try Exercise 43.

6.6 Exercise Set

FOR EXTRA HELP MyMathLab

MathXL PRACTICE WATCH DOWNLOAD READ REVIEW

Concept Reinforcement *In each of Exercises 1–4, complete the sentence.*

1. As a first step when factoring polynomials, always check for a ________.
2. When factoring a trinomial, if two terms are not squares, it cannot be a ________.
3. If a polynomial has four terms and no common factor, it may be possible to factor by ________.
4. It is always possible to check a factorization by ________.

Factor completely. If a polynomial is prime, state this.

5. $10a^2 - 640$

6. $5x^2 - 45$

7. $y^2 + 49 - 14y$

8. $a^2 + 25 + 10a$

9. $2t^2 + 11t + 12$

10. $8t^2 - 18t - 5$

11. $x^3 - 18x^2 + 81x$

12. $x^3 - 24x^2 + 144x$

13. $x^3 - 5x^2 - 25x + 125$

14. $x^3 + 3x^2 - 4x - 12$

15. $27t^3 - 3t$

16. $98t^2 - 18$

17. $9x^3 + 12x^2 - 45x$

18. $20x^3 - 4x^2 - 72x$

19. $t^2 + 25$

20. $x^2 + 4$

21. $6x^2 + 3x - 45$

22. $2t^2 + 6t - 20$

23. $-2a^6 + 8a^5 - 8a^4$

24. $-x^5 - 14x^4 - 49x^3$

25. $5x^5 - 80x$

26. $4x^4 - 64$

27. $t^4 - 9$

28. $9 + t^8$

29. $-x^6 + 2x^5 - 7x^4$

30. $-x^5 + 4x^4 - 3x^3$

31. $x^3 - y^3$

32. $8t^3 + 1$

33. $ax^2 + ay^2$

34. $12n^2 + 24n^3$

35. $80cd^2 - 36c^2d + 4c^3$

36. $a^5 - 4a^4b - 5a^3b^2$

37. $2\pi rh + 2\pi r^2$

38. $4\pi r^2 + 2\pi r$

Aha! 39. $(a + b)5a + (a + b)3b$

40. $5c(a^3 + b) - (a^3 + b)$

41. $x^2 + x + xy + y$

42. $n^2 + 2n + np + 2p$

43. $n^2 - 10n + 25 - 9m^2$

44. $t^2 + 2t + 1 - 81r^2$

45. $3x^2 + 13xy - 10y^2$

46. $-x^2 - y^2 - 2xy$

47. $4b^2 + a^2 - 4ab$

48. $a^2 - 7a - 6$

49. $16x^2 + 24xy + 9y^2$

50. $9w^2 - 9n^2$

51. $t^2 - 8t + 10$

52. $25z^2 + 10zy + y^2$

53. $64t^6 - 1$

54. $m^6 - 1$

55. $8m^3n - 32m^2n^2 + 24mn$

56. $6a^2b^3 + 12a^3b^2 - 3a^4b^2$

57. $3b^2 + 17ab - 6a^2$

58. $2mn - 360n^2 + m^2$

59. $-12 - x^2y^2 - 8xy$

60. $m^2n^2 - 4mn - 32$

61. $t^8 - s^{10} - 12s^5 - 36$

62. $p^4 - 1 + 2x - x^2$

63. $54a^4 + 16ab^3$

64. $54x^3y - 250y^4$

65. $x^6 + x^5y - 2x^4y^2$

66. $2s^6t^2 + 10s^3t^3 + 12t^4$

67. $36a^2 - 15a + \frac{25}{16}$

68. $a^2 + 2a^2bc + a^2b^2c^2$

69. $\frac{1}{81}x^2 - \frac{8}{27}x + \frac{16}{9}$

70. $\frac{1}{4}a^2 + \frac{1}{3}ab + \frac{1}{9}b^2$

71. $1 - 16x^{12}y^{12}$

72. $b^4a - 81a^5$

73. $4a^2b^2 + 12ab + 9$

74. $9c^2 + 6cd + d^2$

75. $a^4 + 8a^2 + 8a^3 + 64a$

76. $t^4 + 7t^2 - 3t^3 - 21t$

TW 77. Kelly factored $16 - 8x + x^2$ as $(x - 4)^2$, while Tony factored it as $(4 - x)^2$. Are they both correct? Why or why not?

TW 78. Describe in your own words or draw a diagram representing a strategy for factoring polynomials.

SKILL REVIEW

To prepare for Section 6.7, review solving problems using the five-step strategy (Section 2.5).

Translate to an algebraic expression. [1.1]

79. The square of the sum of two numbers

80. The sum of the squares of two numbers

81. The product of two consecutive integers

Solve. [2.5]

82. In 2009, shoppers spent \$23.5 billion on gifts for Mother's Day and for Father's Day combined. They spent \$4.7 billion more for Mother's Day than for Father's Day. How much did shoppers spend for each holiday?
Source: National Retail Federation

83. The first angle of a triangle is four times as large as the second. The measure of the third angle is 30° less than that of the second. How large are the angles?

84. A rectangular table top is twice as long as it is wide. The perimeter of the table is 192 in. What are the dimensions of the table?

SYNTHESIS

TW 85. There are third-degree polynomials in x that we are not yet able to factor, despite the fact that they are not prime. Explain how such a polynomial could be created.

TW 86. Describe a method that could be used to find a binomial of degree 16 that can be expressed as the product of prime binomial factors.

Factor.

87. $-(x^5 + 7x^3 - 18x)$

88. $18 + a^3 - 9a - 2a^2$

89. $-3a^4 + 15a^2 - 12$

90. $-x^4 + 7x^2 + 18$

Aha! 91. $y^2(y + 1) - 4y(y + 1) - 21(y + 1)$

92. $y^2(y - 1) - 2y(y - 1) + (y - 1)$

93. $6(x - 1)^2 + 7y(x - 1) - 3y^2$

94. $(y + 4)^2 + 2x(y + 4) + x^2$

95. $2(a + 3)^4 - (a + 3)^3(b - 2) - (a + 3)^2(b - 2)^2$

96. $5(t - 1)^5 - 6(t - 1)^4(s - 1) + (t - 1)^3(s - 1)^2$

97. $49x^4 + 14x^2 + 1 - 25x^6$

Try Exercise Answers: Section 6.6

5. $10(a + 8)(a - 8)$ **7.** $(y - 7)^2$ **21.** $3(x + 3)(2x - 5)$ **41.** $(x + 1)(x + y)$ **43.** $(n - 5 + 3m)(n - 5 - 3m)$ **53.** $(2t + 1)(4t^2 - 2t + 1)(2t - 1)(4t^2 + 2t + 1)$ **55.** $8mn(m^2 - 4mn + 3)$ **59.** $-1(xy + 2)(xy + 6)$, or $-(xy + 2)(xy + 6)$ **73.** $(2ab + 3)^2$

Collaborative Corner

Matching Factorizations*

Focus: Factoring
Time: 20 minutes
Group Size: Begin with the entire class. If there is an odd number of students, the instructor should participate.
Materials: Prepared sheets of paper, pins or tape. On half of the sheets, the instructor writes a polynomial. On the remaining sheets, the instructor writes the factorization of those polynomials. The polynomials and factorizations should be similar; for example,

$x^2 - 2x - 8,\quad (x - 2)(x - 4),$
$x^2 - 6x + 8,\quad (x - 1)(x - 8),$
$x^2 - 9x + 8,\quad (x + 2)(x - 4).$

ACTIVITY

1. As class members enter the room, the instructor pins or tapes either a polynomial or a factorization to the back of each student. Class members are told only whether their sheet of paper contains a polynomial or a factorization.
2. After all students are wearing a sheet of paper, they should mingle with one another, attempting to match up their factorization with the appropriate polynomial or vice versa. They may ask questions of one another that relate to factoring and polynomials. Answers to the questions should be yes or no. For example, a legitimate question might be "Is my last term negative?" or "Do my factors have opposite signs?"
3. The game is over when all factorization/polynomial pairs have "found" one another.

*Thanks to Jann MacInnes of Florida Community College at Jacksonville–Kent Campus for suggesting this activity.

6.7 Applications of Polynomial Equations

- Problem Solving
- The Pythagorean Theorem
- Fitting Polynomial Functions to Data

Polynomial equations and functions occur frequently in problem solving and in data analysis.

PROBLEM SOLVING

Some problems can be translated to polynomial equations that we can now solve. The problem-solving process is the same as that used for other kinds of problems.

EXAMPLE 1 Race Numbers. Terry and Jody each entered a boat in the Lakeport Race. The racing numbers of their boats were consecutive numbers, the product of which was 156. Find the numbers.

SOLUTION

1. **Familiarize.** Consecutive numbers are one apart, like 49 and 50. Let $x =$ the first boat number; then $x + 1 =$ the next boat number.
2. **Translate.** We reword the problem before translating:

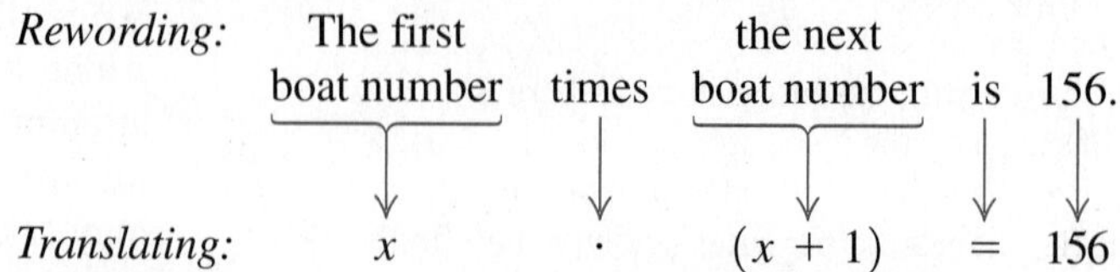

Rewording: The first boat number times the next boat number is 156.

Translating: $x \cdot (x + 1) = 156$

3. **Carry out.** We solve the equation as follows:

$$x(x + 1) = 156$$
$$x^2 + x = 156 \quad \textbf{Multiplying}$$
$$x^2 + x - 156 = 0 \quad \textbf{Subtracting 156 to get 0 on one side}$$
$$(x - 12)(x + 13) = 0 \quad \textbf{Factoring}$$
$$x - 12 = 0 \quad or \quad x + 13 = 0 \quad \textbf{Using the principle of zero products}$$
$$x = 12 \quad or \quad x = -13. \quad \textbf{Solving each equation}$$

4. **Check.** The solutions of the equation are 12 and -13. Since race numbers are not negative, -13 must be rejected. On the other hand, if x is 12, then $x + 1$ is 13 and $12 \cdot 13 = 156$. Thus the solution 12 checks.
5. **State.** The boat numbers for Terry and Jody were 12 and 13.

Try Exercise 3.

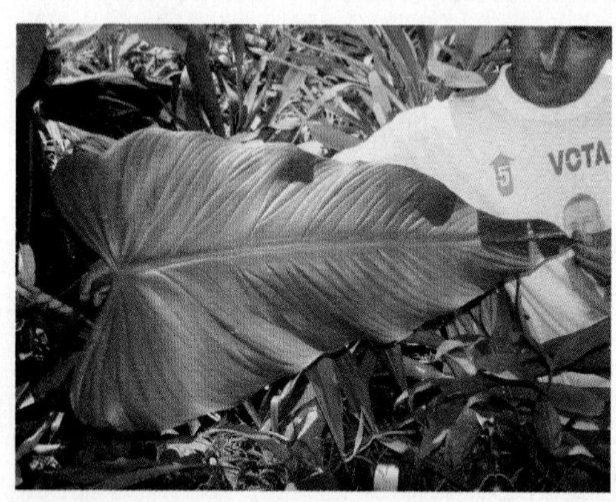

EXAMPLE 2 Dimensions of a Leaf. Each leaf of one particular *Philodendron* species is approximately a triangle. A typical leaf has an area of 320 in^2. If the leaf is 12 in. longer than it is wide, find the length and the width of the leaf.

SOLUTION

1. **Familiarize.** The formula for the area of a triangle is

$$\text{Area} = \tfrac{1}{2} \cdot (\text{base}) \cdot (\text{height}).$$

We let $b =$ the width, in inches, of the triangle's base and $b + 12 =$ the height, in inches.

2. **Translate.** We reword and translate as follows:

Rewording: The area of the leaf is 320 in^2.

Translating: $\frac{1}{2} \cdot b(b + 12) = 320$

3. **Carry out.** We solve the equation as follows:

$$\tfrac{1}{2} \cdot b \cdot (b + 12) = 320$$
$$\tfrac{1}{2}(b^2 + 12b) = 320 \quad \textbf{Multiplying}$$
$$b^2 + 12b = 640 \quad \textbf{Multiplying by 2 to clear fractions}$$
$$b^2 + 12b - 640 = 0 \quad \textbf{Subtracting 640 to get 0 on one side}$$
$$(b + 32)(b - 20) = 0 \quad \textbf{Factoring}$$
$$b + 32 = 0 \quad or \quad b - 20 = 0 \quad \textbf{Using the principle of zero products}$$
$$b = -32 \quad or \quad b = 20.$$

4. **Check.** The width must be positive, so -32 cannot be a solution. Suppose the base is 20 in. The height would be $20 + 12$, or 32 in., and the area $\frac{1}{2}(20)(32)$, or 320 in^2. These numbers check in the original problem.
5. **State.** The leaf is 32 in. long and 20 in. wide.

Try Exercise 9.

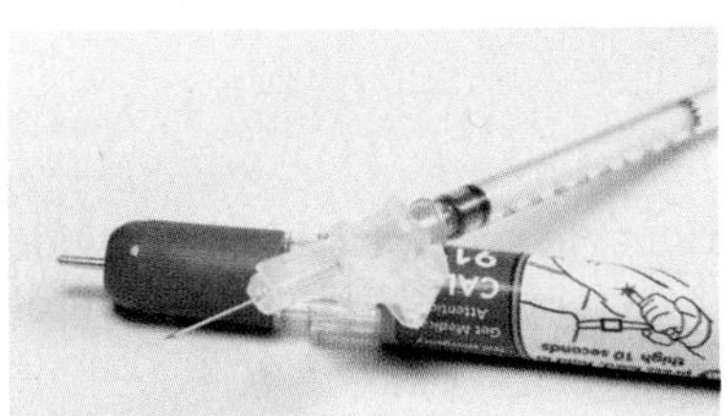

EXAMPLE 3 Medicine. For certain people suffering an extreme allergic reaction, the drug epinephrine (adrenaline) is sometimes prescribed. The number of micrograms $N(t)$ of epinephrine in an adult's bloodstream t minutes after 250 micrograms have been injected can be approximated by

$$N(t) = -10t^2 + 100t.$$

How long after an injection will there be about 210 micrograms of epinephrine in the bloodstream?

Source: Based on information in Chohan, Naina, Rita M. Doyle, and Patricia Nayle (eds.), *Nursing Handbook,* 21st ed. Springhouse, PA: Springhouse Corporation, 2001

SOLUTION

1. **Familiarize.** We make a drawing and label it. Note that t cannot be negative, since it represents the amount of time since the injection.

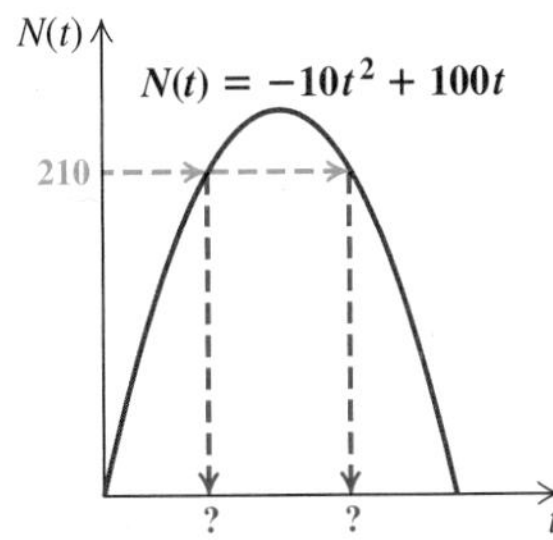

2. **Translate.** To find the length of time after injection when 210 micrograms are in the bloodstream, we replace $N(t)$ with 210 in the formula above:

$$210 = -10t^2 + 100t.$$

3. **Carry out.** We solve both algebraically and graphically, as follows.

ALGEBRAIC APPROACH

We solve by factoring:

$$210 = -10t^2 + 100t$$

$10t^2 - 100t + 210 = 0$ — **Adding $10t^2 - 100t$ to both sides to get 0 on one side**

$10(t^2 - 10t + 21) = 0$
$10(t - 3)(t - 7) = 0$ — **Factoring**

$t - 3 = 0$ *or* $t - 7 = 0$ — **Using the principle of zero products**

$t = 3$ *or* $t = 7.$

GRAPHICAL APPROACH

We graph $y_1 = 210$ and $y_2 = -10x^2 + 100x$ and look for any points of intersection of the graphs. Using the INTERSECT option of the CALC menu, we see that the graphs intersect at $(3, 210)$ and $(7, 210)$. Thus the solutions of the equation are 3 and 7.

$y_1 = 210,$
$y_2 = -10x^2 + 100x$

300
(7, 210)
y_1
(3, 210)
−1
12
−50
y_2
Yscl = 50

4. **Check.** We have

$$N(3) = -10(3)^2 + 100(3) = -10 \cdot 9 + 300 = -90 + 300 = 210;$$
$$N(7) = -10(7)^2 + 100(7) = -10 \cdot 49 + 700 = -490 + 700 = 210.$$

Both 3 and 7 check.

5. **State.** There will be 210 micrograms of epinephrine in the bloodstream approximately 3 minutes and 7 minutes after injection.

Try Exercise 15.

EXAMPLE 4 Display of a Sports Card. A valuable sports card is 4 cm wide and 5 cm long. The card is to be sandwiched by two pieces of Lucite, each of which is $5\frac{1}{2}$ times the area of the card. Determine the dimensions of the Lucite that will ensure a uniform border around the card.

SOLUTION

1. **Familiarize.** We make a drawing and label it, using x to represent the width of the border, in centimeters. Since the border extends uniformly around the entire card, the length of the Lucite must be $5 + 2x$ and the width must be $4 + 2x$.

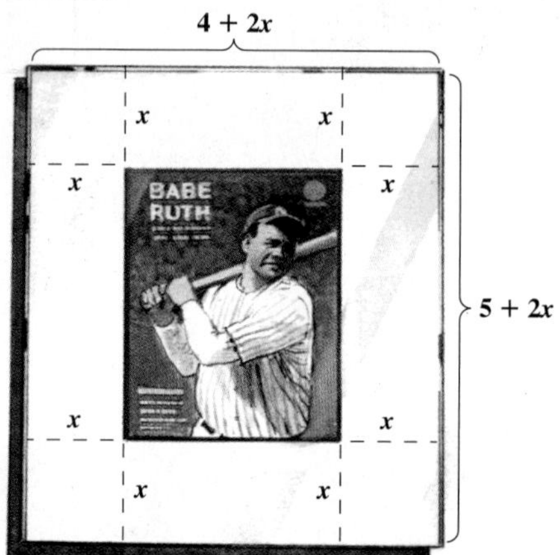

2. **Translate.** We rephrase the information given and translate as follows:

Area of Lucite is $5\frac{1}{2}$ times area of card.

$$(5 + 2x)(4 + 2x) = 5\tfrac{1}{2} \cdot 5 \cdot 4$$

3. **Carry out.** We solve both algebraically and graphically, as follows.

ALGEBRAIC APPROACH

We solve the equation:

$$(5 + 2x)(4 + 2x) = 5\tfrac{1}{2} \cdot 5 \cdot 4$$

$20 + 10x + 8x + 4x^2 = 110$ **Multiplying**

$4x^2 + 18x - 90 = 0$ **Finding standard form**

$2x^2 + 9x - 45 = 0$ **Multiplying by $\frac{1}{2}$ on both sides**

$(2x + 15)(x - 3) = 0$ **Factoring**

$2x + 15 = 0$ *or* $x - 3 = 0$ **Principle of zero products**

$x = -7\frac{1}{2}$ *or* $x = 3$.

GRAPHICAL APPROACH

We graph $y_1 = (5 + 2x)(4 + 2x)$ and $y_2 = 5.5(5)(4)$. Since we have $y_2 = 110$, we use a viewing window with Ymax > 110. Using INTERSECT, we find the points of intersection $(-7.5, 110)$ and $(3, 110)$. The solutions are -7.5 and 3.

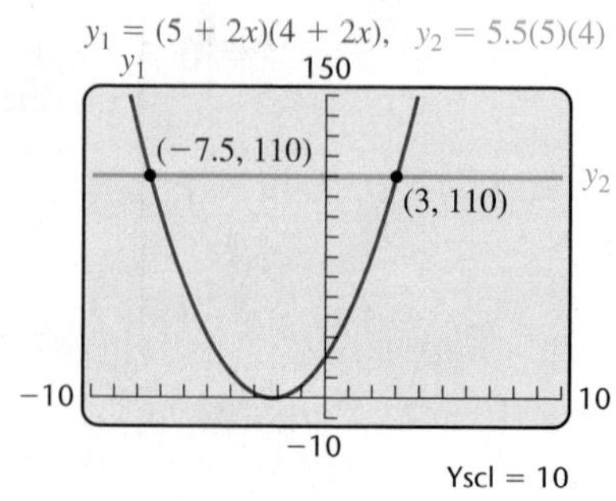

4. **Check.** We check 3 in the original problem. (Note that $-7\frac{1}{2}$ is not a solution because measurements cannot be negative.) If the border is 3 cm wide, the Lucite will have a length of $5 + 2 \cdot 3$, or 11 cm, and a width of $4 + 2 \cdot 3$, or 10 cm. The area of the Lucite is thus $11 \cdot 10$, or 110 cm^2. Since the area of the card is 20 cm^2 and 110 cm^2 is $5\frac{1}{2}$ times 20 cm^2, the number 3 checks.
5. **State.** Each piece of Lucite should be 11 cm long and 10 cm wide.

Try Exercise 25.

THE PYTHAGOREAN THEOREM

The following problem involves the Pythagorean theorem, which relates the lengths of the sides of a *right* triangle. A triangle is a **right triangle** if it has a 90°, or *right*, angle. The side opposite the 90° angle is called the **hypotenuse.** The other sides are called **legs.**

The Pythagorean Theorem

In any right triangle, if a and b are the lengths of the legs and c is the length of the hypotenuse, then

$$a^2 + b^2 = c^2.$$

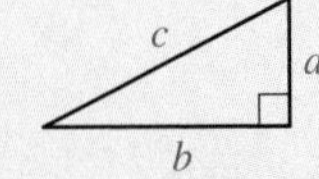

The symbol ⌐ denotes a 90° angle.

The Pythagorean theorem is named for the Greek mathematician Pythagoras (569?–500? B.C.). We can think of this relationship as adding areas.

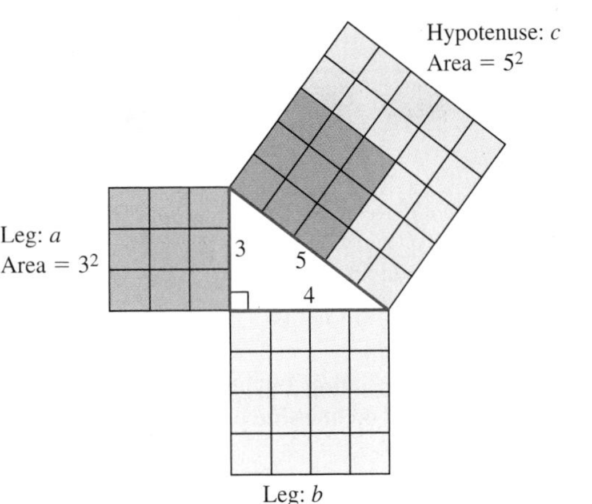

$$a^2 + b^2 = c^2$$
$$3^2 + 4^2 = 5^2$$
$$9 + 16 = 25$$

EXAMPLE 5 Bridge Design. A 50-ft diagonal brace on a bridge connects a support at the center of the bridge to a side support on the bridge. The horizontal distance that it spans is 10 ft longer than the height that it reaches on the side of the bridge. Find both distances.

SOLUTION

1. **Familiarize.** We first make a drawing. The diagonal brace and the missing distances form the hypotenuse and the legs of a right triangle. We let $x =$ the length of the vertical leg. Then $x + 10 =$ the length of the horizontal leg. The hypotenuse has length 50 ft.

2. **Translate.** Since the triangle is a right triangle, we can use the Pythagorean theorem:

$$a^2 + b^2 = c^2$$
$$x^2 + (x + 10)^2 = 50^2. \quad \textbf{Substituting}$$

3. **Carry out.** We solve both algebraically and graphically.

ALGEBRAIC APPROACH

We solve the equation as follows:

$$x^2 + (x + 10)^2 = 50^2$$
$$x^2 + (x^2 + 20x + 100) = 2500 \quad \textbf{Squaring}$$
$$2x^2 + 20x + 100 = 2500 \quad \textbf{Combining like terms}$$
$$2x^2 + 20x - 2400 = 0 \quad \textbf{Subtracting 2500 to get 0 on one side}$$
$$\left.\begin{aligned} 2(x^2 + 10x - 1200) &= 0 \\ 2(x + 40)(x - 30) &= 0 \end{aligned}\right\} \quad \textbf{Factoring}$$
$$x + 40 = 0 \quad or \quad x - 30 = 0 \quad \textbf{Using the principle of zero products}$$
$$x = -40 \quad or \quad x = 30.$$

GRAPHICAL APPROACH

We graph $y_1 = x^2 + (x + 10)^2$ and $y_2 = 50^2$. The points of intersection are $(-40, 2500)$ and $(30, 2500)$, so the solutions are -40 and 30.

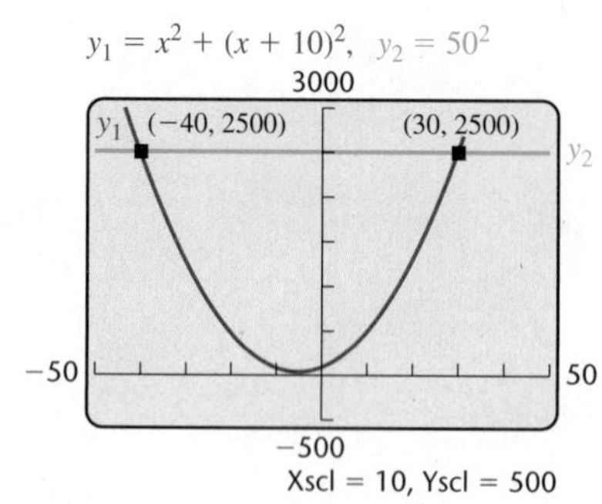

4. **Check.** The integer -40 cannot be a length of a side because it is negative. If the length is 30 ft, $x + 10 = 40$, and $30^2 + 40^2 = 900 + 1600 = 2500$, which is 50^2. So the solution 30 checks.

5. **State.** The height that the brace reaches on the side of the bridge is 30 ft, and the distance that it reaches to the middle of the bridge is 40 ft.

Try Exercise 33.

FITTING POLYNOMIAL FUNCTIONS TO DATA

In Chapter 3, we modeled data using linear functions. Some data that are not linear can be modeled using polynomial functions with higher degrees.

Graphs of quadratic functions are *parabolas*. We will study these graphs in detail in Chapter 11. Data that, when graphed, follow one of these patterns may best be modeled with a quadratic function.

Graphs of quadratic equations

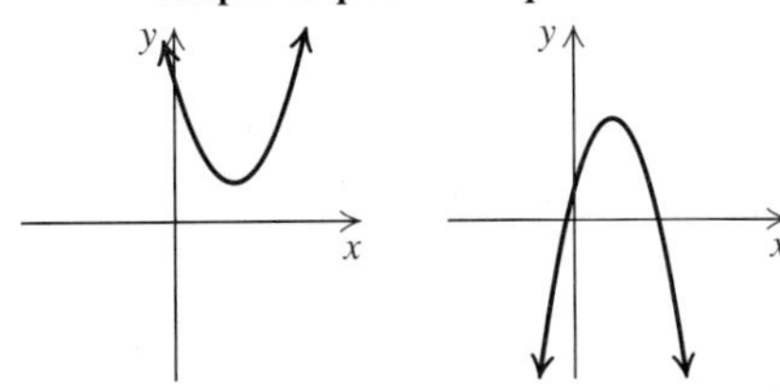

Graphs of cubic functions can follow any of the general shapes below. Graphs of polynomials of higher degree have even more possible shapes. Choosing a model may involve forming the model and noting how well it matches the data. Another consideration is whether estimates and predictions made using the model are reasonable. In this section, we will indicate what type of model to use for each set of data.

Graphs of cubic equations

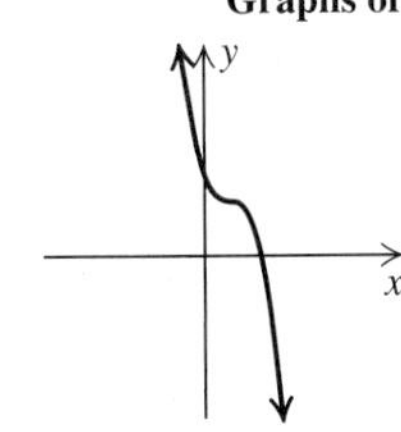

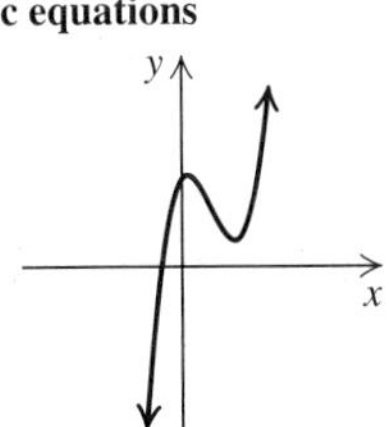

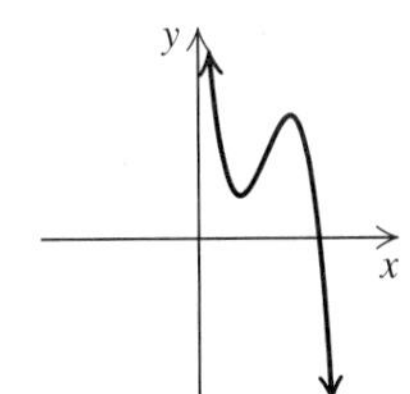

EXAMPLE 6 Athletes' Salaries. The minimum salary for a National Football League (NFL) player depends on the player's seasons of play. The table and graph below show some minimum salaries for 2011.

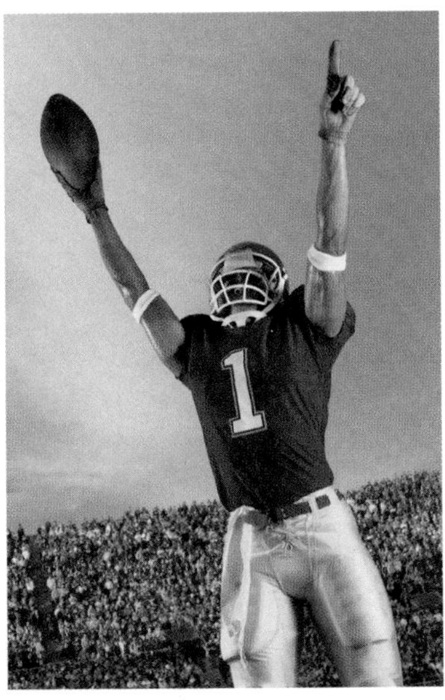

Credited Seasons	Minimum Salary (in thousands)
0	\$340
2	490
4	650
6	650
8	775

Source: "NFL—Minimum Salaries for 2008–2010 Seasons," found on proathletesonly.com

NFL Players' Salaries

NFL minimum salary (in thousands): \$900, 800, 700, 600, 500, 400, 300, 200, 100

Number of credited seasons: 0, 2, 4, 6, 8

a) Fit a cubic (degree 3) polynomial function to the data.

b) Use the function found in part (a) to estimate the minimum salary for a player with three seasons of experience.

c) If a player's minimum salary is \$900,000, for how many seasons has he played?

SOLUTION

a) We enter the data in a graphing calculator. The seasons are entered as L1 and the salary as L2.

L1	L2	L3
0	340	-----
2	490	
4	650	
6	650	
8	775	
-----	-----	

L1(1) = 0

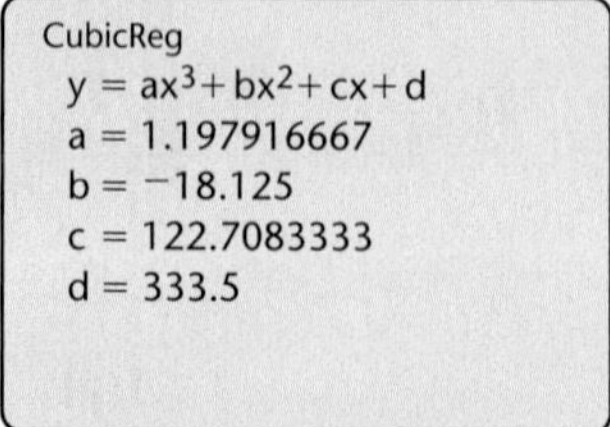
CubicReg
y = ax3+bx2+cx+d
a = 1.197916667
b = −18.125
c = 122.7083333
d = 333.5

To fit a cubic function to the data, we choose the CubicReg option from the STAT CALC menu. Pressing STAT ⟩ 6 VARS ⟩ 1 1 ENTER will calculate the regression equation and copy it to Y1. Rounding coefficients, we see that the cubic function that fits the data is

$$f(x) = 1.198x^3 - 18.125x^2 + 122.708x + 333.5.$$

b) We use a table with Indpnt set to Ask to estimate the minimum salary for a player with three seasons of experience. On the basis of this model, we estimate that the minimum salary for such a player is about \$571,000.

X	Y1	
3	570.84	

c) To determine the number of seasons for which the minimum salary is \$900,000, we solve the equation $f(x) = 900$. We graph $Y1 = f(x)$ and $Y2 = 900$. To view the scatterplot of the data, we turn on Plot1 and check that Xlist is L1 and Ylist is L2. The graphs intersect at (9.5904464, 900). Rounding, we estimate that a player who has a minimum salary of \$900,000 has played 10 seasons.

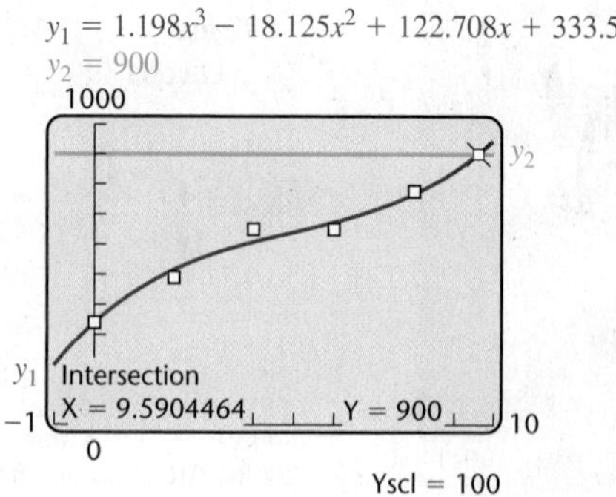

Try Exercise 41.

Visualizing for Success

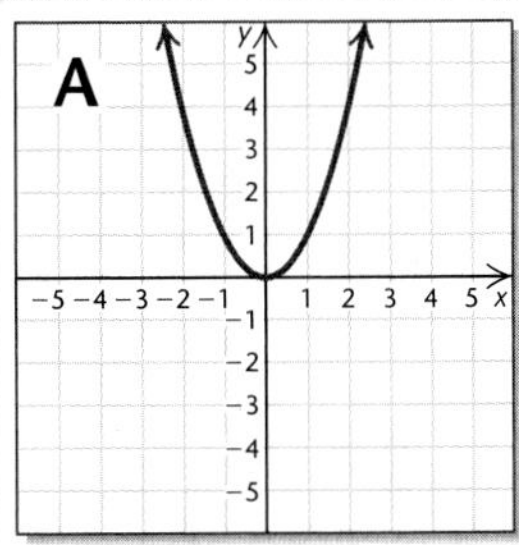

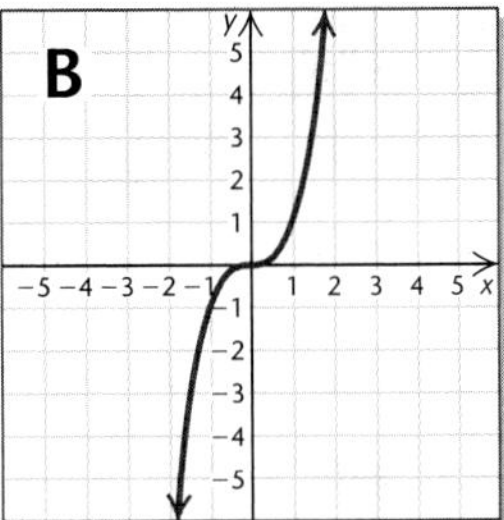

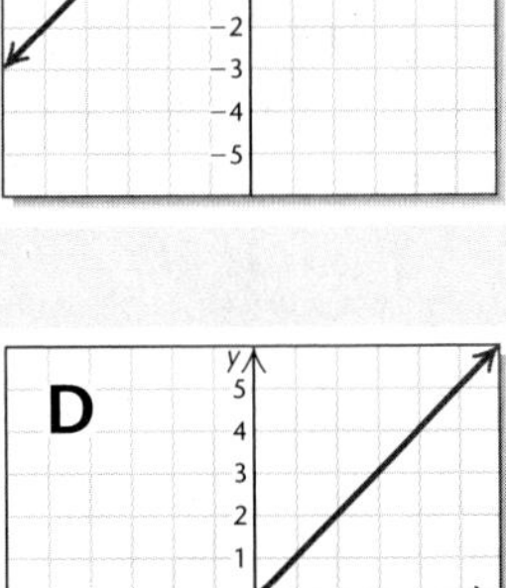

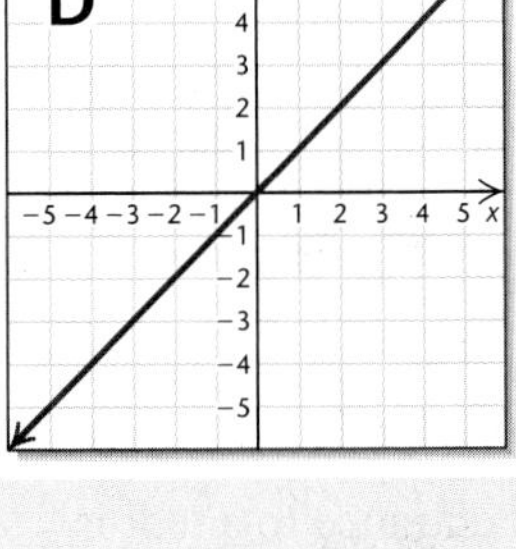

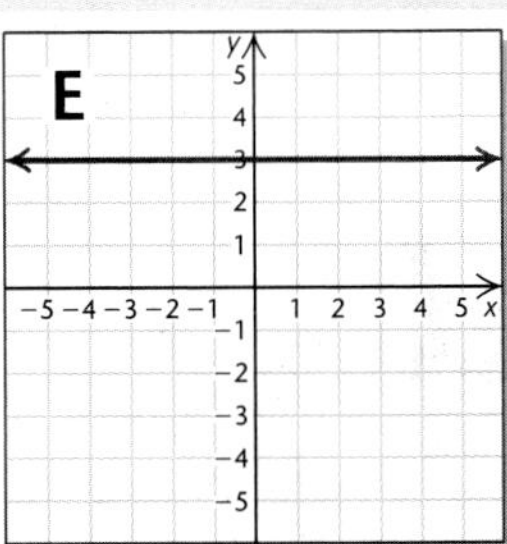

Match each equation with its graph.

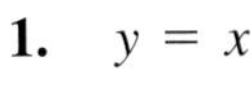
1. $y = x$
2. $y = |x|$
3. $y = x^2$
4. $y = x^3$
5. $y = 3$
6. $y = x + 3$
7. $y = x - 3$
8. $y = 2x$
9. $y = -2x$
10. $y = 2x + 3$

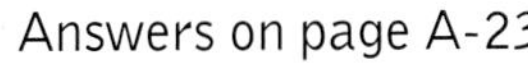
Answers on page A-23

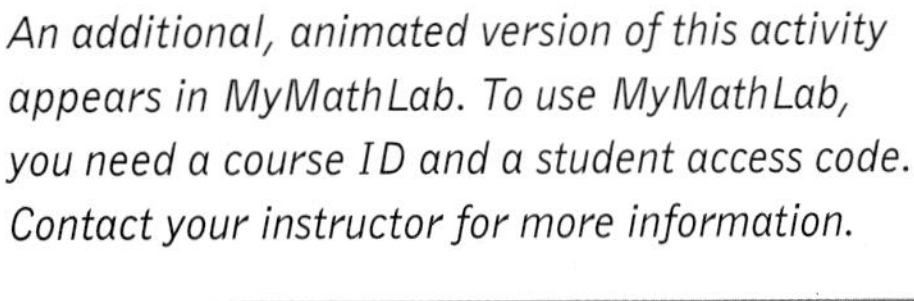
An additional, animated version of this activity appears in MyMathLab. To use MyMathLab, you need a course ID and a student access code. Contact your instructor for more information.

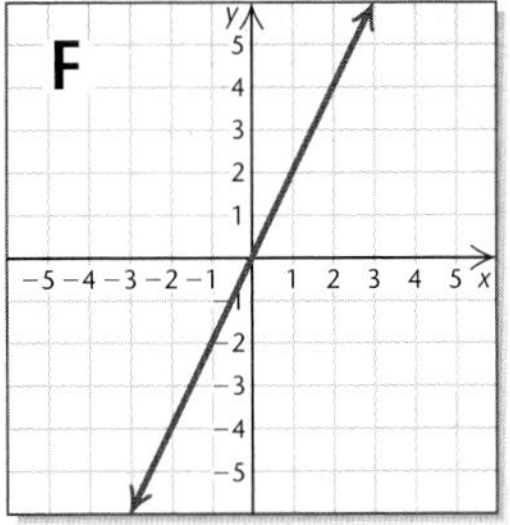

G

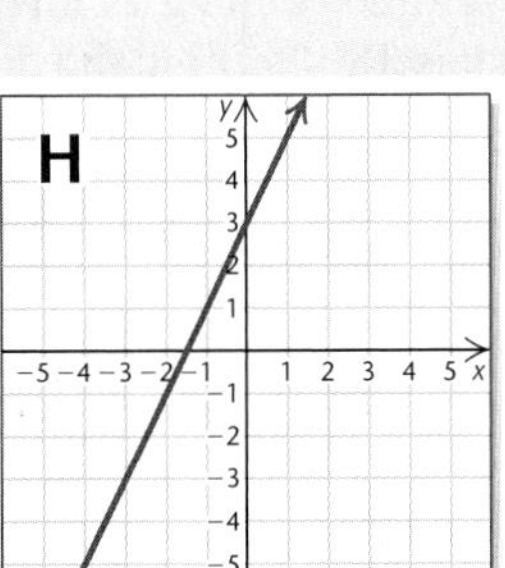

I

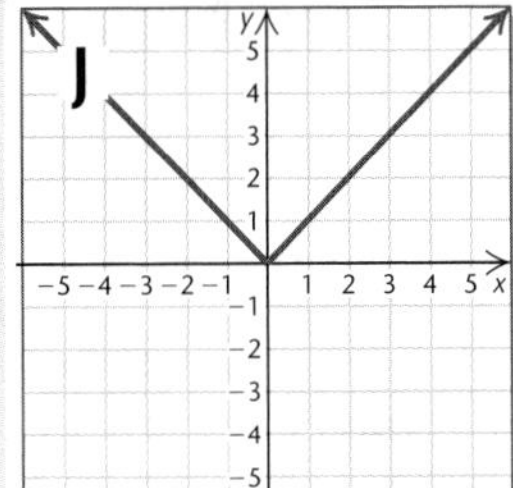

6.7 Exercise Set

FOR EXTRA HELP MyMathLab

Solve.

1. The square of a number plus the number is 132. Find all such numbers.

2. A number is 30 less than its square. Find all such numbers.

3. *Parking-Space Numbers.* The product of two consecutive parking spaces is 110. Find the parking-space numbers.

4. *Page Numbers.* The product of the page numbers on two facing pages of a book is 420. Find the page numbers.

5. *Construction.* The front porch on Trent's new home is five times as long as it is wide. If the area of the porch is 180 ft^2, find the dimensions.

6. *Furnishings.* The work surface of Anita's desk is a rectangle that is twice as long as it is wide. If the area of the desktop is 18 ft^2, find the length and the width of the desk.

7. A photo is 5 cm longer than it is wide. Find the length and the width if the area is 84 cm^2.

8. An envelope is 4 cm longer than it is wide. The area is 96 cm^2. Find the length and the width.

9. *Dimensions of a Sail.* The height of the jib sail on a Lightning sailboat is 5 ft greater than the length of its "foot." If the area of the sail is , find the length of the foot and the height of the sail.

10. *Dimensions of a Triangle.* A triangle is 10 cm wider than it is tall. The area is 48 cm^2. Find the height and the base.

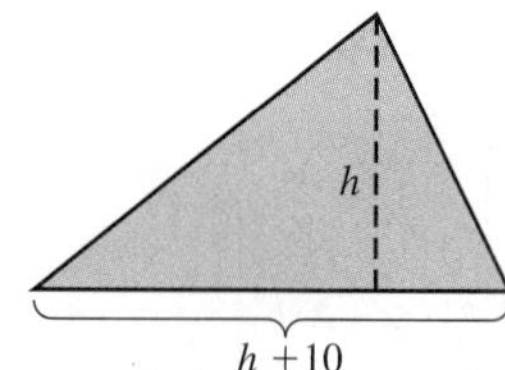

11. *Road Design.* A triangular traffic island has a base half as long as its height. Find the base and the height if the island has an area of 64 ft^2.

12. *Tent Design.* The triangular entrance to a tent is $\frac{2}{3}$ as wide as it is tall. The area of the entrance is 12 ft^2. Find the height and the base.

Games in a League's Schedule. In a sports league of x teams in which all teams play each other twice, the total number N of games played is given by

$$x^2 - x = N.$$

Use this formula for Exercises 13 and 14.

13. The Colchester Youth Soccer League plays a total of 240 games, with all teams playing each other twice. How many teams are in the league?

14. The teams in a women's softball league play each other twice, for a total of 132 games. How many teams are in the league?

15. *Medicine.* For many people suffering from constricted bronchial muscles, the drug Albuterol is prescribed. The number of micrograms A of Albuterol in a person's bloodstream t minutes after 200 micrograms have been inhaled can be approximated by

$$A = -50t^2 + 200t.$$

How long after an inhalation will there be about 150 micrograms of Albuterol in the bloodstream?
Source: Based on information in Chohan, Naina, Rita M. Doyle, and Patricia Nayle (eds.), *Nursing Handbook*, 21st ed. Springhouse, PA: Springhouse Corporation, 2001

16. *Medicine.* For adults with certain heart conditions, the drug Primacor (milrinone lactate) is prescribed. The number of milligrams M of Primacor in the bloodstream of a 132-lb patient t hours after a 3-mg dose has been injected can be approximated by

$$M = -\frac{1}{2}t^2 + \frac{5}{2}t.$$

How long after an injection will there be about 2 mg in the bloodstream?
Source: Based on information in Chohan, Naina, Rita M. Doyle, and Patricia Nayle (eds.), *Nursing Handbook*, 21st ed. Springhouse, PA: Springhouse Corporation, 2001

17. *Wave Height.* The height of waves in a storm depends on the speed of the wind. Assuming the wind has no obstructions for a long distance, the maximum wave height H for a wind speed x can be approximated by

$$H = 0.006x^2 + 0.6x.$$

Here H is in feet and x is in knots (nautical miles per hour). For what wind speed would the maximum wave height be 6.6 ft?
Source: Based on information from cimss.ssec.wisc.edu

18. *Cabinet Making.* Dovetail Woodworking determines that the revenue R, in thousands of dollars, from the sale of x sets of cabinets is given by $R(x) = 2x^2 + x$. If the cost C, in thousands of dollars, of producing x sets of cabinets is given by $C(x) = x^2 - 2x + 10$, how many sets must be produced and sold in order for the company to break even?

19. *Prize Tee Shirts.* During a game's intermission, a team mascot launches tightly rolled tee shirts into the stands. The height $h(t)$, in feet, of an airborne tee shirt t seconds after being launched can be approximated by

$$h(t) = -15t^2 + 75t + 10.$$

After peaking, a rolled-up tee shirt is caught by a fan 70 ft above ground level. For how long was the tee shirt in the air?

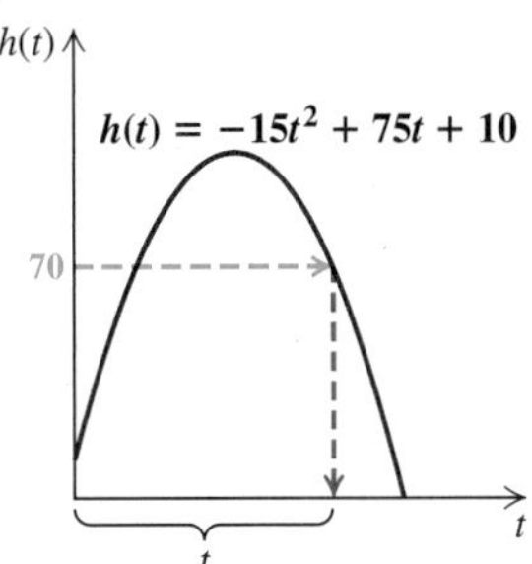

20. *Prize Tee Shirts.* Using the model in Exercise 19, determine how long a tee shirt has been airborne if it is caught on the way *up* by a fan 100 ft above ground level.

21. *Fireworks Displays.* Fireworks are typically launched from a mortar with an upward velocity (initial speed) of about 64 ft/sec. The height $h(t)$, in feet, of a "weeping willow" display, t seconds after having been launched from an 80-ft high rooftop, is given by

$$h(t) = -16t^2 + 64t + 80.$$

After how long will the cardboard shell from the fireworks reach the ground?

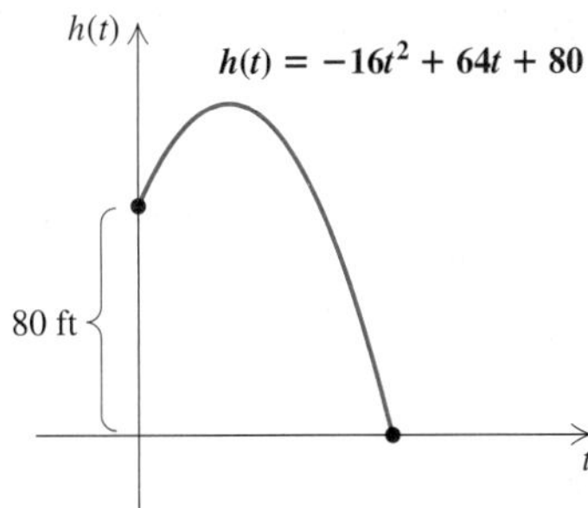

22. *Safety Flares.* Suppose that a flare is launched upward with an initial velocity of 80 ft/sec from a height of 224 ft. Its height $h(t)$, in feet, after t seconds is given by

$$h(t) = -16t^2 + 80t + 224.$$

After how long will the flare reach the ground?

23. *Geometry.* If each of the sides of a square is lengthened by 4 m, the area becomes 49 m^2. Find the length of a side of the original square.

24. *Geometry.* If each of the sides of a square is lengthened by 6 cm, the area becomes 144 cm^2. Find the length of a side of the original square.

25. *Framing a Picture.* A picture frame measures 12 cm by 20 cm, and 84 cm^2 of picture shows. Find the width of the frame.

26. *Framing a Picture.* A picture frame measures 14 cm by 20 cm, and 160 cm^2 of picture shows. Find the width of the frame.

27. *Landscaping.* A rectangular lawn measures 60 ft by 80 ft. Part of the lawn is torn up to install a sidewalk of uniform width around it. The area of the new lawn is 2400 ft^2. How wide is the sidewalk?

28. *Landscaping.* A rectangular garden is 30 ft by 40 ft. Part of the garden is removed in order to install a walkway of uniform width around it. The area of the new garden is one-half the area of the old garden. How wide is the walkway?

29. *Construction.* The diagonal braces in a lookout tower are 15 ft long and span a horizontal distance of 12 ft. How high does each brace reach vertically?

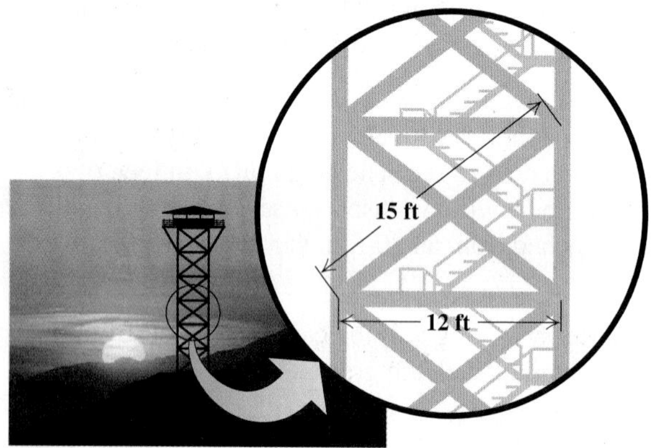

30. *Reach of a Ladder.* Twyla has a 26-ft ladder leaning against her house. If the bottom of the ladder is 10 ft from the base of the house, how high does the ladder reach?

31. *Roadway Design.* Elliott Street is 24 ft wide when it ends at Main Street in Brattleboro, Vermont. A 40-ft long diagonal crosswalk allows pedestrians to cross Main Street to or from either corner of Elliott Street (see the figure). Determine the width of Main Street.

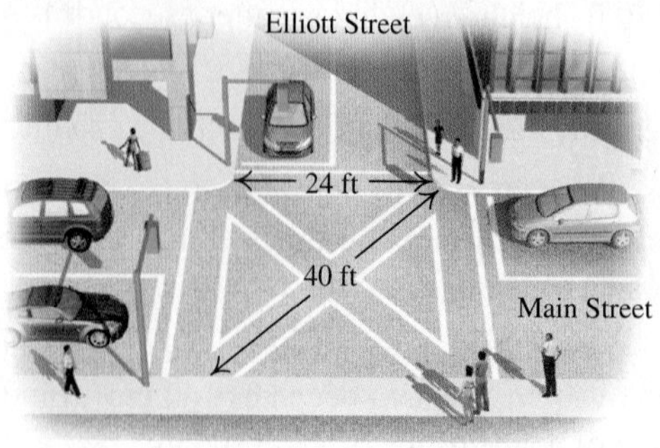

32. *Aviation.* Engine failure forced Robbin to pilot her Cessna 150 to an emergency landing. To land, Robbin's plane glided 17,000 ft over a 15,000-ft stretch of deserted highway. From what altitude did the descent begin?

33. *Archaeology.* Archaeologists have discovered that the 18th-century garden of the Charles Carroll House in Annapolis, Maryland, was a right triangle. One leg of the triangle was formed by a 400-ft long sea wall. The hypotenuse of the triangle was 200 ft longer than the other leg. What were the dimensions of the garden?
Source: www.bsos.umd.edu

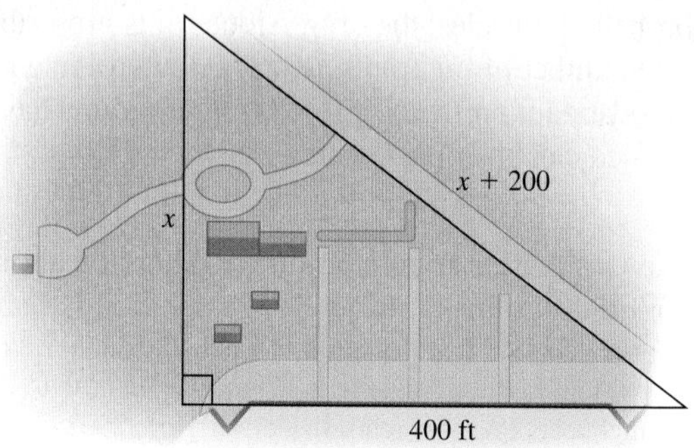

34. One leg of a right triangle is 7 cm shorter than the other leg. The length of the hypotenuse is 13 cm. Find the length of each side.

35. *Antenna Wires.* A wire is stretched from the ground to the top of an antenna tower. The wire is 20 ft long. The height of the tower is 4 ft greater than the distance d from the tower's base to the bottom of the wire. Find the distance d and the height of the tower.

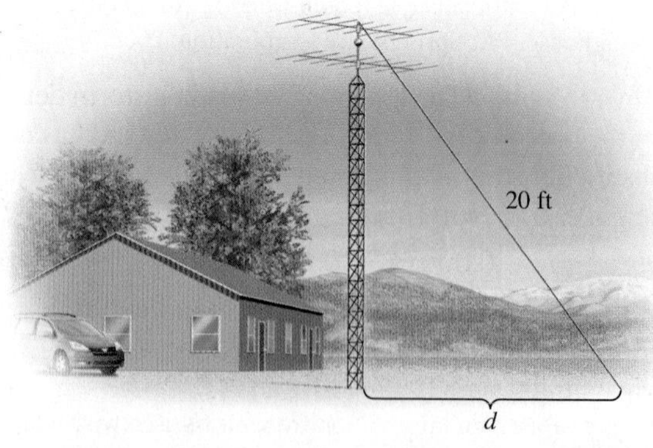

36. *Parking Lot Design.* A rectangular parking lot is 50 ft longer than it is wide. Determine the dimensions of the parking lot if it measures 250 ft diagonally.

37. *Carpentry.* In order to build a deck at a right angle to their house, Lucinda and Felipe place a stake in the ground a precise distance from the back wall of their house. This stake will combine with two marks on the house to form a right triangle. From a course in geometry, Lucinda remembers that there are three consecutive integers that can work as sides of a right triangle. Find the sides of that triangle.

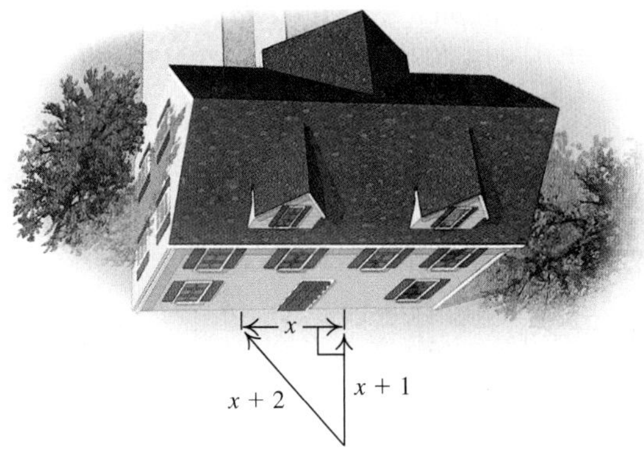

38. *Ladder Location.* The foot of an extension ladder is 9 ft from a wall. The height that the ladder reaches on the wall and the length of the ladder are consecutive integers. How long is the ladder?

39. *Architecture.* An architect has allocated a rectangular space of 264 ft^2 for a square dining room and a 10-ft wide kitchen, as shown in the figure. Find the dimensions of each room.

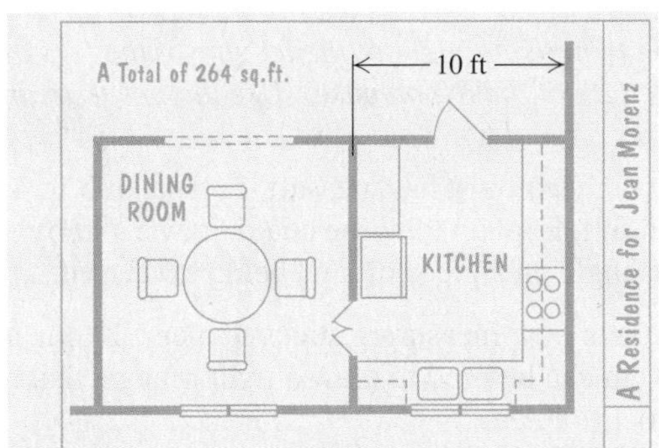

40. *Design.* A window panel for a sun porch consists of a 7-ft high rectangular window stacked above a square window. The windows have the same width. If the total area of the window panel is 18 ft^2, find the dimensions of each window.

41. *Health-Care Costs.* The table below lists the average annual percentage change in national prescription drug expenditures.

Year	Number of Years After 1990, x	Average Annual Percentage Change in Prescription Drug Expenditures, P
1996	6	13
1998	8	14
2000	10	15
2002	12	14
2004	14	8
2006	16	9

Source: Kaiser Family Foundation calculations using National Health Expenditure historical data from Centers for Medicare & Medicaid Services

a) Use regression to find a quartic function P that can be used to estimate the average annual percentage change in prescription drug expenditures x years after 1990. Round coefficients to five decimal places.
b) Estimate the average annual percentage change in prescription drug expenditures in 2005.
c) In what years did the average prescription drug expenditures increase 12%?

42. *Fuel Economy.* The table below lists the average fuel economy of motor vehicles at various speeds.

Speed x (in miles per hour)	Fuel Economy c, (in miles per gallon)
5	11
25	27
35	29
45	30
55	31
65	28
75	23

Source: Based on data from the U.S. Department of Energy

a) Use regression to find a quartic function c that can be used to estimate the average fuel economy at a speed of x miles per hour. Round coefficients to six decimal places.
b) Estimate the average fuel economy at 15 mph.
c) For what speeds is the average fuel economy 25 mpg?

43. *Olympics.* The table below lists the number of female athletes participating in the summer Olympic games for various years.

Year	Number of Years After 1900, *x*	Number of Female Athletes in the Summer Olympic Games, *F*
1960	60	611
1976	76	1260
1992	92	2704
2000	100	4069
2008	108	4746

Source: olympic.org

a) Use regression to find a cubic polynomial function F that can be used to estimate the number of female athletes competing in the summer Olympic games x years after 1900. Round coefficients to five decimal places.
b) Estimate the number of female athletes competing in 2012.
c) In what year will 6000 female athletes compete in the summer Olympic games?

44. *Life Insurance Premiums.* The table below lists monthly life insurance premiums for a \$500,000 policy for females of various ages.

Age	Monthly Premium for \$500,000 Life Insurance Policy for Females
35	\$ 14
40	18
45	24
50	33
55	48
60	70
65	110
70	188

a) Use regression to find a cubic polynomial function L that can be used to estimate the monthly life insurance premium for a female of age x. Round coefficients to six decimal places.
b) Estimate the monthly premium for a female of age 58.
c) For what age is the monthly premium \$100?

TW **45.** Tyler disregards any negative solutions that he finds when solving applied problems. Is this approach correct? Why or why not?

TW **46.** Write a chart of the population of two imaginary cities. Devise the numbers in such a way that one city has linear growth and the other has nonlinear growth.

SKILL REVIEW

To prepare for Chapter 7, review addition, subtraction, multiplication, and division using fraction notation (Sections 1.3, 1.5, 1.6, and 1.7).

Simplify.

47. $-\frac{3}{5} \cdot \frac{4}{7}$ [1.7]

48. $-\frac{3}{5} \div \frac{4}{7}$ [1.7]

49. $-\frac{5}{6} - \frac{1}{6}$ [1.6]

50. $\frac{3}{4} + \left(-\frac{5}{2}\right)$ [1.5]

51. $-\frac{3}{8} \cdot \left(-\frac{10}{15}\right)$ [1.7]

52. $\dfrac{-\frac{8}{15}}{-\frac{2}{3}}$ [1.7]

53. $\frac{5}{24} + \frac{3}{28}$ [1.3]

54. $\frac{5}{6} - \left(-\frac{2}{9}\right)$ [1.6]

SYNTHESIS

The converse of the Pythagorean theorem is also true. That is, if $a^2 + b^2 = c^2$, then the triangle is a right triangle (where a and b are the lengths of the legs and c is the length of the hypotenuse). Use this result to answer Exercises 55 and 56.

TW Aha! **55.** An archaeologist has measuring sticks of 3 ft, 4 ft, and 5 ft. Explain how she could draw a 7-ft by 9-ft rectangle on a piece of land being excavated.

TW **56.** Explain how measuring sticks of 5 cm, 12 cm, and 13 cm can be used to draw a right triangle that has two 45° angles.

57. *Roofing.* A *square* of shingles covers 100 ft^2 of surface area. How many squares will be needed to reshingle the house shown?

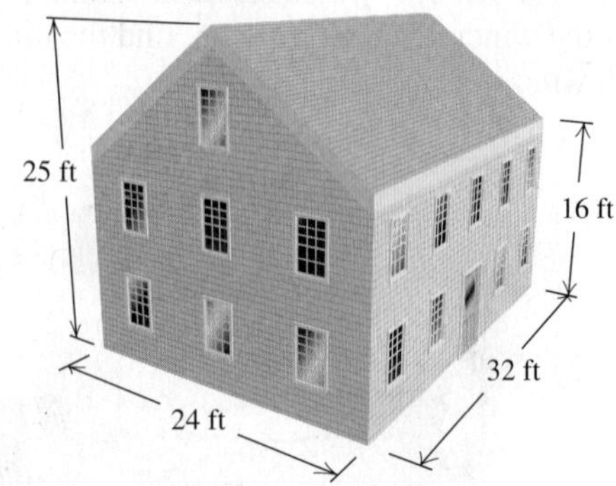

58. Solve for x.

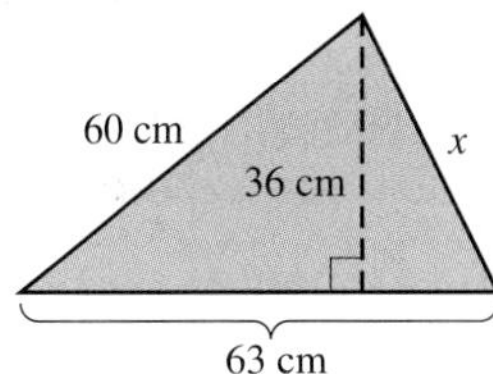

Medicine. For certain people with acid reflux, the drug Pepcid (famotidine) is used. The number of milligrams N of Pepcid in an adult's bloodstream t hours after a 20-mg tablet has been swallowed can be approximated by

$$N(t) = -0.009t(t - 12)^3.$$

Use a graphing calculator with the window $[-1, 13, -1, 25]$ *to answer Exercises 59–61.*

Source: Based on information in Chohan, Naina, Rita M. Doyle, and Patricia Nayle (eds.), *Nursing Handbook,* 21st ed. Springhouse, PA: Springhouse Corporation, 2001

59. Approximately how long after a tablet has been swallowed will there be 18 mg in the bloodstream?

60. Approximately how long after a tablet has been swallowed will there be 10 mg in the bloodstream?

61. Approximately how long after a tablet has been swallowed will the peak dosage in the bloodstream occur?

62. *Folding Sheet Metal.* An open rectangular gutter is made by turning up the sides of a piece of metal 20 in. wide, as shown. The area of the cross section of the gutter is 48 in^2. Find the possible depths of the gutter.

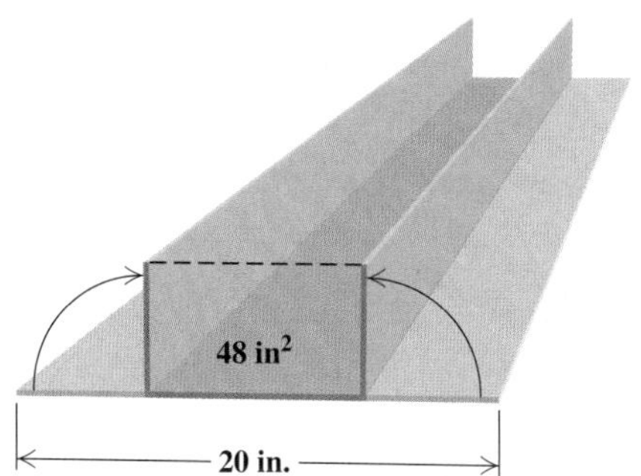

63. *Box Construction.* A rectangular piece of tin is twice as long as it is wide. Squares 2 cm on a side are cut out of each corner, and the ends are turned up to make a box whose volume is 480 cm^3. What are the dimensions of the piece of tin?

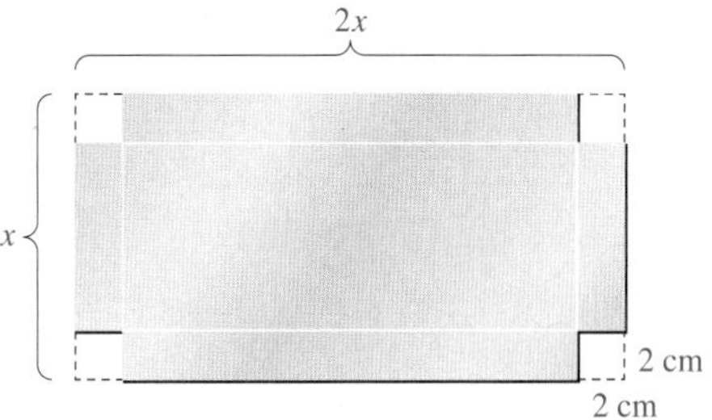

64. *Navigation.* A tugboat and a freighter leave the same port at the same time at right angles. The freighter travels 7 km/h slower than the tugboat. After 4 hr, they are 68 km apart. Find the speed of each boat.

65. *Skydiving.* During the first 13 sec of a jump, a skydiver falls approximately $11.12t^2$ feet in t seconds. A small heavy object (with less wind resistance) falls about $15.4t^2$ feet in t seconds. Suppose that a skydiver jumps from 30,000 ft, and 1 sec later a camera falls out of the airplane. How long will it take the camera to catch up to the skydiver?

Try Exercise Answers: Section 6.7

3. 10, 11 **9.** Foot: 7 ft; height: 12 ft **15.** 1 min, 3 min
25. 3 cm **33.** 300 ft by 400 ft by 500 ft
41. **(a)** $P(x) = 0.01823x^4 - 0.77199x^3 + 11.62153x^2 - 73.65807x + 179.76190$; **(b)** 7%; **(c)** 2003 and 2007

Study Summary

KEY TERMS AND CONCEPTS	EXAMPLES	PRACTICE EXERCISES

SECTION 6.1: INTRODUCTION TO POLYNOMIAL FACTORIZATIONS AND EQUATIONS

To **factor** a polynomial means to write it as a product of polynomials. Always begin by factoring out the **largest common factor.**	$12x^4 - 30x^3 = 6x^3(2x - 5)$	**1.** Factor: $12x^4 - 18x^3 + 30x$.
Some polynomials with four terms can be **factored by grouping.**	$3x^3 - x^2 - 6x + 2 = x^2(3x - 1) - 2(3x - 1)$ $= (3x - 1)(x^2 - 2)$	**2.** Factor: $2x^3 - 6x^2 - x + 3$.
The Principle of Zero Products For any real numbers a and b: If $ab = 0$, then $a = 0$ or $b = 0$. If $a = 0$ or $b = 0$, then $ab = 0$.	Solve: $2x^2 = 8x$. $2x^2 - 8x = 0$ **Getting 0 on one side** $2x(x - 4) = 0$ **Factoring** $2x = 0$ *or* $x - 4 = 0$ **Using the principle of zero products** $x = 0$ *or* $x = 4$ The solutions are 0 and 4.	**3.** Solve: $8x = 6x^2$.

SECTION 6.2: TRINOMIALS OF THE TYPE $x^2 + bx + c$

Some trinomials of the type $x^2 + bx + c$ can be factored by reversing the steps of FOIL.	Factor: $x^2 - 11x + 18$. (see table below) The factorization is $(x - 2)(x - 9)$.	**4.** Factor: $x^2 - 7x - 18$.

Pairs of Factors of 18	Sums of Factors
$-1, -18$	-19
$-2, -9$	-11

SECTION 6.3: TRINOMIALS OF THE TYPE $ax^2 + bx + c$

One method for factoring trinomials of the type $ax^2 + bx + c$ is a FOIL-based method.	Factor: $6x^2 - 5x - 6$. The factors will be in the form $(3x + \quad)(2x + \quad)$ or $(6x + \quad)(x + \quad)$. We list all pairs of factors of -6, and check possible products by multiplying any possibilities that do not contain a common factor. $(3x - 2)(2x + 3) = 6x^2 + 5x - 6$, $(3x + 2)(2x - 3) = 6x^2 - 5x - 6$ ← **This is the correct product, so we stop here.** The factorization is $(3x + 2)(2x - 3)$.	**5.** Factor: $6x^2 + x - 2$.

Another method for factoring trinomials of the type $ax^2 + bx + c$ involves factoring by grouping.	Factor: $6x^2 - 5x - 6$. Multiply the leading coefficient and the constant term: $6(-6) = -36$. Look for factors of -36 that add to -5. Pairs of Factors of -36 / Sums of Factors: $1, -36$ → -35 $2, -18$ → -16 $3, -12$ → -9 $4, -9$ → -5 Rewrite $-5x$ as $4x - 9x$ and factor by grouping: $6x^2 - 5x - 6 = 6x^2 + 4x - 9x - 6$ $= 2x(3x + 2) - 3(3x + 2)$ $= (3x + 2)(2x - 3)$.	**6.** Factor: $8x^2 - 22x + 15$.

Pairs of Factors of -36	Sums of Factors
$1, -36$	-35
$2, -18$	-16
$3, -12$	-9
$4, -9$	-5

SECTION 6.4: PERFECT-SQUARE TRINOMIALS AND DIFFERENCES OF SQUARES

Factoring a Perfect-Square Trinomial $A^2 + 2AB + B^2 = (A + B)^2$; $A^2 - 2AB + B^2 = (A - B)^2$	Factor: $y^2 + 100 - 20y$. $A^2 - 2AB + B^2 = (A - B)^2$ $\downarrow \quad \downarrow \quad \downarrow \quad \downarrow \quad \downarrow$ $y^2 + 100 - 20y = y^2 - 20y + 100 = (y - 10)^2$	**7.** Factor: $100n^2 + 81 + 180n$.
Factoring a Difference of Squares $A^2 - B^2 = (A + B)(A - B)$	Factor: $9t^2 - 1$. $A^2 - B^2 = (A + B)(A - B)$ $\downarrow \quad \downarrow \quad \downarrow \quad \downarrow \quad \downarrow \quad \downarrow$ $9t^2 - 1 = (3t + 1)(3t - 1)$	**8.** Factor: $144t^2 - 25$.

SECTION 6.5: SUMS OR DIFFERENCES OF CUBES

Factoring a Sum or a Difference of Cubes $A^3 + B^3 = (A + B)(A^2 - AB + B^2)$ $A^3 - B^3 = (A - B)(A^2 + AB + B^2)$	$x^3 + 1000 = (x + 10)(x^2 - 10x + 100)$ $z^6 - 8w^3 = (z^2 - 2w)(z^4 + 2wz^2 + 4w^2)$	**9.** Factor: $a^3 - 1$.

SECTION 6.6: FACTORING: A GENERAL STRATEGY

A general strategy for factoring polynomials can be found on p. 471.

Factor: $5x^5 - 80x$.

$$5x^5 - 80x = 5x(x^4 - 16)$$ **$5x$ is the largest common factor.**

$$= 5x(x^2 + 4)(x^2 - 4)$$ **$x^4 - 16$ is a difference of squares.**

$$= 5x(x^2 + 4)(x + 2)(x - 2)$$ **$x^2 - 4$ is also a difference of squares.**

Check:

$$5x(x^2 + 4)(x + 2)(x - 2) = 5x(x^2 + 4)(x^2 - 4)$$
$$= 5x(x^4 - 16)$$
$$= 5x^5 - 80x.$$

Factor: $-x^2y^2 - 3xy + 10$.

$$-x^2y^2 - 3xy + 10 = -(x^2y^2 + 3xy - 10)$$ **Factor out -1 to make the leading coefficient positive.**

$$= -(xy + 5)(xy - 2)$$

Check: $-(xy + 5)(xy - 2) = -(x^2y^2 + 3xy - 10)$
$$= -x^2y^2 - 3xy + 10.$$

10. Factor:

$3x^4 - 24x + 5x^3 - 40$.

SECTION 6.7: APPLICATIONS OF POLYNOMIAL EQUATIONS

Pythagorean Theorem

In any right triangle, if a and b are the lengths of the legs and c is the length of the hypotenuse, then $a^2 + b^2 = c^2$.

Find the lengths of the legs in this triangle.

5, x, $x + 1$

$$x^2 + (x + 1)^2 = 5^2$$
$$x^2 + x^2 + 2x + 1 = 25$$
$$2x^2 + 2x - 24 = 0$$
$$x^2 + x - 12 = 0$$
$$(x + 4)(x - 3) = 0$$
$$x + 4 = 0 \quad or \quad x - 3 = 0$$
$$x = -4 \quad or \quad x = 3$$

Since lengths are not negative, -4 is not a solution. The lengths of the legs are 3 and 4.

11. Find the lengths of the sides in this triangle.

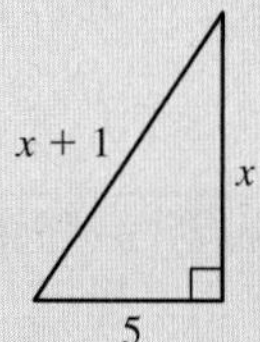

Review Exercises 6

Concept Reinforcement *Classify each of the following statements as either true or false.*

1. Every polynomial has a common factor other than 1 or −1. [6.1]
2. A prime polynomial has no common factor other than 1 or −1. [6.1]
3. Every perfect-square trinomial can be expressed as a binomial squared. [6.4]
4. The last step in factoring is to factor out a common factor. [6.1]
5. Every quadratic equation has two different solutions. [6.1]
6. The principle of zero products can be applied whenever a product equals 0. [6.1]
7. In a right triangle, the hypotenuse is always longer than either leg. [6.7]
8. The Pythagorean theorem can be applied to any triangle that has an angle measuring at least 90°. [6.7]

Factor completely. If a polynomial is prime, state this.

9. $7x^2 + 6x$ [6.1]
10. $18y^4 - 6y^2$ [6.1]
11. $100t^2 - 1$ [6.4]
12. $a^2 - 12a + 27$ [6.2]
13. $3m^2 + 14m + 8$ [6.3]
14. $25x^2 + 20x + 4$ [6.4]
15. $4y^2 - 16$ [6.4]
16. $5x^2 + x^3 - 14x$ [6.2]
17. $ax + 2bx - ay - 2by$ [6.1]
18. $3y^3 + 6y^2 - 5y - 10$ [6.1]
19. $81a^4 - 1$ [6.4]
20. $48t^2 - 28t + 6$ [6.1]
21. $27x^3 - 8$ [6.5]
22. $-t^3 + t^2 + 42t$ [6.2]
23. $a^2b^4 - 64$ [6.4]
24. $3x + x^2 + 5$ [6.2]
25. $2z^8 - 16z^6$ [6.1]
26. $54x^6y - 2y$ [6.5]
27. $75 + 12x^2 - 60x$ [6.4]
28. $6t^2 + 17pt + 5p^2$ [6.3]
29. $x^3 + 2x^2 - 9x - 18$ [6.4]
30. $a^2 - 2ab + b^2 - 4t^2$ [6.4]
31. The graph below is that of a polynomial function $p(x)$. Use the graph to find the zeros of p. [6.1]

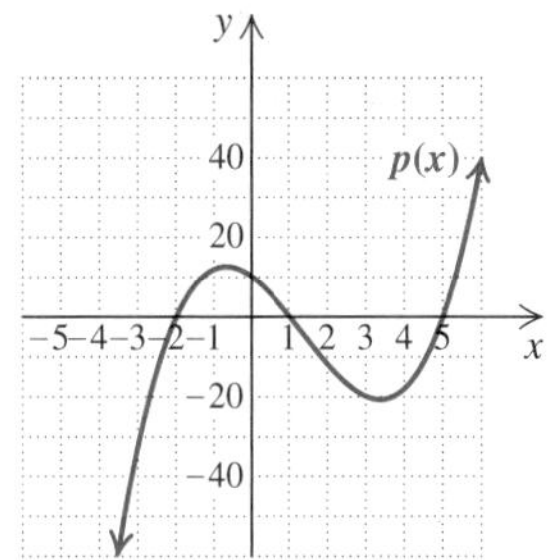

32. Find the zeros of the function given by $f(x) = x^2 - 11x + 28$. [6.2]

Solve. Where appropriate, round answers to the nearest thousandth.

33. $(x - 9)(x + 11) = 0$ [6.1]
34. $6b^2 - 13b + 6 = 0$ [6.3]
35. $8t^2 = 14t$ [6.1]
36. $x^2 - 20x = -100$ [6.4]
37. $r^2 = 16$ [6.4]
38. $a^3 = 4a^2 + 21a$ [6.2]
39. $x(x - 1) = 20$ [6.2]
40. $x^3 - 5x^2 - 16x + 80 = 0$ [6.4]
41. $x^2 + 180 = 27x$ [6.2]
42. $x^2 - 2x = 6$ [6.2]
43. Let $f(x) = x^2 - 7x - 40$. Find a such that $f(a) = 4$. [6.3]
44. Find the domain of the function f given by
$$f(x) = \frac{x - 3}{3x^2 + 19x - 14}.$$ [6.3]
45. The formula $x^2 - x = N$ can be used to determine the total number of games played, N, in a league of x teams in which all teams play each other twice. Serena referees for a soccer league in which all teams play each other twice and a total of 90 games is played. How many teams are in the league? [6.7]

46. The gable of St. Bridget's Convent Ruins in Estonia is $\frac{3}{4}$ as tall as it is wide. Its area is 216 m^2. Find the height and the base. [6.7]

47. A photograph is 3 in. longer than it is wide. When a 2-in. border is placed around the photograph, the total area of the photograph and the border is 108 in^2. Find the dimensions of the photograph. [6.7]

48. Roy is designing a rectangular garden with a width of 8 ft. The path that leads diagonally across the garden is 2 ft longer than the length of the garden. How long is the path? [6.7]

49. *Labor Force.* The table below lists the percent of the U.S. male population age 65 and older participating in the labor force. [6.7]

Year	Year After 1970, x	Labor Force Participation Rate, Men 65 and Older
1970	0	26.8
1980	10	19.0
1990	20	16.3
2000	30	17.7
2007	37	20.5

Source: U.S. Bureau of Labor Statistics

a) Use regression to find a quadratic polynomial function P that can be used to estimate the labor participation rate for men 65 and older x years after 1970. Round coefficients to four decimal places.
b) Estimate the participation rate in 2012.
c) In what year or years is the participation rate 25?

SYNTHESIS

TW **50.** Explain how to find the zeros of a polynomial function from its graph. [6.1]

TW **51.** When the principle of zero products is used to solve a quadratic equation, will there always be two different solutions? Why or why not? [6.1]

Factor. [6.5]

52. $128x^6 - 2y^6$

53. $(x - 1)^3 - (x + 1)^3$

Solve. [6.4]

54. $(x + 1)^3 = x^2(x + 1)$

Aha! **55.** $x^2 + 100 = 0$

Chapter Test

Factor completely. If a polynomial is prime, state this.

1. $x^2 - 10x + 25$

2. $y^3 + 5y^2 - 4y - 20$

3. $p^2 - 12p - 28$

4. $t^7 - 3t^5$

5. $12m^2 + 20m + 3$

6. $9y^2 - 25$

7. $3r^3 - 3$

8. $45x^2 + 20 + 60x$

9. $3x^4 - 48y^4$

10. $y^2 + 8y + 16 - 100t^2$

11. $x^2 + 3x + 6$

12. $20a^2 - 5b^2$

13. $24x^2 - 46x + 10$

14. $3m^2 - 9mn - 30n^2$

15. The graph below is that of the polynomial function $p(x)$. Use the graph to determine the zeros of p.

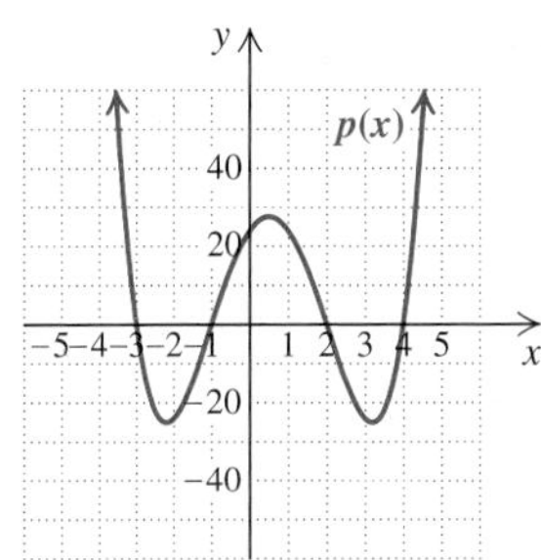

16. Find the zeros of the function given by $f(x) = 2x^2 - 11x - 40$.

Solve. Where appropriate, round solutions to the nearest thousandth.

17. $x^2 - 3x - 18 = 0$

18. $5t^2 = 125$

19. $2x^2 + 21 = -17x$

20. $9x^2 + 3x = 0$

21. $x^2 + 81 = 18x$

22. $x^2(x + 1) = 8x$

23. Let $f(x) = 3x^2 - 15x + 11$. Find a such that $f(a) = 11$.

24. Find the domain of the function f given by

$$f(x) = \frac{3 - x}{x^2 + 2x + 1}.$$

25. A photograph is 3 cm longer than it is wide. Its area is 40 cm^2. Find its length and its width.

26. To celebrate a town's centennial, fireworks are launched over a lake off a dam 36 ft above the water. The height of a display, t seconds after it has been launched, is given by

$$h(t) = -16t^2 + 64t + 36.$$

After how long will the shell from the fireworks reach the water?

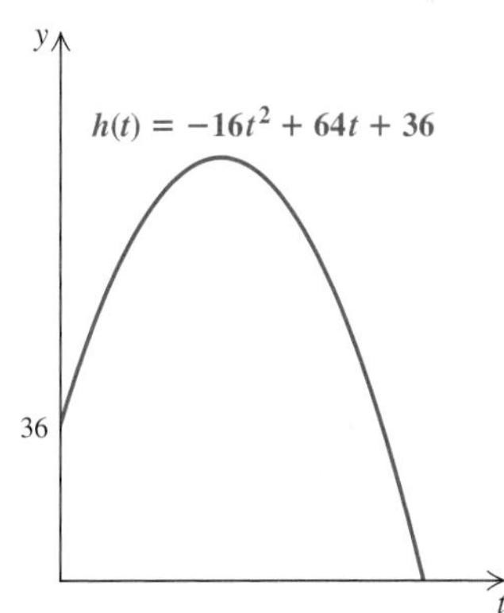

27. *Ladder Location.* The foot of an extension ladder is 10 ft from a wall. The ladder is 2 ft longer than the height that it reaches on the wall. How far up the wall does the ladder reach?

28. *IRS Revenue.* The table below lists the amount of IRS enforcement revenue, in billions of dollars, for various years.

Year	Years After 2000, x	IRS Enforcement Revenue (in billions)
2001	1	$33.8
2003	3	37.6
2005	5	47.3
2007	7	59.2

Source: irs.gov

a) Use regression to find a quadratic polynomial function $E(x)$ that can be used to estimate the IRS enforcement revenue x years after 2000.
b) Estimate the enforcement revenue in 2009.
c) In what year will the enforcement revenue be $100 billion?

SYNTHESIS

29. Factor: $(a + 3)^2 - 2(a + 3) - 35$.

30. Solve: $20x(x + 2)(x - 1) = 5x^3 - 24x - 14x^2$.

Cumulative Review: Chapters 1–6

1. Evaluate $-x$ for $x = -8$. [1.6]

2. Evaluate $-(-x)$ for $x = -8$. [1.6]

3. Find $f(-2)$, for $f(x) = -0.1x^3 - 3x^2 + 10x + 0.5$. [3.8]

Simplify. Do not use negative exponents in the answer.

4. $-2 + (20 \div 4)^2 - 6 \cdot (-1)^3$ [1.8]

5. $(3x^2y^3)^{-2}$ [5.2]

6. $(3x^4 - 2x^2 + x - 7) + (5x^3 + 2x^2 - 3)$ [5.4]

7. $(a^2b - 2ab^2 + 3b^3) - (4a^2b - ab^2 + b^3)$ [5.7]

8. $\dfrac{3t^3s^{-1}}{12t^{-5}s}$ [5.2]

9. $\left(\dfrac{-2x^2y}{3z^4}\right)^3$ [5.1]

Multiply.

10. $-4t^5(t^3 - 2t - 5)$ [5.5]

11. $(6x - 5y)^2$ [5.7]

12. $(10x^5 + 1)(10x^5 - 1)$ [5.6]

13. $(x - 1)(x^2 - x - 1)$ [5.5]

Divide. [5.8]

14. $\dfrac{15x^4 - 12x^3 + 6x^2 + 2x + 18}{3x^2}$

15. $(x^4 + 2x^3 + 6x^2 + 2x + 18) \div (x + 3)$

Factor completely.

16. $c^2 - 1$ [6.4]

17. $5x + 5y + 10x^2 + 10xy$ [6.1]

18. $6x - 2x^2 - 24x^4$ [6.1]

19. $16x^2 - 81$ [6.4]

20. $t^2 - 10t + 24$ [6.2]

21. $8x^2 + 10x + 3$ [6.3]

22. $6x^2 - 28x + 16$ [6.3]

23. $2x^3 + 250$ [6.5]

24. $16x^2 + 40x + 25$ [6.4]

Solve.

25. $5(x - 2) = 40$ [2.2]

26. $-4x < -18$ [2.6]

27. $(x - 1)(x + 3) = 0$ [6.1]

28. $x^2 + 10 = 11x$ [6.2]

29. $\frac{1}{3}x - \frac{2}{9} = \frac{2}{3} + \frac{4}{9}x$ [2.2]

30. $3x + 2y = 5,$
$x - 3y = 9$ [4.2]

31. $2x^2 + 7x = 4$ [6.3]

32. $4(x + 7) < 5(x - 3)$ [2.6]

33. $2y = 4x - 3,$
$4x = 1 - y$ [4.3]

34. Solve $a = bc + dc$ for c. [2.3]

Combine like terms.

35. $x + 2y - 2z + \frac{1}{2}x - z$ [1.6]

36. $2x^3 - 7 + \frac{3}{7}x^2 - 6x^3 - \frac{4}{7}x^2 + 5$ [1.6]

Graph by hand on a plane.

37. $y = 1 - \frac{1}{2}x$ [3.6]

38. $x = -3$ [3.3]

39. $x - 6y = 6$ [3.3]

40. $y = 6$ [3.3]

41. Graph using a graphing calculator: $y = x^2 - 4$. [3.2]

42. Find the slope of the line containing the points $(1, 5)$ and $(2, 3)$. [3.5]

43. Determine the slope and the y-intercept of the graph of $3x + 4y = 8$. [3.6]

44. Find an equation for the line with slope $\frac{1}{2}$ that contains $(0, -7)$. [3.6]

45. Find an equation for the line that is perpendicular to $y = x + 3$ and has the same y-intercept. [3.6]

Solve.

46. From June 2007 to June 2008, the number of U.S. wireless subscribers with 3G devices grew 80% to 64.2 million users. How many U.S. subscribers had a 3G device in June 2007? [2.4]
Source: CTIA—The Wireless Association

47. Emily and Andrew co-wrote a book of advice for new business owners. They agreed that Emily should get twice as much in royalty income as Andrew. If royalties for the first year are \$1260, how much should each person receive? [2.5]

48. The number of people in the United States, in thousands, who are on a waiting list for an organ transplant can be approximated by the polynomial $2.38t + 77.38$,

where t is the number of years since 2000. Estimate the number of people on a waiting list for an organ transplant in 2010. [5.3]
Source: Based on information from The Organ Procurement and Transplantation Network

49. Wilt Chamberlain once scored 100 points, setting a record for points scored in an NBA game. Chamberlain took only two-point shots and (one-point) foul shots and made a total of 64 shots. How many shots of each type did he make? [4.4]

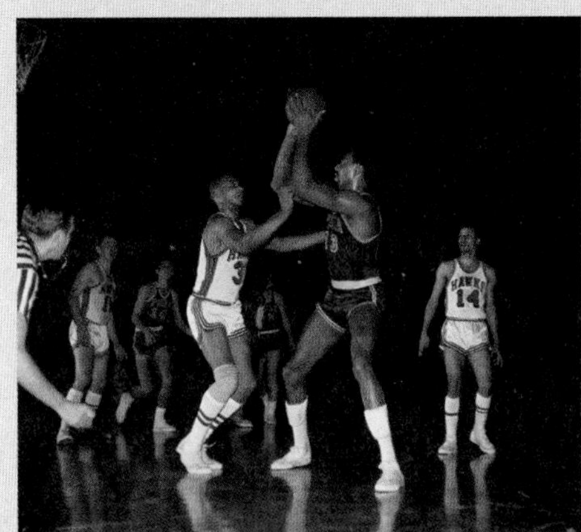

50. A 13-ft ladder is placed against a building in such a way that the distance from the top of the ladder to the ground is 7 ft more than the distance from the bottom of the ladder to the building. Find both distances. [6.7]

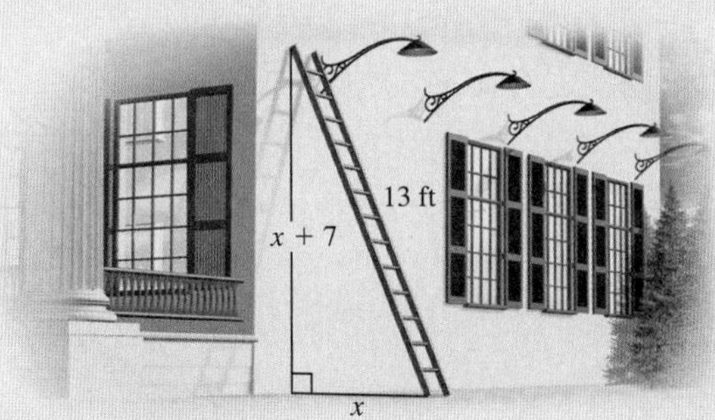

51. A rectangular table in Arlo's House of Tunes is six times as long as it is wide. If the area of the table is 24 ft^2, find the length and the width of the table. [6.7]

52. A large truck traveling 55 mph gets 7.6 miles per gallon of fuel. The same truck traveling 70 mph gets 6.1 miles per gallon.
Source: Kenworth Truck Co.

a) Graph the data and determine an equation that describes the miles per gallon y in terms of the speed x. [3.7]
b) Use the equation to estimate the miles per gallon for a truck traveling 60 mph. [3.7]

Determine whether a linear function could be used to model each set of data. [3.7]

53. **Student Volunteers**

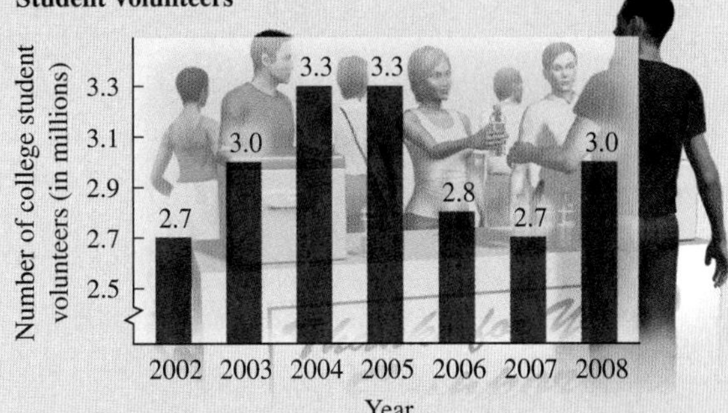

Source: Volunteering in America, Corporation for National and Community Service

54. **College Students**

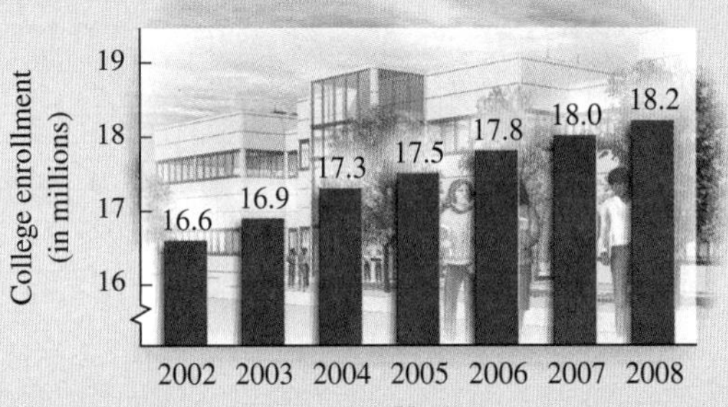

Source: U.S. National Education Statistics, Digest of Education Statistics and Projections of Education Statistics, anual

55. One of the sets of data in Exercises 53 and 54 is linear. Use linear regression to find a function that can be used to model the data. Round coefficients to four decimal places. [3.7]

Estimate the domain and the range of each function from its graph. [3.8]

56.
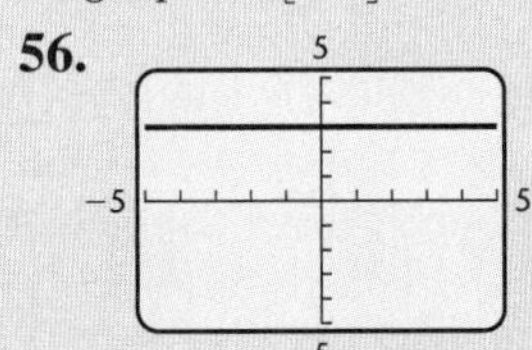

57.
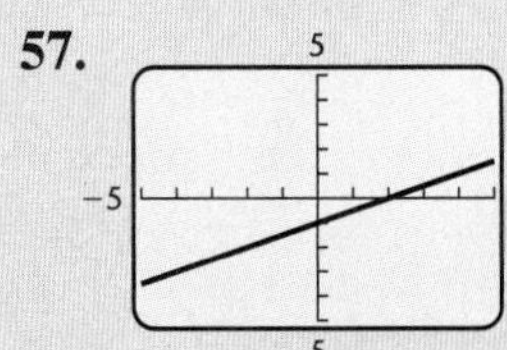

58.
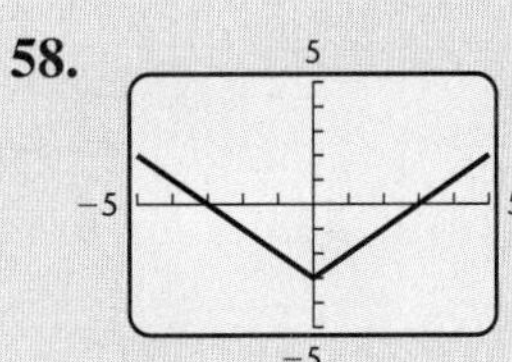

59.
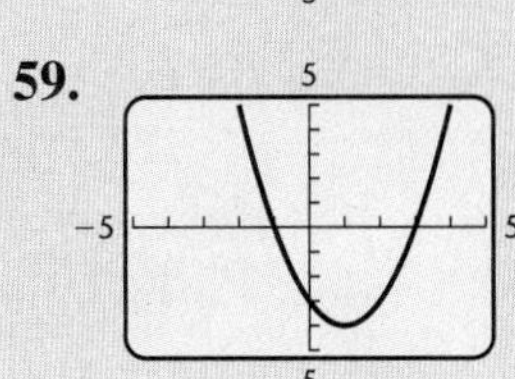

SYNTHESIS

60. Simplify: $(x + 7)(x - 4) - (x + 8)(x - 5)$. [5.4], [5.5]

61. Factor: $2a^{32} - 13{,}122b^{40}$. [6.4]

62. Simplify: $-\left|0.875 - \left(-\frac{1}{8}\right) - 8\right|$. [1.4], [1.6]

63. Find all zeros of $f(x) = x^4 - 34x^2 + 225$. [6.2], [6.4]

7

Rational Expressions, Equations, and Functions

How Long Does a Heart Last?

In general, animals with higher heart rates have shorter life spans, as shown by the graph. In Example 10 of Section 7.8, we use the given data to estimate the average life span of a cat.

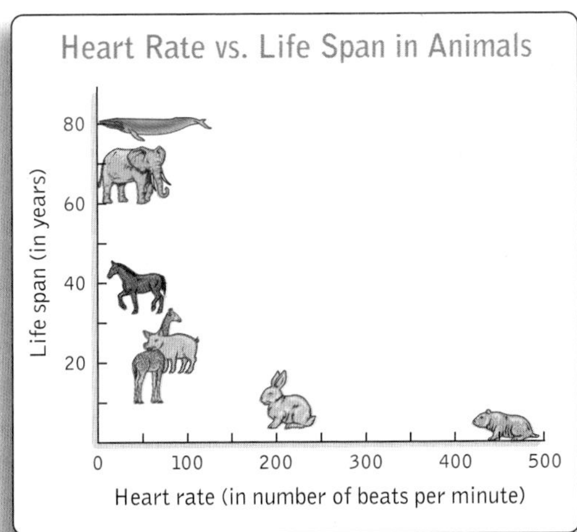

Like fractions in arithmetic, a rational expression is a ratio of two expressions. We now learn to simplify, add, subtract, multiply, and divide rational expressions. We then use these skills to solve equations and applications involving rational expressions.

7.1 Rational Expressions and Functions

- Rational Functions
- Simplifying Rational Expressions
- Factors That Are Opposites
- Vertical Asymptotes

A **rational expression** consists of a polynomial divided by a nonzero polynomial. The following are examples of rational expressions:

$$\frac{3}{4}, \quad \frac{x}{y}, \quad \frac{9}{a+b}, \quad \frac{x^2+7xy-4}{x^3-y^3}.$$

RATIONAL FUNCTIONS

Functions described by rational expressions are called **rational functions.** Following is the graph of the rational function

$$f(x) = \frac{1}{x}.$$

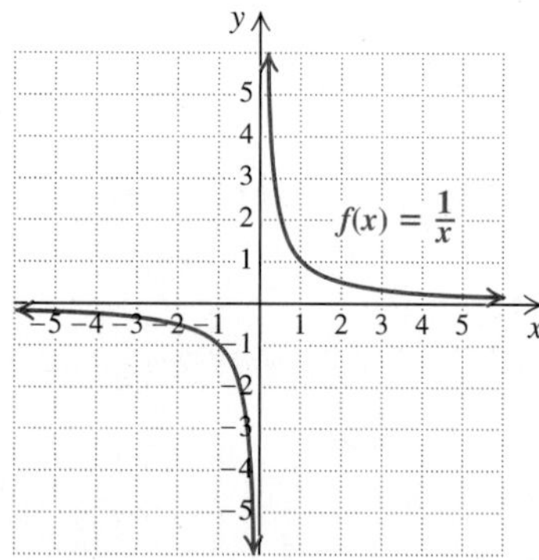

STUDY TIP

Try an Exercise Break

Often the best way to regain your energy or focus is to take a break from your studies in order to exercise. Jogging, biking, or walking briskly are just a few of the activities that can improve your concentration.

Graphs of rational functions vary widely in shape, but some general statements can be drawn from the graph shown above.

1. Graphs of rational functions may not be continuous—that is, there may be a break in the graph. The graph of $f(x) = 1/x$ consists of two unconnected parts, with a break at $x = 0$.
2. The domain of a rational function may not include all real numbers. The domain of $f(x) = 1/x$ does not include 0.

Although we will not consider graphs of rational functions in detail in this course, you should add the graph of $f(x) = 1/x$ to your library of functions.

EXAMPLE 1 The function given by

$$H(t) = \frac{t^2 + 5t}{2t + 5}$$

gives the time, in hours, for two machines, working together, to complete a job that the first machine could do alone in t hours and the other machine could do in $t + 5$ hours. How long will the two machines, working together, require for the job if the first machine alone would take 1 hour?

SOLUTION We find $H(1)$:

$$H(1) = \frac{1^2 + 5 \cdot 1}{2 \cdot 1 + 5} = \frac{1 + 5}{2 + 5} = \frac{6}{7} \text{ hr.}$$

Try Exercise 7.

$y_1 = (x^2 + 5x) / (2x + 5)$

X	Y1	
−2.5	ERROR	

The domain of a rational function must exclude any numbers for which the denominator is 0. For a function like H above, the denominator is 0 when t is $-\frac{5}{2}$, so the domain of H is $\{t \mid t \text{ is a real number } and\ t \neq -\frac{5}{2}\}$. Note from the table at left that H is not defined when $t = -\frac{5}{2}$.

EXAMPLE 2 Find all numbers for which the rational expression

$$\frac{x + 4}{x^2 - 3x - 10}$$

is undefined.

SOLUTION The value of the numerator has no bearing on whether or not a rational expression is defined. To determine which numbers make the rational expression undefined, we set the *denominator* equal to 0 and solve:

$$x^2 - 3x - 10 = 0 \quad \textbf{Setting the denominator equal to 0}$$

$$(x - 5)(x + 2) = 0 \quad \textbf{Factoring}$$

$$x - 5 = 0 \quad or \quad x + 2 = 0 \quad \textbf{Using the principle of zero products}$$

$$x = 5 \quad or \quad x = -2. \quad \textbf{Solving each equation}$$

$y_1 = (x + 4)/(x^2 - 3x - 10)$

X	Y1	
5	ERROR	
−2	ERROR	

To check, we let $y_1 = (x + 4)/(x^2 - 3x - 10)$ and evaluate y_1 for $x = 5$ and for $x = -2$. The table at left confirms that the expression is undefined when $x = 5$ and when $x = -2$.

Try Exercise 13.

In Example 2, we found that the expression

$$\frac{x + 4}{x^2 - 3x - 10}$$

is undefined when $x = 5$ and when $x = -2$. The expression *is* defined for all other values of x. This tells us that the domain of the function given by

$$f(x) = \frac{x + 4}{x^2 - 3x - 10}$$

is $\{x \mid x \text{ is a real number } and\ x \neq -2\ and\ x \neq 5\}$.

SIMPLIFYING RATIONAL EXPRESSIONS

Simplifying rational expressions is similar to simplifying the fraction expressions studied in Section 1.3. We saw, for example, that an expression like $\frac{15}{40}$ can be simplified as follows:

$$\frac{15}{40} = \frac{3 \cdot 5}{8 \cdot 5}$$ **Factoring the numerator and the denominator. Note the common factor, 5.**

$$= \frac{3}{8} \cdot \frac{5}{5}$$ **Rewriting as a product of two fractions**

$$= \frac{3}{8} \cdot 1$$ $\frac{5}{5} = 1$

$$= \frac{3}{8}.$$ **Using the identity property of 1 to remove the factor 1**

Similar steps are followed when simplifying rational expressions: We factor and remove a factor equal to 1, using the fact that

$$\frac{ab}{cb} = \frac{a}{c} \cdot \frac{b}{b}.$$

Student Notes

When using a graphing calculator or tutorial software, use parentheses around the numerator and around the denominator of rational expressions. Note that

$5/3x$ means $\frac{5}{3}x$

and

$5/(3x)$ means $\frac{5}{3x}$.

Some calculators have an n/d option in the MATH NUM menu that will allow you to enter a rational expression without using parentheses around the numerator and around the denominator.

EXAMPLE 3 Simplify: $\frac{8x^2}{24x}$.

SOLUTION

$$\frac{8x^2}{24x} = \frac{8 \cdot x \cdot x}{3 \cdot 8 \cdot x}$$ **Factoring the numerator and the denominator. Note the common factor of $8 \cdot x$.**

$$= \frac{x}{3} \cdot \frac{8x}{8x}$$ **Rewriting as a product of two rational expressions**

$$= \frac{x}{3} \cdot 1$$ $\frac{8x}{8x} = 1$

$$= \frac{x}{3}$$ **Removing the factor 1**

Try Exercise 23.

EXAMPLE 4 Simplify: $\frac{7a^2 + 21a}{14a}$.

SOLUTION We have

$$\frac{7a^2 + 21a}{14a} = \frac{7a(a + 3)}{7 \cdot 2 \cdot a}$$ **Factoring the numerator and the denominator. The greatest common factor is $7a$.**

$$= \frac{7a}{7a} \cdot \frac{a + 3}{2}$$ **Rewriting as a product of two rational expressions**

$$= 1 \cdot \frac{a + 3}{2}$$ $\frac{7a}{7a} = 1$; **try to do this step mentally.**

$$= \frac{a + 3}{2}.$$ **Removing the factor 1**

X	Y1	Y2
−3	0	0
−2	.5	.5
−1	1	1
0	ERROR	1.5
1	2	2
2	2.5	2.5
3	3	3

X = −3

As a partial check, we let $y_1 = (7x^2 + 21x)/(14x)$ and $y_2 = (x + 3)/2$ and compare values in a table. Note that the y-values are the same for any given x-value except 0. We exclude 0 because although $(x + 3)/2$ is defined when $x = 0$, $(7x^2 + 21x)/(14x)$ is not. Both expressions represent the same value when x is replaced with a number that can be used in *either* expression, so they are equivalent.

Try Exercise 27.

A rational expression is said to be **simplified** when no factors equal to 1 can be removed. If the largest common factor is not found, simplifying may require two or more steps. For example, suppose we remove 7/7 instead of $(7a)/(7a)$ in Example 4. We would then have

$$\left.\begin{aligned}\frac{7a^2 + 21a}{14a} &= \frac{7(a^2 + 3a)}{7 \cdot 2a}\\ &= \frac{a^2 + 3a}{2a}.\end{aligned}\right\}$$ **Removing a factor equal to 1: $\frac{7}{7} = 1$**

Here, since 7 is not the *greatest* common factor, we need to simplify further:

$$\left.\begin{aligned}\frac{a^2 + 3a}{2a} &= \frac{a(a + 3)}{a \cdot 2}\\ &= \frac{a + 3}{2}.\end{aligned}\right\}$$ **Removing another factor equal to 1: $a/a = 1$. The rational expression is now simplified.**

Sometimes the common factor has two or more terms.

EXAMPLE 5 Simplify.

a) $\dfrac{6x + 12}{7x + 14}$

b) $\dfrac{x^2 - 1}{x^2 + 3x + 2}$

SOLUTION

a) $\dfrac{6x + 12}{7x + 14} = \dfrac{6(x + 2)}{7(x + 2)}$ **Factoring the numerator and the denominator. The common factor is $x + 2$.**

$= \dfrac{6}{7} \cdot \dfrac{x + 2}{x + 2}$ **Rewriting as a product of two rational expressions**

$= \dfrac{6}{7} \cdot 1$ **$\dfrac{x + 2}{x + 2} = 1$**

$= \dfrac{6}{7}$ **Removing the factor 1**

b) $\dfrac{x^2 - 1}{x^2 + 3x + 2} = \dfrac{(x + 1)(x - 1)}{(x + 1)(x + 2)}$ **Factoring; $x + 1$ is the common factor.**

$= \dfrac{x + 1}{x + 1} \cdot \dfrac{x - 1}{x + 2}$ **Rewriting as a product of two rational expressions**

$= 1 \cdot \dfrac{x - 1}{x + 2}$ **$\dfrac{x + 1}{x + 1} = 1$**

$= \dfrac{x - 1}{x + 2}$ **Removing the factor 1**

Try Exercise 37.

Canceling

Canceling is a shortcut that can be used—and easily *misused*—to simplify rational expressions. Thus canceling must be done with care and understanding. Essentially, canceling streamlines the steps in which we remove a factor equal to 1. Example 5(b) could have been streamlined as follows:

$$\frac{x^2 - 1}{x^2 + 3x + 2} = \frac{\cancel{(x + 1)}(x - 1)}{\cancel{(x + 1)}(x + 2)}$$

When a factor equal to 1 is noted, it is "canceled": $\frac{x+1}{x+1} = 1$.

$$= \frac{x - 1}{x + 2}.$$

Simplifying

CAUTION! Canceling is often performed incorrectly:

$$\cancel{\frac{\cancel{x} + 3}{\cancel{x}} = 3}, \qquad \cancel{\frac{\cancel{4}x + 3}{\cancel{2}} = 2x + 3}, \qquad \cancel{\frac{\cancel{5}}{\cancel{5} + x} = \frac{1}{x}}.$$

Incorrect! Incorrect! Incorrect!

To check that these are not equivalent, substitute a number for x or compare graphs.

None of the above cancellations removes a factor equal to 1. Factors are parts of products. For example, in $x \cdot 3$, x and 3 are factors, but in $x + 3$, x and 3 are terms, *not factors*. Only factors can be canceled.

EXAMPLE 6 Simplify: $\dfrac{3x^2 - 2x - 1}{x^2 - 3x + 2}$.

SOLUTION We factor the numerator and the denominator and look for common factors:

$$\frac{3x^2 - 2x - 1}{x^2 - 3x + 2} = \frac{(3x + 1)\cancel{(x - 1)}}{(x - 2)\cancel{(x - 1)}}$$

Try to visualize this as $\frac{3x+1}{x-2} \cdot \frac{x-1}{x-1}$.

$$= \frac{3x + 1}{x - 2}.$$

Removing a factor equal to 1: $\frac{x-1}{x-1} = 1$

Try Exercise 35.

In Example 4, we found that

$$\frac{7x^2 + 21x}{14x} = \frac{x + 3}{2}.$$

These expressions are equivalent; their values are equal for every value of x except $x = 0$, for which the expression on the left is not defined. However, if

$$f(x) = \frac{7x^2 + 21x}{14x} \quad \text{and} \quad g(x) = \frac{x + 3}{2},$$

we cannot say that $f = g$, because the domains of the functions are not the same. When we simplify a rational expression that defines a function, we must carefully specify the domain.

EXAMPLE 7 Consider the function $f(x) = \dfrac{x^2 - 4}{2x^2 - 3x - 2}$.

a) Find the domain of f.

b) Simplify the rational expression.

c) Write the simplified form of the function, specifying any restrictions on the domain.

SOLUTION

a) To find the domain of f, we set the denominator equal to 0 and solve:

$$2x^2 - 3x - 2 = 0$$

$$(2x + 1)(x - 2) = 0 \qquad \text{Factoring the denominator}$$

$$2x + 1 = 0 \quad \text{or} \quad x - 2 = 0 \qquad \text{Using the principle of zero products}$$

$$x = -\tfrac{1}{2} \quad \text{or} \quad x = 2.$$

The numbers $-\frac{1}{2}$ and 2 are not in the domain of the function. Thus the domain of f is $\left\{x \mid x \text{ is a real number } and\ x \neq -\frac{1}{2}\ and\ x \neq 2\right\}$.

b) To simplify the expression, we begin by factoring:

$$\frac{x^2 - 4}{2x^2 - 3x - 2} = \frac{(x - 2)(x + 2)}{(2x + 1)(x - 2)} \qquad \text{Factoring the numerator and the denominator}$$

$$= \frac{x - 2}{x - 2} \cdot \frac{x + 2}{2x + 1} \qquad \text{Factoring the rational expression; } \frac{x - 2}{x - 2} = 1$$

$$= \frac{x + 2}{2x + 1}. \qquad \text{Removing a factor equal to 1}$$

The simplified form of the expression is $\dfrac{x + 2}{2x + 1}$.

c) When writing the function in simplified form, we list the numbers that are not in the domain of f:

$$f(x) = \frac{x + 2}{2x + 1}, \quad x \neq -\frac{1}{2}, \quad x \neq 2.$$

Try Exercise 63.

FACTORS THAT ARE OPPOSITES

Consider

$$\frac{x - 4}{8 - 2x}, \quad \text{or, equivalently,} \quad \frac{x - 4}{2(4 - x)}.$$

At first glance, the numerator and the denominator do not appear to have any common factors. But $x - 4$ and $4 - x$ are opposites, or additive inverses, of each other. Thus we can find a common factor by factoring out -1 in one expression.

EXAMPLE 8 Simplify $\dfrac{x-4}{8-2x}$ and check by evaluating.

SOLUTION We have

$$\frac{x-4}{8-2x} = \frac{x-4}{2(4-x)} \qquad \text{Factoring}$$

$$= \frac{x-4}{2(-1)(x-4)} \qquad \text{Note that } 4 - x = -(x-4).$$

$$= \frac{x-4}{-2(x-4)} \qquad \text{Had we originally factored out } -2, \text{ we could have gone directly to this step.}$$

$$= \frac{1}{-2}\cdot\frac{x-4}{x-4} = -\frac{1}{2}. \qquad \text{Removing a factor equal to 1: } (x-4)/(x-4) = 1$$

As a partial check, note that for any choice of x other than 4, the value of the rational expression is $-\frac{1}{2}$. For example, if $x = 6$, then

$$\frac{x-4}{8-2x} = \frac{6-4}{8-2\cdot 6} = \frac{2}{-4} = -\frac{1}{2}.$$

Try Exercise 47.

VERTICAL ASYMPTOTES

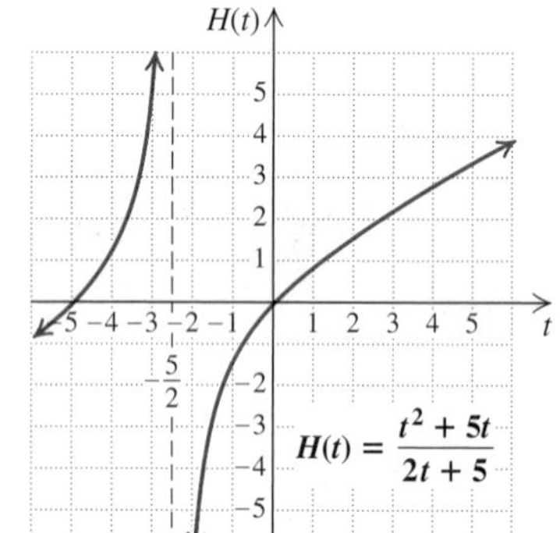

Consider the graph of the function in Example 1,

$$H(t) = \frac{t^2 + 5t}{2t + 5}.$$

Note that the graph consists of two unconnected "branches." Since $-\frac{5}{2}$ is not in the domain of H, there is no point on the graph for $t = -\frac{5}{2}$.

The line $t = -\frac{5}{2}$ is called a **vertical asymptote.** As t gets closer to, or *approaches*, $-\frac{5}{2}$, $H(t)$ approaches the asymptote.

Now consider the same graph drawn using a graphing calculator. The vertical line that appears on the screen on the left below is not part of the graph, nor should it be considered a vertical asymptote. Since a graphing calculator graphs an equation by plotting points and connecting them, the vertical line is actually connecting a point just to the left of the line $x = -\frac{5}{2}$ with a point just to the right of the line $x = -\frac{5}{2}$. Such a vertical line may or may not appear for values not in the domain of a function.

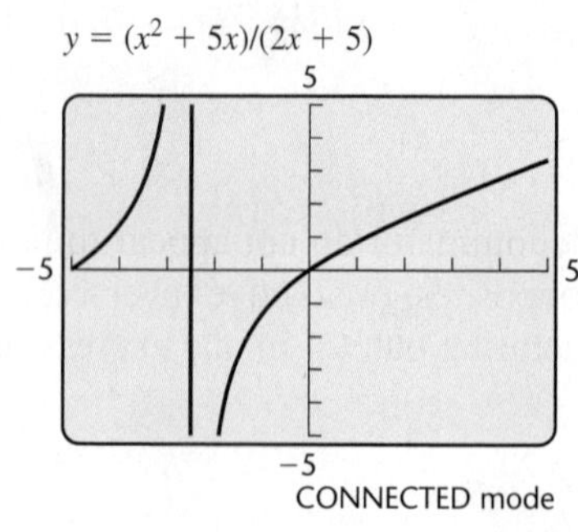

CONNECTED mode

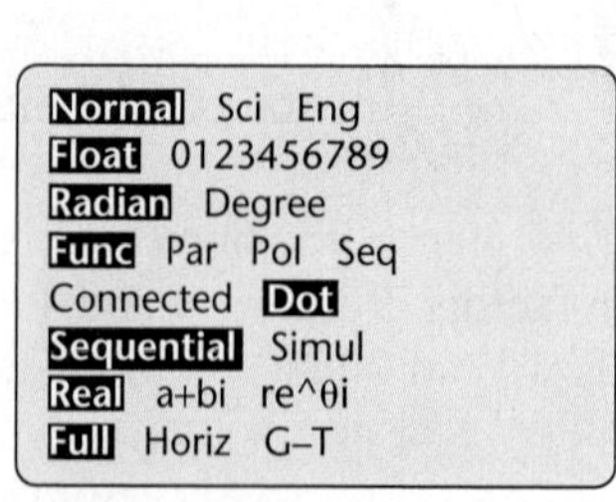

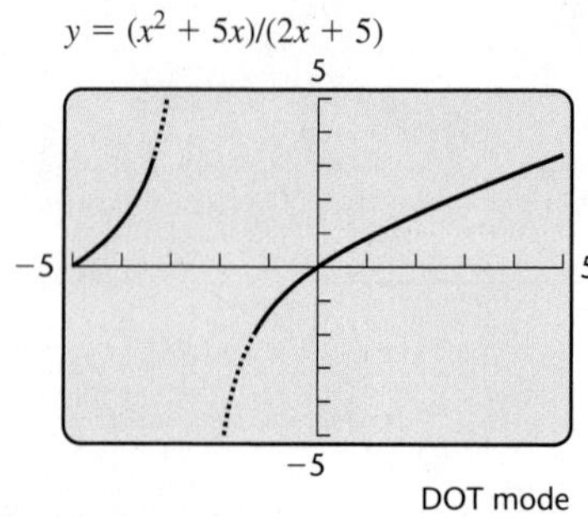

DOT mode

If we tell the calculator not to connect the points that it plots, the vertical line will not appear. Pressing MODE and changing from CONNECTED to DOT gives us the graph on the right above.

Not every number excluded from the domain of a function corresponds to a vertical asymptote.

Interactive Discovery

Let

$$f(x) = \frac{x^2 - 4}{2x^2 - 3x - 2} \quad \text{and} \quad g(x) = \frac{x + 2}{2x + 1}.$$

As we saw in Example 7, the rational expressions describing these functions are equivalent but the domains of the functions are different.

1. Graph $f(x)$. How many vertical asymptotes does the graph appear to have? What is the vertical asymptote?
2. Graph $g(x)$ and determine the vertical asymptote.

As you may have discovered in the Interactive Discovery, the graph of a rational function has the same asymptotes as the graph of the equivalent function in simplified form.

If a function $f(x)$ is described by a *simplified* rational expression, and a is a number that makes the denominator 0, then $x = a$ is a vertical asymptote of the graph of $f(x)$.

EXAMPLE 9 Determine the vertical asymptotes of the graph of

$$f(x) = \frac{9x^2 + 6x - 3}{12x^2 - 12}.$$

SOLUTION We first simplify the rational expression describing the function:

$$\frac{9x^2 + 6x - 3}{12x^2 - 12} = \frac{3(x + 1)(3x - 1)}{3 \cdot 4(x + 1)(x - 1)} \quad \textbf{Factoring the numerator and the denominator}$$

$$= \frac{3(x + 1)}{3(x + 1)} \cdot \frac{3x - 1}{4(x - 1)} \quad \textbf{Factoring the rational expression}$$

$$= \frac{3x - 1}{4(x - 1)}. \quad \textbf{Removing a factor equal to 1}$$

The denominator of the simplified expression, $4(x - 1)$, is 0 when $x = 1$. Thus, $x = 1$ is a vertical asymptote of the graph. Although the domain of the function also excludes -1, there is no asymptote at $x = -1$, only a "hole." This hole should be indicated on a hand-drawn graph but it may not be obvious on a graphing-calculator screen. The vertical asymptote is $x = 1$.

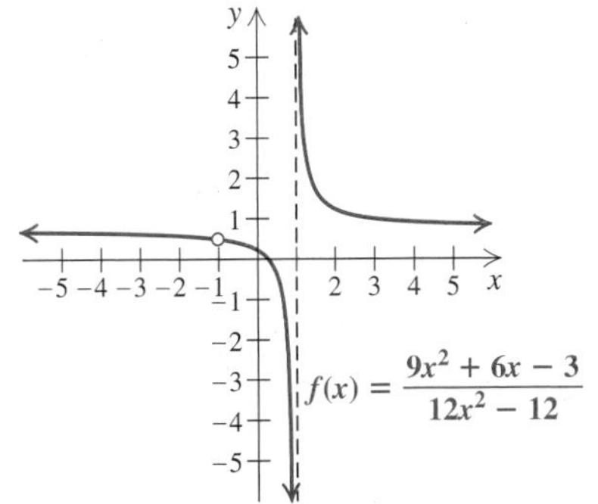

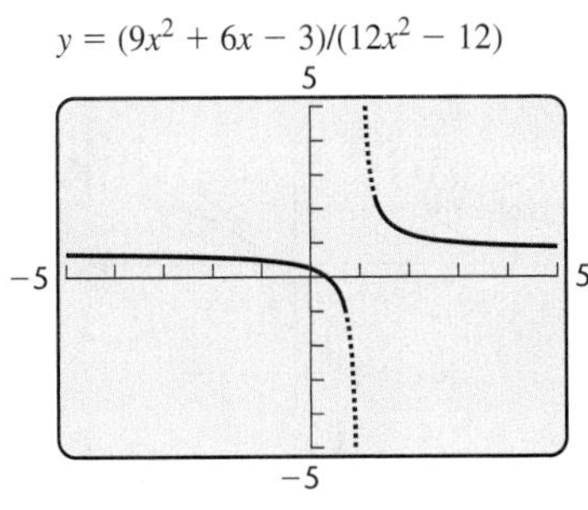

Try Exercise 75.

Visualizing for Success

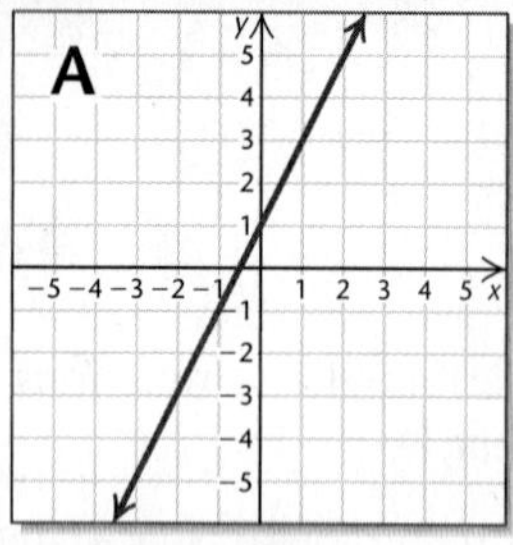

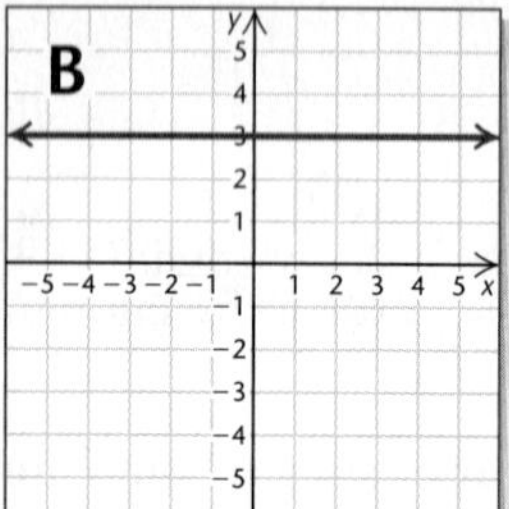

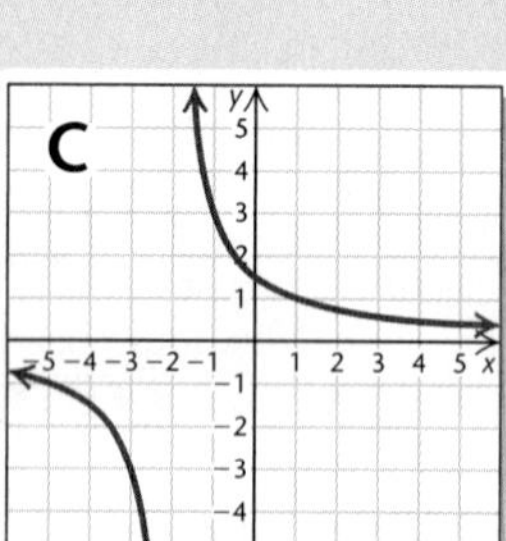

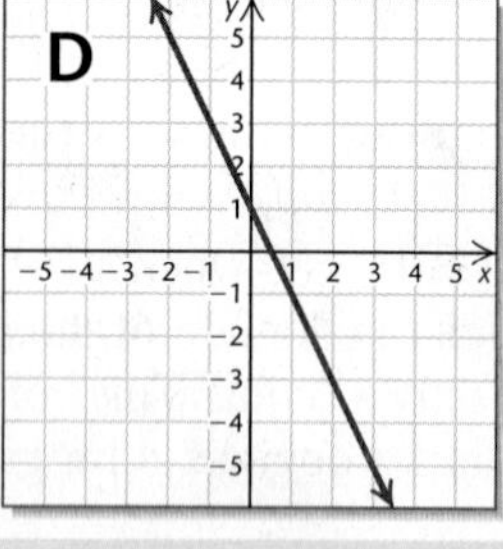

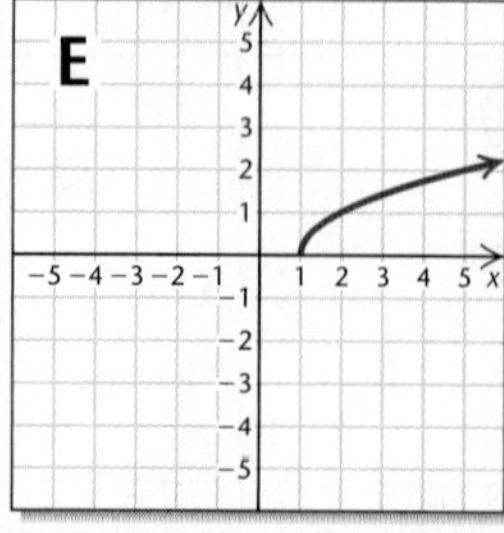

Match each function with its graph.

1. $f(x) = x^2 - 3$

2. $f(x) = 3$

3. $f(x) = \dfrac{1}{x}$

4. $f(x) = 2x + 1$

5. $f(x) = 1 - 2x$

6. $f(x) = \dfrac{3}{x + 2}$

7. $f(x) = \sqrt{x - 1}$

8. $f(x) = \dfrac{2}{x - 1}$

9. $f(x) = x - 1$

10. $f(x) = |x - 1|$

Answers on page A-25

An additional, animated version of this activity appears in MyMathLab. To use MyMathLab, you need a course ID and a student access code. Contact your instructor for more information.

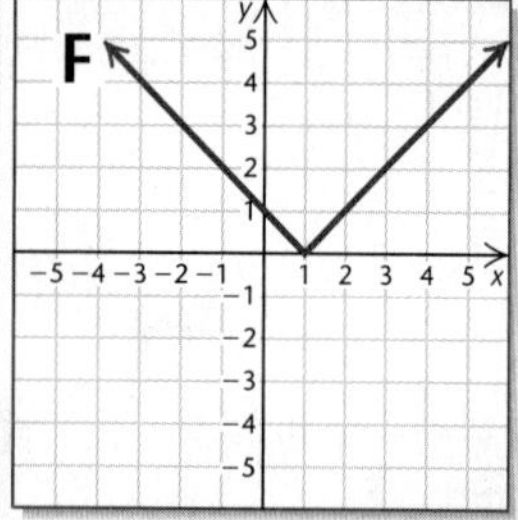

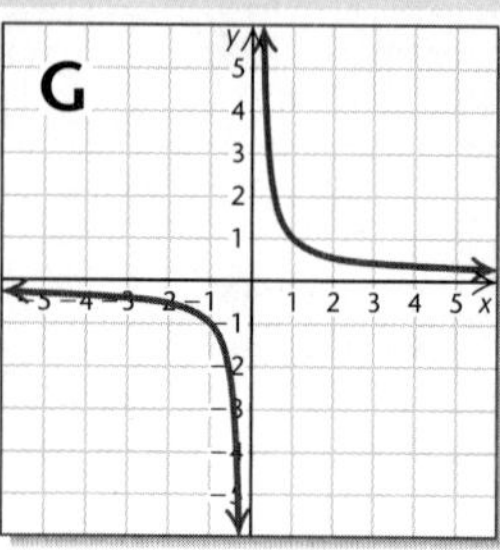

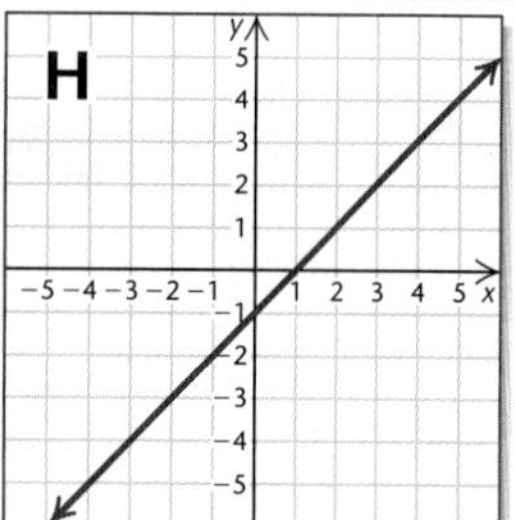

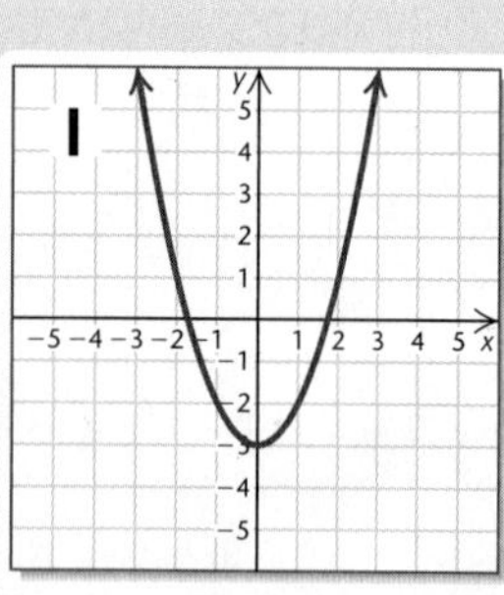

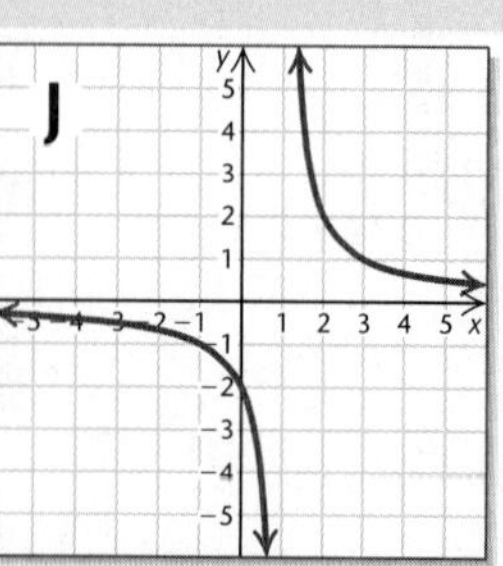

7.1 Exercise Set

FOR EXTRA HELP MyMathLab Math XL PRACTICE  WATCH DOWNLOAD  READ REVIEW

Concept Reinforcement In each of Exercises 1–6, match the function described with the appropriate domain from the column on the right. Some choices of domain will not be used.

1. ___ $f(x) = \dfrac{2 - x}{x - 5}$

2. ___ $f(x) = \dfrac{x + 5}{x + 2}$

3. ___ $g(x) = \dfrac{x - 3}{(x - 2)(x - 5)}$

4. ___ $g(x) = \dfrac{x + 3}{(x + 2)(x - 5)}$

5. ___ $h(x) = \dfrac{(x - 2)(x - 3)}{x + 3}$

6. ___ $f(x) = \dfrac{(x + 2)(x + 3)}{x - 3}$

a) $\{x | x \neq -5, x \neq 2\}$
b) $\{x | x \neq 3\}$
c) $\{x | x \neq -2\}$
d) $\{x | x \neq -3\}$
e) $\{x | x \neq 5\}$
f) $\{x | x \neq -2, x \neq 5\}$
g) $\{x | x \neq 2\}$
h) $\{x | x \neq -2, x \neq -5\}$
i) $\{x | x \neq 2, x \neq 5\}$
j) $\{x | x \neq -5\}$

Processing Orders. *Jasmine usually takes 3 hr more than Molly does to process a day's orders at Books To Go. If Molly takes t hr to process a day's orders, the function given by*

$$H(t) = \frac{t^2 + 3t}{2t + 3}$$

can be used to determine how long it would take if they worked together.

7. How long will it take them, working together, to complete a day's orders if Molly can process the orders alone in 5 hr?

8. How long will it take them, working together, to complete a day's orders if Molly can process the orders alone in 7 hr?

For each rational function, find the function values indicated, provided the value exists.

9. $v(t) = \dfrac{4t^2 - 5t + 2}{t + 3}$; $v(0), v(-2), v(7)$

10. $f(x) = \dfrac{5x^2 + 4x - 12}{6 - x}$; $f(0), f(-1), f(3)$

11. $g(x) = \dfrac{2x^3 - 9}{x^2 - 4x + 4}$; $g(0), g(2), g(-1)$

12. $r(t) = \dfrac{t^2 - 5t + 4}{t^2 - 9}$; $r(1), r(2), r(-3)$

List all numbers for which each rational expression is undefined.

13. $\dfrac{25}{-7x}$

14. $\dfrac{14}{-5y}$

15. $\dfrac{t - 3}{t + 8}$

16. $\dfrac{a - 8}{a + 7}$

17. $\dfrac{a}{3a - 12}$

18. $\dfrac{x^2}{4x - 12}$

19. $\dfrac{x^2 - 16}{x^2 - 3x - 28}$

20. $\dfrac{p^2 - 9}{p^2 - 7p + 10}$

21. $\dfrac{m^3 - 2m}{m^2 - 25}$

22. $\dfrac{7 - 3x + x^2}{49 - x^2}$

Simplify by removing a factor equal to 1.

23. $\dfrac{15x}{5x^2}$

24. $\dfrac{7a^3}{21a}$

25. $\dfrac{18t^3w^2}{27t^7w}$

26. $\dfrac{8y^5z}{4y^9z^3}$

27. $\frac{2a-10}{2}$

28. $\frac{3a+12}{3}$

29. $\frac{3x^2-12x}{3x^2+15x}$

30. $\frac{4y^2-20y}{4y^2+12y}$

31. $\frac{6a^2-3a}{7a^2-7a}$

32. $\frac{3m^2+3m}{6m^2+9m}$

33. $\frac{3a-1}{2-6a}$

34. $\frac{6-5a}{10a-12}$

35. $\frac{3a^2+9a-12}{6a^2-30a+24}$

36. $\frac{2t^2-6t+4}{4t^2+12t-16}$

37. $\frac{x^2+8x+16}{x^2-16}$

38. $\frac{x^2-25}{x^2-10x+25}$

39. $\frac{t^2-1}{t+1}$

40. $\frac{a^2-1}{a-1}$

41. $\frac{y^2+4}{y+2}$

42. $\frac{x^2+1}{x+1}$

43. $\frac{5x^2-20}{10x^2-40}$

44. $\frac{6x^2-54}{4x^2-36}$

45. $\frac{x-8}{8-x}$

46. $\frac{6-x}{x-6}$

47. $\frac{2t-1}{1-4t^2}$

48. $\frac{3a-2}{4-9a^2}$

49. $\frac{a^2-25}{a^2+10a+25}$

50. $\frac{a^2-16}{a^2-8a+16}$

Aha! 51. $\frac{7s^2-28t^2}{28t^2-7s^2}$

52. $\frac{9m^2-4n^2}{4n^2-9m^2}$

53. $\frac{x^3-1}{x^2-1}$

54. $\frac{a^3+8}{a^2-4}$

55. $\frac{3y^3+24}{y^2-2y+4}$

56. $\frac{x^3-27}{5x^2+15x+45}$

Write simplified form for each of the following. Be sure to list all restrictions on the domain, as in Example 7.

57. $f(x)=\frac{3x+21}{x^2+7x}$

58. $f(x)=\frac{5x+20}{x^2+4x}$

59. $g(x)=\frac{x^2-9}{5x+15}$

60. $g(x)=\frac{8x-16}{x^2-4}$

61. $h(x)=\frac{4-x}{5x-20}$

62. $h(x)=\frac{7-x}{3x-21}$

63. $f(t)=\frac{t^2-9}{t^2+4t+3}$

64. $f(t)=\frac{t^2-25}{t^2-6t+5}$

65. $g(t)=\frac{21-7t}{3t-9}$

66. $g(t)=\frac{12-6t}{5t-10}$

67. $h(t)=\frac{t^2+5t+4}{t^2-8t-9}$

68. $h(t)=\frac{t^2-3t-4}{t^2+9t+8}$

69. $f(x)=\frac{9x^2-4}{3x-2}$

70. $f(x)=\frac{4x^2-1}{2x-1}$

Determine the vertical asymptotes of the graph of each function.

71. $f(x)=\frac{3x-12}{3x+15}$

72. $f(x)=\frac{4x-20}{4x+12}$

73. $g(x)=\frac{12-6x}{5x-10}$

74. $r(x)=\frac{21-7x}{3x-9}$

75. $t(x)=\frac{x^3+3x^2}{x^2+6x+9}$

76. $g(x)=\frac{x^2-4}{2x^2-5x+2}$

77. $f(x)=\frac{x^2-x-6}{x^2-6x+8}$

78. $f(x)=\frac{x^2+2x+1}{x^2-2x+1}$

In Exercises 79–84, match each function with one of the following graphs.

a) 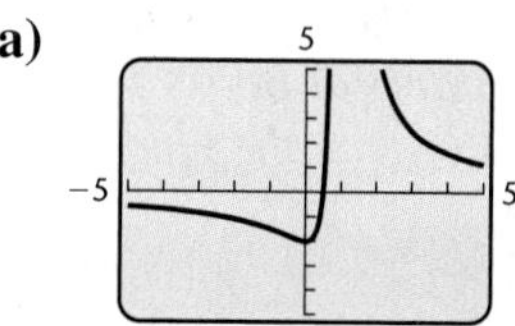

b) 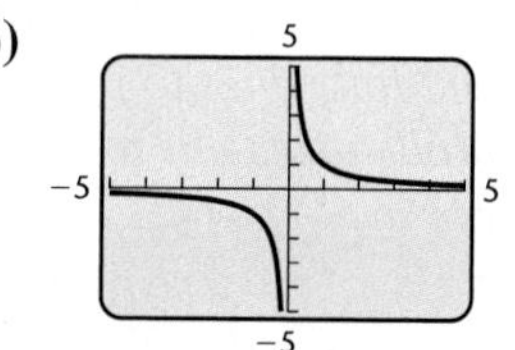

c) 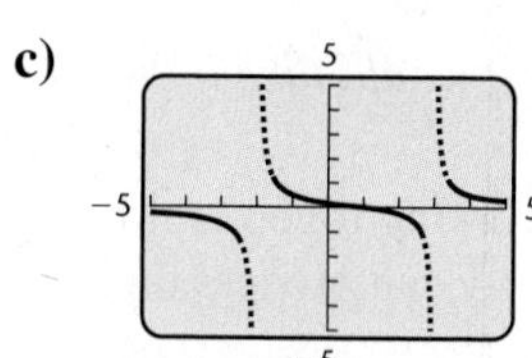

d) 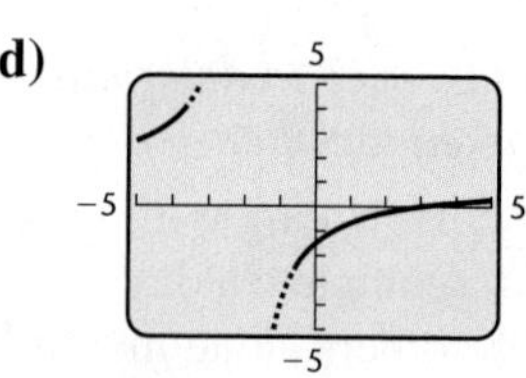

e) 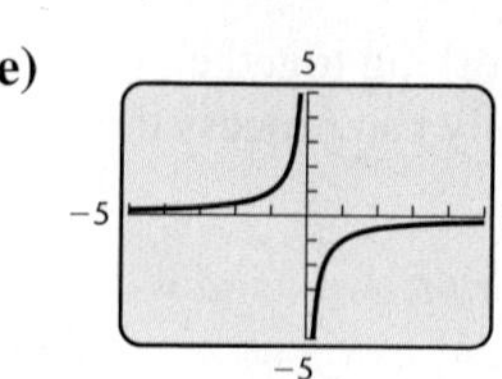

f) 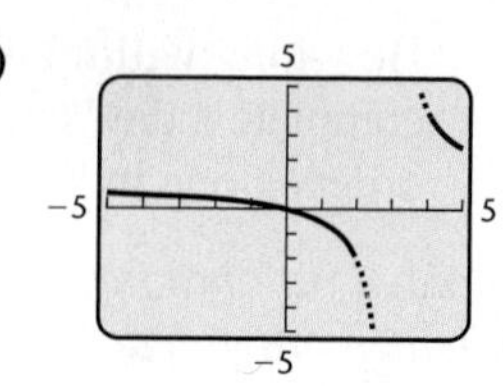

79. $h(x)=\frac{1}{x}$

80. $q(x)=-\frac{1}{x}$

81. $f(x)=\frac{x}{x-3}$

82. $g(x)=\frac{x-3}{x+2}$

83. $r(x)=\frac{4x-2}{x^2-2x+1}$

84. $t(x)=\frac{x-1}{x^2-x-6}$

TW **85.** Explain why the graphs of $f(x) = 5x$ and $g(x) = \frac{5x^2}{x}$ differ.

TW **86.** If a rational expression is undefined for $x = 5$ and $x = -3$, what is the degree of the denominator? Why?

SKILL REVIEW

To prepare for Section 7.2, review multiplication and division using fraction notation (Section 1.7).

Simplify.

87. $-\frac{2}{15} \cdot \frac{10}{7}$ [1.7]

88. $\left(\frac{3}{4}\right)\left(\frac{-20}{9}\right)$ [1.7]

89. $\frac{5}{8} \div \left(-\frac{1}{6}\right)$ [1.7]

90. $\frac{7}{10} \div \left(-\frac{8}{15}\right)$ [1.7]

91. $\frac{7}{9} - \frac{2}{3} \cdot \frac{6}{7}$ [1.8]

92. $\frac{2}{3} - \left(\frac{3}{4}\right)^2$ [1.8]

SYNTHESIS

TW **93.** Keith *incorrectly* simplifies

$$\frac{x^2 + x - 2}{x^2 + 3x + 2} \quad \text{as} \quad \frac{x - 1}{x + 2}.$$

He then checks his simplification by evaluating both expressions for $x = 1$. Use this situation to explain why evaluating is not a foolproof check.

TW **94.** How could you convince someone that $a - b$ and $b - a$ are opposites of each other?

95. Calculate the slope of the line passing through $(a, f(a))$ and $(a + h, f(a + h))$ for the function f given by $f(x) = x^2 + 5$. Be sure your answer is simplified.

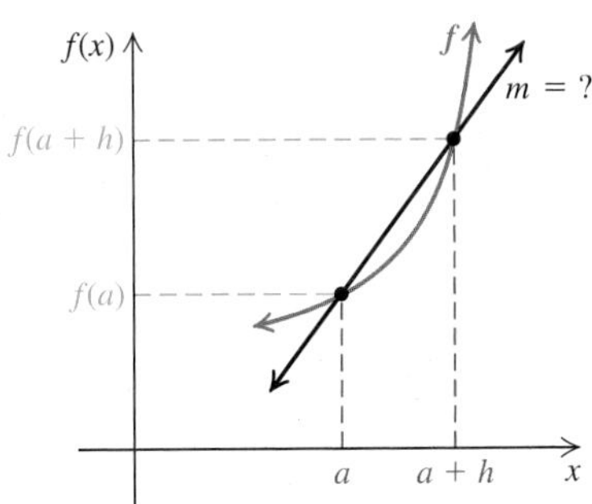

96. Calculate the slope of the line passing through the points $(a, f(a))$ and $(a + h, f(a + h))$ for the function f given by $f(x) = 3x^2$. Be sure your answer is simplified.

Simplify.

97. $\frac{x^4 - y^4}{(y - x)^4}$

98. $\frac{16y^4 - x^4}{(x^2 + 4y^2)(x - 2y)}$

99. $\frac{(x - 1)(x^4 - 1)(x^2 - 1)}{(x^2 + 1)(x - 1)^2(x^4 - 2x^2 + 1)}$

100. $\frac{(t^4 - 1)(t^2 - 9)(t - 9)^2}{(t^4 - 81)(t^2 + 1)(t + 1)^2}$

101. $\frac{a^3 - 2a^2 + 2a - 4}{a^3 - 2a^2 - 3a + 6}$

102. $\frac{x^3 + x^2 - y^3 - y^2}{x^2 - 2xy + y^2}$

103. $\frac{(t + 2)^3(t^2 + 2t + 1)(t + 1)}{(t + 1)^3(t^2 + 4t + 4)(t + 2)}$

104. $\frac{(x^2 - y^2)(x^2 - 2xy + y^2)}{(x + y)^2(x^2 - 4xy - 5y^2)}$

Determine the domain and the range of each function from its graph.

105.

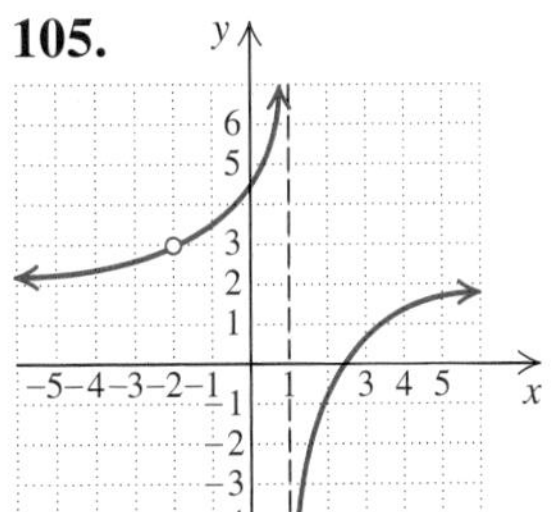

106.

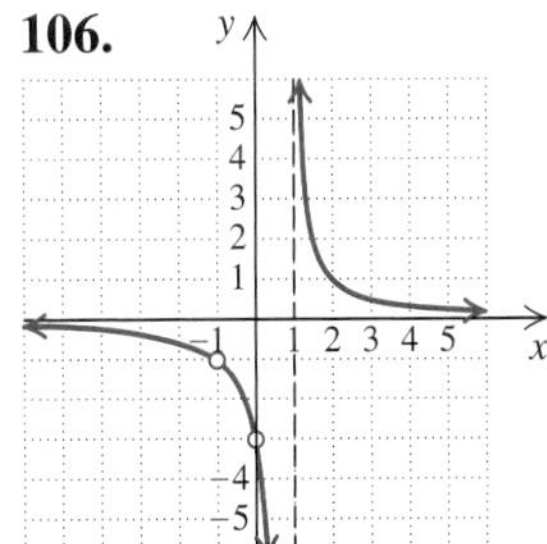

107.

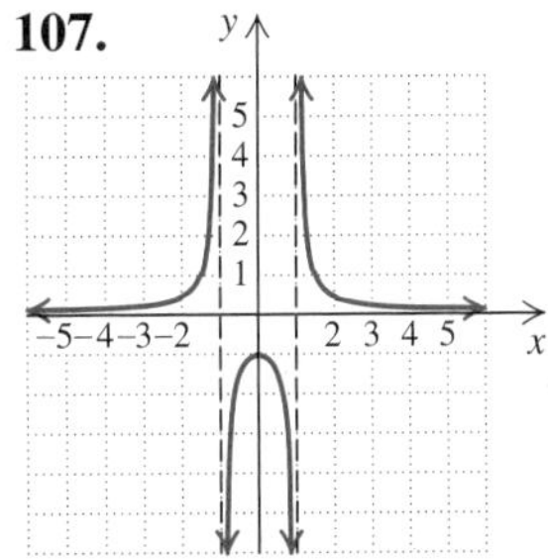

108. Graph the function given by

$$f(x) = \frac{x^2 - 9}{x - 3}.$$

(*Hint*: Determine the domain of f and simplify.)

TW **109.** Select any number x, multiply by 2, add 5, multiply by 5, subtract 25, and divide by 10. What do you get? Explain how this procedure can be used for a number trick.

Try Exercise Answers: Section 7.1

7. $\frac{40}{13}$ hr, or $3\frac{1}{13}$ hr **13.** 0 **23.** $\frac{3}{x}$ **27.** $a - 5$ **35.** $\frac{a + 4}{2(a - 4)}$ **37.** $\frac{x + 4}{x - 4}$ **47.** $-\frac{1}{1 + 2t}$ **63.** $f(t) = \frac{t - 3}{t + 1}, t \neq -3, -1$ **75.** $x = -3$

7.2 Multiplication and Division

- Multiplication
- Division

Multiplication and division of rational expressions are similar to multiplication and division with fractions. In this section, we again assume that all denominators are nonzero.

MULTIPLICATION

Recall that to multiply fractions, we multiply numerator times numerator and denominator times denominator. Rational expressions are multiplied in a similar way.

STUDY TIP

Get a Terrific Seat

To make the most of your class time, try to seat yourself at or near the front of the classroom. This makes it easier to participate and helps minimize distractions.

The Product of Two Rational Expressions To multiply rational expressions, multiply numerators and multiply denominators:

$$\frac{A}{B} \cdot \frac{C}{D} = \frac{AC}{BD}.$$

Then factor and, if possible, simplify the result.

For example,

$$\frac{3}{5} \cdot \frac{8}{11} = \frac{3 \cdot 8}{5 \cdot 11} \quad \text{and} \quad \frac{x}{3} \cdot \frac{x+2}{y} = \frac{x(x+2)}{3y}.$$

Fraction bars are grouping symbols, so parentheses are needed when we are writing some products.

EXAMPLE 1 Multiply. (Write the product as a single rational expression.) Then simplify by removing a factor equal to 1.

a) $\dfrac{5a^3}{4} \cdot \dfrac{2}{5a}$

b) $(x^2 - 3x - 10) \cdot \dfrac{x+4}{x^2 - 10x + 25}$

c) $\dfrac{10x + 20}{2x^2 - 3x + 1} \cdot \dfrac{x^2 - 1}{5x + 10}$

SOLUTION

a) $\dfrac{5a^3}{4} \cdot \dfrac{2}{5a} = \dfrac{5a^3(2)}{4(5a)}$ **Forming the product of the numerators and the product of the denominators**

$= \dfrac{2 \cdot 5 \cdot a \cdot a \cdot a}{2 \cdot 2 \cdot 5 \cdot a}$ **Factoring the numerator and the denominator**

$= \dfrac{\cancel{2} \cdot \cancel{5} \cdot \cancel{a} \cdot a \cdot a}{\cancel{2} \cdot 2 \cdot \cancel{5} \cdot \cancel{a}}$

$= \dfrac{a^2}{2}$ **Removing a factor equal to 1:** $\dfrac{2 \cdot 5 \cdot a}{2 \cdot 5 \cdot a} = 1$

b) $(x^2 - 3x - 10) \cdot \dfrac{x + 4}{x^2 - 10x + 25}$

$= \dfrac{x^2 - 3x - 10}{1} \cdot \dfrac{x + 4}{x^2 - 10x + 25}$ — **Writing $x^2 - 3x - 10$ as a rational expression**

$= \dfrac{(x^2 - 3x - 10)(x + 4)}{1(x^2 - 10x + 25)}$ — **Multiplying the numerators and the denominators**

$= \dfrac{(x - 5)(x + 2)(x + 4)}{(x - 5)(x - 5)}$ — **Factoring the numerator and the denominator**

$= \dfrac{\cancel{(x - 5)}(x + 2)(x + 4)}{\cancel{(x - 5)}(x - 5)}$ — **Removing a factor equal to 1:** $\dfrac{x - 5}{x - 5} = 1$

$= \dfrac{(x + 2)(x + 4)}{x - 5}$

c) $\dfrac{10x + 20}{2x^2 - 3x + 1} \cdot \dfrac{x^2 - 1}{5x + 10}$

$= \dfrac{(10x + 20)(x^2 - 1)}{(2x^2 - 3x + 1)(5x + 10)}$ — Multiply.

$= \dfrac{5(2)(x + 2)(x + 1)(x - 1)}{(x - 1)(2x - 1)5(x + 2)}$ — Factor. **Try to go directly to this step.**

$= \dfrac{\cancel{5}(2)\cancel{(x + 2)}(x + 1)\cancel{(x - 1)}}{\cancel{(x - 1)}(2x - 1)\cancel{5}\cancel{(x + 2)}}$ — Simplify. $\dfrac{5(x + 2)(x - 1)}{5(x + 2)(x - 1)} = 1$

$= \dfrac{2(x + 1)}{2x - 1}$

Try Exercise 19.

Student Notes

Neatness is important for work with rational expressions. Consider using the lines on your notebook paper to indicate where the fraction bars should be drawn in each example. Write the equals sign at the same height as the fraction bars.

Because our results are often used in problems that require factored form, there is no need to multiply out the numerator or the denominator.

DIVISION

Two expressions are reciprocals of each other if their product is 1. As with fractions, to find the reciprocal of a rational expression, we interchange the numerator and the denominator.

The reciprocal of $\dfrac{x}{x^2 + 3}$ is $\dfrac{x^2 + 3}{x}$.

The reciprocal of $y - 8$ is $\dfrac{1}{y - 8}$.

The Quotient of Two Rational Expressions To divide by a rational expression, multiply by its reciprocal:

$$\frac{A}{B} \div \frac{C}{D} = \frac{A}{B} \cdot \frac{D}{C} = \frac{AD}{BC}.$$

Then factor and, if possible, simplify.

EXAMPLE 2 Divide: **(a)** $\frac{x}{5} \div \frac{7}{y}$; **(b)** $(x + 2) \div \frac{x - 1}{x + 3}$.

SOLUTION

a) $\frac{x}{5} \div \frac{7}{y} = \frac{x}{5} \cdot \frac{y}{7}$ **Multiplying by the reciprocal of the divisor**

$= \frac{xy}{35}$ **Multiplying rational expressions**

b) $(x + 2) \div \frac{x - 1}{x + 3} = \frac{x + 2}{1} \cdot \frac{x + 3}{x - 1}$ **Multiplying by the reciprocal of the divisor. Writing $x + 2$ as $\frac{x + 2}{1}$ can be helpful.**

$= \frac{(x + 2)(x + 3)}{x - 1}$

Try Exercise 49.

As usual, we should simplify when possible. Often that will require us to factor one or more polynomials, hoping to discover a common factor that appears in both the numerator and the denominator.

EXAMPLE 3 Divide and, if possible, simplify: $\frac{x + 1}{x^2 - 1} \div \frac{x + 1}{x^2 - 2x + 1}$.

SOLUTION

$$\frac{x + 1}{x^2 - 1} \div \frac{x + 1}{x^2 - 2x + 1} = \frac{x + 1}{x^2 - 1} \cdot \frac{x^2 - 2x + 1}{x + 1}$$ Rewrite as multiplication.

$$= \frac{(x + 1)(x - 1)(x - 1)}{(x + 1)(x - 1)(x + 1)}$$ Multiply and factor.

$$= \frac{\cancel{(x + 1)}\cancel{(x - 1)}(x - 1)}{\cancel{(x + 1)}\cancel{(x - 1)}(x + 1)}$$ Simplify. $\frac{(x + 1)(x - 1)}{(x + 1)(x - 1)} = 1$

$$= \frac{x - 1}{x + 1}$$

Try Exercise 53.

EXAMPLE 4 Divide and, if possible, simplify.

a) $\dfrac{a^2 + 3a + 2}{a^2 + 4} \div (5a^2 + 10a)$

b) $\dfrac{x^2 - 2x - 3}{x^2 - 4} \div \dfrac{x + 1}{x + 5}$

SOLUTION

a) $\dfrac{a^2 + 3a + 2}{a^2 + 4} \div (5a^2 + 10a)$

$$= \frac{a^2 + 3a + 2}{a^2 + 4} \cdot \frac{1}{5a^2 + 10a}$$ **Multiplying by the reciprocal of the divisor**

$$= \frac{(a + 2)(a + 1)}{(a^2 + 4)5a(a + 2)}$$ **Multiplying rational expressions and factoring**

$$= \frac{\cancel{(a + 2)}(a + 1)}{(a^2 + 4)5a\cancel{(a + 2)}}$$

$$= \frac{a + 1}{(a^2 + 4)5a}$$ **Removing a factor equal to 1:** $\dfrac{a + 2}{a + 2} = 1$

b) $\dfrac{x^2 - 2x - 3}{x^2 - 4} \div \dfrac{x + 1}{x + 5} = \dfrac{x^2 - 2x - 3}{x^2 - 4} \cdot \dfrac{x + 5}{x + 1}$ **Multiplying by the reciprocal of the divisor**

$$= \frac{(x - 3)(x + 1)(x + 5)}{(x - 2)(x + 2)(x + 1)}$$ **Multiplying rational expressions and factoring**

$$= \frac{(x - 3)\cancel{(x + 1)}(x + 5)}{(x - 2)(x + 2)\cancel{(x + 1)}}$$

$$= \frac{(x - 3)(x + 5)}{(x - 2)(x + 2)}$$ **Removing a factor equal to 1:** $\dfrac{x + 1}{x + 1} = 1$

We can perform a partial check of the result using a graphing calculator. We enter the original expression as

$$y_1 = ((x^2 - 2x - 3)/(x^2 - 4))/((x + 1)/(x + 5))$$

and the simplified expression as

$$y_2 = ((x - 3)(x + 5))/((x - 2)(x + 2)).$$

Note that parentheses are placed around the entire numerator and around the entire denominator of each expression. Comparing values of y_1 and y_2 in the table shown at left, we see that the simplification is probably correct.

X	Y1	Y2
−5	ERROR	0
−4	−.5833	−.5833
−3	−2.4	−2.4
−2	ERROR	ERROR
−1	ERROR	5.3333
0	3.75	3.75
1	4	4

X = −5

Try Exercise 57.

7.2 Exercise Set

FOR EXTRA HELP MyMathLab MathXL PRACTICE 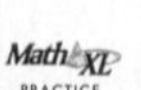 WATCH  DOWNLOAD READ 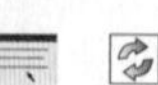 REVIEW

Concept Reinforcement *In Exercises 1–6, match each product or quotient with its result from the column on the right.*

1. ___ $\frac{x}{2} \cdot \frac{5}{y}$ a) $\frac{5x}{y}$
2. ___ $\frac{x}{2} \div \frac{5}{y}$ b) $\frac{xy}{5}$
3. ___ $x \cdot \frac{5}{y}$ c) $\frac{xy}{10}$
4. ___ $\frac{x}{2} \div y$ d) $\frac{5x}{2y}$
5. ___ $x \div \frac{5}{y}$ e) $\frac{x}{2y}$
6. ___ $\frac{5}{y} \div \frac{x}{2}$ f) $\frac{10}{xy}$

Multiply. Leave each answer in factored form.

7. $\frac{7x}{5} \cdot \frac{x-5}{2x+1}$
8. $\frac{3x}{4} \cdot \frac{5x+2}{x-1}$
9. $\frac{a-4}{a+6} \cdot \frac{a+2}{a+6}$
10. $\frac{a+3}{a+6} \cdot \frac{a+3}{a-1}$
11. $\frac{2x+3}{4} \cdot \frac{x+1}{x-5}$
12. $\frac{x+2}{3x-4} \cdot \frac{4}{5x+6}$

Multiply and, if possible, simplify.

13. $\frac{5a^4}{6a} \cdot \frac{2}{a}$
14. $\frac{10}{t^7} \cdot \frac{3t^2}{25t}$
15. $\frac{3c}{d^2} \cdot \frac{8d}{6c^3}$
16. $\frac{3x^2y}{2} \cdot \frac{4}{xy^3}$
17. $\frac{y^2-16}{4y+12} \cdot \frac{y+3}{y-4}$
18. $\frac{m^2-n^2}{4m+4n} \cdot \frac{m+n}{m-n}$
19. $\frac{x^2-3x-10}{(x-2)^2} \cdot \frac{x-2}{x-5}$
20. $\frac{t+2}{t-2} \cdot \frac{t^2-5t+6}{(t+2)^2}$
21. $\frac{y^2-y}{y^2+5y+4} \cdot (y+4)$
22. $(n-3) \cdot \frac{n^2+4n}{n^2-5n+6}$
23. $\frac{a^2-9}{a^2} \cdot \frac{5a}{a^2+a-12}$
24. $\frac{x^2+10x-11}{5x} \cdot \frac{x^3}{x+11}$
25. $\frac{4a^2}{3a^2-12a+12} \cdot \frac{3a-6}{2a}$
26. $\frac{5v+5}{v-2} \cdot \frac{2v^2-8v+8}{v^2-1}$
27. $\frac{t^2+2t-3}{t^2+4t-5} \cdot \frac{t^2-3t-10}{t^2+5t+6}$
28. $\frac{x^2+5x+4}{x^2-6x+8} \cdot \frac{x^2+5x-14}{x^2+8x+7}$
29. $\frac{5a^2-180}{10a^2-10} \cdot \frac{20a+20}{2a-12}$
30. $\frac{2t^2-98}{4t^2-4} \cdot \frac{8t+8}{16t-112}$

Aha! 31. $\frac{x^2+4x+4}{(x-1)^2} \cdot \frac{x^2-2x+1}{(x+2)^2}$

32. $\frac{t^3-4t}{t-t^4} \cdot \frac{t^4-t}{4t-t^3}$
33. $\frac{t^2+8t+16}{(t+4)^3} \cdot \frac{(t+2)^3}{t^2+4t+4}$
34. $\frac{(y-1)^3}{y^2-2y+1} \cdot \frac{y^2-4y+4}{(y-2)^3}$
35. $\frac{7a-14}{4-a^2} \cdot \frac{5a^2+6a+1}{35a+7}$
36. $\frac{a^2-1}{2-5a} \cdot \frac{15a-6}{a^2+5a-6}$
37. $(10x^2-x-2) \cdot \frac{4x^2-8x+3}{10x^2-11x-6}$

38. $\dfrac{2x^2 - 5x + 3}{6x^2 - 5x - 1} \cdot (6x^2 + 13x + 2)$

39. $\dfrac{c^3 + 8}{c^5 - 4c^3} \cdot \dfrac{c^6 - 4c^5 + 4c^4}{c^2 - 2c + 4}$

40. $\dfrac{x^3 - 27}{x^4 - 9x^2} \cdot \dfrac{x^5 - 6x^4 + 9x^3}{x^2 + 3x + 9}$

Find the reciprocal of each expression.

41. $\dfrac{3x}{7}$

42. $\dfrac{3 - x}{x^2 + 4}$

43. $a^3 - 8a$

44. $\dfrac{1}{a^2 - b^2}$

Divide and, if possible, simplify.

45. $\dfrac{5}{8} \div \dfrac{3}{7}$

46. $\dfrac{4}{9} \div \dfrac{5}{7}$

47. $\dfrac{x}{4} \div \dfrac{5}{x}$

48. $\dfrac{5}{x} \div \dfrac{x}{12}$

49. $\dfrac{a^5}{b^4} \div \dfrac{a^2}{b^7}$

50. $\dfrac{c}{d^6} \div \dfrac{c^2}{d}$

51. $\dfrac{y + 5}{4} \div \dfrac{y}{2}$

52. $\dfrac{a + 2}{a - 3} \div \dfrac{a - 1}{a + 3}$

53. $\dfrac{4y - 8}{y + 2} \div \dfrac{y - 2}{y^2 - 4}$

54. $\dfrac{x^2 - 1}{x} \div \dfrac{x + 1}{x - 1}$

55. $\dfrac{a}{a - b} \div \dfrac{b}{b - a}$

56. $\dfrac{x - y}{6} \div \dfrac{y - x}{3}$

57. $(y^2 - 9) \div \dfrac{y^2 - 2y - 3}{y^2 + 1}$

58. $(x^2 - 5x - 6) \div \dfrac{x^2 - 1}{x + 6}$

59. $\dfrac{-3 + 3x}{16} \div \dfrac{x - 1}{5}$

60. $\dfrac{-12 + 4x}{12} \div \dfrac{-6 + 2x}{6}$

61. $\dfrac{a + 2}{a - 1} \div \dfrac{3a + 6}{a - 5}$

62. $\dfrac{t - 3}{t + 2} \div \dfrac{4t - 12}{t + 1}$

63. $\dfrac{25x^2 - 4}{x^2 - 9} \div \dfrac{2 - 5x}{x + 3}$

64. $\dfrac{4a^2 - 1}{a^2 - 4} \div \dfrac{2a - 1}{2 - a}$

65. $(2x - 1) \div \dfrac{2x^2 - 11x + 5}{4x^2 - 1}$

66. $(a + 7) \div \dfrac{3a^2 + 14a - 49}{a^2 + 8a + 7}$

67. $\dfrac{w^2 - 14w + 49}{2w^2 - 3w - 14} \div \dfrac{3w^2 - 20w - 7}{w^2 - 6w - 16}$

68. $\dfrac{2m^2 + 59m - 30}{m^2 - 10m + 25} \div \dfrac{2m^2 - 21m + 10}{m^2 + m - 30}$

69. $\dfrac{c^2 + 10c + 21}{c^2 - 2c - 15} \div (5c^2 + 32c - 21)$

70. $\dfrac{z^2 - 2z + 1}{z^2 - 1} \div (4z^2 - z - 3)$

71. $\dfrac{x^3 - 64}{x^3 + 64} \div \dfrac{x^2 - 16}{x^2 - 4x + 16}$

72. $\dfrac{8y^3 - 27}{64y^3 - 1} \div \dfrac{4y^2 - 9}{16y^2 + 4y + 1}$

73. $\dfrac{8a^3 + b^3}{2a^2 + 3ab + b^2} \div \dfrac{8a^2 - 4ab + 2b^2}{4a^2 + 4ab + b^2}$

74. $\dfrac{x^3 + 8y^3}{2x^2 + 5xy + 2y^2} \div \dfrac{x^3 - 2x^2y + 4xy^2}{8x^2 - 2y^2}$

TW **75.** Why is it important to insert parentheses when multiplying rational expressions such as

$$\frac{x + 2}{5x - 7} \cdot \frac{3x - 1}{x + 4}?$$

TW **76.** As a first step in dividing $\dfrac{x}{3}$ by $\dfrac{7}{x}$, Jan canceled the x's. Explain why this was incorrect, and show the correct division.

SKILL REVIEW

To prepare for Section 7.3, review addition and subtraction with fraction notation (Sections 1.3 and 1.6) and subtraction of polynomials (Section 5.4).

Simplify.

77. $\dfrac{3}{4} + \dfrac{5}{6}$ [1.3]

78. $\dfrac{7}{8} + \dfrac{5}{6}$ [1.3]

79. $\dfrac{2}{9} - \dfrac{1}{6}$ [1.3]

80. $\dfrac{3}{10} - \dfrac{7}{15}$ [1.6]

81. $2x^2 - x + 1 - (x^2 - x - 2)$ [5.4]

82. $3x^2 + x - 7 - (5x^2 + 5x - 8)$ [5.4]

SYNTHESIS

TW **83.** Is the reciprocal of a product the product of the two reciprocals? Why or why not?

TW **84.** Explain why, in the check for Example 4(b), there were three error messages in the Y1-column but only one error in the Y2-column.

85. Find the reciprocal of $2\frac{1}{3}x$.

86. Find the reciprocal of $7.25x$.

Simplify.

Aha! **87.** $\frac{3x-y}{2x+y} \div \frac{3x-y}{2x+y}$

88. $\frac{2a^2-5ab}{c-3d} \div (4a^2-25b^2)$

89. $(x-2a) \div \frac{a^2x^2-4a^4}{a^2x+2a^3}$

90. $\frac{3a^2-5ab-12b^2}{3ab+4b^2} \div (3b^2-ab)^2$

Aha! **91.** $\frac{a^2-3b}{a^2+2b} \cdot \frac{a^2-2b}{a^2+3b} \cdot \frac{a^2+2b}{a^2-3b}$

92. $\frac{z^2-8z+16}{z^2+8z+16} \div \frac{(z-4)^5}{(z+4)^5} \div \frac{3z+12}{z^2-16}$

Perform the indicated operations and simplify.

93. $\left[\frac{r^2-4s^2}{r+2s} \div (r+2s)\right] \cdot \frac{2s}{r-2s}$

94. $\left[\frac{d^2-d}{d^2-6d+8} \cdot \frac{d-2}{d^2+5d}\right] \div \frac{5d}{d^2-9d+20}$

95. Let

$$g(x) = \frac{2x+3}{4x-1}.$$

Determine each of the following.

a) $g(x+h)$
b) $g(2x-2) \cdot g(x)$
c) $g\left(\frac{1}{2}x+1\right) \cdot g(x)$

96. Let

$$f(x) = \frac{4}{x^2-1} \quad \text{and} \quad g(x) = \frac{4x^2+8x+4}{x^3-1}.$$

Find each of the following.

a) $(f \cdot g)(x)$
b) $(f/g)(x)$
c) $(g/f)(x)$

Try Exercise Answers: Section 7.2

19. $\frac{x+2}{x-2}$ **49.** a^3b^3 **53.** $4(y-2)$ **57.** $\frac{(y+3)(y^2+1)}{y+1}$

7.3 Addition, Subtraction, and Least Common Denominators

- Addition When Denominators Are the Same
- Subtraction When Denominators Are the Same
- Least Common Multiples and Denominators

ADDITION WHEN DENOMINATORS ARE THE SAME

Recall that to add fractions having the same denominator, like $\frac{2}{7}$ and $\frac{3}{7}$, we add the numerators and keep the common denominator: $\frac{2}{7} + \frac{3}{7} = \frac{5}{7}$. The same procedure is used when rational expressions share a common denominator. We assume in this section that all denominators are nonzero.

The Sum of Two Rational Expressions To add when the denominators are the same, add the numerators and keep the common denominator:

$$\frac{A}{B} + \frac{C}{B} = \frac{A+C}{B}.$$

X	Y1	Y2
−2	−2.5	−2.5
−1	−6	−6
0	ERROR	ERROR
1	8	8
2	4.5	4.5
3	3.3333	3.3333
4	2.75	2.75
X = −2		

CAUTION! Using a table can indicate that two expressions are equivalent, but it does not verify that one of them is completely simplified.

EXAMPLE 1 Add: $\frac{3 + x}{x} + \frac{4}{x}$.

SOLUTION We have

$$\frac{3 + x}{x} + \frac{4}{x} = \frac{3 + x + 4}{x} = \frac{x + 7}{x}.$$

The denominators are the same, so we add the numerators and keep the common denominator.

To check, we let $y_1 = (3 + x)/x + 4/x$ and $y_2 = (x + 7)/x$. The table at left shows that $y_1 = y_2$ for all x not equal to 0.

Try Exercise 7.

EXAMPLE 2 Add. Simplify the result, if possible.

a) $\frac{2x^2 + 3x - 7}{2x + 1} + \frac{x^2 + x - 8}{2x + 1}$ **b)** $\frac{x - 5}{x^2 - 9} + \frac{2}{x^2 - 9}$

SOLUTION

a) $\frac{2x^2 + 3x - 7}{2x + 1} + \frac{x^2 + x - 8}{2x + 1} = \frac{(2x^2 + 3x - 7) + (x^2 + x - 8)}{2x + 1}$

$= \frac{3x^2 + 4x - 15}{2x + 1}$ **Combining like terms in the numerator**

$= \frac{(3x - 5)(x + 3)}{2x + 1}$ **Factoring. There are no common factors, so we cannot simplify further.**

b) $\frac{x - 5}{x^2 - 9} + \frac{2}{x^2 - 9} = \frac{x - 3}{x^2 - 9}$ **Combining like terms in the numerator: $x - 5 + 2 = x - 3$**

$= \frac{x - 3}{(x - 3)(x + 3)}$ **Factoring**

$= \frac{1 \cdot \cancel{(x - 3)}}{\cancel{(x - 3)}(x + 3)}$

$= \frac{1}{x + 3}$ **Removing a factor equal to 1: $\frac{x - 3}{x - 3} = 1$**

Try Exercise 19.

SUBTRACTION WHEN DENOMINATORS ARE THE SAME

When two fractions have the same denominator, we subtract one numerator from the other and keep the common denominator: $\frac{5}{7} - \frac{2}{7} = \frac{3}{7}$. The same procedure is used with rational expressions.

The Difference of Two Rational Expressions To subtract when the denominators are the same, subtract the second numerator from the first numerator and keep the common denominator:

$$\frac{A}{B} - \frac{C}{B} = \frac{A - C}{B}.$$

STUDY TIP 

If a Question Stumps You

Unless you know the material "cold," don't be surprised if a quiz or test includes a question that you feel unsure about. Should this happen, simply skip the question and continue with the quiz or test, returning to the trouble spot after the other questions have been answered.

CAUTION! The fraction bar under a numerator is a grouping symbol, just like parentheses. Thus, when a numerator is subtracted, it is important to subtract *every* term in that numerator.

EXAMPLE 3 Subtract and, if possible, simplify.

a) $\dfrac{3x}{x+2} - \dfrac{x-5}{x+2}$ b) $\dfrac{x^2}{x-4} - \dfrac{x+12}{x-4}$

SOLUTION

a) $\dfrac{3x}{x+2} - \dfrac{x-5}{x+2} = \dfrac{3x-(x-5)}{x+2}$ **The parentheses are needed to make sure that we subtract both terms.**

$= \dfrac{3x - x + 5}{x+2}$ **Removing the parentheses and changing signs (using the distributive law)**

$= \dfrac{2x+5}{x+2}$ **Combining like terms**

b) $\dfrac{x^2}{x-4} - \dfrac{x+12}{x-4} = \dfrac{x^2-(x+12)}{x-4}$ **Remember the parentheses!**

$= \dfrac{x^2 - x - 12}{x-4}$ **Removing parentheses (using the distributive law)**

$= \dfrac{(x-4)(x+3)}{x-4}$ **Factoring, in hopes of simplifying**

$= \dfrac{\cancel{(x-4)}(x+3)}{\cancel{(x-4)}}$ **Removing a factor equal to 1:** $\dfrac{x-4}{x-4} = 1$

$= x + 3$

Try Exercise 31.

LEAST COMMON MULTIPLES AND DENOMINATORS

Thus far, every pair of rational expressions that we have added or subtracted shared a common denominator. To add or subtract rational expressions that have different denominators, we must first find equivalent rational expressions that *do* have a common denominator.

In algebra, we find a common denominator much as we do in arithmetic. Recall that to add $\frac{1}{12}$ and $\frac{7}{30}$, we first identify the smallest number that contains both 12 and 30 as factors. Such a number, the **least common multiple (LCM)** of the denominators, is then used as the **least common denominator (LCD).**

Let's find the LCM of 12 and 30 using a method that can also be used with polynomials. We begin by writing the prime factorization of 12 and 30:

$$12 = 2 \cdot 2 \cdot 3;$$
$$30 = 2 \cdot 3 \cdot 5.$$

The LCM must include the factors of each number, so it must include each prime factor the greatest number of times that it appears in either of the factorizations. To find the LCM for 12 and 30, we select one factorization, say

$$2 \cdot 2 \cdot 3,$$

and note that because it lacks a factor of 5, it does not contain the entire factorization of 30. If we multiply $2 \cdot 2 \cdot 3$ by 5, every prime factor occurs just often enough to contain both 12 and 30 as factors.

$$\text{LCM} = 2 \cdot 2 \cdot 3 \cdot 5$$

12 is a factor of the LCM.

30 is a factor of the LCM.

Note that each prime factor—2, 3, and 5—is used the greatest number of times that it appears in either of the individual factorizations. The factor 2 occurs twice and the factors 3 and 5 once each.

Student Notes

The terms *least common multiple* and *least common denominator* are similar enough that they may be confusing. We find the LCM of polynomials, and we find the LCD of rational expressions.

To Find the Least Common Denominator (LCD)

1. Write the prime factorization of each denominator.
2. Select one of the factorizations and inspect it to see if it contains the other factorization.
   a) If it does, it represents the LCM of the denominators.
   b) If it does not, multiply that factorization by any factors of the other denominator that it lacks. The final product is the LCM of the denominators.

The LCD is the LCM of the denominators. It should contain each factor the greatest number of times that it occurs in any of the individual factorizations.

EXAMPLE 4 Find the LCD of $\frac{5}{36x^2}$ and $\frac{7}{24x}$.

SOLUTION

1. We begin by writing the prime factorizations of $36x^2$ and $24x$:

$$36x^2 = 2 \cdot 2 \cdot 3 \cdot 3 \cdot x \cdot x;$$
$$24x = 2 \cdot 2 \cdot 2 \cdot 3 \cdot x.$$

2. We select the factorization of $36x^2$. Except for a third factor of 2, this factorization contains the entire factorization of $24x$. Thus we multiply $36x^2$ by a third factor of 2.

$$\text{LCM} = 2 \cdot 2 \cdot 3 \cdot 3 \cdot x \cdot x \cdot 2$$

$36x^2$ is a factor of the LCM.

Note that each factor appears the greatest number of times that it occurs in either of the above factorizations.

$24x$ is a factor of the LCM.

The LCM of the denominators is thus $2^3 \cdot 3^2 \cdot x^2$, or $72x^2$, so the LCD of the expressions is $72x^2$.

Let's add $\frac{1}{12}$ and $\frac{7}{30}$:

$$\frac{1}{12} + \frac{7}{30} = \frac{1}{2 \cdot 2 \cdot 3} + \frac{7}{2 \cdot 3 \cdot 5}.$$

The least common denominator (LCD) is $2 \cdot 2 \cdot 3 \cdot 5$.

To get the LCD, we see that the first denominator needs a factor of 5, and the second denominator needs another factor of 2. Thus we multiply $\frac{1}{12}$ by 1, using $\frac{5}{5}$, and we multiply $\frac{7}{30}$ by 1, using $\frac{2}{2}$. Since $a \cdot 1 = a$, for any number a, the values of the fractions are not changed:

$$\frac{1}{12} + \frac{7}{30} = \frac{1}{2 \cdot 2 \cdot 3} \cdot \frac{5}{5} + \frac{7}{2 \cdot 3 \cdot 5} \cdot \frac{2}{2} \qquad \frac{5}{5} = 1 \text{ and } \frac{2}{2} = 1$$

$$= \frac{5}{60} + \frac{14}{60} \qquad \text{Both denominators are now the LCD.}$$

$$= \frac{19}{60}. \qquad \text{Adding the numerators and keeping the LCD}$$

Expressions like $\frac{5}{36x^2}$ and $\frac{7}{24x}$ are added in much the same manner. In Example 4, we found that the LCD is $2 \cdot 2 \cdot 2 \cdot 3 \cdot 3 \cdot x \cdot x$. To obtain equivalent expressions with this LCD, we multiply each expression by 1, using the missing factors of the LCD to write 1:

$$\frac{5}{36x^2} + \frac{7}{24x} = \frac{5}{2 \cdot 2 \cdot 3 \cdot 3 \cdot x \cdot x} + \frac{7}{2 \cdot 2 \cdot 2 \cdot 3 \cdot x}$$

$$= \frac{5}{2 \cdot 2 \cdot 3 \cdot 3 \cdot x \cdot x} \cdot \frac{2}{2} + \frac{7}{2 \cdot 2 \cdot 2 \cdot 3 \cdot x} \cdot \frac{3 \cdot x}{3 \cdot x}$$

The LCD requires another factor of 2. The LCD requires additional factors of 3 and x.

$$= \frac{10}{72x^2} + \frac{21x}{72x^2} \qquad \text{Both denominators are now the LCD.}$$

$$= \frac{21x + 10}{72x^2}.$$

You now have the "big" picture of why LCMs are needed when adding rational expressions. For the remainder of this section, we will practice finding LCMs and rewriting rational expressions so that they have the LCD as the denominator. In Section 7.4, we will return to the addition and subtraction of rational expressions.

EXAMPLE 5 For each pair of polynomials, find the least common multiple.

a) $15a$ and $35b$

b) $21x^3y^6$ and $7x^5y^2$

c) $x^2 + 5x - 6$ and $x^2 - 1$

SOLUTION

a) We write the prime factorizations and then construct the LCM, starting with the factorization of $15a$.

$$15a = 3 \cdot 5 \cdot a$$
$$35b = 5 \cdot 7 \cdot b$$

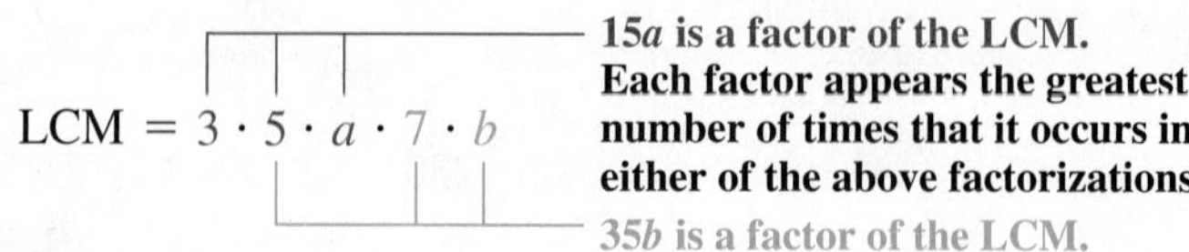

The LCM is $3 \cdot 5 \cdot a \cdot 7 \cdot b$, or $105ab$.

b) $21x^3y^6 = 3 \cdot 7 \cdot x \cdot x \cdot x \cdot y \cdot y \cdot y \cdot y \cdot y \cdot y$ **Try to visualize the factors of x and y mentally.**

$7x^5y^2 = 7 \cdot x \cdot x \cdot x \cdot x \cdot x \cdot y \cdot y$

We start with the factorization of $21x^3y^6$.

$$\text{LCM} = 3 \cdot 7 \cdot x \cdot x \cdot x \cdot y \cdot y \cdot y \cdot y \cdot y \cdot y \cdot x \cdot x$$

We multiply by the factors of $7x^5y^2$ that are lacking.

Note that we used the highest power of each factor in $21x^3y^6$ and $7x^5y^2$. The LCM is $21x^5y^6$.

c) $x^2 + 5x - 6 = (x - 1)(x + 6)$

$x^2 - 1 = (x - 1)(x + 1)$

We start with the factorization of $x^2 + 5x - 6$.

$$\text{LCM} = (x - 1)(x + 6)(x + 1)$$

We multiply by the factor of $x^2 - 1$ that is missing.

The LCM is $(x - 1)(x + 6)(x + 1)$. There is no need to multiply this out.

Try Exercise 49.

The above procedure can be used to find the LCM of three or more polynomials as well. We factor each polynomial and then construct the LCM using each factor the greatest number of times that it appears in any one factorization.

Student Notes

If you prefer, the LCM for a group of three polynomials can be found by finding the LCM of two of them and then finding the LCM of that result and the remaining polynomial.

EXAMPLE 6 For each group of polynomials, find the LCM.

a) $12x$, $16y$, and $8xyz$ **b)** $x^2 + 4$, $x + 1$, and 5

SOLUTION

a) $12x = 2 \cdot 2 \cdot 3 \cdot x$

$16y = 2 \cdot 2 \cdot 2 \cdot 2 \cdot y$

$8xyz = 2 \cdot 2 \cdot 2 \cdot x \cdot y \cdot z$

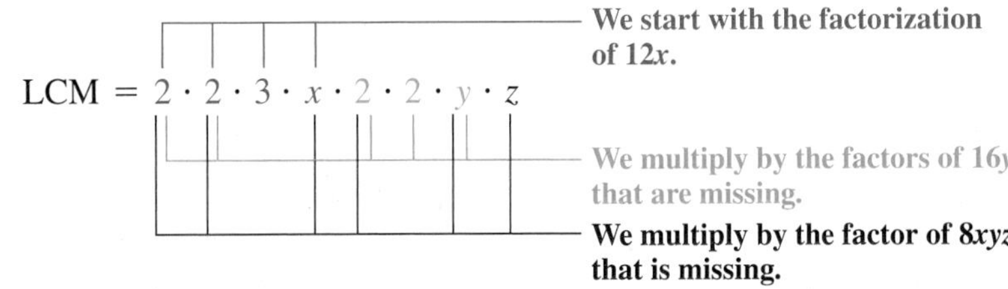

The LCM is $2^4 \cdot 3 \cdot xyz$, or $48xyz$.

b) Since $x^2 + 4$, $x + 1$, and 5 are not factorable, the LCM is their product: $5(x^2 + 4)(x + 1)$.

Try Exercise 55.

To add or subtract rational expressions with different denominators, we first write equivalent expressions that have the LCD. To do this, we multiply each rational expression by a carefully constructed form of 1.

EXAMPLE 7 Find equivalent expressions that have the LCD:

$$\frac{x+3}{x^2+5x-6}, \quad \frac{x+7}{x^2-1}.$$

SOLUTION From Example 5(c), we know that the LCD is

$$(x+6)(x-1)(x+1).$$

Since $x^2 + 5x - 6 = (x+6)(x-1)$, the factor of the LCD that is missing from the first denominator is $x + 1$. We multiply by 1 using $(x+1)/(x+1)$:

$$\left.\begin{aligned}\frac{x+3}{x^2+5x-6} &= \frac{x+3}{(x+6)(x-1)} \cdot \frac{x+1}{x+1}\\ &= \frac{(x+3)(x+1)}{(x+6)(x-1)(x+1)}.\end{aligned}\right\}$$

Finding an equivalent expression that has the least common denominator

For the second expression, we have $x^2 - 1 = (x+1)(x-1)$. The factor of the LCD that is missing is $x + 6$. We multiply by 1 using $(x+6)/(x+6)$:

$$\left.\begin{aligned}\frac{x+7}{x^2-1} &= \frac{x+7}{(x+1)(x-1)} \cdot \frac{x+6}{x+6}\\ &= \frac{(x+7)(x+6)}{(x+1)(x-1)(x+6)}.\end{aligned}\right\}$$

Finding an equivalent expression that has the least common denominator

We leave the results in factored form. In Section 7.4, we will carry out the actual addition and subtraction of such rational expressions.

Try Exercise 69.

7.3 Exercise Set

FOR EXTRA HELP MyMathLab MathXL PRACTICE  WATCH  DOWNLOAD  READ  REVIEW

Concept Reinforcement *Use one or more words to complete each of the following sentences.*

1. To add two rational expressions when the denominators are the same, add __________ and keep the common __________.
2. When a numerator is being subtracted, use parentheses to make sure to subtract every __________ in that numerator.
3. The least common multiple of two denominators is usually referred to as the __________ and is abbreviated __________.
4. The least common denominator of two fractions must contain the prime __________ of both __________.

Perform the indicated operation. Simplify, if possible.

5. $\frac{6}{x} + \frac{4}{x}$

6. $\frac{4}{a^2} + \frac{9}{a^2}$

7. $\frac{x}{12} + \frac{2x+5}{12}$

8. $\frac{a}{7} + \frac{3a-4}{7}$

9. $\frac{4}{a+3} + \frac{5}{a+3}$

10. $\frac{5}{x+2} + \frac{8}{x+2}$

11. $\frac{11}{4x-7} - \frac{3}{4x-7}$

12. $\frac{9}{2x+3} - \frac{5}{2x+3}$

13. $\frac{3y+8}{2y} - \frac{y+1}{2y}$

14. $\frac{5+3t}{4t} - \frac{2t+1}{4t}$

15. $\frac{7x+8}{x+1} + \frac{4x+3}{x+1}$

16. $\frac{3a+13}{a+4} + \frac{2a+7}{a+4}$

17. $\dfrac{7x+8}{x+1} - \dfrac{4x+3}{x+1}$

18. $\dfrac{3a+13}{a+4} - \dfrac{2a+7}{a+4}$

19. $\dfrac{a^2}{a-4} + \dfrac{a-20}{a-4}$

20. $\dfrac{x^2}{x+5} + \dfrac{7x+10}{x+5}$

21. $\dfrac{x^2}{x-2} - \dfrac{6x-8}{x-2}$

22. $\dfrac{a^2}{a+3} - \dfrac{2a+15}{a+3}$

Aha! 23. $\dfrac{t^2-5t}{t-1} + \dfrac{5t-t^2}{t-1}$

24. $\dfrac{y^2+6y}{y+2} + \dfrac{2y+12}{y+2}$

25. $\dfrac{x-6}{x^2+5x+6} + \dfrac{9}{x^2+5x+6}$

26. $\dfrac{x-5}{x^2-4x+3} + \dfrac{2}{x^2-4x+3}$

27. $\dfrac{3a^2+14}{a^2+5a-6} - \dfrac{13a}{a^2+5a-6}$

28. $\dfrac{2a^2+15}{a^2-7a+12} - \dfrac{11a}{a^2-7a+12}$

29. $\dfrac{t^2-3t}{t^2+6t+9} + \dfrac{2t-12}{t^2+6t+9}$

30. $\dfrac{y^2-7y}{y^2+8y+16} + \dfrac{6y-20}{y^2+8y+16}$

31. $\dfrac{2x^2+x}{x^2-8x+12} - \dfrac{x^2-2x+10}{x^2-8x+12}$

32. $\dfrac{2x^2+3}{x^2-6x+5} - \dfrac{3+2x^2}{x^2-6x+5}$

33. $\dfrac{3-2x}{x^2-6x+8} + \dfrac{7-3x}{x^2-6x+8}$

34. $\dfrac{1-2t}{t^2-5t+4} + \dfrac{4-3t}{t^2-5t+4}$

35. $\dfrac{x-9}{x^2+3x-4} - \dfrac{2x-5}{x^2+3x-4}$

36. $\dfrac{5-3x}{x^2-2x+1} - \dfrac{x+1}{x^2-2x+1}$

Find the LCM.

37. 15, 27

38. 10, 15

39. 8, 9

40. 12, 15

41. 6, 12, 15

42. 8, 32, 50

Find the LCM.

43. $12x^2$, $6x^3$

44. $10t^3$, $5t^4$

45. $15a^4b^7$, $10a^2b^8$

46. $6a^2b^7$, $9a^5b^2$

47. $2(y-3)$, $6(y-3)$

48. $4(x-1)$, $8(x-1)$

49. x^2-4, x^2+5x+6

50. x^2+3x+2, x^2-4

51. t^3+4t^2+4t, t^2-4t

52. y^3-y^2, y^4-y^2

53. $10x^2y$, $6y^2z$, $5xz^3$

54. $12s^3t$, $15sv^2$, $6t^4v$

55. $a+1$, $(a-1)^2$, a^2-1

56. x^2-9, $x+3$, $(x-3)^2$

57. $2n^2+n-1$, $2n^2+3n-2$

58. m^2-2m-3, $2m^2+3m+1$

Aha! 59. $t-3$, $t+3$, t^2-9

60. $a-5$, $a^2-10a+25$

61. $6x^3-24x^2+18x$, $4x^5-24x^4+20x^3$

62. $9x^3-9x^2-18x$, $6x^5-24x^4+24x^3$

63. $2t^3-2$, t^2-1

64. $5n+5$, n^3+1

Find equivalent expressions that have the LCD.

65. $\dfrac{5}{6x^5}, \dfrac{y}{12x^3}$

66. $\dfrac{3}{10a^3}, \dfrac{b}{5a^6}$

67. $\dfrac{3}{2a^2b}, \dfrac{7}{8ab^2}$

68. $\dfrac{7}{3x^4y^2}, \dfrac{4}{9xy^3}$

69. $\dfrac{2x}{x^2-4}, \dfrac{4x}{x^2+5x+6}$

70. $\dfrac{5x}{x^2-9}, \dfrac{2x}{x^2+11x+24}$

TW 71. Explain why the product of two numbers is not always their least common multiple.

TW 72. If the LCM of two numbers is their product, what can you conclude about the two numbers?

SKILL REVIEW

To prepare for Section 7.4, review opposites (Sections 1.7 and 1.8).

Write each number in two equivalent forms. [1.7]

73. $-\dfrac{5}{8}$

74. $\dfrac{4}{-11}$

Write an equivalent expression without parentheses. [1.8]

75. $-(x-y)$

76. $-(3-a)$

Multiply and simplify. [1.8]

77. $-1(2x - 7)$

78. $-1(a - b)$

SYNTHESIS

TW 79. If the LCM of a binomial and a trinomial is the trinomial, what relationship exists between the two expressions?

TW 80. If the LCM of two third-degree polynomials is a sixth-degree polynomial, what can be concluded about the two polynomials?

Perform the indicated operations. Simplify, if possible.

81. $\dfrac{6x - 1}{x - 1} + \dfrac{3(2x + 5)}{x - 1} + \dfrac{3(2x - 3)}{x - 1}$

82. $\dfrac{2x + 11}{x - 3} \cdot \dfrac{3}{x + 4} + \dfrac{-1}{4 + x} \cdot \dfrac{6x + 3}{x - 3}$

83. $\dfrac{x^2}{3x^2 - 5x - 2} - \dfrac{2x}{3x + 1} \cdot \dfrac{1}{x - 2}$

84. $\dfrac{x + y}{x^2 - y^2} + \dfrac{x - y}{x^2 - y^2} - \dfrac{2x}{x^2 - y^2}$

African Artistry. In southern Africa, the design of every woven handbag, or gipatsi *(plural,* sipatsi*) is created by repeating two or more geometric patterns. Each pattern encircles the bag, sharing the strands of fabric with any pattern above or below. The length, or period, of each pattern is the number of strands required to construct the pattern. For a gipatsi to be considered beautiful, each individual pattern must fit a whole number of times around the bag.*

Source: Gerdes, Paulus, *Women, Art and Geometry in Southern Africa*. Asmara, Eritrea: Africa World Press, Inc., p. 5

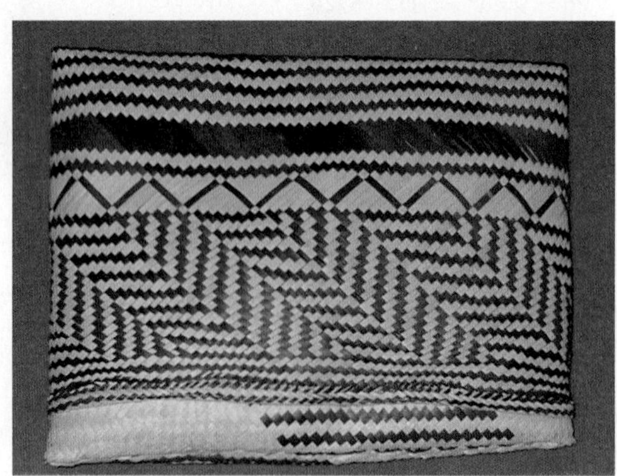

85. A weaver is using two patterns to create a gipatsi. Pattern A is 10 strands long, and pattern B is 3 strands long. What is the smallest number of strands that can be used to complete the gipatsi?

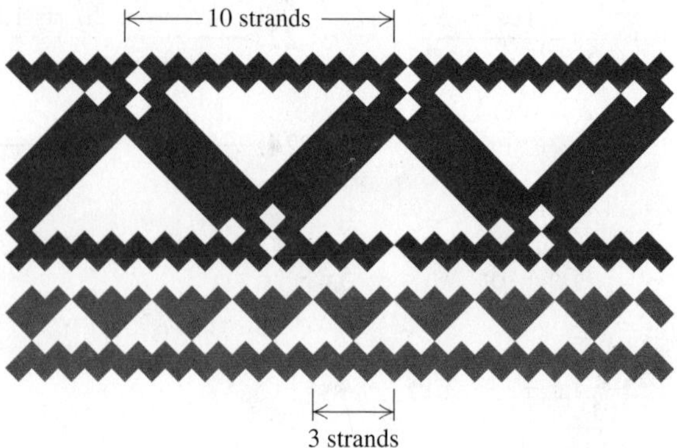

86. A weaver is using a four-strand pattern, a six-strand pattern, and an eight-strand pattern. What is the smallest number of strands that can be used to complete the gipatsi?

87. For technical reasons, the number of strands is generally a multiple of 4. Answer Exercise 85 with this additional requirement in mind.

Find the LCM.

88. 80, 96, 108

89. $4x^2 - 25$, $6x^2 - 7x - 20$, $(9x^2 + 24x + 16)^2$

90. $9n^2 - 9$, $(5n^2 - 10n + 5)^2$, $15n - 15$

91. *Copiers.* The Sharp DX-C400 copier can print 40 pages per minute. The Canon imageRUNNER 1025N can print 25 pages per minute. If both machines begin printing at the same instant, how long will it be until they again begin copying a page at exactly the same time?

Source: Manufacturers' Web sites

92. *Running.* Beth and Todd leave the starting point of a fitness loop at the same time. Beth jogs a lap in 6 min and Todd jogs one in 8 min. Assuming they continue to run at the same pace, when will they next meet at the starting place?

93. *Bus Schedules.* Beginning at 5:00 A.M., a hotel shuttle bus leaves Salton Airport every 25 min, and the downtown shuttle bus leaves the airport every 35 min. What time will it be when both shuttles again leave at the same time?

94. *Appliances.* Dishwashers last an average of 9 years, garbage disposals an average of 12 years, and gas ranges an average of 15 years. If an apartment house is equipped with new dishwashers, garbage disposals, and gas ranges in 2012, in what year will all three appliances need to be replaced at once?
Source: National Association of Home Builders/Bank of America Home Equity Study of Life Expectancy of Home Components

TW **95.** Explain how evaluating can be used to perform a partial check on the result of Example 2(b):

$$\frac{x-5}{x^2-9}+\frac{2}{x^2-9}=\frac{1}{x+3}.$$

TW **96.** On p. 523, the second step in finding an LCD is to select one of the factorizations of the denominators. Does it matter which one is selected? Why or why not?

Try Exercise Answers: Section 7.3

7. $\frac{3x+5}{12}$ **19.** $a+5$ **31.** $\frac{x+5}{x-6}$
49. $(x-2)(x+2)(x+3)$ **55.** $(a+1)(a-1)^2$
69. $\frac{2x(x+3)}{(x-2)(x+2)(x+3)}, \frac{4x(x-2)}{(x-2)(x+2)(x+3)}$

7.4 Addition and Subtraction with Unlike Denominators

- Adding and Subtracting with LCDs
- When Factors Are Opposites

ADDING AND SUBTRACTING WITH LCDS

We now know how to rewrite two rational expressions in equivalent forms that use the LCD. Once rational expressions share a common denominator, they can be added or subtracted just as in Section 7.3.

To Add or Subtract Rational Expressions Having Different Denominators

1. Find the LCD.
2. Multiply each rational expression by a form of 1 made up of the factors of the LCD missing from that expression's denominator.
3. Add or subtract the numerators, as indicated. Write the sum or the difference over the LCD.
4. Simplify, if possible.

STUDY TIP 

Neatness Counts

Make an effort to work neatly. When working with rational expressions, be sure to write the fraction bar long enough to completely separate the numerator and the denominator. Extra time invested in neat work will pay off in fewer errors.

EXAMPLE 1 Add: $\frac{5x^2}{8}+\frac{7x}{12}$.

SOLUTION

1. First, we find the LCD:

$$\left.\begin{aligned} 8 &= 2\cdot 2\cdot 2 \\ 12 &= 2\cdot 2\cdot 3 \end{aligned}\right\} \quad \text{LCD} = 2\cdot 2\cdot 2\cdot 3, \text{ or } 24.$$

2. The denominator 8 must be multiplied by 3 in order to obtain the LCD. The denominator 12 must be multiplied by 2 in order to obtain the LCD. Thus we multiply the first expression by $\frac{3}{3}$ and the second expression by $\frac{2}{2}$ to get the LCD:

$$\frac{5x^2}{8} + \frac{7x}{12} = \frac{5x^2}{2 \cdot 2 \cdot 2} + \frac{7x}{2 \cdot 2 \cdot 3}$$

$$= \frac{5x^2}{2 \cdot 2 \cdot 2} \cdot \frac{3}{3} + \frac{7x}{2 \cdot 2 \cdot 3} \cdot \frac{2}{2}$$ **Multiplying each expression by a form of 1 to get the LCD**

$$= \frac{15x^2}{24} + \frac{14x}{24}.$$

3. Next, we add the numerators:

$$\frac{15x^2}{24} + \frac{14x}{24} = \frac{15x^2 + 14x}{24}.$$

4. Since $15x^2 + 14x$ and 24 have no common factor,

$$\frac{15x^2 + 14x}{24}$$

cannot be simplified any further.

Try Exercise 5.

EXAMPLE 2 Subtract: $\frac{7}{8x} - \frac{5}{12x^2}$.

SOLUTION We follow the four steps shown above. First, we find the LCD:

$$\left.\begin{aligned} 8x &= 2 \cdot 2 \cdot 2 \cdot x \\ 12x^2 &= 2 \cdot 2 \cdot 3 \cdot x \cdot x \end{aligned}\right\} \quad \text{LCD} = 2 \cdot 2 \cdot 3 \cdot x \cdot x \cdot 2, \text{ or } 24x^2.$$

The denominator $8x$ must be multiplied by $3x$ in order to obtain the LCD. The denominator $12x^2$ must be multiplied by 2 in order to obtain the LCD. Thus we multiply by $\frac{3x}{3x}$ and $\frac{2}{2}$ to get the LCD. Then we subtract and, if possible, simplify.

$$\frac{7}{8x} - \frac{5}{12x^2} = \frac{7}{8x} \cdot \frac{3x}{3x} - \frac{5}{12x^2} \cdot \frac{2}{2}$$

$$= \frac{21x}{24x^2} - \frac{10}{24x^2}$$

CAUTION! Do not simplify *these* rational expressions or you will lose the LCD.

$$= \frac{21x - 10}{24x^2}$$ **This cannot be simplified, so we are done.**

Try Exercise 7.

EXAMPLE 3 Add: $\frac{2a}{a^2 - 1} + \frac{1}{a^2 + a}$.

SOLUTION First, we find the LCD:

Find the LCD.

$$\left.\begin{aligned} a^2 - 1 &= (a - 1)(a + 1) \\ a^2 + a &= a(a + 1) \end{aligned}\right\} \quad \text{LCD} = (a - 1)(a + 1)a.$$

We multiply by a form of 1 to get the LCD in each expression:

Write each expression with the LCD.

$$\frac{2a}{a^2-1}+\frac{1}{a^2+a}=\frac{2a}{(a-1)(a+1)}\cdot\frac{a}{a}+\frac{1}{a(a+1)}\cdot\frac{a-1}{a-1}$$

Multiplying by $\frac{a}{a}$ and $\frac{a-1}{a-1}$ to get the LCD

$$=\frac{2a^2}{(a-1)(a+1)a}+\frac{a-1}{a(a+1)(a-1)}$$

Add numerators.

$$=\frac{2a^2+a-1}{a(a-1)(a+1)}$$ **Adding numerators**

Simplify.

$$=\frac{(2a-1)\cancel{(a+1)}}{a(a-1)\cancel{(a+1)}}$$

$$=\frac{2a-1}{a(a-1)}.$$

Simplifying by factoring and removing a factor equal to 1: $\frac{a+1}{a+1}=1$

Try Exercise 31.

Parentheses are important to use when subtracting a numerator with more than one term.

EXAMPLE 4 Subtract: $\frac{x+4}{x-2}-\frac{x-7}{x+5}$.

SOLUTION First, we find the LCD. It is just the product of the denominators:

$$\text{LCD}=(x-2)(x+5).$$

We multiply by a form of 1 to get the LCD in each expression. Then we subtract and try to simplify.

$$\frac{x+4}{x-2}-\frac{x-7}{x+5}=\frac{x+4}{x-2}\cdot\frac{x+5}{x+5}-\frac{x-7}{x+5}\cdot\frac{x-2}{x-2}$$

$$=\frac{x^2+9x+20}{(x-2)(x+5)}-\frac{x^2-9x+14}{(x-2)(x+5)}$$ **Multiplying out numerators (but not denominators)**

$$=\frac{x^2+9x+20-(x^2-9x+14)}{(x-2)(x+5)}$$ **Parentheses are important.**

$$=\frac{x^2+9x+20-x^2+9x-14}{(x-2)(x+5)}$$ **Removing parentheses and subtracting every term**

$$=\frac{18x+6}{(x-2)(x+5)}$$

$$=\frac{6(3x+1)}{(x-2)(x+5)}$$ **We cannot simplify.**

Try Exercise 27.

EXAMPLE 5 Subtract: $\dfrac{x}{x^2 + 5x + 6} - \dfrac{2}{x^2 + 3x + 2}$.

SOLUTION

$$\frac{x}{x^2 + 5x + 6} - \frac{2}{x^2 + 3x + 2}$$

Find the LCD.

$$= \frac{x}{(x + 2)(x + 3)} - \frac{2}{(x + 2)(x + 1)}$$

Factoring denominators. The LCD is $(x + 2)(x + 3)(x + 1)$.

Write each expression with the LCD.

$$= \frac{x}{(x + 2)(x + 3)} \cdot \frac{x + 1}{x + 1} - \frac{2}{(x + 2)(x + 1)} \cdot \frac{x + 3}{x + 3}$$

$$= \frac{x^2 + x}{(x + 2)(x + 3)(x + 1)} - \frac{2x + 6}{(x + 2)(x + 3)(x + 1)}$$

Subtract numerators.

$$= \frac{x^2 + x - (2x + 6)}{(x + 2)(x + 3)(x + 1)}$$

Don't forget the parentheses!

$$= \frac{x^2 + x - 2x - 6}{(x + 2)(x + 3)(x + 1)}$$

Remember to subtract each term in $2x + 6$.

$$= \frac{x^2 - x - 6}{(x + 2)(x + 3)(x + 1)}$$

Combining like terms in the numerator

Simplify.

$$= \frac{\cancel{(x + 2)}(x - 3)}{\cancel{(x + 2)}(x + 3)(x + 1)}$$

$$= \frac{x - 3}{(x + 3)(x + 1)}$$

Factoring and simplifying; $\dfrac{x + 2}{x + 2} = 1$

Try Exercise 47.

WHEN FACTORS ARE OPPOSITES

Expressions of the form $a - b$ and $b - a$ are opposites of each other. When either of these binomials is multiplied by -1, the result is the other binomial:

$$-1(a - b) = -a + b = b + (-a) = b - a;$$
$$-1(b - a) = -b + a = a + (-b) = a - b.$$

Multiplication by -1 reverses the order in which subtraction occurs.

When one denominator is the opposite of the other, we can multiply either expression by 1 using $\dfrac{-1}{-1}$.

EXAMPLE 6 Add: $\dfrac{3}{8a} + \dfrac{1}{-8a}$.

SOLUTION

$$\frac{3}{8a} + \frac{1}{-8a} = \frac{3}{8a} + \frac{-1}{-1} \cdot \frac{1}{-8a}$$

$$= \frac{3}{8a} + \frac{-1}{8a} = \frac{2}{8a}$$

When denominators are opposites, we multiply one rational expression by $-1/-1$ to get the LCD.

$$= \frac{\cancel{2} \cdot 1}{\cancel{2} \cdot 4a} = \frac{1}{4a}$$

Simplifying by removing a factor equal to 1: $\dfrac{2}{2} = 1$

Try Exercise 53.

EXAMPLE 7 Subtract: $\dfrac{5x}{x-7} - \dfrac{3x}{7-x}$.

SOLUTION

$$\frac{5x}{x-7} - \frac{3x}{7-x} = \frac{5x}{x-7} - \frac{-1}{-1}\cdot\frac{3x}{7-x}$$ **Note that $x - 7$ and $7 - x$ are opposites.**

$$= \frac{5x}{x-7} - \frac{-3x}{x-7}$$ **Performing the multiplication. *Note*: $-1(7-x) = -7 + x = x - 7$.**

$$= \frac{5x-(-3x)}{x-7}$$

$$= \frac{5x+3x}{x-7}$$ **Subtracting. The parentheses are important.**

$$= \frac{8x}{x-7}$$ **Simplifying**

Try Exercise 55.

EXAMPLE 8 Find simplified form for the function given by

$$f(x) = \frac{2x}{x^2-4} + \frac{5}{2-x} - \frac{1}{2+x}$$

and list all restrictions on the domain.

SOLUTION We have

$$\frac{2x}{x^2-4} + \frac{5}{2-x} - \frac{1}{2+x}$$

$$= \frac{2x}{(x-2)(x+2)} + \frac{5}{2-x} - \frac{1}{2+x}$$ **Factoring. Note that $x \neq -2, 2$.**

$$= \frac{2x}{(x-2)(x+2)} + \frac{-1}{-1}\cdot\frac{5}{(2-x)} - \frac{1}{x+2}$$ **Multiplying by $-1/-1$ since $2 - x$ is the opposite of $x - 2$**

$$= \frac{2x}{(x-2)(x+2)} + \frac{-5}{x-2} - \frac{1}{x+2}$$ **The LCD is $(x-2)(x+2)$.**

$$= \frac{2x}{(x-2)(x+2)} + \frac{-5}{x-2}\cdot\frac{x+2}{x+2} - \frac{1}{x+2}\cdot\frac{x-2}{x-2}$$ **Multiplying by 1 to get the LCD**

$$= \frac{2x-5(x+2)-(x-2)}{(x-2)(x+2)}$$

$$= \frac{2x-5x-10-x+2}{(x-2)(x+2)}$$

$$= \frac{-4x-8}{(x-2)(x+2)}$$

$$= \frac{-4(x+2)}{(x-2)\,(x+2)}$$

$$= \frac{-4\cancel{(x+2)}}{(x-2)\cancel{(x+2)}}$$ **Removing a factor equal to 1: $\dfrac{x+2}{x+2} = 1, x \neq -2$**

$$= \frac{-4}{x-2}, \text{ or } -\frac{4}{x-2}, x \neq \pm 2.$$

Try Exercise 73.

Student Notes

Your answer may differ slightly from the answer found at the back of the book and still be correct. For example, an equivalent answer to Example 8 is $\dfrac{4}{2-x}$:

$$-\frac{4}{x-2} = \frac{4}{-(x-2)} = \frac{4}{2-x}.$$

Before reworking an exercise, be sure that your answer is indeed different from the correct answer.

Our work in Example 8 indicates that if

$$f(x) = \frac{2x}{x^2 - 4} + \frac{5}{2 - x} - \frac{1}{2 + x} \quad \text{and} \quad g(x) = \frac{-4}{x - 2},$$

then, for $x \neq -2$ and $x \neq 2$, we have $f = g$. Note that whereas the domain of f includes all real numbers except -2 or 2, the domain of g excludes only 2. This is illustrated in the following graphs. Methods for drawing such graphs by hand are discussed in more advanced courses. The graphs here are for visualization only.

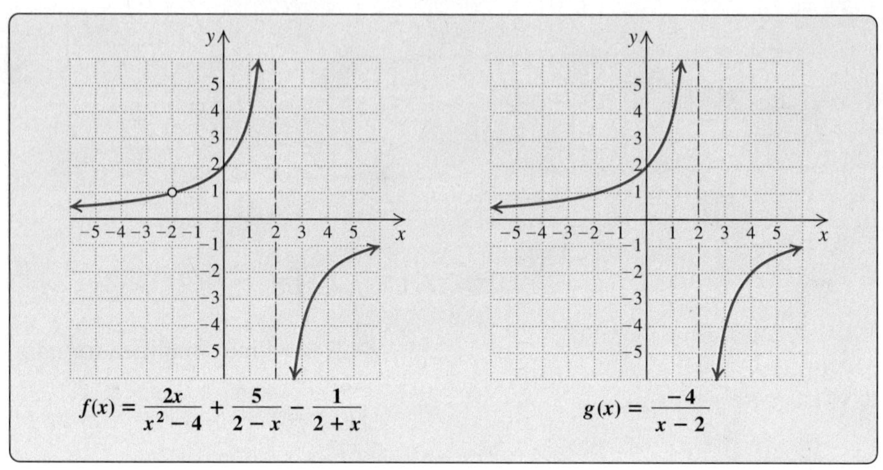

A computer-generated visualization of Example 8

7.4 Exercise Set

FOR EXTRA HELP MyMathLab

PRACTICE WATCH DOWNLOAD READ REVIEW

Concept Reinforcement *In Exercises 1–4, the four steps for adding rational expressions with different denominators are listed. Fill in the missing word or words for each step.*

1. To add or subtract when the denominators are different, first find the __________.

2. Multiply each rational expression by a form of 1 made up of the factors of the LCD that are __________ from that expression's __________.

3. Add or subtract the __________, as indicated. Write the sum or difference over the __________.

4. __________, if possible.

Perform the indicated operation. Simplify, if possible.

5. $\frac{4}{x} + \frac{9}{x^2}$

6. $\frac{5}{x} + \frac{6}{x^2}$

7. $\frac{1}{6r} - \frac{3}{8r}$

8. $\frac{4}{9t} - \frac{7}{6t}$

9. $\frac{2}{c^2d} + \frac{7}{cd^3}$

10. $\frac{4}{xy^2} + \frac{2}{x^2y}$

11. $\frac{-2}{3xy^2} - \frac{6}{x^2y^3}$

12. $\frac{8}{9t^3} - \frac{5}{6t^2}$

13. $\frac{x - 4}{9} + \frac{x + 5}{6}$

14. $\frac{x + 5}{8} + \frac{x - 3}{12}$

15. $\frac{a + 2}{2} - \frac{a - 4}{4}$

16. $\frac{x - 2}{6} - \frac{x + 1}{3}$

17. $\dfrac{2a-1}{3a^2}+\dfrac{5a+1}{9a}$

18. $\dfrac{a+4}{16a}+\dfrac{3a+4}{4a^2}$

19. $\dfrac{x-1}{4x}-\dfrac{2x+3}{x}$

20. $\dfrac{4z-9}{3z}-\dfrac{3z-8}{4z}$

21. $\dfrac{2c-d}{c^2d}+\dfrac{c+d}{cd^2}$

22. $\dfrac{x+y}{xy^2}+\dfrac{3x+y}{x^2y}$

23. $\dfrac{5x+3y}{2x^2y}-\dfrac{3x+4y}{xy^2}$

24. $\dfrac{4x+2t}{3xt^2}-\dfrac{5x-3t}{x^2t}$

25. $\dfrac{5}{x-1}+\dfrac{5}{x+1}$

26. $\dfrac{3}{x-2}+\dfrac{3}{x+2}$

27. $\dfrac{4}{z-1}-\dfrac{2}{z+1}$

28. $\dfrac{5}{x+5}-\dfrac{3}{x-5}$

29. $\dfrac{2}{x+5}+\dfrac{3}{4x}$

30. $\dfrac{3}{x+1}+\dfrac{2}{3x}$

31. $\dfrac{8}{3t^2-15t}-\dfrac{3}{2t-10}$

32. $\dfrac{3}{2t^2-2t}-\dfrac{5}{2t-2}$

33. $\dfrac{4x}{x^2-25}+\dfrac{x}{x+5}$

34. $\dfrac{2x}{x^2-16}+\dfrac{x}{x-4}$

35. $\dfrac{t}{t-3}-\dfrac{5}{4t-12}$

36. $\dfrac{6}{z+4}-\dfrac{2}{3z+12}$

37. $\dfrac{2}{x+3}+\dfrac{4}{(x+3)^2}$

38. $\dfrac{3}{x-1}+\dfrac{2}{(x-1)^2}$

39. $\dfrac{t-3}{t^3-1}-\dfrac{2}{1-t^3}$

40. $\dfrac{1-6m}{1-m^3}-\dfrac{5}{m^3-1}$

41. $\dfrac{3a}{4a-20}+\dfrac{9a}{6a-30}$

42. $\dfrac{4a}{5a-10}+\dfrac{3a}{10a-20}$

Aha! 43. $\dfrac{x}{x-5}+\dfrac{x}{5-x}$

44. $\dfrac{x+4}{x}+\dfrac{x}{x+4}$

45. $\dfrac{6}{a^2+a-2}+\dfrac{4}{a^2-4a+3}$

46. $\dfrac{x}{x^2+2x+1}+\dfrac{1}{x^2+5x+4}$

47. $\dfrac{x}{x^2+9x+20}-\dfrac{4}{x^2+7x+12}$

48. $\dfrac{x}{x^2+5x+6}-\dfrac{2}{x^2+3x+2}$

49. $\dfrac{3z}{z^2-4z+4}+\dfrac{10}{z^2+z-6}$

50. $\dfrac{3}{x^2-9}+\dfrac{2}{x^2-x-6}$

Aha! 51. $\dfrac{-5}{x^2+17x+16}-\dfrac{0}{x^2+9x+8}$

52. $\dfrac{x}{x^2+15x+56}-\dfrac{1}{x^2+13x+42}$

53. $\dfrac{4x}{5}+\dfrac{x+3}{-5}$

54. $\dfrac{x}{4}-\dfrac{3x-5}{-4}$

55. $\dfrac{y^2}{y-3}-\dfrac{9}{3-y}$

56. $\dfrac{t^2}{t-2}+\dfrac{4}{2-t}$

Aha! 57. $\dfrac{c-5}{c^2-64}-\dfrac{5-c}{64-c^2}$

58. $\dfrac{b-4}{b^2-49}+\dfrac{b-4}{49-b^2}$

59. $\dfrac{y+2}{y-7}+\dfrac{3-y}{49-y^2}$

60. $\dfrac{4-p}{25-p^2}+\dfrac{p+1}{p-5}$

61. $\dfrac{x}{x-4}-\dfrac{3}{16-x^2}$

62. $\dfrac{x}{3-x}-\dfrac{2}{x^2-9}$

63. $\dfrac{3x+2}{3x+6}+\dfrac{x}{4-x^2}$

64. $\dfrac{a}{a^2-1}+\dfrac{2a}{a-a^2}$

65. $\dfrac{4-a^2}{a^2-9}-\dfrac{a-2}{3-a}$

66. $\dfrac{4x}{x^2-y^2}-\dfrac{6}{y-x}$

Perform the indicated operations. Simplify, if possible.

67. $\dfrac{x-3}{2-x}-\dfrac{x+3}{x+2}+\dfrac{x+6}{4-x^2}$

68. $\dfrac{t-5}{1-t}-\dfrac{t+4}{t+1}+\dfrac{t+2}{t^2-1}$

69. $\dfrac{x+5}{x+3}+\dfrac{x+7}{x+2}-\dfrac{7x+19}{(x+3)(x+2)}$

70. $\dfrac{2x+5}{x+1}+\dfrac{x+7}{x+5}-\dfrac{5x+17}{(x+1)(x+5)}$

71. $\dfrac{1}{x+y}+\dfrac{1}{x-y}-\dfrac{2x}{x^2-y^2}$

72. $\dfrac{2r}{r^2-s^2}+\dfrac{1}{r+s}-\dfrac{1}{r-s}$

Find simplified form for $f(x)$ and list all restrictions on the domain.

73. $f(x)=2+\dfrac{x}{x-3}-\dfrac{18}{x^2-9}$

74. $f(x)=5+\dfrac{x}{x+2}-\dfrac{8}{x^2-4}$

75. $f(x) = \dfrac{3x - 1}{x^2 + 2x - 3} - \dfrac{x + 4}{x^2 - 16}$

76. $f(x) = \dfrac{3x - 2}{x^2 + 2x - 24} - \dfrac{x - 3}{x^2 - 9}$

Aha! 77. $f(x) = \dfrac{1}{x^2 + 5x + 6} - \dfrac{2}{x^2 + 3x + 2} - \dfrac{1}{x^2 + 5x + 6}$

78. $f(x) = \dfrac{2}{x^2 - 5x + 6} - \dfrac{4}{x^2 - 2x - 3} + \dfrac{2}{x^2 + 4x + 3}$

TW 79. What is the advantage of using the *least* common denominator—rather than just *any* common denominator—when adding or subtracting rational expressions?

TW 80. Describe a procedure that can be used to add any two rational expressions.

SKILL REVIEW

To prepare for Section 7.5, review division of fractions and rational expressions (Sections 1.3, 1.7, and 7.2).

Simplify.

81. $-\dfrac{3}{8} \div \dfrac{11}{4}$ [1.7]

82. $-\dfrac{7}{12} \div \left(-\dfrac{3}{4}\right)$ [1.7]

83. $\dfrac{\frac{3}{4}}{\frac{5}{6}}$ [1.3]

84. $\dfrac{\frac{8}{15}}{\frac{9}{10}}$ [1.3]

85. $\dfrac{2x + 6}{x - 1} \div \dfrac{3x + 9}{x - 1}$ [7.2]

86. $\dfrac{x^2 - 9}{x^2 - 4} \div \dfrac{x^2 + 6x + 9}{x^2 + 4x + 4}$ [7.2]

SYNTHESIS

TW 87. How could you convince someone that

$$\frac{1}{3 - x} \quad \text{and} \quad \frac{1}{x - 3}$$

are opposites of each other?

TW 88. Are parentheses as important for adding rational expressions as they are for subtracting rational expressions? Why or why not?

Write expressions for the perimeter and the area of each rectangle.

89.

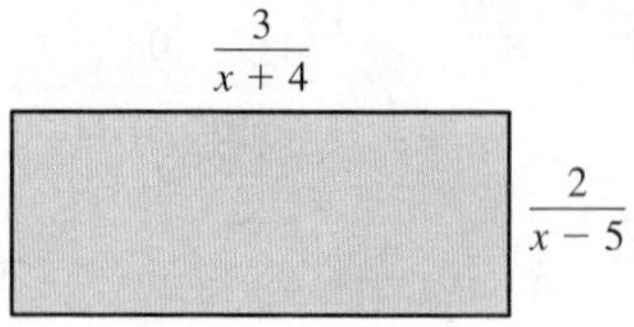

90.

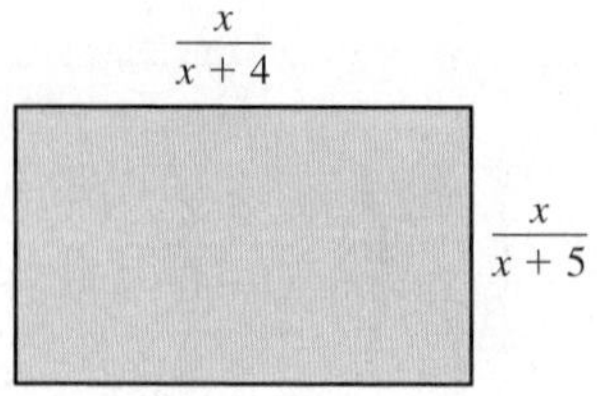

Perform the indicated operations and simplify.

91. $\dfrac{2x + 11}{x - 3} \cdot \dfrac{3}{x + 4} + \dfrac{2x + 1}{4 + x} \cdot \dfrac{3}{3 - x}$

92. $\dfrac{x^2}{3x^2 - 5x - 2} - \dfrac{2x}{3x + 1} \cdot \dfrac{1}{x - 2}$

Aha! 93. $\left(\dfrac{x}{x + 7} - \dfrac{3}{x + 2}\right)\left(\dfrac{x}{x + 7} + \dfrac{3}{x + 2}\right)$

94. $\dfrac{1}{ay - 3a + 2xy - 6x} - \dfrac{xy + ay}{a^2 - 4x^2}\left(\dfrac{1}{y - 3}\right)^2$

95. $\dfrac{2x^2 + 5x - 3}{2x^2 - 9x + 9} + \dfrac{x + 1}{3 - 2x} + \dfrac{4x^2 + 8x + 3}{x - 3} \cdot \dfrac{x + 3}{9 - 4x^2}$

96. $\left(\dfrac{a}{a - b} + \dfrac{b}{a + b}\right)\left(\dfrac{1}{3a + b} + \dfrac{2a + 6b}{9a^2 - b^2}\right)$

97. $5(x - 3)^{-1} + 4(x + 3)^{-1} - 2(x + 3)^{-2}$

98. $4(y - 1)(2y - 5)^{-1} + 5(2y + 3)(5 - 2y)^{-1} + (y - 4)(2y - 5)^{-1}$

99. Express

$$\frac{a - 3b}{a - b}$$

as a sum of two rational expressions with denominators that are opposites of each other. Answers may vary.

100. Determine the domain and the range of the function graphed below.

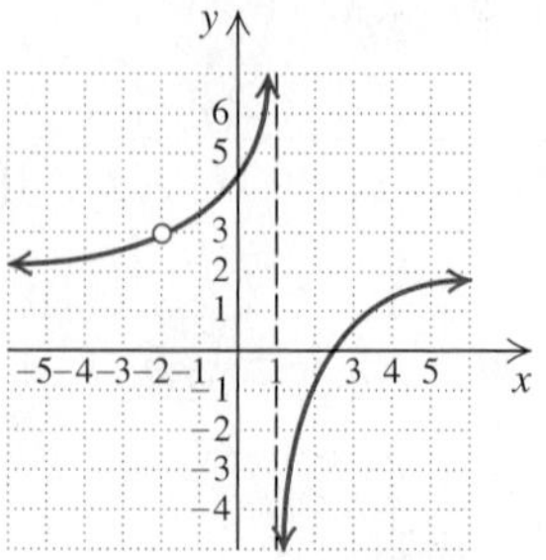

If

$$f(x) = \frac{x^3}{x^2 - 4} \quad and \quad g(x) = \frac{x^2}{x^2 + 3x - 10},$$

find each of the following.

101. $(f + g)(x)$

102. $(f - g)(x)$

103. $(f \cdot g)(x)$

104. $(f/g)(x)$

105. The domain of $f + g$

106. The domain of f/g

Determine the domain and estimate the range of each function.

107. $f(x) = 2 + \frac{x - 3}{x + 1}$

108. $g(x) = \frac{2}{(x + 1)^2} + 5$

109. $r(x) = \frac{1}{x^2} + \frac{1}{(x - 1)^2}$

Try Exercise Answers: Section 7.4

5. $\frac{4x + 9}{x^2}$ **7.** $-\frac{5}{24r}$ **27.** $\frac{2z + 6}{(z - 1)(z + 1)}$ **31.** $\frac{16 - 9t}{6t(t - 5)}$

47. $\frac{x - 5}{(x + 5)(x + 3)}$ **53.** $\frac{3x - 3}{5}$ **55.** $\frac{y^2 + 9}{y - 3}$

73. $f(x) = \frac{3(x + 4)}{x + 3}, x \neq \pm 3$

Mid-Chapter Review

The process of adding and subtracting rational expressions is significantly different from multiplying and dividing them. The first thing you should take note of when combining rational expressions is the operation sign.

Operation	Need Common Denominator?	Procedure	Tips and Cautions
Addition	Yes	Write with a common denominator. Add numerators. Keep denominator.	Do not simplify after writing with the LCD. Instead, simplify after adding the numerators.
Subtraction	Yes	Write with a common denominator. Subtract numerators. Keep denominator.	Use parentheses around the numerator being subtracted. Simplify after subtracting the numerators.
Multiplication	No	Multiply numerators. Multiply denominators.	Do not carry out the multiplications. Instead, factor and try to simplify.
Division	No	Multiply by the reciprocal of the divisor.	Begin by rewriting as a multiplication using the reciprocal of the divisor.

GUIDED SOLUTIONS

1. Divide: $\dfrac{a^2}{a-10} \div \dfrac{a^2+5a}{a^2-100}$. Simplify, if possible. [7.2]

Solution

$$\frac{a^2}{a-10} \div \frac{a^2+5a}{a^2-100}$$

$$= \frac{a^2}{a-10} \cdot \frac{\square}{\square}$$ **Multiplying by the reciprocal of the divisor**

$$= \frac{a \cdot a(a+10) \cdot \square}{(a-10) \cdot a \cdot \square}$$ **Multiplying and factoring**

$$= \frac{\square}{\square} \cdot \frac{a(a+10)}{a+5}$$ **Factoring out a factor equal to 1**

$$= \frac{a(a+10)}{a+5}$$ **Simplifying**

2. Add: $\dfrac{2}{x} + \dfrac{1}{x^2+x}$. Simplify, if possible. [7.4]

Solution

$$\frac{2}{x} + \frac{1}{x^2+x}$$

$$= \frac{2}{x} + \frac{1}{x\square}$$ **Factoring denominators. The LCD is $x(x+1)$.**

$$= \frac{2}{x} \cdot \frac{\square}{\square} + \frac{1}{x(x+1)}$$ **Multiplying by 1 to get the LCD in the first denominator**

$$= \frac{\square}{x(x+1)} + \frac{1}{x(x+1)}$$ **Multiplying**

$$= \frac{\square}{x(x+1)}$$ **Adding numerators. We cannot simplify.**

MIXED REVIEW

Perform the indicated operation and, if possible, simplify.

1. $\dfrac{3}{5x} + \dfrac{2}{x^2}$ [7.4]

2. $\dfrac{3}{5x} \cdot \dfrac{2}{x^2}$ [7.2]

3. $\dfrac{3}{5x} \div \dfrac{2}{x^2}$ [7.2]

4. $\dfrac{3}{5x} - \dfrac{2}{x^2}$ [7.4]

5. $\dfrac{2x-6}{5x+10} \cdot \dfrac{x+2}{6x-12}$ [7.2]

6. $\dfrac{2}{x+3} \cdot \dfrac{3}{x+4}$ [7.2]

7. $\dfrac{2}{x-5} \div \dfrac{6}{x-5}$ [7.2]

8. $\dfrac{x}{x+2} - \dfrac{1}{x-1}$ [7.4]

9. $\dfrac{2}{x+3} + \dfrac{3}{x+4}$ [7.4]

10. $\dfrac{5}{2x-1} + \dfrac{10x}{1-2x}$ [7.4]

11. $\dfrac{3}{x-4} - \dfrac{2}{4-x}$ [7.4]

12. $\dfrac{(x-2)(2x+3)}{(x+1)(x-5)} \div \dfrac{(x-2)(x+1)}{(x-5)(x+3)}$ [7.2]

13. $\dfrac{a}{6a-9b} - \dfrac{b}{4a-6b}$ [7.4]

14. $\dfrac{x^2-16}{x^2-x} \cdot \dfrac{x^2}{x^2-5x+4}$ [7.2]

15. $\dfrac{x+1}{x^2-7x+10} + \dfrac{3}{x^2-x-2}$ [7.4]

16. $\dfrac{3u^2-3}{4} \div \dfrac{4u+4}{3}$ [7.2]

17. $\dfrac{t+2}{10} + \dfrac{2t+1}{15}$ [7.4]

18. $(t^2+t-20) \cdot \dfrac{t+5}{t-4}$ [7.2]

19. $\dfrac{a^2-2a+1}{a^2-4} \div (a^2-3a+2)$ [7.2]

20. $\dfrac{2x-7}{x} - \dfrac{3x-5}{2}$ [7.4]

7.5 Complex Rational Expressions

- Multiplying by the LCD
- Using Division to Simplify

A **complex rational expression** is a rational expression that has one or more rational expressions within its numerator or its denominator. Here are some examples:

$$\frac{x+\frac{5}{x}}{4x}, \quad \frac{\frac{x-y}{x+y}}{\frac{2x-y}{3x+y}}, \quad \frac{\frac{4}{3}+\frac{1}{5}}{\frac{2}{x}-\frac{x}{7}}.$$

The rational expressions within each complex rational expression are red.

STUDY TIP

Multiple Methods

When more than one method is presented, as is the case for simplifying complex rational expressions, it can be helpful to learn both methods. If two different approaches yield the same result, you can be confident that your answer is correct.

When we simplify a complex rational expression, we rewrite it so that it is no longer complex. We will consider two methods for simplifying complex rational expressions. Determining restrictions on variables may now require the solution of equations not yet studied. *Thus for this section we will not state restrictions on variables.*

METHOD 1: MULTIPLYING BY THE LCD

One method of simplifying a complex rational expression is to multiply the entire expression by 1. If we choose the expression for 1 carefully, the multiplication will give an expression that is no longer complex.

To Simplify a Complex Rational Expression by Multiplying by the LCD

1. Find the LCD of all rational expressions *within* the complex rational expression.
2. Multiply the complex rational expression by an expression equal to 1. Write 1 as the LCD over itself (LCD/LCD).
3. Simplify. No fraction expressions should remain within the complex rational expression.
4. Factor and, if possible, simplify.

EXAMPLE 1 Simplify: $\dfrac{\frac{1}{2}+\frac{3}{4}}{\frac{5}{6}-\frac{3}{8}}$.

SOLUTION

1. The LCD of $\frac{1}{2}, \frac{3}{4}, \frac{5}{6}$, and $\frac{3}{8}$ is 24.
2. We multiply by an expression equal to 1:

$$\frac{\frac{1}{2}+\frac{3}{4}}{\frac{5}{6}-\frac{3}{8}} = \frac{\frac{1}{2}+\frac{3}{4}}{\frac{5}{6}-\frac{3}{8}} \cdot \frac{24}{24}.$$

Multiplying by an expression equal to 1, using the LCD: $\frac{24}{24} = 1$

3. Using the distributive law, we perform the multiplication:

$$\frac{\frac{1}{2}+\frac{3}{4}}{\frac{5}{6}-\frac{3}{8}}\cdot\frac{24}{24}=\frac{\left(\frac{1}{2}+\frac{3}{4}\right)24}{\left(\frac{5}{6}-\frac{3}{8}\right)24}$$

← **Multiplying the numerator by 24**
Don't forget the parentheses!
← **Multiplying the denominator by 24**

$$=\frac{\frac{1}{2}(24)+\frac{3}{4}(24)}{\frac{5}{6}(24)-\frac{3}{8}(24)}$$ **Using the distributive law**

$$=\frac{12+18}{20-9},\text{ or }\frac{30}{11}.$$ **Simplifying**

4. The result, $\frac{30}{11}$, cannot be factored or simplified, so we are done.

Try Exercise 5.

Multiplying in such a way effectively clears fractions in both the top and the bottom of the complex rational expression.

EXAMPLE 2 Simplify: $\dfrac{1-\frac{1}{x}}{1-\frac{1}{x^2}}$.

SOLUTION

1. Find the LCD.

2. Multiply by LCD/LCD.

$$\frac{1-\frac{1}{x}}{1-\frac{1}{x^2}}=\frac{1-\frac{1}{x}}{1-\frac{1}{x^2}}\cdot\frac{x^2}{x^2}$$ **The LCD is x^2 so we multiply by 1 using x^2/x^2.**

$$=\frac{1\cdot x^2-\frac{1}{x}\cdot x^2}{1\cdot x^2-\frac{1}{x^2}\cdot x^2}$$ **Using the distributive law**

3. Multiply and simplify.

$$=\frac{x^2-x}{x^2-1}$$ **All fractions have been cleared within the complex rational expression.**

4. Factor and simplify.

$$=\frac{x\cancel{(x-1)}}{(x+1)\cancel{(x-1)}}$$ **Factoring and simplifying: $\frac{x-1}{x-1}=1$**

$$=\frac{x}{x+1}$$

Try Exercise 9.

EXAMPLE 3 Simplify:

$$\frac{\dfrac{3}{2x-2}-\dfrac{1}{x+1}}{\dfrac{1}{x-1}+\dfrac{x}{x^2-1}}.$$

SOLUTION In this case, to find the LCD, we must factor first:

$$\frac{\dfrac{3}{2x-2}-\dfrac{1}{x+1}}{\dfrac{1}{x-1}+\dfrac{x}{x^2-1}}=\frac{\dfrac{3}{2(x-1)}-\dfrac{1}{x+1}}{\dfrac{1}{x-1}+\dfrac{x}{(x-1)(x+1)}}$$ **The LCD is $2(x-1)(x+1)$.**

Student Notes

Writing $2(x-1)(x+1)$ as

$$\frac{2(x-1)(x+1)}{1}$$

may help with lining up numerators and denominators for multiplying.

$$=\frac{\dfrac{3}{2(x-1)}-\dfrac{1}{x+1}}{\dfrac{1}{x-1}+\dfrac{x}{(x-1)(x+1)}}\cdot\frac{2(x-1)(x+1)}{2(x-1)(x+1)}$$ **Multiplying by 1, using the LCD**

$$=\frac{\dfrac{3}{2(x-1)}\cdot 2(x-1)(x+1)-\dfrac{1}{x+1}\cdot 2(x-1)(x+1)}{\dfrac{1}{x-1}\cdot 2(x-1)(x+1)+\dfrac{x}{(x-1)(x+1)}\cdot 2(x-1)(x+1)}$$ **Using the distributive law**

$$=\frac{\dfrac{\cancel{2(x-1)}}{\cancel{2(x-1)}}\cdot 3(x+1)-\dfrac{\cancel{x+1}}{\cancel{x+1}}\cdot 2(x-1)}{\dfrac{\cancel{x-1}}{\cancel{x-1}}\cdot 2(x+1)+\dfrac{\cancel{(x-1)(x+1)}}{\cancel{(x-1)(x+1)}}\cdot 2x}$$ **Removing factors that equal 1**

$$=\frac{3(x+1)-2(x-1)}{2(x+1)+2x}$$ **Simplifying**

$$=\frac{3x+3-2x+2}{2x+2+2x}$$ **Using the distributive law**

$$=\frac{x+5}{4x+2}.$$ **Combining like terms**

Try Exercise 41.

Entering Rational Expressions

To enter a complex rational expression in a calculator, use parentheses around the entire numerator and parentheses around the entire denominator. If necessary, also use parentheses around numerators and denominators of rational expressions within the complex rational expression. Following are some examples of complex rational expressions rewritten with parentheses.

(continued)

Complex Rational Expression	Expression Rewritten with Parentheses
$\dfrac{x + \dfrac{5}{x}}{4x}$	$(x + 5/x)/(4x)$
$\dfrac{\dfrac{3}{2x - 2} - \dfrac{1}{x + 1}}{\dfrac{1}{x - 1} + \dfrac{x}{x^2 - 1}}$	$(3/(2x - 2) - 1/(x + 1))/(1/(x - 1) + x/(x^2 - 1))$

The following may help in placing parentheses properly.

1. Close each set of parentheses: For every left parenthesis, there should be a right parenthesis.
2. Remember the rules for order of operations when deciding how to place parentheses.
3. When in doubt, place parentheses around every numerator and around every denominator.

Your Turn

1. Enter $y_1 = \dfrac{x + \dfrac{1}{x}}{\dfrac{1}{x + 1} - 1}$.
2. Check your entry by evaluating y_1 for $x = 1$ and for $x = 2$. $-4, -3.75$

METHOD 2: USING DIVISION TO SIMPLIFY

Another method for simplifying complex rational expressions involves rewriting the expression as a quotient of two rational expressions.

EXAMPLE 4 Simplify: $\dfrac{\dfrac{x}{x - 3}}{\dfrac{4}{5x - 15}}$.

SOLUTION The numerator and the denominator are single rational expressions. We divide the numerator by the denominator:

$$\frac{\dfrac{x}{x - 3}}{\dfrac{4}{5x - 15}} = \frac{x}{x - 3} \div \frac{4}{5x - 15}$$ **Rewriting with a division symbol**

$$= \frac{x}{x - 3} \cdot \frac{5x - 15}{4}$$ **Multiplying by the reciprocal of the divisor (inverting and multiplying)**

$$= \frac{x}{\cancel{x - 3}} \cdot \frac{5\cancel{(x - 3)}}{4}$$ **Factoring and removing a factor equal to 1: $\dfrac{x - 3}{x - 3} = 1$**

$$= \frac{5x}{4}.$$

$y_1 = (x/(x - 3))/(4/(5x - 15))$,
$y_2 = (5x)/4$

X	Y1	Y2
−2	−2.5	−2.5
−1	−1.25	−1.25
0	0	0
1	1.25	1.25
2	2.5	2.5
3	ERROR	3.75
4	5	5

X = −2

A table of values, as shown at left, indicates that the original expression and the simplified expression are equivalent.

Try Exercise 11.

We can use this method even when the numerator and the denominator are not (yet) written as single rational expressions.

To Simplify a Complex Rational Expression by Dividing

1. Add or subtract, as needed, to get a single rational expression in the numerator.
2. Add or subtract, as needed, to get a single rational expression in the denominator.
3. Divide the numerator by the denominator (invert the divisor and multiply).
4. If possible, simplify by removing a factor equal to 1.

The key here is to express a complex rational expression as one rational expression divided by another.

EXAMPLE 5 Simplify:

$$\frac{1+\dfrac{2}{x}}{1-\dfrac{4}{x^2}}.$$

SOLUTION We have

$$\frac{1+\dfrac{2}{x}}{1-\dfrac{4}{x^2}} = \frac{\dfrac{x}{x}+\dfrac{2}{x}}{\dfrac{x^2}{x^2}-\dfrac{4}{x^2}}$$
Finding a common denominator (numerator)
Finding a common denominator (denominator)

1. Add to get a single rational expression in the numerator.

2. Subtract to get a single rational expression in the denominator.

$$= \frac{\dfrac{x+2}{x}}{\dfrac{x^2-4}{x^2}}$$
Adding in the numerator
Subtracting in the denominator

3. Divide the numerator by the denominator.

$$= \frac{x+2}{x} \div \frac{x^2-4}{x^2}$$
Rewriting with a division symbol

$$= \frac{x+2}{x} \cdot \frac{x^2}{x^2-4}$$
Multiplying by the reciprocal of the divisor

$$= \frac{(x+2)\cdot x\cdot x}{x(x+2)(x-2)}$$
Factoring. Remember to simplify when possible.

4. Simplify.

$$= \frac{\cancel{(x+2)}\cancel{x}\cdot x}{\cancel{x}\cancel{(x+2)}(x-2)}$$
Removing a factor equal to 1: $\frac{(x+2)x}{(x+2)x} = 1$

$$= \frac{x}{x-2}.$$
Simplifying

As a quick, partial check, we select a convenient value for x—say, 1—and evaluate both the original expression and the simplified expression.

Evaluating the Original Expression for $x = 1$

$$\frac{1 + \frac{2}{1}}{1 - \frac{4}{1^2}} = \frac{1 + 2}{1 - 4} = \frac{3}{-3} = -1$$

Evaluating the Simplified Expression for $x = 1$

$$\frac{1}{1 - 2} = \frac{1}{-1} = -1$$

The value of both expressions is -1, so the simplification is probably correct. Evaluating the expression for more values of x would make the check more certain.

Try Exercise 19.

If negative exponents occur, we first find an equivalent expression using positive exponents.

EXAMPLE 6 Simplify: $\dfrac{x^2y^{-1} - 5x^{-1}}{xz}$.

SOLUTION

$$\frac{x^2y^{-1} - 5x^{-1}}{xz} = \frac{\frac{x^2}{y} - \frac{5}{x}}{xz}$$ Rewriting with positive exponents

$$= \frac{\frac{x^2}{y} \cdot \frac{x}{x} - \frac{5}{x} \cdot \frac{y}{y}}{xz}$$ ← Multiplying by 1 to get the LCD, xy, for the numerator of the complex rational expression

$$= \frac{\frac{x^3}{xy} - \frac{5y}{xy}}{xz}$$

$$= \frac{\frac{x^3 - 5y}{xy}}{xz}$$ ← Subtracting; xz ← If you prefer, write xz as $\frac{xz}{1}$.

$$= \frac{x^3 - 5y}{xy} \div (xz)$$ Rewriting with a division symbol

$$= \frac{x^3 - 5y}{xy} \cdot \frac{1}{xz}$$ Multiplying by the reciprocal of the divisor (inverting and multiplying)

$$= \frac{x^3 - 5y}{x^2yz}$$

Try Exercise 21.

There is no one method that is best to use. When it is little or no work to write an expression as a quotient of two rational expressions, the second method is probably easier to use. On the other hand, some expressions require fewer steps if we use the first method. Either method can be used with any complex rational expression.

7.5 Exercise Set

FOR EXTRA HELP MyMathLab   

Concept Reinforcement *Each of Exercises 1–4 shows a complex rational expression and the first step taken to simplify that expression. Indicate for each which method is being used:* **(a)** *multiplying by the LCD (method 1), or* **(b)** *using division to simplify (method 2).*

1. ____ $\dfrac{\frac{1}{x}+\frac{1}{2}}{\frac{1}{3}-\frac{1}{x}} \Rightarrow \dfrac{\frac{1}{x}+\frac{1}{2}}{\frac{1}{3}-\frac{1}{x}}\cdot\dfrac{6x}{6x}$

2. ____ $\dfrac{\frac{1}{x}+\frac{1}{2}}{\frac{1}{3}-\frac{1}{x}} \Rightarrow \dfrac{\frac{1}{x}\cdot\frac{2}{2}+\frac{1}{2}\cdot\frac{x}{x}}{\frac{1}{3}\cdot\frac{x}{x}-\frac{1}{x}\cdot\frac{3}{3}}$

3. ____ $\dfrac{\frac{x-1}{x}}{\frac{x^2}{x^2-1}} \Rightarrow \dfrac{x-1}{x}\div\dfrac{x^2}{x^2-1}$

4. ____ $\dfrac{\frac{x-1}{x}}{\frac{x^2}{x^2-1}} \Rightarrow \dfrac{\frac{x-1}{x}}{\frac{x^2}{x^2-1}}\cdot\dfrac{x(x+1)(x-1)}{x(x+1)(x-1)}$

Simplify. If possible, use a second method, evaluation, or a graphing calculator as a check.

5. $\dfrac{\frac{1}{2}+\frac{1}{3}}{\frac{1}{4}-\frac{1}{6}}$

6. $\dfrac{\frac{2}{5}-\frac{1}{10}}{\frac{7}{20}-\frac{4}{15}}$

7. $\dfrac{1+\frac{1}{4}}{2+\frac{3}{4}}$

8. $\dfrac{3+\frac{1}{4}}{1+\frac{1}{2}}$

9. $\dfrac{\frac{x}{4}+x}{\frac{4}{x}+x}$

10. $\dfrac{\frac{1}{c}+2}{\frac{1}{c}-5}$

11. $\dfrac{\frac{x+2}{x-1}}{\frac{x+4}{x-3}}$

12. $\dfrac{\frac{x-1}{x+3}}{\frac{x-6}{x+2}}$

13. $\dfrac{\frac{5}{a}-\frac{4}{b}}{\frac{2}{a}+\frac{3}{b}}$

14. $\dfrac{\frac{2}{r}-\frac{3}{t}}{\frac{4}{r}+\frac{5}{t}}$

15. $\dfrac{\frac{1}{t}-\frac{1}{5}}{\frac{t-5}{t}}$

16. $\dfrac{\frac{1}{9}-\frac{1}{n}}{\frac{n+9}{9}}$

17. $\dfrac{\frac{a^2-b^2}{ab}}{\frac{a-b}{b}}$

18. $\dfrac{\frac{x^2-y^2}{xy}}{\frac{x-y}{y}}$

19. $\dfrac{1-\frac{2}{3x}}{x-\frac{4}{9x}}$

20. $\dfrac{\frac{3x}{y}-x}{2y-\frac{y}{x}}$

21. $\dfrac{x^{-1}+y^{-1}}{\frac{x^2-y^2}{xy}}$

22. $\dfrac{a^{-1}+b^{-1}}{\frac{a^2-b^2}{ab}}$

23. $\dfrac{\frac{1}{a-h}-\frac{1}{a}}{h}$

24. $\dfrac{\frac{1}{x+h}-\frac{1}{x}}{h}$

25. $\dfrac{\frac{a^2-4}{a^2+3a+2}}{\frac{a^2-5a-6}{a^2-6a-7}}$

26. $\dfrac{\frac{x^2-x-12}{x^2-2x-15}}{\frac{x^2+8x+12}{x^2-5x-14}}$

27. $\dfrac{\frac{x}{x^2+3x-4}-\frac{1}{x^2+3x-4}}{\frac{x}{x^2+6x+8}+\frac{3}{x^2+6x+8}}$

28. $\dfrac{\frac{x}{x^2+5x-6}+\frac{6}{x^2+5x-6}}{\frac{x}{x^2-5x+4}-\frac{2}{x^2-5x+4}}$

29. $\dfrac{y+y^{-1}}{y-y^{-1}}$

30. $\dfrac{x-x^{-1}}{x+x^{-1}}$

31. $\dfrac{\dfrac{x^2}{x^2 - y^2}}{\dfrac{x}{x + y}}$

32. $\dfrac{\dfrac{a^2}{a - 2}}{\dfrac{3a}{a^2 - 4}}$

Aha! 33. $\dfrac{\dfrac{x}{5y^3} + \dfrac{3}{10y}}{\dfrac{3}{10y} + \dfrac{x}{5y^3}}$

34. $\dfrac{\dfrac{a}{6b^3} + \dfrac{4}{9b^2}}{\dfrac{5}{6b} - \dfrac{1}{9b^3}}$

35. $\dfrac{\dfrac{3}{ab^4} + \dfrac{4}{a^3b}}{ab}$

36. $\dfrac{\dfrac{2}{x^2y} + \dfrac{3}{xy^2}}{xy}$

37. $\dfrac{x - y}{\dfrac{1}{x^3} - \dfrac{1}{y^3}}$

38. $\dfrac{\dfrac{1}{a} + \dfrac{1}{b}}{\dfrac{1}{a^3} + \dfrac{1}{b^3}}$

39. $\dfrac{a(a + 3)^{-1} - 2(a - 1)^{-1}}{a(a + 3)^{-1} - (a - 1)^{-1}}$

40. $\dfrac{a(a + 2)^{-1} - 3(a - 3)^{-1}}{a(a + 2)^{-1} - (a - 3)^{-1}}$

41. $\dfrac{\dfrac{2}{a^2 - 1} + \dfrac{1}{a + 1}}{\dfrac{3}{a^2 - 1} + \dfrac{2}{a - 1}}$

42. $\dfrac{\dfrac{3}{a^2 - 9} + \dfrac{2}{a + 3}}{\dfrac{4}{a^2 - 9} + \dfrac{1}{a + 3}}$

43. $\dfrac{\dfrac{y^2}{y^2 - 25} - \dfrac{y}{y - 5}}{\dfrac{y}{y^2 - 25} - \dfrac{1}{y + 5}}$

44. $\dfrac{\dfrac{y^2}{y^2 - 9} - \dfrac{y}{y + 3}}{\dfrac{y}{y^2 - 9} - \dfrac{1}{y - 3}}$

45. $\dfrac{\dfrac{3}{a^2 - 4a + 3} + \dfrac{3}{a^2 - 5a + 6}}{\dfrac{3}{a^2 - 3a + 2} + \dfrac{3}{a^2 + 3a - 10}}$

46. $\dfrac{\dfrac{1}{a^2 + 7a + 10} - \dfrac{2}{a^2 - 7a + 12}}{\dfrac{2}{a^2 - a - 6} - \dfrac{1}{a^2 + a - 20}}$

47. $\dfrac{t + 5 + \dfrac{3}{t}}{t + 2 + \dfrac{1}{t}}$

48. $\dfrac{a + 3 + \dfrac{2}{a}}{a + 2 + \dfrac{5}{a}}$

49. $\dfrac{x - 2 - \dfrac{1}{x}}{x - 5 - \dfrac{4}{x}}$

50. $\dfrac{x - 3 - \dfrac{2}{x}}{x - 4 - \dfrac{3}{x}}$

TW 51. Is it possible to simplify complex rational expressions without knowing how to divide rational expressions? Why or why not?

TW 52. Why is the distributive law important when simplifying complex rational expressions?

SKILL REVIEW

To prepare for Section 7.6, review solving linear and quadratic equations (Section 2.2 and Chapter 6).

Solve.

53. $3x - 5 + 2(4x - 1) = 12x - 3$ [2.2]

54. $(x - 1)7 - (x + 1)9 = 4(x + 2)$ [2.2]

55. $\frac{3}{4}x - \frac{5}{8} = \frac{3}{8}x + \frac{7}{4}$ [2.2]

56. $\frac{5}{9} - \frac{2x}{3} = \frac{5x}{6} + \frac{4}{3}$ [2.2]

57. $x^2 - 7x + 12 = 0$ [6.2]

58. $x^2 + 13x - 30 = 0$ [6.2]

SYNTHESIS

TW 59. Compare the solutions begun in Concept Reinforcement Exercises 1 and 2. Which method is the better method to use in order to simplify the given expression? Also compare the solutions begun in Exercises 3 and 4. Which method is better? Explain your answers.

TW 60. Use algebra to determine the domain of the function given by

$$f(x) = \frac{\dfrac{1}{x - 2}}{\dfrac{x}{x - 2} - \dfrac{5}{x - 2}}.$$

Then explain how a graphing calculator could be used to check your answer.

In Exercises 61–64, find all x-values for which the given expression is undefined.

61. $\dfrac{\dfrac{x - 5}{x - 6}}{\dfrac{x - 7}{x - 8}}$

62. $\dfrac{\dfrac{x + 1}{x + 2}}{\dfrac{x + 3}{x + 4}}$

63. $\dfrac{\dfrac{2x+3}{5x+4}}{\dfrac{3}{7}-\dfrac{x^2}{21}}$

64. $\dfrac{\dfrac{3x-5}{2x-7}}{\dfrac{4x}{5}-\dfrac{8}{15}}$

65. Use multiplication by the LCD (method 1) to show that

$$\frac{A}{B} \div \frac{C}{D} = \frac{A}{B} \cdot \frac{D}{C}.$$

(*Hint*: Begin by forming a complex rational expression.)

66. The formula

$$\frac{P\left(1+\dfrac{i}{12}\right)^2}{\dfrac{\left(1+\dfrac{i}{12}\right)^2-1}{\dfrac{i}{12}}},$$

where P is a loan amount and i is an interest rate, arises in certain business situations. Simplify this expression. (*Hint*: Expand the binomials.)

67. *Astronomy.* When two galaxies are moving in opposite directions at velocities v_1 and v_2, an observer in one of the galaxies would see the other galaxy receding at speed

$$\frac{v_1+v_2}{1+\dfrac{v_1v_2}{c^2}},$$

where c is the speed of light. Determine the observed speed if v_1 and v_2 are both one-fourth the speed of light.

Simplify.

68. $\dfrac{5x^{-2}+10x^{-1}y^{-1}+5y^{-2}}{3x^{-2}-3y^{-2}}$

Aha! **69.** $\left[\dfrac{\dfrac{x-1}{x-1}-1}{\dfrac{x+1}{x-1}+1}\right]^5$

70. $1+\dfrac{1}{1+\dfrac{1}{1+\dfrac{1}{x}}}$

71. $\dfrac{\dfrac{z}{1-\dfrac{z}{2+2z}}-2z}{\dfrac{2z}{5z-2}-3}$

72. Find the simplified form for the reciprocal of

$$\frac{2}{x-1}-\frac{1}{3x-2}.$$

Find and simplify

$$\frac{f(x+h)-f(x)}{h}$$

for each rational function f in Exercises 73 and 74.

73. $f(x)=\dfrac{3}{x}$

74. $f(x)=\dfrac{x}{1-x}$

75. If

$$F(x)=\frac{3+\dfrac{1}{x}}{2-\dfrac{8}{x^2}},$$

find the domain of F.

76. For $f(x)=\dfrac{2}{2+x}$, find $f(f(a))$.

77. *Financial Planning.* Alexis wishes to invest a portion of each month's pay in an account that pays 7.5% interest. If he wants to have \$30,000 in the account after 10 years, the amount invested each month is given by

$$\frac{30{,}000 \cdot \dfrac{0.075}{12}}{\left(1+\dfrac{0.075}{12}\right)^{120}-1}.$$

Find the amount of Alexis' monthly investment.

Try Exercise Answers: Section 7.5

5. 10 **9.** $\dfrac{5x^2}{4(x^2+4)}$ **11.** $\dfrac{(x+2)(x-3)}{(x-1)(x+4)}$ **19.** $\dfrac{3}{3x+2}$ **21.** $\dfrac{1}{x-y}$ **41.** $\dfrac{a+1}{2a+5}$

Collaborative Corner

Which Method Is Better?

Focus: Complex rational expressions
Time: 10–15 minutes
Group Size: 3–4

ACTIVITY

Consider the steps in Examples 2 and 5 for simplifying a complex rational expression by each of the two methods. Then, work as a group to simplify

$$\frac{\dfrac{5}{x+1} - \dfrac{1}{x}}{\dfrac{2}{x^2} + \dfrac{4}{x}}$$

subject to the following conditions.

1. The group should predict which method will more easily simplify this expression.
2. Using the method selected in part (1), one group member should perform the first step in the simplification and then pass the problem on to another member of the group. That person then checks the work, performs the next step, and passes the problem on to another group member. If a mistake is found, the problem should be passed to the person who made the mistake for repair. This process continues until, eventually, the simplification is complete.
3. At the same time that part (2) is being performed, another group member should perform the first step of the solution using the method not selected in part (1). He or she should then pass the problem to another group member and so on, just as in part (2).
4. What method *was* easier? Why? Compare your responses with those of other groups.

7.6 Rational Equations

- Solving Rational Equations

Our study of rational expressions allows us to solve a type of equation that we could not have solved prior to this chapter.

STUDY TIP 

Find All Solutions

Keep in mind that many problems—in math and elsewhere—have more than one solution. When asked to solve an equation, we are expected to find any and all solutions of the equation.

SOLVING RATIONAL EQUATIONS

A **rational equation** is an equation that contains one or more rational expressions. Here are some examples:

$$\frac{2}{3} - \frac{5}{6} = \frac{1}{t}, \qquad \frac{a-1}{a-5} = \frac{4}{a^2-25}, \qquad x^3 + \frac{6}{x} = 5.$$

To solve rational equations, recall that one way to *clear fractions* from an equation is to multiply both sides of the equation by the LCM of the denominators.

To Solve a Rational Equation

1. List any restrictions that exist. Numbers that make a denominator equal 0 can never be solutions.
2. Clear the equation of fractions by multiplying both sides by the LCM of the denominators.
3. Solve the resulting equation using the addition principle, the multiplication principle, and the principle of zero products, as needed.
4. Check the possible solution(s) in the original equation.

When clearing an equation of fractions, we use the terminology LCM instead of LCD because we are *not* adding or subtracting rational expressions.

Recall that the multiplication principle states that $a = b$ is equivalent to $a \cdot c = b \cdot c$, *provided c is not 0*. Because rational equations often have variables in a denominator, clearing fractions will now require us to multiply both sides by a variable expression. Since a variable expression could represent 0, *multiplying both sides of an equation by a variable expression does not always produce an equivalent equation*. Thus checking each solution in the original equation is essential.

EXAMPLE 1 Solve.

a) $\frac{2}{3x} + \frac{1}{x} = 10$

b) $1 + \frac{3x}{x+2} = \frac{-6}{x+2}$

SOLUTION

1. List restrictions.

a) If $x = 0$, both denominators are 0. We list this restriction:

$x \neq 0.$

We now clear the equation of fractions and solve:

2. Clear fractions.

3. Solve.

$$\frac{2}{3x} + \frac{1}{x} = 10$$ **We cannot have $x = 0$. The LCM is $3x$.**

$$3x\left(\frac{2}{3x} + \frac{1}{x}\right) = 3x \cdot 10$$ **Using the multiplication principle to multiply both sides by the LCM.** ***Don't forget the parentheses!***

$$\cancel{3x} \cdot \frac{2}{\cancel{3x}} + 3\cancel{x} \cdot \frac{1}{\cancel{x}} = 3x \cdot 10$$ **Using the distributive law. Removing factors equal to 1: $(3x)/(3x) = 1$ and $x/x = 1$.**

$$2 + 3 = 30x$$ **All fractions are now cleared.**

$$5 = 30x$$

$$\frac{5}{30} = x, \quad \text{so } x = \frac{1}{6}.$$ **Dividing both sides by 30, or multiplying both sides by 1/30**

Since $\frac{1}{6} \neq 0$, and 0 is the only restricted value, $\frac{1}{6}$ *should* check.

4. Check.

Check: $\frac{2}{3x} + \frac{1}{x} = 10$

$\frac{2}{3 \cdot \frac{1}{6}} + \frac{1}{\frac{1}{6}}$	10
$\frac{2}{\frac{1}{2}} + \frac{1}{\frac{1}{6}}$	
$2 \cdot \frac{2}{1} + 1 \cdot \frac{6}{1}$	
$4 + 6$	

$10 \stackrel{?}{=} 10$ TRUE

$y_1 = 2/(3x) + 1/x,$
$y_2 = 10$

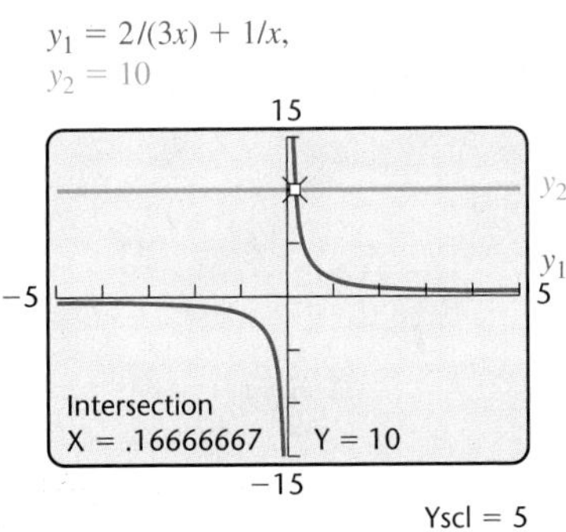

Yscl = 5

The solution is $\frac{1}{6}$. The solution is confirmed by the graph shown at left.

Student Notes

Not all checking is for finding errors in computation. For these equations, the solution process itself can introduce numbers that do not check.

b) The only denominator is $x + 2$. We set this equal to 0 and solve:

$$x + 2 = 0$$
$$x = -2.$$

If $x = -2$, the rational expressions are undefined. We list the restriction:

$$x \neq -2.$$

We clear fractions and solve:

$$1 + \frac{3x}{x + 2} = \frac{-6}{x + 2}$$ **We cannot have $x = -2$. The LCM is $x + 2$.**

$$(x + 2)\left(1 + \frac{3x}{x + 2}\right) = (x + 2)\frac{-6}{x + 2}$$ **Multiplying both sides by the LCM.** ***Don't forget the parentheses!***

$$(x + 2) \cdot 1 + \cancel{(x + 2)}\frac{3x}{\cancel{x + 2}} = \cancel{(x + 2)}\frac{-6}{\cancel{x + 2}}$$ **Using the distributive law; removing a factor equal to 1: $(x + 2)/(x + 2) = 1$**

$$x + 2 + 3x = -6$$
$$4x + 2 = -6$$
$$4x = -8$$
$$x = -2.$$ **Above, we stated that $x \neq -2$.**

Because of the above restriction, -2 must be rejected as a solution. The check below simply confirms this.

Check:

$$\begin{array}{c|c} 1 + \dfrac{3x}{x + 2} & \dfrac{-6}{x + 2} \\ \hline 1 + \dfrac{3(-2)}{-2 + 2} & \dfrac{-6}{-2 + 2} \\ 1 + \dfrac{-6}{0} \overset{?}{=} & \dfrac{-6}{0} \end{array}$$ FALSE

The equation has no solution.

Try Exercise 13.

When we are solving rational equations graphically, it can be difficult to tell whether two graphs actually intersect. Also, it is easy to "miss" seeing a solution. This is especially true where there are two or more branches of the graph. Solving by graphing does make a good check, however, of an algebraic solution.

EXAMPLE 2 Solve: $\dfrac{x^2}{x - 3} = \dfrac{9}{x - 3}$.

ALGEBRAIC APPROACH

Note that $x \neq 3$. We multiply both sides by the LCM of the denominators, $x - 3$:

$$(x-3) \cdot \frac{x^2}{x-3} = (x-3) \cdot \frac{9}{x-3}$$

$$x^2 = 9 \quad \textbf{Simplifying}$$

$$x^2 - 9 = 0 \quad \textbf{Getting 0 on one side}$$

$$(x-3)(x+3) = 0 \quad \textbf{Factoring}$$

$$x = 3 \quad or \quad x = -3. \quad \textbf{Using the principle of zero products}$$

Although 3 is a solution of $x^2 = 9$, it must be rejected as a solution of the rational equation because of the restriction stated in red above. A check shows that -3 is the only solution.

Check:

For 3:

$$\frac{x^2}{x-3} = \frac{9}{x-3}$$

$$\frac{3^2}{3-3} \stackrel{?}{=} \frac{9}{3-3} \quad \text{FALSE} \quad \textbf{Division by 0 is undefined.}$$

For -3:

$$\frac{x^2}{x-3} = \frac{9}{x-3}$$

$$\frac{(-3)^2}{-3-3} \;\Big|\; \frac{9}{-3-3}$$

$$\frac{9}{-6} \stackrel{?}{=} \frac{9}{-6} \quad \text{TRUE}$$

The solution is -3.

GRAPHICAL APPROACH

We graph $y_1 = x^2/(x-3)$ and $y_2 = 9/(x-3)$ and look for any points of intersection of the graphs.

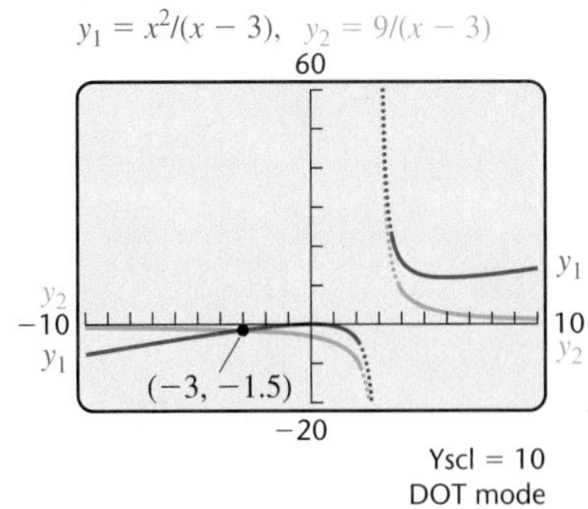

The graphs intersect at $(-3, -1.5)$, so -3 is a solution. Both graphs approach the asymptote $x = 3$, but they do not intersect. This is difficult to determine from the graph. The solution is -3.

Try Exercise 31.

EXAMPLE 3 Solve: $\frac{2}{x+5} + \frac{1}{x-5} = \frac{16}{x^2 - 25}$.

SOLUTION To find all restrictions and to assist in finding the LCM, we factor:

$$\frac{2}{x+5} + \frac{1}{x-5} = \frac{16}{(x+5)(x-5)}. \quad \textbf{Factoring } x^2 - 25$$

Note that $x \neq -5$ and $x \neq 5$. We multiply by the LCM, $(x+5)(x-5)$, and then use the distributive law:

$$(x+5)(x-5)\left(\frac{2}{x+5} + \frac{1}{x-5}\right) = (x+5)(x-5) \cdot \frac{16}{(x+5)(x-5)}$$

$$(x+5)(x-5)\frac{2}{x+5} + (x+5)(x-5)\frac{1}{x-5} = \frac{(x+5)(x-5)16}{(x+5)(x-5)}$$

$$2(x-5) + (x+5) = 16$$

$$2x - 10 + x + 5 = 16$$

$$3x - 5 = 16$$

$$3x = 21$$

$$x = 7.$$

A check will confirm that the solution is 7.

Try Exercise 43.

EXAMPLE 4 Let $f(x) = x + \frac{6}{x}$. Find all values of a for which $f(a) = 5$.

SOLUTION Since $f(a) = a + \frac{6}{a}$, the problem asks that we find all values of a for which

$$a + \frac{6}{a} = 5.$$

ALGEBRAIC APPROACH

First note that $a \neq 0$. To solve for a, we multiply both sides of the equation by the LCM of the denominators, a:

$$a\left(a + \frac{6}{a}\right) = 5 \cdot a \qquad \text{Multiplying both sides by } a. \text{ Parentheses are important.}$$

$$a \cdot a + a \cdot \frac{6}{a} = 5a \qquad \text{Using the distributive law}$$

$$a^2 + 6 = 5a \qquad \text{Simplifying}$$

$$a^2 - 5a + 6 = 0 \qquad \text{Getting 0 on one side}$$

$$(a - 3)(a - 2) = 0 \qquad \text{Factoring}$$

$$a = 3 \quad or \quad a = 2. \qquad \text{Using the principle of zero products}$$

Check:

$$f(3) = 3 + \frac{6}{3} = 3 + 2 = 5;$$

$$f(2) = 2 + \frac{6}{2} = 2 + 3 = 5.$$

The solutions are 2 and 3.

GRAPHICAL APPROACH

We graph $y_1 = x + 6/x$ and $y_2 = 5$, and look for any points of intersection of the graphs.

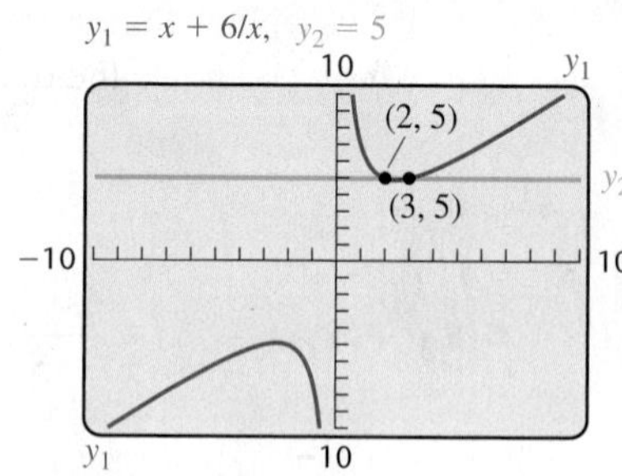

The graphs intersect at $(2, 5)$ and $(3, 5)$. The solutions are 2 and 3.

Try Exercise 49.

Connecting the Concepts

An equation contains an equals sign; an expression does not. Be careful not to confuse simplifying an expression with solving an equation. When expressions are simplified, the result is an equivalent expression. When equations are solved, the result is a solution. Compare the following.

Simplify: $\frac{x-1}{6x} + \frac{4}{9}$.

SOLUTION We have

$$\frac{x-1}{6x} + \frac{4}{9} = \frac{x-1}{6x} \cdot \frac{3}{3} + \frac{4}{9} \cdot \frac{2x}{2x}$$

The equals signs indicate that all the expressions are equivalent.

$$= \frac{3x-3}{18x} + \frac{8x}{18x}$$ Writing with the LCD, $18x$

$$= \frac{11x-3}{18x}.$$ The result is an expression.

The expressions

$$\frac{x-1}{6x} + \frac{4}{9} \quad \text{and} \quad \frac{11x-3}{18x}$$

are equivalent.

Solve: $\frac{x-1}{6x} = \frac{4}{9}$.

SOLUTION We have

$$\frac{x-1}{6x} = \frac{4}{9}$$ **Each line is an equivalent equation.**

$$18x \cdot \frac{x-1}{6x} = 18x \cdot \frac{4}{9}$$ Multiplying by the LCM, $18x$

$$3 \cdot \cancel{6x} \cdot \frac{x-1}{\cancel{6x}} = 2 \cdot \cancel{9} \cdot x \cdot \frac{4}{\cancel{9}}$$

$$3(x-1) = 2x \cdot 4$$

$$3x - 3 = 8x$$

$$-3 = 5x$$

$$-\frac{3}{5} = x.$$ The result is a solution.

The solution of $\frac{x-1}{6x} = \frac{4}{9}$ is $-\frac{3}{5}$.

7.6 Exercise Set

FOR EXTRA HELP MyMathLab    

Concept Reinforcement *Classify each of the following as either an expression or an equation.*

1. $\frac{5x}{x+2} - \frac{3}{x} = 7$
2. $\frac{3}{x-4} + \frac{2}{x+4}$
3. $\frac{4}{t^2-1} + \frac{3}{t+1}$
4. $\frac{2}{t^2-1} + \frac{3}{t+1} = 5$
5. $\frac{2}{x+7} + \frac{6}{5x} = 4$
6. $\frac{5}{2x} - \frac{3}{x^2} = 7$
7. $\frac{t+3}{t-4} = \frac{t-5}{t-7}$
8. $\frac{7t}{2t-3} \div \frac{3t}{2t+3}$
9. $\frac{5x}{x^2-4} \cdot \frac{7}{x^2-5x+4}$
10. $\frac{7x}{2-x} = \frac{3}{4-x}$

Solve. If no solution exists, state this.

11. $\frac{3}{5} - \frac{2}{3} = \frac{x}{6}$
12. $\frac{5}{8} - \frac{3}{5} = \frac{x}{10}$
13. $\frac{1}{8} + \frac{1}{12} = \frac{1}{t}$
14. $\frac{1}{6} + \frac{1}{10} = \frac{1}{t}$
15. $\frac{x}{6} - \frac{6}{x} = 0$
16. $\frac{x}{7} - \frac{7}{x} = 0$
17. $\frac{2}{3} - \frac{1}{t} = \frac{7}{3t}$
18. $\frac{1}{2} - \frac{2}{t} = \frac{3}{2t}$
19. $\frac{n+2}{n-6} = \frac{1}{2}$
20. $\frac{a-4}{a+6} = \frac{1}{3}$
21. $\frac{12}{x} = \frac{x}{3}$
22. $\frac{x}{2} = \frac{18}{x}$

Aha! 23. $\frac{2}{6} + \frac{1}{2x} = \frac{1}{3}$

24. $\frac{12}{15} - \frac{1}{3x} = \frac{4}{5}$

25. $y + \frac{4}{y} = -5$

26. $t + \frac{6}{t} = -5$

27. $x - \frac{12}{x} = 4$

28. $y - \frac{14}{y} = 5$

29. $\frac{y+3}{y-3} = \frac{6}{y-3}$

30. $\frac{4}{8-a} = \frac{4-a}{a-8}$

31. $\frac{x}{x-5} = \frac{25}{x^2-5x}$

32. $\frac{t}{t-6} = \frac{36}{t^2-6t}$

33. $\frac{n+1}{n+2} = \frac{n-3}{n+1}$

34. $\frac{n+2}{n-3} = \frac{n+1}{n-2}$

Aha! 35. $\frac{x^2+4}{x-1} = \frac{5}{x-1}$

36. $\frac{x^2-1}{x+2} = \frac{3}{x+2}$

37. $\frac{6}{a+1} = \frac{a}{a-1}$

38. $\frac{4}{a-7} = \frac{-2a}{a+3}$

39. $\frac{60}{t-5} - \frac{18}{t} = \frac{40}{t}$

40. $\frac{50}{t-2} - \frac{16}{t} = \frac{30}{t}$

41. $\frac{3}{x-3} + \frac{5}{x+2} = \frac{5x}{x^2-x-6}$

42. $\frac{2}{x-2} + \frac{1}{x+4} = \frac{x}{x^2+2x-8}$

43. $\frac{3}{x} + \frac{x}{x+2} = \frac{4}{x^2+2x}$

44. $\frac{x}{x+1} + \frac{5}{x} = \frac{1}{x^2+x}$

45. $\frac{5}{x+2} - \frac{3}{x-2} = \frac{2x}{4-x^2}$

46. $\frac{y+3}{y+2} - \frac{y}{y^2-4} = \frac{y}{y-2}$

47. $\frac{3}{x^2-6x+9} + \frac{x-2}{3x-9} = \frac{x}{2x-6}$

48. $\frac{3-2y}{y+1} - \frac{10}{y^2-1} = \frac{2y+3}{1-y}$

In Exercises 49–54, a rational function f is given. Find all values of a for which $f(a)$ is the indicated value.

49. $f(x) = 2x - \frac{15}{x}$; $f(a) = 7$

50. $f(x) = 2x - \frac{6}{x}$; $f(a) = 1$

51. $f(x) = \frac{x-5}{x+1}$; $f(a) = \frac{3}{5}$

52. $f(x) = \frac{x-3}{x+2}$; $f(a) = \frac{1}{5}$

53. $f(x) = \frac{12}{x} - \frac{12}{2x}$; $f(a) = 8$

54. $f(x) = \frac{6}{x} - \frac{6}{2x}$; $f(a) = 5$

For each pair of functions f and g, find all values of a for which $f(a) = g(a)$.

55. $f(x) = \frac{x+1}{3} - 1$,

$g(x) = \frac{x-1}{2}$

56. $f(x) = \frac{x+4}{3x}$,

$g(x) = 2 - \frac{x+8}{5x}$

57. $f(x) = \frac{12}{x^2-6x+9}$,

$g(x) = \frac{4}{x-3} + \frac{2x}{x-3}$

58. $f(x) = \frac{14}{x^2-25}$,

$g(x) = \frac{x}{x+5} - \frac{5}{x-5}$

TW 59. Below are unlabeled graphs of $f(x) = x + 2$ and $g(x) = (x^2 - 4)/(x - 2)$. How could you determine which graph represents f and which graph represents g?

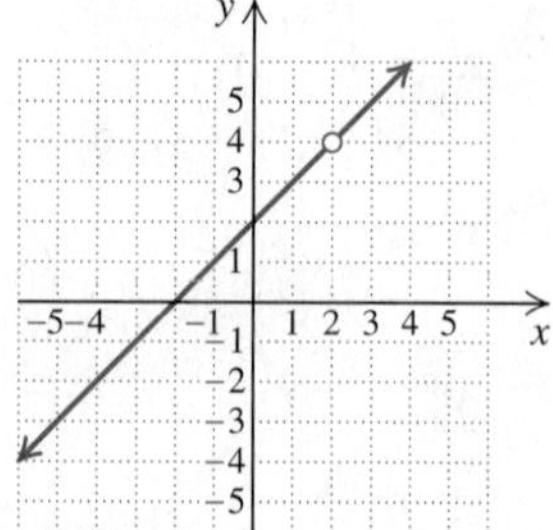

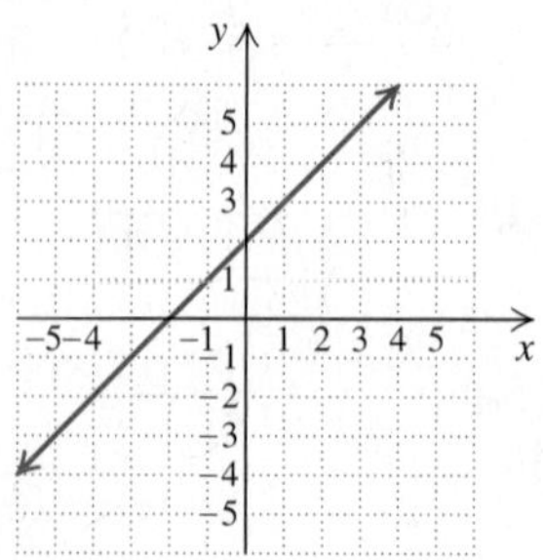

TW 60. Explain the difference between adding rational expressions and solving rational equations.

SKILL REVIEW

To prepare for Section 7.7, review solving applications and rates of change (Sections 2.5, 3.4, and 6.7).

61. The sum of two consecutive odd numbers is 276. Find the numbers. [2.5]

62. The length of a rectangular picture window is 3 yd greater than the width. The area of the rectangle is 10 yd^2. Find the perimeter. [6.7]

63. The height of a triangle is 3 cm longer than its base. If the area of the triangle is 54 cm^2, find the measurements of the base and the height. [6.7]

64. The product of two consecutive even integers is 48. Find the numbers. [6.7]

65. *Human Physiology.* Between June 9 and June 24, Seth's beard grew 0.9 cm. Find the rate at which Seth's beard grows. [3.4]

66. *Gardening.* Between July 7 and July 12, Carla's string beans grew 1.4 in. Find the growth rate of the string beans. [3.4]

SYNTHESIS

TW **67.** Karin and Kurt are working together to solve the equation

$$\frac{x}{x^2 - x - 6} = 2x - 1.$$

Karin obtains the first graph below using the DOT mode of a graphing calculator, while Kurt obtains the second graph using the CONNECTED mode. Which approach is the better method of showing the solutions? Why?

$y_1 = x/(x^2 - x - 6)$, $y_2 = 2x - 1$

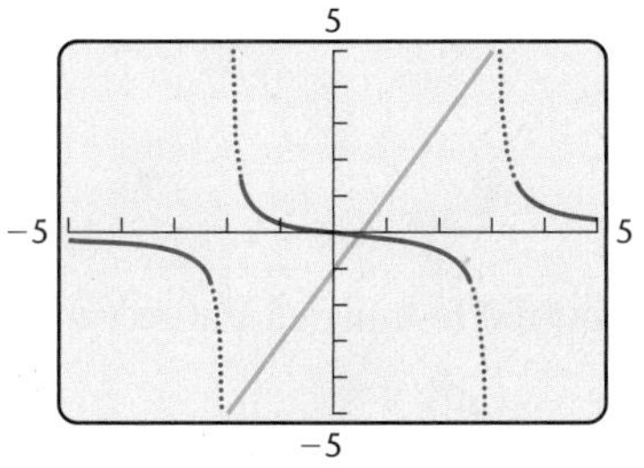

5, −5, 5, −5

$y_1 = x/(x^2 - x - 6)$, $y_2 = 2x - 1$

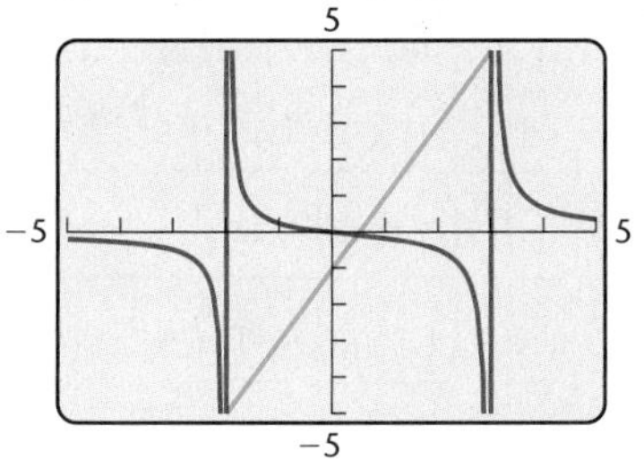

5, −5, 5, −5

TW **68.** Is the following statement true or false: "For any real numbers a, b, and c, if $ac = bc$, then $a = b$"? Explain why you answered as you did.

For each pair of functions f and g, find all values of a for which $f(a) = g(a)$.

69. $f(x) = \dfrac{x - \frac{2}{3}}{x + \frac{1}{2}}$, $g(x) = \dfrac{x + \frac{2}{3}}{x - \frac{3}{2}}$

70. $f(x) = \dfrac{2 - \frac{x}{4}}{2}$, $g(x) = \dfrac{\frac{x}{4} - 2}{\frac{x}{2} + 2}$

71. $f(x) = \dfrac{x + 3}{x + 2} - \dfrac{x + 4}{x + 3}$, $g(x) = \dfrac{x + 5}{x + 4} - \dfrac{x + 6}{x + 5}$

72. $f(x) = \dfrac{1}{1 + x} + \dfrac{x}{1 - x}$, $g(x) = \dfrac{1}{1 - x} - \dfrac{x}{1 + x}$

Solve.

73. $1 + \dfrac{x - 1}{x - 3} = \dfrac{2}{x - 3} - x$

74. $\dfrac{x}{x^2 + 3x - 4} + \dfrac{x + 1}{x^2 + 6x + 8} = \dfrac{2x}{x^2 + x - 2}$

75. $\dfrac{5 - 3a}{a^2 + 4a + 3} - \dfrac{2a + 2}{a + 3} = \dfrac{3 - a}{a + 1}$

76. $\dfrac{\frac{1}{x} + 1}{x} = \dfrac{\frac{1}{x}}{2}$

77. $\dfrac{\frac{1}{3}}{x} = \dfrac{1 - \frac{1}{x}}{x}$

Try Exercise Answers: Section 7.6

13. $\frac{24}{5}$ **31.** -5 **43.** -1 **49.** $-\frac{3}{2}, 5$

7.7 Applications Using Rational Equations and Proportions

- Problems Involving Work
- Problems Involving Motion
- Problems Involving Proportions

Applications involving rates, proportions, or reciprocals translate to rational equations. By using the five steps for problem solving, we can now solve such problems.

PROBLEMS INVOLVING WORK

EXAMPLE 1 Bryn and Reggie volunteer in a community garden. Bryn can mulch the garden alone in 8 hr and Reggie can mulch the garden alone in 10 hr. How long will it take the two of them, working together, to mulch the garden?

SOLUTION

1. **Familiarize.** This *work problem* is a type of problem that we have not yet encountered. Work problems are often *incorrectly* translated to mathematical language in several ways.

   a) Add the times together: 8 hr + 10 hr = 18 hr. ⟵ Incorrect
   This cannot be the correct approach since it should not take longer for Bryn and Reggie to do the job together than it takes either of them working alone.

   b) Average the times: $(8\text{ hr} + 10\text{ hr})/2 = 9\text{ hr}$. ⟵ Incorrect
   Again, this is longer than it would take Bryn to do the job alone.

   c) Assume that each person does half the job. ⟵ Incorrect
   If each person does half the job, Bryn would be finished with her half in 4 hr, and Reggie with his half in 5 hr. Since they are working together, Bryn will continue to help Reggie after completing her half. This does tell us that the job will take between 4 hr and 5 hr.

   Each incorrect approach began with the time it takes each worker to do the job. The correct approach instead focuses on the *rate* of work, or the amount of the job that each person completes in 1 hr.

   Since it takes Bryn 8 hr to mulch the entire garden, in 1 hr she mulches $\frac{1}{8}$ of the garden. Since it takes Reggie 10 hr to mulch the entire garden, in 1 hr he mulches $\frac{1}{10}$ of the garden. Together, they mulch $\frac{1}{8} + \frac{1}{10} = \frac{5}{40} + \frac{4}{40} = \frac{9}{40}$ of the garden per hour. The rates with which we work are thus

   Bryn: $\frac{1}{8}$ garden per hour,
   Reggie: $\frac{1}{10}$ garden per hour,
   Together: $\frac{9}{40}$ garden per hour.

   We are looking for the time required to mulch 1 entire garden.

	Fraction of the Garden Mulched		
Time	**By Bryn**	**By Reggie**	**Together**
1 hr	$\frac{1}{8}$	$\frac{1}{10}$	$\frac{1}{8} + \frac{1}{10}$, or $\frac{9}{40}$
2 hr	$\frac{1}{8} \cdot 2$	$\frac{1}{10} \cdot 2$	$\left(\frac{1}{8} + \frac{1}{10}\right)2$, or $\frac{9}{40} \cdot 2$, or $\frac{9}{20}$
3 hr	$\frac{1}{8} \cdot 3$	$\frac{1}{10} \cdot 3$	$\left(\frac{1}{8} + \frac{1}{10}\right)3$, or $\frac{9}{40} \cdot 3$, or $\frac{27}{40}$
t hr	$\frac{1}{8} \cdot t$	$\frac{1}{10} \cdot t$	$\left(\frac{1}{8} + \frac{1}{10}\right)t$, or $\frac{9}{40} \cdot t$

STUDY TIP

Take Advantage of Free Checking

Always check an answer if it is possible to do so. When an applied problem is being solved, do check an answer in the equation from which it came. However, it is even more important to check the answer with the words of the original problem.

2. **Translate.** From the table, we see that t must be some number for which

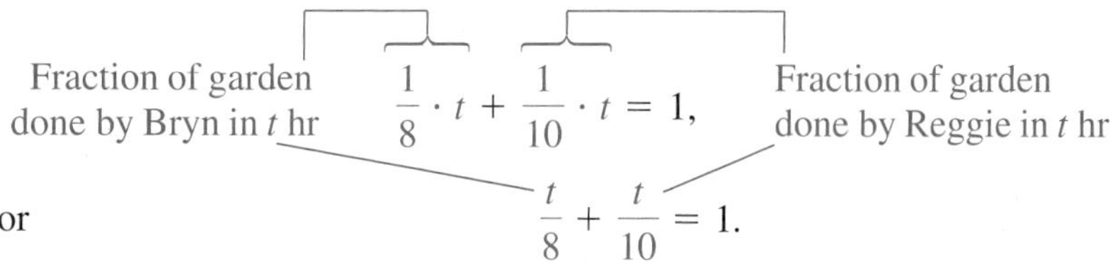

$$\frac{1}{8} \cdot t + \frac{1}{10} \cdot t = 1,$$

or

$$\frac{t}{8} + \frac{t}{10} = 1.$$

3. **Carry out.** We solve the equation both algebraically and graphically.

ALGEBRAIC APPROACH

We solve the equation:

$$\frac{t}{8} + \frac{t}{10} = 1$$

$$40\left(\frac{t}{8} + \frac{t}{10}\right) = 40 \cdot 1 \quad \text{Multiplying by the LCM}$$

$$\frac{40t}{8} + \frac{40t}{10} = 40 \quad \text{Distributing the 40}$$

$$5t + 4t = 40 \quad \text{Simplifying}$$

$$9t = 40$$

$$t = \frac{40}{9}, \text{ or } 4\frac{4}{9}.$$

GRAPHICAL APPROACH

We graph $y_1 = x/8 + x/10$ and $y_2 = 1$.

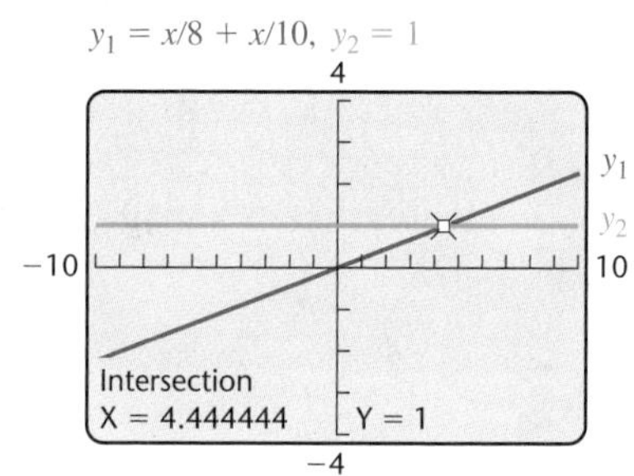

We can convert X to fraction notation and write the point of intersection as $\left(\frac{40}{9}, 1\right)$. The solution is $\frac{40}{9}$, or about 4.4.

4. **Check.** In $\frac{40}{9}$ hr, Bryn mulches $\frac{1}{8} \cdot \frac{40}{9}$, or $\frac{5}{9}$, of the garden and Reggie mulches $\frac{1}{10} \cdot \frac{40}{9}$, or $\frac{4}{9}$, of the garden. Together, they mulch $\frac{5}{9} + \frac{4}{9}$, or 1 garden. The fact that our solution is between 4 hr and 5 hr (see step 1 above) is also a check.
5. **State.** It will take $4\frac{4}{9}$ hr for Bryn and Reggie, working together, to mulch the garden.

Try Exercise 7.

EXAMPLE 2 It takes Jordyn 9 hr longer than it does Manuel to install a wood floor. Working together, they can do the job in 20 hr. How long would it take each, working alone, to install the floor?

SOLUTION

1. **Familiarize.** Unlike Example 1, this problem does not provide us with the times required by the individuals to do the job alone. We let m = the number of hours it would take Manuel working alone and $m + 9$ = the number of hours it would take Jordyn working alone.
2. **Translate.** Using the same reasoning as in Example 1, we see that Manuel completes $\frac{1}{m}$ of the job in 1 hr and Jordyn completes $\frac{1}{m + 9}$ of the job in 1 hr. In 2 hr, Manuel completes $\frac{1}{m} \cdot 2$ of the job and Jordyn completes $\frac{1}{m + 9} \cdot 2$

of the job. We are told that, working together, Manuel and Jordyn can complete the entire job in 20 hr. This gives the following:

Fraction of job done by Manuel in 20 hr; Fraction of job done by Jordyn in 20 hr

$$\frac{1}{m} \cdot 20 + \frac{1}{m+9} \cdot 20 = 1,$$

or

$$\frac{20}{m} + \frac{20}{m+9} = 1.$$

3. Carry out. We solve the equation both algebraically and graphically.

ALGEBRAIC APPROACH

We have

$$\frac{20}{m} + \frac{20}{m+9} = 1$$

$$m(m+9)\left(\frac{20}{m} + \frac{20}{m+9}\right) = m(m+9)1 \quad \textbf{Multiplying by the LCM}$$

$$(m+9)20 + m \cdot 20 = m(m+9) \quad \textbf{Distributing and simplifying}$$

$$40m + 180 = m^2 + 9m$$

$$0 = m^2 - 31m - 180 \quad \textbf{Getting 0 on one side}$$

$$0 = (m-36)(m+5) \quad \textbf{Factoring}$$

$$m - 36 = 0 \quad or \quad m + 5 = 0 \quad \textbf{Principle of zero products}$$

$$m = 36 \quad or \quad m = -5.$$

GRAPHICAL APPROACH

We graph $y_1 = 20/x + 20/(x+9)$ and $y_2 = 1$ and find the points of intersection. From the window $[-8, 8, -10, 10]$, we see that there is a point of intersection near the x-value -5. It also appears that the graphs may intersect at a point off the screen to the right. Thus we adjust the window to $[-8, 50, 0, 5]$ and see that there is indeed another point of intersection.

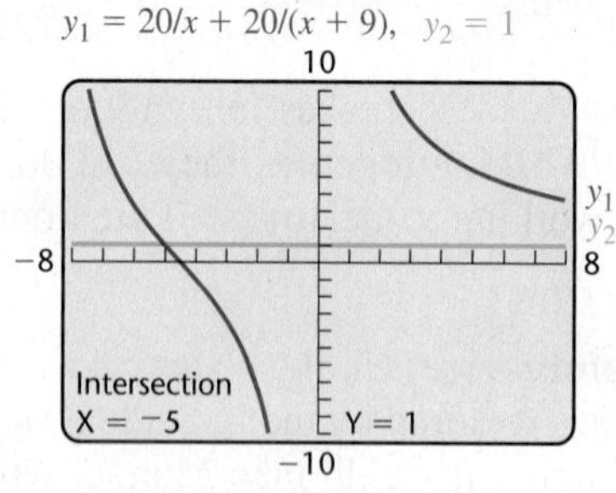

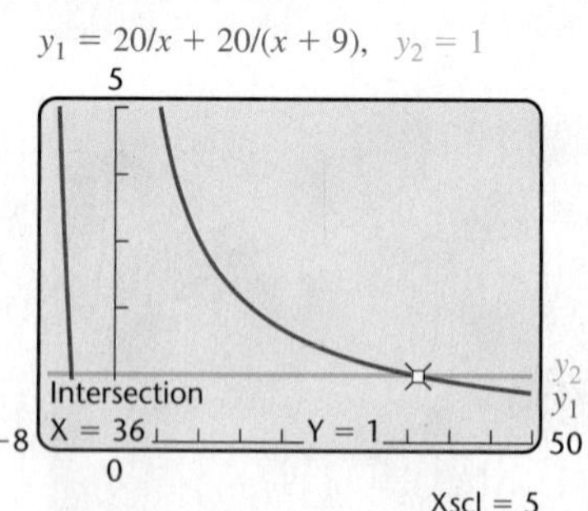

We have $x = 36$ or $x = -5$.

4. Check. Since negative time has no meaning in the problem, -5 is not a solution to the original problem. The number 36 checks since, if Manuel takes

36 hr alone and Jordyn takes $36 + 9 = 45$ hr alone, in 20 hr they would have installed

$$\frac{20}{36} + \frac{20}{45} = \frac{5}{9} + \frac{4}{9} = 1 \text{ complete floor.}$$

5. **State.** It would take Manuel 36 hr to install the floor alone, and Jordyn 45 hr.

Try Exercise 15.

The equations used in Examples 1 and 2 can be generalized as follows.

Modeling Work Problems

If

a = the time needed for A to complete the work alone,
b = the time needed for B to complete the work alone, and
t = the time needed for A and B to complete the work together,

then

$$\frac{t}{a} + \frac{t}{b} = 1.$$

The following are equivalent equations that can also be used:

$$\frac{1}{a} \cdot t + \frac{1}{b} \cdot t = 1 \quad \text{and} \quad \frac{1}{a} + \frac{1}{b} = \frac{1}{t}.$$

Student Notes

You need remember only the motion formula $d = rt$. Then you can divide both sides by t to get $r = d/t$, or you can divide both sides by r to get $t = d/r$.

PROBLEMS INVOLVING MOTION

Problems dealing with distance, rate (or speed), and time are called **motion problems.** To translate them, we use either the basic motion formula, $d = rt$, or the formulas $r = d/t$ or $t = d/r$, which can be derived from $d = rt$.

EXAMPLE 3 On her road bike, Paige bikes 15 km/h faster than Sean does on his mountain bike. In the time it takes Paige to travel 80 km, Sean travels 50 km. Find the speed of each bicyclist.

SOLUTION

1. **Familiarize.** Let's make a guess and check it.

Guess: Sean's speed: 10 km/h,
Paige's speed: $10 + 15 = 25$ km/h,
Sean's time: $50/10 = 5$ hr,
Paige's time: $80/25 = 3.2$ hr
} The times are not the same.

Our guess is wrong, but we can make some observations.

If Sean's speed $= r$, in kilometers/hour, then Paige's speed $= r + 15$.

Sean's travel time t is the same as Paige's travel time.

We make a drawing and construct a table, listing the information we know.

	Distance	Speed	Time
Sean's Mountain Bike	50	r	t
Paige's Road Bike	80	$r + 15$	t

2. **Translate.** By looking at how we checked our guess, we see that in the **Time** column of the table, the t's can be replaced, using the formula $Time = Distance/Rate$, as follows.

	Distance	Speed	Time
Sean's Mountain Bike	50	r	$50/r$
Paige's Road Bike	80	$r + 15$	$80/(r + 15)$

Since we are told that the times are the same, we can write an equation:

$$\frac{50}{r} = \frac{80}{r + 15}.$$

3. **Carry out.** We solve the equation both algebraically and graphically.

ALGEBRAIC APPROACH

We have

$$\frac{50}{r} = \frac{80}{r + 15}$$

$$r(r + 15)\frac{50}{r} = r(r + 15)\frac{80}{r + 15} \quad \textbf{Multiplying by the LCM}$$

$$50r + 750 = 80r \quad \textbf{Simplifying}$$

$$750 = 30r$$

$$25 = r.$$

GRAPHICAL APPROACH

We graph the equations $y_1 = 50/x$ and $y_2 = 80/(x + 15)$, and look for a point of intersection. Using a $[-10, 40, 0, 8]$ window, we see that the graphs intersect at the point $(25, 2)$.

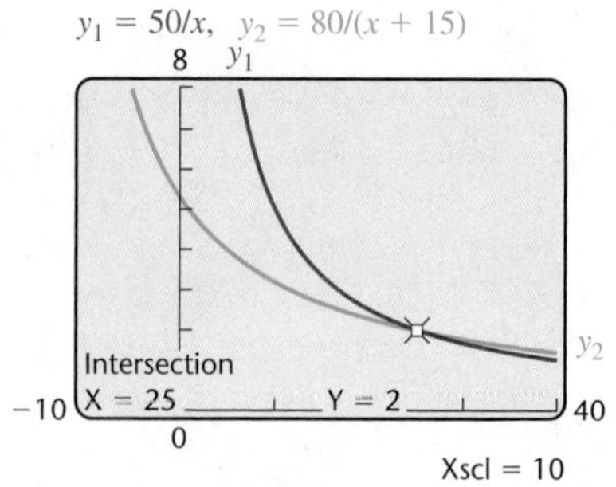

The solution is 25.

4. **Check.** If our answer checks, Sean's mountain bike is going 25 km/h and Paige's road bike is going $25 + 15 = 40$ km/h.

   Traveling 80 km at 40 km/h, Paige is riding for $\frac{80}{40} = 2$ hr. Traveling 50 km at 25 km/h, Sean is riding for $\frac{50}{25} = 2$ hr. Our answer checks since the two times are the same.

5. **State.** Paige's speed is 40 km/h, and Sean's speed is 25 km/h.

Try Exercise 21.

In the following example, although distance is the same in both directions, the key to the translation lies in an additional piece of given information.

EXAMPLE 4 A Hudson River tugboat goes 10 mph in still water. It travels 24 mi upstream and 24 mi back in a total time of 5 hr. What is the speed of the current?

Sources: Based on information from the Department of the Interior, U.S. Geological Survey, and *The Tugboat Captain*, Montgomery County Community College

SOLUTION

1. **Familiarize.** Let's make a guess and check it.

Guess:

Speed of current:	4 mph,	
Tugboat's speed upstream:	$10 - 4 = 6$ mph,	
Tugboat's speed downstream:	$10 + 4 = 14$ mph,	
Travel time upstream:	$\frac{24}{6} = 4$ hr,	The total time
Travel time downstream:	$\frac{24}{14} = 1\frac{5}{7}$ hr	is not 5 hr.

Our guess is wrong, but we can make some observations. If $c =$ the current's rate, in miles per hour, we have the following.

The tugboat's speed upstream is $(10 - c)$ mph.

The tugboat's speed downstream is $(10 + c)$ mph.

The total travel time is 5 hr.

We make a sketch and construct a table, listing the information we know.

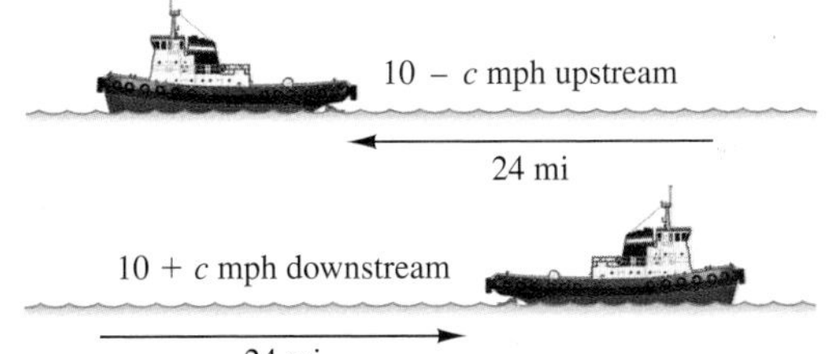

	Distance	Speed	Time
Upstream	24	$10 - c$	
Downstream	24	$10 + c$	

2. **Translate.** From examining our guess, we see that the time traveled can be represented using the formula *Time* = *Distance*/*Rate*:

	Distance	Speed	Time
Upstream	24	$10 - c$	$24/(10 - c)$
Downstream	24	$10 + c$	$24/(10 + c)$

Since the total time upstream and back is 5 hr, we use the last column of the table to form an equation:

$$\frac{24}{10 - c} + \frac{24}{10 + c} = 5.$$

3. **Carry out.** We solve the equation both algebraically and graphically.

ALGEBRAIC APPROACH

We have

$$\frac{24}{10 - c} + \frac{24}{10 + c} = 5$$

$$(10 - c)(10 + c)\left[\frac{24}{10 - c} + \frac{24}{10 + c}\right] = (10 - c)(10 + c)5$$

Multiplying by the LCM

$$24(10 + c) + 24(10 - c) = (100 - c^2)5$$

$$480 = 500 - 5c^2 \quad \textbf{Simplifying}$$

$$5c^2 - 20 = 0$$

$$5(c^2 - 4) = 0$$

$$5(c - 2)(c + 2) = 0$$

$$c = 2 \quad or \quad c = -2.$$

GRAPHICAL APPROACH

We graph

$y_1 = 24/(10 - x) + 24/(10 + x)$ and

$y_2 = 5$

and see that there are two points of intersection.

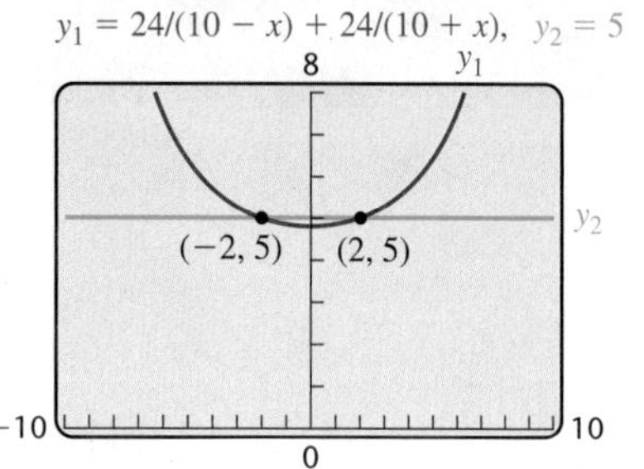

The solutions are -2 and 2.

4. **Check.** Since speed cannot be negative in this problem, -2 cannot be a solution. You should confirm that 2 checks in the original problem.
5. **State.** The speed of the current is 2 mph.

Try Exercise 31.

PROBLEMS INVOLVING PROPORTIONS

A **ratio** of two quantities is their quotient. For example, 37% is the ratio of 37 to 100, or $\frac{37}{100}$. A **proportion** is an equation stating that two ratios are equal.

Proportion An equality of ratios,

$$\frac{A}{B} = \frac{C}{D},$$

is called a *proportion*. The numbers within a proportion are said to be *proportional* to each other.

In geometry, if two triangles are **similar,** then their corresponding angles have the same measure and their corresponding sides are *proportional*. To illustrate, if triangle ABC is similar to triangle RST, then angles A and R have the same measure, angles B and S have the same measure, angles C and T have the same measure, and

$$\frac{a}{r} = \frac{b}{s} = \frac{c}{t}.$$

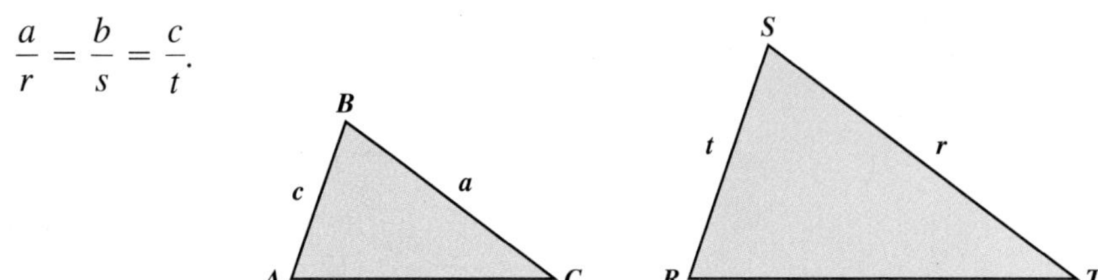

EXAMPLE 5 Similar Triangles. Triangles ABC and XYZ are similar. Solve for z if $x = 10$, $a = 8$, and $c = 5$.

SOLUTION We make a drawing, write a proportion, and then solve. Note that side a is always opposite angle A, side x is always opposite angle X, and so on.

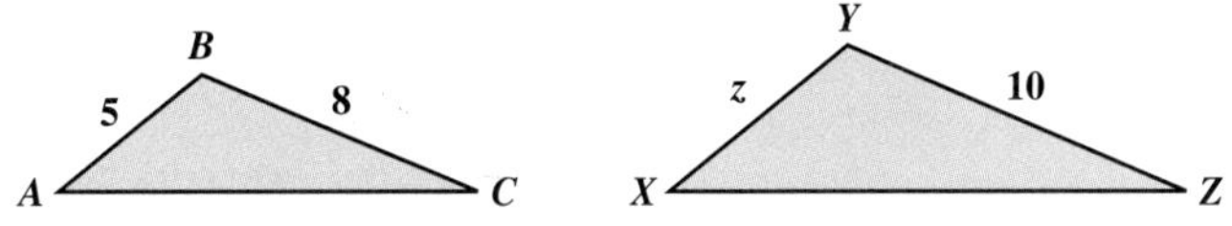

We have

$$\frac{z}{5} = \frac{10}{8}$$ **The proportions $\frac{5}{z} = \frac{8}{10}, \frac{5}{8} = \frac{z}{10}$, or $\frac{8}{5} = \frac{10}{z}$ could also be used.**

$$40 \cdot \frac{z}{5} = 40 \cdot \frac{10}{8}$$ **Multiplying both sides by the LCM, 40**

$$8z = 50$$ **Simplifying**

$$z = \frac{50}{8}, \text{ or } 6.25.$$

Try Exercise 37.

EXAMPLE 6 Architecture. A *blueprint* is a scale drawing of a building representing an architect's plans. Lia is adding 12 ft to the length of an apartment and needs to indicate the addition on an existing blueprint. If a 10-ft long bedroom is represented by $2\frac{1}{2}$ in. on the blueprint, how much longer should Lia make the drawing in order to represent the addition?

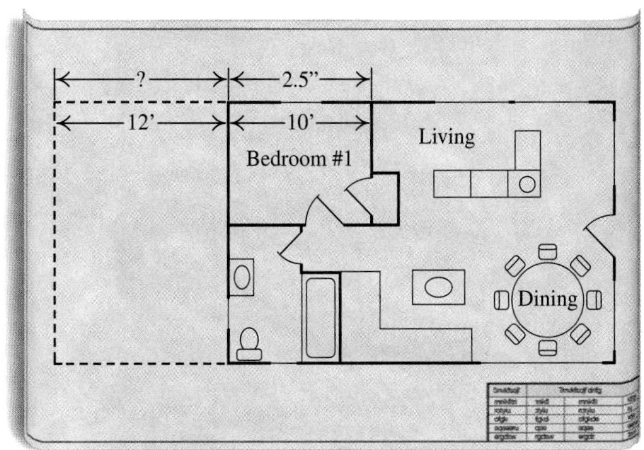

SOLUTION We let w = the width, in inches, of the addition that Lia is drawing. Because the drawing must be to scale, we have

$$\text{Inches on drawing} \rightarrow \frac{w}{12} = \frac{2.5}{10}. \leftarrow \text{Inches on drawing}$$
$$\text{Feet in real life} \rightarrow \qquad \leftarrow \text{Feet in real life}$$

To solve for w, we multiply both sides by the LCM of the denominators, 60:

$$60 \cdot \frac{w}{12} = 60 \cdot \frac{2.5}{10}$$

$$5w = 6 \cdot 2.5 \qquad \textbf{Simplifying}$$

$$w = \frac{15}{5}, \text{ or } 3.$$

Lia should make the blueprint 3 in. longer.

Try Exercise 41.

Proportions can be used to solve a variety of applied problems.

EXAMPLE 7 Wildlife Population. To determine the number of brook trout in River Denys, Cape Breton, Nova Scotia, a team of volunteers and professionals caught and marked 1190 brook trout. Later, they captured 915 brook trout, of which 24 were marked. Estimate the number of brook trout in River Denys.
Source: www.gov.ns.ca

SOLUTION We let T = the brook trout population in River Denys. If we assume that the percentage of marked trout in the second group of trout captured is the same as the percentage of marked trout in the entire river, we can form a proportion in which this percentage is expressed in two ways:

$$\text{Trout originally marked} \rightarrow \frac{1190}{T} = \frac{24}{915}. \leftarrow \text{Marked trout in second group}$$
$$\text{Entire population} \rightarrow \qquad \leftarrow \text{Total trout in second group}$$

To solve for T, we multiply by the LCM, $915T$:

$$915\cancel{T} \cdot \frac{1190}{\cancel{T}} = \cancel{915}T \cdot \frac{24}{\cancel{915}} \qquad \textbf{Multiplying both sides by } 915T$$

$$915 \cdot 1190 = 24T \qquad \textbf{Removing factors equal to 1: } T/T = 1 \textbf{ and } 915/915 = 1$$

$$\frac{915 \cdot 1190}{24} = T \text{ or } T \approx 45{,}369. \qquad \textbf{Dividing both sides by 24}$$

There are about 45,369 brook trout in the river.

Try Exercise 59.

7.7 Exercise Set

FOR EXTRA HELP MyMathLab Math XL PRACTICE  WATCH  DOWNLOAD  READ REVIEW

Concept Reinforcement *Find each rate.*

1. If Sandy can decorate a cake in 2 hr, what is her rate?

2. If Eric can decorate a cake in 3 hr, what is his rate?

3. If Sandy can decorate a cake in 2 hr and Eric can decorate the same cake in 3 hr, what is their rate, working together?

4. If Lisa and Mark can mow a lawn together in 1 hr, what is their rate?

5. If Lisa can mow a lawn by herself in 3 hr, what is her rate?

6. If Lisa and Mark can mow a lawn together in 1 hr, and Lisa can mow the same lawn by herself in 3 hr, what is Mark's rate, working alone?

7. *Home Restoration.* Trey can refinish the floor of an apartment in 8 hr. Matt can refinish the floor in 6 hr. How long will it take them, working together, to refinish the floor?

8. *Custom Embroidery.* Chandra can embroider logos on a team's sweatshirts in 6 hr. Traci, a new employee, needs 9 hr to complete the same job. Working together, how long will it take them to do the job?

9. *Filling a Pool.* The San Paulo community swimming pool can be filled in 12 hr if water enters through a pipe alone or in 30 hr if water enters through a hose alone. If water is entering through both the pipe and the hose, how long will it take to fill the pool?

10. *Filling a Tank.* A community water tank can be filled in 18 hr by the town office well alone and in 22 hr by the high school well alone. How long will it take to fill the tank if both wells are working?

11. *Pumping Water.* A $\frac{1}{4}$ HP Simer 2905 Mark I sump pump can remove water from Martha's flooded basement in 70 min. The $\frac{1}{3}$ HP Wayne SPV 500 sump pump can complete the same job in 30 min. How long would it take the two pumps together to pump out the basement?
Source: Based on data from manufacturers' Web sites

12. *Hotel Management.* The Honeywell 17000 air cleaner can clean the air in a conference room in 12 min. The Allerair 4000 can clean the air in a room of the same size in 10 min. How long would it take the two machines together to clean the air in such a room?
Source: Based on data from manufacturers' Web sites

13. *Copiers.* The Aficio MP C2500 takes three times as long as the MP C7500 to copy Ragheda's grant proposal. If working together the two machines can complete the job in 1.5 min, how long would it take each machine, working alone, to copy the proposal?
Source: Ricol-usa.com

14. *Computer Printers.* The HP Laser Jet P2035 works twice as fast as the Laser Jet P1005. If the machines work together, a university can produce all its staff manuals in 15 hr. Find the time it would take each machine, working alone, to complete the same job.
Source: www.shopping.hp.com

15. *Hotel Management.* The Airgle 750 can purify the air in a conference hall in 20 fewer minutes than it takes the Austin Healthmate 400 to do the same job. Together the two machines can purify the air in the conference hall in 10.5 min. How long would it take each machine, working alone, to purify the air in the room?
Source: Based on information from manufacturers' Web sites

16. *Photo Printing.* It takes the Canon PIXMA iP6310D 15 min longer to print a set of photo proofs than it takes the HP Officejet H470b Mobile Printer. Together it would take them $\frac{180}{7}$, or $25\frac{5}{7}$ min to print the photos. How long would it take each machine, working alone, to print the photos?
Sources: www.shoppinghp.com; www.staples.com

17. *Forest Fires.* The Erickson Air-Crane helicopter can douse a certain forest fire four times as fast as an S-58T helicopter. Working together, the two helicopters can douse the fire in 8 hr. How long would it take each helicopter, working alone, to douse the fire?
Sources: Based on information from www.emergency.com and www.arishelicopters.com

18. *Newspaper Delivery.* Jared can deliver papers three times as fast as Kevin can. If they work together, it takes them 1 hr. How long would it take each to deliver the papers alone?

19. *Sorting Recyclables.* Together, it takes Dawn and Deb 2 hr 55 min to sort recyclables. Alone, Dawn would require 2 more hours than Deb. How long would it take Deb to do the job alone? (*Hint*: Convert minutes to hours or hours to minutes.)

20. *Paving.* Together, Travis and Nick require 4 hr 48 min to pave a driveway. Alone, Travis would require 4 hr longer than Nick. How long would it take Nick to do the job alone? (*Hint*: Convert minutes to hours.)

21. *Train Speeds.* A B&M freight train is traveling 14 km/h slower than an AMTRAK passenger train. The B&M train travels 330 km in the same time that it takes the AMTRAK train to travel 400 km. Find their speeds. Complete the following table as part of the familiarization.

Distance = *Rate* · *Time*

	Distance (in km)	Speed (in km/h)	Time (in hours)
B&M	330		
AMTRAK	400	r	$\frac{400}{r}$

22. *Speed of Travel.* A loaded Roadway truck is moving 40 mph faster than a New York Railways freight train. In the time that it takes the train to travel 150 mi, the truck travels 350 mi. Find their speeds. Complete the following table as part of the familiarization.

Distance = *Rate* · *Time*

	Distance (in miles)	Speed (in miles per hour)	Time (in hours)
Truck	350	r	$\frac{350}{r}$
Train	150		

23. *Kayaking.* The speed of the current in Catamount Creek is 3 mph. Cory can kayak 4 mi upstream in the same time that it takes him to kayak 10 mi downstream. What is the speed of Cory's kayak in still water?

24. *Boating.* The current in the Lazy River moves at a rate of 4 mph. Heather's dinghy motors 6 mi upstream in the same time that it takes to motor 12 mi downstream. What is the speed of the dinghy in still water?

Aha! **25.** *Bus Travel.* A local bus travels 7 mph slower than the express. The express travels 45 mi in the time that it takes the local to travel 38 mi. Find the speed of each bus.

26. *Walking.* Nicole walks 2 mph slower than Simone. In the time that it takes Simone to walk 8 mi, Nicole walks 5 mi. Find the speed of each person.

27. *Moving Sidewalks.* Newark Airport's moving sidewalk moves at a speed of 1.7 ft/sec. Walking on the moving sidewalk, Kaitlyn can travel 120 ft forward in the same time that it takes to travel 52 ft in the opposite direction. How fast would Kaitlyn be walking on a nonmoving sidewalk?

28. *Moving Sidewalks.* The moving sidewalk at O'Hare Airport in Chicago moves 1.8 ft/sec. Walking on the moving sidewalk, Cameron travels 105 ft forward in the same time that it takes to travel 51 ft in the opposite direction. How fast would Cameron be walking on a nonmoving sidewalk?

Aha! **29.** *Tractor Speed.* Manley's tractor is just as fast as Caledonia's. It takes Manley 1 hr more than it takes Caledonia to drive to town. If Manley is 20 mi from town and Caledonia is 15 mi from town, how long does it take Caledonia to drive to town?

30. *Boat Speed.* Tory and Emilio's motorboats travel at the same speed. Tory pilots her boat 40 km before docking. Emilio continues for another 2 hr, traveling a total of 100 km before docking. How long did it take Tory to navigate the 40 km?

31. *Boating.* Destinee's Mercruiser travels 15 km/h in still water. She motors 140 km downstream in the same time that it takes to travel 35 km upstream. What is the speed of the river?

32. *Boating.* Sierra's paddleboat travels 2 km/h in still water. The boat is paddled 4 km downstream in the same time that it takes to go 1 km upstream. What is the speed of the river?

33. *Shipping.* A barge moves 7 km/h in still water. It travels 45 km upriver and 45 km downriver in a total time of 14 hr. What is the speed of the current?

34. *Aviation.* A Citation II Jet travels 350 mph in still air and flies 487.5 mi into the wind and 487.5 mi with the wind in a total of 2.8 hr. Find the wind speed.
Source: Eastern Air Charter

35. *Train Travel.* A freight train covered 120 mi at a certain speed. Had the train been able to travel 10 mph faster, the trip would have been 2 hr shorter. How fast did the train go?

36. *Moped Speed.* Julio's moped travels 8 km/h faster than Ellia's. Julio travels 69 km in the same time that Ellia travels 45 km. Find the speed of each person's moped.

Geometry. *For each pair of similar triangles, find the value of the indicated letter.*

37. b

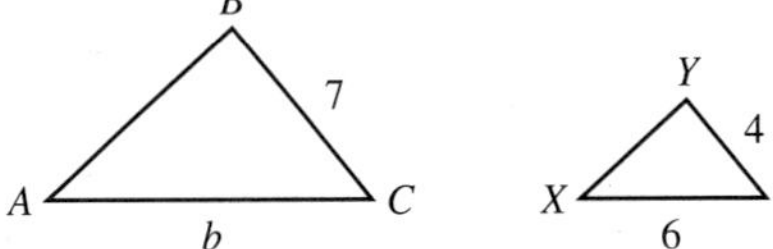

38. a

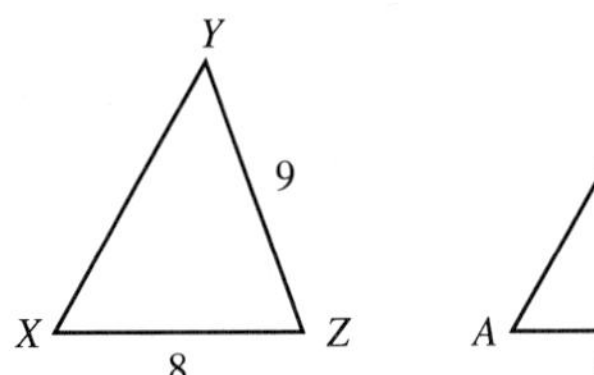

39. f

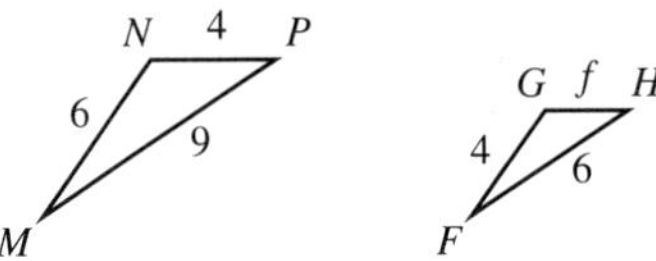

40. r

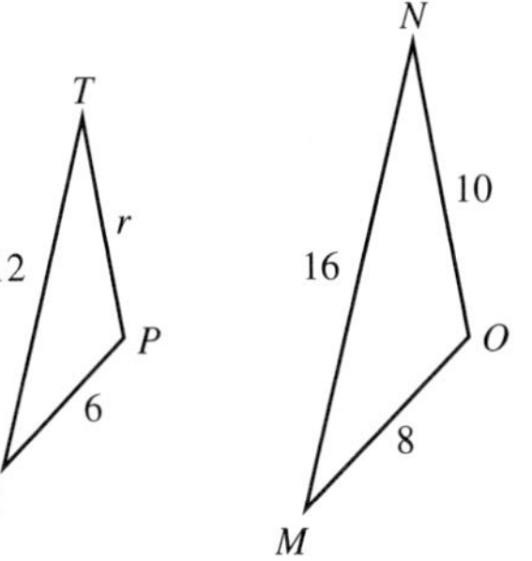

Architecture. *Use the blueprint below to find the indicated length.*

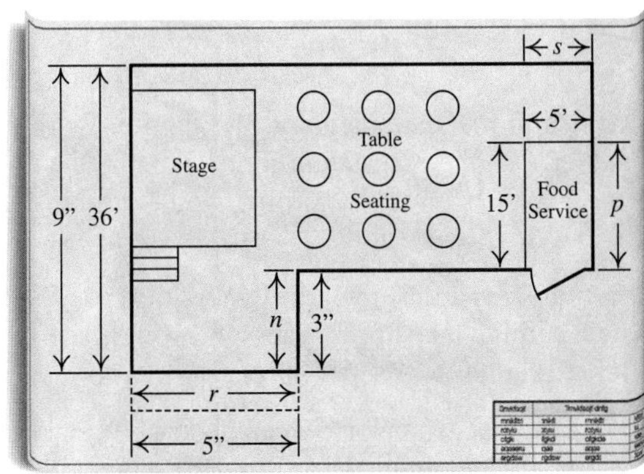

41. p, in inches on blueprint

42. s, in inches on blueprint

43. r, in feet on actual building

44. n, in feet on actual building

Construction. *Find the indicated length.*

45. l

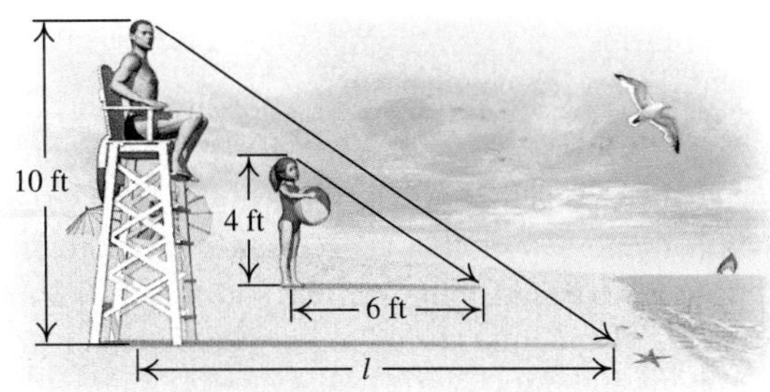

46. h

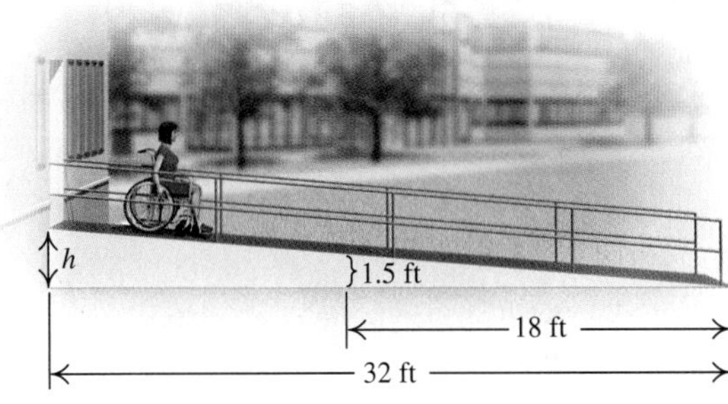

Graphing. *Find the indicated length.*

47. r

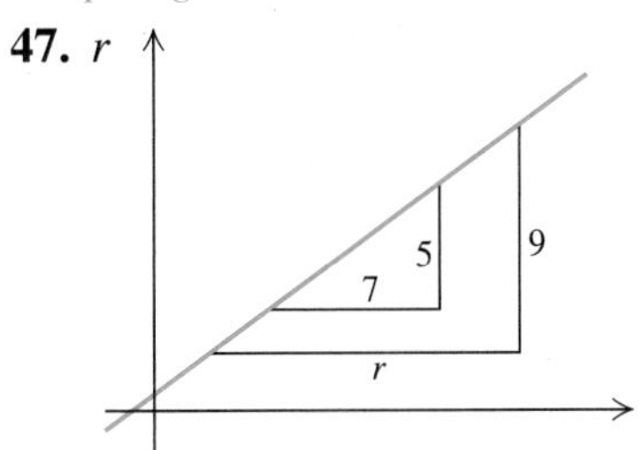

48. s

49. *Text Messaging.* Brett sent or received 384 text messages in 8 days. At this rate, how many text messages would he send or receive in 30 days?

50. *Burning Calories.* The average 140-lb adult burns about 160 calories playing touch football for 20 min. How long should the average 140-lb adult play touch football in order to burn 200 calories?
Source: changingshape.com

Aha! **51.** *Photography.* Rema snapped 234 photos over a period of 14 days. At this rate, how many would she take in 42 days?

52. *Mileage.* The Honda Civic Hybrid is a gasoline–electric car that travels approximately 180 mi on 4 gal of gas. Find the amount of gas required for an 810-mi trip.
Source: www.greenhybrid.com

53. *Wing Aspect Ratio.* The wing aspect ratio for a bird or an airplane is the ratio of the wing span to the wing width. Generally, higher aspect ratios are more efficient during low speed flying. Herons and storks, both waders, have comparable wing aspect ratios. A grey heron has a wing span of 180 cm and a wing width of 24 cm. A white stork has a wing span of 200 cm. What is the wing width of a stork?
Source: birds.ecoport.org

Aha! 54. *Money.* The ratio of the weight of copper to the weight of zinc in a U.S. penny is $\frac{1}{39}$. If 50 kg of zinc is being turned into pennies, how much copper is needed?
Source: United States Mint

55. *Flash Drives.* A sample of 150 flash drives contained 7 defective drives. How many defective flash drives would you expect in a batch of 2700 flash drives?

56. *Light Bulbs.* A sample of 184 compact fluorescent light bulbs contained 6 defective bulbs. How many defective bulbs would you expect in a sample of 1288 bulbs?

57. *Veterinary Science.* The amount of water needed by a small dog depends on its weight. A moderately active 8-lb Shih Tzu needs approximately 12 oz of water per day. How much water does a moderately active 5-lb Bolognese require each day?
Source: www.smalldogsparadise.com

58. *Miles Driven.* Carlos is allowed to drive his leased car for 45,000 mi in 4 years without penalty. In the first $1\frac{1}{2}$ years, Carlos has driven 16,000 mi. At this rate, will he exceed the mileage allowed for 4 years?

59. *Environmental Science.* To determine the number of humpback whales in a pod, a marine biologist, using tail markings, identifies 27 members of the pod. Several weeks later, 40 whales from the pod are randomly sighted. Of the 40 sighted, 12 are from the 27 originally identified. Estimate the number of whales in the pod.

60. *Fox Population.* To determine the number of foxes in King County, a naturalist catches, tags, and then releases 25 foxes. Later, 36 foxes are caught; 4 of them have tags. Estimate the fox population of the county.

61. *Weight on the Moon.* The ratio of the weight of an object on the moon to the weight of that object on Earth is 0.16 to 1.

a) How much would a 12-T rocket weigh on the moon?
b) How much would a 180-lb astronaut weigh on the moon?

62. *Weight on Mars.* The ratio of the weight of an object on Mars to the weight of that object on Earth is 0.4 to 1.

a) How much would a 12-T rocket weigh on Mars?
b) How much would a 120-lb astronaut weigh on Mars?

TW 63. Is it correct to assume that two workers will complete a task twice as quickly as one person working alone? Why or why not?

TW 64. If two triangles are exactly the same shape and size, are they similar? Why or why not?

SKILL REVIEW

To prepare for Section 7.8, review solving a formula for a variable (Section 2.3).

Solve. [2.3]

65. $a = \frac{b}{c}$, for b

66. $a = \frac{b}{c}$, for c

67. $2x - 5y = 10$, for y

68. $12 + 6y = 2x$, for y

69. $an + b = a$, for a

70. $xy + xz = 1$, for x

SYNTHESIS

TW 71. Two steamrollers are paving a parking lot. Working together, will the two steamrollers take less than half as long as the slower steamroller would working alone? Why or why not?

TW 72. Two fuel lines are filling a freighter with oil. Will the faster fuel line take more or less than twice as long to fill the freighter by itself? Why?

73. *Filling a Bog.* The Norwich cranberry bog can be filled in 9 hr and drained in 11 hr. How long will it take to fill the bog if the drainage gate is left open?

74. *Filling a Tub.* Kayla's hot tub can be filled in 10 min and drained in 8 min. How long will it take to empty a full tub if the water is left on?

75. *Grading.* Julia can grade a batch of placement exams in 3 hr. Tristan can grade a batch in 4 hr. If they work together to grade a batch of exams, what percentage of the exams will have been graded by Julia?

76. According to the U.S. Census Bureau, Population Division, in October 2009, there was one birth every 7 sec, one death every 13 sec, and one new international migrant every 36 sec. How many seconds does it take for a net gain of one person?

77. *Escalators.* Together, a 100-cm wide escalator and a 60-cm wide escalator can empty a 1575-person auditorium in 14 min. The wider escalator moves twice as many people as the narrower one. How many people per hour does the 60-cm wide escalator move?
Source: *McGraw-Hill Encyclopedia of Science and Technology*

78. *Aviation.* A Coast Guard plane has enough fuel to fly for 6 hr, and its speed in still air is 240 mph. The plane departs with a 40-mph tailwind and returns to the same airport flying into the same wind. How far from the airport can the plane travel under these conditions? (Assume that the plane can use all its fuel.)

79. *Boating.* Shoreline Travel operates a 3-hr paddleboat cruise on the Missouri River. If the speed of the boat in still water is 12 mph, how far upriver can the pilot travel against a 5-mph current before it is time to turn around?

80. *Travel by Car.* Melissa drives to work at 50 mph and arrives 1 min late. She drives to work at 60 mph and arrives 5 min early. How far does Melissa live from work?

81. *Photocopying.* The printer in an admissions office can print a 500-page document in 50 min, while the printer in the business office can print the same document in 40 min. If the two printers work together to print the document, with the faster machine starting on page 1 and the slower machine working backwards from page 500, at what page will the two machines meet to complete the job?

82. At what time after 4:00 will the minute hand and the hour hand of a clock first be in the same position?

83. At what time after 10:30 will the hands of a clock first be perpendicular?

Average speed is defined as total distance divided by total time.

84. Paloma drove 200 km. For the first 100 km of the trip, she drove at a speed of 40 km/h. For the second half of the trip, she traveled at a speed of 60 km/h. What was the average speed of the entire trip? (It was *not* 50 km/h.)

85. For the first 50 mi of a 100-mi trip, Liam drove 40 mph. What speed would he have to travel for the last half of the trip so that the average speed for the entire trip would be 45 mph?

Try Exercise Answers: Section 7.7

7. $3\frac{3}{7}$ hr **15.** Airgle: 15 min; Austin: 35 min
21. B&M speed: $r - 14$; B&M time: $\frac{330}{r-14}$; AMTRAK: 80 km/h; B&M: 66 km/h **31.** 9 km/h **37.** 10.5
41. $3\frac{3}{4}$ in. **59.** 90 whales

7.8 Formulas, Applications, and Variation

- Formulas
- Direct Variation
- Inverse Variation
- Joint Variation and Combined Variation
- Models

FORMULAS

Formulas occur frequently as mathematical models. Many formulas contain rational expressions, and to solve such formulas for a specified letter, we proceed as when solving rational equations.

EXAMPLE 1 Electronics. The formula

$$\frac{1}{R} = \frac{1}{r_1} + \frac{1}{r_2}$$

is used by electricians to determine the resistance R of two resistors r_1 and r_2 connected in parallel.* Solve for r_1.

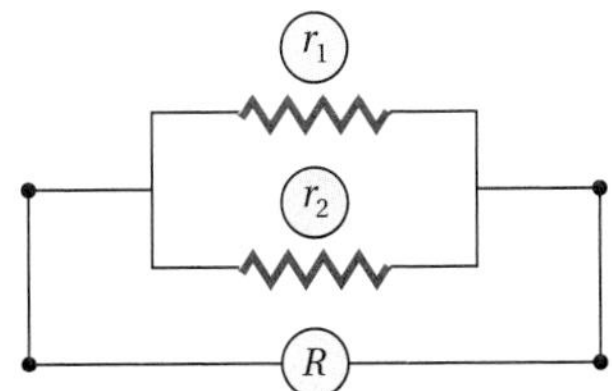

*The subscripts 1 and 2 indicate that r_1 and r_2 are different variables representing similar quantities.

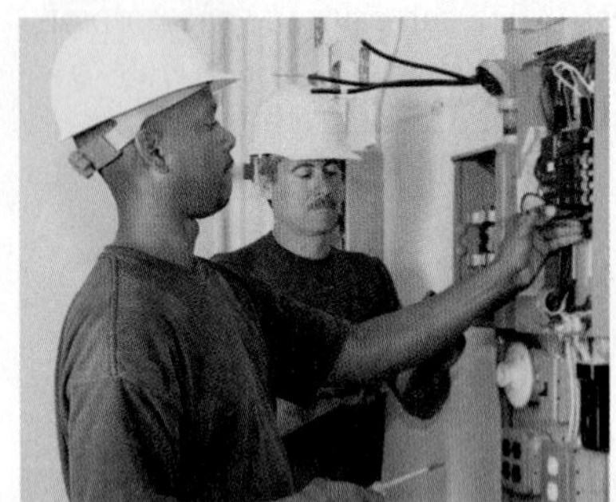

SOLUTION We use the same approach discussed in Section 7.6:

$$Rr_1r_2 \cdot \frac{1}{R} = Rr_1r_2 \cdot \left(\frac{1}{r_1} + \frac{1}{r_2}\right)$$ **Multiplying both sides by the LCM to clear fractions**

$$Rr_1r_2 \cdot \frac{1}{R} = Rr_1r_2 \cdot \frac{1}{r_1} + Rr_1r_2 \cdot \frac{1}{r_2}$$ **Multiplying to remove parentheses**

$$r_1r_2 = Rr_2 + Rr_1.$$ **Simplifying by removing factors equal to 1: $\frac{R}{R} = 1; \frac{r_1}{r_1} = 1; \frac{r_2}{r_2} = 1$**

At this point it is tempting to multiply by $1/r_2$ to get r_1 alone on the left, *but* note that there is an r_1 on the right. We must get all the terms involving r_1 on the *same side* of the equation.

$$r_1r_2 - Rr_1 = Rr_2$$ **Subtracting Rr_1 from both sides**

$$r_1(r_2 - R) = Rr_2$$ **Factoring out r_1 in order to combine like terms**

$$r_1 = \frac{Rr_2}{r_2 - R}$$ **Dividing both sides by $r_2 - R$ to get r_1 alone**

This formula can be used to calculate r_1 whenever R and r_2 are known.

Try Exercise 17.

EXAMPLE 2 Astronomy. The formula

$$\frac{V^2}{R^2} = \frac{2g}{R + h}$$

is used to find a satellite's *escape velocity* V, where R is a planet's radius, h is the satellite's height above the planet, and g is the planet's gravitational constant. Solve for h.

SOLUTION We first multiply by the LCM, $R^2(R + h)$, to clear fractions:

$$\frac{V^2}{R^2} = \frac{2g}{R + h}$$

$$R^2(R + h)\frac{V^2}{R^2} = R^2(R + h)\frac{2g}{R + h}$$ **Multiplying to clear fractions**

$$\frac{R^2(R + h)V^2}{R^2} = \frac{R^2(R + h)2g}{R + h}$$

$$(R + h)V^2 = R^2 \cdot 2g.$$ **Removing factors equal to 1: $\frac{R^2}{R^2} = 1$ and $\frac{R + h}{R + h} = 1$**

Remember: We are solving for h. Although we *could* distribute V^2, since h appears only within the factor $R + h$, it is easier to divide both sides by V^2:

$$\frac{(R + h)V^2}{V^2} = \frac{2R^2g}{V^2}$$ **Dividing both sides by V^2**

$$R + h = \frac{2R^2g}{V^2}$$ **Removing a factor equal to 1: $\frac{V^2}{V^2} = 1$**

$$h = \frac{2R^2g}{V^2} - R.$$ **Subtracting R from both sides**

The last equation can be used to determine the height of a satellite above a planet when the planet's radius and gravitational constant, along with the satellite's escape velocity, are known.

Try Exercise 19.

EXAMPLE 3 Acoustics (the Doppler Effect). The formula

$$f = \frac{sg}{s + v}$$

is used to determine the frequency f of a sound that is moving at velocity v toward a listener who hears the sound as frequency g. Here s is the speed of sound in a particular medium. Solve for s.

SOLUTION We first clear fractions by multiplying by the LCD, $s + v$:

$$f \cdot (s + v) = \frac{sg}{s + v}(s + v)$$

$$fs + fv = sg. \quad \text{The variable for which we are solving, } s\text{, appears on both sides, forcing us to distribute on the left side.}$$

Next, we must get all terms containing s on one side:

$$fv = sg - fs \quad \text{Subtracting } fs \text{ from both sides}$$

$$fv = s(g - f) \quad \text{Factoring out } s$$

$$\frac{fv}{g - f} = s. \quad \text{Dividing both sides by } g - f$$

Since s is isolated on one side, we have solved for s. This last equation can be used to determine the speed of sound whenever f, v, and g are known.

Try Exercise 21.

Student Notes

The steps used to solve equations are precisely the same steps used to solve formulas. If you feel "rusty" in this regard, study the earlier section in which this type of equation first appeared. Then make sure that you can consistently solve those equations before returning to the work with formulas.

To Solve a Rational Equation for a Specified Variable

1. Multiply both sides by the LCM of the denominators to clear fractions, if necessary.
2. Multiply to remove parentheses, if necessary.
3. Get all terms with the specified variable alone on one side.
4. Factor out the specified variable if it is in more than one term.
5. Multiply or divide on both sides to isolate the specified variable.

VARIATION

To extend our study of formulas and functions, we now examine three real-world situations: direct variation, inverse variation, and combined variation.

Direct Variation

A fitness trainer earns \$22 per hour. In 1 hr, \$22 is earned. In 2 hr, \$44 is earned. In 3 hr, \$66 is earned, and so on. From this information, we can form a set of ordered pairs:

$$(1, 22), (2, 44), (3, 66), (4, 88), \quad \text{and so on.}$$

Note that the ratio of earnings E to time t is $\frac{22}{1}$ in every case.

If a situation is modeled by pairs for which the ratio is constant, we say that there is **direct variation.** In this case, earnings *vary directly* as the time. Since $E/t = 22$,

$$E = 22t \quad \text{or, using function notation,} \quad E(t) = 22t.$$

Direct Variation When a situation is modeled by a linear function of the form $f(x) = kx$, or $y = kx$, where k is a nonzero constant, we say that there is *direct variation*, that y *varies directly* as x, or that y *is proportional to* x. The number k is called the *variation constant*, or the *constant of proportionality*.

Note that for $k > 0$, any equation of the form $y = kx$ indicates that as x increases, y increases as well.

EXAMPLE 4 Find the variation constant and an equation of variation if y varies directly as x, and $y = 32$ when $x = 2$.

SOLUTION We know that (2, 32) is a solution of $y = kx$. Therefore,

$$32 = k \cdot 2 \qquad \text{Substituting}$$

$$\frac{32}{2} = k, \quad \text{or} \quad k = 16. \qquad \text{Solving for } k$$

The variation constant is 16. The equation of variation is $y = 16x$. The notation $y(x) = 16x$ or $f(x) = 16x$ is also used.

Try Exercise 43.

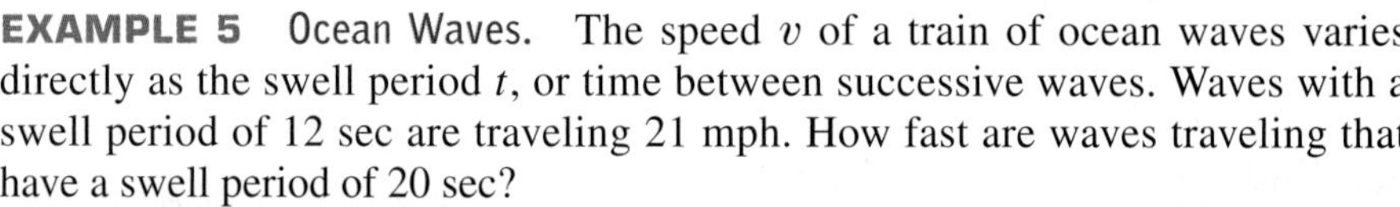

EXAMPLE 5 Ocean Waves. The speed v of a train of ocean waves varies directly as the swell period t, or time between successive waves. Waves with a swell period of 12 sec are traveling 21 mph. How fast are waves traveling that have a swell period of 20 sec?
Source: www.rodntube.com

SOLUTION

1. **Familiarize.** Because of the phrase "v . . . varies directly as . . . t," we express the speed of the wave v, in miles per hour, as a function of the swell period t, in seconds. Thus, $v(t) = kt$, where k is the variation constant. Because we are using ratios, we can use the units "seconds" and "miles per hour" without converting sec to hr or hr to sec. Knowing that waves with a swell period of 12 sec are traveling 21 mph, we have $v(12) = 21$.
2. **Translate.** We find the variation constant using the data and then use it to write the equation of variation:

$$v(t) = kt$$

$$v(12) = k \cdot 12 \qquad \text{Replacing } t \text{ with 12}$$

$$21 = k \cdot 12 \qquad \text{Replacing } v(12) \text{ with 21}$$

$$\frac{21}{12} = k \qquad \text{Solving for } k$$

$$1.75 = k. \qquad \text{This is the variation constant.}$$

The equation of variation is $v(t) = 1.75t$. This is the translation.

3. **Carry out.** To find the speed of waves with a swell period of 20 sec, we compute $v(20)$:

$$\begin{aligned} v(t) &= 1.75t \\ v(20) &= 1.75(20) \qquad \textbf{Substituting 20 for } t \\ &= 35. \end{aligned}$$

4. **Check.** To check, we could reexamine all our calculations. Note that our answer seems reasonable since the ratios $21/12$ and $35/20$ are both 1.75.

5. **State.** Waves with a swell period of 20 sec are traveling 35 mph.

Try Exercise 55.

STUDY TIP

Visualize the Steps

If you have completed all assignments and are studying for a quiz or test, a productive use of your time is to reread the assigned problems, making certain that you can visualize the steps that lead to a solution. When you are unsure of how to solve a problem, redo that problem in its entirety, asking for outside help as needed.

Inverse Variation

Suppose a bus is traveling 20 mi. At 20 mph, the trip takes 1 hr. At 40 mph, it takes $\frac{1}{2}$ hr. At 60 mph, it takes $\frac{1}{3}$ hr, and so on. This gives us pairs of numbers, all having the same product:

$$(20, 1), \left(40, \tfrac{1}{2}\right), \left(60, \tfrac{1}{3}\right), \left(80, \tfrac{1}{4}\right), \quad \text{and so on.}$$

Note that the product of each pair is 20. When a situation is modeled by pairs for which the product is constant, we say that there is **inverse variation.** Since $r \cdot t = 20$,

$$t = \frac{20}{r} \quad \text{or, using function notation,} \quad t(r) = \frac{20}{r}.$$

Inverse Variation When a situation is modeled by a rational function of the form $f(x) = k/x$, or $y = k/x$, where k is a nonzero constant, we say that there is *inverse variation*, that *y varies inversely as x*, or that *y is inversely proportional to x*. The number k is called the *variation constant*, or the *constant of proportionality*.

Note that for $k > 0$, any equation of the form $y = k/x$ indicates that as x increases, y decreases.

EXAMPLE 6 Find the variation constant and an equation of variation if y varies inversely as x, and $y = 32$ when $x = 0.2$.

SOLUTION We know that (0.2, 32) is a solution of

$$y = \frac{k}{x}.$$

Therefore,

$$\begin{aligned} 32 &= \frac{k}{0.2} \qquad \textbf{Substituting} \\ (0.2)32 &= k \\ 6.4 &= k. \qquad \textbf{Solving for } k \end{aligned}$$

The variation constant is 6.4. The equation of variation is

$$y = \frac{6.4}{x}.$$

Try Exercise 49.

Many real-life quantities vary inversely.

EXAMPLE 7 Movie Downloads. The time t that it takes to download a movie file varies inversely as the transfer speed s of the Internet connection. A typical full-length movie file will transfer in 48 min at a transfer speed of 256 KB/s (kilobytes per second). How long will it take to transfer the same movie file at a transfer speed of 32 KB/s?
Source: www.xsvidmovies.com

SOLUTION

1. **Familiarize.** Because of the phrase "... varies inversely as the transfer speed," we express the download time t, in minutes, as a function of the transfer speed s, in kilobytes per second. Thus, $t(s) = k/s$.
2. **Translate.** We use the given information to solve for k. We will then use that result to write the equation of variation:

$$t(s) = \frac{k}{s}$$

$$t(256) = \frac{k}{256} \quad \text{Replacing } s \text{ with } 256$$

$$48 = \frac{k}{256} \quad \text{Replacing } t(256) \text{ with } 48$$

$$12{,}288 = k.$$

The equation of variation is $t(s) = 12{,}288/s$. This is the translation.

3. **Carry out.** To find the download time at a transfer speed of 32 KB/s, we calculate $t(32)$:

$$t(32) = \frac{12{,}288}{32} = 384.$$

4. **Check.** Note that, as expected, as the transfer speed goes *down*, the download time goes *up*. Also, the products $48 \cdot 256$ and $32 \cdot 384$ are both 12,288.
5. **State.** At a transfer speed of 32 KB/s, it will take 384 min, or 6 hr 24 min, to download the movie file.

Try Exercise 57.

Joint Variation and Combined Variation

When a variable varies directly with more than one other variable, we say that there is *joint variation*. For example, in the formula for the volume of a right circular cylinder, $V = \pi r^2 h$, we say that V varies *jointly* as h and the square of r.

Joint Variation y varies *jointly* as x and z if, for some nonzero constant k, $y = kxz$.

EXAMPLE 8 Find an equation of variation if y varies jointly as x and z, and $y = 30$ when $x = 2$ and $z = 3$.

SOLUTION We have

$$y = kxz,$$

so

$$30 = k \cdot 2 \cdot 3$$
$$k = 5. \quad \text{The variation constant is 5.}$$

The equation of variation is $y = 5xz$.

Try Exercise 73.

Joint variation is one form of *combined variation.* In general, when a variable varies directly and/or inversely, at the same time, with more than one other variable, there is **combined variation.** Examples 8 and 9 are both examples of combined variation.

EXAMPLE 9 Find an equation of variation if y varies jointly as x and z and inversely as the square of w, and $y = 105$ when $x = 3$, $z = 20$, and $w = 2$.

SOLUTION The equation of variation is of the form

$$y = k \cdot \frac{xz}{w^2},$$

so, substituting, we have

$$105 = k \cdot \frac{3 \cdot 20}{2^2}$$
$$105 = k \cdot 15$$
$$k = 7.$$

Thus,

$$y = 7 \cdot \frac{xz}{w^2}.$$

Try Exercise 69.

MODELS

A graph like the one on the left below indicates that the quantities represented vary directly. The graph on the right below represents quantities that vary inversely. If we know the type of variation involved, we can choose one data point and calculate an equation of variation.

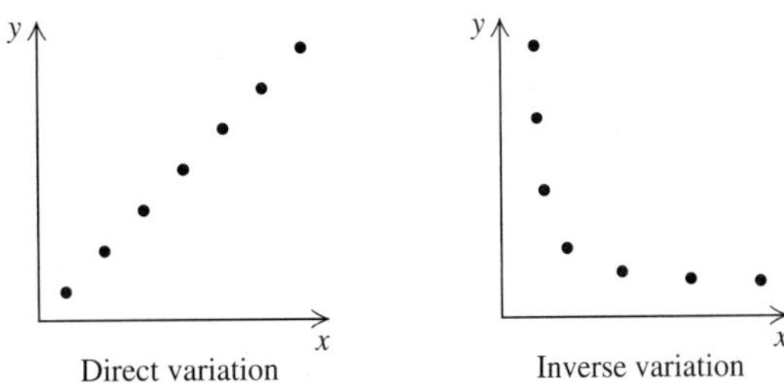

Direct variation   Inverse variation

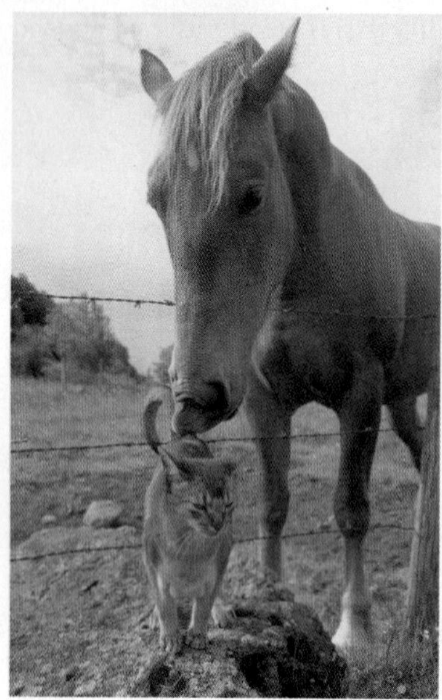

EXAMPLE 10 Longevity and Heart Rate. In general, animals with higher heart rates have shorter life spans. The table below lists average heart rates, in number of beats per minute, and average life spans, in years, for various animals.

Animal	Heart Rate (in number of beats per minute)	Life Span (in years)
Hamster	450	3
Rabbit	205	9
Pig	70	25
Giraffe	65	20
Horse	44	40
Elephant	30	70
Large whale	20	80

Source: Based on data from "Animal Longevity and Scale," found on sjsu.edu/faculty/watkins/longevity.html

a) Determine whether the data indicate direct variation or inverse variation.

b) Use the data point (70, 25) to find an equation of variation that describes the data.

c) Use the equation to estimate the life span of a cat with a heart rate of 150 beats per minute.

SOLUTION

a) We graph the data, letting $x =$ the heart rate, in number of beats per minute, and $y =$ the life span, in years. Since the points approximate the graph of a function of the type $y = k/x$, we see that heart rate varies inversely as life span.

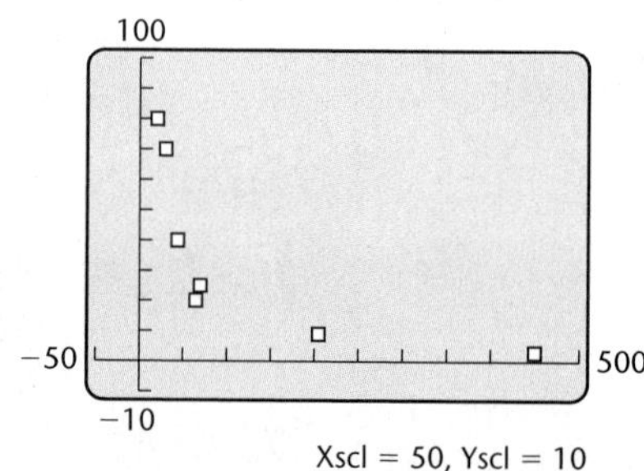

Xscl = 50, Yscl = 10

b) To find an equation of variation, we substitute 70 for x and 25 for y:

$$y = \frac{k}{x} \qquad \text{An equation of inverse variation}$$

$$25 = \frac{k}{70} \qquad \text{Substituting 70 for } x \text{ and 25 for } y$$

$$1750 = k$$

$$y = \frac{1750}{x}. \qquad \text{This is the equation of variation.}$$

The graph at left shows that this equation does approximate the data.

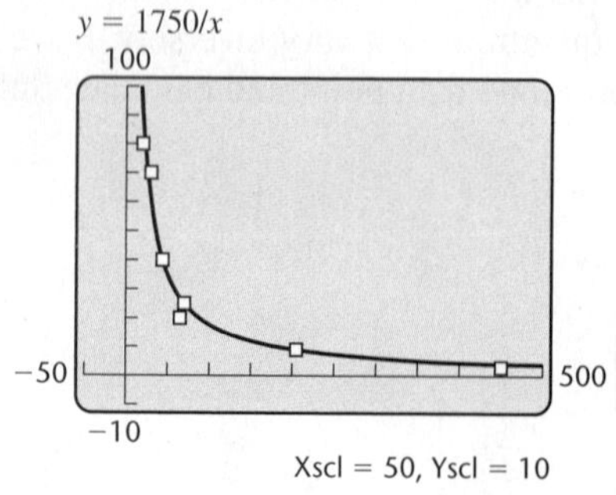

Xscl = 50, Yscl = 10

c) To estimate the life span of a cat, we substitute 150 for x in the equation of variation and calculate y:

$$y = \frac{1750}{x} \quad \textbf{The equation of variation}$$

$$= \frac{1750}{150} \quad \textbf{Substituting 150 for } x$$

$$\approx 11.7.$$

The life span of a cat with a heart rate of 150 beats per minute is approximately 11.7 years.

Try Exercise 83.

7.8 Exercise Set

FOR EXTRA HELP MyMathLab

READ

Concept Reinforcement *Match each statement with the correct term that completes it from the list on the right.*

1. To clear fractions, we can multiply both sides of an equation by the ____.
2. With direct variation, pairs of numbers have a constant ____.
3. With inverse variation, pairs of numbers have a constant ____.
4. If $y = k/x$, then y varies ____ as x.
5. If $y = kx$, then y varies ____ as x.
6. If $y = kxz$, then y varies ____ as x and z.

a) Directly
b) Inversely
c) Jointly
d) LCD
e) Product
f) Ratio

Determine whether each situation represents direct variation or inverse variation.

7. Two painters can scrape a house in 9 hr, whereas three painters can scrape the house in 6 hr.
8. Jayden planted 5 bulbs in 20 min and 7 bulbs in 28 min.
9. Mia swam 2 laps in 7 min and 6 laps in 21 min.
10. It took 2 band members 80 min to set up for a show; with 4 members working, it took 40 min.
11. It took 3 hr for 4 volunteers to wrap the campus' collection of Toys for Tots, but only 1.5 hr with 8 volunteers working.
12. Neveah's air conditioner cooled off 1000 ft^3 in 10 min and 3000 ft^3 in 30 min.

Solve each formula for the specified variable.

13. $f = \frac{L}{d}$; d

14. $\frac{W_1}{W_2} = \frac{d_1}{d_2}$; W_1

15. $s = \frac{(v_1 + v_2)t}{2}$; v_1

16. $s = \frac{(v_1 + v_2)t}{2}$; t

17. $\frac{t}{a} + \frac{t}{b} = 1$; b

18. $\frac{1}{R} = \frac{1}{r_1} + \frac{1}{r_2}$; R

19. $I = \frac{2V}{R + 2r}$; R

20. $I = \frac{2V}{R + 2r}$; r

21. $R = \frac{gs}{g + s}$; g

22. $K = \frac{rt}{r - t}$; t

23. $\frac{1}{p} + \frac{1}{q} = \frac{1}{f}$; q

24. $\frac{1}{p} + \frac{1}{q} = \frac{1}{f}$; p

25. $S = \frac{H}{m(t_1 - t_2)}$; t_1

26. $S = \frac{H}{m(t_1 - t_2)}$; H

27. $\frac{E}{e} = \frac{R + r}{r}$; r

28. $\frac{E}{e} = \frac{R + r}{R}$; R

29. $S = \frac{a}{1 - r}$; r

30. $S = \frac{a - ar^n}{1 - r}$; a

Aha! **31.** $c = \frac{1f}{(a + b)c}$; $a + b$

32. $d = \frac{g}{d(c + f)}$; $c + f$

33. *Interest.* The formula

$$P = \frac{A}{1 + r}$$

is used to determine what principal P should be invested for one year at $(100 \cdot r)\%$ simple interest in order to have A dollars after a year. Solve for r.

34. *Taxable Interest.* The formula

$$I_t = \frac{I_f}{1 - T}$$

gives the *taxable interest rate* I_t equivalent to the *tax-free interest rate* I_f for a person in the $(100 \cdot T)\%$ tax bracket. Solve for T.

35. *Average Speed.* The formula

$$v = \frac{d_2 - d_1}{t_2 - t_1}$$

gives an object's average speed v when that object has traveled d_1 miles in t_1 hours and d_2 miles in t_2 hours. Solve for t_2.

36. *Average Acceleration.* The formula

$$a = \frac{v_2 - v_1}{t_2 - t_1}$$

gives a vehicle's *average acceleration* when its velocity changes from v_1 at time t_1 to v_2 at time t_2. Solve for t_1.

37. *Work Rate.* The formula

$$\frac{1}{t} = \frac{1}{a} + \frac{1}{b}$$

gives the total time t required for two workers to complete a job, if the workers' individual times are a and b. Solve for t.

38. *Planetary Orbits.* The formula

$$\frac{x^2}{a^2} + \frac{y^2}{b^2} = 1$$

can be used to plot a planet's elliptical orbit of width $2a$ and length $2b$. Solve for b^2.

39. *Semester Average.* The formula

$$A = \frac{2Tt + Qq}{2T + Q}$$

gives a student's average A after T tests and Q quizzes, where each test counts as 2 quizzes, t is the test average, and q is the quiz average. Solve for Q.

40. *Astronomy.* The formula

$$L = \frac{dR}{D - d},$$

where D is the diameter of the sun, d is the diameter of the earth, R is the earth's distance from the sun, and L is some fixed distance, is used in calculating when lunar eclipses occur. Solve for D.

41. *Body-Fat Percentage.* The YMCA calculates men's body-fat percentage p using the formula

$$p = \frac{-98.42 + 4.15c - 0.082w}{w},$$

where c is the waist measurement, in inches, and w is the weight, in pounds. Solve for w.
Source: YMCA guide to Physical Fitness Assessment

42. *Preferred Viewing Distance.* Researchers model the distance D from which an observer prefers to watch television in "picture heights"—that is, multiples of the height of the viewing screen. The preferred viewing distance is given by

$$D = \frac{3.55H + 0.9}{H},$$

where D is in picture heights and H is in meters. Solve for H.
Source: www.tid.es, Telefonica Investigación y Desarrollo, S.A. Unipersonal

Find the variation constant and an equation of variation if y varies directly as x and the following conditions apply.

43. $y = 28$ when $x = 4$

44. $y = 5$ when $x = 12$

45. $y = 3.4$ when $x = 2$

46. $y = 2$ when $x = 5$

47. $y = 2$ when $x = \frac{1}{3}$

48. $y = 0.9$ when $x = 0.5$

Find the variation constant and an equation of variation in which y varies inversely as x, and the following conditions exist.

49. $y = 3$ when $x = 20$

50. $y = 16$ when $x = 4$

51. $y = 11$ when $x = 6$

52. $y = 9$ when $x = 5$

53. $y = 27$ when $x = \frac{1}{3}$

54. $y = 81$ when $x = \frac{1}{9}$

55. *Hooke's Law.* Hooke's law states that the distance d that a spring is stretched by a hanging object varies directly as the mass m of the object. If the distance is 20 cm when the mass is 3 kg, what is the distance when the mass is 5 kg?

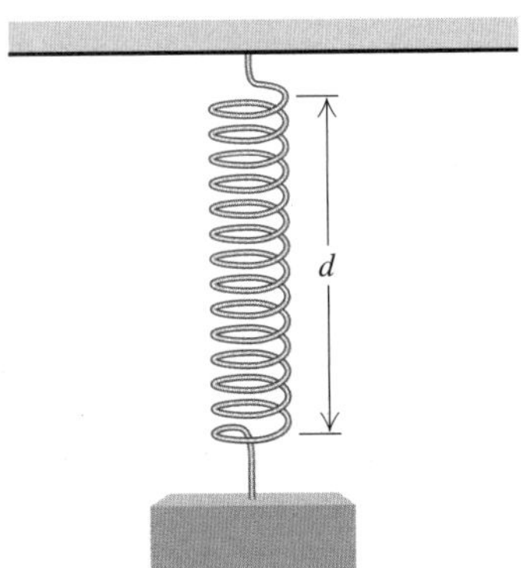

56. *Ohm's Law.* The electric current I, in amperes, in a circuit varies directly as the voltage V. When 15 volts are applied, the current is 5 amperes. What is the current when 18 volts are applied?

57. *Work Rate.* The time T required to do a job varies inversely as the number of people P working. It takes 5 hr for 7 volunteers to pick up rubbish from 1 mi of roadway. How long would it take 10 volunteers to complete the job?

58. *Pumping Rate.* The time t required to empty a tank varies inversely as the rate r of pumping. If a Briggs and Stratton pump can empty a tank in 45 min at the rate of 600 kL/min, how long will it take the pump to empty the tank at 1000 kL/min?

59. *Water from Melting Snow.* The number of centimeters W of water produced from melting snow varies directly as the number of centimeters S of snow. Meteorologists know that under certain conditions, 150 cm of snow will melt to 16.8 cm of water. The average annual snowfall in Alta, Utah, is 500 in. Assuming the above conditions, how much water will replace the 500 in. of snow?

60. *Gardening.* The number of calories burned by a gardener is directly proportional to the time spent gardening. It takes 30 min to burn 180 calories. How long would it take to burn 240 calories when gardening?
Source: www.healthstatus.com

Aha! **61.** *Mass of Water in a Human.* The number of kilograms W of water in a human body varies directly as the mass of the body. A 96-kg person contains 64 kg of water. How many kilograms of water are in a 48-kg person?

62. *Weight on Mars.* The weight M of an object on Mars varies directly as its weight E on Earth. A person who weighs 95 lb on Earth weighs 38 lb on Mars. How much would a 100-lb person weigh on Mars?

63. *String Length and Frequency.* The frequency of a string is inversely proportional to its length. A violin string that is 33 cm long vibrates with a frequency of 260 Hz. What is the frequency when the string is shortened to 30 cm?

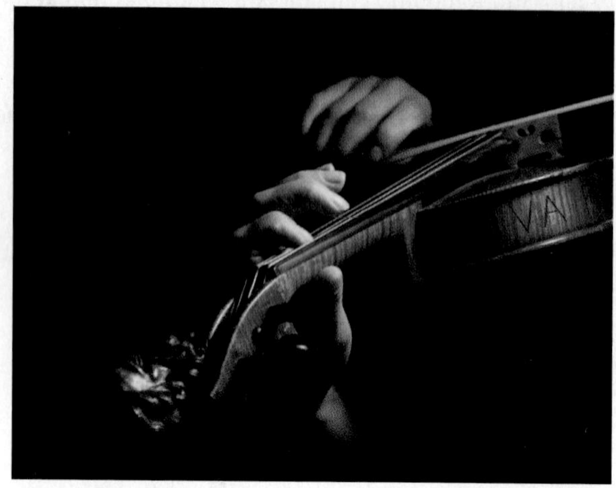

64. *Wavelength and Frequency.* The wavelength W of a radio wave varies inversely as its frequency F. A wave with a frequency of 1200 kilohertz has a length of 300 m. What is the length of a wave with a frequency of 800 kilohertz?

65. *Ultraviolet Index.* At an ultraviolet, or UV, rating of 4, those people who are moderately sensitive to the sun will burn in 70 min. Given that the number of minutes it takes to burn, t, varies inversely with the UV rating, u, how long will it take moderately sensitive people to burn when the UV rating is 14?
Source: *The Electronic Textbook of Dermatology* at www.telemedicine.org

66. *Current and Resistance.* The current I in an electrical conductor varies inversely as the resistance R of the conductor. If the current is $\frac{1}{2}$ ampere when the resistance is 240 ohms, what is the current when the resistance is 540 ohms?

67. *Air Pollution.* The average U.S. household of 2.6 people released 0.65 tons of carbon monoxide into the environment in a recent year. How many tons were released nationally? (Use 308,000,000 as the U.S. population.)
Sources: Based on data from the U.S. Environmental Protection Agency and the U.S. Census Bureau

68. *Fabric Manufacturing.* Knitted fabric is described in terms of wales per inch (for the fabric width) and courses per inch (for the fabric length). The CPI (courses per inch) is inversely proportional to the stitch length. For a specific fabric with a stitch length of 0.166 in., the CPI is 34.85. What would the CPI be if the stitch length were increased to 0.175 in.?
Source: Based on information from "Engineered Knitting Program for 100% Cotton Knit Fabrics" found on cottoninc.com

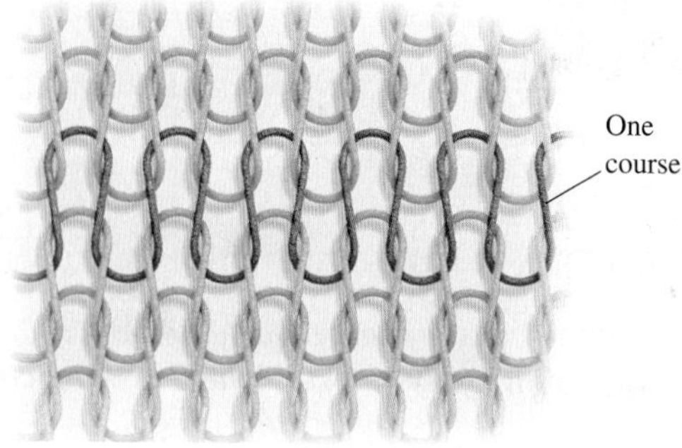

Find an equation of variation in which:

69. y varies directly as the square of x, and $y = 6$ when $x = 3$.

70. y varies directly as the square of x, and $y = 0.15$ when $x = 0.1$.

71. y varies inversely as the square of x, and $y = 6$ when $x = 3$.

72. y varies inversely as the square of x, and $y = 0.15$ when $x = 0.1$.

73. y varies jointly as x and the square of z, and $y = 105$ when $x = 14$ and $z = 5$.

74. y varies jointly as x and z and inversely as w, and $y = \frac{3}{2}$ when $x = 2$, $z = 3$, and $w = 4$.

75. y varies jointly as w and the square of x and inversely as z, and $y = 49$ when $w = 3$, $x = 7$, and $z = 12$.

76. y varies directly as x and inversely as w and the square of z, and $y = 4.5$ when $x = 15$, $w = 5$, and $z = 2$.

77. *Stopping Distance of a Car.* The stopping distance d of a car after the brakes have been applied varies directly as the square of the speed r. Once the brakes are applied, a car traveling 60 mph can stop in 138 ft. What stopping distance corresponds to a speed of 40 mph?
Source: Based on data from Edmunds.com

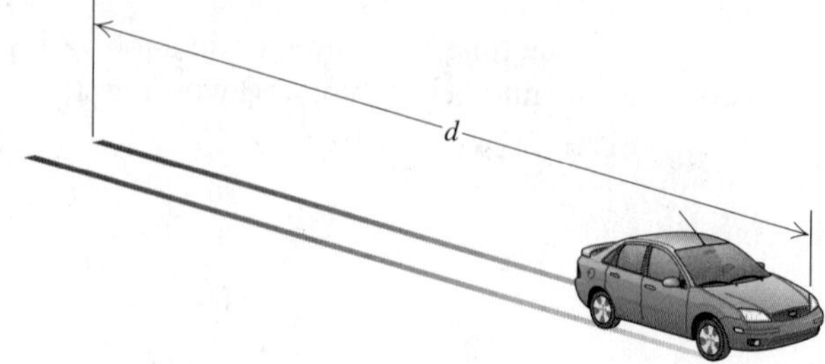

78. *Reverberation Time.* A sound's reverberation time T is the time it takes for the sound level to decrease by 60 dB (decibels) after the sound has been turned off. Reverberation time varies directly as the volume V of a room and inversely as the sound absorption A of the room. A given sound has a reverberation time of 1.5 sec in a room with a volume of 90 m^3 and a sound absorption of 9.6. What is the reverberation time of the same sound in a room with a volume of 84 m^3 and a sound absorption of 10.5?
Source: Based on data from www.isover.co.uk

79. *Volume of a Gas.* The volume V of a given mass of a gas varies directly as the temperature T and inversely as the pressure P. If $V = 231$ cm^3 when $T = 300°$K (Kelvin) and $P = 20$ lb/cm^2, what is the volume when $T = 320°$K and $P = 16$ lb/cm^2?

80. *Intensity of a Signal.* The intensity I of a television signal varies inversely as the square of the distance d from the transmitter. If the intensity is 25 W/m^2 at a distance of 2 km, what is the intensity 6.25 km from the transmitter?

81. *Atmospheric Drag.* Wind resistance, or atmospheric drag, tends to slow down moving objects. Atmospheric drag W varies jointly as an object's surface area A and velocity v. If a car traveling at a speed of 40 mph with a surface area of 37.8 ft^2 experiences a drag of 222 N (Newtons), how fast must a car with 51 ft^2 of surface area travel in order to experience a drag force of 430 N?

82. *Drag Force.* The drag force F on a boat varies jointly as the wetted surface area A and the square of the velocity of the boat. If a boat traveling 6.5 mph experiences a drag force of 86 N when the wetted surface area is 41.2 ft^2, find the wetted surface area of a boat traveling 8.2 mph with a drag force of 94 N.

83. *Commuter Travel.* One factor influencing urban planning is VMT, or vehicle miles traveled. The table below lists the annual VMT per household for various densities for a typical urban area.

Population Density (in number of households per residential acre)	Annual VMT per Household
25	12,000
50	6,000
100	3,000
200	1,500

Source: Based on information from http://www.sflcv.org/density

a) Determine whether the data indicate direct variation or inverse variation.
b) Find an equation of variation that describes the data.
c) Use the equation to estimate the annual VMT per household for areas with 10 households per residential acre.

84. *Ultraviolet Index.* The table below lists the safe exposure time to the sun for people with less sensitive skin.

UV Index	Safe Exposure Time (in minutes)
2	120
4	75
6	50
8	35
10	25

a) Determine whether the data indicate direct variation or inverse variation.
b) Find an equation of variation that approximates the data. Use the data point (6, 50).
c) Use the equation to predict the safe exposure time for people with less sensitive skin when the UV rating is 3.

85. *Internet Auctions.* ListAndSell is an auction drop-off store that accepts items for sale on Internet auctions. For a fee, the items are listed on the auction and shipped when sold. The table below lists the amount that a seller receives from the sale of various items.

Item Selling Price	Amount Seller Receives
\$ 75.00	\$ 41.42
100.00	55.50
200.00	111.85
400.00	240.55

a) Determine whether the amount that the seller receives varies directly or inversely as the selling price.
b) Find an equation of variation that approximates the data. Use the data point (200, 111.85).
c) Use the equation to predict the amount the seller receives if an item sells for \$150.00.

86. *Motor Vehicle Registrations.* The table below lists the number of motor vehicle registrations y and the number of licensed drivers x for various states.

State	Number of Licensed Drivers (in thousands)	Number of Motor Vehicles Registered (in thousands)
Alabama	3,665	4,630
Connecticut	2,805	3,052
Georgia	5,907	8,286
Idaho	1,008	1,275
Maryland	3,694	4,488
Texas	14,907	17,538
Wyoming	391	645

Source: U.S. Federal Highway Administration, *Highway Statistics*

a) Determine whether the number of motor vehicles registered annually varies directly or inversely as the number of licensed drivers.

b) Find an equation of variation that approximates the data. Use the data point $(3694, 4488)$.

c) Use the equation to estimate the number of motor vehicles registered in Hawaii, where there are 867 thousand licensed drivers.

TW **87.** If two quantities vary directly, does this mean that one is "caused" by the other? Why or why not?

TW **88.** If y varies directly as x, does doubling x cause y to be doubled as well? Why or why not?

SKILL REVIEW

To prepare for Chapter 8, review solving inequalities (Section 2.6).

Solve. [2.6]

89. $3 - x < 3x + 5$

90. $\frac{2}{3}x + 1 \leq x + 4$

91. $2(x - 1) \geq 4(x + 1)$

92. $\frac{1}{4}(4x - 3) > \frac{3}{4}(x + 1)$

93. $\dfrac{2x + 3}{4} \leq 5$

94. $\dfrac{5 - x}{2} \geq 1$

SYNTHESIS

TW **95.** Suppose that the number of customer complaints is inversely proportional to the number of employees hired. Will a firm reduce the number of complaints more by expanding from 5 to 10 employees, or from 20 to 25? Explain. Consider using a graph to help justify your answer.

TW **96.** Why do you think subscripts are used in Exercises 15 and 25 but not in Exercises 27 and 28?

97. *Escape Velocity.* A satellite's escape velocity is 6.5 mi/sec, the radius of the earth is 3960 mi, and the earth's gravitational constant is 32.2 ft/sec^2. How far is the satellite from the surface of the earth? (See Example 2.)

98. The *harmonic mean* of two numbers a and b is a number M such that the reciprocal of M is the average of the reciprocals of a and b. Find a formula for the harmonic mean.

99. *Health Care.* Young's rule for determining the size of a particular child's medicine dosage c is

$$c = \frac{a}{a + 12} \cdot d,$$

where a is the child's age and d is the typical adult dosage. If a child's age is doubled, the dosage increases. Find the ratio of the larger dosage to the smaller dosage. By what percent does the dosage increase?

Source: Olsen, June Looby, Leon J. Ablon, and Anthony Patrick Giangrasso, *Medical Dosage Calculations*, 6th ed.

100. Solve for x:

$$x^2\left(1 - \frac{2pq}{x}\right) = \frac{2p^2q^3 - pq^2x}{-q}.$$

101. *Average Acceleration.* The formula

$$a = \frac{\dfrac{d_4 - d_3}{t_4 - t_3} - \dfrac{d_2 - d_1}{t_2 - t_1}}{t_4 - t_2}$$

can be used to approximate average acceleration, where the d's are distances and the t's are the corresponding times. Solve for t_1.

102. If y varies inversely as the cube of x and x is multiplied by 0.5, what is the effect on y?

103. *Intensity of Light.* The intensity I of light from a bulb varies directly as the wattage of W the bulb and inversely as the square of the distance d from the bulb. If the wattage of a light source and its distance from reading matter are both doubled, how does the intensity change?

104. Describe in words the variation represented by

$$W = \frac{km_1M_1}{d^2}.$$

Assume k is a constant.

105. *Tension of a Musical String.* The tension T on a string in a musical instrument varies jointly as the string's mass per unit length m, the square of its length l, and the square of its fundamental frequency f. A 2-m long string of mass 5 gm/m with a fundamental frequency of 80 has a tension of 100 N. How long should the same string be if its tension is going to be changed to 72 N?

106. *Volume and Cost.* A peanut butter jar in the shape of a right circular cylinder is 4 in. high and 3 in. in diameter and sells for \$1.20. If we assume that cost is proportional to volume, how much should a jar 6 in. high and 6 in. in diameter cost?

107. *Golf Distance Finder.* A device used in golf to estimate the distance d to a hole measures the size s that the 7-ft pin *appears* to be in a viewfinder. The viewfinder uses the principle, diagrammed here, that s gets bigger when d gets smaller. If $s = 0.56$ in. when $d = 50$ yd, find an equation of variation that expresses d as a function of s. What is d when $s = 0.40$ in.?

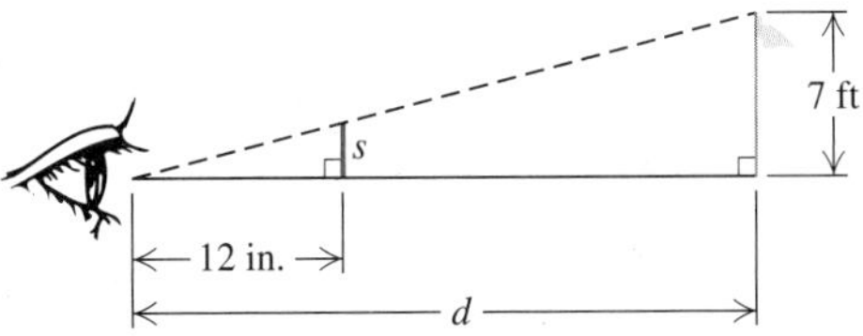

HOW IT WORKS:

Just sight the flagstick through the viewfinder...
fit flag between top dashed line and the solid line below...
...read the distance, 50 – 220 yards.

50 70 90 110 130 150 170 190 210
RANGE YARDS

Try Exercise Answers: Section 7.8

17. $b = \dfrac{at}{a - t}$ **19.** $R = \dfrac{2V}{I} - 2r$, or $\dfrac{2V - 2Ir}{I}$
21. $g = \dfrac{Rs}{s - R}$ **43.** $k = 7; y = 7x$ **49.** $k = 60; y = \dfrac{60}{x}$
55. $33\frac{1}{3}$ cm **57.** 3.5 hr **69.** $y = \frac{2}{3}x^2$ **73.** $y = 0.3xz^2$
83. **(a)** Inverse; **(b)** $y = \dfrac{300{,}000}{x}$; **(c)** 30,000 VMT

Study Summary

KEY TERMS AND CONCEPTS	EXAMPLES	PRACTICE EXERCISES
SECTION 7.1: RATIONAL EXPRESSIONS AND FUNCTIONS		
A **rational expression** can be written as the quotient of two polynomials and is undefined when the denominator is 0.	List all numbers for which the expression is undefined: $\dfrac{x+2}{x^2-3x}$. The rational expression is undefined when the denominator is 0. $x^2 - 3x = 0$ $x(x-3) = 0$ $x = 0 \quad or \quad x - 3 = 0$ $x = 0 \quad or \quad x = 3$ The expression is undefined when $x = 0$ or when $x = 3$.	**1.** List all numbers for which the expression is undefined: $\dfrac{-3}{(t+7)^2}$.
To simplify rational expressions, remove a factor equal to 1.	$\dfrac{x^2-3x-4}{x^2-1} = \dfrac{(x+1)(x-4)}{(x+1)(x-1)}$ **Factoring the numerator and the denominator** $= \dfrac{x-4}{x-1}$ $\quad \dfrac{x+1}{x+1} = 1$	**2.** Simplify: $\dfrac{y^2-5y}{y^2-25}$.
SECTION 7.2: MULTIPLICATION AND DIVISION		
The Product of Two Rational Expressions $\dfrac{A}{B} \cdot \dfrac{C}{D} = \dfrac{AC}{BD}$	$\dfrac{5v+5}{v-2} \cdot \dfrac{2v^2-8v+8}{v^2-1}$ $= \dfrac{5(v+1) \cdot 2(v-2)(v-2)}{(v-2)(v+1)(v-1)}$ **Multiplying numerators, multiplying denominators, and factoring** $= \dfrac{10(v-2)}{v-1}$ $\quad \dfrac{(v+1)(v-2)}{(v+1)(v-2)} = 1$	**3.** Multiply and, if possible, simplify: $\dfrac{6x-12}{2x^2+3x-2} \cdot \dfrac{x^2-4}{8x-8}$.
The Quotient of Two Rational Expressions $\dfrac{A}{B} \div \dfrac{C}{D} = \dfrac{A}{B} \cdot \dfrac{D}{C} = \dfrac{AD}{BC}$	$(x^2-5x-6) \div \dfrac{x^2-1}{x+6}$ $= \dfrac{x^2-5x-6}{1} \cdot \dfrac{x+6}{x^2-1}$ **Multiplying by the reciprocal of the divisor** $= \dfrac{(x-6)(x+1)(x+6)}{(x+1)(x-1)}$ **Multiplying numerators, multiplying denominators, and factoring** $= \dfrac{(x-6)(x+6)}{x-1}$ $\quad \dfrac{x+1}{x+1} = 1$	**4.** Divide and, if possible, simplify: $\dfrac{t-3}{6} \div \dfrac{t+1}{15}$.

SECTION 7.3: ADDITION, SUBTRACTION, AND LEAST COMMON DENOMINATORS

The Sum of Two Rational Expressions

$$\frac{A}{B} + \frac{C}{B} = \frac{A + C}{B}$$

$$\frac{7x + 8}{x + 1} + \frac{4x + 3}{x + 1} = \frac{7x + 8 + 4x + 3}{x + 1}$$ **Adding numerators and keeping the denominator**

$$= \frac{11x + 11}{x + 1}$$

$$= \frac{11(x + 1)}{x + 1}$$ **Factoring**

$$= 11$$ $\frac{x + 1}{x + 1} = 1$

5. Add and, if possible, simplify:

$$\frac{5x + 4}{x + 3} + \frac{4x + 1}{x + 3}.$$

The Difference of Two Rational Expressions

$$\frac{A}{B} - \frac{C}{B} = \frac{A - C}{B}$$

$$\frac{7x + 8}{x + 1} - \frac{4x + 3}{x + 1} = \frac{7x + 8 - (4x + 3)}{x + 1}$$

Subtracting numerators and keeping the denominator. The parentheses are necessary.

$$= \frac{7x + 8 - 4x - 3}{x + 1}$$ **Removing parentheses**

$$= \frac{3x + 5}{x + 1}$$

6. Subtract and, if possible, simplify:

$$\frac{5x + 4}{x + 3} - \frac{4x + 1}{x + 3}.$$

To find the **least common multiple (LCM)** of two or more expressions, write the prime factorizations of the expressions. The LCM contains each factor the greatest number of times that it occurs in any of the individual factorizations.

Find the LCM of $m^2 - 5m + 6$ and $m^2 - 4m + 4$.

$$m^2 - 5m + 6 = (m - 2)(m - 3)$$
$$m^2 - 4m + 4 = (m - 2)(m - 2)$$
Factoring each expression

LCM $= (m - 2)(m - 2)(m - 3)$

7. Find the LCM of $x^2 - 2x - 15$ and $x^2 - 9$.

SECTION 7.4: ADDITION AND SUBTRACTION WITH UNLIKE DENOMINATORS

To add or subtract rational expressions with different denominators, first rewrite the expressions as equivalent expressions with a common denominator.

$$\frac{2x}{x^2 - 16} + \frac{x}{x - 4}$$

$$= \frac{2x}{(x + 4)(x - 4)} + \frac{x}{x - 4}$$ **The LCD is $(x + 4)(x - 4)$.**

$$= \frac{2x}{(x + 4)(x - 4)} + \frac{x}{x - 4} \cdot \frac{x + 4}{x + 4}$$ **Multiplying by 1 to get the LCD in the second expression**

$$= \frac{2x}{(x + 4)(x - 4)} + \frac{x^2 + 4x}{(x + 4)(x - 4)}$$

$$= \frac{x^2 + 6x}{(x + 4)(x - 4)}$$

8. Subtract and, if possible, simplify:

$$\frac{t}{t - 1} - \frac{t - 2}{t + 1}.$$

SECTION 7.5: COMPLEX RATIONAL EXPRESSIONS

Complex rational expressions contain one or more rational expressions within the numerator and/or the denominator. They can be simplified either by using division or by multiplying by a form of 1 to clear the fractions.

Multiplying by the LCD:

$$\frac{\frac{4}{x}}{\frac{3}{x}+\frac{2}{x^2}} = \frac{\frac{4}{x}}{\frac{3}{x}+\frac{2}{x^2}} \cdot \frac{x^2}{x^2}$$

The LCD of all the denominators is x^2; multiplying by $\frac{x^2}{x^2}$

$$= \frac{\frac{4}{x}\cdot\frac{x^2}{1}}{\left(\frac{3}{x}+\frac{2}{x^2}\right)\cdot\frac{x^2}{1}}$$

$$= \frac{\frac{4\cdot x\cdot \cancel{x}}{\cancel{x}}}{\frac{3\cdot x\cdot\cancel{x}}{\cancel{x}}+\frac{2\cdot\cancel{x^2}}{\cancel{x^2}}} = \frac{4x}{3x+2}$$

The fractions are cleared.

9. Simplify: $\dfrac{\frac{4}{x}-4}{\frac{7}{x}-7}$.

Using division to simplify:

$$\frac{\frac{1}{6}-\frac{1}{x}}{\frac{6-x}{6}} = \frac{\frac{1}{6}\cdot\frac{x}{x}-\frac{1}{x}\cdot\frac{6}{6}}{\frac{6-x}{6}} = \frac{\frac{x-6}{6x}}{\frac{6-x}{6}}$$

Subtracting to get a single rational expression in the numerator

$$= \frac{x-6}{6x} \div \frac{6-x}{6} = \frac{x-6}{6x}\cdot\frac{6}{6-x}$$

Dividing the numerator by the denominator

$$= \frac{\cancel{6(x-6)}}{\cancel{6}x(-1)\cancel{(x-6)}} = \frac{1}{-x} = -\frac{1}{x}$$

Factoring and simplifying: $\frac{6(x-6)}{6(x-6)} = 1$

SECTION 7.6: RATIONAL EQUATIONS

To Solve a Rational Equation

1. List any restrictions.
2. Clear the equation of fractions.
3. Solve the resulting equation.
4. Check the possible solution(s) in the original equation.

Solve: $\frac{2}{x+1} = \frac{1}{x-2}$. **The restrictions are $x \neq -1, x \neq 2$.**

$$\frac{2}{x+1} = \frac{1}{x-2}$$

$$\cancel{(x+1)}(x-2)\cdot\frac{2}{\cancel{x+1}} = (x+1)\cancel{(x-2)}\cdot\frac{1}{\cancel{x-2}}$$

$$2(x-2) = x+1$$

$$2x-4 = x+1$$

$$x = 5$$

Check: Since $\frac{2}{5+1} = \frac{1}{5-2}$, the solution is 5.

10. Solve: $\frac{3}{x+4} = \frac{1}{x-1}$.

SECTION 7.7: APPLICATIONS USING RATIONAL EQUATIONS AND PROPORTIONS

The Work Principle

If a = the time needed for A to complete the work alone,

b = the time needed for B to complete the work alone, and

t = the time needed for A and B to complete the work together, then:

$$\frac{t}{a} + \frac{t}{b} = 1;$$

$$\frac{1}{a} \cdot t + \frac{1}{b} \cdot t = 1;$$

$$\frac{1}{a} + \frac{1}{b} = \frac{1}{t}.$$

It takes Jordyn 9 hr longer than Manuel to install a wood floor. Working together, they can do the job in 20 hr. How long would it take each, working alone, to install the floor?

We model the situation using the work principle, with $a = m, b = m + 9$, and $t = 20$:

$$\frac{20}{m} + \frac{20}{m + 9} = 1$$

$$m = 36. \quad \textbf{Solving the equation}$$

It would take Manuel 36 hr to install the floor alone, and Jordyn 45 hr.

See Example 2 in Section 7.7 for a complete solution of this problem.

11. Jackson can sand the oak floors and stairs in a home in 12 hr. Charis can do the same job in 9 hr. How long would it take if they worked together? (Assume that two sanders are available.)

The Motion Formula

$$d = r \cdot t, \quad r = \frac{d}{t}, \quad \text{or}$$

$$t = \frac{d}{r}$$

On her road bike, Paige bikes 15 km/h faster than Sean does on his mountain bike. In the time it takes Paige to travel 80 km, Sean travels 50 km. Find the speed of each bicyclist.

Sean's speed:	r km/h
Sean's time:	$50/r$ hr
Paige's speed:	$(r + 15)$ km/h
Paige's time:	$80/(r + 15)$ hr

The times are equal:

$$\frac{50}{r} = \frac{80}{r + 15}$$

$$r = 25.$$

Paige's speed is 40 km/h, and Sean's speed is 25 km/h.

See Example 3 in Section 7.7 for a complete solution of this problem.

12. The current in the South River is 4 mph. Drew's boat travels 35 mi downstream in the same time that it takes to travel 15 mi upstream. What is the speed of Drew's boat in still water?

In geometry, proportions arise in the study of **similar triangles.**

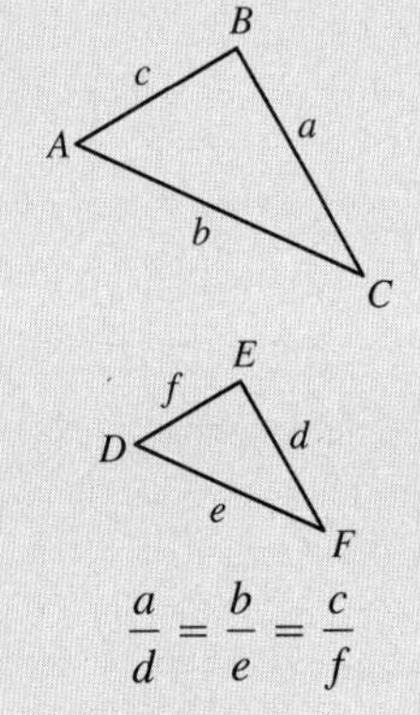

$$\frac{a}{d} = \frac{b}{e} = \frac{c}{f}$$

Triangles DEF and UVW are similar. Solve for u.

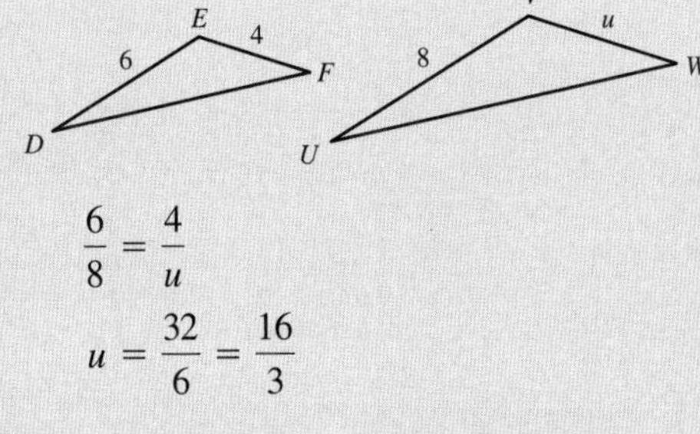

$$\frac{6}{8} = \frac{4}{u}$$

$$u = \frac{32}{6} = \frac{16}{3}$$

13. Triangles LMN and PQR are similar. Find the value of m.

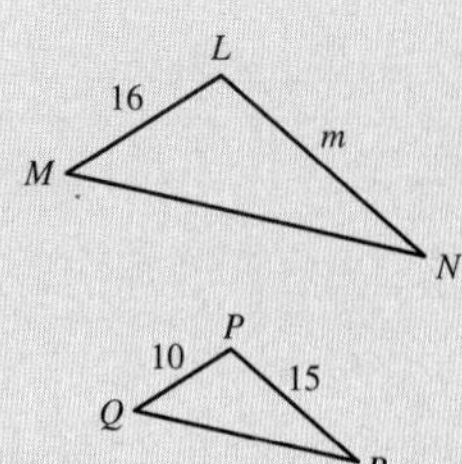

SECTION 7.8: FORMULAS, APPLICATIONS, AND VARIATION

Direct Variation $y = kx$	If y varies directly as x, and $y = 45$ when $x = 0.15$, find the equation of variation. $y = kx$ $45 = k(0.15)$ $300 = k$ The equation of variation is $y = 300x$.	**14.** If y varies directly as x, and $y = 10$ when $x = 0.2$, find the equation of variation.
Inverse Variation $y = \frac{k}{x}$	If y varies inversely as x, and $y = 45$ when $x = 0.15$, find the equation of variation. $y = \frac{k}{x}$ $45 = \frac{k}{0.15}$ $6.75 = k$ The equation of variation is $y = \frac{6.75}{x}$.	**15.** If y varies inversely as x, and $y = 5$ when $x = 8$, find the equation of variation.
Joint Variation $y = kxz$	If y varies jointly as x and z, and $y = 40$ when $x = 5$ and $z = 4$, find the equation of variation. $y = kxz$ $40 = k \cdot 5 \cdot 4$ $2 = k$ The equation of variation is $y = 2xz$.	**16.** If y varies jointly as x and z, and $y = 2$ when $x = 5$ and $z = 4$, find the equation of variation.

Review Exercises

Concept Reinforcement *Classify each of the following statements as either true or false.*

1. Every rational expression can be simplified. [7.1]
2. The expression $(t - 3)/(t^2 - 4)$ is undefined for $t = 2$. [7.1]
3. The expression $(t - 3)/(t^2 - 4)$ is undefined for $t = 3$. [7.1]
4. You need a common denominator in order to multiply rational expressions. [7.2]
5. You need a common denominator in order to divide rational expressions. [7.2]
6. You need a common denominator in order to add rational expressions. [7.3]
7. You need a common denominator in order to subtract rational expressions. [7.3]
8. Every rational equation has at least one solution. [7.6]
9. The number 0 can never be a solution of a rational equation. [7.6]
10. If Carlie swims 5 km/h in still water and heads into a current of 2 km/h, her speed will change to 3 km/h. [7.7]

11. If

$$f(t) = \frac{t^2 - 3t + 2}{t^2 - 9},$$

find the following function values. [7.1]

a) $f(0)$
b) $f(-1)$
c) $f(2)$

List all numbers for which each expression is undefined. [7.1]

12. $\dfrac{17}{-x^2}$
13. $\dfrac{9}{a - 4}$
14. $\dfrac{x - 5}{x^2 - 36}$
15. $\dfrac{x^2 + 3x + 2}{x^2 + x - 30}$

Simplify. [7.1]

16. $\dfrac{4x^2 - 8x}{4x^2 + 4x}$
17. $\dfrac{14x^2 - x - 3}{2x^2 - 7x + 3}$
18. $\dfrac{5x^2 - 20y^2}{2y - x}$

Multiply or divide and, if possible, simplify. [7.2]

19. $\dfrac{a^2 - 36}{10a} \cdot \dfrac{2a}{a + 6}$
20. $\dfrac{8t + 8}{2t^2 + t - 1} \cdot \dfrac{t^2 - 1}{t^2 - 2t + 1}$
21. $\dfrac{16 - 8t}{3} \div \dfrac{t - 2}{12t}$
22. $\dfrac{4x^4}{x^2 - 1} \div \dfrac{2x^3}{x^2 - 2x + 1}$
23. $\dfrac{x^2 + 1}{x - 2} \cdot \dfrac{2x + 1}{x + 1}$
24. $(t^2 + 3t - 4) \div \dfrac{t^2 - 1}{t + 4}$

Find the LCM. [7.3]

25. $10a^3b^8$, $12a^5b$
26. $x^2 - x$, $x^5 - x^3$, x^4
27. $y^2 - y - 2$, $y^2 - 4$

Add or subtract and, if possible, simplify.

28. $\dfrac{x + 6}{x + 3} + \dfrac{9 - 4x}{x + 3}$ [7.3]
29. $\dfrac{6x - 3}{x^2 - x - 12} - \dfrac{2x - 15}{x^2 - x - 12}$ [7.3]
30. $\dfrac{3x - 1}{2x} - \dfrac{x - 3}{x}$ [7.4]
31. $\dfrac{x + 5}{x - 2} - \dfrac{x}{2 - x}$ [7.4]
32. $\dfrac{2a}{a + 1} - \dfrac{4a}{1 - a^2}$ [7.4]
33. $\dfrac{d^2}{d - 2} + \dfrac{4}{2 - d}$ [7.4]
34. $\dfrac{1}{x^2 - 25} - \dfrac{x - 5}{x^2 - 4x - 5}$ [7.4]
35. $\dfrac{2}{5x} + \dfrac{3}{2x + 4}$ [7.4]

Find simplified form for $f(x)$ and list all restrictions on the domain.

36. $f(x) = \dfrac{14x^2 - x - 3}{2x^2 - 7x + 3}$ [7.1]

37. $f(x) = \dfrac{3x}{x + 2} - \dfrac{x}{x - 2} + \dfrac{8}{x^2 - 4}$ [7.4]

Simplify. [7.5]

38. $\dfrac{\dfrac{1}{z} + 1}{\dfrac{1}{z^2} - 1}$

39. $\dfrac{\dfrac{5}{2x^2}}{\dfrac{3}{4x} + \dfrac{4}{x^3}}$

40. $\dfrac{\dfrac{y^2 + 4y - 77}{y^2 - 10y + 25}}{\dfrac{y^2 - 5y - 14}{y^2 - 25}}$

41. $\dfrac{\dfrac{5}{x^2 - 9} - \dfrac{3}{x + 3}}{\dfrac{4}{x^2 + 6x + 9} + \dfrac{2}{x - 3}}$

Solve. [7.6]

42. $\dfrac{3}{x} + \dfrac{7}{x} = 5$

43. $\dfrac{5}{3x + 2} = \dfrac{3}{2x}$

44. $x + \dfrac{6}{x} = -7$

45. $\dfrac{x + 6}{x^2 + x - 6} + \dfrac{x}{x^2 + 4x + 3} = \dfrac{x + 2}{x^2 - x - 2}$

46. If

$$f(x) = \frac{2}{x - 1} + \frac{2}{x + 2},$$

find all a for which $f(a) = 1$. [7.6]

Solve. [7.7]

47. Megan can arrange the books for a book sale in 9 hr. Kelly can set up for the same book sale in 12 hr. How long would it take them, working together, to set up for the book sale?

48. A research company uses employees' computers to process data while the employee is not using the computer. An Intel Core 2 Quad processor can process a data file in 15 sec less time than an Intel Core 2 Duo processor. Working together, the computers can process the file in 18 sec. How long does it take each computer to process the file?

49. The Black River's current is 6 mph. A boat travels 50 mi downstream in the same time that it takes to travel 30 mi upstream. What is the speed of the boat in still water?

50. A car and a motorcycle leave a rest area at the same time, with the car traveling 8 mph faster than the motorcycle. The car then travels 105 mi in the time that it takes the motorcycle to travel 93 mi. Find the speed of each vehicle.

51. To estimate the harbor seal population in Bristol Bay, scientists radio-tagged 33 seals. Several days later, they gathered a sample of 40 seals, and found that 24 of them were tagged. Estimate the seal population of the bay.

52. Triangles *ABC* and *XYZ* are similar. Find the value of x.

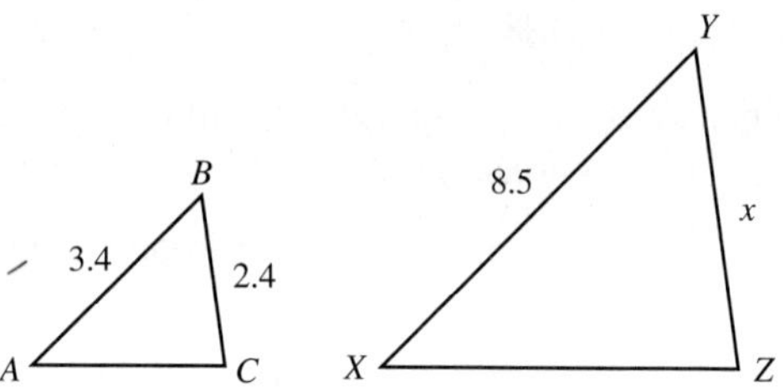

Solve. [7.8]

53. $R = \dfrac{gs}{g + s}$, for s

54. $S = \dfrac{H}{m(t_1 - t_2)}$, for m

55. $\dfrac{1}{ac} = \dfrac{2}{ab} - \dfrac{3}{bc}$, for c

56. $T = \dfrac{A}{v(t_2 - t_1)}$, for t_1

57. The amount of waste generated by a household varies directly as the number of people in the household. The average U.S. family has 2.6 people and generates 11.96 lb of waste daily. How many pounds of waste would be generated daily by a family of 5?
Source: Based on data from the U.S. Environmental Protection Agency

58. A warning dye is used by people in lifeboats to aid search planes. The volume V of the dye used varies directly as the square of the diameter d of the circular patch of water formed by the dye. If 4 L of dye is required for a 10-m wide circle, how much dye is needed for a 40-m wide circle?

59. Find an equation of variation in which y varies inversely as x, and $y = 3$ when $x = \frac{1}{4}$.

60. The table below lists the size y of one serving of breakfast cereal for the corresponding number of servings x obtained from a given box. [7.8]

Number of Servings	Size of Serving (in ounces)
1	24
4	6
6	4
12	2
16	1.5

a) Determine whether the data indicate direct variation or inverse variation.
b) Find an equation of variation that fits the data. Use the data point $(12, 2)$.
c) Use the equation to estimate the size of each serving when 8 servings are obtained from the box.

SYNTHESIS

TW **61.** Discuss at least three different uses of the LCD studied in this chapter. [7.4], [7.5], [7.6]

TW **62.** Explain the difference between a rational expression and a rational equation. [7.1]

Solve.

63. $\dfrac{5}{x-13} - \dfrac{5}{x} = \dfrac{65}{x^2 - 13x}$ [7.6]

64. $\dfrac{\dfrac{x}{x^2-25} + \dfrac{2}{x-5}}{\dfrac{3}{x-5} - \dfrac{4}{x^2-10x+25}} = 1$ [7.5], [7.6]

Simplify.

65. $\dfrac{2a^2+5a-3}{a^2} \cdot \dfrac{5a^3+30a^2}{2a^2+7a-4} \div \dfrac{a^2+6a}{a^2+7a+12}$ [7.2]

Aha! **66.** $\dfrac{5(x-y)}{(x-y)(x+2y)} - \dfrac{5(x-3y)}{(x+2y)(x-3y)}$ [7.3]

Chapter Test 

List all numbers for which each expression is undefined.

1. $\dfrac{2-x}{5x}$

2. $\dfrac{x^2+x-30}{x^2-3x+2}$

3. Simplify: $\dfrac{6x^2-7x-5}{3x^2-2x-5}$.

Multiply or divide and, if possible, simplify.

4. $\dfrac{a^2-25}{9a} \cdot \dfrac{6a}{5-a}$

5. $\dfrac{25y^2-1}{9y^2-6y} \div \dfrac{5y^2+9y-2}{3y^2+y-2}$

6. $\dfrac{4x^2-1}{x^2-2x+1} \div \dfrac{x-2}{x^2+1}$

7. $(x^2+6x+9) \cdot \dfrac{(x-3)^2}{x^2-9}$

8. Find the LCM:
$y^2-9,\quad y^2+10y+21,\quad y^2+4y-21.$

Add or subtract, and, if possible, simplify.

9. $\dfrac{2+x}{x^3} + \dfrac{7-4x}{x^3}$

10. $\dfrac{5-t}{t^2+1} - \dfrac{t-3}{t^2+1}$

11. $\dfrac{x-4}{x-3} + \dfrac{x-1}{3-x}$

12. $\dfrac{x-4}{x-3} - \dfrac{x-1}{3-x}$

13. $\dfrac{7}{t-2} + \dfrac{4}{t}$

14. $\dfrac{4}{x^2-16} - \dfrac{x-1}{x^2+5x+4}$

Find simplified form for $f(x)$ and list all restrictions on the domain.

15. $f(x) = \dfrac{6x^2+17x+7}{2x^2+7x+3}$

16. $f(x) = \dfrac{4}{x+3} - \dfrac{x}{x-2} + \dfrac{x^2+4}{x^2+x-6}$

Simplify.

17. $\dfrac{9 - \dfrac{1}{y^2}}{3 - \dfrac{1}{y}}$

18. $\dfrac{\dfrac{x}{8} - \dfrac{8}{x}}{\dfrac{1}{8} + \dfrac{1}{x}}$

19. $\dfrac{\dfrac{x^2 - 5x - 36}{x^2 - 36}}{\dfrac{x^2 + x - 12}{x^2 - 12x + 36}}$

Solve.

20. $\dfrac{1}{t} + \dfrac{1}{3t} = \dfrac{1}{2}$

21. $\dfrac{4}{2x - 5} = \dfrac{6}{5x + 3}$

22. $\dfrac{15}{x} - \dfrac{15}{x - 2} = -2$

For Exercises 23 and 24, let $f(x) = \dfrac{x + 3}{x - 1}$.

23. Find $f(2)$ and $f(-3)$.

24. Find all a for which $f(a) = 7$.

25. Solve $A = \dfrac{h(b_1 + b_2)}{2}$ for b_1.

26. Emma bicycles 12 mph with no wind. Against the wind, she bikes 8 mi in the same time that it takes to bike 14 mi with the wind. What is the speed of the wind?

27. Kyla can install a vinyl kitchen floor in 3.5 hr. Brock can perform the same job in 4.5 hr. How long will it take them, working together, to install the vinyl?

28. Pe'rez and Ellia work together to mulch the flower beds around an office complex in $2\frac{6}{7}$ hr. Working alone, it would take Pe'rez 6 hr more than it would take Ellia. How long would it take each of them to complete the landscaping working alone?

29. A recipe for pizza crust calls for $3\frac{1}{2}$ cups of whole wheat flour and $1\frac{1}{4}$ cups of warm water. If 6 cups of whole wheat flour are used, how much water should be used?

30. The number of workers n needed to clean a stadium after a game varies inversely as the amount of time t allowed for the cleanup. If it takes 25 workers to clean the stadium when there are 6 hr allowed for the job, how many workers are needed if the stadium must be cleaned in 5 hr?

31. The surface area of a balloon varies directly as the square of its radius. The area is 325 in^2 when the radius is 5 in. What is the surface area when the radius is 7 in.?

SYNTHESIS

32. Solve: $\dfrac{6}{x - 15} - \dfrac{6}{x} = \dfrac{90}{x^2 - 15x}$.

33. Simplify: $1 - \dfrac{1}{1 - \dfrac{1}{1 - \dfrac{1}{a}}}$.

34. One summer, Andy mowed 4 lawns for every 3 lawns mowed by his brother Chad. Together, they mowed 98 lawns. How many lawns did each mow?

8 Inequalities

What Do You Do to Unwind?

As the graph indicates, on average, people are spending less time listening to music and more time playing video games. In Example 3 of Section 8.1, we predict when time spent playing video games will be greater than time spent listening to music.

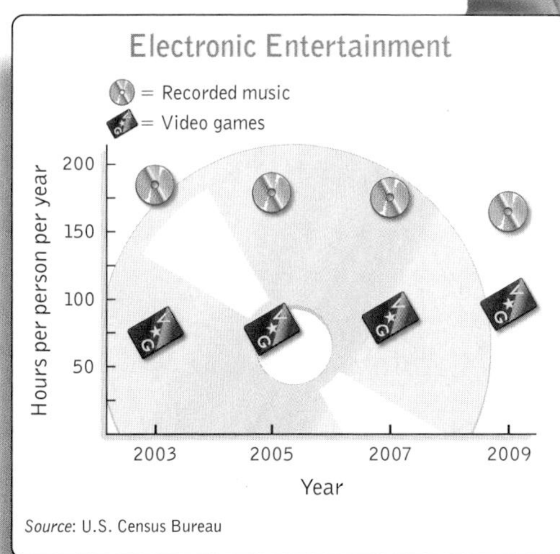

Source: U.S. Census Bureau

Inequalities are mathematical sentences containing symbols such as $<$ (is less than). In this chapter, we use the principles for solving inequalities developed in Chapter 2 to solve various kinds of inequalities. We also combine our knowledge of inequalities and systems of equations to solve systems of inequalities.

8.1 Graphical Solutions and Compound Inequalities

- Solving Inequalities Graphically
- Intersections of Sets and Conjunctions of Sentences
- Unions of Sets and Disjunctions of Sentences
- Interval Notation and Domains

STUDY TIP

How Involved Are You?

Be an active participant in your class. Listen, ask questions when appropriate, and avoid daydreaming. The more you learn in class, the less you will have to learn on your own time outside of class.

Recall from Chapter 1 that an **inequality** is any sentence containing $<, >, \leq, \geq$, or $\neq$ (see Section 1.4)—for example,

$$-2 < a, \quad x > 4, \quad x + 3 \leq 6, \quad 7y \geq 10y - 4, \quad \text{and} \quad 5x \neq 10.$$

Any replacement for the variable that makes an inequality true is called a **solution.** The set of all solutions is called the **solution set.** When all solutions of an inequality are found, we say that we have **solved** the inequality.

We solved inequalities algebraically in Chapter 2, using the addition and multiplication principles for inequalities. These principles are similar to those used to solve equations with one important difference: The direction of the inequality symbol must be reversed when multiplying or dividing by a negative number.

We now look at a graphical method for solving inequalities.

SOLVING INEQUALITIES GRAPHICALLY

Consider the graphs of the functions $f(x) = 2x + 4$ and $g(x) = -x + 1$.

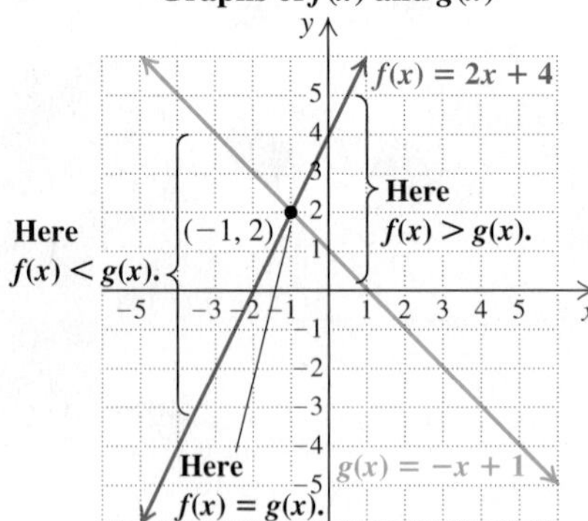

We can see the following from the graph:

- The graphs intersect at $(-1, 2)$. This means that $f(x) = g(x)$ when $x = -1$.
- The graph of f lies above the graph of g when $x > -1$. This is illustrated on the graph for $x = 1$. This means that $f(x) > g(x)$ when $x > -1$.
- The graph of f lies below the graph of g when $x < -1$. This is illustrated on the graph for $x = -4$. This means that $f(x) < g(x)$ when $x < -1$.

In summary, for $f(x) = 2x + 4$ and $g(x) = -x + 1$, we have the following.

Equation/Inequality	Solution Set	Graph of Solution Set
$f(x) = g(x)$	$\{-1\}$	
$f(x) < g(x)$	$(-\infty, -1)$	
$f(x) > g(x)$	$(-1, \infty)$	

EXAMPLE 1 Solve graphically: $-3x + 1 \geq x - 7$.

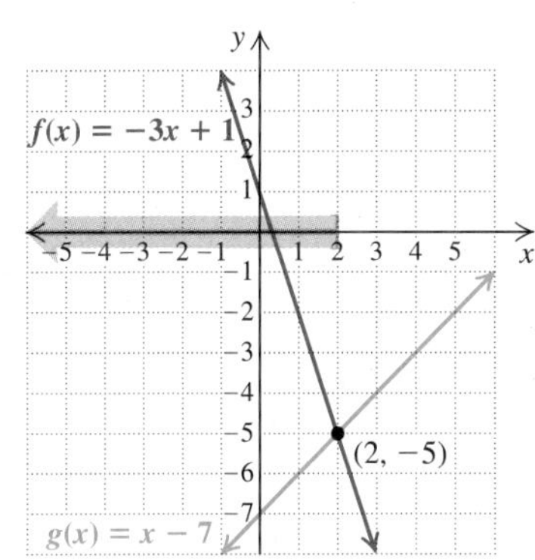

SOLUTION We let $f(x) = -3x + 1$ and $g(x) = x - 7$, and graph both functions. The solution set will consist of the interval for which the graph of f lies on or above the graph of g.

We can confirm that $f(x) = g(x)$ when $x = 2$. The graph of f lies *on* the graph of g when $x = 2$. It lies *above* the graph of g when $x < 2$. Thus the solution of $-3x + 1 \geq x - 7$ is

$$\{x \mid x \leq 2\}, \quad \text{or} \quad (-\infty, 2].$$

This set is indicated by the purple shading on the x-axis.

Try Exercise 17.

EXAMPLE 2 Solve graphically: $16 - 7x \geq 10x - 4$.

SOLUTION We let $y_1 = 16 - 7x$ and $y_2 = 10x - 4$, and graph y_1 and y_2 in the window $[-5, 5, -5, 15]$.

From the graph, we see that $y_1 = y_2$ at the point of intersection, or when $x \approx 1.1765$. To the left of the point of intersection, $y_1 > y_2$, as shown in the graph on the left below.

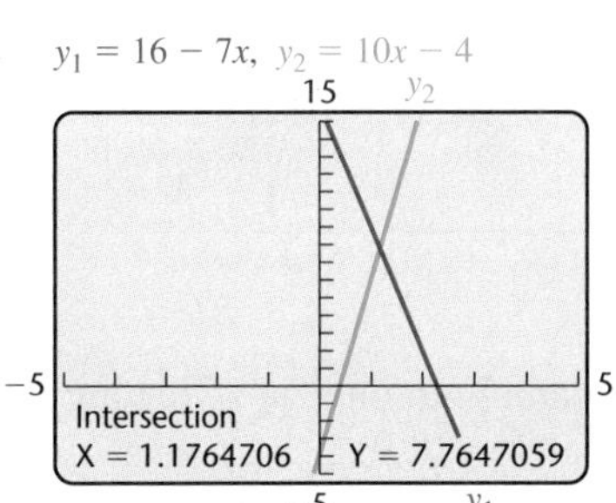

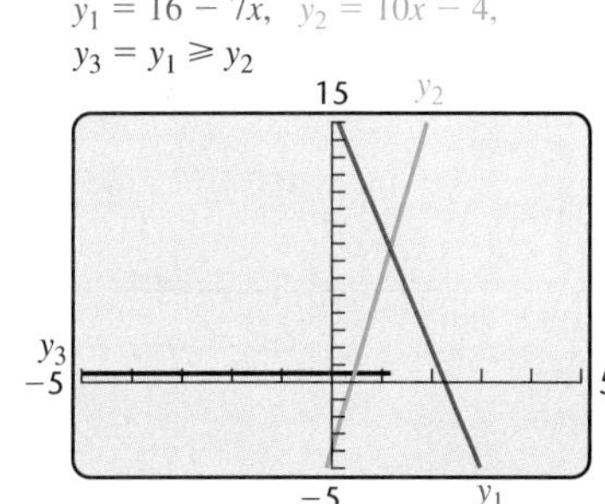

On many graphing calculators, the interval that is the solution set can be indicated by using the VARS and TEST keys. We enter and graph $y_3 = y_1 \geq y_2$. (The "$\geq$" symbol can be found in the TEST menu.) Where this is true, the value of y_3 will be 1, and where it is false, the value will be 0. The solution set is thus displayed as an interval, shown by a horizontal line 1 unit above the x-axis in the graph on the right above. The endpoint of the interval corresponds to the intersection of the graphs of the equations.

The solution set contains all x-values to the left of the point of intersection as well as the x-coordinate of the point of intersection. Thus the solution set is approximately $(-\infty, 1.1765]$, or converted to fraction notation, $\left(-\infty, \frac{20}{17}\right]$.

Try Exercise 25.

EXAMPLE 3 *Media Usage.* The table below lists the number of hours spent, on average, per person per year listening to recorded music and playing video games. Use linear regression to find two linear functions that can be used to estimate the number of hours $m(x)$ spent listening to music and the number of hours $v(x)$ spent playing video games x years after 2000. Then predict those years in which, on average, a person will spend more time playing video games than listening to music.

Year	Recorded Music (in hours per person per year)	Video Games (in hours per person per year)
2003	184	75
2005	179	78
2007	175	86
2009	165	96

Source: U.S. Census Bureau

SOLUTION We enter the number of years after 2000 as L1, the number of hours listening to music as L2, and the number of hours spent playing video games as L3. Graphing the data, we see that both sets appear to be linear, as shown at left.

L1	L2	L3 1
3	184	75
5	179	78
7	175	86
9	165	96

L1(1) = 3

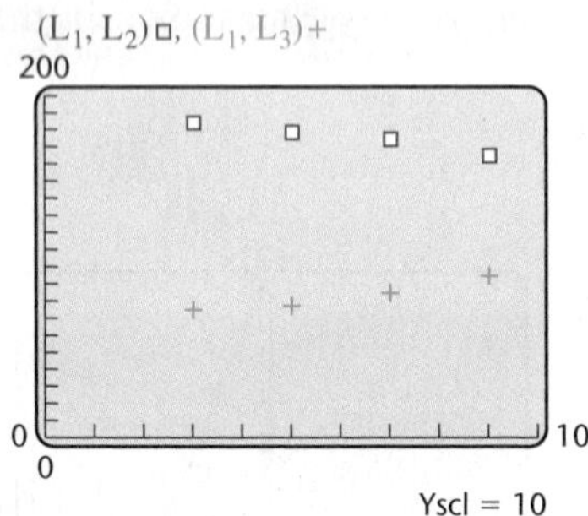

We use linear regression to fit two lines to the data. In the figure on the left below, Y1 is the linear regression line relating L1 and L2, and Y2 is the line relating L1 and L3. Thus we have

$$m(x) = -3.05x + 194.05,$$
$$v(x) = 3.55x + 62.45.$$

```
Plot1  Plot2  Plot3
\Y1= -3.05X+194.0
5
\Y2= 3.55X+62.45
\Y3=
\Y4=
\Y5=
\Y6=
```

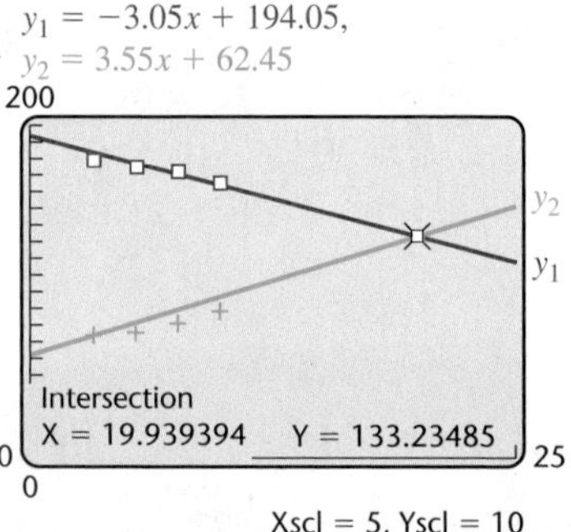

We graph both lines in a viewing window that shows the point of intersection, approximately $(19.9, 133.2)$, as shown in the figure on the right above. From the graph, we can see that when x is greater than 19.9, the number of hours spent playing video games will be greater than the number of hours spent listening to music. Since x is the number of years since 2000, we state that beginning in 2020, on average, a person will spend more time playing video games than listening to music.

Try Exercise 29.

Two inequalities joined by the word "and" or the word "or" are called **compound inequalities.** Thus, "$x < 3$ *or* $x > 4$" and "$x < 9$ *and* $x > -5$" are two examples of compound inequalities. Before discussing how to solve compound inequalities, we must first study ways in which sets can be combined.

INTERSECTIONS OF SETS AND CONJUNCTIONS OF SENTENCES

The **intersection** of two sets A and B is the set of all elements that are common to both A and B. We denote the intersection of sets A and B as

$$A \cap B.$$

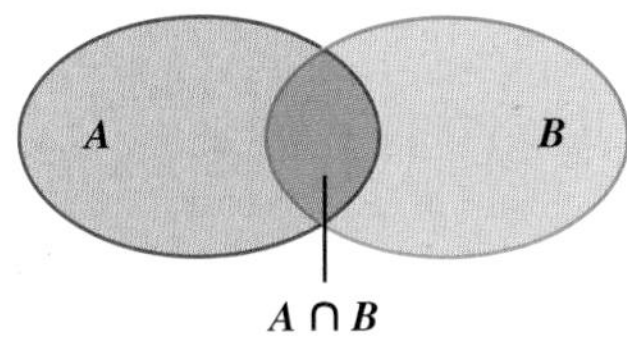

The intersection of two sets is represented by the purple region shown in the figure at left. For example, if $A = \{\text{all students who are taking a math class}\}$ and $B = \{\text{all students who are taking a history class}\}$, then $A \cap B = \{\text{all students who are taking a math class } \textit{and} \text{ a history class}\}$.

EXAMPLE 4 Find the intersection: $\{1, 2, 3, 4, 5\} \cap \{-2, -1, 0, 1, 2, 3\}$.

SOLUTION The numbers 1, 2, and 3 are common to both sets, so the intersection is $\{1, 2, 3\}$.

Try Exercise 35.

When two or more sentences are joined by the word *and* to make a compound sentence, the new sentence is called a **conjunction** of the sentences. The following is a conjunction of inequalities:

$$-2 < x \quad and \quad x < 1.$$

The word *and* means that *both* sentences must be true if the conjunction is to be true.

A number is a solution of a conjunction if it is a solution of *both* of the separate parts. For example, -1 is a solution because it is a solution of $-2 < x$ as well as $x < 1$; that is, -1 is *both* greater than -2 *and* less than 1.

The solution set of a conjunction is the intersection of the solution sets of the individual sentences.

EXAMPLE 5 Graph and write interval notation for the conjunction

$$-2 < x \quad and \quad x < 1.$$

SOLUTION We first graph $-2 < x$, then $x < 1$, and finally the conjunction $-2 < x \text{ and } x < 1$.

$\{x \mid -2 < x\}$ —7 −6 −5 −4 −3 −2 −1 0 1 2 3 4 5 6 7— $(-2, \infty)$

$\{x \mid x < 1\}$ —7 −6 −5 −4 −3 −2 −1 0 1 2 3 4 5 6 7— $(-\infty, 1)$

$\{x \mid -2 < x\} \cap \{x \mid x < 1\}$
$= \{x \mid -2 < x \text{ and } x < 1\}$ —7 −6 −5 −4 −3 −2 −1 0 1 2 3 4 5 6 7— $(-2, 1)$

Because there are numbers that are both greater than -2 and less than 1, the solution set of the conjunction $-2 < x \text{ and } x < 1$ is the interval $(-2, 1)$. In set-builder notation, this is written $\{x \mid -2 < x < 1\}$, the set of all numbers that are *simultaneously* greater than -2 *and* less than 1.

Try Exercise 55.

For $a < b$,

$$a < x \quad and \quad x < b \quad \textbf{can be abbreviated} \quad a < x < b;$$

and, equivalently,

$$b > x \quad and \quad x > a \quad \textbf{can be abbreviated} \quad b > x > a.$$

Mathematical Use of the Word "and" The word "and" corresponds to "intersection" and to the symbol "$\cap$". Any solution of a conjunction must make each part of the conjunction true.

EXAMPLE 6 Solve and graph: $-1 \le 2x + 5 < 13$.

SOLUTION This inequality is an abbreviation for the conjunction

$$-1 \le 2x + 5 \quad \textit{and} \quad 2x + 5 < 13.$$

We solve both algebraically and graphically.

ALGEBRAIC APPROACH

The word *and* corresponds to set *intersection*. To solve the conjunction, we solve each of the two inequalities separately and then find the intersection of the solution sets:

$$-1 \le 2x + 5 \quad \textit{and} \quad 2x + 5 < 13$$

$$-6 \le 2x \quad \textit{and} \quad 2x < 8$$ **Subtracting 5 from both sides of each inequality**

$$-3 \le x \quad \textit{and} \quad x < 4.$$ **Dividing both sides of each inequality by 2**

These steps are sometimes combined as follows:

$$-1 \le 2x + 5 < 13$$

$$-1 - 5 \le 2x + 5 - 5 < 13 - 5$$ **Subtracting 5 from all three regions**

$$-6 \le 2x < 8$$

$$-3 \le x < 4.$$ **Dividing by 2 in all three regions**

The solution set is $\{x \mid -3 \le x < 4\}$, or, in interval notation, $[-3, 4)$. The graph is the intersection of the two separate solution sets.

$\{x \mid -3 \le x\}$, or $[-3, \infty)$

-7 -6 -5 -4 -3 -2 -1 0 1 2 3 4 5 6 7

$\{x \mid x < 4\}$, or $(-\infty, 4)$

-7 -6 -5 -4 -3 -2 -1 0 1 2 3 4 5 6 7

$\{x \mid -3 \le x\} \cap \{x \mid x < 4\}$
$= \{x \mid -3 \le x < 4\}$, or
$[-3, 4)$

-7 -6 -5 -4 -3 -2 -1 0 1 2 3 4 5 6 7

GRAPHICAL APPROACH

We graph the equations $y_1 = -1$, $y_2 = 2x + 5$, and $y_3 = 13$, and determine those x-values for which $y_1 \le y_2$ *and* $y_2 < y_3$.

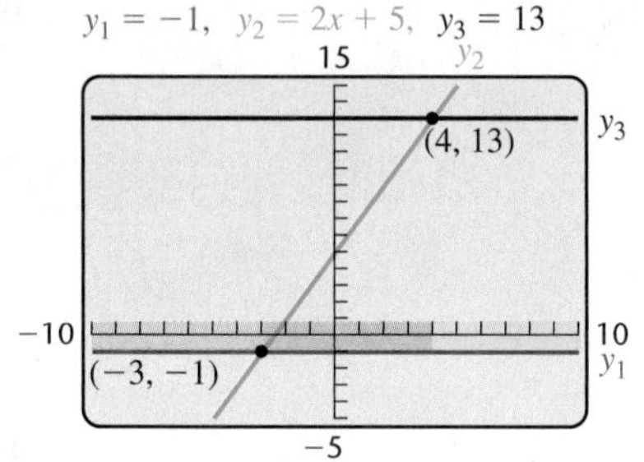

From the graph, we see that $y_1 < y_2$ for x-values greater than -3, as indicated by the purple and blue shading on the x-axis. (The shading does not appear on the calculator, but is added here to illustrate the intervals.) We also see that $y_2 < y_3$ for x-values less than 4, as shown by the purple and red shading on the x-axis. The solution set, indicated by the purple shading, is the intersection of these sets, as well as the number -3. It includes all x-values for which the line $y_2 = 2x + 5$ is both on or above the line $y_1 = -1$ *and* below the line $y_3 = 13$. This can be written $\{x \mid -3 \le x < 4\}$, or $[-3, 4)$.

Try Exercise 69.

CAUTION! The abbreviated form of a conjunction, like $-3 \le x < 4$, can be written only if both inequality symbols point in the same direction.

EXAMPLE 7 Solve and graph: $2x - 5 \ge -3$ *and* $5x + 2 \ge 17$.

SOLUTION We first solve each inequality separately, retaining the word *and*:

$$2x - 5 \ge -3 \quad \textit{and} \quad 5x + 2 \ge 17$$

$$2x \ge 2 \quad \textit{and} \quad 5x \ge 15$$

$$x \ge 1 \quad \textit{and} \quad x \ge 3.$$

$y_1 = 2x - 5$, $y_2 = 5x + 2$,
$y_3 = -3$, $y_4 = 17$

Next, we find the intersection of the two separate solution sets.

$\{x \mid x \geq 1\}$

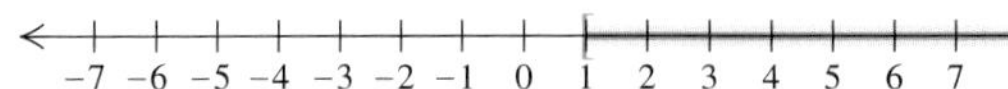

$\{x \mid x \geq 3\}$

$\{x \mid x \geq 1\} \cap \{x \mid x \geq 3\}$
$= \{x \mid x \geq 3\}$

The numbers common to both sets are those greater than or equal to 3. Thus the solution set is $\{x \mid x \geq 3\}$, or, in interval notation, $[3, \infty)$. You should check that any number in $[3, \infty)$ satisfies the conjunction whereas numbers outside $[3, \infty)$ do not. The graph at left serves as another check.

Try Exercise 91.

Sometimes there is no way to solve both parts of a conjunction at once.

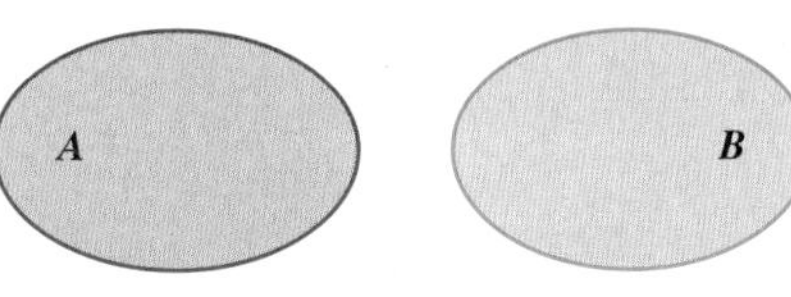

$A \cap B = \varnothing$

When $A \cap B = \varnothing$, A and B are said to be *disjoint*.

EXAMPLE 8 Solve and graph: $2x - 3 > 1$ *and* $3x - 1 < 2$.

SOLUTION We solve each inequality separately:

$$2x - 3 > 1 \quad and \quad 3x - 1 < 2$$
$$2x > 4 \quad and \quad 3x < 3$$
$$x > 2 \quad and \quad x < 1.$$

The solution set is the intersection of the individual inequalities.

$\{x \mid x > 2\}$

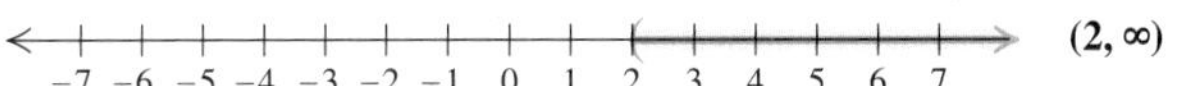

$(2, \infty)$

$\{x \mid x < 1\}$

$(-\infty, 1)$

$\{x \mid x > 2\} \cap \{x \mid x < 1\}$
$= \{x \mid x > 2$ *and* $x < 1\} = \varnothing$

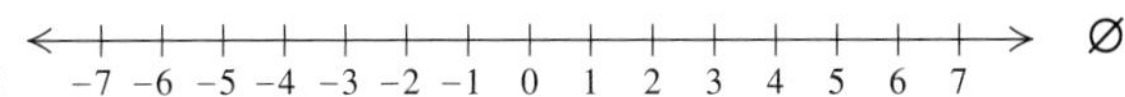

$\varnothing$

$y_1 = 2x - 3$, $y_2 = 3x - 1$,
$y_3 = 1$, $y_4 = 2$

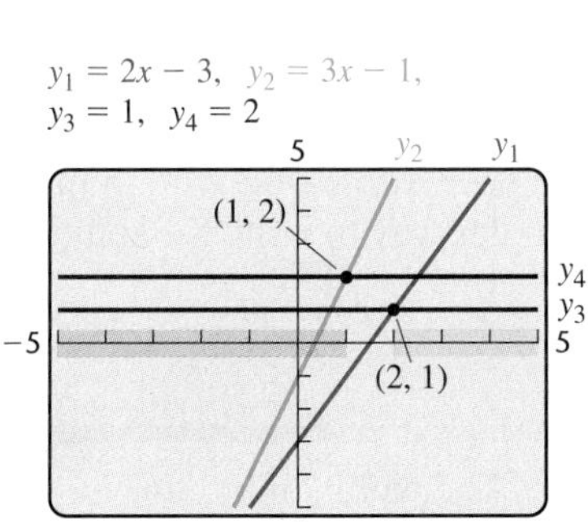

Since no number is both greater than 2 and less than 1, the solution set is the empty set, $\varnothing$. The graph at left confirms that the solution set is empty.

Try Exercise 93.

UNIONS OF SETS AND DISJUNCTIONS OF SENTENCES

The **union** of two sets A and B is the collection of elements belonging to A and/or B. We denote the union of A and B by

$$A \cup B.$$

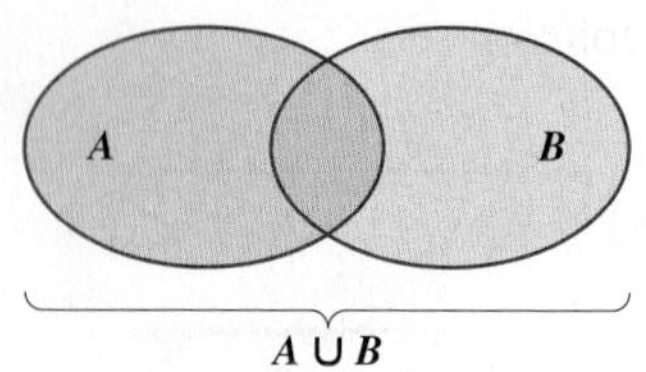

$A \cup B$

The union of two sets is often pictured as shown at left. For example, if $A = \{$all students who are taking a math class$\}$ and $B = \{$all students who are taking a history class$\}$, then $A \cup B = \{$all students who are taking a math class *or* a history class$\}$. Note that this set includes students who are taking a math class *and* a history class.

EXAMPLE 9 Find the union: $\{2, 3, 4\} \cup \{3, 5, 7\}$.

SOLUTION The numbers in either or both sets are 2, 3, 4, 5, and 7, so the union is $\{2, 3, 4, 5, 7\}$.

Try Exercise 37.

Student Notes

Remember that the union or the intersection of two sets is itself a set and should be written with set braces.

When two or more sentences are joined by the word *or* to make a compound sentence, the new sentence is called a **disjunction** of the sentences. Here is an example:

$$x < -3 \quad or \quad x > 3.$$

A number is a solution of a disjunction if it is a solution of at least one of the separate parts. For example, -5 is a solution of the disjunction since -5 is a solution of $x < -3$.

The solution set of a disjunction is the union of the solution sets of the individual sentences.

EXAMPLE 10 Graph and write interval notation for the disjunction

$$x < -3 \quad or \quad x > 3.$$

SOLUTION We first graph $x < -3$, then $x > 3$, and finally the disjunction $x < -3$ *or* $x > 3$.

$\{x \mid x < -3\}$ —6 −5 −4 −3 −2 −1 0 1 2 3 4 5 6— $(-\infty, -3)$

$\{x \mid x > 3\}$ —6 −5 −4 −3 −2 −1 0 1 2 3 4 5 6— $(3, \infty)$

$\{x \mid x < -3\} \cup \{x \mid x > 3\}$
$= \{x \mid x < -3 \text{ or } x > 3\}$ —6 −5 −4 −3 −2 −1 0 1 2 3 4 5 6— $(-\infty, -3) \cup (3, \infty)$

The solution set of $x < -3$ *or* $x > 3$ is $\{x \mid x < -3 \text{ or } x > 3\}$, or, in interval notation, $(-\infty, -3) \cup (3, \infty)$. There is no simpler way to write the solution.

Try Exercise 51.

Mathematical Use of the Word "or" The word "or" corresponds to "union" and to the symbol "$\cup$". For a number to be a solution of a disjunction, it must be in *at least one* of the solution sets of the individual sentences.

EXAMPLE 11 Solve and graph: $7 + 2x < -1$ *or* $13 - 5x \le 3$.

SOLUTION We solve each inequality separately, retaining the word *or*:

$$7 + 2x < -1 \quad or \quad 13 - 5x \le 3$$
$$2x < -8 \quad or \quad -5x \le -10$$

Dividing by a negative number and reversing the symbol

$$x < -4 \quad or \quad x \ge 2.$$

To find the solution set of the disjunction, we consider the individual graphs. We graph $x < -4$ and then $x \geq 2$. Then we take the union of the graphs.

$\{x \mid x < -4\}$ $(-\infty, -4)$

$\{x \mid x \geq 2\}$ $[2, \infty)$

$\{x \mid x < -4\} \cup \{x \mid x \geq 2\}$
$= \{x \mid x < -4 \text{ or } x \geq 2\}$ $(-\infty, -4) \cup [2, \infty)$

$y_1 = 7 + 2x,\ y_2 = 13 - 5x,$
$y_3 = -1,\ y_4 = 3$

The solution set is $\{x \mid x < -4 \text{ or } x \geq 2\}$, or $(-\infty, -4) \cup [2, \infty)$. This is confirmed by the graph at left.

Try Exercise 73.

CAUTION! A compound inequality like

$$x < -4 \quad or \quad x \geq 2,$$

as in Example 11, *cannot* be expressed as $2 \leq x < -4$ because to do so would be to say that x is *simultaneously* less than -4 and greater than or equal to 2. No number is both less than -4 *and* greater than 2, but many are less than -4 *or* greater than 2.

EXAMPLE 12 Solve: $-2x - 5 < -2$ *or* $x - 3 < -10$.

SOLUTION We solve the individual inequalities separately, retaining the word *or*:

$$-2x - 5 < -2 \quad or \quad x - 3 < -10$$
$$-2x < 3 \quad or \quad x < -7$$
$$x > -\tfrac{3}{2} \quad or \quad x < -7.$$

Dividing by a negative number and reversing the symbol

Keep the word "or."

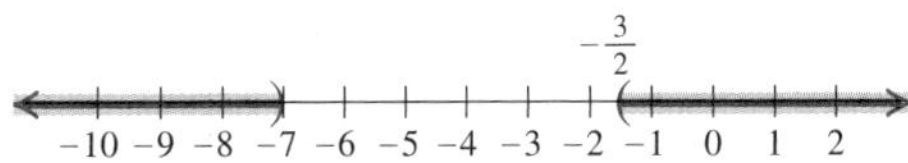

The solution set is $\left\{x \mid x < -7 \text{ or } x > -\frac{3}{2}\right\}$, or $(-\infty, -7) \cup \left(-\frac{3}{2}, \infty\right)$.

Try Exercise 89.

EXAMPLE 13 Solve: $3x - 11 < 4$ *or* $4x + 9 \geq 1$.

SOLUTION We solve the individual inequalities separately, retaining the word *or*:

$$3x - 11 < 4 \quad or \quad 4x + 9 \geq 1$$
$$3x < 15 \quad or \quad 4x \geq -8$$
$$x < 5 \quad or \quad x \geq -2.$$

Keep the word "or."

To find the solution set, we first look at the individual graphs.

$\{x \mid x < 5\}$ $(-\infty, 5)$

$\{x \mid x \geq -2\}$ $[-2, \infty)$

$\{x \mid x < 5\} \cup \{x \mid x \geq -2\}$
$= \{x \mid x < 5 \text{ or } x \geq -2\}$ $(-\infty, \infty) = \mathbb{R}$

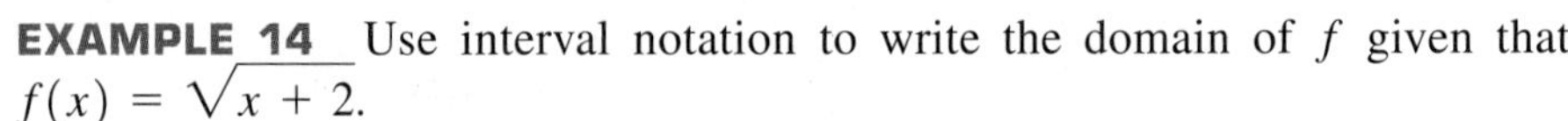

Since *all* numbers are less than 5 or greater than or equal to -2, the two sets fill the entire number line. Thus the solution set is $\mathbb{R}$, the set of all real numbers.

Try Exercise 75.

INTERVAL NOTATION AND DOMAINS

$\left(-\infty, \frac{7}{3}\right) \cup \left(\frac{7}{3}, \infty\right)$

In Section 6.3, we saw that if $g(x) = \dfrac{5x - 2}{3x - 7}$, then the domain of $g = \left\{x \mid x \text{ is a real number } and\ x \neq \frac{7}{3}\right\}$. We can now represent such a set using interval notation:

$$\left\{x \mid x \text{ is a real number } and\ x \neq \tfrac{7}{3}\right\} = \left(-\infty, \tfrac{7}{3}\right) \cup \left(\tfrac{7}{3}, \infty\right).$$

EXAMPLE 14 Use interval notation to write the domain of f given that $f(x) = \sqrt{x + 2}$.

SOLUTION The expression $\sqrt{x + 2}$ is not a real number when $x + 2$ is negative. Thus the domain of f is the set of all x-values for which $x + 2 \geq 0$:

$x + 2 \geq 0$ **$x + 2$ cannot be negative.**

$x \geq -2.$ **Adding -2 to both sides**

$[-2, \infty)$

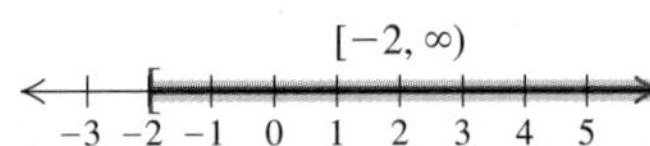

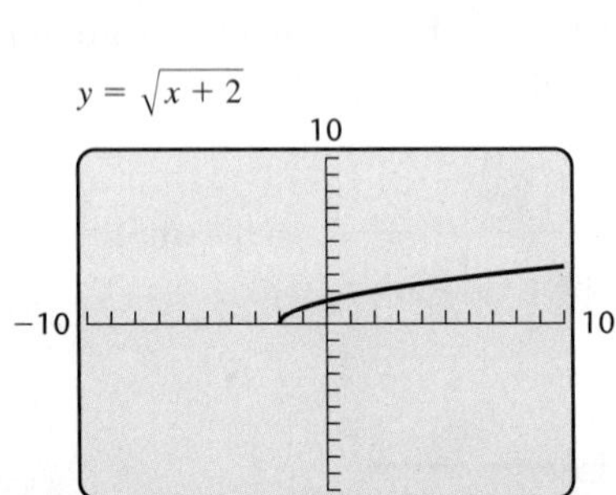

We have the domain of $f = \{x \mid x \geq -2\} = [-2, \infty)$.

We can partially check that $[-2, \infty)$ is the domain of f either by using a table of values or by viewing the graph of $f(x)$, as shown at left. By tracing the curve, we can confirm that no y-value is given for x-values less than -2.

Try Exercise 103.

Interactive Discovery

Consider the functions $f(x) = \sqrt{3 - x}$ and $g(x) = \sqrt{x + 1}$. Enter these in a graphing calculator as $y_1 = \sqrt{}(3 - x)$ and $y_2 = \sqrt{}(x + 1)$.

1. Determine algebraically the domain of f and the domain of g. Then graph y_1 and y_2, and trace each curve to verify the domains.
2. Graph $y_1 + y_2$, $y_1 - y_2$, and $y_1 \cdot y_2$. Use the graphs to determine the domains of $f + g$, $f - g$, and $f \cdot g$.
3. How can the domains of the sum, the difference, and the product of f and g be found algebraically?

Domain of the Sum, Difference, or Product of Functions

The domain of the sum, the difference, or the product of the functions f and g is the intersection of the domains of f and g.

EXAMPLE 15 Find the domain of $f + g$ given that $f(x) = \sqrt{2x - 5}$ and $g(x) = \sqrt{x + 1}$.

SOLUTION We first find the domain of f and the domain of g. The domain of f is the set of all x-values for which $2x - 5 \geq 0$, or $\left\{x \mid x \geq \frac{5}{2}\right\}$, or $\left[\frac{5}{2}, \infty\right)$. Similarly, the domain of g is $\{x \mid x \geq -1\}$, or $[-1, \infty)$. The intersection of the domains is $\left\{x \mid x \geq \frac{5}{2}\right\}$, or $\left[\frac{5}{2}, \infty\right)$. We can confirm this at least approximately by tracing the graph of $f + g$.

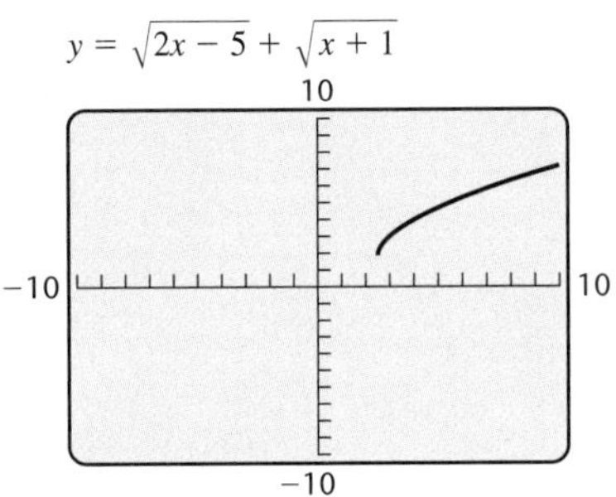

Thus the

$$\text{Domain of } f + g = \left\{x \mid x \geq \tfrac{5}{2}\right\}, \text{ or } \left[\tfrac{5}{2}, \infty\right).$$

Try Exercise 109.

8.1 Exercise Set

FOR EXTRA HELP MyMathLab

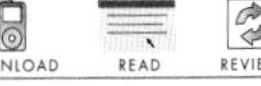

Concept Reinforcement *In each of Exercises 1–10, match the set with the most appropriate choice from the column on the right.*

1. ___ $\{x \mid x < -2 \text{ or } x > 2\}$ a) [number line: −2, 2]
2. ___ $\{x \mid x < -2 \text{ and } x > 2\}$ b) [number line: 2]
3. ___ $\{x \mid x > -2\} \cap \{x \mid x < 2\}$ c) [number line: −2, 2]
4. ___ $\{x \mid x \leq -2\} \cup \{x \mid x \geq 2\}$ d) [number line: −2]
5. ___ $\{x \mid x \leq -2\} \cup \{x \mid x \leq 2\}$ e) [number line: 2]
6. ___ $\{x \mid x \leq -2\} \cap \{x \mid x \leq 2\}$ f) [number line: −2, 2]
7. ___ $\{x \mid x \geq -2\} \cap \{x \mid x \geq 2\}$ g) [number line: −2]
8. ___ $\{x \mid x \geq -2\} \cup \{x \mid x \geq 2\}$ h) [number line: −2, 2]
9. ___ $\{x \mid x \leq 2\}$ *and* $\{x \mid x \geq -2\}$ i) $\mathbb{R}$
10. ___ $\{x \mid x \leq 2\}$ *or* $\{x \mid x \geq -2\}$ j) $\varnothing$

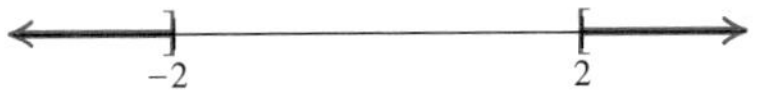

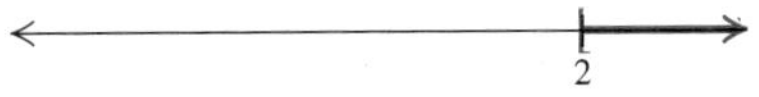

Solve each inequality using the given graph.

11. $f(x) \geq g(x)$

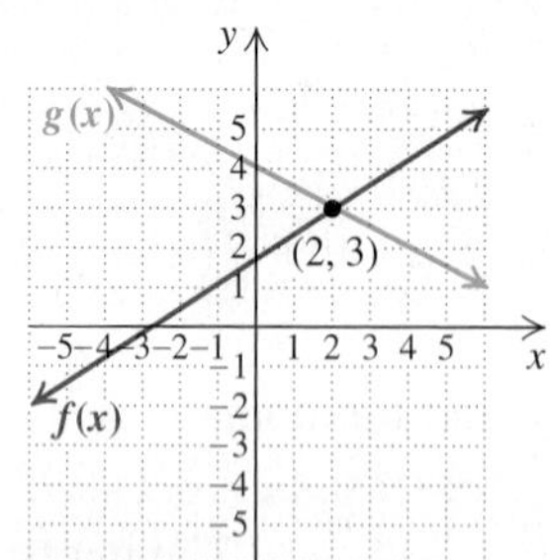

12. $f(x) < g(x)$

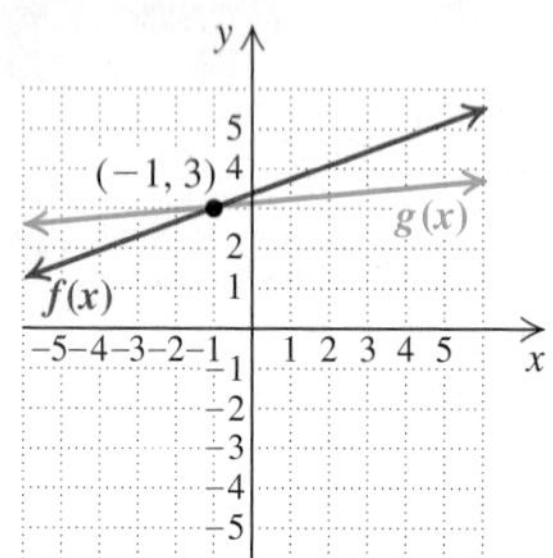

13. $y_1 < y_2$

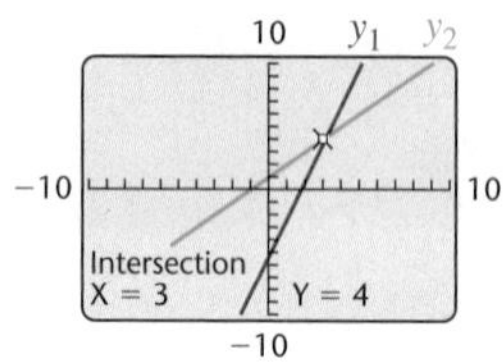

14. $y_1 \geq y_2$

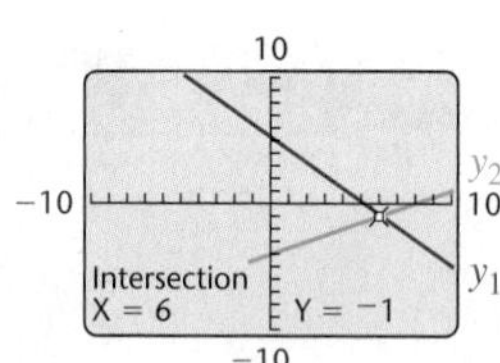

15. The graphs of $f(x) = 2x + 1$, $g(x) = -\frac{1}{2}x + 3$, and $h(x) = x - 1$ are as shown below. Solve each inequality, referring only to the figure.

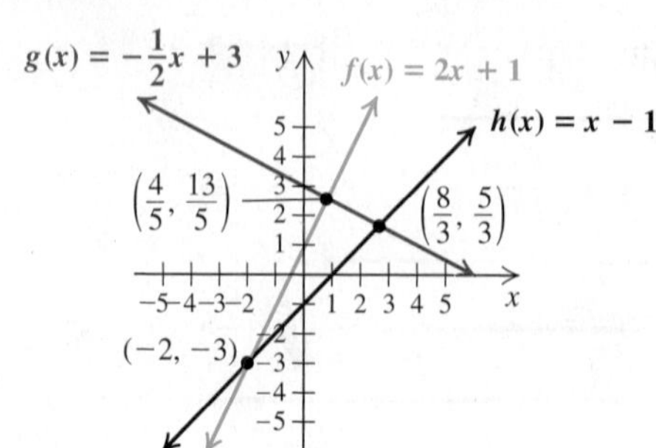

a) $2x + 1 \leq x - 1$
b) $x - 1 > -\frac{1}{2}x + 3$
c) $-\frac{1}{2}x + 3 < 2x + 1$

16. The graphs of $y_1 = -\frac{1}{2}x + 5$, $y_2 = x - 1$, and $y_3 = 2x - 3$ are as shown below. Solve each inequality, referring only to the figure.

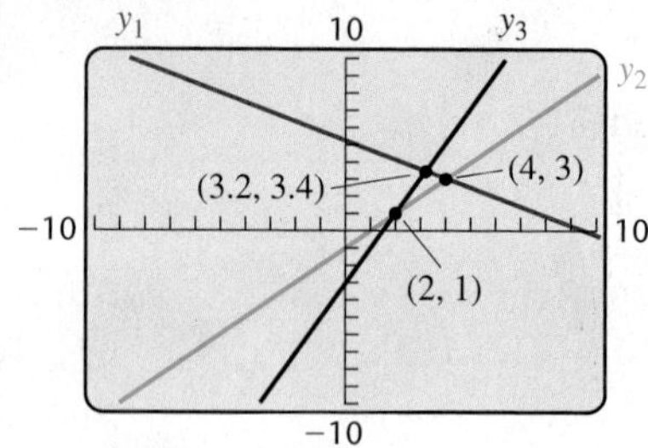

a) $-\frac{1}{2}x + 5 > x - 1$
b) $x - 1 \leq 2x - 3$
c) $2x - 3 \geq -\frac{1}{2}x + 5$

Solve graphically.

17. $x - 3 < 4$

18. $x + 4 \geq 6$

19. $2x - 3 \geq 1$

20. $3x + 1 < 1$

21. $x + 3 > 2x - 5$

22. $3x - 5 \leq 3 - x$

23. $\frac{1}{2}x - 2 \leq 1 - x$

24. $x + 5 > \frac{1}{3}x - 1$

25. $4x + 7 \leq 3 - 5x$

26. $5x + 6 < 8x - 11$

Use a graph to estimate the solution in each of the following. Be sure to use graph paper and a straightedge if graphing by hand.

27. *Show Business.* A band receives \$750 plus 15% of receipts over \$750 for playing a club date. If a club charges a \$6 cover charge, how many people must attend in order for the band to receive at least \$1200?

28. *Temperature Conversion.* The function

$$C(F) = \frac{5}{9}(F - 32)$$

can be used to find the Celsius temperature $C(F)$ that corresponds to $F°$ Fahrenheit.

a) Gold is solid at Celsius temperatures less than 1063°C. Find the Fahrenheit temperatures for which gold is solid.
b) Silver is solid at Celsius temperatures less than 960.8°C. Find the Fahrenheit temperatures for which silver is solid.

29. *Amusement Park Attendance.* The table below lists the estimated annual ridership at amusement parks for various years. Use linear regression to find a linear function that can be used to predict the ridership $r(x)$, in billions, where x is the number of years after 2000. Then predict those years in which there will be fewer than 1.5 billion riders.

Year	Ridership (in billions of people)
2003	1.95
2004	1.81
2005	1.82
2006	1.76
2007	1.78
2008	1.70

Source: Heiden & McGonegal

30. *Smoking.* The table below lists the percent of teenagers in the United States who smoked cigarettes during various years. Use linear regression to find a linear function that can be used to predict the percent $p(x)$ of teenagers who smoked, where x is the number of years after 2000. Then predict those years in which the *Healthy People 2010* goal of no more than 16% teenage smokers will be met.

Year	Percent of Teenagers Who Smoked
2000	31.4%
2001	29.5
2002	26.7
2003	24.4
2004	25
2005	23.2
2007	20

Source: Centers for Disease Control and Prevention

31. *Advertising.* The table below lists advertising revenue for newspapers and the Internet for various years. Use linear regression to find two linear functions that can be used to estimate advertising revenue $n(x)$, in billions of dollars, for newspapers and $t(x)$ for the Internet, where x is the number of years after 2006. Then predict the years for which advertising revenue for the Internet will exceed that for newspapers.

Year	Advertising Revenue, Newspapers (in billions)	Advertising Revenue, Internet (in billions)
2007	\$50	\$15
2008	45	20
2009	35	24
2010	32	26

Source: Based on information from ZenithOptimedia Advertising Forecast

32. *Income.* The table below lists the median adjusted household income for unmarried and married men for various years. Use linear regression to find two linear functions that can be used to estimate the median household earnings $u(x)$ for unmarried men and $m(x)$ for married men, where x is the number of years after 1970. Then estimate the years for which the household earnings for married men exceeds that for unmarried men.

Year	Median Adjusted Household Income, Unmarried Men	Median Adjusted Household Income, Married Men
1970	\$58,000	\$42,000
1980	63,000	56,000
1990	64,000	62,000
2000	65,000	70,000
2007	66,000	74,000

Source: Based on data from Pew Research Center

33. *100-Meter Freestyle.* The table below lists the world record for the women's 100-meter freestyle long-course swim for various years. Use linear regression to find a linear function that can be used to predict the world record x years after 1990. Then predict those years in which the world record will be less than 50 sec.

Year	Time (in seconds)	Athlete
1992	54.48	Jenny Thompson (United States)
1994	54.01	Jingyi Le (China)
2000	53.77	Inge de Bruijn (Netherlands)
2004	53.52	Jodie Henry (Australia)
2006	53.30	Britta Steffen (Germany)
2008	52.88	Lisbeth Trickett (Australia)

Source: www.iaaf.org

34. *Mile Run.* The table below lists the world record for the mile run for various years. Use linear regression to find a linear function that can be used to predict the world record for the mile run x years after 1954. Then predict those years in which the world record will be less than 3.5 min.

Year	Time for Mile (in minutes)	Athlete
1954	3.99	Sir Roger Bannister (Great Britain)
1962	3.9017	Peter Snell (New Zealand)
1975	3.8233	John Walker (New Zealand)
1981	3.7888	Sebastian Coe (Great Britain)
1993	3.7398	Noureddine Morceli (Algeria)
1999	3.7188	Hican El Guerrouj (Morocco)

Source: www.iaaf.org

Find each indicated intersection or union.

35. $\{5, 9, 11\} \cap \{9, 11, 18\}$

36. $\{2, 4, 8\} \cup \{8, 9, 10\}$

37. $\{0, 5, 10, 15\} \cup \{5, 15, 20\}$

38. $\{2, 5, 9, 13\} \cap \{5, 8, 10\}$

39. $\{a, b, c, d, e, f\} \cap \{b, d, f\}$

40. $\{a, b, c\} \cup \{a, c\}$

41. $\{r, s, t\} \cup \{r, u, t, s, v\}$

42. $\{m, n, o, p\} \cap \{m, o, p\}$

43. $\{3, 6, 9, 12\} \cap \{5, 10, 15\}$

44. $\{1, 5, 9\} \cup \{4, 6, 8\}$

45. $\{3, 5, 7\} \cup \varnothing$

46. $\{3, 5, 7\} \cap \varnothing$

Graph and write interval notation for each compound inequality.

47. $3 < x < 7$

48. $0 \le y \le 4$

49. $-6 \le y \le 0$

50. $-9 \le x < -5$

51. $x < -1$ *or* $x > 4$

52. $x < -5$ *or* $x > 1$

53. $x \le -2$ *or* $x > 1$

54. $x \le -5$ or $x > 2$

55. $x > -2$ *and* $x < 4$

56. $x > -7$ *and* $x < -2$

57. $-4 \le -x < 2$

58. $3 > -x \ge -1$

59. $5 > a$ *or* $a > 7$

60. $t \ge 2$ *or* $-3 > t$

61. $x \ge 5$ *or* $-x \ge 4$

62. $-x < 3$ *or* $x < -6$

63. $7 > y$ *and* $y \ge -3$

64. $6 > -x \ge 0$

65. $x < 7$ *and* $x \ge 3$

66. $x \ge -3$ *and* $x < 3$

Aha! **67.** $t < 2$ *or* $t < 5$

68. $t > 4$ *or* $t > -1$

Solve and graph each solution set.

69. $-2 < t + 1 < 8$

70. $-3 < t + 1 \le 5$

71. $4 < x + 4$ *and* $x - 1 < 3$

72. $-1 < x + 2$ *and* $x - 4 < 3$

73. $-7 \le 2a - 3$ *or* $3a + 1 > 7$

74. $-4 \le 3n + 5$ *or* $2n - 3 \le 7$

Aha! **75.** $x + 7 \le -2$ *or* $x + 7 \ge -3$

76. $x + 5 < -3$ *or* $x + 5 \ge 4$

77. $-7 \le 4x + 5 \le 13$

78. $-4 \le 2x + 3 \le 15$

79. $5 > \dfrac{x - 3}{4} > 1$

80. $3 \ge \dfrac{x - 1}{2} \ge -4$

81. $-2 \le \dfrac{x + 2}{-5} \le 6$

82. $-10 \le \dfrac{x + 6}{-3} \le -8$

83. $2 \le f(x) \le 8$, where $f(x) = 3x - 1$

84. $7 \ge g(x) \ge -2$, where $g(x) = 3x - 5$

85. $-21 \le f(x) < 0$, where $f(x) = -2x - 7$

86. $4 > g(t) \geq 2$, where $g(t) = -3t - 8$

87. $f(t) < 3$ *or* $f(t) > 8$, where $f(t) = 5t + 3$

88. $g(x) \leq -2$ *or* $g(x) \geq 10$, where $g(x) = 3x - 5$

89. $6 > 2a - 1$ *or* $-4 \leq -3a + 2$

90. $3a - 7 > -10$ *or* $5a + 2 \leq 22$

91. $a + 3 < -2$ *and* $3a - 4 < 8$

92. $1 - a < -2$ *and* $2a + 1 > 9$

93. $3x + 2 < 2$ *and* $3 - x < 1$

94. $2x - 1 > 5$ *and* $2 - 3x > 11$

95. $2t - 7 \leq 5$ *or* $5 - 2t > 3$

96. $5 - 3a \leq 8$ *or* $2a + 1 > 7$

97. Use the accompanying graph of $f(x) = 2x - 5$ to solve $-7 < 2x - 5 < 7$.

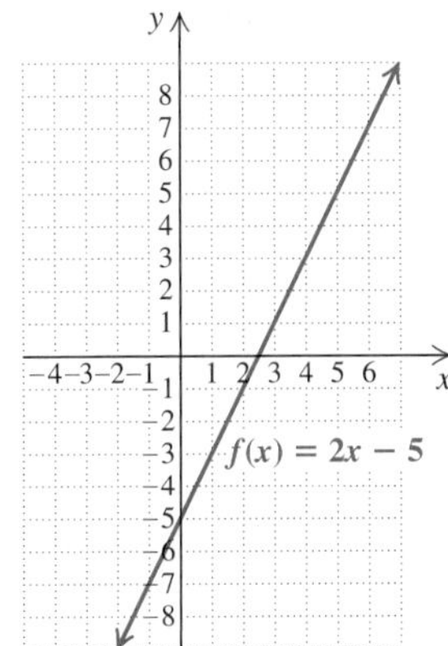

98. Use the accompanying graph of $g(x) = 4 - x$ to solve $4 - x < -2$ *or* $4 - x > 7$.

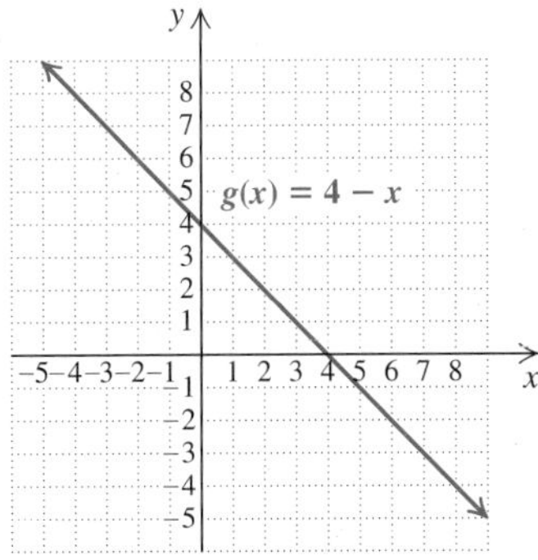

For $f(x)$ as given, use interval notation to write the domain of f.

99. $f(x) = \dfrac{9}{x + 8}$

100. $f(x) = \dfrac{2}{x + 3}$

101. $f(x) = \dfrac{-8}{x}$

102. $f(x) = \dfrac{x + 3}{2x - 8}$

103. $f(x) = \sqrt{x - 6}$

104. $f(x) = \sqrt{x - 2}$

105. $f(x) = \sqrt{2x + 7}$

106. $f(x) = \sqrt{8 - 5x}$

107. $f(x) = \sqrt{8 - 2x}$

108. $f(x) = \sqrt{10 - 2x}$

Use interval notation to write each domain.

109. The domain of $f + g$, if $f(x) = \sqrt{x - 5}$ and $g(x) = \sqrt{\frac{1}{2}x + 1}$

110. The domain of $f - g$, if $f(x) = \sqrt{x + 3}$ and $g(x) = \sqrt{2x - 1}$

111. The domain of $f \cdot g$, if $f(x) = \sqrt{3 - x}$ and $g(x) = \sqrt{3x - 2}$

112. The domain of $f + g$, if $f(x) = \sqrt{3 - 4x}$ and $g(x) = \sqrt{x + 2}$

TW **113.** Why can the conjunction $2 < x$ *and* $x < 5$ be rewritten as $2 < x < 5$, but the disjunction $2 < x$ *or* $x < 5$ cannot be rewritten as $2 < x < 5$?

TW **114.** Can the solution set of a disjunction be empty? Why or why not?

SKILL REVIEW

To prepare for Section 8.2, review graphing and solving equations by graphing (Sections 3.2 and 4.5).

Graph. [3.2]

115. $g(x) = 2x$

116. $f(x) = 4$

117. $g(x) = -3$

118. $f(x) = |x|$

Solve by graphing. [4.5]

119. $x + 4 = 3$

120. $x - 1 = -5$

SYNTHESIS

TW **121.** What can you conclude about a, b, c, and d, if $[a, b] \cup [c, d] = [a, d]$? Why?

TW **122.** What can you conclude about a, b, c, and d, if $[a, b] \cap [c, d] = [a, b]$? Why?

123. *Counseling.* The function given by $s(t) = 500t + 16{,}500$ can be used to estimate the number of student visits to Cornell University's counseling center t years after 2000. For what years is the number of student visits between 18,000 and 21,000?
Source: Based on data from Cornell University

124. *Pressure at Sea Depth.* The function given by $P(d) = 1 + (d/33)$ gives the pressure, in atmospheres (atm), at a depth of d feet in the sea. For what depths d is the pressure at least 1 atm and at most 7 atm?

125. *Converting Dress Sizes.* The function given by $f(x) = 2(x + 10)$ can be used to convert dress sizes x in the United States to dress sizes $f(x)$ in Italy. For what dress sizes in the United States will dress sizes in Italy be between 32 and 46?

126. *Solid-Waste Generation.* The function given by $w(t) = 0.0125t + 4.525$ can be used to estimate the number of pounds of solid waste, $w(t)$, produced daily, on average, by each person in the United States, t years after 2000. For what years will waste production range from 4.6 lb to 4.8 lb per person per day?

127. *Body-Fat Percentage.* The function given by $F(d) = (4.95/d - 4.50) \times 100$ can be used to estimate the body fat percentage $F(d)$ of a person with an average body density d, in kilograms per liter. A woman's body fat percentage is considered acceptable if $25 \leq F(d) \leq 31$. What body densities are considered acceptable for a woman?

128. *Minimizing Tolls.* A \$6.00 toll is charged to cross the bridge from mainland Florida to Sanibel Island. A six-month reduced-fare pass, costing \$50.00, reduces the toll to \$2.00. A six-month unlimited-trip pass costs \$300 and allows free crossings. How many crossings in six months does it take for the reduced-fare pass to be the more economical choice?
Source: www.leewayinfo.com

Solve and graph.

129. $4m - 8 > 6m + 5$ *or* $5m - 8 < -2$

130. $4a - 2 \leq a + 1 \leq 3a + 4$

131. $3x < 4 - 5x < 5 + 3x$

132. $x - 10 < 5x + 6 \leq x + 10$

Determine whether each sentence is true or false for all real numbers a, b, and c.

133. If $-b < -a$, then $a < b$.

134. If $a \leq c$ and $c \leq b$, then $b > a$.

135. If $a < c$ and $b < c$, then $a < b$.

136. If $-a < c$ and $-c > b$, then $a > b$.

For $f(x)$ as given, use interval notation to write the domain of f.

137. $f(x) = \dfrac{\sqrt{3 - 4x}}{x + 7}$

138. $f(x) = \dfrac{\sqrt{5 + 2x}}{x - 1}$

139. On many graphing calculators, the TEST key provides access to inequality symbols, while the LOGIC option of that same key accesses the conjunction *and* and the disjunction *or*. Thus, if $y_1 = x > -2$ and $y_2 = x < 4$, Exercise 55 can be checked by forming the expression $y_3 = y_1$ *and* y_2. The interval(s) in the solution set appears as a horizontal line 1 unit above the x-axis. (Be careful to "deselect" y_1 and y_2 so that only y_3 is drawn.) Use the TEST key to check Exercises 59, 63, 65, and 67.

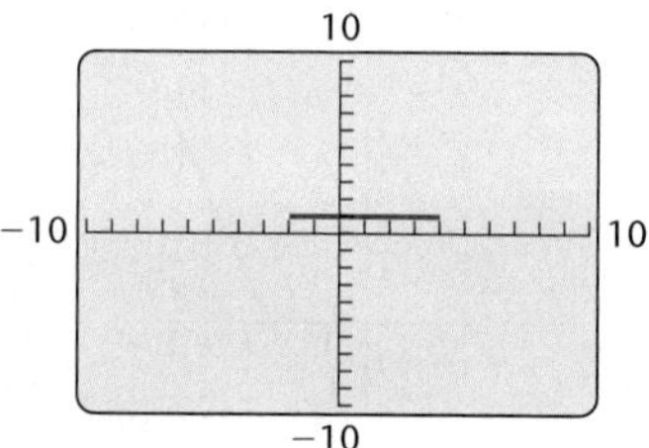

Try Exercise Answers: Section 8.1

17. $\{x \mid x < 7\}$, or $(-\infty, 7)$ **25.** $\{x \mid x \leq -\frac{4}{9}\}$, or $(-\infty, -\frac{4}{9}]$
29. $r(x) = -0.04x + 2.0233$; years after 2013 **35.** $\{9, 11\}$
37. $\{0, 5, 10, 15, 20\}$ **51.** $(-\infty, -1) \cup (4, \infty)$
55. $(-2, 4)$
69. $\{t \mid -3 < t < 7\}$, or $(-3, 7)$
73. $\{a \mid a \leq -2$ *or* $a > 2\}$, or $(-\infty, -2] \cup (2, \infty)$
75. $\mathbb{R}$, or $(-\infty, \infty)$
89. $\{a \mid a < \frac{7}{2}\}$, or $(-\infty, \frac{7}{2})$
91. $\{a \mid a < -5\}$, or $(-\infty, -5)$ **93.** $\varnothing$
103. $[6, \infty)$ **109.** $[5, \infty)$

8.2 Absolute-Value Equations and Inequalities

- Equations with Absolute Value
- Inequalities with Absolute Value

EQUATIONS WITH ABSOLUTE VALUE

Recall from Section 1.4 the definition of absolute value.

Absolute Value The absolute value of x, denoted $|x|$, is defined as

$$|x| = \begin{cases} x, & \text{if } x \geq 0, \\ -x, & \text{if } x < 0. \end{cases}$$

(When x is nonnegative, the absolute value of x is x. When x is negative, the absolute value of x is the opposite of x.)

To better understand this definition, suppose x is -5. Then $|x| = |-5| = 5$, and 5 is the opposite of -5. When x represents a negative number, the absolute value of x is the opposite of x, which is positive.

Since distance is always nonnegative, we can think of a number's absolute value as its distance from zero on the number line.

EXAMPLE 1 Find the solution set.

a) $|x| = 4$ **b)** $|x| = 0$ **c)** $|x| = -7$

SOLUTION

a) We interpret $|x| = 4$ to mean that the number x is 4 units from zero on the number line. There are two such numbers, 4 and -4. Thus the solution set is $\{-4, 4\}$.

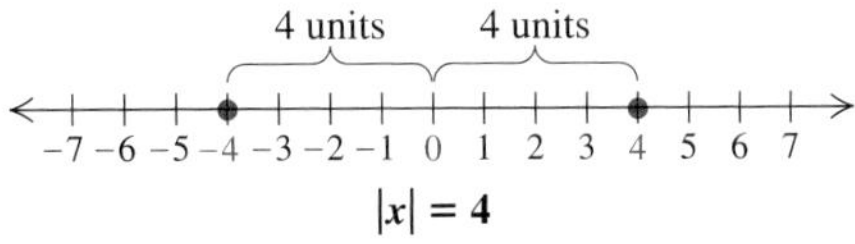

$|x| = 4$

b) We interpret $|x| = 0$ to mean that x is 0 units from zero on the number line. The only number that satisfies this is 0 itself. Thus the solution set is $\{0\}$.

c) Since distance is always nonnegative, it doesn't make sense to talk about a number that is -7 units from zero. *Remember*: The absolute value of a number is never negative. Thus, $|x| = -7$ has no solution; the solution set is $\varnothing$.

Try Exercise 21.

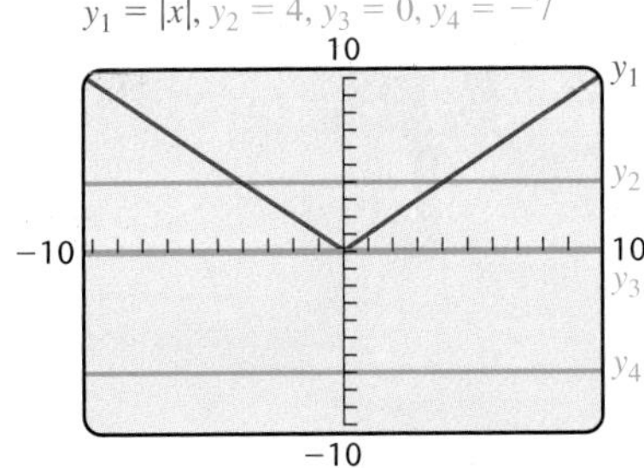

The graph at left illustrates that, in Example 1, there are two solutions of $|x| = 4$, one solution of $|x| = 0$, and no solutions of $|x| = -7$.

Other equations involving absolute value can be solved graphically.

EXAMPLE 2 Solve: $|x - 3| = 5$.

SOLUTION We let $y_1 = \text{abs}(x - 3)$ and $y_2 = 5$. The solution set of the equation consists of the x-coordinates of any points of intersection of the graphs of y_1 and y_2. We see that the graphs intersect at two points. We use INTERSECT twice to find the coordinates of the points of intersection. Each time, we enter a GUESS that is close to the point we are looking for.

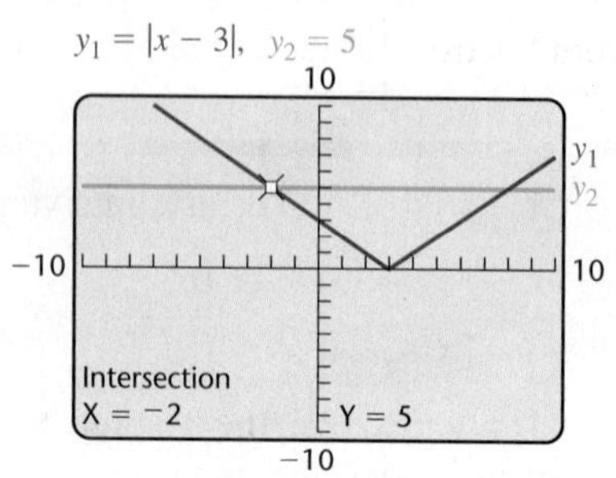

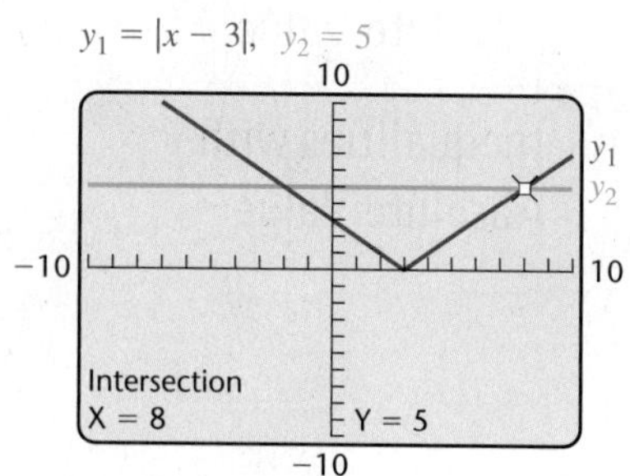

The graphs intersect at $(-2, 5)$ and $(8, 5)$. Thus the solutions are -2 and 8. To check, we substitute each in the original equation.

Check:

For -2:

$\lvert x - 3\rvert$	$= 5$
$\lvert -2 - 3\rvert$	5
$\lvert -5\rvert$	
5	$\overset{?}{=} 5$ TRUE

For 8:

$\lvert x - 3\rvert$	$= 5$
$\lvert 8 - 3\rvert$	5
$\lvert 5\rvert$	
5	$\overset{?}{=} 5$ TRUE

The solution set is $\{-2, 8\}$.

Try Exercise 31.

We can solve equations involving absolute value algebraically using the following principle.

Student Notes

The absolute-value principle will be used with a variety of replacements for X. Make sure that the principle, as stated here, makes sense to you before going further.

The Absolute-Value Principle for Equations

For any positive number p and any algebraic expression X:

a) The solutions of $|X| = p$ are those numbers that satisfy

$$X = -p \quad or \quad X = p.$$

b) The equation $|X| = 0$ is equivalent to the equation $X = 0$.

c) The equation $|X| = -p$ has no solution.

EXAMPLE 3 Find the solution set: **(a)** $|2x + 5| = 13$; **(b)** $|4 - 7x| = -8$.

a)

ALGEBRAIC APPROACH

We use the absolute-value principle:

$$|X| = p$$

$$|2x + 5| = 13 \quad \text{Substituting}$$

$$2x + 5 = -13 \quad or \quad 2x + 5 = 13$$

$$2x = -18 \quad or \quad 2x = 8$$

$$x = -9 \quad or \quad x = 4.$$

The solutions are -9 and 4.

Check: For -9:

$$\begin{array}{c|c} |2x+5| & 13 \\ \hline |2(-9)+5| & 13 \\ |-18+5| & \\ |-13| & \\ \end{array}$$

$13 \stackrel{?}{=} 13$ TRUE

For 4:

$$\begin{array}{c|c} |2x+5| & 13 \\ \hline |2 \cdot 4+5| & 13 \\ |8+5| & \\ |13| & \\ \end{array}$$

$13 \stackrel{?}{=} 13$ TRUE

The number $2x + 5$ is 13 units from zero if x is replaced with -9 or 4. The solution set is $\{-9, 4\}$.

GRAPHICAL APPROACH

We graph $y_1 = \text{abs}(2x + 5)$ and $y_2 = 13$ and use INTERSECT to find any points of intersection. The graphs intersect at $(-9, 13)$ and $(4, 13)$. The x-coordinates, -9 and 4, are the solutions.

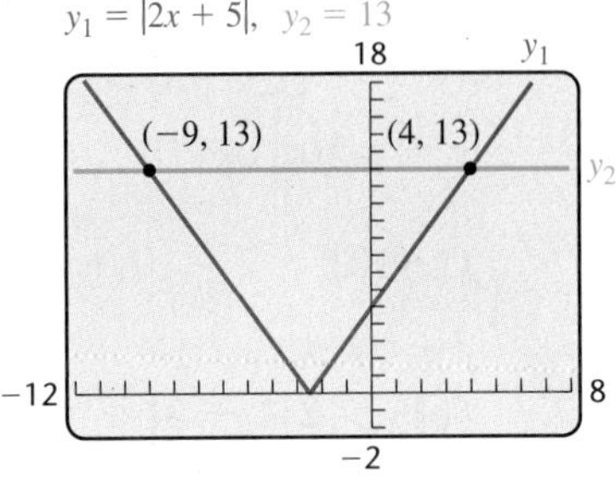

The algebraic solution serves as a check. The solution set is $\{-9, 4\}$.

b)

ALGEBRAIC APPROACH

The absolute-value principle reminds us that absolute value is always nonnegative. Thus the equation $|4 - 7x| = -8$ has no solution. The solution set is $\varnothing$.

GRAPHICAL APPROACH

We graph $y_1 = \text{abs}(4 - 7x)$ and $y_2 = -8$. There are no points of intersection, so the solution set is $\varnothing$.

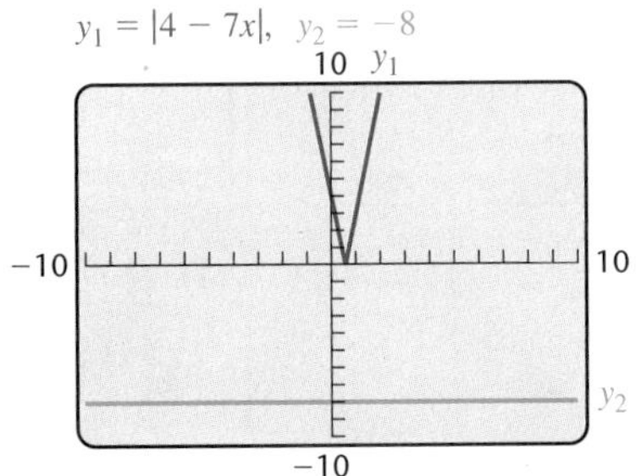

Try Exercise 27.

To use the absolute-value principle, we must be sure that the absolute-value expression is alone on one side of the equation.

EXAMPLE 4 Given that $f(x) = 2|x + 3| + 1$, find all x for which $f(x) = 15$.

ALGEBRAIC APPROACH

Since we are looking for $f(x) = 15$, we substitute:

$$f(x) = 15$$
$$2|x + 3| + 1 = 15 \quad \text{Replacing } f(x) \text{ with } 2|x + 3| + 1$$
$$2|x + 3| = 14 \quad \text{Subtracting 1 from both sides}$$
$$|x + 3| = 7 \quad \text{Dividing both sides by 2}$$
$$x + 3 = -7 \quad or \quad x + 3 = 7 \quad \text{Using the absolute-value principle for equations}$$
$$x = -10 \quad or \quad x = 4.$$

Check:
$$f(-10) = 2|-10 + 3| + 1 = 2|-7| + 1 = 2 \cdot 7 + 1 = 15;$$
$$f(4) = 2|4 + 3| + 1 = 2|7| + 1 = 2 \cdot 7 + 1 = 15$$

The solution set is $\{-10, 4\}$.

GRAPHICAL APPROACH

We graph $y_1 = 2\text{ abs}(x + 3) + 1$ and $y_2 = 15$ and determine the coordinates of any points of intersection.

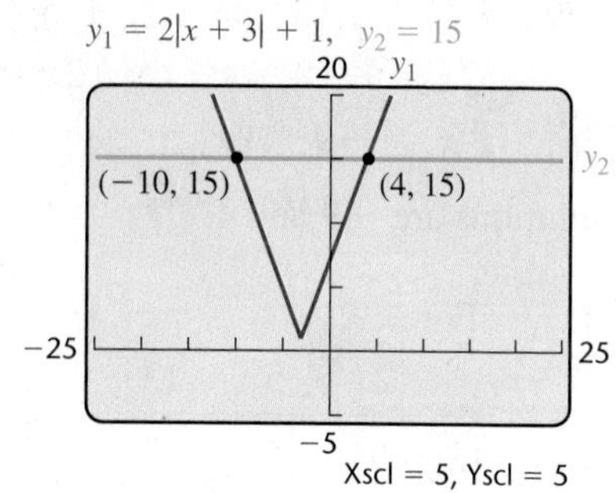

The solutions are the x-coordinates of the points of intersection, -10 and 4. The solution set is $\{-10, 4\}$.

Try Exercise 49.

EXAMPLE 5 Solve: $|x - 2| = 3$.

SOLUTION Because this equation is of the form $|a - b| = c$, it can be solved two ways.

> **CAUTION!** There are two solutions of $|x - 2| = 3$. Simply solving $x - 2 = 3$ will yield only one of those solutions.

Method 1. We interpret $|x - 2| = 3$ as stating that the number $x - 2$ is 3 units from zero. Using the absolute-value principle, we replace X with $x - 2$ and p with 3:

$$|X| = p$$
$$|x - 2| = 3$$
$$x - 2 = -3 \quad or \quad x - 2 = 3 \quad \text{Using the absolute-value principle}$$
$$x = -1 \quad or \quad x = 5.$$

Method 2. This approach is helpful in calculus. The expressions $|a - b|$ and $|b - a|$ can be used to represent the *distance between a* and *b on the number line*. For example, the distance between 7 and 8 is given by $|8 - 7|$ or $|7 - 8|$. From this viewpoint, the equation $|x - 2| = 3$ states that the distance between x and 2 is 3 units. We draw a number line and locate all numbers that are 3 units from 2.

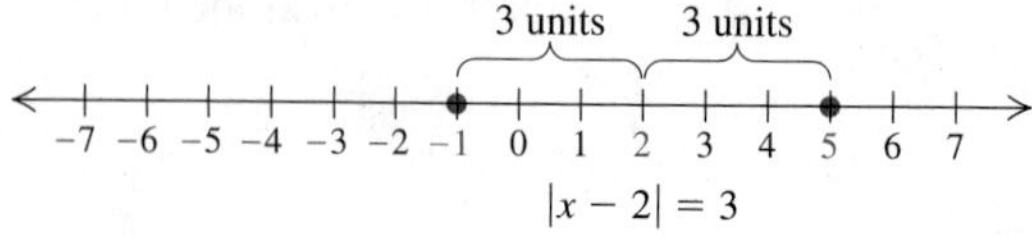

The solutions of $|x - 2| = 3$ are -1 and 5.

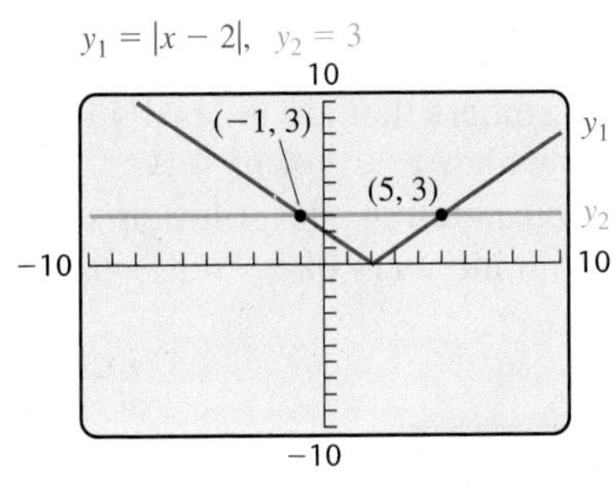

Check: The check consists of noting that both methods give the same solutions. The graphical solution shown at left serves as another check. The solution set is $\{-1, 5\}$.

Try Exercise 33.

Sometimes an equation has two absolute-value expressions. Consider $|a| = |b|$. This means that a and b are the same distance from zero.

If a and b are the same distance from zero, then either they are the same number or they are opposites.

For any algebraic expressions X and Y:

If $|X| = |Y|$, then $X = Y$ or $X = -Y$.

EXAMPLE 6 Solve: $|2x - 3| = |x + 5|$.

ALGEBRAIC APPROACH

The given equation tells us that $2x - 3$ and $x + 5$ are the same distance from zero. This means that they are either the same number or opposites.

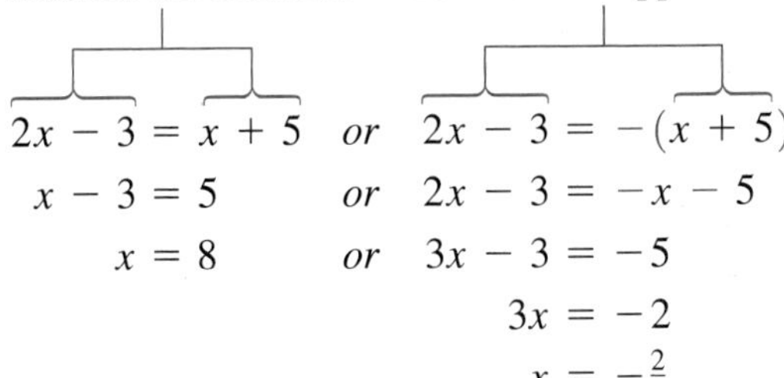

$$\begin{aligned} 2x - 3 &= x + 5 \quad &or \quad 2x - 3 &= -(x + 5) \\ x - 3 &= 5 &or \quad 2x - 3 &= -x - 5 \\ x &= 8 &or \quad 3x - 3 &= -5 \\ & & 3x &= -2 \\ & & x &= -\tfrac{2}{3} \end{aligned}$$

The solutions are 8 and $-\frac{2}{3}$. The solution set is $\left\{-\frac{2}{3}, 8\right\}$.

GRAPHICAL APPROACH

We graph $y_1 = \text{abs}(2x - 3)$ and $y_2 = \text{abs}(x + 5)$ and look for any points of intersection. The x-coordinates of the points of intersection are -0.6666667 and 8.

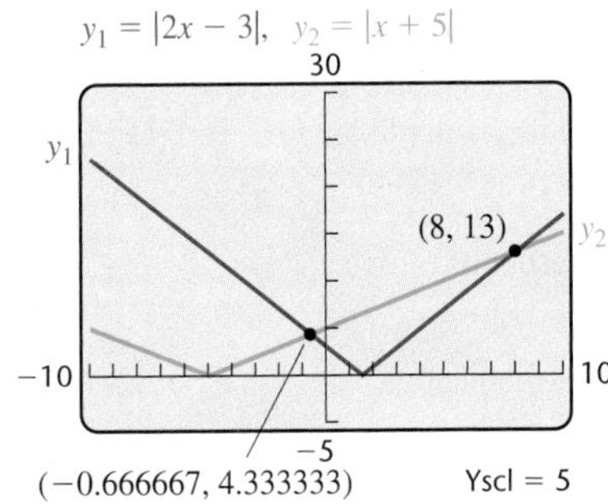

The solution set is $\{-0.6666667, 8\}$. Converted to fraction notation, the solution set is $\left\{-\frac{2}{3}, 8\right\}$.

Try Exercise 53.

INEQUALITIES WITH ABSOLUTE VALUE

Our methods for solving equations with absolute value can be adapted for solving inequalities.

EXAMPLE 7 Solve $|x| < 4$. Then graph.

SOLUTION The solutions of $|x| < 4$ are all numbers whose *distance from zero is less than* 4. By substituting or by looking at the number line, we can see that numbers like $-3, -2, -1, -\frac{1}{2}, -\frac{1}{4}, 0, \frac{1}{4}, \frac{1}{2}$, 1, 2, and 3 are all solutions. In fact, the solutions are all the numbers between -4 and 4. The solution set is $\{x \mid -4 < x < 4\}$. In interval notation, the solution set is $(-4, 4)$. The graph is as follows:

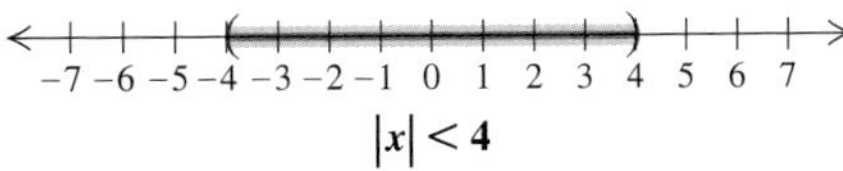

Try Exercise 63.

EXAMPLE 8 Solve $|x| \geq 4$. Then graph.

SOLUTION The solutions of $|x| \geq 4$ are all numbers that are at least 4 units from zero—in other words, those numbers x for which $x \leq -4$ *or* $4 \leq x$. The solution set is $\{x | x \leq -4 \text{ or } x \geq 4\}$. In interval notation, the solution set is $(-\infty, -4] \cup [4, \infty)$. We can check mentally with numbers like -4.1, -5, 4.1, and 5. The graph is as follows:

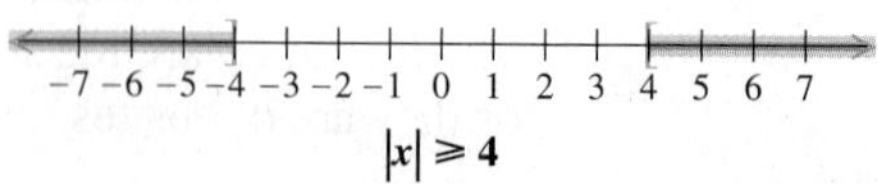

Try Exercise 65.

Examples 1, 7, and 8 illustrate three types of problems in which absolute-value symbols appear. The following is a general principle for solving such problems.

Principles for Solving Absolute-Value Problems For any positive number p and any expression X:

a) The solutions of $|X| = p$ are those numbers that satisfy $X = -p$ *or* $X = p$.

b) The solutions of $|X| < p$ are those numbers that satisfy $-p < X < p$.

c) The solutions of $|X| > p$ are those numbers that satisfy $X < -p$ *or* $p < X$.

Student Notes

Simply ignoring the absolute-value symbol and solving the resulting equation or inequality will not lead to the correct solution.

$|X| = p$ corresponds to two equations.

$|X| < p$ corresponds to a conjunction.

$|X| > p$ corresponds to a disjunction.

The above principles are true for any positive number p.

If p is negative, any value of X will satisfy the inequality $|X| > p$ because absolute value is never negative. Thus, $|2x - 7| > -3$ is true for any real number x, and the solution set is $\mathbb{R}$.

If p is not positive, the inequality $|X| < p$ has no solution. Thus, $|2x - 7| < -3$ has no solution, and the solution set is $\varnothing$.

STUDY TIP

A Place of Your Own

If you can, find a place to study regularly that you do not have to share. If that is not possible, schedule separate study times from others who use that area. When you are ready to study, you should have a place to go.

EXAMPLE 9 Solve: $|3x - 2| < 4$.

ALGEBRAIC APPROACH

The number $3x - 2$ must be less than 4 units from zero. We can visualize this on the number line:

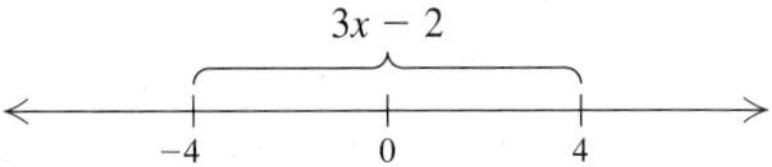

The symbol in the inequality is <, so part (b) of the principles listed above applies.

$$|X| < p$$
$$|3x - 2| < 4 \qquad \textbf{Replacing } X \textbf{ with } 3x - 2 \textbf{ and } p \textbf{ with } 4$$
$$-4 < 3x - 2 < 4$$
$$-2 < 3x < 6 \qquad \textbf{Adding 2}$$
$$-\tfrac{2}{3} < x < 2 \qquad \textbf{Multiplying by } \tfrac{1}{3}$$

The solution set is $\left\{x \mid -\frac{2}{3} < x < 2\right\}$, or, in interval notation, $\left(-\frac{2}{3}, 2\right)$. The graph is as follows:

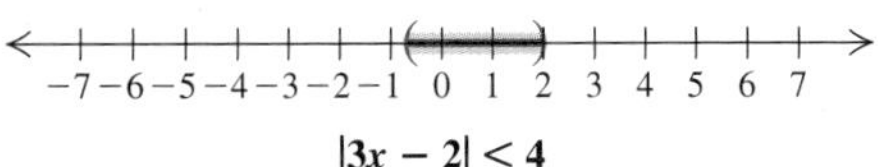

$|3x - 2| < 4$

When x is between $-\frac{2}{3}$ and 2, $3x - 2$ is between -4 and 4.

GRAPHICAL APPROACH

We graph $y_1 = \text{abs}(3x - 2)$ and $y_2 = 4$.

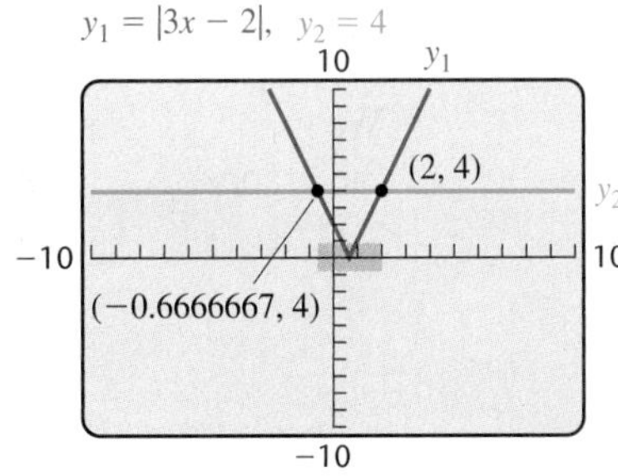

We see that y_1 is less than y_2 between the points $(-0.6666667, 4)$ and $(2, 4)$. The solution set of the inequality is thus the interval $(-0.6666667, 2)$, as indicated by the shading on the x-axis. Converted to fraction notation, the solution set is $\left(-\frac{2}{3}, 2\right)$.

Try Exercise 69.

EXAMPLE 10 Given that $f(x) = |4x + 2|$, find all x for which $f(x) \geq 6$.

ALGEBRAIC APPROACH

We have

$$f(x) \geq 6,$$

or $\quad |4x + 2| \geq 6.$ **Substituting**

The number $4x + 2$ must be 6 or more units from zero. We can visualize this on the number line:

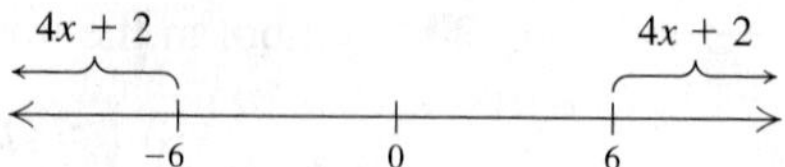

The symbol in the inequality is $\geq$, so we use part (c) of the principles listed above.

$$|X| \geq p$$

$$|4x + 2| \geq 6 \qquad \textbf{Replacing } X \textbf{ with } 4x + 2 \textbf{ and } p \textbf{ with } 6$$

$$4x + 2 \leq -6 \quad or \quad 6 \leq 4x + 2$$

$$4x \leq -8 \quad or \quad 4 \leq 4x \qquad \textbf{Adding } -2$$

$$x \leq -2 \quad or \quad 1 \leq x \qquad \textbf{Multiplying by } \tfrac{1}{4}$$

The solution set is $\{x | x \leq -2 \text{ or } x \geq 1\}$, or, in interval notation, $(-\infty, -2] \cup [1, \infty)$. The graph is as follows:

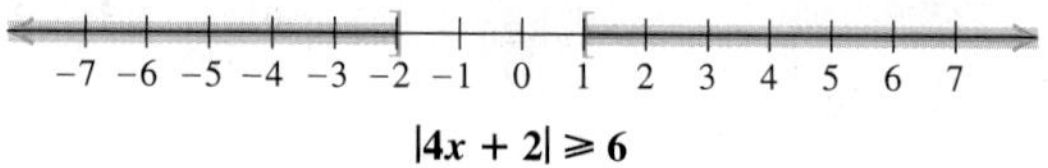

|4x + 2| ≥ 6

When $x \leq -2$ *or* when $x \geq 1$, $f(x) \geq 6$.

GRAPHICAL APPROACH

We graph $y_1 = \text{abs}(4x + 2)$ and $y_2 = 6$ and determine the points of intersection of the graphs.

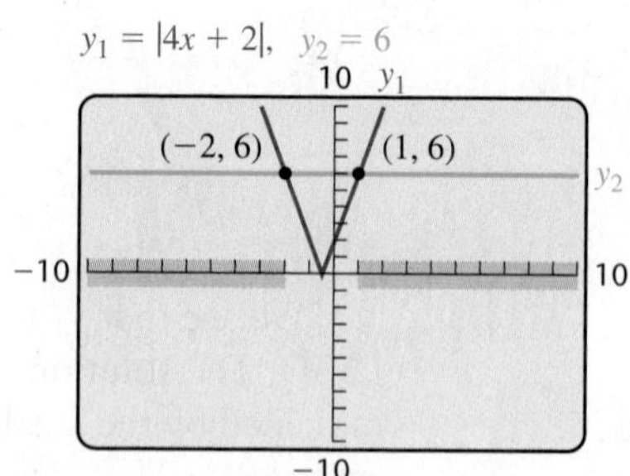

The solution set consists of all x-values for which the graph of $y_1 = |4x + 2|$ is *on or above* the graph of $y_2 = 6$. The graph of y_1 lies *on* the graph of y_2 when $x = -2$ *or* $x = 1$. The graph of y_1 is *above* the graph of y_2 when $x < -2$ *or* $x > 1$. Thus the solution set is $(-\infty, -2] \cup [1, \infty)$, as indicated by the shading on the x-axis.

Try Exercise 93.

Connecting the Concepts

We can visualize Examples 1(a), 7, and 8 by graphing $f(x) = |x|$ and $g(x) = 4$.

Solve: $|x| = 4$. **The symbol is =.**

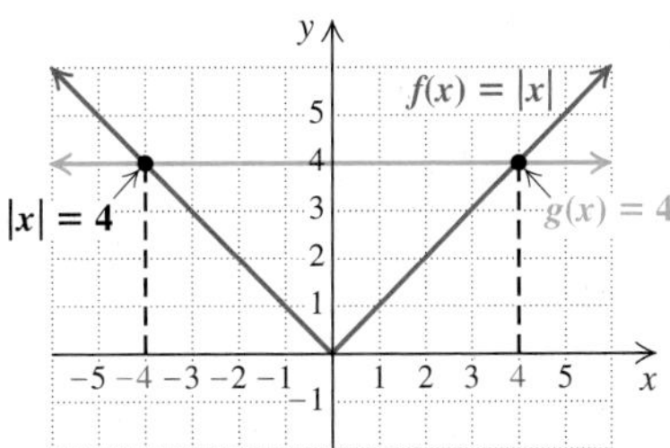

$|x| = 4$ when $x = -4$ *or* $x = 4$.

The solution set of $|x| = 4$ is $\{-4, 4\}$.

Solve: $|x| < 4$. **The symbol is <.**

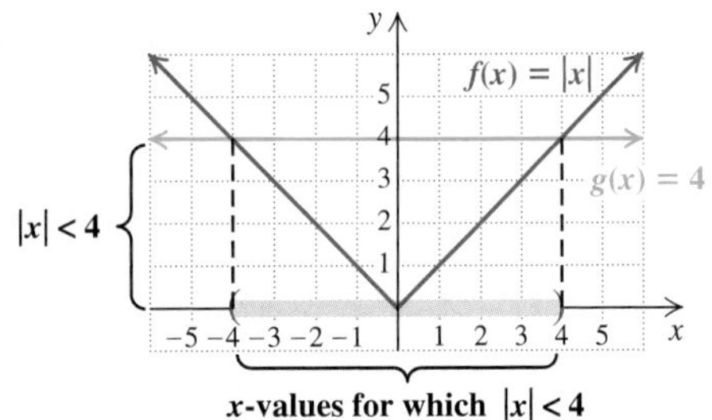

$|x| < 4$ when $-4 < x < 4$.

The solution set of $|x| < 4$ is $(-4, 4)$.

Solve: $|x| \geq 4$. **The symbol is ≥.**

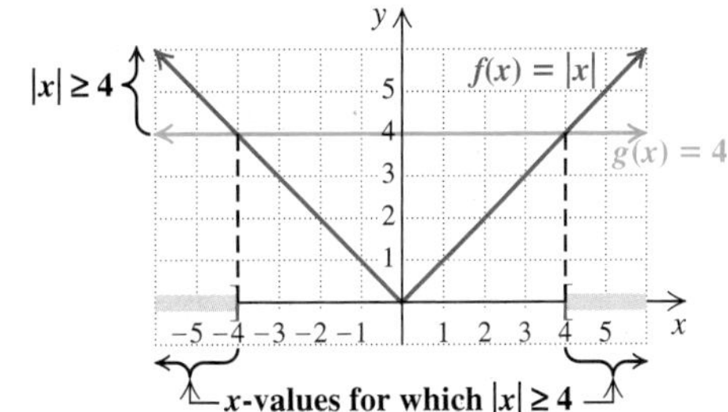

$|x| \geq 4$ when $x \leq -4$ *or* $x \geq 4$.

The solution set of $|x| \geq 4$ is $(-\infty, -4] \cup [4, \infty)$.

8.2 Exercise Set

FOR EXTRA HELP MyMathLab

Concept Reinforcement *Classify each of the following statements as either true or false.*

1. $|x|$ is never negative.
2. $|x|$ is always positive.
3. If x is negative, then $|x| = -x$.
4. The distance between a and b can be expressed as $|a - b|$.
5. The number a is $|a|$ units from 0.
6. There are two solutions of $|3x - 8| = 17$.
7. There is no solution of $|4x + 9| > -5$.
8. All real numbers are solutions of $|2x - 7| < -3$.

Match each equation or inequality with an equivalent statement from the column on the right. Letters may be used more than once or not at all.

9. $\lvert x - 3\rvert = 5$	**a)** The solution set is $\varnothing$.
10. $\lvert x - 3\rvert < 5$	**b)** The solution set is $\mathbb{R}$.
11. $\lvert x - 3\rvert > 5$	**c)** $x - 3 > 5$
12. $\lvert x - 3\rvert < -5$	**d)** $x - 3 < -5$ *or* $x - 3 > 5$
13. $\lvert x - 3\rvert = -5$	**e)** $x - 3 = 5$
14. $\lvert x - 3\rvert > -5$	**f)** $x - 3 < 5$
	g) $x - 3 = -5$ *or* $x - 3 = 5$
	h) $-5 < x - 3 < 5$

Use the following graph to solve Exercises 15–20.

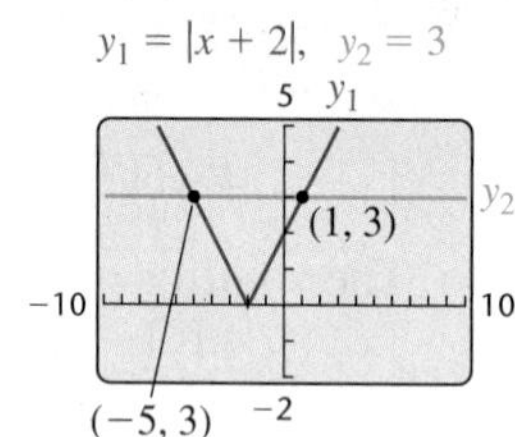

15. $\lvert x + 2\rvert = 3$

16. $\lvert x + 2\rvert \le 3$

17. $\lvert x + 2\rvert < 3$

18. $\lvert x + 2\rvert > 3$

19. $\lvert x + 2\rvert \ge 3$

20. $\lvert x + 2\rvert = -1$

Solve.

21. $\lvert x\rvert = 7$

22. $\lvert x\rvert = 9$

Aha! **23.** $\lvert x\rvert = -6$

24. $\lvert x\rvert = -3$

25. $\lvert p\rvert = 0$

26. $\lvert y\rvert = 7.3$

27. $\lvert 2x - 3\rvert = 4$

28. $\lvert 5x + 2\rvert = 7$

29. $\lvert 3x - 5\rvert = -8$

30. $\lvert 7x - 2\rvert = -9$

31. $\lvert x - 2\rvert = 6$

32. $\lvert x - 3\rvert = 8$

33. $\lvert x - 5\rvert = 3$

34. $\lvert x - 6\rvert = 1$

35. $\lvert t\rvert + 1.1 = 6.6$

36. $\lvert m\rvert + 3 = 3$

37. $\lvert 5x\rvert - 3 = 37$

38. $\lvert 2y\rvert - 5 = 13$

39. $7\lvert q\rvert - 2 = 9$

40. $7\lvert z\rvert + 2 = 16$

41. $\left\lvert \frac{2x - 1}{3} \right\rvert = 4$

42. $\left\lvert \frac{4 - 5x}{6} \right\rvert = 3$

43. $\lvert 5 - m\rvert + 9 = 16$

44. $\lvert t - 7\rvert + 1 = 4$

45. $5 - 2\lvert 3x - 4\rvert = -5$

46. $3\lvert 2x - 5\rvert - 7 = -1$

47. Let $f(x) = \lvert 2x + 6\rvert$. Find all x for which $f(x) = 8$.

48. Let $f(x) = \lvert 2x + 4\rvert$. Find all x for which $f(x) = 10$.

49. Let $f(x) = \lvert x\rvert - 3$. Find all x for which $f(x) = 5.7$.

50. Let $f(x) = \lvert x\rvert + 7$. Find all x for which $f(x) = 18$.

51. Let $f(x) = \left\lvert \frac{3x - 2}{5} \right\rvert$. Find all x for which $f(x) = 2$.

52. Let $f(x) = \left\lvert \frac{1 - 2x}{3} \right\rvert$. Find all x for which $f(x) = 1$.

Solve.

53. $\lvert x + 4\rvert = \lvert 2x - 7\rvert$

54. $\lvert 3x + 2\rvert = \lvert x - 6\rvert$

55. $\lvert x + 4\rvert = \lvert x - 3\rvert$

56. $\lvert x - 9\rvert = \lvert x + 6\rvert$

57. $\lvert 3a - 1\rvert = \lvert 2a + 4\rvert$

58. $\lvert 5t + 7\rvert = \lvert 4t + 3\rvert$

Aha! **59.** $\lvert n - 3\rvert = \lvert 3 - n\rvert$

60. $\lvert y - 2\rvert = \lvert 2 - y\rvert$

61. $\lvert 7 - 4a\rvert = \lvert 4a + 5\rvert$

62. $\lvert 6 - 5t\rvert = \lvert 5t + 8\rvert$

Solve and graph.

63. $\lvert a\rvert \le 9$

64. $\lvert x\rvert < 2$

65. $\lvert t\rvert > 0$

66. $\lvert t\rvert \ge 1$

67. $\lvert x - 1\rvert < 4$

68. $\lvert x - 1\rvert < 3$

69. $\lvert x + 2\rvert \le 6$

70. $\lvert x + 4\rvert \le 1$

71. $\lvert x - 3\rvert + 2 > 7$

72. $\lvert x - 4\rvert + 5 > 2$

Aha! **73.** $\lvert 2y - 9\rvert > -5$

74. $\lvert 3y - 4\rvert > 8$

75. $\lvert 3a - 4\rvert + 2 \ge 8$

76. $\lvert 2a - 5\rvert + 1 \ge 9$

77. $\lvert y - 3\rvert < 12$

78. $\lvert p - 2\rvert < 3$

79. $9 - \lvert x + 4\rvert \le 5$

80. $12 - \lvert x - 5\rvert \le 9$

81. $6 + \lvert 3 - 2x\rvert > 10$

82. $7 + \lvert 4a - 5\rvert \le 26$

Aha! **83.** $\lvert 5 - 4x\rvert < -6$

84. $\lvert 7 - 2y\rvert < -6$

85. $\left\lvert \frac{2 - 5x}{4} \right\rvert \ge \frac{2}{3}$

86. $\left\lvert \frac{1 + 3x}{5} \right\rvert > \frac{7}{8}$

87. $\lvert m + 3\rvert + 8 \le 14$

88. $\lvert t - 7\rvert + 3 \ge 4$

89. $25 - 2\lvert a + 3\rvert > 19$

90. $30 - 4\lvert a + 2\rvert > 12$

91. Let $f(x) = |2x - 3|$. Find all x for which $f(x) \le 4$.

92. Let $f(x) = |5x + 2|$. Find all x for which $f(x) \le 3$.

93. Let $f(x) = 5 + |3x - 4|$. Find all x for which $f(x) \ge 16$.

94. Let $f(x) = |2 - 9x|$. Find all x for which $f(x) \ge 25$.

95. Let $f(x) = 7 + |2x - 1|$. Find all x for which $f(x) < 16$.

96. Let $f(x) = 5 + |3x + 2|$. Find all x for which $f(x) < 19$.

TW **97.** Explain in your own words why -7 is not a solution of $|x| < 5$.

TW **98.** Explain in your own words why $[6, \infty)$ is only part of the solution of $|x| \ge 6$.

SKILL REVIEW

To prepare for Section 8.3, review graphing equations and solving systems of equations (Sections 3.3, 3.6, 4.2, and 4.3).

Graph.

99. $3x - y = 6$ [3.3]

100. $y = \frac{1}{2}x - 1$ [3.6]

101. $x = -2$ [3.3]

102. $y = 4$ [3.3]

Solve using substitution or elimination. [4.2], [4.3]

103. $x - 3y = 8,$
$2x + 3y = 4$

104. $x - 2y = 3,$
$x = y + 4$

105. $y = 1 - 5x,$
$2x - y = 4$

106. $3x - 2y = 4,$
$5x - 3y = 5$

SYNTHESIS

TW **107.** Explain why the inequality $|x + 5| \ge 2$ can be interpreted as "the number x is at least 2 units from -5."

TW **108.** Isabel is using the following graph to solve $|x - 3| < 4$. How can you tell that a mistake has been made?

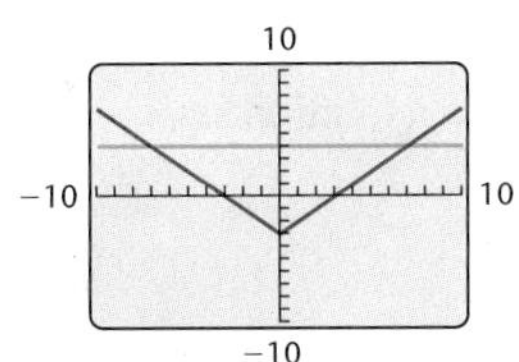

109. From the definition of absolute value, $|x| = x$ only when $x \ge 0$. Solve $|3t - 5| = 3t - 5$ using this same reasoning.

Solve.

110. $|3x - 5| = x$

111. $|x + 2| > x$

112. $2 \le |x - 1| \le 5$

113. $|5t - 3| = 2t + 4$

114. $t - 2 \le |t - 3|$

Find an equivalent inequality with absolute value.

115. $-3 < x < 3$

116. $x \le -6$ *or* $6 \le x$

117. $x < -8$ *or* $2 < x$

118. $-5 < x < 1$

119. x is less than 2 units from 7.

120. x is less than 1 unit from 5.

Write an absolute-value inequality for which the interval shown is the solution.

121. −7 −6 −5 −4 −3 −2 −1 0 1 2 3 4 5 6 7

122. −5 −4 −3 −2 −1 0 1 2 3 4 5 6 7 8 9

123. −7 −6 −5 −4 −3 −2 −1 0 1 2 3 4 5 6 7

124. 0 1 2 3 4 5 6 7 8 9 10 11 12 13 14

125. *Bungee Jumping.* A bungee jumper is bouncing up and down so that her distance d above a river satisfies the inequality $|d - 60 \text{ ft}| \le 10 \text{ ft}$ (see the figure below). If the bridge from which she jumped is 150 ft above the river, how far is the bungee jumper from the bridge at any given time?

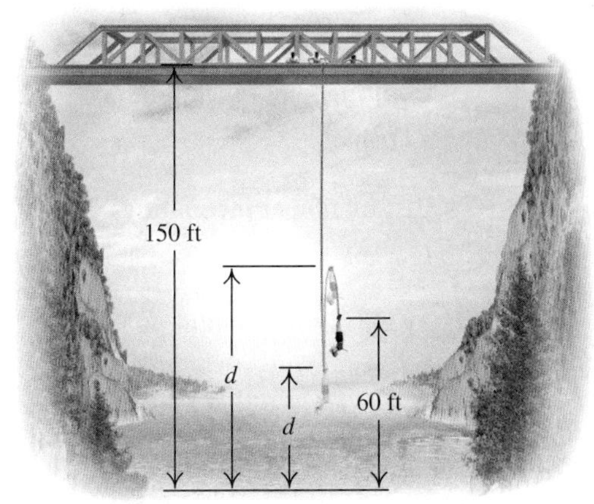

126. *Water Level.* Depending on how dry or wet the weather has been, water in a well will rise and fall. The distance d that a well's water level is below the ground satisfies the inequality $|d - 15| \leq 2.5$ (see the figure below).

a) Solve for d.

b) How tall a column of water is in the well at any given time?

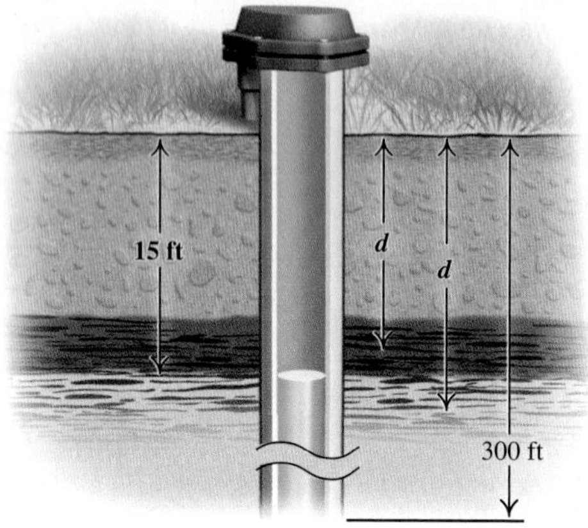

Try Exercise Answers: Section 8.2

21. $\{-7, 7\}$ **27.** $\left\{-\frac{1}{2}, \frac{7}{2}\right\}$ **31.** $\{-4, 8\}$ **33.** $\{2, 8\}$

49. $\{-8.7, 8.7\}$ **53.** $\{1, 11\}$

63. $\{a \mid -9 \leq a \leq 9\}$, or $[-9, 9]$ 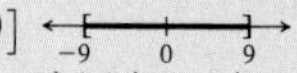

65. $\{t \mid t < 0 \text{ or } t > 0\}$, or $(-\infty, 0) \cup (0, \infty)$ 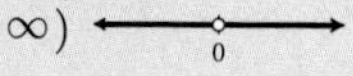

69. $\{x \mid -8 \leq x \leq 4\}$, or $[-8, 4]$ 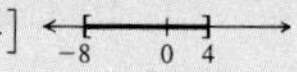

93. $\left\{x \mid x \leq -\frac{7}{3} \text{ or } x \geq 5\right\}$, or $\left(-\infty, -\frac{7}{3}\right] \cup [5, \infty)$

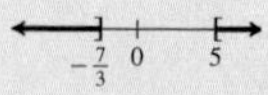

Mid-Chapter Review

The first step in solving the types of equations and inequalities that we have considered so far in this chapter is to determine the correct approach to finding the solution.

Type of Equation or Inequality	Approach
Conjunction of inequalities *Example:* $-3 < x - 5 < 6$	Find the intersection of the separate solution sets.
Disjunction of inequalities *Example:* $x + 8 < 2 \text{ or } x - 4 > 9$	Find the union of the separate solution sets.
Absolute-value equation *Example:* $\lvert x - 4 \rvert = 10$	Translate to two equations: If $\lvert X \rvert = p$, then $X = -p \text{ or } X = p$.
Absolute-value inequality including $<$ *Example:* $\lvert x + 2 \rvert < 5$	Translate to a conjunction: If $\lvert X \rvert < p$, then $-p < X < p$.
Absolute-value inequality including $>$ *Example:* $\lvert x - 1 \rvert > 9$	Translate to a disjunction: If $\lvert X \rvert > p$, then $X < -p \text{ or } p < X$.

GUIDED SOLUTIONS

1. Solve: $-3 < x - 5 < 6$. [8.1]

Solution

$\square < x < \square$ Adding 5

The solution is $(\square, \square)$.

2. Solve: $|x - 1| > 9$. [8.2]

Solution

$x - 1 < \square$ *or* $\square < x - 1$

$x < \square$ *or* $\square < x$ Adding 1

The solution is $(-\infty, \square) \cup (\square, \infty)$.

MIXED REVIEW

Solve.

1. $|x| = 15$ [8.2]
2. $|t| < 10$ [8.2]
3. $|p| > 15$ [8.2]
4. $|2x + 1| = 7$ [8.2]
5. $-3 < x - 5 < 6$ [8.1]
6. $5|t| < 20$ [8.2]
7. $x + 8 < 2$ *or* $x - 4 > 9$ [8.1]
8. $|x + 2| \leq 5$ [8.2]
9. $2 + |3x| = 10$ [8.2]
10. $|x - 3| \leq 10$ [8.2]
11. $-12 < 2n + 6$ *and* $3n - 1 \leq 7$ [8.1]
12. $|t| < 0$ [8.2]
13. $|2x + 5| + 1 \geq 13$ [8.2]
14. $5x + 1 < 1$ *and* $7 - x < 2$ [8.1]
15. $|m + 6| - 8 < 10$ [8.2]
16. $\left|\frac{x + 2}{5}\right| = 8$ [8.2]
17. $4 - |7 - t| \leq 1$ [8.2]
18. $|8x - 11| + 6 < 2$ [8.2]
19. $8 - 5|a + 6| > 3$ [8.2]
20. $|5x + 7| + 9 \geq 4$ [8.2]

8.3 Inequalities in Two Variables

- Graphs of Linear Inequalities
- Systems of Linear Inequalities

We have graphed inequalities in one variable on the number line. Now we graph inequalities in two variables on a plane.

GRAPHS OF LINEAR INEQUALITIES

When the equals sign in a linear equation is replaced with an inequality sign, a **linear inequality** is formed. Solutions of linear inequalities are ordered pairs.

EXAMPLE 1 Determine whether $(-3, 2)$ and $(6, -7)$ are solutions of the inequality $5x - 4y > 13$.

Student Notes

Pay careful attention to the inequality symbol when determining whether an ordered pair is a solution of an inequality. Writing the symbol at the end of the check, as in Example 1, will help you compare the numbers correctly.

SOLUTION Below, on the left, we replace x with -3 and y with 2. On the right, we replace x with 6 and y with -7.

$$\begin{array}{r|l} \multicolumn{2}{c}{5x - 4y > 13} \\ \hline 5(-3) - 4 \cdot 2 & 13 \\ -15 - 8 & \\ -23 \overset{?}{>} 13 & \text{FALSE} \end{array}$$

Since $-23 > 13$ is false, $(-3, 2)$ *is not* a solution.

$$\begin{array}{r|l} \multicolumn{2}{c}{5x - 4y > 13} \\ \hline 5(6) - 4(-7) & 13 \\ 30 + 28 & \\ 58 \overset{?}{>} 13 & \text{TRUE} \end{array}$$

Since $58 > 13$ is true, $(6, -7)$ *is* a solution.

Try Exercise 7.

The graph of a linear equation is a straight line. The graph of a linear inequality is a **half-plane,** with a **boundary** that is a straight line. To find the equation of the boundary, we replace the inequality sign with an equals sign.

EXAMPLE 2 Graph: $y \leq x$.

SOLUTION We first graph the equation of the boundary, $y = x$. Every solution of $y = x$ is an ordered pair, like (3, 3), in which both coordinates are the same. The graph of $y = x$ is shown on the left below. Since the inequality symbol is $\leq$, the line is drawn solid and is part of the graph of $y \leq x$.

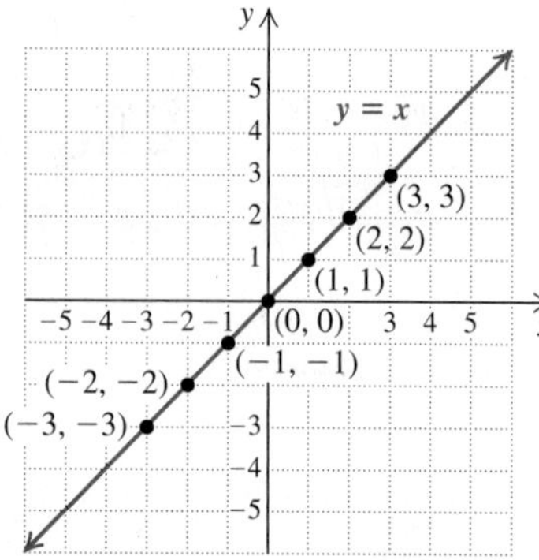

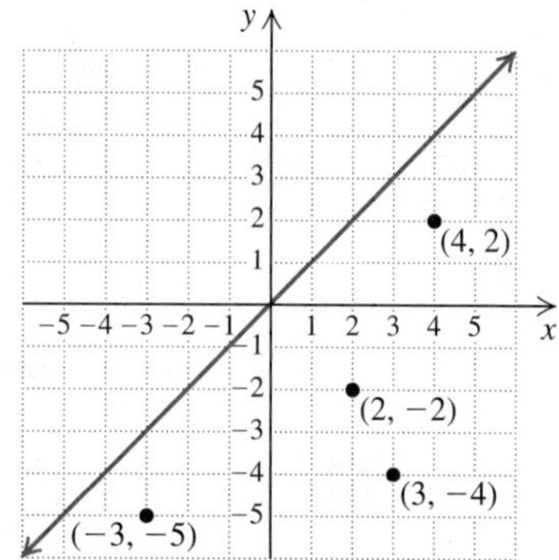

Note that in the graph on the right above each ordered pair on the half-plane below $y = x$ contains a y-coordinate that is less than the x-coordinate. All these pairs represent solutions of $y \leq x$. We check one pair, (4, 2), as follows:

$$\begin{array}{c} y \leq x \\ \hline 2 \overset{?}{\leq} 4 \quad \text{TRUE} \end{array}$$

It turns out that *any* point on the same side of $y = x$ as (4, 2) is also a solution. Thus, if one point in a half-plane is a solution, then *all* points in that half-plane are solutions. We finish drawing the solution set by shading the half-plane below $y = x$. The complete solution set consists of the shaded half-plane as well as the boundary line itself.

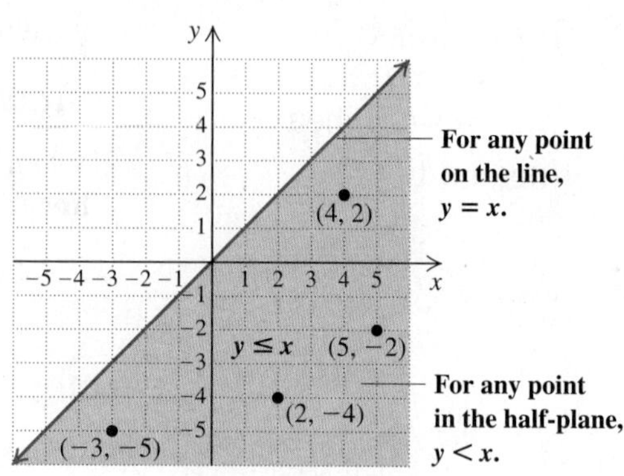

Try Exercise 11.

From Example 2, we see that for any inequality of the form $y \le f(x)$ or $y < f(x)$, we shade *below* the graph of $y = f(x)$.

STUDY TIP

Improve Your Study Skills

The time you spend learning to study better will be returned many times over. Study skills resources such as books and videos are available, your school may offer a class on study skills, or you can find Web sites that offer tips and instruction.

EXAMPLE 3 Graph: $8x + 3y > 24$.

SOLUTION First, we sketch the graph of $8x + 3y = 24$. Since the inequality sign is $>$, points on this line do not represent solutions of the inequality, so the line is drawn dashed. Points representing solutions of $8x + 3y > 24$ are in either the half-plane above the line or the half-plane below the line. To determine which, we select a point that is not on the line and check whether it is a solution of $8x + 3y > 24$. Let's use $(-3, 4)$ as this *test point*:

$8x + 3y > 24$	
$8(-3) + 3 \cdot 4$	24
$-24 + 12$	
$-12 \overset{?}{>} 24$	FALSE

Since $-12 > 24$ is *false*, $(-3, 4)$ is not a solution. Thus no point in the half-plane containing $(-3, 4)$ is a solution. The points in the other half-plane *are* solutions, so we shade that half-plane and obtain the graph shown below.

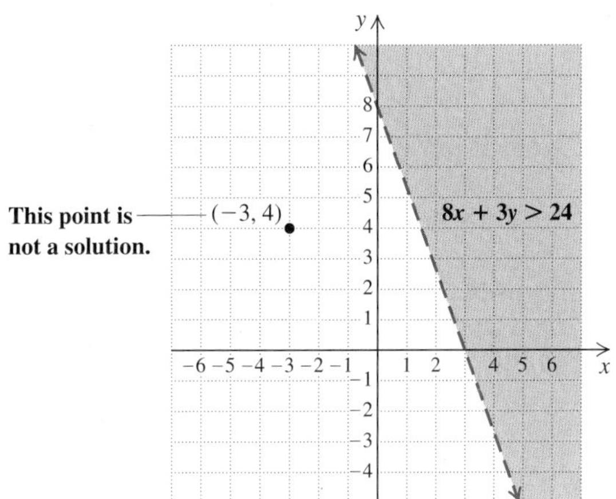

Try Exercise 19.

Steps for Graphing Linear Inequalities

1. Replace the inequality sign with an equals sign and graph this line as the boundary. If the inequality symbol is $<$ or $>$, draw the line dashed. If the symbol is $\le$ or $\ge$, draw the line solid.
2. The graph of the inequality consists of a half-plane on one side of the line and, if the line is solid, the line as well.

   a) If the inequality is of the form $y < mx + b$ or $y \le mx + b$, shade *below* the line.
   If the inequality is of the form $y > mx + b$ or $y \ge mx + b$, shade *above* the line.

   b) If y is not isolated, either solve for y and proceed as in part (a) or select a test point not on the line. If the test point *is* a solution of the inequality, shade the half-plane containing the point. If it is not a solution, shade the other half-plane.

Linear Inequalities

On most graphing calculators, an inequality like $y < \frac{6}{5}x + 3.49$ can be drawn by entering $(6/5)x + 3.49$ as y_1, moving the cursor to the GraphStyle icon just to the left of y_1, pressing ENTER until ◣ appears, and then pressing GRAPH.

Many calculators have an INEQUALZ program that is accessed using the APPS key. Running this program allows us to write inequalities at the Y= screen by pressing ALPHA and then one of the five keys just below the screen, as shown on the left below. With this program, the boundary line will appear dashed when $<$ or $>$ is selected.

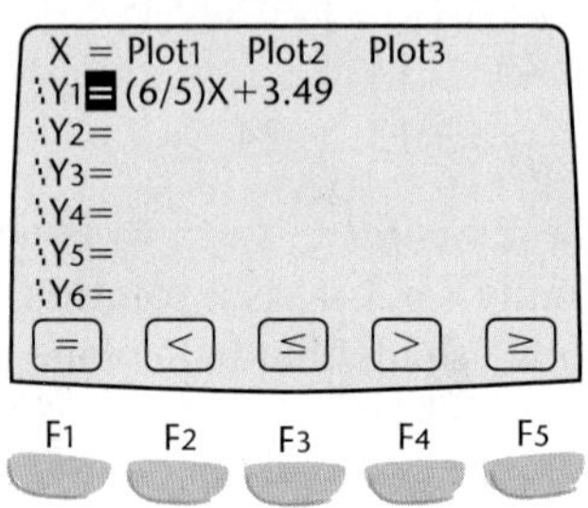

$y_1 < (6/5)x + 3.49$, or
◣ $y_1 = (6/5)x + 3.49$

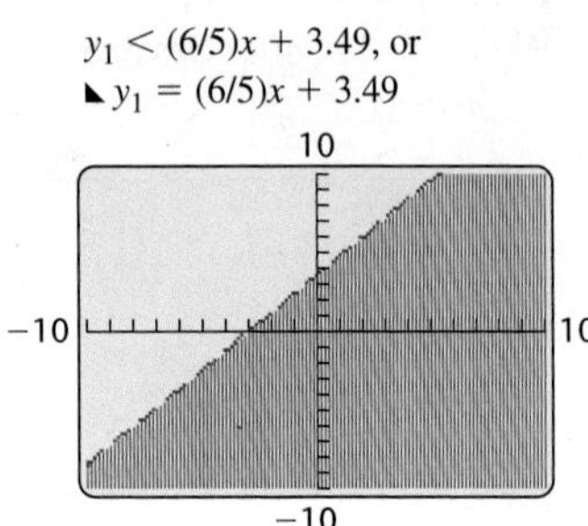

When you are finished with the INEQUALZ application, select it again from the APPS menu and quit the program.

Your Turn

1. Enter $y_1 = 3x - 5$.
2. Move the cursor to the GraphStyle icon to the left of Y1. Press ENTER until ◣ appears, and then press ZOOM 6. You should see the graph of $y < 3x - 5$.
3. If your calculator has an APPS key, press the key and scroll through the options. If INEQUALZ appears, highlight INEQUALZ and press ENTER. From the Y= screen, change $=$ to $<$ for Y1 and graph the inequality. Then select INEQUALZ again and quit the program.

EXAMPLE 4 Graph: $6x - 2y < 12$.

SOLUTION We graph both by hand and using a graphing calculator.

BY HAND

We could graph $6x - 2y = 12$ and use a test point, as in Example 3. Instead, let's solve $6x - 2y < 12$ for y:

$$6x - 2y < 12$$
$$-2y < -6x + 12 \quad \text{Adding } -6x \text{ to both sides}$$
$$y > 3x - 6. \quad \text{Dividing both sides by } -2 \text{ and reversing the } < \text{ symbol}$$

The graph consists of the half-plane above the dashed boundary line $y = 3x - 6$.

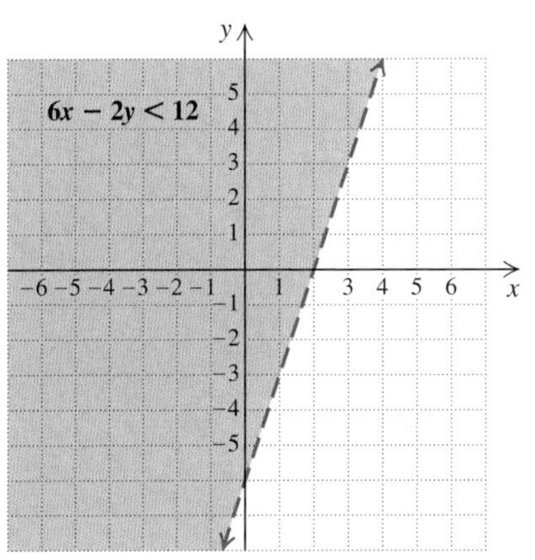

USING A GRAPHING CALCULATOR

We solve the inequality for y:

$$6x - 2y < 12$$
$$-2y < -6x + 12$$
$$y > 3x - 6. \quad \text{Reversing the } < \text{ symbol}$$

We enter the boundary equation, $y = 3x - 6$, and use the INEQUALZ application to graph.

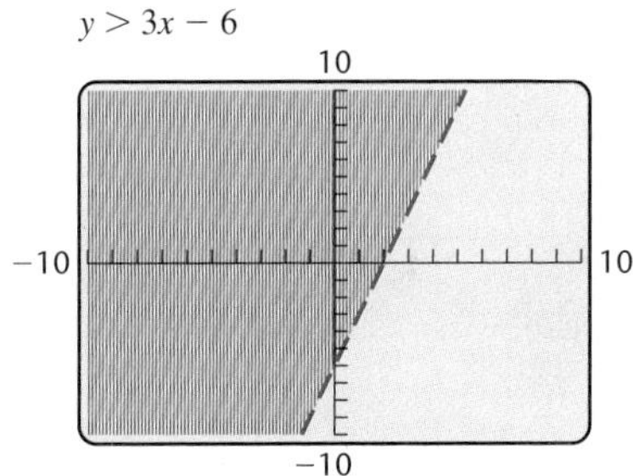

Many calculators do not draw a dashed line, so the graph of $y > 3x - 6$ appears to be the same as the graph of $y \geq 3x - 6$.

Quit the INEQUALZ application when you are finished.

Try Exercise 21.

Although a graphing calculator can be a very useful tool, some equations and inequalities are actually quicker to graph by hand. It is important to be able to quickly sketch basic linear equations and inequalities by hand. Also, when we are using a calculator to graph equations, knowing the basic shape of the graph can serve as a check that the equation was entered correctly.

EXAMPLE 5 Graph $x > -3$ on a plane.

SOLUTION There is only one variable in this inequality. If we graph the inequality on the number line, its graph is as follows:

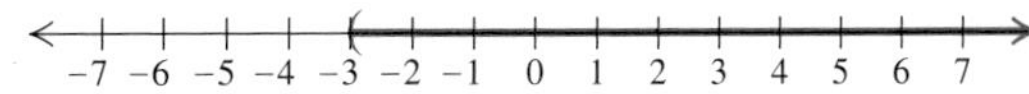

However, we can also write this inequality as $x + 0y > -3$ and graph it on a plane. We can use the same technique as in the examples above. First, we graph the boundary $x = -3$ on a plane. Then we test some point, say, (2, 5):

$$\begin{array}{c|c} x + 0y & > -3 \\ \hline 2 + 0 \cdot 5 & -3 \\ 2 & \overset{?}{>} -3 \end{array} \quad \text{TRUE}$$

Since (2, 5) is a solution, all points in the half-plane containing (2, 5) are solutions. We shade that half-plane. Another approach is to simply note that the solutions of $x > -3$ are all pairs with first coordinates greater than -3.

Although many graphing calculators can graph this inequality from the DRAW menu, it is simpler to graph it by hand.

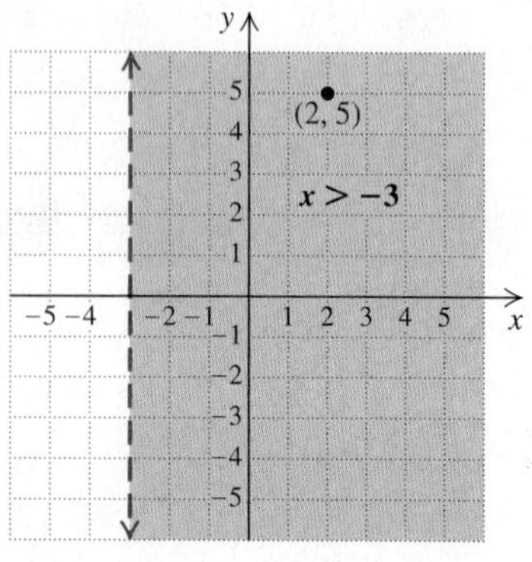

Try Exercise 25.

EXAMPLE 6 Graph $y \leq 4$ on a plane.

SOLUTION The inequality is of the form $y \leq mx + b$ (with $m = 0$), so we shade below the solid horizontal line representing $y = 4$.

This inequality can also be graphed by drawing $y = 4$ and testing a point above or below the line. The half-plane below $y = 4$ should be shaded.

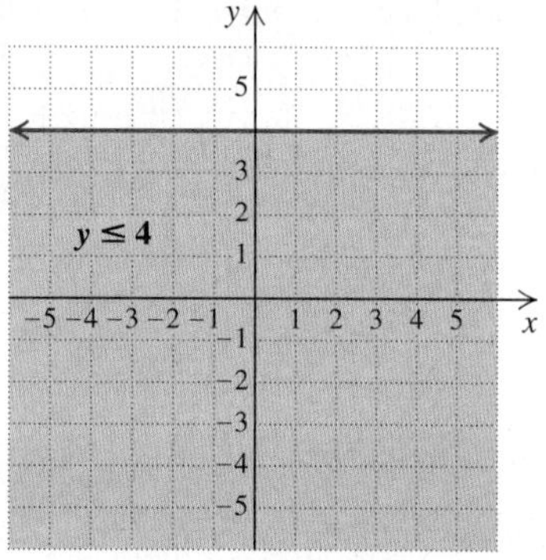

Try Exercise 27.

SYSTEMS OF LINEAR INEQUALITIES

To graph a system of equations, we graph the individual equations and then find the intersection of the graphs. We do the same thing for a system of inequalities: We graph each inequality and find the intersection of the graphs.

EXAMPLE 7 Graph the system

$$x + y \leq 4,$$
$$x - y < 4.$$

SOLUTION To graph $x + y \leq 4$, we graph $x + y = 4$ using a solid line. Since the test point $(0, 0)$ *is* a solution and $(0, 0)$ is below the line, we shade the half-plane below the graph red. The arrows near the ends of the line are another way of indicating the half-plane containing solutions.

Next, we graph $x - y < 4$. We graph $x - y = 4$ using a dashed line and consider $(0, 0)$ as a test point. Again, $(0, 0)$ is a solution, so we shade that side of the line blue. The solution set of the system is the region that is shaded purple (both red and blue) and part of the line $x + y = 4$.

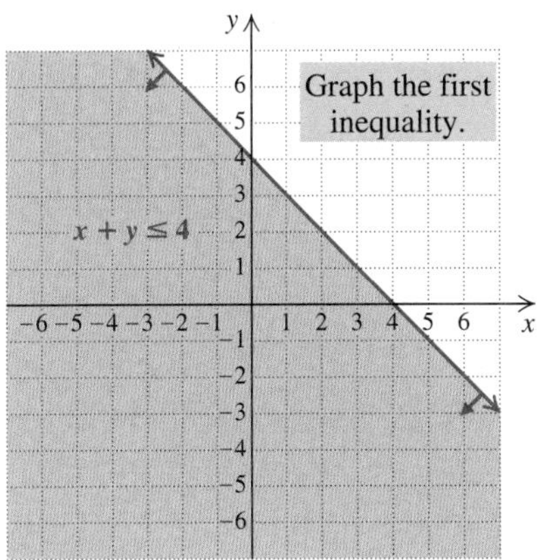

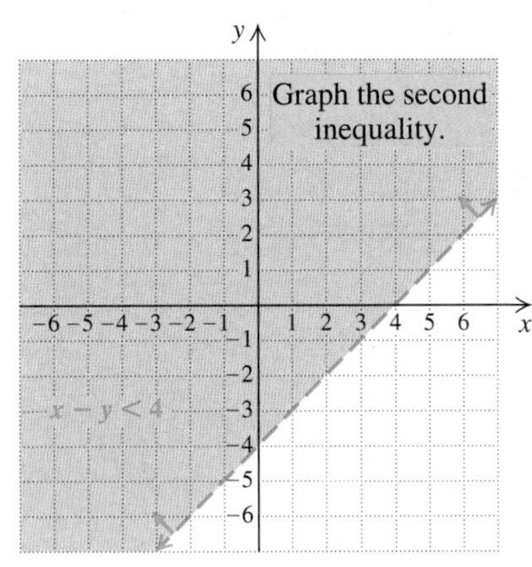

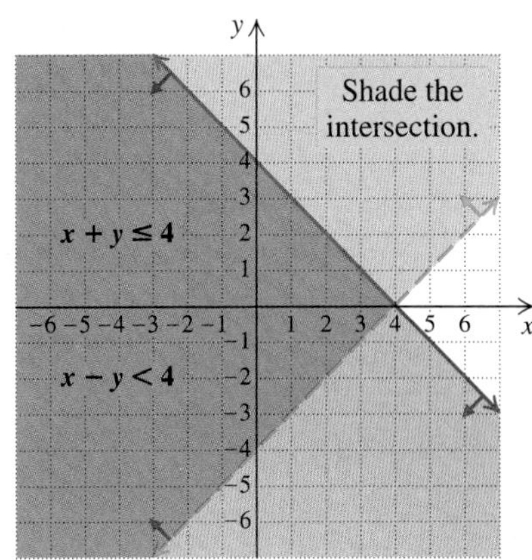

Try Exercise 39.

Student Notes

If you don't use differently colored pencils or pens to shade different regions, consider using a pencil to make slashes that tilt in different directions in each region, similar to the way shading is done on a graphing calculator. You may also find it useful to attach arrows to the lines, as in the examples shown.

EXAMPLE 8 Graph: $-2 < x \leq 3$.

SOLUTION This is a system of inequalities:

$$-2 < x,$$
$$x \leq 3.$$

We graph the equation $-2 = x$, and see that the graph of the first inequality is the half-plane to the right of the boundary $-2 = x$. It is shaded red.

We graph the second inequality, starting with the line $x = 3$, and find that its graph is the line and also the half-plane to its left. It is shaded blue.

The solution set of the system is the region that is the intersection of the individual graphs. Since it is shaded both blue and red, it appears to be purple. All points in this region have x-coordinates that are greater than -2 but do not exceed 3.

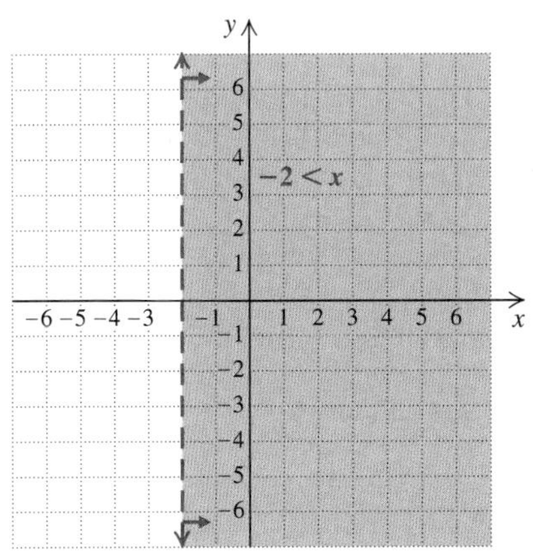

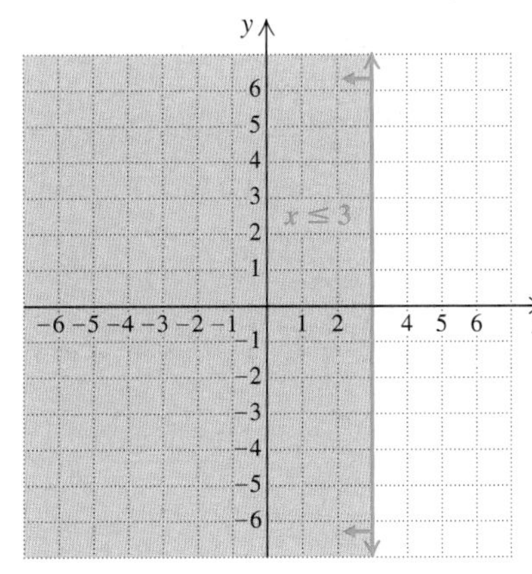

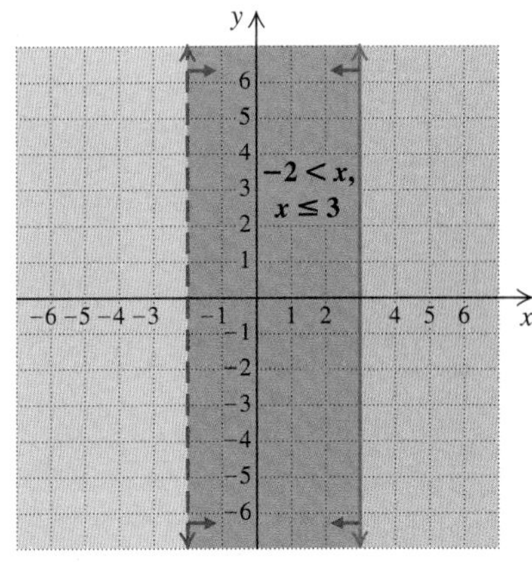

Try Exercise 29.

A system of inequalities may have a graph that consists of a polygon and its interior. The corners of such a graph are called *vertices* (singular, *vertex*).

Systems of Linear Inequalities

Systems of inequalities can be graphed by solving for y and then graphing each inequality. The graph of the system will be the intersection of the shaded regions. To graph systems directly using the INEQUALZ application, enter the separate inequalities, press GRAPH, and then press ALPHA and Shades (F1 or F2). At the SHADES menu, select Ineq Intersection to see the final graph. To find the vertices, or points of intersection, select Pol-Trace from the graph menu.

Your Turn

1. Graph the system

$$y \geq -3,$$
$$y \leq 5.$$

Enter $y_1 = -3$ and $y_2 = 5$, and then indicate the half-plane above y_1 and the half-plane below y_2. Use INEQUALZ if it is available. Press ZOOM 6 when you are ready to graph. The graph consists of the lines $y = -3$ and $y = 5$ and the area between them.

EXAMPLE 9 Graph the system of inequalities. Find the coordinates of any vertices formed.

$$6x - 2y \leq 12, \quad (1)$$
$$y - 3 \leq 0, \quad (2)$$
$$x + y \geq 0. \quad (3)$$

SOLUTION We graph both by hand and using a graphing calculator.

BY HAND

We graph the boundaries

$$6x - 2y = 12,$$
$$y - 3 = 0,$$

and $\quad x + y = 0$

using solid lines. The regions for each inequality are indicated by the arrows near the ends of the lines. We note where the regions overlap and shade the region of solutions purple.

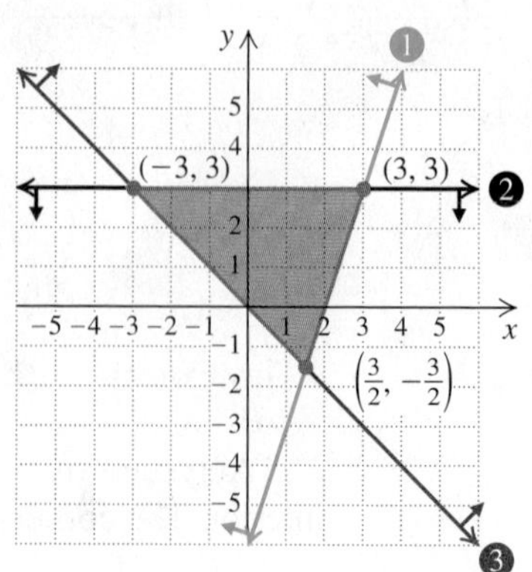

To find the vertices, we solve three different systems of two equations. The system of boundary equations from inequalities (1) and (2) is

$$6x - 2y = 12,$$
$$y - 3 = 0.$$

You can use graphing, substitution, or elimination to solve these systems.

Solving, we obtain the vertex $(3, 3)$.

The system of boundary equations from inequalities (1) and (3) is

$$6x - 2y = 12,$$
$$x + y = 0.$$

Solving, we obtain the vertex $\left(\frac{3}{2}, -\frac{3}{2}\right)$.

The system of boundary equations from inequalities (2) and (3) is

$$y - 3 = 0,$$
$$x + y = 0.$$

Solving, we obtain the vertex $(-3, 3)$.

USING A GRAPHING CALCULATOR

First, we solve each inequality for y and obtain the equivalent system of inequalities

$$y_1 \geq (12 - 6x)/(-2),$$
$$y_2 \leq 3,$$
$$y_3 \geq -x.$$

We run the INEQUALZ application, enter each inequality, and press GRAPH. We then press ALPHA and F1 or F2 to choose Shades. Selecting the Ineq Intersection option results in the graph shown on the left below.

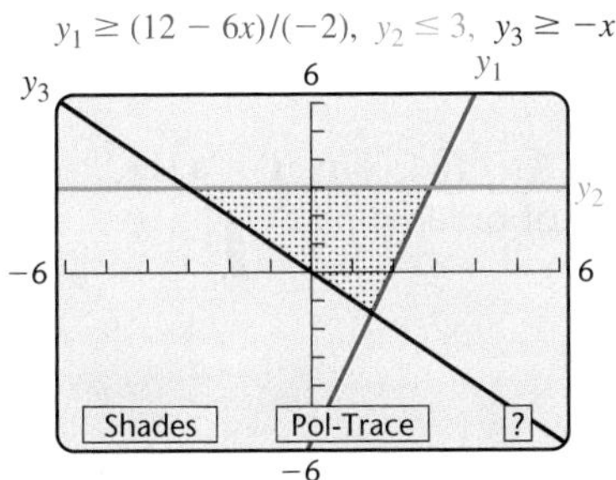

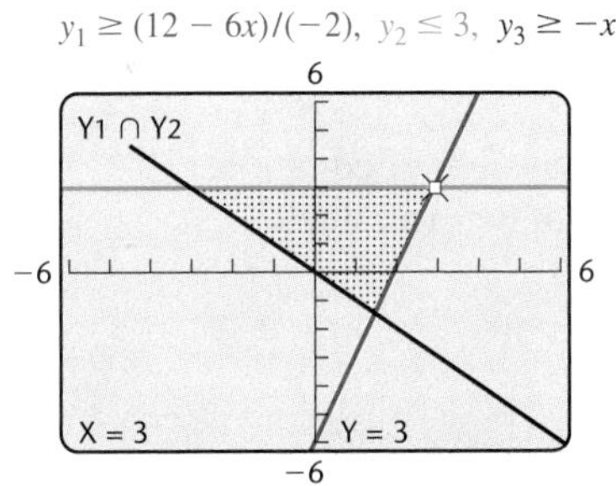

The vertices are the points of intersection of the graphs of the boundary equations y_1 and y_2, y_1 and y_3, and y_2 and y_3. We can find them by choosing Pol-Trace. Pressing the arrow keys moves the cursor to different vertices. The vertices are $(3, 3)$, $(-3, 3)$, and $(1.5, -1.5)$.

Try Exercise 53.

Graphs of systems of inequalities and the coordinates of vertices are used to solve a variety of problems using a branch of mathematics called *linear programming*.

Connecting the Concepts

We have now solved a variety of equations, inequalities, systems of equations, and systems of inequalities. Below is a list of the different types of problems we have solved, illustrations of each type, and descriptions of the solution. Note that a solution set may be empty.

Type	Example	Solution	Graph
Linear equation in one variable	$2x - 8 = 3(x + 5)$	A number	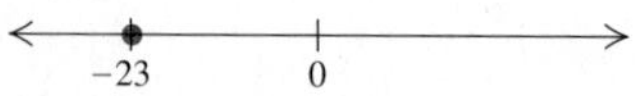 (number line: point at -23; marks -23, 0)
Linear inequality in one variable	$-3x + 5 > 2$	A set of numbers; an interval	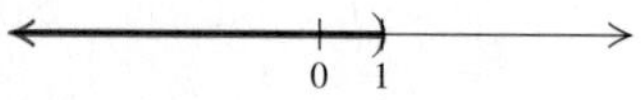 (number line: shaded to the left of 1, open at 1; marks 0, 1)
Linear equation in two variables	$2x + y = 7$	A set of ordered pairs; a line	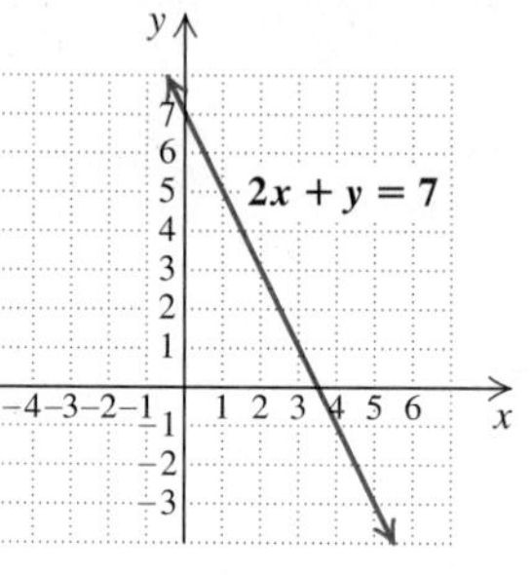
Linear inequality in two variables	$x + y \geq 4$	A set of ordered pairs; a half-plane	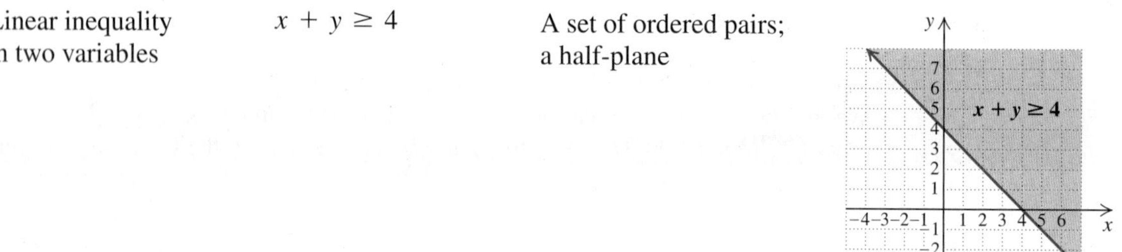 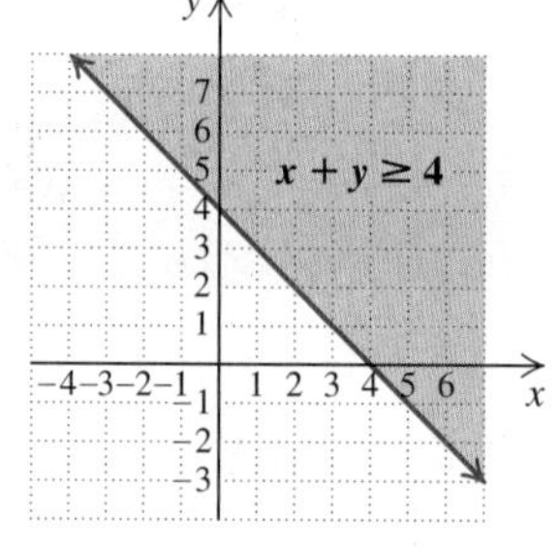
System of equations in two variables	$x + y = 3,$ $5x - y = -27$	An ordered pair or a set of ordered pairs	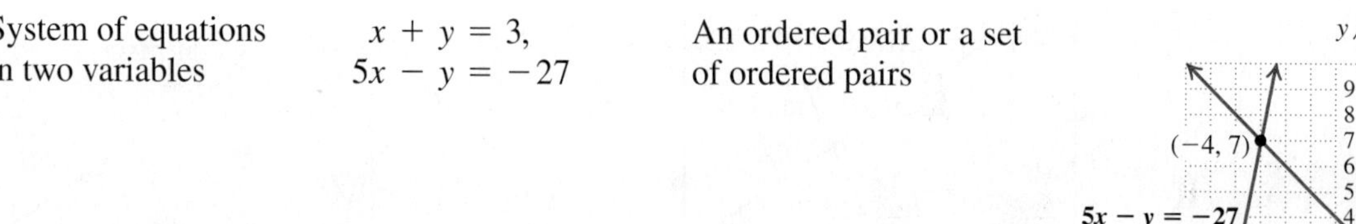 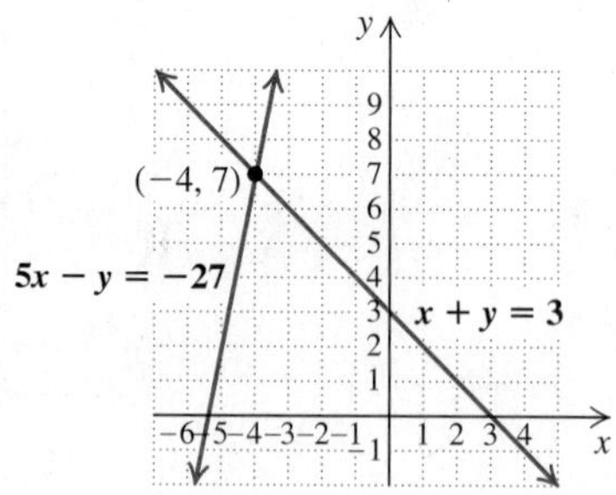
System of inequalities in two variables	$6x - 2y \leq 12,$ $y - 3 \leq 0,$ $x + y \geq 0$	A set of ordered pairs; a region of a plane	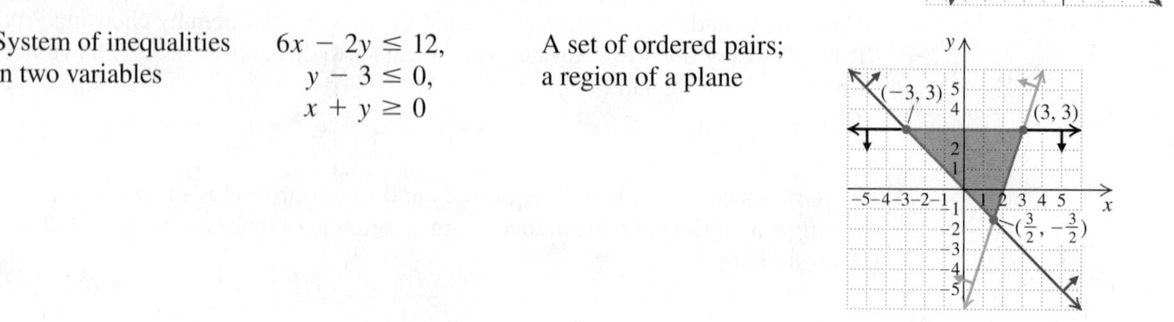

Visualizing for Success

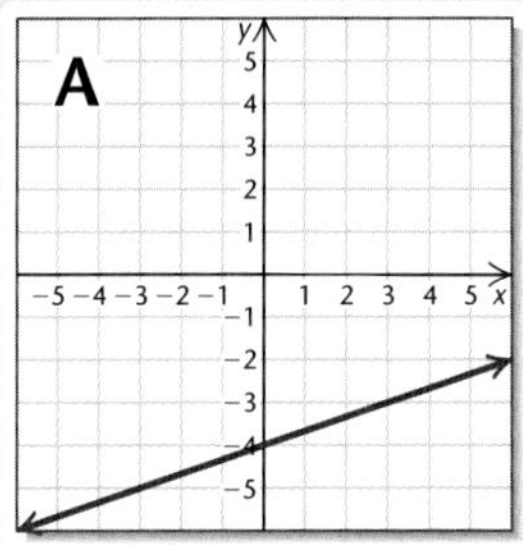

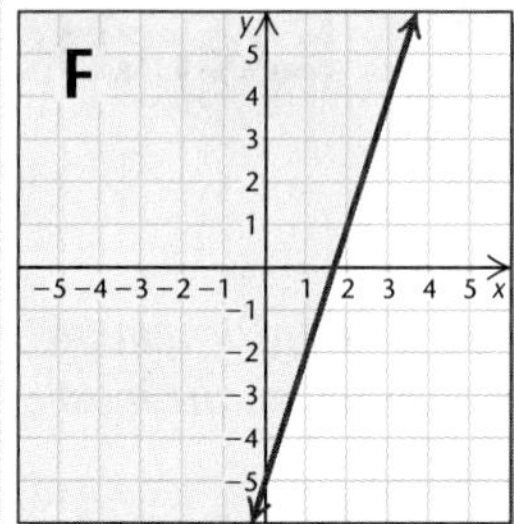

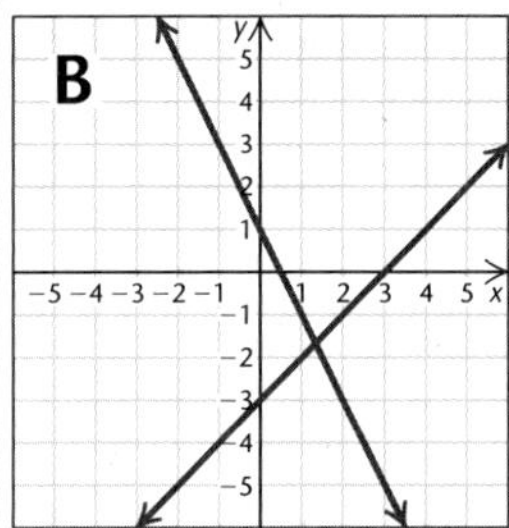

Match each equation, inequality, or system of equations or inequalities to its graph.

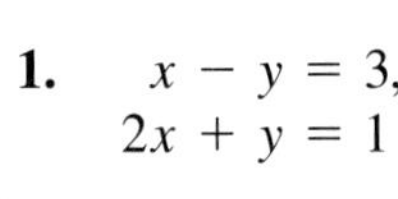

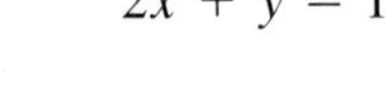

1. $x - y = 3,$
   $2x + y = 1$

2. $3x - y \leq 5$

3. $x > -3$

4. $y = \frac{1}{3}x - 4$

5. $y > \frac{1}{3}x - 4,$
   $y \leq x$

6. $x = y$

7. $y = 2x - 1,$
   $y = 2x - 3$

8. $2x - 5y = 10$

9. $x + y \leq 3,$
   $2y \leq x + 1$

10. $y = \frac{3}{2}$

Answers on page A-30

An additional, animated version of this activity appears in MyMathLab. To use MyMathLab, you need a course ID and a student access code. Contact your instructor for more information.

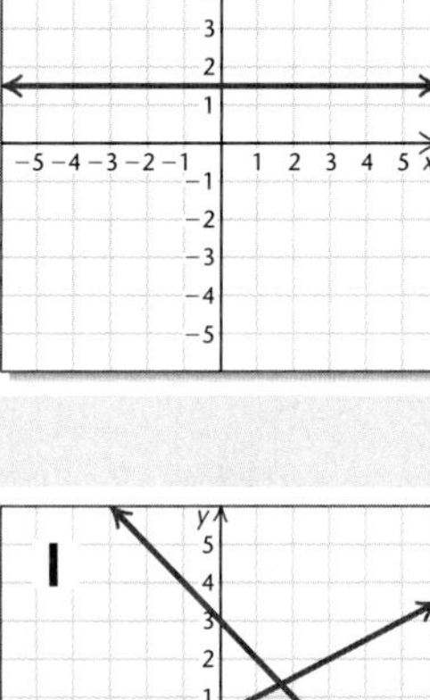

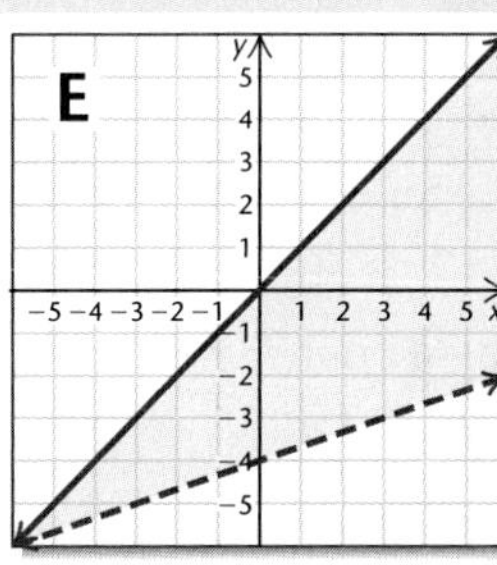

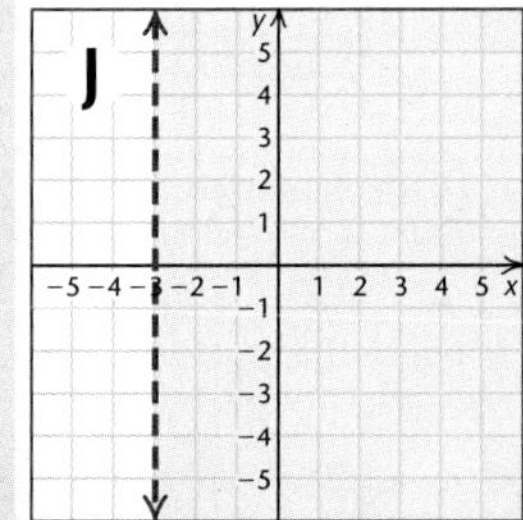

8.3 Exercise Set

FOR EXTRA HELP MyMathLab Math XL PRACTICE  WATCH  DOWNLOAD  READ  REVIEW

Concept Reinforcement *In each of Exercises 1–6, match the phrase with the most appropriate choice from the column on the right.*

1. ____ A solution of a linear inequality
2. ____ The graph of a linear inequality
3. ____ The graph of a system of linear inequalities
4. ____ Often a convenient test point
5. ____ The name for the corners of a graph of a system of linear inequalities
6. ____ A dashed line

a) $(0, 0)$
b) Vertices
c) A half-plane
d) The intersection of two or more half-planes
e) An ordered pair that satisfies the inequality
f) Indicates the line is not part of the solution

Determine whether each ordered pair is a solution of the given inequality.

7. $(-4, 2)$; $2x + 3y < -1$

8. $(3, -6)$; $4x + 2y \leq -2$

9. $(8, 14)$; $2y - 3x \geq 9$

10. $(5, 8)$; $3y - 5x \leq 0$

Graph on a plane.

11. $y \geq \frac{1}{2}x$

12. $y \leq 3x$

13. $y > x - 3$

14. $y < x + 3$

15. $y \leq x + 5$

16. $y > x - 2$

17. $x - y \leq 4$

18. $x + y < 4$

19. $2x + 3y > 6$

20. $3x + 4y \leq 12$

21. $2y - x \leq 4$

22. $2y - 3x > 6$

23. $2x - 2y \geq 8 + 2y$

24. $3x - 2 \leq 5x + y$

25. $x > -2$

26. $x \geq 3$

27. $y \leq 6$

28. $y < -1$

29. $-2 < y < 7$

30. $-4 < y < -1$

31. $-4 \leq x \leq 2$

32. $-3 \leq y \leq 4$

33. $0 \leq y \leq 3$

34. $0 \leq x \leq 6$

Graph using a graphing calculator.

35. $y > x + 3.5$

36. $7y \leq 2x + 5$

37. $8x - 2y < 11$

38. $11x + 13y + 4 \geq 0$

Graph each system.

39. $y > x,$ $y < -x + 3$

40. $y < x,$ $y > -x + 1$

41. $y \leq x,$ $y \leq 2x - 5$

42. $y \geq x,$ $y \leq -x + 4$

43. $y \leq -3,$ $x \geq -1$

44. $y \geq -3,$ $x \geq 1$

45. $x > -4,$ $y < -2x + 3$

46. $x < 3,$ $y > -3x + 2$

47. $y \leq 5,$ $y \geq -x + 4$

48. $y \geq -2,$ $y \geq x + 3$

49. $x + y \leq 6,$ $x - y \leq 4$

50. $x + y < 1,$ $x - y < 2$

51. $y + 3x > 0,$ $y + 3x < 2$

52. $y - 2x \geq 1,$ $y - 2x \leq 3$

Graph each system of inequalities. Find the coordinates of any vertices formed.

53. $y \leq 2x - 3,$ $y \geq -2x + 1,$ $x \leq 5$

54. $2y - x \leq 2,$ $y - 3x \geq -4,$ $y \geq -1$

55. $x + 2y \leq 12,$ $2x + y \leq 12,$ $x \geq 0,$ $y \geq 0$

56. $x - y \leq 2,$ $x + 2y \geq 8,$ $y \leq 4$

57. $8x + 5y \leq 40,$ $x + 2y \leq 8,$ $x \geq 0,$ $y \geq 0$

58. $4y - 3x \geq -12,$ $4y + 3x \geq -36,$ $y \leq 0,$ $x \leq 0$

59. $y - x \geq 2,$ $y - x \leq 4,$ $2 \leq x \leq 5$

60. $3x + 4y \geq 12,$ $5x + 6y \leq 30,$ $1 \leq x \leq 3$

TW **61.** Explain in your own words why a boundary line is drawn dashed for the symbols $<$ and $>$ and why it is drawn solid for the symbols $\leq$ and $\geq$.

TW **62.** When graphing linear inequalities, Ron makes a habit of always shading above the line when the symbol $\geq$ is used. Is this wise? Why or why not?

SKILL REVIEW

To prepare for Section 8.4, review solving polynomial equations and rational equations (Chapters 6 and 7).

Solve.

63. $x^2 - 1 = 0$ [6.4]

64. $5x^2 = 2x + 3$ [6.3]

65. $10x^3 - 30x^2 + 20x = 0$ [6.2]

66. $x^4 - x^3 - 6x^2 = 0$ [6.2]

Solve. [7.6]

67. $\frac{x - 3}{x + 4} = 5$

68. $\frac{x}{x - 1} = 1$

69. $\frac{x}{(x - 3)(x + 7)} = 0$

70. $\frac{(x + 6)(x - 9)}{x + 5} = 0$

SYNTHESIS

TW **71.** Explain how a system of linear inequalities could have a solution set containing exactly one ordered pair.

TW **72.** In Example 7, is the point $(4, 0)$ part of the solution set? Why or why not?

Graph.

73. $x + y > 8,$
$x + y \leq -2$

74. $x + y \geq 1,$
$-x + y \geq 2,$
$x \geq -2,$
$y \geq 2,$
$y \leq 4,$
$x \leq 2$

75. $x - 2y \leq 0,$
$-2x + y \leq 2,$
$x \leq 2,$
$y \leq 2,$
$x + y \leq 4$

76. Write four systems of four inequalities that describe a 2-unit by 2-unit square that has $(0, 0)$ as one of the vertices.

77. *Luggage Size.* Unless an additional fee is paid, most major airlines will not check any luggage for which the sum of the item's length, width, and height exceeds 62 in. The U.S. Postal Service will ship a package only if the sum of the package's length and girth (distance around its midsection) does not exceed 130 in. Video Promotions is ordering several 30-in. long cases that will be both mailed and checked as luggage. Using w and h for width and height (in inches), respectively, write and graph an inequality that represents all acceptable combinations of width and height.

Sources: U.S. Postal Service; www.case2go.com

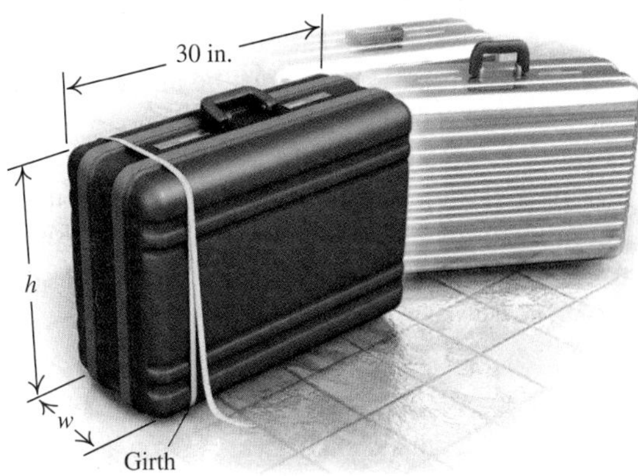

78. *Hockey Wins and Losses.* The Skating Stars figure that they need at least 60 points for the season in order to make the playoffs. A win is worth 2 points and a tie is worth 1 point. Graph a system of inequalities that describes the situation. (*Hint*: Let w = the number of wins and t = the number of ties.)

79. *Waterfalls.* In order for a waterfall to be classified as a classical waterfall, its height must be no more than twice its crest width, and its crest width cannot exceed one-and-a-half times its height. The tallest waterfall in the world is about 3200 ft high. Let h represent a waterfall's height, in feet, and w the crest width, in feet. Write and graph a system of inequalities that represents all possible combinations of heights and crest widths of classical waterfalls.

80. *Widths of a Basketball Floor.* Sizes of basketball floors vary due to building sizes and other constraints such as cost. The length L is to be at most 94 ft and the width W is to be at most 50 ft. Graph a system of inequalities that describes the possible dimensions of a basketball floor.

81. *Graduate-School Admissions.* Students entering the Master of Science program in Computer Science and Engineering at University of Texas Arlington must meet minimum score requirements on the Graduate Records Examination (GRE). The GRE Quantitative score must be at least 700 and the GRE Verbal score must be at least 400. The sum of the GRE Quantitative and Verbal scores must be at least 1150. Both scores have a maximum of 800. Using q for the quantitative score and v for the verbal score, write and graph a system of inequalities that represents all combinations that meet the requirements for entrance into the program.
Source: University of Texas Arlington

82. *Elevators.* Many elevators have a capacity of 1 metric ton (1000 kg). Suppose that c children, each weighing 35 kg, and a adults, each 75 kg, are on an elevator. Graph a system of inequalities that indicates when the elevator is overloaded.

Write a system of inequalities for each region shown.

83.

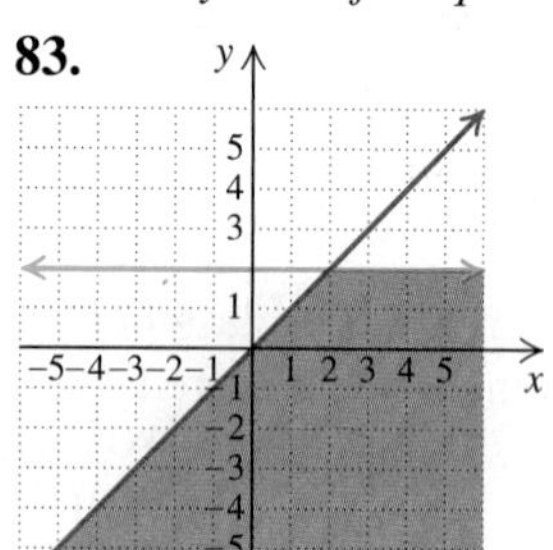

84.

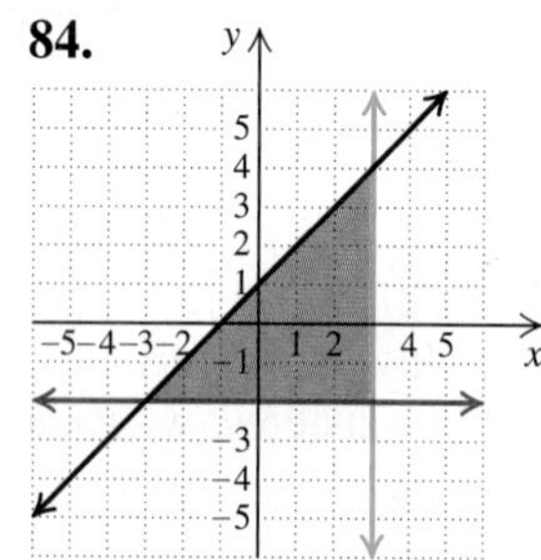

85.

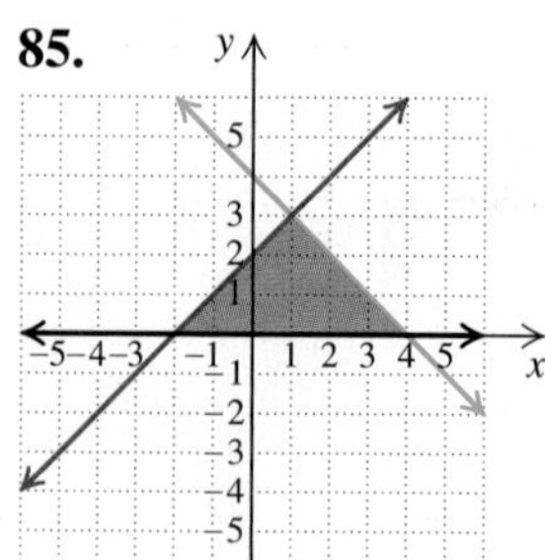

86.

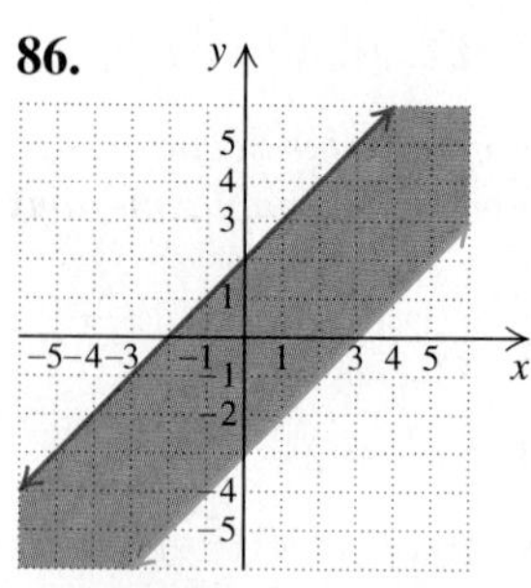

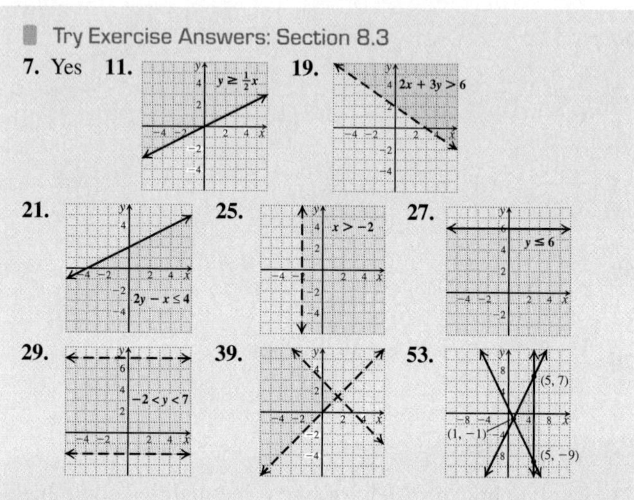

Collaborative Corner

How Old Is Old Enough?

Focus: Linear inequalities
Time: 15–25 minutes
Group Size: 2

It is not unusual for the ages of a bride and groom to differ significantly. Yet is it possible for the difference in age to be too great? In answer to this question, the following rule of thumb has emerged: *The younger spouse's age should be at least seven more than half the age of the older spouse.*
Source: http://home.earthlink.net/~mybrainhurts/2002_06_01_archive.html

ACTIVITY

1. Let b = the age of the bride, in years, and g = the age of the groom, in years. One group member should write an equation for calculating the bride's minimum age if the groom's age is known. The other group member should write an equation for finding the groom's minimum age if the bride's age is known. The equations should look similar.
2. Convert each equation into an inequality by selecting the appropriate symbol from $<, >, \leq$, and $\geq$. Be sure to reflect the rule of thumb stated above.
3. Graph both inequalities from step (2) as a system of linear inequalities. What does the solution set represent?
4. If your group feels that a minimum or maximum age for marriage should exist, adjust your graph accordingly.
5. Compare your finished graph with those of other groups.

8.4 Polynomial Inequalities and Rational Inequalities

- Quadratic and Other Polynomial Inequalities
- Rational Inequalities

QUADRATIC AND OTHER POLYNOMIAL INEQUALITIES

Inequalities like the following are called *polynomial inequalities*:

$$x^3 - 5x > x^2 + 7, \qquad 4x - 3 < 9, \qquad 5x^2 - 3x + 2 \geq 0.$$

Second-degree polynomial inequalities in one variable are called *quadratic inequalities*. To solve polynomial inequalities, we often focus attention on where the outputs of a polynomial function are positive and where they are negative.

STUDY TIP

Map Out Your Day

As the semester winds down and term papers are due, it becomes more critical than ever that you manage your time wisely. If you aren't already doing so, consider writing out an hour-by-hour schedule for each day and then abide by it as much as possible.

EXAMPLE 1 Solve: $x^2 + 3x - 10 > 0$.

SOLUTION Consider the "related" function $f(x) = x^2 + 3x - 10$ and its graph. We are looking for those x-values for which $f(x) > 0$. Graphically, function values are positive when the graph is above the x-axis.

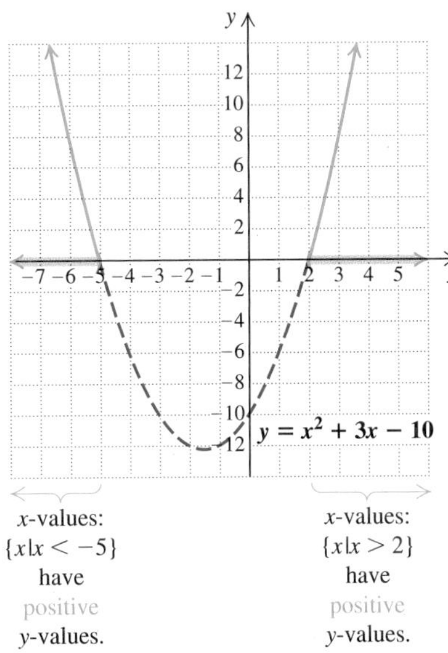

Values of y will be positive to the left and right of the x-intercepts, as shown. To find the intercepts, we set the polynomial equal to 0 and solve:

$$x^2 + 3x - 10 = 0$$
$$(x + 5)(x - 2) = 0$$
$$x + 5 = 0 \quad \text{or} \quad x - 2 = 0$$
$$x = -5 \quad \text{or} \quad x = 2.$$

The x-intercepts are $(-5, 0)$ and $(2, 0)$.

Thus the solution set of the inequality is

$$\{x \mid x < -5 \text{ or } x > 2\}, \quad \text{or} \quad (-\infty, -5) \cup (2, \infty).$$

Try Exercise 7.

Any inequality with 0 on one side can be solved by considering a graph of the related function and finding intercepts as in Example 1.

EXAMPLE 2 Solve: $5x^2 - 2x \le 7$.

SOLUTION We first find standard form with 0 on one side:

$$5x^2 - 2x - 7 \le 0. \quad \textbf{This is equivalent to the original inequality.}$$

If $f(x) = 5x^2 - 2x - 7$, then $5x^2 - 2x - 7 \le 0$ when $f(x)$ is 0 or negative. As shown in the graphs below, values of $f(x)$ are negative for x-values between the x-intercepts. We find the x-intercepts by solving $f(x) = 0$:

$$5x^2 - 2x - 7 = 0$$
$$(5x - 7)(x + 1) = 0$$
$$5x - 7 = 0 \quad or \quad x + 1 = 0$$
$$5x = 7 \quad or \quad x = -1$$
$$x = \frac{7}{5} \quad or \quad x = -1. \quad \textbf{The } x\textbf{-intercepts are } \left(\tfrac{7}{5}, 0\right) \textbf{ and } (-1, 0).$$

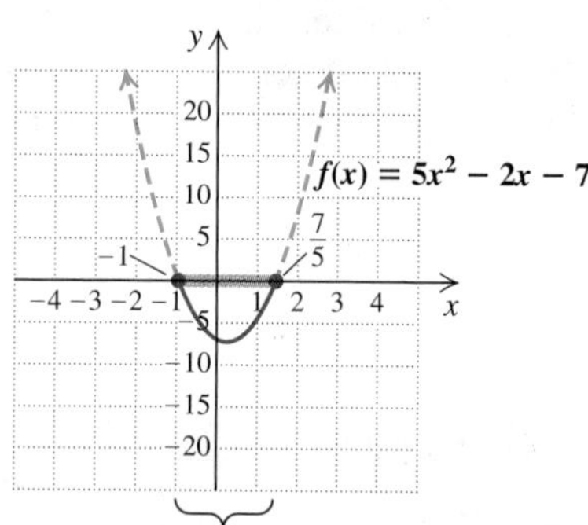

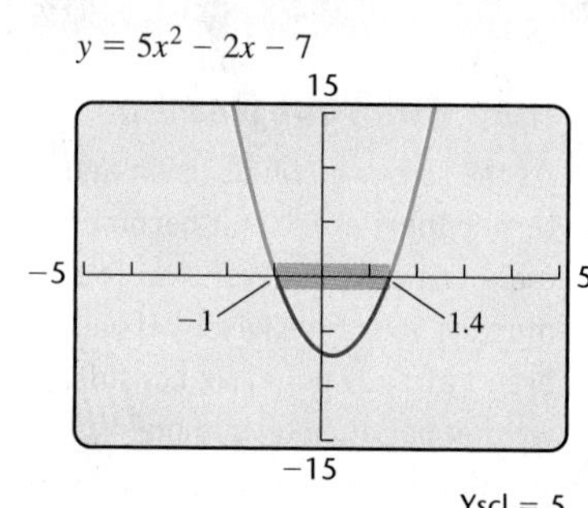

We can also find the x-intercepts using a graphing calculator. At the x-intercepts, the value of $f(x)$ is 0. Thus the solution set of the inequality is

$$\left\{x \middle| -1 \le x \le \frac{7}{5}\right\}, \quad \text{or} \quad \left[-1, \frac{7}{5}\right], \quad \text{or} \quad [-1, 1.4].$$

Try Exercise 21.

In Example 2, it was not essential to draw the graph. The important information came from finding the x-intercepts and the sign of $f(x)$ on each side of those intercepts. We now solve a third-degree polynomial inequality, without graphing, by locating the x-intercepts, or ***zeros***, of f and then using *test points* to determine the sign of $f(x)$ over each interval of the x-axis.

EXAMPLE 3 For $f(x) = 5x^3 + 10x^2 - 15x$, find all x-values for which $f(x) > 0$.

SOLUTION We first solve the related equation:

$$f(x) = 0$$
$$5x^3 + 10x^2 - 15x = 0 \quad \textbf{Substituting}$$
$$5x(x^2 + 2x - 3) = 0$$
$$5x(x + 3)(x - 1) = 0 \quad \textbf{Factoring completely}$$
$$5x = 0 \quad or \quad x + 3 = 0 \quad or \quad x - 1 = 0$$
$$x = 0 \quad or \quad x = -3 \quad or \quad x = 1.$$

The zeros of f are -3, 0, and 1. These zeros divide the number line, or x-axis, into four intervals: A, B, C, and D.

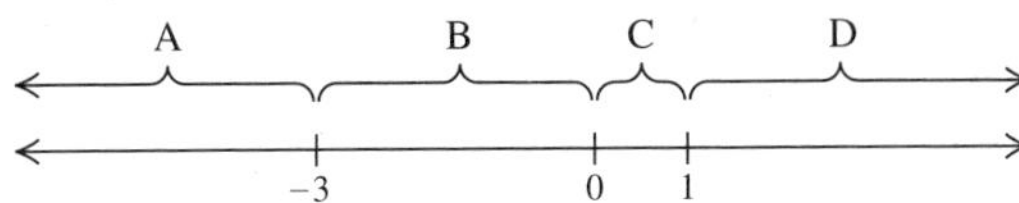

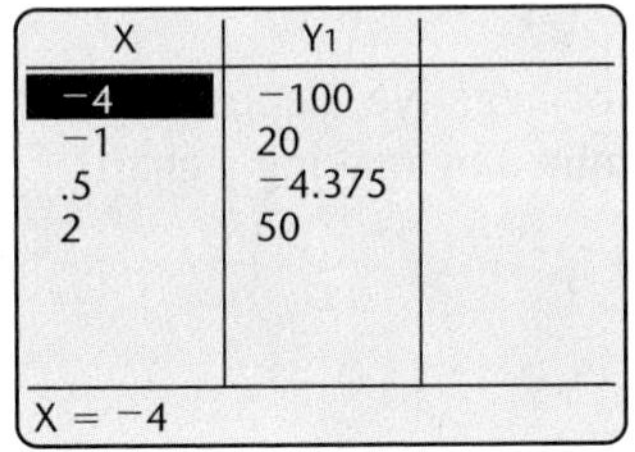

Next, selecting one convenient test value from each interval, we determine the sign of $f(x)$ for that interval. Within each interval, the sign of $f(x)$ cannot change. If it did, there would need to be another zero in that interval.

We choose -4 for a test value from interval A, -1 from interval B, 0.5 from interval C, and 2 from interval D. We enter $y_1 = 5x^3 + 10x^2 - 15x$ and use a table to evaluate the polynomial for each test value.

We are interested only in the signs of each function value. From the table at left, we see that $f(-4)$ and $f(0.5)$ are negative and that $f(-1)$ and $f(2)$ are positive. We indicate on the number line the sign of $f(x)$ in each interval.

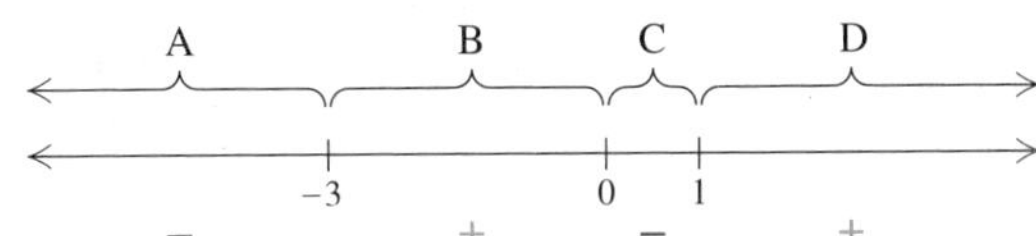

Recall that we are looking for all x for which $5x^3 + 10x^2 - 15x > 0$. The calculations above indicate that $f(x)$ is positive for any number in intervals B and D. The solution set of the original inequality is

$$(-3, 0) \cup (1, \infty), \quad \text{or} \quad \{x \mid -3 < x < 0 \text{ or } x > 1\}.$$

Try Exercise 35.

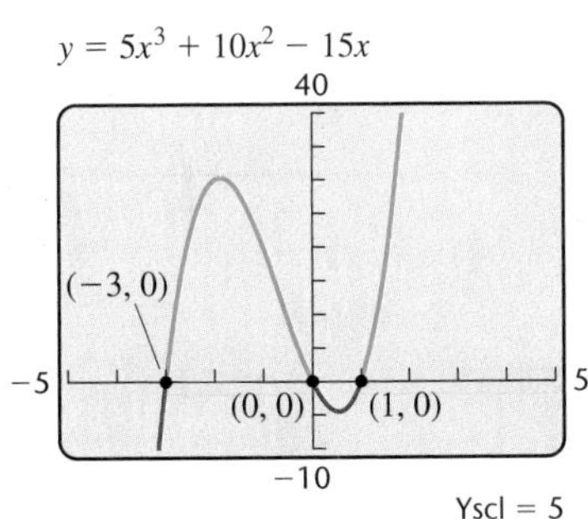

The method of Example 3 works because polynomial function values can change signs only when the graph of the function crosses the x-axis. The graph of $f(x) = 5x^3 + 10x^2 - 15x$ illustrates the solution of Example 3. We can see from the graph at left that the function values are positive in the intervals $(-3, 0)$ and $(1, \infty)$.

One alternative to using a calculator's TABLE feature is to graph the function and determine the appropriate intervals visually.

An efficient algebraic approach involves using a factored form of the polynomial and focusing on only the *sign* of $f(x)$. By looking at how many positive or negative factors are multiplied, we are able to determine the sign of the polynomial function.

EXAMPLE 4 For $f(x) = 4x^4 - 4x^2$, find all x-values for which $f(x) < 0$.

SOLUTION We first solve the related equation:

Solve $f(x) = 0$.

$$\begin{aligned} f(x) &= 0 \\ 4x^4 - 4x^2 &= 0 \\ 4x^2(x^2 - 1) &= 0 \\ 4x^2(x + 1)(x - 1) &= 0 \\ 4x^2 = 0 \quad \text{or} \quad x + 1 &= 0 \quad \text{or} \quad x - 1 = 0 \\ x = 0 \quad \text{or} \quad x &= -1 \quad \text{or} \quad x = 1. \end{aligned}$$

Divide the number line into intervals.

The function f has zeros at -1, 0, and 1, so we divide the number line into four intervals:

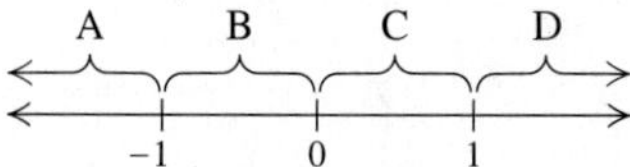

The product $4x^2(x + 1)(x - 1)$ is positive or negative, depending on the signs of $4x^2$, $x + 1$, and $x - 1$. This can be determined by making a chart.

Determine the sign of the function over each interval.

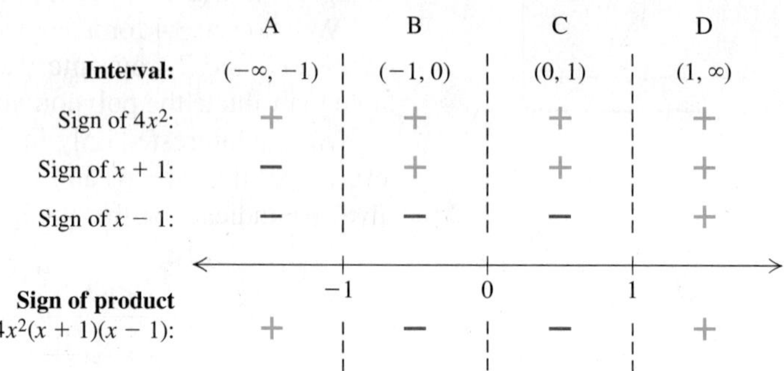

A product is negative when it has an odd number of negative factors. Since the $<$ sign is used, the endpoints -1, 0, and 1 are not solutions. From the chart, we see that the solution set is

Select the interval(s) for which the inequality is satisfied.

$$\{x \mid -1 < x < 0 \text{ or } 0 < x < 1\}, \quad \text{or} \quad (-1, 0) \cup (0, 1).$$

The graph of the function confirms the solution.

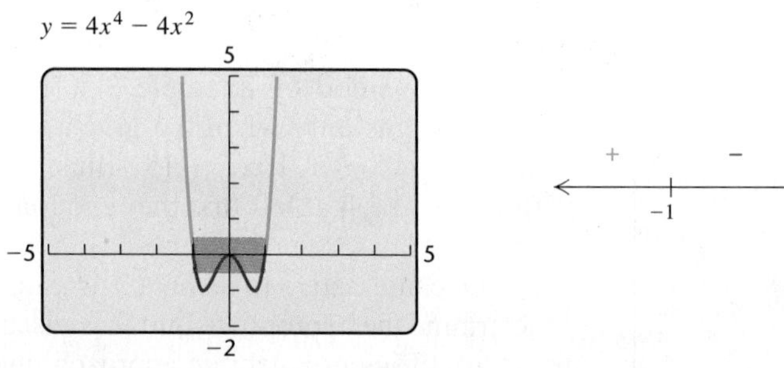

Try Exercise 31.

In Example 4, if the inequality symbol were $\leq$, the endpoints of the intervals would be included in the solution set. The solution of $4x^4 - 4x^2 \leq 0$ is $[-1, 0] \cup [0, 1]$, or simply $[-1, 1]$.

To Solve a Polynomial Inequality

1. Add or subtract to get 0 on one side and solve the related polynomial equation $p(x) = 0$.
2. Use the numbers found in step (1) to divide the number line into intervals.
3. Using a test value from each interval or the graph of the related function, determine the sign of $p(x)$ over each interval.
4. Select the interval(s) for which the inequality is satisfied and write set-builder notation or interval notation for the solution set. Include the endpoints of the intervals when $\leq$ or $\geq$ is used.

RATIONAL INEQUALITIES

Inequalities involving rational expressions are called **rational inequalities.** Like polynomial inequalities, rational inequalities can be solved using test values. Unlike polynomials, however, rational expressions often have values for which the expression is undefined. These values must be used when dividing the number line into intervals.

EXAMPLE 5 Solve: $\frac{x-3}{x+4} \geq 2$.

SOLUTION We write the related equation by changing the $\geq$ symbol to $=$:

$$\frac{x-3}{x+4} = 2. \qquad \textbf{Note that } x \neq -4.$$

We show both algebraic and graphical approaches.

ALGEBRAIC APPROACH

We first solve the related equation:

$$(x+4) \cdot \frac{x-3}{x+4} = (x+4) \cdot 2 \qquad \textbf{Multiplying both sides by the LCD, } x+4$$

$$x - 3 = 2x + 8$$

$$-11 = x. \qquad \textbf{Solving for } x$$

In the case of rational inequalities, we must always find any values that make the denominator 0. As noted at the beginning of the example, $x \neq -4$.

Now we use -11 and -4 to divide the number line into intervals:

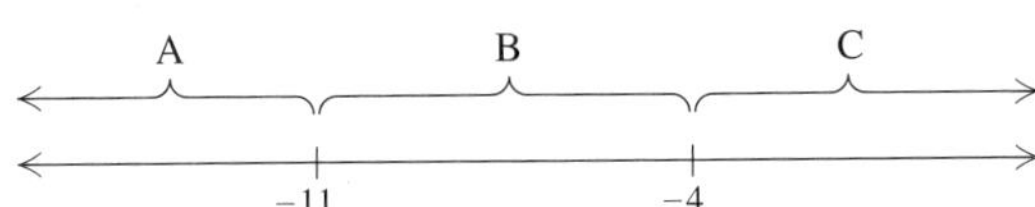

We test a number in each interval to see where the original inequality is satisfied:

$$\frac{x-3}{x+4} \geq 2.$$

If we enter $y_1 = (x-3)/(x+4)$ and $y_2 = 2$, the inequality is satisfied when $y_1 \geq y_2$.

X	Y1	Y2
-15	1.6364	2
-8	2.75	2
1	-.4	2
X =		

$y_1 < y_2$; -15 *is not* a solution.
$y_1 > y_2$; -8 *is* a solution.
$y_1 < y_2$; 1 *is not* a solution.

The solution set includes interval B. The endpoint -11 is included because the inequality symbol is $\geq$ and -11 is a solution of the related equation. The number -4 is *not* included because $(x-3)/(x+4)$ is undefined for $x = -4$. Thus the solution set of the original inequality is

$$[-11, -4), \quad \text{or} \quad \{x \mid -11 \leq x < -4\}.$$

GRAPHICAL APPROACH

We graph $y_1 = (x-3)/(x+4)$ and $y_2 = 2$ using DOT mode. The inequality is true for those values of x for which the graph of y_1 is above or intersects the graph of y_2, or where $y_1 \geq y_2$.

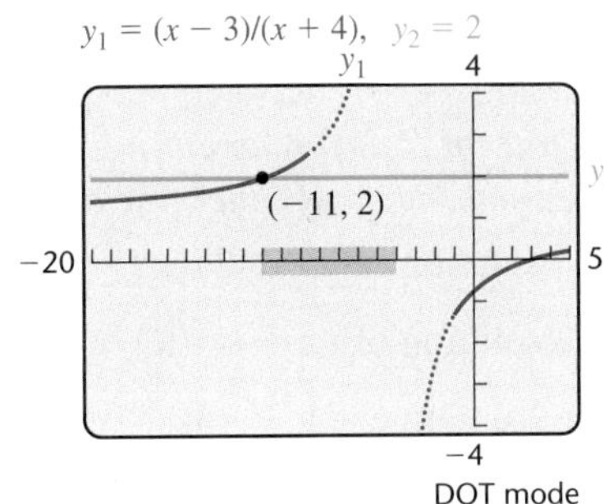

We see that

$$y_1 = y_2 \quad \text{when } x = -11$$

and

$$y_1 > y_2 \quad \text{when } -11 < x < -4.$$

Thus the solution set is $[-11, -4)$.

Try Exercise 41.

To Solve a Rational Inequality

1. Find any replacements for which the rational expression is undefined.
2. Change the inequality symbol to an equals sign and solve the related equation.
3. Use the numbers found in steps (1) and (2) to divide the number line into intervals.
4. Substitute a test value from each interval into the inequality. If the number is a solution, then the interval to which it belongs is part of the solution set. If solving graphically, examine the graph to determine the intervals that satisfy the inequality.
5. Select the interval(s) and any endpoints for which the inequality is satisfied and write set-builder notation or interval notation for the solution set. If the inequality symbol is $\leq$ or $\geq$, then the solutions from step (2) are also included in the solution set. Those numbers found in step (1) should be excluded from the solution set, even if they are solutions from step (2).

8.4 Exercise Set

FOR EXTRA HELP

Concept Reinforcement *Classify each of the following statements as either true or false.*

1. The solution of $(x-3)(x+2) \leq 0$ is $[-2, 3]$.
2. The solution of $(x+5)(x-4) \geq 0$ is $[-5, 4]$.
3. The solution of $(x-1)(x-6) > 0$ is $\{x \mid x < 1 \text{ or } x > 6\}$.
4. The solution of $(x+4)(x+2) < 0$ is $(-4, -2)$.
5. To solve $\dfrac{x+2}{x-3} < 0$ using intervals, we divide the number line into the intervals $(-\infty, -2)$ and $(-2, \infty)$.
6. To solve $\dfrac{x-5}{x+4} \geq 0$ using intervals, we divide the number line into the intervals $(-\infty, -4)$, $(-4, 5)$, and $(5, \infty)$.

Determine the solution set of each inequality from the given graph.

7. $p(x) \leq 0$

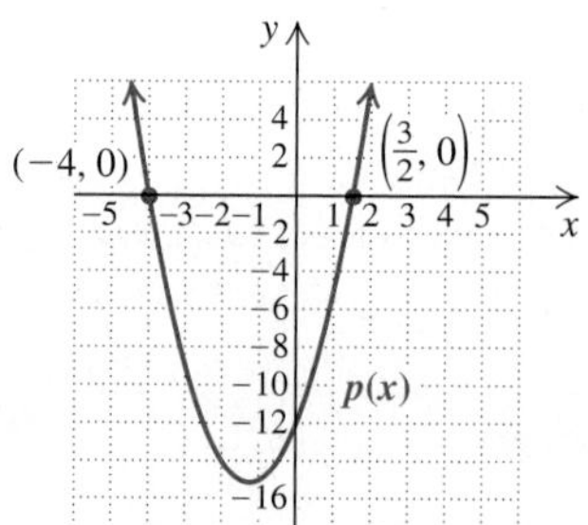

8. $p(x) < 0$

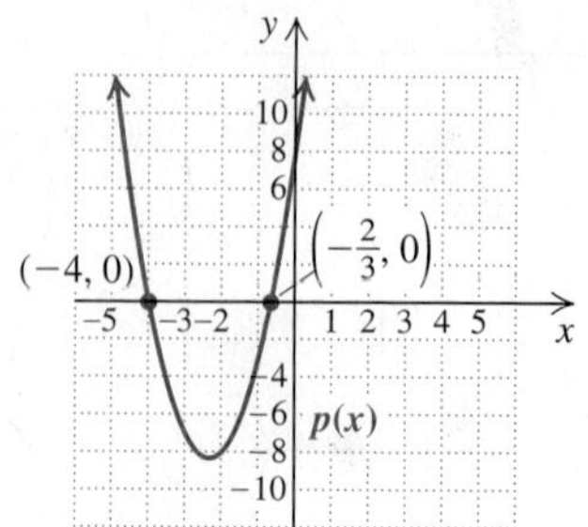

9. $x^4 + 12x > 3x^3 + 4x^2$

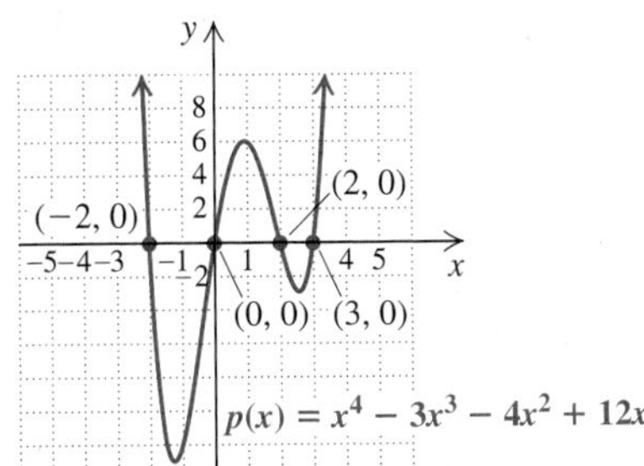

10. $x^4 + x^3 \geq 6x^2$

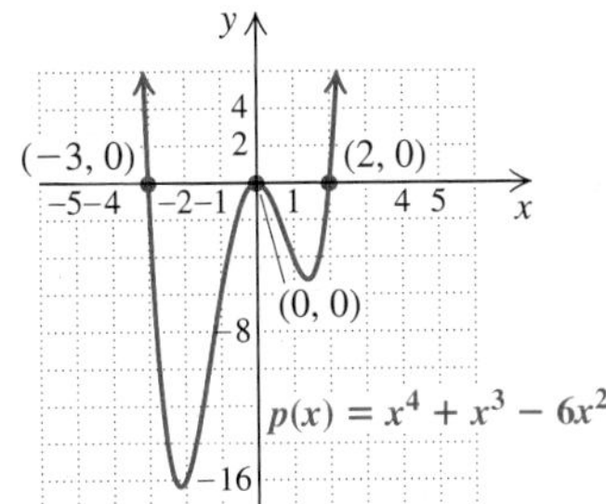

11. $\dfrac{x - 1}{x + 2} < 3$

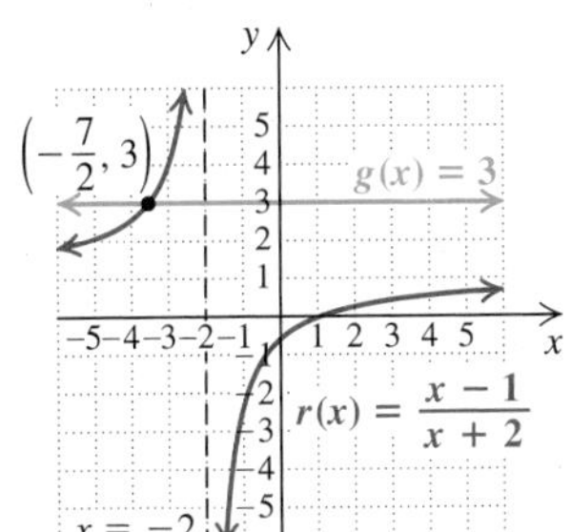

12. $\dfrac{2x - 1}{x - 5} \geq 1$

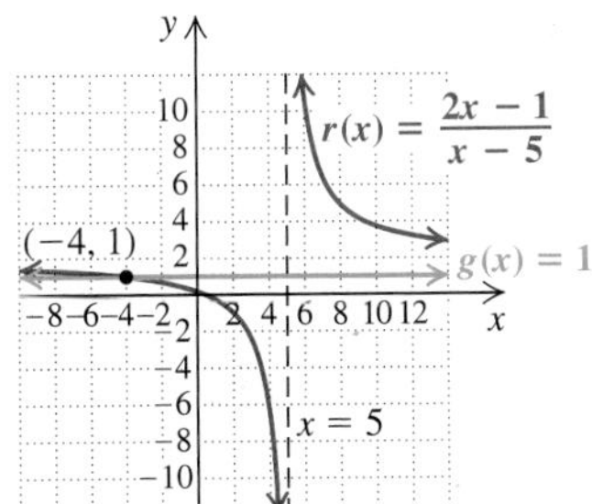

Solve.

13. $(x + 4)(x - 3) < 0$
14. $(x + 5)(x + 2) > 0$
15. $(x + 7)(x - 2) \geq 0$
16. $(x - 1)(x + 4) \leq 0$
17. $x^2 - x - 2 > 0$
18. $x^2 + x - 2 < 0$
19. (Aha!) $x^2 + 4x + 4 < 0$
20. $x^2 + 6x + 9 < 0$
21. $x^2 - 4x < 12$
22. $x^2 + 6x > -8$
23. $3x(x + 2)(x - 2) < 0$
24. $5x(x + 1)(x - 1) > 0$
25. $(x - 1)(x + 2)(x - 4) \geq 0$
26. $(x + 3)(x + 2)(x - 1) < 0$
27. $4.32x^2 - 3.54x - 5.34 \leq 0$
28. $7.34x^2 - 16.55x - 3.89 \geq 0$
29. $x^3 - 2x^2 - 5x + 6 < 0$
30. $\frac{1}{3}x^3 - x + \frac{2}{3} > 0$
31. For $f(x) = 7 - x^2$, find all x-values for which $f(x) \geq 3$.
32. For $f(x) = 14 - x^2$, find all x-values for which $f(x) > 5$.
33. For $g(x) = (x - 2)(x - 3)(x + 1)$, find all x-values for which $g(x) > 0$.
34. For $g(x) = (x + 3)(x - 2)(x + 1)$, find all x-values for which $g(x) < 0$.
35. For $F(x) = x^3 - 7x^2 + 10x$, find all x-values for which $F(x) \leq 0$.
36. For $G(x) = x^3 - 8x^2 + 12x$, find all x-values for which $G(x) \geq 0$.

Solve.

37. $\dfrac{1}{x + 5} < 0$
38. $\dfrac{1}{x + 4} > 0$
39. $\dfrac{x + 1}{x - 3} \geq 0$
40. $\dfrac{x - 2}{x + 4} \leq 0$
41. $\dfrac{x + 1}{x + 6} \geq 1$
42. $\dfrac{x - 1}{x - 2} \leq 1$
43. $\dfrac{(x - 2)(x + 1)}{x - 5} \leq 0$
44. $\dfrac{(x + 4)(x - 1)}{x + 3} \geq 0$
45. $\dfrac{x}{x + 3} \geq 0$
46. $\dfrac{x - 2}{x} \leq 0$
47. $\dfrac{x - 5}{x} < 1$
48. $\dfrac{x}{x - 1} > 2$
49. $\dfrac{x - 1}{(x - 3)(x + 4)} \leq 0$
50. $\dfrac{x + 2}{(x - 2)(x + 7)} \geq 0$

51. For $f(x) = \dfrac{5 - 2x}{4x + 3}$, find all x-values for which $f(x) \geq 0$.

52. For $g(x) = \dfrac{2 + 3x}{2x - 4}$, find all x-values for which $g(x) \geq 0$.

53. For $G(x) = \dfrac{1}{x - 2}$, find all x-values for which $G(x) \leq 1$.

54. For $F(x) = \dfrac{1}{x - 3}$, find all x-values for which $F(x) \leq 2$.

TW 55. Explain how any quadratic inequality can be solved by examining a parabola.

TW 56. Describe a method for creating a quadratic inequality for which there is no solution.

SKILL REVIEW

To prepare for Chapter 9, review solving systems of equations using elimination (Section 4.3).

Solve. [4.3]

57. $3x - 2y = 7,$
$x + 2y = 1$

58. $x + 3y = 5,$
$x - 5y = 7$

59. $3x - 5y = 1,$
$2x + 3y = 7$

60. $10x - 2y = 4,$
$3x + 5y = 2$

61. $y - 5x = 2,$
$3x + y = -1$

62. $4x - 1 = 3y,$
$2y - 4 = 5x$

SYNTHESIS

TW 63. Step (5) on p. 640 states that even when the inequality symbol is $\leq$ or $\geq$, the solutions from step (2) are not always part of the solution set. Why?

TW 64. Describe a method that could be used to create quadratic inequalities that have $(-\infty, a] \cup [b, \infty)$ as the solution set.

Find each solution set.

65. $x^4 + x^2 < 0$

66. $x^4 + 2x^2 \geq 0$

67. $x^4 + 3x^2 \leq 0$

68. $\left|\dfrac{x + 2}{x - 1}\right| \leq 3$

69. *Total Profit.* Derex, Inc., determines that its total-profit function is given by

$$P(x) = -3x^2 + 630x - 6000.$$

a) Find all values of x for which Derex makes a profit.

b) Find all values of x for which Derex loses money.

70. *Height of a Thrown Object.* The function

$$S(t) = -16t^2 + 32t + 1920$$

gives the height S, in feet, of an object thrown from a cliff that is 1920 ft high. Here t is the time, in seconds, that the object is in the air.

a) For what times does the height exceed 1920 ft?

b) For what times is the height less than 640 ft?

71. *Number of Handshakes.* There are n people in a room. The number N of possible handshakes by the people is given by the function

$$N(n) = \frac{n(n - 1)}{2}.$$

For what number of people n is $66 \leq N \leq 300$?

72. *Number of Diagonals.* A polygon with n sides has D diagonals, where D is given by the function

$$D(n) = \frac{n(n - 3)}{2}.$$

Find the number of sides n if $27 \leq D \leq 230$.

Use a graphing calculator to graph each function and find solutions of $f(x) = 0$. Then solve the inequalities $f(x) < 0$ and $f(x) > 0$.

73. $f(x) = x + \dfrac{1}{x}$

74. $f(x) = x - \sqrt{x}, x \geq 0$

75. $f(x) = \dfrac{x^3 - x^2 - 2x}{x^2 + x - 6}$

76. $f(x) = x^4 - 4x^3 - x^2 + 16x - 12$

Find the domain of each function

77. $f(x) = \sqrt{x^2 - 4x - 45}$

78. $f(x) = \sqrt{9 - x^2}$

79. $f(x) = \sqrt{x^2 + 8x}$

80. $f(x) = \sqrt{x^2 + 2x + 1}$

Try Exercise Answers: Section 8.4

7. $\left[-4, \frac{3}{2}\right]$, or $\left\{x \mid -4 \leq x \leq \frac{3}{2}\right\}$
21. $(-2, 6)$, or $\{x \mid -2 < x < 6\}$
31. $[-2, 2]$, or $\{x \mid -2 \leq x \leq 2\}$
35. $(-\infty, 0] \cup [2, 5]$, or $\{x \mid x \leq 0 \text{ or } 2 \leq x \leq 5\}$
41. $(-\infty, -6)$, or $\{x \mid x < -6\}$

Study Summary

KEY TERMS AND CONCEPTS	EXAMPLES	PRACTICE EXERCISES

SECTION 8.1: GRAPHICAL SOLUTIONS AND COMPOUND INEQUALITIES

KEY TERMS AND CONCEPTS	EXAMPLES	PRACTICE EXERCISES
Inequalities can be solved graphically by determining the x-values for which one graph lies above or below another.	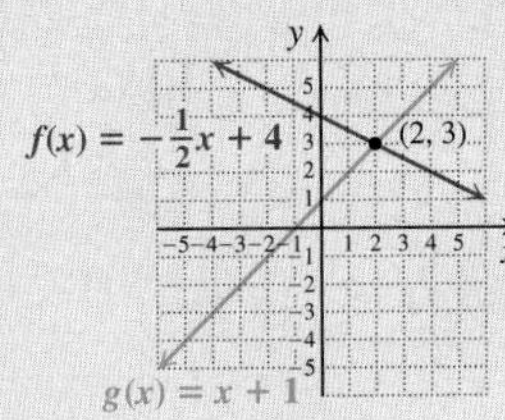 From the graph above, we can read the solution sets of the following inequalities. **Inequality** / **Solution Set** $f(x) < g(x)$ — $\{x \mid x > 2\}$, or $(2, \infty)$ $f(x) \le g(x)$ — $\{x \mid x \ge 2\}$, or $[2, \infty)$ $f(x) > g(x)$ — $\{x \mid x < 2\}$, or $(-\infty, 2)$ $f(x) \ge g(x)$ — $\{x \mid x \le 2\}$, or $(-\infty, 2]$	**1.** Solve graphically: $2x - 1 < x$.
A **conjunction** consists of two or more sentences joined by the word *and*. The solution set of the conjunction is the **intersection** of the solution sets of the individual sentences.	$-4 \le x - 1 \le 5$ $-4 \le x - 1$ *and* $x - 1 \le 5$ $-3 \le x$ *and* $x \le 6$ The solution set is $\{x \mid -3 \le x \le 6\}$, or $[-3, 6]$. $\{x \mid -3 \le x\}$ $\{x \mid x \le 6\}$ $\{x \mid -3 \le x \le 6\}$	**2.** Solve: $-5 < 4x + 3 \le 0$.
A **disjunction** consists of two or more sentences joined by the word *or*. The solution set of the disjunction is the **union** of the solution sets of the individual sentences.	$2x + 9 < 1$ *or* $5x - 2 \ge 3$ $2x < -8$ *or* $5x \ge 5$ $x < -4$ *or* $x \ge 1$ The solution set is $\{x \mid x < -4 \text{ or } x \ge 1\}$, or $(-\infty, -4) \cup [1, \infty)$. $\{x \mid x < -4\}$ $\{x \mid x \ge 1\}$ $\{x \mid x < -4 \text{ or } x \ge 1\}$	**3.** Solve: $x - 3 \le 10$ *or* $25 - x < 3$.

SECTION 8.2: ABSOLUTE-VALUE EQUATIONS AND INEQUALITIES

For any positive number p and any algebraic expression X:

a) The solutions of $|X| = p$ are those numbers that satisfy

$$X = -p \quad or \quad X = p.$$

$$|x + 3| = 4$$
$$x + 3 = 4 \quad or \quad x + 3 = -4 \qquad \textbf{Using part (a)}$$
$$x = 1 \quad or \quad x = -7$$

The solution set is $\{-7, 1\}$.

b) The solutions of $|X| < p$ are those numbers that satisfy

$$-p < X < p.$$

$$|x + 3| < 4$$
$$-4 < x + 3 < 4 \qquad \textbf{Using part (b)}$$
$$-7 < x < 1$$

The solution set is $\{x | -7 < x < 1\}$, or $(-7, 1)$.

c) The solutions of $|X| > p$ are those numbers that satisfy

$$X < -p \quad or \quad p < X.$$

$$|x + 3| \geq 4$$
$$x + 3 \leq -4 \quad or \quad 4 \leq x + 3 \qquad \textbf{Using part (c)}$$
$$x \leq -7 \quad or \quad 1 \leq x$$

The solution set is $\{x | x \leq -7 \text{ or } x \geq 1\}$, or $(-\infty, -7] \cup [1, \infty)$.

If $|X| = 0$, then $X = 0$. If p is negative, then $|X| = p$ and $|X| < p$ have no solution, and any value of X will satisfy $|X| > p$.

Solve.

4. $|4x - 7| = 11$

5. $|x - 12| \leq 1$

6. $|2x + 3| > 7$

SECTION 8.3: INEQUALITIES IN TWO VARIABLES

To graph a linear inequality:

1. Graph the **boundary line.** Draw a dashed line if the inequality symbol is $<$ or $>$, and draw a solid line if the inequality symbol is $\leq$ or $\geq$.
2. Determine which side of the boundary line contains the solution set, and shade that **half-plane.**

Graph: $x + y < -1$.

1. Graph $x + y = -1$ using a dashed line.
2. Choose a test point not on the line: $(0, 0)$.

$$x + y < -1$$
$$0 + 0 \;|\; -1$$
$$0 \overset{?}{<} -1 \quad \text{FALSE}$$

Since $0 < -1$ is false, shade the half-plane that does *not* contain $(0, 0)$.

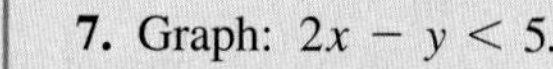

7. Graph: $2x - y < 5$.

SECTION 8.4: POLYNOMIAL INEQUALITIES AND RATIONAL INEQUALITIES

The x-intercepts, or **zeros,** of a function are used to divide the x-axis into intervals when solving a **polynomial inequality.** (See p. 638.)

Solve: $x^2 - 2x - 15 > 0$.

$$x^2 - 2x - 15 = 0 \qquad \textbf{Solving the related equation}$$
$$(x - 5)(x + 3) = 0$$
$$x = 5 \quad or \quad x = -3 \qquad \textbf{−3 and 5 divide the number line into three intervals.}$$

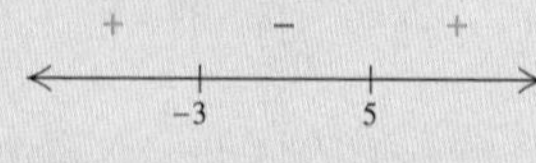

The solutions of a rational equation and any replacements that make a denominator zero are both used to divide the x-axis into intervals when solving a **rational inequality.** (See p. 640.)

For $f(x) = x^2 - 2x - 15 = (x - 5)(x + 3)$:

$f(x)$ is positive for $x < -3$;

$f(x)$ is negative for $-3 < x < 5$;

$f(x)$ is positive for $x > 5$.

Thus, $x^2 - 2x - 15 > 0$ for $(-\infty, -3) \cup (5, \infty)$, or $\{x | x < -3 \text{ or } x > 5\}$.

8. Solve:

$x^2 - 11x - 12 < 0$.

Review Exercises 8

Concept Reinforcement *Classify each of the following statements as either true or false.*

1. The solution of $|3x - 5| \le 8$ is a closed interval. [8.2]
2. The inequality $2 < 5x + 1 < 9$ is equivalent to $2 < 5x + 1$ *or* $5x + 1 < 9$. [8.1]
3. The solution set of a disjunction is the union of two solution sets. [8.1]
4. In mathematics, the word "or" means "one or the other or both." [8.1]
5. The domain of the sum of two functions is the union of the domains of the individual functions. [8.1]
6. The equation $|x| = p$ has no solution when p is negative. [8.2]
7. $|f(x)| > 3$ is equivalent to $f(x) < -3$ *or* $f(x) > 3$. [8.2]
8. A test point is used to determine whether the line in a linear inequality is drawn solid or dashed. [8.3]
9. The graph of a system of linear inequalities is always a half-plane. [8.3]
10. To solve a polynomial inequality, we often must solve a polynomial equation. [8.4]

Solve graphically. [8.1]

11. $4 - 3x > 1$
12. $x - 3 < 3x + 5$
13. $x + 1 \ge \frac{1}{2}x - 2$
14. Let $f(x) = 3x + 2$ and $g(x) = 10 - x$. Find all values of x for which $f(x) \le g(x)$. [8.1]
15. Find the intersection:

$$\{1, 2, 5, 6, 9\} \cap \{1, 3, 5, 9\}. \text{ [8.1]}$$

16. Find the union:

$$\{1, 2, 5, 6, 9\} \cup \{1, 3, 5, 9\}. \text{ [8.1]}$$

Graph and write interval notation. [8.1]

17. $x \le 2$ *and* $x > -3$
18. $x \le 3$ *or* $x > -5$

Solve and graph each solution set. [8.1]

19. $-4 < x + 8 \le 5$
20. $-15 < -4x - 5 < 0$
21. $3x < -9$ *or* $-5x < -5$
22. $2x + 5 < -17$ *or* $-4x + 10 \le 34$
23. $2x + 7 \le -5$ *or* $x + 7 \ge 15$
24. $f(x) < -5$ *or* $f(x) > 5$, where $f(x) = 3 - 5x$

For $f(x)$ as given, use interval notation to write the domain of f. [8.1]

25. $f(x) = \dfrac{2x}{x - 8}$
26. $f(x) = \sqrt{x + 5}$
27. $f(x) = \sqrt{8 - 3x}$

Solve. [8.2]

28. $|x| = 5$
29. $|t| \ge 21$
30. $|x - 3| = 7$
31. $|4a + 3| < 11$
32. $|3x - 4| - 6 \ge 9$
33. $|2x + 5| = |x - 9|$
34. $|5n + 6| = -8$
35. $\left|\dfrac{x + 4}{6}\right| \le 2$
36. $2|x - 5| - 7 > 3$
37. Let $f(x) = |3x - 5|$. Find all x for which $f(x) < 0$. [8.2]
38. Graph $x - 2y \ge 6$ on a plane. [8.3]

Graph each system of inequalities. Find the coordinates of any vertices formed. [8.3]

39. $x + 3y > -1,$
    $x + 3y < 4$
40. $x - 3y \le 3,$
    $x + 3y \ge 9,$
    $y \le 6$

Solve. [8.4]

41. $x^3 - 3x > 2x^2$
42. $\dfrac{x - 5}{x + 3} \le 0$

SYNTHESIS

TW 43. Explain in your own words why $|X| = p$ has two solutions when p is positive and no solution when p is negative. [8.2]

TW 44. Explain why the graph of the solution of a system of linear inequalities is the intersection, not the union, of the individual graphs. [8.3]

45. Solve: $|2x + 5| \le |x + 3|$. [8.2]
46. The Twinrocker Paper Company makes paper by hand. Each sheet is between 18 thousandths and 25 thousandths of an inch thick. Write the thickness t of a sheet of handmade paper as an inequality with absolute value. [8.2]

Chapter Test 8

Solve graphically.

1. $3 - x < 2$

2. $2x - 3 \geq x + 1$

3. Find the intersection:
$\{1, 3, 5, 7, 9\} \cap \{3, 5, 11, 13\}$.

4. Find the union:
$\{1, 3, 5, 7, 9\} \cup \{3, 5, 11, 13\}$.

Write the domain of f using interval notation.

5. $f(x) = \sqrt{6 - 3x}$

6. $f(x) = \dfrac{x}{x - 7}$

Solve and graph each solution set.

7. $-2 < x - 3 < 5$

8. $-11 \leq -5t - 2 < 0$

9. $3x - 2 < 7$ *or* $x - 2 > 4$

10. $-3x > 12$ *or* $4x > -10$

11. $|x| = 13$

12. $|a| > 7$

13. $|3x - 1| < 7$

14. $|-5t - 3| \geq 10$

15. $|2 - 5x| = -12$

16. $g(x) < -3$ *or* $g(x) > 3$, where $g(x) = 4 - 2x$

17. Let $f(x) = |x + 10|$ and $g(x) = |x - 12|$. Find all values of x for which $f(x) = g(x)$.

18. Graph $y \leq 2x + 1$ on a plane.

Graph each system of inequalities. Find the coordinates of any vertices formed.

19. $x + y \geq 3,$
$x - y \geq 5$

20. $2y - x \geq -7,$
$2y + 3x \leq 15,$
$y \leq 0,$
$x \leq 0$

Solve.

21. $x^2 + 5x \leq 6$

22. $x - \dfrac{1}{x} > 0$

SYNTHESIS

Solve. Write the solution set using interval notation.

23. $|2x - 5| \leq 7$ *and* $|x - 2| \geq 2$

24. $7x < 8 - 3x < 6 + 7x$

25. Write an absolute-value inequality for which the interval shown is the solution.

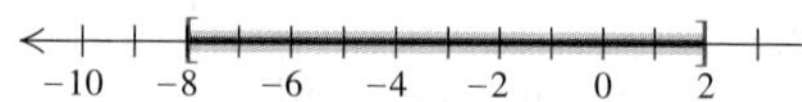

9

More on Systems

Is Silver Rarer Than Gold?

Some analysts think that silver is rarer than gold because silver has so many industrial applications. Products as diverse as toasters, toys, bandages, and clothing often contain silver, and, for most of these uses, silver is consumed and not recoverable. The graph below shows how industrial demand for silver has grown. In Exercise 46 in Section 9.5, we estimate when this demand will equal the above-ground stock of silver.

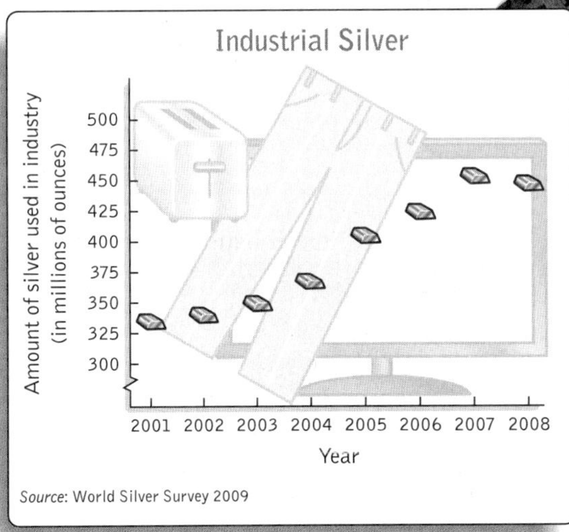

Source: World Silver Survey 2009

In this chapter, we extend the study of systems of equations that we began in Chapter 4. We solve systems of equations in three variables, solve systems using matrices and determinants, and look at some applications from economics.

9.1 Systems of Equations in Three Variables

- Identifying Solutions
- Solving Systems in Three Variables
- Dependency, Inconsistency, and Geometric Considerations

As we saw in Chapter 4, some problems translate directly to a system of two equations. Others call for a translation to three or more equations. In this section, we discuss how to solve systems of three linear equations. Later, we will use such systems in problem-solving situations.

IDENTIFYING SOLUTIONS

A **linear equation in three variables** is an equation equivalent to one in the form $Ax + By + Cz = D$, where A, B, C, and D are real numbers. We refer to the form $Ax + By + Cz = D$ as *standard form* for a linear equation in three variables.

A solution of a system of three equations in three variables is an ordered triple (x, y, z) that makes *all three* equations true. The numbers in an ordered triple correspond to the variables in alphabetical order unless otherwise indicated.

EXAMPLE 1 Determine whether $\left(\frac{3}{2}, -4, 3\right)$ is a solution of the system

$$\begin{aligned} 4x - 2y - 3z &= 5, \\ -8x - y + z &= -5, \\ 2x + y + 2z &= 5. \end{aligned}$$

SOLUTION We substitute $\left(\frac{3}{2}, -4, 3\right)$ into the three equations, using alphabetical order.

BY HAND

We have the following:

$$\begin{array}{r|l} 4x - 2y - 3z & = 5 \\ \hline 4 \cdot \frac{3}{2} - 2(-4) - 3 \cdot 3 & 5 \\ 6 + 8 - 9 & \\ 5 & \stackrel{?}{=} 5 \quad \text{TRUE} \end{array}$$

$$\begin{array}{r|l} -8x - y + z & = -5 \\ \hline -8 \cdot \frac{3}{2} - (-4) + 3 & -5 \\ -12 + 4 + 3 & \\ -5 & \stackrel{?}{=} -5 \quad \text{TRUE} \end{array}$$

$$\begin{array}{r|l} 2x + y + 2z & = 5 \\ \hline 2 \cdot \frac{3}{2} + (-4) + 2 \cdot 3 & 5 \\ 3 - 4 + 6 & \\ 5 & \stackrel{?}{=} 5 \quad \text{TRUE} \end{array}$$

The triple makes all three equations true, so it is a solution.

USING A GRAPHING CALCULATOR

We store $\frac{3}{2}$ as X, -4 as Y, and 3 as Z, using the ALPHA key to enter Y and Z.

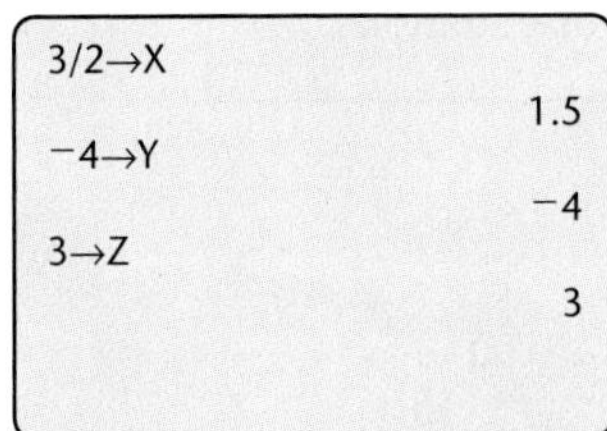

Now we enter the expression on the left side of each equation and press ENTER. If the value is the same as the right side of the equation, the ordered triple makes the equation true.

```
4X−2Y−3Z
                5
−8X−Y+Z
               −5
2X+Y+2Z
                5
```

The triple makes all three equations true, so it is a solution.

Try Exercise 7.

SOLVING SYSTEMS IN THREE VARIABLES

In Chapter 4, we solved systems of two equations using graphing, substitution, and elimination. The graph of a linear equation in three variables is a plane. Because a three-dimensional coordinate system is required, solving systems in three variables graphically is difficult. The substitution method *can* be used but is practical only when one or more of the equations has only two variables. Fortunately, the elimination method works well for a system of three equations in three variables.

EXAMPLE 2 Solve the following system of equations:

$$\begin{aligned} x + y + z &= 4, && (1) \\ x - 2y - z &= 1, && (2) \\ 2x - y - 2z &= -1. && (3) \end{aligned}$$

SOLUTION We select *any* two of the three equations and work to get an equation in two variables. Let's add equations (1) and (2):

$$\begin{aligned} x + y + z &= 4 && (1) \\ x - 2y - z &= 1 && (2) \\ \hline 2x - y \quad &= 5. && (4) \quad \text{Adding to eliminate } z \end{aligned}$$

Next, we select a different pair of equations and eliminate the *same variable* that we did above. Let's use equations (1) and (3) to again eliminate z. Be careful here! A common error is to eliminate a different variable in this step.

CAUTION! Be sure to eliminate the same variable in both pairs of equations.

$$\begin{aligned} x + y + z &= 4, \\ 2x - y - 2z &= -1 \end{aligned} \xrightarrow{\text{Multiplying both sides of equation (1) by 2}} \begin{aligned} 2x + 2y + 2z &= 8 \\ 2x - y - 2z &= -1 \\ \hline 4x + y &= 7 \quad (5) \end{aligned}$$

Now we solve the resulting system of equations (4) and (5). That solution will give us two of the numbers in the solution of the original system.

$$\begin{aligned} 2x - y &= 5 && (4) \\ 4x + y &= 7 && (5) \\ \hline 6x &= 12 && \text{Adding} \\ x &= 2 \end{aligned}$$

Note that we now have two equations in two variables. Had we not eliminated the *same* variable in both of the above steps, this would not be the case.

We can use either equation (4) or (5) to find y. We choose equation (5):

$$\begin{aligned} 4x + y &= 7 && (5) \\ 4 \cdot 2 + y &= 7 && \text{Substituting 2 for } x \text{ in equation (5)} \\ 8 + y &= 7 \\ y &= -1. \end{aligned}$$

We now have $x = 2$ and $y = -1$. To find the value for z, we use any of the original three equations and substitute to find the third number, z. Let's use equation (1) and substitute our two numbers in it:

$$\begin{aligned} x + y + z &= 4 && (1) \\ 2 + (-1) + z &= 4 && \text{Substituting 2 for } x \text{ and } -1 \text{ for } y \\ 1 + z &= 4 \\ z &= 3. \end{aligned}$$

We have obtained the triple $(2, -1, 3)$. It should check in *all three* equations:

$$\begin{array}{c|c} x + y + z & 4 \\ \hline 2 + (-1) + 3 & 4 \\ 4 \stackrel{?}{=} 4 & \text{TRUE} \end{array} \qquad \begin{array}{c|c} x - 2y - z & 1 \\ \hline 2 - 2(-1) - 3 & 1 \\ 1 \stackrel{?}{=} 1 & \text{TRUE} \end{array} \qquad \begin{array}{c|c} 2x - y - 2z & -1 \\ \hline 2 \cdot 2 - (-1) - 2 \cdot 3 & -1 \\ -1 \stackrel{?}{=} -1 & \text{TRUE} \end{array}$$

The solution is $(2, -1, 3)$.

Try Exercise 9.

Student Notes

Because solving systems of three equations can be lengthy, it is important that you use plenty of paper, work in pencil, and double-check each step as you proceed.

Solving Systems of Three Linear Equations

To use the elimination method to solve systems of three linear equations:

1. Write all equations in the standard form $Ax + By + Cz = D$.
2. Clear any decimals or fractions.
3. Choose a variable to eliminate. Then select two of the three equations and work to get one equation in which the selected variable is eliminated.
4. Next, use a different pair of equations and eliminate the same variable that you did in step (3).
5. Solve the system of equations that resulted from steps (3) and (4).
6. Substitute the solution from step (5) into one of the original three equations and solve for the third variable. Then check.

EXAMPLE 3 Solve the system

$$\begin{aligned} 4x - 2y - 3z &= 5, &(1)\\ -8x - y + z &= -5, &(2)\\ 2x + y + 2z &= 5. &(3) \end{aligned}$$

SOLUTION

Write in standard form.

1., 2. The equations are already in standard form with no fractions or decimals.

3. Next, select a variable to eliminate. We decide on y because the y-terms are opposites of each other in equations (2) and (3). We add:

Eliminate a variable. (We choose y.)

$$\begin{aligned} -8x - y + z &= -5 &(2)\\ 2x + y + 2z &= 5 &(3)\\ \hline -6x \quad + 3z &= 0. &(4) \quad \textbf{Adding} \end{aligned}$$

4. We use another pair of equations to create a second equation in x and z. That is, we eliminate the same variable, y, as in step (3). We use equations (1) and (3):

Eliminate the same variable using a different pair of equations.

$$\begin{aligned} 4x - 2y - 3z &= 5,\\ 2x + y + 2z &= 5 \end{aligned} \xrightarrow[\textbf{of equation (3) by 2}]{\textbf{Multiplying both sides}} \begin{aligned} 4x - 2y - 3z &= 5\\ 4x + 2y + 4z &= 10\\ \hline 8x \quad + z &= 15. \quad (5) \end{aligned}$$

5. Now we solve the resulting system of equations (4) and (5). That allows us to find two parts of the ordered triple.

Solve the system of two equations in two variables.

$$\begin{aligned} -6x + 3z &= 0,\\ 8x + z &= 15 \end{aligned} \xrightarrow[\textbf{of equation (5) by } -3]{\textbf{Multiplying both sides}} \begin{aligned} -6x + 3z &= 0\\ -24x - 3z &= -45\\ \hline -30x \quad &= -45\\ x &= \tfrac{-45}{-30} = \tfrac{3}{2} \end{aligned}$$

We use equation (5) to find z:

$$\begin{aligned} 8x + z &= 15 &(5)\\ 8 \cdot \tfrac{3}{2} + z &= 15 &\textbf{Substituting } \tfrac{3}{2} \textbf{ for } x\\ 12 + z &= 15\\ z &= 3. \end{aligned}$$

Solve for the remaining variable.

6. Finally, we use any of the original equations and substitute to find the third number, y. We choose equation (3):

$$\begin{aligned} 2x + y + 2z &= 5 &(3)\\ 2 \cdot \tfrac{3}{2} + y + 2 \cdot 3 &= 5 &\textbf{Substituting } \tfrac{3}{2} \textbf{ for } x \textbf{ and 3 for } z\\ 3 + y + 6 &= 5\\ y + 9 &= 5\\ y &= -4. \end{aligned}$$

Check.

The solution is $\left(\frac{3}{2}, -4, 3\right)$. The check was performed as Example 1.

Try Exercise 23.

Sometimes, certain variables are missing at the outset.

EXAMPLE 4 Solve the system

$$\begin{aligned} x + y + z &= 180, \quad (1)\\ x \qquad - z &= -70, \quad (2)\\ 2y - z &= 0. \quad (3) \end{aligned}$$

SOLUTION

1., 2. The equations appear in standard form with no fractions or decimals.

3., 4. Note that there is no y in equation (2). Thus, at the outset, we already have y eliminated from one equation. We need another equation with y eliminated, so we work with equations (1) and (3):

$$\begin{aligned} x + y + z &= 180, \\ 2y - z &= 0 \end{aligned} \xrightarrow[\text{of equation (1) by } -2]{\text{Multiplying both sides}} \begin{aligned} -2x - 2y - 2z &= -360 \\ 2y - z &= 0 \\ \hline -2x \qquad - 3z &= -360. \quad (4) \end{aligned}$$

5., 6. Now we solve the resulting system of equations (2) and (4):

$$\begin{aligned} x - z &= -70, \\ -2x - 3z &= -360 \end{aligned} \xrightarrow[\text{of equation (2) by 2}]{\text{Multiplying both sides}} \begin{aligned} 2x - 2z &= -140 \\ -2x - 3z &= -360 \\ \hline -5z &= -500 \\ z &= 100. \end{aligned}$$

Continuing as in Examples 2 and 3, we get the solution $(30, 50, 100)$. The check is left to the student.

Try Exercise 29.

DEPENDENCY, INCONSISTENCY, AND GEOMETRIC CONSIDERATIONS

Each equation in Examples 2, 3, and 4 has a graph that is a plane in three dimensions. The solutions are points common to the planes of each system. Since three planes can have an infinite number of points in common or no points at all in common, we need to generalize the concept of *consistency*.

Planes intersect at one point. System is *consistent* and has one solution.

Planes intersect along a common line. System is *consistent* and has an infinite number of solutions.

Three parallel planes. System is *inconsistent;* it has no solution.

Planes intersect two at a time, with no point common to all three. System is *inconsistent;* it has no solution.

Consistency

A system of equations that has at least one solution is said to be **consistent.**

A system of equations that has no solution is said to be **inconsistent.**

EXAMPLE 5 Solve:

$$\begin{aligned} y + 3z &= 4, && (1)\\ -x - y + 2z &= 0, && (2)\\ x + 2y + z &= 1. && (3) \end{aligned}$$

SOLUTION The variable x is missing in equation (1). By adding equations (2) and (3), we can find a second equation in which x is missing:

$$\begin{aligned} -x - y + 2z &= 0 && (2)\\ x + 2y + z &= 1 && (3)\\ \hline y + 3z &= 1. && (4) \quad \textbf{Adding} \end{aligned}$$

Equations (1) and (4) form a system in y and z. We solve as before:

$$\begin{aligned} y + 3z &= 4, \xrightarrow[\text{of equation (1) by } -1]{\text{Multiplying both sides}} & -y - 3z &= -4\\ y + 3z &= 1 & y + 3z &= 1\\ & \text{This is a contradiction.} \longrightarrow & 0 &= -3. \quad \textbf{Adding} \end{aligned}$$

Since we end up with a *false* equation, or contradiction, we know that the system has no solution. It is *inconsistent*.

Try Exercise 15.

STUDY TIP

Guess What Comes Next

If you have at least skimmed over the day's material before you come to class, you will be better able to follow the instructor. As you listen, pay attention to the direction the lecture is taking, and try to predict what topic or idea the instructor will present next. As you take a more active role in listening, you will understand more of the material taught.

The notion of *dependency* from Section 4.1 can also be extended.

EXAMPLE 6 Solve:

$$\begin{aligned} 2x + y + z &= 3, && (1)\\ x - 2y - z &= 1, && (2)\\ 3x + 4y + 3z &= 5. && (3) \end{aligned}$$

SOLUTION Our plan is to first use equations (1) and (2) to eliminate z. Then we will select another pair of equations and again eliminate z:

$$\begin{aligned} 2x + y + z &= 3\\ x - 2y - z &= 1\\ \hline 3x - y \quad &= 4. && (4) \end{aligned}$$

Next, we use equations (2) and (3) to eliminate z again:

$$\begin{aligned} x - 2y - z &= 1, \xrightarrow[\text{of equation (2) by 3}]{\text{Multiplying both sides}} & 3x - 6y - 3z &= 3\\ 3x + 4y + 3z &= 5 & 3x + 4y + 3z &= 5\\ & & 6x - 2y \quad &= 8. && (5) \end{aligned}$$

We now try to solve the resulting system of equations (4) and (5):

$$\begin{aligned} 3x - y &= 4, \xrightarrow[\text{of equation (4) by } -2]{\text{Multiplying both sides}} & -6x + 2y &= -8\\ 6x - 2y &= 8 & 6x - 2y &= 8\\ & & 0 &= 0. && (6) \end{aligned}$$

Equation (6), which is an identity, indicates that equations (1), (2), and (3) are *dependent*. This means that the original system of three equations is equivalent to a system of two equations. Note here that two times equation (1), minus equation (2), is equation (3). Thus removing equation (3) from the system does not affect the solution of the system.* In writing an answer to this problem, we simply state that "the equations are dependent."

Try Exercise 21.

Recall that when dependent equations appeared in Section 4.1, the solution sets were always infinite in size and were written in set-builder notation. There, all systems of dependent equations were *consistent*. This is not always the case for systems of three or more equations. The following figures illustrate some possibilities geometrically.

The planes intersect along a common line. The equations are dependent and the system is consistent. There is an infinite number of solutions.

The planes coincide. The equations are dependent and the system is consistent. There is an infinite number of solutions.

Two planes coincide. The third plane is parallel. The equations are dependent and the system is inconsistent. There is no solution.

9.1 Exercise Set

FOR EXTRA HELP MyMathLab

Concept Reinforcement *Classify each of the following statements as either true or false.*

1. $3x + 5y + 4z = 7$ is a linear equation in three variables.

2. It is not difficult to solve a system of three equations in three unknowns by graphing.

3. Every system of three equations in three unknowns has at least one solution.

4. If, when we are solving a system of three equations, a false equation results from adding a multiple of one equation to another, the system is inconsistent.

5. If, when we are solving a system of three equations, an identity results from adding a multiple of one equation to another, the equations are dependent.

6. Whenever a system of three equations contains dependent equations, there is an infinite number of solutions.

7. Determine whether $(2, -1, -2)$ is a solution of the system

$$\begin{aligned} x + y - 2z &= 5, \\ 2x - y - z &= 7, \\ -x - 2y - 3z &= 6. \end{aligned}$$

8. Determine whether $(-1, -3, 2)$ is a solution of the system

$$\begin{aligned} x - y + z &= 4, \\ x - 2y - z &= 3, \\ 3x + 2y - z &= 1. \end{aligned}$$

Solve each system. If a system's equations are dependent or if there is no solution, state this.

9. $$\begin{aligned} x - y - z &= 0, \\ 2x - 3y + 2z &= 7, \\ -x + 2y + z &= 1 \end{aligned}$$

10. $$\begin{aligned} x + y - z &= 0, \\ 2x - y + z &= 3, \\ -x + 5y - 3z &= 2 \end{aligned}$$

---

*A set of equations is dependent if at least one equation can be expressed as a sum of multiples of other equations in that set.

11. $x - y - z = 1,$
$2x + y + 2z = 4,$
$x + y + 3z = 5$

12. $x + y - 3z = 4,$
$2x + 3y + z = 6,$
$2x - y + z = -14$

13. $3x + 4y - 3z = 4,$
$5x - y + 2z = 3,$
$x + 2y - z = -2$

14. $2x - 3y + z = 5,$
$x + 3y + 8z = 22,$
$3x - y + 2z = 12$

15. $x + y + z = 0,$
$2x + 3y + 2z = -3,$
$-x - 2y - z = 1$

16. $3a - 2b + 7c = 13,$
$a + 8b - 6c = -47,$
$7a - 9b - 9c = -3$

17. $2x - 3y - z = -9,$
$2x + 5y + z = 1,$
$x - y + z = 3$

18. $4x + y + z = 17,$
$x - 3y + 2z = -8,$
$5x - 2y + 3z = 5$

Aha! 19. $a + b + c = 5,$
$2a + 3b - c = 2,$
$2a + 3b - 2c = 4$

20. $u - v + 6w = 8,$
$3u - v + 6w = 14,$
$-u - 2v - 3w = 7$

21. $-2x + 8y + 2z = 4,$
$x + 6y + 3z = 4,$
$3x - 2y + z = 0$

22. $x - y + z = 4,$
$5x + 2y - 3z = 2,$
$4x + 3y - 4z = -2$

23. $2u - 4v - w = 8,$
$3u + 2v + w = 6,$
$5u - 2v + 3w = 2$

24. $4a + b + c = 3,$
$2a - b + c = 6,$
$2a + 2b - c = -9$

25. $r + \frac{3}{2}s + 6t = 2,$
$2r - 3s + 3t = 0.5,$
$r + s + t = 1$

26. $5x + 3y + \frac{1}{2}z = \frac{7}{2},$
$0.5x - 0.9y - 0.2z = 0.3,$
$3x - 2.4y + 0.4z = -1$

27. $4a + 9b = 8,$
$8a + 6c = -1,$
$6b + 6c = -1$

28. $3u + 2w = 11,$
$v - 7w = 4,$
$u - 6v = 1$

29. $x + y + z = 57,$
$-2x + y = 3,$
$x - z = 6$

30. $x + y + z = 105,$
$10y - z = 11,$
$2x - 3y = 7$

31. $a - 3c = 6,$
$b + 2c = 2,$
$7a - 3b - 5c = 14$

32. $2a - 3b = 2,$
$7a + 4c = \frac{3}{4},$
$2c - 3b = 1$

Aha! 33. $x + y + z = 83,$
$y = 2x + 3,$
$z = 40 + x$

34. $l + m = 7,$
$3m + 2n = 9,$
$4l + n = 5$

35. $x + z = 0,$
$x + y + 2z = 3,$
$y + z = 2$

36. $x + y = 0,$
$x + z = 1,$
$2x + y + z = 2$

37. $x + y + z = 1,$
$-x + 2y + z = 2,$
$2x - y = -1$

38. $y + z = 1,$
$x + y + z = 1,$
$x + 2y + 2z = 2$

TW 39. Rondel always begins solving systems of three equations in three variables by using the first two equations to eliminate x. Is this a good approach? Why or why not?

TW 40. Describe a method for writing an inconsistent system of three equations in three variables.

SKILL REVIEW

To prepare for Section 9.2, review translating sentences to equations (Section 1.1).

Translate each sentence to an equation. [1.1]

41. One number is half another.

42. The difference of two numbers is twice the first number.

43. The sum of three consecutive numbers is 100.

44. The sum of three numbers is 100.

45. The product of two numbers is five times a third number.

46. The product of two numbers is twice their sum.

SYNTHESIS

TW 47. Is it possible for a system of three linear equations to have exactly two ordered triples in its solution set? Why or why not?

TW 48. Kadi and Ahmed both correctly solve the system

$$x + 2y - z = 1,$$
$$-x - 2y + z = 3,$$
$$2x + 4y - 2z = 2.$$

Kadi states "the equations are dependent" while Ahmed states "there is no solution." How did each person reach the conclusion?

Solve.

49. $\frac{x+2}{3} - \frac{y+4}{2} + \frac{z+1}{6} = 0,$
$\frac{x-4}{3} + \frac{y+1}{4} + \frac{z-2}{2} = -1,$
$\frac{x+1}{2} + \frac{y}{2} + \frac{z-1}{4} = \frac{3}{4}$

50. $w + x + y + z = 2,$
$w + 2x + 2y + 4z = 1,$
$w - x + y + z = 6,$
$w - 3x - y + z = 2$

51. $w + x - y + z = 0,$
$w - 2x - 2y - z = -5,$
$w - 3x - y + z = 4,$
$2w - x - y + 3z = 7$

For Exercises 52 and 53, let u represent $1/x$, v represent $1/y$, and w represent $1/z$. Solve for u, v, and w, and then solve for x, y, and z.

52. $\frac{2}{x} - \frac{1}{y} - \frac{3}{z} = -1,$
$\frac{2}{x} - \frac{1}{y} + \frac{1}{z} = -9,$
$\frac{1}{x} + \frac{2}{y} - \frac{4}{z} = 17$

53. $\frac{2}{x} + \frac{2}{y} - \frac{3}{z} = 3,$
$\frac{1}{x} - \frac{2}{y} - \frac{3}{z} = 9,$
$\frac{7}{x} - \frac{2}{y} + \frac{9}{z} = -39$

Determine k so that each system is dependent.

54. $x - 3y + 2z = 1,$
$2x + y - z = 3,$
$9x - 6y + 3z = k$

55. $5x - 6y + kz = -5,$
$x + 3y - 2z = 2,$
$2x - y + 4z = -1$

In each case, three solutions of an equation in x, y, and z are given. Find the equation.

56. $Ax + By + Cz = 12;$
$\left(1, \frac{3}{4}, 3\right), \left(\frac{4}{3}, 1, 2\right)$, and $(2, 1, 1)$

57. $z = b - mx - ny;$
$(1, 1, 2), (3, 2, -6)$, and $\left(\frac{3}{2}, 1, 1\right)$

58. Write an inconsistent system of equations that contains dependent equations.

Try Exercise Answers: Section 9.1

7. Yes **9.** $(3, 1, 2)$ **15.** No solution **21.** The equations are dependent. **23.** $\left(3, \frac{1}{2}, -4\right)$ **29.** $(15, 33, 9)$

Collaborative Corner

Finding the Preferred Approach

Focus: Systems of three linear equations
Time: 10–15 minutes
Group Size: 3

Consider the six steps outlined on p. 650 along with the following system:

$$2x + 4y = 3 - 5z,$$
$$0.3x = 0.2y + 0.7z + 1.4,$$
$$0.04x + 0.03y = 0.07 + 0.04z.$$

ACTIVITY

1. Working independently, each group member should solve the system above. One person should begin by eliminating x, one should first eliminate y, and one should first eliminate z. Write neatly so that others can follow your steps.
2. Once all group members have solved the system, compare your answers. If the answers do not check, exchange notebooks and check each other's work. If a mistake is detected, allow the person who made the mistake to make the repair.
3. Decide as a group which of the three approaches above (if any) ranks as easiest and which (if any) ranks as most difficult. Then compare your rankings with the other groups in the class.

9.2 Solving Applications: Systems of Three Equations

- Applications of Three Equations in Three Unknowns

Solving systems of three or more equations is important in many applications. Such systems arise in the natural and social sciences, business, and engineering. To begin, let's look at a purely numerical application.

EXAMPLE 1 The sum of three numbers is 4. The first number minus twice the second, minus the third is 1. Twice the first number minus the second, minus twice the third is -1. Find the numbers.

STUDY TIP

You've Got Mail

If your instructor makes his or her e-mail address available, consider using it to get help. Often, just the act of writing out your question brings some clarity. If you do use e-mail, allow some time for your instructor to reply.

SOLUTION

1. **Familiarize.** There are three statements involving the same three numbers. Let's label these numbers x, y, and z.
2. **Translate.** We can translate directly as follows.

The sum of the three numbers is 4.

$$x + y + z = 4$$

The first number minus twice the second minus the third is 1.

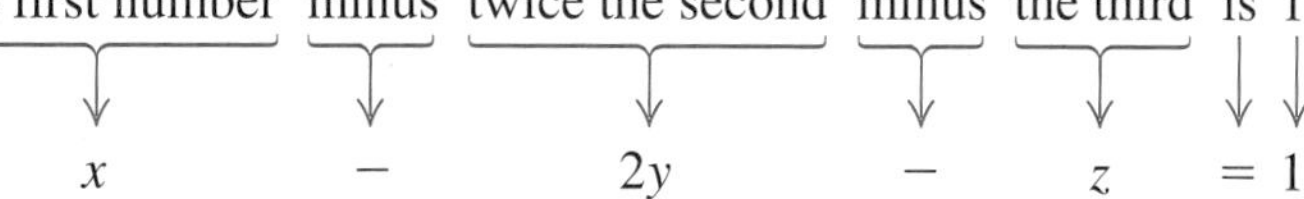

$$x - 2y - z = 1$$

Twice the first number minus the second minus twice the third is -1.

$$2x - y - 2z = -1$$

We now have a system of three equations:

$$\begin{aligned} x + y + z &= 4, \\ x - 2y - z &= 1, \\ 2x - y - 2z &= -1. \end{aligned}$$

3. **Carry out.** We need to solve the system of equations. Note that we found the solution, $(2, -1, 3)$, in Example 2 of Section 9.1.
4. **Check.** The first statement of the problem says that the sum of the three numbers is 4. That checks, because $2 + (-1) + 3 = 4$. The second statement says that the first number minus twice the second, minus the third is 1: $2 - 2(-1) - 3 = 1$. That checks. The check of the third statement is left to the student.
5. **State.** The three numbers are 2, -1, and 3.

Try Exercise 1.

EXAMPLE 2 Architecture. In a triangular cross section of a roof, the largest angle is 70° greater than the smallest angle. The largest angle is twice as large as the remaining angle. Find the measure of each angle.

SOLUTION

1. **Familiarize.** The first thing we do is make a drawing, or a sketch.

Since we don't know the size of any angle, we use x, y, and z to represent the three measures, from smallest to largest. Recall that the measures of the angles in any triangle add up to 180°.

Student Notes

It is quite likely that you are expected to remember that the sum of the measures of the angles in any triangle is 180°. You may want to ask your instructor which other formulas from geometry and elsewhere you are expected to know.

2. **Translate.** This geometric fact about triangles gives us one equation:

$$x + y + z = 180.$$

Two of the statements can be translated almost directly.

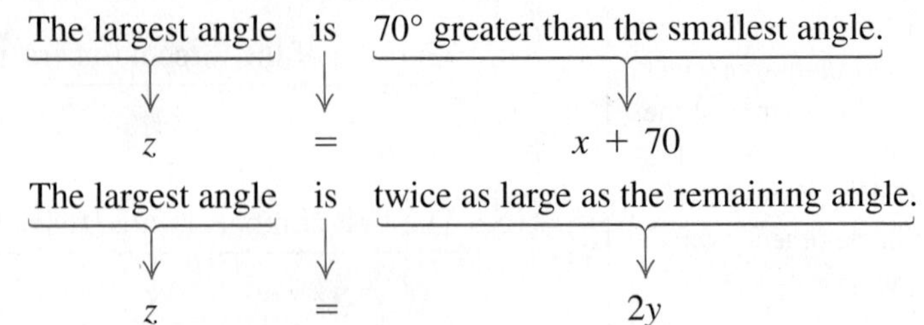

The largest angle is 70° greater than the smallest angle.

$$z = x + 70$$

The largest angle is twice as large as the remaining angle.

$$z = 2y$$

We now have a system of three equations:

$$\begin{aligned} x + y + z &= 180, \\ x + 70 &= z, \\ 2y &= z; \end{aligned} \quad \text{or} \quad \begin{aligned} x + y + z &= 180, \\ x \qquad - z &= -70, \\ 2y - z &= 0, \end{aligned}$$

Rewriting in standard form

3. **Carry out.** The system was solved in Example 4 of Section 9.1. The solution is $(30, 50, 100)$.
4. **Check.** The sum of the numbers is 180, so that checks. The measure of the largest angle, 100°, is 70° greater than the measure of the smallest angle, 30°, so that checks. The measure of the largest angle is also twice the measure of the remaining angle, 50°. Thus we have a check.
5. **State.** The angles in the triangle measure 30°, 50° and 100°.

Try Exercise 5.

EXAMPLE 3 Downloads. Kaya frequently downloads music, TV shows, and iPod games. In January, she downloaded 5 songs, 10 TV shows, and 3 games for a total of \$40. In February, she spent a total of \$135 for 25 songs, 25 TV shows, and 12 games. In March, she spent a total of \$56 for 15 songs, 8 TV shows, and 5 games. Assuming each song is the same price, each TV show is the same price, and each iPod game is the same price, how much does each cost?
Source: Based on information from www.iTunes.com

SOLUTION

1. **Familiarize.** We let s = the cost, in dollars, per song, t = the cost, in dollars, per TV show, and g = the cost, in dollars, per game. Then in January, Kaya spent $5 \cdot s$ for songs, $10 \cdot t$ for TV shows, and $3 \cdot g$ for iPod games. The sum of these amounts was \$40. Each month's downloads will translate to an equation.
2. **Translate.** We can organize the information in a table.

	Cost of Songs	Cost of TV Shows	Cost of iPod Games	Total Cost	
January	$5s$	$10t$	$3g$	40	$\longrightarrow 5s + 10t + 3g = 40$
February	$25s$	$25t$	$12g$	135	$\longrightarrow 25s + 25t + 12g = 135$
March	$15s$	$8t$	$5g$	56	$\longrightarrow 15s + 8t + 5g = 56$

We now have a system of three equations:

$$\begin{aligned} 5s + 10t + 3g &= 40, \quad (1)\\ 25s + 25t + 12g &= 135, \quad (2)\\ 15s + 8t + 5g &= 56. \quad (3) \end{aligned}$$

3. **Carry out.** We begin by using equations (1) and (2) to eliminate s.

$$\begin{aligned} 5s + 10t + 3g &= 40,\\ 25s + 25t + 12g &= 135 \end{aligned} \quad \xrightarrow[\text{of equation (1) by } -5]{\text{Multiplying both sides}} \quad \begin{aligned} -25s - 50t - 15g &= -200\\ 25s + 25t + 12g &= 135\\ \hline -25t - 3g &= -65 \quad (4) \end{aligned}$$

We then use equations (1) and (3) to again eliminate s.

$$\begin{aligned} 5s + 10t + 3g &= 40,\\ 15s + 8t + 5g &= 56 \end{aligned} \quad \xrightarrow[\text{of equation (1) by } -3]{\text{Multiplying both sides}} \quad \begin{aligned} -15s - 30t - 9g &= -120\\ 15s + 8t + 5g &= 56\\ \hline -22t - 4g &= -64 \quad (5) \end{aligned}$$

Now we solve the resulting system of equations (4) and (5).

$$\begin{aligned} -25t - 3g = -65 \quad &\xrightarrow[\text{of equation (4) by } -4]{\text{Multiplying both sides}} \quad 100t + 12g = 260\\ -22t - 4g = -64 \quad &\xrightarrow[\text{of equation (5) by } 3]{\text{Multiplying both sides}} \quad \underline{-66t - 12g = -192}\\ &\qquad\qquad\qquad\qquad\qquad 34t = 68\\ &\qquad\qquad\qquad\qquad\qquad t = 2 \end{aligned}$$

To find g, we use equation (4):

$$\begin{aligned} -25t - 3g &= -65\\ -25 \cdot 2 - 3g &= -65 \quad \text{Substituting 2 for } t\\ -50 - 3g &= -65\\ -3g &= -15\\ g &= 5. \end{aligned}$$

Finally, we use equation (1) to find s:

$$\begin{aligned} 5s + 10t + 3g &= 40\\ 5s + 10 \cdot 2 + 3 \cdot 5 &= 40 \quad \text{Substituting 2 for } t \text{ and 5 for } g\\ 5s + 20 + 15 &= 40\\ 5s + 35 &= 40\\ 5s &= 5\\ s &= 1. \end{aligned}$$

4. **Check.** If a song costs \$1, a TV show costs \$2, and an iPod game costs \$5, then the total cost for each month's downloads is as follows:

January: $5 \cdot \$1 + 10 \cdot \$2 + 3 \cdot \$5 = \$5 + \$20 + \$15 = \$40;$
February: $25 \cdot \$1 + 25 \cdot \$2 + 12 \cdot \$5 = \$25 + \$50 + \$60 = \$135;$
March: $15 \cdot \$1 + 8 \cdot \$2 + 5 \cdot \$5 = \$15 + \$16 + \$25 = \$56.$

This checks with the information given in the problem.

5. **State.** A song costs \$1, a TV show costs \$2, and an iPod game costs \$5.

Try Exercise 9.

9.2 Exercise Set

FOR EXTRA HELP MyMathLab MathXL PRACTICE   WATCH  DOWNLOAD READ 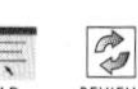 REVIEW

Solve.

1. The sum of three numbers is 57. The second is 3 more than the first. The third is 6 more than the first. Find the numbers.

2. The sum of three numbers is 5. The first number minus the second plus the third is 1. The first minus the third is 3 more than the second. Find the numbers.

3. The sum of three numbers is 26. Twice the first minus the second is 2 less than the third. The third is the second minus three times the first. Find the numbers.

4. The sum of three numbers is 105. The third is 11 less than ten times the second. Twice the first is 7 more than three times the second. Find the numbers.

5. *Geometry.* In triangle ABC, the measure of angle B is three times that of angle A. The measure of angle C is 20° more than that of angle A. Find the angle measures.

6. *Geometry.* In triangle ABC, the measure of angle B is twice the measure of angle A. The measure of angle C is 80° more than that of angle A. Find the angle measures.

7. *Scholastic Aptitude Test.* Many high-school students take the Scholastic Aptitude Test (SAT). Beginning in March 2005, students taking the SAT received three scores: a critical reading score, a mathematics score, and a writing score. The average total score of 2009 high-school seniors who took the SAT was 1509. The average mathematics score exceeded the reading score by 14 points and the average writing score was 8 points less than the reading score. What was the average score for each category?
Source: College Entrance Examination Board

8. *Advertising.* In 2008, U.S. companies spent a total of \$118.2 billion on newspaper, television, and magazine ads. The total amount spent on television ads was \$10.8 billion more than the amount spent on newspaper and magazine ads together. The amount spent on magazine ads was \$3.5 billion more than the amount spent on newspaper ads. How much was spent on each form of advertising?
Source: TNS Media Intelligence

9. *Nutrition.* Most nutritionists now agree that a healthy adult diet should include 25–35 g of fiber each day. A breakfast of 2 bran muffins, 1 banana, and a 1-cup serving of Wheaties® contains 9 g of fiber; a breakfast of 1 bran muffin, 2 bananas, and a 1-cup serving of Wheaties® contains 10.5 g of fiber; and a breakfast of 2 bran muffins and a 1-cup serving of Wheaties® contains 6 g of fiber. How much fiber is in each of these foods?
Sources: usda.gov; InteliHealth.com

10. *Nutrition.* Refer to Exercise 9. A breakfast consisting of 2 pancakes and a 1-cup serving of strawberries contains 4.5 g of fiber, whereas a breakfast of 2 pancakes and a 1-cup serving of Cheerios® contains 4 g of fiber. When a meal consists of 1 pancake, a 1-cup serving of Cheerios®, and a 1-cup serving of strawberries, it contains 7 g of fiber. How much fiber is in each of these foods?
Source: InteliHealth.com

Aha! **11.** *Automobile Pricing.* The basic model of a 2010 Honda Civic Hybrid with a car cover cost \$24,030. When equipped with satellite radio and a car cover, the vehicle's price rose to \$24,340. The price of the basic model with satellite radio was \$24,110. Find the basic price, the price of satellite radio, and the price of a car cover.

12. *Telemarketing.* Sven, Laurie, and Isaiah can process 740 telephone orders per day. Sven and Laurie together can process 470 orders, while Laurie and Isaiah together can process 520 orders per day. How many orders can each person process alone?

13. *Coffee Prices.* Reba works at a Starbucks® coffee shop where a 12-oz cup of coffee costs \$1.65, a 16-oz cup costs \$1.85, and a 20-oz cup costs \$1.95. During one busy period, Reba served 55 cups of coffee, emptying six 144-oz "brewers" while collecting a total of \$99.65. How many cups of each size did Reba fill?

14. *Restaurant Management.* Chick-fil-A® recently sold small lemonades for \$1.29, medium lemonades for \$1.49, and large lemonades for \$1.85. During a lunch-time rush, Chris sold 40 lemonades for a total of \$59.40. The number of small drinks and large drinks, combined, was 10 fewer than the number of medium drinks. How many drinks of each size were sold?

15. *Small-Business Loans.* Chelsea took out three loans for a total of \$120,000 to start an organic orchard. Her bank loan was at an interest rate of 8%, the small-business loan was at an interest rate of 5%, and the mortgage on her house was at an interest rate of 4%. The total simple interest due on the loans in one year was \$5750. The annual simple interest on the mortgage was \$1600 more than the interest on the bank loan. How much did she borrow from each source?

16. *Investments.* A business class divided an imaginary investment of \$80,000 among three mutual funds. The first fund grew by 10%, the second by 6%, and the third by 15%. Total earnings were \$8850. The earnings from the first fund were \$750 more than the earnings from the third. How much was invested in each fund?

17. *Gold Alloys.* Gold used to make jewelry is often a blend of gold, silver, and copper. The relative amounts of the metals determine the color of the alloy. Red gold is 75% gold, 5% silver, and 20% copper. Yellow gold is 75% gold, 12.5% silver, and 12.5% copper. White gold is 37.5% gold and 62.5% silver. If 100 g of red gold costs \$2265.40, 100 g of yellow gold costs \$2287.75, and 100 g of white gold costs \$1312.50, how much do gold, silver, and copper cost?
Source: World Gold Council

18. *Blending Teas.* Verity has recently created three custom tea blends. A 5-oz package of Southern Sandalwood sells for \$13.15 and contains 2 oz of Keemun tea, 2 oz of Assam tea, and 1 oz of a berry blend. A 4-oz package of Golden Sunshine sells for \$12.50 and contains 3 oz of Assam tea and 1 oz of the berry blend. A 6-oz package of Mountain Morning sells for \$12.50 and contains 2 oz of the berry blend, 3 oz of Keemun tea, and 1 oz of Assam tea. What is the price per ounce of Keemun tea, Assam tea, and the berry blend?

19. *Nutrition.* A dietician in a hospital prepares meals under the guidance of a physician. Suppose that for a particular patient a physician prescribes a meal to have 800 calories, 55 g of protein, and 220 mg of vitamin C. The dietician prepares a meal of roast beef, baked potatoes, and broccoli according to the data in the table below.

Serving Size	Calories	Protein (in grams)	Vitamin C (in milligrams)
Roast Beef, 3 oz	300	20	0
Baked Potato, 1	100	5	20
Broccoli, 156 g	50	5	100

How many servings of each food are needed in order to satisfy the doctor's orders?

20. *Nutrition.* Repeat Exercise 19 but replace the broccoli with asparagus, for which a 180-g serving contains 50 calories, 5 g of protein, and 44 mg of vitamin C. Which meal would you prefer eating?

21. *Concert Tickets.* Students in a Listening Responses class bought 40 tickets for a piano concert. The number of tickets purchased for seats either in the first mezzanine or on the main floor was the same as the number purchased for seats in the second mezzanine. First mezzanine seats cost \$52, main floor seats cost \$38, and second mezzanine seats cost \$28. The total cost of the tickets was \$1432. How many of each type of ticket were purchased?

22. *World Population Growth.* The world population is projected to be 9.4 billion in 2050. At that time, there is expected to be approximately 3.5 billion more people in Asia than in Africa. The population for the rest of the world will be approximately 0.3 billion less than two-fifths the population of Asia. Find the projected populations of Asia, Africa, and the rest of the world in 2050.
Source: U.S. Census Bureau

23. *Basketball Scoring.* The New York Knicks recently scored a total of 92 points on a combination of 2-point field goals, 3-point field goals, and 1-point foul shots. Altogether, the Knicks made 50 baskets and 19 more 2-pointers than foul shots. How many shots of each kind were made?

24. *History.* Find the year in which the first U.S. transcontinental railroad was completed. The following are some facts about the number. The sum of the digits in the year is 24. The ones digit is 1 more than the hundreds digit. Both the tens and the ones digits are multiples of 3.

TW **25.** Problems like Exercises 13 and 14 could be classified as total-value problems. How do these problems differ from the total-value problems of Section 4.4?

TW **26.** Write a problem for a classmate to solve. Design the problem so that it translates to a system of three equations in three variables.

SKILL REVIEW

To prepare for Section 9.3, review simplifying expressions (Section 1.8).

Simplify. [1.8]

27. $-2(2x - 3y)$

28. $-(x - 6y)$

29. $-6(x - 2y) + (6x - 5y)$

30. $3(2a + 4b) + (5a - 12b)$

31. $-(2a - b - 6c)$

32. $-10(5a + 3b - c)$

33. $-2(3x - y + z) + 3(-2x + y - 2z)$

34. $(8x - 10y + 7z) + 5(3x + 2y - 4z)$

SYNTHESIS

TW **35.** Consider Exercise 23. Suppose there were no foul shots made. Would there still be a solution? Why or why not?

TW **36.** Consider Exercise 13. Suppose Reba collected $50. Could the problem still be solved? Why or why not?

37. *Health Insurance.* In 2010, United Health One insurance for a 45-year-old and his or her spouse cost $135 per month. That rate increased to $154 per month if a child were included and $173 per month if two children were included. The rate dropped to $102 per month for just the applicant and one child. Find the separate costs for insuring the applicant, the spouse, the first child, and the second child.
Source: United Health One Insurance Company® through www.ehealth.com

38. Find a three-digit positive integer such that the sum of all three digits is 14, the tens digit is 2 more than the ones digit, and if the digits are reversed, the number is unchanged.

39. *Ages.* Tammy's age is the sum of the ages of Carmen and Dennis. Carmen's age is 2 more than the sum of the ages of Dennis and Mark. Dennis's age is four times Mark's age. The sum of all four ages is 42. How old is Tammy?

40. *Ticket Revenue.* A magic show's audience of 100 people consists of adults, students, and children. The ticket prices are $10 for adults, $3 for students, and 50¢ for children. The total amount of money taken in is $100. How many adults, students, and children are in attendance? Does there seem to be some information missing? Do some more careful reasoning.

41. *Sharing Raffle Tickets.* Hal gives Tom as many raffle tickets as Tom first had and Gary as many as Gary first had. In like manner, Tom then gives Hal and Gary as many tickets as each then has. Similarly, Gary gives Hal and Tom as many tickets as each then has. If each finally has 40 tickets, with how many tickets does Tom begin?

42. Find the sum of the angle measures at the tips of the star in this figure.

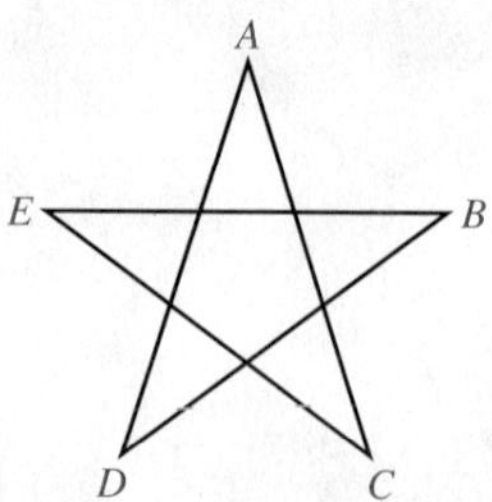

Try Exercise Answers: Section 9.2

1. 16, 19, 22 **5.** 32°, 96°, 52° **9.** Bran muffin: 1.5 g; banana: 3 g; 1 cup of Wheaties: 3 g

9.3 Elimination Using Matrices

- Matrices and Systems
- Row-Equivalent Operations

STUDY TIP

Double-check the Numbers

Solving problems is challenging enough, without miscopying information. Always double-check that you have accurately transferred numbers from the correct exercise in the exercise set.

In solving systems of equations, we perform computations with the constants. If we keep all like terms in the same column, we can simplify writing a system by omitting the variables. For example, the system

$$\begin{aligned} 3x + 4y &= 5, \\ x - 2y &= 1 \end{aligned} \quad \text{simplifies to} \quad \begin{matrix} 3 & 4 & 5 \\ 1 & -2 & 1 \end{matrix}$$

if we do not write the variables, the operation of addition, and the equals signs. Note that the coefficients of the x-terms are written first, the coefficients of the y-terms are written next, and the constant terms are written last. Equations must be written in standard form $Ax + By = C$ before they are written without variables.

MATRICES AND SYSTEMS

In the example above, we have written a rectangular array of numbers. Such an array is called a **matrix** (plural, **matrices**). We ordinarily write brackets around matrices. The following are matrices:

$$\begin{bmatrix} -3 & 1 \\ 0 & 5 \end{bmatrix}, \quad \begin{bmatrix} 2 & 0 & -1 & 3 \\ -5 & 2 & 7 & -1 \\ 4 & 5 & 3 & 0 \end{bmatrix}, \quad \begin{bmatrix} 2 & 3 \\ 7 & 15 \\ -2 & 23 \\ 4 & 1 \end{bmatrix}.$$

The individual numbers are called *elements* or *entries*.

The **rows** of a matrix are horizontal, and the **columns** are vertical.

$$A = \begin{bmatrix} 5 & -2 & -2 \\ 1 & 0 & 1 \\ 4 & -3 & 2 \end{bmatrix} \begin{matrix} \leftarrow \text{row 1} \\ \leftarrow \text{row 2} \\ \leftarrow \text{row 3} \end{matrix}$$

column 1 column 2 column 3

Let's see how matrices can be used to solve a system.

EXAMPLE 1 Solve the system

$$\begin{aligned} 5x - 4y &= -1, \\ -2x + 3y &= 2. \end{aligned}$$

As an aid for understanding, we list the corresponding system in the margin.

$$\begin{aligned} 5x - 4y &= -1, \\ -2x + 3y &= 2 \end{aligned}$$

SOLUTION We write a matrix using only coefficients and constants, listing x-coefficients in the first column and y-coefficients in the second. A dashed line separates the coefficients from the constants:

$$\left[\begin{array}{cc:c} 5 & -4 & -1 \\ -2 & 3 & 2 \end{array}\right].$$

Refer to the notes in the margin for further information.

Our goal is to transform

$$\left[\begin{array}{cc:c} 5 & -4 & -1 \\ -2 & 3 & 2 \end{array}\right] \quad \text{into the form} \quad \left[\begin{array}{cc:c} a & b & c \\ 0 & d & e \end{array}\right].$$

We can then reinsert the variables x and y, form equations, and complete the solution.

Our calculations are similar to those that we would do if we wrote the entire equations. The first step is to multiply and/or interchange the rows so that each number in the first column below the first number is a multiple of that number. Here that means multiplying Row 2 by 5. This corresponds to multiplying both sides of the second equation by 5.

$$5x - 4y = -1,$$
$$-10x + 15y = 10$$

$$\left[\begin{array}{cc:c} 5 & -4 & -1 \\ -10 & 15 & 10 \end{array}\right]$$

New Row 2 = 5(Row 2 from the step above) = 5(−2 3 ⋮ 2) = (−10 15 ⋮ 10)

Next, we multiply the first row by 2, add this to Row 2, and write that result as the "new" Row 2. This corresponds to multiplying the first equation by 2 and adding the result to the second equation in order to eliminate a variable. Write out these computations as necessary.

$$5x - 4y = -1,$$
$$7y = 8$$

$$\left[\begin{array}{cc:c} 5 & -4 & -1 \\ 0 & 7 & 8 \end{array}\right]$$

2(Row 1) = 2(5 −4 ⋮ −1) = (10 −8 ⋮ −2)
New Row 2 = (10 −8 ⋮ −2) + (−10 15 ⋮ 10) = (0 7 ⋮ 8)

If we now reinsert the variables, we have

$$5x - 4y = -1, \quad (1) \quad \textbf{From Row 1}$$
$$7y = 8. \quad (2) \quad \textbf{From Row 2}$$

Solving equation (2) for y gives us

$$7y = 8 \quad (2)$$
$$y = \tfrac{8}{7}.$$

Next, we substitute $\frac{8}{7}$ for y in equation (1):

$$5x - 4y = -1 \quad (1)$$
$$5x - 4 \cdot \tfrac{8}{7} = -1 \quad \textbf{Substituting } \tfrac{8}{7} \textbf{ for } y \textbf{ in equation (1)}$$
$$x = \tfrac{5}{7}. \quad \textbf{Solving for } x$$

The solution is $\left(\frac{5}{7}, \frac{8}{7}\right)$. The check is left to the student.

Try Exercise 7.

All the systems of equations shown in the margin by Example 1 are **equivalent systems of equations;** that is, they all have the same solution.

EXAMPLE 2 Solve the system

$$2x - y + 4z = -3,$$
$$x \quad - 4z = 5,$$
$$6x - y + 2z = 10.$$

SOLUTION We first write a matrix, using only the constants. Where there are missing terms, we must write 0's:

$$2x - y + 4z = -3,$$
$$x \quad - 4z = 5,$$
$$6x - y + 2z = 10$$

$$\left[\begin{array}{ccc:c} 2 & -1 & 4 & -3 \\ 1 & 0 & -4 & 5 \\ 6 & -1 & 2 & 10 \end{array}\right].$$

Note that the x-coefficients are in column 1, the y-coefficients in column 2, the z-coefficients in column 3, and the constant terms in the last column.

Our goal is to transform the matrix to one of the form

$$ax + by + cz = d,$$
$$ey + fz = g,$$
$$hz = i$$

$$\left[\begin{array}{ccc:c} a & b & c & d \\ 0 & e & f & g \\ 0 & 0 & h & i \end{array}\right].$$

This matrix is in a form called *row-echelon form*.

A matrix of this form can be rewritten as a system of equations that is equivalent to the original system, and from which a solution can be easily found.

The first step is to multiply and/or interchange the rows so that each number in the first column is a multiple of the first number in the first row. In this case, we do so by interchanging Rows 1 and 2:

$$\begin{aligned} x \quad\quad - 4z &= 5, \\ 2x - y + 4z &= -3, \\ 6x - y + 2z &= 10 \end{aligned}$$

$$\left[\begin{array}{rrr|r} 1 & 0 & -4 & 5 \\ 2 & -1 & 4 & -3 \\ 6 & -1 & 2 & 10 \end{array}\right].$$

This corresponds to interchanging the first two equations.

Next, we multiply the first row by -2, add it to the second row, and replace Row 2 with the result:

$$\begin{aligned} x \quad\quad - 4z &= 5, \\ -y + 12z &= -13, \\ 6x - y + 2z &= 10 \end{aligned}$$

$$\left[\begin{array}{rrr|r} 1 & 0 & -4 & 5 \\ 0 & -1 & 12 & -13 \\ 6 & -1 & 2 & 10 \end{array}\right].$$

$\mathbf{-2(1 \quad 0 \quad -4 \mid 5) = (-2 \quad 0 \quad 8 \mid -10)}$ **and** $\mathbf{(-2 \quad 0 \quad 8 \mid -10) + (2 \quad -1 \quad 4 \mid -3) = (0 \quad -1 \quad 12 \mid -13)}$

Now we multiply the first row by -6, add it to the third row, and replace Row 3 with the result:

$$\begin{aligned} x \quad\quad - 4z &= 5, \\ -y + 12z &= -13, \\ -y + 26z &= -20 \end{aligned}$$

$$\left[\begin{array}{rrr|r} 1 & 0 & -4 & 5 \\ 0 & -1 & 12 & -13 \\ 0 & -1 & 26 & -20 \end{array}\right].$$

$\mathbf{-6(1 \quad 0 \quad -4 \mid 5) = (-6 \quad 0 \quad 24 \mid -30)}$ **and** $\mathbf{(-6 \quad 0 \quad 24 \mid -30) + (6 \quad -1 \quad 2 \mid 10) = (0 \quad -1 \quad 26 \mid -20)}$

Next, we multiply Row 2 by -1, add it to the third row, and replace Row 3 with the result:

$$\begin{aligned} x \quad\quad - 4z &= 5, \\ -y + 12z &= -13, \\ 14z &= -7 \end{aligned}$$

$$\left[\begin{array}{rrr|r} 1 & 0 & -4 & 5 \\ 0 & -1 & 12 & -13 \\ 0 & 0 & 14 & -7 \end{array}\right].$$

$\mathbf{-1(0 \quad -1 \quad 12 \mid -13) = (0 \quad 1 \quad -12 \mid 13)}$ **and** $\mathbf{(0 \quad 1 \quad -12 \mid 13) + (0 \quad -1 \quad 26 \mid -20) = (0 \quad 0 \quad 14 \mid -7)}$

Reinserting the variables gives us

$$\begin{aligned} x \quad\quad - 4z &= 5, \\ -y + 12z &= -13, \\ 14z &= -7. \end{aligned}$$

We now solve this last equation for z and get $z = -\frac{1}{2}$. Next, we substitute $-\frac{1}{2}$ for z in the preceding equation and solve for y: $-y + 12\left(-\frac{1}{2}\right) = -13$, so $y = 7$. Since there is no y-term in the first equation of this last system, we need only substitute $-\frac{1}{2}$ for z to solve for x: $x - 4\left(-\frac{1}{2}\right) = 5$, so $x = 3$. The solution is $\left(3, 7, -\frac{1}{2}\right)$. The check is left to the student.

Try Exercise 13.

The operations used in the preceding example correspond to those used to produce equivalent systems of equations. We call the matrices **row-equivalent** and the operations that produce them **row-equivalent operations.**

ROW-EQUIVALENT OPERATIONS

Student Notes

Note that row-equivalent matrices are not *equal*. It is the solutions of the corresponding equations that are the same.

Row-Equivalent Operations

Each of the following row-equivalent operations produces a row-equivalent matrix:

a) Interchanging any two rows.
b) Multiplying all elements of a row by a nonzero constant.
c) Replacing a row with the sum of that row and a multiple of another row.

The best overall method for solving systems of equations is by row-equivalent matrices; even computers are programmed to use them. Matrices are part of a branch of mathematics known as linear algebra. They are also studied in many courses in finite mathematics.

We can continue to use row-equivalent operations to write a row-equivalent matrix in **reduced row-echelon form,** from which the solution of the system can often be read directly.

Reduced Row-Echelon Form

A matrix is in reduced row-echelon form if:

1. All rows consisting entirely of zeros are at the bottom of the matrix.
2. The first nonzero number in any nonzero row is 1, called a leading 1.
3. The leading 1 in any row is farther to the left than the leading 1 in any lower row.
4. Each column that contains a leading 1 has zeros everywhere else.

EXAMPLE 3 Solve the system

$$\begin{aligned} 5x - 4y &= -1, \\ -2x + 3y &= 2 \end{aligned}$$

by writing a reduced row-echelon matrix.

SOLUTION We began this solution in Example 1, and used row-equivalent operations to write the row-equivalent matrix:

$$\begin{aligned} 5x - 4y &= -1, \\ 7y &= 8 \end{aligned}$$

$$\left[\begin{array}{cc|c} 5 & -4 & -1 \\ 0 & 7 & 8 \end{array}\right].$$

Before we can write reduced row-echelon form, the first nonzero number in any nonzero row must be 1. We multiply Row 1 by $\frac{1}{5}$ and Row 2 by $\frac{1}{7}$:

$$\begin{aligned} x - \tfrac{4}{5}y &= -\tfrac{1}{5}, \\ y &= \tfrac{8}{7} \end{aligned}$$

$$\left[\begin{array}{cc|c} 1 & -\frac{4}{5} & -\frac{1}{5} \\ 0 & 1 & \frac{8}{7} \end{array}\right].$$

New Row 1 = $\frac{1}{5}$(Row 1 from above)
New Row 2 = $\frac{1}{7}$(Row 2 from above)

There are no zero rows, there is a leading 1 in each row, and the leading 1 in Row 1 is farther to the left than the leading 1 in Row 2. Thus conditions (1)–(3) for reduced row-echelon form have been met.

To satisfy condition (4), we must obtain a 0 in Row 1, Column 2, above the leading 1 in Row 2. We multiply Row 2 by $\frac{4}{5}$ and add it to Row 1:

$$\begin{aligned} x &= \tfrac{5}{7}, \\ y &= \tfrac{8}{7} \end{aligned}$$

$$\left[\begin{array}{cc|c} 1 & 0 & \frac{5}{7} \\ 0 & 1 & \frac{8}{7} \end{array}\right].$$

$\frac{4}{5}\left(0 \quad 1 \mid \frac{8}{7}\right) = \left(0 \quad \frac{4}{5} \mid \frac{32}{35}\right)$ **and**
$\left(0 \quad \frac{4}{5} \mid \frac{32}{35}\right) + \left(1 \quad -\frac{4}{5} \mid -\frac{1}{5}\right) = \left(1 \quad 0 \mid \frac{5}{7}\right)$
New Row 1 = $\frac{4}{5}$(Row 2) + Row 1

This matrix is in reduced row-echelon form. If we reinsert the variables, we have

$$\begin{aligned} x &= \tfrac{5}{7}, \\ y &= \tfrac{8}{7}. \end{aligned}$$

The solution, $\left(\frac{5}{7}, \frac{8}{7}\right)$, can be read directly from the last column of the matrix.

Try Exercise 11.

Finding reduced row-echelon form often involves extensive calculations with fractions or decimals. Thus it is common to use a computer or a graphing calculator to store and manipulate matrices.

Matrix Operations

On many graphing calculators, matrix operations are accessed by pressing MATRIX, often the 2nd feature associated with x^{-1}. The MATRIX menu has three submenus: NAMES, MATH, and EDIT.

The EDIT menu allows matrices to be entered. For example, to enter

$$A = \begin{bmatrix} 1 & 2 & 5 \\ -3 & 0 & 4 \end{bmatrix},$$

choose option [A] from the MATRIX EDIT menu. Enter the **dimensions,** or number of rows and columns, of the matrix first, listing the number of rows before the number of columns. Thus the dimensions of A are 2×3, read "2 by 3." Then enter each element of A by pressing the number and ENTER. The notation $2, 3 = 4$ indicates that the 3rd entry of the 2nd row is 4. Matrices are generally entered row by row rather than column by column.

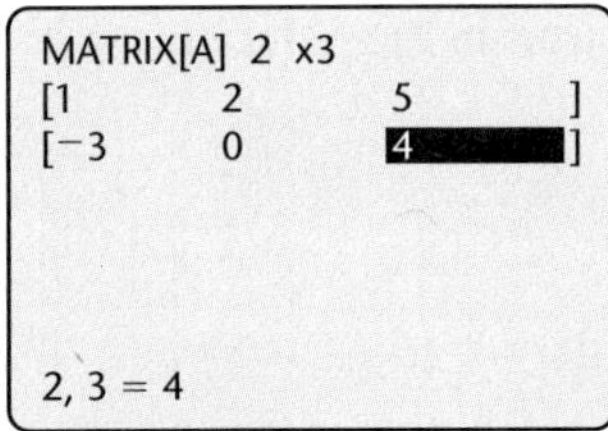

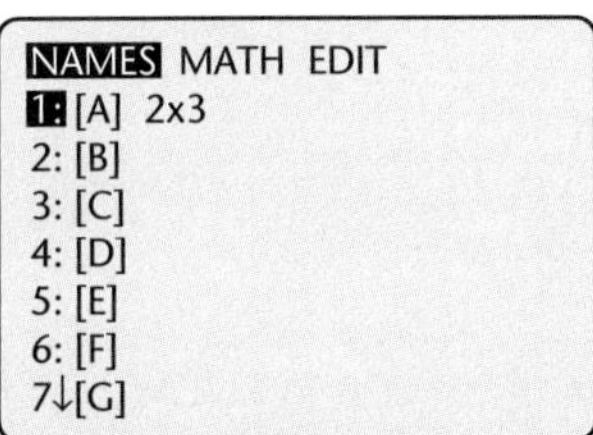

After entering the matrix, exit the matrix editor by pressing 2ND QUIT. To access a matrix once it has been entered, use the MATRIX NAMES submenu as shown above on the right. The brackets around A indicate that it is a matrix.

The submenu MATRIX MATH lists operations that can be performed on matrices.

Your Turn

1. Enter the matrix corresponding to the system in Example 1.
   a) Press MATRIX ▷ ▷ ENTER to choose matrix A from the EDIT menu.
   b) Press 2 ENTER 3 ENTER to enter the dimensions of the matrix.
   c) Type each of the numbers in Row 1, pressing ENTER after each number. Then do the same for Row 2. Press QUIT.
2. Now find the reduced row-echelon form of the matrix.
   a) Press MATRIX ▷ to move to the MATRIX MATH menu. Scroll down the menu to the RREF(option and press ENTER.
   b) Press MATRIX ENTER) to choose matrix A and close parentheses.
   c) Press MATH 1 ENTER to convert entries to fraction notation. You should see the same matrix that appears near the end of Example 3.

EXAMPLE 4 Solve the following system using a graphing calculator:

$$\begin{aligned} 2x + 5y - 8z &= 7, \\ 3x + 4y - 3z &= 8, \\ 5y - 2x &= 9. \end{aligned}$$

SOLUTION Before writing a matrix to represent this system, we rewrite the third equation in the form $ax + by + cz = d$:

$$\begin{aligned} 2x + 5y - 8z &= 7, \\ 3x + 4y - 3z &= 8, \\ -2x + 5y + 0z &= 9. \end{aligned}$$

The following matrix represents this system:

$$\left[\begin{array}{rrr|r} 2 & 5 & -8 & 7 \\ 3 & 4 & -3 & 8 \\ -2 & 5 & 0 & 9 \end{array}\right].$$

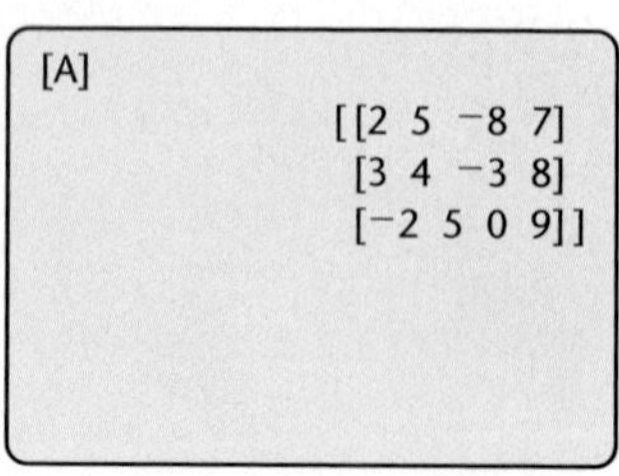

We enter the matrix as A using the MATRIX EDIT menu, noting that its dimensions are 3×4. Once we have entered each element of the matrix, we return to the home screen to perform operations. The contents of matrix A can be displayed on the screen, using the MATRIX NAMES menu.

Each of the row-equivalent operations can be performed using the MATRIX MATH menu, or we can go directly to the reduced row-echelon form. To find this form, from the home screen, we choose the RREF option from the MATRIX MATH menu and then choose [A] from the MATRIX NAMES menu. We press MATH 1 ENTER to write the entries in the reduced row-echelon form using fraction notation.

```
rref ( [A] )▸Frac
   [[1 0 0 1/2]
    [0 1 0 2]
    [0 0 1 1/2]]
```

$$\left[\begin{array}{ccc|c} 1 & 0 & 0 & \frac{1}{2} \\ 0 & 1 & 0 & 2 \\ 0 & 0 & 1 & \frac{1}{2} \end{array}\right].$$

The reduced row-echelon form shown above is equivalent to the system of equations

$$x = \tfrac{1}{2},$$
$$y = 2,$$
$$z = \tfrac{1}{2}.$$

The solution of the system is thus $\left(\frac{1}{2}, 2, \frac{1}{2}\right)$, which can be read directly from the last column of the reduced row-echelon matrix.

Try Exercise 17.

Recall that some systems of equations are inconsistent, and some sets of equations are dependent. When these cases occur, the reduced row-echelon form of the corresponding matrix will have a row or a column of zeros. For example, the matrix

$$\left[\begin{array}{cc|c} 1 & 0 & 4 \\ 0 & 0 & 0 \end{array}\right]$$

is in reduced row-echelon form. The second row translates to the equation $0 = 0$, and indicates that the equations in the system are dependent. The second row of the matrix

$$\left[\begin{array}{cc|c} 1 & 0 & 4 \\ 0 & 0 & 6 \end{array}\right]$$

translates to the equation $0 = 6$. This system is inconsistent.

9.3 Exercise Set

FOR EXTRA HELP MyMathLab   

PRACTICE WATCH DOWNLOAD READ REVIEW

Concept Reinforcement *Complete each of the following statements.*

1. A(n) ______________ is a rectangular array of numbers.

2. The rows of a matrix are ______________ and the ______________ are vertical.

3. Each number in a matrix is called a(n) ______________ or element.

4. The plural of the word matrix is ______________.

5. As part of solving a system using matrices, we can interchange any two ______________.

6. In the final step of solving a system of equations, the leftmost column has zeros in all rows except the ______________ one.

Solve using matrices.

7. $x + 2y = 11,$
$3x - y = 5$

8. $x + 3y = 16,$
$6x + y = 11$

9. $x + 4y = 8,$
$3x + 5y = 3$

10. $x + 4y = 5,$
$-3x + 2y = 13$

11. $6x - 2y = 4,$
$7x + y = 13$

12. $3x + 4y = 7,$
$-5x + 2y = 10$

13. $3x + 2y + 2z = 3,$
$x + 2y - z = 5,$
$2x - 4y + z = 0$

14. $4x - y - 3z = 19,$
$8x + y - z = 11,$
$2x + y + 2z = -7$

15. $a - 2b - 3c = 3,$
$2a - b - 2c = 4,$
$4a + 5b + 6c = 4$

16. $x + 2y - 3z = 9,$
$2x - y + 2z = -8,$
$3x - y - 4z = 3$

17. $3u + 2w = 11,$
$v - 7w = 4,$
$u - 6v = 1$

18. $4a + 9b = 8,$
$8a + 6c = -1,$
$6b + 6c = -1$

19. $2x + 2y - 2z - 2w = -10,$
$w + y + z + x = -5,$
$x - y + 4z + 3w = -2,$
$w - 2y + 2z + 3x = -6$

20. $-w - 3y + z + 2x = -8,$
$x + y - z - w = -4,$
$w + y + z + x = 22,$
$x - y - z - w = -14$

Solve using matrices.

21. *Coin Value.* A collection of 42 coins consists of dimes and nickels. The total value is \$3.00. How many dimes and how many nickels are there?

22. *Coin Value.* A collection of 43 coins consists of dimes and quarters. The total value is \$7.60. How many dimes and how many quarters are there?

23. *Snack Mix.* Bree sells a dried-fruit mixture for \$5.80 per pound and Hawaiian macadamia nuts for \$14.75 per pound. She wants to blend the two to get a 15-lb mixture that she will sell for \$9.38 per pound. How much of each should she use?

24. *Mixing Paint.* Higher quality paint typically contains more solids. Grant has available paint that contains 45% solids and paint that contains 25% solids. How much of each should he use to create 20 gal of paint that contains 39% solids?

25. *Investments.* Elena receives \$212 per year in simple interest from three investments totaling \$2500. Part is invested at 7%, part at 8%, and part at 9%. There is \$1100 more invested at 9% than at 8%. Find the amount invested at each rate.

26. *Investments.* Alejandro receives \$306 per year in simple interest from three investments totaling \$3200. Part is invested at 8%, part at 9%, and part at 10%. There is \$1900 more invested at 10% than at 9%. Find the amount invested at each rate.

TW **27.** Explain how you can recognize dependent equations when solving with matrices.

TW **28.** Explain how you can recognize an inconsistent system when solving with matrices.

SKILL REVIEW

To prepare for Section 9.4, review order of operations (Section 1.8).

Simplify. [1.8]

29. $5(-3) - (-7)4$

30. $8(-5) - (-2)9$

31. $-2(5 \cdot 3 - 4 \cdot 6) - 3(2 \cdot 7 - 15) + 4(3 \cdot 8 - 5 \cdot 4)$

32. $6(2 \cdot 7 - 3(-4)) - 4(3(-8) - 10) + 5(4 \cdot 3 - (-2)7)$

SYNTHESIS

TW **33.** If the systems corresponding to the matrices

$$\begin{bmatrix} a_1 & b_1 & \vdots & c_1 \\ d_1 & e_1 & \vdots & f_1 \end{bmatrix} \quad \text{and} \quad \begin{bmatrix} a_2 & b_2 & \vdots & c_2 \\ d_2 & e_2 & \vdots & f_2 \end{bmatrix}$$

share the same solution, does it follow that the corresponding entries are all equal to each other ($a_1 = a_2, b_1 = b_2$, etc.)? Why or why not?

TW **34.** Explain how the row-equivalent operations make use of the addition, multiplication, and distributive properties.

35. The sum of the digits in a four-digit number is 10. Twice the sum of the thousands digit and the tens digit is 1 less than the sum of the other two digits. The tens digit is twice the thousands digit. The ones digit equals the sum of the thousands digit and the hundreds digit. Find the four-digit number.

36. Solve for x and y:

$$ax + by = c,$$
$$dx + ey = f.$$

Try Exercise Answers: Section 9.3

7. (3, 4) **11.** $\left(\frac{3}{2}, \frac{5}{2}\right)$ **13.** $\left(2, \frac{1}{2}, -2\right)$ **17.** $\left(4, \frac{1}{2}, -\frac{1}{2}\right)$

Mid-Chapter Review

We solve systems of three or more equations using elimination. This process leads to the use of matrices and, eventually, computers for solving systems.

GUIDED SOLUTIONS

1. Solve:

$$x - y + z = 4, \quad (1)$$
$$x + y - 2z = 3, \quad (2)$$
$$2x - y - z = 9. \quad (3) \quad [9.1]$$

Solution

$$x - y + z = 4$$
$$x + y - 2z = 3$$
$$\square = \square$$ **Adding equations (1) and (2) to eliminate *y***

$$x + y - 2z = 3$$
$$2x - y - z = 9$$
$$\square = \square$$ **Adding equations (2) and (3) to again eliminate *y***

$$2x - z = 7$$
$$3x - 3z = 12$$
$$x = \square, \; z = \square$$ **Solving the resulting system of two equations in *x* and *z***

$$x + y - 2z = 3$$
$$\square + y - 2\square = 3$$
$$y = \square$$ **Solving for *y***

The solution is $(\square, \square, \square)$.

2. Solve using matrices:

$$2x + 3y = 6,$$
$$4x - 5y = 1. \quad [9.3]$$

Solution

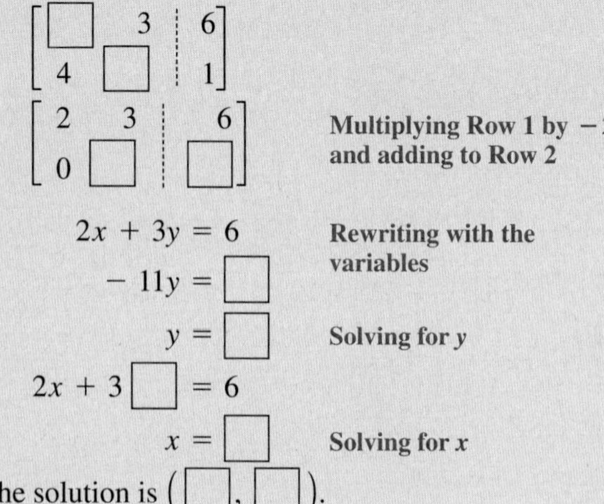

$$\begin{bmatrix} \square & 3 & \vdots & 6 \\ 4 & \square & \vdots & 1 \end{bmatrix}$$

$$\begin{bmatrix} 2 & 3 & \vdots & 6 \\ 0 & \square & \vdots & \square \end{bmatrix}$$ **Multiplying Row 1 by −2 and adding to Row 2**

$$2x + 3y = 6$$
$$-11y = \square$$ **Rewriting with the variables**

$$y = \square$$ **Solving for *y***

$$2x + 3\square = 6$$
$$x = \square$$ **Solving for *x***

The solution is $(\square, \square)$.

MIXED REVIEW

Solve.

1. $x + 2y + z = 4,$
$2x - y - z = 1,$
$x - 2y + z = 8$

2. $x + 2y - z = 3,$
$x - y + z = -7,$
$2x + 3y - z = 2$

3. $5a - 2b - c = 0,$
$3a + 4b + c = 8,$
$9a - 6b - 2c = -1$

4. $2u + v - w = 3,$
$u + 5v + 2w = 2,$
$2u - v - w = 4$

5. $y + 3z = 2,$
$x - 2z = 4,$
$2x - y = 0$

6. $2x - z = 1,$
$x - 2y + 3z = 2,$
$x + 2y - 4z = -2$

Solve using matrices.

7. $2x + 4y = -1,$
$x + 3y = 2$

8. $4x - y = 5,$
$2x + 3y = -1$

9. $2x + y + 2z = 5,$
$x + 2y + 4z = 6,$
$2x + 3y + 5z = 8$

10. $-2a + 8b + c = -2,$
$4a - 2b - 6c = 1,$
$2a - b - c = 2$

Solve.

11. The sum of three numbers is 15. The first number minus twice the second is twice the third. The first minus the second plus the third is 19. Find the numbers.

12. In triangle ABC, the measure of angle B is twice that of angle A. The measure of angle C is 40° more than that of angle B. Find the angle measures.

13. The Golden Eagles sell a brownie for 75¢, a bag of chips for \$1.00, and a hot dog for \$2.00 at their concession stand. At the last game, they sold a total of 125 items and took in \$163.75. The number of hot dogs sold was 25 less than the total number of brownies and bags of chips. How many of each did they sell?

14. Christie owes \$10,300 on three credit cards. On one card the monthly interest rate is 2%, on the second it is 1%, and on the third it is 1.5%. The total interest due one month was \$151. She owes \$3300 more on the 2% card and the 1.5% card, combined, than she does on the 1% card. How much does she owe on each card?

9.4 Determinants and Cramer's Rule

- Determinants of 2×2 Matrices
- Cramer's Rule: 2×2 Systems
- Cramer's Rule: 3×3 Systems

DETERMINANTS OF 2×2 MATRICES

When a matrix has m rows and n columns, it is called an "m by n" matrix. Thus its *dimensions* are denoted by $m \times n$. If a matrix has the same number of rows and columns, it is called a **square matrix**. Associated with every square matrix is a number called its **determinant,** defined as follows for 2×2 matrices.

2 × 2 Determinants The determinant of a two-by-two matrix $\begin{bmatrix} a & c \\ b & d \end{bmatrix}$ is denoted $\begin{vmatrix} a & c \\ b & d \end{vmatrix}$ and is defined as follows:

$$\begin{vmatrix} a & c \\ b & d \end{vmatrix} = ad - bc.$$

STUDY TIP

Catch the Emphasis

Pay attention to phrases from instructors such as "Write this down" and "You will need to know this." Most instructors will place special emphasis on the material they feel is important, and this is the material most likely to appear on a test.

EXAMPLE 1 Evaluate: $\begin{vmatrix} 2 & -5 \\ 6 & 7 \end{vmatrix}$.

SOLUTION We multiply and subtract as follows:

$$\begin{vmatrix} 2 & -5 \\ 6 & 7 \end{vmatrix} = 2 \cdot 7 - 6 \cdot (-5) = 14 + 30 = 44.$$

Try Exercise 7.

CRAMER'S RULE: 2 × 2 SYSTEMS

One of the many uses for determinants is in solving systems of linear equations in which the number of variables is the same as the number of equations and the constants are not all 0. Let's consider a system of two equations:

$$a_1x + b_1y = c_1,$$
$$a_2x + b_2y = c_2.$$

If we use the elimination method, a series of steps can show that

$$x = \frac{c_1b_2 - c_2b_1}{a_1b_2 - a_2b_1} \quad \text{and} \quad y = \frac{a_1c_2 - a_2c_1}{a_1b_2 - a_2b_1}.$$

These fractions can be rewritten using determinants.

Cramer's Rule: 2 × 2 Systems The solution of the system

$$a_1x + b_1y = c_1,$$
$$a_2x + b_2y = c_2,$$

if it is unique, is given by

$$x = \frac{\begin{vmatrix} c_1 & b_1 \\ c_2 & b_2 \end{vmatrix}}{\begin{vmatrix} a_1 & b_1 \\ a_2 & b_2 \end{vmatrix}}, \qquad y = \frac{\begin{vmatrix} a_1 & c_1 \\ a_2 & c_2 \end{vmatrix}}{\begin{vmatrix} a_1 & b_1 \\ a_2 & b_2 \end{vmatrix}}.$$

These formulas apply only if the denominator is not 0. If the denominator *is* 0, then one of two things happens:

1. If the denominator is 0 and the numerators are also 0, then the equations in the system are dependent.
2. If the denominator is 0 and at least one numerator is not 0, then the system is inconsistent.

To use Cramer's rule, we find the determinants and compute x and y as shown above. Note that the denominators are identical and the coefficients of x and y appear in the same position as in the original equations. In the numerator of x, the constants c_1 and c_2 replace a_1 and a_2. In the numerator of y, the constants c_1 and c_2 replace b_1 and b_2.

EXAMPLE 2 Solve using Cramer's rule:

$$2x + 5y = 7,$$
$$5x - 2y = -3.$$

SOLUTION We have

$$x = \frac{\begin{vmatrix} 7 & 5 \\ -3 & -2 \end{vmatrix}}{\begin{vmatrix} 2 & 5 \\ 5 & -2 \end{vmatrix}}$$

← **The constants $\begin{matrix} 7 \\ -3 \end{matrix}$ form the first column.**

← **The columns are the coefficients of the variables.**

$$= \frac{7(-2) - (-3)5}{2(-2) - 5 \cdot 5} = -\frac{1}{29}$$

and

$$y = \frac{\begin{vmatrix} 2 & 7 \\ 5 & -3 \end{vmatrix}}{\begin{vmatrix} 2 & 5 \\ 5 & -2 \end{vmatrix}}$$

← **The constants $\begin{matrix} 7 \\ -3 \end{matrix}$ form the second column.**

← **The denominator is the same as in the expression for x.**

$$= \frac{2(-3) - 5 \cdot 7}{-29} = \frac{41}{29}.$$

The solution is $\left(-\frac{1}{29}, \frac{41}{29}\right)$. The check is left to the student.

Try Exercise 15.

CRAMER'S RULE: 3 × 3 SYSTEMS

Cramer's rule can be extended for systems of three linear equations. However, before doing so, we must define what a 3 × 3 determinant is.

3 × 3 Determinants The determinant of a three-by-three matrix is defined as follows:

Subtract. Add.

$$\begin{vmatrix} a_1 & b_1 & c_1 \\ a_2 & b_2 & c_2 \\ a_3 & b_3 & c_3 \end{vmatrix} = a_1\begin{vmatrix} b_2 & c_2 \\ b_3 & c_3 \end{vmatrix} - a_2\begin{vmatrix} b_1 & c_1 \\ b_3 & c_3 \end{vmatrix} + a_3\begin{vmatrix} b_1 & c_1 \\ b_2 & c_2 \end{vmatrix}.$$

Student Notes

Cramer's rule and the evaluation of determinants rely on patterns. Recognizing and remembering the patterns will help you understand and use the definitions.

Note that the a's come from the first column. Note too that the 2 × 2 determinants above can be obtained by crossing out the row and the column in which the a occurs.

For a_1: $\begin{vmatrix} \cancel{a_1} & \cancel{b_1} & \cancel{c_1} \\ \cancel{a_2} & b_2 & c_2 \\ \cancel{a_3} & b_3 & c_3 \end{vmatrix}$ For a_2: $\begin{vmatrix} \cancel{a_1} & b_1 & c_1 \\ \cancel{a_2} & \cancel{b_2} & \cancel{c_2} \\ \cancel{a_3} & b_3 & c_3 \end{vmatrix}$ For a_3: $\begin{vmatrix} \cancel{a_1} & b_1 & c_1 \\ \cancel{a_2} & b_2 & c_2 \\ \cancel{a_3} & \cancel{b_3} & \cancel{c_3} \end{vmatrix}$

EXAMPLE 3 Evaluate:

$$\begin{vmatrix} -1 & 0 & 1 \\ -5 & 1 & -1 \\ 4 & 8 & 1 \end{vmatrix}.$$

SOLUTION We have

Subtract. Add.

$$\begin{vmatrix} -1 & 0 & 1 \\ -5 & 1 & -1 \\ 4 & 8 & 1 \end{vmatrix} = -1\begin{vmatrix} 1 & -1 \\ 8 & 1 \end{vmatrix} - (-5)\begin{vmatrix} 0 & 1 \\ 8 & 1 \end{vmatrix} + 4\begin{vmatrix} 0 & 1 \\ 1 & -1 \end{vmatrix}$$

$$= -1(1 + 8) + 5(0 - 8) + 4(0 - 1)$$ **Evaluating the three determinants**

$$= -9 - 40 - 4 = -53.$$

Try Exercise 11.

Cramer's Rule: 3 × 3 Systems The solution of the system

$$a_1x + b_1y + c_1z = d_1,$$
$$a_2x + b_2y + c_2z = d_2,$$
$$a_3x + b_3y + c_3z = d_3$$

can be found using the following determinants:

$$D = \begin{vmatrix} a_1 & b_1 & c_1 \\ a_2 & b_2 & c_2 \\ a_3 & b_3 & c_3 \end{vmatrix}, \quad D_x = \begin{vmatrix} d_1 & b_1 & c_1 \\ d_2 & b_2 & c_2 \\ d_3 & b_3 & c_3 \end{vmatrix},$$

D contains only coefficients. In D_x, the d's replace the a's.

$$D_y = \begin{vmatrix} a_1 & d_1 & c_1 \\ a_2 & d_2 & c_2 \\ a_3 & d_3 & c_3 \end{vmatrix}, \quad D_z = \begin{vmatrix} a_1 & b_1 & d_1 \\ a_2 & b_2 & d_2 \\ a_3 & b_3 & d_3 \end{vmatrix}.$$

In D_y, the d's replace the b's. In D_z, the d's replace the c's.

If a unique solution exists, it is given by

$$x = \frac{D_x}{D}, \quad y = \frac{D_y}{D}, \quad z = \frac{D_z}{D}.$$

EXAMPLE 4 Solve using Cramer's rule:

$$x - 3y + 7z = 13,$$
$$x + y + z = 1,$$
$$x - 2y + 3z = 4.$$

SOLUTION We compute D, D_x, D_y, and D_z:

$$D = \begin{vmatrix} 1 & -3 & 7 \\ 1 & 1 & 1 \\ 1 & -2 & 3 \end{vmatrix} = -10; \qquad D_x = \begin{vmatrix} 13 & -3 & 7 \\ 1 & 1 & 1 \\ 4 & -2 & 3 \end{vmatrix} = 20;$$

$$D_y = \begin{vmatrix} 1 & 13 & 7 \\ 1 & 1 & 1 \\ 1 & 4 & 3 \end{vmatrix} = -6; \qquad D_z = \begin{vmatrix} 1 & -3 & 13 \\ 1 & 1 & 1 \\ 1 & -2 & 4 \end{vmatrix} = -24.$$

Then

$$x = \frac{D_x}{D} = \frac{20}{-10} = -2;$$

$$y = \frac{D_y}{D} = \frac{-6}{-10} = \frac{3}{5};$$

$$z = \frac{D_z}{D} = \frac{-24}{-10} = \frac{12}{5}.$$

The solution is $\left(-2, \frac{3}{5}, \frac{12}{5}\right)$. The check is left to the student.

Try Exercise 19.

In Example 4, we need not have evaluated D_z. Once x and y were found, we could have substituted them into one of the equations to find z.

To use Cramer's rule, we divide by D, provided $D \neq 0$. If $D = 0$ and at least one of the other determinants is not 0, then the system is inconsistent. If *all* the determinants are 0, then the equations in the system are dependent.

Determinants

Determinants can be evaluated using the DET(option of the MATRIX MATH menu. After entering the matrix, we go to the home screen and select the determinant operation. Then we enter the name of the matrix using the MATRIX NAMES menu. The graphing calculator will return the value of the determinant of the matrix. For example, if

$$\mathbf{A} = \begin{bmatrix} 1 & 6 & -1 \\ -3 & -5 & 3 \\ 0 & 4 & 2 \end{bmatrix},$$

we have

```
det([A])
          26
```
.

Your Turn

1. Enter the matrix $\begin{bmatrix} 1 & -3 \\ 4 & -2 \end{bmatrix}$ as matrix A.
2. From the home screen, press MATRIX ▷ 1 MATRIX 1) ENTER.
   $\det([A]) = 10$

9.4 Exercise Set

FOR EXTRA HELP MyMathLab MathXL PRACTICE  WATCH 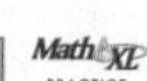 DOWNLOAD  READ  REVIEW

Concept Reinforcement *Classify each of the following statements as either true or false.*

1. A square matrix has the same number of rows and columns.
2. A 3×4 matrix has 3 rows and 4 columns.
3. A determinant is a number.
4. Cramer's rule exists only for 2×2 systems.
5. Whenever Cramer's rule yields a denominator that is 0, the system has no solution.
6. Whenever Cramer's rule yields a numerator that is 0, the equations are dependent.

Evaluate.

7. $\begin{vmatrix} 5 & 1 \\ 2 & 4 \end{vmatrix}$

8. $\begin{vmatrix} 3 & 2 \\ 2 & -3 \end{vmatrix}$

9. $\begin{vmatrix} 10 & 8 \\ -5 & -9 \end{vmatrix}$

10. $\begin{vmatrix} 3 & 2 \\ -7 & 5 \end{vmatrix}$

11. $\begin{vmatrix} 1 & 4 & 0 \\ 0 & -1 & 2 \\ 3 & -2 & 1 \end{vmatrix}$

12. $\begin{vmatrix} 2 & 4 & -2 \\ 1 & 0 & 2 \\ 0 & 1 & 3 \end{vmatrix}$

13. $\begin{vmatrix} -4 & -2 & 3 \\ -3 & 1 & 2 \\ 3 & 4 & -2 \end{vmatrix}$

14. $\begin{vmatrix} 2 & -1 & 1 \\ 1 & 2 & -1 \\ 3 & 4 & -3 \end{vmatrix}$

Solve using Cramer's rule.

15. $5x + 8y = 1,$
$3x + 7y = 5$

16. $3x - 4y = 6,$
$5x + 9y = 10$

17. $5x - 4y = -3,$
$7x + 2y = 6$

18. $-2x + 4y = 3,$
$3x - 7y = 1$

19. $3x - y + 2z = 1,$
$x - y + 2z = 3,$
$-2x + 3y + z = 1$

20. $3x + 2y - z = 4,$
$3x - 2y + z = 5,$
$4x - 5y - z = -1$

21. $2x - 3y + 5z = 27,$
$x + 2y - z = -4,$
$5x - y + 4z = 27$

22. $x - y + 2z = -3,$
$x + 2y + 3z = 4,$
$2x + y + z = -3$

23. $r - 2s + 3t = 6,$
$2r - s - t = -3,$
$r + s + t = 6$

24. $a - 3c = 6,$
$b + 2c = 2,$
$7a - 3b - 5c = 14$

TW 25. Describe at least one of the patterns in Cramer's rule.

TW 26. Which version of Cramer's rule do you find more useful: the version for 2×2 systems or the version for 3×3 systems? Why?

SKILL REVIEW

To prepare for Section 9.5, review functions (Sections 3.8 and 5.9).

Find each of the following, given $f(x) = 80x + 2500$ and $g(x) = 150x$.

27. $f(90)$ [3.8]

28. $(g - f)(x)$ [5.9]

29. $(g - f)(10)$ [5.9]

30. $(g - f)(100)$ [5.9]

31. All values of x for which $f(x) = g(x)$ [3.8]

32. All values of x for which $(g - f)(x) = 0$ [5.9]

SYNTHESIS

TW 33. Cramer's rule states that if $a_1x + b_1y = c_1$ and $a_2x + b_2y = c_2$ are dependent, then

$$\begin{vmatrix} a_1 & b_1 \\ a_2 & b_2 \end{vmatrix} = 0.$$

Explain why this will always happen.

TW 34. Under what conditions can a 3×3 system of linear equations be consistent but unable to be solved using Cramer's rule?

Solve.

35. $\begin{vmatrix} y & -2 \\ 4 & 3 \end{vmatrix} = 44$

36. $\begin{vmatrix} 2 & x & -1 \\ -1 & 3 & 2 \\ -2 & 1 & 1 \end{vmatrix} = -12$

37. $\begin{vmatrix} m+1 & -2 \\ m-2 & 1 \end{vmatrix} = 27$

38. Show that an equation of the line through (x_1, y_1) and (x_2, y_2) can be written

$$\begin{vmatrix} x & y & 1 \\ x_1 & y_1 & 1 \\ x_2 & y_2 & 1 \end{vmatrix} = 0.$$

Try Exercise Answers: Section 9.4

7. 18 11. 27 15. $(-3, 2)$ 19. $\left(-1, -\frac{6}{7}, \frac{11}{7}\right)$

9.5 Business and Economics Applications

- Break-Even Analysis
- Supply and Demand

BREAK-EVEN ANALYSIS

The money that a business spends to manufacture a product is its *cost.* The **total cost** of production can be thought of as a function C, where $C(x)$ is the cost of producing x units. When the company sells the product, it takes in money. This is *revenue* and can be thought of as a function R, where $R(x)$ is the **total revenue** from the sale of x units. **Total profit** is the money taken in less the money spent, or total revenue minus total cost. Total profit from the production and sale of x units is a function P given by

$$\textbf{Profit} = \textbf{Revenue} - \textbf{Cost} \quad \text{or} \quad P(x) = R(x) - C(x).$$

If $R(x)$ is greater than $C(x)$, there is a gain and $P(x)$ is positive. If $C(x)$ is greater than $R(x)$, there is a loss and $P(x)$ is negative. When $R(x) = C(x)$, the company breaks even.

There are two kinds of costs. First, there are costs like rent, insurance, machinery, and so on. These costs, which must be paid regardless of how many items are produced, are called *fixed costs.* Second, costs for labor, materials, marketing, and so on are called *variable costs*, because they vary according to the amount being produced. The sum of the fixed cost and the variable cost gives the **total cost.**

STUDY TIP 

Keep It Relevant

Finding applications of math in your everyday life is a great study aid. Try to extend this idea to the newspapers, magazines, and books that you read. Look with a critical eye at graphs and their labels. Not only will this help with your math, it will make you a more informed citizen.

CAUTION! Do not confuse "cost" with "price." When we discuss the *cost* of an item, we are referring to what it costs to produce the item. The *price* of an item is what a consumer pays to purchase the item and is used when calculating revenue.

EXAMPLE 1 Manufacturing Chairs. Renewable Designs is planning to make a new chair. Fixed costs will be \$90,000, and it will cost \$150 to produce each chair (variable costs). Each chair sells for \$400.

a) Find the total cost $C(x)$ of producing x chairs.

b) Find the total revenue $R(x)$ from the sale of x chairs.

c) Find the total profit $P(x)$ from the production and sale of x chairs.

d) What profit will the company realize from the production and sale of 300 chairs? of 800 chairs?

e) Graph the total-cost, total-revenue, and total-profit functions using the same set of axes. Determine the break-even point.

SOLUTION

a) Total cost, in dollars, is given by

$$C(x) = (\text{Fixed costs}) \text{ plus } (\text{Variable costs}),$$
$$\text{or } C(x) = 90{,}000 + 150x$$

where x is the number of chairs produced.

b) Total revenue, in dollars, is given by

$$R(x) = 400x.$$ **$400 times the number of chairs sold. We assume that every chair produced is sold.**

c) Total profit, in dollars, is given by

$$P(x) = R(x) - C(x)$$ **Profit is revenue minus cost.**
$$= 400x - (90{,}000 + 150x)$$
$$= 250x - 90{,}000.$$

d) Profits will be

$y_1 = 250x - 90{,}000$

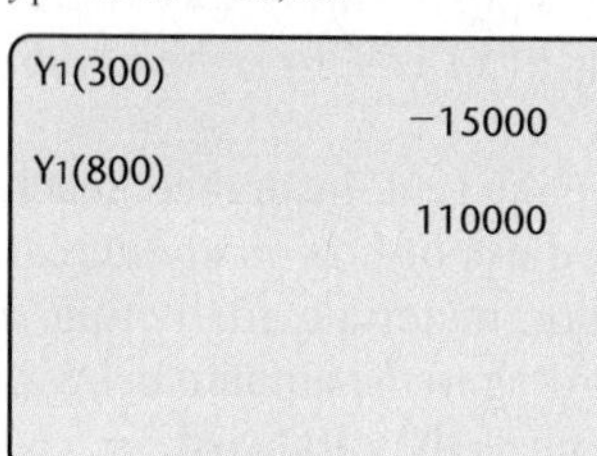

$$P(300) = 250 \cdot 300 - 90{,}000 = -\$15{,}000$$

when 300 chairs are produced and sold, and

$$P(800) = 250 \cdot 800 - 90{,}000 = \$110{,}000$$

when 800 chairs are produced and sold. These values can also be found using a graphing calculator, as shown in the figure at left. Thus the company loses money if only 300 chairs are sold, but makes money if 800 are sold.

e) The graphs of each of the three functions are shown below:

$C(x) = 90{,}000 + 150x,$ **This represents the cost function.**
$R(x) = 400x,$ **This represents the revenue function.**
$P(x) = 250x - 90{,}000.$ **This represents the profit function.**

$C(x)$, $R(x)$, and $P(x)$ are all in dollars.

The revenue function has a graph that goes through the origin and has a slope of 400. The cost function has an intercept on the \$-axis of 90,000 and has a slope of 150. The profit function has an intercept on the \$-axis of −90,000 and has a slope of 250. It is shown by the red and black dashed line. The red portion of the dashed line shows a "negative" profit, which is a loss. (That is what is known as "being in the red.") The black portion of the dashed line shows a "positive" profit, or gain. (That is what is known as "being in the black.")

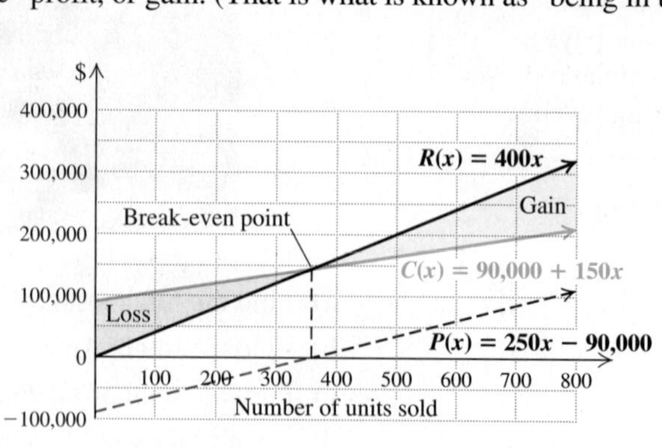

Gains occur where revenue exceeds cost. Losses occur where the revenue is less than cost. The **break-even point** occurs where the graphs of R and C cross. Thus to find the break-even point, we solve a system:

$$R(x) = 400x,$$
$$C(x) = 90{,}000 + 150x.$$

Since both revenue and cost are in *dollars* and they are equal at the break-even point, the system can be rewritten as

$$d = 400x, \qquad (1)$$
$$d = 90{,}000 + 150x \qquad (2)$$

and solved using substitution:

$$400x = 90{,}000 + 150x \qquad \textbf{Substituting } 400x \textbf{ for } d \textbf{ in equation (2)}$$
$$250x = 90{,}000$$
$$x = 360.$$

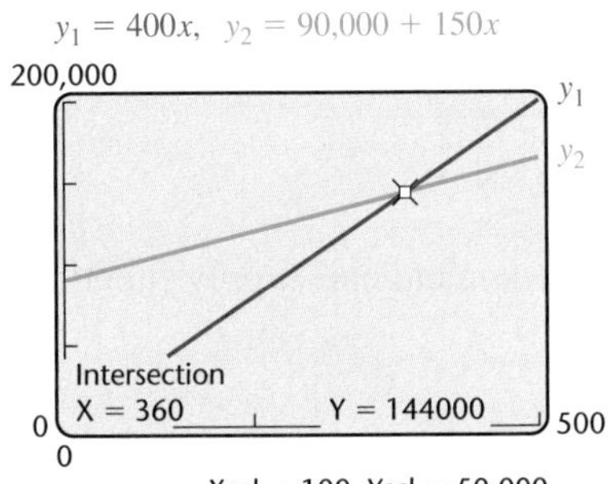

The figure at left shows the solution found using a graphing calculator.

The firm will break even if it produces and sells 360 chairs and takes in a total of $R(360) = \$400(360) = \$144{,}000$ in revenue. Note that the x-coordinate of the break-even point can also be found by solving $P(x) = 0$. The break-even point is (360 chairs, \$144,000).

Try Exercise 9.

Student Notes

If you plan to study business or economics, expect to see these topics arise in other courses.

SUPPLY AND DEMAND

As the price of coffee varies, so too does the amount sold. The table and graph below show that *consumers will demand less as the price goes up.*

Demand function, D

Price, p, per Kilogram	Quantity, $D(p)$ (in millions of kilograms)
\$ 8.00	25
9.00	20
10.00	15
11.00	10
12.00	5

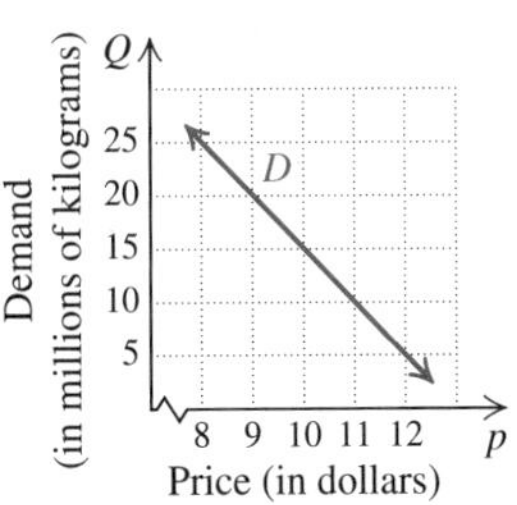

As the price of coffee varies, the amount made available varies as well. The table and graph below show that *sellers will supply more as the price goes up.*

Supply function, S

Price, p, per Kilogram	Quantity, $S(p)$ (in millions of kilograms)
\$ 9.00	5
9.50	10
10.00	15
10.50	20
11.00	25

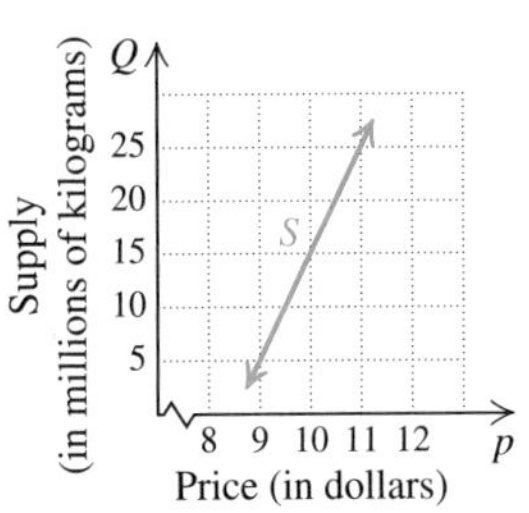

Considering demand and supply together, we see that as price increases, demand decreases. As price increases, supply increases. The point of intersection is called the **equilibrium point.** At that price, the amount that the seller will supply is the same amount that the consumer will buy. The situation is similar to that of a buyer and a seller negotiating the price of an item. The equilibrium point is the price and the quantity that they finally agree on.

Any ordered pair of coordinates from the graph is (price, quantity), because the horizontal axis is the price axis and the vertical axis is the quantity axis. If D is a demand function and S is a supply function, then the equilibrium point is where demand equals supply:

$$D(p) = S(p).$$

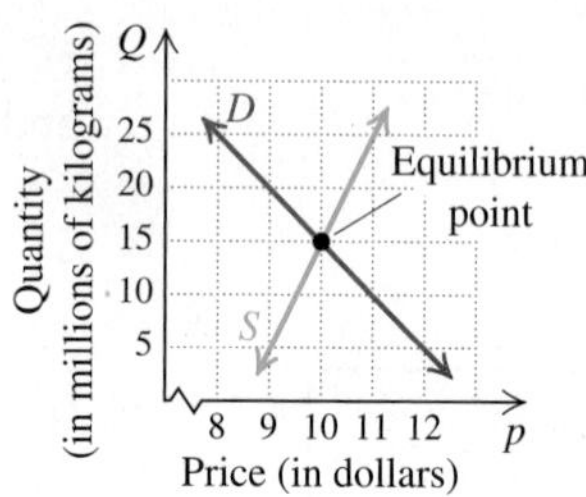

EXAMPLE 2 Find the equilibrium point for the demand and supply functions given:

$$D(p) = 1000 - 60p, \quad (1)$$
$$S(p) = 200 + 4p. \quad (2)$$

SOLUTION Since both demand and supply are *quantities* and they are equal at the equilibrium point, we rewrite the system as

$$q = 1000 - 60p, \quad (1)$$
$$q = 200 + 4p. \quad (2)$$

We substitute $200 + 4p$ for q in equation (1) and solve:

$$200 + 4p = 1000 - 60p \quad \textbf{Substituting } 200 + 4p \textbf{ for } q \textbf{ in equation (1)}$$
$$200 + 64p = 1000 \quad \textbf{Adding } 60p \textbf{ to both sides}$$
$$64p = 800 \quad \textbf{Adding } -200 \textbf{ to both sides}$$
$$p = \tfrac{800}{64} = 12.5.$$

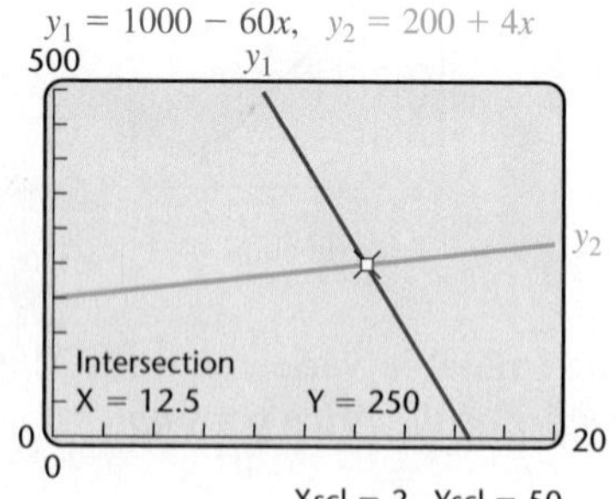

Thus the equilibrium price is \$12.50 per unit.

To find the equilibrium quantity, we substitute \$12.50 into either $D(p)$ or $S(p)$. We use $S(p)$:

$$S(12.5) = 200 + 4(12.5) = 200 + 50 = 250.$$

Thus the equilibrium quantity is 250 units, and the equilibrium point is (\$12.50, 250). The graph at left confirms the solution.

Try Exercise 19.

Visualizing for Success

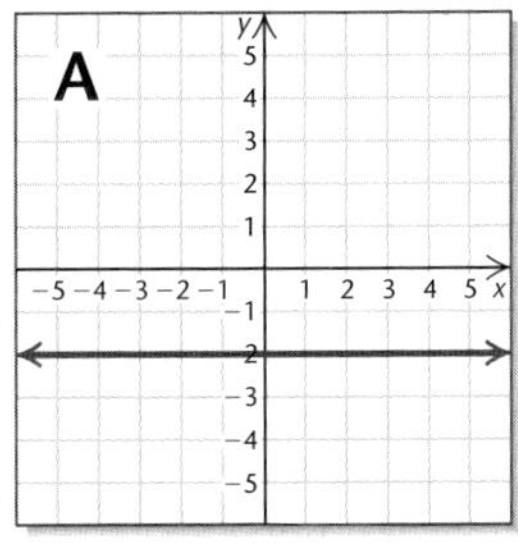

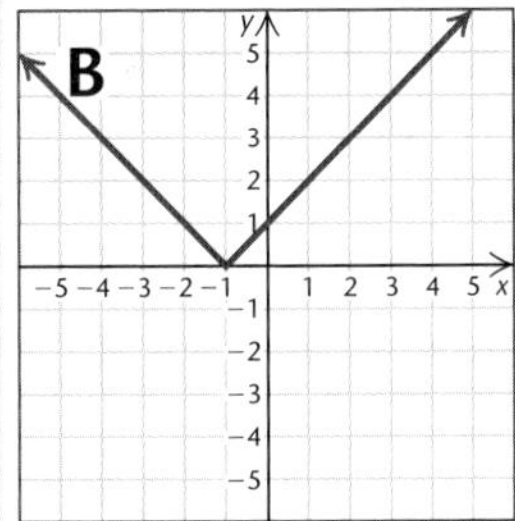

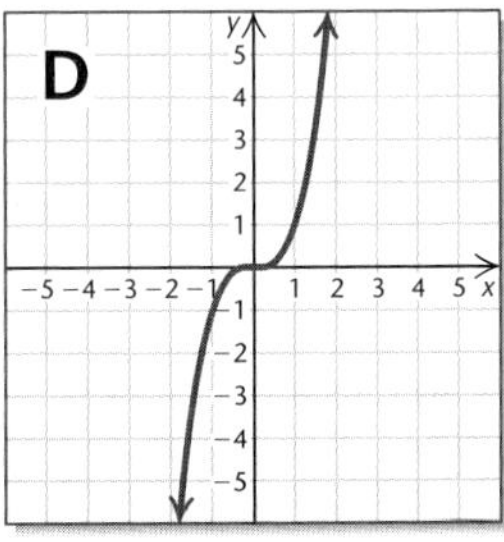

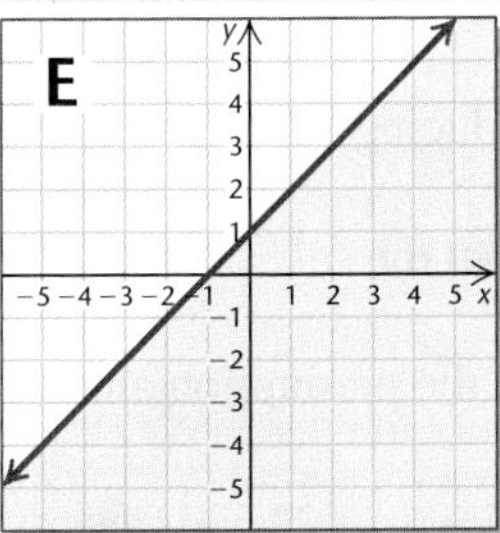

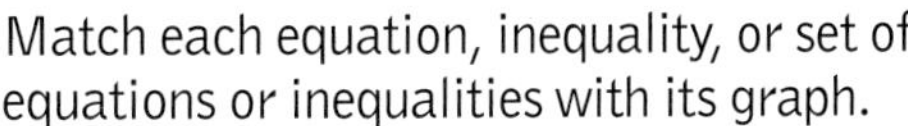

Match each equation, inequality, or set of equations or inequalities with its graph.

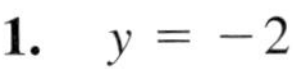

1. $y = -2$

2. $y = x^3$

3. $y = x^2$

4. $y = \dfrac{1}{x}$

5. $y = x + 1$

6. $y = |x + 1|$

7. $y \leq x + 1$

8. $y > 2x - 3$

9. $y = x + 1,$
$y = 2x - 3$

10. $y \leq x + 1,$
$y > 2x - 3$

Answers on page A-34

An additional, animated version of this activity appears in MyMathLab. To use MyMathLab, you need a course ID and a student access code. Contact your instructor for more information.

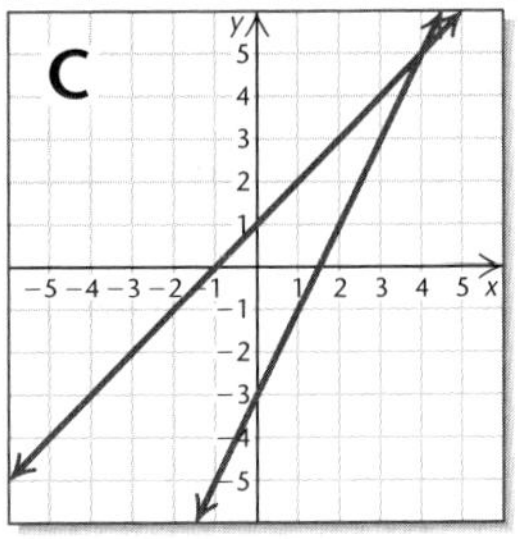

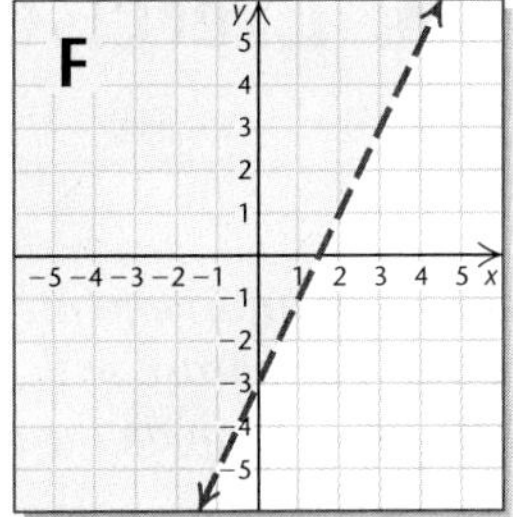

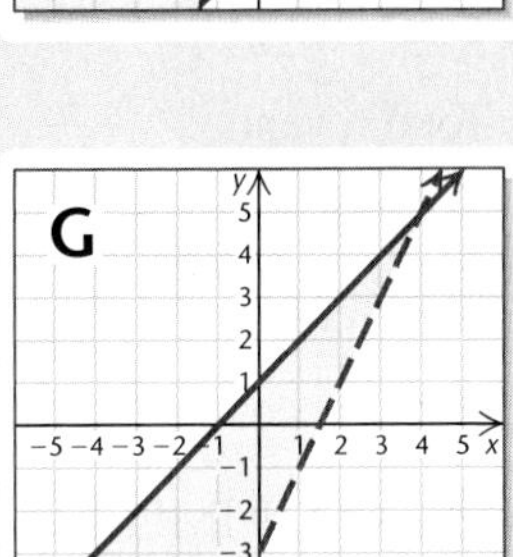

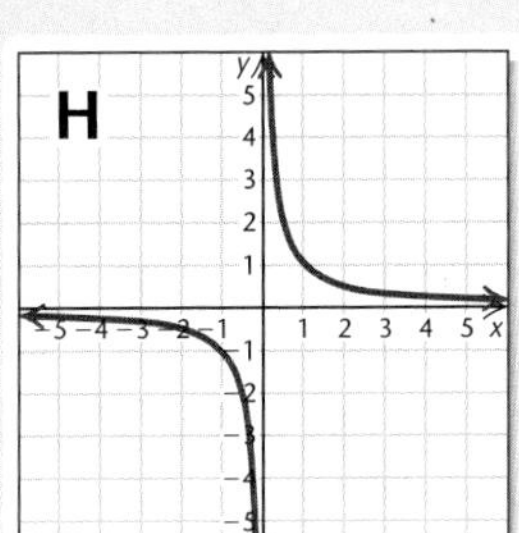

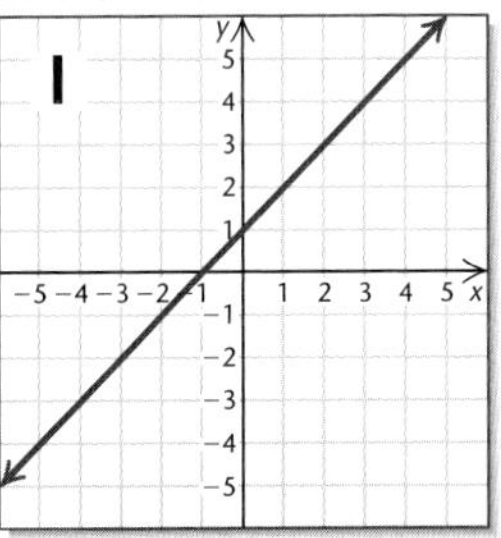

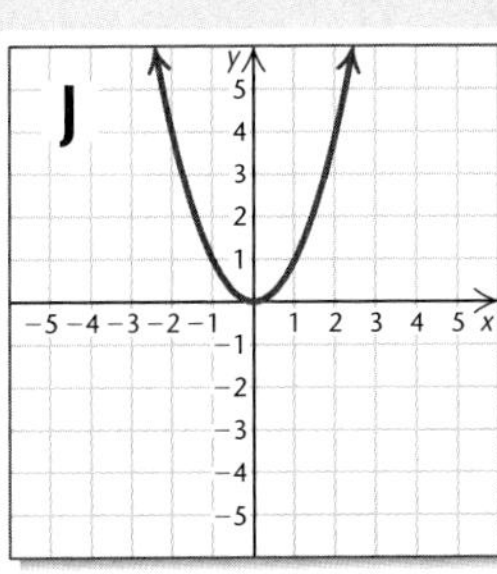

9.5 Exercise Set

FOR EXTRA HELP MyMathLab Math XL PRACTICE  WATCH  DOWNLOAD 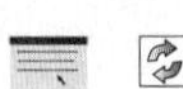 READ REVIEW

Concept Reinforcement *In each of Exercises 1–8, match the word or phrase with the most appropriate choice from the column on the right.*

1. ____ Total cost
2. ____ Total revenue
3. ____ Total profit
4. ____ Fixed costs
5. ____ Variable costs
6. ____ Break-even point
7. ____ Equilibrium point
8. ____ Price

a) The amount of money that a company takes in
b) The sum of fixed costs and variable costs
c) The point at which total revenue equals total cost
d) What consumers pay per item
e) The difference between total revenue and total cost
f) What companies spend whether or not a product is produced
g) The point at which supply equals demand
h) The costs that vary according to the number of items produced

For each of the following pairs of total-cost and total-revenue functions, find **(a)** *the total-profit function and* **(b)** *the break-even point.*

9. $C(x) = 45x + 300{,}000;$ $R(x) = 65x$

10. $C(x) = 25x + 270{,}000;$ $R(x) = 70x$

11. $C(x) = 15x + 3100;$ $R(x) = 40x$

12. $C(x) = 30x + 49{,}500;$ $R(x) = 85x$

13. $C(x) = 40x + 22{,}500;$ $R(x) = 85x$

14. $C(x) = 20x + 10{,}000;$ $R(x) = 100x$

15. $C(x) = 24x + 50{,}000;$ $R(x) = 40x$

16. $C(x) = 40x + 8010;$ $R(x) = 58x$

Aha! **17.** $C(x) = 75x + 100{,}000;$ $R(x) = 125x$

18. $C(x) = 20x + 120{,}000;$ $R(x) = 50x$

Find the equilibrium point for each of the following pairs of demand and supply functions.

19. $D(p) = 1000 - 10p,$ $S(p) = 230 + p$

20. $D(p) = 2000 - 60p,$ $S(p) = 460 + 94p$

21. $D(p) = 760 - 13p,$ $S(p) = 430 + 2p$

22. $D(p) = 800 - 43p,$ $S(p) = 210 + 16p$

23. $D(p) = 7500 - 25p,$ $S(p) = 6000 + 5p$

24. $D(p) = 8800 - 30p,$ $S(p) = 7000 + 15p$

25. $D(p) = 1600 - 53p,$ $S(p) = 320 + 75p$

26. $D(p) = 5500 - 40p,$ $S(p) = 1000 + 85p$

Solve.

27. *Manufacturing MP3 Players.* SoundGen, Inc., is planning to manufacture a new type of MP3 player/cell phone. The fixed costs for production are \$45,000. The variable costs for producing each unit are estimated to be \$40. The revenue from each unit is to be \$130. Find the following.

a) The total cost $C(x)$ of producing x MP3/cell phones
b) The total revenue $R(x)$ from the sale of x MP3/cell phones
c) The total profit $P(x)$ from the production and sale of x MP3/cell phones
d) The profit or loss from the production and sale of 3000 MP3/cell phones; of 400 MP3/cell phones
e) The break-even point

28. *Computer Manufacturing.* Current Electronics is planning to introduce a new laptop computer. The fixed costs for production are \$125,300. The variable costs for producing each computer are \$450. The revenue from each computer is \$800. Find the following.

a) The total cost $C(x)$ of producing x computers
b) The total revenue $R(x)$ from the sale of x computers
c) The total profit $P(x)$ from the production and sale of x computers
d) The profit or loss from the production and sale of 100 computers; of 400 computers
e) The break-even point

29. *Pet Safety.* Christine designed and is now producing a pet car seat. The fixed costs for setting up production are \$10,000. The variable costs for producing each seat are \$30. The revenue from each seat is to be \$80. Find the following.

a) The total cost $C(x)$ of producing x seats
b) The total revenue $R(x)$ from the sale of x seats
c) The total profit $P(x)$ from the production and sale of x seats
d) The profit or loss from the production and sale of 2000 seats; of 50 seats
e) The break-even point

30. *Manufacturing Caps.* Martina's Custom Printing is planning to add painter's caps to its product line. For the first year, the fixed costs for setting up production are \$16,404. The variable costs for producing a dozen caps are \$6.00. The revenue on each dozen caps will be \$18.00. Find the following.

a) The total cost $C(x)$ of producing x dozen caps
b) The total revenue $R(x)$ from the sale of x dozen caps
c) The total profit $P(x)$ from the production and sale of x dozen caps
d) The profit or loss from the production and sale of 3000 dozen caps; of 1000 dozen caps
e) The break-even point

31. *Dog Food Production.* Puppy Love, Inc., will soon begin producing a new line of puppy food. The marketing department predicts that the demand function will be $D(p) = -14.97p + 987.35$ and the supply function will be $S(p) = 98.55p - 5.13$.

a) To the nearest cent, what price per unit should be charged in order to have equilibrium between supply and demand?
b) The production of the puppy food involves \$87,985 in fixed costs and \$5.15 per unit in variable costs. If the price per unit is the value you found in part (a), how many units must be sold in order to break even?

32. *Computer Production.* Brushstroke Computers is planning a new line of computers, each of which will sell for \$970. The fixed costs in setting up production are \$1,235,580 and the variable costs for each computer are \$697.

a) What is the break-even point? (Round to the nearest whole number.)
b) The marketing department at Brushstroke is not sure that \$970 is the best price. Their demand function for the new computers is given by $D(p) = -304.5p + 374{,}580$ and their supply function is given by $S(p) = 788.7p - 576{,}504$. What price p would result in equilibrium between supply and demand?
c) If the computers are sold for the equilibrium price found in part (b), what is the break-even point?

TW **33.** In Example 1, the slope of the line representing Revenue is the sum of the slopes of the other two lines. This is not a coincidence. Explain why.

TW **34.** Variable costs and fixed costs are often compared to the slope and the y-intercept, respectively, of an equation for a line. Explain why you feel this analogy is or is not valid.

SKILL REVIEW

Review function notation and domains of functions (Sections 3.8, 5.9, 8.1, and 8.2).

35. If $f(x) = 4x - 7$, find $f(a) + h$. [3.8]

36. If $f(x) = 4x - 7$, find $f(a + h)$. [5.9]

Find the domain of f.

37. $f(x) = \dfrac{x - 5}{2x + 1}$ [3.8], [8.2]

38. $f(x) = \dfrac{3x}{x^2 + 1}$ [8.2]

39. $f(x) = \sqrt{2x + 8}$ [8.1]

40. $f(x) = \dfrac{3x}{x^2 - 1}$ [3.8], [8.2]

SYNTHESIS

TW **41.** Les claims that since his fixed costs are $3000, he need sell only 10 custom birdbaths at $300 each in order to break even. Does this sound plausible? Why or why not?

TW **42.** In this section, we examined supply and demand functions for coffee. Does it seem realistic to you for the graph of D to have a constant slope? Why or why not?

43. *Yo-yo Production.* Bing Boing Hobbies is willing to produce 100 yo-yo's at $2.00 each and 500 yo-yo's at $8.00 each. Research indicates that the public will buy 500 yo-yo's at $1.00 each and 100 yo-yo's at $9.00 each. Find the equilibrium point.

44. *Loudspeaker Production.* Sonority Speakers, Inc., has fixed costs of $15,400 and variable costs of $100 for each pair of speakers produced. If the speakers sell for $250 per pair, how many pairs of speakers must be produced (and sold) in order to have enough profit to cover the fixed costs of two additional facilities? Assume that all fixed costs are identical.

45. *Peanut Butter.* The table below lists the data for supply and demand of an 18-oz jar of peanut butter at various prices.

Price	Supply (in millions)	Demand (in millions)
$1.59	23.4	22.5
1.29	19.2	24.8
1.69	26.8	22.2
1.19	18.4	29.7
1.99	30.7	19.3

a) Use linear regression to find the supply function $S(p)$ for suppliers of peanut butter at price p.
b) Use linear regression to find the demand function $D(p)$ for consumers of peanut butter at price p.
c) Find the equilibrium point.

46. *Silver.* Because silver is a good conductor of electricity and heat, and is both antimicrobial and antibacterial, it is used in many industrial applications. For instance, silver is used in batteries, toasters, cell phones, flat-screen televisions, ink, and bandages for wounds, in clothing to minimize odor, and in wood to resist mold. Because of such a wide variety of applications, industrial demand for silver is growing, as shown in the table below.

Year	Amount of Silver Used in Industrial Applications (in millions of ounces)
2001	335.2
2002	339.1
2003	349.7
2004	367.1
2005	405.1
2006	424.5
2007	453.5
2008	447.2

Source: World Silver Survey 2009

a) Find a linear function that can be used to estimate the industrial demand $S(t)$, in millions of ounces, for silver t years after 2000.
b) In 2010, there were about one billion ounces of silver in above-ground supply. In what year will industrial demand of silver be one billion ounces?

Try Exercise Answers: Section 9.5

9. **(a)** $P(x) = 20x - 300{,}000$; **(b)** (15,000 units, $975,000)
19. ($70, 300)

Study Summary

KEY TERMS AND CONCEPTS	EXAMPLES	PRACTICE EXERCISES
SECTION 9.1: SYSTEMS OF EQUATIONS IN THREE VARIABLES		
Systems of three equations in three variables are usually easiest to solve using elimination.	Solve: $x+y-z=3$, (1) $-x+y+2z=-5$, (2) $2x-y-3z=9$. (3) *Eliminate x using two equations:* $x+y-z=3$ (1); $-x+y+2z=-5$ (2); $2y+z=-2$ *Eliminate x again using two different equations:* $-2x-2y+2z=-6$ (1); $2x-y-3z=9$ (3); $-3y-z=3$ *Solve the system of two equations for y and z:* $2y+z=-2$; $-3y-z=3$; $-y=1$; $y=-1$; $2(-1)+z=-2$; $z=0$. *Substitute and solve for x:* $x+y-z=3$; $x+(-1)-0=3$; $x=4$. The solution is $(4,-1,0)$.	**1.** Solve: $x-2y-z=8$, $2x+2y-z=8$, $x-8y+z=1$.
SECTION 9.2: SOLVING APPLICATIONS: SYSTEMS OF THREE EQUATIONS		
Many problems with three unknowns can be solved after translating to a system of three equations.	In a triangular cross section of a roof, the largest angle is 70° greater than the smallest angle, and the largest angle is twice as large as the remaining angle. Find the measure of each angle. The angles in the triangle measure 30°, 50°, and 100°. (See Example 2 on pp. 657–658 for a complete solution.)	**2.** The sum of three numbers is 9. The third number is half the sum of the first and second numbers. The second number is 2 less than the sum of the first and third numbers. Find the numbers.
SECTION 9.3: ELIMINATION USING MATRICES		
A **matrix** (plural, **matrices**) is a rectangular array of numbers. The individual numbers are called **entries** or **elements.** By using **row-equivalent** operations, we can solve systems of equations using matrices.	Solve: $x+4y=1$, $2x-y=3$. Write as a matrix in row-echelon form: $\left[\begin{array}{cc\|c} 1 & 4 & 1 \\ 2 & -1 & 3 \end{array}\right] \longrightarrow \left[\begin{array}{cc\|c} 1 & 4 & 1 \\ 0 & -9 & 1 \end{array}\right]$. Rewrite as equations and solve: $-9y=1$, $y=-\frac{1}{9}$ $\longrightarrow$ $x+4\left(-\frac{1}{9}\right)=1$, $x=\frac{13}{9}$. The solution is $\left(\frac{13}{9},-\frac{1}{9}\right)$.	**3.** Solve using matrices: $3x-2y=10$, $x+y=5$.

SECTION 9.4: DETERMINANTS AND CRAMER'S RULE

A **determinant** is a number associated with a square matrix.

Determinant of a 2 × 2 Matrix

$$\begin{vmatrix} a & c \\ b & d \end{vmatrix} = ad - bc$$

$$\begin{vmatrix} 2 & 3 \\ -1 & 5 \end{vmatrix} = 2 \cdot 5 - (-1)(3) = 13$$

Determinant of a 3 × 3 Matrix

$$\begin{vmatrix} a_1 & b_1 & c_1 \\ a_2 & b_2 & c_2 \\ a_3 & b_3 & c_3 \end{vmatrix} = a_1\begin{vmatrix} b_2 & c_2 \\ b_3 & c_3 \end{vmatrix} - a_2\begin{vmatrix} b_1 & c_1 \\ b_3 & c_3 \end{vmatrix} + a_3\begin{vmatrix} b_1 & c_1 \\ b_2 & c_2 \end{vmatrix}$$

$$\begin{vmatrix} 2 & 3 & 2 \\ 0 & 1 & 0 \\ -1 & 5 & -4 \end{vmatrix}$$
$$= 2\begin{vmatrix} 1 & 0 \\ 5 & -4 \end{vmatrix} - 0\begin{vmatrix} 3 & 2 \\ 5 & -4 \end{vmatrix} + (-1)\begin{vmatrix} 3 & 2 \\ 1 & 0 \end{vmatrix}$$
$$= 2(-4 - 0) - 0 - 1(0 - 2)$$
$$= -8 + 2 = -6$$

Evaluate.

4. $\begin{vmatrix} 3 & -5 \\ 2 & 6 \end{vmatrix}$

5. $\begin{vmatrix} 1 & 2 & -1 \\ 2 & 0 & 3 \\ 0 & 1 & 5 \end{vmatrix}$

We can use matrices and **Cramer's rule** to solve systems of equations.

Cramer's rule for 2 × 2 matrices is given on p. 672.

Cramer's rule for 3 × 3 matrices is given on p. 674.

Solve:

$$x - 3y = 7,$$
$$2x + 5y = 4.$$

$$x = \frac{\begin{vmatrix} 7 & -3 \\ 4 & 5 \end{vmatrix}}{\begin{vmatrix} 1 & -3 \\ 2 & 5 \end{vmatrix}}; \qquad y = \frac{\begin{vmatrix} 1 & 7 \\ 2 & 4 \end{vmatrix}}{\begin{vmatrix} 1 & -3 \\ 2 & 5 \end{vmatrix}}$$

$$x = \tfrac{47}{11} \qquad y = \tfrac{-10}{11}$$

The solution is $\left(\frac{47}{11}, -\frac{10}{11}\right)$.

6. Solve using Cramer's rule:

$$3x - 5y = 12,$$
$$2x + 6y = 1.$$

SECTION 9.5: BUSINESS AND ECONOMICS APPLICATIONS

The **break-even point** occurs where the **revenue** equals the **cost**, or where **profit** is 0.

Find **(a)** the total-profit function and **(b)** the break-even point for the total-cost and total-revenue functions

$$C(x) = 38x + 4320 \quad \text{and} \quad R(x) = 62x.$$

a) Profit = Revenue − Cost

$$P(x) = R(x) - C(x)$$
$$P(x) = 62x - (38x + 4320)$$
$$P(x) = 24x - 4320$$

b) $C(x) = R(x)$ **At the break-even point, revenue = cost.**

$38x + 4320 = 62x$

$180 = x$ **Solving for x**

$R(180) = 11{,}160$ **Finding the revenue (or cost) at the break-even point**

The break-even point is (180, $11,160).

7. Find **(a)** the total-profit function and **(b)** the break-even point for the total-cost and total-profit functions

$$C(x) = 15x + 9000,$$
$$R(x) = 90x.$$

An **equilibrium point** occurs where the **supply** equals the **demand**.	Find the equilibrium point for the demand and supply functions $S(p) = 60 + 7p$ and $D(p) = 90 - 13p$. $S(p) = D(p)$ **At the equilibrium point, supply = demand.** $60 + 7p = 90 - 13p$ $20p = 30$ $p = 1.5$ **Solving for p** $S(1.5) = 70.5$ **Finding the supply (or demand) at the equilibrium point** The equilibrium point is $(\$1.50, 70.5)$.	**8.** Find the equilibrium point for the supply and demand functions $S(p) = 60 + 9p,$ $D(p) = 195 - 6p.$

Review Exercises 9

Concept Reinforcement *Complete each of the following sentences.*

1. Systems of three equations in three variables are usually solved using the ________ method. [9.1]
2. If a system has at least one solution, it is said to be ________. [9.1]
3. The sum of the measures of the angles in any triangle is ________. [9.2]
4. When a matrix is in reduced row-echelon form, the first nonzero number in any nonzero row is ________. [9.3]
5. When a matrix has the same number of rows and columns, it is said to be ________. [9.4]
6. Cramer's rule is a formula in which the numerator and the denominator of each fraction is a(n) ________. [9.4]
7. Total revenue minus total cost is ________. [9.5]
8. ________ costs must be paid whether a product is produced or not. [9.5]
9. At the ________, the amount that the seller will supply is the same as the amount that the consumer will buy. [9.5]
10. At the break-even point, the value of the profit function is ________. [9.5]

Solve. If a system's equations are dependent or if there is no solution, state this. [9.1]

11. $x + 4y + 3z = 2,$
$2x + y + z = 10,$
$-x + y + 2z = 8$

12. $4x + 2y - 6z = 34,$
$2x + y + 3z = 3,$
$6x + 3y - 3z = 37$

13. $2x - 5y - 2z = -4,$
$7x + 2y - 5z = -6,$
$-2x + 3y + 2z = 4$

14. $2x - 3y + z = 1,$
$x - y + 2z = 5,$
$3x - 4y + 3z = -2$

15. $3x + y = 2,$
$x + 3y + z = 0,$
$x + z = 2$

Solve.

16. In triangle *ABC*, the measure of angle *A* is four times the measure of angle *C*, and the measure of angle *B* is 45° more than the measure of angle *C*. What are the measures of the angles of the triangle? [9.2]

17. The sum of the average number of times a man, a woman, and a one-year-old child cry each month is 56.7. A woman cries 3.9 more times than a man. The average number of times a one-year-old cries per month is 43.3 more than the average number of times combined that a man and a woman cry. What is the average number of times per month that each cries? [9.2]

Solve using matrices. Show your work. [9.3]

18. $3x + 4y = -13,$
$5x + 6y = 8$

19. $3x - y + z = -1,$
$2x + 3y + z = 4,$
$5x + 4y + 2z = 5$

Evaluate. [9.4]

20. $\begin{vmatrix} -2 & -5 \\ 3 & 10 \end{vmatrix}$

21. $\begin{vmatrix} 2 & 3 & 0 \\ 1 & 4 & -2 \\ 2 & -1 & 5 \end{vmatrix}$

Solve using Cramer's rule. Show your work. [9.4]

22. $2x + 3y = 6,$
$x - 4y = 14$

23. $2x + y + z = -2,$
$2x - y + 3z = 6,$
$3x - 5y + 4z = 7$

24. Find **(a)** the total-profit function and **(b)** the break-even point for the total-cost and total-revenue functions

$$C(x) = 30x + 15{,}800,$$
$$R(x) = 50x. \text{ [9.5]}$$

25. Find the equilibrium point for the demand and supply functions

$$S(p) = 60 + 7p,$$
$$D(p) = 120 - 13p. \text{ [9.5]}$$

26. Danae is beginning to produce organic honey. For the first year, the fixed costs for setting up production are \$54,000. The variable costs for producing each pint of honey are \$4.75. The revenue from each pint of honey is \$9.25. Find the following. [9.5]

a) The total cost $C(x)$ of producing x pints of honey
b) The total revenue $R(x)$ from the sale of x pints of honey
c) The total profit $P(x)$ from the production and sale of x pints of honey
d) The profit or loss from the production and sale of 5000 pints of honey; of 15,000 pints of honey
e) The break-even point

SYNTHESIS

TW **27.** How would you go about solving a problem that involves four variables? [9.2]

TW **28.** Explain how a system of equations can be both dependent and inconsistent. [9.1]

29. Danae is leaving a job that pays \$36,000 per year to make honey (see Exercise 26). How many pints of honey must she produce and sell in order to make as much money as she earned at her previous job? [9.5]

30. The graph of $f(x) = ax^2 + bx + c$ contains the points $(-2, 3)$, $(1, 1)$, and $(0, 3)$. Find *a, b,* and *c* and give a formula for the function. [9.2]

Chapter Test 9

Solve. If a system's equations are dependent or if there is no solution, state this.

1. $-3x + y - 2z = 8,$
$-x + 2y - z = 5,$
$2x + y + z = -3$

2. $6x + 2y - 4z = 15,$
$-3x - 4y + 2z = -6,$
$4x - 6y + 3z = 8$

3. $2x + 2y = 0,$
$4x + 4z = 4,$
$2x + y + z = 2$

4. $3x + 3z = 0,$
$2x + 2y = 2,$
$3y + 3z = 3$

Solve using matrices. Show your work.

5. $4x + y = 12,$
$3x + 2y = 2$

6. $x + 3y - 3z = 12,$
$3x - y + 4z = 0,$
$-x + 2y - z = 1$

Evaluate.

7. $\begin{vmatrix} 4 & -2 \\ 3 & -5 \end{vmatrix}$

8. $\begin{vmatrix} 3 & 4 & 2 \\ -2 & -5 & 4 \\ 0 & 5 & -3 \end{vmatrix}$

9. Solve using Cramer's rule:
$3x + 4y = -1,$
$5x - 2y = 4.$

10. In triangle ABC, the measure of angle B is 5° less than three times that of angle A. The measure of angle C is 5° more than three times that of angle B. Find the angle measures.

11. An electrician, a carpenter, and a plumber are hired to work on a house. The electrician earns \$30 per hour, the carpenter \$28.50 per hour, and the plumber \$34 per hour. The first day on the job, they worked a total of 21.5 hr and earned a total of \$673.00. If the plumber worked 2 more hours than the carpenter did, how many hours did each work?

12. Find the equilibrium point for the demand and supply functions

$D(p) = 79 - 8p$ and $S(p) = 37 + 6p,$

where p is the price, in dollars, $D(p)$ is the number of units demanded, and $S(p)$ is the number of units supplied.

13. Kick Back, Inc., is producing a new hammock. For the first year, the fixed costs for setting up production are \$44,000. The variable costs for producing each hammock are \$25. The revenue from each hammock is \$80. Find the following.

a) The total cost $C(x)$ of producing x hammocks
b) The total revenue $R(x)$ from the sale of x hammocks
c) The total profit $P(x)$ from the production and sale of x hammocks
d) The profit or loss from the production and sale of 300 hammocks; of 900 hammocks
e) The break-even point

SYNTHESIS

14. Solve:
$2w - x + 5y - z = 1,$
$w + x - 2y + z = 2,$
$w + 3x - y + z = 0,$
$3w + x + y - z = 0.$

15. At a county fair, an adult's ticket sold for \$5.50, a senior citizen's ticket for \$4.00, and a child's ticket for \$1.50. On opening day, the number of adults' and senior citizens' tickets sold was 30 more than the number of children's tickets sold. The number of adults' tickets sold was 6 more than four times the number of senior citizens' tickets sold. Total receipts from the ticket sales were \$11,219.50. How many of each type of ticket were sold?

Cumulative Review: Chapters 1–9

Simplify. Do not leave negative exponents in your answers.

1. $3 + 24 \div 2^2 \cdot 3 - (6 - 7)$ [5.2]
2. $3c - [8 - 2(1 - c)]$ [5.2]
3. -10^{-2} [5.2]
4. $(3xy^{-4})(-2x^3y)$ [5.2]
5. $\left(\dfrac{18a^2b^{-1}}{12a^{-1}b}\right)^2$ [5.2]
6. $\dfrac{2x - 10}{x^3 - 125}$ [7.1]

Perform the indicated operations.

7. $(x - 5)(x + 5)$ [5.6]
8. $(3n - 2)(5n + 7)$ [5.6]
9. $\dfrac{1}{x + 5} - \dfrac{1}{x - 5}$ [7.4]
10. $\dfrac{x^2 - 3x}{2x^2 - x - 3} \div \dfrac{x^3}{x^2 - 2x - 3}$ [7.2]
11. $y + \dfrac{2}{3y}$ [7.4]

Factor.

12. $4x^3 + 18x^2$ [6.1]
13. $x^2 + 8x - 84$ [6.2]
14. $16y^2 - 81$ [6.4]
15. $64x^3 + 8$ [6.5]
16. $t^2 - 16t + 64$ [6.4]
17. $x^6 - x^2$ [6.4]
18. $0.027b^3 - 0.008c^3$ [6.5]
19. $20x^2 + 7x - 3$ [6.3]

Solve.

20. $3(x - 2) = 14 - x$ [2.2]
21. $x - 2 < 6$ *or* $2x + 1 > 5$ [8.1]
22. $x^2 - 2x - 3 = 5$ [6.2]
23. $\dfrac{3}{x + 1} = \dfrac{x}{4}$ [7.6]
24. $y = \frac{1}{2}x - 7$,
$2x - 4y = 3$ [4.2]
25. $x + 3y = 8$,
$2x - 3y = 7$ [4.3]
26. $-3x + 4y + z = -5$,
$x - 3y - z = 6$,
$2x + 3y + 5z = -8$ [9.1]
27. $|2x - 1| = 8$ [8.2]
28. $9(x - 3) - 4x < 2 - (3 - x)$ [2.6]
29. $|4t| > 12$ [8.2]
30. $|3x - 2| \leq 8$ [8.2]

In Exercises 31–34, match each equation with one of the following graphs.

a)

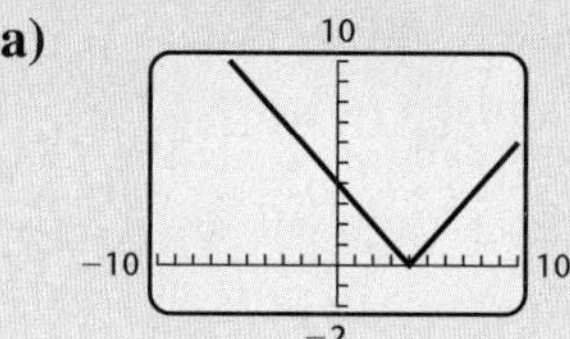

b)

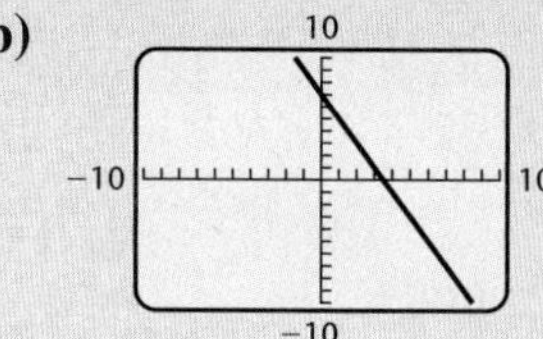

c)

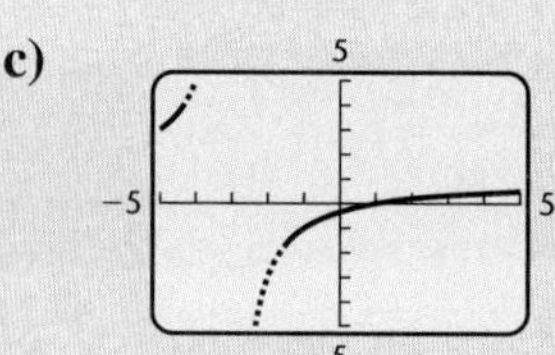

d)

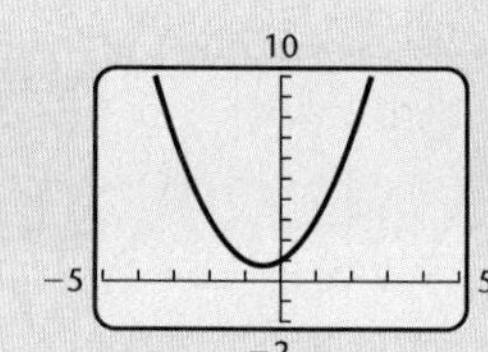

31. $y = -2x + 7$ [3.6]
32. $y = |x - 4|$ [8.2]
33. $y = x^2 + x + 1$ [6.1]
34. $y = \dfrac{x - 1}{x + 3}$ [7.1]

Graph on a plane.

35. $y = \frac{2}{3}x - 4$ [3.6]
36. $x = -3$ [3.3]
37. $3x - y = 3$ [3.3]
38. $x + y \geq -2$ [8.3]
39. $f(x) = -x + 1$ [3.8]
40. $x - 2y > 4$,
$x + 2y \geq -2$ [8.3]
41. Find the slope and the y-intercept of the line given by $4x - 9y = 18$. [3.6]
42. Write a slope–intercept equation for the line with slope -7 and containing the point $(-3, -4)$. [3.7]

43. Find an equation of the line with y-intercept $(0, 4)$ and perpendicular to the line given by $3x + 2y = 1$. [3.6]

44. For the graph of f shown, determine the domain and the range of f. [3.8]

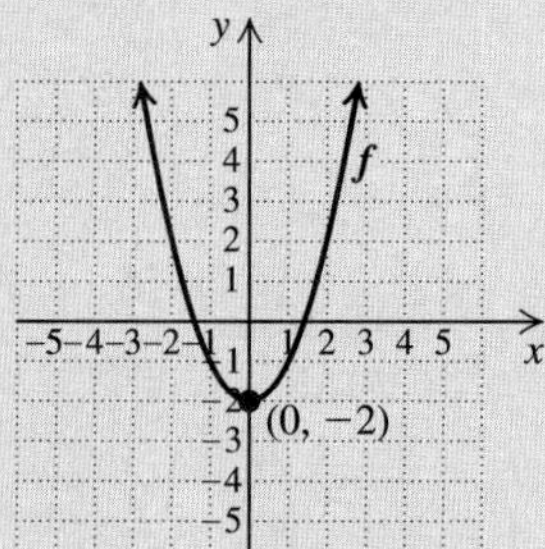

45. Determine the domain of the function given by

$$f(x) = \frac{3}{2x + 5}. \quad [4.1]$$

46. Find $g(-2)$ if $g(x) = 3x^2 - 5x$. [3.8]

47. Find $(f - g)(x)$ if $f(x) = x^2 + 3x$ and $g(x) = 9 - 3x$. [5.9]

48. Graph the solution set of $-3 \leq f(x) \leq 2$, where $f(x) = 1 - x$. [8.1]

49. Find the domain of h/g if $h(x) = \frac{1}{x}$ and $g(x) = 3x - 1$. [5.9]

50. Solve for t: $at - dt = c$. [2.3]

Find the solutions of each equation, inequality, or system from the given graph.

51. $10 - 5x = 3x + 2$ [4.5]

52. $f(x) \leq g(x)$ [8.1]

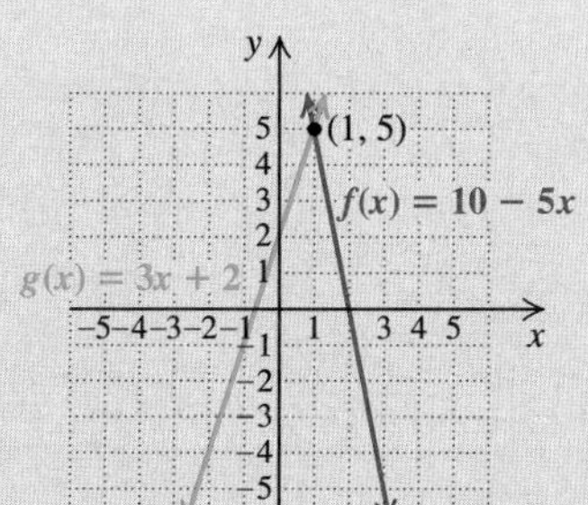

53. $y = x - 5,$
$y = 1 - 2x$ [4.2]

54. $x - 5 = 1 - 2x$ [4.5]

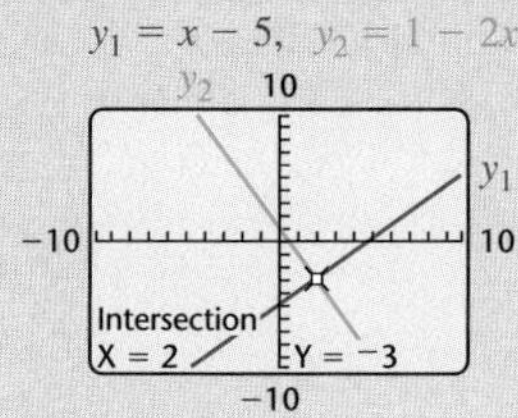

55. $x^2 - 2x + 1 = 0$ [6.1]

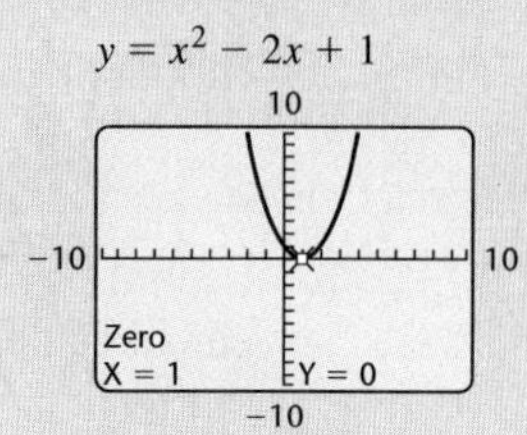

56. The Baqueira, a resort in Spain, uses 549 snow cannons to make snow for its 4344 acres of ski runs. How many snow cannons should a resort containing 1448 acres of runs use in order to make a comparable amount of snow? [7.7]
Source: www.bluebookski.com

57. *Water Usage.* In dry climates, it takes about 11,600 gal of water to produce a pound of beef and a pound of wheat. The pound of beef requires 7000 more gallons of water than the pound of wheat. How much water does it take to produce each? [4.4]
Source: *The Wall Street Journal*, 1/28/08

58. *Fundraising.* Michelle is planning a fundraising dinner for Happy Hollow Children's Camp. The banquet facility charges a rental fee of \$1500, but will waive the rental fee if more than \$6000 is spent for catering. Michelle knows that 150 people will attend the dinner. [2.7]

a) How much should each dinner cost in order for the rental fee to be waived?

b) For what costs per person will the total cost (including the rental fee) exceed \$6000?

59. *Perimeter of a Rectangle.* The perimeter of a rectangle is 32 cm. If five times the width equals three times the length, what are the dimensions of the rectangle? [4.4]

60. *Utility Bills.* One month Lori and Tony spent \$920 for electricity, rent, and cell phone. The electric bill was $\frac{1}{4}$ the amount of the rent, and the phone bill was \$40 less than the electric bill. How much was the rent? [9.2]

61. *Fast-Food Sales.* In 2006, 18- to 34-year-olds ate fast food an average of 19 times per month. By 2009, this figure had dropped to 14 times per month. [3.7], [3.8]
Source: Based on data from Sandelman & Associates

a) Let $f(t)$ = the average number of times per month that 18- to 34-year-olds ate fast food, where t is the number of years after 2006. Find a linear function that fits the data.

b) Use the function from part (a) to predict the average number of visits in 2012.

c) In what year will 18- to 34-year-olds eat fast food an average of 4 times per month?

62. *Theft.* The table below lists the number of truckloads of cargo stolen in the United States for various years. [3.7], [3.8]

Year	Number of Truckloads Stolen
2007	672
2008	767
2009	859

Source: FreightWatch International

a) Use linear regression to find a function $c(t)$ that can be used to estimate the number of truckloads stolen t years after 2007.

b) Use the function of part (a) to estimate the number of truckloads of cargo stolen in 2011.

SYNTHESIS

63. Use interval notation to write the domain of the function given by

$$f(x) = \frac{\sqrt{x+4}}{x}. \quad [8.1]$$

64. Simplify: $\dfrac{2^{a-1} \cdot 2^{4a}}{2^{3(-2a+5)}}$. [5.2]

65. Find all roots for $f(x) = x^4 - 34x^2 + 225$. [6.4]

Solve.

66. $4 \le |3 - x| \le 6$ [8.1], [8.2]

67. $\dfrac{18}{x-9} + \dfrac{10}{x+5} = \dfrac{28x}{x^2 - 4x - 45}$ [7.6]

Exponents and Radical Functions

How Much Does Health Insurance Cost?

The graph below shows the average total cost to both employer and employee for family health insurance from 2001 through 2009. In Example 14 of Section 10.1, we use these data to estimate the average total cost for 2010.

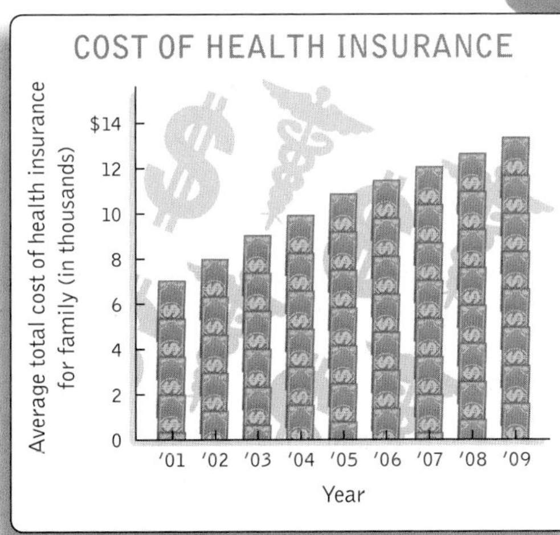

In this chapter, we learn about square roots, cube roots, fourth roots, and so on. These roots can be written using radical notation and appear in both radical expressions and radical equations. Exponents that are fractions are also studied and are used to ease some of our work with radicals. The chapter closes with an introduction to the complex-number system.

10.1 Radical Expressions, Functions, and Models

- Square Roots and Square-Root Functions
- Expressions of the Form $\sqrt{a^2}$
- Cube Roots
- Odd and Even nth Roots
- Radical Functions and Models

In this section, we consider roots, such as square roots and cube roots, and the *radical* expressions involving such roots.

SQUARE ROOTS AND SQUARE-ROOT FUNCTIONS

When a number is multiplied by itself, we say that the number is squared. If we can find a number that was squared in order to produce some value a, we call that number a *square root* of a.

Square Root The number c is a *square root* of a if $c^2 = a$.

For example,

9 has -3 and 3 as square roots because $(-3)^2 = 9$ and $3^2 = 9$.

25 has -5 and 5 as square roots because $(-5)^2 = 25$ and $5^2 = 25$.

-4 does not have a real-number square root because there is no real number c such that $c^2 = -4$.

Every positive number has two square roots, and 0 has only itself as a square root. Negative numbers do not have real-number square roots, although later in this chapter we introduce the *complex-number system* in which such square roots do exist.

EXAMPLE 1 Find the two square roots of 64.

SOLUTION The square roots of 64 are 8 and -8, because $8^2 = 64$ and $(-8)^2 = 64$.

Try Exercise 9.

Student Notes

It is important to remember the difference between *the* square root of 9 and *a* square root of 9. *A* square root of 9 means either 3 or -3, but *the* square root of 9, or $\sqrt{9}$, means the principal square root of 9, or 3.

Whenever we refer to *the* square root of a number, we mean the nonnegative square root of that number. This is often referred to as the *principal square root* of the number.

Principal Square Root The *principal square root* of a nonnegative number is its nonnegative square root. The symbol $\sqrt{}$ is called a *radical sign* and is used to indicate the principal square root of the number over which it appears.

EXAMPLE 2 Simplify each of the following.

a) $\sqrt{25}$ **b)** $\sqrt{\frac{25}{64}}$

c) $-\sqrt{64}$ **d)** $\sqrt{0.0049}$

SOLUTION

a) $\sqrt{25} = 5$ — **$\sqrt{\ }$ indicates the principal square root. Note that $\sqrt{25} \neq -5$.**

b) $\sqrt{\frac{25}{64}} = \frac{5}{8}$ — **Since $\left(\frac{5}{8}\right)^2 = \frac{25}{64}$**

c) $-\sqrt{64} = -8$ — **Since $\sqrt{64} = 8$, $-\sqrt{64} = -8$.**

d) $\sqrt{0.0049} = 0.07$ — **$(0.07)(0.07) = 0.0049$. Note too that $\sqrt{0.0049} = \sqrt{\frac{49}{10{,}000}} = \frac{7}{100} = 0.07$.**

Try Exercise 17.

STUDY TIP

Advance Planning Pays Off

The best way to prepare for a final exam is to do so over a period of at least two weeks. First review each chapter, studying the terms, formulas, problems, properties, and procedures in the Study Summaries. Then retake your quizzes and tests. If you miss any questions, spend extra time reviewing the corresponding topics. Watch the videos that accompany the text or use the accompanying tutorial resources. Also consider participating in a study group or attending a tutoring or review session.

In addition to being read as "the principal square root of a," $\sqrt{a}$ is also read as "the square root of a," or simply "root a" or "radical a." Any expression in which a radical sign appears is called a *radical expression*. The following are radical expressions:

$$\sqrt{5}, \quad \sqrt{a}, \quad -\sqrt{3x}, \quad \sqrt{\frac{y^2+7}{y}}, \quad \sqrt{x} + 8.$$

The expression under the radical sign is called the **radicand.** In the expressions above, the radicands are 5, a, $3x$, $(y^2 + 7)/y$, and x, respectively.

To calculate $\sqrt{5}$ on most graphing calculators, we press [√] [5] [)] [ENTER] if a left parenthesis is automatically supplied. On some calculators, [√] is pressed after 5 has been entered. A calculator will display an approximation like

2.23606798

for $\sqrt{5}$. The exact value of $\sqrt{5}$ is not given by any repeating or terminating decimal. In general, for any whole number a that is not a perfect square, $\sqrt{a}$ *is* a nonterminating, nonrepeating decimal, or an *irrational number*.

The square-root function, given by

$$f(x) = \sqrt{x},$$

has $[0, \infty)$ as its domain and $[0, \infty)$ as its range. We can draw its graph by selecting convenient values for x and calculating the corresponding outputs. Once these ordered pairs have been graphed, a smooth curve can be drawn.

$f(x) = \sqrt{x}$

x	$\sqrt{x}$	$(x, f(x))$
0	0	$(0, 0)$
1	1	$(1, 1)$
4	2	$(4, 2)$
9	3	$(9, 3)$

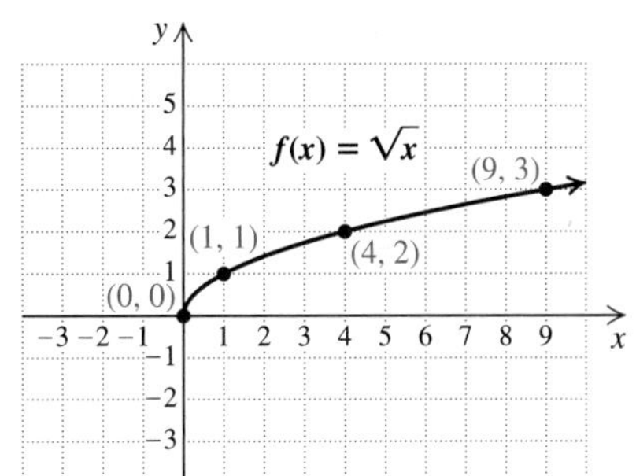

EXAMPLE 3 For each function, find the indicated function value.

a) $f(x) = \sqrt{3x - 2};\ f(1)$
b) $g(z) = -\sqrt{6z + 4};\ g(3)$

SOLUTION

a) $f(1) = \sqrt{3 \cdot 1 - 2}$ **Substituting**
$= \sqrt{1} = 1$ **Simplifying**

b) $g(3) = -\sqrt{6 \cdot 3 + 4}$ **Substituting**
$= -\sqrt{22}$ **Simplifying. This answer is exact.**
≈ -4.69041576 **Using a calculator to approximate $\sqrt{22}$**

Try Exercise 33.

EXPRESSIONS OF THE FORM $\sqrt{a^2}$

As the next example shows, $\sqrt{a^2}$ does not always simplify to a.

EXAMPLE 4 Evaluate $\sqrt{a^2}$ for the following values: **(a)** 5; **(b)** 0; **(c)** -5.

SOLUTION

a) $\sqrt{5^2} = \sqrt{25} = 5$ — **Same**

b) $\sqrt{0^2} = \sqrt{0} = 0$ — **Same**

c) $\sqrt{(-5)^2} = \sqrt{25} = 5$ — **Opposites** **Note that $\sqrt{(-5)^2} \neq -5$.**

You may have noticed that evaluating $\sqrt{a^2}$ is just like evaluating $|a|$.

Interactive Discovery

Use graphs or tables to determine which of the following are identities. Be sure to enclose the entire radicand in parentheses, if your calculator requires them.

1. $\sqrt{x^2} = x$
2. $\sqrt{x^2} = -x$
3. $\sqrt{x^2} = |x|$
4. $\sqrt{(x + 3)^2} = x + 3$
5. $\sqrt{(x + 3)^2} = |x + 3|$
6. $\sqrt{x^8} = x^4$

In general, we cannot say that $\sqrt{x^2} = x$. However, we can simplify $\sqrt{x^2}$ using absolute value.

Simplifying $\sqrt{a^2}$ For any real number a,

$$\sqrt{a^2} = |a|.$$

(The principal square root of a^2 is the absolute value of a.)

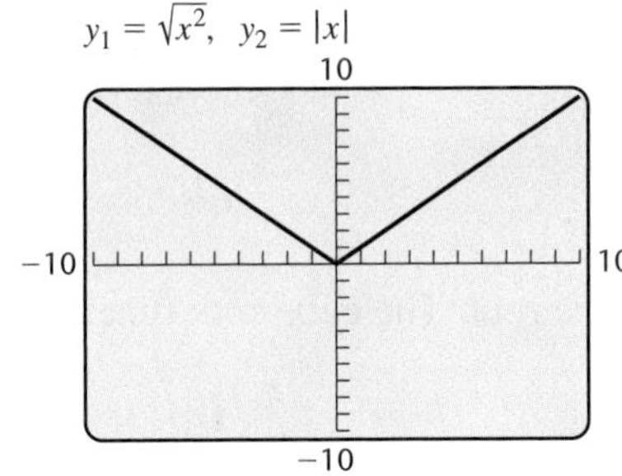

If we graph $f(x) = \sqrt{x^2}$ and $g(x) = |x|$, we find that the graphs of the functions coincide, as illustrated at left.

When a radicand is the square of a variable expression, like $(x + 1)^2$ or $36t^2$, absolute-value signs are needed when simplifying. We use absolute-value signs unless we know that the expression being squared is nonnegative. This ensures that our result is never negative.

EXAMPLE 5 Simplify each expression. Assume that the variable can represent any real number.

a) $\sqrt{36t^2}$ **b)** $\sqrt{(x + 1)^2}$ **c)** $\sqrt{x^2 - 8x + 16}$
d) $\sqrt{a^8}$ **e)** $\sqrt{t^6}$

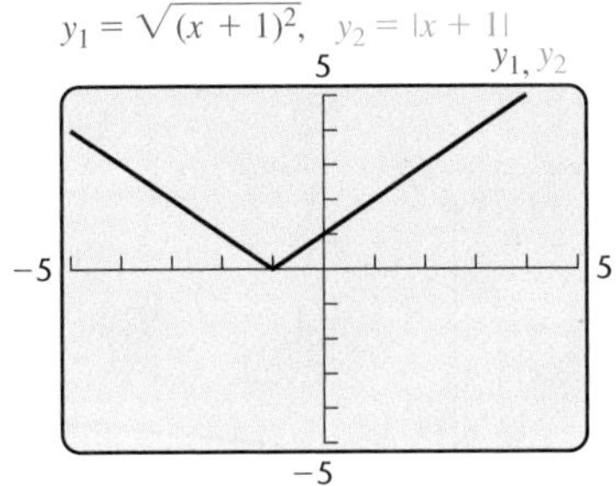

SOLUTION

a) $\sqrt{36t^2} = \sqrt{(6t)^2} = |6t|$, or $6|t|$ **Since t might be negative, absolute-value notation is necessary.**

b) $\sqrt{(x + 1)^2} = |x + 1|$ **Since $x + 1$ might be negative (for example, if $x = -3$), absolute-value notation is necessary.**

The graph at left confirms the identity.

c) $\sqrt{x^2 - 8x + 16} = \sqrt{(x - 4)^2} = |x - 4|$ **Since $x - 4$ might be negative, absolute-value notation is necessary.**

d) Note that $(a^4)^2 = a^8$ and that a^4 is never negative. Thus,

$\sqrt{a^8} = |a^4| = a^4.$ **Absolute-value notation is unnecessary here.**

e) Note that $(t^3)^2 = t^6$. Thus,

$\sqrt{t^6} = |t^3|.$ **Since t^3 can be negative, absolute-value notation is necessary.**

Try Exercise 39.

Student Notes

Some absolute-value notation can be simplified.

- $|ab| = |a| \cdot |b|$, so an expression like $|3x|$ can be written

  $|3x| = |3| \cdot |x| = 3|x|.$

- Even powers of real numbers are always positive, so

  $|x^2| = x^2,$
  $|x^4| = x^4,$
  $|x^6| = x^6,$ and so on.

If we assume that the expression being squared is nonnegative, then absolute-value notation is not necessary.

EXAMPLE 6 Simplify each expression. Assume that no radicands were formed by squaring negative quantities.

a) $\sqrt{y^2}$ **b)** $\sqrt{a^{10}}$ **c)** $\sqrt{9x^2 - 6x + 1}$

SOLUTION

a) $\sqrt{y^2} = y$ **We are assuming that y is nonnegative, so no absolute-value notation is necessary. When y is negative, $\sqrt{y^2} \neq y$.**

b) $\sqrt{a^{10}} = a^5$ **Assuming that a^5 is nonnegative. Note that $(a^5)^2 = a^{10}$.**

c) $\sqrt{9x^2 - 6x + 1} = \sqrt{(3x - 1)^2} = 3x - 1$ **Assuming that $3x - 1$ is nonnegative**

Try Exercise 69.

CUBE ROOTS

We often need to know what number cubed produces a certain value. When such a number is found, we say that we have found a *cube root*. For example,

2 is the cube root of 8 because $2^3 = 2 \cdot 2 \cdot 2 = 8$;
-4 is the cube root of -64 because $(-4)^3 = (-4)(-4)(-4) = -64$.

Cube Root The number c is the *cube root* of a if $c^3 = a$. In symbols, we write $\sqrt[3]{a}$ to denote the cube root of a.

Each real number has only one real-number cube root. The cube-root function, given by

$$f(x) = \sqrt[3]{x},$$

has $\mathbb{R}$ as its domain and $\mathbb{R}$ as its range. We can draw its graph by selecting convenient values for x and calculating the corresponding outputs. Once these ordered pairs have been graphed, a smooth curve can be drawn. Note that the cube root of a positive number is positive, the cube root of a negative number is negative, and the cube root of 0 is 0.

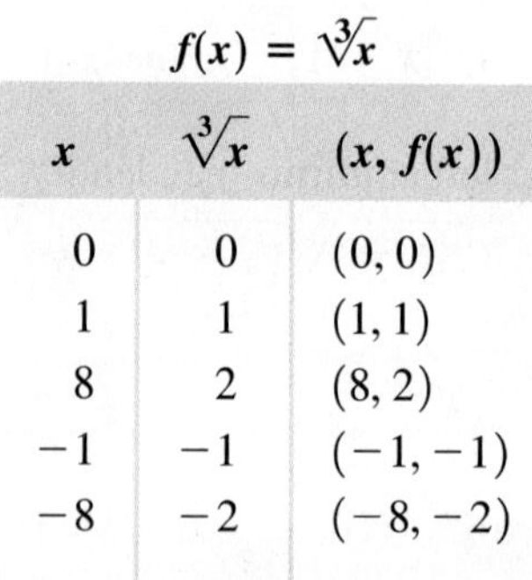

$f(x) = \sqrt[3]{x}$

x	$\sqrt[3]{x}$	$(x, f(x))$
0	0	(0, 0)
1	1	(1, 1)
8	2	(8, 2)
−1	−1	(−1, −1)
−8	−2	(−8, −2)

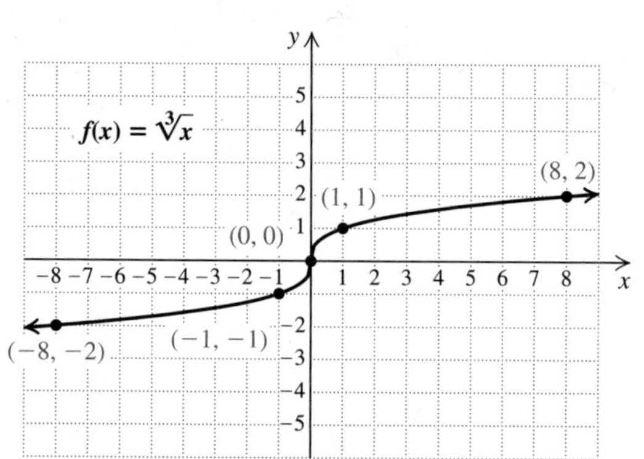

The following graphs illustrate that

$$\sqrt{x^2} = |x| \quad \text{and} \quad \sqrt[3]{x^3} = x.$$

$y = \sqrt{x^2}$

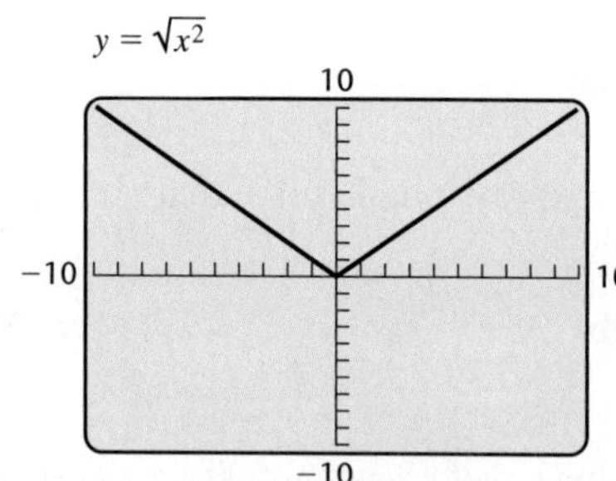

The graphs of $y = \sqrt{x^2}$ and $y = |x|$ are the same. Note that $y = \sqrt{x^2}$ is never negative.

$y = \sqrt[3]{x^3}$

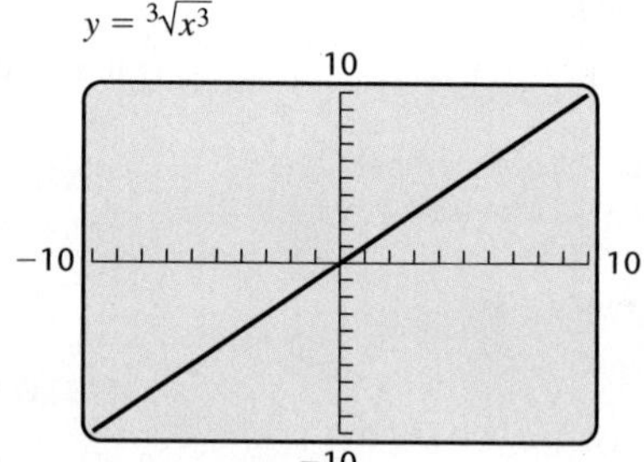

The graphs of $y = \sqrt[3]{x^3}$ and $y = x$ are the same. Note that $y = \sqrt[3]{x^3}$ can be negative.

EXAMPLE 7 For each function, find the indicated function value.

a) $f(y) = \sqrt[3]{y}$; $f(125)$

b) $g(x) = \sqrt[3]{x - 3}$; $g(-24)$

SOLUTION

a) $f(125) = \sqrt[3]{125} = 5$ **Since $5 \cdot 5 \cdot 5 = 125$**

b) $g(-24) = \sqrt[3]{-24 - 3}$

$= \sqrt[3]{-27}$

$= -3$ **Since $(-3)(-3)(-3) = -27$**

Try Exercise 89.

EXAMPLE 8 Simplify: $\sqrt[3]{-8y^3}$.

SOLUTION

$$\sqrt[3]{-8y^3} = -2y \qquad \text{Since } (-2y)(-2y)(-2y) = -8y^3$$

Try Exercise 83.

ODD AND EVEN nTH ROOTS

The 4th root of a number a is the number c for which $c^4 = a$. There are also 5th roots, 6th roots, and so on. We write $\sqrt[n]{a}$ for the principal nth root. The number n is called the **index** (plural, **indices**). When the index is 2, we do not write it.

When the index is odd, we are taking an *odd root.*

> Every number has exactly one real root when n is odd. Odd roots of positive numbers are positive and odd roots of negative numbers are negative.
>
> Absolute-value signs are not used when finding odd roots.

EXAMPLE 9 Simplify each expression.

a) $\sqrt[5]{32}$ **b)** $\sqrt[5]{-32}$ **c)** $-\sqrt[5]{32}$

d) $-\sqrt[5]{-32}$ **e)** $\sqrt[7]{x^7}$ **f)** $\sqrt[9]{(t-1)^9}$

SOLUTION

a) $\sqrt[5]{32} = 2$ Since $2^5 = 32$

b) $\sqrt[5]{-32} = -2$ Since $(-2)^5 = -32$

c) $-\sqrt[5]{32} = -2$ Taking the opposite of $\sqrt[5]{32}$

d) $-\sqrt[5]{-32} = -(-2) = 2$ Taking the opposite of $\sqrt[5]{-32}$

e) $\sqrt[7]{x^7} = x$ No absolute-value signs are needed.

f) $\sqrt[9]{(t-1)^9} = t - 1$

Try Exercise 57.

When the index n is even, we are taking an *even root.*

> Every positive real number has two real nth roots when n is even—one positive and one negative. Negative numbers do not have real nth roots when n is even.
>
> When n is even, the notation $\sqrt[n]{a}$ indicates the nonnegative nth root. Thus, when finding nth roots, absolute-value signs are often necessary.

Compare the following.

Odd Root	Even Root
$\sqrt[3]{8} = 2$	$\sqrt[4]{16} = 2$
$\sqrt[3]{-8} = -2$	$\sqrt[4]{-16}$ is not a real number.
$\sqrt[3]{x^3} = x$	$\sqrt[4]{x^4} = \lvert x\rvert$

EXAMPLE 10 Simplify each expression, if possible. Assume that variables can represent any real number.

a) $\sqrt[4]{16}$ b) $-\sqrt[4]{16}$ c) $\sqrt[4]{-16}$

d) $\sqrt[4]{81x^4}$ e) $\sqrt[6]{(y+7)^6}$

SOLUTION

a) $\sqrt[4]{16} = 2$ **Since $2^4 = 16$**

b) $-\sqrt[4]{16} = -2$ **Taking the opposite of $\sqrt[4]{16}$**

c) $\sqrt[4]{-16}$ cannot be simplified. **$\sqrt[4]{-16}$ is not a real number.**

d) $\sqrt[4]{81x^4} = |3x|$, or $3|x|$ **Using absolute-value notation since x could represent a negative number**

e) $\sqrt[6]{(y+7)^6} = |y+7|$ **Using absolute-value notation since $y + 7$ is negative for $y < -7$**

Try Exercise 63.

We summarize as follows.

Simplifying nth roots

n	a	$\sqrt[n]{a}$	$\sqrt[n]{a^n}$
Even	Positive	Positive	$\lvert a \rvert$
	Negative	Not a real number	
Odd	Positive	Positive	a
	Negative	Negative	

RADICAL FUNCTIONS AND MODELS

A **radical function** is a function that can be described by a radical expression. When the index of the expression is odd, the domain is the set of all real numbers. When the index is even, the radicand must be nonnegative. This often means that some numbers are not in the domain of a radical function.

EXAMPLE 11 Determine the domain of the function given by

$$g(x) = \sqrt[6]{7 - 3x}.$$

SOLUTION Since the index is even, the radicand, $7 - 3x$, must be nonnegative. We solve the inequality:

Set the radicand greater than or equal to 0.

$7 - 3x \geq 0$ **We cannot find the 6th root of a negative number.**

$-3x \geq -7$

Solve.

$x \leq \frac{7}{3}$. **Multiplying both sides by $-\frac{1}{3}$ and reversing the inequality**

Thus the domain of g is $\left\{x \middle| x \leq \frac{7}{3}\right\}$, or $\left(-\infty, \frac{7}{3}\right]$.

Try Exercise 93.

Radical Expressions and Functions

When entering radical expressions on a graphing calculator, care must be taken to place parentheses properly. For example, we enter

$$f(x) = \frac{-2 + \sqrt{1 - 6x}}{5}$$

on the Y = screen as ((−) 2 + 2ND √ 1 − 6 X,T,Θ,n)) ÷ 5. The outer parentheses enclose the numerator of the expression. The first right parenthesis) indicates the end of the radicand; the left parenthesis of the radicand is supplied by the calculator when √ is pressed.

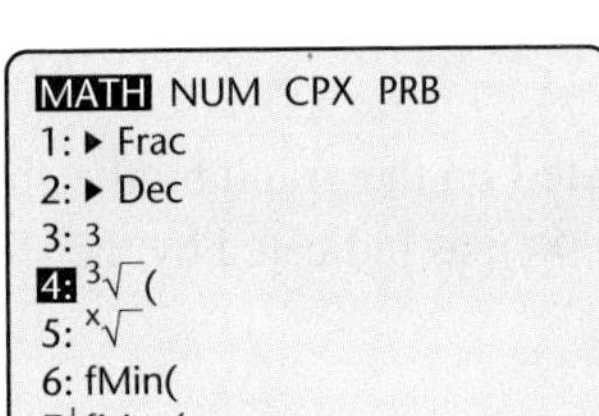

Cube roots can be entered using the $\sqrt[3]{}$(option in the MATH MATH menu, as shown at left. For an index other than 2 or 3, first enter the index, then choose the $\sqrt[x]{}$ option from the MATH MATH menu, and then enter the radicand. The $\sqrt[x]{}$ option may not supply the left parenthesis.

If we can determine the domain of the function algebraically, we can use that information to choose an appropriate viewing window. For example, compare the following graphs of the function $f(x) = 2\sqrt{15 - x}$. Note that the domain of f is $\{x \mid x \leq 15\}$, or $(-\infty, 15]$.

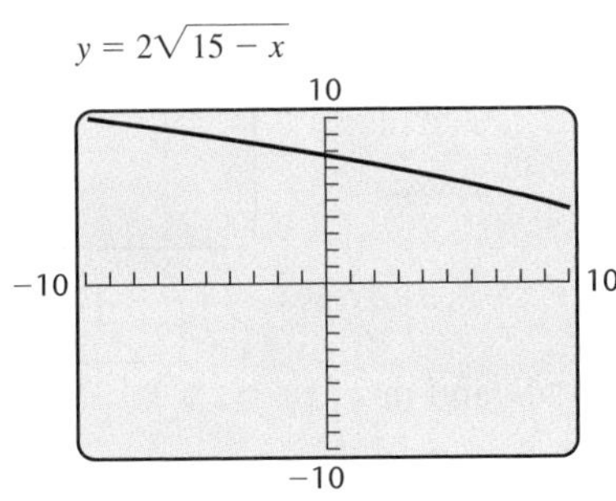

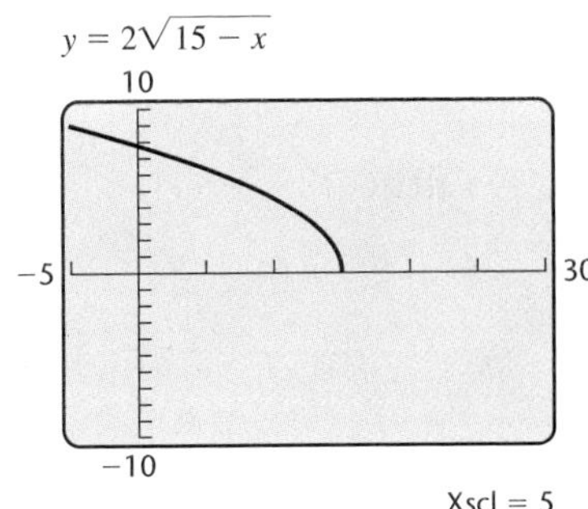

The graph on the left looks almost like a straight line. The viewing window on the right includes the endpoint, 15, on the x-axis. Knowing the domain of the function helps us choose Xmin and Xmax.

Student Notes

Some calculators display radical expressions as they would be written by hand, without the need for additional parentheses. On such a calculator, the expression

$$\sqrt{x + 2} - 3$$

is entered by pressing √ X,T,Θ,n + 2 ▷ − 3.

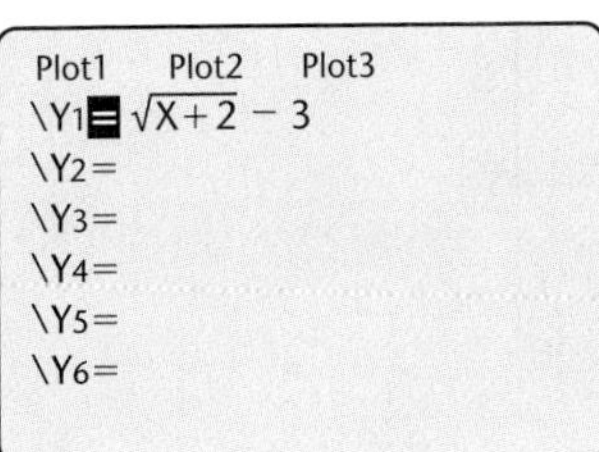

Your Turn

1. Enter $f(x) = \sqrt{x + 2} - 3$. Check your work by finding $f(7)$ and $f(-1)$ using a TABLE set to ASK. $f(7) = 0, f(-1) = -2$
2. Determine the domain of $g(x) = \sqrt{x + 30}$. Which of the following windows would best show the graph? (b)
   a) $[-10, 50, -10, 10]$
   b) $[-50, 10, -10, 10]$
   Check by graphing.

EXAMPLE 12 Find the domain of the function given by each of the following equations. Check by graphing the function. Then, from the graph, estimate the range of the function.

a) $t(x) = \sqrt{2x - 5} - 3$

b) $f(x) = \sqrt{x^2 + 1}$

SOLUTION

a) The function $t(x) = \sqrt{2x - 5} - 3$ is defined when $2x - 5 \geq 0$:

Set the radicand greater than or equal to 0.

$2x - 5 \geq 0$ The radicand must be nonnegative.

$2x \geq 5$ Adding 5

Solve.

$x \geq \frac{5}{2}$. Dividing by 2

The solution gives the domain.

The domain is $\{x \mid x \geq \frac{5}{2}\}$, or $[\frac{5}{2}, \infty)$, as indicated on the graph by the shading on the x-axis. The range appears to be $[-3, \infty)$, as indicated by the shading on the y-axis.

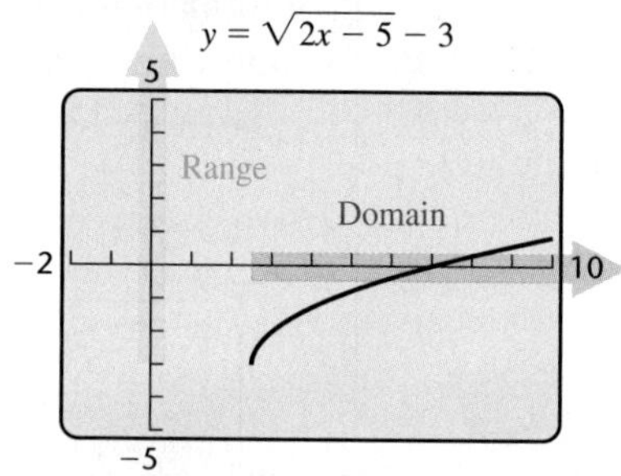

b) The radicand in $f(x) = \sqrt{x^2 + 1}$ is $x^2 + 1$. We must have

$x^2 + 1 \geq 0$ The radicand must be nonnegative.

$x^2 \geq -1$.

Since x^2 is nonnegative for all real numbers x, the inequality is true for all real numbers. The domain is $(-\infty, \infty)$. The range appears to be $[1, \infty)$.

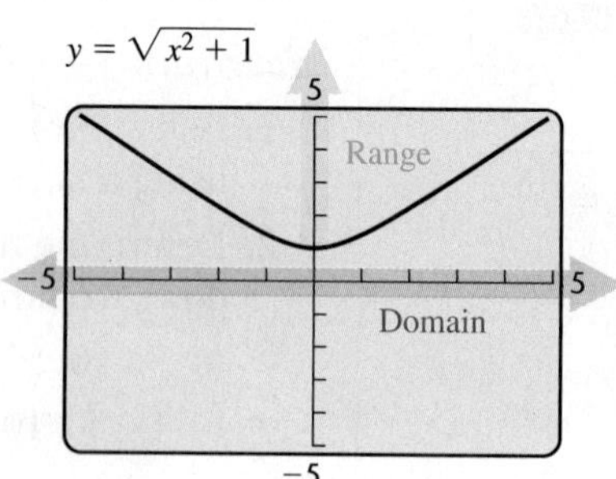

Try Exercise 105.

Some situations can be modeled using radical functions. The graphs of radical functions can have many different shapes. However, radical functions given by equations of the form

$$r(x) = \sqrt{ax + b}$$

will have the general shape of the graph of $f(x) = \sqrt{x}$.

We can determine whether a radical function might fit a set of data by plotting the points.

EXAMPLE 13 Health Insurance. The average total cost of a family health insurance policy through an employer has risen every year from 2001 through 2009, as shown in the table below. Graph the data and determine whether a radical function can be used to model average total cost of family health insurance.

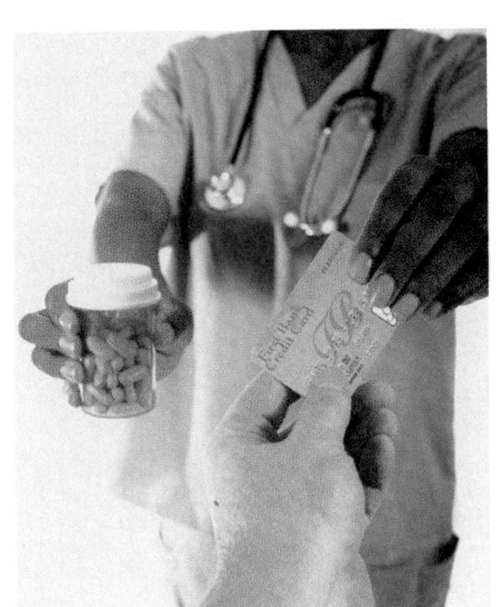

Year	Years Since 2000	Average Total Cost of Family Health Insurance (in thousands)
2001	1	\$ 7.061
2002	2	8.003
2003	3	9.068
2004	4	9.950
2005	5	10.880
2006	6	11.480
2007	7	12.106
2008	8	12.680
2009	9	13.375

Source: Kaiser Family Foundation; Health Research & Educational Trust

SOLUTION To make the graph easier to read, we use the number of years since 2000 instead of the actual year, as shown in the second column of the table. We graph the data, entering the years since 2000 as L1 and the costs as L2. The data appear to follow the pattern of the graph of a radical function. Thus we determine that a radical function could be used to model family health insurance cost.

L1	L2	L3 1
1	7.061	-----
2	8.003	
3	9.068	
4	9.950	
5	10.880	
6	11.480	
7	12.106	

L1(1) = 1

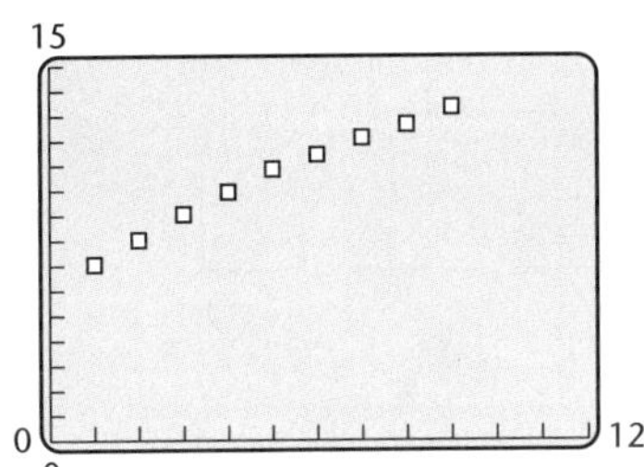

Try Exercise 115.

$y_1 = (0.3 + \sqrt{0.282 + 0.32x})/0.16$

15

0 12

0

The function in Example 14 is a good model of the data in Example 13.

EXAMPLE 14 Health Insurance. The average total cost of family health insurance, through an employer, can be modeled by the radical function

$$(0.3 + \sqrt{0.282 + 0.32x})/0.16,$$

where x is the number of years since 2000. Estimate the average total cost of family health insurance in 2010.

SOLUTION We enter the equation and use a table with Indpnt set to Ask. For 2010, $x = 10$, and the health insurance cost is estimated to be \$13,538.

X	Y1	Y2
10	13.538	

Try Exercise 121.

Visualizing for Success

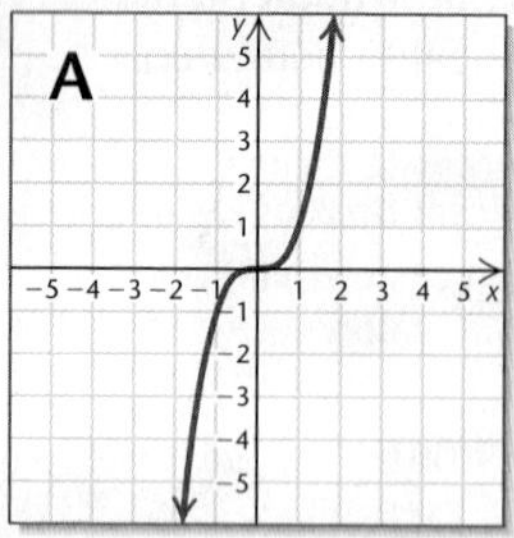

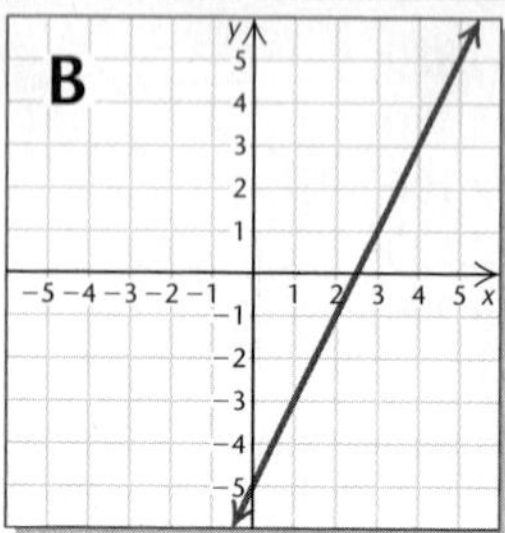

C

D

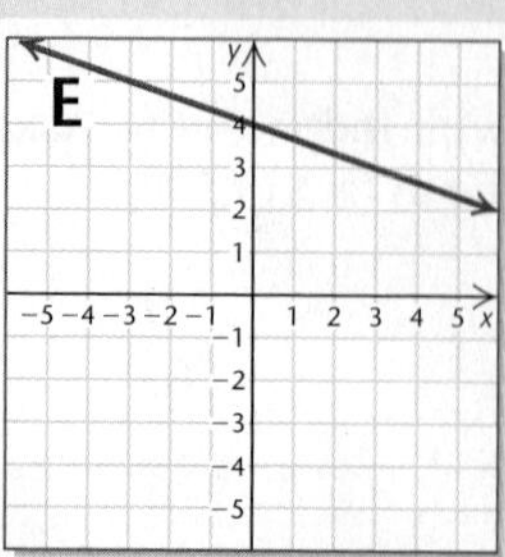

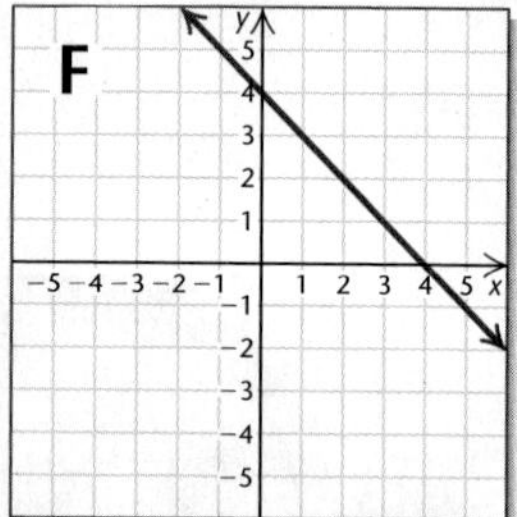

G

H

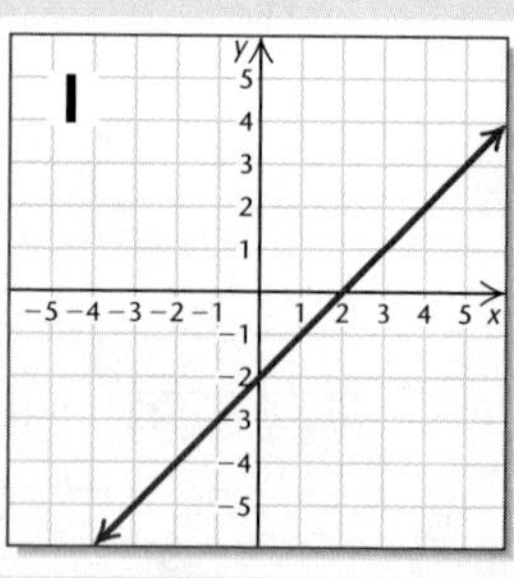

J

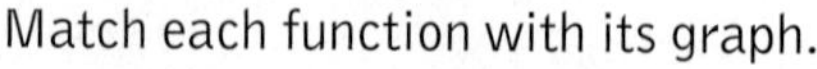

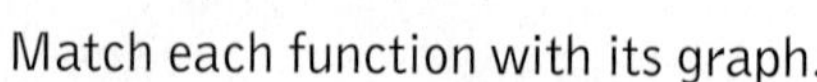

Match each function with its graph.

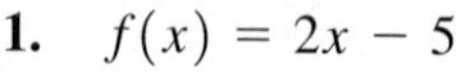

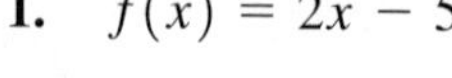

1. $f(x) = 2x - 5$

2. $f(x) = x^2 - 1$

3. $f(x) = \sqrt{x - 2}$

4. $f(x) = x - 2$

5. $f(x) = -\frac{1}{3}x$

6. $f(x) = x^3$

7. $f(x) = 4 - x$

8. $f(x) = |2x - 5|$

9. $f(x) = -2$

10. $f(x) = -\frac{1}{3}x + 4$

Answers on page A-35

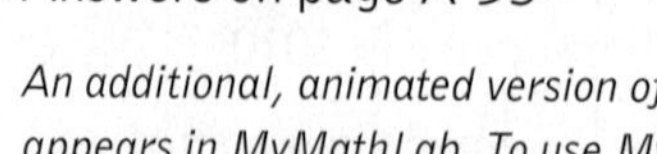

An additional, animated version of this activity appears in MyMathLab. To use MyMathLab, you need a course ID and a student access code. Contact your instructor for more information.

10.1 Exercise Set

FOR EXTRA HELP MyMathLab MathXL PRACTICE WATCH  DOWNLOAD  READ 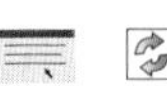 REVIEW

Concept Reinforcement *Select the appropriate word to complete each of the following.*

1. Every positive number has ______________ (one/two) square root(s).
2. The principal square root is never ______________ (negative/positive).
3. For any ______________ (negative/positive) number a, we have $\sqrt{a^2} = a$.
4. For any ______________ (negative/positive) number a, we have $\sqrt{a^2} = -a$.
5. If a is a whole number that is not a perfect square, then $\sqrt{a}$ is a(n) ______________ (irrational/rational) number.
6. The domain of the function f given by $f(x) = \sqrt[3]{x}$ is the set of all ______________ (whole/real/positive) numbers.
7. If $\sqrt[4]{x}$ is a real number, then x must be ______________ (negative/positive/nonnegative).
8. If $\sqrt[3]{x}$ is negative, then x must be ______________ (negative/positive).

For each number, find all of its square roots.

9. 49 **10.** 81

11. 144 **12.** 9

13. 400 **14.** 2500

15. 900 **16.** 225

Simplify.

17. $\sqrt{49}$ **18.** $\sqrt{144}$

19. $-\sqrt{16}$ **20.** $-\sqrt{100}$

21. $\sqrt{\frac{36}{49}}$ **22.** $\sqrt{\frac{4}{9}}$

23. $-\sqrt{\frac{16}{81}}$ **24.** $-\sqrt{\frac{81}{144}}$

25. $\sqrt{0.04}$ **26.** $\sqrt{0.36}$

27. $\sqrt{0.0081}$ **28.** $\sqrt{0.0016}$

Identify the radicand and the index for each expression.

29. $5\sqrt{p^2 + 4}$ **30.** $-7\sqrt{y^2 - 8}$

31. $xy\sqrt[5]{\frac{x}{y + 4}}$ **32.** $\frac{a}{b}\sqrt[6]{a^2 + 1}$

For each function, find the specified function value, if it exists. If it does not exist, state this.

33. $f(t) = \sqrt{5t - 10}$; $f(3), f(2), f(1), f(-1)$

34. $g(x) = \sqrt{x^2 - 25}$; $g(-6), g(3), g(6), g(13)$

35. $t(x) = -\sqrt{2x^2 - 1}$; $t(5), t(0), t(-1), t\left(-\frac{1}{2}\right)$

36. $p(z) = \sqrt{2z - 20}$; $p(4), p(10), p(12), p(0)$

37. $f(t) = \sqrt{t^2 + 1}$; $f(0), f(-1), f(-10)$

38. $g(x) = -\sqrt{(x + 1)^2}$; $g(-3), g(4), g(-5)$

Simplify. Remember to use absolute-value notation when necessary. If a root cannot be simplified, state this.

39. $\sqrt{64x^2}$ **40.** $\sqrt{25t^2}$

41. $\sqrt{(-4b)^2}$ **42.** $\sqrt{(-7c)^2}$

43. $\sqrt{(8 - t)^2}$ **44.** $\sqrt{(a + 3)^2}$

45. $\sqrt{y^2 + 16y + 64}$ **46.** $\sqrt{x^2 - 4x + 4}$

47. $\sqrt{4x^2 + 28x + 49}$ **48.** $\sqrt{9x^2 - 30x + 25}$

49. $-\sqrt[4]{256}$ **50.** $-\sqrt[4]{625}$

51. $\sqrt[5]{-1}$ **52.** $-\sqrt[3]{-1000}$

53. $-\sqrt[5]{-\frac{32}{243}}$ **54.** $\sqrt[5]{-\frac{1}{32}}$

55. $\sqrt[6]{x^6}$ **56.** $\sqrt[8]{y^8}$

57. $\sqrt[9]{t^9}$ **58.** $\sqrt[5]{a^5}$

59. $\sqrt[4]{(6a)^4}$ **60.** $\sqrt[4]{(8y)^4}$

61. $\sqrt[10]{(-6)^{10}}$ **62.** $\sqrt[12]{(-10)^{12}}$

63. $\sqrt[414]{(a + b)^{414}}$ **64.** $\sqrt[1976]{(2a + b)^{1976}}$

65. $\sqrt{a^{22}}$ **66.** $\sqrt{x^{10}}$

67. $\sqrt{-25}$ **68.** $\sqrt{-16}$

Simplify. Assume that no radicands were formed by raising negative quantities to even powers.

69. $\sqrt{16x^2}$ **70.** $\sqrt{25t^2}$

71. $-\sqrt{(3t)^2}$ **72.** $-\sqrt{(7c)^2}$

73. $\sqrt{(a+1)^2}$

74. $\sqrt{(5+b)^2}$

75. $\sqrt{9t^2-12t+4}$

76. $\sqrt{25t^2-20t+4}$

77. $\sqrt[3]{27a^3}$

78. $-\sqrt[3]{64y^3}$

79. $\sqrt[4]{16x^4}$

80. $\sqrt[4]{81x^4}$

81. $\sqrt[5]{(x-1)^5}$

82. $-\sqrt[7]{(1-x)^7}$

83. $-\sqrt[3]{-125y^3}$

84. $-\sqrt[3]{-64x^3}$

85. $\sqrt{t^{18}}$

86. $\sqrt{a^{14}}$

87. $\sqrt{(x-2)^8}$

88. $\sqrt{(x+3)^{10}}$

For each function, find the specified function value, if it exists. If it does not exist, state this.

89. $f(x)=\sqrt[3]{x+1}$; $f(7), f(26), f(-9), f(-65)$

90. $g(x)=-\sqrt[3]{2x-1}$; $g(0), g(-62), g(-13), g(63)$

91. $g(t)=\sqrt[4]{t-3}$; $g(19), g(-13), g(1), g(84)$

92. $f(t)=\sqrt[4]{t+1}$; $f(0), f(15), f(-82), f(80)$

Determine the domain of each function described.

93. $f(x)=\sqrt{x-6}$

94. $g(x)=\sqrt{x+8}$

95. $g(t)=\sqrt[4]{t+8}$

96. $f(x)=\sqrt[4]{x-9}$

97. $g(x)=\sqrt[4]{2x-10}$

98. $g(t)=\sqrt[3]{2t-6}$

99. $f(t)=\sqrt[5]{8-3t}$

100. $f(t)=\sqrt[6]{4-3t}$

101. $h(z)=-\sqrt[6]{5z+2}$

102. $d(x)=-\sqrt[4]{7x-5}$

Aha! 103. $f(t)=7+\sqrt[8]{t^8}$

104. $g(t)=9+\sqrt[6]{t^6}$

Determine algebraically the domain of each function described. Then use a graphing calculator to confirm your answer and to estimate the range.

105. $f(x)=\sqrt{5-x}$

106. $g(x)=\sqrt{2x+1}$

107. $f(x)=1-\sqrt{x+1}$

108. $g(x)=2+\sqrt{3x-5}$

109. $g(x)=3+\sqrt{x^2+4}$

110. $f(x)=5-\sqrt{3x^2+1}$

In Exercises 111–114, match each function with one of the following graphs without using a calculator.

a)

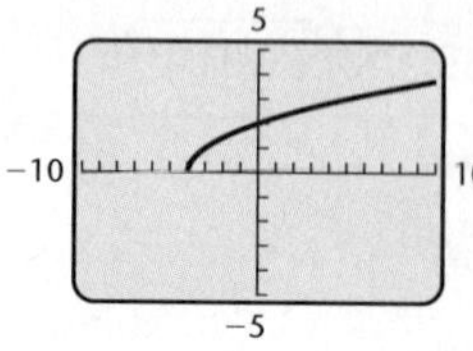

b)

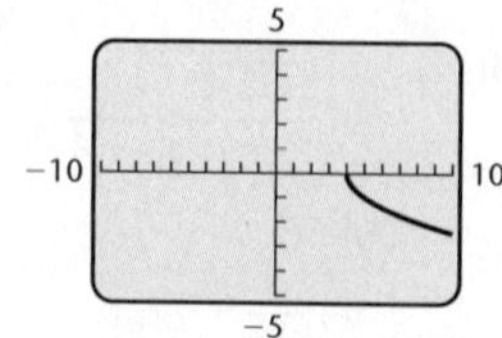

c)

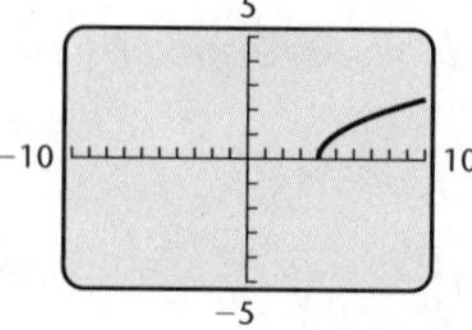

d)

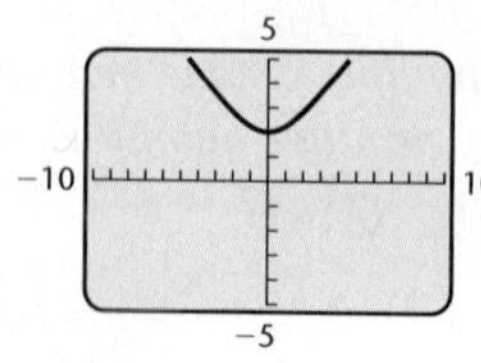

111. $f(x)=\sqrt{x-4}$

112. $g(x)=\sqrt{x+4}$

113. $h(x)=\sqrt{x^2+4}$

114. $f(x)=-\sqrt{x-4}$

In Exercises 115–120, determine whether a radical function would be a good model of the given situation.

115. *Sports Salaries.* The table below lists the average salary of a major-league baseball player, based on the number of years in his contract.

Length of Contract (in years)	Average Salary (in millions)
1	\$ 2.1
2	3.6
3	7.0
4	9.6
5	10.5
6	12.5
7	13.8
8	14.3
9	14.6

Source: "Long-term contracts" by Dave Studeman, 12/06/07, on www.hardballtimes.com

116. *Firefighting.* The number of gallons per minute discharged from a fire hose depends on the diameter of the hose and the nozzle pressure. The table below lists the amount of water flow for a 2-in. diameter solid bore nozzle at various nozzle pressures.

Nozzle Pressure (in pounds per square inch, psi)	Water Flow (in gallons per minute, GPM)
40	752
60	921
80	1063
100	1188
120	1302
150	1455
200	1681

Source: www.firetactics.com

117. *Koi Growth.* Koi, a popular fish for backyard pools, grow from $\frac{1}{40}$ cm when newly hatched to an average length of 80 cm. The table below lists the length of a koi at various ages.

Age (in months)	Length (in centimeters)
1	2.9
2	5.0
4	9.1
13	24.9
16	29.3
30	45.8
36	51.1
48	59.3
60	65.2
72	69.4

Source: www.coloradokoi.com

118. *Farm Size.* The table below lists the average size of United States' farms for various years from 1960 to 2007.

Year	Average Size of Farm (in acres)
1960	303
1980	426
1997	431
2002	441
2007	418

Source: U.S. Department of Agriculture

119. *Cancer Research.* The table below lists the amount of federal funds allotted to the National Cancer Institute for cancer research in the United States from 2005 to 2010.

Year	Funds (in billions)
2005	$4.83
2006	4.79
2007	4.70
2008	4.93
2009	4.97
2010*	5.15

*Requested
Source: National Cancer Institute

120. *Cable Television.* The table below lists the percent of households with basic cable television service for various years.

Year	Percent of Households Served by Basic Cable Television
1985	46.2
1990	59.0
1995	65.7
2000	67.9
2005	66.8
2007	58.0

Source: Nielsen Media Research

121. *Firefighting.* The water flow, in number of gallons per minute (GPM), for a 2-in. diameter solid bore nozzle is given by

$$f(x) = 118.8\sqrt{x},$$

where x is the nozzle pressure, in pounds per square inch (psi). (See Exercise 116.) Use the function to estimate the water flow when the nozzle pressure is 50 psi and when it is 175 psi.

122. *Koi Growth.* The length, in centimeters, of a koi of age x months can be estimated using the function

$$f(x) = 0.27 + \sqrt{71.94x - 164.41}.$$

(See Exercise 117.) Use the function to estimate the length of a koi at 8 months and at 20 months.

TW **123.** Explain how to write the negative square root of a number using radical notation.

TW **124.** Does the square root of a number's absolute value always exist? Why or why not?

SKILL REVIEW

To prepare for Section 10.2, review exponents (Sections 5.1 and 5.2).

Simplify. Do not use negative exponents in your answer. [5.1], [5.2]

125. $(a^2b)(a^4b)$

126. $(3xy^8)(5x^2y)$

127. $(5x^2y^{-3})^3$

128. $(2a^{-1}b^2c)^{-3}$

129. $\left(\dfrac{10x^{-1}y^5}{5x^2y^{-1}}\right)^{-1}$

130. $\left(\dfrac{8x^3y^{-2}}{2xz^4}\right)^{-2}$

SYNTHESIS

TW **131.** If the domain of $f = [1, \infty)$ and the range of $f = [2, \infty)$, find a possible expression for $f(x)$ and explain how such an expression is formulated.

TW **132.** Natasha obtains the following graph of

$$f(x) = \sqrt{x^2 - 4x - 12}$$

and concludes that the domain of f is $(-\infty, -2]$. Is she correct? If not, what mistake is she making?

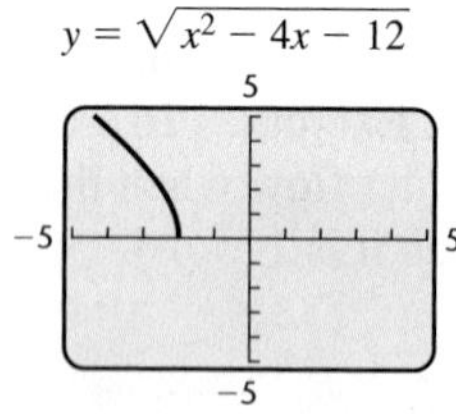

TW **133.** Could the following situation possibly be modeled using a radical function? Why or why not? "For each year, the yield increases. The amount of increase is smaller each year."

TW **134.** Could the following situation possibly be modeled using a radical function? Why or why not? "For each year, the costs increase. The amount of increase is the same each year."

135. *Spaces in a Parking Lot.* A parking lot has attendants to park the cars. The number N of stalls needed for waiting cars before attendants can get to them is given by the formula $N = 2.5\sqrt{A}$, where A is the number of arrivals in peak hours. Find the number of spaces needed for the given number of arrivals in peak hours: **(a)** 25; **(b)** 36; **(c)** 49; **(d)** 64.

Determine the domain of each function described. Then draw the graph of each function.

136. $g(x) = \sqrt{x} + 5$

137. $f(x) = \sqrt{x + 5}$

138. $f(x) = \sqrt{x - 2}$

139. $g(x) = \sqrt{x} - 2$

140. Find the domain of f if

$$f(x) = \frac{\sqrt{x+3}}{\sqrt[4]{2-x}}.$$

141. Find the domain of g if

$$g(x) = \frac{\sqrt[4]{5-x}}{\sqrt[6]{x+4}}.$$

142. Find the domain of F if $F(x) = \dfrac{x}{\sqrt{x^2 - 5x - 6}}$.

143. Examine the graph of the data in Exercise 119. What type of function could be used to model the data?

TW **144.** In Exercise 120, a radical function could be used to model the data through 2000. What might have caused the number of basic cable subscribers to change its pattern of growth?

Try Exercise Answers: Section 10.1

9. $7, -7$ **17.** 7 **33.** $\sqrt{5}$; 0; does not exist; does not exist **39.** $|8x|$, or $8|x|$ **57.** t **63.** $|a + b|$ **69.** $4x$ **83.** $5y$ **89.** $2; 3; -2; -4$ **93.** $\{x \mid x \geq 6\}$, or $[6, \infty)$ **105.** Domain: $\{x \mid x \leq 5\}$, or $(-\infty, 5]$; range: $\{y \mid y \geq 0\}$, or $[0, \infty)$ **115.** Yes **121.** Approximately 840 GPM; approximately 1572 GPM

10.2 Rational Numbers as Exponents

- Rational Exponents
- Negative Rational Exponents
- Laws of Exponents
- Simplifying Radical Expressions

STUDY TIP

It Looks Familiar

Sometimes a topic seems familiar and students are tempted to assume that they already know the material. Try to avoid this tendency. Often new extensions or applications are included when a topic reappears.

We have already considered natural-number exponents and integer exponents. We now expand the study of exponents further to include all rational numbers. This will give meaning to expressions like $a^{1/3}$, $7^{-1/2}$, and $(3x)^{4/5}$. Such notation will help us simplify certain radical expressions.

RATIONAL EXPONENTS

When defining rational exponents, we want the rules for exponents to hold for rational exponents just as they do for integer exponents. In particular, we still want to add exponents when multiplying.

If $a^{1/2} \cdot a^{1/2} = a^{1/2+1/2} = a^1$, then $a^{1/2}$ should mean $\sqrt{a}$.

If $a^{1/3} \cdot a^{1/3} \cdot a^{1/3} = a^{1/3+1/3+1/3} = a^1$, then $a^{1/3}$ should mean $= \sqrt[3]{a}$.

$a^{1/n} = \sqrt[n]{a}$ $\quad a^{1/n}$ means $\sqrt[n]{a}$.

When a is nonnegative, n can be any natural number greater than 1. When a is negative, n can be any odd natural number greater than 1.

Thus, $a^{1/5} = \sqrt[5]{a}$ and $a^{1/10} = \sqrt[10]{a}$. Note that the denominator of the exponent becomes the index and the base becomes the radicand.

EXAMPLE 1 Write an equivalent expression using radical notation and, if possible, simplify.

a) $x^{1/2}$ **b)** $(-8)^{1/3}$
c) $(abc)^{1/5}$ **d)** $(25x^{16})^{1/2}$

SOLUTION

a) $x^{1/2} = \sqrt{x}$
b) $(-8)^{1/3} = \sqrt[3]{-8} = -2$
c) $(abc)^{1/5} = \sqrt[5]{abc}$

The denominator of the exponent becomes the index. The base becomes the radicand. Recall that for square roots, the index 2 is understood without being written.

d) $(25x^{16})^{1/2} = \sqrt{25x^{16}} = 5x^8$

Try Exercise 9.

EXAMPLE 2 Write an equivalent expression using exponential notation.

a) $\sqrt[5]{7ab}$ **b)** $\sqrt[7]{\dfrac{x^3y}{4}}$ **c)** $\sqrt{5x}$

SOLUTION Parentheses are required to indicate the base.

a) $\sqrt[5]{7ab} = (7ab)^{1/5}$
b) $\sqrt[7]{\dfrac{x^3y}{4}} = \left(\dfrac{x^3y}{4}\right)^{1/7}$

The index becomes the denominator of the exponent. The radicand becomes the base.

c) $\sqrt{5x} = (5x)^{1/2}$ The index 2 is understood without being written. We assume $x \geq 0$.

Try Exercise 39.

> **CAUTION!** When we are converting from radical notation to exponential notation, parentheses are necessary to indicate the base.
>
> $\sqrt{5x} = (5x)^{1/2}$

Rational Exponents

We can enter a radical expression using rational exponents.

To use rational exponents, enter the radicand enclosed in parentheses, then press [^], and then enter the rational exponent, also enclosed in parentheses. If the rational exponent is written as a single decimal number, the parentheses around the exponent are unnecessary. If decimal notation for a rational exponent must be rounded, fraction notation should be used.

Your Turn

1. Use a calculator to find $256^{1/8}$. 2
2. Use rational exponents to find $\sqrt[4]{0.00390625}$. 0.25

EXAMPLE 3 Graph: $f(x) = \sqrt[4]{2x - 7}$.

SOLUTION We can enter the equation in radical notation using $\sqrt[x]{\ }$. Alternatively, we can rewrite the radical expression using a rational exponent:

$$f(x) = (2x - 7)^{1/4}, \quad \text{or} \quad f(x) = (2x - 7)^{0.25}.$$

Then we let $y = (2x - 7)^\wedge(1/4)$ or $y = (2x - 7)^\wedge.25$. The screen on the left below shows all three forms of the equation; they are equivalent.

Knowing the domain of the function can help us determine an appropriate viewing window. Since the index is even, the domain is the set of all x for which the radicand is nonnegative, or $\left[\frac{7}{2}, \infty\right)$. We choose a viewing window of $[1, 10, -1, 5]$. This will show the axes and the first quadrant. The graph is shown on the right below.

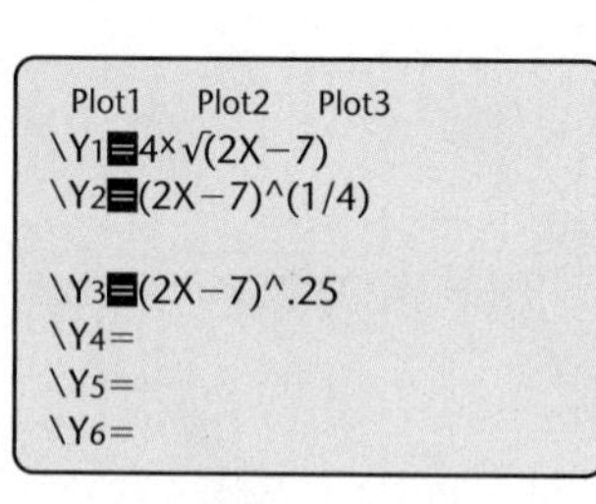

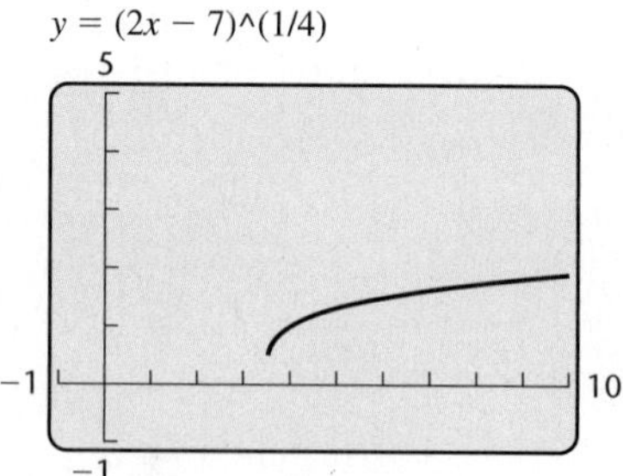

Try Exercise 65.

How shall we define $a^{2/3}$? If the property for multiplying exponents is to hold, we must have $a^{2/3} = (a^{1/3})^2$ and $a^{2/3} = (a^2)^{1/3}$. This would suggest that $a^{2/3} = (\sqrt[3]{a})^2$ and $a^{2/3} = \sqrt[3]{a^2}$. We make our definition accordingly.

> **Positive Rational Exponents** For any natural numbers m and n ($n \neq 1$) and any real number a for which $\sqrt[n]{a}$ exists,
>
> $a^{m/n}$ means $(\sqrt[n]{a})^m$, or $\sqrt[n]{a^m}$.

Student Notes

It is important to remember both meanings of $a^{m/n}$. When the root of the base a is known, $(\sqrt[n]{a})^m$ is generally easier to work with. When it is not known, $\sqrt[n]{a^m}$ is often more convenient.

EXAMPLE 4 Write an equivalent expression using radical notation and simplify.

a) $27^{2/3}$

b) $25^{3/2}$

SOLUTION

a) $27^{2/3}$ means $(\sqrt[3]{27})^2$ or, equivalently, $\sqrt[3]{27^2}$. Let's see which is easier to simplify:

$$(\sqrt[3]{27})^2 = 3^2 = 9; \qquad \sqrt[3]{27^2} = \sqrt[3]{729} = 9.$$

The simplification on the left is probably easier for most people.

b) $25^{3/2}$ means $(\sqrt[2]{25})^3$ or, equivalently, $\sqrt[2]{25^3}$ (the index 2 is normally omitted). Since $\sqrt{25}$ is more commonly known than $\sqrt{25^3}$, we use that form:

$$25^{3/2} = (\sqrt{25})^3 = 5^3 = 125.$$

Try Exercise 23.

EXAMPLE 5 Write an equivalent expression using exponential notation.

a) $\sqrt[3]{9^4}$

b) $(\sqrt[4]{7xy})^5$

SOLUTION

a) $\sqrt[3]{9^4} = 9^{4/3}$

b) $(\sqrt[4]{7xy})^5 = (7xy)^{5/4}$

The index becomes the denominator of the fraction that is the exponent.

Try Exercise 43.

Rational roots of numbers can be approximated on a calculator.

EXAMPLE 6 Approximate $\sqrt[5]{(-23)^3}$. Round to the nearest thousandth.

SOLUTION We first rewrite the expression using a rational exponent:

$$\sqrt[5]{(-23)^3} = (-23)^{3/5}.$$

Using a calculator, we have

$$(-23)\wedge(3/5) \approx -6.562.$$

Try Exercise 71.

NEGATIVE RATIONAL EXPONENTS

Recall that $x^{-2} = \frac{1}{x^2}$. Negative rational exponents behave similarly.

Negative Rational Exponents For any rational number m/n and any nonzero real number a for which $a^{m/n}$ exists,

$$a^{-m/n} \quad \text{means} \quad \frac{1}{a^{m/n}}.$$

CAUTION! A negative exponent does not indicate that the expression in which it appears is negative.

EXAMPLE 7 Write an equivalent expression with positive exponents and, if possible, simplify.

a) $9^{-1/2}$ b) $(5xy)^{-4/5}$ c) $64^{-2/3}$

d) $4x^{-2/3}y^{1/5}$ e) $\left(\frac{3r}{7s}\right)^{-5/2}$

SOLUTION

a) $9^{-1/2} = \frac{1}{9^{1/2}}$ **$9^{-1/2}$ is the reciprocal of $9^{1/2}$.**

Since $9^{1/2} = \sqrt{9} = 3$, the answer simplifies to $\frac{1}{3}$.

b) $(5xy)^{-4/5} = \frac{1}{(5xy)^{4/5}}$ **$(5xy)^{-4/5}$ is the reciprocal of $(5xy)^{4/5}$.**

c) $64^{-2/3} = \frac{1}{64^{2/3}}$ **$64^{-2/3}$ is the reciprocal of $64^{2/3}$.**

Since $64^{2/3} = (\sqrt[3]{64})^2 = 4^2 = 16$, the answer simplifies to $\frac{1}{16}$.

d) $4x^{-2/3}y^{1/5} = 4 \cdot \frac{1}{x^{2/3}} \cdot y^{1/5} = \frac{4y^{1/5}}{x^{2/3}}$

e) In Section 5.2, we found that $(a/b)^{-n} = (b/a)^n$. This property holds for *any* negative exponent:

$$\left(\frac{3r}{7s}\right)^{-5/2} = \left(\frac{7s}{3r}\right)^{5/2}.$$ **Writing the reciprocal of the base and changing the sign of the exponent**

Try Exercise 49.

LAWS OF EXPONENTS

The same laws hold for rational exponents as for integer exponents.

Laws of Exponents For any real numbers a and b and any rational exponents m and n for which a^m, a^n, and b^m are defined:

1. $a^m \cdot a^n = a^{m+n}$ When multiplying, add exponents if the bases are the same.
2. $\frac{a^m}{a^n} = a^{m-n}$ When dividing, subtract exponents if the bases are the same. (Assume $a \neq 0$.)
3. $(a^m)^n = a^{m \cdot n}$ To raise a power to a power, multiply the exponents.
4. $(ab)^m = a^m b^m$ To raise a product to a power, raise each factor to the power and multiply.

EXAMPLE 8 Use the laws of exponents to simplify.

a) $3^{1/5} \cdot 3^{3/5}$ b) $\frac{a^{1/4}}{a^{1/2}}$

c) $(7.2^{2/3})^{3/4}$ d) $(a^{-1/3}b^{2/5})^{1/2}$

SOLUTION

a) $3^{1/5} \cdot 3^{3/5} = 3^{1/5+3/5} = 3^{4/5}$ — **Adding exponents**

b) $\dfrac{a^{1/4}}{a^{1/2}} = a^{1/4-1/2} = a^{1/4-2/4}$ — **Subtracting exponents after finding a common denominator**

$= a^{-1/4}$, or $\dfrac{1}{a^{1/4}}$ — **$a^{-1/4}$ is the reciprocal of $a^{1/4}$.**

c) $(7.2^{2/3})^{3/4} = 7.2^{(2/3)(3/4)} = 7.2^{6/12}$ — **Multiplying exponents**

$= 7.2^{1/2}$ — **Using arithmetic to simplify the exponent**

d) $(a^{-1/3}b^{2/5})^{1/2} = a^{(-1/3)(1/2)} \cdot b^{(2/5)(1/2)}$ — **Raising a product to a power and multiplying exponents**

$= a^{-1/6}b^{1/5}$, or $\dfrac{b^{1/5}}{a^{1/6}}$

Try Exercise 77.

SIMPLIFYING RADICAL EXPRESSIONS

Many radical expressions can be simplified using rational exponents.

To Simplify Radical Expressions

1. Convert radical expressions to exponential expressions.
2. Use arithmetic and the laws of exponents to simplify.
3. Convert back to radical notation as needed.

EXAMPLE 9 Use rational exponents to simplify. Do not use exponents that are fractions in the final answer.

a) $\sqrt[6]{(5x)^3}$

b) $\sqrt[5]{t^{20}}$

c) $\left(\sqrt[3]{ab^2c}\right)^{12}$

d) $\sqrt{\sqrt[3]{x}}$

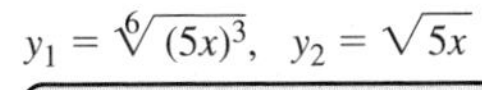

$y_1 = \sqrt[6]{(5x)^3}, \quad y_2 = \sqrt{5x}$

X	Y1	Y2
0	0	0
1	2.2361	2.2361
2	3.1623	3.1623
3	3.873	3.873
4	4.4721	4.4721
5	5	5
6	5.4772	5.4772
X = 0		

SOLUTION

a) $\sqrt[6]{(5x)^3} = (5x)^{3/6}$ — **Converting to exponential notation**

$= (5x)^{1/2}$ — **Simplifying the exponent**

$= \sqrt{5x}$ — **Returning to radical notation**

To check on a graphing calculator, we let $y_1 = \sqrt[6]{(5x)^3}$ and $y_2 = \sqrt{5x}$ and compare values in a table. If we scroll through the table (see the figure at left), we see that $y_1 = y_2$, so our simplification is probably correct.

b) $\sqrt[5]{t^{20}} = t^{20/5}$ — **Converting to exponential notation**

$= t^4$ — **Simplifying the exponent**

c) $\left(\sqrt[3]{ab^2c}\right)^{12} = (ab^2c)^{12/3}$ — **Converting to exponential notation**

$= (ab^2c)^4$ — **Simplifying the exponent**

$= a^4b^8c^4$ — **Using the laws of exponents**

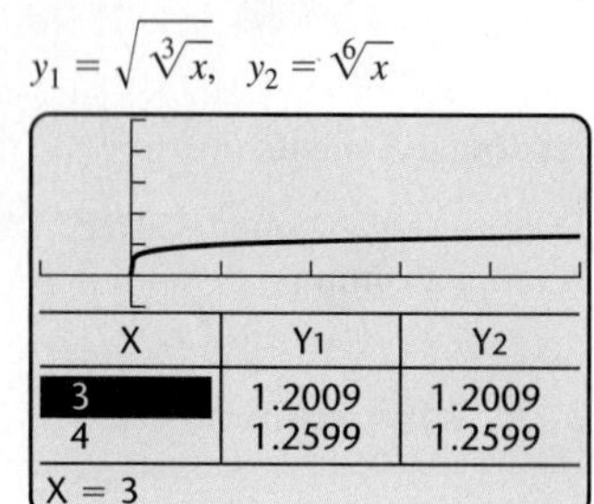

d) $\sqrt{\sqrt[3]{x}} = \sqrt{x^{1/3}}$ **Converting the radicand to exponential notation**

$= (x^{1/3})^{1/2}$

$= x^{1/6}$ **Using the laws of exponents**

$= \sqrt[6]{x}$ **Returning to radical notation**

We can check by graphing $y_1 = \sqrt{\sqrt[3]{x}}$ and $y_2 = \sqrt[6]{x}$. The graphs coincide, as we also see by scrolling through the table of values shown at left.

Try Exercise 91.

10.2 Exercise Set

FOR EXTRA HELP MyMathLab   

Concept Reinforcement *In each of Exercises 1–8, match the expression with the equivalent expression from the column on the right.*

1. ___ $x^{2/5}$ a) $x^{3/5}$
2. ___ $x^{5/2}$ b) $(\sqrt[5]{x})^4$
3. ___ $x^{-5/2}$ c) $\sqrt{x^5}$
4. ___ $x^{-2/5}$ d) $x^{1/2}$
5. ___ $x^{1/5} \cdot x^{2/5}$ e) $\dfrac{1}{(\sqrt{x})^5}$
6. ___ $(x^{1/5})^{5/2}$ f) $\sqrt[4]{x^5}$
7. ___ $\sqrt[5]{x^4}$ g) $\sqrt[5]{x^2}$
8. ___ $(\sqrt[4]{x})^5$ h) $\dfrac{1}{(\sqrt[5]{x})^2}$

Note: Assume for all exercises that even roots are of nonnegative quantities and that all denominators are nonzero.

Write an equivalent expression using radical notation and, if possible, simplify.

9. $x^{1/6}$
10. $y^{1/5}$
11. $16^{1/2}$
12. $8^{1/3}$
13. $32^{1/5}$
14. $64^{1/6}$
15. $9^{1/2}$
16. $25^{1/2}$
17. $(xyz)^{1/2}$
18. $(ab)^{1/4}$
19. $(a^2b^2)^{1/5}$
20. $(x^3y^3)^{1/4}$
21. $t^{2/5}$
22. $b^{3/2}$
23. $16^{3/4}$
24. $4^{7/2}$
25. $27^{4/3}$
26. $9^{5/2}$
27. $(81x)^{3/4}$
28. $(125a)^{2/3}$
29. $(25x^4)^{3/2}$
30. $(9y^6)^{3/2}$

Write an equivalent expression using exponential notation.

31. $\sqrt[3]{20}$
32. $\sqrt[3]{19}$
33. $\sqrt{17}$
34. $\sqrt{6}$
35. $\sqrt{x^3}$
36. $\sqrt{a^5}$
37. $\sqrt[5]{m^2}$
38. $\sqrt[5]{n^4}$
39. $\sqrt[4]{cd}$
40. $\sqrt[5]{xy}$
41. $\sqrt[5]{xy^2z}$
42. $\sqrt[7]{x^3y^2z^2}$
43. $(\sqrt{3mn})^3$
44. $(\sqrt[3]{7xy})^4$
45. $(\sqrt[7]{8x^2y})^5$
46. $(\sqrt[6]{2a^5b})^7$
47. $\dfrac{2x}{\sqrt[3]{z^2}}$
48. $\dfrac{3a}{\sqrt[5]{c^2}}$

Write an equivalent expression with positive exponents and, if possible, simplify.

49. $8^{-1/3}$
50. $10{,}000^{-1/4}$
51. $(2rs)^{-3/4}$
52. $(5xy)^{-5/6}$
53. $\left(\dfrac{1}{16}\right)^{-3/4}$
54. $\left(\dfrac{1}{8}\right)^{-2/3}$
55. $\dfrac{2c}{a^{-3/5}}$
56. $\dfrac{3b}{a^{-5/7}}$
57. $5x^{-2/3}y^{4/5}z$
58. $2ab^{-1/2}c^{2/3}$
59. $3^{-5/2}a^3b^{-7/3}$
60. $2^{-1/3}x^4y^{-2/7}$

61. $\left(\frac{2ab}{3c}\right)^{-5/6}$

62. $\left(\frac{7x}{8yz}\right)^{-3/5}$

63. $\frac{6a}{\sqrt[4]{b}}$

64. $\frac{7x}{\sqrt[3]{z}}$

Graph using a graphing calculator.

65. $f(x) = \sqrt[4]{x + 7}$

66. $g(x) = \sqrt[5]{4 - x}$

67. $r(x) = \sqrt[7]{3x - 2}$

68. $q(x) = \sqrt[6]{2x + 3}$

69. $f(x) = \sqrt[6]{x^3}$

70. $g(x) = \sqrt[8]{x^2}$

Approximate. Round to the nearest thousandth.

71. $\sqrt[5]{9}$

72. $\sqrt[6]{13}$

73. $\sqrt[4]{10}$

74. $\sqrt[7]{-127}$

75. $\sqrt[3]{(-3)^5}$

76. $\sqrt[10]{(1.5)^6}$

Use the laws of exponents to simplify. Do not use negative exponents in any answers.

77. $7^{3/4} \cdot 7^{1/8}$

78. $11^{2/3} \cdot 11^{1/2}$

79. $\frac{3^{5/8}}{3^{-1/8}}$

80. $\frac{8^{7/11}}{8^{-2/11}}$

81. $\frac{5.2^{-1/6}}{5.2^{-2/3}}$

82. $\frac{2.3^{-3/10}}{2.3^{-1/5}}$

83. $(10^{3/5})^{2/5}$

84. $(5^{5/4})^{3/7}$

85. $a^{2/3} \cdot a^{5/4}$

86. $x^{3/4} \cdot x^{1/3}$

Aha! 87. $(64^{3/4})^{4/3}$

88. $(27^{-2/3})^{3/2}$

89. $(m^{2/3}n^{-1/4})^{1/2}$

90. $(x^{-1/3}y^{2/5})^{1/4}$

Use rational exponents to simplify. Do not use fraction exponents in the final answer.

91. $\sqrt[8]{x^4}$

92. $\sqrt[6]{a^2}$

93. $\sqrt[4]{a^{12}}$

94. $\sqrt[3]{x^{15}}$

95. $\sqrt[12]{y^8}$

96. $\sqrt[10]{t^6}$

97. $(\sqrt[7]{xy})^{14}$

98. $(\sqrt[3]{ab})^{15}$

99. $\sqrt[4]{(7a)^2}$

100. $\sqrt[8]{(3x)^2}$

101. $(\sqrt[8]{2x})^6$

102. $(\sqrt[10]{3a})^5$

103. $\sqrt{\sqrt[5]{m}}$

104. $\sqrt[4]{\sqrt{x}}$

105. $\sqrt[4]{(xy)^{12}}$

106. $\sqrt{(ab)^6}$

107. $(\sqrt[5]{a^2b^4})^{15}$

108. $(\sqrt[3]{x^2y^5})^{12}$

109. $\sqrt[3]{\sqrt[4]{xy}}$

110. $\sqrt[5]{\sqrt{2a}}$

TW 111. If $f(x) = (x + 5)^{1/2}(x + 7)^{-1/2}$, find the domain of f. Explain how you found your answer.

TW 112. Let $f(x) = 5x^{-1/3}$. Under what condition will we have $f(x) > 0$? Why?

SKILL REVIEW

To prepare for Section 10.3, review multiplying and factoring polynomials (Sections 4.5 and 5.4).

Multiply. [4.5]

113. $(x + 5)(x - 5)$

114. $(x - 2)(x^2 + 2x + 4)$

Factor. [5.4]

115. $4x^2 + 20x + 25$

116. $9a^2 - 24a + 16$

117. $5t^2 - 10t + 5$

118. $3n^2 + 12n + 12$

SYNTHESIS

TW 119. Explain why $\sqrt[3]{x^6} = x^2$ for any value of x, whereas $\sqrt[2]{x^6} = x^3$ only when $x \geq 0$.

TW 120. If $g(x) = x^{3/n}$, in what way does the domain of g depend on whether n is odd or even?

Use rational exponents to simplify.

121. $\sqrt{x\sqrt[3]{x^2}}$

122. $\sqrt[4]{\sqrt[3]{8x^3y^6}}$

123. $\sqrt[12]{p^2 + 2pq + q^2}$

124. *Herpetology.* The daily number of calories c needed by a reptile of weight w pounds can be approximated by $c = 10w^{3/4}$. Find the daily calorie requirement of a green iguana weighing 16 lb.
Source: www.anapsid.org

Music. The function given by $f(x) = k2^{x/12}$ can be used to determine the frequency, in cycles per second, of a musical note that is x half-steps above a note with frequency k. Use this information for Exercises 125–127.*

125. The frequency of concert A for a trumpet is 440 cycles per second. Find the frequency of the A that is two octaves (24 half-steps) above concert A. (Few trumpeters can reach this note!)

*This application was inspired by information provided by Dr. Homer B. Tilton of Pima Community College East.

126. Show that the G that is 7 half-steps (a "perfect fifth") above middle C (262 cycles per second) has a frequency that is about 1.5 times that of middle C.

127. Show that the C sharp that is 4 half-steps (a "major third") above concert A (see Exercise 125) has a frequency that is about 25% greater than that of concert A.

128. *Baseball.* The statistician Bill James has found that a baseball team's winning percentage P can be approximated by

$$P = \frac{r^{1.83}}{r^{1.83} + \sigma^{1.83}},$$

where r is the total number of runs scored by that team and σ is the total number of runs scored by their opponents. During a recent season, the San Francisco Giants scored 799 runs and their opponents scored 749 runs. Use James's formula to predict the Giants' winning percentage. (The team actually won 55.6% of their games.)

Source: M. Bittinger, *One Man's Journey Through Mathematics*. Boston: Addison-Wesley, 2004

129. *Road Pavement Messages.* In a psychological study, it was determined that the proper length L of the letters of a word printed on pavement is given by

$$L = \frac{0.000169d^{2.27}}{h},$$

where d is the distance of a car from the lettering and h is the height of the eye above the surface of the road. All units are in meters. This formula says that if a person is h meters above the surface of the road and is to be able to recognize a message d meters away, that message will be the most recognizable if the length of the letters is L. Find L to the nearest tenth of a meter, given d and h.

a) $h = 1$ m, $d = 60$ m
b) $h = 0.9906$ m, $d = 75$ m
c) $h = 2.4$ m, $d = 80$ m
d) $h = 1.1$ m, $d = 100$ m

130. *Physics.* The equation

$$m = m_0(1 - v^2c^{-2})^{-1/2},$$

developed by Albert Einstein, is used to determine the mass m of an object that is moving v meters per second and has mass m_0 before the motion begins. The constant c is the speed of light, approximately 3×10^8 m/sec. Suppose that a particle with mass 8 mg is accelerated to a speed of $\frac{9}{5} \times 10^8$ m/sec. Without using a calculator, find the new mass of the particle.

131. *Forestry.* The total wood volume T, in cubic feet, in a California black oak can be estimated using the formula

$$T = 0.936d^{1.97}h^{0.85},$$

where d is the diameter of the tree at breast height and h is the total height of the tree. How much wood is in a California black oak that is 3 ft in diameter at breast height and 80 ft high?

Source: Norman H. Pillsbury and Michael L. Kirkley, 1984. Equations for total, wood, and saw-log volume for thirteen California hardwoods, USDA Forest Service PNW Research Note No. 414: 52 p.

132. A person's body surface area (BSA) can be approximated by the DuBois formula

$$\text{BSA} = 0.007184w^{0.425}h^{0.725},$$

where w is mass, in kilograms, h is height, in centimeters, and BSA is in square meters. What is the BSA of a child who is 122 cm tall and has a mass of 29.5 kg?

Source: www.halls.md

133. Using a graphing calculator and a $[-10, 10, -1, 8]$ window, select the MODE SIMUL and the FORMAT EXPROFF. Then graph

$$y_1 = x^{1/2}, \quad y_2 = 3x^{2/5},$$
$$y_3 = x^{4/7}, \quad \text{and} \quad y_4 = \tfrac{1}{5}x^{3/4}.$$

Looking only at coordinates, match each graph with its equation.

Try Exercise Answers: Section 10.2

9. $\sqrt[6]{x}$ **23.** 8 **39.** $(cd)^{1/4}$ **43.** $(3mn)^{3/2}$ **49.** $\frac{1}{2}$
65.

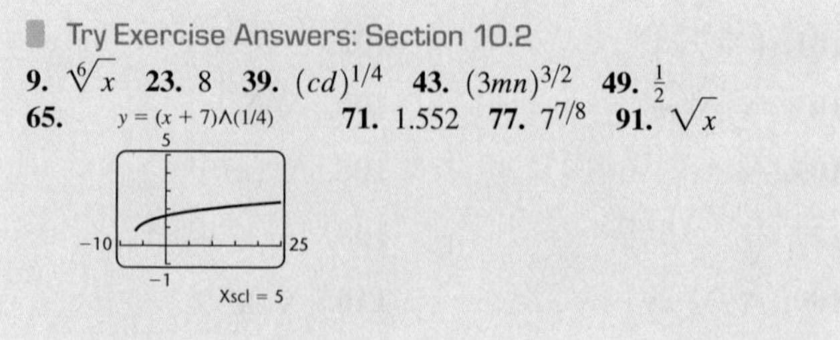

71. 1.552 **77.** $7^{7/8}$ **91.** $\sqrt{x}$

Collaborative Corner

Are Equivalent Fractions Equivalent Exponents?

Focus: Functions and rational exponents
Time: 10–20 minutes
Group Size: 3
Materials: Graph paper

In arithmetic, we have seen that $\frac{1}{3}$, $\frac{1}{6} \cdot 2$, and $2 \cdot \frac{1}{6}$ all represent the same number. Interestingly,

$$f(x) = x^{1/3},$$
$$g(x) = (x^{1/6})^2, \quad \text{and}$$
$$h(x) = (x^2)^{1/6}$$

represent three *different* functions.

ACTIVITY

1. Selecting a variety of values for x and using the definition of positive rational exponents, one group member should graph f, a second group member should graph g, and a third group member should graph h. Be sure to check whether negative x-values are in the domain of the function.
2. Compare the three graphs and check each other's work. How and why do the graphs differ?
3. Decide as a group which graph, if any, would best represent the graph of $k(x) = x^{2/6}$. Then be prepared to explain your reasoning to the entire class. (*Hint*: Study the definition of $a^{m/n}$ on p. 710 carefully.)

10.3 Multiplying Radical Expressions

- Multiplying Radical Expressions
- Simplifying by Factoring
- Multiplying and Simplifying

Interested in Something Extra?

Students interested in extra-credit projects are often pleasantly surprised to find that their instructor will not only provide a possible project for them to pursue, but might even tailor that project to the student's interests. Remember that if you do an extra-credit project, you must still complete the regular course work.

MULTIPLYING RADICAL EXPRESSIONS

Note that $\sqrt{4}\sqrt{25} = 2 \cdot 5 = 10$. Also $\sqrt{4 \cdot 25} = \sqrt{100} = 10$. Likewise,

$$\sqrt[3]{27}\sqrt[3]{8} = 3 \cdot 2 = 6 \quad \text{and} \quad \sqrt[3]{27 \cdot 8} = \sqrt[3]{216} = 6.$$

These examples suggest the following.

The Product Rule for Radicals For any real numbers $\sqrt[n]{a}$ and $\sqrt[n]{b}$,

$$\sqrt[n]{a} \cdot \sqrt[n]{b} = \sqrt[n]{a \cdot b}.$$

(The product of two nth roots is the nth root of the product of the two radicands.)

Rational exponents can be used to derive this rule:

$$\sqrt[n]{a} \cdot \sqrt[n]{b} = a^{1/n} \cdot b^{1/n} = (a \cdot b)^{1/n} = \sqrt[n]{a \cdot b}.$$

EXAMPLE 1 Multiply.

a) $\sqrt{3} \cdot \sqrt{5}$

b) $\sqrt{x + 3}\sqrt{x - 3}$

c) $\sqrt[3]{4} \cdot \sqrt[3]{5}$

d) $\sqrt[4]{\frac{y}{5}} \cdot \sqrt[4]{\frac{7}{x}}$

SOLUTION

a) When no index is written, roots are understood to be square roots with an unwritten index of two. We apply the product rule:

$$\sqrt{3} \cdot \sqrt{5} = \sqrt{3 \cdot 5} = \sqrt{15}.$$

CAUTION!
$\sqrt{x^2 - 9} \neq \sqrt{x^2} - \sqrt{9}$.
To see this, note that
$\sqrt{5^2 - 9} = \sqrt{16} = 4$, but
$\sqrt{5^2} - \sqrt{9} = 5 - 3 = 2$.

CAUTION! The product rule for radicals applies only when radicals have the same index:

$$\sqrt[n]{a} \cdot \sqrt[m]{b} \neq \sqrt[nm]{a \cdot b}.$$

b) $\sqrt{x+3}\,\sqrt{x-3} = \sqrt{(x+3)(x-3)}$ **The product of two square roots is the square root of the product.**

$= \sqrt{x^2 - 9}$

c) Both $\sqrt[3]{4}$ and $\sqrt[3]{5}$ have indices of three, so to multiply we can use the product rule:

$$\sqrt[3]{4} \cdot \sqrt[3]{5} = \sqrt[3]{4 \cdot 5} = \sqrt[3]{20}.$$

d) $\sqrt[4]{\frac{y}{5}} \cdot \sqrt[4]{\frac{7}{x}} = \sqrt[4]{\frac{y}{5} \cdot \frac{7}{x}} = \sqrt[4]{\frac{7y}{5x}}$ **In Section 10.4, we discuss other ways to write answers like this.**

Try Exercise 7.

Connecting the Concepts

We saw in Example 1(b) that

$$\sqrt{x+3}\,\sqrt{x-3} = \sqrt{x^2 - 9}.$$

Is it then true that

$$f(x) = \sqrt{x+3}\,\sqrt{x-3} \quad \text{and} \quad g(x) = \sqrt{x^2 - 9}$$

represent the same function? To answer this, let's examine the domains of f and g.

For $f(x) = \sqrt{x+3}\,\sqrt{x-3}$, we must have

$$x + 3 \geq 0 \quad \textit{and} \quad x - 3 \geq 0$$
$$x \geq -3 \quad \textit{and} \quad x \geq 3.$$

The domain of f is thus $\{x \mid x \geq -3 \textit{ and } x \geq 3\}$, or $[3, \infty)$.

For $g(x) = \sqrt{x^2 - 9}$, we must have

$$x^2 - 9 \geq 0$$
$$x^2 \geq 9.$$

This is true for $x \geq 3$ or for $x \leq -3$. Thus the domain of g is $\{x \mid x \leq -3 \textit{ or } x \geq 3\}$, or $(-\infty, -3] \cup [3, \infty)$.

The domains are not the same, as confirmed by the following graphs. Note that the graphs do coincide where the domains coincide. We say that the expressions are equivalent because they represent the same number for all possible replacements for *both* expressions.

$y = \sqrt{x+3}\,\sqrt{x-3}$

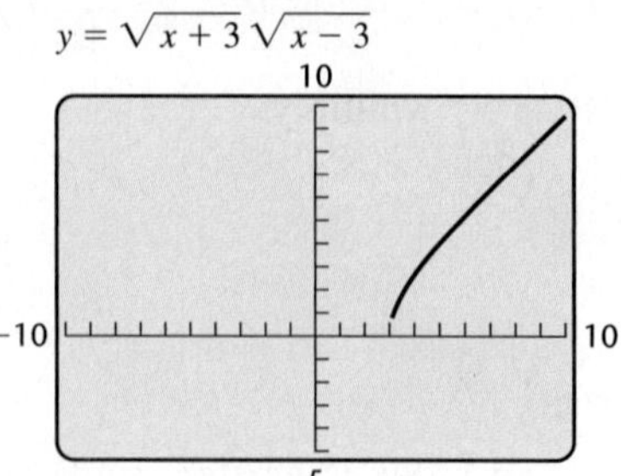

$y = \sqrt{x^2 - 9}$

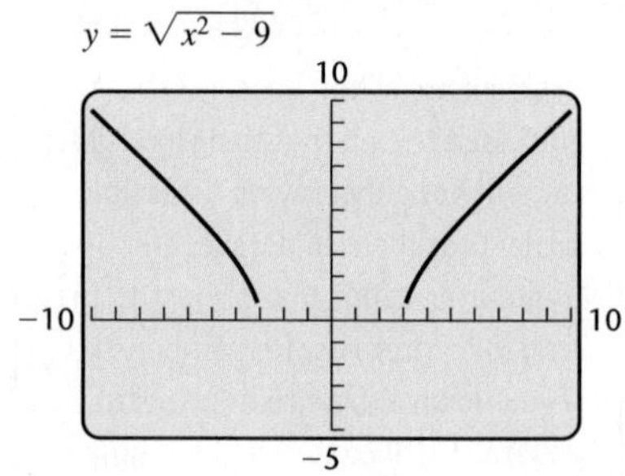

SIMPLIFYING BY FACTORING

The number p is a *perfect square* if there exists a rational number q for which $q^2 = p$. We say that p is a *perfect cube* if $q^3 = p$ for some rational number q. In general, p is a *perfect nth power* if $q^n = p$ for some rational number q. The product rule allows us to simplify $\sqrt[n]{ab}$ when a or b is a perfect nth power.

Using the Product Rule to Simplify

$\sqrt[n]{ab} = \sqrt[n]{a} \cdot \sqrt[n]{b}.$

$\left(\sqrt[n]{a} \text{ and } \sqrt[n]{b} \text{ must both be real numbers.}\right)$

To illustrate, suppose we wish to simplify $\sqrt{20}$. Since this is a *square* root, we check to see if there is a factor of 20 that is a perfect square. There is one, 4, so we express 20 as $4 \cdot 5$ and use the product rule.

$$\sqrt{20} = \sqrt{4 \cdot 5} \quad \text{Factoring the radicand (4 is a perfect square)}$$
$$= \sqrt{4} \cdot \sqrt{5} \quad \text{Factoring into two radicals}$$
$$= 2\sqrt{5} \quad \text{Taking the square root of 4}$$

To Simplify a Radical Expression with Index n by Factoring

1. Express the radicand as a product in which one factor is the largest perfect nth power possible.
2. Rewrite the expression as the nth root of each factor.
3. Simplify the expression containing the perfect nth power.
4. Simplification is complete when no radicand has a factor that is a perfect nth power.

It is often safe to assume that a radicand does not represent a negative number raised to an even power. We will henceforth make this assumption—unless functions are involved—and discontinue use of absolute-value notation when taking even roots.

EXAMPLE 2 Simplify by factoring: **(a)** $\sqrt{200}$; **(b)** $\sqrt{18x^2y}$; **(c)** $\sqrt[3]{-72}$; **(d)** $\sqrt[4]{162x^6}$.

SOLUTION

a) $\sqrt{200} = \sqrt{100 \cdot 2}$ 100 is the largest perfect-square factor of 200.

$= \sqrt{100} \cdot \sqrt{2} = 10\sqrt{2}$

Express the radicand as a product.

Rewrite as the nth root of each factor.

Simplify.

b) $\sqrt{18x^2y} = \sqrt{9 \cdot 2 \cdot x^2 \cdot y}$ $9x^2$ is the largest perfect-square factor of $18x^2y$.

$= \sqrt{9x^2} \cdot \sqrt{2y}$ Factoring into two radicals

$= 3x\sqrt{2y}$ Taking the square root of $9x^2$

c) $\sqrt[3]{-72} = \sqrt[3]{-8 \cdot 9}$ -8 is a perfect-cube (third-power) factor of -72.

$= \sqrt[3]{-8} \cdot \sqrt[3]{9} = -2\sqrt[3]{9}$

d) $\sqrt[4]{162x^6} = \sqrt[4]{81 \cdot 2 \cdot x^4 \cdot x^2}$ $81 \cdot x^4$ is the largest perfect fourth-power factor of $162x^6$.

$= \sqrt[4]{81x^4} \cdot \sqrt[4]{2x^2}$ Factoring into two radicals

$= 3x\sqrt[4]{2x^2}$ Taking fourth roots

Let's look at this example another way. We write a complete factorization and look for quadruples of factors. Each quadruple makes a perfect fourth power:

$$\sqrt[4]{162x^6} = \sqrt[4]{3 \cdot 3 \cdot 3 \cdot 3 \cdot 2 \cdot x \cdot x \cdot x \cdot x \cdot x \cdot x} \qquad 3 \cdot 3 \cdot 3 \cdot 3 = 3^4 \text{ and } x \cdot x \cdot x \cdot x = x^4$$
$$= 3 \cdot x \cdot \sqrt[4]{2 \cdot x \cdot x} = 3x\sqrt[4]{2x^2}.$$

Try Exercise 31.

EXAMPLE 3 If $f(x) = \sqrt{3x^2 - 6x + 3}$, find a simplified form for $f(x)$. Because we are working with a function, assume that x can be any real number.

SOLUTION

$$f(x) = \sqrt{3x^2 - 6x + 3}$$
$$= \sqrt{3(x^2 - 2x + 1)}$$
$$= \sqrt{(x - 1)^2 \cdot 3} \qquad \text{Factoring the radicand; } x^2 - 2x + 1 \text{ is a perfect square.}$$
$$= \sqrt{(x - 1)^2} \cdot \sqrt{3} \qquad \text{Factoring into two radicals}$$
$$= |x - 1|\sqrt{3} \qquad \text{Taking the square root of } (x - 1)^2$$

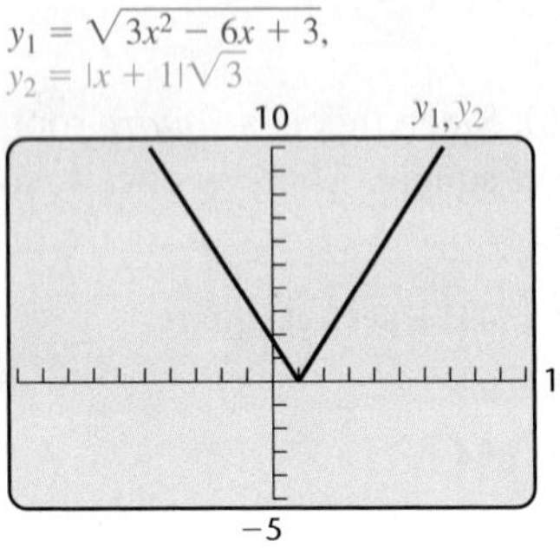

We can check by graphing $y_1 = \sqrt{3x^2 - 6x + 3}$ and $y_2 = |x - 1|\sqrt{3}$, as shown in the graph at left. It appears that the graphs coincide, and scrolling through a table of values would show us that this is indeed the case.

Try Exercise 45.

EXAMPLE 4 Simplify: **(a)** $\sqrt{x^7y^{11}z^9}$; **(b)** $\sqrt[3]{16a^7b^{14}}$.

SOLUTION

a) There are many ways to factor $x^7y^{11}z^9$. Because of the square root (index of 2), we identify the largest exponents that are multiples of 2:

$$\sqrt{x^7y^{11}z^9} = \sqrt{x^6 \cdot x \cdot y^{10} \cdot y \cdot z^8 \cdot z} \qquad \text{The largest perfect-square factor is } x^6y^{10}z^8.$$
$$= \sqrt{x^6}\sqrt{y^{10}}\sqrt{z^8}\sqrt{xyz} \qquad \text{Factoring into several radicals}$$
$$= x^{6/2}\ y^{10/2}\ z^{8/2}\sqrt{xyz} \qquad \text{Converting to rational exponents}$$
$$= x^3y^5z^4\sqrt{xyz}.$$

Check: $(x^3y^5z^4\sqrt{xyz})^2 = (x^3)^2(y^5)^2(z^4)^2(\sqrt{xyz})^2$
$= x^6 \cdot y^{10} \cdot z^8 \cdot xyz = x^7y^{11}z^9$

Our check shows that $x^3y^5z^4\sqrt{xyz}$ is the square root of $x^7y^{11}z^9$.

b) There are many ways to factor $16a^7b^{14}$. Because of the cube root (index of 3), we identify factors with the largest exponents that are multiples of 3:

$$\sqrt[3]{16a^7b^{14}} = \sqrt[3]{8 \cdot 2 \cdot a^6 \cdot a \cdot b^{12} \cdot b^2} \qquad \text{The largest perfect-cube factor is } 8a^6b^{12}.$$
$$= \sqrt[3]{8a^6b^{12}}\sqrt[3]{2ab^2} \qquad \text{Rewriting as a product of cube roots}$$
$$= 2a^2b^4\sqrt[3]{2ab^2}. \qquad \text{Simplifying the expression containing the perfect cube}$$

As a check, let's redo the problem using a complete factorization of the radicand:

$$\sqrt[3]{16a^7b^{14}} = \sqrt[3]{2 \cdot 2 \cdot 2 \cdot 2 \cdot a \cdot a \cdot a \cdot a \cdot a \cdot a \cdot a \cdot b \cdot b \cdot b \cdot b \cdot b \cdot b \cdot b \cdot b \cdot b \cdot b \cdot b \cdot b \cdot b \cdot b}$$

Each triple of factors makes a cube.

$$= 2 \cdot a \cdot a \cdot b \cdot b \cdot b \cdot b \cdot \sqrt[3]{2 \cdot a \cdot b \cdot b}$$
$$= 2a^2b^4\sqrt[3]{2ab^2}. \qquad \text{Our answer checks.}$$

Try Exercise 51.

To simplify an nth root, identify factors in the radicand with exponents that are multiples of n.

MULTIPLYING AND SIMPLIFYING

We have used the product rule for radicals to find products and also to simplify radical expressions. For some radical expressions, it is possible to do both: First find a product and then simplify.

EXAMPLE 5 Multiply and simplify.

a) $\sqrt{15}\sqrt{6}$ **b)** $3\sqrt[3]{25} \cdot 2\sqrt[3]{5}$ **c)** $\sqrt[4]{8x^3y^5}\sqrt[4]{4x^2y^3}$

SOLUTION

a) $\sqrt{15}\sqrt{6} = \sqrt{15 \cdot 6}$ **Multiplying radicands**

$= \sqrt{90} = \sqrt{9 \cdot 10}$ **9 is a perfect square.**

$= 3\sqrt{10}$

b) $3\sqrt[3]{25} \cdot 2\sqrt[3]{5} = 3 \cdot 2 \cdot \sqrt[3]{25 \cdot 5}$ **Using a commutative law; multiplying radicands**

$= 6 \cdot \sqrt[3]{125}$ **125 is a perfect cube.**

$= 6 \cdot 5$, or 30

c) $\sqrt[4]{8x^3y^5}\sqrt[4]{4x^2y^3} = \sqrt[4]{32x^5y^8}$ **Multiplying radicands**

$= \sqrt[4]{16x^4y^8 \cdot 2x}$ **Identifying the largest perfect fourth-power factor**

$= \sqrt[4]{16x^4y^8}\sqrt[4]{2x}$ **Factoring into radicals**

$= 2xy^2\sqrt[4]{2x}$ **Finding the fourth roots; assume $x \geq 0$.**

The checks are left to the student.

Try Exercise 65.

Student Notes

To multiply $\sqrt{x} \cdot \sqrt{x}$, remember what $\sqrt{x}$ represents and go directly to the product, x. For $x \geq 0$,

$\sqrt{x} \cdot \sqrt{x} = x$,

$(\sqrt{x})^2 = x$, and

$\sqrt{x^2} = x$.

10.3 Exercise Set

FOR EXTRA HELP MyMathLab

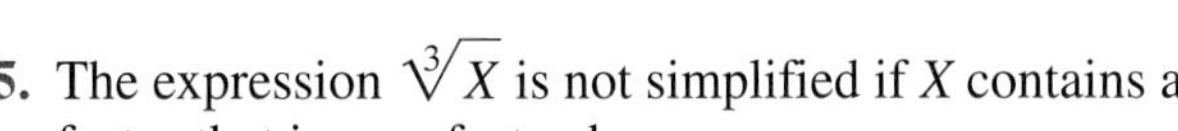

Concept Reinforcement Classify each of the following statements as either true or false.

1. For any real numbers $\sqrt[n]{a}$ and $\sqrt[n]{b}$, $\sqrt[n]{a} \cdot \sqrt[n]{b} = \sqrt[n]{ab}$.

2. For any real numbers $\sqrt[n]{a}$ and $\sqrt[n]{b}$, $\sqrt[n]{a} + \sqrt[n]{b} = \sqrt[n]{a + b}$.

3. For any real numbers $\sqrt[n]{a}$ and $\sqrt[m]{b}$, $\sqrt[n]{a} \cdot \sqrt[m]{b} = \sqrt[nm]{ab}$.

4. For $x > 0$, $\sqrt{x^2 - 9} = x - 3$.

5. The expression $\sqrt[3]{X}$ is not simplified if X contains a factor that is a perfect cube.

6. It is often possible to simplify $\sqrt{A \cdot B}$ even though $\sqrt{A}$ and $\sqrt{B}$ cannot be simplified.

Multiply.

7. $\sqrt{5}\sqrt{7}$
8. $\sqrt{10}\sqrt{3}$
9. $\sqrt[3]{3}\sqrt[3]{2}$
10. $\sqrt[3]{2}\sqrt[3]{5}$
11. $\sqrt[4]{6}\sqrt[4]{3}$
12. $\sqrt[4]{8}\sqrt[4]{9}$

13. $\sqrt{2x}\sqrt{13y}$

14. $\sqrt{5a}\sqrt{6b}$

15. $\sqrt[5]{8y^3}\sqrt[5]{10y}$

16. $\sqrt[5]{9t^2}\sqrt[5]{2t}$

17. $\sqrt{y-b}\sqrt{y+b}$

18. $\sqrt{x-a}\sqrt{x+a}$

19. $\sqrt[3]{0.7y}\sqrt[3]{0.3y}$

20. $\sqrt[3]{0.5x}\sqrt[3]{0.2x}$

21. $\sqrt[5]{x-2}\sqrt[5]{(x-2)^2}$

22. $\sqrt[4]{x-1}\sqrt[4]{x^2+x+1}$

23. $\sqrt{\dfrac{3}{t}}\sqrt{\dfrac{7s}{11}}$

24. $\sqrt{\dfrac{5x}{6}}\sqrt{\dfrac{11}{y}}$

25. $\sqrt[7]{\dfrac{x-3}{4}}\sqrt[7]{\dfrac{5}{x+2}}$

26. $\sqrt[6]{\dfrac{a}{b-2}}\sqrt[6]{\dfrac{3}{b+2}}$

Simplify by factoring.

27. $\sqrt{18}$

28. $\sqrt{50}$

29. $\sqrt{27}$

30. $\sqrt{45}$

31. $\sqrt{8x^9}$

32. $\sqrt{75y^5}$

33. $\sqrt{120}$

34. $\sqrt{350}$

35. $\sqrt{36a^4b}$

36. $\sqrt{175y^8}$

37. $\sqrt[3]{8x^3y^2}$

38. $\sqrt[3]{27ab^6}$

39. $\sqrt[3]{-16x^6}$

40. $\sqrt[3]{-32a^6}$

Find a simplified form of $f(x)$. Assume that x can be any real number.

41. $f(x) = \sqrt[3]{125x^5}$

42. $f(x) = \sqrt[3]{16x^6}$

43. $f(x) = \sqrt{49(x-3)^2}$

44. $f(x) = \sqrt{81(x-1)^2}$

45. $f(x) = \sqrt{5x^2 - 10x + 5}$

46. $f(x) = \sqrt{2x^2 + 8x + 8}$

Simplify. Assume that no radicands were formed by raising negative numbers to even powers.

47. $\sqrt{a^6b^7}$

48. $\sqrt{x^6y^9}$

49. $\sqrt[3]{x^5y^6z^{10}}$

50. $\sqrt[3]{a^6b^7c^{13}}$

51. $\sqrt[4]{16x^5y^{11}}$

52. $\sqrt[5]{-32a^7b^{11}}$

53. $\sqrt[5]{x^{13}y^8z^{17}}$

54. $\sqrt[5]{a^6b^8c^9}$

55. $\sqrt[3]{-80a^{14}}$

56. $\sqrt[4]{810x^9}$

Multiply and simplify. Assume that no radicands were formed by raising negative numbers to even powers.

57. $\sqrt{6}\sqrt{3}$

58. $\sqrt{15}\sqrt{5}$

59. $\sqrt{10}\sqrt{14}$

60. $\sqrt{6}\sqrt{33}$

61. $\sqrt[3]{9}\sqrt[3]{3}$

62. $\sqrt[3]{2}\sqrt[3]{4}$

Aha! 63. $\sqrt{18a^3}\sqrt{18a^3}$

64. $\sqrt{75x^7}\sqrt{75x^7}$

65. $\sqrt[3]{5a^2}\sqrt[3]{2a}$

66. $\sqrt[3]{7x}\sqrt[3]{3x^2}$

67. $3\sqrt{2x^5}\cdot 4\sqrt{10x^2}$

68. $3\sqrt{5a^7}\,2\sqrt{15a^3}$

69. $\sqrt[3]{s^2t^4}\sqrt[3]{s^4t^6}$

70. $\sqrt[3]{x^2y^4}\sqrt[3]{x^2y^6}$

71. $\sqrt[3]{(x+5)^2}\sqrt[3]{(x+5)^4}$

72. $\sqrt[3]{(a-b)^5}\sqrt[3]{(a-b)^7}$

73. $\sqrt[4]{20a^3b^7}\sqrt[4]{4a^2b^5}$

74. $\sqrt[4]{9x^7y^2}\sqrt[4]{9x^2y^9}$

75. $\sqrt[5]{x^3(y+z)^6}\sqrt[5]{x^3(y+z)^4}$

76. $\sqrt[5]{a^3(b-c)^4}\sqrt[5]{a^7(b-c)^4}$

TW 77. Explain how you could convince a friend that $\sqrt{x^2-16} \neq \sqrt{x^2} - \sqrt{16}$.

TW 78. Why is it incorrect to say that, in general, $\sqrt{x^2} = x$?

SKILL REVIEW

Review simplifying rational expressions (Sections 7.1–7.5).

Perform the indicated operation and, if possible, simplify.

79. $\dfrac{15a^2x}{8b} \cdot \dfrac{24b^2x}{5a}$ [7.2]

80. $\dfrac{x^2-1}{x^2-4} \div \dfrac{x^2-x-2}{x^2+x-2}$ [7.2]

81. $\dfrac{x-3}{2x-10} - \dfrac{3x-5}{x^2-25}$ [7.4]

82. $\dfrac{6x}{25y^2} + \dfrac{3y}{10x}$ [7.4]

83. $\dfrac{a^{-1}+b^{-1}}{ab}$ [7.5]

84. $\dfrac{\dfrac{1}{x+1} - \dfrac{2}{x}}{\dfrac{3}{x} + \dfrac{1}{x+1}}$ [7.5]

SYNTHESIS

TW 85. Abdul is puzzled. When he uses a graphing calculator to graph $y = \sqrt{x}\cdot\sqrt{x}$, he gets the following screen. Explain why Abdul did not get the complete line $y = x$.

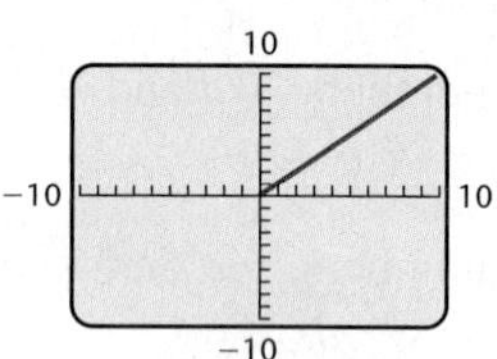

TW **86.** Is the equation $\sqrt{(2x + 3)^8} = (2x + 3)^4$ always, sometimes, or never true? Why?

87. *Radar Range.* The function given by

$$R(x) = \frac{1}{2}\sqrt[4]{\frac{x \cdot 3.0 \times 10^6}{\pi^2}}$$

can be used to determine the maximum range $R(x)$, in miles, of an ARSR-3 surveillance radar with a peak power of x watts. Determine the maximum radar range when the peak power is 5×10^4 watts.
Source: Introduction to RADAR Techniques, Federal Aviation Administration, 1988

88. *Speed of a Skidding Car.* Under certain conditions, police can estimate the speed at which a car was traveling by measuring its skid marks. The function given by

$$r(L) = 2\sqrt{5L}$$

can be used, where L is the length of a skid mark, in feet, and $r(L)$ is the speed, in miles per hour. Find the exact speed and an estimate (to the nearest tenth mile per hour) for the speed of a car that left skid marks **(a)** 20 ft long; **(b)** 70 ft long; **(c)** 90 ft long. See also Exercise 101.

89. *Wind Chill Temperature.* When the temperature is T degrees Celsius and the wind speed is v meters per second, the *wind chill temperature*, T_w, is the temperature (with no wind) that it feels like. Here is a formula for finding wind chill temperature:

$$T_w = 33 - \frac{(10.45 + 10\sqrt{v} - v)(33 - T)}{22}.$$

Estimate the wind chill temperature (to the nearest tenth of a degree) for the given actual temperatures and wind speeds.

a) $T = 7°C,\ v = 8\ m/sec$
b) $T = 0°C,\ v = 12\ m/sec$
c) $T = -5°C,\ v = 14\ m/sec$
d) $T = -23°C,\ v = 15\ m/sec$

Simplify. Assume that all variables are nonnegative.

90. $(\sqrt{r^3 t})^7$

91. $(\sqrt[3]{25x^4})^4$

92. $(\sqrt[3]{a^2b^4})^5$

93. $(\sqrt{a^3b^5})^7$

Draw and compare the graphs of each group of functions.

94. $f(x) = \sqrt{x^2 + 2x + 1}$,
$g(x) = x + 1$,
$h(x) = |x + 1|$

95. $f(x) = \sqrt{x^2 - 2x + 1}$,
$g(x) = x - 1$,
$h(x) = |x - 1|$

96. If $f(t) = \sqrt{t^2 - 3t - 4}$, what is the domain of f?

97. What is the domain of g, if $g(x) = \sqrt{x^2 - 6x + 8}$?

Solve.

98. $\sqrt[3]{5x^{k+1}}\,\sqrt[3]{25x^k} = 5x^7$, for k

99. $\sqrt[5]{4a^{3k+2}}\,\sqrt[5]{8a^{6-k}} = 2a^4$, for k

100. Use a graphing calculator to check your answers to Exercises 21 and 41.

TW **101.** Does a car traveling twice as fast as another car leave a skid mark that is twice as long? (See Exercise 88.) Why or why not?

Try Exercise Answers: Section 10.3
7. $\sqrt{35}$ **31.** $2x^4\sqrt{2x}$ **45.** $f(x) = |x - 1|\sqrt{5}$
51. $2xy^2\sqrt[4]{xy^3}$ **65.** $a\sqrt[3]{10}$

10.4 Dividing Radical Expressions

- Dividing and Simplifying
- Rationalizing Denominators or Numerators with One Term

DIVIDING AND SIMPLIFYING

Just as the root of a product can be expressed as the product of two roots, the root of a quotient can be expressed as the quotient of two roots. For example,

$$\sqrt[3]{\frac{27}{8}} = \frac{3}{2} \quad \text{and} \quad \frac{\sqrt[3]{27}}{\sqrt[3]{8}} = \frac{3}{2}.$$

This example suggests the following.

The Quotient Rule for Radicals For any real numbers $\sqrt[n]{a}$ and $\sqrt[n]{b}$, $b \neq 0$,

$$\sqrt[n]{\frac{a}{b}} = \frac{\sqrt[n]{a}}{\sqrt[n]{b}}.$$

STUDY TIP

Professors Are Human

Even the best professors sometimes make mistakes. If, as you review your notes, you find that something doesn't make sense, it may be due to your instructor having made a mistake. If, after double-checking, you still perceive a mistake, politely ask him or her about it. Your instructor will welcome the opportunity to correct any errors.

Remember that an nth root is simplified when its radicand has no factors that are perfect nth powers. Recall too that we assume that no radicands represent negative quantities raised to an even power.

EXAMPLE 1 Simplify by taking the roots of the numerator and the denominator.

a) $\sqrt[3]{\frac{27}{125}}$

b) $\sqrt{\frac{25}{y^2}}$

SOLUTION

a) $\sqrt[3]{\frac{27}{125}} = \frac{\sqrt[3]{27}}{\sqrt[3]{125}} = \frac{3}{5}$ **Taking the cube roots of the numerator and the denominator**

b) $\sqrt{\frac{25}{y^2}} = \frac{\sqrt{25}}{\sqrt{y^2}} = \frac{5}{y}$ **Taking the square roots of the numerator and the denominator. Assume $y > 0$.**

Try Exercise 9.

Any radical expressions appearing in the answers should be simplified as much as possible.

EXAMPLE 2 Simplify: **(a)** $\sqrt{\frac{16x^3}{y^8}}$; **(b)** $\sqrt[3]{\frac{27y^{14}}{8x^3}}$.

SOLUTION

a) $\sqrt{\dfrac{16x^3}{y^8}} = \dfrac{\sqrt{16x^3}}{\sqrt{y^8}}$

$= \dfrac{\sqrt{16x^2 \cdot x}}{\sqrt{y^8}} = \dfrac{4x\sqrt{x}}{y^4}$ **Simplifying the numerator and the denominator**

b) $\sqrt[3]{\dfrac{27y^{14}}{8x^3}} = \dfrac{\sqrt[3]{27y^{14}}}{\sqrt[3]{8x^3}}$

$= \dfrac{\sqrt[3]{27y^{12}y^2}}{\sqrt[3]{8x^3}} = \dfrac{\sqrt[3]{27y^{12}}\sqrt[3]{y^2}}{\sqrt[3]{8x^3}} = \dfrac{3y^4\sqrt[3]{y^2}}{2x}$ **Simplifying the numerator and the denominator**

Try Exercise 15.

If we read from right to left, the quotient rule tells us that to divide two radical expressions that have the same index, we can divide the radicands.

EXAMPLE 3 Divide and, if possible, simplify.

a) $\dfrac{\sqrt{80}}{\sqrt{5}}$ b) $\dfrac{5\sqrt[3]{32}}{\sqrt[3]{2}}$ c) $\dfrac{\sqrt{72xy}}{2\sqrt{2}}$ d) $\dfrac{\sqrt[4]{18a^9b^5}}{\sqrt[4]{3b}}$

Student Notes

When writing radical signs, pay careful attention to what is included as the radicand. Each of the following represents a *different* number:

$$\sqrt{\frac{5 \cdot 2}{3}}, \quad \frac{\sqrt{5 \cdot 2}}{3}, \quad \frac{\sqrt{5} \cdot 2}{3}.$$

SOLUTION

a) $\dfrac{\sqrt{80}}{\sqrt{5}} = \sqrt{\dfrac{80}{5}} = \sqrt{16} = 4$ **Because the indices match, we can divide the radicands.**

b) $\dfrac{5\sqrt[3]{32}}{\sqrt[3]{2}} = 5\sqrt[3]{\dfrac{32}{2}} = 5\sqrt[3]{16}$

$= 5\sqrt[3]{8 \cdot 2}$ **8 is the largest perfect-cube factor of 16.**

$= 5\sqrt[3]{8}\sqrt[3]{2} = 5 \cdot 2\sqrt[3]{2}$

$= 10\sqrt[3]{2}$

c) $\dfrac{\sqrt{72xy}}{2\sqrt{2}} = \dfrac{1}{2}\sqrt{\dfrac{72xy}{2}}$ **Because the indices match, we can divide the radicands.**

$= \dfrac{1}{2}\sqrt{36xy} = \dfrac{1}{2} \cdot 6\sqrt{xy} = 3\sqrt{xy}$

d) $\dfrac{\sqrt[4]{18a^9b^5}}{\sqrt[4]{3b}} = \sqrt[4]{\dfrac{18a^9b^5}{3b}}$

$= \sqrt[4]{6a^9b^4} = \sqrt[4]{a^8b^4}\sqrt[4]{6a}$ **Note that 8 is the largest power less than 9 that is a multiple of the index 4.**

$= a^2b\sqrt[4]{6a}$

Try Exercise 27.

RATIONALIZING DENOMINATORS OR NUMERATORS WITH ONE TERM*

The expressions

$$\frac{1}{\sqrt{2}} \quad \text{and} \quad \frac{\sqrt{2}}{2}$$

are equivalent, but the second expression does not have a radical expression in the denominator.† We can **rationalize the denominator** of a radical expression if we multiply by 1 in either of two ways.

One way is to multiply by 1 *under* the radical to make the denominator of the radicand a perfect power.

EXAMPLE 4 Rationalize each denominator.

a) $\sqrt{\frac{7}{3}}$ **b)** $\sqrt[3]{\frac{5}{16}}$

SOLUTION

a) We multiply by 1 under the radical, using $\frac{3}{3}$. We do this so that the denominator of the radicand will be a perfect square:

$$\sqrt{\frac{7}{3}} = \sqrt{\frac{7}{3} \cdot \frac{3}{3}} \quad \text{Multiplying by 1 under the radical}$$

$$= \sqrt{\frac{21}{9}} \quad \text{The denominator, 9, is now a perfect square.}$$

$$= \frac{\sqrt{21}}{\sqrt{9}} = \frac{\sqrt{21}}{3}.$$

b) Note that $16 = 4^2$. Thus, to make the denominator a perfect cube, we multiply under the radical by $\frac{4}{4}$:

$$\sqrt[3]{\frac{5}{16}} = \sqrt[3]{\frac{5}{4 \cdot 4} \cdot \frac{4}{4}} \quad \text{Since the index is 3, we need 3 identical factors in the denominator.}$$

$$= \sqrt[3]{\frac{20}{4^3}} \quad \text{The denominator is now a perfect cube.}$$

$$= \frac{\sqrt[3]{20}}{\sqrt[3]{4^3}} = \frac{\sqrt[3]{20}}{4}.$$

Try Exercise 41.

Another way to rationalize a denominator is to multiply by 1 *outside* the radical.

EXAMPLE 5 Rationalize each denominator.

a) $\sqrt{\frac{4}{5b}}$ **b)** $\frac{\sqrt[3]{a}}{\sqrt[3]{9x}}$

*Denominators and numerators with two terms are rationalized in Section 10.5.

†See Exercise 73 on p. 729.

SOLUTION

a) We rewrite the expression as a quotient of two radicals. Then we simplify and multiply by 1:

$$\sqrt{\frac{4}{5b}} = \frac{\sqrt{4}}{\sqrt{5b}} = \frac{2}{\sqrt{5b}} \qquad \text{We assume } b > 0.$$

$$= \frac{2}{\sqrt{5b}} \cdot \frac{\sqrt{5b}}{\sqrt{5b}} \qquad \text{Multiplying by 1}$$

$$= \frac{2\sqrt{5b}}{\left(\sqrt{5b}\right)^2} \qquad \text{Try to do this step mentally.}$$

$$= \frac{2\sqrt{5b}}{5b}.$$

b) To rationalize the denominator $\sqrt[3]{9x}$, note that $9x$ is $3 \cdot 3 \cdot x$. In order for this radicand to be a cube, we need another factor of 3 and two more factors of x. Thus we multiply by 1, using $\sqrt[3]{3x^2}/\sqrt[3]{3x^2}$:

$$\frac{\sqrt[3]{a}}{\sqrt[3]{9x}} = \frac{\sqrt[3]{a}}{\sqrt[3]{9x}} \cdot \frac{\sqrt[3]{3x^2}}{\sqrt[3]{3x^2}} \qquad \text{Multiplying by 1}$$

$$= \frac{\sqrt[3]{3ax^2}}{\sqrt[3]{27x^3}} \longleftarrow \text{This radicand is now a perfect cube.}$$

$$= \frac{\sqrt[3]{3ax^2}}{3x}.$$

Try Exercise 49.

Sometimes in calculus it is necessary to rationalize a numerator. To do so, we multiply by 1 to make the radicand in the *numerator* a perfect power.

EXAMPLE 6 Rationalize each numerator.

a) $\sqrt{\frac{7}{5}}$

b) $\frac{\sqrt[3]{4a^2}}{\sqrt[3]{5b}}$

SOLUTION

a)
$$\sqrt{\frac{7}{5}} = \sqrt{\frac{7}{5} \cdot \frac{7}{7}} \qquad \text{Multiplying by 1 under the radical. We also could have multiplied by } \sqrt{7}/\sqrt{7} \text{ outside the radical.}$$

$$= \sqrt{\frac{49}{35}} \qquad \text{The numerator is now a perfect square.}$$

$$= \frac{\sqrt{49}}{\sqrt{35}} = \frac{7}{\sqrt{35}}$$

b)
$$\frac{\sqrt[3]{4a^2}}{\sqrt[3]{5b}} = \frac{\sqrt[3]{4a^2}}{\sqrt[3]{5b}} \cdot \frac{\sqrt[3]{2a}}{\sqrt[3]{2a}} \qquad \text{Multiplying by 1}$$

$$= \frac{\sqrt[3]{8a^3}}{\sqrt[3]{10ba}} \longleftarrow \text{This radicand is now a perfect cube.}$$

$$= \frac{2a}{\sqrt[3]{10ab}}$$

Try Exercise 59.

10.4 Exercise Set

FOR EXTRA HELP MyMathLab MathXL PRACTICE  WATCH  DOWNLOAD READ  REVIEW

Concept Reinforcement *In each of Exercises 1–8, match the expression with an equivalent expression from the column on the right. Assume $a, b > 0$.*

1. ___ $\sqrt[3]{\frac{a^2}{b^6}}$
2. ___ $\frac{\sqrt[3]{a^6}}{\sqrt[3]{b^9}}$
3. ___ $\sqrt[5]{\frac{a^6}{b^4}}$
4. ___ $\sqrt{\frac{a}{b^3}}$
5. ___ $\frac{\sqrt[5]{a^2}}{\sqrt[5]{b^2}}$
6. ___ $\frac{\sqrt{5a^4}}{\sqrt{5a^3}}$
7. ___ $\frac{\sqrt[5]{a^2}}{\sqrt[5]{b^3}}$
8. ___ $\sqrt[4]{\frac{16a^6}{a^2}}$

a) $\frac{\sqrt[5]{a^2}\,\sqrt[5]{b^2}}{\sqrt[5]{b^5}}$

b) $\frac{a^2}{b^3}$

c) $\sqrt{\frac{a \cdot b}{b^3 \cdot b}}$

d) $\sqrt{a}$

e) $\frac{\sqrt[3]{a^2}}{b^2}$

f) $\sqrt[5]{\frac{a^6 b}{b^4 \cdot b}}$

g) $2a$

h) $\frac{\sqrt[5]{a^2 b^3}}{\sqrt[5]{b^5}}$

Simplify by taking the roots of the numerator and the denominator. Assume that all variables represent positive numbers.

9. $\sqrt{\frac{36}{25}}$
10. $\sqrt{\frac{100}{81}}$
11. $\sqrt[3]{\frac{64}{27}}$
12. $\sqrt[3]{\frac{343}{1000}}$
13. $\sqrt{\frac{49}{y^2}}$
14. $\sqrt{\frac{121}{x^2}}$
15. $\sqrt{\frac{36y^3}{x^4}}$
16. $\sqrt{\frac{25a^5}{b^6}}$
17. $\sqrt[3]{\frac{27a^4}{8b^3}}$
18. $\sqrt[3]{\frac{64x^7}{216y^6}}$
19. $\sqrt[4]{\frac{32a^4}{2b^4c^8}}$
20. $\sqrt[4]{\frac{81x^4}{y^8z^4}}$
21. $\sqrt[4]{\frac{a^5b^8}{c^{10}}}$
22. $\sqrt[4]{\frac{x^9y^{12}}{z^6}}$
23. $\sqrt[5]{\frac{32x^6}{y^{11}}}$
24. $\sqrt[5]{\frac{243a^9}{b^{13}}}$
25. $\sqrt[6]{\frac{x^6y^8}{z^{15}}}$
26. $\sqrt[6]{\frac{a^9b^{12}}{c^{13}}}$

Divide and, if possible, simplify. Assume that all variables represent positive numbers.

27. $\frac{\sqrt{18y}}{\sqrt{2y}}$
28. $\frac{\sqrt{700x}}{\sqrt{7x}}$
29. $\frac{\sqrt[3]{26}}{\sqrt[3]{13}}$
30. $\frac{\sqrt[3]{35}}{\sqrt[3]{5}}$
31. $\frac{\sqrt{40xy^3}}{\sqrt{8x}}$
32. $\frac{\sqrt{56ab^3}}{\sqrt{7a}}$
33. $\frac{\sqrt[3]{96a^4b^2}}{\sqrt[3]{12a^2b}}$
34. $\frac{\sqrt[3]{189x^5y^7}}{\sqrt[3]{7x^2y^2}}$
35. $\frac{\sqrt{100ab}}{5\sqrt{2}}$
36. $\frac{\sqrt{75ab}}{3\sqrt{3}}$
37. $\frac{\sqrt[4]{48x^9y^{13}}}{\sqrt[4]{3xy^{-2}}}$
38. $\frac{\sqrt[5]{64a^{11}b^{28}}}{\sqrt[5]{2ab^{-2}}}$
39. $\frac{\sqrt[3]{x^3 - y^3}}{\sqrt[3]{x - y}}$
40. $\frac{\sqrt[3]{r^3 + s^3}}{\sqrt[3]{r + s}}$

***Hint*: Factor and then simplify.**

Rationalize each denominator. Assume that all variables represent positive numbers.

41. $\sqrt{\frac{3}{2}}$
42. $\sqrt{\frac{6}{7}}$
43. $\frac{2\sqrt{5}}{7\sqrt{3}}$
44. $\frac{3\sqrt{5}}{2\sqrt{2}}$
45. $\sqrt[3]{\frac{5}{4}}$
46. $\sqrt[3]{\frac{2}{9}}$
47. $\frac{\sqrt[3]{3a}}{\sqrt[3]{5c}}$
48. $\frac{\sqrt[3]{7x}}{\sqrt[3]{3y}}$

49. $\dfrac{\sqrt[4]{5y^6}}{\sqrt[4]{9x}}$

50. $\dfrac{\sqrt[5]{3a^4}}{\sqrt[5]{2b^7}}$

51. $\sqrt[3]{\dfrac{2}{x^2y}}$

52. $\sqrt[3]{\dfrac{5}{ab^2}}$

53. $\sqrt{\dfrac{7a}{18}}$

54. $\sqrt{\dfrac{3x}{10}}$

55. $\sqrt{\dfrac{9}{20x^2y}}$

56. $\sqrt{\dfrac{7}{32a^2b}}$

Aha! 57. $\sqrt{\dfrac{10ab^2}{72a^3b}}$

58. $\sqrt{\dfrac{21x^2y}{75xy^5}}$

Rationalize each numerator. Assume that all variables represent positive numbers.

59. $\sqrt{\dfrac{5}{11}}$

60. $\sqrt{\dfrac{2}{3}}$

61. $\dfrac{2\sqrt{6}}{5\sqrt{7}}$

62. $\dfrac{3\sqrt{10}}{2\sqrt{3}}$

63. $\dfrac{\sqrt{8}}{2\sqrt{3x}}$

64. $\dfrac{\sqrt{12}}{\sqrt{5y}}$

65. $\dfrac{\sqrt[3]{7}}{\sqrt[3]{2}}$

66. $\dfrac{\sqrt[3]{5}}{\sqrt[3]{4}}$

67. $\sqrt{\dfrac{7x}{3y}}$

68. $\sqrt{\dfrac{7a}{6b}}$

69. $\sqrt[3]{\dfrac{2a^5}{5b}}$

70. $\sqrt[3]{\dfrac{2a^4}{7b}}$

71. $\sqrt{\dfrac{x^3y}{2}}$

72. $\sqrt{\dfrac{ab^5}{3}}$

TW 73. Explain why it is easier to approximate

$$\frac{\sqrt{2}}{2} \quad \text{than} \quad \frac{1}{\sqrt{2}}$$

if no calculator is available and we know that $\sqrt{2} \approx 1.414213562$.

TW 74. A student *incorrectly* claims that

$$\frac{5+\sqrt{2}}{\sqrt{18}} = \frac{5+\sqrt{1}}{\sqrt{9}} = \frac{5+1}{3}.$$

How could you convince the student that a mistake has been made? How would you explain the correct way of rationalizing the denominator?

SKILL REVIEW

To prepare for Section 10.5, review factoring expressions and multiplying polynomials (Sections 5.6 and 6.1).

Factor. [6.1]

75. $3x - 8xy + 2xz$

76. $4a^2c + 9ac - 3a^3c$

Multiply. [5.6]

77. $(a + b)(a - b)$

78. $(a^2 - 2y)(a^2 + 2y)$

79. $(8 + 3x)(7 - 4x)$

80. $(2y - x)(3a - c)$

SYNTHESIS

TW 81. Is the quotient of two irrational numbers always an irrational number? Why or why not?

TW 82. Is it possible to understand how to rationalize a denominator without knowing how to multiply rational expressions? Why or why not?

83. *Pendulums.* The *period* of a pendulum is the time it takes to complete one cycle, swinging to and fro. For a pendulum that is L centimeters long, the period T is given by the formula

$$T = 2\pi\sqrt{\frac{L}{980}},$$

where T is in seconds. Find, to the nearest hundredth of a second, the period of a pendulum of length **(a)** 65 cm; **(b)** 98 cm; **(c)** 120 cm. Use a calculator's (π) key if possible.

Perform the indicated operations.

84. $\dfrac{7\sqrt{a^2b}\sqrt{25xy}}{5\sqrt{a^{-4}b^{-1}}\sqrt{49x^{-1}y^{-3}}}$

85. $\dfrac{(\sqrt[3]{81mn^2})^2}{(\sqrt[3]{mn})^2}$

86. $\dfrac{\sqrt{44x^2y^9z}\sqrt{22y^9z^6}}{(\sqrt{11xy^8z^2})^2}$

87. $\sqrt{a^2 - 3} - \dfrac{a^2}{\sqrt{a^2 - 3}}$

88. $5\sqrt{\dfrac{x}{y}} + 4\sqrt{\dfrac{y}{x}} - \dfrac{3}{\sqrt{xy}}$

89. Provide a reason for each step in the following derivation of the quotient rule:

$$\sqrt[n]{\frac{a}{b}} = \left(\frac{a}{b}\right)^{1/n} \qquad \_\_\_\_\_\_\_\_\_\_$$
$$= \frac{a^{1/n}}{b^{1/n}} \qquad \_\_\_\_\_\_\_\_\_\_$$
$$= \frac{\sqrt[n]{a}}{\sqrt[n]{b}} \qquad \_\_\_\_\_\_\_\_\_\_$$

90. Show that $\dfrac{\sqrt[n]{a}}{\sqrt[n]{b}}$ is the nth root of $\dfrac{a}{b}$ by raising it to the nth power and simplifying.

91. Let $f(x) = \sqrt{18x^3}$ and $g(x) = \sqrt{2x}$. Find $(f/g)(x)$ and specify the domain of f/g.

92. Let $f(t) = \sqrt{2t}$ and $g(t) = \sqrt{50t^3}$. Find $(f/g)(t)$ and specify the domain of f/g.

93. Let $f(x) = \sqrt{x^2 - 9}$ and $g(x) = \sqrt{x - 3}$. Find $(f/g)(x)$ and specify the domain of f/g.

Try Exercise Answers: Section 10.4

9. $\dfrac{6}{5}$ **15.** $\dfrac{6y\sqrt{y}}{x^2}$ **27.** 3 **41.** $\dfrac{\sqrt{6}}{2}$ **49.** $\dfrac{y\sqrt[4]{45x^3y^2}}{3x}$

59. $\dfrac{5}{\sqrt{55}}$

10.5 Expressions Containing Several Radical Terms

- Adding and Subtracting Radical Expressions
- Products and Quotients of Two or More Radical Terms
- Rationalizing Denominators and Numerators with Two Terms
- Terms with Differing Indices

Radical expressions like $6\sqrt{7} + 4\sqrt{7}$ or $\left(\sqrt{a} + \sqrt{b}\right)\left(\sqrt{a} - \sqrt{b}\right)$ contain more than one *radical term* and can sometimes be simplified.

ADDING AND SUBTRACTING RADICAL EXPRESSIONS

When two radical expressions have the same indices and radicands, they are said to be **like radicals.** Like radicals can be combined (added or subtracted) in much the same way that we combine like terms.

EXAMPLE 1 Simplify by combining like radical terms.

a) $6\sqrt{7} + 4\sqrt{7}$

b) $\sqrt[3]{2} - 7x\sqrt[3]{2} + 5\sqrt[3]{2}$

c) $6\sqrt[5]{4x} + 3\sqrt[5]{4x} - \sqrt[3]{4x}$

SOLUTION

a) $6\sqrt{7} + 4\sqrt{7} = (6 + 4)\sqrt{7}$ — Using the distributive law (factoring out $\sqrt{7}$)

$= 10\sqrt{7}$ — You can think: 6 square roots of 7 plus 4 square roots of 7 results in 10 square roots of 7.

b) $\sqrt[3]{2} - 7x\sqrt[3]{2} + 5\sqrt[3]{2} = (1 - 7x + 5)\sqrt[3]{2}$ — Factoring out $\sqrt[3]{2}$

$= (6 - 7x)\sqrt[3]{2}$ — These parentheses are important!

Student Notes

Combining like radicals is similar to combining like terms. Recall the following:

$$3x + 8x = (3 + 8)x = 11x$$

and

$$6x^2 - 7x^2 = (6 - 7)x^2 = -x^2.$$

c) $6\sqrt[5]{4x} + 3\sqrt[5]{4x} - \sqrt[3]{4x} = (6 + 3)\sqrt[5]{4x} - \sqrt[3]{4x}$ **Try to do this step mentally.**

$= 9\sqrt[5]{4x} - \sqrt[3]{4x}$ **The indices are different. We cannot combine these terms.**

Try Exercise 7.

Our ability to simplify radical expressions can help us to find like radicals even when, at first, it may appear that none exists.

EXAMPLE 2 Simplify by combining like radical terms, if possible.

a) $3\sqrt{8} - 5\sqrt{2}$ **b)** $9\sqrt{5} - 4\sqrt{3}$

c) $\sqrt[3]{2x^6y^4} + 7\sqrt[3]{2y}$

SOLUTION

a) $3\sqrt{8} - 5\sqrt{2} = 3\sqrt{4 \cdot 2} - 5\sqrt{2}$
$= 3\sqrt{4} \cdot \sqrt{2} - 5\sqrt{2}$ **Simplifying $\sqrt{8}$**
$= 3 \cdot 2 \cdot \sqrt{2} - 5\sqrt{2}$
$= 6\sqrt{2} - 5\sqrt{2}$
$= \sqrt{2}$ **Combining like radicals**

b) $9\sqrt{5} - 4\sqrt{3}$ cannot be simplified. **The radicands are different.**

c) $\sqrt[3]{2x^6y^4} + 7\sqrt[3]{2y} = \sqrt[3]{x^6y^3 \cdot 2y} + 7\sqrt[3]{2y}$
$= \sqrt[3]{x^6y^3} \cdot \sqrt[3]{2y} + 7\sqrt[3]{2y}$ **Simplifying $\sqrt[3]{2x^6y^4}$**
$= x^2y \cdot \sqrt[3]{2y} + 7\sqrt[3]{2y}$
$= (x^2y + 7)\sqrt[3]{2y}$ **Factoring to combine like radical terms**

Try Exercise 17.

PRODUCTS AND QUOTIENTS OF TWO OR MORE RADICAL TERMS

Radical expressions often contain factors that have more than one term. Multiplying such expressions is similar to finding products of polynomials. Some products will yield like radical terms, which we can now combine.

STUDY TIP

Review Material on Your Own

Never hesitate to review earlier material in which you feel a lack of confidence. For example, if you feel unsure about how to multiply with fraction notation, be sure to review that material before studying any new material that involves multiplication of fractions. Doing (and checking) some practice problems from that section is also a good way to sharpen any skills that you may not have used for a while.

EXAMPLE 3 Multiply.

a) $\sqrt{3}(x - \sqrt{5})$ **b)** $\sqrt[3]{y}(\sqrt[3]{y^2} + \sqrt[3]{2})$

c) $(4 - \sqrt{7})^2$ **d)** $(4\sqrt{3} + \sqrt{2})(\sqrt{3} - 5\sqrt{2})$

e) $(\sqrt{a} + \sqrt{b})(\sqrt{a} - \sqrt{b})$

SOLUTION

a) $\sqrt{3}(x - \sqrt{5}) = \sqrt{3} \cdot x - \sqrt{3} \cdot \sqrt{5}$ **Using the distributive law**
$= x\sqrt{3} - \sqrt{15}$ **Multiplying radicals**

b) $\sqrt[3]{y}(\sqrt[3]{y^2} + \sqrt[3]{2}) = \sqrt[3]{y} \cdot \sqrt[3]{y^2} + \sqrt[3]{y} \cdot \sqrt[3]{2}$ **Using the distributive law**
$= \sqrt[3]{y^3} + \sqrt[3]{2y}$ **Multiplying radicals**
$= y + \sqrt[3]{2y}$ **Simplifying $\sqrt[3]{y^3}$**

c) $\left(4-\sqrt{7}\right)^2=\left(4-\sqrt{7}\right)\left(4-\sqrt{7}\right)$ **We could also use the pattern $(A-B)^2=A^2-2AB+B^2$.**

$$\begin{aligned}&\quad\;\; \text{F} \qquad \text{O} \qquad\quad \text{I} \qquad\quad \text{L}\\ &=4^2-4\sqrt{7}-4\sqrt{7}+\left(\sqrt{7}\right)^2\\ &=16-8\sqrt{7}+7 \qquad \textbf{Squaring and combining like terms}\\ &=23-8\sqrt{7} \qquad \textbf{Adding 16 and 7}\end{aligned}$$

d) $$\begin{aligned}\left(4\sqrt{3}+\sqrt{2}\right)\left(\sqrt{3}-5\sqrt{2}\right)&\overset{\text{F}\quad\text{O}\quad\text{I}\quad\text{L}}{=}4\left(\sqrt{3}\right)^2-20\sqrt{3}\cdot\sqrt{2}+\sqrt{2}\cdot\sqrt{3}-5\left(\sqrt{2}\right)^2\\ &=4\cdot3-20\sqrt{6}+\sqrt{6}-5\cdot2 \qquad \textbf{Multiplying radicals}\\ &=12-20\sqrt{6}+\sqrt{6}-10\\ &=2-19\sqrt{6} \qquad \textbf{Combining like terms}\end{aligned}$$

e) $$\begin{aligned}\left(\sqrt{a}+\sqrt{b}\right)\left(\sqrt{a}-\sqrt{b}\right)&=\left(\sqrt{a}\right)^2-\sqrt{a}\sqrt{b}+\sqrt{a}\sqrt{b}-\left(\sqrt{b}\right)^2 \qquad \textbf{Using FOIL}\\ &=a-b \qquad \textbf{Combining like terms}\end{aligned}$$

Try Exercise 33.

In Example 3(e) above, you may have noticed that since the outer and inner products in FOIL are opposites, the result, $a-b$, is not itself a radical expression. Pairs of radical expressions like $\sqrt{a}+\sqrt{b}$ and $\sqrt{a}-\sqrt{b}$ are called **conjugates.**

RATIONALIZING DENOMINATORS AND NUMERATORS WITH TWO TERMS

The use of conjugates allows us to rationalize denominators or numerators with two terms.

EXAMPLE 4 Rationalize each denominator: **(a)** $\dfrac{4}{\sqrt{3}+x}$; **(b)** $\dfrac{4+\sqrt{2}}{\sqrt{5}-\sqrt{2}}$.

SOLUTION

a) $$\begin{aligned}\frac{4}{\sqrt{3}+x}&=\frac{4}{\sqrt{3}+x}\cdot\frac{\sqrt{3}-x}{\sqrt{3}-x} &&\textbf{Multiplying by 1, using the conjugate of } \sqrt{3}+x\textbf{, which is } \sqrt{3}-x\\ &=\frac{4\left(\sqrt{3}-x\right)}{\left(\sqrt{3}+x\right)\left(\sqrt{3}-x\right)} &&\textbf{Multiplying numerators and denominators}\\ &=\frac{4\left(\sqrt{3}-x\right)}{\left(\sqrt{3}\right)^2-x^2} &&\textbf{Using FOIL in the denominator}\\ &=\frac{4\sqrt{3}-4x}{3-x^2} &&\textbf{Simplifying. No radicals remain in the denominator.}\end{aligned}$$

b) $$\frac{4+\sqrt{2}}{\sqrt{5}-\sqrt{2}} = \frac{4+\sqrt{2}}{\sqrt{5}-\sqrt{2}} \cdot \frac{\sqrt{5}+\sqrt{2}}{\sqrt{5}+\sqrt{2}}$$ Multiplying by 1, using the conjugate of $\sqrt{5}-\sqrt{2}$, which is $\sqrt{5}+\sqrt{2}$

$$= \frac{(4+\sqrt{2})(\sqrt{5}+\sqrt{2})}{(\sqrt{5}-\sqrt{2})(\sqrt{5}+\sqrt{2})}$$ Multiplying numerators and denominators

$$= \frac{4\sqrt{5}+4\sqrt{2}+\sqrt{2}\sqrt{5}+(\sqrt{2})^2}{(\sqrt{5})^2-(\sqrt{2})^2}$$ Using FOIL

$$= \frac{4\sqrt{5}+4\sqrt{2}+\sqrt{10}+2}{5-2}$$ Squaring in the denominator and the numerator

$$= \frac{4\sqrt{5}+4\sqrt{2}+\sqrt{10}+2}{3}$$ No radicals remain in the denominator.

```
(4+√(2))/(√(5)-√
(2))
                 6.587801273
(4√(5)+4√(2)+√(1
0)+2)/3
                 6.587801273
```

We can check by approximating the value of the original expression and the value of the rationalized expression, as shown at left. Care must be taken to place the parentheses properly. The approximate values are the same, so we have a check.

Try Exercise 61.

To rationalize a numerator with two terms, we use the conjugate of the numerator.

EXAMPLE 5 Rationalize the numerator: $\dfrac{4+\sqrt{2}}{\sqrt{5}-\sqrt{2}}$.

SOLUTION We have

$$\frac{4+\sqrt{2}}{\sqrt{5}-\sqrt{2}} = \frac{4+\sqrt{2}}{\sqrt{5}-\sqrt{2}} \cdot \frac{4-\sqrt{2}}{4-\sqrt{2}}$$ Multiplying by 1, using the conjugate of $4+\sqrt{2}$, which is $4-\sqrt{2}$

$$= \frac{16-(\sqrt{2})^2}{4\sqrt{5}-\sqrt{5}\sqrt{2}-4\sqrt{2}+(\sqrt{2})^2}$$

$$= \frac{14}{4\sqrt{5}-\sqrt{10}-4\sqrt{2}+2}.$$

```
(4+√(2))/(√(5)-√
(2))
                 6.587801273
14/(4√(5)-√(10)-
4√(2)+2)
                 6.587801273
```

We check by comparing the values of the original expression and the expression with a rationalized numerator, as shown at left.

Try Exercise 71.

TERMS WITH DIFFERING INDICES

To multiply or divide radical terms with different indices, we can often convert to exponential notation, use the rules for exponents, and then convert back to radical notation.

Student Notes

Expressions similar to the one in Example 6 are most easily simplified by rewriting the expression using exponents in place of radicals. After simplifying, remember to write your final result in radical notation. In general, if a problem is presented in one form, it is expected that the final result be presented in the same form.

EXAMPLE 6 Divide and, if possible, simplify: $\dfrac{\sqrt[4]{(x+y)^3}}{\sqrt{x+y}}$.

SOLUTION

$$\frac{\sqrt[4]{(x+y)^3}}{\sqrt{x+y}} = \frac{(x+y)^{3/4}}{(x+y)^{1/2}}$$ Converting to exponential notation

$$= (x+y)^{3/4-1/2}$$ Since the bases are identical, we can subtract exponents: $\frac{3}{4} - \frac{1}{2} = \frac{3}{4} - \frac{2}{4} = \frac{1}{4}$.

$$= (x+y)^{1/4}$$
$$= \sqrt[4]{x+y}$$ Converting back to radical notation

Try Exercise 89.

To Simplify Products or Quotients with Differing Indices

1. Convert all radical expressions to exponential notation.
2. When the bases are identical, subtract exponents to divide and add exponents to multiply. This may require finding a common denominator.
3. Convert back to radical notation and, if possible, simplify.

EXAMPLE 7 Multiply and simplify: $\sqrt{x^3}\,\sqrt[3]{x}$.

SOLUTION

$$\sqrt{x^3}\,\sqrt[3]{x} = x^{3/2} \cdot x^{1/3}$$ Converting to exponential notation

$$= x^{11/6}$$ Adding exponents: $\frac{3}{2} + \frac{1}{3} = \frac{9}{6} + \frac{2}{6}$

$$= \sqrt[6]{x^{11}}$$ Converting back to radical notation

$$= \sqrt[6]{x^6}\,\sqrt[6]{x^5}$$
$$= x\sqrt[6]{x^5}$$ Simplifying

Try Exercise 79.

EXAMPLE 8 If $f(x) = \sqrt[3]{x^2}$ and $g(x) = \sqrt{x} + \sqrt[4]{x}$, find $(f \cdot g)(x)$.

SOLUTION Recall from Section 5.9 that $(f \cdot g)(x) = f(x) \cdot g(x)$. Thus,

$$(f \cdot g)(x) = \sqrt[3]{x^2}\left(\sqrt{x} + \sqrt[4]{x}\right)$$ x is assumed to be nonnegative.

$$= x^{2/3}(x^{1/2} + x^{1/4})$$ Converting to exponential notation

$$= x^{2/3} \cdot x^{1/2} + x^{2/3} \cdot x^{1/4}$$ Using the distributive law

$$= x^{2/3+1/2} + x^{2/3+1/4}$$ Adding exponents

$$= x^{7/6} + x^{11/12}$$ $\frac{2}{3} + \frac{1}{2} = \frac{4}{6} + \frac{3}{6}$; $\frac{2}{3} + \frac{1}{4} = \frac{8}{12} + \frac{3}{12}$

$$= \sqrt[6]{x^7} + \sqrt[12]{x^{11}}$$ Converting back to radical notation

$$= \sqrt[6]{x^6}\,\sqrt[6]{x} + \sqrt[12]{x^{11}}$$
$$= x\sqrt[6]{x} + \sqrt[12]{x^{11}}$$ Simplifying

Try Exercise 103.

We can often write a result as a single radical expression by finding a common denominator in the exponents.

EXAMPLE 9 Divide and, if possible, simplify: $\dfrac{\sqrt[3]{a^2b^4}}{\sqrt{ab}}$.

SOLUTION

$$\begin{aligned}\frac{\sqrt[3]{a^2b^4}}{\sqrt{ab}} &= \frac{(a^2b^4)^{1/3}}{(ab)^{1/2}} && \text{Converting to exponential notation}\\ &= \frac{a^{2/3}b^{4/3}}{a^{1/2}b^{1/2}} && \text{Using the product and power rules}\\ &= a^{2/3-1/2}b^{4/3-1/2} && \text{Subtracting exponents}\\ &= a^{1/6}b^{5/6}\\ &= \sqrt[6]{a}\,\sqrt[6]{b^5} && \text{Converting to radical notation}\\ &= \sqrt[6]{ab^5} && \text{Using the product rule for radicals}\end{aligned}$$

Try Exercise 93.

10.5 Exercise Set

FOR EXTRA HELP MyMathLab   

Concept Reinforcement For each of Exercises 1–6, fill in the blanks by selecting from the following words (which may be used more than once):

radicand(s), indices, conjugate(s), base(s), denominator(s), numerator(s).

1. To add radical expressions, the ____________ and the ____________ must be the same.
2. To multiply radical expressions, the ____________ must be the same.
3. To find a product by adding exponents, the ____________ must be the same.
4. To add rational expressions, the ____________ must be the same.
5. To rationalize the ____________ of $\dfrac{\sqrt{c}-\sqrt{a}}{5}$, we multiply by a form of 1, using the ____________ of $\sqrt{c}-\sqrt{a}$, or $\sqrt{c}+\sqrt{a}$, to write 1.
6. To find a quotient by subtracting exponents, the ____________ must be the same.

Add or subtract. Simplify by combining like radical terms, if possible. Assume that all variables and radicands represent positive real numbers.

7. $2\sqrt{5} + 7\sqrt{5}$
8. $4\sqrt{7} + 2\sqrt{7}$
9. $7\sqrt[3]{4} - 5\sqrt[3]{4}$
10. $14\sqrt[5]{2} - 6\sqrt[5]{2}$
11. $\sqrt[3]{y} + 9\sqrt[3]{y}$
12. $9\sqrt[4]{t} - \sqrt[4]{t}$
13. $8\sqrt{2} - \sqrt{2} + 5\sqrt{2}$
14. $\sqrt{6} - 8\sqrt{6} + 2\sqrt{6}$
15. $9\sqrt[3]{7} - \sqrt{3} + 4\sqrt[3]{7} + 2\sqrt{3}$
16. $5\sqrt{7} - 8\sqrt[4]{11} + \sqrt{7} + 9\sqrt[4]{11}$
17. $4\sqrt{27} - 3\sqrt{3}$
18. $9\sqrt{50} - 4\sqrt{2}$
19. $3\sqrt{45} - 8\sqrt{20}$
20. $5\sqrt{12} + 16\sqrt{27}$

21. $3\sqrt[3]{16} + \sqrt[3]{54}$

22. $\sqrt[3]{27} - 5\sqrt[3]{8}$

23. $\sqrt{a} + 3\sqrt{16a^3}$

24. $2\sqrt{9x^3} - \sqrt{x}$

25. $\sqrt[3]{6x^4} - \sqrt[3]{48x}$

26. $\sqrt[3]{54x} - \sqrt[3]{2x^4}$

27. $\sqrt{4a-4} + \sqrt{a-1}$

28. $\sqrt{9y+27} + \sqrt{y+3}$

29. $\sqrt{x^3-x^2} + \sqrt{9x-9}$

30. $\sqrt{4x-4} - \sqrt{x^3-x^2}$

Multiply. Assume that all variables represent nonnegative real numbers.

31. $\sqrt{3}(4+\sqrt{3})$

32. $\sqrt{7}(3-\sqrt{7})$

33. $3\sqrt{5}(\sqrt{5}-\sqrt{2})$

34. $4\sqrt{2}(\sqrt{3}-\sqrt{5})$

35. $\sqrt{2}(3\sqrt{10}-2\sqrt{2})$

36. $\sqrt{3}(2\sqrt{5}-3\sqrt{4})$

37. $\sqrt[3]{3}(\sqrt[3]{9}-4\sqrt[3]{21})$

38. $\sqrt[3]{2}(\sqrt[3]{4}-2\sqrt[3]{32})$

39. $\sqrt[3]{a}(\sqrt[3]{a^2}+\sqrt[3]{24a^2})$

40. $\sqrt[3]{x}(\sqrt[3]{3x^2}-\sqrt[3]{81x^2})$

41. $(2+\sqrt{6})(5-\sqrt{6})$

42. $(4-\sqrt{5})(2+\sqrt{5})$

43. $(\sqrt{2}+\sqrt{7})(\sqrt{3}-\sqrt{7})$

44. $(\sqrt{7}-\sqrt{2})(\sqrt{5}+\sqrt{2})$

45. $(3-\sqrt{5})(3+\sqrt{5})$

46. $(2+\sqrt{11})(2-\sqrt{11})$

47. $(\sqrt{6}+\sqrt{8})(\sqrt{6}-\sqrt{8})$

48. $(\sqrt{12}-\sqrt{7})(\sqrt{12}+\sqrt{7})$

49. $(3\sqrt{7}+2\sqrt{5})(2\sqrt{7}-4\sqrt{5})$

50. $(4\sqrt{5}-3\sqrt{2})(2\sqrt{5}+4\sqrt{2})$

51. $(2+\sqrt{3})^2$

52. $(3+\sqrt{7})^2$

53. $(\sqrt{3}-\sqrt{2})^2$

54. $(\sqrt{5}-\sqrt{3})^2$

55. $(\sqrt{2t}+\sqrt{5})^2$

56. $(\sqrt{3x}-\sqrt{2})^2$

57. $(3-\sqrt{x+5})^2$

58. $(4+\sqrt{x-3})^2$

59. $(2\sqrt[4]{7}-\sqrt[4]{6})(3\sqrt[4]{9}+2\sqrt[4]{5})$

60. $(4\sqrt[3]{3}+\sqrt[3]{10})(2\sqrt[3]{7}+5\sqrt[3]{6})$

Rationalize each denominator.

61. $\dfrac{6}{3-\sqrt{2}}$

62. $\dfrac{3}{4-\sqrt{7}}$

63. $\dfrac{2+\sqrt{5}}{6+\sqrt{3}}$

64. $\dfrac{1+\sqrt{2}}{3+\sqrt{5}}$

65. $\dfrac{\sqrt{a}}{\sqrt{a}+\sqrt{b}}$

66. $\dfrac{\sqrt{z}}{\sqrt{x}-\sqrt{z}}$

Aha! 67. $\dfrac{\sqrt{7}-\sqrt{3}}{\sqrt{3}-\sqrt{7}}$

68. $\dfrac{\sqrt{7}+\sqrt{5}}{\sqrt{5}+\sqrt{2}}$

69. $\dfrac{3\sqrt{2}-\sqrt{7}}{4\sqrt{2}+2\sqrt{5}}$

70. $\dfrac{5\sqrt{3}-\sqrt{11}}{2\sqrt{3}-5\sqrt{2}}$

Rationalize each numerator. If possible, simplify your result.

71. $\dfrac{\sqrt{5}+1}{4}$

72. $\dfrac{\sqrt{3}+1}{4}$

73. $\dfrac{\sqrt{6}-2}{\sqrt{3}+7}$

74. $\dfrac{\sqrt{10}+4}{\sqrt{2}-3}$

75. $\dfrac{\sqrt{x}-\sqrt{y}}{\sqrt{x}+\sqrt{y}}$

76. $\dfrac{\sqrt{a}+\sqrt{b}}{\sqrt{a}-\sqrt{b}}$

77. $\dfrac{\sqrt{a+h}-\sqrt{a}}{h}$

78. $\dfrac{\sqrt{x-h}-\sqrt{x}}{h}$

Perform the indicated operation and simplify. Assume that all variables represent positive real numbers.

79. $\sqrt[3]{a}\,\sqrt[6]{a}$

80. $\sqrt[10]{a}\,\sqrt[5]{a^2}$

81. $\sqrt[5]{b^2}\,\sqrt{b^3}$

82. $\sqrt[4]{a^3}\,\sqrt[3]{a^2}$

83. $\sqrt{xy^3}\,\sqrt[3]{x^2y}$

84. $\sqrt[5]{a^3b}\,\sqrt{ab}$

85. $\sqrt[4]{9ab^3}\,\sqrt{3a^4b}$

86. $\sqrt{2x^3y^3}\,\sqrt[3]{4xy^2}$

87. $\sqrt{a^4b^3c^4}\,\sqrt[3]{ab^2c}$

88. $\sqrt[3]{xy^2z}\,\sqrt{x^3yz^2}$

89. $\dfrac{\sqrt[3]{a^2}}{\sqrt[4]{a}}$

90. $\dfrac{\sqrt[3]{x^2}}{\sqrt[5]{x}}$

91. $\dfrac{\sqrt[4]{x^2y^3}}{\sqrt[3]{xy}}$

92. $\dfrac{\sqrt[5]{a^4b}}{\sqrt[3]{ab}}$

93. $\dfrac{\sqrt{ab^3}}{\sqrt[5]{a^2b^3}}$

94. $\dfrac{\sqrt[5]{x^3y^4}}{\sqrt{xy}}$

95. $\dfrac{\sqrt{(7-y)^3}}{\sqrt[3]{(7-y)^2}}$

96. $\dfrac{\sqrt[5]{(y-9)^3}}{\sqrt{y-9}}$

97. $\dfrac{\sqrt[4]{(5+3x)^3}}{\sqrt[3]{(5+3x)^2}}$

98. $\dfrac{\sqrt[3]{(2x+1)^2}}{\sqrt[5]{(2x+1)^2}}$

99. $\sqrt[3]{x^2y}\left(\sqrt{xy} - \sqrt[5]{xy^3}\right)$

100. $\sqrt[4]{a^2b}\left(\sqrt[3]{a^2b} - \sqrt[5]{a^2b^2}\right)$

101. $\left(m + \sqrt[3]{n^2}\right)\left(2m + \sqrt[4]{n}\right)$

102. $\left(r - \sqrt[4]{s^3}\right)\left(3r - \sqrt[5]{s}\right)$

In Exercises 103–106, $f(x)$ and $g(x)$ are as given. Find $(f \cdot g)(x)$. Assume that all variables represent nonnegative real numbers.

103. $f(x) = \sqrt[4]{x},\ g(x) = 2\sqrt{x} - \sqrt[3]{x^2}$

104. $f(x) = \sqrt[4]{x} + 5\sqrt{x},\ g(x) = \sqrt[3]{x^2}$

105. $f(x) = x + \sqrt{7},\ g(x) = x - \sqrt{7}$

106. $f(x) = x - \sqrt{2},\ g(x) = x + \sqrt{6}$

Let $f(x) = x^2$. Find each of the following.

107. $f\left(5 + \sqrt{2}\right)$

108. $f\left(7 + \sqrt{3}\right)$

109. $f\left(\sqrt{3} - \sqrt{5}\right)$

110. $f\left(\sqrt{6} - \sqrt{3}\right)$

TW 111. In what way(s) is combining like radical terms similar to combining like terms that are monomials?

TW 112. Why do we need to know how to multiply radical expressions before learning how to add them?

SKILL REVIEW

To prepare for Section 10.6, review solving equations (Sections 2.2 and 7.6 and Chapter 6).

Solve.

113. $3x - 1 = 125$ [2.2]

114. $x + 5 - 2x = 3x + 6 - x$ [2.2]

115. $x^2 + 2x + 1 = 22 - 2x$ [6.2]

116. $9x^2 - 6x + 1 = 7 + 5x - x^2$ [6.3]

117. $\dfrac{1}{x} + \dfrac{1}{2} = \dfrac{1}{6}$ [7.6]

118. $\dfrac{x}{x-4} + \dfrac{2}{x+4} = \dfrac{x-2}{x^2-16}$ [7.6]

SYNTHESIS

TW 119. Ramon *incorrectly* writes

$$\sqrt[5]{x^2} \cdot \sqrt{x^3} = x^{2/5} \cdot x^{3/2} = \sqrt[5]{x^3}.$$

What mistake do you suspect he is making?

TW 120. After examining the expression $\sqrt[4]{25xy^3}\,\sqrt{5x^4y}$, Marika (correctly) concludes that both x and y are nonnegative. Explain how she could reach this conclusion.

Find a simplified form for $f(x)$. Assumex ≥ 0.

121. $f(x) = \sqrt{x^3 - x^2} + \sqrt{9x^3 - 9x^2} - \sqrt{4x^3 - 4x^2}$

122. $f(x) = \sqrt{20x^2 + 4x^3} - 3x\sqrt{45 + 9x} + \sqrt{5x^2 + x^3}$

123. $f(x) = \sqrt[4]{x^5 - x^4} + 3\sqrt[4]{x^9 - x^8}$

124. $f(x) = \sqrt[4]{16x^4 + 16x^5} - 2\sqrt[4]{x^8 + x^9}$

Simplify.

125. $7x\sqrt{(x+y)^3} - 5xy\sqrt{x+y} - 2y\sqrt{(x+y)^3}$

126. $\sqrt{27a^5(b+1)}\,\sqrt[3]{81a(b+1)^4}$

127. $\sqrt{8x(y+z)^5}\,\sqrt[3]{4x^2(y+z)^2}$

128. $\frac{1}{2}\sqrt{36a^5bc^4} - \frac{1}{2}\sqrt[3]{64a^4bc^6} + \frac{1}{6}\sqrt{144a^3bc^6}$

129. $\dfrac{\dfrac{1}{\sqrt{w}} - \sqrt{w}}{\dfrac{\sqrt{w}+1}{\sqrt{w}}}$

130. $\dfrac{1}{4+\sqrt{3}} + \dfrac{1}{\sqrt{3}} + \dfrac{1}{\sqrt{3}-4}$

Express each of the following as the product of two radical expressions.

131. $x - 5$

132. $y - 7$

133. $x - a$

Multiply.

134. $\sqrt{9 + 3\sqrt{5}}\,\sqrt{9 - 3\sqrt{5}}$

135. $\left(\sqrt{x+2} - \sqrt{x-2}\right)^2$

Try Exercise Answers: Section 10.5

7. $9\sqrt{5}$ 17. $9\sqrt{3}$ 33. $15 - 3\sqrt{10}$ 61. $\dfrac{18 + 6\sqrt{2}}{7}$

71. $\dfrac{1}{\sqrt{5}-1}$ 79. $\sqrt{a}$ 89. $\sqrt[12]{a^5}$ 93. $\sqrt[10]{ab^9}$

103. $2\sqrt[4]{x^3} - \sqrt[12]{x^{11}}$

Mid-Chapter Review

Many radical expressions can be simplified. It is important to know under which conditions radical expressions can be multiplied and divided and radical terms can be combined.

Multiplication and division: The indices must be the same.

$$\frac{\sqrt{50t^5}}{\sqrt{2t^{11}}} = \sqrt{\frac{50t^5}{2t^{11}}} = \sqrt{\frac{25}{t^6}} = \frac{5}{t^3}; \qquad \sqrt[4]{8x^3} \cdot \sqrt[4]{2x} = \sqrt[4]{16x^4} = 2x$$

Combining like terms: The indices and the radicands must both be the same.

$$\sqrt{75x} + \sqrt{12x} - \sqrt{3x} = 5\sqrt{3x} + 2\sqrt{3x} - \sqrt{3x} = 6\sqrt{3x}$$

Radical expressions with differing indices can sometimes be simplified using rational exponents.

$$\sqrt[3]{x^2}\,\sqrt{x} = x^{2/3}x^{1/2} = x^{4/6}x^{3/6} = x^{7/6} = \sqrt[6]{x^7} = x\sqrt[6]{x}$$

GUIDED SOLUTIONS

1. Multiply and simplify: $\sqrt{6x^9} \cdot \sqrt{2xy}$. [10.3]

Solution

$$\begin{aligned}\sqrt{6x^9} \cdot \sqrt{2xy} &= \sqrt{6x^9 \cdot \square} \\ &= \sqrt{12x^{10}y} \\ &= \sqrt{\square \cdot 3y} \\ &= \sqrt{\square} \cdot \sqrt{3y} \\ &= \square\sqrt{3y} \quad \textbf{Taking the square root}\end{aligned}$$

2. Combine like terms: $\sqrt{12} - 3\sqrt{75} + \sqrt{8}$. [10.5]

Solution

$$\begin{aligned}&\sqrt{12} - 3\sqrt{75} + \sqrt{8} \\ &= \square - 3 \cdot 5\sqrt{3} + \square \quad \textbf{Simplifying each term} \\ &= 2\sqrt{3} - \square + 2\sqrt{2} \quad \textbf{Multiplying} \\ &= \square + 2\sqrt{2} \quad \textbf{Combining like terms}\end{aligned}$$

MIXED REVIEW

Simplify. Assume that variables can represent any real number.

1. $\sqrt{81}$ [10.1]
2. $-\sqrt{\frac{9}{100}}$ [10.1]
3. $\sqrt{64t^2}$ [10.1]
4. $\sqrt[5]{x^5}$ [10.1]
5. Find $f(-5)$ if $f(x) = \sqrt[3]{12x - 4}$. [10.1]
6. Determine the domain of g if $g(x) = \sqrt[4]{10 - x}$. [10.1]
7. Write an equivalent expression using radical notation and simplify: $8^{2/3}$. [10.2]

Simplify. Assume for the remainder of the exercises that no radicands were formed by raising negative numbers to even powers.

8. $\sqrt[6]{\sqrt{a}}$ [10.2]
9. $\sqrt[3]{y^{24}}$ [10.2]
10. $\sqrt{(t + 5)^2}$ [10.1]
11. $\sqrt[3]{-27a^{12}}$ [10.1]
12. $\sqrt{6x}\,\sqrt{15x}$ [10.3]
13. $\frac{\sqrt{20y}}{\sqrt{45y}}$ [10.4]
14. $\sqrt{15t} + 4\sqrt{15t}$ [10.5]
15. $\sqrt[5]{a^5b^{10}c^{11}}$ [10.1]
16. $\sqrt{6}(\sqrt{10} - \sqrt{33})$ [10.5]
17. $\frac{\sqrt{t}}{\sqrt[8]{t^3}}$ [10.5]
18. $\sqrt[5]{\frac{3a^{12}}{96a^2}}$ [10.4]
19. $2\sqrt{3} - 5\sqrt{12}$ [10.5]
20. $(\sqrt{5} + 3)(\sqrt{5} - 3)$ [10.5]
21. $(\sqrt{15} + \sqrt{10})^2$ [10.5]
22. $\sqrt{25x - 25} - \sqrt{9x - 9}$ [10.5]
23. $\sqrt{x^3y}\,\sqrt[5]{xy^4}$ [10.5]
24. $\sqrt[3]{5000} + \sqrt[3]{625}$ [10.5]
25. $\sqrt[3]{12x^2y^5}\,\sqrt[3]{18x^7y}$ [10.3]

10.6 Solving Radical Equations

- The Principle of Powers
- Equations with Two Radical Terms

In Sections 10.1–10.5, we learned how to manipulate radical expressions as well as expressions containing rational exponents. We performed this work to find *equivalent expressions.*

Now that we know how to work with radicals and rational exponents, we can learn how to solve a new type of equation.

THE PRINCIPLE OF POWERS

A **radical equation** is an equation in which the variable appears in a radicand. Examples are

$$\sqrt[3]{2x} + 1 = 5, \quad \sqrt{a - 2} = 7, \quad \text{and} \quad 4 - \sqrt{3x + 1} = \sqrt{6 - x}.$$

To solve such equations, we need a new principle. Suppose $a = b$ is true. If we square both sides, we get another true equation: $a^2 = b^2$. This can be generalized.

The Principle of Powers If $a = b$, then $a^n = b^n$ for any exponent n.

Note that the principle of powers is an "if–then" statement. If we interchange the two parts of the sentence, then we have the statement

"If $a^n = b^n$ for some exponent n, then $a = b$."

This statement is *not* always true. For example, "if $x = 3$, then $x^2 = 9$" is true, but the statement "if $x^2 = 9$, then $x = 3$" is *not* true when x is replaced with -3.

Interactive Discovery

Solve each pair of equations graphically, and compare the solution sets.

1. $x = 3$; $x^2 = 9$
2. $x = -2$; $x^2 = 4$
3. $\sqrt{x} = 5$; $x = 25$
4. $\sqrt{x} = -3$; $x = 9$

Every solution of $x = a$ is a solution of $x^2 = a^2$, but not every solution of $x^2 = a^2$ is necessarily a solution of $x = a$. A similar statement can be made for any even exponent n; for example, the solution sets of $x^4 = 3^4$ and $x = 3$ are not the same.

When we raise both sides of an equation to an even exponent, it is essential that we check the answer in the *original equation.*

EXAMPLE 1 Solve: $\sqrt{x} - 3 = 4$.

ALGEBRAIC APPROACH

Before using the principle of powers, we must isolate the radical term:

$$\sqrt{x} - 3 = 4$$

$$\sqrt{x} = 7 \quad \text{Isolating the radical by adding 3 to both sides}$$

$$(\sqrt{x})^2 = 7^2 \quad \text{Using the principle of powers}$$

$$x = 49.$$

Check:

$\sqrt{x} - 3 = 4$	
$\sqrt{49} - 3$	4
$7 - 3$	
$4 \stackrel{?}{=} 4$	TRUE

The solution is 49.

GRAPHICAL APPROACH

We let $f(x) = \sqrt{x} - 3$ and $g(x) = 4$. We use a viewing window of $[-10, 100, -5, 5]$ and find the point of intersection. (It may require some trial and error to find an appropriate viewing window.)

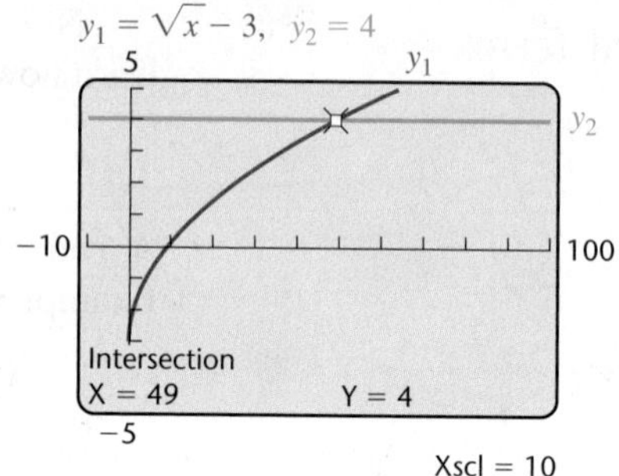

We have a solution of 49, which we can check by substituting into the original equation, as shown in the algebraic approach.

The solution is 49.

Try Exercise 7.

EXAMPLE 2 Solve: $\sqrt{x} + 5 = 3$.

ALGEBRAIC APPROACH

We have

$$\sqrt{x} + 5 = 3$$

$$\sqrt{x} = -2 \quad \text{Isolating the radical by subtracting 5 from both sides}$$

The equation $\sqrt{x} = -2$ has no solution because the principal square root of a number is never negative. We continue as in Example 1 for comparison.

$$(\sqrt{x})^2 = (-2)^2 \quad \text{Using the principle of powers (squaring)}$$

$$x = 4.$$

Check:

$\sqrt{x} + 5 = 3$	
$\sqrt{4} + 5$	3
$2 + 5$	
$7 \stackrel{?}{=} 3$	FALSE

The number 4 does not check. Thus, $\sqrt{x} + 5 = 3$ has no solution.

GRAPHICAL APPROACH

We let $f(x) = \sqrt{x} + 5$ and $g(x) = 3$ and graph.

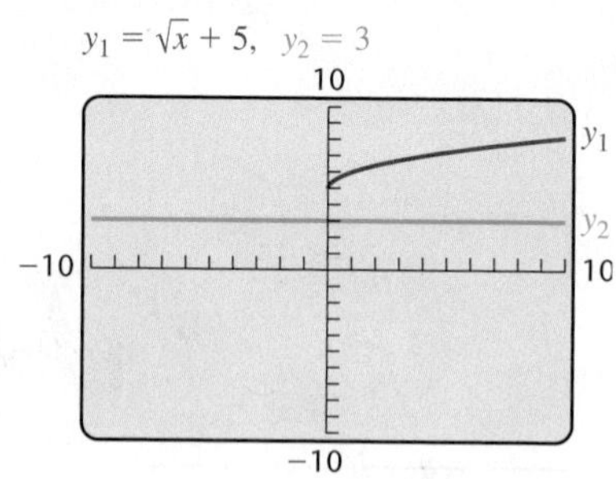

The graphs do not intersect. There is no real-number solution.

Try Exercise 27.

CAUTION! Raising both sides of an equation to an even power may not produce an equivalent equation. In this case, a check is essential.

Note in Example 2 that $x = 4$ has solution 4 but $\sqrt{x} + 5 = 3$ has *no* solution. Thus the equations $x = 4$ and $\sqrt{x} + 5 = 3$ are *not* equivalent.

To Solve an Equation with a Radical Term

1. Isolate the radical term on one side of the equation.
2. Use the principle of powers and solve the resulting equation.
3. Check any possible solution in the original equation.

EXAMPLE 3 Solve: $x = \sqrt{x + 7} + 5$.

ALGEBRAIC APPROACH

$$x = \sqrt{x + 7} + 5$$

$$x - 5 = \sqrt{x + 7}$$ **Isolating the radical by subtracting 5 from both sides**

$$\left.\begin{aligned}(x - 5)^2 &= \left(\sqrt{x + 7}\right)^2 \\ x^2 - 10x + 25 &= x + 7\end{aligned}\right\}$$ **Using the principle of powers; squaring both sides**

$$x^2 - 11x + 18 = 0$$ **Adding $-x - 7$ to both sides to write the quadratic equation in standard form**

$$(x - 9)(x - 2) = 0$$ **Factoring**

$$x = 9 \quad or \quad x = 2$$ **Using the principle of zero products**

The possible solutions are 9 and 2. Let's check.

Check:

For 9:

$$x = \sqrt{x + 7} + 5$$
$$9 \;|\; \sqrt{9 + 7} + 5$$
$$9 \stackrel{?}{=} 9 \qquad \text{TRUE}$$

For 2:

$$x = \sqrt{x + 7} + 5$$
$$2 \;|\; \sqrt{2 + 7} + 5$$
$$2 \stackrel{?}{=} 8 \qquad \text{FALSE}$$

Since 9 checks but 2 does not, the solution is 9.

GRAPHICAL APPROACH

We graph $y_1 = x$ and $y_2 = \sqrt{x + 7} + 5$ and determine any points of intersection.

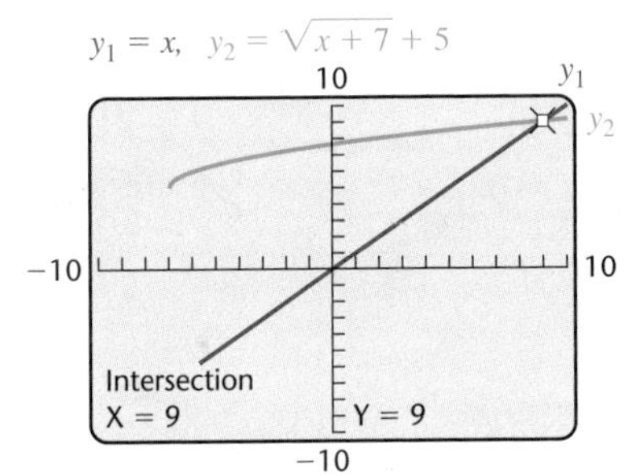

We obtain a solution of 9.

Try Exercise 39.

It is important to isolate a radical term before using the principle of powers. Suppose in Example 3 that both sides of the equation were squared *before* isolating the radical. We then would have had the expression

$$\left(\sqrt{x + 7} + 5\right)^2 \quad \text{or} \quad x + 7 + 10\sqrt{x + 7} + 25$$

in the third step of the algebraic approach, and the radical would have remained in the problem.

EXAMPLE 4 Solve: $(2x + 1)^{1/3} + 5 = 0$.

ALGEBRAIC APPROACH

We can use exponential notation to solve:

$$\begin{aligned} (2x + 1)^{1/3} + 5 &= 0 \\ (2x + 1)^{1/3} &= -5 && \text{Subtracting 5 from both sides} \\ [(2x + 1)^{1/3}]^3 &= (-5)^3 && \text{Cubing both sides} \\ (2x + 1)^1 &= (-5)^3 && \text{Multiplying exponents} \\ 2x + 1 &= -125 \\ 2x &= -126 && \text{Subtracting 1 from both sides} \\ x &= -63. \end{aligned}$$

Because both sides were raised to an *odd* power, we need check only the accuracy of our work.

Check:

$(2x + 1)^{1/3} + 5 = 0$	
$[2(-63) + 1]^{1/3} + 5$	0
$[-126 + 1]^{1/3} + 5$	
$(-125)^{1/3} + 5$	
$\sqrt[3]{-125} + 5$	
$-5 + 5$	
$0 \stackrel{?}{=} 0$ TRUE	

GRAPHICAL APPROACH

We let $f(x) = (2x + 1)^{1/3} + 5$ and determine any values of x for which $f(x) = 0$. Using a viewing window of $[-100, 100, -10, 10]$, we find that $f(x) = 0$ when $x = -63$.

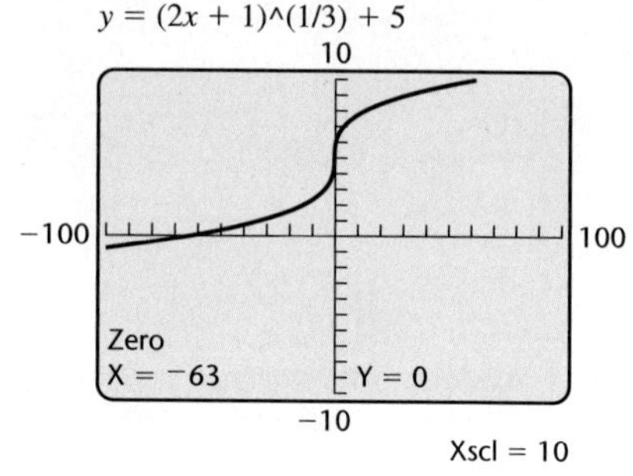

Try Exercise 33.

STUDY TIP

Plan Your Future

As you register for next semester's courses, be careful to consider your work and family commitments. Speak to faculty and other students to estimate how demanding each course is before signing up. If in doubt, it is usually better to take one fewer course than one too many.

EQUATIONS WITH TWO RADICAL TERMS

A strategy for solving equations with two or more radical terms is as follows.

To Solve an Equation with Two or More Radical Terms

1. Isolate one of the radical terms.
2. Use the principle of powers.
3. If a radical remains, perform steps (1) and (2) again.
4. Solve the resulting equation.
5. Check possible solutions in the original equation.

EXAMPLE 5 Solve: $\sqrt{2x - 5} = 1 + \sqrt{x - 3}$.

ALGEBRAIC APPROACH

We have

$$\sqrt{2x-5} = 1 + \sqrt{x-3}$$

$$\left(\sqrt{2x-5}\right)^2 = \left(1 + \sqrt{x-3}\right)^2 \qquad \text{One radical is already isolated. We square both sides.}$$

This is like squaring a binomial. We square 1, then find twice the product of 1 and $\sqrt{x - 3}$, and then the square of $\sqrt{x - 3}$.

$$2x - 5 = 1 + 2\sqrt{x-3} + \left(\sqrt{x-3}\right)^2$$

$$2x - 5 = 1 + 2\sqrt{x-3} + (x-3)$$

$$x - 3 = 2\sqrt{x-3} \qquad \text{Isolating the remaining radical term}$$

$$(x-3)^2 = \left(2\sqrt{x-3}\right)^2 \qquad \text{Squaring both sides}$$

$$x^2 - 6x + 9 = 4(x-3) \qquad \text{Remember to square both the 2 and the } \sqrt{x-3} \text{ on the right side.}$$

$$x^2 - 6x + 9 = 4x - 12$$

$$x^2 - 10x + 21 = 0$$

$$(x-7)(x-3) = 0 \qquad \text{Factoring}$$

$$x = 7 \quad or \quad x = 3. \qquad \text{Using the principle of zero products}$$

The possible solutions are 7 and 3. Both 7 and 3 check in the original equation and are the solutions.

GRAPHICAL APPROACH

We let $f(x) = \sqrt{2x - 5}$ and $g(x) = 1 + \sqrt{x - 3}$. The domain of f is $\left[\frac{5}{2}, \infty\right)$, and the domain of g is $[3, \infty)$. Since the intersection of their domains is $[3, \infty)$, any point of intersection will occur in that interval. We choose a viewing window of $[-1, 10, -1, 5]$.

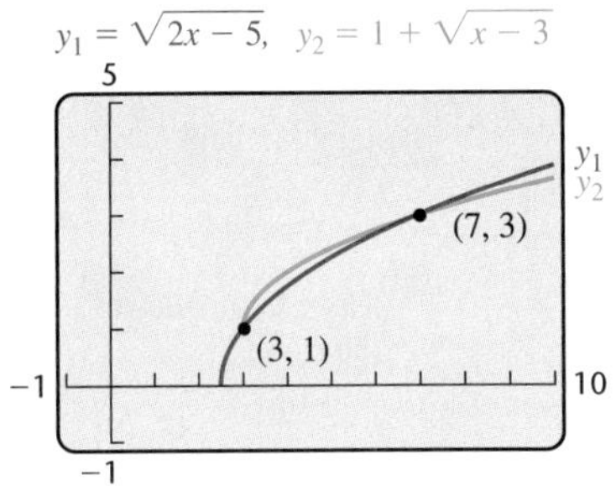

The graphs appear to intersect when x is approximately 3 and again when x is approximately 7. The points of intersection are $(3, 1)$ and $(7, 3)$. The solutions are 3 and 7.

Try Exercise 41.

CAUTION! A common error in solving equations like

$$\sqrt{2x-5} = 1 + \sqrt{x-3}$$

is to obtain $1 + (x - 3)$ as the square of the right side. This is wrong because $(A + B)^2 \neq A^2 + B^2$. Placing parentheses around each side when squaring serves as a reminder to square the entire expression.

EXAMPLE 6 Let $f(x) = \sqrt{x+5} - \sqrt{x-7}$. Find all x-values for which $f(x) = 2$.

SOLUTION We must have $f(x) = 2$, or

$$\sqrt{x+5} - \sqrt{x-7} = 2. \qquad \textbf{Substituting for } f(x)$$

To solve, we isolate one radical term and square both sides:

Isolate a radical term.

$$\sqrt{x+5} = 2 + \sqrt{x-7}$$ **Adding $\sqrt{x-7}$ to both sides. This isolates one of the radical terms.**

Raise both sides to the same power.

$$\left(\sqrt{x+5}\right)^2 = \left(2 + \sqrt{x-7}\right)^2$$ **Using the principle of powers (squaring both sides)**

$$x + 5 = 4 + 4\sqrt{x-7} + (x-7)$$ **Using $(A+B)^2 = A^2 + 2AB + B^2$**

$$5 = 4\sqrt{x-7} - 3$$ **Adding $-x$ to both sides and combining like terms**

Isolate a radical term.

$$8 = 4\sqrt{x-7}$$ **Isolating the remaining radical term**

$$2 = \sqrt{x-7}$$

Raise both sides to the same power.

$$2^2 = \left(\sqrt{x-7}\right)^2$$ **Squaring both sides**

$$4 = x - 7$$

Solve.

$$11 = x.$$

Check.

Check: $f(11) = \sqrt{11+5} - \sqrt{11-7}$
$= \sqrt{16} - \sqrt{4}$
$= 4 - 2 = 2.$

We have $f(x) = 2$ when $x = 11$.

Try Exercise 49.

10.6 Exercise Set

FOR EXTRA HELP MyMathLab Math XL PRACTICE WATCH DOWNLOAD READ REVIEW

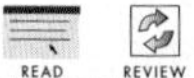

Concept Reinforcement *Classify each of the following statements as either true or false.*

1. If $x^2 = 25$, then $x = 5$.
2. If $t = 7$, then $t^2 = 49$.
3. If $\sqrt{x} = 3$, then $\left(\sqrt{x}\right)^2 = 3^2$.
4. If $x^2 = 36$, then $x = 6$.
5. $\sqrt{x} - 8 = 7$ is equivalent to $\sqrt{x} = 15$.
6. $\sqrt{t} + 5 = 8$ is equivalent to $\sqrt{t} = 3$.

Solve.

7. $\sqrt{5x+1} = 4$
8. $\sqrt{7x-3} = 5$
9. $\sqrt{3x} + 1 = 6$
10. $\sqrt{2x} - 1 = 2$
11. $\sqrt{y+1} - 5 = 8$
12. $\sqrt{x-2} - 7 = -4$
13. $\sqrt{8-x} + 7 = 10$
14. $\sqrt{y+4} + 6 = 7$
15. $\sqrt[3]{x+5} = 2$
16. $\sqrt[3]{x-2} = 3$
17. $\sqrt[4]{y-1} = 3$
18. $\sqrt[4]{x+3} = 2$
19. $3\sqrt{x} = x$
20. $8\sqrt{y} = y$

21. $2y^{1/2} - 13 = 7$

22. $3x^{1/2} + 12 = 9$

23. $\sqrt[3]{x} = -3$

24. $\sqrt[3]{y} = -4$

25. $z^{1/4} + 8 = 10$

26. $x^{1/4} - 2 = 1$

Aha! 27. $\sqrt{n} = -2$

28. $\sqrt{a} = -1$

29. $\sqrt[4]{3x + 1} - 4 = -1$

30. $\sqrt[4]{2x + 3} - 5 = -2$

31. $(21x + 55)^{1/3} = 10$

32. $(5y + 31)^{1/4} = 2$

33. $\sqrt[3]{3y + 6} + 7 = 8$

34. $\sqrt[3]{6x + 9} + 5 = 2$

35. $\sqrt{3t + 4} = \sqrt{4t + 3}$

36. $\sqrt{2t - 7} = \sqrt{3t - 12}$

37. $3(4 - t)^{1/4} = 6^{1/4}$

38. $2(1 - x)^{1/3} = 4^{1/3}$

39. $3 + \sqrt{5 - x} = x$

40. $x = \sqrt{x - 1} + 3$

41. $\sqrt{4x - 3} = 2 + \sqrt{2x - 5}$

42. $3 + \sqrt{z - 6} = \sqrt{z + 9}$

43. $\sqrt{20 - x} + 8 = \sqrt{9 - x} + 11$

44. $4 + \sqrt{10 - x} = 6 + \sqrt{4 - x}$

45. $\sqrt{x + 2} + \sqrt{3x + 4} = 2$

46. $\sqrt{6x + 7} - \sqrt{3x + 3} = 1$

47. If $f(x) = \sqrt{x} + \sqrt{x - 9}$, find any x for which $f(x) = 1$.

48. If $g(x) = \sqrt{x} + \sqrt{x - 5}$, find any x for which $g(x) = 5$.

49. If $f(t) = \sqrt{t - 2} - \sqrt{4t + 1}$, find any t for which $f(t) = -3$.

50. If $g(t) = \sqrt{2t + 7} - \sqrt{t + 15}$, find any t for which $g(t) = -1$.

51. If $f(x) = \sqrt{2x - 3}$ and $g(x) = \sqrt{x + 7} - 2$, find any x for which $f(x) = g(x)$.

52. If $f(x) = 2\sqrt{3x + 6}$ and $g(x) = 5 + \sqrt{4x + 9}$, find any x for which $f(x) = g(x)$.

53. If $f(t) = 4 - \sqrt{t - 3}$ and $g(t) = (t + 5)^{1/2}$, find any t for which $f(t) = g(t)$.

54. If $f(t) = 7 + \sqrt{2t - 5}$ and $g(t) = 3(t + 1)^{1/2}$, find any t for which $f(t) = g(t)$.

TW 55. Explain in your own words why it is important to check your answers when using the principle of powers.

TW 56. The principle of powers is an "if–then" statement that becomes false when the sentence parts are interchanged. Give an example of another such if–then statement from everyday life (answers will vary).

SKILL REVIEW

To prepare for Section 10.7, review finding dimensions of triangles and rectangles (Sections 2.5 and 6.7).

Solve.

57. *Sign Dimensions.* The largest sign in the United States is a rectangle with a perimeter of 430 ft. The length of the rectangle is 5 ft longer than thirteen times the width. Find the dimensions of the sign. [2.5]
Source: Florida Center for Instructional Technology

58. *Sign Dimensions.* The base of a triangular sign is 4 in. longer than twice the height. The area of the sign is 255 in^2. Find the dimensions of the sign. [6.7]

59. *Photograph Dimensions.* A rectangular family photo is 4 in. longer than it is wide. The area of the photo is 140 in^2. Find the dimensions of the photograph. [6.7]

60. *Sidewalk Length.* The length of a rectangular lawn between classroom buildings is 2 yd less than twice the width of the lawn. A path that is 34 yd long stretches diagonally across the area. What are the dimensions of the lawn? [6.7]

61. The sides of a right triangle are consecutive even integers. Find the length of each side. [6.7]

62. One leg of a right triangle is 5 cm long. The hypotenuse is 1 cm longer than the other leg. Find the length of the hypotenuse. [6.7]

SYNTHESIS

TW 63. Describe a procedure that could be used to create radical equations that have no solution.

TW 64. Is checking essential when the principle of powers is used with an odd power n? Why or why not?

65. *Firefighting.* The velocity of water flow, in feet per second, from a nozzle is given by

$$v(p) = 12.1\sqrt{p},$$

where p is the nozzle pressure, in pounds per square inch (psi). Find the nozzle pressure if the water flow is 100 feet per second.
Source: Houston Fire Department Continuing Education

66. *Firefighting.* The velocity of water flow, in feet per second, from a water tank that is h feet high is given by

$$v(h) = 8\sqrt{h}.$$

Find the height of a water tank that provides a water flow of 60 feet per second.
Source: Houston Fire Department Continuing Education

67. *Music.* The frequency of a violin string varies directly with the square root of the tension on the string. A violin string vibrates with a frequency of 260 Hz when the tension on the string is 28 N. What is the frequency when the tension is 32 N?

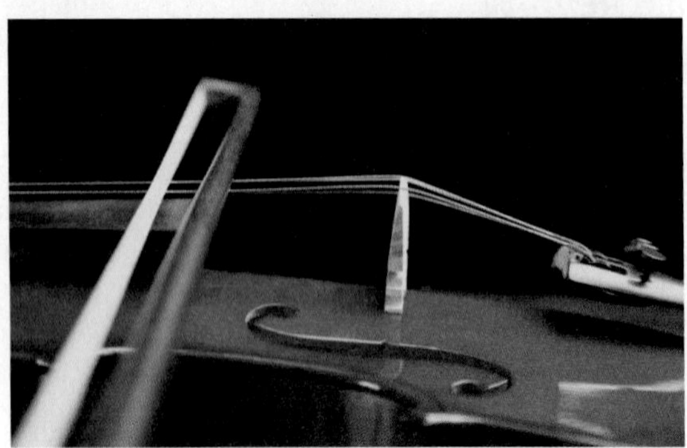

68. *Music.* The frequency of a violin string varies inversely with the square root of the density of the string. A nylon violin string with a density of 1200 kg/m^3 vibrates with a frequency of 250 Hz. What is the frequency of a silk violin string with a density of 1300 kg/m^3?
Source: www.speech.kth.se

Steel Manufacturing. In the production of steel and other metals, the temperature of the molten metal is so great that conventional thermometers melt. Instead, sound is transmitted across the surface of the metal to a receiver on the far side and the speed of the sound is measured. The formula

$$S(t) = 1087.7\sqrt{\frac{9t + 2617}{2457}}$$

gives the speed of sound $S(t)$, in feet per second, at a temperature of t degrees Celsius.

69. Find the temperature of a blast furnace where sound travels 1880 ft/sec.

70. Find the temperature of a blast furnace where sound travels 1502.3 ft/sec.

71. Solve the above equation for t.

Automotive Repair. For an engine with a displacement of 2.8 L, the function given by

$$d(n) = 0.75\sqrt{2.8n}$$

can be used to determine the diameter size of the carburetor's opening, in millimeters. Here n is the number of rpm's at which the engine achieves peak performance.
Source: macdizzy.com

72. If a carburetor's opening is 81 mm, for what number of rpm's will the engine produce peak power?

73. If a carburetor's opening is 84 mm, for what number of rpm's will the engine produce peak power?

Escape Velocity. A formula for the escape velocity v of a satellite is

$$v = \sqrt{2gr}\sqrt{\frac{h}{r + h}},$$

where g is the force of gravity, r is the planet or star's radius, and h is the height of the satellite above the planet or star's surface.

74. Solve for h.

75. Solve for r.

Solve.

76. $\left(\frac{z}{4} - 5\right)^{2/3} = \frac{1}{25}$

77. $\frac{x + \sqrt{x + 1}}{x - \sqrt{x + 1}} = \frac{5}{11}$

78. $\sqrt{\sqrt{y} + 49} = 7$

79. $(z^2 + 17)^{3/4} = 27$

80. $x^2 - 5x - \sqrt{x^2 - 5x - 2} = 4$
($Hint$: Let $u = x^2 - 5x - 2$.)

81. $\sqrt{8 - b} = b\sqrt{8 - b}$

Without graphing, determine the x-intercepts of the graphs given by each of the following.

82. $f(x) = \sqrt{x - 2} - \sqrt{x + 2} + 2$

83. $g(x) = 6x^{1/2} + 6x^{-1/2} - 37$

84. $f(x) = (x^2 + 30x)^{1/2} - x - (5x)^{1/2}$

Try Exercise Answers: Section 10.6
7. 3 **27.** No solution **33.** $-\frac{5}{3}$ **39.** 4 **41.** 3, 7 **49.** 2, 6

Collaborative Corner

Tailgater Alert

Focus: Radical equations and problem solving
Time: 15–25 minutes
Group size: 2–3
Materials: Calculators

The faster a car is traveling, the more distance it needs to stop. Thus it is important for drivers to allow sufficient space between their vehicle and the vehicle in front of them. Police recommend that for each 10 mph of speed, a driver allow 1 car length. Thus a driver going 30 mph should have at least 3 car lengths between his or her vehicle and the one in front.

In Exercise Set 10.3, the function $r(L) = 2\sqrt{5L}$ was used to find the speed, in miles per hour, that a car was traveling when it left skid marks L feet long.

ACTIVITY

1. Each group member should estimate the length of a car in which he or she frequently travels. (Each should use a different length, if possible.)
2. Using a calculator as needed, each group member should complete the table at right. Column 1 gives a car's speed s, column 2 lists the minimum amount of space between cars traveling s miles per hour, as recommended by police. Column 3 is the speed that a vehicle *could* travel were it forced to stop in the distance listed in column 2, using the above function.

Column 1 s (in miles per hour)	Column 2 $L(s)$ (in feet)	Column 3 $r(L)$ (in miles per hour)
20		
30		
40		
50		
60		
70		

3. Determine whether there are any speeds at which the "1 car length per 10 mph" guideline might not suffice. On what reasoning do you base your answer? Compare tables to determine how car length affects the results. What recommendations would your group make to a new driver?

10.7 The Distance Formula, the Midpoint Formula, and Other Applications

- Using the Pythagorean Theorem
- Two Special Triangles
- The Distance Formula and the Midpoint Formula

USING THE PYTHAGOREAN THEOREM

There are many kinds of problems that involve powers and roots. Many also involve right triangles and the Pythagorean theorem, which we studied in Section 6.7 and restate here.

The Pythagorean Theorem* In any right triangle, if a and b are the lengths of the legs and c is the length of the hypotenuse, then

$$a^2 + b^2 = c^2.$$

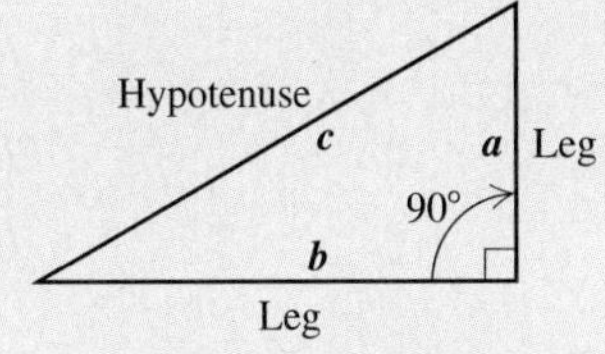

*The converse of the Pythagorean theorem also holds. That is, if a, b, and c are the lengths of the sides of a triangle and $a^2 + b^2 = c^2$, then the triangle is a right triangle.

In using the Pythagorean theorem, we often make use of the following principle.

The Principle of Square Roots For any nonnegative real number n,

If $x^2 = n$, then $x = \sqrt{n}$ or $x = -\sqrt{n}$.

For most real-world applications involving length or distance, $-\sqrt{n}$ is not needed.

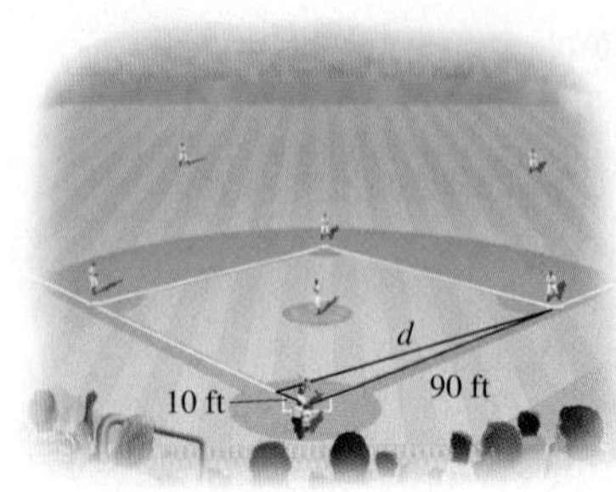

EXAMPLE 1 Baseball. A baseball diamond is actually a square 90 ft on a side. Suppose a catcher fields a ball while standing on the third-base line 10 ft from home plate, as shown in the figure at left. How far is the catcher's throw to first base? Give an exact answer and an approximation to three decimal places.

SOLUTION We make a drawing and let $d =$ the distance, in feet, to first base. Note that a right triangle is formed in which the leg from home plate to first base measures 90 ft and the leg from home plate to where the catcher fields the ball measures 10 ft.

We substitute these values into the Pythagorean theorem to find d:

$$d^2 = 90^2 + 10^2$$
$$d^2 = 8100 + 100$$
$$d^2 = 8200.$$

We now use the principle of square roots: If $d^2 = 8200$, then $d = \sqrt{8200}$ or $d = -\sqrt{8200}$. Since d represents a length, it follows that d is the positive square root of 8200:

$d = \sqrt{8200}$ ft **This is an exact answer.**

$d \approx 90.554$ ft. **Using a calculator for an approximation**

Try Exercise 19.

STUDY TIP

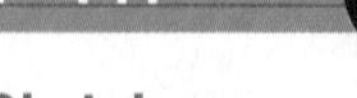
Making Sketches

One need not be an artist to make highly useful mathematical sketches. That said, it is important to make sure that your sketches are drawn accurately enough to represent the relative sizes within each shape. For example, if one side of a triangle is clearly the longest, make sure your drawing reflects this.

EXAMPLE 2 Guy Wires. The base of a 40-ft long guy wire is located 15 ft from the telephone pole that it is anchoring. How high up the pole does the guy wire reach? Give an exact answer and an approximation to three decimal places.

SOLUTION We make a drawing and let $h =$ the height, in feet, to which the guy wire reaches. A right triangle is formed in which one leg measures 15 ft and the hypotenuse measures 40 ft. Using the Pythagorean theorem, we have

$$h^2 + 15^2 = 40^2$$
$$h^2 + 225 = 1600$$
$$h^2 = 1375$$
$$h = \sqrt{1375}.$$ **Using the positive square root**

Exact answer:

$h = \sqrt{1375}$ ft

Approximation:

$h \approx 37.081$ ft **Using a calculator**

Try Exercise 15.

TWO SPECIAL TRIANGLES

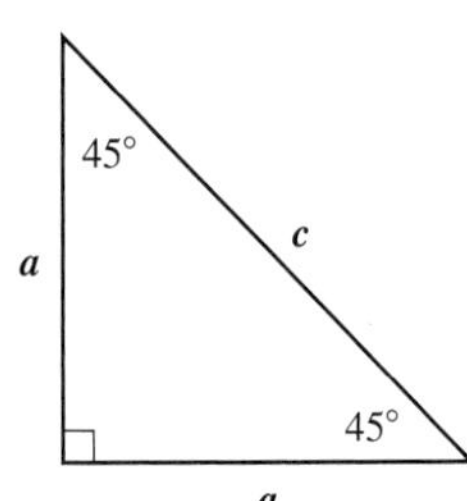

When both legs of a right triangle are the same size, as shown at left, we call the triangle an *isosceles right triangle*. If one leg of an isosceles right triangle has length a, we can find a formula for the length of the hypotenuse as follows:

$$c^2 = a^2 + b^2$$

$$c^2 = a^2 + a^2 \quad \text{Because the triangle is isosceles, both legs are the same size: } a = b.$$

$$c^2 = 2a^2. \quad \text{Combining like terms}$$

Next, we use the principle of square roots. Because a, b, and c are lengths, there is no need to consider negative square roots or absolute values. Thus,

$$c = \sqrt{2a^2} \quad \text{Using the principle of square roots}$$

$$= \sqrt{a^2 \cdot 2} = a\sqrt{2}.$$

EXAMPLE 3 One leg of an isosceles right triangle measures 7 cm. Find the length of the hypotenuse. Give an exact answer and an approximation to three decimal places.

SOLUTION We substitute:

$$c = a\sqrt{2} \quad \text{This equation is worth memorizing.}$$

$$= 7\sqrt{2}.$$

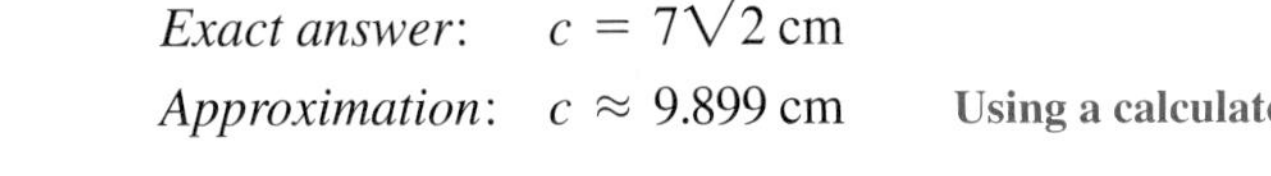

Exact answer: $c = 7\sqrt{2}$ cm

Approximation: $c \approx 9.899$ cm Using a calculator

Try Exercise 29.

When the hypotenuse of an isosceles right triangle is known, the lengths of the legs can be found.

EXAMPLE 4 The hypotenuse of an isosceles right triangle is 5 ft long. Find the length of a leg. Give an exact answer and an approximation to three decimal places.

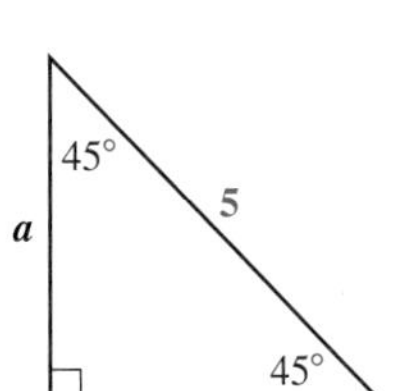

SOLUTION We replace c with 5 and solve for a:

$$5 = a\sqrt{2} \quad \text{Substituting 5 for } c \text{ in } c = a\sqrt{2}$$

$$\frac{5}{\sqrt{2}} = a \quad \text{Dividing both sides by } \sqrt{2}$$

$$\frac{5\sqrt{2}}{2} = a. \quad \text{Rationalize the denominator if desired.}$$

Exact answer: $a = \dfrac{5}{\sqrt{2}}$ ft, or $\dfrac{5\sqrt{2}}{2}$ ft

Approximation: $a \approx 3.536$ ft Using a calculator

Try Exercise 35.

A second special triangle is known as a 30°–60°–90° right triangle, so named because of the measures of its angles. Note that in an equilateral triangle, all sides have the same length and all angles are 60°. An altitude, drawn dashed in the figure, bisects, or splits in half, one angle and one side. Two 30°–60°–90° right triangles are thus formed.

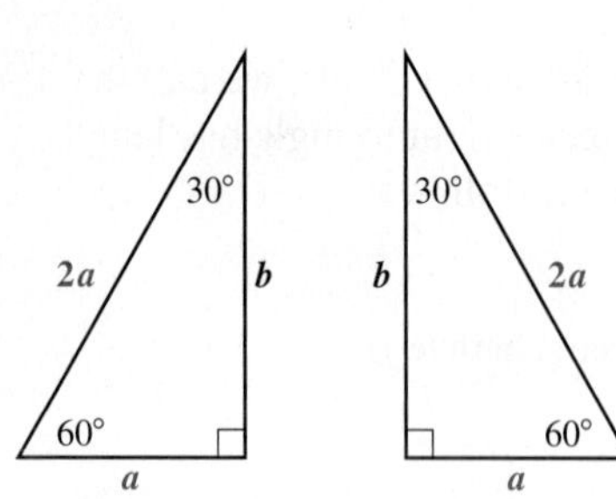

If we let a represent the length of the shorter leg in a 30°–60°–90° right triangle, then $2a$ represents the length of the hypotenuse. We have

$$a^2 + b^2 = (2a)^2 \quad \textbf{Using the Pythagorean theorem}$$
$$a^2 + b^2 = 4a^2$$
$$b^2 = 3a^2 \quad \textbf{Subtracting } a^2 \textbf{ from both sides}$$
$$b = \sqrt{3a^2} \quad \textbf{Considering only the positive square root}$$
$$b = \sqrt{a^2 \cdot 3}$$
$$b = a\sqrt{3}.$$

EXAMPLE 5 The shorter leg of a 30°–60°–90° right triangle measures 8 in. Find the lengths of the other sides. Give exact answers and, where appropriate, an approximation to three decimal places.

SOLUTION The hypotenuse is twice as long as the shorter leg, so we have

$$c = 2a \quad \textbf{This relationship is worth memorizing.}$$
$$= 2 \cdot 8 = 16 \text{ in.} \quad \textbf{This is the length of the hypotenuse.}$$

The length of the longer leg is the length of the shorter leg times $\sqrt{3}$. This gives us

$$b = a\sqrt{3} \quad \textbf{This is also worth memorizing.}$$
$$= 8\sqrt{3} \text{ in.} \quad \textbf{This is the length of the longer leg.}$$

Exact answer: $c = 16$ in., $b = 8\sqrt{3}$ in.

Approximation: $b \approx 13.856$ in.

Try Exercise 37.

EXAMPLE 6 The length of the longer leg of a 30°–60°–90° right triangle is 14 cm. Find the length of the hypotenuse. Give an exact answer and an approximation to three decimal places.

SOLUTION The length of the hypotenuse is twice the length of the shorter leg. We first find a, the length of the shorter leg, by using the length of the longer leg:

$$14 = a\sqrt{3} \quad \textbf{Substituting 14 for } b \textbf{ in } b = a\sqrt{3}$$
$$\frac{14}{\sqrt{3}} = a. \quad \textbf{Dividing by } \sqrt{3}$$

Since the hypotenuse is twice as long as the shorter leg, we have

$$c = 2a$$
$$= 2 \cdot \frac{14}{\sqrt{3}} \quad \textbf{Substituting}$$
$$= \frac{28}{\sqrt{3}} \text{ cm.}$$

Exact answer: $c = \dfrac{28}{\sqrt{3}}$ cm, or $\dfrac{28\sqrt{3}}{3}$ cm if the denominator is rationalized.

Approximation: $c \approx 16.166$ cm

Try Exercise 33.

Student Notes

Perhaps the easiest way to remember the important results listed in the adjacent box is to write out, on your own, the derivations shown on pp. 749 and 750.

Lengths Within Isosceles Triangles and 30°–60°–90° Right Triangles The length of the hypotenuse in an isosceles right triangle is the length of a leg times $\sqrt{2}$.

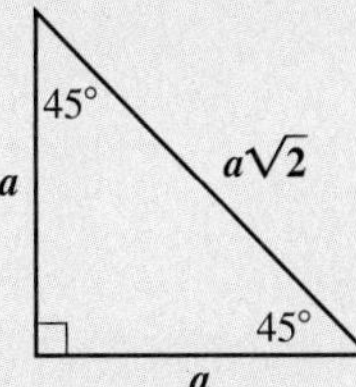

The length of the longer leg in a 30°–60°–90° right triangle is the length of the shorter leg times $\sqrt{3}$. The hypotenuse is twice as long as the shorter leg.

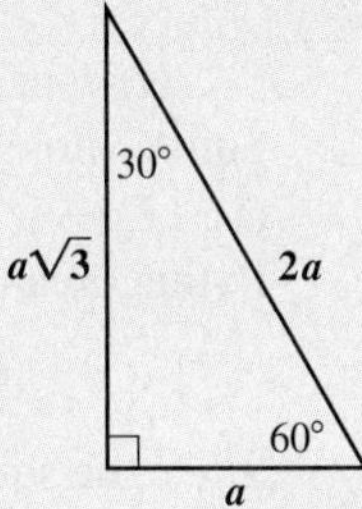

THE DISTANCE FORMULA AND MIDPOINT FORMULA

We can use the Pythagorean theorem to find the distance between two points on a plane.

To find the distance between two points on the number line, we subtract. Depending on the order in which we subtract, the difference may be positive or negative. However, if we take the absolute value of the difference, we always obtain a positive value for the distance:

7 units

−4 −3 −2 −1 0 1 2 3 4

$$|4 - (-3)| = |7| = 7$$
$$|-3 - 4| = |-7| = 7$$

If two points are on a horizontal line, they have the same second coordinate. We can find the distance between them by subtracting their first coordinates and taking the absolute value of that difference.

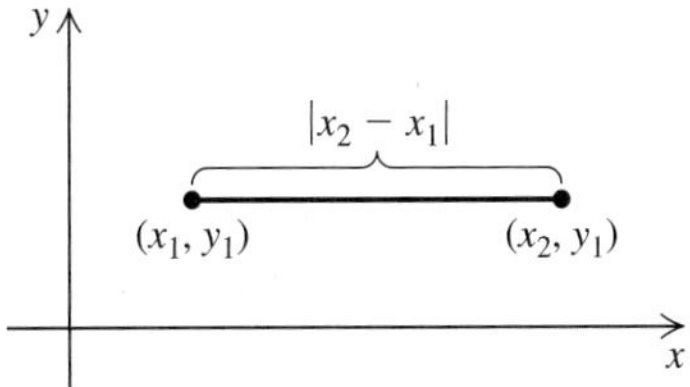

The distance between the points (x_1, y_1) and (x_2, y_1) on a horizontal line is thus $|x_2 - x_1|$. Similarly, the distance between the points (x_2, y_1) and (x_2, y_2) on a vertical line is $|y_2 - y_1|$.

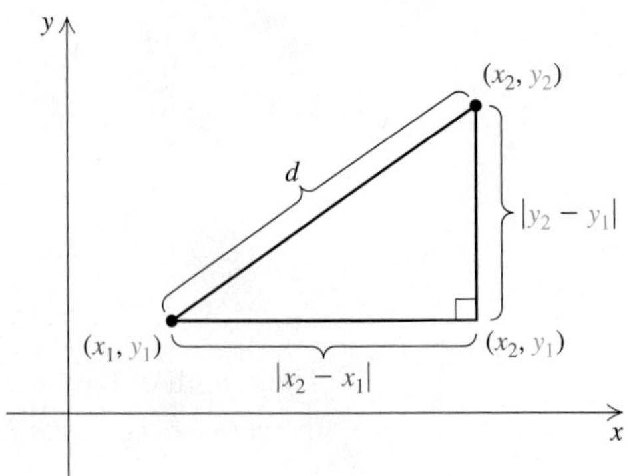

Now consider two points (x_1, y_1) and (x_2, y_2). If $x_1 \neq x_2$ and $y_1 \neq y_2$, these points, along with the point (x_2, y_1), describe a right triangle. The lengths of the legs are $|x_2 - x_1|$ and $|y_2 - y_1|$. We find d, the length of the hypotenuse, by using the Pythagorean theorem:

$$d^2 = |x_2 - x_1|^2 + |y_2 - y_1|^2.$$

Since the square of a number is the same as the square of its opposite, we can replace the absolute-value signs with parentheses:

$$d^2 = (x_2 - x_1)^2 + (y_2 - y_1)^2.$$

Taking the principal square root, we have a formula for distance.

The Distance Formula The distance d between any two points (x_1, y_1) and (x_2, y_2) is given by

$$d = \sqrt{(x_2 - x_1)^2 + (y_2 - y_1)^2}.$$

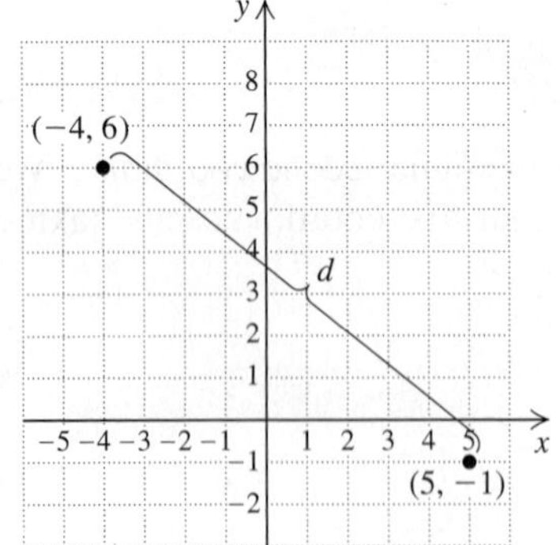

EXAMPLE 7 Find the distance between $(5, -1)$ and $(-4, 6)$. Find an exact answer and an approximation to three decimal places.

SOLUTION We substitute into the distance formula:

$$d = \sqrt{(-4 - 5)^2 + [6 - (-1)]^2}$$ **Substituting. A drawing is optional.**

$$= \sqrt{(-9)^2 + 7^2}$$

$$= \sqrt{130}$$ **This is exact.**

$$\approx 11.402.$$ **Using a calculator for an approximation**

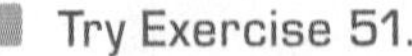

Try Exercise 51.

The distance formula is needed to verify a formula for the coordinates of the *midpoint* of a segment connecting two points. We state the midpoint formula and leave its proof to the exercises.

Student Notes

To help remember the formulas correctly, note that the distance formula (a variation on the Pythagorean theorem) involves both subtraction and addition, whereas the midpoint formula does not include any subtraction.

The Midpoint Formula If the endpoints of a segment are (x_1, y_1) and (x_2, y_2), then the coordinates of the midpoint are

$$\left(\frac{x_1 + x_2}{2}, \frac{y_1 + y_2}{2}\right).$$

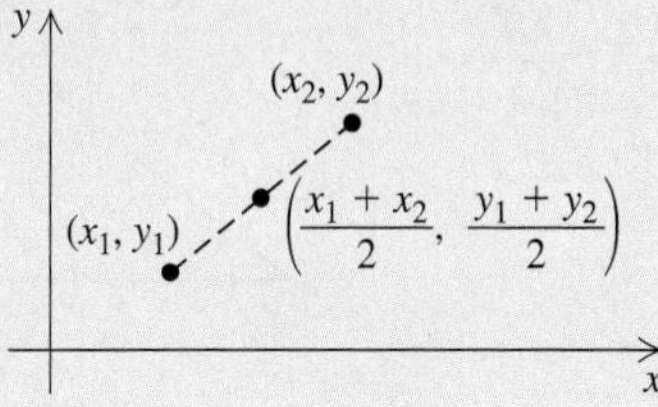

(To locate the midpoint, average the x-coordinates and average the y-coordinates.)

EXAMPLE 8 Find the midpoint of the segment with endpoints $(-2, 3)$ and $(4, -6)$.

SOLUTION Using the midpoint formula, we obtain

$$\left(\frac{-2 + 4}{2}, \frac{3 + (-6)}{2}\right), \text{ or } \left(\frac{2}{2}, \frac{-3}{2}\right), \text{ or } \left(1, -\frac{3}{2}\right).$$

Try Exercise 65.

10.7 Exercise Set

FOR EXTRA HELP

Concept Reinforcement *Complete each sentence with the best choice from the column on the right.*

1. In any ___ triangle, the square of the length of the hypotenuse is the sum of the squares of the lengths of the legs.
2. The shortest side of a right triangle is always one of the two ___.
3. The principle of ___ states that if $x^2 = n$, then $x = \sqrt{n}$ or $x = -\sqrt{n}$.
4. In a(n) ___ right triangle, both legs have the same length.
5. In a(n) ___ right triangle, the hypotenuse is twice as long as the shorter leg.
6. If both legs in a right triangle have measure a, then the ___ measures $a\sqrt{2}$.

a) Hypotenuse
b) Isosceles
c) Legs
d) Right
e) Square roots
f) $30°-60°-90°$

In a right triangle, find the length of the side not given. Give an exact answer and, where appropriate, an approximation to three decimal places.

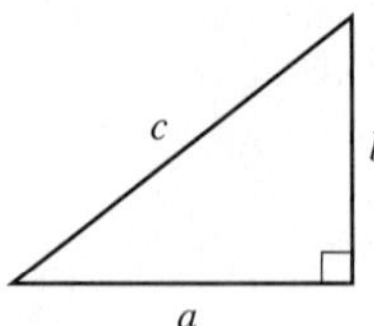

7. $a = 5, b = 3$
8. $a = 8, b = 10$
9. Aha! $a = 9, b = 9$
10. $a = 10, b = 10$
11. $b = 12, c = 13$
12. $a = 7, c = 25$

In Exercises 13–28, give an exact answer and, where appropriate, an approximation to three decimal places.

13. A right triangle's hypotenuse is 8 m and one leg is $4\sqrt{3}$ m. Find the length of the other leg.
14. A right triangle's hypotenuse is 6 cm and one leg is $\sqrt{5}$ cm. Find the length of the other leg.
15. The hypotenuse of a right triangle is $\sqrt{20}$ in. and one leg measures 1 in. Find the length of the other leg.
16. The hypotenuse of a right triangle is $\sqrt{15}$ ft and one leg measures 2 ft. Find the length of the other leg.
17. Aha! One leg of a right triangle is 1 m and the hypotenuse measures $\sqrt{2}$ m. Find the length of the other leg.
18. One leg of a right triangle is 1 yd and the hypotenuse measures 2 yd. Find the length of the other leg.
19. *Bicycling.* Amanda routinely bicycles across a rectangular parking lot on her way to class. If the lot is 200 ft long and 150 ft wide, how far does Amanda travel when she rides across the lot diagonally?

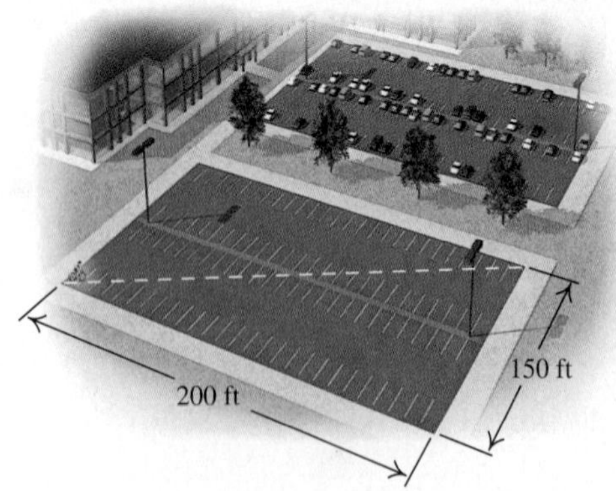

20. *Guy Wire.* How long is a guy wire if it reaches from the top of a 15-ft pole to a point on the ground 10 ft from the pole?
21. *Softball.* A slow-pitch softball diamond is actually a square 65 ft on a side. How far is it from home plate to second base?
22. *Baseball.* Suppose the catcher in Example 1 makes a throw to second base from the same location. How far is that throw?
23. *Television Sets.* What does it mean to refer to a 51-in. TV set? Such units refer to the diagonal of the screen. A 51-in. TV set has a width of 45 in. What is its height?

24. *Television Sets.* A 53-in. TV set has a screen with a height of 28 in. What is its width? (See Exercise 23.)
25. *Speaker Placement.* A stereo receiver is in a corner of a 12-ft by 14-ft room. Wire will run under a rug, diagonally, to a speaker in the far corner. If 4 ft of slack is required on each end, how long a piece of wire should be purchased?
26. *Distance over Water.* To determine the width of a pond, a surveyor locates two stakes at either end of the pond and uses instrumentation to place a third stake so that the distance across the pond is the length of a hypotenuse. If the third stake is 90 m from one stake and 70 m from the other, what is the distance across the pond?

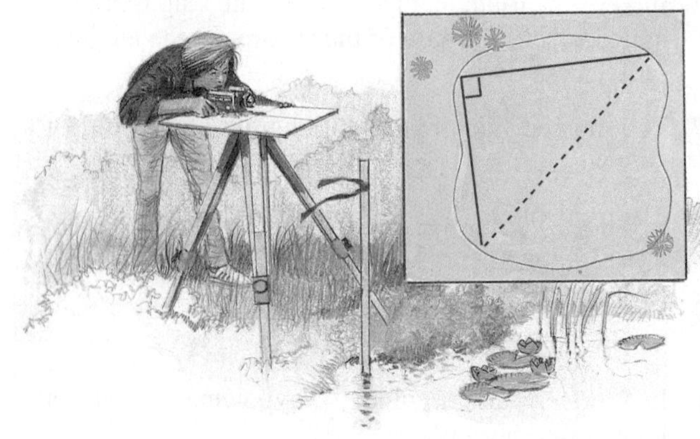

27. *Walking.* Students at Mossway Community College have worn a path that cuts diagonally across the campus "quad." If the quad is actually a rectangle that Angie measured to be 70 paces long and 40 paces wide, how many paces will Angie save by using the diagonal path?

28. *Crosswalks.* The diagonal crosswalk at the intersection of State St. and Main St. is the hypotenuse of a triangle in which the crosswalks across State St. and Main St. are the legs. If State St. is 28 ft wide and Main St. is 40 ft wide, how much shorter is the distance traveled by pedestrians using the diagonal crosswalk?

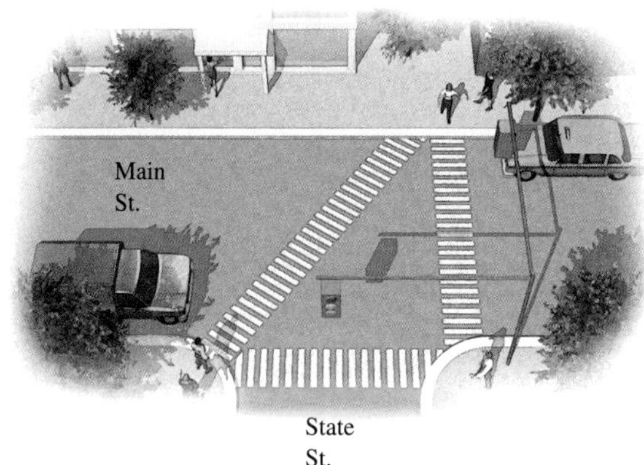

For each triangle, find the missing length(s). Give an exact answer and, where appropriate, an approximation to three decimal places.

29.

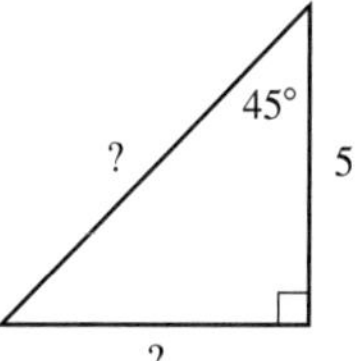

30.

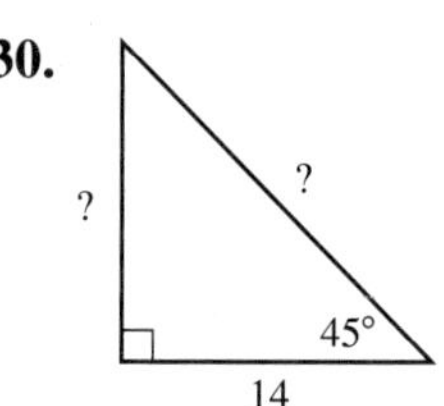

31.

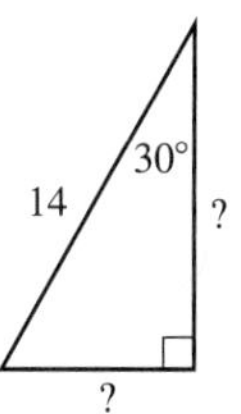

32.

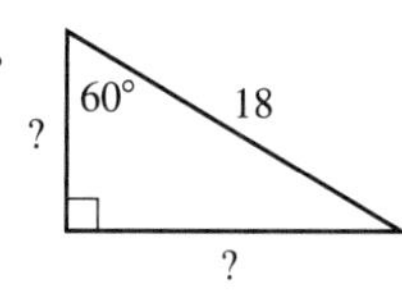

33.

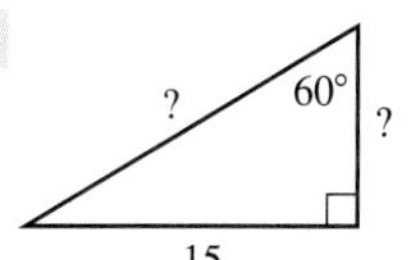

34.

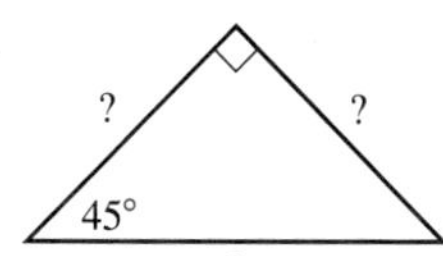

35.

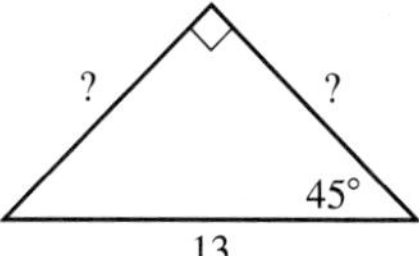

36.

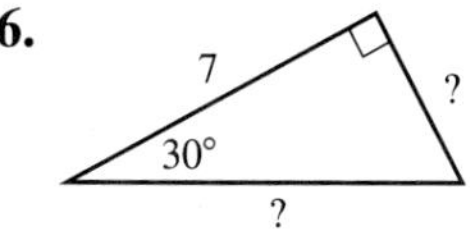

37.

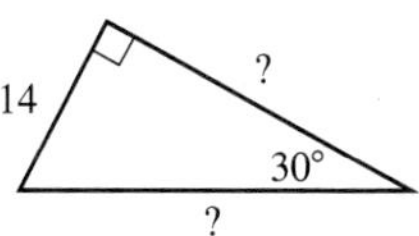

38.

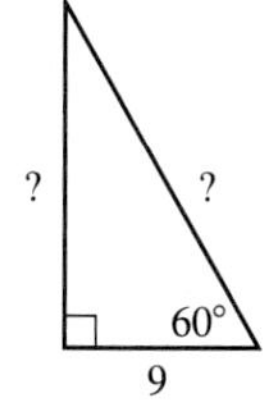

39.

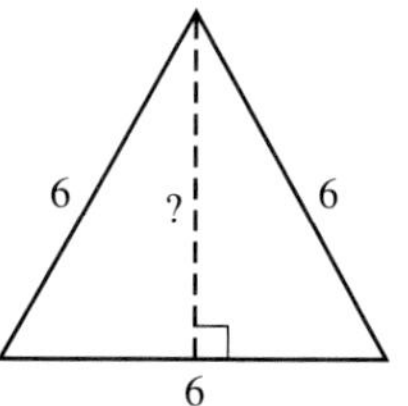

40.

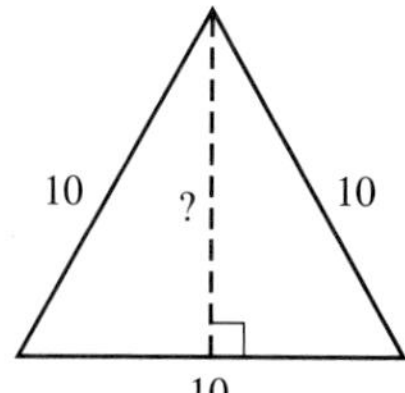

41.

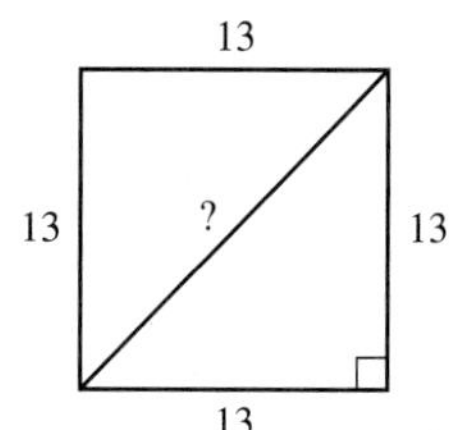

42.

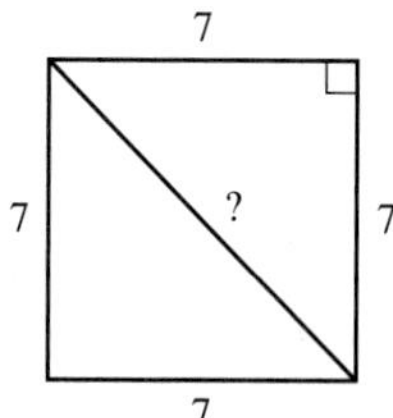

43.

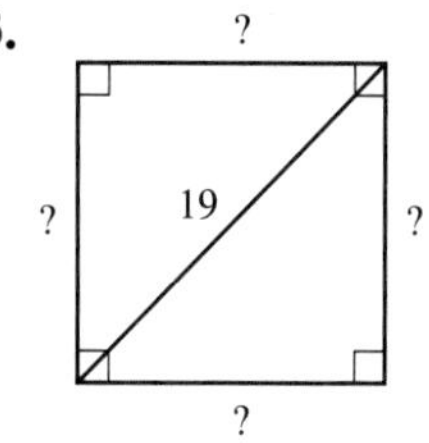

44.

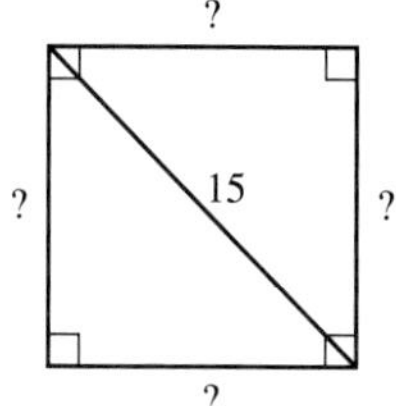

In Exercises 45–50, give an exact answer and, where appropriate, an approximation to three decimal places.

45. *Bridge Expansion.* During the summer heat, a 2-mi bridge expands 2 ft in length. If we assume that the bulge occurs straight up the middle, how high is the bulge? (The answer may surprise you. Most bridges have expansion spaces to avoid such buckling.)

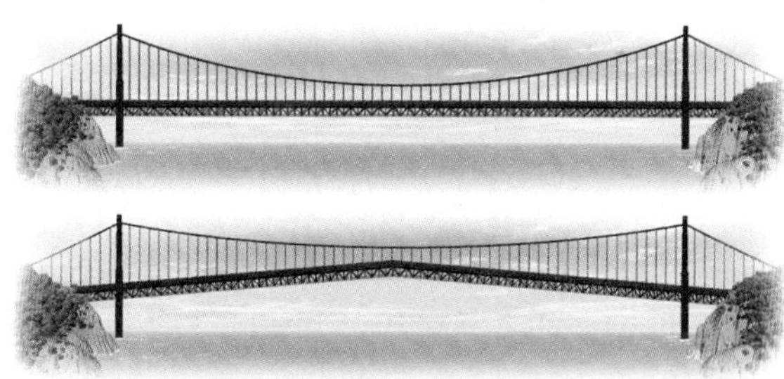

46. Triangle ABC has sides of lengths 25 ft, 25 ft, and 30 ft. Triangle PQR has sides of lengths 25 ft, 25 ft, and 40 ft. Which triangle, if either, has the greater area and by how much?

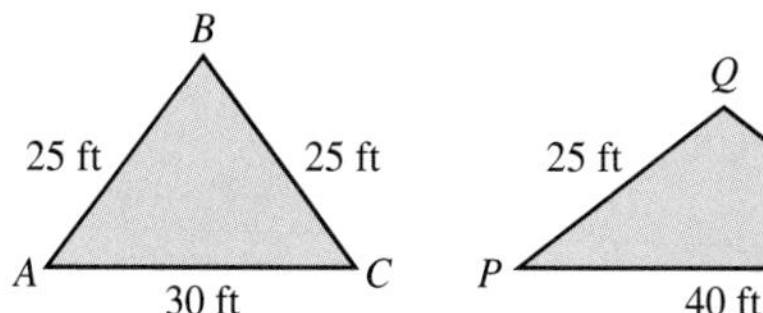

47. *Architecture.* The Rushton Triangular Lodge in Northamptonshire, England, was designed and constructed by Sir Thomas Tresham between 1593 and 1597. The building is in the shape of an equilateral triangle with walls of length 33 ft. How many square feet of land is covered by the lodge?
Source: The Internet Encyclopedia of Science

48. *Antenna Length.* As part of an emergency radio communication station, Rik sets up an "Inverted-V" antenna. He stretches a copper wire from one point on the ground to a point on a tree and then back down to the ground, forming two 30°–60°–90° triangles. If the wire is fastened to the tree 34 ft above the ground, how long is the copper wire?

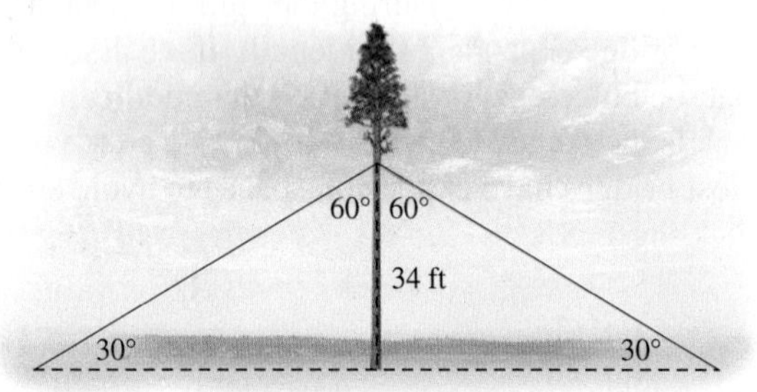

49. Find all points on the y-axis of a Cartesian coordinate system that are 5 units from the point $(3, 0)$.

50. Find all points on the x-axis of a Cartesian coordinate system that are 5 units from the point $(0, 4)$.

Find the distance between each pair of points. Where appropriate, find an approximation to three decimal places.

51. $(4, 5)$ and $(7, 1)$

52. $(0, 8)$ and $(6, 0)$

53. $(0, -5)$ and $(1, -2)$

54. $(-1, -4)$ and $(-3, -5)$

55. $(-4, 4)$ and $(6, -6)$

56. $(5, 21)$ and $(-3, 1)$

Aha! 57. $(8.6, -3.4)$ and $(-9.2, -3.4)$

58. $(5.9, 2)$ and $(3.7, -7.7)$

59. $\left(\frac{1}{2}, \frac{1}{3}\right)$ and $\left(\frac{5}{6}, -\frac{1}{6}\right)$

60. $\left(\frac{5}{7}, \frac{1}{14}\right)$ and $\left(\frac{1}{7}, \frac{11}{14}\right)$

61. $\left(-\sqrt{6}, \sqrt{6}\right)$ and $(0, 0)$

62. $\left(\sqrt{5}, -\sqrt{3}\right)$ and $(0, 0)$

63. $(-1, -30)$ and $(-2, -40)$

64. $(0.5, 100)$ and $(1.5, -100)$

Find the midpoint of each segment with the given endpoints.

65. $(-2, 5)$ and $(8, 3)$

66. $(1, 4)$ and $(9, -6)$

67. $(2, -1)$ and $(5, 8)$

68. $(-1, 2)$ and $(1, -3)$

69. $(-8, -5)$ and $(6, -1)$

70. $(8, -2)$ and $(-3, 4)$

71. $(-3.4, 8.1)$ and $(4.8, -8.1)$

72. $(4.1, 6.9)$ and $(5.2, -8.9)$

73. $\left(\frac{1}{6}, -\frac{3}{4}\right)$ and $\left(-\frac{1}{3}, \frac{5}{6}\right)$

74. $\left(-\frac{4}{5}, -\frac{2}{3}\right)$ and $\left(\frac{1}{8}, \frac{3}{4}\right)$

75. $\left(\sqrt{2}, -1\right)$ and $\left(\sqrt{3}, 4\right)$

76. $\left(9, 2\sqrt{3}\right)$ and $\left(-4, 5\sqrt{3}\right)$

TW 77. Are there any right triangles, other than those with sides measuring 3, 4, and 5, that have consecutive numbers for the lengths of the sides? Why or why not?

TW 78. If a 30°–60°–90° triangle and an isosceles right triangle have the same perimeter, which will have the greater area? Why?

SKILL REVIEW

Review graphing (Sections 3.2, 3.3, 3.6, and 9.4).

Graph on a plane.

79. $y = 2x - 3$ [3.6]

80. $y < x$ [9.4]

81. $8x - 4y = 8$ [3.3]

82. $2y - 1 = 7$ [3.3]

83. $x \geq 1$ [9.4]

84. $x - 5 = 6 - 2y$ [3.2]

SYNTHESIS

TW 85. Describe a procedure that uses the distance formula to determine whether three points, (x_1, y_1), (x_2, y_2), and (x_3, y_3), are vertices of a right triangle.

TW 86. Outline a procedure that uses the distance formula to determine whether three points, (x_1, y_1), (x_2, y_2), and (x_3, y_3), are collinear (lie on the same line).

87. The perimeter of a regular hexagon is 72 cm. Determine the area of the shaded region shown.

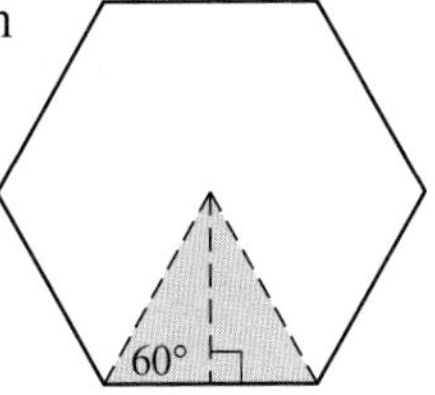

88. If the perimeter of a regular hexagon is 120 ft, what is its area? (*Hint*: See Exercise 87.)

89. Each side of a regular octagon has length s. Find a formula for the distance d between the parallel sides of the octagon.

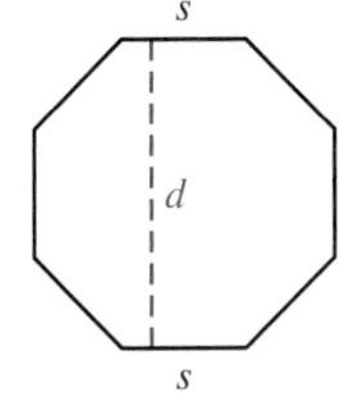

90. *Roofing.* Kit's home, which is 24 ft wide and 32 ft long, needs a new roof. By counting clapboards that are 4 in. apart, Kit determines that the peak of the roof is 6 ft higher than the sides. If one packet of shingles covers 100 ft^2, how many packets will the job require?

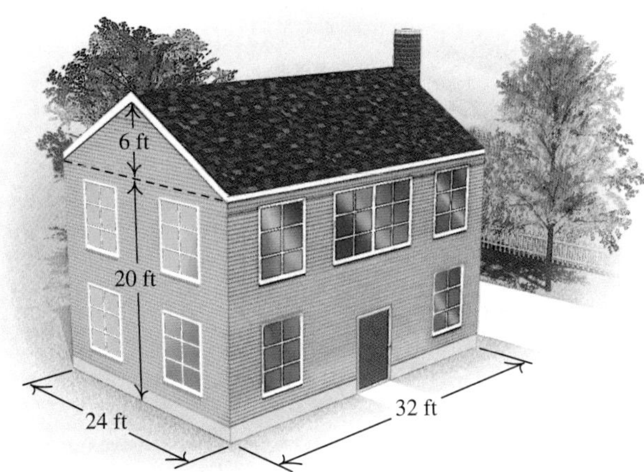

91. *Painting.* (Refer to Exercise 90.) A gallon of Benjamin Moore® exterior acrylic paint covers 450–500 ft^2. If Kit's house has dimensions as shown above, how many gallons of paint should be bought to paint the house? What assumption(s) is made in your answer?

92. *Contracting.* Oxford Builders has an extension cord on their generator that permits them to work, with electricity, anywhere in a circular area of 3850 ft^2. Find the dimensions of the largest square room they could work on without having to relocate the generator to reach each corner of the floor plan.

93. *Contracting.* Cleary Construction has a hose attached to their insulation blower that permits them to work, with electricity, anywhere in a circular area of 6160 ft^2. Find the dimensions of the largest square room with 12-ft ceilings in which they could reach all corners with the hose while leaving the blower centrally located. Assume that the blower sits on the floor.

94. A cube measures 5 cm on each side. How long is the diagonal that connects two opposite corners of the cube? Give an exact answer.

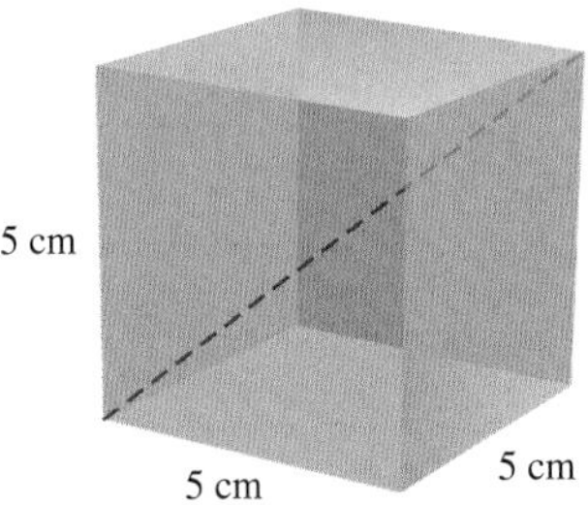

95. Prove the midpoint formula by showing that

i) the distance from (x_1, y_1) to

$$\left(\frac{x_1 + x_2}{2}, \frac{y_1 + y_2}{2}\right)$$

equals the distance from (x_2, y_2) to

$$\left(\frac{x_1 + x_2}{2}, \frac{y_1 + y_2}{2}\right);$$

and

ii) the points

$$(x_1, y_1), \left(\frac{x_1 + x_2}{2}, \frac{y_1 + y_2}{2}\right),$$

and

$$(x_2, y_2)$$

lie on the same line (see Exercise 86).

Try Exercise Answers: Section 10.7

15. $\sqrt{19}$ in.; 4.359 in. **19.** 250 ft
29. Leg = 5; hypotenuse = $5\sqrt{2} \approx 7.071$
33. Leg = $5\sqrt{3} \approx 8.660$; hypotenuse = $10\sqrt{3} \approx 17.321$
35. Both legs = $\frac{13\sqrt{2}}{2} \approx 9.192$
37. Leg = $14\sqrt{3} \approx 24.249$; hypotenuse = 28
51. 5 **65.** (3, 4)

10.8 The Complex Numbers

- Imaginary Numbers and Complex Numbers
- Addition and Subtraction
- Multiplication
- Conjugates and Division
- Powers of i

IMAGINARY NUMBERS AND COMPLEX NUMBERS

Negative numbers do not have square roots in the real-number system. However, a larger number system that contains the real-number system is designed so that negative numbers *do* have square roots. That system is called the **complex-number system,** and it will allow us to solve equations like $x^2 + 1 = 0$. The complex-number system makes use of i, a number that is, by definition, a square root of -1.

The Number i i is the unique number for which $i = \sqrt{-1}$ and $i^2 = -1$.

We can now define the square root of a negative number as follows:

$$\sqrt{-p} = \sqrt{-1}\sqrt{p} = i\sqrt{p} \text{ or } \sqrt{p}i, \quad \text{for any positive number } p.$$

EXAMPLE 1 Express in terms of i: **(a)** $\sqrt{-7}$; **(b)** $\sqrt{-16}$; **(c)** $-\sqrt{-13}$; **(d)** $-\sqrt{-50}$.

SOLUTION

i is *not* under the radical.

a) $\sqrt{-7} = \sqrt{-1 \cdot 7} = \sqrt{-1} \cdot \sqrt{7} = i\sqrt{7}$, or $\sqrt{7}i$

b) $\sqrt{-16} = \sqrt{-1 \cdot 16} = \sqrt{-1} \cdot \sqrt{16} = i \cdot 4 = 4i$

c) $-\sqrt{-13} = -\sqrt{-1 \cdot 13} = -\sqrt{-1} \cdot \sqrt{13} = -i\sqrt{13}$, or $-\sqrt{13}i$

d) $-\sqrt{-50} = -\sqrt{-1} \cdot \sqrt{25} \cdot \sqrt{2} = -i \cdot 5 \cdot \sqrt{2}$

$= -5i\sqrt{2}$, or $-5\sqrt{2}i$

Try Exercise 9.

Imaginary Numbers An *imaginary number* is a number that can be written in the form $a + bi$, where a and b are real numbers and $b \neq 0$.

Don't let the name "imaginary" fool you. Imaginary numbers appear in fields such as engineering and the physical sciences. The following are examples of imaginary numbers:

$5 + 4i$, **Here $a = 5, b = 4$.**

$\sqrt{3} - \pi i$, **Here $a = \sqrt{3}, b = -\pi$.**

$\sqrt{7}i$. **Here $a = 0, b = \sqrt{7}$.**

The union of the set of all imaginary numbers and the set of all real numbers is the set of all **complex numbers.**

Complex Numbers A *complex number* is any number that can be written in the form $a + bi$, where a and b are real numbers. (Note that a and b both can be 0.)

STUDY TIP

Studying Together by Phone

Working with a classmate over the telephone can be a very effective way to receive or give help. Because you cannot point to figures on paper, you must verbalize the mathematics. This tends to improve your understanding of the material. In some cases it may be more effective to study with a classmate over the phone than in person.

The following are examples of complex numbers:

$7 + 3i$ (here $a \neq 0, b \neq 0$); $4i$ (here $a = 0, b \neq 0$);
8 (here $a \neq 0, b = 0$); 0 (here $a = 0, b = 0$).

Complex numbers like $17i$ or $4i$, in which $a = 0$ and $b \neq 0$, are called *pure imaginary numbers.* For $b = 0$, we have $a + 0i = a$, so every real number is a complex number. The relationships among various real numbers and complex numbers are shown below.

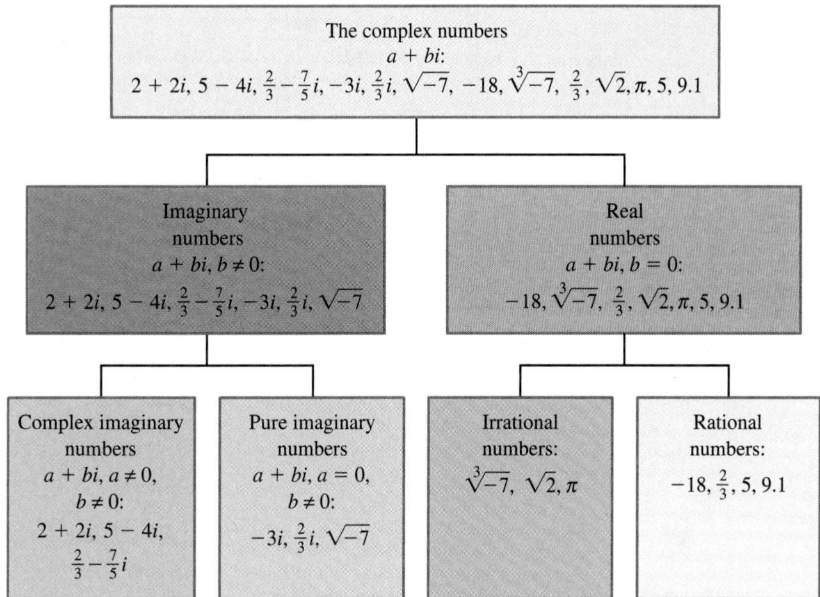

Note that although $\sqrt{-7}$ and $\sqrt[3]{-7}$ are both complex numbers, $\sqrt{-7}$ is imaginary whereas $\sqrt[3]{-7}$ is real.

ADDITION AND SUBTRACTION

The commutative, associative, and distributive laws hold for complex numbers. Thus we can add and subtract them as we do binomials.

EXAMPLE 2 Add or subtract and simplify.

a) $(8 + 6i) + (3 + 2i)$ **b)** $(4 + 5i) - (6 - 3i)$

SOLUTION

a) $(8 + 6i) + (3 + 2i) = (8 + 3) + (6i + 2i)$ **Combining the real parts and the imaginary parts**

$= 11 + (6 + 2)i = 11 + 8i$

b) $(4 + 5i) - (6 - 3i) = (4 - 6) + [5i - (-3i)]$ **Note that the 6 and the $-3i$ are *both* being subtracted.**

$= -2 + 8i$

Try Exercise 27.

Student Notes

The rule developed in Section 10.3, $\sqrt[n]{a} \cdot \sqrt[n]{b} = \sqrt[n]{a \cdot b}$, does *not* apply when n is 2 and either a or b is negative. Indeed this condition is stated on p. 717 when it is specified that $\sqrt[n]{a}$ and $\sqrt[n]{b}$ are both *real* numbers.

MULTIPLICATION

To multiply square roots of negative real numbers, we first express them in terms of i. For example,

$$\begin{aligned} \sqrt{-2} \cdot \sqrt{-5} &= \sqrt{-1} \cdot \sqrt{2} \cdot \sqrt{-1} \cdot \sqrt{5} \\ &= i \cdot \sqrt{2} \cdot i \cdot \sqrt{5} \\ &= i^2 \cdot \sqrt{10} \qquad \text{Since } i \text{ is a square root of } -1, i^2 = -1. \\ &= -1\sqrt{10} = -\sqrt{10} \text{ is correct!} \end{aligned}$$

CAUTION! With complex numbers, simply multiplying radicands is *incorrect* when both radicands are negative:

$$\sqrt{-2} \cdot \sqrt{-5} \neq \sqrt{10}.$$

With this in mind, we can now multiply complex numbers.

EXAMPLE 3 Multiply and simplify. When possible, write answers in the form $a + bi$.

a) $\sqrt{-4} \cdot \sqrt{-25}$ **b)** $\sqrt{-5} \cdot \sqrt{-7}$ **c)** $-3i \cdot 8i$
d) $-4i(3 - 5i)$ **e)** $(1 + 2i)(4 + 3i)$

SOLUTION

a) $$\begin{aligned} \sqrt{-4} \cdot \sqrt{-25} &= \sqrt{-1} \cdot \sqrt{4} \cdot \sqrt{-1} \cdot \sqrt{25} \\ &= i \cdot 2 \cdot i \cdot 5 \\ &= i^2 \cdot 10 \\ &= -1 \cdot 10 \qquad i^2 = -1 \\ &= -10 \end{aligned}$$

b) $$\begin{aligned} \sqrt{-5} \cdot \sqrt{-7} &= \sqrt{-1} \cdot \sqrt{5} \cdot \sqrt{-1} \cdot \sqrt{7} \qquad \text{Try to do this step mentally.} \\ &= i \cdot \sqrt{5} \cdot i \cdot \sqrt{7} \\ &= i^2 \cdot \sqrt{35} \\ &= -1 \cdot \sqrt{35} \qquad i^2 = -1 \\ &= -\sqrt{35} \end{aligned}$$

c) $$\begin{aligned} -3i \cdot 8i &= -24 \cdot i^2 \\ &= -24 \cdot (-1) \qquad i^2 = -1 \\ &= 24 \end{aligned}$$

d) $$\begin{aligned} -4i(3 - 5i) &= -4i \cdot 3 + (-4i)(-5i) \qquad \text{Using the distributive law} \\ &= -12i + 20i^2 \\ &= -12i - 20 \qquad i^2 = -1 \\ &= -20 - 12i \qquad \text{Writing in the form } a + bi \end{aligned}$$

e) $$\begin{aligned} (1 + 2i)(4 + 3i) &= 4 + 3i + 8i + 6i^2 \qquad \text{Multiplying each term of } 4 + 3i \text{ by every term of } 1 + 2i \text{ (FOIL)} \\ &= 4 + 3i + 8i - 6 \qquad i^2 = -1 \\ &= -2 + 11i \qquad \text{Combining like terms} \end{aligned}$$

Try Exercises 39 and 45.

CONJUGATES AND DIVISION

Recall that the conjugate of $4 + \sqrt{2}$ is $4 - \sqrt{2}$.

Conjugates of complex numbers are defined in a similar manner.

Conjugate of a Complex Number The *conjugate* of a complex number $a + bi$ is $a - bi$, and the *conjugate* of $a - bi$ is $a + bi$.

EXAMPLE 4 Find the conjugate of each number.

a) $-3 - 7i$ **b)** $4i$

SOLUTION

a) $-3 - 7i$ The conjugate is $-3 + 7i$.

b) $4i$ The conjugate is $-4i$. Note that $4i = 0 + 4i$.

The product of a complex number and its conjugate is a real number.

EXAMPLE 5 Multiply: $(5 + 7i)(5 - 7i)$.

SOLUTION

$$\begin{aligned}(5 + 7i)(5 - 7i) &= 5^2 - (7i)^2 && \text{Using } (A + B)(A - B) = A^2 - B^2\\ &= 25 - 49i^2\\ &= 25 - 49(-1) && i^2 = -1\\ &= 25 + 49 = 74\end{aligned}$$

Try Exercise 55.

Conjugates are used when dividing complex numbers. The procedure is much like that used to rationalize denominators with two terms.

EXAMPLE 6 Divide and simplify to the form $a + bi$.

a) $\dfrac{-5 + 9i}{1 - 2i}$ **b)** $\dfrac{7 + 3i}{5i}$

SOLUTION

a) To divide and simplify $(-5 + 9i)/(1 - 2i)$, we multiply by 1, using the conjugate of the denominator to form 1:

$$\begin{aligned}\frac{-5 + 9i}{1 - 2i} &= \frac{-5 + 9i}{1 - 2i} \cdot \frac{1 + 2i}{1 + 2i} && \text{Multiplying by 1 using the conjugate of the denominator in the symbol for 1}\\ &= \frac{-5 - 10i + 9i + 18i^2}{1^2 - 4i^2} && \text{Multiplying using FOIL}\\ &= \frac{-5 - i - 18}{1 - 4(-1)} && i^2 = -1\\ &= \frac{-23 - i}{5}\\ &= -\frac{23}{5} - \frac{1}{5}i. && \text{Writing in the form } a + bi. \text{ Note that } \frac{X + Y}{Z} = \frac{X}{Z} + \frac{Y}{Z}.\end{aligned}$$

b) The conjugate of $5i$ is $-5i$, so we *could* multiply by $-5i/(-5i)$. However, when the denominator is a pure imaginary number, it is easiest if we multiply by i/i:

$$\frac{7+3i}{5i} = \frac{7+3i}{5i}\cdot\frac{i}{i}$$ **Multiplying by 1 using i/i. We can also use the conjugate of $5i$ to write $-5i/(-5i)$.**

$$= \frac{7i+3i^2}{5i^2}$$ **Multiplying**

$$= \frac{7i+3(-1)}{5(-1)}$$ $i^2 = -1$

$$= \frac{7i-3}{-5} = \frac{-3}{-5} + \frac{7}{-5}i, \text{ or } \frac{3}{5} - \frac{7}{5}i.$$ **Writing in the form $a + bi$**

Try Exercise 67.

POWERS OF *i*

Answers to problems involving complex numbers are generally written in the form $a + bi$. In the following discussion, we show why there is no need to use powers of i (other than 1) when writing answers.

Recall that -1 raised to an *even* power is 1, and -1 raised to an *odd* power is -1. Simplifying powers of i can then be done by using the fact that $i^2 = -1$ and expressing the given power of i in terms of i^2. Consider the following:

$$\begin{aligned} i^2 &= -1, \\ i^3 = i^2\cdot i &= (-1)i = -i, \\ i^4 = (i^2)^2 &= (-1)^2 = 1, \\ i^5 = i^4\cdot i = (i^2)^2\cdot i &= (-1)^2\cdot i = i, \\ i^6 = (i^2)^3 &= (-1)^3 = -1. \end{aligned}$$

The pattern is now repeating.

The powers of i cycle themselves through the values $i, -1, -i$, and 1. Even powers of i are -1 or 1 whereas odd powers of i are i or $-i$.

EXAMPLE 7 Simplify: **(a)** i^{18}; **(b)** i^{24}.

SOLUTION

a) $i^{18} = (i^2)^9$ **Using the power rule**

$= (-1)^9 = -1$ **Raising -1 to a power**

b) $i^{24} = (i^2)^{12} = (-1)^{12} = 1$

Try Exercise 85.

To simplify i^n when n is odd, we rewrite i^n as $i^{n-1}\cdot i$.

EXAMPLE 8 Simplify: **(a)** i^{29}; **(b)** i^{75}.

SOLUTION

a) $i^{29} = i^{28}i^1$ **Using the product rule. This is a key step when i is raised to an odd power.**

$= (i^2)^{14}i$ **Using the power rule**

$= (-1)^{14}i = 1\cdot i = i$

b) $i^{75} = i^{74}i^1$ **Using the product rule**

$= (i^2)^{37}i$ **Using the power rule**

$= (-1)^{37}i = -1\cdot i = -i$

Try Exercise 83.

Complex Numbers

To perform operations with complex numbers using a calculator, you may need to first choose the a + bi setting in the MODE screen.

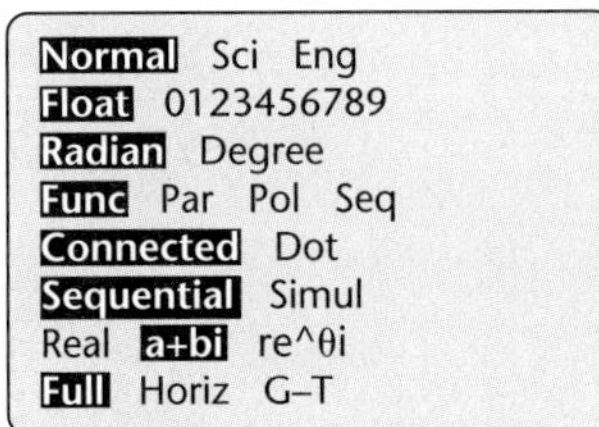

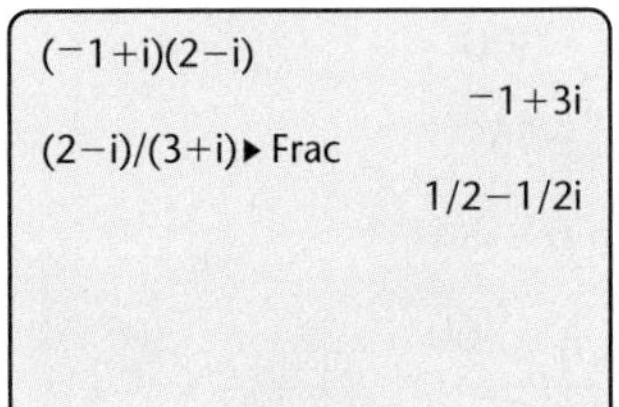

Press (i) for the complex number i. ((i) is often the 2nd option associated with the (.) key.) For example, to calculate $(-1 + i)(2 - i)$, press ((−) 1 + i) (2 − i) ENTER. As seen in the screen on the right above, the result is $-1 + 3i$. To calculate $\frac{2 - i}{3 + i}$ and write the result using fraction notation, press (2 − i) ÷ (3 + i) MATH 1 ENTER. Note the need for parentheses around the numerator and the denominator of the expression. The result appears in the form $a + bi$; i is not in the denominator of the fraction. The result is $\frac{1}{2} - \frac{1}{2}i$.

Your Turn

1. Set your calculator to complex mode.
2. Calculate $(1 - i)(3 + i)$ and $\frac{3 - 2i}{4 + i}$. $4 - 2i$; $\frac{10}{17} - \frac{11}{17}i$
3. Return your calculator to REAL mode.

10.8 Exercise Set

FOR EXTRA HELP MyMathLab

Concept Reinforcement *Classify each of the following statements as either true or false.*

1. Imaginary numbers are so named because they have no real-world applications.
2. Every real number is imaginary, but not every imaginary number is real.
3. Every imaginary number is a complex number, but not every complex number is imaginary.
4. Every real number is a complex number, but not every complex number is real.
5. We multiply complex numbers by multiplying real parts and multiplying imaginary parts.
6. The product of a complex number and its conjugate is always a real number.
7. The square of a complex number is always a real number.
8. The quotient of two complex numbers is always a complex number.

Express in terms of i.

9. $\sqrt{-100}$

10. $\sqrt{-25}$

11. $\sqrt{-13}$

12. $\sqrt{-19}$

13. $\sqrt{-8}$

14. $\sqrt{-12}$

15. $-\sqrt{-3}$
16. $-\sqrt{-17}$
17. $-\sqrt{-81}$
18. $-\sqrt{-49}$
19. $-\sqrt{-300}$
20. $-\sqrt{-75}$
21. $6-\sqrt{-84}$
22. $4-\sqrt{-60}$
23. $-\sqrt{-76}+\sqrt{-125}$
24. $\sqrt{-4}+\sqrt{-12}$
25. $\sqrt{-18}-\sqrt{-100}$
26. $\sqrt{-72}-\sqrt{-25}$

Perform the indicated operation and simplify. Write each answer in the form a + bi.

27. $(6+7i)+(5+3i)$
28. $(4-5i)+(3+9i)$
29. $(9+8i)-(5+3i)$
30. $(9+7i)-(2+4i)$
31. $(7-4i)-(5-3i)$
32. $(5-3i)-(9+2i)$
33. $(-5-i)-(7+4i)$
34. $(-2+6i)-(-7+i)$
35. $7i\cdot 6i$
36. $6i\cdot 9i$
37. $(-4i)(-6i)$
38. $7i\cdot(-8i)$
39. $\sqrt{-36}\sqrt{-9}$
40. $\sqrt{-49}\sqrt{-25}$
41. $\sqrt{-5}\sqrt{-2}$
42. $\sqrt{-6}\sqrt{-7}$
43. $\sqrt{-6}\sqrt{-21}$
44. $\sqrt{-15}\sqrt{-10}$
45. $5i(2+6i)$
46. $2i(7+3i)$
47. $-7i(3-4i)$
48. $-4i(6-5i)$
49. $(1+i)(3+2i)$
50. $(4+i)(2+3i)$
51. $(6-5i)(3+4i)$
52. $(5-6i)(2+5i)$
53. $(7-2i)(2-6i)$
54. $(-4+5i)(3-4i)$
55. $(3+8i)(3-8i)$
56. $(1+2i)(1-2i)$
57. $(-7+i)(-7-i)$
58. $(-4+5i)(-4-5i)$
59. $(4-2i)^2$
60. $(1-2i)^2$
61. $(2+3i)^2$
62. $(3+2i)^2$
63. $(-2+3i)^2$
64. $(-5-2i)^2$
65. $\dfrac{10}{3+i}$
66. $\dfrac{26}{5+i}$
67. $\dfrac{2}{3-2i}$
68. $\dfrac{4}{2-3i}$
69. $\dfrac{2i}{5+3i}$
70. $\dfrac{3i}{4+2i}$
71. $\dfrac{5}{6i}$
72. $\dfrac{4}{7i}$
73. $\dfrac{5-3i}{4i}$
74. $\dfrac{2+7i}{5i}$

Aha! 75. $\dfrac{7i+14}{7i}$

76. $\dfrac{6i+3}{3i}$
77. $\dfrac{4+5i}{3-7i}$
78. $\dfrac{5+3i}{7-4i}$
79. $\dfrac{2+3i}{2+5i}$
80. $\dfrac{3+2i}{4+3i}$
81. $\dfrac{3-2i}{4+3i}$
82. $\dfrac{5-2i}{3+6i}$

Simplify.

83. i^7
84. i^{11}
85. i^{32}
86. i^{38}
87. i^{42}
88. i^{64}
89. i^9
90. $(-i)^{71}$
91. $(-i)^6$
92. $(-i)^4$
93. $(5i)^3$
94. $(-3i)^5$
95. i^2+i^4
96. $5i^5+4i^3$

TW 97. Is the product of two imaginary numbers always an imaginary number? Why or why not?

TW 98. In what way(s) are conjugates of complex numbers similar to the conjugates used in Section 10.5?

SKILL REVIEW

To prepare for Section 11.1, review solving quadratic equations (Chapter 6).

Solve.

99. $x^2 - x - 6 = 0$ [6.2]

100. $(x - 5)^2 = 0$ [6.1]

101. $t^2 = 100$ [6.4]

102. $2t^2 - 50 = 0$ [6.4]

103. $15x^2 = 14x + 8$ [6.3]

104. $6x^2 = 5x + 6$ [6.3]

SYNTHESIS

TW **105.** Is the set of real numbers a subset of the set of complex numbers? Why or why not?

TW **106.** Is the union of the set of imaginary numbers and the set of real numbers the set of complex numbers? Why or why not?

Complex numbers are often graphed on a plane. The horizontal axis is the real axis and the vertical axis is the imaginary axis. A complex number such as $5 - 2i$ then corresponds to 5 on the real axis and -2 on the imaginary axis.

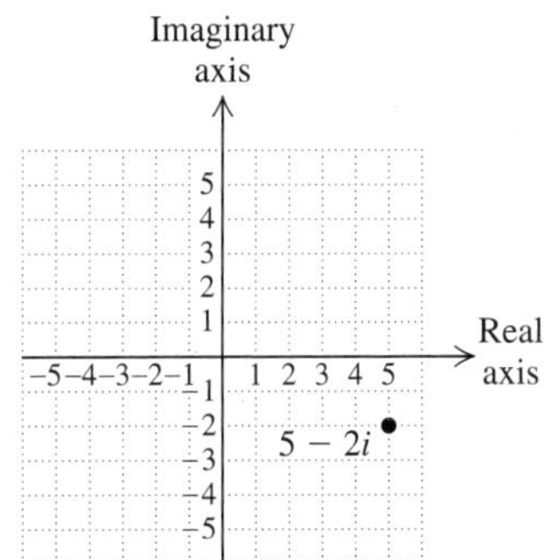

107. Graph each of the following.

a) $3 + 2i$ **b)** $-1 + 4i$
c) $3 - i$ **d)** $-5i$

108. Graph each of the following.

a) $1 - 4i$ **b)** $-2 - 3i$
c) i **d)** 4

The absolute value of a complex number $a + bi$ is its distance from the origin. Using the distance formula, we have $|a + bi| = \sqrt{a^2 + b^2}$. Find the absolute value of each complex number.

109. $|3 + 4i|$

110. $|8 - 6i|$

111. $|-1 + i|$

112. $|-3 - i|$

A function g is given by

$$g(z) = \frac{z^4 - z^2}{z - 1}.$$

113. Find $g(3i)$.

114. Find $g(1 + i)$.

115. Find $g(5i - 1)$.

116. Find $g(2 - 3i)$.

117. Evaluate

$$\frac{1}{w - w^2} \quad \text{for} \quad w = \frac{1 - i}{10}.$$

Simplify.

118. $\dfrac{i^5 + i^6 + i^7 + i^8}{(1 - i)^4}$

119. $(1 - i)^3(1 + i)^3$

120. $\dfrac{5 - \sqrt{5}i}{\sqrt{5}i}$

121. $\dfrac{6}{1 + \dfrac{3}{i}}$

122. $\left(\frac{1}{2} - \frac{1}{3}i\right)^2 - \left(\frac{1}{2} + \frac{1}{3}i\right)^2$

123. $\dfrac{i - i^{38}}{1 + i}$

Try Exercise Answers: Section 10.8

9. $10i$ **27.** $11 + 10i$ **39.** -18 **45.** $-30 + 10i$ **55.** 73 **67.** $\frac{6}{13} + \frac{4}{13}i$ **83.** $-i$ **85.** 1

Study Summary

KEY TERMS AND CONCEPTS	EXAMPLES	PRACTICE EXERCISES
SECTION 10.1: RADICAL EXPRESSIONS, FUNCTIONS, AND MODELS		
c is a **square root** of a if $c^2 = a$. c is a **cube root** of a if $c^3 = a$. $\sqrt{a}$ indicates the **principal** square root of a. $\sqrt[n]{a}$ indicates the ***n*th root** of a. **index** — $\sqrt[n]{a}$ — **radicand** **radical symbol**	The square roots of 25 are -5 and 5. The cube root of -8 is -2. $\sqrt{25} = 5$ $\sqrt[3]{-8} = -2$	*Simplify.* **1.** $-\sqrt{81}$ **2.** $\sqrt[3]{-1}$
For all a, $\sqrt[n]{a^n} = \lvert a \rvert$ when n is even; $\sqrt[n]{a^n} = a$ when n is odd. If a represents a nonnegative number, $\sqrt[n]{a^n} = a$.	Assume that x can represent any real number. $\sqrt{(3+x)^2} = \lvert 3 + x \rvert$ Assume that x represents a nonnegative number. $\sqrt{(7x)^2} = 7x$	**3.** Simplify. Assume that x can represent any real number. $\sqrt{36x^2}$ **4.** Simplify. Assume that x represents a nonnegative number. $\sqrt[4]{x^4}$
SECTION 10.2: RATIONAL NUMBERS AS EXPONENTS		
$a^{1/n}$ means $\sqrt[n]{a}$. $a^{m/n}$ means $\left(\sqrt[n]{a}\right)^m$ or $\sqrt[n]{a^m}$. $a^{-m/n}$ means $\frac{1}{a^{m/n}}$.	$64^{1/2} = \sqrt{64} = 8$ $125^{2/3} = \left(\sqrt[3]{125}\right)^2 = 5^2 = 25$ $8^{-1/3} = \frac{1}{8^{1/3}} = \frac{1}{2}$	**5.** Simplify: $100^{-1/2}$.
SECTION 10.3: MULTIPLYING RADICAL EXPRESSIONS		
The Product Rule for Radicals For any real numbers $\sqrt[n]{a}$ and $\sqrt[n]{b}$, $\sqrt[n]{a} \cdot \sqrt[n]{b} = \sqrt[n]{a \cdot b}$.	$\sqrt[3]{4x} \cdot \sqrt[3]{5y} = \sqrt[3]{20xy}$	**6.** Multiply: $\sqrt{7x} \cdot \sqrt{3y}$.
Using the Product Rule to Simplify For any real numbers $\sqrt[n]{a}$ and $\sqrt[n]{b}$, $\sqrt[n]{a \cdot b} = \sqrt[n]{a} \cdot \sqrt[n]{b}$.	$\sqrt{75x^8y^{11}} = \sqrt{25 \cdot x^8 \cdot y^{10} \cdot 3 \cdot y}$ $= \sqrt{25} \cdot \sqrt{x^8} \cdot \sqrt{y^{10}} \cdot \sqrt{3y}$ $= 5x^4y^5\sqrt{3y}$ **Assuming y is nonnegative**	**7.** Simplify: $\sqrt{200x^5y^{18}}$.

SECTION 10.4: DIVIDING RADICAL EXPRESSIONS

The Quotient Rule for Radicals For any real numbers $\sqrt[n]{a}$ and $\sqrt[n]{b}$, $b \neq 0$, $\sqrt[n]{\frac{a}{b}} = \frac{\sqrt[n]{a}}{\sqrt[n]{b}}$.	$\sqrt[3]{\frac{8y^4}{125}} = \frac{\sqrt[3]{8y^4}}{\sqrt[3]{125}} = \frac{2y\sqrt[3]{y}}{5}$ $\frac{\sqrt{18a^9}}{\sqrt{2a^3}} = \sqrt{\frac{18a^9}{2a^3}} = \sqrt{9a^6} = 3a^3$ **Assuming a is positive**	**8.** Simplify: $\sqrt{\frac{12x^3}{25}}$.
We can **rationalize a denominator** by multiplying by 1.	$\frac{1}{\sqrt{2}} = \frac{1}{\sqrt{2}} \cdot \frac{\sqrt{2}}{\sqrt{2}} = \frac{\sqrt{2}}{2}$	**9.** Rationalize the denominator: $\sqrt{\frac{2x}{3y^2}}$.

SECTION 10.5: EXPRESSIONS CONTAINING SEVERAL RADICAL TERMS

Like radicals have the same indices and radicands.	$\sqrt{12} + 5\sqrt{3} = \sqrt{4 \cdot 3} + 5\sqrt{3} = 2\sqrt{3} + 5\sqrt{3} = 7\sqrt{3}$	**10.** Simplify: $5\sqrt{8} - 3\sqrt{50}$.
Radical expressions are multiplied in much the same way that polynomials are multiplied.	$(1 + 5\sqrt{6})(4 - \sqrt{6})$ $= 1 \cdot 4 - 1\sqrt{6} + 4 \cdot 5\sqrt{6} - 5\sqrt{6} \cdot \sqrt{6}$ $= 4 - \sqrt{6} + 20\sqrt{6} - 5 \cdot 6$ $= -26 + 19\sqrt{6}$	**11.** Simplify: $(2 - \sqrt{3})(5 - 7\sqrt{3})$.
To rationalize a denominator containing two terms, we use the **conjugate** of the denominator to write a form of 1.	$\frac{2}{1 - \sqrt{3}} = \frac{2}{1 - \sqrt{3}} \cdot \frac{1 + \sqrt{3}}{1 + \sqrt{3}}$ **$1 + \sqrt{3}$ is the conjugate of $1 - \sqrt{3}$.** $= \frac{2(1 + \sqrt{3})}{-2} = -1 - \sqrt{3}$	**12.** Rationalize the denominator: $\frac{\sqrt{15}}{3 + \sqrt{5}}$.
When terms have different indices, we can often use rational exponents to simplify.	$\sqrt[3]{p} \cdot \sqrt[4]{q^3} = p^{1/3} \cdot q^{3/4}$ $= p^{4/12} \cdot q^{9/12}$ **Finding a common denominator** $= \sqrt[12]{p^4q^9}$	**13.** Simplify: $\frac{\sqrt{x^5}}{\sqrt[3]{x}}$.

SECTION 10.6: SOLVING RADICAL EQUATIONS

The Principle of Powers If $a = b$, then $a^n = b^n$. To solve a radical equation, use the principle of powers and the steps on pp. 741 and 742. Solutions found using the principle of powers must be checked in the original equation.	$x - 7 = \sqrt{x - 5}$ $(x - 7)^2 = (\sqrt{x - 5})^2$ $x^2 - 14x + 49 = x - 5$ $x^2 - 15x + 54 = 0$ $(x - 6)(x - 9) = 0$ $x = 6$ *or* $x = 9$ Only 9 checks and is the solution. $2 + \sqrt{t} = \sqrt{t + 8}$ $(2 + \sqrt{t})^2 = (\sqrt{t + 8})^2$ $4 + 4\sqrt{t} + t = t + 8$ $4\sqrt{t} = 4$ $\sqrt{t} = 1$ $(\sqrt{t})^2 = (1)^2$ $t = 1$ 1 checks and is the solution.	**14.** Solve: $\sqrt{2x + 3} = x$.

SECTION 10.7: THE DISTANCE FORMULA, THE MIDPOINT FORMULA, AND OTHER APPLICATIONS

The Pythagorean Theorem

In any right triangle, if a and b are the lengths of the legs and c is the length of the hypotenuse, then

$$a^2 + b^2 = c^2.$$

Find the length of the hypotenuse of a right triangle with legs of lengths 4 and 7. Give an exact answer in radical notation, as well as a decimal approximation to three decimal places.

$$a^2 + b^2 = c^2$$
$$4^2 + 7^2 = c^2 \quad \textbf{Substituting}$$
$$16 + 49 = c^2$$
$$65 = c^2$$
$$\sqrt{65} = c \quad \textbf{This is exact.}$$
$$8.062 \approx c \quad \textbf{This is approximate.}$$

15. The hypotenuse of a right triangle is 10 m long, and one leg is 7 m long. Find the length of the other leg. Give an exact answer in radical notation, as well as a decimal approximation to three decimal places.

Special Triangles

The length of the hypotenuse in an isosceles right triangle is the length of a leg times $\sqrt{2}$.

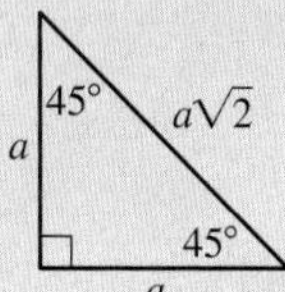

The length of the longer leg in a 30°–60°–90° triangle is the length of the shorter leg times $\sqrt{3}$. The hypotenuse is twice as long as the shorter leg.

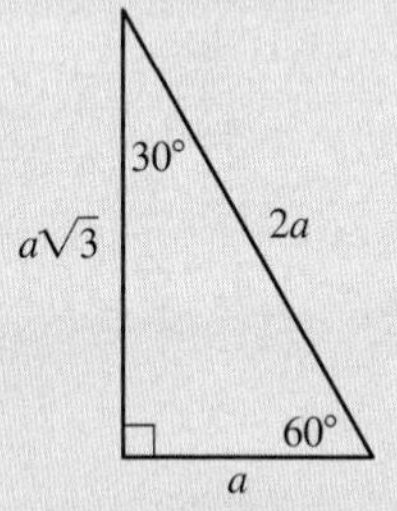

Find the missing lengths. Give an exact answer and, where appropriate, an approximation to three decimal places.

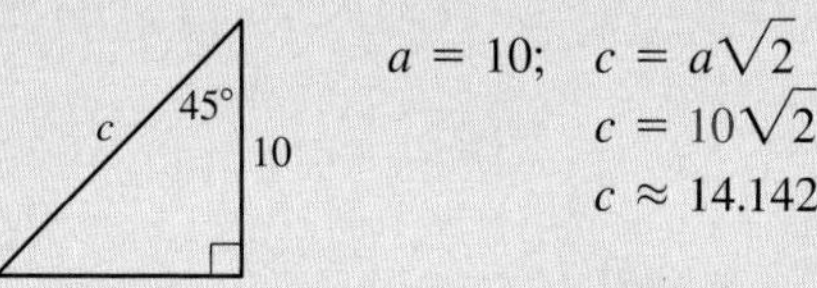

$$a = 10; \quad c = a\sqrt{2}$$
$$c = 10\sqrt{2}$$
$$c \approx 14.142$$

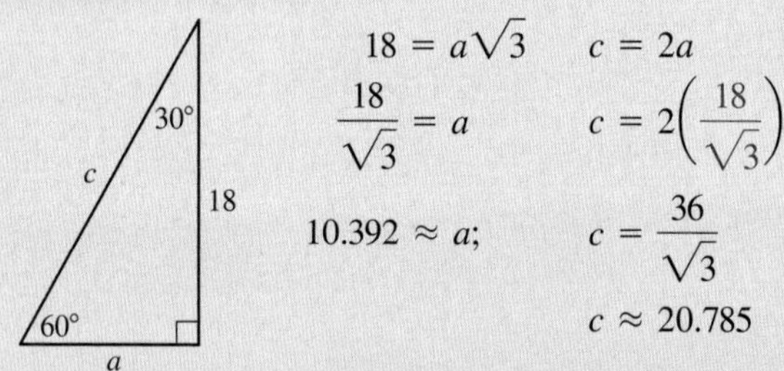

$$18 = a\sqrt{3} \qquad c = 2a$$
$$\frac{18}{\sqrt{3}} = a \qquad c = 2\left(\frac{18}{\sqrt{3}}\right)$$
$$10.392 \approx a; \qquad c = \frac{36}{\sqrt{3}}$$
$$c \approx 20.785$$

Find the missing lengths. Give an exact answer and, where appropriate, an approximation to three decimal places.

16.

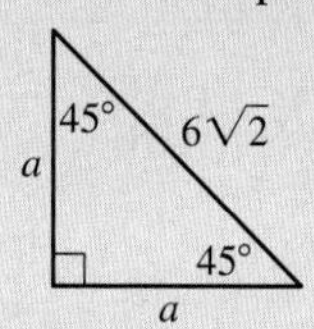

17.

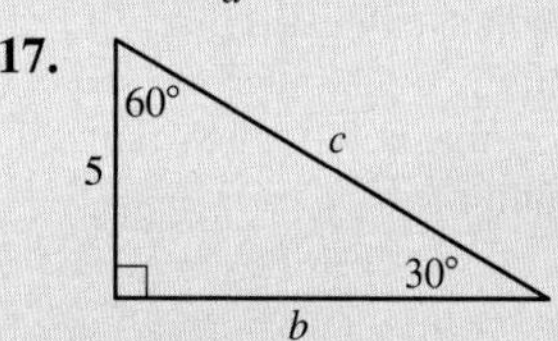

The Distance Formula

The distance d between any two points (x_1, y_1) and (x_2, y_2) is given by $d = \sqrt{(x_2 - x_1)^2 + (y_2 - y_1)^2}$.

Find the distance between $(3, -5)$ and $(-1, -2)$.

$$d = \sqrt{(-1 - 3)^2 + (-2 - (-5))^2}$$
$$= \sqrt{(-4)^2 + (3)^2}$$
$$= \sqrt{16 + 9} = \sqrt{25} = 5$$

18. Find the distance between $(-2, 1)$ and $(6, -10)$. Give an exact answer and an approximation to three decimal places.

The Midpoint Formula

If the endpoints of a segment are (x_1, y_1) and (x_2, y_2), then the coordinates of the midpoint are

$$\left(\frac{x_1 + x_2}{2}, \frac{y_1 + y_2}{2}\right).$$

Find the midpoint of the segment with endpoints $(3, -5)$ and $(-1, -2)$.

$$\left(\frac{3 + (-1)}{2}, \frac{-5 + (-2)}{2}\right), \text{ or } \left(1, -\frac{7}{2}\right)$$

19. Find the midpoint of the segment with endpoints $(-2, 1)$ and $(6, -10)$.

SECTION 10.8: THE COMPLEX NUMBERS

A **complex number** is any number that can be written in the form $a + bi$, where a and b are real numbers, $i = \sqrt{-1}$, and $i^2 = -1$.	$(3 + 2i) + (4 - 7i) = 7 - 5i$; $(8 + 6i) - (5 + 2i) = 3 + 4i$; $(2 + 3i)(4 - i) = 8 - 2i + 12i - 3i^2 = 8 + 10i - 3(-1) = 11 + 10i$; $\frac{1 - 4i}{3 - 2i} = \frac{1 - 4i}{3 - 2i} \cdot \frac{3 + 2i}{3 + 2i} = \frac{3 + 2i - 12i - 8i^2}{9 + 6i - 6i - 4i^2} = \frac{3 - 10i - 8(-1)}{9 - 4(-1)} = \frac{11 - 10i}{13} = \frac{11}{13} - \frac{10}{13}i$	**20.** Add: $(5 - 3i) + (-8 - 9i)$. **21.** Subtract: $(2 - i) - (-1 + i)$. **22.** Multiply: $(1 - 7i)(3 - 5i)$. **23.** Divide: $\frac{1 + i}{1 - i}$.

Review Exercises

Concept Reinforcement *Classify each of the following statements as either true or false.*

1. $\sqrt{ab} = \sqrt{a} \cdot \sqrt{b}$ for any real numbers $\sqrt{a}$ and $\sqrt{b}$. [10.3]
2. $\sqrt{a + b} = \sqrt{a} + \sqrt{b}$ for any real numbers $\sqrt{a}$ and $\sqrt{b}$. [10.5]
3. $\sqrt{a^2} = a$, for any real number a. [10.1]
4. $\sqrt[3]{a^3} = a$, for any real number a. [10.1]
5. $x^{2/5}$ means $\sqrt[5]{x^2}$ and $(\sqrt[5]{x})^2$. [10.2]
6. The hypotenuse of a right triangle is never shorter than either leg. [10.7]
7. Some radical equations have no solution. [10.6]
8. If $f(x) = \sqrt{x - 5}$, then the domain of f is the set of all nonnegative real numbers. [10.1]

Simplify. [10.1]

9. $\sqrt{\frac{49}{9}}$

10. $-\sqrt{0.25}$

Let $f(x) = \sqrt{2x - 7}$. *Find the following.* [10.1]

11. $f(16)$

12. The domain of f

13. *Dairy Farming.* As a calf grows, it needs more milk for nourishment. The number of pounds of milk, M, required by a calf weighing x pounds can be estimated using the formula

$$M = -5 + \sqrt{6.7x - 444}.$$

Estimate the number of pounds of milk required by a calf of the given weight: **(a)** 300 lb; **(b)** 100 lb; **(c)** 200 lb; **(d)** 400 lb. [10.1]

Source: www.ext.vt.edu

Simplify. Assume that each variable can represent any real number.

14. $\sqrt{25t^2}$ [10.1]

15. $\sqrt{(c + 8)^2}$ [10.1]

16. $\sqrt{4x^2 + 4x + 1}$ [10.1]

17. $\sqrt[5]{-32}$ [10.1]

18. Write an equivalent expression using exponential notation: $(\sqrt[3]{5ab})^4$. [10.2]

19. Write an equivalent expression using radical notation: $(16a^6)^{3/4}$. [10.2]

Use rational exponents to simplify. Assume x, y ≥ 0. [10.2]

20. $\sqrt{x^6y^{10}}$

21. $(\sqrt[6]{x^2y})^2$

Simplify. Do not use negative exponents in the answers. [10.2]

22. $(x^{-2/3})^{3/5}$

23. $\frac{7^{-1/3}}{7^{-1/2}}$

24. If $f(x) = \sqrt{25(x - 6)^2}$, find a simplified form for $f(x)$. [10.3]

Simplify. Write all answers using radical notation. Assume that all variables represent nonnegative numbers.

25. $\sqrt[4]{16x^{20}y^8}$ [10.3]

26. $\sqrt{250x^3y^2}$ [10.3]

27. $\sqrt{2x}\,\sqrt{3y}$ [10.3]

28. $\sqrt[3]{3x^4b}\,\sqrt[3]{9xb^2}$ [10.3]

29. $\sqrt[3]{-24x^{10}y^8}\,\sqrt[3]{18x^7y^4}$ [10.3]

30. $\dfrac{\sqrt[3]{60xy^3}}{\sqrt[3]{10x}}$ [10.4]

31. $\dfrac{\sqrt{75x}}{2\sqrt{3}}$ [10.4]

32. $\sqrt[4]{\dfrac{48a^{11}}{c^8}}$ [10.4]

33. $5\sqrt[3]{x} + 2\sqrt[3]{x}$ [10.5]

34. $2\sqrt{75} - 9\sqrt{3}$ [10.5]

35. $\sqrt[3]{8x^4} + \sqrt[3]{xy^6}$ [10.5]

36. $\sqrt{50} + 2\sqrt{18} + \sqrt{32}$ [10.5]

37. $(3 + \sqrt{10})(3 - \sqrt{10})$ [10.5]

38. $(\sqrt{3} - 3\sqrt{8})(\sqrt{5} + 2\sqrt{8})$ [10.5]

39. $\sqrt[4]{x}\,\sqrt{x}$ [10.5]

40. $\dfrac{\sqrt[3]{x^2}}{\sqrt[4]{x}}$ [10.5]

41. If $f(x) = x^2$, find $f(a - \sqrt{2})$. [10.5]

42. Rationalize the denominator:

$$\sqrt{\frac{x}{8y}}.\ [10.4]$$

43. Rationalize the denominator:

$$\frac{4\sqrt{5}}{\sqrt{2} + \sqrt{3}}.\ [10.5]$$

44. Rationalize the numerator of the expression in Exercise 43. [10.5]

Solve. [10.6]

45. $\sqrt{y + 6} - 2 = 3$

46. $\sqrt{x} = x - 6$

47. $(x + 1)^{1/3} = -5$

48. $1 + \sqrt{x} = \sqrt{3x - 3}$

49. If $f(x) = \sqrt[4]{x + 2}$, find a such that $f(a) = 2$. [10.6]

Solve. Give an exact answer and, where appropriate, an approximation to three decimal places. [10.7]

50. The diagonal of a square has length 10 cm. Find the length of a side of the square.

51. A skate-park jump has a ramp that is 6 ft long and is 2 ft high. How long is its base?

52. Find the missing lengths. Give exact answers and, where appropriate, an approximation to three decimal places.

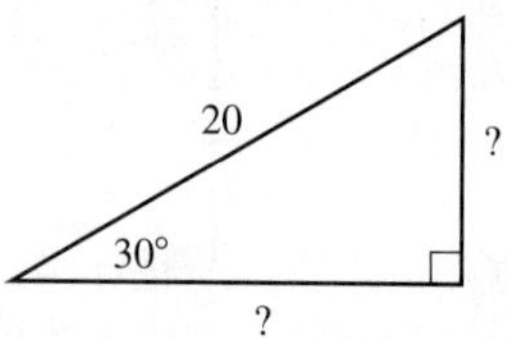

53. Find the distance between $(-6, 4)$ and $(-1, 5)$. Give an exact answer and an approximation to three decimal places. [10.7]

54. Find the midpoint of the segment with endpoints $(-7, -2)$ and $(3, -1)$. [10.7]

55. Express in terms of i and simplify: $\sqrt{-45}$. [10.8]

56. Add: $(-4 + 3i) + (2 - 12i)$. [10.8]

57. Subtract: $(9 - 7i) - (3 - 8i)$. [10.8]

Simplify. [10.8]

58. $(2 + 5i)(2 - 5i)$

59. i^{18}

60. $(6 - 3i)(2 - i)$

61. Divide and simplify to the form $a + bi$:

$$\frac{7 - 2i}{3 + 4i}.\ [10.8]$$

SYNTHESIS

TW **62.** What makes some complex numbers real and others imaginary? [10.8]

TW **63.** Explain why $\sqrt[n]{x^n} = |x|$ when n is even, but $\sqrt[n]{x^n} = x$ when n is odd. [10.1]

64. Solve:

$$\sqrt{11x + \sqrt{6 + x}} = 6.\ [10.6]$$

65. Simplify:

$$\frac{2}{1 - 3i} - \frac{3}{4 + 2i}.\ [10.8]$$

66. Write a quotient of two imaginary numbers that is a real number. (Answers may vary.) [10.8]

67. Don's Discount Shoes has two locations. The sign at the original location is shaped like an isosceles right triangle. The sign at the newer location is shaped like a 30°–60°–90° triangle. The hypotenuse of each sign measures 6 ft. Which sign has the greater area and by how much? (Round to three decimal places.) [10.7]

Chapter Test 10

Simplify. Assume that variables can represent any real number.

1. $\sqrt{50}$

2. $\sqrt[3]{-\frac{8}{x^6}}$

3. $\sqrt{81a^2}$

4. $\sqrt{x^2 - 8x + 16}$

5. Write an equivalent expression using exponential notation: $\sqrt{7xy}$.

6. Write an equivalent expression using radical notation: $(4a^3b)^{5/6}$.

7. If $f(x) = \sqrt{2x - 10}$, determine the domain of f.

8. If $f(x) = x^2$, find $f(5 + \sqrt{2})$.

Simplify. Write all answers using radical notation. Assume that all variables represent positive numbers.

9. $\sqrt[5]{32x^{16}y^{10}}$

10. $\sqrt[3]{4w}\,\sqrt[3]{4v^2}$

11. $\sqrt{\frac{100a^4}{9b^6}}$

12. $\frac{\sqrt[5]{48x^6y^{10}}}{\sqrt[5]{16x^2y^9}}$

13. $\sqrt[4]{x^3}\,\sqrt{x}$

14. $\frac{\sqrt{y}}{\sqrt[10]{y}}$

15. $8\sqrt{2} - 2\sqrt{2}$

16. $\sqrt{x^4y} + \sqrt{9y^3}$

17. $(7 + \sqrt{x})(2 - 3\sqrt{x})$

18. Rationalize the denominator:

$$\frac{\sqrt[3]{x}}{\sqrt[3]{4y}}.$$

Solve.

19. $6 = \sqrt{x - 3} + 5$

20. $x = \sqrt{3x + 3} - 1$

21. $\sqrt{2x} = \sqrt{x + 1} + 1$

Solve. For Exercises 22–24, give exact answers and approximations to three decimal places.

22. A referee jogs diagonally from one corner of a 50-ft by 90-ft basketball court to the far corner. How far does she jog?

23. The hypotenuse of a 30°–60°–90° triangle is 10 cm long. Find the lengths of the legs.

24. Find the distance between the points $(3, 7)$ and $(-1, 8)$.

25. Find the midpoint of the segment with endpoints $(2, -5)$ and $(1, -7)$.

26. Express in terms of i and simplify: $\sqrt{-50}$.

27. Subtract: $(9 + 8i) - (-3 + 6i)$.

28. Multiply: $\sqrt{-16}\,\sqrt{-36}$.

29. Multiply. Write the answer in the form $a + bi$.

$$(4 - i)^2$$

30. Divide and simplify to the form $a + bi$:

$$\frac{-2 + i}{3 - 5i}.$$

31. Simplify: i^{37}.

SYNTHESIS

32. Solve:

$$\sqrt{2x - 2} + \sqrt{7x + 4} = \sqrt{13x + 10}.$$

33. Simplify:

$$\frac{1 - 4i}{4i(1 + 4i)^{-1}}.$$

34. The function $D(h) = 1.2\sqrt{h}$ can be used to approximate the distance D, in miles, that a person can see to the horizon from a height h, in feet. How far above sea level must a pilot fly in order to see a horizon that is 180 mi away?

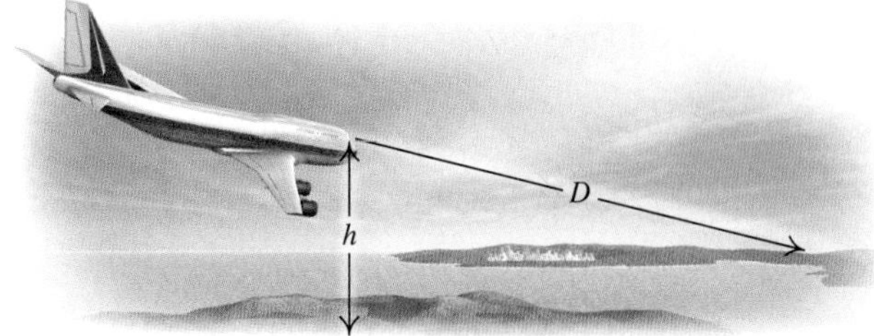

Quadratic Functions and Equations

How Great Is America's "Sweet Tooth"?

As the graph indicates, Americans ate more candy in 2006 than in 2004, but less in 2008 than in 2006. In Example 4 of Section 11.8, we estimate how much candy Americans will have consumed in 2009 if the trend continues.

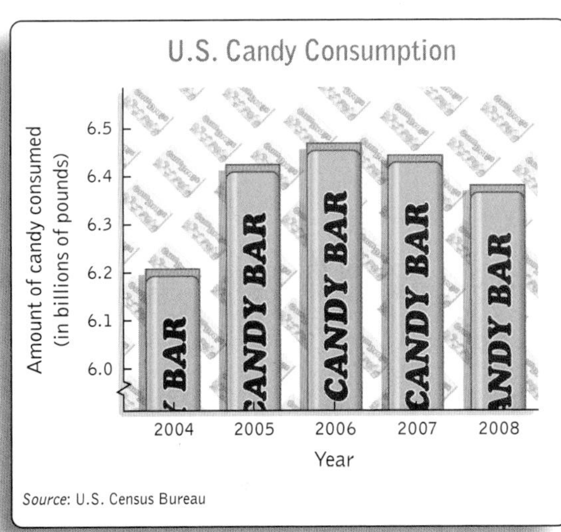

The mathematical translation of a problem is often a function or an equation containing a second-degree polynomial in one variable. Such functions or equations are said to be *quadratic*. In this chapter, we examine a variety of ways to solve quadratic equations and look at graphs and applications of quadratic functions.

11.1 Quadratic Equations

- The Principle of Square Roots
- Completing the Square
- Problem Solving

The general form of a quadratic function is

$$f(x) = ax^2 + bx + c, \quad \text{with } a \neq 0.$$

The graph of a quadratic function is a *parabola*. Such graphs open up or down and can have 0, 1, or 2 x-intercepts. We learn to graph quadratic functions later in this chapter.

The graph of the quadratic function

$$f(x) = x^2 + 6x + 8$$

is shown at right. Note that $(-4, 0)$ and $(-2, 0)$ are the x-intercepts of the graph of $f(x) = x^2 + 6x + 8$.

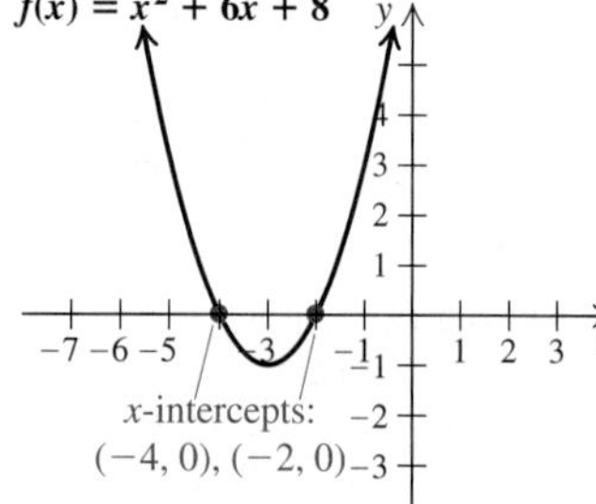

In Chapter 6, we solved equations like $x^2 + 6x + 8 = 0$ by factoring:

$$x^2 + 6x + 8 = 0$$

$$(x + 4)(x + 2) = 0 \quad \textbf{Factoring}$$

$$x + 4 = 0 \quad \text{or} \quad x + 2 = 0 \quad \textbf{Using the principle of zero products}$$

$$x = -4 \quad \text{or} \quad x = -2.$$

Note that -4 and -2 are the first coordinates of the x-intercepts of the graph of $f(x)$ above.

Interactive Discovery

Graph each function and determine the number of x-intercepts.

1. $f(x) = x^2$
2. $g(x) = -x^2$
3. $h(x) = (x - 2)^2$
4. $p(x) = 2x^2 + 1$
5. $f(x) = -1.5x^2 + x - 3$
6. $g(x) = 4x^2 - 2x - 7$
7. Describe the shape of the graph of a quadratic function.

The graph of a quadratic function can have no, one, or two x-intercepts, as illustrated below. Thus a quadratic equation can have no, one, or two real-number roots. Quadratic equations can have nonreal-number roots. These must be found using algebraic methods.

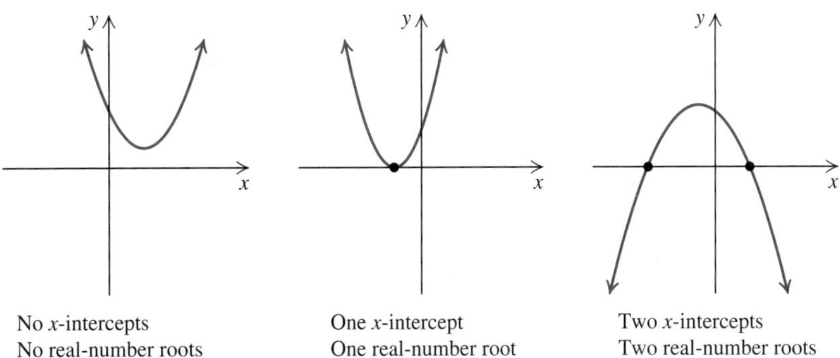

To solve a quadratic equation by factoring, we write the equation in the *standard form* $ax^2 + bx + c = 0$, factor, and use the principle of zero products.

EXAMPLE 1 Solve: $x^2 = 25$.

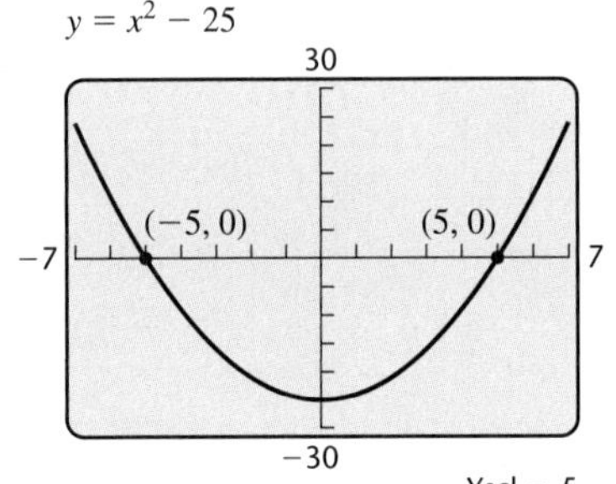

SOLUTION We have

$$x^2 = 25$$

$$x^2 - 25 = 0 \qquad \text{Writing in standard form}$$

$$(x - 5)(x + 5) = 0 \qquad \text{Factoring}$$

$$x - 5 = 0 \quad or \quad x + 5 = 0 \qquad \text{Using the principle of zero products}$$

$$x = 5 \quad or \quad x = -5.$$

The solutions are 5 and -5. The graph at left confirms the solutions.

Try Exercise 13.

In this section and the next, we develop algebraic methods for solving any quadratic equation, whether it is factorable or not.

THE PRINCIPLE OF SQUARE ROOTS

Consider the equation $x^2 = 25$ again. We know from Chapter 10 that the number 25 has two real-number square roots, 5 and -5, the solutions of the equation in Example 1. Thus we see that square roots provide quick solutions for equations of the type $x^2 = k$.

The Principle of Square Roots For any real number k, if $x^2 = k$, then

$$x = \sqrt{k} \quad \text{or} \quad x = -\sqrt{k}.$$

EXAMPLE 2 Solve: $3x^2 = 6$. Give exact solutions and approximations to three decimal places.

SOLUTION We have

$$3x^2 = 6$$
$$x^2 = 2 \qquad \textbf{Isolating } x^2$$
$$x = \sqrt{2} \quad or \quad x = -\sqrt{2}. \qquad \textbf{Using the principle of square roots}$$

We often use the symbol $\pm\sqrt{2}$ to represent the two numbers $\sqrt{2}$ and $-\sqrt{2}$.

CAUTION! The symbol $\pm\sqrt{2}$ represents *two* solutions: $\sqrt{2}$ and $-\sqrt{2}$.

Check:

For $\sqrt{2}$:

$3x^2$	$= 6$
$3(\sqrt{2})^2$	6
$3 \cdot 2$	
$6 \stackrel{?}{=} 6$	TRUE

For $-\sqrt{2}$:

$3x^2$	$= 6$
$3(-\sqrt{2})^2$	6
$3 \cdot 2$	
$6 \stackrel{?}{=} 6$	TRUE

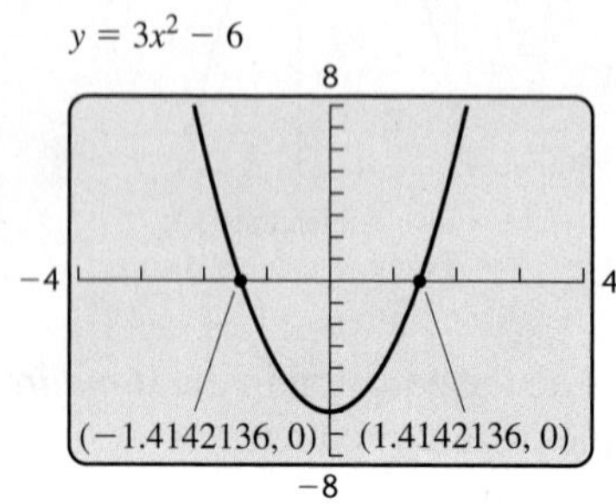

The solutions are $\sqrt{2}$ and $-\sqrt{2}$, or $\pm\sqrt{2}$, which round to 1.414 and -1.414. A graphical solution, shown at left, yields approximate solutions, since $\sqrt{2}$ is irrational. The solutions found graphically correspond to those found algebraically.

Try Exercise 17.

EXAMPLE 3 Solve: $-5x^2 + 2 = 0$.

SOLUTION We have

$$-5x^2 + 2 = 0$$
$$x^2 = \frac{2}{5} \qquad \textbf{Isolating } x^2$$
$$x = \sqrt{\frac{2}{5}} \quad or \quad x = -\sqrt{\frac{2}{5}}. \qquad \textbf{Using the principle of square roots}$$

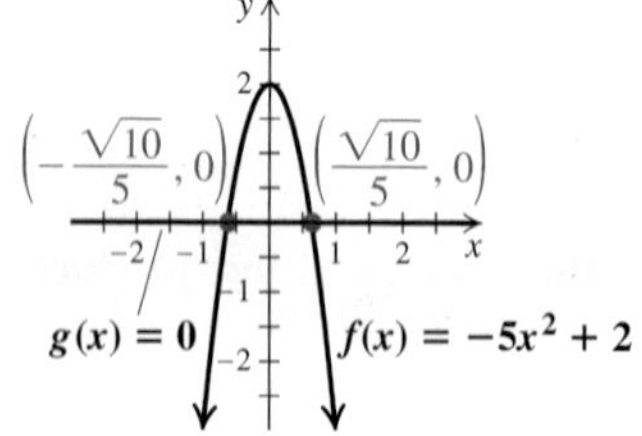

A visualization of Example 3

The solutions are $\sqrt{\frac{2}{5}}$ and $-\sqrt{\frac{2}{5}}$. This can also be written as $\pm\sqrt{\frac{2}{5}}$. If we rationalize the denominator, the solutions are written $\pm\frac{\sqrt{10}}{5}$. The checks are left to the student.

Try Exercise 23.

Sometimes we get solutions that are imaginary numbers.

EXAMPLE 4 Solve: $4x^2 + 9 = 0$.

SOLUTION We have

$$4x^2 + 9 = 0$$
$$x^2 = -\tfrac{9}{4} \qquad \textbf{Isolating } x^2$$
$$x = \sqrt{-\tfrac{9}{4}} \quad or \quad x = -\sqrt{-\tfrac{9}{4}} \qquad \textbf{Using the principle of square roots}$$
$$x = \sqrt{\tfrac{9}{4}}\sqrt{-1} \quad or \quad x = -\sqrt{\tfrac{9}{4}}\sqrt{-1}$$
$$x = \tfrac{3}{2}i \quad or \quad x = -\tfrac{3}{2}i. \qquad \textbf{Recall that } \sqrt{-1} = i.$$

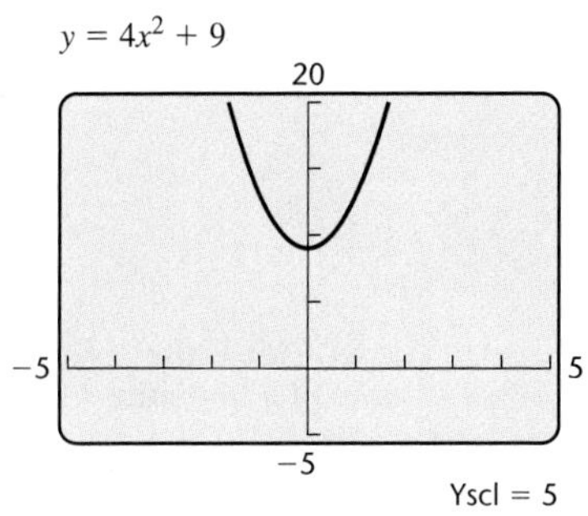

Check: For $\frac{3}{2}i$:

$$\begin{array}{c|c} 4x^2 + 9 & = 0 \\ \hline 4\left(\frac{3}{2}i\right)^2 + 9 & 0 \\ 4 \cdot \frac{9}{4} \cdot i^2 + 9 & \\ 9(-1) + 9 & \\ 0 & \stackrel{?}{=} 0 \quad \text{TRUE} \end{array}$$

For $-\frac{3}{2}i$:

$$\begin{array}{c|c} 4x^2 + 9 & = 0 \\ \hline 4\left(-\frac{3}{2}i\right)^2 + 9 & 0 \\ 4 \cdot \frac{9}{4} \cdot i^2 + 9 & \\ 9(-1) + 9 & \\ 0 & \stackrel{?}{=} 0 \quad \text{TRUE} \end{array}$$

The solutions are $\frac{3}{2}i$ and $-\frac{3}{2}i$, or $\pm\frac{3}{2}i$. The graph at left confirms that there are no real-number solutions, because there are no x-intercepts.

Try Exercise 25.

The principle of square roots can be restated in a more general form.

The Principle of Square Roots (Generalized Form)

For any real number k and any algebraic expression X,

If $X^2 = k$, then $X = \sqrt{k}$ or $X = -\sqrt{k}$.

EXAMPLE 5 Let $f(x) = (x - 2)^2$. Find all x-values for which $f(x) = 7$.

SOLUTION We are asked to find all x-values for which

$$f(x) = 7,$$

or

$$(x - 2)^2 = 7. \quad \textbf{Substituting } (x - 2)^2 \textbf{ for } f(x)$$

We solve both algebraically and graphically.

ALGEBRAIC APPROACH

The generalized principle of square roots gives us

$$x - 2 = \sqrt{7} \quad \text{or} \quad x - 2 = -\sqrt{7}$$
$$x = 2 + \sqrt{7} \quad \text{or} \quad x = 2 - \sqrt{7}.$$

Check: $f(2 + \sqrt{7}) = (2 + \sqrt{7} - 2)^2 = (\sqrt{7})^2 = 7.$

Similarly,

$$f(2 - \sqrt{7}) = (2 - \sqrt{7} - 2)^2 = (-\sqrt{7})^2 = 7.$$

Rounded to the nearest thousandth,

$$2 + \sqrt{7} \approx 4.646 \quad \text{and} \quad 2 - \sqrt{7} \approx -0.646,$$

so the graphical solutions match the algebraic solutions.

The solutions are $2 + \sqrt{7}$ and $2 - \sqrt{7}$, or simply $2 \pm \sqrt{7}$.

GRAPHICAL APPROACH

We graph the equations $y_1 = (x - 2)^2$ and $y_2 = 7$. If there are any real-number x-values for which $f(x) = 7$, they will be the x-coordinates of the points of intersection of the graphs.

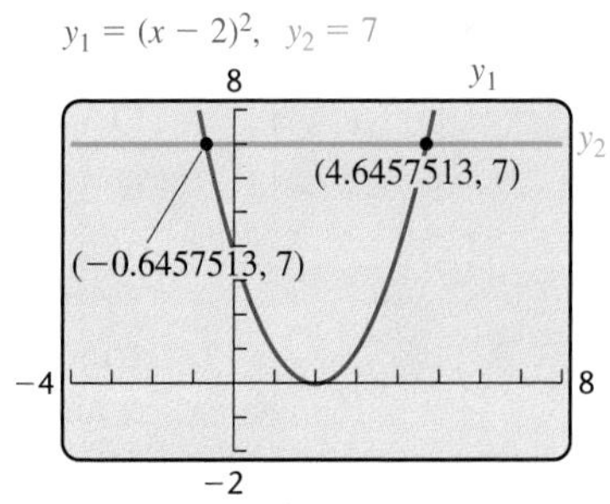

Rounded to the nearest thousandth, the solutions are approximately −0.646 and 4.646.

We can check the algebraic solutions using the VALUE option of the CALC menu. If we enter $2 + \sqrt{7}$ or $2 - \sqrt{7}$ for x, the corresponding value of y_1 will be 7, so the solutions check.

Try Exercise 41.

Example 5 is of the form $(x - a)^2 = c$, where a and c are constants. Sometimes we must factor to obtain this form.

EXAMPLE 6 Solve: $x^2 + 6x + 9 = 2$.

SOLUTION We have

$$x^2 + 6x + 9 = 2 \quad \text{The left side is the square of a binomial.}$$

$$(x + 3)^2 = 2 \quad \text{Factoring}$$

$$x + 3 = \sqrt{2} \quad or \quad x + 3 = -\sqrt{2} \quad \text{Using the principle of square roots}$$

$$x = -3 + \sqrt{2} \quad or \quad x = -3 - \sqrt{2}. \quad \text{Adding } -3 \text{ to both sides}$$

Y1(−3 + √(2))
2
Y1(−3 − √(2))
2

One way to check the solutions on a graphing calculator is to enter $y_1 = x^2 + 6x + 9$ and then evaluate $y_1(-3 + \sqrt{2})$ and $y_1(-3 - \sqrt{2})$, as shown at left. The second calculation can be entered quickly by copying the first entry and editing it.

The solutions are $-3 + \sqrt{2}$ and $-3 - \sqrt{2}$, or $-3 \pm \sqrt{2}$.

Try Exercise 35.

COMPLETING THE SQUARE

By using a method called *completing the square*, we can use the principle of square roots to solve *any* quadratic equation.

EXAMPLE 7 Solve: $x^2 + 6x + 4 = 0$.

SOLUTION We have

$$x^2 + 6x + 4 = 0$$

$$x^2 + 6x = -4 \quad \text{Subtracting 4 from both sides}$$

$$x^2 + 6x + 9 = -4 + 9 \quad \text{Adding 9 to both sides. We explain this shortly.}$$

$$(x + 3)^2 = 5 \quad \text{Factoring the perfect-square trinomial}$$

$$x + 3 = \pm\sqrt{5} \quad \text{Using the principle of square roots. Remember that } \pm\sqrt{5} \text{ represents two numbers.}$$

$$x = -3 \pm \sqrt{5}. \quad \text{Adding } -3 \text{ to both sides}$$

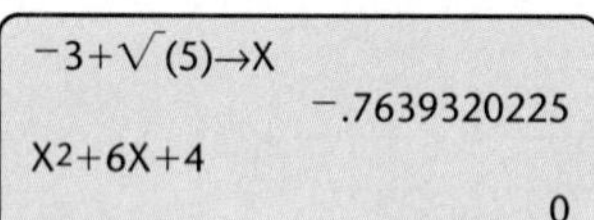

We check by storing $-3 + \sqrt{5}$ as x and evaluating $x^2 + 6x + 4$, as shown at left. We can repeat the process for $-3 - \sqrt{5}$ to check the second solution. The solutions are $-3 + \sqrt{5}$ and $-3 - \sqrt{5}$.

In Example 7, we chose to add 9 to both sides because it creates a perfect-square trinomial on the left side. The 9 was determined by taking half of the coefficient of x and squaring it.

To understand why this procedure works, examine the following drawings.

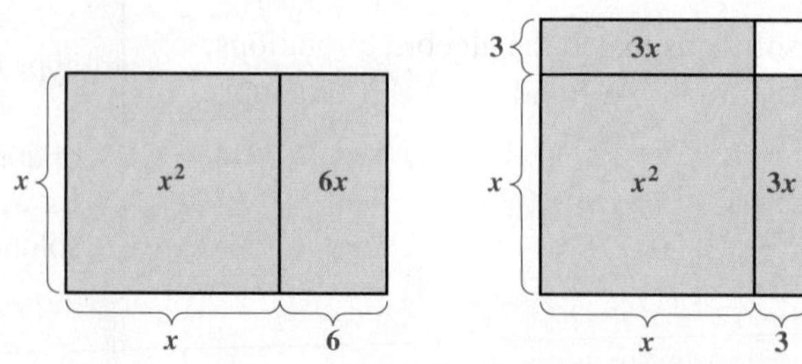

Note that the shaded areas in both figures represent the same area, $x^2 + 6x$. However, only the figure on the right, in which the $6x$ is halved, can be converted into a square with the addition of a constant term. The constant 9 is the "missing" piece that *completes* the square.

To complete the square for $x^2 + bx$, we add $\left(\frac{b}{2}\right)^2$.

Example 8, which follows, provides practice in finding numbers that complete the square. We will then use this skill to solve equations.

EXAMPLE 8 Replace the blanks in each equation with constants to form a true equation.

a) $x^2 + 14x + \_\_\_ = (x + \_\_\_)^2$

b) $x^2 - 5x + \_\_\_ = (x - \_\_\_)^2$

c) $x^2 + \frac{3}{4}x + \_\_\_ = (x + \_\_\_)^2$

SOLUTION

a) Take half of the coefficient of x: Half of 14 is 7.

Square this number: $7^2 = 49$.

Add 49 to complete the square: $x^2 + 14x + 49 = (x + 7)^2$.

b) Take half of the coefficient of x: Half of -5 is $-\frac{5}{2}$.

Square this number: $\left(-\frac{5}{2}\right)^2 = \frac{25}{4}$.

Add $\frac{25}{4}$ to complete the square: $x^2 - 5x + \frac{25}{4} = \left(x - \frac{5}{2}\right)^2$.

c) Take half of the coefficient of x: Half of $\frac{3}{4}$ is $\frac{3}{8}$.

Square this number: $\left(\frac{3}{8}\right)^2 = \frac{9}{64}$.

Add $\frac{9}{64}$ to complete the square: $x^2 + \frac{3}{4}x + \frac{9}{64} = \left(x + \frac{3}{8}\right)^2$.

Student Notes

In problems like Examples 8(b) and (c), it is best to avoid decimal notation. Most students have an easier time recognizing $\frac{9}{64}$ as $\left(\frac{3}{8}\right)^2$ than seeing 0.140625 as 0.375^2.

Try Exercise 45.

We can now use the method of completing the square to solve equations.

EXAMPLE 9 Solve: $x^2 - 8x - 7 = 0$.

SOLUTION We begin by adding 7 to both sides:

$$x^2 - 8x - 7 = 0$$

$$x^2 - 8x = 7$$ **Adding 7 to both sides. We can now complete the square on the left side.**

$$x^2 - 8x + 16 = 7 + 16$$ **Adding 16 to both sides to complete the square: $\frac{1}{2}(-8) = -4$, and $(-4)^2 = 16$**

$$(x - 4)^2 = 23$$ **Factoring and simplifying**

$$x - 4 = \pm\sqrt{23}$$ **Using the principle of square roots**

$$x = 4 \pm \sqrt{23}.$$ **Adding 4 to both sides**

Check: For $4 + \sqrt{23}$:

$$\begin{array}{r|l} x^2 - 8x - 7 = 0 & \\ \hline (4 + \sqrt{23})^2 - 8(4 + \sqrt{23}) - 7 & 0 \\ 16 + 8\sqrt{23} + 23 - 32 - 8\sqrt{23} - 7 & \\ 16 + 23 - 32 - 7 + 8\sqrt{23} - 8\sqrt{23} & \\ 0 \stackrel{?}{=} 0 & \text{TRUE} \end{array}$$

For $4 - \sqrt{23}$:

$$\begin{array}{r|l} x^2 - 8x - 7 = 0 & \\ \hline (4 - \sqrt{23})^2 - 8(4 - \sqrt{23}) - 7 & 0 \\ 16 - 8\sqrt{23} + 23 - 32 + 8\sqrt{23} - 7 & \\ 16 + 23 - 32 - 7 - 8\sqrt{23} + 8\sqrt{23} & \\ 0 \stackrel{?}{=} 0 & \text{TRUE} \end{array}$$

The solutions are $4 + \sqrt{23}$ and $4 - \sqrt{23}$, or $4 \pm \sqrt{23}$.

Try Exercise 57.

Recall that the value of $f(x)$ must be 0 at any x-intercept of the graph of f. If $f(a) = 0$, then $(a, 0)$ is an x-intercept of the graph.

EXAMPLE 10 Find the x-intercepts of the graph of $f(x) = x^2 + 5x - 3$.

SOLUTION We set $f(x)$ equal to 0 and solve:

$$f(x) = 0$$

$$x^2 + 5x - 3 = 0 \quad \text{Substituting}$$

$$x^2 + 5x = 3 \quad \text{Adding 3 to both sides}$$

$$x^2 + 5x + \frac{25}{4} = 3 + \frac{25}{4} \quad \text{Completing the square: } \tfrac{1}{2} \cdot 5 = \tfrac{5}{2}, \text{ and } \left(\tfrac{5}{2}\right)^2 = \tfrac{25}{4}$$

$$\left(x + \frac{5}{2}\right)^2 = \frac{37}{4} \quad \text{Factoring and simplifying}$$

$$x + \frac{5}{2} = \pm\frac{\sqrt{37}}{2} \quad \text{Using the principle of square roots and the quotient rule for radicals}$$

$$x = -\frac{5}{2} \pm \frac{\sqrt{37}}{2}, \text{ or } \frac{-5 \pm \sqrt{37}}{2}. \quad \text{Adding } -\tfrac{5}{2} \text{ to both sides}$$

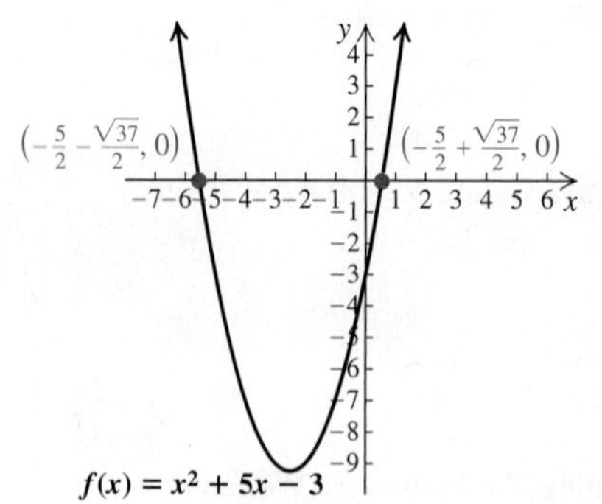

A visualization of Example 10

The x-intercepts are

$$\left(-\frac{5}{2} - \frac{\sqrt{37}}{2}, 0\right) \text{ and } \left(-\frac{5}{2} + \frac{\sqrt{37}}{2}, 0\right), \text{ or}$$

$$\left(\frac{-5 - \sqrt{37}}{2}, 0\right) \text{ and } \left(\frac{-5 + \sqrt{37}}{2}, 0\right).$$

The checks are left to the student.

Try Exercise 65.

Before we complete the square in a quadratic equation, the leading coefficient must be 1. When it is not 1, we divide both sides of the equation by whatever that coefficient may be.

To Solve a Quadratic Equation in x by Completing the Square

1. Isolate the terms with variables on one side of the equation, and arrange them in descending order.
2. Divide both sides by the coefficient of x^2 if that coefficient is not 1.
3. Complete the square by taking half of the coefficient of x and adding its square to both sides.
4. Express the trinomial as the square of a binomial (factor the trinomial) and simplify the other side of the equation.
5. Use the principle of square roots (find the square roots of both sides).
6. Solve for x by adding or subtracting on both sides.

EXAMPLE 11 Solve: $3x^2 + 7x - 2 = 0$.

SOLUTION We follow the steps listed above:

$$3x^2 + 7x - 2 = 0$$

Isolate the variable terms.

$$3x^2 + 7x = 2 \qquad \text{Adding 2 to both sides}$$

Divide both sides by the x^2-coefficient.

$$x^2 + \frac{7}{3}x = \frac{2}{3} \qquad \text{Dividing both sides by 3}$$

Complete the square.

$$x^2 + \frac{7}{3}x + \frac{49}{36} = \frac{2}{3} + \frac{49}{36} \qquad \text{Completing the square: } \left(\frac{1}{2}\cdot\frac{7}{3}\right)^2 = \frac{49}{36}$$

Factor the trinomial.

$$\left(x + \frac{7}{6}\right)^2 = \frac{73}{36} \qquad \text{Factoring and simplifying}$$

Use the principle of square roots.

$$x + \frac{7}{6} = \pm\frac{\sqrt{73}}{6} \qquad \text{Using the principle of square roots and the quotient rule for radicals}$$

Solve for x.

$$x = -\frac{7}{6} \pm \frac{\sqrt{73}}{6}, \text{ or } \frac{-7 \pm \sqrt{73}}{6}. \qquad \text{Adding } -\frac{7}{6} \text{ to both sides}$$

The checks are left to the student. The solutions are $-\frac{7}{6} \pm \frac{\sqrt{73}}{6}$, or $\frac{-7 \pm \sqrt{73}}{6}$. This can be written as

$$-\frac{7}{6} + \frac{\sqrt{73}}{6} \text{ and } -\frac{7}{6} - \frac{\sqrt{73}}{6}, \text{ or } \frac{-7 + \sqrt{73}}{6} \text{ and } \frac{-7 - \sqrt{73}}{6}.$$

Try Exercise 71.

Any quadratic equation can be solved by completing the square. The procedure is also useful when graphing quadratic equations and will be used in the next section to develop a formula for solving quadratic equations.

PROBLEM SOLVING

After one year, an amount of money P, invested at 4% per year, is worth 104% of P, or $P(1.04)$. If that amount continues to earn 4% interest per year, after the second year the investment will be worth 104% of $P(1.04)$, or $P(1.04)^2$. This is called **compounding interest** since after the first time period, interest is earned on both the initial investment *and* the interest from the first time period. Continuing the above pattern, we see that after the third year, the investment will be worth 104% of $P(1.04)^2$. Generalizing, we have the following.

STUDY TIP 

Choosing Tunes

Some students prefer working while listening to music. Whether listening for pleasure or to block out nearby noise, you may find that music without lyrics is most conducive to focusing your concentration on your studies. At least one study has shown that classical background music can improve one's ability to concentrate.

The Compound-Interest Formula If an amount of money P is invested at interest rate r, compounded annually, then in t years, it will grow to the amount A given by

$$A = P(1 + r)^t. \qquad (r \text{ is written in decimal notation.})$$

We can use quadratic equations to solve certain interest problems.

EXAMPLE 12 Investment Growth. Clare invested \$4000 at interest rate r, compounded annually. In 2 years, it had grown to \$4410. What was the interest rate?

SOLUTION

1. **Familiarize.** We are already familiar with the compound-interest formula. If we were not, we would need to consult an outside source.
2. **Translate.** The translation consists of substituting into the formula:

$$A = P(1 + r)^t$$
$$4410 = 4000(1 + r)^2. \qquad \textbf{Substituting}$$

3. **Carry out.** We solve for r both algebraically and graphically.

ALGEBRAIC APPROACH

We have

$$4410 = 4000(1 + r)^2$$
$$\frac{4410}{4000} = (1 + r)^2 \qquad \text{Dividing both sides by 4000}$$
$$\frac{441}{400} = (1 + r)^2 \qquad \text{Simplifying}$$
$$\pm\sqrt{\frac{441}{400}} = 1 + r \qquad \text{Using the principle of square roots}$$
$$\pm\frac{21}{20} = 1 + r \qquad \text{Simplifying}$$
$$-\frac{20}{20} \pm \frac{21}{20} = r \qquad \text{Adding } -1, \text{ or } -\tfrac{20}{20}, \text{ to both sides}$$
$$\frac{1}{20} = r \quad or \quad -\frac{41}{20} = r.$$

Since r represents an interest rate, we convert to decimal notation. We now have

$$r = 0.05 \quad or \quad r = -2.05.$$

GRAPHICAL APPROACH

We let $y_1 = 4410$ and $y_2 = 4000(1 + x)^2$. Using the window $[-3, 1, 0, 5000]$, we see that the graphs intersect at two points. The first coordinates of the points of intersection are -2.05 and 0.05.

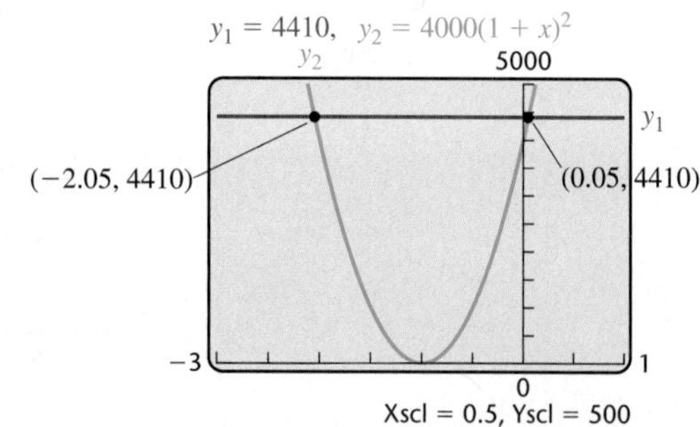

4. **Check.** Since the interest rate cannot be negative, we need only check 0.05, or 5%. If \$4000 were invested at 5% interest, compounded annually, then in 2 years it would have grown to $4000(1.05)^2$, or \$4410. The rate 5% checks.
5. **State.** The interest rate was 5%.

Try Exercise 81.

EXAMPLE 13 Free-Falling Objects. The formula $s = 16t^2$ is used to approximate the distance s, in feet, that an object falls freely from rest in t seconds. The highest point of the Mike O'Callaghan–Pat Tillman Memorial Bridge is 890 ft above the Colorado River. How long will it take a stone to fall from the bridge to the river? Round to the nearest tenth of a second.
Source: www.desertusa.com

SOLUTION

1. **Familiarize.** We agree to disregard air resistance and use the given formula.
2. **Translate.** We substitute into the formula:

$$s = 16t^2$$
$$890 = 16t^2.$$

3. **Carry out.** We solve for t:

$$890 = 16t^2$$
$$55.625 = t^2$$
$$\sqrt{55.625} = t \quad \text{Using the principle of square roots; rejecting the negative square root since } t \text{ cannot be negative in this problem}$$
$$7.5 \approx t. \quad \text{Using a calculator and rounding to the nearest tenth}$$

4. **Check.** Since $16(7.5)^2 = 900 \approx 890$, our answer checks.
5. **State.** It takes about 7.5 sec for a stone to fall freely from the bridge to the river.

Try Exercise 85.

11.1 Exercise Set

FOR EXTRA HELP MyMathLab

Concept Reinforcement *Complete each of the following to form a true statement.*

1. The principle of square roots states that if $x^2 = k$, then $x =$ ___ or $x =$ ___.

2. If $(x + 5)^2 = 49$, then $x + 5 =$ ___ or $x + 5 =$ ___.

3. If $t^2 + 6t + 9 = 17$, then $(\_\_)^2 = 17$ and ___ $= \pm\sqrt{17}$.

4. The equations $x^2 + 8x +$ ___ $= 23$ and $x^2 + 8x = 7$ are equivalent.

5. The expressions $t^2 + 10t +$ ___ and $(t +$ ___$)^2$ are equivalent.

6. The expressions $x^2 - 6x +$ ___ and $(x -$ ___$)^2$ are equivalent.

Determine the number of real-number solutions of each equation from the given graph.

7. $x^2 + x - 12 = 0$

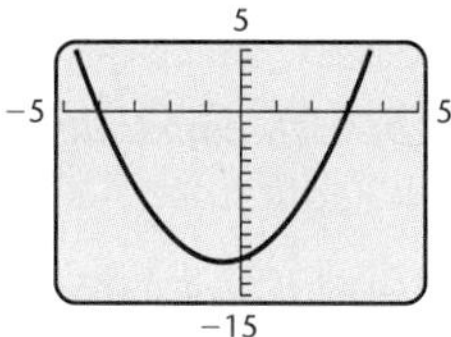

8. $-3x^2 - x - 7 = 0$

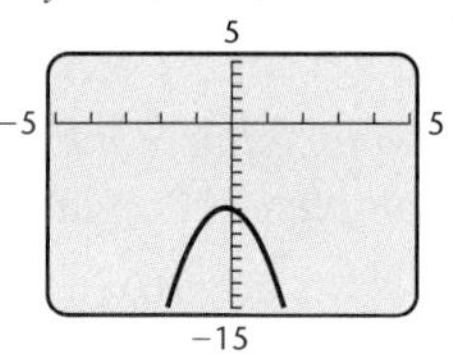

9. $4x^2 + 9 = 12x$

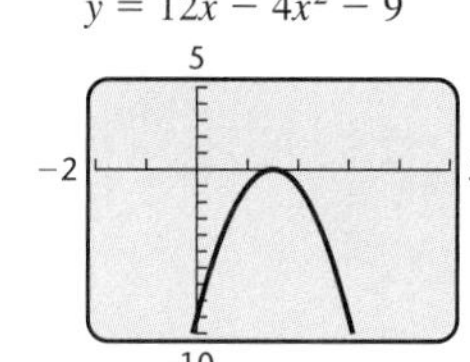

10. $2x^2 + 3 = 6x$

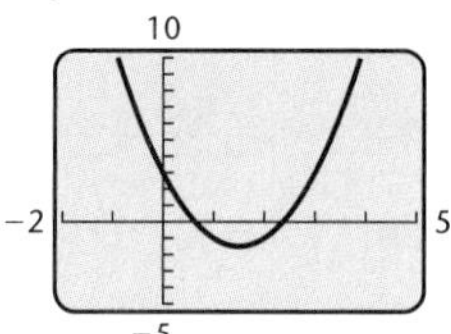

11. $f(x) = 0$

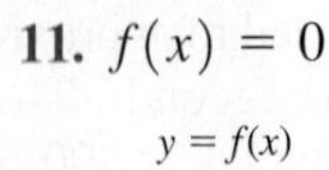

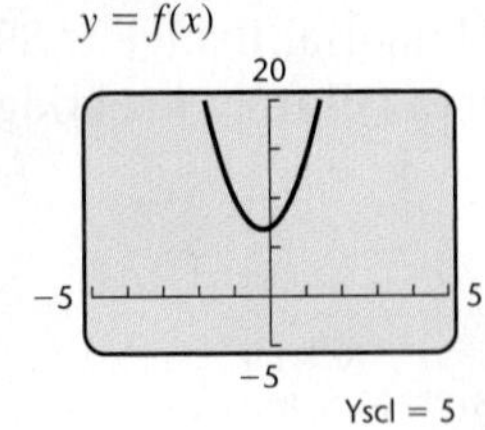

12. $f(x) = 0$

Solve.

13. $x^2 = 100$

14. $t^2 = 144$

15. $p^2 - 50 = 0$

16. $c^2 - 8 = 0$

17. $4x^2 = 20$

18. $7x^2 = 21$

19. $x^2 = -4$

20. $x^2 = -9$

21. $9x^2 - 16 = 0$

22. $25x^2 - 4 = 0$

23. $5t^2 - 3 = 4$

24. $3t^2 - 1 = 6$

25. $4d^2 + 81 = 0$

26. $25y^2 + 16 = 0$

27. $(x - 1)^2 = 49$

28. $(x + 2)^2 = 25$

29. $(a - 13)^2 = 18$

30. $(a + 5)^2 = 8$

31. $(x + 1)^2 = -9$

32. $(x - 1)^2 = -49$

33. $\left(y + \frac{3}{4}\right)^2 = \frac{17}{16}$

34. $\left(t + \frac{3}{2}\right)^2 = \frac{7}{2}$

35. $x^2 - 10x + 25 = 64$

36. $x^2 - 6x + 9 = 100$

37. Let $f(x) = x^2$. Find x such that $f(x) = 19$.

38. Let $f(x) = x^2$. Find x such that $f(x) = 11$.

39. Let $f(x) = (x - 5)^2$. Find x such that $f(x) = 16$.

40. Let $g(x) = (x - 2)^2$. Find x such that $g(x) = 25$.

41. Let $F(t) = (t + 4)^2$. Find t such that $F(t) = 13$.

42. Let $f(t) = (t + 6)^2$. Find t such that $f(t) = 15$.

Aha! 43. Let $g(x) = x^2 + 14x + 49$. Find x such that $g(x) = 49$.

44. Let $F(x) = x^2 + 8x + 16$. Find x such that $F(x) = 9$.

Replace the blanks in each equation with constants to complete the square and form a true equation.

45. $x^2 + 16x +$ ___ $= (x +$ ___$)^2$

46. $x^2 + 8x +$ ___ $= (x +$ ___$)^2$

47. $t^2 - 10t +$ ___ $= (t -$ ___$)^2$

48. $t^2 - 6t +$ ___ $= (t -$ ___$)^2$

49. $t^2 - 2t +$ ___ $= (t -$ ___$)^2$

50. $x^2 + 2x +$ ___ $= (x +$ ___$)^2$

51. $x^2 + 3x +$ ___ $= (x +$ ___$)^2$

52. $t^2 - 9t +$ ___ $= (t -$ ___$)^2$

53. $x^2 + \frac{2}{5}x +$ ___ $= (x +$ ___$)^2$

54. $x^2 + \frac{2}{3}x +$ ___ $= (x +$ ___$)^2$

55. $t^2 - \frac{5}{6}t +$ ___ $= (t -$ ___$)^2$

56. $t^2 - \frac{5}{3}t +$ ___ $= (t -$ ___$)^2$

Solve by completing the square. Show your work.

57. $x^2 + 6x = 7$

58. $x^2 + 8x = 9$

59. $t^2 - 10t = -23$

60. $t^2 - 4t = -1$

61. $x^2 + 12x + 32 = 0$

62. $x^2 + 16x + 15 = 0$

63. $t^2 + 8t - 3 = 0$

64. $t^2 + 6t - 5 = 0$

Complete the square to find the x-intercepts of each function given by the equation listed.

65. $f(x) = x^2 + 6x + 7$

66. $f(x) = x^2 + 10x - 2$

67. $g(x) = x^2 + 9x - 25$

68. $g(x) = x^2 + 5x + 2$

69. $f(x) = x^2 - 10x - 22$

70. $f(x) = x^2 - 8x - 10$

Solve by completing the square. Remember to first divide, as in Example 11, to make sure that the coefficient of x^2 is 1.

71. $9x^2 + 18x = -8$

72. $4x^2 + 8x = -3$

73. $3x^2 - 5x - 2 = 0$

74. $2x^2 - 5x - 3 = 0$

75. $5x^2 + 4x - 3 = 0$

76. $4x^2 + 3x - 5 = 0$

77. Find the x-intercepts of the function given by $f(x) = 4x^2 + 2x - 3$.

78. Find the x-intercepts of the function given by $f(x) = 3x^2 + x - 5$.

79. Find the x-intercepts of the function given by $g(x) = 2x^2 - 3x - 1$.

80. Find the x-intercepts of the function given by $g(x) = 3x^2 - 5x - 1$.

Interest. *Use $A = P(1 + r)^t$ to find the interest rate in Exercises 81–84. Refer to Example 12.*

81. \$2000 grows to \$2420 in 2 years

82. \$1000 grows to \$1440 in 2 years

83. \$6250 grows to \$6760 in 2 years

84. \$6250 grows to \$7290 in 2 years

Free-Falling Objects. *Use $s = 16t^2$ for Exercises 85–88. Refer to Example 13 and neglect air resistance.*

85. The Grand Canyon skywalk is 4000 ft above the Colorado River. How long will it take a stone to fall from the skywalk to the river?
Source: www.grandcanyonskywalk.com

86. The Sears Tower in Chicago is 1454 ft tall. How long would it take an object to fall freely from the top?

87. At 2063 ft, the KVLY-TV tower in North Dakota is the tallest supported tower in the United States. How long would it take an object to fall freely from the top?
Source: North Dakota Tourism Division

88. El Capitan in Yosemite National Park is 3593 ft high. How long would it take a carabiner to fall freely from the top?
Source: *Guinness World Records* 2008

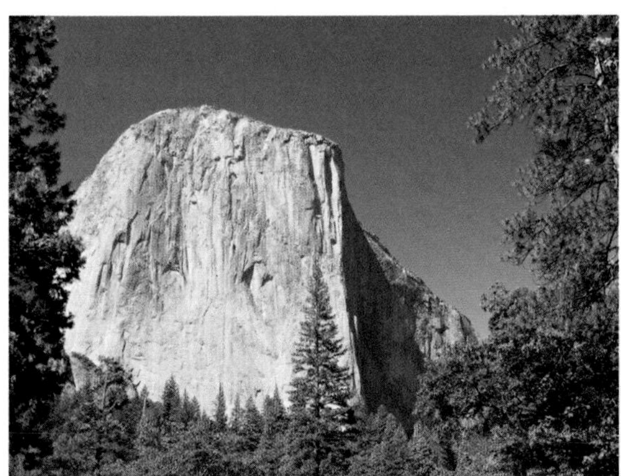

TW **89.** Explain in your own words a sequence of steps that can be used to solve any quadratic equation in the quickest way.

TW **90.** Describe how to write a quadratic equation that can be solved algebraically but not graphically.

SKILL REVIEW

To prepare for Section 11.2, review evaluating expressions and simplifying radical expressions (Sections 1.8, 10.3, and 10.8).

Evaluate. [1.8]

91. $b^2 - 4ac$, for $a = 3$, $b = 2$, and $c = -5$

92. $b^2 - 4ac$, for $a = 1$, $b = -1$, and $c = 4$

Simplify. [10.3], [10.8]

93. $\sqrt{200}$

94. $\sqrt{96}$

95. $\sqrt{-4}$

96. $\sqrt{-25}$

97. $\sqrt{-8}$

98. $\sqrt{-24}$

SYNTHESIS

TW **99.** What would be better: to receive 3% interest every 6 months or to receive 6% interest every 12 months? Why?

TW **100.** Example 12 was solved with a graphing calculator by graphing each side of

$$4410 = 4000(1 + r)^2.$$

How could you determine, from a reading of the problem, a suitable viewing window?

Find b such that each trinomial is a square.

101. $x^2 + bx + 81$

102. $x^2 + bx + 49$

103. If $f(x) = 2x^5 - 9x^4 - 66x^3 + 45x^2 + 280x$ and $x^2 - 5$ is a factor of $f(x)$, find all a for which $f(a) = 0$.

104. If $f(x) = \left(x - \frac{1}{3}\right)(x^2 + 6)$ and $g(x) = \left(x - \frac{1}{3}\right)\left(x^2 - \frac{2}{3}\right)$, find all a for which $(f + g)(a) = 0$.

105. *Boating.* A barge and a fishing boat leave a dock at the same time, traveling at a right angle to each other. The barge travels 7 km/h slower than the fishing boat. After 4 hr, the boats are 68 km apart. Find the speed of each boat.

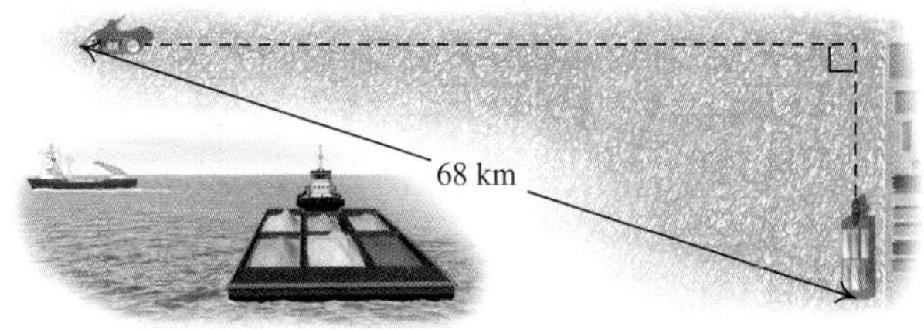

106. Find three consecutive integers such that the square of the first plus the product of the other two is 67.

Try Exercise Answers: Section 11.1

13. ± 10 **17.** $\pm\sqrt{5}$ **23.** $\pm\sqrt{\frac{7}{5}}$, or $\pm\frac{\sqrt{35}}{5}$ **25.** $\pm\frac{9}{2}i$
35. $-3, 13$ **41.** $-4 \pm \sqrt{13}$ **45.** $x^2 + 16x + 64 = (x + 8)^2$
57. $-7, 1$ **65.** $\left(-3 - \sqrt{2}, 0\right), \left(-3 + \sqrt{2}, 0\right)$ **71.** $-\frac{4}{3}, -\frac{2}{3}$
81. 10% **85.** About 15.8 sec

11.2 The Quadratic Formula

- Solving Using the Quadratic Formula
- Approximating Solutions

We can use the process of completing the square to develop a general formula for solving quadratic equations.

SOLVING USING THE QUADRATIC FORMULA

Each time we solve by completing the square, the procedure is the same. When a procedure is repeated many times, we can often develop a formula to speed up our work.

We begin with a quadratic equation in standard form,

$$ax^2 + bx + c = 0,$$

with $a > 0$. For $a < 0$, a slightly different derivation is needed (see Exercise 60), but the result is the same. Let's solve by completing the square. As the steps are performed, compare them with Example 11 on p. 781.

$$ax^2 + bx = -c \quad \textbf{Adding } -c \textbf{ to both sides}$$

$$x^2 + \frac{b}{a}x = -\frac{c}{a} \quad \textbf{Dividing both sides by } a$$

Half of $\frac{b}{a}$ is $\frac{b}{2a}$ and $\left(\frac{b}{2a}\right)^2$ is $\frac{b^2}{4a^2}$. We add $\frac{b^2}{4a^2}$ to both sides:

$$x^2 + \frac{b}{a}x + \frac{b^2}{4a^2} = -\frac{c}{a} + \frac{b^2}{4a^2} \quad \textbf{Adding } \frac{b^2}{4a^2} \textbf{ to complete the square}$$

$$\left(x + \frac{b}{2a}\right)^2 = -\frac{4ac}{4a^2} + \frac{b^2}{4a^2} \quad \textbf{Factoring on the left side; finding a common denominator on the right side}$$

$$\left(x + \frac{b}{2a}\right)^2 = \frac{b^2 - 4ac}{4a^2}$$

$$x + \frac{b}{2a} = \pm\frac{\sqrt{b^2 - 4ac}}{2a} \quad \textbf{Using the principle of square roots and the quotient rule for radicals; since } a > 0, \sqrt{4a^2} = 2a$$

$$x = \frac{-b \pm \sqrt{b^2 - 4ac}}{2a}. \quad \textbf{Adding } -\frac{b}{2a} \textbf{ to both sides}$$

It is important to remember the quadratic formula and know how to use it.

The Quadratic Formula The solutions of $ax^2 + bx + c = 0$, $a \neq 0$, are given by

$$x = \frac{-b \pm \sqrt{b^2 - 4ac}}{2a}.$$

STUDY TIP 

Know It "By Heart"

When memorizing something like the quadratic formula, try to first understand and write out the derivation. Doing so will help you remember the formula.

Student Notes

To avoid common errors when using the quadratic formula, use the following suggestions.

- Read "$-b$" as "the opposite of b." If b is negative, $-b$ will be positive.
- Write the fraction bar under the entire expression $-b \pm \sqrt{b^2 - 4ac}$.
- If a, b, or c is negative, use parentheses when substituting in the formula.

EXAMPLE 1 Solve $5x^2 + 8x = -3$ using the quadratic formula.

SOLUTION We first find standard form and determine a, b, and c:

$$5x^2 + 8x + 3 = 0; \quad \textbf{Adding 3 to both sides to get 0 on one side}$$

$$a = 5, \quad b = 8, \quad c = 3.$$

Next, we use the quadratic formula:

$$x = \frac{-b \pm \sqrt{b^2 - 4ac}}{2a}$$

$$x = \frac{-8 \pm \sqrt{8^2 - 4 \cdot 5 \cdot 3}}{2 \cdot 5}$$ **Substituting**

Be sure to write the fraction bar all the way across.

$$x = \frac{-8 \pm \sqrt{64 - 60}}{10}$$

$$x = \frac{-8 \pm \sqrt{4}}{10} = \frac{-8 \pm 2}{10}$$

$$x = \frac{-8 + 2}{10} \quad or \quad x = \frac{-8 - 2}{10}$$ **The symbol $\pm$ indicates that there are two solutions.**

$$x = \frac{-6}{10} \quad or \quad x = \frac{-10}{10}$$

$$x = -\frac{3}{5} \quad or \quad x = -1.$$

The solutions are $-\frac{3}{5}$ and -1. The checks are left to the student.

Try Exercise 7.

Because $5x^2 + 8x + 3$ can be factored as $(5x + 3)(x + 1)$, the quadratic formula may not have been the fastest way to solve Example 1. However, because the quadratic formula works for *any* quadratic equation, we need not spend too much time struggling to solve a quadratic equation by factoring.

To Solve a Quadratic Equation

1. If the equation can be easily written in the form $ax^2 = p$ or $(x + k)^2 = d$, use the principle of square roots as in Section 11.1.
2. If step (1) does not apply, write the equation in the form $ax^2 + bx + c = 0$.
3. Try factoring and using the principle of zero products.
4. If factoring seems difficult or impossible, use the quadratic formula. Completing the square can also be used.

The solutions of a quadratic equation can always be found using the quadratic formula. They cannot always be found by factoring.

Recall that a second-degree polynomial in one variable is said to be quadratic. Similarly, a second-degree polynomial function in one variable is said to be a **quadratic function.**

EXAMPLE 2 For the quadratic function given by $f(x) = 3x^2 - 6x - 4$, find all x for which $f(x) = 0$.

SOLUTION We substitute and solve for x:

$$f(x) = 0$$

$$3x^2 - 6x - 4 = 0$$ **Substituting. We cannot factor $3x^2 - 6x - 4$.**

$$a = 3, \quad b = -6, \quad c = -4.$$

We then substitute into the quadratic formula:

$$x = \frac{-(-6) \pm \sqrt{(-6)^2 - 4 \cdot 3 \cdot (-4)}}{2 \cdot 3}$$ **Use parentheses when substituting negative numbers.**

$$= \frac{6 \pm \sqrt{36 + 48}}{6}$$ $(-6)^2 - 4 \cdot 3 \cdot (-4) = 36 - (-48) = 36 + 48$

$$= \frac{6 \pm \sqrt{84}}{6}$$

$$= \frac{6}{6} \pm \frac{\sqrt{84}}{6}$$ **Note that 4 is a perfect-square factor of 84.**

$$= 1 \pm \frac{\sqrt{4}\sqrt{21}}{6}$$ $84 = 4 \cdot 21$

$$= 1 \pm \frac{2\sqrt{21}}{6}$$

$$= 1 \pm \frac{\sqrt{21}}{3}.$$ **Simplifying**

$y = 3x^2 - 6x - 4$

10, −5, 5, −10

X = 2.5275252 Y = 0

To check, we graph $y_1 = 3x^2 - 6x - 4$, and press TRACE. When we enter $1 + \sqrt{\ }(21)/3$, a rational approximation and the y-value 0 appear, as shown in the graph at left. The number $1 - \sqrt{21}/3$ also checks.

The solutions are $1 - \frac{\sqrt{21}}{3}$ and $1 + \frac{\sqrt{21}}{3}$.

Try Exercise 39.

Some quadratic equations have solutions that are imaginary numbers.

EXAMPLE 3 Solve: $x(x + 5) = 2(2x - 1)$.

SOLUTION We first find standard form:

$$x^2 + 5x = 4x - 2$$ **Multiplying**

$$x^2 + x + 2 = 0.$$ **Subtracting $4x$ and adding 2 to both sides**

Since we cannot factor $x^2 + x + 2$, we use the quadratic formula with $a = 1$, $b = 1$, and $c = 2$:

$$x = \frac{-1 \pm \sqrt{1^2 - 4 \cdot 1 \cdot 2}}{2 \cdot 1}$$ **Substituting**

$$= \frac{-1 \pm \sqrt{1 - 8}}{2}$$

$$= \frac{-1 \pm \sqrt{-7}}{2}$$

$$= \frac{-1 \pm i\sqrt{7}}{2}, \text{ or } -\frac{1}{2} \pm \frac{\sqrt{7}}{2}i.$$

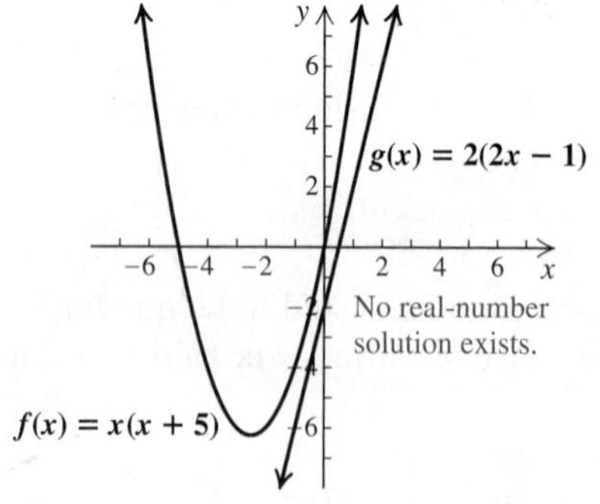

A visualization of Example 3

The solutions are $-\frac{1}{2} - \frac{\sqrt{7}}{2}i$ and $-\frac{1}{2} + \frac{\sqrt{7}}{2}i$. The checks are left to the student.

Try Exercise 35.

The quadratic formula can be used to solve certain rational equations.

EXAMPLE 4 If

$$f(x) = 2 + \frac{7}{x} \quad \text{and} \quad g(x) = \frac{4}{x^2},$$

find all x for which $f(x) = g(x)$.

SOLUTION We set $f(x)$ equal to $g(x)$ and solve:

$$f(x) = g(x)$$

$$2 + \frac{7}{x} = \frac{4}{x^2}. \qquad \text{Substituting. Note that } x \neq 0.$$

This is a rational equation similar to those in Section 7.6. To solve, we multiply both sides by the LCD, x^2:

$$x^2\left(2 + \frac{7}{x}\right) = x^2 \cdot \frac{4}{x^2}$$

$$2x^2 + 7x = 4 \qquad \text{Simplifying}$$

$$2x^2 + 7x - 4 = 0. \qquad \text{Subtracting 4 from both sides}$$

We have

$$a = 2, \quad b = 7, \quad \text{and} \quad c = -4.$$

Substituting then gives us

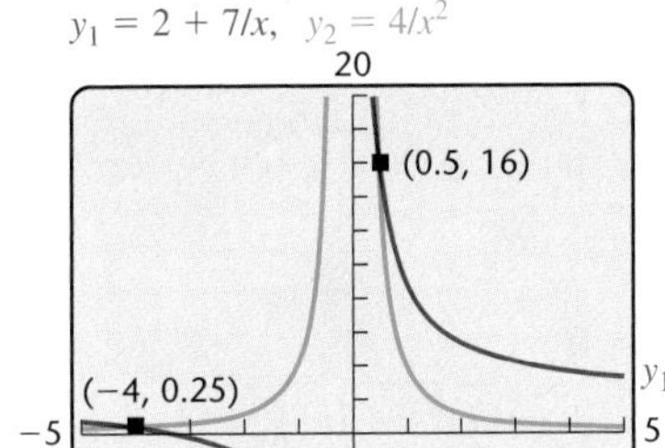

$$x = \frac{-7 \pm \sqrt{7^2 - 4 \cdot 2 \cdot (-4)}}{2 \cdot 2}$$

$$= \frac{-7 \pm \sqrt{49 + 32}}{4} = \frac{-7 \pm \sqrt{81}}{4} = \frac{-7 \pm 9}{4}$$

$$x = \frac{-7 + 9}{4} = \frac{1}{2} \quad or \quad x = \frac{-7 - 9}{4} = -4. \qquad \text{Both answers should check since } x \neq 0.$$

You can confirm that $f\left(\frac{1}{2}\right) = g\left(\frac{1}{2}\right)$ and $f(-4) = g(-4)$. The graph at left also serves as a check. The solutions are $\frac{1}{2}$ and -4.

Try Exercise 43.

APPROXIMATING SOLUTIONS

When the solution of an equation is irrational, a rational-number approximation is often useful. This is often the case in real-world applications similar to those found in Section 11.4.

EXAMPLE 5 Use a calculator to approximate, to three decimal places, the solutions of Example 2.

SOLUTION On many graphing calculators, the following sequence of keystrokes can be used to approximate $1 + \sqrt{21}/3$:

Similar keystrokes can be used to approximate $1 - \sqrt{21}/3$.

The solutions are approximately 2.527525232 and −0.5275252317. Rounded to three decimal places, the solutions are approximately 2.528 and −0.528.

Try Exercise 45.

Student Notes

It is important that you understand both the rules for order of operations *and* the manner in which your calculator applies those rules.

Connecting the Concepts

We have studied four different ways of solving quadratic equations. Each method has advantages and disadvantages, as outlined below. Note that although the quadratic formula can be used to solve *any* quadratic equation, the other methods are sometimes faster and easier to use.

Method	Advantages	Disadvantages	Example
Factoring	Can be very fast.	Can be used only on certain equations. Many equations are difficult or impossible to solve by factoring.	$x^2 - x - 6 = 0$ $(x - 3)(x + 2) = 0$ $x = 3$ *or* $x = -2$
The principle of square roots	Fastest way to solve equations of the form $X^2 = k$. Can be used to solve *any* quadratic equation.	Can be slow when original equation is not written in the form $X^2 = k$.	$(x - 5)^2 = 2$ $x - 5 = \pm\sqrt{2}$ $x = 5 \pm \sqrt{2}$
Completing the square	Works well on equations of the form $x^2 + bx = -c$, when b is even. Can be used to solve *any* quadratic equation.	Can be complicated when $a \neq 1$ or when b is not even in $x^2 + bx = -c$.	$x^2 + 14x = -2$ $x^2 + 14x + 49 = -2 + 49$ $(x + 7)^2 = 47$ $x + 7 = \pm\sqrt{47}$ $x = -7 \pm \sqrt{47}$
The quadratic formula	Can be used to solve *any* quadratic equation.	Can be slower than factoring or the principle of square roots for certain equations.	$x^2 - 2x - 5 = 0$ $x = \frac{-(-2) \pm \sqrt{(-2)^2 - 4(1)(-5)}}{2 \cdot 1}$ $= \frac{2 \pm \sqrt{24}}{2}$ $= \frac{2}{2} \pm \frac{2\sqrt{6}}{2} = 1 \pm \sqrt{6}$

11.2 Exercise Set

FOR EXTRA HELP MyMathLab Math XL PRACTICE   WATCH  DOWNLOAD  READ REVIEW

Concept Reinforcement *Classify each of the following statements as either true or false.*

1. The quadratic formula can be used to solve *any* quadratic equation.
2. The steps used to derive the quadratic formula are the same as those used when solving by completing the square.
3. The quadratic formula does not work if solutions are imaginary numbers.
4. Solving by factoring is always slower than using the quadratic formula.
5. A quadratic equation can have as many as four solutions.
6. It is possible for a quadratic equation to have no real-number solutions.

Solve.

7. $2x^2 + 3x - 5 = 0$

8. $3x^2 - 7x + 2 = 0$

9. $u^2 + 2u - 4 = 0$

10. $u^2 - 2u - 2 = 0$

11. $3p^2 = 18p - 6$

12. $3u^2 = 8u - 5$

13. $h^2 + 4 = 6h$

14. $t^2 + 4t = 1$

15. $x^2 = 3x + 5$

16. $x^2 + 5x = -3$

17. $3t(t + 2) = 1$

18. $2t(t + 2) = 1$

19. $\frac{1}{x^2} - 3 = \frac{8}{x}$

20. $\frac{9}{x} - 2 = \frac{5}{x^2}$

21. $t^2 + 10 = 6t$

22. $t^2 + 10t + 26 = 0$

23. $x^2 + 4x + 6 = 0$

24. $x^2 + 11 = 6x$

25. $12t^2 + 17t = 40$

26. $15t^2 + 7t = 2$

27. $25x^2 - 20x + 4 = 0$

28. $36x^2 + 84x + 49 = 0$

29. $7x(x + 2) + 5 = 3x(x + 1)$

30. $5x(x - 1) - 7 = 4x(x - 2)$

31. $14(x - 4) - (x + 2) = (x + 2)(x - 4)$

32. $11(x - 2) + (x - 5) = (x + 2)(x - 6)$

33. $5x^2 = 13x + 17$

34. $25x = 3x^2 + 28$

35. $x(x - 3) = x - 9$

36. $x(x - 1) = 2x - 7$

37. $x^3 - 8 = 0$ (*Hint*: Factor the difference of cubes. Then use the quadratic formula.)

38. $x^3 + 1 = 0$

39. Let $g(x) = 4x^2 - 2x - 3$. Find x such that $g(x) = 0$.

40. Let $f(x) = 6x^2 - 7x - 20$. Find x such that $f(x) = 0$.

41. Let

$$g(x) = \frac{2}{x} + \frac{2}{x + 3}.$$

Find all x for which $g(x) = 1$.

42. Let

$$f(x) = \frac{7}{x} + \frac{7}{x + 4}.$$

Find all x for which $f(x) = 1$.

43. Let

$$F(x) = \frac{x + 3}{x} \quad \text{and} \quad G(x) = \frac{x - 4}{3}.$$

Find all x for which $F(x) = G(x)$.

44. Let

$$f(x) = \frac{3 - x}{4} \quad \text{and} \quad g(x) = \frac{1}{4x}.$$

Find all x for which $f(x) = g(x)$.

Solve. Use a calculator to approximate, to three decimal places, the solutions as rational numbers.

45. $x^2 + 4x - 7 = 0$

46. $x^2 + 6x + 4 = 0$

47. $x^2 - 6x + 4 = 0$

48. $x^2 - 4x + 1 = 0$

49. $2x^2 - 3x - 7 = 0$

50. $3x^2 - 3x - 2 = 0$

TW **51.** Are there any equations that can be solved by the quadratic formula but not by completing the square? Why or why not?

TW **52.** Suppose you are solving a quadratic equation with no constant term ($c = 0$). Would you use factoring or the quadratic formula to solve? Why?

SKILL REVIEW

To prepare for Section 11.3, review multiplying and simplifying radical expressions and complex-number expressions (Sections 10.3, 10.5, and 10.8).

Multiply and simplify.

53. $(x - 2i)(x + 2i)$ [10.8]

54. $(x - 6\sqrt{5})(x + 6\sqrt{5})$ [10.5]

55. $(x - (2 - \sqrt{7}))(x - (2 + \sqrt{7}))$ [10.5]

56. $(x - (-3 + 5i))(x - (-3 - 5i))$ [10.8]

Simplify.

57. $\dfrac{-6 \pm \sqrt{(-4)^2 - 4(2)(2)}}{2(2)}$ [10.3]

58. $\dfrac{-(-1) \pm \sqrt{(6)^2 - 4(3)(5)}}{2(3)}$ [10.8]

SYNTHESIS

TW **59.** Explain how you could use the quadratic formula to help factor a quadratic polynomial.

TW **60.** If $a < 0$ and $ax^2 + bx + c = 0$, then $-a$ is positive and the equivalent equation, $-ax^2 - bx - c = 0$, can be solved using the quadratic formula.

a) Find this solution, replacing a, b, and c in the formula with $-a$, $-b$, and $-c$ from the equation.

b) How does the result of part (a) indicate that the quadratic formula "works" regardless of the sign of a?

For Exercises 61–63, let

$$f(x) = \frac{x^2}{x - 2} + 1 \quad \textit{and} \quad g(x) = \frac{4x - 2}{x - 2} + \frac{x + 4}{2}.$$

61. Find the x-intercepts of the graph of f.

62. Find the x-intercepts of the graph of g.

63. Find all x for which $f(x) = g(x)$.

Solve. Approximate the solutions to three decimal places.

64. $x^2 - 0.75x - 0.5 = 0$

65. $z^2 + 0.84z - 0.4 = 0$

Solve.

66. $(1 + \sqrt{3})x^2 - (3 + 2\sqrt{3})x + 3 = 0$

67. $\sqrt{2}x^2 + 5x + \sqrt{2} = 0$

68. $ix^2 - 2x + 1 = 0$

69. One solution of $kx^2 + 3x - k = 0$ is -2. Find the other.

TW **70.** Can a graph be used to solve *any* quadratic equation? Why or why not?

TW **71.** Solve Example 2 graphically and compare with the algebraic solution. Which method is faster? Which method is more precise?

TW **72.** Solve Example 4 graphically and compare with the algebraic solution. Which method is faster? Which method is more precise?

Try Exercise Answers: Section 11.1

7. $-\frac{5}{2}, 1$ **35.** $2 \pm \sqrt{5}i$ **39.** $\frac{1}{4} \pm \frac{\sqrt{13}}{4}$ **43.** $\frac{7}{2} \pm \frac{\sqrt{85}}{2}$ **45.** $-5.317, 1.317$

11.3 Studying Solutions of Quadratic Equations

- The Discriminant
- Writing Equations from Solutions

THE DISCRIMINANT

It is sometimes enough to know what *type* of number a solution will be, without actually solving the equation. Suppose we want to know if $4x^2 + 7x - 15 = 0$ has rational solutions (and thus can be solved by factoring). Using the quadratic formula, we would have

$$x = \frac{-b \pm \sqrt{b^2 - 4ac}}{2a}$$

$$= \frac{-7 \pm \sqrt{7^2 - 4 \cdot 4 \cdot (-15)}}{2 \cdot 4}.$$

Since $7^2 - 4 \cdot 4 \cdot (-15) = 49 - 16(-15) = 289$, and since 289 is a perfect square $\left(\sqrt{289} = 17\right)$, the solutions of the equation will be two rational numbers. This means that $4x^2 + 7x - 15 = 0$ *can* be solved by factoring. Note that the radicand, 289, determines what type of number the solutions will be.

The radicand $b^2 - 4ac$ is known as the **discriminant.** If a, b, and c are rational, then we can make the following observations on the basis of the value of the discriminant.

Discriminant	Observation	Example
$b^2 - 4ac = 0$	We get the same solution twice. There is one *repeated* solution and it is rational.	$9x^2 + 6x + 1 = 0$ $b^2 - 4ac = 6^2 - 4 \cdot 9 \cdot 1 = 0$ Solving, we have $x = \frac{-6 \pm \sqrt{0}}{2 \cdot 9}$. The (repeated) solution is $-\frac{1}{3}$.
$b^2 - 4ac$ is positive. **1.** $b^2 - 4ac$ *is a* perfect square. **2.** $b^2 - 4ac$ *is not* a perfect square.	There are two different real-number solutions. **1.** The solutions are rational numbers. **2.** The solutions are irrational conjugates.	**1.** $6x^2 + 5x + 1 = 0$ $b^2 - 4ac = 5^2 - 4 \cdot 6 \cdot 1 = 1$ Solving, we have $x = \frac{-5 \pm \sqrt{1}}{2 \cdot 6}$. The solutions are $-\frac{1}{3}$ and $-\frac{1}{2}$. **2.** $x^2 + 4x + 2 = 0$ $b^2 - 4ac = 4^2 - 4 \cdot 1 \cdot 2 = 8$ Solving, we have $x = \frac{-4 \pm \sqrt{8}}{2 \cdot 1}$. The solutions are $-2 + \sqrt{2}$ and $-2 - \sqrt{2}$.
$b^2 - 4ac$ is negative.	There are two different imaginary-number solutions. They are complex conjugates.	$x^2 + 4x + 5 = 0$ $b^2 - 4ac = 4^2 - 4 \cdot 1 \cdot 5 = -4$ Solving, we have $x = \frac{-4 \pm \sqrt{-4}}{2 \cdot 1}$. The solutions are $-2 + i$ and $-2 - i$.

Note that all quadratic equations have one or two solutions. They can always be found algebraically; only real-number solutions can be found graphically.

EXAMPLE 1 For each equation, determine what type of number the solutions are and how many solutions exist.

a) $9x^2 - 12x + 4 = 0$

b) $x^2 + 5x + 8 = 0$

c) $2x^2 + 7x - 3 = 0$

SOLUTION

a) For $9x^2 - 12x + 4 = 0$, we have

$$a = 9, \quad b = -12, \quad c = 4.$$

We substitute and compute the discriminant:

$$\begin{aligned} b^2 - 4ac &= (-12)^2 - 4 \cdot 9 \cdot 4 \\ &= 144 - 144 = 0. \end{aligned}$$

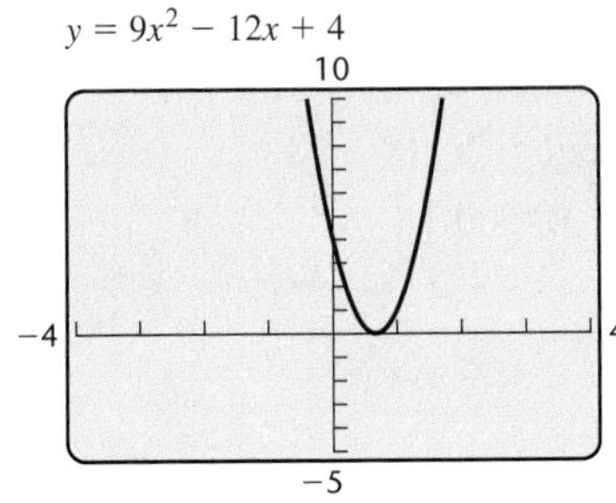

There is exactly one solution, and it is rational. This indicates that the equation $9x^2 - 12x + 4 = 0$ can be solved by factoring. The graph at left confirms that $9x^2 - 12x + 4 = 0$ has just one solution.

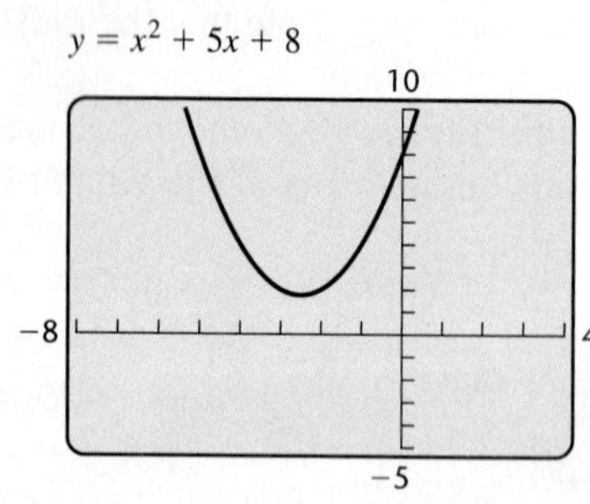

b) For $x^2 + 5x + 8 = 0$, we have

$$a = 1, \quad b = 5, \quad c = 8.$$

We substitute and compute the discriminant:

$$b^2 - 4ac = 5^2 - 4 \cdot 1 \cdot 8$$
$$= 25 - 32 = -7.$$

Since the discriminant is negative, there are two different imaginary-number solutions that are complex conjugates of each other. As the graph at left shows, there are no x-intercepts of the graph of $y = x^2 + 5x + 8$, and thus no real solutions of the equation $x^2 + 5x + 8 = 0$.

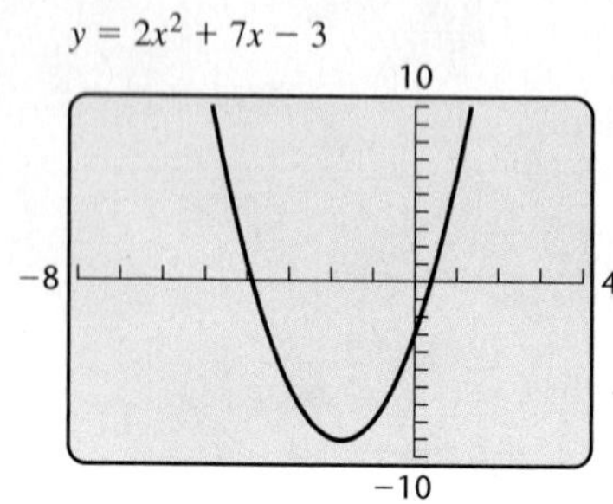

c) For $2x^2 + 7x - 3 = 0$, we have

$$a = 2, \quad b = 7, \quad c = -3;$$
$$b^2 - 4ac = 7^2 - 4 \cdot 2(-3)$$
$$= 49 - (-24) = 73.$$

The discriminant is a positive number that is not a perfect square. Thus there are two different irrational solutions that are conjugates of each other. From the graph at left, we see that there are two solutions of the equation. We cannot tell from simply observing the graph whether the solutions are rational or irrational.

Try Exercise 7.

Since the discriminant tells us the nature of the solutions of a quadratic equation, it also tells us the number of x-intercepts of the graph of a quadratic equation.

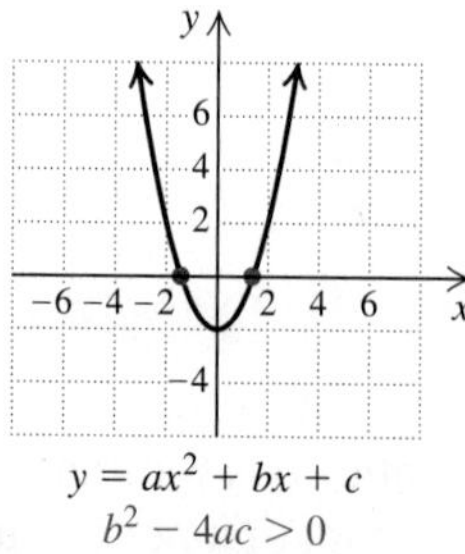

$y = ax^2 + bx + c$
$b^2 - 4ac > 0$
Two real solutions of $ax^2 + bx + c = 0$
Two x-intercepts

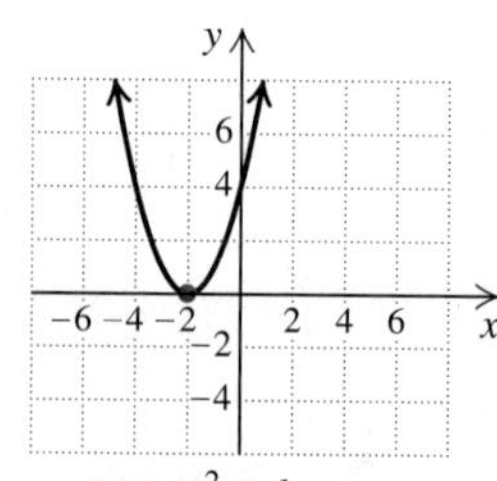

$y = ax^2 + bx + c$
$b^2 - 4ac = 0$
One real solution of $ax^2 + bx + c = 0$
One x-intercept

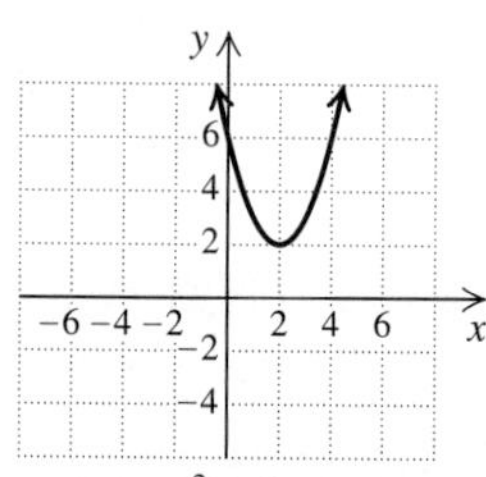

$y = ax^2 + bx + c$
$b^2 - 4ac < 0$
No real solutions of $ax^2 + bx + c = 0$
No x-intercept

WRITING EQUATIONS FROM SOLUTIONS

We know by the principle of zero products that $(x - 2)(x + 3) = 0$ has solutions 2 and -3. If we know the solutions of an equation, we can write an equation, using the principle in reverse.

EXAMPLE 2 Find an equation for which the given numbers are solutions.

a) 3 and $-\frac{2}{5}$ **b)** $2i$ and $-2i$

c) $5\sqrt{7}$ and $-5\sqrt{7}$ **d)** -4, 0, and 1

STUDY TIP

When Something Seems Familiar

Every so often, you may encounter a lesson that you remember from a previous math course. When this occurs, make sure that *all* of that lesson is understood. Take time to review any new concepts and to sharpen any old skills.

SOLUTION

a)

$$\begin{aligned} x = 3 \quad &or \quad x = -\tfrac{2}{5} \\ x - 3 = 0 \quad &or \quad x + \tfrac{2}{5} = 0 && \text{Getting 0's on one side} \\ (x - 3)\left(x + \tfrac{2}{5}\right) &= 0 && \text{Using the principle of zero products (multiplying)} \\ x^2 + \tfrac{2}{5}x - 3x - 3 \cdot \tfrac{2}{5} &= 0 && \text{Multiplying} \\ x^2 - \tfrac{13}{5}x - \tfrac{6}{5} &= 0 && \text{Combining like terms} \\ 5x^2 - 13x - 6 &= 0 && \text{Multiplying both sides by 5 to clear fractions} \end{aligned}$$

Note that multiplying both sides by the LCD, 5, clears the equations of fractions. Had we preferred, we could have multiplied $x + \frac{2}{5} = 0$ by 5, thus clearing fractions *before* using the principle of zero products.

b)

$$\begin{aligned} x = 2i \quad &or \quad x = -2i \\ x - 2i = 0 \quad &or \quad x + 2i = 0 && \text{Getting 0's on one side} \\ (x - 2i)(x + 2i) &= 0 && \text{Using the principle of zero products (multiplying)} \\ x^2 - (2i)^2 &= 0 && \text{Finding the product of a sum and a difference} \\ x^2 - 4i^2 &= 0 \\ x^2 + 4 &= 0 && i^2 = -1 \end{aligned}$$

c)

$$\begin{aligned} x = 5\sqrt{7} \quad &or \quad x = -5\sqrt{7} \\ x - 5\sqrt{7} = 0 \quad &or \quad x + 5\sqrt{7} = 0 && \text{Getting 0's on one side} \\ \left(x - 5\sqrt{7}\right)\left(x + 5\sqrt{7}\right) &= 0 && \text{Using the principle of zero products} \\ x^2 - \left(5\sqrt{7}\right)^2 &= 0 && \text{Finding the product of a sum and a difference} \\ x^2 - 25 \cdot 7 &= 0 \\ x^2 - 175 &= 0 \end{aligned}$$

d)

$$\begin{aligned} x = -4 \quad or \quad x = 0 \quad &or \quad x = 1 \\ x + 4 = 0 \quad or \quad x = 0 \quad &or \quad x - 1 = 0 && \text{Getting 0's on one side} \\ (x + 4)x(x - 1) &= 0 && \text{Using the principle of zero products} \\ x(x^2 + 3x - 4) &= 0 && \text{Multiplying} \\ x^3 + 3x^2 - 4x &= 0 \end{aligned}$$

Try Exercise 29.

11.3 Exercise Set

Concept Reinforcement *Complete each of the following statements.*

1. In the quadratic formula, the expression $b^2 - 4ac$ is called the ______________.

2. When $b^2 - 4ac$ is 0, there is/are ______________ solution(s).

3. When $b^2 - 4ac$ is positive, there is/are ______________ solution(s).

4. When $b^2 - 4ac$ is negative, there is/are ______________ solution(s).

5. When $b^2 - 4ac$ is a perfect square, the solutions are ______________ numbers.

6. When $b^2 - 4ac$ is negative, the solutions are ______________ numbers.

For each equation, determine what type of number the solutions are and how many solutions exist.

7. $x^2 - 7x + 5 = 0$
8. $x^2 - 5x + 3 = 0$
9. $x^2 + 3 = 0$
10. $x^2 + 5 = 0$
11. $x^2 - 5 = 0$
12. $x^2 - 3 = 0$
13. $4x^2 + 8x - 5 = 0$
14. $4x^2 - 12x + 9 = 0$
15. $x^2 + 4x + 6 = 0$
16. $x^2 - 2x + 4 = 0$
17. $9t^2 - 48t + 64 = 0$
18. $6t^2 - 19t - 20 = 0$
19. Aha! $9t^2 - 3t = 0$
20. $4m^2 + 7m = 0$
21. $x^2 + 4x = 8$
22. $x^2 + 5x = 9$
23. $2a^2 - 3a = -5$
24. $3a^2 + 5 = 7a$
25. $7x^2 = 19x$
26. $5x^2 = 48x$
27. $y^2 + \frac{9}{4} = 4y$
28. $x^2 = \frac{1}{2}x - \frac{3}{5}$

Write a quadratic equation having the given numbers as solutions.

29. $-7, 3$
30. $-6, 4$
31. 3, only solution (*Hint*: It must be a repeated solution.)
32. -5, only solution
33. $-1, -3$
34. $-2, -5$
35. $5, \frac{3}{4}$
36. $4, \frac{2}{3}$
37. $-\frac{1}{4}, -\frac{1}{2}$
38. $\frac{1}{2}, \frac{1}{3}$
39. $2.4, -0.4$
40. $-0.6, 1.4$
41. $-\sqrt{3}, \sqrt{3}$
42. $-\sqrt{7}, \sqrt{7}$
43. $2\sqrt{5}, -2\sqrt{5}$
44. $3\sqrt{2}, -3\sqrt{2}$
45. $4i, -4i$
46. $3i, -3i$
47. $2 - 7i, 2 + 7i$
48. $5 - 2i, 5 + 2i$
49. $3 - \sqrt{14}, 3 + \sqrt{14}$
50. $2 - \sqrt{10}, 2 + \sqrt{10}$
51. $1 - \frac{\sqrt{21}}{3}, 1 + \frac{\sqrt{21}}{3}$
52. $\frac{5}{4} - \frac{\sqrt{33}}{4}, \frac{5}{4} + \frac{\sqrt{33}}{4}$

Write a third-degree equation having the given numbers as solutions.

53. $-2, 1, 5$
54. $-5, 0, 2$
55. $-1, 0, 3$
56. $-2, 2, 3$

57. TW Explain why there are not two different solutions when the discriminant is 0.

58. TW While solving a quadratic equation of the form $ax^2 + bx + c = 0$ with a graphing calculator, Amberley gets the following screen. How could the sign of the discriminant help her check the graph?

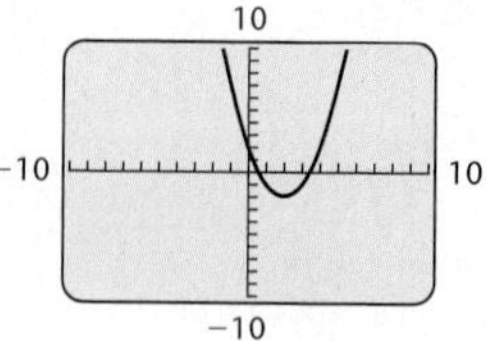

SKILL REVIEW

To prepare for Section 11.4, review solving formulas and solving motion problems (Sections 4.4, 7.7, and 7.8).

Solve each formula for the specified variable. [7.8]

59. $\frac{c}{d} = c + d$, for c
60. $\frac{p}{q} = \frac{a + b}{b}$, for b
61. $x = \frac{3}{1 - y}$, for y

Solve.

62. *Boating.* Kiara's motorboat took 4 hr to make a trip downstream with a 2-mph current. The return trip against the same current took 6 hr. Find the speed of the boat in still water. [4.4]
63. *Walking.* Jamal walks 1.5 mph faster than Kade. In the time it takes Jamal to walk 7 mi, Kade walks 4 mi. Find the speed of each person. [7.7]
64. *Aviation.* Taryn's Cessna travels 120 mph in still air. She flies 140 mi into the wind and 140 mi with the wind in a total of 2.4 hr. Find the wind speed. [7.7]

SYNTHESIS

TW **65.** If we assume that a quadratic equation has integers for coefficients, will the product of the solutions always be a real number? Why or why not?

TW **66.** Can a fourth-degree equation have exactly three irrational solutions? Why or why not?

67. The graph of an equation of the form

$$y = ax^2 + bx + c$$

is a curve similar to the one shown below. Determine a, b, and c from the information given.

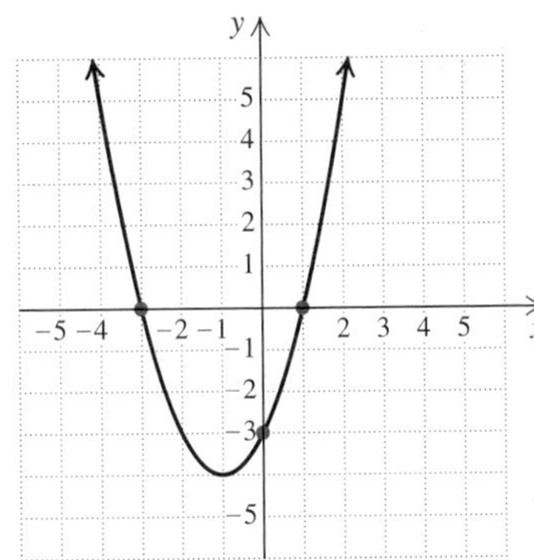

68. Show that the product of the solutions of $ax^2 + bx + c = 0$ is c/a.

For each equation under the given condition, **(a)** *find k and* **(b)** *find the other solution.*

69. $kx^2 - 2x + k = 0$; one solution is -3

70. $x^2 - kx + 2 = 0$; one solution is $1 + i$

71. $x^2 - (6 + 3i)x + k = 0$; one solution is 3

72. Show that the sum of the solutions of $ax^2 + bx + c = 0$ is $-b/a$.

73. Show that whenever there is just one solution of $ax^2 + bx + c = 0$, that solution is of the form $-b/(2a)$.

74. Find h and k, where $3x^2 - hx + 4k = 0$, the sum of the solutions is -12, and the product of the solutions is 20. (*Hint*: See Exercises 68 and 72.)

75. Suppose that $f(x) = ax^2 + bx + c$, with $f(-3) = 0$, $f\left(\frac{1}{2}\right) = 0$, and $f(0) = -12$. Find a, b, and c.

76. Find an equation for which $2 - \sqrt{3}$, $2 + \sqrt{3}$, $5 - 2i$, and $5 + 2i$ are solutions.

Aha! **77.** Write a quadratic equation with integer coefficients for which $-\sqrt{2}$ is one solution.

78. Write a quadratic equation with integer coefficients for which $10i$ is one solution.

79. Find an equation with integer coefficients for which $1 - \sqrt{5}$ and $3 + 2i$ are two of the solutions.

TW **80.** A discriminant that is a perfect square indicates that factoring can be used to solve the quadratic equation. Why?

Try Exercise Answers: Section 11.3

7. Two irrational **29.** $x^2 + 4x - 21 = 0$

11.4 Applications Involving Quadratic Equations

- Solving Problems
- Solving Formulas

SOLVING PROBLEMS

As we found in Section 7.7, some problems translate to rational equations. The solution of such rational equations can involve quadratic equations.

EXAMPLE 1 Motorcycle Travel. Keisha rode her motorcycle 300 mi at a certain average speed. Had she averaged 10 mph more, the trip would have taken 1 hr less. Find Keisha's average speed.

SOLUTION

1. **Familiarize.** We make a drawing, labeling it with the information provided. As in Section 7.7, we can create a table. We let $r =$ the rate, in miles per hour, and t the time, in hours, for Keisha's trip.

STUDY TIP

Keep Your Book Handy

You can supplement your regular studying by using smaller blocks of time for reviewing material. By keeping your text and notebook nearby (in your car or bag), you can review course material in a doctor's waiting room, while sitting through a child's soccer practice, while riding a bus, or whenever you find yourself with some free time.

Distance	Speed	Time	
300	r	t	$\longrightarrow r = \dfrac{300}{t}$
300	$r + 10$	$t - 1$	$\longrightarrow r + 10 = \dfrac{300}{t-1}$

Recall that the definition of speed, $r = d/t$, relates the three quantities.

2. **Translate.** From the first two lines of the table, we obtain

$$r = \frac{300}{t} \quad \text{and} \quad r + 10 = \frac{300}{t - 1}.$$

3. **Carry out.** A system of equations has been formed. We substitute for r from the first equation into the second and solve the resulting equation:

$$\frac{300}{t} + 10 = \frac{300}{t - 1}$$ **Substituting 300/t for r**

$$t(t-1) \cdot \left[\frac{300}{t} + 10\right] = t(t-1) \cdot \frac{300}{t-1}$$ **Multiplying by the LCD**

$$\cancel{t}(t-1) \cdot \frac{300}{\cancel{t}} + t(t-1) \cdot 10 = t\cancel{(t-1)} \cdot \frac{300}{\cancel{t-1}}$$ **Using the distributive law and removing factors that equal 1: $\frac{t}{t} = 1$; $\frac{t-1}{t-1} = 1$**

$$300(t-1) + 10(t^2 - t) = 300t$$

$$300t - 300 + 10t^2 - 10t = 300t$$

$$10t^2 - 10t - 300 = 0$$ **Rewriting in standard form**

$$t^2 - t - 30 = 0$$ **Multiplying by $\frac{1}{10}$ or dividing by 10**

$$(t - 6)(t + 5) = 0$$ **Factoring**

$$t = 6 \quad or \quad t = -5.$$ **Principle of zero products**

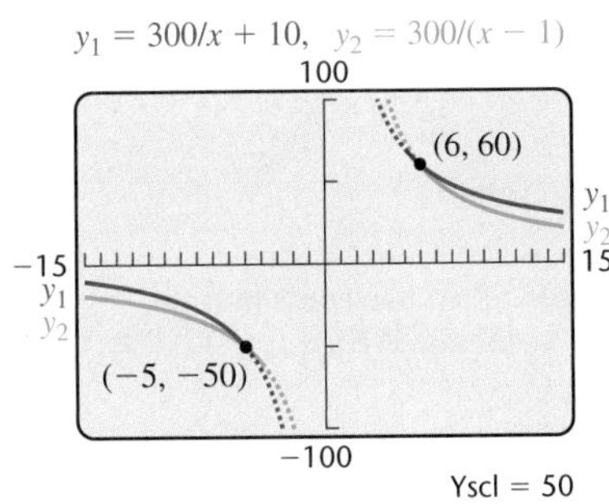

4. **Check.** As a partial check, we graph the equations $y_1 = 300/x + 10$ and $y_2 = 300/(x - 1)$. The graph at left shows that the solutions are -5 and 6.

Note that we have solved for t, not r as required. Since negative time has no meaning here, we disregard the -5 and use 6 hr to find r:

$$r = \frac{300 \text{ mi}}{6 \text{ hr}} = 50 \text{ mph}.$$

CAUTION! Always make sure that you find the quantity asked for in the problem.

To see if 50 mph checks, we increase the speed 10 mph to 60 mph and see how long the trip would have taken at that speed:

$$t = \frac{d}{r} = \frac{300 \text{ mi}}{60 \text{ mph}} = 5 \text{ hr}.$$

Note that mi/mph = mi ÷ $\frac{\text{mi}}{\text{hr}}$ = $\cancel{\text{mi}} \cdot \frac{\text{hr}}{\cancel{\text{mi}}}$ = hr.

This is 1 hr less than the trip actually took, so the answer checks.

5. **State.** Keisha traveled at an average speed of 50 mph.

Try Exercise 1.

SOLVING FORMULAS

Recall that to solve a formula for a certain letter, we use the principles for solving equations to get that letter alone on one side.

EXAMPLE 2 Period of a Pendulum. The time T required for a pendulum of length l to swing back and forth (complete one period) is given by the formula $T = 2\pi\sqrt{l/g}$, where g is the earth's gravitational constant. Solve for l.

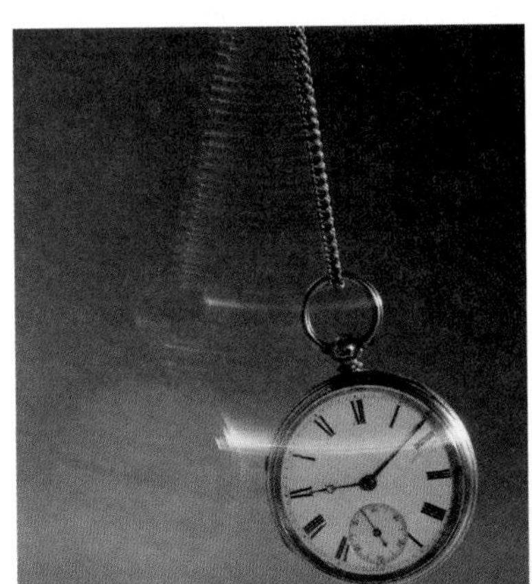

SOLUTION We have

$$T = 2\pi\sqrt{\frac{l}{g}} \qquad \text{This is a radical equation (see Section 10.6).}$$

$$T^2 = \left(2\pi\sqrt{\frac{l}{g}}\right)^2 \qquad \text{Principle of powers (squaring both sides)}$$

$$T^2 = 2^2\pi^2\frac{l}{g}$$

$$gT^2 = 4\pi^2 l \qquad \text{Multiplying both sides by } g \text{ to clear fractions}$$

$$\frac{gT^2}{4\pi^2} = l. \qquad \text{Dividing both sides by } 4\pi^2$$

We now have l alone on one side and l does not appear on the other side, so the formula is solved for l.

Try Exercise 21.

In formulas for which variables represent nonnegative numbers, there is no need for absolute-value signs when taking square roots.

EXAMPLE 3 Hang Time.* An athlete's *hang time* is the amount of time that the athlete can remain airborne when jumping. A formula relating an athlete's vertical leap V, in inches, to hang time T, in seconds, is $V = 48T^2$. Solve for T.

*This formula is taken from an article by Peter Brancazio, "The Mechanics of a Slam Dunk," *Popular Mechanics*, November 1991. Courtesy of Professor Peter Brancazio, Brooklyn College.

SOLUTION We have

$$48T^2 = V$$

$$T^2 = \frac{V}{48} \quad \text{Dividing by 48 to isolate } T^2$$

$$T = \frac{\sqrt{V}}{\sqrt{48}} \quad \text{Using the principle of square roots and the quotient rule for radicals. We assume } V, T \geq 0.$$

$$= \frac{\sqrt{V}}{\sqrt{16}\sqrt{3}}$$

$$= \frac{\sqrt{V}}{4\sqrt{3}}$$

$$= \frac{\sqrt{V}}{4\sqrt{3}} \cdot \frac{\sqrt{3}}{\sqrt{3}} = \frac{\sqrt{3V}}{12}. \quad \text{Rationalizing the denominator}$$

Try Exercise 15.

EXAMPLE 4 Falling Distance. An object tossed downward with an initial speed (velocity) of v_0 will travel a distance of s meters, where $s = 4.9t^2 + v_0t$ and t is measured in seconds. Solve for t.

SOLUTION Since t is squared in one term and raised to the first power in the other term, the equation is quadratic in t.

$$4.9t^2 + v_0t = s$$

$$4.9t^2 + v_0t - s = 0 \quad \text{Writing standard form}$$

$$a = 4.9, \quad b = v_0, \quad c = -s$$

$$t = \frac{-v_0 \pm \sqrt{(v_0)^2 - 4(4.9)(-s)}}{2(4.9)} \quad \text{Using the quadratic formula}$$

Since the negative square root would yield a negative value for t, we use only the positive root:

$$t = \frac{-v_0 + \sqrt{(v_0)^2 + 19.6s}}{9.8}.$$

Try Exercise 17.

The following list of steps should help you when solving formulas for a given letter. Try to remember that when solving a formula, you use the same approach that you would to solve an equation.

To Solve a Formula for a Letter—Say, *h*

1. Clear fractions and use the principle of powers, as needed. Perform these steps until radicals containing h are gone and h is not in any denominator.
2. Combine all like terms.
3. If the only power of h is h^1, the equation can be solved as in Sections 2.3 and 7.8. (See Example 2.)
4. If h^2 appears but h does not, solve for h^2 and use the principle of square roots to solve for h. (See Example 3.)
5. If there are terms containing both h and h^2, put the equation in standard form and use the quadratic formula. (See Example 4.)

11.4 Exercise Set

FOR EXTRA HELP MyMathLab    

Solve.

1. *Car Trips.* During the first part of a trip, Jaclyn's Honda traveled 120 mi at a certain speed. Jaclyn then drove another 100 mi at a speed that was 10 mph slower. If the total time of Jaclyn's trip was 4 hr, what was her speed on each part of the trip?

2. *Canoeing.* During the first part of a canoe trip, Terrell covered 60 km at a certain speed. He then traveled 24 km at a speed that was 4 km/h slower. If the total time for the trip was 8 hr, what was the speed on each part of the trip?

3. *Car Trips.* Franklin's Ford travels 200 mi averaging a certain speed. If the car had gone 10 mph faster, the trip would have taken 1 hr less. Find Franklin's average speed.

4. *Car Trips.* Mallory's Mazda travels 280 mi averaging a certain speed. If the car had gone 5 mph faster, the trip would have taken 1 hr less. Find Mallory's average speed.

5. *Air Travel.* A Cessna flies 600 mi at a certain speed. A Beechcraft flies 1000 mi at a speed that is 50 mph faster, but takes 1 hr longer. Find the speed of each plane.

6. *Air Travel.* A turbo-jet flies 50 mph faster than a super-prop plane. If a turbo-jet goes 2000 mi in 3 hr less time than it takes the super-prop to go 2800 mi, find the speed of each plane.

7. *Bicycling.* Naoki bikes the 40 mi to Hillsboro averaging a certain speed. The return trip is made at a speed that is 6 mph slower. Total time for the round trip is 14 hr. Find Naoki's average speed on each part of the trip.

8. *Car Speed.* On a sales trip, Jay drives the 600 mi to Richmond averaging a certain speed. The return trip is made at an average speed that is 10 mph slower. Total time for the round trip is 22 hr. Find Jay's average speed on each part of the trip.

9. *Navigation.* The Hudson River flows at a rate of 3 mph. A patrol boat travels 60 mi upriver and returns in a total time of 9 hr. What is the speed of the boat in still water?

10. *Navigation.* The current in a typical Mississippi River shipping route flows at a rate of 4 mph. In order for a barge to travel 24 mi upriver and then return in a total of 5 hr, approximately how fast must the barge be able to travel in still water?

11. *Filling a Pool.* A well and a spring are filling a swimming pool. Together, they can fill the pool in 4 hr. The well, working alone, can fill the pool in 6 hr less time than the spring. How long would the spring take, working alone, to fill the pool?

12. *Filling a Tank.* Two pipes are connected to the same tank. Working together, they can fill the tank in 2 hr. The larger pipe, working alone, can fill the tank in 3 hr less time than the smaller one. How long would the smaller one take, working alone, to fill the tank?

13. *Paddleboats.* Antonio paddles 1 mi upstream and 1 mi back in a total time of 1 hr. The speed of the river is 2 mph. Find the speed of Antonio's paddleboat in still water.

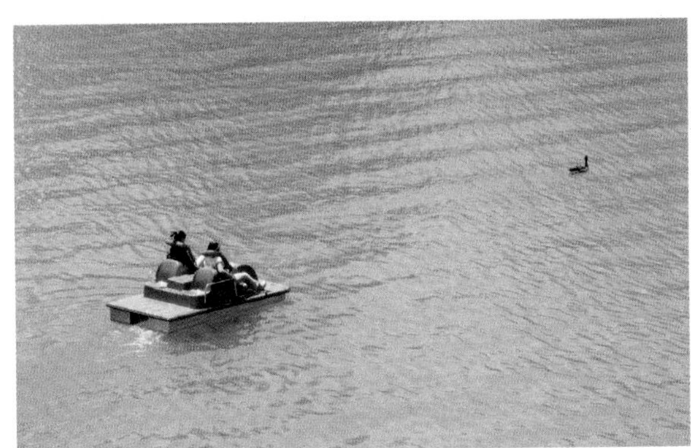

14. *Rowing.* Sydney rows 10 km upstream and 10 km back in a total time of 3 hr. The speed of the river is 5 km/h. Find Sydney's speed in still water.

Solve each formula for the indicated letter. Assume that all variables represent nonnegative numbers.

15. $A = 4\pi r^2$, for r
(Surface area of a sphere of radius r)

16. $A = 6s^2$, for s
(Surface area of a cube with sides of length s)

17. $A = 2\pi r^2 + 2\pi rh$, for r
(Surface area of a right cylindrical solid with radius r and height h)

18. $N = \frac{k^2 - 3k}{2}$, for k
(Number of diagonals of a polygon with k sides)

19. $F = \frac{Gm_1m_2}{r^2}$, for r
(Law of gravity)

20. $N = \frac{kQ_1Q_2}{s^2}$, for s
(Number of phone calls between two cities)

21. $c = \sqrt{gH}$, for H
(Velocity of an ocean wave)

22. $V = 3.5\sqrt{h}$, for h
(Distance to the horizon from a height)

23. $w = \frac{lg^2}{800}$, for g
(An ancient fisherman's formula)

24. $V = \pi r^2 h$, for r
(Volume of a cylinder)

25. $a^2 + b^2 = c^2$, for b
(Pythagorean formula in two dimensions)

26. $a^2 + b^2 + c^2 = d^2$, for c
(Pythagorean formula in three dimensions)

27. $s = v_0t + \frac{gt^2}{2}$, for t
(A motion formula)

28. $A = \pi r^2 + \pi rs$, for r
(Surface area of a cone)

29. $N = \frac{1}{2}(n^2 - n)$, for n
(Number of games if n teams play each other once)

30. $A = A_0(1 - r)^2$, for r
(A business formula)

31. $T = 2\pi\sqrt{\frac{l}{g}}$, for g
(A pendulum formula)

32. $W = \sqrt{\frac{1}{LC}}$, for L
(An electricity formula)

Aha! **33.** $at^2 + bt + c = 0$, for t
(An algebraic formula)

34. $A = P_1(1 + r)^2 + P_2(1 + r)$, for r
(Amount in an account when P_1 is invested for 2 years and P_2 for 1 year at interest rate r)

Solve.

35. *Falling Distance.* (Use $4.9t^2 + v_0t = s$.)

a) A bolt falls off an airplane at an altitude of 500 m. Approximately how long does it take the bolt to reach the ground?

b) A ball is thrown downward at a speed of 30 m/sec from an altitude of 500 m. Approximately how long does it take the ball to reach the ground?

c) Approximately how far will an object fall in 5 sec, when thrown downward at an initial velocity of 30 m/sec from a plane?

36. *Falling Distance.* (Use $4.9t^2 + v_0t = s$.)

a) A ring is dropped from a helicopter at an altitude of 75 m. Approximately how long does it take the ring to reach the ground?

b) A coin is tossed downward with an initial velocity of 30 m/sec from an altitude of 75 m. Approximately how long does it take the coin to reach the ground?

c) Approximately how far will an object fall in 2 sec, if thrown downward at an initial velocity of 20 m/sec from a helicopter?

37. *Bungee Jumping.* Jaime is tied to one end of a 40-m elasticized (bungee) cord. The other end of the cord is tied to the middle of a bridge. If Jaime jumps off the bridge, for how long will he fall before the cord begins to stretch? (Use $4.9t^2 = s$.)

38. *Bungee Jumping.* Mariah is tied to a bungee cord (see Exercise 37) and falls for 2.5 sec before her cord begins to stretch. How long is the bungee cord?

39. *Hang Time.* The NBA's LeBron James reportedly has a vertical leap of 44 in. What is his hang time? (Use $V = 48T^2$.)
Source: www.vertcoach.com

40. *League Schedules.* In a bowling league, each team plays each of the other teams once. If a total of 66 games is played, how many teams are in the league? (See Exercise 29.)

For Exercises 41 and 42, use $4.9t^2 + v_0t = s$.

41. *Downward Speed.* An object thrown downward from a 100-m cliff travels 51.6 m in 3 sec. What was the initial velocity of the object?

42. *Downward Speed.* An object thrown downward from a 200-m cliff travels 91.2 m in 4 sec. What was the initial velocity of the object?

For Exercises 43 and 44, use

$$A = P_1(1 + r)^2 + P_2(1 + r).$$

(See Exercise 34.)

43. *Compound Interest.* A firm invests \$3000 in a savings account for 2 years. At the beginning of the second year, an additional \$1700 is invested. If a total of \$5253.70 is in the account at the end of the second year, what is the annual interest rate?

44. *Compound Interest.* A business invests \$10,000 in a savings account for 2 years. At the beginning of the second year, an additional \$3500 is invested. If a total of \$15,569.75 is in the account at the end of the second year, what is the annual interest rate?

TW 45. Marti is tied to a bungee cord that is twice as long as the cord tied to Tivon. Will Marti's fall take twice as long as Tivon's before their cords begin to stretch? Why or why not? (See Exercises 37 and 38.)

TW 46. Under what circumstances would a negative value for t, time, have meaning?

SKILL REVIEW

To prepare for Section 11.5, review raising a power to a power and solving rational equations and radical equations (Sections 5.2, 7.6, 10.2, and 10.6).

Simplify.

47. $(m^{-1})^2$ [5.2]

48. $(t^{1/3})^2$ [10.2]

49. $(y^{1/6})^2$ [10.2]

50. $(z^{1/4})^2$ [10.2]

Solve.

51. $t^{-1} = \frac{1}{2}$ [7.6]

52. $x^{1/4} = 3$ [10.6]

SYNTHESIS

TW 53. Write a problem for a classmate to solve. Devise the problem so that (**a**) the solution is found after solving a rational equation and (**b**) the solution is "The express train travels 90 mph."

TW 54. In what ways do the motion problems in this section (like Example 1) differ from the motion problems in Section 7.7?

55. *Biochemistry.* The equation

$$A = 6.5 - \frac{20.4t}{t^2 + 36}$$

is used to calculate the acid level A in a person's blood t minutes after sugar is consumed. Solve for t.

56. *Special Relativity.* Einstein found that an object with initial mass m_0 and traveling velocity v has mass

$$m = \frac{m_0}{\sqrt{1 - \frac{v^2}{c^2}}},$$

where c is the speed of light. Solve the formula for c.

57. Find a number for which the reciprocal of 1 less than the number is the same as 1 more than the number.

58. *Purchasing.* A discount store bought a quantity of paperback books for \$250 and sold all but 15 at a profit of \$3.50 per book. With the total amount received, the manager could buy 4 more than twice as many as were bought before. Find the cost per book.

59. *Art and Aesthetics.* For over 2000 years, artists, sculptors, and architects have regarded the proportions of a "golden" rectangle as visually appealing. A rectangle of width w and length l is considered "golden" if

$$\frac{w}{l} = \frac{l}{w + l}.$$

Solve for l.

60. *Diagonal of a Cube.* Find a formula that expresses the length of the three-dimensional diagonal of a cube as a function of the cube's surface area.

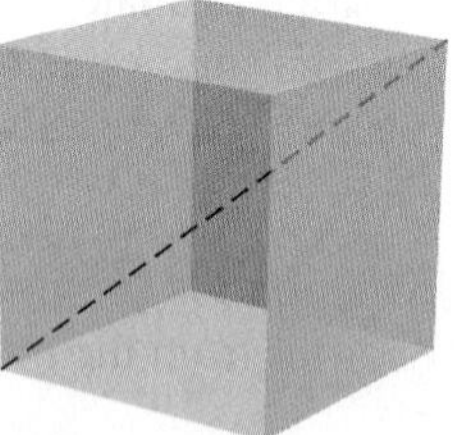

61. Solve for n:

$$mn^4 - r^2pm^3 - r^2n^2 + p = 0.$$

62. *Surface Area.* Find a formula that expresses the diameter of a right cylindrical solid as a function of its surface area and its height. (See Exercise 17.)

63. A sphere is inscribed in a cube as shown in the figure below. Express the surface area of the sphere as a function of the surface area S of the cube. (See Exercise 15.)

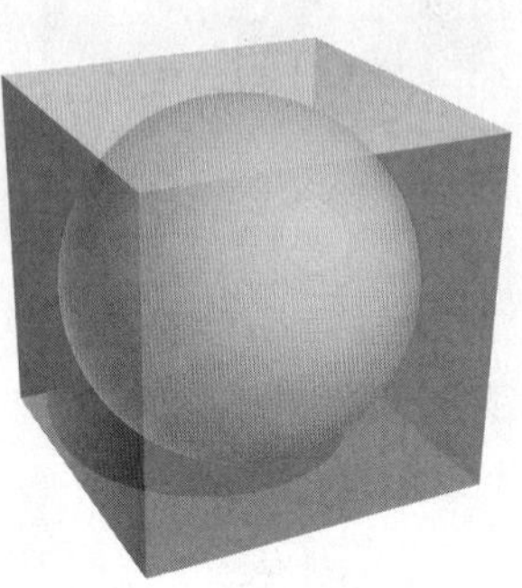

Try Exercise Answers: Section 11.4

1. First part: 60 mph; second part: 50 mph
15. $r = \frac{1}{2}\sqrt{\frac{A}{\pi}}$, or $\frac{\sqrt{A\pi}}{2\pi}$ **17.** $r = \frac{-\pi h + \sqrt{\pi^2h^2 + 2\pi A}}{2\pi}$
21. $H = \frac{c^2}{g}$

11.5 Equations Reducible to Quadratic

- Recognizing Equations in Quadratic Form
- Radical Equations and Rational Equations

RECOGNIZING EQUATIONS IN QUADRATIC FORM

Certain equations that are not really quadratic can be thought of in such a way that they can be solved as quadratic. For example, because the square of x^2 is x^4, the equation $x^4 - 9x^2 + 8 = 0$ is said to be "quadratic in x^2":

$$x^4 - 9x^2 + 8 = 0$$

$$\downarrow$$

$$(x^2)^2 - 9(x^2) + 8 = 0 \qquad \text{Thinking of } x^4 \text{ as } (x^2)^2$$

$$\downarrow$$

$$u^2 - 9u + 8 = 0. \qquad \text{To make this clearer, write } u \text{ instead of } x^2.$$

The equation $u^2 - 9u + 8 = 0$ can be solved by factoring or by the quadratic formula. Then, remembering that $u = x^2$, we can solve for x. Equations that can be solved like this are *reducible to quadratic* and are said to be *in quadratic form*.

EXAMPLE 1 Solve: $x^4 - 9x^2 + 8 = 0$.

SOLUTION We solve both algebraically and graphically.

ALGEBRAIC APPROACH

Let $u = x^2$. Then $u^2 = x^4$. We solve by substituting u for x^2 and u^2 for x^4:

$$u^2 - 9u + 8 = 0$$
$$(u - 8)(u - 1) = 0 \quad \text{Factoring}$$
$$u - 8 = 0 \quad or \quad u - 1 = 0 \quad \text{Principle of zero products}$$
$$u = 8 \quad or \quad u = 1.$$

We replace u with x^2 and solve these equations:

$$x^2 = 8 \quad or \quad x^2 = 1$$
$$x = \pm\sqrt{8} \quad or \quad x = \pm 1$$
$$x = \pm 2\sqrt{2} \quad or \quad x = \pm 1.$$

To check, note that for both $x = 2\sqrt{2}$ and $-2\sqrt{2}$, we have $x^2 = 8$ and $x^4 = 64$. Similarly, for both $x = 1$ and -1, we have $x^2 = 1$ and $x^4 = 1$. Thus instead of making four checks, we need make only two.

Check:

For $\pm 2\sqrt{2}$:

$x^4 - 9x^2 + 8 = 0$	
$(\pm 2\sqrt{2})^4 - 9(\pm 2\sqrt{2})^2 + 8$	0
$64 - 9 \cdot 8 + 8$	
$0 \stackrel{?}{=} 0$	TRUE

For ± 1:

$x^4 - 9x^2 + 8 = 0$	
$(\pm 1)^4 - 9(\pm 1)^2 + 8$	0
$1 - 9 + 8$	
$0 \stackrel{?}{=} 0$	TRUE

The solutions are $1, -1, 2\sqrt{2}$, and $-2\sqrt{2}$.

GRAPHICAL APPROACH

We let $f(x) = x^4 - 9x^2 + 8$ and determine the zeros of the function.

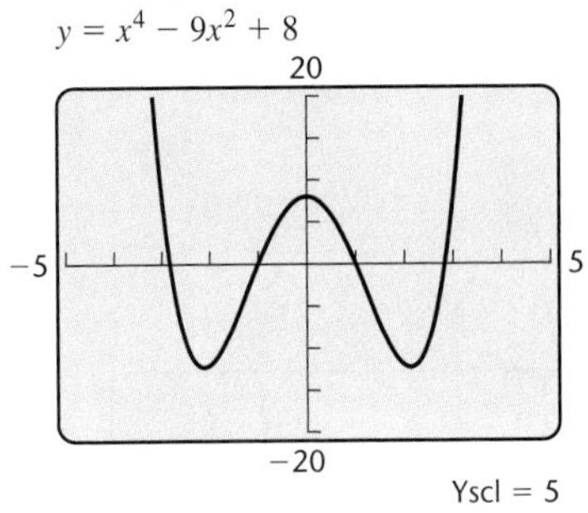

It appears as though the graph has 4 x-intercepts. Recall from Chapter 6 that a fourth-degree polynomial function has at most 4 zeros, so we know that we have not missed any zeros. Using ZERO four times, we obtain the solutions -2.8284271, -1, 1, and 2.8284271. The solutions are ± 1 and, approximately, ± 2.8284271. Note that since $2\sqrt{2} \approx 2.8284271$, the solutions found graphically are the same as those found algebraically.

Try Exercise 15.

Equations like those in Example 1 can be solved directly by factoring:

$$x^4 - 9x^2 + 8 = 0$$
$$(x^2 - 1)(x^2 - 8) = 0$$
$$x^2 - 1 = 0 \quad or \quad x^2 - 8 = 0$$
$$x^2 = 1 \quad or \quad x^2 = 8$$
$$x = \pm 1 \quad or \quad x = \pm 2\sqrt{2}.$$

However, it becomes difficult to solve some equations without first making a substitution.

EXAMPLE 2 Find the x-intercepts of the graph of

$$f(x) = (x^2 - 1)^2 - (x^2 - 1) - 2.$$

SOLUTION The x-intercepts occur where $f(x) = 0$ so we must have

$$(x^2 - 1)^2 - (x^2 - 1) - 2 = 0. \quad \textbf{Setting } f(x) \textbf{ equal to 0}$$

If we identify $x^2 - 1$ as u, then the equation can be written in quadratic form:

$$u = x^2 - 1$$
$$u^2 = (x^2 - 1)^2.$$

Substituting, we have

$$u^2 - u - 2 = 0 \quad \textbf{Substituting in } (x^2 - 1)^2 - (x^2 - 1) - 2 = 0$$
$$(u - 2)(u + 1) = 0$$
$$u = 2 \quad or \quad u = -1. \quad \textbf{Using the principle of zero products}$$

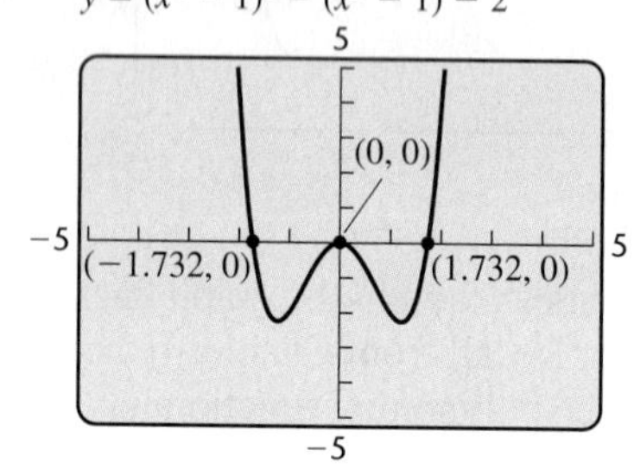

Next, we replace u with $x^2 - 1$ and solve these equations:

$$x^2 - 1 = 2 \quad or \quad x^2 - 1 = -1$$
$$x^2 = 3 \quad or \quad x^2 = 0 \quad \textbf{Adding 1 to both sides}$$
$$x = \pm\sqrt{3} \quad or \quad x = 0. \quad \textbf{Using the principle of square roots}$$

The x-intercepts occur at $(-\sqrt{3}, 0)$, $(0, 0)$, and $(\sqrt{3}, 0)$. These solutions are confirmed by the graph at left.

Try Exercise 49.

RADICAL EQUATIONS AND RATIONAL EQUATIONS

Sometimes rational equations, radical equations, or equations containing exponents that are fractions are reducible to quadratic. It is especially important that answers to these equations be checked in the original equation.

EXAMPLE 3 Solve: $x - 3\sqrt{x} - 4 = 0$.

SOLUTION This radical equation could be solved using the method discussed in Section 10.6. However, if we note that the square of $\sqrt{x}$ is x, we can regard the equation as "quadratic in $\sqrt{x}$."

We determine u and u^2:

$$u = \sqrt{x}$$
$$u^2 = x.$$

Substituting, we have

$$x - 3\sqrt{x} - 4 = 0$$
$$u^2 - 3u - 4 = 0 \quad \textbf{Substituting}$$
$$(u - 4)(u + 1) = 0$$
$$u = 4 \quad or \quad u = -1. \quad \textbf{Using the principle of zero products}$$

CAUTION! A common error is to solve for u but forget to solve for x. Remember to solve for the *original* variable!

Next, we replace u with $\sqrt{x}$ and solve these equations:

$$\sqrt{x} = 4 \quad or \quad \sqrt{x} = -1.$$

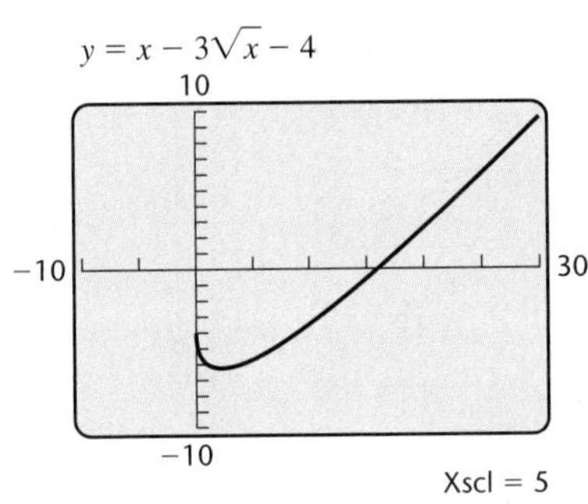

Squaring gives us $x = 16$ or $x = 1$ and also makes checking essential.

Check: For 16:

$$\begin{array}{c|c} x - 3\sqrt{x} - 4 & 0 \\ \hline 16 - 3\sqrt{16} - 4 & 0 \\ 16 - 3 \cdot 4 - 4 & \\ 0 \stackrel{?}{=} 0 & \text{TRUE} \end{array}$$

For 1:

$$\begin{array}{c|c} x - 3\sqrt{x} - 4 & 0 \\ \hline 1 - 3\sqrt{1} - 4 & 0 \\ 1 - 3 \cdot 1 - 4 & \\ -6 \stackrel{?}{=} 0 & \text{FALSE} \end{array}$$

The number 16 checks, but 1 does not. Had we noticed that $\sqrt{x} = -1$ has no solution (since principal square roots are never negative), we could have solved only the equation $\sqrt{x} = 4$. The graph at left provides further evidence that although 16 is a solution, -1 is not. The solution is 16.

Try Exercise 21.

The following tips may prove useful.

To Solve an Equation That Is Reducible to Quadratic

1. The equation is quadratic in form if the variable factor in one term is the square of the variable factor in the other variable term.
2. Write down any substitutions that you are making.
3. Whenever you make a substitution, be sure to solve for the variable that is used in the original equation.
4. Check possible answers in the original equation.

EXAMPLE 4 Solve: $2m^{-2} + m^{-1} - 15 = 0$.

SOLUTION Note that the square of m^{-1} is $(m^{-1})^2$, or m^{-2}. We let $u = m^{-1}$:

Determine u and u^2.

$$u = m^{-1}$$
$$u^2 = m^{-2}.$$

Substituting, we have

Substitute.

$$2u^2 + u - 15 = 0 \qquad \textbf{Substituting in } 2m^{-2} + m^{-1} - 15 = 0$$

$$(2u - 5)(u + 3) = 0$$

$$2u - 5 = 0 \quad or \quad u + 3 = 0 \qquad \textbf{Using the principle of zero products}$$

$$2u = 5 \quad or \quad u = -3$$

Solve for u.

$$u = \frac{5}{2} \quad or \quad u = -3.$$

Now we replace u with m^{-1} and solve:

$$m^{-1} = \frac{5}{2} \quad or \quad m^{-1} = -3$$

$$\frac{1}{m} = \frac{5}{2} \quad or \quad \frac{1}{m} = -3 \qquad \textbf{Recall that } m^{-1} = \frac{1}{m}.$$

$$1 = \frac{5}{2}m \quad or \quad 1 = -3m \qquad \textbf{Multiplying both sides by } m$$

Solve for the original variable.

$$\frac{2}{5} = m \quad or \quad -\frac{1}{3} = m. \qquad \textbf{Solving for } m$$

Check.

Check:

For $\frac{2}{5}$:

$$\begin{array}{r|l} 2m^{-2} + m^{-1} - 15 = 0 & \\ \hline 2\left(\frac{2}{5}\right)^{-2} + \left(\frac{2}{5}\right)^{-1} - 15 & 0 \\ 2\left(\frac{5}{2}\right)^{2} + \left(\frac{5}{2}\right) - 15 & \\ 2\left(\frac{25}{4}\right) + \frac{5}{2} - 15 & \\ \frac{25}{2} + \frac{5}{2} - 15 & \\ \frac{30}{2} - 15 & \\ 0 \stackrel{?}{=} 0 & \text{TRUE} \end{array}$$

For $-\frac{1}{3}$:

$$\begin{array}{r|l} 2m^{-2} + m^{-1} - 15 = 0 & \\ \hline 2\left(-\frac{1}{3}\right)^{-2} + \left(-\frac{1}{3}\right)^{-1} - 15 & 0 \\ 2\left(-\frac{3}{1}\right)^{2} + \left(-\frac{3}{1}\right) - 15 & \\ 2(9) + (-3) - 15 & \\ 18 - 3 - 15 & \\ 0 \stackrel{?}{=} 0 & \text{TRUE} \end{array}$$

STUDY TIP

Use Your Words

When a new word or phrase (such as *reducible to quadratic*) arises, try to use it in conversation with classmates, your instructor, or a study partner. Although it may seem awkward at first, doing so will help you in your reading, deepen your level of understanding, and increase your confidence.

We could also check the solutions using a calculator by entering $y_1 = 2x^{-2} + x^{-1} - 15$ and evaluating y_1 for both numbers, as shown on the left below.

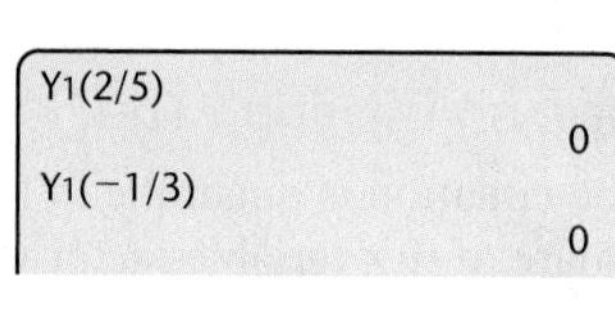

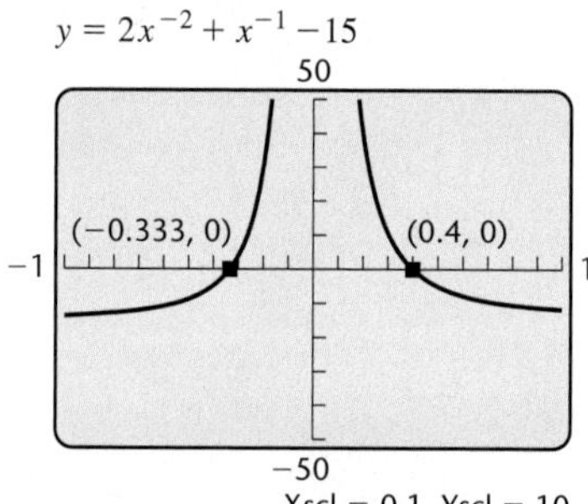

As another check, we solve the equation graphically. The first coordinates of the x-intercepts of the graph on the right above give the solutions.

Both numbers check. The solutions are $-\frac{1}{3}$ and $\frac{2}{5}$, or approximately -0.333 and 0.4.

Try Exercise 31.

Note that Example 4 can also be written

$$\frac{2}{m^2} + \frac{1}{m} - 15 = 0.$$

It can then be solved by letting $u = 1/m$ and $u^2 = 1/m^2$ or by clearing fractions as in Section 7.6.

EXAMPLE 5 Solve: $t^{2/5} - t^{1/5} - 2 = 0$.

SOLUTION Note that the square of $t^{1/5}$ is $(t^{1/5})^2$, or $t^{2/5}$. The equation is therefore quadratic in $t^{1/5}$, so we let $u = t^{1/5}$:

$$u = t^{1/5}$$
$$u^2 = t^{2/5}.$$

Substituting, we have

$$u^2 - u - 2 = 0 \quad \textbf{Substituting in } t^{2/5} - t^{1/5} - 2 = 0$$
$$(u - 2)(u + 1) = 0$$
$$u = 2 \quad or \quad u = -1. \quad \textbf{Using the principle of zero products}$$

Now we replace u with $t^{1/5}$ and solve:

$$t^{1/5} = 2 \quad or \quad t^{1/5} = -1$$
$$t = 32 \quad or \quad t = -1. \qquad \textbf{Principle of powers; raising to the 5th power}$$

Check:

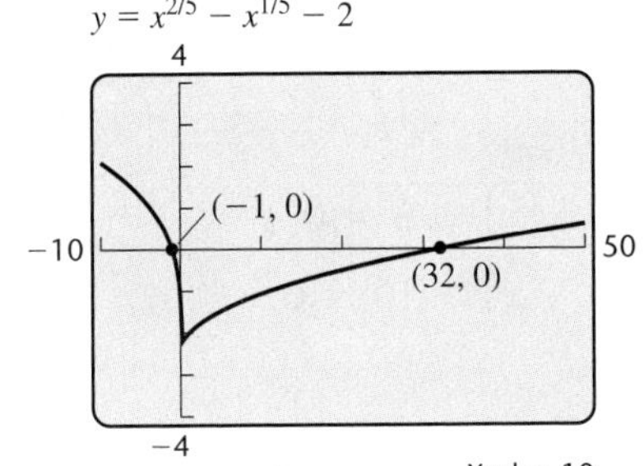

For 32:

$$\begin{array}{l|l} t^{2/5} - t^{1/5} - 2 = 0 & \\ \hline 32^{2/5} - 32^{1/5} - 2 & 0 \\ (32^{1/5})^2 - 32^{1/5} - 2 & \\ 2^2 - 2 - 2 & \\ 0 \stackrel{?}{=} 0 & \text{TRUE} \end{array}$$

For -1:

$$\begin{array}{l|l} t^{2/5} - t^{1/5} - 2 = 0 & \\ \hline (-1)^{2/5} - (-1)^{1/5} - 2 & 0 \\ [(-1)^{1/5}]^2 - (-1)^{1/5} - 2 & \\ (-1)^2 - (-1) - 2 & \\ 0 \stackrel{?}{=} 0 & \text{TRUE} \end{array}$$

Both numbers check and are confirmed by the graph at left. The solutions are 32 and -1.

Try Exercise 33.

EXAMPLE 6 Solve: $(5 + \sqrt{r})^2 + 6(5 + \sqrt{r}) + 2 = 0$.

SOLUTION We determine u and u^2:

$$u = 5 + \sqrt{r}$$
$$u^2 = (5 + \sqrt{r})^2.$$

Substituting, we have

$$u^2 + 6u + 2 = 0$$

$$u = \frac{-6 \pm \sqrt{6^2 - 4 \cdot 1 \cdot 2}}{2 \cdot 1} \qquad \textbf{Using the quadratic formula}$$

$$\left.\begin{aligned} &= \frac{-6 \pm \sqrt{28}}{2} \\ &= \frac{-6}{2} \pm \frac{2\sqrt{7}}{2} \\ &= -3 \pm \sqrt{7}. \end{aligned}\right\} \qquad \textbf{Simplifying; } \sqrt{28} = \sqrt{4}\sqrt{7}$$

Now we replace u with $5 + \sqrt{r}$ and solve for r:

$$5 + \sqrt{r} = -3 + \sqrt{7} \quad or \quad 5 + \sqrt{r} = -3 - \sqrt{7} \qquad u = -3 + \sqrt{7}$$
$$\sqrt{r} = -8 + \sqrt{7} \quad or \quad \sqrt{r} = -8 - \sqrt{7}. \qquad or\ u = -3 - \sqrt{7}$$

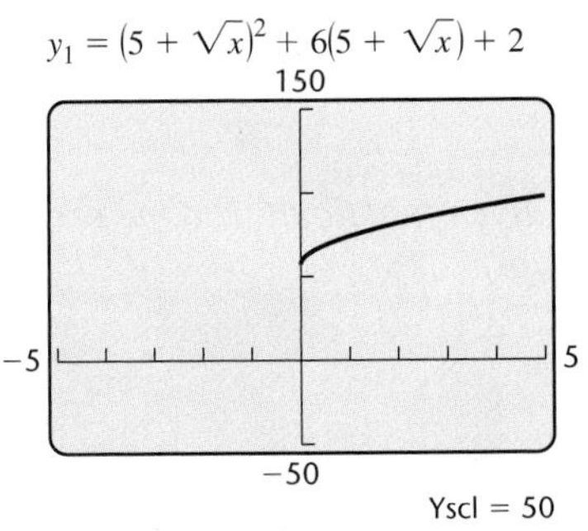

We could now solve for r and check possible solutions, but first let's examine $-8 + \sqrt{7}$ and $-8 - \sqrt{7}$. Since $\sqrt{7} \approx 2.6$, both $-8 + \sqrt{7}$ and $-8 - \sqrt{7}$ are negative. Since the principal square root of r is never negative, both values of $\sqrt{r}$ must be rejected. Note too that in the original equation, $(5 + \sqrt{r})^2$, $6(5 + \sqrt{r})$, and 2 are all positive. Thus it is impossible for their sum to be 0.

As another check, using the graph at left, we see that the domain of the function $f(x) = (5 + \sqrt{x})^2 + 6(5 + \sqrt{x}) + 2$ is $(0, \infty)$, so neither value found is an allowable input. Also, there are no x-intercepts of the graph, and thus no solution of $f(x) = 0$.

The original equation has no solution.

Try Exercise 27.

11.5 Exercise Set

FOR EXTRA HELP MyMathLab MathXL PRACTICE  WATCH  DOWNLOAD 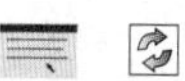 READ REVIEW

Concept Reinforcement *In each of Exercises 1–8, match the equation with a substitution from the column on the right that could be used to reduce the equation to quadratic form.*

1. ___ $4x^6 - 2x^3 + 1 = 0$ **a)** $u = x^{-1/3}$
2. ___ $3x^4 + 4x^2 - 7 = 0$ **b)** $u = x^{1/3}$
3. ___ $5x^8 + 2x^4 - 3 = 0$ **c)** $u = x^{-2}$
4. ___ $2x^{2/3} - 5x^{1/3} + 4 = 0$ **d)** $u = x^2$
5. ___ $3x^{4/3} + 4x^{2/3} - 7 = 0$ **e)** $u = x^{-2/3}$
6. ___ $2x^{-2/3} + x^{-1/3} + 6 = 0$ **f)** $u = x^3$
7. ___ $4x^{-4/3} - 2x^{-2/3} + 3 = 0$ **g)** $u = x^{2/3}$
8. ___ $3x^{-4} + 4x^{-2} - 2 = 0$ **h)** $u = x^4$

Write the substitution that could be used to make each equation quadratic in u.

9. For $3p - 4\sqrt{p} + 6 = 0$, use $u =$ ________.
10. For $x^{1/2} - x^{1/4} - 2 = 0$, use $u =$ ________.
11. For $(x^2 + 3)^2 + (x^2 + 3) - 7 = 0$, use $u =$ ________.
12. For $t^{-6} + 5t^{-3} - 6 = 0$, use $u =$ ________.
13. For $(1 + t)^4 + (1 + t)^2 + 4 = 0$, use $u =$ ________.
14. For $w^{1/3} - 3w^{1/6} + 8 = 0$, use $u =$ ________.

Solve.

15. $x^4 - 5x^2 + 4 = 0$
16. $x^4 - 10x^2 + 9 = 0$
17. $x^4 - 9x^2 + 20 = 0$
18. $x^4 - 12x^2 + 27 = 0$
19. $4t^4 - 19t^2 + 12 = 0$
20. $9t^4 - 14t^2 + 5 = 0$
21. $w + 4\sqrt{w} - 12 = 0$
22. $s + 3\sqrt{s} - 40 = 0$
23. $(x^2 - 7)^2 - 3(x^2 - 7) + 2 = 0$
24. $(x^2 - 2)^2 - 12(x^2 - 2) + 20 = 0$
25. $r - 2\sqrt{r} - 6 = 0$
26. $s - 4\sqrt{s} - 1 = 0$
27. $(1 + \sqrt{x})^2 + 5(1 + \sqrt{x}) + 6 = 0$
28. $(3 + \sqrt{x})^2 + 3(3 + \sqrt{x}) - 10 = 0$
29. $x^{-2} - x^{-1} - 6 = 0$
30. $2x^{-2} - x^{-1} - 1 = 0$
31. $4y^{-2} - 3y^{-1} - 1 = 0$
32. $m^{-2} + 9m^{-1} - 10 = 0$
33. $t^{2/3} + t^{1/3} - 6 = 0$
34. $w^{2/3} - 2w^{1/3} - 8 = 0$
35. $y^{1/3} - y^{1/6} - 6 = 0$
36. $t^{1/2} + 3t^{1/4} + 2 = 0$
37. $t^{1/3} + 2t^{1/6} = 3$
38. $m^{1/2} + 6 = 5m^{1/4}$
39. $(10 - \sqrt{x})^2 - 2(10 - \sqrt{x}) - 35 = 0$
40. $(5 + \sqrt{x})^2 - 12(5 + \sqrt{x}) + 33 = 0$
41. $16\left(\frac{x - 1}{x - 8}\right)^2 + 8\left(\frac{x - 1}{x - 8}\right) + 1 = 0$
42. $9\left(\frac{x + 2}{x + 3}\right)^2 - 6\left(\frac{x + 2}{x + 3}\right) + 1 = 0$
43. $x^4 + 5x^2 - 36 = 0$
44. $x^4 + 5x^2 + 4 = 0$
45. $(n^2 + 6)^2 - 7(n^2 + 6) + 10 = 0$
46. $(m^2 + 7)^2 - 6(m^2 + 7) - 16 = 0$

Find all x-intercepts of the given function f. If none exists, state this.

47. $f(x) = 5x + 13\sqrt{x} - 6$
48. $f(x) = 3x + 10\sqrt{x} - 8$
49. $f(x) = (x^2 - 3x)^2 - 10(x^2 - 3x) + 24$
50. $f(x) = (x^2 - 6x)^2 - 2(x^2 - 6x) - 35$
51. $f(x) = x^{2/5} + x^{1/5} - 6$
52. $f(x) = x^{1/2} - x^{1/4} - 6$

Aha! **53.** $f(x) = \left(\frac{x^2 + 2}{x}\right)^4 + 7\left(\frac{x^2 + 2}{x}\right)^2 + 5$

54. $f(x) = \left(\frac{x^2 + 1}{x}\right)^4 + 4\left(\frac{x^2 + 1}{x}\right)^2 + 12$

TW **55.** To solve $25x^6 - 10x^3 + 1 = 0$, Margaret lets $u = 5x^3$ and Murray lets $u = x^3$. Can they both be correct? Why or why not?

TW **56.** While trying to solve $0.05x^4 - 0.8 = 0$ with a graphing calculator, Carmela gets the following screen. Can Carmela solve this equation with a graphing calculator? Why or why not?

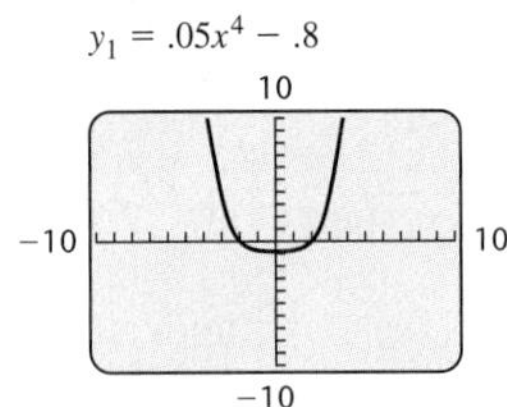

SKILL REVIEW

To prepare for Section 11.6, review graphing functions (Section 7.3).

Graph. [7.3]

57. $f(x) = x$

58. $g(x) = x + 2$

59. $h(x) = x - 2$

60. $f(x) = x^2$

61. $g(x) = x^2 + 2$

62. $h(x) = x^2 - 2$

SYNTHESIS

TW **63.** Describe a procedure that could be used to solve any equation of the form $ax^4 + bx^2 + c = 0$.

TW **64.** Describe a procedure that could be used to write an equation that is quadratic in $3x^2 - 1$. Then explain how the procedure could be adjusted to write equations that are quadratic in $3x^2 - 1$ and have no real-number solution.

Solve.

65. $5x^4 - 7x^2 + 1 = 0$

66. $3x^4 + 5x^2 - 1 = 0$

67. $(x^2 - 4x - 2)^2 - 13(x^2 - 4x - 2) + 30 = 0$

68. $(x^2 - 5x - 1)^2 - 18(x^2 - 5x - 1) + 65 = 0$

69. $\frac{x}{x - 1} - 6\sqrt{\frac{x}{x - 1}} - 40 = 0$

70. $\left(\sqrt{\frac{x}{x - 3}}\right)^2 - 24 = 10\sqrt{\frac{x}{x - 3}}$

71. $a^5(a^2 - 25) + 13a^3(25 - a^2) + 36a(a^2 - 25) = 0$

72. $a^3 - 26a^{3/2} - 27 = 0$

73. $x^6 - 28x^3 + 27 = 0$

74. $x^6 + 7x^3 - 8 = 0$

Try Exercise Answers: Section 11.5

15. $\pm 1, \pm 2$ **21.** 4 **27.** No solution **31.** $-4, 1$ **33.** $-27, 8$
49. $\left(\frac{3}{2} + \frac{\sqrt{33}}{2}, 0\right), \left(\frac{3}{2} - \frac{\sqrt{33}}{2}, 0\right), (4, 0), (-1, 0)$

Mid-Chapter Review

We have discussed four methods of solving quadratic equations:

- factoring and the principle of zero products;
- the principle of square roots;
- completing the square;
- the quadratic formula.

Any of these may also be appropriate when solving an applied problem or an equation that is reducible to quadratic form.

GUIDED SOLUTIONS

1. Solve: $(x - 7)^2 = 5$.

Solution

$x - 7 = \square$ **Using the principle of square roots**

$x = \square$ **Adding 7 to both sides**

The solutions are $7 + \square$ and $7 - \square$.

2. Solve: $x^2 - 2x - 1 = 0$.

Solution

$a = \square,\ b = \square,\ c = \square$

$$x = \frac{-(\square) \pm \sqrt{(\square)^2 - 4 \cdot 1 \cdot (\square)}}{2 \cdot \square}$$

$$x = \frac{\square \pm \sqrt{\square}}{\square}$$

$$x = \frac{2}{2} \pm \frac{\square\sqrt{2}}{2}$$

The solutions are $1 + \square$ and $1 - \square$.

MIXED REVIEW

Solve. Examine each exercise carefully, and try to solve using the easiest method.

1. $x^2 - 3x - 10 = 0$ [11.1]

2. $x^2 = 121$ [11.1]

3. $x^2 + 6x = 10$ [11.1], [11.2]

4. $x^2 + x - 3 = 0$ [11.2]

5. $(x + 1)^2 = 2$ [11.1]

6. $x^2 - 10x + 25 = 0$ [11.1]

7. $4t^2 = 11$ [11.1]

8. $2t^2 + 1 = 3t$ [11.1]

9. $16c^2 = 7c$ [11.1]

10. $y^2 - 2y + 8 = 0$ [11.2]

11. $x^4 - 10x^2 + 9 = 0$ [11.5]

12. $x^4 - 8x^2 - 9 = 0$ [11.5]

13. $(t + 4)(t - 3) = 18$ [11.1]

14. $m^{-4} - 5m^{-2} + 6 = 0$ [11.5]

For each equation, determine what type of number the solutions are and how many solutions exist. [11.3]

15. $x^2 - 8x + 1 = 0$

16. $3x^2 = 4x + 7$

17. $5x^2 - x + 6 = 0$

Solve each formula for the indicated letter. Assume that all variables represent nonnegative numbers. [11.4]

18. $F = \dfrac{Av^2}{400}$, for v
(Force of wind on a sail)

19. $D^2 - 2Dd - 2hd = 0$, for D
(Dynamic load)

20. Sophie drove 225 mi south through the Smoky Mountains. It was snowing on her return trip, so her average speed on the return trip was 30 mph slower. The total driving time was 8 hr. Find Sophie's average speed on each part of the trip. [11.4]

11.6 Quadratic Functions and Their Graphs

- The Graph of $f(x) = ax^2$
- The Graph of $f(x) = a(x - h)^2$
- The Graph of $f(x) = a(x - h)^2 + k$

The graph of any *linear* function $f(x) = mx + b$ is a straight line. In this section and the next, we will see that the graph of any *quadratic* function $f(x) = ax^2 + bx + c$ is a *parabola*.

THE GRAPH OF $f(x) = ax^2$

The most basic quadratic function is $f(x) = x^2$.

EXAMPLE 1 Graph: $f(x) = x^2$.

SOLUTION We choose some values for x and compute $f(x)$ for each. Then we plot the ordered pairs and connect them with a smooth curve.

x	$f(x) = x^2$	$(x, f(x))$
-3	9	$(-3, 9)$
-2	4	$(-2, 4)$
-1	1	$(-1, 1)$
0	0	$(0, 0)$
1	1	$(1, 1)$
2	4	$(2, 4)$
3	9	$(3, 9)$

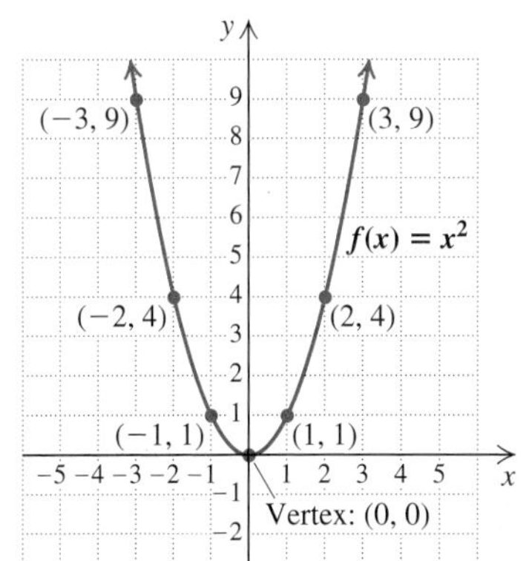

Try Exercise 15.

All quadratic functions have graphs similar to the one in Example 1. Such curves are called **parabolas.** They are U-shaped and symmetric with respect to a vertical line known as the parabola's *axis of symmetry*. For the graph of $f(x) = x^2$, the y-axis (or the line $x = 0$) is the axis of symmetry. Were the paper folded on this line, the two halves of the curve would match. The point $(0, 0)$ is known as the *vertex* of this parabola.

Since x^2 is a polynomial, we know that the domain of $f(x) = x^2$ is the set of all real numbers, or $(-\infty, \infty)$. We see from the graph and the table in Example 1 that the range of f is $\{y \mid y \geq 0\}$, or $[0, \infty)$. The function has a *minimum value* of 0. This minimum value occurs when $x = 0$.

Now let's consider a quadratic function of the form $g(x) = ax^2$. How does the constant a affect the graph of the function?

Student Notes

By paying attention to the symmetry of each parabola and the location of the vertex, you save yourself considerable work. Note too that when we are graphing ax^2, the x-values 1 unit to the right or left of the vertex are paired with the y-value a units above the vertex. Thus the graph of $y = \frac{3}{2}x^2$ includes the points $\left(-1, \frac{3}{2}\right)$ and $\left(1, \frac{3}{2}\right)$.

Interactive Discovery

Graph each of the following equations along with $y_1 = x^2$ in a $[-5, 5, -10, 10]$ window. If your calculator has a Transfrm application, run that application and graph $y_2 = Ax^2$. Then enter the values for A indicated in Exercises 1–6. For each equation, answer the following questions.

a) What is the vertex of the graph of y_2?

b) What is the axis of symmetry of the graph of y_2?

c) Does the graph of y_2 open upward or downward?

d) Is the parabola narrower or wider than the parabola given by $y_1 = x^2$?

1. $y_2 = 3x^2$ **2.** $y_2 = \frac{1}{3}x^2$ **3.** $y_2 = 0.2x^2$

4. $y_2 = -x^2$ **5.** $y_2 = -4x^2$ **6.** $y_2 = -\frac{2}{3}x^2$

7. Describe the effect of multiplying x^2 by a when $a > 1$ and when $0 < a < 1$.

8. Describe the effect of multiplying x^2 by a when $a < -1$ and when $-1 < a < 0$.

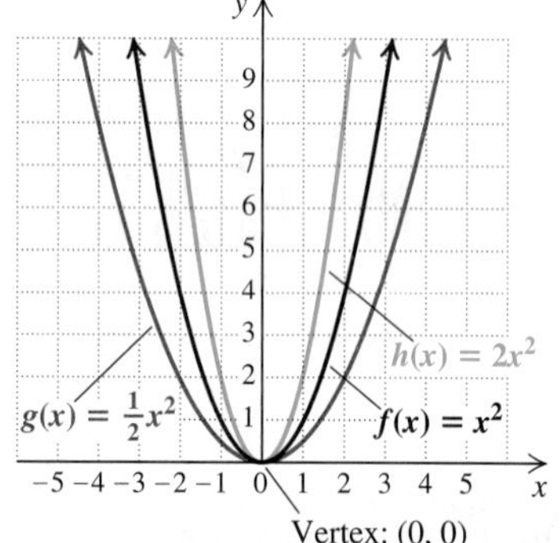

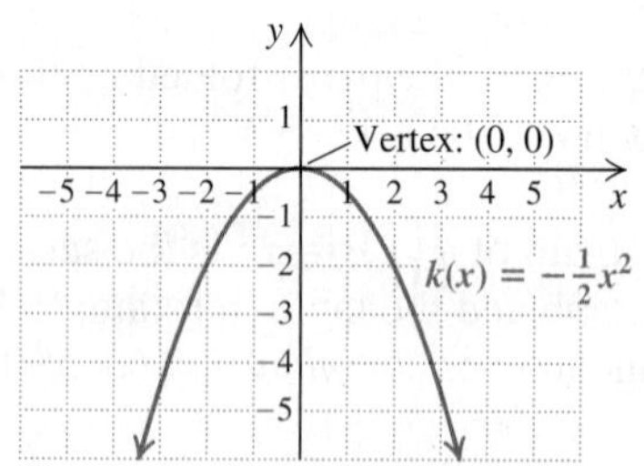

You may have noticed that for $f(x) = ax^2$, the constant a, when $|a| \neq 1$, makes the graph wider or narrower than the graph of $g(x) = x^2$. When a is negative, the graph opens downward.

Graphing $f(x) = ax^2$

The graph of $f(x) = ax^2$ is a parabola with $x = 0$ as its axis of symmetry. Its vertex is the origin. The domain of f is $(-\infty, \infty)$.

For $a > 0$, the parabola opens upward. The range of the function is $[0, \infty)$.

For $a < 0$, the parabola opens downward. The range of the function is $(-\infty, 0]$.

If $|a|$ is greater than 1, the parabola is narrower than $y = x^2$.

If $|a|$ is between 0 and 1, the parabola is wider than $y = x^2$.

EXAMPLE 2 Graph: $g(x) = \frac{1}{2}x^2$.

SOLUTION The function is in the form $g(x) = ax^2$, where $a = \frac{1}{2}$. The vertex is $(0, 0)$ and the axis of symmetry is $x = 0$. Since $\frac{1}{2} > 0$, the parabola opens upward. The parabola is wider than $y = x^2$ since $|\frac{1}{2}|$ is between 0 and 1.

To graph the function, we calculate and plot points. Because we know the general shape of the graph, we can sketch the graph using only a few points.

STUDY TIP

Know Yourself

Not everyone learns in the same way. Use a learning center on campus, a book, or the Internet to find out what your learning style is. Then study in a way that fits your learning style best.

x	$g(x) = \frac{1}{2}x^2$
-3	$\frac{9}{2}$
-2	2
-1	$\frac{1}{2}$
0	0
1	$\frac{1}{2}$
2	2
3	$\frac{9}{2}$

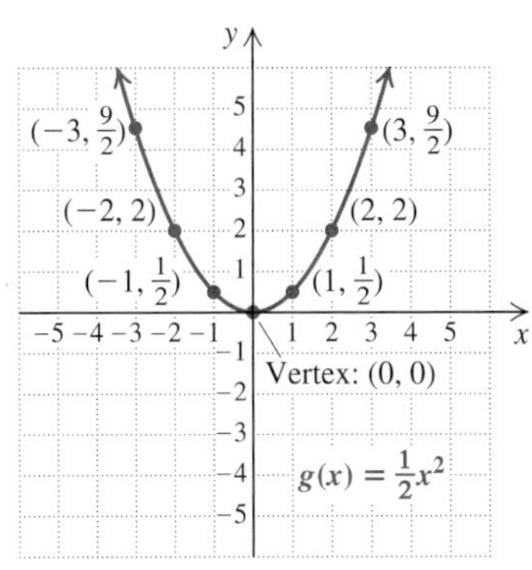

Try Exercise 19.

Note in Example 2 that since a parabola is symmetric, we can calculate function values for inputs to one side of the vertex and then plot the mirror images of those points on the other "half" of the graph.

THE GRAPH OF $f(x) = a(x - h)^2$

The width of a parabola and whether it opens upward or downward are determined by the coefficient a in $f(x) = ax^2 + bx + c$. To determine the vertex, we look first at functions of the form $f(x) = a(x - h)^2$.

Interactive Discovery

Graph each of the following pairs of equations in a $[-5, 5, -10, 10]$ window. For each pair, answer the following questions.

a) What is the vertex of the graph of y_2?

b) What is the axis of symmetry of the graph of y_2?

c) Is the graph of y_2 narrower, wider, or the same shape as the graph of y_1?

1. $y_1 = x^2$; $y_2 = (x - 3)^2$
2. $y_1 = 2x^2$; $y_2 = 2(x + 1)^2$
3. $y_1 = -\frac{1}{2}x^2$; $y_2 = -\frac{1}{2}\left(x - \frac{3}{2}\right)^2$
4. $y_1 = -3x^2$; $y_2 = -3(x + 2)^2$
5. Describe the effect of h on the graph of $g(x) = a(x - h)^2$.

The graph of $g(x) = a(x - h)^2$ looks just like the graph of $f(x) = ax^2$, except that it is moved, or *translated*, $|h|$ units horizontally.

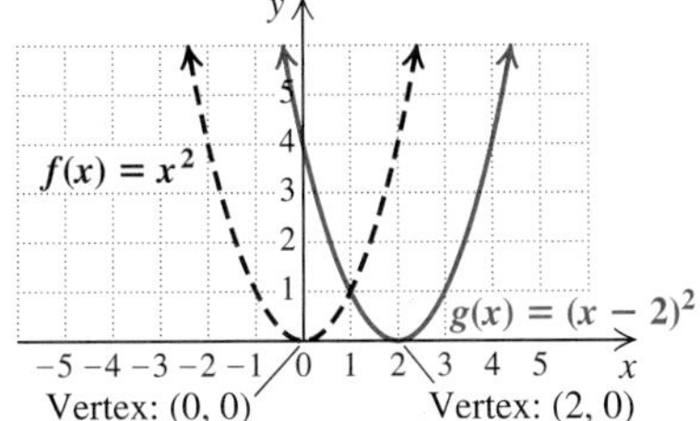

Graphing $f(x) = a(x - h)^2$

The graph of $f(x) = a(x - h)^2$ has the same shape as the graph of $y = ax^2$.
The domain of f is $(-\infty, \infty)$.

If h is positive, the graph of $y = ax^2$ is shifted h units to the right.

If h is negative, the graph of $y = ax^2$ is shifted $|h|$ units to the left.

The vertex is $(h, 0)$, and the axis of symmetry is $x = h$.

EXAMPLE 3 Graph $g(x) = -2(x + 3)^2$. Label the vertex, and draw the line of symmetry.

SOLUTION We rewrite the equation as $g(x) = -2[x - (-3)]^2$. In this case, $a = -2$ and $h = -3$, so the graph looks like that of $y = 2x^2$ translated 3 units to the left and, since $-2 < 0$, it opens downward. The vertex is $(-3, 0)$, and the axis of symmetry is $x = -3$. Plotting points as needed, we obtain the graph shown below. We say that g is a *reflection* of the graph of $f(x) = 2(x + 3)^2$ across the x-axis.

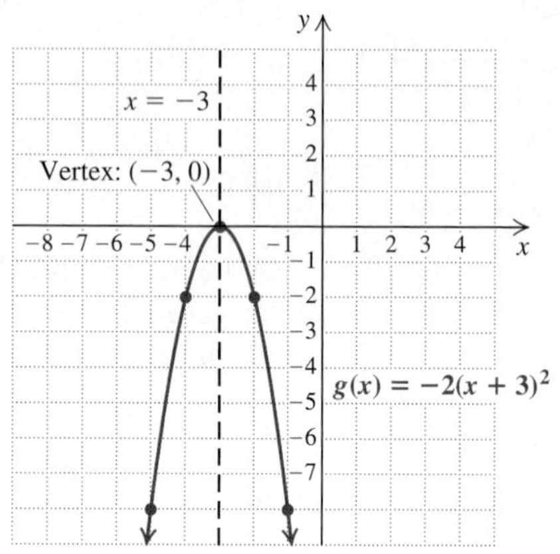

Try Exercise 25.

THE GRAPH OF $f(x) = a(x - h)^2 + k$

If we add a positive constant k to $f(x) = a(x - h)^2$, the graph of $f(x)$ is moved up. If we add a negative constant k, the graph is moved down. The axis of symmetry for the parabola remains at $x = h$, but the vertex will be at (h, k).

Because of the shape of their graphs, quadratic functions have either a *minimum* value or a *maximum* value. Many real-world applications involve finding that value. For example, a business owner is concerned with minimizing cost and maximizing profit. If a parabola opens upward $(a > 0)$, the function value, or y-value, at the vertex is a least, or minimum, value. That is, it is less than the y-value at any other point on the graph. If the parabola opens downward $(a < 0)$, the function value at the vertex is a greatest, or maximum, value.

Graphs of $f(x) = a(x - h)^2 + k$

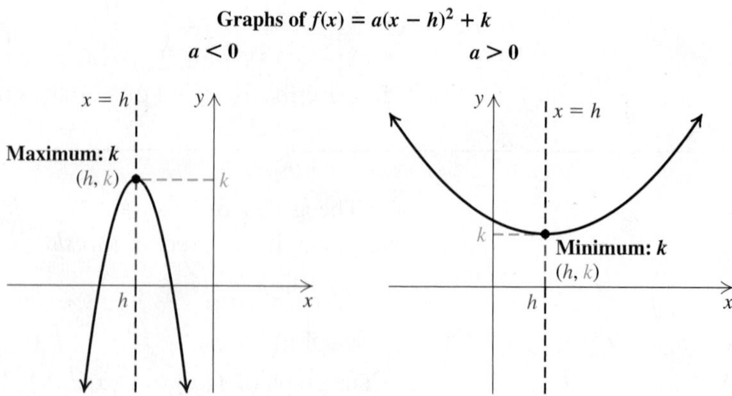

Graphing $f(x) = a(x - h)^2 + k$

The graph of $f(x) = a(x - h)^2 + k$ has the same shape as the graph of $y = a(x - h)^2$.

If k is positive, the graph of $y = a(x - h)^2$ is shifted k units up.

If k is negative, the graph of $y = a(x - h)^2$ is shifted $|k|$ units down.

The vertex is (h, k), and the axis of symmetry is $x = h$.

The domain of f is $(-\infty, \infty)$.

For $a > 0$, the range of f is $[k, \infty)$. The minimum function value is k, which occurs when $x = h$.

For $a < 0$, the range of f is $(-\infty, k]$. The maximum function value is k, which occurs when $x = h$.

EXAMPLE 4 Graph $g(x) = (x - 3)^2 - 5$, and find the minimum function value and the range of g.

SOLUTION The graph will look like that of $f(x) = (x - 3)^2$ but shifted 5 units down. You can confirm this by plotting some points.

The vertex is now $(3, -5)$, and the minimum function value is -5. The range of g is $[-5, \infty)$.

x	$g(x) = (x - 3)^2 - 5$
0	4
1	-1
2	-4
3	-5
4	-4
5	-1
6	4

← Vertex

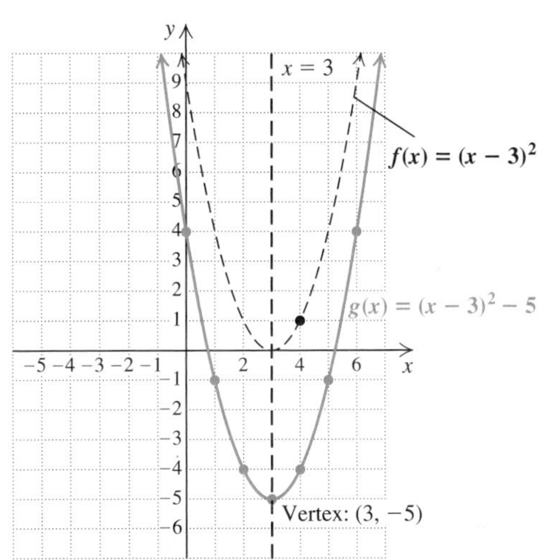

Try Exercise 45.

EXAMPLE 5 Graph $h(x) = \frac{1}{2}(x - 3)^2 + 4$, and find the minimum function value and the range of h.

SOLUTION The graph looks just like that of $f(x) = \frac{1}{2}x^2$ but moved 3 units to the right and 4 units up. The vertex is $(3, 4)$, and the axis of symmetry is $x = 3$. We draw $f(x) = \frac{1}{2}x^2$ and then shift the curve over and up.

By plotting some points, we have a check. The minimum function value is 4, and the range is $[4, \infty)$.

x	$h(x) = \frac{1}{2}(x - 3)^2 + 4$
0	$8\frac{1}{2}$
1	6
3	4
5	6
6	$8\frac{1}{2}$

← Vertex

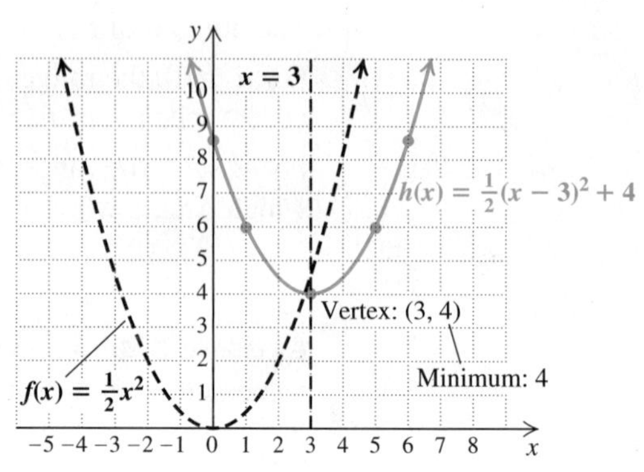

Try Exercise 49.

EXAMPLE 6 Graph $f(x) = -2(x + 3)^2 + 5$. Find the vertex, the axis of symmetry, the maximum or minimum function value, and the range of f.

SOLUTION We first express the equation in the equivalent form

$$f(x) = -2[x - (-3)]^2 + 5. \quad \text{This is in the form } y = a(x - h)^2 + k.$$

The graph looks like that of $y = -2x^2$ translated 3 units to the left and 5 units up. The vertex is $(-3, 5)$, and the axis of symmetry is $x = -3$. Since -2 is negative, we know that 5, the second coordinate of the vertex, is the maximum function value. The range of f is $(-\infty, 5]$.

We compute a few points as needed, selecting convenient x-values on either side of the vertex. The graph is shown here.

x	$f(x) = -2(x + 3)^2 + 5$
−4	3
−3	5
−2	3

← Vertex

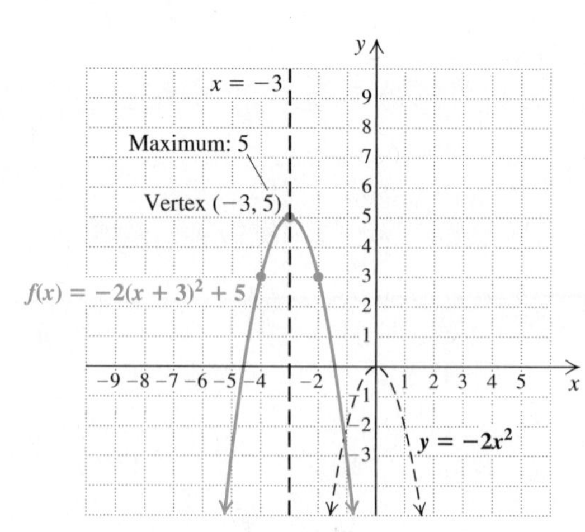

Try Exercise 59.

Connecting the Concepts

We can read information about the graph of a quadratic function directly from the constants a, h, and k in $f(x) = a(x - h)^2 + k$.

$$f(x) = a(x - h)^2 + k$$

- a: Determines the width of the parabola; Tells whether the graph opens up or down; Tells whether the function has a maximum or a minimum
- h: Gives the axis of symmetry of the graph
- k: Gives the maximum or minimum function value
- h and k: Together, give the vertex of the graph

11.6 Exercise Set

FOR EXTRA HELP MyMathLab   Math XL PRACTICE WATCH DOWNLOAD READ REVIEW

Concept Reinforcement *In each of Exercises 1–8, match the equation with the corresponding graph from those shown.*

1. ___ $f(x) = 2(x - 1)^2 + 3$
2. ___ $f(x) = -2(x - 1)^2 + 3$
3. ___ $f(x) = 2(x + 1)^2 + 3$
4. ___ $f(x) = 2(x - 1)^2 - 3$
5. ___ $f(x) = -2(x + 1)^2 + 3$
6. ___ $f(x) = -2(x + 1)^2 - 3$
7. ___ $f(x) = 2(x + 1)^2 - 3$
8. ___ $f(x) = -2(x - 1)^2 - 3$

a) 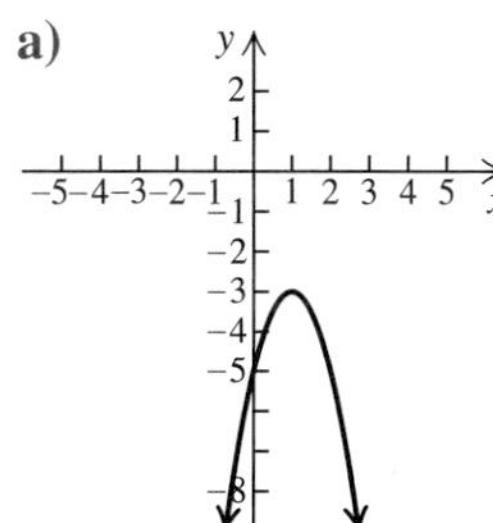

b) 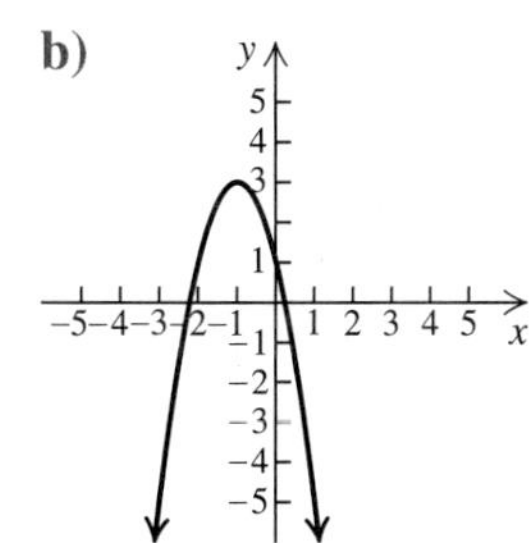

c) 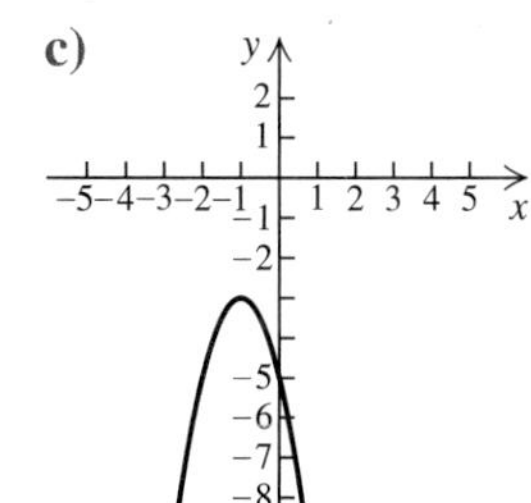

d) 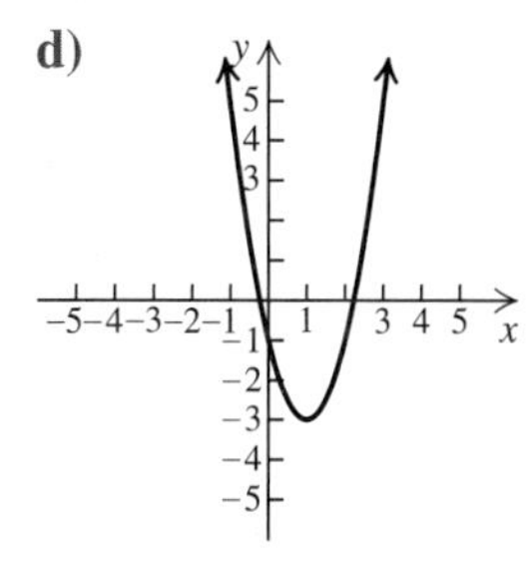

e) 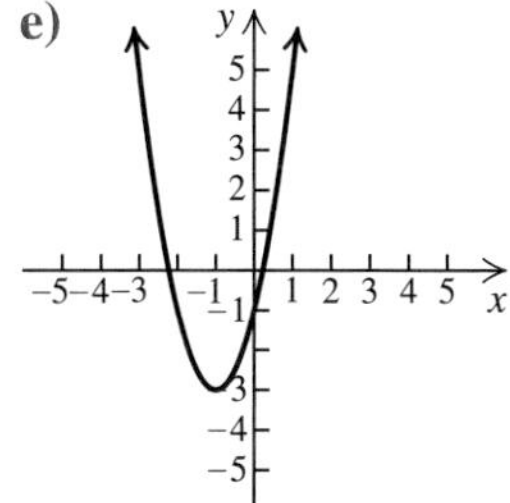

f) 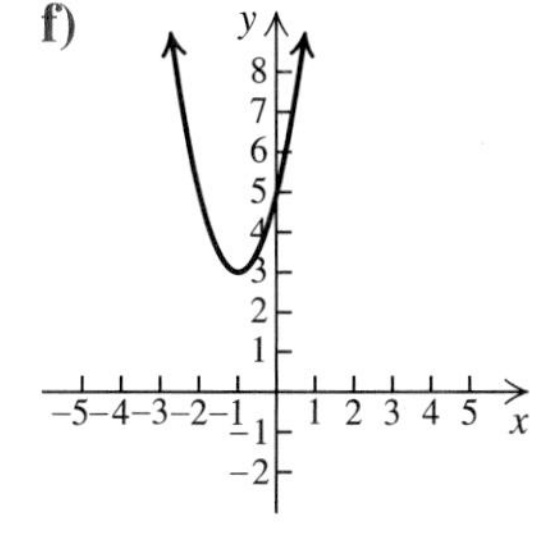

g) 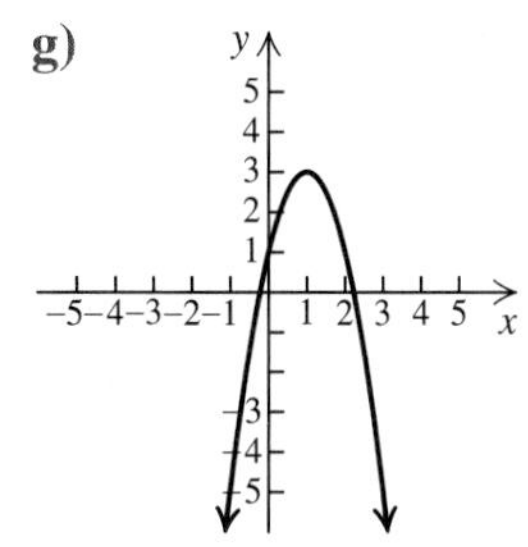

h) 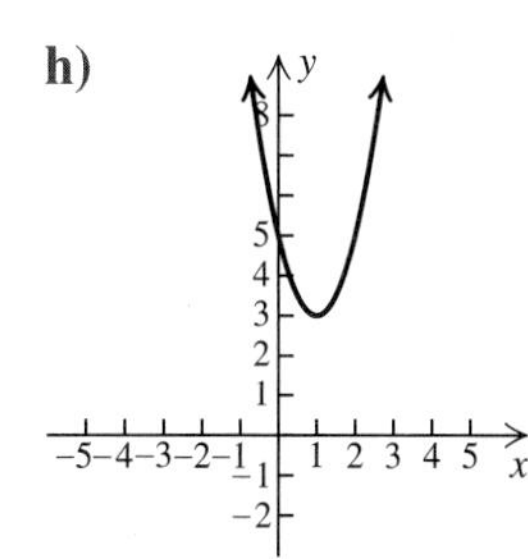

For each graph of a quadratic function

$$f(x) = a(x - h)^2 + k$$

in Exercises 9–14:

a) *Tell whether a is positive or negative.*
b) *Determine the vertex.*
c) *Determine the axis of symmetry.*
d) *Determine the range.*

9.

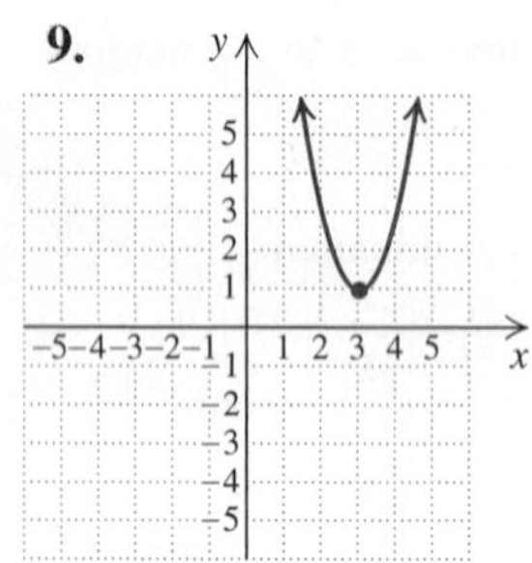

10.

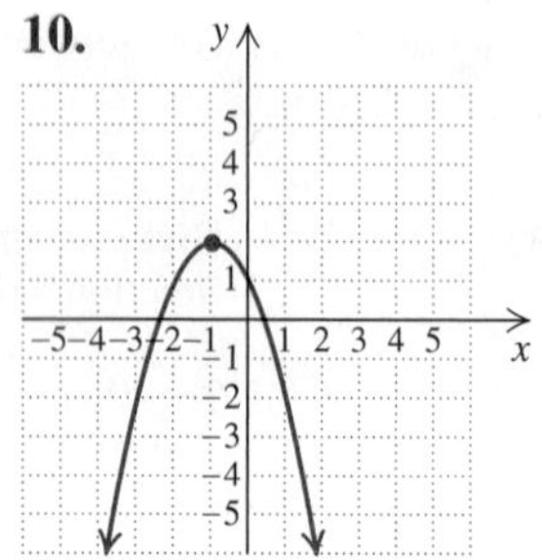

11.

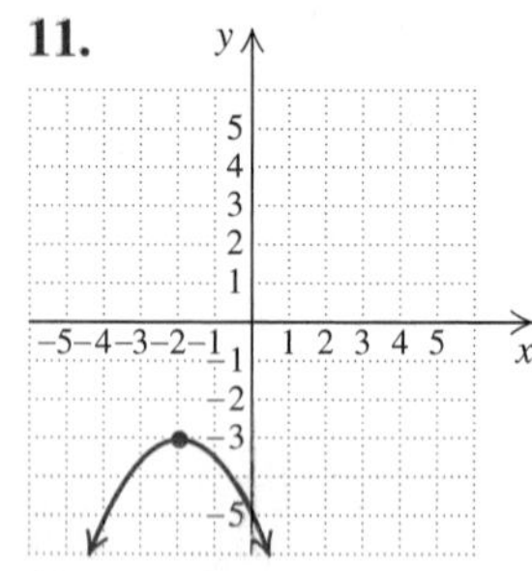

12.

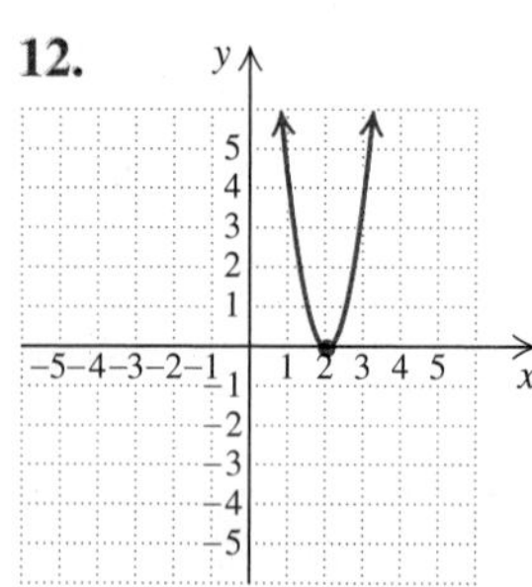

13.

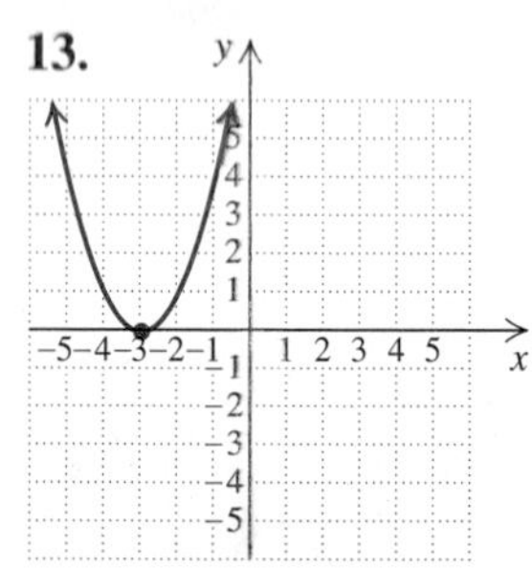

14.

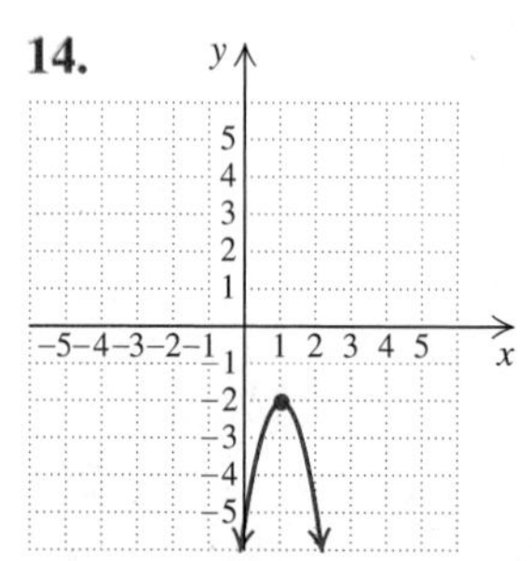

Graph.

15. $f(x) = x^2$

16. $f(x) = -x^2$

17. $f(x) = -2x^2$

18. $f(x) = -3x^2$

19. $g(x) = \frac{1}{3}x^2$

20. $g(x) = \frac{1}{4}x^2$

Aha! **21.** $h(x) = -\frac{1}{3}x^2$

22. $h(x) = -\frac{1}{4}x^2$

23. $f(x) = \frac{5}{2}x^2$

24. $f(x) = \frac{3}{2}x^2$

For each of the following, graph the function, label the vertex, and draw the axis of symmetry.

25. $g(x) = (x + 1)^2$

26. $g(x) = (x + 4)^2$

27. $f(x) = (x - 2)^2$

28. $f(x) = (x - 1)^2$

29. $f(x) = -(x + 1)^2$

30. $f(x) = -(x - 1)^2$

31. $g(x) = -(x - 2)^2$

32. $g(x) = -(x + 4)^2$

33. $f(x) = 2(x + 1)^2$

34. $f(x) = 2(x + 4)^2$

35. $g(x) = 3(x - 4)^2$

36. $g(x) = 3(x - 5)^2$

37. $h(x) = -\frac{1}{2}(x - 4)^2$

38. $h(x) = -\frac{3}{2}(x - 2)^2$

39. $f(x) = \frac{1}{2}(x - 1)^2$

40. $f(x) = \frac{1}{3}(x + 2)^2$

41. $f(x) = -2(x + 5)^2$

42. $f(x) = 2(x + 7)^2$

43. $h(x) = -3\left(x - \frac{1}{2}\right)^2$

44. $h(x) = -2\left(x + \frac{1}{2}\right)^2$

For each of the following, graph the function and find the maximum value or the minimum value and the range of the function.

45. $f(x) = (x - 5)^2 + 2$

46. $f(x) = (x + 3)^2 - 2$

47. $f(x) = -(x + 2)^2 - 1$

48. $f(x) = -(x - 1)^2 + 3$

49. $g(x) = \frac{1}{2}(x + 4)^2 + 3$

50. $g(x) = 2(x - 4)^2 - 1$

51. $h(x) = -2(x - 1)^2 - 3$

52. $h(x) = -\frac{1}{2}(x + 2)^2 + 1$

For each of the following, graph the function and find the vertex, the axis of symmetry, the maximum value or the minimum value, and the range of the function.

53. $f(x) = (x + 1)^2 - 3$

54. $f(x) = (x - 1)^2 + 2$

55. $g(x) = -(x + 3)^2 + 5$

56. $g(x) = -(x - 2)^2 - 4$

57. $f(x) = \frac{1}{2}(x - 2)^2 + 1$

58. $f(x) = -\frac{1}{2}(x + 1)^2 - 1$

59. $h(x) = -2(x - 1)^2 - 3$

60. $h(x) = -2(x + 1)^2 + 4$

61. $f(x) = 2(x + 4)^2 + 1$

62. $f(x) = 2(x - 5)^2 - 3$

63. $g(x) = -\frac{3}{2}(x - 1)^2 + 4$

64. $g(x) = \frac{3}{2}(x + 2)^2 - 3$

Without graphing, find the vertex, the axis of symmetry, and the maximum value or the minimum value.

65. $f(x) = 6(x - 8)^2 + 7$

66. $f(x) = 4(x + 5)^2 - 6$

67. $h(x) = -\frac{2}{7}(x + 6)^2 + 11$

68. $h(x) = -\frac{3}{11}(x - 7)^2 - 9$

69. $f(x) = \left(x - \frac{7}{2}\right)^2 - \frac{29}{4}$

70. $f(x) = -\left(x + \frac{3}{4}\right)^2 + \frac{17}{16}$

71. $f(x) = \sqrt{2}(x + 4.58)^2 + 65\pi$

72. $f(x) = 4\pi(x - 38.2)^2 - \sqrt{34}$

TW **73.** While trying to graph $y = -\frac{1}{2}x^2 + 3x + 1$, Ibrahim gets the following screen. How can Ibrahim tell at a glance that a mistake has been made?

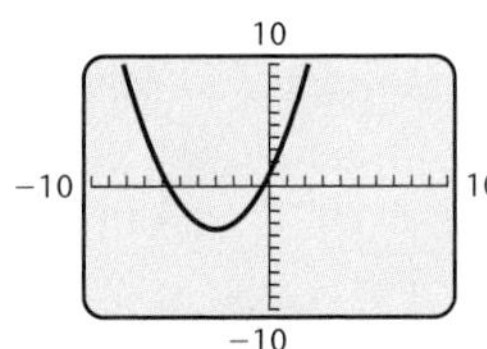

TW **74.** Explain, without plotting points, why the graph of $y = (x + 2)^2$ looks like the graph of $y = x^2$ translated 2 units to the left.

SKILL REVIEW

To prepare for Section 11.7, review finding intercepts and completing the square (Sections 3.3, 6.2, 6.3, and 11.1).

Find the x-intercept and the y-intercept. [3.3]

75. $8x - 6y = 24$

76. $3x + 4y = 8$

Find the x-intercepts.

77. $y = x^2 + 8x + 15$ [6.2]

78. $y = 2x^2 - x - 3$ [6.3]

Replace the blanks with constants to form a true equation. [11.1]

79. $x^2 - 14x + \_\_\_ = (x - \_\_\_)^2$

80. $x^2 + 7x + \_\_\_ = \left(x + \_\_\_\right)^2$

SYNTHESIS

TW **81.** Before graphing a quadratic function, Cassandra always plots five points. First, she calculates and plots the coordinates of the vertex. Then she plots *four* more points after calculating *two* more ordered pairs. How is this possible?

TW **82.** If the graphs of $f(x) = a_1(x - h_1)^2 + k_1$ and $g(x) = a_2(x - h_2)^2 + k_2$ have the same shape, what, if anything, can you conclude about the a's, the h's, and the k's? Why?

Write an equation for a function having a graph with the same shape as the graph of $f(x) = \frac{3}{5}x^2$, but with the given point as the vertex.

83. (4, 1)	**84.** (2, 6)	**85.** (3, −1)
86. (5, −6)	**87.** (−2, −5)	**88.** (−4, −2)

For each of the following, write the equation of the parabola that has the shape of $f(x) = 2x^2$ or $g(x) = -2x^2$ and has a maximum value or a minimum value at the specified point.

89. Minimum: (2, 0)

90. Minimum: (−4, 0)

91. Maximum: (0, 3)

92. Maximum: (3, 8)

Use the following graph of $f(x) = a(x - h)^2 + k$ for Exercises 93–96.

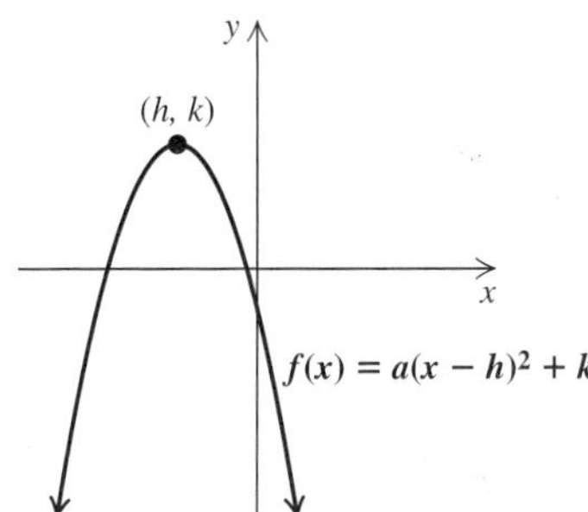

93. Describe what will happen to the graph if h is increased.

94. Describe what will happen to the graph if k is decreased.

95. Describe what will happen to the graph if a is replaced with $-a$.

96. Describe what will happen to the graph if $(x - h)$ is replaced with $(x + h)$.

Find an equation for a quadratic function F that satisfies the following conditions.

97. The graph of F is the same shape as the graph of f, where $f(x) = 3(x + 2)^2 + 7$, and $F(x)$ is a minimum at the same point that $g(x) = -2(x - 5)^2 + 1$ is a maximum.

98. The graph of F is the same shape as the graph of f, where $f(x) = -\frac{1}{3}(x - 2)^2 + 7$, and $F(x)$ is a maximum at the same point that $g(x) = 2(x + 4)^2 - 6$ is a minimum.

Functions other than parabolas can be translated. When calculating $f(x)$, if we replace x with $x - h$, where h is a constant, the graph will be moved horizontally. If we replace $f(x)$ with $f(x) + k$, the graph will be moved vertically. Use the graph below for Exercises 99–104.

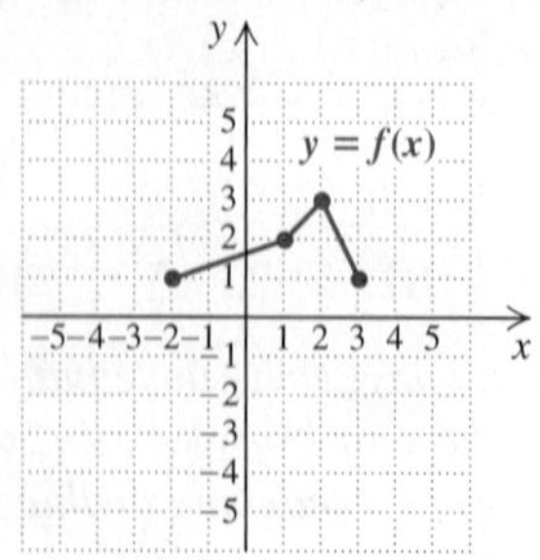

Draw a graph of each of the following.

99. $y = f(x - 1)$ **100.** $y = f(x + 2)$

101. $y = f(x) + 2$ **102.** $y = f(x) - 3$

103. $y = f(x + 3) - 2$ **104.** $y = f(x - 3) + 1$

Try Exercise Answers: Section 11.6

15. **19.**

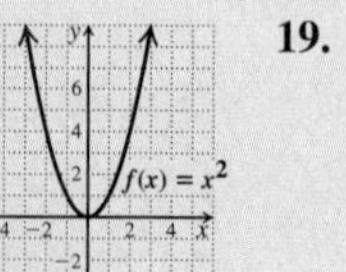

25. Vertex: $(-1, 0)$; axis of symmetry: $x = -1$

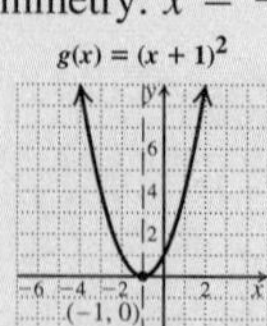

45. Minimum: 2; range: $[2, \infty)$

49. Minimum: 3; range: $[3, \infty)$

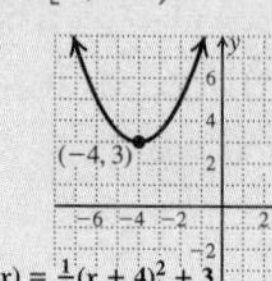

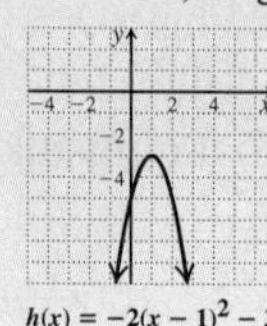

59. Vertex: $(1, -3)$; axis of symmetry: $x = 1$; maximum: -3; range: $(-\infty, -3]$

$h(x) = -2(x - 1)^2 - 3$

Collaborative Corner

Match the Graph

Focus: Graphing quadratic functions
Time: 15–20 minutes
Group size: 6
Materials: Index cards

ACTIVITY

1. On each of six index cards, write one of the following equations:

$$y = \tfrac{1}{2}(x - 3)^2 + 1; \quad y = \tfrac{1}{2}(x - 1)^2 + 3;$$
$$y = \tfrac{1}{2}(x + 1)^2 - 3; \quad y = \tfrac{1}{2}(x + 3)^2 + 1;$$
$$y = \tfrac{1}{2}(x + 3)^2 - 1; \quad y = \tfrac{1}{2}(x + 1)^2 + 3.$$

2. Fold each index card and mix up the six cards in a hat or a bag. Then, one by one, each group member should select one of the equations. Do not let anyone see your equation.
3. Each group member should carefully graph the equation selected. Make the graph large enough so that when it is finished, it can be easily viewed by the rest of the group. Be sure to scale the axes and label the vertex, but **do not label the graph with the equation used.**
4. When all group members have drawn a graph, place the graphs in a pile. The group should then match and agree on the correct equation for each graph *with no help from the person who drew the graph.* If a mistake has been made and a graph has no match, determine what its equation *should* be.
5. Compare your group's labeled graphs with those of other groups to reach consensus within the class on the correct label for each graph.

11.7 More About Graphing Quadratic Functions

- Finding the Vertex
- Finding Intercepts

FINDING THE VERTEX

By *completing the square* (see Section 11.1), we can rewrite any polynomial $ax^2 + bx + c$ in the form $a(x - h)^2 + k$. Once that has been done, the procedures discussed in Section 11.6 will enable us to graph any quadratic function.

EXAMPLE 1 Graph: $g(x) = x^2 - 6x + 4$. Label the vertex and the axis of symmetry.

SOLUTION We have

$$g(x) = x^2 - 6x + 4$$
$$= (x^2 - 6x) + 4.$$

To complete the square inside the parentheses, we take half of the x-coefficient, $\frac{1}{2} \cdot (-6) = -3$, and square it to get $(-3)^2 = 9$. Then we add $9 - 9$ inside the parentheses:

$$g(x) = (x^2 - 6x + 9 - 9) + 4$$ The effect is of adding 0.

$$= (x^2 - 6x + 9) + (-9 + 4)$$ Using the associative law of addition to regroup

$$= (x - 3)^2 - 5.$$ Factoring and simplifying

This equation appeared as Example 4 of Section 11.6. The graph is that of $f(x) = x^2$ translated 3 units to the right and 5 units down. The vertex is $(3, -5)$, and the axis of symmetry is $x = 3$.

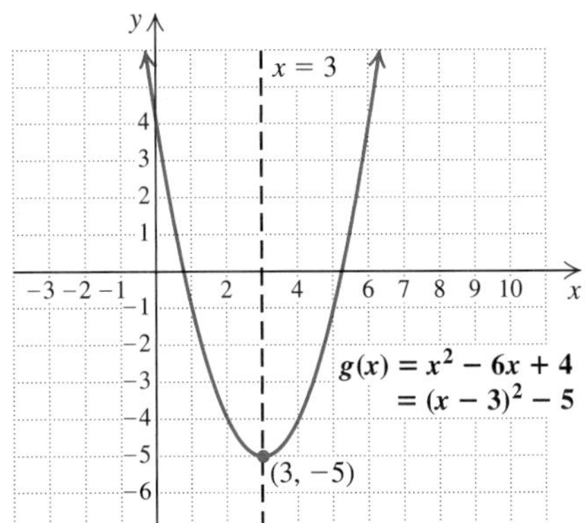

Try Exercise 19.

STUDY TIP 

Be Prepared

Before sitting down to study, take time to get ready for the study session. Collect your notes, assignments, textbook, paper, pencils, and eraser. Get a drink if you are thirsty, turn off your phone, and plan to spend some uninterrupted study time.

When the leading coefficient is not 1, we factor out that number from the first two terms. Then we complete the square and use the distributive law.

EXAMPLE 2 Graph: $f(x) = 3x^2 + 12x + 13$. Label the vertex and the axis of symmetry.

SOLUTION Since the coefficient of x^2 is not 1, we need to factor out that number—in this case, 3—from the first two terms. Remember that we want the form $f(x) = a(x - h)^2 + k$:

$$f(x) = 3x^2 + 12x + 13 = 3(x^2 + 4x) + 13.$$

Now we complete the square as before. We take half of the x-coefficient, $\frac{1}{2} \cdot 4 = 2$, and square it: $2^2 = 4$. Then we add $4 - 4$ inside the parentheses:

$$f(x) = 3(x^2 + 4x + 4 - 4) + 13.$$ **Adding 4 − 4, or 0, inside the parentheses**

The distributive law allows us to separate the -4 from the perfect-square trinomial so long as it is multiplied by 3:

The −4 was added inside the parentheses. To separate it, we *must* multiply it by 3.

$$f(x) = 3(x^2 + 4x + 4) + 3(-4) + 13$$ **This leaves a perfect-square trinomial inside the parentheses.**

$$= 3(x + 2)^2 + 1.$$ **Factoring and simplifying**

The vertex is $(-2, 1)$, and the axis of symmetry is $x = -2$. The coefficient of x^2 is 3, so the graph is narrow and opens upward. We choose a few x-values on either side of the vertex, compute y-values, and then graph the parabola.

x	$f(x) = 3(x + 2)^2 + 1$
-2	1 ← Vertex
-3	4
-1	4

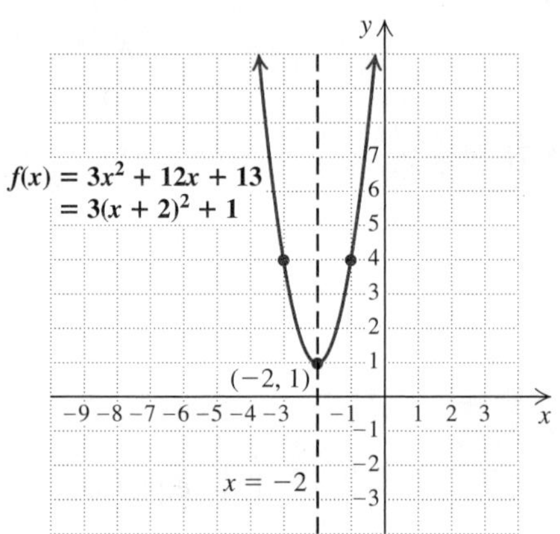

Try Exercise 23.

EXAMPLE 3 Graph $f(x) = -2x^2 + 10x - 7$, and find the maximum or minimum function value.

SOLUTION We first find the vertex by completing the square. To do so, we factor out -2 from the first two terms of the expression. This makes the coefficient of x^2 inside the parentheses 1:

$$f(x) = -2x^2 + 10x - 7$$

Factor out a from both variable terms.

$$= -2(x^2 - 5x) - 7.$$

Now we complete the square as before. We take half of the x-coefficient and square it to get $\frac{25}{4}$. Then we add $\frac{25}{4} - \frac{25}{4}$ inside the parentheses:

Complete the square inside the parentheses.

$$f(x) = -2\left(x^2 - 5x + \frac{25}{4} - \frac{25}{4}\right) - 7$$

Regroup.

$$= -2\left(x^2 - 5x + \frac{25}{4}\right) + (-2)\left(-\frac{25}{4}\right) - 7$$ **Multiplying by −2, using the distributive law, and regrouping**

Factor.

$$= -2\left(x - \frac{5}{2}\right)^2 + \frac{11}{2}.$$ **Factoring and simplifying**

The vertex is $\left(\frac{5}{2}, \frac{11}{2}\right)$, and the axis of symmetry is $x = \frac{5}{2}$. The coefficient of x^2, -2, is negative, so the graph opens downward and the second coordinate of the vertex, $\frac{11}{2}$, is the maximum function value. We plot a few points on either side of the vertex, including the y-intercept, $f(0)$, and graph the parabola.

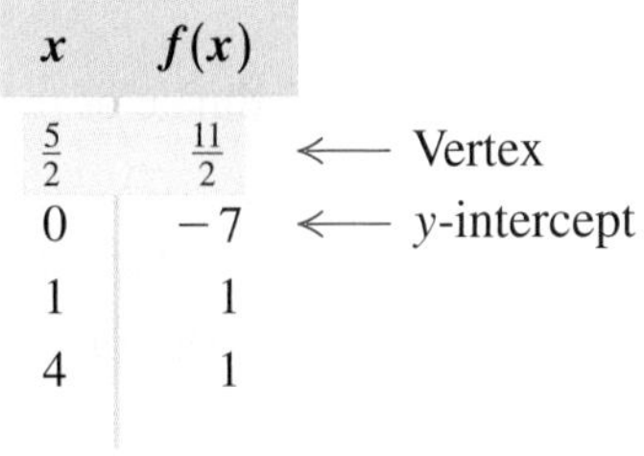

x	$f(x)$	
$\frac{5}{2}$	$\frac{11}{2}$	← Vertex
0	-7	← y-intercept
1	1	
4	1	

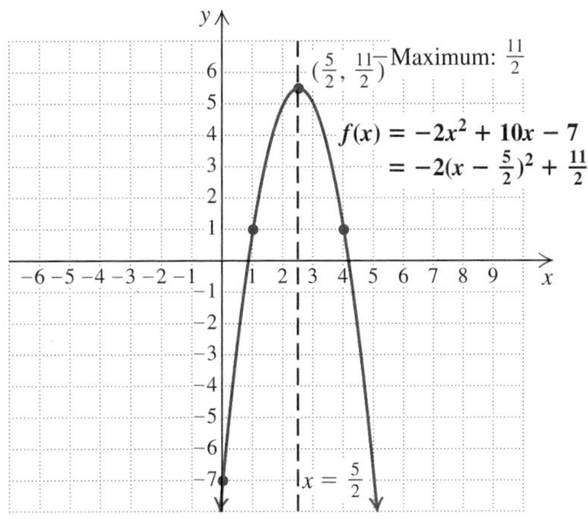

Try Exercise 39.

The method used in Examples 1–3 can be generalized to find a formula for locating the vertex. We complete the square as follows:

$$f(x) = ax^2 + bx + c$$

$$= a\left(x^2 + \frac{b}{a}x\right) + c. \qquad \textbf{Factoring } a \textbf{ out of the first two terms. Check by multiplying.}$$

Half of the x-coefficient, $\frac{b}{a}$, is $\frac{b}{2a}$. We square it to get $\frac{b^2}{4a^2}$ and add $\frac{b^2}{4a^2} - \frac{b^2}{4a^2}$ inside the parentheses. Then we distribute the a and regroup terms:

$$f(x) = a\left(x^2 + \frac{b}{a}x + \frac{b^2}{4a^2} - \frac{b^2}{4a^2}\right) + c$$

$$= a\left(x^2 + \frac{b}{a}x + \frac{b^2}{4a^2}\right) + a\left(-\frac{b^2}{4a^2}\right) + c \qquad \textbf{Using the distributive law}$$

$$= a\left(x + \frac{b}{2a}\right)^2 + \frac{-b^2}{4a} + \frac{4ac}{4a} \qquad \textbf{Factoring and finding a common denominator}$$

$$= a\left[x - \left(-\frac{b}{2a}\right)\right]^2 + \frac{4ac - b^2}{4a}.$$

Thus we have the following.

Student Notes

It is easier to remember a formula when you understand its derivation. Check with your instructor to determine what formulas you will be expected to remember.

The Vertex of a Parabola The vertex of the parabola given by $f(x) = ax^2 + bx + c$ is

$$\left(-\frac{b}{2a}, f\left(-\frac{b}{2a}\right)\right), \text{ or } \left(-\frac{b}{2a}, \frac{4ac - b^2}{4a}\right).$$

- The x-coordinate of the vertex is $-b/(2a)$.
- The axis of symmetry is $x = -b/(2a)$.
- The second coordinate of the vertex is most commonly found by computing $f\left(-\frac{b}{2a}\right)$.

Let's reexamine Example 3 to see how we could have found the vertex directly. From the formula above, we see that

$$\text{the } x\text{-coordinate of the vertex is } -\frac{b}{2a} = -\frac{10}{2(-2)} = \frac{5}{2}.$$

Substituting $\frac{5}{2}$ into $f(x) = -2x^2 + 10x - 7$, we find the second coordinate of the vertex:

$$\begin{aligned} f\left(\tfrac{5}{2}\right) &= -2\left(\tfrac{5}{2}\right)^2 + 10\left(\tfrac{5}{2}\right) - 7 \\ &= -2\left(\tfrac{25}{4}\right) + 25 - 7 \\ &= -\tfrac{25}{2} + 18 \\ &= -\tfrac{25}{2} + \tfrac{36}{2} = \tfrac{11}{2}. \end{aligned}$$

The vertex is $\left(\frac{5}{2}, \frac{11}{2}\right)$. The axis of symmetry is $x = \frac{5}{2}$.

A quadratic function has a maximum value or a minimum value at the vertex of its graph. Thus determining the maximum value or the minimum value of the function allows us to find the vertex of the graph.

Maximums and Minimums

On most graphing calculators, a MAXIMUM or MINIMUM feature is found in the CALC menu. (CALC is the 2nd option associated with the TRACE key.)

To find a maximum or a minimum of a quadratic function, enter and graph the function, choosing a viewing window that will show the vertex. Next, press CALC and choose either the MAXIMUM or MINIMUM option in the menu. The graphing calculator will find the maximum or minimum function value over a specified interval, so the left and right endpoints, or bounds, of the interval must be entered as well as a guess near where the maximum or minimum occurs. The calculator will return the coordinates of the point for which the function value is a maximum or minimum within the interval.

Your Turn

1. Enter and graph $y_1 = x^2 - 3x - 5$, using a standard viewing window. The function is a *minimum* at the vertex.
2. Open the CALC menu and choose the minimum option.
3. The vertex occurs near $x = 1$, so enter a number less than 1 (say, 0) for the left bound.
4. Enter a number greater than 1 (say, 4) for the right bound.
5. Enter 1 as a guess. The coordinates shown are the vertex. The coordinates should be approximately $(1.5, -7.25)$.

Student Notes

It is easy to press an incorrect key when using a calculator. Always check to see if your answer is reasonable. For example, if the MINIMUM option were chosen in Example 4 instead of MAXIMUM, one endpoint of the interval would have been returned instead of the maximum. In this case, noting that the highest point on the graph is marked is a quick check that we did indeed find the maximum.

EXAMPLE 4 Use a graphing calculator to determine the vertex of the graph of the function given by $f(x) = -2x^2 + 10x - 7$.

SOLUTION The coefficient of x^2 is negative, so the parabola opens downward and the function has a maximum value. We enter and graph the function, choosing a window that will show the vertex.

We choose the MAXIMUM option from the CALC menu. For a left bound, we visually locate the vertex and either move the cursor to a point left of the vertex using the left and right arrow keys or type a value of x that is less than the x-value at the vertex. (See the graph on the left below.) After pressing ENTER, we choose a right bound in a similar manner (as shown in the graph on the right below), and press ENTER again.

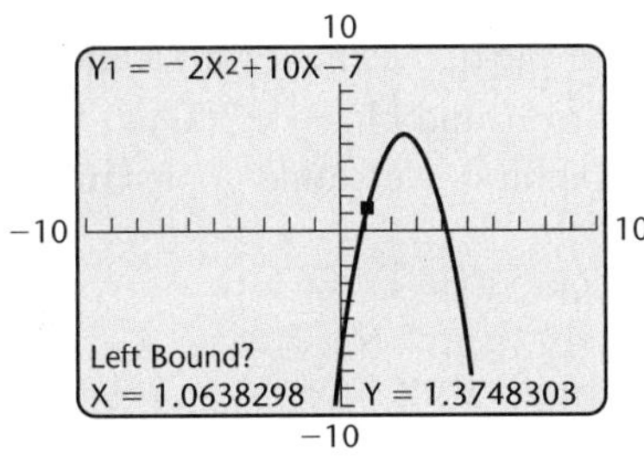

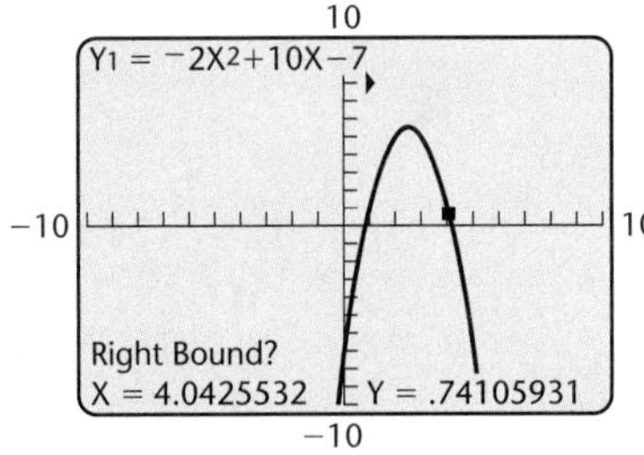

Finally, we enter a guess that is close to the vertex (as in the graph on the left below), and press ENTER. The calculator indicates that the coordinates of the vertex are (2.5, 5.5).

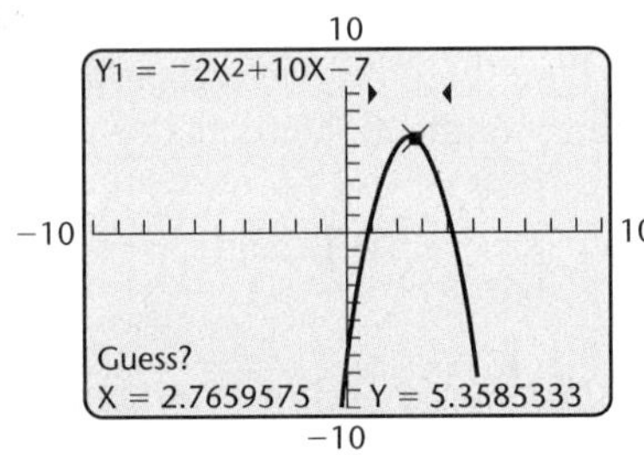

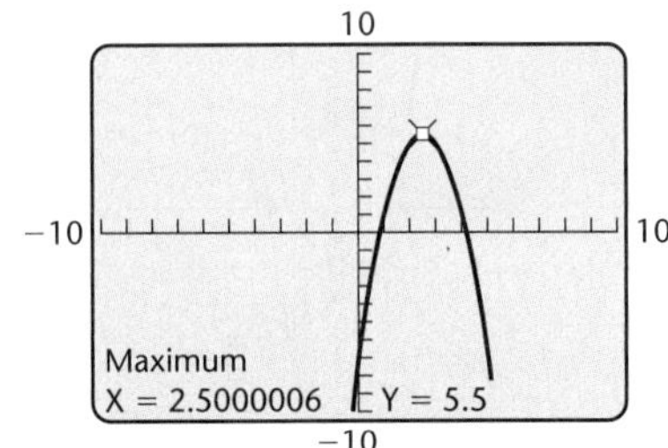

Try Exercise 43.

In Example 3, we found the coordinates of the vertex of the graph of $f(x) = -2x^2 + 10x - 7$, $\left(\frac{5}{2}, \frac{11}{2}\right)$, by completing the square. On p. 826, we also found these coordinates using the formula for the vertex of a parabola. The coordinates found using a calculator in Example 4 were in decimal notation. Since $2.5000006 \approx \frac{5}{2}$, and since $5.5 = \frac{11}{2}$, the coordinates check.

We have actually developed three methods for finding the vertex. One is by completing the square, the second is by using a formula, and the third is by using a graphing calculator. You should check with your instructor about which method to use.

FINDING INTERCEPTS

All quadratic functions have a y-intercept and 0, 1, or 2 x-intercepts. For $f(x) = ax^2 + bx + c$, the y-intercept is $(0, f(0))$, or $(0, c)$. To find x-intercepts, we look for points where $y = 0$ or $f(x) = 0$. Thus, for $f(x) = ax^2 + bx + c$, the x-intercepts occur at those x-values for which $ax^2 + bx + c = 0$.

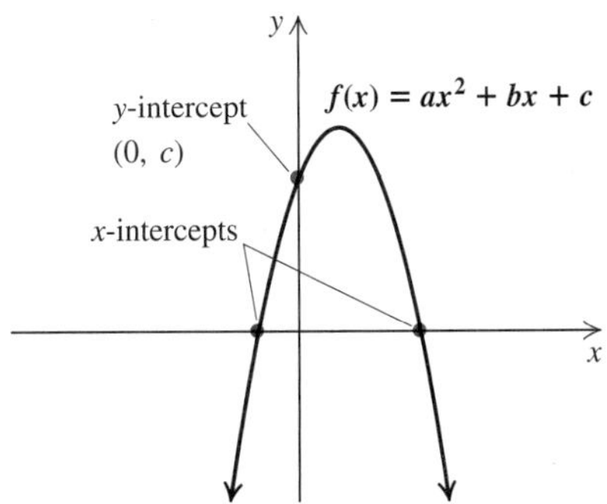

EXAMPLE 5 Find any x-intercepts and the y-intercept of the graph of $f(x) = x^2 - 2x - 2$.

SOLUTION The y-intercept is $(0, f(0))$, or $(0, -2)$. To find the x-intercepts, we solve the equation $0 = x^2 - 2x - 2$. We are unable to factor $x^2 - 2x - 2$, so we use the quadratic formula and get $x = 1 \pm \sqrt{3}$. Thus the x-intercepts are $(1 - \sqrt{3}, 0)$ and $(1 + \sqrt{3}, 0)$.

If graphing, we would approximate, to get $(-0.7, 0)$ and $(2.7, 0)$.

Try Exercise 49.

If the solutions of $f(x) = 0$ are imaginary, the graph of f has no x-intercepts.

Connecting the Concepts

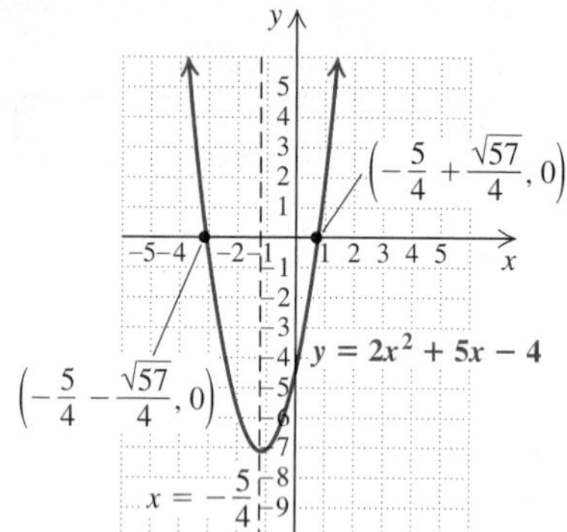

Because the graph of a quadratic equation is symmetric, the x-intercepts of the graph, if they exist, will be symmetric with respect to the axis of symmetry. This symmetry can be seen directly if the x-intercepts are found using the quadratic formula.

For example, the x-intercepts of the graph of $y = 2x^2 + 5x - 4$ are

$$\left(-\frac{5}{4} + \frac{\sqrt{57}}{4}, 0\right) \text{ and } \left(-\frac{5}{4} - \frac{\sqrt{57}}{4}, 0\right).$$

For this equation, the axis of symmetry is $x = -\frac{5}{4}$ and the x-intercepts are $\frac{\sqrt{57}}{4}$ units to the left and right of $-\frac{5}{4}$ on the x-axis.

The Graph of a Quadratic Function Given by $f(x) = ax^2 + bx + c$ or $f(x) = a(x - h)^2 + k$

The graph is a parabola.

The vertex is (h, k) or $\left(-\frac{b}{2a}, f\left(-\frac{b}{2a}\right)\right)$.

The axis of symmetry is $x = h$.

The y-intercept of the graph is $(0, c)$.

The x-intercepts can be found by solving $ax^2 + bx + c = 0$.

If $b^2 - 4ac > 0$, there are two x-intercepts.
If $b^2 - 4ac = 0$, there is one x-intercept.
If $b^2 - 4ac < 0$, there are no x-intercepts.

The domain of the function is $(-\infty, \infty)$.

If a is positive: The graph opens upward.
The function has a minimum value, given by k.
This occurs when $x = h$.
The range of the function is $[k, \infty)$.

If a is negative: The graph opens downward.
The function has a maximum value, given by k.
This occurs when $x = h$.
The range of the function is $(-\infty, k]$.

Visualizing for Success

Match each function with its graph.

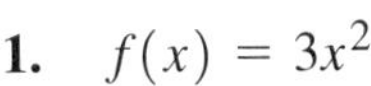
1. $f(x) = 3x^2$

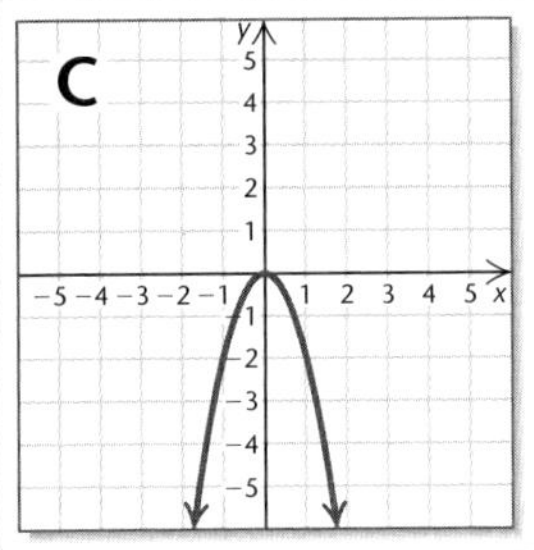

2. $f(x) = x^2 - 4$

3. $f(x) = (x - 4)^2$

4. $f(x) = x - 4$

5. $f(x) = -2x^2$

6. $f(x) = x + 3$

7. $f(x) = |x + 3|$

8. $f(x) = (x + 3)^2$

9. $f(x) = \sqrt{x + 3}$

10. $f(x) = (x + 3)^2 - 4$

Answers on page A-43

An additional, animated version of this activity appears in MyMathLab. To use MyMathLab, you need a course ID and a student access code. Contact your instructor for more information.

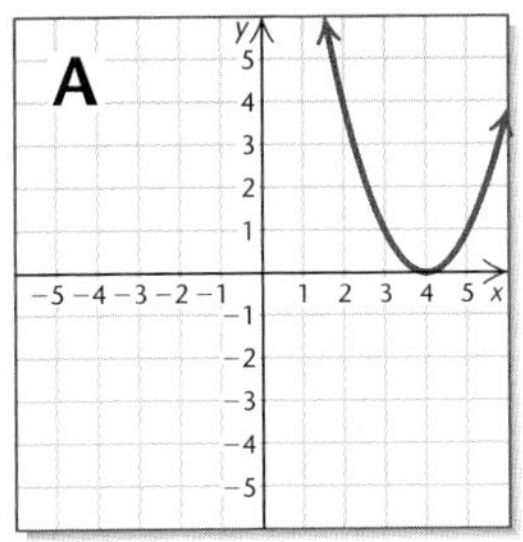

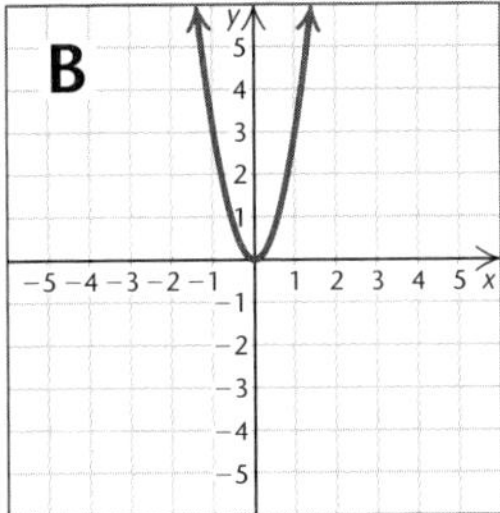

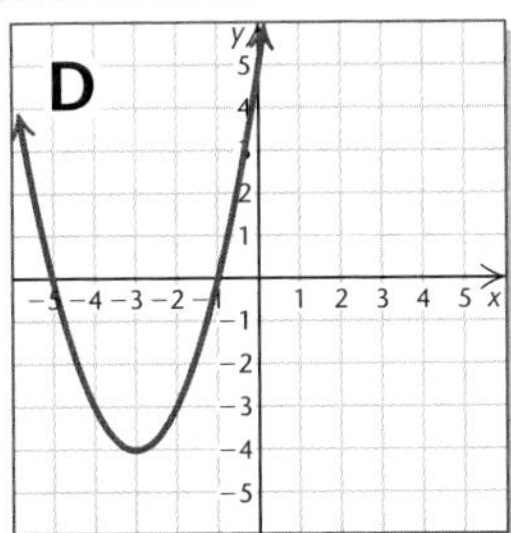

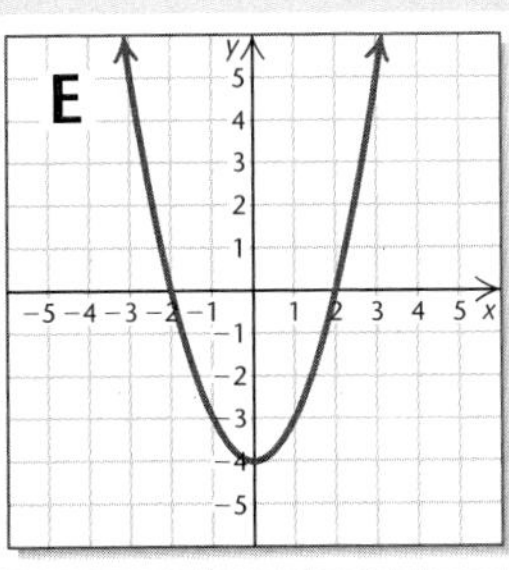

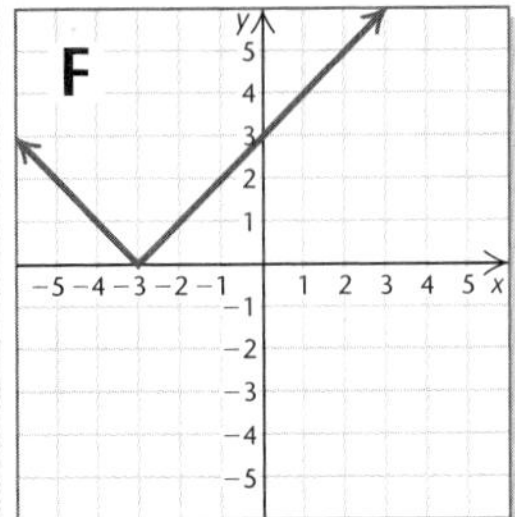

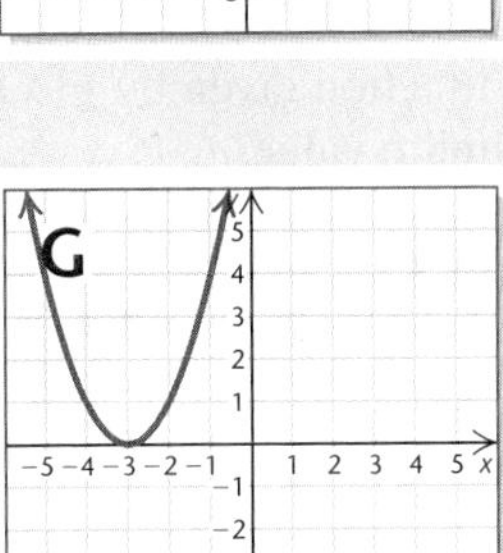

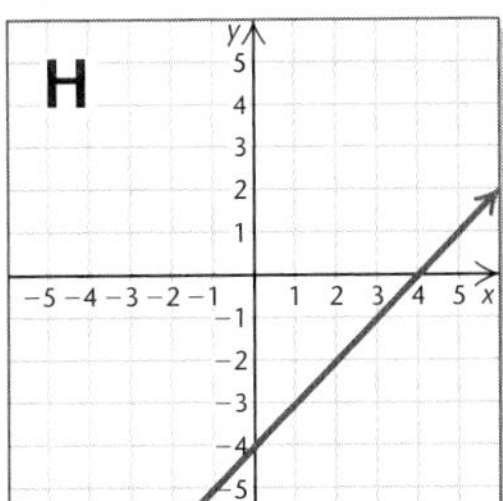

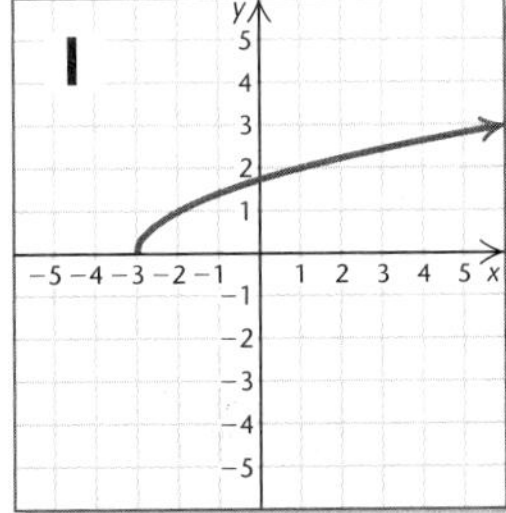

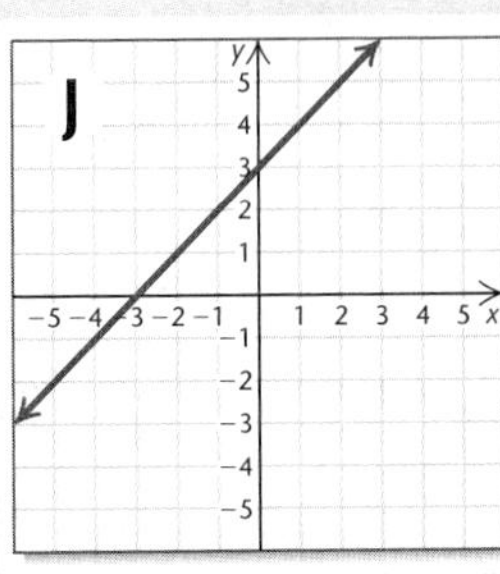

Exercise Set

FOR EXTRA HELP MyMathLab  PRACTICE WATCH  DOWNLOAD 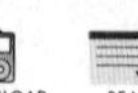 READ REVIEW

Concept Reinforcement *Classify each of the following statements as either true or false.*

1. The graph of $f(x) = 3x^2 - x + 6$ opens upward.
2. The function given by $g(x) = -x^2 + 3x + 1$ has a minimum value.
3. The graph of $f(x) = -2(x - 3)^2 + 7$ has its vertex at $(3, 7)$.
4. The graph of $g(x) = 4(x + 6)^2 - 2$ has its vertex at $(-6, -2)$.
5. The graph of $g(x) = \frac{1}{2}(x - \frac{3}{2})^2 + \frac{1}{4}$ has $x = \frac{1}{4}$ as its axis of symmetry.
6. The function given by $f(x) = (x - 2)^2 - 5$ has a minimum value of -5.
7. The y-intercept of the graph of $f(x) = 2x^2 - 6x + 7$ is $(7, 0)$.
8. If the graph of a quadratic function f opens upward and has a vertex of $(1, 5)$, then the graph has no x-intercepts.

Complete the square to write each function in the form $f(x) = a(x - h)^2 + k$.

9. $f(x) = x^2 - 8x + 2$
10. $f(x) = x^2 - 6x - 1$
11. $f(x) = x^2 + 3x - 5$
12. $f(x) = x^2 + 5x + 3$
13. $f(x) = 3x^2 + 6x - 2$
14. $f(x) = 2x^2 - 20x - 3$
15. $f(x) = -x^2 - 4x - 7$
16. $f(x) = -2x^2 - 8x + 4$
17. $f(x) = 2x^2 - 5x + 10$
18. $f(x) = 3x^2 + 7x - 3$

*For each quadratic function, **(a)** find the vertex and the axis of symmetry and **(b)** graph the function.*

19. $f(x) = x^2 + 4x + 5$
20. $f(x) = x^2 + 2x - 5$
21. $f(x) = x^2 + 8x + 20$
22. $f(x) = x^2 - 10x + 21$
23. $h(x) = 2x^2 - 16x + 25$
24. $h(x) = 2x^2 + 16x + 23$
25. $f(x) = -x^2 + 2x + 5$
26. $f(x) = -x^2 - 2x + 7$
27. $g(x) = x^2 + 3x - 10$
28. $g(x) = x^2 + 5x + 4$
29. $h(x) = x^2 + 7x$
30. $h(x) = x^2 - 5x$
31. $f(x) = -2x^2 - 4x - 6$
32. $f(x) = -3x^2 + 6x + 2$

*For each quadratic function, **(a)** find the vertex, the axis of symmetry, and the maximum or minimum function value and **(b)** graph the function.*

33. $g(x) = x^2 - 6x + 13$
34. $g(x) = x^2 - 4x + 5$
35. $g(x) = 2x^2 - 8x + 3$
36. $g(x) = 2x^2 + 5x - 1$
37. $f(x) = 3x^2 - 24x + 50$
38. $f(x) = 4x^2 + 16x + 13$
39. $f(x) = -3x^2 + 5x - 2$
40. $f(x) = -3x^2 - 7x + 2$
41. $h(x) = \frac{1}{2}x^2 + 4x + \frac{19}{3}$
42. $h(x) = \frac{1}{2}x^2 - 3x + 2$

Use a graphing calculator to find the vertex of the graph of each function.

43. $f(x) = x^2 + x - 6$
44. $f(x) = x^2 + 2x - 5$
45. $f(x) = 5x^2 - x + 1$
46. $f(x) = -4x^2 - 3x + 7$
47. $f(x) = -0.2x^2 + 1.4x - 6.7$
48. $f(x) = 0.5x^2 + 2.4x + 3.2$

Find any x-intercepts and the y-intercept. If no x-intercepts exist, state this.

49. $f(x) = x^2 - 6x + 3$

50. $f(x) = x^2 + 5x + 2$

51. $g(x) = -x^2 + 2x + 3$

52. $g(x) = x^2 - 6x + 9$

Aha! **53.** $f(x) = x^2 - 9x$

54. $f(x) = x^2 - 7x$

55. $h(x) = -x^2 + 4x - 4$

56. $h(x) = -2x^2 - 20x - 50$

57. $g(x) = x^2 + x - 5$

58. $g(x) = 2x^2 + 3x - 1$

59. $f(x) = 2x^2 - 4x + 6$

60. $f(x) = x^2 - x + 2$

TW **61.** The graph of a quadratic function f opens downward and has no x-intercepts. In what quadrant(s) must the vertex lie? Explain your reasoning.

TW **62.** Is it possible for the graph of a quadratic function to have only one x-intercept if the vertex is off the x-axis? Why or why not?

SKILL REVIEW

To prepare for Section 11.8, review solving systems of three equations in three unknowns (Section 9.1).

Solve. [9.1]

63. $x + y + z = 3,$
$x - y + z = 1,$
$-x - y + z = -1$

64. $x - y + z = -6,$
$2x + y + z = 2,$
$3x + y + z = 0$

65. $z = 8,$
$x + y + z = 23,$
$2x + y - z = 17$

66. $z = -5,$
$2x - y + 3z = -27,$
$x + 2y + 7z = -26$

67. $1.5 = c,$
$52.5 = 25a + 5b + c,$
$7.5 = 4a + 2b + c$

68. $\frac{1}{2} = c,$
$5 = 9a + 6b + 2c,$
$29 = 81a + 9b + c$

SYNTHESIS

TW **69.** If the graphs of two quadratic functions have the same x-intercepts, will they also have the same vertex? Why or why not?

TW **70.** Suppose that the graph of $f(x) = ax^2 + bx + c$ has $(x_1, 0)$ and $(x_2, 0)$ as x-intercepts. Explain why the graph of $g(x) = -ax^2 - bx - c$ will also have $(x_1, 0)$ and $(x_2, 0)$ as x-intercepts.

For each quadratic function, find **(a)** *the maximum or minimum value and* **(b)** *the x- and y-intercepts. Round to the nearest hundredth.*

71. $f(x) = 2.31x^2 - 3.135x - 5.89$

72. $f(x) = -18.8x^2 + 7.92x + 6.18$

73. $g(x) = -1.25x^2 + 3.42x - 2.79$

74. $g(x) = 0.45x^2 - 1.72x + 12.92$

75. Graph the function
$$f(x) = x^2 - x - 6.$$
Then use the graph to approximate solutions to each of the following equations.

a) $x^2 - x - 6 = 2$
b) $x^2 - x - 6 = -3$

76. Graph the function
$$f(x) = \frac{x^2}{2} + x - \frac{3}{2}.$$
Then use the graph to approximate solutions to each of the following equations.

a) $\frac{x^2}{2} + x - \frac{3}{2} = 0$
b) $\frac{x^2}{2} + x - \frac{3}{2} = 1$
c) $\frac{x^2}{2} + x - \frac{3}{2} = 2$

Find an equivalent equation of the type
$$f(x) = a(x - h)^2 + k.$$

77. $f(x) = mx^2 - nx + p$

78. $f(x) = 3x^2 + mx + m^2$

79. A quadratic function has $(-1, 0)$ as one of its intercepts and $(3, -5)$ as its vertex. Find an equation for the function.

80. A quadratic function has $(4, 0)$ as one of its intercepts and $(-1, 7)$ as its vertex. Find an equation for the function.

Graph.

81. $f(x) = |x^2 - 1|$

82. $f(x) = |x^2 - 3x - 4|$

83. $f(x) = |2(x - 3)^2 - 5|$

Try Exercise Answers: Section 11.7

19. (a) Vertex: $(-2, 1)$; axis of symmetry: $x = -2$; **(b)**

$f(x) = x^2 + 4x + 5$

23. (a) Vertex: $(4, -7)$; axis of symmetry: $x = 4$; **(b)**

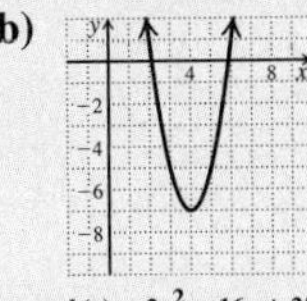

$h(x) = 2x^2 - 16x + 25$

39. (a) Vertex: $\left(\frac{5}{6}, \frac{1}{12}\right)$; axis of symmetry: $x = \frac{5}{6}$; maximum: $\frac{1}{12}$; **(b)**

$f(x) = -3x^2 + 5x - 2$

43. $(-0.5, -6.25)$ **49.** $(3 - \sqrt{6}, 0), (3 + \sqrt{6}, 0)$; $(0, 3)$

11.8 Problem Solving and Quadratic Functions

- Maximum and Minimum Problems
- Fitting Quadratic Functions to Data

Let's look now at some of the many situations in which quadratic functions are used for problem solving.

MAXIMUM AND MINIMUM PROBLEMS

We have seen that for any quadratic function f, the value of $f(x)$ at the vertex is either a maximum or a minimum. Thus problems in which a quantity must be maximized or minimized can often be solved by finding the coordinates of a vertex, assuming the problem can be modeled with a quadratic function.

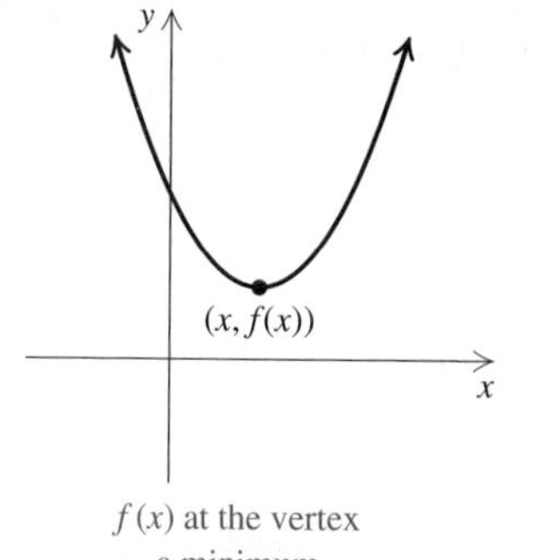

$f(x)$ at the vertex a minimum

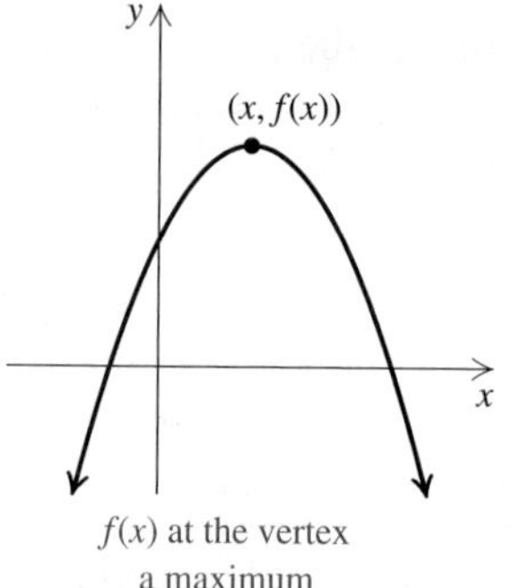

$f(x)$ at the vertex a maximum

STUDY TIP 

Don't Fight Sleep

Study time is too precious to waste. If you find yourself with heavy eyelids, don't continue studying even if you're nearly finished. A problem that may require 20 minutes to complete when you are sleepy often takes 5 minutes when you are fresh and rested.

EXAMPLE 1 Sonoma Sunshine. The percent of days each month during which the sun shines in Sonoma, California, can be modeled by the quadratic function

$$s(n) = -1.625n^2 + 21.7n + 24.925,$$

where $n = 1$ represents January, $n = 2$ represents February, and so on. What month has the most sunshine, and during what percent of that month's days does the sun shine?

Source: Based on information from www.city-data.com

SOLUTION

1., 2. Familiarize and **Translate.** We are given the function that models the percent of days with sunshine in Sonoma. Since the coefficient of n^2 is negative, the quadratic function has a maximum value at the vertex of its graph.

3. Carry out. We find the vertex using the formula giving the first coordinate of the vertex:

$$n = -\frac{b}{2a} = -\frac{21.7}{2(-1.625)} = -\frac{21.7}{-3.25} \approx 6.7.$$

If we round this number, we have $n \approx 7$, which represents the month of July. The percent of days in July with sunshine would then be

$$s(7) = -1.625 \cdot 7^2 + 21.7 \cdot 7 + 24.925 = 97.2.$$

Thus the sun shines during approximately 97% of the days in July.

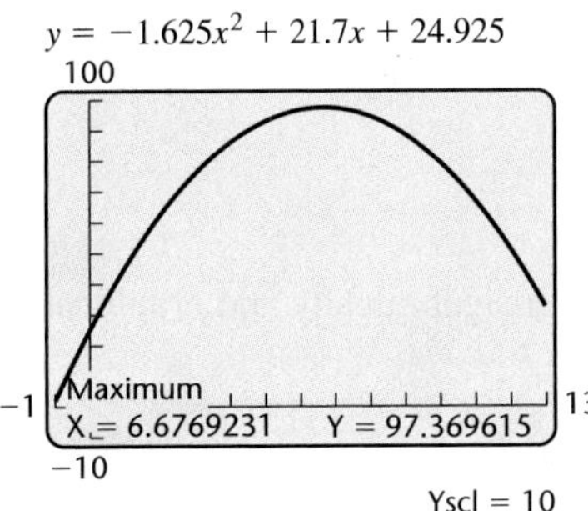

4. **Check.** Using the MAXIMUM feature of a graphing calculator, as shown at left, we find that the vertex of the graph of $s(n)$ is approximately $(6.677, 97.370)$. This is very close to our answer, allowing for rounding error.
5. **State.** In Sonoma, the sun shines during approximately 97% of the days in July, making July the month with the most sunshine.

Try Exercise 7.

EXAMPLE 2 Swimming Area. A lifeguard has 100 m of roped-together flotation devices with which to cordon off a rectangular swimming area at Lakeside Beach. If the shoreline forms one side of the rectangle, what dimensions will maximize the size of the area for swimming?

SOLUTION

1. **Familiarize.** We make a drawing and label it, letting w = the width of the rectangle, in meters, and l = the length of the rectangle, in meters.

Recall that Area $= l \cdot w$ and Perimeter $= 2w + 2l$. Since the beach forms one length of the rectangle, the flotation devices form three sides. Thus,

$$2w + l = 100.$$

To get a better feel for the problem, we can look at some possible dimensions for a rectangular area that can be enclosed with 100 m of flotation devices. All possibilities are chosen so that $2w + l = 100$.

We can make a table by hand or use a graphing calculator. To use a calculator, we let x represent the width of the swimming area. Solving $2w + l = 100$ for l, we have $l = 100 - 2w$. Thus, if $y_1 = 100 - 2x$, then the area is $y_2 = x \cdot y_1$.

$y_1 = 100 - 2x,\ y_2 = x \cdot y_1$

X	Y1	Y2
20	60	1200
22	56	1232
24	52	1248
26	48	1248
28	44	1232
30	40	1200
32	36	1152
X = 20		

What choice of X will maximize Y2?

l	w	Rope Length	Area
40 m	30 m	100 m	1200 m^2
30 m	35 m	100 m	1050 m^2
20 m	40 m	100 m	800 m^2
.	.	.	.
.	.	.	.
.	.	.	.

What choice of l and w will maximize A?

2. **Translate.** We have two equations: One guarantees that all 100 m of flotation devices are used; the other expresses area in terms of length and width.

$$2w + l = 100,$$
$$A = l \cdot w$$

3. **Carry out.** We solve the system of equations both algebraically and graphically.

ALGEBRAIC APPROACH

We need to express A as a function of l or w but not both. To do so, we solve for l in the first equation to obtain $l = 100 - 2w$. Substituting for l in the second equation, we get a quadratic function:

$$A(w) = (100 - 2w)w$$ **Substituting for l**

$$= 100w - 2w^2.$$ **This represents a parabola opening downward, so a maximum exists.**

Factoring and completing the square, we get

$$A(w) = -2(w^2 - 50w + 625 - 625)$$ **We could also use the vertex formula.**

$$= -2(w - 25)^2 + 1250.$$ **This suggests a maximum of 1250 m^2 when $w = 25$ m.**

There is a maximum value of 1250 when $w = 25$.

GRAPHICAL APPROACH

As in the *Familiarize* step, we let $x =$ the width. Then

length $= y_1 = 100 - 2x$, and

area $= y_2 = x \cdot y_1 = 100x - 2x^2$.

We find the maximum area.

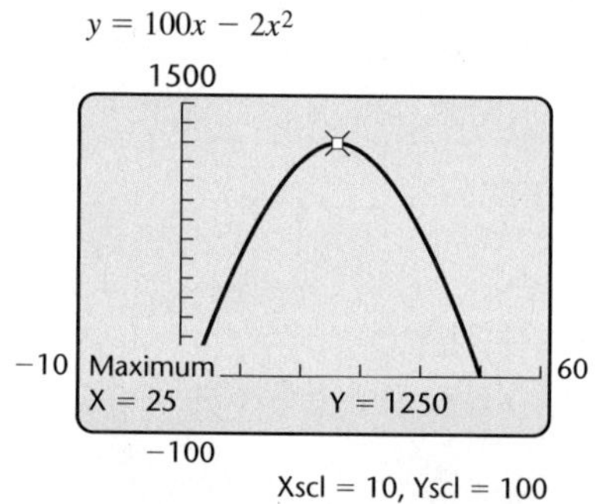

The maximum is 1250 when the width, x, is 25. If $w = 25$ m, then $l = 100 - 2 \cdot 25 = 50$ m. These dimensions give an area of 1250 m^2.

4. **Check.** Note that 1250 m^2 is greater than any of the values for A found in the *Familiarize* step. To be more certain, we could check values other than those used in that step. For example, if $w = 26$ m, then $l = 48$ m, and $A = 26 \cdot 48 = 1248$ m^2. Since 1250 m^2 is greater than 1248 m^2, it appears that we have a maximum.

5. **State.** The largest rectangular area for swimming that can be enclosed is 25 m by 50 m.

Try Exercise 11.

FITTING QUADRATIC FUNCTIONS TO DATA

We can now model some real-world situations using quadratic functions. As always, before attempting to fit an equation to data, we should graph the data and compare the result to the shape of the graphs of different types of functions and ask what type of function seems appropriate.

Connecting the Concepts

The general shapes of the graphs of many functions that we have studied are shown below.

Linear function: $f(x) = mx + b$

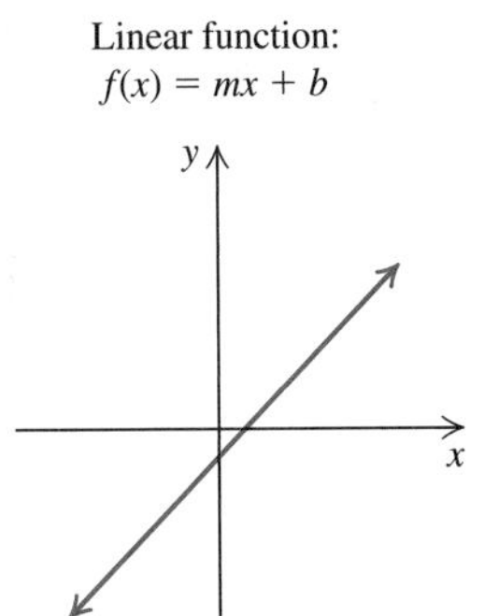

Absolute-value function: $f(x) = |x|$

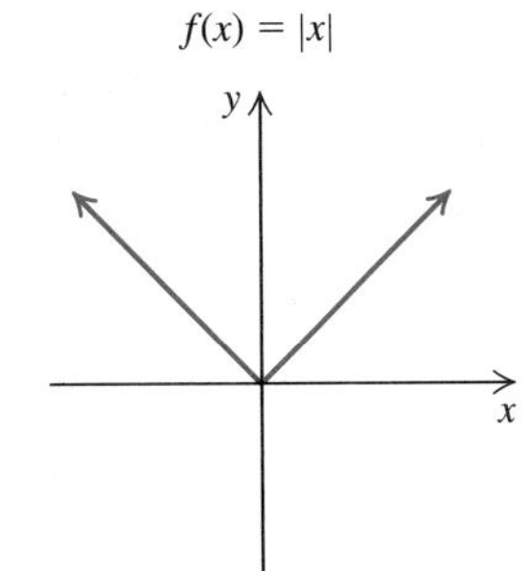

Rational function: $f(x) = \frac{1}{x}$

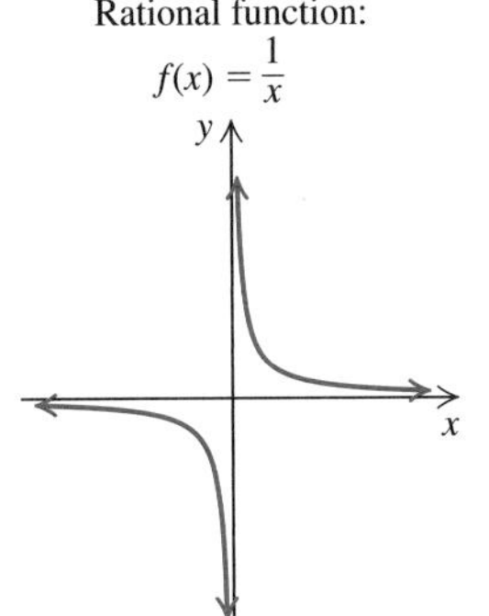

Radical function: $f(x) = \sqrt{x}$

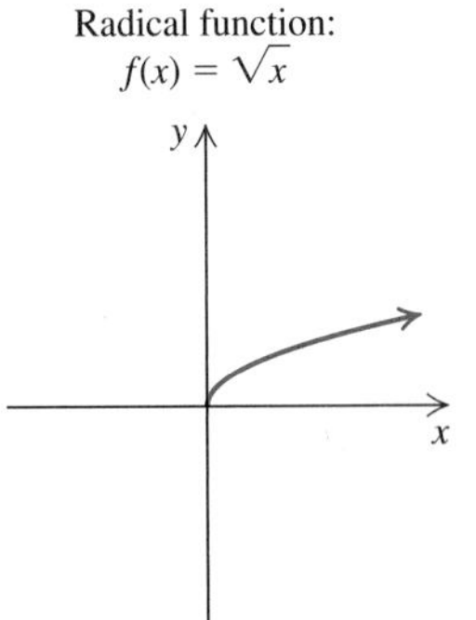

In order for a quadratic function to fit a set of data, the graph of the data must approximate the shape of a parabola. Data that resemble one half of a parabola might also be modeled using a quadratic function with a restricted domain.

Quadratic function: $f(x) = ax^2 + bx + c, a > 0$

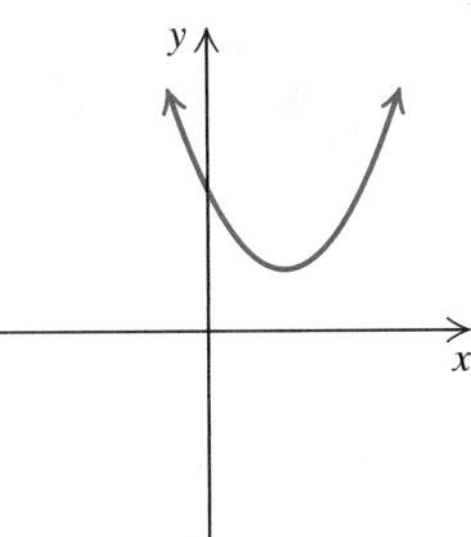

Quadratic function: $f(x) = ax^2 + bx + c, a < 0$

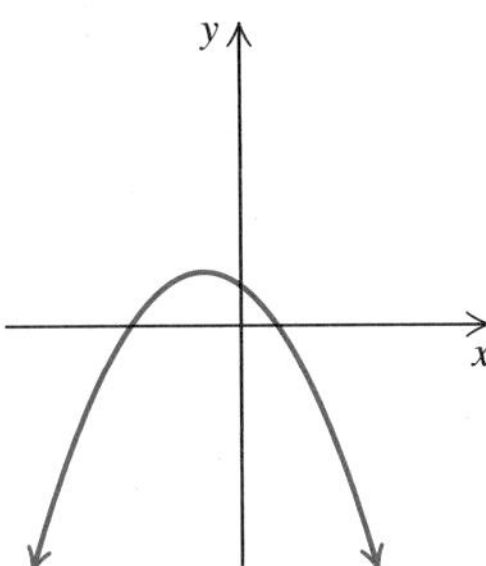

Quadratic function: $f(x) = ax^2 + bx + c,$ $a > 0,\ x \geq 0$

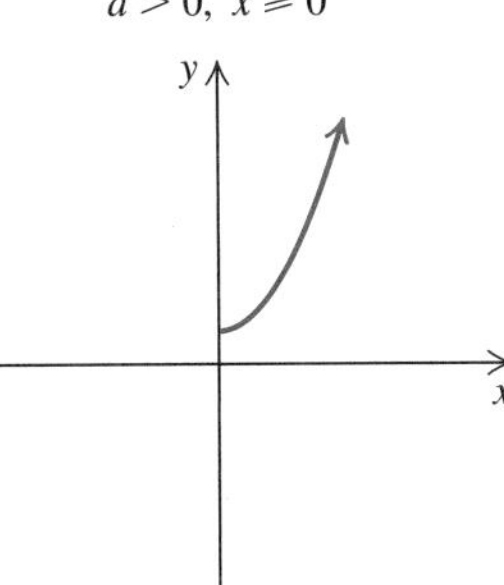

Quadratic function: $f(x) = ax^2 + bx + c,$ $a < 0,\ x \geq 0$

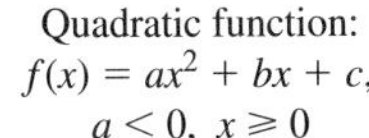

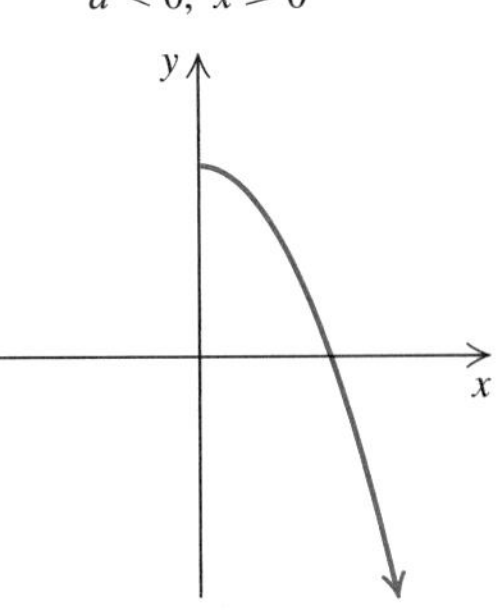

EXAMPLE 3 Determine whether a quadratic function can be used to model each of the following situations.

a) The quantity of candy consumed each year in the United States

U.S. Candy Consumption

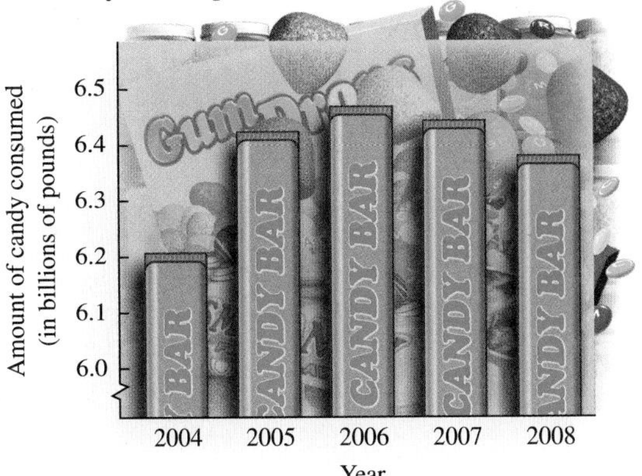

Source: U.S. Census Bureau

b) The amount of precipitation each month in Sonoma, California

Sonoma Precipitation

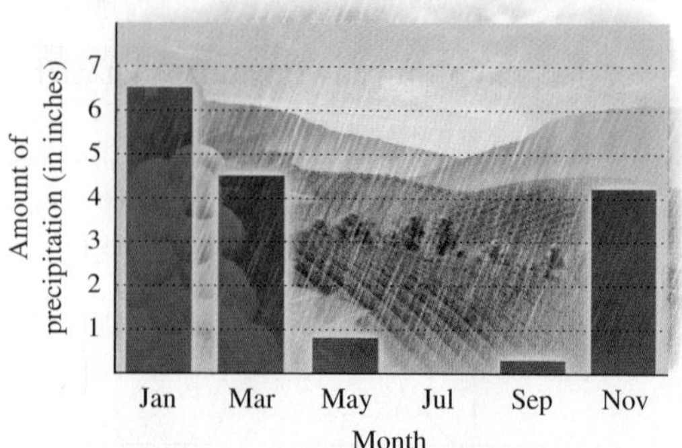

Source: www.city-data.com

c) The amount of precipitation each month in Atlanta, Georgia

Atlanta Precipitation

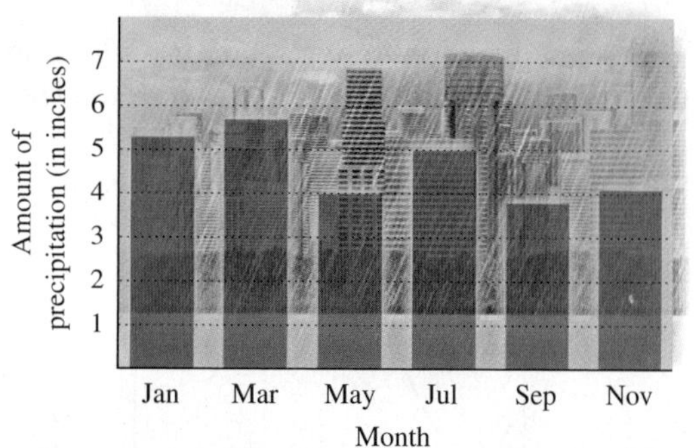

Source: www.city-data.com

d) The number of digital theater screens in the world

Digital Theater Screens

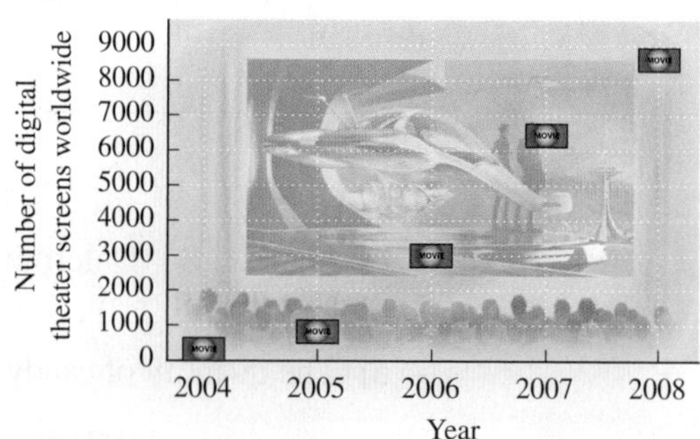

Source: Motion Picture Association of America, *Theatrical Market Statistics, 2008*

e) The average cinema admission price in the United States

Cinema Admission Price

Source: U.S. Bureau of Labor Statistics

SOLUTION

a) The data rise and then fall, resembling a parabola that opens downward. The situation could be modeled by a quadratic function $f(x) = ax^2 + bx + c, a < 0$.

b) The data fall and then rise, resembling a parabola that opens upward. A quadratic function $f(x) = ax^2 + bx + c, a > 0$, might be used as a model for this situation.

c) The data do not resemble a parabola. A quadratic function would not be a good model for this situation.

d) The data resemble the right half of a parabola that opens upward. We could use a quadratic function $f(x) = ax^2 + bx + c, a > 0, x \geq 0$, to model the situation.

e) The data appear nearly linear. A linear function is a better model for this situation than a quadratic function.

Try Exercise 23.

EXAMPLE 4 Candy Consumption. The amount of candy consumed each year in the United States can be modeled by a quadratic function, as seen in Example 3(a).

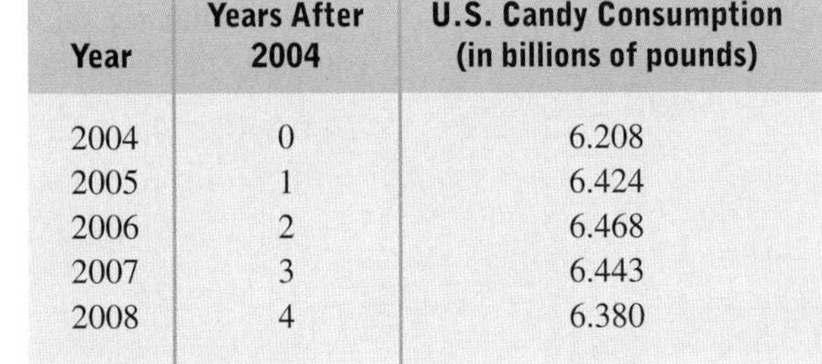

Year	Years After 2004	U.S. Candy Consumption (in billions of pounds)
2004	0	6.208
2005	1	6.424
2006	2	6.468
2007	3	6.443
2008	4	6.380

a) Use the data points (0, 6.208), (2, 6.468), and (4, 6.380) to fit a quadratic function $f(x)$ to the data.

b) Use the function from part (a) to estimate the candy consumption in 2009.

c) Use the REGRESSION feature of a graphing calculator to fit a quadratic function $g(x)$ to the data.

d) Use the function from part (c) to estimate the candy consumption in 2009, and compare the estimate with the estimate from part (b).

SOLUTION

a) Only one parabola will go through any three points that are not on a straight line. We are looking for a function of the form $f(x) = ax^2 + bx + c$, where $f(x)$ is the candy consumption x years after 2004. From the given information, we can write three equations. To find a, b, and c, we will solve the system of equations.

We begin by substituting the values of x and $f(x)$ from the three data points listed:

$$6.208 = a(0)^2 + b(0) + c, \quad \text{Using the data point } (0, 6.208)$$
$$6.468 = a(2)^2 + b(2) + c, \quad \text{Using the data point } (2, 6.468)$$
$$6.380 = a(4)^2 + b(4) + c. \quad \text{Using the data point } (4, 6.380)$$

After simplifying, we have the system of three equations

$$6.208 = c,$$
$$6.468 = 4a + 2b + c,$$
$$6.380 = 16a + 4b + c.$$

From the first equation, we know that $c = 6.208$. We substitute 6.208 for c in the second and third equations, giving us a system of two equations:

$$6.468 = 4a + 2b + 6.208,$$
$$6.380 = 16a + 4b + 6.208.$$

Now we subtract 6.208 from both sides of each equation, giving us

$$0.260 = 4a + 2b,$$
$$0.172 = 16a + 4b.$$

We can solve this system using elimination.

$0.260 = 4a + 2b,$ Multiplying in order to eliminate b → $-0.520 = -8a - 4b$

$0.172 = 16a + 4b$ → $0.172 = 16a + 4b$

$-0.348 = 8a$ **Adding**

$-0.0435 = a$ **Solving for a**

Now that we know $a = -0.0435$, we can solve for b, using one of the equations that contain a and b:

$$0.260 = 4a + 2b$$
$$0.260 = 4(-0.0435) + 2b \quad \textbf{Substituting for } a$$
$$0.260 = -0.174 + 2b$$
$$0.434 = 2b$$
$$0.217 = b. \quad \textbf{Solving for } b$$

We now have values for a, b, and c, and can write the quadratic function:

$$f(x) = ax^2 + bx + c \quad a = -0.0435, b = 0.217, c = 6.208$$
$$f(x) = -0.0435x^2 + 0.217x + 6.208.$$

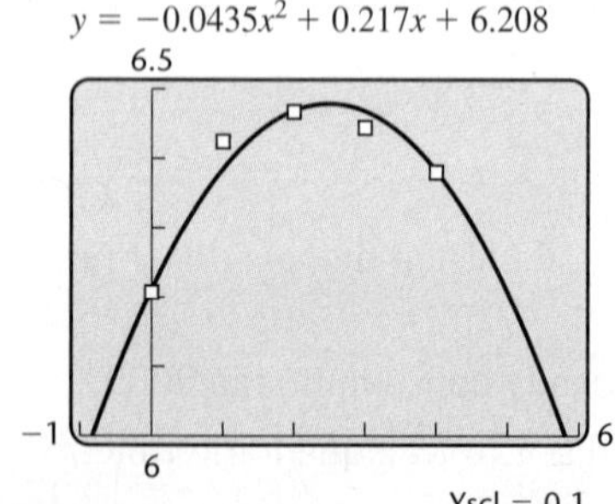

As a partial check, we graph the function along with the data, as shown at left. The graph goes through the three given data points.

b) Since 2009 is 5 years after 2004, we find $f(5)$ in order to estimate the candy consumption in that year:

$$f(5) = -0.0435(5)^2 + 0.217(5) + 6.208 = 6.2055.$$

According to this model, Americans will consume 6.2055 billion pounds of candy in 2009.

c) Using regression allows us to consider all the data given in the problem, not just three points. After entering the data, we choose the QUADREG option in the

STAT CALC menu. The figure on the left below gives the coefficients of the equation. In the figure on the right, the graph of g is shown along with the data. Rounding coefficients, we obtain

$$g(x) = -0.0448x^2 + 0.2154x + 6.2224.$$

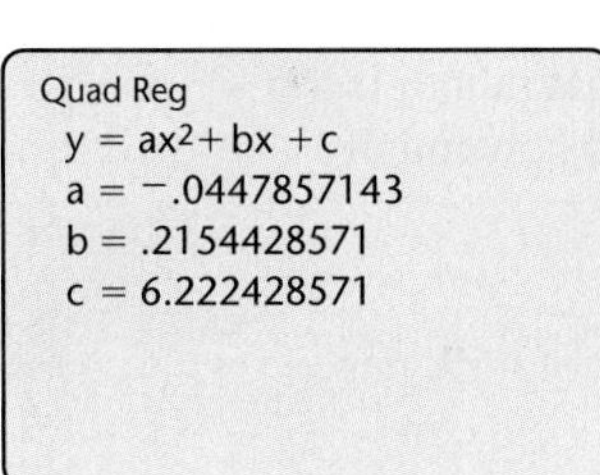

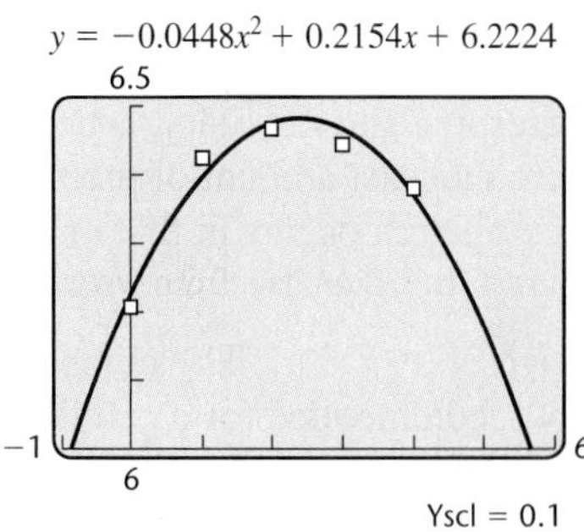

d) To estimate candy consumption in 2009, we evaluate $g(5)$ and obtain

$$g(5) = 6.1794.$$

This estimate of 6.1794 billion pounds of candy is lower than the estimate obtained in part (b). Since the second model took into consideration all the data, it is a better fit. However, without further information, it is difficult to tell which model more accurately describes the pattern of candy consumption in the United States.

Try Exercise 37.

11.8 Exercise Set

FOR EXTRA HELP

Concept Reinforcement *In each of Exercises 1–6, match the description with the graph that displays that characteristic.*

1. ____ A minimum value of $f(x)$ exists.
2. ____ A maximum value of $f(x)$ exists.
3. ____ No maximum or minimum value of $f(x)$ exists.
4. ____ The data points appear to suggest a linear model.
5. ____ The data points appear to suggest a quadratic model with a maximum.
6. ____ The data points appear to suggest a quadratic model with a minimum.

a)

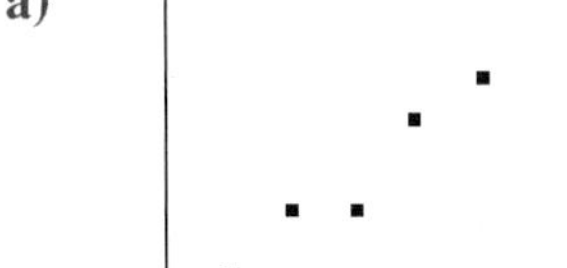

b)

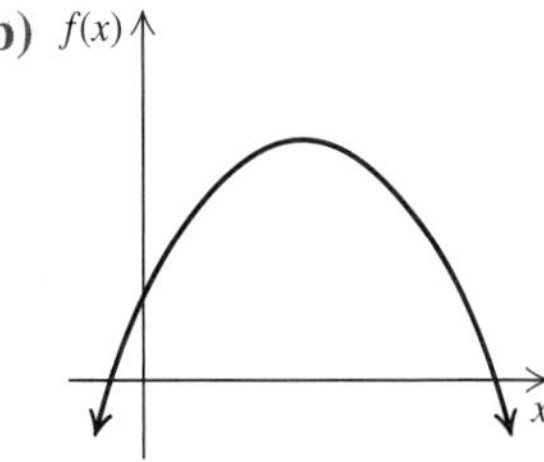

c)

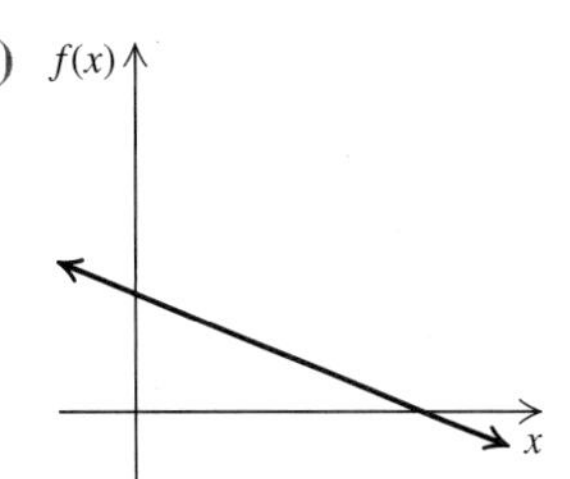

d)

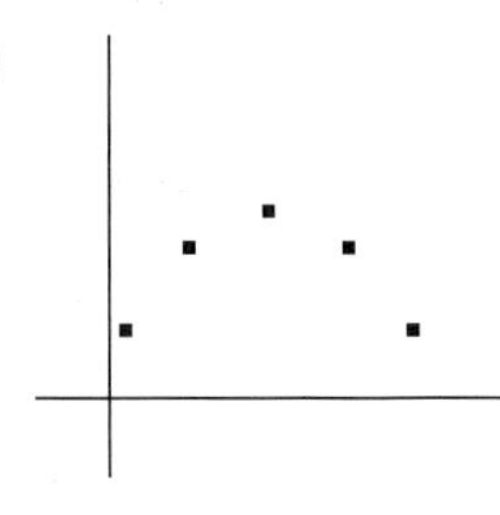

e)

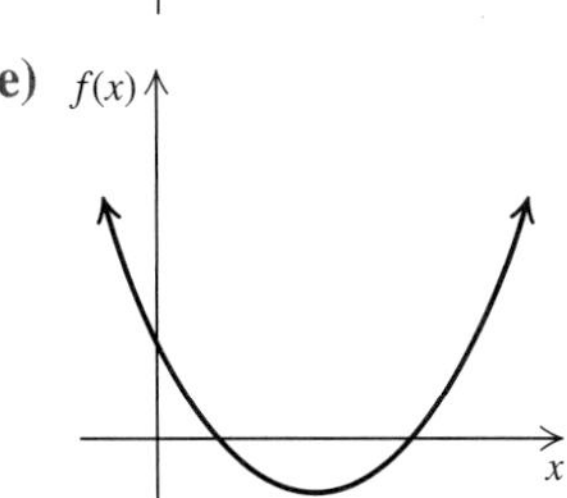

f)

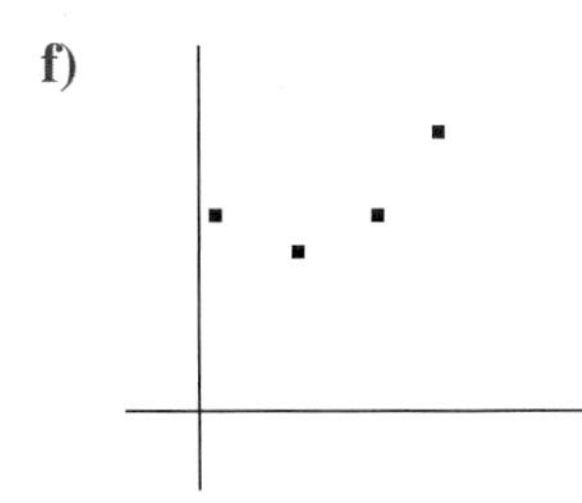

Solve.

7. *Sonoma Precipitation.* The amount of precipitation in Sonoma, California, can be approximated by

$$P(x) = 0.2x^2 - 2.8x + 9.8,$$

where $P(x)$ is in inches and x is the number of the month ($x = 1$ corresponds to January, $x = 2$ corresponds to February, and so on). In what month is there the least amount of precipitation, and how much precipitation occurs in that month?
Source: Based on data from www.city-data.com

8. *Newborn Calves.* The number of pounds of milk per day recommended for a calf that is x weeks old can be approximated by $p(x)$, where

$$p(x) = -0.2x^2 + 1.3x + 6.2.$$

When is a calf's milk consumption greatest and how much milk does it consume at that time?
Source: C. Chaloux, University of Vermont, 1998

9. *Maximizing Profit.* Recall that total profit P is the difference between total revenue R and total cost C. Given

$$R(x) = 1000x - x^2$$

and

$$C(x) = 3000 + 20x,$$

find the total profit, the maximum value of the total profit, and the value of x at which it occurs.

10. *Minimizing Cost.* Sweet Harmony Crafts has determined that when x hundred Dobros are built, the average cost per Dobro can be estimated by

$$C(x) = 0.1x^2 - 0.7x + 2.425,$$

where $C(x)$ is in hundreds of dollars. What is the minimum average cost per Dobro and how many Dobros should be built in order to achieve that minimum?

11. *Furniture Design.* A furniture builder is designing a rectangular end table with a perimeter of 128 in. What dimensions will yield the maximum area?

12. *Architecture.* An architect is designing an atrium for a hotel. The atrium is to be rectangular with a perimeter of 720 ft of brass piping. What dimensions will maximize the area of the atrium?

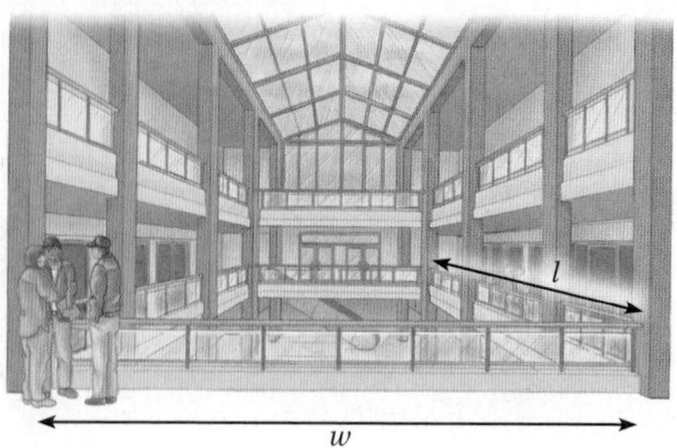

13. *Patio Design.* A stone mason has enough stones to enclose a rectangular patio with 60 ft of perimeter, assuming that the attached house forms one side of the rectangle. What is the maximum area that the mason can enclose? What should the dimensions of the patio be in order to yield this area?

14. *Garden Design.* Ginger is fencing in a rectangular garden, using the side of her house as one side of the rectangle. What is the maximum area that she can enclose with 40 ft of fence? What should the dimensions of the garden be in order to yield this area?

15. *Molding Plastics.* Economite Plastics plans to produce a one-compartment vertical file by bending the long side of an 8-in. by 14-in. sheet of plastic along two lines to form a U shape. How tall should the file be in order to maximize the volume that the file can hold?

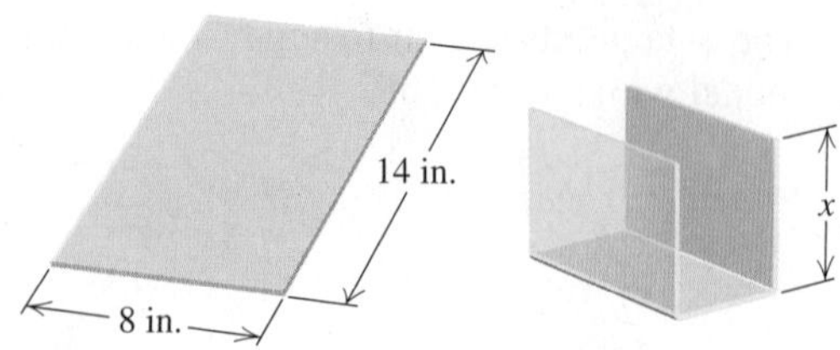

16. *Composting.* A rectangular compost container is to be formed in a corner of a fenced yard, with 8 ft of chicken wire completing the other two sides of the rectangle. If the chicken wire is 3 ft high, what dimensions of the base will maximize the container's volume?

17. What is the maximum product of two numbers that add to 18? What numbers yield this product?

18. What is the maximum product of two numbers that add to 26? What numbers yield this product?

19. What is the minimum product of two numbers that differ by 8? What are the numbers?

20. What is the minimum product of two numbers that differ by 7? What are the numbers?

Aha! 21. What is the maximum product of two numbers that add to -10? What numbers yield this product?

22. What is the maximum product of two numbers that add to -12? What numbers yield this product?

Choosing Models. *For the scatterplots and graphs in Exercises 23–32, determine which, if any, of the following functions might be used as a model for the data: Linear, with $f(x) = mx + b$; quadratic, with $f(x) = ax^2 + bx + c, a > 0$; quadratic, with $f(x) = ax^2 + bx + c, a < 0$; neither quadratic nor linear.*

23. **Dorm Expense**

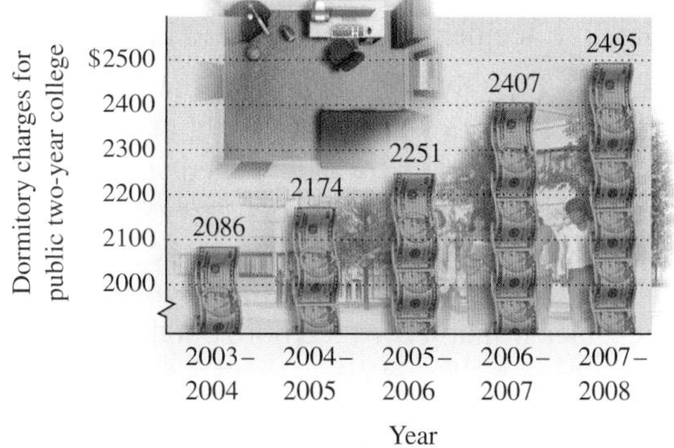

Source: National Center for Education Statistics

24. **Restaurant Industry Sales**

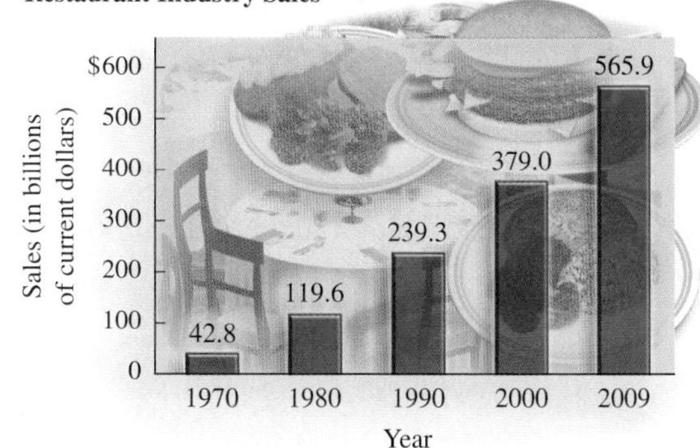

Source: National Restaurant Association

25.

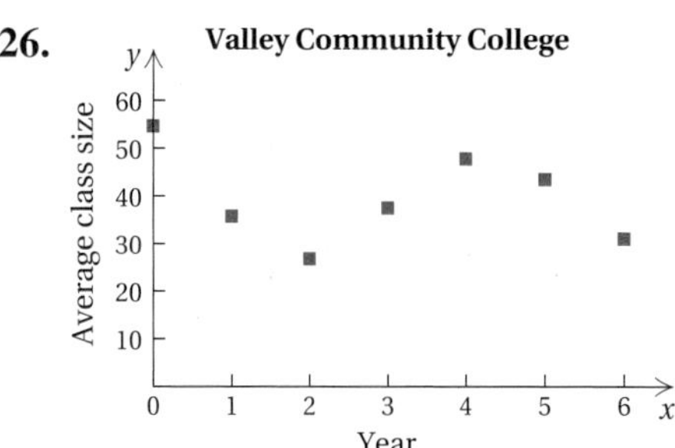

26.

27. **Seniors in the Work Force**

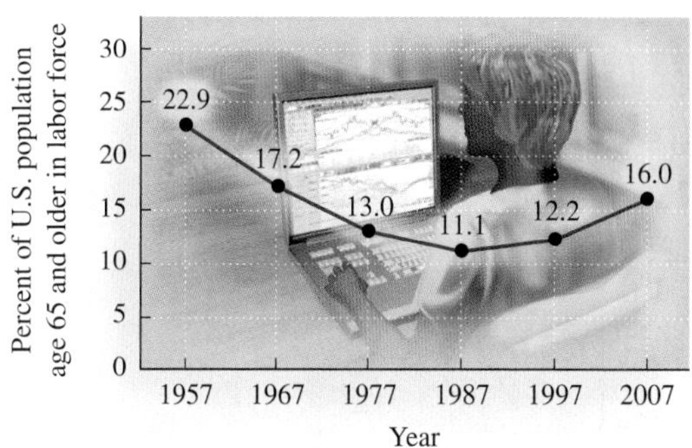

Source: U.S. Bureau of Labor Statistics

28. **Doctor Visits**

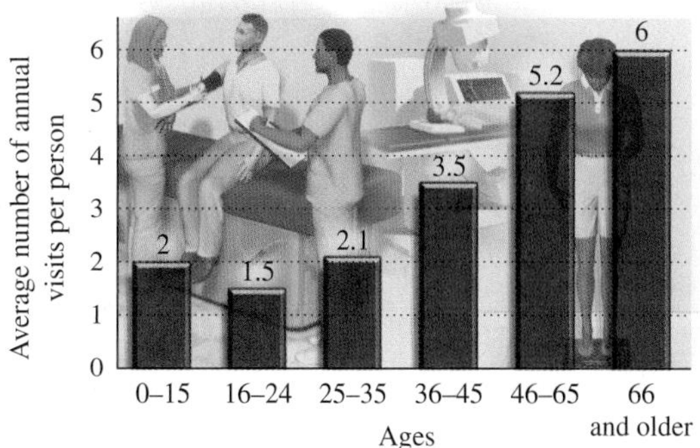

Sources: National Ambulatory Health Care Administration; Merritt, Hawkins & Associates; Council on Graduate Medical Education

29. **Recycling**

Source: Environmental Protection Agency

30. **Average Number of Live Births per 1000 Women**

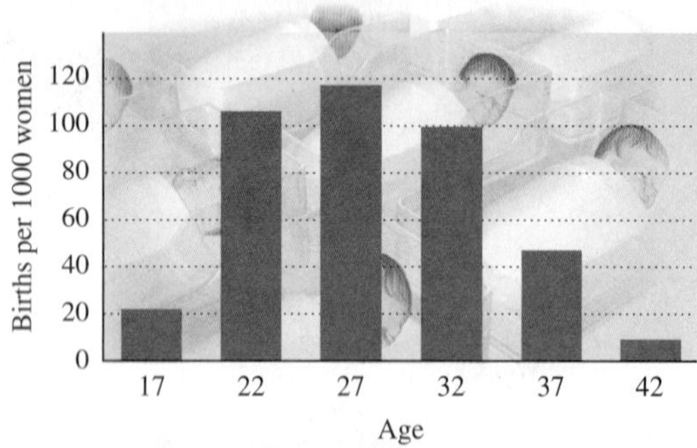

Source: U.S. Centers for Disease Control

31. **Credit Cards**

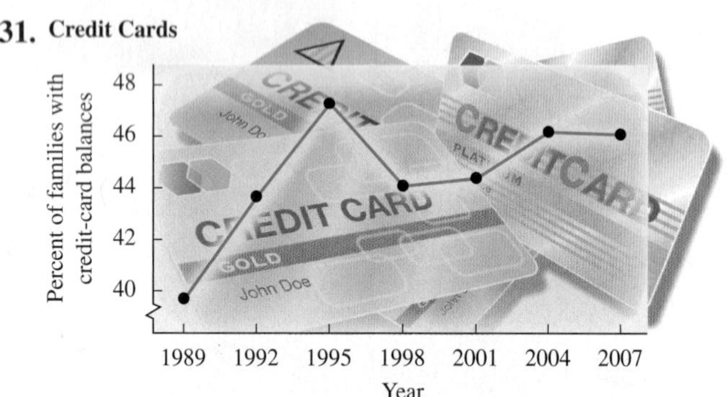

Source: http://www.federalreserve.gov/pubs/oss/oss2/2007/2007%20SCF%20Chartbook.pdf

32. **Wind Power**

Source: Energy Information Administration

Find a quadratic function that fits the set of data points.

33. $(1, 4), (-1, -2), (2, 13)$

34. $(1, 4), (-1, 6), (-2, 16)$

35. $(2, 0), (4, 3), (12, -5)$

36. $(-3, -30), (3, 0), (6, 6)$

37. **a)** Find a quadratic function that fits the following data.

Travel Speed (in kilometers per hour)	Number of Nighttime Accidents (for every 200 million kilometers driven)
60	400
80	250
100	250

b) Use the function to estimate the number of nighttime accidents that occur at 50 km/h.

38. **a)** Find a quadratic function that fits the following data.

Travel Speed (in kilometers per hour)	Number of Daytime Accidents (for every 200 million kilometers driven)
60	100
80	130
100	200

b) Use the function to estimate the number of daytime accidents that occur at 50 km/h.

39. *Archery.* The Olympic flame tower at the 1992 Summer Olympics was lit at a height of about 27 m by a flaming arrow that was launched about 63 m from the base of the tower. If the arrow landed about 63 m beyond the tower, find a quadratic function that expresses the height h of the arrow as a function of the distance d that it traveled horizontally.

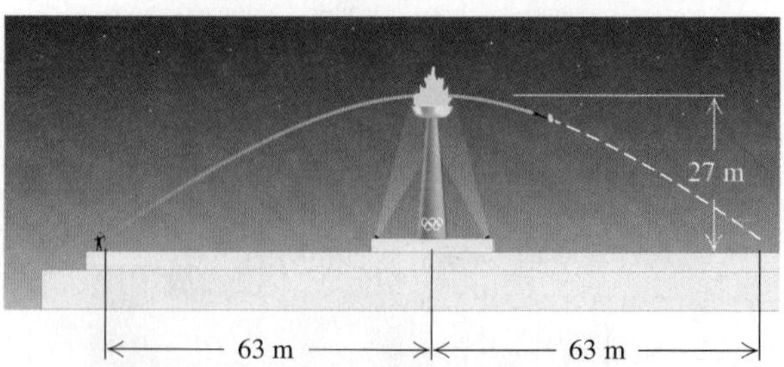

40. *Pizza Prices.* Pizza Unlimited has the following prices for pizzas.

Diameter	Price
8 in.	$ 6.00
12 in.	$ 8.50
16 in.	$11.50

Is price a quadratic function of diameter? It probably should be, because the price should be proportional to the area, and the area is a quadratic function of the diameter. (The area of a circular region is given by $A = \pi r^2$ or $(\pi/4) \cdot d^2$.)

a) Express price as a quadratic function of diameter using the data points $(8, 6)$, $(12, 8.50)$, and $(16, 11.50)$.
b) Use the function to find the price of a 14-in. pizza.

41. *Hydrology.* The drawing below shows the cross section of a river. Typically rivers are deepest in the middle, with the depth decreasing to 0 at the edges. A hydrologist measures the depths D, in feet, of a river at distances x, in feet, from one bank. The results are listed in the table below.

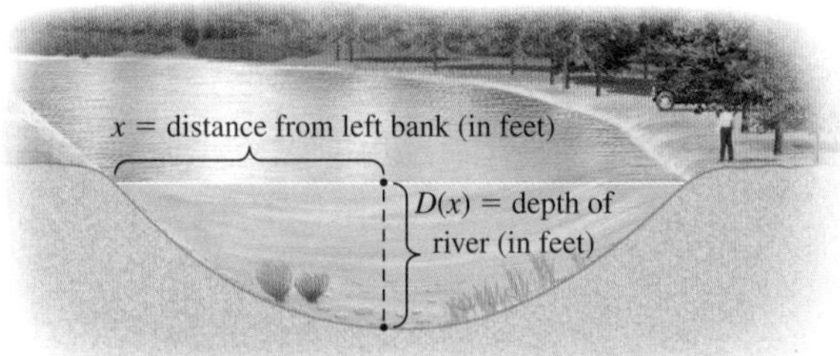

Distance x, from the Left Bank (in feet)	Depth, D, of the River (in feet)
0	0
15	10.2
25	17
50	20
90	7.2
100	0

a) Use regression to find a quadratic function that fits the data.
b) Use the function to estimate the depth of the river 70 ft from the left bank.

42. *Work Force.* The graph in Exercise 27 indicates that the percent of the U.S. population age 65 and older that is in the work force is increasing after several decades of decrease.

a) Use regression to find a quadratic function that can be used to estimate the percent $p(x)$ of the population age 65 and older that is in the work force x years after 1957.
b) Use the function found in part (a) to predict the percent of those age 65 and older who will be in the work force in 2010.

43. *Teacher Shortages.* The estimated shortage of math and science teachers each year is shown in the table below.

Year	Math and Science Teacher Shortage
2005	24,546
2007	25,029
2009	25,049
2011	25,781
2013	26,391
2015	27,214

Source: Based on data from National Center for Education Statistics and Council of Chief of State of School Officers

a) Use regression to find a quadratic function $t(x)$ that can be used to estimate the teacher shortage x years after 2005.
b) Use the function found in part (a) to predict the math and science teacher shortage in 2017.

44. *Restaurant Sales.* The graph in Exercise 24 indicates that restaurant sales can be modeled as a quadratic function of the year.

a) Use regression to find a quadratic function $r(x)$ that can be used to estimate sales, in billions of dollars, x years after 1970.
b) Use the function found in part (a) to estimate restaurant sales in 2015.

TW 45. Does every nonlinear function have a minimum value or a maximum value? Why or why not?

TW 46. Explain how the leading coefficient of a quadratic function can be used to determine if a maximum or a minimum function value exists.

SKILL REVIEW

To prepare for Section 12.1, review function notation (Section 3.8).

Graph each function. [3.1], [3.8]

47. $f(x) = x^3 - 2$

48. $g(x) = \frac{2}{x}$

49. If $f(x) = x + 7$, find $f\left(\frac{1}{a^2}\right)$. [3.8]

50. If $g(x) = x^2 - 3$, find $g\left(\sqrt{a - 5}\right)$. [3.8], [10.1]

51. If $g(x) = x^2 + 2$, find $g(2a + 5)$. [3.8], [5.5]

52. If $f(x) = \sqrt{4x + 1}$, find $g(3a - 5)$. [3.8]

SYNTHESIS

The following graphs can be used to compare the baseball statistics of pitcher Roger Clemens with the 31 other pitchers since 1968 who started at least 10 games in at least 15 seasons and pitched at least 3000 innings. Use the graphs to answer questions 53 and 54.

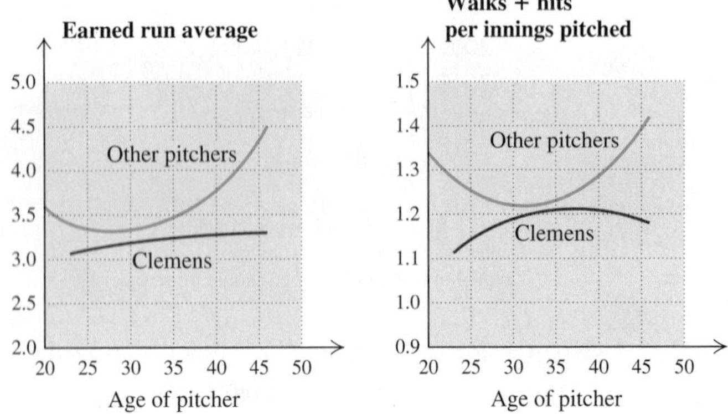

Source: *The New York Times*, February 10, 2008; Eric Bradlow, Shane Jensen, Justin Wolfers and Adi Wyner

TW **53.** The earned run average describes how many runs a pitcher has allowed per game. The lower the earned run average, the better the pitcher. Compare, in terms of maximums or minimums, the earned run average of Roger Clemens with that of other pitchers. Is there any reason to suspect that the aging process was unusual for Clemens? Explain.

TW **54.** The statistic "Walks + hits per innings pitched" is related to how often a pitcher allows a batter to reach a base. The lower this statistic, the better. Compare, in terms of maximums or minimums, the "walks + hits" statistic of Roger Clemens with that of other pitchers.

55. *Bridge Design.* The cables supporting a straight-line suspension bridge are nearly parabolic in shape. Suppose that a suspension bridge is being designed with concrete supports 160 ft apart and with vertical cables 30 ft above road level at the midpoint of the bridge and 80 ft above road level at a point 50 ft from the midpoint of the bridge. How long are the longest vertical cables?

56. *Trajectory of a Flare.* The height above the ground of a launched object is a quadratic function of the time that it is in the air. Suppose that a flare is launched from a cliff 64 ft above sea level. If 3 sec after being launched the flare is again level with the cliff, and if 2 sec after that it lands in the sea, what is the maximum height that the flare will reach?

57. *Norman Window.* A *Norman window* is a rectangle with a semicircle on top. Reid is designing a Norman window that will require 24 ft of trim. What dimensions will give the maximum area of glass?

58. *Crop Yield.* An orange grower finds that she gets an average yield of 40 bushels (bu) per tree when she plants 20 trees on an acre of ground. Each time she adds a tree to an acre, the yield per tree decreases by 1 bu, due to congestion. How many trees per acre should she plant for maximum yield?

59. *Cover Charges.* When the owner of Sweet Sounds charges a \$10 cover charge, an average of 80 people will attend a show. For each 25¢ increase in admission price, the average number attending decreases by 1. What should the owner charge in order to make the most money?

60. *Minimizing Area.* A 36-in. piece of string is cut into two pieces. One piece is used to form a circle while the other is used to form a square. How should the string be cut so that the sum of the areas is a minimum?

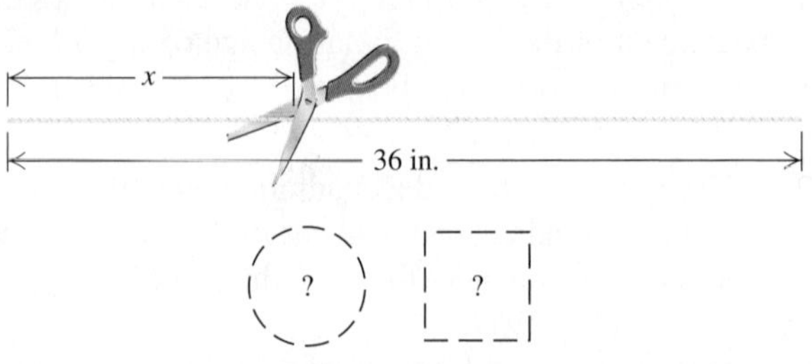

Try Exercise Answers: Section 11.8

7. July; 0 in. **11.** 32 in. by 32 in. **23.** $f(x) = mx + b$
37. (a) $A(s) = \frac{3}{16}s^2 - \frac{135}{4}s + 1750$; **(b)** about 531 accidents

Study Summary

KEY TERMS AND CONCEPTS	EXAMPLES	PRACTICE EXERCISES
SECTION 11.1: QUADRATIC EQUATIONS		
A **quadratic equation in standard form** is written $ax^2 + bx + c = 0$, with a, b, and c constant and $a \neq 0$. Some quadratic equations can be solved by factoring.	$x^2 - 3x - 10 = 0$ $(x + 2)(x - 5) = 0$ $x + 2 = 0$ *or* $x - 5 = 0$ $x = -2$ *or* $x = 5$	**1.** Solve: $x^2 - 12x + 11 = 0.$
The Principle of Square Roots For any real number k, if $X^2 = k$, then $X = \sqrt{k}$ *or* $X = -\sqrt{k}$.	$x^2 - 8x + 16 = 25$ $(x - 4)^2 = 25$ $x - 4 = -5$ *or* $x - 4 = 5$ $x = -1$ *or* $x = 9$	**2.** Solve: $x^2 - 18x + 81 = 5.$
Any quadratic equation can be solved by **completing the square.**	$x^2 + 6x = 1$ $x^2 + 6x + \left(\frac{6}{2}\right)^2 = 1 + \left(\frac{6}{2}\right)^2$ $x^2 + 6x + 9 = 1 + 9$ $(x + 3)^2 = 10$ $x + 3 = \pm\sqrt{10}$ $x = -3 \pm \sqrt{10}$	**3.** Solve by completing the square: $x^2 + 20x = 21.$
SECTION 11.2: THE QUADRATIC FORMULA		
The Quadratic Formula The solutions of $ax^2 + bx + c = 0$ are given by $x = \dfrac{-b \pm \sqrt{b^2 - 4ac}}{2a}$.	$3x^2 - 2x - 5 = 0 \quad a = 3, b = -2, c = -5$ $x = \dfrac{-(-2) \pm \sqrt{(-2)^2 - 4 \cdot 3(-5)}}{2 \cdot 3}$ $x = \dfrac{2 \pm \sqrt{4 + 60}}{6}$ $x = \dfrac{2 \pm \sqrt{64}}{6}$ $x = \dfrac{2 \pm 8}{6}$ $x = \dfrac{10}{6} = \dfrac{5}{3}$ *or* $x = \dfrac{-6}{6} = -1$	**4.** Solve: $2x^2 - 3x - 9 = 0.$
SECTION 11.3: STUDYING SOLUTIONS OF QUADRATIC EQUATIONS		
The **discriminant** of the quadratic formula is $b^2 - 4ac$.		**5.** Use the discriminant to determine the number and type of solutions of $2x^2 + 5x + 9 = 0.$
$b^2 - 4ac = 0 \rightarrow$ One solution; a rational number	For $4x^2 - 12x + 9 = 0$, $b^2 - 4ac = (-12)^2 - 4(4)(9) = 144 - 144 = 0$. Thus, $4x^2 - 12x + 9 = 0$ has one rational solution.	
$b^2 - 4ac > 0 \rightarrow$ Two real solutions; both are rational if $b^2 - 4ac$ is a perfect square.	For $x^2 + 6x - 2 = 0$, $b^2 - 4ac = (6)^2 - 4(1)(-2) = 36 + 8 = 44$. Thus, $x^2 + 6x - 2 = 0$ has two irrational real-number solutions.	

$b^2 - 4ac < 0 \rightarrow$ Two imaginary-number solutions	For $2x^2 - 3x + 5 = 0$, $b^2 - 4ac = (-3)^2 - 4(2)(5) = 9 - 40 = -31$. Thus, $2x^2 - 3x + 5 = 0$ has two imaginary-number solutions.	

SECTION 11.4: APPLICATIONS INVOLVING QUADRATIC EQUATIONS

To solve a formula for a letter, use the same principles used to solve equations.	Solve $y = pn^2 + dn$ for n. $pn^2 + dn - y = 0 \quad a = p, b = d, c = -y$ $n = \dfrac{-d \pm \sqrt{d^2 - 4p(-y)}}{2 \cdot p}$ $n = \dfrac{-d \pm \sqrt{d^2 + 4py}}{2p}$	**6.** Solve $a = n^2 + 1$ for n.

SECTION 11.5: EQUATIONS REDUCIBLE TO QUADRATIC

Equations that are **reducible to quadratic** or in **quadratic form** can be solved by making an appropriate substitution.	$x^4 - 10x^2 + 9 = 0$ **Let $u = x^2$. Then $u^2 = x^4$.** $u^2 - 10u + 9 = 0$ **Substituting** $(u - 9)(u - 1) = 0$ $u - 9 = 0$ *or* $u - 1 = 0$ $u = 9$ *or* $u = 1$ **Solving for u** $x^2 = 9$ *or* $x^2 = 1$ $x = \pm 3$ *or* $x = \pm 1$ **Solving for x**	**7.** Solve: $x - \sqrt{x} - 30 = 0.$

SECTION 11.6: QUADRATIC FUNCTIONS AND THEIR GRAPHS
SECTION 11.7: MORE ABOUT GRAPHING QUADRATIC FUNCTIONS

The graph of a quadratic function $f(x) = ax^2 + bx + c$ $= a(x - h)^2 + k$ is a **parabola.** The graph opens upward for $a > 0$ and downward for $a < 0$. The **vertex** is (h, k) and the **axis of symmetry** is $x = h$. If $a > 0$, the function has a **minimum** value of k, and if $a < 0$, the function has a **maximum** value of k. The vertex and the axis of symmetry occur at $x = -\dfrac{b}{2a}$.	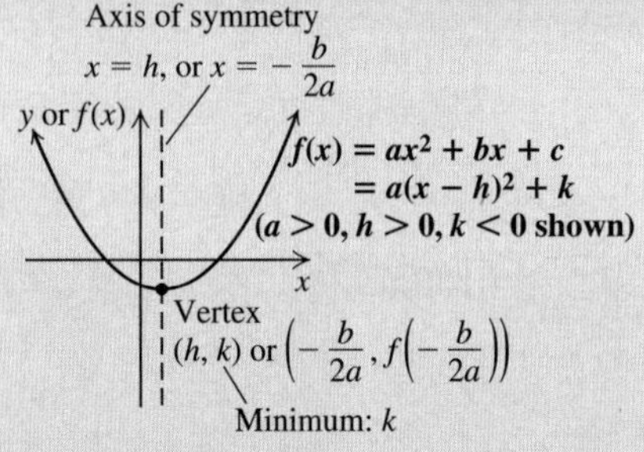	**8.** Graph $f(x) = 2x^2 - 12x + 3$. Label the vertex and the axis of symmetry, and identify the minimum or maximum function value.

SECTION 11.8: PROBLEM SOLVING AND QUADRATIC FUNCTIONS

Some problem situations can be **modeled** using quadratic functions. For those problems, a quantity can often be **maximized** or **minimized** by finding the coordinates of a vertex.	A lifeguard has 100 m of roped-together flotation devices with which to cordon off a rectangular swimming area at Lakeside Beach. If the shoreline forms one side of the rectangle, what dimensions will maximize the size of the area for swimming? This problem and its solution appear as Example 2 on pp. 833–834.	**9.** Loretta is putting fencing around a rectangular vegetable garden. She can afford to buy 120 ft of fencing. What dimensions should she plan for the garden in order to maximize its area?

Review Exercises 11

Concept Reinforcement *Classify each of the following statements as either true or false.*

1. Every quadratic equation has two different solutions. [11.3]
2. Every quadratic equation has at least one real-number solution. [11.3]
3. If an equation cannot be solved by completing the square, it cannot be solved by the quadratic formula. [11.2]
4. A negative discriminant indicates two imaginary-number solutions of a quadratic equation. [11.3]
5. Certain radical or rational equations can be written in quadratic form. [11.5]
6. The graph of $f(x) = 2(x + 3)^2 - 4$ has its vertex at $(3, -4)$. [11.6]
7. The graph of $g(x) = 5x^2$ has $x = 0$ as its axis of symmetry. [11.6]
8. The graph of $f(x) = -2x^2 + 1$ has no minimum value. [11.6]
9. The zeros of $g(x) = x^2 - 9$ are -3 and 3. [11.6]
10. The graph of every quadratic function has at least one x-intercept. [11.7]

11. Consider the following graph of $f(x) = ax^2 + bx + c$.
    a) State the number of real-number solutions of $ax^2 + bx + c = 0$. [11.1]
    b) State whether a is positive or negative. [11.6]
    c) Determine the minimum value of f. [11.6]

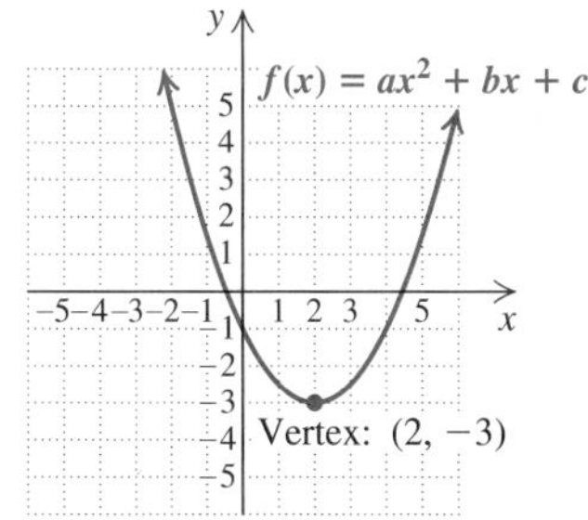

Solve.

12. $9x^2 - 2 = 0$ [11.1]
13. $8x^2 + 6x = 0$ [11.1]
14. $x^2 - 12x + 36 = 9$ [11.1]
15. $x^2 - 4x + 8 = 0$ [11.2]
16. $x(3x + 4) = 4x(x - 1) + 15$ [11.2]
17. $x^2 + 9x = 1$ [11.2]
18. $x^2 - 5x - 2 = 0$. Use a calculator to approximate the solutions with rational numbers. Round to three decimal places. [11.2]
19. Let $f(x) = 4x^2 - 3x - 1$. Find x such that $f(x) = 0$. [11.2]

Replace the blanks with constants to form a true equation. [11.1]

20. $x^2 - 12x +$ ___ $= (x -$ ___$)^2$
21. $x^2 + \frac{3}{5}x +$ ___ $= (x +$ ___$)^2$
22. Solve by completing the square. Show your work.
    $$x^2 - 6x + 1 = 0 \quad [11.1]$$
23. \$2500 grows to \$2704 in 2 years. Use the formula $A = P(1 + r)^t$ to find the interest rate. [11.1]
24. The Singapore Flyer Observation Wheel is 541 ft tall. Use $s = 16t^2$ to approximate how long it would take an object to fall from the top. [11.1]

For each equation, determine whether the solutions are real or imaginary. If they are real, specify whether they are rational or irrational. [11.3]

25. $x^2 + 3x - 6 = 0$
26. $x^2 + 2x + 5 = 0$
27. Write a quadratic equation having the solutions $3i$ and $-3i$. [11.3]
28. Write a quadratic equation having -4 as its only solution. [11.3]

Solve. [11.4]

29. Horizons has a manufacturing plant located 300 mi from company headquarters. Their corporate pilot must fly from headquarters to the plant and back in 4 hr. If there is a 20-mph headwind going and a 20-mph tailwind returning, how fast must the plane be able to travel in still air?

30. Working together, Erica and Shawna can answer a day's worth of customer-service e-mails in 4 hr. Working alone, Erica takes 6 hr longer than Shawna. How long would it take Shawna to answer the e-mails alone?

31. Find all x-intercepts of the graph of

$$f(x) = x^4 - 13x^2 + 36. \text{ [11.5]}$$

Solve. [11.5]

32. $15x^{-2} - 2x^{-1} - 1 = 0$

33. $(x^2 - 4)^2 - (x^2 - 4) - 6 = 0$

34. a) Graph: $f(x) = -3(x + 2)^2 + 4$. [11.6]
b) Label the vertex.
c) Draw the axis of symmetry.
d) Find the maximum or the minimum value.

35. For the function given by $f(x) = 2x^2 - 12x + 23$: [11.7]
a) find the vertex and the axis of symmetry;
b) graph the function.

36. Find any x-intercepts and the y-intercept of the graph of

$$f(x) = x^2 - 9x + 14. \text{ [11.7]}$$

37. Solve $N = 3\pi\sqrt{\frac{1}{p}}$ for p. [11.4]

38. Solve $2A + T = 3T^2$ for T. [11.4]

Determine which, if any, of the following functions might be used as a model for the data: Linear, with $f(x) = mx + b$; quadratic, with $f(x) = ax^2 + bx + c, a > 0$; quadratic, with $f(x) = ax^2 + bx + c, a < 0$; neither quadratic nor linear. [11.8]

39. Recreational Boating

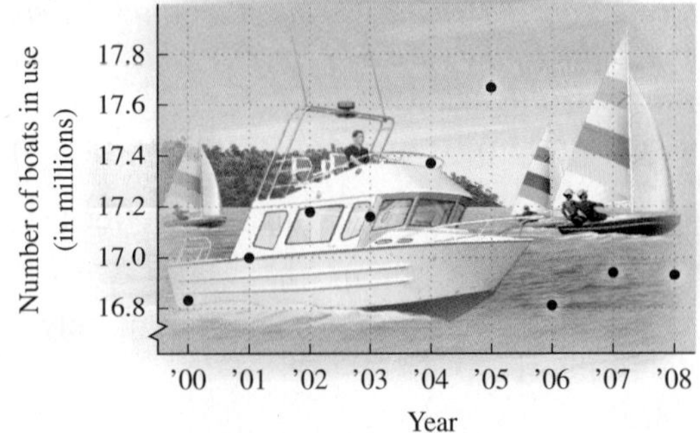

40. McDonalds Restaurants

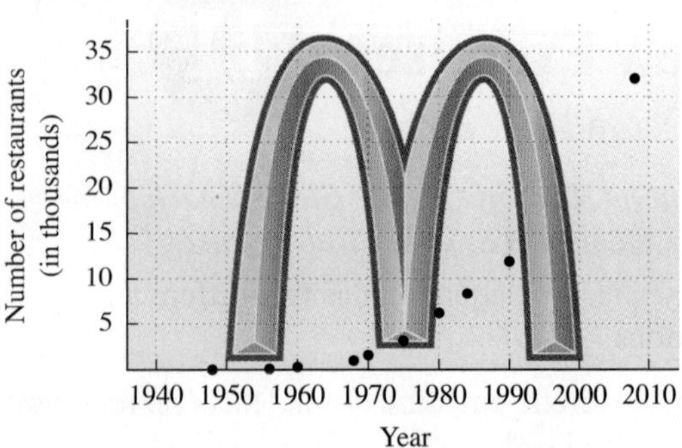

Source: McDonalds

41. Cost of Cable Television

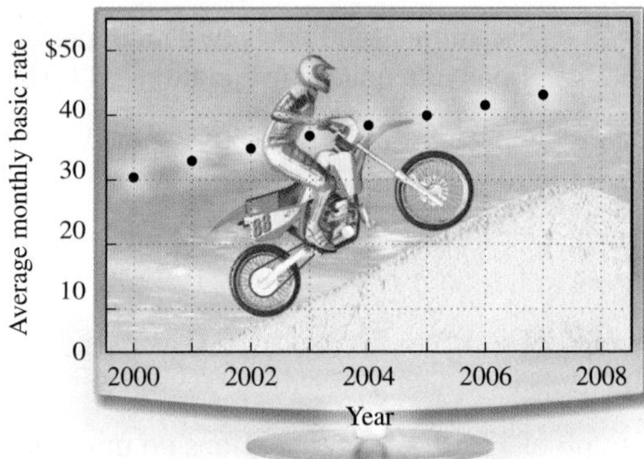

Source: SNL Kagan

42. Eastgate Consignments wants to build a rectangular area in a corner for children to play in while their parents shop. They have 30 ft of low fencing. What is the maximum area that they can enclose? What dimensions will yield this area? [11.8]

43. *McDonalds.* The table below lists the number of McDonalds restaurants for various years (see Exercise 40). [11.8]

Year	Number of McDonalds Restaurants
1948	1
1956	14
1960	228
1968	1,000
1970	1,600
1975	3,076
1980	6,263
1984	8,300
1990	11,800
2008	32,000

a) Use the data points $(0, 1)$, $(20, 1000)$, and $(60, 32{,}000)$ to find a quadratic function $M(x)$ that can be used to estimate the number of McDonalds restaurants x years after 1948.

b) Use the function to estimate the number of McDonalds restaurants in 2020.

44. a) Use the REGRESSION feature of a graphing calculator and all the data in Exercise 43 to find a quadratic function that can be used to estimate the number $M(x)$ of McDonalds restaurants x years after 1948. [11.8]

b) Use the function to estimate the number of McDonalds restaurants in 2020. [11.8]

SYNTHESIS

TW **45.** Discuss two ways in which completing the square was used in this chapter. [11.1], [11.2], [11.7]

TW **46.** Compare the results of Exercises 43(b) and 44(b). Which model do you think gives a better prediction? [11.8]

TW **47.** What is the greatest number of solutions that an equation of the form $ax^4 + bx^2 + c = 0$ can have? Why? [11.5]

48. A quadratic function has x-intercepts at -3 and 5. If the y-intercept is at -7, find an equation for the function. [11.7]

49. Find h and k if, for $3x^2 - hx + 4k = 0$, the sum of the solutions is 20 and the product of the solutions is 80. [11.3]

50. The average of two positive integers is 171. One of the numbers is the square root of the other. Find the integers. [11.5]

Chapter Test 11

1. Consider the following graph of $f(x) = ax^2 + bx + c$.

a) State the number of real-number solutions of $ax^2 + bx + c = 0$.

b) State whether a is positive or negative.

c) Determine the maximum function value of f.

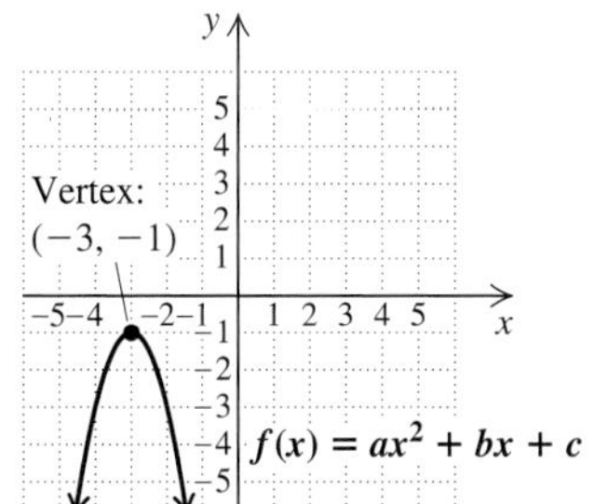

Solve.

2. $25x^2 - 7 = 0$

3. $4x(x - 2) - 3x(x + 1) = -18$

4. $x^2 + 2x + 3 = 0$

5. $2x + 5 = x^2$

6. $x^{-2} - x^{-1} = \frac{3}{4}$

7. $x^2 + 3x = 5$. Use a calculator to approximate the solutions with rational numbers. Round to three decimal places.

8. Let $f(x) = 12x^2 - 19x - 21$. Find x such that $f(x) = 0$.

Replace the blanks with constants to form a true equation.

9. $x^2 - 20x + \_\_\_ = (x - \_\_\_)^2$

10. $x^2 + \frac{2}{7}x + \_\_\_ = \left(x + \_\_\_\right)^2$

11. Solve by completing the square. Show your work.

$$x^2 + 10x + 15 = 0$$

12. Determine the type of number that the solutions of $x^2 + 2x + 5 = 0$ will be.

13. Write a quadratic equation having solutions $\sqrt{11}$ and $-\sqrt{11}$.

Solve.

14. The Connecticut River flows at a rate of 4 km/h for the length of a popular scenic route. In order for a cruiser to travel 60 km upriver and then return in a total of 8 hr, how fast must the boat be able to travel in still water?

15. Dal and Kim can assemble a swing set in $1\frac{1}{2}$ hr. Working alone, it takes Kim 4 hr longer than Dal to assemble the swing set. How long would it take Dal, working alone, to assemble the swing set?

16. Find all x-intercepts of the graph of
$$f(x) = x^4 - 15x^2 - 16.$$

17. a) Graph: $f(x) = 4(x - 3)^2 + 5$.
b) Label the vertex.
c) Draw the axis of symmetry.
d) Find the maximum or the minimum function value.

18. For the function $f(x) = 2x^2 + 4x - 6$:
a) find the vertex and the axis of symmetry;
b) graph the function.

19. Find any x-intercepts and the y-intercept of
$$f(x) = x^2 - x - 6.$$

20. Solve $V = \frac{1}{3}\pi(R^2 + r^2)$ for r. Assume that all variables are positive.

21. State whether the graph appears to represent a linear function, a quadratic function, or neither.

Chicago Air Quality

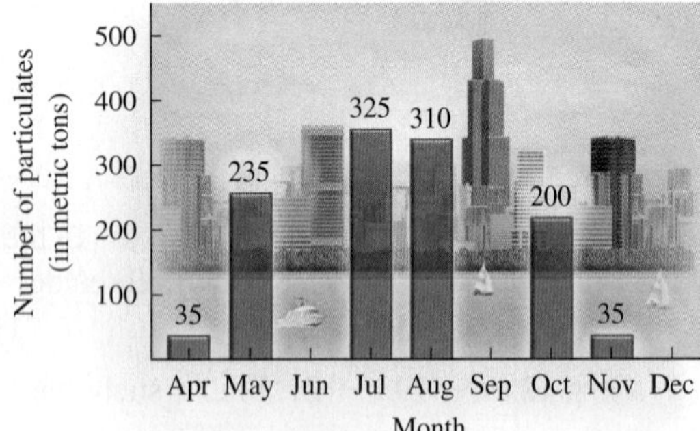

Source: The National Arbor Day Foundation

22. Jay's Metals has determined that when x hundred storage cabinets are built, the average cost per cabinet is given by
$$C(x) = 0.2x^2 - 1.3x + 3.4025,$$
where $C(x)$ is in hundreds of dollars. What is the minimum cost per cabinet and how many cabinets should be built in order to achieve that minimum?

23. Trees improve air quality in part by retaining airborne particles, called particulates, from the air. The table below shows the number of metric tons of particulates retained by trees in Chicago during the spring, summer, and autumn months. (See Exercise 21.)

Month	Amount of Particulates Retained (in metric tons)
April (0)	35
May (1)	235
July (3)	325
August (4)	310
October (6)	200
November (7)	35

Source: The National Arbor Day Foundation

Use the data points (0, 35), (4, 310), and (6, 200) to find a quadratic function that can be used to estimate the number of metric tons p of particulates retained x months after April.

24. Use regression to find a quadratic function that fits all the data in Exercise 23.

SYNTHESIS

25. One solution of $kx^2 + 3x - k = 0$ is -2. Find the other solution.

26. Find a fourth-degree polynomial equation, with integer coefficients, for which $-\sqrt{3}$ and $2i$ are solutions.

27. Solve: $x^4 - 4x^2 - 1 = 0$.

12

Exponential Functions and Logarithmic Functions

Do You Text More Than You Talk?

Beginning in 2008, Americans sent more text messages than they made phone calls. They still talk on the phone just as often, but the number of text messages has increased *exponentially*. In Example 9 of Section 12.7, we use the data in the graph shown here to estimate the number of text messages in 2011.

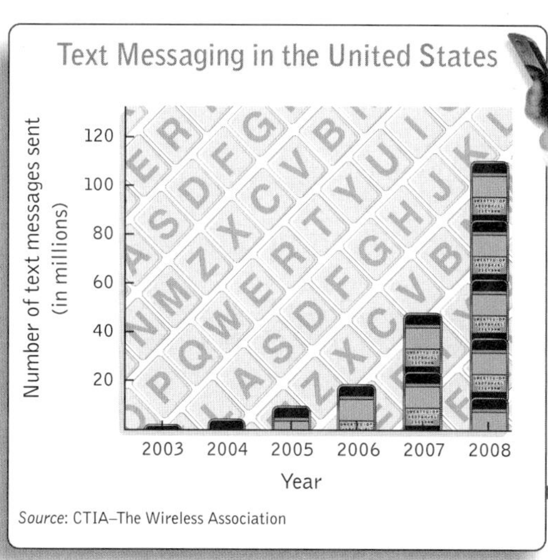

The functions that we consider in this chapter have rich applications in many fields such as finance, epidemiology (the study of the spread of disease), and marketing. The theory centers on functions with variable exponents (*exponential functions*). We study these functions, their inverses, and their properties.

12.1 Composite Functions and Inverse Functions

- Composite Functions
- Inverses and One-to-One Functions
- Finding Formulas for Inverses
- Graphing Functions and Their Inverses
- Inverse Functions and Composition

COMPOSITE FUNCTIONS

In the real world, functions frequently occur in which some quantity depends on a variable that, in turn, depends on another variable. For instance, a firm's profits may depend on the number of items the firm produces, which may in turn depend on the number of employees hired. Such functions are called **composite functions.**

For example, the function g that gives a correspondence between women's shoe sizes in the United States and those in Britain is given by $g(x) = x - 2$, where x is the U.S. size and $g(x)$ is the British size. Thus a U.S. size 4 corresponds to a shoe size of $g(4) = 4 - 2$, or 2, in Britain.

A second function converts women's shoe sizes in Britain to those in Italy. This particular function is given by $f(x) = 2x + 28$, where x is the British size and $f(x)$ is the corresponding Italian size. Thus a British size 2 corresponds to an Italian size $f(2) = 2 \cdot 2 + 28$, or 32.

It is correct to conclude that a U.S. size 4 corresponds to an Italian size 32 and that some function h describes this correspondence.

g: $g(x) = x - 2$ (U.S. → Britain)

f: $f(x) = 2x + 28$ (Britain → Italy)

U.S.	Britain	Italy
4	2	32
5	3	34
6	4	36
7	5	38

h: $h(x) = ?$ (U.S. → Italy)

Size x shoes in the United States correspond to size $g(x)$ shoes in Britain, where

$$g(x) = x - 2.$$

STUDY TIP

Divide and Conquer

In longer sections of reading, there are almost always subsections. Rather than feel obliged to read the entire section at once, use the subsections as natural resting points. Taking a break between subsections can increase your comprehension and is thus an efficient use of your time.

Student Notes

Throughout this chapter, keep in mind that equations such as $g(x) = x - 2$ and $g(t) = t - 2$ describe the same function g. Both equations tell us to find a function value by subtracting 2 from the input.

Size n shoes in Britain correspond to size $f(n)$ shoes in Italy. Similarly, size $g(x)$ shoes in Britain correspond to size $f(g(x))$ shoes in Italy. Since the x in the expression $f(g(x))$ represents a U.S. shoe size, we can find the Italian shoe size that corresponds to a U.S. size x as follows:

$$f(g(x)) = f(x - 2) = 2(x - 2) + 28 \quad \text{Using } g(x) \text{ as an input}$$
$$= 2x - 4 + 28 = 2x + 24.$$

This gives a formula for h: $h(x) = 2x + 24$. Thus U.S. size 4 corresponds to Italian size $h(4) = 2(4) + 24$, or 32. We call h the *composition* of f and g and denote it by $f \circ g$ (read "the composition of f and g," "f composed with g," or "f circle g").

Composition of Functions The *composite function* $f \circ g$, the *composition* of f and g, is defined as

$$(f \circ g)(x) = f(g(x)).$$

We can visualize the composition of functions as follows.

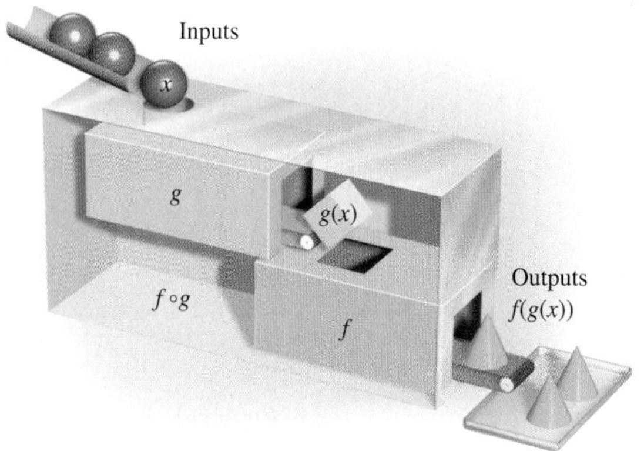

EXAMPLE 1 Given $f(x) = 3x$ and $g(x) = 1 + x^2$:

a) Find $(f \circ g)(5)$ and $(g \circ f)(5)$.
b) Find $(f \circ g)(x)$ and $(g \circ f)(x)$.

SOLUTION Consider each function separately:

$f(x) = 3x$ — **This function multiplies each input by 3.**

and $g(x) = 1 + x^2.$ — **This function adds 1 to the square of each input.**

a) To find $(f \circ g)(5)$, we first find $g(5)$ and then use that as an input for f:

$$(f \circ g)(5) = f(g(5)) = f(1 + 5^2) \quad \text{Using } g(x) = 1 + x^2$$
$$= f(26) = 3 \cdot 26 = 78. \quad \text{Using } f(x) = 3x$$

To find $(g \circ f)(5)$, we first find $f(5)$ and then use that as an input for g:

$$(g \circ f)(5) = g(f(5)) = g(3 \cdot 5) \quad \text{Note that } f(5) = 3 \cdot 5 = 15.$$
$$= g(15) = 1 + 15^2 = 1 + 225 = 226.$$

b) We find $(f \circ g)(x)$ by substituting $g(x)$ for x in the equation for $f(x)$:

$$(f \circ g)(x) = f(g(x)) = f(1 + x^2) \quad \text{Using } g(x) = 1 + x^2$$
$$= 3 \cdot (1 + x^2) = 3 + 3x^2. \quad \text{Using } f(x) = 3x$$

To find $(g \circ f)(x)$, we substitute $f(x)$ for x in the equation for $g(x)$:

$$(g \circ f)(x) = g(f(x)) = g(3x) \quad \text{Substituting } 3x \text{ for } f(x)$$
$$= 1 + (3x)^2 = 1 + 9x^2.$$

We can now find the function values of part (a) using the functions of part (b):

$$(f \circ g)(5) = 3 + 3(5)^2 = 3 + 3 \cdot 25 = 78;$$
$$(g \circ f)(5) = 1 + 9(5)^2 = 1 + 9 \cdot 25 = 226.$$

Try Exercise 9.

Example 1 shows that, in general, $(f \circ g)(x) \neq (g \circ f)(x)$.

EXAMPLE 2 Given $f(x) = \sqrt{x}$ and $g(x) = x - 1$, find $(f \circ g)(x)$ and $(g \circ f)(x)$.

SOLUTION We have

$$(f \circ g)(x) = f(g(x)) = f(x - 1) = \sqrt{x - 1}; \quad \text{Using } g(x) = x - 1$$
$$(g \circ f)(x) = g(f(x)) = g(\sqrt{x}) = \sqrt{x} - 1. \quad \text{Using } f(x) = \sqrt{x}$$

To check using a graphing calculator, let

$$y_1 = f(x) = \sqrt{x},$$
$$y_2 = g(x) = x - 1,$$
$$y_3 = \sqrt{x - 1}, \quad \text{This is the expression we found for } (f \circ g)(x).$$
and $$y_4 = y_1(y_2(x)). \quad \text{This is the notation for } f(g(x)).$$

If our work is correct, y_4 will be equivalent to y_3. We form a table of values, selecting only y_3 and y_4. The table on the left below indicates that the functions are probably equivalent.

X	Y3	Y4
1	0	0
1.5	.70711	.70711
2	1	1
2.5	1.2247	1.2247
3	1.4142	1.4142
3.5	1.5811	1.5811
4	1.7321	1.7321
X = 1		

X	Y5	Y6
1	0	0
1.5	.22474	.22474
2	.41421	.41421
2.5	.58114	.58114
3	.73205	.73205
3.5	.87083	.87083
4	1	1
X = 1		

Next, we let $y_5 = \sqrt{x} - 1$ and $y_6 = y_2(y_1(x))$ and select only y_5 and y_6. The table on the right above indicates that the functions are probably equivalent. Thus we conclude that

$$(f \circ g)(x) = \sqrt{x - 1} \quad \text{and} \quad (g \circ f)(x) = \sqrt{x} - 1.$$

Try Exercise 15.

In order for us to find $f(g(a))$, the output $g(a)$ must be in the domain of f.

EXAMPLE 3 Use the following table to find each value, if possible.

a) $(y_2 \circ y_1)(3)$ **b)** $(y_1 \circ y_2)(3)$

x	y_1	y_2
0	−2	3
1	2	5
2	−1	7
3	1	9
4	0	11
5	0	13
6	−3	15

SOLUTION

a) Since $(y_2 \circ y_1)(3) = y_2(y_1(3))$, we first find $y_1(3)$. We locate 3 in the x-column and then move across to the y_1-column to find $y_1(3) = 1$.

x	y_1	y_2	
0	−2	3	
1	2	**5**	← $y_2(1)$
2	−1	7	
3	1 ←	9	
4	0	11	
5	0	13	
6	−3	15	
		$y_1(3)$	

The output 1 now becomes an input for the function y_2:

$$y_2(y_1(3)) = y_2(1).$$

To find $y_2(1)$, we locate 1 in the x-column and then move across to the y_2-column to find $y_2(1) = \mathbf{5}$. Thus, $(y_2 \circ y_1)(3) = \mathbf{5}$.

b) Since $(y_1 \circ y_2)(3) = y_1(y_2(3))$, we find $y_2(3)$ by locating 3 in the x-column and moving across to the y_2-column. We have $y_2(3) = 9$, so

$$(y_1 \circ y_2)(3) = y_1(y_2(3)) = y_1(9).$$

However, y_1 is not defined for $x = 9$, so $(y_1 \circ y_2)(3)$ is undefined.

Try Exercise 21.

In fields ranging from chemistry to geology to economics, we must recognize how a function can be regarded as the composition of two "simpler" functions. This is sometimes called *de*composition.

EXAMPLE 4 If $h(x) = (7x + 3)^2$, find f and g such that $h(x) = (f \circ g)(x)$.

SOLUTION We can think of $h(x)$ as the result of first forming $7x + 3$ and then squaring. This suggests that $g(x) = 7x + 3$ and $f(x) = x^2$. We check by forming the composition:

$$(f \circ g)(x) = f(g(x)) = f(7x + 3) = (7x + 3)^2 = h(x), \text{ as desired.}$$

There are other less "obvious" answers. For example, if

$$f(x) = (x - 1)^2 \quad \text{and} \quad g(x) = 7x + 4,$$

then

$$\begin{aligned}(f \circ g)(x) &= f(g(x)) = f(7x + 4)\\ &= (7x + 4 - 1)^2 = (7x + 3)^2 = h(x).\end{aligned}$$

Try Exercise 31.

INVERSES AND ONE-TO-ONE FUNCTIONS

Let's view the following two functions as relations, or correspondences.

Countries and Their Capitals

Domain (Set of Inputs)		Range (Set of Outputs)
Australia	→	Canberra
China	→	Beijing
Germany	→	Berlin
Madagascar	→	Antananaviro
Turkey	→	Ankara
United States	→	Washington, D.C.

Phone Keys

Domain (Set of Inputs)		Range (Set of Outputs)
a	→	2
b	→	2
c	→	2
d	→	3
e	→	3
f	→	3

Suppose we reverse the arrows. We obtain what is called the **inverse relation.** Are these inverse relations functions?

Countries and Their Capitals

Range (Set of Outputs)		Domain (Set of Inputs)
Australia	←	Canberra
China	←	Beijing
Germany	←	Berlin
Madagascar	←	Antananaviro
Turkey	←	Ankara
United States	←	Washington, D.C.

Phone Keys

Range (Set of Outputs)		Domain (Set of Inputs)
a	←	2
b	←	2
c	←	2
d	←	3
e	←	3
f	←	3

Recall that for each input, a function provides exactly one output. However, it is possible for different inputs to correspond to the same output. Only when this possibility is *excluded* will the inverse be a function. For the functions listed above, this means the inverse of the "Capitals" correspondence is a function, but the inverse of the "Phone Keys" correspondence is not.

In the Capitals function, each input has its own output, so it is a **one-to-one function.** In the Phone Keys function, a and b are both paired with 2. Thus the Phone Keys function is not a one-to-one function.

One-To-One Function A function f is *one-to-one* if different inputs have different outputs. That is, if for a and b in the domain of f with $a \neq b$, we have $f(a) \neq f(b)$, then f is one-to-one. If a function is one-to-one, then its inverse correspondence is also a function.

How can we tell graphically whether a function is one-to-one?

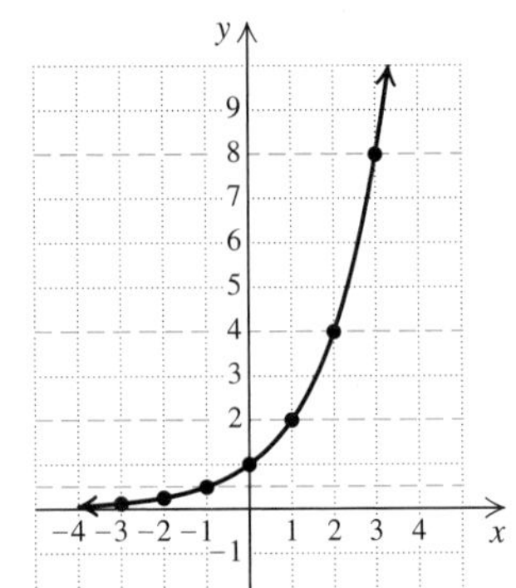

EXAMPLE 5 Shown at left is the graph of a function similar to those we will study in Section 12.2. Determine whether the function is one-to-one and thus has an inverse that is a function.

SOLUTION A function is one-to-one if different inputs have different outputs—that is, if no two x-values have the same y-value. For this function, we cannot find two x-values that have the same y-value. Note that this means that no horizontal line can be drawn so that it crosses the graph more than once. The function is one-to-one so its inverse is a function.

The graph of every function must pass the vertical-line test. In order for a function to have an inverse that is a function, it must pass the *horizontal-line test* as well.

The Horizontal-Line Test If it is impossible to draw a horizontal line that intersects a function's graph more than once, then the function is one-to-one. For every one-to-one function, an inverse function exists.

EXAMPLE 6 Determine whether the function $f(x) = x^2$ is one-to-one and thus has an inverse that is a function.

SOLUTION The graph of $f(x) = x^2$ is shown here. Many horizontal lines cross the graph more than once. For example, the line $y = 4$ crosses where the first coordinates are -2 and 2. Although these are different inputs, they have the same output. That is, $-2 \neq 2$, but $f(-2) = f(2) = 4$.

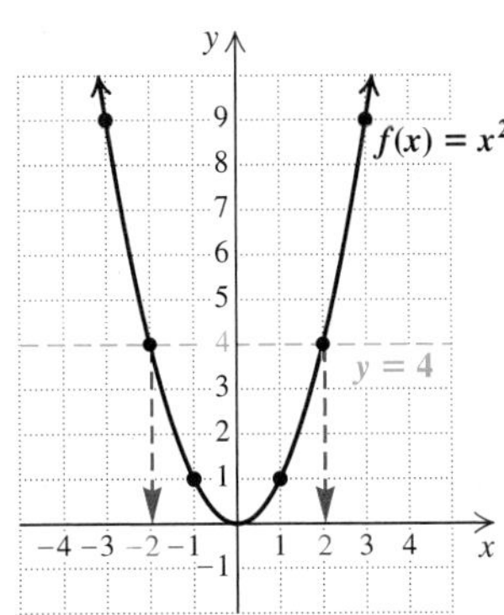

Thus the function is not one-to-one and no inverse function exists.

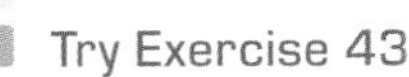

FINDING FORMULAS FOR INVERSES

When the inverse of f is also a function, it is denoted f^{-1} (read "f-inverse").

CAUTION! The -1 in f^{-1} is *not* an exponent!

Suppose a function is described by a formula. If its inverse is a function, how do we find a formula for that inverse? For any equation in two variables, if we interchange the variables, we form an equation of the inverse correspondence. If it is a function, we proceed as follows to find a formula for f^{-1}.

Student Notes

Since we interchange x and y to find an equation for an inverse, the domain of the function is the range of the inverse, and the range of the function is the domain of the inverse.

To Find a Formula for f^{-1}

First make sure that f is one-to-one. Then:

1. Replace $f(x)$ with y.
2. Interchange x and y. (This gives the inverse function.)
3. Solve for y.
4. Replace y with $f^{-1}(x)$. (This is inverse function notation.)

EXAMPLE 7 Determine whether each function is one-to-one and if it is, find a formula for $f^{-1}(x)$.

a) $f(x) = x + 2$ **b)** $f(x) = 2x - 3$

SOLUTION

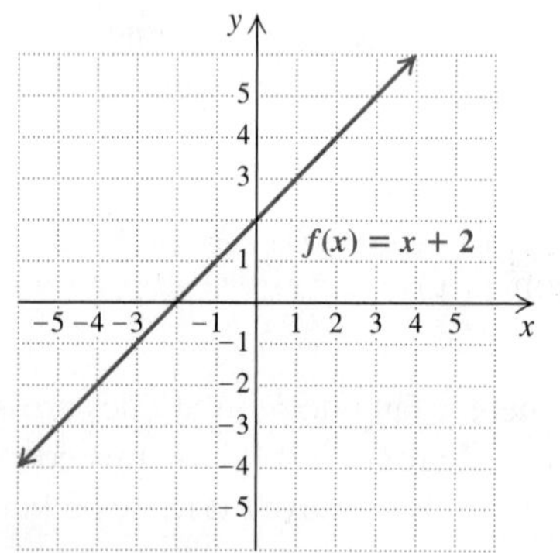

a) The graph of $f(x) = x + 2$ is shown at left. It passes the horizontal-line test, so it is one-to-one. Thus its inverse is a function.

1. Replace $f(x)$ with y: $y = x + 2$.
2. Interchange x and y: $x = y + 2$. **This gives the inverse function.**
3. Solve for y: $x - 2 = y$.
4. Replace y with $f^{-1}(x)$: $f^{-1}(x) = x - 2$. **We also "reversed" the equation.**

In this case, the function f adds 2 to all inputs. Thus, to "undo" f, the function f^{-1} must subtract 2 from its inputs.

b) The function $f(x) = 2x - 3$ is also linear. Any linear function that is not constant will pass the horizontal-line test. Thus, f is one-to-one.

1. Replace $f(x)$ with y: $y = 2x - 3$.
2. Interchange x and y: $x = 2y - 3$.
3. Solve for y: $x + 3 = 2y$

$$\frac{x+3}{2} = y.$$

4. Replace y with $f^{-1}(x)$: $f^{-1}(x) = \dfrac{x+3}{2}$.

In this case, the function f doubles all inputs and then subtracts 3. Thus, to "undo" f, the function f^{-1} adds 3 to each input and then divides by 2.

Try Exercise 49.

GRAPHING FUNCTIONS AND THEIR INVERSES

How do the graphs of a function and its inverse compare?

EXAMPLE 8 Graph $f(x) = 2x - 3$ and $f^{-1}(x) = (x + 3)/2$ on the same set of axes. Then compare.

SOLUTION The graph of each function follows. Note that the graph of f^{-1} can be drawn by reflecting the graph of f across the line $y = x$. That is, if we graph $f(x) = 2x - 3$ in wet ink and fold the paper along the line $y = x$, the graph of $f^{-1}(x) = (x + 3)/2$ will appear as the impression made by f.

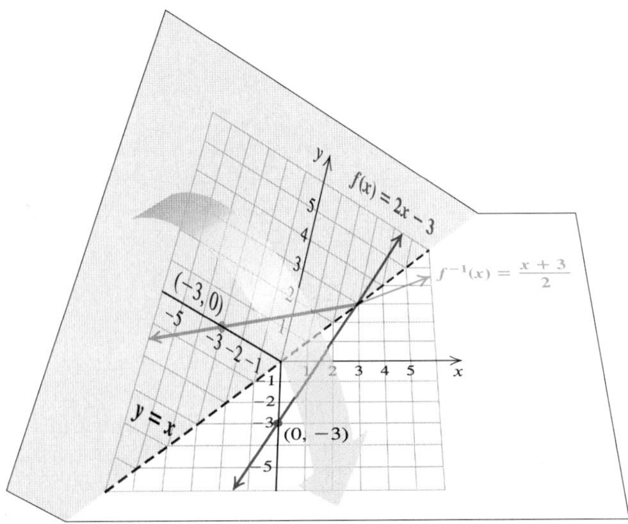

When x and y are interchanged to find a formula for the inverse, we are, in effect, reflecting or flipping the graph of $f(x) = 2x - 3$ across the line $y = x$. For example, when the coordinates of the y-intercept of the graph of f, $(0, -3)$, are reversed, we get $(-3, 0)$, the x-intercept of the graph of f^{-1}.

Try Exercise 73.

> Visualizing Inverses The graph of f^{-1} is a reflection of the graph of f across the line $y = x$.

EXAMPLE 9 Consider $g(x) = x^3 + 2$.

a) Determine whether the function is one-to-one.

b) If it is one-to-one, find a formula for its inverse.

c) Graph the inverse, if it exists.

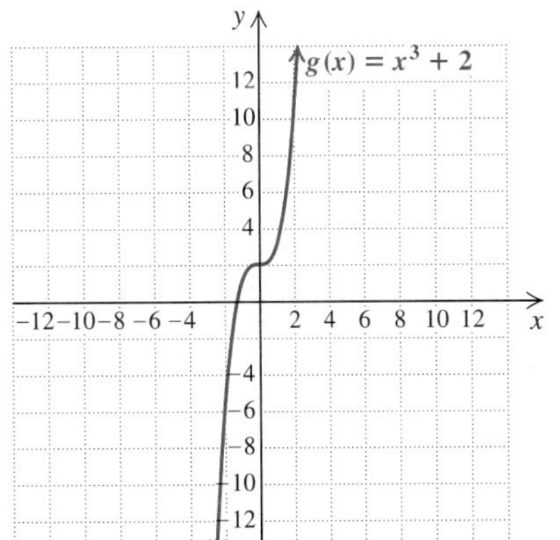

SOLUTION

a) The graph of $g(x) = x^3 + 2$ is shown at left. It passes the horizontal-line test and thus has an inverse that is a function.

b) 1. Replace $g(x)$ with y: $\quad y = x^3 + 2$. **Using $g(x) = x^3 + 2$**

2. Interchange x and y: $\quad x = y^3 + 2$.

3. Solve for y: $\quad x - 2 = y^3$

$\sqrt[3]{x - 2} = y$. **Each number has only one cube root, so we can solve for y.**

4. Replace y with $g^{-1}(x)$: $\quad g^{-1}(x) = \sqrt[3]{x - 2}$.

c) To graph g^{-1}, we can reflect the graph of $g(x) = x^3 + 2$ across the line $y = x$, as we did in Example 8. We can also graph $g^{-1}(x) = \sqrt[3]{x - 2}$ by plotting points. Note that $(2, 10)$ is on the graph of g, whereas $(10, 2)$ is on the graph of g^{-1}. The graphs of g and g^{-1} are shown together below.

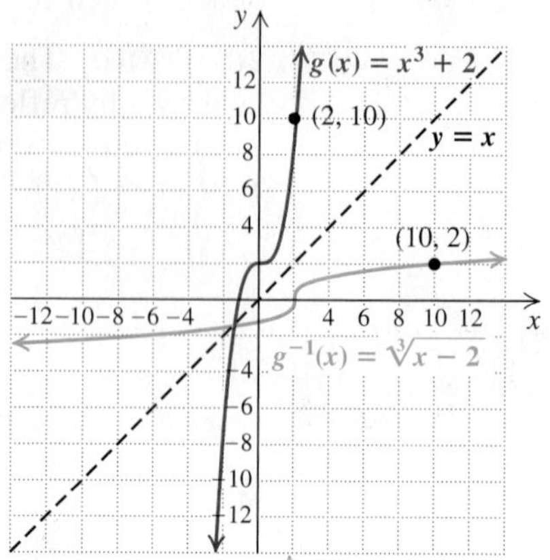

Try Exercise 75.

INVERSE FUNCTIONS AND COMPOSITION

Let's consider inverses of functions in terms of function machines. Suppose that a one-to-one function f is programmed into a machine. If the machine is run in reverse, it performs the inverse function f^{-1}. Inputs then enter at the opposite end, and the entire process is reversed.

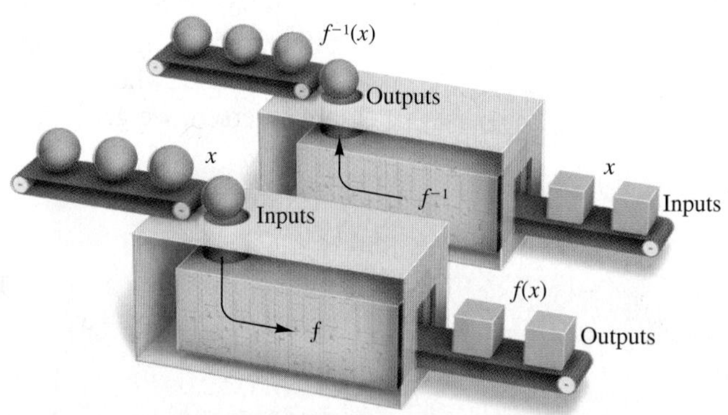

Consider $g(x) = x^3 + 2$ and $g^{-1}(x) = \sqrt[3]{x - 2}$ from Example 9. For the input 3,

$$g(3) = 3^3 + 2 = 27 + 2 = 29.$$

The output is 29. Now we use 29 for the input in the inverse:

$$g^{-1}(29) = \sqrt[3]{29 - 2} = \sqrt[3]{27} = 3.$$

The function g takes 3 to 29. The inverse function g^{-1} takes the number 29 back to 3.

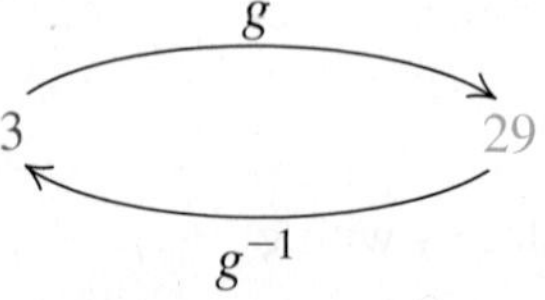

In general, if f is one-to-one, then f^{-1} takes the output $f(x)$ back to x. Similarly, f takes the output $f^{-1}(x)$ back to x.

Composition and Inverses If a function f is one-to-one, then f^{-1} is the unique function for which

$$(f^{-1} \circ f)(x) = f^{-1}(f(x)) = x \quad \text{and} \quad (f \circ f^{-1})(x) = f(f^{-1}(x)) = x.$$

EXAMPLE 10 Let $f(x) = 2x + 1$. Show that

$$f^{-1}(x) = \frac{x - 1}{2}.$$

SOLUTION We find $(f^{-1} \circ f)(x)$ and $(f \circ f^{-1})(x)$ and check to see that each is x.

$$(f^{-1} \circ f)(x) = f^{-1}(f(x)) = f^{-1}(2x + 1)$$
$$= \frac{(2x + 1) - 1}{2}$$
$$= \frac{2x}{2} = x \qquad \text{Thus, } (f^{-1} \circ f)(x) = x.$$

$$(f \circ f^{-1})(x) = f(f^{-1}(x)) = f\left(\frac{x - 1}{2}\right)$$
$$= 2 \cdot \frac{x - 1}{2} + 1$$
$$= x - 1 + 1 = x \qquad \text{Thus, } (f \circ f^{-1})(x) = x.$$

Try Exercise 83.

Inverse Functions: Graphing and Composition

In Example 9, we found that the inverse of $y_1 = x^3 + 2$ is $y_2 = \sqrt[3]{x - 2}$. There are several ways to check this result using a graphing calculator.

1. We can check visually by graphing both functions, along with the line $y = x$, on a "squared" set of axes. The graph of y_2 *appears* to be the reflection of the graph of y_1 across the line $y = x$. This is only a partial check since we are comparing graphs visually.

$y_1 = x^3 + 2,\ y_2 = \sqrt[3]{x - 2}$

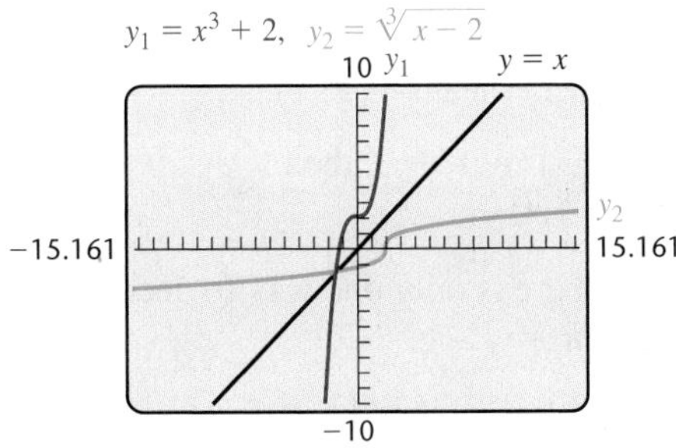

(continued)

2. A second, more precise, visual check begins with graphing $y_1 = x^3 + 2$ and $y_2 = \sqrt[3]{x - 2}$ using a squared viewing window. Then press DRAW 8 VARS ▷ 1 1 ENTER to select the DrawInv option of the DRAW menu and graph the inverse. The resulting graph should coincide with the graph of y_2.
3. For a third check, note that if y_2 is the inverse of y_1, then $(y_2 \circ y_1)(x) = x$ and $(y_1 \circ y_2)(x) = x$. Enter $y_3 = y_2(y_1(x))$ and $y_4 = y_1(y_2(x))$, and form a table to compare x, y_3, and y_4. Note that for the values shown, $y_3 = y_4 = x$.

TblStart = −3, ΔTbl = 1

X	Y3	Y4
−3	−3	−3
−2	−2	−2
−1	−1	−1
0	0	0
1	1	1
2	2	2
3	3	3

X = −3

Your Turn

1. Enter $y_1 = 2x + 1$ and $y_2 = x/2 - 1$. These are *not* inverse functions, as the following three activities will demonstrate.
2. Graph y_1 and y_2 in a squared viewing window. These should not appear to be reflections across the line $y = x$.
3. From the home screen, press DRAW 8 VARS ▷ 1 1 ENTER. The graph drawn should be different from the graph of y_2.
4. Enter $y_3 = y_2(y_1(x))$ and $y_4 = y_1(y_2(x))$ and form a table. Compare x_1, y_3, and y_4. The columns should *not* be the same.

12.1 Exercise Set

FOR EXTRA HELP MyMathLab

Concept Reinforcement *Classify each of the following statements as either true or false.*

1. The composition of two functions f and g is written $f \circ g$.
2. The notation $(f \circ g)(x)$ means $f(g(x))$.
3. If $f(x) = x^2$ and $g(x) = x + 3$, then $(g \circ f)(x) = (x + 3)^2$.
4. For any function h, there is only one way to decompose the function as $h = f \circ g$.
5. The function f is one-to-one if $f(1) = 1$.
6. The -1 in f^{-1} is an exponent.
7. The function f is the inverse of f^{-1}.
8. If g and h are inverses of each other, then $(g \circ h)(x) = x$.

For each pair of functions, find **(a)** $(f \circ g)(1)$; **(b)** $(g \circ f)(1)$; **(c)** $(f \circ g)(x)$; *and* **(d)** $(g \circ f)(x)$.

9. $f(x) = x^2 + 1$; $g(x) = x - 3$

10. $f(x) = x + 4$; $g(x) = x^2 - 5$

11. $f(x) = 5x + 1$; $g(x) = 2x^2 - 7$

12. $f(x) = 3x^2 + 4$; $g(x) = 4x - 1$

13. $f(x) = x + 7;\ g(x) = 1/x^2$

14. $f(x) = 1/x^2;\ g(x) = x + 2$

15. $f(x) = \sqrt{x};\ g(x) = x + 3$

16. $f(x) = 10 - x;\ g(x) = \sqrt{x}$

17. $f(x) = \sqrt{4x};\ g(x) = 1/x$

18. $f(x) = \sqrt{x + 3};\ g(x) = 13/x$

19. $f(x) = x^2 + 4;\ g(x) = \sqrt{x - 1}$

20. $f(x) = x^2 + 8;\ g(x) = \sqrt{x + 17}$

Use the following table to find each value, if possible.

X	Y1	Y2
−3	−4	1
−2	−1	−2
−1	2	−3
0	5	−2
1	8	1
2	11	6
3	14	11

X =

21. $(y_1 \circ y_2)(-3)$

22. $(y_2 \circ y_1)(-3)$

23. $(y_1 \circ y_2)(-1)$

24. $(y_2 \circ y_1)(-1)$

25. $(y_2 \circ y_1)(1)$

26. $(y_1 \circ y_2)(1)$

Use the table below to find each value, if possible.

x	$f(x)$	$g(x)$
1	0	1
2	3	5
3	2	8
4	6	5
5	4	1

27. $(f \circ g)(2)$

28. $(g \circ f)(4)$

29. $f(g(3))$

30. $g(f(5))$

Find $f(x)$ and $g(x)$ such that $h(x) = (f \circ g)(x)$. Answers may vary.

31. $h(x) = (3x - 5)^4$

32. $h(x) = (2x + 7)^3$

33. $h(x) = \sqrt{2x + 7}$

34. $h(x) = \sqrt[3]{4x - 5}$

35. $h(x) = \dfrac{2}{x - 3}$

36. $h(x) = \dfrac{3}{x} + 4$

37. $h(x) = \dfrac{1}{\sqrt{7x + 2}}$

38. $h(x) = \sqrt{x - 7} - 3$

39. $h(x) = \dfrac{1}{\sqrt{3x}} + \sqrt{3x}$

40. $h(x) = \dfrac{1}{\sqrt{2x}} - \sqrt{2x}$

Determine whether each function is one-to-one.

41. $f(x) = x - 5$

42. $f(x) = 5 - 2x$

43. $f(x) = x^2 + 1$

44. $f(x) = 1 - x^2$

45.

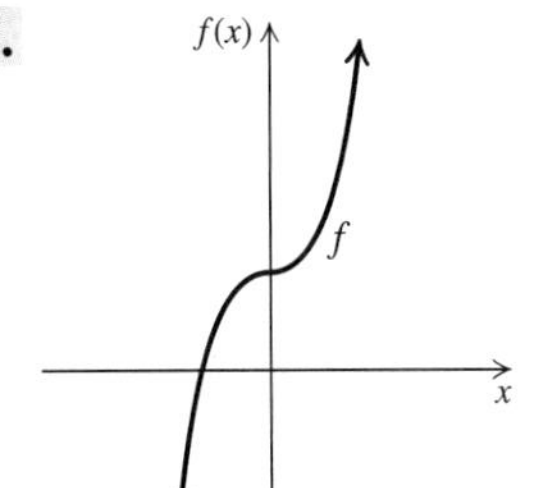

46.

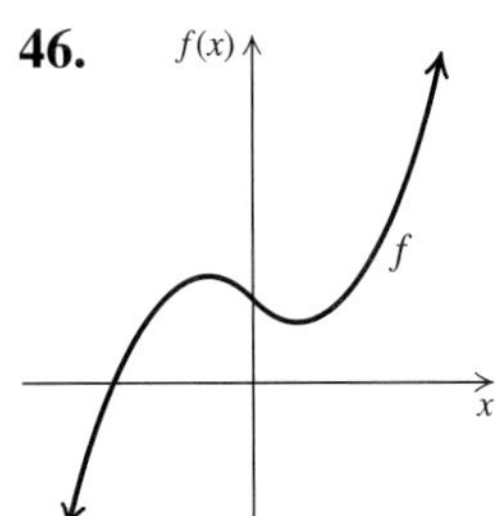

47.

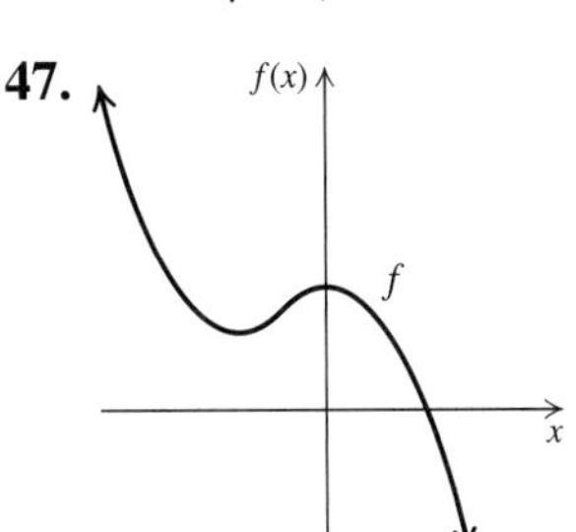

48.

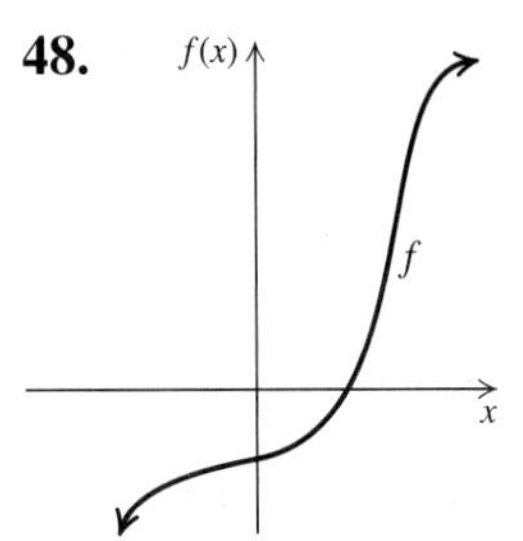

*For each function, **(a)** determine whether it is one-to-one and **(b)** if it is one-to-one, find a formula for the inverse.*

49. $f(x) = x + 4$

50. $f(x) = x + 2$

51. $f(x) = 2x$

52. $f(x) = 3x$

53. $g(x) = 3x - 1$

54. $g(x) = 2x - 5$

55. $f(x) = \frac{1}{2}x + 1$

56. $f(x) = \frac{1}{3}x + 2$

57. $g(x) = x^2 + 5$

58. $g(x) = x^2 - 4$

59. $h(x) = -10 - x$

60. $h(x) = 7 - x$

Aha! 61. $f(x) = \dfrac{1}{x}$

62. $f(x) = \dfrac{3}{x}$

63. $G(x) = 4$

64. $H(x) = 2$

65. $f(x) = \dfrac{2x + 1}{3}$

66. $f(x) = \dfrac{3x + 2}{5}$

67. $f(x) = x^3 - 5$

68. $f(x) = x^3 + 7$

69. $g(x) = (x - 2)^3$

70. $g(x) = (x + 7)^3$

71. $f(x) = \sqrt{x}$

72. $f(x) = \sqrt{x - 1}$

Graph each function and its inverse using the same set of axes.

73. $f(x) = \frac{2}{3}x + 4$

74. $g(x) = \frac{1}{4}x + 2$

75. $f(x) = x^3 + 1$

76. $f(x) = x^3 - 1$

77. $g(x) = \frac{1}{2}x^3$

78. $g(x) = \frac{1}{3}x^3$

79. $F(x) = -\sqrt{x}$

80. $f(x) = \sqrt{x}$

81. $f(x) = -x^2, x \geq 0$

82. $f(x) = x^2 - 1, x \leq 0$

83. Let $f(x) = \sqrt[3]{x - 4}$. Use composition to show that $f^{-1}(x) = x^3 + 4$.

84. Let $f(x) = 3/(x + 2)$. Use composition to show that
$$f^{-1}(x) = \frac{3}{x} - 2.$$

85. Let $f(x) = (1 - x)/x$. Use composition to show that
$$f^{-1}(x) = \frac{1}{x + 1}.$$

86. Let $f(x) = x^3 - 5$. Use composition to show that $f^{-1}(x) = \sqrt[3]{x + 5}$.

Use a graphing calculator to help determine whether or not the given pairs of functions are inverses of each other.

87. $f(x) = 0.75x^2 + 2;\ g(x) = \sqrt{\dfrac{4(x - 2)}{3}}$

88. $f(x) = 1.4x^3 + 3.2;\ g(x) = \sqrt[3]{\dfrac{x - 3.2}{1.4}}$

89. $f(x) = \sqrt{2.5x + 9.25}$;
$g(x) = 0.4x^2 - 3.7,\ x \geq 0$

90. $f(x) = 0.8x^{1/2} + 5.23$;
$g(x) = 1.25(x^2 - 5.23),\ x \geq 0$

In Exercises 91 and 92, match the graph of each function in Column A with the graph of its inverse in Column B.

91. Column A

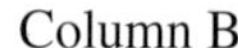
Column B

(1)

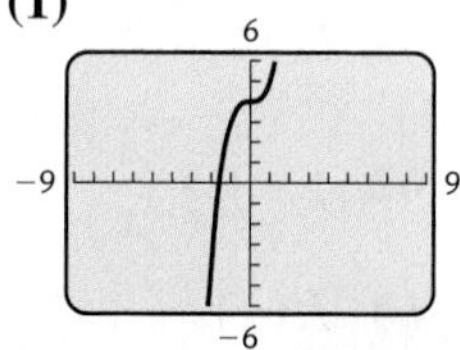

A.

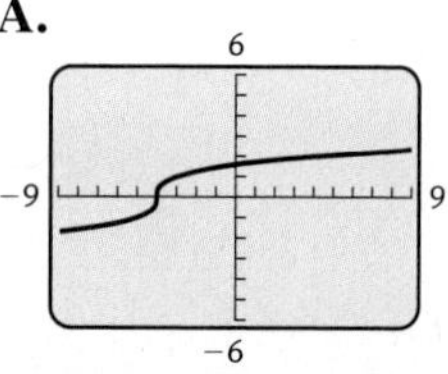

(2)

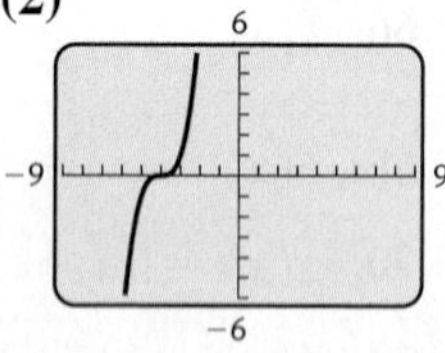

B.

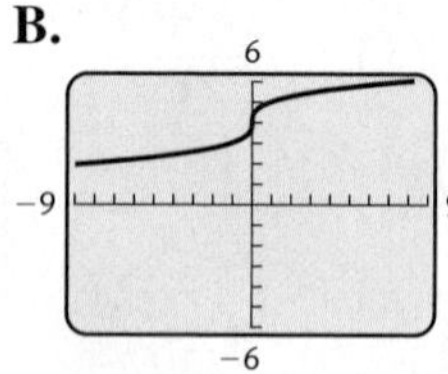

(3)

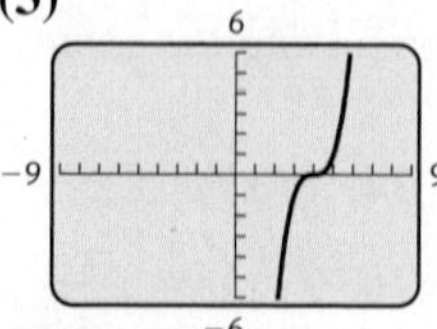

C.

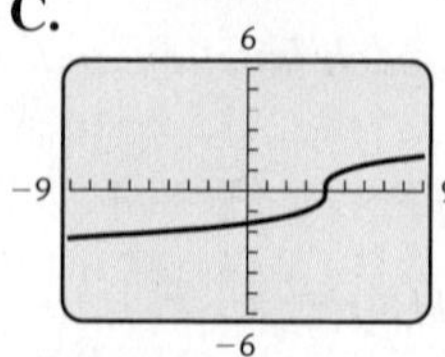

(4)

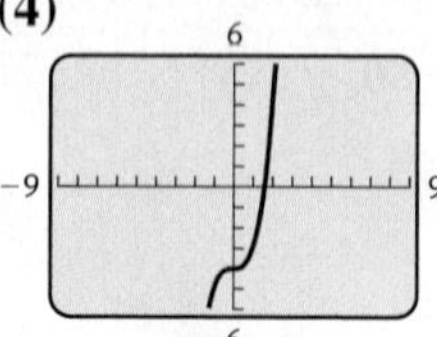

D.

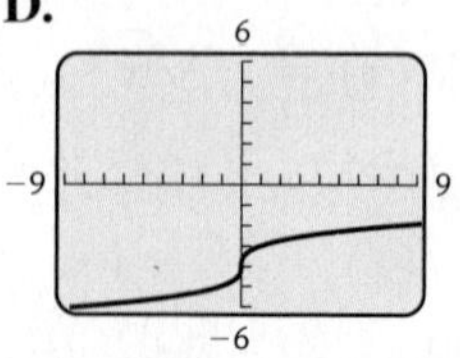

92. Column A

Column B

(1)

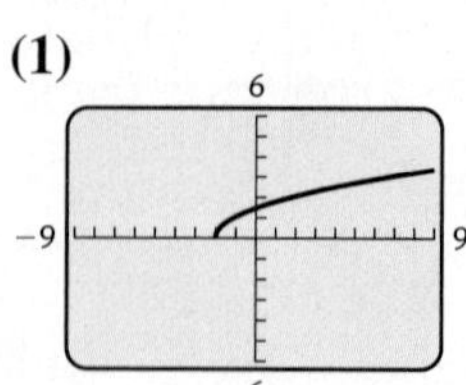

A.

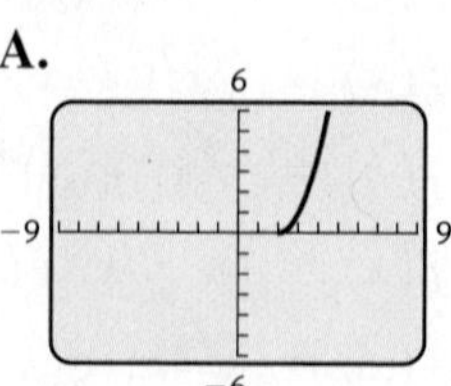

(2)

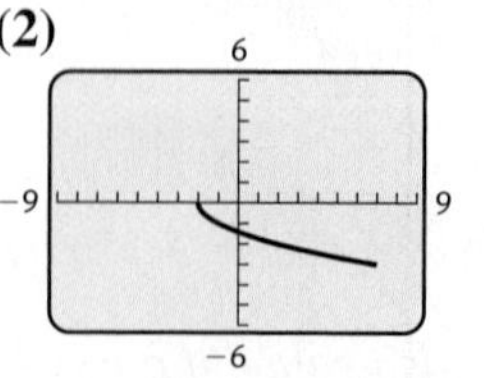

B.

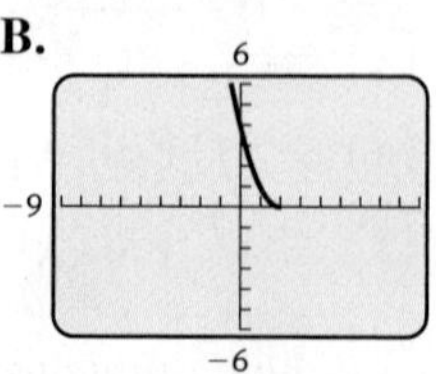

(3)

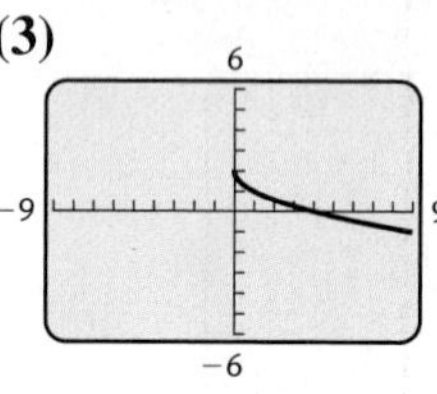

C.

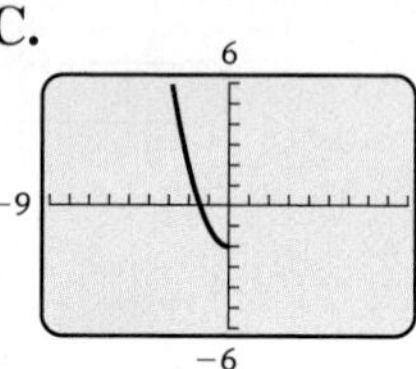

(4)

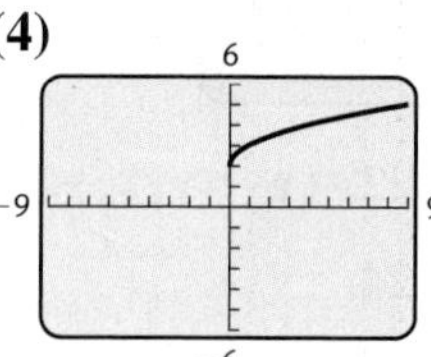

D.

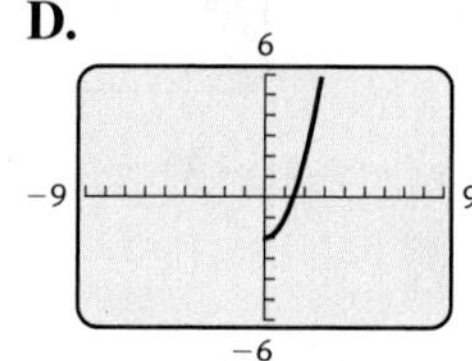

93. *Dress Sizes in the United States and France.* A size-6 dress in the United States is size 38 in France. A function that converts dress sizes in the United States to those in France is
$$f(x) = x + 32.$$

a) Find the dress sizes in France that correspond to sizes 8, 10, 14, and 18 in the United States.

b) Determine whether this function has an inverse that is a function. If so, find a formula for the inverse.

c) Use the inverse function to find dress sizes in the United States that correspond to sizes 40, 42, 46, and 50 in France.

94. *Dress Sizes in the United States and Italy.* A size-6 dress in the United States is size 36 in Italy. A function that converts dress sizes in the United States to those in Italy is
$$f(x) = 2(x + 12).$$

a) Find the dress sizes in Italy that correspond to sizes 8, 10, 14, and 18 in the United States.
b) Determine whether this function has an inverse that is a function. If so, find a formula for the inverse.
c) Use the inverse function to find dress sizes in the United States that correspond to sizes 40, 44, 52, and 60 in Italy.

TW **95.** Is there a one-to-one relationship between items in a store and the price of each of those items? Why or why not?

TW **96.** Mathematicians usually try to select "logical" words when forming definitions. Does the term "one-to-one" seem logical? Why or why not?

SKILL REVIEW

To prepare for Section 12.2, review simplifying exponential expressions and graphing equations (Sections 3.2, 5.2, and 10.2).

Simplify.

97. 2^{-3} [5.2]

98. $5^{(1-3)}$ [5.2]

99. $4^{5/2}$ [10.2]

100. $3^{7/10}$ [10.2]

Graph. [3.2]

101. $y = x^3$

102. $x = y^3$

SYNTHESIS

TW **103.** The function $V(t) = 750(1.2)^t$ is used to predict the value, $V(t)$, of a certain rare stamp t years from 2005. Do not calculate $V^{-1}(t)$, but explain how V^{-1} could be used.

TW **104.** An organization determines that the cost per person of chartering a bus is given by the function

$$C(x) = \frac{100 + 5x}{x},$$

where x is the number of people in the group and $C(x)$ is in dollars. Determine $C^{-1}(x)$ and explain how this inverse function could be used.

For Exercises 105 and 106, graph the inverse of f.

105.

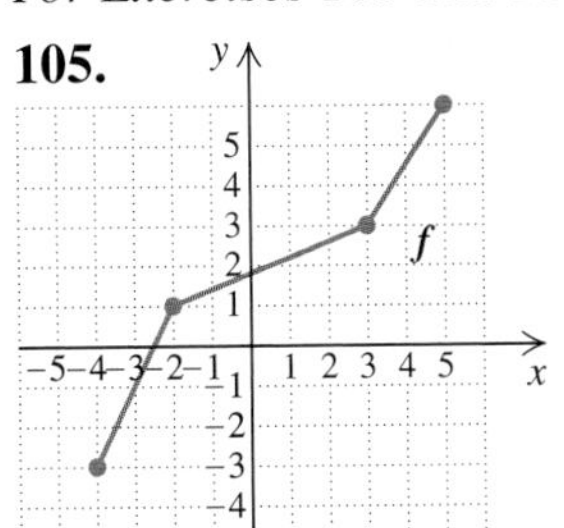

106.

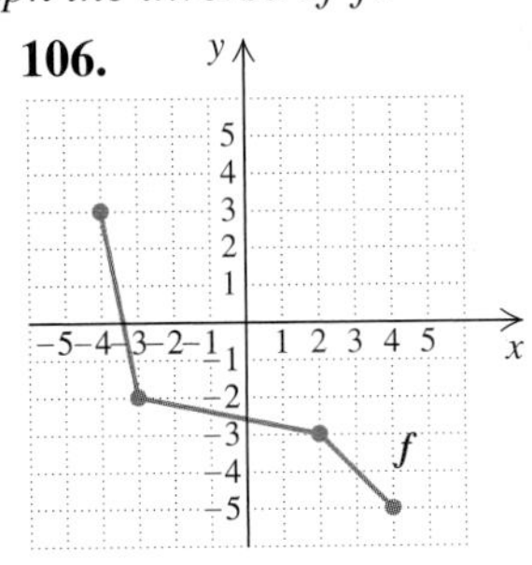

107. *Dress Sizes in France and Italy.* Use the information in Exercises 93 and 94 to find a function for the French dress size that corresponds to a size x dress in Italy.

108. *Dress Sizes in Italy and France.* Use the information in Exercises 93 and 94 to find a function for the Italian dress size that corresponds to a size x dress in France.

TW **109.** What relationship exists between the answers to Exercises 107 and 108? Explain how you determined this.

110. Show that function composition is associative by showing that $((f \circ g) \circ h)(x) = (f \circ (g \circ h))(x)$.

111. Show that if $h(x) = (f \circ g)(x)$, then $h^{-1}(x) = (g^{-1} \circ f^{-1})(x)$. (*Hint*: Use Exercise 110.)

112. Match each function in Column A with its inverse from Column B.

Column A	Column B
(1) $y = 5x^3 + 10$	A. $y = \dfrac{\sqrt[3]{x} - 10}{5}$
(2) $y = (5x + 10)^3$	B. $y = \sqrt[3]{\dfrac{x}{5}} - 10$
(3) $y = 5(x + 10)^3$	C. $y = \sqrt[3]{\dfrac{x - 10}{5}}$
(4) $y = (5x)^3 + 10$	D. $y = \dfrac{\sqrt[3]{x - 10}}{5}$

TW **113.** Examine the following table. Is it possible that f and g could be inverses of each other? Why or why not?

x	$f(x)$	$g(x)$
6	6	6
7	6.5	8
8	7	10
9	7.5	12
10	8	14
11	8.5	16
12	9	18

114. Assume in Exercise 113 that f and g are both linear functions. Find equations for $f(x)$ and $g(x)$ Are f and g inverses of each other?

115. Let $c(w)$ represent the cost of mailing a package that weighs w pounds. Let $f(n)$ represent the weight, in pounds, of n copies of a certain book. Explain what $(c \circ f)(n)$ represents.

116. Let $g(a)$ represent the number of gallons of sealant needed to seal a bamboo floor with area a. Let $c(s)$ represent the cost of s gallons of sealant. Which composition makes sense: $(c \circ g)(a)$ or $(g \circ c)(s)$? What does it represent?

The following graphs show the rate of flow R, in liters per minute, of blood from the heart in a man who bicycles for 20 min, and the pressure P, in millimeters of mercury, in the artery leading to the lungs for a rate of blood flow R from the heart. (This problem was suggested by Kandace Kling, Portland Community College, Sylvania, Oregon.)

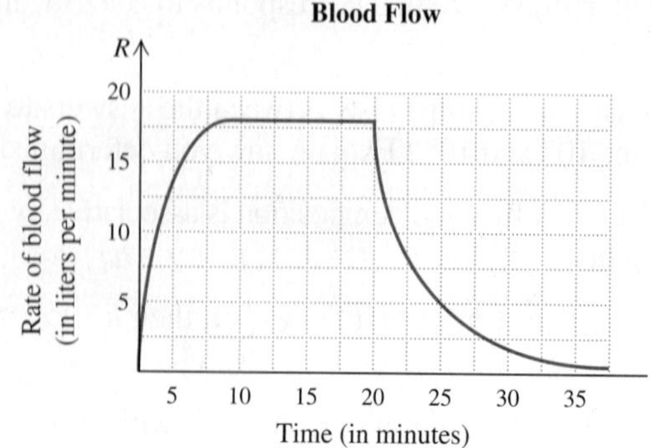

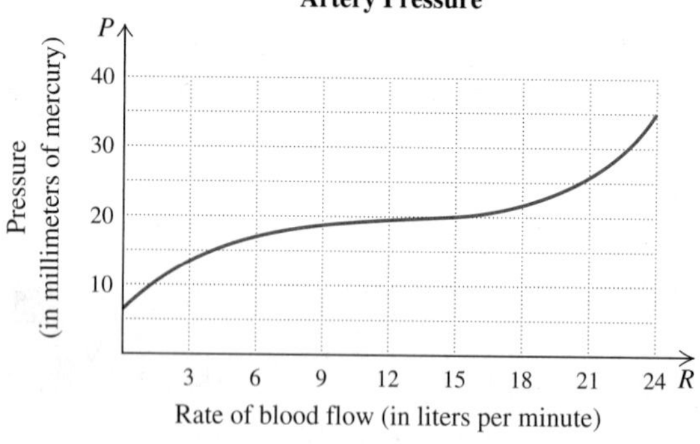

117. Estimate $P(R(10))$.

118. Explain what $P(R(10))$ represents.

119. Estimate $P^{-1}(20)$.

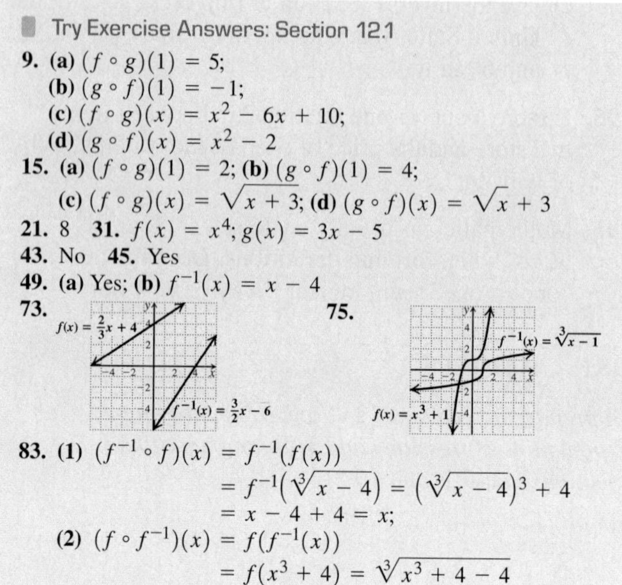

Try Exercise Answers: Section 12.1

9. **(a)** $(f \circ g)(1) = 5$; **(b)** $(g \circ f)(1) = -1$; **(c)** $(f \circ g)(x) = x^2 - 6x + 10$; **(d)** $(g \circ f)(x) = x^2 - 2$
15. **(a)** $(f \circ g)(1) = 2$; **(b)** $(g \circ f)(1) = 4$; **(c)** $(f \circ g)(x) = \sqrt{x+3}$; **(d)** $(g \circ f)(x) = \sqrt{x} + 3$
21. 8 **31.** $f(x) = x^4$; $g(x) = 3x - 5$
43. No **45.** Yes
49. **(a)** Yes; **(b)** $f^{-1}(x) = x - 4$
73. $f(x) = \frac{2}{3}x + 4$; $f^{-1}(x) = \frac{3}{2}x - 6$ **75.** $f^{-1}(x) = \sqrt[3]{x-1}$; $f(x) = x^3 + 1$
83. **(1)** $(f^{-1} \circ f)(x) = f^{-1}(f(x))$
$= f^{-1}(\sqrt[3]{x-4}) = (\sqrt[3]{x-4})^3 + 4$
$= x - 4 + 4 = x$;
(2) $(f \circ f^{-1})(x) = f(f^{-1}(x))$
$= f(x^3 + 4) = \sqrt[3]{x^3 + 4 - 4}$
$= \sqrt[3]{x^3} = x$

12.2 Exponential Functions

- Graphing Exponential Functions
- Equations with *x* and *y* Interchanged
- Applications of Exponential Functions

In this section, we introduce a new type of function, the *exponential function.* These functions and their inverses, called *logarithmic functions*, have applications in many fields.

Consider the graph below. The rapidly rising curve approximates the graph of an *exponential function.* We now consider such functions and some of their applications.

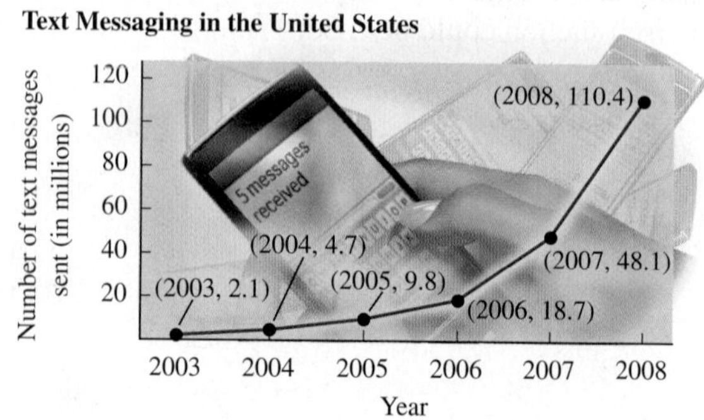

Source: U.S. Census Bureau

STUDY TIP

Know Your Machine

Whether you use a scientific calculator or a graphing calculator, it is a wise investment of time to study the user's manual. If you cannot find a paper manual to consult, an electronic version can usually be found, online, at the manufacturer's Web site. Experimenting by pressing various combinations of keystrokes can also be useful.

GRAPHING EXPONENTIAL FUNCTIONS

In Chapter 10, we studied exponential expressions with rational-number exponents, such as

$$5^{1/4}, \quad 3^{-3/4}, \quad 7^{2.34}, \quad 5^{1.73}.$$

For example, $5^{1.73}$, or $5^{173/100}$, represents the 100th root of 5 raised to the 173rd power. What about expressions with irrational exponents, such as $5^{\sqrt{3}}$ or $7^{-\pi}$? To attach meaning to $5^{\sqrt{3}}$, consider a rational approximation, r, of $\sqrt{3}$. As r gets closer to $\sqrt{3}$, the value of 5^r gets closer to some real number p.

r closes in on $\sqrt{3}$.	5^r closes in on some real number p.
$1.7 < r < 1.8$	$15.426 \approx 5^{1.7} < p < 5^{1.8} \approx 18.119$
$1.73 < r < 1.74$	$16.189 \approx 5^{1.73} < p < 5^{1.74} \approx 16.452$
$1.732 < r < 1.733$	$16.241 \approx 5^{1.732} < p < 5^{1.733} \approx 16.267$

We define $5^{\sqrt{3}}$ to be the number p. To eight decimal places,

$$5^{\sqrt{3}} \approx 16.24245082.$$

Any positive irrational exponent can be interpreted in a similar way. Negative irrational exponents are then defined using reciprocals. Thus, so long as a is positive, a^x has meaning for *any* real number x. All of the laws of exponents still hold, but we will not prove that here. We can now define an *exponential function.*

Exponential Function The function $f(x) = a^x$, where a is a positive constant, $a \neq 1$, and x is any real number, is called the *exponential function,* base a.

We require the base a to be positive to avoid imaginary numbers that would result from taking even roots of negative numbers. The restriction $a \neq 1$ is made to exclude the constant function $f(x) = 1^x$, or $f(x) = 1$.

The following are examples of exponential functions:

$$f(x) = 2^x, \quad f(x) = \left(\tfrac{1}{3}\right)^x, \quad f(x) = 5^{-3x}.$$

Note that $5^{-3x} = (5^{-3})^x$.

Like polynomial functions, the domain of an exponential function is the set of all real numbers. Unlike polynomial functions, exponential functions have a variable exponent. Because of this, graphs of exponential functions either rise or fall dramatically.

EXAMPLE 1 Graph the exponential function given by $y = f(x) = 2^x$.

SOLUTION We compute some function values, thinking of y as $f(x)$, and list the results in a table. It is a good idea to start by letting $x = 0$.

$f(0) = 2^0 = 1;$ $\quad f(-1) = 2^{-1} = \frac{1}{2^1} = \frac{1}{2};$

$f(1) = 2^1 = 2;$

$f(2) = 2^2 = 4;$ $\quad f(-2) = 2^{-2} = \frac{1}{2^2} = \frac{1}{4};$

$f(3) = 2^3 = 8;$ $\quad f(-3) = 2^{-3} = \frac{1}{2^3} = \frac{1}{8}$

x	y, or $f(x)$
0	1
1	2
2	4
3	8
-1	$\frac{1}{2}$
-2	$\frac{1}{4}$
-3	$\frac{1}{8}$

Next, we plot these points and connect them with a smooth curve.

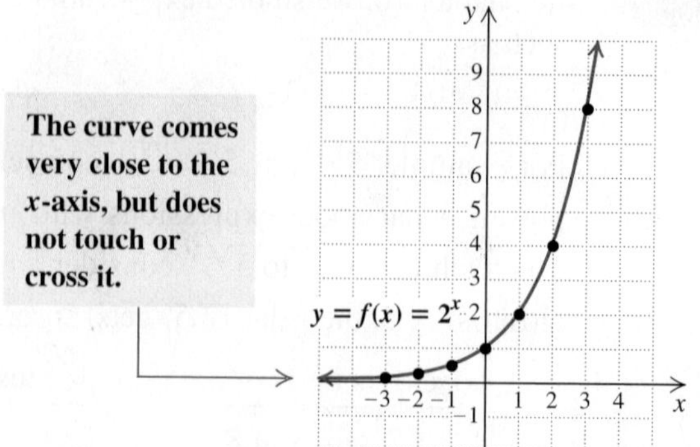

Note that as x increases, the function values increase without bound. As x decreases, the function values decrease, getting very close to 0. The x-axis, or the line $y = 0$, is a horizontal *asymptote*, meaning that the curve gets closer and closer to this line the further we move to the left.

Try Exercise 11.

EXAMPLE 2 Graph: $y = f(x) = \left(\frac{1}{2}\right)^x$.

SOLUTION We compute some function values, thinking of y as $f(x)$, and list the results in a table. Before we do this, note that

$$y = f(x) = \left(\tfrac{1}{2}\right)^x = (2^{-1})^x = 2^{-x}.$$

Then we have

$$f(0) = 2^{-0} = 1;$$
$$f(1) = 2^{-1} = \frac{1}{2^1} = \frac{1}{2};$$
$$f(2) = 2^{-2} = \frac{1}{2^2} = \frac{1}{4};$$
$$f(3) = 2^{-3} = \frac{1}{2^3} = \frac{1}{8};$$
$$f(-1) = 2^{-(-1)} = 2^1 = 2;$$
$$f(-2) = 2^{-(-2)} = 2^2 = 4;$$
$$f(-3) = 2^{-(-3)} = 2^3 = 8.$$

x	y, or $f(x)$
0	1
1	$\frac{1}{2}$
2	$\frac{1}{4}$
3	$\frac{1}{8}$
-1	2
-2	4
-3	8

Next, we plot these points and connect them with a smooth curve. This curve is a mirror image, or *reflection*, of the graph of $y = 2^x$ (see Example 1) across the y-axis. The line $y = 0$ is again the horizontal asymptote.

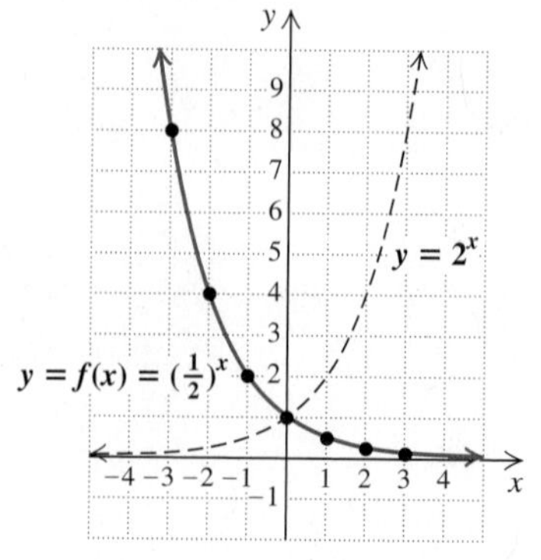

Try Exercise 23.

Interactive Discovery

If the values of $f(x)$ increase as x increases, we say that f is an *increasing function.* The function in Example 1, $f(x) = 2^x$, is increasing. If the values of $f(x)$ decrease as x increases, we say that f is a *decreasing function.* The function in Example 2, $f(x) = \left(\frac{1}{2}\right)^x$, is decreasing.

1. Graph each of the following functions, and determine whether the graph increases or decreases from left to right.

a) $f(x) = 3^x$

b) $g(x) = 4^x$

c) $h(x) = 1.5^x$

d) $r(x) = \left(\frac{1}{3}\right)^x$

e) $t(x) = 0.75^x$

2. How can you tell from the number a in $f(x) = a^x$ whether the graph of f increases or decreases?

3. Compare the graphs of f, g, and h. Which curve is steepest?

4. Compare the graphs of r and t. Which curve is steeper?

We can make the following observations.

- For $a > 1$, the graph of $f(x) = a^x$ increases from left to right. The greater the value of a, the steeper the curve. (See the figure on the left below.)

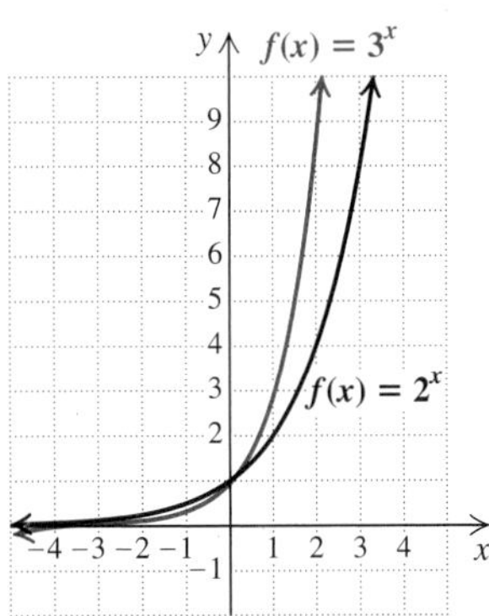

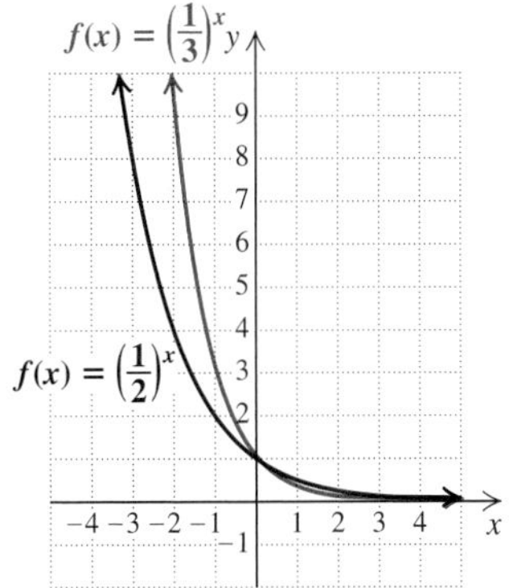

- For $0 < a < 1$, the graph of $f(x) = a^x$ decreases from left to right. For smaller values of a, the curve becomes steeper. (See the figure on the right above.)
- All graphs of $f(x) = a^x$ go through the y-intercept $(0, 1)$.
- All graphs of $f(x) = a^x$ have the x-axis as the horizontal asymptote.
- If $f(x) = a^x$, with $a > 0$, $a \neq 1$, the domain of f is all real numbers, and the range of f is all positive real numbers.
- For $a > 0$, $a \neq 1$, the function given by $f(x) = a^x$ is one-to-one. Its graph passes the horizontal-line test.

EXAMPLE 3 Graph: $y = f(x) = 2^{x-2}$.

SOLUTION We construct a table of values. Then we plot the points and connect them with a smooth curve. Here $x - 2$ is the *exponent.*

Student Notes

When using translations, make sure that you are shifting in the correct direction. When in doubt, substitute a value for x and make some calculations.

$$f(0) = 2^{0-2} = 2^{-2} = \tfrac{1}{4};$$
$$f(1) = 2^{1-2} = 2^{-1} = \tfrac{1}{2};$$
$$f(2) = 2^{2-2} = 2^{0} = 1;$$
$$f(3) = 2^{3-2} = 2^{1} = 2;$$
$$f(4) = 2^{4-2} = 2^{2} = 4;$$
$$f(-1) = 2^{-1-2} = 2^{-3} = \tfrac{1}{8};$$
$$f(-2) = 2^{-2-2} = 2^{-4} = \tfrac{1}{16}$$

x	y, or $f(x)$
0	$\frac{1}{4}$
1	$\frac{1}{2}$
2	1
3	2
4	4
-1	$\frac{1}{8}$
-2	$\frac{1}{16}$

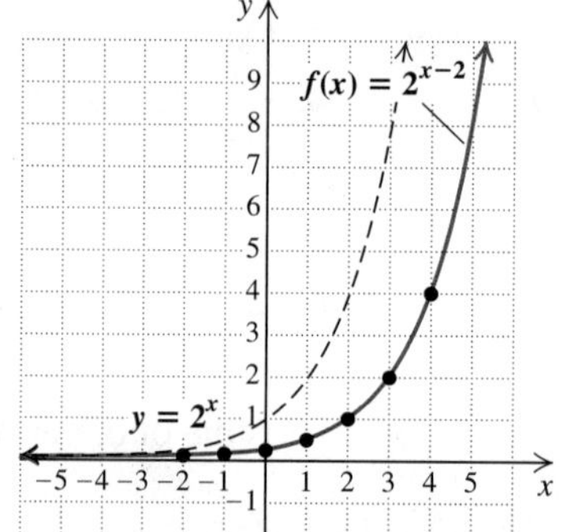

The graph looks just like the graph of $y = 2^x$, but it is translated 2 units to the right. The y-intercept of $y = 2^x$ is $(0, 1)$. The y-intercept of $y = 2^{x-2}$ is $\left(0, \frac{1}{4}\right)$. The line $y = 0$ is again the horizontal asymptote.

Try Exercise 19.

In general, the graph of $f(x) = 2^{x-h}$ will look like the graph of $y = 2^x$ translated right or left. Similarly, the graph of $f(x) = 2^x + k$ will look like the graph of 2^x translated up or down.

This observation is true in general.

The graph of $f(x) = a^{x-h} + k$ looks like the graph of $y = a^x$ translated $|h|$ units left or right and $|k|$ units up or down.

- If $h > 0$, $y = a^x$ is translated h units right.
- If $h < 0$, $y = a^x$ is translated $|h|$ units left.
- If $k > 0$, $y = a^x$ is translated k units up.
- If $k < 0$, $y = a^x$ is translated $|k|$ units down.

EQUATIONS WITH x AND y INTERCHANGED

It will be helpful in later work to be able to graph an equation in which the x and the y in $y = a^x$ are interchanged.

EXAMPLE 4 Graph: $x = 2^y$.

SOLUTION Note that x is alone on one side of the equation. To find ordered pairs that are solutions, we choose values for y and then compute values for x:

For $y = 0$, $x = 2^0 = 1$.

For $y = 1$, $x = 2^1 = 2$.

For $y = 2$, $x = 2^2 = 4$.

For $y = 3$, $x = 2^3 = 8$.

For $y = -1$, $x = 2^{-1} = \frac{1}{2}$.

For $y = -2$, $x = 2^{-2} = \frac{1}{4}$.

For $y = -3$, $x = 2^{-3} = \frac{1}{8}$.

x	y
1	0
2	1
4	2
8	3
$\frac{1}{2}$	-1
$\frac{1}{4}$	-2
$\frac{1}{8}$	-3

(1) Choose values for y.

(2) Compute values for x.

We plot the points and connect them with a smooth curve.

The curve does not touch or cross the y-axis, which serves as a vertical asymptote.

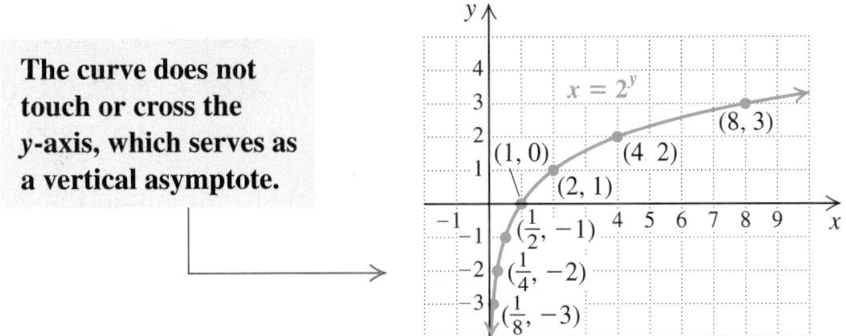

Note too that this curve looks just like the graph of $y = 2^x$, except that it is reflected across the line $y = x$, as shown here.

We have graphed $y = 2^x$ and its inverse, $x = 2^y$.

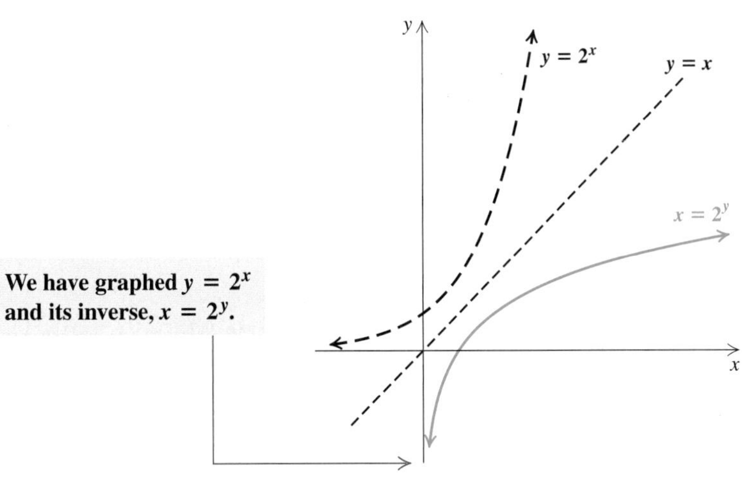

Try Exercise 33.

APPLICATIONS OF EXPONENTIAL FUNCTIONS

EXAMPLE 5 Interest Compounded Annually. The amount of money A that a principal P will be worth after t years at interest rate i, compounded annually, is given by the formula

$$A = P(1 + i)^t.$$ **You might review Example 12 in Section 11.1.**

Suppose that \$100,000 is invested at 8% interest, compounded annually.

a) Find a function for the amount of money in the account after t years.

b) Find the amount of money in the account at $t = 0$, $t = 4$, $t = 8$, and $t = 10$.

c) Graph the function.

SOLUTION

a) If $P = \$100{,}000$ and $i = 8\% = 0.08$, we can substitute these values and form the following function:

$$A(t) = \$100{,}000(1 + 0.08)^t \qquad \text{Using } A = P(1+i)^t$$
$$= \$100{,}000(1.08)^t.$$

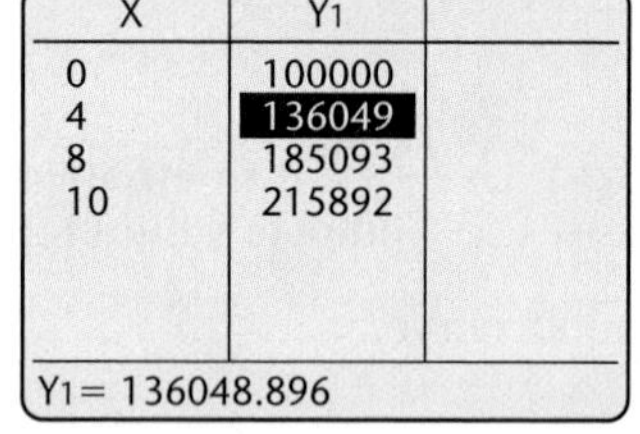

X	Y1	
0	100000	
4	136049	
8	185093	
10	215892	

Y1= 136048.896

b) To find the function values, we let $y_1 = 100{,}000(1.08)^x$ and use the TABLE feature with Indpnt set to Ask, as shown at left. We highlight a table entry if we want to view the function value to more decimal places than is shown in the table. Large values in a table will be written in scientific notation.

Alternatively, we can use $y_1(\ )$ notation to evaluate the function for each value. Using either procedure gives us

$$A(0) = \$100{,}000,$$
$$A(4) \approx \$136{,}048.90,$$
$$A(8) \approx \$185{,}093.02,$$
and $$A(10) \approx \$215{,}892.50.$$

c) We can use the function values computed in part (b), and others if we wish, to draw the graph by hand. Note that the axes will be scaled differently because of the large function values.

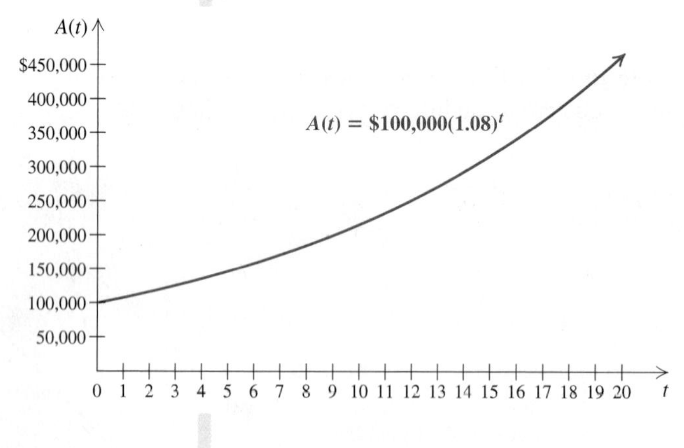

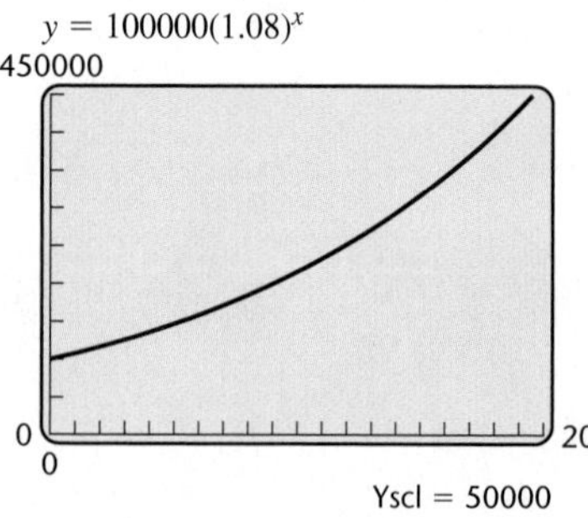

Try Exercise 51.

Connecting the Concepts

Linear and Exponential Growth

Linear functions increase (or decrease) by a constant amount. The rate of change (slope) of a linear graph is constant. The rate of change of an exponential graph is not constant.

To illustrate the difference between linear growth and exponential growth, compare the salaries for two hypothetical jobs, shown in the table and the graph below. Note that the starting salaries are the same, but the salary for job A increases by a constant amount (linearly) and the salary for job B increases by a constant percent (exponentially).

Note that the salary for job A is larger for the first few years only. The rate of change of the salary for job B increases each year, since the increase is based on a larger salary each year.

	Job A	Job B
Starting salary	\$50,000	\$50,000
Guaranteed raise	\$5000 per year	8% per year
Salary function	$f(x) = 50{,}000 + 5000x$	$g(x) = 50{,}000(1.08)^x$

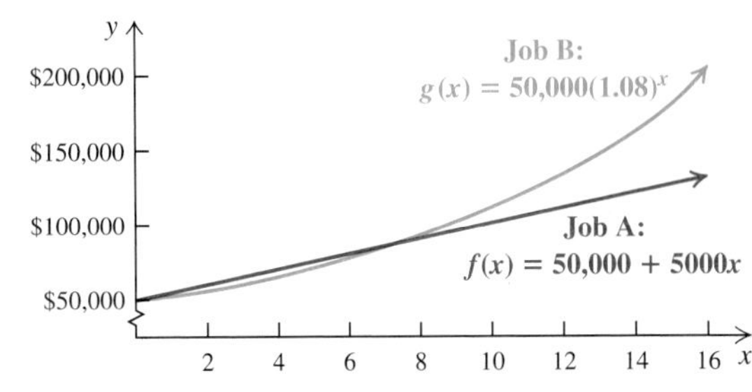

12.2 Exercise Set

FOR EXTRA HELP MyMathLab

REVIEW

Concept Reinforcement *Classify each of the following statements as either true or false.*

1. The graph of $f(x) = a^x$ always passes through the point $(0, 1)$.
2. The graph of $g(x) = \left(\frac{1}{2}\right)^x$ gets closer and closer to the x-axis as x gets larger and larger.
3. The graph of $f(x) = 2^{x-3}$ looks just like the graph of $y = 2^x$, but it is translated 3 units to the right.
4. The graph of $g(x) = 2^x - 3$ looks just like the graph of $y = 2^x$, but it is translated 3 units up.
5. The graph of $y = 3^x$ gets close to, but never touches, the y-axis.
6. The graph of $x = 3^y$ gets close to, but never touches, the y-axis.

Each of Exercises 7–10 shows the graph of a function $f(x) = a^x$. Determine from the graph whether $a > 1$ or $0 < a < 1$.

7. 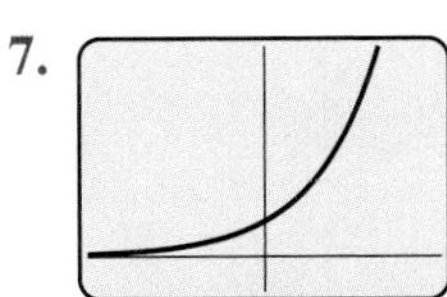

8. 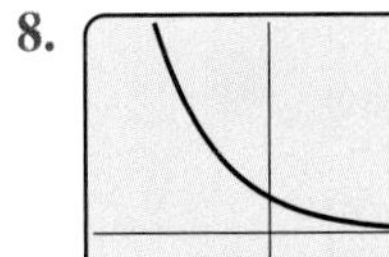

9. 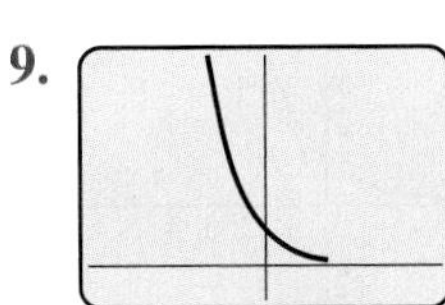

10. 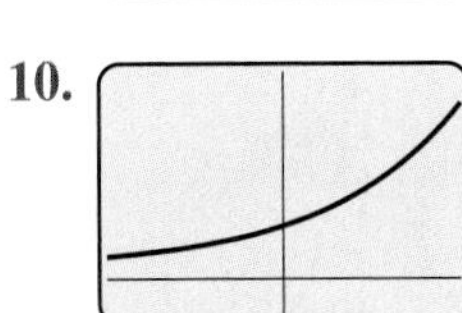

Graph.

11. $y = f(x) = 3^x$

12. $y = f(x) = 4^x$

13. $y = 5^x$

14. $y = 6^x$

15. $y = 2^x + 3$

16. $y = 2^x + 1$

17. $y = 3^x - 1$

18. $y = 3^x - 2$

19. $y = 2^{x-3}$

20. $y = 2^{x-1}$

21. $y = 2^{x+3}$

22. $y = 2^{x+1}$

23. $y = \left(\frac{1}{5}\right)^x$

24. $y = \left(\frac{1}{4}\right)^x$

25. $y = \left(\frac{1}{10}\right)^x$

26. $y = \left(\frac{1}{3}\right)^x$

27. $y = 2^{x-3} - 1$

28. $y = 2^{x+1} - 3$

29. $y = 1.7^x$

30. $y = 4.8^x$

31. $y = 0.15^x$

32. $y = 0.98^x$

33. $x = 3^y$

34. $x = 6^y$

35. $x = 2^{-y}$

36. $x = 3^{-y}$

37. $x = 5^y$

38. $x = 4^y$

39. $x = \left(\frac{3}{2}\right)^y$

40. $x = \left(\frac{4}{3}\right)^y$

Graph each pair of equations using the same set of axes.

41. $y = 3^x,\ x = 3^y$

42. $y = 4^x,\ x = 4^y$

43. $y = \left(\frac{1}{2}\right)^x,\ x = \left(\frac{1}{2}\right)^y$

44. $y = \left(\frac{1}{4}\right)^x,\ x = \left(\frac{1}{4}\right)^y$

Aha! *In Exercises 45–50, match each equation with one of the following graphs.*

a)

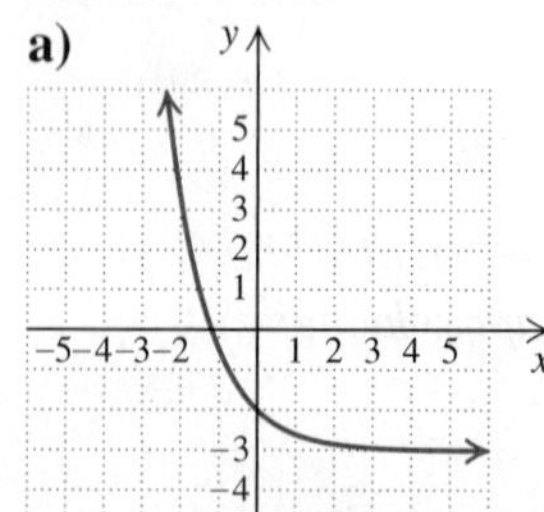

b)

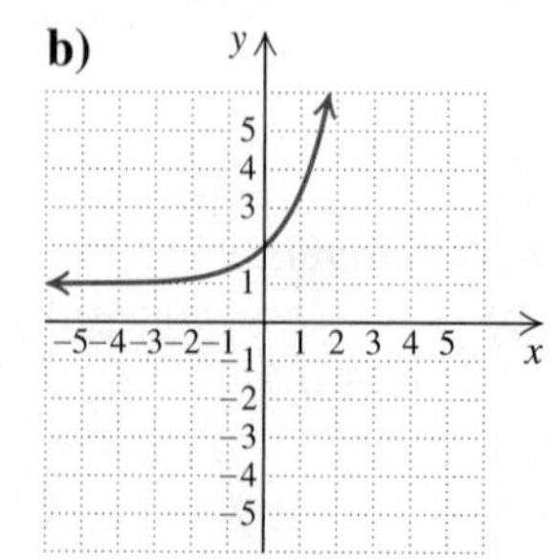

c)

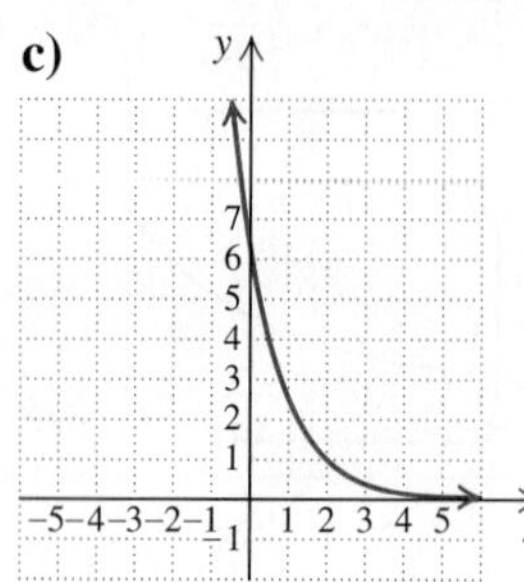

d)

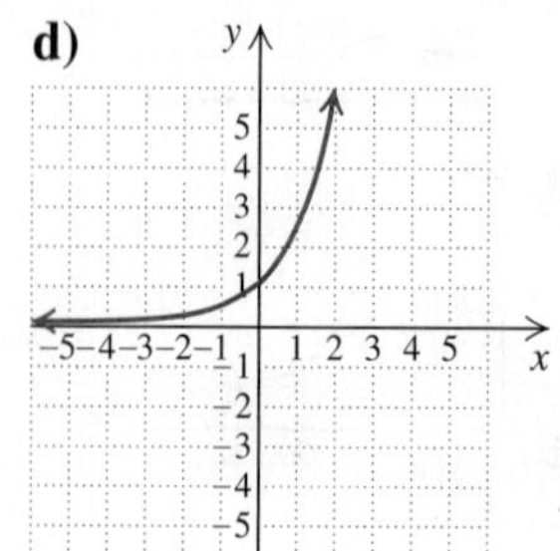

e)

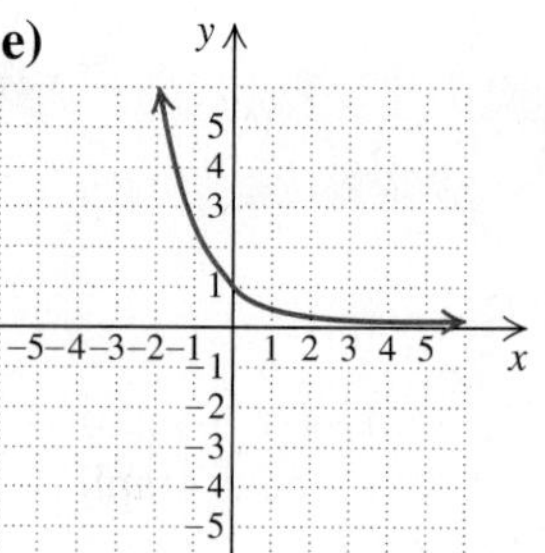

f)

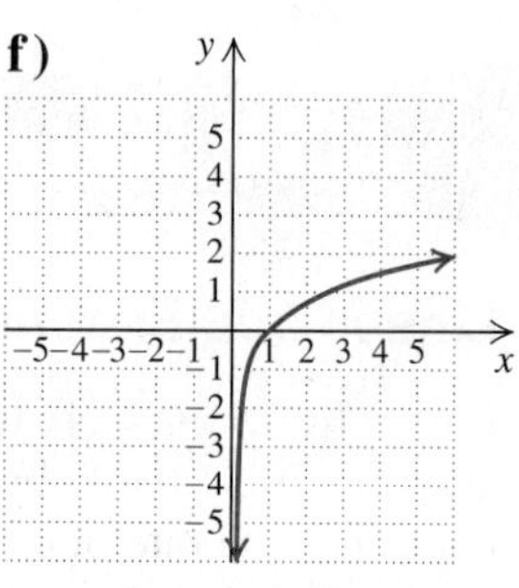

45. $y = \left(\frac{5}{2}\right)^x$

46. $y = \left(\frac{2}{5}\right)^x$

47. $x = \left(\frac{5}{2}\right)^y$

48. $y = \left(\frac{2}{5}\right)^x - 3$

49. $y = \left(\frac{2}{5}\right)^{x-2}$

50. $y = \left(\frac{5}{2}\right)^x + 1$

Solve.

51. *Music Downloads.* The number $M(t)$ of single tracks downloaded, in billions, t years after 2003 can be approximated by

$$M(t) = 0.353(1.244)^t.$$

Source: Based on data from International Federation of the Phonographic Industry

a) Estimate the number of single tracks downloaded in 2006, in 2008, and in 2012.

b) Graph the function.

52. *Growth of Bacteria.* The bacteria *Escherichia coli* are commonly found in the human bladder. Suppose that 3000 of the bacteria are present at time $t = 0$. Then t minutes later, the number of bacteria present can be approximated by

$$N(t) = 3000(2)^{t/20}.$$

a) How many bacteria will be present after 10 min? 20 min? 30 min? 40 min? 60 min?

b) Graph the function.

53. *Smoking Cessation.* The percent of smokers $P(t)$ who, with telephone counseling to quit smoking, are still successful t months later can be approximated by

$$P(t) = 21.4(0.914)^t.$$

Sources: *New England Journal of Medicine*; data from California's Smokers' Hotline

a) Estimate the percent of smokers receiving telephone counseling who are successful in quitting for 1 month, 3 months, and 1 year.

b) Graph the function.

54. *Smoking Cessation.* The percent of smokers $P(t)$ who, without telephone counseling, have successfully quit smoking for t months (see Exercise 53) can be approximated by

$$P(t) = 9.02(0.93)^t.$$

Sources: *New England Journal of Medicine*; data from California's Smokers' Hotline

a) Estimate the percent of smokers not receiving telephone counseling who are successful in quitting for 1 month, 3 months, and 1 year.
b) Graph the function.

55. *Marine Biology.* Due to excessive whaling prior to the mid 1970s, the humpback whale is considered an endangered species. The worldwide population of humpbacks $P(t)$, in thousands, t years after 1900 $(t < 70)$ can be approximated by

$$P(t) = 150(0.960)^t.$$

Source: Based on information from the American Cetacean Society, 2006, and the ASK Archive, 1998

a) How many humpback whales were alive in 1930? in 1960?
b) Graph the function.

56. *Salvage Value.* A laser printer is purchased for \$1200. Its value each year is about 80% of the value of the preceding year. Its value, in dollars, after t years is given by the exponential function

$$V(t) = 1200(0.8)^t.$$

a) Find the value of the printer after 0 year, 1 year, 2 years, 5 years, and 10 years.
b) Graph the function.

57. *Marine Biology.* As a result of preservation efforts in most countries in which whaling was common, the humpback whale population has grown since the 1970s. The worldwide population of humpbacks $P(t)$, in thousands, t years after 1982 can be approximated by

$$P(t) = 5.5(1.08)^t.$$

Source: Based on information from the American Cetacean Society, 2006, and the ASK Archive, 1998

a) How many humpback whales were alive in 1992? in 2006?
b) Graph the function.

58. *Recycling Aluminum Cans.* About $\frac{1}{2}$ of all aluminum cans will be recycled. A beverage company distributes 250,000 cans. The number in use after t years is given by the exponential function

$$N(t) = 250{,}000\left(\tfrac{1}{2}\right)^t.$$

Source: The Aluminum Association, Inc., 2009

a) How many cans are still in use after 0 year? 1 year? 4 years? 10 years?
b) Graph the function.

59. *Invasive Species.* Ruffe is a species of freshwater fish that is considered invasive where it is not native. The function

$$R(t) = 2(1.75)^t$$

can be used to estimate the number of ruffe in a lake t years after 2 fish have been introduced to the lake.
Source: Based on information from invasivespeciesireland.com

a) How many fish will be in the lake 10 years after 2 ruffe have been introduced? after 15 years?
b) Graph the function.

60. *mp3 Players.* The number $m(t)$ of mp3 players sold per year in the United States, in millions, can be estimated by

$$m(t) = 34.3(1.25)^t,$$

where t is the number of years after 2006.
Source: Forrester Research, Inc.

a) Estimate the number of mp3 players sold in 2006, in 2010, and in 2012.
b) Graph the function.

TW **61.** Without using a calculator, explain why 2^{π} must be greater than 8 but less than 16.

TW **62.** Suppose that \$1000 is invested for 5 years at 7% interest, compounded annually. In what year will the most interest be earned? Why?

SKILL REVIEW

Review factoring polynomials (Sections 6.1–6.6).

Factor.

63. $3x^2 - 48$ [6.4]

64. $x^2 - 20x + 100$ [6.4]

65. $6x^2 + x - 12$ [6.3]

66. $8x^6 - 64y^6$ [6.5]

67. $6y^2 + 36y - 240$ [6.2]

68. $5x^4 - 10x^3 - 3x^2 + 6x$ [6.1]

SYNTHESIS

TW **69.** Examine Exercise 60. Do you believe that the equation for the number of mp3 players sold in the United States will be accurate 20 years from now? Why or why not?

TW **70.** Why was it necessary to discuss irrational exponents before graphing exponential functions?

Determine which of the two given numbers is larger. Do not use a calculator.

71. $\pi^{1.3}$ or $\pi^{2.4}$

72. $\sqrt{8^3}$ or $8^{\sqrt{3}}$

Graph.

73. $y = 2^x + 2^{-x}$

74. $y = \left|\left(\frac{1}{2}\right)^x - 1\right|$

75. $y = |2^x - 2|$

76. $y = 2^{-(x-1)^2}$

77. $y = |2^{x^2} - 1|$

78. $y = 3^x + 3^{-x}$

Graph both equations using the same set of axes.

79. $y = 3^{-(x-1)}$, $x = 3^{-(y-1)}$

80. $y = 1^x$, $x = 1^y$

81. *Navigational Devices.* The number of GPS navigational devices in use in the United States has grown from 0.5 million in 2000 to 4 million in 2004 to 50 million in 2008. After pressing STAT and entering the data, use the ExpReg option in the STAT CALC menu to find an exponential function that models the number of navigational devices in use t years after 2000. Then use that function to predict the total number of devices in use in 2012.
Source: Telematics Research Group

82. *Keyboarding Speed.* Trey is studying keyboarding. After he has studied for t hours, Trey's speed, in words per minute, is given by the exponential function

$$S(t) = 200[1 - (0.99)^t].$$

Use a graph and/or table of values to predict Trey's speed after studying for 10 hr, 40 hr, and 80 hr.

TW **83.** The graph below shows growth in the height of ocean waves over time, assuming a steady surface wind.
Source: magicseaweed.com

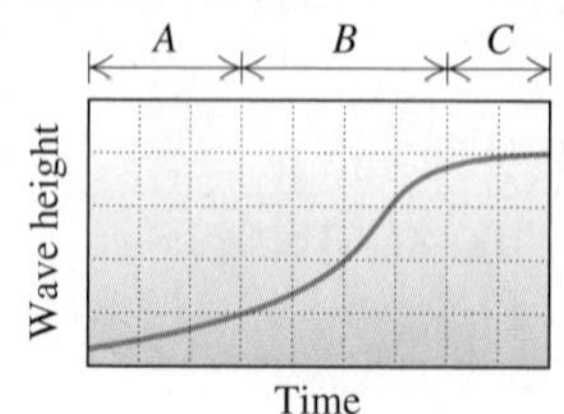

Source: magicseaweed.com

a) Consider the portions of the graph marked A, B, and C. Suppose that each portion can be labeled Exponential Growth, Linear Growth, or Saturation. How would you label each portion?

b) Small vertical movements in wind, surface roughness of water, and gravity are three forces that create waves. How might these forces be related to the shape of the wave-height graph?

TW **84.** Consider any exponential function of the form $f(x) = a^x$ with $a > 1$. Will it always follow that $f(3) - f(2) > f(2) - f(1)$, and, in general, $f(n+2) - f(n+1) > f(n+1) - f(n)$? Why or why not? (*Hint*: Think graphically.)

85. On many graphing calculators, it is possible to enter and graph $y_1 = A\wedge(X - B) + C$ after first pressing APPS Transfrm. Use this application to graph $f(x) = 2.5^{x-3} + 2$, $g(x) = 2.5^{x+3} + 2$, $h(x) = 2.5^{x-3} - 2$, and $k(x) = 2.5^{x+3} - 2$.

Try Exercise Answers: Section 12.2

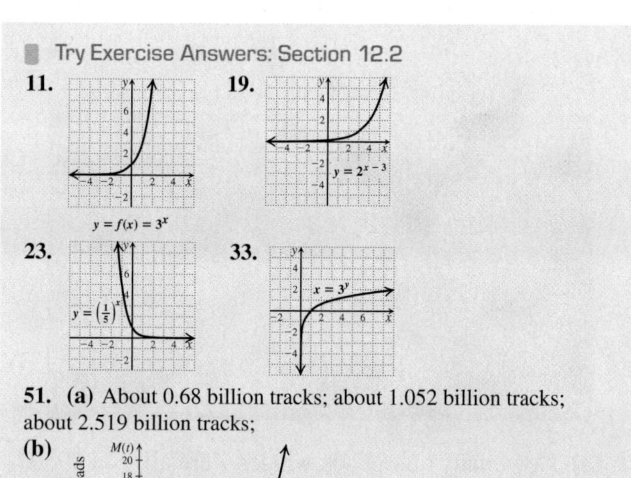

51. **(a)** About 0.68 billion tracks; about 1.052 billion tracks; about 2.519 billion tracks;
(b)

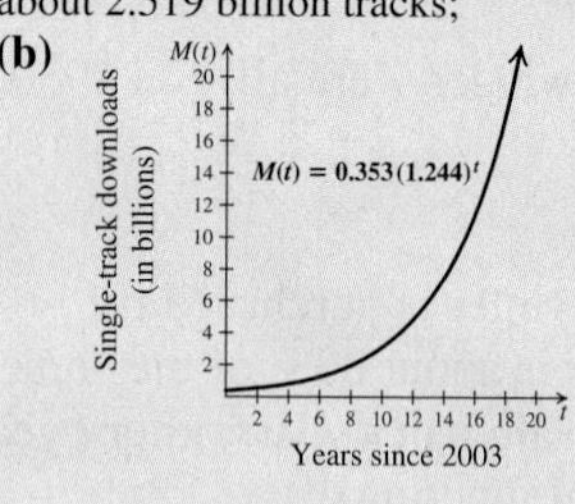

Collaborative Corner

The True Cost of a New Car

Focus: Car loans and exponential functions
Time: 30 minutes
Group Size: 2
Materials: Calculators with exponentiation keys

The formula

$$M = \frac{Pr}{1 - (1 + r)^{-n}}$$

is used to determine the payment size, M, when a loan of P dollars is to be repaid in n equally sized monthly payments. Here r represents the monthly interest rate. Loans repaid in this fashion are said to be *amortized* (spread out equally) over a period of n months.

ACTIVITY

1. Suppose one group member is selling the other a car for \$2600, financed at 1% interest per month for 24 months. What should be the size of each monthly payment?
2. Suppose both group members are shopping for the same model new car. Each group member visits a different dealer. One dealer offers the car for \$13,000 at 10.5% interest (0.00875 monthly interest) for 60 months (no down payment). The other dealer offers the same car for \$12,000, but at 12% interest (0.01 monthly interest) for 48 months (no down payment)
   - **a)** Determine the monthly payment size for each offer. Then determine the total amount paid for the car under each offer. How much of each total is interest?
   - **b)** Work together to find the annual interest rate for which the total cost of 60 monthly payments for the \$13,000 car would equal the total cost of 48 monthly payments for the \$12,000 car (as found in part (a) above).

12.3 Logarithmic Functions

- Graphs of Logarithmic Functions
- Common Logarithms
- Equivalent Equations
- Solving Certain Logarithmic Equations

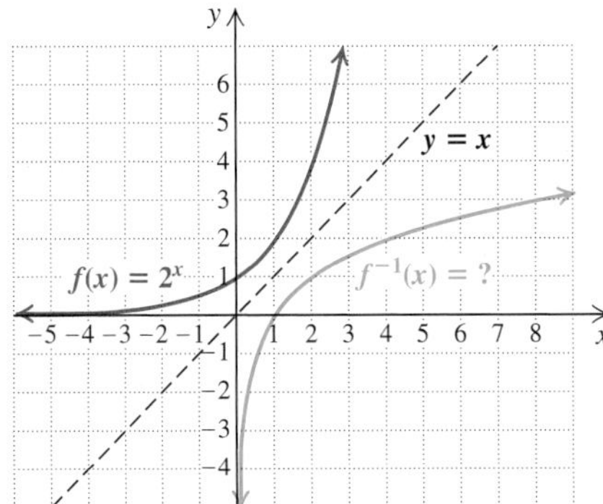

We are now ready to study inverses of exponential functions. These functions have many applications and are referred to as *logarithm*, or *logarithmic, functions*.

GRAPHS OF LOGARITHMIC FUNCTIONS

Consider the exponential function $f(x) = 2^x$. Like all exponential functions, f is one-to-one. Can a formula for f^{-1} be found? To answer this, we use the method of Section 12.1:

1. Replace $f(x)$ with y: $\quad y = 2^x$.
2. Interchange x and y: $\quad x = 2^y$.
3. Solve for y: $\quad y =$ the power to which we raise 2 to get x.
4. Replace y with $f^{-1}(x)$: $\quad f^{-1}(x) =$ the power to which we raise 2 to get x.

We now define a new symbol to replace the words "the power to which we raise 2 to get x":

$\log_2 x$, read "the logarithm, base 2, of x," or "log, base 2, of x," means "the exponent to which we raise 2 to get x."

Thus if $f(x) = 2^x$, then $f^{-1}(x) = \log_2 x$. Note that $f^{-1}(8) = \log_2 8 = 3$, because 3 is *the exponent to which we raise 2 to get* 8.

EXAMPLE 1 Simplify: **(a)** $\log_2 32$; **(b)** $\log_2 1$; **(c)** $\log_2 \frac{1}{8}$.

SOLUTION

a) Think of $\log_2 32$ as the exponent to which we raise 2 to get 32. That exponent is 5. Therefore, $\log_2 32 = 5$.

b) We ask ourselves: "To what exponent do we raise 2 in order to get 1?" That exponent is 0 (recall that $2^0 = 1$). Thus, $\log_2 1 = 0$.

c) To what exponent do we raise 2 in order to get $\frac{1}{8}$? Since $2^{-3} = \frac{1}{8}$, we have $\log_2 \frac{1}{8} = -3$.

Try Exercise 9.

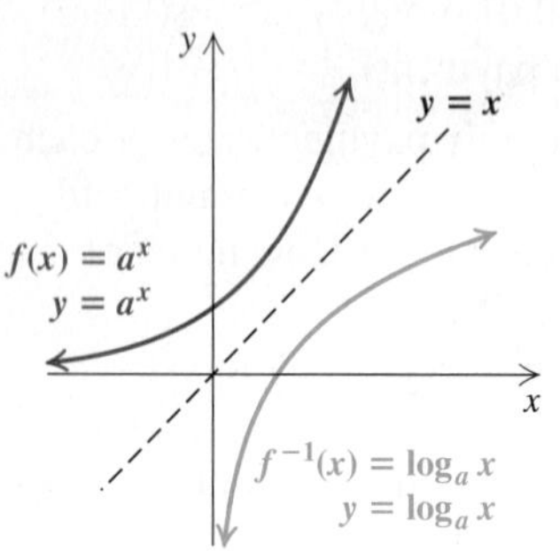

Although numbers like $\log_2 13$ can be only approximated, we must remember that $\log_2 13$ represents *the exponent to which we raise 2 to get* 13. That is, $2^{\log_2 13} = 13$. Using an approximation, we see that $2^{3.7} \approx 13$. Therefore, $\log_2 13 \approx 3.7$.

For any exponential function $f(x) = a^x$, the inverse is called a **logarithmic function, base *a*.** The graph of the inverse can be drawn by reflecting the graph of $f(x) = a^x$ across the line $y = x$. It will be helpful to remember that the inverse of $f(x) = a^x$ is given by $f^{-1}(x) = \log_a x$.

The Meaning of $\log_a x$ For $x > 0$ and a a positive constant other than 1, $\log_a x$ is the exponent to which a must be raised in order to get x. Thus,

$$\log_a x = m \text{ means } a^m = x$$

or, equivalently,

$$\log_a x \text{ is that unique exponent for which } a^{\log_a x} = x.$$

Student Notes

As an aid in remembering what $\log_a x$ means, note that a is called the *base*, just as it is the base in $a^y = x$.

It is important to remember that *a logarithm is an exponent.* It might help to repeat several times: "The logarithm, base a, of a number x is the exponent to which a must be raised in order to get x."

EXAMPLE 2 Simplify: $7^{\log_7 85}$.

SOLUTION Remember that $\log_7 85$ is the exponent to which 7 is raised to get 85. Raising 7 to that exponent, we have

$$7^{\log_7 85} = 85.$$

Try Exercise 35.

Because logarithmic functions and exponential functions are inverses of each other, the result in Example 2 should come as no surprise: If $f(x) = \log_7 x$, then

$$\text{for} \quad f(x) = \log_7 x, \text{ we have } f^{-1}(x) = 7^x$$
$$\text{and} \quad f^{-1}(f(x)) = f^{-1}(\log_7 x) = 7^{\log_7 x} = x.$$

Thus, $f^{-1}(f(85)) = 7^{\log_7 85} = 85$.

The following is a comparison of exponential and logarithmic functions.

Exponential Function	Logarithmic Function
$y = a^x$	$x = a^y$
$f(x) = a^x$	$g(x) = \log_a x$
$a > 0, a \neq 1$	$a > 0, a \neq 1$
The domain is $\mathbb{R}$.	The range is $\mathbb{R}$.
The range is $(0, \infty)$.	The domain is $(0, \infty)$.
$f^{-1}(x) = \log_a x$	$g^{-1}(x) = a^x$
The graph of $f(x)$ contains the points $(0, 1)$ and $(1, a)$.	The graph of $g(x)$ contains the points $(1, 0)$ and $(a, 1)$.

Student Notes

Since $g(x) = \log_a x$ is the inverse of $f(x) = a^x$, the domain of f is the range of g, and the range of f is the domain of g.

EXAMPLE 3 Graph: $y = f(x) = \log_5 x$.

SOLUTION If $y = \log_5 x$, then $5^y = x$. We can find ordered pairs that are solutions by choosing values for y and computing the x-values.

For $y = 0$, $x = 5^0 = 1$.
For $y = 1$, $x = 5^1 = 5$.
For $y = 2$, $x = 5^2 = 25$.
For $y = -1$, $x = 5^{-1} = \frac{1}{5}$.
For $y = -2$, $x = 5^{-2} = \frac{1}{25}$.

(1) Select y.
(2) Compute x.

x, or 5^y	y
1	0
5	1
25	2
$\frac{1}{5}$	-1
$\frac{1}{25}$	-2

This table shows the following:

$$\left.\begin{aligned} \log_5 1 &= 0; \\ \log_5 5 &= 1; \\ \log_5 25 &= 2; \\ \log_5 \tfrac{1}{5} &= -1; \\ \log_5 \tfrac{1}{25} &= -2. \end{aligned}\right\}$$

These can all be checked using the equations above.

We plot the set of ordered pairs and connect the points with a smooth curve. The graphs of $y = 5^x$ and $y = x$ are shown only for reference.

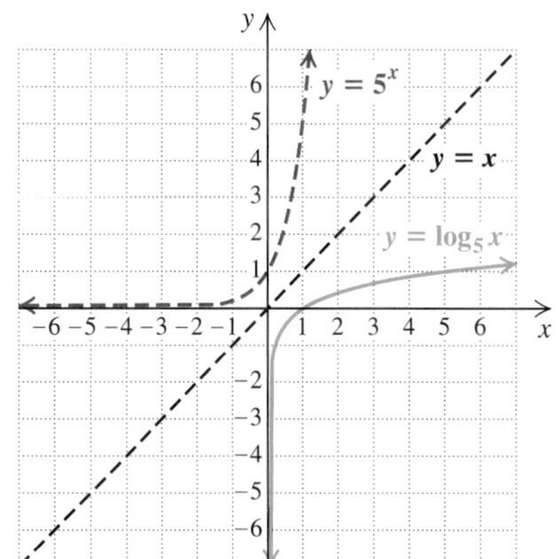

Try Exercise 37.

COMMON LOGARITHMS

Any positive number other than 1 can serve as the base of a logarithmic function. However, some logarithm bases fit into certain applications more naturally than others.

Base-10 logarithms, called **common logarithms,** are useful because they have the same base as our "commonly" used decimal system. Before calculators became so widely available, common logarithms helped with tedious calculations. In fact, that is why logarithms were developed. At first, printed tables listed common logarithms. Today we find common logarithms using calculators.

The abbreviation **log,** with no base written, is generally understood to mean logarithm base 10—that is, a common logarithm. Thus,

$\log 17$ means $\log_{10} 17$. **It is important to remember this abbreviation.**

Common Logarithms

The key for common logarithms is usually marked LOG. Some calculators automatically supply the left parenthesis when LOG is pressed. You may have to supply a right parenthesis to close the expression. Even if a calculator does not require an ending parenthesis, it is a good idea to include one.

For example, if we are using a calculator that supplies the left parenthesis, log 5 can be found by pressing LOG 5) ENTER. Note from the screen on the left below that $\log 5 \approx 0.6990$.

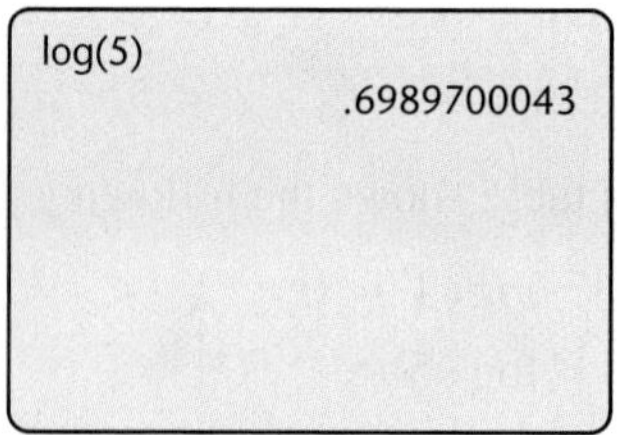

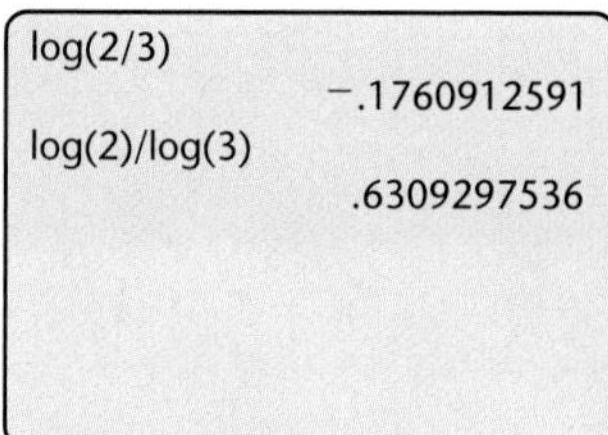

Parentheses must be placed carefully when evaluating expressions containing logarithms. For example, to evaluate

$$\log \frac{2}{3},$$

press LOG 2 ÷ 3) ENTER. To evaluate

$$\frac{\log 2}{\log 3},$$

press LOG 2) ÷ LOG 3) ENTER. Note from the screen on the right above that $\log \frac{2}{3} \approx -0.1761$ and that $\frac{\log 2}{\log 3} \approx 0.6309$.

The inverse of $f(x) = \log_{10} x$ is $g(x) = 10^x$. Because of this, on many calculators the LOG key serves as the 10^x key after the 2ND key is pressed. As with logarithms, some calculators automatically supply a left parenthesis before the exponent. Using such a calculator, we find $10^{3.417}$ by pressing 10^x 3 . 4 1 7) ENTER. The result is approximately 2612.1614. If we then

When a Turn Is Trouble

Occasionally a page turn can interrupt your thoughts as you work through a section. You may find it helpful to rewrite (in pencil) the last equation or sentence appearing on one page at the very top of the next page.

press LOG ANS) ENTER, the result is 3.417. This illustrates that $f(x) = \log x$ and $g(x) = 10^x$ are inverse functions.

```
10^(3.417)
                2612.161354
log(Ans)
                      3.417
```

Your Turn

1. Find log 100. 2
2. Find $\log \frac{15}{8}$. About 0.2730
3. Find $\dfrac{\log 15}{\log 8}$. About 1.3023
4. Find $10^{1.2}$. About 15.8489
5. Find log 96. Then press 10^x ANS. The result should be 96.

It is important to place parentheses properly when entering logarithmic expressions in a graphing calculator.

EXAMPLE 4 Graph: $f(x) = \log \dfrac{x}{5} + 1$.

SOLUTION We enter $y = \log(x/5) + 1$. Since logarithms of negative numbers are not defined, we set the window accordingly. The graph is shown below. Note that although on the calculator the graph appears to touch the y-axis, a table of values or TRACE would show that the y-axis is indeed an asymptote.

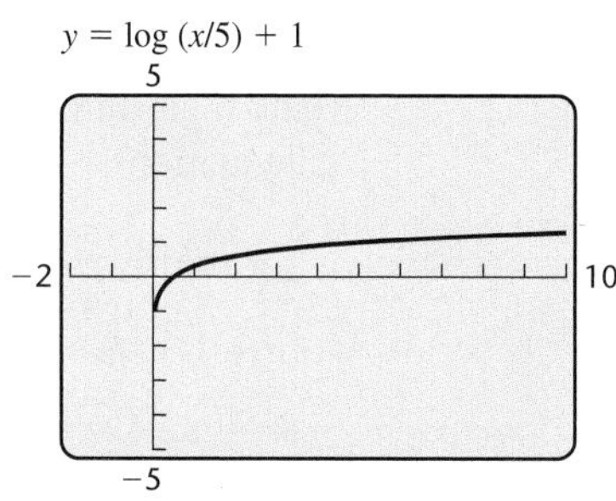

Try Exercise 59.

EQUIVALENT EQUATIONS

We use the definition of logarithm to rewrite a *logarithmic equation* as an equivalent *exponential equation* or the other way around:

$$m = \log_a x \quad \textbf{is equivalent to} \quad a^m = x.$$

CAUTION! **Do not forget this relationship!** It is probably the most important definition in the chapter. Many times this definition will be used to justify a property we are considering.

EXAMPLE 5 Rewrite each as an equivalent exponential equation.

a) $y = \log_3 5$ **b)** $-2 = \log_a 7$

c) $a = \log_b d$

SOLUTION

a) $y = \log_3 5$ is equivalent to $3^y = 5$ **The logarithm is the exponent.**

The base remains the base.

b) $-2 = \log_a 7$ is equivalent to $a^{-2} = 7$

c) $a = \log_b d$ is equivalent to $b^a = d$

Try Exercise 65.

EXAMPLE 6 Rewrite each as an equivalent logarithmic equation: **(a)** $8 = 2^x$; **(b)** $y^{-1} = 4$; **(c)** $a^b = c$.

SOLUTION

a) $8 = 2^x$ is equivalent to $x = \log_2 8$ **The exponent is the logarithm.**

The base remains the base.

b) $y^{-1} = 4$ is equivalent to $-1 = \log_y 4$

c) $a^b = c$ is equivalent to $b = \log_a c$

Try Exercise 81.

SOLVING CERTAIN LOGARITHMIC EQUATIONS

Many logarithmic equations can be solved by rewriting them as equivalent exponential equations.

EXAMPLE 7 Solve: **(a)** $\log_2 x = -3$; **(b)** $\log_x 16 = 2$.

SOLUTION

a) $\log_2 x = -3$

$2^{-3} = x$ **Rewriting as an exponential equation**

$\frac{1}{8} = x$ **Computing 2^{-3}**

Check: $\log_2 \frac{1}{8}$ is the exponent to which 2 is raised to get $\frac{1}{8}$. Since that exponent is -3, we have a check. The solution is $\frac{1}{8}$.

b) $\log_x 16 = 2$

$x^2 = 16$ **Rewriting as an exponential equation**

$x = 4$ *or* $x = -4$ **Principle of square roots**

Check: $\log_4 16 = 2$ because $4^2 = 16$. Thus, 4 is a solution of $\log_x 16 = 2$. Because all logarithmic bases must be positive, -4 cannot be a solution. Logarithmic bases must be positive because logarithms are defined using exponential functions that require positive bases. The solution is 4.

Try Exercise 97.

One method for solving certain logarithmic equations and exponential equations relies on the following property, which results from the fact that exponential functions are one-to-one.

The Principle of Exponential Equality For any real number b, where $b \neq -1, 0$, or 1,

$b^x = b^y$ is equivalent to $x = y$.

(Powers of the same base are equal if and only if the exponents are equal.)

EXAMPLE 8 Solve: **(a)** $\log_{10} 1000 = x$; **(b)** $\log_4 1 = t$.

SOLUTION

a) We rewrite $\log_{10} 1000 = x$ in exponential form and solve:

$10^x = 1000$	Rewriting as an exponential equation
$10^x = 10^3$	Writing 1000 as a power of 10
$x = 3.$	Equating exponents

Check: This equation can also be solved directly by determining the exponent to which we raise 10 in order to get 1000. In both cases, we find that $\log_{10} 1000 = 3$, so we have a check. The solution is 3.

b) We rewrite $\log_4 1 = t$ in exponential form and solve:

$4^t = 1$	Rewriting as an exponential equation
$4^t = 4^0$	Writing 1 as a power of 4. This can be done mentally.
$t = 0.$	Equating exponents

Check: As in part (a), this equation can be solved directly by determining the exponent to which we raise 4 in order to get 1. In both cases, we find that $\log_4 1 = 0$, so we have a check. The solution is 0.

Try Exercise 99.

Example 8 illustrates an important property of logarithms.

$\log_a 1$

The logarithm, base a, of 1 is always 0: $\log_a 1 = 0$.

This follows from the fact that $a^0 = 1$ is equivalent to the logarithmic equation $\log_a 1 = 0$. Thus, $\log_{10} 1 = 0$, $\log_7 1 = 0$, and so on.

Another property results from the fact that $a^1 = a$. This is equivalent to the equation $\log_a a = 1$.

$\log_a a$

The logarithm, base a, of a is always 1: $\log_a a = 1$.

Thus, $\log_{10} 10 = 1$, $\log_8 8 = 1$, and so on.

12.3 Exercise Set

FOR EXTRA HELP MyMathLab    

Concept Reinforcement *In each of Exercises 1–8, match the expression or equation with an equivalent expression or equation from the column on the right.*

1. ___ $\log_5 25$ a) 1
2. ___ $2^5 = x$ b) x
3. ___ $\log_5 5$ c) $x^5 = 27$
4. ___ $\log_2 1$ d) $\log_2 x = 5$
5. ___ $\log_5 5^x$ e) $\log_2 5 = x$
6. ___ $\log_x 27 = 5$ f) $\log_x 5 = -2$
7. ___ $5 = 2^x$ g) 2
8. ___ $x^{-2} = 5$ h) 0

Simplify.

9. $\log_{10} 1000$
10. $\log_{10} 100$
11. $\log_2 16$
12. $\log_2 8$
13. $\log_3 81$
14. $\log_3 27$
15. $\log_4 \frac{1}{16}$
16. $\log_4 \frac{1}{4}$
17. $\log_7 \frac{1}{7}$
18. $\log_7 \frac{1}{49}$
19. $\log_5 625$
20. $\log_5 125$
21. $\log_8 8$
22. $\log_7 1$
23. $\log_8 1$
24. $\log_8 8$
25. Aha! $\log_9 9^5$
26. $\log_9 9^{10}$
27. $\log_{10} 0.01$
28. $\log_{10} 0.1$
29. $\log_9 3$
30. $\log_{16} 4$
31. $\log_9 27$
32. $\log_{16} 64$
33. $\log_{1000} 100$
34. $\log_{27} 9$
35. $5^{\log_5 7}$
36. $6^{\log_6 13}$

Graph by hand.

37. $y = \log_{10} x$
38. $y = \log_2 x$
39. $y = \log_3 x$
40. $y = \log_7 x$
41. $f(x) = \log_6 x$
42. $f(x) = \log_4 x$
43. $f(x) = \log_{2.5} x$
44. $f(x) = \log_{1/2} x$

Graph both functions using the same set of axes.

45. $f(x) = 3^x,\ f^{-1}(x) = \log_3 x$
46. $f(x) = 4^x,\ f^{-1}(x) = \log_4 x$

Use a calculator to find each of the following rounded to four decimal places.

47. $\log 4$
48. $\log 5$
49. $\log 13{,}400$
50. $\log 93{,}100$
51. $\log 0.527$
52. $\log 0.493$

Use a calculator to find each of the following rounded to four decimal places.

53. $10^{2.3}$
54. $10^{0.173}$
55. $10^{-2.9523}$
56. $10^{4.8982}$
57. $10^{0.0012}$
58. $10^{-3.89}$

Graph using a graphing calculator.

59. $\log(x + 2)$
60. $\log(x - 5)$
61. $\log(2x) - 3$
62. $\log(3x) + 2$
63. $\log(x^2)$
64. $\log(x^2 + 1)$

Rewrite each of the following as an equivalent exponential equation. Do not solve.

65. $x = \log_{10} 8$
66. $h = \log_7 10$
67. $\log_9 9 = 1$
68. $\log_6 6 = 1$
69. $\log_{10} 0.1 = -1$
70. $\log_{10} 0.01 = -2$
71. $\log_{10} 7 = 0.845$
72. $\log_{10} 3 = 0.4771$
73. $\log_c m = 8$
74. $\log_b n = 23$
75. $\log_t Q = r$
76. $\log_m P = a$
77. $\log_e 0.25 = -1.3863$
78. $\log_e 0.989 = -0.0111$
79. $\log_r T = -x$
80. $\log_c M = -w$

Rewrite each of the following as an equivalent logarithmic equation. Do not solve.

81. $10^2 = 100$
82. $10^4 = 10{,}000$
83. $4^{-5} = \frac{1}{1024}$
84. $5^{-3} = \frac{1}{125}$
85. $16^{3/4} = 8$
86. $8^{1/3} = 2$
87. $10^{0.4771} = 3$
88. $10^{0.3010} = 2$

89. $z^m = 6$

90. $m^n = r$

91. $p^m = V$

92. $Q^t = x$

93. $e^3 = 20.0855$

94. $e^2 = 7.3891$

95. $e^{-4} = 0.0183$

96. $e^{-2} = 0.1353$

Solve.

97. $\log_3 x = 2$

98. $\log_4 x = 3$

99. $\log_5 125 = x$

100. $\log_4 64 = x$

101. $\log_x 16 = 4$

102. $\log_x 81 = 2$

103. $\log_x 7 = 1$

104. $\log_x 8 = 1$

105. $\log_x 9 = \frac{1}{2}$

106. $\log_x 11 = \frac{1}{2}$

107. $\log_3 x = -2$

108. $\log_2 x = -1$

109. $\log_{32} x = \frac{2}{5}$

110. $\log_8 x = \frac{2}{3}$

TW 111. Explain why we say that "a logarithm is an exponent."

TW 112. Is it easier to find x given $x = \log_9 \frac{1}{3}$ or given $9^x = \frac{1}{3}$? Explain your reasoning.

SKILL REVIEW

To prepare for Section 12.4, review rules for working with exponents (Section 5.1).

Simplify. [5.1]

113. $a^{12} \cdot a^6$

114. $x^4(x^5)$

115. $\dfrac{x^{12}}{x^4}$

116. $\dfrac{a^{15}}{a^3}$

117. $(y^3)^5$

118. $(n^{15})^2$

119. $x^2 \cdot x^3$

120. $(x^2)^3$

SYNTHESIS

TW 121. Would a manufacturer be pleased or unhappy if sales of a product grew logarithmically? Why?

TW 122. Explain why the number $\log_{10} 70$ must be between 1 and 2.

123. Graph both equations using the same set of axes:

$$y = \left(\tfrac{3}{2}\right)^x, \qquad y = \log_{3/2} x.$$

Graph by hand.

124. $y = \log_2 (x - 1)$

125. $y = \log_3 |x + 1|$

Solve.

126. $|\log_3 x| = 2$

127. $\log_4 (3x - 2) = 2$

128. $\log_8 (2x + 1) = -1$

129. $\log_{10} (x^2 + 21x) = 2$

Simplify.

130. $\log_{1/4} \frac{1}{64}$

131. $\log_{1/5} 25$

132. $\log_{81} 3 \cdot \log_3 81$

133. $\log_{10} (\log_4 (\log_3 81))$

134. $\log_2 (\log_2 (\log_4 256))$

135. Show that $b^x = b^y$ is *not* equivalent to $x = y$ for $b = 0$ or $b = 1$.

TW 136. If $\log_b a = x$, does it follow that $\log_a b = 1/x$? Why or why not?

Try Exercise Answers: Section 12.3

9. 3 **35.** 7 **37.**

$y = \log_{10} x$

59.

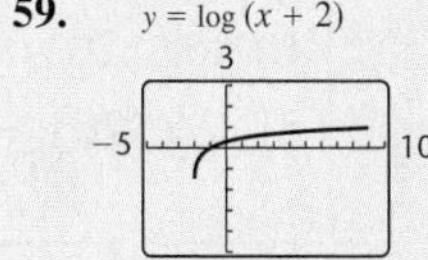

65. $10^x = 8$ **81.** $2 = \log_{10} 100$ **97.** 9 **99.** 3

12.4 Properties of Logarithmic Functions

- Logarithms of Products
- Logarithms of Powers
- Logarithms of Quotients
- Using the Properties Together

Logarithmic functions are important in many applications and in more advanced mathematics. We now establish some basic properties that are useful in manipulating expressions involving logarithms.

The following Interactive Discovery explores some of these properties. In this section, we will state and prove the results you may observe. As their proofs reveal, the properties of logarithms are related to the properties of exponents.

Interactive Discovery

For each of the following, use a calculator to determine which is the equivalent logarithmic expression.

1. $\log(20 \cdot 5)$

a) $\log(20) \cdot \log(5)$

b) $5 \cdot \log(20)$

c) $\log(20) + \log(5)$

2. $\log(20^5)$

a) $\log(20) \cdot \log(5)$

b) $5 \cdot \log(20)$

c) $\log(20) - \log(5)$

3. $\log\left(\frac{20}{5}\right)$

a) $\log(20) \cdot \log(5)$

b) $\log(20)/\log(5)$

c) $\log(20) - \log(5)$

4. $\log\left(\frac{1}{20}\right)$

a) $-\log(20)$

b) $\log(1)/\log(20)$

c) $\log(1) \cdot \log(20)$

LOGARITHMS OF PRODUCTS

The first property we discuss is related to the product rule for exponents: $a^m \cdot a^n = a^{m+n}$. Its proof appears immediately after Example 2.

The Product Rule for Logarithms For any positive numbers M, N, and a $(a \neq 1)$,

$$\log_a(MN) = \log_a M + \log_a N.$$

(The logarithm of a product is the sum of the logarithms of the factors.)

EXAMPLE 1 Express as an equivalent expression that is a sum of logarithms: $\log_2(4 \cdot 16)$.

SOLUTION We have

$$\log_2(4 \cdot 16) = \log_2 4 + \log_2 16. \qquad \text{Using the product rule for logarithms}$$

As a check, note that

$$\log_2 (4 \cdot 16) = \log_2 64 = 6 \qquad 2^6 = 64$$

and that

$$\log_2 4 + \log_2 16 = 2 + 4 = 6. \qquad 2^2 = 4 \text{ and } 2^4 = 16$$

Try Exercise 7.

EXAMPLE 2 Express as an equivalent expression that is a single logarithm: $\log_b 7 + \log_b 5$.

SOLUTION We have

$$\begin{aligned} \log_b 7 + \log_b 5 &= \log_b (7 \cdot 5) && \text{Using the product rule for logarithms} \\ &= \log_b 35. \end{aligned}$$

Try Exercise 13.

STUDY TIP

Find Your Pace

When new material seems challenging to you, do not focus on how quickly or slowly you absorb the concepts. We each move at our own speed when it comes to digesting new material.

A Proof of the Product Rule. Let $\log_a M = x$ and $\log_a N = y$. Converting to exponential equations, we have $a^x = M$ and $a^y = N$.

Now we multiply the left sides of both equations and the right sides of both equations to obtain

$$MN = a^x \cdot a^y, \quad \text{or} \quad MN = a^{x+y}.$$

Converting back to a logarithmic equation, we get

$$\log_a (MN) = x + y.$$

Recalling what x and y represent, we get

$$\log_a (MN) = \log_a M + \log_a N.$$

LOGARITHMS OF POWERS

The second basic property is related to the power rule for exponents: $(a^m)^n = a^{mn}$. Its proof follows Example 3.

The Power Rule for Logarithms For any positive numbers M and a $(a \neq 1)$, and any real number p,

$$\log_a M^p = p \cdot \log_a M.$$

(The logarithm of a power of M is the exponent times the logarithm of M.)

To better understand the power rule, note that

$$\log_a M^3 = \log_a (M \cdot M \cdot M) = \log_a M + \log_a M + \log_a M = 3 \log_a M.$$

EXAMPLE 3 Use the power rule for logarithms to write an equivalent expression that is a product: **(a)** $\log_a 9^{-5}$; **(b)** $\log_7 \sqrt[3]{x}$.

SOLUTION

a) $\log_a 9^{-5} = -5 \log_a 9$ — Using the power rule for logarithms

b) $\log_7 \sqrt[3]{x} = \log_7 x^{1/3}$ — Writing exponential notation

$= \frac{1}{3} \log_7 x$ — Using the power rule for logarithms

Try Exercise 17.

Student Notes

Without understanding and *remembering* the rules of this section, it will be extremely difficult to solve the equations of Section 12.6.

A Proof of the Power Rule. Let $x = \log_a M$. The equivalent exponential equation is $a^x = M$. Raising both sides to the pth power, we get

$$(a^x)^p = M^p, \quad \text{or} \quad a^{xp} = M^p. \qquad \textbf{Multiplying exponents}$$

Converting back to a logarithmic equation gives us

$$\log_a M^p = xp.$$

But $x = \log_a M$, so substituting, we have

$$\log_a M^p = (\log_a M)p = p \cdot \log_a M.$$

LOGARITHMS OF QUOTIENTS

The third property that we study is similar to the quotient rule for exponents: $\frac{a^m}{a^n} = a^{m-n}$. Its proof follows Example 5.

The Quotient Rule for Logarithms For any positive numbers M, N, and a $(a \neq 1)$,

$$\log_a \frac{M}{N} = \log_a M - \log_a N.$$

(The logarithm of a quotient is the logarithm of the dividend minus the logarithm of the divisor.)

To better understand the quotient rule, note that

$$\log_2 \tfrac{8}{32} = \log_2 \tfrac{1}{4} = -2$$

and $\log_2 8 - \log_2 32 = 3 - 5 = -2.$

EXAMPLE 4 Express as an equivalent expression that is a difference of logarithms: $\log_t (6/U)$.

SOLUTION

$$\log_t \frac{6}{U} = \log_t 6 - \log_t U \qquad \textbf{Using the quotient rule for logarithms}$$

Try Exercise 23.

EXAMPLE 5 Express as an equivalent expression that is a single logarithm: $\log_b 17 - \log_b 27$.

SOLUTION

$$\log_b 17 - \log_b 27 = \log_b \frac{17}{27} \qquad \textbf{Using the quotient rule for logarithms "in reverse"}$$

Try Exercise 27.

A Proof of the Quotient Rule. Our proof uses both the product rule and the power rule:

$$\log_a \frac{M}{N} = \log_a MN^{-1} \quad \text{Rewriting } \frac{M}{N} \text{ as } MN^{-1}$$
$$= \log_a M + \log_a N^{-1} \quad \text{Using the product rule for logarithms}$$
$$= \log_a M + (-1)\log_a N \quad \text{Using the power rule for logarithms}$$
$$= \log_a M - \log_a N.$$

USING THE PROPERTIES TOGETHER

EXAMPLE 6 Express as an equivalent expression, using the individual logarithms of x, y, and z.

a) $\log_b \frac{x^3}{yz}$ **b)** $\log_a \sqrt[4]{\frac{xy}{z^3}}$

SOLUTION

a) $\log_b \frac{x^3}{yz} = \log_b x^3 - \log_b yz$ **Using the quotient rule for logarithms**

$= 3\log_b x - \log_b yz$ **Using the power rule for logarithms**

$= 3\log_b x - (\log_b y + \log_b z)$ **Using the product rule for logarithms. Because of the subtraction, parentheses are essential.**

$= 3\log_b x - \log_b y - \log_b z$ **Using the distributive law**

b) $\log_a \sqrt[4]{\frac{xy}{z^3}} = \log_a \left(\frac{xy}{z^3}\right)^{1/4}$ **Writing exponential notation**

$= \frac{1}{4} \cdot \log_a \frac{xy}{z^3}$ **Using the power rule for logarithms**

$= \frac{1}{4}(\log_a xy - \log_a z^3)$ **Using the quotient rule for logarithms. Parentheses are important.**

$= \frac{1}{4}(\log_a x + \log_a y - 3\log_a z)$ **Using the product rule and the power rule for logarithms**

CAUTION! Because the product rule and the quotient rule replace one term with two, parentheses are often necessary, as in Example 6.

Try Exercise 39.

EXAMPLE 7 Express as an equivalent expression that is a single logarithm.

a) $\frac{1}{2}\log_a x - 7\log_a y + \log_a z$ **b)** $\log_a \frac{b}{\sqrt{x}} + \log_a \sqrt{bx}$

SOLUTION

a) $\frac{1}{2}\log_a x - 7\log_a y + \log_a z$

$= \log_a x^{1/2} - \log_a y^7 + \log_a z$ **Using the power rule for logarithms**

$= \left(\log_a \sqrt{x} - \log_a y^7\right) + \log_a z$ **Using parentheses to emphasize the order of operations; $x^{1/2} = \sqrt{x}$**

$= \log_a \frac{\sqrt{x}}{y^7} + \log_a z$ **Using the quotient rule for logarithms**

$= \log_a \frac{z\sqrt{x}}{y^7}$ **Using the product rule for logarithms**

b) $\log_a \frac{b}{\sqrt{x}} + \log_a \sqrt{bx} = \log_a \frac{b \cdot \sqrt{bx}}{\sqrt{x}}$ **Using the product rule for logarithms**

$= \log_a b\sqrt{b}$ **Removing a factor equal to 1: $\frac{\sqrt{x}}{\sqrt{x}} = 1$**

$= \log_a b^{3/2}$, or $\frac{3}{2}\log_a b$ **Since $b\sqrt{b} = b^1 \cdot b^{1/2}$**

Try Exercise 51.

If we know the logarithms of two different numbers (with the same base), the properties allow us to calculate other logarithms.

EXAMPLE 8 Given $\log_a 2 = 0.431$ and $\log_a 3 = 0.683$, use the properties of logarithms to calculate a numerical value for each of the following, if possible.

a) $\log_a 6$ **b)** $\log_a \frac{2}{3}$ **c)** $\log_a 81$

d) $\log_a \frac{1}{3}$ **e)** $\log_a (2a)$ **f)** $\log_a 5$

SOLUTION

a) $\log_a 6 = \log_a (2 \cdot 3) = \log_a 2 + \log_a 3$ **Using the product rule for logarithms**

$= 0.431 + 0.683 = 1.114$

Check: $a^{1.114} = a^{0.431} \cdot a^{0.683} = 2 \cdot 3 = 6$

b) $\log_a \frac{2}{3} = \log_a 2 - \log_a 3$ **Using the quotient rule for logarithms**

$= 0.431 - 0.683 = -0.252$

c) $\log_a 81 = \log_a 3^4 = 4\log_a 3$ **Using the power rule for logarithms**

$= 4(0.683) = 2.732$

d) $\log_a \frac{1}{3} = \log_a 1 - \log_a 3$ **Using the quotient rule for logarithms**

$= 0 - 0.683 = -0.683$

e) $\log_a (2a) = \log_a 2 + \log_a a$ **Using the product rule for logarithms**

$= 0.431 + 1 = 1.431$

f) $\log_a 5$ *cannot be found using these properties.* ($\log_a 5 \neq \log_a 2 + \log_a 3$)

Try Exercise 55.

A final property follows from the product rule: Since $\log_a a^k = k \log_a a$, and $\log_a a = 1$, we have $\log_a a^k = k$.

The Logarithm of the Base to an Exponent For any base a,

$$\log_a a^k = k.$$

(The logarithm, base a, of a to an exponent is the exponent.)

This property also follows from the definition of logarithm: k is the exponent to which you raise a in order to get a^k.

EXAMPLE 9 Simplify: **(a)** $\log_3 3^7$; **(b)** $\log_{10} 10^{-5.2}$.

SOLUTION

a) $\log_3 3^7 = 7$ **7 is the exponent to which you raise 3 in order to get 3^7.**

b) $\log_{10} 10^{-5.2} = -5.2$

Try Exercise 65.

We summarize the properties covered in this section as follows.

For any positive numbers M, N, and a $(a \neq 1)$:

$$\log_a (MN) = \log_a M + \log_a N; \qquad \log_a M^p = p \cdot \log_a M;$$

$$\log_a \frac{M}{N} = \log_a M - \log_a N; \qquad \log_a a^k = k.$$

CAUTION! Keep in mind that, in general,

$\log_a (M + N) \neq \log_a M + \log_a N,$
$\log_a (M - N) \neq \log_a M - \log_a N,$
$\log_a (MN) \neq (\log_a M)(\log_a N),$
$\log_a \frac{M}{N} \neq \frac{\log_a M}{\log_a N}.$

12.4 Exercise Set

FOR EXTRA HELP MyMathLab MathXL PRACTICE  WATCH  DOWNLOAD  READ 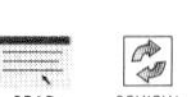 REVIEW

Concept Reinforcement *In each of Exercises 1–6, match the expression with an equivalent expression from the column on the right.*

1. ___ $\log_7 20$		**a)** $\log_7 5 - \log_7 4$
2. ___ $\log_7 5^4$		**b)** 1
3. ___ $\log_7 \frac{5}{4}$		**c)** 0
4. ___ $\log_7 7$		**d)** $\log_7 30$
5. ___ $\log_7 1$		**e)** $\log_7 5 + \log_7 4$
6. ___ $\log_7 5 + \log_7 6$		**f)** $4 \log_7 5$

Express as an equivalent expression that is a sum of logarithms.

7. $\log_3 (81 \cdot 27)$ **8.** $\log_2 (16 \cdot 32)$

9. $\log_4 (64 \cdot 16)$ **10.** $\log_5 (25 \cdot 125)$

11. $\log_c (rst)$ **12.** $\log_t (3ab)$

Express as an equivalent expression that is a single logarithm.

13. $\log_a 5 + \log_a 14$ **14.** $\log_b 65 + \log_b 2$

15. $\log_c t + \log_c y$ **16.** $\log_t H + \log_t M$

Express as an equivalent expression that is a product.

17. $\log_a r^8$ **18.** $\log_b t^5$

19. $\log_2 y^{1/3}$ **20.** $\log_{10} x^{1/2}$

21. $\log_b C^{-3}$ **22.** $\log_c M^{-5}$

Express as an equivalent expression that is a difference of two logarithms.

23. $\log_2 \frac{25}{13}$ **24.** $\log_3 \frac{23}{9}$

25. $\log_b \frac{m}{n}$ **26.** $\log_a \frac{y}{x}$

Express as an equivalent expression that is a single logarithm.

27. $\log_a 17 - \log_a 6$ **28.** $\log_b 32 - \log_b 7$

29. $\log_b 36 - \log_b 4$ **30.** $\log_a 26 - \log_a 2$

31. $\log_a x - \log_a y$ **32.** $\log_b c - \log_b d$

Express as an equivalent expression, using the individual logarithms of w, x, y, and z.

33. $\log_a (xyz)$ **34.** $\log_a (wxy)$

35. $\log_a (x^3z^4)$ **36.** $\log_a (x^2y^5)$

37. $\log_a (x^2y^{-2}z)$ **38.** $\log_a (xy^2z^{-3})$

39. $\log_a \frac{x^4}{y^3z}$ **40.** $\log_a \frac{x^4}{yz^2}$

41. $\log_b \frac{xy^2}{wz^3}$ **42.** $\log_b \frac{w^2x}{y^3z}$

43. $\log_a \sqrt{\frac{x^7}{y^5z^8}}$ **44.** $\log_c \sqrt[3]{\frac{x^4}{y^3z^2}}$

45. $\log_a \sqrt[3]{\frac{x^6y^3}{a^2z^7}}$ **46.** $\log_a \sqrt[4]{\frac{x^8y^{12}}{a^3z^5}}$

Express as an equivalent expression that is a single logarithm and, if possible, simplify.

47. $8 \log_a x + 3 \log_a z$

48. $2 \log_b m + \frac{1}{2} \log_b n$

49. $\log_a x^2 - 2 \log_a \sqrt{x}$

50. $\log_a \frac{a}{\sqrt{x}} - \log_a \sqrt{ax}$

51. $\frac{1}{2} \log_a x + 5 \log_a y - 2 \log_a x$

52. $\log_a (2x) + 3(\log_a x - \log_a y)$

53. $\log_a (x^2 - 4) - \log_a (x + 2)$

54. $\log_a (2x + 10) - \log_a (x^2 - 25)$

Given $\log_b 3 = 0.792$ and $\log_b 5 = 1.161$. If possible, use the properties of logarithms to calculate numerical values for each of the following.

55. $\log_b 15$ **56.** $\log_b \frac{5}{3}$

57. $\log_b \frac{3}{5}$ **58.** $\log_b \frac{1}{3}$

59. $\log_b \frac{1}{5}$ **60.** $\log_b \sqrt{b}$

61. $\log_b \sqrt{b^3}$ **62.** $\log_b (3b)$

63. $\log_b 8$ **64.** $\log_b 45$

Simplify.

65. $\log_t t^7$ **66.** $\log_p p^4$

67. $\log_e e^m$ **68.** $\log_Q Q^{-2}$

Use the properties of logarithms to find each of the following.

Aha! **69.** $\log_5 (125 \cdot 625)$ **70.** $\log_3 (9 \cdot 81)$

71. $\log_2 16^5$ **72.** $\log_3 27^7$

TW **73.** Explain the difference between the phrases "the logarithm of a quotient" and "a quotient of logarithms."

TW **74.** How could you convince someone that

$$\log_a c \neq \log_c a?$$

SKILL REVIEW

To prepare for Section 12.5, review graphing functions and finding domains of functions.

Graph.

75. $f(x) = \sqrt{x} - 3$ [10.1]

76. $g(x) = \sqrt[3]{x} + 1$ [10.1]

77. $g(x) = x^3 + 2$ [3.2], [3.8]

78. $f(x) = 1 - x^2$ [11.7]

Find the domain of each function.

79. $f(x) = \frac{x - 3}{x + 7}$ [9.1]

80. $f(x) = \frac{x}{(x - 2)(x + 3)}$ [9.1]

81. $g(x) = \sqrt{10 - x}$ [10.1]

82. $g(x) = |x^2 - 6x + 7|$ [3.2], [3.8]

SYNTHESIS

TW **83.** A student *incorrectly* reasons that

$$\log_b \frac{1}{x} = \log_b \frac{x}{xx}$$
$$= \log_b x - \log_b x + \log_b x = \log_b x.$$

What mistake has the student made?

TW **84.** Is it true that $\log_a x + \log_b x = \log_{ab} x$? Why or why not?

Express as an equivalent expression that is a single logarithm and, if possible, simplify.

85. $\log_a (x^8 - y^8) - \log_a (x^2 + y^2)$

86. $\log_a (x + y) + \log_a (x^2 - xy + y^2)$

Express as an equivalent expression that is a sum or a difference of logarithms and, if possible, simplify.

87. $\log_a \sqrt{1 - s^2}$ **88.** $\log_a \frac{c - d}{\sqrt{c^2 - d^2}}$

89. If $\log_a x = 2$, $\log_a y = 3$, and $\log_a z = 4$, what is

$$\log_a \frac{\sqrt[3]{x^2 z}}{\sqrt[3]{y^2 z^{-2}}}?$$

90. If $\log_a x = 2$, what is $\log_a (1/x)$?

91. If $\log_a x = 2$, what is $\log_{1/a} x$?

Solve.

92. $\log_{10} 2000 - \log_{10} x = 3$

93. $\log_2 80 + \log_2 x = 5$

Classify each of the following as true or false. Assume a, x, P, and Q > 0, a ≠ 1.

94. $\log_a \left(\frac{P}{Q}\right)^x = x \log_a P - \log_a Q$

95. $\log_a (Q + Q^2) = \log_a Q + \log_a (Q + 1)$

96. Use graphs to show that

$$\log x^2 \neq \log x \cdot \log x.$$

(*Note*: log means $\log_{10}$.)

Try Exercise Answers: Section 12.4

7. $\log_3 81 + \log_3 27$ **13.** $\log_a (5 \cdot 14)$, or $\log_a 70$ **17.** $8 \log_a r$ **23.** $\log_2 25 - \log_2 13$ **27.** $\log_a \frac{17}{6}$ **39.** $4 \log_a x - 3 \log_a y - \log_a z$ **51.** $\log_a \frac{y^5}{x^{3/2}}$ **55.** 1.953 **65.** 7

Mid-Chapter Review

We use the following properties to simplify expressions and to rewrite equivalent logarithmic and exponential equations.

$\log_a x = m$ means $x = a^m$.	$\log_a a^k = k$
$\log_a (MN) = \log_a M + \log_a N$	$\log_a a = 1$
$\log_a \frac{M}{N} = \log_a M - \log_a N$	$\log_a 1 = 0$
$\log_a M^p = p \cdot \log_a M$	$\log x = \log_{10} x$

GUIDED SOLUTIONS

1. Find a formula for the inverse of $f(x) = 2x - 5$. [12.1]

Solution

$y = 2x - 5$

$\square = 2\square - 5$ **Interchanging x and y**

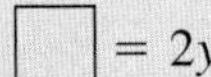

$\square = 2y$

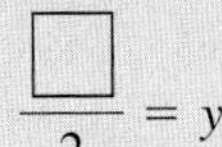

$\frac{\square}{2} = y$ **Solving for y**

$f^{-1}(x) = \square$

2. Solve: $\log_4 x = 1$. [12.3]

Solution

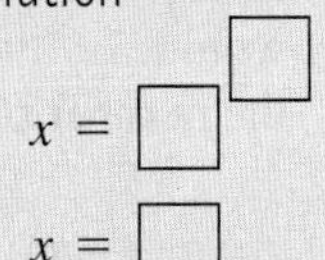

$x = \square^{\square}$ **Rewriting as an exponential equation**

$x = \square$

MIXED REVIEW

1. Find $(f \circ g)(x)$ if $f(x) = x^2 + 1$ and $g(x) = x - 5$. [12.1]

2. If $h(x) = \sqrt{5x - 3}$, find $f(x)$ and $g(x)$ such that $h(x) = (f \circ g)(x)$. Answers may vary. [12.1]

3. Find a formula for the inverse of $g(x) = 6 - x$. [12.1]

4. Graph by hand: $f(x) = 2^x + 3$. [12.2]

Simplify.

5. $\log_4 16$ [12.3]

6. $\log_5 \frac{1}{5}$ [12.3]

7. $\log_{100} 10$ [12.3]

8. $\log_b b$ [12.4]

9. $\log_8 8^{19}$ [12.4]

10. $\log_t 1$ [12.4]

Rewrite each of the following as an equivalent exponential equation.

11. $\log_x 3 = m$ [12.3]

12. $\log_2 1024 = 10$ [12.3]

Rewrite each of the following as an equivalent logarithmic equation.

13. $e^t = x$ [12.3]

14. $64^{2/3} = 16$ [12.3]

15. Express as an equivalent expression using $\log x$, $\log y$, and $\log z$:

$$\log \sqrt{\frac{x^2}{yz^3}}. \text{ [12.4]}$$

16. Express as an equivalent expression that is a single logarithm:

$$\log a - 2 \log b - \log c. \text{ [12.4]}$$

Solve. [12.3]

17. $\log_x 64 = 3$

18. $\log_3 x = -1$

19. $\log x = 5$

20. $\log_x 2 = \frac{1}{2}$

12.5 Natural Logarithms and Changing Bases

- The Base *e* and Natural Logarithms
- Changing Logarithmic Bases
- Graphs of Exponential Functions and Logarithmic Functions, Base *e*

We have already looked at base-10 logarithms, or common logarithms. Another logarithm base widely used today is an irrational number named *e*.

THE BASE *e* AND NATURAL LOGARITHMS

When interest is computed n times per year, the compound interest formula is

$$A = P\left(1 + \frac{r}{n}\right)^{nt},$$

where A is the amount that an initial investment P will be worth after t years at interest rate r. Suppose that \$1 is invested at 100% interest for 1 year (no bank would pay this). The preceding formula becomes a function A defined in terms of the number of compounding periods n:

$$A(n) = \left(1 + \frac{1}{n}\right)^n.$$

Interactive Discovery

1. What happens to the function values of $A(n) = \left(1 + \frac{1}{n}\right)^n$ as n gets larger?

Use the TABLE feature with Indpnt set to Ask to fill in the table below. Round each entry to six decimal places.

n	$A(n) = \left(1 + \frac{1}{n}\right)^n$
1 (compounded annually)	$\left(1 + \frac{1}{1}\right)^1$, or \$2.00
2 (compounded semiannually)	$\left(1 + \frac{1}{2}\right)^2$, or \$?
3	
4 (compounded quarterly)	
12 (compounded monthly)	
52 (compounded weekly)	
365 (compounded daily)	
8760 (compounded hourly)	

2. Which of the following statements appears to be true?

a) $A(n)$ gets very large as n gets very large.

b) $A(n)$ gets very small as n gets very large.

c) $A(n)$ approaches a certain number as n gets very large.

The numbers in the table approach a very important number in mathematics, called e. Because e is irrational, its decimal representation does not terminate or repeat.

The Number e $e \approx 2.7182818284\ldots$

Logarithms base e are called **natural logarithms,** or **Napierian logarithms,** in honor of John Napier (1550–1617), who first "discovered" logarithms.

The abbreviation "ln" is generally used with natural logarithms. Thus,

$\ln 53$ means $\log_e 53$.

Remember: $\log x$ means $\log_{10} x$, and $\ln x$ means $\log_e x$.

STUDY TIP

Is Your Answer Reasonable?

It is always a good idea—especially when using a calculator—to check that your answer is reasonable. It is easy for an incorrect calculation or keystroke to result in an answer that is clearly too big or too small.

Natural Logarithms

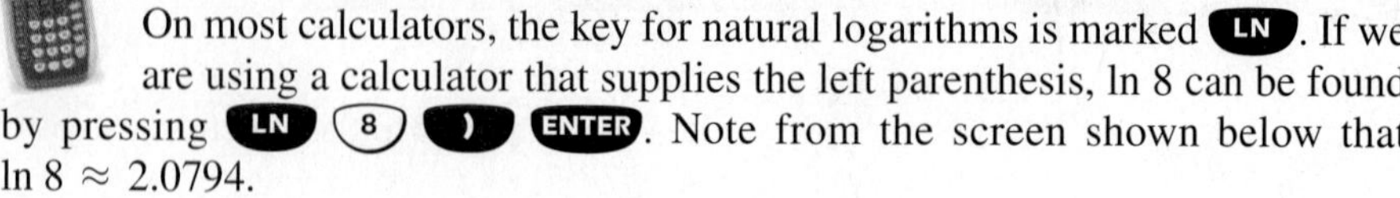

On most calculators, the key for natural logarithms is marked LN. If we are using a calculator that supplies the left parenthesis, ln 8 can be found by pressing LN 8) ENTER. Note from the screen shown below that $\ln 8 \approx 2.0794$.

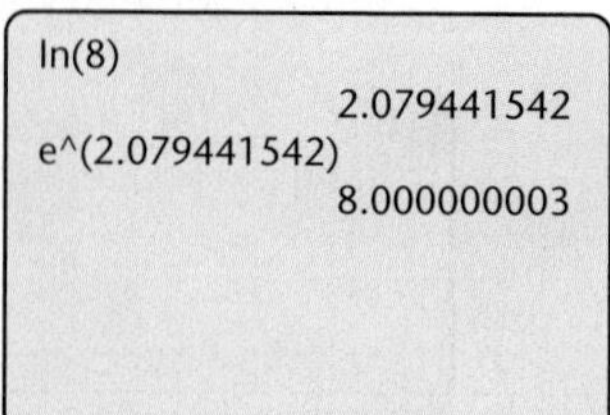

On many calculators, the LN key serves as the e^x key after the 2ND key is pressed. Some calculators automatically supply a left parenthesis before the exponent. If such a calculator is used,

$$e^{2.079441542}$$

is found by pressing e^x 2 . 0 7 9 4 4 1 5 4 2) ENTER. As shown on the screen above, $e^{2.079441542} \approx 8$, as expected, since $f(x) = e^x$ is the inverse function of $g(x) = \ln x$.

Your Turn

1. Approximate ln 5 to four decimal places. 1.6094
2. Approximate $\ln(5) + 1$ to four decimal places. Be sure that only 5 is in parentheses. 2.6094
3. Approximate $e^{2/3}$ to four decimal places. 1.9477
4. Approximate $\dfrac{e^2}{3}$ to four decimal places. Be sure that only 2 is in parentheses. 2.4630

Student Notes

Some calculators can find logarithms with any base, often through a logBASE(option in the MATH MATH submenu. You can fill in the blanks shown below, moving between blanks using the left and right arrow keys, and then press ENTER to find the logarithm.

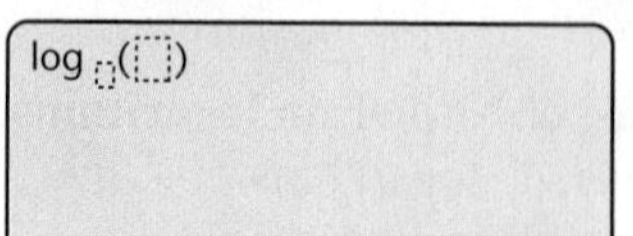

CHANGING LOGARITHMIC BASES

Most calculators can find both common logarithms and natural logarithms. To find a logarithm with some other base, a conversion formula is often needed.

The Change-of-Base Formula For any logarithmic bases a and b, and any positive number M,

$$\log_b M = \frac{\log_a M}{\log_a b}.$$

(To find the log, base b, we typically compute $\log M/\log b$ or $\ln M/\ln b$.)

A Proof of the Change-of-Base Formula. Let $x = \log_b M$. Then,

$$b^x = M \qquad \log_b M = x \text{ is equivalent to } b^x = M.$$

$$\log_a b^x = \log_a M \qquad \text{Taking the logarithm, base } a\text{, on both sides}$$

$$x \log_a b = \log_a M \qquad \text{Using the power rule for logarithms}$$

$$x = \frac{\log_a M}{\log_a b}. \qquad \text{Dividing both sides by } \log_a b$$

But at the outset we stated that $x = \log_b M$. Thus, by substitution, we have

$$\log_b M = \frac{\log_a M}{\log_a b}. \qquad \text{This is the change-of-base formula.}$$

EXAMPLE 1 Find $\log_5 8$ using the change-of-base formula.

SOLUTION We use the change-of-base formula with $a = 10$, $b = 5$, and $M = 8$:

$$\log_5 8 = \frac{\log_{10} 8}{\log_{10} 5} \qquad \text{Substituting into } \log_b M = \frac{\log_a M}{\log_a b}$$

$$\approx \frac{0.903089987}{0.6989700043} \qquad \text{Using LOG twice}$$

$$\approx 1.2920. \qquad$$ **When using a calculator, it is best not to round before dividing.**

Student Notes

The choice of the logarithm base a in the change-of-base formula should be either 10 or e so that the logarithms can be found using a calculator. Either choice will yield the same end result.

The figure below shows the computation using a graphing calculator. To check, note that $\ln 8/\ln 5 \approx 1.2920$. We can also use a calculator to verify that $5^{1.2920} \approx 8$.

```
log(8)/log(5)
                1.292029674
```

Try Exercise 29.

EXAMPLE 2 Find $\log_4 31$.

SOLUTION As shown in the check of Example 1, base e can also be used.

$$\log_4 31 = \frac{\log_e 31}{\log_e 4} \qquad \text{Substituting into } \log_b M = \frac{\log_a M}{\log_a b}$$

$$= \frac{\ln 31}{\ln 4} \approx \frac{3.433987204}{1.386294361} \qquad \text{Using LN twice}$$

$$\approx 2.4771 \qquad \textit{Check: } 4^{2.4771} \approx 31$$

```
ln(31)/ln(4)
                2.477098155
log(31)/log(4)
                2.477098155
4^Ans
                         31
```

The screen at left illustrates that we find the same solution using either natural logarithms or common logarithms.

Try Exercise 31.

GRAPHS OF EXPONENTIAL FUNCTIONS AND LOGARITHMIC FUNCTIONS, BASE *e*

EXAMPLE 3 Graph $f(x) = e^x$ and $g(x) = e^{-x}$ and state the domain and the range of f and g.

SOLUTION We use a calculator with an e^x key to find approximate values of e^x and e^{-x}. Using these values, we can graph the functions.

x	e^x	e^{-x}
0	1	1
1	2.7	0.4
2	7.4	0.1
−1	0.4	2.7
−2	0.1	7.4

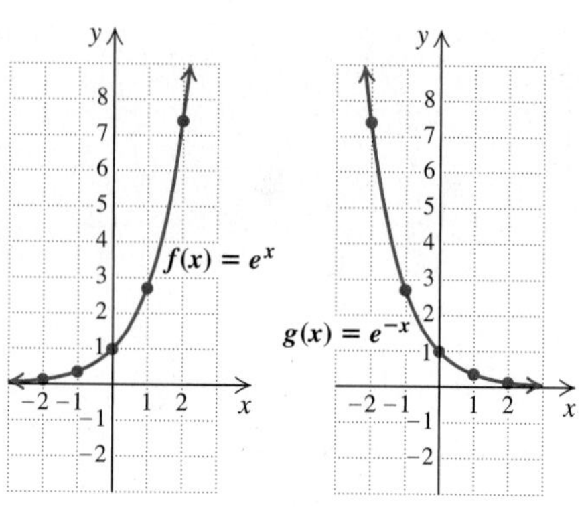

The domain of each function is ℝ and the range of each function is $(0, \infty)$.

Try Exercise 53.

EXAMPLE 4 Graph $f(x) = e^{-x} + 2$ and state the domain and the range of f.

SOLUTION To graph by hand, we find some solutions with a calculator, plot them, and then draw the graph. For example, $f(2) = e^{-2} + 2 \approx 2.1$. The graph is exactly like the graph of $g(x) = e^{-x}$, but is translated 2 units up.

To graph using a graphing calculator, we enter $y = e^{-x} + 2$ and choose a viewing window that shows more of the y-axis than the x-axis, since the function is exponential.

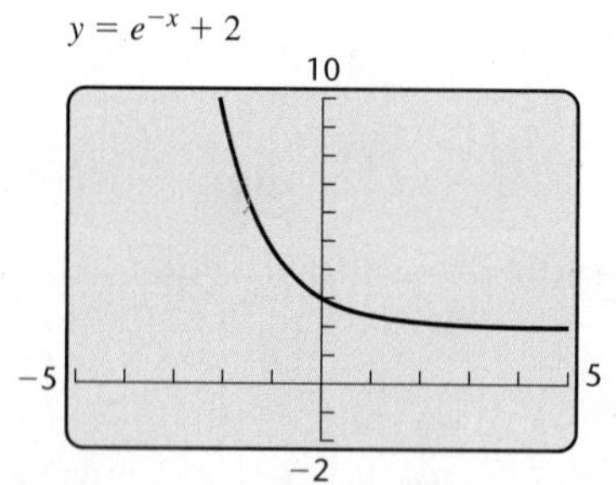

x	$e^{-x} + 2$
0	3
1	2.4
2	2.1
−1	4.7
−2	9.4

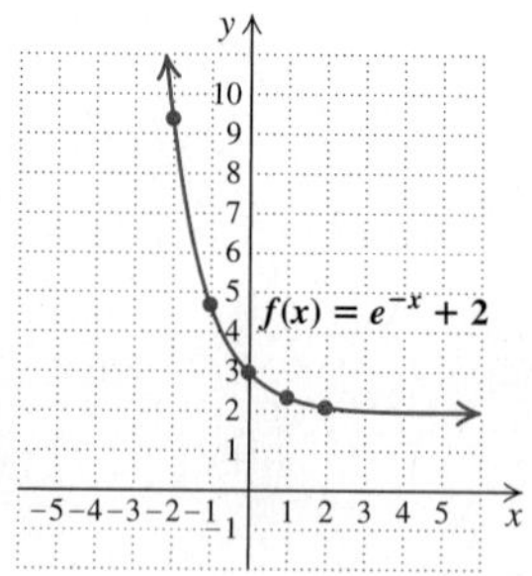

The domain of f is ℝ and the range is $(2, \infty)$.

Try Exercise 41.

EXAMPLE 5 Graph and state the domain and the range of each function.

a) $g(x) = \ln x$ **b)** $f(x) = \ln(x + 3)$

SOLUTION

a) To graph by hand, we find some solutions with a calculator and then draw the graph. As expected, the graph is a reflection across the line $y = x$ of the graph of $y = e^x$. We can also graph $y = \ln(x)$ using a graphing calculator. Since a logarithmic function is the inverse of an exponential function, we choose a viewing window that shows more of the x-axis than the y-axis.

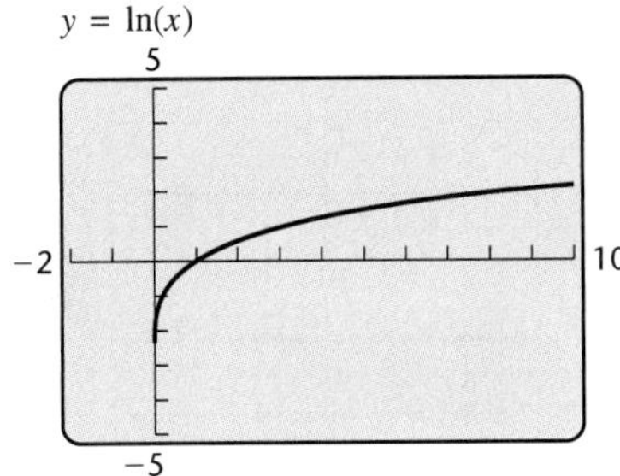

x	$\ln x$
1	0
4	1.4
7	1.9
0.5	−0.7

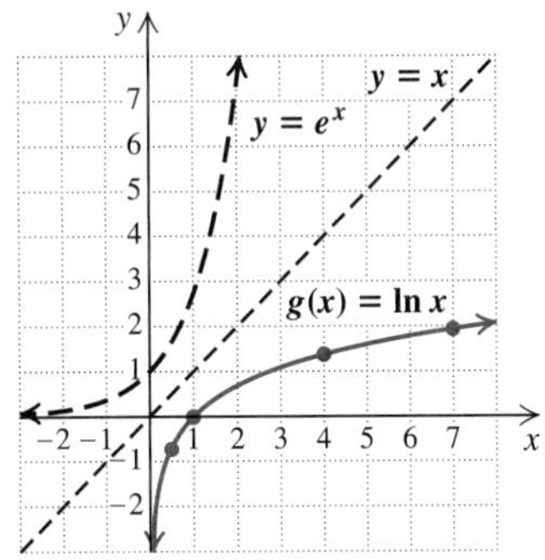

The domain of g is $(0, \infty)$ and the range is $\mathbb{R}$.

b) We find some solutions with a calculator, plot them, and draw the graph by hand. Note that the graph of $y = \ln(x + 3)$ is the graph of $y = \ln x$ translated 3 units to the left. To graph using a graphing calculator, we enter $y = \ln(x + 3)$.

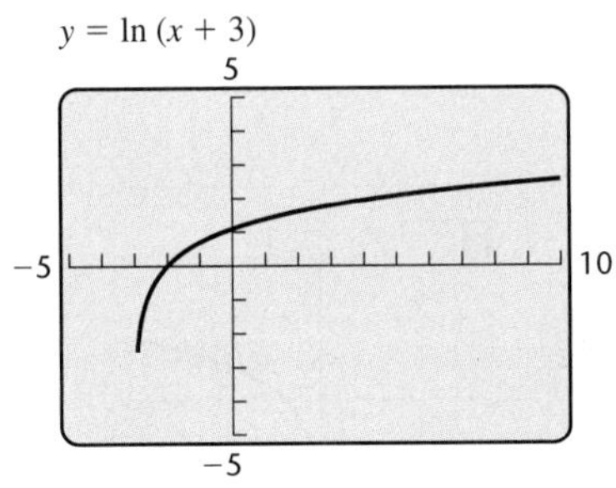

x	$\ln(x + 3)$
0	1.1
1	1.4
2	1.6
3	1.8
4	1.9
−1	0.7
−2	0
−2.5	−0.7

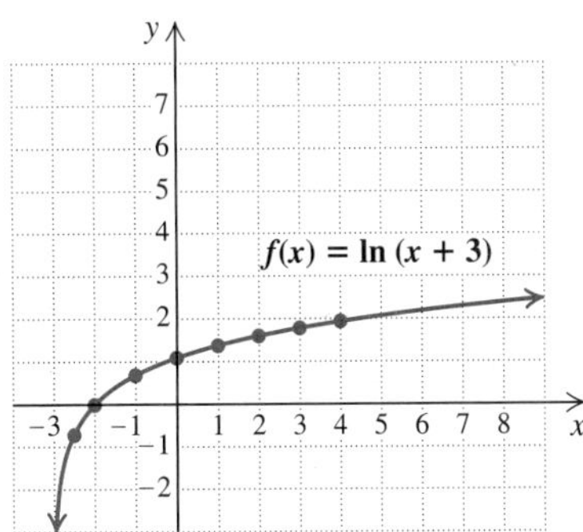

Since $x + 3$ must be positive, the domain of f is $(-3, \infty)$ and the range is $\mathbb{R}$.

Try Exercise 55.

Logarithmic functions with bases other than 10 or e can be graphed on a graphing calculator using the change-of-base formula.

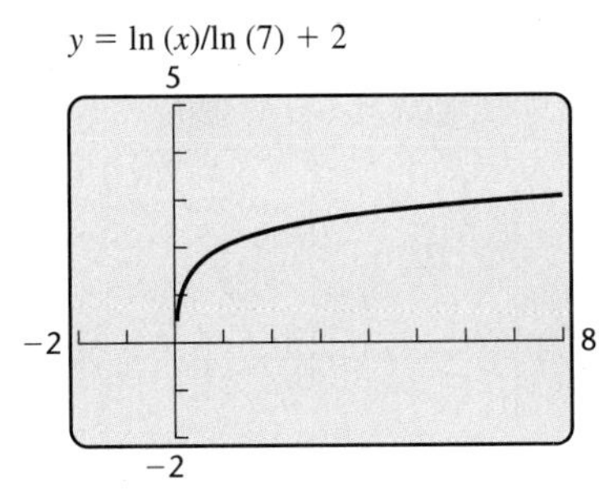

EXAMPLE 6 Graph: $f(x) = \log_7 x + 2$.

SOLUTION We use the change-of-base formula with natural logarithms. (We would get the same graph if we used common logarithms.)

$$f(x) = \log_7 x + 2 \qquad \text{Note that this is not } \log_7 (x + 2).$$

$$= \frac{\ln x}{\ln 7} + 2 \qquad \text{Using the change-of-base formula for } \log_7 x$$

We graph $y = \ln(x)/\ln(7) + 2$.

Try Exercise 67.

Visualizing for Success

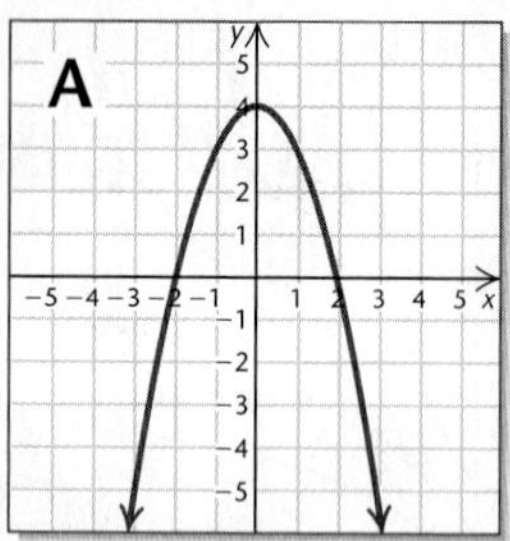

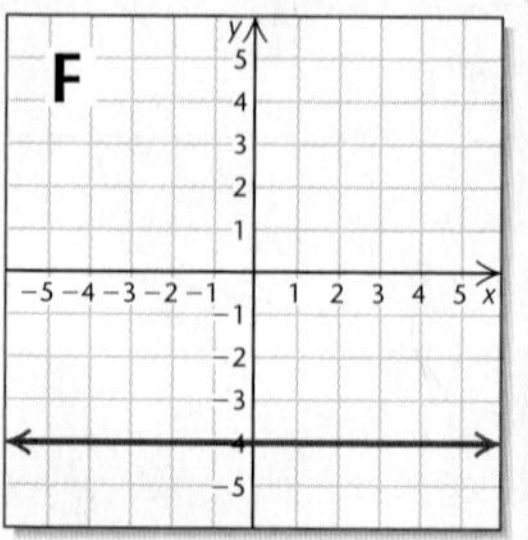

B

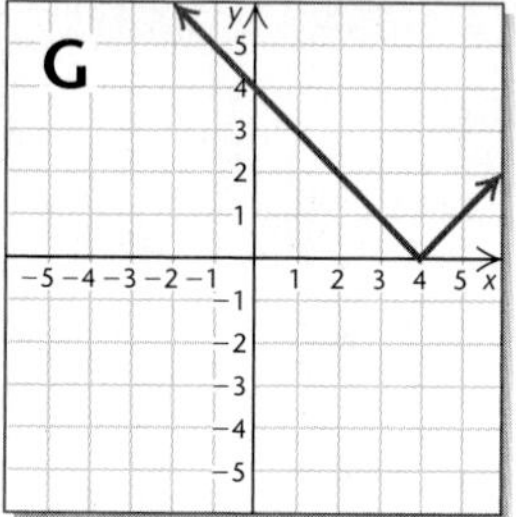

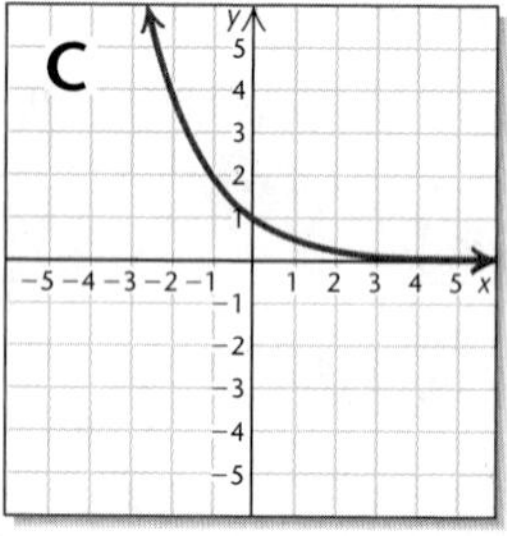

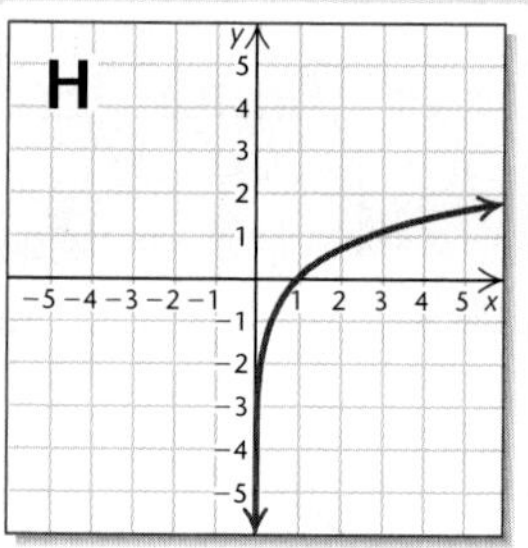

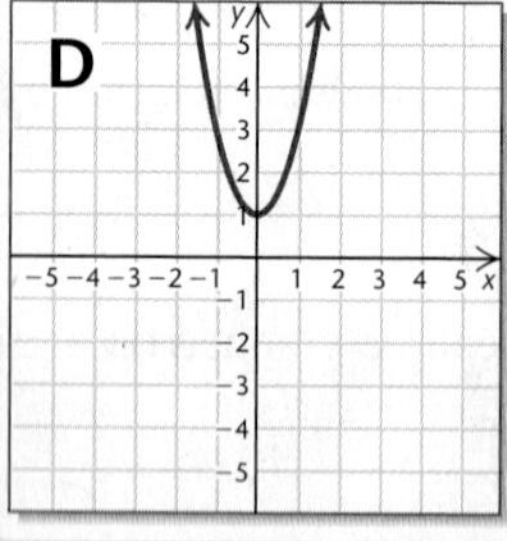

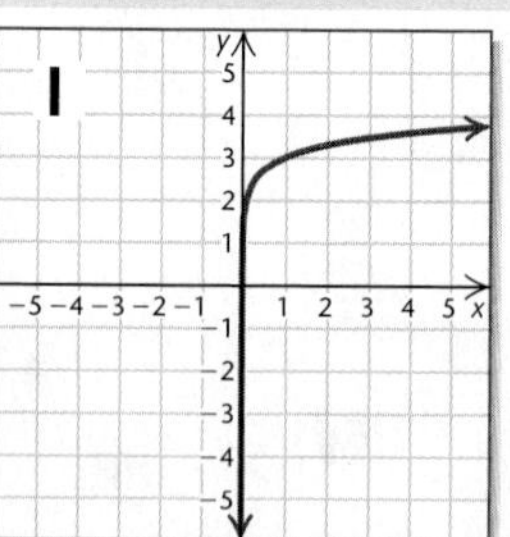

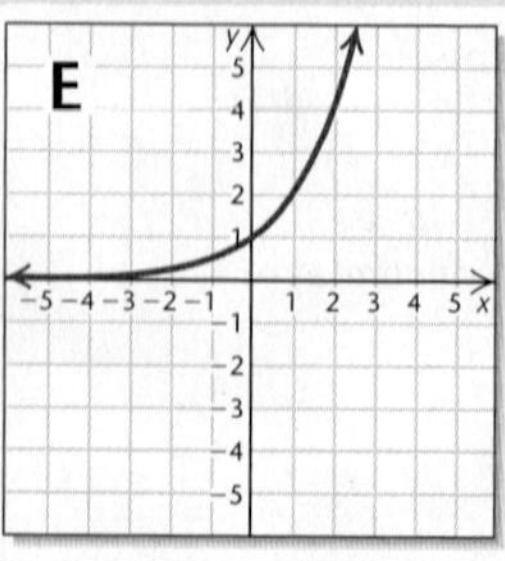

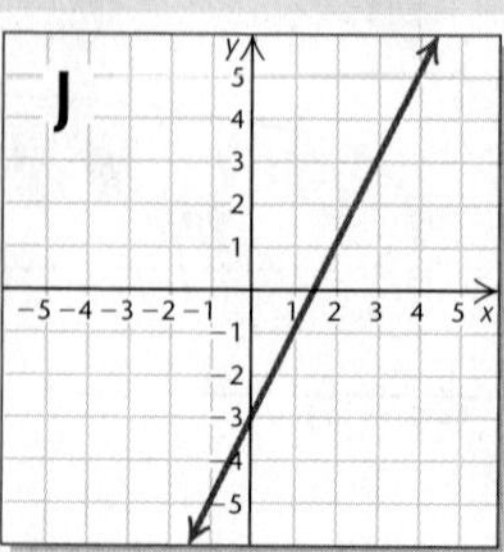

Match each function with its graph.

1. $f(x) = 2x - 3$

2. $f(x) = 2x^2 + 1$

3. $f(x) = \sqrt{x + 5}$

4. $f(x) = |4 - x|$

5. $f(x) = \ln x$

6. $f(x) = 2^{-x}$

7. $f(x) = -4$

8. $f(x) = \log x + 3$

9. $f(x) = 2^x$

10. $f(x) = 4 - x^2$

Answers on page A-49

An additional, animated version of this activity appears in MyMathLab. To use MyMathLab, you need a course ID and a student access code. Contact your instructor for more information.

12.5 Exercise Set

FOR EXTRA HELP MyMathLab MathXL PRACTICE  WATCH  DOWNLOAD  READ REVIEW

Concept Reinforcement *Classify each of the following statements as either true or false.*

1. The expression log 23 means $\log_{10} 23$.
2. The expression ln 7 means $\log_e 7$.
3. The number e is approximately 2.7.
4. The expressions log 9 and log 18/log 2 are equivalent.
5. The expressions log 9 and log 18 − log 2 are equivalent.
6. The expressions $\log_2 9$ and ln 9/ln 2 are equivalent.
7. The expressions ln 81 and 2 ln 9 are equivalent.
8. The domain of the function given by $f(x) = \ln(x + 2)$ is $(-2, \infty)$.
9. The range of the function given by $g(x) = e^x$ is $(0, \infty)$.
10. The range of the function given by $f(x) = \ln x$ is $(-\infty, \infty)$.

Use a calculator to find each of the following to four decimal places.

11. ln 5
12. ln 2
13. ln 0.0062
14. ln 0.00073
15. $\dfrac{\ln 2300}{0.08}$
16. $\dfrac{\ln 1900}{0.07}$
17. $e^{2.71}$
18. $e^{3.06}$
19. $e^{-3.49}$
20. $e^{-2.64}$
21. log 7
22. log 2
23. $\dfrac{\log 8200}{\log 2}$
24. $\dfrac{\log 5700}{\log 5}$
25. $\log \dfrac{3}{8}$
26. $\ln \frac{2}{3}$
27. ln (7) + 3
28. log (6) − 2

Find each of the following logarithms using the change-of-base formula. Round answers to the nearest ten-thousandth.

29. $\log_6 92$
30. $\log_3 78$
31. $\log_2 100$
32. $\log_7 100$
33. $\log_{0.5} 5$
34. $\log_{0.1} 3$
35. $\log_2 0.2$
36. $\log_2 0.08$
37. $\log_\pi 58$
38. $\log_\pi 200$

Graph by hand or using a graphing calculator and state the domain and the range of each function.

39. $f(x) = e^x$
40. $f(x) = e^{-x}$
41. $f(x) = e^x + 3$
42. $f(x) = e^x + 2$
43. $f(x) = e^x - 2$
44. $f(x) = e^x - 3$
45. $f(x) = 0.5e^x$
46. $f(x) = 2e^x$
47. $f(x) = 0.5e^{2x}$
48. $f(x) = 2e^{-0.5x}$
49. $f(x) = e^{x-3}$
50. $f(x) = e^{x-2}$
51. $f(x) = e^{x+2}$
52. $f(x) = e^{x+3}$
53. $f(x) = -e^x$
54. $f(x) = -e^{-x}$
55. $g(x) = \ln x + 1$
56. $g(x) = \ln x + 3$
57. $g(x) = \ln x - 2$
58. $g(x) = \ln x - 1$
59. $g(x) = 2 \ln x$
60. $g(x) = 3 \ln x$
61. $g(x) = -2 \ln x$
62. $g(x) = -\ln x$
63. $g(x) = \ln(x + 2)$
64. $g(x) = \ln(x + 1)$
65. $g(x) = \ln(x - 1)$
66. $g(x) = \ln(x - 3)$

Write an equivalent expression for the function that could be graphed using a graphing calculator. Then graph the function.

67. $f(x) = \log_5 x$
68. $f(x) = \log_3 x$
69. $f(x) = \log_2(x - 5)$
70. $f(x) = \log_5(2x + 1)$
71. $f(x) = \log_3 x + x$
72. $f(x) = \log_2 x - x + 1$

TW 73. Using a calculator, Aden *incorrectly* says that log 79 is between 4 and 5. How could you convince him, without using a calculator, that he is mistaken?

TW 74. Examine Exercise 73. What mistake do you believe Aden made?

SKILL REVIEW

To prepare for Section 12.6, review solving equations.

Solve.

75. $x^2 - 3x - 28 = 0$ [6.2]
76. $5x^2 - 7x = 0$ [6.1]
77. $17x - 15 = 0$ [2.2]
78. $\frac{5}{3} = 2t$ [2.2]

79. $(x - 5) \cdot 9 = 11$ [2.2]

80. $\dfrac{x+3}{x-3} = 7$ [7.6]

81. $x^{1/2} - 6x^{1/4} + 8 = 0$ [11.5]

82. $2y - 7\sqrt{y} + 3 = 0$ [11.5]

SYNTHESIS

TW **83.** In an attempt to solve $\ln x = 1.5$, Emma gets the following graph. How can Emma tell at a glance that she has made a mistake?

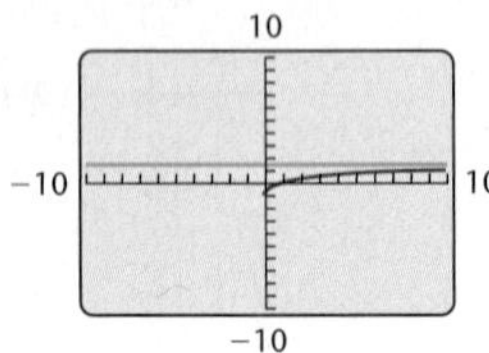

TW **84.** How would you explain to a classmate why $\log_2 5 = \log 5/\log 2$ *and* $\log_2 5 = \ln 5/\ln 2$?

Knowing only that $\log 2 \approx 0.301$ *and* $\log 3 \approx 0.477$, *find each of the following.*

85. $\log_6 81$

86. $\log_9 16$

87. $\log_{12} 36$

88. Find a formula for converting common logarithms to natural logarithms.

89. Find a formula for converting natural logarithms to common logarithms.

Solve for x.

90. $\log (275x^2) = 38$

91. $\log (492x) = 5.728$

92. $\dfrac{3.01}{\ln x} = \dfrac{28}{4.31}$

93. $\log 692 + \log x = \log 3450$

For each function given below, **(a)** *determine the domain and the range,* **(b)** *set an appropriate window, and* **(c)** *draw the graph.*

94. $f(x) = 7.4e^x \ln x$

95. $f(x) = 3.4 \ln x - 0.25e^x$

96. $f(x) = x \ln (x - 2.1)$

97. $f(x) = 2x^3 \ln x$

Try Exercise Answers: Section 12.5

29. 2.5237 **31.** 6.6439

41. Domain: $\mathbb{R}$; range: $(3, \infty)$

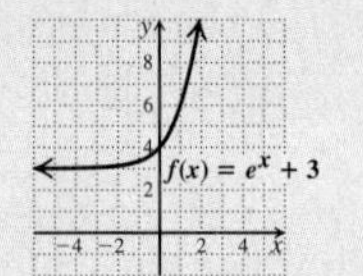

53. Domain: $\mathbb{R}$; range: $(-\infty, 0)$

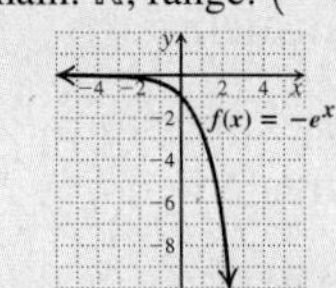

55. Domain: $(0, \infty)$; range: $\mathbb{R}$

67. $f(x) = \log (x)/\log (5)$, or $f(x) = \ln (x)/\ln (5)$

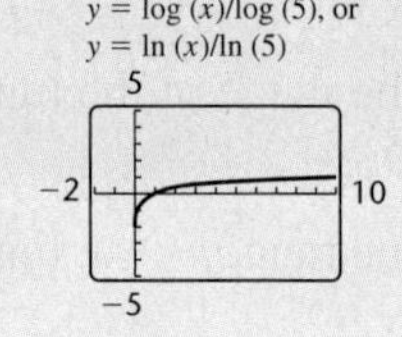

12.6 Solving Exponential and Logarithmic Equations

- Solving Exponential Equations
- Solving Logarithmic Equations

SOLVING EXPONENTIAL EQUATIONS

Equations with variables in exponents, such as $5^x = 12$ and $2^{7x} = 64$, are called **exponential equations.** In Section 12.3, we solved certain equations by using the principle of exponential equality. We restate that principle below.

The Principle of Exponential Equality For any real number b, where $b \neq -1, 0,$ or 1,

$$b^x = b^y \quad \text{is equivalent to} \quad x = y.$$

(Powers of the same base are equal if and only if the exponents are equal.)

EXAMPLE 1 Solve: $4^{3x} = 16$.

SOLUTION Note that $16 = 4^2$. Thus we can write each side as a power of the same base:

$$4^{3x} = 4^2 \quad \text{Rewriting 16 as a power of 4}$$
$$3x = 2 \quad \text{Since the base on each side is 4, the exponents are equal.}$$
$$x = \tfrac{2}{3}. \quad \text{Solving for } x$$

Since $4^{3x} = 4^{3(2/3)} = 4^2 = 16$, the answer checks. The solution is $\frac{2}{3}$.

Try Exercise 9.

In Example 1, we wrote both sides of the equation as powers of 4. When it seems impossible to write both sides of an equation as powers of the same base, we use the following principle and write an equivalent logarithmic equation.

The Principle of Logarithmic Equality For any logarithmic base a, and for $x, y > 0$,

$$x = y \quad \text{is equivalent to} \quad \log_a x = \log_a y.$$

(Two expressions are equal if and only if the logarithms of those expressions are equal.)

Because calculators can generally find only common or natural logarithms (without resorting to the change-of-base formula), we usually take the common or natural logarithm on both sides of the equation.

The principle of logarithmic equality, used together with the power rule for logarithms, allows us to solve equations with a variable in an exponent.

EXAMPLE 2 Solve: $7^{x-2} = 60$.

SOLUTION We have

$$7^{x-2} = 60$$

Take the logarithm of both sides.

$$\log 7^{x-2} = \log 60 \quad \text{Using the principle of logarithmic equality to take the common logarithm on both sides. Natural logarithms also would work.}$$

Use the power rule for logarithms.

$$(x - 2) \log 7 = \log 60 \quad \text{Using the power rule for logarithms}$$

$$x - 2 = \frac{\log 60}{\log 7}$$

CAUTION! This is *not* $\log 60 - \log 7$.

$$x = \frac{\log 60}{\log 7} + 2 \quad \text{Adding 2 to both sides}$$

Solve for x.

$$x \approx 4.1041. \quad \text{Using a calculator and rounding to four decimal places}$$

Check.

Since $7^{4.1041-2} \approx 60.0027$, we have a check. We can also note that since $7^{4-2} = 49$, we expect a solution greater than 4. The solution is $\dfrac{\log 60}{\log 7} + 2$, or approximately 4.1041.

Try Exercise 17.

EXAMPLE 3 Solve: $e^{0.06t} = 1500$.

ALGEBRAIC APPROACH

Since one side is a power of e, we take the *natural logarithm* on both sides:

$\ln e^{0.06t} = \ln 1500$ — **Taking the natural logarithm on both sides**

$0.06t = \ln 1500$ — **Finding the logarithm of the base to a power: $\log_a a^k = k$**

$t = \dfrac{\ln 1500}{0.06}$ — **Dividing both sides by 0.06**

$\approx 121.887.$ — **Using a calculator and rounding to three decimal places**

The solution is 121.887.

GRAPHICAL APPROACH

We graph $y_1 = e^{0.06x}$ and $y_2 = 1500$. Since $y_2 = 1500$, we choose a value for Ymax that is greater than 1500. It may require trial and error to choose appropriate units for the x-axis.

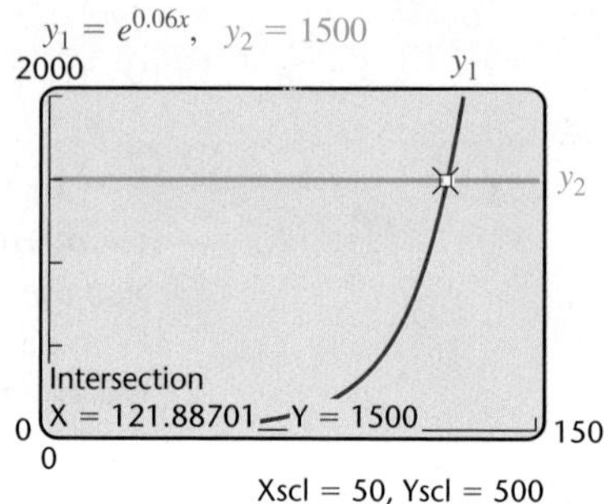

Rounded to three decimal places, the solution is 121.887.

Try Exercise 21.

To Solve an Equation of the Form $a^t = b$ for t

1. Take the logarithm (either natural or common) of both sides.
2. Use the power rule for exponents so that the variable is no longer written as an exponent.
3. Divide both sides by the coefficient of the variable to isolate the variable.
4. If appropriate, use a calculator to find an approximate solution in decimal form.

Some equations, like the one in Example 3, are readily solved algebraically. There are other exponential equations for which we do not have the tools to solve algebraically, but we can nevertheless solve graphically.

EXAMPLE 4 Solve: $xe^{3x-1} = 5$.

SOLUTION We graph $y_1 = xe^{3x-1}$ and $y_2 = 5$ and determine the coordinates of any points of intersection.

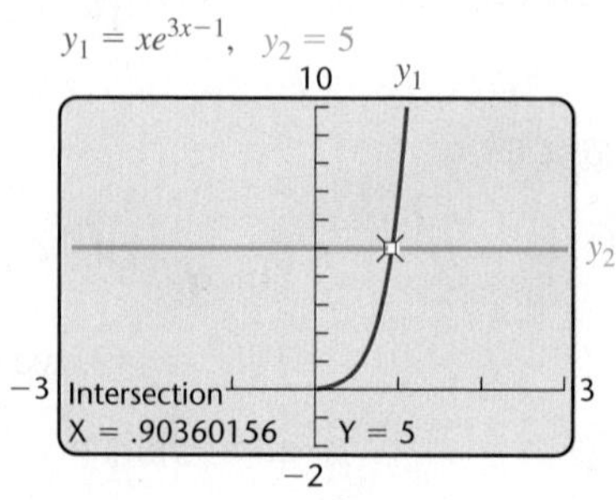

The x-coordinate of the point of intersection is approximately 0.90360156. Thus the solution is approximately 0.904.

Try Exercise 69.

STUDY TIP

Abbrevs. Cn Help U Go Fst

If you take notes and have trouble keeping up with your instructor, use abbreviations to speed up your work. Consider standard abbreviations like "Ex" for "Example," " ≈ " for "is approximately equal to," " ∴ " for "therefore," and " ⇒ " for "implies." Feel free to create your own abbreviations as well.

SOLVING LOGARITHMIC EQUATIONS

Equations containing logarithmic expressions are called **logarithmic equations.** We saw in Section 12.3 that certain logarithmic equations can be solved by writing an equivalent exponential equation.

EXAMPLE 5 Solve: **(a)** $\log_4 (8x - 6) = 3$; **(b)** $\ln (5x) = 27$.

SOLUTION

a) $\log_4 (8x - 6) = 3$

$4^3 = 8x - 6$ **Writing the equivalent exponential equation**

$64 = 8x - 6$

$70 = 8x$ **Adding 6 to both sides**

$x = \frac{70}{8}$, or $\frac{35}{4}$

Check:

$\log_4 (8x - 6) = 3$	
$\log_4 \left(8 \cdot \frac{35}{4} - 6\right)$	3
$\log_4 (2 \cdot 35 - 6)$	
$\log_4 64$	

$3 \stackrel{?}{=} 3$ TRUE

The solution is $\frac{35}{4}$.

b) $\ln (5x) = 27$ **Remember: $\ln (5x)$ means $\log_e (5x)$.**

$e^{27} = 5x$ **Writing the equivalent exponential equation**

$\frac{e^{27}}{5} = x$ **This is a very large number.**

The solution is $\frac{e^{27}}{5}$. The check is left to the student.

Student Notes

It is essential that you remember the properties of logarithms from Section 12.4. Consider reviewing the properties before attempting to solve equations similar to those in Examples 6–8.

Try Exercise 47.

Often the properties for logarithms are needed. The goal is to first write an equivalent equation in which the variable appears in just one logarithmic expression. We then isolate that expression and solve as in Example 5. Since logarithms of negative numbers are not defined, all possible solutions must be checked.

EXAMPLE 6 Solve: $\log x + \log (x - 3) = 1$.

SOLUTION To increase understanding, we write in the base, 10.

Find a single logarithm.

$\log_{10} x + \log_{10} (x - 3) = 1$

$\log_{10} [x(x - 3)] = 1$ **Using the product rule for logarithms to obtain a single logarithm**

Write an equivalent exponential equation.

$x(x - 3) = 10^1$ **Writing an equivalent exponential equation**

$x^2 - 3x = 10$

$x^2 - 3x - 10 = 0$

$(x + 2)(x - 5) = 0$ **Factoring**

$x + 2 = 0$ *or* $x - 5 = 0$ **Using the principle of zero products**

Solve.

$x = -2$ *or* $x = 5$

Check.

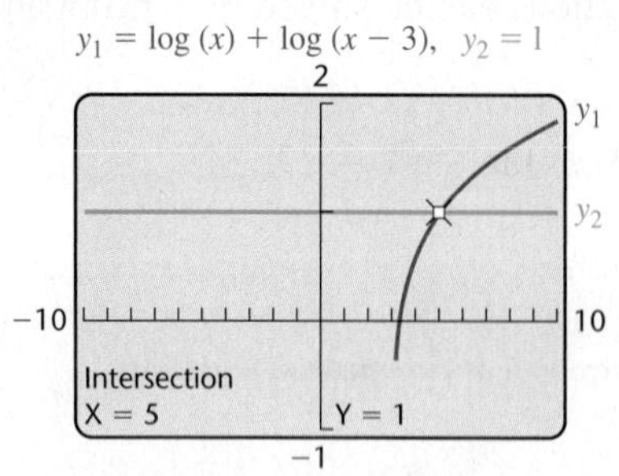

Check:

For -2:

$$\frac{\log x + \log(x-3) = 1}{\log(-2) + \log(-2-3) \stackrel{?}{=} 1} \quad \text{FALSE}$$

For 5:

$$\begin{array}{r|l} \log x + \log(x-3) & = 1 \\ \hline \log 5 + \log(5-3) & 1 \\ \log 5 + \log 2 & \\ \log 10 & \\ 1 & \stackrel{?}{=} 1 \quad \text{TRUE} \end{array}$$

The number -2 *does not check* because the logarithm of a negative number is undefined. Note from the graph at left that there is only one point of intersection. The solution is 5.

Try Exercise 49.

EXAMPLE 7 Solve: $\log_2(x+7) - \log_2(x-7) = 3$.

ALGEBRAIC APPROACH

We have

$$\log_2(x+7) - \log_2(x-7) = 3$$

$$\log_2 \frac{x+7}{x-7} = 3 \qquad \textbf{Using the quotient rule for logarithms to obtain a single logarithm}$$

$$\frac{x+7}{x-7} = 2^3 \qquad \textbf{Writing an equivalent exponential equation}$$

$$\frac{x+7}{x-7} = 8$$

$$x + 7 = 8(x-7) \qquad \textbf{Multiplying by the LCD, } x - 7$$

$$x + 7 = 8x - 56 \qquad \textbf{Using the distributive law}$$

$$63 = 7x$$

$$9 = x. \qquad \textbf{Dividing by 7}$$

GRAPHICAL APPROACH

We first use the change-of-base formula to write the base-2 logarithms using common logarithms. Then we graph and determine the coordinates of any points of intersection.

$y_1 = \log(x+7)/\log(2) - \log(x-7)/\log(2)$, $y_2 = 3$

10

y_2
y_1
−5
15
Intersection
X = 9
Y = 3
−1

The graphs intersect at $(9, 3)$. The solution is 9.

Check:

$$\begin{array}{r|l} \log_2(x+7) - \log_2(x-7) & = 3 \\ \hline \log_2(9+7) - \log_2(9-7) & 3 \\ \log_2 16 - \log_2 2 & \\ 4 - 1 & \\ 3 & \stackrel{?}{=} 3 \quad \text{TRUE} \end{array}$$

The solution is 9.

Try Exercise 65.

EXAMPLE 8 Solve: $\log_7(x+1) + \log_7(x-1) = \log_7 8$.

ALGEBRAIC APPROACH

We have

$$\log_7(x+1) + \log_7(x-1) = \log_7 8$$

$$\log_7[(x+1)(x-1)] = \log_7 8 \quad \text{Using the product rule for logarithms}$$

$$\log_7(x^2-1) = \log_7 8 \quad \text{Multiplying. Note that both sides are base-7 logarithms.}$$

$$x^2 - 1 = 8 \quad \text{Using the principle of logarithmic equality. Study this step carefully.}$$

$$x^2 - 9 = 0$$

$$(x-3)(x+3) = 0 \quad \text{Solving the quadratic equation}$$

$$x = 3 \quad or \quad x = -3.$$

The student should confirm that 3 checks but -3 does not. The solution is 3.

GRAPHICAL APPROACH

Using the change-of-base formula, we graph

$$y_1 = \log(x+1)/\log(7) + \log(x-1)/\log(7)$$

and

$$y_2 = \log(8)/\log(7).$$

$y_1 = \log(x+1)/\log(7) + \log(x-1)/\log(7)$,
$y_2 = \log(8)/\log(7)$

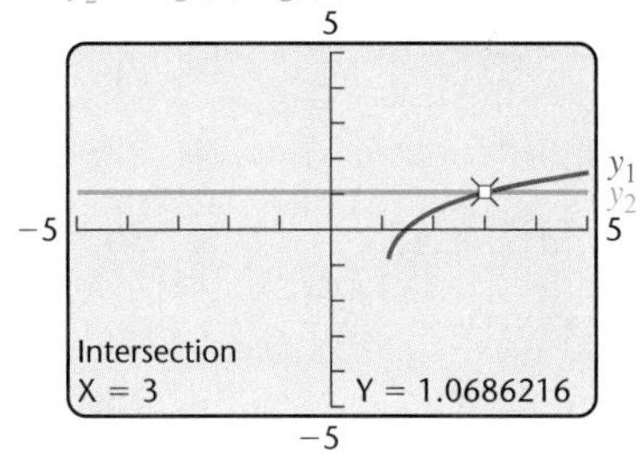

The graphs intersect at $(3, 1.0686216)$. The solution is 3.

Try Exercise 57.

12.6 Exercise Set

FOR EXTRA HELP **MyMathLab**

Concept Reinforcement *In each of Exercises 1–8, match the equation with an equivalent equation from the column on the right that could be the next step in the solution process.*

1. ____ $5^x = 3$
2. ____ $e^{5x} = 3$
3. ____ $\ln x = 3$
4. ____ $\log_x 5 = 3$
5. ____ $\log_5 x + \log_5(x-2) = 3$
6. ____ $\log_5 x - \log_5(x-2) = 3$
7. ____ $\ln x - \ln(x-2) = 3$
8. ____ $\log x + \log(x-2) = 3$

a) $\ln e^{5x} = \ln 3$

b) $\log_5(x^2 - 2x) = 3$

c) $\log(x^2 - 2x) = 3$

d) $\log_5 \dfrac{x}{x-2} = 3$

e) $\log 5^x = \log 3$

f) $e^3 = x$

g) $\ln \dfrac{x}{x-2} = 3$

h) $x^3 = 5$

Solve. Where appropriate, include approximations to three decimal places.

9. $3^{2x} = 81$
10. $2^{3x} = 64$
11. $4^x = 32$
12. $9^x = 27$
13. $2^x = 10$
14. $2^x = 24$
15. $2^{x+5} = 16$
16. $2^{x-1} = 8$
17. $8^{x-3} = 19$
18. $5^{x+2} = 15$
19. $e^t = 50$
20. $e^t = 20$
21. $e^{-0.02t} = 8$
22. $e^{-0.01t} = 100$
23. $5 = 3^{x+1}$
24. $7 = 3^{x-1}$
25. $4.9^x - 87 = 0$
26. $7.2^x - 65 = 0$
27. $19 = 2e^{4x}$
28. $29 = 3e^{2x}$
29. $7 + 3e^{5x} = 13$
30. $4 + 5e^{4x} = 9$

Aha! 31. $\log_3 x = 4$

32. $\log_2 x = 6$
33. $\log_2 x = -3$
34. $\log_5 x = 3$
35. $\ln x = 5$
36. $\ln x = 4$
37. $\ln (4x) = 3$
38. $\ln (3x) = 2$
39. $\log x = 2.5$
40. $\log x = 0.5$
41. $\ln (2x + 1) = 4$
42. $\ln (4x - 2) = 3$

Aha! 43. $\ln x = 1$

44. $\log x = 1$
45. $5 \ln x = -15$
46. $3 \ln x = -3$
47. $\log_2 (8 - 6x) = 5$
48. $\log_5 (2x - 7) = 3$
49. $\log (x - 9) + \log x = 1$
50. $\log (x + 9) + \log x = 1$
51. $\log x - \log (x + 3) = 1$
52. $\log x - \log (x + 7) = -1$

Aha! 53. $\log (2x + 1) = \log 5$

54. $\log (x + 1) - \log x = 0$
55. $\log_4 (x + 3) = 2 + \log_4 (x - 5)$
56. $\log_2 (x + 3) = 4 + \log_2 (x - 3)$
57. $\log_7 (x + 1) + \log_7 (x + 2) = \log_7 6$
58. $\log_6 (x + 3) + \log_6 (x + 2) = \log_6 20$
59. $\log_5 (x + 4) + \log_5 (x - 4) = \log_5 20$
60. $\log_4 (x + 2) + \log_4 (x - 7) = \log_4 10$
61. $\ln (x + 5) + \ln (x + 1) = \ln 12$
62. $\ln (x - 6) + \ln (x + 3) = \ln 22$
63. $\log_2 (x - 3) + \log_2 (x + 3) = 4$
64. $\log_3 (x - 4) + \log_3 (x + 4) = 2$
65. $\log_{12} (x + 5) - \log_{12} (x - 4) = \log_{12} 3$
66. $\log_6 (x + 7) - \log_6 (x - 2) = \log_6 5$
67. $\log_2 (x - 2) + \log_2 x = 3$
68. $\log_4 (x + 6) - \log_4 x = 2$
69. $e^{0.5x} - 7 = 2x + 6$
70. $e^{-x} - 3 = x^2$
71. $\ln (3x) = 3x - 8$
72. $\ln (x^2) = -x^2$
73. Find the value of x for which the natural logarithm is the same as the common logarithm.
74. Find all values of x for which the common logarithm of the square of x is the same as the square of the common logarithm of x.

TW 75. Christina finds that the solution of $\log_3 (x + 4) = 1$ is -1, but rejects -1 as an answer because it is negative. What mistake is she making?

TW 76. Could Example 2 have been solved by taking the natural logarithm on both sides? Why or why not?

SKILL REVIEW

To prepare for Section 12.7, review using the five-step problem-solving strategy.

Solve.

77. A rectangle is 6 ft longer than it is wide. Its perimeter is 26 ft. Find the length and the width. [2.5]

78. Under one health insurance plan offered in California, the maximum co-pay for an individual is \$3000 per calendar year. The co-pay for each visit to a specialist is \$40, and the co-pay for a hospitalization is \$1000. With hospitalizations and specialist visits, Marguerite reached the maximum co-pay in 2010. If she was hospitalized twice, how many visits to specialists did she make? [8.1]
Source: ehealthinsurance.com

79. Joanna wants to mix Golden Days bird seed containing 25% sunflower seeds with Snowy Friends bird seed containing 40% sunflower seeds. She wants 50 lb of a mixture containing 33% sunflower seeds. How much of each type should she use? [3.4]

80. The outside edge of a picture frame measures 12 cm by 19 cm, and 144 cm^2 of picture shows. Find the width of the frame. [6.7]

81. Max can key in a musical score in 2 hr. Miles takes 3 hr to key in the same score. How long would it take them, working together, to key in the score? [7.7]

82. A sign is in the shape of a right triangle. The hypotenuse is 3 ft long, and the base and the height of the triangle are equal. Find the length of the base and the height. Round to the nearest tenth of a foot. [10.7]

SYNTHESIS

TW **83.** Can the principle of logarithmic equality be expanded to include all functions? That is, is the statement "$m = n$ is equivalent to $f(m) = f(n)$" true for any function f? Why or why not?

TW **84.** Explain how Exercises 39 and 40 could be solved using the graph of $f(x) = \log x$.

Solve. If no solution exists, state this.

85. $27^x = 81^{2x-3}$

86. $8^x = 16^{3x+9}$

87. $\log_x(\log_3 27) = 3$

88. $\log_6(\log_2 x) = 0$

89. $x \log \frac{1}{8} = \log 8$

90. $\log_5 \sqrt{x^2 - 9} = 1$

91. $2^{x^2+4x} = \frac{1}{8}$

92. $\log(\log x) = 5$

93. $\log_5 |x| = 4$

94. $\log x^2 = (\log x)^2$

95. $\log \sqrt{2x} = \sqrt{\log 2x}$

96. $1000^{2x+1} = 100^{3x}$

97. $3^{x^2} \cdot 3^{4x} = \frac{1}{27}$

98. $3^{3x} \cdot 3^{x^2} = 81$

99. $\log x^{\log x} = 25$

100. $3^{2x} - 8 \cdot 3^x + 15 = 0$

101. $(81^{x-2})(27^{x+1}) = 9^{2x-3}$

102. $3^{2x} - 3^{2x-1} = 18$

103. Given that $2^y = 16^{x-3}$ and $3^{y+2} = 27^x$, find the value of $x + y$.

104. If $x = (\log_{125} 5)^{\log_5 125}$, what is the value of $\log_3 x$?

Try Exercise Answers: Section 12.6

9. 2 **17.** $\frac{\log 19}{\log 8} + 3 \approx 4.416$ **21.** $\frac{\ln 8}{-0.02} \approx -103.972$
47. -4 **49.** 10 **57.** 1 **65.** $\frac{17}{2}$ **69.** $-6.480, 6.519$

12.7 Applications of Exponential and Logarithmic Functions

- Applications of Logarithmic Functions
- Applications of Exponential Functions

We now consider applications of exponential and logarithmic functions.

APPLICATIONS OF LOGARITHMIC FUNCTIONS

EXAMPLE 1 Sound Levels. To measure the volume, or "loudness," of a sound, the *decibel* scale is used. The loudness L, in decibels (dB), of a sound is given by

$$L = 10 \cdot \log \frac{I}{I_0},$$

where I is the intensity of the sound, in watts per square meter (W/m^2), and $I_0 = 10^{-12}\ \text{W/m}^2$. ($I_0$ is approximately the intensity of the softest sound that can be heard by the human ear.)

a) The average maximum intensity of sound in a New York subway car is about $3.2 \times 10^{-3}\ \text{W/m}^2$. How loud, in decibels, is the sound level?
Source: Columbia University Mailman School of Public Health

b) The Occupational Safety and Health Administration (OSHA) considers sustained sound levels of 90 dB and above unsafe. What is the intensity of such sounds?

SOLUTION

a) To find the loudness, in decibels, we use the above formula:

$$
\begin{aligned}
L &= 10 \cdot \log \frac{I}{I_0} \\
&= 10 \cdot \log \frac{3.2 \times 10^{-3}}{10^{-12}} && \textbf{Substituting} \\
&= 10 \cdot \log (3.2 \times 10^{9}) && \textbf{Subtracting exponents} \\
&= 10(\log 3.2 + \log 10^{9}) && \log MN = \log M + \log N \\
&= 10(\log 3.2 + 9) && \log_{10} 10^9 = 9 \\
&\approx 10(0.5051 + 9) && \textbf{Approximating log 3.2} \\
&= 10(9.5051) && \textbf{Adding within the parentheses} \\
&\approx 95. && \textbf{Multiplying and rounding}
\end{aligned}
$$

The volume of the sound in a subway car is about 95 decibels.

b) We substitute and solve for I:

$$
\begin{aligned}
L &= 10 \cdot \log \frac{I}{I_0} \\
90 &= 10 \cdot \log \frac{I}{10^{-12}} && \textbf{Substituting} \\
9 &= \log \frac{I}{10^{-12}} && \textbf{Dividing both sides by 10} \\
9 &= \log I - \log 10^{-12} && \textbf{Using the quotient rule for logarithms} \\
9 &= \log I - (-12) && \log 10^a = a \\
-3 &= \log I && \textbf{Adding } -12 \textbf{ to both sides} \\
10^{-3} &= I. && \textbf{Converting to an exponential equation}
\end{aligned}
$$

Sustained sounds with intensities exceeding $10^{-3}\ \text{W/m}^2$ are considered unsafe.

Try Exercise 15.

EXAMPLE 2 Chemistry: pH of Liquids. In chemistry, the pH of a liquid is a measure of its acidity. We calculate pH as follows:

$$\text{pH} = -\log [\text{H}^+],$$

where $[\text{H}^+]$ is the hydrogen ion concentration, in moles per liter.

a) The hydrogen ion concentration of human blood is normally about 3.98×10^{-8} moles per liter. Find the pH.
Source: www.merck.com

b) The pH of seawater is about 8.2. Find the hydrogen ion concentration.
Source: www.seafriends.org.nz

SOLUTION

a) To find the pH of blood, we use the above formula:

$$
\begin{aligned}
\text{pH} &= -\log [\text{H}^+] \\
&= -\log [3.98 \times 10^{-8}] \\
&\approx -(-7.400117) && \textbf{Using a calculator} \\
&\approx 7.4.
\end{aligned}
$$

The pH of human blood is normally about 7.4.

b) We substitute and solve for $[\mathrm{H}^+]$:

$$\begin{aligned} 8.2 &= -\log[\mathrm{H}^+] && \textbf{Using pH} = -\log[\mathrm{H}^+] \\ -8.2 &= \log[\mathrm{H}^+] && \textbf{Dividing both sides by } -1 \\ 10^{-8.2} &= [\mathrm{H}^+] && \textbf{Converting to an exponential equation} \\ 6.31 \times 10^{-9} &\approx [\mathrm{H}^+]. && \textbf{Using a calculator; writing scientific notation} \end{aligned}$$

The hydrogen ion concentration of seawater is about 6.31×10^{-9} moles per liter.

Try Exercise 11.

APPLICATIONS OF EXPONENTIAL FUNCTIONS

EXAMPLE 3 Interest Compounded Annually. Suppose that \$25,000 is invested at 4% interest, compounded annually. (See Example 5 in Section 12.2.) In t years, it will grow to the amount A given by the function

$$A(t) = 25{,}000(1.04)^t.$$

a) How long will it take to accumulate \$80,000 in the account?

b) Find the amount of time it takes for the \$25,000 to double itself.

SOLUTION

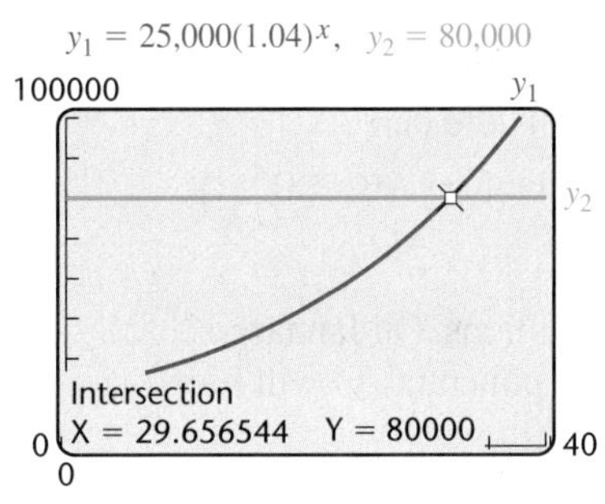

a) We set $A(t) = 80{,}000$ and solve for t:

$$\begin{aligned} 80{,}000 &= 25{,}000(1.04)^t \\ \frac{80{,}000}{25{,}000} &= 1.04^t && \textbf{Dividing both sides by 25,000} \\ 3.2 &= 1.04^t \\ \log 3.2 &= \log 1.04^t && \textbf{Taking the common logarithm on both sides} \\ \log 3.2 &= t\log 1.04 && \textbf{Using the power rule for logarithms} \\ \frac{\log 3.2}{\log 1.04} &= t && \textbf{Dividing both sides by log 1.04} \\ 29.7 &\approx t. && \textbf{Using a calculator} \end{aligned}$$

Remember that when doing a calculation like this on a calculator, it is best to wait until the end to round off. We check by solving graphically, as shown at left. At an interest rate of 4% per year, it will take about 29.7 years for \$25,000 to grow to \$80,000.

Student Notes

Study the different steps in the solution of Example 3(b). Note that if 50,000 and 25,000 are replaced with 8000 and 4000, the doubling time is unchanged.

b) To find the *doubling time*, we replace $A(t)$ with 50,000 and solve for t:

$$\begin{aligned} 50{,}000 &= 25{,}000(1.04)^t \\ 2 &= (1.04)^t && \textbf{Dividing both sides by 25,000} \\ \log 2 &= \log(1.04)^t && \textbf{Taking the common logarithm on both sides} \\ \log 2 &= t\log 1.04 && \textbf{Using the power rule for logarithms} \\ t &= \frac{\log 2}{\log 1.04} \approx 17.7. && \textbf{Dividing both sides by log 1.04 and using a calculator} \end{aligned}$$

At an interest rate of 4% per year, the doubling time is about 17.7 years.

Try Exercise 5.

Like investments, populations often grow exponentially.

Exponential Growth

An **exponential growth model** is a function of the form

$$P(t) = P_0 e^{kt}, \quad k > 0,$$

where P_0 is the population at time 0, $P(t)$ is the population at time t, and k is the **exponential growth rate** for the situation. The **doubling time** is the amount of time necessary for the population to double in size.

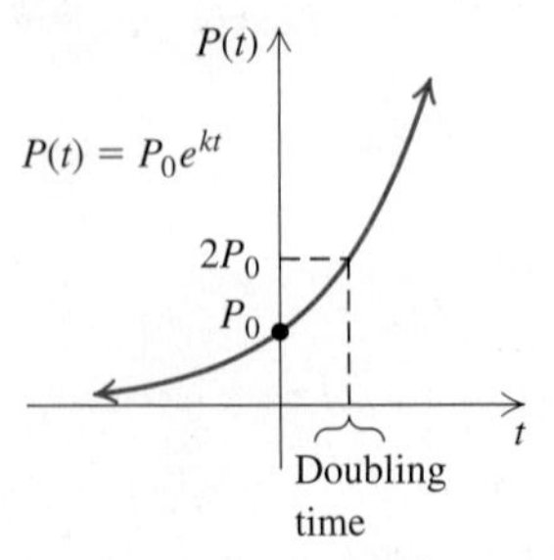

The exponential growth rate is the rate of growth of a population at any *instant* in time. Since the population is continually growing, the percent of total growth after one year will exceed the exponential growth rate.

EXAMPLE 4 Spread of a Computer Virus. The number of computers infected by a virus t hours after it first appears usually increases exponentially. On January 12, 2009, the "Conflicker" worm had infected 2.4 million PCs. This number was increasing exponentially at a rate of 33% per day.
Source: Based on data from *PC World*

a) Find the exponential growth function that models the data.

b) Estimate the number of infected computers on January 16, 2009.

SOLUTION

a) We let $N(t)$ = the number of infected PCs, in millions. On January 12, at $t = 0$, there were 2.4 million infected PCs. Using the exponential growth function

$$N(t) = N_0 e^{kt},$$

we substitute 2.4 for N_0 and 33%, or 0.33, for k:

$$N(t) = 2.4e^{0.33t}.$$

b) On January 16, we have $t = 4$, since 4 days have passed since January 12. We compute $N(4)$:

$$\begin{aligned} N(4) &= 2.4e^{0.33(4)} \\ &= 2.4e^{1.32} \\ &\approx 9.0. \quad \text{Using a calculator} \end{aligned}$$

On January 16, 2009, the worm had infected about 9.0 million PCs.

Try Exercise 23.

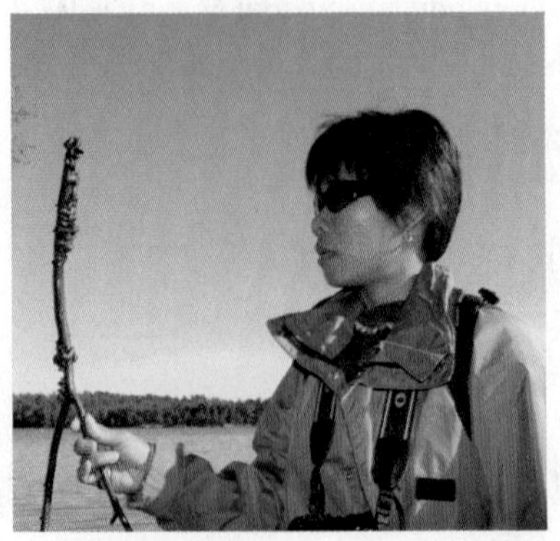

EXAMPLE 5 Growth of Zebra Mussel Populations. Zebra mussels, inadvertently imported from Europe, began fouling North American waters in 1988. These mussels are so prolific that lake and river bottoms, as well as water intake pipes, can become blanketed with them, altering an entire ecosystem. The number of zebra mussels at Prescott, Minnesota, increased exponentially from $72/\text{m}^2$ in 2005 to $574/\text{m}^2$ in 2007.
Source: *Annual Report: Quantitative Assessment of Zebra Mussels Associated with Native Mussels Beds in the Lower St. Croix River*, 2007; prepared for U.S. Army Corps of Engineers, St. Paul District St. Paul, MN, by National Park Service, St. Croix National Scenic Riverway St. Croix Falls, WI, March 2008

a) Find the exponential growth rate and the exponential growth function.

b) Assuming exponential growth, how long will it take for the density to reach 5000 mussels/m^2?

SOLUTION

a) We use $P(t) = P_0e^{kt}$, where t is the number of years since 2005. Substituting 72 for P_0 gives

$$P(t) = 72e^{kt}.$$

To find the exponential growth function, we use the fact that after 2 years, the density was 574/m^2:

$$\left.\begin{aligned} P(2) &= 72e^{k \cdot 2} \\ 574 &= 72e^{2k} \end{aligned}\right\} \quad \textbf{Substituting 2 for } t \textbf{ and 574 for } P(2)$$

$$\frac{574}{72} = e^{2k} \quad \textbf{Dividing both sides by 72}$$

$$\ln\left(\frac{574}{72}\right) = \ln e^{2k} \quad \textbf{Taking the natural logarithm on both sides}$$

$$\ln\left(\frac{574}{72}\right) = 2k \quad \ln e^a = a$$

$$\frac{\ln\left(\frac{574}{72}\right)}{2} = k \quad \textbf{Dividing both sides by 2}$$

$$1.038 \approx k. \quad \textbf{Using a calculator and rounding}$$

The exponential growth rate is 103.8%, and the exponential growth function is given by $P(t) = 72e^{1.038t}$.

b) To estimate how long it will take for the density to reach 5000 mussels/m^2, we replace $P(t)$ with 5000 and solve for t:

$$5000 = 72e^{1.038t}$$

$$\frac{5000}{72} = e^{1.038t} \quad \textbf{Dividing both sides by 72}$$

$$\ln\left(\frac{5000}{72}\right) = \ln e^{1.038t} \quad \textbf{Taking the natural logarithm on both sides}$$

$$\ln\left(\frac{5000}{72}\right) = 1.038t \quad \ln e^a = a$$

$$\frac{\ln\left(\frac{5000}{72}\right)}{1.038} = t \quad \textbf{Dividing both sides by 1.038}$$

$$4.1 \approx t. \quad \textbf{Using a calculator}$$

It will take about 4.1 years for the density to grow to 5000 zebra mussels/m^2.

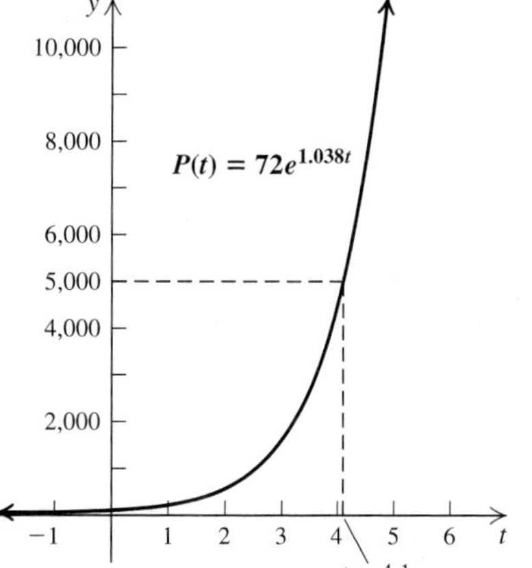

A graphical solution of Example 5

Try Exercise 31.

EXAMPLE 6 Interest Compounded Continuously. When an amount of money P_0 is invested at interest rate k, compounded *continuously*, interest is computed every "instant" and added to the original amount. The balance $P(t)$, after t years, is given by the exponential growth model

$$P(t) = P_0e^{kt}.$$

a) Suppose that \$30,000 is invested and grows to \$44,754.75 in 5 years. Find the exponential growth function.

b) What is the doubling time?

SOLUTION

a) We have $P(0) = 30{,}000$. Thus the exponential growth function is

$$P(t) = 30{,}000e^{kt}, \quad \text{where } k \text{ must still be determined.}$$

We substitute 5 for t and 44,754.75 for $P(5)$:

$$44{,}754.75 = 30{,}000e^{k(5)} = 30{,}000e^{5k}$$

$$\frac{44{,}754.75}{30{,}000} = e^{5k} \qquad \textbf{Dividing both sides by 30,000}$$

$$1.491825 = e^{5k}$$

$$\ln 1.491825 = \ln e^{5k} \qquad \textbf{Taking the natural logarithm on both sides}$$

$$\ln 1.491825 = 5k \qquad \ln e^a = a$$

$$\frac{\ln 1.491825}{5} = k \qquad \textbf{Dividing both sides by 5}$$

$$0.08 \approx k. \qquad \textbf{Using a calculator and rounding}$$

The interest rate is about 0.08, or 8%, compounded continuously. Because interest is being compounded continuously, the yearly interest rate is a bit more than 8%. The exponential growth function is

$$P(t) = 30{,}000e^{0.08t}.$$

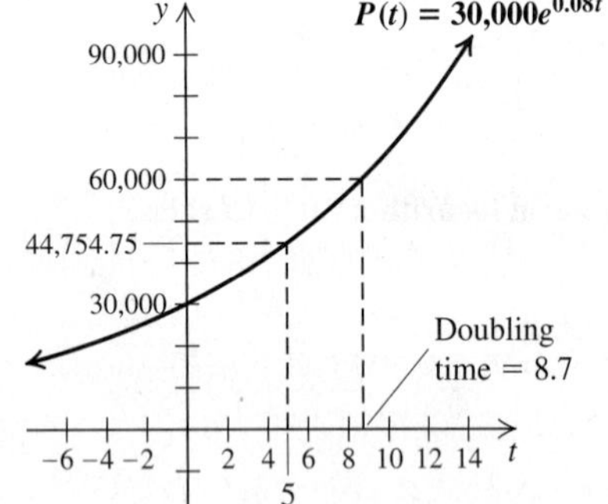

A graphical solution of Example 6

b) To find the doubling time T, we replace $P(T)$ with 60,000 and solve for T:

$$60{,}000 = 30{,}000e^{0.08T}$$

$$2 = e^{0.08T} \qquad \textbf{Dividing both sides by 30,000}$$

$$\ln 2 = \ln e^{0.08T} \qquad \textbf{Taking the natural logarithm on both sides}$$

$$\ln 2 = 0.08T \qquad \ln e^a = a$$

$$\frac{\ln 2}{0.08} = T \qquad \textbf{Dividing both sides by 0.08}$$

$$8.7 \approx T. \qquad \textbf{Using a calculator and rounding}$$

Thus the original investment of \$30,000 will double in about 8.7 years.

Try Exercise 41.

For any specified interest rate, continuous compounding gives the highest yield and the shortest doubling time.

In some real-life situations, a quantity or population is *decreasing* or *decaying* exponentially.

Exponential Decay

An **exponential decay model** is a function of the form

$$P(t) = P_0 e^{-kt}, \quad k > 0,$$

where P_0 is the quantity present at time 0, $P(t)$ is the amount present at time t, and k is the **decay rate.** The **half-life** is the amount of time necessary for half of the quantity to decay.

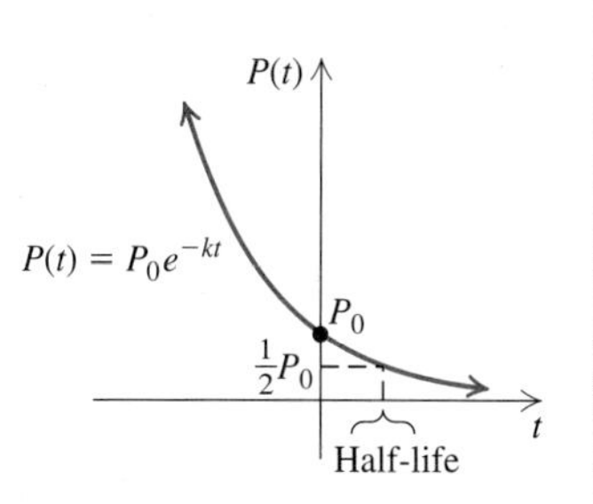

STUDY TIP

Sorting by Type

When a section contains a variety of problems, try to sort them out by type. For instance, interest compounded continuously, population growth, and the spread of a virus can all be regarded as one type of problem: exponential growth. Once you know how to solve this type of problem, you can focus on determining which problems fall into this category. The solution should then follow in a rather straightforward manner.

EXAMPLE 7 Carbon Dating. The radioactive element carbon-14 has a half-life of 5750 years. The percentage of carbon-14 in the remains of organic matter can be used to determine the age of that matter. Recently, while digging near Patuxent River, Maryland, archaeologists found charcoal that had lost 8.1% of its carbon-14. The age of this charcoal indicates that this is the oldest dwelling ever discovered in Maryland. What was the age of the charcoal?

Source: Based on information from *The Baltimore Sun*, "Digging Where Indians Camped Before Columbus," by Frank D. Roylance, July 2, 2009

SOLUTION We first find k. To do so, we use the concept of half-life. When $t = 5750$ (the half-life), $P(t)$ will be half of P_0. Then

$$0.5P_0 = P_0 e^{-k(5750)} \qquad \textbf{Substituting in } P(t) = P_0 e^{-kt}$$
$$0.5 = e^{-5750k} \qquad \textbf{Dividing both sides by } P_0$$
$$\ln 0.5 = \ln e^{-5750k} \qquad \textbf{Taking the natural logarithm on both sides}$$
$$\ln 0.5 = -5750k \qquad \ln e^a = a$$
$$\frac{\ln 0.5}{-5750} = k \qquad \textbf{Dividing}$$
$$0.00012 \approx k. \qquad \textbf{Using a calculator and rounding}$$

Now we have a function for the decay of carbon-14:

$$P(t) = P_0 e^{-0.00012t}. \longleftarrow \textbf{This completes the first part of our solution.}$$

If the charcoal has lost 8.1% of its carbon-14 from an initial amount P_0, then 100% − 8.1%, or 91.9%, of P_0 is still present. To find the age t of the charcoal, we solve this equation for t:

$$0.919P_0 = P_0 e^{-0.00012t} \qquad \textbf{We want to find } t \textbf{ for which } P(t) = 0.919P_0.$$
$$0.919 = e^{-0.00012t} \qquad \textbf{Dividing both sides by } P_0$$
$$\ln 0.919 = \ln e^{-0.00012t} \qquad \textbf{Taking the natural logarithm on both sides}$$
$$\ln 0.919 = -0.00012t \qquad \ln e^a = a$$
$$\frac{\ln 0.919}{-0.00012} = t \qquad \textbf{Dividing}$$
$$700 \approx t. \qquad \textbf{Using a calculator}$$

The charcoal is about 700 years old.

Try Exercise 35.

Student Notes

Like linear functions, two points determine an exponential function. It is important to determine whether data follow a linear model or an exponential model (if either) before fitting a function to the data.

The equation

$$P(t) = P_0 e^{-0.00012t}$$

can be used for any subsequent carbon-dating problem.

By looking at the graph of a set of data, we can tell whether a population or other quantity is growing or decaying exponentially.

EXAMPLE 8 For each of the following graphs, determine whether an exponential function might fit the data.

a) Time to Hearing Loss

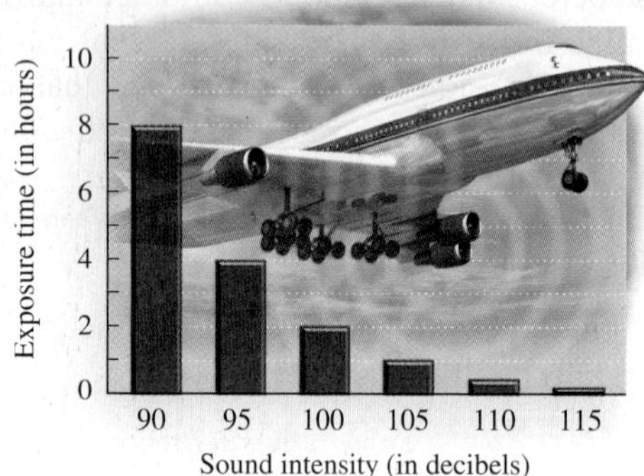

Source: American Hearing Research Foundation

b) Health-Care Providers and Social Workers

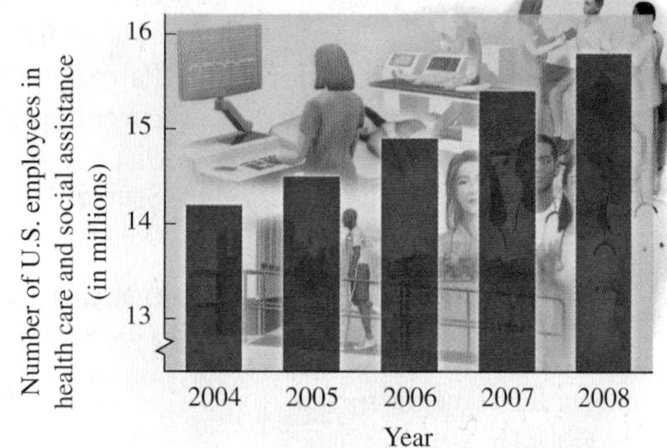

Source: U.S. Bureau of Labor Statistics

c) Atlantic Basin Hurricanes

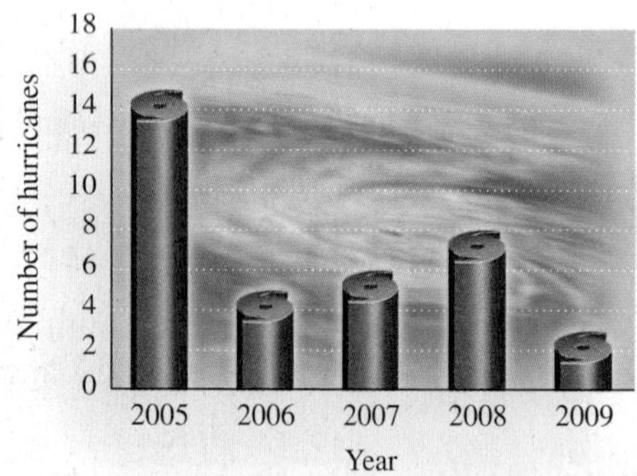

Source: NOAA

d) Text Messaging in the United States

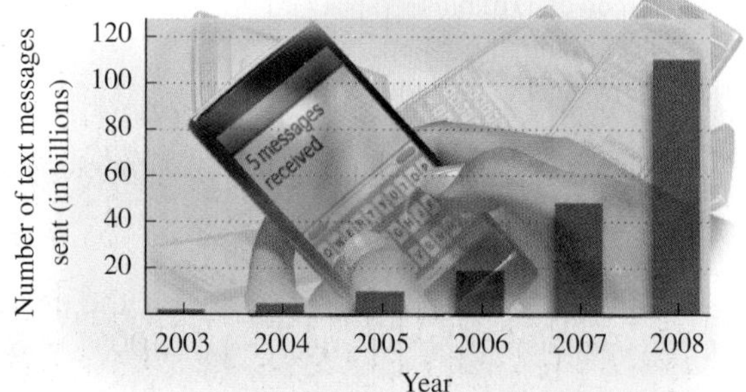

Source: CTIA—The Wireless Association

SOLUTION

a) As the sound intensity increases, the safe exposure time decreases. The amount of decrease gets smaller as the intensity increases. It appears that an exponential decay function might fit the data.

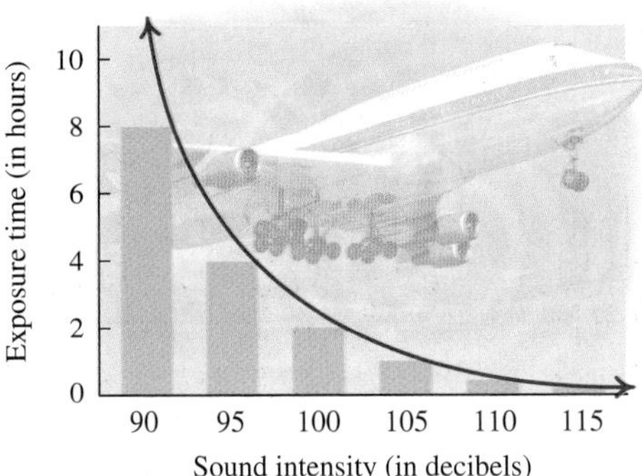

Source: American Hearing Research Foundation

b) The number of health-care providers and social workers increased between 2004 and 2008 at approximately the same rate each year. It does not appear that an exponential function models the data; instead, a linear function might be appropriate.

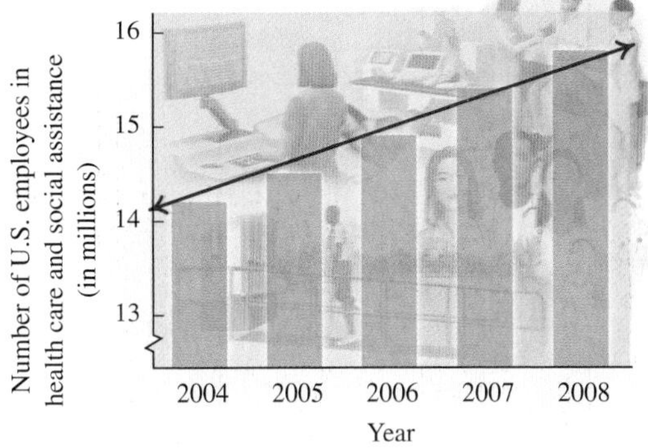

Source: U.S. Bureau of Labor Statistics

c) The number of hurricanes first fell, then rose, and then fell again. This does not fit an exponential model.

d) The number of text messages increased from 2003 to 2008, and the amount of yearly growth also increased during that time. This suggests that an exponential growth function could be used to model this situation.

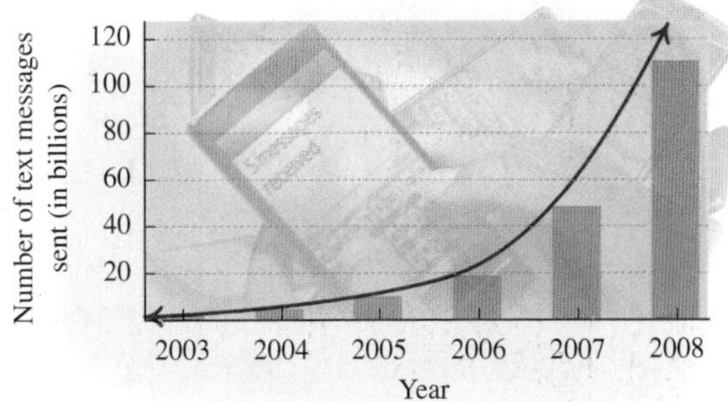

Source: CTIA—The Wireless Association

Try Exercise 45.

Exponential Regression

To fit an exponential function to a set of data, we use the ExpReg option in the STAT CALC menu. After entering the data, with the independent variable as L1 and the dependent variable as L2, choose the ExpReg option. For the data shown on the left below, the calculator will return a screen like that shown on the right below.

L1	L2	L3 1
1	12	------
2	60	
3	300	
4	1500	
5	7500	
6	37500	
------	------	

L1(1)=1

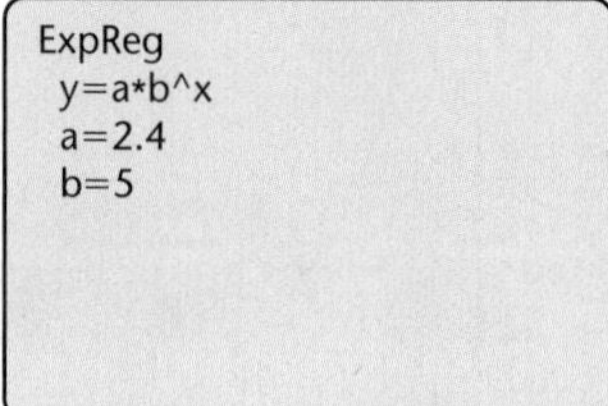

The exponential function found is of the form $f(x) = ab^x$.

We can use the function in this form or convert it to an exponential function of the form $f(x) = ae^{kx}$. The number a remains unchanged. To find k, we solve the equation $b^x = e^{kx}$ for k:

$$b^x = e^{kx}$$

$$b^x = (e^k)^x \quad \textbf{Writing both sides as a power of } x$$

$$b = e^k \quad \textbf{The bases must be equal.}$$

$$\ln b = \ln e^k \quad \textbf{Taking the natural logarithm on both sides}$$

$$\ln b = k. \quad \ln e^k = k$$

Thus to write $f(x) = 2.4(5)^x$ in the form $f(x) = ae^{kx}$, we can write

$$f(x) = 2.4e^{(\ln 5)x}, \quad \text{or} \quad f(x) = 2.4e^{1.609x}.$$

Your Turn

1. Enter the data $(0, 6), (1, 24), (2, 96), (3, 384)$.
2. Plot the points in a $[-1, 4, 0, 500]$ window, with Yscl $= 50$. Verify that the pattern appears to fit an exponential model.
3. Fit an exponential equation to the data. $y = 6(4)^x$
4. Write the equation in the form $y = ae^{kx}$. $a = 6, k = \ln 4; y = 6e^{1.386x}$

EXAMPLE 9 Text Messaging. In 2003, approximately 2.1 million text messages were sent in the United States. This number increased exponentially to 110.4 million in 2008.

Year	Number of Text Messages (in billions)
2003	2.1
2004	4.7
2005	9.8
2006	18.7
2007	48.1
2008	110.4

Source: CTIA—The Wireless Association

a) Use regression to fit an exponential function to the data and graph the function.

b) Determine the exponential growth rate.

c) At this rate, how many text messages will be sent in 2011?

SOLUTION

a) We let $x =$ the number of years since 2003, and enter the data, as shown on the left below. Then we use regression to find the function. Rounding the coefficients, we have

$$f(x) = 2.0545(2.1899)^x.$$

The graph of the function, along with the data points, is shown on the right below.

L1	L2	L3 1
0	2.1	----------
1	4.7	
2	9.8	
3	18.7	
4	48.1	
5	110.4	
----------	----------	

L1(1) = 0

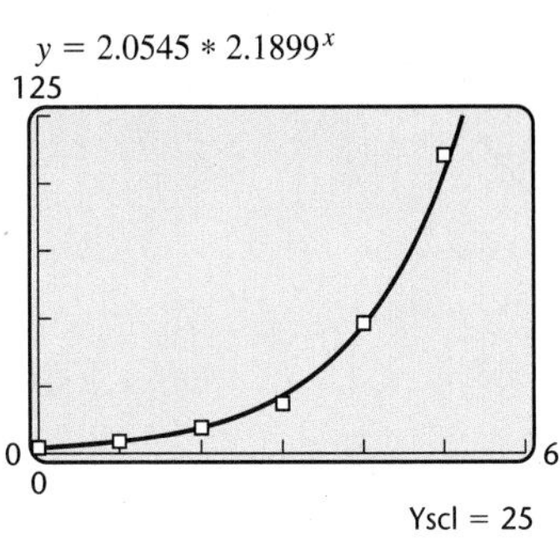

b) The exponential growth rate is the number k in the equation $f(x) = ae^{kx}$. To write $f(x) = 2.0545(2.1899)^x$ in the form $f(x) = ae^{kx}$, we find k:

$$k = \ln b \approx \ln 2.1899 \approx 0.7839.$$

Thus the exponential growth rate is approximately 0.784, or 78.4%. The exponential function written in the form $f(x) = ae^{kx}$ is thus

$$f(x) = 2.0545e^{0.784x}.$$

This form is equivalent to the form $f(x) = 2.0545(2.1899)^x$.

c) Since 2011 is 8 years after 2003, we find $f(8)$:

$$f(8) = 2.0545(2.1899)^8 \approx 1086.7.$$

Using the function, we estimate that there will be approximately 1.1 trillion text messages sent in the United States in 2011.

Try Exercise 47.

Connecting the Concepts

We can now add the exponential functions $f(t) = P_0e^{kt}$ and $f(t) = P_0e^{-kt}$, $k > 0$, and the logarithmic function $f(x) = \log_b x$, $b > 1$, to our library of functions.

Linear function: $f(x) = mx + b$

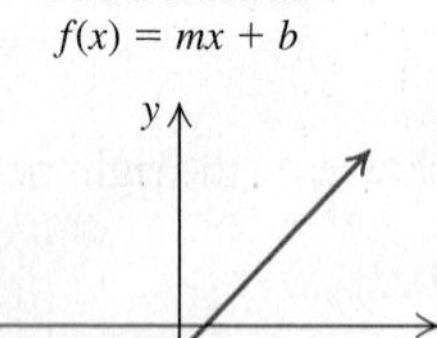

Absolute-value function: $f(x) = |x|$

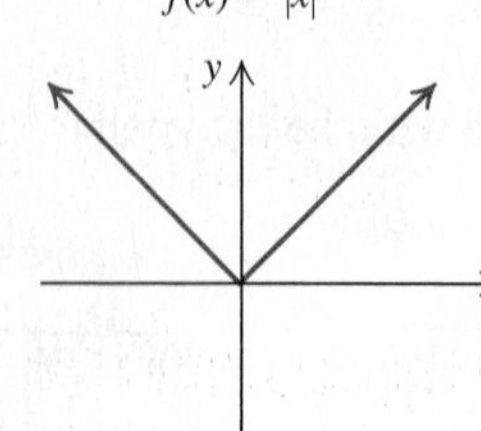

Rational function: $f(x) = \frac{1}{x}$

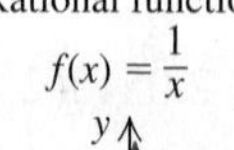

Radical function: $f(x) = \sqrt{x}$

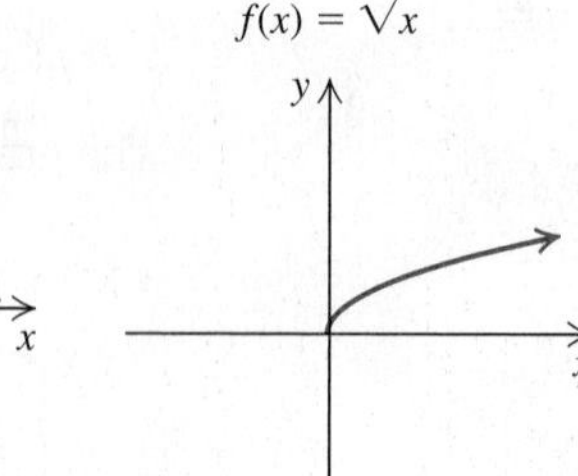

Quadratic function: $f(x) = ax^2 + bx + c$, $a > 0$

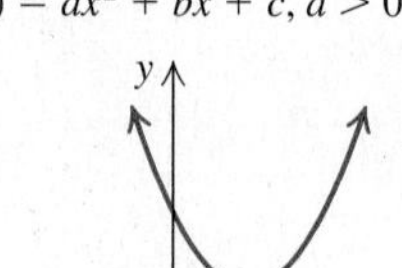

Quadratic function: $f(x) = ax^2 + bx + c$, $a < 0$

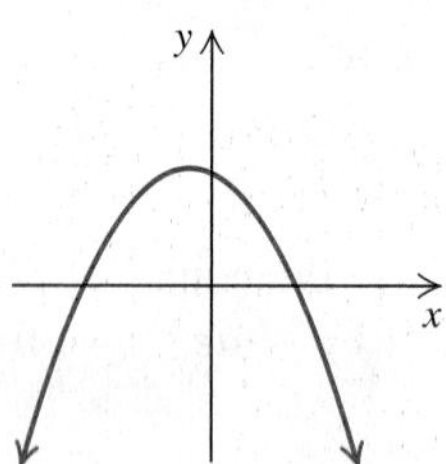

Exponential growth function: $f(t) = P_0e^{kt}$, $k > 0$

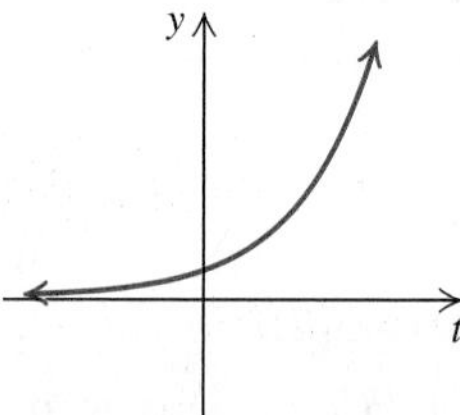

Exponential decay function: $f(t) = P_0e^{-kt}$, $k > 0$

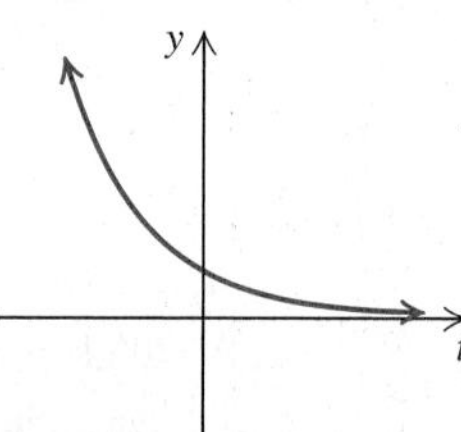

Logarithmic function: $f(x) = \log_b x$, $b > 1$

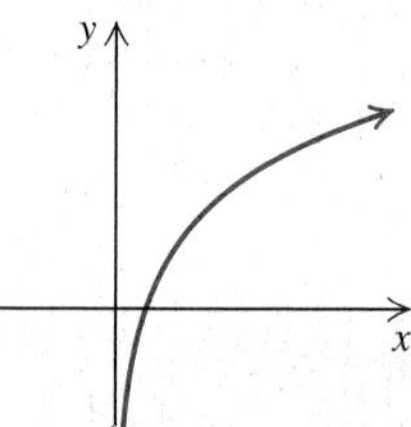

12.7 Exercise Set

FOR EXTRA HELP

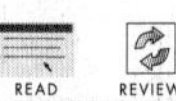

Solve.

1. *Asteroids.* The total number $A(t)$ of known asteroids t years after 1990 can be estimated by

$$A(t) = 77(1.283)^t.$$

Source: Based on data from NASA

a) Determine the year in which the number of known asteroids first reached 4000.

b) What is the doubling time for the number of known asteroids?

2. *Microfinancing.* In India, the number of microfinance institutions, or firms offering small loans, increased rapidly from 2004 to 2009. The total amount of these loans, in millions of dollars, can be estimated by

$$m(t) = 2.05(2.2)^t,$$

where t is the number of years since 2004.

Source: Based on data from Sa-Dhan

a) In what year will the total amount of these loans reach \$1 billion?

b) Find the doubling time.

3. *Health.* The rate of the number of deaths due to stroke in the United States can be estimated by

$$S(t) = 180(0.97)^t,$$

where $S(t)$ is the number of deaths per 100,000 people and t is the number of years since 1960.

Source: Based on data from Centers for Disease Control and Prevention

a) In what year was the death rate due to stroke 100 per 100,000 people?

b) In what year will the death rate due to stroke be 25 per 100,000 people?

4. *Recycling Aluminum Cans.* Approximately one-half of all aluminum cans distributed will be recycled each year. A beverage company distributes 250,000 cans. The number still in use after t years is given by the function

$$N(t) = 250{,}000\left(\tfrac{1}{2}\right)^t.$$

Source: The Aluminum Association, Inc., May 2004

a) After how many years will 60,000 cans still be in use?

b) After what amount of time will only 1000 cans still be in use?

5. *Student Loan Repayment.* A college loan of $29,000 is made at 3% interest, compounded annually. After t years, the amount due, A, is given by the function

$$A(t) = 29{,}000(1.03)^t.$$

a) After what amount of time will the amount due reach $35,000?

b) Find the doubling time.

6. *Spread of a Rumor.* The number of people who have heard a rumor increases exponentially. If all who hear a rumor repeat it to two people per day, and if 20 people start the rumor, the number N of people who have heard the rumor after t days is given by

$$N(t) = 20(3)^t.$$

a) After what amount of time will 1000 people have heard the rumor?

b) What is the doubling time for the number of people who have heard the rumor?

7. *Food Storage.* The maximum storage time, in months, for shelled corn with 13% moisture content can be estimated by

$$m(t) = 1507(0.94)^t,$$

where t is the storage temperature, in degrees Fahrenheit, $t \geq 40°$.

Source: Based on data from University of Minnesota Extension Service

a) At what temperature can corn be stored for 4 years (48 months)?

b) At what temperature can corn be stored for 15 months?

8. *Smoking.* The percentage of smokers who received telephone counseling and had successfully quit smoking for t months is given by

$$P(t) = 21.4(0.914)^t.$$

Sources: *New England Journal of Medicine*; data from California's Smoker's Hotline

a) In what month will 15% of those who quit and used telephone counseling still be smoke-free?

b) In what month will 5% of those who quit and used telephone counseling still be smoke-free?

9. *E-book Sales.* The net amount of e-book sales, in millions of dollars, can be estimated by

$$S(t) = 2.05(1.8)^t,$$

where t is the number of years since 2002.

Source: Based on data from Association of American Publishers

a) In what year will there be $1 billion in e-book net sales?

b) Find the doubling time.

10. *World Population.* The world population $P(t)$, in billions, t years after 2010 can be approximated by

$$P(t) = 6.9(1.011)^t.$$

Sources: Based on data from U.S. Census Bureau; International Data Base

a) In what year will the world population reach 10 billion?

b) Find the doubling time.

Use $\text{pH} = -\log\,[\text{H}^+]$ *for Exercises 11–14.*

11. *Chemistry.* The hydrogen ion concentration of fresh-brewed coffee is about 1.3×10^{-5} moles per liter. Find the pH.

12. *Chemistry.* The hydrogen ion concentration of milk is about 1.6×10^{-7} moles per liter. Find the pH.

13. *Medicine.* When the pH of a patient's blood drops below 7.4, a condition called *acidosis* sets in. Acidosis can be deadly when the patient's pH reaches 7.0. What would the hydrogen ion concentration of the patient's blood be at that point?

14. *Medicine.* When the pH of a patient's blood rises above 7.4, a condition called *alkalosis* sets in. Alkalosis can be deadly when the patient's pH reaches 7.8. What would the hydrogen ion concentration of the patient's blood be at that point?

Use the formula in Example 1 for Exercises 15–18.

15. *Racing.* The intensity of sound from a race car in full throttle is about 10 W/m^2. How loud in decibels is this sound level?
Source: nascar.about.com

16. *Audiology.* The intensity of sound in normal conversation is about 3.2×10^{-6} W/m^2. How loud in decibels is this sound level?

17. *Concerts.* The crowd at a Hearsay concert at Wembley Arena in London cheered at a sound level of 128.8 dB. What is the intensity of such a sound?
Source: www.peterborough.gov.uk

18. *City Ordinances.* In Albuquerque, New Mexico, the maximum allowable sound level from a car's exhaust is 96 dB. What is the intensity of such a sound?
Source: www.cabq.gov

19. *E-mail Volume.* The SenderBase® Security Network ranks e-mail volume using a logarithmic scale. The magnitude M of a network's daily e-mail volume is given by

$$M = \log \frac{v}{1.34},$$

where v is the number of e-mail messages sent each day. How many e-mail messages are sent each day by a network that has a magnitude of 7.5?
Source: forum.spamcop.net

20. *Richter Scale.* The Richter scale, developed in 1935, has been used for years to measure earthquake magnitude. The Richter magnitude of an earthquake $m(A)$ is given by the formula

$$m(A) = \log \frac{A}{A_0},$$

where A is the maximum amplitude of the earthquake and A_0 is a constant. What is the magnitude on the Richter scale of an earthquake with an amplitude that is a million times A_0?

Use $P(t) = P_0 e^{kt}$ for Exercises 21 and 22.

21. *Interest Compounded Continuously.* Suppose that P_0 is invested in a savings account where interest is compounded continuously at 2.5% per year.

a) Express $P(t)$ in terms of P_0 and 0.025.
b) Suppose that \$5000 is invested. What is the balance after 1 year? after 2 years?
c) When will an investment of \$5000 double itself?

22. *Interest Compounded Continuously.* Suppose that P_0 is invested in a savings account where interest is compounded continuously at 3.1% per year.

a) Express $P(t)$ in terms of P_0 and 0.031.
b) Suppose that \$1000 is invested. What is the balance after 1 year? after 2 years?
c) When will an investment of \$1000 double itself?

23. *Population Growth.* In 2010, the population of the United States was 310 million, and the exponential growth rate was 1.0% per year.
Source: U.S. Census Bureau

a) Find the exponential growth function.
b) Predict the U.S. population in 2016.
c) When will the U.S. population reach 350 million?

24. *World Population Growth.* In 2010, the world population was 6.9 billion, and the exponential growth rate was 1.1% per year.
Source: U.S. Census Bureau

a) Find the exponential growth function.
b) Predict the world population in 2014.
c) When will the world population be 8.0 billion?

25. *Online College Studies.* In 2007, the number of college students studying online was 3.9 million, and the exponential growth rate was 12%. What is the doubling time for the number of online college students?
Source: Based on information from "Staying the Course," Sloan Consortium, 2008

26. *Population Growth.* The exponential growth rate of the population of United Arab Emirates is 3.7% per year (one of the highest in the world). What is the doubling time?
Source: Central Intelligence Agency, *The World Factbook*

27. *World Population.* The function

$$Y(x) = 87 \ln \frac{x}{6.1}$$

can be used to estimate the number of years $Y(x)$ after 2000 required for the world population to reach x billion people.
Sources: Based on data from U.S. Census Bureau; International Data Base

a) In what year will the world population reach 10 billion?
b) In what year will the world population reach 12 billion?
c) Graph the function.

28. *Marine Biology.* The function

$$Y(x) = 13 \ln \frac{x}{5.5}$$

can be used to estimate the number of years $Y(x)$ after 1982 required for the world's humpback whale population to reach x thousand whales.

a) In what year will the whale population reach 40,000?

b) In what year will the whale population reach 50,000?

c) Graph the function.

29. *Forgetting.* Students in an English class took a final exam. They took equivalent forms of the exam at monthly intervals thereafter. The average score $S(t)$, in percent, after t months was found to be given by

$$S(t) = 68 - 20 \log (t + 1), \quad t \geq 0.$$

a) What was the average score when they initially took the test $(t = 0)$?

b) What was the average score after 4 months? after 24 months?

c) Graph the function.

d) After what time t was the average score 50%?

30. *Health Insurance.* The amount spent each year by the U.S. government for health insurance for children in low-income families can be estimated by

$$h(t) = 2.6 \ln t,$$

where $h(t)$ is in billions of dollars and t is the number of years after 1998.

Source: Based on data from the Congressional Budget Office

a) How much was spent on health insurance for low-income children in 2007?

b) Graph the function.

c) In what year will \$7 billion be spent on health insurance for low-income children?

31. *Wind Power.* U.S. wind-power capacity has grown exponentially from about 2000 megawatts in 1990 to 35,000 megawatts in 2009.

Source: American Wind Energy Association

a) Find the exponential growth rate k, and write an equation for an exponential function that can be used to predict U.S. wind-power capacity t years after 1990.

b) Estimate the year in which wind-power capacity will reach 50,000 megawatts.

32. *Spread of a Computer Virus.* The number of computers infected by a virus t hours after it first appears usually increases exponentially. In 2004, the "MyDoom" worm spread from 100 computers to about 100,000 computers in 24 hr.

Source: Based on data from IDG News Service

a) Find the exponential growth rate k, and write an equation for an exponential function that can be used to predict the number of computers infected t hours after the virus first appeared in 100 computers.

b) Assuming exponential growth, estimate how long it took the MyDoom worm to infect 9000 computers.

33. *Cable Costs.* In 1997, the cost to construct communication cables under the ocean was approximately \$8200 per gigabit per second per mile. This cost for subsea cables dropped exponentially to \$500 by 2007.

Source: Based on information from TeleGeography

a) Find the exponential decay rate k, and write an equation for an exponential function that can be used to predict the cost of subsea cables t years after 1997.

b) Estimate the cost of subsea cables in 2010.

c) In what year (theoretically) will it cost only \$1 per gigabit per second per mile to construct subsea cables?

34. *Decline in Farmland.* The number of acres of farmland in the United States has decreased from 945 million acres in 2000 to 920 million acres in 2008. Assume the number of acres of farmland is decreasing exponentially.

Source: *Statistical Abstract of the United States*

a) Find the value k, and write an equation for an exponential function that can predict the number of acres of U.S. farmland t years after 2000.

b) Predict the number of acres of farmland in 2015.

c) In what year (theoretically) will there be only 800 million acres of U.S. farmland remaining?

35. *Archaeology.* A date palm seedling is growing in Kibbutz Ketura, Israel, from a seed found in King Herod's palace at Masada. The seed had lost 21% of its carbon-14. How old was the seed? (See Example 7.)

Source: Based on information from www.sfgate.com

36. *Archaeology.* Soil from beneath the Kish Church in Azerbaijan was found to have lost 12% of its carbon-14. How old was the soil? (See Example 7.)
Source: Based on information from www.azer.com

37. *Chemistry.* The exponential decay rate of iodine-131 is 9.6% per day. What is its half-life?

38. *Chemistry.* The decay rate of krypton-85 is 6.3% per year. What is its half-life?

39. *Caffeine.* The half-life of caffeine in the human body for a healthy adult is approximately 5 hr.

a) What is the exponential decay rate?

b) How long will it take 95% of the caffeine consumed to leave the body?

40. *Home Construction.* The chemical urea formaldehyde was found in some insulation used in houses built during the mid to late 1960s. Unknown at the time was the fact that urea formaldehyde emitted toxic fumes as it decayed. The half-life of urea formaldehyde is 1 year.

a) What is its decay rate?

b) How long will it take 95% of the urea formaldehyde present to decay?

41. *Art Masterpieces.* As of August 2010, the highest auction price for a sculpture was \$104.3 million, paid for Alberto Giacometti's bronze sculpture *Walking Man I*. The same sculpture was purchased for about \$9 million in 1990.
Source: Based on information from *The New York Times*, 02/02/10

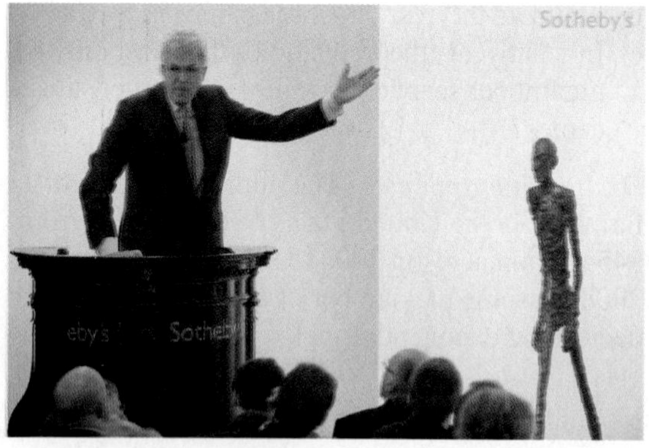

a) Find the exponential growth rate k, and determine the exponential growth function that can be used to estimate the sculpture's value $V(t)$, in millions of dollars, t years after 1990.

b) Estimate the value of the sculpture in 2020.

c) What is the doubling time for the value of the sculpture?

d) How long after 1990 will the value of the sculpture be \$1 billion?

42. *Value of a Sports Card.* Legend has it that because he objected to teenagers smoking, and because his first baseball card was issued in cigarette packs, the great shortstop Honus Wagner halted production of his card before many were produced. One of these cards was purchased in 1991 by hockey great Wayne Gretzky (and a partner) for \$451,000. The same card was sold in 2007 for \$2.8 million. For the following questions, assume that the card's value increases exponentially, as it has for many years.

a) Find the exponential growth rate k, and determine an exponential function that can be used to estimate the dollar value, $V(t)$, of the card t years after 1991.

b) Predict the value of the card in 2012.

c) What is the doubling time for the value of the card?

d) In what year will the value of the card first exceed \$4,000,000?

In Exercises 43–46, determine whether an exponential function might fit the data.

43.

Microprocessor Chip	Year Introduced	Number of Transistors
8080	1974	6,000
8086	1978	29,000
80286	1982	134,000
486 DX	1989	1,200,000
Pentium	1993	3,300,000
Pentium III	1999	9,500,000
Core2 Quad	2006	291,000,000

Sources: "Microprocessors," found on computinghistorymuseum.org; www.neoseeker.com

44. **Loan Payment**

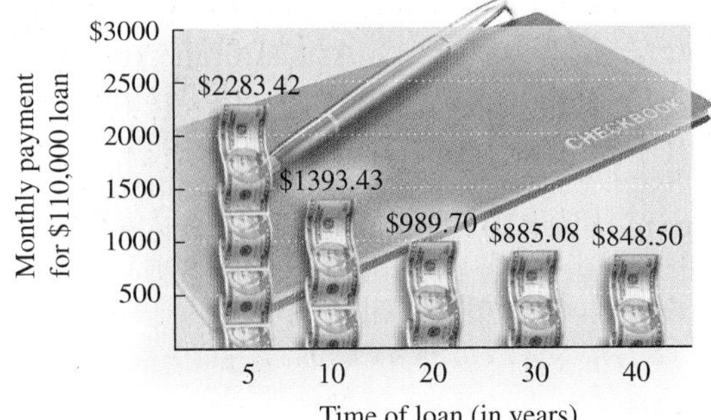

45. **Motor Vehicle Accidents**

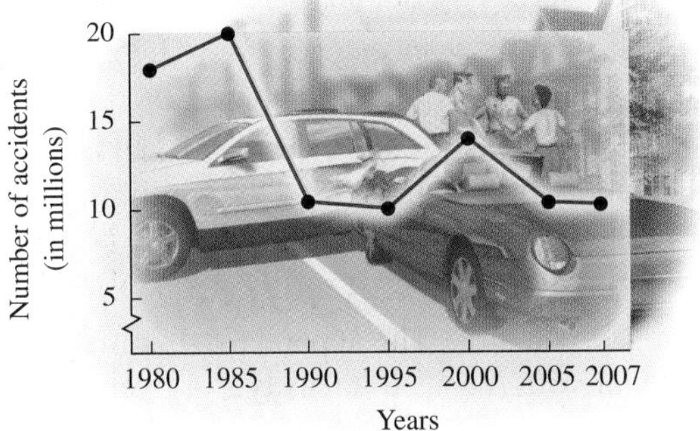

Source: National Safety Council

46. *World automobile production*

Year	Number of Automobiles Produced (in millions)
1950	11
1960	16
1970	29
1980	39
1990	49
2000	60
2008	68

Sources: American Automobile Manufacturers Association, Automotive News Data Center; R.L. Polk Marketing Systems

47. *Technology.* The number of transistors in a microchip has been increasing exponentially. The table in Exercise 43 lists the number of transistors in Intel chips introduced in various years.

a) Use regression to find an exponential function that can be used to estimate the number of transistors $n(t)$, in thousands, in a microchip t years after 1974.

b) Determine the exponential growth rate.

c) Predict the number of transistors in a microchip introduced in 2010.

48. *Stamp Collecting.* There are only two known 1¢ Z Grill stamps. The table below lists the prices paid for one of these stamps as it changed hands.

Year	Price
1975	$ 42,500
1977	90,000
1986	418,000
1998	935,000
2005	2,970,000

Source: Siegel Auctions

a) Use regression to find an exponential function that can be used to estimate the value of the stamp $v(t)$, in thousands of dollars, t years after 1975.

b) Determine the exponential growth rate.

c) Predict the value of the stamp in 2012.

49. *Hearing Loss.* Prolonged exposure to high noise levels can lead to hearing loss. The table below lists the safe exposure time, in hours, for different sound intensities.

Sound Intensity (in decibels)	Safe Exposure Time (in hours)
90	8
100	2
105	1
110	0.5
115	0.25

Source: American Hearing Research Foundation

a) Use regression to fit an exponential function of the form $f(x) = ab^x$ to the data.

b) Estimate the safe exposure time for a sound intensity of 95 dB.

50. *Loan Repayment.* The size of a monthly loan repayment decreases exponentially as the time of the loan increases. The table in Exercise 44 lists the size of monthly loan payments for a \$110,000 loan at a fixed interest rate for various lengths of loans.

a) Use regression to fit an exponential function of the form $f(x) = ab^x$ to the data.

b) Estimate the monthly loan payment for a 15-year \$110,000 loan.

TW **51.** Will the model used to predict the number of text messages in Example 9 still be realistic in 2030? Why or why not?

TW **52.** Examine the restriction on t in Exercise 29.

a) What upper limit might be placed on t?

b) In practice, would this upper limit ever be enforced? Why or why not?

SKILL REVIEW

To prepare for Section 13.1, review the distance and midpoint formulas, completing the square, and graphing parabolas (Sections 10.7, 11.1, and 11.7).

Find the distance between each pair of points. [10.7]

53. $(-3, 7)$ and $(-2, 6)$

54. $(1, 5)$ and $(4, 1)$

Find the coordinates of the midpoint of the segment connecting each pair of points. [10.7]

55. $(3, -8)$ and $(5, -6)$

56. $(2, -11)$ and $(-9, -8)$

Solve by completing the square. [11.1]

57. $x^2 + 8x = 1$

58. $x^2 - 10x = 15$

Graph. [11.7]

59. $y = x^2 - 5x - 6$

60. $g(x) = 2x^2 - 6x + 3$

SYNTHESIS

TW **61.** *Atmospheric Pressure.* Atmospheric pressure P at altitude a is given by

$$P(a) = P_0e^{-0.00005a},$$

where P_0 is the pressure at sea level $\approx 14.7\ \text{lb/in}^2$ (pounds per square inch). Explain how a barometer, or some other device for measuring atmospheric pressure, can be used to find the height of a skyscraper.

TW **62.** Write a problem for a classmate to solve in which information is provided and the classmate is asked to find an exponential growth function. Make the problem as realistic as possible.

63. *Sports Salaries.* As of February 2010, Alex Rodriguez of the New York Yankees had the largest contract in sports history. As part of the 10-year \$275-million deal, he will receive \$20 million in 2016. How much money would need to be invested in 2008 at 4% interest, compounded continuously, in order to have \$20 million for Rodriguez in 2016? (This is much like determining what \$20 million in 2016 is worth in 2008 dollars.)

Source: *The San Francisco Chronicle*

64. *Supply and Demand.* The supply and demand for the sale of stereos by Sound Ideas are given by

$$S(x) = e^x \quad \text{and} \quad D(x) = 162{,}755e^{-x},$$

where $S(x)$ is the price at which the company is willing to supply x stereos and $D(x)$ is the demand price for a quantity of x stereos. Find the equilibrium point. (For reference, see Section 9.5.)

65. *Stellar Magnitude.* The apparent stellar magnitude m of a star with received intensity I is given by

$$m(I) = -(19 + 2.5 \cdot \log I),$$

where I is in watts per square meter (W/m^2). The smaller the apparent stellar magnitude, the brighter the star appears.

Source: The Columbus Optical SETI Observatory

a) The intensity of light received from the sun is $1390\ \text{W/m}^2$. What is the apparent stellar magnitude of the sun?

b) The 5-m diameter Hale telescope on Mt. Palomar can detect a star with magnitude +23. What is the received intensity of light from such a star?

66. *Growth of Bacteria.* The bacteria *Escherichia coli* (*E. coli*) are commonly found in the human bladder. Suppose that 3000 of the bacteria are present at time $t = 0$. Then t minutes later, the number of bacteria present is

$$N(t) = 3000(2)^{t/20}.$$

If 100,000,000 bacteria accumulate, a bladder infection can occur. If, at 11:00 A.M., a patient's bladder contains 25,000 *E. coli* bacteria, at what time can infection occur?

67. Show that for exponential growth at rate k, the doubling time T is given by $T = \dfrac{\ln 2}{k}$.

68. Show that for exponential decay at rate k, the half-life T is given by $T = \dfrac{\ln 2}{k}$.

Logistic Curves. Realistically, most quantities that are growing exponentially eventually level off. The quantity may continue to increase, but at a decreasing rate. This pattern of growth can be modeled by a logistic function

$$f(x) = \frac{c}{1 + ae^{-bx}}.$$

The general shape of this family of functions is shown by the graph below.

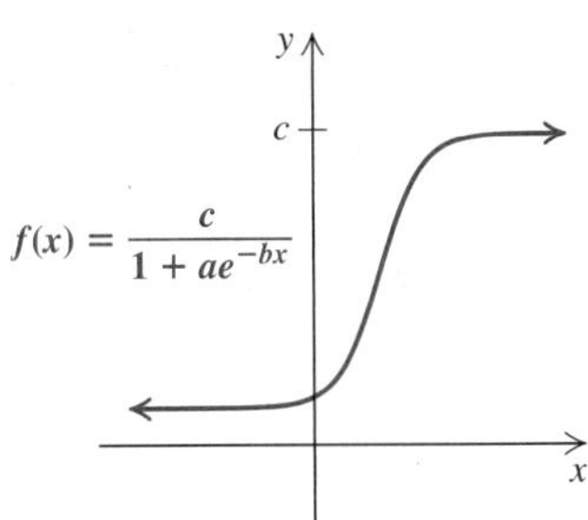

Many graphing calculators can fit a logistic function to a set of data.

69. *Internet Access.* The percent of U.S. households with Internet access for various years is listed in the table below.

Year	Percent of Households with Internet Access
1997	18.6
1998	26.2
2000	41.5
2003	54.6
2007	61.7

Source: www.ntia.doc.gov

a) Find a logistic function

$$f(x) = \frac{c}{1 + ae^{-bx}}$$

that could be used to estimate the percent of U.S. households with Internet access x years after 1997.

b) Use the function found in part (a) to predict the percent of U.S. households with Internet access in 2010.

TW **70.** *Heart Transplants.* In 1967, Dr. Christiaan Barnard of South Africa stunned the world by performing the first heart transplant. Since that time, the operation's popularity has both grown and declined, as shown in the table below.

Year	Number of Heart Transplants Worldwide
1982	189
1985	1189
1988	3157
1991	4186
1994	4402
1997	4039
2000	3246
2003	3020
2006	2200
2008	2927

Source: International Society for Heart & Lung Transplantation

a) Using 1982 as $t = 0$, graph the data.

b) Does it appear that an exponential function might have ever served as an appropriate model for these data? If so, for what years would this have been the case?

c) Considering *all* the data points on your graph, which would be the most appropriate model: a linear, quadratic, polynomial, or exponential function? Why?

Try Exercise Answers: Section 12.7

5. **(a)** 6.4 yr; **(b)** 23.4 yr **11.** 4.9 **15.** 130 dB
23. **(a)** $P(t) = 310e^{0.01t}$; **(b)** 329 million; **(c)** about 2022
31. **(a)** $k \approx 0.151$; $P(t) = 2000e^{0.151t}$; **(b)** 2011
35. About 1964 yr **41.** **(a)** $k \approx 0.123$; $V(t) = 9e^{0.123t}$; **(b)** \$360.4 million; **(c)** 5.6 yr; **(d)** 38.3 yr **45.** No
47. **(a)** $n(t) = 7.8(1.3725)^t$; **(b)** 31.7%; **(c)** about 696,000,000 transistors

Study Summary

KEY TERMS AND CONCEPTS	EXAMPLES	PRACTICE EXERCISES
SECTION 12.1: COMPOSITE FUNCTIONS AND INVERSE FUNCTIONS		
The **composition** of f and g is defined as $(f \circ g)(x) = f(g(x))$.	If $f(x) = \sqrt{x}$ and $g(x) = 2x - 5$, then $(f \circ g)(x) = f(g(x)) = f(2x - 5) = \sqrt{2x - 5}$.	**1.** Find $(f \circ g)(x)$ if $f(x) = 1 - 6x$ and $g(x) = x^2 - 3$.
A function f is **one-to-one** if different inputs have different outputs. The graph of a one-to-one function passes the **horizontal-line test.**	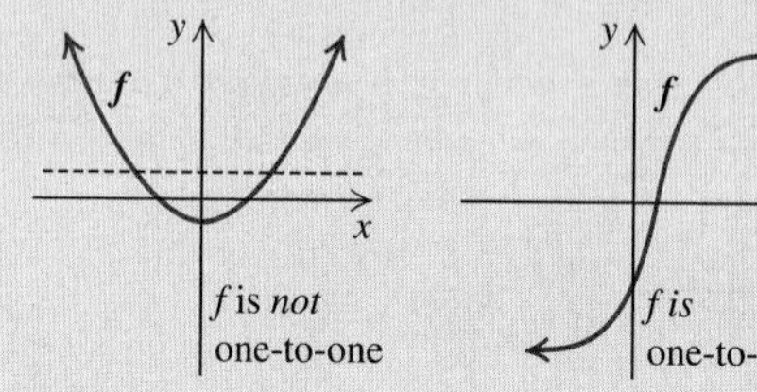	**2.** Determine whether $f(x) = 5x - 7$ is one-to-one.
If f is one-to-one, it is possible to find its inverse: **1.** Replace $f(x)$ with y. **2.** Interchange x and y. **3.** Solve for y. **4.** Replace y with $f^{-1}(x)$.	If $f(x) = 2x - 3$, find $f^{-1}(x)$. **1.** $y = 2x - 3$ **2.** $x = 2y - 3$ **3.** $x + 3 = 2y$ $\frac{x+3}{2} = y$ **4.** $\frac{x+3}{2} = f^{-1}(x)$	**3.** If $f(x) = 5x + 1$, find $f^{-1}(x)$.
SECTION 12.2: EXPONENTIAL FUNCTIONS **SECTION 12.3: LOGARITHMIC FUNCTIONS**		
For an **exponential function** f: $f(x) = a^x$; $a > 0, a \neq 1$; Domain: $\mathbb{R}$; $f^{-1}(x) = \log_a x$. For a **logarithmic function** g: $g(x) = \log_a x$; $a > 0, a \neq 1$; Domain: $(0, \infty)$; $g^{-1}(x) = a^x$.	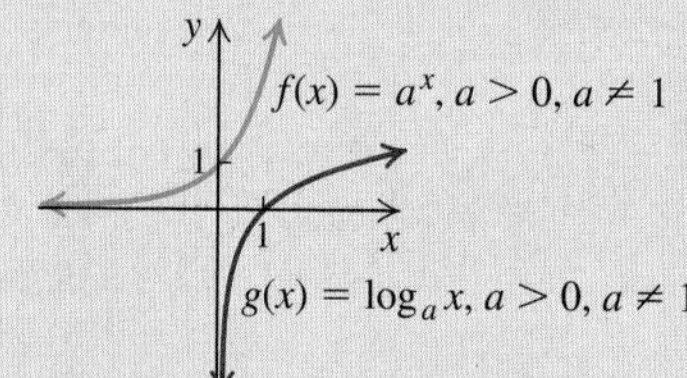	**4.** Graph by hand: $f(x) = 2^x$. **5.** Graph by hand: $f(x) = \log x$.
$\log_a x = m$ means $a^m = x$.	Solve: $\log_8 x = 2$. $\log_8 x = 2$ $8^2 = x$ **Rewriting as an exponential equation** $64 = x$	**6.** Rewrite as an equivalent logarithmic equation: $5^4 = 625$.

SECTION 12.4: PROPERTIES OF LOGARITHMIC FUNCTIONS
SECTION 12.5: NATURAL LOGARITHMS AND CHANGING BASES

Properties of Logarithms

$\log_a (MN) = \log_a M + \log_a N$	$\log_7 10 = \log_7 5 + \log_7 2$
$\log_a \frac{M}{N} = \log_a M - \log_a N$	$\log_5 \frac{14}{3} = \log_5 14 - \log_5 3$
$\log_a M^p = p \cdot \log_a M$	$\log_8 5^{12} = 12 \log_8 5$
$\log_a 1 = 0$	$\log_9 1 = 0$
$\log_a a = 1$	$\log_4 4 = 1$
$\log_a a^k = k$	$\log_3 3^8 = 8$
$\log M = \log_{10} M$	$\log 43 = \log_{10} 43$
$\ln M = \log_e M$	$\ln 37 = \log_e 37$
$\log_b M = \frac{\log_a M}{\log_a b}$	$\log_6 31 = \frac{\log 31}{\log 6} = \frac{\ln 31}{\ln 6}$

7. Express as an equivalent expression that is a sum of logarithms: $\log_9 xy$.

8. Express as an equivalent expression that is a difference of logarithms: $\log_6 \frac{7}{10}$.

9. Express as an equivalent expression that is a product: $\log 7^5$.

10. Simplify: $\log_8 1$.

11. Simplify: $\log_7 7$.

12. Simplify: $\log_t t^{12}$.

13. Simplify: $\log 100$.

14. Simplify: $\ln e$.

15. Use the change-of-base formula to find $\log_2 5$.

SECTION 12.6: SOLVING EXPONENTIAL AND LOGARITHMIC EQUATIONS

The Principle of Exponential Equality

For any real number b, $b \neq -1, 0$, or 1:

$b^x = b^y$ is equivalent to $x = y$.

Solve: $25 = 5^x$.

$$25 = 5^x$$
$$5^2 = 5^x$$
$$2 = x$$

16. Solve: $2^{3x} = 16$.

The Principle of Logarithmic Equality

For any logarithm base a, and for $x, y > 0$:

$x = y$ is equivalent to $\log_a x = \log_a y$.

Solve: $83 = 7^x$.

$$83 = 7^x$$
$$\log 83 = \log 7^x$$
$$\log 83 = x \log 7$$
$$\frac{\log 83}{\log 7} = x$$

17. Solve: $e^{0.1x} = 10$.

SECTION 12.7: APPLICATIONS OF EXPONENTIAL AND LOGARITHMIC FUNCTIONS

Exponential Growth Model

$$P(t) = P_0 e^{kt}, \quad k > 0$$

P_0 is the population at time 0.

$P(t)$ is the population at time t.

k is the **exponential growth rate**.

The **doubling time** is the amount of time necessary for the population to double in size.

If \$1000 is invested at 5%, compounded continuously, then:

- $P_0 = 1000$;
- $k = 0.05$;
- $P(t) = 1000e^{0.05t}$;
- The doubling time is 13.9 years.

18. The population of a town in 2010 was 15,000, and it was increasing at an exponential growth rate of 2.3% per year.

a) Write an exponential function that describes the population $P(t)$, where t is the number of years after 2010.

b) Find the doubling time.

Exponential Decay Model $P(t) = P_0e^{-kt}, \quad k > 0$ P_0 is the quantity present at time 0. $P(t)$ is the amount present at time t. k is the **exponential decay rate.** The **half-life** is the amount of time necessary for half of the quantity to decay.	If the population of a town is 2000, and the population is decreasing exponentially at a rate of 1.5% per year, then: • $P_0 = 2000$; • $k = 0.015$; • $P(t) = 2000e^{-0.015t}$; • The half-life is 46.2 years.	**19.** Argon-37 has an exponential decay rate of 1.98% per day. Find the half-life.

Review Exercises 

Concept Reinforcement *Classify each of the following statements as either true or false.*

1. The functions given by $f(x) = 3^x$ and $g(x) = \log_3 x$ are inverses of each other. [12.5]
2. A function's doubling time is the amount of time t for which $f(t) = 2f(0)$. [12.7]
3. A radioactive isotope's half-life is the amount of time t for which $f(t) = \frac{1}{2}f(0)$. [12.7]
4. $\ln(ab) = \ln a - \ln b$ [12.4]
5. $\log x^a = x \ln a$ [12.4]
6. $\log_a \frac{m}{n} = \log_a m - \log_a n$ [12.4]
7. For $f(x) = 3^x$, the domain of f is $[0, \infty)$. [12.2]
8. For $g(x) = \log_2 x$, the domain of g is $[0, \infty)$. [12.3]
9. The function F is not one-to-one if $F(-2) = F(5)$. [12.1]
10. The function g is one-to-one if it passes the vertical-line test. [12.1]

11. Find $(f \circ g)(x)$ and $(g \circ f)(x)$ if $f(x) = x^2 + 1$ and $g(x) = 2x - 3$. [12.1]

12. If $h(x) = \sqrt{3 - x}$, find $f(x)$ and $g(x)$ such that $h(x) = (f \circ g)(x)$. Answers may vary. [12.1]

13. Determine whether $f(x) = 4 - x^2$ is one-to-one. [12.1]

Find a formula for the inverse of each function. [12.1]

14. $f(x) = x - 8$

15. $g(x) = \dfrac{3x + 1}{2}$

16. $f(x) = 27x^3$

Graph by hand.

17. $f(x) = 3^x + 1$ [12.2]

18. $x = \left(\frac{1}{4}\right)^y$ [12.2]

19. $y = \log_5 x$ [12.3]

Simplify. [12.3]

20. $\log_3 9$

21. $\log_{10} \frac{1}{100}$

22. $\log_5 5^7$

23. $\log_9 3$

Rewrite as an equivalent logarithmic equation. [12.3]

24. $10^{-2} = \frac{1}{100}$

25. $25^{1/2} = 5$

Rewrite as an equivalent exponential equation. [12.3]

26. $\log_4 16 = x$

27. $\log_8 1 = 0$

Express as an equivalent expression using the individual logarithms of x, y, and z. [12.4]

28. $\log_a x^4y^2z^3$

29. $\log_a \dfrac{x^5}{yz^2}$

30. $\log \sqrt[4]{\dfrac{z^2}{x^3y}}$

Express as an equivalent expression that is a single logarithm and, if possible, simplify. [12.4]

31. $\log_a 7 + \log_a 8$

32. $\log_a 72 - \log_a 12$

33. $\frac{1}{2}\log a - \log b - 2\log c$

34. $\frac{1}{3}[\log_a x - 2\log_a y]$

Simplify. [12.4]

35. $\log_m m$

36. $\log_m 1$

37. $\log_m m^{17}$

Given $\log_a 2 = 1.8301$ *and* $\log_a 7 = 5.0999$, *find each of the following.* [12.4]

38. $\log_a 14$

39. $\log_a \frac{2}{7}$

40. $\log_a 28$

41. $\log_a 3.5$

42. $\log_a \sqrt{7}$

43. $\log_a \frac{1}{4}$

Use a calculator to find each of the following to the nearest ten-thousandth. [12.3], [12.5]

44. $\log 75$

45. $10^{1.789}$

46. $\ln 0.05$

47. $e^{-0.98}$

Find each of the following logarithms using the change-of-base formula. Round answers to the nearest ten-thousandth. [12.5]

48. $\log_5 2$

49. $\log_{12} 70$

Graph and state the domain and the range of each function. [12.5]

50. $f(x) = e^x - 1$

51. $g(x) = 0.6 \ln x$

Solve. Where appropriate, include approximations to the nearest ten-thousandth. [12.6]

52. $2^x = 32$

53. $3^{2x} = \frac{1}{9}$

54. $\log_3 x = -4$

55. $\log_x 16 = 4$

56. $\log x = -3$

57. $3 \ln x = -6$

58. $4^{2x-5} = 19$

59. $2^x = 12$

60. $e^{-0.1t} = 0.03$

61. $2 \ln x = -6$

62. $\log (2x - 5) = 1$

63. $\log_4 x - \log_4 (x - 15) = 2$

64. $\log_3 (x - 4) = 3 - \log_3 (x + 4)$

65. In a business class, students were tested at the end of the course with a final exam. They were then tested again 6 months later. The forgetting formula was determined to be

$$S(t) = 82 - 18 \log (t + 1),$$

where t is the time, in months, after taking the final exam. [12.7]

a) Determine the average score when they first took the exam (when $t = 0$).

b) What was the average score after 6 months?

c) After what time was the average score 54?

66. A laptop computer is purchased for \$1500. Its value each year is about 80% of its value in the preceding year. Its value in dollars after t years is given by the exponential function

$$V(t) = 1500(0.8)^t. \text{ [12.7]}$$

a) After what amount of time will the computer's value be \$900?

b) After what amount of time will the computer's value be half the original value?

67. U.S. companies spent \$1.2 billion in e-mail marketing in 2007. This amount was predicted to grow exponentially to \$2.1 billion in 2012. [12.7]
Source: Jupiter Research

a) Find the exponential growth rate k, and write a function that describes the amount $A(t)$, in billions of dollars, spent on e-mail marketing t years after 2007.

b) Estimate the amount spent on e-mail marketing in 2015.

c) In what year will U.S. companies spend \$4 billion on e-mail marketing?

d) Find the doubling time.

68. In 2005, consumers received, on average, 3253 spam messages. The volume of spam messages per consumer is decreasing exponentially at an exponential decay rate of 13.7% per year. [12.7]

a) Find the exponential decay function that can be used to predict the average number of spam messages, $M(t)$, t years after 2005.

b) Predict the number of spam messages received per consumer in 2012.

c) In what year, theoretically, will the average consumer receive 100 spam messages?

69. *Diseases.* The number of hepatitis A cases in the United States has decreased exponentially since 1995. The number of cases for various years are listed in the table below. [12.7]

Year	Number of Hepatitis A Cases (in thousands)
1995	31.6
2000	13.4
2003	7.7
2004	5.7
2005	4.5
2006	3.6
2007	3.0

Source: U.S. Centers for Disease Control

a) Use regression to find an exponential function of the form $f(x) = ab^x$ that can be used to estimate the number of hepatitis A cases x years after 1995.

b) Use the function of part (a) to estimate the number of cases of hepatitis A in 2010.

c) Find the exponential decay rate.

70. The value of Aret's stock market portfolio doubled in 6 years. What was the exponential growth rate? [12.7]

71. How long will it take \$7600 to double itself if it is invested at 4.2%, compounded continuously? [12.7]

72. How old is a skull that has lost 34% of its carbon-14? (Use $P(t) = P_0 e^{-0.00012t}$.) [12.7]

73. What is the pH of coffee if its hydrogen ion concentration is 7.9×10^{-6} moles per liter? (Use $\text{pH} = -\log[\text{H}^+]$.) [12.7]

74. *Nuclear Energy.* Plutonium-239 (Pu-239) is used in nuclear energy plants. The half-life of Pu-239 is 24,360 years. How long will it take for a fuel rod of Pu-239 to lose 90% of its radioactivity? [12.7]
Source: *Microsoft Encarta 97 Encyclopedia*

75. The roar of a lion can reach a sound intensity of $2.5 \times 10^{-1}\ \text{W/m}^2$. How loud in decibels is this sound level? $\left(\text{Use } L = 10 \cdot \log \dfrac{I}{10^{-12}\ \text{W/m}^2}.\right)$ [12.7]
Source: en.allexperts.com

SYNTHESIS

TW **76.** Explain why negative numbers do not have logarithms. [12.3]

TW **77.** Explain why taking the natural logarithm or common logarithm on each side of an equation produces an equivalent equation. [12.6]

Solve. [12.6]

78. $\ln(\ln x) = 3$

79. $2^{x^2+4x} = \frac{1}{8}$

80. Solve the system

$$5^{x+y} = 25,$$
$$2^{2x-y} = 64.\ [12.6]$$

81. *Size of the Internet.* [12.7] Chinese researchers claim that the Internet doubles in size every 5.32 years. What is the exponential growth rate?
Source: Zhang, Guo-Qing, Guo-Qiang Zhang, Qing-Feng Yang, Su-Qi Cheng, and Tao Zhou. "Evolution of the Internet and its Cores," *New Journal of Physics* **10** (2008) 123027.

Chapter Test 12

1. Find $(f \circ g)(x)$ and $(g \circ f)(x)$ if $f(x) = x + x^2$ and $g(x) = 2x + 1$.

2. If

$$h(x) = \frac{1}{2x^2 + 1},$$

find $f(x)$ and $g(x)$ such that $h(x) = (f \circ g)(x)$. Answers may vary.

3. Determine whether $f(x) = |x - 3|$ is one-to-one.

Find a formula for the inverse of each function.

4. $f(x) = 3x + 4$

5. $g(x) = (x + 1)^3$

Graph by hand.

6. $f(x) = 2^x - 3$

7. $g(x) = \log_7 x$

Simplify.

8. $\log_5 125$

9. $\log_{100} 10$

10. $3^{\log_3 18}$

11. $\log_n n$

12. $\log_c 1$

13. $\log_a a^{19}$

14. Rewrite as an equivalent logarithmic equation:
$5^{-4} = \frac{1}{625}$.

15. Rewrite as an equivalent exponential equation:
$m = \log_2 \frac{1}{2}$.

16. Express as an equivalent expression using the individual logarithms of a, b, and c:
$$\log \frac{a^3 b^{1/2}}{c^2}.$$

17. Express as an equivalent expression that is a single logarithm:
$\frac{1}{3} \log_a x + 2 \log_a z$.

Given $\log_a 2 = 0.301$, $\log_a 6 = 0.778$, *and* $\log_a 7 = 0.845$, *find each of the following.*

18. $\log_a 14$

19. $\log_a 3$

20. $\log_a 16$

Use a calculator to find each of the following to the nearest ten-thousandth.

21. $\log 12.3$

22. $10^{-0.8}$

23. $\ln 0.4$

24. $e^{4.8}$

25. Find $\log_3 14$ using the change-of-base formula. Round to the nearest ten-thousandth.

Graph and state the domain and the range of each function.

26. $f(x) = e^x + 3$

27. $g(x) = \ln(x - 4)$

Solve. Where appropriate, include approximations to the nearest ten-thousandth.

28. $2^x = \frac{1}{32}$

29. $\log_4 x = \frac{1}{2}$

30. $\log x = 4$

31. $5^{4-3x} = 87$

32. $7^x = 1.2$

33. $\ln x = 3$

34. $\log(x - 3) + \log(x + 1) = \log 5$

35. The average walking speed R of people living in a city of population P is given by $R = 0.37 \ln P + 0.05$, where R is in feet per second and P is in thousands.

a) The population of Tulsa, Oklahoma, is 383,000. Find the average walking speed.

b) San Diego, California, has an average walking speed of about 3 ft/sec. Find the population.

36. The population of Austria was about 8.2 million in 2009, and the exponential growth rate was 0.052% per year.
Source: Central Intelligence Agency, *The World Factbook*

a) Write an exponential function describing the population of Austria.

b) What will the population be in 2020? in 2050?

c) When will the population be 9 million?

d) What is the doubling time?

37. The average cost of a year at a private four-year college grew exponentially from \$21,855 in 2001 to \$35,600 in 2010.
Source: National Center for Education Statistics

a) Find the exponential growth rate k, and write an exponential function that approximates the cost $C(t)$ of a year of college t years after 2001.

b) Predict the cost of a year of college in 2015.

c) In what year will the average cost of a year of college be \$50,000?

38. *E-Book Readers.* The number of Kindle unit sales grew exponentially from 2008 to 2010, as listed in the table below.

Year	Kindle Unit Sales (in thousands)
2008	500
2009	1027
2010	3533

Source: Citi Investment Research, estimated

a) Use regression to find an exponential function of the form $f(x) = ab^x$ that can be used to estimate Kindle unit sales x years after 2008.

b) Use the function to estimate Kindle unit sales in 2012.

39. An investment with interest compounded continuously doubled itself in 15 years. What is the interest rate?

40. How old is an animal bone that has lost 43% of its carbon-14? (Use $P(t) = P_0 e^{-0.00012t}$.)

41. The sound of fireworks averages 140 dB. What is the intensity of such a sound?
$$\left(\text{Use } L = 10 \cdot \log \frac{I}{I_0}, \text{ where } I_0 = 10^{-12}\ \text{W/m}^2.\right)$$

42. The hydrogen ion concentration of water is 1.0×10^{-7} moles per liter. What is the pH? (Use $\text{pH} = -\log[\text{H}^+]$.)

SYNTHESIS

43. Solve: $\log_5 |2x - 7| = 4$.

44. If $\log_a x = 2$, $\log_a y = 3$, and $\log_a z = 4$, find
$$\log_a \frac{\sqrt[3]{x^2 z}}{\sqrt[3]{y^2 z^{-1}}}.$$

Cumulative Review: Chapters 1–12

1. Evaluate $\dfrac{x^0 + y}{-z}$ for $x = 6, y = 9$, and $z = -5$. [1.8], [5.1]

Simplify.

2. $(-2x^2y^{-3})^{-4}$ [5.2]

3. $(-5x^4y^{-3}z^2)(-4x^2y^2)$ [5.2]

4. $\dfrac{3x^4y^6z^{-2}}{-9x^4y^2z^3}$ [5.2]

5. $(1.5 \times 10^{-3})(4.2 \times 10^{-12})$ [5.2]

6. $3^3 + 2^2 - (32 \div 4 - 16 \div 8)$ [1.8]

Solve.

7. $3(2x - 3) = 9 - 5(2 - x)$ [2.2]

8. $4x - 3y = 15,$
   $3x + 5y = 4$ [4.3]

9. $x + y - 3z = -1,$
   $2x - y + z = 4,$
   $-x - y + z = 1$ [9.1]

10. $x(x - 3) = 70$ [6.2]

11. $\dfrac{7}{x^2 - 5x} - \dfrac{2}{x - 5} = \dfrac{4}{x}$ [7.6]

12. $\sqrt{4 - 5x} = 2x - 1$ [10.6]

13. $\sqrt[3]{2x} = 1$ [10.6]

14. $3x^2 + 48 = 0$ [11.1]

15. $x^4 - 13x^2 + 36 = 0$ [11.5]

16. $\log_x 81 = 2$ [12.3]

17. $3^{5x} = 7$ [12.6]

18. $\ln x - \ln (x - 8) = 1$ [12.6]

19. $x^2 + 4x > 5$ [8.4]

20. If $f(x) = x^2 + 6x$, find a such that $f(a) = 11$. [11.2]

21. If $f(x) = |2x - 3|$, find all x for which $f(x) \geq 7$. [8.2]

Solve.

22. $D = \dfrac{ab}{b + a}$, for a [7.8]

23. $d = ax^2 + vx$, for x [11.4]

24. Find the domain of the function f given by
$$f(x) = \frac{x + 4}{3x^2 - 5x - 2}.$$ [6.3]

Perform the indicated operations and simplify.

25. $(5p^2q^3 + 6pq - p^2 + p) - (2p^2q^3 + p^2 - 5pq - 9)$ [5.7]

26. $(3x^2 - z^3)^2$ [5.7]

27. $\dfrac{1 + \dfrac{3}{x}}{x - 1 - \dfrac{12}{x}}$ [7.5]

28. $\dfrac{a^2 - a - 6}{a^3 - 27} \cdot \dfrac{a^2 + 3a + 9}{6}$ [7.2]

29. $\dfrac{3}{x + 6} - \dfrac{2}{x^2 - 36} + \dfrac{4}{x - 6}$ [7.4]

30. $\dfrac{\sqrt[3]{24xy^8}}{\sqrt[3]{3xy}}$ [10.4]

31. $\sqrt{x + 5}\,\sqrt[5]{x + 5}$ [10.5]

32. $\left(2 - i\sqrt{3}\right)\left(6 + i\sqrt{3}\right)$ [10.8]

33. $(x^4 - 8x^3 + 15x^2 + x - 3) \div (x - 3)$ [5.8]

Factor.

34. $xy + 2xz - xw$ [6.1]

35. $6x^2 + 8xy - 8y^2$ [6.3]

36. $x^4 - 4x^3 + 7x - 28$ [6.1]

37. $2m^2 + 12mn + 18n^2$ [6.4]

38. $x^4 - 16y^4$ [6.4]

39. Rationalize the denominator:
$$\frac{3 - \sqrt{y}}{2 - \sqrt{y}}.$$ [10.5]

40. Find the inverse of f if $f(x) = 9 - 2x$. [12.1]

41. Find a linear equation with a graph that contains the points $(0, -8)$ and $(-1, 2)$. [3.7]

42. Find an equation of the line whose graph has a y-intercept of $(0, 5)$ and is perpendicular to the line given by $2x + y = 6$. [3.6]

Graph by hand.

43. $5x = 15 + 3y$ [3.3]

44. $y = \log_3 x$ [12.3]

45. $-2x - 3y \leq 12$ [8.4]

46. Graph: $f(x) = 2x^2 + 12x + 19$. [11.7]
   a) Label the vertex.
   b) Draw the axis of symmetry.
   c) Find the maximum or minimum value.

47. Graph $f(x) = 2e^x$ and determine the domain and the range. [12.5]

48. Express as a single logarithm:
$$3 \log x - \tfrac{1}{2} \log y - 2 \log z. \quad [12.4]$$

In Exercises 49–52, match each function with one of the following graphs.

a)
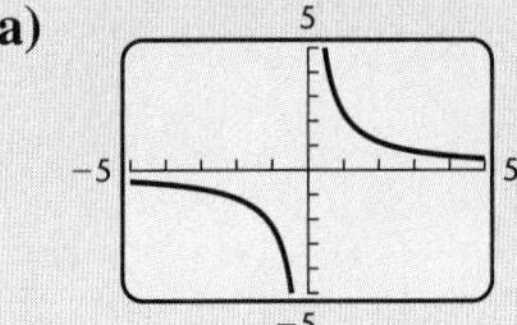

b)
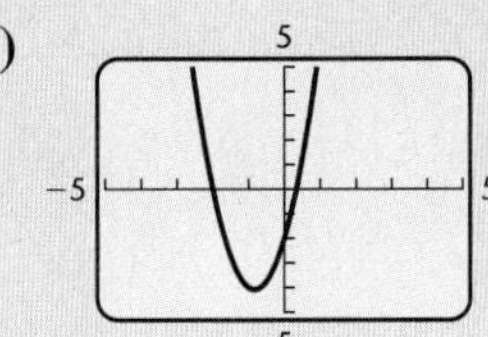

c)
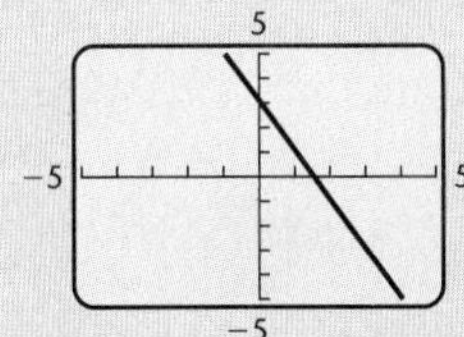

d)
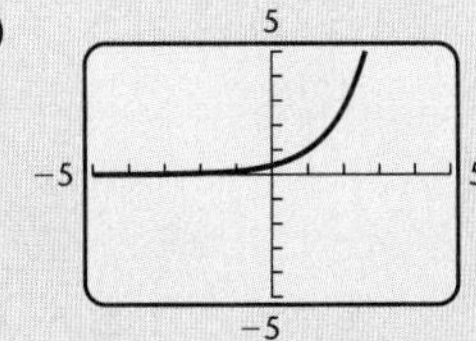

49. $f(x) = -2x + 3.1$ [3.6], [3.8]

50. $p(x) = 3x^2 + 5x - 2$ [11.6]

51. $h(x) = \dfrac{2.3}{x}$ [7.1]

52. $g(x) = e^{x-1}$ [12.5]

Solve.

53. *Colorado River.* The Colorado River delivers 1.5 million acre-feet of water to Mexico each year. This is only 10% of the volume of the river; the remainder is diverted at an earlier time for agricultural use. How much water is diverted each year from the Colorado River? [2.5]
Source: www.sierraclub.org

54. *Desalination.* An increasing number of cities are supplying some of their fresh water through desalination, the process of removing the salt from ocean water. The worldwide desalination capacity has grown exponentially from 15 million m^3 per day in 1990 to 55 million m^3 per day in 2007. [12.7]
Source: Global Water Intelligence

   a) Find the exponential growth rate k, and write an equation for an exponential function that can be used to predict the worldwide desalination capacity $D(t)$, in millions of cubic meters per day, t years after 1990.
   b) Predict the worldwide desalination capacity in 2012.
   c) In what year will the worldwide desalination capacity reach 100 million m^3 per day?

55. Good's Candies of Indiana makes all their chocolates by hand. It takes Abi 10 min to coat a tray of candies in chocolate. It takes Brad 12 min to coat a tray of candies. How long would it take Abi and Brad, working together, to coat the candies? [7.7]

56. Joe's Thick and Tasty salad dressing gets 45% of its calories from fat. The Light and Lean dressing gets 20% of its calories from fat. How many ounces of each should be mixed in order to get 15 oz of dressing that gets 30% of its calories from fat? [4.4]

57. A fishing boat with a trolling motor can move at a speed of 5 km/h in still water. The boat travels 42 km downriver in the same time that it takes to travel 12 km upriver. What is the speed of the river? [7.7]

Travel. *Prices for a massage at Passport Travel Spa in an airport waiting area are shown in the graph below. Use the graph for Exercises 58–62.*

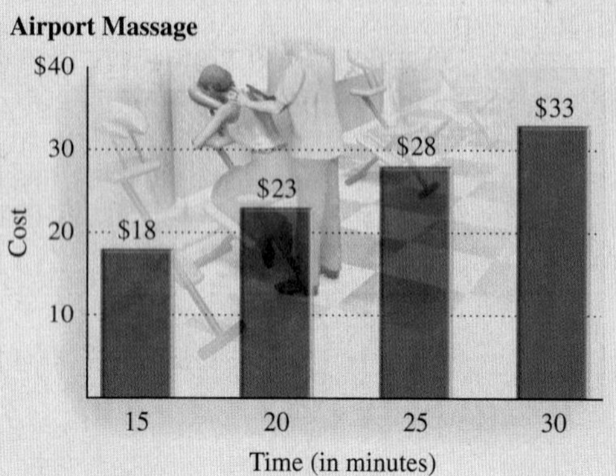

Source: Passport Travel Spa, Indianapolis International Airport

58. Determine whether the data can be modeled by a linear, quadratic, or exponential function. [3.7]

59. Find the rate of change, in dollars per minute. [3.4]

60. Find a linear function $f(x) = mx + b$ that fits the data, where f is the cost of the massage and x is the length, in minutes. [3.7], [3.8]

61. Use the function in Exercise 60 to estimate the cost of a 10-min massage. [3.8]

62. For the function in Exercise 60, what do the values of m and b signify? [3.6]

Life Insurance. *The monthly premium for a term life insurance policy increases as the age of the insured person increases. The graph below shows some premiums for a $250,000 policy for a male.*

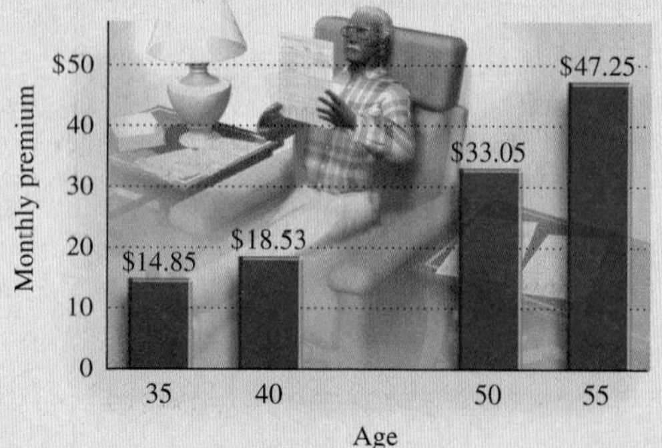

Source: Insurance company advertisement

63. Use regression to find an exponential function of the form $f(x) = ab^x$ that can be used to estimate the monthly insurance premium m for a male who is x years old. [12.7]

64. Use the function in Exercise 63 to estimate the monthly premium for a 45-year-old male. [12.7]

SYNTHESIS

Solve.

65. $\dfrac{5}{3x - 3} + \dfrac{10}{3x + 6} = \dfrac{5x}{x^2 + x - 2}$ [6.6]

66. $\log \sqrt{3x} = \sqrt{\log 3x}$ [12.6]

67. The Danville Express travels 280 mi at a certain speed. If the speed had been increased by 5 mph, the trip could have been made in 1 hr less time. Find the actual speed. [11.3]

Elementary Algebra Review

This chapter is a review of the first six chapters of this text. Each section corresponds to a chapter of the text. For further explanation of the topics in this chapter, refer to the section or pages referenced in the margin.

R.1 Introduction to Algebraic Expressions

- The Real Numbers
- Operations on Real Numbers
- Algebraic Expressions

THE REAL NUMBERS

Sets of real numbers (Section 1.4)

Sets of Numbers

Natural numbers:	$\{1, 2, 3, \ldots\}$
Whole numbers:	$\{0, 1, 2, 3, \ldots\}$
Integers:	$\{\ldots, -3, -2, -1, 0, 1, 2, 3, \ldots\}$
Rational numbers:	$\left\{\frac{a}{b} \,\middle|\, a \text{ and } b \text{ are integers and } b \neq 0\right\}$

Terminating decimals (p. 34)

Repeating decimals (p. 34)

Irrational numbers (p. 35)

Real numbers (p. 36)

Rational numbers can always be written as **terminating** decimals or **repeating** decimals. **Irrational numbers,** like $\sqrt{2}$ or π, can be thought of as nonterminating, nonrepeating decimals. The set of **real numbers** consists of all rational numbers and irrational numbers, taken together.

Real numbers can be represented by points on the number line. For any two numbers, the one to the left is less than the one to the right.

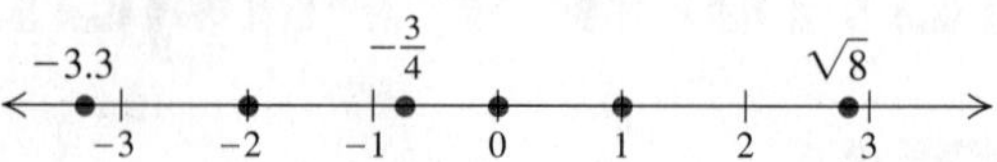

Equation (p. 7)

Inequality (p. 37)

Sentences like $\frac{1}{4} = 0.25$, containing an equals sign, are called **equations.** An **inequality** is a sentence containing $>$ (is greater than), $<$ (is less than), $\geq$ (is greater than or equal to), or $\leq$ (is less than or equal to). Equations and inequalities can be true or false.

EXAMPLE 1 Write true or false for each equation or inequality.

a) $-2\frac{1}{3} = -\frac{7}{3}$ **b)** $1 = -1$ **c)** $-5 < -2$
d) $-3 \geq 2$ **e)** $1.1 \leq 1.1$

SOLUTION

a) $-2\frac{1}{3} = -\frac{7}{3}$ is *true* because $-2\frac{1}{3}$ and $-\frac{7}{3}$ represent the same number.

b) $1 = -1$ is a *false* equation.

c) $-5 < -2$ is *true* because -5 is to the left of -2 on the number line.

d) $-3 \geq 2$ is *false* because neither $-3 > 2$ nor $-3 = 2$ is true.

e) $1.1 \leq 1.1$ is *true* because $1.1 = 1.1$ is true.

Try Exercise 1.

Absolute value (p. 38)

The distance of a number from 0 is called the **absolute value** of the number. The notation $|-4|$ represents the absolute value of -4. The absolute value of a number is never negative.

EXAMPLE 2 Find the absolute value: **(a)** $|-4|$; **(b)** $|\frac{11}{3}|$; **(c)** $|0|$.

SOLUTION

a) $|-4| = 4$ since -4 is 4 units from 0.
b) $|\frac{11}{3}| = \frac{11}{3}$ since $\frac{11}{3}$ is $\frac{11}{3}$ units from 0.
c) $|0| = 0$ since 0 is 0 units from itself.

Try Exercise 7.

OPERATIONS ON REAL NUMBERS

Addition (Section 1.5)

Rules for Addition of Real Numbers

1. *Positive numbers*: Add as usual. The answer is positive.
2. *Negative numbers*: Add absolute values and make the answer negative.
3. *A positive number and a negative number*: Subtract absolute values. Then:
   - **a)** If the positive number has the greater absolute value, the answer is positive.
   - **b)** If the negative number has the greater absolute value, the answer is negative.
   - **c)** If the numbers have the same absolute value, the answer is 0.
4. *One number is zero*: The sum is the other number.

Opposite (p. 48)

Every real number has an **opposite.** The opposite of -6 is 6, the opposite of 3.7 is -3.7, and the opposite of 0 is itself. When opposites are added, the result is 0. Finding the opposite of a number is often called "changing its sign."

Subtraction of real numbers is defined in terms of addition and opposites.

Subtraction (Section 1.6)

Subtraction of Real Numbers To subtract, add the opposite of the number being subtracted.

The rules for multiplication and division can be stated together.

Multiplication and division (Section 1.7)

Rules for Multiplication and Division To multiply or divide two real numbers:

1. Using the absolute values, multiply or divide, as indicated.
2. If the signs are the same, the answer is positive.
3. If the signs are different, the answer is negative.

EXAMPLE 3 Perform the indicated operations: **(a)** $-13 + (-9)$; **(b)** $-\frac{4}{5} + \frac{1}{10}$; **(c)** $-6 - (-7.3)$; **(d)** $3(-1.5)$; **(e)** $\left(-\frac{4}{9}\right) \div \left(-\frac{2}{5}\right)$.

SOLUTION

a) $-13 + (-9) = -22$

Two negatives. *Think*: Add the absolute values, 13 and 9, to get 22. Make the answer *negative*, −22.

b) $-\frac{4}{5} + \frac{1}{10} = -\frac{8}{10} + \frac{1}{10} = -\frac{7}{10}$

A negative and a positive. *Think*: The difference of absolute values is $\frac{8}{10} - \frac{1}{10}$, or $\frac{7}{10}$. The negative number has the larger absolute value, so the answer is *negative*, $-\frac{7}{10}$.

c) $-6 - (-7.3) = -6 + 7.3 = 1.3$

Change the subtraction to addition and add the opposite.

d) $3(-1.5) = -4.5$

***Think*: $3(1.5) = 4.5$. The signs are different, so the answer is negative.**

e) $\left(-\frac{4}{9}\right) \div \left(-\frac{2}{5}\right) = \left(-\frac{4}{9}\right) \cdot \left(-\frac{5}{2}\right)$

$= \frac{20}{18} = \frac{10}{9} \cdot \frac{2}{2} = \frac{10}{9}$

Multiplying by the reciprocal. The answer is positive.

Try Exercises 11 and 25.

Addition, subtraction, and multiplication are defined for all real numbers, but we cannot divide by 0. For example, $\frac{0}{3} = 0 \div 3 = 0$, but $\frac{3}{0} = 3 \div 0$ is *undefined.*

Exponential notation (p. 63)

A product like $2 \cdot 2 \cdot 2 \cdot 2$, in which the factors are the same, is called a **power.** Powers are often written using **exponential notation:**

$$2 \cdot 2 \cdot 2 \cdot 2 = 2^4.$$

← **There are 4 factors; 4 is the *exponent*.**

↑ **2 is the *base*.**

A number raised to the exponent of 1 is the number itself; for example, $3^1 = 3$.

An expression containing a series of operations is not necessarily evaluated from left to right. Instead, we perform the operations according to the following rules.

Rules for Order of Operations

1. Calculate within the innermost grouping symbols, ( ), [ ], { }, | |, and above or below fraction bars.
2. Simplify all exponential expressions.
3. Perform all multiplication and division, working from left to right.
4. Perform all addition and subtraction, working from left to right.

EXAMPLE 4 Simplify: $3 - [(4 \times 5) + 12 \div 2^3 \times 6] + 5$.

SOLUTION

$3 - [(4 \times 5) + 12 \div 2^3 \times 6] + 5$	
$= 3 - [20 + 12 \div 2^3 \times 6] + 5$	**Doing the calculations in the innermost parentheses first**
$= 3 - [20 + 12 \div 8 \times 6] + 5$	**Working inside the brackets; evaluating 2^3**
$= 3 - [20 + 1.5 \times 6] + 5$	**$12 \div 8$ is the first multiplication or division working from left to right.**
$= 3 - [20 + 9] + 5$	**Multiplying**
$= 3 - 29 + 5$	**Completing the calculations within the brackets**
$= -26 + 5$ $= -21$	**Adding and subtracting from left to right**

Try Exercise 45.

ALGEBRAIC EXPRESSIONS

Algebraic expression (p. 2)

Constant (p. 2)

Variable (p. 2)

Substitute (p. 2)

Evaluate (p. 2)

In an **algebraic expression** like $2xt^3$, the number 2 is a **constant** and x and t are **variables.** Algebraic expressions containing variables can be **evaluated** by **substituting** a number for each variable in the expression and following the rules for order of operations.

EXAMPLE 5 The perimeter P of a rectangle of length l and width w is given by the formula $P = 2l + 2w$. Find the perimeter when l is 16 in. and w is 7.5 in.

SOLUTION We evaluate, substituting 16 in. for l and 7.5 in. for w and carrying out the operations:

$$P = 2l + 2w$$

Substitute. $= 2 \cdot 16 + 2 \cdot 7.5$

Carry out the operations. $= 32 + 15$

$= 47$ in.

w

l

Try Exercise 55.

Expressions that represent the same number are said to be **equivalent.** The laws that follow provide methods for writing equivalent expressions.

Laws and Properties of Real Numbers

Commutative laws:	$a + b = b + a$; $ab = ba$
Associative laws:	$a + (b + c) = (a + b) + c$; $a(bc) = (ab)c$
Distributive law:	$a(b + c) = ab + ac$
Identity property of 1:	$1 \cdot a = a \cdot 1 = a$
Identity property of 0:	$a + 0 = 0 + a = a$
Law of opposites:	$a + (-a) = 0$
Multiplicative property of 0:	$0 \cdot a = a \cdot 0 = 0$
Property of -1:	$-1 \cdot a = -a$
Opposite of a sum:	$-(a + b) = -a + (-b)$
$\frac{-a}{b} = \frac{a}{-b} = -\frac{a}{b}$,	$\frac{-a}{-b} = \frac{a}{b}$

Factor (p. 19)

The distributive law can be used to multiply and to **factor** expressions. We factor an expression when we write an equivalent expression that is a product.

EXAMPLE 6 Write an equivalent expression as indicated.

a) Multiply: $-2(5x - 3)$.

b) Factor: $5x + 10y + 5$.

SOLUTION

a)
$$\begin{aligned} -2(5x - 3) &= -2(5x + (-3)) && \text{Adding the opposite} \\ &= -2 \cdot 5x + (-2) \cdot (-3) && \text{Using the distributive law} \\ &= (-2 \cdot 5)x + 6 && \text{Using the associative law for multiplication} \\ &= -10x + 6 \end{aligned}$$

b)
$$\begin{aligned} 5x + 10y + 5 &= 5 \cdot x + 5 \cdot 2y + 5 \cdot 1 && \text{The common factor is 5.} \\ &= 5(x + 2y + 1) && \text{Using the distributive law} \end{aligned}$$

Factoring can be checked by multiplying:

$$5(x + 2y + 1) = 5 \cdot x + 5 \cdot 2y + 5 \cdot 1 = 5x + 10y + 5.$$

Try Exercises 61 and 69.

Terms (p. 18)

Combine like terms (p. 45)

The **terms** of an algebraic expression are separated by plus signs. When two terms have variable factors that are exactly the same, the terms are called **like,** or **similar, terms.** The distributive law enables us to **combine,** or **collect, like terms.**

EXAMPLE 7 Combine like terms: $-5m + 3n - 4n + 10m$.

SOLUTION

$-5m + 3n - 4n + 10m$

$= -5m + 3n + (-4n) + 10m$ **Rewriting as addition**

$= -5m + 10m + 3n + (-4n)$ **Using the commutative law of addition**

$= (-5 + 10)m + (3 + (-4))n$ **Using the distributive law**

$= 5m + (-n)$

$= 5m - n$ **Rewriting as subtraction**

Try Exercise 77.

We can also use the distributive law to help simplify algebraic expressions containing parentheses.

EXAMPLE 8 Simplify: **(a)** $4x - (y - 2x)$; **(b)** $3(t + 2) - 6(t - 1)$.

SOLUTION

a) $4x - (y - 2x) = 4x - y + 2x$ **Removing parentheses and changing the sign of every term**

$= 6x - y$ **Combining like terms**

b) $3(t + 2) - 6(t - 1) = 3t + 6 - 6t + 6$ **Multiplying each term of $t + 2$ by 3 and each term of $t - 1$ by -6**

$= -3t + 12$ **Combining like terms**

Try Exercise 83.

Value (p. 2)

Solution (p. 7)

If the expressions on each side of an equation have the same **value** for a given number, then that number is a **solution** of the equation.

EXAMPLE 9 Determine whether each number is a solution of $x - 2 = -5$.

a) 3 **b)** -3

SOLUTION

a) We have

$x - 2 = -5$ **Writing the equation**

$3 - 2 \mid -5$ **Substituting 3 for x**

$1 \stackrel{?}{=} -5$ **$1 = -5$ is FALSE**

Since $3 - 2 = -5$ is false, 3 is not a solution of $x - 2 = -5$.

b) We have

$x - 2 = -5$

$-3 - 2 \mid -5$

$-5 \stackrel{?}{=} -5$ TRUE

Since $-3 - 2 = -5$ is true, -3 is a solution of $x - 2 = -5$.

Try Exercise 89.

Translating to algebraic expressions (p. 5)

Certain word phrases can be translated to algebraic expressions. These in turn can often be used to translate problems to equations.

Translating to equations (p. 7)

EXAMPLE 10 Energy Use. Translate this problem to an equation. Do not solve.

On average, a home spa costs about \$192 per year to operate. This is 16 times as much as the average annual operating cost of a home computer. How much does it cost to operate a home computer for a year?
Source: U.S. Department of Energy

SOLUTION We let c represent the annual cost of operating a home computer. We then reword the problem to make the translation more direct.

Rewording:	16	times	home computer cost	is	192
	↓	↓	↓	↓	↓
Translating:	16	$\cdot$	c	$=$	192

Try Exercise 95.

R.1 Exercise Set

FOR EXTRA HELP MyMathLab   
PRACTICE WATCH DOWNLOAD READ REVIEW

Classify each equation or inequality as true or false.

1. $2.3 = 2.31$
2. $-3 \geq -3$
3. $-10 < -1$
4. $0 \leq -1$
5. $0 > -5$
6. $\frac{1}{10} = 0.1$

Find each absolute value.

7. $|4|$
8. $\left|\frac{11}{4}\right|$
9. $|-1.3|$
10. $|-105|$

Simplify.

11. $(-13) + (-12)$
12. $3 - (-2)$
13. $-\frac{1}{3} - \frac{2}{5}$
14. $\frac{3}{8} \div \frac{3}{5}$
15. $4.2 - 10.7$
16. $(-1.3)(2.8)$
17. $-15 + 0$
18. $\left(-\frac{1}{2}\right) + \frac{1}{8}$
19. $0 \div (-10)$
20. $0 - 32$
21. $\left(-\frac{3}{10}\right) + \left(-\frac{1}{5}\right)$
22. $\left(-\frac{4}{7}\right)\left(\frac{7}{4}\right)$
23. $-3.8 + 9.6$
24. $-0.01 + 1$
25. $(-12) \div 4$
26. $(-87)(0)$
27. $32 - (-7)$
28. $-100 + 35$
29. $(-10)(-17.5)$
30. $-10 - 2.68$
31. $(-68) + 36$
32. $175 \div (-25)$
33. $2 + (-3) + 7 + 10$
34. $-5 + (-15) + 13 + (-1)$
35. $3 \cdot (-2) \cdot (-1) \cdot (-1)$
36. $(-6) \cdot (-5) \cdot (-4) \cdot (-3) \cdot (-2) \cdot (-1)$
37. $(-1)^4 + 2^3$
38. $(-1)^5 + 2^4$
39. $2 \cdot 6 - 3 \cdot 5$
40. $12 \div 4 + 15 \div 3$
41. $3 - (2 \cdot 4 + 11)$
42. $3 - 11 + 2 \cdot 4$
43. $4 \cdot 5^2$
44. $7 \cdot 2^3$
45. $25 - 8 \cdot 3 + 1$
46. $12 - 16 \cdot 5 + 4$
47. $2 - (3^3 + 16 \div (-2)^3)$
48. $-7 - (8 + 10 \cdot 2^2)$

49. $|6(-3)| + |(-2)(-9)|$

50. $3 - |2 - 7 + 4|$

51. $\dfrac{7000 + (-10)^3}{10^2 \cdot (2 + 4)}$

52. $\dfrac{3 - 2 \cdot 6 - 5}{2(3 + 7)^2}$

53. $2 + 8 \div 2 \cdot 2$

54. $2 + 8 \div (2 \cdot 2)$

Evaluate.

55. $y - x$, for $x = 10$ and $y = 3$

56. $n - 2m$, for $m = 6$ and $n = 11$

57. $-3 - x^2 + 12x$, for $x = 5$

58. $14 + (y - 5)^2 - 12 \div y$, for $y = -2$

59. The area of a parallelogram with base b and height h is bh. Find the area of the parallelogram when the height is 3.5 cm and the base is 8 cm.

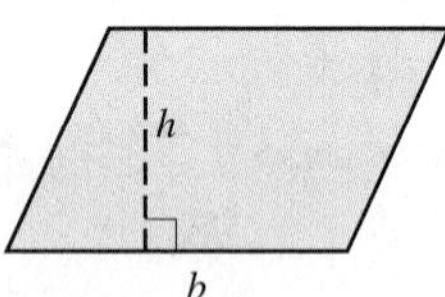

60. The area of a triangle with base b and height h is $\frac{1}{2}bh$. Find the area of the triangle when the height is 2 in. and the base is 6.2 in.

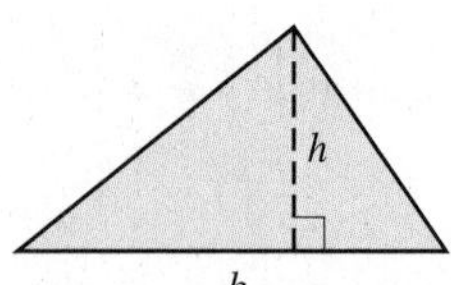

Multiply.

61. $4(2x + 7)$

62. $3(5y + 1)$

63. $-2(15 - 3x)$

64. $-7(3x - 5)$

65. $2(4a + 6b - 3c)$

66. $5(8p + q - 5r)$

67. $-3(2x - y + z)$

68. $-10(-6 - y - z)$

Factor.

69. $8x + 6y$

70. $7p + 14q$

71. $3 + 3w$

72. $4x + 4y$

73. $10x + 50y + 100$

74. $81p + 27q + 9$

Combine like terms.

75. $3p - 2p$

76. $4x + 3x$

77. $4m + 10 - 5m + 12$

78. $3a - 4b - 5b - 6a$

79. $-6x + 7 + 9x$

80. $16r + (-7r) + 3s$

Remove parentheses and simplify.

81. $2p - (7 - 4p)$

82. $4r - (3r + 5)$

83. $6x + 5y - 7(x - y)$

84. $14m - 6(2n - 3m) + n$

85. $6[2a + 4(a - 2b)]$

86. $2[2a + 1 - (3a - 6)]$

87. $3 - 2[5(x - 10y) - (3 + 2y)]$

88. $7 - 4[2(3 - 2x) - 5(4x - 3)]$

Determine whether the given number is a solution of the given equation.

89. 4; $3x - 2 = 10$

90. 12; $100 = 4x + 50$

91. -3; $4 - x = 1$

92. -1; $2 = 5 + 3x$

93. 4.6; $\dfrac{x}{2} = 2.3$

94. 144; $\dfrac{x}{9} = 16$

Translate each problem to an equation. Do not solve.

95. Three times what number is 348?

96. What number added to 256 is 113?

97. *Fast-Food Calories.* A McDonald's Big Mac® contains 500 calories. This is 69 more calories than a Taco Bell Beef Burrito® provides. How many calories are in a Taco Bell Beef Burrito?

98. *Coca-Cola® Consumption.* The average U.S. citizen consumes 296 servings of Coca-Cola each year. This is 7.4 times the international average. What is the international average per capita consumption of Coke?

99. *Vegetable Production.* It takes 42 gal of water to produce 1 lb of broccoli. This is twice the amount of water used to produce 1 lb of lettuce. How many gallons of water does it take to produce 1 lb of lettuce?

100. *Sports Costs.* The average annual cost for badminton is \$12. This is \$458 less than the average annual cost for scuba diving. What is the average annual cost for scuba diving?

Try Exercise Answers: Section R.1

1. False **7.** 4 **11.** -25 **25.** -3 **45.** 2 **55.** -7 **61.** $8x + 28$ **69.** $2(4x + 3y)$ **77.** $-m + 22$ **83.** $-x + 12y$ **89.** Yes **95.** Let n represent the number; $3n = 348$

R.2 Equations, Inequalities, and Problem Solving

- Solving Equations and Formulas
- Solving Inequalities
- Problem Solving

SOLVING EQUATIONS AND FORMULAS

Any replacement for the variable in an equation that makes the equation true is called a *solution* of the equation. To **solve** an equation means to find all of its solutions.

We use the following principles to write **equivalent equations,** or equations with the same solutions.

Equivalent equations (p. 83)

The addition principle (p. 83)

The multiplication principle (p. 85)

The Addition and Multiplication Principles for Equations

The Addition Principle

For any real numbers a, b, and c,

$a = b$ is equivalent to $a + c = b + c$.

The Multiplication Principle

For any real numbers a, b, and c, with $c \neq 0$,

$a = b$ is equivalent to $a \cdot c = b \cdot c$.

To solve $x + a = b$ for x, we add $-a$ to (or subtract a from) both sides. To solve $ax = b$ for x, we multiply both sides by $\frac{1}{a}$ (or divide both sides by a).

To solve an equation like $-3x - 10 = 14$, we first isolate the variable term, $-3x$, using the addition principle. Then we use the multiplication principle to get the variable by itself.

EXAMPLE 1 Solve: $-3x - 10 = 14$.

SOLUTION

$$-3x - 10 = 14$$

$$-3x - 10 + 10 = 14 + 10$$ **Using the addition principle: Adding 10 to both sides**

Isolate the x-term.

$$-3x = 24$$ **Simplifying**

$$\frac{-3x}{-3} = \frac{24}{-3}$$ **Dividing both sides by -3**

Isolate x.

$$x = -8$$ **Simplifying**

Check:

$$\begin{array}{r|l} -3x - 10 & 14 \\ \hline -3(-8) - 10 & 14 \\ 24 - 10 & \\ 14 & \stackrel{?}{=} 14 \end{array}$$ TRUE

Always check in the original equation.

The solution is -8.

Try Exercise 13.

Clearing fractions (p. 94)

Equations are generally easier to solve when they do not contain fractions. The easiest way to clear an equation of fractions is to multiply *every term on both sides* of the equation by the least common denominator.

EXAMPLE 2 Solve: $\frac{5}{2} - \frac{1}{6}t = \frac{2}{3}$.

SOLUTION The number 6 is the least common denominator, so we multiply both sides by 6.

$$6\left(\frac{5}{2} - \frac{1}{6}t\right) = 6 \cdot \frac{2}{3}$$ **Multiplying both sides by 6**

$$6 \cdot \frac{5}{2} - 6 \cdot \frac{1}{6}t = 6 \cdot \frac{2}{3}$$ **Using the distributive law. Be sure to multiply every term by 6.**

$$15 - t = 4$$ **The fractions are cleared.**

$$15 - t - 15 = 4 - 15$$ **Subtracting 15 from both sides**

$$-t = -11$$

$$(-1)(-t) = (-1)(-11)$$ **Multiplying both sides by −1 to change the sign**

$$t = 11$$

Check:

$$\frac{5}{2} - \frac{1}{6}t = \frac{2}{3}$$

$\frac{5}{2} - \frac{1}{6}(11)$	$\frac{2}{3}$
$\frac{5}{2} - \frac{11}{6}$	
$\frac{15}{6} - \frac{11}{6}$	
$\frac{2}{3} \overset{?}{=} \frac{2}{3}$	TRUE

The solution is 11.

Try Exercise 21.

To solve equations that contain parentheses, we can use the distributive law to first remove the parentheses. If like terms appear in an equation, we combine them and then solve.

EXAMPLE 3 Solve: $1 - 3(4 - x) = 2(x + 5) - 3x$.

SOLUTION

$$1 - 3(4 - x) = 2(x + 5) - 3x$$

$$1 - 12 + 3x = 2x + 10 - 3x$$ **Using the distributive law**

$$-11 + 3x = -x + 10$$ **Combining like terms; $1 - 12 = -11$ and $2x - 3x = -x$**

$$-11 + 3x + x = 10$$ **Adding x to both sides to get all x-terms on one side**

$$-11 + 4x = 10$$ **Combining like terms**

$$4x = 10 + 11$$ **Adding 11 to both sides to isolate the x-term**

$$4x = 21$$ **Simplifying**

$$x = \frac{21}{4}$$ **Dividing both sides by 4**

```
1-3(4-21/4)
                    4.75
2(21/4+5)-3*21/4
                    4.75
```

Check:

$$\begin{array}{r|l} \multicolumn{2}{c}{1 - 3(4 - x) = 2(x + 5) - 3x} \\ \hline 1 - 3\left(4 - \frac{21}{4}\right) & 2\left(\frac{21}{4} + 5\right) - 3\left(\frac{21}{4}\right) \\ 1 - 3\left(-\frac{5}{4}\right) & 2\left(\frac{41}{4}\right) - \frac{63}{4} \\ 1 + \frac{15}{4} & \frac{82}{4} - \frac{63}{4} \\ \frac{19}{4} \stackrel{?}{=} & \frac{19}{4} \end{array} \qquad \text{TRUE}$$

We can also check using a calculator, as shown at left. The solution is $\frac{21}{4}$.

Try Exercise 29.

Conditional equation (p. 96)

Identity (p. 96)

Contradiction (p. 96)

Formulas (Section 2.3)

The equations in Examples 1–3 are **conditional**—that is, they are true for some values of x and false for other values of x. An **identity** is an equation like $x + 1 = x + 1$ that is true for all values of x. A **contradiction** is an equation like $x + 1 = x + 2$ that is never true.

A **formula** is an equation using two or more letters that represents a relationship between two or more quantities. A formula can be solved for a specified letter using the principles for solving equations.

EXAMPLE 4 The formula

$$A = \frac{a + b + c + d}{4}$$

gives the average A of four test scores a, b, c, and d. Solve for d.

SOLUTION We have

$$A = \frac{a + b + c + d}{4} \qquad \text{We want the letter } d \text{ alone.}$$

$$4A = a + b + c + d \qquad \text{Multiplying by 4 to clear the fraction}$$

$$4A - a - b - c = d. \qquad \text{Subtracting } a + b + c \text{ from (or adding } -a - b - c \text{ to) both sides. The letter } d \text{ is now isolated.}$$

We can also write this as $d = 4A - a - b - c$. This formula can be used to determine the test score needed to obtain a specified average if three tests have already been taken.

Try Exercise 39.

SOLVING INEQUALITIES

Solutions of inequalities (p. 132)

Graphs of inequalities (p. 133)

A **solution of an inequality** is a replacement of the variable that makes the inequality true. The solutions of an inequality in one variable can be **graphed,** or represented by a drawing, on the number line. All points that are solutions are shaded. A parenthesis indicates an endpoint that is not a solution and a bracket indicates an endpoint that is a solution.

For example, the graph of the inequality $m \leq 2$ is shown below. The solutions of $m \leq 2$ are shown on the number line by shading points to the left of 2 as well as the point at 2. The bracket at 2 indicates that 2 is a part of the graph (that is, it is a solution of $m \leq 2$).

Also shown below is the graph of $-1 \leq x < 4$. In order to be a solution of this inequality, a number must be a solution of both $-1 \leq x$ and $x < 4$. The solutions are shaded on the number line, with a parenthesis indicating that 4 is not a solution and a bracket indicating that -1 is a solution.

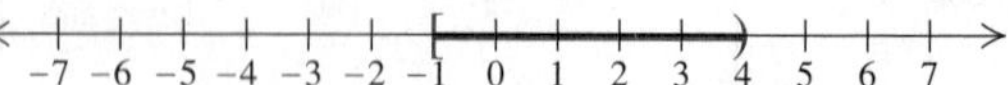

Set-builder notation (p. 133)

Since it is impossible to list all the solutions of most inequalities, we use **set-builder notation** or **interval notation** to write such sets.

Using set-builder notation, we write the solution set of $m \leq 2$ as

$$\{m|m \leq 2\}.$$

This is read "the set of all m such that m is less than or equal to 2."

Interval notation uses parentheses, ( ), and brackets, [ ], to describe a set of real numbers.

Interval notation (p. 134)

Interval Notation	Set-builder Notation	Graph
(a, b) open interval	$\{x\|a < x < b\}$	(a, b); a, b
$[a, b]$ closed interval	$\{x\|a \leq x \leq b\}$	$[a, b]$; a, b
$(a, b]$ half-open interval	$\{x\|a < x \leq b\}$	$(a, b]$; a, b
$[a, b)$ half-open interval	$\{x\|a \leq x < b\}$	$[a, b)$; a, b
(a, ∞)	$\{x\|x > a\}$	a
$[a, \infty)$	$\{x\|x \geq a\}$	a
$(-\infty, a)$	$\{x\|x < a\}$	a
$(-\infty, a]$	$\{x\|x \leq a\}$	a

Thus, for example, the solution set of $m \leq 2$, in interval notation, is $(-\infty, 2]$.

Equivalent inequalities (p. 135)

To solve inequalities, we use principles that enable us to write **equivalent inequalities**—inequalities having the same solution set. The addition principle is similar to the addition principle for equations; the multiplication principle contains an important difference.

The Addition and Multiplication Principles for Inequalities

The addition principle for inequalities (p. 135)

The Addition Principle

For any real numbers a, b, and c,

$a < b$ is equivalent to $a + c < b + c$, and

$a > b$ is equivalent to $a + c > b + c$.

The multiplication principle for inequalities (p. 137)

The Multiplication Principle

For any real numbers a and b, and for any *positive* number c,

$a < b$ is equivalent to $ac < bc$, and

$a > b$ is equivalent to $ac > bc$.

For any real numbers a and b, and for any *negative* number c,

$a < b$ is equivalent to $ac > bc$, and

$a > b$ is equivalent to $ac < bc$.

Similar statements hold for $\leq$ and $\geq$.

Note that when we multiply both sides of an inequality by a negative number, we must reverse the direction of the inequality symbol in order to have an equivalent inequality.

EXAMPLE 5 Solve $-2x \geq 5$ and then graph the solution.

SOLUTION We have

$$-2x \geq 5$$

$$\frac{-2x}{-2} \leq \frac{5}{-2}$$ **Multiplying by $-\frac{1}{2}$ or dividing by -2**

The symbol must be reversed!

$$x \leq -\frac{5}{2}.$$

Any number less than or equal to $-\frac{5}{2}$ is a solution. The graph is as follows.

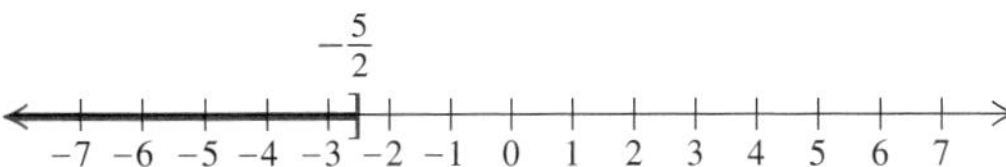

The solution set is $\left\{x \middle| x \leq -\frac{5}{2}\right\}$, or $\left(-\infty, -\frac{5}{2}\right]$.

Try Exercise 51.

We can use the addition and multiplication principles together to solve inequalities. We can also combine like terms, remove parentheses, and clear fractions and decimals.

EXAMPLE 6 Solve: $2 - 3(x + 5) > 4 - 6(x - 1)$.

SOLUTION We have

$$2 - 3(x + 5) > 4 - 6(x - 1)$$

$$2 - 3x - 15 > 4 - 6x + 6$$ **Using the distributive law to remove parentheses**

$$-3x - 13 > -6x + 10$$ **Simplifying**

$$-3x + 6x > 10 + 13$$ **Adding $6x$ and 13, to get all x-terms on one side and the remaining terms on the other side**

$$3x > 23$$ **Combining like terms**

$$x > \frac{23}{3}.$$ **Multiplying by $\frac{1}{3}$. The inequality symbol stays the same because $\frac{1}{3}$ is positive.**

The solution set is $\left\{x \middle| x > \frac{23}{3}\right\}$, or $\left(\frac{23}{3}, \infty\right)$.

Try Exercise 57.

PROBLEM SOLVING

Problem solving (Section 2.5)

One of the most important uses of algebra is as a tool for problem solving. The following five steps can be used to help solve problems of many types.

Five Steps for Problem Solving in Algebra

1. *Familiarize* yourself with the problem.
2. *Translate* to mathematical language. (This often means write an equation.)
3. *Carry out* some mathematical manipulation. (This often means *solve* an equation.)
4. *Check* your possible answer in the original problem.
5. *State* the answer clearly.

EXAMPLE 7 Kitchen Cabinets. Cherry kitchen cabinets cost 10% more than oak cabinets. Shelby Custom Cabinets designs a kitchen using \$7480 worth of cherry cabinets. How much would the same kitchen cost using oak cabinets?

SOLUTION

Familiarize step (p. 117)

Percent (Section 2.4)

1. **Familiarize.** Be sure to allow sufficient time for the *Familiarize* step; it is often the most important step of the five. Listed below are several helpful strategies.
   - List known (and unknown) information.
   - Define a variable to represent the unknown.
   - Make a drawing or a table.
   - Make a guess and check it.
   - Look up further information.

   For this problem, let's make a guess and check it.

   Guess: The oak cabinets cost \$6500.

   Check: The cherry cabinets cost 10% more, or an additional \$650. The total cost of the cherry cabinets would be \$6500 + \$650 = \$7150.

Our guess does *not* check, since $\$7150 \neq \$7480$, but we see that the price of the cherry cabinets is the price of the oak cabinets plus 10% of the price of the oak cabinets.

We let $c =$ the cost of the oak cabinets.

2. **Translate.** What we learned in the *Familiarize* step leads to the translation of the problem to an equation.

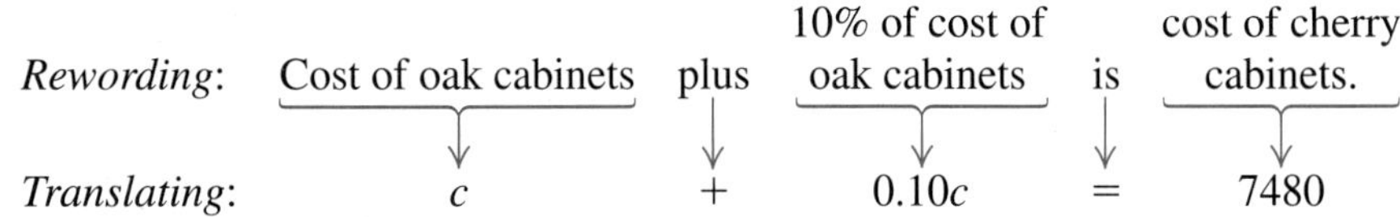

3. **Carry out.** We solve the equation:

$$c + 0.10c = 7480$$
$$1c + 0.10c = 7480 \quad \textbf{Writing } c \textbf{ as } 1c \textbf{ before combining terms}$$
$$1.10c = 7480 \quad \textbf{Combining like terms}$$
$$c = \frac{7480}{1.10} \quad \textbf{Dividing by 1.10}$$
$$c = 6800.$$

4. **Check.** We check in the wording of the stated problem.

Additional cost of cherry cabinets $= 0.10(\$6800) = \$680.$

Total cost of cherry cabinets $= \$6800 + \$680 + \$7480.$

Our answer checks.

5. **State.** The oak cabinets would cost \$6800.

Try Exercise 67.

Sometimes the translation of a problem is an inequality.

EXAMPLE 8 Phone Bills. Elyse pays 10¢ per text message. The monthly charge for her wireless service is \$39.50. How many text messages can she send or receive in a month and not exceed her phone budget of \$50?

SOLUTION

1. **Familiarize.** Suppose that Elyse sends or receives 200 text messages one month. Then her bill would be the monthly fee plus the texting charges, or $\$39.50 + \$0.10(200) = \$59.50$. This amount exceeds \$50, so we know that the number of text messages must be fewer than 200. We let $t =$ the number of text messages sent or received in a month.

Solving applications with inequalities (Section 2.7)

2. **Translate.** The *Familiarize* step helps us reword and translate.

Rewording:	The monthly charge	plus	the texting charge	cannot exceed	\$50.
Translating:	39.50	$+$	$0.10t$	$\leq$	50

3. **Carry out.** We solve the inequality:

$$39.50 + 0.10t \leq 50$$
$$0.10t \leq 10.50 \quad \textbf{Subtracting 39.50 from both sides}$$
$$t \leq 105. \quad \textbf{Dividing by 0.10. The inequality symbol stays the same.}$$

4. **Check.** As a partial check, we determine the bill for 105 texts:

$$\$39.50 + \$0.10(105) = \$50.$$

Since this does not exceed the \$50 budget, and fewer texts will cost even less, our answer checks. We also note that 105 is less than 200, as noted in the *Familiarize* step.

5. **State.** Elyse will not exceed her budget if she sends or receives no more than 105 text messages.

Try Exercise 77.

R.2 Exercise Set

FOR EXTRA HELP MyMathLab   

Solve.

1. $-6 + x = 10$
2. $y + 7 = -3$
3. $t + \frac{1}{3} = \frac{1}{4}$
4. $-\frac{2}{3} + p = \frac{1}{6}$
5. $-1.9 = x - 1.1$
6. $x + 4.6 = 1.7$
7. $-x = \frac{5}{3}$
8. $-y = -\frac{2}{5}$
9. $-\frac{2}{7}x = -12$
10. $-\frac{1}{4}x = 3$
11. $\frac{-t}{5} = 1$
12. $\frac{2}{3} = -\frac{z}{8}$
13. $3x + 7 = 13$
14. $4x + 3 = -1$
15. $3y - 10 = 15$
16. $12 = 5y - 18$
17. $4x + 7 = 3 - 5x$
18. $2x = 5 + 7x$
19. $2x - 7 = 5x + 1 - x$
20. $a + 7 - 2a = 14 + 7a - 10$
21. $\frac{2}{5} + \frac{1}{3}t = 5$
22. $-\frac{5}{6} + t = \frac{1}{2}$
23. $x + 0.45 = 2.6x$
24. $1.8x + 0.16 = 4.2 - 0.05x$
25. $8(3 - m) + 7 = 47$
26. $2(5 - m) = 5(6 + m)$
27. $4 - (6 + x) = 13$
28. $18 = 9 - (3 - x)$
29. $2 + 3(4 + c) = 1 - 5(6 - c)$
30. $b + (b + 5) - 2(b - 5) = 18 + b$
31. $0.1(a - 0.2) = 1.2 + 2.4a$
32. $\frac{2}{3}\left(\frac{1}{2} - x\right) + \frac{5}{6} = \frac{3}{2}\left(\frac{2}{3}x + 1\right)$
33. $A = lw$, for l
34. $A = lw$, for w
35. $p = 30q$, for q
36. $d = 20t$, for t
37. $I = \frac{P}{V}$, for P
38. $b = \frac{A}{h}$, for A
39. $q = \frac{p + r}{2}$, for p
40. $q = \frac{p - r}{2}$, for r
41. $A = \pi r^2 + \pi r^2 h$, for π
42. $ax + by = c$, for a

Determine whether each number is a solution of the given inequality.

43. $x \leq -5$
    a) 5 b) -5
    c) 0 d) -10
44. $y > 0$
    a) -1 b) 1
    c) 0 d) 100

Solve and graph. Write each answer in set-builder notation and in interval notation.

45. $x + 3 \leq 15$
46. $y + 7 < -10$
47. $m - 17 > -5$
48. $x + 9 \geq -8$
49. $2x \geq -3$
50. $-\frac{1}{2}n \leq 4$
51. $-5t > 15$
52. $3x > 10$

Solve. Write each answer in set-builder notation and in interval notation.

53. $2y - 7 > 13$

54. $2 - 6y \leq 18$

55. $6 - 5a \leq a$

56. $4b + 7 > 2 - b$

57. $2(3 + 5x) \geq 7(10 - x)$

58. $2(x + 5) < 8 - 3x$

59. $\frac{2}{3}(6 - x) < \frac{1}{4}(x + 3)$

60. $\frac{2}{3}t + \frac{8}{9} \geq \frac{4}{6} - \frac{1}{4}t$

61. $0.7(2 + x) \geq 1.1x + 5.75$

62. $0.4x + 5.7 \leq 2.6 - 3(1.2x - 7)$

Solve. Use the five-step problem-solving process.

63. Three less than the sum of 2 and some number is 6. What is the number?

64. Five times some number is 10 less than the number. What is the number?

65. The sum of two consecutive even integers is 34. Find the numbers.

66. The sum of three consecutive integers is 195. Find the numbers.

67. *Reading.* Leisa is reading a 500-page book. She has twice as many pages to read as she has already finished. How many pages has she already read?

68. *Mowing.* It takes Caleb 50 min to mow his lawn. It will take him three times as many minutes to finish as he has already spent mowing. How long has he already spent mowing?

69. *Perimeter of a Rectangle.* The perimeter of a rectangle is 28 cm. The width is 5 cm less than the length. Find the width and the length.

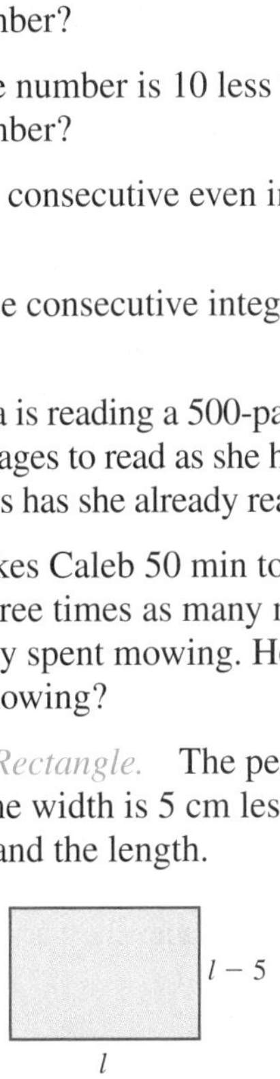

70. *Triangles.* The second angle of a triangle is one-third as large as the first. The third angle is 5° more than the first. Find the measure of the second angle.

71. *Water Usage.* Rural Water Company charges a monthly service fee of \$9.70 plus a volume charge of \$2.60 for every hundred cubic feet of water used. How much water was used if the monthly bill is \$33.10?

72. *Phone Bills.* Brandon pays \$4.95 per month for a long-distance phone service that offers a flat rate of 7¢ per minute. One month his total long-distance phone bill was \$10.69. How many minutes of long-distance phone calls were made that month?

73. *Sale Price.* A can of tomatoes is on sale at 20% off for 64¢. What is the normal selling price of the tomatoes?

74. *Plywood.* The price of a piece of plywood rose 5% to \$42. What was the original price of the plywood?

75. *Practice.* Dierdre's basketball coach requires each team member to average at least 15 min per day shooting baskets. One week Dierdre spent 10 min, 20 min, 5 min, 0 min, 25 min, and 15 min shooting baskets. How long must she practice shooting baskets on the seventh day if she is to meet the requirement?

76. *Perimeter of a Garden.* The perimeter of Garry's rectangular garden cannot exceed the 100 ft of fencing that he purchased. He wants the length to be twice the width. What widths of the garden will meet these conditions?

77. *Meeting Costs.* The Winds charges a \$75 cleaning fee plus \$45 per hour for the use of its meeting room. Complete Consultants has budgeted \$200 to rent a room for a seminar. For how many hours can they rent the meeting room at The Winds?

78. *Meeting Costs.* Spring Haven charges a \$15 setup fee, a \$30 cleanup fee, and \$50 per hour for the use of its meeting room. For what lengths of time will Spring Haven's room be less expensive than the room at The Winds? (See Exercise 77.)

Try Exercise Answers: Section R.2

13. 2 **21.** $\frac{69}{5}$ **29.** $\frac{43}{2}$ **39.** $p = 2q - r$
51. $\{t \mid t < -3\}$, or $(-\infty, -3)$ 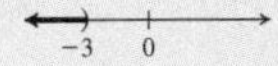

57. $\left\{x \mid x \geq \frac{64}{17}\right\}$, or $\left[\frac{64}{17}, \infty\right)$ **67.** $166\frac{2}{3}$ pages
77. For $2\frac{7}{9}$ hr or less

R.3 Introduction to Graphing and Functions

- Points and Ordered Pairs
- Graphs and Slope
- Linear Equations
- Functions

POINTS AND ORDERED PAIRS

To graph pairs of numbers such as $(2, -5)$ on a plane, we use two perpendicular number lines called **axes.** The axes cross at a point called the **origin.** Arrows on the axes show the positive directions.

Coordinates (p. 163)

The order of the **coordinates,** or numbers in a pair, is important. The **first coordinate** indicates horizontal position and the **second coordinate** indicates vertical position. Such pairs of numbers are called **ordered pairs.** Thus the ordered pairs $(1, -2)$ and $(-2, 1)$ correspond to different points, as shown in the figure below.

Quadrants (p. 165)

The axes divide the plane into four regions, or **quadrants,** as indicated by Roman numerals in the figure below. Points on the axes are not considered to be in any quadrant. The horizontal axis is often labeled the *x*-axis, and the vertical axis the *y*-axis.

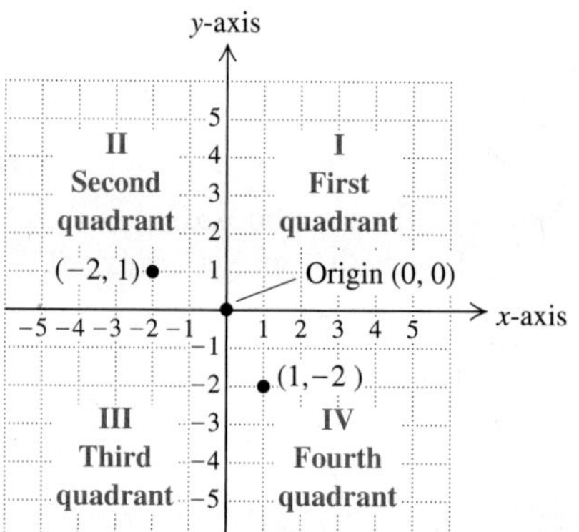

Solutions of equations (p. 173)

GRAPHS AND SLOPE

When an equation contains two variables, solutions must be ordered pairs. Unless stated otherwise, the first number in each pair replaces the variable that occurs first alphabetically.

EXAMPLE 1 Determine whether $(1, 4)$ is a solution of $y - x = 3$.

SOLUTION We substitute 1 for x and 4 for y since x occurs first alphabetically:

$$\begin{array}{c|c} y - x & = 3 \\ \hline 4 - 1 & 3 \\ 3 & \stackrel{?}{=} 3 \end{array} \quad \text{TRUE}$$

Since $3 = 3$ is true, the pair $(1, 4)$ *is* a solution.

Try Exercise 9.

A curve or a line that represents all the solutions of an equation is called its **graph.**

EXAMPLE 2 Graph: $y = -2x + 1$.

SOLUTION We select a value for x, calculate the corresponding value of y, and form an ordered pair.

If $x = 0$, then $y = -2 \cdot 0 + 1 = 1$, and $(0, 1)$ is a solution. Repeating this step, we find other ordered pairs and list the results in a table. We then plot the points corresponding to the pairs. They appear to form a straight line, so we draw a line through the points.

$y = -2x + 1$

x	y	(x, y)
0	1	$(0, 1)$
-1	3	$(-1, 3)$
3	-5	$(3, -5)$
1	-1	$(1, -1)$

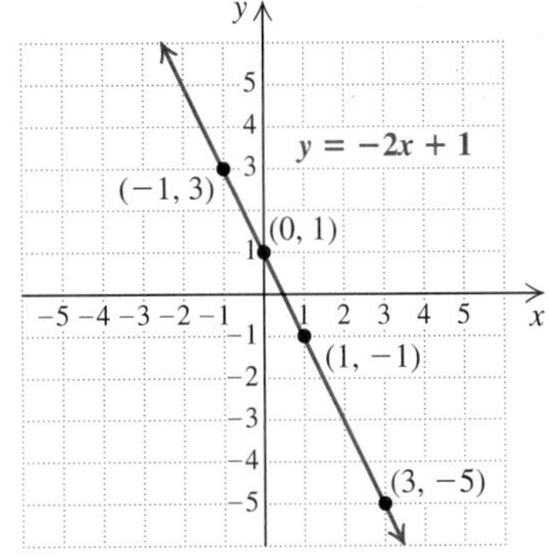

$y = -2x + 1$
10
−10 10
−10

To graph using a graphing calculator, we let $y_1 = -2x + 1$, as shown at left.

Try Exercise 13.

Slope (p. 204)

The graph in Example 2 is a straight line. An equation whose graph is a straight line is a **linear equation.** The *rate of change* of y with respect to x is called the **slope** of a graph. A linear graph has a constant slope that can be found using any two points on a line.

Slope The *slope* of the line containing points (x_1, y_1) and (x_2, y_2) is given by

$$m = \frac{\text{change in } y}{\text{change in } x} = \frac{\text{rise}}{\text{run}} = \frac{y_2 - y_1}{x_2 - x_1}.$$

EXAMPLE 3 Find the slope of the line containing the points $(-2, 1)$ and $(3, -4)$.

SOLUTION From $(-2, 1)$ to $(3, -4)$, the change in y, or the rise, is $-4 - 1$, or -5. The change in x, or the run, is $3 - (-2)$, or 5. Thus,

$$\text{Slope} = \frac{\text{change in } y}{\text{change in } x} = \frac{\text{rise}}{\text{run}} = \frac{-4 - 1}{3 - (-2)} = \frac{-5}{5} = -1.$$

Try Exercise 23.

The slope of a line indicates the direction and the steepness of its slant. The larger the absolute value of the slope, the steeper the line. The direction of the slant is indicated by the sign of the slope, as shown in the figures below.

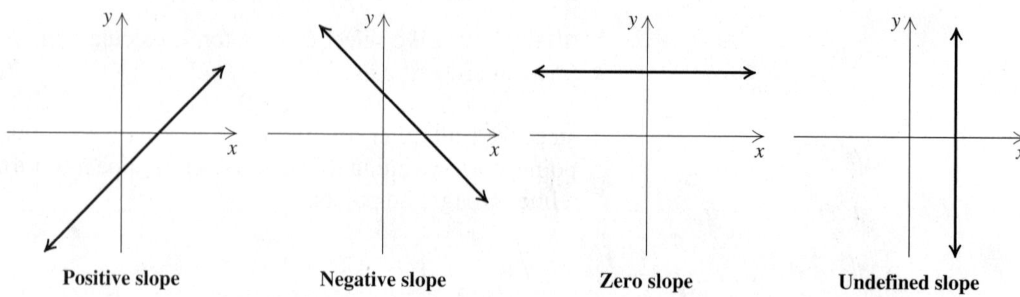

x-intercept (p. 186)

y-intercept (p. 186)

The **x-intercept** of a line, if it exists, is the point at which the graph crosses the x-axis. To find an x-intercept, we replace y with 0 and calculate x.

The **y-intercept** of a line, if it exists, is the point at which the graph crosses the y-axis. To find a y-intercept, we replace x with 0 and calculate y.

LINEAR EQUATIONS

Any equation that can be written in the **standard form** $Ax + By = C$ is linear. Linear equations can also be written in other forms.

Forms of Linear Equations

Standard form:	$Ax + By = C$
Slope–intercept form:	$y = mx + b$
Point–slope form:	$y - y_1 = m(x - x_1)$

The slope and the y-intercept of a line can be read from the slope–intercept form of the line's equation.

Slope and y-intercept For the graph of any equation $y = mx + b$,

- the slope is m, and
- the y-intercept is $(0, b)$.

EXAMPLE 4 Find the slope and the y-intercept of the line given by the equation $4x - 3y = 9$.

SOLUTION We write the equation in slope–intercept form $y = mx + b$:

$$4x - 3y = 9 \qquad \text{We must solve for } y.$$
$$-3y = -4x + 9 \qquad \text{Adding } -4x \text{ to both sides}$$
$$y = \tfrac{4}{3}x - 3. \qquad \text{Dividing both sides by } -3$$

The slope is $\frac{4}{3}$ and the y-intercept is $(0, -3)$.

Try Exercise 29.

If we know that the graph of an equation is a straight line, we can plot two points on the line and draw the line through those points. The intercepts are often convenient points to use.

EXAMPLE 5 Graph $2x - 5y = 10$ using intercepts.

SOLUTION To find the x-intercept, we let $y = 0$ and solve for x:

$$2x - 5 \cdot 0 = 10 \qquad \textbf{Replacing } y \textbf{ with 0}$$
$$2x = 10$$
$$x = 5.$$

To find the y-intercept, we let $x = 0$ and solve for y:

$$2 \cdot 0 - 5y = 10 \qquad \textbf{Replacing } x \textbf{ with 0}$$
$$-5y = 10$$
$$y = -2.$$

Thus the x-intercept is $(5, 0)$ and the y-intercept is $(0, -2)$. The graph is a line, since $2x - 5y = 10$ is in the form $Ax + By = C$. It passes through these two points.

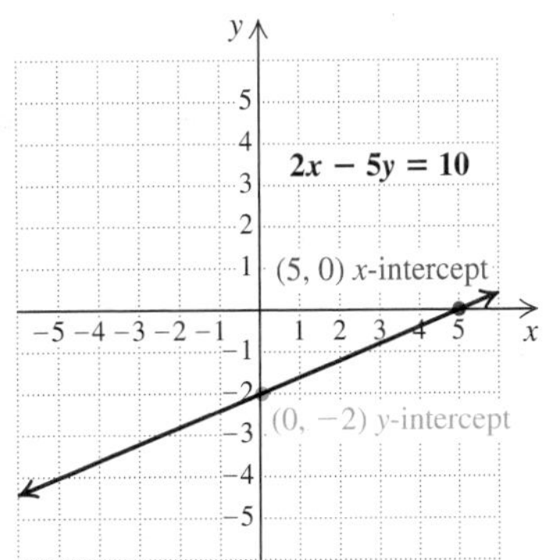

Try Exercise 33.

Alternatively, if we know a point on the line and its slope, we can plot the point and "count off" its slope to locate another point on the line.

EXAMPLE 6 Graph: $y = -\frac{1}{2}x + 3$.

SOLUTION The equation is in slope–intercept form, so we can read the slope and the y-intercept directly from the equation.

Slope: $-\frac{1}{2}$

y-intercept: $(0, 3)$

We plot the y-intercept and use the slope to find another point.

Another way to write the slope is $\frac{-1}{2}$. This means that for a run of 2 units, there is a negative rise, or a fall, of 1 unit. Starting at $(0, 3)$, we move 2 units in the positive horizontal direction and then 1 unit down, to locate the point $(2, 2)$. Then we draw the graph. A third point can be calculated and plotted as a check.

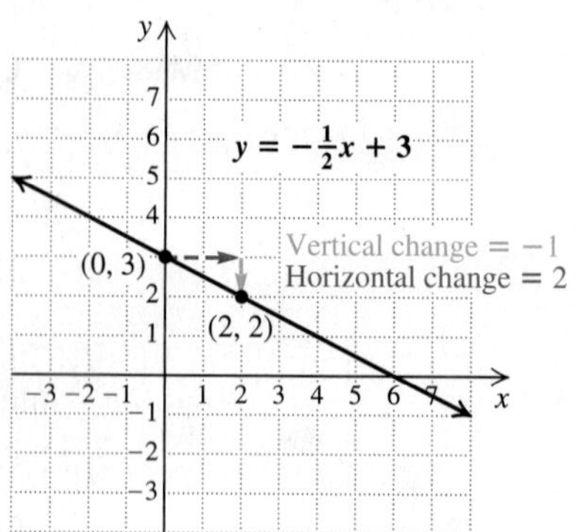

Try Exercise 39.

Horizontal lines and vertical lines intersect only one axis.

Horizontal line (p. 189)

Vertical line (p. 189)

Horizontal Lines and Vertical Lines

Horizontal Line	***Vertical Line***
$y = b$	$x = a$
y-intercept $(0, b)$	x-intercept $(a, 0)$
Slope is 0	Undefined slope
Example: $y = -3$	Example: $x = 2$

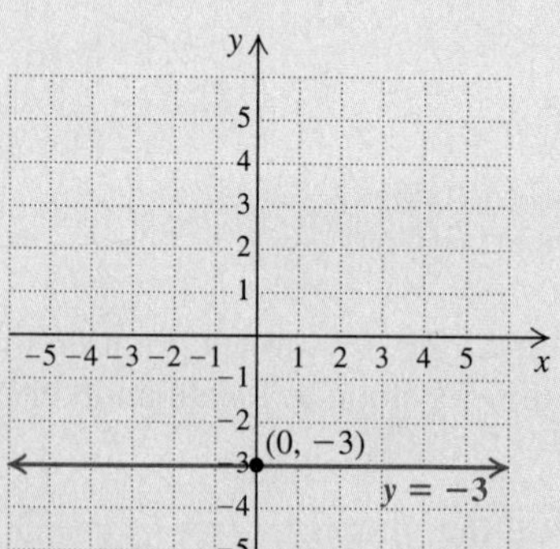

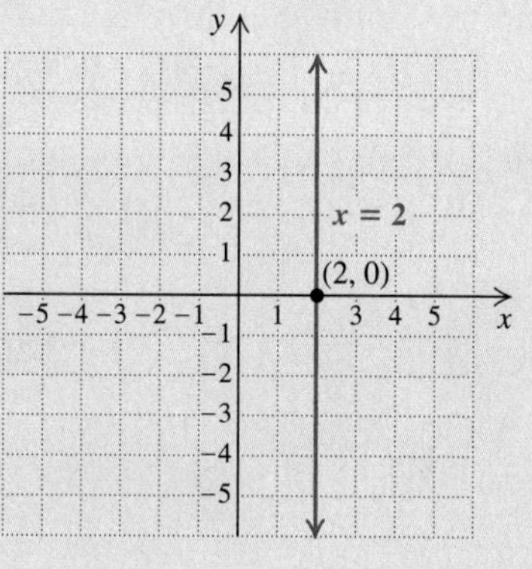

If we know the slope of a line and the coordinates of a point on the line, we can find an equation of the line, using either the slope–intercept equation $y = mx + b$, or the point–slope equation $y - y_1 = m(x - x_1)$.

EXAMPLE 7 Find the slope–intercept equation of a line given the following:

a) The slope is 2, and the y-intercept is $(0, -5)$.

b) The graph contains the points $(-2, 1)$ and $(3, -4)$.

SOLUTION

a) Since the slope and the y-intercept are given, we use the slope–intercept equation:

$$y = mx + b$$

$$y = 2x - 5. \qquad \textbf{Substituting 2 for } m \textbf{ and } -5 \textbf{ for } b$$

b) To use the point–slope equation, we need a point on the line and its slope. The slope can be found from the points given:

Find the slope.

$$m = \frac{1 - (-4)}{-2 - 3} = \frac{5}{-5} = -1.$$

Either point can be used for (x_1, y_1). Using $(-2, 1)$, we have

Find a point–slope equation for the line.

$$y - y_1 = m(x - x_1)$$

$$y - 1 = -1(x - (-2))$$ **Substituting -2 for x_1, 1 for y_1, and -1 for m**

$$y - 1 = -(x + 2)$$

$$y - 1 = -x - 2$$

Find the slope–intercept equation for the line.

$$y = -x - 1.$$ **This is in slope–intercept form.**

Try Exercise 47.

We can tell from the slopes of two lines whether they are parallel (never intersect) or perpendicular (intersect at right angles).

Parallel lines (p. 219)

Perpendicular lines (p. 221)

Parallel Lines and Perpendicular Lines

Two lines are parallel if they have the same slope.

Two lines are perpendicular if the product of the slopes is -1.

EXAMPLE 8 Tell whether the graphs of each pair of lines are parallel, perpendicular, or neither.

a) $2x - y = 7,$
$y = 2x + 3$

b) $4x - y = 8,$
$x + 4y = 8$

SOLUTION

a) The slope of $y = 2x + 3$ is 2. To find the slope of $2x - y = 7$, we solve for y:

$$2x - y = 7$$

$$-y = -2x + 7$$

$$y = 2x - 7.$$ **The slope of $2x - y = 7$ is 2.**

Since the slopes are equal, the lines are parallel.

b) We solve both equations for y in order to determine the slopes of the lines:

$$4x - y = 8$$

$$-y = -4x + 8$$

$$y = 4x - 8.$$ **The slope of $4x - y = 8$ is 4.**

For the second line, we have

$$x + 4y = 8$$

$$4y = -x + 8$$

$$y = -\tfrac{1}{4}x + 2.$$ **The slope of $x + 4y = 8$ is $-\frac{1}{4}$.**

We multiply the slopes 4 and $-\frac{1}{4}$. Since $4 \cdot \left(-\frac{1}{4}\right) = -1$, the lines are perpendicular.

Try Exercise 51.

FUNCTIONS

A correspondence from one set to another is a pairing of one element of the first set with one element of a second set. Such a match can be written as an ordered pair. One type of correspondence used extensively in mathematics is called a **function.**

Function (p. 244)

Domain (p. 244)

Range (p. 244)

Function A *function* is a correspondence between a first set, called the *domain*, and a second set, called the *range*, such that each member of the domain corresponds to *exactly one* member of the range.

A function is often labeled with a letter, such as f or g. The notation $f(x)$—read "f of x," "f at x," or "the value of f at x"—represents the **output** that corresponds to the **input** x. For example, if a function f contains the ordered pair $(0, -5)$, we can write

Function notation (p. 249)

Output (p. 249)

Input (p. 249)

$$f(0) = -5.$$

Many functions are described by equations. To find a function value, we substitute a given number or expression for the variable in the equation.

EXAMPLE 9 For $f(x) = 2x^2 - x + 4$, find $f(-1)$.

SOLUTION We have

$$\begin{aligned} f(-1) &= 2(-1)^2 - (-1) + 4 \\ &= 2(1) + 1 + 4 \\ &= 7. \end{aligned}$$

Try Exercise 55.

When the domain and the range of a function are sets of numbers, we can graph the function. If a graph contains two or more points with the same first coordinate, that graph cannot represent a function.

Vertical-line test (p. 249)

The Vertical-Line Test A graph represents a function if it is not possible to draw a vertical line that intersects the graph more than once.

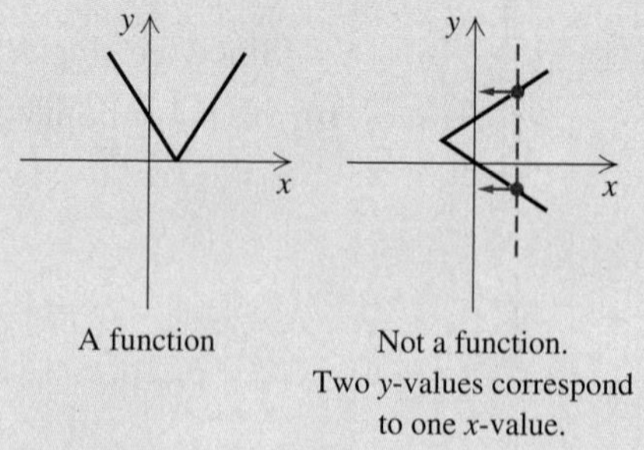

A function

Not a function. Two y-values correspond to one x-value.

If a domain of a function is not specified, we assume that the domain is the set of all real numbers for which the function value can be computed.

EXAMPLE 10 Determine the domain of each of the following functions.

a) $f(x) = 2x^2 + x - 10$ **b)** $g(x) = \dfrac{1}{x + 2}$

c) $h(x) = \sqrt{7 - x}$

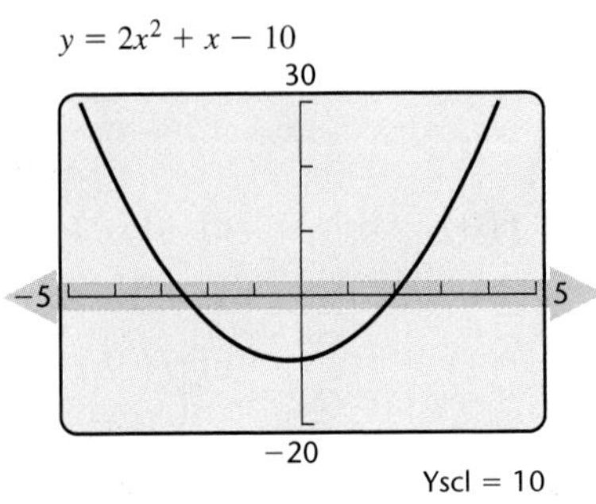

The domain is the set of all possible x-values.

SOLUTION We look for any restrictions on the domain—that is, any numbers for which the function value cannot be computed.

a) We can compute $2x^2 + x - 10$ for any number that replaces x. Therefore, the domain of f is $(-\infty, \infty)$, or $\{x | x \text{ is a real number}\}$, or $\mathbb{R}$. We can confirm this by examining a graph of $f(x) = 2x^2 + x - 10$, as shown at left.

b) The expression $1/(x + 2)$ cannot be computed when the denominator, $x + 2$, is 0. We set $x + 2$ equal to 0 and solve:

$$x + 2 = 0$$
$$x = -2.$$

Thus, -2 is *not* in the domain of g, but all other real numbers are. The domain of g is $\{x | x \text{ is a real number } \textit{and } x \neq -2\}$.

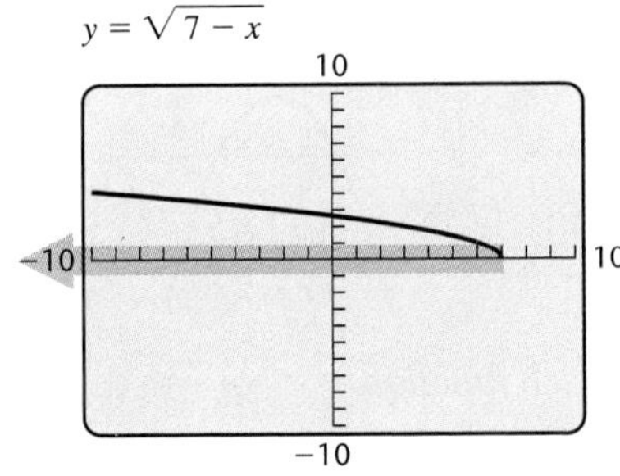

c) Since the square root of a negative number is not a real number, only inputs for which $7 - x$ is nonnegative are in the domain of the function. We solve an inequality:

$$7 - x \geq 0$$
$$7 \geq x.$$

We see that all numbers less than or equal to 7 will be in the domain. The domain of h is $(-\infty, 7]$, or $\{x | x \leq 7\}$. The graph of $h(x) = \sqrt{7 - x}$, as shown at left, confirms this result.

Try Exercise 63.

R.3 Exercise Set

FOR EXTRA HELP MyMathLab    

1. Plot these points.
$(2, -3), (5, 1), (0, 2), (-1, 0),$
$(0, 0), (-2, -5), (-1, 1), (1, -1)$

2. Plot these points.
$(0, -4), (-4, 0), (5, -2), (2, 5),$
$(3, 3), (-3, -1), (-1, 4), (0, 1)$

In which quadrant is each point located?

3. $(2, 1)$

4. $(-2, 5)$

5. $(3, -2.6)$

6. $(-1.7, -5.9)$

7. First coordinates are positive in quadrants ______ and ______.

8. Second coordinates are negative in quadrants ______ and ______.

Determine whether each equation has the given ordered pair as a solution.

9. $y = 2x - 5$; $(1, 3)$

10. $4x + 3y = 8$; $(-1, 4)$

11. $a - 5b = -3$; $(2, 1)$

12. $c = d + 1$; $(1, 2)$

Graph by hand.

13. $y = \frac{1}{3}x + 3$

14. $y = -x - 2$

15. $y = -4x$

16. $y = \frac{3}{4}x + 1$

Graph using a graphing calculator.

17. $y = x^2 - 7$

18. $y = 2x^2 - x + 3$

19. $y = |x + 3|$

20. $y = \sqrt{3 - x}$

21. $y = \sqrt{x^2 + 1}$

22. $y = |x^2 - x - 1|$

Find the slope of the line containing each given pair of points.

23. $(3, 6)$ and $(2, 5)$

24. $(-1, 7)$ and $(-3, 4)$

25. $(0, 3)$ and $(1, -2)$

26. $(-3, 8)$ and $(2, -7)$

27. $\left(-2, -\frac{1}{2}\right)$ and $\left(5, -\frac{1}{2}\right)$

28. $(6.8, 7.5)$ and $(6.8, -3.2)$

Find the slope and the y-intercept of each equation.

29. $y = 2x - 5$

30. $y = 4 - x$

31. $2x + 7y = 1$

32. $x - 2y = 3$

Find the intercepts. Then graph.

33. $2x - y = 4$

34. $2x + 5y = 10$

35. $y = x + 5$

36. $y = -x + 7$

37. $3 - y = 2x$

38. $2y + 1 = x$

Determine the coordinates of the y-intercept of each equation. Then graph the equation.

39. $y = 2x - 5$

40. $y = -\frac{5}{4}x - 3$

41. $2y + 4x = 6$

42. $3y + x = 4$

Find the slope of each line, and graph.

43. $y = 4$

44. $x = -5$

45. $x = 3$

46. $y = -1$

Find the slope–intercept equation of a line given the conditions.

47. The slope is $\frac{1}{3}$ and the y-intercept is $(0, 1)$.

48. The slope is -1 and the y-intercept is $(0, -5)$.

49. The graph contains the points $(0, 3)$ and $(-1, 4)$.

50. The graph contains the points $(5, 1)$ and $(8, 0)$.

Determine whether each pair of lines is parallel, perpendicular, or neither.

51. $x + y = 5,$
$x - y = 1$

52. $2x + y = 3,$
$y = 4 - 2x$

53. $2x + 3y = 1,$
$2x - 3y = 5$

54. $y = \frac{1}{3}x - 7,$
$y + 3x = 1$

Find the function values.

55. $g(x) = \frac{1}{3}x + 7$

a) $g(-3)$ **b)** $g(4)$ **c)** $g(a + 3)$

56. $h(x) = \dfrac{1}{x + 2}$

a) $h(0)$ **b)** $h(2)$ **c)** $h(-2)$

57. $f(x) = 5x^2 + 2x + 3$

a) $f(-1)$ **b)** $f(0)$ **c)** $f(2a)$

58. $f(t) = 6 - t^2$

a) $f(-1)$ **b)** $f(1)$ **c)** $f(3a)$

Determine whether each of the following is the graph of a function.

59.

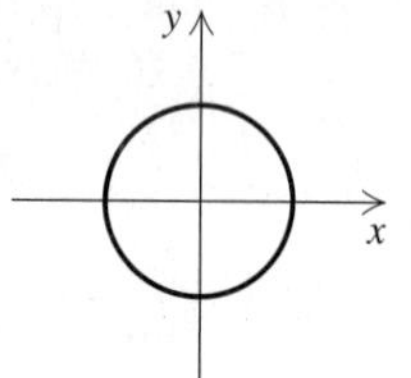

60.

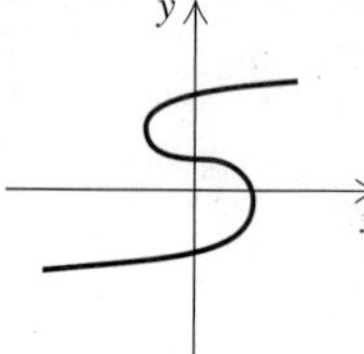

61.

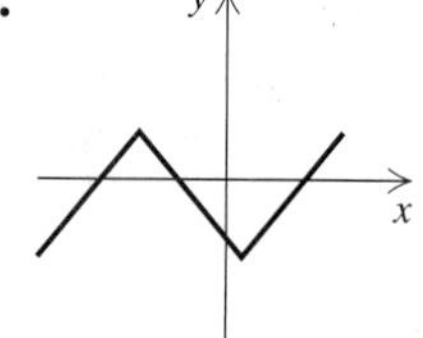

62.

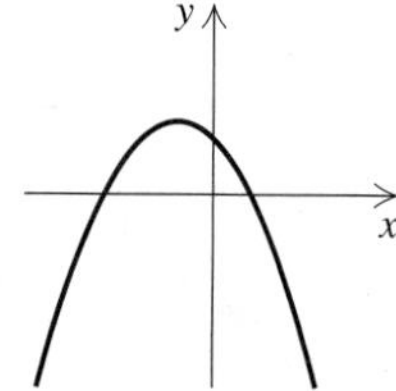

Determine the domain of each function.

63. $f(x) = \dfrac{x}{x - 3}$

64. $g(x) = \sqrt{x}$

65. $g(x) = 2x + 3$

66. $f(x) = |x + 3|$

67. $f(x) = \sqrt{\frac{1}{2}x + 3}$

68. $g(x) = \dfrac{1}{2x + 1}$

Try Exercise Answers: Section R.3

9. No **13.**

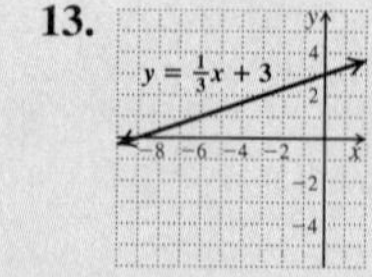

23. 1 **29.** Slope: 2; y-intercept: $(0, -5)$

33.

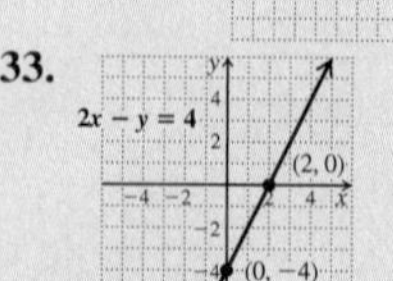

39. (0, −5) $y = 2x - 5$

47. $y = \frac{1}{3}x + 1$

51. Perpendicular **55.** (a) 6; (b) $8\frac{1}{3}$; (c) $\frac{1}{3}a + 8$

63. $\{x \mid x \text{ is a real number } and\ x \neq 3\}$

R.4 Systems of Equations in Two Variables

A **system of equations** is a set of two or more equations that are to be solved simultaneously. An ordered pair is a solution of a system of equations in two variables if it makes *all* of the equations in the system true. We can solve systems of equations in two variables graphically or algebraically.

SOLVING SYSTEMS GRAPHICALLY

The solution of a system of equations is given by the point of intersection of the graphs of the equations.

EXAMPLE 1 Solve by graphing:

$$2x + y = 1,$$
$$x + y = -2.$$

SOLUTION

BY HAND

We graph each equation. The point of intersection has coordinates that make *both* equations true. Apparently, $(3, -5)$ is the solution. Because solving by graphing is not always accurate, we must check. As shown below, $(3, -5)$ is indeed the solution.

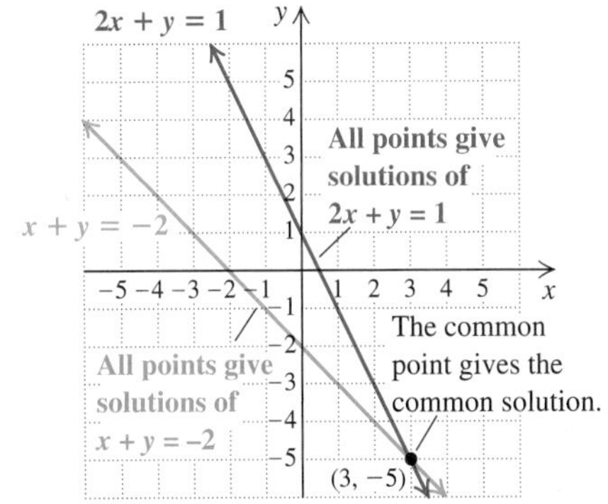

Check:

$$\begin{array}{c|c} 2x + y & 1 \\ \hline 2 \cdot 3 + (-5) & 1 \\ 1 & \stackrel{?}{=} 1 \end{array} \quad \text{TRUE}$$

$$\begin{array}{c|c} x + y & -2 \\ \hline 3 + (-5) & -2 \\ -2 & \stackrel{?}{=} -2 \end{array} \quad \text{TRUE}$$

WITH A GRAPHING CALCULATOR

We first solve each equation for y:

$$2x + y = 1, \qquad x + y = -2$$
$$y = 1 - 2x; \qquad y = -2 - x.$$

Then we enter and graph both equations using the same viewing window.

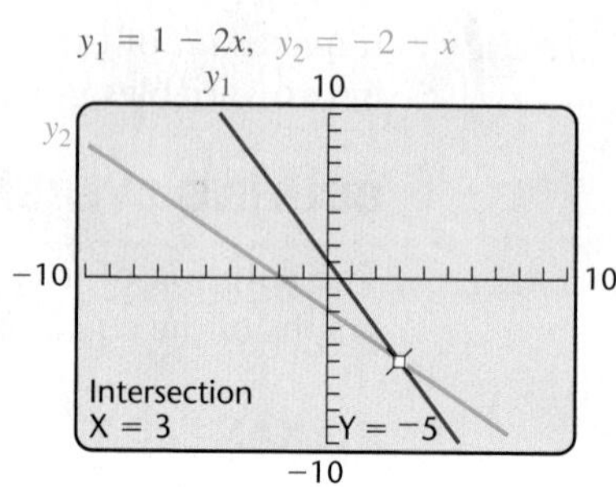

The point of intersection is $(3, -5)$. This ordered pair makes both equations true, so it is the solution of the system.

Try Exercise 1.

The graphs of two linear equations can intersect at one point, can be parallel with no points of intersection, or can have the same graph. Thus systems of two linear equations can have one solution, no solution, or an infinite number of solutions.

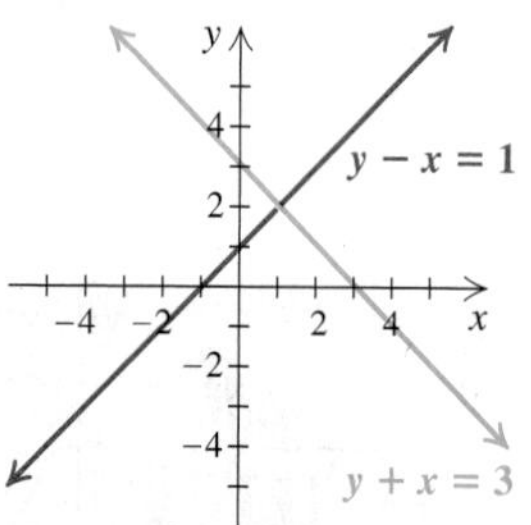

Graphs intersect at one point. The system is *consistent* and has one solution. Since neither equation is a multiple of the other, they are *independent*. The solution is (1, 2).

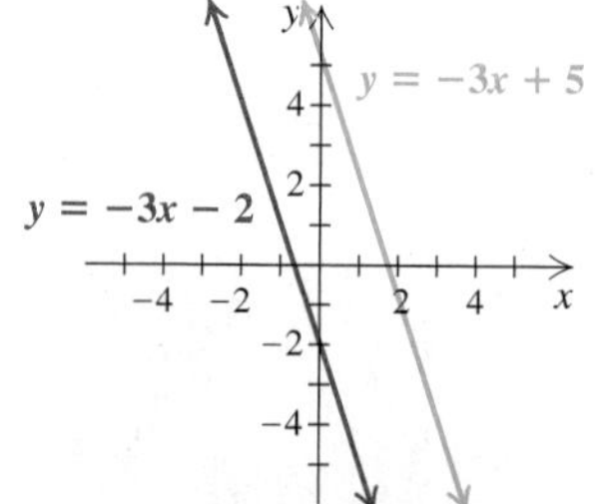

Graphs are parallel. The system is *inconsistent* because there is no solution. Since the equations are not equivalent, they are *independent*. The solution set is $\varnothing$.

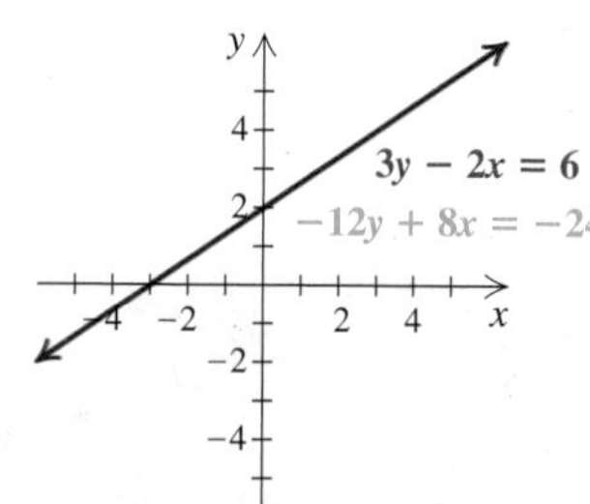

Equations have the same graph. The system is *consistent* and has an infinite number of solutions. The equations are *dependent* since they are equivalent. The solution set is $\{(x, y) \mid 3y - 2x = 6\}$.

Consistent (p. 277)

Independent (p. 277)

SOLVING SYSTEMS ALGEBRAICALLY

Solving by substitution (Section 4.2)

Algebraic (nongraphical) methods for solving systems yield exact answers. The **substitution method** relies on having a variable isolated.

EXAMPLE 2 Solve the system

$$x + y = 4, \qquad (1)$$
$$x - y = 1. \qquad (2)$$

We have numbered the equations for easy reference.

SOLUTION First, we select an equation and solve for one variable. Here we choose to solve for x in equation (2):

Solve for x in terms of y.

$$x - y = 1 \qquad (2)$$
$$x = y + 1. \qquad (3) \quad \textbf{Adding } y \textbf{ to both sides}$$

Equation (3) says that x and $y + 1$ name the same number. Thus we can substitute $y + 1$ for x in equation (1):

Substitute, using parentheses.

$$x + y = 4 \qquad \textbf{Equation (1)}$$
$$(y + 1) + y = 4. \qquad \textbf{Substituting } y + 1 \textbf{ for } x$$

We solve this last equation, using methods learned earlier:

$$(y + 1) + y = 4$$
$$2y + 1 = 4 \qquad \textbf{Removing parentheses and combining like terms}$$
$$2y = 3 \qquad \textbf{Subtracting 1 from both sides}$$
$$y = \tfrac{3}{2}. \qquad \textbf{Dividing by 2}$$

Solve for y.

We now return to the original pair of equations and substitute $\frac{3}{2}$ for y in either equation so that we can solve for x. For this problem, calculations are slightly easier if we use equation (3):

$$x = y + 1 \qquad \textbf{Equation (3)}$$
$$= \tfrac{3}{2} + 1 \qquad \textbf{Substituting } \tfrac{3}{2} \textbf{ for } y$$
$$= \tfrac{3}{2} + \tfrac{2}{2} = \tfrac{5}{2}.$$

Solve for x.

We obtain the ordered pair $\left(\frac{5}{2}, \frac{3}{2}\right)$. A check ensures that it is a solution.

Check.

Check:

$$\begin{array}{c|c} x + y & 4 \\ \hline \frac{5}{2} + \frac{3}{2} & 4 \\ \frac{8}{2} & \\ 4 \stackrel{?}{=} 4 & \text{TRUE} \end{array} \qquad \begin{array}{c|c} x - y & 1 \\ \hline \frac{5}{2} - \frac{3}{2} & 1 \\ \frac{2}{2} & \\ 1 \stackrel{?}{=} 1 & \text{TRUE} \end{array}$$

Since $\left(\frac{5}{2}, \frac{3}{2}\right)$ checks, it is the solution.

Try Exercise 13.

Solving by elimination (Section 4.3)

A second algebraic method, the **elimination method,** makes use of the addition principle: If $a = b$, then $a + c = b + c$. To use the elimination method, we multiply, if necessary, so that the coefficients of one variable are opposites. Then we add the equations, eliminating that variable.

EXAMPLE 3 Solve the system

$$2x + 3y = 17, \qquad (1)$$
$$5x + 7y = 29. \qquad (2)$$

SOLUTION We multiply so that the x-terms are eliminated.

$$2x + 3y = 17, \xrightarrow{\text{Multiplying both sides by 5}} 10x + 15y = 85$$
$$5x + 7y = 29 \xrightarrow{\text{Multiplying both sides by } -2} -10x - 14y = -58$$
$$0 + y = 27$$
$$y = 27$$

The coefficients of x, 10 and -10, are opposites. Adding

Next, we substitute to find x:

$$2x + 3 \cdot 27 = 17 \quad \text{Substituting 27 for } y \text{ in equation (1)}$$
$$2x + 81 = 17$$
$$\left.\begin{aligned} 2x &= -64 \\ x &= -32. \end{aligned}\right\} \quad \text{Solving for } x$$

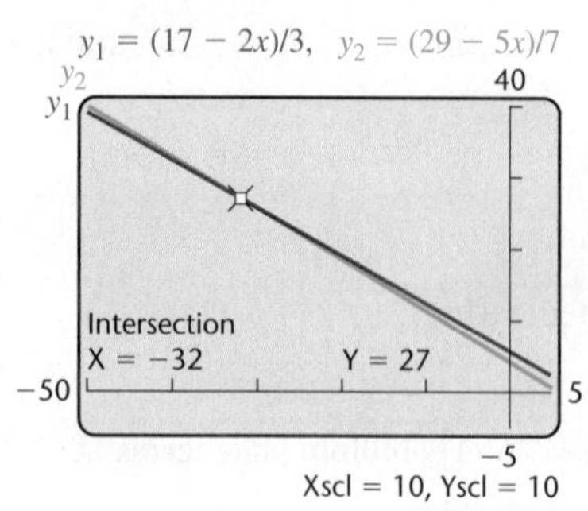

Check:

$2x + 3y = 17$	
$2(-32) + 3(27)$	17
$-64 + 81$	
$17 \stackrel{?}{=} 17$	TRUE

$5x + 7y = 29$	
$5(-32) + 7(27)$	29
$-160 + 189$	
$29 \stackrel{?}{=} 29$	TRUE

We can also check using a graphing calculator, as shown in the figure at left. We obtain $(-32, 27)$, or $x = -32$, $y = 27$, as the solution.

Try Exercise 25.

When solving algebraically, we can tell whether a system has no solutions or an infinite number of solutions.

Rules for Special Cases When solving a system of two linear equations in two variables:

1. If an identity is obtained, such as $0 = 0$, then the system has an infinite number of solutions. The equations are dependent and, since a solution exists, the system is consistent.
2. If a contradiction is obtained, such as $0 = 7$, then the system has no solution. The system is inconsistent.

APPLICATIONS

Many problems that can be translated using two variables can be solved using systems of equations.

EXAMPLE 4 Blending Coffees. The Java Joint wants to mix organic Mexican Chiapas beans that sell for \$8.25 per pound with organic Dark Ethiopian Yirgacheffe beans that sell for \$9.50 per pound in order to form a 50-lb batch of Morning Blend that sells for \$9.00 per pound. How many pounds of each type of bean should go into the blend?

SOLUTION

Mixture problems (p. 305)

1. **Familiarize.** This problem is an example of a **mixture problem.** We can find two relationships in the problem:

   The *weight* of the separate beans must equal the weight of the mixture.

   The *value* of the separate beans must equal the value of the mixture.

   We let c = the number of pounds of organic Mexican coffee used and y = the number of pounds of organic Ethiopian coffee used.

2. **Translate.** Since a 50-lb batch is being made, we must have

$$c + y = 50.$$ **This is the *weight* relationship.**

To find a second equation, we consider the total value of the 50-lb batch. That value must be the same as the value of the Mexican beans and the value of the Ethiopian beans that go into the blend.

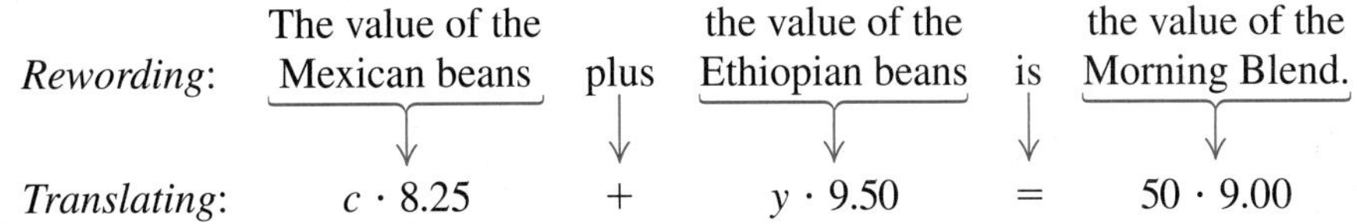

Translating: $c \cdot 8.25 \quad + \quad y \cdot 9.50 \quad = \quad 50 \cdot 9.00$

This information can be presented in a table.

	Mexican	Ethiopian	Morning Blend	
Number of Pounds	c	y	50	→ $c + y = 50$
Price per Pound	\$8.25	\$9.50	\$9.00	
Value of Beans	$8.25c$	$9.50y$	$50 \cdot 9$, or 450	→ $8.25c + 9.50y = 450$

We have translated to a system of equations:

$$c + y = 50, \qquad (1)$$
$$8.25c + 9.50y = 450. \qquad (2)$$

3. **Carry out.** When equation (1) is solved for c, we have $c = 50 - y$. We then substitute $50 - y$ for c in equation (2):

$8.25(50 - y) + 9.50y = 450$ **Solving by substitution**

$412.50 - 8.25y + 9.50y = 450$ **Using the distributive law**

$1.25y = 37.50$ **Combining like terms; subtracting 412.50 from both sides**

$y = 30.$ **Dividing both sides by 1.25**

If $y = 30$, we see from equation (1) that $c = 20$.

4. **Check.** If 20 lb of Mexican beans and 30 lb of Ethiopian beans are mixed, a 50-lb blend will result. The value of 20 lb of Mexican beans is 20(\$8.25), or \$165. The value of 30 lb of Ethiopian beans is 30(\$9.50), or \$285, so the value of the blend is \$165 + \$285 = \$450. A 50-lb blend priced at \$9.00 per pound is also worth \$450, so our answer checks.

5. **State.** The Morning Blend should be made by combining 20 lb of Mexican Chiapas beans with 30 lb of Dark Ethiopian Yirgacheffe beans.

Try Exercise 47.

When a problem deals with distance, speed (rate), and time, we use the following.

Distance, Rate, and Time Equations If r represents rate, t represents time, and d represents distance, then

$$d = rt, \qquad r = \frac{d}{t}, \quad \text{and} \quad t = \frac{d}{r}.$$

EXAMPLE 5 Marine Travel. A Coast-Guard patrol boat travels 4 hr on a trip downstream with a 6-mph current. The return trip against the same current takes 5 hr. Find the speed of the boat in still water.

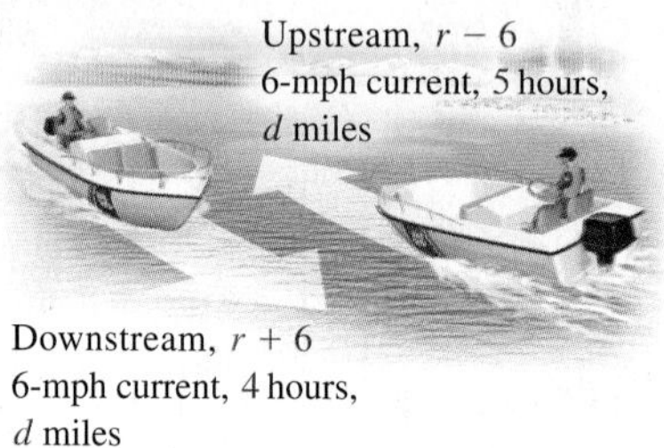

1. **Familiarize.** We first make a drawing and note that the distances are the same. We let $d =$ the distance, in miles, and $r =$ the speed of the boat in still water, in miles per hour. Then, when the boat is traveling downstream, its speed is $r + 6$ (the current helps the boat along). When it is traveling upstream, its speed is $r - 6$ (the current holds the boat back). We can organize the information in a table, using the formula $d = rt$.

$$d \;=\; r \;\cdot\; t$$

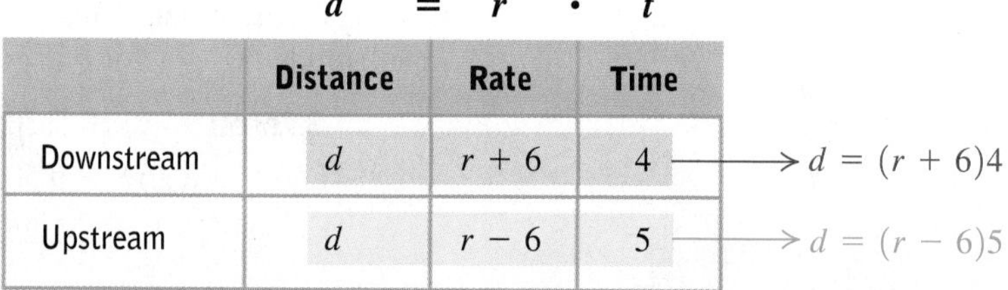

	Distance	Rate	Time	
Downstream	d	$r + 6$	4	→ $d = (r + 6)4$
Upstream	d	$r - 6$	5	→ $d = (r - 6)5$

2. **Translate.** From each row of the table, we get an equation, $d = rt$:

$$d = 4r + 24, \qquad (1)$$
$$d = 5r - 30. \qquad (2)$$

3. **Carry out.** We solve the system by the substitution method:

$$4r + 24 = 5r - 30 \qquad \textbf{Substituting } 4r + 24 \textbf{ for } d \textbf{ in equation (2)}$$
$$\left.\begin{aligned} 24 &= r - 30 \\ 54 &= r. \end{aligned}\right\} \qquad \textbf{Solving for } r$$

4. **Check.** If $r = 54$, then $r + 6 = 60$; and $60 \cdot 4 = 240$, the distance traveled downstream. If $r = 54$, then $r - 6 = 48$; and $48 \cdot 5 = 240$, the distance traveled upstream. The speed of the boat in still water, 54 mph, checks.

5. **State.** The speed of the boat in still water is 54 mph.

Try Exercise 51.

SOLVING EQUATIONS BY GRAPHING

Solving equations graphically (Section 4.5)

When two graphs intersect, that point of intersection represents a solution of *both* equations. We can use this fact to solve equations graphically.

EXAMPLE 6 Solve by graphing: $\frac{1}{2}x - 3 = 2(x + 7)$.

SOLUTION We graph two equations, each corresponding to one side of the equation we wish to solve. For this equation, we graph $y_1 = \frac{1}{2}x - 3$ and $y_2 = 2(x + 7)$ using the same viewing window. We may need to adjust the dimensions of the viewing window in order to see the point of intersection.

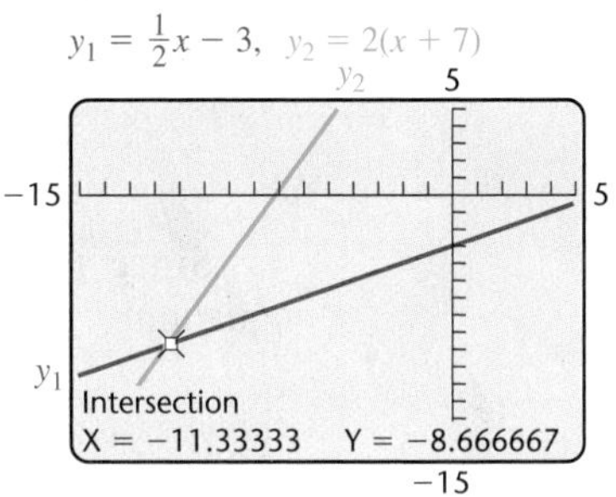

Using INTERSECT, we see that the point of intersection is approximately $(-11.33333, -8.666667)$. The x-coordinate of the point of intersection is the solution of the original equation. Returning to the home screen, we can convert x to fraction notation if we wish. The solution is $-\frac{34}{3}$, or approximately -11.33333.

Try Exercise 59.

Intersect method (p. 314)

Zero method (p. 317)

Zero of a function (p. 317)

We solved the equation in Example 6 using the **Intersect method.** To use the **Zero method,** we find the **zeros of a function,** or inputs that make an output zero.

EXAMPLE 7 Find any zeros of the function given by $f(x) = 3 - 2x$.

SOLUTION A zero is an input x such that $f(x) = 0$:

$$f(x) = 0 \quad \text{The output } f(x) \text{ must be 0.}$$
$$3 - 2x = 0 \quad \text{Substituting}$$
$$\left.\begin{aligned} 3 &= 2x \\ \tfrac{3}{2} &= x. \end{aligned}\right\} \quad \text{Solving for } x$$

Since $f\left(\frac{3}{2}\right) = 3 - 2 \cdot \frac{3}{2} = 0$, $\frac{3}{2}$ is the zero of f.

Try Exercise 55.

EXAMPLE 8 Solve by graphing: $3x + 1 = 5x - 7$.

SOLUTION We use the Zero method, first getting 0 on one side of the equation:

$$3x + 1 = 5x - 7$$
$$-2x + 1 = -7 \quad \text{Subtracting } 5x \text{ from both sides}$$
$$-2x + 8 = 0. \quad \text{Adding 7 to both sides}$$

We let $f(x) = -2x + 8$, and find the zero of the function by locating the x-intercept of its graph.

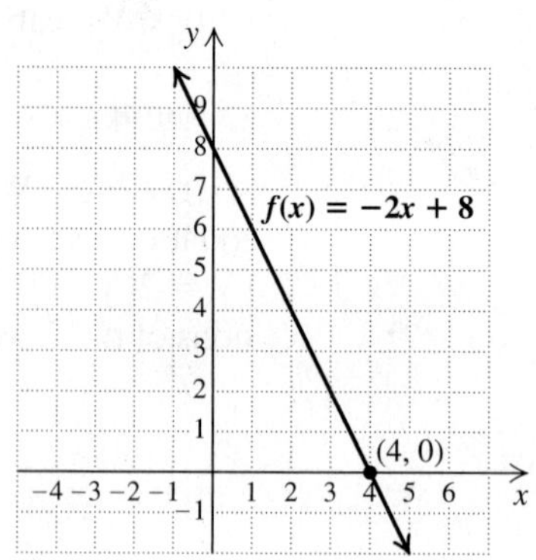

Since $f(4) = 0$, the zero of the function is 4. We check 4 in the original equation.

Check:

$$\begin{array}{r|l} 3x + 1 & = 5x - 7 \\ \hline 3 \cdot 4 + 1 & 5 \cdot 4 - 7 \\ 12 + 1 & 20 - 7 \\ 13 & \stackrel{?}{=} 13 \end{array} \quad \text{TRUE}$$

The solution is 4.

Try Exercise 61.

R.4 Exercise Set

FOR EXTRA HELP MyMathLab

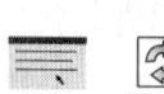

Solve each system using the indicated method. Be sure to check your solution. If a system has an infinite number of solutions, use set-builder notation to write the solution set, as illustrated in the third graph after Example 1. If a system has no solution, state this.

Solve each system graphically.

1. $x + y = 7,$
   $x - y = 1$

2. $x - y = 1,$
   $x + y = 3$

3. $y = -2x + 5,$
   $x + y = 4$

4. $y = 2x - 5,$
   $x + y = 4$

5. $y = x - 3,$
   $y = -2x + 3$

6. $y = -3x + 2,$
   $y = x - 2$

7. $4x - 20 = 5y,$
   $8x - 10y = 12$

8. $6x + 12 = 2y,$
   $6 - y = -3x$

9. $x = 6,$
   $y = -1$

10. $x = -2,$
    $y = 5$

11. $y = \frac{1}{5}x + 4,$
    $2y = \frac{2}{5}x + 8$

12. $y = \frac{1}{3}x + 2,$
    $y = \frac{1}{3}x - 7$

Solve each system using the substitution method.

13. $y = 5 - 4x,$
    $2x + y = 1$

14. $2y + x = 9,$
    $x = 3y - 3$

15. $3x + 5y = 3,$
    $x = 8 - 4y$

16. $9x + 2y = 3,$
    $3x - 6 = y$

17. $3s - 4t = 14,$
    $t = 8 - 5s$

18. $m - 2n = 16,$
    $n = 1 - 4m$

19. $4x - 2y = 6,$
    $2x - 3 = y$

20. $t = 4 - 2s,$
    $t + 2s = 6$

21. $x - 4y = 3,$
$5x + 3y = 4$

22. $2a + 2b = 5,$
$3a - b = 7$

23. $2x - 3 = y,$
$y - 2x = 1$

24. $a - 2b = 3,$
$3a = 6b + 9$

Solve each system using the elimination method.

25. $x + 3y = 7,$
$-x + 4y = 7$

26. $2x + y = 6,$
$x - y = 3$

27. $2x - y = -3,$
$x + y = 9$

28. $x - 2y = 6,$
$-x + 3y = -4$

29. $3x + y = -1,$
$2x - 3y = -8$

30. $x + 3y = 2,$
$2x - y = 11$

31. $4x + 3y = 6,$
$2x - 2y = 1$

32. $3x + 2y = 3,$
$9x - 8y = -2$

33. $5r - 3s = 24,$
$3r + 5s = 28$

34. $5x - 7y = -16,$
$2x + 8y = 26$

35. $6s + 9t = 12,$
$4s + 6t = 5$

36. $10a + 6b = 8,$
$5a + 3b = 2$

37. $12x - 6y = -15,$
$-4x + 2y = 5$

38. $8s + 12t = 16,$
$6s + 9t = 12$

Solve using a system of equations.

39. The sum of two numbers is 89. One number is 3 more than the other. Find the numbers.

40. Find two numbers for which the sum is 108 and the difference is 50.

41. *Supplementary Angles.* Supplementary angles are angles for which the sum of their measures is 180°. Two angles are supplementary. One angle is 33° more than twice the other. Find the measure of each angle.

42. *Complementary Angles.* Complementary angles are angles for which the sum of their measures is 90°. Two angles are complementary. Their difference is 12°. Find the measure of each angle.

43. *Basketball Scoring.* In winning the Western Conference finals, the San Antonio Spurs once scored 76 of their points on a combination of 36 two- and three-pointers. How many shots of each type did they make?
Source: National Basketball Association.

44. *Zoo Admissions.* From April through November, the Bronx Zoo charges $15 per adult and $11 per child for a Total Experience admission. One May day, a total of $8428 was collected from 604 adults and children. How many adult admissions were there?

45. *Dance Lessons.* Jean charges $20 for a private tap lesson and $12 for a group class. One Wednesday, Jean earned $216 from 14 students. How many students of each type did Jean teach?

46. *Used Books.* A Book in Hand sells used paperback books for $1.99 and used hardbacks for $4.99. Holly recently purchased a total of 14 books for a total of $42.86 (before tax). How many paperbacks and how many hardbacks did she buy?

47. *Mixed Nuts.* A grocer wishes to mix peanuts worth $3.20 per pound with Brazil nuts worth $6.40 per pound to make 480 lb of a mixture worth $5.50 per pound. How much of each nut should be used?

48. *Mixed Nuts.* The Nuthouse has 10 kg of mixed cashews and pecans worth $8.40 per kilogram. Cashews alone sell for $8.00 per kilogram, and pecans sell for $9.00 per kilogram. How many kilograms of each nut are in the mixture?

49. *Production.* Streakfree window cleaner is 12% alcohol and Sunstream window cleaner is 30% alcohol. How much of each cleaner should be used to make 90 oz of a cleaner that is 20% alcohol? Complete the following table as part of the *Translate* step.

Type of Solution	Streakfree	Sunstream	Mixture
Amount of Solution	x	y	
Percent Alcohol	12%		20%
Amount of Alcohol in Solution		$0.3y$	

50. *Chemistry.* E-Chem Testing has a solution that is 80% base and another that is 30% base. A technician needs 200 L of a solution that is 62% base. The 200 L will be prepared by mixing the two solutions on hand. How much of each should be used?

51. *Car Travel.* Two cars leave Denver traveling in opposite directions. One car travels at a speed of 75 km/h and the other at 88 km/h. In how many hours will they be 489 km apart?

52. *Canoeing.* Darren paddled for 4 hr with a 6-km/h current to reach a campsite. The return trip against the same current took 16 hr. Find the speed of Darren's canoe in still water.

53. *Air Travel.* Rod is a pilot for Crossland Airways. He computes his flight time against a headwind for a trip of 2900 mi at 5 hr. The flight would take 4 hr and 50 min if the headwind were half as great. Find the headwind and the plane's air speed.

54. *Motorcycle Travel.* Sally and Rocky travel on motorcycles toward each other from Chicago and Indianapolis, which are about 350 km apart, and they are biking at rates of 110 km/h and 90 km/h. They started at the same time. In how many hours will they meet?

Find the zero of each function.

55. $f(x) = 7 - x$

56. $f(x) = \frac{1}{2}x + 2$

57. $f(x) = 2x + 9$

58. $f(x) = 3 - 4x$

Solve by graphing.

59. $2x - 5 = 3x + 1$

60. $x - 7 = 2x + 5$

61. $3x = x + 6$

62. $4x - 7 = x + 2$

63. $5x + 1 = 11x - 3$

64. $2x - 5 = 8x$

Try Exercise Answers: Section R.4

1. (4, 3) **13.** (2, −3) **25.** (1, 2) **47.** Peanuts: 135 lb; Brazil nuts: 345 lb **51.** 3 hr **55.** 7 **59.** −6 **61.** 3

R.5 Polynomials

- Exponents
- Polynomials
- Addition and Subtraction of Polynomials
- Multiplication of Polynomials
- Division of Polynomials

EXPONENTS

We know that x^4 means $x \cdot x \cdot x \cdot x$ and that x^1 means x. Exponential notation is also defined for zero and negative exponents.

Zero and Negative Exponents For any real number a, $a \neq 0$,

$$a^0 = 1 \quad \text{and} \quad a^{-n} = \frac{1}{a^n}.$$

EXAMPLE 1 Simplify: **(a)** $(36)^0$; **(b)** $(-2x)^0$.

SOLUTION

The exponent zero (p. 332)

a) $(36)^0 = 1$ since any number (other than 0 itself) raised to the 0 power is 1.

b) $(-2x)^0 = 1$ for any $x \neq 0$.

Try Exercise 1.

EXAMPLE 2 Write an equivalent expression using positive exponents.

a) x^{-2}

b) $\frac{1}{x^{-2}}$

c) $7y^{-1}$

SOLUTION

Negative exponents (p. 338)

a) $x^{-2} = \frac{1}{x^2}$ **x^{-2} is the reciprocal of x^2.**

b) $\frac{1}{x^{-2}} = x^{-(-2)} = x^2$ **The reciprocal of x^{-2} is $x^{-(-2)}$, or x^2.**

c) $7y^{-1} = 7\left(\frac{1}{y^1}\right) = \frac{7}{y}$ **y^{-1} is the reciprocal of y^1.**

Try Exercise 5.

The following properties hold for any integers m and n and any real numbers a and b, provided no denominators are 0 and 0^0 is not considered.

Properties of Exponents

The Product Rule: $a^m \cdot a^n = a^{m+n}$

The Quotient Rule: $\frac{a^m}{a^n} = a^{m-n}$

The Power Rule: $(a^m)^n = a^{mn}$

Raising a product to a power: $(ab)^n = a^n b^n$

Raising a quotient to a power: $\left(\frac{a}{b}\right)^n = \frac{a^n}{b^n}$

These properties are often used to simplify exponential expressions.

EXAMPLE 3 Simplify.

a) $(x^2y^{-1})(xy^{-3})$ **b)** $\frac{(3p)^3}{(3p)^{-2}}$ **c)** $\left(\frac{ab^2}{3c^3}\right)^{-4}$

SOLUTION

a)
$$\begin{aligned}(x^2y^{-1})(xy^{-3}) &= x^2y^{-1}xy^{-3} && \text{Using an associative law}\\ &= x^2x^1y^{-1}y^{-3} && \text{Using a commutative law; } x = x^1\\ &= x^{2+1}y^{-1+(-3)} && \text{Using the product rule: Adding exponents}\\ &= x^3y^{-4}, \text{ or } \frac{x^3}{y^4}\end{aligned}$$

The product rule (p. 330)

b)
$$\begin{aligned}\frac{(3p)^3}{(3p)^{-2}} &= (3p)^{3-(-2)} && \text{Using the quotient rule: Subtracting exponents}\\ &= (3p)^5\\ &= 3^5p^5 && \text{Raising each factor to the fifth power}\\ &= 243p^5\end{aligned}$$

The quotient rule (p. 331)

The power rule (p. 333)

c)
$$\begin{aligned}\left(\frac{ab^2}{3c^3}\right)^{-4} &= \frac{(ab^2)^{-4}}{(3c^3)^{-4}} && \text{Raising the numerator and the denominator to the } -4 \text{ power}\\ &= \frac{a^{-4}(b^2)^{-4}}{3^{-4}(c^3)^{-4}} && \text{Raising each factor to the } -4 \text{ power}\\ &= \frac{a^{-4}b^{-8}}{3^{-4}c^{-12}} && \text{Multiplying exponents}\\ &= \frac{3^4c^{12}}{a^4b^8}, \text{ or } \frac{81c^{12}}{a^4b^8} && \text{Rewriting without negative exponents}\end{aligned}$$

Raising a quotient to a power (p. 334)

Raising a product to a power (p. 334)

Try Exercise 25.

POLYNOMIALS

Polynomials (Section 5.3)

Terms (p. 350)

Algebraic expressions like

$$2x^3 + 3x - 5, \quad 4x, \quad -7, \quad \text{and} \quad 2a^3b^2 + ab^3$$

are all examples of **polynomials.** All variables in a polynomial are raised to whole-number powers, and there are no variables in a denominator. The **terms** of a

Degree of a term (p. 351)

Leading term (p. 351)

Degree of a polynomial (p. 351)

Coefficient (p. 351)

Leading coefficient (p. 351)

polynomial are separated by addition signs. The **degree of a term** is the number of variable factors in that term. The **leading term** of a polynomial is the term of highest degree. The **degree of a polynomial** is the degree of the leading term. A polynomial is written in *descending order* when the leading term appears first, followed by the term of next highest degree, and so on.

The number -2 in the term $-2y^3$ is called the **coefficient** of that term. The coefficient of the leading term is the **leading coefficient** of the polynomial. To illustrate this terminology, consider the polynomial

$$4y^2 - 8y^5 + y^3 - 6y + 7.$$

The *terms* are $4y^2$, $-8y^5$, y^3, $-6y$, and 7.

The *coefficients* are 4, -8, 1, -6, and 7.

The *degree of each term* is 2, 5, 3, 1, and 0.

The *leading term* is $-8y^5$ and the *leading coefficient* is -8.

The *degree of the polynomial* is 5.

Polynomials are classified by the number of terms and by degree.

A **monomial** has one term. *Example*: $-2x^3y$

A **binomial** has two terms. *Example*: $1.4x^2 - 10$

A **trinomial** has three terms. *Example*: $x^2 - 3x - 6$

A **constant** polynomial has degree 0. *Example*: 7

A **linear** polynomial has degree 1. *Example*: $3x + 5$

A **quadratic** polynomial has degree 2. *Example*: $5x^2 - x$

A **cubic** polynomial has degree 3. *Example*: $x^3 + 2x^2 - \frac{1}{3}$

A **quartic** polynomial has degree 4. *Example*: $-6x^4 - 2x^2 + 19$

Like, or *similar*, *terms* are either constant terms or terms containing the same variable(s) raised to the same power(s). Polynomials containing like terms can be simplified by *combining* those terms.

EXAMPLE 4 Combine like terms: $4x^2y + 2xy - x^2y + xy^2$.

SOLUTION The like terms are $4x^2y$ and $-x^2y$. Thus we have

$$\begin{aligned} 4x^2y + 2xy - x^2y + xy^2 &= 4x^2y - x^2y + 2xy + xy^2 \\ &= 3x^2y + 2xy + xy^2. \end{aligned}$$

Try Exercise 39.

A polynomial can be evaluated by replacing the variable or variables with a number or numbers. We use parentheses when substituting negative numbers.

EXAMPLE 5 Evaluate $-a^2 + 2ab + 5b^2$ for $a = -1$ and $b = 3$.

SOLUTION We replace a with -1 and b with 3 and calculate the value using the rules for order of operations:

$$\begin{aligned} -a^2 + 2ab + 5b^2 &= -(-1)^2 + 2 \cdot (-1) \cdot 3 + 5 \cdot 3^2 \\ &= -1 - 6 + 45 = 38. \end{aligned}$$

Try Exercise 43.

Polynomial function (p. 352)

A function described by a polynomial is called a **polynomial function.**

EXAMPLE 6 Find $P(-2)$ for the polynomial function given by $P(x) = -x^3 + 3x - 7$.

SOLUTION We substitute -2 for x and carry out the operations using the rules for order of operations:

$$\begin{aligned} P(-2) &= -(-2)^3 + 3(-2) - 7 \\ &= -(-8) + (-6) - 7 \qquad \textbf{We cube the input before taking the opposite.} \\ &= 8 - 6 - 7 \\ &= -5. \end{aligned}$$

Try Exercise 49.

ADDITION AND SUBTRACTION OF POLYNOMIALS

Addition of polynomials (Section 5.4)

To add two polynomials, we write a plus sign between them and combine like terms.

EXAMPLE 7 Add: $(4x^3 + 3x^2 + 2x - 7) + (-5x^2 + x - 10)$.

SOLUTION

$$\begin{aligned} &(4x^3 + 3x^2 + 2x - 7) + (-5x^2 + x - 10) \\ &\quad = 4x^3 + (3 - 5)x^2 + (2 + 1)x + (-7 - 10) \\ &\quad = 4x^3 - 2x^2 + 3x - 17 \end{aligned}$$

Try Exercise 53.

Opposite of a polynomial (p. 366)

To find the **opposite of a polynomial,** we replace each term with its opposite. This process is also called *changing the sign* of each term. For example, the opposite of

$$3y^4 - 7y^2 - \tfrac{1}{3}y + 17$$

is

$$-(3y^4 - 7y^2 - \tfrac{1}{3}y + 17) = -3y^4 + 7y^2 + \tfrac{1}{3}y - 17.$$

Subtraction of polynomials (Section 5.4)

To subtract polynomials, we add the opposite of the polynomial being subtracted.

EXAMPLE 8 Subtract: $(3a^4 - 2a + 7) - (-a^3 + 5a - 1)$.

SOLUTION

$$\begin{aligned} &(3a^4 - 2a + 7) - (-a^3 + 5a - 1) \\ &\quad = 3a^4 - 2a + 7 + a^3 - 5a + 1 \qquad \textbf{Adding the opposite} \\ &\quad = 3a^4 + a^3 - 7a + 8 \qquad \textbf{Combining like terms} \end{aligned}$$

Try Exercise 55.

MULTIPLICATION OF POLYNOMIALS

Multiplication of polynomials (Section 5.5)

To multiply two monomials, we multiply coefficients and then multiply variables using the product rule for exponents. To multiply a monomial and a polynomial, we multiply each term of the polynomial by the monomial, using the distributive property.

EXAMPLE 9 Multiply: $4x^3(3x^4 - 2x^3 + 7x - 5)$.

SOLUTION

$$\textit{Think: } \underbrace{4x^3 \cdot 3x^4} - \underbrace{4x^3 \cdot 2x^3} + \underbrace{4x^3 \cdot 7x} - \underbrace{4x^3 \cdot 5}$$

$$4x^3(3x^4 - 2x^3 + 7x - 5) = 12x^7 - 8x^6 + 28x^4 - 20x^3$$

Try Exercise 59.

To multiply any two polynomials P and Q, we select one of the polynomials—say, P. We then multiply each term of P by every term of Q and combine like terms.

EXAMPLE 10 Multiply: $(2a^3 + 3a - 1)(a^2 - 4a)$.

SOLUTION It is often helpful to use columns for a long multiplication. We multiply each term at the top by every term at the bottom, write like terms in columns, and add the results.

$$\begin{array}{rrrrrl}
 & & 2a^3 & + 3a & - 1 & \\
 & & & a^2 & - 4a & \\
\hline
 & -8a^4 & & -12a^2 & + 4a & \quad \textbf{Multiplying the top row by } -4a \\
2a^5 & & + 3a^3 & - a^2 & & \quad \textbf{Multiplying the top row by } a^2 \\
\hline
2a^5 & - 8a^4 & + 3a^3 & - 13a^2 & + 4a & \quad \textbf{Combining like terms. Be sure that like terms are lined up in columns.}
\end{array}$$

Try Exercise 63.

By observing the pattern of the products formed when multiplying two binomials, we can develop a more efficient method to find their product.

The FOIL Method To multiply two binomials, $A + B$ and $C + D$, multiply the First terms AC, the Outer terms AD, the Inner terms BC, and then the Last terms BD. Then combine like terms, if possible.

$$(A + B)(C + D) = AC + AD + BC + BD$$

1. Multiply First terms: AC.
2. Multiply Outer terms: AD.
3. Multiply Inner terms: BC.
4. Multiply Last terms: BD.

↓

FOIL

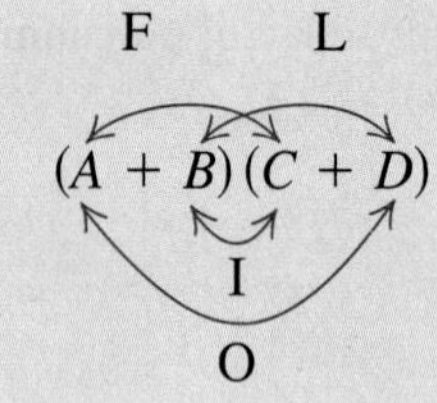

EXAMPLE 11 Multiply: $(3x + 4)(x - 2)$.

FOIL (p. 379)

SOLUTION

F L

I

O

$$\begin{aligned}
(3x + 4)(x - 2) &= \overset{F}{3x^2} - \overset{O}{6x} + \overset{I}{4x} - \overset{L}{8} \\
&= 3x^2 - 2x - 8 \qquad \textbf{Combining like terms}
\end{aligned}$$

Try Exercise 61.

Special products occur so often that specific formulas or methods for computing them have been developed.

Multiplying sums and differences of two terms (p. 381)

Squaring binomials (p. 382)

Special Products The product of a sum and a difference of the same two terms:

$$(A + B)(A - B) = \underbrace{A^2 - B^2}$$

This is called a *difference of squares*.

The square of a binomial:

$$(A + B)^2 = A^2 + 2AB + B^2$$
$$(A - B)^2 = A^2 - 2AB + B^2$$

EXAMPLE 12 Multiply: **(a)** $(x + 3y)(x - 3y)$; **(b)** $(x^3 + 2)^2$.

SOLUTION

$$(A + B)(A - B) = A^2 - B^2$$

a) $(x + 3y)(x - 3y) = x^2 - (3y)^2$ **$A = x$ and $B = 3y$**

$= x^2 - 9y^2$

$$(A + B)^2 = A^2 + 2 \cdot A \cdot B + B^2$$

b) $(x^3 + 2)^2 = (x^3)^2 + 2 \cdot x^3 \cdot 2 + 2^2$ **$A = x^3$ and $B = 2$**

$= x^6 + 4x^3 + 4$

Try Exercise 65.

DIVISION OF POLYNOMIALS

Division of polynomials (Section 5.8)

Polynomial division is similar to division in arithmetic. To divide a polynomial by a monomial, we divide each term by the monomial.

EXAMPLE 13 Divide: $(3x^5 + 8x^3 - 12x) \div (4x)$.

SOLUTION This division can be written

$$\frac{3x^5 + 8x^3 - 12x}{4x} = \frac{3x^5}{4x} + \frac{8x^3}{4x} - \frac{12x}{4x}$$ **Dividing each term by 4x**

$$= \frac{3}{4}x^{5-1} + \frac{8}{4}x^{3-1} - \frac{12}{4}x^{1-1}$$ **Dividing coefficients and subtracting exponents**

$$= \frac{3}{4}x^4 + 2x^2 - 3.$$

To check, we multiply the quotient by $4x$:

$$\left(\tfrac{3}{4}x^4 + 2x^2 - 3\right)4x = 3x^5 + 8x^3 - 12x.$$ **The answer checks.**

Try Exercise 75.

To use long division, we write polynomials in descending order, including terms with 0 coefficients for missing terms. As shown below in Example 14, the procedure ends when the degree of the remainder is less than the degree of the divisor.

EXAMPLE 14 Divide: $(4x^3 - 7x + 1) \div (2x + 1)$.

SOLUTION The polynomials are already written in descending order, but there is no x^2-term in the dividend. We fill in $0x^2$ for that term.

$$\begin{array}{r|l}
 & 2x^2 \\ \hline
2x + 1 & 4x^3 + 0x^2 - 7x + 1 \\
 & \underline{4x^3 + 2x^2} \\
 & -2x^2
\end{array}$$

Divide the first term of the dividend, $4x^3$, by the first term in the divisor, $2x$: $4x^3/(2x) = 2x^2$.

Multiply $2x^2$ by the divisor, $2x + 1$.

Subtract: $(4x^3 + 0x^2) - (4x^3 + 2x^2) = -2x^2$.

Then we bring down the next term of the dividend, $-7x$.

$$\begin{array}{r|l}
 & 2x^2 - x \\ \hline
2x + 1 & 4x^3 + 0x^2 - 7x + 1 \\
 & \underline{4x^3 + 2x^2} \\
 & \quad -2x^2 - 7x \\
 & \quad \underline{-2x^2 - x} \\
 & \qquad\quad -6x
\end{array}$$

Divide the first term of $-2x^2 - 7x$ by the first term in the divisor: $-2x^2/(2x) = -x$.

The $-7x$ has been "brought down."

Multiply $-x$ by the divisor, $2x + 1$.

Subtract: $(-2x^2 - 7x) - (-2x^2 - x) = -6x$.

Since the degree of the remainder, $-6x$, is *not* less than the degree of the divisor, we must continue dividing.

$$\begin{array}{r|l}
 & 2x^2 - x - 3 \\ \hline
2x + 1 & 4x^3 + 0x^2 - 7x + 1 \\
 & \underline{4x^3 + 2x^2} \\
 & \quad -2x^2 - 7x \\
 & \quad \underline{-2x^2 - x} \\
 & \qquad\quad -6x + 1 \\
 & \qquad\quad \underline{-6x - 3} \\
 & \qquad\qquad\quad 4
\end{array}$$

Divide the first term of $-6x + 1$ by the first term in the divisor: $-6x/(2x) = -3$.

The 1 has been "brought down."

Multiply -3 by $2x + 1$.

Subtract.

The answer is $2x^2 - x - 3$ with R 4, or

$$\textbf{Quotient} \longrightarrow 2x^2 - x - 3 + \frac{4 \longleftarrow \textbf{Remainder}}{2x + 1 \longleftarrow \textbf{Divisor}}.$$

Check: We multiply by the divisor and add the remainder:

$$\begin{aligned}
(2x + 1)(2x^2 - x - 3) + 4 &= 4x^3 - 7x - 3 + 4 \\
&= 4x^3 - 7x + 1.
\end{aligned}$$

Try Exercise 77.

R.5 Exercise Set

FOR EXTRA HELP MyMathLab 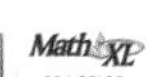  

Simplify.

1. a^0, for $a = -25$
2. y^0, for $y = 6.97$
3. $4^0 - 4^1$
4. $8^1 - 8^0$

Write an equivalent expression using positive exponents. Then, if possible, simplify.

5. 8^{-2}
6. $(-2)^{-3}$
7. $10x^{-5}$
8. $-16y^{-3}$
9. $(ab)^{-2}$
10. ab^{-2}
11. $\dfrac{1}{y^{-10}}$
12. $\dfrac{1}{x^{-t}}$

Write an equivalent expression using negative exponents.

13. $\dfrac{1}{y^4}$
14. $\dfrac{1}{a^2b^3}$
15. $\dfrac{1}{x^t}$
16. $\dfrac{1}{n}$

Simplify. Write answers with positive exponents.

17. $x^5 \cdot x^8$
18. $a^4 \cdot a^{-2}$
19. $\dfrac{a}{a^{-5}}$
20. $\dfrac{p^{-3}}{p^{-8}}$
21. $\dfrac{(4x)^{10}}{(4x)^2}$
22. $\dfrac{a^2b^9}{a^9b^2}$
23. $(7^8)^5$
24. $(x^3)^{-7}$
25. $(x^{-2}y^{-3})^{-4}$
26. $(-2a^2)^3$
27. $\left(\dfrac{y^2}{4}\right)^3$
28. $\left(\dfrac{ab^2}{c^3}\right)^4$
29. $\left(\dfrac{2p^3}{3q^4}\right)^{-2}$
30. $\left(\dfrac{2}{x}\right)^{-5}$

Identify the terms of each polynomial.

31. $8x^3 - 6x^2 + x - 7$
32. $-a^2b + 4a^2 - 8b + 17$

Determine the coefficient and the degree of each term in each polynomial. Then find the degree of each polynomial.

33. $18x^3 + 36x^9 - 7x + 3$
34. $-8y^7 + y + 19$
35. $-x^2y + 4y^3 - 2xy$
36. $8 - x^2y^4 + y^7$

Determine the leading term and the leading coefficient of each polynomial.

37. $-p^2 + 4 + 8p^4 - 7p$
38. $13 + 20t - 30t^2 - t^3$

Combine like terms. Write each answer in descending order.

39. $3x^3 - x^2 + x^4 + x^2$
40. $5t - 8t^2 + 4t^2$
41. $3 - 2t^2 + 8t - 3t - 5t^2 + 7$
42. $8x^5 - \frac{1}{3} + \frac{4}{5}x + 1 - \frac{1}{2}x$

Evaluate each polynomial for the given replacements of the variables.

43. $3x^2 - 7x + 10$, for $x = -2$
44. $-y + 3y^2 + 2y^3$, for $y = 3$
45. $a^2b^3 + 2b^2 - 6a$, for $a = 2$ and $b = -1$
46. $2pq^3 - 5q^2 + 8p$, for $p = -4$ and $q = -2$

The distance s, in feet, traveled by a body falling freely from rest in t seconds is approximated by

$$s = 16t^2.$$

47. A pebble is dropped into a well and takes 3 sec to hit the water. How far down is the surface of the water?
48. An acorn falls from the top of an oak tree and takes 2 sec to hit the ground. How high is the tree?

Find each function value.

49. $f(3)$, if $f(x) = x^2 - x + 5$
50. $g(7)$, if $g(x) = 3x^2 + 2x$
51. $p(-1)$, if $p(x) = -x^3 - x^2 + 4x$
52. $F(-2)$, if $F(x) = 12 - x + 3x^2$

Add or subtract, as indicated.

53. $(3x^3 + 2x^2 + 8x) + (x^3 - 5x^2 + 7)$
54. $(-6x^4 + 3x^2 - 16) + (4x^2 + 4x - 7)$
55. $(8y^2 - 2y - 3) - (9y^2 - 7y - 1)$
56. $(4t^2 + 6t - 7) - (t + 5)$
57. $(-x^2y + 2y^2 + y) - (3y^2 + 2x^2y - 7y)$
58. $(ab + x^2y^2) + (2ab - x^2y^2)$

Multiply.

59. $4x^2(3x^3 - 7x + 7)$
60. $a^2b(a^3 + b^2 - ab - 2b)$
61. $(2a + y)(4a + b)$
62. $(x + 7y)(y - 3x)$
63. $(x + 7)(x^2 - 3x + 1)$
64. $(2x - 3)(x^2 - x - 1)$
65. $(x + 7)(x - 7)$
66. $(2x + 1)^2$
67. $(x + y)^2$
68. $(xy + 1)(xy - 1)$
69. $(2x^2 + 7)(3x^2 - 2)$
70. $(x^2 + 2)^2$
71. $(a - 3b)^2$
72. $(1.1x^2 + 5)(0.1x^2 - 2)$
73. $(6a - 5y)(7a + 3y)$
74. $(3p^2 - q^3)^2$

Divide and check.

75. $(3t^5 + 9t^3 - 6t^2 + 15t) \div (-3t)$
76. $(4x^5 + 10x^4 - 16x^2) \div (4x^2)$
77. $(15x^2 - 16x - 15) \div (3x - 5)$
78. $(x^3 - 2x^2 - 14x + 1) \div (x - 5)$
79. $(2x^3 - x^2 + 1) \div (x + 1)$
80. $(2x^3 + 3x^2 - 50) \div (2x - 5)$
81. $(5x^3 + 3x^2 - 5x) \div (x^2 - 1)$
82. $(2x^3 + 3x^2 + 6x + 10) \div (x^2 + 3)$

Try Exercise Answers: Section R.5

1. 1 **5.** $\frac{1}{8^2} = \frac{1}{64}$ **25.** x^8y^{12} **39.** $x^4 + 3x^3$ **43.** 36
49. 11 **53.** $4x^3 - 3x^2 + 8x + 7$ **55.** $-y^2 + 5y - 2$
59. $12x^5 - 28x^3 + 28x^2$ **61.** $8a^2 + 2ab + 4ay + by$
63. $x^3 + 4x^2 - 20x + 7$ **65.** $x^2 - 49$
75. $-t^4 - 3t^2 + 2t - 5$ **77.** $5x + 3$

R.6 Polynomial Factorizations and Equations

- Common Factors and Factoring by Grouping
- Factoring Trinomials
- Factoring Special Forms
- Solving Polynomial Equations

COMMON FACTORS AND FACTORING BY GROUPING

To *factor* a polynomial is to find an equivalent expression that is a product. For example, three *factorizations* of $50x^6$ are $5 \cdot 10x^6$, $5x^3 \cdot 10x^3$, and $2x \cdot 25x^5$.

If all the terms in a polynomial share a common factor, that factor can be "factored out" of the polynomial. Whenever you are factoring a polynomial with two or more terms, try to first find the largest common factor of the terms, if one exists.

Common factor (p. 429)

EXAMPLE 1 Factor: $3x^6 + 15x^4 - 9x^3$.

SOLUTION The largest factor common to 3, 15, and -9 is 3. The largest power of x common to x^6, x^4, and x^3 is x^3. Thus the largest common factor of the terms of the polynomial is $3x^3$. We factor as follows:

$$3x^6 + 15x^4 - 9x^3 = 3x^3 \cdot x^3 + 3x^3 \cdot 5x - 3x^3 \cdot 3 \quad \text{Factoring each term}$$

$$= 3x^3(x^3 + 5x - 3). \quad \text{Factoring out } 3x^3$$

Factorizations can always be checked by multiplying:

$$3x^3(x^3 + 5x - 3) = 3x^6 + 15x^4 - 9x^3.$$

Try Exercise 1.

A polynomial with two or more terms can be a common factor.

EXAMPLE 2 Factor: $3x^2(x - 2) + 5(x - 2)$.

SOLUTION The binomial $x - 2$ is a factor of both $3x^2(x - 2)$ and $5(x - 2)$. Thus we have

$$3x^2(x - 2) + 5(x - 2) = (x - 2)(3x^2 + 5).$$ **Factoring out the common factor, $x - 2$**

Try Exercise 5.

Factoring by grouping (p. 431)

If a polynomial with four terms can be split into two groups of terms, and both groups share a common binomial factor, the polynomial can be factored. This method is known as **factoring by grouping.**

EXAMPLE 3 Factor by grouping: $2x^3 + 6x^2 - x - 3$.

SOLUTION First, we consider the polynomial as two groups of terms, $2x^3 + 6x^2$ and $-x - 3$. Then we factor each group separately:

$$2x^3 + 6x^2 - x - 3 = 2x^2(x + 3) - 1(x + 3)$$ **Factoring out $2x^2$ and -1 to give the common binomial factor, $x + 3$**

$$= (x + 3)(2x^2 - 1).$$

The check is left to the student.

Try Exercise 23.

Prime polynomial (p. 430)

Not every polynomial with four terms is factorable by grouping. A polynomial that is not factorable is said to be **prime.**

FACTORING TRINOMIALS

Factoring trinomials of the type $x^2 + bx + c$ (Section 6.2)

Many trinomials that have no common factor can be written as the product of two binomials. We look first at trinomials of the form $x^2 + bx + c$, for which the leading coefficient is 1.

Factoring trinomials involves a trial-and-error process. In order for the product of two binomials to be $x^2 + bx + c$, the binomials must look like

$$(x + p)(x + q),$$

where p and q are constants that must be determined. We look for two numbers p and q whose product is c and whose sum is b.

EXAMPLE 4 Factor.

a) $x^2 + 10x + 16$ **b)** $x^2 - 8x + 15$

c) $x^2 - 2x - 24$ **d)** $3t^2 - 33st + 84s^2$

SOLUTION

Constant term positive (p. 438)

a) The factorization is of the form

$$(x + \quad)(x + \quad).$$

To find the constant terms, we need a pair of factors whose product is 16 and whose sum is 10. Since 16 is positive, its factors will have the same sign as 10—that is, we need consider only positive factors of 16.

We list the possible factorizations in a table and calculate the sum of each pair of factors.

Pairs of Factors of 16	Sums of Factors
1, 16	17
2, 8	10 ← The numbers we seek are 2 and 8.
4, 4	8

The factorization of $x^2 + 10x + 16$ is $(x + 2)(x + 8)$. To check, we multiply.

Check: $(x + 2)(x + 8) = x^2 + 8x + 2x + 16 = x^2 + 10x + 16.$

b) For $x^2 - 8x + 15$, c is positive and b is negative. Therefore, the factors of 15 will be negative. Again, we list the possible factorizations in a table.

Pairs of Factors of 15	Sums of Factors
$-1, -15$	-16
$-3, -5$	-8 ← The numbers we need are -3 and -5.

The factorization is $(x - 3)(x - 5)$. The check is left to the student.

Constant term negative (p. 440)

c) For $x^2 - 2x - 24$, c is negative, so one factor of -24 will be negative and one will be positive. Since b is also negative, the negative factor must have the larger absolute value.

Pairs of Factors of -24	Sums of Factors
$1, -24$	-23
$2, -12$	-10
$3, -8$	-5
$4, -6$	-2 ← The numbers we need are 4 and -6.

The factorization is $(x + 4)(x - 6)$.

Check: $(x + 4)(x - 6) = x^2 - 6x + 4x - 24 = x^2 - 2x - 24.$

d) Always look first for a common factor. There is a common factor, 3, which we factor out first:

$$3t^2 - 33st + 84s^2 = 3(t^2 - 11st + 28s^2).$$

Now we consider $t^2 - 11st + 28s^2$. Think of $28s^2$ as the "constant" term c and $-11s$ as the "coefficient" b of the middle term. We try to express $28s^2$ as the product of two factors whose sum is $-11s$. These factors are $-4s$ and $-7s$. Thus the factorization of $t^2 - 11st + 28s^2$ is

$$(t - 4s)(t - 7s).$$ **This is not the entire factorization of $3t^2 - 33st + 84s^2$.**

We now include the common factor, 3, and write

$$3t^2 - 33st + 84s^2 = 3(t - 4s)(t - 7s). \quad \text{This is the factorization.}$$

Check: $3(t - 4s)(t - 7s) = 3(t^2 - 11st + 28s^2)$
$= 3t^2 - 33st + 84s^2.$

Try Exercise 11.

Factoring trinomials of the type $ax^2 + bx + c$ (Section 6.3)

When the leading coefficient of a trinomial is not 1, the number of trials needed to find a factorization can increase dramatically. We will consider two methods for factoring trinomials of the type $ax^2 + bx + c$: factoring by reversing FOIL and the ac-method.

To Factor $ax^2 + bx + c$ Using FOIL

1. Make certain that all common factors have been removed. If any remain, factor out the largest common factor.
2. Find two First terms whose product is ax^2:

$$(\square x + \quad)(\square x + \quad) = ax^2 + bx + c.$$
FOIL

3. Find two Last terms whose product is c:

$$(\quad x + \square)(\quad x + \square) = ax^2 + bx + c.$$
FOIL

4. Check by multiplying to see if the sum of the Outer and Inner products is bx. If necessary, repeat steps (2) and (3) until the correct combination is found.

$$(\square x + \square)(\square x + \square) = ax^2 + bx + c.$$
I O FOIL

If no correct combination exists, state that the polynomial is prime.

EXAMPLE 5 Factor: $20x^3 - 22x^2 - 12x$.

SOLUTION

1. First, we factor out the largest common factor, $2x$:

$$20x^3 - 22x^2 - 12x = 2x(10x^2 - 11x - 6).$$

2. Next, in order to factor the trinomial $10x^2 - 11x - 6$, we search for two terms whose product is $10x^2$. The possibilities are

$$(x + \quad)(10x + \quad) \quad \text{or} \quad (2x + \quad)(5x + \quad).$$

3. There are four pairs of factors of -6. Since the first terms of the binomials are different, the order of the factors is important. So there are eight possibilities for the last terms:

$1, -6$		$-6, 1$
$-1, 6$	and	$6, -1$
$2, -3$		$-3, 2$
$-2, 3$		$3, -2.$

4. Since each of the eight possibilities from step (3) could be used in either of the two possibilities from step (2), there are $2 \cdot 8$, or 16, possible factorizations. We check the possibilities systematically until we find one that gives the correct factorization. Let's first try factors with $(2x + \quad)(5x + \quad)$.

Pair of Factors	Corresponding Trial	Product	
$1, -6$	$(2x + 1)(5x - 6)$	$10x^2 - 7x - 6$	← **Wrong middle term**
$-1, 6$	$(2x - 1)(5x + 6)$	$10x^2 + 7x - 6$	← **Wrong middle term. Changing the signs in the binomials changed the sign of the middle term in the product.**
$2, -3$	$(2x + 2)(5x - 3)$	$10x^2 + 4x - 6$	← **Wrong middle term. We need not consider $(2x - 2)(5x + 3)$.**
$-6, 1$	$(2x - 6)(5x + 1)$	$10x^2 - 28x - 6$	← **Wrong middle term. We need not consider $(2x + 6)(5x - 1)$.**
$-3, 2$	$(2x - 3)(5x + 2)$	$10x^2 - 11x - 6$	← **Correct middle term**

Student Note

Both the third and the fourth trial factorizations in Example 5 contain a binomial with a common factor: $2x + 2 = 2(x + 1)$ and $2x - 6 = 2(x - 3)$. If all common factors are factored out as a first step, you can immediately eliminate any trial factorization containing a binomial with a common factor.

We can stop when we find a correct factorization. Including the common factor $2x$, we now have

$$20x^3 - 22x^2 - 12x = 2x(2x - 3)(5x + 2).$$

This can be checked by multiplying.

Try Exercise 27.

With practice, some of the trials can be skipped or performed mentally.

The second method of factoring trinomials of the type $ax^2 + bx + c$ involves factoring by grouping.

The *ac*-method (p. 451)

To Factor $ax^2 + bx + c$ Using the *ac*-Method

1. Factor out the largest common factor, if one exists.
2. Multiply the leading coefficient a and the constant c.
3. Find a pair of factors of ac whose sum is b.
4. Rewrite the middle term, bx, as a sum or a difference using the factors found in step (3).
5. Factor by grouping.
6. Include any common factor from step (1) and check by multiplying.

EXAMPLE 6 Factor: $7x^2 + 31x + 12$.

SOLUTION

1. There is no common factor (other than 1 or -1).
2. We multiply the leading coefficient, 7, and the constant, 12:

$$7 \cdot 12 = 84.$$

3. We look for a pair of factors of 84 whose sum is 31. Since both 84 and 31 are positive, we need consider only positive factors.

Pairs of Factors of 84	Sums of Factors
1, 84	85
2, 42	44
3, 28	31 ← $3 + 28 = 31$

4. Next, we rewrite $31x$ using the factors 3 and 28:

$$31x = 3x + 28x.$$

5. We now factor by grouping:

$$\begin{aligned} 7x^2 + 31x + 12 &= 7x^2 + 3x + 28x + 12 && \text{Substituting } 3x + 28x \text{ for } 31x \\ &= x(7x + 3) + 4(7x + 3) \\ &= (7x + 3)(x + 4). && \text{Factoring out the common factor, } 7x + 3 \end{aligned}$$

6. *Check:* $(7x + 3)(x + 4) = 7x^2 + 31x + 12$.

Try Exercise 15.

FACTORING SPECIAL FORMS

We can factor certain types of polynomials directly, without using trial and error.

Factoring Formulas

Perfect-square trinomial: $A^2 + 2AB + B^2 = (A + B)^2$,
$A^2 - 2AB + B^2 = (A - B)^2$

Difference of squares: $A^2 - B^2 = (A + B)(A - B)$

Sum of cubes: $A^3 + B^3 = (A + B)(A^2 - AB + B^2)$

Difference of cubes: $A^3 - B^3 = (A - B)(A^2 + AB + B^2)$

Before using the factoring formulas, it is important to check carefully that the expression being factored is indeed in one of the forms listed. Note that there is no factoring formula for the sum of two squares.

EXAMPLE 7 Factor.

a) $2x^2 - 2$

b) $x^2y^2 + 20xy + 100$

c) $p^3 - 64$

d) $3y^2 + 27$

SOLUTION

a) We first factor out a common factor, 2:

$$2x^2 - 2 = 2(x^2 - 1).$$

Recognizing and factoring differences of squares (p. 460)

Looking at $x^2 - 1$, we see that it is a difference of squares, with $A = x$ and $B = 1$. The factorization is thus

$$2x^2 - 2 = \underset{A^2 - B^2}{2(x^2 - 1)} = \underset{(A + B)(A - B)}{2(x + 1)(x - 1)}.$$

Recognizing and factoring perfect-square trinomials (pp. 458, 459)

b) First, we check for a common factor; there is none. The polynomial is a perfect-square trinomial, since x^2y^2 and 100 are squares; there is no minus sign before either square; and $20xy$ is $2 \cdot xy \cdot 10$, where xy and 10 are square roots of x^2y^2 and 100, respectively. The factorization is thus

$$x^2y^2 + 20xy + 100 = (xy)^2 + 2 \cdot xy \cdot 10 + 10^2 = (xy + 10)^2.$$
$$A^2 + 2 \cdot A \cdot B + B^2 = (A + B)^2$$

Factoring sums or differences of cubes (Section 6.5)

c) This is a difference of cubes, with $A = p$ and $B = 4$:

$$\begin{aligned} p^3 - 64 &= (p)^3 - (4)^3 \\ &= (p - 4)(p^2 + 4p + 16). \end{aligned}$$

d) We factor out the common factor, 3:

$$3y^2 + 27 = 3(y^2 + 9).$$

Since $y^2 + 9$ is a sum of squares, no further factorization is possible.

Try Exercise 29.

A polynomial is said to be *factored completely* when no factor can be factored further.

EXAMPLE 8 Factor completely: $x^4 - 1$.

SOLUTION

$$x^4 - 1 = (x^2 + 1)(x^2 - 1)$$ **Factoring a difference of squares**

$$= (x^2 + 1)(x + 1)(x - 1)$$ **The factor $x^2 - 1$ is itself a difference of squares.**

Try Exercise 13.

SOLVING POLYNOMIAL EQUATIONS

Polynomial equation (p. 424)

Intersect method (p. 424)

Zero method (p. 425)

Whenever two polynomials are set equal to each other, we have a **polynomial equation.** We can solve polynomial equations graphically using the Intersect method or the Zero method.

EXAMPLE 9 Solve: $3x^3 = 5x^2 + 2x$.

INTERSECT METHOD

We graph $y_1 = 3x^3$ and $y_2 = 5x^2 + 2x$ and look for any points of intersection. The x-coordinates of those points will be the solutions of the equation.

From the first graph, we see that one point of intersection is $(2, 24)$. In the second graph, we enlarge the section of the graph near the origin and see that $(-0.33333, -0.11111)$ and $(0, 0)$ are also points of intersection.

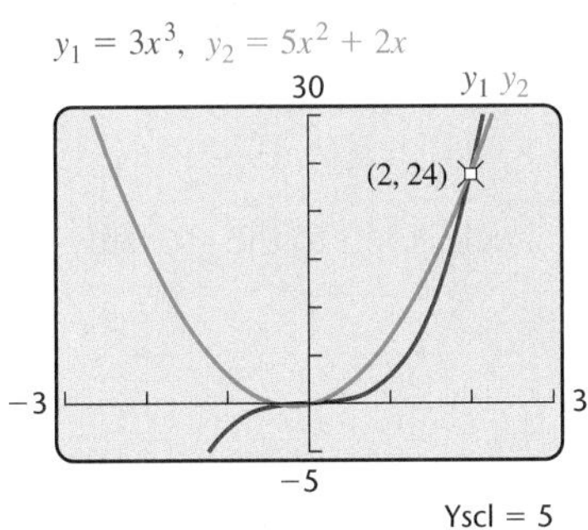

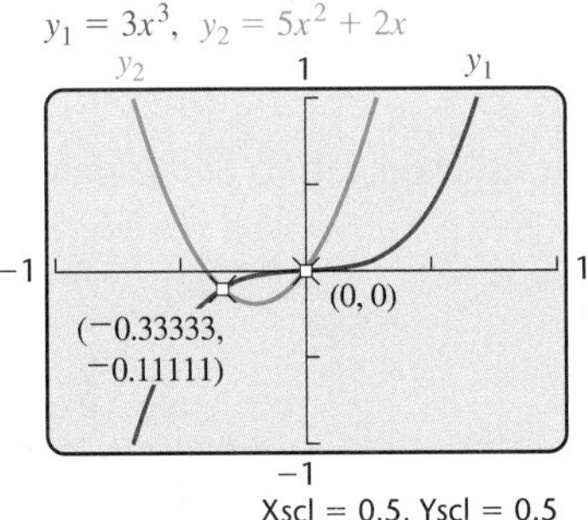

The solutions of the equation are -0.33333, 0, and 2.

ZERO METHOD

We rewrite the equation as $3x^3 - 5x^2 - 2x = 0$. We graph $f(x) = 3x^3 - 5x^2 - 2x$ and find the zeros of the function. Those zeros will be the solutions of the equation.

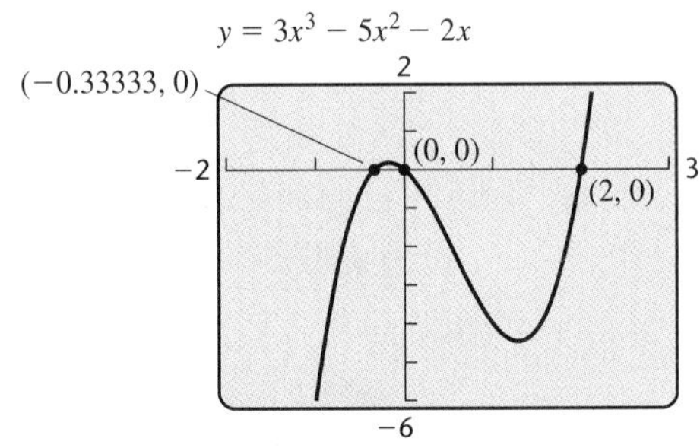

We see that $f(x) = 0$ when $x = -0.33333$, 0, or 2. Thus the solutions of the equation are -0.33333, 0, and 2.

Try Exercise 37.

Note in Example 9 that one of the solutions found is an approximation. If we can factor the polynomial that describes the function, we can find exact solutions using the following principle.

The Principle of Zero Products An equation $ab = 0$ is true if and only if $a = 0$ or $b = 0$, or both. (A product is 0 if and only if at least one factor is 0.)

If we can write an equation as a product that equals 0, we can try to use the principle of zero products to solve the equation.

EXAMPLE 10 Solve.

a) $x^2 - 11x = 12$ **b)** $5x^2 + 10x + 5 = 0$

c) $9x^2 = 1$

SOLUTION

a) We must have 0 on one side of the equation before using the principle of zero products:

$$x^2 - 11x = 12$$

Get 0 on one side. $x^2 - 11x - 12 = 0$ **Subtracting 12 from both sides**

Factor. $(x - 12)(x + 1) = 0$ **Factoring**

Use the principle of zero products. $x - 12 = 0 \quad or \quad x + 1 = 0$ **Using the principle of zero products**

Solve. $x = 12 \quad or \quad x = -1.$

The solutions are 12 and -1. The check is left to the student.

b) We have

$$5x^2 + 10x + 5 = 0$$

$5(x^2 + 2x + 1) = 0$ **Factoring out a common factor**

$5(x + 1)(x + 1) = 0$ **Factoring completely**

$x + 1 = 0 \quad or \quad x + 1 = 0$ **Using the principle of zero products**

$$x = -1 \quad or \quad x = -1.$$

There is only one solution, -1. The check is left to the student.

c) We have

$$9x^2 = 1$$

$9x^2 - 1 = 0$ **Subtracting 1 from both sides to get 0 on one side**

$(3x + 1)(3x - 1) = 0$ **Factoring a difference of squares**

$3x + 1 = 0 \quad or \quad 3x - 1 = 0$ **Using the principle of zero products**

$$3x = -1 \quad or \quad 3x = 1$$

$$x = -\tfrac{1}{3} \quad or \quad x = \tfrac{1}{3}.$$

The solutions are $\frac{1}{3}$ and $-\frac{1}{3}$. The check is left to the student.

Try Exercise 35.

Quadratic equations can be used to solve problems. One important result that uses squared quantities is the Pythagorean theorem. It relates the lengths of the sides of a **right triangle**—that is, a triangle with a 90° angle. The side opposite the 90° angle is called the **hypotenuse,** and the other sides are called the **legs.**

The Pythagorean Theorem The sum of the squares of the legs of a right triangle is equal to the square of the hypotenuse:

$$a^2 + b^2 = c^2.$$

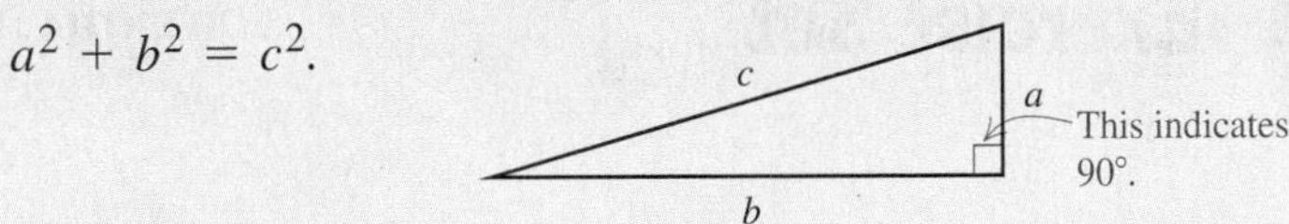

EXAMPLE 11 Hiking. Cheri hiked 500 ft up a steep incline. Her global positioning unit indicated that her horizontal position had changed by 100 ft more than her vertical position had changed. What was the change in altitude?

SOLUTION

1. **Familiarize.** We make a drawing and let $x =$ the change in altitude, in feet. We know then that the horizontal change is $x + 100$, since Cheri's horizontal position has changed by 100 ft more than her vertical position. The hypotenuse has length 500 ft.

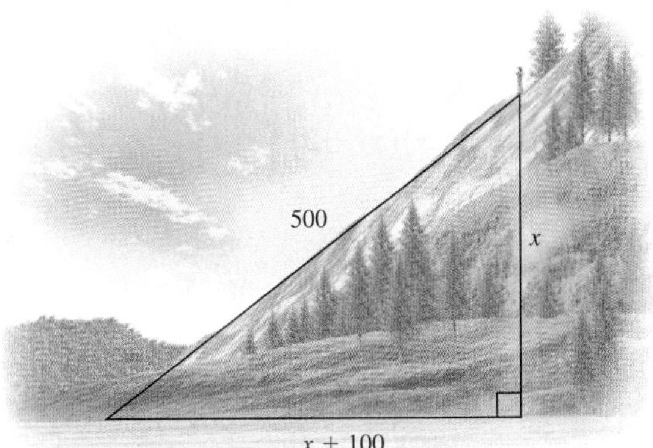

2. **Translate.** Applying the Pythagorean theorem gives us

$$a^2 + b^2 = c^2$$
$$x^2 + (x + 100)^2 = 500^2.$$

3. **Carry out.** We solve the equation:

$$x^2 + (x + 100)^2 = 500^2$$
$$x^2 + x^2 + 200x + 10{,}000 = 250{,}000 \quad \text{Squaring } x + 100\text{; squaring 500}$$
$$2x^2 + 200x + 10{,}000 = 250{,}000 \quad \text{Combining like terms}$$
$$2x^2 + 200x - 240{,}000 = 0 \quad \text{Getting 0 on one side}$$
$$2(x^2 + 100x - 120{,}000) = 0 \quad \text{Factoring out a common factor}$$
$$2(x + 400)(x - 300) = 0 \quad \text{Factoring a trinomial}$$
$$x + 400 = 0 \quad or \quad x - 300 = 0 \quad \text{Using the principle of zero products}$$
$$x = -400 \quad or \quad x = 300.$$

4. **Check.** We know that the change in altitude is positive, so -400 cannot be a solution. When $x = 300$, the horizontal change is 400. Since $300^2 + 400^2 = 500^2$, the solution checks.

5. **State.** Cheri's change in altitude was 300 ft.

Try Exercise 49.

R.6 Exercise Set

FOR EXTRA HELP MyMathLab  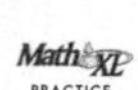  

Factor completely. If a polynomial is prime, state this.

1. $18t^5 - 12t^4 + 6t^3$
2. $x^2y^4 - 2xy^5 + 3x^3y^6$
3. $y^2 - 6y + 9$
4. $4z^2 - 25$
5. $2p^3(p + 2) + (p + 2)$
6. $6y^2 + y - 1$
7. $x^2 + 100$
8. $y^3 - 1$
9. $8t^3 + 27$
10. $a^2b^2 + 24ab + 144$
11. $m^2 + 13m + 42$
12. $2x^3 - 6x^2 + x - 3$
13. $x^4 - 81$
14. $x^2 + x + 1$
15. $8x^2 + 22x + 15$
16. $4x^2 - 40x + 100$
17. $x^3 + 2x^2 - x - 2$
18. $(x + 2y)(x - 1) + (x + 2y)(x - 2)$
19. $0.001t^6 - 0.008$
20. $x^2 - 20 - x$
21. $-\frac{1}{16} + x^4$
22. $5x^8 - 5z^{16}$
23. $mn - 2m + 3n - 6$
24. $t^6 - p^6$
25. $5mn + m^2 - 150n^2$
26. $\frac{1}{27} + x^3$
27. $24x^2y - 6y - 10xy$
28. $-3y^2 - 12y - 12$
29. $y^2 - 121$
30. $t^3 - 2t^2 - 5t + 10$

Solve.

31. $(x - 2)(x + 7) = 0$
32. $(3x - 5)(7 - 4x) = 0$
33. $8x(11 - x) = 0$
34. $(x - 3)(x + 1)(2x - 9) = 0$
35. $x^2 = 100$
36. $8x^2 = 2x$
37. $4x^2 - 18x = 70$
38. $x^2 + 2x + 1 = 0$
39. $2x^3 - 10x^2 = 0$
40. $100x^2 = 81$
41. $(a + 1)(a - 5) = 7$
42. $d(d - 3) = 40$
43. $x^2 + 6x - 55 = 0$
44. $x^2 + 7x - 60 = 0$
45. $\frac{1}{2}x^2 + 5x + \frac{25}{2} = 0$
46. $3 + 10x^2 = 11x$
47. *Design.* The base of a triangular fountain is 2 ft longer than its height. The area of the fountain base is 24 ft^2. What are the dimensions of the fountain?

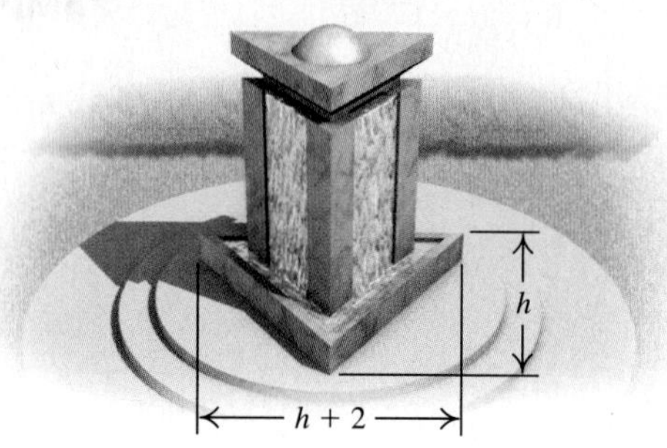

48. *Page Numbers.* The product of the page numbers on two facing pages of a book is 156. Find the page numbers.
49. *Right Triangles.* The hypotenuse of a right triangle is 17 ft. One leg is 1 ft shorter than twice the length of the other leg. Find the length of the legs.

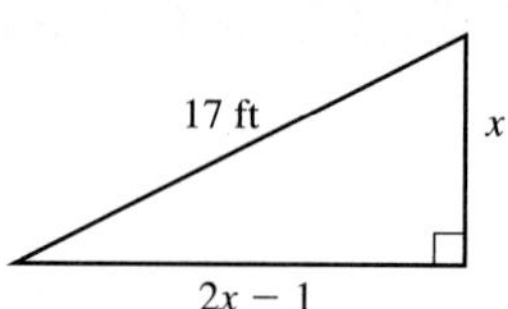

50. *Swing Sets.* The length of a slide on a swing set is 10 ft. The distance from the base of the ladder to the base of the slide is 2 ft more than the height of the ladder. Find the height of the ladder.

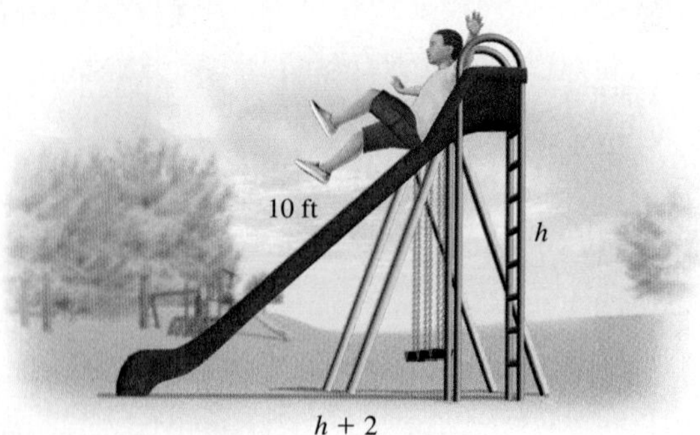

Try Exercise Answers: Section R.6

1. $6t^3(3t^2 - 2t + 1)$ 5. $(p + 2)(2p^3 + 1)$ 11. $(m + 6)(m + 7)$ 13. $(x^2 + 9)(x + 3)(x - 3)$ 15. $(2x + 3)(4x + 5)$ 23. $(n - 2)(m + 3)$ 27. $2y(3x + 1)(4x - 3)$ 29. $(y + 11)(y - 11)$ 35. $-10, 10$ 37. $-\frac{5}{2}, 7$ 49. 8 ft, 15 ft

Answers

Chapter 1

Translating for Success, p. 10

1. H **2.** E **3.** K **4.** B **5.** O **6.** L **7.** M **8.** C **9.** D **10.** F

Exercise Set 1.1, pp. 11–14

1. Expression **2.** Equation **3.** Equation **4.** Expression **5.** Equation **6.** Equation **7.** Expression **8.** Equation **9.** Equation **10.** Expression **11.** Expression **12.** Expression **13.** 27 **15.** 8 **17.** 4 **19.** 7 **21.** 3 **23.** 3438 **25.** 21 ft^2 **27.** 15 cm^2 **29.** 804 ft^2 **31.** Let r represent Ron's age; $r + 5$, or $5 + r$ **33.** $4a$, or $a \cdot 4$ **35.** $c - 9$ **37.** $6 + q$, or $q + 6$ **39.** $m - n$ **41.** $y - x$ **43.** $x \div w$, or $\frac{x}{w}$ **45.** Let l represent the length of the box and h the height; $l + h$, or $h + l$ **47.** Let p represent Panya's speed and w the wind speed; $p - 2w$ **49.** Let y represent "some number"; $\frac{1}{4}y - 12$, or $\frac{y}{4} - 12$ **51.** Let a and b represent the numbers; $8(a - b)$ **53.** Let w represent the number of women attending; 64% of w, or $0.64w$ **55.** Yes **57.** No **59.** Yes **61.** Yes **63.** Let x represent the unknown number; $7x = 1596$ **65.** Let x represent the unknown number; $42x = 2352$ **67.** Let s represent the number of unoccupied squares; $s + 19 = 64$ **69.** Let w represent the amount of solid waste generated, in millions of tons; 33.4% of $w = 85$, or $0.334w = 85$ **71.** (f) **73.** (d) **75.** (g) **77.** (e) **79.** $f = a + 5$ **81.** $n = m + 2.21$ **83.** $v = 10{,}000d$ **85.** TW **87.** TW **89.** \$450 **91.** 9 **93.** $w + 4$ **95.** $l + w + l + w$, or $2l + 2w$ **97.** $t + 8$ **99.** TW

Exercise Set 1.2, pp. 20–22

1. Commutative **2.** Associative **3.** Associative **4.** Commutative **5.** Distributive **6.** Associative **7.** Associative **8.** Commutative **9.** Commutative **10.** Distributive **11.** $x + 7$ **13.** $3y + x$ **15.** $c + ab$ **17.** $5(1 + a)$ **19.** $a \cdot 2$ **21.** ts **23.** $5 + ba$ **25.** $(a + 1)5$ **27.** $a + (5 + b)$ **29.** $(r + t) + 7$ **31.** $ab + (c + d)$ **33.** $7(mn)$ **35.** $(2a)b$ **37.** $(3 \cdot 2)(a + b)$ **39.** $2 + (6 + t)$; $(2 + 6) + t$; $8 + t$ **41.** $(a \cdot 3) \cdot 7$; $a(3 \cdot 7)$; $a \cdot 21$, or $21a$

43.

$(5 + x) + 2$	$= (x + 5) + 2$	Commutative law
	$= x + (5 + 2)$	Associative law
	$= x + 7$	Simplifying

45.

$(m \cdot 3)7$	$= m(3 \cdot 7)$	Associative law
	$= m \cdot 21$	Simplifying
	$= 21m$	Commutative law

47. $4a + 12$ **49.** $6 + 6x$ **51.** $2n + 10$ **53.** $24x + 40y$ **55.** $18x + 54$ **57.** $5r + 10 + 15t$ **59.** $2a + 2b$ **61.** $5x + 5y + 10$ **63.** x, xyz, 19 **65.** $2a$, $\frac{a}{b}$, $5b$ **67.** $2(a + b)$ **69.** $7(1 + y)$ **71.** $3(6x + 1)$ **73.** $5(x + 2 + 3y)$ **75.** $3(4x + 3)$ **77.** $3(a + 3b)$ **79.** $22(2x + 4y + 3z)$ **81.** s, t **83.** 3, $(x + y)$ **85.** 7, a, b **87.** $(a - b)$, $(x - y)$ **89.** TW **91.** Let k represent Kylie's salary; $\frac{1}{2}k$, or $\frac{k}{2}$ **92.** $2(m + 7)$, or $2(7 + m)$ **93.** TW **95.** Yes; distributive law **97.** No; for example, let $m = 1$. Then $7 \div 3 \cdot 1 = \frac{7}{3}$ and $1 \cdot 3 \div 7 = \frac{3}{7}$. **99.** No; for example, let $x = 1$ and $y = 2$. Then $30 \cdot 2 + 1 \cdot 15 = 60 + 15 = 75$ and $5[2(1 + 3 \cdot 2)] = 5[2(7)] = 5 \cdot 14 = 70$. **101.** TW

Exercise Set 1.3, pp. 30–32

1. (b) **2.** (c) **3.** (d) **4.** (a) **5.** $1 \cdot 50$; $2 \cdot 25$; $5 \cdot 10$; 1, 2, 5, 10, 25, 50 **7.** $1 \cdot 42$; $2 \cdot 21$; $3 \cdot 14$; $6 \cdot 7$; 1, 2, 3, 6, 7, 14, 21, 42 **9.** Composite **11.** Prime **13.** Composite **15.** Prime **17.** Neither **19.** Composite **21.** $2 \cdot 13$ **23.** $2 \cdot 3 \cdot 5$ **25.** $3 \cdot 3 \cdot 3$ **27.** $2 \cdot 2 \cdot 2 \cdot 5$ **29.** Prime **31.** $2 \cdot 3 \cdot 5 \cdot 7$ **33.** $5 \cdot 23$ **35.** $\frac{2}{3}$ **37.** $\frac{2}{7}$ **39.** $\frac{1}{8}$ **41.** 4 **43.** $\frac{1}{4}$ **45.** 6 **47.** $\frac{21}{25}$ **49.** $\frac{60}{41}$ **51.** $\frac{15}{7}$ **53.** $\frac{3}{14}$ **55.** $\frac{8}{3}$ **57.** $\frac{1}{2}$ **59.** $\frac{7}{6}$ **61.** $\frac{3b}{7a}$ **63.** $\frac{7}{a}$ **65.** $\frac{5}{6}$ **67.** 1 **69.** $\frac{5}{18}$ **71.** 0 **73.** $\frac{35}{18}$ **75.** $\frac{10}{3}$ **77.** 28 **79.** 1 **81.** $\frac{6}{35}$ **83.** 18 **85.** TW **87.** $5(3 + x)$; answers may vary **88.** $7 + (b + a)$, or $(a + b) + 7$ **89.** TW **91.** Row 1: 7, 2, 36, 14, 8, 8; row 2: 9, 18, 2, 10, 12, 21 **93.** $\frac{2}{5}$ **95.** $\frac{3q}{t}$ **97.** $\frac{6}{25}$ **99.** $\frac{5ap}{2cm}$ **101.** $\frac{23r}{18t}$ **103.** $\frac{28}{45}\,\text{m}^2$ **105.** $14\frac{2}{9}$ m **107.** $27\frac{3}{5}$ cm

Exercise Set 1.4, pp. 39–40

1. Repeating **2.** Terminating **3.** Integer **4.** Whole number **5.** Rational number **6.** Irrational number **7.** Natural number **8.** Absolute value **9.** 100, −80 **11.** −777.68, 936.42 **13.** −12,500, 5000 **15.** 8, −5 **17.** −10, 235 **19.**

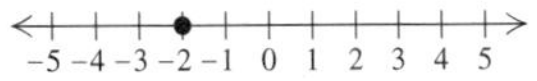

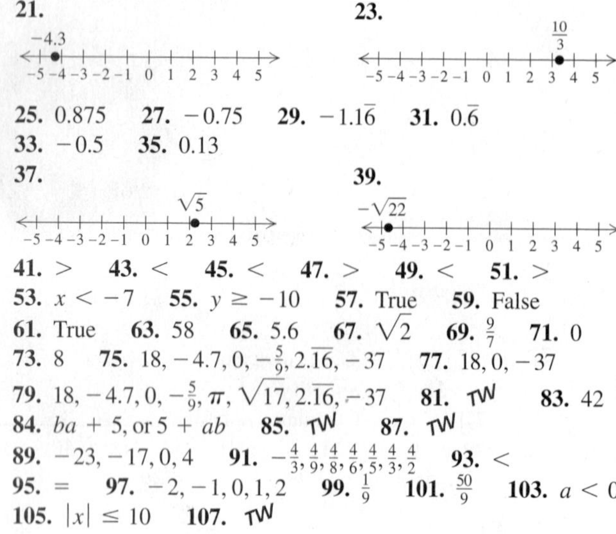

21.

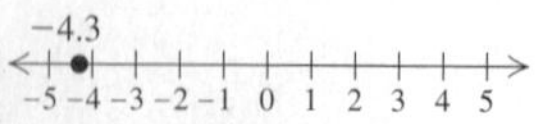

23.

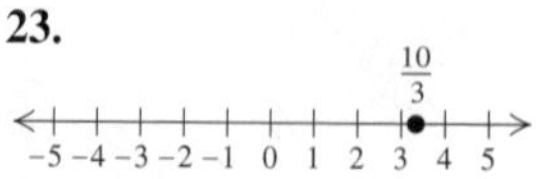

25. 0.875 **27.** −0.75 **29.** $-1.1\overline{6}$ **31.** $0.\overline{6}$
33. −0.5 **35.** 0.13
37.

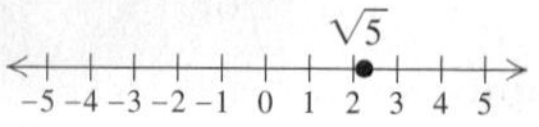

39.

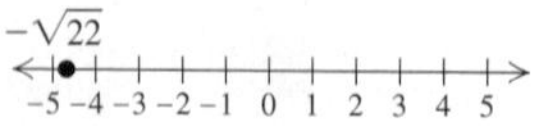

41. > **43.** < **45.** < **47.** > **49.** < **51.** >
53. $x < -7$ **55.** $y \geq -10$ **57.** True **59.** False
61. True **63.** 58 **65.** 5.6 **67.** $\sqrt{2}$ **69.** $\frac{9}{7}$ **71.** 0
73. 8 **75.** $18, -4.7, 0, -\frac{5}{9}, 2.\overline{16}, -37$ **77.** $18, 0, -37$
79. $18, -4.7, 0, -\frac{5}{9}, \pi, \sqrt{17}, 2.\overline{16}, -37$ **81.** TW **83.** 42
84. $ba + 5$, or $5 + ab$ **85.** TW **87.** TW
89. $-23, -17, 0, 4$ **91.** $-\frac{4}{3}, \frac{4}{9}, \frac{4}{8}, \frac{4}{6}, \frac{4}{5}, \frac{4}{3}, \frac{4}{2}$ **93.** <
95. = **97.** $-2, -1, 0, 1, 2$ **99.** $\frac{1}{9}$ **101.** $\frac{50}{9}$ **103.** $a < 0$
105. $|x| \leq 10$ **107.** TW

Mid-Chapter Review: Chapter 1, p. 41

Guided Solutions

1. $\frac{x - y}{3} = \frac{22 - 10}{3} = \frac{12}{3} = 4$
2. $40x + 8 = 8 \cdot 5x + 8 \cdot 1 = 8(5x + 1)$

Mixed Review

1. 15 **2.** 4 **3.** $d - 10$ **4.** Let h represent the number of hours worked; $8h$ **5.** Let s represent the number of students originally enrolled; $27 = s - 5$ **6.** No **7.** $10x + 7$
8. $(3a)b$ **9.** $8x + 32$ **10.** $6m + 15n + 30$
11. $9(2x + 1)$ **12.** $4(2a + 6y + 5)$ **13.** $2 \cdot 2 \cdot 3 \cdot 7$
14. $\frac{6}{5}$ **15.** $\frac{3}{7}$ **16.** $\frac{13}{24}$ **17.** $\frac{44}{45}$
18.

19. −0.15 **20.** >
21. < **22.** $9 \leq x$ **23.** True **24.** 5.6 **25.** 0

Exercise Set 1.5, pp. 46–48

1. (f) **2.** (d) **3.** (e) **4.** (a) **5.** (b) **6.** (c)
7. −3 **9.** 4 **11.** −4 **13.** −8 **15.** −11 **17.** −5
19. 0 **21.** −41 **23.** 0 **25.** 7 **27.** −36 **29.** 11
31. −33 **33.** 0 **35.** 18 **37.** −45 **39.** 0 **41.** 20
43. −1.7 **45.** −9.1 **47.** $\frac{1}{5}$ **49.** $\frac{-6}{7}$ **51.** $-\frac{1}{15}$ **53.** $\frac{2}{9}$
55. −3 **57.** 0 **59.** The price dropped 1¢. **61.** The lake dropped $\frac{1}{20}$ ft. **63.** The total gain was 22 yd. **65.** His new balance was $95. **67.** The elevation of the peak is 13,796 ft.
69. $-3a$ **71.** $9x$ **73.** $-7a$ **75.** $1 - 2x$ **77.** $-4m$
79. $-5x - 3.9$ **81.** $12x + 17$ **83.** $7r + 8t + 16$
85. $18n + 16$ **87.** TW **89.** $21z + 7y + 14$ **90.** $\frac{28}{3}$
91. TW **93.** $451.70 **95.** $-5y$ **97.** $-7m$
99. $-7t, -23$ **101.** 1 under par

Exercise Set 1.6, pp. 53–55

1. (d) **2.** (g) **3.** (f) **4.** (h) **5.** (a) **6.** (c) **7.** (b)
8. (e) **9.** Four minus ten **11.** Two minus negative nine
13. The opposite of x minus y **15.** Negative three minus the opposite of n **17.** −39 **19.** $\frac{11}{2}$ **21.** 3.14 **23.** 45
25. $\frac{14}{3}$ **27.** −0.101 **29.** 72 **31.** $-\frac{2}{5}$ **33.** 1 **35.** −7
37. −2 **39.** −5 **41.** −7 **43.** −6 **45.** 0 **47.** −10
49. 2 **51.** 0 **53.** 0 **55.** 8 **57.** −11 **59.** 16 **61.** −1
63. 17 **65.** 10 **67.** −3 **69.** −21 **71.** −8 **73.** −11
75. −23 **77.** −7.3 **79.** 0.6 **81.** −5.5 **83.** −0.928
85. $-\frac{7}{11}$ **87.** $-\frac{4}{5}$ **89.** $\frac{5}{17}$ **91.** −58 **93.** 34
95. $3.8 - (-5.2); 9$ **97.** $114 - (-79); 193$ **99.** 41
101. −62 **103.** −139 **105.** 0 **107.** $-7x, -4y$
109. $9, -5t, -3st$ **111.** $-3x$ **113.** $-5a + 4$
115. $-n - 9$ **117.** $-5t - 7$ **119.** $-12x + 3y + 9$
121. $8x + 66$ **123.** 138,677 ft **125.** 47°F **127.** 8.9 points
129. TW **131.** 432 ft^2 **132.** $2 \cdot 2 \cdot 2 \cdot 2 \cdot 2 \cdot 3 \cdot 3 \cdot 3$
133. TW **135.** 11:00 P.M., on August 14 **137.** False. For example, let $m = -3$ and $n = -5$. Then $-3 > -5$, but $-3 + (-5) = -8 \not> 0$. **139.** True. For example, for $m = 4$ and $n = -4$, $4 = -(-4)$ and $4 + (-4) = 0$; for $m = -3$ and $n = 3$, $-3 = -3$ and $-3 + 3 = 0$.
141. (−) 9 − (−) 7 ENTER

Exercise Set 1.7, pp. 61–63

1. 1 **2.** 0 **3.** 0 **4.** 1 **5.** 0 **6.** 1 **7.** 1 **8.** 0
9. 1 **10.** 0 **11.** −24 **13.** −56 **15.** −24 **17.** 42
19. −190 **21.** −144 **23.** 1200 **25.** −126 **27.** 11.5
29. 0 **31.** $-\frac{2}{5}$ **33.** $\frac{1}{12}$ **35.** −11.13 **37.** $-\frac{5}{12}$ **39.** 252
41. $\frac{1}{28}$ **43.** 150 **45.** 0 **47.** −720 **49.** −30,240
51. −7 **53.** 2 **55.** −10 **57.** 5.1 **59.** $\frac{100}{11}$ **61.** −8
63. Undefined **65.** −4 **67.** 0 **69.** 0 **71.** $-\frac{8}{3}; \frac{8}{-3}$
73. $-\frac{29}{35}; \frac{-29}{35}$ **75.** $\frac{-7}{3}; \frac{7}{-3}$ **77.** $-\frac{x}{2}; \frac{x}{-2}$ **79.** $-\frac{5}{4}$
81. $-\frac{10}{51}$ **83.** $-\frac{1}{10}$ **85.** $\frac{1}{4.3}$, or $\frac{10}{43}$ **87.** −4
89. Does not exist **91.** $\frac{21}{20}$ **93.** −1 **95.** 1 **97.** $\frac{3}{11}$
99. $-\frac{7}{4}$ **101.** −12 **103.** −3 **105.** 1 **107.** $-\frac{2}{9}$
109. $\frac{1}{10}$ **111.** $-\frac{7}{6}$ **113.** Undefined **115.** $-\frac{14}{15}$ **117.** TW
119. $\frac{22}{39}$ **120.** No **121.** TW **123.** $\frac{1}{a + b}$ **125.** $-(a + b)$
127. $x = -x$ **129.** For 2 and 3, the reciprocal of the sum is $1/(2 + 3)$, or $1/5$. But $1/5 \neq 1/2 + 1/3$. **131.** Negative
133. Negative **135.** Negative **137.** **(a)** m and n have different signs; **(b)** either m or n is zero; **(c)** m and n have the same sign.
139. Distributive law; law of opposites; multiplicative property of zero

Interactive Discovery, p. 65

1. Multiplication **2.** $4 + (2 \times 5)$

Exercise Set 1.8, pp. 70–73

1. **(a)** Division; **(b)** subtraction; **(c)** addition; **(d)** multiplication; **(e)** subtraction; **(f)** multiplication
2. **(a)** Multiplication; **(b)** subtraction; **(c)** addition; **(d)** subtraction; **(e)** division; **(f)** multiplication **3.** x^7 **5.** $(-5)^3$
7. $(3t)^5$ **9.** $2n^4$ **11.** 9 **13.** 16 **15.** −16
17. 64 **19.** 625 **21.** 7 **23.** −32 **25.** $81t^4$
27. $-343x^3$ **29.** 26 **31.** 86 **33.** 5 **35.** 1 **37.** 298
39. 11 **41.** −36 **43.** 13 **45.** 152 **47.** 24 **49.** 1
51. −26 **53.** −2 **55.** −5 **57.** (a) **59.** (d) **61.** 21.563
63. −1.026 **65.** −11 **67.** −3 **69.** −15 **71.** 9 **73.** 30

75. 6 **77.** -17 **79.** 15 **81.** 25.125 **83.** 13,778 **85.** $-9x - 1$ **87.** $-5 + 6x$ **89.** $-4a + 3b - 7c$ **91.** $-3x^2 - 5x + 1$ **93.** $2x - 7$ **95.** $-9x + 6$ **97.** $21t + r$ **99.** $9y - 25z$ **101.** $x^2 + 2$ **103.** $-t^3 + 4t$ **105.** $37a^2 - 23ab + 35b^2$ **107.** $-22t^3 - t^2 + 9t$ **109.** $2x - 25$ **111.** TW **113.** Let n represent the number; $2n + 9$, or $9 + 2n$ **114.** Let m and n represent the two numbers; $\frac{1}{2}(m + n)$ **115.** TW **117.** $-6r - 5t + 21$ **119.** $-2x - f$ **121.** TW **123.** True **125.** False **127.** 0 **129.** 17 **131.** 39,000 **133.** $44x^3$

Study Summary: Chapter 1, pp. 74–77

1. 8 **2.** 4 ft^2 **3.** Let n represent some number; $78 = n - 92$ **4.** No **5.** $10n + 6$ **6.** $(3a)b$ **7.** $50m + 90n + 10$ **8.** $13(2x + 1)$ **9.** 7, 0 **10.** Composite **11.** $2 \cdot 2 \cdot 3 \cdot 7$ **12.** 1 **13.** $\frac{9}{10}$ **14.** $\frac{3}{2}$ **15.** $\frac{9}{20}$ **16.** $\frac{25}{6}$ **17.** 25 **18.** $0, -15, \frac{30}{3}$ **19.** $-1.\overline{1}$ **20.** False **21.** 1.5 **22.** -5 **23.** -2.9 **24.** -12 **25.** 15 **26.** $-7c + 9d - 2$ **27.** 21 **28.** -4 **29.** -100 **30.** -21 **31.** $a - 2b + 3c$ **32.** $5m + 7n - 27$

Review Exercises: Chapter 1, pp. 78–79

1. True **2.** True **3.** False **4.** True **5.** False **6.** False **7.** True **8.** False **9.** False **10.** True **11.** 15 **12.** -16 **13.** -15 **14.** $z - 7$ **15.** $xz + 10$, or $10 + xz$ **16.** Let b represent Brent's speed and w the wind speed; $15(b - w)$ **17.** No **18.** Let b represent the number of calories per hour burned when backpacking; $b = 2 \cdot 237$ **19.** $c = 200t$ **20.** $t \cdot 3 + 5$ **21.** $2x + (y + z)$ **22.** $(4x)y, 4(yx), (4y)x$; answers may vary **23.** $18x + 30y$ **24.** $40x + 24y + 16$ **25.** $3(7x + 5y)$ **26.** $7(5x + 11y + 1)$ **27.** $2 \cdot 2 \cdot 13$ **28.** $\frac{5}{12}$ **29.** $\frac{9}{4}$ **30.** $\frac{31}{36}$ **31.** $\frac{3}{16}$ **32.** $\frac{3}{5}$ **33.** $\frac{72}{25}$ **34.** $172, -820$ **35.** $\frac{-1}{3}$ −5 −4 −3 −2 −1 0 1 2 3 4 5 **36.** $x > -3$ **37.** False **38.** -0.875 **39.** 1 **40.** -9 **41.** -10 **42.** $-\frac{7}{12}$ **43.** 0 **44.** -5 **45.** 5 **46.** $-\frac{2}{5}$ **47.** -7.9 **48.** 54 **49.** -9.18 **50.** $-\frac{2}{7}$ **51.** -140 **52.** -7 **53.** -3 **54.** $\frac{3}{4}$ **55.** 92 **56.** 18 **57.** 48 **58.** 168 **59.** $\frac{21}{8}$ **60.** $\frac{103}{17}$ **61.** $7a - 3b$ **62.** $-2x + 5y$ **63.** 7 **64.** $-\frac{1}{7}$ **65.** $(2x)^4$ **66.** $-125x^3$ **67.** $-3a + 9$ **68.** $3x^4 + 10x$ **69.** $17n^2 + m^2 + 20mn$ **70.** $5x + 28$ **71.** TW The value of a constant never varies. A variable can represent a variety of numbers. **72.** TW A term is one of the parts of an expression that is separated from the other parts by plus signs. A factor is part of a product. **73.** TW The distributive law is used in factoring algebraic expressions, multiplying algebraic expressions, combining like terms, finding the opposite of a sum, and subtracting algebraic expressions. **74.** TW A negative number raised to an even exponent is positive; a negative number raised to an odd exponent is negative. **75.** 25,281 **76.** **(a)** $\frac{3}{11}$; **(b)** $\frac{10}{11}$ **77.** $-\frac{5}{8}$ **78.** -2.1 **79.** (i) **80.** (j) **81.** (a) **82.** (h) **83.** (k) **84.** (b) **85.** (c) **86.** (e) **87.** (d) **88.** (f) **89.** (g)

Test: Chapter 1, p. 80

1. [1.1] 4 **2.** [1.1] Let x and y represent the numbers; $xy - 9$ **3.** [1.1] 240 ft^2 **4.** [1.2] $q + 3p$ **5.** [1.2] $(x \cdot 4) \cdot y$ **6.** [1.1] No **7.** [1.1] Let g represent the number of golden lion tamarins in zoos; $1500 = g + 1050$ **8.** [1.2] $35 + 7x$ **9.** [1.7] $-5y + 10$ **10.** [1.2] $11(1 + 4x)$ **11.** [1.2] $7(x + 1 + 2y)$ **12.** [1.3] $2 \cdot 2 \cdot 3 \cdot 5 \cdot 5$ **13.** [1.3] $\frac{2}{7}$ **14.** [1.4] $<$ **15.** [1.4] $>$ **16.** [1.4] $\frac{9}{4}$ **17.** [1.4] 3.8 **18.** [1.6] $\frac{2}{3}$ **19.** [1.7] $-\frac{7}{4}$ **20.** [1.6] 10 **21.** [1.4] $-5 \geq x$ **22.** [1.6] 7.8 **23.** [1.5] -8 **24.** [1.6] -2.5 **25.** [1.6] $-\frac{7}{8}$ **26.** [1.7] -48 **27.** [1.7] $\frac{2}{9}$ **28.** [1.7] -6 **29.** [1.7] $\frac{3}{4}$ **30.** [1.7] -9.728 **31.** [1.8] 20 **32.** [1.8] -64 **33.** [1.8] 448 **34.** [1.6] $21a - 14y$ **35.** [1.8] $16x^4$ **36.** [1.8] $x + 7$ **37.** [1.8] $9a - 12b - 7$ **38.** [1.8] $-y - 16$ **39.** [1.1] 5 **40.** [1.8] $9 - (3 - 4) + 5 = 15$ **41.** [1.8] 15 **42.** [1.8] $4a$ **43.** [1.8] False

Chapter 2

Exercise Set 2.1, pp. 89–91

1. (c) **2.** (b) **3.** (f) **4.** (a) **5.** (d) **6.** (e) **7.** (d) **8.** (b) **9.** (c) **10.** (a) **11.** No **13.** Yes **15.** Yes **17.** Yes **19.** 17 **21.** -11 **23.** -31 **25.** 13 **27.** 19 **29.** -4 **31.** $\frac{7}{3}$ **33.** $-\frac{1}{10}$ **35.** $-\frac{1}{20}$ **37.** 1.5 **39.** -5 **41.** 14 **43.** 12 **45.** -23 **47.** 8 **49.** $-\frac{5}{2}$ **51.** 8 **53.** -88 **55.** 20 **57.** -54 **59.** $\frac{5}{9}$ **61.** 1 **63.** $\frac{9}{2}$ **65.** -7.6 **67.** -2.5 **69.** -15 **71.** -6 **73.** -128 **75.** $-\frac{1}{2}$ **77.** -15 **79.** 12 **81.** 310.756 **83.** TW **85.** -6 **86.** 2 **87.** 1 **88.** -16 **89.** TW **91.** 9.4 **93.** 2 **95.** $-13, 13$ **97.** 9000 **99.** 250

Exercise Set 2.2, pp. 97–98

1. (c) **2.** (e) **3.** (a) **4.** (f) **5.** (b) **6.** (d) **7.** 8 **9.** -4 **11.** 5 **13.** 14 **15.** -11 **17.** -4 **19.** 19 **21.** -2.8 **23.** $\frac{10}{9}$ **25.** 3 **27.** 15 **29.** -4 **31.** $-\frac{28}{3}$ **33.** All real numbers; identity **35.** -3 **37.** -6 **39.** 2 **41.** 0 **43.** 6 **45.** No solution; contradiction **47.** $-\frac{1}{2}$ **49.** 10 **51.** 4 **53.** 0 **55.** No solution; contradiction **57.** 2 **59.** -8 **61.** All real numbers; identity **63.** 2 **65.** $\frac{16}{3}$ **67.** $\frac{2}{5}$ **69.** 3 **71.** -4 **73.** $1.\overline{6}$ **75.** $-\frac{40}{37}$ **77.** 11 **79.** 6 **81.** $\frac{16}{15}$ **83.** $-\frac{51}{31}$ **85.** 2 **87.** TW **89.** -7 **90.** 15 **91.** -15 **92.** -28 **93.** TW **95.** $\frac{1136}{909}$, or $1.\overline{2497}$ **97.** No solution; contradiction **99.** $\frac{2}{3}$ **101.** 0 **103.** $\frac{52}{45}$ **105.** All real numbers; identity **107.** $-4, 4$

Exercise Set 2.3, pp. 104–107

1. 1423 students **3.** 3450 watts **5.** 8.4734 **7.** 255 mg **9.** $b = \dfrac{A}{h}$ **11.** $P = \dfrac{I}{rt}$ **13.** $m = 65 - H$ **15.** $l = \dfrac{P - 2w}{2}$, or $l = \dfrac{P}{2} - w$ **17.** $\pi = \dfrac{A}{r^2}$ **19.** $h = \dfrac{2A}{b}$ **21.** $m = \dfrac{E}{c^2}$ **23.** $d = 2Q - c$ **25.** $q = p + r - 2$

27. $r = wf$ **29.** $T = \frac{550H}{V}$ **31.** $C = \frac{5}{9}(F - 32)$
33. $y = 2x - 1$ **35.** $y = -\frac{2}{5}x + 2$ **37.** $y = \frac{4}{3}x - 2$
39. $y = -\frac{9}{8}x + \frac{1}{2}$ **41.** $y = \frac{3}{5}x - \frac{8}{5}$ **43.** $t = \frac{A}{a + b}$
45. $h = \frac{2A}{a + b}$ **47.** $x = \frac{z - 13}{2} - y$, or $x = \frac{z - 13 - 2y}{2}$
49. $l = w + 4(t - 27)$ **51.** $L = W - \frac{N(R - r)}{400}$, or $L = \frac{400W - NR + Nr}{400}$ **53.** TW **55.** -10 **56.** -196
57. 0 **58.** -32 **59.** -13 **60.** 65 **61.** TW **63.** 40 yr
65. 27 in^3 **67.** $y = \frac{z^2}{t}$ **69.** $t = \frac{rs}{q - r}$ **71.** $S = 20a$, where S is the number of Btu's saved

Mid-Chapter Review: Chapter 2, p. 108

Guided Solutions

1. $2x + 3 - 3 = 10 - 3$
$2x = 7$
$\frac{1}{2} \cdot 2x = \frac{1}{2} \cdot 7$
$x = \frac{7}{2}$

2. $6 \cdot \frac{1}{2}(x - 3) = 6 \cdot \frac{1}{3}(x - 4)$
$3(x - 3) = 2(x - 4)$
$3x - 9 = 2x - 8$
$3x - 9 + 9 = 2x - 8 + 9$
$3x = 2x + 1$
$3x - 2x = 2x + 1 - 2x$
$x = 1$

Mixed Review

1. 1 **2.** 3 **3.** $\frac{5}{3}$ **4.** -8 **5.** 48 **6.** 0.5 **7.** -5
8. $\frac{8}{3}$ **9.** 6 **10.** 3 **11.** 0 **12.** $\frac{49}{9}$ **13.** $-\frac{4}{11}$ **14.** $\frac{23}{7}$
15. $A = \frac{E}{w}$ **16.** $w = \frac{V}{lh}$ **17.** $y = \frac{C - Ax}{B}$ **18.** $a = \frac{m}{t + p}$
19. $a = \frac{F}{m}$ **20.** $d_1 = d_2 - vt$

Exercise Set 2.4, pp. 113–116

1. (d) **2.** (c) **3.** (e) **4.** (b) **5.** (c) **6.** (d) **7.** (f)
8. (a) **9.** (b) **10.** (e) **11.** 0.67 **13.** 0.02 **15.** 0.035
17. 0.4 **19.** 0.6258 **21.** 0.007 **23.** 1.25 **25.** 13%
27. 1.4% **29.** 32.6% **31.** 90% **33.** 0.49% **35.** 108%
37. 230% **39.** 80% **41.** 32% **43.** 25% **45.** 24%
47. $46\frac{2}{3}$, or $\frac{140}{3}$ **49.** 2.5 **51.** 84 **53.** 125%
55. 0.8 **57.** 50% **59.** $33.\overline{3}\%$, or $33\frac{1}{3}\%$ **61.** \$456
63. \$71.25 **65.** 75 credits **67.** 600 at-bats **69.** (a) 16%; (b) \$29 **71.** About 34.8%; about 65.2% **73.** \$210
75. 285 women **77.** \$19.20 per hour **79.** About 718%
81. \$36 **83.** \$148.50 **85.** About 31.5 lb **87.** About 2.45 billion pieces of mail **89.** 150 calories **91.** TW **93.** Let l represent the length and w represent the width; $2l + 2w$
94. $0.05 \cdot 180$ **95.** Let p represent the number of points Tino scored; $p - 5$ **96.** $15 + 1.5x$ **97.** $10(\frac{1}{2}a)$ **98.** Let n represent the number; $3n + 10$ **99.** Let l represent the length and w represent the width; $w = l - 2$ **100.** Let x represent the first number and y represent the second number; $x = 4y$ **101.** TW
103. 18,500 people **105.** About 5 ft 6 in. **107.** About 3.7% per year; 84 per thousand in 2008, 81 per thousand in 2009 **109.** TW

Exercise Set 2.5, pp. 126–132

1. 8 **3.** 11 **5.** \$150 **7.** \$85 **9.** Approximately $144\frac{1}{3}$ km
11. 160 mi **13.** 1204 and 1205 **15.** 396 and 398
17. 19, 20, 21 **19.** Bride: 102 yr; groom: 83 yr
21. Spam: 175 billion messages; non-spam: 35 billion messages
23. 140 and 141 **25.** Width: 100 ft; length: 160 ft; area: 16,000 ft^2
27. Width: 50 ft; length: 84 ft **29.** $1\frac{1}{2}$ in. by $3\frac{1}{2}$ in.
31. 30°, 90°, 60° **33.** 95° **35.** Bottom: 144 ft; middle: 72 ft; top: 24 ft **37.** $8\frac{3}{4}$ mi **39.** $128\frac{1}{3}$ mi **41.** 65°, 25°
43. Length: 27.9 cm; width: 21.6 cm **45.** \$6600 **47.** 725 points
49. \$1250 **51.** 160 chirps per minute **53.** 150 gal
55. August 19 **57.** 600 ft per minute **59.** 30 mph
61. 1.2 hr, or 72 min **63.** TW **65.** $<$ **66.** $<$ **67.** $<$
68. $>$ **69.** $-4 \le x$ **70.** $5 > x$ **71.** $y < 5$ **72.** $t \ge -10$
73. TW **75.** \$37 **77.** 20 **79.** Half-dollars: 5; quarters: 10; dimes: 20; nickels: 60 **81.** \$95.99 **83.** 30 games **85.** 76
87. TW **89.** Width: 23.31 cm; length: 27.56 cm

Exercise Set 2.6, pp. 140–142

1. $\ge$ **2.** $\le$ **3.** $<$ **4.** $>$ **5.** Equivalent **6.** Equivalent
7. Equivalent **8.** Not equivalent **9.** (a) Yes; (b) yes; (c) no
11. (a) Yes; (b) no; (c) yes **13.** (a) Yes; (b) no; (c) yes
15. $x \le 7$ **17.** $t > -2$
19. $1 \le m$ **21.** $-3 < x \le 5$
23. $0 < x < 3$
25. 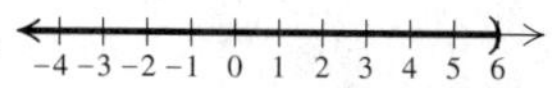 $\{y|y < 6\}, (-\infty, 6)$
27. 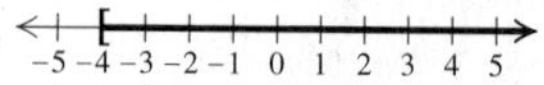 $\{x|x \ge -4\}, [-4, \infty)$
29. 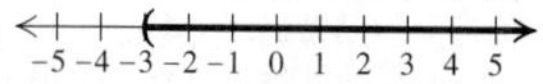 $\{t|t > -3\}, (-3, \infty)$
31. 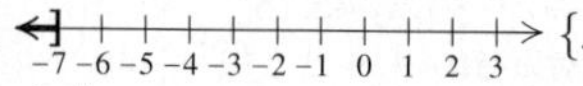 $\{x|x \le -7\}, (-\infty, -7]$
33. $\{x|x > -4\}$, or $(-4, \infty)$ **35.** $\{x|x \le 2\}$, or $(-\infty, 2]$
37. $\{x|x < -1\}$, or $(-\infty, -1)$ **39.** $\{x|x \ge 0\}$, or $[0, \infty)$
41. $\{y|y > 7\}$, or $(7, \infty)$, 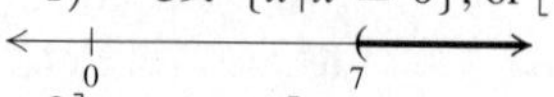
43. $\{x|x \le -2\}$, or $(-\infty, -2]$, 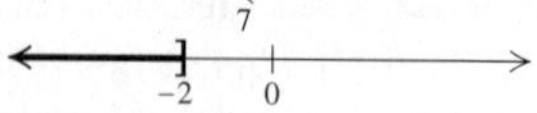
45. $\{t|t \ge -3\}$, or $[-3, \infty)$,
47. $\{x|x \le 5\}$, or $(-\infty, 5]$, 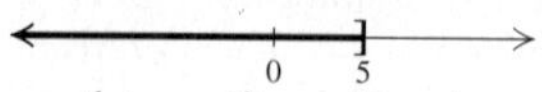
49. $\{y|y \le \frac{1}{2}\}$, or $(-\infty, \frac{1}{2}]$ **51.** $\{t|t > \frac{5}{8}\}$, or $(\frac{5}{8}, \infty)$
53. $\{x|x < 0\}$, or $(-\infty, 0)$ **55.** $\{t|t < 23\}$, or $(-\infty, 23)$
57. $\{x|x < 7\}$, or $(-\infty, 7)$,
59. $\{t|t < -3\}$, or $(-\infty, -3)$,
61. $\{n|n \ge -1.5\}$, $[-1.5, \infty)$,
63. $\{y|y \ge -\frac{1}{10}\}$, or $[-\frac{1}{10}, \infty)$ **65.** $\{x|x > \frac{4}{5}\}$, or $(\frac{4}{5}, \infty)$
67. $\{x|x < 9\}$, or $(-\infty, 9)$ **69.** $\{t|t \le 7\}$, or $(-\infty, 7]$
71. $\{a|a < -12\}$, or $(-\infty, -12)$ **73.** $\{y|y \le 0\}$, or $(-\infty, 0]$
75. $\{x|x > -10\}$, or $(-10, \infty)$ **77.** $\{y|y < 2\}$, or $(-\infty, 2)$
79. $\{y|y \ge 3\}$, or $[3, \infty)$ **81.** $\{x|x > -4\}$, or $(-4, \infty)$

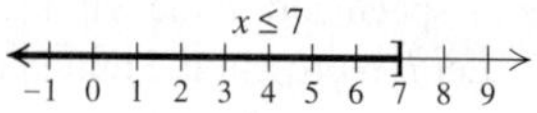

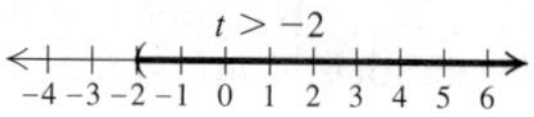

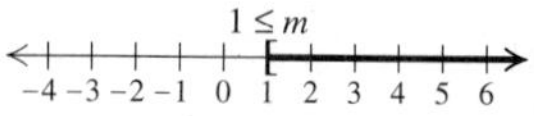

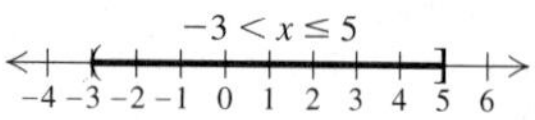

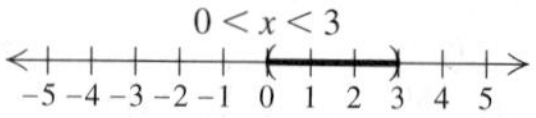

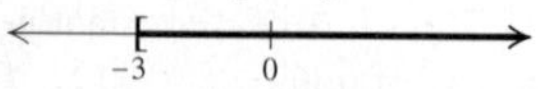

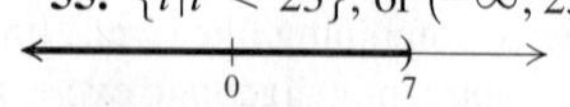

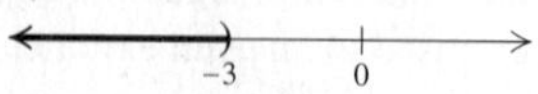

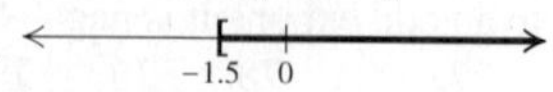

83. $\{n|n \geq 70\}$, or $[70, \infty)$ **85.** $\{x|x \leq 15\}$, or $(-\infty, 15]$ **87.** $\{t|t < 14\}$, or $(-\infty, 14)$ **89.** $\{y|y \leq -4\}$, or $(-\infty, -4]$ **91.** $\{t|t \leq -4\}$, or $(-\infty, -4]$ **93.** $\{r|r > -3\}$, or $(-3, \infty)$ **95.** $\{x|x \geq 8\}$, or $[8, \infty)$ **97.** $\left\{x|x < \frac{11}{18}\right\}$, or $\left(-\infty, \frac{11}{18}\right)$ **99.** TW **101.** 1 **102.** $\{x|x > 1\}$, or $(1, \infty)$ **103.** $\frac{19}{5}$ **104.** $\left\{x|x \geq \frac{19}{5}\right\}$, or $\left[\frac{19}{5}, \infty\right)$ **105.** TW **107.** $\{x|x$ is a real number$\}$, or $(-\infty, \infty)$ **109.** $\left\{x|x \leq \frac{5}{6}\right\}$, or $\left(-\infty, \frac{5}{6}\right]$ **111.** $\{x|x > 7\}$, or $(7, \infty)$ **113.** $\left\{x\middle|x < \frac{y-b}{a}\right\}$, or $\left(-\infty, \frac{y-b}{a}\right)$ **115.** $\{x|x$ is a real number$\}$, or $(-\infty, \infty)$

Translating for Success, p. 147

1. F **2.** I **3.** C **4.** E **5.** D **6.** J **7.** O **8.** M **9.** B **10.** L

Exercise Set 2.7, pp. 148–152

1. $b \leq a$ **2.** $b < a$ **3.** $a \leq b$ **4.** $a < b$ **5.** $b \leq a$ **6.** $a \leq b$ **7.** $b < a$ **8.** $a < b$ **9.** Let n represent the number; $n < 10$ **11.** Let t represent the temperature; $t \leq -3$ **13.** Let a represent the age of the altar; $a > 1200$ **15.** Let d represent the distance to Normandale Community College; $d \leq 15$ **17.** Let d represent the number of years of driving experience; $d \geq 5$ **19.** Let c represent the cost of production; $c \leq 12{,}500$ **21.** More than 1.125 hr **23.** At least 2.25 **25.** Scores greater than or equal to 97 **27.** 8 credits or more **29.** At least 3 plate appearances **31.** Lengths greater than 6 cm **33.** Depths less than 437.5 ft **35.** Blue-book value is greater than or equal to \$10,625 **37.** Lengths greater than or equal to 5 in. **39.** Temperatures greater than 37°C **41.** Heights at least 4 ft **43.** A serving contains at least 6 g of fat. **45.** Dates after September 16 **47.** No more than 134 text messages **49.** Years after 2012 **51.** Mileages less than or equal to 225 **53.** More than 25 checks **55.** Gross sales greater than \$7000 **57.** About 6.8 gal or less **59.** TW **61.** -14 **62.** $-\frac{2}{3}$ **63.** -60 **64.** -11.1 **65.** 0 **66.** 5 **67.** -2 **68.** -1 **69.** TW **71.** Parties of more than 80 guests **73.** More than 6 hr **75.** Lengths less than or equal to 8 cm **77.** They contain at least 7.5 g of fat per serving. **79.** At least \$42 **81.** TW

Study Summary: Chapter 2, pp. 153–155

1. 5 **2.** 4.8 **3.** -1 **4.** $-\frac{1}{10}$ **5.** $c = \frac{d}{a-b}$ **6.** 80 **7.** $47\frac{1}{2}$ mi; $72\frac{1}{2}$ mi **8.** $(-\infty, 0]$ **9.** $\{x|x > 7\}$, or $(7, \infty)$ **10.** $\left\{x|x \geq -\frac{1}{4}\right\}$, or $\left[-\frac{1}{4}, \infty\right)$ **11.** Let d represent the distance Luke runs, in miles; $d \geq 3$

Review Exercises: Chapter 2, pp. 156–157

1. True **2.** False **3.** True **4.** True **5.** True **6.** False **7.** True **8.** True **9.** -25 **10.** 7 **11.** -65 **12.** 3 **13.** -20 **14.** 1.11 **15.** $\frac{1}{2}$ **16.** $\frac{38}{5}$ **17.** -8 **18.** -5 **19.** $-\frac{1}{3}$ **20.** 4 **21.** No solution; contradiction **22.** 4 **23.** 16 **24.** -7 **25.** $-\frac{7}{5}$ **26.** 12 **27.** All real numbers; identity **28.** $d = \frac{C}{\pi}$ **29.** $B = \frac{3V}{h}$ **30.** $y = \frac{5}{2}x - 5$ **31.** $x = \frac{b}{t-a}$ **32.** 0.009 **33.** 44% **34.** 70% **35.** 140 **36.** No **37.** Yes **38.** No **39.**

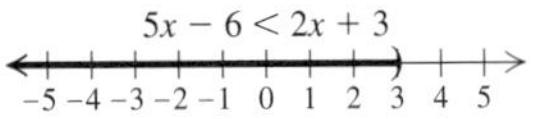

40.

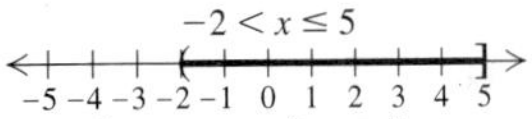

41.

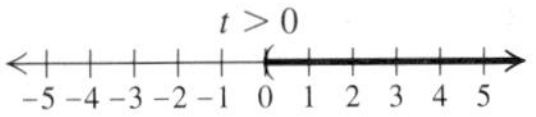

42. $\left\{t|t \geq -\frac{1}{2}\right\}$, or $\left[-\frac{1}{2}, \infty\right)$ **43.** $\{x|x \geq 7\}$, or $[7, \infty)$ **44.** $\{y|y > 3\}$, or $(3, \infty)$ **45.** $\{y|y \leq -4\}$, or $(-\infty, -4]$ **46.** $\{x|x < -11\}$, or $(-\infty, -11)$ **47.** $\{y|y > -7\}$, or $(-7, \infty)$ **48.** $\{x|x > -6\}$, or $(-6, \infty)$ **49.** $\left\{x|x > -\frac{9}{11}\right\}$, or $\left(-\frac{9}{11}, \infty\right)$ **50.** $\{t|t \leq -12\}$, or $(-\infty, -12]$ **51.** $\{x|x \leq -8\}$, or $(-\infty, -8]$ **52.** About \$305.4 billion **53.** 15 ft, 17 ft **54.** 57, 59 **55.** Width: 11 cm; length: 17 cm **56.** \$160 **57.** 16 hr **58.** 35°, 85°, 60° **59.** 7 subscriptions **60.** \$105 or less **61.** Widths greater than 17 cm **62.** TW Multiplying both sides of an equation by any nonzero number results in an equivalent equation. When multiplying on both sides of an inequality, the sign of the number being multiplied by must be considered. If the number is positive, the direction of the inequality symbol remains unchanged; if the number is negative, the direction of the inequality symbol must be reversed to produce an equivalent inequality. **63.** TW The solutions of an equation can usually each be checked. The solutions of an inequality are normally too numerous to check. Checking a few numbers from the solution set found cannot guarantee that the answer is correct, although if any number does not check, the answer found is incorrect. **64.** About 1 hr 36 min **65.** Nile: 4160 mi; Amazon: 4225 mi **66.** \$18,600 **67.** $-23, 23$ **68.** $-20, 20$ **69.** $a = \frac{y-3}{2-b}$

Test: Chapter 2, p. 158

1. [2.1] 9 **2.** [2.1] 15 **3.** [2.1] -3 **4.** [2.1] 49 **5.** [2.1] -12 **6.** [2.2] 2 **7.** [2.1] -8 **8.** [2.2] $-\frac{23}{67}$ **9.** [2.2] 7 **10.** [2.2] $-\frac{5}{3}$ **11.** [2.2] $\frac{23}{3}$ **12.** [2.2] All real numbers; identity **13.** [2.6] $\{x|x > -5\}$, or $(-5, \infty)$ **14.** [2.6] $\{x|x > -13\}$, or $(-13, \infty)$ **15.** [2.6] $\{y|y \geq -13\}$, or $[-13, \infty)$ **16.** [2.6] $\{y|y \leq -8\}$, or $(-\infty, -8]$ **17.** [2.6] $\{n|n < -5\}$, or $(-\infty, -5)$ **18.** [2.6] $\{t|t \geq -1\}$, or $[-1, \infty)$ **19.** [2.6] $\{x|x \leq -1\}$, or $(-\infty, -1]$ **20.** [2.3] $r = \frac{A}{2\pi h}$ **21.** [2.3] $l = 2w - P$ **22.** [2.4] 2.3 **23.** [2.4] 5.4% **24.** [2.4] 16 **25.** [2.4] 44% **26.** [2.6]

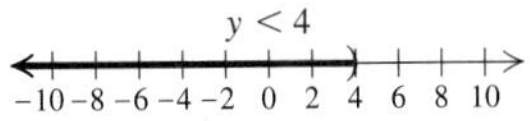

27. [2.6]

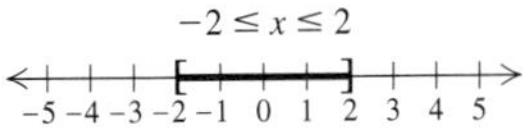

28. [2.5] Width: 7 cm; length: 11 cm **29.** [2.5] 60 mi **30.** [2.5] 81 mm, 83 mm, 85 mm **31.** [2.4] \$65 **32.** [2.7] Mileages less than or equal to 398.3 mi **33.** [2.3] $d = \frac{a}{3}$ **34.** [1.4], [2.2] $-15, 15$ **35.** [2.7] Let h = the number of hours of sun each day; $4 \leq h \leq 6$ **36.** [2.5] 60 tickets

Chapter 3

Exercise Set 3.1, pp. 169–172

1. (a) **2.** (c) **3.** (b) **4.** (d) **5.** 2 drinks **7.** The person weighs more than 140 lb. **9.** About \$4883 **11.** \$6,022,800,000 **13.** About 13.5 million tons **15.** About 3.2 million tons **17.** About 9 million households **19.** 2005 and 2006

21.

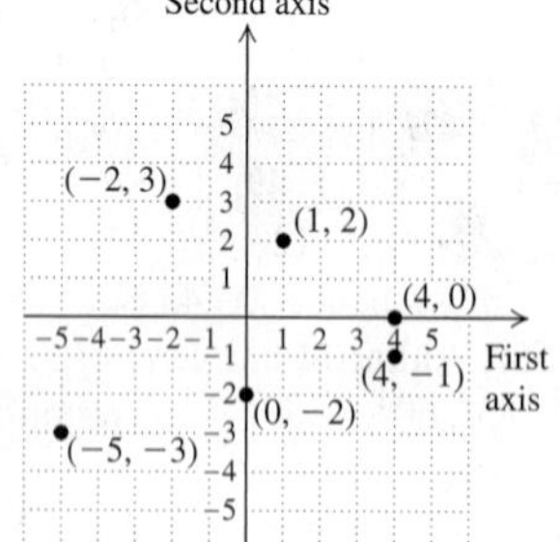

23.

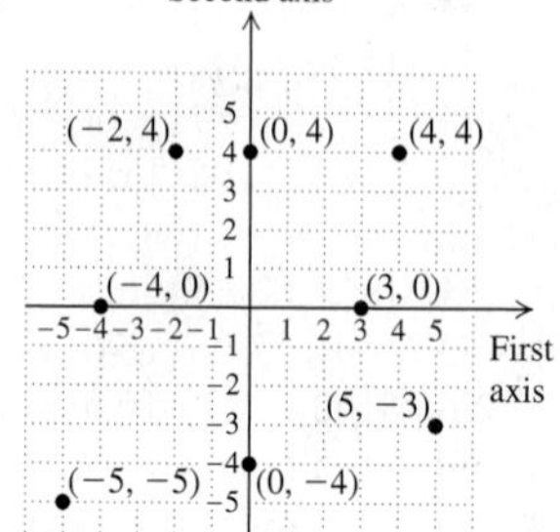

25.

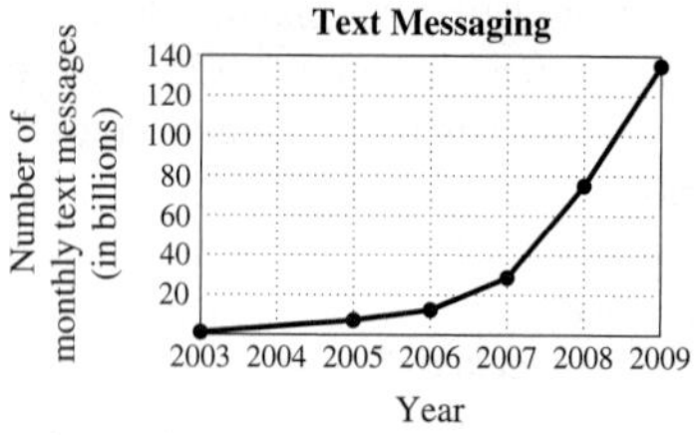

Text Messaging

Number of monthly text messages (in billions)

140, 120, 100, 80, 60, 40, 20

2003 2004 2005 2006 2007 2008 2009

Year

27. $A(-4, 5); B(-3, -3); C(0, 4); D(3, 4); E(3, -4)$

29. $A(4, 1); B(0, -5); C(-4, 0); D(-3, -2); E(3, 0)$

31.

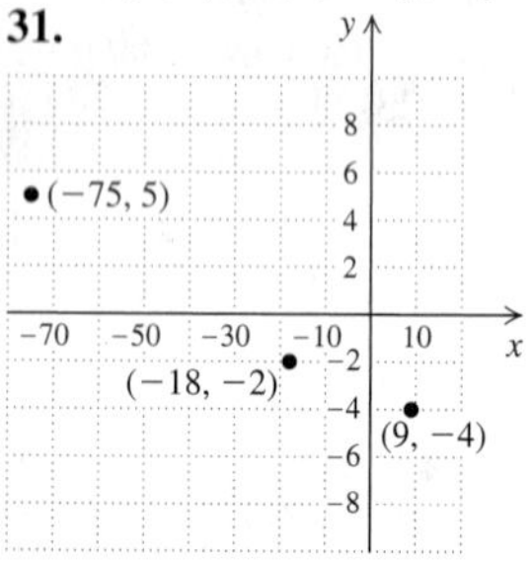

(−75, 5) (−18, −2) (9, −4)

33.

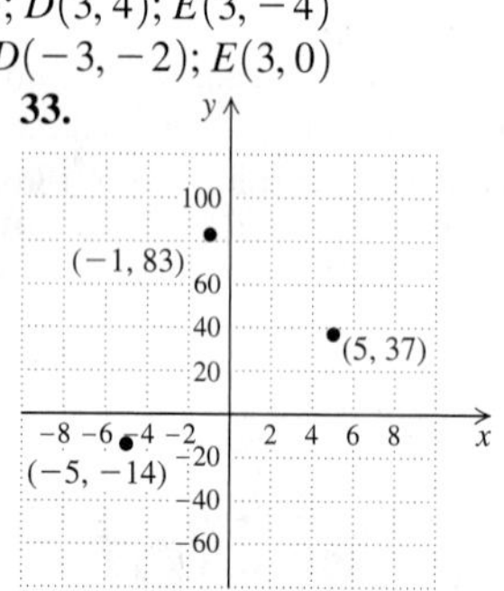

35.

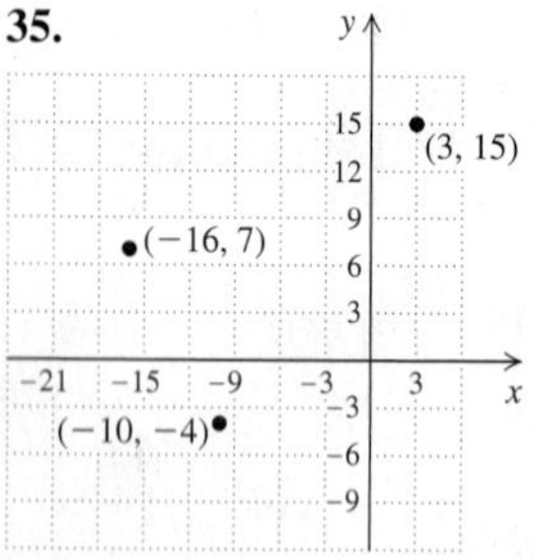

(3, 15) (−16, 7) (−10, −4)

37.

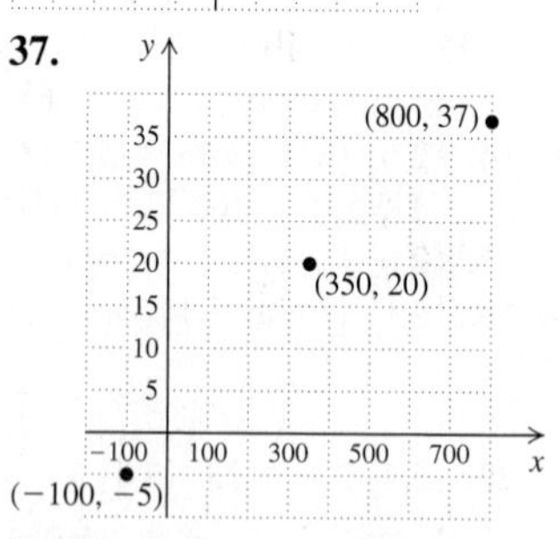

39.

(−83, 491) (−124, −95) (54, −238)

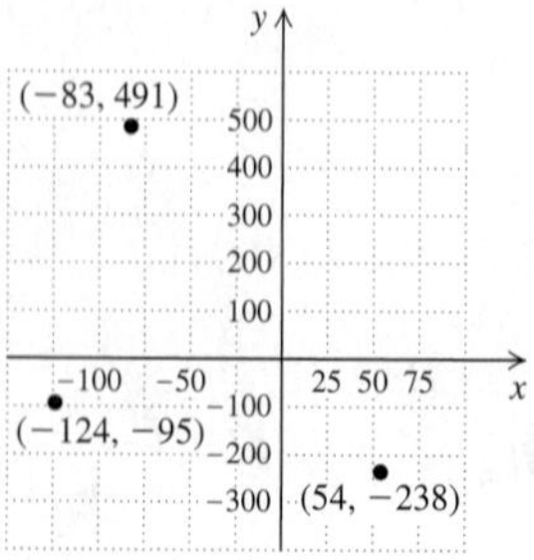

41. IV **43.** III **45.** y-axis **47.** II **49.** x-axis **51.** I **53.** I and IV **55.** I and III

57.

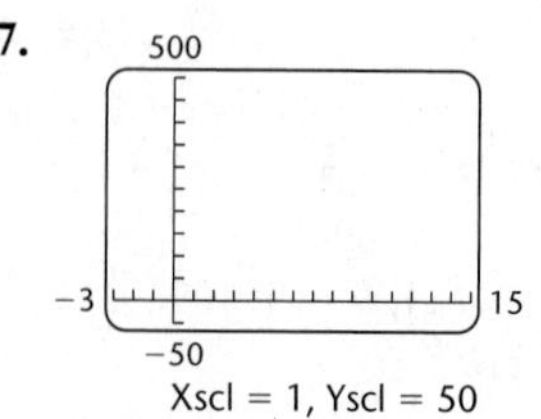

59.

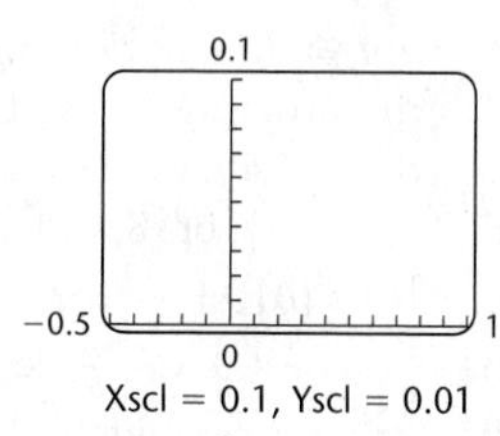

61. TW **63.** $y = \frac{2x}{5}$, or $y = \frac{2}{5}x$ **64.** $y = \frac{-3x}{2}$, or $y = -\frac{3}{2}x$ **65.** $y = x - 8$ **66.** $y = -\frac{2}{5}x + 2$ **67.** $y = -\frac{2}{3}x + \frac{5}{3}$ **68.** $y = \frac{5}{8}x - \frac{1}{8}$ **69.** TW **71.** II or IV **73.** $(-1, -5)$

75.

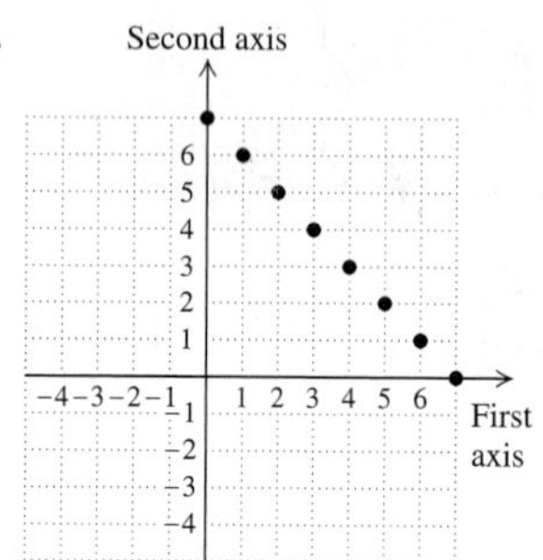

77. $\frac{65}{2}$ sq units **79.** Latitude 27° North; longitude 81° West

Exercise Set 3.2, pp. 181–184

1. False **2.** True **3.** True **4.** True **5.** True **6.** False **7.** Yes **9.** No **11.** No

13.
$$\begin{array}{c|l} y = x + 3 & \\ \hline 2 & -1 + 3 \\ 2 \stackrel{?}{=} 2 & \text{True} \end{array} \qquad \begin{array}{c|l} y = x + 3 & \\ \hline 7 & 4 + 3 \\ 7 \stackrel{?}{=} 7 & \text{True} \end{array}$$
(2, 5); answers may vary

15.
$$\begin{array}{c|l} y = \frac{1}{2}x + 3 & \\ \hline 5 & \frac{1}{2}\cdot 4 + 3 \\ & 2 + 3 \\ 5 \stackrel{?}{=} 5 & \text{True} \end{array} \qquad \begin{array}{c|l} y = \frac{1}{2}x + 3 & \\ \hline 2 & \frac{1}{2}(-2) + 3 \\ & -1 + 3 \\ 2 \stackrel{?}{=} 2 & \text{True} \end{array}$$
(0, 3); answers may vary

17.
$$\begin{array}{l|c} y + 3x = 7 & \\ \hline 1 + 3\cdot 2 & 7 \\ 1 + 6 & \\ 7 \stackrel{?}{=} 7 & \text{True} \end{array} \qquad \begin{array}{l|c} y + 3x = 7 & \\ \hline -5 + 3\cdot 4 & 7 \\ -5 + 12 & \\ 7 \stackrel{?}{=} 7 & \text{True} \end{array}$$
(1, 4); answers may vary

19.
$$\begin{array}{l|c} 4x - 2y = 10 & \\ \hline 4\cdot 0 - 2(-5) & 10 \\ 0 + 10 & \\ 10 \stackrel{?}{=} 10 & \text{True} \end{array} \qquad \begin{array}{l|c} 4x - 2y = 10 & \\ \hline 4\cdot 4 - 2\cdot 3 & 10 \\ 16 - 6 & \\ 10 \stackrel{?}{=} 10 & \text{True} \end{array}$$
(2, −1); answers may vary

21.

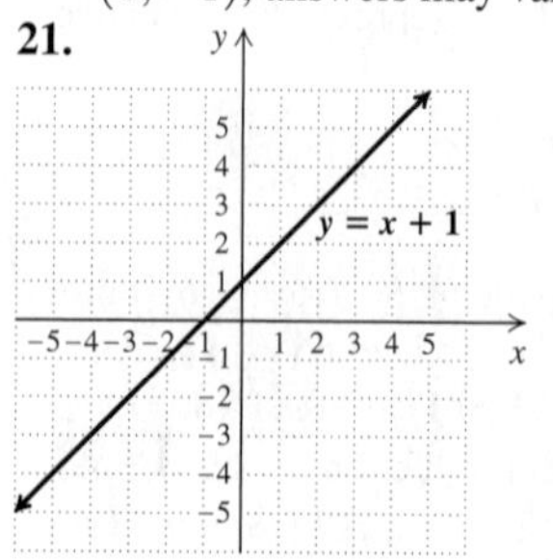

23.

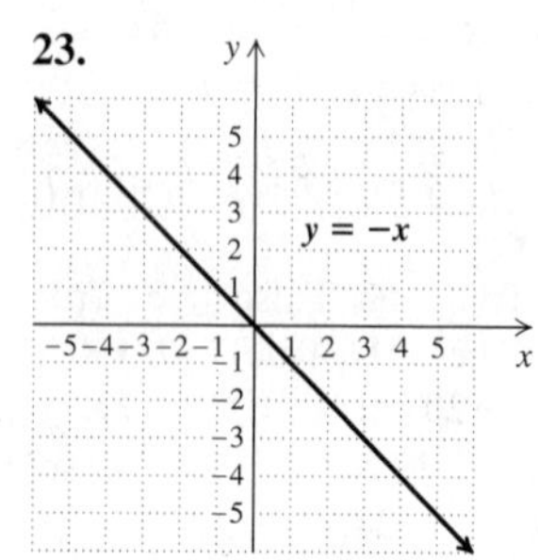

25.

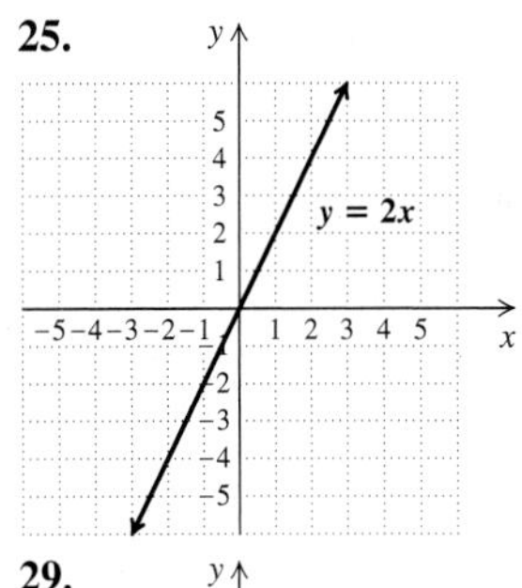

27.

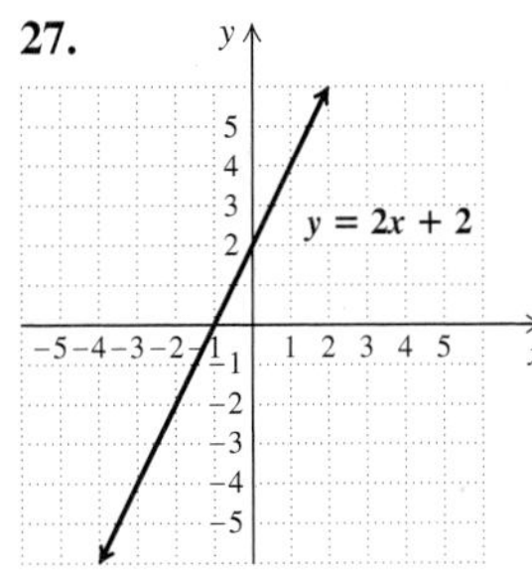

29.

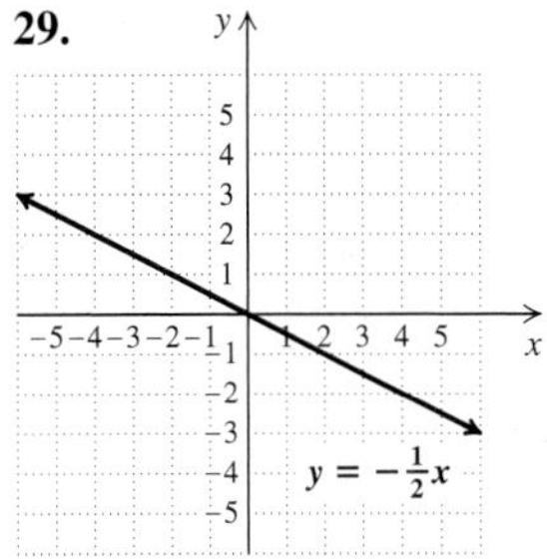

31.

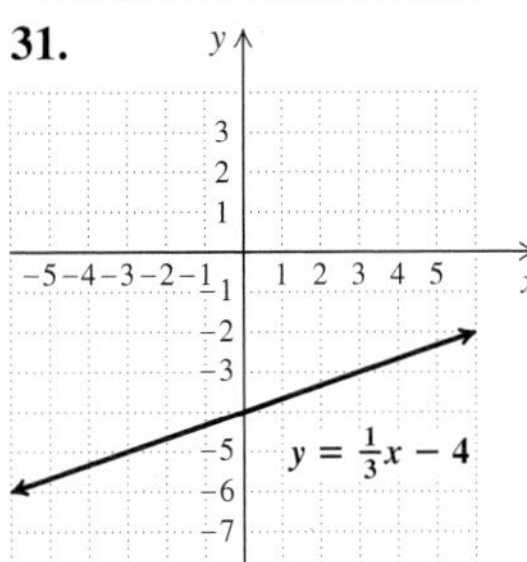

33.

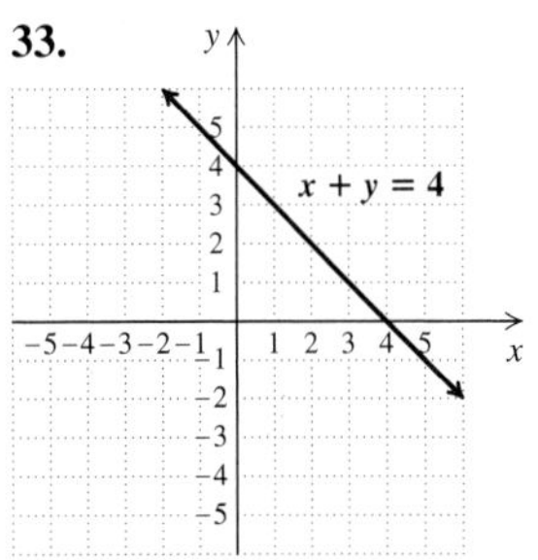

35.

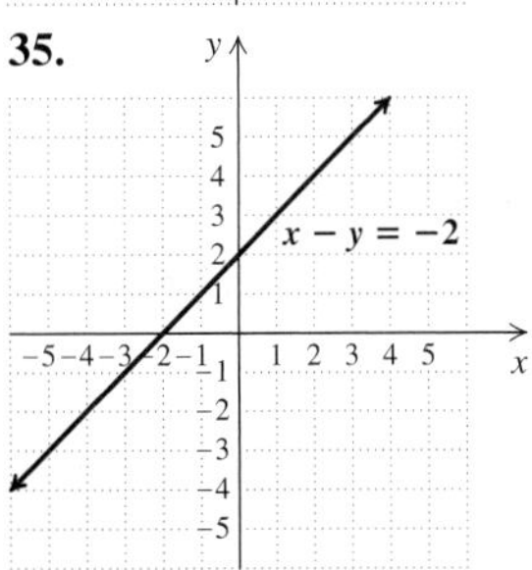

37.

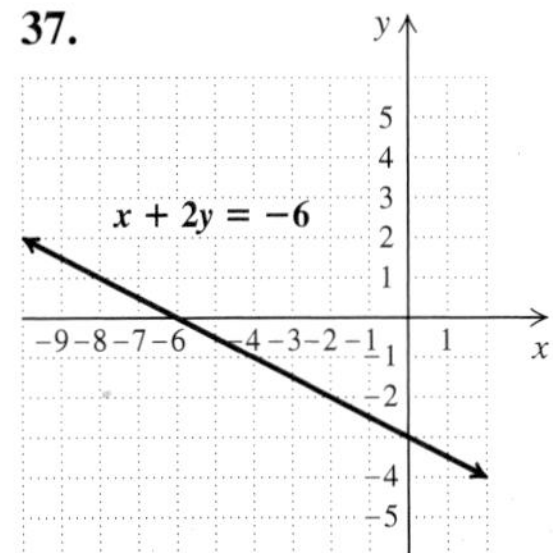

39.

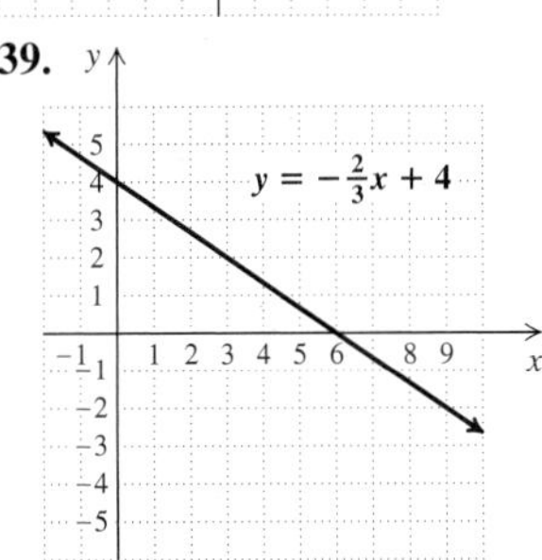

41.

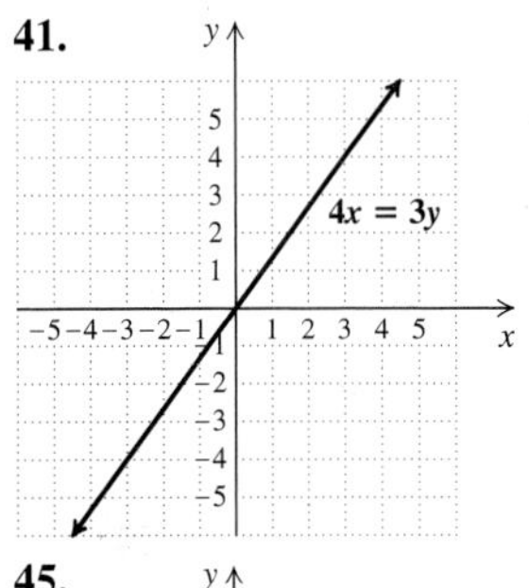

43.

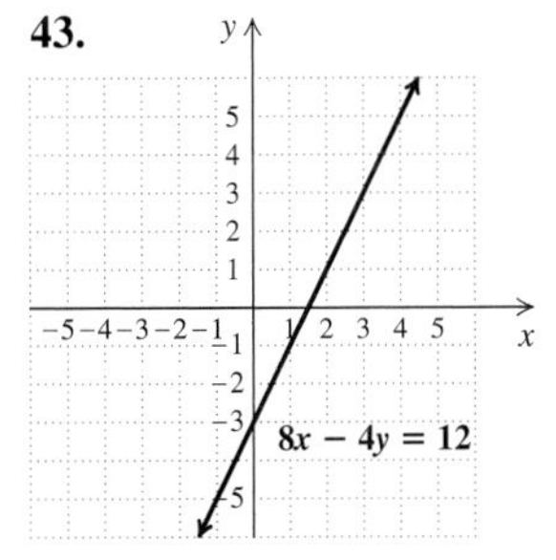

45.

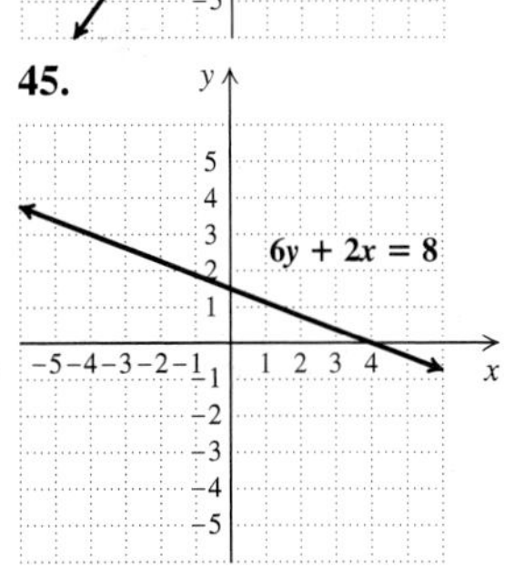

47.

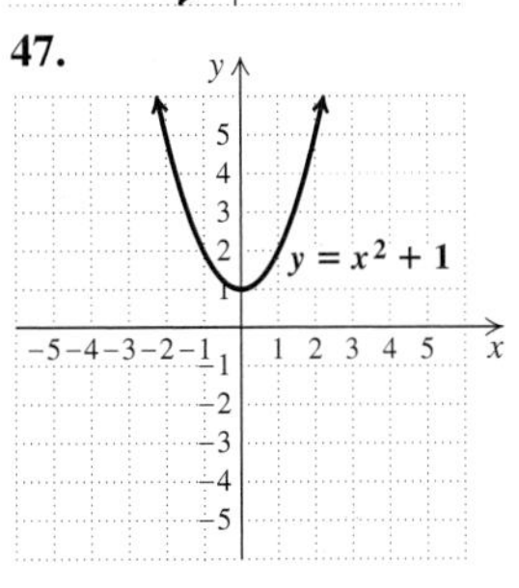

49.

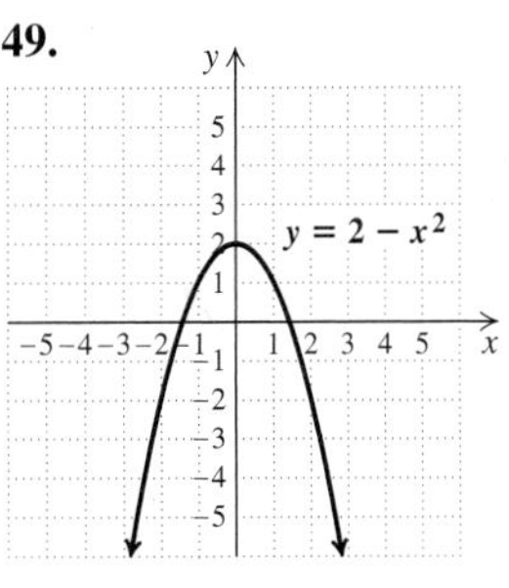

51. (a) **(b)**

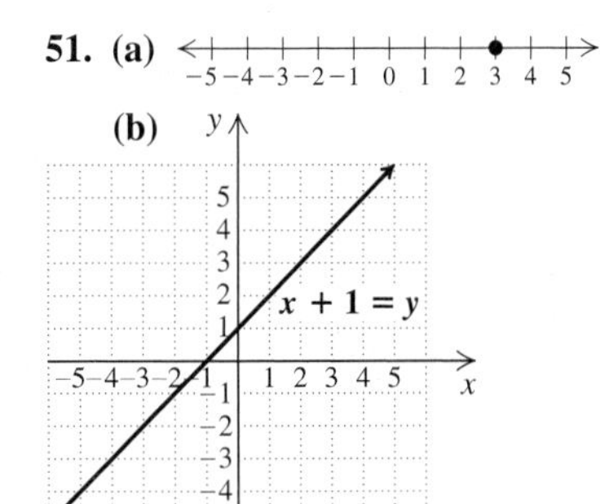

(b)

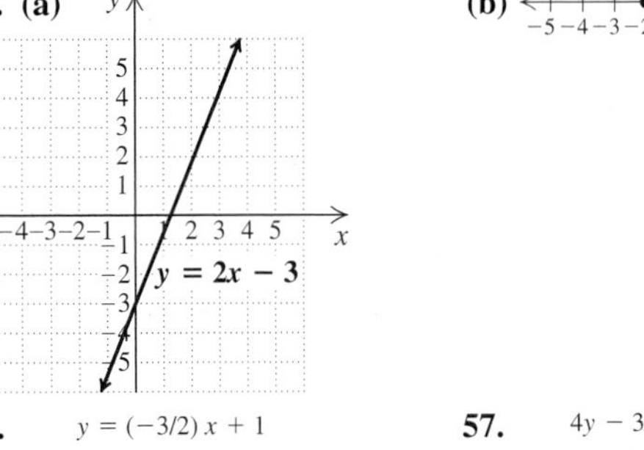

53. (a)

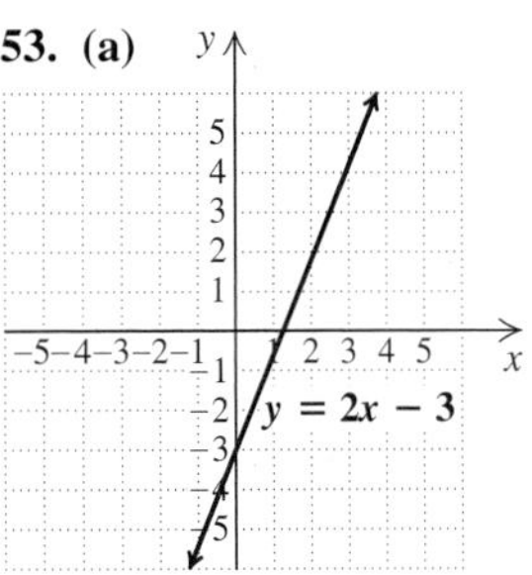

55. $y = (-3/2)x + 1$

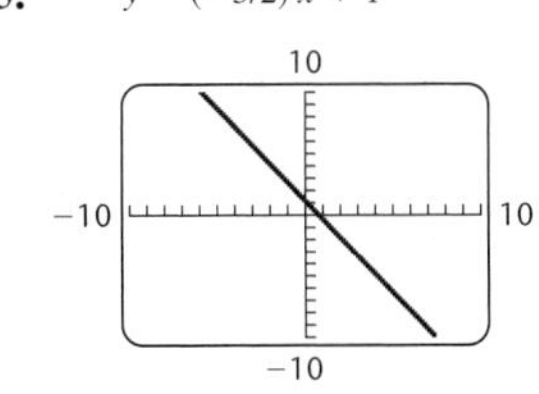

57. $4y - 3x = 1$, or $y = (3x + 1)/4$

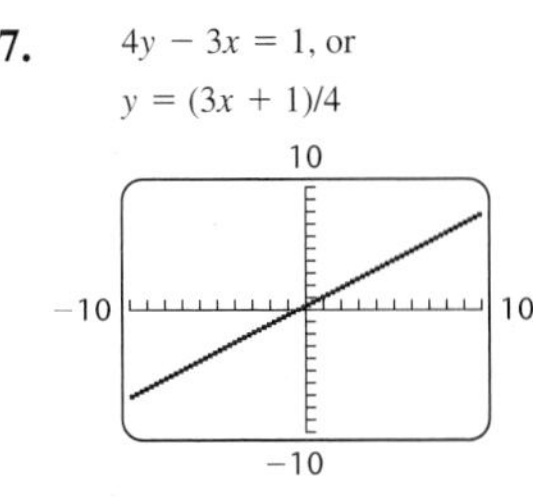

59. $y = -2$

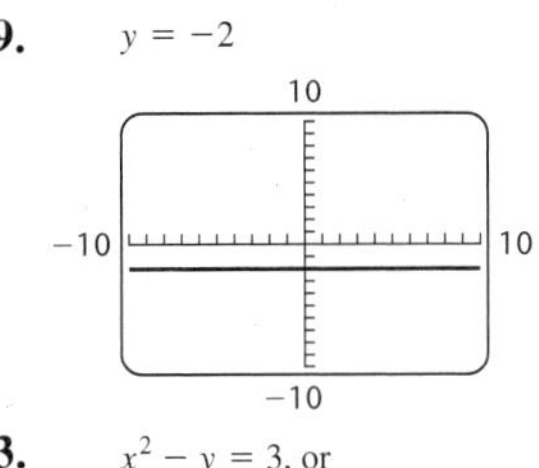

61. $y = -x^2$

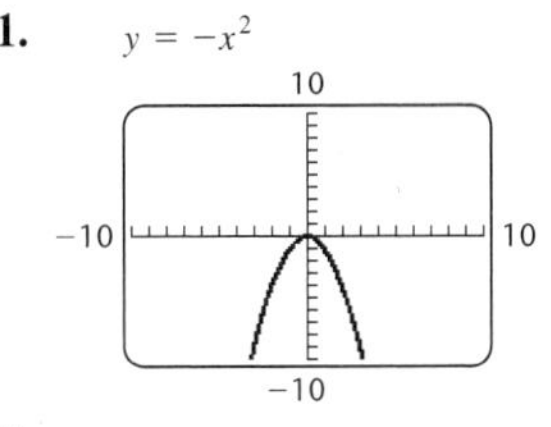

63. $x^2 - y = 3$, or $y = x^2 - 3$

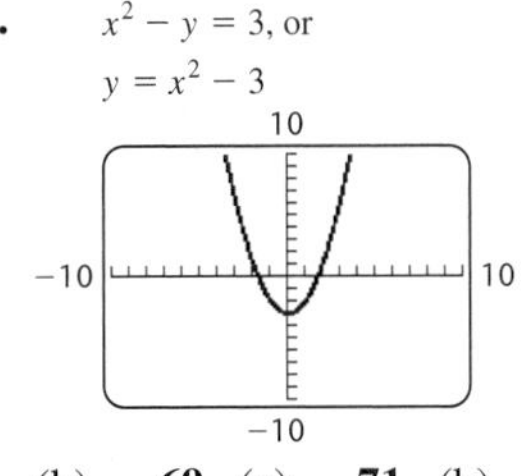

65. $y = x\wedge 3$ $y = x\wedge 3$

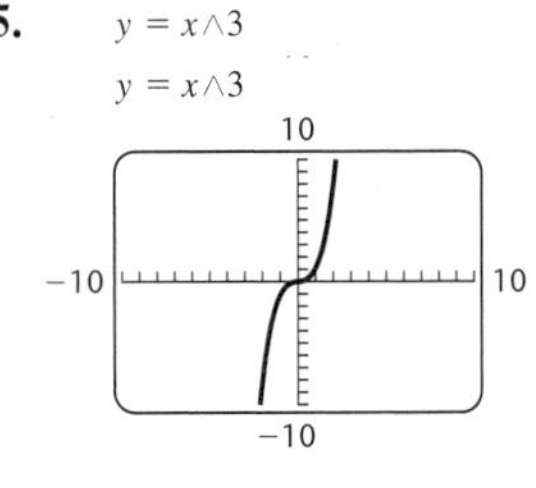

67. (b) **69.** (a) **71.** (b)

73. About \$4000

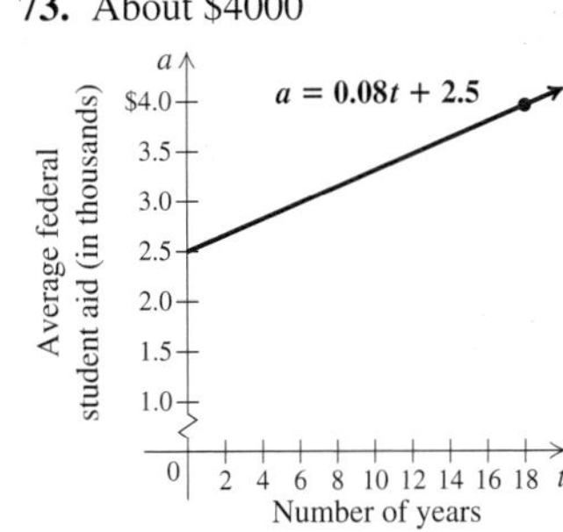

75. About \$49

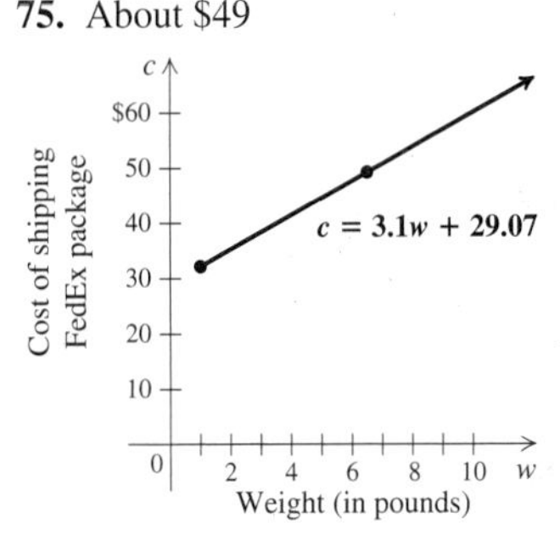

77. About \$96

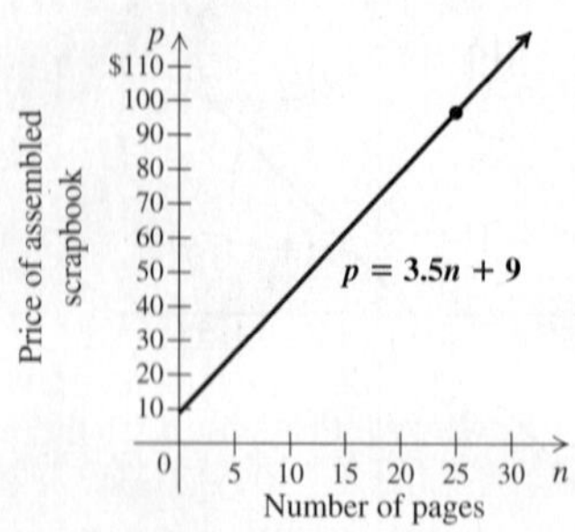

79. About \$1700

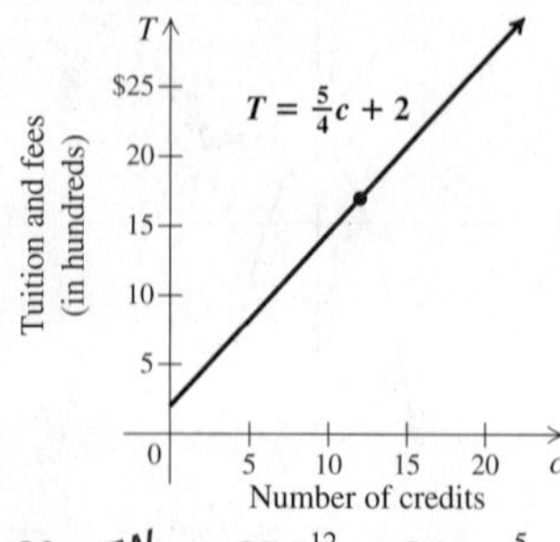

81. About 24°F

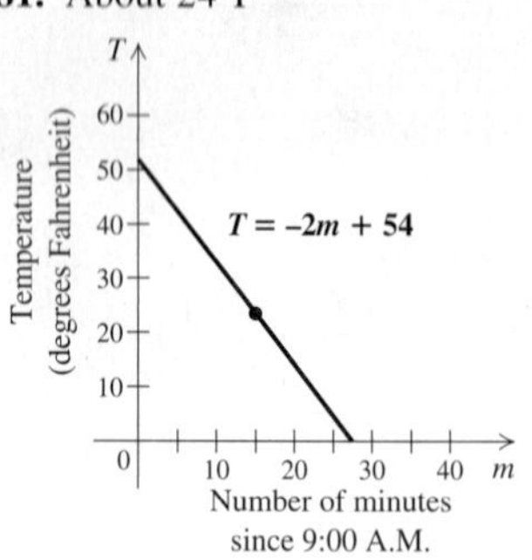

83. TW **85.** $\frac{12}{5}$ **86.** $-\frac{5}{2}$ **87.** 3 **88.** $-\frac{21}{5}$ **89.** $p = \dfrac{w}{t+1}$ **90.** $y = \dfrac{C - Ax}{B}$

91. $Q = 2A - T$ **92.** $y = m(x - h) + k$ **93.** TW **95.** $s + n = 18$

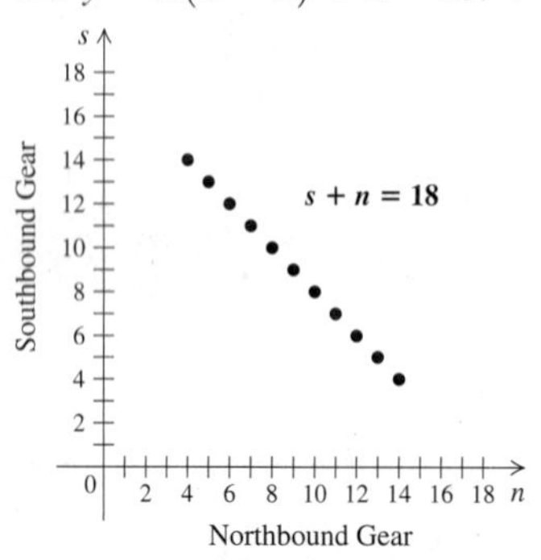

97. $x + y = 2$, or $y = -x + 2$
99. $5x - 3y = 15$, or $y = \frac{5}{3}x - 5$
101.

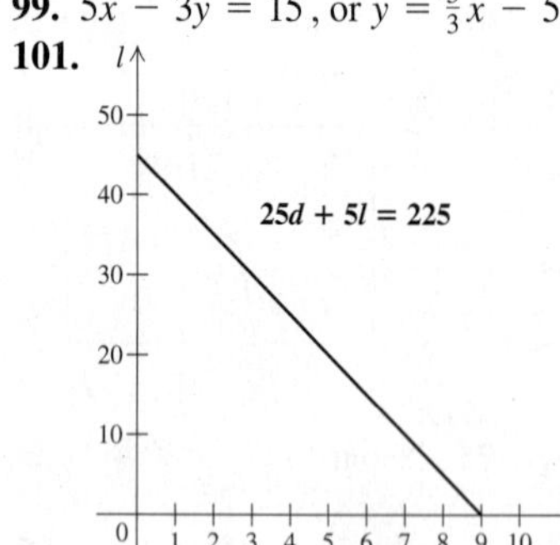

Answers may vary. 1 dinner, 40 lunches; 5 dinners, 20 lunches; 8 dinners, 5 lunches

103.

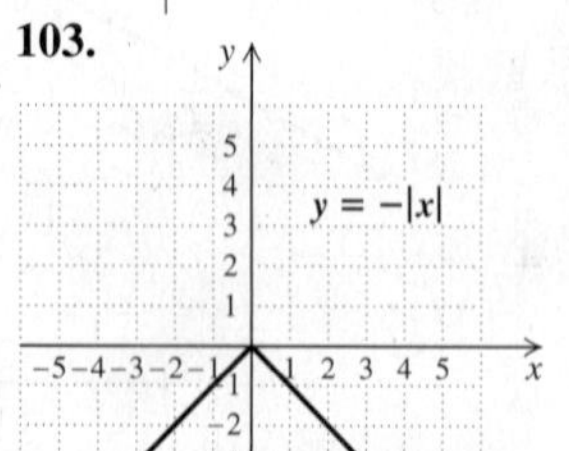

105.

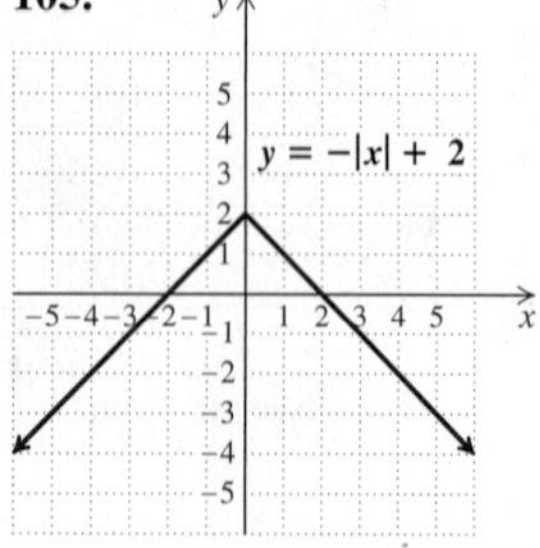

107. \$56.62; 16.2 gal

Exercise Set 3.3, pp. 191–193

1. (f) **2.** (e) **3.** (d) **4.** (c) **5.** (b) **6.** (a) **7.** Linear **9.** Linear **11.** Not linear **13.** Linear **15.** Not linear **17.** **(a)** $(0, 5)$; **(b)** $(2, 0)$ **19.** **(a)** $(0, -4)$; **(b)** $(3, 0)$ **21.** **(a)** $(0, -2)$; **(b)** $(-3, 0)$, $(3, 0)$ **23.** **(a)** $(0, 4)$; **(b)** $(-3, 0)$, $(3, 0)$, $(5, 0)$ **25.** **(a)** $(0, 5)$; **(b)** $(3, 0)$ **27.** **(a)** $(0, -14)$; **(b)** $(4, 0)$ **29.** **(a)** $(0, 50)$; **(b)** $\left(-\frac{75}{2}, 0\right)$ **31.** **(a)** $(0, 9)$; **(b)** none **33.** **(a)** None; **(b)** $(-7, 0)$

35.

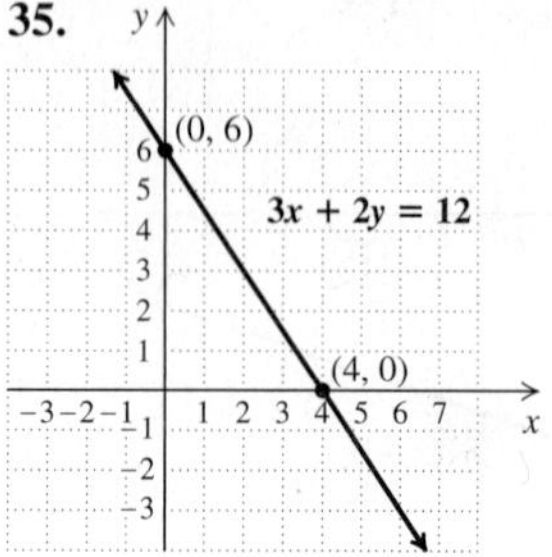

37.

39.

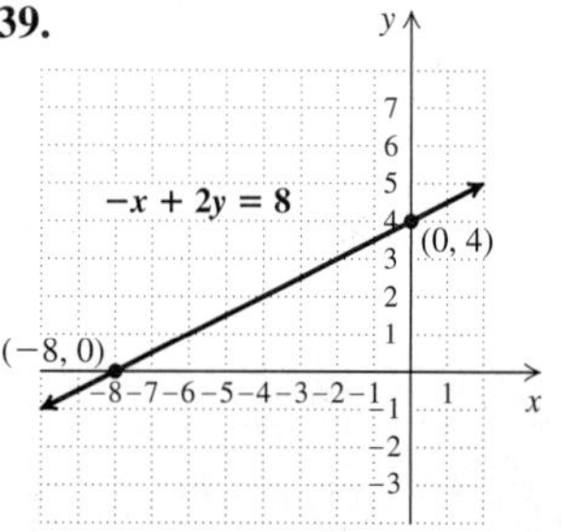

41.

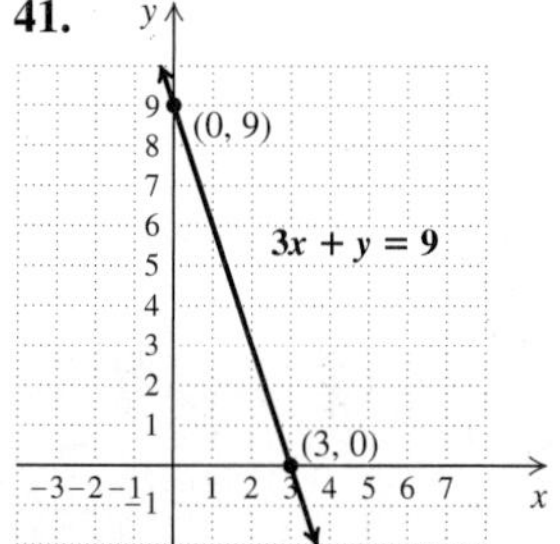

43.

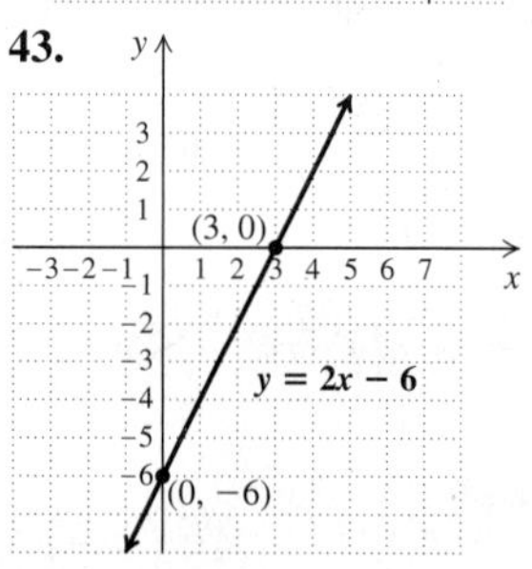

45.

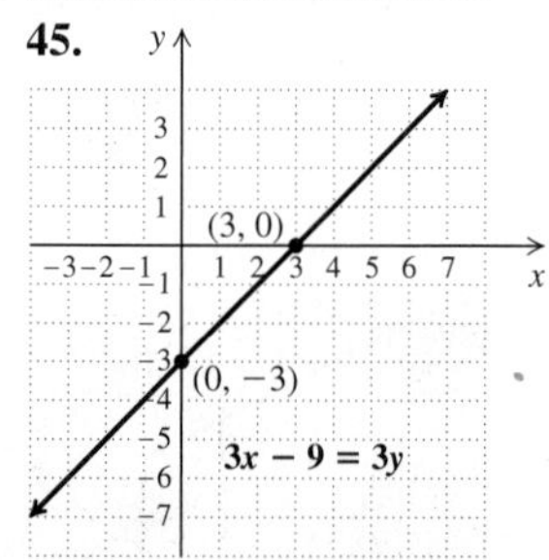

47.

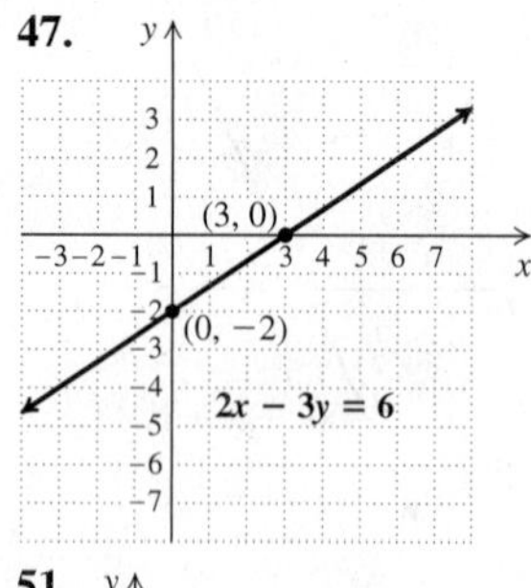

49.

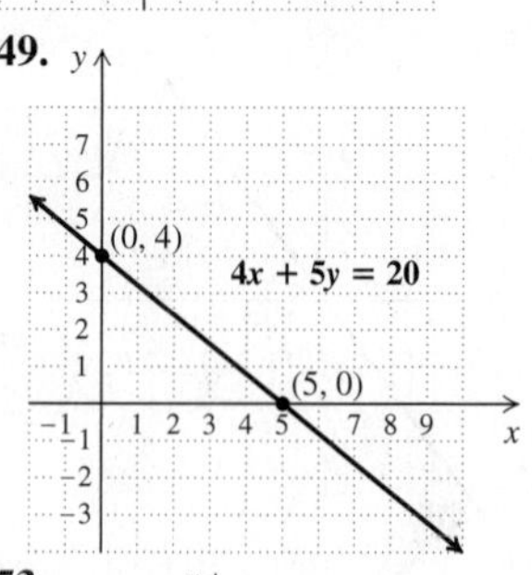

51.

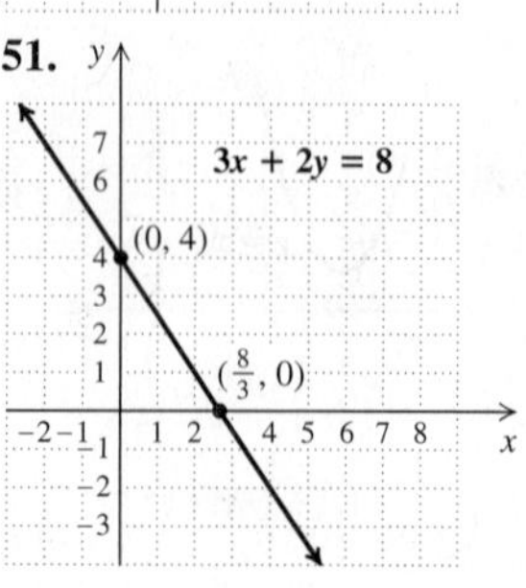

53.

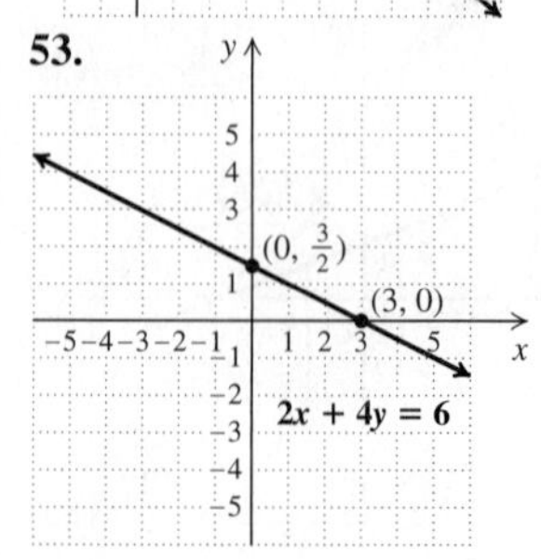

55.
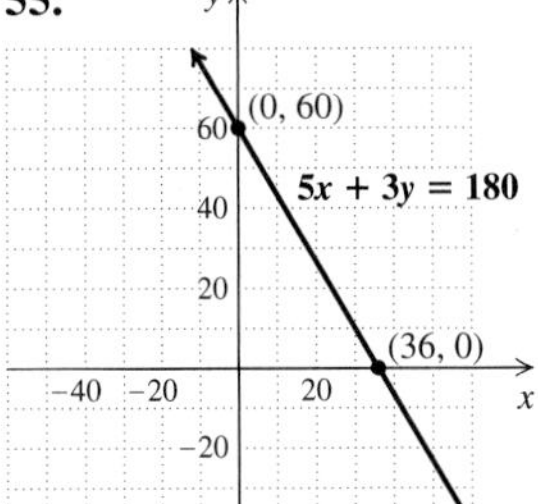

57.
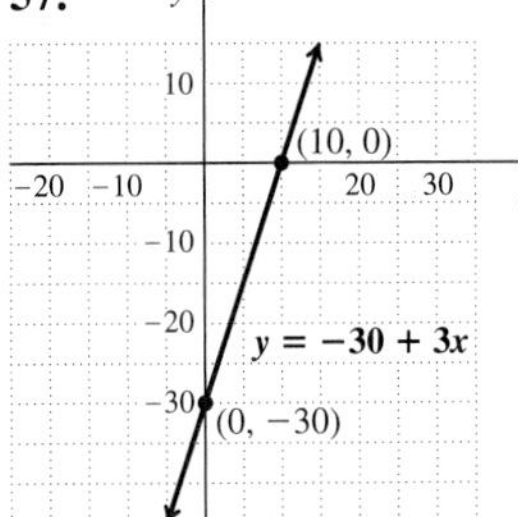

59.
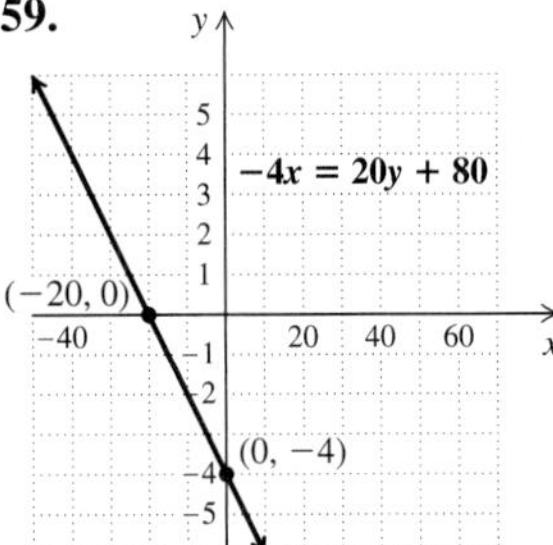

61.
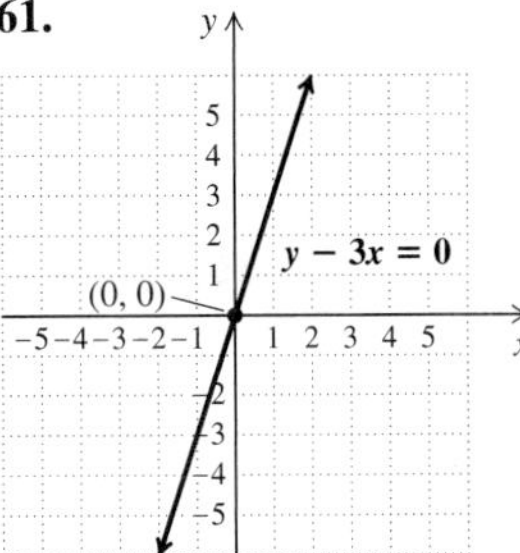

63.
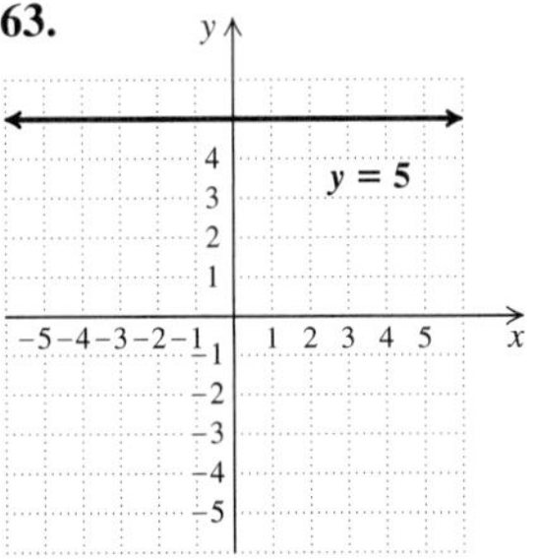

65.
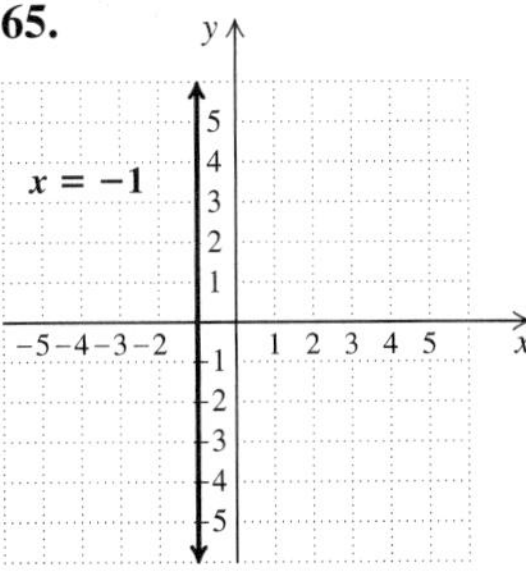

67.
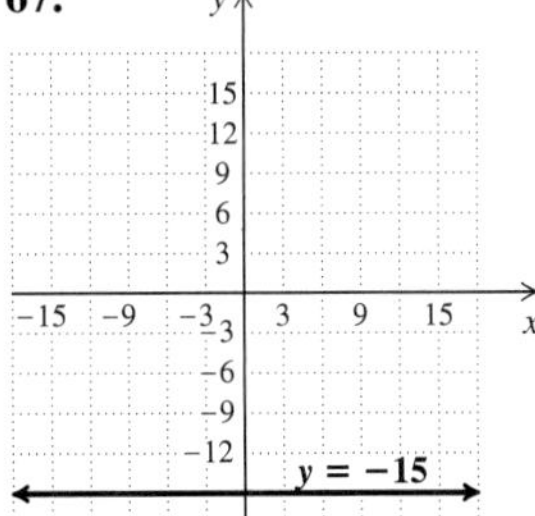

69.
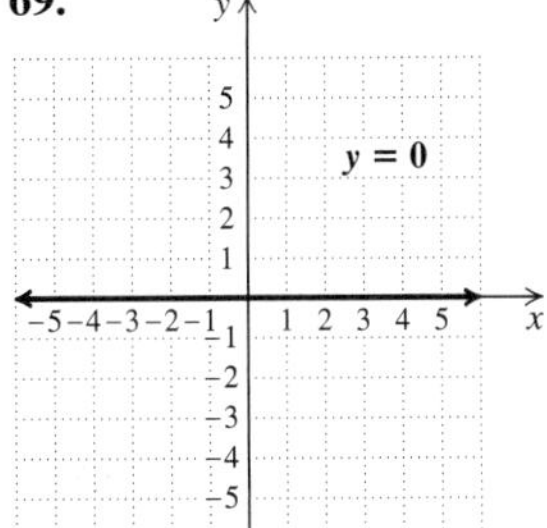

71.
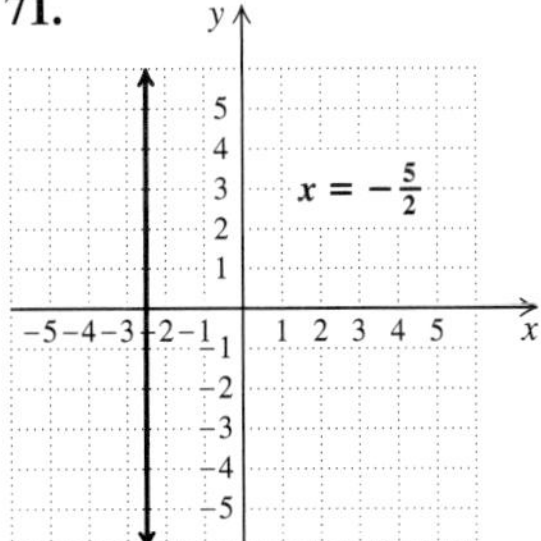

73.
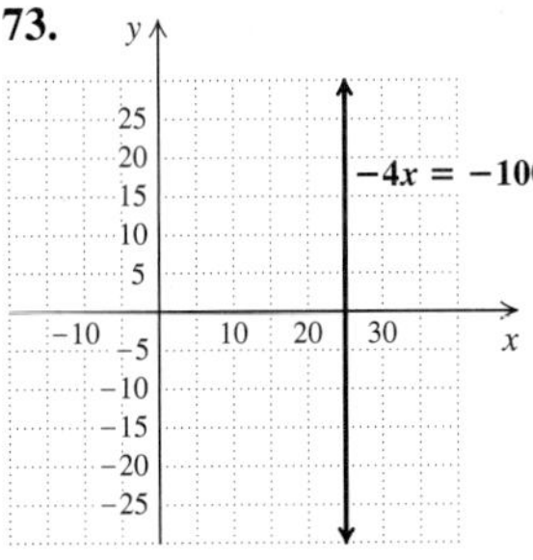

75.
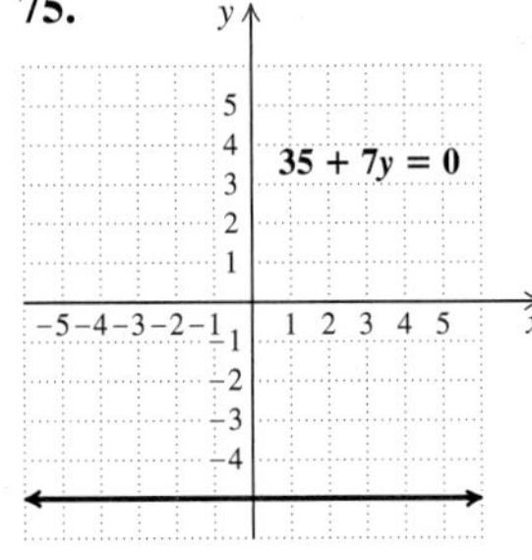

77. $y = -1$ **79.** $x = 4$
81. $y = 0$
83. $(5, 0)$, $(0, 20)$; (c)
85. $(200, 0)$, $(0, 7000)$; (d)

87. $y = -0.72x - 15$

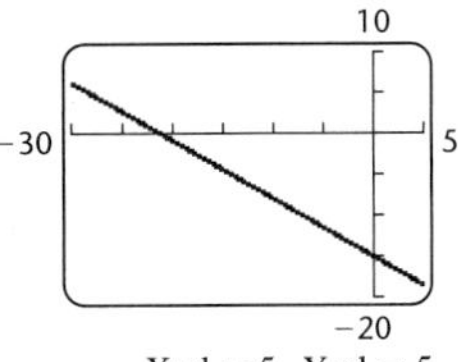

89. $5x + 6y = 84$, or $y = (84 - 5x)/6$

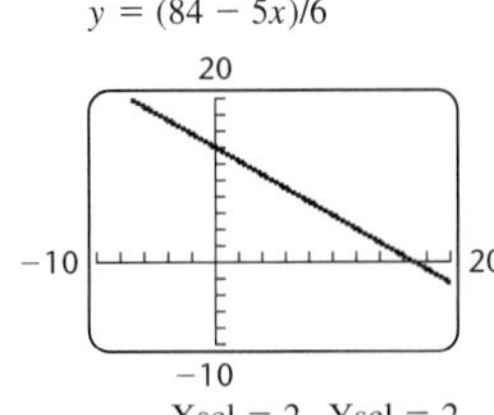

91. $19x - 17y = 200$, or $y = (19x - 200)/17$

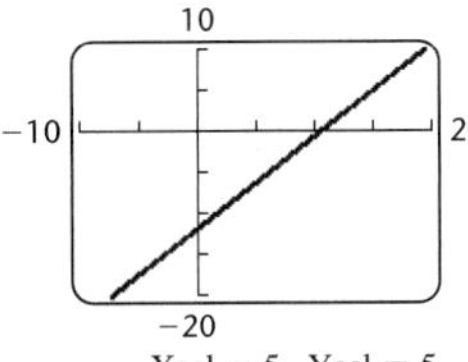

93. TW **95.** $d - 7$ **96.** $5 + w$, or $w + 5$ **97.** Let n represent the number; $7 + 4n$, or $4n + 7$ **98.** Let n represent the number; $3n$ **99.** Let x and y represent the numbers; $2(x + y)$ **100.** Let a and b represent the numbers; $\frac{1}{2}(a + b)$ **101.** TW **103.** $y = 0$ **105.** $x = -2$ **107.** $(-3, -3)$ **109.** $-5x + 3y = 15$, or $y = \frac{5}{3}x + 5$ **111.** -24

Exercise Set 3.4, pp. 197–201

1. Miles per hour, or $\frac{\text{miles}}{\text{hour}}$

2. Hours per chapter, or $\frac{\text{hours}}{\text{chapter}}$

3. Dollars per mile, or $\frac{\text{dollars}}{\text{mile}}$

4. Petunias per foot, or $\frac{\text{petunias}}{\text{foot}}$

5. Minutes per errand, or $\frac{\text{minutes}}{\text{errand}}$

6. Cups of flour per cake, or $\frac{\text{cups of flour}}{\text{cake}}$

7. **(a)** 21 mpg; **(b)** \$39.33/day; **(c)** 91 mi/day; **(d)** 43¢/mile **9.** **(a)** 6 mph; **(b)** \$4/hr; **(c)** \$0.67/mi **11.** **(a)** \$22/hr; **(b)** 20.6 pages/hr; **(c)** \$1.07/page **13.** \$13.33/month **15.** **(a)** 14.5 floors/min; **(b)** 4.14 sec/floor **17.** **(a)** 23.43 ft/min; **(b)** 0.04 min/ft

19.
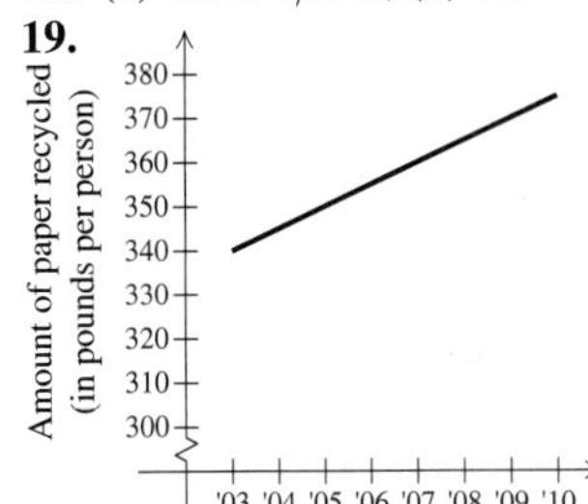

21.
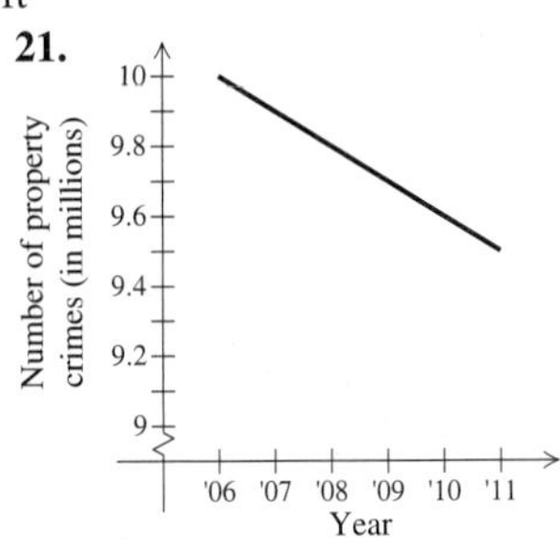

23.

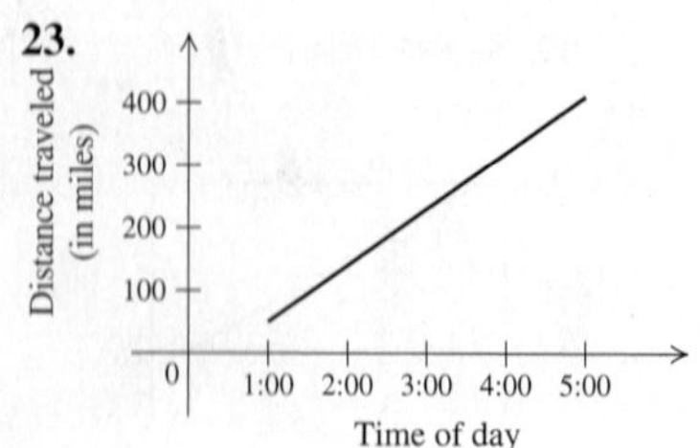

25.

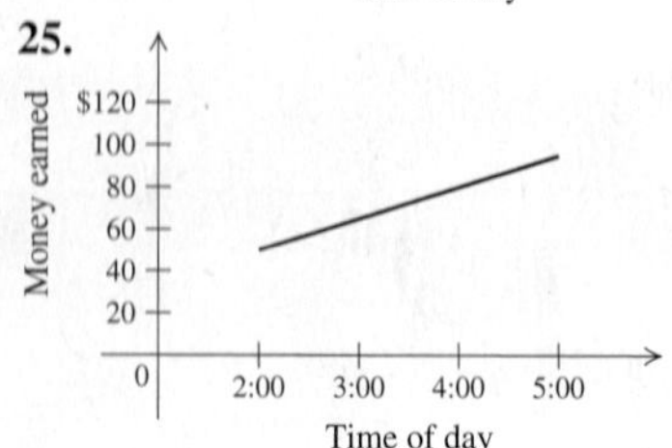

27.

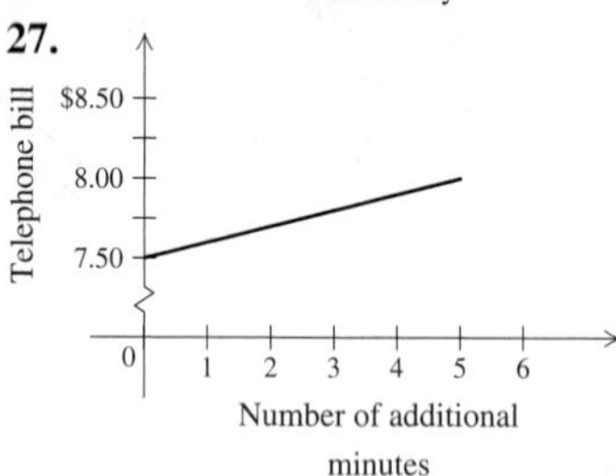

29. 20 calls/hr **31.** 75 mi/hr **33.** 12¢/min
35. -2000 people/yr **37.** 0.025 gal/min **39.** (e) **41.** (d)
43. (b) **45.** TW **47.** 5 **48.** -6 **49.** -1 **50.** $-\frac{4}{3}$
51. $-\frac{4}{3}$ **52.** $-\frac{4}{5}$ **53.** 0 **54.** Undefined **55.** TW

57.

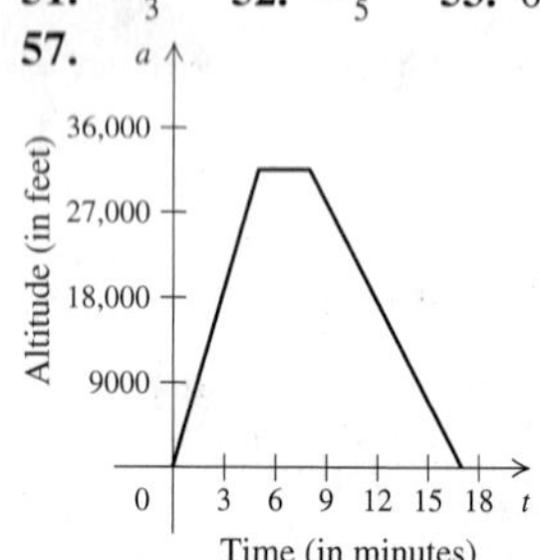

59.

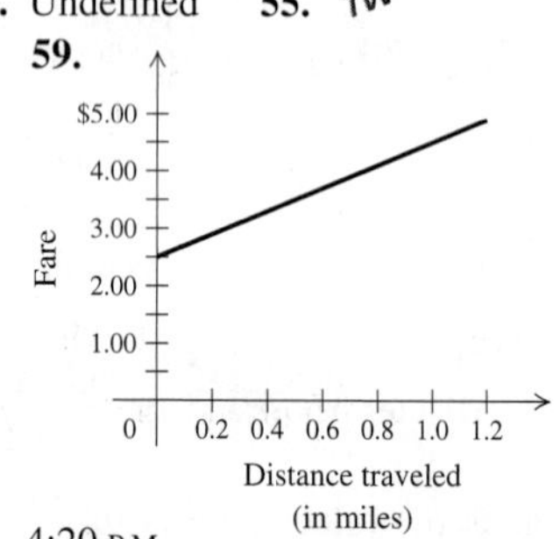

61. **63.** About 41.7 min **65.** 4:20 P.M.

Exercise Set 3.5, pp. 207–213

1. Positive **2.** Negative **3.** Negative **4.** Positive
5. Positive **6.** Negative **7.** Zero **8.** Positive
9. Negative **10.** Zero **11.** 3 million people/yr
13. -0.0075 min/yr, or -0.45 sec/yr **15.** $\frac{3}{4}$ point/$1000 income
17. About $-2.1°$/min **19.** $\frac{3}{4}$ **21.** $\frac{1}{3}$ **23.** -1 **25.** 0 **27.** -2
29. Undefined **31.** $-\frac{1}{4}$ **33.** 5 **35.** $\frac{2}{5}$ **37.** $\frac{3}{2}$ **39.** $-\frac{4}{5}$
41. $\frac{1}{2}$ **43.** -1 **45.** $-\frac{1}{2}$ **47.** 0 **49.** 1 **51.** Undefined
53. 0 **55.** Undefined **57.** Undefined **59.** 0 **61.** 8%
63. 5% **65.** $\frac{29}{98}$, or about 30% **67.** About 5.1%; yes
69. **(a)** II; **(b)** IV; **(c)** I; **(d)** III **71.** TW **73.** $y = \frac{c - ax}{b}$
74. $r = \frac{p + mn}{x}$ **75.** $y = \frac{ax - c}{b}$ **76.** $t = \frac{q - rs}{n}$

77.

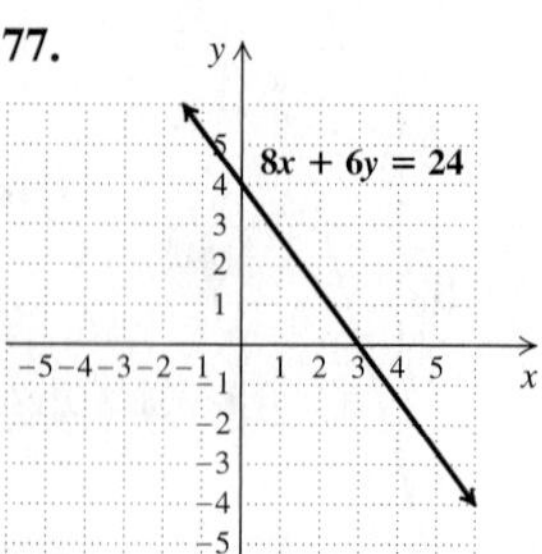

78.

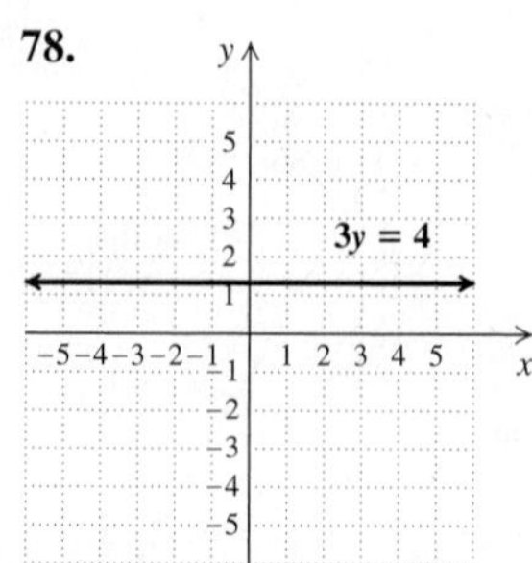

79. TW **81.** $\{m|-\frac{7}{4} \le m \le 0\}$ **83.** $\frac{18 - x}{x}$ **85.** $\frac{1}{4}$
87. 0.364, or 36.4%

Mid-Chapter Review: Chapter 3, p. 214

Guided Solutions

1. *y-intercept*: $y - 3 \cdot 0 = 6$
$y = 6$
The y-intercept is $(0, 6)$.
x-intercept: $0 - 3x = 6$
$-3x = 6$
$x = -2$
The x-intercept is $(-2, 0)$.

2. $m = \frac{y_2 - y_1}{x_2 - x_1} = \frac{-1 - 5}{3 - 1} = \frac{-6}{2} = -3$

Mixed Review

1.

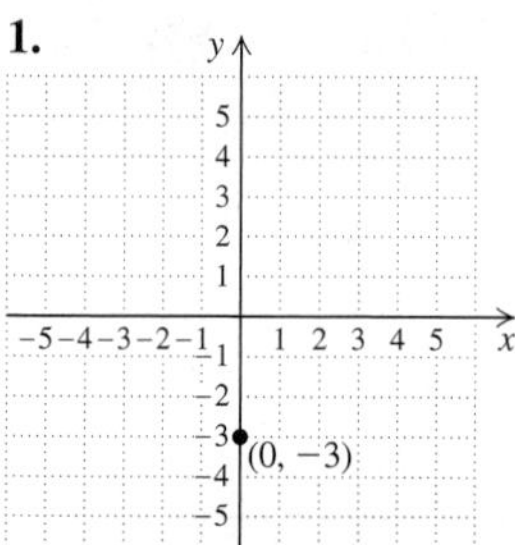

2. IV **3.** No

4.

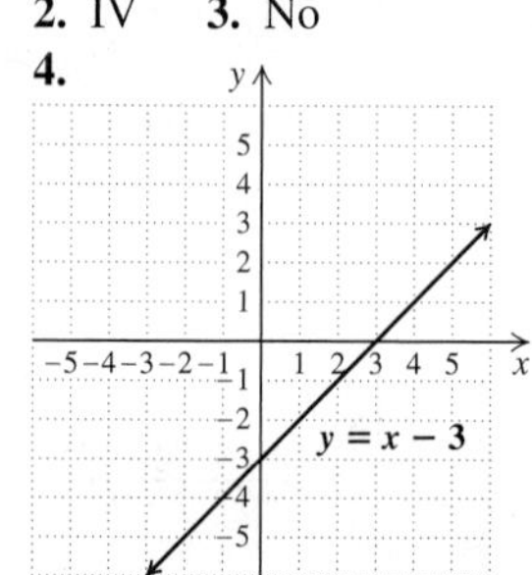

5.

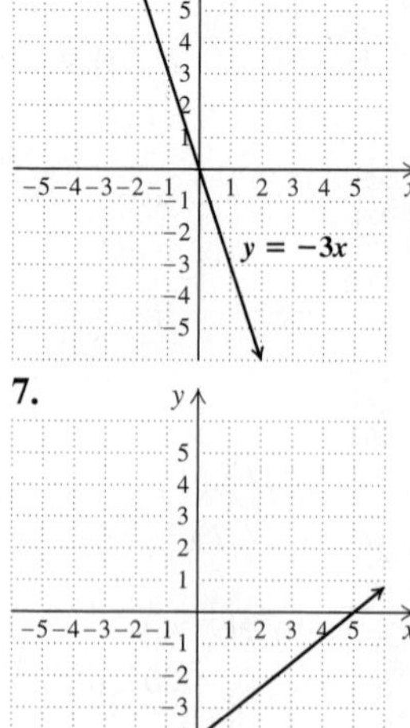

6.

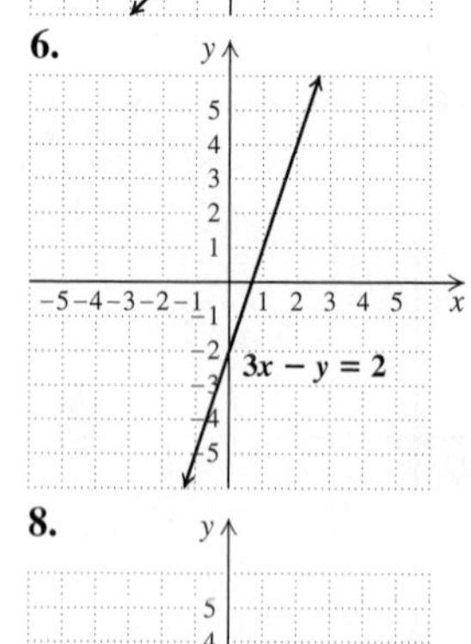

7.

8.

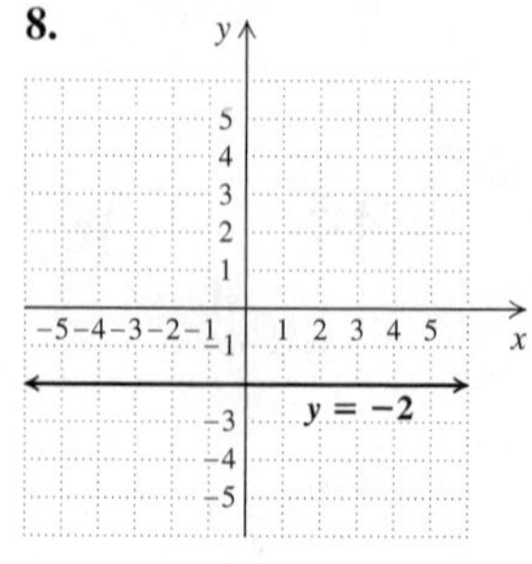

9.

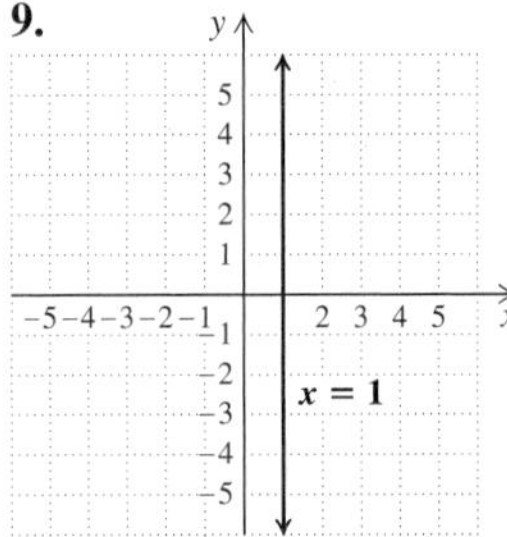

10. $y = 2x^2 + x$

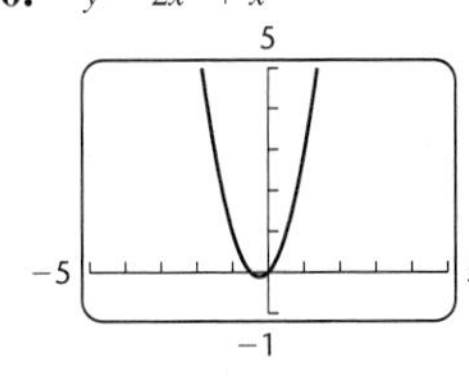

11. Linear 12. Nonlinear 13. 14 homes/month
14. About 29% 15. $\frac{5}{3}$ 16. -3 17. Undefined
18. 0 19. Undefined 20. $(-4, 0)$; $(0, 6)$

Interactive Discovery, p. 216

1. The slope changes as the coefficient of the x-term changes. The y-intercept does not change. 2. The y-intercept changes as the constant term changes. The slope does not change.

Exercise Set 3.6, pp. 223–226

1. (f) 2. (b) 3. (d) 4. (c) 5. (e) 6. (a)

7.

9.

11.

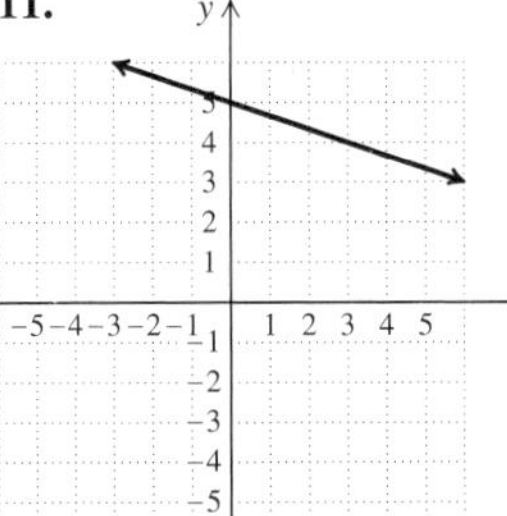

13.

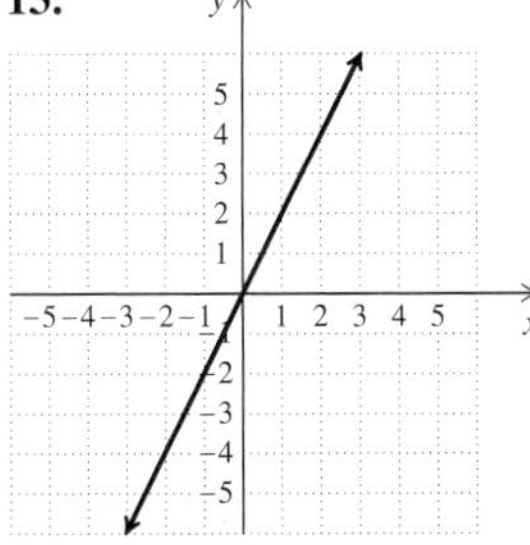

15.

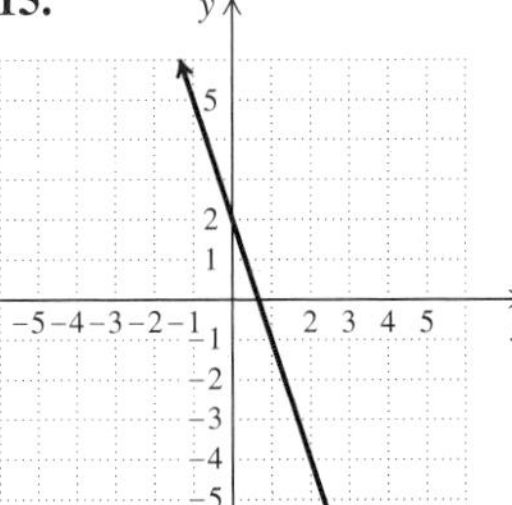

17.

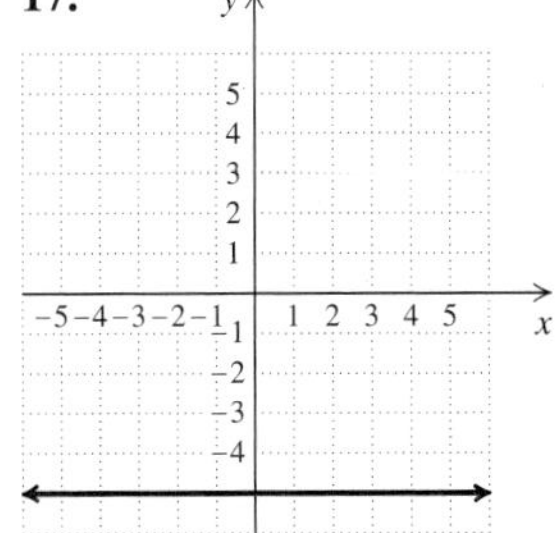

19. $-\frac{2}{7}$; $(0, 5)$ 21. $\frac{1}{3}$; $(0, 7)$ 23. $\frac{9}{5}$; $(0, -4)$ 25. 3; $(0, 7)$
27. -2; $(0, 4)$ 29. 0; $(0, 4)$ 31. $\frac{2}{5}$; $\left(0, \frac{8}{5}\right)$ 33. $\frac{9}{8}$; $(0, 0)$
35. (a) II; (b) IV; (c) III; (d) I 37. $y = 3x + 7$
39. $y = \frac{7}{8}x - 1$ 41. $y = -\frac{5}{3}x - 8$ 43. $y = \frac{1}{3}$
45. $y = 3x + 70$, where y is the number of jobs, in thousands, and x is the number of years since 2006 47. $y = 0.0375x + 2.2$, where y is the number of registered nurses, in millions, and x is the number of years since 2000

49.

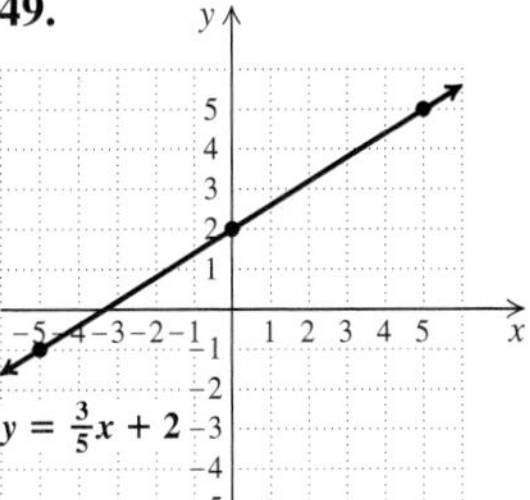

51.

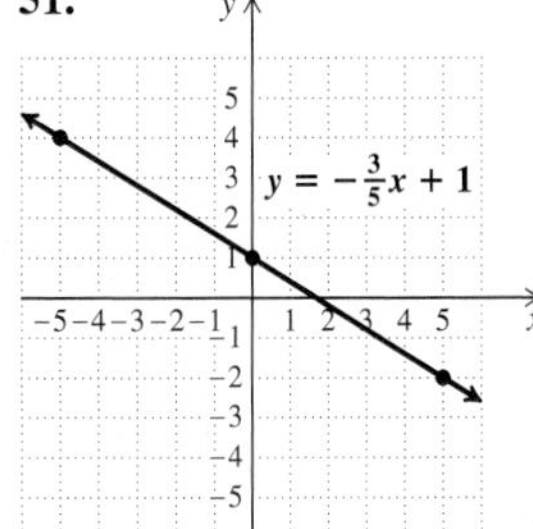

53.

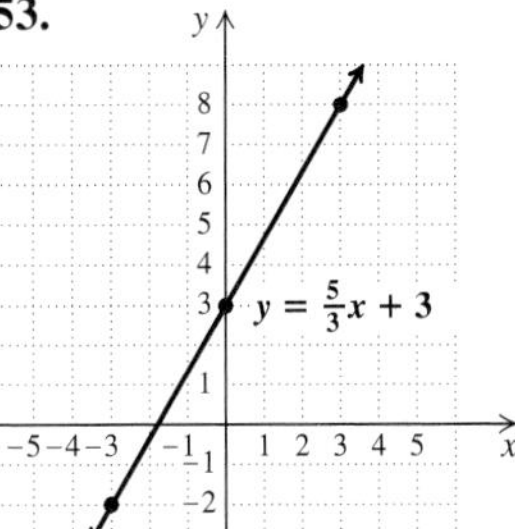

55.

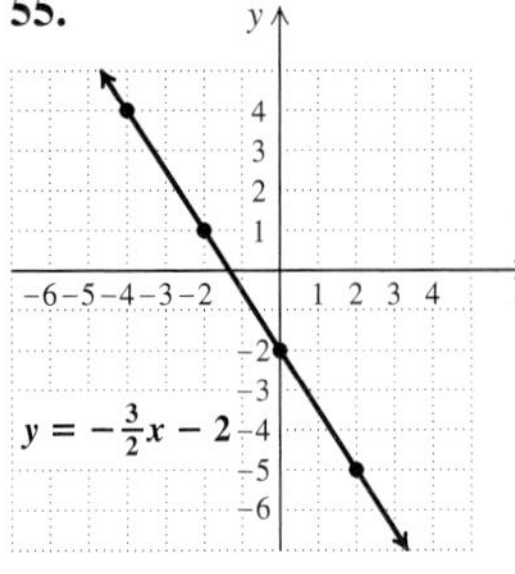

57.

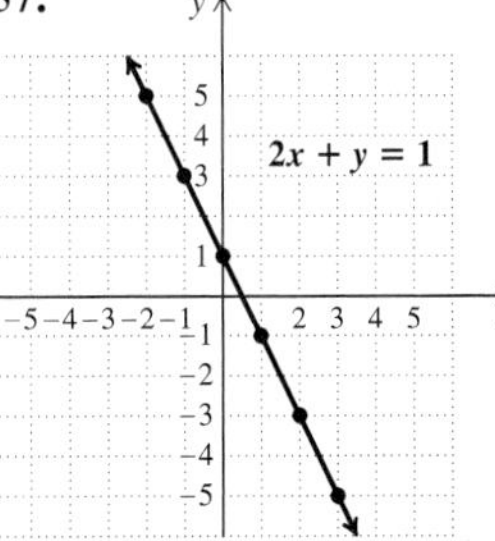

59.

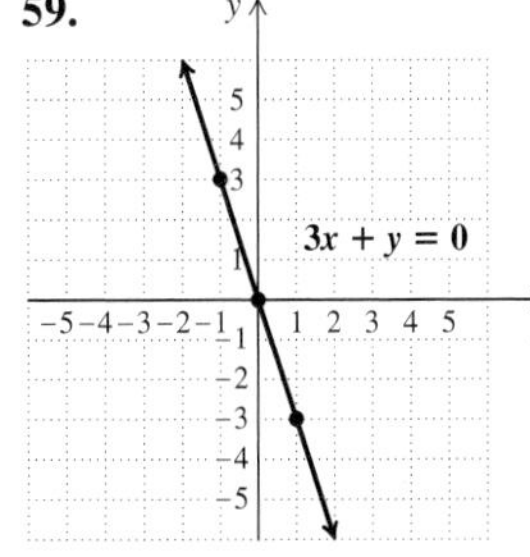

61.

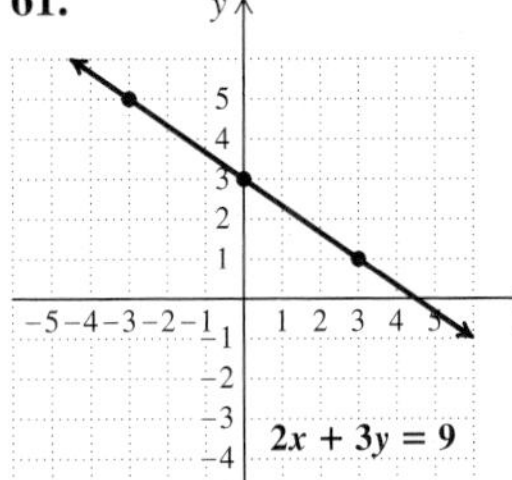

63.

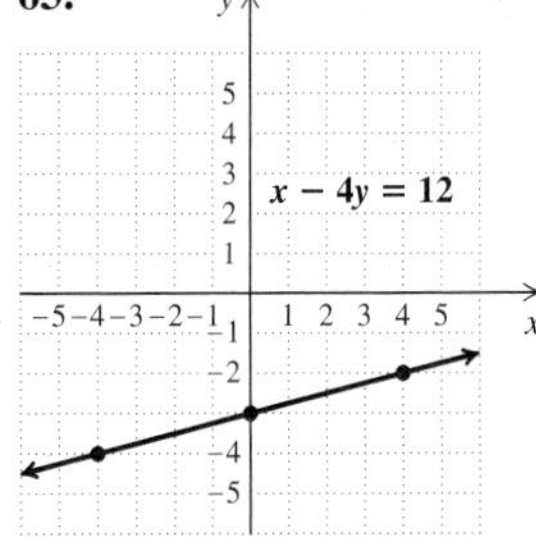

65. Yes 67. No 69. Yes 71. Yes 73. No 75. Yes
77. (a) $\frac{7}{8}$; (b) $-\frac{8}{7}$ 79. (a) $-\frac{1}{4}$; (b) 4 81. (a) 20; (b) $-\frac{1}{20}$
83. (a) -1; (b) 1 85. $y = 5x + 11$ 87. $y = \frac{1}{2}x$
89. $y = x + 3$ 91. $y = x - 4$ 93. TW
95. $y = m(x - h) + k$ 96. $y = -2(x + 4) + 9$ 97. -7
98. 13 99. -9 100. -11 101. TW 103. When $x = 0$, $y = b$, so $(0, b)$ is on the line. When $x = 1$, $y = m + b$, so $(1, m + b)$ is on the line. Then

$$\text{slope} = \frac{(m + b) - b}{1 - 0} = m.$$

105. $y = 1.5x + 16$ 107. Slope: $-\frac{r}{p}$; y-intercept: $\left(0, \frac{s}{p}\right)$

109. TW

Visualizing for Success, p. 237

1. C 2. G 3. F 4. B 5. D 6. A 7. I 8. H
9. J 10. E

Exercise Set 3.7, pp. 238–243

1. (g) **2.** (d) **3.** (e) **4.** (a) **5.** (b) **6.** (h) **7.** (f) **8.** (c) **9.** (c) **10.** (b) **11.** (d) **12.** (a) **13.** $y - 2 = 5(x - 6)$ **15.** $y - 1 = -4(x - 3)$ **17.** $y - (-4) = \frac{3}{2}(x - 5)$ **19.** $y - 6 = -\frac{5}{4}(x - (-2))$ **21.** $y - (-1) = -2(x - (-4))$ **23.** $y - 8 = 1(x - (-2))$ **25.** $\frac{2}{7}$; $(8, 9)$ **27.** -5; $(7, -2)$ **29.** $-\frac{5}{3}$; $(-2, 4)$ **31.** $\frac{4}{7}$; $(0, 0)$ **33.** $y = 2x - 3$ **35.** $y = \frac{7}{4}x - 9$ **37.** $y = -3x + 3$ **39.** $y = -4x - 9$ **41.** $y = -\frac{5}{6}x + 4$ **43.** $y = -\frac{1}{2}x + 9$ **45.** $y = 2x - 7$ **47.** $y = \frac{5}{3}x - \frac{28}{3}$ **49.** $x = 5$ **51.** $y = -2x + 4$ **53.** $y = x + 6$ **55.** $y = -\frac{1}{2}x + 6$ **57.** $x = -3$ **59.** $y = -x + 6$ **61.** $y = \frac{2}{3}x + 3$ **63.** $y = \frac{2}{5}x - 2$ **65.** $y = \frac{3}{4}x - \frac{5}{2}$

67.

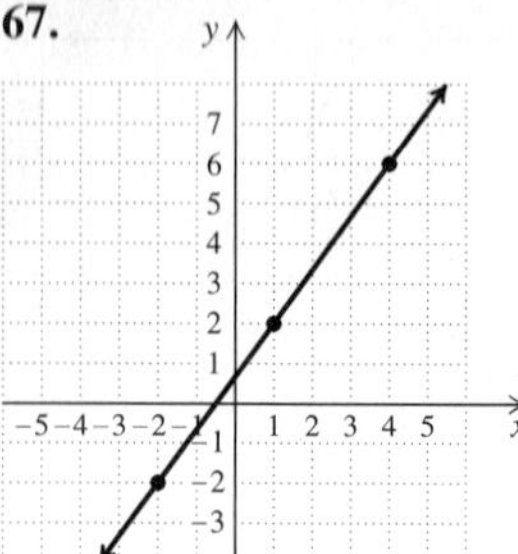
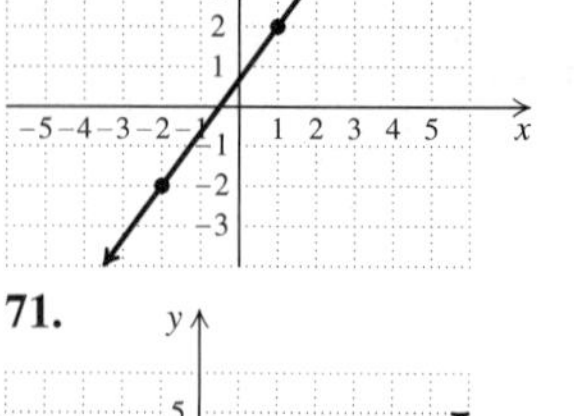

69.

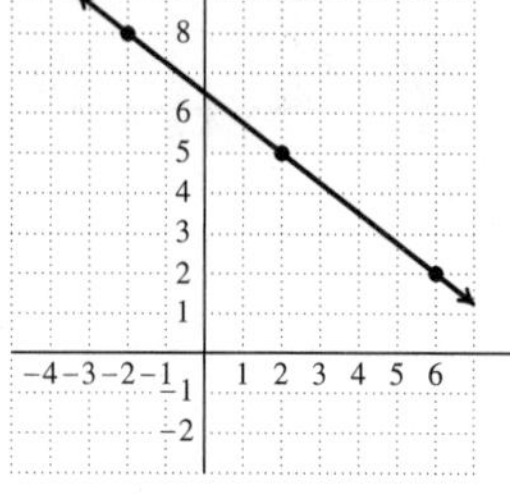

71.

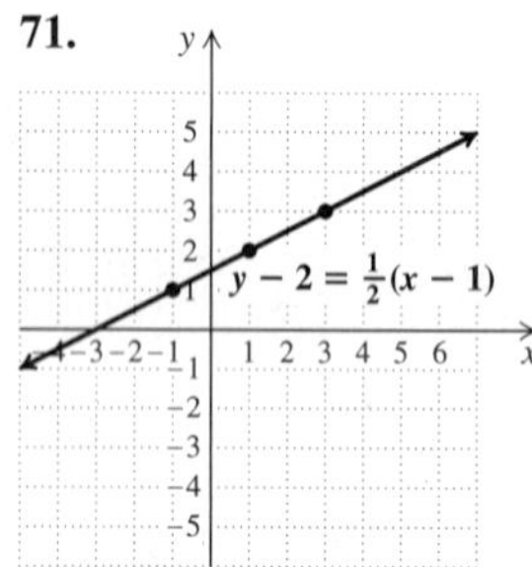

73.

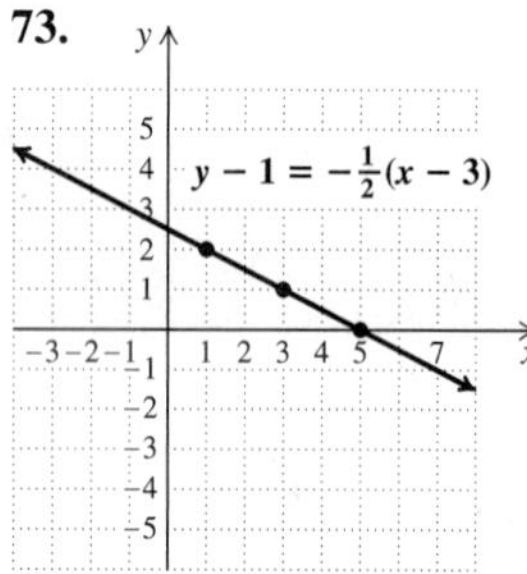

75.

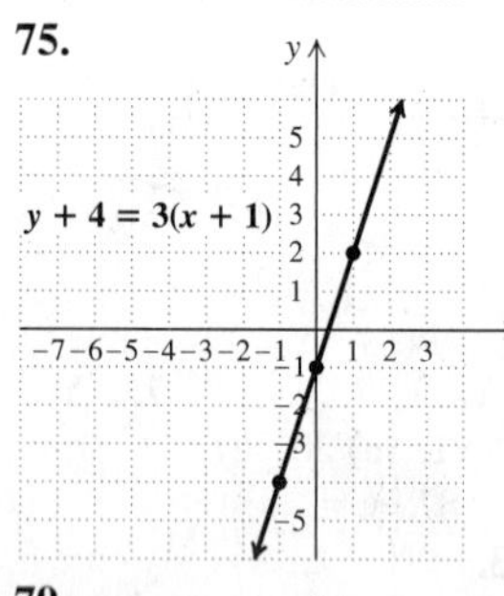

77.

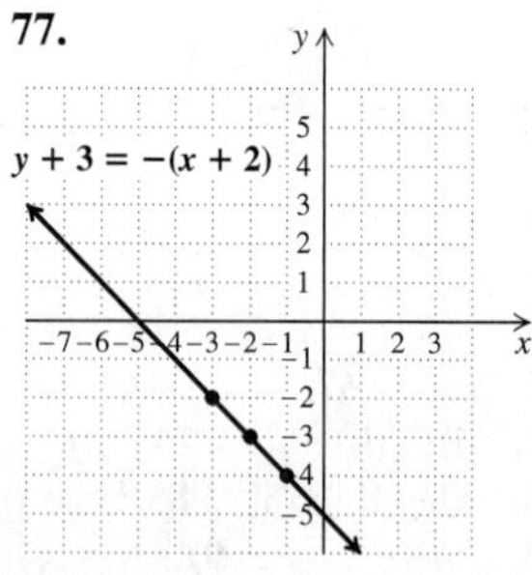

79.

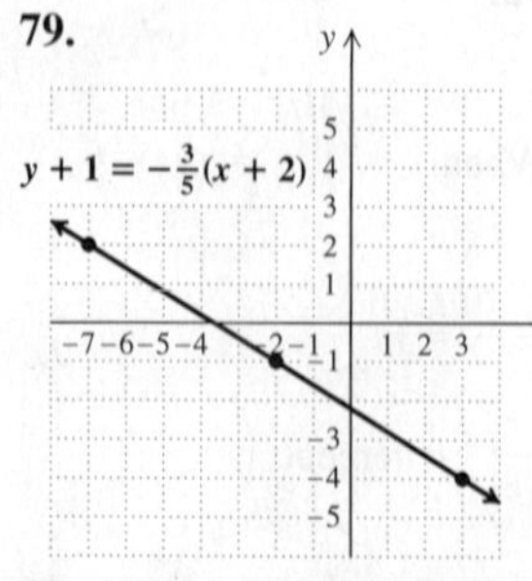

81. **(a)** $y = 37.85x + 354.65$; **(b)** 430.35 million students; **(c)** 619.6 million students **83.** **(a)** $A = 0.0857t + 78.2$; **(b)** about 78.5 million acres; **(c)** about 79.1 million acres **85.** **(a)** $C = 9.5t + 28.5$; **(b)** 85.5%; **(c)** by 2014

87. **(a)** $E = 0.1077t + 78.4769$; **(b)** 80.6 yr **89.** **(a)** $N = 15.43t + 217.8$; **(b)** \$433.82 million **91.** Linear **93.** Not linear **95.** Linear **97.** **(a)** $W = 0.1611x + 63.6983$; **(b)** 81.4; this estimate is 0.8 yr higher **99.** **(a)** $N = 0.0433t + 2.1678$; **(b)** 2.69 million registered nurses **101.** TW

103. -17 **104.** 5 **105.** 6 **106.** 0 **107.** $-\frac{1}{8}$ **108.** 0 **109.** TW **111.**

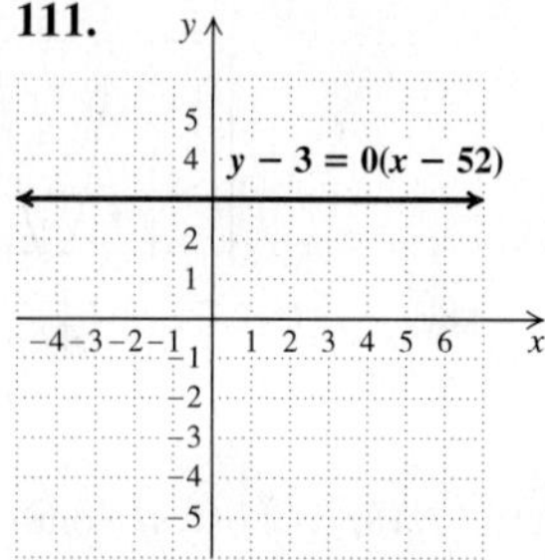

113. $y = 2x - 9$ **115.** $y = \frac{1}{5}x - 2$ **117.** $y = \frac{1}{2}x + 1$ **119.** 21.1°C **121.** \$11,000

Exercise Set 3.8, pp. 255–260

1. Correspondence **2.** Exactly **3.** Domain **4.** Range **5.** Horizontal **6.** Vertical **7.** "f of 3," "f at 3," or "the value of f at 3" **8.** Vertical **9.** Yes **11.** Yes **13.** No **15.** Yes **17.** Function **19.** Function **21.** **(a)** -2; **(b)** $\{x \mid -2 \le x \le 5\}$, or $[-2, 5]$; **(c)** 4; **(d)** $\{y \mid -3 \le y \le 4\}$, or $[-3, 4]$ **23.** **(a)** 3; **(b)** $\{x \mid -1 \le x \le 4\}$, or $[-1, 4]$; **(c)** 3; **(d)** $\{y \mid 1 \le y \le 4\}$, or $[1, 4]$ **25.** **(a)** -2; **(b)** $\{x \mid -4 \le x \le 2\}$, or $[-4, 2]$; **(c)** -2; **(d)** $\{y \mid -3 \le y \le 3\}$, or $[-3, 3]$ **27.** **(a)** 3; **(b)** $\{x \mid -4 \le x \le 3\}$, or $[-4, 3]$; **(c)** -3; **(d)** $\{y \mid -2 \le y \le 5\}$, or $[-2, 5]$ **29.** **(a)** 1; **(b)** $\{-3, -1, 1, 3, 5\}$; **(c)** 3; **(d)** $\{-1, 0, 1, 2, 3\}$ **31.** **(a)** 4; **(b)** $\{x \mid -3 \le x \le 4\}$, or $[-3, 4]$; **(c)** $-1, 3$; **(d)** $\{y \mid -4 \le y \le 5\}$, or $[-4, 5]$ **33.** **(a)** 1; **(b)** $\{x \mid -4 < x \le 5\}$, or $(-4, 5]$; **(c)** $\{x \mid 2 < x \le 5\}$, or $(2, 5]$; **(d)** $\{-1, 1, 2\}$ **35.** Domain: $\mathbb{R}$; range: $\mathbb{R}$ **37.** Domain: $\mathbb{R}$; range: $\{4\}$ **39.** Domain: $\mathbb{R}$; range: $\{y \mid y \ge 1\}$, or $[1, \infty)$ **41.** Domain: $\{x \mid x$ is a real number *and* $x \ne -2\}$; range: $\{y \mid y$ is a real number *and* $y \ne -4\}$ **43.** Domain: $\{x \mid x \ge 0\}$, or $[0, \infty)$; range: $\{y \mid y \ge 0\}$, or $[0, \infty)$ **45.** Yes **47.** Yes **49.** No **51.** No **53.** **(a)** 3; **(b)** -5; **(c)** -11; **(d)** 19; **(e)** $2a + 7$; **(f)** $2a + 5$ **55.** **(a)** 0; **(b)** 1; **(c)** 57; **(d)** $5t^2 + 4t$; **(e)** $20a^2 + 8a$; **(f)** $10a^2 + 8a$ **57.** **(a)** $\frac{3}{5}$; **(b)** $\frac{1}{3}$; **(c)** $\frac{4}{7}$; **(d)** 0; **(e)** $\frac{x - 1}{2x - 1}$ **59.** **(a)** 29; **(b)** 3.59 **61.** **(a)** 2.8; **(b)** 12.25 **63.** $4\sqrt{3}\text{ cm}^2 \approx 6.93\text{ cm}^2$ **65.** $36\pi\text{ in}^2 \approx 113.10\text{ in}^2$ **67.** $1\frac{20}{33}$ atm; $1\frac{10}{11}$ atm; $4\frac{1}{33}$ atm **69.** 11 **71.** 0 **73.** $-\frac{21}{2}$ **75.** $\frac{25}{6}$ **77.** -3 **79.** -25 **81.** $\{x \mid x$ is a real number *and* $x \ne 3\}$ **83.** $\{x \mid x$ is a real number *and* $x \ne \frac{1}{2}\}$ **85.** $\mathbb{R}$ **87.** $\mathbb{R}$ **89.** $\{x \mid x$ is a real number *and* $x \ne 9\}$ **91.** $\mathbb{R}$ **93.** $\mathbb{R}$ **95.** **(a)** -5; **(b)** 1; **(c)** 21 **97.** **(a)** 0; **(b)** 2; **(c)** 7 **99.** **(a)** 100; **(b)** 100; **(c)** 131 **101.** 25 signifies that the cost per person is \$25; 75 signifies that the setup cost for the party is \$75. **103.** $\frac{2}{3}$ signifies that consumption of renewable energy increases $\frac{2}{3}$ quadrillion Btu's per year, for years after 1960; $\frac{10}{3}$ signifies that the consumption was $\frac{10}{3}$ quadrillion Btu's in 1960. **105.** $\frac{1}{8}$ signifies that the grass grows $\frac{1}{8}$ in. per day; 2 signifies that the grass is 2 in. long when cut. **107.** 0.21 signifies that the average price of a movie ticket increases by \$0.21 per year, for years after 2000; 5.43 signifies that the average price was \$5.43 in 2000. **109.** **(a)** -5000 signifies that the depreciation is \$5000 per year; 90,000 signifies that the original value of the truck was \$90,000; **(b)** 18 yr; **(c)** $\{t \mid 0 \le t \le 18\}$

111. **(a)** 46.8 signifies that the record in the 400-m run was 46.8 sec in 1930; -0.075 signifies that the record decreases 0.075 sec per year, for years after 1930; **(b)** 108 yr after 1930, or in 2038; **(c)** $\{t \mid 0 \le t < 624\}$, assuming the record can never reach 0 **113.** TW **115.** -125 **116.** 64 **117.** -64 **118.** 8 **119.** 28 **120.** -4 **121.** TW **123.** 26; 99 **125.** Worm **127.** About 2 min 50 sec **129.** 1 every 3 min **131.** $g(x) = \frac{15}{4}x - \frac{13}{4}$ **133.** False **135.** False

Study Summary: Chapter 3, pp. 261–263

Practice Exercises

1.

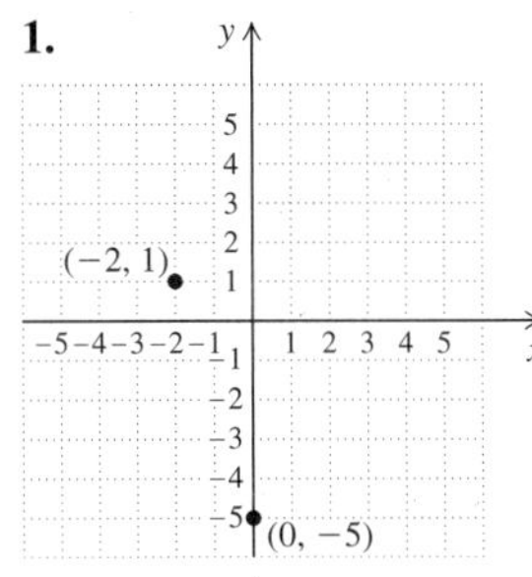

2. III

3.

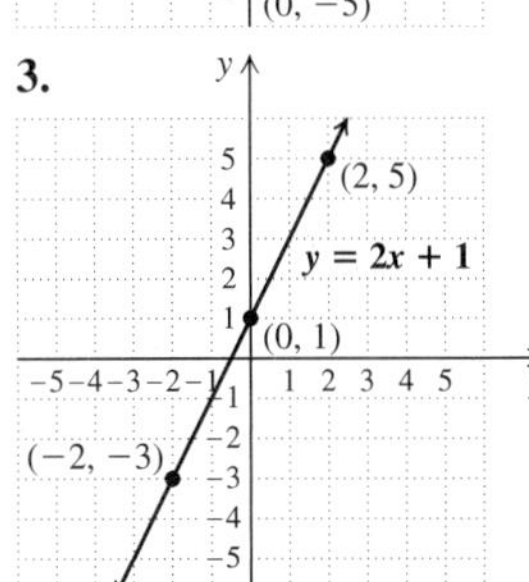

4. x-intercept: $(1, 0)$; y-intercept: $(0, -10)$

5.

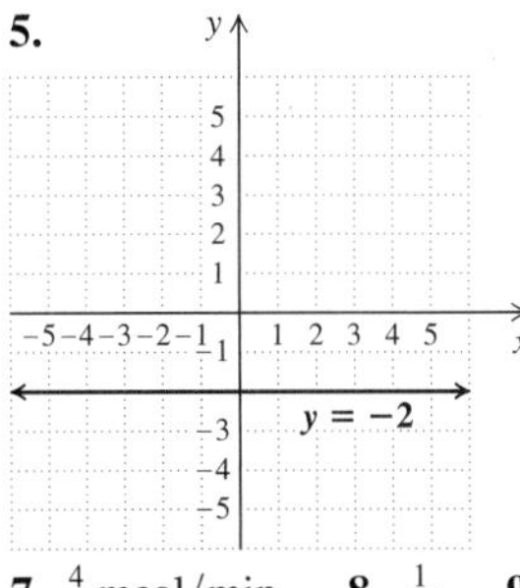

6. $x = 3$

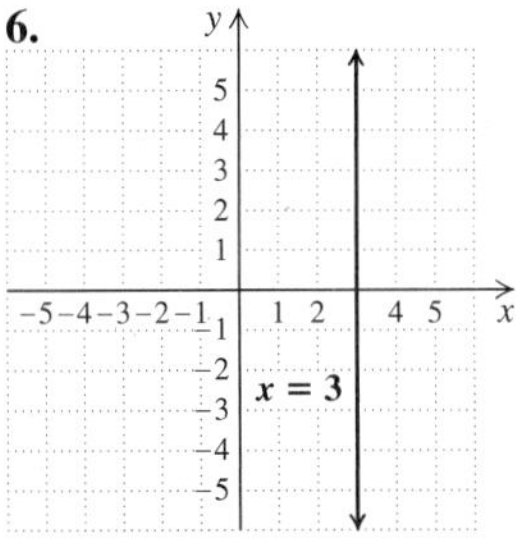

7. $\frac{4}{9}$ meal/min **8.** $\frac{1}{10}$ **9.** 0 **10.** Slope: -4; y-intercept: $\left(0, \frac{2}{5}\right)$

11.

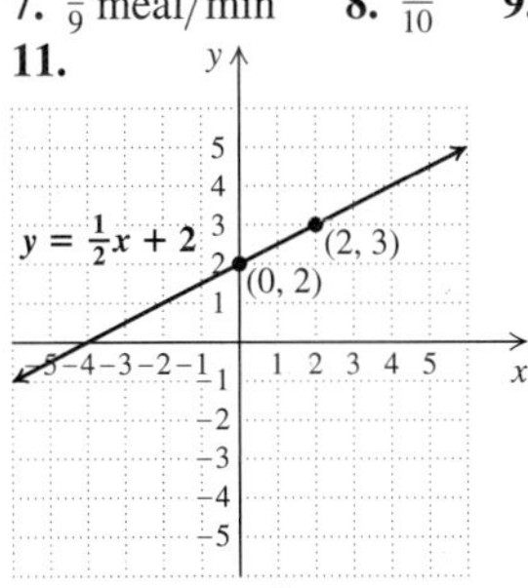

12. No **13.** Yes

14. $y - 6 = \frac{1}{4}(x - (-1))$ **15.** 5 **16.** Yes **17.** Domain: $\mathbb{R}$; range: $\{y \mid y \ge -2\}$, or $[-2, \infty)$ **18.** $\mathbb{R}$

Review Exercises: Chapter 3, pp. 264–267

1. False **2.** True **3.** False **4.** False **5.** True **6.** True **7.** True **8.** False **9.** True **10.** True **11.** About 1.7 billion searches **12.** About 1492 searches

13.–15. (2, 3); (−4, 0); (5, −1)

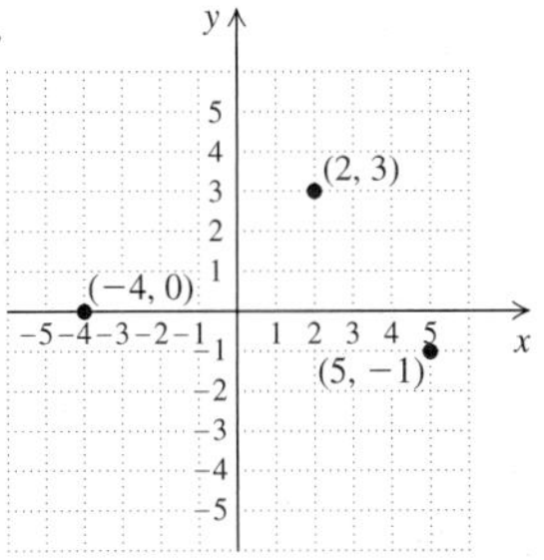

16. IV **17.** III

18. x-axis **19.** $(-5, -1)$ **20.** $(-2, 5)$ **21.** $(3, 0)$

22.

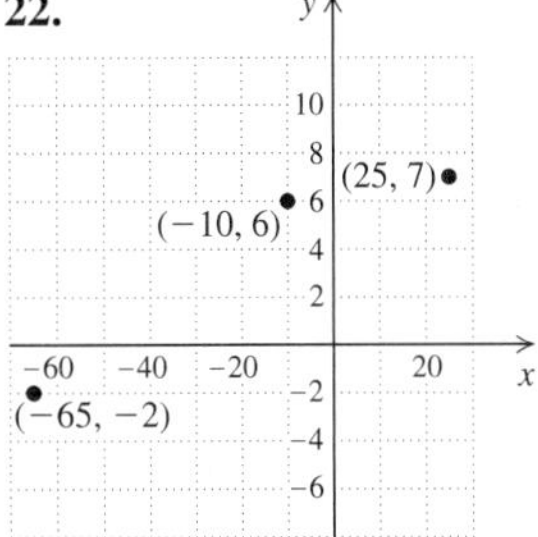

23. **(a)** Yes; **(b)** no

24.

$2x - y = 3$	
$2 \cdot 0 - (-3)$	3
$0 + 3$	
$3 \stackrel{?}{=} 3$	True

$2x - y = 3$	
$2 \cdot 2 - 1$	3
$4 - 1$	
$3 \stackrel{?}{=} 3$	True

$(-1, -5)$; answers may vary

25.

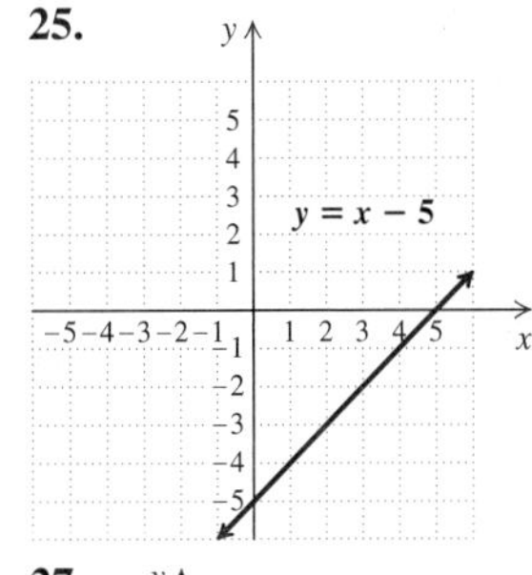

26. $y = -\frac{1}{4}x$

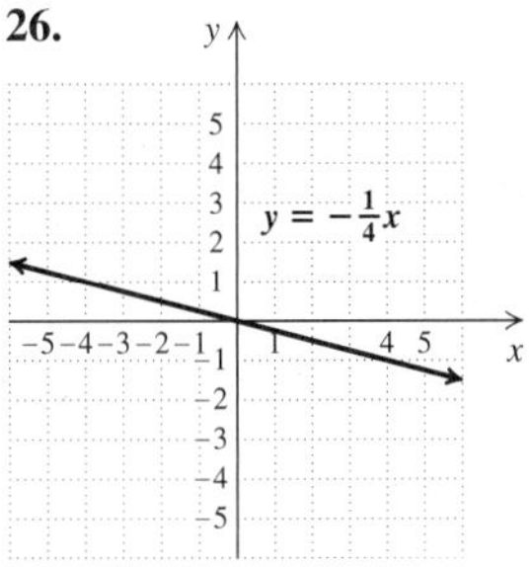

27.

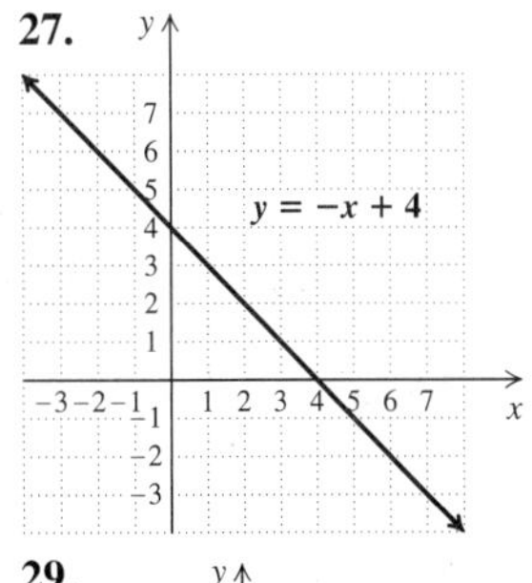

28.

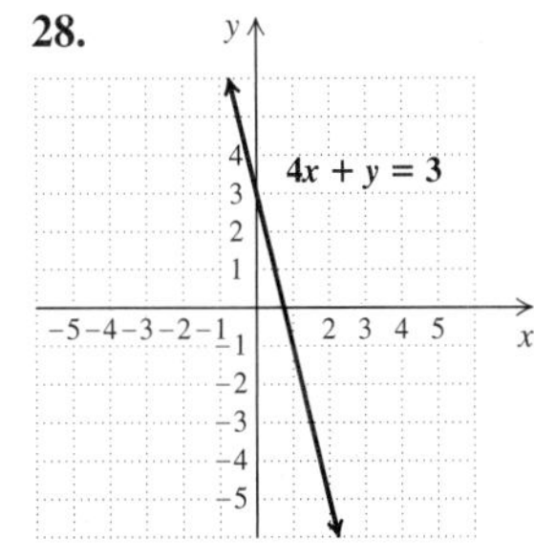

29.

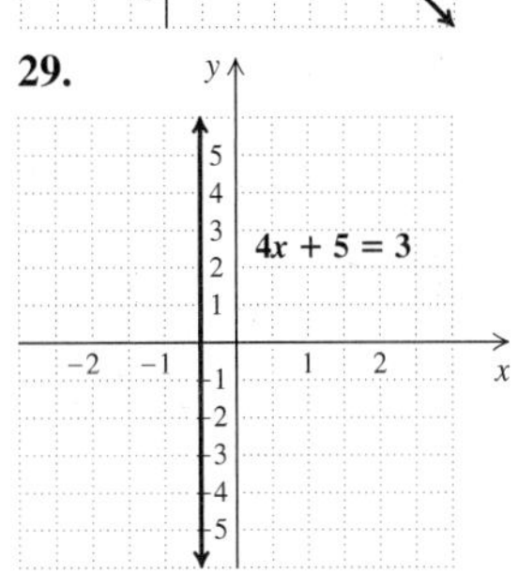

30.

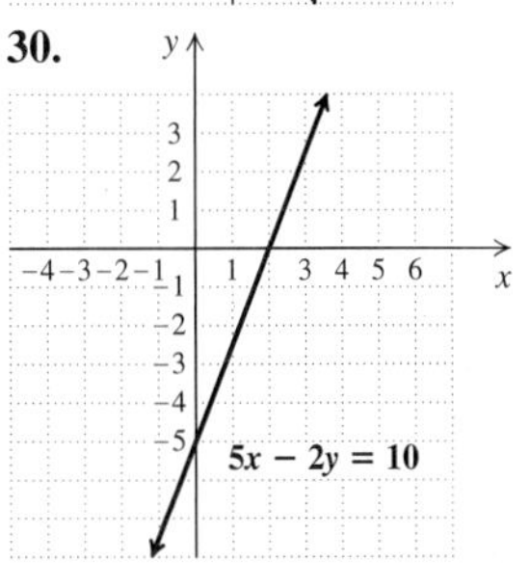

31. $y = x^2 + 1$

10

−10 10

−10

32.

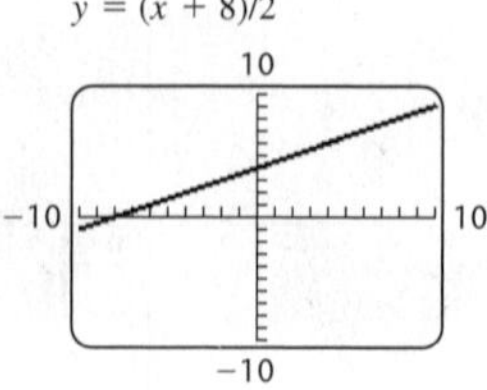

19 million households

33.

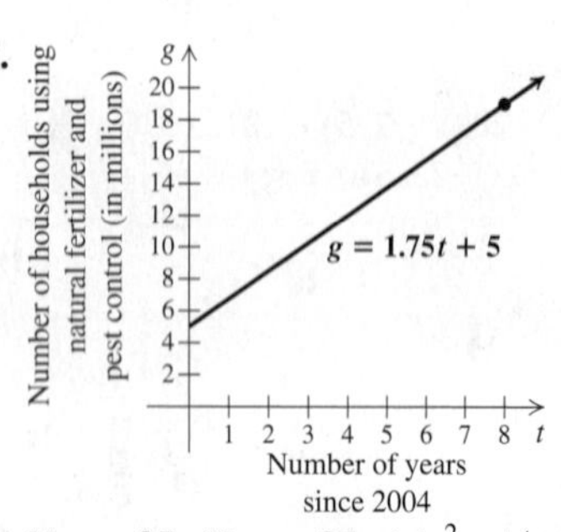

34. No **35.** Yes **36.** **(a)** $\frac{2}{15}$ mi/min; **(b)** 7.5 min/mi **37.** $\frac{1}{12}$ gal/mi **38.** **(a)** $(0, -2)$; **(b)** none; **(c)** 0 **39.** **(a)** $(0, 2)$; **(b)** $(4, 0)$; **(c)** $-\frac{1}{2}$ **40.** **(a)** $(0, -3)$; **(b)** $(2, 0)$; **(c)** $\frac{3}{2}$ **41.** $-\frac{6}{5}$ **42.** 0 **43.** Undefined **44.** 2 **45.** $8.\overline{3}\%$ **46.** $-\frac{7}{10}$ **47.** 0 **48.** Undefined **49.** $\frac{3}{2}$ **50.** x-intercept: $(6, 0)$; y-intercept: $(0, -30)$ **51.** $-\frac{1}{2}$; $(0, 5)$ **52.** Perpendicular **53.** Parallel **54.** $y = -\frac{3}{4}x + 6$ **55.** $y - 6 = -\frac{1}{2}(x - 3)$ **56.** $y = \frac{5}{4}x - \frac{13}{4}$ **57.** $y = -\frac{5}{3}x - \frac{5}{3}$ **58.** **(a)** $a = 0.2t + 10.8$; **(b)** 11.6 million; **(c)** 13.2 million

59.

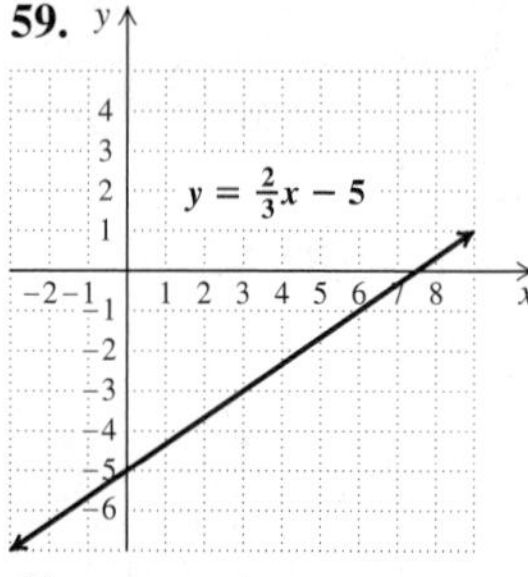

60.

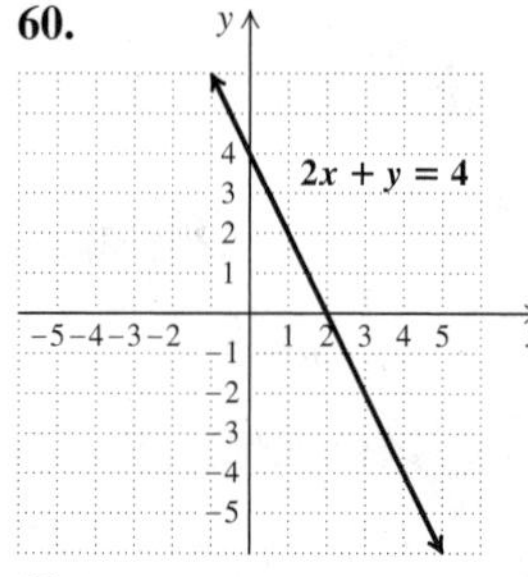

61.

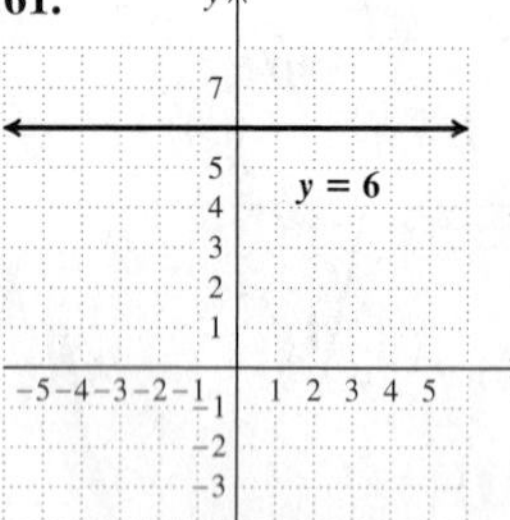

62.

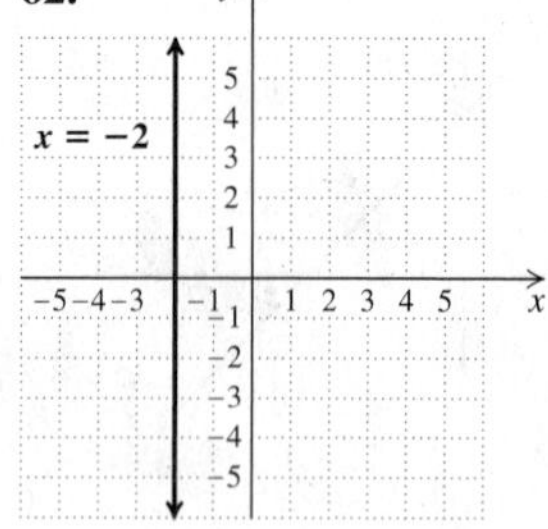

63.

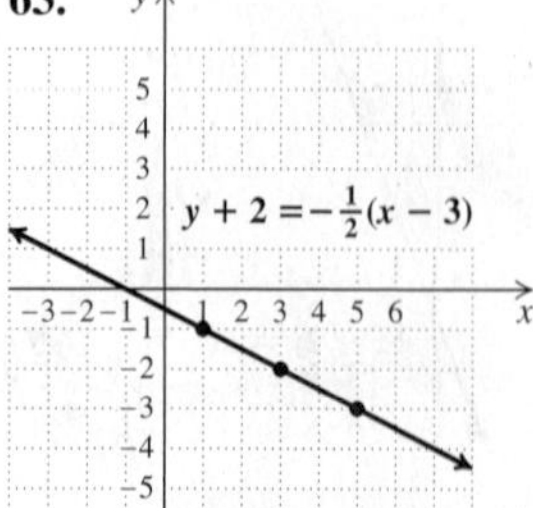

64. Linear

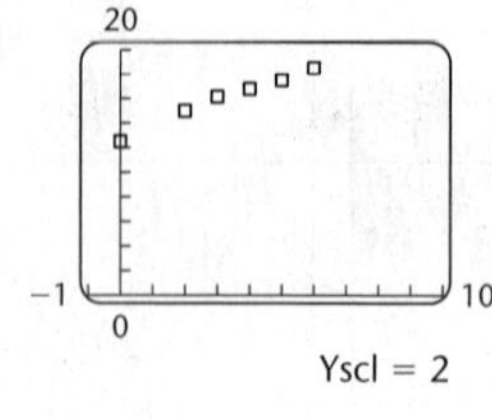

65. $F = 0.96t + 12.85$ **66.** About \$24.4 billion **67.** **(a)** 3; **(b)** $\{x|-2 \leq x \leq 4\}$, or $[-2, 4]$; **(c)** -1; **(d)** $\{y|1 \leq y \leq 5\}$, or $[1, 5]$ **68.** 39 **69.** $4a^2 + 4a - 3$

70. **(a)** 10.1 yr; **(b)** 0.11 signifies that the median age of cars increases 0.11 yr per year; 7.9 signifies that the median age of cars was 7.9 in 1995. **71.** **(a)** Yes; **(b)** domain: $\mathbb{R}$; range: $\{y|y \geq 0\}$, or $[0, \infty)$ **72.** **(a)** No **73.** **(a)** No **74.** **(a)** Yes; **(b)** domain: $\mathbb{R}$; range: $\{-2\}$ **75.** $\mathbb{R}$ **76.** $\{x|x$ is a real number *and* $x \neq 1\}$ **77.** **(a)** 5; **(b)** 4; **(c)** 16; **(d)** 35 **78.** TW Two perpendicular lines share the same y-intercept if their point of intersection is on the y-axis. **79.** TW Two functions that have the same domain and range are not necessarily identical. For example, the functions f: $\{(-2, 1), (-3, 2)\}$ and g: $\{(-2, 2), (-3, 1)\}$ have the same domain and range but are different functions. **80.** -1 **81.** 19 **82.** Area: 45 sq units; perimeter: 28 units **83.** $(0, 4)$, $(1, 3)$, $(-1, 3)$; answers may vary **84.** Domain: $\{x|x \geq -4$ *and* $x \neq 2\}$; range: $\{y|y \geq 0$ *and* $y \neq 3\}$

Test: Chapter 3, pp. 267–268

1. [3.1] About 83 students **2.** [3.1] About 114 students **3.** [3.1] y-axis **4.** [3.1] II **5.** [3.1] $(3, 4)$ **6.** [3.1] $(0, -4)$ **7.** [3.1] $(-5, 2)$

8. [3.2]

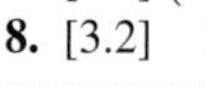

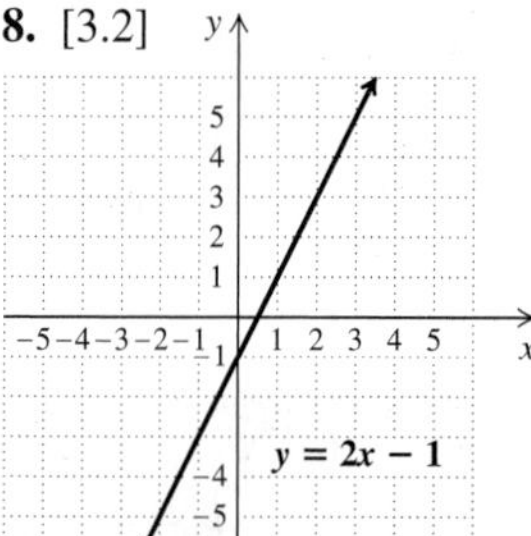

9. [3.3]

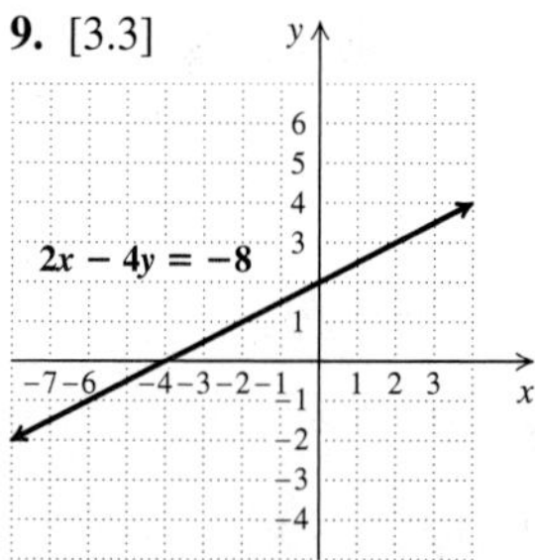

10. [3.7]

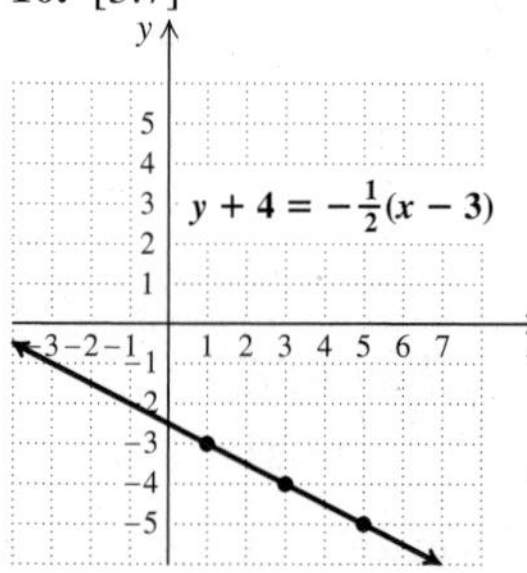

11. [3.2]

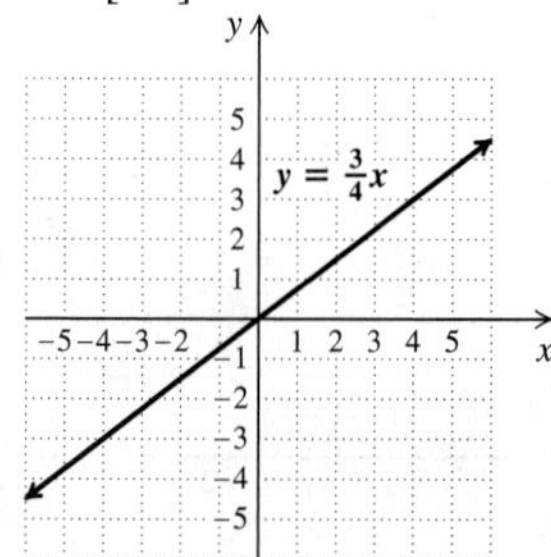

12. [3.2]

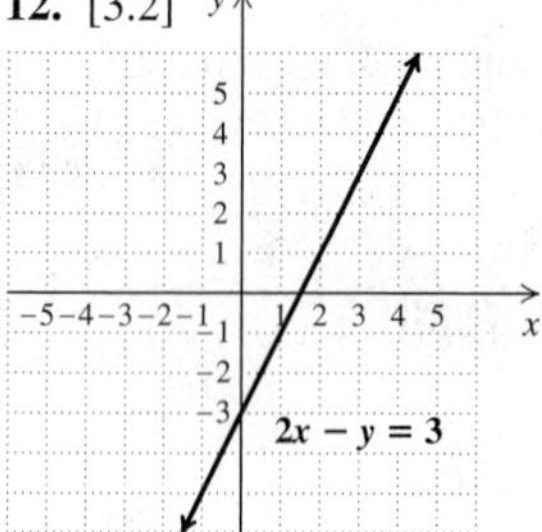

13. [3.3]

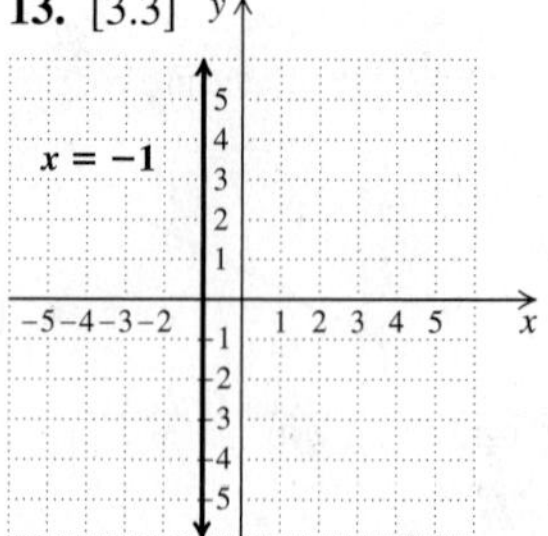

14. [3.2] $1.2x - y = 5$, or $y = 1.2x - 5$

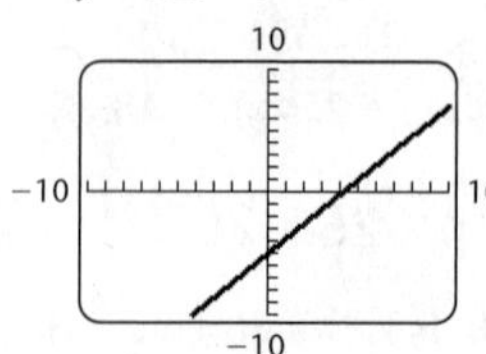

15. [3.3] x-intercept: $(16, 0)$; y-intercept: $(0, -8)$ **16.** [3.5] $\frac{9}{2}$ **17.** [3.4] $\frac{1}{3}$ km/min **18.** [3.5] 31.5% **19.** [3.6] 8; $(0, 10)$

20. [3.6] Parallel **21.** [3.6] Perpendicular
22. [3.7] $y - 8 = -3(x - 6)$ **23.** [3.7] $y = -x + 10$
24. [3.7] **(a)** $c = t + 16$; **(b)** 34 hr; **(c)** 46 hr
25. [3.7] $B = 2.6718t + 65.0212$
26. [3.7] Approximately 151,000 births **27.** [3.8] **(a)** Yes; **(b)** domain: $\{x \mid -4 \le x \le 5\}$, or $[-4, 5]$; range: $\{y \mid -2 \le y \le 4\}$, or $[-2, 4]$ **28.** [3.8] **(a)** Yes; **(b)** domain: $\mathbb{R}$; range: $\{y \mid y \ge 1\}$, or $[1, \infty)$ **29.** [3.8] **(a)** No
30. [3.8] **(a)** 0; **(b)** -4 **31.** [3.8] **(a)** $\frac{4}{3}$; **(b)** $\{x \mid x \text{ is a real number and } x \neq -\frac{1}{2}\}$ **32.** [3.8] **(a)** -5; **(b)** 10 **33.** [3.6] $y = \frac{2}{5}x + 9$
34. [3.1] Area: 25 sq units; perimeter: 20 units

Cumulative Review: Chapters 1–3, pp. 269–270

1. 7 **2.** $12a - 6b + 18$ **3.** $4(2x - y + 1)$ **4.** $2 \cdot 3^3$
5. -0.15 **6.** 37 **7.** $\frac{1}{10}$ **8.** -10 **9.** 0.367 **10.** $\frac{11}{60}$
11. 7.28 **12.** $-\frac{5}{12}$ **13.** -239 **14.** -3 **15.** 27
16. $-2y - 7$ **17.** $5x + 11$ **18.** -2.6 **19.** -27 **20.** 16
21. -6 **22.** 2 **23.** $\frac{7}{9}$ **24.** -17 **25.** 2 **26.** $\{x \mid x < 16\}$, or $(-\infty, 16)$ **27.** $\{x \mid x \le -\frac{11}{8}\}$, or $\left(-\infty, -\frac{11}{8}\right]$
28. $h = \dfrac{A - \pi r^2}{2\pi r}$ **29.** IV

30.
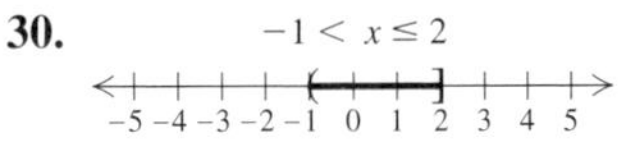

31.
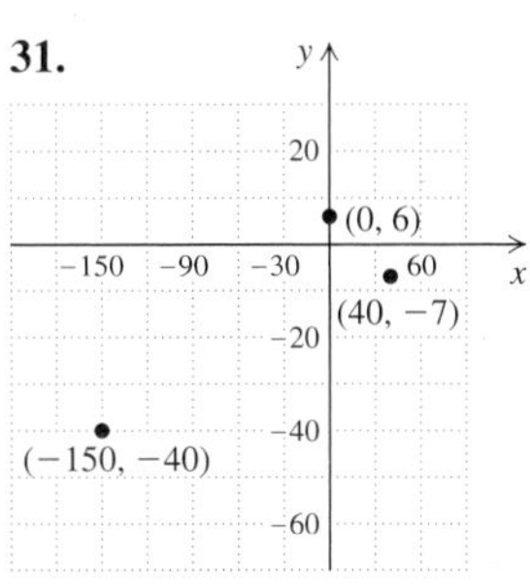

32.
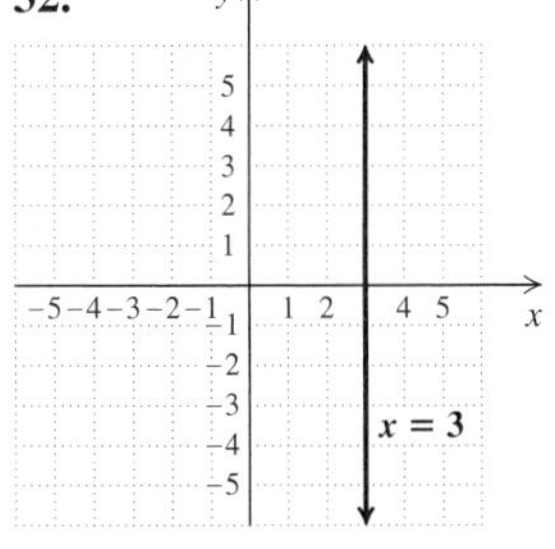

33.
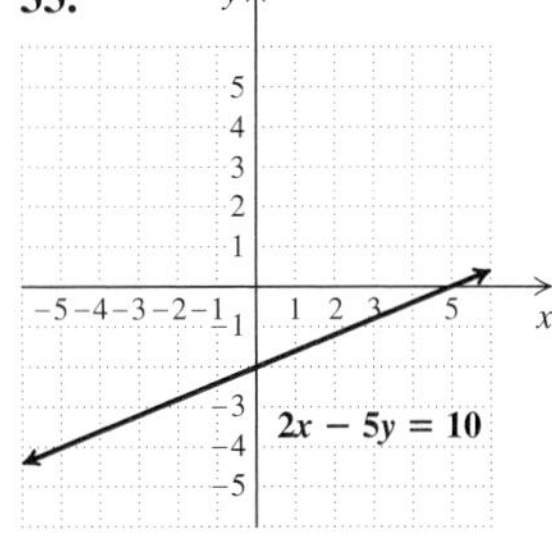

34.
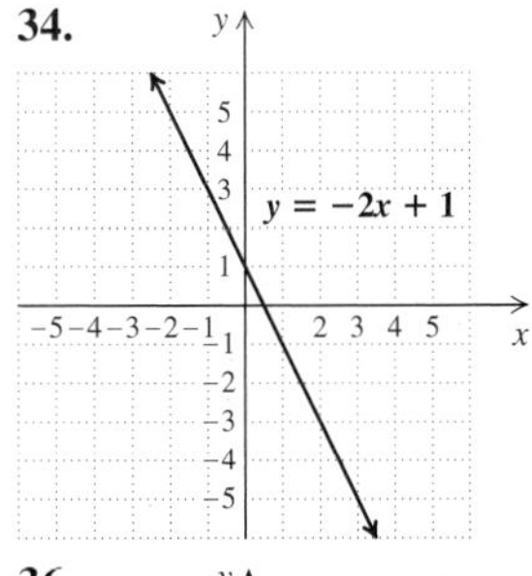

35.
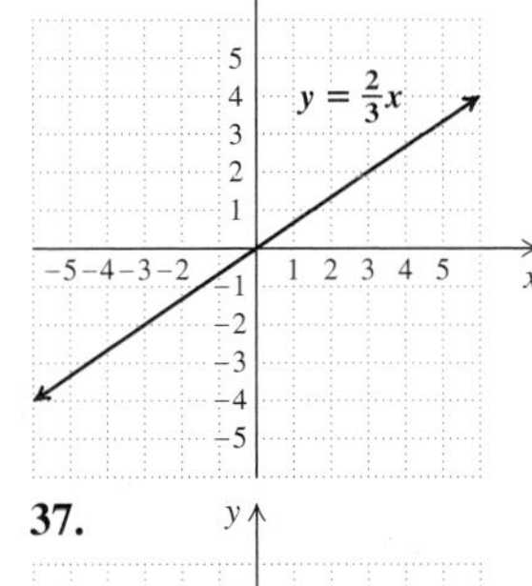

36.
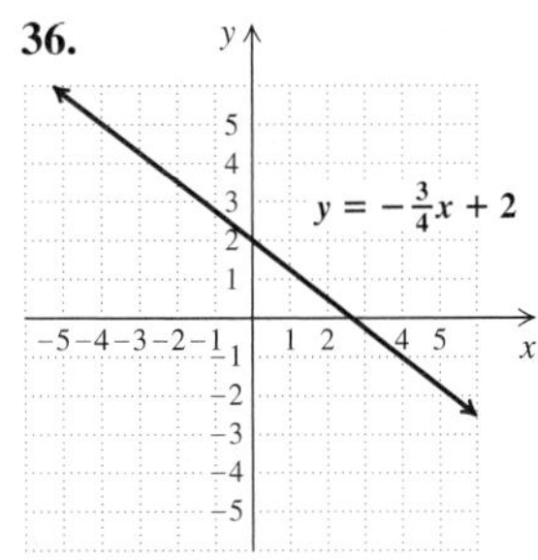

37.
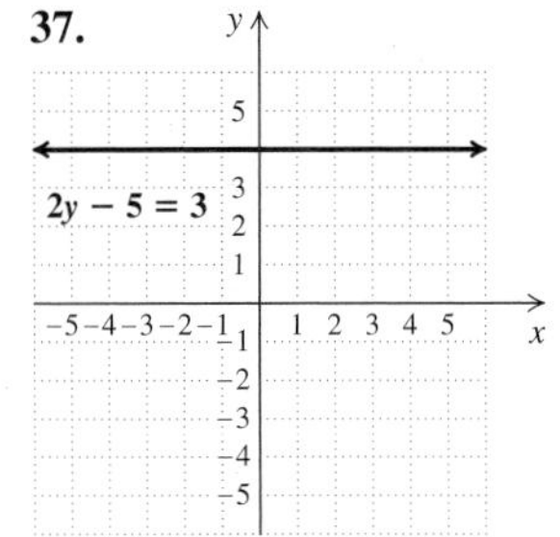

38. $\left(\frac{21}{2}, 0\right), (0, -3)$ **39.** $\left(-\frac{5}{4}, 0\right), (0, 5)$ **40.** 3; $(0, -2)$
41. $-\frac{1}{3}$ **42.** $y = \frac{2}{7}x - 4$ **43.** $y - 4 = -\frac{3}{8}(x - (-6))$
44. $y = -\frac{3}{8}x + \frac{7}{4}$ **45.** $y = 2x + 1$ **46.** $\{-5, -3, -1, 1, 3\}$; $\{-3, -2, 1, 4, 5\}$; -2; 3 **47.** **(a)** -7; **(b)** $\{x \mid x \text{ is a real number and } x \neq \frac{1}{2}\}$
48. **(a)**

$c = \frac{85}{3}r + 70$; Number of calories burned per hour; Cycling rate (in miles per hour)

(b) about 353 calories per hour

49. **(a)** $c = 2.64w + 8.\overline{6}$; **(b)** approximately 365 calories per hour **50.** \$57,196 **51.** 10.5 million Americans **52.** \$120
53. 50 m, 53 m, 40 m **54.** 4 hr **55.** \$25,000 **56.** $-4, 4$
57. 3 **58.** No solution **59.** $Q = \dfrac{2 - pm}{p}$
60. $y = -\frac{7}{3}x + 7$; $y = -\frac{7}{3}x - 7$; $y = \frac{7}{3}x - 7$; $y = \frac{7}{3}x + 7$

Chapter 4

Visualizing for Success, p. 280

1. C **2.** H **3.** J **4.** G **5.** D **6.** I **7.** A **8.** F
9. E **10.** B

Exercise Set 4.1, pp. 281–283

1. True **2.** False **3.** True **4.** True **5.** True **6.** False
7. False **8.** True **9.** Yes **11.** No **13.** Yes **15.** Yes
17. $(4, 1)$ **19.** $(2, -1)$ **21.** $(4, 3)$ **23.** $(-3, -2)$
25. $(-3, 2)$ **27.** $(3, -7)$ **29.** $(7, 2)$ **31.** $(4, 0)$
33. No solution **35.** $\{(x, y) \mid y = 3 - x\}$ **37.** Approximately $(1.53, 2.58)$ **39.** No solution **41.** Approximately $(-6.37, -18.77)$
43. All except 33 and 39 **45.** 35 **47.** Full-time faculty: $y = 9.0524x + 430.6778$; part-time faculty: $y = 14.7175x + 185.3643$; y is in thousands and x is the number of years after 1980; in about 2023 **49.** Independent advisers: $y = 2.7x + 20.4$; financial advisers at firms: $y = -1.5x + 61.6$; y is in thousands and x is the number of years after 2004; in about 2014
51. TW **53.** 15 **54.** $\frac{19}{12}$ **55.** $\frac{9}{20}$ **56.** $\frac{13}{3}$
57. $y = -\frac{3}{4}x + \frac{7}{4}$ **58.** $y = \frac{2}{5}x - \frac{9}{5}$ **59.** TW **61.** Answers may vary. **(a)** $x + y = 6, x - y = 4$; **(b)** $x + y = 1, 2x + 2y = 3$; **(c)** $x + y = 1, 2x + 2y = 2$ **63.** $A = -\frac{17}{4}, B = -\frac{12}{5}$
65. $(0, 0), (1, 1)$ **67.** (c) **69.** (b) **71.** **(a)** $n = \frac{68}{3}t + 46$; $n = 4t + 140$; **(b)** 2009

Exercise Set 4.2, pp. 288–290

1. False **2.** True **3.** True **4.** True **5.** $(1, 4)$ **7.** $(2, 1)$
9. $(4, 3)$ **11.** $(-20, 5)$ **13.** $(-2, 4)$ **15.** No solution
17. $(-1, -3)$ **19.** $\left(\frac{17}{3}, \frac{2}{3}\right)$ **21.** $\{(x, y) \mid x - 2y = 7\}$
23. $\{(x, y) \mid y = 2x + 5\}$ **25.** $\left(\frac{25}{8}, -\frac{11}{4}\right)$ **27.** $(6, 0)$
29. $(10, 5)$ **31.** $\{(x, y) \mid y = 3 - 4x\}$ **33.** $(-3, -4)$
35. No solution **37.** 39, 44 **39.** 42, 51 **41.** 12, 28
43. 55°, 125° **45.** 36°, 54° **47.** Length: 28 in.; width: 22 in.

49. Length: 380 mi; width: 270 mi **51.** Length: 90 yd; width: 50 yd **53.** Height: 20 ft; width: 5 ft **55.** TW **57.** $-5x - 3y$ **58.** $-11x$ **59.** $-11y$ **60.** $-6y - 25$ **61.** $-11y$ **62.** $23x$ **63.** TW **65.** $(7, -1)$ **67.** (4.38, 4.33) **69.** 34 yr **71.** (30, 50, 100) **73.** TW

Exercise Set 4.3, pp. 296–297

1. False **2.** True **3.** True **4.** True **5.** (9, 3) **7.** (5, 1) **9.** (2, 7) **11.** $(-1, 3)$ **13.** $\left(-1, \frac{1}{5}\right)$ **15.** $\{(a, b)|3a - 6b = 8\}$ **17.** $(-3, -5)$ **19.** (4, 5) **21.** (3, 10) **23.** No solution **25.** $\left(\frac{1}{2}, -\frac{1}{2}\right)$ **27.** $(-3, -1)$ **29.** No solution **31.** (50, 18) **33.** $(-2, 2)$ **35.** $(2, -1)$ **37.** $\left(\frac{231}{202}, \frac{117}{202}\right)$ **39.** 40 mi **41.** 26°, 64° **43.** $13\frac{1}{3}$ min **45.** 56°, 124° **47.** Riesling: 340 acres; Chardonnay: 480 acres **49.** Length: 6 ft; width: 3 ft **51.** TW **53.** 0.122 **54.** 0.005 **55.** 40% **56.** 3.06 **57.** Let l = the number of liters; $0.12l$ **58.** Let n = the number of pounds; $0.105n$ **59.** TW **61.** (2, 5) **63.** $\left(\frac{1}{2}, -\frac{1}{2}\right)$ **65.** (0, 3) **67.** $x = \dfrac{c - b}{a - 1}$; $y = \dfrac{ac - b}{a - 1}$ **69.** Rabbits: 12; pheasants: 23 **71.** Man: 45 yr; his daughter: 10 yr

Mid-Chapter Review: Chapter 4, pp. 298–299

Guided Solutions

1.
$$\begin{aligned} 2x - 3(x - 1) &= 5 \\ 2x - 3x + 3 &= 5 \\ -x + 3 &= 5 \\ -x &= 2 \\ x &= -2 \end{aligned}$$

$$\begin{aligned} y &= x - 1 \\ y &= -2 - 1 \\ y &= -3 \end{aligned}$$

The solution is $(-2, -3)$.

2.
$$\begin{aligned} 2x - 5y &= 1 \\ x + 5y &= 8 \\ \hline 3x &= 9 \\ x &= 3 \end{aligned}$$

$$\begin{aligned} x + 5y &= 8 \\ 3 + 5y &= 8 \\ 5y &= 5 \\ y &= 1 \end{aligned}$$

The solution is (3, 1).

Mixed Review

1. (1, 1) **2.** (9, 1) **3.** (4, 3) **4.** (5, 7) **5.** (5, 10) **6.** $\left(2, \frac{2}{5}\right)$ **7.** No solution **8.** $\{(x, y)|x = 2 - y\}$ **9.** (1, 1) **10.** (0, 0) **11.** $(6, -1)$ **12.** No solution **13.** (3, 1) **14.** $\left(\frac{95}{71}, -\frac{1}{142}\right)$ **15.** (1, 1) **16.** $(11, -3)$ **17.** $\{(x, y)|x - 2y = 5\}$ **18.** $\left(1, -\frac{1}{19}\right)$ **19.** $\left(\frac{201}{23}, -\frac{18}{23}\right)$ **20.** $\left(\frac{40}{9}, \frac{10}{3}\right)$

Exercise Set 4.4, pp. 309–313

1. Plant species: 601; animal species: 409 **3.** Facebook: 104 million users; MySpace: 56 million users **5.** 119°, 61° **7.** 3-credit courses: 37; 4-credit courses: 11 **9.** 5-cent bottles or cans: 336; 10-cent bottles or cans: 94 **11.** 2225 motorcycles **13.** Nonrecycled sheets: 38; recycled sheets: 112 **15.** Gemstone beads: 28; silver beads: 52 **17.** HP cartridges: 15; Epson cartridges: 35 **19.** Mexican: 14 lb; Peruvian: 14 lb **21.** Sumac: 8 oz; thyme: 12 oz **23.** 50%-acid solution: 80 mL; 80%-acid solution: 120 mL

x	y	200
50%	80%	68%
$0.5x$	$0.8y$	136

25. 50% chocolate: 7.5 lb; 10% chocolate: 12.5 lb **27.** \$7500 at 6.5%; \$4500 at 7.2% **29.** Steady State: 12.5 L; Even Flow: 7.5 L **31.** 87-octane: 2.5 gal; 95-octane: 7.5 gal **33.** Whole milk: $169\frac{3}{13}$ lb; cream: $30\frac{10}{13}$ lb **35.** 375 km **37.** 14 km/h **39.** About 1489 mi **41.** Foul shots: 28; two-pointers: 36 **43.** Landline: 85 min; wireless: 315 min **45.** Quarters: 17; fifty-cent pieces: 13 **47.** TW

49.

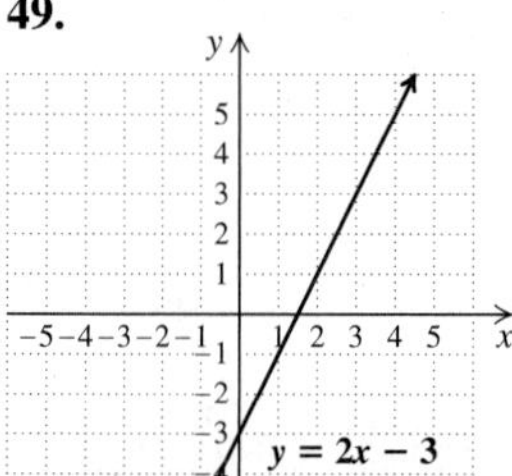

50.

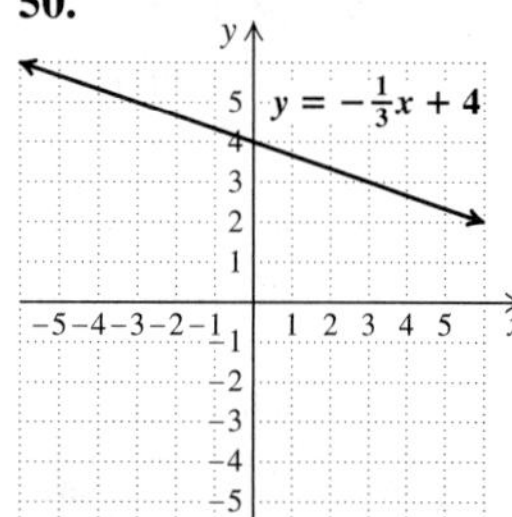

51.

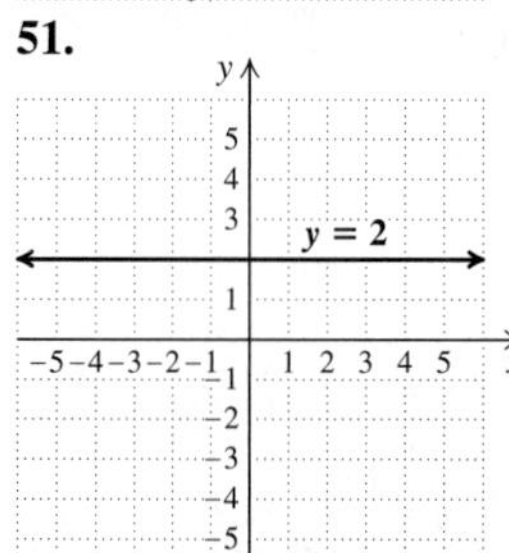

52.

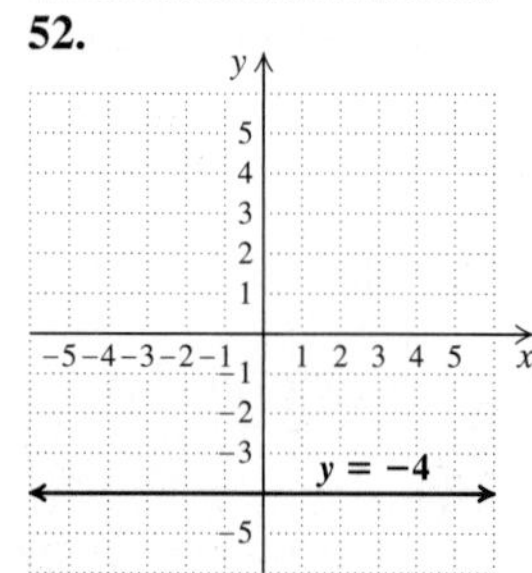

53.

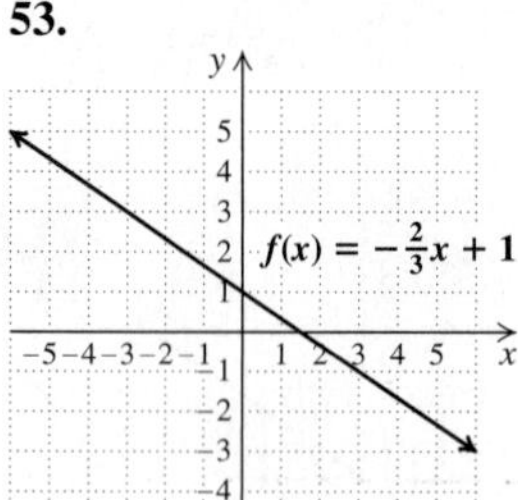

54.

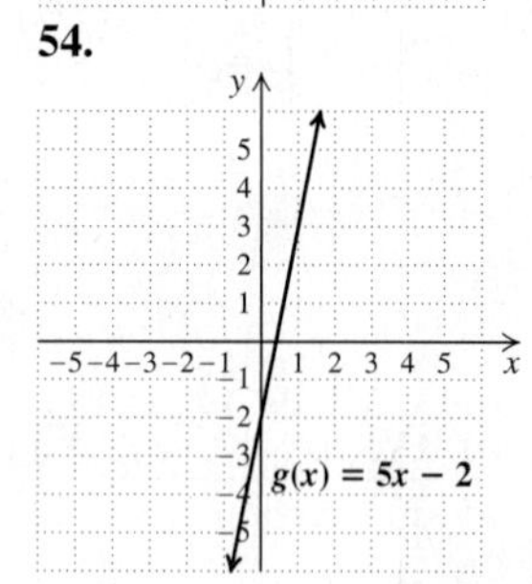

55. TW **57.** $10\frac{2}{3}$ oz **59.** 1.8 L **61.** 12 sets **63.** Brown: 0.8 gal; neutral: 0.2 gal **65.** City: 261 mi; highway: 204 mi **67.** Round Stic: 15 packs; Matic Grip: 9 packs

Exercise Set 4.5, pp. 320–323

1. (e) **2.** (d) **3.** (f) **4.** (b) **5.** (a) **6.** (c) **7.** -2 **9.** -4 **11.** 8 **13.** 0 **15.** 5 **17.** -2 **19.** None **21.** $-2, 2$ **23.** 5 **25.** -20 **27.** 2.7 **29.** $-\frac{7}{3}$ **31.** 7 **33.** 3 **35.** 9 **37.** 1 **39.** 6 **41.** $1\frac{1}{3}$ **43.** \$9500, or \$4500 over \$5000 **45.** 5 months **47.** 2 hr 15 min **49.** 250 lb **51.** TW **53.** -125 **54.** 64 **55.** -64 **56.** 8 **57.** 28

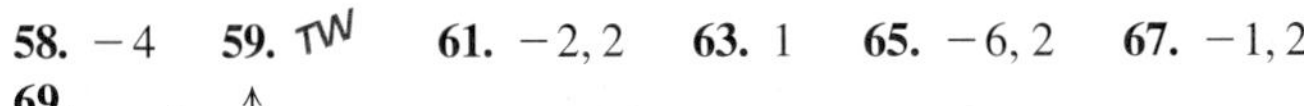
58. -4 **59.** TW **61.** $-2, 2$ **63.** 1 **65.** $-6, 2$ **67.** $-1, 2$
69.

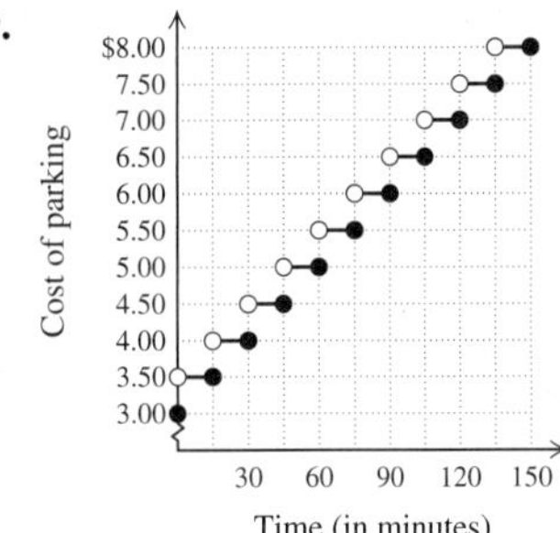

Study Summary: Chapter 4, pp. 324–326

1. $(2, -1)$ **2.** $\left(-\frac{1}{2}, \frac{1}{2}\right)$ **3.** $\left(\frac{16}{7}, -\frac{3}{7}\right)$ **4.** Roller Grip: 32 boxes; eGEL: 88 boxes **5.** 40%-acid: 0.8 L; 15%-acid: 1.2 L **6.** 8 mph **7.** $\frac{1}{8}$ **8.** -1

Review Exercises: Chapter 4, pp. 326–327

1. Substitution **2.** Elimination **3.** Graphical **4.** Dependent **5.** Inconsistent **6.** Contradiction **7.** Parallel **8.** Zero **9.** Alphabetical **10.** x-coordinate **11.** $(4, 1)$ **12.** $(2, 1)$ **13.** $(-5, -13)$ **14.** No solution **15.** $\left(-\frac{4}{5}, \frac{2}{5}\right)$ **16.** $\left(-\frac{9}{2}, -4\right)$ **17.** $\left(\frac{76}{17}, -\frac{2}{119}\right)$ **18.** $(-2, -3)$ **19.** $\{(x, y) \mid 3x + 4y = 6\}$ **20.** Length: 265 ft; width: 165 ft **21.** Oil: $21\frac{1}{3}$ oz; lemon juice: $10\frac{2}{3}$ oz **22.** Private lessons: 7 students; group lessons: 5 students **23.** 163°, 17° **24.** 4 hr **25.** 8% juice: 10 L; 15% juice: 4 L **26.** 1300-word pages: 7; 1850-word pages: 5 **27.** -2 **28.** 1 **29.** $\frac{4}{7}$ **30.** 3 **31.** TW A solution of a system of two equations is an ordered pair that makes both equations true. The graph of an equation represents all ordered pairs that make that equation true. So in order for an ordered pair to make *both* equations true, it must be on both graphs. **32.** TW Both methods involve finding the coordinates of the point of intersection of two graphs. The solution of a system of equations is the ordered pair at the point of intersection; the solution of an equation is the x-coordinate of the point of intersection. **33.** $(0, 2), (1, 3)$ **34.** $C = 1, D = 3$ **35.** 24 **36.** $1800

Test: Chapter 4, p. 328

1. [4.1] $(2, 4)$ **2.** [4.1] No solution **3.** [4.2] $\left(3, -\frac{11}{3}\right)$ **4.** [4.2] $\left(\frac{15}{7}, -\frac{18}{7}\right)$ **5.** [4.2] $\{(x, y) \mid x = 5y - 10\}$ **6.** [4.3] $(2, -1)$ **7.** [4.3] No solution **8.** [4.3] $\left(-\frac{3}{2}, -\frac{3}{2}\right)$ **9.** [4.3] $(0, 1)$ **10.** [4.2] Length: 25 ft; width: 8 ft **11.** [4.2] 38°, 52° **12.** [4.2] Wii game machines: 3.63 million; PlayStation 3 consoles: 1.21 million **13.** [4.4] Hardbacks: 11; paperbacks: 12 **14.** [4.4] Pepperidge Farm Goldfish: 120 g; Rold Gold Pretzels: 500 g **15.** [4.4] 20 mph **16.** [4.4] Nickels: 10; quarters: 3 **17.** [4.5] -2 **18.** [4.5] 10 **19.** [4.2] $(2, 0)$ **20.** [4.3] $(12, -6)$ **21.** [4.4] 9 people **22.** [3.8], [4.3] $m = 7, b = 10$

Chapter 5

Exercise Set 5.1, pp. 336–337

1. (e) **2.** (f) **3.** (b) **4.** (h) **5.** (g) **6.** (a) **7.** (c) **8.** (d) **9.** Base: $5x$; exponent: 7 **11.** Base: n; exponent: 0 **13.** Base: y; exponent: 3 **15.** d^{13} **17.** a^7 **19.** 8^{11} **21.** $(3y)^{12}$ **23.** $7p$ **25.** $(x + 1)^{12}$ **27.** a^5b^9 **29.** r^{10} **31.** m^4n^9 **33.** 7^3 **35.** t^4 **37.** $5a$ **39.** 1 **41.** $(r + s)^8$ **43.** $4a^7b^6$ **45.** $\frac{4}{5}d^7$ **47.** m^9n^4 **49.** 1 **51.** 5 **53.** 2 **55.** -4 **57.** x^{28} **59.** 5^{16} **61.** t^{80} **63.** $49x^2$ **65.** $-8a^3$ **67.** $25n^{14}$ **69.** $a^{14}b^7$ **71.** x^8y^7 **73.** $24x^{19}$ **75.** $\frac{a^3}{64}$ **77.** $\frac{49}{25a^2}$ **79.** $\frac{a^{20}}{b^{15}}$ **81.** $\frac{x^8y^4}{z^{12}}$ **83.** $\frac{a^{12}}{16b^{20}}$ **85.** $-\frac{125x^{21}y^3}{8z^{12}}$ **87.** 1 **89.** TW **91.** -24 **92.** -15 **93.** -11 **94.** 16 **95.** 80 **96.** -14 **97.** TW **99.** TW **101.** Let $a = 1$; then $(a + 5)^2 = 36$, but $a^2 + 5^2 = 26$. **103.** Let $a = 0$; then $\frac{a + 7}{7} = 1$, but $a = 0$. **105.** a^{8k} **107.** $\frac{16}{375}$ **109.** 13 **111.** $<$ **113.** $<$ **115.** $>$ **117.** 4,000,000; 4,194,304; 194,304 **119.** 2,000,000,000; 2,147,483,648; 147,483,648 **121.** 1,536,000 bytes, or approximately 1,500,000 bytes

Exercise Set 5.2, pp. 346–349

1. Positive power of 10 **2.** Negative power of 10 **3.** Negative power of 10 **4.** Positive power of 10 **5.** Positive power of 10 **6.** Negative power of 10 **7.** (c) **8.** (d) **9.** (a) **10.** (b) **11.** $\frac{1}{7^2}$, or $\frac{1}{49}$ **13.** $\frac{1}{(-2)^6}$, or $\frac{1}{64}$ **15.** $\frac{1}{a^3}$ **17.** 5^3, or 125 **19.** $\frac{1}{7}$ **21.** $\frac{8}{x^3}$ **23.** $\frac{3a^8}{b^6}$ **25.** $\frac{1}{3x^5z^4}$ **27.** $\frac{5y^7z^4}{x^2}$ **29.** $\left(\frac{2}{a}\right)^3$, or $\frac{8}{a^3}$ **31.** 8^{-4} **33.** x^{-1} **35.** $\frac{1}{x^{-5}}$ **37.** 8^{-6}, or $\frac{1}{8^6}$ **39.** b^{-3}, or $\frac{1}{b^3}$ **41.** a^2 **43.** $10a^{-6}b^{-2}$, or $\frac{10}{a^6b^2}$ **45.** y^9 **47.** 2^{-3}, or $\frac{1}{2^3}$, or $\frac{1}{8}$ **49.** $-3a^{-1}$, or $-\frac{3}{a}$ **51.** $2a^{-4}b^3$, or $\frac{2b^3}{a^4}$ **53.** $\frac{1}{4}x^3y^{-2}z^{11}$, or $\frac{x^3z^{11}}{4y^2}$ **55.** n^{-15}, or $\frac{1}{n^{15}}$ **57.** t^{40} **59.** $m^{-7}n^{-7}$, or $\frac{1}{m^7n^7}$ **61.** $25r^{-8}t^6$, or $\frac{25t^6}{r^8}$ **63.** $3^{-2}m^{-10}n^6$, or $\frac{n^6}{9m^{10}}$ **65.** $a^{-8}b^5c^4$, or $\frac{b^5c^4}{a^8}$ **67.** $9a^{-8}$, or $\frac{9}{a^8}$ **69.** $\frac{m^{-3}}{n^{-12}}$, or $\frac{n^{12}}{m^3}$ **71.** $\frac{5^4}{(-4)^4}x^{-20}y^{24}$, or $\frac{5^4y^{24}}{4^4x^{20}}$ **73.** 1 **75.** $32a^{-4}$, or $\frac{32}{a^4}$ **77.** -4096 **79.** 0.0625 **81.** 0.648 **83.** 492,000 **85.** 0.00802 **87.** 0.000003497 **89.** 90,300,000,000 **91.** 4.7×10^{10} **93.** 5.83×10^{-3} **95.** 4.07×10^{11} **97.** 6.03×10^{-7} **99.** 5.02×10^{18} **101.** -3.05×10^{-10} **103.** 9.7×10^{-5} **105.** 1.3×10^{-11} **107.** 6.0 **109.** 1.5×10^3 **111.** 3.0×10^{-5} **113.** 1.79×10^{20} **115.** 1.2×10^{24} **117.** 8×10^2 megabytes of information per person **119.** 4.2×10^{10} min **121.** 1×10^5 light years **123.** $3.5\ \text{m}^3$ **125.** 1.4×10^{21} bacteria **127.** Approximately $5 \times 10^2\ \text{in}^3$, or $3 \times 10^{-1}\ \text{ft}^3$ **129.** TW **131.** $8x$ **132.** $-3a - 6b$ **133.** $-2x - 7$ **134.** $-4t - r - 5$ **135.** 1004 **136.** 9 **137.** TW **139.** TW **141.** $1 - \frac{1}{5} = \frac{4}{5}$ **143.** 7 **145.** 7.0×10^{23} **147.** 3^{11} **149.** 1.25×10^{22} **151.** 8×10^{18} grains

Interactive Discovery, p. 356

Answers may include: The graph of a polynomial function has no sharp corners; there are no holes or breaks; the domain is the set of all real numbers.

Visualizing for Success, p. 358

1. C **2.** B **3.** A **4.** D **5.** J **6.** E **7.** I **8.** F **9.** G **10.** H

Exercise Set 5.3, pp. 359–363

1. (b) **2.** (f) **3.** (h) **4.** (d) **5.** (g) **6.** (e) **7.** (a) **8.** (c) **9.** Yes **11.** No **13.** Yes **15.** $7x^4, x^3, -5x, 8$ **17.** $-t^6, 7t^3, -3t^2, 6$ **19.** Coefficients: 4, 7; degrees: 5, 1 **21.** Coefficients: 9, −3, 4; degrees: 2, 1, 0 **23.** Coefficients: 1, −1, 4, −3; degrees: 4, 3, 1, 0 **25.** **(a)** 3, 5, 2; **(b)** $7a^5$, 7; **(c)** 5 **27.** **(a)** 4, 2, 7, 0; **(b)** x^7, 1; **(c)** 7 **29.** **(a)** 1, 4, 0, 3; **(b)** $-a^4$, −1; **(c)** 4

31.

Term	Coefficient	Degree of the Term	Degree of the Polynomial
$8x^5$	8	5	
$-\frac{1}{2}x^4$	$-\frac{1}{2}$	4	
$-4x^3$	−4	3	5
$7x^2$	7	2	
6	6	0	

33. Trinomial **35.** Polynomial with no special name **37.** Binomial **39.** Monomial **41.** $11x^2 + 3x$ **43.** $4a^4$ **45.** $9t^3 + t^2 - 6t$ **47.** $11b^3 + b^2 - b$ **49.** $-x^4 - x^3$ **51.** $\frac{1}{15}x^4 + 10$ **53.** $3.4x^2 + 1.3x + 5.5$ **55.** −17; 25 **57.** 16; 34 **59.** −67; 17 **61.** −27; 81 **63.** −39; 21 **65.** 47; 7 **67.** −45; $-8\frac{19}{27}$ **69.** 9 **71.** \$1.93 billion **73.** 1112 ft **75.** 62.8 cm **77.** 153.86 m^2 **79.** About 135 ft **81.** About 55 million Web sites **83.** 175 million **85.** 14; 55 oranges **87.** About 2.3 mcg/mL **89.** $[0, 10]$ **91.** $(-\infty, 3]$ **93.** $(-\infty, \infty)$ **95.** $[-4, \infty)$ **97.** $[-65, \infty)$ **99.** $[0, \infty)$ **101.** $(-\infty, 5]$ **103.** $(-\infty, \infty)$ **105.** $[-6.7, \infty)$ **107.** TW **109.** $2x + 4$ **110.** $-t + 1$ **111.** $6a + 20$ **112.** $-x - 1$ **113.** $t^4 - 6t^2$ **114.** $0.4n^2 - n + 11$ **115.** TW **117.** $2x^5 + 4x^4 + 6x^2 + 8$; answers may vary **119.** 10 **121.** $5x^9 + 4x^8 + x^2 + 5x$ **123.** $x^3 - 2x^2 - 6x + 3$ **125.** 85.0

127.

t	$-t^2 + 10t - 18$
3	3
4	6
5	7
6	6
7	3

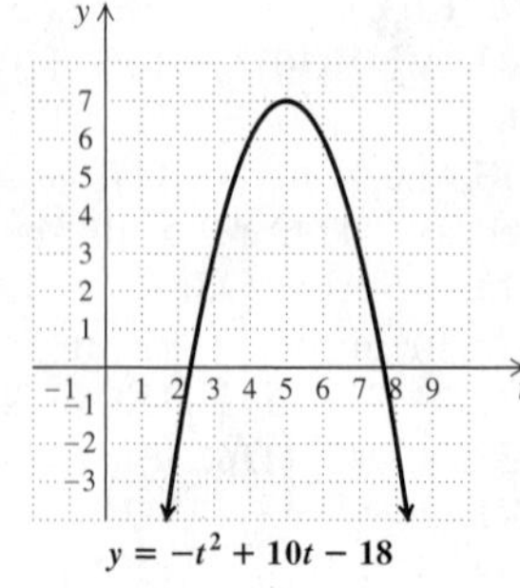

129.

d	$-0.0064d^2 + 0.8d + 2$
0	2
30	20.24
60	26.96
90	22.16
120	5.84

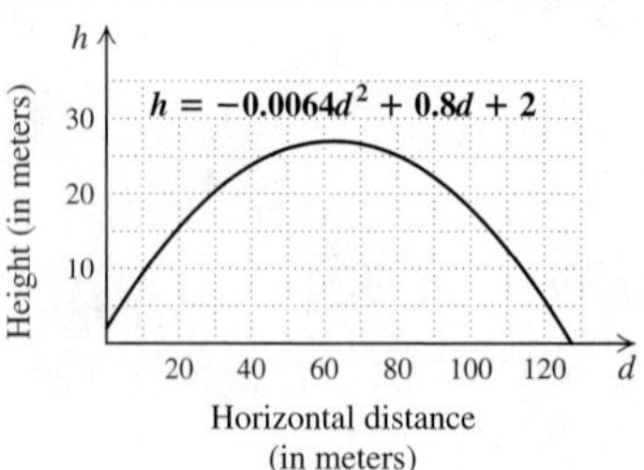

Exercise Set 5.4, pp. 369–372

1. x^2 **2.** −6 **3.** − **4.** + **5.** $4x + 9$ **7.** $-7y - 2$ **9.** $x^2 - 5x - 1$ **11.** $9t^2 + 5t - 3$ **13.** $6m^3 - 3m - 8$ **15.** $7 + 13a - a^2 + 7a^3$ **17.** $9x^8 + 8x^7 - 3x^4 + 2x^2 - 2x + 5$ **19.** $-\frac{1}{2}x^4 + \frac{2}{3}x^3 + x^2$ **21.** $4.2t^3 + 3.5t^2 - 6.4t - 1.8$ **23.** $-3x^4 + 3x^2 + 4x$ **25.** $1.05x^4 + 0.36x^3 + 14.22x^2 + x + 0.97$ **27.** $-(-t^3 + 4t^2 - 9)$; $t^3 - 4t^2 + 9$ **29.** $-(12x^4 - 3x^3 + 3)$; $-12x^4 + 3x^3 - 3$ **31.** $-8x + 9$ **33.** $-3a^4 + 5a^2 - 9$ **35.** $4x^4 - 6x^2 - \frac{3}{4}x + 8$ **37.** $5x + 3$ **39.** $-t^2 - 8t + 7$ **41.** $6y^3 + 16y^2 + y - 14$ **43.** $4.6x^3 + 9.2x^2 - 3.8x - 23$ **45.** 0 **47.** $1 + 2a + 7a^2 - 3a^3$ **49.** $\frac{3}{4}x^3 - \frac{1}{2}x$ **51.** $0.05t^3 - 0.07t^2 + 0.05t$ **53.** $3x + 5$ **55.** $11x^4 + 12x^3 - x^2$ **57.** **(a)** $5x^2 + 4x$; **(b)** 145; 273 **59.** $14y + 25$ **61.** $(r + 11)(r + 9)$; $9r + 99 + 11r + r^2$ **63.** $(x + 3)^2$; $x^2 + 3x + 9 + 3x$ **65.** $\pi r^2 - 25\pi$ **67.** $18z - 64$ **69.** $(x^2 - 12)$ ft^2 **71.** $(z^2 - 36\pi)$ ft^2 **73.** $\left(144 - \frac{d^2}{4}\pi\right)$ m^2 **75.** Not correct **77.** Correct **79.** Not correct **81.** TW **83.** $2x^2 - 2x + 6$ **84.** $-15x^2 + 10x + 35$ **85.** t^{13} **86.** y^7 **87.** $2n^7$ **88.** $-9n^{12}$ **89.** TW **91.** $9t^2 - 20t + 11$ **93.** $-6x + 14$ **95.** $250.591x^3 + 2.812x$ **97.** $20w + 42$ **99.** $2x^2 + 20x$ **101.** **(a)** $P(x) = -x^2 + 175x - 5000$; **(b)** \$2500; **(c)** \$1600

Exercise Set 5.5, pp. 377–379

1. (c) **2.** (d) **3.** (d) **4.** (a) **5.** (c) **6.** (b) **7.** $36x^3$ **9.** x^3 **11.** $-x^8$ **13.** $28t^8$ **15.** $-0.02x^{10}$ **17.** $\frac{1}{15}x^4$ **19.** $19t^2$ **21.** $72y^{10}$ **23.** $4x^2 + 4x$ **25.** $4a^2 - 28a$ **27.** $x^5 + x^2$ **29.** $-6n^3 + 24n^2 - 3n$ **31.** $-15t^5 - 30t^3$ **33.** $4a^9 - 8a^7 - \frac{5}{12}a^4$ **35.** $x^2 + 7x + 6$ **37.** $x^2 + 3x - 10$ **39.** $a^2 - 1.3a + 0.42$ **41.** $x^2 - 9$ **43.** $25 - 15x + 2x^2$ **45.** $t^2 + \frac{17}{6}t + 2$ **47.** $\frac{3}{16}a^2 + \frac{5}{4}a - 2$

49.

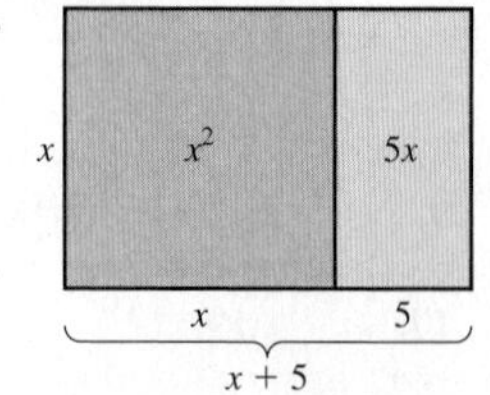

51.

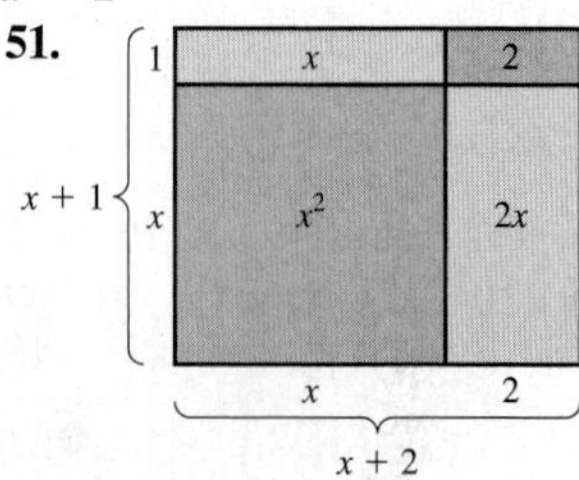

53.

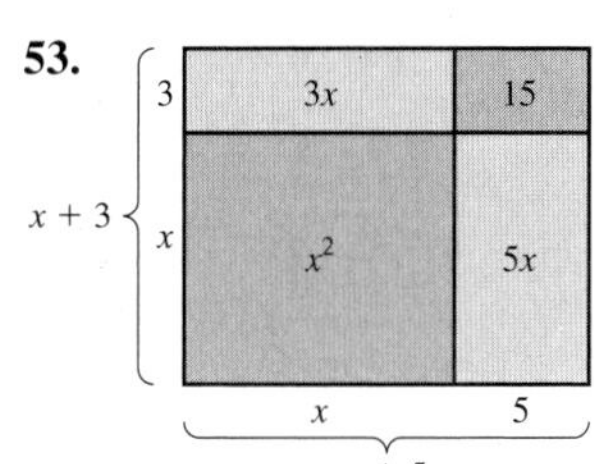

55. $x^3 + 4x + 5$
57. $2a^3 - a^2 - 11a + 10$
59. $2y^5 - 13y^3 + y^2 - 7y - 7$ **61.** $27x^2 + 39x + 14$
63. $x^4 - 2x^3 - x + 2$ **65.** $6t^4 - 17t^3 - 5t^2 - t - 4$
67. $x^4 + 8x^3 + 12x^2 + 9x + 4$ **69.** TW **71.** 0
72. 0 **73.** 8 **74.** 7 **75.** 32 **76.** 50 **77.** TW
79. $75y^2 - 45y$ **81.** 5 **83.** $V = (4x^3 - 48x^2 + 144x)\text{ in}^3$; $S = (-4x^2 + 144)\text{ in}^2$ **85.** $(x^3 - 5x^2 + 8x - 4)\text{ cm}^3$
87. $(x^3 + 2x^2 - 210)\text{ m}^3$ **89.** 0 **91.** $x^3 + x^2 - 22x - 40$
93. 0

Interactive Discovery, p. 381

1. Identity **2.** Not an identity **3.** Not an identity
4. Not an identity

Interactive Discovery, p. 382

1. Not an identity **2.** Identity **3.** Not an identity
4. Not an identity **5.** Identity

Exercise Set 5.6, pp. 385–387

1. True **2.** False **3.** False **4.** True
5. $x^3 + 3x^2 + 5x + 15$ **7.** $t^5 + 7t^4 - 2t - 14$
9. $y^2 - y - 6$ **11.** $9x^2 + 21x + 10$ **13.** $5x^2 + 4x - 12$
15. $2 + 3t - 9t^2$ **17.** $x^4 - 4x^2 - 21$ **19.** $p^2 - \frac{1}{16}$
21. $x^2 - 0.2x + 0.01$ **23.** $-3n^2 - 19n + 14$
25. $a^2 + 18a + 81$ **27.** $1 - 3t + 5t^2 - 15t^3$
29. $x^5 + 3x^3 - x^2 - 3$ **31.** $3x^6 - 2x^4 - 6x^2 + 4$
33. $4t^6 + 16t^3 + 15$ **35.** $8x^5 + 16x^3 + 5x^2 + 10$
37. $100x^4 - 9$ **39.** $x^2 - 49$ **41.** $4x^2 - 1$
43. $25m^4 - 81$ **45.** $36a^6 - 1$ **47.** $x^8 - 0.01$
49. $t^2 - \frac{9}{16}$ **51.** $x^2 + 4x + 4$ **53.** $9x^{10} - 6x^5 + 1$
55. $a^2 - \frac{4}{5}a + \frac{4}{25}$ **57.** $x^4 - x^3 + 3x^2 - x + 2$
59. $4 - 12x^4 + 9x^8$ **61.** $25 + 60t^2 + 36t^4$
63. $49x^2 - 4.2x + 0.09$ **65.** $10a^5 - 5a^3$
67. $a^3 - a^2 - 10a + 12$ **69.** $49 - 42x^4 + 9x^8$
71. $-4x^3 - 24x^2 + 12x$ **73.** $t^6 - 2t^3 + 1$
75. $15t^5 - 3t^4 + 3t^3$ **77.** $36x^8 - 36x^5 + 9x^2$
79. $18a^4 + 0.8a^3 + 4.5a + 0.2$ **81.** $\frac{1}{25} - 36x^8$
83. $a^3 + 1$ **85.** $a^2 + 2a + 1$ **87.** $x^2 + 7x + 10$
89. $x^2 + 14x + 49$ **91.** $t^2 + 10t + 24$
93. $t^2 + 13t + 36$ **95.** $9x^2 + 24x + 16$

97.

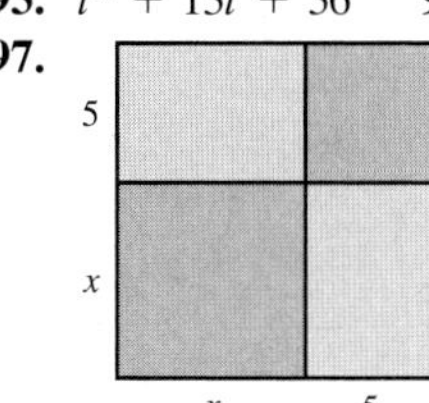

99.

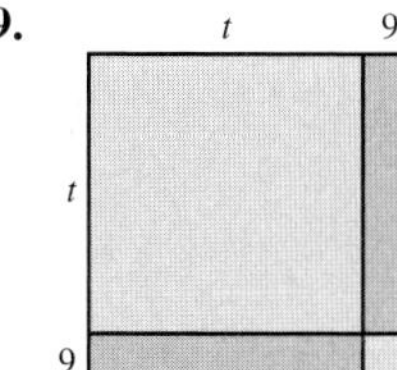

101.

103. TW
105. Washing machine: 9 kWh/mo; refrigerator: 189 kWh/mo; freezer: 99 kWh/mo
106. 10 billion searches
107. $y = \dfrac{8}{5x}$ **108.** $a = \dfrac{c}{3b}$ **109.** $x = \dfrac{by + c}{a}$
110. $y = \dfrac{ax - c}{b}$ **111.** TW **113.** 396 **115.** $16x^4 - 81$
117. $81t^4 - 72t^2 + 16$ **119.** $t^{24} - 4t^{18} + 6t^{12} - 4t^6 + 1$
121. -7 **123.** $y^2 - 4y + 4$ **125.** $l^3 - l$
127. $Q(Q - 14) - 5(Q - 14)$, $(Q - 5)(Q - 14)$; other equivalent expressions are possible. **129.** $(y + 1)(y - 1)$, $y(y + 1) - y - 1$; other equivalent expressions are possible.

Mid-Chapter Review: Chapter 5, p. 388

Guided Solutions

1. $$(2x^2y^{-5})^{-10} = 2^{-10}(x^2)^{-10}(y^{-5})^{-10} = 2^{-10}x^{-20}y^{50} = \frac{y^{50}}{2^{10}x^{20}}$$

2. $$(x^2 + 7x)(x^2 - 7x) = (x^2)^2 - (7x)^2 = x^4 - 49x^2$$

Mixed Review

1. $x^{16}y^{40}$ **2.** 1 **3.** $\dfrac{1}{d^{10}}$ **4.** $\frac{1}{4}a^{10}$ **5.** $\dfrac{m^3}{y}$ **6.** $-\dfrac{8}{3a^{10}b}$
7. $3x^2 + 3x + 3$ **8.** $7x + 7$ **9.** $48x^5 - 42x^3$
10. $6x^2 + x - 2$ **11.** $-6x^2 + 2x - 12$
12. $9x^2 + 45x + 56$ **13.** $t^9 + 5t^7$ **14.** $4m^2 - 4m + 1$
15. $x^3 - 1$ **16.** $c^2 - 9$ **17.** $16y^6 + 56y^3 + 49$
18. $3a^4 - 13a^3 - 13a^2 - 4$ **19.** $16t^4 - 25$
20. $a^8 - 5a^4 - 24$

Exercise Set 5.7, pp. 394–397

1. Coefficient **2.** Degree **3.** Degree **4.** Leading coefficient **5.** Binomial **6.** Three variables **7.** Like terms
8. Binomial **9.** -7 **11.** -92 **13.** 2.97 L
15. About 2494 calories **17.** 73.005 in^2 **19.** 66.4 m
21. Coefficients: 1, −2, 3, −5; degrees: 4, 2, 2, 0; 4
23. Coefficients: 11, −1, 1, 0.5; degrees: 0, 3, 3, 3; 3
25. $3a - 2b$ **27.** $3x^2y - 2xy^2 + x^2 + 5x$
29. $8u^2v - 5uv^2 + 7u^2$ **31.** $6a^2c - 7ab^2 + a^2b$
33. $3x^2 - 4xy + 3y^2$ **35.** $-6a^4 - 8ab + 7ab^2$
37. $-6r^2 - 5rt - t^2$ **39.** $3x^3 - x^2y + xy^2 - 3y^3$
41. $10y^4x^2 - 8y^3x$ **43.** $-8x + 8y$ **45.** $6z^2 + 7uz - 3u^2$
47. $x^2y^2 + 3xy - 28$ **49.** $4a^2 - b^2$ **51.** $15r^2t^2 - rt - 2$
53. $m^6n^2 + 2m^3n - 48$ **55.** $30x^2 - 28xy + 6y^2$
57. $0.01 - a^2w^2$ **59.** $x^2 + 2xh + h^2$
61. $16a^2 - 40ab + 25b^2$ **63.** $a^2b^2 - c^2d^4$
65. $2x^2y^2 + x^3y^2 + 2xy^3 + x^2y^3 + 3xy + 3y^2$
67. $a^2 + 2ab + b^2 - c^2$ **69.** $a^2 - b^2 - 2bc - c^2$
71. $a^2 + ab + ac + bc$ **73.** $x^2 - z^2$
75. $x^2 + y^2 + z^2 + 2xy + 2xz + 2yz$ **77.** $\frac{1}{2}x^2 + \frac{1}{2}xy - y^2$

79. We draw a rectangle with dimensions $r + s$ by $u + v$.

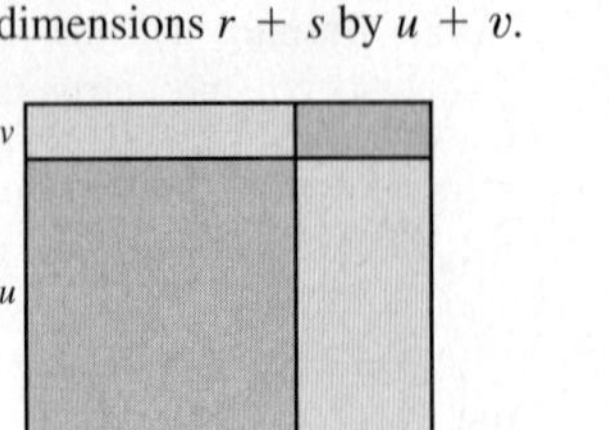

81.

83. **(a)** $t^2 - 2t + 6$; **(b)** $2ah + h^2$; **(c)** $2ah - h^2$ **85.** TW **87.** $x^2 - 8x - 4$ **88.** $2x^3 - x^2 - x + 4$ **89.** $-2x + 5$ **90.** $5x^2 + x$ **91.** $13x^2 + 1$ **92.** $-x - 3$ **93.** TW **95.** $2\pi ab - \pi b^2$ **97.** $a^2 - 4b^2$ **99.** $x^3 + 2y^3 + x^2y + xy^2$ **101.** $2\pi nh + 2\pi mh + 2\pi n^2 - 2\pi m^2$ **103.** TW **105.** 40 **107.** $P + 2Pr + Pr^2$ **109.** \$15,638.03

Exercise Set 5.8, pp. 404–406

1. Quotient **2.** Divisor **3.** Dividend **4.** Remainder **5.** $4x^5 - 3x$ **7.** $1 - 2u + u^6$ **9.** $5t^2 - 8t + 2$ **11.** $6t^2 - 10t + \frac{3}{2}$ **13.** $-5x^5 + 7x^2 + 1$ **15.** $4x - 5 + \dfrac{1}{2x}$ **17.** $-3rs - r + 2s$ **19.** $2x^3y + 3y - 1$ **21.** $x + 3$ **23.** $a - 12 + \dfrac{32}{a + 4}$ **25.** $2x - 1 + \dfrac{1}{x + 6}$ **27.** $y - 5$ **29.** $a^2 - 2a + 4$ **31.** $t + 4 + \dfrac{3}{t - 4}$ **33.** $t^2 - 3t + 1$ **35.** $x - 3 + \dfrac{3}{5x + 1}$ **37.** $t^2 - 2t + 3 + \dfrac{-4}{t + 1}$ **39.** $t^2 - 1 + \dfrac{3t - 1}{t^2 + 5}$ **41.** $2x^2 + 1 + \dfrac{-x}{2x^2 - 3}$ **43.** $x^2 - 3x + 5 + \dfrac{-12}{x + 1}$ **45.** $a + 5 + \dfrac{-4}{a + 3}$ **47.** $x^2 - 5x - 23 + \dfrac{-43}{x - 2}$ **49.** $3x^2 - 2x + 2 + \dfrac{-3}{x + 3}$ **51.** $x^4 + 2x^3 + 4x^2 + 8x + 16$ **53.** $3x^2 + 6x - 3 + \dfrac{2}{x + \frac{1}{3}}$ **55.** TW

57.

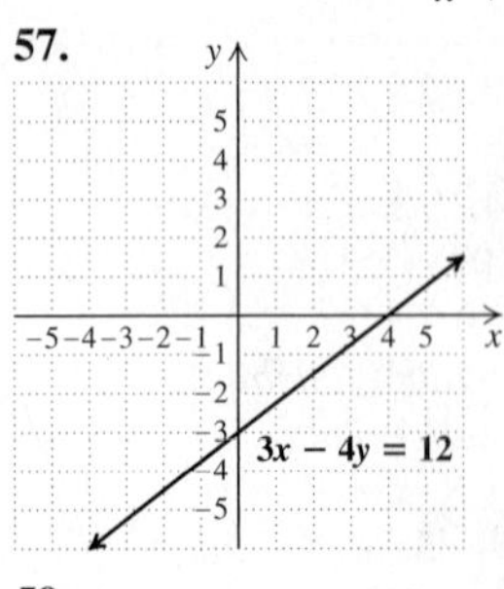

58.

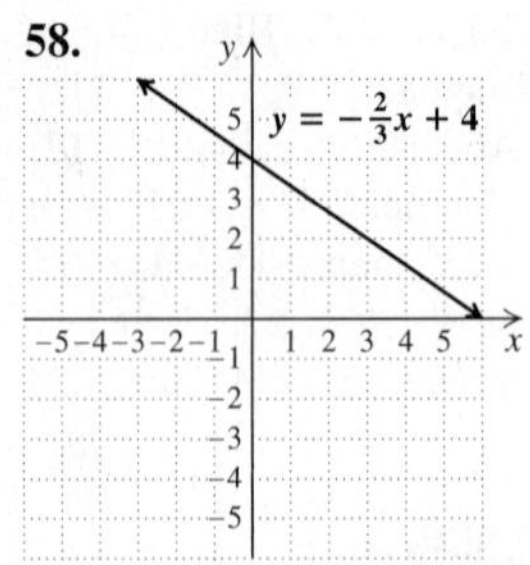

59.

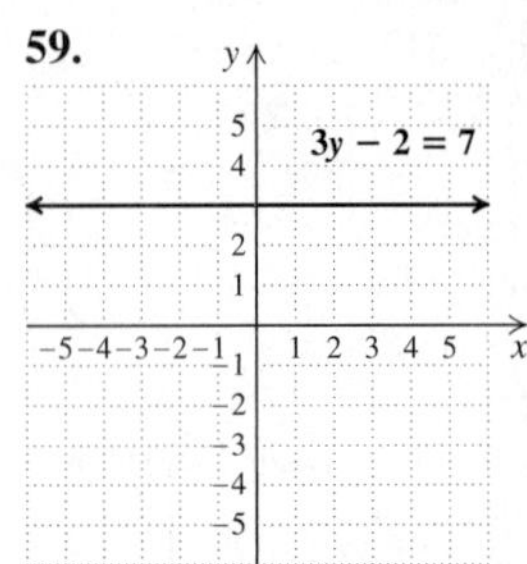

60.

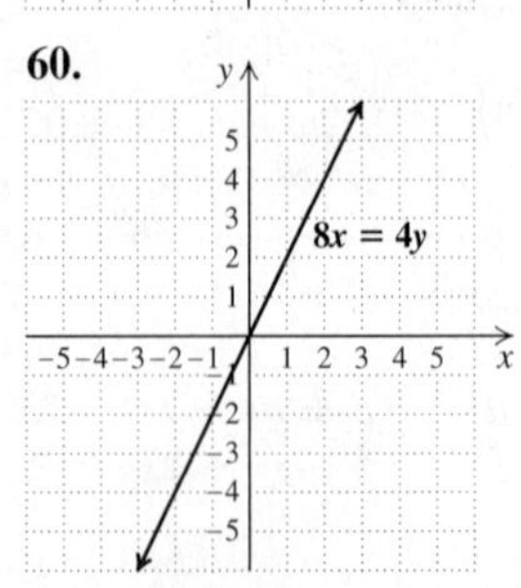

61. $-\frac{3}{10}$ **62.** Slope: 4; y-intercept: $\left(0, \frac{7}{2}\right)$ **63.** $y = -5x - 10$ **64.** $y = \frac{5}{4}x - \frac{9}{2}$ **65.** TW **67.** $5x^{6k} - 16x^{3k} + 14$ **69.** $3t^{2h} + 2t^h - 5$ **71.** $a + 3 + \dfrac{5}{5a^2 - 7a - 2}$ **73.** $2x^2 + x - 3$ **75.** 3 **77.** -1

Interactive Discovery, p. 409

1. 0 **2.** -1 **3.** $-1, 0$ **4.** Yes **5.** Yes **6.** No; 3 is not in the domain of f/g.

Exercise Set 5.9, pp. 412–415

1. Difference **2.** Subtract **3.** Evaluate **4.** Common to **5.** Excluding **6.** Sum **7.** 1 **9.** -41 **11.** 12 **13.** $\frac{13}{18}$ **15.** 5 **17.** $x^2 - 3x + 3$ **19.** $-3x^3 + x^2 - 6x + 2$ **21.** $x^2 - x + 3$ **23.** 23 **25.** 5 **27.** 56 **29.** $\dfrac{x^2 - 2}{5 - x}, x \neq 5$ **31.** $x^2 + x - 7$ **33.** $\frac{2}{7}$ **35.** 4% **37.** $1.2 + 2.9 = 4.1$ million births **39.** About 95 million; the number of tons of municipal solid waste that was composted or recycled in 2005 **41.** About 215 million; the number of tons of municipal solid waste in 1996 **43.** About 230 million; the number of tons of municipal solid waste that was not composted in 2004 **45.** $\mathbb{R}$ **47.** $\{x \mid x$ is a real number *and* $x \neq 3\}$ **49.** $\{x \mid x$ is a real number *and* $x \neq 0\}$ **51.** $\{x \mid x$ is a real number *and* $x \neq 1\}$ **53.** $\{x \mid x$ is a real number *and* $x \neq \frac{9}{2}$ *and* $x \neq 1\}$ **55.** $\{x \mid x$ is a real number *and* $x \neq 3\}$ **57.** $\{x \mid x$ is a real number *and* $x \neq 4\}$ **59.** $\{x \mid x$ is a real number *and* $x \neq 4$ *and* $x \neq 5\}$ **61.** $\left\{x \mid x\right.$ is a real number *and* $\left. x \neq -1 \text{ and } x \neq -\frac{5}{2}\right\}$ **63.** $4x^2 - 6x + 9, x \neq -\frac{3}{2}$ **65.** $2x - 5, x \neq -\frac{2}{3}$ **67.** 4; 3 **69.** 5; -1 **71.** $\{x \mid 0 \leq x \leq 9\}$; $\{x \mid 3 \leq x \leq 10\}$; $\{x \mid 3 \leq x \leq 9\}$; $\{x \mid 3 \leq x \leq 9\}$

73.

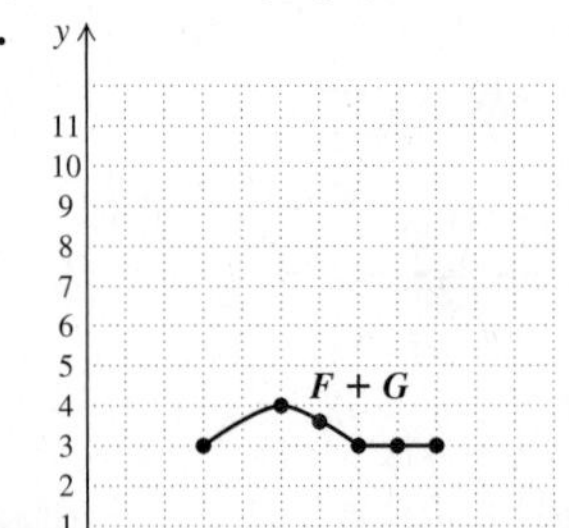

75. TW **77.** $5(3x + 4y + 1)$ **78.** $4(3m + n + 2)$ **79.** -5 **80.** $-\frac{9}{4}$ **81.** $\frac{1}{3}$ **82.** 0 **83.** TW **85.** $\{x \mid x$ is a real number *and* $x \neq -\frac{5}{2}$ *and* $x \neq -3$ *and* $x \neq 1$ *and* $x \neq -1\}$ **87.** Answers may vary.

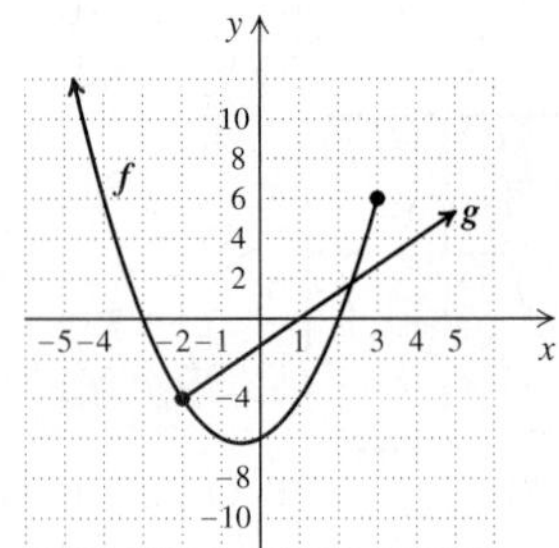

89. $\left\{x \mid x\right.$ is a real number *and* $\left. -1 < x < 5 \text{ and } x \neq \frac{3}{2}\right\}$ **91.** Answers may vary. $f(x) = \dfrac{1}{x + 2}, g(x) = \dfrac{1}{x - 5}$

93. The domain of $y_3 = \dfrac{2.5x + 1.5}{x - 3}$ is $\{x \mid x$ is a real number *and* $x \neq 3\}$. The CONNECTED mode graph crosses the line $x = 3$, whereas the DOT mode graph contains no points having 3 as the first coordinate. Thus the DOT mode graph represents y_3 more accurately.

Study Summary: Chapter 5, pp. 416–419

1. 6 **2.** 1 **3.** x^{16} **4.** 8^7 **5.** y^{15} **6.** $x^{30}y^{10}$ **7.** $\dfrac{x^{10}}{7^5}$ **8.** $\frac{1}{10}$ **9.** $\dfrac{y^3}{x}$ **10.** 9.04×10^{-4} **11.** 690,000 **12.** $x^2, -10, 5x, -8x^6$ **13.** 1 **14.** 1 **15.** $-8x^6$ **16.** -8 **17.** 6 **18.** Trinomial **19.** $3x^2 - 4x$ **20.** 4 **21.** $8x^2 + x$ **22.** $10x^2 - 7x$ **23.** $x^3 - 2x^2 - x + 2$ **24.** $2x^2 + 11x + 12$ **25.** $25 - 9x^2$ **26.** $x^2 + 18x + 81$ **27.** $64x^2 - 16x + 1$ **28.** -7 **29.** 6 **30.** $3cd^2 + 4cd - 7c$ **31.** $-2p^2w$ **32.** $49x^2y^2 - 14x^3y + x^4$ **33.** $y^3 - 2y + 4$ **34.** $x - 2 + \dfrac{6}{x + 1}$ **35.** $2x - 9$ **36.** 5 **37.** $x^2 - 9x + 14$ **38.** $\dfrac{x - 2}{x - 7}, x \neq 7$

Review Exercises: Chapter 5, pp. 419–421

1. False **2.** True **3.** True **4.** False **5.** True **6.** False **7.** True **8.** True **9.** y^{11} **10.** $(3x)^{14}$ **11.** t^8 **12.** 4^3, or 64 **13.** 1 **14.** $-2m^3n^2$ **15.** $-8x^3y^6$ **16.** $18x^5$ **17.** a^7b^6 **18.** $\dfrac{9x^4}{4y^6}$ **19.** $\dfrac{1}{m^7}$ **20.** $\dfrac{1}{7^2}$, or $\dfrac{1}{49}$ **21.** $\dfrac{1}{a^{13}b^7}$ **22.** $\dfrac{1}{x^{12}}$ **23.** $\dfrac{x^6}{4y^2}$ **24.** $\dfrac{y^3}{8x^3}$ **25.** 8,300,000 **26.** 3.28×10^{-5} **27.** 2.1×10^4 **28.** 5.1×10^{-5} **29.** 2.3×10^{12} red blood cells **30.** $3x^2, 6x, \frac{1}{2}$ **31.** $-4y^5, 7y^2, -3y, -2$ **32.** $7, -1, 7$ **33.** $4, 1, -5, \frac{5}{3}$ **34.** **(a)** 2, 0, 3; **(b)** $15t^3$, 15; **(c)** 3 **35.** **(a)** 5, 4, 2, 1; **(b)** $-2x^5, -2$; **(c)** 5 **36.** Binomial **37.** Polynomial with no special name **38.** Monomial **39.** $-x^2 + 9x$ **40.** $-\frac{1}{4}x^3 + 4x^2 + 7$ **41.** $-3x^5 + 25$ **42.** $10x^4 - 7x^2 - x - \frac{1}{2}$ **43.** 10 **44.** 28 **45.** $(-\infty, 2]$ **46.** About 340 mg **47.** About 185 mg **48.** $[0, 345]$ **49.** $[0, 6]$ **50.** $x^5 + 3x^4 + 6x^3 - 2x - 9$ **51.** $2x^2 - 4x - 6$ **52.** $x^5 - 3x^3 - 2x^2 + 8$ **53.** $\frac{3}{4}x^4 + \frac{1}{4}x^3 - \frac{1}{3}x^2 - \frac{7}{4}x + \frac{3}{8}$ **54.** $-x^5 + x^4 - 5x^3 - 2x^2 + 2x$ **55.** **(a)** $4w + 6$; **(b)** $w^2 + 3w$ **56.** $-12x^3$ **57.** $49x^2 + 14x + 1$ **58.** $a^2 - 3a - 28$ **59.** $m^2 - 25$ **60.** $12x^3 - 23x^2 + 13x - 2$ **61.** $x^2 - 18x + 81$ **62.** $15t^5 - 6t^4 + 12t^3$ **63.** $9a^2 - 64$ **64.** $x^2 - 0.8x + 0.15$ **65.** $x^7 + x^5 - 3x^4 + 3x^3 - 2x^2 + 5x - 3$ **66.** $9x^8 - 30x^4 + 25$ **67.** $2t^4 - 11t^2 - 21$ **68.** $a^2 + \frac{1}{6}a - \frac{1}{3}$ **69.** $4n^2 - 49$ **70.** 49 **71.** Coefficients: $1, -7, 9, -8$; degrees: 6, 2, 2, 0; 6 **72.** Coefficients: $1, -1, 1$; degrees: 16, 40, 23; 40 **73.** $-y + 9w - 5$ **74.** $6m^3 + 4m^2n - mn^2$ **75.** $-x^2 - 10xy$ **76.** $11x^3y^2 - 8x^2y - 6x^2 - 6x + 6$ **77.** $p^3 - q^3$ **78.** $25a^2b^2 - 10abcd^2 + c^2d^4$ **79.** $\frac{1}{2}x^2 - \frac{1}{2}y^2$ **80.** $5x^2 - \frac{1}{2}x + 3$ **81.** $3x^2 - 7x + 4 + \dfrac{1}{2x + 3}$ **82.** $t^3 + 2t - 3$ **83.** 102 **84.** -17 **85.** $-\frac{9}{2}$ **86.** $x^2 + 3x - 5$ **87.** $3x^3 - 6x^2 + 3x - 6$ **88.** $\dfrac{x^2 + 1}{3x - 6}, x \neq 2$ **89.** $\mathbb{R}$ **90.** $\{x \mid x$ is a real number *and* $x \neq 2\}$ **91.** TW In the expression $5x^3$, the exponent refers only to the x. In the expression $(5x)^3$, the entire expression within the parentheses is cubed. **92.** TW Using the equation $(A - B)^2 = A^2 - 2AB + B^2$ in reverse, it is possible to determine that $x^2 - 6x + 9$ is $(x - 3)^2$ or $(3 - x)^2$ (if written $9 - 6x + x^2$). Without further information, we cannot determine whether the binomial squared was $x - 3$ or $3 - x$. **93.** **(a)** 12; **(b)** 15 **94.** $-28x^8$ **95.** $8x^4 + 4x^3 + 5x - 2$ **96.** $-16x^6 + x^2 - 10x + 25$ **97.** $\frac{94}{13}$

Test: Chapter 5, p. 422

1. [5.1] t^8 **2.** [5.1] x^{36} **3.** [5.1] 3^3, or 27 **4.** [5.1] 1 **5.** [5.1] $-40x^{19}y^4$ **6.** [5.1] $-3a^5b^3$ **7.** [5.2] $\dfrac{1}{5^3}$ **8.** [5.2] $\dfrac{1}{t^6}$ **9.** [5.2] $\dfrac{4y^5}{5x^5}$ **10.** [5.2] $\dfrac{b^4}{16a^{12}}$ **11.** [5.2] $\dfrac{c^3}{a^3b^3}$ **12.** [5.2] 3.06×10^9 **13.** [5.2] 0.0005 **14.** [5.2] 1.8×10^{17} **15.** [5.2] 1.3×10^{22} **16.** About 1.4×10^7 hr **17.** [5.3] Binomial **18.** [5.3] $\frac{1}{3}, -1, 7$ **19.** [5.3] Degrees of terms: 3, 1, 5, 0; leading term: $7t^5$; leading coefficient: 7; degree of polynomial: 5 **20.** [5.3] -7 **21.** [5.3] $5a^2 - 6$ **22.** [5.3] $\frac{7}{4}y^2 - 4y$ **23.** [5.3] $x^5 + 2x^3 + 4x^2 - 8x + 3$ **24.** [5.3] $(-\infty, \infty)$ **25.** [5.4] $4x^5 + x^4 + 5x^3 - 8x^2 + 2x - 7$ **26.** [5.4] $-4x^4 + x^3 - 8x - 3$ **27.** [5.4] $-t^4 + 2.5t^3 - 0.6t^2 - 9$ **28.** [5.5] $-12x^4 + 9x^3 + 15x^2$ **29.** [5.6] $x^2 - \frac{2}{3}x + \frac{1}{9}$ **30.** [5.6] $25t^2 - 49$ **31.** [5.6] $3b^2 - 4b - 15$ **32.** [5.6] $x^{14} - 4x^8 + 4x^6 - 16$ **33.** [5.6] $48 + 34y - 5y^2$ **34.** [5.5] $6x^3 - 7x^2 - 11x - 3$ **35.** [5.6] $64a^2 + 48a + 9$ **36.** [5.7] $-5x^3y - x^2y^2 + xy^3 - y^3 + 19$ **37.** [5.7] $8a^2b^2 + 6ab + 6ab^2 + ab^3 - 4b^3$ **38.** [5.7] $9x^{10} - 16y^2$ **39.** [5.8] $4x^2 + 3x - 5$ **40.** [5.8] $6x^2 - 20x + 26 + \dfrac{-39}{x + 2}$ **41.** [5.9] $\frac{7}{3}$ **42.** [5.9] $\dfrac{1}{x} + 2x + 1$ **43.** [5.9] $\{x \mid x$ is a real number *and* $x \neq 0$ *and* $x \neq -\frac{1}{2}\}$ **44.** [5.5], [5.6] $V = l(l - 2)(l - 1) = l^3 - 3l^2 + 2l$ **45.** [5.2] $\frac{1}{2} - \frac{1}{4} = \frac{1}{4}$

Chapter 6

Interactive Discovery, p. 426

1. 1 **2.** 0 **3.** 2 **4.** 3 **5.** 2 **6.** 1 **7.** The number of real-number zeros is less than or equal to the degree.

Exercise Set 6.1, pp. 434–438

1. False **2.** False **3.** True **4.** True **5.** True **6.** False **7.** True **8.** False **9.** Expression **10.** Equation

11. Equation **12.** Expression **13.** Equation **14.** Expression
15. $-3, 5$ **17.** $-2, 0$ **19.** $-3, 1$ **21.** $-4, 2$ **23.** $0, 5$
25. $1, 3$ **27.** $10, 15$ **29.** $0, 1, 2$ **31.** $-15, 6, 12$
33. $-0.42857, 0.33333$ **35.** $-5, 9$ **37.** $-0.5, 7$ **39.** $-1, 0, 3$
41. III **43.** I **45.** $2t(t + 4)$ **47.** $y^2(9y - 1)$
49. $5x(3x - x^3 + 1)$ **51.** $4xy(x - 3y)$ **53.** $3(y^2 - y - 3)$
55. $2a(3b - 2d + 6c)$ **57.** $12x(6x^2 - 3x + 2)$
59. $xy^2(x^4y^3 + x^3y + x^2y - 1)$
61. $3x^2y^4z^2(3xy^2 - 4x^2z^2 + 5yz)$ **63.** $-5(x - 7)$
65. $-2(x^2 - 2x + 6)$ **67.** $-3(-y + 8x)$, or $-3(8x - y)$
69. $-(x^2 - 5x + 9)$ **71.** $-a(a^3 - 2a^2 + 13)$
73. $(b - 5)(a + c)$ **75.** $(x + 7)(2x - 3)$
77. $(x - y)(a^2 - 5)$ **79.** $(c + d)(a + b)$
81. $(b - 1)(b^2 + 2)$ **83.** Not factorable by grouping
85. $(a - 3)(a^2 - 2)$ **87.** $x^3(x - 1)(x^2 + 1)$
89. $(y^2 + 3)(2y^2 + 5)$ **91. (a)** $h(t) = -8t(2t - 9)$;
(b) $h(1) = 56$ ft **93.** $R(n) = n(n - 1)$ **95.** $P(x) = x(x - 3)$
97. $R(x) = 0.4x(700 - x)$ **99.** $N(x) = \frac{1}{6}(x^3 + 3x^2 + 2x)$
101. $H(n) = \frac{1}{2}n(n - 1)$ **103.** $-3, 4$ **105.** $-1, 0$ **107.** $0, 3$
109. $-3, 0$ **111.** $-\frac{1}{3}, 0$ **113.** $-7, 3$ **115.** $0, -\frac{9}{5}$ **117.** $0, 3$
119. TW **121.** $x^2 + 9x + 14$ **122.** $x^2 - 9x + 14$
123. $x^2 - 5x - 14$ **124.** $x^2 + 5x - 14$ **125.** $a^2 - 4a + 3$
126. $t^2 + 8t + 15$ **127.** $t^2 + 5t - 50$ **128.** $a^2 - 2a - 24$
129. TW **131.** $(x + 4)(x - 2)$
133. $x^5y^4 + x^4y^6 = x^4y^4(x + y^2)$ **135.** $x^{-9}(x^3 + 1 + x^6)$
137. $x^{1/3}(1 - 5x^{1/6} + 3x^{5/12})$ **139.** $(x - 1)(5x^4 + x^2 + 3)$
141. $2x^a(x^{2a} + 4 + 2x^a)$ **143.** $2x^2(4 - \pi)$

Exercise Set 6.2, pp. 445–447

1. True **2.** True **3.** False **4.** True **5.** True **6.** False
7. True **8.** True **9.** $(x + 2)(x + 6)$ **11.** $(t + 3)(t + 5)$
13. $(a - 3)(a - 4)$ **15.** $(x - 5)(x + 3)$
17. $(x + 5)(x - 3)$ **19.** $2(n - 5)(n - 5)$, or $2(n - 5)^2$
21. $a(a + 8)(a - 9)$ **23.** $(x + 9)(x + 5)$
25. $(x + 5)(x - 2)$ **27.** $3(x - 2)(x - 3)$
29. $-(x - 8)(x + 7)$, or $(-x + 8)(x + 7)$, or $(8 - x)(7 + x)$
31. $-y(y - 8)(y + 4)$, or $y(-y + 8)(y + 4)$, or $y(8 - y)(4 + y)$
33. $x^2(x + 16)(x - 5)$ **35.** Prime **37.** $(p - 8q)(p + 3q)$
39. $(y + 4z)(y + 4z)$, or $(y + 4z)^2$ **41.** $p^2(p - 1)(p - 79)$
43. $-6, -2$ **45.** 5 **47.** $-5, 1$ **49.** $-3, 2$ **51.** $-5, 9$
53. $-3, -1, 0$ **55.** $-9, 5$ **57.** $0, 9$ **59.** $0, 5, 8$ **61.** $-4, 5$
63. $-7, 5$ **65.** $(x + 22)(x - 12)$ **67.** $(x + 24)(x + 16)$
69. $(x + 64)(x - 38)$
71. $f(x) = (x + 1)(x - 2)$, or $f(x) = x^2 - x - 2$
73. $f(x) = (x + 7)(x + 10)$, or $f(x) = x^2 + 17x + 70$
75. $f(x) = x(x - 1)(x - 2)$, or $f(x) = x^3 - 3x^2 + 2x$
77. TW **79.** $6x^2 + 17x + 12$ **80.** $6x^2 + x - 12$
81. $6x^2 - x - 12$ **82.** $6x^2 - 17x + 12$ **83.** $5x^2 - 36x + 7$
84. $3x^2 + 13x - 30$ **85.** TW
87. $\{-1, 3\}$; $(-2, 4)$, or $\{x \mid -2 < x < 4\}$
89. $f(x) = 5x^3 - 20x^2 + 5x + 30$; answers may vary
91. 6.90 **93.** 3.48 **95.** $\left(x + \frac{3}{4}\right)\left(x - \frac{1}{4}\right)$
97. $(x^a + 8)(x^a - 3)$ **99.** $(a + 1)(x + 2)(x + 1)$
101. $(x - 4)(x + 8)$ **103.** $31, -31, 14, -14, 4, -4$
105. $(x + 4)(x + 5)$

Exercise Set 6.3, pp. 455–456

1. (f) **2.** (c) **3.** (e) **4.** (a) **5.** (g) **6.** (d) **7.** (h)
8. (b) **9.** $(2x - 1)(x + 4)$ **11.** $(3x + 1)(x - 6)$
13. $(5a - 3)(3a - 1)$ **15.** $(2t + 1)(3t + 7)$
17. $2(3x + 1)(x - 2)$ **19.** $4(x - 4)(2x + 1)$
21. $x^2(2x - 3)(7x + 1)$ **23.** $(4x - 5)(3x - 2)$
25. $(3x + 4)(3x + 1)$ **27.** $(4x + 3)(x + 3)$
29. Prime **31.** $-2(2t - 3)(2t + 5)$
33. $-1(3z - 2)(3z + 4)$, or $(4 + 3z)(2 - 3z)$
35. $xy(6y + 5)(3y - 2)$ **37.** $(x - 2)(24x + 1)$
39. $3x(7x + 3)(3x + 4)$ **41.** $2x^2(4x - 3)(6x + 5)$
43. $(4a - 3b)(3a - 2b)$ **45.** $(2x - 3y)(x + 2y)$
47. $2(s + t)(4s + 7t)$
49. $(3x - 5y)(3x - 5y)$, or $(3x - 5y)^2$ **51.** $(9xy - 4)(xy + 1)$
53. $-\frac{4}{3}, \frac{2}{3}$ **55.** $-\frac{4}{3}, -\frac{3}{7}, 0$ **57.** $\frac{2}{3}, 2$ **59.** $-2, -\frac{3}{4}, 0$
61. $-\frac{1}{3}, \frac{5}{2}$ **63.** $-\frac{7}{4}, \frac{4}{3}$ **65.** $-\frac{1}{2}, 7$ **67.** $-8, -4$ **69.** $-4, \frac{3}{2}$
71. $\{x \mid x$ is a real number *and* $x \neq 5$ *and* $x \neq -1\}$
73. $\{x \mid x$ is a real number *and* $x \neq 0$ *and* $x \neq \frac{1}{2}\}$
75. $\{x \mid x$ is a real number *and* $x \neq \frac{1}{2}$ *and* $x \neq 4\}$
77. $\{x \mid x$ is a real number *and* $x \neq 0$ *and* $x \neq 2$ *and* $x \neq 5\}$
79. TW **81.** $x^2 - 4x + 4$ **82.** $x^2 + 4x + 4$
83. $x^2 - 4$ **84.** $25t^2 - 30t + 9$ **85.** $16a^2 + 8a + 1$
86. $4n^2 - 49$ **87.** $9c^2 - 60c + 100$ **88.** $1 - 10a + 25a^2$
89. $64n^2 - 9$ **90.** $81 - y^2$ **91.** TW **93.** $(2x + 15)(2x + 45)$
95. $3x(x + 68)(x - 18)$ **97.** $-\frac{11}{8}, -\frac{1}{4}, \frac{2}{3}$ **99.** 1
101. $(3ab + 2)(6ab - 5)$ **103.** Prime **105.** $(5t^5 - 1)^2$
107. $(10x^n + 3)(2x^n + 1)$ **109.** $[7(t - 3)^n - 2][(t - 3)^n + 1]$
111. $(2a^2b^3 + 5)(a^2b^3 - 4)$ **113.** Since $ax^2 + bx + c = (mx + r)(nx + s)$, from FOIL we know that $a = mn$, $c = rs$, and $b = ms + rn$. If $P = ms$ and $Q = rn$, then $b = P + Q$. Since $ac = mnrs = msrn$, we have $ac = PQ$.

Mid-Chapter Review: Chapter 6, p. 457

Guided Solutions

1. $12x^3y - 8xy^2 + 24x^2y = 4xy(3x^2 - 2y + 6x)$
2. $3a^3 - 3a^2 - 90a = 3a(a^2 - a - 30)$
$= 3a(a - 6)(a + 5)$

Mixed Review

1. $6x^2(x^3 - 3)$ **2.** $(x + 2)(x + 8)$ **3.** $(x + 7)(2x - 1)$
4. $(x + 3)(x^2 + 2)$ **5.** $5(x - 2)(x + 10)$ **6.** Prime
7. $7y(x - 4)(x + 1)$ **8.** $3a^2(5a^2 - 9b^2 + 7b)$
9. $(b - 7)(b - 7)$, or $(b - 7)^2$ **10.** $(3x - 1)(4x + 1)$
11. $(xy - 2)(xy + 1)$ **12.** $2(x - 5)(x + 20)$ **13.** Prime
14. $15(d^2 - 2d + 5)$ **15.** $(3p + 2x)(5p + 2x)$
16. $-2t(t + 2)(t + 3)$ **17.** $(x + 11)(x - 7)$
18. $10(c + 1)(c + 1)$, or $10(c + 1)^2$ **19.** $-1(2x - 5)(x + 1)$
20. $2n(m - 5)(m^2 - 3)$

Exercise Set 6.4, pp. 464–466

1. Difference of two squares **2.** Perfect-square trinomial
3. Perfect-square trinomial **4.** Difference of two squares
5. None of these **6.** Prime polynomial **7.** Prime polynomial

8. None of these **9.** Yes **11.** No **13.** No **15.** Yes
17. $(t + 3)^2$ **19.** $(a - 7)^2$ **21.** $4(a - 2)^2$
23. $(t - 1)^2$, or $(1 - t)^2$ **25.** $a(a + 12)^2$ **27.** $5(2x + 5)^2$
29. $(4d^3 + 1)^2$ **31.** $-y(y - 4)^2$ **33.** $(0.5x + 0.3)^2$
35. $(x - y)^2$ **37.** $(5a^3 + 3b^3)^2$ **39.** $5(a - b)^2$ **41.** Yes
43. No **45.** Yes **47.** $(y + 10)(y - 10)$
49. $(m + 8)(m - 8)$ **51.** $(7 + t)(-7 + t)$, or $(t + 7)(t - 7)$
53. $8(x + y)(x - y)$ **55.** $-5(4a^3 + 3)(4a^3 - 3)$ **57.** Prime
59. $(t^2 + 1)(t + 1)(t - 1)$ **61.** $a^2(3a + 5b^2)(3a - 5b^2)$
63. $(4x^2 + y^2)(2x + y)(2x - y)$ **65.** $\left(\frac{1}{7} + x\right)\left(\frac{1}{7} - x\right)$
67. $(a + b + 3)(a + b - 3)$ **69.** $(x - 3 + y)(x - 3 - y)$
71. $(t + 8)(t + 1)(t - 1)$ **73.** $(r - 3)^2(r + 3)$
75. $(m - n + 5)(m - n - 5)$ **77.** $(6 + x + y)(6 - x - y)$
79. $(4 + a + b)(4 - a - b)$ **81.** $(a - 2)(a + b)(a - b)$
83. 1 **85.** 6 **87.** $-3, 3$ **89.** $-\frac{1}{5}, \frac{1}{5}$ **91.** $-\frac{1}{2}, \frac{1}{2}$
93. $-1, 1, 3$ **95.** $-1.541, 4.541$ **97.** $-3.871, -0.129$
99. $-2.414, -1, 0.414$ **101.** 6 **103.** $-4, 4$ **105.** -1
107. $-1, 1, 2$ **109.** TW **111.** $8x^6y^{12}$ **112.** $-125x^6y^3$
113. $x^3 + 3x^2 + 3x + 1$ **114.** $x^3 - 3x^2 + 3x - 1$
115. $m^3 + 3m^2n + 3mn^2 + n^3$ **116.** $m^3 - 3m^2n + 3mn^2 - n^3$
117. TW **119.** $(x^4 + 2^4)(x^2 + 2^2)(x + 2)(x - 2)$, or $(x^4 + 16)(x^2 + 4)(x + 2)(x - 2)$ **121.** $3\left(x + \frac{1}{3}\right)\left(x - \frac{1}{3}\right)$
123. $(0.3x^4 + 0.8)^2$, or $\frac{1}{100}(3x^4 + 8)^2$
125. $(r + s + 1)(r - s - 9)$ **127.** $(x^{2a} + 7y^a)(x^{2a} - 7y^a)$
129. $3(x + 3)^2$ **131.** $(3x^n - 1)^2$ **133.** $(s - 2t + 2)^2$
135. $h(2a + h)(2a^2 + 2ah + h^2)$

Exercise Set 6.5, pp. 470–471

1. Difference of cubes **2.** Sum of cubes **3.** Difference of squares **4.** None of these **5.** Sum of cubes **6.** Difference of cubes **7.** None of these **8.** Difference of squares
9. Difference of cubes **10.** None of these
11. $(x + 4)(x^2 - 4x + 16)$ **13.** $(z - 1)(z^2 + z + 1)$
15. $(t - 10)(t^2 + 10t + 100)$ **17.** $(3x + 1)(9x^2 - 3x + 1)$
19. $(4 - 5x)(16 + 20x + 25x^2)$ **21.** $8(y + 2)(y^2 - 2y + 4)$
23. $(x - y)(x^2 + xy + y^2)$ **25.** $\left(a + \frac{1}{2}\right)\left(a^2 - \frac{1}{2}a + \frac{1}{4}\right)$
27. $8(t - 1)(t^2 + t + 1)$ **29.** $\left(y - \frac{1}{10}\right)\left(y^2 + \frac{1}{10}y + \frac{1}{100}\right)$
31. $a(b + 5)(b^2 - 5b + 25)$ **33.** $5(x - 2z)(x^2 + 2xz + 4z^2)$
35. $(x + 0.1)(x^2 - 0.1x + 0.01)$
37. $8(2x^2 - t^2)(4x^4 + 2x^2t^2 + t^4)$
39. $2y(y - 4)(y^2 + 4y + 16)$
41. $(z + 1)(z^2 - z + 1)(z - 1)(z^2 + z + 1)$
43. $(t^2 + 4y^2)(t^4 - 4t^2y^2 + 16y^4)$
45. $(x^4 - yz^4)(x^8 + x^4yz^4 + y^2z^8)$
47. -1 **49.** $\frac{3}{2}$ **51.** 10 **53.** TW **55.** $-\frac{1}{5}$ **56.** $\frac{1}{4}$

57.

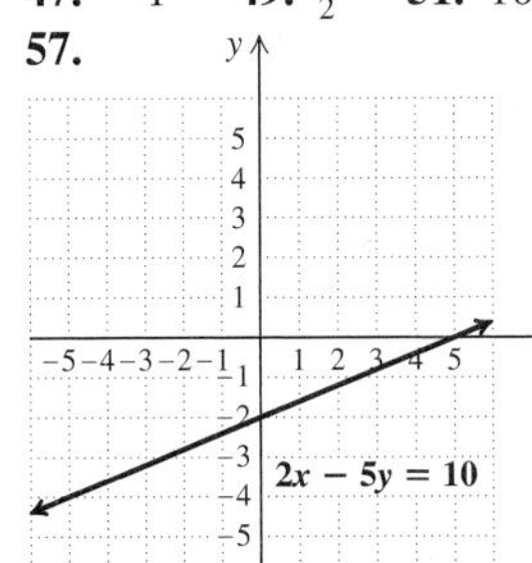

58.

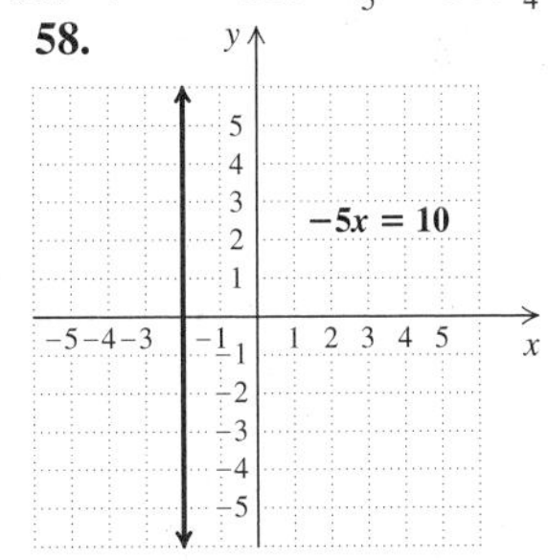

59.

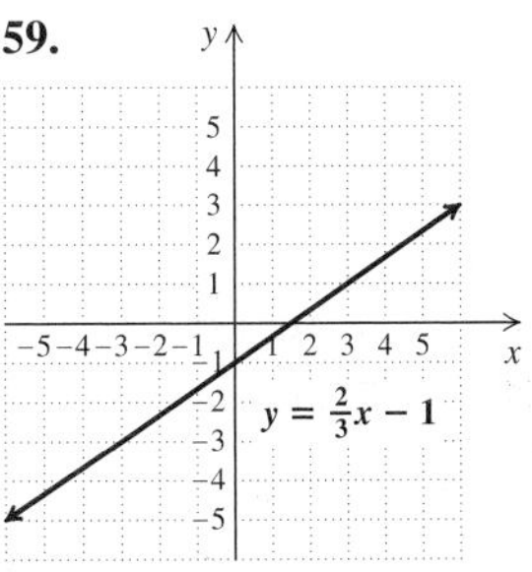

60.

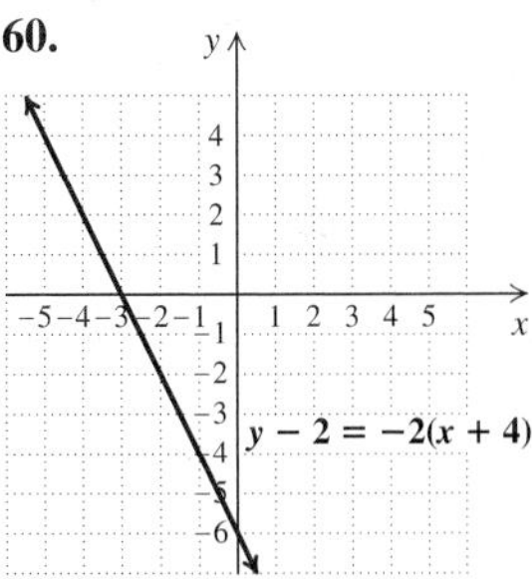

61. TW **63.** $(x^{2a} - y^b)(x^{4a} + x^{2a}y^b + y^{2b})$
65. $2x(x^2 + 75)$ **67.** $5\left(xy^2 - \frac{1}{2}\right)\left(x^2y^4 + \frac{1}{2}xy^2 + \frac{1}{4}\right)$
69. $-(3x^{4a} + 3x^{2a} + 1)$ **71.** $(t - 8)(t - 1)(t^2 + t + 1)$
73. $h(2a + h)(a^2 + ah + h^2)(3a^2 + 3ah + h^2)$

Exercise Set 6.6, pp. 475–476

1. Common factor **2.** Perfect-square trinomial **3.** Grouping
4. Multiplying **5.** $10(a + 8)(a - 8)$ **7.** $(y - 7)^2$
9. $(2t + 3)(t + 4)$ **11.** $x(x - 9)^2$ **13.** $(x - 5)^2(x + 5)$
15. $3t(3t + 1)(3t - 1)$ **17.** $3x(3x - 5)(x + 3)$ **19.** Prime
21. $3(x + 3)(2x - 5)$ **23.** $-2a^4(a - 2)^2$
25. $5x(x^2 + 4)(x + 2)(x - 2)$ **27.** $(t^2 + 3)(t^2 - 3)$
29. $-x^4(x^2 - 2x + 7)$ **31.** $(x - y)(x^2 + xy + y^2)$
33. $a(x^2 + y^2)$ **35.** $4c(4d - c)(5d - c)$ **37.** $2\pi r(h + r)$
39. $(a + b)(5a + 3b)$ **41.** $(x + 1)(x + y)$
43. $(n - 5 + 3m)(n - 5 - 3m)$ **45.** $(3x - 2y)(x + 5y)$
47. $(a - 2b)^2$ **49.** $(4x + 3y)^2$ **51.** Prime
53. $(2t + 1)(4t^2 - 2t + 1)(2t - 1)(4t^2 + 2t + 1)$
55. $8mn(m^2 - 4mn + 3)$ **57.** $(3b - a)(b + 6a)$
59. $-1(xy + 2)(xy + 6)$, or $-(xy + 2)(xy + 6)$
61. $(t^4 + s^5 + 6)(t^4 - s^5 - 6)$
63. $2a(3a + 2b)(9a^2 - 6ab + 4b^2)$ **65.** $x^4(x + 2y)(x - y)$
67. $\left(6a - \frac{5}{4}\right)^2$ **69.** $\left(\frac{1}{9}x - \frac{4}{3}\right)^2$
71. $(1 + 4x^6y^6)(1 + 2x^3y^3)(1 - 2x^3y^3)$
73. $(2ab + 3)^2$ **75.** $a(a^2 + 8)(a + 8)$ **77.** TW
79. Let m and n represent the numbers; $(m + n)^2$ **80.** Let m and n represent the numbers; $m^2 + n^2$ **81.** Let x represent the first integer; then $x + 1$ represents the second integer; $x(x + 1)$
82. Mother's Day: \$14.1 billion; Father's Day: \$9.4 billion
83. 140°, 35°, 5° **84.** Length: 64 in.; width: 32 in. **85.** TW
87. $-x(x^2 + 9)(x^2 - 2)$
89. $-3(a + 1)(a - 1)(a + 2)(a - 2)$
91. $(y + 1)(y - 7)(y + 3)$ **93.** $(2x - 2 + 3y)(3x - 3 - y)$
95. $(a + 3)^2(2a + b + 4)(a - b + 5)$
97. $(7x^2 + 1 + 5x^3)(7x^2 + 1 - 5x^3)$

Visualizing for Success, p. 485

1. D **2.** J **3.** A **4.** B **5.** E **6.** C **7.** I **8.** F
9. G **10.** H

Exercise Set 6.7, pp. 486–491

1. $-12, 11$ **3.** 10, 11 **5.** Length: 30 ft; width: 6 ft
7. Length: 12 cm; width: 7 cm **9.** Foot: 7 ft; height: 12 ft
11. Base: 8 ft; height: 16 ft **13.** 16 teams **15.** 1 min, 3 min
17. 10 knots **19.** 4 sec **21.** 5 sec **23.** 3 m **25.** 3 cm

27. 10 ft **29.** 9 ft **31.** 32 ft **33.** 300 ft by 400 ft by 500 ft
35. Distance d: 12 ft; tower height: 16 ft **37.** 3, 4, 5
39. Dining room: 12 ft by 12 ft; kitchen: 12 ft by 10 ft
41. **(a)** $P(x) = 0.01823x^4 - 0.77199x^3 + 11.62153x^2 - 73.65807x + 179.76190$; **(b)** 7%; **(c)** 2003 and 2007
43. **(a)** $F(x) = -0.03587x^3 + 10.35169x^2 - 871.97543x + 23423.45189$; **(b)** 5220 athletes; **(c)** 2024; $x = 122$, which gives 2022; the first games after 2022 will be in 2024 **45.** TW **47.** $-\frac{12}{35}$
48. $-\frac{21}{20}$ **49.** -1 **50.** $-\frac{7}{4}$ **51.** $\frac{1}{4}$ **52.** $\frac{4}{5}$ **53.** $\frac{53}{168}$
54. $\frac{19}{18}$ **55.** TW **57.** 10 squares **59.** 2 hr, 4.2 hr **61.** 3 hr
63. Length: 28 cm; width: 14 cm **65.** About 5.7 sec

Study Summary: Chapter 6, pp. 492–494

1. $6x(2x^3 - 3x^2 + 5)$ **2.** $(x - 3)(2x^2 - 1)$ **3.** $0, \frac{4}{3}$
4. $(x - 9)(x + 2)$ **5.** $(3x + 2)(2x - 1)$
6. $(2x - 3)(4x - 5)$ **7.** $(10n + 9)^2$ **8.** $(12t + 5)(12t - 5)$
9. $(a - 1)(a^2 + a + 1)$ **10.** $(x - 2)(x^2 + 2x + 4)(3x + 5)$
11. 12, 13

Review Exercises: Chapter 6, pp. 495–496

1. False **2.** True **3.** True **4.** False **5.** False **6.** True
7. True **8.** False **9.** $x(7x + 6)$ **10.** $6y^2(3y^2 - 1)$
11. $(10t + 1)(10t - 1)$ **12.** $(a - 9)(a - 3)$
13. $(3m + 2)(m + 4)$ **14.** $(5x + 2)^2$ **15.** $4(y + 2)(y - 2)$
16. $x(x - 2)(x + 7)$ **17.** $(a + 2b)(x - y)$
18. $(y + 2)(3y^2 - 5)$ **19.** $(9a^2 + 1)(3a + 1)(3a - 1)$
20. $2(24t^2 - 14t + 3)$ **21.** $(3x - 2)(9x^2 + 6x + 4)$
22. $-t(t + 6)(t - 7)$ **23.** $(ab^2 + 8)(ab^2 - 8)$
24. Prime **25.** $2z^6(z^2 - 8)$
26. $2y(3x^2 - 1)(9x^4 + 3x^2 + 1)$ **27.** $3(2x - 5)^2$
28. $(3t + p)(2t + 5p)$ **29.** $(x + 3)(x - 3)(x + 2)$
30. $(a - b + 2t)(a - b - 2t)$ **31.** $-2, 1, 5$ **32.** 4, 7
33. $-11, 9$ **34.** $\frac{2}{3}, \frac{3}{2}$ **35.** $0, \frac{7}{4}$ **36.** 10 **37.** $-4, 4$
38. $-3, 0, 7$ **39.** $-4, 5$ **40.** $-4, 4, 5$ **41.** 12, 15
42. $-1.646, 3.646$ **43.** $-4, 11$ **44.** $\{x \mid x$ is a real number *and* $x \neq -7$ *and* $x \neq \frac{2}{3}\}$ **45.** 10 teams **46.** Height: 18 m; base: 24 m **47.** Length: 8 in.; width: 5 in. **48.** 17 ft
49. **(a)** $P(x) = 0.0212x^2 - 0.9442x + 26.6508$; **(b)** 24.4; **(c)** approximately 1972 and 2013 **50.** TW The roots of a polynomial function are the x-coordinates of the points at which the graph of the function crosses or touches the x-axis. **51.** TW If the factors of the quadratic polynomial are different, there will be two different solutions. If the polynomial is a perfect square, the factors will be the same, and there will not be two different solutions.
52. $2(2x - y)(4x^2 + 2xy + y^2)(2x + y)(4x^2 - 2xy + y^2)$
53. $-2(3x^2 + 1)$ **54.** $-1, -\frac{1}{2}$ **55.** No real solution

Test: Chapter 6, pp. 496–497

1. [6.4] $(x - 5)^2$ **2.** [6.4] $(y + 5)(y + 2)(y - 2)$
3. [6.2] $(p - 14)(p + 2)$ **4.** [6.1] $t^5(t^2 - 3)$
5. [6.3] $(6m + 1)(2m + 3)$ **6.** [6.4] $(3y + 5)(3y - 5)$
7. [6.5] $3(r - 1)(r^2 + r + 1)$ **8.** [6.4] $5(3x + 2)^2$
9. [6.4] $3(x^2 + 4y^2)(x + 2y)(x - 2y)$
10. [6.4] $(y + 4 + 10t)(y + 4 - 10t)$ **11.** [6.2] Prime
12. [6.4] $5(2a - b)(2a + b)$ **13.** [6.3] $2(4x - 1)(3x - 5)$
14. [6.2] $3(m - 5n)(m + 2n)$ **15.** [6.1] $-3, -1, 2, 4$
16. [6.3] $-\frac{5}{2}, 8$ **17.** [6.2] $-3, 6$ **18.** [6.4] $-5, 5$
19. [6.3] $-7, -\frac{3}{2}$ **20.** [6.1] $-\frac{1}{3}, 0$ **21.** [6.4] 9
22. [6.2] $-3.372, 0, 2.372$ **23.** [6.3] 0, 5 **24.** [6.3] $\{x \mid x$ is a real number *and* $x \neq -1\}$ **25.** [6.7] Length: 8 cm; width: 5 cm
26. [6.7] $4\frac{1}{2}$ sec **27.** [6.7] 24 ft **28.** [6.7] **(a)** $E(x) = 0.50625x^2 + 0.245x + 32.86375$; **(b)** about \$76.1 billion; **(c)** in about 2011 **29.** [6.2] $(a - 4)(a + 8)$ **30.** [6.3] $-\frac{8}{3}, 0, \frac{2}{5}$

Cumulative Review: Chapters 1–6, pp. 498–499

1. 8 **2.** -8 **3.** -30.7 **4.** 29 **5.** $\dfrac{1}{9x^4y^6}$

6. $3x^4 + 5x^3 + x - 10$ **7.** $-3a^2b - ab^2 + 2b^3$ **8.** $\dfrac{t^8}{4s^2}$

9. $-\dfrac{8x^6y^3}{27z^{12}}$ **10.** $-4t^8 + 8t^6 + 20t^5$ **11.** $36x^2 - 60xy + 25y^2$
12. $100x^{10} - 1$ **13.** $x^3 - 2x^2 + 1$
14. $5x^2 - 4x + 2 + \dfrac{2}{3x} + \dfrac{6}{x^2}$
15. $x^3 - x^2 + 9x - 25 + \dfrac{93}{x + 3}$ **16.** $(c + 1)(c - 1)$
17. $5(x + y)(1 + 2x)$ **18.** $2x(3 - x - 12x^3)$
19. $(4x + 9)(4x - 9)$ **20.** $(t - 4)(t - 6)$
21. $(4x + 3)(2x + 1)$ **22.** $2(3x - 2)(x - 4)$
23. $2(x + 5)(x^2 - 5x + 25)$ **24.** $(4x + 5)^2$ **25.** 10
26. $\left\{x \mid x > \frac{9}{2}\right\}$, or $\left(\frac{9}{2}, \infty\right)$ **27.** $-3, 1$ **28.** 1, 10 **29.** -8
30. $(3, -2)$ **31.** $-4, \frac{1}{2}$ **32.** $\{x \mid x > 43\}$, or $(43, \infty)$
33. $\left(\frac{5}{12}, -\frac{2}{3}\right)$ **34.** $c = \dfrac{a}{b + d}$ **35.** $\frac{3}{2}x + 2y - 3z$
36. $-4x^3 - \frac{1}{7}x^2 - 2$

37.
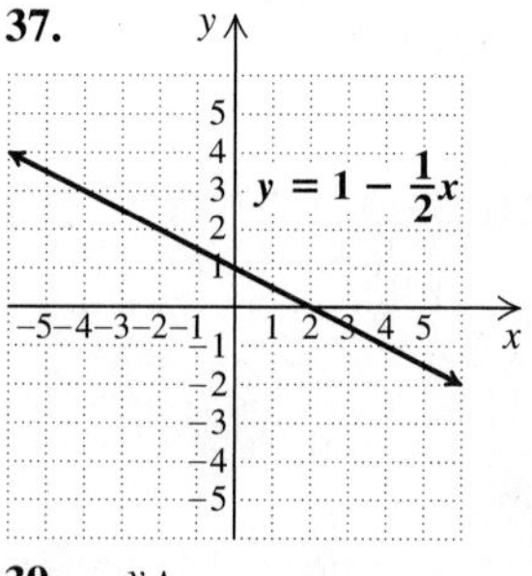

38.
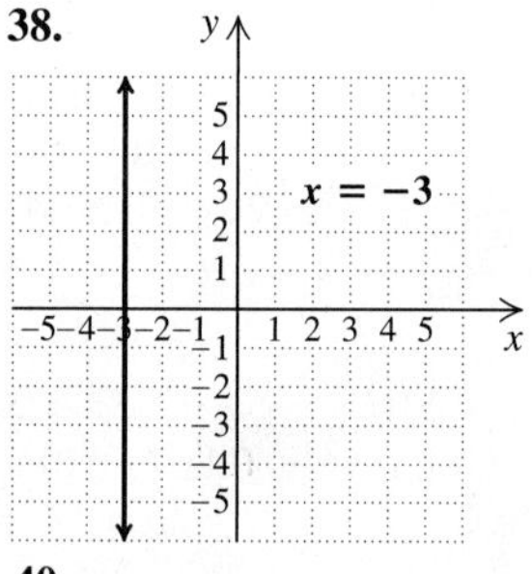

39.
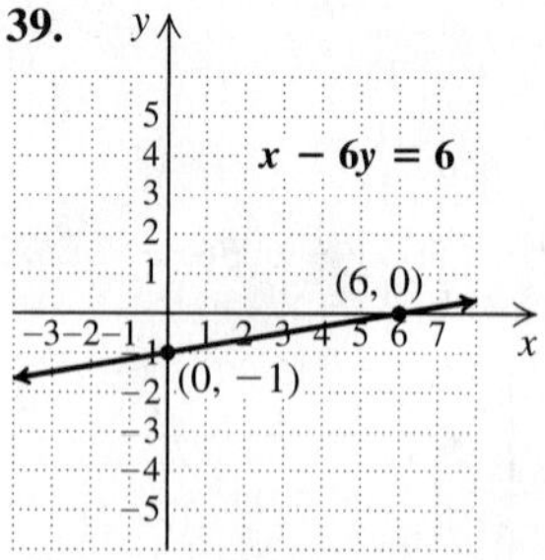

40.
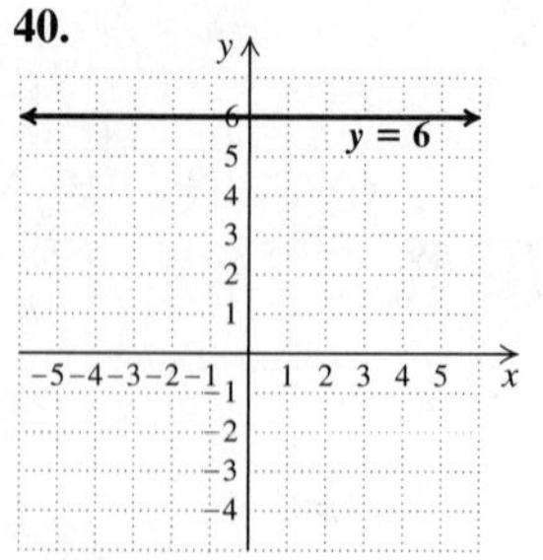

41.
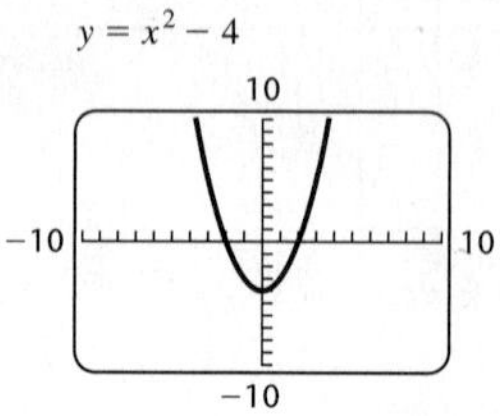

42. -2

43. Slope: $-\frac{3}{4}$; y-intercept: $(0, 2)$ **44.** $y = \frac{1}{2}x - 7$
45. $y = -x + 3$ **46.** About 35.7 million subscribers
47. Emily: \$840; Andrew: \$420 **48.** 101,180 people
49. 36 two-point shots; 28 foul shots **50.** Bottom of ladder to building: 5 ft; top of ladder to ground: 12 ft **51.** Length: 12 ft; width: 2 ft **52.** **(a)** $y = -0.1x + 13.1$; **(b)** 7.1 miles per gallon
53. Not linear **54.** Linear **55.** $f(x) = 0.2679x + 16.6679$, where x is the number of years after 2002
56. Domain: $\mathbb{R}$; range: $\{3\}$ **57.** Domain: $\mathbb{R}$; range: $\mathbb{R}$
58. Domain: $\mathbb{R}$; range: $\{y | y \geq -3\}$, or $[-3, \infty)$
59. Domain: $\mathbb{R}$; range: $\{y | y \geq -4\}$, or $[-4, \infty)$ **60.** 12
61. $2(a^{16} + 81b^{20})(a^8 + 9b^{10})(a^4 + 3b^5)(a^4 - 3b^5)$
62. -7 **63.** $-5, -3, 3, 5$

Chapter 7

Interactive Discovery, p. 509

1. One vertical asymptote: $x = -\frac{1}{2}$ **2.** $x = -\frac{1}{2}$

Visualizing for Success, p. 510

1. I **2.** B **3.** G **4.** A **5.** D **6.** C **7.** E
8. J **9.** H **10.** F

Exercise Set 7.1, pp. 511–513

1. (e) **2.** (c) **3.** (i) **4.** (f) **5.** (d) **6.** (b)
7. $\frac{40}{13}$ hr, or $3\frac{1}{13}$ hr **9.** $\frac{2}{3}$; 28; $\frac{163}{10}$ **11.** $-\frac{9}{4}$; does not exist; $-\frac{11}{9}$
13. 0 **15.** -8 **17.** 4 **19.** $-4, 7$ **21.** $-5, 5$
23. $\frac{3}{x}$ **25.** $\frac{2w}{3t^4}$ **27.** $a - 5$ **29.** $\frac{x - 4}{x + 5}$ **31.** $\frac{3(2a - 1)}{7(a - 1)}$
33. $-\frac{1}{2}$ **35.** $\frac{a + 4}{2(a - 4)}$ **37.** $\frac{x + 4}{x - 4}$ **39.** $t - 1$
41. $\frac{y^2 + 4}{y + 2}$ **43.** $\frac{1}{2}$ **45.** -1 **47.** $-\frac{1}{1 + 2t}$ **49.** $\frac{a - 5}{a + 5}$
51. -1 **53.** $\frac{x^2 + x + 1}{x + 1}$ **55.** $3(y + 2)$
57. $f(x) = \frac{3}{x}, x \neq -7, 0$ **59.** $g(x) = \frac{x - 3}{5}, x \neq -3$
61. $h(x) = -\frac{1}{5}, x \neq 4$ **63.** $f(t) = \frac{t - 3}{t + 1}, t \neq -3, -1$
65. $g(t) = -\frac{7}{3}, t \neq 3$ **67.** $h(t) = \frac{t + 4}{t - 9}, t \neq -1, 9$
69. $f(x) = 3x + 2, x \neq \frac{2}{3}$ **71.** $x = -5$
73. No vertical asymptotes **75.** $x = -3$ **77.** $x = 2, x = 4$
79. (b) **81.** (f) **83.** (a) **85.** TW **87.** $-\frac{4}{21}$
88. $-\frac{5}{3}$ **89.** $-\frac{15}{4}$ **90.** $-\frac{21}{16}$ **91.** $\frac{13}{63}$ **92.** $\frac{5}{48}$
93. TW **95.** $2a + h$ **97.** $\frac{(x^2 + y^2)(x + y)}{(x - y)^3}$
99. $\frac{1}{x - 1}$ **101.** $\frac{a^2 + 2}{a^2 - 3}$ **103.** 1
105. Domain: $\{x | x$ is a real number *and* $x \neq -2$ *and* $x \neq 1\}$; range: $\{y | y$ is a real number *and* $y \neq 2$ *and* $y \neq 3\}$
107. Domain: $\{x | x$ is a real number *and* $x \neq -1$ *and* $x \neq 1\}$; range: $\{y | y \leq -1$ *or* $y > 0\}$ **109.** TW

Exercise Set 7.2, pp. 518–520

1. (d) **2.** (c) **3.** (a) **4.** (e) **5.** (b) **6.** (f)
7. $\frac{7x(x - 5)}{5(2x + 1)}$ **9.** $\frac{(a - 4)(a + 2)}{(a + 6)^2}$ **11.** $\frac{(2x + 3)(x + 1)}{4(x - 5)}$
13. $\frac{5a^2}{3}$ **15.** $\frac{4}{c^2 d}$ **17.** $\frac{y + 4}{4}$ **19.** $\frac{x + 2}{x - 2}$ **21.** $\frac{y(y - 1)}{y + 1}$
23. $\frac{5(a + 3)}{a(a + 4)}$ **25.** $\frac{2a}{a - 2}$ **27.** $\frac{t - 5}{t + 5}$ **29.** $\frac{5(a + 6)}{a - 1}$
31. 1 **33.** $\frac{t + 2}{t + 4}$ **35.** $-\frac{a + 1}{2 + a}$ **37.** $(2x - 1)^2$
39. $c(c - 2)$ **41.** $\frac{7}{3x}$ **43.** $\frac{1}{a^3 - 8a}$ **45.** $\frac{35}{24}$ **47.** $\frac{x^2}{20}$
49. a^3b^3 **51.** $\frac{y + 5}{2y}$ **53.** $4(y - 2)$ **55.** $-\frac{a}{b}$
57. $\frac{(y + 3)(y^2 + 1)}{y + 1}$ **59.** $\frac{15}{16}$ **61.** $\frac{a - 5}{3(a - 1)}$ **63.** $-\frac{5x + 2}{x - 3}$
65. $\frac{(2x - 1)(2x + 1)}{x - 5}$ **67.** $\frac{(w - 7)(w - 8)}{(2w - 7)(3w + 1)}$
69. $\frac{1}{(c - 5)(5c - 3)}$ **71.** $\frac{x^2 + 4x + 16}{(x + 4)^2}$ **73.** $\frac{(2a + b)^2}{2(a + b)}$
75. TW **77.** $\frac{19}{12}$ **78.** $\frac{41}{24}$ **79.** $\frac{1}{18}$ **80.** $-\frac{1}{6}$ **81.** $x^2 + 3$
82. $-2x^2 - 4x + 1$ **83.** TW **85.** $\frac{3}{7x}$ **87.** 1
89. 1 **91.** $\frac{a^2 - 2b}{a^2 + 3b}$ **93.** $\frac{2s}{r + 2s}$
95. **(a)** $\frac{2x + 2h + 3}{4x + 4h - 1}$; **(b)** $\frac{2x + 3}{8x - 9}$; **(c)** $\frac{x + 5}{4x - 1}$

Exercise Set 7.3, pp. 526–529

1. Numerators; denominator **2.** Term
3. Least common denominator; LCD
4. Factorizations; denominators **5.** $\frac{10}{x}$ **7.** $\frac{3x + 5}{12}$
9. $\frac{9}{a + 3}$ **11.** $\frac{8}{4x - 7}$ **13.** $\frac{2y + 7}{2y}$ **15.** 11
17. $\frac{3x + 5}{x + 1}$ **19.** $a + 5$ **21.** $x - 4$ **23.** 0 **25.** $\frac{1}{x + 2}$
27. $\frac{(3a - 7)(a - 2)}{(a + 6)(a - 1)}$ **29.** $\frac{t - 4}{t + 3}$ **31.** $\frac{x + 5}{x - 6}$
33. $-\frac{5}{x - 4}$, or $\frac{5}{4 - x}$ **35.** $-\frac{1}{x - 1}$, or $\frac{1}{1 - x}$
37. 135 **39.** 72 **41.** 60 **43.** $12x^3$ **45.** $30a^4b^8$
47. $6(y - 3)$ **49.** $(x - 2)(x + 2)(x + 3)$
51. $t(t - 4)(t + 2)^2$ **53.** $30x^2y^2z^3$
55. $(a + 1)(a - 1)^2$ **57.** $(2n - 1)(n + 1)(n + 2)$
59. $t^2 - 9$ **61.** $12x^3(x - 5)(x - 3)(x - 1)$
63. $2(t + 1)(t - 1)(t^2 + t + 1)$ **65.** $\frac{10}{12x^5}, \frac{x^2y}{12x^5}$
67. $\frac{12b}{8a^2b^2}, \frac{7a}{8a^2b^2}$ **69.** $\frac{2x(x + 3)}{(x - 2)(x + 2)(x + 3)}$, $\frac{4x(x - 2)}{(x - 2)(x + 2)(x + 3)}$ **71.** TW **73.** $\frac{-5}{8}, \frac{5}{-8}$ **74.** $\frac{-4}{11}, -\frac{4}{11}$

75. $-x + y$, or $y - x$ **76.** $-3 + a$, or $a - 3$
77. $-2x + 7$, or $7 - 2x$ **78.** $-a + b$, or $b - a$ **79.** TW
81. $\frac{18x + 5}{x - 1}$ **83.** $\frac{x}{3x + 1}$ **85.** 30 strands **87.** 60 strands
89. $(2x + 5)(2x - 5)(3x + 4)^4$ **91.** 12 sec
93. 7:55 A.M. **95.** TW

Exercise Set 7.4, pp. 534–537

1. LCD **2.** Missing; denominator **3.** Numerators; LCD
4. Simplify **5.** $\frac{4x + 9}{x^2}$ **7.** $-\frac{5}{24r}$ **9.** $\frac{2d^2 + 7c}{c^2d^3}$
11. $\frac{-2xy - 18}{3x^2y^3}$ **13.** $\frac{5x + 7}{18}$ **15.** $\frac{a + 8}{4}$
17. $\frac{5a^2 + 7a - 3}{9a^2}$ **19.** $\frac{-7x - 13}{4x}$ **21.** $\frac{c^2 + 3cd - d^2}{c^2d^2}$
23. $\frac{3y^2 - 3xy - 6x^2}{2x^2y^2}$ **25.** $\frac{10x}{(x - 1)(x + 1)}$
27. $\frac{2z + 6}{(z - 1)(z + 1)}$ **29.** $\frac{11x + 15}{4x(x + 5)}$ **31.** $\frac{16 - 9t}{6t(t - 5)}$
33. $\frac{x^2 - x}{(x - 5)(x + 5)}$ **35.** $\frac{4t - 5}{4(t - 3)}$ **37.** $\frac{2x + 10}{(x + 3)^2}$
39. $\frac{1}{t^2 + t + 1}$ **41.** $\frac{9a}{4(a - 5)}$ **43.** 0
45. $\frac{10}{(a - 3)(a + 2)}$ **47.** $\frac{x - 5}{(x + 5)(x + 3)}$
49. $\frac{3z^2 + 19z - 20}{(z - 2)^2(z + 3)}$ **51.** $\frac{-5}{x^2 + 17x + 16}$
53. $\frac{3x - 3}{5}$ **55.** $\frac{y^2 + 9}{y - 3}$ **57.** 0 **59.** $\frac{y^2 + 10y + 11}{(y - 7)(y + 7)}$
61. $\frac{(x + 1)(x + 3)}{(x - 4)(x + 4)}$ **63.** $\frac{3x^2 - 7x - 4}{3(x - 2)(x + 2)}$
65. $\frac{a - 2}{(a - 3)(a + 3)}$ **67.** $\frac{2x - 3}{2 - x}$ **69.** 2
71. 0 **73.** $f(x) = \frac{3(x + 4)}{x + 3}, x \neq \pm 3$
75. $f(x) = \frac{(x - 7)(2x - 1)}{(x - 4)(x - 1)(x + 3)}, x \neq -4, -3, 1, 4$
77. $f(x) = \frac{-2}{(x + 1)(x + 2)}, x \neq -3, -2, -1$ **79.** TW
81. $-\frac{3}{22}$ **82.** $\frac{7}{9}$ **83.** $\frac{9}{10}$ **84.** $\frac{16}{27}$ **85.** $\frac{2}{3}$
86. $\frac{(x - 3)(x + 2)}{(x - 2)(x + 3)}$ **87.** TW
89. Perimeter: $\frac{2(5x - 7)}{(x - 5)(x + 4)}$; area: $\frac{6}{(x - 5)(x + 4)}$
91. $\frac{30}{(x - 3)(x + 4)}$ **93.** $\frac{x^4 + 4x^3 - 5x^2 - 126x - 441}{(x + 2)^2(x + 7)^2}$
95. $\frac{-x^2 - 3}{(2x - 3)(x - 3)}$ **97.** $\frac{9x^2 + 28x + 15}{(x - 3)(x + 3)^2}$
99. $\frac{a}{a - b} + \frac{3b}{b - a}$; Answers may vary.
101. $\frac{x^4 + 6x^3 + 2x^2}{(x + 2)(x - 2)(x + 5)}$ **103.** $\frac{x^5}{(x^2 - 4)(x^2 + 3x - 10)}$
105. $\{x \mid x \text{ is a real number and } x \neq -5 \text{ and } x \neq -2 \text{ and } x \neq 2\}$
107. Domain: $\{x \mid x \text{ is a real number and } x \neq -1\}$; range: $\{y \mid y \text{ is a real number and } y \neq 3\}$
109. Domain: $\{x \mid x \text{ is a real number and } x \neq 0 \text{ and } x \neq 1\}$; range: $\{y \mid y > 0\}$

Mid-Chapter Review: Chapter 7, pp. 537–538

Guided Solutions

1.
$$\frac{a^2}{a - 10} \div \frac{a^2 + 5a}{a^2 - 100} = \frac{a^2}{a - 10} \cdot \frac{a^2 - 100}{a^2 + 5a}$$
$$= \frac{a \cdot a \cdot (a + 10) \cdot (a - 10)}{(a - 10) \cdot a \cdot (a + 5)}$$
$$= \frac{a(a - 10)}{a(a - 10)} \cdot \frac{a(a + 10)}{a + 5}$$
$$= \frac{a(a + 10)}{a + 5}$$

2.
$$\frac{2}{x} + \frac{1}{x^2 + x} = \frac{2}{x} + \frac{1}{x(x + 1)}$$
$$= \frac{2}{x} \cdot \frac{x + 1}{x + 1} + \frac{1}{x(x + 1)}$$
$$= \frac{2x + 2}{x(x + 1)} + \frac{1}{x(x + 1)}$$
$$= \frac{2x + 3}{x(x + 1)}$$

Mixed Review

1. $\frac{3x + 10}{5x^2}$ **2.** $\frac{6}{5x^3}$ **3.** $\frac{3x}{10}$ **4.** $\frac{3x - 10}{5x^2}$ **5.** $\frac{x - 3}{15(x - 2)}$
6. $\frac{6}{(x + 3)(x + 4)}$ **7.** $\frac{1}{3}$ **8.** $\frac{x^2 - 2x - 2}{(x - 1)(x + 2)}$
9. $\frac{5x + 17}{(x + 3)(x + 4)}$ **10.** -5 **11.** $\frac{5}{x - 4}$
12. $\frac{(2x + 3)(x + 3)}{(x + 1)^2}$ **13.** $\frac{1}{6}$ **14.** $\frac{x(x + 4)}{(x - 1)^2}$
15. $\frac{x + 7}{(x - 5)(x + 1)}$ **16.** $\frac{9(u - 1)}{16}$ **17.** $\frac{7t + 8}{30}$
18. $(t + 5)^2$ **19.** $\frac{a - 1}{(a + 2)(a - 2)^2}$ **20.** $\frac{-3x^2 + 9x - 14}{2x}$

Exercise Set 7.5, pp. 545–547

1. (a) **2.** (b) **3.** (b) **4.** (a) **5.** 10 **7.** $\frac{5}{11}$
9. $\frac{5x^2}{4(x^2 + 4)}$ **11.** $\frac{(x + 2)(x - 3)}{(x - 1)(x + 4)}$ **13.** $\frac{5b - 4a}{2b + 3a}$
15. $-\frac{1}{5}$ **17.** $\frac{a + b}{a}$ **19.** $\frac{3}{3x + 2}$ **21.** $\frac{1}{x - y}$
23. $\frac{1}{a(a - h)}$ **25.** $\frac{(a - 2)(a - 7)}{(a + 1)(a - 6)}$ **27.** $\frac{x + 2}{x + 3}$
29. $\frac{y^2 + 1}{y^2 - 1}$ **31.** $\frac{x}{x - y}$ **33.** 1 **35.** $\frac{3a^2 + 4b^3}{a^4b^5}$
37. $\frac{-x^3y^3}{y^2 + xy + x^2}$ **39.** $\frac{a^2 - 3a - 6}{a^2 - 2a - 3}$ **41.** $\frac{a + 1}{2a + 5}$
43. $-y$ **45.** $\frac{(2a - 3)(a + 5)}{2(a - 3)(a + 2)}$ **47.** $\frac{t^2 + 5t + 3}{(t + 1)^2}$

49. $\dfrac{x^2 - 2x - 1}{x^2 - 5x - 4}$ **51.** TW **53.** -4 **54.** -4 **55.** $\frac{19}{3}$
56. $-\frac{14}{27}$ **57.** 3, 4 **58.** $-15, 2$ **59.** TW **61.** 6, 7, 8
63. $-3, -\frac{4}{5}, 3$ **65.** $\dfrac{A}{B} \div \dfrac{C}{D} = \dfrac{\frac{A}{B}}{\frac{C}{D}} = \dfrac{\frac{A}{B}}{\frac{C}{D}} \cdot \dfrac{BD}{BD} = \dfrac{AD}{BC} = \dfrac{A}{B} \cdot \dfrac{D}{C}$
67. $\dfrac{8c}{17}$ **69.** 0 **71.** $\dfrac{2z(5z - 2)}{(z + 2)(13z - 6)}$ **73.** $\dfrac{-3}{x(x + h)}$
75. $\{x \mid x$ is a real number *and* $x \neq 0$ *and* $x \neq -2$ *and* $x \neq 2\}$
77. \$168.61

Exercise Set 7.6, pp. 553–555

1. Equation **2.** Expression **3.** Expression **4.** Equation
5. Equation **6.** Equation **7.** Equation **8.** Expression
9. Expression **10.** Equation **11.** $-\frac{2}{5}$ **13.** $\frac{24}{5}$ **15.** $-6, 6$
17. 5 **19.** -10 **21.** $-6, 6$ **23.** No solution **25.** $-4, -1$
27. $-2, 6$ **29.** No solution **31.** -5 **33.** $-\frac{7}{3}$ **35.** -1
37. 2, 3 **39.** -145 **41.** No solution **43.** -1 **45.** 4
47. $-6, 5$ **49.** $-\frac{3}{2}, 5$ **51.** 14 **53.** $\frac{3}{4}$ **55.** -1
57. $-3, 4$ **59.** TW **61.** 137, 139 **62.** 14 yd
63. Base: 9 cm; height: 12 cm **64.** $-8, -6$; 6, 8
65. 0.06 cm per day **66.** 0.28 in. per day **67.** TW
69. $\frac{1}{5}$ **71.** $-\frac{7}{2}$ **73.** -2 **75.** -6 **77.** $\frac{3}{2}$

Exercise Set 7.7, pp. 565–569

1. $\frac{1}{2}$ cake per hour **2.** $\frac{1}{3}$ cake per hour **3.** $\frac{5}{6}$ cake per hour
4. 1 lawn per hour **5.** $\frac{1}{3}$ lawn per hour **6.** $\frac{2}{3}$ lawn per hour
7. $3\frac{3}{7}$ hr **9.** $8\frac{4}{7}$ hr **11.** 21 min
13. MP C2500: 6 min; MP C7500: 2 min
15. Airgle: 15 min; Austin: 35 min
17. Erickson Air-Crane: 10 hr; S-58T: 40 hr
19. 300 min, or 5 hr
21. B&M speed: $r - 14$; B&M time: $\dfrac{330}{r - 14}$;
AMTRAK: 80 km/h; B&M: 66 km/h **23.** 7 mph
25. Express: 45 mph; local: 38 mph **27.** 4.3 ft/sec
29. 3 hr **31.** 9 km/h **33.** 2 km/h **35.** 20 mph
37. 10.5 **39.** $\frac{8}{3}$ **41.** $3\frac{3}{4}$ in. **43.** 20 ft **45.** 15 ft
47. 12.6 **49.** 1440 messages **51.** 702 photos
53. $26\frac{2}{3}$ cm **55.** 126 flash drives **57.** $7\frac{1}{2}$ oz
59. 90 whales **61.** **(a)** 1.92 T; **(b)** 28.8 lb **63.** TW
65. $b = ac$ **66.** $c = \dfrac{b}{a}$ **67.** $y = \frac{2}{5}x - 2$ **68.** $y = \frac{1}{3}x - 2$
69. $a = \dfrac{b}{1 - n}$ **70.** $x = \dfrac{1}{y + z}$ **71.** TW **73.** $49\frac{1}{2}$ hr
75. About 57% **77.** 10,125 people per hour **79.** $14\frac{7}{8}$ mi
81. Page 278 **83.** $8\frac{2}{11}$ min after 10:30 **85.** $51\frac{3}{7}$ mph

Exercise Set 7.8, pp. 577–583

1. (d) **2.** (f) **3.** (e) **4.** (b) **5.** (a) **6.** (c)
7. Inverse **8.** Direct **9.** Direct **10.** Inverse
11. Inverse **12.** Direct **13.** $d = \dfrac{L}{f}$
15. $v_1 = \dfrac{2s}{t} - v_2$, or $\dfrac{2s - tv_2}{t}$ **17.** $b = \dfrac{at}{a - t}$
19. $R = \dfrac{2V}{I} - 2r$, or $\dfrac{2V - 2Ir}{I}$ **21.** $g = \dfrac{Rs}{s - R}$
23. $q = \dfrac{pf}{p - f}$ **25.** $t_1 = \dfrac{H}{Sm} + t_2$, or $\dfrac{H + Smt_2}{Sm}$
27. $r = \dfrac{Re}{E - e}$ **29.** $r = 1 - \dfrac{a}{S}$, or $\dfrac{S - a}{S}$
31. $a + b = \dfrac{f}{c^2}$ **33.** $r = \dfrac{A}{P} - 1$, or $\dfrac{A - P}{P}$
35. $t_2 = \dfrac{d_2 - d_1}{v} + t_1$, or $\dfrac{d_2 - d_1 + t_1 v}{v}$ **37.** $t = \dfrac{ab}{b + a}$
39. $Q = \dfrac{2Tt - 2AT}{A - q}$ **41.** $w = \dfrac{4.15c - 98.42}{p + 0.082}$
43. $k = 7; y = 7x$ **45.** $k = 1.7; y = 1.7x$
47. $k = 6; y = 6x$ **49.** $k = 60; y = \dfrac{60}{x}$
51. $k = 66; y = \dfrac{66}{x}$ **53.** $k = 9; y = \dfrac{9}{x}$ **55.** $33\frac{1}{3}$ cm
57. 3.5 hr **59.** 56 in. **61.** 32 kg **63.** 286 Hz
65. 20 min **67.** 77,000,000 tons **69.** $y = \frac{2}{3}x^2$
71. $y = \dfrac{54}{x^2}$ **73.** $y = 0.3xz^2$ **75.** $y = \dfrac{4wx^2}{z}$
77. 61.3 ft **79.** 308 cm^3 **81.** About 57.42 mph
83. **(a)** Inverse; **(b)** $y = \dfrac{300{,}000}{x}$; **(c)** 30,000 VMT
85. **(a)** Directly; **(b)** $y \approx 0.56x$; **(c)** \$84 **87.** TW
89. $\left\{x \mid x > -\frac{1}{2}\right\}$, or $\left(-\frac{1}{2}, \infty\right)$ **90.** $\{x \mid x \geq -9\}$, or $[-9, \infty)$
91. $\{x \mid x \leq -3\}$, or $(-\infty, -3]$ **92.** $\{x \mid x > 6\}$, or $(6, \infty)$
93. $\left\{x \mid x \leq \frac{17}{2}\right\}$, or $\left(-\infty, \frac{17}{2}\right]$ **94.** $\{x \mid x \leq 3\}$, or $(-\infty, 3]$
95. TW **97.** 567 mi
99. Ratio is $\dfrac{a + 12}{a + 6}$; percent increase is $\dfrac{6}{a + 6} \cdot 100\%$, or $\dfrac{600}{a + 6}\%$
101. $t_1 = t_2 + \dfrac{(d_2 - d_1)(t_4 - t_3)}{a(t_4 - t_2)(t_4 - t_3) + d_3 - d_4}$
103. The intensity is halved. **105.** About 1.697 m
107. $d(s) = \dfrac{28}{s}$; 70 yd

Study Summary: Chapter 7, pp. 584–588

1. -7 **2.** $\dfrac{y}{y + 5}$ **3.** $\dfrac{3(x - 2)^2}{4(x - 1)(2x - 1)}$ **4.** $\dfrac{5(t - 3)}{2(t + 1)}$
5. $\dfrac{9x + 5}{x + 3}$ **6.** 1 **7.** $(x - 5)(x + 3)(x - 3)$
8. $\dfrac{2(2t - 1)}{(t - 1)(t + 1)}$ **9.** $\frac{4}{7}$ **10.** $\frac{7}{2}$ **11.** $5\frac{1}{7}$ hr
12. 10 mph **13.** 24 **14.** $y = 50x$ **15.** $y = \dfrac{40}{x}$
16. $y = \frac{1}{10}xz$

Review Exercises: Chapter 7, pp. 589–591

1. False **2.** True **3.** False **4.** False **5.** False
6. True **7.** True **8.** False **9.** False **10.** True

11. (a) $-\frac{2}{9}$; (b) $-\frac{3}{4}$; (c) 0 **12.** 0 **13.** 4 **14.** $-6, 6$
15. $-6, 5$ **16.** $\dfrac{x-2}{x+1}$ **17.** $\dfrac{7x+3}{x-3}$ **18.** $-5(x+2y)$
19. $\dfrac{a-6}{5}$ **20.** $\dfrac{8(t+1)}{(2t-1)(t-1)}$ **21.** $-32t$
22. $\dfrac{2x(x-1)}{x+1}$ **23.** $\dfrac{(x^2+1)(2x+1)}{(x-2)(x+1)}$ **24.** $\dfrac{(t+4)^2}{t+1}$
25. $60a^5b^8$ **26.** $x^4(x-1)(x+1)$
27. $(y-2)(y+2)(y+1)$ **28.** $\dfrac{15-3x}{x+3}$ **29.** $\dfrac{4}{x-4}$
30. $\dfrac{x+5}{2x}$ **31.** $\dfrac{2x+5}{x-2}$ **32.** $\dfrac{2a}{a-1}$ **33.** $d+2$
34. $\dfrac{-x^2+x+26}{(x+1)(x-5)(x+5)}$ **35.** $\dfrac{19x+8}{10x(x+2)}$
36. $f(x)=\dfrac{7x+3}{x-3}, x \neq \frac{1}{2}, x \neq 3$
37. $f(x)=\dfrac{2(x-2)}{x+2}, x \neq -2, x \neq 2$ **38.** $\dfrac{z}{1-z}$
39. $\dfrac{10x}{3x^2+16}$ **40.** $\dfrac{(y+11)(y+5)}{(y-5)(y+2)}$ **41.** $\dfrac{(14-3x)(x+3)}{2x^2+16x+6}$
42. 2 **43.** 6 **44.** $-6, -1$ **45.** 0 **46.** $-1, 4$ **47.** $5\frac{1}{7}$ hr
48. Core 2 Duo: 45 sec; Core 2 Quad: 30 sec **49.** 24 mph
50. Motorcycle: 62 mph; car: 70 mph **51.** 55 seals **52.** 6
53. $s=\dfrac{Rg}{g-R}$ **54.** $m=\dfrac{H}{S(t_1-t_2)}$ **55.** $c=\dfrac{b+3a}{2}$
56. $t_1=\dfrac{-A}{vT}+t_2$, or $\dfrac{-A+vTt_2}{vT}$ **57.** About 23 lb **58.** 64 L
59. $y=\dfrac{\frac{3}{4}}{x}$ **60.** (a) Inverse; (b) $y=\dfrac{24}{x}$; (c) 3 oz
61. TW The least common denominator was used to add and subtract rational expressions, to simplify complex rational expressions, and to solve rational equations. **62.** TW A rational *expression* is a quotient of two polynomials. Expressions can be simplified, multiplied, or added, but they cannot be solved for a variable. A rational *equation* is an equation containing rational expressions. In a rational equation, we often can solve for a variable.
63. All real numbers except 0 and 13
64. 45 **65.** $\dfrac{5(a+3)^2}{a}$ **66.** 0

Test: Chapter 7, pp. 591–592

1. [7.1] 0 **2.** [7.1] 1, 2 **3.** [7.1] $\dfrac{2x+1}{x+1}$
4. [7.2] $\dfrac{-2(a+5)}{3}$ **5.** [7.2] $\dfrac{(5y+1)(y+1)}{3y(y+2)}$
6. [7.2] $\dfrac{(2x+1)(2x-1)(x^2+1)}{(x-1)^2(x-2)}$
7. [7.2] $(x+3)(x-3)$ **8.** [7.3] $(y-3)(y+3)(y+7)$
9. [7.3] $\dfrac{-3x+9}{x^3}$ **10.** [7.3] $\dfrac{-2t+8}{t^2+1}$ **11.** [7.4] $\dfrac{3}{3-x}$
12. [7.4] $\dfrac{2x-5}{x-3}$ **13.** [7.4] $\dfrac{11t-8}{t(t-2)}$
14. [7.4] $\dfrac{-x(x-9)}{(x+4)(x-4)(x+1)}$
15. [7.1] $f(x)=\dfrac{3x+7}{x+3}, x \neq -3, -\frac{1}{2}$
16. [7.4] $f(x)=\dfrac{x-4}{(x+3)(x-2)}, x \neq -3, 2$
17. [7.5] $\dfrac{3y+1}{y}$ **18.** [7.5] $x-8$ **19.** [7.5] $\dfrac{(x-9)(x-6)}{(x+6)(x-3)}$
20. [7.6] $\frac{8}{3}$ **21.** [7.6] $-\frac{21}{4}$ **22.** [7.6] $-3, 5$ **23.** [7.1] 5; 0
24. [7.6] $\frac{5}{3}$ **25.** [7.8] $b_1=\dfrac{2A}{h}-b_2$, or $\dfrac{2A-b_2h}{h}$
26. [7.7] $3\frac{3}{11}$ mph **27.** [7.7] $1\frac{31}{32}$ hr
28. [7.7] Ellia: 4 hr; Pe'rez: 10 hr **29.** [7.7] $2\frac{1}{7}$ c
30. [7.8] 30 workers **31.** [7.8] 637 in^2
32. [7.6] $\{x \mid x \text{ is a real number and } x \neq 0 \text{ and } x \neq 15\}$
33. [7.5] a **34.** [7.7] Andy: 56 lawns; Chad: 42 lawns

Chapter 8

Interactive Discovery, p. 602

1. Domain of $f=(-\infty, 3]$; domain of $g=[-1, \infty)$
2. Domain of $f+g$ = domain of $f-g$ = domain of $f \cdot g=[-1, 3]$ **3.** By finding the intersection of the domains of f and g

Exercise Set 8.1, pp. 603–608

1. (h) **2.** (j) **3.** (f) **4.** (a) **5.** (e) **6.** (d)
7. (b) **8.** (g) **9.** (c) **10.** (i)
11. $\{x \mid x \geq 2\}$, or $[2, \infty)$ **13.** $\{x \mid x < 3\}$, or $(-\infty, 3)$
15. (a) $\{x \mid x \leq -2\}$, or $(-\infty, -2]$; (b) $\{x \mid x > \frac{8}{3}\}$, or $(\frac{8}{3}, \infty)$; (c) $\{x \mid x > \frac{4}{5}\}$, or $(\frac{4}{5}, \infty)$ **17.** $\{x \mid x < 7\}$, or $(-\infty, 7)$
19. $\{x \mid x \geq 2\}$, or $[2, \infty)$ **21.** $\{x \mid x < 8\}$, or $(-\infty, 8)$
23. $\{x \mid x \leq 2\}$, or $(-\infty, 2]$ **25.** $\{x \mid x \leq -\frac{4}{9}\}$, or $(-\infty, -\frac{4}{9}]$
27. At least 625 people
29. $r(x)=-0.04x+2.0233$; years after 2013
31. $n(x)=-6.4x+56.5, t(x)=3.7x+12$; years after 2010
33. $f(x)=-0.0826x+54.541$; years after 2045
35. $\{9, 11\}$ **37.** $\{0, 5, 10, 15, 20\}$ **39.** $\{b, d, f\}$
41. $\{r, s, t, u, v\}$ **43.** $\varnothing$ **45.** $\{3, 5, 7\}$
47. $(3, 7)$
49. $[-6, 0]$
51. $(-\infty, -1) \cup (4, \infty)$
53. $(-\infty, -2] \cup (1, \infty)$
55. $(-2, 4)$
57. $(-2, 4]$
59. $(-\infty, 5) \cup (7, \infty)$
61. $(-\infty, -4] \cup [5, \infty)$
63. $[-3, 7)$
65. $[3, 7)$

67. $(-\infty, 5)$

69. $\{t \mid -3 < t < 7\}$, or $(-3, 7)$

71. $\{x \mid 0 < x < 4\}$, or $(0, 4)$

73. $\{a \mid a \le -2 \text{ or } a > 2\}$, or $(-\infty, -2] \cup (2, \infty)$

75. $\mathbb{R}$, or $(-\infty, \infty)$

77. $\{x \mid -3 \le x \le 2\}$, or $[-3, 2]$

79. $\{x \mid 7 < x < 23\}$, or $(7, 23)$

81. $\{x \mid -32 \le x \le 8\}$, or $[-32, 8]$

83. $\{x \mid 1 \le x \le 3\}$, or $[1, 3]$

85. $\{x \mid -\frac{7}{2} < x \le 7\}$, or $(-\frac{7}{2}, 7]$

87. $\{t \mid t < 0 \text{ or } t > 1\}$, or $(-\infty, 0) \cup (1, \infty)$

89. $\{a \mid a < \frac{7}{2}\}$, or $(-\infty, \frac{7}{2})$

91. $\{a \mid a < -5\}$, or $(-\infty, -5)$

93. $\varnothing$ **95.** $\{t \mid t \le 6\}$, or $(-\infty, 6]$

97. $(-1, 6)$ **99.** $(-\infty, -8) \cup (-8, \infty)$

101. $(-\infty, 0) \cup (0, \infty)$ **103.** $[6, \infty)$ **105.** $[-\frac{7}{2}, \infty)$

107. $(-\infty, 4]$ **109.** $[5, \infty)$ **111.** $[\frac{2}{3}, 3]$ **113.** TW

115.

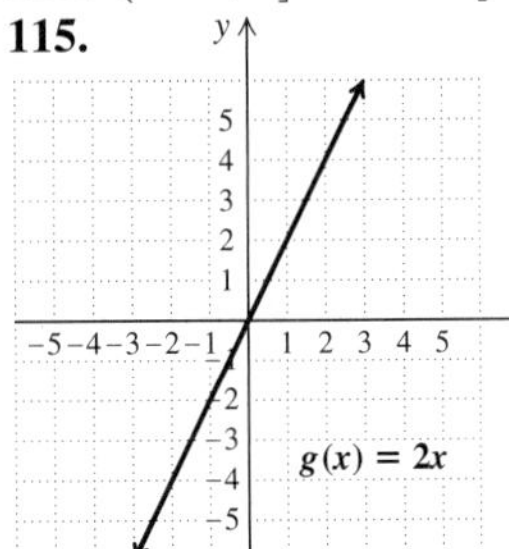

116.

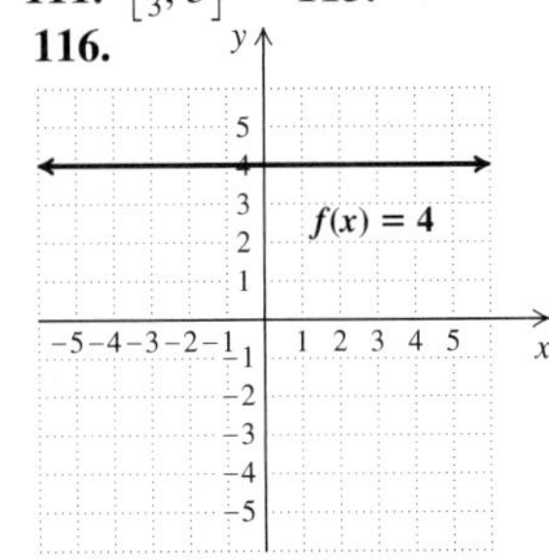

117.

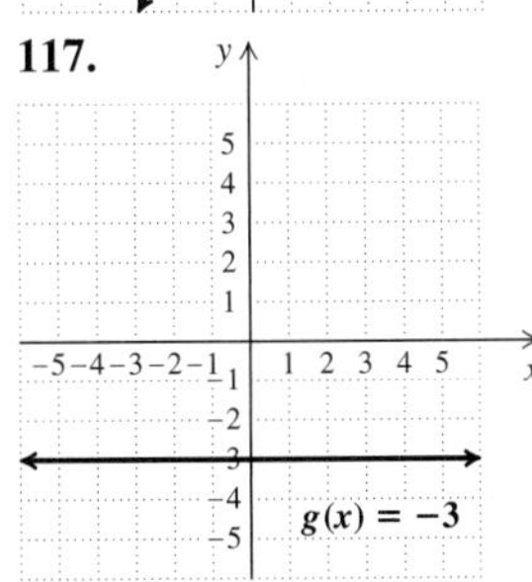

118.

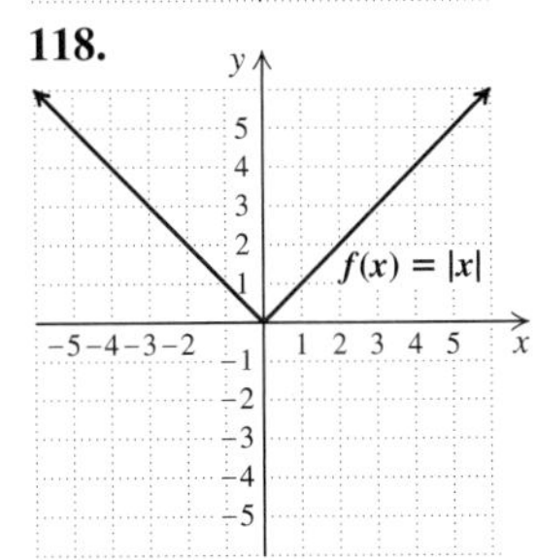

119. -1 **120.** -4 **121.** TW

123. Between 2003 and 2009 **125.** Sizes between 6 and 13

127. Densities between 1.03 kg/L and 1.04 kg/L

129. $\{m \mid m < \frac{6}{5}\}$, or $(-\infty, \frac{6}{5})$

131. $\{x \mid -\frac{1}{8} < x < \frac{1}{2}\}$, or $(-\frac{1}{8}, \frac{1}{2})$

133. True **135.** False **137.** $(-\infty, -7) \cup (-7, \frac{3}{4}]$

139.

Exercise Set 8.2, pp. 617–620

1. True **2.** False **3.** True **4.** True **5.** True **6.** True **7.** False **8.** False **9.** (g) **10.** (h) **11.** (d) **12.** (a) **13.** (a) **14.** (b) **15.** $\{-5, 1\}$ **16.** $[-5, 1]$ **17.** $(-5, 1)$ **18.** $(-\infty, -5) \cup (1, \infty)$ **19.** $(-\infty, -5] \cup [1, \infty)$ **20.** $\varnothing$ **21.** $\{-7, 7\}$ **23.** $\varnothing$ **25.** $\{0\}$ **27.** $\{-\frac{1}{2}, \frac{7}{2}\}$ **29.** $\varnothing$ **31.** $\{-4, 8\}$ **33.** $\{2, 8\}$ **35.** $\{-5.5, 5.5\}$ **37.** $\{-8, 8\}$ **39.** $\{-\frac{11}{7}, \frac{11}{7}\}$ **41.** $\{-\frac{11}{2}, \frac{13}{2}\}$ **43.** $\{-2, 12\}$ **45.** $\{-\frac{1}{3}, 3\}$ **47.** $\{-7, 1\}$ **49.** $\{-8.7, 8.7\}$ **51.** $\{-\frac{8}{3}, 4\}$ **53.** $\{1, 11\}$ **55.** $\{-\frac{1}{2}\}$ **57.** $\{-\frac{3}{5}, 5\}$ **59.** $\mathbb{R}$ **61.** $\{\frac{1}{4}\}$

63. $\{a \mid -9 \le a \le 9\}$, or $[-9, 9]$

65. $\{t \mid t < 0 \text{ or } t > 0\}$, or $(-\infty, 0) \cup (0, \infty)$

67. $\{x \mid -3 < x < 5\}$, or $(-3, 5)$

69. $\{x \mid -8 \le x \le 4\}$, or $[-8, 4]$

71. $\{x \mid x < -2 \text{ or } x > 8\}$, or $(-\infty, -2) \cup (8, \infty)$

73. $\mathbb{R}$, or $(-\infty, \infty)$

75. $\{a \mid a \le -\frac{2}{3} \text{ or } a \ge \frac{10}{3}\}$, or $(-\infty, -\frac{2}{3}] \cup [\frac{10}{3}, \infty)$

77. $\{y \mid -9 < y < 15\}$, or $(-9, 15)$

79. $\{x \mid x \le -8 \text{ or } x \ge 0\}$, or $(-\infty, -8] \cup [0, \infty)$

81. $\{x \mid x < -\frac{1}{2} \text{ or } x > \frac{7}{2}\}$, or $(-\infty, -\frac{1}{2}) \cup (\frac{7}{2}, \infty)$

83. $\varnothing$ **85.** $\{x \mid x \le -\frac{2}{15} \text{ or } x \ge \frac{14}{15}\}$, or $(-\infty, -\frac{2}{15}] \cup [\frac{14}{15}, \infty)$

87. $\{m \mid -9 \le m \le 3\}$, or $[-9, 3]$

89. $\{a \mid -6 < a < 0\}$, or $(-6, 0)$

91. $\{x \mid -\frac{1}{2} \le x \le \frac{7}{2}\}$, or $[-\frac{1}{2}, \frac{7}{2}]$

93. $\{x \mid x \le -\frac{7}{3} \text{ or } x \ge 5\}$, or $(-\infty, -\frac{7}{3}] \cup [5, \infty)$

95. $\{x \mid -4 < x < 5\}$, or $(-4, 5)$

97. TW

99.

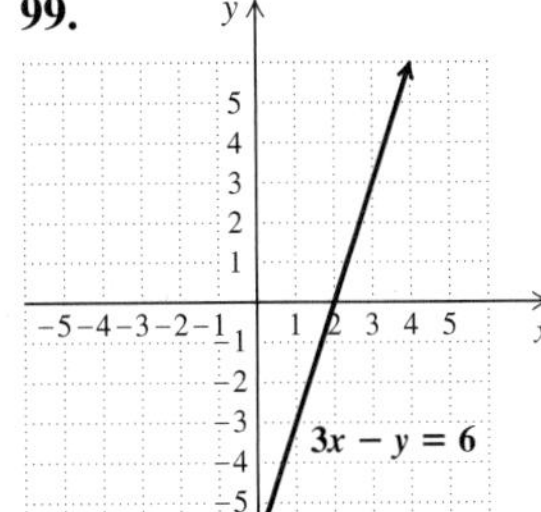

100.

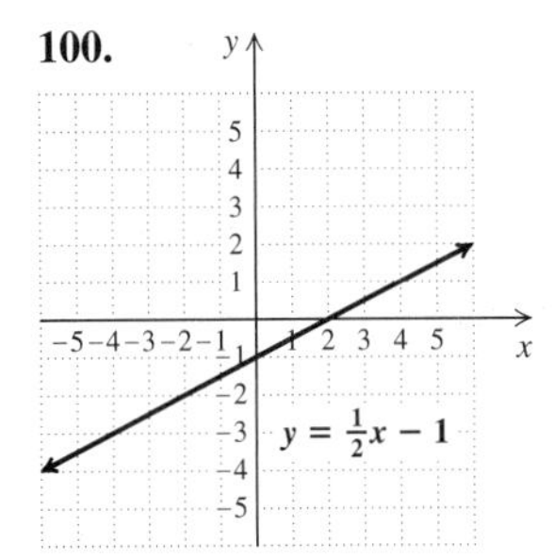

101.

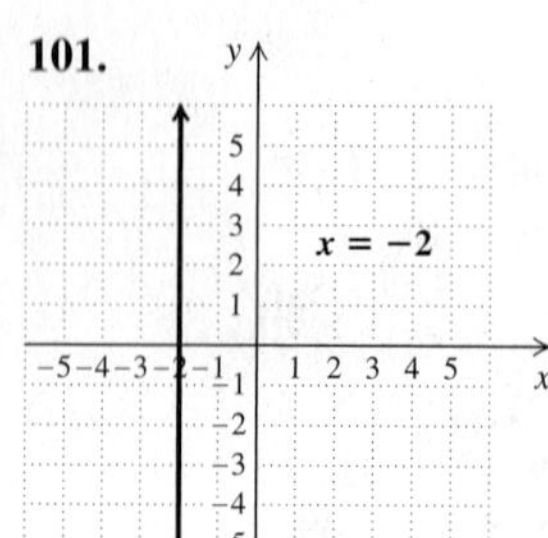

102.

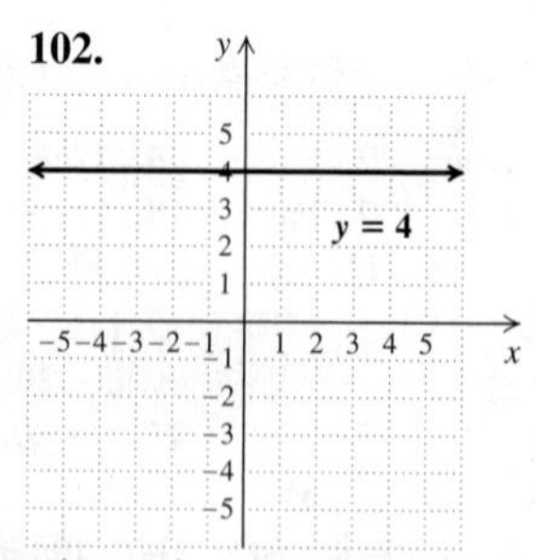

103. $\left(4, -\frac{4}{3}\right)$ **104.** $(5, 1)$ **105.** $\left(\frac{5}{7}, -\frac{18}{7}\right)$ **106.** $(-2, -5)$
107. TW **109.** $\left\{t \mid t \geq \frac{5}{3}\right\}$, or $\left[\frac{5}{3}, \infty\right)$ **111.** $\mathbb{R}$, or $(-\infty, \infty)$
113. $\left\{-\frac{1}{7}, \frac{7}{3}\right\}$ **115.** $|x| < 3$ **117.** $|x + 3| > 5$
119. $|x - 7| < 2$, or $|7 - x| < 2$ **121.** $|x - 3| \leq 4$
123. $|x + 4| < 3$ **125.** Between 80 ft and 100 ft

Mid-Chapter Review: Chapter 8, pp. 620–621

Guided Solutions

1. $2 < x < 11$
The solution is $(2, 11)$.

2. $x - 1 < -9$ *or* $9 < x - 1$
$x < -8$ *or* $10 < x$
The solution is $(-\infty, -8) \cup (10, \infty)$.

Mixed Review

1. $\{-15, 15\}$ **2.** $\{t \mid -10 < t < 10\}$, or $(-10, 10)$
3. $\{p \mid p < -15 \text{ or } p > 15\}$, or $(-\infty, -15) \cup (15, \infty)$
4. $\{-4, 3\}$ **5.** $\{x \mid 2 < x < 11\}$, or $(2, 11)$
6. $\{t \mid -4 < t < 4\}$, or $(-4, 4)$
7. $\{x \mid x < -6 \text{ or } x > 13\}$, or $(-\infty, -6) \cup (13, \infty)$
8. $\{x \mid -7 \leq x \leq 3\}$, or $[-7, 3]$ **9.** $\left\{-\frac{8}{3}, \frac{8}{3}\right\}$
10. $\{x \mid -7 \leq x \leq 13\}$, or $[-7, 13]$
11. $\left\{n \mid -9 < n \leq \frac{8}{3}\right\}$, or $\left(-9, \frac{8}{3}\right]$ **12.** $\varnothing$
13. $\left\{x \mid x \leq -\frac{17}{2} \text{ or } x \geq \frac{7}{2}\right\}$, or $\left(-\infty, -\frac{17}{2}\right] \cup \left[\frac{7}{2}, \infty\right)$
14. $\varnothing$ **15.** $\{m \mid -24 < m < 12\}$, or $(-24, 12)$
16. $\{-42, 38\}$
17. $\{t \mid t \leq 4 \text{ or } t \geq 10\}$, or $(-\infty, 4] \cup [10, \infty)$
18. $\varnothing$ **19.** $\{a \mid -7 < a < -5\}$, or $(-7, -5)$
20. $\mathbb{R}$, or $(-\infty, \infty)$

Visualizing for Success, p. 631

1. B **2.** F **3.** J **4.** A **5.** E **6.** G **7.** C
8. D **9.** I **10.** H

Exercise Set 8.3, pp. 632–634

1. (e) **2.** (c) **3.** (d) **4.** (a) **5.** (b)
6. (f) **7.** Yes **9.** No

11.

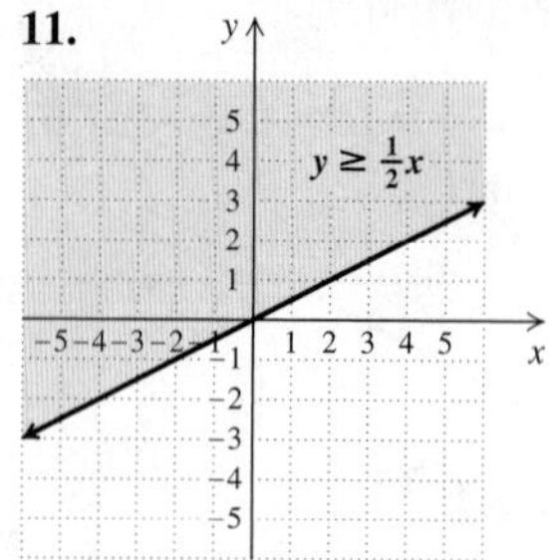

13.

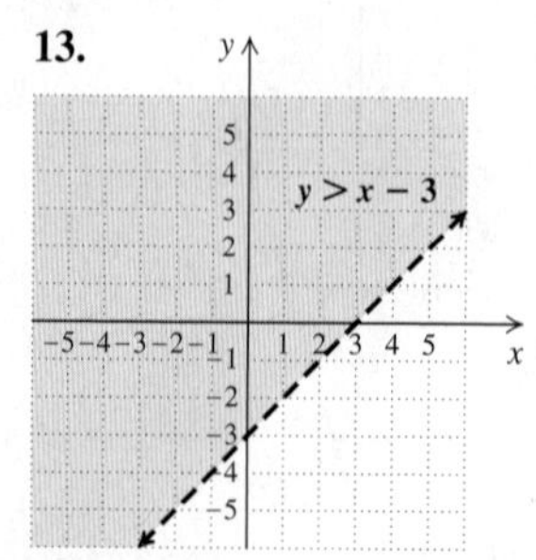

15.

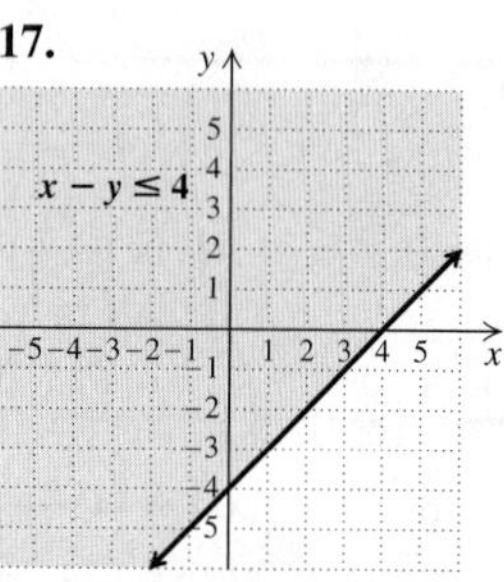

17.

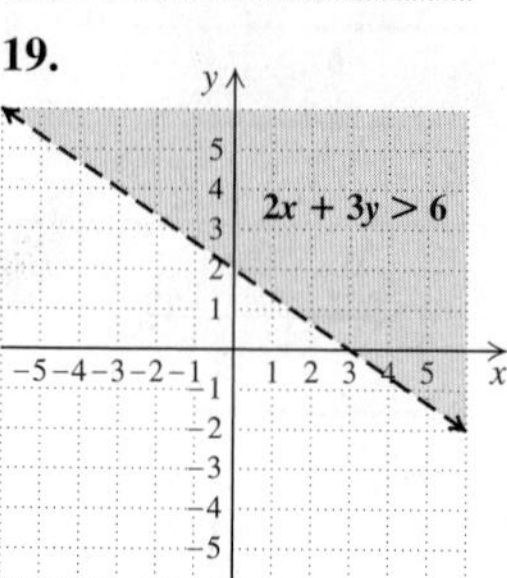

19.

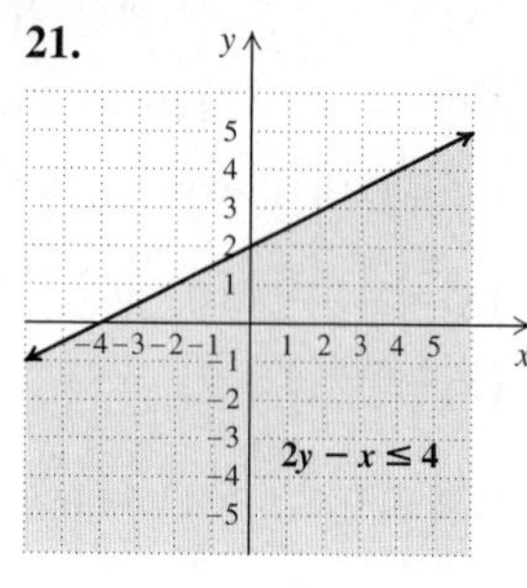

21.

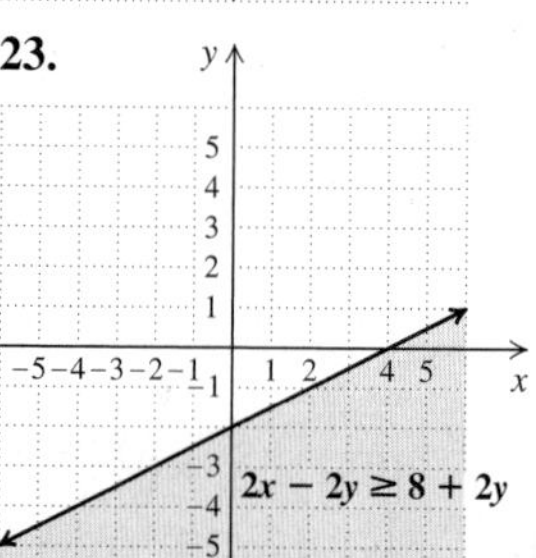

23.

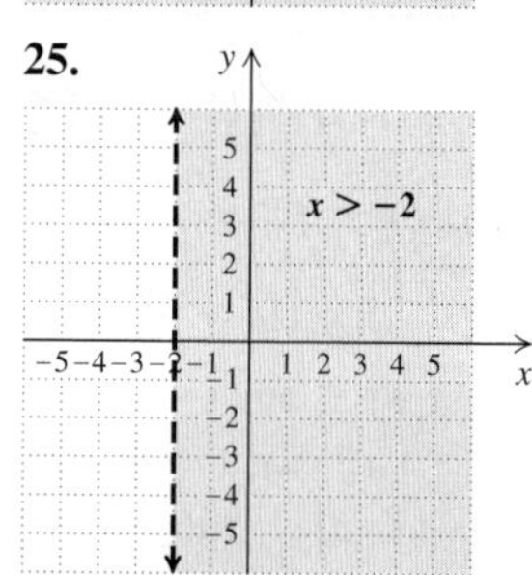

25.

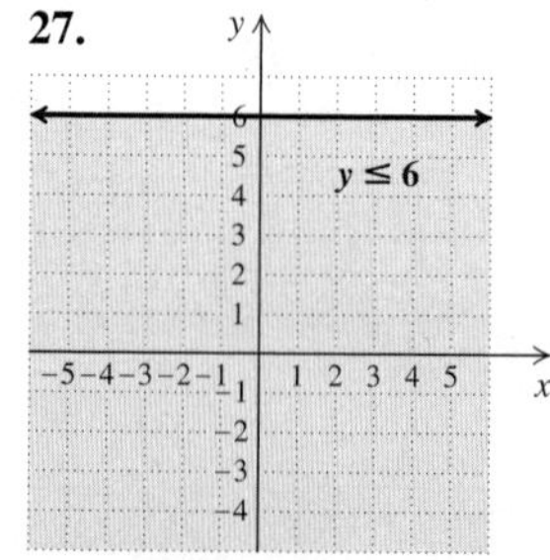

27.

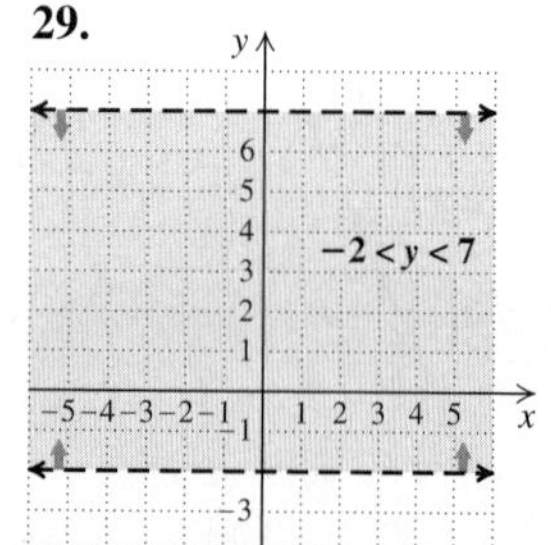

29.

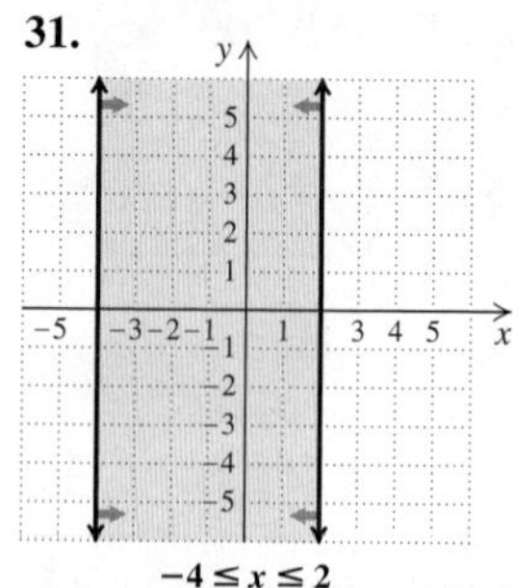

31.

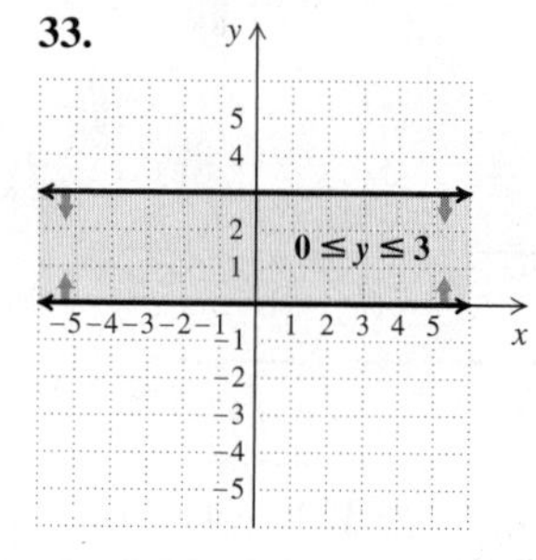

33.

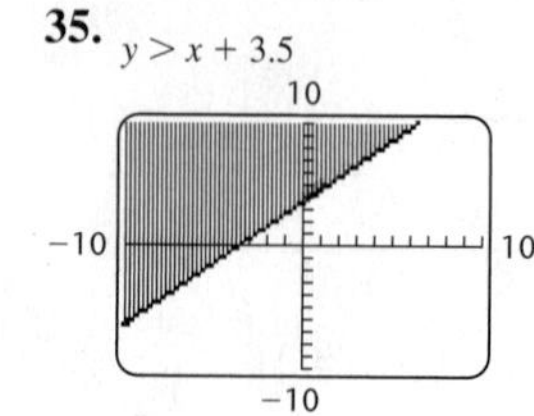

35. $y > x + 3.5$

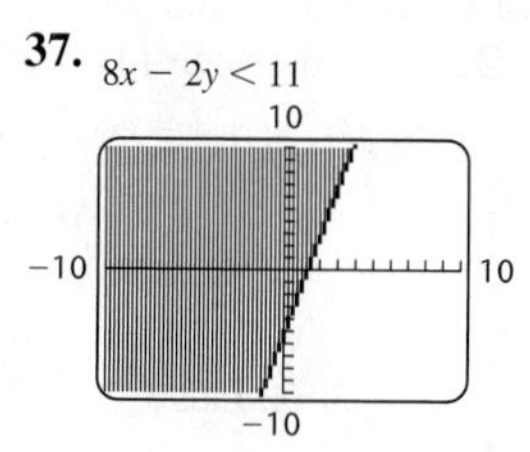

37. $8x - 2y < 11$

39.
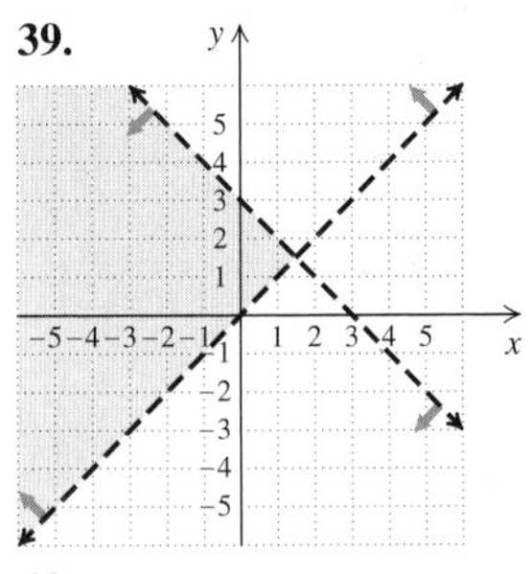

41.
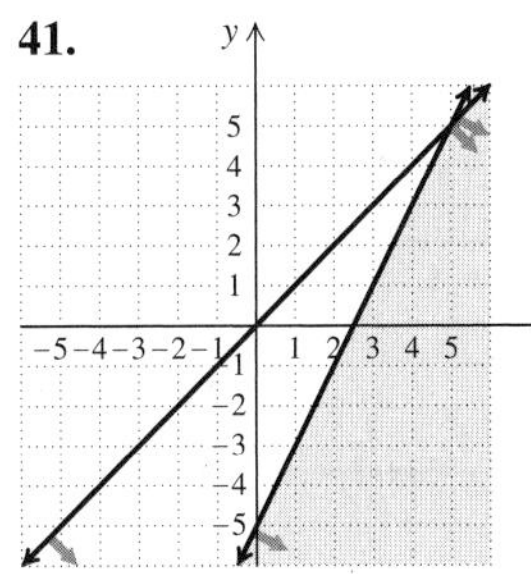

43.
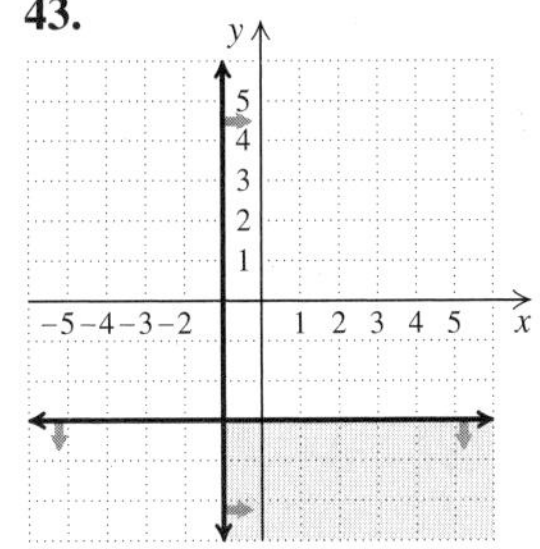

45.
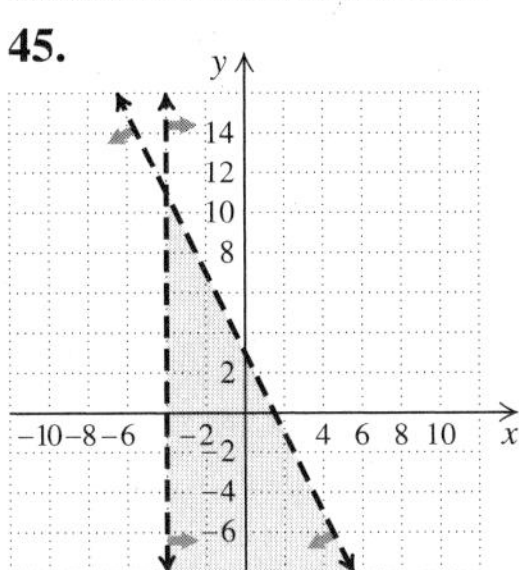

47.
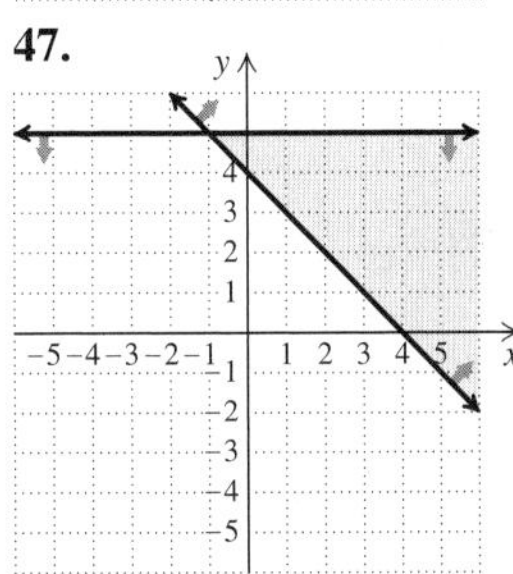

49.
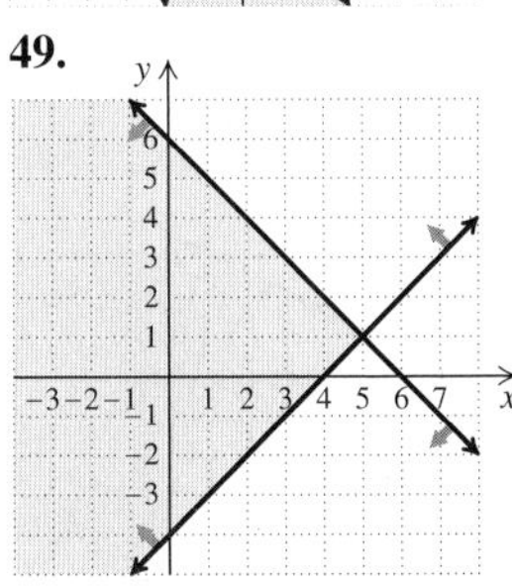

51.
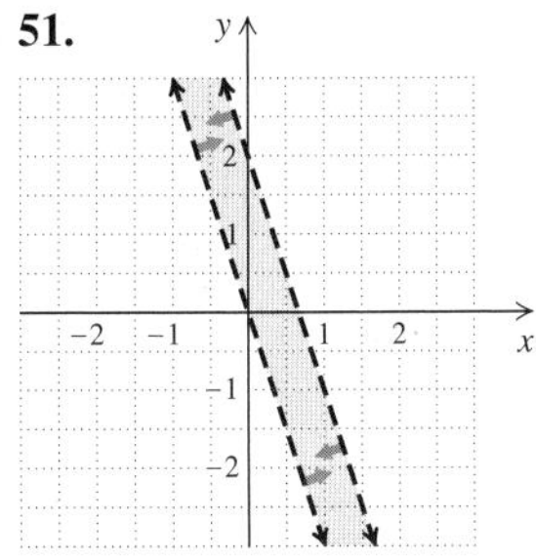

53.
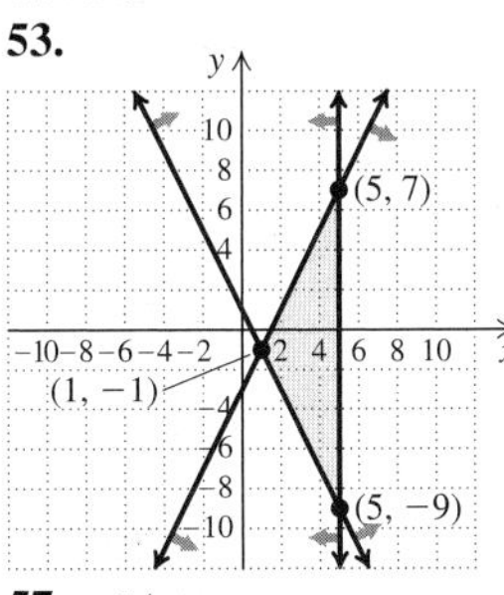

55.
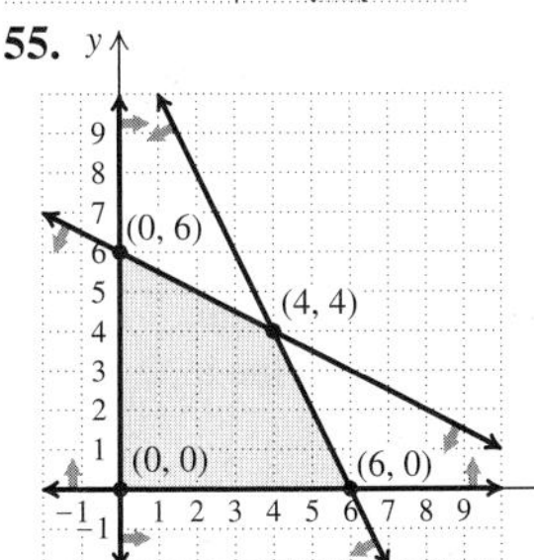

57.
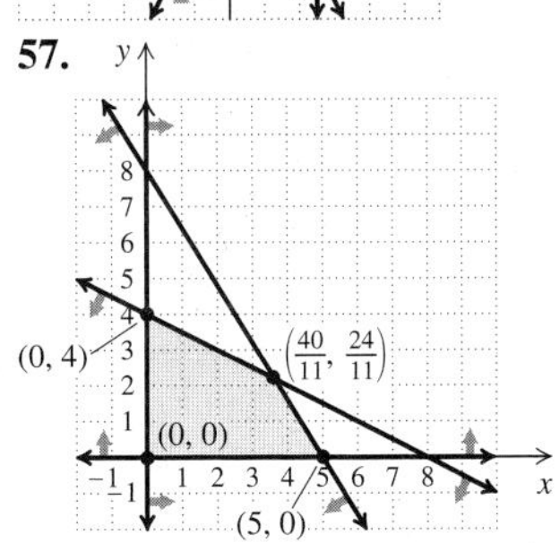

59.
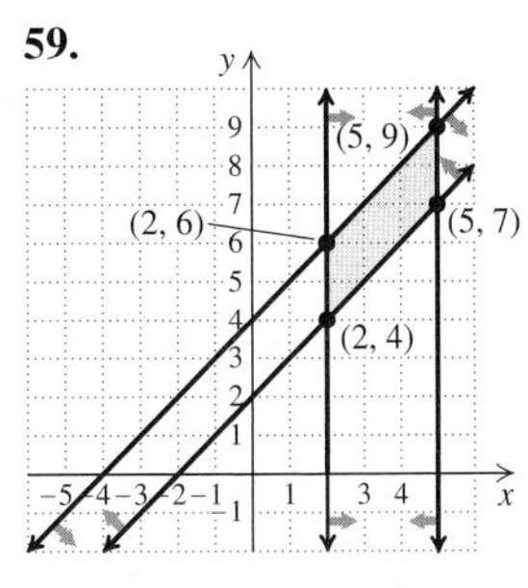

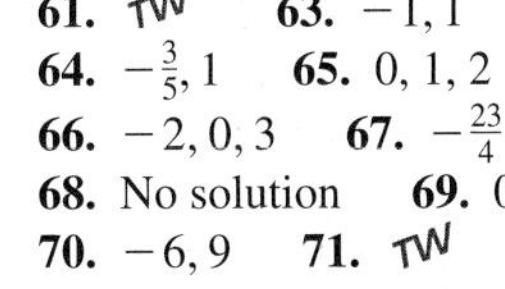
61. TW **63.** $-1, 1$
64. $-\frac{3}{5}, 1$ **65.** 0, 1, 2
66. $-2, 0, 3$ **67.** $-\frac{23}{4}$
68. No solution **69.** 0
70. $-6, 9$ **71.** TW

73.
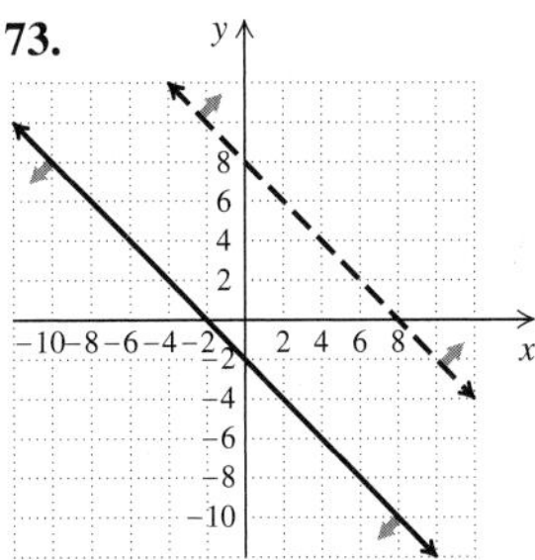

75.
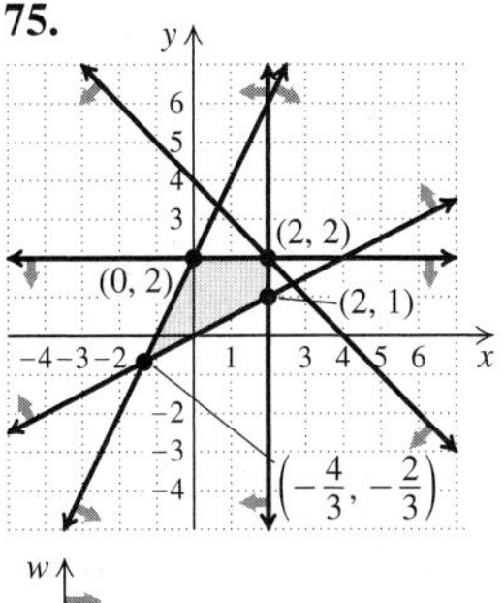

77.
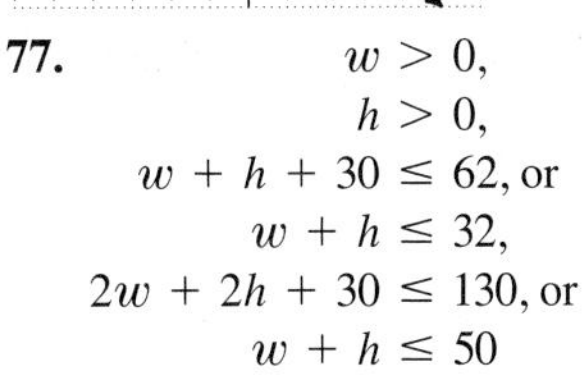
$w > 0,$
$h > 0,$
$w + h + 30 \le 62$, or
$w + h \le 32,$
$2w + 2h + 30 \le 130$, or
$w + h \le 50$

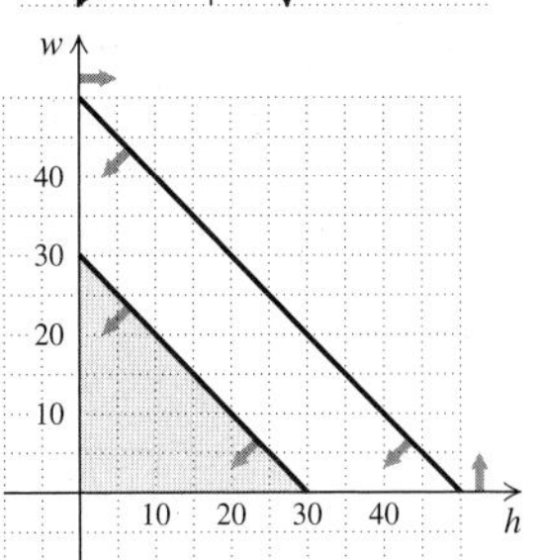

79. $h \le 2w,$
$w \le 1.5h,$
$h \le 3200,$
$h \ge 0,$
$w \ge 0$

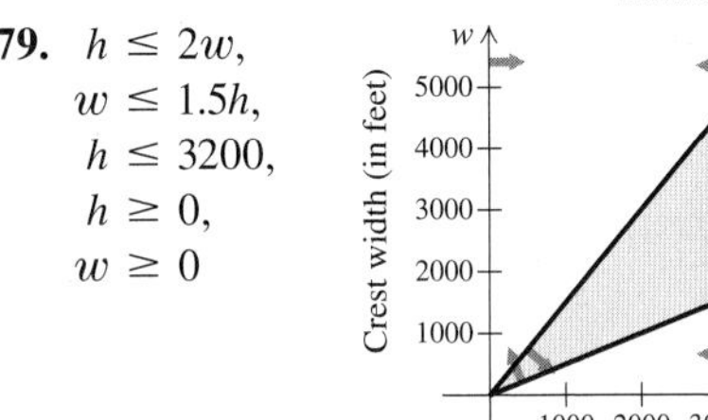

81. $q + v \le 1150,$
$q \ge 700,$
$q \le 800,$
$v \ge 400,$
$v \le 800$

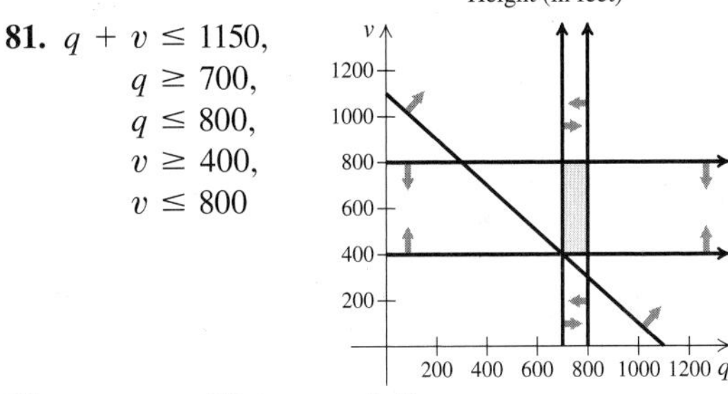

83. $y \le x,$
$y \le 2$

85. $y \le x + 2,$
$y \le -x + 4,$
$y \ge 0$

Exercise Set 8.4, pp. 640–642

1. True **2.** False **3.** True **4.** True **5.** False
6. True **7.** $\left[-4, \frac{3}{2}\right]$, or $\left\{x \mid -4 \le x \le \frac{3}{2}\right\}$
9. $(-\infty, -2) \cup (0, 2) \cup (3, \infty)$, or $\{x \mid x < -2 \text{ or } 0 < x < 2 \text{ or } x > 3\}$
11. $\left(-\infty, -\frac{7}{2}\right) \cup (-2, \infty)$, or $\left\{x \mid x < -\frac{7}{2} \text{ or } x > -2\right\}$
13. $(-4, 3)$, or $\{x \mid -4 < x < 3\}$
15. $(-\infty, -7] \cup [2, \infty)$, or $\{x \mid x \le -7 \text{ or } x \ge 2\}$
17. $(-\infty, -1) \cup (2, \infty)$, or $\{x \mid x < -1 \text{ or } x > 2\}$
19. $\varnothing$ **21.** $(-2, 6)$, or $\{x \mid -2 < x < 6\}$
23. $(-\infty, -2) \cup (0, 2)$, or $\{x \mid x < -2 \text{ or } 0 < x < 2\}$
25. $[-2, 1] \cup [4, \infty)$, or $\{x \mid -2 \le x \le 1 \text{ or } x \ge 4\}$
27. $[-0.78, 1.59]$, or $\{x \mid -0.78 \le x \le 1.59\}$
29. $(-\infty, -2) \cup (1, 3)$, or $\{x \mid x < -2 \text{ or } 1 < x < 3\}$
31. $[-2, 2]$, or $\{x \mid -2 \le x \le 2\}$
33. $(-1, 2) \cup (3, \infty)$, or $\{x \mid -1 < x < 2 \text{ or } x > 3\}$
35. $(-\infty, 0] \cup [2, 5]$, or $\{x \mid x \le 0 \text{ or } 2 \le x \le 5\}$

37. $(-\infty, -5)$, or $\{x \mid x < -5\}$
39. $(-\infty, -1] \cup (3, \infty)$, or $\{x \mid x \le -1 \text{ or } x > 3\}$
41. $(-\infty, -6)$, or $\{x \mid x < -6\}$
43. $(-\infty, -1] \cup [2, 5)$, or $\{x \mid x \le -1 \text{ or } 2 \le x < 5\}$
45. $(-\infty, -3) \cup [0, \infty)$, or $\{x \mid x < -3 \text{ or } x \ge 0\}$
47. $(0, \infty)$, or $\{x \mid x > 0\}$
49. $(-\infty, -4) \cup [1, 3)$, or $\{x \mid x < -4 \text{ or } 1 \le x < 3\}$
51. $\left(-\frac{3}{4}, \frac{5}{2}\right]$, or $\left\{x \mid -\frac{3}{4} < x \le \frac{5}{2}\right\}$
53. $(-\infty, 2) \cup [3, \infty)$, or $\{x \mid x < 2 \text{ or } x \ge 3\}$ **55.** TW
57. $\left(2, -\frac{1}{2}\right)$ **58.** $\left(\frac{23}{4}, -\frac{1}{4}\right)$ **59.** $(2, 1)$ **60.** $\left(\frac{3}{7}, \frac{1}{7}\right)$
61. $\left(-\frac{3}{8}, \frac{1}{8}\right)$ **62.** $(-2, -3)$ **63.** TW **65.** $\varnothing$ **67.** $\{0\}$
69. (a) $(10, 200)$, or $\{x \mid 10 < x < 200\}$;
(b) $[0, 10) \cup (200, \infty)$, or $\{x \mid 0 \le x < 10 \text{ or } x > 200\}$
71. $\{n \mid n \text{ is an integer and } 12 \le n \le 25\}$ **73.** $f(x)$ has no zeros; $f(x) < 0$ for $(-\infty, 0)$, or $\{x \mid x < 0\}$; $f(x) > 0$ for $(0, \infty)$, or $\{x \mid x > 0\}$ **75.** $f(x) = 0$ for $x = -1, 0$; $f(x) < 0$ for $(-\infty, -3) \cup (-1, 0)$, or $\{x \mid x < -3 \text{ or } -1 < x < 0\}$; $f(x) > 0$ for $(-3, -1) \cup (0, 2) \cup (2, \infty)$, or $\{x \mid -3 < x < -1 \text{ or } 0 < x < 2 \text{ or } x > 2\}$
77. $(-\infty, -5] \cup [9, \infty)$, or $\{x \mid x \le -5 \text{ or } x \ge 9\}$
79. $(-\infty, -8] \cup [0, \infty)$, or $\{x \mid x \le -8 \text{ or } x \ge 0\}$

Study Summary: Chapter 8, pp. 643–644

1. $\{x \mid x < 1\}$, or $(-\infty, 1)$
2. $\left\{x \mid -2 < x \le -\frac{3}{4}\right\}$, or $\left(-2, -\frac{3}{4}\right]$
3. $\{x \mid x \le 13 \text{ or } x > 22\}$, or $(-\infty, 13] \cup (22, \infty)$
4. $\left\{-1, \frac{9}{2}\right\}$ **5.** $\{x \mid 11 \le x \le 13\}$, or $[11, 13]$
6. $\{x \mid x < -5 \text{ or } x > 2\}$, or $(-\infty, -5) \cup (2, \infty)$
7.

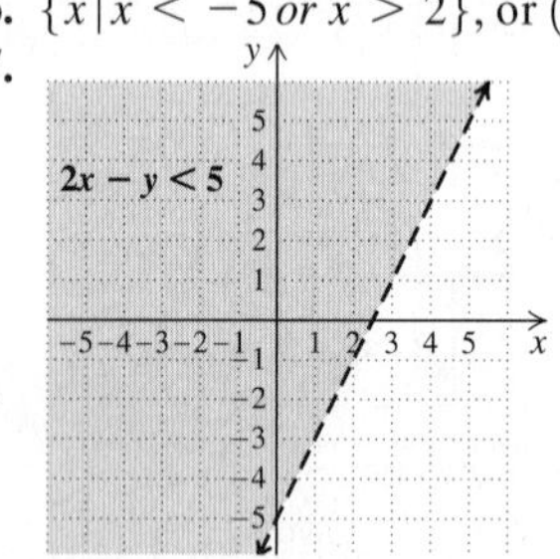

8. $\{x \mid -1 < x < 12\}$, or $(-1, 12)$

Review Exercises: Chapter 8, p. 645

1. True **2.** False **3.** True **4.** True **5.** False
6. True **7.** True **8.** False **9.** False **10.** True
11. $\{x \mid x < 1\}$, or $(-\infty, 1)$ **12.** $\{x \mid x > -4\}$, or $(-4, \infty)$
13. $\{x \mid x \ge -6\}$, or $[-6, \infty)$ **14.** $\{x \mid x \le 2\}$, or $(-\infty, 2]$
15. $\{1, 5, 9\}$ **16.** $\{1, 2, 3, 5, 6, 9\}$
17. $(-3, 2]$
18. $(-\infty, \infty)$
19. $\{x \mid -12 < x \le -3\}$, or $(-12, -3]$
20. $\left\{x \mid -\frac{5}{4} < x < \frac{5}{2}\right\}$, or $\left(-\frac{5}{4}, \frac{5}{2}\right)$
21. $\{x \mid x < -3 \text{ or } x > 1\}$, or $(-\infty, -3) \cup (1, \infty)$
22. $\{x \mid x < -11 \text{ or } x \ge -6\}$, or $(-\infty, -11) \cup [-6, \infty)$
23. $\{x \mid x \le -6 \text{ or } x \ge 8\}$, or $(-\infty, -6] \cup [8, \infty)$
24. $\left\{x \mid x < -\frac{2}{5} \text{ or } x > \frac{8}{5}\right\}$, or $\left(-\infty, -\frac{2}{5}\right) \cup \left(\frac{8}{5}, \infty\right)$
25. $(-\infty, 8) \cup (8, \infty)$ **26.** $[-5, \infty)$ **27.** $\left(-\infty, \frac{8}{3}\right]$
28. $\{-5, 5\}$ **29.** $\{t \mid t \le -21 \text{ or } t \ge 21\}$, or $(-\infty, -21] \cup [21, \infty)$ **30.** $\{-4, 10\}$
31. $\left\{a \mid -\frac{7}{2} < a < 2\right\}$, or $\left(-\frac{7}{2}, 2\right)$
32. $\left\{x \mid x \le -\frac{11}{3} \text{ or } x \ge \frac{19}{3}\right\}$, or $\left(-\infty, -\frac{11}{3}\right] \cup \left[\frac{19}{3}, \infty\right)$
33. $\left\{-14, \frac{4}{3}\right\}$ **34.** $\varnothing$
35. $\{x \mid -16 \le x \le 8\}$, or $[-16, 8]$
36. $\{x \mid x < 0 \text{ or } x > 10\}$, or $(-\infty, 0) \cup (10, \infty)$ **37.** $\varnothing$
38.

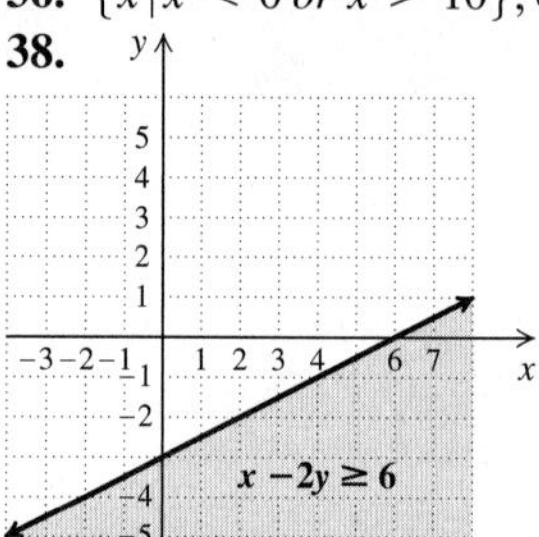

39.

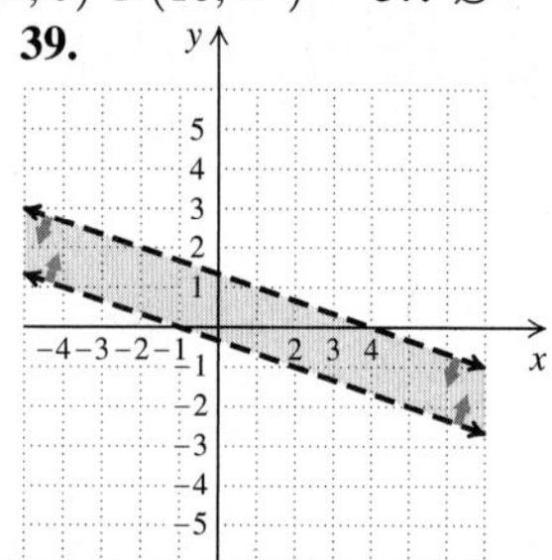

40.

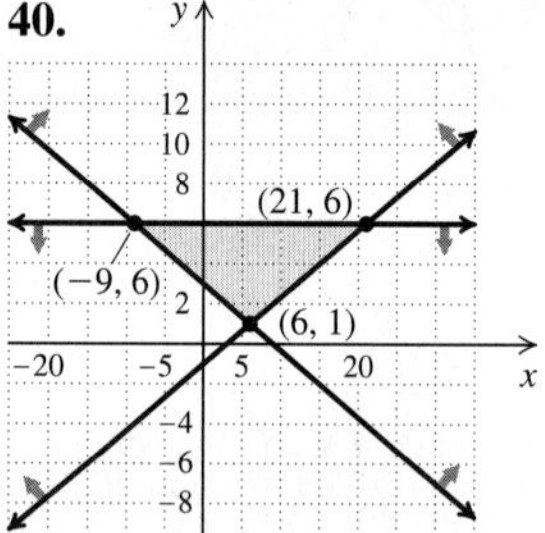

41. $(-1, 0) \cup (3, \infty)$, or $\{x \mid -1 < x < 0 \text{ or } x > 3\}$
42. $(-3, 5]$, or $\{x \mid -3 < x \le 5\}$
43. TW The equation $|X| = p$ has two solutions when p is positive because X can be either p or $-p$. The same equation has no solution when p is negative because no number has a negative absolute value. **44.** TW The solution set of a system of inequalities is all ordered pairs that make *all* the individual inequalities true. This consists of ordered pairs that are common to all the individual solution sets, or the intersection of the graphs.
45. $\left\{x \mid -\frac{8}{3} \le x \le -2\right\}$, or $\left[-\frac{8}{3}, -2\right]$
46. $|t - 21.5| \le 3.5$, where t is in thousandths of an inch

Test: Chapter 8, p. 646

1. [8.1] $\{x \mid x > 1\}$, or $(1, \infty)$ **2.** [8.1] $\{x \mid x \ge 4\}$, or $[4, \infty)$
3. [8.1] $\{3, 5\}$ **4.** [8.1] $\{1, 3, 5, 7, 9, 11, 13\}$
5. [8.1] $(-\infty, 2]$ **6.** [8.1] $(-\infty, 7) \cup (7, \infty)$
7. [8.1] $\{x \mid 1 < x < 8\}$, or $(1, 8)$
8. [8.1] $\left\{t \mid -\frac{2}{5} < t \le \frac{9}{5}\right\}$, or $\left(-\frac{2}{5}, \frac{9}{5}\right]$

9. [8.1] $\{x \mid x < 3 \text{ or } x > 6\}$, or $(-\infty, 3) \cup (6, \infty)$

10. [8.1] $\{x \mid x < -4 \text{ or } x > -\frac{5}{2}\}$, or $(-\infty, -4) \cup (-\frac{5}{2}, \infty)$

11. [8.2] $\{-13, 13\}$

12. [8.2] $\{a \mid a < -7 \text{ or } a > 7\}$, or $(-\infty, -7) \cup (7, \infty)$

13. [8.2] $\{x \mid -2 < x < \frac{8}{3}\}$, or $(-2, \frac{8}{3})$

14. [8.2] $\{t \mid t \le -\frac{13}{5} \text{ or } t \ge \frac{7}{5}\}$, or $(-\infty, -\frac{13}{5}] \cup [\frac{7}{5}, \infty)$

15. [8.2] $\varnothing$

16. [8.1] $\{x \mid x < \frac{1}{2} \text{ or } x > \frac{7}{2}\}$, or $(-\infty, \frac{1}{2}) \cup (\frac{7}{2}, \infty)$

17. [8.2] $\{1\}$

18. [8.3]

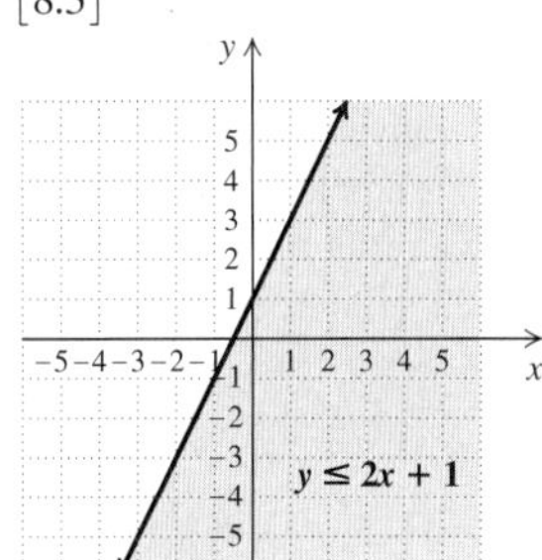

19. [8.3]

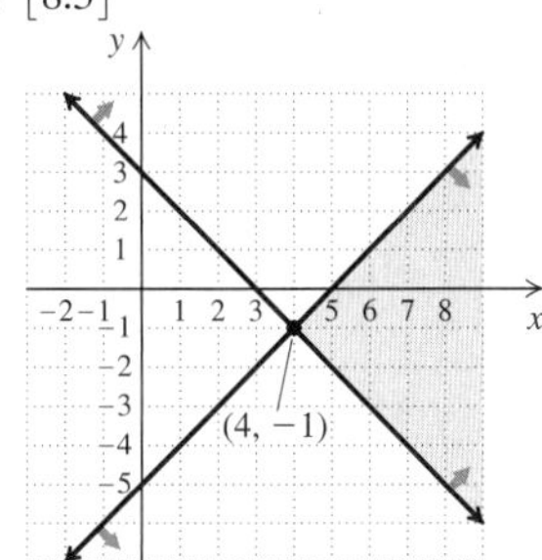

20. [8.3]

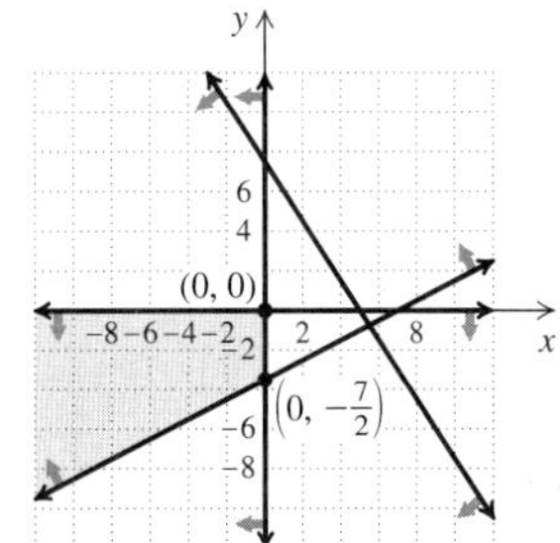

21. [8.4] $[-6, 1]$, or $\{x \mid -6 \le x \le 1\}$

22. [8.4] $(-1, 0) \cup (1, \infty)$, or $\{x \mid -1 < x < 0 \text{ or } x > 1\}$

23. [8.2] $[-1, 0] \cup [4, 6]$ **24.** [8.1] $(\frac{1}{5}, \frac{4}{5})$

25. [8.2] $|x + 3| \le 5$

Chapter 9

Exercise Set 9.1, pp. 654–656

1. True **2.** False **3.** False **4.** True **5.** True **6.** False **7.** Yes **9.** $(3, 1, 2)$ **11.** $(1, -2, 2)$ **13.** $(2, -5, -6)$ **15.** No solution **17.** $(-2, 0, 5)$ **19.** $(21, -14, -2)$ **21.** The equations are dependent. **23.** $(3, \frac{1}{2}, -4)$ **25.** $(\frac{1}{2}, \frac{1}{3}, \frac{1}{6})$ **27.** $(\frac{1}{2}, \frac{2}{3}, -\frac{5}{6})$ **29.** $(15, 33, 9)$ **31.** $(3, 4, -1)$ **33.** $(10, 23, 50)$ **35.** No solution **37.** The equations are dependent. **39.** TW **41.** Let x and y represent the numbers; $x = \frac{1}{2}y$ **42.** Let x and y represent the numbers; $x - y = 2x$ **43.** Let x represent the first number; $x + (x + 1) + (x + 2) = 100$ **44.** Let x, y, and z represent the numbers; $x + y + z = 100$ **45.** Let x, y, and z represent the numbers; $xy = 5z$ **46.** Let x and y represent the numbers; $xy = 2(x + y)$ **47.** TW **49.** $(1, -1, 2)$ **51.** $(-3, -1, 0, 4)$ **53.** $(-\frac{1}{2}, -1, -\frac{1}{3})$ **55.** 14 **57.** $z = 8 - 2x - 4y$

Exercise Set 9.2, pp. 660–662

1. 16, 19, 22 **3.** 8, 21, −3 **5.** 32°, 96°, 52° **7.** Reading: 501; mathematics: 515; writing: 493 **9.** Bran muffin: 1.5 g; banana: 3 g; 1 cup Wheaties: 3 g **11.** Basic price: \$23,800; satellite radio: \$310; car cover: \$230 **13.** 12-oz cups: 17; 16-oz cups: 25; 20-oz cups: 13 **15.** Bank loan: \$15,000; small-business loan: \$35,000; mortgage: \$70,000 **17.** Gold: \$30/g; silver: \$3/g; copper: \$0.02/g **19.** Roast beef: 2 servings; baked potato: 1 serving; broccoli: 2 servings **21.** First mezzanine: 8 tickets; main floor: 12 tickets; second mezzanine: 20 tickets **23.** Two-point field goals: 32; three-point field goals: 5; foul shots: 13 **25.** TW **27.** $-4x + 6y$ **28.** $-x + 6y$ **29.** $7y$ **30.** $11a$ **31.** $-2a + b + 6c$ **32.** $-50a - 30b + 10c$ **33.** $-12x + 5y - 8z$ **34.** $23x - 13z$ **35.** TW **37.** Applicant: \$83; spouse: \$52; first child: \$19; second child: \$19 **39.** 20 yr **41.** 35 tickets

Exercise Set 9.3, pp. 669–670

1. Matrix **2.** Horizontal; columns **3.** Entry **4.** Matrices **5.** Rows **6.** First **7.** $(3, 4)$ **9.** $(-4, 3)$ **11.** $(\frac{3}{2}, \frac{5}{2})$ **13.** $(2, \frac{1}{2}, -2)$ **15.** $(2, -2, 1)$ **17.** $(4, \frac{1}{2}, -\frac{1}{2})$ **19.** $(1, -3, -2, -1)$ **21.** Dimes: 18; nickels: 24 **23.** Dried fruit: 9 lb; macadamia nuts: 6 lb **25.** \$400 at 7%; \$500 at 8%; \$1600 at 9% **27.** TW **29.** 13 **30.** −22 **31.** 37 **32.** 422 **33.** TW **35.** 1324

Mid-Chapter Review: Chapter 9, pp. 670–671

Guided Solutions

1.

$$\begin{array}{r} x - y + z = 4 \\ x + y - 2z = 3 \\ \hline 2x \quad\quad - z = 7 \end{array}$$

$$\begin{array}{r} x + y - 2z = 3 \\ 2x - y - z = 9 \\ \hline 3x \quad\quad - 3z = 12 \end{array}$$

$$\begin{aligned} 2x - z &= 7 \\ 3x - 3z &= 12 \end{aligned}$$

$x = 3, z = -1$

$$\begin{aligned} x + y - 2z &= 3 \\ 3 + y - 2(-1) &= 3 \\ y &= -2 \end{aligned}$$

The solution is $(3, -2, -1)$.

2. $\left[\begin{array}{rr|r} 2 & 3 & 6 \\ 4 & -5 & 1 \end{array}\right]$

$\left[\begin{array}{rr|r} 2 & 3 & 6 \\ 0 & -11 & -11 \end{array}\right]$

$2x + 3y = 6$
$-11y = -11$
$y = 1$

$2x + 3(1) = 6$
$x = \frac{3}{2}$

The solution is $\left(\frac{3}{2}, 1\right)$.

Mixed Review

1. $(2, -1, 4)$ **2.** $(-3, 2, -2)$ **3.** $(1, 0, 5)$ **4.** $\left(\frac{23}{10}, -\frac{1}{2}, \frac{11}{10}\right)$ **5.** $\left(\frac{16}{7}, \frac{32}{7}, -\frac{6}{7}\right)$ **6.** No solution **7.** $\left(-\frac{11}{2}, \frac{5}{2}\right)$ **8.** $(1, -1)$ **9.** $\left(\frac{4}{3}, -1, \frac{5}{3}\right)$ **10.** $\left(\frac{11}{8}, 0, \frac{3}{4}\right)$ **11.** 10, -2,7 **12.** 28°, 56°, 96° **13.** Brownies: 45; bags of chips: 30; hot dogs: 50 **14.** 2%: \$2800; 1%: \$3500; 1.5%: \$4000

Exercise Set 9.4, p. 676

1. True **2.** True **3.** True **4.** False **5.** False **6.** False **7.** 18 **9.** -50 **11.** 27 **13.** -5 **15.** $(-3, 2)$ **17.** $\left(\frac{9}{19}, \frac{51}{38}\right)$ **19.** $\left(-1, -\frac{6}{7}, \frac{11}{7}\right)$ **21.** $(2, -1, 4)$ **23.** $(1, 2, 3)$ **25.** TW **27.** 9700 **28.** $70x - 2500$ **29.** -1800 **30.** 4500 **31.** $\frac{250}{7}$ **32.** $\frac{250}{7}$ **33.** TW **35.** 12 **37.** 10

Visualizing for Success, p. 681

1. A **2.** D **3.** J **4.** H **5.** I **6.** B **7.** E **8.** F **9.** C **10.** G

Exercise Set 9.5, pp. 682–684

1. (b) **2.** (a) **3.** (e) **4.** (f) **5.** (h) **6.** (c) **7.** (g) **8.** (d) **9. (a)** $P(x) = 20x - 300{,}000$; **(b)** (15,000 units, \$975,000) **11. (a)** $P(x) = 25x - 3100$; **(b)** (124 units, \$4960) **13. (a)** $P(x) = 45x - 22{,}500$; **(b)** (500 units, \$42,500) **15. (a)** $P(x) = 16x - 50{,}000$; **(b)** (3125 units, \$125,000) **17. (a)** $P(x) = 50x - 100{,}000$; **(b)** (2000 units, \$250,000) **19.** (\$70, 300) **21.** (\$22, 474) **23.** (\$50, 6250) **25.** (\$10, 1070) **27. (a)** $C(x) = 45{,}000 + 40x$; **(b)** $R(x) = 130x$; **(c)** $P(x) = 90x - 45{,}000$; **(d)** \$225,000 profit, \$9000 loss **(e)** (500 phones, \$65,000) **29. (a)** $C(x) = 10{,}000 + 30x$; **(b)** $R(x) = 80x$; **(c)** $P(x) = 50x - 10{,}000$; **(d)** \$90,000 profit, \$7500 loss; **(e)** (200 seats, \$16,000) **31. (a)** \$8.74; **(b)** 24,509 units **33.** TW **35.** $4a - 7 + h$ **36.** $4a + 4h - 7$ **37.** $\{x \mid x$ is a real number *and* $x \neq -\frac{1}{2}\}$, or $\left(-\infty, -\frac{1}{2}\right) \cup \left(-\frac{1}{2}, \infty\right)$ **38.** $\mathbb{R}$ **39.** $\{x \mid x \geq -4\}$, or $[-4, \infty)$ **40.** $\{x \mid x$ is a real number *and* $x \neq -1$ *and* $x \neq 1\}$, or $(-\infty, -1) \cup (-1, 1) \cup (1, \infty)$ **41.** TW **43.** (\$5, 300 yo-yo's) **45. (a)** $S(p) = 15.97p - 1.05$; **(b)** $D(p) = -11.26p + 41.16$; **(c)** (\$1.55, 23.7 million jars)

Study Summary: Chapter 9, pp. 685–687

1. $\left(2, -\frac{1}{2}, -5\right)$ **2.** $(2.5, 3.5, 3)$ **3.** $(4, 1)$ **4.** 28 **5.** -25 **6.** $\left(\frac{11}{4}, -\frac{3}{4}\right)$ **7. (a)** $P(x) = 75x - 9000$; **(b)** (120, \$10,800) **8.** (\$9, 141)

Review Exercises: Chapter 9, pp. 687–688

1. Elimination **2.** Consistent **3.** 180° **4.** 1 **5.** Square **6.** Determinant **7.** Total profit **8.** Fixed **9.** Equilibrium point **10.** Zero **11.** $(4, -8, 10)$ **12.** The equations are dependent. **13.** $(2, 0, 4)$ **14.** No solution **15.** $\left(\frac{8}{9}, -\frac{2}{3}, \frac{10}{9}\right)$ **16.** A: 90°; B: 67.5°; C: 22.5° **17.** Man: 1.4; woman: 5.3; one-year-old child: 50 **18.** $\left(55, -\frac{89}{2}\right)$ **19.** $(-1, 1, 3)$ **20.** -5 **21.** 9 **22.** $(6, -2)$ **23.** $(-3, 0, 4)$ **24. (a)** $P(x) = 20x - 15{,}800$; **(b)** (790 units, \$39,500) **25.** (\$3, 81) **26. (a)** $C(x) = 4.75x + 54{,}000$; **(b)** $R(x) = 9.25x$; **(c)** $P(x) = 4.5x - 54{,}000$; **(d)** \$31,500 loss; \$13,500 profit; **(e)** (12,000 pints of honey, \$111,000) **27.** TW To solve a problem involving four variables, go through the *Familiarize* and *Translate* steps as usual. The resulting system of equations can be solved using the elimination method just as for three variables but likely with more steps. **28.** TW A system of equations can be both dependent and inconsistent if it is equivalent to a system with fewer equations that has no solution. An example is a system of three equations in three unknowns in which two of the equations represent the same plane, and the third represents a parallel plane. **29.** 20,000 pints **30.** $a = -\frac{2}{3}, b = -\frac{4}{3}, c = 3$; $f(x) = -\frac{2}{3}x^2 - \frac{4}{3}x + 3$

Test: Chapter 9, p. 689

1. [9.1] The equations are dependent. **2.** [9.1] $\left(2, -\frac{1}{2}, -1\right)$ **3.** [9.1] No solution **4.** [9.1] $(0, 1, 0)$ **5.** [9.3] $\left(\frac{22}{5}, -\frac{28}{5}\right)$ **6.** [9.3] $(3, 1, -2)$ **7.** [9.4] -14 **8.** [9.4] -59 **9.** [9.4] $\left(\frac{7}{13}, -\frac{17}{26}\right)$ **10.** [9.2] A: 15°; B: 40°; C: 125° **11.** [9.2] Electrician: 3.5 hr; carpenter: 8 hr; plumber: 10 hr **12.** [9.5] (\$3, 55) **13.** [9.5] **(a)** $C(x) = 25x + 44{,}000$; **(b)** $R(x) = 80x$; **(c)** $P(x) = 55x - 44{,}000$; **(d)** \$27,500 loss, \$5500 profit; **(e)** (880 hammocks, \$64,000) **14.** [9.1] $(1, -1, 0, 2)$ **15.** [9.2] Adults' tickets: 1346; senior citizens' tickets: 335; children's tickets: 1651

Cumulative Review: Chapters 1–9, pp. 690–692

1. 22 **2.** $c - 6$ **3.** $-\frac{1}{100}$ **4.** $-\dfrac{6x^4}{y^3}$ **5.** $\dfrac{9a^6}{4b^4}$ **6.** $\dfrac{2}{x^2 + 5x + 25}$ **7.** $x^2 - 25$ **8.** $15n^2 + 11n - 14$ **9.** $\dfrac{-10}{(x + 5)(x - 5)}$ **10.** $\dfrac{(x - 3)^2}{x^2(2x - 3)}$ **11.** $\dfrac{3y^2 + 2}{3y}$ **12.** $2x^2(2x + 9)$ **13.** $(x - 6)(x + 14)$ **14.** $(4y - 9)(4y + 9)$ **15.** $8(2x + 1)(4x^2 - 2x + 1)$ **16.** $(t - 8)^2$ **17.** $x^2(x - 1)(x + 1)(x^2 + 1)$ **18.** $(0.3b - 0.2c)(0.09b^2 + 0.06bc + 0.04c^2)$

19. $(4x-1)(5x+3)$ **20.** 5 **21.** $\mathbb{R}$ **22.** $-2, 4$
23. $-4, 3$ **24.** No solution **25.** $(5, 1)$ **26.** $(-2, -3, 1)$
27. $\left\{-\frac{7}{2}, \frac{9}{2}\right\}$ **28.** $\left\{x \mid x < \frac{13}{2}\right\}$, or $\left(-\infty, \frac{13}{2}\right)$
29. $\{t \mid t < -3 \text{ or } t > 3\}$, or $(-\infty, -3) \cup (3, \infty)$
30. $\left\{x \mid -2 \le x \le \frac{10}{3}\right\}$, or $\left[-2, \frac{10}{3}\right]$ **31.** (b) **32.** (a)
33. (d) **34.** (c)

35.
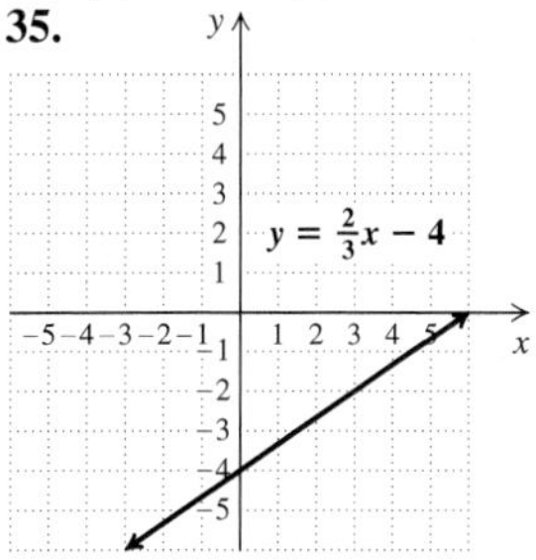

36.
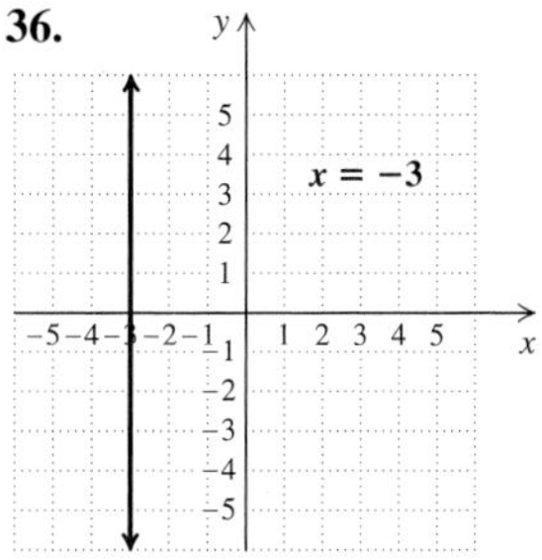

37.
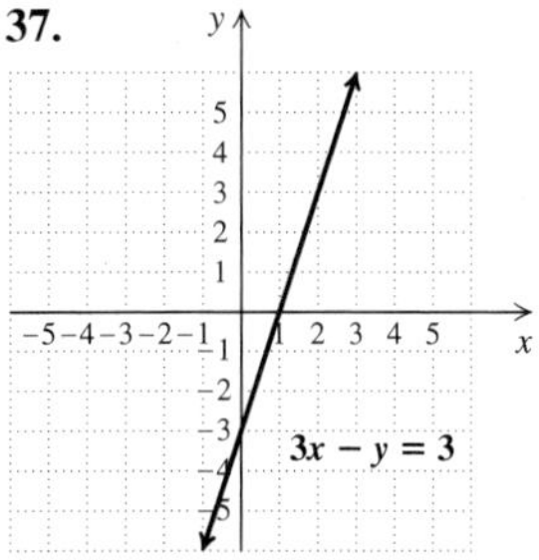

38.
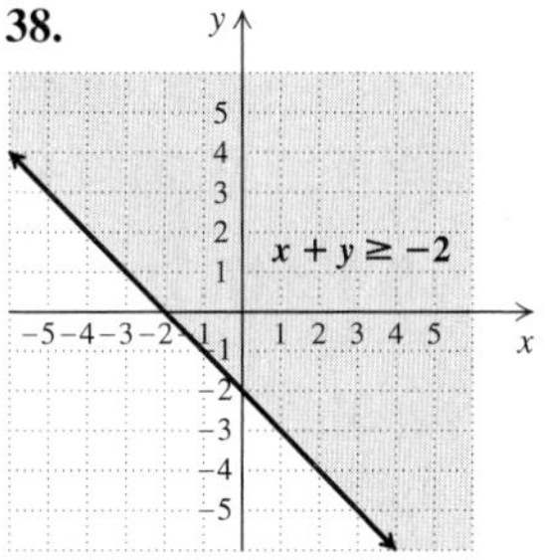

39.
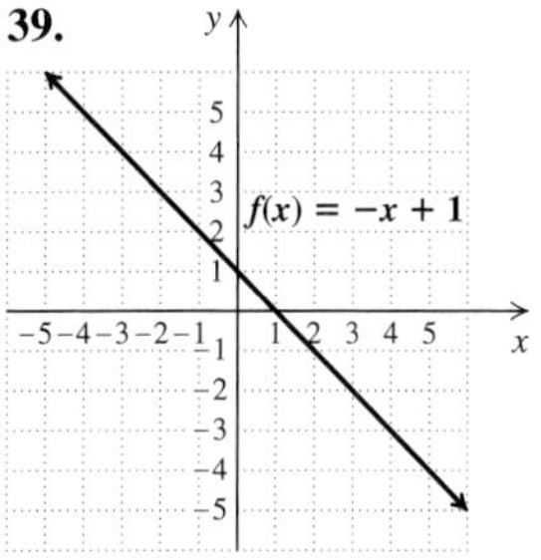

40.
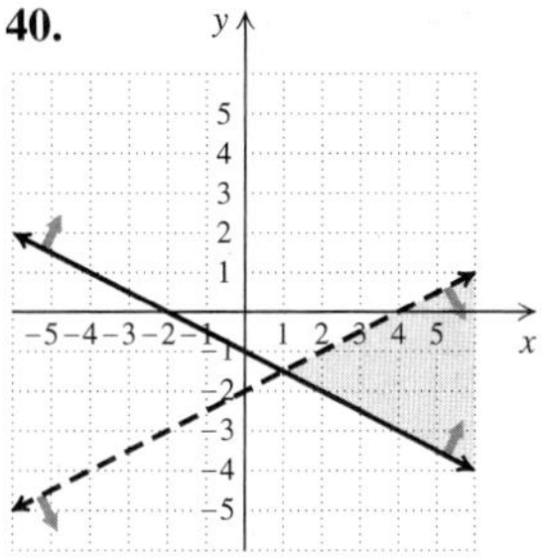

41. Slope: $\frac{4}{9}$; y-intercept: $(0, -2)$ **42.** $y = -7x - 25$
43. $y = \frac{2}{3}x + 4$ **44.** Domain: $\mathbb{R}$; range: $\{y \mid y \ge -2\}$, or $[-2, \infty)$ **45.** $\left\{x \mid x \text{ is a real number } and\ x \ne -\frac{5}{2}\right\}$, or $\left(-\infty, -\frac{5}{2}\right) \cup \left(-\frac{5}{2}, \infty\right)$ **46.** 22 **47.** $x^2 + 6x - 9$
48.
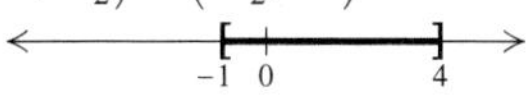

49. $\left\{x \mid x \text{ is a real number } and\ x \ne 0 \ and\ x \ne \frac{1}{3}\right\}$
50. $t = \dfrac{c}{a - d}$ **51.** 1 **52.** $\{x \mid x \ge 1\}$, or $[1, \infty)$
53. $(2, -3)$ **54.** 2 **55.** 1 **56.** 183 cannons
57. Beef: 9300 gal; wheat: 2300 gal
58. **(a)** More than \$40; **(b)** costs greater than \$30
59. Length: 10 cm; width: 6 cm **60.** \$640
61. **(a)** $f(t) = -\frac{5}{3}t + 19$; **(b)** 9 times per month; **(c)** 2015
62. **(a)** $c(t) = 93.5t + 672.5$; **(b)** about 1047 truckloads
63. $[-4, 0) \cup (0, \infty)$ **64.** 2^{11a-16} **65.** $-3, 3, -5, 5$
66. $\{x \mid -3 \le x \le -1 \text{ or } 7 \le x \le 9\}$, or $[-3, -1] \cup [7, 9]$
67. All real numbers except 9 and -5

Chapter 10

Interactive Discovery, p. 696

1. Not an identity **2.** Not an identity **3.** Identity
4. Not an identity **5.** Identity **6.** Identity

Visualizing for Success, p. 704

1. B **2.** H **3.** C **4.** I **5.** D **6.** A **7.** F
8. J **9.** G **10.** E

Exercise Set 10.1, pp. 705–708

1. Two **2.** Negative **3.** Positive **4.** Negative **5.** Irrational
6. Real **7.** Nonnegative **8.** Negative **9.** $7, -7$
11. $12, -12$ **13.** $20, -20$ **15.** $30, -30$ **17.** 7 **19.** -4
21. $\frac{6}{7}$ **23.** $-\frac{4}{9}$ **25.** 0.2 **27.** 0.09 **29.** p^2; 2
31. $\dfrac{x}{y+4}$; 5 **33.** $\sqrt{5}$; 0; does not exist; does not exist
35. -7; does not exist; -1; does not exist **37.** 1; $\sqrt{2}$; $\sqrt{101}$
39. $|8x|$, or $8|x|$ **41.** $|-4b|$, or $4|b|$ **43.** $|8 - t|$
45. $|y + 8|$ **47.** $|2x + 7|$ **49.** -4 **51.** -1 **53.** $\frac{2}{3}$
55. $|x|$ **57.** t **59.** $|6a|$, or $6|a|$ **61.** 6 **63.** $|a + b|$
65. $|a^{11}|$ **67.** Cannot be simplified **69.** $4x$ **71.** $-3t$
73. $a + 1$ **75.** $3t - 2$ **77.** $3a$ **79.** $2x$ **81.** $x - 1$
83. $5y$ **85.** t^9 **87.** $(x-2)^4$ **89.** 2; 3; -2; -4
91. 2; does not exist; does not exist; 3
93. $\{x \mid x \ge 6\}$, or $[6, \infty)$ **95.** $\{t \mid t \ge -8\}$, or $[-8, \infty)$
97. $\{x \mid x \ge 5\}$, or $[5, \infty)$ **99.** $\mathbb{R}$
101. $\left\{z \mid z \ge -\frac{2}{5}\right\}$, or $\left[-\frac{2}{5}, \infty\right)$ **103.** $\mathbb{R}$
105. Domain: $\{x \mid x \le 5\}$, or $(-\infty, 5]$; range: $\{y \mid y \ge 0\}$, or $[0, \infty)$
107. Domain: $\{x \mid x \ge -1\}$, or $[-1, \infty)$; range: $\{y \mid y \le 1\}$, or $(-\infty, 1]$
109. Domain: $\mathbb{R}$; range: $\{y \mid y \ge 5\}$, or $[5, \infty)$ **111.** (c)
113. (d) **115.** Yes **117.** Yes **119.** No
121. Approximately 840 GPM; approximately 1572 GPM
123. TW **125.** a^6b^2 **126.** $15x^3y^9$ **127.** $\dfrac{125x^6}{y^9}$ **128.** $\dfrac{a^3}{8b^6c^3}$
129. $\dfrac{x^3}{2y^6}$ **130.** $\dfrac{y^4z^8}{16x^4}$ **131.** TW **133.** TW
135. **(a)** 13; **(b)** 15; **(c)** 18; **(d)** 20
137. $\{x \mid x \ge -5\}$, or $[-5, \infty)$ **139.** $\{x \mid x \ge 0\}$, or $[0, \infty)$

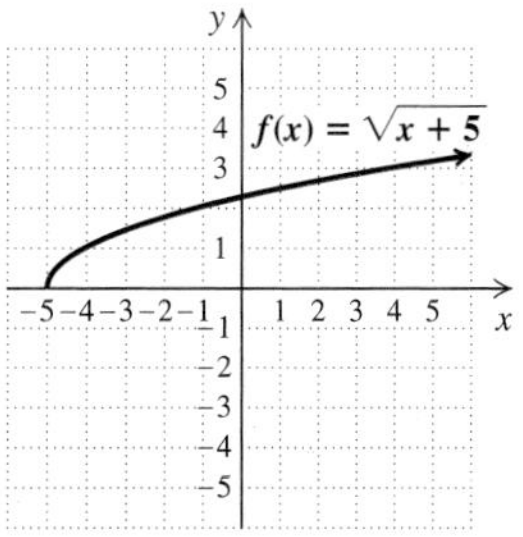

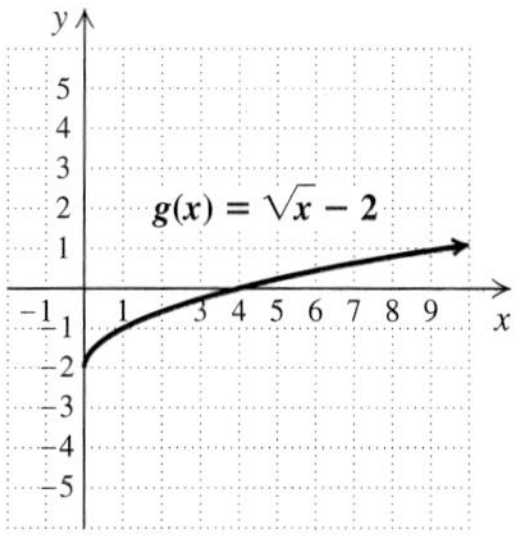

141. $\{x \mid -4 < x \le 5\}$, or $(-4, 5]$ **143.** Cubic

Exercise Set 10.2, pp. 714–716

1. (g) **2.** (c) **3.** (e) **4.** (h) **5.** (a) **6.** (d) **7.** (b) **8.** (f) **9.** $\sqrt[6]{x}$ **11.** 4 **13.** 2 **15.** 3 **17.** $\sqrt{xyz}$ **19.** $\sqrt[5]{a^2b^2}$ **21.** $\sqrt[5]{t^2}$ **23.** 8 **25.** 81 **27.** $27\sqrt[4]{x^3}$ **29.** $125x^6$ **31.** $20^{1/3}$ **33.** $17^{1/2}$ **35.** $x^{3/2}$ **37.** $m^{2/5}$ **39.** $(cd)^{1/4}$ **41.** $(xy^2z)^{1/5}$ **43.** $(3mn)^{3/2}$ **45.** $(8x^2y)^{5/7}$ **47.** $\dfrac{2x}{z^{2/3}}$ **49.** $\frac{1}{2}$ **51.** $\dfrac{1}{(2rs)^{3/4}}$ **53.** 8 **55.** $2a^{3/5}c$ **57.** $\dfrac{5y^{4/5}z}{x^{2/3}}$ **59.** $\dfrac{a^3}{3^{5/2}b^{7/3}}$ **61.** $\left(\dfrac{3c}{2ab}\right)^{5/6}$ **63.** $\dfrac{6a}{b^{1/4}}$

65. y = (x + 7)∧(1/4)

5

−10 25

−1

Xscl = 5

67. y = (3x − 2)∧(1/7)

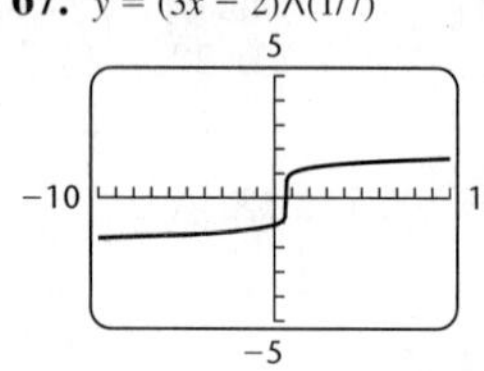

69. y = x∧(3/6)

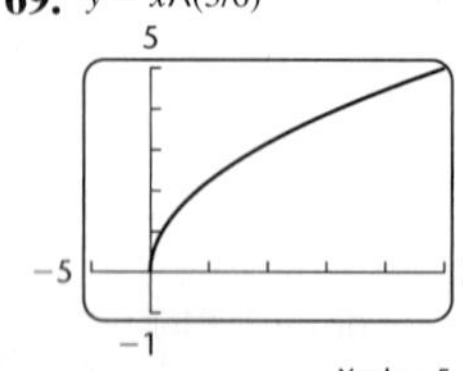

71. 1.552 **73.** 1.778 **75.** −6.240 **77.** $7^{7/8}$ **79.** $3^{3/4}$ **81.** $5.2^{1/2}$ **83.** $10^{6/25}$ **85.** $a^{23/12}$ **87.** 64 **89.** $\dfrac{m^{1/3}}{n^{1/8}}$ **91.** $\sqrt{x}$ **93.** a^3

95. $\sqrt[3]{y^2}$ **97.** x^2y^2 **99.** $\sqrt{7a}$ **101.** $\sqrt[4]{8x^3}$ **103.** $\sqrt[10]{m}$ **105.** x^3y^3 **107.** a^6b^{12} **109.** $\sqrt[12]{xy}$ **111.** TW **113.** $x^2 - 25$ **114.** $x^3 - 8$ **115.** $(2x + 5)^2$ **116.** $(3a - 4)^2$ **117.** $5(t - 1)^2$ **118.** $3(n + 2)^2$ **119.** TW **121.** $\sqrt[6]{x^5}$ **123.** $\sqrt[6]{p + q}$
125. 1760 cycles per second
127. $2^{4/12} \approx 1.2599 \approx 1.25$, which is 25% greater than 1.
129. **(a)** 1.8 m; **(b)** 3.1 m; **(c)** 1.5 m; **(d)** 5.3 m
131. 338 cubic feet **133.**

Exercise Set 10.3, pp. 721–723

1. True **2.** False **3.** False **4.** False **5.** True **6.** True **7.** $\sqrt{35}$ **9.** $\sqrt[3]{6}$ **11.** $\sqrt[4]{18}$ **13.** $\sqrt{26xy}$ **15.** $\sqrt[5]{80y^4}$ **17.** $\sqrt{y^2 - b^2}$ **19.** $\sqrt[3]{0.21y^2}$ **21.** $\sqrt[5]{(x - 2)^3}$ **23.** $\sqrt{\dfrac{21s}{11t}}$ **25.** $\sqrt[7]{\dfrac{5x - 15}{4x + 8}}$ **27.** $3\sqrt{2}$ **29.** $3\sqrt{3}$ **31.** $2x^4\sqrt{2x}$ **33.** $2\sqrt{30}$ **35.** $6a^2\sqrt{b}$ **37.** $2x\sqrt[3]{y^2}$ **39.** $-2x^2\sqrt[3]{2}$ **41.** $f(x) = 5x\sqrt[3]{x^2}$ **43.** $f(x) = |7(x - 3)|$, or $7|x - 3|$ **45.** $f(x) = |x - 1|\sqrt{5}$ **47.** $a^3b^3\sqrt{b}$ **49.** $xy^2z^3\sqrt[3]{x^2z}$ **51.** $2xy^2\sqrt[4]{xy^3}$ **53.** $x^2yz^3\sqrt[5]{x^3y^3z^2}$ **55.** $-2a^4\sqrt[3]{10a^2}$ **57.** $3\sqrt{2}$ **59.** $2\sqrt{35}$ **61.** 3 **63.** $18a^3$ **65.** $a\sqrt[3]{10}$ **67.** $24x^3\sqrt{5x}$ **69.** $s^2t^3\sqrt[3]{t}$ **71.** $(x + 5)^2$ **73.** $2ab^3\sqrt[4]{5a}$ **75.** $x(y + z)^2\sqrt[5]{x}$ **77.** TW **79.** $9abx^2$ **80.** $\dfrac{(x - 1)^2}{(x - 2)^2}$ **81.** $\dfrac{x + 1}{2(x + 5)}$ **82.** $\dfrac{3(4x^2 + 5y^3)}{50xy^2}$ **83.** $\dfrac{b + a}{a^2b^2}$ **84.** $\dfrac{-x - 2}{4x + 3}$ **85.** TW, **87.** 175.6 mi
89. **(a)** −3.3°C; **(b)** −16.6°C; **(c)** −25.5°C; **(d)** −54.0°C
91. $25x^5\sqrt[3]{25x}$ **93.** $a^{10}b^{17}\sqrt{ab}$
95. $f(x) = h(x); f(x) \neq g(x)$

y
h(x)
f(x)
g(x)
−5 −4 −3 −2 −1 1 2 3 4 5 x

97. $\{x | x \leq 2 \text{ or } x \geq 4\}$, or $(-\infty, 2] \cup [4, \infty)$ **99.** 6
101. TW

Exercise Set 10.4, pp. 728–730

1. (e) **2.** (b) **3.** (f) **4.** (c) **5.** (h) **6.** (d) **7.** (a) **8.** (g) **9.** $\frac{6}{5}$ **11.** $\frac{4}{3}$ **13.** $\dfrac{7}{y}$ **15.** $\dfrac{6y\sqrt{y}}{x^2}$ **17.** $\dfrac{3a\sqrt[3]{a}}{2b}$ **19.** $\dfrac{2a}{bc^2}$ **21.** $\dfrac{ab^2}{c^2}\sqrt[4]{\dfrac{a}{c^2}}$ **23.** $\dfrac{2x}{y^2}\sqrt[5]{\dfrac{x}{y}}$ **25.** $\dfrac{xy}{z^2}\sqrt[6]{\dfrac{y^2}{z^3}}$ **27.** 3 **29.** $\sqrt[3]{2}$ **31.** $y\sqrt{5y}$ **33.** $2\sqrt[3]{a^2b}$ **35.** $\sqrt{2ab}$ **37.** $2x^2y^3\sqrt[4]{y^3}$ **39.** $\sqrt[3]{x^2 + xy + y^2}$ **41.** $\dfrac{\sqrt{6}}{2}$ **43.** $\dfrac{2\sqrt{15}}{21}$ **45.** $\dfrac{\sqrt[3]{10}}{2}$ **47.** $\dfrac{\sqrt[3]{75ac^2}}{5c}$ **49.** $\dfrac{y\sqrt[4]{45x^3y^2}}{3x}$ **51.** $\dfrac{\sqrt[3]{2xy^2}}{xy}$ **53.** $\dfrac{\sqrt{14a}}{6}$ **55.** $\dfrac{3\sqrt{5y}}{10xy}$ **57.** $\dfrac{\sqrt{5b}}{6a}$ **59.** $\dfrac{5}{\sqrt{55}}$ **61.** $\dfrac{12}{5\sqrt{42}}$ **63.** $\dfrac{2}{\sqrt{6x}}$ **65.** $\dfrac{7}{\sqrt[3]{98}}$ **67.** $\dfrac{7x}{\sqrt{21xy}}$ **69.** $\dfrac{2a^2}{\sqrt[3]{20ab}}$ **71.** $\dfrac{x^2y}{\sqrt{2xy}}$ **73.** TW **75.** $x(3 - 8y + 2z)$ **76.** $ac(4a + 9 - 3a^2)$ **77.** $a^2 - b^2$ **78.** $a^4 - 4y^2$ **79.** $56 - 11x - 12x^2$ **80.** $6ay - 2cy - 3ax + cx$ **81.** TW **83.** **(a)** 1.62 sec; **(b)** 1.99 sec; **(c)** 2.20 sec **85.** $9\sqrt[3]{9n^2}$ **87.** $\dfrac{-3\sqrt{a^2 - 3}}{a^2 - 3}$, or $\dfrac{-3}{\sqrt{a^2 - 3}}$
89. Step 1: $\sqrt[n]{a} = a^{1/n}$, by definition; Step 2: $\left(\dfrac{a}{b}\right)^n = \dfrac{a^n}{b^n}$, raising a quotient to a power; Step 3: $a^{1/n} = \sqrt[n]{a}$, by definition
91. $(f/g)(x) = 3x$, where x is a real number and $x > 0$
93. $(f/g)(x) = \sqrt{x + 3}$, where x is a real number and $x > 3$

Exercise Set 10.5, pp. 735–737

1. Radicands; indices **2.** Indices **3.** Bases **4.** Denominators **5.** Numerator; conjugate **6.** Bases **7.** $9\sqrt{5}$ **9.** $2\sqrt[3]{4}$ **11.** $10\sqrt[3]{y}$ **13.** $12\sqrt{2}$ **15.** $13\sqrt[3]{7} + \sqrt{3}$ **17.** $9\sqrt{3}$ **19.** $-7\sqrt{5}$ **21.** $9\sqrt[3]{2}$ **23.** $(1 + 12a)\sqrt{a}$ **25.** $(x - 2)\sqrt[3]{6x}$ **27.** $3\sqrt{a - 1}$ **29.** $(x + 3)\sqrt{x - 1}$ **31.** $4\sqrt{3} + 3$ **33.** $15 - 3\sqrt{10}$ **35.** $6\sqrt{5} - 4$ **37.** $3 - 4\sqrt[3]{63}$ **39.** $a + 2a\sqrt[3]{3}$ **41.** $4 + 3\sqrt{6}$ **43.** $\sqrt{6} - \sqrt{14} + \sqrt{21} - 7$ **45.** 4 **47.** −2 **49.** $2 - 8\sqrt{35}$ **51.** $7 + 4\sqrt{3}$ **53.** $5 - 2\sqrt{6}$

55. $2t + 5 + 2\sqrt{10t}$ **57.** $14 + x - 6\sqrt{x + 5}$
59. $6\sqrt[4]{63} + 4\sqrt[4]{35} - 3\sqrt[4]{54} - 2\sqrt[4]{30}$ **61.** $\frac{18 + 6\sqrt{2}}{7}$
63. $\frac{12 - 2\sqrt{3} + 6\sqrt{5} - \sqrt{15}}{33}$ **65.** $\frac{a - \sqrt{ab}}{a - b}$
67. -1 **69.** $\frac{12 - 3\sqrt{10} - 2\sqrt{14} + \sqrt{35}}{6}$ **71.** $\frac{1}{\sqrt{5} - 1}$
73. $\frac{2}{14 + 2\sqrt{3} + 3\sqrt{2} + 7\sqrt{6}}$ **75.** $\frac{x - y}{x + 2\sqrt{xy} + y}$
77. $\frac{1}{\sqrt{a + h} + \sqrt{a}}$ **79.** $\sqrt{a}$ **81.** $b\sqrt[10]{b^9}$
83. $xy\sqrt[6]{xy^5}$ **85.** $3a^2b\sqrt[4]{ab}$ **87.** $a^2b^2c^2\sqrt[6]{a^2bc^2}$
89. $\sqrt[12]{a^5}$ **91.** $\sqrt[12]{x^2y^5}$ **93.** $\sqrt[10]{ab^9}$ **95.** $\sqrt[6]{(7 - y)^5}$
97. $\sqrt[12]{5 + 3x}$ **99.** $x\sqrt[6]{xy^5} - \sqrt[15]{x^{13}y^{14}}$
101. $2m^2 + m\sqrt[4]{n} + 2m\sqrt[3]{n^2} + \sqrt[12]{n^{11}}$
103. $2\sqrt[4]{x^3} - \sqrt[12]{x^{11}}$ **105.** $x^2 - 7$ **107.** $27 + 10\sqrt{2}$
109. $8 - 2\sqrt{15}$ **111.** TW **113.** 42 **114.** $-\frac{1}{3}$
115. $-7, 3$ **116.** $-\frac{2}{5}, \frac{3}{2}$ **117.** -3 **118.** $-6, 1$ **119.** TW
121. $f(x) = 2x\sqrt{x - 1}$ **123.** $f(x) = (x + 3x^2)\sqrt[4]{x - 1}$
125. $(7x^2 - 2y^2)\sqrt{x + y}$ **127.** $4x(y + z)^3\sqrt[6]{2x(y + z)}$
129. $1 - \sqrt{w}$ **131.** $(\sqrt{x} + \sqrt{5})(\sqrt{x} - \sqrt{5})$
133. $(\sqrt{x} + \sqrt{a})(\sqrt{x} - \sqrt{a})$ **135.** $2x - 2\sqrt{x^2 - 4}$

Mid-Chapter Review: Chapter 10, p. 738

Guided Solutions

1. $\sqrt{6x^9} \cdot \sqrt{2xy} = \sqrt{6x^9 \cdot 2xy}$
$= \sqrt{12x^{10}y}$
$= \sqrt{4x^{10} \cdot 3y}$
$= \sqrt{4x^{10}} \cdot \sqrt{3y}$
$= 2x^5\sqrt{3y}$

2. $\sqrt{12} - 3\sqrt{75} + \sqrt{8} = 2\sqrt{3} - 3 \cdot 5\sqrt{3} + 2\sqrt{2}$
$= 2\sqrt{3} - 15\sqrt{3} + 2\sqrt{2}$
$= -13\sqrt{3} + 2\sqrt{2}$

Mixed Review

1. 9 **2.** $-\frac{3}{10}$ **3.** $|8t|$, or $8|t|$ **4.** x **5.** -4
6. $\{x|x \le 10\}$, or $(-\infty, 10]$ **7.** 4 **8.** $\sqrt[12]{a}$
9. y^8 **10.** $t + 5$ **11.** $-3a^4$ **12.** $3x\sqrt{10}$ **13.** $\frac{2}{3}$
14. $5\sqrt{15t}$ **15.** $ab^2c^2\sqrt[5]{c}$ **16.** $2\sqrt{15} - 3\sqrt{22}$
17. $\sqrt[8]{t}$ **18.** $\frac{a^2}{2}$ **19.** $-8\sqrt{3}$ **20.** -4 **21.** $25 + 10\sqrt{6}$
22. $2\sqrt{x - 1}$ **23.** $xy\sqrt[10]{x^7y^3}$ **24.** $15\sqrt[3]{5}$ **25.** $6x^3y^2$

Interactive Discovery, p. 739

1. $\{3\}; \{-3, 3\}$ **2.** $\{-2\}; \{-2, 2\}$ **3.** $\{25\}; \{25\}$
4. $\varnothing; \{9\}$

Exercise Set 10.6, pp. 744–746

1. False **2.** True **3.** True **4.** False **5.** True **6.** True
7. 3 **9.** $\frac{25}{3}$ **11.** 168 **13.** -1 **15.** 3 **17.** 82 **19.** 0, 9
21. 100 **23.** -27 **25.** 16 **27.** No solution **29.** $\frac{80}{3}$
31. 45 **33.** $-\frac{5}{3}$ **35.** 1 **37.** $\frac{106}{27}$ **39.** 4 **41.** 3, 7
43. $\frac{80}{9}$ **45.** -1 **47.** No solution **49.** 2, 6 **51.** 2 **53.** 4
55. TW **57.** Length: 200 ft; width: 15 ft **58.** Base: 34 in.; height: 15 in. **59.** Length: 14 in.; width: 10 in. **60.** Length: 30 yd; width: 16 yd **61.** 6, 8, 10 **62.** 13 cm **63.** TW
65. About 68 psi **67.** About 278 Hz **69.** 524.8°C
71. $t = \frac{1}{9}\left(\frac{S^2 \cdot 2457}{1087.7^2} - 2617\right)$ **73.** 4480 rpm
75. $r = \frac{v^2h}{2gh - v^2}$ **77.** $-\frac{8}{9}$ **79.** $-8, 8$ **81.** 1, 8
83. $\left(\frac{1}{36}, 0\right), (36, 0)$

Exercise Set 10.7, pp. 753–757

1. (d) **2.** (c) **3.** (e) **4.** (b) **5.** (f) **6.** (a)
7. $\sqrt{34}$; 5.831 **9.** $9\sqrt{2}$; 12.728 **11.** 5 **13.** 4 m
15. $\sqrt{19}$ in.; 4.359 in. **17.** 1 m **19.** 250 ft
21. $\sqrt{8450}$, or $65\sqrt{2}$ ft; 91.924 ft **23.** 24 in.
25. $(\sqrt{340} + 8)$ ft; 26.439 ft
27. $(110 - \sqrt{6500})$ paces; 29.377 paces
29. Leg = 5; hypotenuse = $5\sqrt{2} \approx 7.071$
31. Shorter leg = 7; longer leg = $7\sqrt{3} \approx 12.124$
33. Leg = $5\sqrt{3} \approx 8.660$; hypotenuse = $10\sqrt{3} \approx 17.321$
35. Both legs = $\frac{13\sqrt{2}}{2} \approx 9.192$
37. Leg = $14\sqrt{3} \approx 24.249$; hypotenuse = 28
39. $3\sqrt{3} \approx 5.196$ **41.** $13\sqrt{2} \approx 18.385$
43. $\frac{19\sqrt{2}}{2} \approx 13.435$ **45.** $\sqrt{10{,}561}$ ft ≈ 102.767 ft
47. $\frac{1089}{4}\sqrt{3}$ ft^2 ≈ 471.551 ft^2 **49.** $(0, -4), (0, 4)$
51. 5 **53.** $\sqrt{10} \approx 3.162$ **55.** $\sqrt{200} \approx 14.142$ **57.** 17.8
59. $\frac{\sqrt{13}}{6} \approx 0.601$ **61.** $\sqrt{12} \approx 3.464$ **63.** $\sqrt{101} \approx 10.050$
65. $(3, 4)$ **67.** $\left(\frac{7}{2}, \frac{7}{2}\right)$ **69.** $(-1, -3)$ **71.** $(0.7, 0)$
73. $\left(-\frac{1}{12}, \frac{1}{24}\right)$ **75.** $\left(\frac{\sqrt{2} + \sqrt{3}}{2}, \frac{3}{2}\right)$ **77.** TW

79.

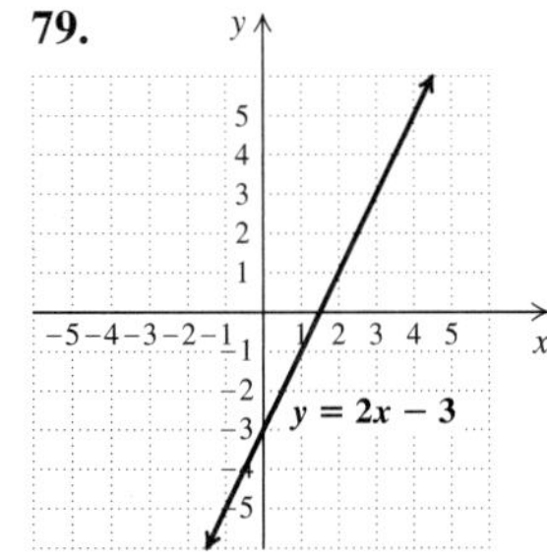

80.

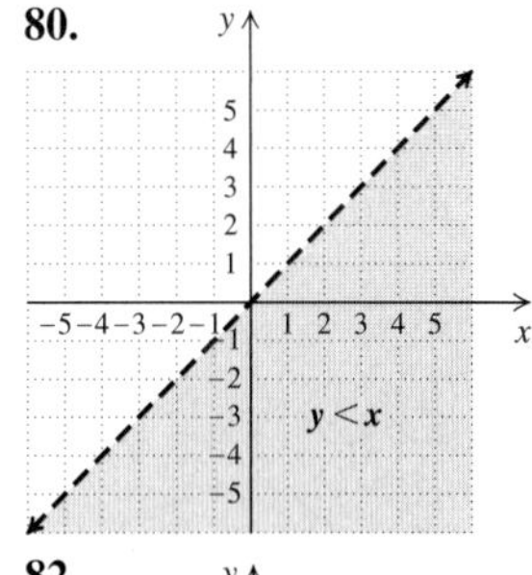

81.

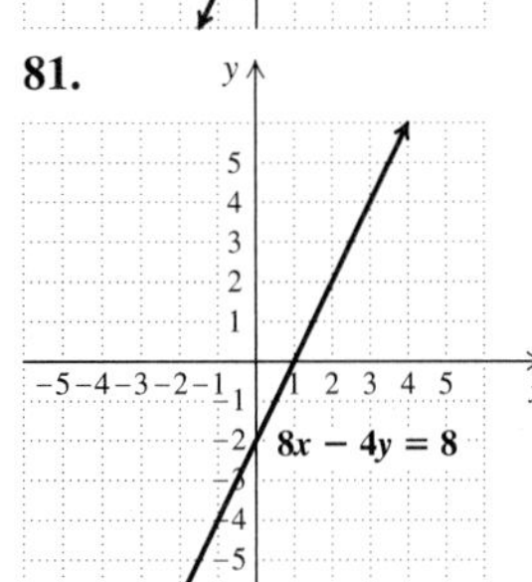

82.

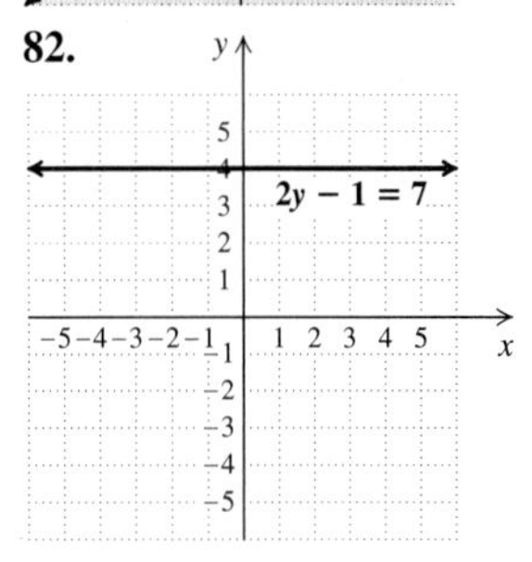

83.

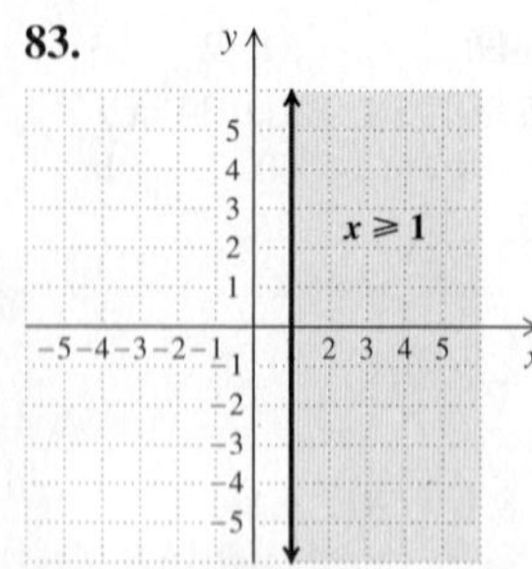

84.

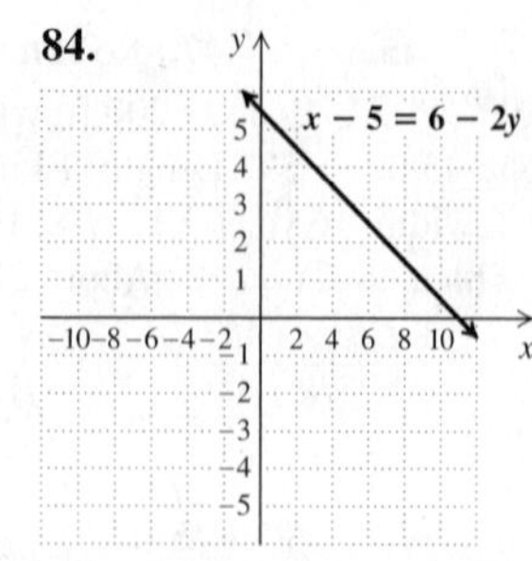

85. TW **87.** $36\sqrt{3}\text{ cm}^2$; 62.354 cm^2 **89.** $d = s + s\sqrt{2}$
91. 5 gal. The total area of the doors and windows is 134 ft^2 or more.
93. 60.28 ft by 60.28 ft
95. Let $P_1 = (x_1, y_1)$, $P_2 = (x_2, y_2)$, and

$$M = \left(\frac{x_1 + x_2}{2}, \frac{y_1 + y_2}{2}\right).$$

Let $d(AB)$ denote the distance from point A to point B.

$$\text{(i) } d(P_1M) = \sqrt{\left(\frac{x_1 + x_2}{2} - x_1\right)^2 + \left(\frac{y_1 + y_2}{2} - y_1\right)^2} = \frac{1}{2}\sqrt{(x_2 - x_1)^2 + (y_2 - y_1)^2};$$

$$d(P_2M) = \sqrt{\left(\frac{x_1 + x_2}{2} - x_2\right)^2 + \left(\frac{y_1 + y_2}{2} - y_2\right)^2} = \frac{1}{2}\sqrt{(x_1 - x_2)^2 + (y_1 - y_2)^2} = \frac{1}{2}\sqrt{(x_2 - x_1)^2 + (y_2 - y_1)^2} = d(P_1M).$$

$$\text{(ii) } d(P_1M) + d(P_2M) = \frac{1}{2}\sqrt{(x_2 - x_1)^2 + (y_2 - y_1)^2} + \frac{1}{2}\sqrt{(x_2 - x_1)^2 + (y_2 - y_1)^2} = \sqrt{(x_2 - x_1)^2 + (y_2 - y_1)^2} = d(P_1P_2).$$

Exercise Set 10.8, pp. 763–765

1. False **2.** False **3.** True **4.** True **5.** False **6.** True **7.** False **8.** True **9.** $10i$ **11.** $i\sqrt{13}$, or $\sqrt{13}i$ **13.** $2i\sqrt{2}$, or $2\sqrt{2}i$ **15.** $-i\sqrt{3}$, or $-\sqrt{3}i$ **17.** $-9i$ **19.** $-10i\sqrt{3}$, or $-10\sqrt{3}i$ **21.** $6 - 2i\sqrt{21}$, or $6 - 2\sqrt{21}i$ **23.** $(-2\sqrt{19} + 5\sqrt{5})i$ **25.** $(3\sqrt{2} - 10)i$ **27.** $11 + 10i$ **29.** $4 + 5i$ **31.** $2 - i$ **33.** $-12 - 5i$ **35.** -42 **37.** -24 **39.** -18 **41.** $-\sqrt{10}$ **43.** $-3\sqrt{14}$ **45.** $-30 + 10i$ **47.** $-28 - 21i$ **49.** $1 + 5i$ **51.** $38 + 9i$ **53.** $2 - 46i$ **55.** 73 **57.** 50 **59.** $12 - 16i$ **61.** $-5 + 12i$ **63.** $-5 - 12i$ **65.** $3 - i$ **67.** $\frac{6}{13} + \frac{4}{13}i$ **69.** $\frac{3}{17} + \frac{5}{17}i$ **71.** $-\frac{5}{6}i$ **73.** $-\frac{3}{4} - \frac{5}{4}i$ **75.** $1 - 2i$ **77.** $-\frac{23}{58} + \frac{43}{58}i$ **79.** $\frac{19}{29} - \frac{4}{29}i$ **81.** $\frac{6}{25} - \frac{17}{25}i$ **83.** $-i$ **85.** 1 **87.** -1 **89.** i **91.** -1 **93.** $-125i$ **95.** 0 **97.** TW **99.** $-2, 3$ **100.** 5 **101.** $-10, 10$ **102.** $-5, 5$ **103.** $-\frac{2}{5}, \frac{4}{3}$ **104.** $-\frac{2}{3}, \frac{3}{2}$ **105.** TW

107.

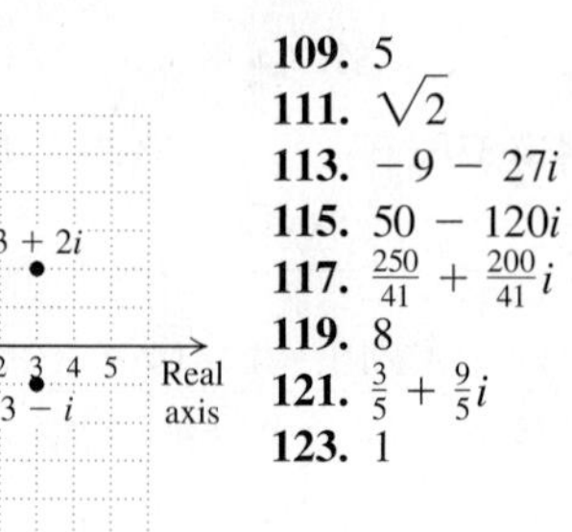

109. 5
111. $\sqrt{2}$
113. $-9 - 27i$
115. $50 - 120i$
117. $\frac{250}{41} + \frac{200}{41}i$
119. 8
121. $\frac{3}{5} + \frac{9}{5}i$
123. 1

Study Summary: Chapter 10, pp. 766–769

1. -9 **2.** -1 **3.** $|6x|$, or $6|x|$ **4.** x **5.** $\frac{1}{10}$ **6.** $\sqrt{21xy}$ **7.** $10x^2y^9\sqrt{2x}$ **8.** $\frac{2x\sqrt{3x}}{5}$ **9.** $\frac{\sqrt{6x}}{3y}$ **10.** $-5\sqrt{2}$ **11.** $31 - 19\sqrt{3}$ **12.** $\frac{3\sqrt{15} - 5\sqrt{3}}{4}$ **13.** $\sqrt[6]{x^{13}}$ **14.** 3 **15.** $\sqrt{51}\text{ m} \approx 7.141\text{ m}$ **16.** $a = 6$ **17.** $b = 5\sqrt{3} \approx 8.660$; $c = 10$ **18.** $\sqrt{185} \approx 13.601$ **19.** $\left(2, -\frac{9}{2}\right)$ **20.** $-3 - 12i$ **21.** $3 - 2i$ **22.** $-32 - 26i$ **23.** i

Review Exercises: Chapter 10, pp. 769–770

1. True **2.** False **3.** False **4.** True **5.** True **6.** True **7.** True **8.** False **9.** $\frac{7}{3}$ **10.** -0.5 **11.** 5 **12.** $\{x|x \geq \frac{7}{2}\}$, or $\left[\frac{7}{2}, \infty\right)$ **13.** **(a)** 34.6 lb; **(b)** 10.0 lb; **(c)** 24.9 lb; **(d)** 42.3 lb **14.** $|5t|$, or $5|t|$ **15.** $|c + 8|$ **16.** $|2x + 1|$ **17.** -2 **18.** $(5ab)^{4/3}$ **19.** $8a^4\sqrt{a}$ **20.** x^3y^5 **21.** $\sqrt[3]{x^2y}$ **22.** $\frac{1}{x^{2/5}}$ **23.** $7^{1/6}$ **24.** $f(x) = 5|x - 6|$ **25.** $2x^5y^2$ **26.** $5xy\sqrt{10x}$ **27.** $\sqrt{6xy}$ **28.** $3xb\sqrt[3]{x^2}$ **29.** $-6x^5y^4\sqrt[3]{2x^2}$ **30.** $y\sqrt[3]{6}$ **31.** $\frac{5\sqrt{x}}{2}$ **32.** $\frac{2a^2\sqrt[4]{3a^3}}{c^2}$ **33.** $7\sqrt[3]{x}$ **34.** $\sqrt{3}$ **35.** $(2x + y^2)\sqrt[3]{x}$ **36.** $15\sqrt{2}$ **37.** -1 **38.** $\sqrt{15} + 4\sqrt{6} - 6\sqrt{10} - 48$ **39.** $\sqrt[4]{x^3}$ **40.** $\sqrt[12]{x^5}$ **41.** $a^2 - 2a\sqrt{2} + 2$ **42.** $\frac{\sqrt{2xy}}{4y}$ **43.** $-4\sqrt{10} + 4\sqrt{15}$ **44.** $\frac{20}{\sqrt{10} + \sqrt{15}}$ **45.** 19 **46.** 9 **47.** -126 **48.** 4 **49.** 14 **50.** $5\sqrt{2}$ cm; 7.071 cm **51.** $\sqrt{32}$ ft; 5.657 ft **52.** Short leg = 10; long leg = $10\sqrt{3} \approx 17.321$ **53.** $\sqrt{26} \approx 5.099$ **54.** $\left(-2, -\frac{3}{2}\right)$ **55.** $3i\sqrt{5}$, or $3\sqrt{5}i$ **56.** $-2 - 9i$ **57.** $6 + i$ **58.** 29 **59.** -1 **60.** $9 - 12i$ **61.** $\frac{13}{25} - \frac{34}{25}i$ **62.** TW A complex number $a + bi$ is real when $b = 0$. It is imaginary when $b \neq 0$. **63.** TW An absolute-value sign must be used to simplify $\sqrt[n]{x^n}$ when n is even, since x may be negative. If x is negative while n is even, the radical expression cannot be simplified to x, since $\sqrt[n]{x^n}$ represents the principal, or positive, root. When n is odd, there is only one root, and it will be positive or negative depending on the sign of x. Thus there is no absolute-value sign when n is odd. **64.** 3 **65.** $-\frac{2}{5} + \frac{9}{10}i$ **66.** $\frac{2i}{3i}$; answers may vary **67.** The isosceles triangle is larger by about 1.206 ft^2.

Test: Chapter 10, p. 771

1. [10.3] $5\sqrt{2}$ **2.** [10.4] $-\frac{2}{x^2}$ **3.** [10.1] $|9a|$, or $9|a|$
4. [10.1] $|x - 4|$ **5.** [10.2] $(7xy)^{1/2}$ **6.** [10.2] $\sqrt[6]{(4a^3b)^5}$
7. [10.1] $\{x|x \geq 5\}$, or $[5, \infty)$ **8.** [10.5] $27 + 10\sqrt{2}$
9. [10.3] $2x^3y^2\sqrt[3]{x}$ **10.** [10.3] $2\sqrt[3]{2wv^2}$
11. [10.4] $\frac{10a^2}{3b^3}$ **12.** [10.4] $\sqrt[5]{3x^4y}$ **13.** [10.5] $x\sqrt[4]{x}$
14. [10.5] $\sqrt[5]{y^2}$ **15.** [10.5] $6\sqrt{2}$ **16.** [10.5] $(x^2 + 3y)\sqrt{y}$
17. [10.5] $14 - 19\sqrt{x} - 3x$ **18.** [10.4] $\frac{\sqrt[3]{2xy^2}}{2y}$
19. [10.6] 4 **20.** [10.6] $-1, 2$ **21.** [10.6] 8
22. [10.7] $\sqrt{10{,}600}$ ft $\approx$ 102.956 ft
23. [10.7] 5 cm; $5\sqrt{3}$ cm $\approx$ 8.660 cm
24. [10.7] $\sqrt{17} \approx 4.123$ **25.** [10.7] $\left(\frac{3}{2}, -6\right)$
26. [10.8] $5i\sqrt{2}$, or $5\sqrt{2}i$ **27.** [10.8] $12 + 2i$
28. [10.8] -24 **29.** [10.8] $15 - 8i$ **30.** [10.8] $-\frac{11}{34} - \frac{7}{34}i$
31. [10.8] i **32.** [10.6] 3 **33.** [10.8] $-\frac{17}{4}i$
34. [10.6] 22,500 ft

Chapter 11

Interactive Discovery, p. 774

1. 1 **2.** 1 **3.** 1 **4.** 0 **5.** 0 **6.** 2
7. A cup-shaped curve opening up or down

Exercise Set 11.1, pp. 783–785

1. $\sqrt{k}; -\sqrt{k}$ **2.** $7; -7$ **3.** $t + 3; t + 3$ **4.** 16
5. 25; 5 **6.** 9; 3 **7.** 2 **8.** 0 **9.** 1 **10.** 2 **11.** 0
12. 1 **13.** ± 10 **15.** $\pm 5\sqrt{2}$ **17.** $\pm\sqrt{5}$ **19.** $\pm 2i$
21. $\pm\frac{4}{3}$ **23.** $\pm\sqrt{\frac{7}{5}}$, or $\pm\frac{\sqrt{35}}{5}$ **25.** $\pm\frac{9}{2}i$ **27.** $-6, 8$
29. $13 \pm 3\sqrt{2}$ **31.** $-1 \pm 3i$ **33.** $-\frac{3}{4} \pm \frac{\sqrt{17}}{4}$, or $\frac{-3 \pm \sqrt{17}}{4}$
35. $-3, 13$ **37.** $\pm\sqrt{19}$ **39.** 1, 9 **41.** $-4 \pm \sqrt{13}$
43. $-14, 0$ **45.** $x^2 + 16x + 64 = (x + 8)^2$
47. $t^2 - 10t + 25 = (t - 5)^2$ **49.** $t^2 - 2t + 1 = (t - 1)^2$
51. $x^2 + 3x + \frac{9}{4} = \left(x + \frac{3}{2}\right)^2$ **53.** $x^2 + \frac{2}{5}x + \frac{1}{25} = \left(x + \frac{1}{5}\right)^2$
55. $t^2 - \frac{5}{6}t + \frac{25}{144} = \left(t - \frac{5}{12}\right)^2$ **57.** $-7, 1$ **59.** $5 \pm \sqrt{2}$
61. $-8, -4$ **63.** $-4 \pm \sqrt{19}$
65. $\left(-3 - \sqrt{2}, 0\right), \left(-3 + \sqrt{2}, 0\right)$
67. $\left(-\frac{9}{2} - \frac{\sqrt{181}}{2}, 0\right), \left(-\frac{9}{2} + \frac{\sqrt{181}}{2}, 0\right)$, or $\left(\frac{-9 - \sqrt{181}}{2}, 0\right), \left(\frac{-9 + \sqrt{181}}{2}, 0\right)$
69. $\left(5 - \sqrt{47}, 0\right), \left(5 + \sqrt{47}, 0\right)$ **71.** $-\frac{4}{3}, -\frac{2}{3}$ **73.** $-\frac{1}{3}, 2$
75. $-\frac{2}{5} \pm \frac{\sqrt{19}}{5}$, or $\frac{-2 \pm \sqrt{19}}{5}$
77. $\left(-\frac{1}{4} - \frac{\sqrt{13}}{4}, 0\right), \left(-\frac{1}{4} + \frac{\sqrt{13}}{4}, 0\right)$, or $\left(\frac{-1 - \sqrt{13}}{4}, 0\right), \left(\frac{-1 + \sqrt{13}}{4}, 0\right)$
79. $\left(\frac{3}{4} - \frac{\sqrt{17}}{4}, 0\right), \left(\frac{3}{4} + \frac{\sqrt{17}}{4}, 0\right)$, or $\left(\frac{3 - \sqrt{17}}{4}, 0\right), \left(\frac{3 + \sqrt{17}}{4}, 0\right)$ **81.** 10% **83.** 4%
85. About 15.8 sec **87.** About 11.4 sec **89.** TW **91.** 64
92. -15 **93.** $10\sqrt{2}$ **94.** $4\sqrt{6}$ **95.** $2i$ **96.** $5i$
97. $2i\sqrt{2}$, or $2\sqrt{2}i$ **98.** $2i\sqrt{6}$, or $2\sqrt{6}i$ **99.** TW
101. ± 18 **103.** $-\frac{7}{2}, -\sqrt{5}, 0, \sqrt{5}, 8$
105. Barge: 8 km/h; fishing boat: 15 km/h

Exercise Set 11.2, pp. 791–792

1. True **2.** True **3.** False **4.** False **5.** False **6.** True
7. $-\frac{5}{2}, 1$ **9.** $-1 \pm \sqrt{5}$ **11.** $3 \pm \sqrt{7}$ **13.** $3 \pm \sqrt{5}$
15. $\frac{3}{2} \pm \frac{\sqrt{29}}{2}$ **17.** $-1 \pm \frac{2\sqrt{3}}{3}$ **19.** $-\frac{4}{3} \pm \frac{\sqrt{19}}{3}$
21. $3 \pm i$ **23.** $-2 \pm \sqrt{2}i$ **25.** $-\frac{8}{3}, \frac{5}{4}$ **27.** $\frac{2}{5}$
29. $-\frac{11}{8} \pm \frac{\sqrt{41}}{8}$ **31.** 5, 10 **33.** $\frac{13}{10} \pm \frac{\sqrt{509}}{10}$
35. $2 \pm \sqrt{5}i$ **37.** $2, -1 \pm \sqrt{3}i$ **39.** $\frac{1}{4} \pm \frac{\sqrt{13}}{4}$ **41.** $-2, 3$
43. $\frac{7}{2} \pm \frac{\sqrt{85}}{2}$ **45.** $-5.317, 1.317$ **47.** 0.764, 5.236
49. $-1.266, 2.766$ **51.** TW **53.** $x^2 + 4$ **54.** $x^2 - 180$
55. $x^2 - 4x - 3$ **56.** $x^2 + 6x + 34$ **57.** $-\frac{3}{2}$
58. $\frac{1}{6} \pm \frac{\sqrt{6}}{3}i$ **59.** TW **61.** $(-2, 0), (1, 0)$
63. $4 - 2\sqrt{2}, 4 + 2\sqrt{2}$ **65.** $-1.179, 0.339$
67. $\frac{-5\sqrt{2}}{4} \pm \frac{\sqrt{34}}{4}$ **69.** $\frac{1}{2}$ **71.** TW

Exercise Set 11.3, pp. 795–797

1. Discriminant **2.** One **3.** Two **4.** Two **5.** Rational
6. Imaginary **7.** Two irrational **9.** Two imaginary
11. Two irrational **13.** Two rational **15.** Two imaginary
17. One rational **19.** Two rational **21.** Two irrational
23. Two imaginary **25.** Two rational **27.** Two irrational
29. $x^2 + 4x - 21 = 0$ **31.** $x^2 - 6x + 9 = 0$
33. $x^2 + 4x + 3 = 0$ **35.** $4x^2 - 23x + 15 = 0$
37. $8x^2 + 6x + 1 = 0$ **39.** $x^2 - 2x - 0.96 = 0$
41. $x^2 - 3 = 0$ **43.** $x^2 - 20 = 0$ **45.** $x^2 + 16 = 0$
47. $x^2 - 4x + 53 = 0$ **49.** $x^2 - 6x - 5 = 0$
51. $3x^2 - 6x - 4 = 0$ **53.** $x^3 - 4x^2 - 7x + 10 = 0$
55. $x^3 - 2x^2 - 3x = 0$ **57.** TW **59.** $c = \frac{d^2}{1 - d}$
60. $b = \frac{aq}{p - q}$ **61.** $y = \frac{x - 3}{x}$, or $1 - \frac{3}{x}$ **62.** 10 mph
63. Jamal: 3.5 mph; Kade: 2 mph **64.** 20 mph **65.** TW
67. $a = 1, b = 2, c = -3$ **69.** **(a)** $-\frac{3}{5}$; **(b)** $-\frac{1}{3}$
71. **(a)** $9 + 9i$; **(b)** $3 + 3i$
73. The solutions of $ax^2 + bx + c = 0$ are $x = \frac{-b \pm \sqrt{b^2 - 4ac}}{2a}$. When there is just one solution, $b^2 - 4ac$ must be 0, so $x = \frac{-b \pm 0}{2a} = \frac{-b}{2a}$.

75. $a = 8, b = 20, c = -12$ **77.** $x^2 - 2 = 0$
79. $x^4 - 8x^3 + 21x^2 - 2x - 52 = 0$

Exercise Set 11.4, pp. 801–804

1. First part: 60 mph; second part: 50 mph **3.** 40 mph
5. Cessna: 150 mph, Beechcraft: 200 mph; or Cessna: 200 mph, Beechcraft: 250 mph **7.** To Hillsboro: 10 mph; return trip: 4 mph
9. About 14 mph **11.** 12 hr **13.** About 3.24 mph
15. $r = \frac{1}{2}\sqrt{\frac{A}{\pi}}$, or $\frac{\sqrt{A\pi}}{2\pi}$ **17.** $r = \frac{-\pi h + \sqrt{\pi^2h^2 + 2\pi A}}{2\pi}$
19. $r = \sqrt{\frac{Gm_1m_2}{F}}$, or $\frac{\sqrt{FGm_1m_2}}{F}$ **21.** $H = \frac{c^2}{g}$
23. $g = \sqrt{\frac{800w}{l}}$, or $\frac{20\sqrt{2lw}}{l}$ **25.** $b = \sqrt{c^2 - a^2}$
27. $t = \frac{-v_0 + \sqrt{(v_0)^2 + 2gs}}{g}$ **29.** $n = \frac{1 + \sqrt{1 + 8N}}{2}$
31. $g = \frac{4\pi^2 l}{T^2}$ **33.** $t = \frac{-b \pm \sqrt{b^2 - 4ac}}{2a}$
35. **(a)** 10.1 sec; **(b)** 7.49 sec; **(c)** 272.5 m **37.** 2.9 sec
39. 0.957 sec **41.** 2.5 m/sec **43.** 7% **45.** TW
47. m^{-2}, or $\frac{1}{m^2}$ **48.** $t^{2/3}$ **49.** $y^{1/3}$ **50.** $z^{1/2}$ **51.** 2
52. 81 **53.** TW **55.** $t = \frac{-10.2 + 6\sqrt{-A^2 + 13A - 39.36}}{A - 6.5}$
57. $\pm\sqrt{2}$ **59.** $l = \frac{w + w\sqrt{5}}{2}$
61. $n = \pm\sqrt{\frac{r^2 \pm \sqrt{r^4 + 4m^4r^2p - 4mp}}{2m}}$ **63.** $A(S) = \frac{\pi S}{6}$

Exercise Set 11.5, pp. 810–811

1. (f) **2.** (d) **3.** (h) **4.** (b) **5.** (g) **6.** (a) **7.** (e)
8. (c) **9.** $\sqrt{p}$ **10.** $x^{1/4}$ **11.** $x^2 + 3$ **12.** t^{-3}
13. $(1 + t)^2$ **14.** $w^{1/6}$ **15.** $\pm 1, \pm 2$ **17.** $\pm\sqrt{5}, \pm 2$
19. $\pm\frac{\sqrt{3}}{2}, \pm 2$ **21.** 4 **23.** $\pm 2\sqrt{2}, \pm 3$ **25.** $8 + 2\sqrt{7}$
27. No solution **29.** $-\frac{1}{2}, \frac{1}{3}$ **31.** $-4, 1$ **33.** $-27, 8$
35. 729 **37.** 1 **39.** 9, 225 **41.** $\frac{12}{5}$ **43.** $\pm 2, \pm 3i$
45. $\pm i, \pm 2i$ **47.** $\left(\frac{4}{25}, 0\right)$
49. $\left(\frac{3}{2} + \frac{\sqrt{33}}{2}, 0\right), \left(\frac{3}{2} - \frac{\sqrt{33}}{2}, 0\right), (4, 0), (-1, 0)$
51. $(-243, 0), (32, 0)$ **53.** No x-intercepts **55.** TW
57.

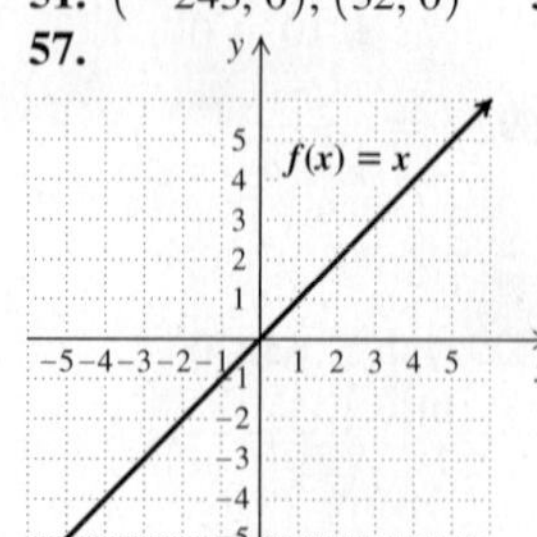

58.

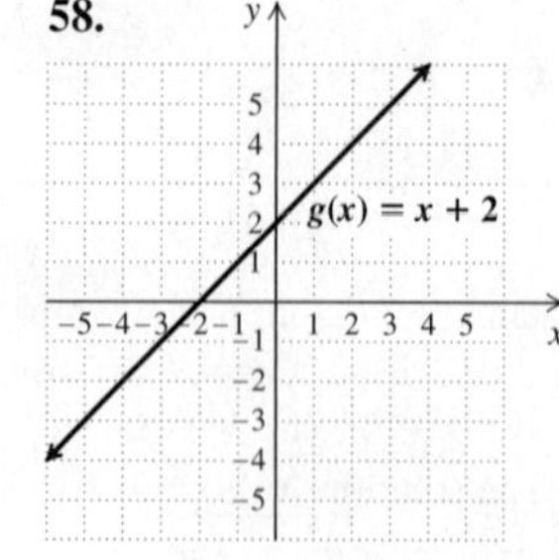

59.

60.

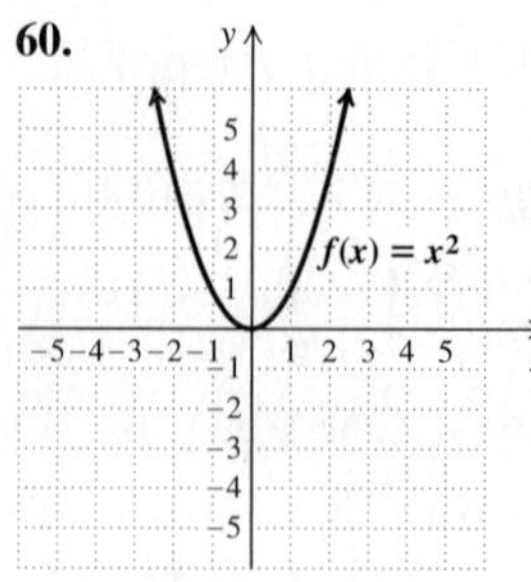

61.

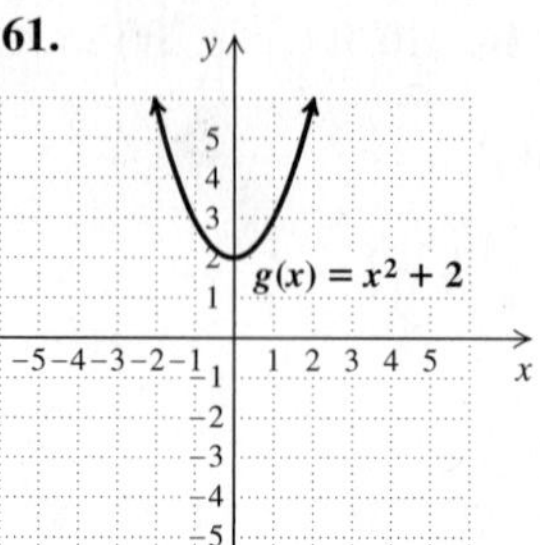

62.

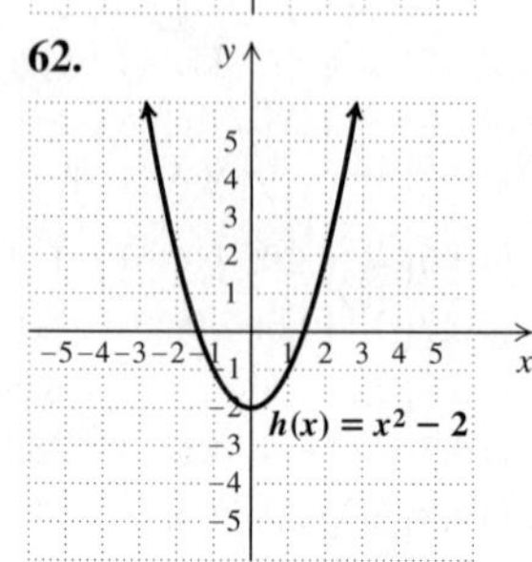

63. TW **65.** $\pm\sqrt{\frac{7 \pm \sqrt{29}}{10}}$ **67.** $-2, -1, 5, 6$
69. $\frac{100}{99}$ **71.** $-5, -3, -2, 0, 2, 3, 5$
73. $1, 3, -\frac{1}{2} + \frac{\sqrt{3}}{2}i, -\frac{1}{2} - \frac{\sqrt{3}}{2}i, -\frac{3}{2} + \frac{3\sqrt{3}}{2}i, -\frac{3}{2} - \frac{3\sqrt{3}}{2}i$

Mid-Chapter Review: Chapter 11, pp. 811–812

Guided Solutions

1. $x - 7 = \pm\sqrt{5}$
$x = 7 \pm \sqrt{5}$
The solutions are $7 + \sqrt{5}$ and $7 - \sqrt{5}$.

2. $a = 1, b = -2, c = -1$
$$x = \frac{-(-2) \pm \sqrt{(-2)^2 - 4 \cdot 1 \cdot (-1)}}{2 \cdot 1}$$
$$x = \frac{2 \pm \sqrt{8}}{2}$$
$$x = \frac{2}{2} \pm \frac{2\sqrt{2}}{2}$$
The solutions are $1 + \sqrt{2}$ and $1 - \sqrt{2}$.

Mixed Review

1. $-2, 5$ **2.** ± 11 **3.** $-3 \pm \sqrt{19}$ **4.** $-\frac{1}{2} \pm \frac{\sqrt{13}}{2}$
5. $-1 \pm \sqrt{2}$ **6.** 5 **7.** $\pm\frac{\sqrt{11}}{2}$ **8.** $\frac{1}{2}, 1$ **9.** $0, \frac{7}{16}$
10. $1 \pm \sqrt{7}i$ **11.** $\pm 1, \pm 3$ **12.** $\pm 3, \pm i$ **13.** $-6, 5$
14. $\pm\frac{\sqrt{3}}{3}, \pm\frac{\sqrt{2}}{2}$ **15.** Two irrational
16. Two rational **17.** Two imaginary **18.** $v = 20\sqrt{\frac{F}{A}}$, or $\frac{20\sqrt{FA}}{A}$ **19.** $D = d + \sqrt{d^2 + 2hd}$
20. South: 75 mph; north: 45 mph

Interactive Discovery, p. 814

1. **(a)** $(0, 0)$; **(b)** $x = 0$; **(c)** upward; **(d)** narrower
2. **(a)** $(0, 0)$; **(b)** $x = 0$; **(c)** upward; **(d)** wider
3. **(a)** $(0, 0)$; **(b)** $x = 0$; **(c)** upward; **(d)** wider
4. **(a)** $(0, 0)$; **(b)** $x = 0$; **(c)** downward; **(d)** neither narrower nor wider **5.** **(a)** $(0, 0)$; **(b)** $x = 0$; **(c)** downward; **(d)** narrower **6.** **(a)** $(0, 0)$; **(b)** $x = 0$; **(c)** downward; **(d)** wider **7.** When $a > 1$, the graph of $y = ax^2$ is narrower than the graph of $y = x^2$. When $0 < a < 1$, the graph of $y = ax^2$ is wider than the graph of $y = x^2$. **8.** When $a < -1$, the graph of $y = ax^2$ is narrower than the graph of $y = x^2$ and the graph opens downward. When $-1 < a < 0$, the graph of $y = ax^2$ is wider than the graph of $y = x^2$ and the graph opens downward.

Interactive Discovery, p. 815

1. **(a)** $(3, 0)$; **(b)** $x = 3$; **(c)** same shape **2.** **(a)** $(-1, 0)$; **(b)** $x = -1$; **(c)** same shape **3.** **(a)** $\left(\frac{3}{2}, 0\right)$; **(b)** $x = \frac{3}{2}$; **(c)** same shape **4.** **(a)** $(-2, 0)$; **(b)** $x = -2$; **(c)** same shape **5.** The graph of $g(x) = a(x - h)^2$ looks like the graph of $f(x) = ax^2$, except that it is moved left or right.

Exercise Set 11.6, pp. 819–822

1. (h) **2.** (g) **3.** (f) **4.** (d) **5.** (b) **6.** (c) **7.** (e) **8.** (a) **9.** **(a)** Positive; **(b)** $(3, 1)$; **(c)** $x = 3$; **(d)** $[1, \infty)$ **10.** **(a)** Negative; **(b)** $(-1, 2)$; **(c)** $x = -1$; **(d)** $(-\infty, 2]$ **11.** **(a)** Negative; **(b)** $(-2, -3)$; **(c)** $x = -2$; **(d)** $(-\infty, -3]$ **12.** **(a)** Positive; **(b)** $(2, 0)$; **(c)** $x = 2$; **(d)** $[0, \infty)$ **13.** **(a)** Positive; **(b)** $(-3, 0)$; **(c)** $x = -3$; **(d)** $[0, \infty)$ **14.** **(a)** Negative; **(b)** $(1, -2)$; **(c)** $x = 1$; **(d)** $(-\infty, -2]$

15. $f(x) = x^2$

17. $f(x) = -2x^2$

19. $g(x) = \frac{1}{3}x^2$

21.

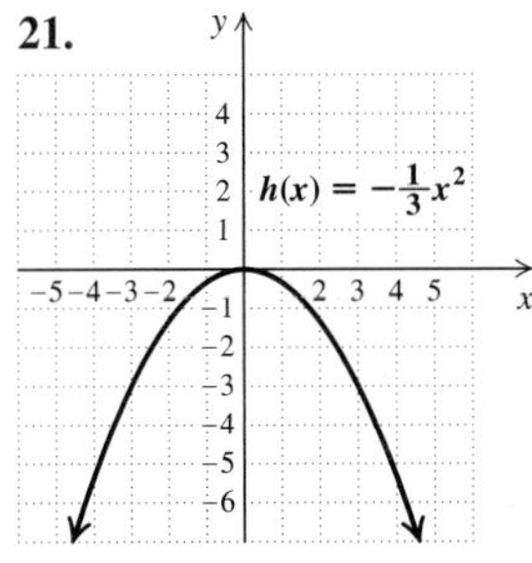

23.

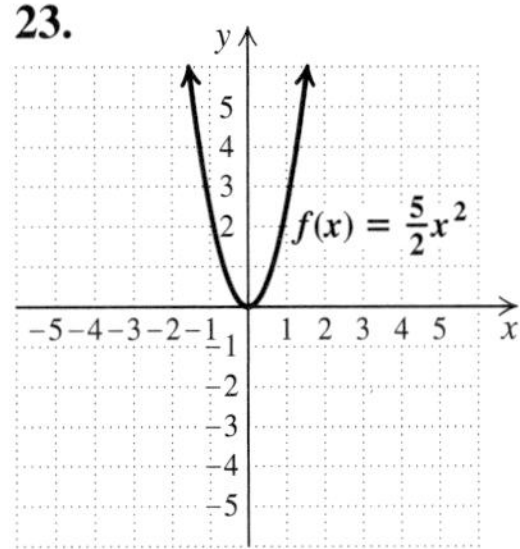

25. Vertex: $(-1, 0)$; axis of symmetry: $x = -1$

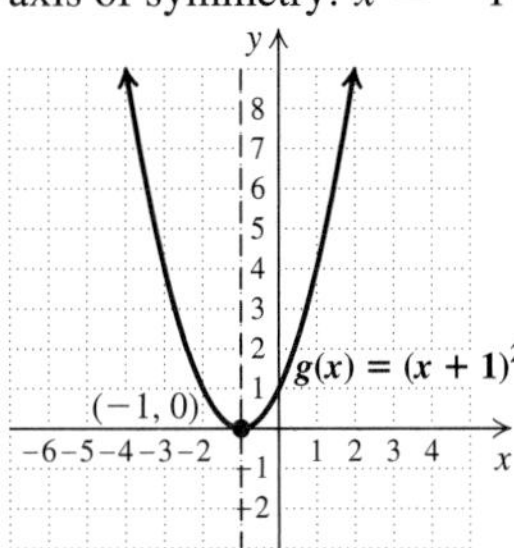

27. Vertex: $(2, 0)$; axis of symmetry: $x = 2$

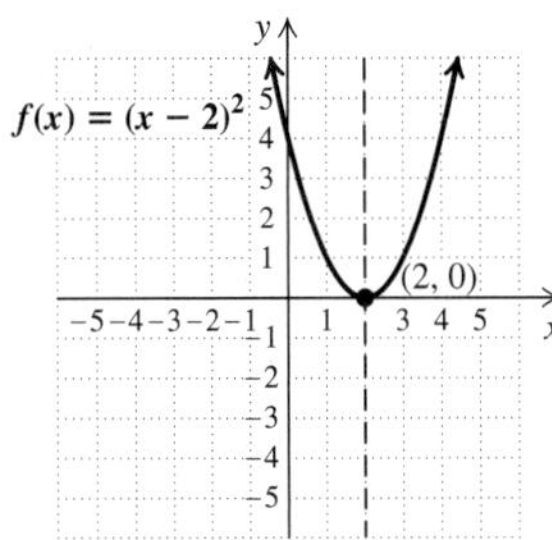

29. Vertex: $(-1, 0)$; axis of symmetry: $x = -1$

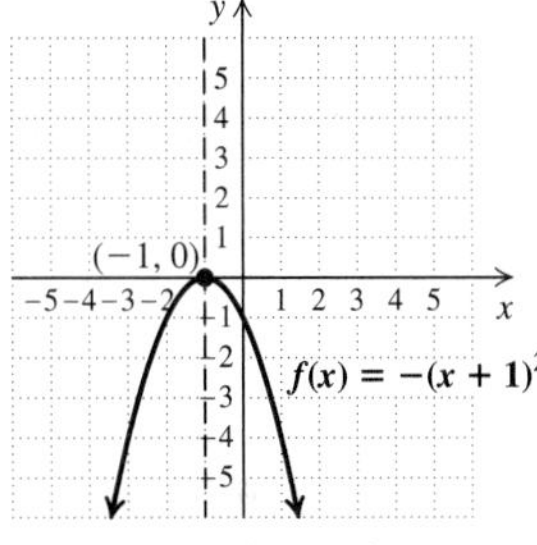

31. Vertex: $(2, 0)$; axis of symmetry: $x = 2$

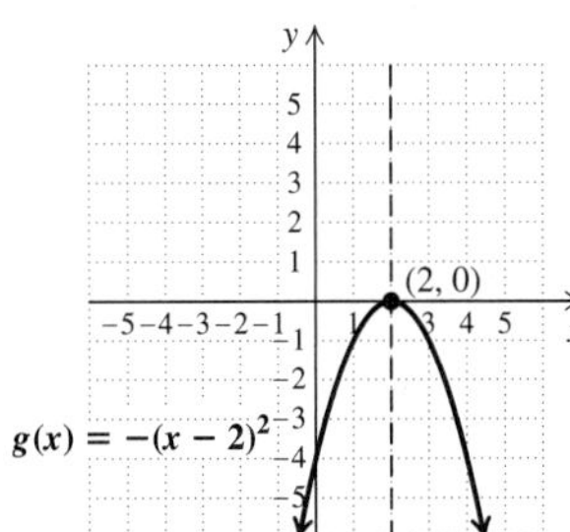

33. Vertex: $(-1, 0)$; axis of symmetry: $x = -1$

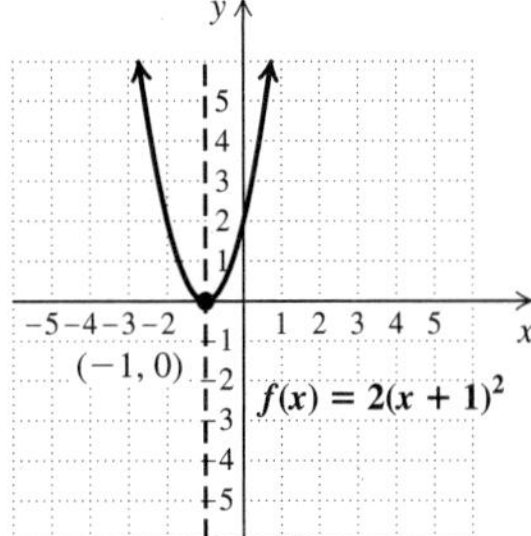

35. Vertex: $(4, 0)$; axis of symmetry: $x = 4$

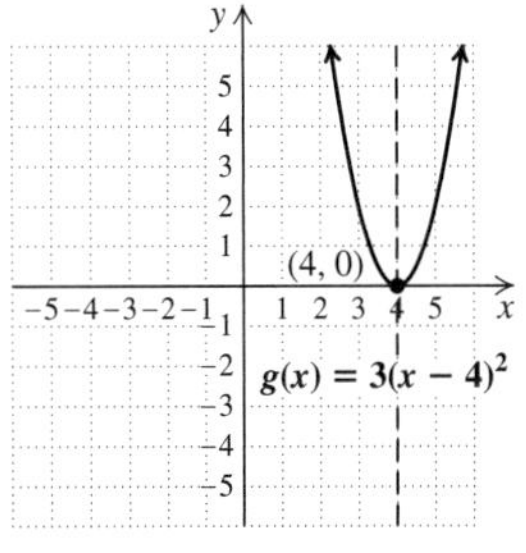

37. Vertex: $(4, 0)$; axis of symmetry: $x = 4$

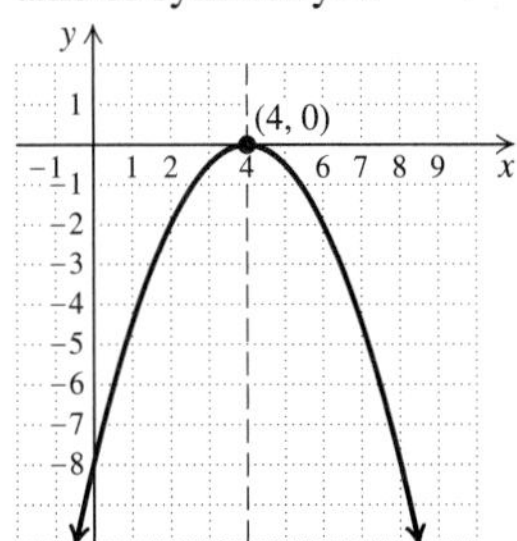

$h(x) = -\frac{1}{2}(x - 4)^2$

39. Vertex: $(1, 0)$; axis of symmetry: $x = 1$

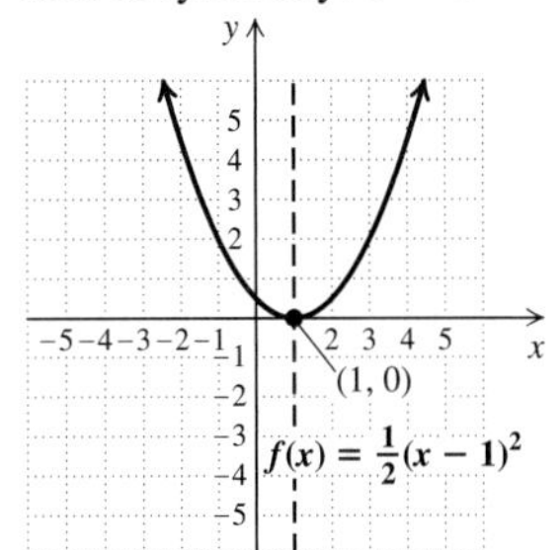

41. Vertex: $(-5, 0)$; axis of symmetry: $x = -5$

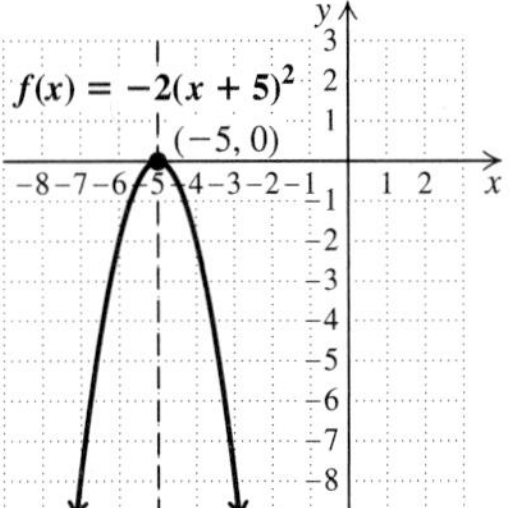

43. Vertex: $\left(\frac{1}{2}, 0\right)$; axis of symmetry: $x = \frac{1}{2}$

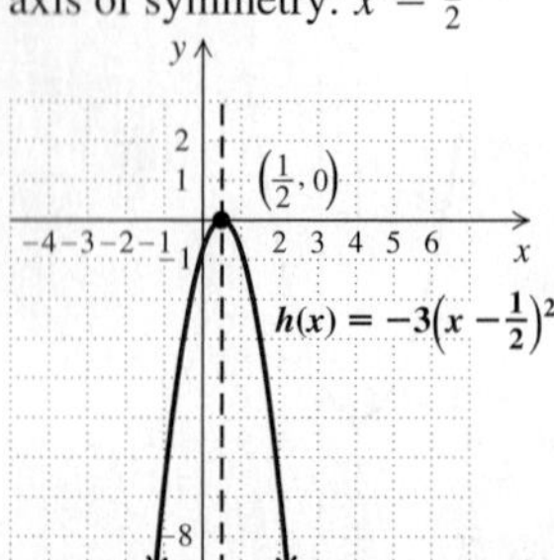

45. Minimum: 2; range: $[2, \infty)$

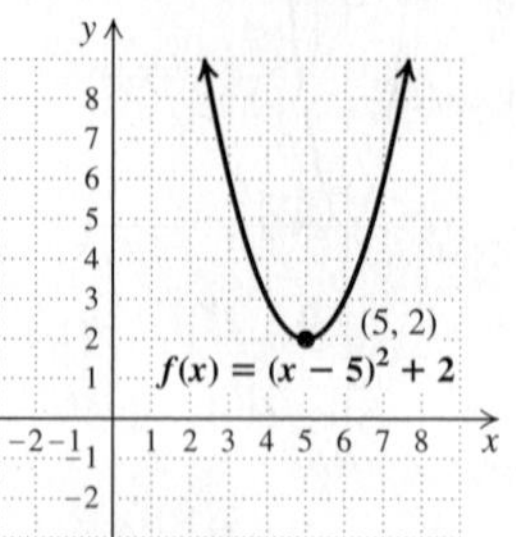

47. Maximum: -1; range: $(-\infty, -1]$

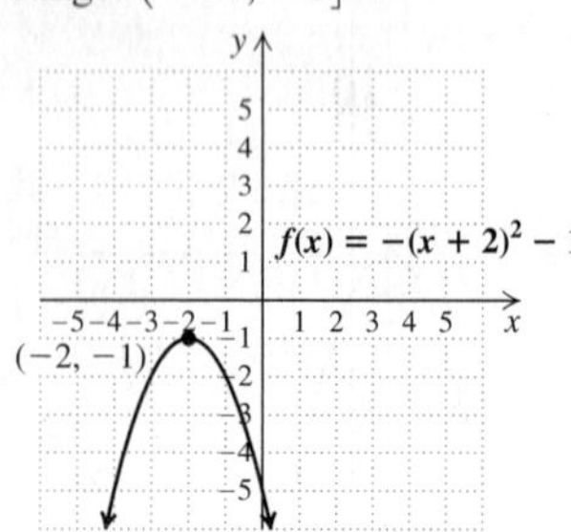

49. Minimum: 3; range: $[3, \infty)$

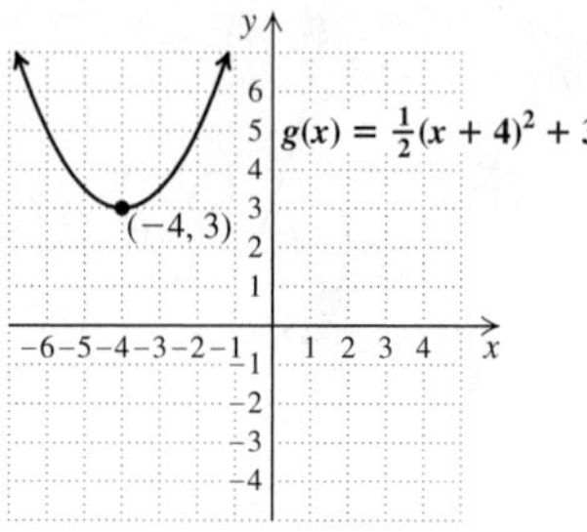

51. Maximum: -3; range: $(-\infty, -3]$

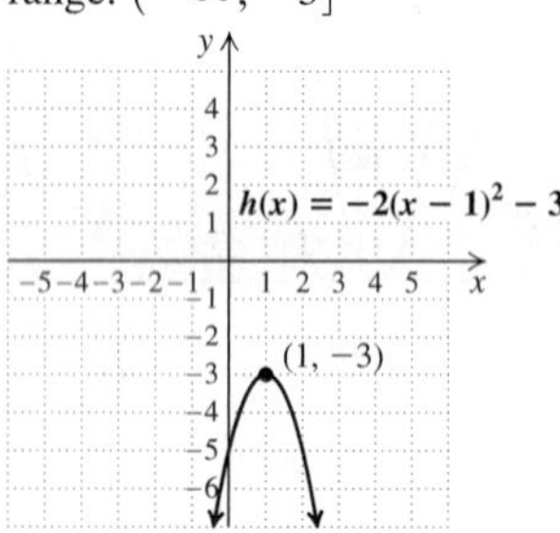

53. Vertex: $(-1, -3)$; axis of symmetry: $x = -1$; minimum: -3; range: $[-3, \infty)$

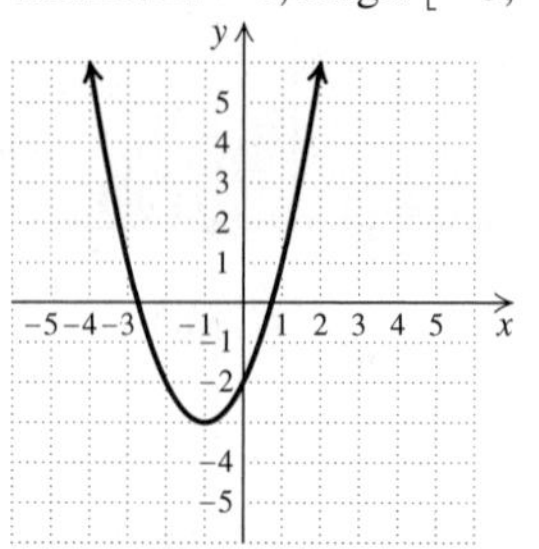

55. Vertex: $(-3, 5)$; axis of symmetry: $x = -3$; maximum: 5; range: $(-\infty, 5]$

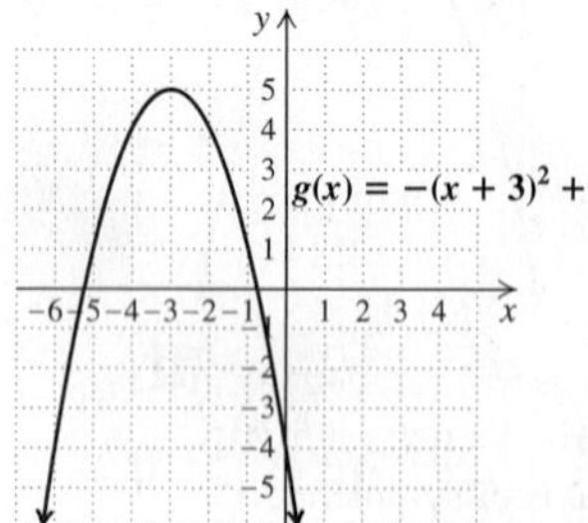

57. Vertex: $(2, 1)$; axis of symmetry: $x = 2$; minimum: 1; range: $[1, \infty)$

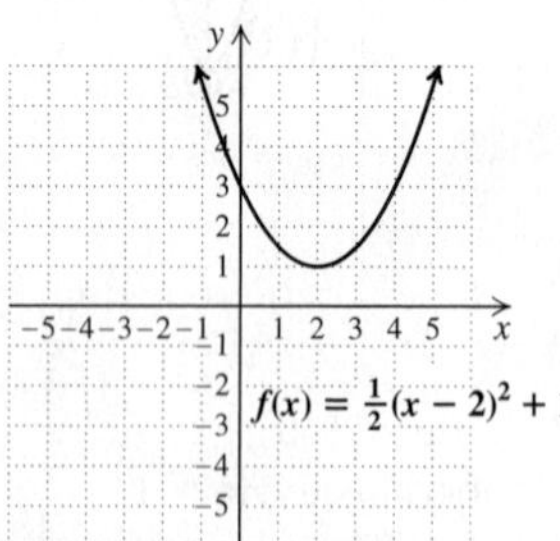

59. Vertex: $(1, -3)$; axis of symmetry: $x = 1$; maximum: -3; range: $(-\infty, -3]$

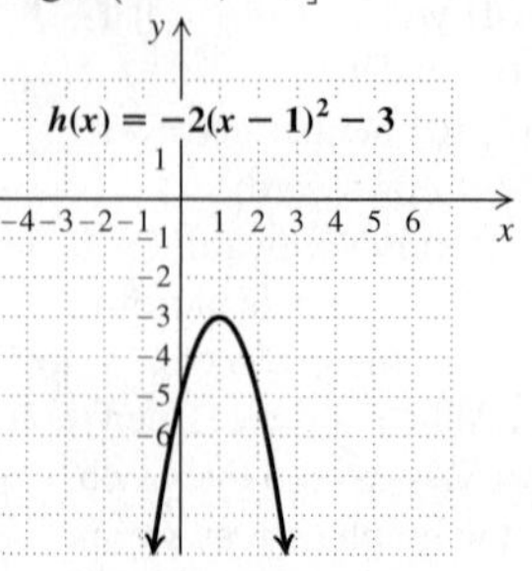

61. Vertex: $(-4, 1)$; axis of symmetry: $x = -4$; minimum: 1; range: $[1, \infty)$

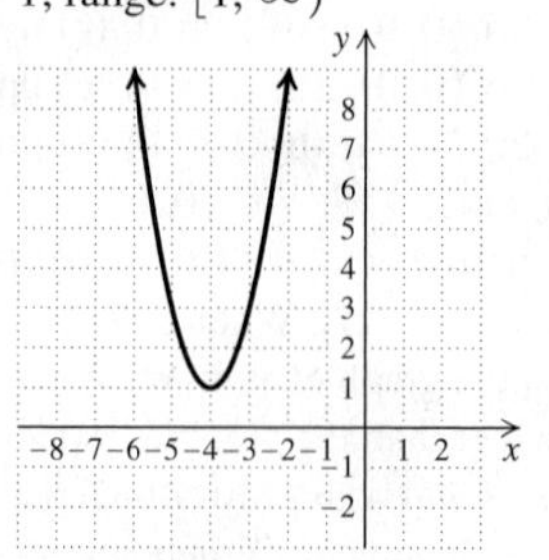

63. Vertex: axis of symmetry: maximum: 4; range:

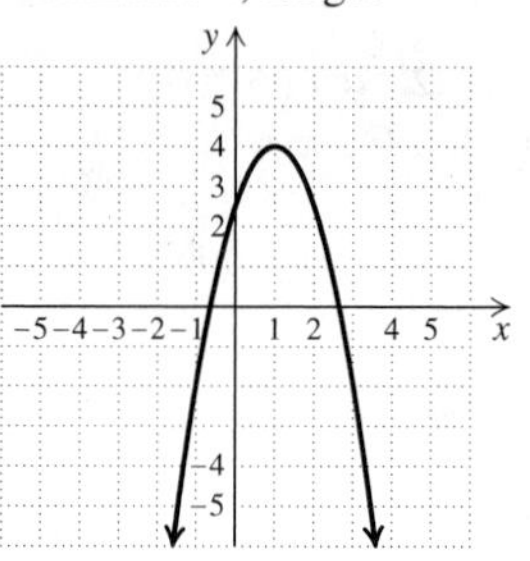

65. Vertex: $(8, 7)$; axis of symmetry: $x = 8$; minimum: 7
67. Vertex: $(-6, 11)$; axis of symmetry: $x = -6$; maximum: 11
69. Vertex: $\left(\frac{7}{2}, -\frac{29}{4}\right)$; axis of symmetry: $x = \frac{7}{2}$; minimum: $-\frac{29}{4}$
71. Vertex: $(-4.58, 65\pi)$; axis of symmetry: $x = -4.58$; minimum: 65π **73.** TW **75.** x-intercept: $(3, 0)$; y-intercept: $(0, -4)$ **76.** x-intercept: $\left(\frac{8}{3}, 0\right)$; y-intercept: $(0, 2)$
77. $(-5, 0), (-3, 0)$ **78.** $(-1, 0), \left(\frac{3}{2}, 0\right)$
79. $x^2 - 14x + 49 = (x - 7)^2$
80. $x^2 + 7x + \frac{49}{4} = \left(x + \frac{7}{2}\right)^2$ **81.** TW
83. $f(x) = \frac{3}{5}(x - 4)^2 + 1$ **85.** $f(x) = \frac{3}{5}(x - 3)^2 - 1$
87. $f(x) = \frac{3}{5}(x + 2)^2 - 5$ **89.** $f(x) = 2(x - 2)^2$
91. $g(x) = -2x^2 + 3$ **93.** The graph will move to the right.
95. The graph will be reflected across the x-axis.
97. $F(x) = 3(x - 5)^2 + 1$

99.

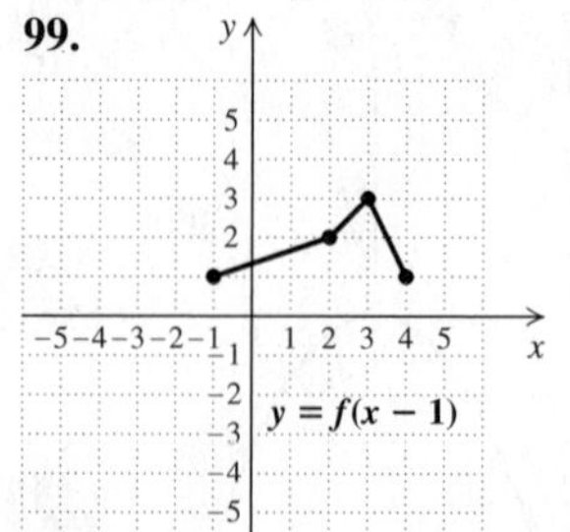

101.

$y = f(x) + 2$

103.

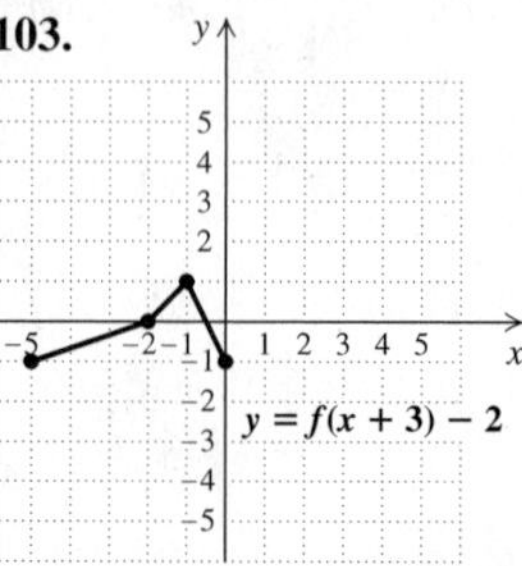

Visualizing for Success, p. 829

1. B **2.** E **3.** A **4.** H **5.** C **6.** J **7.** F **8.** G **9.** I **10.** D

Exercise Set 11.7, pp. 830–831

1. True **2.** False **3.** True **4.** True **5.** False **6.** True **7.** False **8.** True
9. $f(x) = (x - 4)^2 + (-14)$
11. $f(x) = \left(x - \left(-\frac{3}{2}\right)\right)^2 + \left(-\frac{29}{4}\right)$
13. $f(x) = 3(x - (-1))^2 + (-5)$
15. $f(x) = -(x - (-2))^2 + (-3)$
17. $f(x) = 2\left(x - \frac{5}{4}\right)^2 + \frac{55}{8}$
19. **(a)** Vertex: $(-2, 1)$; axis of symmetry: $x = -2$;
(b)

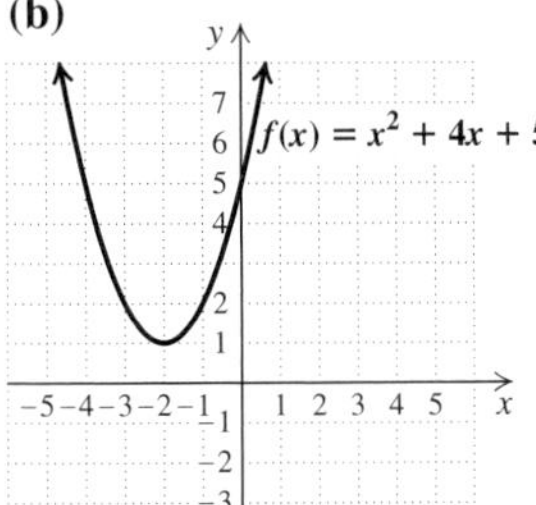

$f(x) = x^2 + 4x + 5$

21. **(a)** Vertex: $(-4, 4)$; axis of symmetry: $x = -4$;
(b)

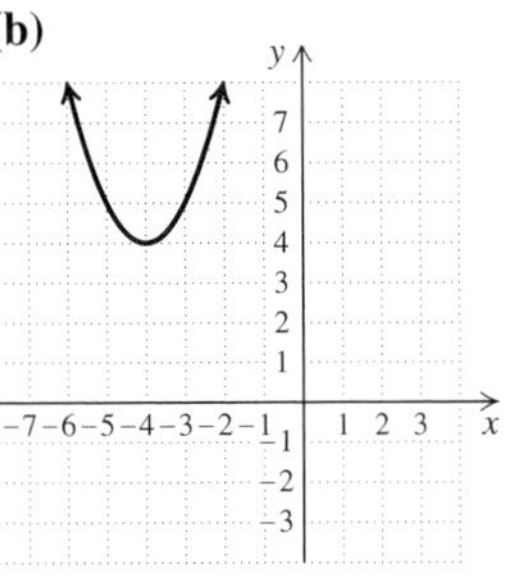

$f(x) = x^2 + 8x + 20$

23. **(a)** Vertex: $(4, -7)$; axis of symmetry: $x = 4$;
(b)

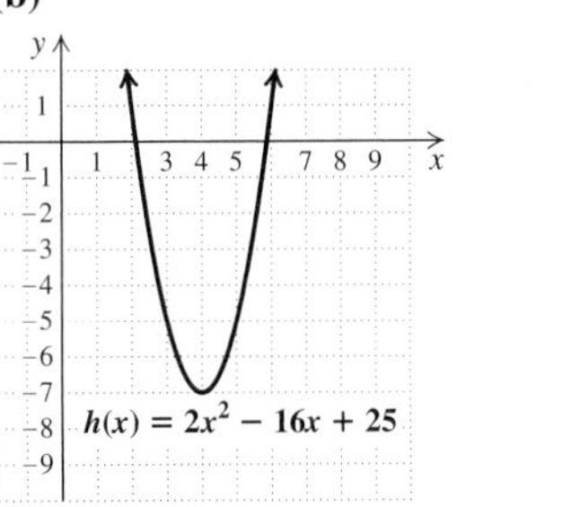

$h(x) = 2x^2 - 16x + 25$

25. **(a)** Vertex: $(1, 6)$; axis of symmetry: $x = 1$;
(b)

$f(x) = -x^2 + 2x + 5$

27. **(a)** Vertex: $\left(-\frac{3}{2}, -\frac{49}{4}\right)$; axis of symmetry: $x = -\frac{3}{2}$;
(b)

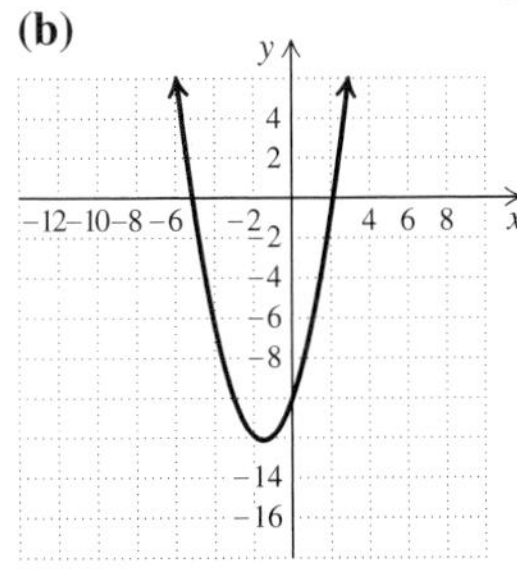

$g(x) = x^2 + 3x - 10$

29. **(a)** Vertex: $\left(-\frac{7}{2}, -\frac{49}{4}\right)$; axis of symmetry: $x = -\frac{7}{2}$;
(b)

$h(x) = x^2 + 7x$

31. **(a)** Vertex: $(-1, -4)$; axis of symmetry: $x = -1$;
(b)

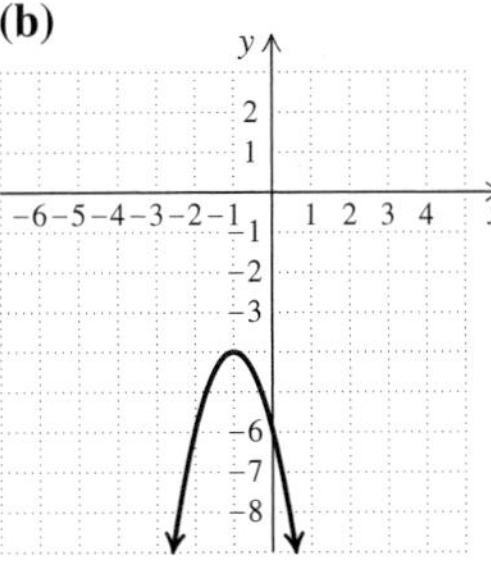

$f(x) = -2x^2 - 4x - 6$

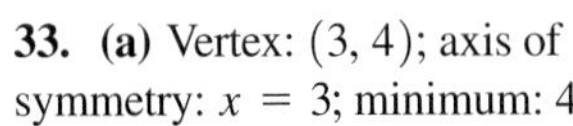

33. **(a)** Vertex: $(3, 4)$; axis of symmetry: $x = 3$; minimum: 4;
(b)

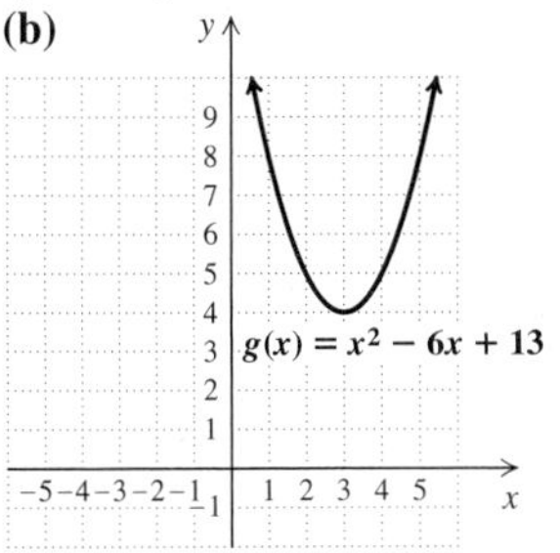

$g(x) = x^2 - 6x + 13$

35. **(a)** Vertex: $(2, -5)$; axis of symmetry: $x = 2$; minimum: -5;
(b)

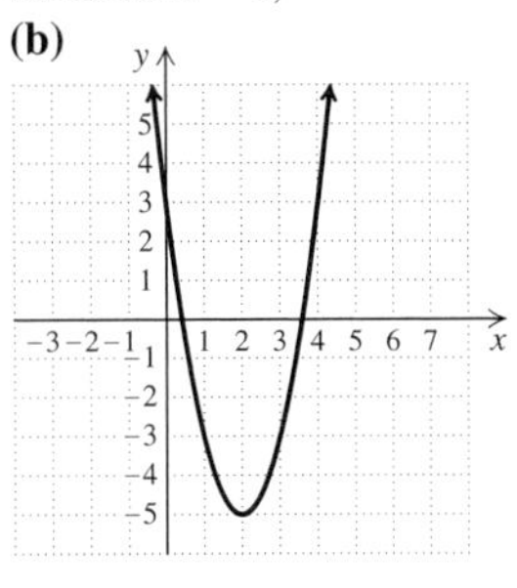

$g(x) = 2x^2 - 8x + 3$

37. **(a)** Vertex: $(4, 2)$; axis of symmetry: $x = 4$; minimum: 2;
(b)

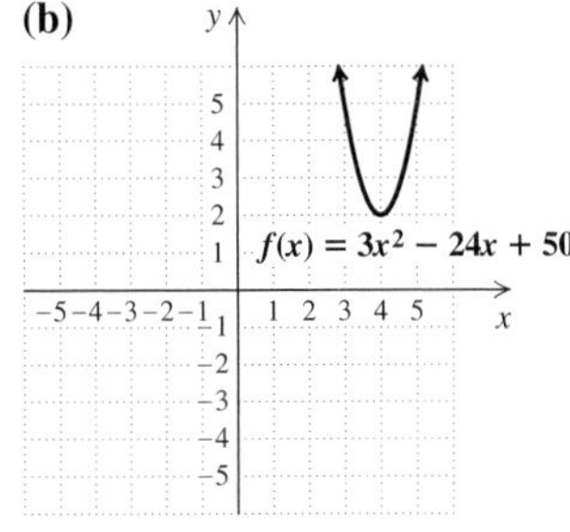

$f(x) = 3x^2 - 24x + 50$

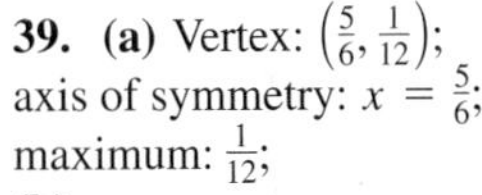

39. **(a)** Vertex: $\left(\frac{5}{6}, \frac{1}{12}\right)$; axis of symmetry: $x = \frac{5}{6}$; maximum: $\frac{1}{12}$;
(b)

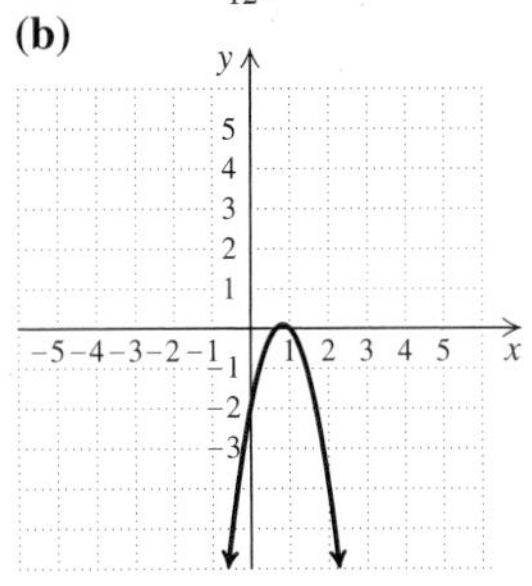

$f(x) = -3x^2 + 5x - 2$

41. **(a)** Vertex: $\left(-4, -\frac{5}{3}\right)$; axis of symmetry: $x = -4$; minimum: $-\frac{5}{3}$;
(b)

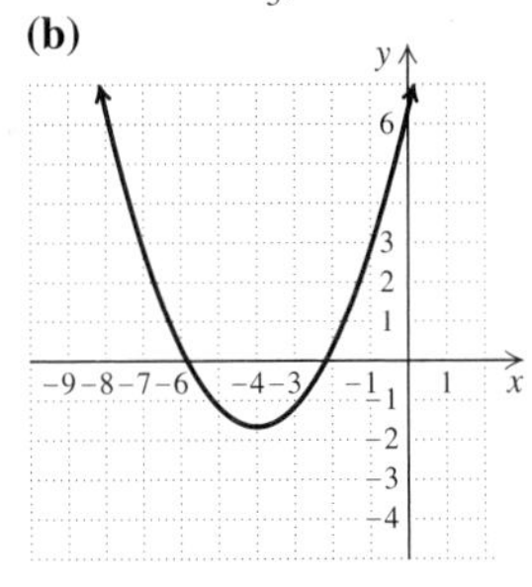

$h(x) = \frac{1}{2}x^2 + 4x + \frac{19}{3}$

43. $(-0.5, -6.25)$ **45.** $(0.1, 0.95)$ **47.** $(3.5, -4.25)$
49. $(3 - \sqrt{6}, 0), (3 + \sqrt{6}, 0); (0, 3)$
51. $(-1, 0), (3, 0); (0, 3)$ **53.** $(0, 0), (9, 0); (0, 0)$
55. $(2, 0); (0, -4)$
57. $\left(-\frac{1}{2} - \frac{\sqrt{21}}{2}, 0\right), \left(-\frac{1}{2} + \frac{\sqrt{21}}{2}, 0\right); (0, -5)$
59. No x-intercept; $(0, 6)$ **61.** TW **63.** $(1, 1, 1)$
64. $(-2, 5, 1)$ **65.** $(10, 5, 8)$ **66.** $(-3, 6, -5)$
67. $(2.4, -1.8, 1.5)$ **68.** $\left(\frac{1}{3}, \frac{1}{6}, \frac{1}{2}\right)$ **69.** TW
71. **(a)** Minimum: -6.95; **(b)** $(-1.06, 0), (2.41, 0); (0, -5.89)$

73. **(a)** Maximum: -0.45; **(b)** no x-intercept; $(0, -2.79)$
75. **(a)** $-2.4, 3.4$; **(b)** $-1.3, 2.3$
77. $f(x) = m\left(x - \frac{n}{2m}\right)^2 + \frac{4mp - n^2}{4m}$
79. $f(x) = \frac{5}{16}x^2 - \frac{15}{8}x - \frac{35}{16}$, or $f(x) = \frac{5}{16}(x - 3)^2 - 5$

81.

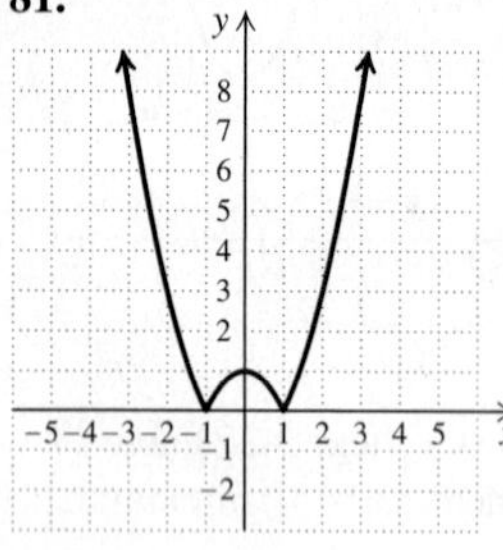

$f(x) = |x^2 - 1|$

83.

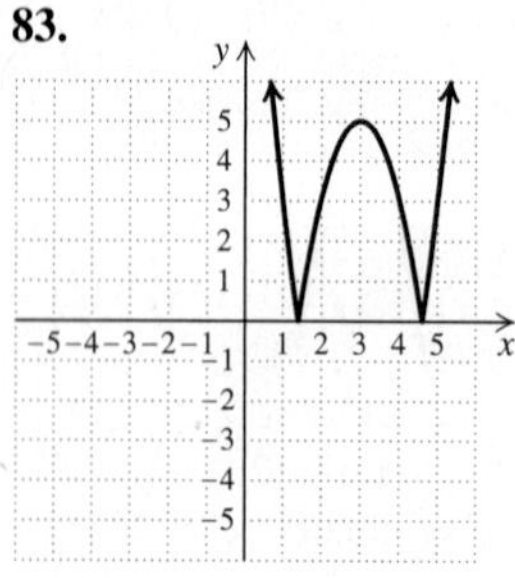

$f(x) = |2(x - 3)^2 - 5|$

Exercise Set 11.8, pp. 839–844

1. (e) **2.** (b) **3.** (c) **4.** (a) **5.** (d) **6.** (f)
7. July; 0 in.
9. $P(x) = -x^2 + 980x - 3000$; \$237,100 at $x = 490$
11. 32 in. by 32 in. **13.** 450 ft^2; 15 ft by 30 ft (The house serves as a 30-ft side.) **15.** 3.5 in. **17.** 81; 9 and 9
19. -16; 4 and -4 **21.** 25; -5 and -5
23. $f(x) = mx + b$ **25.** $f(x) = ax^2 + bx + c, a < 0$
27. $f(x) = ax^2 + bx + c, a > 0$
29. $f(x) = ax^2 + bx + c, a > 0$
31. Neither quadratic nor linear **33.** $f(x) = 2x^2 + 3x - 1$
35. $f(x) = -\frac{1}{4}x^2 + 3x - 5$
37. **(a)** $A(s) = \frac{3}{16}s^2 - \frac{135}{4}s + 1750$; **(b)** about 531 accidents
39. $h(d) = -0.0068d^2 + 0.8571d$
41. **(a)** $D(x) = -0.0083x^2 + 0.8243x + 0.2122$; **(b)** 17.325 ft
43. **(a)** $t(x) = 18.125x^2 + 78.15x + 24{,}613$; 28,161 teachers
45. TW

47.

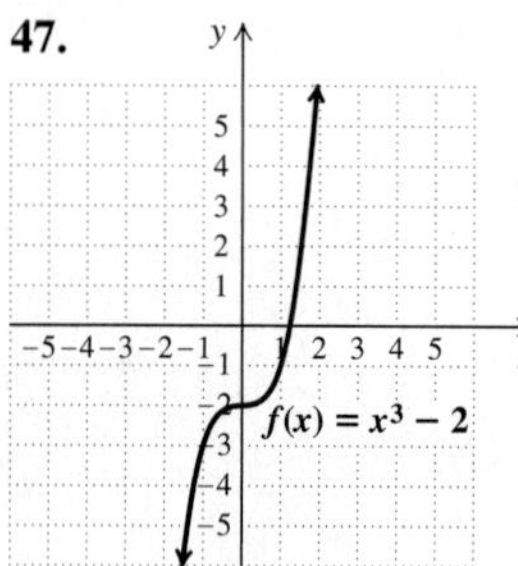

48.

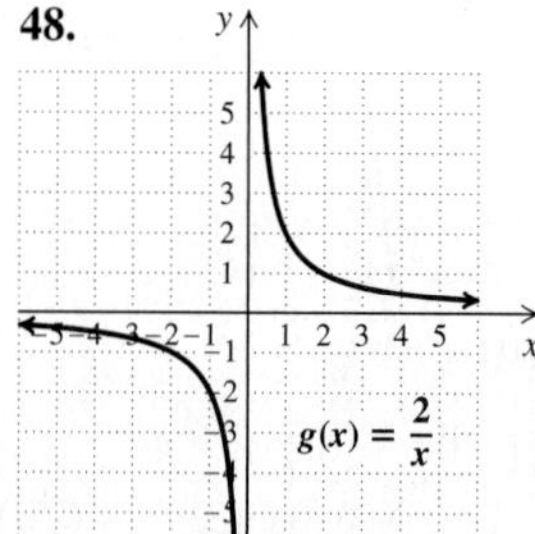

49. $\frac{1}{a^2} + 7$ **50.** $a - 8$ **51.** $4a^2 + 20a + 27$
52. $\sqrt{12a - 19}$ **53.** TW **55.** 158 ft **57.** The radius of the circular portion of the window and the height of the rectangular portion should each be $\frac{24}{\pi + 4}$ ft. **59.** \$15

Study Summary: Chapter 11, pp. 845–846

1. 1, 11 **2.** $9 \pm \sqrt{5}$ **3.** $-21, 1$ **4.** $-\frac{3}{2}, 3$ **5.** Two imaginary-number solutions **6.** $n = \pm\sqrt{a - 1}$ **7.** 36

8.

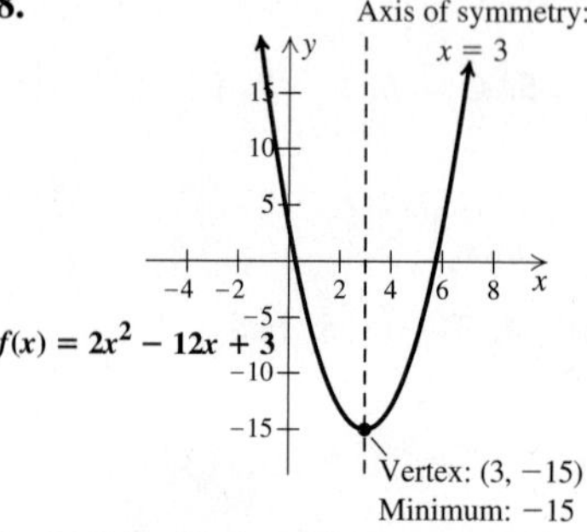

9. 30 ft by 30 ft

Review Exercises: Chapter 11, pp. 847–849

1. False **2.** True **3.** True **4.** True **5.** True
6. False **7.** True **8.** True **9.** True **10.** False
11. **(a)** 2; **(b)** positive; **(c)** -3 **12.** $\pm\frac{\sqrt{2}}{3}$ **13.** $0, -\frac{3}{4}$
14. 3, 9 **15.** $2 \pm 2i$ **16.** 3, 5 **17.** $-\frac{9}{2} \pm \frac{\sqrt{85}}{2}$
18. $-0.372, 5.372$ **19.** $-\frac{1}{4}, 1$
20. $x^2 - 12x + 36 = (x - 6)^2$
21. $x^2 + \frac{3}{5}x + \frac{9}{100} = \left(x + \frac{3}{10}\right)^2$ **22.** $3 \pm 2\sqrt{2}$ **23.** 4%
24. 5.8 sec **25.** Two irrational real numbers
26. Two imaginary numbers **27.** $x^2 + 9 = 0$
28. $x^2 + 8x + 16 = 0$ **29.** About 153 mph **30.** 6 hr
31. $(-3, 0), (-2, 0), (2, 0), (3, 0)$ **32.** $-5, 3$
33. $\pm\sqrt{2}, \pm\sqrt{7}$
34.

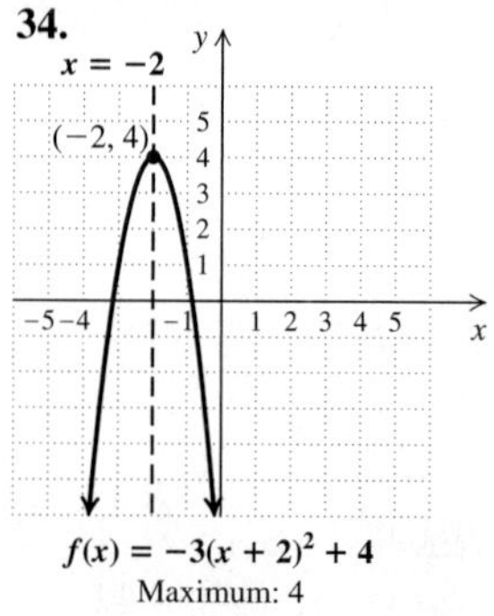

$x = -2$; $(-2, 4)$

$f(x) = -3(x + 2)^2 + 4$
Maximum: 4

35. **(a)** Vertex: $(3, 5)$; axis of symmetry: $x = 3$;
(b)

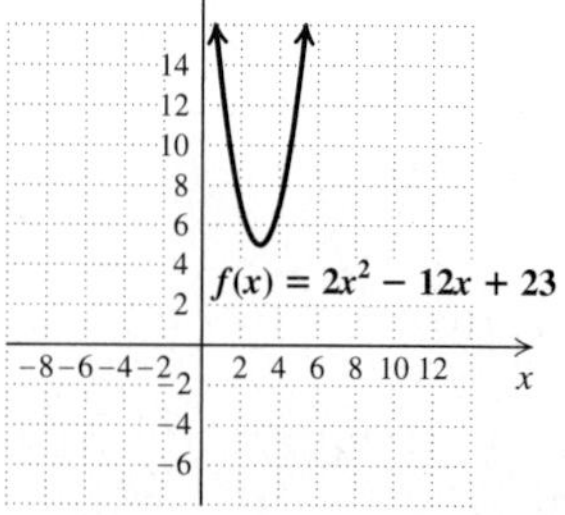

$f(x) = 2x^2 - 12x + 23$

36. $(2, 0), (7, 0); (0, 14)$ **37.** $p = \frac{9\pi^2}{N^2}$
38. $T = \frac{1 \pm \sqrt{1 + 24A}}{6}$ **39.** Neither quadratic nor linear
40. $f(x) = ax^2 + bx + c, a > 0$ **41.** $f(x) = mx + b$
42. 225 ft^2; 15 ft by 15 ft
43. **(a)** $M(x) = \frac{14{,}501}{1200}x^2 - \frac{2876}{15}x + 1$; **(b)** 48,841 restaurants
44. **(a)** $M(x) = 12.6207x^2 - 242.2557x + 706.6461$; **(b)** 48,690 restaurants **45.** TW Completing the square was used to solve quadratic equations and to graph quadratic functions by rewriting the function in the form $f(x) = a(x - h)^2 + k$.
46. TW The model found in Exercise 44 predicts 151 fewer restaurants in 2020 than the model from Exercise 43. Since the

function in Exercise 44 considers all the data, we would expect it to be a better model. **47.** TW The equation can have at most four solutions. If we substitute u for x^2, we see that $au^2 + bu + c = 0$ can have at most two solutions. Suppose $u = m$ or $u = n$. Then $x^2 = m$ or $x^2 = n$, and $x = \pm\sqrt{m}$ or $x = \pm\sqrt{n}$. When $m \neq n$, this gives four solutions, the maximum number possible. **48.** $f(x) = \frac{7}{15}x^2 - \frac{14}{15}x - 7$ **49.** $h = 60, k = 60$ **50.** 18, 324

Test: Chapter 11, pp. 849–850

1. **(a)** [11.1] 0; **(b)** [11.6] negative; **(c)** [11.6] -1
2. [11.1] $\pm\frac{\sqrt{7}}{5}$ **3.** [11.2] 2, 9 **4.** [11.2] $-1 \pm \sqrt{2}i$
5. [11.2] $1 \pm \sqrt{6}$ **6.** [11.5] $-2, \frac{2}{3}$ **7.** [11.2] $-4.193, 1.193$
8. [11.2] $-\frac{3}{4}, \frac{7}{3}$ **9.** [11.1] $x^2 - 20x + 100 = (x - 10)^2$
10. [11.1] $x^2 + \frac{2}{7}x + \frac{1}{49} = \left(x + \frac{1}{7}\right)^2$ **11.** [11.1] $-5 \pm \sqrt{10}$
12. [11.3] Two imaginary numbers **13.** [11.3] $x^2 - 11 = 0$
14. [11.4] 16 km/h **15.** [11.4] 2 hr **16.** [11.5] $(-4, 0), (4, 0)$

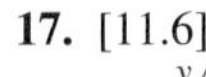

17. [11.6]

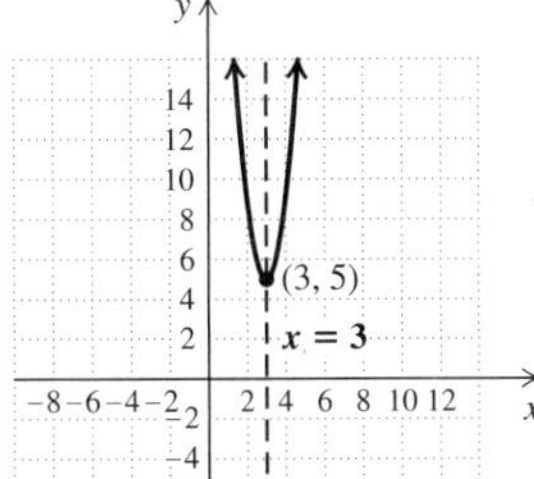

$f(x) = 4(x - 3)^2 + 5$
Minimum: 5

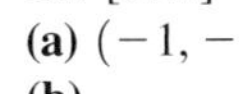

18. [11.7]
(a) $(-1, -8), x = -1$;
(b)

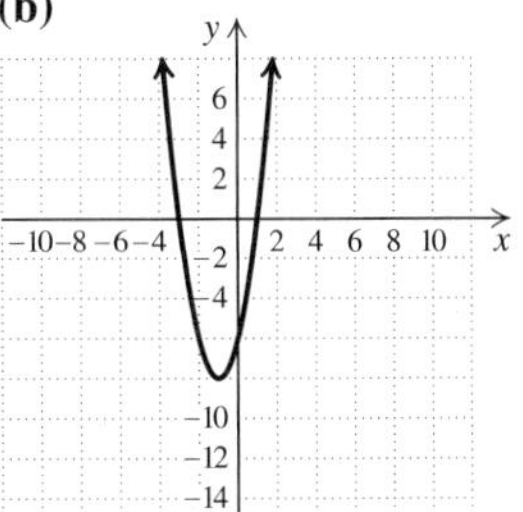

$f(x) = 2x^2 + 4x - 6$

19. [11.7] $(-2, 0), (3, 0); (0, -6)$
20. [11.4] $r = \sqrt{\frac{3V}{\pi} - R^2}$ **21.** [11.8] Quadratic; the data approximate a parabola opening downward.
22. [11.8] Minimum \$129 per cabinet when 325 cabinets are built
23. [11.8] $p(x) = -\frac{165}{8}x^2 + \frac{605}{4}x + 35$
24. [11.8] $p(x) = -23.5417x^2 + 162.2583x + 57.6167$
25. [11.3] $\frac{1}{2}$ **26.** [11.3] $x^4 + x^2 - 12 = 0$; answers may vary
27. [11.5] $\pm\sqrt{\sqrt{5} + 2}, \pm\sqrt{\sqrt{5} - 2}i$

Chapter 12

Exercise Set 12.1, pp. 862–866

1. True **2.** True **3.** False **4.** False **5.** False
6. False **7.** True **8.** True
9. **(a)** $(f \circ g)(1) = 5$; **(b)** $(g \circ f)(1) = -1$;
(c) $(f \circ g)(x) = x^2 - 6x + 10$; **(d)** $(g \circ f)(x) = x^2 - 2$
11. **(a)** $(f \circ g)(1) = -24$; **(b)** $(g \circ f)(1) = 65$;
(c) $(f \circ g)(x) = 10x^2 - 34$; **(d)** $(g \circ f)(x) = 50x^2 + 20x - 5$
13. **(a)** $(f \circ g)(1) = 8$; **(b)** $(g \circ f)(1) = \frac{1}{64}$;
(c) $(f \circ g)(x) = \frac{1}{x^2} + 7$; **(d)** $(g \circ f)(x) = \frac{1}{(x + 7)^2}$
15. **(a)** $(f \circ g)(1) = 2$; **(b)** $(g \circ f)(1) = 4$;
(c) $(f \circ g)(x) = \sqrt{x + 3}$; **(d)** $(g \circ f)(x) = \sqrt{x} + 3$
17. **(a)** $(f \circ g)(1) = 2$; **(b)** $(g \circ f)(1) = \frac{1}{2}$;
(c) $(f \circ g)(x) = \sqrt{\frac{4}{x}}$; **(d)** $(g \circ f)(x) = \frac{1}{\sqrt{4x}}$
19. **(a)** $(f \circ g)(1) = 4$; **(b)** $(g \circ f)(1) = 2$;
(c) $(f \circ g)(x) = x + 3$; **(d)** $(g \circ f)(x) = \sqrt{x^2 + 3}$ **21.** 8
23. -4 **25.** Not defined **27.** 4 **29.** Not defined
31. $f(x) = x^4; g(x) = 3x - 5$ **33.** $f(x) = \sqrt{x}$;
$g(x) = 2x + 7$ **35.** $f(x) = \frac{2}{x}; g(x) = x - 3$
37. $f(x) = \frac{1}{\sqrt{x}}; g(x) = 7x + 2$ **39.** $f(x) = \frac{1}{x} + x$;
$g(x) = \sqrt{3x}$ **41.** Yes **43.** No **45.** Yes **47.** No
49. **(a)** Yes; **(b)** $f^{-1}(x) = x - 4$ **51.** **(a)** Yes;
(b) $f^{-1}(x) = \frac{x}{2}$ **53.** **(a)** Yes; **(b)** $g^{-1}(x) = \frac{x + 1}{3}$
55. **(a)** Yes; **(b)** $f^{-1}(x) = 2x - 2$ **57.** **(a)** No **59.** **(a)** Yes;
(b) $h^{-1}(x) = -10 - x$ **61.** **(a)** Yes; **(b)** $f^{-1}(x) = \frac{1}{x}$
63. **(a)** No **65.** **(a)** Yes; **(b)** $f^{-1}(x) = \frac{3x - 1}{2}$
67. **(a)** Yes; **(b)** $f^{-1}(x) = \sqrt[3]{x + 5}$ **69.** **(a)** Yes;
(b) $g^{-1}(x) = \sqrt[3]{x} + 2$ **71.** **(a)** Yes; **(b)** $f^{-1}(x) = x^2, x \geq 0$

73.

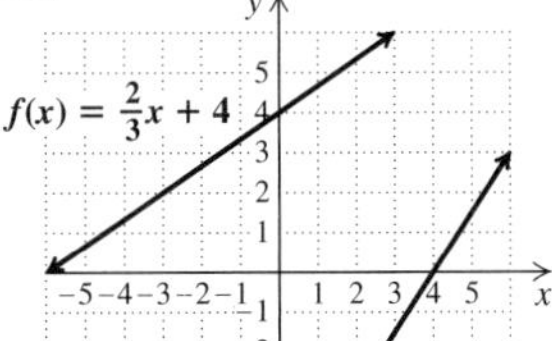

75.

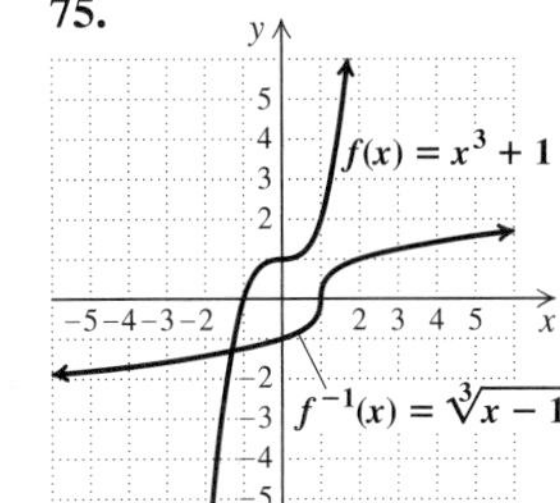

77.

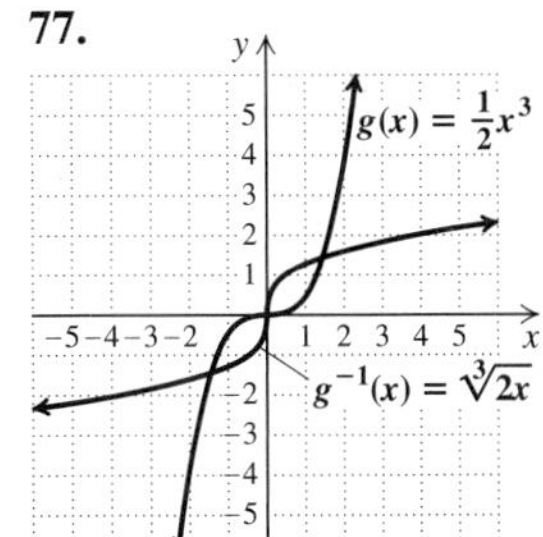

79.

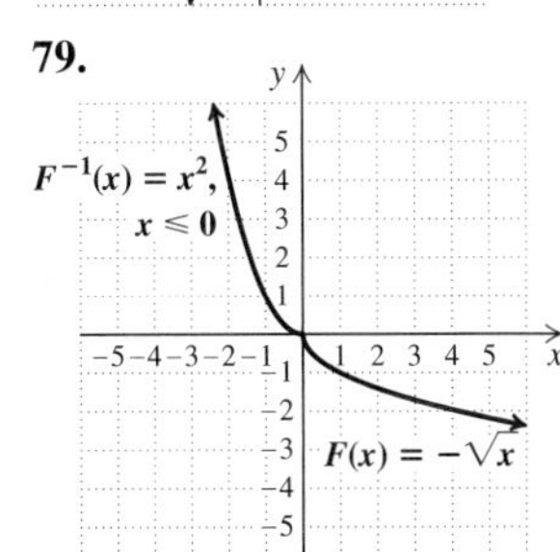

81.

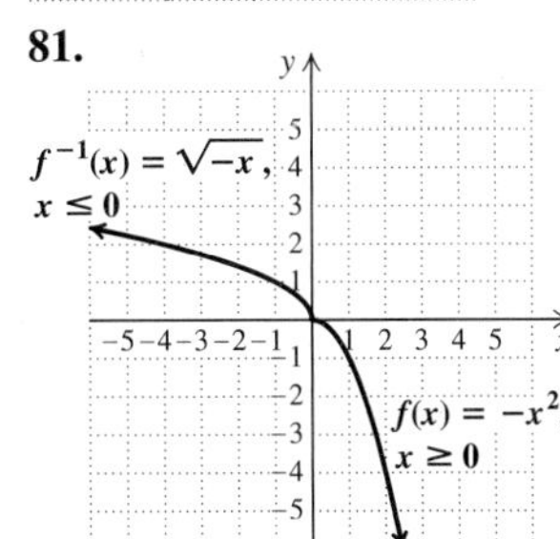

83. **(1)** $(f^{-1} \circ f)(x) = f^{-1}(f(x))$
$= f^{-1}(\sqrt[3]{x-4}) = (\sqrt[3]{x-4})^3 + 4$
$= x - 4 + 4 = x;$
(2) $(f \circ f^{-1})(x) = f(f^{-1}(x))$
$= f(x^3 + 4) = \sqrt[3]{x^3 + 4 - 4}$
$= \sqrt[3]{x^3} = x$

85. **(1)** $(f^{1} \circ f)(x) = f^{-1}(f(x)) = f^{-1}\left(\frac{1-x}{x}\right)$

$$= \frac{1}{\left(\frac{1-x}{x}\right) + 1}$$

$$= \frac{1}{\frac{1-x+x}{x}}$$

$= x;$

(2) $(f \circ f^{-1})(x) = f(f^{-1}(x)) = f\left(\frac{1}{x+1}\right)$

$$= \frac{1 - \left(\frac{1}{x+1}\right)}{\left(\frac{1}{x+1}\right)}$$

$$= \frac{\frac{x+1-1}{x+1}}{\frac{1}{x+1}} = x$$

87. No **89.** Yes **91.** **(1)** C; **(2)** D; **(3)** B; **(4)** A
93. **(a)** 40, 42, 46, 50; **(b)** yes; $f^{-1}(x) = x - 32$;
(c) 8, 10, 14, 18 **95.** TW **97.** $\frac{1}{8}$ **98.** $\frac{1}{25}$
99. 32 **100.** Approximately 2.1577

101.

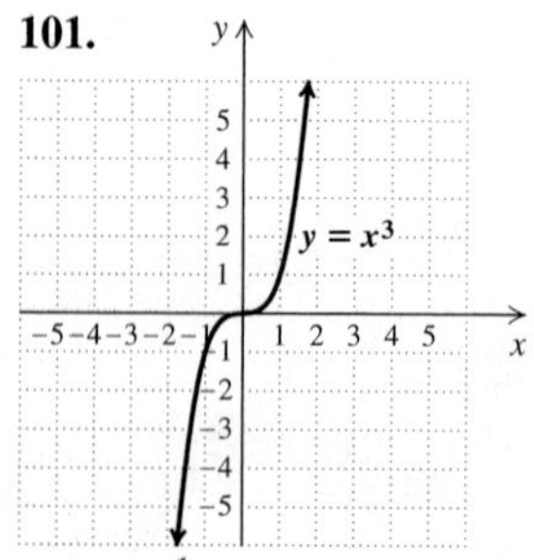

102.

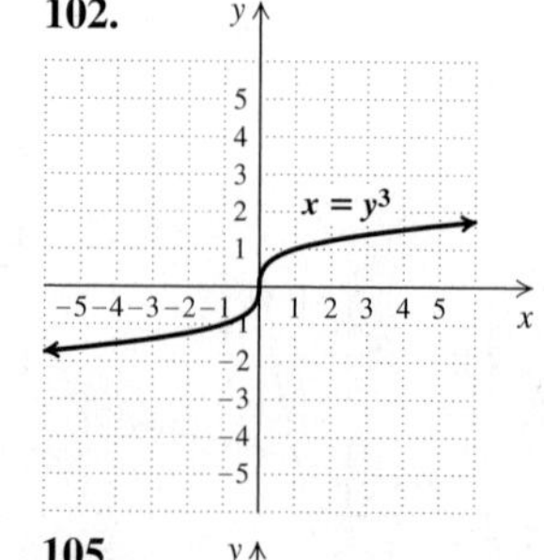

103. TW

105.

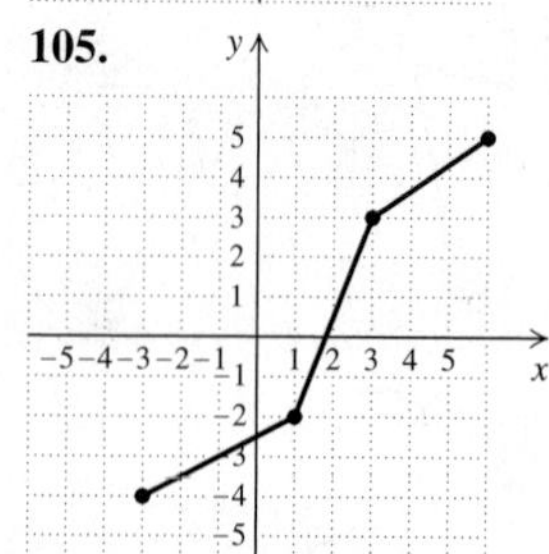

107. $g(x) = \frac{x}{2} + 20$ **109.** TW

111. Suppose that $h(x) = (f \circ g)(x)$. First, note that for $I(x) = x$, $(f \circ I)(x) = f(I(x)) = f(x)$ for any function f.
(i) $((g^{-1} \circ f^{-1}) \circ h)(x) = ((g^{-1} \circ f^{-1}) \circ (f \circ g))(x)$
$= ((g^{-1} \circ (f^{-1} \circ f)) \circ g)(x)$
$= ((g^{-1} \circ I) \circ g)(x)$
$= (g^{-1} \circ g)(x) = x$
(ii) $(h \circ (g^{-1} \circ f^{-1}))(x) = ((f \circ g) \circ (g^{-1} \circ f^{-1}))(x)$
$= ((f \circ (g \circ g^{-1})) \circ f^{-1})(x)$
$= ((f \circ I) \circ f^{-1})(x)$
$= (f \circ f^{-1})(x) = x.$
Therefore, $(g^{-1} \circ f^{-1})(x) = h^{-1}(x)$.
113. TW **115.** The cost of mailing n copies of the book
117. 22 mm **119.** 15 L/min

Interactive Discovery, p. 869

1. **(a)** Increases; **(b)** increases; **(c)** increases; **(d)** decreases; **(e)** decreases **2.** If $a > 1$, the graph increases; if $0 < a < 1$, the graph decreases. **3.** g **4.** r

Exercise Set 12.2, pp. 873–876

1. True **2.** True **3.** True **4.** False **5.** False
6. True **7.** $a > 1$ **9.** $0 < a < 1$

11.

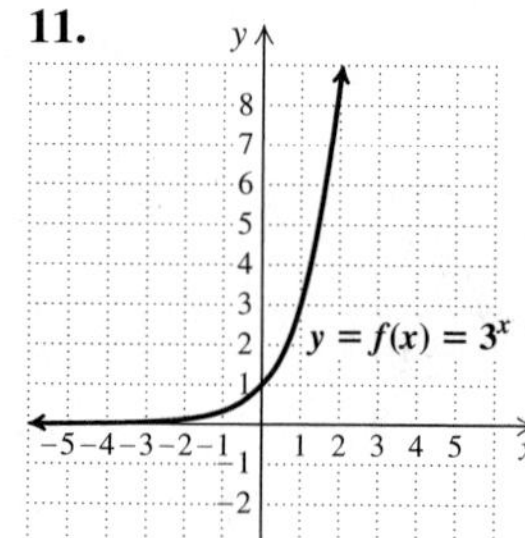

13.

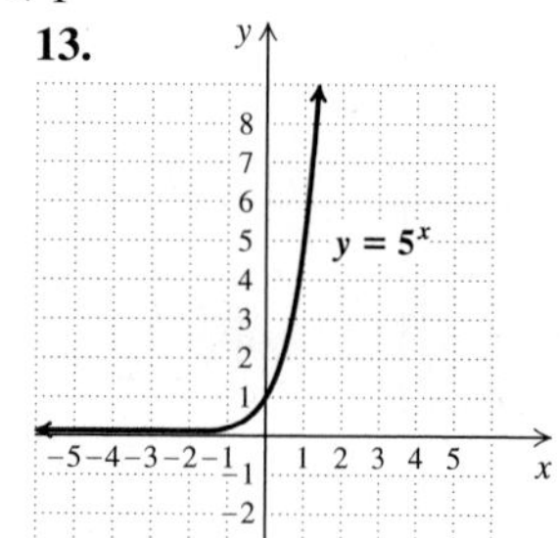

15.

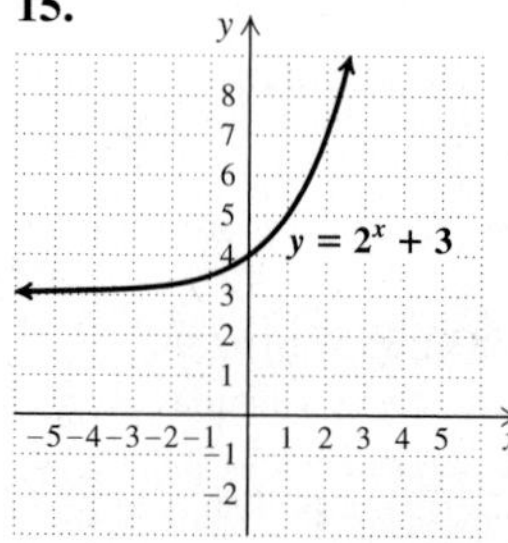

17.

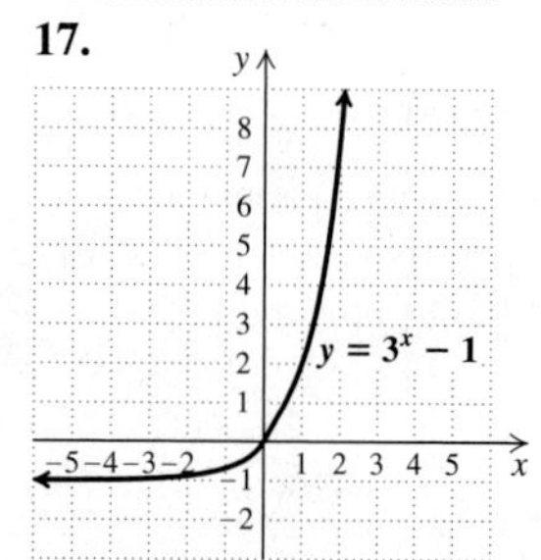

19.

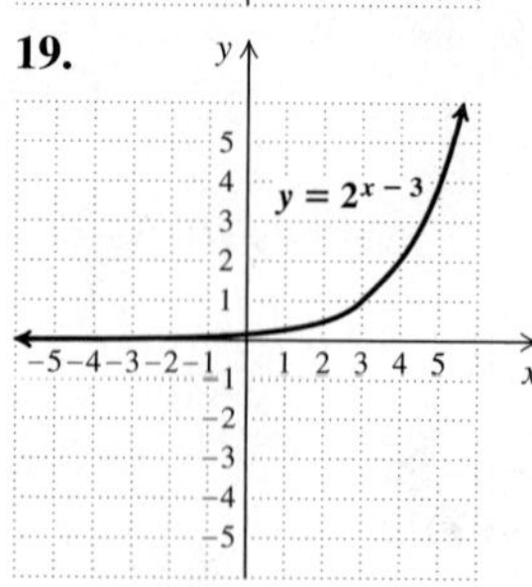

21.

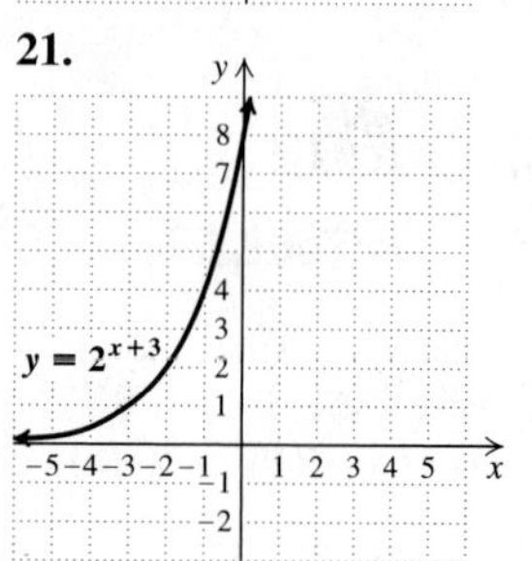

23.

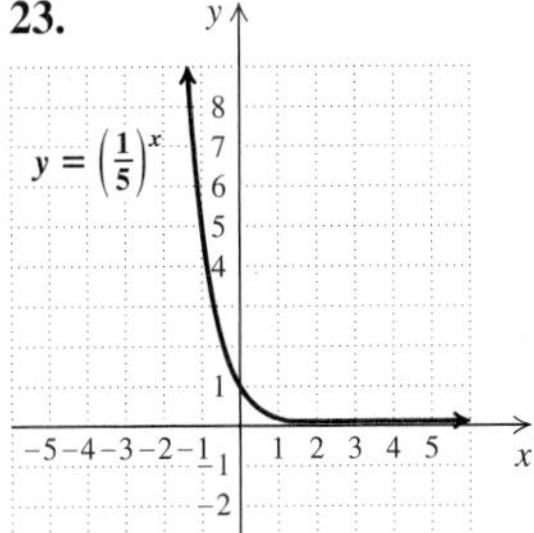

25.

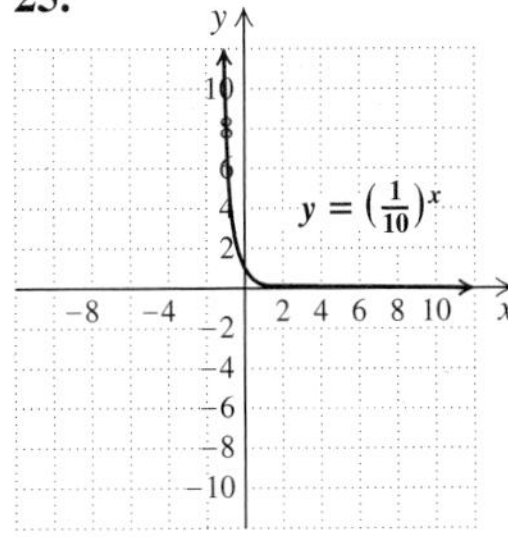

27.

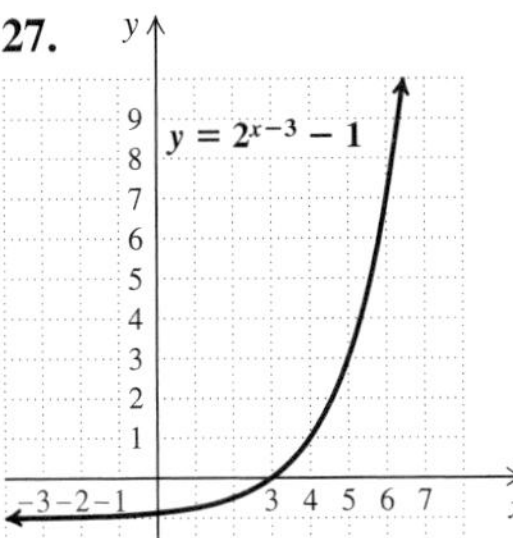

29.

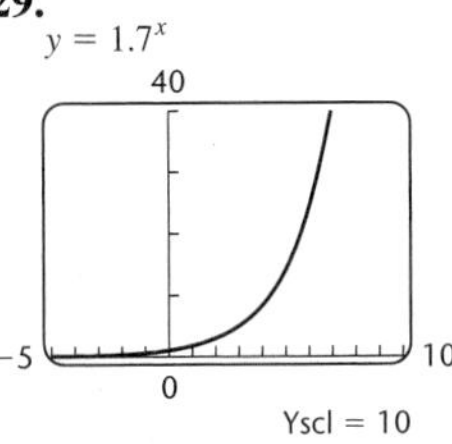

31.

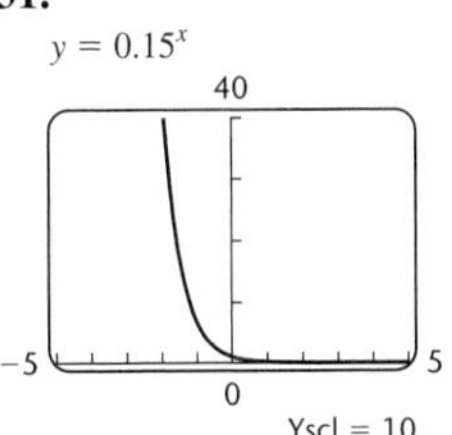

33.

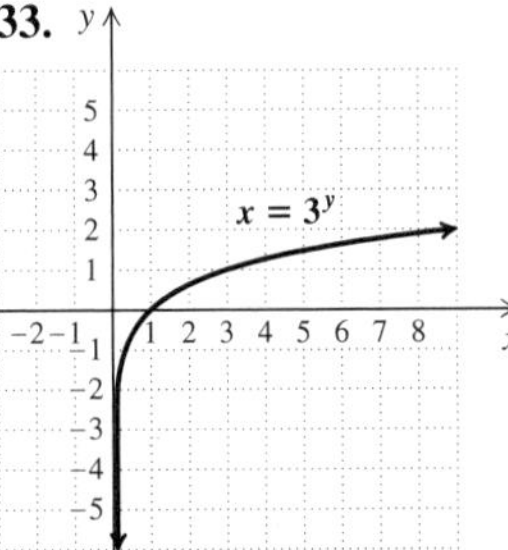

35.

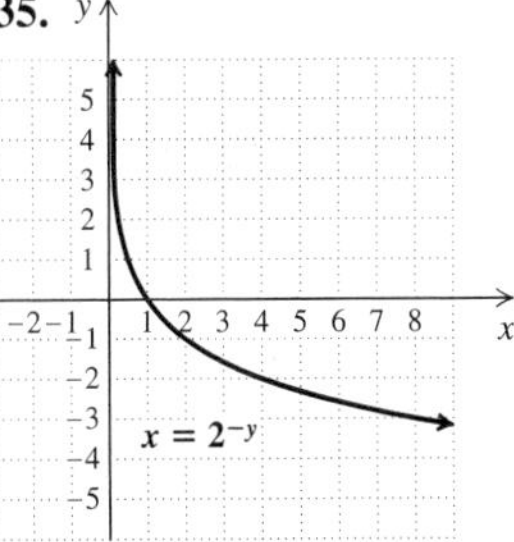

37.

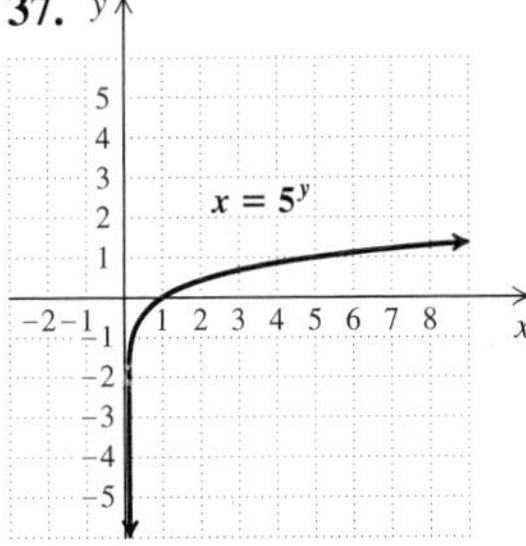

39.

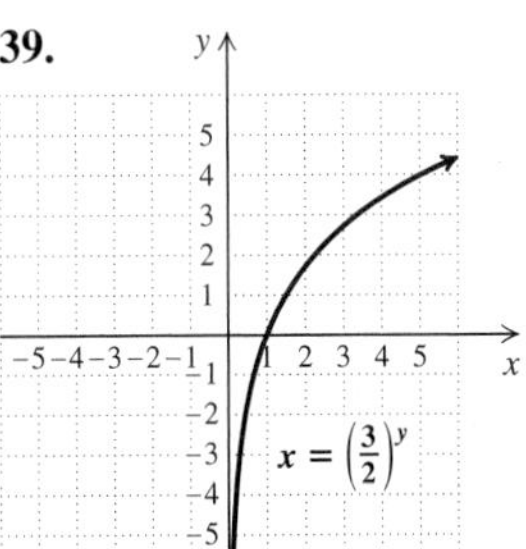

41.

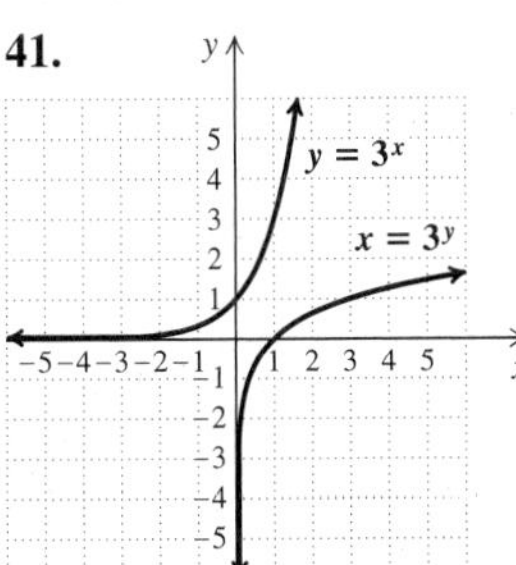

43.

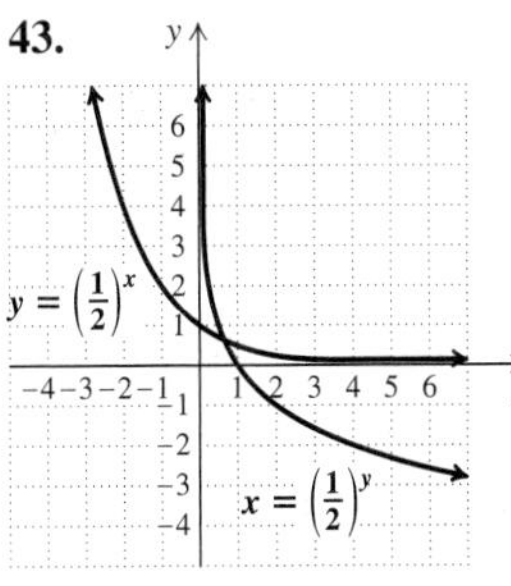

45. (d)
47. (f)
49. (c)

51. (a) About 0.68 billion tracks; about 1.052 billion tracks; about 2.519 billion tracks;
(b)

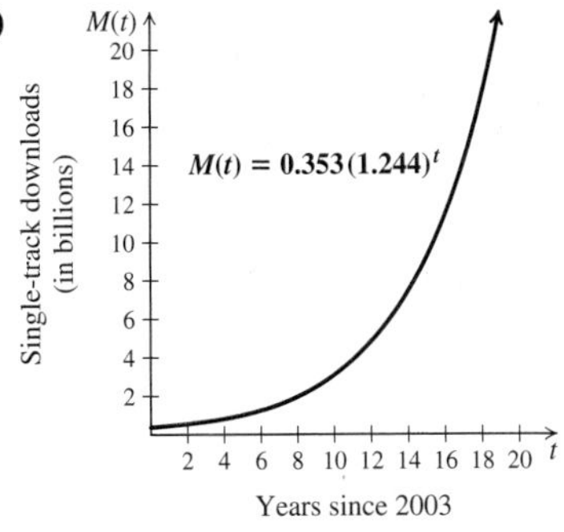

53. (a) 19.6%; 16.3%; 7.3%;
(b)

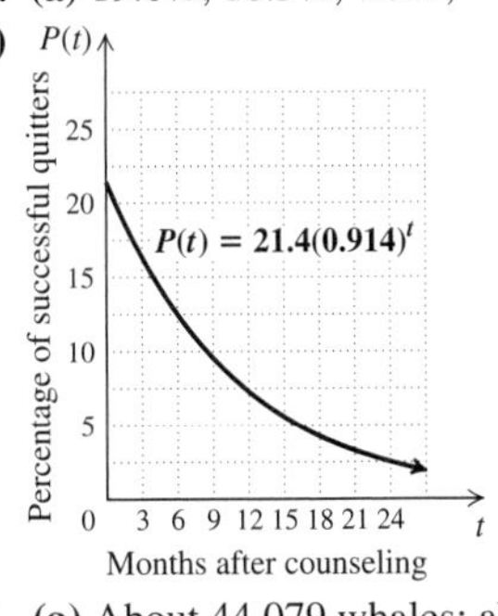

55. (a) About 44,079 whales; about 12,953 whales;
(b)

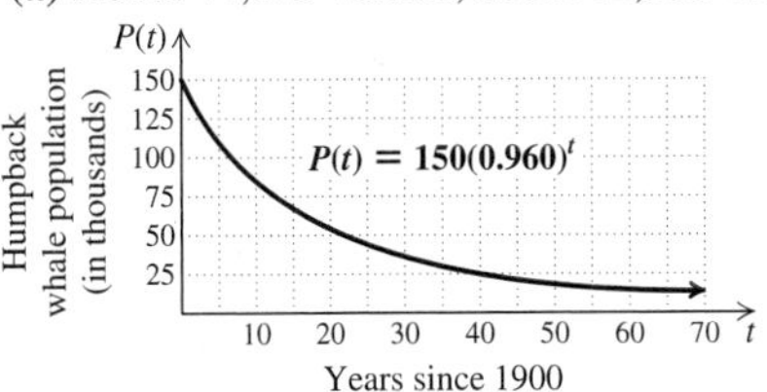

57. (a) About 11,874 whales; about 34,876 whales;
(b)

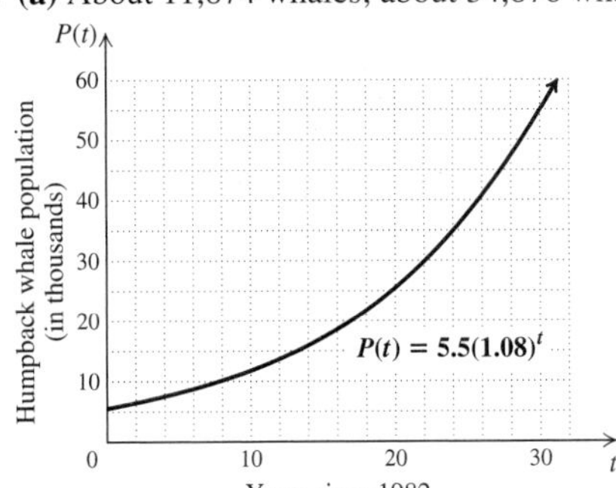

59. (a) About 539 fish; about 8843 fish;
(b)

$R(t) = 2(1.75)^t$

Number of ruffe in lake

Years after introduction

61. TW **63.** $3(x + 4)(x - 4)$ **64.** $(x - 10)^2$
65. $(2x + 3)(3x - 4)$
66. $8(x^2 - 2y^2)(x^4 + 2x^2y^2 + 4y^4)$ **67.** $6(y - 4)(y + 10)$
68. $x(x - 2)(5x^2 - 3)$ **69.** TW **71.** $\pi^{2.4}$

73.

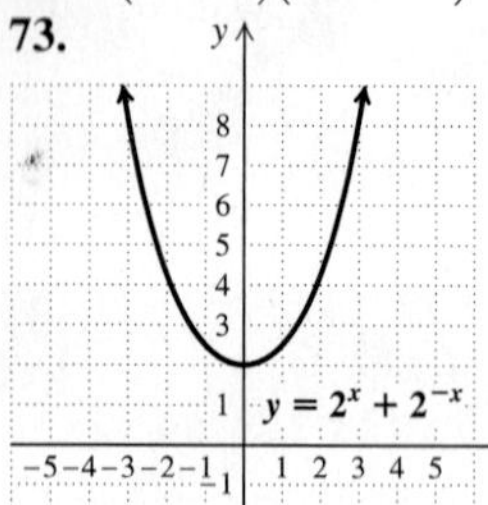

75.

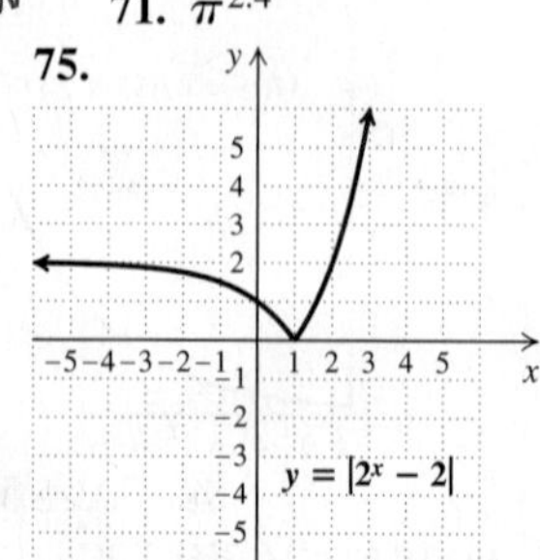

77.

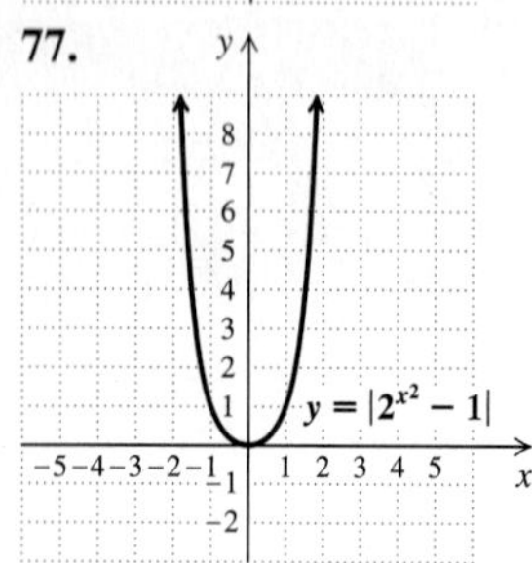

79.

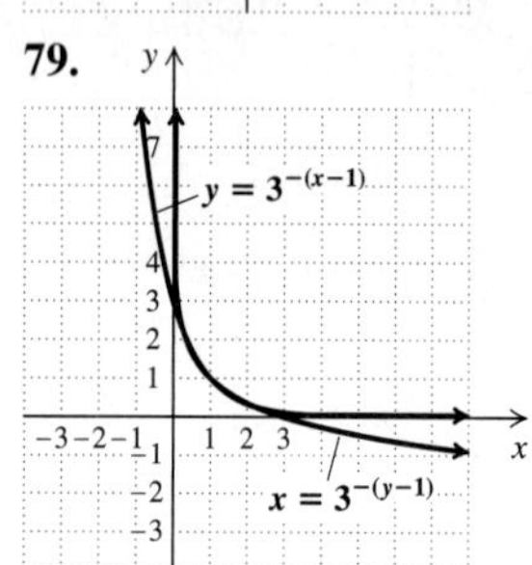

81. $N(t) = 0.464(1.778)^t$; about 464 million devices
83. TW **85.**

Exercise Set 12.3, pp. 884–885

1. (g) **2.** (d) **3.** (a) **4.** (h) **5.** (b) **6.** (c)
7. (e) **8.** (f) **9.** 3 **11.** 4 **13.** 4 **15.** -2
17. -1 **19.** 4 **21.** 1 **23.** 0 **25.** 5 **27.** -2
29. $\frac{1}{2}$ **31.** $\frac{3}{2}$ **33.** $\frac{2}{3}$ **35.** 7

37.

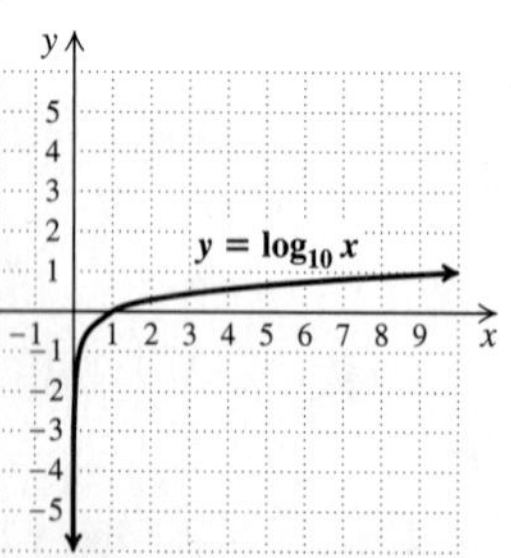

39.

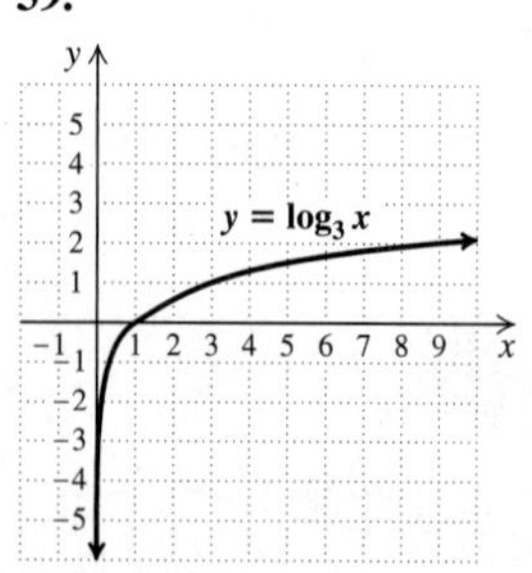

41.

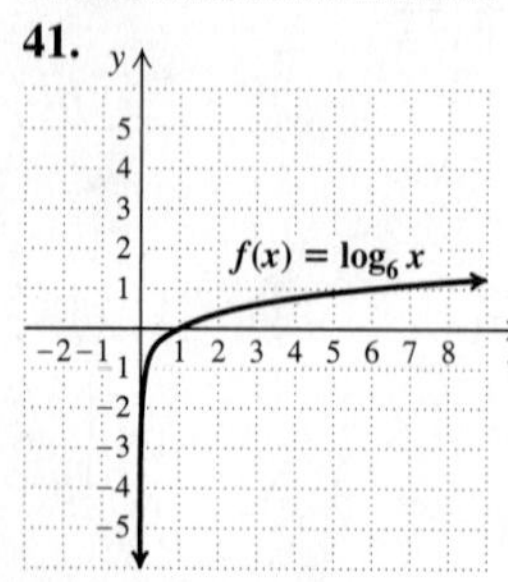

43.

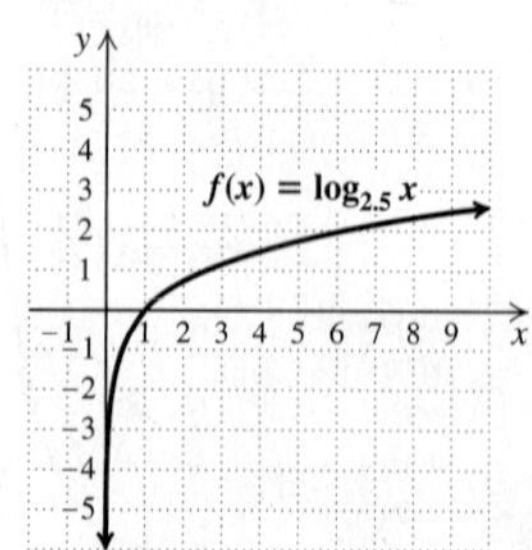

45.

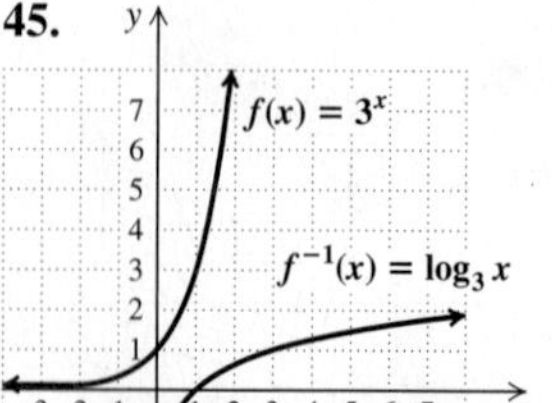
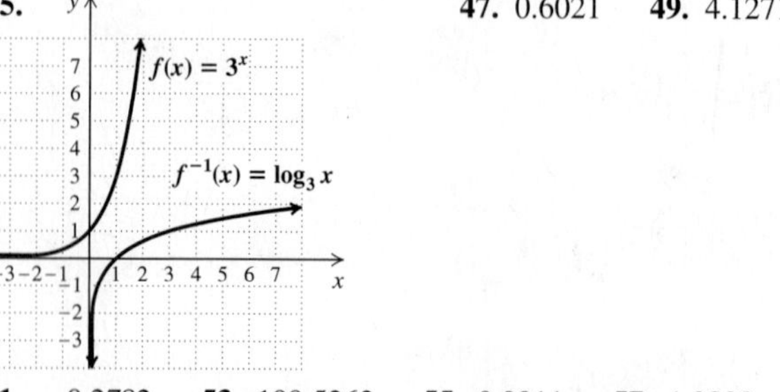

47. 0.6021 **49.** 4.1271
51. -0.2782 **53.** 199.5262 **55.** 0.0011 **57.** 1.0028

59.

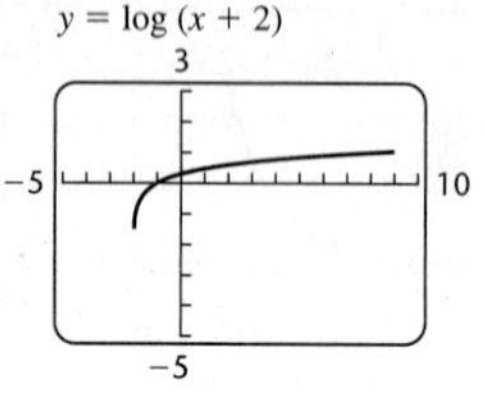

61.

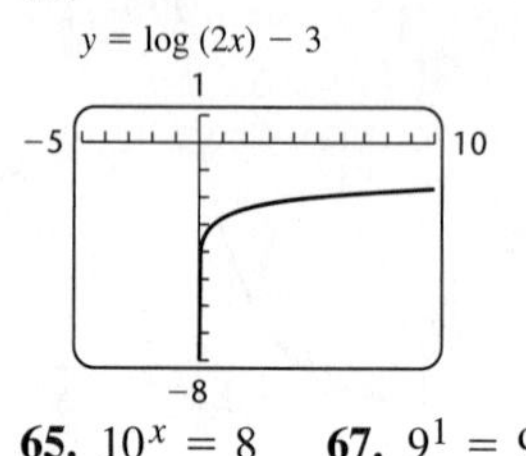

63. $y = \log(x^2)$

65. $10^x = 8$ **67.** $9^1 = 9$
69. $10^{-1} = 0.1$ **71.** $10^{0.845} = 7$ **73.** $c^8 = m$
75. $t^r = Q$ **77.** $e^{-1.3863} = 0.25$ **79.** $r^{-x} = T$
81. $2 = \log_{10} 100$ **83.** $-5 = \log_4 \frac{1}{1024}$ **85.** $\frac{3}{4} = \log_{16} 8$
87. $0.4771 = \log_{10} 3$ **89.** $m = \log_z 6$ **91.** $m = \log_p V$
93. $3 = \log_e 20.0855$ **95.** $-4 = \log_e 0.0183$ **97.** 9
99. 3 **101.** 2 **103.** 7 **105.** 81 **107.** $\frac{1}{9}$ **109.** 4
111. TW **113.** a^{18} **114.** x^9 **115.** x^8 **116.** a^{12}
117. y^{15} **118.** n^{30} **119.** x^5 **120.** x^6 **121.** TW

123.

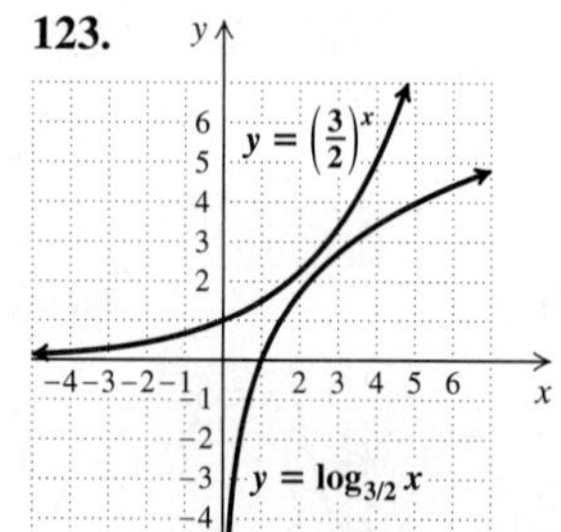

125.

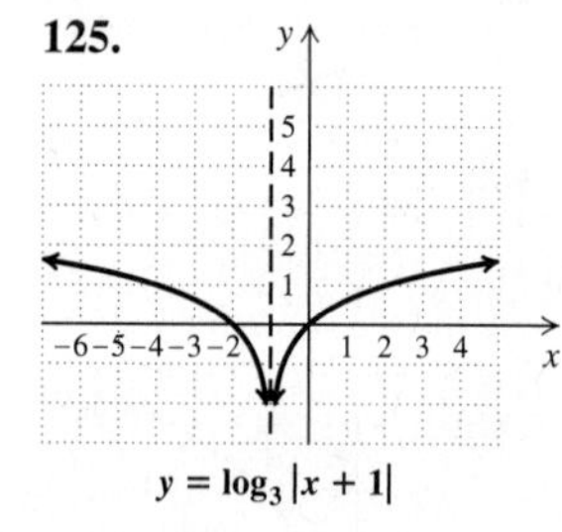

127. 6 **129.** $-25, 4$ **131.** -2 **133.** 0
135. Let $b = 0$, and suppose that $x = 1$ and $y = 2$. Then $0^1 = 0^2$, but $1 \neq 2$. Then let $b = 1$, and suppose that $x = 1$ and $y = 2$. Then $1^1 = 1^2$, but $1 \neq 2$.

Interactive Discovery, p. 886

1. (c) **2.** (b) **3.** (c) **4.** (a)

Exercise Set 12.4, pp. 891–893

1. (e) **2.** (f) **3.** (a) **4.** (b) **5.** (c) **6.** (d)
7. $\log_3 81 + \log_3 27$ **9.** $\log_4 64 + \log_4 16$
11. $\log_c r + \log_c s + \log_c t$ **13.** $\log_a (5 \cdot 14)$, or $\log_a 70$

15. $\log_c(t \cdot y)$ **17.** $8\log_a r$ **19.** $\frac{1}{3}\log_2 y$
21. $-3\log_b C$ **23.** $\log_2 25 - \log_2 13$
25. $\log_b m - \log_b n$ **27.** $\log_a \frac{17}{6}$ **29.** $\log_b \frac{36}{4}$, or $\log_b 9$
31. $\log_a \frac{x}{y}$ **33.** $\log_a x + \log_a y + \log_a z$
35. $3\log_a x + 4\log_a z$ **37.** $2\log_a x - 2\log_a y + \log_a z$
39. $4\log_a x - 3\log_a y - \log_a z$
41. $\log_b x + 2\log_b y - \log_b w - 3\log_b z$
43. $\frac{1}{2}(7\log_a x - 5\log_a y - 8\log_a z)$
45. $\frac{1}{3}(6\log_a x + 3\log_a y - 2 - 7\log_a z)$ **47.** $\log_a(x^8z^3)$
49. $\log_a x$ **51.** $\log_a \frac{y^5}{x^{3/2}}$ **53.** $\log_a(x - 2)$ **55.** 1.953
57. -0.369 **59.** -1.161 **61.** $\frac{3}{2}$ **63.** Cannot be found
65. 7 **67.** m **69.** $\log_5 125 + \log_5 625 = 7$
71. $5 \cdot \log_2 16 = 20$ **73.** TW

75.

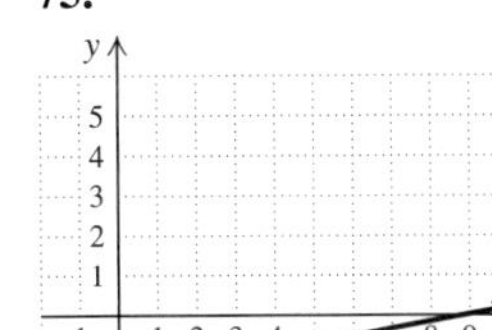

76.

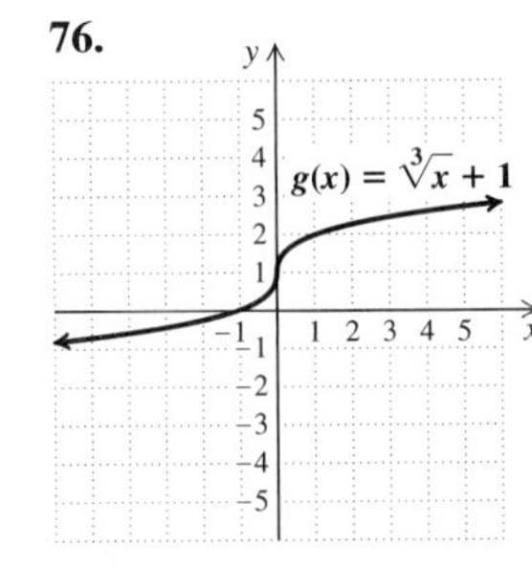

77.

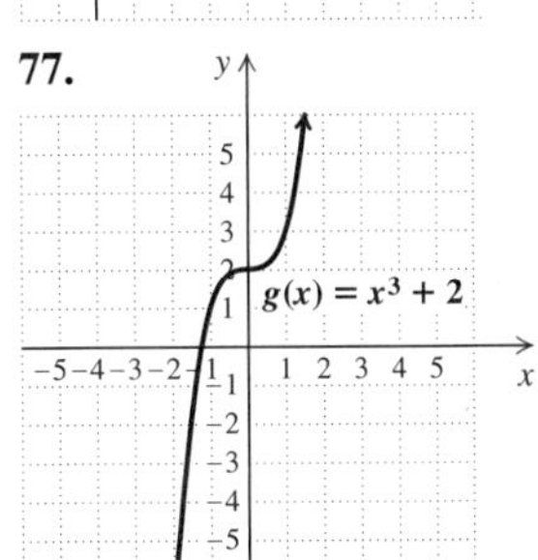

78.

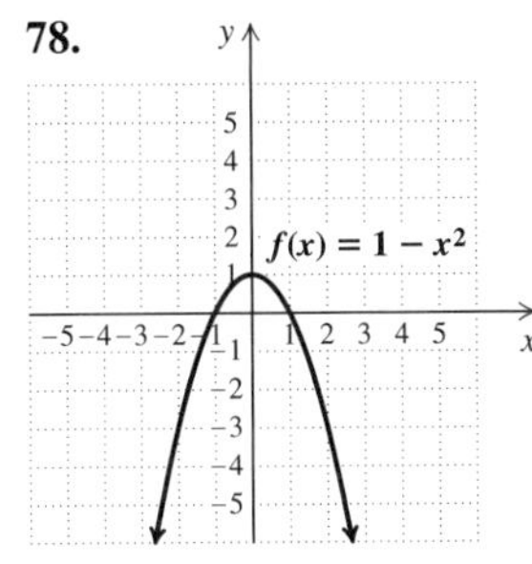

79. $(-\infty, -7) \cup (-7, \infty)$, or $\{x | x \text{ is a real number } and\ x \neq -7\}$
80. $(-\infty, -3) \cup (-3, 2) \cup (2, \infty)$, or $\{x | x \text{ is a real number } and\ x \neq -3\ and\ x \neq 2\}$
81. $(-\infty, 10]$, or $\{x | x \leq 10\}$ **82.** $(-\infty, \infty)$, or $\mathbb{R}$
83. TW **85.** $\log_a(x^6 - x^4y^2 + x^2y^4 - y^6)$
87. $\frac{1}{2}\log_a(1 - s) + \frac{1}{2}\log_a(1 + s)$ **89.** $\frac{10}{3}$ **91.** -2
93. $\frac{2}{5}$ **95.** True

Mid-Chapter Review: Chapter 12, pp. 893–894

Guided Solutions

1.
$$y = 2x - 5$$
$$x = 2y - 5$$
$$x + 5 = 2y$$
$$\frac{x + 5}{2} = y$$
$$f^{-1}(x) = \frac{x + 5}{2}$$

2. $x = 4^1$
$x = 4$

Mixed Review

1. $x^2 - 10x + 26$ **2.** $f(x) = \sqrt{x}; g(x) = 5x - 3$
3. $g^{-1}(x) = 6 - x$ **4.**

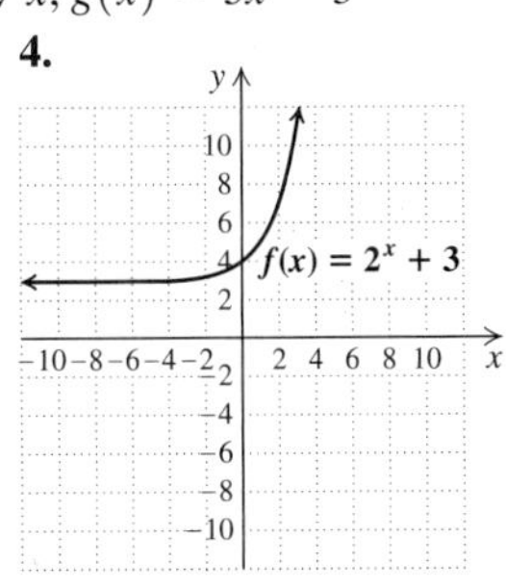

5. 2 **6.** -1 **7.** $\frac{1}{2}$ **8.** 1 **9.** 19 **10.** 0 **11.** $x^m = 3$
12. $2^{10} = 1024$ **13.** $t = \log_e x$ **14.** $\frac{2}{3} = \log_{64} 16$
15. $\log x - \frac{1}{2}\log y - \frac{3}{2}\log z$ **16.** $\log \frac{a}{b^2c}$ **17.** 4
18. $\frac{1}{3}$ **19.** 100,000 **20.** 4

Interactive Discovery, p. 895

1. \$2.25; \$2.370370; \$2.441406; \$2.613035; \$2.692597; \$2.714567; \$2.718127 **2.** (c)

Visualizing for Success, p. 900

1. J **2.** D **3.** B **4.** G **5.** H **6.** C **7.** F
8. I **9.** E **10.** A

Exercise Set 12.5, pp. 901–902

1. True **2.** True **3.** True **4.** False **5.** True
6. True **7.** True **8.** True **9.** True **10.** True
11. 1.6094 **13.** -5.0832 **15.** 96.7583 **17.** 15.0293
19. 0.0305 **21.** 0.8451 **23.** 13.0014 **25.** -0.4260
27. 4.9459 **29.** 2.5237 **31.** 6.6439 **33.** -2.3219
35. -2.3219 **37.** 3.5471

39. Domain: $\mathbb{R}$; range: $(0, \infty)$

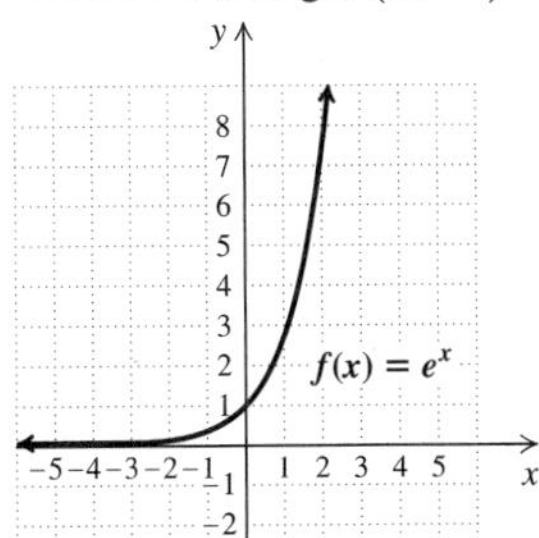

41. Domain: $\mathbb{R}$; range: $(3, \infty)$

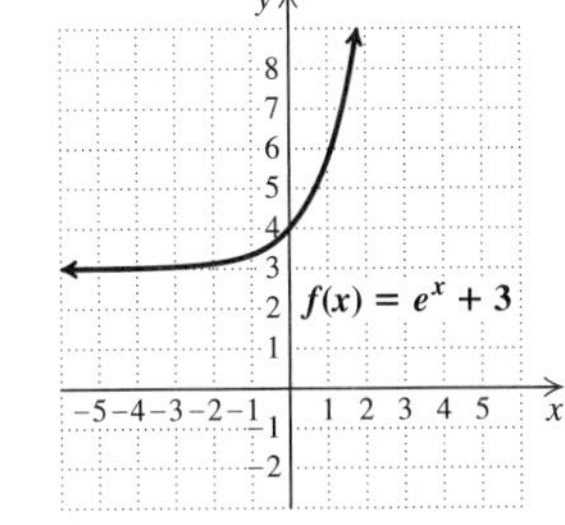

43. Domain: $\mathbb{R}$; range: $(-2, \infty)$

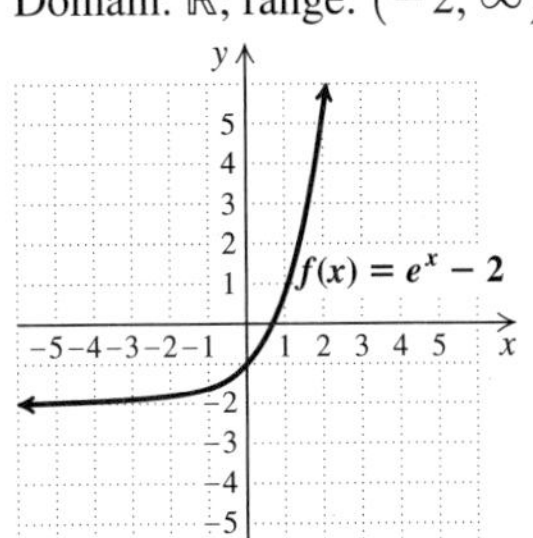

45. Domain: $\mathbb{R}$; range: $(0, \infty)$

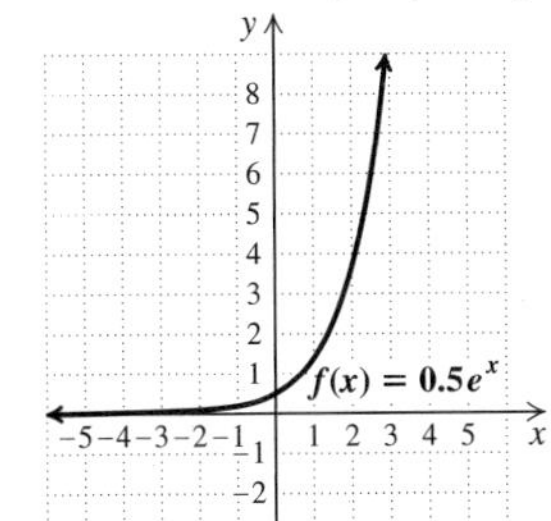

47. Domain: $\mathbb{R}$; range: $(0, \infty)$

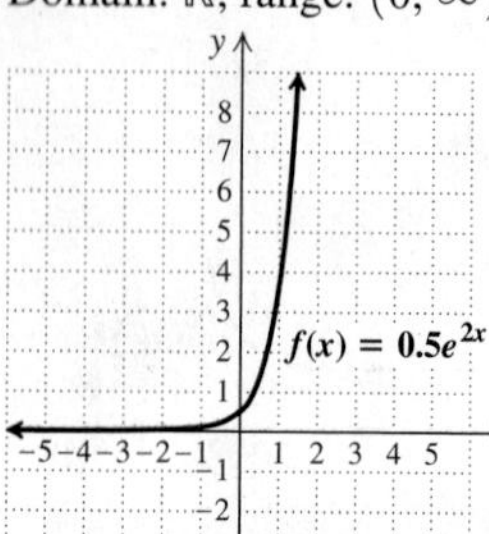

49. Domain: $\mathbb{R}$; range: $(0, \infty)$

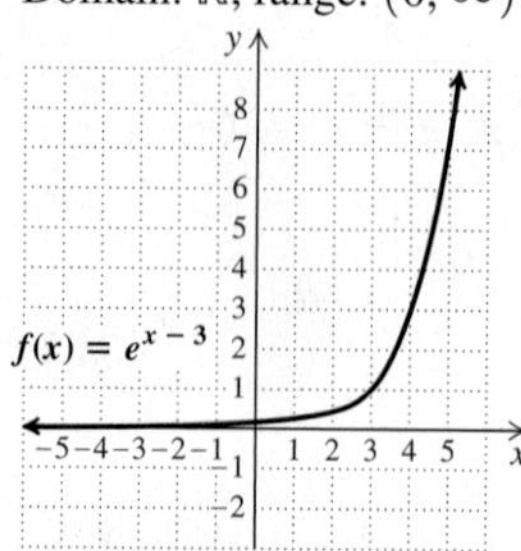

51. Domain: $\mathbb{R}$; range: $(0, \infty)$

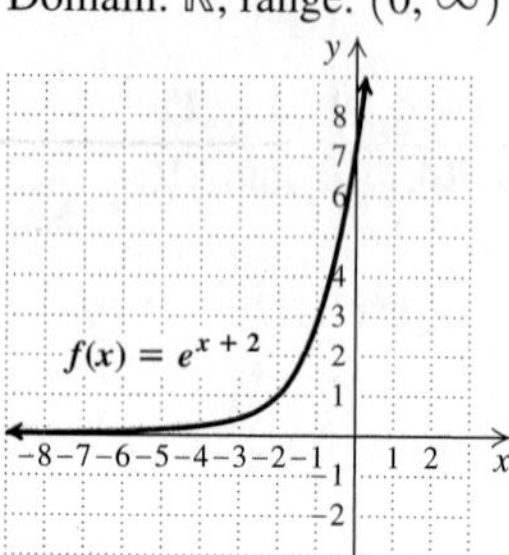

53. Domain: $\mathbb{R}$; range: $(-\infty, 0)$

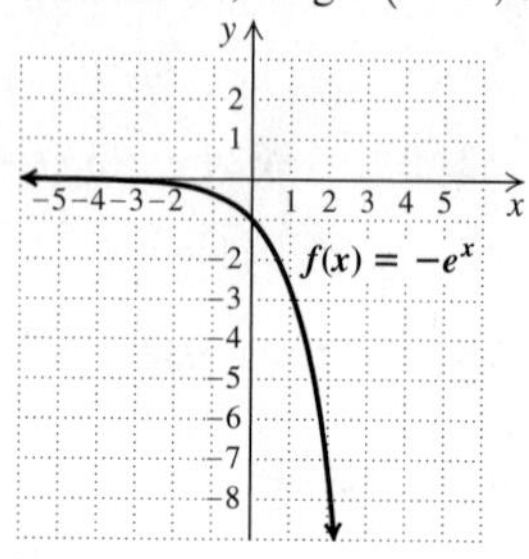

55. Domain: $(0, \infty)$; range: $\mathbb{R}$

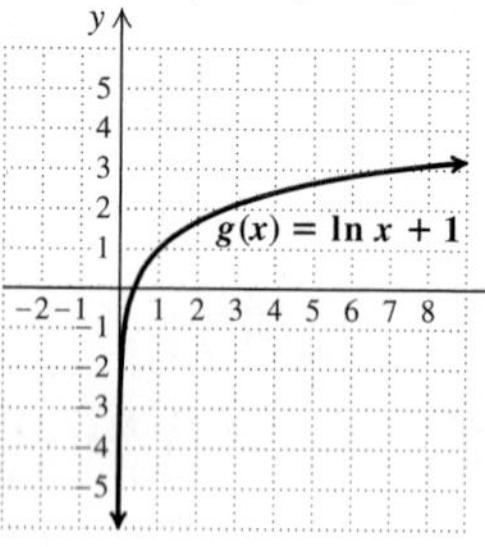

57. Domain: $(0, \infty)$; range: $\mathbb{R}$

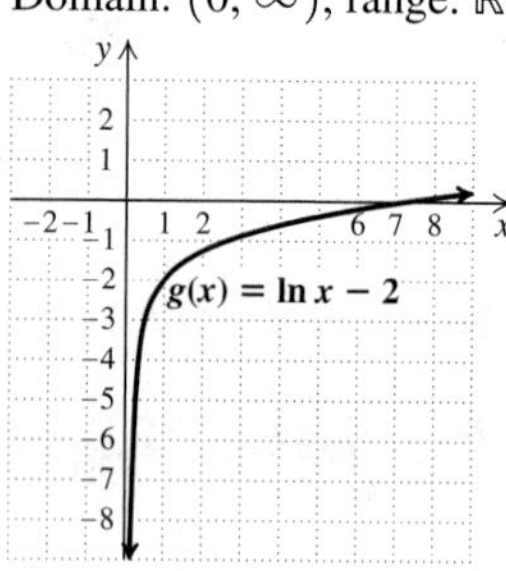

59. Domain: $(0, \infty)$; range: $\mathbb{R}$

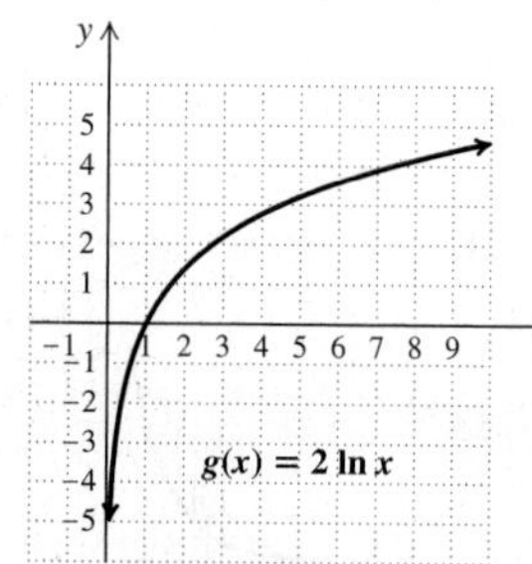

61. Domain: $(0, \infty)$; range: $\mathbb{R}$

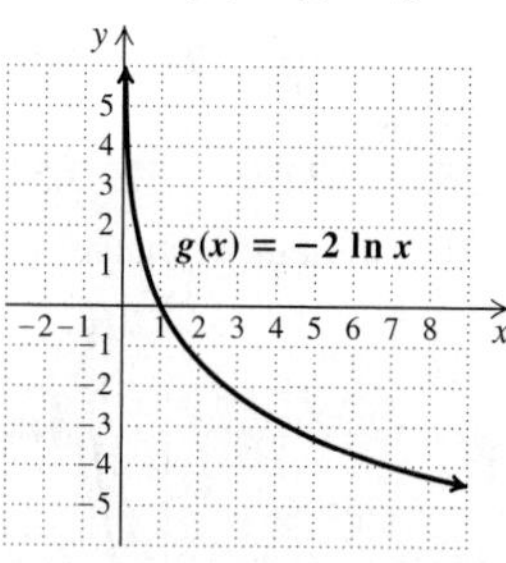

63. Domain: $(-2, \infty)$; range: $\mathbb{R}$

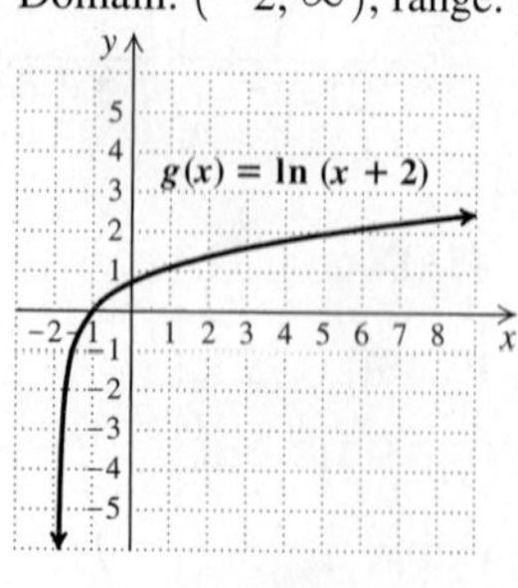

65. Domain: $(1, \infty)$; range: $\mathbb{R}$

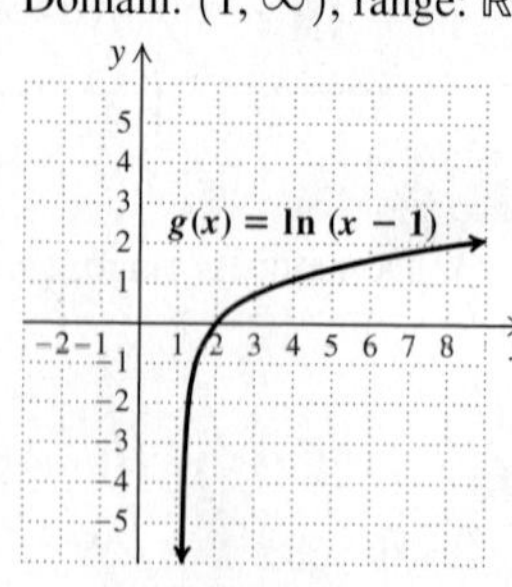

67. $f(x) = \log(x)/\log(5)$, or $f(x) = \ln(x)/\ln(5)$

$y = \log(x)/\log(5)$, or $y = \ln(x)/\ln(5)$

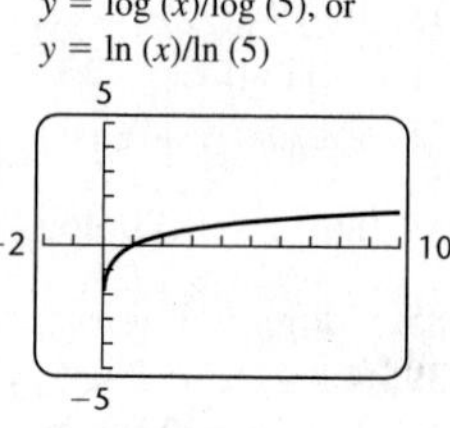

69. $f(x) = \log(x - 5)/\log(2)$, or $f(x) = \ln(x - 5)/\ln(2)$

$y = \log(x - 5)/\log(2)$, or $y = \ln(x - 5)/\ln(2)$

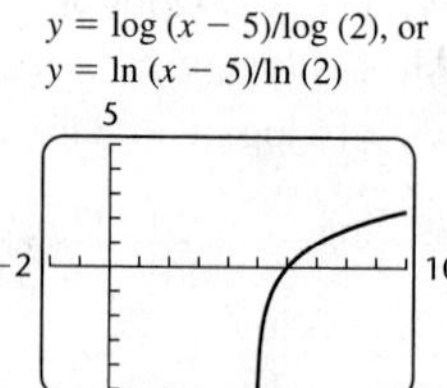

71. $f(x) = \log(x)/\log(3) + x$, or $f(x) = \ln(x)/\ln(3) + x$

$y = \log(x)/\log(3) + x$, or $y = \ln(x)/\ln(3) + x$

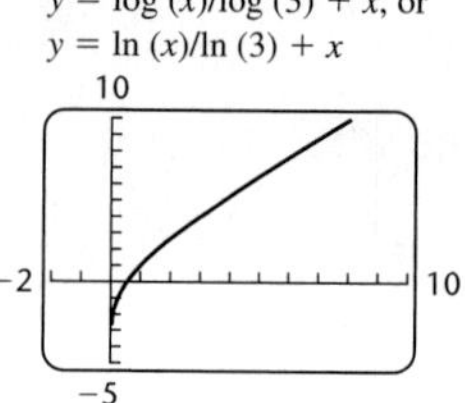

73. TW **75.** $-4, 7$ **76.** $0, \frac{7}{5}$ **77.** $\frac{15}{17}$ **78.** $\frac{5}{6}$

79. $\frac{56}{9}$ **80.** 4 **81.** 16, 256 **82.** $\frac{1}{4}, 9$ **83.** TW

85. 2.452 **87.** 1.442 **89.** $\log M = \dfrac{\ln M}{\ln 10}$

91. 1086.5129 **93.** 4.9855

95. **(a)** Domain: $\{x \mid x > 0\}$, or $(0, \infty)$; range: $\{y \mid y < 0.5135\}$, or $(-\infty, 0.5135)$; **(b)** $[-1, 5, -10, 5]$;
(c)

$y = 3.4 \ln x - 0.25e^x$

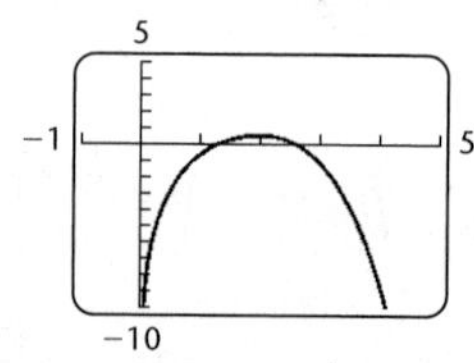

97. **(a)** Domain: $\{x \mid x > 0\}$, or $(0, \infty)$; range: $\{y \mid y > -0.2453\}$, or $(-0.2453, \infty)$; **(b)** $[-1, 5, -1, 10]$;
(c)

$y = 2x^3 \ln x$

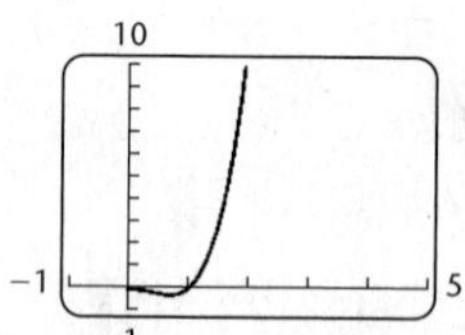

Exercise Set 12.6, pp. 907–909

1. (e) **2.** (a) **3.** (f) **4.** (h) **5.** (b) **6.** (d)

7. (g) **8.** (c) **9.** 2 **11.** $\frac{5}{2}$ **13.** $\dfrac{\log 10}{\log 2} \approx 3.322$

15. -1 **17.** $\dfrac{\log 19}{\log 8} + 3 \approx 4.416$ **19.** $\ln 50 \approx 3.912$

21. $\frac{\ln 8}{-0.02} \approx -103.972$ **23.** $\frac{\log 5}{\log 3} - 1 \approx 0.465$

25. $\frac{\log 87}{\log 4.9} \approx 2.810$ **27.** $\frac{\ln\left(\frac{19}{2}\right)}{4} \approx 0.563$ **29.** $\frac{\ln 2}{5} \approx 0.139$

31. 81 **33.** $\frac{1}{8}$ **35.** $e^5 \approx 148.413$ **37.** $\frac{e^3}{4} \approx 5.021$

39. $10^{2.5} \approx 316.228$ **41.** $\frac{e^4 - 1}{2} \approx 26.799$ **43.** $e \approx 2.718$

45. $e^{-3} \approx 0.050$ **47.** -4 **49.** 10 **51.** No solution
53. 2 **55.** $\frac{83}{15}$ **57.** 1 **59.** 6 **61.** 1 **63.** 5
65. $\frac{17}{2}$ **67.** 4 **69.** $-6.480, 6.519$ **71.** 0.000112, 3.445
73. 1 **75.** TW **77.** Length: 9.5 ft; width: 3.5 ft
78. 25 visits or more **79.** Golden Days; $23\frac{1}{3}$ lb; Snowy Friends: $26\frac{2}{3}$ lb **80.** 1.5 cm **81.** $1\frac{1}{5}$ hr **82.** Approximately 2.1 ft
83. TW **85.** $\frac{12}{5}$ **87.** $\sqrt[3]{3}$ **89.** -1 **91.** $-3, -1$
93. $-625, 625$ **95.** $\frac{1}{2}, 5000$ **97.** $-3, -1$
99. $\frac{1}{100{,}000}$, 100,000 **101.** $-\frac{1}{3}$ **103.** 38

Exercise Set 12.7, pp. 920–927

1. **(a)** About 2006; **(b)** 2.8 yr **3.** **(a)** About 1979; **(b)** about 2025 **5.** **(a)** 6.4 yr; **(b)** 23.4 yr **7.** **(a)** About 56°F; **(b)** about 75°F **9.** **(a)** About 2013; **(b)** 1.2 yr **11.** 4.9
13. 10^{-7} moles per liter **15.** 130 dB **17.** 7.6 W/m^2
19. Approximately 42.4 million messages per day
21. **(a)** $P(t) = P_0e^{0.025t}$; **(b)** \$5126.58; \$5256.36; **(c)** 27.7 yr
23. **(a)** $P(t) = 310e^{0.01t}$; **(b)** 329 million; **(c)** about 2022
25. 5.8 yr **27.** **(a)** About 2043; **(b)** about 2059;
(c)

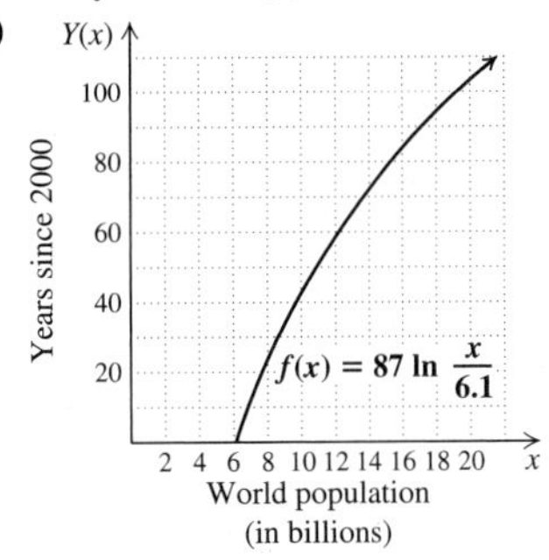

29. **(a)** 68%; **(b)** 54%; 40%
(c)

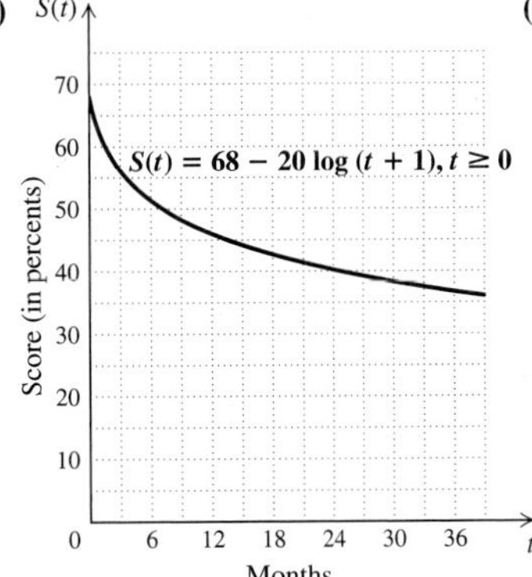

(d) 6.9 months

31. **(a)** $k \approx 0.151$; $P(t) = 2000e^{0.151t}$; **(b)** 2011
33. **(a)** $k \approx 0.280$; $P(t) = 8200e^{-0.280t}$; **(b)** \$215 per gigabit per second per mile; **(c)** 2029
35. About 1964 yr **37.** About 7.2 days **39.** **(a)** 13.9% per hour; **(b)** 21.6 hr **41.** **(a)** $k \approx 0.123$; $V(t) = 9e^{0.123t}$; **(b)** \$360.4 million; **(c)** 5.6 yr; **(d)** 38.3 yr **43.** Yes **45.** No
47. **(a)** $n(t) = 7.8(1.3725)^t$; **(b)** 31.7%; **(c)** about 696,000,000 transistors **49.** **(a)** $f(x) = 2{,}097{,}152(0.8706)^x$; **(b)** 4 hr
51. TW **53.** $\sqrt{2}$ **54.** 5 **55.** $(4, -7)$
56. $\left(-\frac{7}{2}, -\frac{19}{2}\right)$ **57.** $-4 \pm \sqrt{17}$ **58.** $5 \pm 2\sqrt{10}$
59.

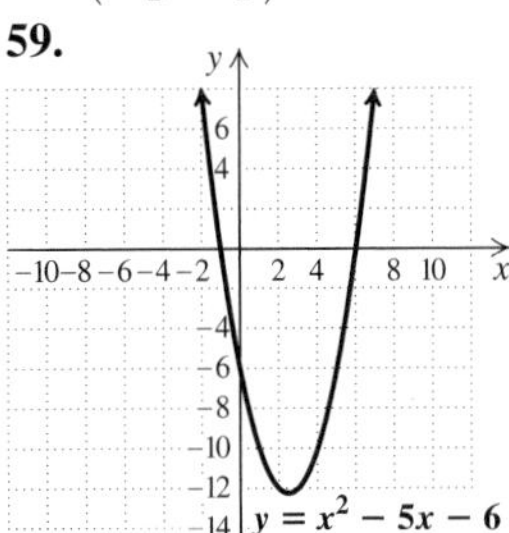

60.

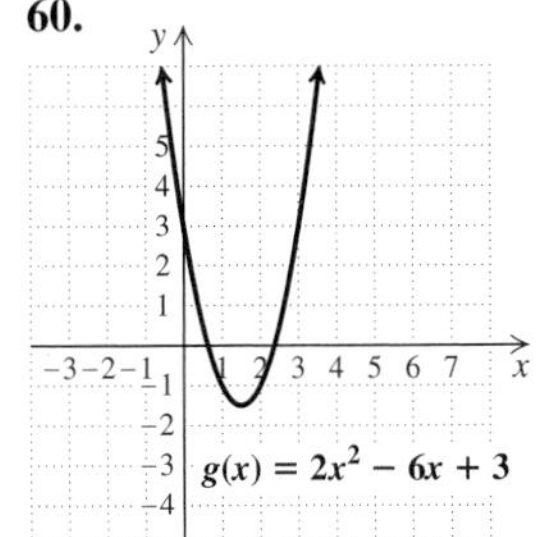

61. TW **63.** \$14.5 million
65. **(a)** -26.9; **(b)** 1.58×10^{-17} W/m^2
67. Consider an exponential growth function $P(t) = P_0e^{kt}$. At time T, $P(T) = 2P_0$.
Solve for T:

$$2P_0 = P_0e^{kT}$$
$$2 = e^{kT}$$
$$\ln 2 = kT$$
$$\frac{\ln 2}{k} = T.$$

69. **(a)** $f(x) = \frac{62.2245}{1 + 2.2661e^{-0.4893x}}$; **(b)** 62.0%

Study Summary: Chapter 12, pp. 928–930

1. $(f \circ g)(x) = 19 - 6x^2$ **2.** Yes **3.** $f^{-1}(x) = \frac{x - 1}{5}$
4.

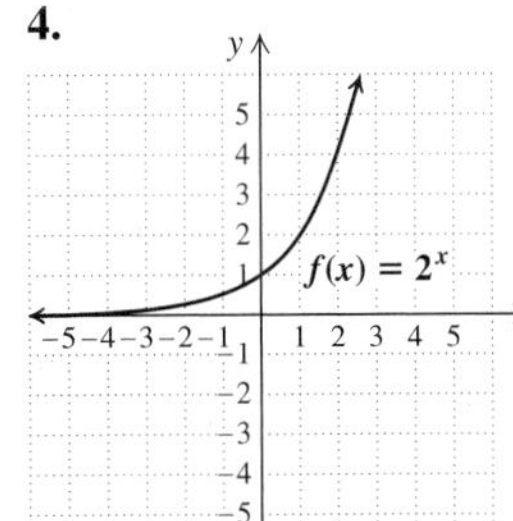

5.

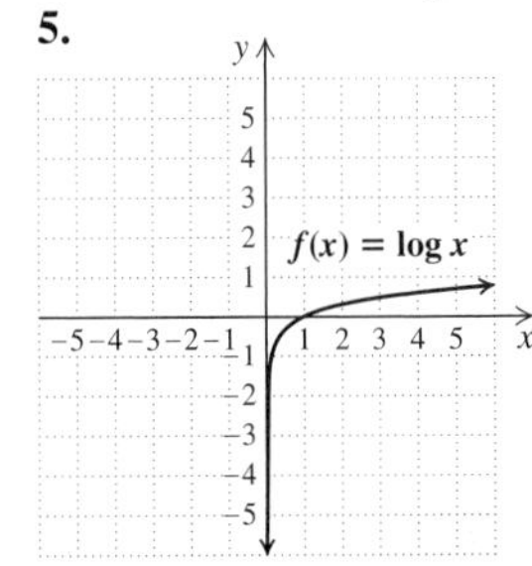

6. $\log_5 625 = 4$ **7.** $\log_9 x + \log_9 y$ **8.** $\log_6 7 - \log_6 10$
9. $5 \log 7$ **10.** 0 **11.** 1 **12.** 12 **13.** 2 **14.** 1
15. 2.3219 **16.** $\frac{4}{3}$ **17.** $\frac{\ln 10}{0.1} \approx 23.0259$
18. **(a)** $P(t) = 15{,}000e^{0.023t}$; **(b)** 30.1 yr **19.** 35 days

Review Exercises: Chapter 12, pp. 930–932

1. True **2.** True **3.** True **4.** False **5.** False
6. True **7.** False **8.** False **9.** True **10.** False
11. $(f \circ g)(x) = 4x^2 - 12x + 10$; $(g \circ f)(x) = 2x^2 - 1$

12. $f(x) = \sqrt{x}$; $g(x) = 3 - x$ **13.** No
14. $f^{-1}(x) = x + 8$ **15.** $g^{-1}(x) = \dfrac{2x - 1}{3}$
16. $f^{-1}(x) = \dfrac{\sqrt[3]{x}}{3}$
17.

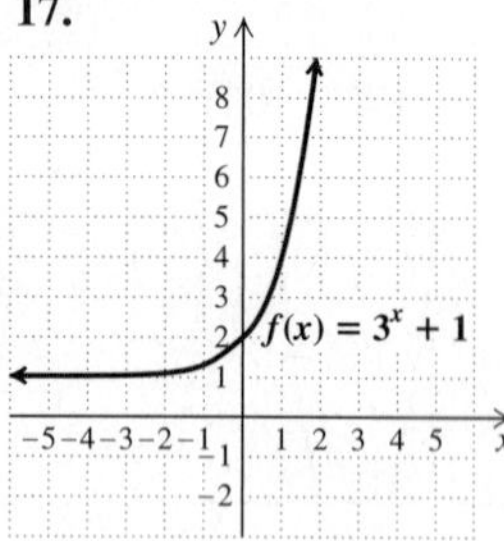

18.

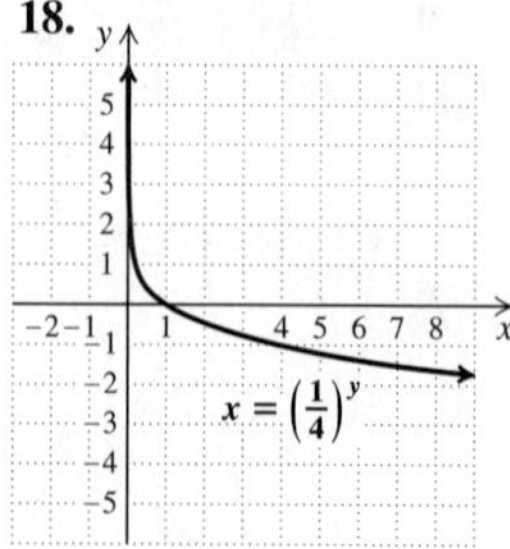

19.

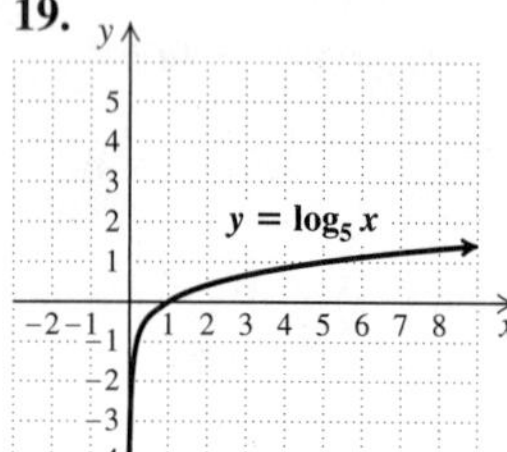

20. 2
21. -2
22. 7
23. $\frac{1}{2}$
24. $\log_{10} \frac{1}{100} = -2$
25. $\log_{25} 5 = \frac{1}{2}$
26. $16 = 4^x$
27. $1 = 8^0$
28. $4 \log_a x + 2 \log_a y + 3 \log_a z$
29. $5 \log_a x - (\log_a y + 2 \log_a z)$, or $5 \log_a x - \log_a y - 2 \log_a z$
30. $\frac{1}{4}(2 \log z - 3 \log x - \log y)$ **31.** $\log_a (7 \cdot 8)$, or $\log_a 56$
32. $\log_a \frac{72}{12}$, or $\log_a 6$ **33.** $\log \dfrac{a^{1/2}}{bc^2}$ **34.** $\log_a \sqrt[3]{\dfrac{x}{y^2}}$
35. 1 **36.** 0 **37.** 17 **38.** 6.93 **39.** -3.2698
40. 8.7601 **41.** 3.2698 **42.** 2.54995 **43.** -3.6602
44. 1.8751 **45.** 61.5177 **46.** -2.9957 **47.** 0.3753
48. 0.4307 **49.** 1.7097
50. Domain: $\mathbb{R}$; range: $(-1, \infty)$ **51.** Domain: $(0, \infty)$; range: $\mathbb{R}$

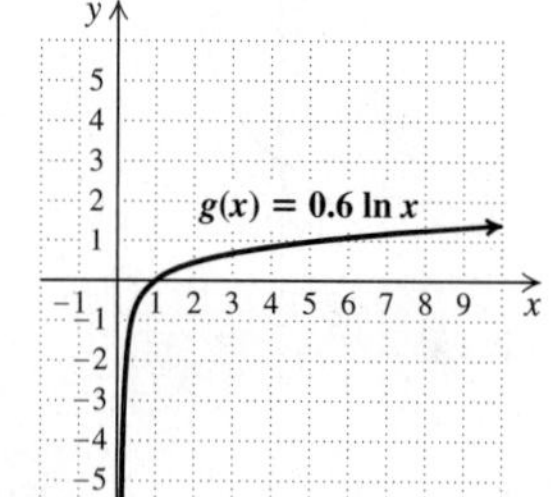

52. 5 **53.** -1 **54.** $\frac{1}{81}$ **55.** 2 **56.** $\frac{1}{1000}$
57. $e^{-2} \approx 0.1353$ **58.** $\dfrac{1}{2}\left(\dfrac{\log 19}{\log 4} + 5\right) \approx 3.5620$
59. $\dfrac{\log 12}{\log 2} \approx 3.5850$ **60.** $\dfrac{\ln 0.03}{-0.1} \approx 35.0656$
61. $e^{-3} \approx 0.0498$ **62.** $\frac{15}{2}$ **63.** 16 **64.** $\sqrt{43}$
65. **(a)** 82; **(b)** 66.8; **(c)** 35 months **66.** **(a)** 2.3 yr; **(b)** 3.1 yr
67. **(a)** $k \approx 0.112$; $A(t) = 1.2e^{0.112t}$; **(b)** \$2.9 billion; **(c)** 2018; **(d)** 6.2 yr **68.** **(a)** $M(t) = 3253e^{-0.137t}$; **(b)** 1247 spam messages per consumer; **(c)** 2030
69. **(a)** $f(x) = 33.8684(0.8196)^x$; **(b)** about 1700 cases; **(c)** 19.9% **70.** 11.553% per year **71.** 16.5 yr
72. 3463 yr **73.** 5.1 **74.** About 80,922 yr, or with rounding of k, about 80,792 yr **75.** About 114 dB
76. TW Negative numbers do not have logarithms because logarithm bases are positive, and there is no exponent to which a positive number can be raised to yield a negative number.
77. TW Taking the logarithm on each side of an equation produces an equivalent equation because the logarithm function is one-to-one. If two quantities are equal, their logarithms must be equal, and if the logarithms of two quantities are equal, the quantities must be the same.
78. e^{e^3} **79.** $-3, -1$ **80.** $\left(\frac{8}{3}, -\frac{2}{3}\right)$ **81.** 13.03%

Test: Chapter 12, pp. 932–933

1. [12.1] $(f \circ g)(x) = 2 + 6x + 4x^2$; $(g \circ f)(x) = 2x^2 + 2x + 1$
2. [12.1] $f(x) = \dfrac{1}{x}$; $g(x) = 2x^2 + 1$ **3.** [12.1] No
4. [12.1] $f^{-1}(x) = \dfrac{x - 4}{3}$ **5.** [12.1] $g^{-1}(x) = \sqrt[3]{x} - 1$
6. [12.2]

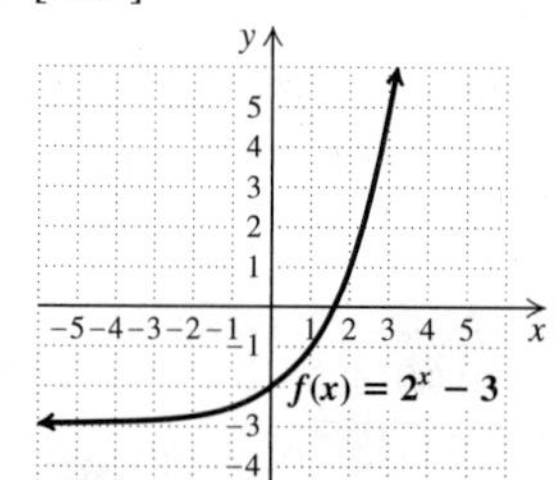

7. [12.3]

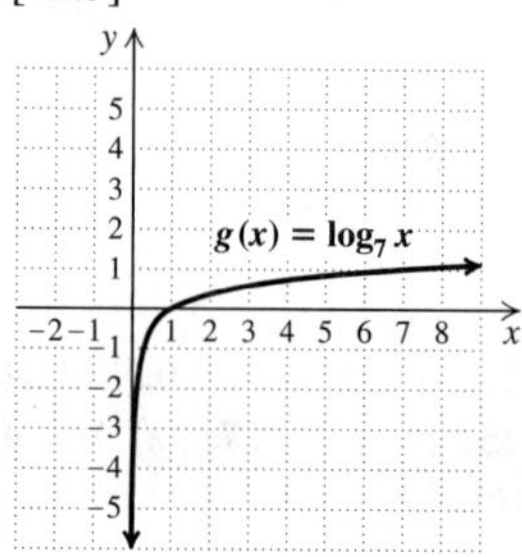

8. [12.3] 3 **9.** [12.3] $\frac{1}{2}$ **10.** [12.3] 18 **11.** [12.4] 1
12. [12.4] 0 **13.** [12.4] 19 **14.** [12.3] $\log_5 \frac{1}{625} = -4$
15. [12.3] $2^m = \frac{1}{2}$ **16.** [12.4] $3 \log a + \frac{1}{2} \log b - 2 \log c$
17. [12.4] $\log_a \left(z^2 \sqrt[3]{x}\right)$ **18.** [12.4] 1.146 **19.** [12.4] 0.477
20. [12.4] 1.204 **21.** [12.3] 1.0899 **22.** [12.3] 0.1585
23. [12.5] -0.9163 **24.** [12.5] 121.5104 **25.** [12.5] 2.4022
26. [12.5] Domain: $\mathbb{R}$; range: $(3, \infty)$

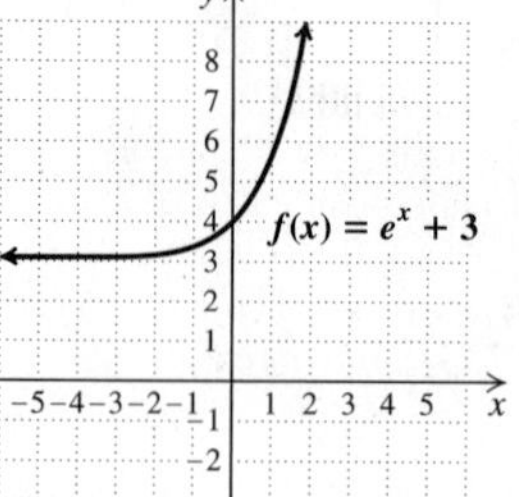

27. [12.5] Domain: $(4, \infty)$; range: $\mathbb{R}$

28. [12.6] -5 **29.** [12.6] 2 **30.** [12.6] 10,000
31. [12.6] $-\dfrac{1}{3}\left(\dfrac{\log 87}{\log 5} - 4\right) \approx 0.4084$
32. [12.6] $\dfrac{\log 1.2}{\log 7} \approx 0.0937$ **33.** [12.6] $e^3 \approx 20.0855$
34. [12.6] 4 **35.** [12.7] **(a)** 2.25 ft/sec; **(b)** 2,901,000
36. [12.7] **(a)** $P(t) = 8.2e^{0.00052t}$, where t is the number of years after 2009 and $P(t)$ is in millions; **(b)** 8.25 million; 8.38 million; **(c)** 2188; **(d)** 1333 yr **37.** [12.7] **(a)** $k \approx 0.054$; $C(t) = 21{,}855e^{0.054t}$; **(b)** \$46,545; **(c)** 2016 **38.** [12.7] **(a)** $f(x) = 458.8188(2.6582)^x$, where $f(x)$ is in thousands; **(b)** about 22,908,000 units

39. [12.7] 4.6% **40.** [12.7] About 4684 yr **41.** [12.7] 10^2 W/m^2 **42.** [12.7] 7.0 **43.** [12.6] $-309, 316$ **44.** [12.4] 2

Cumulative Review: Chapters 1–12, pp. 934–936

1. 2 **2.** $\dfrac{y^{12}}{16x^8}$ **3.** $\dfrac{20x^6z^2}{y}$ **4.** $-\dfrac{y^4}{3z^5}$ **5.** 6.3×10^{-15}
6. 25 **7.** 8 **8.** $(3, -1)$ **9.** $(1, -2, 0)$ **10.** $-7, 10$
11. $\frac{9}{2}$ **12.** $\frac{3}{4}$ **13.** $\frac{1}{2}$ **14.** $\pm 4i$ **15.** $\pm 2, \pm 3$ **16.** 9
17. $\dfrac{\log 7}{5 \log 3} \approx 0.3542$ **18.** $\dfrac{8e}{e-1} \approx 12.6558$
19. $(-\infty, -5) \cup (1, \infty)$, or $\{x \mid x < -5 \text{ or } x > 1\}$
20. $-3 \pm 2\sqrt{5}$ **21.** $\{x \mid x \le -2 \text{ or } x \ge 5\}$, or $(-\infty, -2] \cup [5, \infty)$ **22.** $a = \dfrac{Db}{b - D}$
23. $x = \dfrac{-v \pm \sqrt{v^2 + 4ad}}{2a}$
24. $\{x \mid x \text{ is a real number and } x \neq -\frac{1}{3} \text{ and } x \neq 2\}$, or $\left(-\infty, -\frac{1}{3}\right) \cup \left(-\frac{1}{3}, 2\right) \cup (2, \infty)$
25. $3p^2q^3 + 11pq - 2p^2 + p + 9$ **26.** $9x^4 - 6x^2z^3 + z^6$
27. $\dfrac{1}{x-4}$ **28.** $\dfrac{a+2}{6}$ **29.** $\dfrac{7x+4}{(x+6)(x-6)}$ **30.** $2y^2\sqrt[3]{y}$
31. $\sqrt[10]{(x+5)^7}$ **32.** $15 - 4\sqrt{3}i$ **33.** $x^3 - 5x^2 + 1$
34. $x(y + 2z - w)$ **35.** $2(3x - 2y)(x + 2y)$
36. $(x - 4)(x^3 + 7)$ **37.** $2(m + 3n)^2$
38. $(x - 2y)(x + 2y)(x^2 + 4y^2)$ **39.** $\dfrac{6 + \sqrt{y} - y}{4 - y}$
40. $f^{-1}(x) = \dfrac{x-9}{-2}$, or $f^{-1}(x) = \dfrac{9-x}{2}$
41. $f(x) = -10x - 8$ **42.** $y = \frac{1}{2}x + 5$

43.

44.

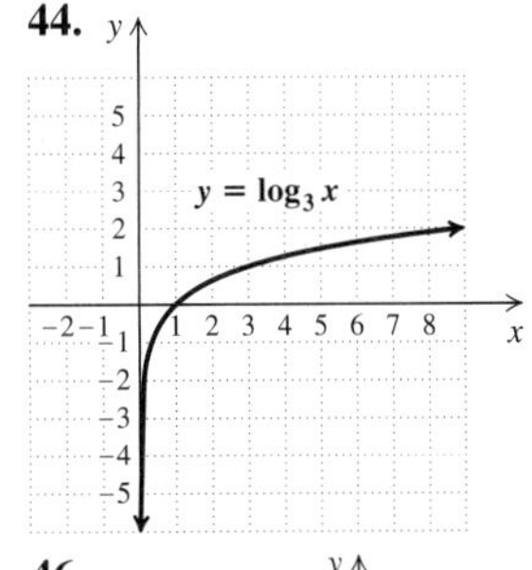

45.

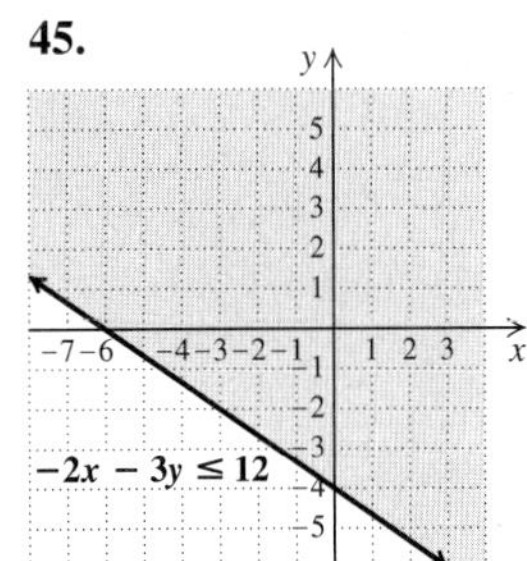

46.

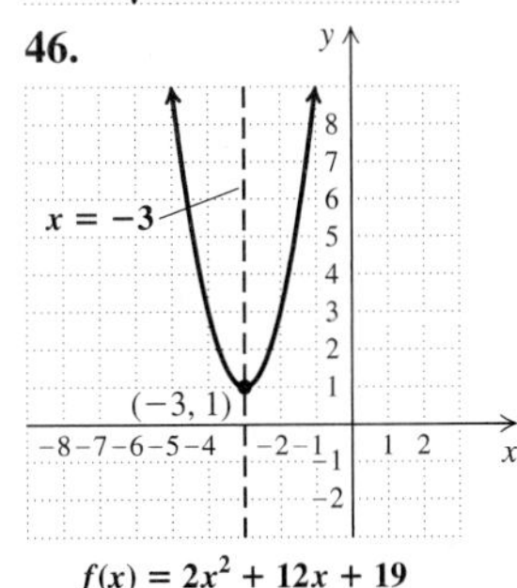

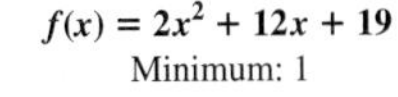
$f(x) = 2x^2 + 12x + 19$
Minimum: 1

47. Domain: $\mathbb{R}$; range: $(0, \infty)$

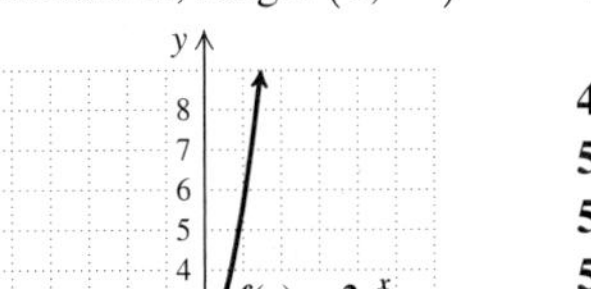

48. 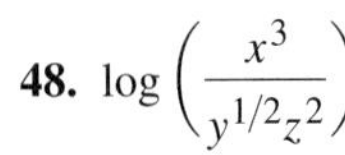 $\log\left(\dfrac{x^3}{y^{1/2}z^2}\right)$

49. (c)
50. (b)
51. (a)
52. (d)
53. 13.5 million acre-feet

54. **(a)** $k \approx 0.076$; $D(t) = 15e^{0.076t}$; **(b)** 79.8 million cubic meters per day; **(c)** 2015 **55.** $5\frac{5}{11}$ min **56.** Thick and Tasty: 6 oz; Light and Lean: 9 oz **57.** $2\frac{7}{9}$ km/h **58.** Linear
59. \$1/min **60.** $f(x) = x + 3$ **61.** \$13
62. $m = 1$ signifies the cost per minute; $b = 3$ signifies the startup cost of each massage **63.** $m(x) = 1.8937(1.0596)^x$
64. \$25.63 **65.** All real numbers except 1 and -2
66. $\frac{1}{3}, \frac{10{,}000}{3}$ **67.** 35 mph

Chapter 13

Exercise Set 13.1, pp. 945–947

1. (f) **2.** (e) **3.** (g) **4.** (h) **5.** (c)
6. (b) **7.** (d) **8.** (a)

9.

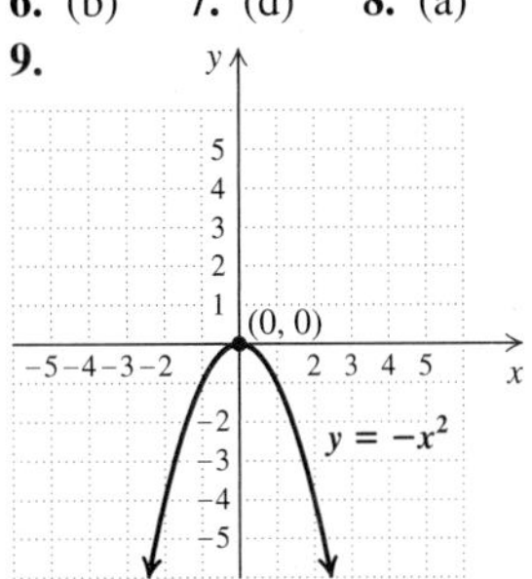

11.

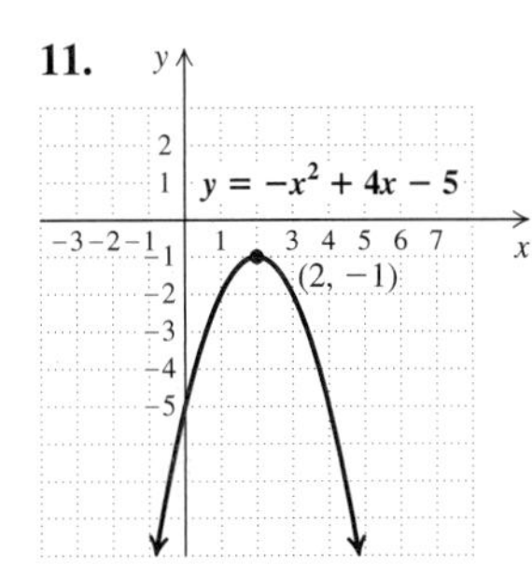

13.

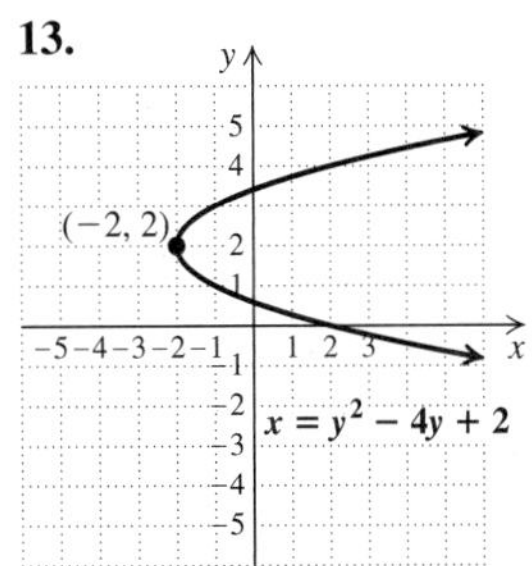

15.

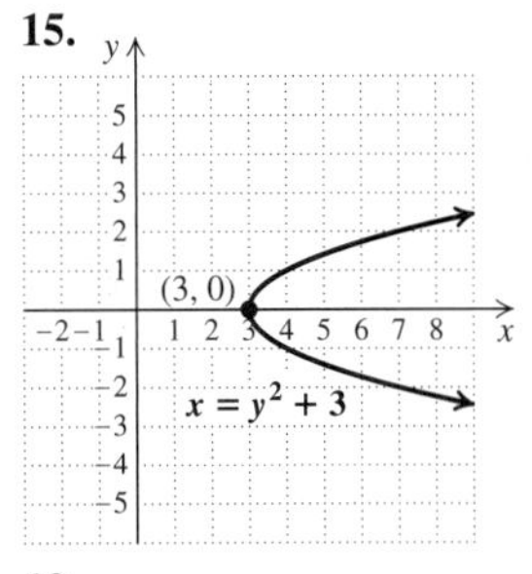

17.

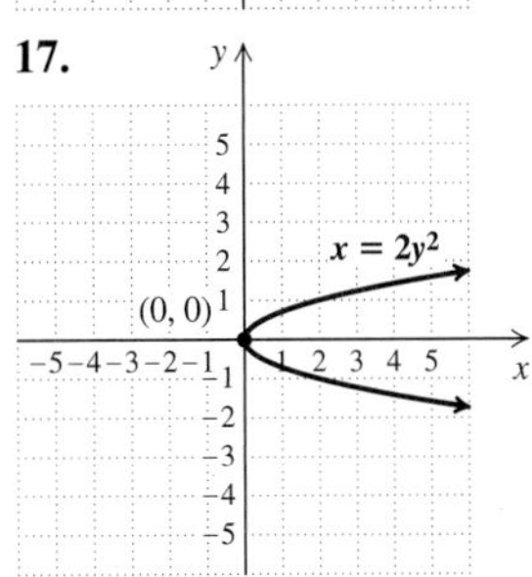

19.

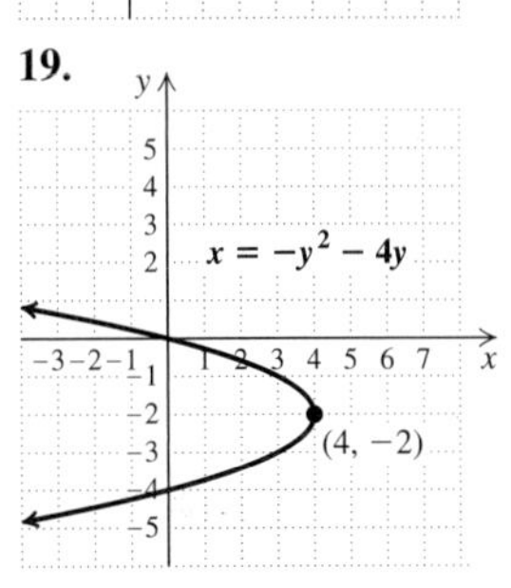

21.

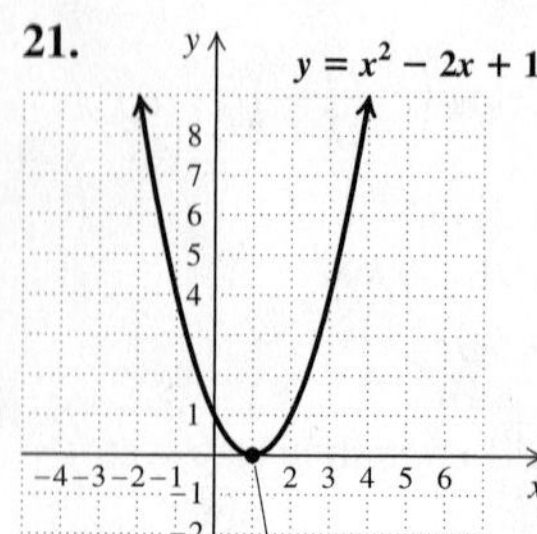

23.

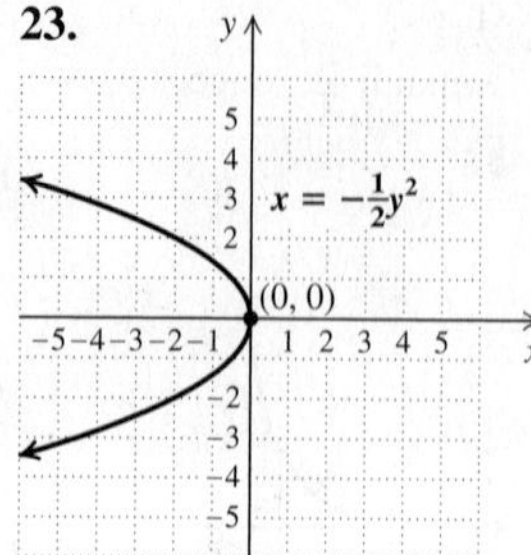

25.

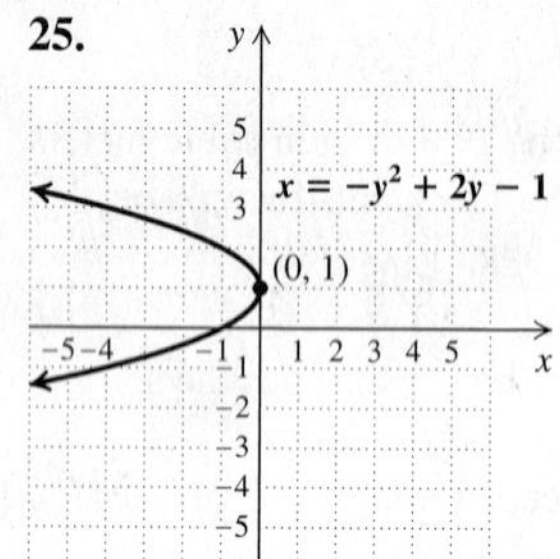

27.

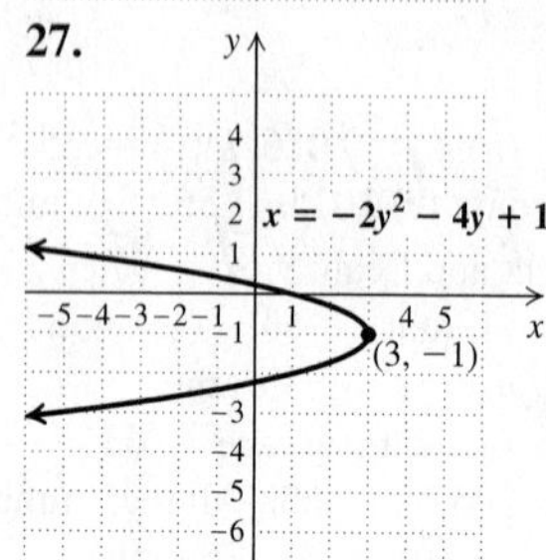

29. $x^2 + y^2 = 36$ **31.** $(x - 7)^2 + (y - 3)^2 = 5$
33. $(x + 4)^2 + (y - 3)^2 = 48$
35. $(x + 7)^2 + (y + 2)^2 = 50$
37. $x^2 + y^2 = 25$ **39.** $(x + 4)^2 + (y - 1)^2 = 20$
41. $(0, 0); 8$ **43.** $(-1, -3); 6$

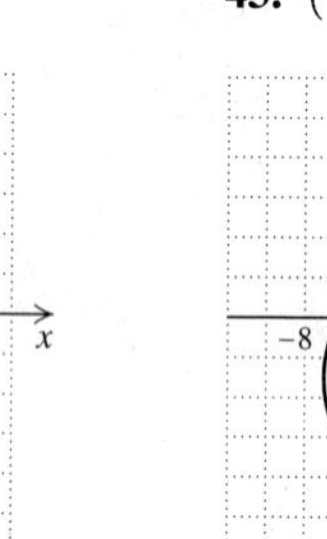

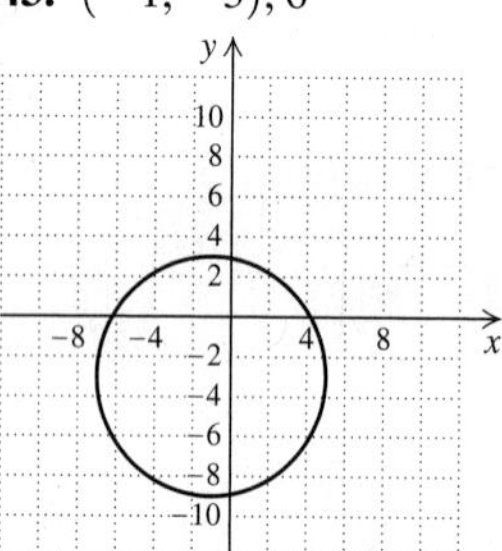

$(x + 1)^2 + (y + 3)^2 = 36$

45. $(4, -3); \sqrt{10}$ **47.** $(0, 0); \sqrt{10}$

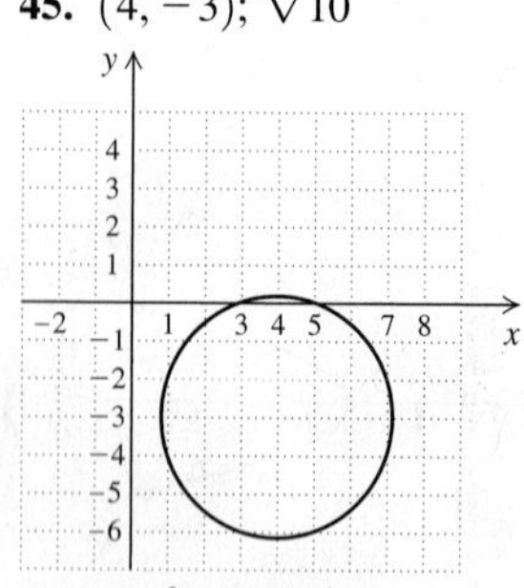

$(x - 4)^2 + (y + 3)^2 = 10$

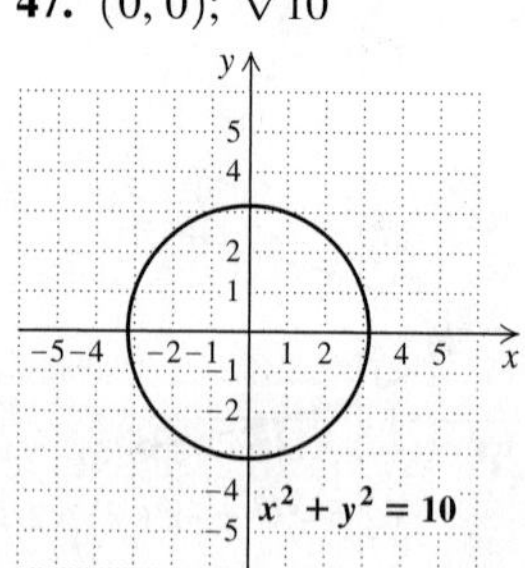

49. $(5, 0); \frac{1}{2}$ **51.** $(-4, 3); \sqrt{40}$, or $2\sqrt{10}$

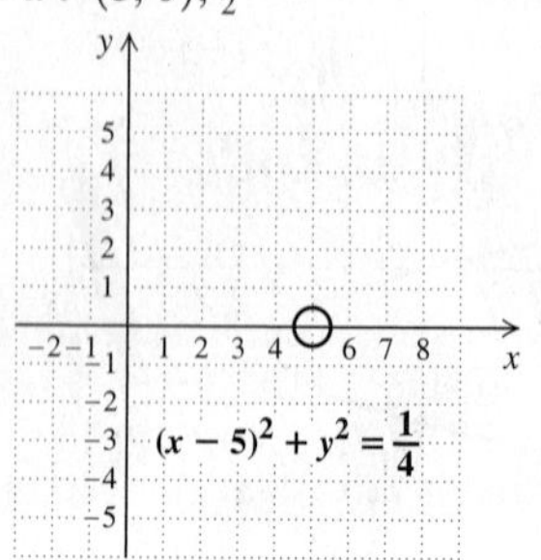

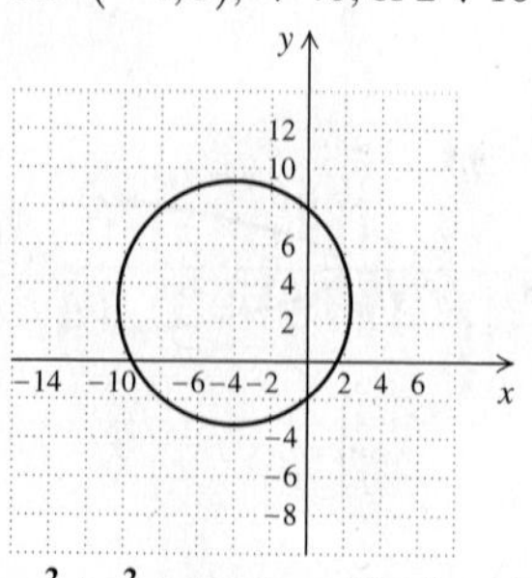

$x^2 + y^2 + 8x - 6y - 15 = 0$

53. $(4, -1); 2$ **55.** $(0, -5); 10$

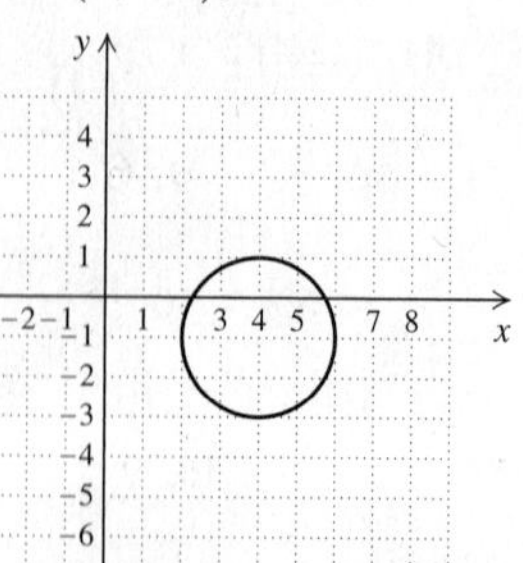

$x^2 + y^2 - 8x + 2y + 13 = 0$

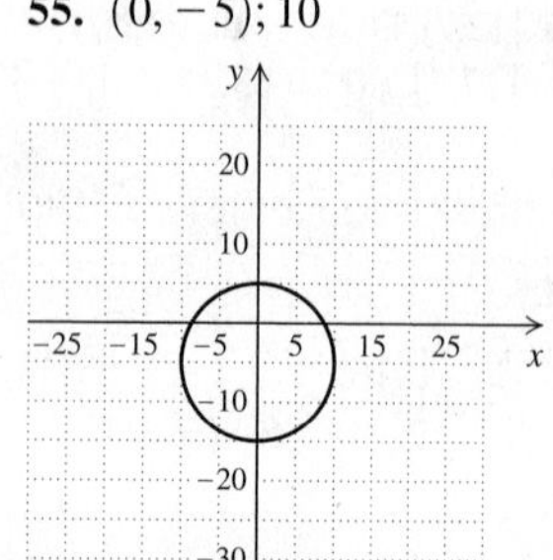

$x^2 + y^2 + 10y - 75 = 0$

57. $\left(-\frac{7}{2}, \frac{3}{2}\right); \sqrt{\frac{98}{4}}$, or $\frac{7\sqrt{2}}{2}$ **59.** $(0, 0); \frac{1}{6}$

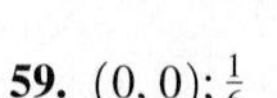

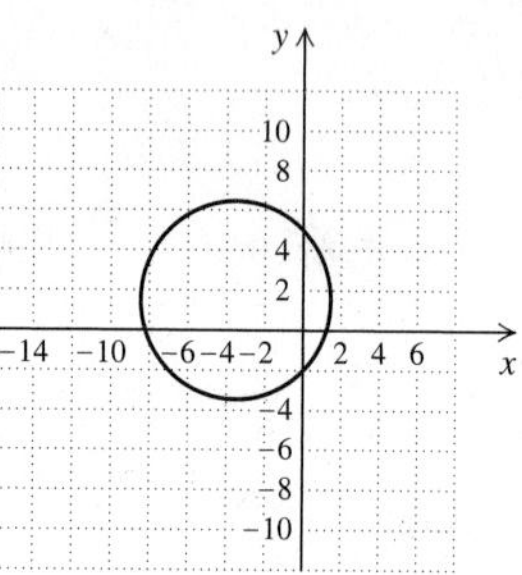

$x^2 + y^2 + 7x - 3y - 10 = 0$

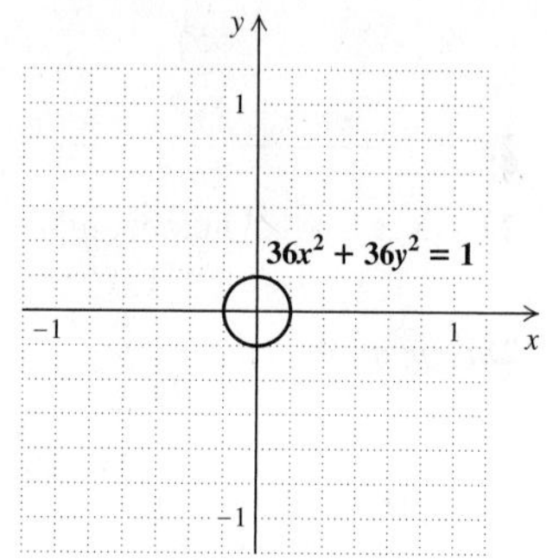

61. $x^2 + y^2 - 16 = 0$ **63.** $x^2 + y^2 + 14x - 16y + 54 = 0$

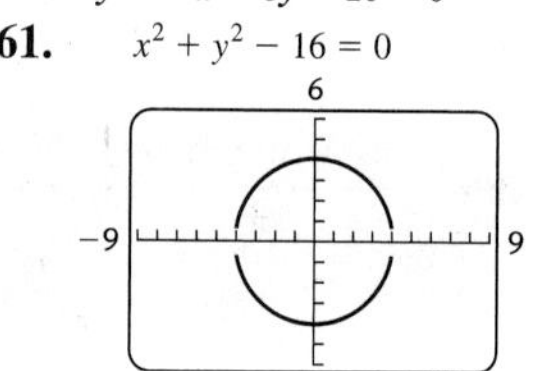

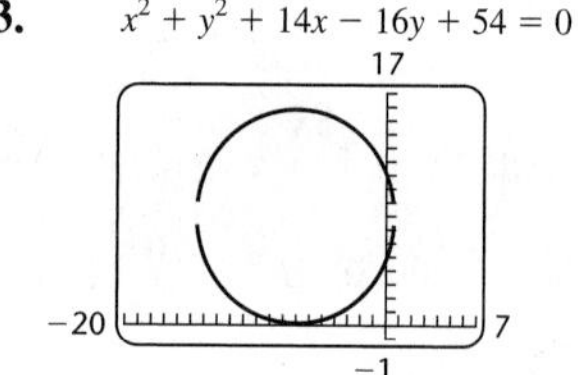

65. TW **67.** ± 4 **68.** $\pm a$ **69.** $-4, 6$ **70.** $-5 \pm 2\sqrt{3}$
71. $-3 \pm 3\sqrt{3}$ **72.** $2 \pm \frac{4\sqrt{2}}{3}$ **73.** TW
75. $(x - 3)^2 + (y + 5)^2 = 9$ **77.** $(x - 3)^2 + y^2 = 25$
79. $(0, 4)$ **81.** $\frac{17}{4}\pi$ m^2, or approximately 13.4 m^2
83. 7169 mm **85.** **(a)** $(0, -3)$; **(b)** 5 ft
87. $x^2 + (y - 30.6)^2 = 590.49$

Exercise Set 13.2, pp. 952–955

1. True **2.** False **3.** False **4.** False
5. True **6.** True **7.** True **8.** True

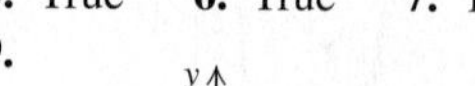

9. **11.**

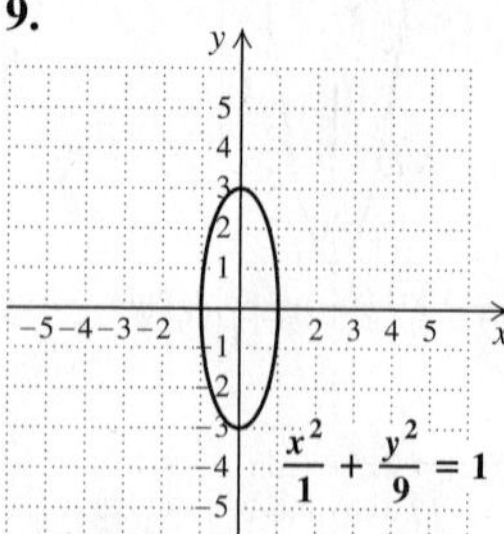

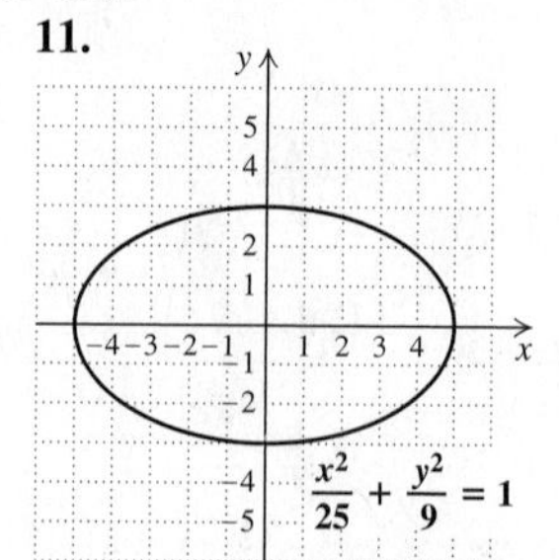

13.

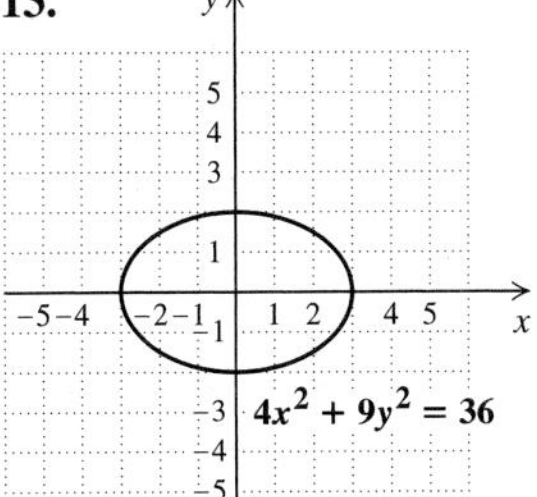

$4x^2 + 9y^2 = 36$

15.

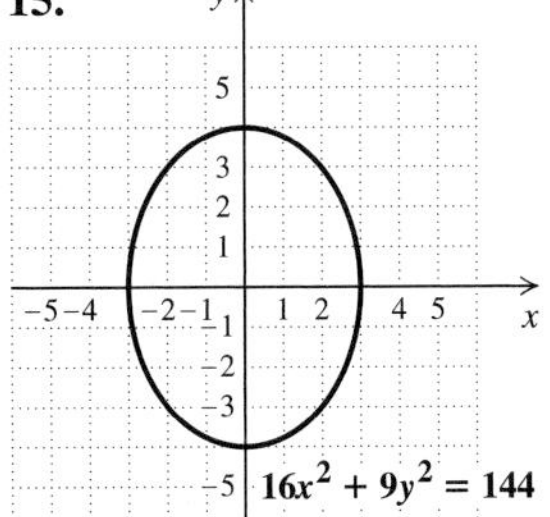

$16x^2 + 9y^2 = 144$

17.

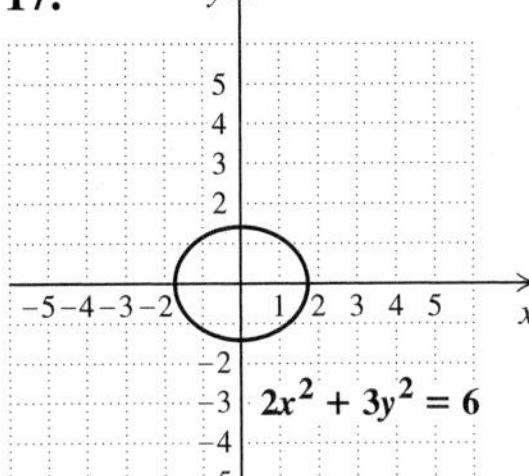

$2x^2 + 3y^2 = 6$

19.

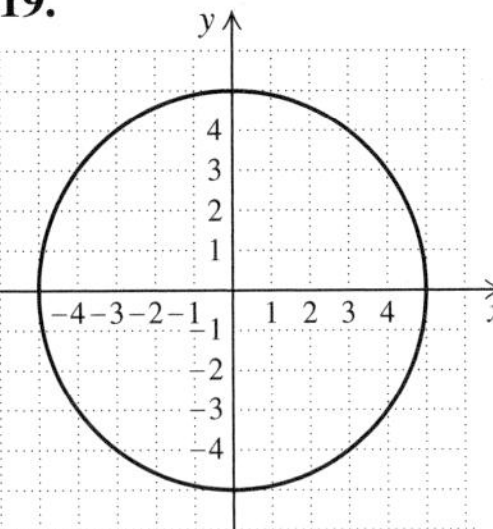

$5x^2 + 5y^2 = 125$

21.

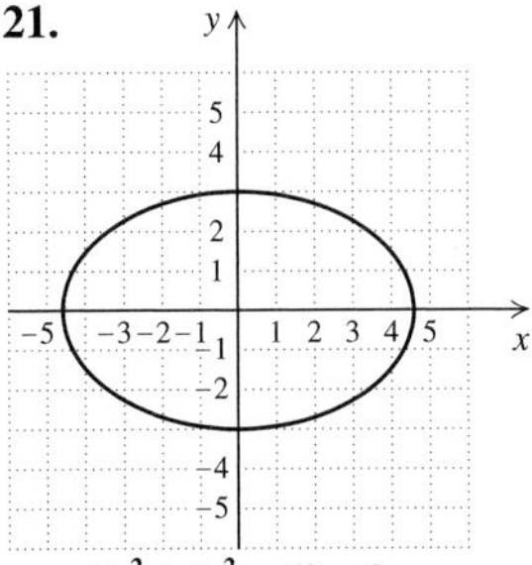

$3x^2 + 7y^2 - 63 = 0$

23.

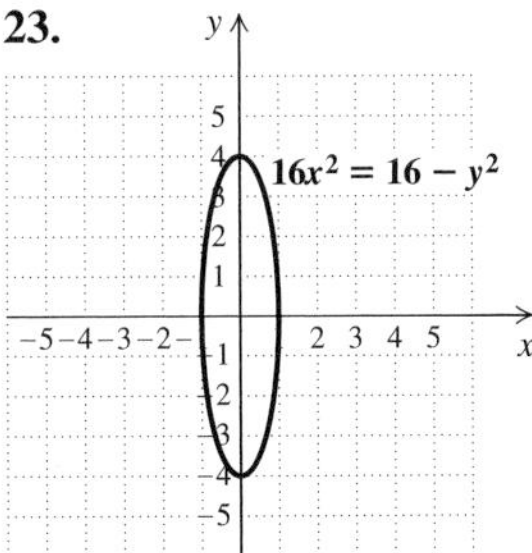

$16x^2 = 16 - y^2$

25.

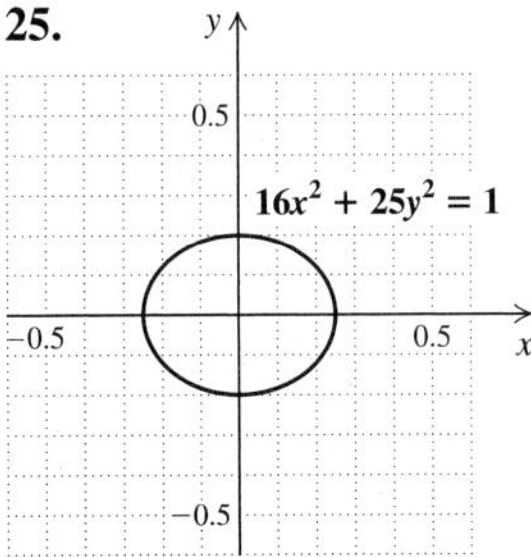

$16x^2 + 25y^2 = 1$

27.

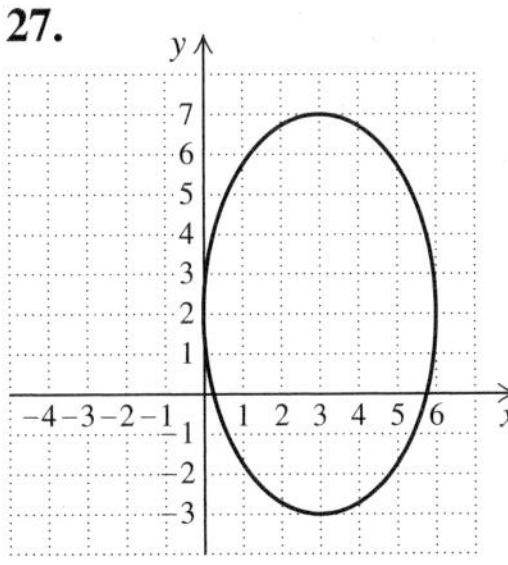

$\frac{(x-3)^2}{9} + \frac{(y-2)^2}{25} = 1$

29.

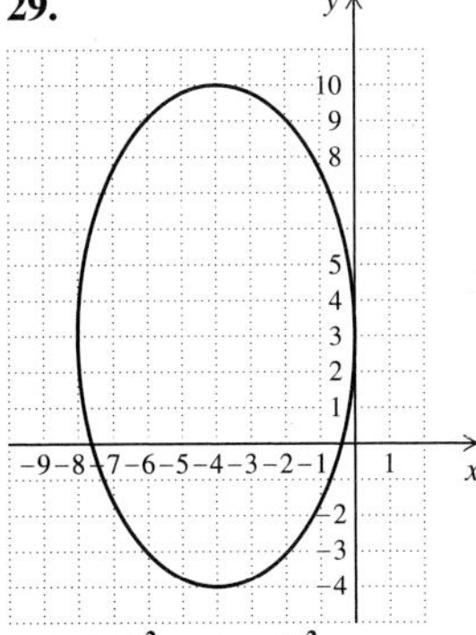

$\frac{(x+4)^2}{16} + \frac{(y-3)^2}{49} = 1$

31.

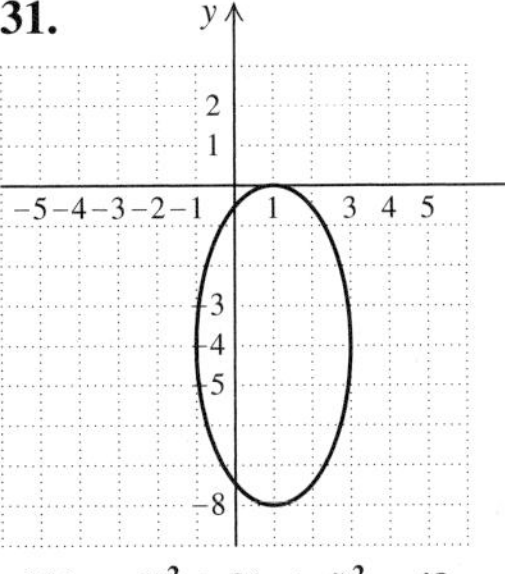

$12(x-1)^2 + 3(y+4)^2 = 48$

33.

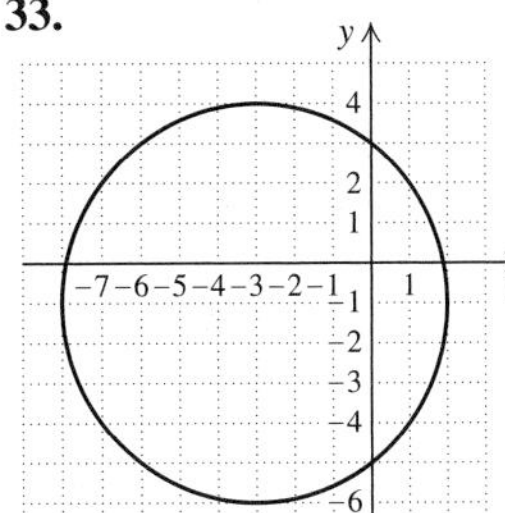

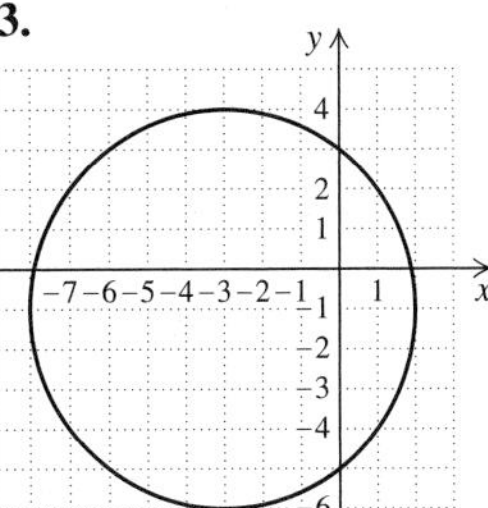

$4(x+3)^2 + 4(y+1)^2 - 10 = 90$

35. TW

37. $\frac{5}{2} \pm \frac{\sqrt{13}}{2}$

38. 3

39. $-\frac{3}{4}, 2$

40. $\frac{5}{2}$

41. $-\sqrt{11}, \sqrt{11}$

42. $-10, 6$

43. TW

45. $\frac{x^2}{81} + \frac{y^2}{121} = 1$

47. $\frac{(x-2)^2}{16} + \frac{(y+1)^2}{9} = 1$

49. $\frac{x^2}{9} + \frac{y^2}{25} = 1$

51. **(a)** Let $F_1 = (-c, 0)$ and $F_2 = (c, 0)$. Then the sum of the distances from the foci to P is $2a$. By the distance formula,

$$\sqrt{(x+c)^2 + y^2} + \sqrt{(x-c)^2 + y^2} = 2a, \text{ or}$$
$$\sqrt{(x+c)^2 + y^2} = 2a - \sqrt{(x-c)^2 + y^2}.$$

Squaring, we get

$$(x+c)^2 + y^2 = 4a^2 - 4a\sqrt{(x-c)^2 + y^2} + (x-c)^2 + y^2,$$

or

$$x^2 + 2cx + c^2 + y^2 = 4a^2 - 4a\sqrt{(x-c)^2 + y^2} + x^2 - 2cx + c^2 + y^2.$$

Thus

$$-4a^2 + 4cx = -4a\sqrt{(x-c)^2 + y^2}$$
$$a^2 - cx = a\sqrt{(x-c)^2 + y^2}.$$

Squaring again, we get

$$a^4 - 2a^2cx + c^2x^2 = a^2(x^2 - 2cx + c^2 + y^2)$$
$$a^4 - 2a^2cx + c^2x^2 = a^2x^2 - 2a^2cx + a^2c^2 + a^2y^2,$$

or

$$x^2(a^2 - c^2) + a^2y^2 = a^2(a^2 - c^2)$$
$$\frac{x^2}{a^2} + \frac{y^2}{a^2 - c^2} = 1.$$

(b) When P is at $(0, b)$, it follows that $b^2 = a^2 - c^2$. Substituting, we have

$$\frac{x^2}{a^2} + \frac{y^2}{b^2} = 1.$$

53. 5.66 ft

55.

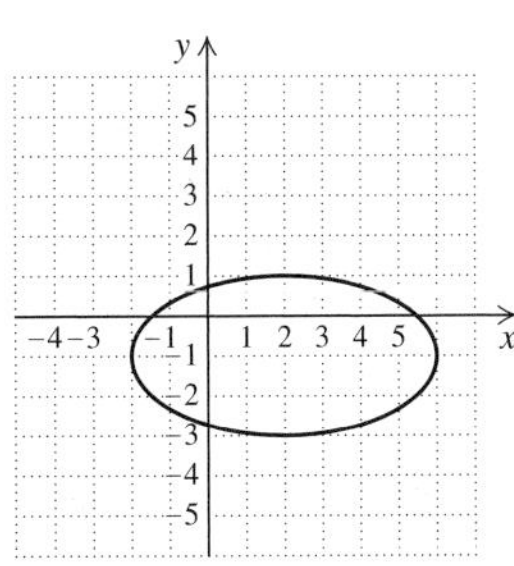

$\frac{(x-2)^2}{16} + \frac{(y+1)^2}{4} = 1$

57. 152.1 million km

Visualizing for Success, p. 963

1. C **2.** A **3.** F **4.** B **5.** J **6.** D **7.** H
8. I **9.** G **10.** E

Exercise Set 13.3, pp. 964–965

1. (d) **2.** (f) **3.** (h) **4.** (a) **5.** (g) **6.** (b)
7. (c) **8.** (e)

9.

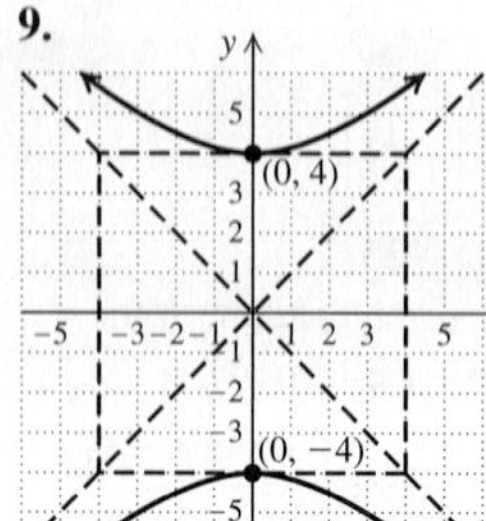

$\frac{y^2}{16} - \frac{x^2}{16} = 1$

11.

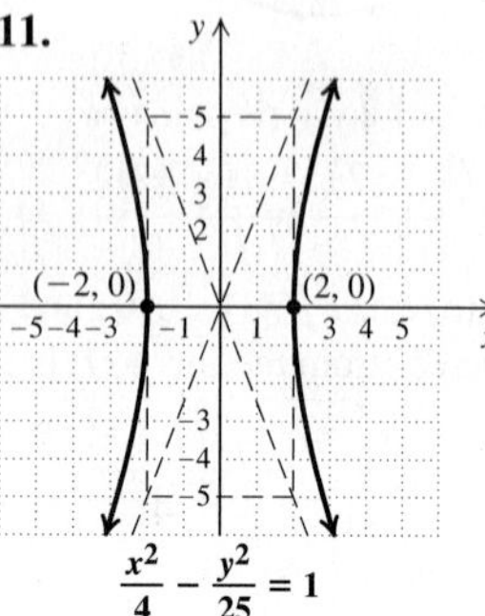

$\frac{x^2}{4} - \frac{y^2}{25} = 1$

13.

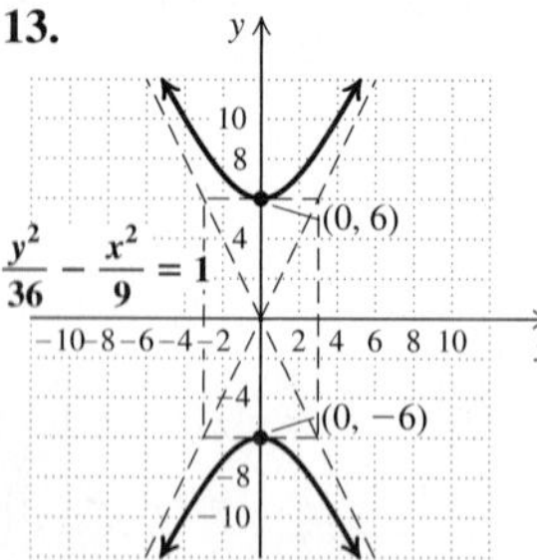

$\frac{y^2}{36} - \frac{x^2}{9} = 1$

15.

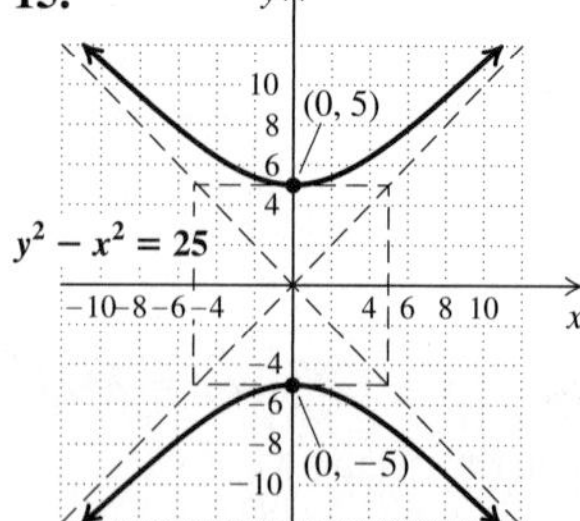

$y^2 - x^2 = 25$

17.

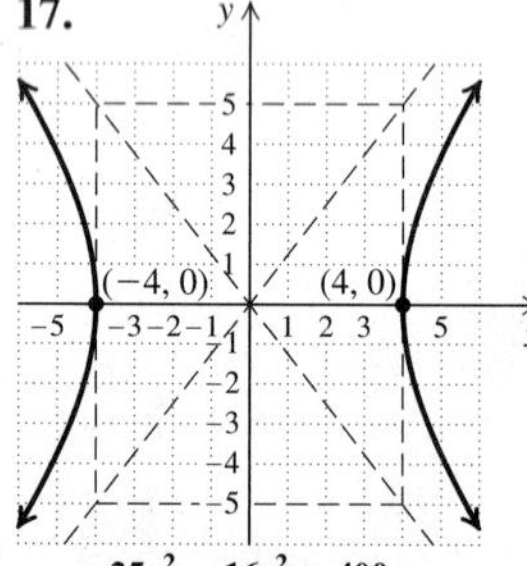

$25x^2 - 16y^2 = 400$

19.

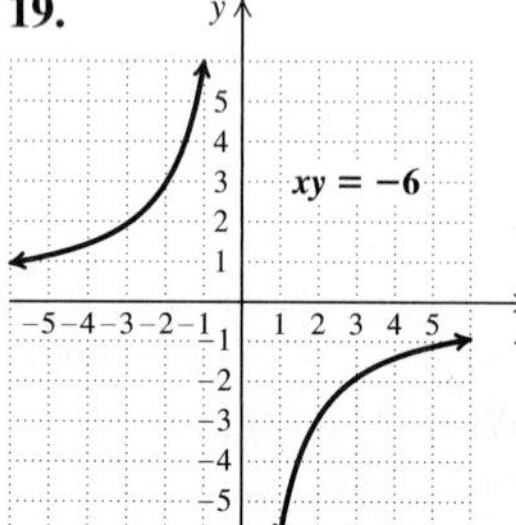

$xy = -6$

21.

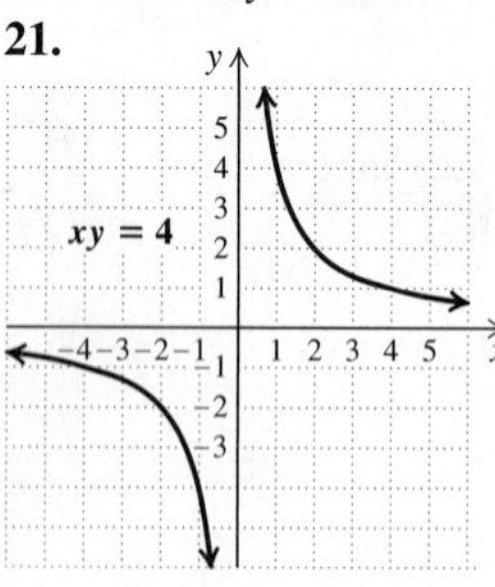

$xy = 4$

23.

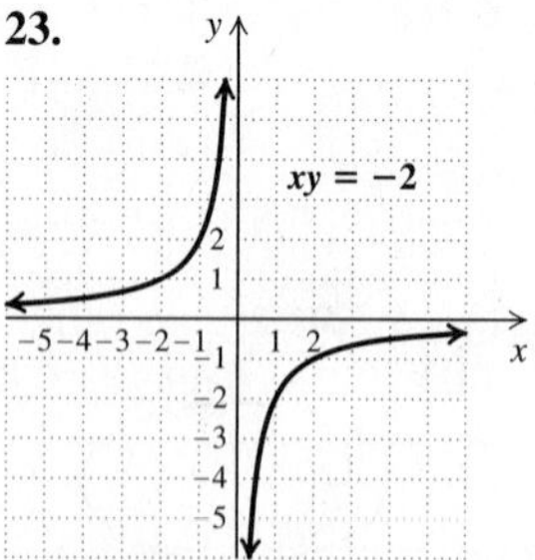

$xy = -2$

25.

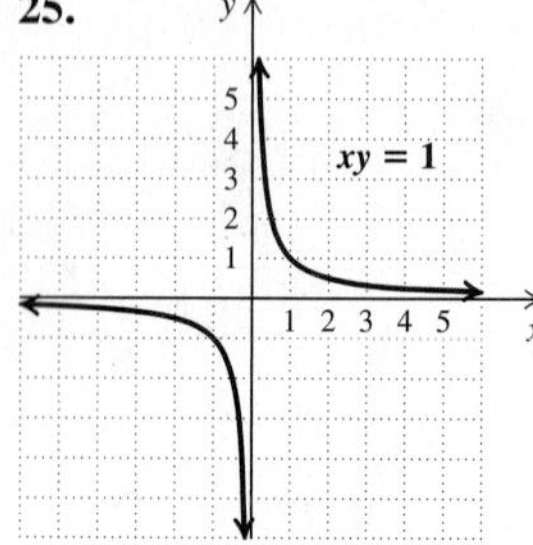

$xy = 1$

27. Circle
29. Ellipse
31. Hyperbola
33. Circle
35. Parabola
37. Hyperbola
39. Parabola
41. Hyperbola
43. Circle
45. Ellipse

47. TW **49.** $(-3, 6)$ **50.** $\left(\frac{1}{2}, -\frac{3}{2}\right)$ **51.** $-2, 2$

52. $-4, \frac{2}{3}$ **53.** $\frac{3}{2} \pm \frac{\sqrt{13}}{2}$ **54.** $\pm 1, \pm 5$ **55.** TW

57. $\frac{y^2}{36} - \frac{x^2}{4} = 1$ **59.** C: $(5, 2)$; V: $(-1, 2), (11, 2)$; asymptotes: $y - 2 = \frac{5}{6}(x - 5), y - 2 = -\frac{5}{6}(x - 5)$

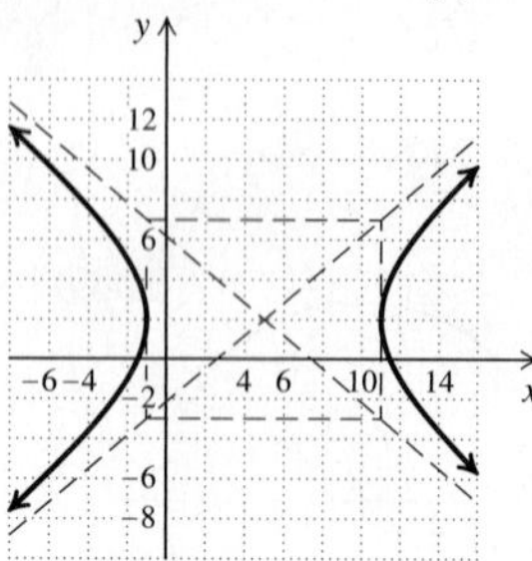

$\frac{(x-5)^2}{36} - \frac{(y-2)^2}{25} = 1$

61. $\frac{(y+3)^2}{4} - \frac{(x-4)^2}{16} = 1$; C: $(4, -3)$; V: $(4, -5), (4, -1)$; asymptotes: $y + 3 = \frac{1}{2}(x - 4), y + 3 = -\frac{1}{2}(x - 4)$

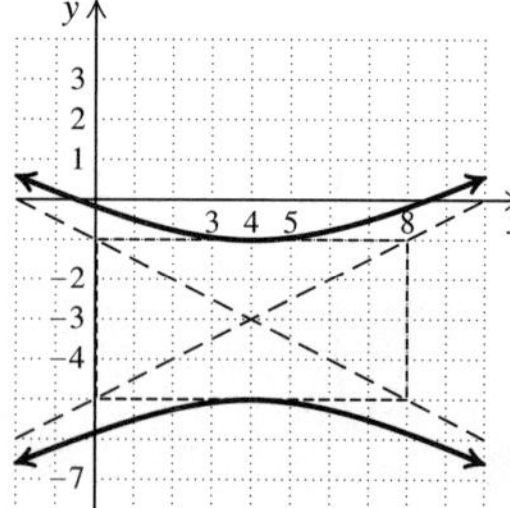

$8(y+3)^2 - 2(x-4)^2 = 32$

63. $\frac{(x+3)^2}{1} - \frac{(y-2)^2}{4} = 1$; C: $(-3, 2)$; V: $(-4, 2), (-2, 2)$; asymptotes: $y - 2 = 2(x + 3), y - 2 = -2(x + 3)$

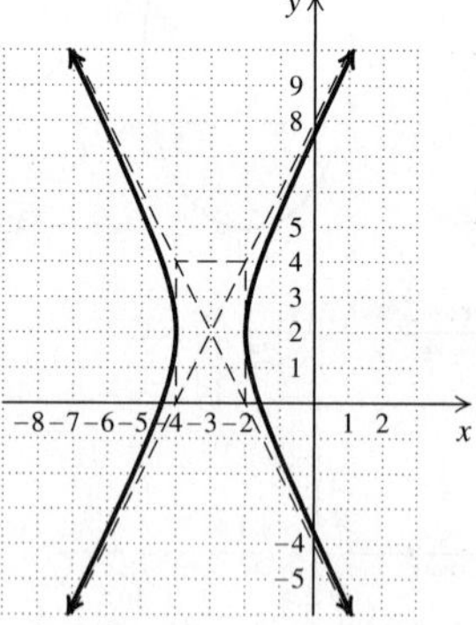

$4x^2 - y^2 + 24x + 4y + 28 = 0$

Mid-Chapter Review: Chapter 13, pp. 966–967

Guided Solutions

1.

$$(x^2 - 4x) + (y^2 + 2y) = 6$$
$$(x^2 - 4x + 4) + (y^2 + 2y + 1) = 6 + 4 + 1$$
$$(x - 2)^2 + (y + 1)^2 = 11$$

The center of the circle is $(2, -1)$. The radius is $\sqrt{11}$.

2. **1.** Is there both an x^2-term and a y^2-term? Yes
2. Do both the x^2-term and the y^2-term have the same sign? No
3. The graph of the equation is a hyperbola.

Mixed Review

1. $(4, 1); x = 4$ **2.** $(2, -1); y = -1$ **3.** $(3, 2)$
4. $(-3, -5)$ **5.** $(-12, 0), (12, 0), (0, -9), (0, 9)$
6. $(-3, 0), (3, 0)$ **7.** $(0, -1), (0, 1)$ **8.** $y = \frac{3}{2}x, y = -\frac{3}{2}x$

9. Circle

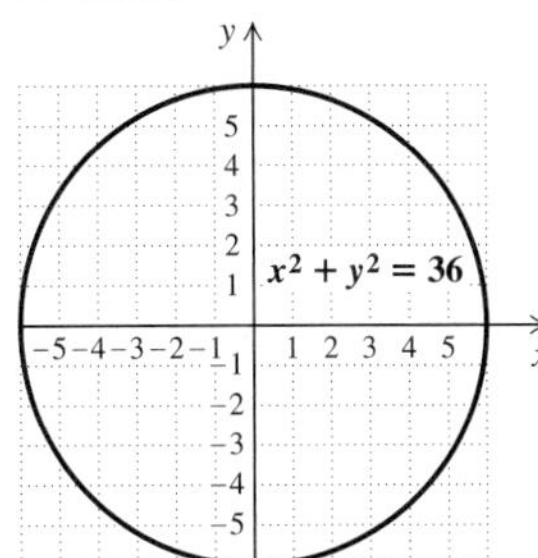

10. Parabola

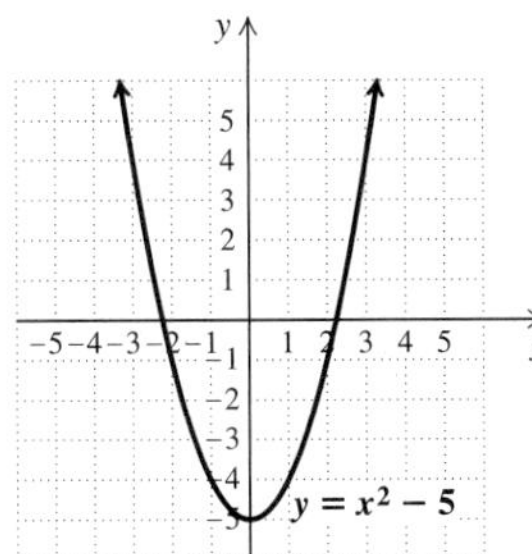

11. Ellipse

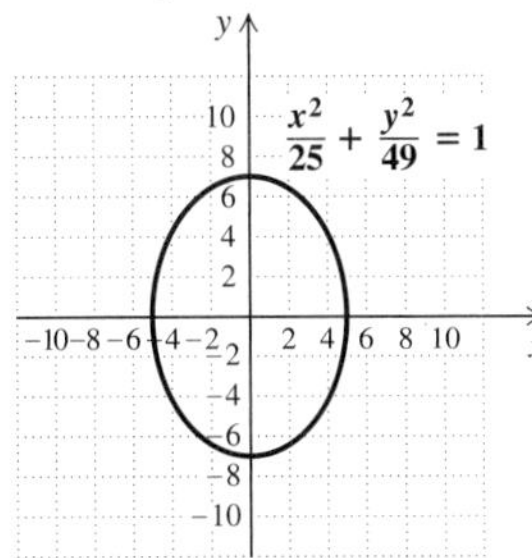

12. Hyperbola

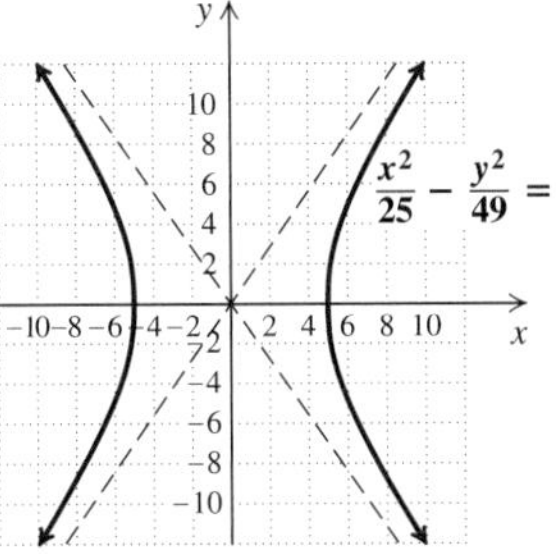

13. Parabola

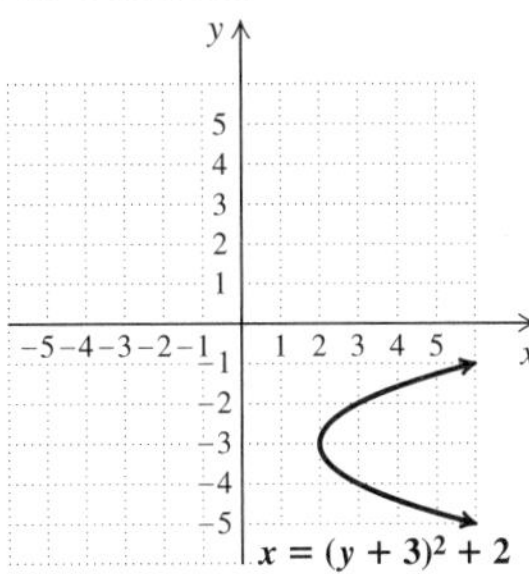

14. Ellipse

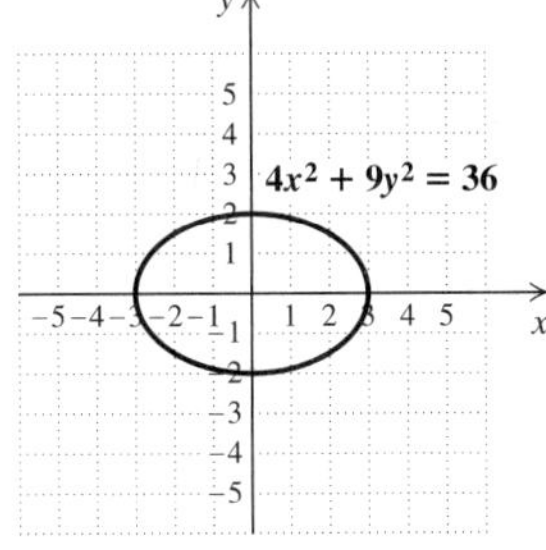

15. Hyperbola

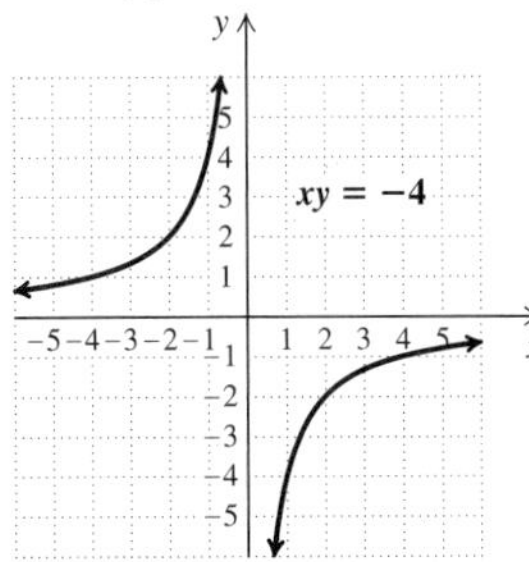

16. Circle

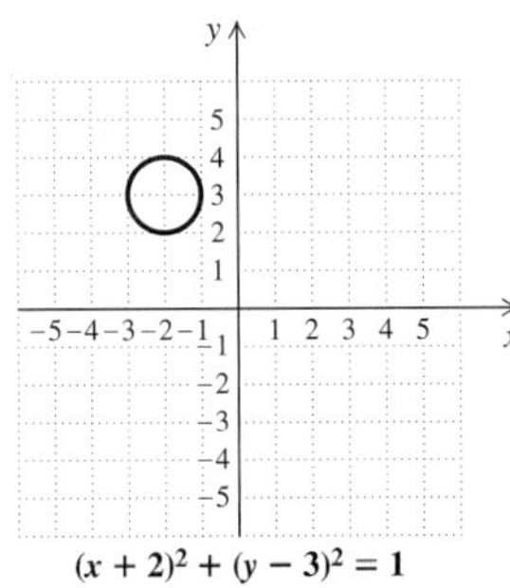

17. Circle

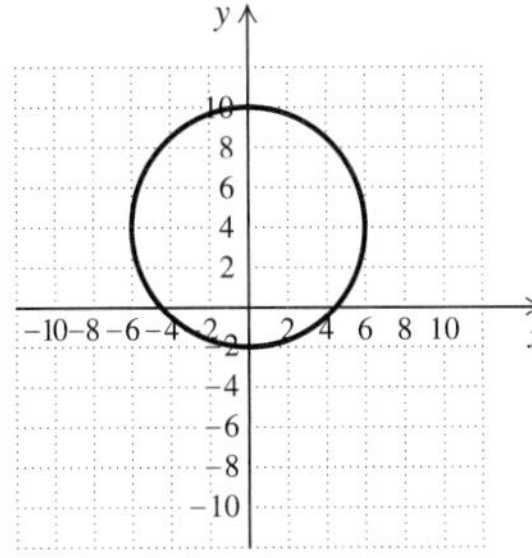

18. Parabola

$x = y^2 + 2y$

19. Hyperbola

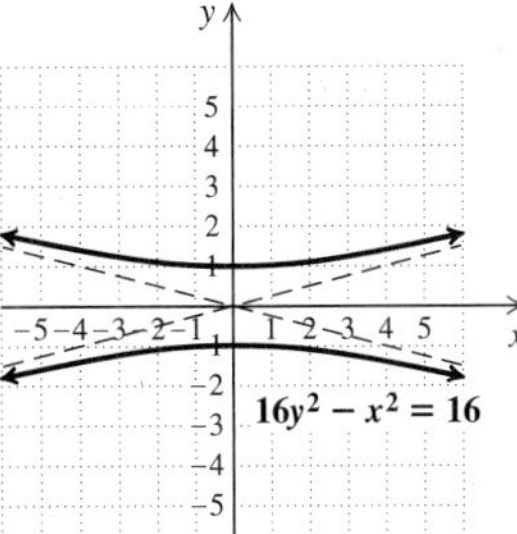

20. Hyperbola

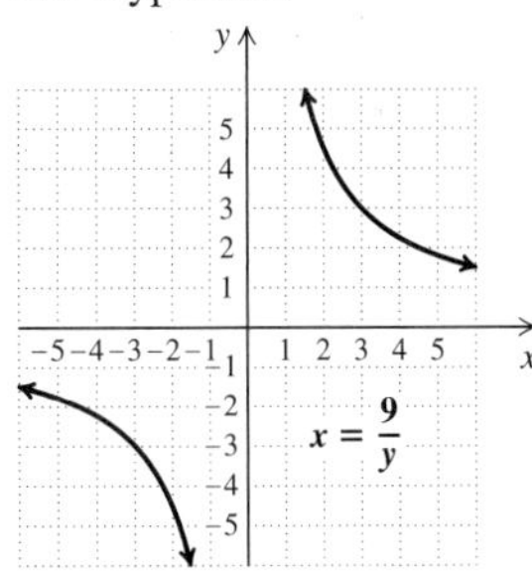

Exercise Set 13.4, pp. 973–975

1. True **2.** True **3.** False **4.** False **5.** True
6. True **7.** $(-4, -3), (3, 4)$ **9.** $(0, 2), (3, 0)$ **11.** $(-2, 1)$
13. $\left(\frac{5 + \sqrt{70}}{3}, \frac{-1 + \sqrt{70}}{3}\right), \left(\frac{5 - \sqrt{70}}{3}, \frac{-1 - \sqrt{70}}{3}\right)$
15. $\left(4, \frac{3}{2}\right), (3, 2)$ **17.** $\left(\frac{7}{3}, \frac{1}{3}\right), (1, -1)$ **19.** $\left(\frac{11}{4}, -\frac{5}{4}\right), (1, 4)$
21. $(2, 4), (4, 2)$ **23.** $(3, -5), (-1, 3)$ **25.** $(-5, -8), (8, 5)$
27. $(0, 0), (1, 1), \left(-\frac{1}{2} + \frac{\sqrt{3}}{2}i, -\frac{1}{2} - \frac{\sqrt{3}}{2}i\right),$
$\left(-\frac{1}{2} - \frac{\sqrt{3}}{2}i, -\frac{1}{2} + \frac{\sqrt{3}}{2}i\right)$ **29.** $(-3, 0), (3, 0)$
31. $(-4, -3), (-3, -4), (3, 4), (4, 3)$
33. $\left(\frac{16}{3}, \frac{5\sqrt{7}}{3}i\right), \left(\frac{16}{3}, -\frac{5\sqrt{7}}{3}i\right), \left(-\frac{16}{3}, \frac{5\sqrt{7}}{3}i\right),$
$\left(-\frac{16}{3}, -\frac{5\sqrt{7}}{3}i\right)$ **35.** $(-3, -\sqrt{5}), (-3, \sqrt{5}), (3, -\sqrt{5}),$
$(3, \sqrt{5})$ **37.** $(4, 2), (-4, -2), (2, 4), (-2, -4)$
39. $(4, 1), (-4, -1), (2, 2), (-2, -2)$ **41.** $(2, 1), (-2, -1)$
43. $\left(2, -\frac{4}{5}\right), \left(-2, -\frac{4}{5}\right), (5, 2), (-5, 2)$
45. $(-\sqrt{2}, \sqrt{2}), (\sqrt{2}, -\sqrt{2})$ **47.** Length: 8 cm; width: 6 cm
49. Length: 2 yd; width: 1 yd **51.** Length: 12 ft; width: 5 ft
53. 6 and 15; -6 and -15 **55.** 24 ft, 16 ft
57. Length: $\sqrt{3}$ m; width: 1 m **59.** TW **61.** -9 **62.** -27
63. -1 **64.** $\frac{1}{5}$ **65.** 77 **66.** $\frac{21}{2}$ **67.** TW
69. $(x + 2)^2 + (y - 1)^2 = 4$
71. $(-2, 3), (2, -3), (-3, 2), (3, -2)$ **73.** Length: 55 ft; width: 45 ft **75.** 10 in. by 7 in. by 5 in. **77.** Length: 63.6 in.; height: 35.8 in. **79.** $(-1.50, -1.17); (3.50, 0.50)$

Study Summary: Chapter 13, pp. 976–977

1.

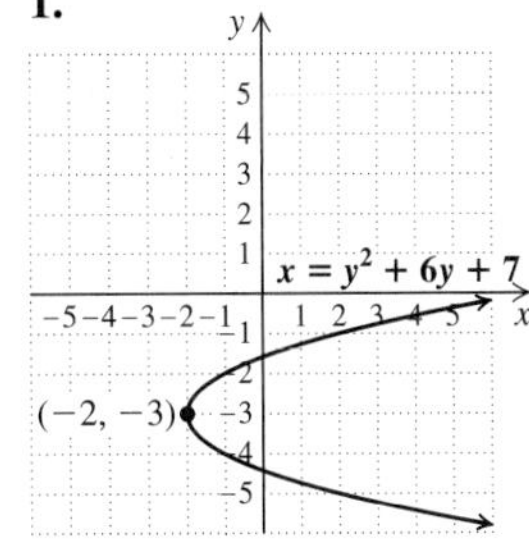

2. Center: $(3, 0)$; radius 2

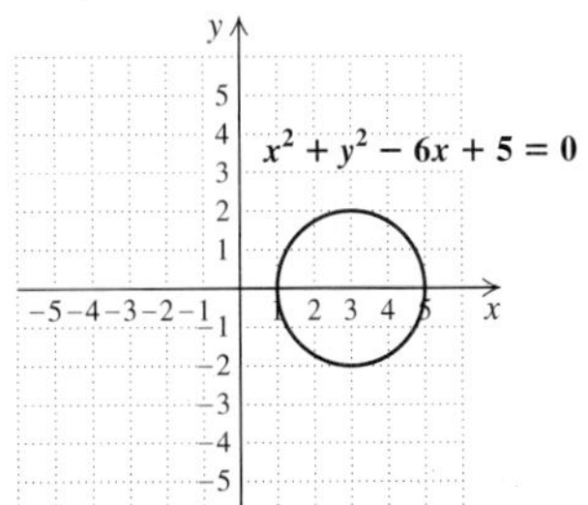

3.

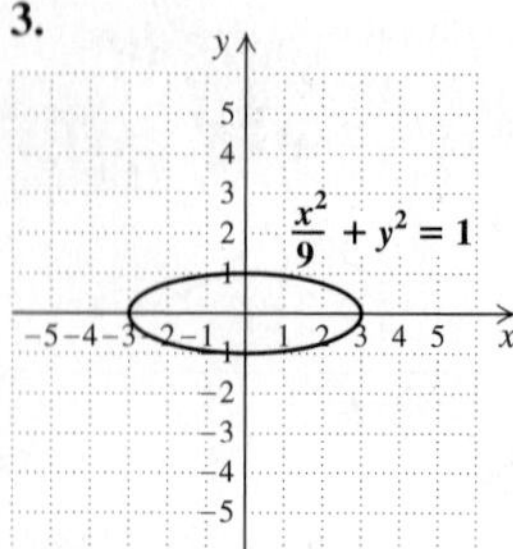

4.

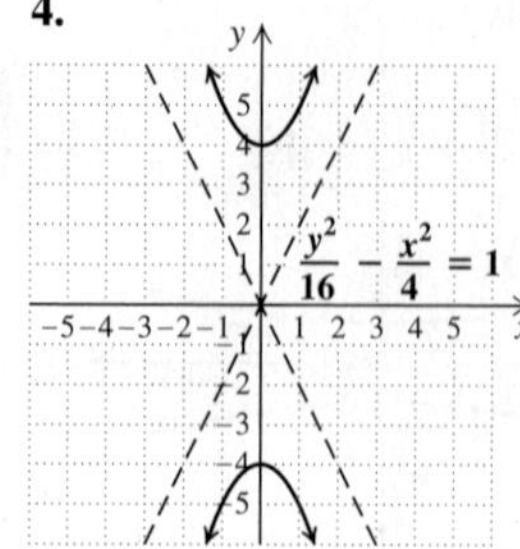

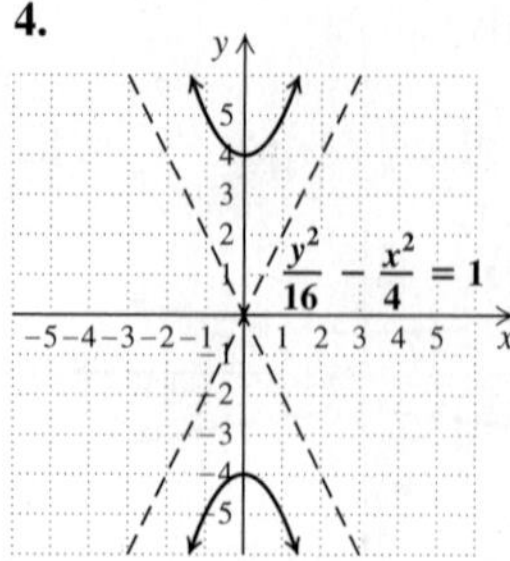

5. $(-5, -4), (4, 5)$

Review Exercises: Chapter 13, pp. 978–979

1. True **2.** False **3.** False **4.** True **5.** True **6.** True **7.** False **8.** True **9.** $(-3, 2), 4$ **10.** $(5, 0), \sqrt{11}$ **11.** $(3, 1), 3$ **12.** $(-4, 3), 3\sqrt{5}$ **13.** $(x + 4)^2 + (y - 3)^2 = 16$ **14.** $(x - 7)^2 + (y + 2)^2 = 20$

15. Circle

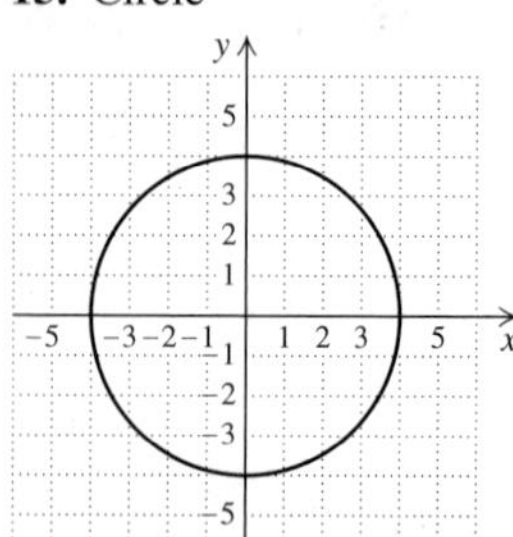

$5x^2 + 5y^2 = 80$

16. Ellipse

$9x^2 + 2y^2 = 18$

17. Parabola

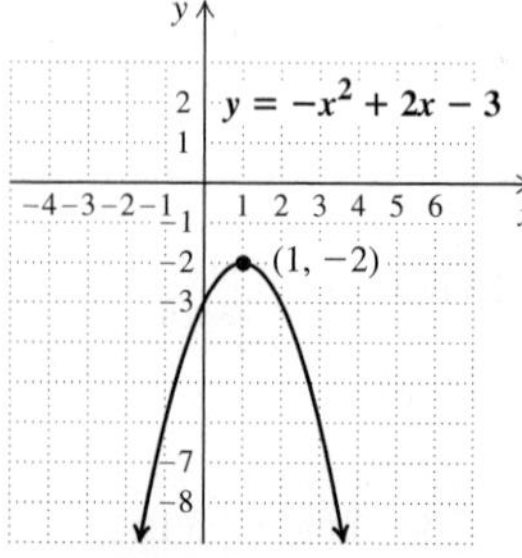

18. Hyperbola

$\frac{y^2}{9} - \frac{x^2}{4} = 1$

19. Hyperbola

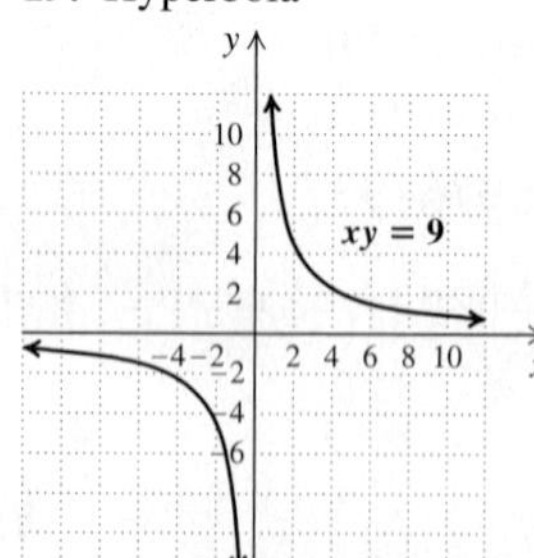

20. Parabola

$x = y^2 + 2y - 2$

21. Ellipse

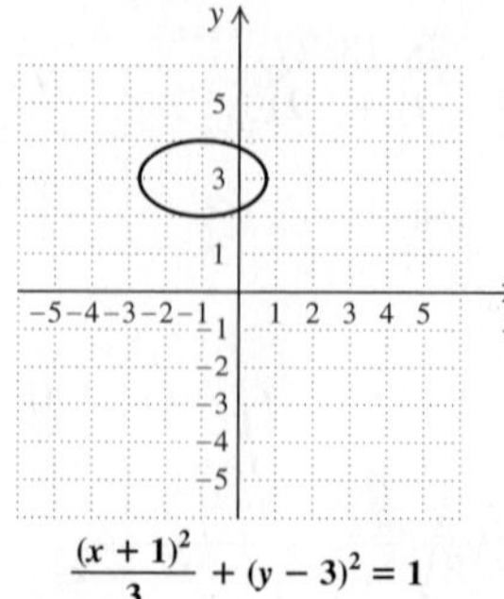

$\frac{(x + 1)^2}{3} + (y - 3)^2 = 1$

22. Circle

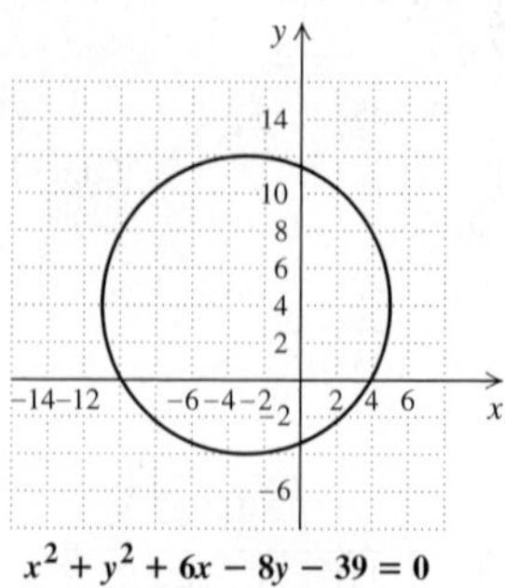

$x^2 + y^2 + 6x - 8y - 39 = 0$

23. $(5, -2)$ **24.** $(2, 2), \left(\frac{32}{9}, -\frac{10}{9}\right)$ **25.** $(0, -5), (2, -1)$ **26.** $(4, i), (4, -i), (-4, i), (-4, -i)$ **27.** $(2, 1), (\sqrt{3}, 0), (-2, 1), (-\sqrt{3}, 0)$ **28.** $(3, -3), \left(-\frac{3}{5}, \frac{21}{5}\right)$ **29.** $(6, 8), (6, -8), (-6, 8), (-6, -8)$ **30.** $(2, 2), (-2, -2), (2\sqrt{2}, \sqrt{2}), (-2\sqrt{2}, -\sqrt{2})$ **31.** Length: 12 m; width: 7 m **32.** Length: 12 in.; width: 9 in. **33.** 32 cm, 20 cm **34.** 3 ft, 11 ft **35.** TW The graph of a parabola has one branch whereas the graph of a hyperbola has two branches. A hyperbola has asymptotes, but a parabola does not. **36.** TW Function notation rarely appears in this chapter because many of the relations are not functions. Function notation could be used for vertical parabolas and for hyperbolas that have the axes as asymptotes. **37.** $(-5, -4\sqrt{2}), (-5, 4\sqrt{2}), (3, -2\sqrt{2}), (3, 2\sqrt{2})$ **38.** $(0, 6), (0, -6)$ **39.** $(x - 2)^2 + (y + 1)^2 = 25$ **40.** $\frac{x^2}{81} + \frac{y^2}{25} = 1$ **41.** $\left(\frac{9}{4}, 0\right)$

Test: Chapter 13, p. 979

1. [13.1] $(x - 3)^2 + (y + 4)^2 = 12$ **2.** [13.1] $(4, -1), \sqrt{5}$ **3.** [13.1] $(-2, 3), 3$

4. [13.1], [13.3] Parabola

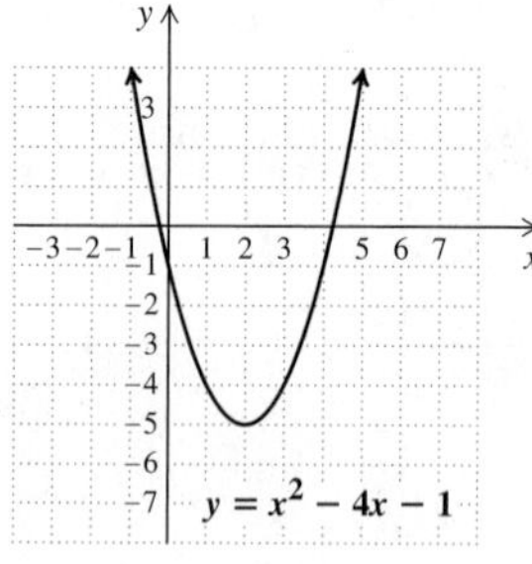

5. [13.1], [13.3] Circle

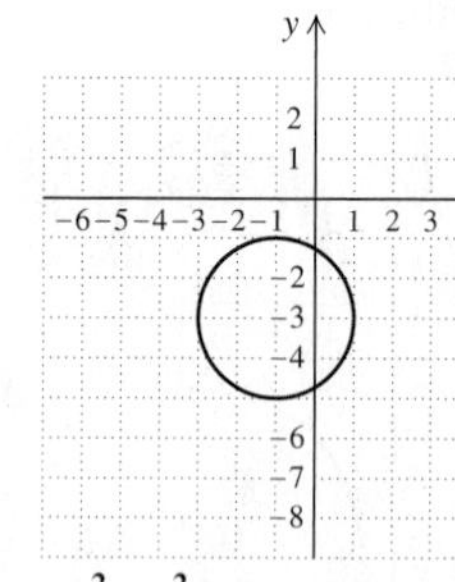

$x^2 + y^2 + 2x + 6y + 6 = 0$

6. [13.3] Hyperbola

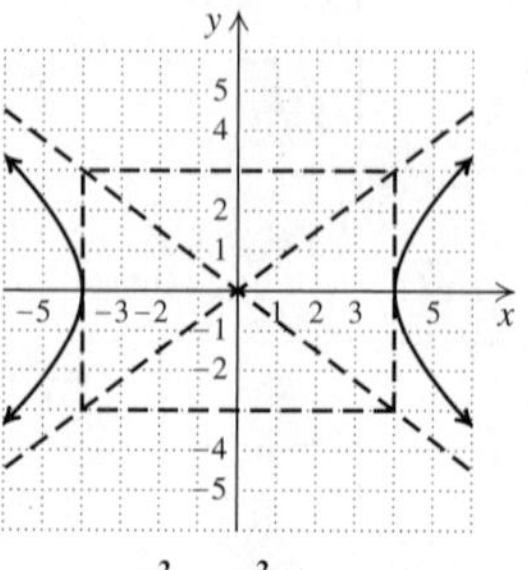

$\frac{x^2}{16} - \frac{y^2}{9} = 1$

7. [13.2], [13.3] Ellipse

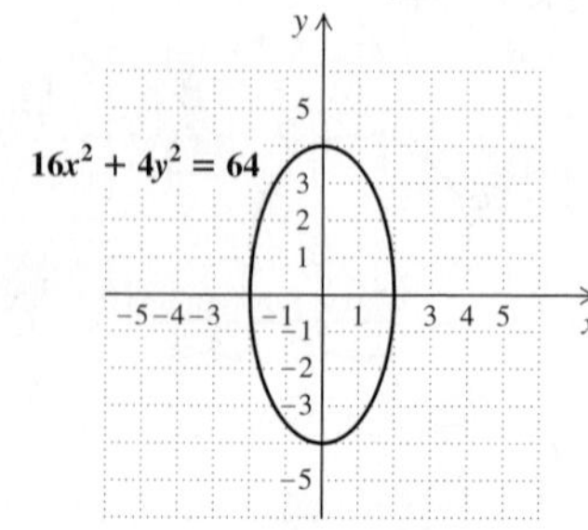

8. [13.3] Hyperbola

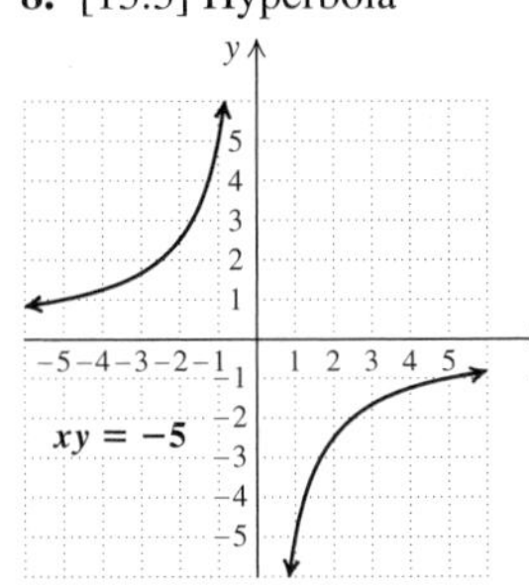

9. [13.1], [13.3] Parabola

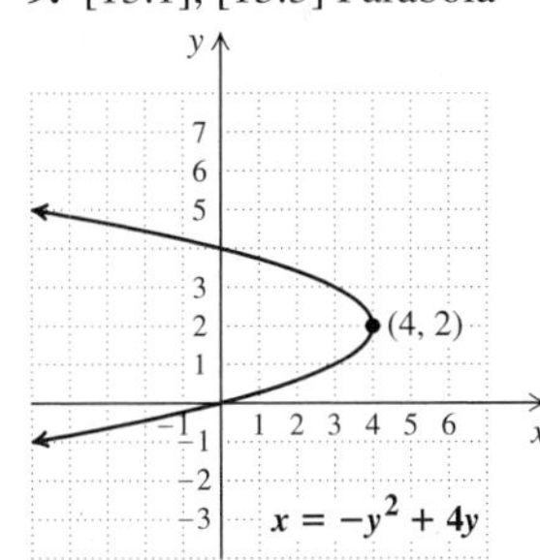

10. [13.4] $(0, 6), \left(\frac{144}{25}, \frac{42}{25}\right)$ **11.** [13.4] $(-4, 13), (2, 1)$
12. [13.4] $(2, 1), (-2, -1), (i, -2i), (-i, 2i)$
13. [13.4] $(\sqrt{6}, 2), (\sqrt{6}, -2), (-\sqrt{6}, 2), (-\sqrt{6}, -2)$
14. [13.4] 2 in. by 11 in. **15.** [13.4] $\sqrt{5}$ m, $\sqrt{3}$ m
16. [13.4] Length: 32 ft; width: 24 ft **17.** [13.4] \$1200, 6%
18. [13.2] $\dfrac{(x-6)^2}{25} + \dfrac{(y-3)^2}{9} = 1$ **19.** [13.1] $\left(0, -\frac{31}{4}\right)$
20. [13.4] 9 **21.** [13.2] $\dfrac{x^2}{16} + \dfrac{y^2}{49} = 1$

Chapter 14

Exercise Set 14.1, pp. 988–990

1. (f) **2.** (a) **3.** (d) **4.** (b) **5.** (c) **6.** (e)
7. 13 **9.** 364 **11.** -23.5 **13.** -363 **15.** $\dfrac{441}{400}$
17. 5, 7, 9, 11; 23; 33 **19.** 3, 6, 11, 18; 102; 227
21. $\frac{1}{2}, \frac{2}{3}, \frac{3}{4}, \frac{4}{5}; \frac{10}{11}; \frac{15}{16}$ **23.** $1, -\frac{1}{2}, \frac{1}{4}, -\frac{1}{8}; -\frac{1}{512}; \frac{1}{16{,}384}$
25. $-1, \frac{1}{2}, -\frac{1}{3}, \frac{1}{4}; \frac{1}{10}; -\frac{1}{15}$ **27.** 0, 7, -26, 63; 999; -3374
29. $-3, -1, 1, 3, 5$ **31.** $-1, -1, 1, 5, 11$ **33.** $\frac{1}{8}, \frac{4}{25}, \frac{1}{6}, \frac{8}{49}, \frac{5}{32}$
35. $2n$ **37.** $(-1)^{n+1}$ **39.** $(-1)^n \cdot n$ **41.** $2n + 1$
43. $n^2 - 1$, or $(n+1)(n-1)$ **45.** $\dfrac{n}{n+1}$ **47.** 5^n
49. $(-1)^n \cdot n^2$ **51.** 4 **53.** 30 **55.** $2, 3, \frac{11}{3}, \frac{25}{6}$
57. $-1, 3, -6, 10$ **59.** $\frac{1}{2} + \frac{1}{4} + \frac{1}{6} + \frac{1}{8} + \frac{1}{10} = \frac{137}{120}$
61. $10^0 + 10^1 + 10^2 + 10^3 + 10^4 = 11{,}111$
63. $2 + \frac{3}{2} + \frac{4}{3} + \frac{5}{4} + \frac{6}{5} + \frac{7}{6} + \frac{8}{7} = \frac{1343}{140}$
65. $(-1)^2 2^1 + (-1)^3 2^2 + (-1)^4 2^3 + (-1)^5 2^4 + (-1)^6 2^5 + (-1)^7 2^6 + (-1)^8 2^7 + (-1)^9 2^8 = -170$
67. $(0^2 - 2 \cdot 0 + 3) + (1^2 - 2 \cdot 1 + 3) + (2^2 - 2 \cdot 2 + 3) + (3^2 - 2 \cdot 3 + 3) + (4^2 - 2 \cdot 4 + 3) + (5^2 - 2 \cdot 5 + 3) = 43$
69. $\dfrac{(-1)^3}{3 \cdot 4} + \dfrac{(-1)^4}{4 \cdot 5} + \dfrac{(-1)^5}{5 \cdot 6} = -\dfrac{1}{15}$
71. $\sum_{k=1}^{5} \dfrac{k+1}{k+2}$ **73.** $\sum_{k=1}^{6} k^2$ **75.** $\sum_{k=2}^{n} (-1)^k k^2$
77. $\sum_{k=1}^{\infty} 5k$ **79.** $\sum_{k=1}^{\infty} \dfrac{1}{k(k+1)}$

81. $u = 3n + 1$

Yscl = 5

83. $u = (-1)^\wedge n(n^2)$

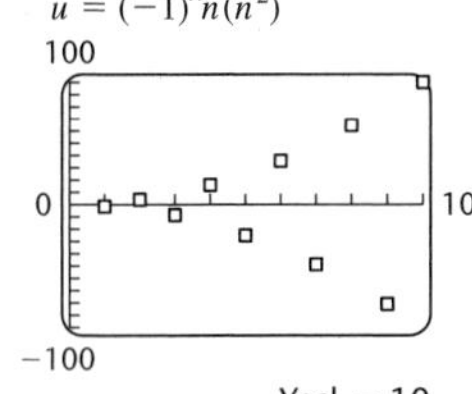

Yscl = 10

85. $u = 1/n$

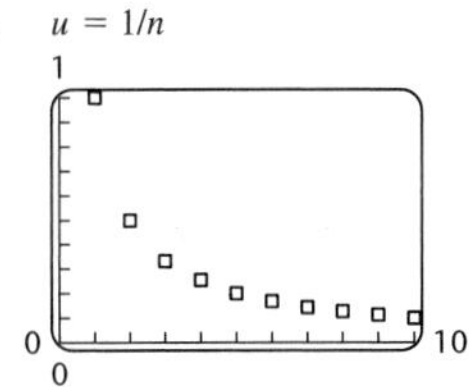

Yscl = 0.1

87. $u = (-1)^\wedge n/(n + 2)$

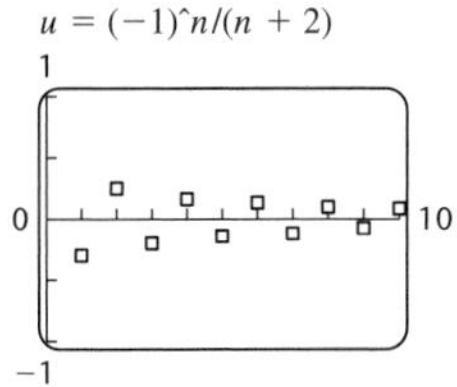

Yscl = 0.5

89. TW **91.** 98 **92.** -15 **93.** $a_1 + 4d$ **94.** $a_1 + a_n$
95. $3(a_1 + a_n)$, or $3a_1 + 3a_n$ **96.** d **97.** TW
99. 1, 3, 13, 63, 313, 1563 **101.** \$2500, \$2000, \$1600, \$1280, \$1024, \$819.20, \$655.36, \$524.29, \$419.43, \$335.54
103. $S_{100} = 0$; $S_{101} = -1$ **105.** $i, -1, -i, 1, i; i$
107. 11th term

Interactive Discovery, p. 993

1. The points lie on a straight line with positive slope.
2. The points lie on a straight line with positive slope.
3. The points lie on a straight line with negative slope.
4. The points lie on a straight line with negative slope.

Exercise Set 14.2, pp. 997–1000

1. True **2.** True **3.** False **4.** False **5.** True
6. True **7.** False **8.** False **9.** $a_1 = 2, d = 4$
11. $a_1 = 7, d = -4$ **13.** $a_1 = \frac{3}{2}, d = \frac{3}{4}$
15. $a_1 = \$5.12, d = \$0.12$ **17.** 49 **19.** -94
21. $-\$1628.16$ **23.** 26th **25.** 57th **27.** 82 **29.** 5
31. 28 **33.** $a_1 = 8; d = -3; 8, 5, 2, -1, -4$
35. $a_1 = 1; d = 1$ **37.** 780 **39.** 31,375 **41.** 2550
43. 918 **45.** 1030 **47.** 35 musicians; 315 musicians
49. 180 stones **51.** \$49.60 **53.** 722 seats **55.** TW
57. $y = \frac{1}{3}x + 10$ **58.** $y = -4x + 11$ **59.** $y = -2x + 10$
60. $y = -\frac{4}{3}x - \frac{16}{3}$ **61.** $x^2 + y^2 = 16$
62. $(x+2)^2 + (y-1)^2 = 20$ **63.** TW **65.** 33 jumps
67. \$8760, \$7961.77, \$7163.54, \$6365.31, \$5567.08; \$4768.85, \$3970.62, \$3172.39, \$2374.16, \$1575.93 **69.** Let d = the common difference. Since p, m, and q form an arithmetic sequence, $m = p + d$ and $q = p + 2d$. Then
$$\frac{p+q}{2} = \frac{p + (p + 2d)}{2} = p + d = m.$$
71. 156,375 **73.** Arithmetic; $a_n = 0.7n + 68.3$, where $n = 1$ corresponds to a ball speed of 100 mph, $n = 2$ corresponds to a ball speed of 101 mph, and so on **75.** Not arithmetic

Interactive Discovery, p. 1002

1. The points lie on an exponential curve with $a > 1$.
2. The points lie on an exponential curve with $a > 1$.
3. The points lie on an exponential curve with $0 < a < 1$.
4. The points lie on an exponential curve with $0 < a < 1$.

Interactive Discovery, p. 1005

1. (a) $\frac{1}{3}$; **(b)** 1, 1.33333, 1.44444, 1.48148, 1.49383, 1.49794, 1.49931, 1.49977, 1.49992, 1.49997 (sums are rounded to 5 decimal places); **(c)** 1.5 **2. (a)** $-\frac{7}{2}$; **(b)** 1, −2.5, 9.75, −33.125, 116.9375, −408.28125, 1429.984375, −5003.945313, 17514.80859, −61300.83008; **(c)** does not exist **3. (a)** 2; **(b)** 4, 12, 28, 60, 124, 252, 508, 1020, 2044, 4092; **(c)** does not exist **4. (a)** −0.4; **(b)** 1.3, 0.78, 0.988, 0.9048, 0.93808, 0.92477, 0.93009, 0.92796, 0.92881, 0.92847 (sums are rounded to 5 decimal places); **(c)** 0.928 **5.** S_∞ does not exist if $|r| > 1$.

Visualizing for Success, p. 1009

1. J **2.** G **3.** A **4.** H **5.** I **6.** B **7.** E **8.** D **9.** F **10.** C

Exercise Set 14.3, pp. 1010–1012

1. Geometric sequence **2.** Arithmetic sequence **3.** Arithmetic sequence **4.** Geometric sequence **5.** Geometric series **6.** Arithmetic series **7.** Geometric series **8.** None of these **9.** 2 **11.** −0.1 **13.** $-\frac{1}{2}$ **15.** $\frac{1}{5}$ **17.** $\frac{6}{m}$ **19.** 192 **21.** 243 **23.** 52,488 **25.** \$2331.64 **27.** $a_n = 5^{n-1}$ **29.** $a_n = (-1)^{n-1}$, or $a_n = (-1)^{n+1}$ **31.** $a_n = \frac{1}{x^n}$, or $a_n = x^{-n}$ **33.** 3066 **35.** $\frac{547}{18}$ **37.** $\frac{1-x^8}{1-x}$, or $(1+x)(1+x^2)(1+x^4)$ **39.** \$5134.51 **41.** 27 **43.** $\frac{49}{4}$ **45.** No **47.** No **49.** $\frac{43}{99}$ **51.** \$25,000 **53.** $\frac{7}{9}$ **55.** $\frac{830}{99}$ **57.** $\frac{5}{33}$ **59.** $\frac{5}{1024}$ ft **61.** 155,797 **63.** 2710 flies **65.** Approximately 179.9 billion coffees **67.** 3100.35 ft **69.** 20.48 in. **71.** Arithmetic **73.** Geometric **75.** Geometric **77.** TW **79.** $x^2 + 2xy + y^2$ **80.** $x^3 + 3x^2y + 3xy^2 + y^3$ **81.** $x^3 - 3x^2y + 3xy^2 - y^3$ **82.** $x^4 - 4x^3y + 6x^2y^2 - 4xy^3 + y^4$ **83.** $8x^3 + 12x^2y + 6xy^2 + y^3$ **84.** $8x^3 - 12x^2y + 6xy^2 - y^3$ **85.** TW **87.** 54 **89.** $\frac{x^2[1-(-x)^n]}{1+x}$ **91.** 512 cm^2

Mid-Chapter Review: Chapter 14, pp. 1013–1014

Guided Solutions

1. $a_n = a_1 + (n-1)d$
$n = 14, a_1 = -6, d = 5$
$a_{14} = -6 + (14-1)5$
$a_{14} = 59$

2. $a_n = a_1r^{n-1}$
$n = 7, a_1 = \frac{1}{9}, r = -3$
$a_7 = \frac{1}{9} \cdot (-3)^{7-1}$
$a_7 = 81$

Mixed Review

1. 300 **2.** $\frac{1}{n+1}$ **3.** 78 **4.** $2^2 + 3^2 + 4^2 + 5^2 = 54$ **5.** $\sum_{k=1}^{6} (-1)^{k+1} \cdot k$ **6.** −3 **7.** 110 **8.** 61st **9.** −39 **10.** 21 **11.** 11 **12.** 4410 **13.** $-\frac{1}{2}$ **14.** 640 **15.** $2(-1)^{n+1}$ **16.** \$1146.39 **17.** 1 **18.** No **19.** \$465 **20.** \$1,073,741,823

Exercise Set 14.4, pp. 1021–1023

1. 2^5, or 32 **2.** 8 **3.** 9 **4.** 4! **5.** $\binom{8}{5}$, or $\binom{8}{3}$ **6.** 1 **7.** x^7y^2 **8.** 10 choose 4 **9.** 24 **11.** 39,916,800 **13.** 56 **15.** 126 **17.** 35 **19.** 1 **21.** 435 **23.** 780 **25.** $a^4 - 4a^3b + 6a^2b^2 - 4ab^3 + b^4$ **27.** $p^7 + 7p^6q + 21p^5q^2 + 35p^4q^3 + 35p^3q^4 + 21p^2q^5 + 7pq^6 + q^7$ **29.** $2187c^7 - 5103c^6d + 5103c^5d^2 - 2835c^4d^3 + 945c^3d^4 - 189c^2d^5 + 21cd^6 - d^7$ **31.** $t^{-12} + 12t^{-10} + 60t^{-8} + 160t^{-6} + 240t^{-4} + 192t^{-2} + 64$ **33.** $x^5 - 5x^4y + 10x^3y^2 - 10x^2y^3 + 5xy^4 - y^5$ **35.** $19{,}683s^9 + \frac{59{,}049s^8}{t} + \frac{78{,}732s^7}{t^2} + \frac{61{,}236s^6}{t^3} + \frac{30{,}618s^5}{t^4} + \frac{10{,}206s^4}{t^5} + \frac{2268s^3}{t^6} + \frac{324s^2}{t^7} + \frac{27s}{t^8} + \frac{1}{t^9}$ **37.** $x^{15} - 10x^{12}y + 40x^9y^2 - 80x^6y^3 + 80x^3y^4 - 32y^5$ **39.** $125 + 150\sqrt{5}t + 375t^2 + 100\sqrt{5}t^3 + 75t^4 + 6\sqrt{5}t^5 + t^6$ **41.** $x^{-3} - 6x^{-2} + 15x^{-1} - 20 + 15x - 6x^2 + x^3$ **43.** $15a^4b^2$ **45.** $-64{,}481{,}508a^3$ **47.** $1120x^{12}y^2$ **49.** $1{,}959{,}552u^5v^{10}$ **51.** y^8 **53.** TW

55.

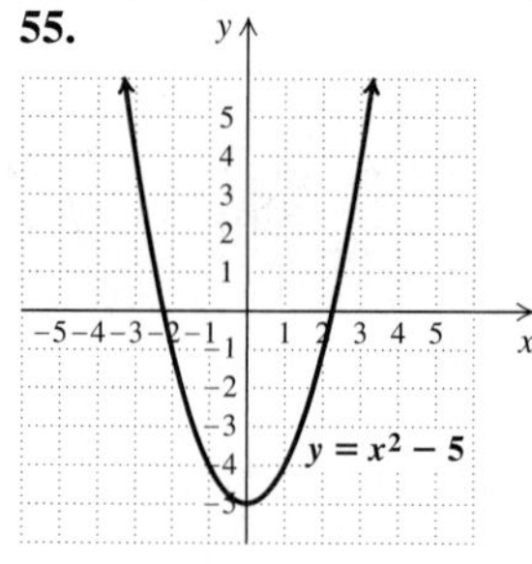

56.

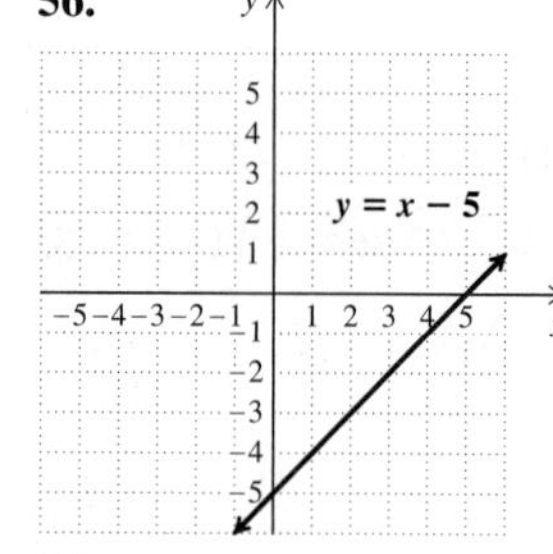

57.

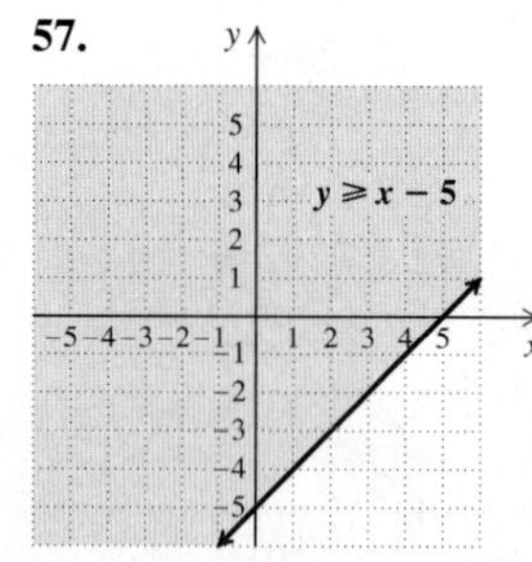

58.

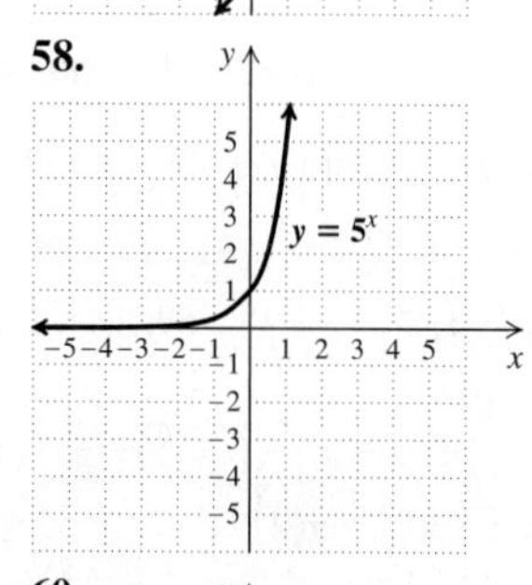

59.

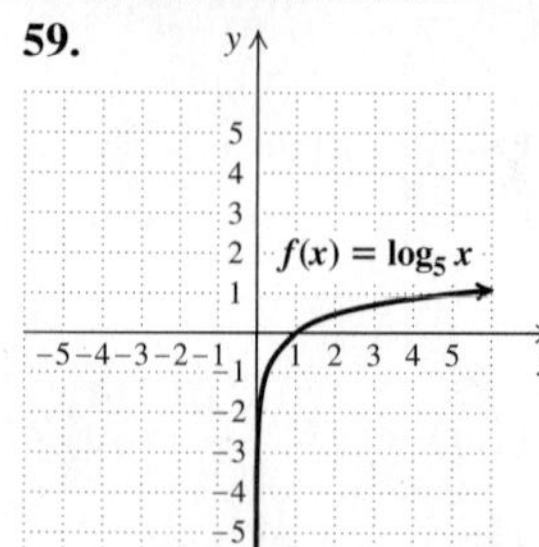

60.

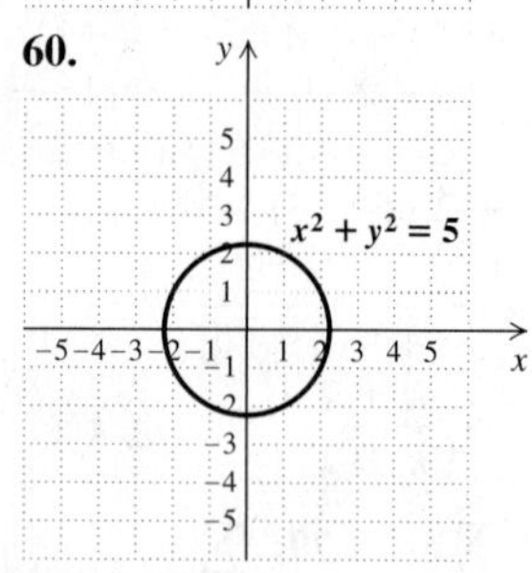

61. TW **63.** List all the subsets of size 3: $\{a, b, c\}$, $\{a, b, d\}$, $\{a, b, e\}$, $\{a, c, d\}$, $\{a, c, e\}$, $\{a, d, e\}$, $\{b, c, d\}$, $\{b, c, e\}$, $\{b, d, e\}$, $\{c, d, e\}$. There are exactly 10 subsets of size 3 and $\binom{5}{3} = 10$, so there are exactly $\binom{5}{3}$ ways of forming a subset of size 3 from $\{a, b, c, d, e\}$.

65. $\binom{8}{5}(0.15)^3(0.85)^5 \approx 0.084$

67. $\binom{8}{6}(0.15)^2(0.85)^6 + \binom{8}{7}(0.15)(0.85)^7 + \binom{8}{8}(0.85)^8 \approx 0.89$

69. $\binom{n}{n-r} = \dfrac{n!}{[n-(n-r)]!\,(n-r)!} = \dfrac{n!}{r!\,(n-r)!} = \binom{n}{r}$

71. $\dfrac{-\sqrt[3]{q}}{2p}$ **73.** $x^7 + 7x^6y + 21x^5y^2 + 35x^4y^3 + 35x^3y^4 + 21x^2y^5 + 7xy^6 + y^7$

Study Summary: Chapter 14, pp. 1024–1025

1. 143 **2.** -25 **3.** $5 \cdot 0 + 5 \cdot 1 + 5 \cdot 2 + 5 \cdot 3 = 30$ **4.** 15.5 **5.** 215 **6.** -640 **7.** $-20{,}475$ **8.** 16 **9.** 39,916,800 **10.** 84 **11.** $x^{10} - 10x^8 + 40x^6 - 80x^4 + 80x^2 - 32$ **12.** $3240t^7$

Review Exercises: Chapter 14, pp. 1025–1026

1. False **2.** True **3.** True **4.** False **5.** False **6.** True **7.** False **8.** False **9.** 1, 5, 9, 13; 29; 45 **10.** $0, \frac{1}{5}, \frac{1}{5}, \frac{3}{17}; \frac{7}{65}; \frac{11}{145}$ **11.** $a_n = -5n$ **12.** $a_n = (-1)^n(2n-1)$ **13.** $-2 + 4 + (-8) + 16 + (-32) = -22$ **14.** $-3 + (-5) + (-7) + (-9) + (-11) + (-13) = -48$ **15.** $\sum_{k=1}^{5} 4k$ **16.** $\sum_{k=1}^{5} \frac{1}{(-2)^k}$ **17.** 85 **18.** $\frac{8}{3}$ **19.** $a_1 = -15, d = 5$ **20.** -544 **21.** 25,250 **22.** $1024\sqrt{2}$ **23.** $\frac{3}{4}$ **24.** $a_n = 2(-1)^n$ **25.** $a_n = 3\left(\frac{x}{4}\right)^{n-1}$ **26.** 4095 **27.** $-4095x$ **28.** 12 **29.** $\frac{49}{11}$ **30.** No **31.** No **32.** \$40,000 **33.** $\frac{5}{9}$ **34.** $\frac{46}{33}$ **35.** \$24.30 **36.** 903 poles **37.** \$15,791.18 **38.** 6 m **39.** 5040 **40.** 56 **41.** $190a^{18}b^2$ **42.** $x^4 - 8x^3y + 24x^2y^2 - 32xy^3 + 16y^4$ **43.** TW For a geometric sequence with $|r| < 1$, as n gets larger, the absolute value of the terms gets smaller, since $|r^n|$ gets smaller. **44.** TW The first form of the binomial theorem draws the coefficients from Pascal's triangle; the second form uses factorial notation. The second form avoids the need to compute all preceding rows of Pascal's triangle, and is generally easier to use when only one term of an expansion is needed. When several terms of an expansion are needed and n is not large (say, $n \leq 8$), it is often easier to use Pascal's triangle. **45.** $\dfrac{1-(-x)^n}{x+1}$ **46.** $x^{-15} + 5x^{-9} + 10x^{-3} + 10x^3 + 5x^9 + x^{15}$

Test: Chapter 14, pp. 1026–1027

1. [14.1] $\frac{1}{2}, \frac{1}{5}, \frac{1}{10}, \frac{1}{17}, \frac{1}{26}; \frac{1}{145}$ **2.** [14.1] $a_n = 4\left(\frac{1}{3}\right)^n$ **3.** [14.1] $-3 + (-7) + (-15) + (-31) = -56$ **4.** [14.1] $\sum_{k=1}^{5} (-1)^{k+1}k^3$ **5.** [14.2] $\frac{13}{2}$ **6.** [14.2] -3 **7.** [14.2] $a_1 = 32; d = -4$ **8.** [14.2] 2508 **9.** [14.3] 1536 **10.** [14.3] $\frac{2}{3}$ **11.** [14.3] 3^n **12.** [14.3] 5621 **13.** [14.3] 1 **14.** [14.3] No **15.** [14.3] $\frac{\$25{,}000}{23} \approx \$1086.96$ **16.** [14.3] $\frac{85}{99}$ **17.** [14.2] 63 seats **18.** [14.2] \$17,100 **19.** [14.3] \$5987.37 **20.** [14.3] 36 m **21.** [14.4] 220 **22.** [14.4] $x^5 - 15x^4y + 90x^3y^2 - 270x^2y^3 + 405xy^4 - 243y^5$ **23.** [14.4] $220a^9x^3$ **24.** [14.2] $n(n+1)$ **25.** [14.3] $\dfrac{1-\left(\frac{1}{x}\right)^n}{1-\frac{1}{x}}$, or $\dfrac{x^n - 1}{x^{n-1}(x-1)}$

Cumulative Review/Final Exam: Chapters 1–14, pp. 1027–1030

1. $\frac{7}{15}$ **2.** $-4y + 17$ **3.** 280 **4.** 8.4×10^{-15} **5.** $\frac{7}{6}$ **6.** $3a^2 - 8ab - 15b^2$ **7.** $4a^2 - 1$ **8.** $9a^4 - 30a^2y + 25y^2$ **9.** $\dfrac{4}{x+2}$ **10.** $\dfrac{x-4}{4(x+2)}$ **11.** $\dfrac{(x+y)(x^2+xy+y^2)}{x^2+y^2}$ **12.** $x - a$ **13.** $12a^2\sqrt{b}$ **14.** $-27x^{10}y^{-2}$, or $-\dfrac{27x^{10}}{y^2}$ **15.** $25x^4y^{1/3}$ **16.** $y\sqrt[12]{x^5y^2}, y \geq 0$ **17.** $14 + 8i$ **18.** $(2x-3)^2$ **19.** $(3a-2)(9a^2+6a+4)$ **20.** $12(s^2+2t)(s^2-2t)$ **21.** $3(y^2+3)(5y^2-4)$ **22.** $7x^3 + 9x^2 + 19x + 38 + \dfrac{72}{x-2}$ **23.** 20 **24.** $[4, \infty)$, or $\{x|x \geq 4\}$ **25.** $(-\infty, 5) \cup (5, \infty)$, or $\{x|x \neq 5\}$ **26.** $\dfrac{1-2\sqrt{x}+x}{1-x}$ **27.** $y = 3x - 8$ **28.** $x^2 - 50 = 0$ **29.** $(2, -3); 6$ **30.** $\log_a \dfrac{\sqrt[3]{x^2} \cdot z^5}{\sqrt{y}}$ **31.** $a^5 = c$ **32.** 2.0792 **33.** 0.6826 **34.** 5 **35.** -121 **36.** 875 **37.** $16\left(\frac{1}{4}\right)^{n-1}$ **38.** $13{,}440a^4b^6$ **39.** $\frac{19{,}171}{64}$, or 299.546875 **40.** $\frac{3}{5}$ **41.** $-\frac{6}{5}, 4$ **42.** $\mathbb{R}$, or $(-\infty, \infty)$ **43.** $\left(-1, \frac{1}{2}\right)$ **44.** $(2, -1, 1)$ **45.** 2 **46.** $\pm 2, \pm 5$ **47.** $(\sqrt{5}, \sqrt{3}), (\sqrt{5}, -\sqrt{3}), (-\sqrt{5}, \sqrt{3}), (-\sqrt{5}, -\sqrt{3})$ **48.** 1.7925 **49.** 1005 **50.** $\frac{1}{25}$ **51.** $-\frac{1}{2}$ **52.** $\{x|-2 \leq x \leq 3\}$, or $[-2, 3]$ **53.** $\pm i\sqrt{3}$ **54.** $-2 \pm \sqrt{7}$ **55.** $\{y|y < -5 \text{ or } y > 2\}$, or $(-\infty, -5) \cup (2, \infty)$ **56.** $-8, 10$ **57.** 3 **58.** $r = \dfrac{V-P}{-Pt}$, or $\dfrac{P-V}{Pt}$ **59.** $R = \dfrac{Ir}{1-I}$ **60.** **(a)** Linear; **(b)** 2 **61.** **(a)** Logarithmic; **(b)** -1 **62.** **(a)** Quadratic; **(b)** $-1, 4$ **63.** **(a)** Exponential; **(b)** no real zeros

64.

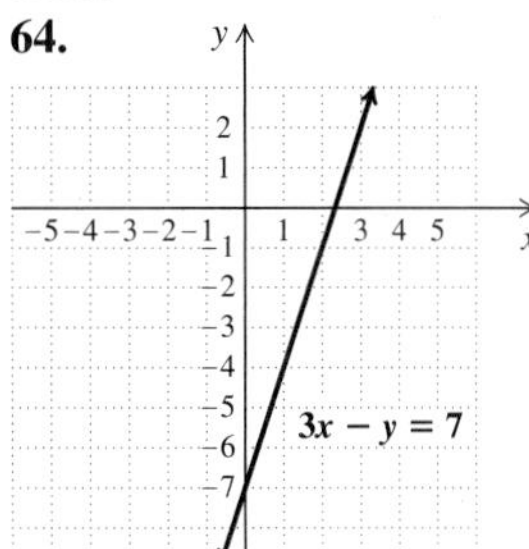

65.

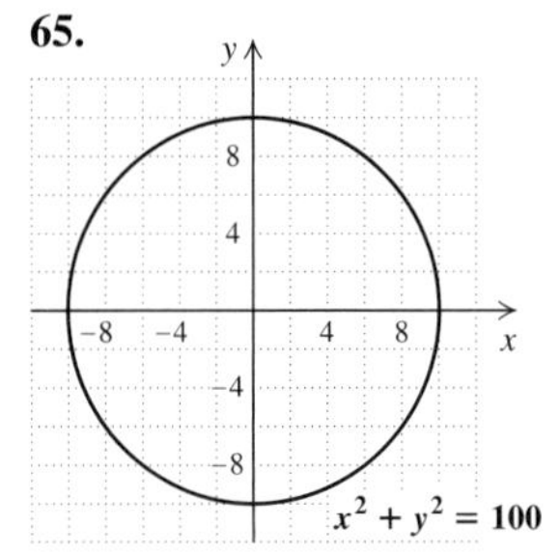

66.

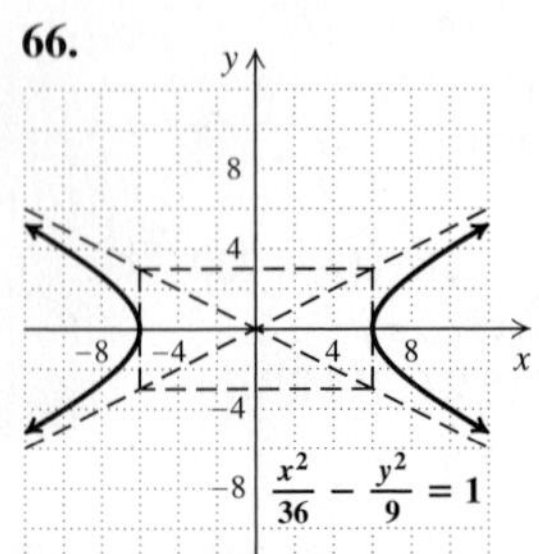

67.

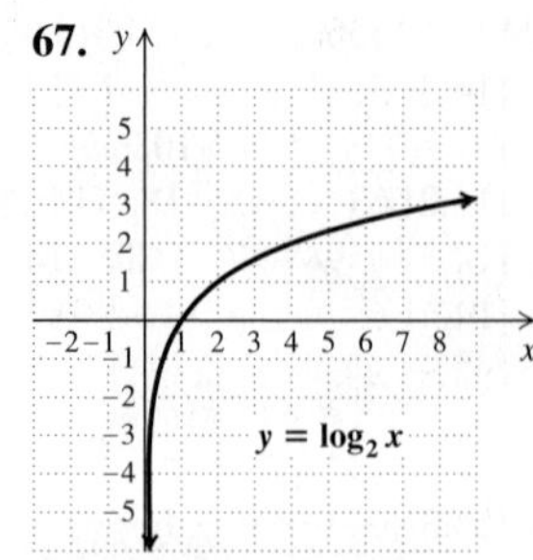

68.

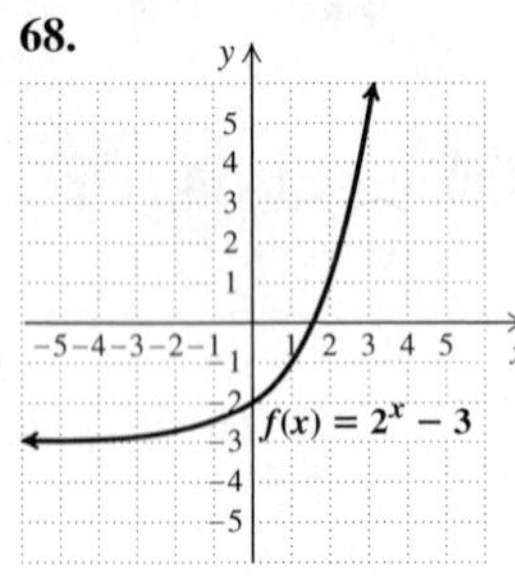

69.

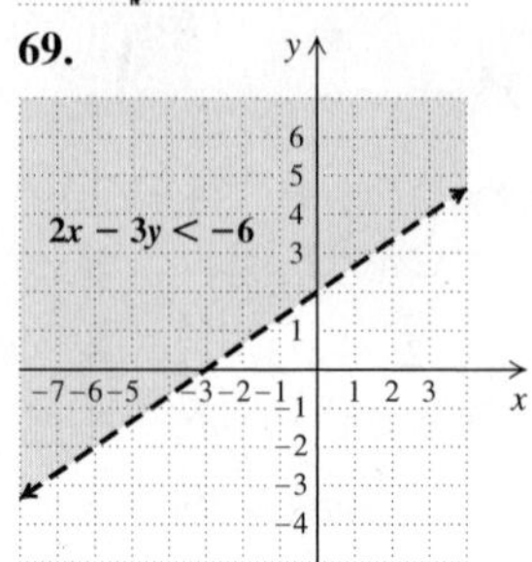

70.

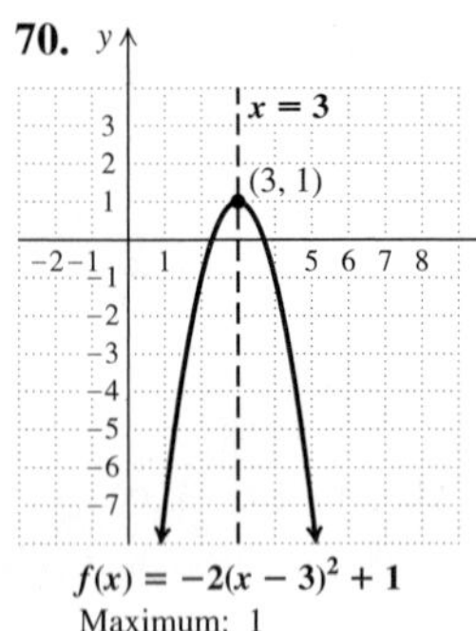

71.

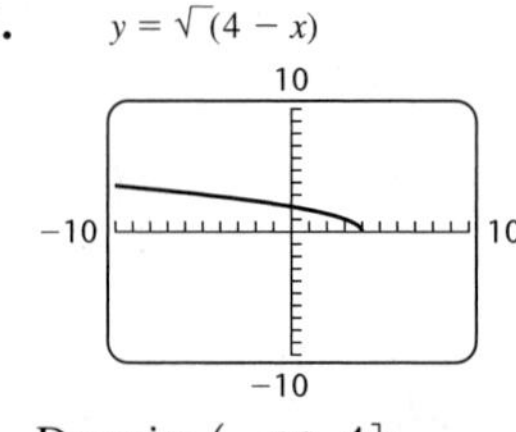

Domain: $(-\infty, 4]$; range: $[0, \infty)$

72. 5000 ft^2 **73.** 5 ft by 12 ft **74.** More than 25 rentals **75.** \$2.68 herb: 10 oz; \$4.60 herb: 14 oz **76.** 350 mph **77.** $8\frac{2}{5}$ hr, or 8 hr 24 min **78.** 20 **79.** **(a)** 2 million card holders/year; **(b)** $c = 2t + 173$; **(c)** 191 million card holders; **(d)** 2044 **80.** **(a)** $k \approx 0.092$; $P(t) = 5e^{0.092t}$, where $P(t)$ is in billions of dollars; **(b)** \$104 billion; **(c)** 2019 **81.** \$14,079.98 **82.** All real numbers except 0 and -12 **83.** 81 **84.** y gets divided by 8 **85.** 84 yr

Chapter R

Exercise Set R.1, pp. R-7–R-8

1. False **3.** True **5.** True **7.** 4 **9.** 1.3 **11.** -25 **13.** $-\frac{11}{15}$ **15.** -6.5 **17.** -15 **19.** 0 **21.** $-\frac{1}{2}$ **23.** 5.8 **25.** -3 **27.** 39 **29.** 175 **31.** -32 **33.** 16 **35.** -6 **37.** 9 **39.** -3 **41.** -16 **43.** 100 **45.** 2 **47.** -23 **49.** 36 **51.** 10 **53.** 10 **55.** -7 **57.** 32 **59.** 28 cm^2 **61.** $8x + 28$ **63.** $-30 + 6x$ **65.** $8a + 12b - 6c$ **67.** $-6x + 3y - 3z$ **69.** $2(4x + 3y)$ **71.** $3(1 + w)$ **73.** $10(x + 5y + 10)$ **75.** p **77.** $-m + 22$ **79.** $3x + 7$ **81.** $6p - 7$ **83.** $-x + 12y$ **85.** $36a - 48b$ **87.** $-10x + 104y + 9$ **89.** Yes **91.** No **93.** Yes **95.** Let n represent the number; $3n = 348$ **97.** Let c represent the number of calories in a Taco Bell Beef Burrito; $c + 69 = 500$ **99.** Let l represent the amount of water used to produce 1 lb of lettuce; $42 = 2l$

Exercise Set R.2, pp. R-16–R-17

1. 16 **3.** $-\frac{1}{12}$ **5.** -0.8 **7.** $-\frac{5}{3}$ **9.** 42 **11.** -5 **13.** 2 **15.** $\frac{25}{3}$ **17.** $-\frac{4}{9}$ **19.** -4 **21.** $\frac{69}{5}$ **23.** $\frac{9}{32}$ **25.** -2 **27.** -15 **29.** $\frac{43}{2}$ **31.** $-\frac{61}{115}$ **33.** $l = \frac{A}{w}$ **35.** $q = \frac{p}{30}$ **37.** $P = IV$ **39.** $p = 2q - r$ **41.** $\pi = \frac{A}{r^2 + r^2h}$ **43.** **(a)** No; **(b)** yes; **(c)** no; **(d)** yes **45.** $\{x \mid x \leq 12\}$, or $(-\infty, 12]$ **47.** $\{m \mid m > 12\}$, or $(12, \infty)$ **49.** $\{x \mid x \geq -\frac{3}{2}\}$, or $[-\frac{3}{2}, \infty)$ **51.** $\{t \mid t < -3\}$, or $(-\infty, -3)$ **53.** $\{y \mid y > 10\}$, or $(10, \infty)$ **55.** $\{a \mid a \geq 1\}$, or $[1, \infty)$ **57.** $\{x \mid x \geq \frac{64}{17}\}$, or $[\frac{64}{17}, \infty)$ **59.** $\{x \mid x > \frac{39}{11}\}$, or $(\frac{39}{11}, \infty)$ **61.** $\{x \mid x \leq -10.875\}$, or $(-\infty, -10.875]$ **63.** 7 **65.** 16, 18 **67.** $166\frac{2}{3}$ pages **69.** 4.5 cm, 9.5 cm **71.** 900 cubic feet **73.** 80¢ **75.** 30 min or more **77.** For $2\frac{7}{9}$ hr or less

Exercise Set R.3, pp. R-25–R-26

1.

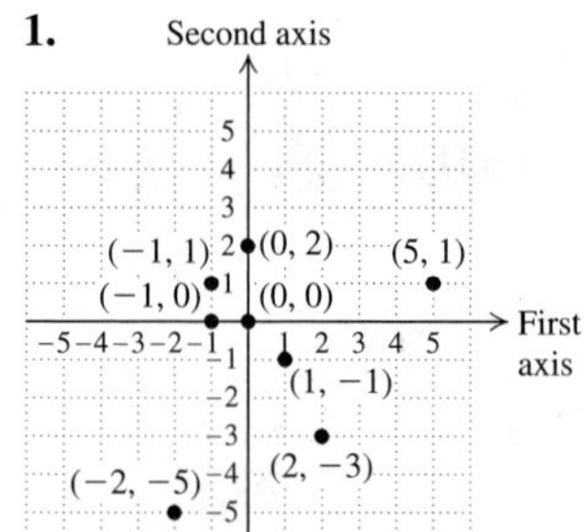

3. I **5.** IV **7.** I, IV **9.** No **11.** Yes

13.

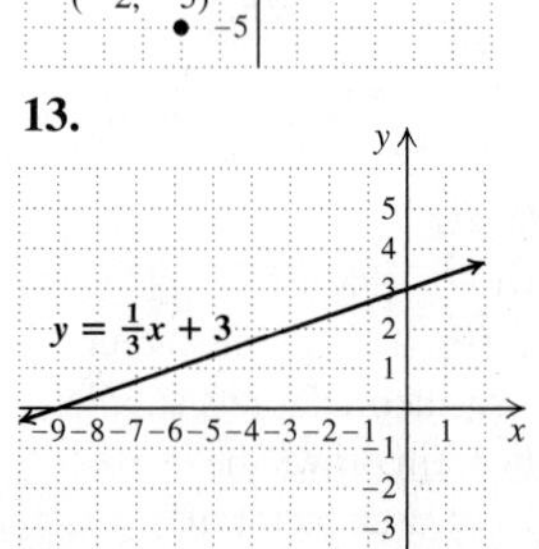

15.

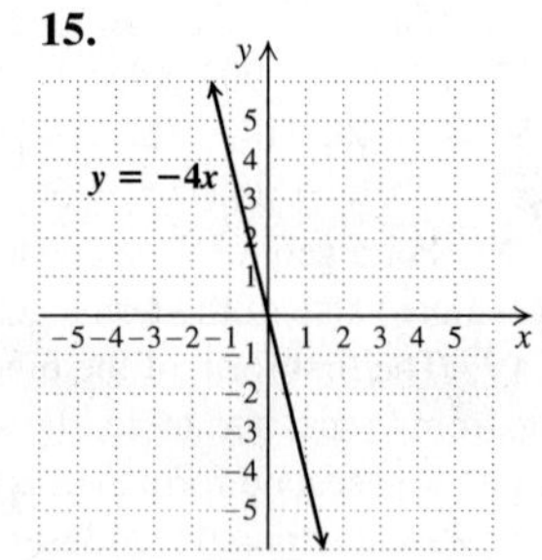

17.

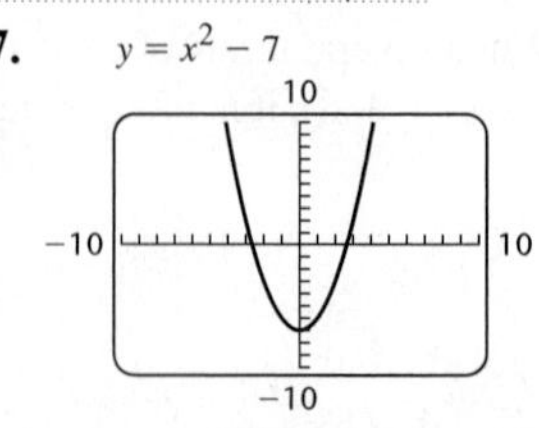

19.

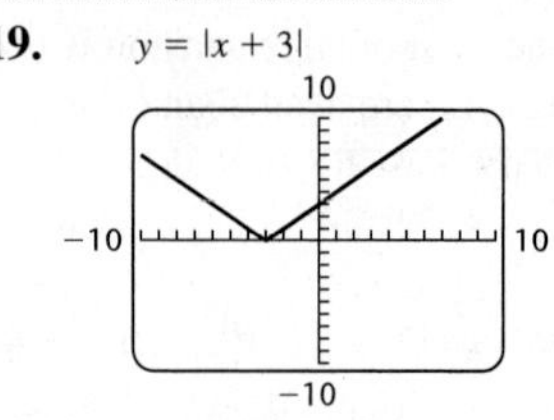

21. $y = \sqrt{x^2 + 1}$

23. 1 **25.** -5 **27.** 0

29. Slope: 2; y-intercept: $(0, -5)$

31. Slope: $-\frac{2}{7}$; y-intercept: $\left(0, \frac{1}{7}\right)$

33.

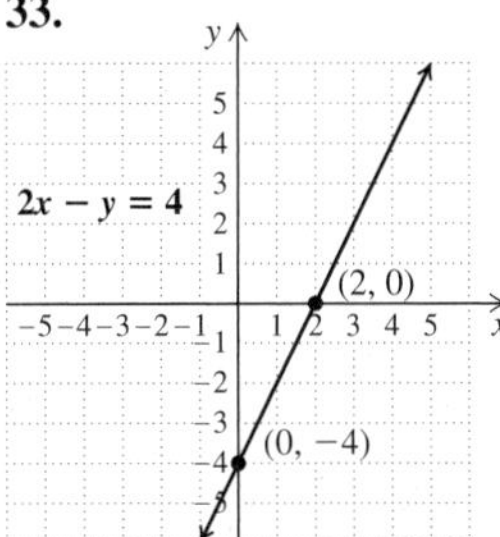

35.

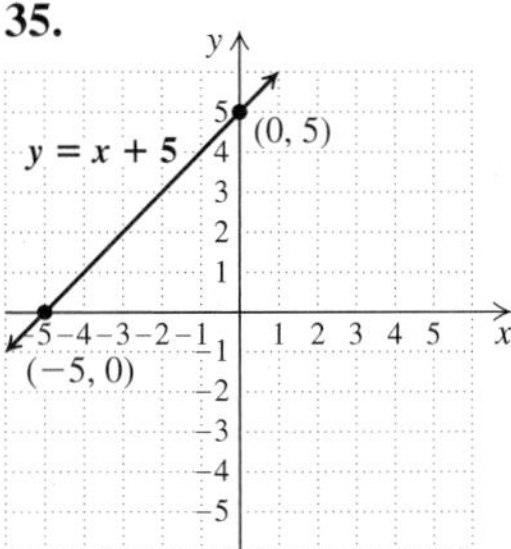

37.

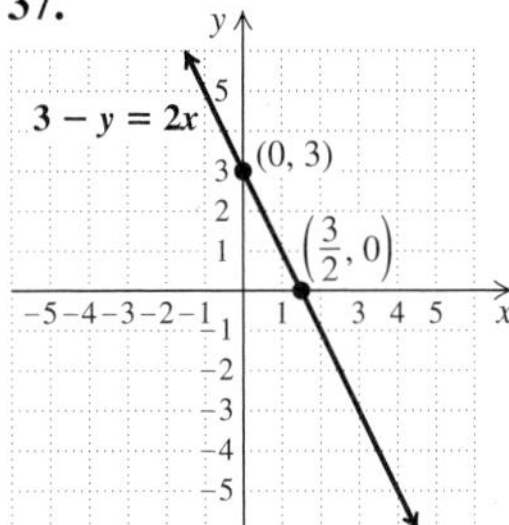

39.

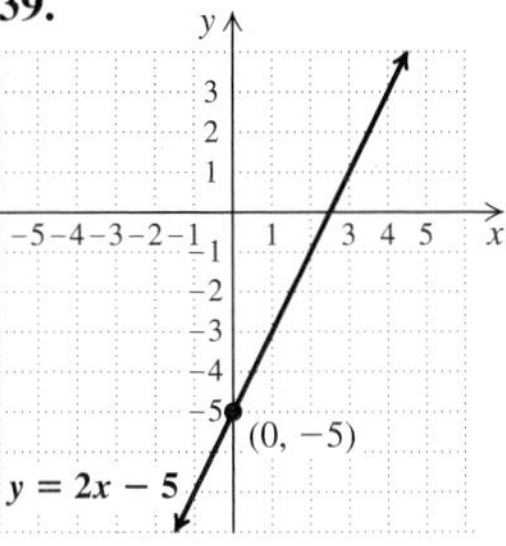

41.

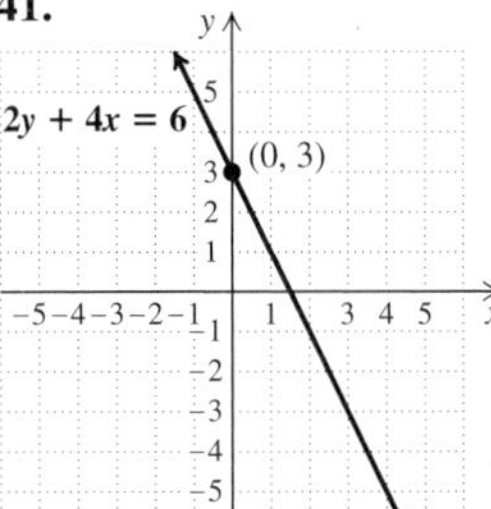

43. 0

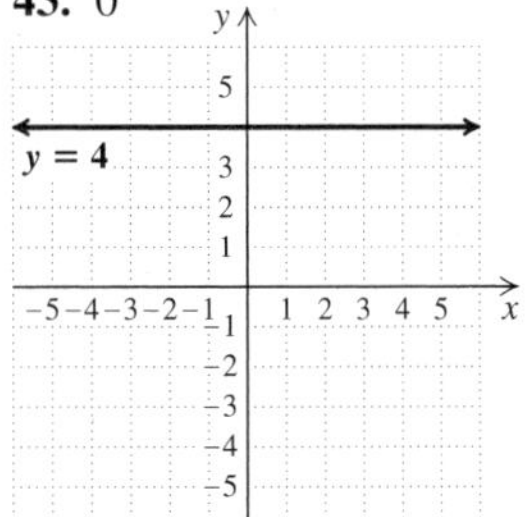

45. Undefined

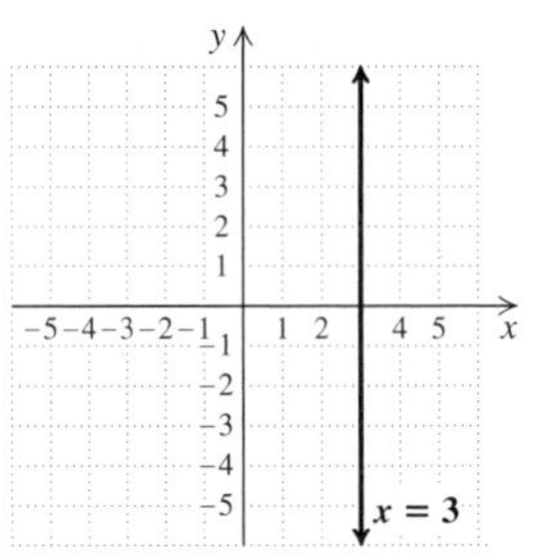

47. $y = \frac{1}{3}x + 1$ **49.** $y = -x + 3$ **51.** Perpendicular
53. Neither **55. (a)** 6; **(b)** $8\frac{1}{3}$; **(c)** $\frac{1}{3}a + 8$
57. (a) 6; **(b)** 3; **(c)** $20a^2 + 4a + 3$ **59.** No **61.** Yes
63. $\{x \mid x \text{ is a real number } and\ x \neq 3\}$ **65.** $\mathbb{R}$
67. $\{x \mid x \geq -6\}$, or $[-6, \infty)$

Exercise Set R.4, pp. R-34–R-36

1. $(4, 3)$ **3.** $(1, 3)$ **5.** $(2, -1)$ **7.** No solution
9. $(6, -1)$ **11.** $\left\{(x, y) \mid y = \frac{1}{5}x + 4\right\}$ **13.** $(2, -3)$
15. $(-4, 3)$ **17.** $(2, -2)$ **19.** $\{(x, y) \mid 2x - 3 = y\}$
21. $\left(\frac{25}{23}, -\frac{11}{23}\right)$ **23.** No solution **25.** $(1, 2)$ **27.** $(2, 7)$
29. $(-1, 2)$ **31.** $\left(\frac{15}{14}, \frac{4}{7}\right)$ **33.** $(6, 2)$ **35.** No solution
37. $\{(x, y) \mid -4x + 2y = 5\}$ **39.** 43, 46 **41.** 49°, 131°
43. Two-pointers: 32; three-pointers: 4
45. Private lessons: 6 students; group lessons: 8 students
47. Peanuts: 135 lb; Brazil nuts: 345 lb

49.

x	y	90
12%	30%	20%
$0.12x$	$0.3y$	$0.2(90)$

Streakfree: 50 oz;
Sunstream: 40 oz

51. 3 hr **53.** Headwind: 40 mph; plane: 620 mph **55.** 7
57. $-\frac{9}{2}$ **59.** -6 **61.** 3 **63.** 0.666667, or $\frac{2}{3}$

Exercise Set R.5, pp. R-43–R-44

1. 1 **3.** -3 **5.** $\dfrac{1}{8^2} = \dfrac{1}{64}$ **7.** $\dfrac{10}{x^5}$ **9.** $\dfrac{1}{(ab)^2}$
11. y^{10} **13.** y^{-4} **15.** x^{-t} **17.** x^{13} **19.** a^6 **21.** $(4x)^8$
23. 7^{40} **25.** x^8y^{12} **27.** $\dfrac{y^6}{64}$ **29.** $\dfrac{9q^8}{4p^6}$
31. $8x^3, -6x^2, x, -7$ **33.** 18, 36, −7, 3; 3, 9, 1, 0; 9
35. −1, 4, −2; 3, 3, 2; 3 **37.** $8p^4$; 8 **39.** $x^4 + 3x^3$
41. $-7t^2 + 5t + 10$ **43.** 36 **45.** -14 **47.** 144 ft **49.** 11
51. -4 **53.** $4x^3 - 3x^2 + 8x + 7$ **55.** $-y^2 + 5y - 2$
57. $-3x^2y - y^2 + 8y$ **59.** $12x^5 - 28x^3 + 28x^2$
61. $8a^2 + 2ab + 4ay + by$ **63.** $x^3 + 4x^2 - 20x + 7$
65. $x^2 - 49$ **67.** $x^2 + 2xy + y^2$ **69.** $6x^4 + 17x^2 - 14$
71. $a^2 - 6ab + 9b^2$ **73.** $42a^2 - 17ay - 15y^2$
75. $-t^4 - 3t^2 + 2t - 5$ **77.** $5x + 3$
79. $2x^2 - 3x + 3 + \dfrac{-2}{x + 1}$ **81.** $5x + 3 + \dfrac{3}{x^2 - 1}$

Exercise Set R.6, p. R-54

1. $6t^3(3t^2 - 2t + 1)$ **3.** $(y - 3)^2$ **5.** $(p + 2)(2p^3 + 1)$
7. Prime **9.** $(2t + 3)(4t^2 - 6t + 9)$
11. $(m + 6)(m + 7)$ **13.** $(x^2 + 9)(x + 3)(x - 3)$
15. $(2x + 3)(4x + 5)$ **17.** $(x + 2)(x + 1)(x - 1)$
19. $(0.1t^2 - 0.2)(0.01t^4 + 0.02t^2 + 0.04)$
21. $\left(x^2 + \frac{1}{4}\right)\left(x + \frac{1}{2}\right)\left(x - \frac{1}{2}\right)$ **23.** $(n - 2)(m + 3)$
25. $(m + 15n)(m - 10n)$ **27.** $2y(3x + 1)(4x - 3)$
29. $(y + 11)(y - 11)$ **31.** $-7, 2$ **33.** 0, 11 **35.** $-10, 10$
37. $-\frac{5}{2}, 7$ **39.** 0, 5 **41.** $-2, 6$ **43.** $-11, 5$ **45.** -5
47. Base: 8 ft; height: 6 ft **49.** 8 ft, 15 ft

Glossary

Absolute value [1.4] The distance that a number is from 0 on the number line

***ac*-method** [6.3] A method for factoring a trinomial that uses factoring by grouping; also called *grouping method*

Additive identity [1.5] The number 0

Additive inverse [1.6] A number's opposite; two numbers are additive inverses of each other if their sum is zero.

Algebraic expression [1.1] An expression consisting of variables and/or numerals, often with operation signs and grouping symbols

Arithmetic sequence (or **progression)** [14.2] A sequence in which adding the same number to any term gives the next term in the sequence

Arithmetic series [14.2] A series for which the associated sequence is arithmetic

Ascending order [5.4] A polynomial written with the terms arranged according to degree, from least to greatest

Associative law for addition [1.2] The statement that when three numbers are added, regrouping the addends gives the same sum

Associative law for multiplication [1.2] The statement that when three numbers are multiplied, regrouping the factors gives the same product

Asymptote [13.3] A line that a graph approaches more and more closely as x increases or as x decreases

Average [2.7] Most commonly, the sum of a set of numbers divided by the number of addends; also called the *mean*

Axes [3.1] Two perpendicular number lines used to identify points in a plane

Axis of symmetry [11.6] A line that can be drawn through a graph such that the part of the graph on one side of the line is an exact reflection of the part on the opposite side

Bar graph [3.1] A graphic display of data using bars proportional in length to the numbers represented

Base [1.8] In exponential notation, the number being raised to a power

Binomial [5.3] A polynomial composed of two terms

Branches [7.1], [13.3] The two or more curves that comprise the graph of some relations

Break-even point [9.5] In business, the point of intersection of the revenue function and the cost function

Circle [13.1] A set of points in a plane that are a fixed distance r, called the radius, from a fixed point (h, k), called the center

Circle graph [3.1] A graphic display of data using sectors of a circle to represent percents

Circumference [2.3] The distance around a circle

Closed interval $[a, b]$ [2.6] The set of all numbers x for which $a \leq x \leq b$; thus, $[a, b] = \{x|a \leq x \leq b\}$

Coefficient [2.1] The numerical multiplier of a variable

Columns of a matrix [9.3] The vertical elements of a matrix

Combined variation [7.8] A mathematical relationship in which a variable varies directly and/or inversely, at the same time, with more than one other variable

Common logarithm [12.3] A logarithm with base 10

Commutative law for addition [1.2] The statement that when two numbers are added, the order in which the numbers are added does not affect the sum

Commutative law for multiplication [1.2] The statement that when two numbers are multiplied, the order in which the numbers are multiplied does not affect the product

Completing the square [11.1] A method of adding a particular constant to an expression so that the resulting sum is a perfect square

Complex number [10.8] Any number that can be written in the form $a + bi$, where a and b are real numbers and $i = \sqrt{-1}$

Complex rational expression [7.5] A rational expression that has one or more rational expressions within its numerator and/or its denominator

Complex-number system [10.8] A number system that contains the real-number system and is designed so that negative numbers have defined square roots

Composite function [12.1] A function in which a quantity depends on a variable that, in turn, depends on another variable

Composite number [1.3] A natural number other than 1 that is not prime

Compound inequality [8.1] A statement in which two or more inequalities are joined by the word "and" or the word "or"
Compound interest [11.1] Interest computed on the sum of an original principal and the interest previously accrued by that principal
Conditional equation [2.2] An equation that is true for some replacements of a variable and false for others
Conic section [13.1] A curve formed by the intersection of a plane and a cone
Conjugates [10.5] Pairs of radical expressions, like $\sqrt{a} + \sqrt{b}$ and $\sqrt{a} - \sqrt{b}$, for which the product does not have a radical term
Conjugate of a complex number [10.8] The *conjugate* of a complex number $a + bi$ is $a - bi$, and the *conjugate* of $a - bi$ is $a + bi$.
Conjunction [8.1] A sentence in which two or more sentences are joined by the word "and"
Consecutive even integers [2.5] Integers that are even and two units apart
Consecutive integers [2.5] Integers that are one unit apart
Consecutive odd integers [2.5] Integers that are odd and two units apart
Consistent system of equations [4.1] A system of equations that has at least one solution
Constant [1.1] A known number that never changes
Constant function [3.8] A function given by an equation of the form $f(x) = b$, where b is a real number
Constant of proportionality [7.8] The constant, k, in an equation of variation; also called *variation constant*
Contradiction [2.2] An equation that is never true
Coordinates [3.1] The numbers in an ordered pair
Cube root [10.1] The number c is the *cube root* of a if $c^3 = a$.
Cubic function [5.3] A polynomial function in one variable of degree 3
Curve fitting [3.7] The process of understanding and interpreting data to determine if it has a pattern and if it does, fitting an equation to the data

Data [1.1] Numerical information
Degree of a polynomial [5.3] The degree of the term of highest degree in a polynomial
Degree of a term [5.3] The number of variable factors in a term
Demand function [9.5] A function modeling the relationship between the price of a good and the quantity of that good demanded
Denominator [1.3] The bottom number or expression in a fraction
Dependent equations [4.1], [9.1] Equations in a system are dependent if one equation can be removed without changing the solution set.
Dependent variable [2.3] In an equation with two variables, the variable that *depends* on the value of the other
Descending order [5.3] A polynomial written with the terms arranged according to degree, from greatest to least
Determinant [9.4] A descriptor of a square matrix. The determinant of a two-by-two matrix $\begin{bmatrix} a & c \\ b & d \end{bmatrix}$ is denoted by $\begin{vmatrix} a & c \\ b & d \end{vmatrix}$ and is defined as $ad - bc$.
Difference of squares [6.4] An expression that can be written in the form $A^2 - B^2$
Difference of two cubes [6.5] An expression that can be written in the form $A^3 - B^3$
Direct variation [7.8] A situation that is modeled by a linear function of the form $f(x) = kx$, or $y = kx$, where k is a nonzero constant
Discriminant [11.3] The radicand $b^2 - 4ac$ from the quadratic formula
Disjunction [8.1] A sentence in which two or more sentences are joined by the word "or"
Distributive law [1.2] The statement that multiplying a factor by the sum of two numbers gives the same result as multiplying the factor by each of the two numbers and then adding
Domain [3.8] The set of all first coordinates of the ordered pairs in a function
Double root [6.4] A repeated root that appears twice
Doubling time [12.7] The amount of time necessary for a population to double in size

Elements of a matrix [9.3] The individual numbers in a matrix; also called *entries*
Elimination method [4.3] An algebraic method that uses the addition principle to solve a system of equations
Ellipse [13.2] The set of all points in a plane for which the sum of the distances from two fixed points F_1 and F_2 is constant
Empty set [2.2] The set containing no elements, denoted $\varnothing$ or $\{ \}$
Equation [1.1] A number sentence with the verb =
Equilibrium point [9.5] The point of intersection between the demand function and the supply function
Equivalent equations [2.1] Equations with the same solutions
Equivalent expressions [1.2] Expressions that have the same value for all allowable replacements
Equivalent inequalities [2.6] Inequalities that have the same solution set

Equivalent systems of equations [9.3] Systems of equations with the same solutions
Evaluate [1.1] To substitute a number for each variable in the expression and calculate the result
Even root [10.1] A root with an even index
Exponent [1.8] In an expression of the form b^n, the number n is an exponent. For n a natural number, b^n represents n factors of b.
Exponential decay [12.7] A decrease in quantity over time that can be modeled by an exponential function of the form $P(t) = P_0e^{-kt}$, $k > 0$, where P_0 is the quantity present at time 0, $P(t)$ is the amount present at time t, and k is the decay rate
Exponential equation [12.6] An equation in which a variable appears as an exponent
Exponential function [12.2] A function that can be described by an exponential equation
Exponential growth [12.7] An increase in quantity over time that can be modeled by an exponential function of the form $P(t) = P_0e^{kt}$, $k > 0$, where P_0 is the population at time 0, $P(t)$ is the population at time t, and k is the exponential growth rate
Exponential notation [1.8] A representation of a number using a base raised to a power
Extrapolation [3.7] The process of predicting a future value on the basis of given data

Factor [1.2] *Verb*: to write an equivalent expression that is a product; *noun*: a multiplier
Factoring [1.2] The process of writing an expression as a product
Finite sequence [14.1] A function having for its domain a set of natural numbers: $\{1, 2, 3, 4, 5, \ldots, n\}$, for some natural number n
Finite series [14.1] The sum of the first n terms of a sequence: $a_1 + a_2 + a_3 + \cdots + a_n$; also called *partial sum*
Fixed costs [9.5] In business, costs that are incurred whether or not a product is produced
Foci (plural of focus) [13.2] Two fixed points that determine the points of an ellipse
FOIL method [5.6] To multiply two binomials $A + B$ and $C + D$, multiply the First terms AC, the Outside terms AD, the Inner terms BC, and then the Last terms BD. Then combine terms if possible.
Formula [2.3] An equation that uses two or more letters to represent a relationship between two or more quantities
Fraction notation [1.3] A number written using a numerator and a denominator
Function [3.8] A correspondence between a first set, called the *domain*, and a second set, called the *range*, such that each member of the domain corresponds to *exactly one* member of the range

General term of a sequence [14.1] The nth term, denoted a_n
Geometric sequence [14.3] A sequence in which multiplying a certain fixed number by any term gives the next term in the sequence
Geometric series [14.3] A series for which the associated sequence is geometric
Grade [3.5] The ratio of the vertical distance that a road rises over the horizontal distance that it runs, expressed as a percent
Graph [2.6] A picture or diagram of the data in a table; a line, curve, or collection of points that represents all the solutions of an equation
Greatest common factor [6.1] The greatest common factor of a polynomial is the largest common factor of the coefficients times the greatest common factor of the variable(s) in all of the terms.
Grouping method [6.3] A method for factoring a trinomial that uses factoring by grouping; also called *ac-method*

Half-life [12.7] The amount of time necessary for half of a quantity to decay
Half-open interval [2.6] An interval that contains one endpoint and not the other
Half-plane [8.3] A region of a plane consisting of all points on one side of a straight line
Horizontal-line test [12.1] If it is impossible to draw a horizontal line that intersects a function's graph more than once, then the function is one-to-one.
Hyperbola [13.3] The set of all points P in a plane such that the difference of the distance from P to two fixed points is constant
Hypotenuse [6.7] In a right triangle, the side opposite the 90° angle

i [10.8] The square root of -1; $i = \sqrt{-1}$ and $i^2 = -1$
Identity [2.2] An equation that is always true
Identity property of 0 [1.5] The statement that the sum of a number and 0 is always the original number
Identity property of 1 [1.3] The statement that the product of a number and 1 is always the original number
Imaginary number [10.8] A number that can be written in the form $a + bi$, where a and b are real numbers and $b \neq 0$
Inconsistent system of equations [4.1] A system of equations for which there is no solution
Independent equations [4.1] Equations that are not dependent
Independent variable [2.3] In an equation with two variables, the variable that does *not depend* on the other
Index [10.1] In the radical $\sqrt[n]{a}$, the number n is called the index.

Inequality [1.4] A mathematical sentence using $<, >, \leq, \geq$, or $\neq$
Infinite geometric series [14.3] The sum of the terms of an infinite geometric sequence
Infinite sequence [14.1] A function having for its domain the set of natural numbers: $\{1, 2, 3, 4, 5, \ldots\}$
Infinite series [14.1] Given the infinite sequence $a_1, a_2, a_3, a_4, \ldots a_n, \ldots$, the sum of the terms $a_1 + a_2 + a_3 + a_4 + \cdots + a_n + \cdots$ is called an infinite series.
Input [3.8] An element of the domain of a function
Integers [1.4] The whole numbers and their opposites: $\{\ldots, -4, -3, -2, -1, 0, 1, 2, 3, 4, \ldots\}$
Interpolation [3.7] The process of estimating a value between given values
Intersection of *A* and *B* [8.1] The set of all elements that are common to *both* A and B; denoted $A \cap B$
Interval notation [2.6] The use of a pair of numbers inside parentheses and brackets to represent the set of all numbers between those two numbers; see also *closed, open,* and *half-open intervals*
Inverse relation [12.1] The relation formed by interchanging the members of the domain and the range of a relation
Inverse variation [7.8] A situation that is modeled by a rational function of the form $f(x) = k/x$, or $y = k/x$, where k is a nonzero constant
Irrational number [1.4] A real number that cannot be named as a ratio of two integers
Isosceles right triangle [10.7] A right triangle in which both legs have the same length

Joint variation [7.8] A situation that is modeled by an equation of the form $y = kxz$, where k is a nonzero constant

Leading coefficient [5.3] The coefficient of the term of highest degree in a polynomial
Leading term [5.3] The term of highest degree in a polynomial
Least common denominator (LCD) [7.3] The least common multiple of the denominators of two or more fractions
Least common multiple (LCM) [7.3] The smallest number that is a multiple of two or more numbers
Legs [6.7] In a right triangle, the two sides that form the 90° angle
Like radicals [10.5] Radical expressions that have the same indices and radicands
Like terms [1.5] Terms that have exactly the same variable factors
Line graph [3.1] A graph in which quantities are represented as points connected by straight-line segments
Linear equation in two variables [2.2] Any equation whose graph is a straight line, and can be written in the form $y = mx + b$, or $Ax + By = C$, where x and y are variables
Linear equation in three variables [9.1] An equation equivalent to one in the form $Ax + By + Cz = D$, where x, y, and z are variables
Linear function [3.8] A function whose graph is a straight line and can be written in the form $f(x) = mx + b$
Linear inequality [8.3] An inequality whose related equation is a linear equation
Linear regression [3.7] A method of fitting a line to a set of data
Logarithmic equation [12.6] An equation containing a logarithmic expression
Logarithmic function, base *a* [12.3] The inverse of an exponential function $f(x) = a^x$

Mathematical model [1.1] A mathematical representation of a real-world situation
Matrix [9.3] A rectangular array of numbers
Maximum value [11.6] The greatest function value (output) achieved by a function
Mean [2.7] The sum of a set of numbers divided by the number of addends; also called *average*
Minimum value [11.6] The least function value (output) achieved by a function
Monomial [5.3] A constant, a variable, or a product of a constant and/or variables
Motion problem [4.4] A problem that deals with distance, speed (rate), and time
Multiplicative identity [1.3] The number 1
Multiplicative inverses [1.3] Reciprocals; two numbers are multiplicative inverses if their product is 1.
Multiplicative property of zero [1.7] The statement that the product of 0 and any real number is 0

Natural logarithm [12.5] A logarithm with base e; also called *Napierian logarithm*
Natural numbers [1.3] The counting numbers: 1, 2, 3, 4, 5, . . .
Nonlinear equation [3.2] An equation whose graph is not a straight line
***n*th root** [10.1] A number c is called the nth root of a if $c^n = a$.
Numerator [1.3] The top number or expression in a fraction

Odd root [10.1] A root with an odd index
One-to-one function [12.1] A function for which different inputs have different outputs
Open interval (*a*, *b*) [2.6] The set of all numbers x for which $a < x < b$; thus, $(a, b) = \{x | a < x < b\}$

Opposite [1.6] The opposite, or additive inverse, of a number a is written $-a$; opposites are the same distance from 0 on the number line but on different sides of 0.

Opposite of a polynomial [5.4] Two polynomials are opposites if their sum is 0.

Ordered pair [3.1] A pair of numbers of the form (a, b) for which the order in which the numbers are listed is important

Origin [3.1] The point (0, 0) on a coordinate plane where the two axes intersect

Output [3.8] An element of the range of a function

Parabola [11.6] A graph of a second-degree polynomial in one variable

Parallel lines [3.6] Lines in the same plane that never intersect; two lines are parallel if they have the same slope.

Partial sum [14.1] The sum of the first n terms of a sequence: $a_1 + a_2 + a_3 + \cdots + a_n$; also called *finite series*

Pascal's triangle [14.4] A triangular array of coefficients of the expansion $(a + b)^n$ for $n = 0, 1, 2, \ldots$

Perfect-square trinomial [6.4] A trinomial that is the square of a binomial

Perpendicular lines [3.6] Two lines that intersect to form a right angle; two lines are perpendicular if the product of their slopes is -1.

Point–slope equation [3.7] An equation of a line of the form $y - y_1 = m(x - x_1)$, where x and y are variables, m is the slope, and (x_1, y_1) is a point through which the line passes

Polynomial [5.3] A monomial or a sum of monomials

Polynomial equation [6.1] An equation in which two polynomials are set equal to each other

Polynomial function [5.3] A function in which outputs are determined by evaluating a polynomial

Polynomial inequality [8.4] An inequality that is equivalent to an inequality with a polynomial as one side and 0 as the other

Prime factorization [1.3] The factorization of a composite number into a product of prime numbers

Prime number [1.3] A natural number that has exactly two different factors: the number itself and 1

Prime polynomial [6.1] A polynomial that cannot be factored

Principal square root [10.1] The nonnegative square root of a number

Product [1.2] The result of multiplication

Proportion [7.7] An equation stating that two ratios are equal

Pure imaginary number [10.8] A complex number of the form $a + bi$, in which $a = 0$ and $b \neq 0$

Pythagorean theorem [6.7] In any right triangle, if a and b are the lengths of the legs and c is the length of the hypotenuse, then $a^2 + b^2 = c^2$.

Quadrants [3.1] The four regions into which the axes divide a plane

Quadratic equation [11.1] An equation equivalent to one of the form $ax^2 + bx + c = 0$, where $a \neq 0$

Quadratic formula [11.2] The solutions of $ax^2 + bx + c = 0, a \neq 0$, are given by the equation $x = \left(-b \pm \sqrt{b^2 - 4ac}\right)/(2a)$

Quadratic function [11.2] A second-degree polynomial function in one variable

Quadratic inequality [8.4] A second-degree polynomial inequality in one variable

Quartic function [5.3] A polynomial function in one variable of degree 4

Radical equation [10.6] An equation in which a variable appears in a radicand

Radical expression [10.1] An algebraic expression in which a radical sign appears

Radical function [10.1] A function that can be described by a radical equation

Radical sign [10.1] The symbol $\sqrt{}$

Radical term [10.5] A term in which a radical sign appears

Radicand [10.1] The expression under the radical sign

Radius [13.1] The distance from the center of a circle to a point on the circle; a segment connecting a point on the circle to the center of the circle

Range [3.8] The set of all second coordinates of the ordered pairs in a function

Rate [3.4] A ratio that indicates how two quantities change with respect to each other

Ratio [7.7] The ratio of a to b is a/b, also written $a : b$

Rational equation [7.6] An equation that contains one or more rational expressions

Rational expression [7.1] A polynomial divided by a nonzero polynomial

Rational function [7.1] A function described by a rational equation

Rational inequality [8.4] An inequality involving rational expressions

Rational number [1.4] A number that can be written in the form a/b, where a and b are integers and $b \neq 0$

Rationalizing the denominator [10.4] A procedure for finding an equivalent expression without a radical in the denominator

Rationalizing the numerator [10.4] A procedure for finding an equivalent expression without a radical in the numerator

Real number [1.4] Any number that is either rational or irrational

Reciprocal [1.3] A multiplicative inverse; two numbers are reciprocals if their product is 1.
Reflection [11.6] The mirror image of a graph
Relation [3.8] A correspondence between a first set, called the *domain*, and a second set, called the *range*, such that each member of the domain corresponds to *at least one* member of the range
Repeated root [6.4] A root resulting from a factor that appears two or more times in a factorization
Repeating decimal [1.4] A decimal in which a block of digits repeats indefinitely
Right triangle [6.7] A triangle that includes a 90° angle
Roots of an equation [6.1] The x-values for which an equation such as $f(x) = 0$ is true
Root of multiplicity two [6.4] A repeated root that appears twice
Row-equivalent operations [9.3] Operations used to produce equivalent systems of equations
Rows of a matrix [9.3] The horizontal elements of a matrix

Scientific notation [5.2] A number written in the form $N \times 10^m$, where N is at least 1 but less than 10 $(1 \leq N \leq 10)$, N is expressed in decimal notation, and m is an integer
Sequence [14.1] A function for which the domain is a set of consecutive positive integers beginning with 1
Series [14.1] The sum of specified terms in a sequence
Set [1.4] A collection of objects
Set-builder notation [2.6] The naming of a set by describing basic characteristics of the elements in the set
Sigma notation [14.1] The naming of a sum using the Greek letter Σ (sigma) as part of an abbreviated form; also called *summation notation*
Significant digits [5.2] When working with decimals in science, significant digits are used to determine how accurate a measurement is.
Similar triangles [7.7] Triangles in which corresponding angles have the same measure and corresponding sides are proportional
Simplify [1.3] To rewrite an expression in an equivalent, abbreviated form
Slope [3.5] The ratio of the rise to the run for any two points on a line
Slope–intercept equation [3.6] An equation of a line of the form $y = mx + b$, where x and y are variables, m is the slope, and $(0, b)$ is the y-intercept
Solution [1.1] A replacement or a substitution that makes an equation or an inequality or a system of equations or inequalities true
Solution set [2.2] The set of all solutions of an equation, an inequality, or a system of equations or inequalities
Solve [2.1] To find all solutions of an equation, an inequality, or a system of equations or inequalities; to find the solution(s) of a problem
Speed [3.4] The ratio of distance traveled to the time required to travel that distance
Square matrix [9.4] A matrix with the same number of rows and columns
Square root [10.1] The number c is a *square root* of a if $c^2 = a$.
Standard form of a linear equation [3.3] An equation of the form $Ax + By = C$, where A, B, and C are real numbers and A and B are not both 0
Substitute [1.1] To replace a variable with a number
Substitution method [4.2] An algebraic method for solving a system of equations
Sum [1.2] The result of addition
Sum of two cubes [6.5] An expression that can be written in the form $A^3 + B^3$
Supply function [9.5] A function modeling the relationship between the price of a good and the quantity of that good supplied
Synthetic division [5.8] A method used to divide a polynomial by a binomial of the type $x - a$
System of equations [4.1] A set of two or more equations, in two or more variables, for which we seek a common solution

Term [1.2] A number, a variable, or a product or a quotient of numbers and/or variables
Terminating decimal [1.4] A decimal that can be written using a finite number of decimal places
Total cost [9.5] The money spent to produce a product
Total profit [9.5] The money taken in less the money spent, or total revenue minus total cost
Total revenue [9.5] The amount taken in from the sale of a product
Trinomial [5.3] A polynomial composed of three terms

Undefined [1.7] An expression that has no meaning attached to it
Union of A and B [8.1] The collection of elements belonging to A and/or B; denoted $A \cup B$

Value [1.1] The numerical result after a number has been substituted into an expression
Variable [1.1] A letter that represents an unknown number
Variable costs [9.5] In business, costs that vary according to the amount being produced
Variable expression [1.1] An expression containing a variable

Variation constant [7.8] The constant, k, in an equation of variation

Vertex [11.7], [13.1], [13.2], [13.3] The point at which a parabola, an ellipse, or a hyperbola crosses its axis of symmetry

Vertical asymptote [7.1] A vertical line that a graph approaches, but never touches

Vertical-line test [3.8] The statement that a graph represents a function if it is impossible to draw a vertical line that intersects the graph more than once

Whole numbers [1.4] The set of natural numbers and 0: $\{0, 1, 2, 3, 4, 5, \ldots\}$

***x*-axis** [3.1] The horizontal axis in a coordinate plane

***x*-intercept** [3.3] A point at which a graph crosses the x-axis

***y*-axis** [3.1] The vertical axis in a coordinate plane

***y*-intercept** [3.3] A point at which a graph crosses the y-axis

Zero of a function [4.5], [6.1] An input whose corresponding output is 0; if a is a zero of the function f, then $f(a) = 0$.

Photo Credits

1, Choon Boon/210 MyLoupe, Inc. **8 (top),** Radius Images/Corbis **8 (bottom),** Choon Boon/2010 MyLoupe, Inc. **12,** *Spring Arcs* (2004), Edward Tufte. Solid stainless steel, footprint 12′ × 67′/www.tufte.com **33,** Alex Farnsworth/The Image Works **39,** Richard Drew/Associated Press **45,** SuperStock Royalty-Free/SuperStock **47,** Amwell/Photodisc/Getty Images **52,** Ivern Photo/Royalty-Free/Pixtal/AGE Fotostock **55,** Peter Turnley/Corbis **80,** Ikuru Kuwajima/Syracuse Newspapers/The Image Works **82,** Kristian Sekulic/Shutterstock **99,** Pixel4Images/Shutterstock **109,** VanHart/Shutterstock **112,** Radius Images/JupiterImages **113,** V. J. Matthew/Shutterstock **115,** Gary C. Caskey/UPI Photo/Landov **118,** Appalachian Trail Conservancy **122,** David Sacks/Lifesize/Thinkstock **123,** Martina Ebel/Shutterstock **127,** AFP/Getty Images **129,** Ed Quinn/Corbis **130 (left),** Annabelle Breakey/DigitalVision/JupiterImages **130 (right),** Don Mason/Blend Images/JupiterImages **143,** Papa Roux, Inc./papa@papaRouxIndy.com **144,** JupiterImages **148,** Liv friis-larsen/Shutterstock **150,** Terry Davis/Shutterstock **159,** Getty Images Entertainment **195 (top),** iStockphoto/Thinkstock **195 (bottom),** Corbis/JupiterImages **198,** Shutterstock **202,** David Muir/Getty Images **212,** AbbiOrca.com **226,** Michael Pettigrew/Shutterstock **230,** Getty Images/Thinkstock **234,** Shutterstock **240 (top),** James M. Phelps, Jr./Shutterstock **240 (bottom),** Richard B. Levine/Newscom **259,** Rafa Irusta/Shutterstock **267,** AGE Fotostock/SuperStock **270,** Gorilla/Shutterstock **271,** Tim Matsui/Aurora Open/Corbis **278,** Mike Norton/Shutterstock **295,** Orla/Shutterstock **296,** Hans Huber/Westend61/JupiterImages **299,** Avava/Shutterstock **302,** TeaPots n Treasures **309,** Joe White/Shutterstock **310,** Tony Campbell/Shutterstock **311 (left),** Vladimir Wrangel/Shutterstock **311 (bottom),** Kesu/Shutterstock **312,** Alex Mares-Manton/Getty Images **329,** Nice One Productions/Corbis **353,** NASA **354,** Nice One Productions/Corbis **360,** Joggie Botma/Shutterstock **361,** Barcroft/Fame **423,** Pete Saloutos/Shutterstock **476,** Ivonne Wierink/Shutterstock **478,** Ron Kaufmann, University of San Diego **479,** Gila Photography/Shutterstock **483,** Pete Saloutos/Shutterstock **496,** Ragne Kabanova/Shutterstock **499,** Bettmann/Corbis **501,** Teresa Kaufman/Photolibrary New York **528,** Dr. Paulus Gerdes **556,** Golden Pixels LLC/Shutterstock **557,** Getty Images/JupiterImages **561,** George Green/Shutterstock **566,** Sarah Scott/Shutterstock **568,** Nice_Pictures/Shutterstock **570 (top),** iStockphoto **570 (bottom),** Neo Edmund/Shutterstock **572,** SVLumagraphica/Shutterstock **574,** Andresr/Shutterstock **576,** Teresa Kaufman/Photolibrary New York **580,** Silver-john/Shutterstock **583,** Joe Belanger **593, 596,** Paparazzit/Shutterstock **605,** Cheryl Casey/Shutterstock **608,** Bill Gozansky/Alamy **633,** Leksele/Shutterstock **647,** Westend61 GmbH/Alamy **660,** Marylooo/Shutterstock **661 (left),** Vladoleg/Shutterstock **661 (right),** Mihai Simonia/Shutterstock **677,** Andresr/Shutterstock **683,** FidoProducts **684 (left),** Corbis **684 (right),** Trombax/Shutterstock **691,** Sergei Butorin/Shutterstock **693, 703,** Dewayne Flowers/Shutterstock **707,** Mishella/Shutterstock **715,** Ljupco Smokovski/Shutterstock **723,** Timothy Large/Shutterstock **729,** Nikolais/Dreamstime **746 (top),** Perkus/Shutterstock **746 (bottom),** Hardtmuth/Shutterstock **773,** Rob Marmion/Shutterstock **783,** iStockphoto **785 (top),** Alamy **785 (bottom),** Eric Ricks/Shutterstock **796,** iStockphoto **799,** Thinkstock **800,** Ohn G. Mabanglo/epa/Landov **801,** Tony Wear/Shutterstock **803,** Ron Jenkins/Fort Worth Star-Telegram/MCT **832,** Ciapix/Shutterstock **837,** Rob Marmion/Shutterstock **847,** Allan Szeto/Shutterstock **851,** Chris Curtis/Shutterstock **852,** AGE Fotostock/SuperStock **864,** Bhowe/Shutterstock **875,** AndiPu/Shutterstock **910,** Emin Kuliyev/Shutterstock **912,** Photawa/Dreamstime **918,** Chris Curtis/Shutterstock **921 (left),** Iofoto/Shutterstock **921 (right),** Claudio Bravo/Shutterstock **923,** Bobby Wjr/Shutterstock **924 (left),** Anthony Devlin/PA Photos/Landov **924 (right), 925,** Reuters/Landov **932 (left),** Chris Kruger/Shutterstock **932 (right),** Sergej Khakimullin/Shutterstock **935,** Irabel8/Shutterstock **937,** Olinchuk/Shutterstock **951,** Phanie/Photo Researchers, Inc. **954 (left),** Chuck Aghoian/Shutterstock **954 (right),** NASA **974,** Courtesy of Miniature Art Society of Florida **981,** Reaktori/Dreamstime **1000,** Tilo/Dreamstime **1011,** fotogiunta/Shutterstock **1030,** Joseppi/Dreamstime **R-17,** Pétur Ásgeirsson

Index

*Page numbers followed by an asterisk pertain to Chapters 13 and 14.

E

H

I

S

T

U

V

Index of Applications

*Page numbers followed by an asterisk pertain to Chapters 13 and 14.

Chemistry

Construction

Consumer

Economics

Education

Engineering

Environment

Finance

Geometry

Health/Medicine

Social Sciences

Sports/Entertainment

Statistics/Demographics

Technology

Transportation

Frequently Used Symbols and Formulas

A Key to the Icons Appearing in the Exercise Sets

Icon	Description
	Concept reinforcement exercises, indicated by purple exercise numbers, provide basic practice with the new concepts and vocabulary.
1	Following most examples, students are directed to Try Exercise. These selected exercises are identified with a color block around the exercise numbers.
Aha!	Exercises labeled Aha! can often be solved quickly with the proper insight.
	Calculator exercises are designed to be worked using a scientific calculator or a graphing calculator.
	Graphing calculator exercises are designed to be worked using a graphing calculator.
TW	Writing exercises are designed to be answered using one or more complete sentences.

Symbols

Symbol	Meaning
$=$	Is equal to
$\approx$	Is approximately equal to
$>$	Is greater than
$<$	Is less than
$\geq$	Is greater than or equal to
$\leq$	Is less than or equal to
$\in$	Is an element of
$\subseteq$	Is a subset of
$\lvert x \rvert$	The absolute value of x
$\{x \mid x \ldots\}$	The set of all x such that $x \ldots$
$-x$	The opposite of x
$\sqrt{x}$	The square root of x
$\sqrt[n]{x}$	The nth root of x
LCM	Least Common Multiple
LCD	Least Common Denominator
π	Pi
i	$\sqrt{-1}$
$f(x)$	f of x, or f at x
$f^{-1}(x)$	f inverse of x
$(f \circ g)(x)$	$f(g(x))$
e	Approximately 2.7
Σ	Summation
$n!$	Factorial notation

Formulas

Formula	Description
$m = \dfrac{y_2 - y_1}{x_2 - x_1}$	Slope of a line
$y = mx + b$	Slope–intercept form of a linear equation
$y - y_1 = m(x - x_1)$	Point–slope form of a linear equation
$(A + B)(A - B) = A^2 - B^2$	Product of the sum and the difference of the same two terms
$(A + B)^2 = A^2 + 2AB + B^2$, $(A - B)^2 = A^2 - 2AB + B^2$	Square of a binomial
$d = rt$	Formula for distance traveled
$\dfrac{1}{a} \cdot t + \dfrac{1}{b} \cdot t = 1$	Work principle
$s = 16t^2$	Free-fall distance
$y = kx$	Direct variation
$y = \dfrac{k}{x}$	Inverse variation
$x = \dfrac{-b \pm \sqrt{b^2 - 4ac}}{2a}$	Quadratic formula
$P(t) = P_0 e^{kt}, k > 0$	Exponential growth
$P(t) = P_0 e^{-kt}, k > 0$	Exponential decay
$d = \sqrt{(x_2 - x_1)^2 + (y_2 - y_1)^2}$	Distance formula
$\dbinom{n}{r} = \dfrac{n!}{(n - r)!\,r!}$	$\dbinom{n}{r}$ notation

A Library of Functions

Constant Function
$y = b$

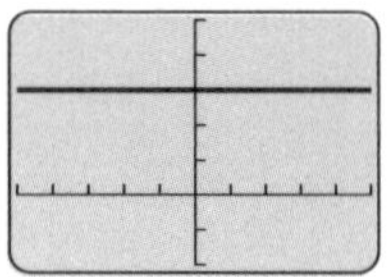

Linear Function
$y = mx + b$

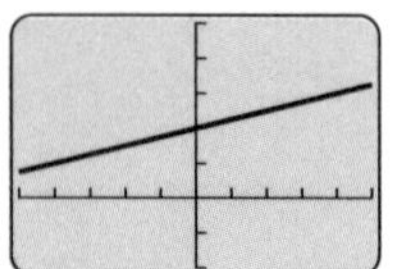

Squaring Function
$y = x^2$

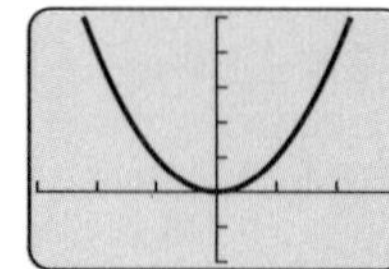

Quadratic Function
$y = ax^2 + bx + c, a > 0$

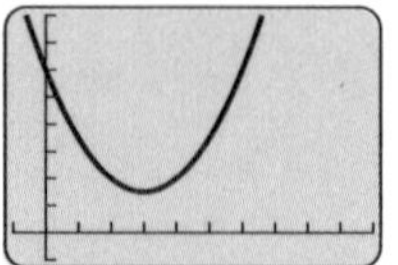

Quadratic Function
$y = ax^2 + bx + c, a < 0$

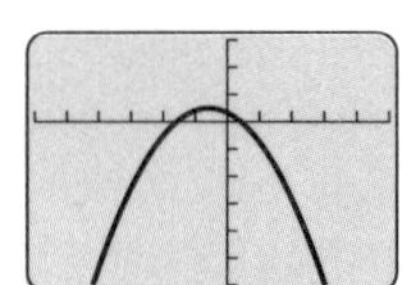

Square Root Function
$y = \sqrt{x}$

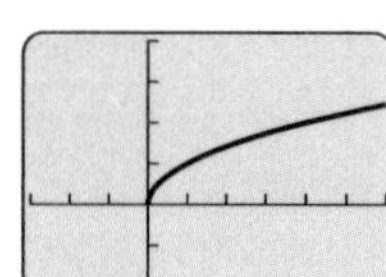

Cubing Function
$y = x^3$

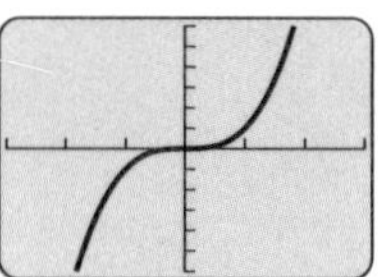

Cube Root Function
$y = \sqrt[3]{x}$

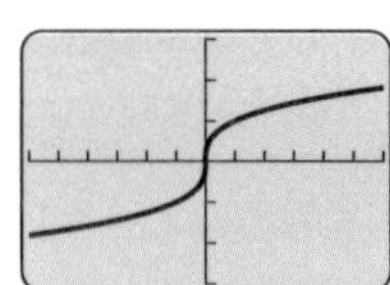

Reciprocal Function
$y = \frac{1}{x}$

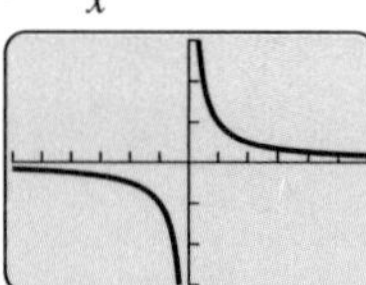

Absolute Value Function
$y = |x|$

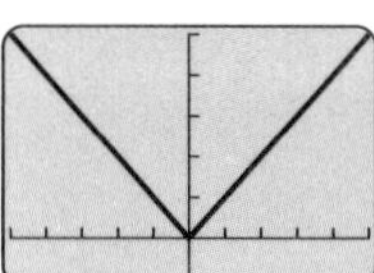

Exponential Function
$y = ab^x$, or ae^{kx}
$a, b > 0, k > 0$

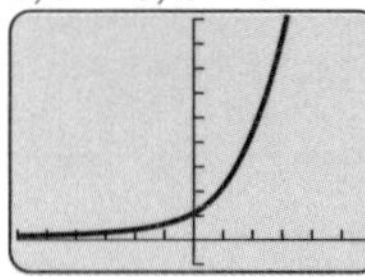

Exponential Function
$y = ab^{-x}$, or ae^{-kx}
$a, b > 0, k > 0$

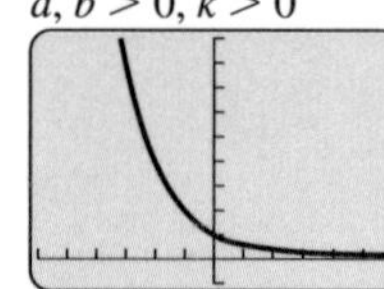

Logarithmic Function
$y = a + b \ln x$

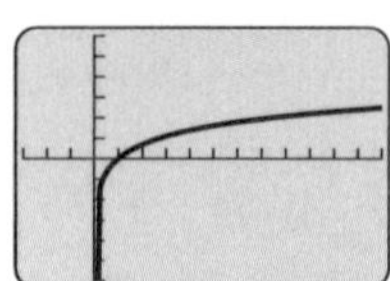